Plant Cell Biology

Plant Cell Biology

From Astronomy to Zoology

Second Edition

Randy Wayne

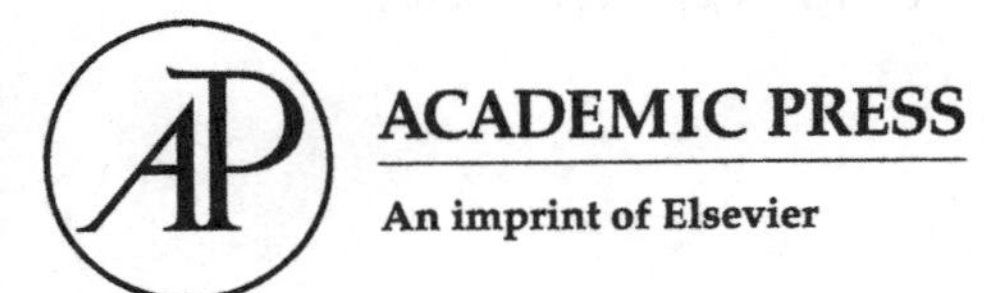

Academic Press is an imprint of Elsevier
125 London Wall, London EC2Y 5AS, United Kingdom
525 B Street, Suite 1650, San Diego, CA 92101, United States
50 Hampshire Street, 5th Floor, Cambridge, MA 02139, United States
The Boulevard, Langford Lane, Kidlington, Oxford OX5 1GB, United Kingdom

Notices

Knowledge and best practice in this field are constantly changing. As new research and experience broaden our understanding, changes in research methods, professional practices, or medical treatment may become necessary.

Practitioners and researchers must always rely on their own experience and knowledge in evaluating and using any information, methods, compounds, or experiments described herein. In using such information or methods they should be mindful of their own safety and the safety of others, including parties for whom they have a professional responsibility.

To the fullest extent of the law, neither the Publisher nor the authors, contributors, or editors, assume any liability for any injury and/or damage to persons or property as a matter of products liability, negligence or otherwise, or from any use or operation of any methods, products, instructions, or ideas contained in the material herein.

Library of Congress Cataloging-in-Publication Data
A catalog record for this book is available from the Library of Congress

British Library Cataloguing-in-Publication Data
A catalogue record for this book is available from the British Library

ISBN: 978-0-12-814371-1

For information on all Academic Press publications visit our website at https://www.elsevier.com/books-and-journals

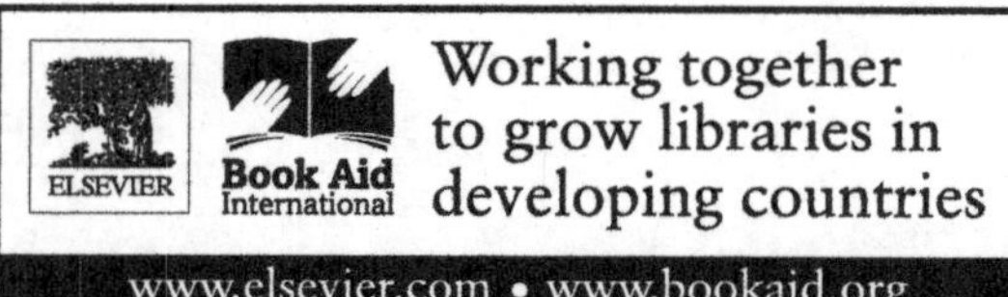

Publisher: Andre G. Wolff
Acquisition Editor: Mary Preap
Editorial Project Manager: Mary Preap
Production Project Manager: Mohanapriyan Rajendran
Cover Designer: Mark Rogers

Cover pictures: Photomicrograph of diatoms arranged on a microscope slide taken with a differential interference contrast microscope (top, left); Photomicrograph of an onion epidermal cell taken with a phase contrast microscope (top, center); Photomicrograph of *Paramecium bursaria* taken with a differential interference microscope (top, right); Photomicrograph of cotton hair cells taken with a polarized light microscope (bottom).

Typeset by TNQ Technologies

This book is dedicated to Amy, my wife and soulmate, and to Erwin Chargaff, an Apollonian and Dionysian scientific hero who first isolated and characterized DNA as an informational macromolecule, suggested that we should sequence it, and understood where science was going.

Contents

Preface to the First Edition

This book is in essence the lectures I give in my plant cell biology course at Cornell University. Heretofore, the lecture notes have gone by various titles, including "Cell La Vie," "The Book Formerly Known as Cell La Vie," "Molecular Theology of the Cell," "Know Thy Cell" (with apologies to Socrates), "Cell This Book" (with apologies to Abbie Hoffman), and "Impressionistic Plant Cell Biology." I would like to take this opportunity to describe this course. It is a semester-long course for undergraduate and graduate students. Because the undergraduate biology majors are required to take genetics, biochemistry, and evolution as well as 1 year each of mathematics and physics and 2 years of chemistry, I have done my best to integrate these disciplines into my teaching. Moreover, many of the students also take plant anatomy, plant physiology, plant growth and development, plant taxonomy, plant biochemistry, plant molecular biology, and a variety of courses that end with the suffix "-omics"; I have tried to show the connections between these courses and plant cell biology. Nonbotanists can find a good introduction to plant biology in Mauseth (2009) and Taiz and Zeiger (2006).

Much of the content has grown over the past 20 years from the questions and insights of the students and teaching assistants who have participated in the class. The students' interest has been sparked by the imaginative and insightful studies done by the worldwide community of cell biologists, which I had the honor of presenting.

I have taken the approach that real divisions not only exist between subject areas taught in a university but only in the state of mind of the teachers and researchers. With this approach, I hope that my students do not see plant cell biology as an isolated subject area, but as an entry into every aspect of human endeavor. One of the goals of my course is to try to reestablish the connections that once existed between mathematics, astronomy, physics, chemistry, geology, philosophy, and biology. It is my own personal attempt, and it is an ongoing process. Consequently, it is far from complete. Even so, I try to provide the motivation and resources for my students to weave together the threads of these disciplines to create their own personal tapestry of the cell from the various lines of research.

Recognizing the basic similarities between all living eukaryotic cells (Quekett, 1852, 1854; Huxley, 1893), I discuss both animal and plant cells in my course. Although the examples are biased toward plants (as they should be in a plant cell biology course), I try to present the best example to illustrate a process and sometimes the best examples are from animal cells. I take the approach used by August Krogh (1929), that is, there are many organisms in the treasure house of nature and if one respects this treasure, one can find an organism created to best illuminate each principle! I try to present my course in a balanced manner, covering all aspects of plant cell biology without emphasizing any one plant, organelle, molecule, or technique. I realize, however, that the majority of papers in plant cell biology today are using a few model organisms and "-omic" techniques. My students can learn about the successes gained though this approach in a multitude of other courses. I teach them that there are other approaches.

Pythagoras believed in the power of numbers, and I believe that the power of numbers is useful for understanding the nature of the cell. In my class, I apply the power of numbers to help relate quantities that one wishes to know to things that can be easily measured (Hobson, 1923; Whitehead, 1925; Hardy, 1940; Synge, 1951, 1970; Feynman, 1965a; Schrödinger, 1996). For example, the area of a rectangle is difficult to measure. However, if one knows its length and width, and the relation that area is the product of length and width, the area can be calculated from the easily measurable quantities. Likewise, the circumference or area of a circle is relatively difficult to measure. However, if one measures the diameter and multiplies it by π, or the square of the diameter by $\pi/4$, one can easily obtain the circumference and area, respectively. In the same way, one can easily estimate the height of a tree from easily measurable quantities if one understands trigonometry and the definition of tangent.

My teaching was greatly influenced by a story that Hans Bethe told at a meeting at Cornell University commemorating the 50th anniversary of the chain reaction produced by Enrico Fermi. Bethe spoke about the difference between his graduate adviser, Arnold Sommerfeld, and his postdoctoral adviser, Enrico Fermi. He said that, in the field of atomic physics, Sommerfeld was a genius at creating a mathematical theory to describe the available data. Sommerfeld's skill, however, depended on the presence of data. Fermi, on the other hand, could come up with theories even if the relevant data were not apparent. He would make estimates of the data from first principles. For example, he estimated the force of the first atomic bomb by measuring the distance small pieces of paper flew as they fell to the ground during the blast in Alamogordo. Knowing that the force of the blast diminished with the square of the distance from the bomb, Fermi estimated the force of the bomb relative to the force of gravity. Within seconds of the blast, he calculated the force of the bomb to be approximately 20 kilotons, similar to which the expensive machines recorded (Fermi, 1954; Lamont, 1965).

To train his students to estimate things that they did not know, Fermi would ask them, "How many piano tuners are there in Los Angeles?" After they looked befuddled, he would say, "You can estimate the number of piano tuners from first principles! For example, how many people are there in Los Angeles? One million? What percentage has pianos? Five percent? Then there are 50,000 pianos in Los Angeles. How often does a piano need to be tuned? About once a year? Then 50,000 pianos need to be tuned in a year. How many pianos can a piano tuner tune in a day? Three? Then one tuner must spend 16,667 days a year tuning pianos. But since there are not that many days in a year, and he or she probably only works 250 days a year, then there must be around 67 piano tuners in Los Angeles."

My students apply the power of numbers to the study of cellular processes, including membrane transport, photosynthesis, and respiration, to get a feel for these processes and the interconversions that occur during these processes between different forms of energy. My students apply the power of numbers to the study of cell growth, chromosome motion, and membrane trafficking to postulate and evaluate the potential mechanisms involved in these processes, and the relationships between these processes and the bioenergetic events that power them. Becoming facile with numbers allows the students to understand, develop, and critique theories. "As the Greek origin of the word [theory] implies, the Theory is the true seeing of things—the insight that should come with healthy sight" (Adams and Whicher, 1949).

Using the power of numbers to relate seemingly unrelated processes, my students are able to try to analyze all their conclusions in terms of first principles. They also learn to make predictions based on first principles. The students must be explicit in terms of what they are considering to be facts, what they are considering to be the relationship between facts, and where they are making assumptions. This provides a good entry into research because the facts must be refined and the assumptions must be tested (East, 1923).

I do not try to introduce any more terminology in my class than is necessary, and I try to explain the origin of each term. Some specialized terms are essential for precise communication in science just as it is in describing love and beauty. However, some terms are created to hide our ignorance, and consequently prevent further inquiry, because something with an official-sounding name seems well understood (Locke, 1824; Hayakawa, 1941; Rapoport, 1975). In Goethe's (1808) "Faust Part One," Mephistopheles says: "For at the point where concepts fail. At the right time a word is thrust in there. With words we fitly can our foes assail." Francis Bacon (1620) referred to this problem as the "Idols of the Marketplace." Often we think we are great thinkers when we answer a question with a Greek or Latin word. For example, if I am asked, "Why are leaves green?" I quickly retort, "Because they have chlorophyll." The questioner is satisfied, and says "Oh." The conversation ends. However, chlorophyll is just the Greek word for green leaf. Thus, I really answered the question with a tautology. I really said "Leaves are green because leaves are green" and did not answer the question at all. It was as if I was reciting a sentence from scripture, which I had committed to memory without giving it much thought. However, I gave the answer in Greek, and with authority … so it was a scientific answer.

In "An Essay Concerning Human Understanding," John Locke (1824) admonished that words are often used in a nonintellectual manner. He wrote,

> *… he would not be much better than the Indian before-mentioned, who, saying that the world was supported by a great elephant, was asked what the elephant rested on; to which his answer was, a great tortoise. But being again pressed to know what gave support to the broad-backed tortoise, replied, something he knew not what. And thus here, as in all other cases where we use words without having clear and distinct ideas, we talk like children; who being questioned what such a thing is, which they know not, readily give the satisfactory answer, that it is something; which in truth signifies no more, when so used either by children or men, but that they know not what; and that the thing that they pretend to know and talk of is what they have no distinct idea of at all, and so are perfectly ignorant of it, and in the dark.*

Sometimes terms are created to become the shibboleths of a field, and sometimes they are created for political reasons, financial reasons, or to transfer credit from someone who discovers something to someone who renames it

(Agre et al., 1995). Joseph Fruton (1992) recounted (and translated) a story of a conversation with a famous chemist in Honoré de Balzac's La Peau de Chagrin:

> *"Well, my old friend," said Planchette upon seeing Japhet seated in an armchair and examining a precipitate, "How goes it in chemistry?"*
>
> *It is asleep. Nothing new. The Académie has in the meantime recognized the existence of salicine. But salicine, asparagine, vauqueline, digitaline are not new discoveries.*
>
> *"If one is unable to produce new things," said Raphael, "it seems that you are reduced to inventing new names."*
>
> *That is indeed true, young man.*

I teach plant cell biology with a historical approach and teach "not only of the fruits but also of the trees which have borne them, and of those who planted these trees" (Lenard, 1906). This approach also allows them to understand the origins and meanings of terms; to capture the excitement of the moment of discovery; to elucidate how we, as a scientific community, know what we know; and it emphasizes the unity and continuity of human thought (Haldane, 1985). I want my students to become familiar with the great innovators in science and to learn their way of doing science (Wayne and Staves, 1998, 2008). I want my students to learn how the scientists we learn about choose and pose questions, and how they go about solving them. I do not want my students to know just the results and regurgitate those results on a test (Szent-Györgyi, 1964; Farber, 1969). I do not want my students to become scientists who merely repeat on another organism the work of others. I want my students to become like the citizens of Athens, who according to Pericles "do not imitate—but are a model to others." Whether or not my students become professional cell biologists, I hope they forever remain amateurs and dilettantes in terms of cell biology. That is, I hope that I have helped them become "one who loves cell biology" and "one who delights in cell biology" (Chargaff, 1986)—not someone who cannot recognize the difference between a pile of bricks and an edifice (Forscher, 1963), not someone who sells "buyology" (Wayne and Staves, 2008), and not someone who sells his or her academic freedom (Rabounski, 2006; Apostol, 2007).

Often people think that a science course should teach what is new, but I answer this with an amusing anecdote said by Erwin Chargaff (1986): "Kaiser Wilhelm I of Germany, Bismark's old emperor, visited the Bonn Observatory and asked the director: 'Well, dear Argelander, what's new in the starry sky?' The director answered promptly: 'Does your Majesty already know the old?' The emperor reportedly shook with laughter every time he retold the story."

According to R. John Ellis (1996),

> *It is useful to consider the origins of a new subject for two reasons. First, it can be instructive; the history of science provides sobering take-home messages about the importance of not ignoring observations that do not fit the prevailing conceptual paradigm, and about the value of thinking laterally, in case apparently unrelated phenomena conceal common principles. Second, once a new idea has become accepted there is often a tendency to believe that it was obvious all along—hindsight is a wonderful thing, but the problem is that it is never around when you need it!*

The historical approach is necessary, in the words of George Palade (1963), "to indicate that recent findings and present concepts are only the last approximation in a long series of similar attempts which, of course, is not ended."

I teach my students that it is important to be skeptical when considering old as well as new ideas. According to Thomas Gold (1989),

> *New ideas in science are not always right just because they are new. Nor are the old ideas always wrong just because they are old. A critical attitude is clearly required of every scientist. But what is required is to be equally critical to the old ideas as to the new. Whenever the established ideas are accepted uncritically, but conflicting new evidence is brushed aside and not reported because it does not fit, then that particular science is in deep trouble—and it has happened quite often in the historical past.*

To emphasize the problem of scientists unquestioningly accepting the conventional wisdom, Conrad H. Waddington (1977) proposed the acronym COWDUNG to signify the Conventional Wisdom of the Dominant Group.

In teaching in a historical manner, I recognize the importance of Thomas H. Huxley's (1853b) warnings that "Truth often has more than one Avatar, and whatever the forgetfulness of men, history should be just, and not allow those who had the misfortune to be before their time to pass for that reason into oblivion" and "The world, always too happy to join in toadying the rich, and taking away the 'one ewe lamb' from the poor." Indeed, it is often difficult to determine who makes a discovery (Djerassi and Hoffmann, 2001). I try to the best of my ability to give a fair and accurate account of the historical aspects of cell biology.

My course includes a laboratory section and my students perform experiments to acquire personal experience in understanding the living cell and how it works (Hume, 1748; Wilson, 1952; Ramón y Cajal, 1999). Justus von Liebig (1840) described the importance of the experimental approach this way:

> *Nature speaks to us in a peculiar language, in the language of phenomena; she answers at all times the questions which are put to her; and such questions are experiments. An experiment is the expression of a thought: we are near the truth when the phenomenon, elicited by the experiment, corresponds to the thought; while the opposite result shows that the question was falsely stated, and that the conception was erroneous.*

My students cannot wait to get into the laboratory. In fact, they often come in on nights and weekends to use the microscopes to take photomicrographs. At the end of the semester, the students come over to my house for dinner (I worked my way through college as a cook) and bring their best photomicrographs. After dinner, they vote on the 12 best, and those are incorporated into a class calendar. The calendars are beautiful and the students often make extra to give as gifts.

In 1952, Edgar Bright Wilson Jr. wrote in *An Introduction to Scientific Research*, "There is no excuse for doing a given job in an expensive way when it can be carried through equally effectively with less expenditure." Today, with an emphasis on research that can garner significant money for a college or university through indirect costs, there is an emphasis on the first use of expensive techniques to answer cell biological questions and often questions that have already been answered. However, the very expense of the techniques often prevents one from performing the preliminary experiments necessary to learn how to do the experiment so that meaningful and valuable data and not just lists are generated. Unfortunately, the lists generated with expensive techniques often require statisticians and computer programmers, who are far removed from experiencing the living cells through observation and measurement, to tell the scientist which entries on the list are meaningful. Thus, there is a potential for the distinction between meaningful science and meaningless science to become a blur. I use John Synge's (1951) essay on vicious circles to help my students realize that there is a need to distinguish for themselves what is fundamental and what is derived.

By contrast, this book emphasizes the importance of the scientists who have made the great discoveries in cell biology using relatively low-tech quantitative and observational methods. But—and this is a big but—these scientists also treated their brains, eyes, and hands as highly developed scientific instruments. I want my students to have the ability to get to know these great scientists. I ask them to name who they think are the 10 best scientists who ever lived. Then I ask if they have ever read any of their original work. In the majority of the cases, they have never read a single work by the people who they consider to be the best scientists. This is a shame. They read the work of others … but not the best. Interestingly, they usually are well read when it comes to reading the best writers (e.g., Shakespeare, Faulkner, etc.).

Typically, the people on my students' lists of best scientists have written books for the layperson or an autobiography (Wayne and Staves, 1998). Even Isaac Newton wrote a book for the layperson! I give my class these references and encourage them to become familiar with their favorite scientists first hand. The goal of my lectures and this book is to facilitate my students' personal and continual journey in the study of life.

My goal in teaching plant cell biology is not only to help my students understand the mechanisms of the cell and its organelles in converting energy and material matter into a living organism that performs all the functions we ascribe to life. I also hope to deepen my students' ideas of the meaning, beauty, and value of life and the value in searching for meaning and understanding in all processes involved in living.

I thank Mark Staves and my family, Michelle, Katherine, Zack, Beth, Scott, my mother and father, and aunts and uncles for their support over the years. I also thank my colleagues at Cornell University and teachers at the Universities of Massachusetts, Georgia, and California at Los Angeles, and especially Peter Hepler and Masashi Tazawa, who taught me how to see the universe in a living cell.

Randy Wayne
Department of Plant Biology, Cornell University

Preface to the Second Edition

The first edition of this book emphasized the great ideas regarding what makes life possible on the cellular level and the great observations and experiments that stimulated, supported, and challenged those ideas. That is, the first edition emphasized the questions asked by great thinkers to understand what life is and the observations and experiments performed by great observers and experimentalists to provide answers to those questions. It was an inquiry into the origin, certainty, and extent of our cell biological knowledge and left room for further questioning and wonder. Scientific inquiry not only begins with wonder but also usually ends with it too. The questions are eternal and the answers are provisional—becoming fleshed out and sometimes changed over time. Extensive references were given so that the reader could read the original papers and develop a personal relationship with the scientists who came up with the great observations, experiments, and ideas. My goal was to facilitate the prospect that the reader would be inspired to join a century-long collaboration with the scientists who wrote the original papers. Even if the reader did not join a century-long collaboration, he or she would not only be informed from reading the original references, meaning that the reader would know that something was the answer; but also, the reader would be enlightened, knowing why something was the answer. This goal was the scientific equivalent of Mortimer Adler's (See Adler and van Doren, 1967) philosophy that drove him to edit the "Great Books of the Western World" series and write "How to Read a Book."

Biology has now become systems biology. The approach of systems biology (Cullis, 2008) "*is to accumulate information and biological resources relating to as much of the genome as possible and then determining which parts are of importance*." Moreover, the questions and answers obtained in the past are often considered biased and perhaps unworthy of study. According to Sheth and Thaker (2014), "*the main challenge confronting the field is not to look back (incorporating previous findings is critical but will be comparatively easy) but to look forward to how one might plan and interpret the enormous new data that soon will be generated*." Technology is being developed to obtain big data at the systems biology level that is unbiased. However, recognizing what bias is and what wisdom and knowledge are depends on an ability to read the relevant papers, the ability to work with one's eyes and hands to make observations and do experiments, and the ability to understand the ideas, observations, and experiments. This edition maintains and updates the philosophical outlook of the first edition, emphasizing how we know what we know, so that the reader can come to his or her own conclusion about the value of the work.

This edition also includes a new chapter entitled, Omic Science: Platforms and Pipelines, which emphasizes the great ideas and experiments that provided the basis for the technologies used in systems biology. I emphasize the development of the technologies rather than the results obtained from the technologies, as a study of the development of the technologies reveals real scientific achievement and creativity, and the explanation of their development has real pedagogical value. On the other hand, the results produced by the technologies are by necessity wanting given that the goal is to characterize everything in an unbiased manner (Chory et al., 2000), and, at the present time, we are as close to knowing nothing new from the systems approach as we are to knowing everything.

According to Bruce Alberts, Marc Kirschner, Shirley Tilghman, and Harold Varmus (2012) "*the system now favors those who can guarantee results rather than those with potentially path-breaking ideas that, by definition, cannot promise success. Young investigators are discouraged from departing too far from their postdoctoral work, when they should instead be posing new questions and inventing new approaches. Seasoned investigators are inclined to stick to their tried-and-true formulas for success rather than explore new fields*." Where is the freedom to think? Not only are trained scientists working like automatons but also being replaced by automatons. According to Steve Strogatz (Manjoo, 2011), "*Our time is limited. As thinking machines, they have a lot of advantages over us—this is obvious...We're not going to be the best players in town. I do think we'll be put out of business. This is really going to happen*."

More importantly, I'd like to ask the question: What are the goals of science? Are the goals best done by computers? Or should the goals be based on the proposition that science is a field of human endeavor that promotes freedom by training people to think. John Dewey (1910) realized, "*Genuine freedom, in short, is intellectual; it rests in the trained power of thought, in ability to 'turn things over,' to look at matters deliberately, to judge whether the amount and kind of evidence requisite for decision is at hand, and if not, to tell where and how to seek such evidence. If a man's actions are not guided by thoughtful conclusions, then they are guided by inconsiderate impulse, unbalanced appetite, caprice, or the circumstances of the moment. To cultivate unhindered, unreflective external activity is to foster enslavement, for it leaves the person at the mercy of appetite, sense, and circumstance*." We have to take seriously what Marcus Garvey (1938) said in a speech given in Nova Scotia, "*We are going to emancipate ourselves from mental slavery because whilst others might free the body, none but ourselves can free the mind*." Bob Marley immortalized these words in *Redemption Song*.

The culture of scientific research is in danger of eliminating what is necessary to cultivate the intelligent and creative use of technology. Both Jacobus van't Hoff (1967), who viewed chemicals in three-dimensional space, and Peter Mitchell (1980), who viewed metabolism in three-dimensional space as well as time, quoted Henry Thomas Buckle's (1872) description of the type of imagination needed to make great scientific discoveries: "*there is a spiritual, a poetic, and for aught we know a spontaneous and uncaused element in the human mind, which ever and anon, suddenly and without warning, gives us a glimpse and a forecast of the future, and urges us to seize the truth as it were by anticipation*." Yes, serendipity favors a prepared mind.

This book is written for the students who are craving to be free to do curiosity-driven science that requires thinking, imagination, and skill. It is also written for the taxpayer who expects their hard-earned money to be invested wisely. It is my hope that this book will help turn the scientific pendulum back so that science can once again encourage a way of understanding nature that is fueled by the intelligent use of technology and the scientific imagination (Holton, 1978).

I wrote the following paragraph in the Preface to the First Edition:

In 1952, Edgar Bright Wilson Jr. wrote in *An Introduction to Scientific Research*, "*There is no excuse for doing a given job in an expensive way when it can be carried through equally effectively with less expenditure*." Today, with an emphasis on research that can garner significant money for a college or university through indirect costs, there is an emphasis on the first use of expensive techniques to answer cell biological questions and often questions that have already been answered. However, the very expense of the techniques often prevents one from performing the preliminary experiments necessary to learn how to do the experiment so that meaningful and valuable data and not just lists are generated. Unfortunately, the lists generated with expensive techniques often require statisticians and computer programmers, who are far removed from experiencing the living cells through observation and measurement, to tell the scientist which entries on the list are meaningful. Thus, there is a potential for the distinction between meaningful science and meaningless science to become a blur.

In this preface, I would like to update the paragraph by substituting "Facilities & Administrative (F&A) costs" for "indirect costs."

I would like to heartily thank Mohanapriyan Rajendran for finding so many of my mistakes, typos, and omissions. Lastly, I would like to thank Karl Niklas, a scholar and a gentleman, for the fascinating and thought-provoking conversations we have about biology and teaching every morning.

Chapter 1

On the Nature of Cells

The world globes itself in a drop of dew. The microscope cannot find the animalcule which is less perfect for being little. Eyes, ears, taste, smell, motion, resistance, appetite, and organs of reproduction that take hold on eternity—all find room to consist in the small creature. So do we put our life into every act. The true doctrine of omnipresence is that God reappears with all His parts in every moss and cobweb.

Ralph Waldo Emerson, "Compensation."

The path of modern organ physiology is straight and clear, and we are not far from a complete understanding of life as an association of organs. But the organ is an assembly of cells and its properties and activities are dependent on the properties and activities of its component cells. Organ physiology has therefore, so to speak, begun its study from the midst of life; the beginning, the basis of life is in the cell.

Ivan Pavlov (cited in Heilbrunn, 1952).

1.1 INTRODUCTION: WHAT IS A CELL?

In the introduction to his book, *Grundzüge der Botanik*, Matthias Schleiden (1842), often considered the cofounder of the cell theory, admonished, "Anyone who has an idea of learning botany from the present book, may just as well put it at once aside unread; for from books botany is not learnt" (quoted in Goebel, 1926). Likewise, I would like to stress that an understanding of plant cell biology, and what a plant cell is, comes from direct experience. I hope that this book helps facilitate your own personal journey into the world of the cell.

Dom Pérignon, according to André Simon (1934), "did not discover, invent or create sparkling Champagne. He never claimed to have done so, nor did any of his contemporaries claim any such honour for him." In fact, Champagne was already being made in England where the oiled hemp rag that was traditionally used to stopper the bottle had just been replaced by cork (see history of Vintners' Hall at http://www.vintnershall.co.uk/?page=history_of_the_hall). While the oiled hemp rag kept the dust out, the cork allowed the build up of carbon dioxide in the wine. Exploring the world made accessible by the invention of the microscope, Robert Hooke (1665) took a look at the cork that may have been used by William Russell, the first Duke and fifth Earl of Bedford to stopper champagne bottles (Simon, 1934; Taber, 2007, 2009). Serendipitously, Hooke discovered a regular, repeating structure in cork that he called a cell. The word *cell* comes from the Latin *celle*, which in Hooke's time meant "a small apartment, *esp*. one of several such in the same building, used e.g., for a store-closet, slave's room, prison cell; also cell of a honeycomb; … also a monk's or hermit's cell" (Oxford English Dictionary, 1933). Hooke used the word *cell* to denote the stark appearance of the air-filled pores he saw in the honeycomb-like pattern in the cork that he viewed with his microscope (Fig. 1.1). Hooke's perspective of the emptiness of cells was propagated by Nehemiah Grew (1682), who compared the cells of the pith of asparagus to the froth of beer (Fig. 1.2), and is still implied in words with the prefix *cytos*, which in Greek means "hollow place." Hooke, however, did realize that there might be more to a cell than he could see. He wrote,

Now, though I have with great diligence endeavoured to find whether there be any such thing in those microscopical pores of wood or piths, as the valves in the heart, veins and other passages of animals, that open and give passage to the contained fluid juices one way, and shut themselves, and impede the passage of such liquors back again, yet have I not hitherto been able to say anything positive in it; … but … some diligent observer, if helped with better microscopes, may in time, detect [them].

Hugo von Mohl (1852) pointed out in *Principles of the Anatomy and Physiology of the Vegetable Cell*, the first textbook devoted to plant cell biology, that indeed plant cells are not vacuous when viewed with optically corrected microscopes but contain a nucleus and "an opake, viscid fluid of a white colour, having granules intermingled in it, which fluid I call protoplasm." Von Mohl, echoing the conclusions of Henri Dutrochet (1824) and John Quekett (1852), further revealed through his developmental studies that cells have a variety of shapes (Fig. 1.3) and give rise to all structures in the plant including the phloem and xylem. This was contrary to the earlier opinions of de Candolle and Sprengel (1821), who believed that there were three elementary forms in plants—dodecahedral-shaped cells, noncellular tubes, and noncellular spirals (Fig. 1.4). By focusing on mature plants, de Candolle and Sprengel had

Plant Cell Biology. https://doi.org/10.1016/B978-0-12-814371-1.00001-1

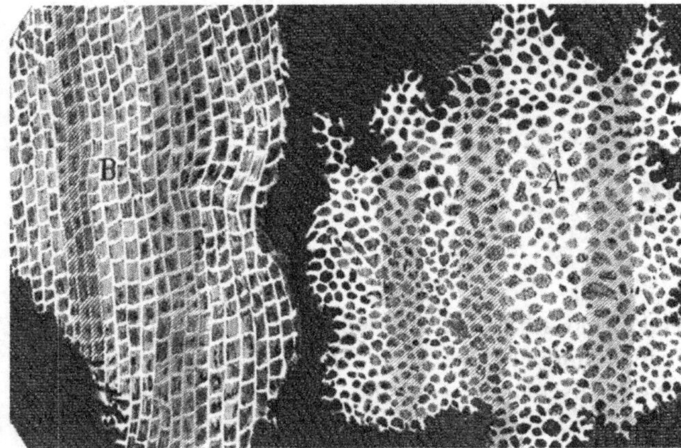

FIGURE 1.1 Cells of cork. *From Hooke, R., [1665], 1961. Micrographia or Some Physiological Descriptions of Minute Bodies Made by Manifying Glasses with Observations and Inquiries thereupon. Dover Publications, New York.*

FIGURE 1.3 Stellate cells from the petiole of a banana. *From von Mohl, H., 1852. Principles of the Anatomy and Physiology of the Vegetable Cell. Translated by Henfrey, A. John van Voorst, London.*

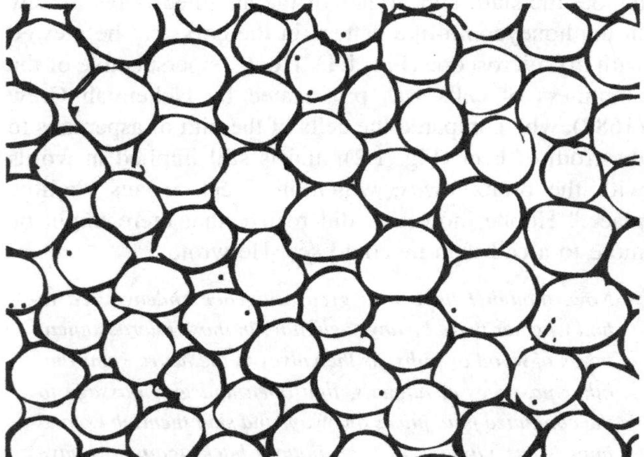

FIGURE 1.2 The cortical cells of a small root of asparagus. *From Grew, N., [1682], 1965. The Anatomy of Plants with an Idea of a Philosophical History of Plants and Several other Lectures Read before The Royal Society. Johnson Reprint Corp., New York.*

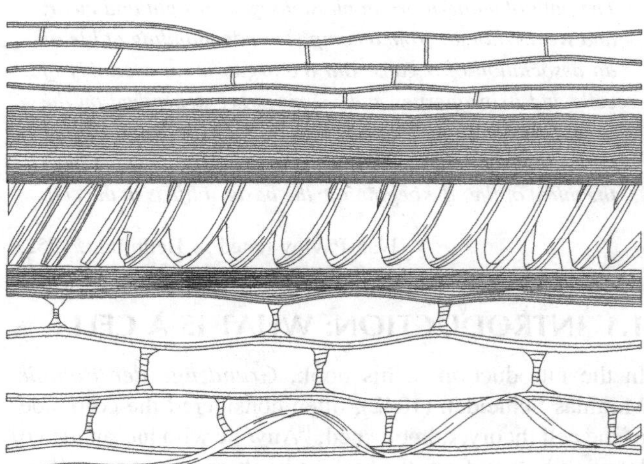

FIGURE 1.4 Spiral vessels, sap tubes, and cells of *Maranatha lutea*. *From deCandolle, A.P., Sprengel, K., 1821. Elements of the Philosophy of Plants. William Blackwood, Edinburgh.*

not realized that the tubelike vessels and the spiral-like protoxylem developed from dodecahedral-shaped cells. To further emphasize the vitality of cells, von Mohl also stressed that cells were endowed with the ability to perform all kinds of movements.

In the world of the living cell, the only thing that is certain is change—movement occurs at all levels, from the molecular to the whole cell. While I was taught that plants, unlike animals, do not move, some plants can constantly change their position. Get a drop of pond water and look at it under the microscope. Watch a single-celled alga such as *Dunaliella* under the microscope (Fig. 1.5). See it swim? These plant cells are Olympic-class swimmers: they swim about 50 μm/s—equivalent to five body lengths per second. Not only can the cells swim but also change their motile behavior in response to external stimuli. When a bright flash of light (from the sun or a photographic flash) strikes swimming *Dunaliella* cells, like synchronous swimmers, they all swim backward for about a half second. From this observation, even a casual observer will conclude that individual cells have well-developed sensory systems that can sense and respond to external stimuli (Wayne et al., 1991).

In contrast to *Dunaliella*, some cells, particularly those of higher plants, remain static within an immobile cell wall. Yet, if you look inside the cell, you are again faced with movement. You see that the protoplasm dramatically flows throughout a plant cell, a phenomenon known as *cytoplasmic streaming* (Kamiya, 1959). Look at the giant internodal cell of *Chara* (Fig. 1.6). The cytoplasm rotates around the cell at about 100 μm/s. If you electrically stimulate the

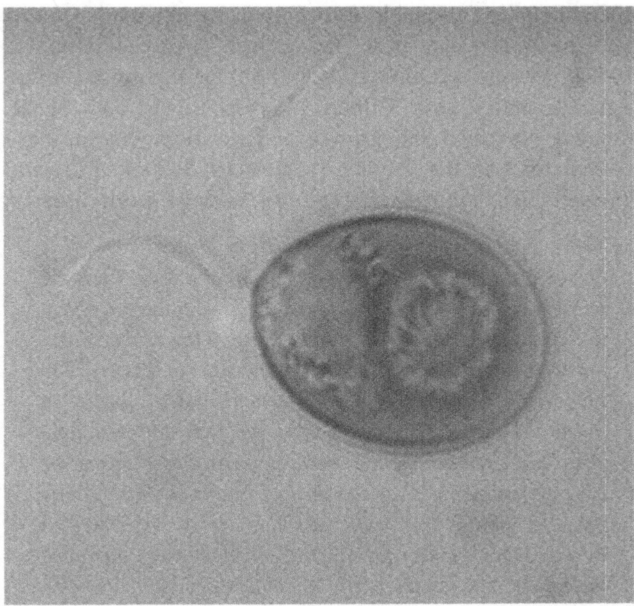

FIGURE 1.5 Photomicrograph of a swimming *Dunaliella* cell taken with Nomarski differential interference contrast optics.

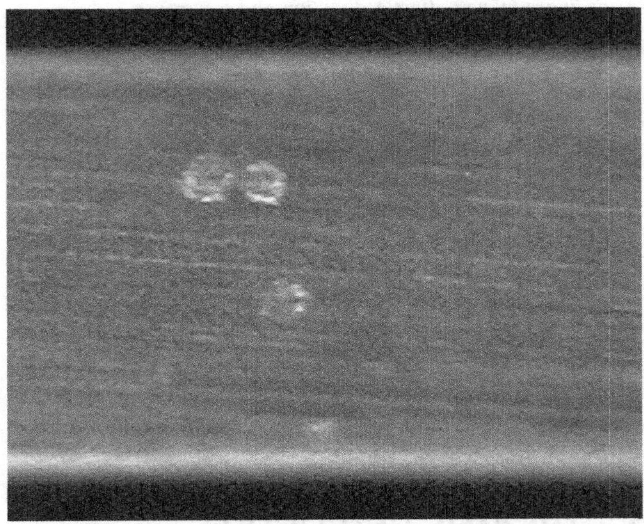

FIGURE 1.6 Photomicrograph of a portion of a giant internodal cell of *Chara* showing several nuclei being carried by cytoplasmic streaming.

cell, the cytoplasmic streaming ceases instantly. As the neurobiologists say, the cell is excitable and responds to external stimuli. In fact, action potentials were observed in characean internodal cells before they were observed in the nerve cells of animals (Cole and Curtis, 1938, 1939). The events that occur between electrical stimulation and the cessation of streaming are relatively well understood, and I discuss these throughout the book.

Lastly, take a look at the large single-celled plasmodium of the slime mold *Physarum* (Fig. 1.7; Coman, 1940; Kamiya, 1959; Carlisle, 1970; Konijn and Koevenig, 1971; Ueda et al., 1975; Durham and Ridgway, 1976; Chet et al., 1977; Kincaid and Mansour, 1978a,b; Hato, 1979; Dove and Rusch, 1980; Sauer, 1982; Dove et al., 1986; Bailey, 1997; Bozzone and Martin, 1998). Its cytoplasm streams at about 2000 μm/s. The force exerted by the streaming causes the plasmodium to migrate about 0.1 μm/s. Why does it move so slowly when streaming is so rapid? Notice that the cytoplasmic streaming changes direction in a rhythmic manner. The velocity in one direction is slightly greater than the velocity in the opposite direction. This causes the cell to migrate in the direction of the more rapid streaming. Because the plasmodium migrates toward food, the velocity of cytoplasmic streaming in each direction is probably affected by the gradient of nutrients. Nobody knows how this cell perceives the direction of food and how this signal is converted into directions for migration. Will you find out?

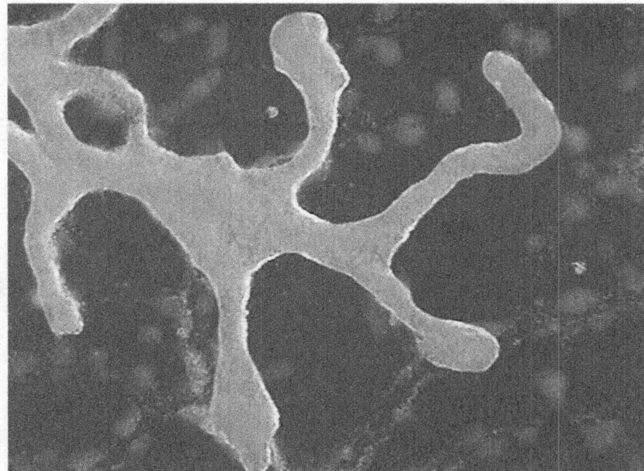

FIGURE 1.7 Dark-field photomicrograph of the slime mold *Physarum polycephalum*.

While looking at *Physarum*, notice that the protoplasm is not homogeneous but is full of relatively large round bodies rushing through the cell (Fig. 1.8). Is what you see the true nature of protoplasm, or are there smaller entities, which are invisible in a light microscope, that are also important in the understanding of cells? Edmund B. Wilson (1923) describes the power and the limitations of the light microscope in studying protoplasm:

> *When viewed under a relatively low magnification … only the larger bodies are seen; but as … we increase the magnification … we see smaller and smaller bodies coming into view, at every stage graduating down to the limit of vision … which in round numbers is not less than 200 submicrons. … Such an order of magnitude seems to be far greater than that of the molecules of proteins and other inorganic substances. … Therefore an immense gap remains between the smallest bodies visible with the microscope and the molecules of even the most complex organic*

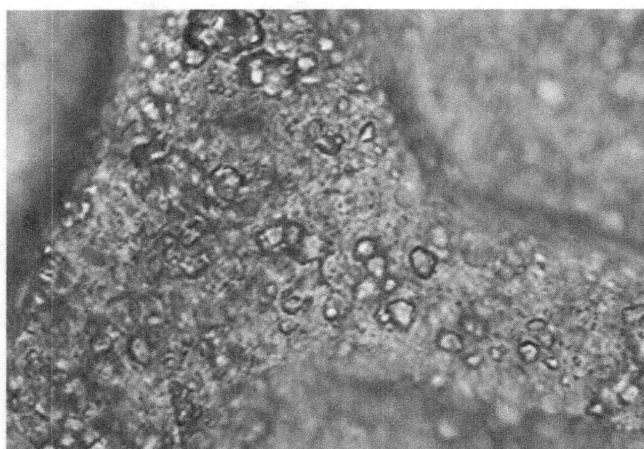

FIGURE 1.8 Bright-field photomicrograph of the streaming cytoplasm of the slime mold *Physarum polycephalum.*

substances. For these reasons alone … we should be certain that below the horizon of our present high-power microscopes there exists an invisible realm peopled by a multitude of suspended or dispersed particles, and one that is perhaps quite as complex as the visible region of the system with which the cytologist is directly occupied.

We have now arrived at a borderland, where the cytologist and the colloidal chemist are almost within hailing distance of each other—a region, it must be added, where both are treading on dangerous ground. Some of our friends seem disposed to think that the cytologist should halt at the artificial boundary set by the existing limits of microscopical vision and hand over his inquiry to the biochemist and biophysicist with a farewell greeting. The cytologist views the matter somewhat differently. Unless he is afflicted with complete paralysis of his cerebral protoplasm he can not stop at the artificial boundary set up by the existing limits of microscopical vision.

Looking at the streaming plasmodium of *Physarum* inspires a sense of awe and wonder about life. How is that single cell able to sense the presence of the oatmeal flake and move toward it? How does it generate the force to move from within? What kind of endogenous timekeeper is in the cell that allows the streaming cytoplasm to move back and forth with the rhythm and regularity of a beating heart (Time, 1937, 1940)? We will explore these and other questions about living cells. However, to cross the "artificial boundaries" and comprehend the nature of the living cell, it is necessary to develop knowledge of mathematics, chemistry, and physics as well as cytology, anatomy, physiology, genetics, and developmental biology. The practice of cell biology that incorporates these various disciplines is still in its adolescent period and is "treading on dangerous ground." As in any developing science, observations and measurements contain a given amount of uncertainty or "probable error," and the exactness of the measurements, and thus the science, evolves (Hubble, 1954). Perhaps cell biology is at the stage thermodynamics was a century ago. Gilbert Newton Lewis and Merle Randall described the growth and development of thermodynamics in the Preface to their 1923 book, *Thermodynamics and the Free Energy of Chemical Substances*:

There are ancient cathedrals which, apart from their consecrated purpose, inspire solemnity and awe. Even the curious visitor speaks of serious things, with hushed voice, and as each whisper reverberates through the vaulted nave, the returning echo seems to bear a message of mystery. The labor of generations of architects and artisans has been forgotten, the scaffolding erected for their toil has long since been removed, their mistakes have been erased, or have become hidden by the dust of centuries. Seeing only the perfection of the completed whole, we are impressed as by some superhuman agency. But sometimes we enter such an edifice that is still partly under construction; then the sound of hammers, the reek of tobacco, the trivial jests bandied from workman to workman, enable us to realize that these great structures are but the result of giving to ordinary human effort a direction and a purpose.

Science has its cathedrals.

Cell biology is a young, vibrant, growing science, the beginnings of which took place in the early part of the 19th century when scientists, including Schleiden (1853), pondered what regular element may underlie the vast array of plant forms from "the slender palm, waving its elegant crown in the refreshing breezes … to the delicate moss, barely an inch in length, which clothes our damp grottos with its phosphorescent verdue." Schleiden felt that "we may never expect to be enabled to spy into the mysteries of nature, until we are guided by our researches to very simple relations … the simple element, the regular basis of all the various forms."

1.2 THE BASIC UNIT OF LIFE

Prior to 1824, organic particles or a vegetative force that organized organic particles were considered by some prominent scientists including Gottfried Leibniz, Comte de Buffon, and John Needham to be the basic unit of life (Roger, 1997). In fact, John Needham (1749) and John Bywater (1817, 1824) observed these living particles in infusions of plant and animal material that they placed under the microscope. Bywater observed that they writhed about in a very active manner and conjectured that the immediate source of the movement was thermal energy, which originated from the "particles of [sun]light which come in contact with the earth, and have lost their rapid momentum." Bywater considered sunlight to carry the vital force, and concluded "that the particles of which bodies are composed, are not merely inert matter, but have received from the Deity certain qualities, which render them actively instrumental in promoting the physical economy of the world."[1]

Henri Dutrochet (1824) emphasized the importance of the cell, as opposed to living particles or the whole organism, as the basic unit of life. Dutrochet came to this conclusion from his microscopical observations, by which he observed "plants are derived entirely from cells, or of organs which are obviously derived from cells." He extended his observations to animals and concluded that all organic beings are "composed of an infinite number of microscopic parts, which are only related by their proximity" (quoted in Rich, 1926). More than a decade later, Dutrochet's cell theory was promoted by Schleiden and Schwann. Schleiden (1838), a botanist, wrote

> *Every plant developed in any higher degree, is an aggregate of fully individualized, independent, separate beings, even the cells themselves. Each cell leads a double life: an independent one pertaining to its own development alone, and another incidental, in so far as it has become an integral part of a plant. It is, however, easy to perceive that the vital process of the individual cells must form the very first, absolutely indispensable fundamental basis.*

Likewise, Schwann (1838), a zoologist, concluded that "the whole animal body, like that of plants, is thus composed of cells and does not differ fundamentally in its structure from plant tissue." Thanks to the extensive research, and active promotion by Schleiden and Schwann; by the end of the 1830s, Dutrochet's concept that the cell is the basic unit of all life became well established, accepted, and extended to emphasize the interrelationships between cells. The expanded cell theory provided a framework to understand the nature of life and its origin and continuity.

We often divide various objects on Earth into two categories: the living and the lifeless. Therefore, the investigation of cells may provide us with a method to understand the question, "What is life?" We often characterize life as something that possesses attributes that the lifeless lack (Beale, 1892; Blackman, 1906; Tashiro, 1917; Osterhout, 1924; Harold, 2001). The power of movement is a distinctive aspect of living matter, where the movement has an internal rather than an external origin. Living matter generates electricity. Living matter also takes up nutrients from the external environment and, by performing synthetic reactions at ambient temperatures, converts the inorganic elements into living matter. Living matter also expels the matter that would be toxic to it. The ability to synthesize macromolecules from inorganic elements allows growth, another characteristic of living matter. Living matter also contains information, and thus has the ability to reproduce itself, with near-perfect fidelity. Lastly, living matter is self-regulating. It is capable of sensing and responding to environmental signals to maintain a homeostasis (Cannon, 1932, 1941) or to adjust to new conditions by entering metastable states, or other states, in a process known as allostasis (Spencer, 1864; Emerson, 1954; Sapolsky, 1998).

The above-mentioned properties are characteristic of living things and their possession defines a living thing. Mathews (1916) notes, "When we speak of life we mean this peculiar group of phenomena; and when we speak of explaining life, we mean the explanation of these phenomena in the terms of better known processes in the nonliving." There are entities like viruses that exhibit some, but not all, of the characteristics of life. Are viruses the smallest living organism as the botanist Martinus Beijerinck thought when he isolated the tobacco mosaic virus in 1898 or are they the largest molecules as the chemist Wendell Stanley thought when he crystallized the tobacco mosaic virus in 1935 (Stanley and Valens, 1961)? While the distinction between nonliving and living is truly blurred (Anonymous, 1905; Twain, 1923; Pirie, 1938; Baitsell, 1940), the cell in general is the smallest unit capable of performing all the processes associated with life.

For centuries, people believed that the difference between living and nonliving matter arises from the fact that living matter possesses a vital force, also known as the *vis vitalis*, a purpose, a soul, Maxwell's demon, a spirit, an archeus, or an entelechy (Reil, 1796; Loew, 1896; Lovejoy, 1911; Ritter, 1911; Driesch, 1914, 1929; Frankl, 1973; Waddington, 1977). According to the view of the "vitalists and dualists," the laws of physics and chemistry used to describe inorganic nature are, in principle, incapable of describing living things. By contrast, mechanists, materialists, mechanical materialists, and monists believe that there is a unity of nature and a continuum between the nonliving and the living—and all things, whether living or not, are made of the same material and are subject to the same physical laws and mechanisms (Bernard, 1865; Dutrochet, 1824; von Helmholtz, 1903; Koenigsberger, 1906; Rich, 1926; Brooks and Cranefield, 1959). We will ask along with Erwin Schrödinger (1946), "How can the events in space and time which take place within the spatial boundary of a living organism be accounted for by physics and chemistry?"

Mary Shelley (1818) wrote about the potential of the materialistic/mechanical view and the ethics involved in experimentation on the nature of life when she described how Victor Frankenstein discovered that life could emerge spontaneously when he puts together the right combination of matter and activated it with electrical energy. In the

1. Robert Brown (1828, 1829) independently observed the movement of particles. However, Brown, in contrast to Bywater, did not consider the movement of the particles to be a sign of vitality or life, but just a physical process.

materialist/mechanical view, living matter is merely a complex arrangement of atoms and molecules, performing chemical reactions and following physical laws. Thus, according to this view, the laws of chemistry and physics are not only applicable but also essential to the understanding of life (Belfast Address, Tyndall, 1898). Claude Bernard (1865) believed that "the term 'vital properties' is only provisional; because we call properties vital which we have not yet been able to reduce to physico-chemical terms; but in that we shall doubtless succeed some day." An understanding of the relationship between nonliving matter and living matter underlies the understanding of the relationship between the body and the soul and the definition of personal identity, free will, and immortality (Dennett, 1978; Perry, 1978; Popper and Eccles, 1977; Eccles, 1979).

Thomas H. Huxley (1890) explains

The existence of the matter of life depends on the pre-existence of certain compounds; namely, carbonic acid, water and ammonia. Withdraw any one of these three from the world, and all vital phenomena come to an end. They are related to the protoplasm of the plant, as the protoplasm of the plant is to that of the animal. Carbon, hydrogen, oxygen, and nitrogen are all lifeless bodies. Of these, carbon and oxygen unite, in certain proportions and under certain conditions, to give rise to carbonic acid; hydrogen and oxygen produce water; nitrogen and hydrogen give rise to ammonia. These new compounds, like the elementary bodies of which they are composed, are lifeless. But when they are brought together, under certain conditions they give rise to the still more complex body, protoplasm, and this protoplasm exhibits the phenomena of life.

When hydrogen and oxygen are mixed in a certain proportion, and an electric spark is passed through them, they disappear, and a quantity of water … appears in their place. … At 32° Fahrenheit and far below that temperature, oxygen and hydrogen are elastic gaseous bodies. … Water, at the same temperature, is a strong though brittle solid. … Nevertheless, … we do not hesitate to believe that … [the properties of water] result from the properties of the component elements of the water. We do not assume that a something called 'aquosity' entered into and took possession of the oxide of hydrogen as soon as it was formed. … On the contrary, we live in the hope and in the faith that, by the advance of molecular physics, we shall by and by be able to see our way clearly from the constituents of water to the properties of water, as we are now able to deduce the operations of a watch from the form of its parts and the manner in which they are put together.

Is the case in any way changed when carbonic acid, water, and ammonium disappear, and in their place, under the influence of pre-existing living protoplasm, an equivalent weight of the matter of life makes its appearance? … What better philosophical status has 'vitality' than 'aquosity'?

With a like mind, Edmund B. Wilson (1923) concluded his essay on "The Physical Basis of Life" by saying

I do not in the least mean by this that our faith in mechanistic methods and conceptions is shaken. It is by following precisely these methods and conceptions that observation and experiment are every day enlarging our knowledge of colloidal systems, lifeless and living. Who will set a limit to their future progress? But I am not speaking of tomorrow but of today; and the mechanist should not deceive himself in regard to the magnitude of the task that still lies before him. Perhaps, indeed, a day may come (and here I use the words of Professor Troland) when we may be able 'to show how in accordance with recognized principles of physics a complex of specific, autocatalytic, colloidal particles in the germ-cell can engineer the construction of a vertebrate organism'; but assuredly that day is not yet within sight. … Shall we then join hands with the neo-vitalists in referring the unifying and regulatory principle to the operation of an unknown power …? … No, a thousand times, if we hope really to advance our understanding of the living organism.

In the spirit of E. B. Wilson and many others, we will begin our study of the cell by becoming familiar with its chemical and physical nature. During our journey, I will neither take the extreme perspective of Richard Dawkins (1995, 2006) and Edward O. Wilson (1998) that life can be reduced to the laws of physics nor will I take the extreme perspective of the electrophysiologist Emil du Bois-Reymond (1872), who proclaimed that there are absolute limits to our knowledge of nature, and moreover, he would not try to find these limits using science ("*Ignoramus et ignorabimus*"). I will also not take the perspective offered by the Copenhagen School of Physics that states that "*We now know that we shall never know*" (Seifriz, 1943a) and blurs the distinction between living and nonliving when it states that until you observe a cell that has been kept from view, that cell is both living and dead according to the rules of quantum superposition. This view was ridiculed by Erwin Schrödinger in his story of the cat in a box (Gribbin, 1984, 1995). I will try to take a middle ground (Heitler, 1963), looking at the cell physicochemically without losing sight of the miracle, value, and meaning of life (Bischof, 1996; Berry, 2000).

Max Planck wrote, "In my opinion every philosophy has the task of developing an understanding of the meaning of life, and in setting up this task one supposes that life really has a meaning. Therefore whoever denies the meaning of life at the same time denies the precondition of every ethics and of every philosophy that penetrates to fundamentals" (quoted in Heilbron, 1986). As discoveries made by cell biologists become techniques used by biotechnologists to create new choices for humanity, we realize that our own discoveries can have profound effects on the meaning of life.

1.3 THE CHEMICAL COMPOSITION OF CELLS

Living cells are made out of the same elements found in the inorganic world. However, out of the more than 100 elements available on Earth, cells are primarily made out of carbon, hydrogen, and oxygen (Mulder, 1849; see also Table 1.1). According to Lawrence Henderson (1917), it is the special physicochemical properties of these elements and their compounds that allow life, as we know it, to exist.

The vast majority of the oxygen and hydrogen in the cells exists in the cell as water, which provides the milieu in which the other chemicals exist (Ball, 2000; Franks, 2000). The large numbers of atoms of carbon, oxygen, hydrogen, nitrogen, sulfur, and phosphorous found in cells are for the most part combined into macromolecules. The macromolecular composition of a "typical" bacterial cell calculated by Albert Lehninger in his book *Bioenergetics* (1965) is shown in Table 1.2.

The cell uses these various macromolecules to build the machinery of the cell. A cell has various components that help it to transform information into structure, and it has various structures to help it convert mass and energy into work so it can maintain a homeostasis, move, grow, and reproduce. We will begin discussing the organization of the cell in Chapter 2. For now, let us get a sense of scale.

TABLE 1.1 Atomic Composition of the Large Spore Cells of *Onoclea*

Element	Percent Dry Weight	nmol/mg Dry Weight	Atoms/Cell
C	58.59	48,784	4×10^{15}
O	21.25	13,281	1×10^{15}
H	7.76	76,942	6×10^{15}
N	4.59	3277	2×10^{14}
P	0.82	255	2×10^{13}
K	0.70	179	1×10^{13}
S	0.53	164	1×10^{13}
Mg	0.34	140	1×10^{13}
Na	0.23	100	8×10^{12}
Ca	0.20	50	4×10^{12}
Cl	0.11	31	2×10^{12}
Co	0.04	7	5×10^{11}
Fe	0.02	4	3×10^{11}
Ni	0.01	2	2×10^{11}
Mn	0.01	1	8×10^{10}
Zn	0.01	1	8×10^{10}
Cu	0.01	1	8×10^{10}

From Wayne, R., Hepler, P.K., 1985b. The atomic composition of *Onoclea sensibilis* spores. Amer. Fern J. 75, 12–18.

TABLE 1.2 Macromolecular Composition of a Bacterial Cell

Chemical Component	Percent of Dry Weight	Number of Molecules per Cell
DNA	5	4
RNA	10	15,000
Protein	70	1,700,000
Lipid	10	15,000,000
Polysaccharides	5	39,000

From Lehninger, A.L., 1965. Bioenergetics. W. A. Benjamin, New York.

Before we discuss the scale of living cells, let us discuss an experiment described by Irving Langmuir to get a feeling for the size of a macromolecule, for example, a lipid (Langmuir, 1917; Taylor et al., 1942; see also Appendix 1). When you place a drop (10^{-7} m^3) of lipid such as olive oil on the surface of a trough full of water, the olive oil will spread out and form a monolayer. As the lipid is amphiphilic, in that it has both a hydrophilic end (glycerol) and a hydrophobic or lipophilic end (the hydrocarbon derived from oleic acid), the hydrophilic glycerol end will dissolve in the water and the hydrophobic hydrocarbon end will stick into the air. We can use this observation to determine the size of the lipid molecules—but how?

If we know the volume of oil we started with and the area of the monolayer, we can estimate the thickness of the oil molecules. For example, Benjamin Franklin found that a teaspoonful[2] of oil covers a surface of about half an acre (Tanford, 1989). As a teaspoonful of oil contains approximately 2×10^{-6} m^3 of oil and a half acre is approximately 2000 m^2, the thickness of the monolayer and thus the length of the molecule, obtained by dividing the volume by the area, is approximately 1 nm (Laidler, 1993).

Franklin never made this calculation, probably because at the time the concept of molecules had not been developed. However, now that we understand the molecular organization of matter, we can go even further in our analysis. For example, if we know the density (ρ) and molecular mass (M_r) of the oil (e.g., $\rho = 900$ kg/m^3 and $M_r = 0.282$ kg/mol for olive oil), we can calculate the number of molecules in the drop using dimensional

2. For reference, 1 mL is one-millionth of a cubic meter, and 1 L is one-thousandth of a cubic meter.

analysis and Avogadro's number (6.02×10^{23} molecules/mol; Avogadro, 1837; Deslattes, 1980):

$$\left(2 \times 10^{-6}\text{m}^3\right)\left(900\text{ kg m}^{-3}\right)\left(0.282\text{ kg mol}^{-1}\right)^{-1}$$
$$\left(6.02 \times 10^{23}\text{ molecules mol}^{-1}\right) = 3.8 \times 10^{21}\text{ molecules}$$

Because we know how many molecules we applied to the water and the area the oil takes up, we can calculate the cross-sectional area of each molecule. We obtain the cross-sectional area of each molecule (5.3×10^{-19} m^2) by dividing the area of the monolayer by the number of molecules in it. If we assume that the molecules have circular cross sections, we can estimate their diameter ($2r$) from their area (πr^2). We get a diameter of approximately 0.8 nm. We can do the experiment more rigorously using pipettes and a Langmuir trough, but the answers are not so different.

It is amazing how much you can learn with a teaspoon and a ruler if you apply a little algebra! Insightful scientific results do not necessarily require expensive commercial equipment. Ernest Rutherford, who discovered the nucleus of the atom, was known to say (Andrade, 1964), "We've got no money, so we've got to think." You have just deduced the size of a molecule from first principles using dimensional analysis! Lipids are important in the structure of cellular membranes. However, as membranes are exposed to aqueous solutions on both sides, the lipids form double layers also known as *bilayers*. Membranes are also composed of proteins that have characteristic lengths on the order of 5 nm. As I will discuss in Chapter 2, the diameters of proteins can be determined from studies on their rate/of diffusion. Can you estimate the thickness of a membrane composed of proteins inserted in a single lipid bilayer?

1.4 A SENSE OF CELLULAR SCALE

To understand cells we must get a grasp of their dimensions because, while there are many similarities between the living processes of cells and multicellular organisms such as ourselves, of which we are most familiar, we will find that there are limits to the similarities between single cell and multicellular organisms that must be taken into consideration (Hill, 1926).

How small can a cell be? The lower size limit of a cell is determined by the minimal number and size of the components that are necessary for an autonomous existence (Koch, 1996). To live autonomously, a cell has to perform approximately 100 metabolic reactions involved with primary metabolism (e.g., the biosynthesis of amino acids, nucleotides, sugars, and lipids, as well as the polymers of these molecules) and transport. Therefore, about 100 different enzymes, with an average diameter of 5 nm, and the corresponding amount of substrate molecules must be present. In addition, 1 DNA molecule, 100 mRNA molecules, 20 tRNA molecules, and several rRNA molecules are needed to synthesize these enzymes. If we assume that there is one copy of each molecule, we can estimate the volume of the molecules and the water needed to dissolve them. To keep the enzymes together, the cell must have a limiting membrane. If we add the dimensions of a plasma membrane (10 nm thick), we find that the minimum cell diameter is about 65 nm. The smallest known organisms are *Rickettsia* (Bovarnick, 1955), ultramicrobacteria (Levin and Angert, 2015; Luef et al., 2015), and various mycoplasmas (Maniloff and Morowitz, 1972; Hutchison et al., 1999), which have diameters of approximately 100 nm.

There is a limit as to how big a cell can be. Assume that a cell is spherical. The surface area of a cell with radius r will be given by $4\pi r^2$ and its volume will be given by (4/3) πr^3. Thus, its surface-to-volume ratio will be $3/r$, and as the cell gets larger and larger, its surface-to-volume ratio will decrease exponentially. This limits the cell's ability to take up nutrients and to eliminate wastes (Table 1.3).

Some cells are very large. For example, an ostrich egg can be 10.5 cm in diameter. In this case, a large portion of the intracellular volume is occupied by the yolk. The yolk is "inert" relative to the cytoplasm. In the case of large plant cells, the vacuole functions as an inert space filler. Haldane (1985) illustrates the bridge between mathematics and biology beautifully in his essay "On Being the Right Size." In it he writes, "Comparative anatomy is largely the story of the struggle to increase surface in proportion to volume."

TABLE 1.3 Relationship Between Surface and Volume of a Sphere

Radius (r, in m)	Surface Area (A, in m^2) ($4\pi r^2$)	Volume (V, in m^3) ($(4/3)\pi r^3$)	Surface-to-Volume Ratio (A/V, in m^{-1}) = ($3/r$)
0.1	0.126	0.0042	30.0
1	12.56	4.19	3.0
10	1256.64	4188.79	0.3
100	125663.71	4188790.21	0.03
1000	12566370.61	4,188,790,205	0.003

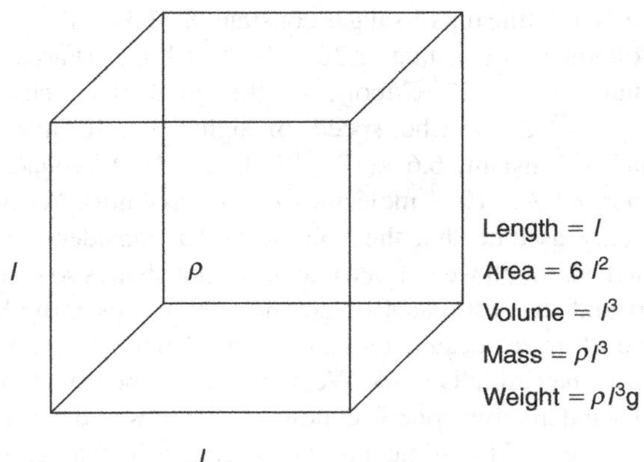

FIGURE 1.9 A geometrical model of a cell.

How long is a typical plant cell? While their lengths vary from a few micrometers in meristematic cells to 1.5 mm in root hairs and 25 cm in phloem fibers (Haberlandt, 1914; Esau, 1965; Ridge and Emons, 2000; Bhaskar, 2003), for the present we will assume that a typical plant cell is a cube where each side has a length of 10^{-5} m. Such a typical cell has a surface area of 6×10^{-10} m^2 and a volume of 10^{-15} m^3.

How much does a cell weigh? We can estimate its weight from "first principles." A cell is composed mostly of water, so let us assume that it is made totally out of water, which has a density (ρ) of 10^3 kg/m^3. Using dimensional analysis and multiplying the volume of the cell by its density, we see that the mass of the cell is 1×10^{-12} kg or 1 ng (Fig. 1.9). Multiplying its mass by the acceleration due to gravity (*g*), we find that it weighs 9.8×10^{-12} N (or 9.8 pN). As the actual density of the protoplasm is about 1015 kg/m^3, the weight of a single cell is 9.95 pN. Our approximation was not so bad, was it?

We often talk about the importance of pH in enzyme reactions and the energetics of cells. The pH is a measure of the concentration of protons, which are ionized hydrogen atoms. Concentration is a measure of the amount of a substance in moles divided by the volume. Usually we do not realize how small that volume is when we talk about cells. So, to get a feel for cellular volumes, let us calculate how many protons there are in a mitochondrion, an organelle that is involved in molecular free energy (*E*, in Joules [J]) transduction. A mitochondrion has a volume of approximately $(10^{-6}\ m)^3$ or 10^{-18} m^3, a value that is about the size of a prokaryotic cell and one-thousandth the size of a typical eukaryotic cell.

Consider that the mitochondrion has an internal pH of 7. Because pH is -log [H^+], at pH 7 there are 10^{-7} mol H^+/l, which is equal to 10^{-4} mol/m^3. Now we will need to use Avogadro's number as a conversion factor that relates the number of particles to the number of moles of that particle. Now that all the units match, we will use dimensional analysis to calculate how many protons are there in the mitochondrion (Fig. 1.10):

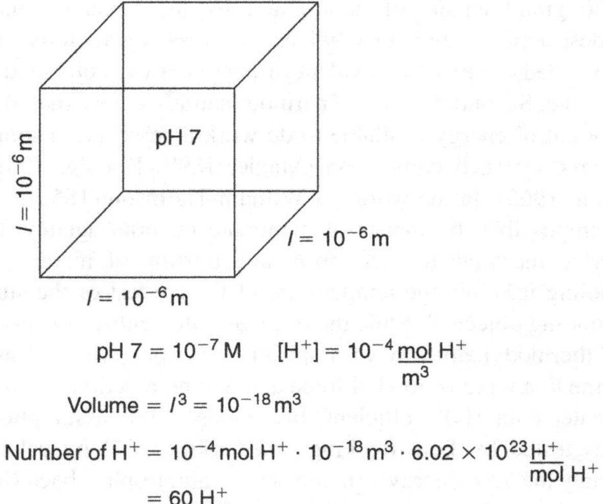

FIGURE 1.10 A calculation of the number of H^+ in a mitochondrion.

$$10^{-4}\ \text{mol m}^{-3}\left(10^{-6}\ \text{m}\right)^3\left(6.02\ \times 10^{23}\text{protons m}^{-1}\right)$$
$$= 60\ \text{protons}$$

If the pH of the mitochondrion is raised to 8, how many protons are now in the mitochondrion?

$$10^{-5}\ \text{mol m}^{-3}\left(10^{-6}\ \text{m}\right)^3\left(6.02\ \times 10^{23}\text{protons m}^{-1}\right)$$
$$= 54\ \text{protons}$$

Thus, 54 protons would have to leave the mitochondrion to raise the pH from pH 7 to pH 8. Interestingly, while it is common knowledge to every introductory biology student that energy conversion in the mitochondrion involves the movement of protons, have you ever realized how few protons actually move? Now we are beginning to understand the scale of the cell (Peters, 1929; McLaren and Babcock, 1959; Chance, 1967; McCabe, 1967; Mitchell, 1967).

1.5 THE ENERGETICS OF CELLS

The molecular free energy (*E*, in J) is the cellular currency, and all cellular processes can be considered as free energy—transduction mechanisms that convert one form of free energy to another according to the First Law of Thermodynamics proposed by the physician Julius Robert Mayer and demonstrated by the brewer James Joule. That is, while energy can be converted from one form to another in various processes, it is conserved and thus cannot be created or destroyed (Joule, 1852, 1892; Grove et al., 1867; Maxwell, 1897; Lenard, 1933; Lindsay, 1973; Fenn, 1982; Müller, 2007). In the words of James Joule (1843), "the grand agents of nature are, by the Creator's fiat, indestructible; and that whatever mechanical force is expended, an exact equivalent of heat is always obtained."

The Second Law of Thermodynamics states that the amount of energy available to do work is lessened to some degree by each conversion (Magie, 1899; Koenig, 1959; Bent, 1965). In the words of William Thomson (1852), "It is impossible, by means of inanimate material agency, to derive mechanical effect from any portion of matter by cooling it below the temperature of the coldest of the surrounding objects." While the original statements of the laws of thermodynamics have a spiritual overtone, we will assume that there is no vital force and that no reactions can be greater than 100% efficient. Interestingly, this assumption was tested by Baas-Becking and Parks (1927) by calculating the free-energy efficiencies of autotrophic bacteria. They never found thermodynamic efficiencies greater than 100% and concluded that the laws of thermodynamics apply to living systems.

That the First Law of Thermodynamics applies to living things should be of no surprise. Indeed, the First Law of Thermodynamics, like many other physical principles we will discuss throughout this book (e.g., Fick's Law, Poiseuille's Law, Brownian motion, sound waves involved in hearing, light waves involved in vision), has its roots in biological observations. Mayer, while spending the summer of 1840 in Java, noticed that the venous blood of the people was bright red and not bluish, as it was in people of temperate regions. He concluded that the venous blood was so bright because less oxidation was needed to maintain the body temperature in hot climates compared with cold ones, and as a result, the excess oxygen remained in the venous blood. Mayer also realized that people not only generate heat inside their bodies but also outside as well by performing work, and he postulated that there is a fixed relationship between the amount of food oxidized and the total amount of heat generated by a body. He wrote: "I count, therefore, upon your agreement with me when I state as an axiomatic truth, that during vital processes, the conversion only and never the creation of matter or force occurs" (quoted in Tyndall, 1898).

Using a thermometer, James Joule observed that electrical energy, mechanical energy, and chemical energy produced heat, and then he developed the quantitative relationships between the different forms of energy in terms of the equivalent amount of heat generated. Energy is a particularly convenient measure to compare various seemingly unrelated things because energy, unlike force and velocity, is a scalar quantity and not a vector quantity. Thus, the difference in energy over time and space can be determined with simple algebra. Thus, we will typically convert measurements of force, the electric field, concentration, etc. into energy units (Joules) by using a number of coefficients that transform numbers with given units into numbers with energy units. These include g the acceleration due to gravity ($9.8\ m/s^2$), R (the universal gas constant, $8.31\ J\ mol^{-1}\ K^{-1}$), k (Boltzmann's constant, 1.38×10^{-23} J/K), F (Faraday's constant, 9.65×10^4 C/mol), e (the elementary charge, 1.6×10^{-19} C), c (the speed of light, 3×10^8 m/s), h (Planck's constant, 6.6×10^{-34} J s), and N_A (Avogadro's number, 6.02×10^{-3} molecules (or atoms)/mol). We will implicitly assume that the volume under consideration is defined, although we will see that this is not always so simple to do and that estimates of geometrical values provide a source of error because they are more difficult to estimate than one may initially think. We will also assume that all cells are at standard atmospheric conditions of 298K and 0.1 MPa of pressure, and for all intents and purposes, the temperature and pressure remain constant. We will also assume that the reactions take place at equilibrium. Because life is best approximated by steady-state conditions, where there is a constant input of energy, and equilibrium only occurs at death, this is clearly a first order and limited approximation (Katchalsky and Curran, 1965; Schmitt and Livingston, 1972, 1973). Using these assumptions, in later chapters, we will determine the minimum energy capable of performing mechanical work to move a vesicle, chromosome, or cell; osmotic work to move a solute; or biosynthetic work to form new chemical bonds.

The potential energy of a given mass equals the product of force and distance. The gravitational potential energy of a protoplast settling inside a static extracellular matrix can be converted into the potential energy of a stretched springlike protein in the extracellular matrix if a helical, springlike region of the protein is attached to both the plasma membrane and the extracellular matrix of the settling protoplast. Let us determine the potential energy of the falling of a protoplast. The potential energy equals *force* × *distance*, so if a cell that weighs 9.95×10^{-12} N falls 1 nm in a gravitational field (i.e., changes its position by -1 nm), it makes available 9.95×10^{-21} J of energy that can be used to do work. Some of the potential energy will be degraded as a result of friction, and thus the potential energy in the springlike protein will be somewhat less than the gravitational energy of the protoplast. The potential energy released by the falling protoplast is used for the perception of gravity (Fig. 1.11; Wayne and Staves, 1997).

What are the minimum and maximum values for molecular free energies in cellular processes (Fig. 1.12)? The unitary processes that utilize the greatest quantity of energy are typically light-activated processes. One such process is photosynthesis, which uses the radiant energy of sunlight to convert water and carbon dioxide to carbohydrates. The energy in a photon of light depends on its wavelength (λ) and is given by the equation: $E = hc/\lambda$. Photosynthesis utilizes both blue and red light. These colors represent photons with the highest and lowest energy contents, respectively. Because blue light has a wavelength of 450 nm and red light has a wavelength of 650 nm, the

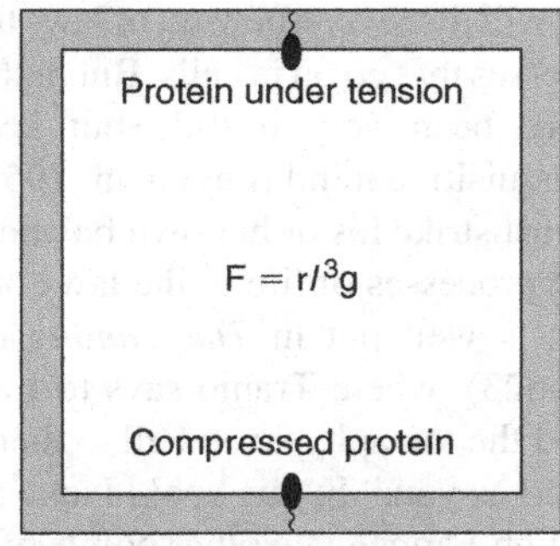

FIGURE 1.11 Potential energy of a protoplast in a gravitational field.

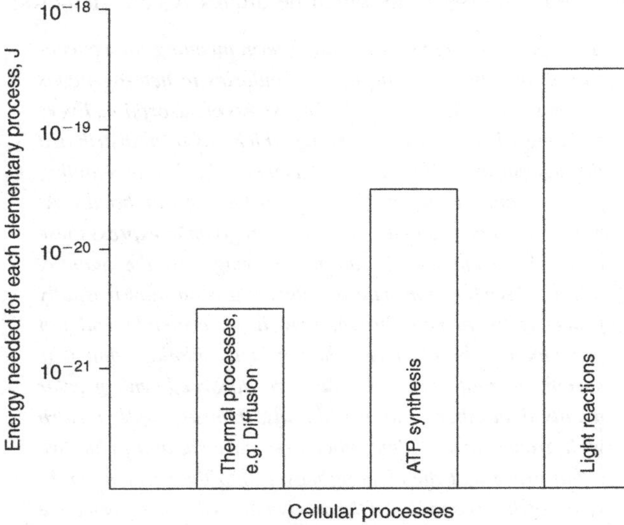

FIGURE 1.12 A comparison of the energetics of some cellular processes.

energy of a photon of blue and red light is 4.4×10^{-19} and 3.0×10^{-19} J, respectively. Because light-driven processes are high-energy reactions in cells, we might expect a typical single reaction to require or release free energy on the order of less than 4×10^{-19} J.

What is the minimum free energy that may be involved in a cellular reaction? The free energy generated by the collisions of molecules in the cell at the ambient temperature is approximately equal to kT, which is (1.38×10^{-23} J/K) (298 K) $= 4.1 \times 10^{-21}$ J at room temperature. An input of free energy lower than this cannot be utilized by a receptor in a cell to do work because the effect of such small energies will be overshadowed by random changes in the state of the receptor due to thermal collisions between the receptor and the water or lipid molecules that surround it.

The free energy of single reactions in a cell thus falls between 4×10^{-21} and 4×10^{-19} J. For a reference, let us look at adenosine triphosphate (ATP), a molecule involved in the activation of many molecules in the cell (Lipmann, 1941). The hydrolysis of one ATP molecule liberates a maximum of 8×10^{-20} J of free energy, which, if coupled to other processes, is capable of doing work (Rosing and Slater, 1972; Shikama and Nakamura, 1973; Jencks, 1975). This is only an order of magnitude greater than the energy of thermal motion. Because many reactions that require an input of free energy (i.e., endergonic reactions) are coupled to the hydrolysis of ATP, many unitary, endergonic cellular reactions will require energies on the order of 8×10^{-20} J to proceed. I am calculating the free energies per molecule to stress the small number of molecules found in cells compared with the number found in experiments with ideal gases and to help us visualize the possible mechanisms of cellular reactions. I am assuming that the average energy of any molecule is equal to the average energy of all the molecules. The free energy in a molecule is related to the free energy in a mole of molecules by Avogadro's number, as Boltzmann's constant, k, is equal to R, the universal gas constant, divided by N_A. Therefore, RT gives the free energy in a mole of molecules, and kT gives the free energy in one molecule.

1.6 ARE THERE LIMITS TO THE MECHANISTIC VIEW?

Many people have applied the laws of thermodynamics to cells. These laws are extremely helpful in all aspects of cell biology from calculating the permeability of molecules passing through the membrane to calculating the free energy liberated from the hydrolysis of ATP. Thermodynamics allows us to calculate equilibrium, affinity, and dissociation constants. Thermodynamics provides the boundary conditions, which the reactions of the cell must obey, independent of the detailed physical mechanisms. However, thermodynamics does not tell us anything about the mechanisms of the processes. In our everyday experience, kinetic theory and statistical mechanics provide a model to explain thermodynamics (Clausius, 1879; Maxwell, 1897; Loeb, 1961; Jeans, 1962; Boltzmann, 1964; Brush, 1983; Tucker, 1983; Garber et al., 1986; Schroeder, 2000). However, the assumptions that the models on which statistical mechanics are based may not be met in the cell (Schrödinger, 1946, 1964). According to Albert Szent-Györgyi von Nagyrapolt (1960; Bendiner, 1982)

> *There is a basic difference between physics and biology. Physics is the science of probabilities. ... Biology is the science of the improbable and I think it is on principle that the body works only with reactions that are statistically improbable. ... I do not mean to say that biological reactions do not obey physics. In the last instance it is physics which has to explain them, only over a detour which may seem entirely improbable on first sight.*

According to Erwin Schrödinger (1946), there should be about 10^{20} molecules or ions present before the

TABLE 1.4 Relationship Between Number of Molecules and Statistical Noise

Number of Molecules (n)	Noise ($n^{1/2}$)	Proportion of Noise ($n^{1/2}/n$)
10^{20}	10^{10}	10^{-10}
10^{10}	10^{5}	10^{-5}
10^{6}	10^{3}	10^{-3}
10^{3}	31.6	0.03
10^{2}	10	0.1
30	5.5	0.18
10	3.16	0.32

predictions based on the laws of statistical mechanics are accurate. The need for large numbers results from the fact that the statistical noise is equal to $\sqrt{n}$, where n is the number of molecules or ions (Table 1.4). That is, if there were on the average 1,000,000 molecules in a given sample volume, on sampling that volume you may find between 999,000 and 1,001,000 molecules, and thus the relative error is 0.1%. Likewise, if there were on the average 100 molecules in a given sample volume, on sampling you would find 90–110, and the relative error would be 10%. We can see from these calculations that the number of protons in a cell or mitochondrion is small compared with the number required for accurate predictions using statistical mechanics (Guye, n.d.). Even in the large spore cell of *Onoclea*, if we count all the atoms, there are 10,000 times too few to use reliably the laws of statistical mechanics.

Can we use statistical mechanics to understand cells? Yes and no. Perhaps it is possible that cells function on a statistical basis where the noise level is typically 10%. We should consider statistical mechanics to be a first approximation, as the assumptions on which it is based do not take into consideration the scale of a single cell. Furthermore, the cell is not just a reaction vessel but a polyphasic system composed of a number of compartments, solid-state supports, and transport systems (e.g., membranes and cytoskeletal elements) that facilitate biochemical reactions in cells (Peters, 1929, 1937; Needham, 1936). Because of the complex structure and small numbers of atoms or molecules within each compartment, we may need a solid-state, quantum mechanical model to fully understand the nature of the living cell (Donnan, 1928, 1937). The Planck distribution as opposed to the Boltzmann distribution has the advantage of describing steady-state processes, consistent with life rather than equilibrium processes, consistent with death (Wayne, 2016a). According to Niels Bohr (1950), mechanistic and vitalistic arguments are complementary and must be reconciled to understand life.

Perhaps you will discover a new set of laws that will better predict the processes that go on in cells. But first, learn the old laws—they have been very useful—but keep an open, skeptical, and inquisitive mind (Feynman, 1955, 1969).

Everyone must strike his or her own balance in reducing the complicated processes of life to the laws of physics and chemistry. This is well put in *The Taming of the Shrew* (Shakespeare, 1623), where Tranio says to Lucentio, "The mathematics and the metaphysics—Fall to them as you find your stomach serves you." In this book, I take a reductionist approach, although I appreciate other points of view (Clark, 1890; Stokes, 1891, 1893; Duncan and Eakin, 1981; Campbell, 2014). The absurdity of blindly applying the laws of physics to complicated situations is well described by Needham (1930), in which he quotes Albert Mathews:

> *Adsorption is a physico-chemical term meaning the concentration of substances at phase-boundaries in heterogeneous systems. Dressing can be called a process of adsorption. Every morning when we dress, clothing which has been distributed throughout our environment—dispersed in the surrounding phase—concentrates itself at the surface of our bodies. At night the process is reversed. We might go on to express these events by a curve or isotherm, showing how the quantity adsorbed is a function of the amount in the room, how it usually proceeds to an equilibrium, how it is reversible and not accompanied by chemical change in the clothes, that it is specific in that certain clothes are adsorbed with greater avidity than others, that certain adsorbants (people) adsorb with greater avidity than others, or more so, and finally we could prove that the clothing moved into the surface film in virtue of the second law of thermodynamics and in consonance with the principle of Willard Gibbs.*

1.7 THE MECHANISTIC VIEWPOINT AND GOD

In general, there seems to be a war between science and religion (White, 1877, 1913; Draper, 1898; Alexander, 1972, 2001; Alexander and White, 2004a,b; McGrath, 2005; Dawkins, 2006; McGrath and McGrath, 2007; Numbers, 2007; Stenger, 2007; Berlinski, 2008; Spencer, 2009), but this does not need to occur (Ray, 1691; Howey, 1948; Lack, 1957; Jaki, 1978; Kutschera and Nick, 2017). In studying mechanisms, one must deconstruct the whole into its parts and determine the relationships between the parts as well as the relationships between the parts and the whole. Each community has words or a word to describe "the whole." Throughout civilization, *Homo sapiens* have strived to live up to our specific epithet by struggling to understand the relationship between the parts and the whole in terms of understanding, among other things, our place in the universe, our relation to other people, our relationship to other species, and our relationship to the environment

(Leopold, 1949). Science and religion have been guides throughout this struggle to understand (Power, 1664; Kneller, 1911; Griffiths, 2008; Lerner and Griffiths, 2008; Seifriz, 2008; Wayne and Staves, 2008). Science and religion may be two sides of the same coin of understanding, each with a measure of truth, and each complementing the other. Herbert Spencer (1880) writes,

> *Assuming, then, that since these two great realities are constituents of the same mind and respond to different aspects of the same universe, there must be a fundamental harmony between them; we see good reason to conclude that the most abstract truth contained in religion and the most abstract truth contained in science must be the one in which the two coalesce. … Uniting these positive and negative poles of human thought, it must be the ultimate fact in our intelligence.*

If there is only one right answer, it is the answer to the metaphysical question: Is essence prior to existence (Plato, 1892; Gilson, 1956), or is existence prior to essence (Sartre, 1946, 2007; Heidegger, 2010).

It is often thought that a mechanistic viewpoint of nature excludes God. Philosophers have discussed the relationship between God and mechanics (Planck, 1932), and many scientists, including Kepler, Galileo, Boyle, Newton, Schleiden, Planck, Einstein, and Millikan, believed that the study of nature led to an understanding of God. For example, while imprisoned by the forces of the Inquisition, Galileo wrote (quoted in Gamow, 1988),

> *When I ask: whose work is the Sun, the Moon, the Earth, the Stars, their motions and dispositions, I shall probably be told that they are God's work. When I continue to ask whose work is Holy Scripture, I shall certainly be told that it is the work of the Holy Ghost, i.e. God's work also. If now I ask if the Holy Ghost uses words which are manifest contradictions of the truth as to satisfy the understanding of the generally uneducated masses, I am convinced that I shall be told, with many citations from all the sanctified writers, that this is indeed the custom that taken literally would be nothing but heresy and blasphemy, for in them God appears as a Being full of hatred, guilt and forgetfulness. If now I ask whether God, so as to be understood by the masses, had ever altered His works, or else if Nature, unchangeable and inaccessible as it is to human desires, has always retained the same kinds of motion, forms and divisions of the Universe, I am certain to be told that the Moon has always been round, even though it was long considered to be flat. To condense all this into one phrase: Nobody will maintain that Nature has ever changed in order to make its works palatable to men. If this be the case, then I ask why it is that, in order to arrive at an understanding of the different parts of the world, we must begin with the investigation of the Words of God, rather than of His Works. Is then the Work less venerable than the Word? If someone had held it to be heresy to say that the Earth moves, and if later verification and experiments were to show us that it does indeed do so, what difficulties would the church not encounter! If, on the contrary, whenever the Works and the Word cannot be made to agree, we consider Holy Scripture as secondary, no harm will befall it, for it has often been modified to suit the masses and has frequently attributed false qualities to God. Therefore I must ask why it is that we insist that whenever it speaks of the Sun or of the Earth, Holy Scripture is considered quite infallible?*

In this book, I will not base any mechanisms on the existence of God, and at the same time, I will not conclude that the discovery of a mechanism precludes the existence of a God. After all the Works of Nature are not identical with the Words from scientists who study nature.

1.8 WHAT IS CELL BIOLOGY?

First, let me define biology. According to Thomas Beddoes (1799) and G.R. Treviranus (1802), who along with J.B. Lamarck (1802) gave us the term *biology*: "The subject of our researches will be the different forms and phenomena of life, the conditions and laws under which this state occurs, and the causes which produce it. We shall designate the science which is occupied with these things as biology or the theory of life" (quoted in Driesch, 1914). By the end of the 19th century, the Roman Catholic priest, Jean Baptiste Carnoy (1884) stressed the importance of establishing a field of cellular biology to understand all aspects of biology. He envisioned cell biology as a multidisciplinary field, saying, "To be complete it is necessary to envision the cell from all of its facets, from the point of view of its morphology, its anatomy, its physiology and its biochemistry." Carnoy (1884) also stressed that scientific observations that were valuable require good instruments; good material; and a good spirit that possesses specific knowledge of the instrument and the material and a general knowledge of the cell, a spirit that is calm, patient, tenacious, perseverant, positive, and sometimes skeptical. By 1939, Lorande Woodruff wrote that when it comes to biology, the study of life, the cell has become "a sort of half-way house through which biological problems must pass, going or coming before they complete their destiny."

We will center our study of biology on the cell—the basic unit of life. We will try to understand the processes that contribute to our definition of life from first principles, that is, with the fewest assumptions possible (Northrop, 1931). Peter Mitchell (1960) noted that "when *Homo sapiens* is confronted by a thing or a phenomenon which is too large or too complex for him to investigate as it stands, he slices it up arbitrarily (or perhaps one should say homocentrically) into parts of more manageable dimensions and proceed to investigate the parts. Thus, biology has been sliced one way into Botany, Zoology,

Microbiology, another way into Systematics, Ecology, Behaviour, Anatomy, Physiology, Pharmacology, Biochemistry, again into Digestion, Metabolism, Synthesis, Cytology, Biomolecular structure, Biochemical cytology, and again and again into smaller and smaller pieces especially as the chemists and physicists have become hungry for the knowledge of just another little slice." One of the aims of this book is to put the pieces back together—or as Mitchell would say, "to sympose them." When symposing, it is valuable to treat each discipline non-hierarchically (Wayne, 2016b). In our search, we will use the techniques and tenets of biochemistry, biophysics, microscopy, immunology, physiology, genetics, and the various "-omics." By studying the basic unit of life, we will try to understand the nature of life and its unity.

Enjoy your search into the nature of the cell and remember what Albert Szent-Györgyi (1960) said about research: "The basic texture of research consists of dreams into which the threads of reasoning, measurement, and calculation are woven."

1.9 SUMMARY

Life consists of the ability to move and generate electricity; to take up nutrients and expel wastes; to perform chemical syntheses of organic molecules at ambient temperatures and pressures, and therefore grow; to reproduce itself with near-perfect fidelity; and to sense and respond to changes in the external environment to maintain itself. The cell is the lowest level of organization that has the ability to perform all these processes and thus is the basic unit of life (Table 1.5). Our endeavor is to understand the vital processes that are made possible by cells from a physicochemical standpoint.

TABLE 1.5 Cell as the Basic Unit of Life in Context

Numbers and constants (mathematics)—e, π, 21, 0, 1, etc., k, h, c, G Elementary particles (physics)—quarks, antiquarks, leptons Free lifeless particles Elements (chemistry)—H, C, N, O, P, S, etc. Molecules—H_2O, CO_2, NO_3, PO_4, etc. Minerals (mineralogy)—(e.g., clays, which are able to grow and reproduce themselves in an ionic solution) Simple organic molecules (organic chemistry)—CH_4, NH_3, H_2S, HCN, etc. (organic chemistry) $C(H_2O)$, $C(OOH)C(HR)NH_3$, fatty acids, adenine, etc. Organic macromolecules (biochemistry, physical chemistry, molecular biology, genomics and other -omics)—carbohydrates, proteins, lipids, nucleic acids
All the above levels of lifeless particles show passive translational motion (i.e., diffusion) and move passively in response to pressure and thermal gradients and electromagnetic and gravitational fields. Radiant energy causes a change in their electronic structure.
Viruses (proteins and nucleic acids; virology): Viruses are able to reproduce, adapt, and evolve in a living environment created by other organisms. *Cells—living particles (cell biology)*: Cells are able to take up nutrients, grow, synthesize compounds at body temperature and 1 atm of pressure, degrade compounds at body temperature and 1 atm of pressure, cause conversion of kinetically stable compounds into kinetically unstable compounds to be used as a ready supply of energy to perform endergonic reactions, expel wastes, regulate the biosynthetic and degradative processes, sense and respond to the environment in an adaptive manner, and move actively and reproduce with near-perfect fidelity to allow for the continuity of life as well as adaptation by natural selection. *Bacteria and protoctists (single-celled prokaryotic and eukaryotic organisms)*: Able to perform all the functions of life (microbiology). *Colonies (psychology, invertebrate biology)* *Multicellular organisms*: Animals, Fungi, Plants (zoology, mycology, botany, biology, anatomy, morphology, physiology, developmental biology, taxonomy, systematics, biogeography, biomechanics, biophysics, etc.) Soul (neurobiology, behavior, psychology, psychiatry, philosophy, theology) Mind (neurobiology, behavior, psychology, psychiatry, philosophy) Thinking (neurobiology, behavior, psychology, psychiatry, philosophy) Personality (neurobiology, behavior, psychology, psychiatry, philosophy) Individuality (neurobiology, behavior, psychology, psychiatry, philosophy, political theory) Spirituality (neurobiology, behavior, psychology, psychiatry, philosophy, theology) *Cells interact within an organism to make possible highly specialized cells, tissues, organs, and the processes that they perform.* *Multiorganism level*: Relationship between the organism and other organisms (political science, ecology, psychology, sociology, philosophy, theology) Relationship between the organism and the environment (political science, ecology, psychology, sociology, philosophy, theology) Relationship between the organism and the universe (theology, astronomy, cosmology) *Each organism does not live in isolation. For example, it may be the predator and/or the prey. Or it may be a symbiont. It may be a member of a pioneer species, or it may come into an environment after the way is prepared, etc. And unbelievably, Homo sapiens have the ability to know their place in the universe.*

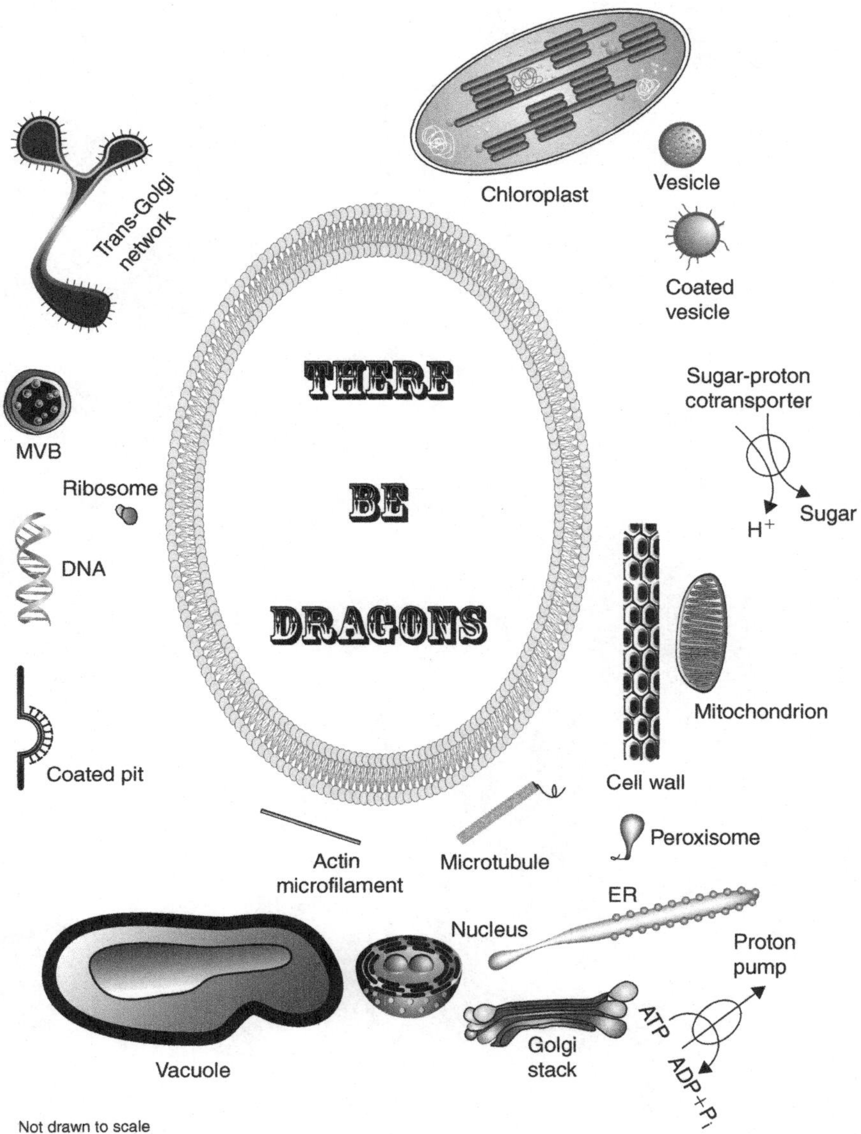

FIGURE 1.13 Build your own map of a cell. Make a copy of this figure and place the organelles in the cell as you learn about them. What are the similarities between cells? What are the differences? How do the organelles change their positions? Which organelles are derived from other organelles?

Your own personal map of the cell is provided in Fig. 1.13. Throughout your journey through the cell, add the landmarks you discover. Keep in mind that the quantity, composition, and arrangement of any of the landmarks may change during cell development, following a change in the cell's environment, and as you travel from cell to cell. Develop an idea of which landmarks and which of their characteristics are fundamental, which are important in specialized systems, and which may be ephemeral.

1.10 QUESTIONS

1.1. What is life and how can the study of cells help us bring meaning and value to life?

1.2. How can mathematics help us understand living processes?

The references for this chapter can be found in the references at the end of the book.

Chapter 2

Plasma Membrane

By fate, not option, frugal Nature gave …
It was her stern necessity: all things
Are of one pattern made …
Deceive us, seeming to be many things,
And are but one …
To know one element, explore another,
And in the second reappears the first …

Ralph Waldo Emerson, "Xenophanes."

2.1 THE CELL BOUNDARY

The plasma membrane provides a barrier that separates the living cellular protoplasm from the external environment and, as such, resists the attainment of equilibrium prophesied by the kinetic theory of molecules and the Second Law of Thermodynamics (Clausius, 1879; Bünning, 1989). For this reason, the plasma membrane is a *sine qua non* for life (Just, 1939). However, the boundary must not be absolutely impassable but must allow the entrance of nutrients into and the excretion of waste products out of the cell. That is, the plasma membrane, like the border of a country, must neither be open nor closed but differentially permeable. Moreover, the plasma membrane, by virtue of its position at the frontier of the protoplasmic substance, is particularly suited to sense changes in the external environment so that the cell can act appropriately. Of course, the external environment for a cell in a multicellular plant also includes all the other cells in the same plant that are connected through chemical signals or physical forces! As the interface of the cell, the plasma membrane is involved in every cellular process that depends on the cell's ability to respond to external stimuli, including light, gravity, hormones, salinity changes, pollination, and pathogen attack.

The honeycomb-like appearance of the thick walls found in wood and cork inspired Hooke (1665) to name the compartments *cells*. However, confusion concerning the definition of a cell ensued when it was discovered that animal cells lack such a wall even though they contain a nucleus and protoplasm (Baker, 1988). As a result of this discovery, the cell was redefined by Franz von Leydig (1857).

Leydig (1857) defined the cell as a soft substance containing a nucleus and surrounded by a plasma membrane. Johannes von Hanstein (1880) used the term *protoplast* for the soft substance containing a nucleus and surrounded by a plasma membrane, to avoid the confusion that could be caused by having two definitions of a cell. However, the term never caught on (see Sachs, 1882a,b). The great plant physiologist Julius Sachs (1892) rejected the use of the word *cell* for a wall-less protoplast, and said sarcastically, that, if it were correct to call the protoplast a cell, then a bee should be called a cell and the honeycomb should be called the capsule of the cell! Unbelievably, this argument over terminology was resurrected 100 years later (Robinson, 1991; Sack, 1991; Staehelin, 1991; Stafford, 1991; Connolly and Berlyn, 1996)! In this book, I will use the words *protoplast* and *cell* interchangeably to emphasize that the plasma membrane and not the wall provides the major functional division between the living matter and the external environment.

2.2 TOPOLOGY OF THE CELL

The plasma membrane divides the volumes inside and outside the cell topologically into two compartments: the external space or E-space, which is the volume external to the plasma membrane; and the protoplasmic space or P-space, which is the volume immediately inside the plasma membrane (Fig. 2.1; Mitchell, 1959). That is, the plasma membrane separates the "living" space from the "lifeless" space. In most bacteria, which typically have only one membrane, these are the only compartments, although this too is a generalization beecause there can be more than one aqueous phase in a single membrane-enclosed compartment (Tehei et al., 2007). At first glance, the situation appears unreasonably complicated in eukaryotic cells, which contain many membranous compartments. However, we will see that keeping track of the topology of each compartment in the cell will help us to develop some generalizations about the cell. For example, the P-space will generally have a lower Ca^{2+} concentration than the E-space.

The E-space is not totally lifeless or abiotic. For example, surrounding the epidermal cells in many roots, symbiotic mycorrhizal fungi live, which secrete a protein known as glomalin that allows better root development by structuring the soil so that it is more permeable to air and water (Wright and Upadhyaya, 1996, 1998, 1999).

Plant Cell Biology. https://doi.org/10.1016/B978-0-12-814371-1.00002-3

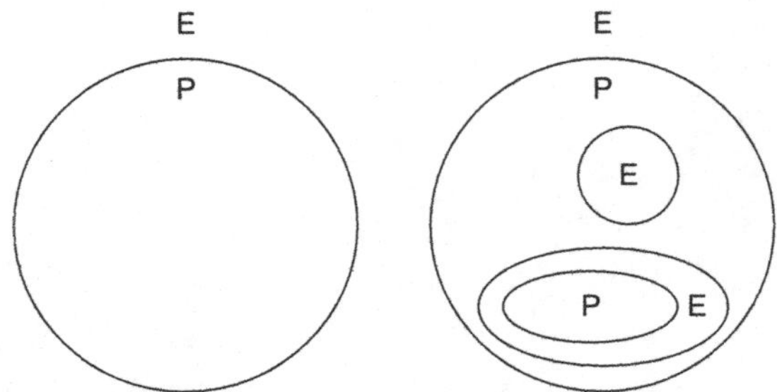

FIGURE 2.1 The topology of the cell. The external space (E-space) and protoplasmic space (P-space) are separated by a membrane. The figure on the right shows the topology a cell with many compartments.

Gilbert Ling (1984, 2001) proposed an alternative to the membrane theory of separation between the E- and P-spaces. He proposed that the colloidal nature of the proteins in the P-space differentially binds ions and causes a matrix or Donnan potential similar in magnitude to the observed "membrane" potential. If Ling were correct, the high electrical resistance of the plasma membrane (discussed in the following section) would be superfluous because the proteins in the cytosol would bind the ions so tightly that the proteins would resist the movement of ions more than the membrane would. While there may be some truth in the membrane and matrix theories, evidence including the high electrical resistance of the plasma membrane (Walker, 1960) and the rapid diffusion of dyes in the cytoplasm but not across the membrane (Chambers, 1922; Plowe, 1931) suggests that the membrane theory is the best first approximation of reality.

2.3 EVIDENCE FOR THE EXISTENCE OF A PLASMA MEMBRANE

In 1844, Carl von Nägeli noticed that the protoplasts of various algae, including *Nitella* and *Bryopsis*, pull away from the wall when the cells are exposed to various concentrated solutions, and that they return to their normal size when the concentrated solutions are replaced by dilute solutions (Fig. 2.2). Realizing that protoplasts exhibited the same osmotic properties that Jean-Antoine Nollet and Henri Dutrochet described for animal bladders, Nägeli and Nathanael Pringsheim (1854) concluded that there must be a differentially permeable membrane around the protoplast. After investigating the osmoticum-induced contraction of the protoplast of many plant cells that could be easily visualized as a consequence of their anthocyanin content, von Nägeli and Cramer (1855) concluded that a cell membrane was a typical characteristic of plant cells.

In 1867, Wilhelm Hofmeister found that the protoplasts that make up beetroots shrink in concentrated NaCl solutions. However, Hofmeister turned his attention to the shrinking of the easily visible red-colored vacuole and concluded that the entire protoplast, and not an invisible surface layer, is responsible for the osmotic properties of the cell. Hofmeister proposed that the osmotic movement of water into and out of the protoplast is primarily responsible for plant movements, including the touch-sensitive and sleep movements of the leaves of *Mimosa* (de Mairan, 1729), the temperature-induced opening and closing of tulip flowers, and the light- and/or gravity-induced bending of plant organs (Goebel, 1926).

Hugo de Vries (1885, 1888a), like Hofmeister, performed similar plasmolytic experiments on the violet epidermal cells of *Tradescantia discolor*. He also noticed that the protoplast detached from the cell wall and it was de Vries who named this phenomenon *plasmolysis*. However, due to the obvious shrinkage of the violet vacuole, he believed that the vacuolar membrane, which he termed the *tonoplast* because of its putative role in tonicity or turgor, and the protoplasm surrounding the tonoplast were responsible for the osmotic properties of the cell. While, in reality, these botanists demonstrated the differential permeability of the plasma membrane, they did not think that the surface layer was an important regulator of the osmotic properties of the cell (Briggs and Robertson, 1957). Many influential botanists held on to this view up until the 1960s, thus impeding the advancement of plant membrane biology (Dainty, 1962; Hope and Walker, 1975; Wayne, 1994).

Wilhelm Pfeffer, a botanist influenced by the physicochemical philosophy of Hermann von Helmholtz, turned to the study of cellular mechanisms to satisfy his curiosity

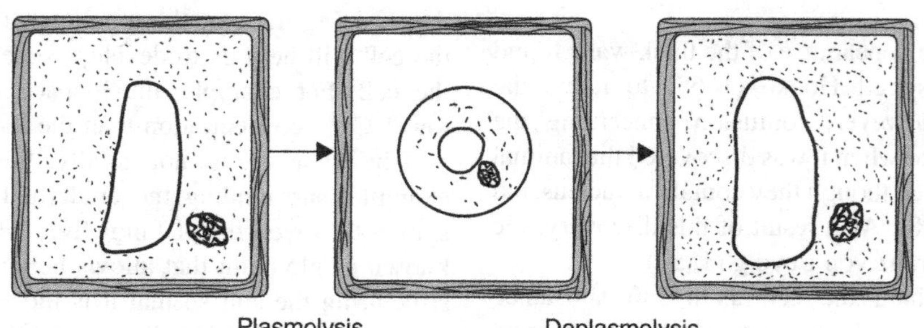

FIGURE 2.2 Plasmolysis and deplasmolysis of a plant cell.

to understand the thigmonastic or touch-induced leaf movements of the sensitive plant, *Mimosa pudica,* and the rapid movements of the stamen of the knapweed, *Centaurea jacea* (Bünning, 1989). To understand osmosis,[1] the basis of the pushing force that causes the leaf movement, Wilhelm Pfeffer (1877) turned to the model membranes that Justus Liebig's student, Moritz Traube, designed out of copper ferrocyanide.[2] Traube (1867) designed these membranes to create "artificial cells" so he could study the processes of living cells, including growth and osmosis (Fig. 2.3). The artificial cells could expand and bud like living cells; however, the artificial membranes were not strong enough to withstand the osmotic pressure that developed within them and consequently broke easily. To overcome this problem, Pfeffer deposited the copper ferrocyanide membranes on a porous clay pot, imagining that the pot would protect the artificial membrane from lysing, much like a plant cell wall prevents the rupturing of the protoplast. The membrane-covered porous pot was connected to a thin capillary tube (Fig. 2.4). When he added solutions of sucrose to the inside of the copper ferrocyanide membrane, the water moved from the outside of the membrane into the sucrose solution and caused the sucrose solution to rise in the capillary. He defined osmotic pressure as the pressure that must be added to the top of the capillary to prevent the water from entering the sucrose solution surrounded by the differentially permeable membrane. Pfeffer's membranes were strong enough to perform repeated measurements, and he was able to get quantitative results (see Chapter 7).

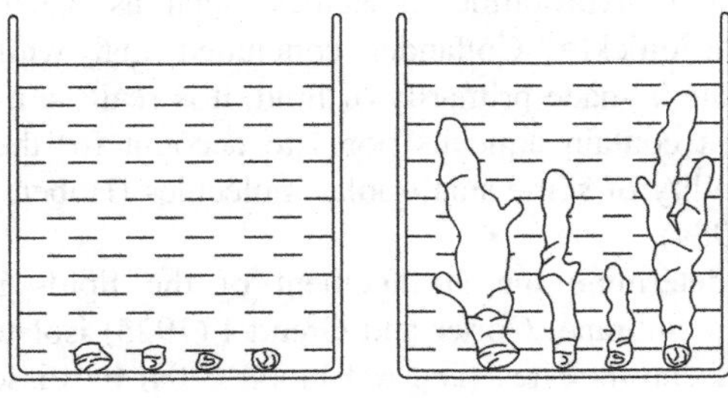

FIGURE 2.3 A Traube cell. The diagram on the left shows the initial situation and the diagram on the right shows "cell" growth after approximately 15 minutes.

1. From *osmos*, the Greek word meaning "to push."
2. To make artificial cells, fill a beaker three-quarters full with a 5% copper sulfate solution ($CuSO_4$). Use forceps to drop a small crystal of ferrocyanide ($K_4Fe(CN)_6$) into the solution. Do not disturb. Observe the formation of a precipitation membrane of copper ferrocyanide. Note the growth of the "cell" and the different colored solutions that are outside and inside the precipitation membrane. For other work on artificial membranes, see Collander (1924, 1925) and Michaelis (1926a).

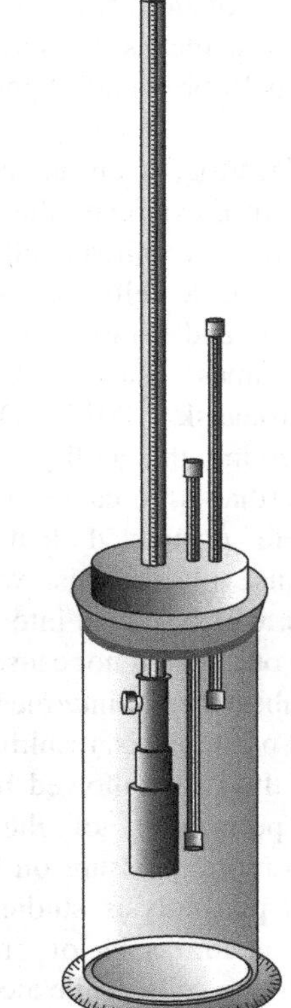

FIGURE 2.4 Wilhelm Pfeffer's osmometer.

Noticing that thin artificial membranes exhibited the same osmotic phenomena as protoplasts, Pfeffer postulated that a thin plasma membrane surrounded the entire protoplast and regulated the osmotic properties of the cell (Bünning, 1988). Indeed, he also mentioned that the analogous behavior of cells and model membranes indicates that a vital force is not responsible for the permeability properties of cells.

Pfeffer also used classical anatomical techniques to understand the nature of the plasma membrane. He noticed that the plasma membrane could be stained by iodine or mercury, and thus concluded that it was composed, at least in part, of proteins. Pfeffer (1877) also believed that physiological experiments would provide a method for understanding the structure of the plasma membrane. He said

> *Though our understanding of the experimental results does not necessarily compel us to presuppose a very definite conception of the molecular structure of precipitation membranes, we feel even more an intellectual need for deeper insight because only on the basis of such insight can our thoughts follow the path of a solute through the membrane.*

Pfeffer coined the term *plasma membrane* to emphasize its differentially permeable nature. Others, who did not believe that there was a functional differentially permeable boundary at the surface of the cell, called this region the *plasmalemma*, which just means the surface of the protoplast (Mast, 1924; MacRobbie and Dainty, 1957; Dainty, 1960).

In 1899, Ernest Overton, a distant cousin of Charles Darwin (Collander, 1965), examined the permeability of a medley of living plant and animal cells, including root hairs, algal filaments, muscle cells, and red blood cells, to about 500 compounds and came up with a series of generalizations—sometimes called Overton's Rules (Overton, 1899b; Perouansky, 2015). Overton measured permeability by observing the ability of a substance to induce cell shrinkage (that is, to cause osmotic water flow out of the cell). He postulated that less permeable substances would cause plasmolysis, whereas the more permeable substances, which moved into the cell as fast as the water could move out, would not cause plasmolysis. He also postulated that substances of intermediate permeability would cause an initial plasmolysis, resulting from the initial flow of water out of the cell, followed by deplasmolysis, resulting from the permeation of the solute and an equilibration of the osmotic pressure on both sides of the membrane. From his plasmolysis studies, he found that sugars, amino acids, neutral salts of organic acids, and glycerol barely enter living cells, while alcohols, aldehydes, ketones, and hydrocarbons permeated rapidly. He concluded that the possession of a charged or polar functional group (COO^-, OH, NH_2) in a chemical substance decreased its ability to permeate living cells.

The polarity of a functional group can be estimated from the polar nature of the bonds that make up the functional group. The greater the difference between the electronegativities of the two atoms that make up a bond, the greater the ionic or electrical dipole nature of the bond. A functional group composed of primarily ionic bonds will be polar. On the other hand, the smaller the difference between the electronegativities of the two atoms that make up a bond, the more covalent the bond is. Because a pure covalent bond has a vanishingly small electrical dipole, functional groups composed primarily of covalent bonds are nonpolar. Linus Pauling (1954a) created the electronegativity scale to characterize the electrical nature of bonds. From Fig. 2.5, one sees why functional groups composed primarily of OH bonds are more polar than functional groups composed of CH bonds. Functional groups composed of NH bonds are intermediate.

Consistent with the electrical interpretation of the chemical bonds just described, Overton (1900) also found that the ability of chemicals to permeate living cells increased as the length of their hydrocarbon chains increased. His observations indicated a positive correlation between lipid solubility or lipophilicity and permeability. He postulated that the plasma membrane, which regulated the permeability of the cell, must be composed of lipid.

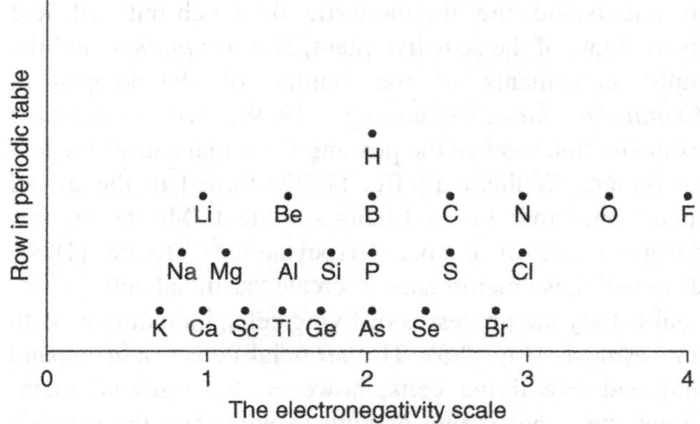

FIGURE 2.5 Linus Pauling's electronegativity scale. The greater the difference between the electronegativities of the two atoms that make up a bond, the greater the polarity of the bond and the greater the polarity of the functional group (e.g., OH) that is composed of these bonded atoms. The smaller the difference between the electronegativities of the two atoms that make up a bond, the more covalent the bond and the less polar the functional group (e.g., CH) that is composed of these bonded atoms. *Data from Pauling, L., 1954a. General Chemistry, second ed. W. H. Freeman and Co., San Francisco.*

Overton went on to show that dyes that were soluble in lipid permeated the cell faster than those that were not.

Considerably more support for Overton's theory came from the work of Collander and Bärlund (1933), who analyzed chemically the amount of a given substance that appeared in the cell sap of characean cells a certain time after the cells were placed in a given concentration of the substance. Runar Collander (1937, 1959) found that, in general, the more lipophilic a substance is, the greater its ability to permeate the membrane (Fig. 2.6). However, some small hydrophilic molecules, such as water, also permeate quickly. Collander concluded that while the membrane is made primarily of lipid, it is really a mosaic, and must contain aqueous pores to account for the high permeability of some small polar molecules (Höber, 1945; Ling, 1984).

To determine the arrangement of the lipids in the plasma membrane, Gorter and Grendel (1925) isolated the lipids of chromocytes (i.e., red blood cells) by dissolving them in acetone. Mammalian red blood cells are a favorite material of plasma membrane biologists because, unlike most eukaryotic cells, the plasma membrane in mammalian red blood cells is the only membrane in the cell (Bretscher and Raff, 1975). Gorter and Grendel floated the isolated lipids on the surface of a Langmuir trough (Adam, 1941; Taylor et al., 1942; Becher, 1965) and decreased the surface area of the trough until the lipids formed a monolayer (Fig. 2.7). They measured the surface area of the monolayer and calculated the surface area of the original red blood cells from measurements of the cell radius (r). The area of the monolayer was twice as large as the area of the red blood cells, and so they concluded that the plasma

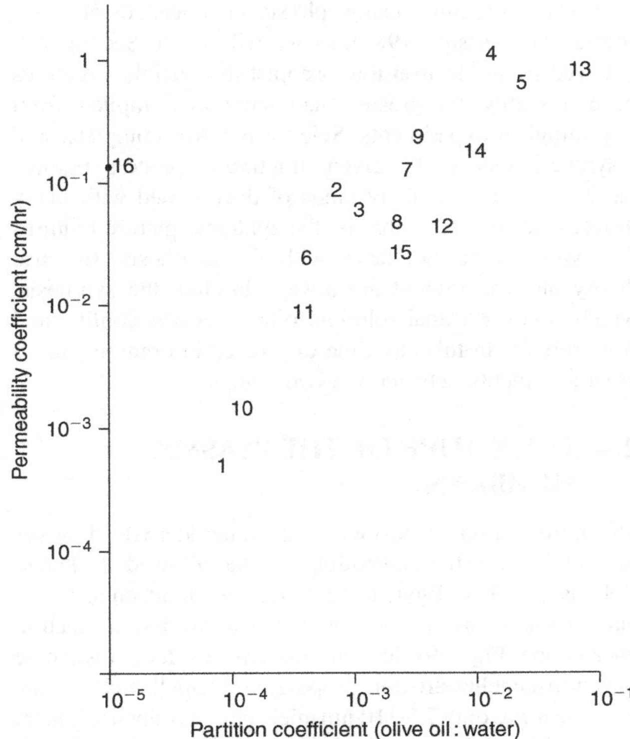

FIGURE 2.6 The relationship between the permeability coefficient and olive oil:water partition coefficient. 1, glycerol; 2, formamide; 3, acetamide; 4, methanol; 5, ethanol; 6, ethylene glycol; 7, propylene glycol; 8, succinamide; 9, cyanamide; 10, urea; 11, methylurea; 12, dimethylurea; 13, urethane; 14, n-butyramide; 15, dimethylurea; 16, DHO. *Data from Collander, R., 1937. The permeability of plant protoplasts to nonelectrolytes. Trans. Faraday Soc. 33, 985–990; Collander, R., 1959. Cell membranes: their resistance to penetration and their capacity for transport. In: Stewart, F.C. (Ed.), Plant Physiology. A treatise. Plants in Relation to Water and Solutes, vol. 2. Academic Press, New York, London, pp. 3–102.*

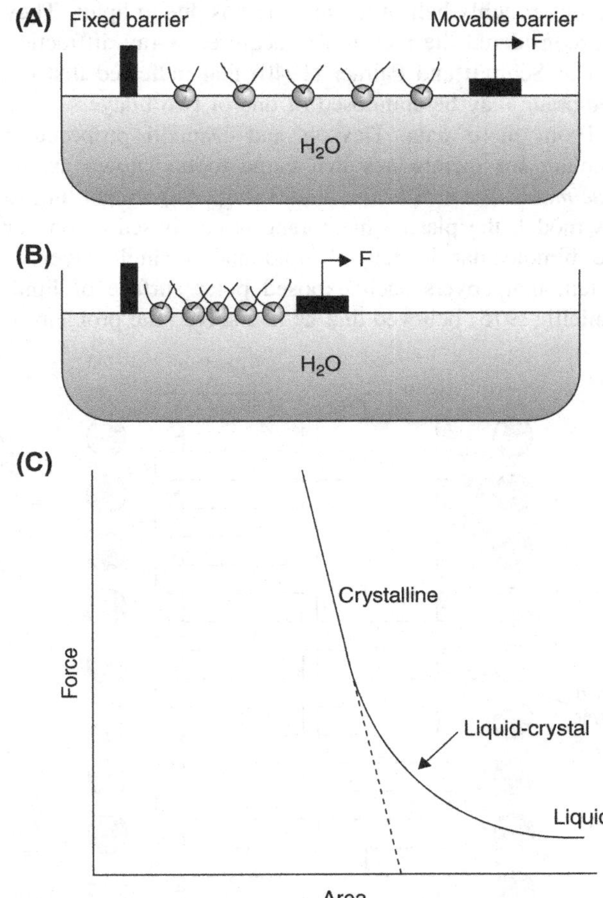

FIGURE 2.7 Using a Langmuir trough to determine the area taken up by a known amount of lipids. The lipids are placed on the surface of the water (A) of the trough and a movable barrier attached to a force transducer is moved toward the lipids (B). As the lipids are compressed, the force increases as the lipids are transformed from a liquid state to a liquid crystalline state to a crystalline state (C). The area of the lipids in the liquid crystalline state is determined by extrapolating the force-area curve to zero force.

membrane is composed of a lipid bilayer, with the hydrocarbon tails oriented inward and the polar head groups facing the outside.

Actually, Gorter and Grendel's conclusion was not justified because the acetone did not extract all the lipids from the membrane. Luckily, they also made a mistake in calculating the surface area of the cells, which they underestimated to be equal to $8r^2$, so they could conclude, serendipitously, that the lipids form a bilayer around the cell. Newer experiments show that there are only enough lipids to cover approximately 1.5 cell surface areas, not two (Bar et al., 1966). It is possible that an overlapping of the hydrocarbon tails comprising each leaflet of the membrane could increase the surface covered by the lipids somewhat, but we now know that much of the surface of the membrane is taken up by proteins.

Hugh Davson and James Danielli (1943, 1952), two physical chemists, put together the known physicochemical data on plasma membranes to deduce the structure of the plasma membrane. They included Overton's observations on permeability; the determination by Parpart and Dziemian (1940) that the plasma membrane is composed of lipids and proteins with a ratio (w/w) of 1:1.7 (Jain, 1972); Hugo Fricke's (1925) and McClendon's (1927) measurements on the electrical impedance of red blood cells, which showed that the membrane had a high resistance and a capacitance reminiscent of a lipid bilayer (see Section 2.8); the measurements of the tension at the surface of the plasma membrane that indicated that the low value for the surface tension may be due to a coating of protein (Danielli and Harvey, 1935); the observations by Schmitt et al. (1936, 1938; Liu, 2018) that plasma membranes have radially positive intrinsic birefringence and negative-form birefringence when viewed with a polarization microscope (Wayne, 2009); and the observations by Waugh and Schmitt (1940) using interference microscopy that showed that the plasma membrane is approximately 20 nm thick and that roughly half of the thickness is due to lipids. They also considered the then newly acquired X-ray diffraction data of Schmitt and Palmer (1940) that indicated that the membrane may be composed of one or two bilayers.

From these data, Davson and Danielli proposed a structure for membranes that came to be known as the *paucimolecular membrane model* (Fig. 2.8). According to this model, the plasma membrane is composed of one or two bimolecular leaflets of lipid and a single layer of protein that covers each exposed polar surface of lipid. Danielli (1975) believed that he had solved the problem of membrane structure using physicochemical theory. By contrast, Robertson (1987), as we will see in Section 2.4, believed as an electron microscopist that visible structures have a reality far greater than structures implied from physiological experiments. Science is never complete, and a synthetic theory of a given structure or process requires the combination of many kinds of data gained with many different techniques—that is, the synthetic picture requires the synthesis of the thesis with the antithesis. Because theory and experiment are always limited, the synthesis, which is never a final solution, often involves conflict and compromise. In this case, the difference in opinion caused great arguments between the two camps.

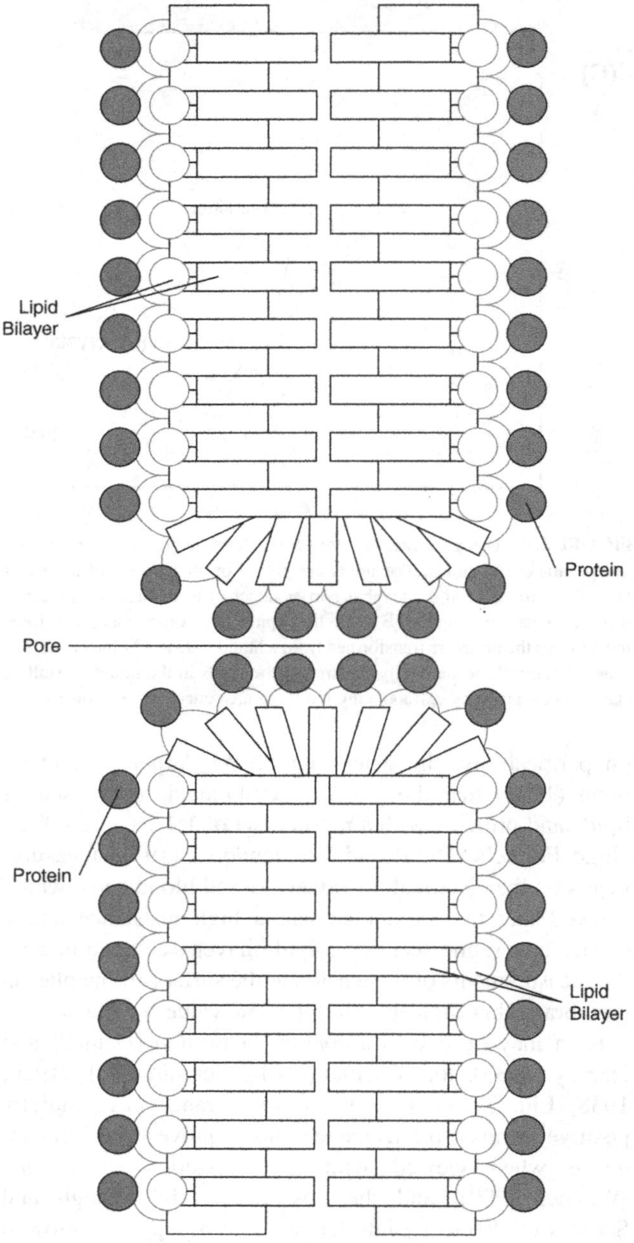

FIGURE 2.8 Danielli and Davson's paucimolecular model of a membrane.

2.4 STRUCTURE OF THE PLASMA MEMBRANE

The introduction of $KMnO_4$ and glutaraldehyde fixatives as well as plastic embedding media allowed J. David Robertson (1959, 1964) to visualize the architecture of the plasma membrane of cells with the transmission electron microscope (Fig. 2.9). Imagine Robertson's delight when he saw the trilamellar structure proposed by Danielli and Davson. Because it was only 7.5–10 nm thick, he determined that there was only one 3.5-nm lipid bilayer, coated on both sides with approximately 2-nm thick protein layers. Moreover, due to the differential fixation and staining properties of the two protein layers, Robertson concluded that the membrane was asymmetric. Robertson also stressed that the membrane had no pores because he could not visualize them at the level of resolution attainable at the time. He proposed that all membranes have the same structure and proclaimed the concept of the "unit membrane." This model became unfashionable, not because it is so wrong, but because it emphasized the static, constant properties of membranes at a time when biochemistry was showing that each membrane was chemically distinct, enzymatically unique, and dynamically active (Korn, 1966; Stoeckenius and Engelman, 1969; Robertson, 1987).

Moor et al. (1961) and Moor and Mühlethaler (1963) introduced a new method called *freeze fracture* or *freeze etching* for visualizing cells that prepared the way for a dynamic view of the plasma membrane. With this technique, shown in Fig. 2.10, frozen cells are nicked with a razor blade in such a manner that the membrane splits between the two leaflets of the lipid bilayer, disclosing face views of the membrane that show the distribution of particles that are approximately 8–10 nm in diameter (Branton, 1966; see Fig. 2.11). Every membrane has characteristic particles, which are not uniformly distributed over the membrane but are restricted to either the external surface (ES), the protoplasmic surface (PS), or the fracture face of either the leaflet on the external side (EF) or the leaflet on the protoplasmic side (PF).

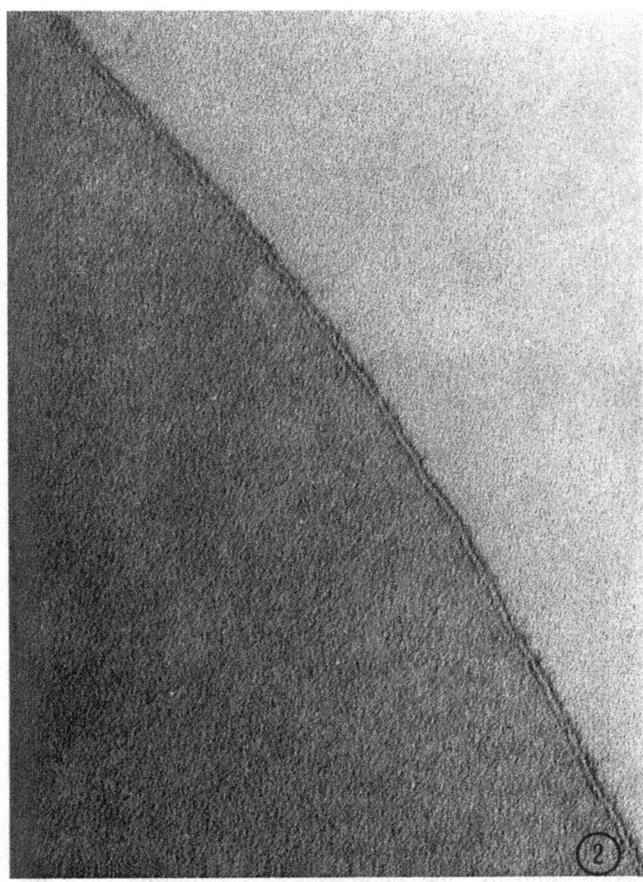

FIGURE 2.9 Electron micrograph of the plasma membrane of a red blood cell showing the structure of a unit membrane. *From Robertson, J.D., 1964. Unit membranes: A review with recent new studies of experimental alterations and a new subunit structure in synaptic membranes. In: Locke, M. (Ed.), Cellular Membranes in Development. Academic Press, New York, pp. 1–81.*

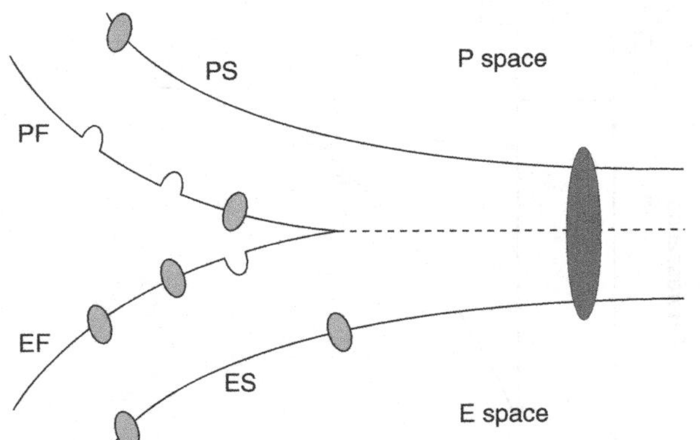

FIGURE 2.10 Diagram of a freeze-fractured membrane showing the PS, PF, EF, and ES.

The present model of the architecture of the plasma membrane, known as the *fluid mosaic model*, explains elegantly the structure of the membrane observed with freeze-fracture electron microscopy (Singer and Nicolson,

FIGURE 2.11 Freeze-fracture micrograph of the plasma membrane of a mesophyll cell. The *arrows* point to hexagonally arranged depressions from which particles have been pulled away during the freeze-fracturing process. Bar, 100 nm. *From Schnabl, H., Vienken, J., Zimmermann, U., 1980. Regular arrays of intramembranous particles in the plasmalemma of guard cells and mesophyll cell protoplasts of Vicia faba. Planta 148, 231–237.*

1972; Engelman, 2005; Nicholson, 2014; Goñi, 2014; Lombard, 2014). In this model, shown in Fig. 2.12, the particles are considered to be proteins. The fluid mosaic model arose from the curiosity of S. Jonathan Singer (1975, 1992), who as a protein physical chemist, wondered why some proteins are soluble in the cytoplasm, while others are associated with membranes. He suggested that an accurate model of membrane structure must be able to provide an explanation for the specific association of proteins with membranes. Singer (1990) grouped membrane proteins into two classes: peripheral or extrinsic proteins, which are primarily hydrophilic and can be removed by mild treatments such as raising or lowering the ionic strength, changing the pH, or using bivalent ion chelators such as ethylenediaminetetraacetic acid or ethylene glycol tetraacetic acid; and integral or intrinsic proteins, which are primarily hydrophobic and can be removed only by detergents or organic solvents. Singer realized that the Davson–Danielli–Robertson model of the membrane was thermodynamically unsound because it would require hydrophobic membrane proteins to be in contact with aqueous solutions. One can get a feel for the high amounts of free energy required to mix hydrophobic and hydrophilic molecules by trying to mix an oil and vinegar–type salad dressing using various amounts of shaking.

Singer visualized that the intrinsic membrane proteins were globular and amphipathic; that is, they had both hydrophilic and hydrophobic ends. He proposed that the hydrophobic portions of proteins would be embedded in the hydrocarbons derived from fatty acids in the lipid bilayer, and either one or two hydrophilic portions would extend into the polar head groups and out into the

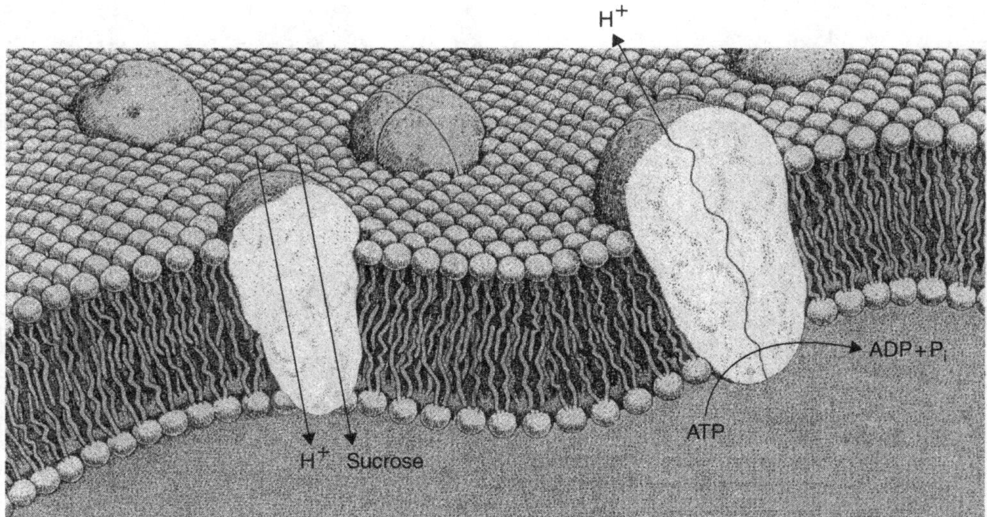

FIGURE 2.12 Singer and Nicholson's fluid mosaic model of a membrane.

aqueous media. He visualized the proteins as icebergs floating in a sea of lipid and imagined that the structure of the membrane is determined primarily by the various hydrophobic and hydrophilic interactions between the proteins and lipids. Some of the proteins, he guessed, traversed the entire thickness of the membrane. This was confirmed by Mark Bretscher's (1971a,b, 1973) study in which he labeled a protein (later named band 3, or the anion transporter) in the intact erythrocyte membrane with a radioactive probe. Presumably, the protein would be labeled only if it was exposed to the E-surface of the plasma membrane. When Bretscher previously treated the intact erythrocytes with pronase, an impermeable enzyme that digests the parts of the proteins on the outside of the membrane, the band 3 protein in the intact cell could not be labeled.

Subsequently, Bretscher made membrane ghosts, which are either inside out or right side out. This allowed the probe to label the inside and the outside. Because a peptide fragment of the same protein becomes labeled in ghosts made from cells that were previously treated with pronase, Bretscher concluded that the band 3 protein spanned the entire width of the membrane.

If the membrane proteins coated the lipid bilayer, as suggested by the Davson–Danielli–Robertson model, it would be unlikely that the membrane proteins would be mobile. However, experiments involving the fusion of human and mouse cells showed that membrane proteins from both cell types become uniformly distributed in the heterokaryon, indicating that proteins can translate through the plasma membrane (Frye and Edidin, 1970). This observation was extended by Singer's thermodynamic calculations that showed that membrane molecules should be able to move in the plane of the membrane. Thus, according to the fluid mosaic model, lateral movements of proteins and lipids in the plane of the membrane are possible, and in fact do occur. Movements of membrane molecules can be observed with a technique known as *fluorescence redistribution after photobleaching* (FRAP; Axelrod et al., 1976). With this technique, shown in Fig. 2.13, membrane molecules are selectively labeled with fluorescent probes. The distribution of the probe in the membrane is then observed with a fluorescence microscope. Initially, the fluorescence is uniform. Then the fluorescent molecules in an area of the membrane are destroyed with a laser and the fluorescence decreases. The fluorescence begins to recover over time as fluorescent molecules diffuse back into the bleached area.

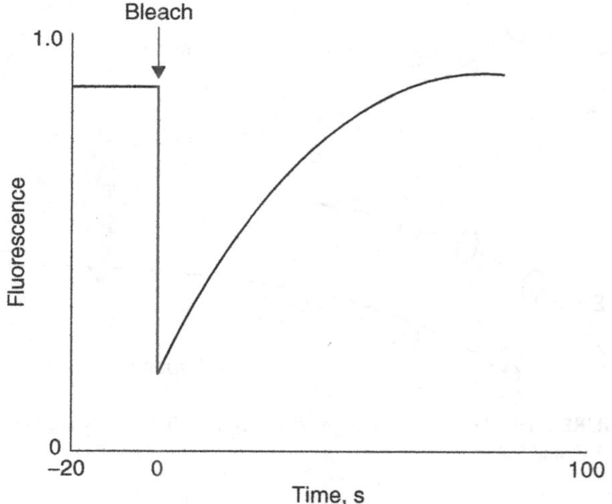

FIGURE 2.13 Results from a fluorescence redistribution after photobleaching experiment. The sample is initially fluorescent. Then the fluorescence is bleached at time zero. The fluorescence recovers over time and the diffusion coefficient is calculated from the slope of the recovery curve.

The rate of fluorescence recovery is an indication of the rate of diffusion of the molecules in the plane of the membrane. The diffusion coefficients for proteins in plant plasma membranes fall between 5×10^{-15} and 3×10^{-14} m^2/s (Metcalf et al., 1986a; Dugas et al., 1989). While the proteins are able to diffuse within the plane of the membrane, they are not free to diffuse anywhere. In fact, the values obtained for the diffusion coefficients of proteins vary in part because peripheral proteins known as the *membrane skeleton* may cause a compartmentalization of the plasma membrane into domains that are approximately 0.1–1 μm^2. The membrane skeletal proteins "corral" the intrinsic proteins, and consequently, the smaller the domain tested with FRAP, the larger the diffusion coefficient appears to be (Kusumi and Sako, 1996).

The mobility of individual plasma membrane proteins can also be observed using single particle–tracking techniques that depend on computer-enhanced microscopy (Saxton and Jacobson, 1997; Tomishige et al., 1998; Smith et al., 1999; Tomishige and Kusumi, 1999; Mirchev and Golan, 2001; Douglass and Vale, 2005). The diffusion-restricted domains in the plasma membrane can be evanescent or stable, and they may represent regions of the plasma membrane with specialized functions (Edidin, 1992, 2001). As we will see later, the diffusion coefficients for protein in water are between 10^{-11} and 10^{-10} m^2/s, indicating that the lipid bilayer is really a very viscous solution through which the proteins diffuse.

The lipids also diffuse through the leaflet of the bilayer; although, due to their small size, they diffuse 10–100 times faster than the proteins. Their diffusion coefficients fall between 3×10^{-14} and 10^{-12} m^2/s (Kornberg and McConnell, 1971b; Metcalf et al., 1986b; Walko and Nothnagel, 1989). As we will see, the ability of molecules to move translationally in the membrane does not mean that they necessarily become randomly arranged. Indeed, like the proteins, the lipids can form Triton X-100 insoluble microdomains, known as *lipid rafts* (Simons and Ikonen, 1997; Simons and Toomre, 2000; Mongrand et al., 2004; Bhat and Panstruga, 2005; Borner et al., 2005; Martin et al., 2005; Lefebvre et al., 2007; Cacas et al., 2016).

In contrast to the array of movements that take place in the plane of the membrane, it is thermodynamically unlikely for a lipid or protein to flip-flop from one leaflet of the bilayer to the other as a consequence of the high energies that are required to make contact between hydrophilic and hydrophobic molecules. This kind of movement can be monitored by labeling membrane molecules with a molecule that contains an unpaired electron (i.e., a spin label), which acts like a magnet that can be aligned in a magnetic field (Compton, 1921a,b; Uhlenbeck and Goudsmit, 1926; McMillan, 1968; Goudsmit, 1971). Some spins are aligned parallel to the magnetic field and others are aligned antiparallel to the magnetic field. The greater the number of spin labels in the system, the greater the amount of microwave energy that can be absorbed by the system to flip the spin label from a lower-energy spin parallel to the field to the higher-energy spin antiparallel to the field.

The spin label is inserted into one side of the membrane and observed with an electron spin resonance spectrometer. If the labeled lipids flip-flop to the other side of the membrane, the signal will not be able to be quenched with ascorbic acid, a quencher of free radicals. The percentage of lipid molecules that flip across the membrane during a given time can be estimated by determining the proportion of the signal that is protected from being quenched by ascorbic acid. These measurements indicate that 10^{-6}–10^{-4} lipids flip across the membrane every second, or putting it another way, it takes 10^4–10^6 seconds (i.e., several days) for a single lipid to flip-flop across the membrane (Kornberg and McConnell, 1971a; Contreras et al., 2010). While the rate of lipid flip-flop is slow, the rate of protein flip-flop is even slower and has not yet been observed. The high energies required for lipids and proteins to flip-flop between the two sides of the membrane assures that if the two leaflets of the membrane are synthesized asymmetrically, they will maintain their asymmetry.

All the data obtained to date are consistent with the fluid mosaic model. Moreover, this model of a dynamic membrane helps us to conceptualize many cellular processes, including energy generation, nutrient uptake, and those processes involved in responding to the external environment.

2.5 ISOLATION OF THE PLASMA MEMBRANE

In most cells, the plasma membrane accounts for less than 5% of the cellular membranes. Thus, if we want to be sure that the plasma membrane itself is involved in a given process, we must isolate it from all the other membranes in the cell. However, one of the rules of cell biology is to analyze the properties of the plasma membrane in vitro in light of the known properties in vivo, and once information about the properties of this membrane is gathered in vitro, apply it right away to understand the function of the plasma membrane in the living cell. That is, we must always keep in mind the relationship of the parts to the whole, and in the words of Lester Sharp (1934), "a true conception of the organism can be approached only when analysis into physico-chemical components is followed by resynthesis into a biological whole."

Eduard Buchner (1907) successfully developed a method to observe cellular processes in cell-free systems. In his Nobel Lecture, Buchner described the principles of isolating subcellular objects: "For the chemical investigation of the cell contents, it was necessary to remove the membrane and the plasma envelope by crushing them to pieces. Furthermore, all chemically active solvents and the use of higher temperatures had to be avoided. Finally, it was important that the process should reach completion in the shortest possible time, which would exclude any change whilst the operation was proceeding." Christian de Duve (1975) added the power of centrifugation to isolating subcellular objects and described his journey through the cell with the aid of a centrifuge. In the process, he described the history, theory, and practice of cell fractionation. Anyone interested in isolating organelles should read his Nobel Lecture. In each chapter of this book, I describe a general procedure for isolating a given organelle, although it must be realized that the procedure usually has to be modified for each tissue and species. Furthermore, it is likely that techniques will be developed that further maximize the yield, minimize the contamination due to other cellular components, and minimize the loss of molecules from the organelle that are present in vivo. With these and other (Hillman, 2001) caveats in mind, I describe the isolation of the plasma membrane.

To isolate the plasma membrane, the tissue must first be homogenized. The homogenate is then passed through a filter to remove the walls and whole cells. The filtrate is then centrifuged at 10,000 g for 15 minutes to get a supernatant free of nuclei, plastids, and mitochondria. This differential centrifugation separates organelles solely on the basis of their differential rates of sedimentation. In general, because the densities of particles are similar, the larger the particle, the faster it sediments. The supernatant from the differential centrifugation step contains small particles. This supernatant is centrifuged at 100,000 g for 30 minutes to separate the membranes from the majority of cytosolic proteins. The supernatant is discarded and the pellet is then resuspended. At this stage, the plasma membrane can be isolated from the other cellular membranes in one of three ways. One way of isolating the plasma membrane is based on equilibrium density-gradient sedimentation, where the membranes are separated on the basis of their densities. Before the membranes are applied to the centrifuge tube, the tube is filled with sucrose or a polymer and centrifuged to establish a gradient of densities. The gradient is maintained indefinitely due to the opposing actions of diffusion and sedimentation (Meselson et al., 1957). Once the gradient is established, the membranes are loaded on the top of the gradient and centrifuged at approximately 100,000 g for a few hours.

Each type of membrane forms a band at its particular density (Fig. 2.14A). In general, the greater the proportion of protein ($\rho \approx$ 1.33 g/mL) to lipid ($\rho \approx$ 0.9 g/mL), the greater the density of the membrane. The density of the plasma membrane is typically about 1.16 g/mL. The identity of the plasma membrane is confirmed by assaying whether or not it has high vanadate-sensitive,

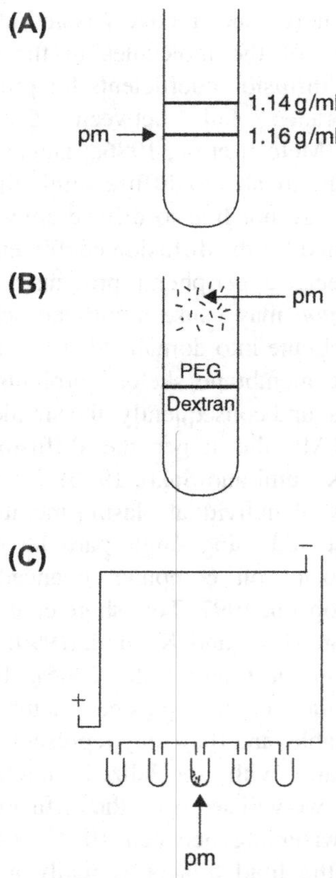

FIGURE 2.14 Techniques used to isolate plasma membranes: (A) equilibrium density-gradient centrifugation, (B) aqueous two-phase partitioning, and (C) free-flow electrophoresis.

K^+-stimulated H^+-ATPase activity for the amount of protein in the sample. Each membrane has a specific enzyme that can be used to identify it. These enzymes are called *marker enzymes* (Quail, 1979). The plasma membrane fraction should also be assayed for the marker enzymes of the other membranes to determine the amount of contamination in the fraction.

The plasma membrane can also be isolated from the other membranes on the basis of its surface properties, including charge and hydrophobicity using an aqueous two-phase partitioning technique (see Fig. 2.14B; Kjellbom and Larsson, 1984; Asenjo and Andrews, 2011). To perform this technique, the membranes are mixed with a solution of 6.4% (w/w) dextran T500, which is more hydrophilic, and 6.4% (w/w) polyethylene glycol 3400, which is more hydrophobic. The solutions are then centrifuged so that the polyethylene glycol forms a layer above and the dextran forms a layer below. Right-side-out plasma membrane vesicles end up in the upper phase. To purify the right-side-out plasma membranes further, the partitioning step can be repeated by collecting the upper phase and mixing it with fresh lower phase solution (Faraday and Spanswick, 1992; Faraday et al., 1996).

Plasma membranes can also be purified on the basis of their charge densities with a technique known as free-flow electrophoresis (see Fig. 2.14C; Sandelius et al., 1986). With this technique, a mixture of membranes is introduced into a separation buffer flowing perpendicular to an electric field. Membranes bearing different electrical charge densities will migrate different distances along the separation chamber. Each type of membrane flows into a different collecting tube, which is then centrifuged at 110,000 g for 30 minutes to concentrate the membranes. This technique can resolve vesicles of different sidedness.

How can we determine the sidedness of the membranes? There is a good trick (see Fig. 2.15; Canut et al., 1987). Do you remember when we talked about the E- and P-spaces? Adenosine triphosphate (ATP) is almost always in the P-space. Therefore, the portions of membrane proteins that bind ATP must be on the P-side of the membrane. Thus, if all the membranes are tightly sealed and right side out, and we add ATP to assay the VO_4-sensitive, K^+-stimulated, H^+-ATPase activity, we should see no activity because the membrane prevents a large hydrophilic molecule like ATP from getting to the P-space. If we then add a detergent such as Triton X-100 to permeabilize the membrane so that ATP can enter the P-space and bind to the enzyme on the P-side of the plasma membrane, we should see an enhancement of the ATPase activity. If we see a detergent enhancement, which is known as *latency*, we say that the membrane vesicles are tightly sealed and are right side out. If the ATPase exhibits high activity both with and without the detergent, the membranes are either inside out or leaky.

Once we have isolated right-side-out plasma membrane vesicles, we can use them to characterize the function of the lipids and proteins that determine the permeability of the membrane. We can also use the isolated plasma membranes to determine the chemical profile of the lipids and proteins that reside in the membrane.

2.6 CHEMICAL COMPOSITION OF THE PLASMA MEMBRANE

Because most nutrients and metabolites are polar, the nonpolar lipids in the membrane bilayer act as a barrier to their passage. On the other hand, once a given molecule gets to the right place in or out of the cell, the lipid bilayer will help it stay there. As we will see in Section 2.8, the lipid bilayer also forms a nonconducting layer that gives the plasma membrane the capacity to store charge (i.e., act as a capacitor), and the stored charge can be used to do work. The lipids also provide the fluid environment where the integral membrane proteins involved in transporting polar molecules reside.

To characterize the lipid fraction, the plasma membranes must be extracted with nonpolar organic solvents (e.g., chloroform, methanol, HCl). The lipids in the extract are then separated by polarity on silica Sep-Pak cartridges, and the fractions are identified by thin-layer chromatography, gas chromatography, and mass spectrometry (Schneiter et al., 1999). Lipids have been characterized in the plasma membrane of a variety of plants (Sheffer et al., 1986; Lynch and Steponkus, 1987; Sandstrom and Cleland, 1989; Brown and DuPont, 1989; Navari-Izzo et al., 1989; Peeler et al., 1989; Cahoon and Lynch, 1991; Bohn et al., 2001; Kerkeb et al., 2001; Quartacci et al., 2002; Lin et al., 2003; Welti and Wang, 2004; Cacas et al., 2016). Omic approaches are now being applied to the study of lipids in plants, although the resolution is currently at the whole-cell level (Welti and Wang, 2004; Welti et al., 2007; Shulaev and Chapman, 2017).

The membrane lipids make up about 30% of the weight of the plasma membrane. The membrane lipids of the plasma membrane are mainly represented by phospholipids, glycolipids, and sterols. Glycolipids possess either a glycerol backbone and are known as glyceroglycolipids or they possess a sphingosine (18 carbon amino alcohol)

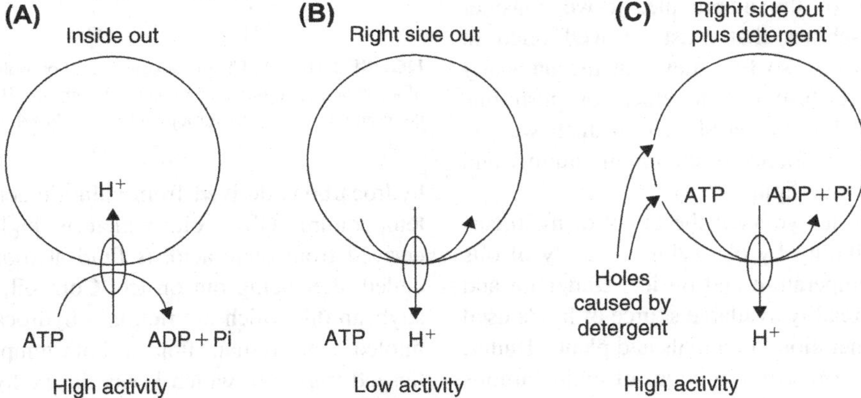

FIGURE 2.15 Determination of the orientation of a membrane vesicle. (A) intact inside-out vesicle; (B) intact right-side-out vesicle; (C) detergent-permeabilized right-side-out vesicle.

backbone and are known as sphingolipids. Sterols are neutral and nonpolar lipids. In plants, unlike animals where cholesterol is the only sterol, the sterols form a complex mixture, which may include sitosterol, stigmasterol, cholesterol, and brassicasterol (Nes, 1977; Hartmann and Benveniste, 1987; Gachotte et al., 1995; Hartmann, 1998). The glycolipids are polar lipids that contain one or more sugar groups. The phospholipids are polar lipids that contain a glycerol molecule with a phosphate group attached through an ester bond to one carbon and two hydrocarbon tails, which result from the esterification of fatty acids to hydroxyl groups attached to the remaining two carbons of the glycerol molecule (Fig. 2.16). A polar group, such as choline, serine, inositol, glycerol, or ethanolamine, is attached to the phosphate group through an ester bond. One molecule of water is eliminated in the formation of each ester bond. The combination of the glycerol phosphate and the additional group is known as the *polar head group*. Careful workers determine the true molecular species of each lipid, which involves identifying the polar head group and the fatty acids that are derived from the hydrocarbon tails. They also determine the position of each hydrocarbon tail in each lipid. Less analytically precise workers analyze the polar head groups and hydrocarbons separately. The polar head group composition of a variety of plasma membranes are given in Table 2.1.

Notice that there is a basic similarity between different plasma membranes. On the average, the noted membranes are composed of (in mole percent) 40.7% phospholipids, 27.3% glycolipids, and 25.4 free sterols—values very similar to those found in the plasma membranes of animal cells.

By contrast, the hydrocarbon compositions of plant and animal plasma membranes are somewhat different. Mammalian membranes are enriched in hydrocarbons derived from stearic acid (18:0) and oleic acid (18:1), whereas plant plasma membranes are enriched in hydrocarbons derived from palmitic (16:0), linoleic (18:2), and linolenic (18:3) acids. This is particularly intriguing because the hydrocarbon composition may have a large effect on the fluidity of the membrane. If we consider enzymes as little machines that must be well oiled to change their conformation so that they can mechanically split a molecule apart, fuse two together, or push one through a membrane (Johnson et al., 1974), then we can visualize the possible importance of membrane fluidity and the diversity in membrane lipids.

You can easily see for yourself the effect of hydrocarbon composition in fluidity by observing a variety of oils and butter at room temperature and on ice. Butter fat and vegetable oils are the readily available source of lipids used to nourish the next generation of animals and plants. Butter, an animal product, is primarily a mixture of hydrocarbons derived from palmitic and oleic acids and is solid both at room temperature and on ice. Palm oil, which contains

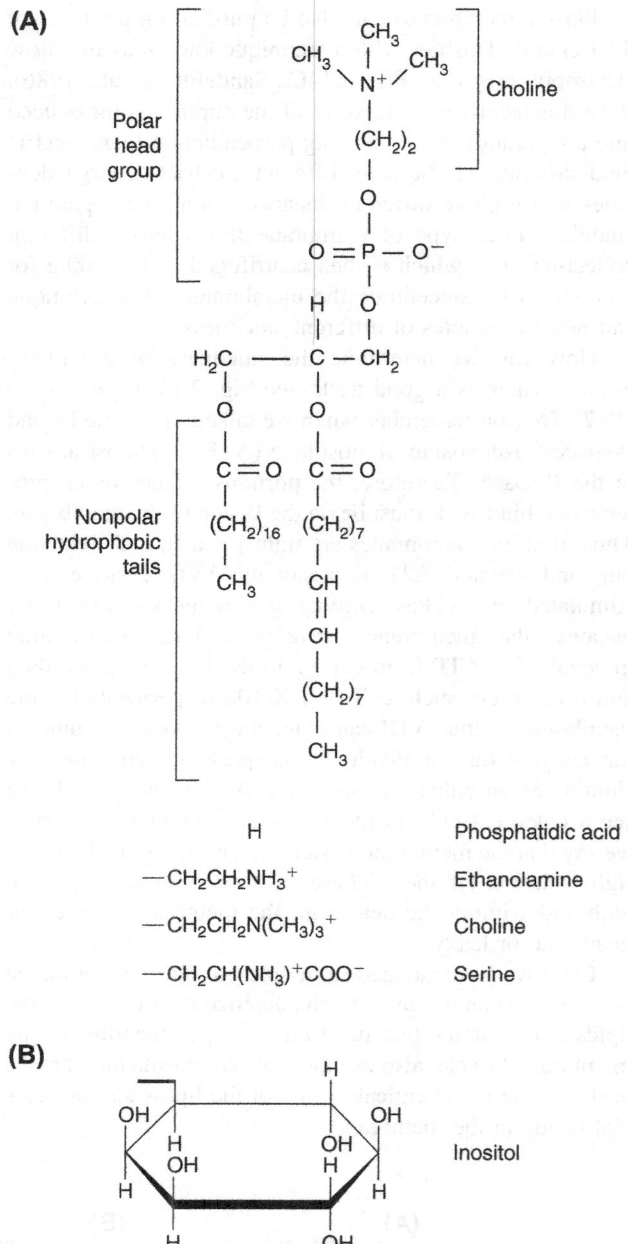

FIGURE 2.16 (A) Chemical structure of phosphatidylcholine composed of a stearic acyl group and an oleic acyl group. (B) Chemical structures of the common polar head groups found in phospholipids.

hydrocarbons derived from palmitic acid, is solid at room temperature. Olive oil, which is high in hydrocarbons derived from oleic acid, is fluid at room temperature and gelled after being put on ice. Corn oil, sunflower oil, and soybean oil, which are rich in a hydrocarbon derived from linoleic acid, remain fluid at both temperatures. Likewise, linseed (flax) oil, which is rich in the hydrocarbon derived from linolenic acid (omega-3 fatty acid), remains liquid at both temperatures.

TABLE 2.1 Lipid Composition of Plasma Membranes (Polar Head Groups)

	Oat Coleoptile	Oat Root	*Dunaliella*	Barley Root	Corn Shoot	Rye Leaves
Total phospholipid	41.7	50.1	29.2	43.3	47.2	31.7/41.9
LPC	1.0	1.7				
PI	2.3	1.5				0.7/0.5
PS	3.2	4.2				1.5/1.0
PC	9.6	14.3	13.2			14.8/19.5
PG	1.8	1.3	5.3			1.8/2.1
PA	14.8	11.8				1.7/2.3
PE	9.0	15.3	10.7			10.9/15.7
Total glycolipid	38.9	25.8	7.7	25.5	44.2	35.6/13.3
ASG	5.	4.5				4.3/1.1
SG	7.3	10.6				15.1/5.7
GC	26.1	10.1				16.2/6.8
Total free sterol	19.4	24.6	26.4	25.5	4.6	32.7/44.4
Cholesterol	3.1	0.1				0.5/0.4
Brassicasterol	2.0	n.d.				
Campesterol	1.9	2.0			3.5/1.3	
Stigmasterol	1.6	12.1			0.6/<0.1	
β-Sitosterol	9.2	5.1				20.8/32.6
Unknown	1.5	5.2				

ASG, acylated sterol glycoside; *GC*, glycocerebroside; *LPC*, lysophosphatidylcholine; *PA*, phosphatidic acid; *PC*, phosphatidylcholine; *PE*, phosphatidylethanolamine; *PG*, phosphatidylglycerol; *PI*, phosphatidylinositol; *PS*, phosphatidylserine; *SG*, steryl glycoside.

The diversity in chemical composition of lipids is probably adaptive, as butter comes from warm-blooded animals, palms live in the tropics, olive trees live in warm Mediterranean climates, the crops rich in the hydrocarbons derived from linoleic acid grow in temperate regions, and crops rich in hydrocarbons derived from linolenic acid come from frost-resistant flax grown in the colder regions of the temperate zone. Cold-water fish also contain large amounts of linolenic acid.

The membrane fluidity is affected by the chain length of the hydrocarbons. As the chain length increases, the hydrocarbons interact with each other to a greater extent and form a more gel-like membrane. This causes the melting point to increase. The melting point decreases as the number of double bonds in the hydrocarbons increase. This is because the double bonds cause a kink in the hydrocarbons, which prevents them from interacting with each other. This decreases their gel-like properties and thus decreases the melting point. Therefore, the relative proportion of each hydrocarbon determines the fluidity of the membrane at a given temperature (Table 2.2). Indeed, the hydrocarbon composition of some phospholipids changes after cold acclimation or osmotic stress, indicating

TABLE 2.2 Names and Melting Points of Various Fatty Acids

Symbol[a]	Common Name	Melting Point (°C)
14:0	Myristic	53.9
16:0	Palmitic	63.1
18:0	Stearic	69.6
20:0	Arachidic	76.5
$16:1^{9}$	Palmitoleic	−0.5
$18:1^{9}$	Oleic	13.4
$18:2^{9,12}$	Linoleic	−5
$18:3^{9,12,15}$	Linolenic	−11
$20:4^{5,8,1,14}$	Arachidonic	−49.5

[a]*The numbers in the superscript represent the positions of the double bond(s) counting from the carbon that occurs in the carboxylic acid group. For example, $18:2^{9,12}$ indicates that there are 18 carbons in the fatty acyl chain with two double bonds. One double bond is in the 9th position and the other is in the 12th position from the carboxylic end.*

that there is a relationship between the lipid composition of a membrane and its function (Lynch and Steponkus, 1987; Peeler et al., 1989; Uemura, 2001).

Why are there approximately 100 different kinds of lipids that coexist in the plasma membrane? We really do not know. It is possible that a variety of lipids is required to maintain the membrane fluidity in the correct range throughout the day and throughout the seasons. Interestingly enough, just the presence of many types of lipids will maintain the fluidity of the bilayer because each one acts like an impurity for the others and prevents crystallization. Each lipid has a different conformational shape, and we may find that different lipids are necessary to maintain a tight barrier against the free diffusion of polar molecules or ions across the membrane in curved areas and in flat areas of the membrane—or in regions of the membrane with various intermediate curvatures. We may find that many lipids are necessary because each molecular species of lipid performs a specific function. For example, some proteins, including the proton ATPase on the plasma membrane, require certain lipids for activation (Kasamo and Sakakibara, 1995; Kasamo, 2003), and some proteins require glycosylphosphatidylinositol (GPI) to anchor them in the membrane. Some lipids, including phosphatidylinositol and sphingolipids, participate in cell signaling.

The plasma membrane is also composed of proteins, which make up approximately two-thirds of the weight of the membrane. Proteins are composed of one or more polypeptide chains of amino acids. Amino acids are bifunctional molecules with the following structure: $H_2N{-}CRH{-}COOH$ (Fig. 2.17). The properties of the proteins are due, in part, to the properties of their amino acids, the properties of the amino acids are due to the various radical or R groups, and the properties of the R groups depend on the properties of the atoms that constitute them. The salient property of the atoms is their electronegativity, which is a semiquantitative value that can be given in Joules (J) and represents their affinities for electrons (Pauling, 1932, 1940, 1954a, 1970). Higher electronegativities represent greater affinities for electrons. In general, carbon (2.5) and hydrogen (2.1) atoms have approximately equal electronegativities. Consequently, R groups that contain many hydrocarbon (CH) bonds are typically nonpolar. By contrast, the oxygen (3.5) atom has a greater electronegativity than the hydrogen atom, and thus the OH bond has a partial ionic character and acts like an electrical dipole, which makes R groups with an OH group polar.

R groups with more than one polar group may be charged as a result of the combined action of the polar groups on a given atom. In complex R groups, nitrogen becomes positively charged and oxygen becomes negatively charged. As a result of the different electronegativities of the C, H, O, N, and S that make up the R groups, the amino acids that make up the proteins can be nonpolar, polar, or charged. The charged and polar amino acids will be soluble in water and are thus called *hydrophilic*. The relative hydrophilicities of the 20 amino acids that make up proteins are given in Table 2.3. The amino acids with the lowest hydrophilicities are the most likely to be in contact with the hydrocarbons derived from fatty acids in the lipid bilayer.

The various polypeptides that reside in the plasma membrane can be observed using polyacrylamide gel electrophoresis (PAGE) after dissolving the isolated plasma membrane in an ionic detergent like sodium dodecyl sulfate (SDS). The SDS unfolds and coats each integral and peripheral protein so that the number of SDS molecules attached is proportional to the mass of the protein. The SDS molecules are negatively charged and this overwhelmingly negative charge masks the intrinsic charge of the protein so that each protein travels through the gel with essentially the same charge-to-mass ratio.

The SDS—protein complexes are put on top of an acrylamide gel that has a given pore size. When an electric field is placed across the gel, the proteins are separated by mass because the smaller proteins migrate through the gel faster than the larger ones as they move toward the positive pole (anode). There are approximately 100 polypeptides in the plasma membrane as determined by SDS PAGE (see Fig. 2.18) and over 200 by proteomic analysis using mass spectrometry (Alexandersson et al., 2004; Lefebvre et al., 2007; Marmagne et al., 2004, 2007; Tang et al., 2008; Zhang et al., 2008; Han et al., 2010; Paul et al., 2016). Out of these, only a few have been well characterized. As we learned from growing pollen tubes, the protein composition of a given membrane may change during the development of the cell (Pertl et al., 2009; Pertl-Obermeyer and Obermeyer, 2013; Pertl-Obermeyer, 2017).

While the lipid bilayer prevents water-soluble solutes from entering or leaving the cell, membranes contain special transport mechanisms made of proteins and coded by genes (Sze et al., 2014) that contain an aqueous pathway that facilitates the movement of hydrophilic solutes that permeate the membrane (Bethe, 1930). Depending on the tortuosity or openness of the pathway, the proteins are classified into two groups: carrier proteins and channel proteins. As with any system of classification, the division is convenient, but artificial, and intermediates between carriers and channels exist (Eisenberg, 1990; Läuger, 1991).

Carrier proteins act like enzymes in that their ability to transport a solute increases as the solute concentration increases, but eventually saturates. Carrier proteins can also be characterized by their maximal velocity and their affinity for the transported solute (Mitchell, 1954a,b; Epstein et al., 1963; Welch and Epstein, 1968; Weiss, 1996). Because carrier proteins undergo a conformational change to move

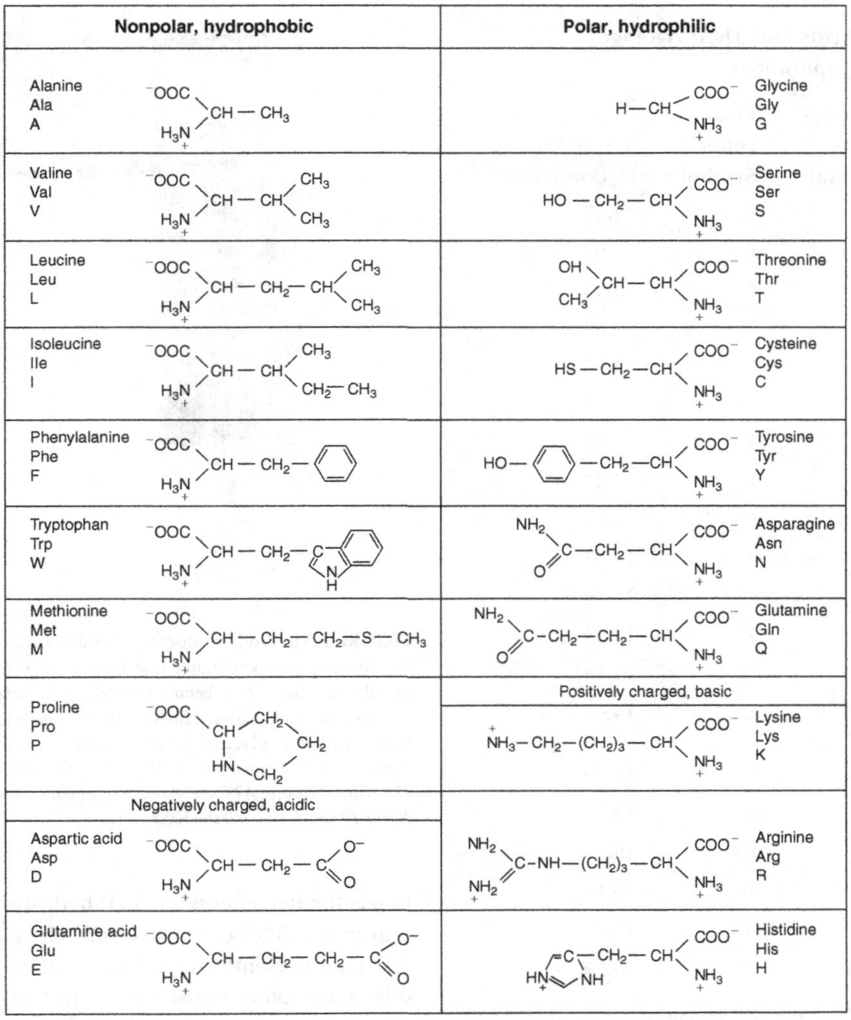

Nonpolar, hydrophobic		Polar, hydrophilic	
Alanine Ala A	$^-OOC, H_3\overset{+}{N}$ >CH — CH_3	H—CH< COO^-, $\overset{+}{N}H_3$	Glycine Gly G
Valine Val V	$^-OOC, H_3\overset{+}{N}$ >CH — CH< CH_3, CH_3	HO — CH_2— CH< COO^-, $\overset{+}{N}H_3$	Serine Ser S
Leucine Leu L	$^-OOC, H_3\overset{+}{N}$ >CH — CH_2— CH< CH_3, CH_3	OH, CH_3 >CH — CH< COO^-, $\overset{+}{N}H_3$	Threonine Thr T
Isoleucine Ile I	$^-OOC, H_3\overset{+}{N}$ >CH — CH< CH_3, CH_2— CH_3	HS — CH_2—CH< COO^-, $\overset{+}{N}H_3$	Cysteine Cys C
Phenylalanine Phe F	$^-OOC, H_3\overset{+}{N}$ >CH — CH_2—(benzene ring)	HO—(benzene ring)—CH_2—CH< COO^-, $\overset{+}{N}H_3$	Tyrosine Tyr Y
Tryptophan Trp W	$^-OOC, H_3\overset{+}{N}$ >CH — CH_2—C(indole ring, N H)	NH_2, O >C—CH_2— CH< COO^-, $\overset{+}{N}H_3$	Asparagine Asn N
Methionine Met M	$^-OOC, H_3\overset{+}{N}$ >CH — CH_2— CH_2—S— CH_3	NH_2, O >C-CH_2—CH_2—CH< COO^-, $\overset{+}{N}H_3$	Glutamine Gln Q
		Positively charged, basic	
Proline Pro P	^-OOC CH (ring: CH_2, CH_2, CH_2, HN)	$\overset{+}{N}H_3$— CH_2—$(CH_2)_3$— CH< COO^-, $\overset{+}{N}H_3$	Lysine Lys K
Negatively charged, acidic			
Aspartic acid Asp D	$^-OOC, H_3\overset{+}{N}$ >CH — CH_2— C< O^-, O	NH_2, $\overset{+}{N}H_2$ >C-NH—$(CH_2)_3$—CH< COO^-, $\overset{+}{N}H_3$	Arginine Arg R
Glutamine acid Glu E	$^-OOC, H_3\overset{+}{N}$ >CH — CH_2— CH_2— C< O^-, O	(imidazole ring: C, $H\overset{+}{N}$, NH)—CH_2— CH< COO^-, $\overset{+}{N}H_3$	Histidine His H

FIGURE 2.17 Structure of amino acids with their three-letter and one-letter abbreviations.

solutes through a relatively tortuous pathway, they are relatively slow and can transport only 10^2–10^4 solutes/s. Peter Mitchell (1978) named carriers in terms of their chemiosmotic functions (see Chapters 13 and 14). A carrier protein can act as a uniporter, which transports one type of ion in one direction (e.g., H^+-ATPase), or it can be involved in coupled transport, where it transports two types of solutes in the same direction (e.g., H^+/amino acid symporter; Borstlap and Schuurmans, 2001) or in opposite directions (e.g., Ca^{2+}/H^+ antiporter). When carrier proteins facilitate the movement of solutes down an electrochemical difference, in a process known as *facilitated diffusion*, the process is passive. That is, it does not require any additional inputs of free energy. The evolution of carrier proteins are being determined by analyzing the sequence of nucleotides that encode them (Chanroj et al., 2012; Emery et al., 2012).

Carrier protein transport can also be coupled to ATP hydrolysis. When this occurs, the carriers can facilitate the movement of a solute against its electrochemical difference (Ussing, 1947, 1949). This process is known as *active transport* because a free-energy input in the form of ATP is required. When a carrier protein requires the hydrolysis of a phosphoanhydride bond (e.g., ATP or pyrophosphate), thereby converting the free energy of chemical bonds into the free energy of an electrochemical difference, the carrier is considered to be a *primary transporter* or pump. In plant cells, primary pumps typically transport protons. By contrast, *secondary transporters* transport ions and organic molecules by using the free energy inherent in the electrochemical difference across the membrane formed by a primary transporter. In plant cells, secondary transport dissipates the electrochemical difference of protons that is established by the primary transporter. In secondary transport, the free energy made available by the movement of protons down their electrochemical difference is used to move another type of ion or uncharged organic molecule

TABLE 2.3 Amino Acids and Their Average Hydrophilicity-Hydrophobicity

Amino Acid	Three-Letter Symbol	One-Letter Symbol	Hydrophilicity[a]
Tryptophan	Trp	W	2.57
Phenylalanine	Phe	F	2.64
Leucine	Leu	L	3.29
Isoleucine	Ile	I	3.64
Tyrosine	Tyr	Y	4.57
Methionine	Met	M	6.57
Valine	Val	V	7.50
Proline	Pro	P	7.57
Cysteine	Cys	C	8.29
Alanine	Ala	A	12.07
Histidine	His	H	12.79
Threonine	Thr	T	13.64
Glutamine	Gln	Q	14.36
Glutamic acid	Glu	E	14.64
Glycine	Gly	G	14.79
Serine	Ser	S	14.93
Arginine	Arg	R	15.93
Asparagine	Asn	N	16.14
Lysine	Lys	K	16.21
Aspartic acid	Asp	D	16.29

[a]*The greater the hydrophilicity number, the more hydrophilic the amino acid.*

against its electrochemical or concentration difference, respectively. While most primary pumps transport ions using the free energy made available from ATP hydrolysis, some primary pumps transport protons and other ions at the expense of free energy made available by electron transport chains (Conway, 1953; Møller and Lin, 1986; Rubinstein and Luster, 1993; Trost, 2003).

Channel proteins contain relatively large aqueous pores that have a diameter of about 0.6 nm (Hille, 1992). As a reference, ions have a diameter of about 0.2 nm. Because the pores are so large, they can pass solutes at a rate of 10^7-10^8 particles/s. Eisenberg (1990) suggests that channel proteins should also be considered as enzymes that catalyze the flow of current. However, for a channel protein, the maximal velocity is higher and the affinity for a solute is lower than it would be for a carrier protein that binds a solute and transports it through a tortuous path. However, unlike carrier proteins, channel proteins cannot be coupled

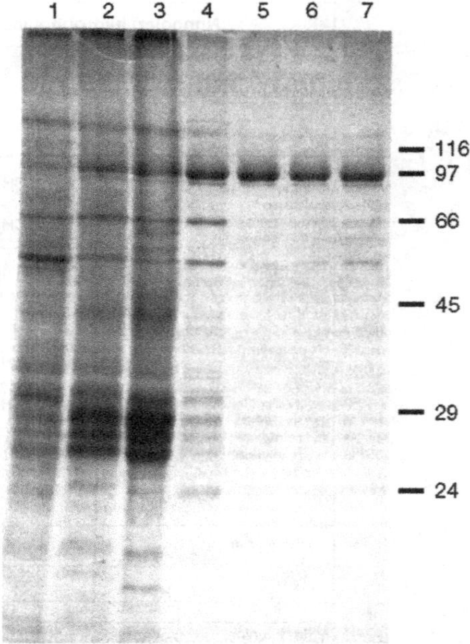

FIGURE 2.18 sodium dodecyl sulfate polyacrylamide gel of (1) crude membranes, (2) membranes that have been washed with Triton/KBr, (3) membranes that have been extracted with octyl glucoside plus deoxycholate, (4) the lysolecithin-solubilized supernatant. (5), (6), and (7) are three different glycerol-gradient purified protein preparations. *From Anthon, G.E., Spanswick, R.M., 1986. Purification and properties of the H^+-translocating ATPase from the plasma membrane of tomato roots. Plant Physiol. 81, 1080–1085.*

to use the free energy of ATP hydrolysis, and thus can only transport solutes down an electrochemical gradient and thus dissipate it. Some channels pass only a given solute, while others are rather nonselective. In fact, the selectivity for a particular channel can be changed by mutating a single amino acid (Yang et al., 1993a,b; Tang et al., 1993; Catterall, 1994, 1995; Jan and Jan, 1994; Uozumi et al., 1995; Mäser et al., 2001, 2002).

If the pores of channel proteins were always open, the movement of solutes would continue until the free energies inherent in the electrochemical gradients of various solutes were dissipated. Thus, the pores must be opened and closed in a regulated manner. This is called *gating*. Channel gates can be opened and closed by the mechanical energy of compression, tension and stretch, the electromagnetic energy of light, chemically by the binding of a ligand (e.g., hormone or nucleotide), or electrically by the change in the electrical potential across the membrane. These channels are referred to as *mechanically gated* (Falke et al., 1988; Cosgrove and Hedrich, 1991; Dutta and Robinson, 2004; Qi et al., 2004; Haswell, 2007; Haswell et al., 2008; Martinac et al., 2008; Hamilton et al., 2015), light-gated (Li et al., 2005; Zhang et al., 2006; Hegemann and Tsunoda, 2007; Zhang and Oertner, 2007; Hegemann, 2008), ligand-gated (Dietrich et al., 2010; Zelman et al., 2012; DeFalco et al., 2016), and voltage-gated channels, respectively. The gates regulate the communication between the living protoplasm and the lifeless environment—the P- and E-spaces.

Many of the transport proteins may have related domains, which have joined together in various ways to make related yet unique proteins (Doolittle, 1995). By comparing the sequence of nucleotides of genes that code for the diverse transport proteins, one can make predictions about the function of a given domain. Currently, the structures of various plasma membrane proteins are studied bioinformatically in silico, using software that predicts the structure and function of the protein from its amino acid sequence inferred from the nucleotide sequence of the gene that encodes the protein. Algorithms are available online that use the nucleotide sequence of genes that encode ion channels to predict the properties of the channel protein, including transmembrane domains, the beta-barrel regions, the presence of a signal peptide, the presence of a myristoylation site, the presence of a glycosylphosphatidylinositol (GPI) anchoring site, and the subcellular localization of the protein.

According to Ullrich Lüttge (2016), Whole-plant physiology shows that transport is the basis of the functioning of entire plants. Transport is the pathway for interaction and integration creating plant's individuality as unitary organisms. The integration of modules via transport leads to the emergence of holistic systems across a large range of scalar levels from compartments within cells to cells and eventually the whole biosphere. Comprehending emergence of holism leads to understanding life beyond mechanistic modularity. Now that I have discussed the general nature of membrane transport, I will discuss two proteins that function in transporting solutes across the lipid bilayer, but first let us become familiar with the ways to quantify transport.

2.7 TRANSPORT PHYSIOLOGY

Plasmolysis studies show that the plasma membrane is readily permeable to water but not to the salts and sugars. Yet these solutes, which are so necessary for life, must be taken up and the plasma membrane has the mechanisms necessary to take up and eliminate each solute in a regulated manner. Differential membrane permeability and the mechanisms that cause it can be observed in certain cells and organs, including the root hair and endodermal cells of roots and the cells of the brush border of the intestines of animals that specialize in nutrient uptake. Likewise, differential membrane permeability and the mechanisms that cause it can be observed in certain cells and organs, including the salt glands of plants and the kidneys of animals that specialize in the elimination of wastes. The differential ionic permeability and the mechanisms that cause it can be observed in the guard cells of the stomatal complexes of plants, which are important in regulating the balance between carbon dioxide uptake and water loss. Differential membrane permeability and the mechanisms that regulate it can be observed in pollen tubes and root hairs in plants and myoblasts and neurons in animals, cells that grow in a polarized manner (Lund, 1947; Jaffe and Poo, 1979; Hinkle et al., 1981; Nuccitelli, 1983; Patel and Poo, 1984; Robinson, 1985). The differential membrane permeability and the mechanisms that control it can be observed in the neurons of animals and in some plant cells responsible for the coordination between parts of large organisms (Cole and Curtis, 1938, 1939; Eccles, 1963; Hodgkin, 1963; Huxley, 1963; Erickson and Nuccitelli, 1982; Nuccitelli and Erickson, 1983; Cooper and Keller, 1984; Cooper and Schliwa, 1985). Differential membrane permeability and the mechanisms that cause it can also be observed in pathogens that use electric fields to target roots (Morris and Gow, 1993; Robinson and Messerli, 2002; van West et al., 2002). Differential membrane permeability and the mechanisms that cause it are not static but change during cell development and aging (Laties, 1964). It is interesting to think about all the plasma membranes that a nutrient crosses as it is transported through the food chain from the soil, to the plant, to the herbivore, carnivore, or omnivore, and back to the soil (Weiss, 1996).

As important as membrane permeability to a particular solute is, it cannot be measured directly (Jost, 1907). Rather, we must postulate a relationship between the permeability of the membrane for a given solute and the measurable quantities, including the amount of the solute transported, time, area, and driving force. Consequently, according to Jack Dainty (1962), "no permeability coefficient should ever be quoted without stating the theory according to which it was calculated." As no theory of membrane permeability was available to Adolf Fick, a physiologist who was searching for a physical description of how kidneys work, he had to come up with his own theory. Using the concept of analogy, Fick (1855) considered that a

> *law for the diffusion of a salt in its solvent must be identical with that … [which describes] the diffusion of heat … and … the diffusion of electricity. According to this law, the transfer of salt, and water occurring in a unit of time, between two elements of space filled with differently concentrated solutions of the same salt, must be cœteris paribus, directly proportional to the difference in the concentration, and inversely proportional to the distance of the elements from one another.*

When nonelectrolytes are the permeators, the driving force is expressed as the concentration difference divided by the distance between the high and low concentrations, according to Fick's First Law

$$\left(\frac{ds}{dt}\right)A = J = -D\left(\frac{dC}{dx}\right) \quad (2.1)$$

where *ds* is the amount of solute (in mol) passing through the membrane in a given time (*dt*, in s); *A* is the area of the membrane (in m^2); and [(*ds*/*dt*) (1/*A*)] is defined as the flux in mol $m^{-2}s^{-1}$ and is often denoted by *J*. *dC* is the concentration difference (in mol/m^3) over distance *dx* (in m), and it is defined as the low concentration minus the high concentration (i.e., *dC*/*dx* is the concentration gradient, which is defined as the negative of the concentration drop). *D* is the diffusion coefficient (in m^2/s) that relates the flux to the concentration difference. When the concentration differences of two different substances are identical but the fluxes are different, the substance with the greater flux will have a greater diffusion coefficient. The diffusion coefficient is related to velocity. Alternative definitions of the diffusion coefficient are given in Eqs. (2.5) and (2.8).

Fick (1855) investigated the diffusion of salt in water or across aqueous porous membranes and did not have to take into consideration the presence of a membrane barrier that could hinder the passage of the solute on the basis of the lack of solubility of the substance in the membrane. However, when we apply Fick's Law to biological membranes we have to account for the fact that each solute must enter and leave the membrane, and the solubility of the solute in the membrane material will be a factor that will also determine its flow. We use the dimensionless partition coefficient, *K*, introduced by the chemist Marcellin Pierre Berthelot, as an estimate of how soluble a given solute is in the membrane relative to its solubility in an aqueous solution:

$$J = -DK\left(\frac{dC}{dx}\right) \quad (2.2)$$

Because we do not know the actual partition coefficient of the solute in the membrane compared to water, we estimate it by measuring the relative distribution of that solute in olive oil, or any other solvent (e.g., octanol) that mimics the hydrophobic properties of the membrane, and water (Stein, 1986).

In the case of plasma membranes, *dx* is usually not a measured or measurable quantity and consequently it is almost impossible to determine the diffusion coefficient. Thus, we use Runnström's (1911) modification of Fick's Law to relate the flux of nonelectrolytes to the magnitude of measurable quantities:

$$J = -P(dC) \quad (2.3)$$

where *P* is equal to (*D*)(*K*)/*dx* and is called the *permeability coefficient*. It is given in units of m/s. Consequently, the permeability coefficient gives us an idea of the velocity with which a given substance will permeate a given membrane.

Let us do an example. Assume that the concentration of sucrose outside the cell is 100 mol/m^3, the concentration inside is 1 mol/m^3, and the cell is a cube, the sides of which have a length (10^{-5} m). Using a radioactive tracer, we measure how many moles of sucrose are in the cell after a given time and we find that the flow (*ds*/*dt*) is 1.2×10^{-15} mol/s. Now we will calculate the permeability coefficient:

$$P = -\frac{\left(\frac{ds}{dt}\right)}{A\,dC}$$

$$A = 6(10 \times 10^{-6}\text{m})^2 = 6 \times 10^{-10}\text{ m}^2$$

$$dC = 1\text{ mol/m}^3 - 100\text{ mol/m}^3 = -99\text{ mol/m}^3$$

$$P = -(1.2 \times 10^{-15}\text{ mol/s})(6 \times 10^{-10}\text{ m}^2)^{-1}(-99\text{ mol/m}^3)^{-1}$$

$$P = 2 \times 10^{-8}\text{ m/s}$$

Remember that the use of these equations depends on the validity of the assumptions. First, we assume that the concentration difference does not change during the experiment; therefore, we must use short transport times. We also assume that the concentration of the bulk solution is identical to the concentration at the membrane and therefore there are no unstirred layers (Dainty, 1990). Lastly, it is the absolute activity of the solute, and not the concentration, that is important. The absolute activity is less than the concentration because some of the molecules in question may be bound to each other or to other molecules. The absolute activity is equal to the concentration times the activity coefficient, which is the proportion of the free to the total solute. Thus, we are assuming that the activity coefficients on both sides of the membrane are equal to 1.

The equations can only be used for calculating the permeability coefficients of nonelectrolytes because they assume that the only driving force is the concentration difference. While Fick studied the movement of electrolytes, he used membranes that had little or no capacitance (see Section 2.8) and thus could not develop an electrical potential across them. Without an electrical potential difference, there can be no electrical driving force. The flux of electrolytes is not only affected by the concentration difference but also by the electrical properties of the membrane (Michaelis, 1925; Höber, 1930), including the electrical driving force, which can develop when ions diffuse across a differentially permeable membrane that has capacitance. Thus, the equation used to determine the permeability coefficient of an electrolyte must be an expanded form of Fick's Law. Such an expansion was done by Walther Nernst (1888). The Nernst equation will be used for many applications, which will allow us to determine the contribution of a given ion to the resting membrane potential; to determine if transport is active or passive; to determine the driving force for ion movements; and to determine the specificity of the ionic channels that are observed in patch clamp studies. Remember that all movements require energy. Active transport requires metabolic energy, and passive transport requires thermal energy.

Like any robust equation, the Nernst equation can be derived from many starting points. I will derive the Nernst equation by considering two basic principles behind the movement of ions: Fick's Law and Ohm's Law.

As I already discussed, Fick's Law describes the movement of uncharged solutes. To get a better understanding of Fick's Law, consider two groups of solutes in communication with each other. As a consequence of the thermal motion of the particles, there will be a tendency for the two groups to mix, and a net flux (J_{dif}) of particles will occur from the dense to the sparse group. The magnitude of the flux will depend on the concentration gradient (dC/dx) and the diffusion coefficient (D) in the following manner:

$$J_{dif} = -D\left(\frac{dC}{dx}\right) \tag{2.4}$$

where dC is the concentration difference in molecules/m^3.

The average flux is also proportional to temperature; thus the diffusion coefficient is proportional to the absolute temperature (T, in K). This seems reasonable because diffusion is a consequence of the thermal motion of the solutes (Fig. 2.19).

Boltzmann's constant ($k = 1.38 \times 10^{-23}$ J/K) relates the free energy of the solute to the temperature. The higher the temperature, the more free energy the solute has and the faster it moves. The coefficient of proportionality that relates the diffusion coefficient to kT is called the mobility (u, in m^2/J s). The relation is given by the Nernst–Einstein equation (Einstein, 1956):

$$D = ukT \tag{2.5}$$

where u (in m^2/Js) relates the diffusion coefficient to the thermal energy. In a more graphic sense because Joules/meter = Newton, u (in (m/s)/N) is a measure of the velocity (v, in m/s) a given solute travels when subjected to a given force (F, in N). Therefore, Fick's Law can be rewritten as

$$J_{dif} = -ukT\left(\frac{dC}{dx}\right) \tag{2.6}$$

Eq. (2.6) describes the movement of solutes as a result of diffusion. The concentration difference produces the force needed for movement, and u is a coefficient that relates the velocity of movement to the applied force. That is, $u = v/F$. Note that this equation assumes that velocity and not acceleration, as is found in Newton's Second Law, is proportional to force. That is, a molecule does not accelerate in response to a force because it constantly collides with other molecules. In the microscopic world, molecular resistance is so great that Newton's Second Law does not apply.

Because the mobility is a coefficient that relates the velocity of a particle to the force that causes it to move, we can consider it in terms of Stokes' Law, which states that a force causes a spherical particle to move with a given velocity. However, as a consequence of friction, the velocity of the particle is inversely proportional to the hydrodynamic radius of the particle (r_H, in m) and the viscosity of the medium through which it moves (η, in Pa s). While this relationship was worked out by George Stokes (1922) for the macroscopic movement of pendulums, it is also applicable to the microscopic movement of atoms and molecules, although one can still introduce higher-order corrections (Millikan, 1935). Stokes' Law is

$$v = \frac{F}{6\pi r_H \eta} \tag{2.7}$$

and because $v/F = u$, then $u = 1/(6\pi r_H \eta)$ and

$$D = \frac{kT}{6\pi r_H \eta} \tag{2.8}$$

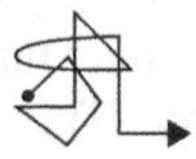

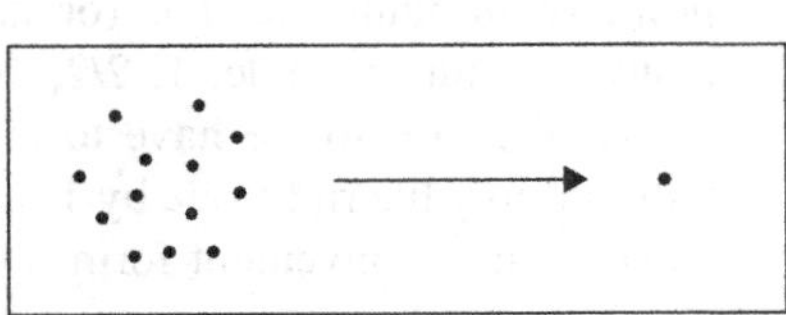

FIGURE 2.19 Thermal motion of a single particle. Due to the uneven initial distribution of particles (i.e., chemical potential), a particle will typically diffuse along an axis from regions where the initial concentration is high to regions where the initial concentration is low. The speed of the particles will remain constant in an isolated system. However, the average velocity vector of the particles, which will initially be finite and pointing from high concentration to low concentration, will vanish as the distribution equalizes. The vanishing of the velocity vector makes diffusion irreversible, and contrary to Poincare's recurrence theorem, the initial condition can only be achieved with an input of energy (Wayne, 2012a).

Eq. (2.8) is known as the Einstein–Stokes (or Stokes–Einstein) equation (Weiss, 1996). It describes the diffusion coefficient as the ratio of the amount of free energy in a spherical particle at a given temperature (kT) to the friction experienced by that particle (with a hydrodynamic radius of r_H) moving through a solution with a viscosity, η. Larger solutes experience more friction than smaller solutes. Thus, Ca^{2+} with a radius of 99 pm will have a smaller diffusion coefficient than Mg^{2+} with a radius of 65 pm. Likewise, a protein with a hydrodynamic radius of 2.5 nm will have a smaller diffusion coefficient than a glucose molecule that has a hydrodynamic radius of 350 pm ceteris paribus. In general, the diffusion coefficient for "typical" small molecules like glucose ($r_H = 0.35$ nm) in water ($\eta = 0.001$ Pa s) at room temperature (298K) is about 6×10^{-10} m²/s.

To model the electrical effects on solute movement we will use the relation discovered by Georg Ohm that describes the flow of current in wires. Ohm's Law (1827) describes the current (in A) that results when a potential, ψ (in V), is put across a given resistance, R (in Ω):

$$I = -\frac{\psi}{R} \tag{2.9}$$

where 1 A = 1 C/s, 1 V = 1 J/C, and 1 Ω = 1 V s/C = 1 Js/C². The negative sign indicates that a positive current moves away from a positive voltage source. The negative sign is not used in most forms of Ohm's Law but is used here because it is consistent with the use of negative signs in all other flux equations.

The transport of electricity can take place in two different ways: with or without the simultaneous transport of atomic nuclei (Nernst, 1923). In wires (i.e., metallic conductors), electricity is transported without the simultaneous transport of atomic nuclei. In fact, in wires, electricity is carried by the movement of electrons from one potential to a more positive potential. Unfortunately, Ben Franklin defined the movement of electricity as the movement of positive charge from one potential to a more negative potential. In electrolyte solutions, electricity is transported by atomic nuclei and thus can be described by fluxes. To convert Ohm's Law into a flux equation, I will write Ohm's Law in a form that describes the simultaneous transport of electrons and matter (be it atomic nuclei and electrons or electrons alone).

Ohm's Law can also be used to describe the net motion of charged solutes in an electric field ($d\psi/dx$) where $d\psi$ is the electrical potential difference (in V) across the distance x (in m).

The flux of monovalent cations in an electric field, J_{el} (in mol/m² s), is related to the current (in A or C/s) by the following relation:

$$J_{el} = \frac{I}{FA} \tag{2.10}$$

where F is the Faraday (9.65×10^4 C/mol) and A is the area perpendicular to the electric field through which the solutes move (in m²).

To derive the Nernst equation, I will start with Ohm's Law. However, to determine the movement of electrolytes in solution instead of electrons in a wire, both sides of the equation must be divided by FA. This converts the current (I) into a flux (J_{el}).

$$J_{el} = \frac{I}{FA} = -\frac{\psi}{RFA} \tag{2.11}$$

Eq. (2.11) is only true for the flux of monovalent cations, yet the flux of ions for a given current density (in A/m²) depends on the valence of the ion (z). Thus, for a given current density, the flux of a bivalent ion is one-half of the flux of a monovalent ion, and the general equation is

$$J_{el} = \frac{I}{zFA} = -\frac{\psi}{zRFA} \tag{2.12}$$

Eq. (2.12) is actually a typical flux equation because $I/(zFA)$ is equal to J_{el} and $\psi/RzFA$ is equal to $uzeC(d\psi/dx)$. To show this, I will use dimensional analysis; that is, I will write out the units of each term.

$$\frac{\mathrm{C/s}}{(\mathrm{C/mol})(\mathrm{m}^2)} = \frac{-\mathrm{V}}{(\mathrm{Js/C}^2)(\mathrm{C/mol})(\mathrm{m}^2)}$$

After canceling C on both sides, we get

$$\frac{1/\mathrm{s}}{(1/\mathrm{mol})(\mathrm{m}^2)} = \frac{-\mathrm{V}}{(\mathrm{Js/C})(1/\mathrm{mol})(\mathrm{m}^2)}$$

We can see that the left side is in units of flux and thus is equal to J_{el}, which is the flux of ions in response to an electric field.

$$J_{el} = \frac{-\mathrm{V}}{(\mathrm{Js/C})(1/\mathrm{mol})(\mathrm{m}^2)}$$

Now we have to use a mathematical trick. The first time you use it, it is a trick. The second time you use it, it seems clever. By the third time you use it, it will seem self-evident and will become one of your mathematical skills. Now for the trick: We are going to multiply the right side by 1. We know that multiplying anything by 1 (or adding 0) does not change its value, but it will help us to put the equation in a simple form. This trick takes advantage of the identity property of multiplication (or addition). Now, there are many 1s—for example, 1, 2/2, 100/100, m/m, and m²/m² all equal to 1—so we have to make the right choice. We will multiply the right side by 1 = m²/m² to end up with the equation in a convenient form.

$$J_{el} = \frac{-\mathrm{V}(\mathrm{m}^2)}{(\mathrm{Js/C})(1/\mathrm{mol})(\mathrm{m}^2)(\mathrm{m}^2)}$$

After rearranging terms and converting C/J into V^{-1}, we get

$$J_{el} = \frac{-V}{m}\frac{m^2}{Vs}\frac{mol}{m^3}$$

Replacing the units with symbols, we get

$$J_{el} = \frac{-d\psi}{dx}u'C$$

where $d\psi$ is the electrical potential difference (in V) across a membrane of thickness x (in m) and u' is the electrical mobility (in m^2/V s). Again, even though an electric field accelerates a charged particle, the particle collides with other particles. On collision, the acceleration stops and must start anew after each collision. Consequently, the velocity, and not the acceleration, is proportional to the electrical force. According to Robinson and Stokes (1959), u' is equal to uze by definition. Thus,

$$J_{el} = -uzeC\frac{d\psi}{dx} \tag{2.13}$$

and we see that the electrical flux equation is really nothing more than Ohm's law for ions. The total flux due to diffusion and electric forces is

$$J_{tot} = J_{dif} + J_{el} \tag{2.14}$$

and the total flux of a given solute in response to the driving forces, (dC/dx) and $(d\psi/dx)$, will be a function of the mobility of the solute.

$$J_{tot} = -ukT\left(\frac{dC}{dx}\right) - uzeC\frac{d\psi}{dx} \tag{2.15}$$

Eq. (2.15) also assumes that the solubility of the solute in the membrane and the solutions on either side of the membrane are the same. This is not a valid assumption for biological membranes, so we must account for this by including the partition coefficient:

$$J_{tot} = -KukT\left(\frac{dC}{dx}\right) - KuzeC\frac{d\psi}{dx} \tag{2.16}$$

Now consider two solutions of monovalent ions (e.g., KCl) separated by a membrane. Imagine that the membrane passes only the positively charged cation (K^+) but not the negatively charged anion (Cl^-), which is a good assumption for a plasma membrane. On each side of the membrane, the concentrations of ions are different; although, according to the rule of electroneutrality, on the macroscopic level there must be almost the same number of cations and anions. Let us consider a situation where there is no net flux (i.e., $J_{tot} = 0$).

$$J_{tot} = -KukT\left(\frac{dC}{dx}\right) - KuzeC\frac{d\psi}{dx} = 0 \tag{2.17}$$

This is an equilibrium situation, so any term that involves a rate should cancel out. First, let us rearrange the terms:

$$KuzeC\frac{d\psi}{dx} = -KukT\left(\frac{dC}{dx}\right) \tag{2.18}$$

Now let us solve for $d\psi/dx$:

$$\frac{d\psi}{dx} = -\frac{KukT}{KuzeC}\left(\frac{dC}{dx}\right) \tag{2.19}$$

After canceling u and K, we get

$$\frac{d\psi}{dx} = -\frac{kT}{ze}\left(\frac{dC/C}{dx}\right) \tag{2.20}$$

To eliminate the derivatives and get a simple, easy-to-use, powerful equation, we must integrate the equation. I will integrate between the two limits of the membrane thickness from outside to inside and assume k, T, z, and e are constant to get a simple yet powerful equation (Lakshminarayanaiah, 1965, 1969).

We take the constants out of the integral and then integrate. To integrate, we must remember that $\int_o^i \frac{dC}{C}dx = (\ln C_i - \ln C_o)$.

I set up this and every equation in this book to reflect a change from the initial state (outside the cell) to the final state (inside the cell). According to the fundamental law of calculus, we must subtract the initial state (outside the membrane) from the final state (inside the membrane). Thus, on integration, we get

$$\psi_i - \psi_o = -\frac{kT}{ze}(\ln C_i - \ln C_o) \tag{2.21}$$

where C_o and C_i are the concentration outside and inside the membrane, respectively, and ψ_i and ψ_o are the electrical potentials inside and outside the membrane, respectively. Remember that $\ln C_i - \ln C_o = \ln \frac{C_i}{C_o} = -\ln \frac{C_o}{C_i}$; thus

$$\psi_i - \psi_o = \frac{kT}{ze}\ln\frac{C_o}{C_i} \tag{2.22}$$

Boltzmann's constant (k) and the elementary charge (e) are related to the universal gas constant (R) and the Faraday (F), respectively, through Avogadro's number (Perrin, 1926). Because $R = kN_A$ and $F = eN_A$, kT/e is also equal to RT/F. Thus,

$$\psi_i - \psi_o = \frac{RT}{zF}\ln\frac{C_o}{C_i} \tag{2.23}$$

And because, by convention $\psi_o = 0$, which means that practically, we zero the potential measuring electrode outside the cell:

$$\psi_i = \frac{RT}{zF}\ln\frac{C_o}{C_i} = \frac{kT}{ze}\ln\frac{C_o}{C_i} \tag{2.24}$$

which are the familiar forms of the Nernst equation.

Let us use the Nernst equation right away to calculate the electrical potential across a membrane that has a concentration of 100 mol/m^3 K$^+$ inside and 1 mol/m^3 K$^+$ outside (Fig. 2.20).

$$\psi_i = \frac{RT}{zF}\ln\frac{C_o}{C_i}$$

$$\psi_i = \frac{(8.31\ \text{J/mol K})(298\text{K})}{9.65\times10^4\ \text{C/mol}}\ln\frac{1}{100}$$

$$\psi_i = \frac{(8.31\ \text{J/mol K})(298\text{K})}{9.65\times10^4\ \text{C/mol}}(-4.6)$$

$$\psi_i = -0.118\ \text{V} = -118\ \text{mV}$$

The result from the Nernst equation tells us that leakage of K$^+$ down its concentration difference creates an electrical difference across the membrane, such that the inside of the membrane becomes more and more negatively charged until the electrical potential difference that develops exactly balances the concentration difference. At equilibrium, the resulting voltage difference is −0.118 V.

To maintain an electrical potential across the plasma membrane, the membrane must have a property known as capacitance, which is the ability to store charge or resist changes in voltages. A capacitor results when two conductors are separated by a nonconductor (Fig. 2.21). The plasma membrane is a capacitor because the lipid bilayer serves as a nonconductor that separates the aqueous conducting fluids on both sides of the membrane. The specific capacitance (C_{sp}, in F/m^2, where F = C/V) is the proportionality coefficient that relates the charge per unit area (q/A in C/m^2) that is produced on either side of a nonconductor to a given electrical potential difference (ψ, in V). The capacitance (C) is defined as q/ψ, and the specific capacitance (C_{sp}) is defined as $q/(A\psi)$.

The capacitance of the membrane determines how many K$^+$ ions have to move across the membrane to obtain the membrane potential predicted by the Nernst equation. How many K$^+$ ions have to move across the membrane to obtain a membrane potential of −0.118 V? Assume the cell is a cube where the length of each edge is 10^{-5} m and the specific capacitance of the membrane is 10^{-2} F/m^2.

$$\frac{q}{A} = C_{sp}\psi_i \tag{2.25}$$

After plugging in the above values for the specific capacitance and the membrane potential, we get

$$\frac{q}{A} = (10^{-2}\ \text{CV}^{-1}\ \text{m}^{-2})(-0.118\ \text{V})$$

$$= -1.18\times10^{-3}\ \text{Cm}^{-2}$$

The number of K$^+$ per unit area needed to charge the membrane is obtained by dividing the charge per unit area (on either side of the membrane) by the elementary charge (e):

$$\frac{-1.18\times10^{-3}\ \text{C/m}^2}{1.6\times10^{-19}\ \text{C/K}^+}$$

Because the surface area = 6(10^{-5} m)^2 = 6 × 10^{-10} m^2, then (−7.36 × 10^{15} K$^+$/m^2 × 6 × 10^{-10} m^2) = −4.4 × 10^6 K$^+$ must cross the membrane to charge it to a voltage of −0.118 V. Because the sign of the membrane potential was determined by the fact that we integrated from outside the membrane to inside, the negative sign for the number of K$^+$ that must cross the membrane to charge it to −0.118 V means that the ions must move from inside the cell to outside

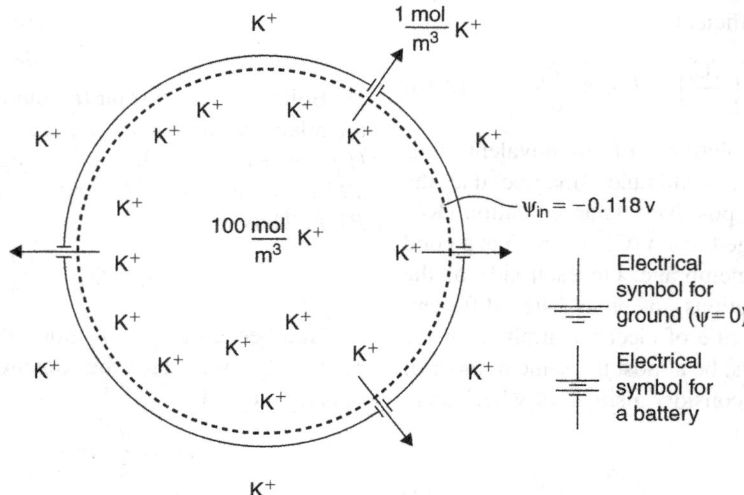

FIGURE 2.20 A membrane potential (e.g., battery) develops as a result of the unequal distribution of ions across the membrane and transport proteins that allow the ions to permeate across the membrane.

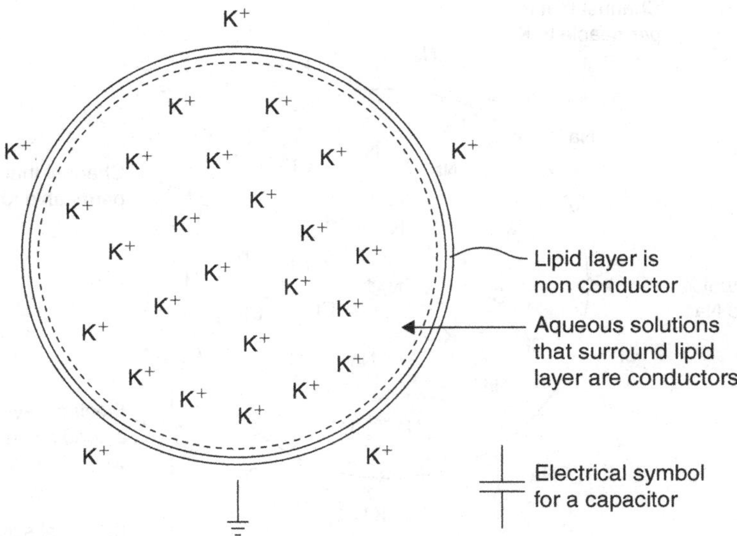

FIGURE 2.21 The membrane potential developed by the permeation of ions across the membrane would dissipate if the membrane did not have the electrical capacity (capacitance) to store electrical charge. The lipid bilayer functions as the nonconducting or dielectric layer between the two conducting aqueous layers. The three layers form a capacitor.

the cell, which as we know is down their concentration difference.

When approximately four million ions leave the cell, what is left? Or, put another way, how does this change the initial conditions in a cell with a volume of 10^{-15} m^3?

Because the initial concentration of K^+ in the cell was 10^2 mol/m^3 and the volume of the cell is 10^{-15} m^3, then the number of K^+ initially in the cell was

$$\left(10^2 \text{ mol/m}^3\right)\left(10^{-15} \text{ m}^3\right)\left(6.02 \times 10^{23} \text{ K}^+/\text{mol}\right) = 6.02 \times 10^{10} \text{ K}^+$$

Thus, only $\frac{4.4\times10^6 \text{ K}^+}{6.02\times10^{10} \text{ K}^+}$ 100%, or 0.007%, of the K^+ leaves the cell due to its concentration difference. After this trivial loss, the membrane potential is charged to −0.118 V, which prevents any additional net loss of K^+. Therefore, K^+ diffuses out of the cell until the membrane potential becomes negative enough to balance the driving force due to diffusion. If the membrane capacitance were zero, it would be impossible to develop an electrical potential across it. It would run down over time. Because the capacitance is due to the lipid bilayer, the membrane potential is, in part, due to the lipids, as well as the transport proteins and the concentration differences of the various ions.

The creation of a membrane potential due to the passive movement of ions across a membrane was just described. However, the membrane potential created by the diffusion of ions depends on the ability of that ion to dissolve in and diffuse across the membrane. Therefore, if a membrane is completely impermeable to an ion, the ion will not be able to diffuse and leave behind the opposite charge and establish a membrane potential. In reality, membranes are usually much more permeable to K^+ than to any other abundant ion due to the large proportion and high conductance of K^+ channels. For this reason, K^+ is the ion that contributes the most to the resting diffusion potential (Fig. 2.22).

Nernst determined the relationship between the diffusion of a single type of ion and the electrical potential at equilibrium where the net flux equals zero ($J_{tot} = 0$). Max Planck integrated the flux equation for situations where $J_{tot} \neq 0$. Planck's form of the equation can be used to determine the permeability coefficients of a membrane for ions from flux experiments. Later, David Goldman, Alan Hodgkin, and Bernard Katz derived an equation to account for the diffusion of multiple ions at equilibrium. We can calculate the resting membrane potential due to passive diffusion of all the abundant monovalent ions using the Nernst potential for these ions and the permeability coefficients for each ion. These are all combined into the Goldman–Hodgkin–Katz equation:

$$\psi_i = \frac{RT}{F} \frac{\left(P_K C_K^o + P_{Na} C_{Na}^o + P_{Cl} C_{Cl}^i\right)}{\left(P_K C_K^i + P_{Na} C_{Na}^i + P_{Cl} C_{Cl}^o\right)} \tag{2.26}$$

of which the complete derivation is given in Wayne (1994). Only the monovalent ions are considered in this equation because they are the most abundant and their mobilities

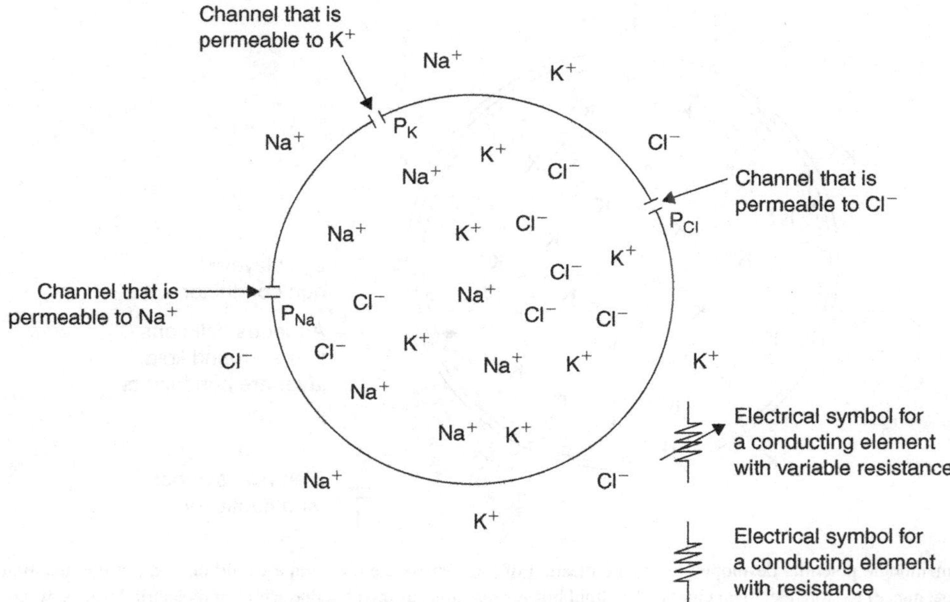

FIGURE 2.22 The magnitude of the membrane potential depends, in part, on the relative permeability of the membrane to various ions. The permeability of a channel to a given ion is a variable, not a constant.

are greater than the mobilities of the abundant bivalent cations due to their low charge density (Nernst, 1923).

Once the permeabilities are known, we can simplify the equation by using relative permeabilities:

$$\psi_i = \frac{RT}{F}\frac{(C^o_K + \alpha C^o_{Na} + \beta C^i_{Cl})}{(C^i_K + \alpha C^i_{Na} + \beta C^o_{Cl})} \tag{2.27}$$

where $\alpha = P_{Na}/P_K \approx 0.1$ and $\beta = P_{Cl}/P_K \approx 0.01$ in the resting state. When solving Eq. (2.27), remember the order of operations in arithmetic. Add the concentration terms in the parentheses before dividing the two sums.

When $\alpha = 0$ and $\beta = 0$, the Goldman–Hodgkin–Katz equation reduces to the Nernst equation for K^+. The permeability coefficients vary from about 10^{-4} m/s for water to 10^{-6} m/s for auxin (Rutschow et al., 2014) to about 10^{-11} m/s for Cl^-. Moreover, the permeability coefficients are not constant but depend on such factors as the age of the cell, the light quality, pH, and Ca^{2+}. Indeed, fluctuations in the permeability coefficients of ions lead to dramatic changes in cell physiology and development (Jaffe, 1980, 1981, 2006, 2007; Harold, 1986, 1990; Raschke et al., 1988).

Typically, the membrane potential of plant cells is far more negative (hyperpolarized) than would be predicted by the Goldman–Hodgkin–Katz equation. When the membrane potential is greater than that accounted for by the passive diffusion of ions, then active transport must be taking place. Active transport can be diagnosed by treating the cells with metabolic inhibitors and seeing whether the membrane potential rapidly and reversibly depolarizes.

In animal cells, the membrane potential is only slightly greater than the diffusion potential and a Na^+/K^+-ATPase is the most common electrogenic pump (Ussing and Zerahn, 1951; Kerkut and York, 1971). While there are also Na^+-ATPases in plant cells, fungi, and bacteria, in these organisms, the H^+-ATPase is the most common electrogenic pump (Spanswick, 2006).

We can already determine the Nernst potential for K^+, Na^+, and Cl^-. Now we will determine whether or not the uptake or effluxes of these ions are active or passive. Assume that the membrane potential is −0.25 V, $C^o_K = 0.1$ mol/m^3, $C^o_{Na} = 0.1$ mol/m^3, $C^i_{Cl} = 20$ mol/m^3, $C^i_K = 100$ mol/m^3, $C^i_{Na} = 10$ mol/m^3, $C^o_{Cl} = 0.2$ mol/m^3, and $T = 298$K. We will determine if an ion is at equilibrium by determining if the Nernst potential for that ion is equal to the membrane potential. We will determine whether the movement is active or passive by calculating the driving force for an ion and multiplying that value by ze. The free energy (in J) needed to move an ion is given by the following formula:

$$\Delta E = (\psi_m - \psi_{ion})ze \tag{2.28}$$

where ψ_m is the membrane potential (in V = J/C), ψ_{ion} is the Nernst potential for the ion (in V = J/C), z is the valence of the ion (dimensionless), and e is the elementary charge (in C). Because we integrated the Nernst equation from outside to inside, that set all the signs. If the free energy obtained is negative (exergonic), the flux into the cell is spontaneous, or passive, requiring only thermal energy. If the free energy is positive (endergonic), the flux into the cell is active and requires metabolic energy.

The difference between the membrane potential and the Nernst potential for an ion provides the driving force[3] for the uptake of that ion. In terms of K^+, $(\psi_m - \psi_{ion})$ is given by -0.250 V − (-0.177 V) = -0.073 V. The free energy involved in moving a K^+ from outside the cell to the inside is (-0.073 V)$ze = -1.2 \times 10^{-20}$ J. Because the free energy is negative, the inward movement is passive. The inward movement of positive charge results in an inward current.

In terms of Na^+, $(\psi_m - \psi_{ion})$ is given by -0.250 V − (-0.118 V) = -0.132 V, and the free energy of transport is given by (-0.132 V) $ze = -2.1 \times 10^{-20}$ J. Because the free energy is negative, there is a passive inward movement of positive charge.

In terms of Cl^-, $(\psi_m - \psi_{ion})$ is given by -0.250 V − (0.118 V) = -0.368 V, and the free energy of transport is given by (-0.368 V)$ze = +5.9 \times 10^{-20}$ J. Because the free energy is positive, the inward movement of ions requires active transport. However, the outward movement of Cl^- is passive. The outward movement of negative charges is also called an inward current because it behaves as if positive charges move into the cell.

In conclusion, if the product of $(\psi_m - \psi_{ion})$ and ze is negative, it means that ions will move into the cell passively. If the product of $(\psi_m - \psi_{ion})$ and ze is positive, it means that ions will move out of the cell passively. If the product of $(\psi_m - \psi_{ion})$ and ze is zero, there will be no net movement of anions or cations and the ion is at equilibrium. If the net movement in one direction is spontaneous (i.e., passive), it must be active in the other direction. When we measure the distribution of all ions on both sides of the membrane, we see that none of the ions are in equilibrium. That is, maintaining the normal distribution of ions requires a constant input of energy.

A change in the membrane potential (ψ_m) can determine whether uptake is active or passive. An environmental stimulus (e.g., touch, light, or hormones) can often cause a membrane depolarization and thus a change in influx and efflux. Given the above situations, if the membrane potential is depolarized to -0.1 V, would the flux change from passive to active or from active to passive for any of the ions?

2.8 ELECTRICAL PROPERTIES OF THE PLASMA MEMBRANE

As we have already seen, the electrical properties of the plasma membrane influence the transport of electrolytes across the membrane. The membrane electrical potential has a great influence, and it can be readily measured. Our study of permeability and transport physiology is bringing us into the field of electrophysiology. Although this field is unfamiliar to many botanists, historically the study of electricity began with plants. Thales of Miletus discovered that amber (e.g., fossilized pine resin from which succinic acid was first extracted), when rubbed with fur, attracts little pieces of pith and cork. William Gilbert (1600), named this attraction electricity, after *electron*, the Greek work for "amber" (Laidler, 1993).

It turns out that there are two kinds of electricity with opposite properties. Rubbed amber contains resinous electricity (i.e., an excess of electrons) and is negatively charged, whereas rubbed glass repels the things that amber attracts and contains vitreous electricity (i.e., a dearth of electrons) and is positively charged. Electricity is dynamic and current flows from positively charged substances to negatively charged substances. In the late 18th century, animal biologists played a role in the development of galvanic electricity from static electricity when Luigi Galvani noticed that when two different metals touched a frog's leg electricity was generated (Galvani, 1953a,b). This work was followed up by Alessandro Volta, who found that he could still generate electricity without the frog as long as the two different metals were placed in a solution more conductive than water. These observations led Volta to invent the battery (Conant, 1947). In the early 18th century, the voltaic pile was used to separate chemical compounds into their constituent elements (Nicolson and Carlisle, 1800; Davy, 1821; Urban, 1829; Arrhenius, 1902; Nernst, 1923; Ostwald, 1980). At this time, it was believed in some circles that electricity could also be used to animate matter and to create life (Aldini, 1803; Shelley, 1818; Ure, 1819). The importance of electricity in living organisms was further established when Emil du Bois-Reymond (1848) showed that electricity was the "nervous principle" transmitted by nerves, and Guillaume Duchenne (1862, 1871, 1949) showed that electrical stimulation was involved in the contraction of all muscles—including the ones that give rise to smiles. The electrical nature of the nervous system of animals was further characterized by Sherrington (1906), Lucas (1917), Langley (1921), Creed et al. (1932), and Eccles (1964). There has been a call for a resurgence in studies of the electrical nature of plant communication (Staves et al., 2008). The techniques involved in electrophysiology are described briefly in the next section. A fuller discussion on the techniques can be found in Bures et al. (1967), Hille (1992), Weiss (1997), and Volkov (2006).

To measure the membrane potential, two electrodes are connected to an electrometer (see Fig. 2.23; Walker, 1955). Then the two electrodes are placed in the solution bathing the cell and the electrometer is zeroed. This is why the

3. The driving force, like any force, is properly given in Newtons. Thus, the proper driving force is equal to the negative spatial derivative (or gradient, in vector calculus notation) of the electrochemical potential. That is, the proper driving force is the difference between the high electrochemical potential and the low electrochemical potential divided by the distance between the high and low electrochemical potentials.

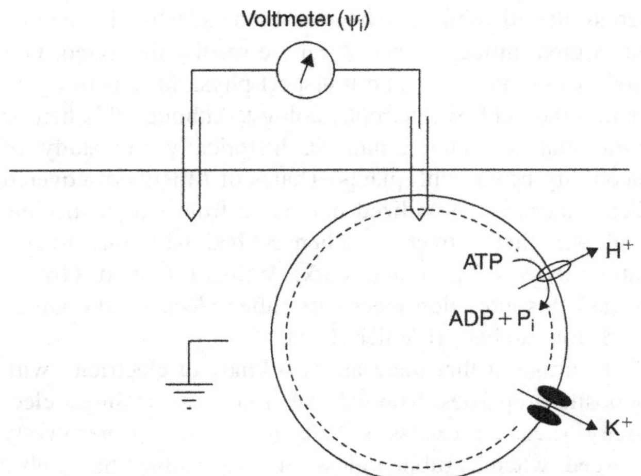

FIGURE 2.23 Measuring the membrane potential with two microcapillary electrodes connected to a voltmeter in the current clamp mode. By convention, the electrical potential difference between the two electrodes is set to zero before the microcapillary electrode is inserted into the cell.

external electrical potential is considered zero. Then the glass microcapillary electrode, filled with 3M KCl, is inserted into the cell with the help of a micromanipulator and ψ_i is measured (in V). The membrane potentials of plant cells typically range from −0.12 to −0.25 V (Etherton and Higinbotham, 1960; Etherton, 1963). Using their wits, Wright and Fisher (1981) measured the membrane potential of narrow and difficult-to-access sieve tubes by using aphid stylets as microcapillaries. Membrane potentials found in animal cells range from −0.06 to −0.1 V.

The resistance of the membrane (in Ω) is the property of the membrane that determines the relationship between the current and the electrical potential. The greater the resistance, the smaller the change in current for a given change in voltage. Alternatively, the greater the resistance, the greater the change in voltage across a resistor for a given change in current. To measure the membrane resistance, a tiny current is intermittently passed through the membrane while the membrane potential is being measured (see Fig. 2.24; Blinks, 1930, 1939; Walker, 1960a,b). The membrane potential changes when the current flows. The membrane resistance is then obtained from Ohm's Law, where the membrane resistance is calculated by dividing the change in membrane potential by the change in membrane current. Because the resistance is a function of cell size, we usually use the specific resistance (in $\Omega\ m^2$) to characterize the membrane. The specific resistance is obtained by multiplying the membrane resistance by the surface area of the cell. The reciprocal of the specific resistance is the specific conductance (in S/m^2). The specific conductance of the membrane is a measure of its permeability to all ions and is determined by the quantity and type of transport proteins embedded in the lipid bilayer.

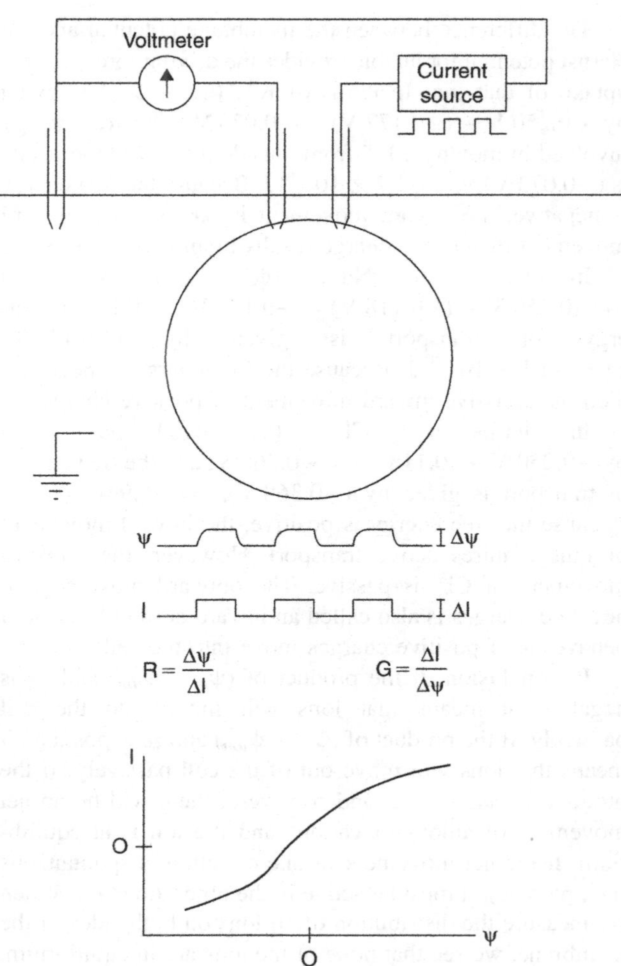

FIGURE 2.24 A voltage clamp measures the amount of current that passes through a membrane when an electrical potential difference is established artificially across that membrane. The current is plotted against the membrane potential. The conductance of the membrane at a given voltage is calculated from the slope of the *I*-ψ curve.

Capacitance is the property of a membrane that resists changes in voltage when a current is applied. When a current is applied to the cell, the voltage does not instantaneously attain the value predicted by the resistance but rises logarithmically to that value. Once the resistance is known, the capacitance is measured by determining the time it takes for the membrane potential to reach 63% of the maximal value it reaches at infinite time (Fig. 2.25). The time it takes to reach this value is equal to the product of the resistance and the capacitance. The time needed to reach 63% of the maximal value is known as the *time constant* and in general is around 10 ms. Capacitance, like resistance, also depends on the surface area of the cell, and thus we usually talk about the specific capacitance (in F/m^2), which is given by the capacitance divided by the surface area.

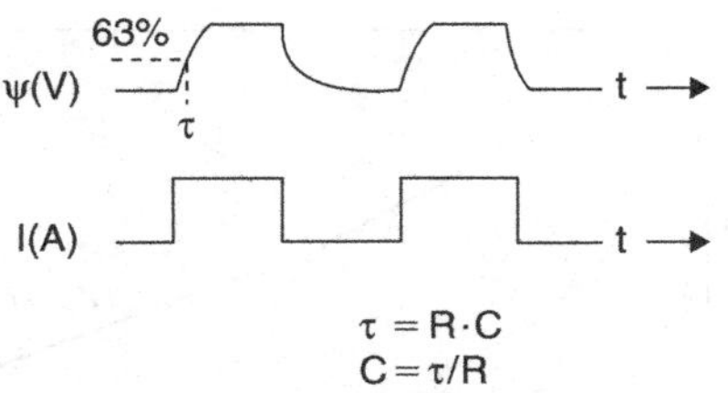

FIGURE 2.25 When a current is applied to a membrane in a *square wave*, the membrane potential follows a sawtooth pattern. The shape of the sawtooth depends on the membrane resistance and the membrane capacitance. The voltage change reaches 63% of the maximum in a time (τ) equal to the product of the resistance (*R*) and the capacitance (*C*). The membrane capacitance is equal to τ/*R*.

Hugo Fricke (1925) measured the specific capacitance of the plasma membrane of red blood cells to be 0.01 F/m^2. The specific membrane capacitance depends on an electrical property of the membrane known as the dielectric constant or the *relative permittivity* (ε, dimensionless) and the thickness of the nonconducting layer (*dx*). Fricke assumed that the nonconducting layer was made of lipids and guessed that the relative permittivity (or dielectric constant) of the membrane was the same as it is for lipids (ε = 3). By assuming that the relative permittivity of the plasma membrane was 3, the thickness of the lipid layer can be calculated from the specific capacitance of the membrane. Plugging these values into the following equation, which is used to calculate the specific capacitance of a parallel plate capacitor, Fricke estimated that the thickness of the lipid layer was approximately 3 nm.

$$C_{sp} = \varepsilon_o\varepsilon/dx \quad \text{or} \quad dx = \varepsilon_o\varepsilon/C_{sp} \qquad (2.29)$$

where ε_o, the permittivity of a vacuum, is 8.85×10^{-12} F/m.

2.9 CHARACTERIZATION OF TWO TRANSPORT PROTEINS OF THE PLASMA MEMBRANE

The plasma membrane of plants contains a diverse array of transport proteins including a H^+-ATPase, a Ca^{2+}-ATPase, a Na^+-ATPase (Benito and Rodriguez-Navarro, 2003), a Cl^--ATPase (Gradmann and Klemke, 1974; Mummert et al., 1981), an ATP-binding cassette-type transporter (Jasinski et al., 2001; Sanchez-Fernandez et al., 2001; Kobae et al., 2006), an amino acid symporter (Etherton and Rubinstein, 1978; Kinraide and Etherton, 1980), a Cl^-/H^+ symporter (Beilby, 2016), an auxin efflux carrier (Hertel and Leopold, 1963; Christie and Leopold, 1965a,b; Goldsmith, 1966a,b; Leopold and Hall, 1966; Goldsmith and Ray, 1973; Goldsmith et al., 1981; Chen et al., 1998; Gälweiler et al., 1998; Petrášek et al., 2006; Wiśiewska et al., 2006; Křeček et al., 2009; Petrášek and Friml, 2009; Yang and Murphy, 2009; Carraro et al., 2012; Forestan et al., 2012; Luschnig and Vert, 2014), a Ca^{2+}/H^+ antiporter, a sucrose/H^+ antiporter, a silicon influx (Ma et al., 2006), and a silicon efflux transporter (Ma et al., 2007), as well as Cl^-, K^+, and Ca^{2+} channels (Mäser et al., 2001; Axelsen and Palmgren, 2001; Hedrich and Marten, 2006; Ward et al., 2009). There are also channel proteins that pass small polar and nonpolar molecules, including H_2O and CO_2 (Wayne and Tazawa, 1990; Wayne et al., 1994; Tyerman et al., 2002; Terashima and Ono, 2002). The approaches used to characterize two of the major and ubiquitous transport proteins—the H^+-pumping ATPase and the K^+ channel—are discussed next.

2.9.1 Proton-Pumping ATPase

The H^+-ATPase is one of the best characterized proteins in the plasma membrane of plants. Its presence in the plasma membrane was first inferred by H. Kitasato in 1968 when he noticed that, contrary to the predictions of the Goldman–Hodgkin–Katz equation, the plasma membranes of characean cells were relatively insensitive to changes in the external K^+ at concentrations below 1 mol/m^3. He did observe, however, that the membrane potential was sensitive to changes in the external H^+ concentration (Kitasato, 2003).

Using the Nernst equation, Kitasato calculated that if the protons were distributed passively, the internal pH should be < 3, given the external pH and the observed membrane potential. Because the internal pH is approximately 7, Kitasato proposed that H^+ was actively pumped out of the cell. He found that dinitrophenol, a protonophore, reduced the membrane potential. Later, Roger Spanswick (1972, 1974a) and Keifer and Spanswick (1978, 1979) provided evidence that H^+ was the ion pumped because dicyclohexylcarbodiimide (DCCD), an inhibitor of H^+ transport in mitochondria and chloroplasts, decreased the membrane potential to the value predicted by the Goldman–Hodgkin–Katz equation. DCCD also increased the membrane resistance, indicating that there is a conductance in the plasma membrane for H^+. Teruo Shimmen and Masashi Tazawa (1977) perfused the inside of characean internodal cells with ATP and demonstrated that the membrane potential and the efflux of H^+ were dependent of the intracellular ATP concentration (Tazawa, 2011). Takeshige et al. (1986) have shown that the extrusion of H^+ can be quantitatively accounted for by the action of the proton pump by showing the equivalence between the proton efflux (J^{H+} in mol/m^2 s) and the pump current density (I/A in A/m^2). They used the following equation:

$$J^{H^+} = \frac{I}{zFA} \qquad (2.30)$$

Not every segment of the plasma membrane is identical. This can be visualized easily and elegantly in the large internodal cells of *Chara* (Dorn and Weisenseel, 1984; Shimmen and Wakabayashi, 2008). The plasma membrane is differentiated into regions that have a net proton efflux and regions that have net proton influx. These regions are known as the acid and alkaline bands, respectively. The bands can be beautifully visualized by placing cells on nutrient agar containing phenol red. The phenol red will turn yellow where the pH is acidic and red where the pH is basic (Spear et al., 1969).

At the same time that work with whole cells or cell models was advancing, investigations at the biochemical level were also making progress. Knowing that ion transport is often dependent on respiration (Briggs and Petrie, 1931; Steward, 1933, 1941; Lundegardh, 1955; Laties, 1959), Tom Hodges and his colleagues (Fisher and Hodges, 1969; Fisher et al., 1970; Hodges et al., 1972; Leonard et al., 1973; Hodges and Leonard, 1974; Balke and Hodges, 1975; Sze and Hodges, 1977; Briskin et al., 1987; Leonard and Hodges, 2000) began searching for the molecular mechanism that converts respiratory energy into the work of ion transport. They purified plasma membranes from roots and discovered that purified plasma membranes had the ability to hydrolyze ATP, a product of respiration. Moreover, the purified ATPase contained an amino acid sequence in the phosphorylation site that was similar to the amino acid sequences in the phosphorylation sites of mammalian and prokaryotic ATPases (Walderhaug et al., 1985). The activity of the plasma membrane ATPase must be regulated because if it ran continuously, it would consume 25%–50% of the cellular ATP (Felle, 1982).

The H^+-ATPase has been purified from many plants and accounts for approximately 1% of the plasma membrane protein and approximately 0.01% of the total cellular protein (Sussman and Harper, 1989; Sussman, 1994). Anthon and Spanswick (1986) purified the H^+-ATPase from the plasma membranes of tomato. They washed a crude membrane fraction with high salt and 0.1% Triton to remove the peripheral and loosely held integral membrane proteins. The membranes were further extracted with octyl glucoside/deoxycholate, a detergent that removes other integral proteins, but not the H^+-ATPase. The ATPase was finally solubilized with lysolecithin and released into the supernatant fraction. The supernatant was centrifuged through a glycerol gradient, and the H^+-ATPase was collected in the 37% glycerol fraction. It is possible to follow the purification of the H^+-ATPase because the specific activity increases as the protein is purified.

Purity can also be estimated from SDS PAGE (see Fig. 2.18), and the activity of the purified protein can be measured with functional assays of its ATPase activity and its ability to pump protons. As the specific activity increases, a single band becomes more and more prominent and this is assumed to be the H^+-ATPase polypeptide.

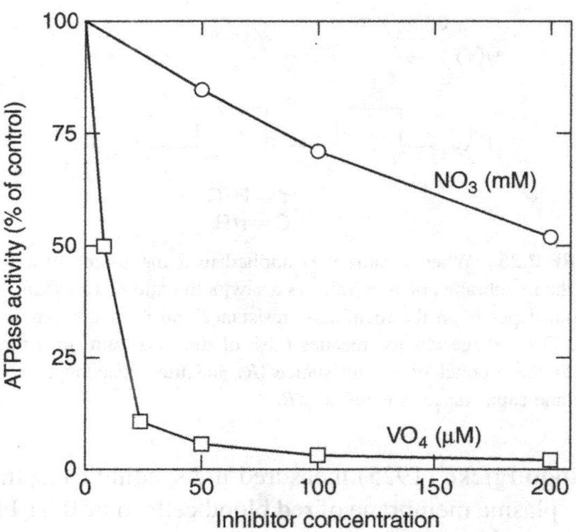

FIGURE 2.26 Inhibition of plasma membrane ATPase activity by vanadate but not by nitrate. *From Anthon, G.E., Spanswick, R.M., 1986. Purification and properties of the H^+-translocating ATPase from the plasma membrane of tomato roots. Plant Physiol. 81, 1080–1085.*

The H^+-ATPase can be characterized based on the nature of the compounds that inhibit it (Fig. 2.26). For example, the plasma membrane proton ATPase activity is inhibited by micromolar concentrations of vanadate, which inhibits all ATPases that form an inorganic phosphate (P_i)–enzyme intermediate. By contrast, it is not inhibited by millimolar concentrations of nitrate, which is an inhibitor of the vacuolar membrane proton ATPase. The plasma membrane ATPase is also inhibited by DCCD, a compound that depolarizes the membrane potential.

The purified ATPase is able to pump H^+ after it is inserted into proteoliposomes filled with a fluorescent dye, quinacrine, the fluorescence of which depends on the pH of its environment, and the fluorescence of quinacrine decreases on the accumulation of H^+. H^+ pumping requires ATP and is inhibited by DCCD and vanadate (Fig. 2.27). Carbonyl cyanide-p-trifluoromethoxyphenylhydrazone, a proton ionophore, increases the fluorescence by releasing protons, providing evidence that the fluorescence decrease is due to H^+ pumping and that the ATPase is an H^+ pump.

The functions of the various segments of the H^+-ATPase are becoming clear (Portillo, 2000; Bukrinsky et al., 2001; Kühlbrandt et al., 2002; Wurtele et al., 2003). For example, when inside-out vesicles are challenged with ATP, they pump protons at the expense of the ATP. If the vesicles are treated with trypsin, so that a 7-kDa polypeptide is removed from the carboxy-terminus of the H^+-ATPase, both the ATP hydrolyzing and proton-pumping activities are stimulated. This stimulation can again be inhibited by the addition of the 7-kDa fragment, indicating that the carboxy-terminal end of the ATPase regulates the activity of the rest of the protein (Palmgren et al., 1991).

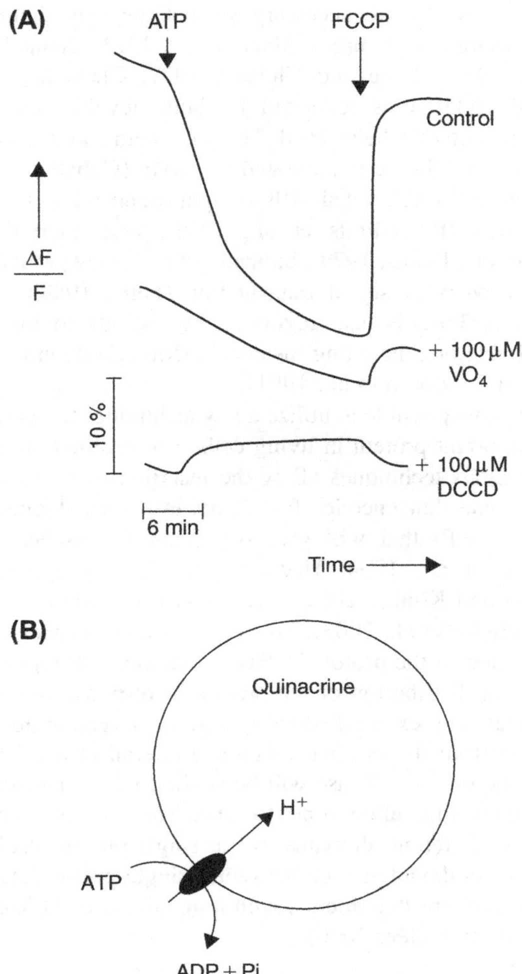

FIGURE 2.27 (A) Inhibition of plasma membrane ATP-dependent proton-pumping activity in proteoliposomes by vanadate and dicyclohexylcarbodiimide. (B) Only the activities of the proton pumps in the "inside out" orientation are measured. *From Anthon, G.E., Spanswick, R.M., 1986. Purification and properties of the H^+-translocating ATPase from the plasma membrane of tomato roots. Plant Physiol. 81, 1080–1085.*

The H^+-ATPase is regulated by its phosphorylation state. The carboxy-terminal end has a threonine residue that can be phosphorylated. On phosphorylation, a regulatory protein known as 14-3-3 binds to the carboxy-terminal end and activates the H^+-ATPase (Jahn et al., 1997; Svennelid et al., 1999). The carboxy-terminal end also has a serine residue, which on phosphorylation inhibits the binding of the 14-3-3 protein and thus inactivates the H^+-ATPase (Fulsang et al., 2007). The activated protein complex consists of six phosphorylated proton ATPase molecules and six 14-3-3 molecules assembled in a hexameric structure (Kanczewska et al., 2005). The kinases that phosphorylate the H^+-ATPase are being identified (Falhof et al., 2016).

The mechanism of how the H^+-ATPase pumps protons across the membrane has been postulated by looking for analogies with the better known Na,K-ATPase and Ca-ATPase of animal cells (Jørgensen and Pedersen, 2001; Toyoshima and Nomura, 2002; Buch-Pedersen and Palmgren, 2003). The phosphorylation of the enzyme, which results from the incorporation of the phosphate from the ATP used to power the enzyme, probably induces a conformational change in the protein that moves the H^+-binding site of the protein from the protoplasmic side of the plasma membrane to the extracellular side of the plasma membrane. Concurrently, the affinity of the H^+-binding site for H^+ decreases. These changes result in the release of the H^+ to the external space and the net transport of protons across the membrane. Dephosphorylation of the H^+-transporting ATPase returns it to its initial conformation where it can again bind an H^+ on the protoplasmic side of the membrane.

The powers of electrophysiological techniques and biochemical techniques have been combined to study the proton ATPase by reconstituting the proton ATPase into a planer lipid bilayer. In this way, the electrical and chemical environments on both sides of the proton ATPase can be regulated at the same time (Briskin et al., 1995).

In the 1970s and 1980s, immunologists were cloning genes based on their knowledge of the sequence of the proteins that they encoded (Hood, 2008). In 1989, the gene for the H^+-ATPase was cloned and identified by its ability to hybridize to a synthetic DNA oligonucleotide containing the nucleotide sequence that coded for the conserved phosphorylated site in yeast, animal, and bacterial ATPases (Serrano et al., 1986; Serrano, 1988, 1989; Harper et al., 1989; Pardo and Serrano, 1989; Sussman and Harper, 1989). The gene, which was then sequenced, was more similar to the yeast H^+-ATPase than the other ATPases, suggesting that it too was a H^+-ATPase. The gene also contained the nucleotide sequences predicted from the amino acid sequences of tryptic peptides obtained from H^+-ATPase purified from oats (Schaller and Sussman, 1988). Using the hydrophilic and hydrophobic properties of the amino acids that are encoded by the sequence, a first approximation of the structure and topography of this H^+-ATPase was made, and the structure suggested that this canonical H^+-ATPase is a multipass integral membrane protein. Southern blots showed that there were at least three isoforms of the H^+-ATPase (Harper et al., 1990).

Molecular genetics has taught us that there are typically multiple and distinguishable copies of genes that encode transport proteins like the plasma membrane H^+-ATPase, and that the first transport protein characterized will most likely turn out to be just one example of a class of transport proteins that may vary from cell to cell or during the life of a cell in a single organism. Thus, we must not be too dogmatic and we must be sure to remember that the canonical protein with its characteristics may just be one example of the range of possible transport proteins that may differ in their kinetics, their sensitivity to inhibitors, their regulation, and their transport selectivity, as a result of being encoded by genes of which the domains have been joined together in various ways though evolutionary time to code for related yet unique proteins (Doolittle, 1995).

Genetic analysis shows that the proton ATPase is a member of a class of ion-translocating ATPases called the *P-type ion-translocating ATPases*. P-type ATPases are characterized by the formation of a phosphorylated intermediate in its reaction cycle and thus are inhibited by vanadate. In plants, the plasma membrane–bound and endoplasmic reticulum (ER)–bound Ca^{2+}-ATPases are also P-type ATPases and are related to the Na^+/K^+-ATPase and the Ca^{2+}-ATPase of animal cells, the H^+-ATPase of fungal cells, and the K^+-ATPase of bacterial cells (Wimmers et al., 1990; Pedersen et al., 2012). Information about P-type ATPases can be found at *http://www.traplabs.dk/patbase/*.

The H^+ pump has been characterized, purified, and cloned because it is so important to the life of the cell (Obermeyer et al., 1992; Felle, 2002; Tazawa, 2003; Pertl et al., 2010; Pertl-Obermeyer et al., 2014; Falhof et al., 2016). The H^+-ATPase creates an electrochemical difference of protons. As long as the plasma membrane is not freely permeable to the protons being pumped out, the free energy stored in the electrochemical difference of protons set up by the proton pump can be used to drive a number of secondary transport processes, including sugar and amino acid transport, and the passive transport of K^+ (Vreugdenhil and Spanswick, 1987; Raschke et al., 1988; Sanders, 1990). The proton ATPase is present in all cell types (Michelet et al., 1994) and is abundant in cells specialized for the transport of nutrients, including root epidermal cells, phloem companion cells, and transfer cells (Jahn and Palmgren, 2002).

The proton pump uses the energy of ATP to pump protons out of the cell in an electrogenic manner. The membrane potential, which becomes negative inside as a result of the activity of the pump, drives the inward flux of K^+. The increased osmotic pressure due to the increased K^+ concentration within the cell causes water to move into the cell, which in turn causes an increase in the turgor pressure. This turgor pressure, which follows indirectly from the activity of the electrogenic proton pump, is necessary for cell and plant growth as well as movements, including tropisms, leaflet movement, and stomatal opening/closure. In the cells of many tissues, the proton ATPases is not uniformly distributed throughout the plasma membrane but is differentially and/or asymmetrically localized (Bouchè-Pillon et al., 1994; Jahn et al., 1998; Jahn and Palmgren, 2002; Certal et al., 2008).

The H^+ pump is involved in another aspect of growth. It acidifies the wall, thus activating the wall-loosening enzymes, which are necessary so that the wall yields to the pressure due to turgor (Hager et al., 1971; Cleland and Rayle, 1977; Rayle and Cleland, 1977; Cleland, 2002). The H^+-ATPase is regulated by both development and environment (Michelet et al., 1994)—nature and nurture. The proton ATPase is regulated by auxin (Gabathuler and Cleland, 1985; Frias et al., 1996), light (Spanswick, 1974a), salt stress (Perez-Prats et al., 1994), and internal pH (Vesper and Evans, 1979), indicating that it may participate in all aspects of signal transduction (Felle, 1989b). The proton ATPase is also regulated by various toxins and fungal elicitors, including fusicoccin (Rasi-Caldogno et al., 1986; Hagendoorn et al., 1991).

It is now possible to utilize a few techniques to visualize this important protein in living cells. For example, recombinant DNA techniques allow the insertion of a sequence into a gene that encodes for a protein (green fluorescent protein, GFP) that will give off green fluorescent light (Chalfie et al., 1994; Hadjantonakis and Nagy, 2001; Hanson and Köhler, 2001; van Roessel and Brand, 2002; Luby-Phelps et al., 2003). This allows one to visualize the distribution of the proton ATPase in various cell types, the change in distribution of the protein in response to developmental and external stimuli, and its targeting to and removal from the plasma membrane (Certal et al., 2008). Soon, the proton ATPase will be studied with microscopic techniques that allow one to visualize the interactions between different domains of a single-proton ATPase molecule or the interaction between a single-proton ATPase and the proteins that interact with it in a living cell (Gadella et al., 1999; Uhlén, 2006).

2.9.2 The K^+ Channel

K^+, which gets its name from being the major constituent of potash (Davy, 1808), is an essential macronutrient that accounts for 1%–10% of the dry mass of a plant and, as the major ionic contributor to cell turgor, plays a role in cell growth and other turgor-dependent cell movements (Epstein, 1972; Epstein and Bloom, 2005; Moran, 2007; Britto and Kronzucker, 2008; Wang and Wu, 2013). Although water-selective channels known as aquaporins exist and are important in plants for a variety of functions (Maurel, 1997; Anderberg et al., 2012), including pollen hydration on the stigma (Ikeda et al., 1997; Sommer et al., 2008; Kayum et al., 2017) and pollen tube growth (Obermeyer, 2017), by virtue of their aqueous pore, K^+ channels also serve as water channels (Wayne and Tazawa, 1990; Tazawa et al., 2001). The K^+ channels of the plasma membrane of plants, particularly those found in guard cells, are becoming well understood as a consequence of the introduction of the patch clamp technique (Schroeder, 1988, 1989; Schroeder et al., 1987, 1994; Cao et al., 1995; Schachtman et al., 1992).

Patch clamping is an electrophysiological technique. However, unlike classical electrophysiological methods, where a microcapillary electrode is inserted into the cell, with patch clamping, a microcapillary electrode is pressed against a clean membrane surface, and suction is applied to make a tight seal (Hedrich, 1995). The high-resistance seal that is formed between the pipette and the membrane allows the recording of tiny currents, including those that pass through single channels.

There are several configurations used in the patch-clamp technique (Hamill et al., 1981), one of which is the whole-cell configuration, in which the current through the whole membrane is studied.

Using the whole-cell configuration, Julian Schroeder and his colleagues (Schroeder et al., 1987; Schroeder, 1988) discovered that the activity of K^+ channels is controlled by membrane potential (Fig. 2.28). To perform these experiments, the electrical potential across the membrane is varied and the steady-state current that flows at each potential is measured. The steady-state current is then plotted with respect to the membrane potential used to elicit those currents. Such a plot is referred to as an *I*-ψ curve or, more commonly, an *I*-V curve where *I* represents current (a variable) and V stands for voltage (a unit of measurement). We can determine if K^+ is the ion that flows through the channel by calculating the equilibrium potential for K^+ using the Nernst equation and the concentrations of K^+ on both sides of the membrane. Because the external and internal concentrations of K^+ are 11 and 105 mol/m^3, respectively, the equilibrium potential is approximately −0.058 V. If K^+ is the ion moving through the channels, there should be no current at the applied potential that is equal to the equilibrium potential for K^+. We see in the figure that the curve in Fig. 2.28D intercepts the *x*-axis at approximately the voltage equal to the equilibrium potential of K^+ and there is no current flow. Moreover, the direction and magnitude of the currents passing through the channels depend on the deviation of the electrical potential from the equilibrium potential for K^+ and are consistent with the movement of positive charge. The currents can be interpreted as an uptake of K^+ at potentials more negative than the equilibrium potential and a release at more positive potentials. If the electrical potential only influenced the K^+ currents by changing the displacement of the membrane potential from the equilibrium potential for K^+, then the I-ψ curve would be linear with an *x*-intercept at the K^+ equilibrium potential. However, because the curve is nonlinear, the electrical potential must be activating voltage-gated channels, which are virtually closed when the membrane potential lies between −0.05 and −0.08 *V*. The channels are activated at both hyperpolarized and depolarized potentials, and the conductances of the membrane at hyperpolarized or depolarized potentials can be calculated from the slope of the curve.

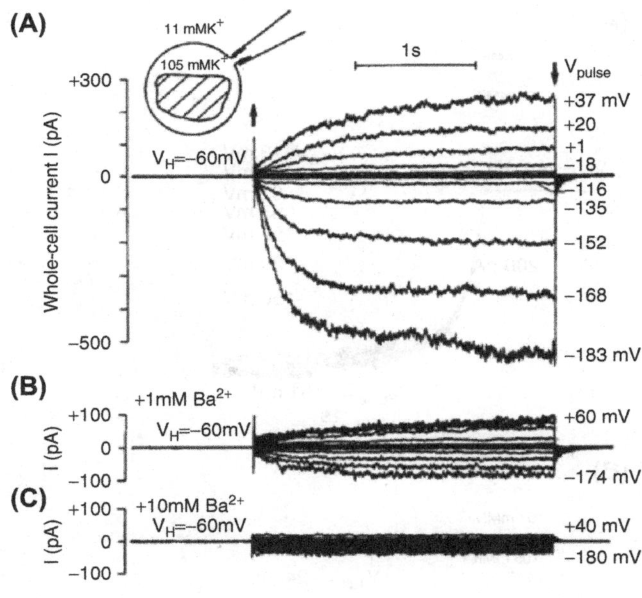

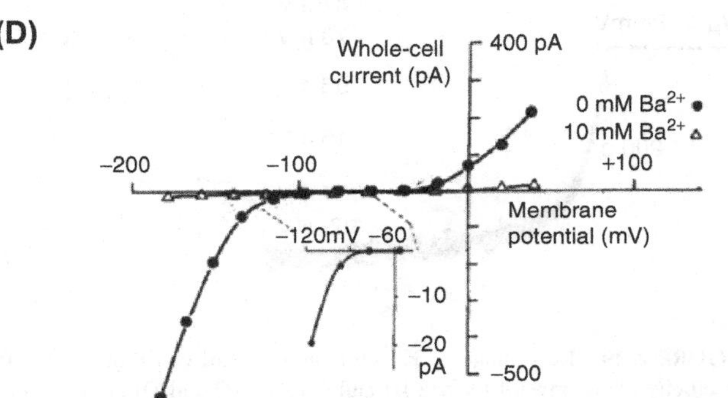

FIGURE 2.28 Recordings of K^+ channel currents in guard cell protoplasts of *Vicia faba* using the whole-cell configuration of recording. (A) Current versus time curves at the indicated pulsed voltages (right) (B and C) Current versus time curves at pulsed voltages when the cells are treated with Ba^{2+}. (D) Current voltage or *I*-ψ curve that represents the data shown in (A) and (C). Note that the channels pass K^+ and not Cl^- because the x-intercept is closer to the equilibrium potential for K^+ (−0.058 V) than it is to the equilibrium potential for Cl^- (+0.058 V). *From Schroeder, J.I., Raschke, K., Neher, E., 1987. Voltage dependence of K^+ channels in guard-cell protoplasts. Proc. Natl. Acad. Sci. USA 84, 4108–4112.*

We can determine the selectivity of the channels for K^+ by doing the following experiment (Fig. 2.29). First, the channel in question is activated by applying either a hyperpolarizing or a depolarizing pulse. Then the voltage is rapidly changed to various values to see where there is neither inward nor outward "tail" current flow. The potential that causes neither an inward nor an outward current represents the reversal potential (ψ_{rev}). If there are only two permeant ions used at a time and one is on one side of the membrane and one is on the other, the relative permeabilities can be calculated from the reversal potential obtained using an exponentiated form of a simplified

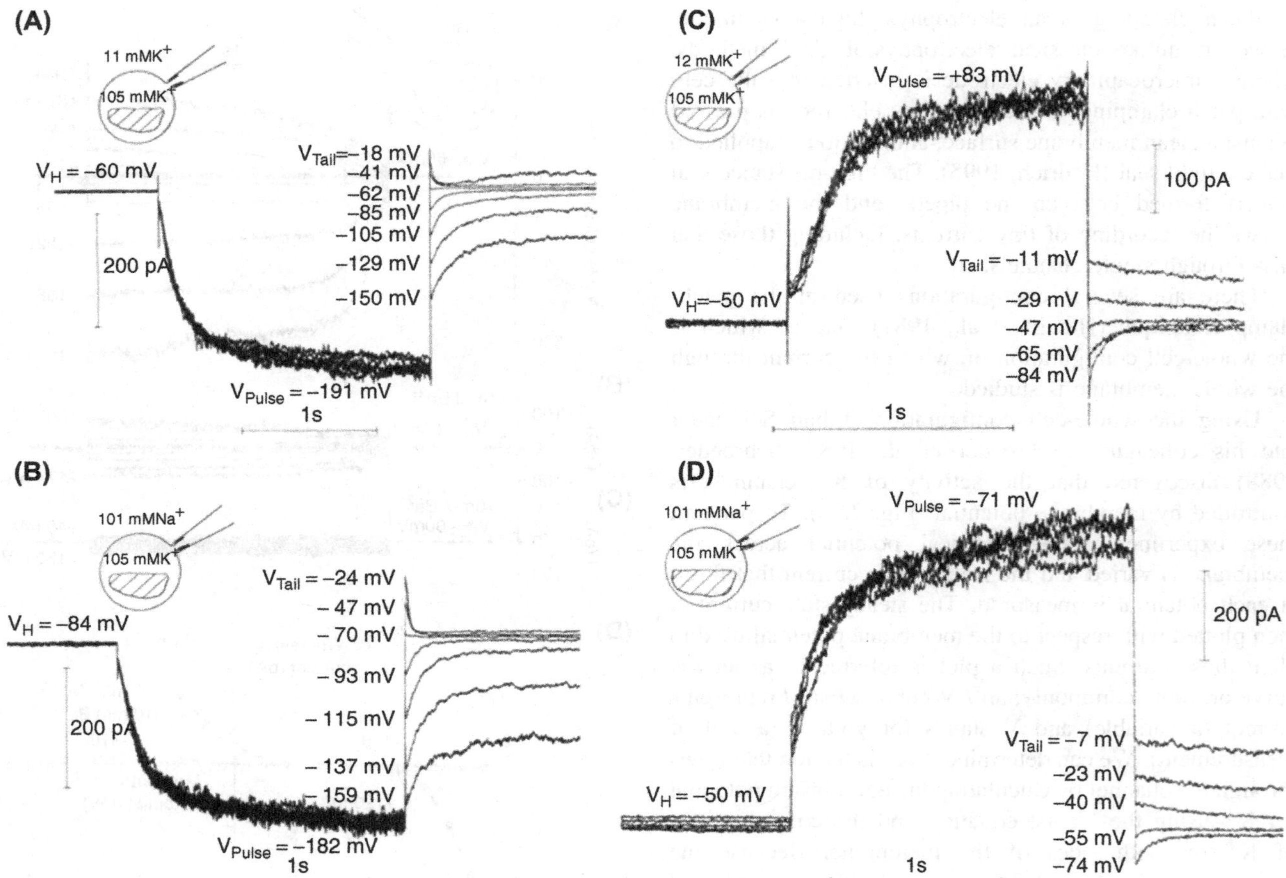

FIGURE 2.29 Recordings of K^+ currents in guard cell protoplasts of *Vicia faba* using the whole-cell configuration of recording to determine the specificity of the inward (A and B) and outward (C and D) currents for K^+. Notice that the concentrations of Na^+ and K^+ have been varied in each experiment. *From Schroeder, J.I., Raschke, K., Neher, E., 1987. Voltage dependence of K⁺ channels in guard-cell protoplasts. Proc. Natl. Acad. Sci. USA 84, 4108–4112.*

version of the Goldman equation, where everything cancels except the following terms:

$$\psi_{rev} = \frac{kT}{e} \ln \frac{P_{Na}}{P_K} \text{ or } \frac{P_{Na}}{P_K} = e^{\frac{e\psi_{rev}}{kT}} \qquad (2.31)$$

These experiments show that the permeability sequence for the inward-rectifying channel is $K^+ >$. $Rb^+ > Na^+ > Li^+ >> Cs^+$. P_{Na}/P_K for the inward-rectifying channel is 0.06. It is 0.132 for the outward-rectifying channel.

Ba^{2+} blocks the inward and outward current (Fig. 2.28), whereas Al^{3+} only blocks the inward current but not the outward current. This provides evidence that there are two distinct classes of channels, one that allows K^+ to move into the cell (e.g., inward rectifying) and one that facilitates the movement of K^+ out of the cell (e.g., outward rectifying).

Currents that pass through a single channel can be visualized with the patch clamp technique even though they may be less than 1 pA in magnitude (Fig. 2.30). To accomplish this, a single patch of membrane approximately 1 μm in diameter must be removed from the cell. This is done by pulling the tightly attached patch-clamp pipette away from the cell. The single-channel currents consist of rectangular pulses of random duration. The upward and downward spikes represent small conformational changes in the channel-gating polypeptide. Each upward step represents the closing and each downward step represents the opening of a single inward-rectifying cation (e.g., K^+) channel. By convention, inward current, which is defined as the movement of positive charge from an E-space to a P-space, is presented as a downward deflection from zero, and outward current is presented as an upward deflection. The height of the opening is a measure of the current that passes through the channel. As long as the channel is open, ions pass through it being driven by their electrochemical difference. The current amplitude indicates how many ions pass through the channel in a given time because the number of ions/s passing through a channel times *ze* equals the current.

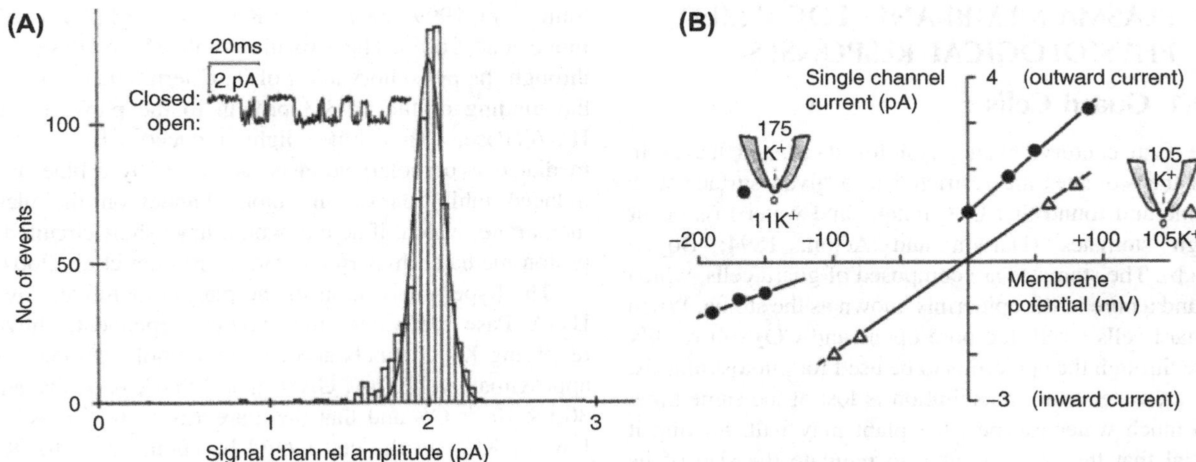

FIGURE 2.30 Recordings of K^+ currents through single channels in guard cell protoplasts of *Vicia faba* using the patch-clamp configuration of recording. (A) Shows the current recording of an individual channel and a histogram showing the distribution of amplitudes through a channel. (B) Shows current-voltage curves for channels in outside-out patches. *From Schroeder, J.I., Raschke, K., Neher, E., 1987. Voltage dependence of* K^+ *channels in guard-cell protoplasts. Proc. Natl. Acad. Sci. USA 84, 4108–4112.*

$$\text{Single channel flow} = \text{ions/s}$$

$$\text{ions/s} = \text{single channel current}/ze$$

$$\text{ions/s} = 2 \times 10^{-12}\text{A}/1.6 \times 10^{-19}\text{C/K}^+$$

$$\text{ions/s} = 1.25 \times 10^7\ \text{K}^+/\text{s}$$

We can estimate the conductance of the K^+ channels from the current (I) and electrical potential (ψ), or I-ψ curves, for the membrane patch because the slope of the curve is equal to the conductance ($-I/\psi$). At hyperpolarizing potentials, the single-channel conductance is 10 pS, and at depolarizing potentials, the single-channel conductance is 25 pS, where 1 S = 1 A/V. The difference in conductance supports the contention that there are two distinct types of channels on the plasma membrane of guard cells, one inwardly rectifying and the other outwardly rectifying. Notice that the observed conductance depends on the concentration of K^+ on each side of the membrane.

How many inward-rectifying K^+ channels are there on the plasma membrane? If the whole-cell current is approximately 300 pA and each channel passes 2 pA, then there are 150 open channels/cell. If the area of the cell is $6 \times 10^{-10}\ \text{m}^2$, then there are 2.5×10^{11} channels/m^2 or 0.25 channels/μm^2. That is, about one channel can be found for every four patches made. The inward-rectifying K^+ can be observed in the plasma membrane using GFP fusion proteins (Hurst et al., 2004).

The genes for K^+ channels have been cloned by transforming yeast that is unable to grow on low K^+ with cDNA from *Arabidopsis* (Anderson et al., 1992; Sentenac et al., 1992). If the transformed yeast can grow on low K^+ with a given DNA, then that DNA used to transform it is likely to be some kind of K^+ transporter. Subsequently, many families of genes that encode K^+ channels have been discovered using other cloning strategies (Mäser et al., 2001; Véry and Sentenac, 2003; Hosy et al., 2003; Gierth and Mäser, 2007; Grabov, 2007; Ward et al., 2009; Wang and Wu, 2013). There are many different kinds of K^+ channels that are expressed throughout the plant, consistent with the idea that DNA is promiscuous, and thus regions that code for properties such as K^+ selectivity, K^+ affinity, certain gating characteristics, channel regulation, and kinetics, as well as placement in the membrane with respect to space and time can be mixed and matched through evolutionary time in a number of ways to result in a great variety of adaptive channels. The various polypeptides that confer the ability to transport K^+ come together as subunits to form dimers or heterotetrameric K^+ channels with diverse properties that depend on the relative composition of the subunits (Hoth et al., 2001; Véry and Sentenac, 2003; Xicluna et al., 2007). Molecular genetics has taught us that a single transporter is often made up of the polypeptides encoded by several members of a gene family (Hedrich and Marten, 2006).

Using recombinant DNA technology, Julian Schroeder and his coworkers (Cao et al., 1995; Rubio et al., 1995; Uozumi et al., 1995; Mäser et al., 2002) have identified the amino acids or peptide regions of various K^+ channels that confer ion selectivity and the ability to act as a rectifier, and this work has been extended by others (Marten and Hoshi, 1998; Hoth et al., 2001). The properties of the K^+ channels can be regulated through the action of many regulatory proteins, including protein kinases, protein phosphatases, and 14-3-3 proteins (Véry and Sentenac, 2003; Jones et al., 2014).

2.10 PLASMA MEMBRANE—LOCALIZED PHYSIOLOGICAL RESPONSES

2.10.1 Guard Cells

In the 19th century, plant physiologists coated leaves in which the stomates are restricted to a given surface with Vaseline and found that CO_2 uptake and water loss occur through stomates (Darwin and Acton, 1894; Boyer, 2015a,b). The stomates are composed of guard cells, which surround a pore in the epidermis known as the stoma. When the guard cells swell, the pore opens and CO_2 can readily diffuse through the epidermis to be used for photosynthesis; however, water from transpiration is lost at the same time. If too much water escapes, the plant may wilt, making it essential that the plant be able to regulate the size of its stoma. The stoma closes when the guard cells shrink, and this closure not only prevents the loss of water but also prevents the influx of CO_2 necessary for photosynthesis. The swelling and shrinkage of the guard cells are consequences of their water uptake or loss, respectively. The guard cells act as osmometers; water moves in and out of them passively depending to a large extent on the difference in the osmotic pressure on both sides of the plasma membrane and to a smaller extent on the elasticity of the guard cell wall (Roelfsema and Hedrich, 2002). In the main, the osmotic pressure in the guard cells is due to K^+ and Cl^-. Thus, the channels involved in K^+ transport across the plasma membrane provide the molecular mechanism for regulating many aspects of whole-plant physiology, including photosynthesis, transpiration, thermoregulation, and the ascent of sap due to transpiration (Dixon and Joly, 1895; Larmor, 1905; Ewart, 1906; Dixon, 1938; Nobel, 1983, 1991, 2005). The properties of the K^+ channels that I discussed above can account for the known properties of guard cells that were obtained from physiological studies. That is, the properties of these channels can account for guard cell swelling, which requires an increase in the $[K^+]$ of about 400 mol/m^3.

Assume that at rest, the membrane potential of guard cells is equal to the Nernst potential of K^+ (-0.058 V), so that there is no net movement of K^+ into or out of the cell (Saftner and Raschke, 1981). The opening and closing of the stomatal pore are regulated by a myriad of environmental signals, which are perceived and integrated by the guard cells themselves (Schroeder et al., 2001; Merlot et al., 2007; Jones et al., 2014a,b,c; Murata et al., 2015). Blue light, which signals the beginning of the day, is absorbed by phototropins and causes guard cells to swell as a result of the blue light—induced activation of the H^+-ATPase in the plasma membrane (Wang et al., 2014). The activation of the H^+-ATPase results in the hyperpolarization of the membrane potential to approximately -0.16 V (Zeiger et al., 1977; Shimazaki et al., 1992; Kinoshita and Shimazaki, 1999, 2001, 2002; Kinoshita et al., 2001, 2003; Inoue et al., 2008). The activation of the H^+-ATPase occurs through the phosphorylation of its C-terminus followed by the binding of the 14-3-3 protein to the phosphorylated H^+-ATPase. The blue light—induced, H^+-ATPase—mediated hyperpolarization is facilitated by a blue light—induced inhibition of an anion channel on the plasma membrane, which, if active, would have short-circuited the proton-mediated hyperpolarization (Marten et al., 2007).

The hyperpolarization of the plasma membrane by the H^+-ATPase activates the voltage-dependent, inward-rectifying K^+ channels and causes a whole-cell current of approximately 300 pA. Given that 300 pA is equivalent to 300×10^{-12} C/s and that there are 1.6×10^{-19} C/K^+, the flow of K^+ into the cell would be about 1.9×10^9 K^+/s. Given that there are approximately 150 inward-rectifying channels per cell, then approximately 1.25×10^7 K^+ must pass through each channel every second. If the volume of a guard cell is approximately 10^{-14} m^3 and the $[K^+]$ in the guard cell must increase by 400 mol/m^3, then the channels must pass 400×10^{-14} mol of K^+, which, using Avogadro's number as a conversion factor, is equivalent to 2.4×10^{12} K^+. Thus, if all the channels were activated it would take (2.4×10^{12} K^+)/(1.9×10^9 K^+/s) or approximately 1260 seconds (21 minutes) for the guard cell to swell. The opening of some stoma, like those from corn, takes only a few minutes, whereas those from the broad bean take hours (Blatt, 1991; Tallman, 1992; Assmann, 1993). Variation in the kinetics of opening depends not only in variations in the properties and abundance of potassium channels and guard cell size but also in part on the fact that opening can result from the accumulation of solutes besides potassium, including chloride, malate, and sucrose (Schroeder et al., 2001). There are many environmental factors, hormones, nucleotides, enzymes, ions, and pathogens that regulate stomatal opening and closure in the treasure house of plant species (Kim et al., 1995; Eun and Lee, 1997, 2000; Li et al., 1998, 2000; Li and Assmann, 2000; Mori et al., 2000; Eun et al., 2001; Hwang and Lee, 2001; Schroeder et al., 2001; Jung et al., 2002; Taiz and Zeiger, 2006; Kim et al., 2010; Kollist et al., 2014; McLachlan et al., 2014; Murata et al., 2015; Marom et al., 2017).

2.10.2 Motor Organs

Many legumes, including *Mimosa*, the sensitive plant, *Neptunia*, *Albizia*, and *Samanea*, show leaflet movements. The leaflet movements result from changes in turgor. The changes in turgor result from water movement that is controlled by ion movements across the plasma membrane of specialized cells in organs known as pulvini (Brücke, 1898a; Toriyama, 1955, 1962, 1974; Dutt, 1957; Datta, 1957; Jaffe and Galston, 1967; Toriyama and

Satô, 1968a,b, 1970; Allen, 1969; Toriyama and Jaffe, 1972; Campbell and Thomson, 1977; Campbell et al., 1979; Moran et al., 1988; Satter et al., 1988; Hollins and Jaffe, 1997; Leopold et al., 2000; Suh et al., 2000; Moshelion and Moran, 2000; Yu et al., 2001; Moshelion et al., 2002a,b; Okazaki, 2002).

2.10.3 Action Potentials

The necessity of cells to osmoregulate rapidly resulted in the evolution of ion channels. Once these channels evolved, they could be used to communicate electrical signals within a cell or from cell to cell in the form of action potentials (Di Palma et al., 1961; Sibaoka, 1962, 1966; Cole, 1968, 1979; Kishimoto, 1968; Goldsworthy, 1983; Huxley, 1992, 1994; Wayne, 1994; Shimmen, 2001a,b, 2003, 2008; Johnson et al., 2002; Baudenbacher et al., 2005; Iwabuchi et al., 2005, 2008; Kaneko et al., 2005; Fromm and Lautner, 2007; Visnovitz et al., 2007; Volkov et al., 2010a,b; Basir et al., 2015). An action potential is a transient depolarization of the plasma membrane that is propagated along the length of the cell. In characean cells, a mechanical or electrical stimulus transiently activates a mechanosensitive calcium channel. The resulting influx of Ca^{2+} causes an increase in cytosolic Ca^{2+} that activates Cl^- channels on the plasma membrane. The efflux of Cl^- along its electrochemical gradient through the channels depolarizes the adjacent membrane, which opens more Ca^{2+} channels, and the cycle repeats as the depolarization propagates along the cell. In the case of characean cells, the action potential results in an electrically or mechanically induced cessation of cytoplasmic streaming (see Chapter 12).

2.10.4 Cell Polarization

Many cells, including rhizoids, protonema, and pollen tubes, exhibit a polarized distribution of ionic currents, which most likely result from the unequal distribution of pumps and channels in the plasma membrane. These currents are involved in many aspects of cell polarization that occur during development (Jaffe, 1979, 1981; Saunders, 1986a,b; Harold, 1990; Feijó et al., 1995, 2001; Holdaway-Clarke et al., 1997; Messerli and Robinson, 1997; Feijó et al., 1999; Franklin-Tong, 1999; Messerli et al., 1999, 2000; Hepler et al., 2001; Griessner and Obermeyer, 2003; Certal et al., 2008). Indeed, cells can also generate electric fields that may participate in localizing the proteins of the plasma membrane in a polar manner by electrophoresis in the plane of the membrane (Jaffe, 1977; Poo and Robinson, 1977; Poo, 1981; see also Chapter 19).

The electrical polarization of individual cells may play a role in the growth of whole plants (Goldsworthy and Rathore, 1985; Desrosiers and Bandurski, 1988). Moreover, the polarized distribution of plasma membrane proteins is important for the transport of auxin and for the formation of Casparian strips in the endodermis.

Auxin is a plant hormone that promotes cell elongation and thus influences nearly every aspect of plant growth and development (Okada et al., 1991; Abel and Theologis, 2010). As a result of the acidic pH of the cell exterior and the neutral pH of the cytoplasm, auxin permeates the plasma membrane of the cell in its neutral form but becomes ionized within the cell. Because auxin becomes ionized, it cannot permeate the lipid bilayer of the plasma membrane and must exit through an auxin efflux carrier protein. The polarized distribution of auxin efflux carriers in the plasma membrane of each cell determines the pathway of auxin transport within the plant, and the polarity of auxin transport is correlated with many developmental and physiological responses (Gälweiler et al., 1998; Geldner, 2009; Luschnig and Vert, 2014; Adamowski and Friml, 2015).

Water and nutrients move from protoplast to protoplast through the root through the symplast or through the wall around the cells through the apoplast. The cells of the endodermis have specialized thickenings of lignin and suberin in the anticlinal cell walls that inhibit the free diffusion of polar molecules and ions through the apoplast (Van Fleet, 1961; Bonnett, 1968; Karahara and Shibaoka, 1988, 1992; Naseer et al., 2012; Roppolo and Geldner, 2012; Geldner, 2013a; Andersen et al., 2015; Kamiya et al., 2015; Barberon et al., 2016; Pauluzzi and Bailey-Serres, 2016; Foster and Miklavcic, 2017). The specialized thickening, known as the Casparian strip, ensures that in the root, the water and nutrients that enter the stele pass through the symplast—specifically through the periclinal surfaces of the plasma membrane of the endodermal cells, which are specialized for ion transport (Ma et al., 2006, 2007; Alassimone et al., 2010; Takano et al., 2010; Barberon, 2017). In rice endodermal cells, the silicon influx transporter is localized to the plasma membrane facing the soil and the silicon efflux transporter is localized to the plasma membrane facing the stele (Ma et al., 2006, 2007). Prior to the formation of the specialized wall thickenings, plasma membrane–localized proteins involved in Casparian strip formation exhibit a polarized distribution that predicts the appearance of the specialized thickenings (Grebe, 2011; Roppolo et al., 2011; Alassimone et al., 2012; Geldner, 2013b).

2.11 STRUCTURAL SPECIALIZATIONS OF THE PLASMA MEMBRANE

Invaginations of the plasma membrane, analogous to brush borders in intestines, increase the surface area in a variety of plant cells. *Dunaliella*, an alga that lives in the Dead Sea, increases the surface area of its plasma membrane through a rapid and continuous process of endocytosis and exocytosis (Ginzburg et al., 1999). Other organisms increase the area of the plasma membrane by forming apparently less dynamic invaginations known as lomasomes, charasomes, or plasmalemmasomes (Moore and McAlean, 1961; Chau et al., 1994; Foissner et al., 2016). In some cells, which occur at bottlenecks in solute transport pathways, invaginations of the extracellular matrix that increase the surface area of the plasma membrane occur. Such cells are known as *transfer cells* (Pate and Gunning, 1972; Gunning and Pate, 1974; Offler et al., 2003; Royo et al., 2007). Invaginations of the plasma membrane are observed in gland cells of carnivorous plants that secrete digestive enzymes (Robins and Juniper, 1983; Scala et al., 1968; Schwab et al., 1969), in the cells of flowering plants bordering mycorrhizal fungi (Allaway et al., 1985; Ashford and Allaway, 1985), in cells at the interface of two generations (Offler et al., 2003), and in the salt glands of *Limonium* (Faraday and Thomson, 1986a,b,c; see Fig. 2.31). In fact, when limnologist Robert Lauterborn declared that the self-purification of fresh water necessary to prevent eutrophication is directly proportional to the surface area of the flora, he essentially realized the relationship between the area of the plasma membrane and its ability to take up nutrients.

The facilitated transport of solutes across the plasma membrane does not only depend solely on transport proteins, including pumps, channels, and carriers, but can also occur as a result of exocytosis and endocytosis, which are discussed in Chapter 8. Ed Etxeberria and his colleagues (Etxeberria et al., 2005a,b,c, 2007; Baroja-Fernandez et al., 2006; Pozueta-Romero et al., 2008) have demonstrated that the uptake of apoplastic sugars into plant cells depend on two parallel pathways—one consisting of a carrier protein and the other consisting of endocytotic vesicles. Transport due to the carrier protein has rectangular hyperbolic uptake kinetics and is inhibited by phloridzin (Bush, 1993), whereas transport due to the endocytotic mechanism is linear and inhibited by wortmannin-A (Emans et al., 2002) and latrunculin-B (Baluska et al., 2004).

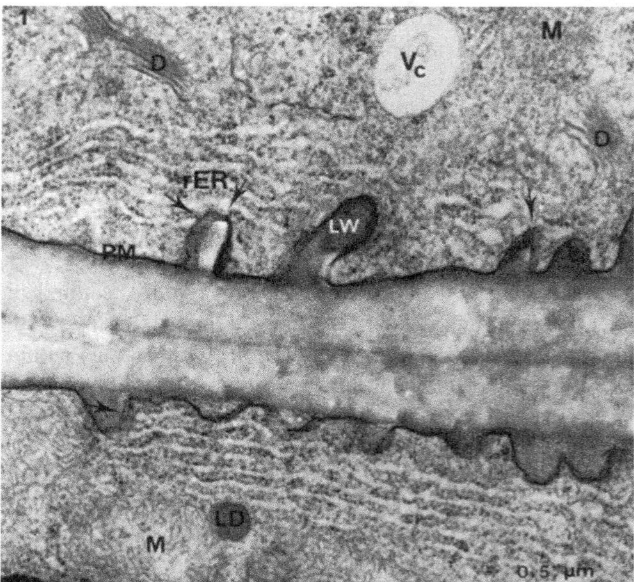

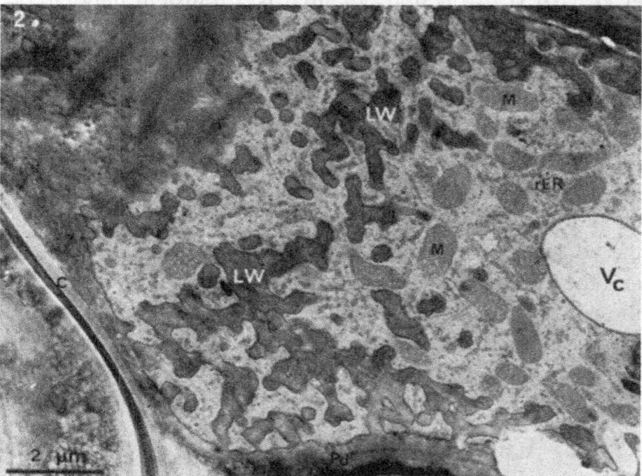

FIGURE 2.31 The surface area of the plasma membrane of the glandular cells of *Dionaea muscipula* is increased due to the labyrinthine invaginations in the extracellular matrix (LW). (1) Cross-section, (2) tangential section. *c*, cuticle; *D*, Golgi stack; *M*, mitochondrion; *PM*, plasma membrane; *rER*, rough endoplasmic reticulum; V_c, vacuole. *From Robins and Juniper (1980).*

2.12 THE CYTOSKELETON—PLASMA MEMBRANE—EXTRACELLULAR MATRIX CONTINUUM

While investigating plasmolysis in a number of plants, Bower (1883) noticed that the protoplasm does not detach uniformly from the cell wall as was often shown in studies of plasmolysis, but does so in a nonuniform manner, as if the protoplasm adhered to the cell wall in a number of places. The thin strands of protoplasm that adhere to the cell wall are now know as Hechtian strands, named after Hecht (1912), who observed them while studying plasmolysis in onion cells (Fig. 2.32; Küster, 1929; Oparka, 1994; Lang-Pauluzzi, 2000; Buer et al., 2000; Lang-Pauluzzi and Gunning, 2000). In both plant and animal cells, the plasma membrane does not exist in isolation but is intimately attached to the extracellular matrix on the outside and the cytoskeleton on the inside. The plasma membrane proteins that connect the extracellular matrix proteins to the cytoskeleton are usually called *integrins* (McDonald, 1988; Ruoslahti, 1988; Burridge et al., 1988; Pennell et al., 1989; Schindler et al., 1989; Roberts, 1990; Humphries, 1990; Kaminskyj and Heath, 1995; Canut et al., 1998; Laval et al., 1999; Nagpal and Quatrano, 1999; Swatzell et al., 1999; Sun et al., 2000; Sonobe et al., 2001; Sakurai et al., 2004; Gens et al., 2000; Pickard and

FIGURE 2.32 Hechtian strands in onion epidermal cells that demonstrate the connections between the protoplast and the extracellular matrix. Bar, 20 μm. *From Lang-Pauluzzi, I., Gunning, B.E.S., 2000. A plasmolytic cycle: the fate of cytoskeletal elements. Protoplasma 212, 174–185.*

Fujiki, 2005; Pickard, 2007; Knepper et al., 2011). The cytoskeleton and extracellular matrix are discussed in Chapters 10, 11, and 20.

The attachment of the plasma membrane to the extracellular matrix can be seen easily by plasmolyzing the cells (see Fig. 2.32; Cholodny and Sankewitsch, 1933; Lang-Pauluzzi and Gunning, 2000). Interestingly, the attachments are not uniform but may show a distinct polarity within the cell (Strugger, 1935; Stebbins and Jain, 1960). The plasma membrane is attached structurally and functionally to an underlying skeleton known as the membrane skeleton (Bennett and Gilligan, 1993). The membrane skeleton is composed of proteins, including spectrin, ankyrin, etc., and ankyrin may bind directly to some of the transport proteins (Faraday and Spanswick, 1993). The membrane skeleton may also attach directly to the cytoskeleton. The integrin-like proteins that connect the extracellular matrix with the cytoskeleton can act as mechanosensors (Eychmans et al., 2011) and appear to be involved in the ability of cells to sense gravity (Wayne et al., 1990, 1992; Hemmersbach and Braun, 2006) and touch-induced responses (Haberlandt, 1914; Junker, 1977; Bünning, 1989; Jaffe et al., 2002). The cytoskeleton–plasma membrane–extracellular matrix continuum also appears to be the way pathogenic fungi sense the epidermal cell pattern on leaves that facilitates the directional growth of the mycelium and the formation of an appressorium (Johnson, 1934; Hoch et al., 1987; Correa et al., 1996).

2.13 SUMMARY

The plasma membrane is at the frontier of the plant cell and not only separates the living protoplasm from the external medium but also coordinates the relationships between the protoplasm and the external world. In general, the lipids in the plasma membrane provide a barrier to mixing, while the membrane proteins facilitate the transport of polar substances across the membrane. In this chapter, I have discussed how to quantify phenomena that are not directly measurable by postulating relationships between the desired quantities and measurable quantities. Remember that these relationships are postulates and must be changed to accommodate newly discovered relationships and interactions. I have also discussed the techniques used to characterize ion fluxes and visualize the movement of ions through a single channel. Imagine how delighted Wilhelm Pfeffer would be to know that we now have the theoretical and technical tools to understand leaflet movement in *Mimosa*, the phenomenon that started Pfeffer in his investigations of the plasma membrane.

2.14 QUESTIONS

2.1. What is the evidence that the plasma membrane provides a barrier between the living and nonliving world of a cell?

2.2. What are the mechanisms by which the plasma membrane and its components regulate transport between the inside and outside of the cell?

2.3. What are the limitations of thinking about the plasma membrane as a barrier or as the sole barrier?

The references for this chapter can be found in the references at the end of the book.

Chapter 3

Plasmodesmata

TO understand political power right, and derive it from its original, we must consider, what state all men are naturally in, and that is, a state of perfect freedom to order their actions, and dispose of their possessions and persons, as they think fit, within the bounds of the law of nature, without asking leave, or depending upon the will of any other man….The state of nature has a law of nature to govern it, which obliges every one: and reason, which is that law, teaches all mankind, who will but consult it, that being all equal and independent, no one ought to harm another in his life, health, liberty, or possessions: for men being all the workmanship of one omnipotent, and infinitely wise maker.

John Locke (The Second Treatise of Civil Government).

3.1 THE RELATIONSHIP BETWEEN CELLS AND THE ORGANISM

Cells in multicellular organisms are both autonomous and interdependent (Huxley, 1912; Canguilhem, 1969; Andrews, 2002). Biologists have argued about the relationship between cells and the organism—or the part to the whole—with as much passion as those who argue about the relationship of the individual to the state (Donne, 1634; Hobbes, 1651; Locke, 1690; Hume, 1748; Rousseau, 1762; Priestley, 1771; de Lafayette and Jefferson, 1789; Stanton, 1848; Thoreau, 1849; Spencer, 1860; Roberts, 1938; Hamilton et al., 1961; Jahn, 1964; King, 1964; Hanser, 1979; Metaxas, 2007, 2010; McDonough, 2009; Clinton and Kaine, 2016; Trump, 2015; Sanders, 2016) or the individual to the rest of the world (Taylor, 2008a,b). That is, who or what controls the borders and their permeability or openness (Stephenne et al., 2009). Proponents of the organismal theory of plant development and the cellular theory of plant development still argue vehemently about the respective importance of each level of organization in plant development, although, like many arguments, there are elements of truth in both views (Weiss, 1940; Kaplan and Hagemann, 1991; Kaplan, 1992; Baluska et al., 2004a,b).

The organismal theory of plant development arose after botanists, including Charles-François Brisseau de Mirbel (1808) and Augustin de Candolle and Kurt Sprengel (1821), studied static sections of plants and concluded that there were three elementary components of plants: cells, tubes, and spirals. Consequently, the whole plant was considered the single most elementary form of vegetable life. By contrast, Dutrochet (1824) took a dynamic developmental approach and noticed that all the structures in plants, including the tubes and spirals, developed from cells. Dutrochet not only championed the view that the cell is the fundamental element in multicellular organisms but also emphasized that cells were independent entities. Dutrochet (1824, in Buvat, 1969) wrote,

> *I may repeat here what I have revealed previously about the organic texture of plants. We have seen that these organisms were entirely composed of cells, or of organs obviously derived from cells. We have seen that these hollow organs were simply contiguous, and held to each other by a cohesive force, but that such an assembly of cells did not really form one continuous tissue. Thus it seemed to us that an organic creature consists of an infinite number of microscopic components, which have no relationship to each other beyond that of being adjacent.*

The cell view was later supported by the evolutionary interpretation of the trends in both the plant and animal kingdoms to form more and more elaborate organisms. We see such trends vividly in the green algae where some organisms, like *Chlamydomonas*, are composed of only a single cell, whereas others are organized loosely into colonies that show no (e.g., *Gonium*) or minimal (e.g., *Volvox*) differentiation. Still others, for example, *Ulva*, are even differentiated into leaflike and rhizoidal tissues (Bold and Wynne, 1978). This phylogenetic series implies that multicellular organisms are *cell republics*, which result from the assemblage of a large number of independent units.

The organismal view was supported by Julius Sachs (1887), who wrote,

> *That plants consist of cells is now known to every well-informed man; yet the true meaning of the word cell may be quite clear to but a few, the less so since biologists themselves, even now, hold and discuss the most different opinions upon it. To many, the cell is always an independent living being, which sometimes exists for itself alone, and sometimes "becomes joined with" others—millions of its like in order to form a cell-colony, or, as Häckel has named it for the plant particularly, a cell republic. To others again, to whom the author of this book also belongs, cell-formation is a phenomenon very general, it is true, in*

Plant Cell Biology. https://doi.org/10.1016/B978-0-12-814371-1.00003-5

organic life, but still only of secondary significance; at all events, it is merely one of the numerous expressions of the formative forces which reside in all matter, in the highest degree, however, in organic substances.

T.H. Huxley (1853a,b) also felt that cells "are not instruments, but indications—that they are no more the producers of the vital phenomena than the shells scattered in orderly lines along the sea-beach are the instruments by which the gravitative force of the moon acts upon the ocean. Like these, the cells mark only where the vital tides have been, and how they have acted." Anton de Bary put it more succinctly: "The plant forms cells, not cells the plant" (quoted in Barlow, 1982).

The organismal theory was further supported by Whitman (1894) and Lester Sharp (1934), who wrote in his book, *Introduction to Cytology*, "The body is not an aggregation of elementary organisms, but a single organism which has evolved a multicellular structure." He noted that many plants, particularly the gymnosperms and *Paeonia*, pass through a coenocytic stage during early embryogenesis (Bierhorst, 1971), and indeed, the differentiation of the organism into cells is not necessary for complex development because there are large organisms, including *Caulerpa* and *Bryopsis*, that consist of only one cell, yet differentiate into leaflike, stemlike, and rootlike structures (Coneva and Chitwood, 2015; Ranjan et al., 2015). However, Sharp went on to say

The presence of cell partitions allows a more effective segregation of functionally specialized regions and a fuller play to those important physico-chemical processes which depend on surfaces and thin films for their action. Furthermore, it permits the development of larger plant bodies by furnishing an ideal basis for the more effective operation of turgor and for the deposition of supporting materials… The evolution of higher organisms has unquestionably been very largely conditioned by the multicellular state, but we should think of such organisms primarily as highly differentiated protoplasmic individuals rather than cell republics.

Is this old and ever-recurrent problem of cell theory versus organismal theory a moot question? According to Wilhelm Ostwald (1910), we can determine whether a question is moot by asking ourselves, "What would be the difference empirically if the one or the other view were correct?" I think that both theories have elements of truth that help understand plant development. I concur with the organismal view of multicellular organization and believe that it is erroneous to work on the assumption that an organism is only equal to the sum of its parts and has no greater level of organization and coordination.

Multicellular organisms have emergent properties that the individual cells themselves lack (Heitler, 1963). Even water has a higher level of organization and integration than the oxygen and hydrogen of which it is composed! A purely cellular view could hinder research on higher levels of integration. However, it will become clear as we continue our journey that a purely organismal view could lead to erroneous experimental results. Each organism is made up of many different cell types, each of which is surrounded by a differentially permeable membrane that determines the degree of autonomy of each cell. Some of these cells may be undergoing different processes at a given time than others. Thus, when breaking the organism up into its parts to understand it physiology, misleading results and unjust interpretations may occur unless one separates and studies individual cell types (Wayne, 1994). On the other hand, no cell in a multicellular organism is completely autonomous, and when we isolate cells, we must be aware of the mechanical, electrical, and chemical influences we are severing (Lintilhac, 1999; Roelfsema and Hedrich, 2002). Indeed, the enucleate sieve tube elements are completely dependent on their companion cells for a continuous supply of protein (Parthasarathy, 1974; Esau and Thorsch, 1985; Lough and Lucas, 2006).

In Chapter 2, I spoke about cells as if they existed in isolation, protected by the plasma membrane from an ever-changing and sometimes hostile environment. However, the cells in multicellular plants are not only physically touching but also often connected by small structures called plasmodesmata (singular, plasmodesma), which allow direct cell-to-cell communication (Tangl, 1879; Elsberg, 1883; Goebel, 1926; Pickard and Beachy, 1999; Ehlers and Kollmann, 2001; Oparka and Roberts, 2001; Roberts, 2005; Niklas and Newman, 2013; Niklas, 2014; Sager and Lee, 2014; Brunkard et al., 2015; Heinlein, 2015; Brunkard and Zambryski, 2017). Indeed, the presence of functioning plasmodesmata is correlated with the ability of cells to divide synchronously (Ehlers and Kollmann, 2000), and the loss of plasmodesmatal function is correlated with programmed cell death (Zhu and Rost, 2000). Moreover, as a result of the presence of plasmodesmata, the plasma membrane of one cell is continuous with the plasma membrane of the adjoining cell, thus forming a continuum of P-spaces, known as the *symplast*, and a continuum of E-spaces outside the plasma membrane, known as the *apoplast*.

The cells in multicellular animals are often connected by structures analogous to plasmodesmata, known as *gap junctions* (Sjöstrand et al., 1958; Revel and Karnovsky, 1967; McNutt and Weinstein, 1973; Cox, 1974). Intercellular connections between animal cells, 50–200 nm in diameter and several cell diameters long, are known as *cytonemes, nanotubular structures*, or *tunneling nanotubes* (TNT; Ramirez-Weber and Kornberg, 1999; Rustom et al., 2004). The plasma membranes of the connected cells are in direct communication and can be seen to exchange fluorescently labeled fusion proteins (Rustom et al., 2004).

3.2 DISCOVERY AND OCCURRENCE OF PLASMODESMATA

Eduard Tangl (1879) was the first person to observe connections between cells, while observing with the light microscope the endosperm of a variety of plants (Fig. 3.1; Witztum and Wayne, 2012). He serendipitously discovered these connections while investigating cell walls with organic dyes (Köhler and Carr, 2006). Tangl proposed that these connections were important for transport between cells. These intercellular connections, which pass through the surrounding extracellular matrix, came to be known as *plasmodesmata* (Livingston, 1933; Meeuse, 1957). Plasmodesmata occur in all the major groups of plants from algae to higher plants, and although the structure of the plasmodesmata in all these groups is remarkably similar (Robards and Lucas, 1990), there is some variation at the microscopic (Franceschi et al., 1994; Botha et al., 2005) and nanoscopic (Beebe and Turgeon, 1991; Waigmann et al., 1997) levels. Evolutionary studies of plasmodesmata are ongoing (Cooke et al., 1997; Raven, 1997, 2005; Cooke and Graham, 1999; van Bel and van Kesteren, 1999; Brecknock et al., 2011).

Plasmodesmata between two sister cells are typically formed during cytokinesis and are called *primary plasmodesmata*. However, plasmodesmata formation can take place between any two adjacent cells, forming new symplastic pathways. Plasmodesmata that are formed between two cells that are already separated by an extracellular matrix are called *secondary plasmodesmata*. In terms of primary and secondary plasmodesmata, plasmodesmatal formation in *Chara*, a genus of algae on the evolutionary line that gave rise to higher plants, is of interest. While *Chara zeylanica* produces both primary and secondary plasmodesmata (Cooke et al., 1997), *Chara corallina* produces only secondary plasmodesmata (Franceschi et al., 1994). The secondary plasmodesmata may have different transport characteristics from the primary plasmodesmata in the same cell (Itaya et al., 1998).

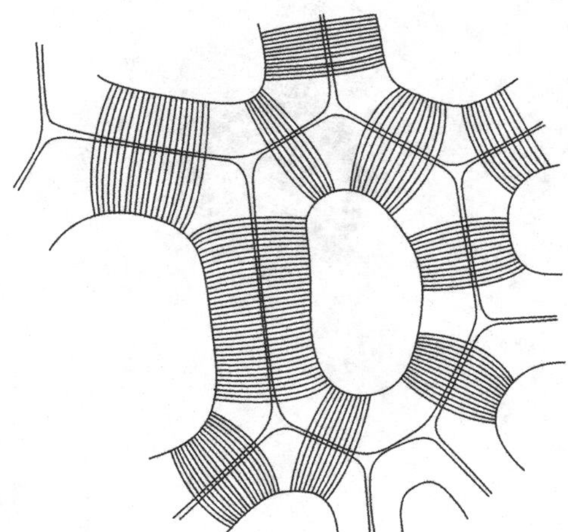

FIGURE 3.1 Intercellular connections (plasmodesmata) between endosperm cells of *Strychnos nuxvomica*. The neuromuscular poisons, strychine and brucine, come from this plant. *From Tangl, E., 1879. Ueber oftene Communicationen zwichen den Zellen des Endosperms einiger Samen. Jahrb. wiss. Bot. 12, 170–190.*

The biogenesis of the primary plasmodesmata will be discussed in Chapter 19. Secondary plasmodesmata, however, begin their formation when the extracellular matrix thins in regions where the endoplasmic reticulum (ER) is abutting the plasma membrane. As the extracellular matrix dissolves in this localized area, the endoplasmic reticula of the two adjoining cells, as well as the bordering plasma membranes, fuse to form a plasmodesma (Kollmann and Glockmann, 1991). Both primary and secondary plasmodesmata are initially simple in structure but can form complex structures through branching and/or fusion of existing plasmodesmata or the fusion of established and newly formed plasmodesmata (Oparka et al., 1999; Ehlers and Kollmann, 2001; Roberts et al., 2001; Faulkner et al., 2008).

There are generally between 1 and 15 plasmodesmata/πm^2, although as many as 41 plasmodesmata/πm^2 have been observed (Livingston, 1933). Plasmodesmata can either be uniformly distributed around the cell or occur in aggregates. In a given cell, at a given time, the number and density of plasmodesmata are precisely determined (Tilney et al., 1990b). The density and permeability of plasmodesmata determine the extent of symplastically connected cells or symplastic fields (Kim et al., 2005b). However, the density of plasmodesmata, their structure, and/or their unitary conductance can change over time (Palevitz and Hepler, 1985; Zambryski and Crawford, 2000; Kwiatkowska, 2003). At maturity, guard cells and tracheary elements lose all plasmodesmatal connections to neighboring cells (Wille and Lucas, 1984; Erwee et al., 1985; Palevitz and Hepler, 1985; Lachaud and Maurousset, 1996). The frequency of plasmodesmata is influenced by daylength and cytokinin application (Ormenese et al., 2006).

3.3 STRUCTURE OF PLASMODESMATA

Based on electron microscopic evidence, López-Sáez et al. (1966a) proposed a model for plasmodesmatal structure (Fig. 3.2). Although this model has been contested (Gunning and Robards, 1976), it is still widely accepted (Overall et al., 1982; Hepler, 1982). Electron micrographs show that a plasmodesma is a cylindrical, membrane-lined, mostly aqueous canal that is 20–40 nm in diameter and can be hundreds to thousands of nanometers long, depending on the thickness of the intervening extracellular matrix (Fig. 3.3). In the center of the canal is a cylindrical structure. It was originally called the *axial component* and

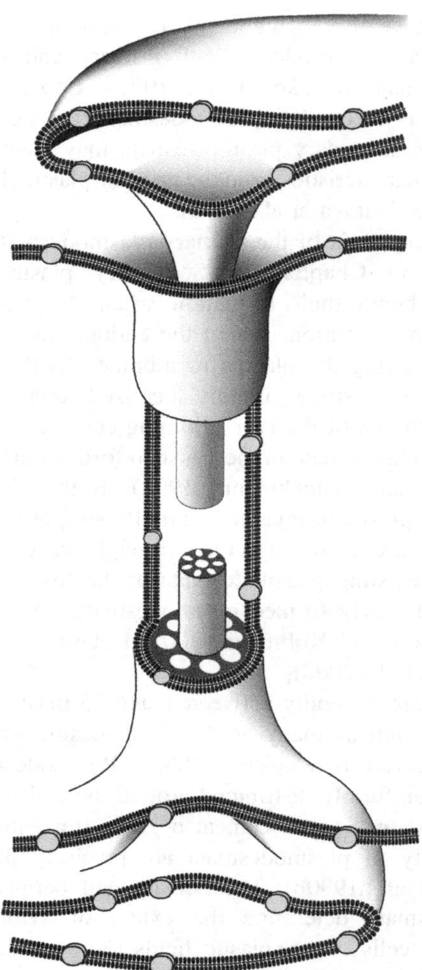

FIGURE 3.2 Diagram of a plasmodesmata showing the three-dimensional relationship between the endoplasmic reticulum, plasma membrane, and desmotubule. *From Gunning, B.E.S., Overall, R.L., 1983. Plasmodesmata and cell-to-cell transport in plants. Bioscience 33, 260–265.*

now is commonly called the *desmotubule* (Fig. 3.4). The desmotubule is continuous with the ER (see Chapter 4). A cytoplasmic pathway, called the *cytoplasmic annulus*, surrounds the desmotubule and is continuous from cell to cell. The ends of the cytoplasmic annulus often seem to be constricted. These constrictions may regulate the flux of substances through the cytoplasmic annulus, although currently there is no evidence for this.

Electron microscopic images of plasmodesmata are shown in Figs. 3.3 and 3.4. The plasma membrane shows up as a tripartite structure that is 7.2 nm wide, and the dense central rod is 1.4 nm in radius. The width of the pale ring that surrounds the dense central rod is 2.2 nm. This is consistent with the hypothesis that the desmotubule is made of the membrane of the ER without any lumen. The central rod represents the polar head groups of two oppressed inner leaflets of the ER membrane that are close-packed, and the

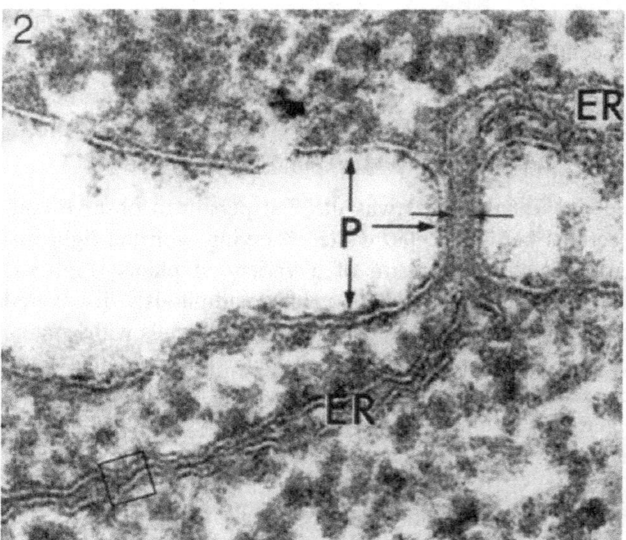

FIGURE 3.3 Longitudinal view of a plasmodesma in *Azolla pinnata* root cells. The arrow points to the desmotubule. *ER*, endoplasmic reticulum; *P*, plasma membrane, ×175,000. *From Overall, R.L., Wolfe, J., Gunning, B.E.S., 1982. Intercellular communication in Azolla roots I. Ultrastructure of plasmodesmata. Protoplasma 111, 134–150.*

clear ring represents the fatty acyl groups of the bilayer. The layer between the inner leaflet of the plasma membrane and the hydrocarbon ring of the desmotubule is called the *cytoplasmic annulus*. The cytoplasmic annulus appears as a

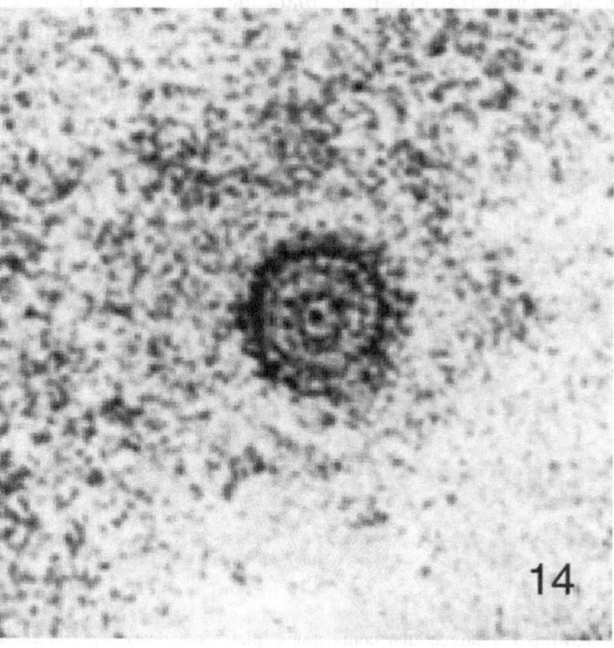

FIGURE 3.4 Transverse view of a plasmodesma in a lettuce root tip cell, ×210,000. *From Hepler, P.K., 1982. Endoplasmic reticulum in the formation of the cell plate and plasmodesmata. Protoplasma 111, 121–133.*

densely stained region, approximately 4.5 nm wide, and shows some substructure. The lumen of the ER is not continuous within a plasmodesma between cells, as evidenced by the discontinuity in staining by a lumen-filling stain (Fig. 3.5), as well as the lack of cell-to-cell transport of green fluorescent protein (GFP) that targeted the lumen of the ER (Oparka et al., 1999).

The neck region often appears different from the rest of the plasmodesmata (Robards and Lucas, 1990; Badelt et al., 2016). An extracellular ring of large particles appears to surround the outer part of the neck construction (Taiz and Jones, 1973; Olesen, 1979; Mollenhauer and Morré, 1987). It is possible that these extracellular particles regulate the size of the cytoplasmic annulus. However, the extracellular particles are not seen in rapidly freeze-fixed tissues, indicating that they may be wound-induced localized formations of callose, which under natural wounding conditions would serve to isolate the wounded cell (Ding et al., 1992b; Radford et al., 1998; Levy et al., 2007; Simpson et al., 2009; Zavaliev et al., 2011), particularly following pathogen attack (Lee and Lu, 2011; Lee et al., 2011a,b; Lee, 2015). This is a fair amount of evidence that shows that the large particles are composed of callose and that callose deposition controls the closing of plasmodesmata and callose degradation controls their opening (Han et al., 2014; Amsbury et al., 2018).

Freeze fixation followed by freeze substitution has allowed a more detailed knowledge of plasmodesmatal structure compared with chemical fixation because with freeze fixation, the cells are killed and the structures are fixed within milliseconds. With chemical fixation, cells take several seconds to die due to the relatively slow penetration of chemicals compared to the rate in which heat can be dissipated (Mersey and McCully, 1978). Thus, during chemical fixation, there is sufficient time for wound processes to occur and for cellular structures to become modified (Buvat, 1969).

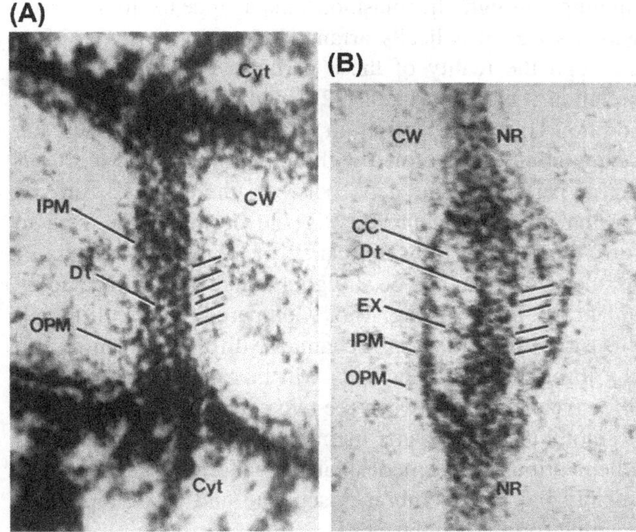

FIGURE 3.6 Longitudinal sections through plasmodesmata of tobacco cells that have been prepared by freeze fixation and freeze substitution: (A) a plasmodesma between phloem parenchyma cells; (B) a plasmodesma between a phloem parenchyma cell and a bundle sheath cell. *CC, central cavity; CW*, extracellular matrix; *cyt*, cytoplasm; *Dt*, desmotubule; *EX*, spoke-like extensions; *IPM*, inner leaflet of the plasma membrane; *NR*, neck region; *OPM*, outer leaflet of the plasma membrane. *From Ding, B., Turgeon, R., Parthasarathy, M.V., 1992b. Substructure of freeze-substituted plasmodesmata. Protoplasma 169, 28–41.*

Freeze fixation is done by plunging a cell or small tissue into liquid propane. Then the vitrified water in the sample is removed with organic solvents. Then chemical fixatives are added to stabilize the cellular structures. The samples are then warmed to room temperature, embedded in plastic, sectioned, stained, and viewed with an electron microscope.

The general structure of plasmodesmata in a freeze-substituted tobacco leaf is similar to that seen in chemically fixed materials. However, new details in the substructure can be seen (Ding et al., 1992b; Ding et al., 1999; see Fig. 3.6). The inner leaflet of the plasma membrane

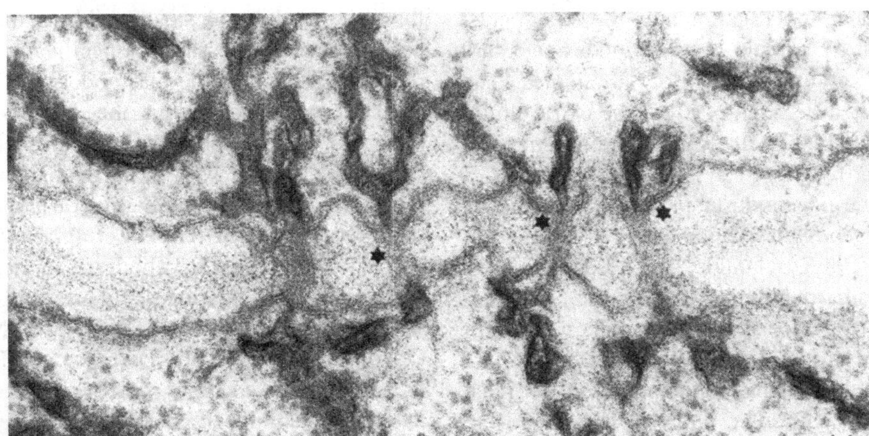

FIGURE 3.5 Longitudinal view of plasmodesmata in lettuce root tip cells. The lumen of the endoplasmic reticulum (ER) is stained with OsFeCN. The cisternal space is constricted where the ER enters the plasmodesmata (asterisks) ×100,000. *From Hepler, P.K., 1982. Endoplasmic reticulum in the formation of the cell plate and plasmodesmata. Protoplasma 111, 121–133.*

running through the plasmodesmata appears to be lined with a series of helically arranged electron-dense particles, although the reality of these particles is subject to interpretation (Overall et al., 1982; Brecknock et al., 2011). In addition, the outer leaflet of the ER that makes up the desmotubule is also lined with helically arranged electron-dense particles. The gaps between the particles on the plasma membrane inner leaflet and desmotubule seem to form the aqueous transport canals of the plasmodesmata. If so, the canals may not be straight but helical as indicated by unlabeled lines in Fig. 3.6. Compared with cell-to-cell diffusion through straight channels, diffusion from cell to cell through helical channels will take longer because the effective distance between the two cells will be longer. The desmotubule and plasma membrane are initially appressed when primary plasmodesmata are formed, and subsequently the two membranes separate and the cytoplasmic annulus forms (Nicolas et al., 2017).

A variety of intercellular connections that range from large simple holes to elaborate structures can be found in the fungi (Reichle and Alexander, 1965; Carroll, 1967; Brenner and Carroll, 1968; Carroll, 1972; Furtado, 1971; Beckett et al., 1974), red algae (Bold and Wynne, 1978), and the brown algae (Schmitz and Kühn, 1982; Terauchi et al., 2015). Each structure represents a compromise between cell individuality and the organismal whole.

3.4 ISOLATION AND COMPOSITION OF PLASMODESMATA

Pure and intact plasmodesmata can be isolated (Kotlizsky et al., 1992; Epel et al., 1996; Bayer et al., 2004; Fernandez-Calvino et al., 2011; Grison et al., 2015a,b). To isolate plasmodesmata, plants are frozen and pulverized to a fine powder. The powder is further homogenized in a buffer and passed through a nylon mesh that retains the plasmodesmata embedded in the extracellular matrix. The extracellular matrix fraction is then passed through a valve under pressure to shear the fraction into tiny fragments. These fragments, which contain the plasmodesmata, are collected by centrifugation at 600 g for 10 minutes. The plasmodesmata can then be released from the extracellular matrix through the action of cell wall degrading enzymes (Grison et al., 2015a,b).

The proteins of the plasmodesmata are then characterized by solubilizing them in sodium dodecyl sulfate and subjecting them to polyacrylamide gel electrophoresis. Interestingly, the plasma membrane H^+-ATPase, an aquaporin, and cellulose synthase are not present in plasmodesmata (Fleurat-Lessard et al., 1995; Grison et al., 2015a,b). While there are many polypeptides in the plasmodesmata, one is of particular interest. It is a 26- to 27-kDa protein that cross-reacts with antibodies made against connexin (Meiners and Schindler, 1987, 1989; Meiners et al., 1991b; Yahalom et al., 1991), which is a component of the intercellular connections (i.e., gap junctions) of animal cells.

Yahalom et al. (1991), using immunolocalization electron microscopy, found that a connexin-like protein is present in the plasmodesmata along the entire length, including the cytoplasmic annulus and the neck region. Immunolocalization electron microscopy involves treating thin sections with an antibody that is specific for an antigen, which in this case is a connexin-like protein. After washing away the loosely bound antibodies, the sections are treated with a secondary antibody attached to 12- to 15-nm particles of gold. This secondary antibody recognizes the primary antibody. The antigen can be localized because the electron-dense gold is precipitated nearby. A calcium-dependent protein kinase (Yahalom et al., 1998), centrin (Blackman et al., 1999), calreticulin (Baluska et al., 1999; Bayer et al., 2004), myosin (Radford and White, 1998; Reichelt et al., 1999), actin (White et al., 1994; Blackman and Overall, 1998), a reversibly glycosylated polypeptide (Sagi et al., 2005), a protein kinase (Lee et al., 2005), a callose binding protein (Simpson et al., 2009), and a β-1,3-glucanase (Levy et al., 2007; Grison et al., 2015a,b) have also been localized in the plasmodesmata. More than 1000, as yet unidentified, proteins have been observed to be associated with plasmodesmata through proteomic analysis (Faulkner et al., 2005; Faulkner and Maule, 2011; Fernandez-Calvino et al., 2011; Salmon and Bayer, 2012), although assignment of function by gene ontology indicates that 49% of the proteins are signaling receptor kinases (Fernandez-Calvino et al., 2011).

Plasmodesmatal proteins are being identified by fusing sequences that encode GFP with random stretches of cDNA, and then after transient expression, looking for those proteins that localize to the plasmodesmata (Escobar et al., 2003). Thomas et al. (2008) have discovered a protein that is capable of influencing the transport of GFP through the plasmodesmata and have discovered the amino acid sequence necessary to specifically target this plasmodesmatal protein to the plasmodesmata.

The phospholipids in the plasmodesmata are more saturated than the phospholipids in the rest of the plasma membrane, which was isolated using aqueous two-phase partitioning after the wall fraction containing plasmodesmata was discarded (Grison et al., 2015a,b). The lipids in the plasmodesmata are modestly enriched in sterols and sphingolipids, indicating that these lipids in plasmodesmata may, in part, be present as lipid rafts (Naulin et al., 2014; Grison et al., 2015a,b; Iswanto and Kim, 2017). Some plasmodesmata-specific proteins are associated with the detergent-insoluble lipid rafts (Farquharson, 2015; Grison et al., 2015a,b).

3.5 PERMEABILITY OF PLASMODESMATA

The fundamental significance of plasmodesmata is that they form a low-resistance pathway between two cells through which large hydrophilic molecules can travel faster than they would if they had to pass through the plasma membrane to leave a cell and through another plasma membrane to enter the next cell. To calculate the permeability coefficient of plasmodesmata, Goodwin et al. (1990) injected fluorescent dyes into cells of *Egeria* and measured the rate in which the dyes diffused into the next cell. They also calculated the permeability coefficient for the plasma membrane by measuring the rate in which the dye diffused into the cell from the extracellular medium. The permeabilities of the plasma membrane and plasmodesmata are shown in Table 3.1.

The plasmodesmata are approximately 10,000 times more permeable than the plasma membrane to the dyes with molecular masses less than 700 Da. For dye molecules greater than 1000 Da, the permeability coefficients of the plasmodesmata become indistinguishable from those of the plasma membrane.

The plasmodesmatal permeability coefficients (P) are obtained by assuming that the dyes move from cell 1 (C_1) to cell 2 (C_2) by diffusion during time t and can thus be modeled by Runnström's (1911) modification of Fick's Law (see Chapter 2):

$$ds_2/Adt = -P(C_2 - C_1) \tag{3.1}$$

where A is the area between cell 1 and cell 2 and ds_2/dt is the flow of solute (in mol/s) from cell 1 to cell 2.

The volumes of the cells (V_1 and V_2) remain constant during the experiment. The amount of dye that diffuses into cell 2 is equal to the change in concentration (dC_2) in cell 2 times the volume (V_2) of cell 2. That is, because $ds_2 = V_2 dC_2$, then

$$\frac{V_2}{A}\frac{dC_2}{dt} = -P(C_2 - C_1) \tag{3.2}$$

and

$$\frac{dC_2}{dt} = -P\frac{A}{V_2}(C_2 - C_1) \tag{3.3}$$

After dividing both sides by ($C_2 - C_1$) and multiplying both sides by dt, we get

$$\frac{dC_2}{(C_2 - C_1)} = -P\frac{A}{V_2}dt \tag{3.4}$$

To calculate P, we must integrate Eq. (3.4). To integrate easily, we must assume that P, A, V_2, and C_1 remain constant. Because we know that C_1 will decrease with time, we must do the experiment over short periods of time.

$$\int_{t=0}^{t=t} \frac{dC_2}{(C_2 - C_1)} = -P\frac{A}{V_2}\int_{t=0}^{t=t} dt \tag{3.5}$$

First, let us integrate the left side. Let $u = C_2 - C_1$, thus, if C_1 is constant, then $du = dC_2$ and, $\int_{t=0}^{t=t} \frac{dC_2}{(C_2-C_1)} = \int_{t=0}^{t=t} \frac{du}{u}$ which according to the Fundamental Theorem of Calculus is equal to $\ln(u_t/u_0)$, which after substitution is equal to $\ln[(C_2 - C_1)_t/(C_2 - C_1)_0]$.

Now let us integrate the right side

$$-P\frac{A}{V_2}\int_{t=0}^{t=t} dt = -P\frac{A}{V_2}t \tag{3.6}$$

Thus,

$$\ln\frac{(C_2 - C_1)_t}{(C_2 - C_1)_0} = -P\frac{A}{V_2}t \tag{3.7}$$

At $t = 0$, $C_2 = 0$, thus $(C_2 - C_1)_0 = (-C_1)_0$, and

$$\ln\frac{(C_2 - C_1)_t}{(-C_1)_0} = -P\frac{A}{V_2}t \tag{3.8}$$

If the experiment is done for short times and C_1 barely changes, and $(C_1)_0 = (C_1)_t$, then

$$\ln\left(\frac{(-C_2)_t}{(C_1)_0} + 1\right) = \ln\left(1 - \frac{(C_2)_t}{(C_1)_0}\right) = -P\frac{A}{V_2}t \tag{3.9}$$

TABLE 3.1 Permeability Coefficients of Plasmodesmata and Plasma Membrane of *Egeria*

	Permeability Coefficient (m/second)	
Molecule (M_r)	**Plasmodesmata**	**Plasma Membrane**
6-CF (376)	112×10^{-8}	2.4×10^{-12}
FITC + glutamic acid (536)	15.5×10^{-8}	1.8×10^{-12}
FITC + glutamyl-glutamic acid (665)	11.4×10^{-8}	1.3×10^{-12}
FITC + hexaglycine (744)	0.66×10^{-8}	2.3×10^{-12}
FITC + leucyl diglutamyl-leucine (874)	0.009×10^{-8}	1.6×10^{-12}

From Goodwin, P.B., Shepherd, V., Erwee, M.G., 1990. Compartmentation of fluorescent tracers injected into the epidermal cells of *Egeria densa* leaves. Planta 181, 129–136.

Because C_2, C_1, A, V_2, and t are all measurable quantities, we can calculate P from the slope of an experimentally derived curve that relates $-\ln[1 - (C_2)_t/(C_1)_0]$ to t. P is equal to the slope (in s^{-1}) times (V_2/A). The permeability of the plasmodesmata is influenced to some extent on the tissue preparation technique (Radford and White, 2001).

Dye movement experiments have been performed on filaments of soybean culture cells using fluorescence redistribution after photobleaching (FRAP; Baron-Epel et al., 1988b). With this technique, the hydrophobic form of carboxyfluorescein (i.e., carboxyfluorescein diacetate) is added to the external medium. The dye is passively taken up across the plasma membrane in the ester form. Esterases then cleave the hydrophilic portion of the dye from the hydrophobic acetates. The cell then glows from the dye. Then a laser beam bleaches the dye in one cell, and the movement of dye into this cell from neighboring cells is monitored over time. A rate constant (1/time) is obtained from these data. The rate constant can be transformed into a permeability coefficient if we postulate that the rate (K) that the dye moves into the cell is proportional to the area (A) on two sides of the cell because the plasmodesmata are only on two sides of soybean culture cells. We must also assume that the rate in which the cell gets brighter is inversely proportional to the volume (V) of the cell. Lastly, we must define the permeability coefficient (P) as the conversion factor that relates the rate to the area and volume. Thus,

$$K = P\frac{A}{V} \tag{3.10}$$

Baron-Epel et al. (1988a,b) obtained a rate of 0.0015 second^{-1}. Because for soybean culture cells, $A/V = 1.7 \times 10^5\ \text{m}^{-1}$, then $P = .9 \times 10^{-8}$ m/s, which is the ballpark of the values found by Goodwin et al. (1990) for *Egeria*.

The diameter of the aqueous canals of the plasmodesmata can be estimated from dye injection experiments by using dyes of various sizes. The diameters depend on the cell type tested. These experiments show that plasmodesmata can pass molecules that have a molecular mass of less than 376–800 Da in *Elodea* (Goodwin, 1983; Erwee and Goodwin, 1985), 700–800 Da in *Setcreasea* stamen hairs (Tucker, 1982), 850–900 Da in bundle sheath cells of C4 plants (Weiner et al., 1988) and molecules as large as 1090 Da in the nectary trichome cells of *Abutilon* (Terry and Robards, 1987; Fisher, 1999), and 20,000 Da in the internodal cells of *Nitella* (Kikuyama et al., 1992).

Dye permeation experiments are being done more cleanly in plants that have been transformed to express a molecule that fluoresces only after it is photoactivated in the cell of interest. Using this method, the difference in apically directed and basally directed plasmodesmatal transport from a given cell along a moss protonema has been determined (Kitagawa and Fujita, 2013, 2015).

TABLE 3.2 Relationship Between Molecular Mass and Hydrodynamic Radius

Molecular Mass (M_r, Da) x	Hydrodynamic Radius (r_H) nm
100	0.26
200	0.35
400	0.51
700	0.76
800	0.85
900	0.93
1000	1.01
1090	1.09

Dye permeation experiments can help us determine the size of the plasmodesmatal canals because there is a direct relationship between molecular mass and the hydrodynamic radius for small organic molecules (Table 3.2). The hydrodynamic radius of a molecule can be determined from diffusion or viscosity measurements with molecules of known molecular mass (Schultz and Solomon, 1961).

The hydrodynamic radius (r_H) of a spherical molecule can be calculated from the Stokes–Einstein equation presented in Chapter 2, as long as one knows the diffusion coefficient of the molecule and the viscosity of the solution:

$$r_H = \frac{kT}{6\pi D\eta} \tag{3.11}$$

Using the measurements of the hydrodynamic radius determined by using the Stokes–Einstein equation and the molecular masses of the solutes, I have come up with the following empirical formula to express the relationship between the hydrodynamic radius (in nm) and the molecular mass (M_r, in Da):

$$r_H = 0.00083327\ (M_r) + 0.18 \tag{3.12}$$

Thus, the dye permeation studies indicate that the cytoplasmic annuli have size exclusion limits that typically vary between 0.7 and 4 nm, depending on the cell type. These estimates are compatible with what would be expected from structural studies. Movement through the plasmodesmata is not restricted to hydrophilic molecules. Hydrophobic molecules may also pass from cell to cell by translation through the lipid bilayers in the membranes that make up the plasmodesmata (Baron-Epel et al., 1988b; Grabski et al., 1993; Fisher, 1999).

To test the influence of particular amino acid sequences on plasmodesmatal transport, the biolistic bombardment technique is used to quantify transport (Oparka and Boevink, 2005). With this transient expression technique, genes that are engineered to express proteins that are fluorescent, have various enzymatic or regulatory activities, and plasmodesmatal targeting sequences are shot into a cell using the gene gun. Once the protein encoded by the engineered gene is expressed, the movement of the fluorescent protein with the engineered sequences to neighboring cells is observed and quantified.

Classical electrophysiological techniques similar to those used to characterize the plasma membrane show that the plasmodesmata provide a high-conductance pathway for the movement of ions between cells (Spanswick and Costerton, 1967; Overall and Gunning, 1982; van Bel and Ehlers, 2005; Beilby, 2016). The plasmodesmata have a specific conductance approximately 50 times greater than that of the plasma membrane (Spanswick, 1974b).

The permeability of plasmodesmata can be regulated. For example, Ding and Tazawa (1989) and Oparka and Prior (1992) have shown that pressure can regulate plasmodesmatal conductivity, and Baron-Epel et al. (1988b) and Tucker (1990) have shown that increasing the intracellular Ca^{2+} concentration inhibits intercellular movement of dyes. Holdaway-Clarke et al. (2000) have shown that elevated cytosolic concentrations of Ca^{2+} increase the resistance of the plasmodesmata, providing further evidence that the plasmodesmata close in response to Ca^{2+}. This is particularly interesting because the $[Ca^{2+}]$ outside the cell is typically high (~ 1 mol/m^3), whereas it is low in the cell ($\sim 10^{-4}$ mol/m^3), thus a high intracellular $[Ca^{2+}]$ is a sign of a damaged cell (e.g., the plasma membrane is lysed). Thus, the decreased conductance of the plasmodesmata due to high Ca^{2+} may isolate a damaged cell from its healthy neighbors. External stimuli, including red light, can also influence plasmodesmatal conductance (Racusen, 1976). Plasmodesmatal permeability is also regulated by actin microfilaments (Ding et al., 1996; Su et al., 2010) and can change during cell development (Gisel et al., 1999, 2001; Oparka and Turgeon, 1999; Ruan et al., 2001; Kim et al., 2002a,b; 2005a,b). The search to identify the genes involved in plasmodesmata structure or the regulation of plasmodesmatal permeability is currently being studied in mutants (Kim et al., 2002a,b; Burch-Smith and Zambryski, 2010; Burch-Smith et al., 2011; Xu et al., 2012a,b). The severity of the mutations makes it hard to determine the direct effect of the gene itself.

Plasmodesmatal permeability is not only regulated by physiological and developmental signals but also increased by some of the proteins that are trafficked through them. Viruses can travel through plasmodesmata (Bennett, 1956; Esau, 1956, 1961), and plant virologists wanted to know how globular viruses 18–80 nm in diameter, or helical or filamentous rods 10–25 nm in diameter and up to 2.5 μm in length, pass through plasmodesmata (Lazarowitz, 1999; Lazarowitz and Beachy, 1999; Niehl and Heinlein, 2011). Some viruses, such as the dahlia mosaic virus and cauliflower mosaic virus, somehow drastically modify the structure of the plasmodesmata, getting rid of the desmotubule and expanding the diameter of the cytoplasmic annulus to 60–80 nm. These two viruses are commonly found within the plasmodesmata in transmission electron micrographs, indicating that the viruses move through the plasmodesmata to attack the host everywhere.

By contrast, the tobacco mosaic virus (TMV; Zaitlin, 1998) is never observed in plasmodesmata. It is possible that only the small RNA genome passes through the plasmodesmata so that the plasmodesmata structure is only minimally affected. Through genetic studies of a temperature-sensitive mutant of this virus, Nishiguchi et al. (1980) found the gene that coded for the ability of the virus to move through the plant. They found the gene by obtaining a mutant virus that was able to replicate at 32°C but was unable to move through the plant at this temperature. However, the virus was able to also move through the plant, if the temperature was lowered to the permissive level of 22°C.

Leonard and Zaitlin (1982) discovered the protein involved in virus movement when they found that tryptic digests of the in vitro translation products of the mutant and wild type differed only for a 30-kDa protein. They concluded that this protein is involved in virus movement. The genes of the wild type and mutant have been sequenced, and they differ only in one amino acid at position 154. The wild type has serine, whereas the mutant protein has proline (Ohno et al., 1983). Tomenius et al. (1987) have used immunogold cytochemistry to localize the 30-kDa protein in infected tobacco leaves and find it in the plasmodesmata. The plasmodesmatal proteins that interact with the movement protein are being identified (Kishi-Kaboshi et al., 2005).

A breakthrough in plasmodesmata research occurred when Deom et al. (1987, 1990, 1991) combined techniques of plant biotechnology and virology to construct a chimeric gene that encoded the 30-kDa movement protein and introduced it into tobacco plants. This allowed the study of the function of the 30-kDa gene product in the absence of all the other TMV gene products. They found that the 30-kDa protein was associated with the extracellular matrix fraction (see Chapter 20). Furthermore, in a type of complementation study it was found that mutant viruses could move through the transgenic plant at nonpermissive temperatures.

Wolf et al. (1989, 1991) showed with dye movement experiments that while control tobacco plants have a size exclusion limit of c750 Da for cell-to-cell transport, transgenic tobacco plants that are expressing the movement protein have a size exclusion limit between 9400 and 17,200 Da. Thus, the movement protein is capable of regulating the size of the plasmodesmatal canals. Wolf et al. (1989) postulated that the exclusion limit of control plants was approximately 0.73 nm, whereas the transgenic plants had a size exclusion limit of 2.4–3.1 nm. This is still too small to pass the 8 × 300-nm virus or its approximately 10-nm RNA. Thus, it is likely that the movement protein acts as a chaperone to facilitate the movement of the viral RNA through the plasmodesmata (Lucas et al., 1993; Wolf and Lucas, 1994; Ghoshroy et al., 1997).

Movement proteins can also facilitate the movement of viral DNA through the plasmodesmata. Plant cells injected with movement protein (from bean dwarf mosaic geminivirus) and fluorescently labeled viral DNA show that the movement protein causes the movement of viral DNA from cell to cell (Noueiry et al., 1994). By contrast, red clover necrotic virus movement protein enhances the movement of RNA, but not DNA (Fujiwara et al., 1993).

It is likely that specific amino acid sequences are necessary for proteins to bind to and pass through the plasmodesmata. This hypothesis is supported by the observation that a fusion protein made by combining the targeting sequence from the viral movement protein with the sequence for the GFP enhances the cell-to-cell movement of the GFP (Crawford and Zambryski, 2000; Zambryski and Crawford, 2000; Liarzi and Epel, 2005). The low activation energy for the transport of GFP and other proteins, not normally targeted to the plasmodesmata through the plasmodesmata, indicates that any conformational changes of the plasmodesmata necessary to allow the movement of this large molecule must be minimal (Schönknecht et al., 2008). The movement of other proteins through plasmodesmata may require the transported protein to unfold to enter the plasmodesmata and refold when they exit. Moreover, the movement of proteins through plasmodesmata may require the plasmodesmatal proteins to change their conformation to increase the size-exclusion limit. The folding and unfolding may be facilitated by molecular chaperone proteins, including heat shock proteins, protein disulfide isomerases, and peptidyl-proyl *cis-trans* isomerases.

The search for native plant polypeptides that interact with the plasmodesmata and facilitate the movement of themselves or other proteins through the plasmodesmata is ongoing. Some proteins specifically target themselves or other proteins to the plasmodesmata and others unfold the proteins so that they can fit through the plasmodesmata and refold them on passage or interact directly with the plasmodesmata to increase the size-exclusion limit (Kragler et al., 2000; Zambryski and Crawford, 2000; Haywood et al., 2002; Kragler, 2005). Recently, Gottschalk et al. (2008) have shown that the chaperone peptidyl-proyl *cis-trans* isomerase, which is also known as cyclophilin, is able to increase the size-exclusion limit of the plasmodesmata between mesophyll cells so that a 10-kDa fluorescent dextran can pass from the injected cell to other cells.

In many plants, the concentration of sucrose is greater in the mesophyll cells where it is produced by photosynthesis than in the cells of the phloem. Consequently, the sucrose formed in the mesophyll cells is transported by diffusion through the plasmodesmata connecting the cells between the mesophyll and the phloem (Turgeon and Medville, 2004). To move by diffusion, in these plants, the sucrose concentration must be higher in the mesophyll cells than in the sieve tube elements. In many other plants, however, the concentration of sugar is greater in the sieve tube elements than in the mesophyll cells. In these cases, a mechanism must exist to actively load the sugar into the phloem (Roberts and Oparka, 2003; Turgeon, 2010). There are two major hypotheses to describe how sugar is transported into the phloem against its concentration gradient. Data obtained by Robert Turgeon show that it is not an either/or situation. Some plants use one mechanism for phloem loading, others use the second mechanism for phloem loading exclusively, and still others use additional mechanisms.

According to the canonical apoplastic hypothesis of phloem loading, sugars pass through plasmodesmata from the mesophyll cells until they reach the phloem. At this point, the plasmodesmata are occluded and thus the sugars are unloaded into the apoplast (Beebe and Evert, 1992). According to the apoplastic hypothesis, the sugar is then loaded into the phloem against its concentration gradient in an ATP-dependent manner (Geiger et al., 1973, 1974; Sovonick et al., 1974; Giaquinta, 1976; Maynard and Lucas, 1982). Specifically, the sugars are then taken up through the plasma membranes of the sieve tube element–companion cell complex by sucrose/H^+ symporters that use the free energy inherent in the electrochemical difference of protons across the membrane formed by the H^+-ATPase.

According to the canonical symplastic–loading hypothesis, the sugar stays within the symplast. The major problem with the symplastic-loading hypothesis is being able to explain how sugars can move by diffusion through plasmodesmata against their concentration gradient (Turgeon and Hepler, 1989; van Bel, 1989). Robert Turgeon and Ester Gowan (1990) and Turgeon (1991) propose that special cells in the phloem known as intermediary cells act as a "molecular size–discrimination trap." In this model, sucrose and galactinol synthesized by the photosynthesizing mesophyll cells diffuse down their concentration gradients through the plasmodesmata between the bundle sheath cells and the intermediary cells. At this point, an enzyme combines the two small molecules into the larger raffinose, which is too big to diffuse back through the plasmodesmata that are thinner on the intermediary cell side (Fig. 3.7). In this way, small molecules diffuse down their concentration gradient to the phloem, where they are converted to raffinose. The raffinose, unable to move back into the bundle sheath cell, then diffuses into the sieve tubes (Turgeon and Beebe, 1991; Turgeon, 2000). Advection, or the transfer of matter horizontally as a result of

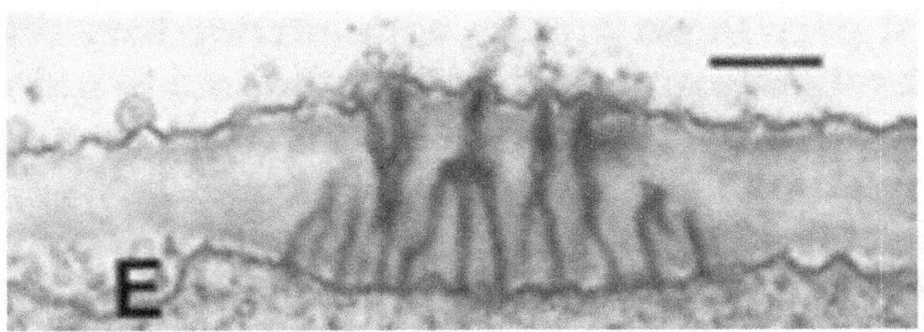

FIGURE 3.7 Plasmodesmata between an intermediary cell and a bundle sheath cell of *Alonsoa warscewiczii*. The portion of the plasmodesmata on the intermediary cell side is extensively branched, and the branches are narrower than those on the bundle sheath cell side. Bar, 250 nm. *From Turgeon, R., Beebe, D.U., Gowan, E., 1993. The intermediary cell: minor-vein anatomy and raffinose oligosaccharide synthesis in the Scrophulariaceae. Planta 191, 446–456.*

water flow, has been added to the polymer trap model to account for the observed selectivity of the plasmodesmata for the passage of sucrose over raffinose (Comtet et al., 2017a,b). This polymer trap model has also been used to explain oligofructan transport (Wang and Nobel, 1998).

The proteins needed by the sieve tube elements, which do not contain a nucleus, are synthesized in the companion cells. The proteins then pass from the companion cells through the plasmodesmata to the sieve tube elements (Fisher et al., 1992). The large cytoplasmic pathway through these plasmodesmata can be visualized by following the movement of GFP, which is a cylinder 2.1 nm in diameter and 4.2 nm long (Imlau et al., 1999). There is a protein in the companion cells that increases the permeability of the plasmodesmata so that proteins can move from the companion cells into the sieve tube elements. This protein is a plant homolog of the viral movement protein (Xoconostle-Cázares et al., 1999). It is a member of the cytochrome b_5 reductase family and must be processed by a protease in the companion cell before it can pass through the plasmodesmata to the sieve tube elements (Xoconostle-Cázares et al., 2000).

I opened this chapter by discussing whether the organism has a level of coordination that is greater than that of cells. While physical and readily diffusible hormonal factors certainly are important in communication within an organism (D'Arcy Thompson, 1959; Turing, 1992a,b,c), plasmodesmata must also be important in integrating the parts with the whole (Goebel, 1926; Sharp, 1934). Patterns of morphogenesis are related to the ability of plasmodesmata to transport certain macromolecules, including RNA and transcription factors that are able to influence cell differentiation (Lucas et al., 1995; van der Shoot et al., 1995; Bergmans et al., 1997; Ding, 1998; Lucas, 1999; Zambryski and Crawford, 2000; Kim et al., 2001; Nakajima et al., 2001; Itaya et al., 2002; Haywood et al., 2002; Kim et al., 2002a,b; Wu et al., 2002; Cilia and Jackson, 2004, 2005; Kim et al., 2003; Heinlein and Epel, 2004; Qi et al., 2004a; Ryabov et al., 2004; Yoo et al., 2004; Zambryski, 2004; Heinlein, 2005; Kobayashi et al., 2005; Ding and Itaya, 2007; Zhong et al., 2007; Zhong and Ding, 2008; Rim et al., 2011; Wu and Gallagher, 2011, 2012; Benitez-Alfonso et al., 2013; Brunkard et al., 2013; Yadav et al., 2014). Because plasmodesmata allow the passage of important informational macromolecules that affect gene expression such as RNA and transcription factors from cell to cell, plasmodesmata ensure that cells are not autonomous units and that there are higher levels of coordination within a plant.

3.6 SUMMARY

Plasmodesmata are structures that provide a pathway for the transport of information in the form of molecules from cell to cell. Along with other positional influences that determine development, the distribution and unitary conductance of plasmodesmata will determine the degree in which a given cell will act as an individual or as a member of the whole organism.

3.7 QUESTIONS

3.1. What is the evidence that the plasmodesmata provide a mechanism by which cells communicate with each other?

3.2. What are the mechanisms by which the plasmodesmata can facilitate cell-to-cell communication?

3.3. What are the limitations of thinking about the plasmodesmata as the sole mechanism of cell-to-cell communication?

The references for this chapter can be found in the references at the end of the book.

Chapter 4

Endoplasmic Reticulum

The simplest plants, such as the green algae growing in stagnant water or on the bark of trees, are mere round cells. The higher plants increase their surface by putting out leaves and roots. Comparative anatomy is largely the story of the struggle to increase surface in proportion to volume.

J.B.S. Haldane (On Being the Right Size, 1929).

4.1 SIGNIFICANCE AND EVOLUTION OF THE ENDOPLASMIC RETICULUM

In Chapter 2, I discussed cells as if their only membrane was the plasma membrane. Perhaps this is just what the precursors of the first eukaryotic cells were like. The plasma membrane of the precursor cell, like those of present-day prokaryotic cells, probably performed all of the membrane-dependent functions. It is likely that the precursor prokaryotic cell perhaps had a volume of 10^{-18} m^3 and a surface-to-volume ratio of 10^6 m^{-1}, whereas a modern eukaryotic plant cell has a volume of 10^{-15} m^3 or more and a surface-to-volume ratio of 10^5 m^{-1} or less. That is, the volume of a eukaryotic plant cell is approximately 1000 times greater than the volume of the putative precursor. Because the surface-to-volume ratio decreases as the radius increases ($A/V = 3/r$ for spherical cells), it may have been impossible for a large eukaryotic cell to perform all the required membrane-dependent processes on the plasma membrane alone.

As larger cells evolved, the plasma membrane may have invaginated and pinched off, forming membrane-bound vesicles, a process that would maintain a high surface-to-volume ratio. The inside of such a vesicle is called the lumen and is topologically an E-space. Indeed, *Epulopiscium fishelsoni*, the largest prokaryote, has a highly invaginated plasma membrane (Angert et al., 1993, 1996; Bresler et al., 1998; Robinow and Angert, 1998; Levin and Angert, 2015). In eukaryotes today, the internal membranes, known collectively as the endomembrane system, are differentiated into the endoplasmic reticulum (ER), the Golgi apparatus, and the vacuolar compartment along with all the adjoining membranes. Each compartment has its own function (Lunn, 2006).

The ER is a highly convoluted, netlike meshwork that extends throughout the cytoplasm (Staehelin, 1997). It is composed of a single membrane and constitutes more than half of the total membrane of the cell. It contributes to a surface-to-volume ratio of approximately 10^6 m^{-1} in root cells and 10^7 m^{-1} in tapetal cells (Gunning and Steer, 1996). The ER can be stationary or motile (Ueda et al., 2010). I will discuss the ER in terms of how it complements the plasma membrane in performing transport activities and how it is involved in the synthesis of many membranes, including the plasma membrane (Brandizzi et al., 2002b,c; Saint-Jore et al., 2002; Stefano et al., 2014a,b). As Günter Blobel (1999) said in his must-read Nobel lecture, "omnis membrane e membrane (each membrane comes from a pre-existing membrane). Membranes and compartments are not created de novo, but recreate themselves."

4.2 DISCOVERY OF THE ENDOPLASMIC RETICULUM

The introduction of electron microscopy into the study of cells opened up a whole new world that was approximately 100 times smaller than that which had been previously visualized. According to Jan-Erik Edström (1974), "Although the light microscope certainly opened a door to a new world during the 19th century, it had obvious limitations. The components of the cell are so small that it was not possible to study their inner structure, their mutual relations or their different roles. To take a metaphor from an earlier Prize Winner, the cell was like a mother's work basket, in that it contained objects strewn about in no discernible order and evidently, for him, with no recognizable functions." In 1945, Keith Porter, Albert Claude, and Ernest Fullam first observed a lacelike reticulum in cultured chick embryo cells (Figs. 4.1–4.3; Palade, 1977; Peachey and Brinkley, 1983; Satir, 1997a,b; Moberg, 2012; Peachey, 2013). They used cultured cells because they were thin enough to be penetrated by an electron beam. This was important as the ultramicrotome had not yet been invented. Imagine their excitement when they saw this beautiful lacelike structure revealed by the electron microscope. Approximately 100 years earlier, Félix Dujardin (1835, quoted in Buvat, 1969) had described protoplasm viewed

Plant Cell Biology. https://doi.org/10.1016/B978-0-12-814371-1.00004-7

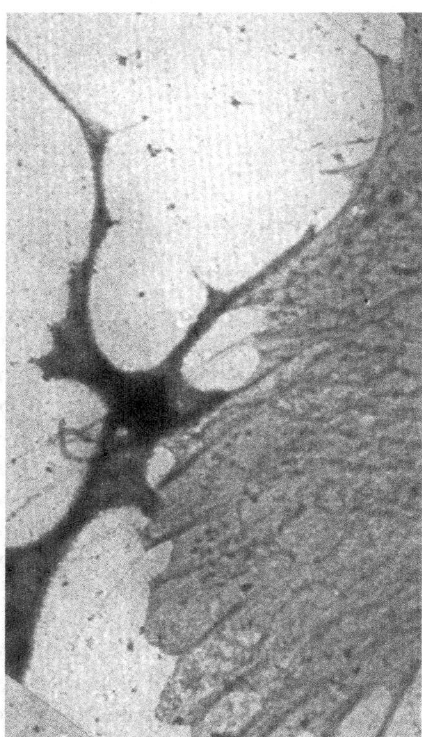

FIGURE 4.1 Cytoplasmic reticulum in a fibroblast-like cell cultured from chick embryo tissue. *From Porter, K.R., Claude, A., Fullam, E., 1945. A study of tissue culture cells by electron microscopy. J. Exp. Med. 81, 233–241.*

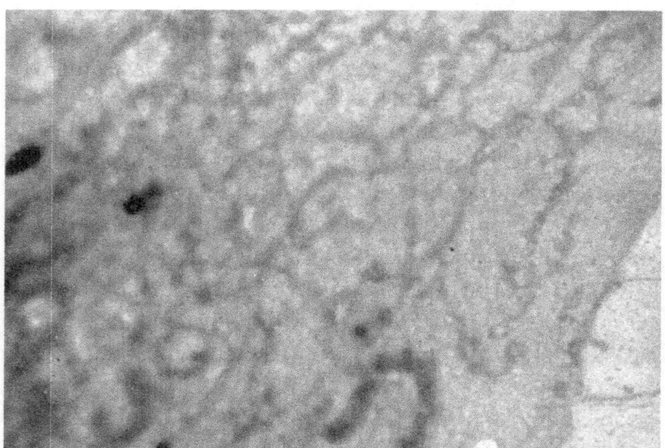

FIGURE 4.2 Lacelike reticulum in a cultured chick fibroblast cell that in places appears to be made up of chains of vesicles. *From Porter, K.R., Claude, A., Fullam, E., 1945. A study of tissue culture cells by electron microscopy. J. Exp. Med. 81, 233–241.*

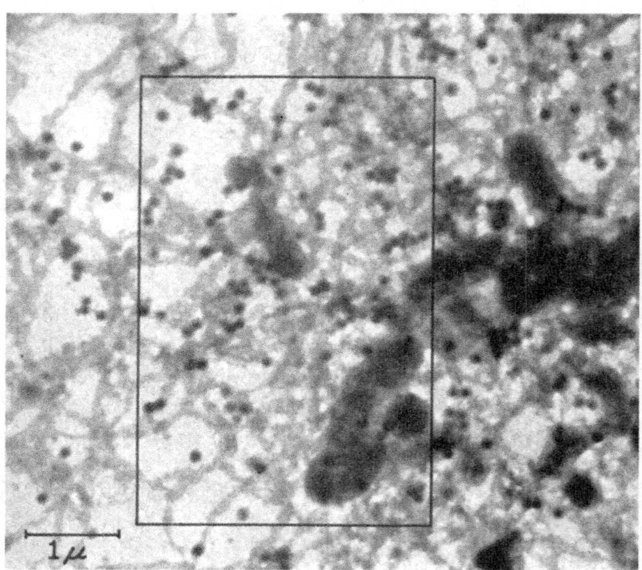

FIGURE 4.3 The endoplasmic reticulum of an epithelial tumor cell. The preparation was dried directly on a wire mesh and observed with the electron microscope. *From Porter, K.R., Thompson, H.P., 1948. A particulate body associated with epithelial cells cultured from mammary carcinomas of mice of a milk-factor strain. J. Exp. Med. 88, 15–24.*

with a light microscope as a substance that has "absolutely no trace of any organization … neither fibres, nor membranes, nor any sign of cellular structure."

Immediately following the discovery of the lacelike reticulum, Albert Claude (1943a,b; 1946a,b; 1948) isolated it from zymogen particle–containing liver and pancreatic cells using the technique of differential centrifugation developed by Bensley and Hoerr (1934). Previously Claude (1938, 1939, 1940, 1941) had used centrifugation to investigate the RNA-rich subcellular particles in normal and tumor cells, which he originally thought were mitochondria because they lacked DNA and thus could not be nuclei. However, differences in chemical composition, behavior in the centrifuge, and stainability showed that these particles differed from mitochondria and secretory granules and were perhaps part of a heretofore unknown cellular structure. Brachet and Jeener (1944) showed that these RNA-rich particles could be isolated from all cell types tested, and moreover the RNA-rich particles contained the proteins that the cells were known to secrete. While the RNA-rich particles could be detected with a dark field microscope, the resolving power was limited. With the introduction of the electron microscope, morphology and biochemistry was combined in a new and complementary way (Mazia, 1975), and Claude could clearly see that the RNA-rich subcellular particles were not mitochondria, but a new organelle. Claude called the nonmitochondrial membranes he isolated *microsomes*, a term originally coined by Johannes Hanstein to mean the unidentified vesicles he saw in plant cells. Claude used *microsome* as a noncommittal term emphasizing only the size. Claude chemically analyzed the microsomes and found that they contained approximately 9% N, 2.5% P, 40%–45% lipid, 0.75% S, 0.01% Cu, and 0.03% Fe. A few years later, the lacelike reticulum, which was visible in the electron microscope, and which did not breach into the cortical region

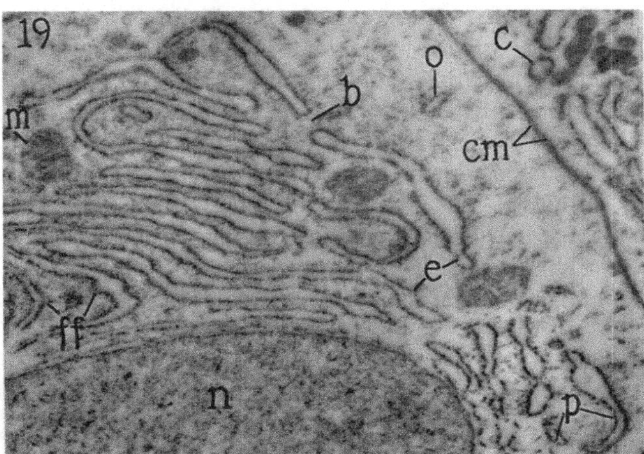

FIGURE 4.4 Electron micrograph of a thin section through the endoplasmic reticulum in a paratid gland cell. *From Palade and Porter, 1954).*

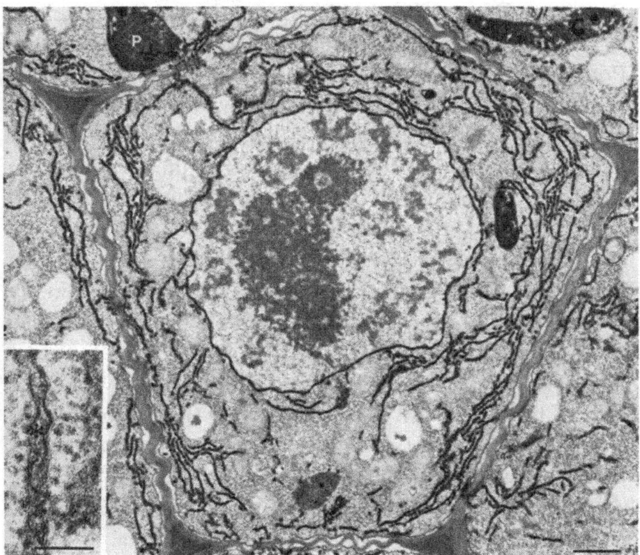

FIGURE 4.5 Electron micrograph of the endoplasmic reticulum of a lettuce root cell that has been fixed in OsFeCN. Bar, 1 μm. (*Inset*) High-magnification electron micrograph of a segment of endoplasmic reticulum showing that the inner leaflet is stained more darkly than the outer leaflet. Bar, 100 nm. *From Hepler, P.K., 1981. The structure of the endoplasmic reticulum revealed by osmium tetroxide-potassium ferricyanide staining. Eur. J. Cell Biol. 26, 102–110.*

of the cytoplasm, was renamed the *endoplasmic reticulum* by Porter and Thompson (1948).

With the advent of the ultramicrotome- and methacrylate-embedding procedures (Pease and Porter, 1981), the ER was first seen with high resolution by George Palade and Keith Porter in 1954 (Fig. 4.4). With thin sections, it was possible to see that the ER was composed of membranes that were 5.5–6.5 nm thick. Perhaps it was lucky that the lacelike reticulum had been discovered before the invention of the ultramicrotome because it is possible that the three-dimensional arrangement of the ER may not have been deduced from 20- to 40-nm thick sections (Palade, 1956). By 1956, Palade and Siekevitz began an integrated study combining electron microscopy and biochemistry, a combination that led to the award of the Nobel Prize to Palade (1975). Buvat and Carasso (1957) contributed to the notion that the ER was a fundamental part of the protoplasm of eukaryotic cells by showing that it is present in the cells of the plant kingdom and those of the animal kingdom.

4.3 STRUCTURE OF THE ENDOPLASMIC RETICULUM

The architecture of the ER is dynamic; it varies from cell to cell and changes throughout the cell cycle (Haguenau, 1958; Hepler, 1989a,b; Stefano and Brandizzi, 2018). The form of the ER can be seen best by treating the cells with stains that fill the luminal space of the ER and consequently contrast the ER against the rest of the cell (see Fig. 4.5; Hepler, 1981; Stephenson and Hawes, 1986). The ER exists both as tubules and as lamellae. The ER is continuous with the nuclear envelope, and some of the lamellae of the ER have pores or fenestrations that are reminiscent of nuclear pores (see Chapter 16 and Fig. 4.6). In addition, the outer membrane of the nuclear envelope contains ribosomes

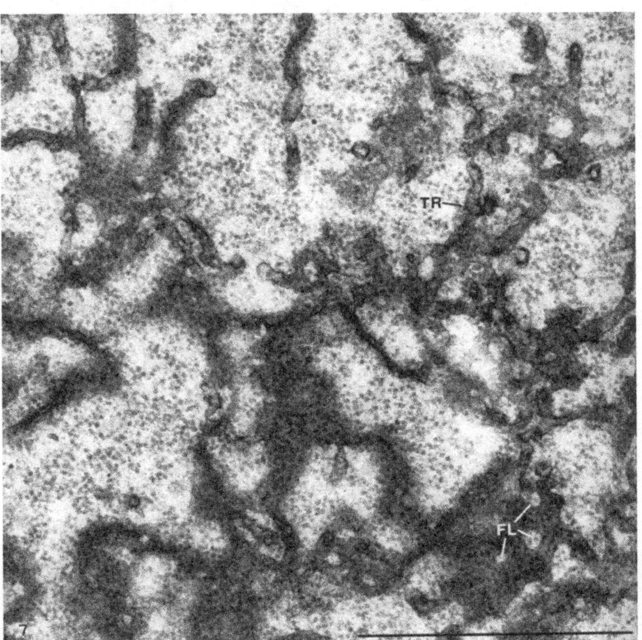

FIGURE 4.6 Electron micrograph of the endoplasmic reticulum of a lettuce root cell that has been fixed in OsFeCN. Notice the fenestrated lamellae (FL). The tubular elements (TRs) intergrade with the cisternal elements. Bar, 1 μm. *From Hepler, P.K., 1981. The structure of the endoplasmic reticulum revealed by osmium tetroxide-potassium ferricyanide staining. Eur. J. Cell Biol. 26, 102–110.*

(Porter and Bonneville, 1968). Indeed the nuclear envelope has been described as perinuclear cisternae of the ER (Porter and Moses, 1958). Focal arrays of ER can also be seen, and these may be the sites of active membrane growth (Hepler, 1981). Zheng and Staehelin (2001) call similar focal arrays *nodal ER* and suggest that the nodal ER in columella cells are involved in gravity sensing. The ER also forms contact sites where it is appressed to the plasma membrane (Hepler et al., 1990; Lancelle and Hepler, 1992; Saheki and De Camilli, 2017), thus possibly forming a continuum from the plasma membrane to the nucleus (Porter and Moses, 1958).

It is also possible to visualize the exquisitely delicate form of the ER in the light microscope (Url, 1964; Lichtscheidl and Url, 1990). The three-dimensional arrangement of ER is particularly clear after staining the cells with the lipophilic, anionic, fluorescent dye, $DiOC_6(3)$ (see Fig. 4.7; Terasaki et al., 1984, 1986; Quader and Schnepf, 1986; Quader et al., 1987; Terasaki, 1989), ER-directed green fluorescent protein (GFP) (Boevink et al., 1996; Hawes et al., 2001; Brandizzi et al., 2002a; Goodin et al., 2007; Noguchi et al., 2014), or other fluorescent proteins (Held et al., 2008).

There are various architectural classes of ER, which are interconnected. One class consists of thin, flat, variably sized cisternae that are connected by thin tubular elements that are approximately 100–400 nm in diameter when viewed with a light microscope and about 60 nm in diameter when viewed with superresolution microscopy (Griffing, 2010). This form of *ER*, which has a lacelike appearance, is found in the thin cytoplasm adjacent to and parallel with the plasma membrane (Lancelle and Hepler, 1992). Ironically, it is found in the *ecto*plasm or cortical cytoplasm of plant cells! Another type of ER consists of bundles of long thin tubular elements that run away from or toward the nucleus through transvacuolar strands. A third class, not only rediscovered in GFP-transformed cells but also found in wild-type cells, consists of fusiform bodies several micrometers long and a few micrometers wide (Bonnett and Newcomb, 1965; Hawes et al., 2001; Matsushima et al., 2003; Gunning, 1998; Noguchi et al., 2014). The distribution of the ER is cell type specific, and distinct forms of ER are found in various cells, including sieve tube elements (Sjolund and Shih, 1983; Schulz, 1992), the tip of *Chara* rhizoids (Bartnik and Sievers, 1988), and the statocytes of *Lepidium* (Hensel, 1987).

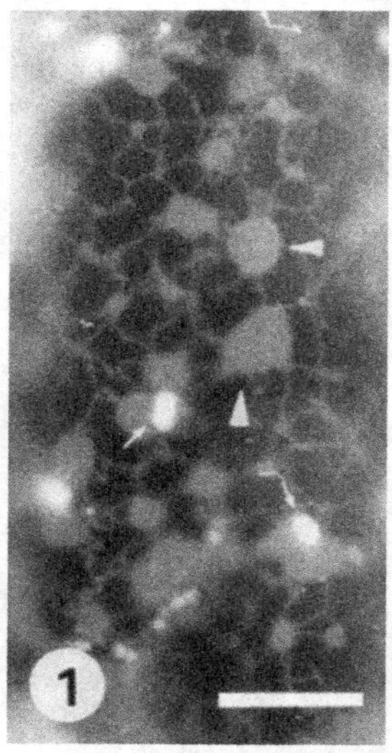

FIGURE 4.7 Fluorescence light micrograph of the endoplasmic reticulum in an onion bulb scale cell stained with $DiOC_6(3)$. The arrowheads indicate cisternal endoplasmic reticulum and the arrow indicates a mitochondrion. Bar, 20 μm. *From Quader, H., Schnepf., E., 1986. Endoplasmic reticulum and cytoplasmic streaming: fluorescence microscopical observations in adaxial epidermis cells of onion bulb scales. Protoplasma 131, 250–252.*

One advantage of light microscopy is that the ER can be observed in living cells and one can see that it is not a static organelle but a dynamic one that exhibits constant movement and undergoes dramatic transformations (Goodin et al., 2007; Goyal and Blackstone, 2013; Lin et al., 2014a,b,c). For example, some tubules grow and shrink at a rate of about 10 μm/second, while other sites do not move (Knebel et al., 1990). The shape of the ER is controlled by temperature, Ca^{2+}, and pH (Quader, 1990; Quader and Fast, 1990), and a class of proteins known as reticulons (Nziengui et al., 2007; Tolley et al., 2008, 2010; Chen et al., 2012; English and Voeltz, 2013; Lee et al., 2013; Stefano et al., 2014b; Hawes et al., 2015). The shape and position of the ER also depend on cytoplasmic structures, known as microfilaments and microtubules, which are discussed in Chapters 10 and 11 (Quader et al., 1987; Lancelle and Hepler, 1988; Allen and Brown, 1988; Lee et al., 1989; Quader, 1990; Lichtscheidl et al., 1990; Knebel et al., 1990; Lancelle and Hepler, 1992; Liebe and Quader, 1994; Yokota et al., 2009; Griffing, 2010; Sparkes et al., 2010; Gao, 2014; Griffing et al., 2014).

4.4 STRUCTURAL SPECIALIZATIONS THAT RELATE TO FUNCTION

The first step in membrane biosynthesis begins on the ER, where the component proteins and lipids are synthesized. Proteins that are destined to become integral membrane proteins are synthesized on polyribosomes that are attached to the ER. Ribosomes, originally called *Palade's small particles*, are 15- to 20-nm complexes that are composed of ribonucleic acid and protein. They provide the workbench for protein synthesis, which will be discussed in Chapter 17. Because the ribosomes cover the P-surface of the ER, they give the ER a "rough" appearance, and these regions of the ER are called the *rough endoplasmic reticulum* or RER (see Fig. 4.8; Palade, 1955). Cells that are active in

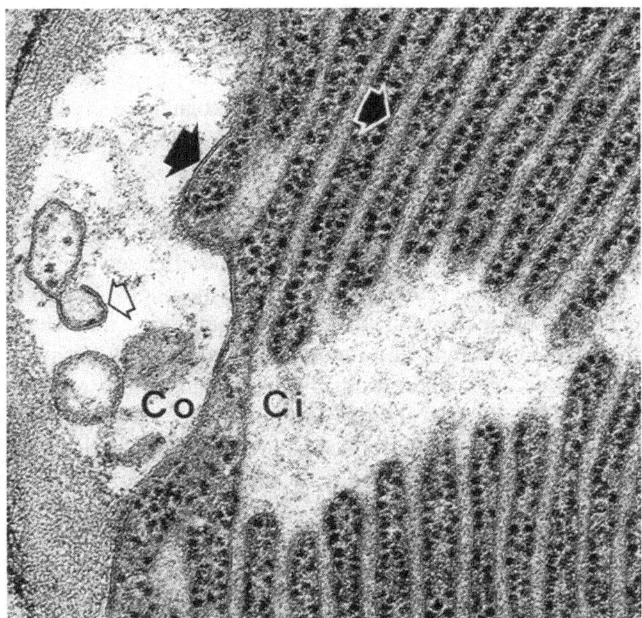

FIGURE 4.8 Rough endoplasmic reticulum in the trichomes of *Coleus blumei*. The ribosome-studded tubular endoplasmic reticulum is connected by a cisterna (Ci). The plasma membrane (*black arrow*) is thicker than the endoplasmic reticulum membrane (*black-and-white arrow*). ×75,000. *From Gunning, B.E.S., Steer, M.W., 1996. Plant Cell Biology. Structure and Function. Jones and Bartlett Publishers, Sudbury, MA.*

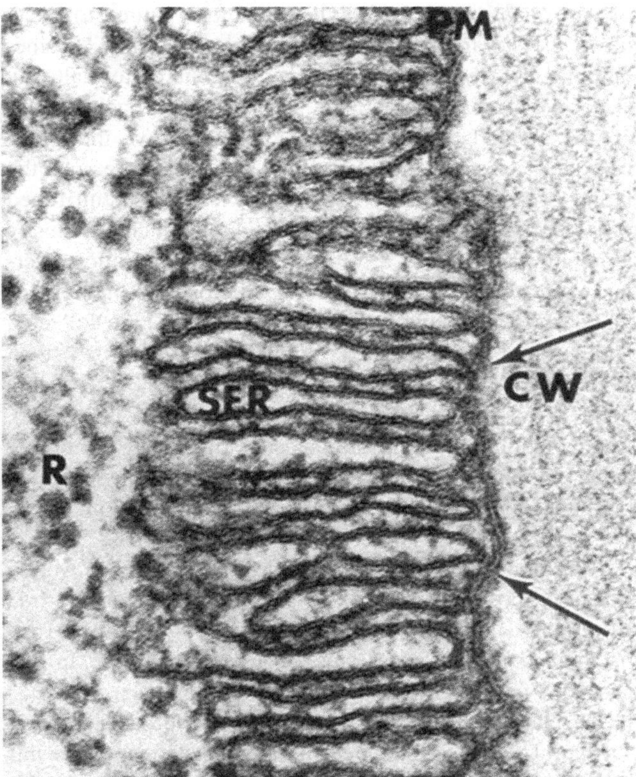

FIGURE 4.9 Smooth endoplasmic reticulum (SER)in the periphery of the sieve tube elements of *Streptanthus tortuosus*. *CW*, extracellular matrix; *PM*, plasma membrane; *R*, ribosomes. ×165,000. *From Sjolund, R.D., Shih, C.Y., 1983. Freeze-fracture analysis of phloem structure in plant tissue cultures. I. The sieve element reticulum. J. Ultrastruc. Res. 82, 111–121.*

secreting proteins are rich in RER, indicating that the RER is involved in protein synthesis. The proteins synthesized by the ribosomes that are attached to the ER are imported into the ER as they are synthesized. Because the protein is translocated into the ER as the linear mRNA sequence is being translated into a linear sequence of amino acids, the import of the nascent proteins is called *cotranslational import*. While the majority of proteins that enter the ER are imported cotranslationally in a guanosine triphosphate (GTP)–dependent manner (Nyathi et al., 2013), some are synthesized on cytosolic ribosomes and enter the ER *posttranslationally* in an ATP- or GTP-dependent manner (Mueckler and Lodish, 1986; Waters and Blobel, 1986; Blobel, 1999; Johnson et al., 2013).

Some regions of the ER lack ribosomes and appear smooth (Fig. 4.9). These regions are called the *smooth endoplasmic reticulum* or SER. Cells that have abundant SER are specialized for lipid production, indicating that the SER may be responsible for lipid biosynthesis. The oil glands of *Arctium* or the stigmatic cells of *Petunia* have an extensive network of SER needed for the synthesis and secretion of lipophilic molecules (Konar and Linskins, 1966; Schnepf, 1969a,b,c), perhaps at the contact sites between the ER and the plasma membrane (Hepler et al., 1990; Lancelle and Hepler, 1992). It is possible that the ER also delivers lipids to the chloroplast by means of contact sites (Wang and Benning, 2012; Hurlock et al., 2014). The SER of plant cells functions in detoxification much as it does in liver cells (Kreuz et al., 1996). There are regions of ER that are partly smooth and partly rough and are called *transitional elements* (Paulik et al., 1987; Morré et al., 1989a). Some transitional elements are specialized regions involved in producing the vesicles that transport newly synthesized proteins and lipids to the Golgi apparatus. Other transitional elements produce osmotically active lipid bodies and their associated proteins (Wu et al., 1997; Thompson et al., 1998; Murphy and Vance, 1999; Hsieh and Huang, 2004; Lersten et al., 2006; Horn and Benning, 2016). Such lipid bodies may serve as a novel source of biofuel (Chisti, 2007, 2008; Fortman et al., 2008; Li et al., 2008).

4.5 ISOLATION OF RER AND SER

The aleurone layer is a tissue that surrounds the endosperm in cereal grains and has been a favorite material for the study of ER because it contains a large amount of ER (Jones, 1969a,b; Vigil and Ruddat, 1973). The ER in the cells of this tissue is involved in the synthesis and secretion of vast quantities of hydrolytic enzymes required to break down the storage products of the endosperm into the metabolites used by the beer industry (i.e., starch to maltose).

To isolate ER membranes, aleurone layers are homogenized, filtered through cheesecloth to remove the extracellular matrix, and then centrifuged at 100 g to remove the large organelles. The supernatant is then centrifuged (70,000 g; 2.5 hours) on a discontinuous sucrose density gradient consisting of a 50% (w/w) sucrose cushion overlaid with 13% (w/w) sucrose. The microsomal membranes that accumulate between the 50/13% interface are collected and layered on a sucrose density gradient, and then centrifuged at 70,000 g for 14 hours. The ER forms a defined peak near 30% sucrose (density = 1.127 g/mL). The isolation is done in the presence of the Mg^{2+}-binding agent ethylenediaminetetraacetic acid (EDTA) to "capture" both the RER and the SER in the same fraction (Lord, 1983; Bush et al., 1989a,b; Sticher et al., 1990).

Polyribosome binding to the ER requires Mg^{2+}. Because ribosome-studded ER is denser than ribosome-free ER, the RER membranes undergo an Mg^{2+}-dependent shift in their densities on sucrose density gradients. In the absence of Mg^{2+}, the ER forms a sharp band at 1.12 g/mL; in the presence of Mg^{2+}, the ER forms a broader band at 1.16 g/mL (Lord, 1983). Because the plasma membrane has a peak between 1.14 and 1.17 g/mL (Hall, 1983), the inclusion of EDTA to chelate the Mg^{2+} ions helps to isolate pure ER membranes from sucrose density gradients. The ER can also be isolated using aqueous two-phase partitioning. With this procedure, the ER membranes are preferentially accumulated in the lower phase (Walker et al., 1993; Gilroy and Jones, 1993).

During isolation of the ER, its presence and purity are determined with the help of marker enzymes. The ER contains a number of enzymes that are endemic to it. Some of these enzymes, including NADH- and NADPH-dependent cytochrome c reductases, are involved in oxidation–reduction reactions and can be readily assayed spectrophotometrically. Consequently, these enzymes are often used for marker enzymes (Martin and Morton, 1956). The ER also contains a number of cytochromes that can be identified spectrophotometrically by their difference spectra. The oxidized minus dithionite-reduced difference spectrum of ER membranes has peaks at 555, 527, and 410 nm, which are typical of cytochrome b_5.

4.6 COMPOSITION OF THE ENDOPLASMIC RETICULUM

The membrane and lumen of the ER contain proteins that are involved in lipid synthesis, protein synthesis and processing, and ionic regulation. An auxin-binding protein also appears to be localized in the ER (Hesse et al., 1989; Inohara et al., 1989; Friml and Jones, 2010). Moreover, there are specific proteins that allow the attachment of the ribosomes to the ER. I will discuss some of these proteins individually. The CRISPER/Cas9 system of gene editing is being used to identify proteins in the ER that are involved in the signal peptide processing pathway (see Chapter 21; Zhang et al., 2016).

The lipid composition of the ER is similar, although not identical to the lipid composition of the plasma membrane (Philipp et al., 1976; Donaldson and Beevers, 1977; Coughlan et al., 1996). In fact, all membranes in the endomembrane system have a basic similarity related to their common origin and function as permeability barriers. The differences may result from specializations of the various membranes. However, unlike the case of yeast (Schneiter et al., 1999), it must be noted that the lipids of all the membranes from a single plant cell type of a single species have not yet been characterized. This observation, combined with the fact that the composition of the ER lipids varies depending on the environmental conditions (Holden et al., 1994), makes comparisons between different membranes somewhat tenuous (Table 4.1).

TABLE 4.1 Lipid Composition of Endoplasmic Reticulum Membrane

Lipid	Onion	Castor Bean
Phospholipids	(% of lipid phosphorus)	
Phosphatidylcholine	30.3	45.3
Lysophosphatidylcholine	4.4	—
Phosphatidic acid	24.6	—
Phosphatidylethanolamine	21.4	28.7
Lysophosphatidylethanolamine	2.5	—
Phosphatidylinositol	7.4	13.1
Phosphatidylserine	2.1	2.3
Phosphatidylglycerol	6.0	3.6
Cardiolipin	1.3	2.5
Major Sterols	(% wt of sterols)	
b-sitosterol	81.0	
Campesterol	6.6	
Fatty Acid Precursors of Phospholipids		(%wt)
16:0 stearic acid	32	
18:1 oleic acid	6	
18:2 linoleic acid	53	
18:3 linolenic acid	7	

Lysophospholipids contain a single acyl chain.
From Philipp et al. (1967) and Donaldson, R.P., Beevers, H., 1977. Lipid composition of organelles from germinating castor bean endosperm. Plant Physiol. 59, 259–263.

4.7 FUNCTION OF THE ENDOPLASMIC RETICULUM

4.7.1 Lipid Synthesis

The ER produces most of the lipids needed for membrane synthesis (Moore, 1982, 1987; Chapman and Trelease, 1991a,b; Vance and Vance, 2008). One representative biosynthetic pathway involves the formation of phosphatidylcholine from a glycerol-3-phosphate molecule, a cytidine diphosphate-choline molecule, and two fatty acids that have been activated by coenzyme A (CoA)(Lipmann, 1971). The A stands for acylation. CoA is a derivative of the vitamin pantothenic acid and is involved in the activation of acyl groups. This activation process is necessary to make the acetic acid groups reactive enough to participate in fatty acid elongation and attachment to the glycerol-3-phosphate molecule, which is produced in the cytosol by glycolysis (see Chapter 14). The fatty acids used in lipid synthesis are made in the cytosol of animal cells and in the plastids of plant cells (Ohlrogge et al., 1979). The physical pathway defining how lipids get from the plastid to the ER is being investigated (Andersson et al., 2007a,b; Schattat et al., 2011a,b; Mehrshahi et al., 2013, 2014). To initiate the synthesis of lipids on the ER, an acyl transferase combines the glycerol-3-phosphate with the two fatty acyl CoAs in a dehydration reaction to form phosphatidic acid and releases two CoA and two water molecules in the process (Fig. 4.10). Subsequently a phosphatase cleaves the phosphate from phosphatidic acid, thus producing diacylglycerol. Then choline phosphotransferase catalyzes the exchange of choline phosphate from cytidinediphosphatecholine (CDP-choline) to diacylglycerol, thus producing phosphatidylcholine and cytidine monophosphate. Phosphatidylethanolamine and phosphatidylserine are synthesized in a similar manner.

In addition, phosphatidylethanolamine can be converted to phosphatidylcholine by a methylation reaction, and phosphatidylserine can be converted to phosphatidylethanolamine by a decarboxylation reaction. Exchange reactions

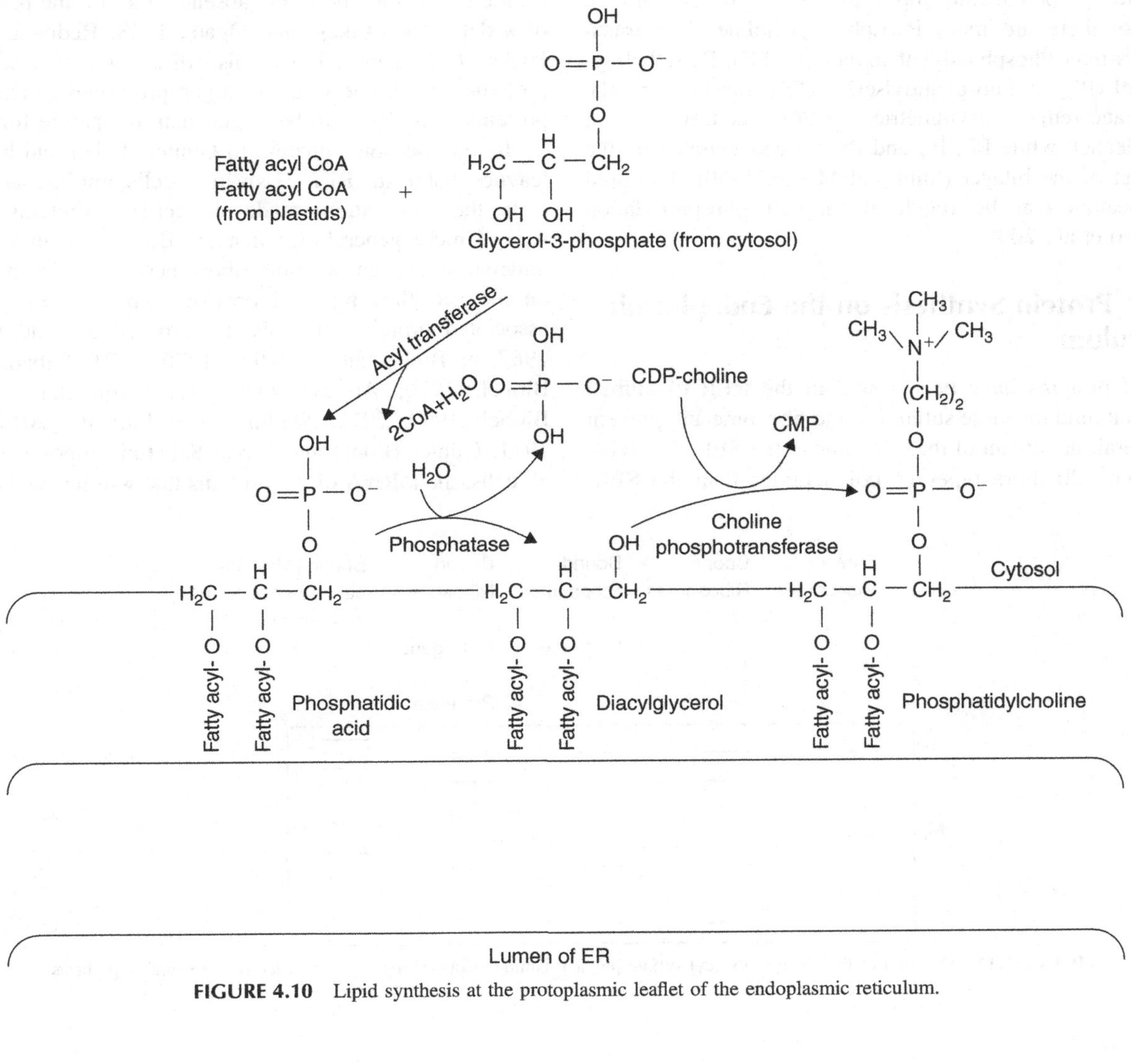

FIGURE 4.10 Lipid synthesis at the protoplasmic leaflet of the endoplasmic reticulum.

also take place, in which serine replaces the ethanolamine in phosphatidylethanolamine to form phosphatidylserine or ethanolamine replaces the serine in phosphatidylserine to form phosphatidylethanolamine. There are numerous enzymes and pathways involved in synthesizing the various lipids. It would be wonderful to know why nature goes to such lengths to form the lipid bilayer.

It is not always the head group that is activated by CDP. In the case of phosphatidylinositol synthesis, CDP activates the diacylglycerol molecule, which then attaches to an inositol molecule to form phosphatidylinositol.

The addition of the phosphatidic acid to the membrane results in membrane growth. Each step in lipid biosynthesis occurs on the cytoplasmic leaflet of the ER membrane. If this was kept up, a monolayer would be formed. However, a bilayer results not just due to thermodynamics but because the ER has head group–specific phospholipid translocators, which flip-flop the lipid across the membrane at a rate of 10^{-2} second^{-1}. This means it takes a lipid approximately 10^{2} seconds to be translocated across the membrane. The translocator-facilitated rate is 100–10,000 times greater than the rate of spontaneous flip-flops (10^{-4}–10^{-6} second^{-1}). Because there are more Phosphatidylcholine (PC) translocators than Phosphatidylethanolamine (PE), Phosphatidylinositol (PI), or Phosphatidylserine (PS) translocators, the membrane remains asymmetric and PC is concentrated on the E-leaflet, while PE, PI, and PS are concentrated on the P-leaflet of the bilayer (Shin and Moore, 1990). The lipid translocators can be regulated through phosphorylation (Nakano et al., 2008).

4.7.2 Protein Synthesis on the Endoplasmic Reticulum

Special proteins have been found in the RER of animal cells that bind the large subunit of the ribosome and prevent the lateral movement of the ribosome to the SER. The RER has about 20 more types of polypeptides than the SER. Some of these polypeptides may be involved in ribosome anchoring; others may be involved in maintaining the shape of the flattened cisternae.

The mechanism of protein synthesis is discussed in Chapter 17. For now, let us accept the fact that ribosomes contain the means to synthesize proteins, which was demonstrated by measuring the incorporation of radioactive amino acids into proteins in the presence of isolated ribosomes. Comparative cytochemical studies in secretory cells suggested that the free ribosomes synthesize proteins that were used by the cell, whereas bound ribosomes synthesize secreted proteins (Siekevitz and Palade, 1960a,b). Subsequent biochemical work using free and ER-bound ribosomes that had been separated from each other and allowed to translate the mRNA in them into proteins confirmed that the two populations of ribosomes in a single cell type produce different proteins (Fig. 4.11). Subsequent work with the in vitro translation system showed that the secretory proteins were probably inserted into the lumen of the ER, as experiments using isolated microsomes showed that the newly formed proteins were protected from proteolysis by the ER membrane in the absence, but not the presence, of a detergent (Takagi and Ogata, 1968; Redman, 1969; Hicks et al., 1969). It was also discovered that secreted proteins are synthesized as larger proproteins. The proproteins were found to be larger than the mature form.

It soon became apparent to Günter Blobel and his colleagues that as the RER exists in all cells, not just secretory cells, the observations made on secretory proteins might have a more general significance. But before they could understand the reason some ribosomes synthesize proteins on the ER, they repeated previous work on membrane-associated protein synthesis in vitro (Blobel and Potter, 1967a,b; Blobel and Sabatini, 1970, 1971; Sabatini and Blobel, 1970; Anonymous, 1999; Birmingham, 1999; Blobel, 1999, 2012; Matlin, 2013; Labonte, 2017). By 1971, Günter Blobel and David Sabatini proposed that in all cells, the mRNA of the proteins that will be synthesized

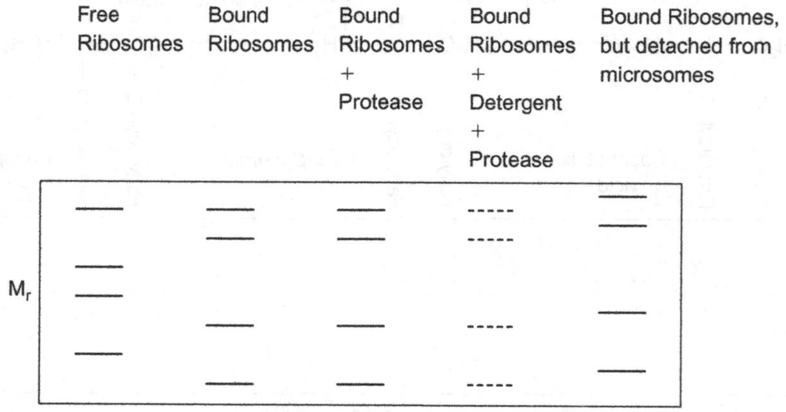

FIGURE 4.11 Diagram of the sodium dodecyl sulfate polyacrylamide electrophoresis gels that led to the signal hypothesis.

on the RER, unlike those that are synthesized on free ribosomes, would prove to have a certain sequence at the 5′ end of a gene, which would result in a certain amino acid sequence at the amino-terminus. They predicted that this sequence would cause the ribosomes that have bound that particular mRNA to be delivered to the ER. Protein synthesis would then continue on the ER where the nascent polypeptide would be vectorially transported into the lumen and the signal peptide would be removed.

This proposal, which came to be known as the *signal hypothesis*, was directly tested by Blobel and Dobberstein (1975a,b). They found that ribosomes detached from microsomes produce longer proteins than do ribosomes in the presence of microsomal membranes (see Fig. 4.11). They also found that proteins made in the absence of microsomes were degraded by an added protease, whereas those made in the presence of microsomes were protected, indicating that the newly synthesized proteins were in the ER lumen. They also found that the protein produced by free ribosomes had an amino-terminal leader peptide that was cleaved in the presence of microsomes to make a protein of the correct size, while the rest of the protein was still being synthesized. They named this ER-localized peptidase the *signal peptidase*. Reconstitution experiments using free ribosomes and mRNA that encoded a secreted protein showed that the mRNA has the information necessary to deliver the ribosome to the ER.

The importance of the signal sequence is dramatically shown in experiments in which the DNA that codes for this sequence is inserted in front of a DNA sequence that encodes a protein that is typically translated on free ribosomes. Instead of being translated on free ribosomes and ending up in the cytosol, the fusion protein is translated by bound ribosomes and inserted into the ER! Moreover, when recombinant DNA technology is used to delete the signal sequence from proteins typically synthesized on membrane-bound ribosomes, these proteins are synthesized on free cytosolic ribosomes. Much is known about the structure of signal peptides (von Heijne, 1990). They have a three-domain structure that includes an amino-terminal positively charged region that is 1–5 amino acids long, a central hydrophobic region that is 7–15 amino acids long, and a more polar carboxy-terminal domain that is 3–7 amino acids long. Beyond this pattern, there is no precise sequence conservation. Site-directed mutagenesis shows that the amino and central regions are required for translocation, while the carboxy region contains the sequence that is recognized by the signal peptidase and thus specifies the cleavage site. The amino acid sequences that target nascent proteins to the correct organelle are known as *molecular zip codes* and are discussed in Chapter 17.

The signal hypothesis has been very powerful in providing a theoretical framework to understand how a given protein ends up in the appropriate organelle, which is surrounded by an otherwise impermeable lipid bilayer that ensures the physical and functional separation of biochemical functions. The problem of delivering a protein to the correct organelle in a cell is analogous to the problem of sending a package to the correct address in a large city. "Naturally, to avoid chaos, each product requires a clearly labeled address tag. Günter Blobel [was] awarded [the] Nobel Prize in Physiology or Medicine for having shown that newly synthesized proteins, analogous to the products manufactured in the factory, contain built-in signals, or address tags, that direct them to their proper cellular destination." (Pettersson, 1999). According to the general theory of protein targeting and translocation (Blobel, 1980; Simon and Blobel, 1991)

1. A protein that is translocated across or integrated into a distinct membrane must contain a signal sequence.
2. The signal sequence is specific for each membrane.
3. A signal sequence—specific recognition factor and its receptor on the correct membrane are needed for successful targeting.
4. Translocation across the membrane occurs through a proteinaceous channel.
5. The nascent protein has a series of amino acids that form an alpha helix, the outer surface of which is hydrophobic and functions as a start- or stop-transfer sequence, depending on its position in the polypeptide.
6. If the protein is to be integrated into the membrane, a start- or stop-transfer sequence in the polypeptide opens the protein-conducting channel and displaces the polypeptide from the aqueous environment of the channel into the lipid bilayer.

The signal peptide of a protein that is destined to be synthesized on the ER is guided into the ER by a signal recognition particle (SRP; see Fig. 4.12; Ng and Walter, 1994). The SRP is composed of six different polypeptide chains bound to a single molecule of 7S RNA. Just as in ribosomes, here is another example where proteins and RNA function together in a complex. The SRP binds to the signal peptide as soon as it emerges from the ribosome. Actually the 54-kDa polypeptide of the SRP is methionine rich and forms a hydrophobic pocket and binds to the signal peptide (High and Dobberstein, 1991). The binding of the SRP to the nascent polypeptide somehow causes a halt in the synthesis of that protein, thus allowing time for the large subunit of the ribosome to bind to the ER. Protein synthesis is reinitiated once the SRP binds to an SRP receptor on the ER (Gilmore et al., 1982a,b). This occurs because the SRP receptor displaces the SRP from the nascent polypeptide.

The association of the SRP with the nascent protein synthesized by the ribosome, and the association of the SRP receptor with the protein-translocating channel, requires GTP (Mandon et al., 2003; Shan and Walter, 2005).

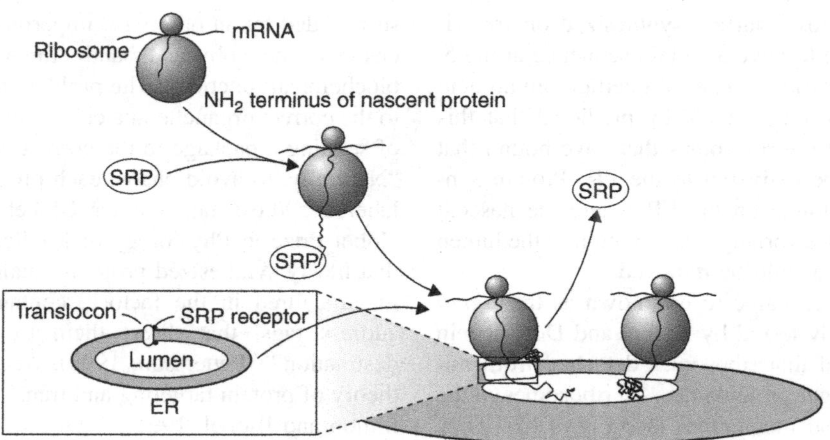

FIGURE 4.12 The delivery of a nascent protein to the endoplasmic reticulum. *SRP*, Signal Recognition Particle.

The SRP receptor then brings the ribosome and its nascent SRP-binding polypeptide in contact with the protein-translocating channel (Walter, 1997). Cross-linking studies performed at various times following the interaction of the 54-kDa subunit of the SRP with the ribosome and ending with the binding of the ribosome to the protein-translocating channel have allowed the identification of a number of polypeptides involved in these processes (Krieg et al., 1989).

In vitro studies, where detergent-treated ER is depleted of the SRP receptor, show that the SRP receptor is essential for protein translocation across the ER membrane (Migliaccio et al., 1992). Both the SRP and the SRP receptor were originally identified as components needed to reconstitute in vitro protein translocation into the ER. Again, we see the importance of a functional assay, involving reconstitution to identify the proteins involved and their functions. The SRP receptor is an integral membrane protein that contains four polypeptide chains. The function of each polypeptide has been determined by reconstitution experiments (Görlich and Rapoport, 1993). Genetic studies of yeast mutants that are unable to secrete proteins have identified the secretory, or sec genes that encode proteins involved in protein translocation into the ER. We are just beginning to find that plants use the same protein-targeting and protein-translocation mechanisms (Thoyts et al., 1995; Beaudoin et al., 2000; Shy et al., 2001; Jang et al., 2005).

Exciting work has begun on determining the mechanism of how proteins can pass through the ER membrane (Simon, 1993, 2002; Schatz and Dobberstein, 1996). Simon and Blobel (1991) and Simon et al. (1989) have identified protein-translocating channels using electrophysiological techniques. They isolated vesicles of the RER and incorporated them into one side of a planar black lipid membrane (BLM). They then applied an electrical potential (ψ) across the two sides of the bilayer and measured the resulting current (I). They calculated the conductance (G) of the protein-translocating channels using Ohm's Law ($G = -I/\psi$).

Initially, the conductance is approximately 0 pS. However, after adding 100 μM of puromycin, an adenosine derivative that uncouples a nascent polypeptide from its ribosome-bound peptidyl-tRNA, a large increase in conductance occurs. When a low concentration of puromycin is added, so that elongation of one chain is stopped at a time, discrete changes in conductance of 220-pS steps is seen (Fig. 4.13). The conductance results from the fact that the nascent polypeptide no longer occludes the channel, and now K^+ can move through the channel and produce a current in response to the applied voltage. The protein-translocating channel probably remains open until the ribosome moves away from the membrane because the high-conductance state is stable until the ribosomes are washed off with high salt (Fig. 4.14). Electron microscopy of the protein-translocating channel indicates that the pore has a diameter of 4–6 nm (Hanein et al., 1996; Hamman et al., 1997).

Crowley et al. (1993, 1994) have created a great technique to monitor the polarity of the channel that the nascent polypeptide goes through. They incorporated a fluorescent probe into the nascent polypeptide. The probe was chosen so that its fluorescence lifetime depends on the environment immediately surrounding it. The lifetime is short when it is in an aqueous environment and long when it is in a lipid environment.

When attached to the signal sequence, the dye has a short fluorescence lifetime, indicating that the signal sequence goes through an aqueous channel. However, because the fluorescence cannot be eliminated by adding an aqueous-quenching agent, the aqueous pore is not continuous with the aqueous environment surrounding the membrane when the polypeptide is in the pore. Cross-linking studies show that the hydrophobic portion of the signal sequence is in contact

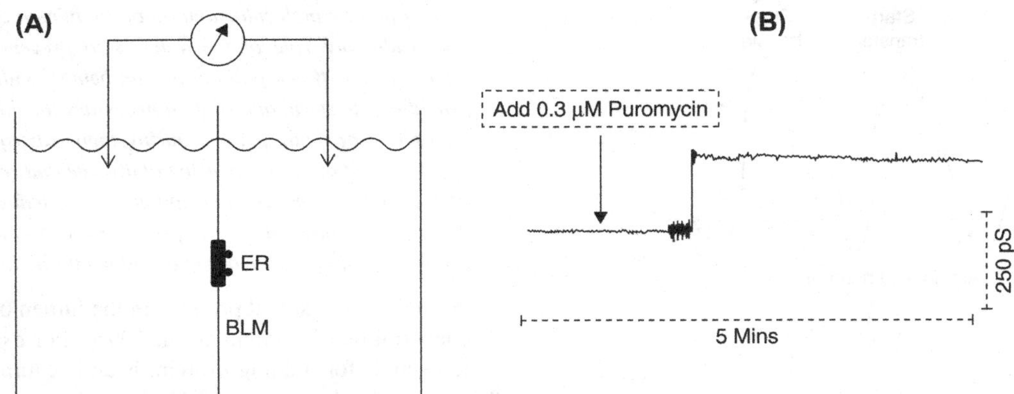

FIGURE 4.13 (A) An electrophysiological experiment demonstrating a single protein-translocating channel revealed by the application of puromycin. (B) The protein-translocating channel has a conductance of approximately 220 pS. This is much greater than the conductance of ion channels, the activation of which is shown by the small variations in the conductance trace. *BLM*, Black Lipid Membrane. *From Simon, S.M., Blobel, G., 1991. A protein-conducting channel in the endoplasmic reticulum. Cell 65, 371–380.*

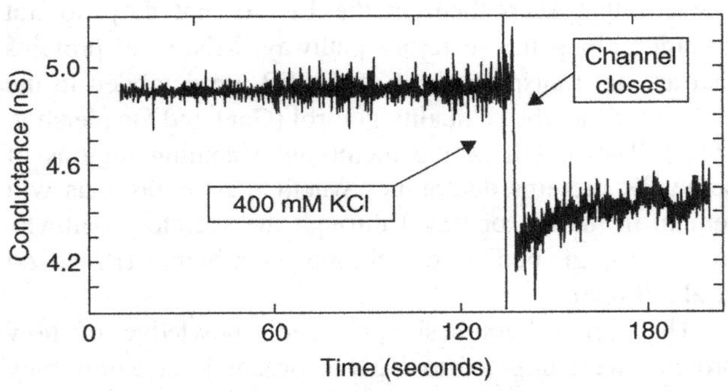

FIGURE 4.14 An electrophysiological experiment demonstrating the closure of one of four protein-translocating channels following the addition of 400 mM KCl, which probably washed off the ribosome on the channel that closed. *From Simon, S.M., Blobel, G., 1991. A protein-conducting channel in the endoplasmic reticulum. Cell 65, 371–380.*

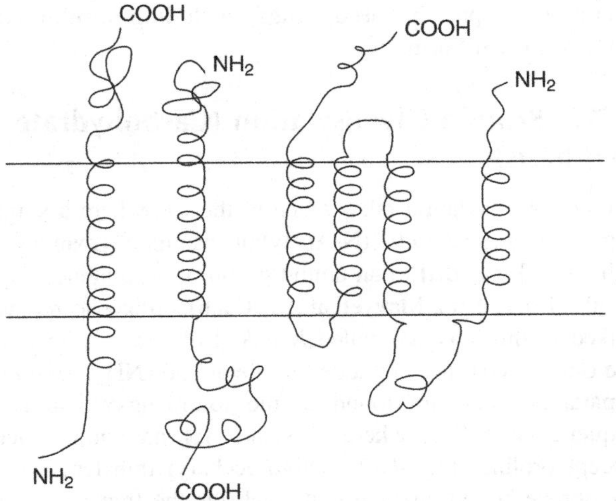

FIGURE 4.15 Membrane proteins in the endoplasmic reticulum with various orientations.

with lipids in the bilayer during an early stage of protein insertion, indicating that the protein-translocating channel can open laterally, at least during an early stage of protein insertion (Martoglio et al., 1995).

How is an integral protein inserted into the ER membrane (High and Dobberstein, 1992)? The actual mechanism still needs to be elucidated, but we assume that a sequence of approximately 7–21 nonpolar amino acids are long enough to span the membrane and encourage the nascent polypeptide to partition out of the translocon and into the lipid bilayer, while a series of polar amino acids encourages the polypeptide to remain where the hydrophilic sequence begins—either on the cytosolic side or on the luminal side of the ER (Fig. 4.15).

Proteins are synthesized from the amino-terminus to the carboxy-terminus. The first sequence of hydrophobic amino acids acts as a start-transfer sequence and the subsequent sequences of hydrophobic amino acid act alternately as stop- and start-transfer sequences. If the membrane protein is to be inserted such that only the amino-terminus is to remain in the lumen, there must be a start-transfer sequence at the amino-terminus followed by a stop-transfer sequence. If the carboxy-terminus is to remain in the lumen, the start-transfer sequence occurs distal to the amino-terminus and there is no stop-transfer sequence. In the case of multipass transmembrane proteins, there are many start- and stop-transfer sequences that allow polypeptide chains to repeatedly pass the membrane, leaving loops in the cytoplasmic space and the luminal space (Singer, 1990).

While little is known about how the start- and stop-transfer sequences function, in general they are probably determined by the hydrophobicity and charge of the amino acids (see Fig. 4.16; High and Dobberstein, 1992). The chemical properties of the amino acid sequences may influence the partition of that segment of the polypeptide

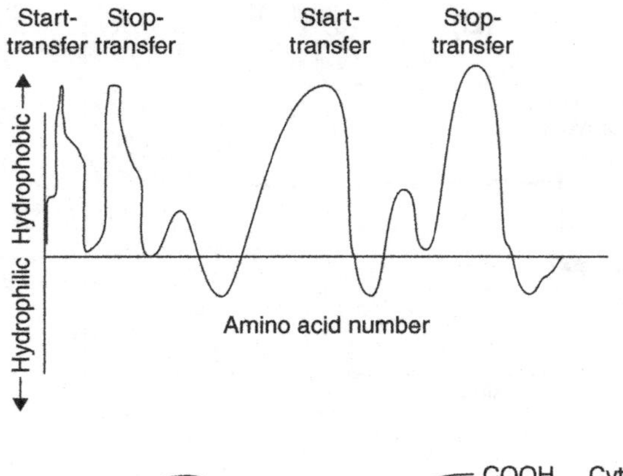

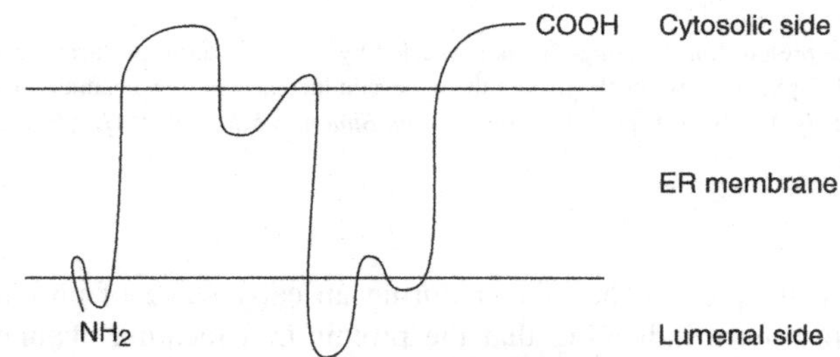

FIGURE 4.16 Comparison between a Kyte-Doolittle plot, which characterizes the hydrophobicity of regions of a protein with the localization of those regions in the membrane.

between the aqueous channel, the lipid bilayer, the cytosol, and the lumen. The start- and stop-transfer sequences must influence the axial and lateral gating behavior of the protein-translocating channel. The gating is not only influenced by the amino acid sequence in the nascent protein that is going through the channel but also by the adjoining transmembrane amino acid sequences. In addition, there are also long-range allosteric effects on translocon gating determined by the amino acid sequence still in the ribosome (Liao et al., 1997).

The protein conducting channel (PCC) is, as Blobel (1999) said, "clearly one of the marvels of nature. Unlike an ion-conducting channel that opens and closes in only one dimension, the PCC opens and closes in two dimensions, across the lipid bilayer and in the plane of the bilayer. The PCC is not merely a passive conduit but it scans the unfolded nascent chain as it passes across it, responding to a passing stop-transfer sequence by lateral opening. Moreover, the PCC is constructed in such a way that it does not leak significant amounts of small molecules."

What allows the translocating protein to move through the pore? It probably reptates back and forth though the pore as a result of thermal energy (Simon et al., 1992). However, it is also possible that the energy and direction of the vectorial transport may be influenced by the binding of sugars to the nascent polypeptide (Nicchitta and Blobel, 1993), or the binding of other proteins, known as chaperonins, that help fold a polypeptide into its mature conformation. A chaperonin, according to R.J. Ellis (1996),

> *... is a precise molecular analog of the human chaperone. The traditional role of the latter is to prevent incorrect interactions between pairs of human beings, without either providing the steric information necessary for their correct interaction or being present during their subsequent married life—but often reappearing at divorce and remarriage! So the term is a precise description of an essential function that we now know all cells require in order to increase the probability of correct macromolecular interactions.*

Some of the resident proteins in the lumen of the ER act as chaperonins (Coughlan et al., 1996). For example, BiP, which stands for *bi*nding *p*rotein, is an ER lumenal protein that is involved in protein folding, perhaps by acting as a detergent that helps solubilize the polypeptide so that it can be properly folded (Pelham, 1989a,b; Jones and Bush, 1991; Fontes et al., 1991; Li et al., 1993a,b; Anderson et al., 1994; Donohoe et al., 2013). BiP is an unusual protein because it utilizes ATP in the E-space. The proteins that will remain in the lumen of the ER have specific sequences that keep them in the ER so that they do not continue along the secretory pathway. Misfolded proteins that are not transported out of the ER are degraded in the ER itself as a type of quality control (Gant and Hendershot, 1993). The length of the membrane-spanning regions of some ER proteins determines whether those proteins will remain in the ER or travel through the secretory pathway to the Golgi bodies or plasma membrane (Brandizzi et al., 2002c).

The signal hypothesis provided knowledge of how proteins were targeted to a given organelle and how they were translocated through the membrane. In addition to this, it provided a framework to understand the mechanisms underlying hereditary diseases in which specific proteins were mislocalized. It also provided the guiding principles necessary for the pharmaceutical industry to turn cultured cells into efficient minifactories for the production and secretion of protein-based drugs, including insulin and human growth hormone.

4.7.3 Protein Glycosylation (Carbohydrate Synthesis)

Protein glycosylation takes place in the ER where a single type of oligosaccharide [(N-Acetylglucosamine)$_2$(mannose)$_9$ (glucose)$_3$] is added to an amino group of asparagine (Faye et al., 1989; Ruiz-May et al., 2012a,b). Oligosaccharides linked in this way are called *N*-linked oligosaccharides, as the oligosaccharides are added to the amino (NH_2) group of asparagines that are found in the following amino acid sequence N-X-S/T, where X stands for any amino acid except proline (Fig. 4.17). Oligosaccharyl transferase is a membrane-bound enzyme that catalyzes the transfer of the entire oligosaccharide from dolichyl pyrophosphate to the asparagine group.

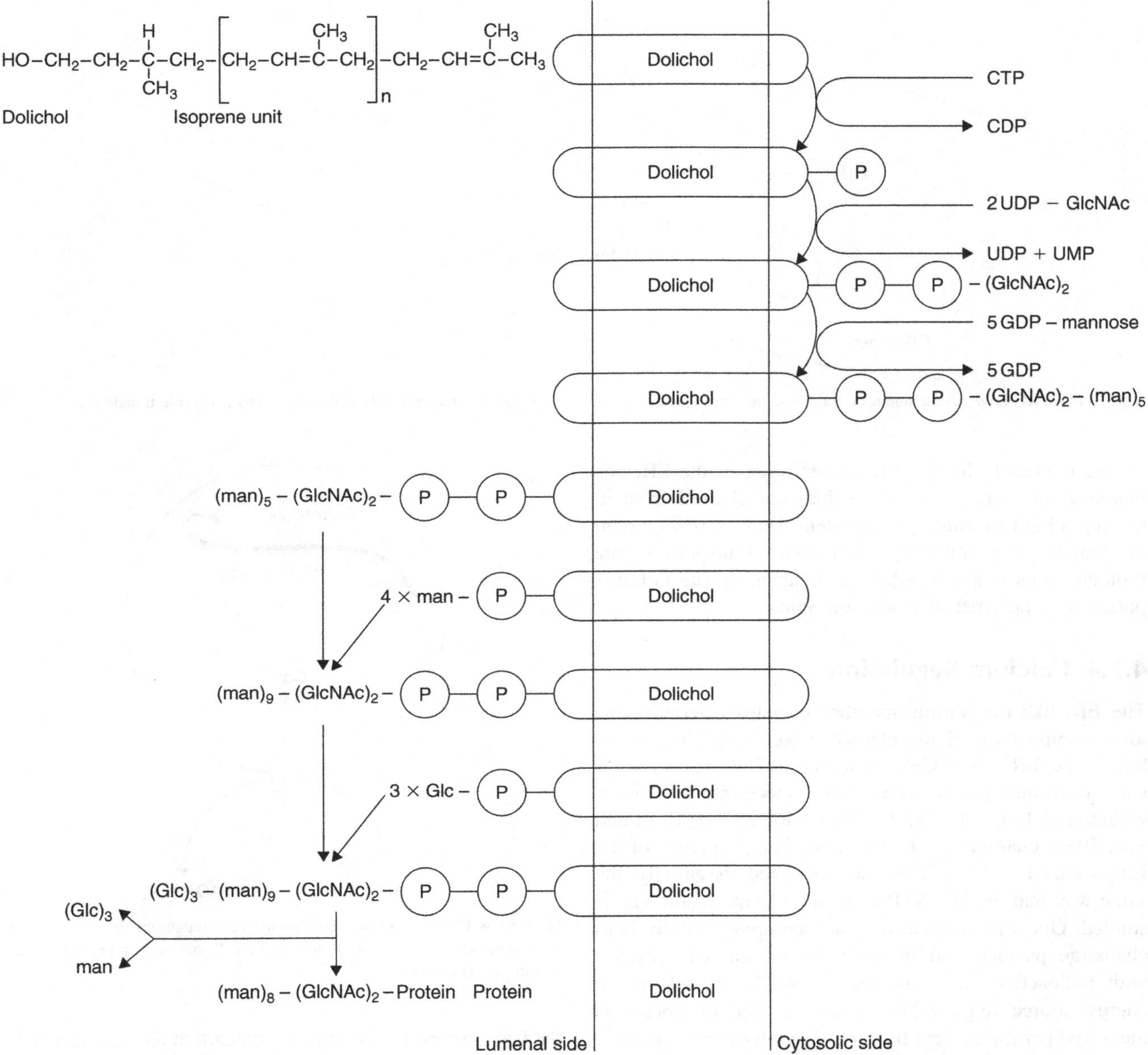

FIGURE 4.17 Synthesis of the carbohydrate group of *N*-linked glycoproteins in the membrane.

Dolichols are long-chain unsaturated alcohols made up of 16–21 isoprenoid units (–$CH_2C[CH_3]$=CH–CH_2–). Dolichols are found in the ER and the Golgi apparatus and serve as intermediates in the formation of oligosaccharides (Fig. 4.18; Elbein, 1979). The dolichol that is in the ER membrane must be charged by GTP to make dolichyl phosphate. This process takes place on the cytosolic side of the membrane or in the P-space. Subsequently, two UDP-*N*-acetyl-glucosamines are added to form dolichol pyrophosphate (*N*-acetyl-glucosamine)$_2$, uridine monophosphate, and uridine diphosphate. In cells, these nucleoside phosphates are specific for transferring sugar molecules much like CoA is specific for transferring acyl groups. Tunicamycin is a specific inhibitor of this step, and thus a good way of testing the importance of glycosylation in a given process.

Five guanosine diphosphate mannoses (GDP-mannoses) are added to the *N*-acetyl-glucosamine and five GDPs are released. At this point, the lipid–sugar complex flips to the other side of the membrane where it faces the lumen. From this point on, four mannose and three glucose residues are added to the oligosaccharide from dolichyl-P-mannose and dolichyl-P-glucose, which were originally formed on the cytosolic surface of the ER membrane and then flipped across into the lumenal side.

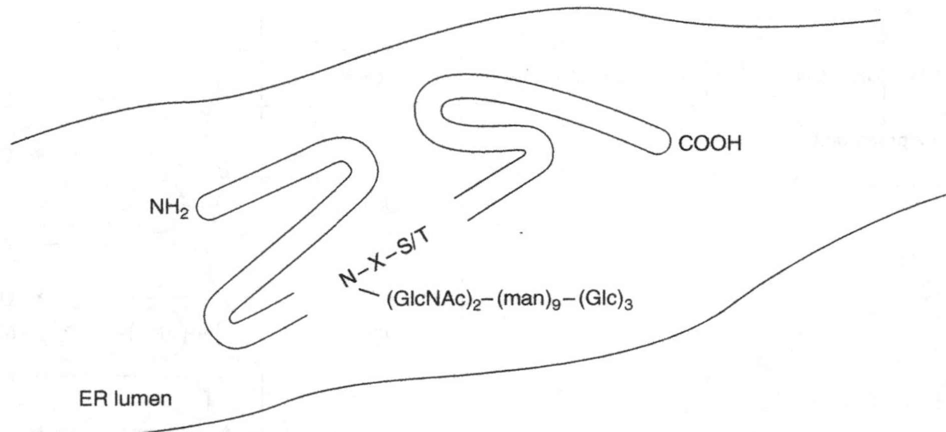

FIGURE 4.18 Attachment of carbohydrate groups to the asparagine residue of a nascent polypeptide by oligosaccharide transferase.

Before the newly formed glycoprotein leaves the ER, one mannose and three glucose residues are cleaved from it, leaving a high-mannose glycoprotein. Here, it is important to realize that not all membrane transporters are proteins—this is a case where a sugar-phosphate is transported by a polymer of isoprenoid units.

4.7.4 Calcium Regulation

The ER, like the plasma membrane, helps to control the ionic composition of the cytosol. Like the plasma membrane, the ER has a Ca^{2+} pumping ATPase that pumps Ca^{2+} from the cytosol, which is a P-space into the lumen, which is an E-space (Fig. 4.19; Buckhout, 1984; Bush and Sze, 1986; Giannini et al., 1987a,b). The properties of the ER-localized Ca^{2+}-ATPase are analyzed essentially the same way that the H^+-ATPase of the plasma membrane is studied. One way to measure Ca^{2+} pumping activity is to challenge purified and intact, right-side-out ER vesicles with radioactive Ca^{2+} and initiate uptake by adding an energy source (e.g., ATP). Then the rate of uptake is measured per mg protein by capturing the membranes on a filter. Using such an assay, Williams et al. (1990) found that the ER Ca^{2+}-ATPase is almost identical to the plasma membrane–localized Ca^{2+}-ATPase except that the plasma membrane–bound ATPase can use GTP as a substrate, whereas the ER one cannot. The ER Ca^{2+}-ATPase, like the plasma membrane Ca^{2+}- and H^+-ATPases, is inhibited by vanadate and dicyclohexylcarbodiimide. Ca^{2+} uptake into the ER of barley aleurone cells requires 0.07 mol/m^3 ATP and 0.0005 mol/m^3 Ca^{2+} for half-maximal activity (Bush et al., 1989a). A temperature-sensitive fluorescent dye that targets the ER shows that the temperature of the ER increases by 1.7 C when the Ca^{2+}-ATPase is working (Arai et al., 2014).

The sarcoplasmic reticulum of muscle cells is a highly developed form of ER that specializes in storing and releasing

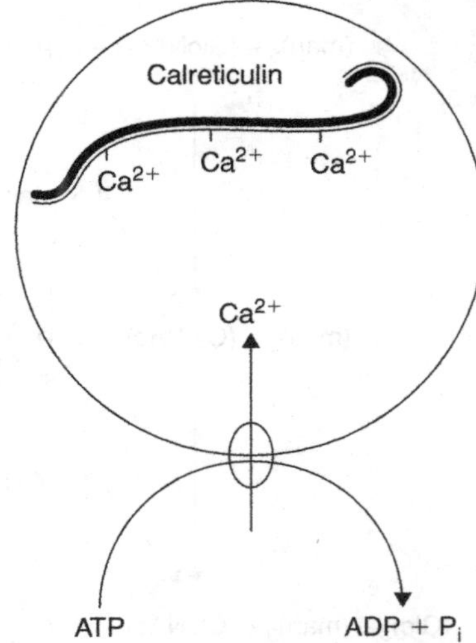

FIGURE 4.19 A right-side-out endoplasmic reticulum vesicle showing the orientation of a Ca^{2+}-ATPase and the calcium-binding protein, calreticulin, in the E-space.

the Ca^{2+} necessary for muscle contraction and relaxation. It contains a low-affinity, high-capacity Ca^{2+}-binding protein called *calsequestrin* in its lumen. The affinity constant of calsequestrin for Ca^{2+} is approximately 1 m^3/mol, and it can bind approximately 50 Ca^{2+} ions per 51-kDa molecule (Ebashi, 1985). Calsequestrin also occurs in the ER of plant cells (Krause et al., 1989; Chou et al., 1989).

Calreticulin, another low-affinity, high-capacity Ca^{2+}-binding protein, also occurs in the ER of plant cells (Opas et al., 1996). What evidence should we look for to determine whether calreticulin or calsequestrin are integral membrane proteins or localized in the lumen? First, they can be isolated and purified from osmotically shocked microsomes without detergents; second, the purified protein always stays in the hydrophilic phase and never goes into the hydrophobic phase during two-phase partitioning; third, immunogold labeling shows that the protein is localized in the lumen; last, the protein is protected by the ER and not degraded when the ER is treated with trypsin, a protease that cannot cross membranes.

How would you measure the binding capacity and affinity of a Ca^{2+} binding protein? Put the protein in a dialysis membrane surrounded by solutions with various concentrations of Ca^{2+}. At equilibrium, remove the protein, measure the amount of protein spectrophotometrically and the Ca^{2+} content of the protein using atomic absorption spectrophotometry, and then determine how many moles of Ca^{2+} are bound to each mole of protein at each Ca^{2+} concentration. Then plot this ratio versus the Ca^{2+} concentration. The Ca^{2+} concentration that gives the half-maximal saturation of the protein is an estimate of the dissociation constant (K_d, in M) of the protein for Ca^{2+}. The dissociation constant is the reciprocal of the affinity constant (K_a, in M^{-1}) of the protein for Ca^{2+}. The capacity and affinity of typical lumenal Ca^{2+}-binding proteins are 20–50 mol Ca^{2+}/mol protein and 1 m^3/mol, respectively (Ebashi, 1985; Macer and Koch, 1988; Michalak et al., 1992). The Ca^{2+} is typically bound to the negatively charged, acidic amino acids.

We are now ready to learn a general principle. In a particular location, if an enzyme has the function for which it is named, then the substrate in its local environment, particularly around the active site, must be present in concentrations around the enzyme's binding constants (K_m, $1/K_a$ or K_d) for that substrate. That is, we can use the values of the binding constants to estimate the concentrations of molecules and ions in various compartments in the cell. This assumes that the binding constants were determined with intact enzymes, under the correct physiological conditions. Given the binding constants found for the proteins involved in Ca^{2+} regulation, we can assume that the Ca^{2+} concentration in the lumen of the ER (E-space) is approximately 1 mol/m^3 and the Ca^{2+} concentration in the cytosol (P-space) is approximately 10^{-4} mol/m^3. As I will discuss in Chapter 12, Ca^{2+} is a cytotoxin, and this metastable, nonequilibrium distribution must be strongly regulated or cell death will result.

4.7.5 Phenylpropanoid and Flavonoid Synthesis

The phenylpropanoid pathway is part of the plant aromatic pathway, and one of the major enzymes of this pathway (cinnamate 4-hydroxylase) is embedded in the ER as part of a multienzyme complex known as a metabolon (Wagner and Hrazdina, 1984; Hrazdina and Wagner, 1985; Hrazdina and Jensen, 1992; Burbulis and Winkel-Shirley, 1999; Winkel-Shirley, 2002; Winkel, 2004). The phenylpropanoid pathway begins with the amino acid phenylalanine, which is the end product of the shikimic acid pathway. Phenylalanine is the precursor for the synthesis of phytoalexins, which are involved in defense mechanisms; flavonoids, which in part cause the colors of plants that attract pollinators; lignin, which is the major molecule participating in the incrustation and waterproofing of walls; and coumarin, the molecule that gives us the smell of freshly cut grass.

The key enzyme in the branch of the phenylpropanoid pathway involved in flavonoid synthesis (chalcone synthase; malonyl-CoA:4 coumaroyl-CoA malonyltransferase) has been localized in the ER fraction (Hrazdina et al., 1987). Furthermore, immunocytochemistry with colloidal gold particles shows that the enzyme is associated with the cytosolic leaflet of the ER membrane. The flavonoids synthesized by this enzyme are stored in the vacuole (see Chapter 7).

4.8 SUMMARY

The plasma membrane of small prokaryotic cells perform all the functions that depend on membranes, including ion transport, protein secretion, and perhaps the anchoring of DNA during replication (Jacob et al., 1963; Funnell, 1993; Toro and Shapiro, 2010; Badrinarayanan et al., 2015). The plasma membrane of eukaryotic cells does not have a sufficient surface area to perform these functions. The ER functions in maintaining a surface-to-volume ratio in large cells of approximately 10^6 m^{-1} and thus duplicates many of the transport functions of the plasma membrane, particularly those involved with Ca^{2+} transport. The ER also serves as the workbench of the cell for building itself and other membranes, and thus is endowed with the enzymes necessary for lipid synthesis and the ribosomes necessary for protein synthesis. We have learned that proteins that are synthesized on the ER go there because they contain a signal peptide. As we will see, the signal hypothesis describes a general mechanism of how specific proteins are targeted to each organelle. The ER is also continuous with the nuclear envelope and has fenestrated lamellae that appear to be nuclear pores without the nuclear pore complex.

4.9 QUESTIONS

4.1. How is the ER similar to the plasma membrane, and how is it different?

4.2. Why is the surface-to-volume ratio important in biology?

4.3. How is the ER involved in the synthesis of the plasma membrane?

The references for this chapter can be found in the references at the end of the book.

Chapter 5

Peroxisomes

The cell, too, has a geography, and its reactions occur in colloidal apparatus, of which the form, and the catalytic activity of its manifold surfaces, must efficiently contribute to the due guidance of chemical reactions.

Frederick Gowland Hopkins (The dynamic side of biochemistry, 1913).

5.1 DISCOVERY OF MICROBODIES

Microbodies are usually considered to be single membrane–enclosed organelles approximately 0.2–1.5 μm in diameter. While they are often round in thin sections, they can also appear ellipsoidal or dumb bell shaped (Frederick et al., 1968; Gruber et al., 1972). As discussed later, they may actually form a peroxisomal reticulum. Microbodies have a limiting membrane, which is approximately 6.5 nm thick and surrounds a matrix that can appear amorphous, granular, fibrillar, or paracrystalline (Vigil, 1983).

While microbodies were first seen in electron micrographs by Johannes Rhodin (1954), who coined the term *microbody*, Rouiller and Bernhard (1956) presented the first widely available picture of microbodies in liver cells (Fig. 5.1). Approximately a decade later, Christian de Duve et al. isolated microbodies from rat liver cells as a contaminant of the mitochondrial fraction (Baudhuin et al., 1965; de Duve, 2012.). In plants, microbodies were first isolated from castor bean seedlings by Breidenbach and Beevers (1967) and from spinach leaves by Tolbert et al. (1968). The papers written by the pioneers in microbody research are very exciting because biochemists and electron microscopists were meeting at the borderlands of their sciences and providing cellular biology with the depth originally envisioned by Jean Baptiste Carnoy when he coined the term *cellular biology* in 1884. Now biochemists could "see" the organelles that contained the enzymes of interest and electron microscopists could assign a function to the structures and significance to the relative positions of organelles.

De Duve and Baudhuin (1966) did not like the term *microbody* because it was so general and strictly morphological, but they did not want to name it something else until they knew more about its true function. These were the days when objects were given functional names only after the functions were understood. However, after learning more about the function of microbodies, they named them *peroxisomes* (Parcharoenwattana and Smith, 2008). They defined the peroxisome as a single membrane–enclosed organelle that contains at least one oxidase that forms the toxic molecule H_2O_2, as well as catalase, an enzyme that breaks the H_2O_2 down into nontoxic oxygen and water. Because peroxisomes contain catalase, they are easily identified in electron microscopical sections that have been stained with diaminobenzidine, as this agent forms an electron-dense deposit in the presence of catalase (Vigil, 1970). Functioning peroxisomes are necessary for normal plant growth and development (Zolman et al., 2001; Hu et al., 2002; Zolman and Bartel, 2004; Woodward and Bartel, 2005) and are involved in auxin metabolism and the synthesis of jasmonic acid (Cruz Castillo et al., 2004; Zolman et al., 2007; Baker and Paudyal, 2014).

5.2 ISOLATION OF PEROXISOMES

To isolate peroxisomes, the tissue is homogenized and filtered to remove the extracellular matrix. Then the filtrate is centrifuged at about 500 g for 10 min to remove the nuclei, starch, plastids, and fat. The fat-free supernatant is then centrifuged at 10,000 g for 20 min to get the mitochondrial fraction, which contains the peroxisomes. The resuspended pellet is layered on a linear sucrose gradient (30%–60% sucrose) and centrifuged at 62,000 g for 4–5 hours. The peroxisomes show up as a very dense (1.25 g/mL) fraction (Vigil, 1983), which indicates that they are high in protein.

The peroxisomes can be further fractionated by osmotic shock. The membrane can be separated from the matrix by placing the peroxisomes into a dilute buffer and then recentrifuging them at 100,000 g for 60 min to pellet the membranes. The matrix proteins remain in the supernatant.

Peroxisomes have also been purified using free-flow electrophoresis (Eubel et al., 2008).

5.3 COMPOSITION OF PEROXISOMES

The proteins of the peroxisome have been named peroxins (Pool et al., 1998a,b; Tugal et al., 1999; Lopez-Huertas et al., 1999; Nito et al., 2007). Almost all peroxisomes contain catalase, but depending on their function, they have a variety of oxidases, which are discussed in the next

Plant Cell Biology. https://doi.org/10.1016/B978-0-12-814371-1.00005-9

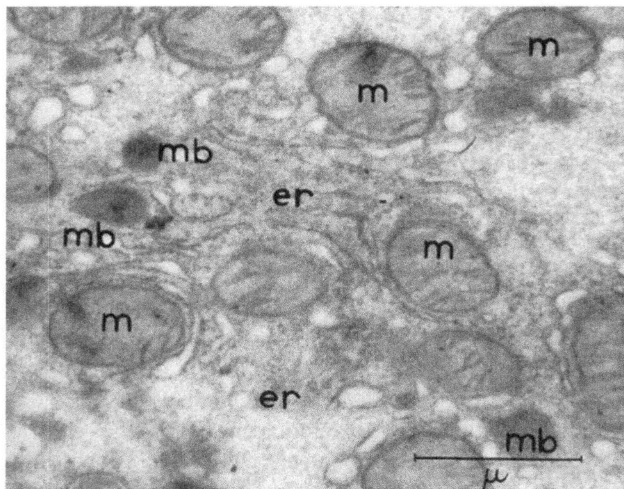

FIGURE 5.1 Microbodies (mb) in the cytoplasm of a rat liver cell. *er*, ergastoplasm; *m*, mitochondria. *From Rouiller and Bernhard*, 1956. J. Biophys. Biochem. Cytol. 2 (no. 4) Suppl. 355.

TABLE 5.1 Lipid Composition of the Castor Bean Peroxisomal Membrane

(mol%) of Lipid Phosphorus in Membrane	
Phospholipid	
Phosphatidylcholine	51.4
Phosphatidylethanolamine	27.2
Phosphatidylinositol	9.0
Phosphatidylserine	1.5
Phosphatidylglycerol	2.7
Cardiolipin	2.3
Other	2.0

From Donaldson, R.P., Beevers, H., 1977. Lipid composition of organelles from germinating castor bean endosperm. Plant Physiol. 59, 259–263.

section. Proteomic analysis of the matrix suggests that there are approximately 100 proteins (Kaur and Hu, 2011; Reumann, 2011; Hu et al., 2012). Porins have been identified as part of the peroxisomal membrane (Reumann et al., 1995, 1998; Corpas et al., 2000).

Although the phospholipid content of peroxisomes is particularly high, the lipid composition is similar to that of the endoplasmic reticulum (ER) and plasma membrane (Donaldson and Beevers, 1977; Donaldson et al., 1972, 1981; Chapman and Trelease, 1991a). Again, the similar composition reflects similar function, and the rationale for the differences is not yet known. The lipid compositions of peroxisomes are given in Table 5.1.

5.4 FUNCTION OF PEROXISOMES

Peroxisomes participate in a diverse set of biochemical reactions, including H_2O_2-based respiration, the β-oxidation of fatty acyl chains, the initial reactions in ether glycerolipid biosynthesis, cholesterol and dolichol synthesis, the glyoxylate cycle, photorespiration, jasmonic acid synthesis, alcohol oxidation, transaminations, purine and polyamine catabolism, ureide anabolism, nitric oxide synthesis, and the formation of glycine betaine, an osmoprotectant (Frederick et al., 1975; Tolbert, 1981; Huang et al., 1983; Masters and Crane, 1995; Corpas et al., 2001, 2004; Minorsky, 2002; Emanuelsson et al., 2003; Lin et al., 2004; Reumann, 2004; Reumann et al., 2004; Travis, 2004; del Rio, 2011; Rőszer, 2012). Interestingly, the β-oxidation of fatty acyl chains was first found to occur in the peroxisomes of castor bean cells (Cooper and Beevers, 1969) and only later found in the peroxisomes of animal cells (Lazarow and de Duve, 1976). Two of the pathways, including the β-oxidation and the glycolate pathways, are discussed in the following sections to demonstrate where marker enzymes fit into the picture and to see how the various organelles cooperate in the realization of complete biochemical pathways. I also show that duplication or functional redundancy exists in cells in that many of the same enzymes (e.g., malate dehydrogenase and citrate synthase) exist in more than one organelle.

5.4.1 β-Oxidation

In general, β-oxidation is involved in the formation of sucrose from fatty acyl chains and is very important during the germination of oil-rich seeds and spores (Kornberg and Beevers, 1957a,b; Kornberg and Madsen, 1957; DeMaggio et al., 1980; Hayashi et al., 2001; Kresge et al., 2010) and in heterotrophic cells living on a lipid food source (Binns et al., 2006). In these cases, the peroxisomes are found in close proximity with lipid bodies and mitochondria (Fig. 5.2). Peroxisomes can and do move in cells, apparently to where they need to be. In animal cells, they tend to move along microtubules (Rapp et al., 1996), whereas in plant cells, they tend to move along microfilaments (Collings et al., 2002; Mano et al., 2002; Mathur et al., 2002; Jedd and Chua, 2002; Sparkes, 2010). Catalase, isocitrate lyase (Harrop and Kornberg, 1966), and malate synthase are some of the enzymes that function in β-oxidation and are used as marker enzymes for peroxisomes.

Peroxisomes in fatty cells contain lipases that break down the stored lipids into their constitutive fatty acyl chains. Fatty acyl CoA enters peroxisomes by way of an ABC (ATP-binding cassette) transport protein (Zolman et al., 2001; Footitt et al., 2002; Hayashi et al., 2002; Theodoulou et al., 2005). The fatty acyl chains then undergo β-oxidation, a series of reactions that lead to the breakdown of long fatty acyl chains into many acetyl-CoA

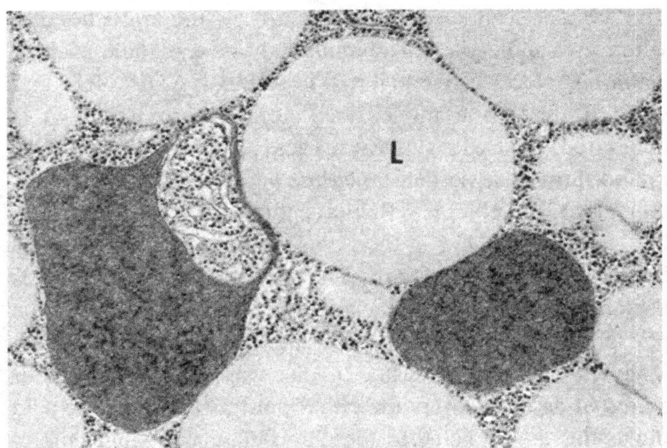

FIGURE 5.2 Peroxisome (glyoxysome) next to lipid bodies (L) in a tomato cotyledon cell. ×29,000. Notice how the peroxisome on the left encloses a mitochondrion. *From Frederick, S.E., Gruber, P.J., Newcomb, E.H., 1975. Plant microbodies, Protoplasma 84, 1–29.*

molecules (Fig. 5.3). CoA is involved in the activation and transfer of acetate groups between molecules (Lipmann, 1971). The enzymes involved in β-oxidation include fatty acyl–CoA synthetase and fatty acyl–CoA oxidase. The fatty acyl–CoA oxidase generates hydrogen peroxide. The hydrogen peroxide would be deadly to the cell, but it is broken down by catalase to $\frac{1}{2}O_2 + H_2O$, both of which are nontoxic. Subsequently, the acetyl-CoA molecules formed during β-oxidation are joined by malate synthase to glyoxylic acid molecules to make malic acid. This is one way acetyl-CoA enters the glyoxylic acid cycle.

The malic acid is oxidized to oxaloacetic acid by malate dehydrogenase. Another acetyl-CoA is attached to the oxaloacetic acid by citrate synthase to form citric acid. The citric acid is converted to isocitric acid by aconitase. Then isocitrate lyase splits the isocitric acid into glyoxylic acid, which is used to recharge the glyoxylic acid cycle, and succinic acid, which leaves the peroxisome. In essence, the glyoxylic acid cycle in the peroxisome converts two acetic acid molecules into one succinic acid molecule.

The succinic acid moves from the peroxisome to the mitochondrion, where it is converted to malic acid. The malic acid may exit the mitochondrion on the same transporter that allows the entrance of succinic acid so that a 1:1 stoichiometry is maintained. The malic acid that leaves the mitochondrion is converted to oxaloacetic acid in the cytosol by malate dehydrogenase. Then the oxaloacetic acid is converted into phosphoenolpyruvic acid by phosphoenolpyruvate carboxykinase. In seeds, the phosphoenolpyruvic acid is converted to glucose via the gluconeogenesis pathway. Because glucose is a reducing sugar, which makes it highly reactive, it must be converted to a relatively nonreactive form such as sucrose or starch. The hydrophilic sucrose molecules are then translocated out of the seed where they nourish the growing regions of the plant.

Hans Kornberg originally discovered the glyoxylic acid cycle when he investigated how certain bacteria can grow on acetate as the sole carbon source (Kornberg and Krebs, 1957). Almost immediately Hans Kornberg and Harry Beevers together discovered that the glyoxylic acid cycle was also involved in β-oxidation in plants. Hans Kornberg (2003) reminisced, "*There are, all too rarely, occasions when a person [Kornberg] with a solution but no idea to which problem it might apply meets a person [Beevers] with a problem but no knowledge of the solution. This happy conjunction occurred in 1957 when Harry Beevers was spending a sabbatical in Oxford and told Krebs of his interest in the possible mechanisms that might account for the almost stoichiometric conversion of oil to storage carbohydrates in germinating castor bean seeds. Harry and I met and immediately recognized that the glyoxylate cycle might well provide that route. The next day, Harry appeared with castor beans; we disrupted them in a blender and by late afternoon had demonstrated the presence of a highly active isocitrate lyase (as isocitritase is now more properly called). Incubation of the cell-free extract with [^{14}C]acetate, ATP, coenzyme A, and isocitrate yielded malate as the sole labeled product; clearly, the sought for mechanism had been found. This laid the foundation for Harry's later distinguished work that identified a novel intracellular particle (the "glyoxysome") as the location of these enzymes in plants.*"

5.4.2 Photorespiration

John Decker (1955, 1959) discovered a light-dependent respiratory pathway in plants that came to be known as *photorespiration*. His discovery was not widely accepted because recognition of photorespiration would mean that scientists who studied photosynthesis would have to reinterpret much of their data (Zelitch, 2001). In fact, Decker's name is barely mentioned in the photorespiration literature (Goldsworthy, 1976). Photorespiration turned out to be real and it involves the capture of glycolic acid produced in the chloroplast (Ludwig and Canvin, 1971; Tolbert, 1971; Chollet and Ogren, 1975; Zelitch, 1964, 1971, 1975, 2001; Ogren, 2003). Glycolic acid is formed from 2-phosphoglycolic acid, one of the products formed when ribulose bisphosphate carboxylase–oxygenase (i.e., Rubisco) catalyzes the addition of oxygen instead of carbon dioxide to ribulose bisphosphate (see Chapter 13). Thus, photorespiration is the cause of the "Warburg effect," which is the apparent inhibition of photosynthesis by oxygen discovered by Otto Warburg (1920). In green leaves, the enzymes involved in photorespiration are found in the peroxisomes and the peroxisomes are found in close proximity to the chloroplasts and mitochondria (Frederick and Newcomb, 1969; Oikawa et al., 2015; see Fig. 5.4). Catalase, serine:glyoxylate aminotransferase, and hydroxypyruvate reductase are three important enzymes involved in photorespiration (Fig. 5.5).

The 2-phosphoglycolic acid formed by Rubisco is dephosphorylated to glycolic acid. The glycolic acid diffuses to the peroxisome where it is converted to glyoxylic acid and then to glycine. The enzyme that causes the formation of glyoxylic acid from glycolic acid is glycolate oxidase. This reaction also produces hydrogen peroxide, which is then broken down by catalase. The glyoxylic acid is converted to glycine by glutamate:glyoxylate aminotransferase. An aminotransferase is an enzyme that transfers an amino group from an amino acid to an α-keto acid. Peroxisomes contain aminotransferases that convert glyoxylate to glycine. At the same time, some of the aminotransferases convert glutamate to α-ketoglutarate, serine to hydroxypyruvate, aspartate to oxaloacetate, or alanine to pyruvate. The glycine formed by the aminotransferase enters the cytosol where it can participate in protein synthesis. The glycine can also be taken up by the mitochondria where two molecules of glycine and one molecule each of water and NAD^+ are converted to serine, NADH, NH_3, and CO_2. If the serine moves back to the peroxisome, it is converted to hydroxypyruvic acid by serine:glyoxylate aminotransferase (at the same time another glyoxylic acid is converted to glycine; Raghavendra et al., 1998). The hydroxypyruvic acid is then converted to glyceric acid by hydroxypyruvate reductase. The glyceric acid then moves back to the chloroplast, where it is phosphorylated. The phosphorylated form then enters the Calvin cycle to participate in starch metabolism. In this way, three out of four atoms of carbon lost by the chloroplast as two glycolic acid molecules are recycled to the chloroplast as a single molecule of the three-carbon glyceric acid (Berry et al., 1978).

To deduce the pathways involved in β-oxidation and photorespiration, experiments using radioactive carbon compounds are performed in vivo and in vitro to determine

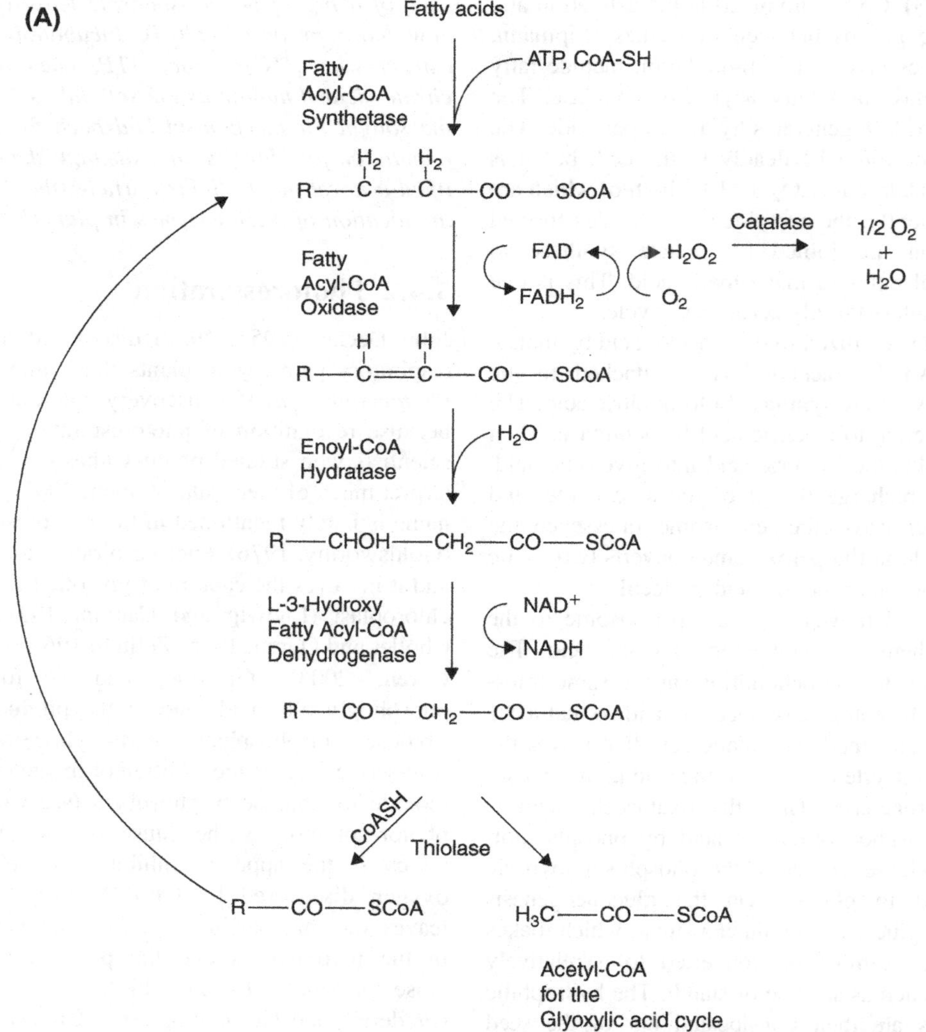

FIGURE 5.3 Pathway for the breakdown of fatty acyl chains: (A) β-oxidation of fatty acyl chains.

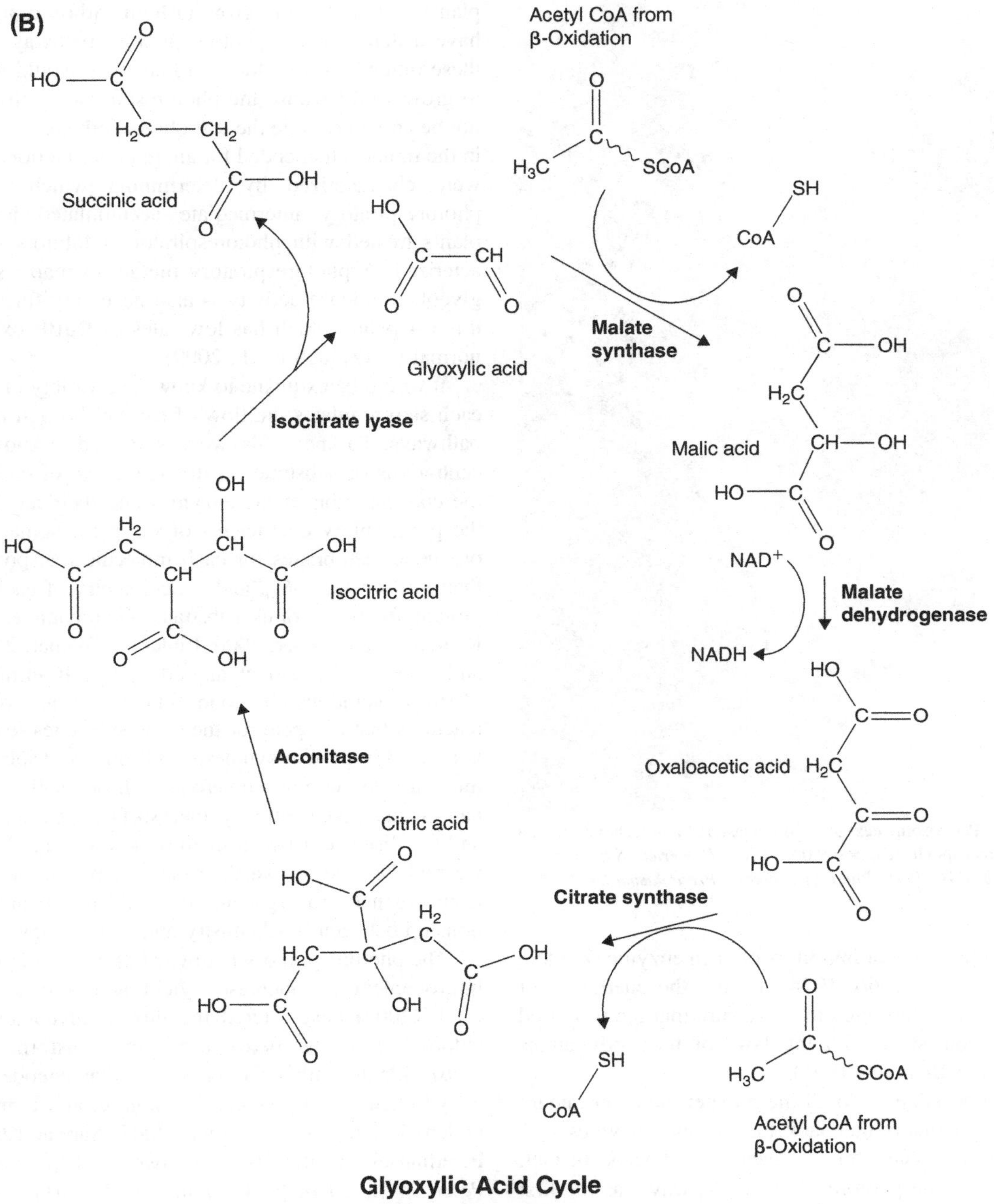

(B)

Glyoxylic Acid Cycle

FIGURE 5.3 cont'd (B) the glyoxylic acid cycle.

the temporal sequence in which intermediates become labeled. Radioactive labeling experiments also provide information about the rates that carbon moves through the pathway. These rates are then compared with the rates of each enzyme reaction estimated from the maximal velocity of that reaction (v_{max}) and the concentration of substrate [S] that is needed to achieve the half-maximal rate (K_m; Zelitch and Ochoa, 1953; Zelitch, 1953, 1955). The rate or velocity of an enzyme reaction (v) is estimated with the following equation that will be derived in Chapter 12.

$$v = \frac{v_{max}}{\frac{k_m}{[S]} + 1} \tag{5.1}$$

To estimate the extent in which each enzyme reaction is rate-limiting, the flux through the pathway is tested in the

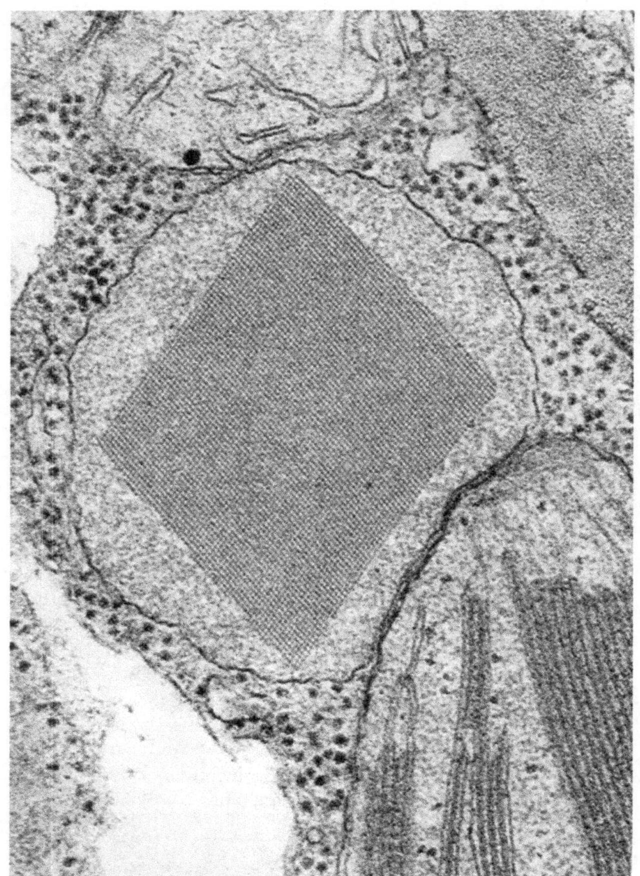

FIGURE 5.4 Peroxisome next to a chloroplast and a mitochondrion in a tobacco leaf mesophyll cell. ×109,000. *From Frederick, S.E., Gruber, P.J., Newcomb, E.H., 1975. Plant microbodies, Protoplasma 84, 1–29.*

absence and presence of inhibitors of each enzyme (Zelitch, 1957, 1959, 1965, 1966, 1974). Lastly, the energetics of each reaction is determined to make sure that the proposed pathway is consistent with the laws of thermodynamics (Anderson and Beardall, 1991).

It is also possible to do all these experiments in mutant or transformed plants that lack the normal enzymes presumed to be necessary for the reaction pathways. In fact, confirmation of the photorespiratory pathway was the first use of *Arabidopsis* as a model system (Somerville and Ogren, 1979, 1980, 1982; Somerville, 1986, 2001; Ogren, 2003). The confirmation was accomplished by screening chemically induced mutants for their ability to grow in 1% carbon dioxide but not in natural air, which contains 0.03% carbon dioxide. Because the oxygenase activity of Rubisco would be low in a high—carbon dioxide environment, the activity of the photorespiratory pathway would also be low and consequently the growth of photorespiratory mutants would not differ significantly compared with the growth of wild type plants under high carbon dioxide but would differ under low carbon dioxide conditions. Presumably, the plants that could only grow in high carbon dioxide would have a defect in the photorespiratory pathway, and when these mutants were returned to air, they would not be able to grow well because the photorespiratory pathway would not be able to recycle the lost photosynthetic carbon. Genes in the mutants that coded for altered nonfunctional enzymes were characterized by determining which ^{14}C-labeled photorespiratory intermediate accumulated just as the plants treated with photorespiratory inhibitors were characterized. A photorespiratory mutant in maize shows that glycolate oxidase activity is also necessary for survival of this C4 plant, which has low rates of RuBP oxidation, in normal air (Zelitch et al., 2009).

It would be exquisite to know, in a variety of cells, how each step regulates the flow of carbon through the various pathways. To know this, we would need to know the concentrations of substrates at the active site of each enzyme, the concentration of the enzymes and their regulators, and the permeability coefficients of the peroxisomal and other organelle membranes for each molecule transported across them, or the v_{max}, K_m, and stoichiometry of each transport protein for the various substrates (Reumann et al., 1998; Reumann and Weber, 2006; Linka and Weber, 2010; Linka and Esser, 2012; Hu et al., 2012; van Roermund et al., 2016). It would also be good to know the rates of the other reactions that compete for the same substrates (e.g., glycine is utilized in protein synthesis and in photorespiration). We must also know how numerous each organelle that participates is in a given pathway, their surface areas and volumes, and the distance between them (Oikawa et al., 2015). Then we will be able to visualize the flux of carbon molecules from organelle to organelle and truly understand the relationship between biochemistry and cell biology.

The photorespiratory pathway has been a target for crop improvement for increased yield and biomass. Kebeish et al. (2007) have altered the flux of glycolate from the chloroplast to the peroxisomes by transforming *Arabidopsis* plants with bacterial genes that encode glycolate dehydrogenase, glyoxylate carboligase, and tartronic semialdehyde reductase (Leegood, 2007; Sarwar Khan, 2007; Peterhänsel et al., 2008; Bauwe, 2010; Maurino and Peterhansel, 2010; de Carvalho et al., 2011; Peterhänsel and Maurino, 2011; Hagemann and Bauwe, 2016). The inserted genes also have a chloroplast-targeting sequence so that the enzymes they encode end up in the stroma of the chloroplast. In the transformed plants, the flux of glycolate from the chloroplast to the peroxisome is reduced and the chloroplastic glycolate is converted directly to glycerate. This results in faster-growing plants that produce more biomass and soluble sugars. When applied to crop plants, such technology could increase the yield of plants without diminishing their taste. While the photorespiratory pathway is usually considered to be wasteful of fixed carbon and thus a detriment to plants, it is also beneficial in promoting

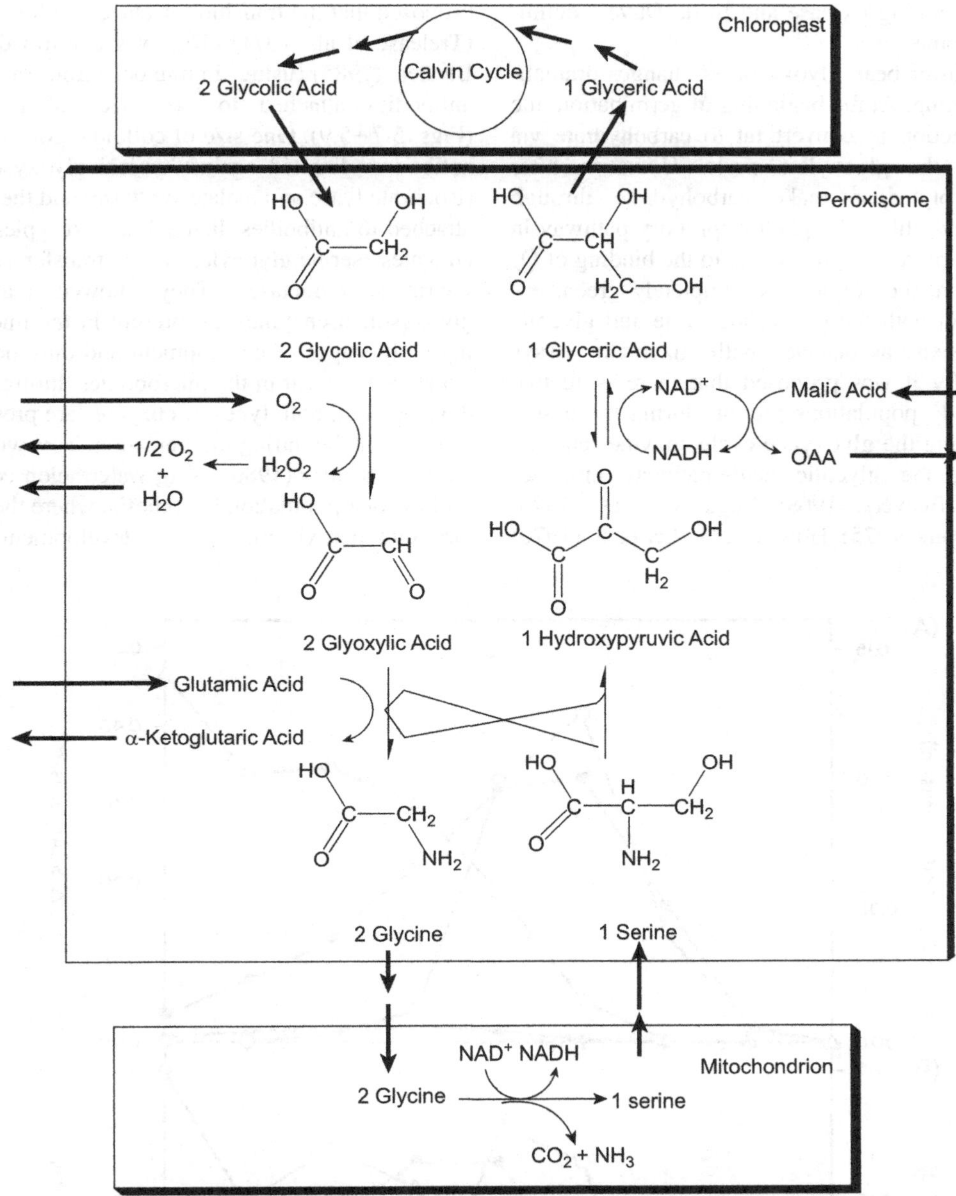

FIGURE 5.5 The glycolic acid pathway. Glycolic acid is converted to glyoxylic acid by glycollate oxidase. The breakdown of H_2O_2 to ½O_2 and H_2O is catalyzed by catalase. Glyoxylic acid is converted to glycine by glutamate:glyoxylate aminotransferase. The conversion of serine to hydroxypyruvic acid is catalyzed by serine:glyoxylate aminotransferase. The conversion of hydroxypyruvic acid to glyceric acid is catalyzed by hydroxypyruvate reductase.

nitrate assimilation (Rachmilevitch et al., 2004; Bloom et al., 2010), in reducing photoinhibition of photosystem II (see Chapter 13; Somerville and Ogren, 1979; Kozaki and Takeba, 1996; Takahashi et al., 2007; Cui et al., 2016), and in maintaining the dynamic opening response of guard cells to reduced CO_2 (Eisenhut et al., 2017). By altering the expression of the genes introduced by Kebeish et al. (2007), one could titrate the flux of glycolate from the chloroplast to the peroxisome and thereby possibly maximize the benefits and minimize the demerits of the photorespiratory pathway in the peroxisomes.

5.5 RELATIONSHIP BETWEEN GLYOXYSOMES AND PEROXISOMES

The organelle to which Rhodin gave the name *microbody* has been called *cytosome, phragmosome, spherosome*, and *unidentified cytoplasmic organelle* by morphologists (Huang et al., 1983). When microbodies were first isolated from castor beans, they were given the name *glyoxysomes* because they contained the enzymes of the glyoxylate cycle. However, later it was found that they contain catalase and an H_2O_2-generating oxidase and fit de Duve's definition of a peroxisome.

The fate of castor bean glyoxysomes changes dramatically during greening. At the beginning of germination, the glyoxysomes function to convert fat to carbohydrate via β-oxidation and the glyoxylate cycle. However, after greening, the cotyledons make carbohydrate through photosynthesis and utilize the photorespiratory pathway in peroxisomes to capture carbon lost due to the binding of O_2 to Rubisco. Before the cotyledons completely green, enzymes involved in both the glyoxylate cycle and glycolic acid pathway coexist as detected with enzymatic assays (Fig. 5.6). Initially it was proposed that there were two separate microbody populations present during the transition: one containing the glyoxylate-cycle enzymes and the other containing the glycolic acid–pathway enzymes (McGregor and Beevers, 1969; Kagawa et al., 1973; Kagawa and Beevers, 1975). However, Gruber et al. (1970) proposed that the transition occurred within one population (Trelease et al., 1971). This was confirmed by Titus and Becker (1985) using immunoelectron microscopy with antibodies attached to two sizes of protein A gold (Figs. 5.7–5.9). One size of colloidal gold was attached to antibodies directed against typical glyoxysomal enzymes (isocitrate lyase and malate synthase) and the other size was attached to antibodies directed against typical peroxisomal enzymes (serine:glyoxylate aminotransferase and hydroxypyruvate reductase). They showed that while only glyoxysomal enzymes are present in the microbodies during early stages of development and only peroxisomal enzymes are present in the microbodies during later stages of development, both types of enzymes are present within the same organelle during the transition. This work and similar work by Sautter (1986) using watermelon cotyledons support the one-population hypothesis, where the glyoxysomes turn into peroxisomes during development and they are

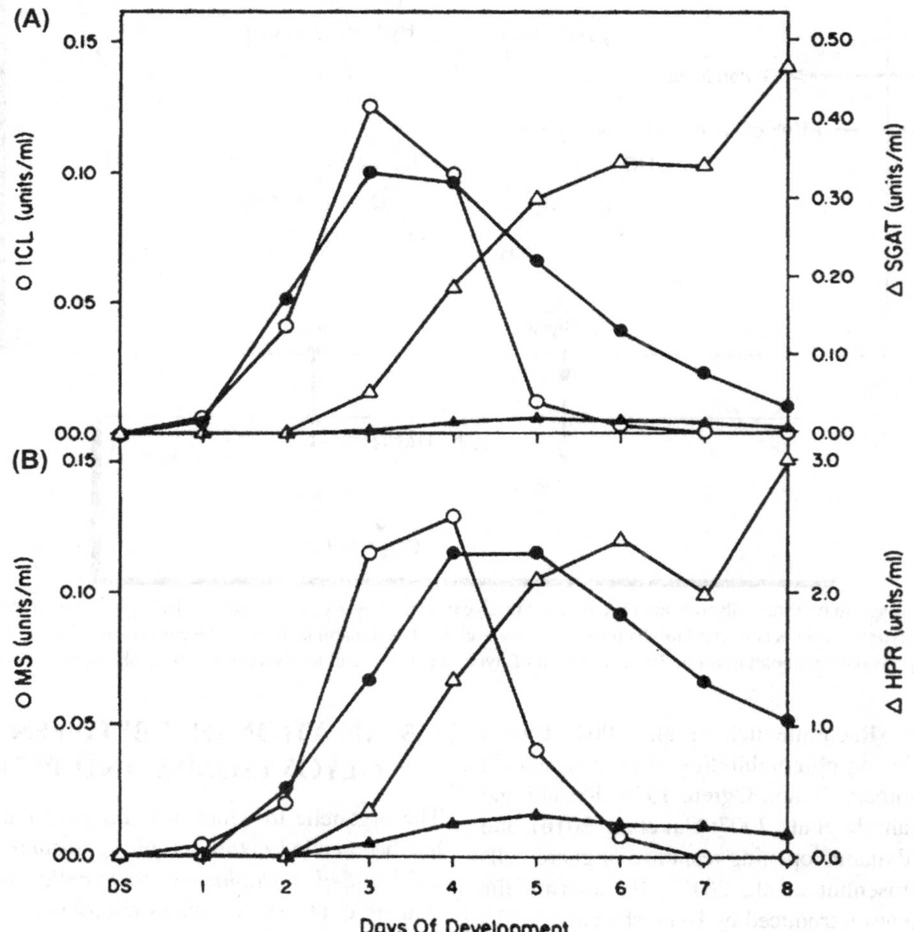

FIGURE 5.6 The activities of glyoxysomal and peroxisomal enzymes in cotyledons of light-grown (empty symbols) and dark-grown (filled symbols) plants. Note that the activities of serine:glyoxylate aminotransferase (SGAT) and hydroxypyruvate reductase (HPR) are light dependent. *From Titus, D.E., Becker, W.M., 1985. Investigation of the glyoxysome- peroxisome transition in germinating cucumber cotyledons using double-label immunoelectron microscopy. J. Cell Biol. 101, 1288–1299.*

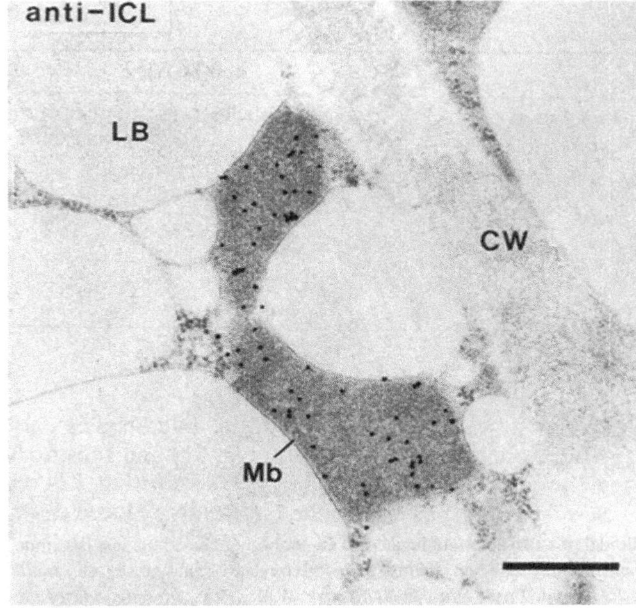

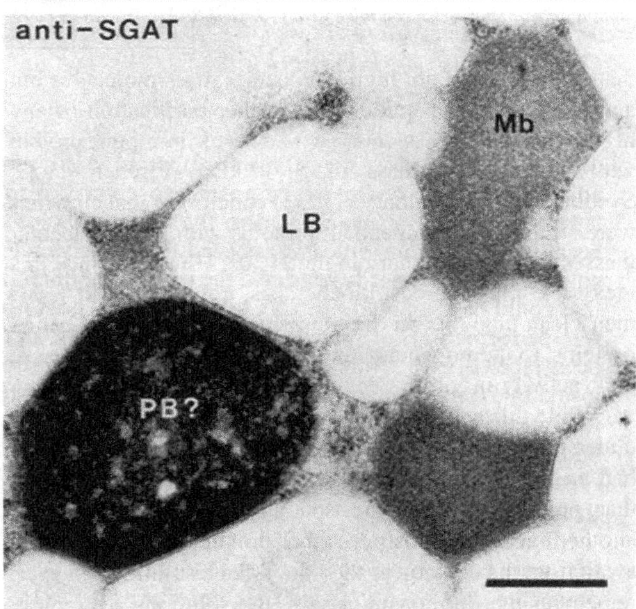

FIGURE 5.7 Cell of a 2-day-old cucumber cotyledon stained with colloidal gold particles conjugated to an antibody directed against isocitrate lyase (ICL) but not by antibodies directed against serine:glyoxylate. *CW*, Cell Wall; *LB*, Lipid body; *Mb*, Microbody; *PB*, Protein body.

really one and the same organelle that can have two specialized functions.

Interestingly, during leaf senescence that accompanies the beautiful autumn colors (Matzke, 1942; Thimann, 1950), peroxisomes transform back to glyoxysomes; that is, green leaves have peroxisomes with high catalase and hydroxypyruvate activities, whereas senescent leaves have peroxisomes with high malate synthase and isocitrate lyase activities (Nishimura et al., 1986, 1993). The peroxisomes in senescing cells may be responsible for recycling the fatty

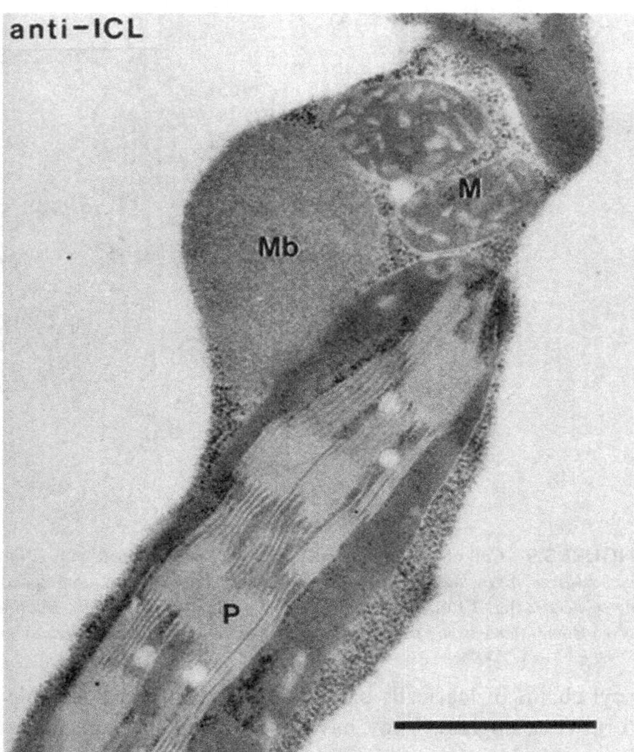

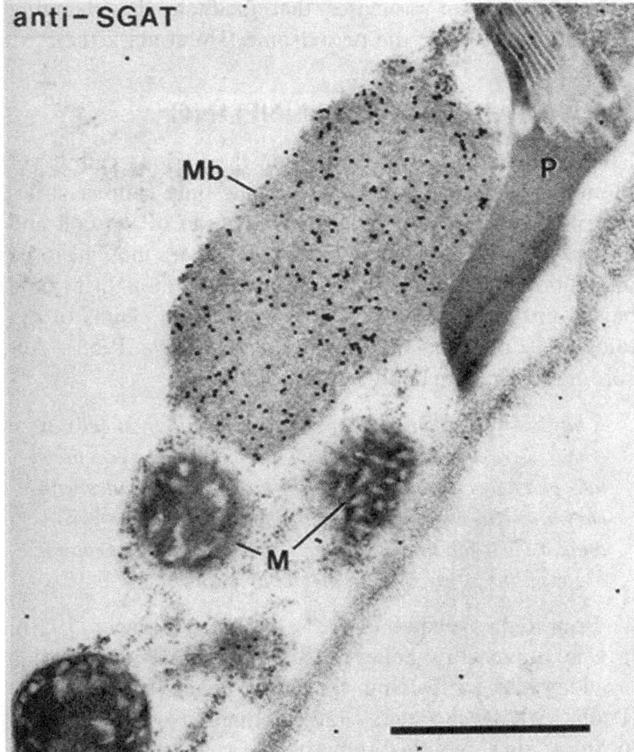

FIGURE 5.8 Cell of an 8-day-old cucumber cotyledon stained with colloidal gold particles conjugated to an antibody directed against serine:glyoxylate aminotransferase (SGAT) but not by antibodies directed against isocitrate lyase (ICL). Bar, 1 μm. *ICL*, Isocitrate lyase; *M*, Mitochondrion; *Mb*, Microbody; *P*, Plastid; *SGAT*, Serine:glyoxylate aminotransferase. *From Titus and Becker, 1985*

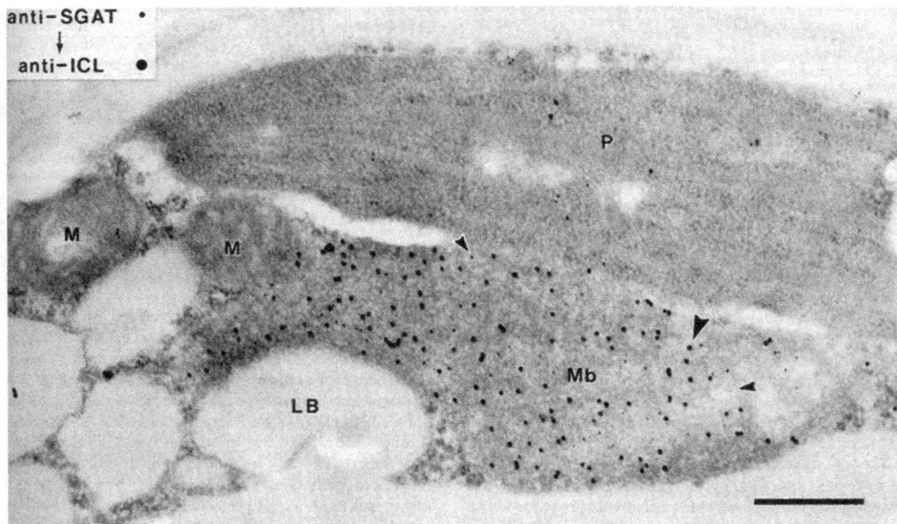

FIGURE 5.9 Cell of a 4-day-old cucumber cotyledon stained with 20-nm colloidal gold particles conjugated to an antibody directed against isocitrate lyase (*large arrowhead*) and 10-nm colloidal gold particles conjugated to an antibody directed against serine:glyoxylate aminotransferase (*small arrowhead*). Bar, 1 μm. *LB*, Lipid Body; *M*, Mitochondrion; *Mb*, Microbody; *P*, Plastid. *From Titus, D.E., Becker, W.M., 1985. Investigation of the glyoxysome- peroxisome transition in germinating cucumber cotyledons using double-label immunoelectron microscopy. J. Cell Biol. 101, 1288–1299.*

acyl chains of leaf cells back to the plant in translocatable form (i.e., sugars). They have been given the name *gerontosomes* by Vincentini and Matile (1993). Little is known about the role of proteases that facilitate the changing biochemical roles of the peroxisome (Hu et al., 2012).

5.6 METABOLITE CHANNELING

How do metabolites move through the various pathways? Membranes are usually considered the only cellular structure involved in the compartmentalization of the cell and the enzymes enclosed by the membranes are thought to be in solution. That is, the cell was typically thought to be a bag of enzymes (Asimov, 1954) despite the claims of cytologists (Ephrussi, 1953; Porter and Moses, 1958). According to Hrazdina and Jensen (1992).

> *It has been both convenient and productive to treat cells as if they were simply a bag full of enzymes where reactions take place by chance encounter of substrate molecules with enzymes. However, improved methodology is now generating a basis for suggesting the existence of a strict spatial organization of enzymes in metabolic pathways.*

Prior to the isolation of urease by James Sumner (1926), it was universally believed that proteinaceous colloids provided the scaffolding for nonproteinaceous enzymes (Pauli, 1907; Willstätter, 1927). Perhaps it was reasonable to think that proteins only acted as structural entities because at that time they were known to be the major constituent of hair, nails, horns, and hooves (Astbury and Street, 1932; Astbury, 1933; Astbury and Woods, 1934; Astbury and Sisson, 1935). Furthermore, it was believed that proteins were not high–molecular mass molecules but aggregates of polypeptides, and further purification of any high–molecular mass entity would result in a pure protein with a molecular mass of 5–17 kDa (Fischer, 1923; Svedberg, 1937). Sumner's (1933) conclusion that enzymes were structurally independent proteins with high molecular masses was revolutionary in the 1930s. However, once this idea took hold, the best biochemists, not wanting to waste their clean thoughts on dirty enzymes, customarily purified proteins to homogeneity and then began to characterize their activity in vitro.

While this approach has contributed substantially to our understanding of cell metabolism, unfortunately, it has also had an undesirable side effect. That is, it led to the belief that proteins in vivo are spatially independent of one another, and the substrates and products diffuse to and away from the enzyme as they do in the test tube. However, remember that in enzyme assays, the solutions are rapidly stirred, which causes convection so that the substrate concentration around the active site does not become depleted due to the limited speed of diffusion. Thus, while it was accepted that proteins could have a primary, secondary, tertiary, and quaternary structure, thoughts of a quintinary structure, where successive proteins in a complex form a real unit, were forbidden (Ovádi, 1991; Kühn-Velten, 1993; Mathews, 1993; Cascante et al., 1994). Thus, any observed protein–protein interactions that could potentially channel the product of one enzyme to the next enzyme in the pathway were considered an artifact of isolation. However, to paraphrase George Baitsell (1940), may it not be possible that the matrix of the peroxisome is a protein crystal in which a number of proteins are solidified into an ultramicroscopic pattern to perform the function of the organelle? Even in the inorganic world, there are many familiar examples of supramolecular structures, where the whole has different and emergent properties than the parts. I have already discussed T.H. Huxley's comments on the properties of water, which differ markedly from those of its constituents, oxygen, and hydrogen. Irving Langmuir (1916) wrote, "it had been taken for granted that crystals were built up of molecules. But ... it is clear that in crystals of this type [NaCl] the identity of the molecules is wholly lost, except in so far as we may look upon the whole crystal as composing a *single molecule*."

Recent work suggests that in some cases, the protein matrix in the cell (e.g., cytoskeleton) or an organelle (e.g., peroxisomal matrix) may provide a structure on which some enzymes reside, so that the enzymes form a highly efficient supramolecular structure in which the product of one enzyme is immediately channeled to another enzyme of which it is a substrate (Heupel and Heldt, 1994; Reumann et al., 1994; Reumann, 2000). It is also possible that the structure is composed either entirely or in part by the enzymes themselves.

Heupel et al. (1991) isolated peroxisomes from spinach leaves using Percoll density gradient centrifugation. They then assayed six peroxisomal enzymes for activity in the presence and absence of detergent (Triton X-100). In the case of malate dehydrogenase, hydroxypyruvate reductase, serine:glyoxylate aminotransferase, catalase, and glutamate:glyoxylate aminotransferase, the enzyme activity of each enzyme was higher after Triton X-100 treatment than before (Fig. 5.10). Significantly, the activity of the first enzyme in the pathway, glycolate oxidase, shows no increase in activity following detergent treatment, indicating that its active site is always exposed to its substrate. The increase in enzyme activity following detergent treatment is known as the latency of the enzyme. Because the detergent permeabilizes membranes so that the substrates can enter the peroxisomes, the barrier to the diffusion of the substrate is usually thought to be due exclusively to the membrane.

Heupel et al. (1991) broke the membrane by osmotically shocking the peroxisomes. Electron microscopy showed that the osmotically shocked peroxisomes remained compact but no longer had a continuous boundary membrane. Heupel et al. (1991) found that even after the membrane was lysed, the enzymes still showed increased activity when treated with Triton X-100, indicating that the diffusion barrier may not be due exclusively to the membrane but may also be due to the specific positioning of enzymes into a complex. They concluded that Triton X-100 increased the availability of the active sites of the enzymes to the substrates by solubilizing the enzymes from the complex. They believe that in vivo, this multiprotein organization allows the movement of the

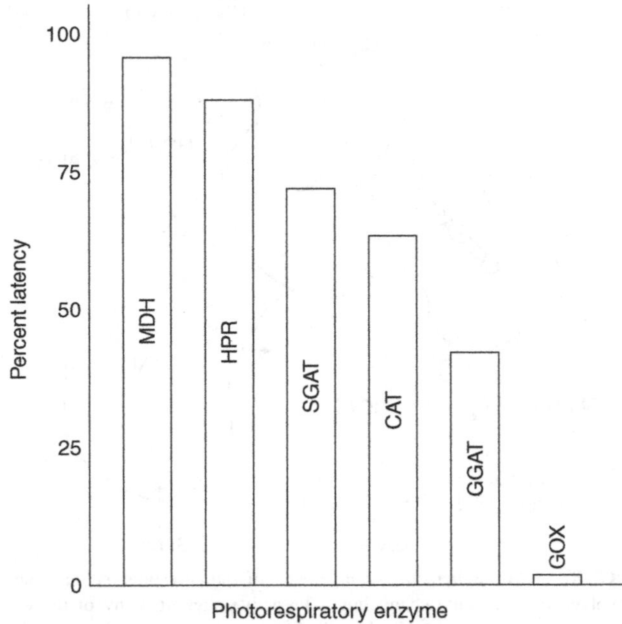

FIGURE 5.10 The latency of various photorespiratory enzymes. *Data from Heupel, R., Markgraf, T., Robinson, D.G., Heldt, H.W., 1991. Compartmentation studies on spinach leaf peroxisomes. Evidence for channeling of photorespiratory metabolites in peroxisomes devoid of intact boundary membrane. Plant Physiol. 96, 971–979.*

product of one enzyme directly to the active site of the next enzyme of which it is a substrate and prevents competing substrates from getting to the active site (Fig. 5.11). The ordered arrangement of enzymes leads to a process known as metabolite channeling.

Now let us look at the whole pathway. Heupel and Heldt (1994) measured the synthesis of glyceric acid from glycolic acid, glutamic acid, serine, and malic acid and found that it is inhibited in detergent-treated peroxisomes compared with intact or osmotically shocked peroxisomes (Fig. 5.12). Likewise, the synthesis of glycine from glycolic acid, glutamic acid, and serine in detergent-treated peroxisomes occurs at a much reduced rate compared to either intact or osmotically shocked peroxisomes. They concluded that detergent treatment solubilized the enzymes so that the substrates had to diffuse long distances to the next enzyme in the pathway and there was no longer metabolite channeling. According to Einstein's (1906) random walk equation

$$t = \frac{x^2}{2D} \tag{5.2}$$

the time (t) it takes a molecule with a diffusion coefficient $D = [kT/(6\pi r_h \eta)]$ to diffuse from one place to another is proportional to the square of the distance (x). Thus, the rate of a diffusion-limited reaction will be 10,000 times faster if a substrate only has to diffuse 1 nm instead of 100 nm. If an enzymatic reaction is slow and not limited

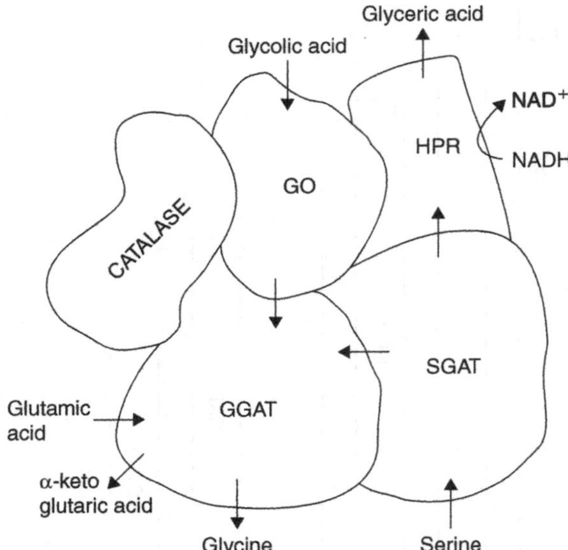

FIGURE 5.11 Schematic arrangement of photorespiratory enzymes involved in metabolite channeling. The active sites of many of the enzymes are blocked by protein–protein interactions. For example, glyoxylic acid can only get to the active site of GGAT by entering the complex at GO as glycolic acid, being converted to glyoxylic acid, and exiting GO as glyoxylic acid. In this way, glyoxylic acid is concentrated at the active site of GGAT, where its concentration will be close to the K_m of GGAT for glyoxylic acid. However, the average concentration of glyoxylic acid in the peroxisome will be low. Even with high concentrations of glyoxylic acid, the activity of GGAT will be low because the active site is not available. The active site becomes available following detergent treatment and the enzyme activity increases. *GGAT*, glutamate:glyoxylate aminotransferase; *GO*, glycolate oxidase; *SGAT*, serine:glyoxylate aminotransferase; *HPR*, hydroxypyruvate reductase.

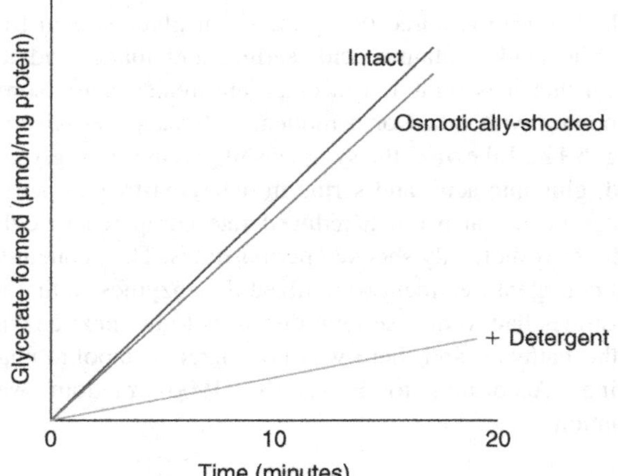

FIGURE 5.12 Time course of glycerate formation in intact, osmotically shocked, and detergent-treated peroxisomes.

by diffusion, then metabolite channeling will not be helpful. However, if the enzyme reactions are limited by diffusion, metabolite channeling can be important. Consequently, the reduced rate of the metabolic pathway in detergent-treated peroxisomes is strong evidence for the importance of metabolite channeling by multienzyme complexes.

Interestingly, in intact or osmotically shocked peroxisomes, the intermediate, glyoxylic acid, is barely detectable (<1 μM). However, in detergent-lysed peroxisomes, glyoxylic acid accumulates to high levels. This is because in detergent-lysed peroxisomes, the concentration of glyoxylic acid at the active site of the enzyme that converts glyoxylate to glycine (glutamate:glyoxylate aminotransferase) is too low to be transformed by the enzyme. For this enzyme to work, the concentration of glyoxylic acid must be approximately 150 μM at the active site. Presumably, in intact or osmotically shocked peroxisomes, the concentration of glyoxylic acid at the active site is approximately 150 μM, and the enzyme actively converts glyoxylic acid to glycine. We are beginning to understand the relationships between local concentrations and enzyme affinities so that we can interpret the measured fluxes through biochemical pathways at the level of molecular dimensions $(1-10\ \text{nm})^3$.

We can deduce the presence and/or contribution of metabolite channeling through multienzyme complexes based on the following criteria:

1. The flux through an entire pathway is faster when the enzymes are in a complex rather than when they are separate.
2. In multienzyme complexes that depend on metabolite channeling, the local concentration of an intermediate may be very high, whereas the overall concentration is low.
3. It should be possible to reconstitute a multienzyme complex from the component enzymes and regain the rapid flux due to metabolite channeling.

5.7 OTHER FUNCTIONS

The peroxisomes are multifunctional organelles that are capable of adapting to various cell types by adding or deleting enzymes involved in a variety of pathways (Baker and Graham, 2002). In animal cells, and perhaps some plant cells too, the peroxisomes are important in the catabolism of purines. The peroxisomes in nitrogen-fixing nodules may be specialized for ureide formation (Huang et al., 1983). In some fungi, they also participate in the biosynthesis of antibiotics. Using immunogold cytochemistry and cell fractionation, vandenBosch et al. (1992) have shown that the final enzyme involved in penicillin biosynthesis is localized in the peroxisomes of *Penicillium chrysogenum*. The peroxisomes of plants contain a Ca^{2+}-dependent nitric oxide synthase (Barroso et al., 1999; del Rio et al., 2002) and a sulfite oxidase (Eilers et al., 2001; Nakamura et al., 2002).

5.8 BIOGENESIS OF PEROXISOMES

When I was learning cell biology in the 1970s and 1980s, I was taught that peroxisomes are formed directly from the ER by a budding process. This was based on electron micrographs made by Novikoff and Shin (1964; see Fig. 5.13). They interpreted these micrographs to reveal connections between the peroxisomes and the ER. Their interpretation was based on the work of Higashi and Peters (1963), which showed that newly synthesized catalase is found in the ER fraction. Gonzalez (1982) and Gonzalez and Beevers (1976) later obtained similar results in plants. However, with the introduction of in vitro translation techniques, something seemed amiss with the interpretation that the peroxisomes formed from the budding of the ER (Beevers, 1979; Tolbert, 1981; Kindl, 1982a,b; Vigil, 1983; Trelease, 1984; Lazarow and Fujiki, 1985). That is, if the peroxisomal proteins were synthesized on the ER, they should have a signal peptide, and thus the protein formed in vitro in the absence of microsomes should have a greater molecular mass than those synthesized in vivo. However, it was found that peroxisomal enzymes, including isocitrate lyase, glycolate oxidase, bifunctional enoyl-CoA hydratase/β-hydroxyacyl-CoA dehydrogenase, and catalase, are synthesized in vitro in a cell-free system at the same size that they are found in vivo (Frevert et al., 1980; Yamaguchi and Nishimura, 1984).

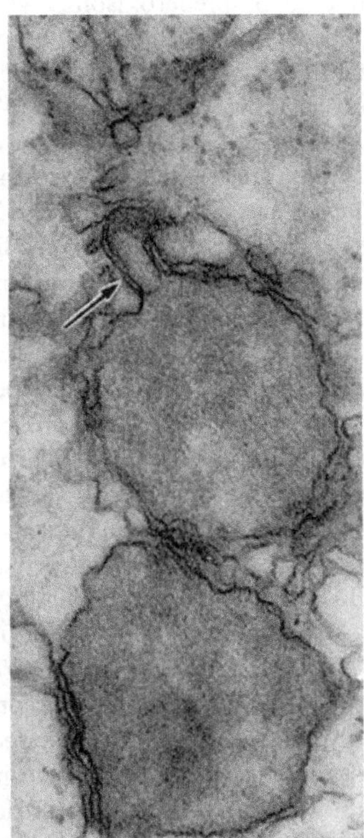

FIGURE 5.13 Microbody that appears to be continuous with the smooth endoplasmic reticulum (*arrow*). ×56,000. *From Novikoff, A.B., Shin, W.-Y., 1964. The endoplasmic reticulum in the Golgi zone and its relations to microbodies, Golgi apparatus and autophagic vacuoles in rat liver cells. Journal de Microscopie 3, 187–206.*

The majority of the peroxisomal peptides are produced on cytosolic ribosomes and lack ER-specific signal peptides (Walk and Hoch, 1978; Riezman et al., 1980; Kruse et al., 1981; Lord and Roberts, 1982; Gietl, 1990). Thus, how can the peroxisomes be produced by budding off of the ER? They cannot! Then how can we reinterpret the data that support the budding hypothesis (Hu et al., 2012; Cross et al., 2016; Reumann and Bartel, 2016)?

First, perhaps as a consequence of their ability to form multienzyme complexes, peroxisomal proteins form aggregates that artifactually cosediment with ER membranes (Kruse and Kindl, 1983). Second, serial sections of cells show that peroxisomes actually exist as a "peroxisomal reticulum" (Gorgas, 1984; Ferreira et al., 1989; see Fig. 5.14) and the so-called attachments to the ER are interconnections between the peroxisomes themselves. At this point, it is worthwhile to remind ourselves that a single electron micrograph provides a static two-dimensional view of a dynamic three-dimensional cell. Thus, if we want to make three-dimensional interpretations, we should reconstruct images from serial sections, and if we wish to make dynamic interpretations, we should fix many cells at various points of time. The time interval between when we fix each sample should be at least half as long as the time resolution we would like to achieve in understanding the biological process in question. This sampling theorem is known as the Nyquist Theorem (Nyquist, 1924, 1928; Horowitz and Hill, 1989).

It is important to realize that while it is convenient to represent a three-dimensional structure in two dimensions,

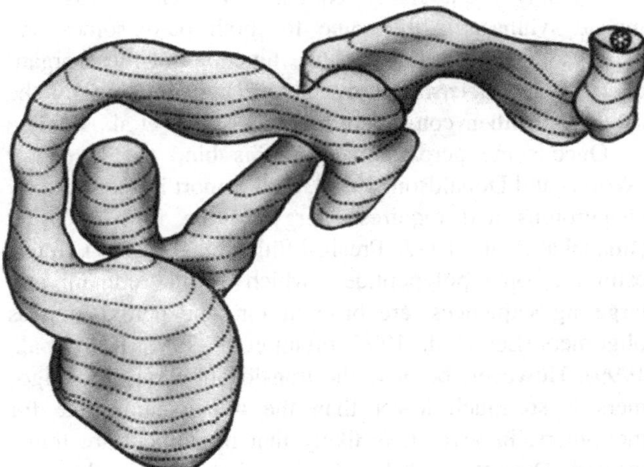

FIGURE 5.14 Three-dimensional reconstruction of a peroxisome from serial sections. *From Gorgas, K., 1984. Peroxisomes in sebaceous glands. V. Complex peroxisomes in the mouse preputial gland: serial sectioning and three-dimensional reconstruction studies. Anat. Embryol. 169, 261–270.*

we must not think of a cell or organelle as being two dimensional. This is the lesson that Jacobus van't Hoff taught organic chemists in the past century when he introduced the field of stereochemistry to explain the mechanism of stereoisomerism. Indeed, Hermann Kolbe thought van't Hoff was crazy for introducing a three-dimensional aspect to chemicals, the structure of which could be written on a two-dimensional piece of paper. Kolbe wrote about his distaste for van't Hoff and his stereochemistry in no uncertain terms (van't Hoff, 1967)! We must remember that cells live in a three-dimensional world, and consequently, we must fight the temptation to look at the cell as an inhabitant of "Flatland" (Square, 1899).

If peroxisomes do not come from the ER, where do they come from? Lazarow and Fujiki (1985) and Lazarow (2003) propose that peroxisomes originate from preexisting peroxisomes. Peroxisome growth occurs by the incorporation of new protein and lipid into preexisting peroxisomes. Peroxisomes, like plastids and mitochondria, seem to increase in number by fission (Dinis and Mesquita, 1994) mediated by a dynamin-like protein (Koch et al., 2003; Li and Gould, 2003; Mano et al., 2004; see Chapters 13 and 14).

How do proteins get into the peroxisomes? As we will discuss in Chapter 17, the peroxisomal proteins do not have the ER signal peptide, but they do have another specific amino acid sequence that targets proteins to the peroxisome (Trelease et al., 1996; Mullen, 2002; Reumann, 2004). Isolated peroxisomes take up polypeptides that contain a peroxisomal targeting sequence. The peroxisomal targeting sequence is either SKL (serine-lysine-leucine) on the carboxy-terminal end of matrix proteins (Gould et al., 1990; Keller et al., 1991) or arg-leu/gln/ile-X5-his-leu on the amino-terminus (Gietl, 1990). This indicates that there may be multiple translocator pathways (Olsen and Harada, 1995). Interestingly, the targeting sequence for isocitrate lyase and malate synthase is the same for both peroxisomes and glyoxysomes, indicating that the functions of these organelles are not determined by protein targeting but by the synthesis of their constituent proteins (Olsen et al., 1993).

Once some peroxisomal proteins bind to a receptor (Wolins and Donaldson, 1994), their import is enhanced by chaperonins and requires energy in the form of ATP (Imanaka et al., 1987; Presig-Müller et al., 1994). Interestingly, some polypeptides, which do not contain any targeting sequences, are brought into the peroxisome as oligomers (Lee et al., 1997; Flynn et al., 1998; Kato et al., 1999). However, because the translocation rate for oligomers is so much lower than the translocation rate for monomers, in vivo, it is likely that monomers are translocated (Dias et al., 2016). Nevertheless, the translocation of oligomers indicates that the protein-translocating pore is relatively large. The protein translocator in the peroxisomal membrane, unlike those in the chloroplast or mitochondrial membranes (see Chapters 13, 14, and 17), can transport proteins in their native form. In fact, proteins containing the SKL targeting sequence will even bring 4- to 9-nm gold particles into the peroxisome (Walton et al., 1995). The porous peroxisomal protein translocators, or peroxisomal translocons, are suitable in a membrane like the peroxisomal membrane, which does not maintain an electrical membrane potential difference. Such a large ungated pore, which may allow the substrates and products of peroxisomal pathways, would be incompatible in a membrane that must maintain an electrical potential difference.

Just when the "independent growth and division" model of peroxisome biogenesis seemed to win over the "ER-vesiculation" model, Robert Mullen et al. (Mullen et al., 1999, 2001; Mullen and Trelease, 2000; Lisenbee et al., 2003; Titorenko and Mullen, 2006; Hu et al., 2012; van der Zand et al., 2012) discovered that ascorbate peroxidase, a peroxisomal membrane protein, is posttranslationally inserted into the ER. This step requires ATP and chaperonins. The protein is localized in a distinct region of the ER. Inhibition of ER vesicle blebbing by brefeldin A prevents the movement of this protein into peroxisomes.

Interestingly, in yeast, two biochemically distinct types of vesicles bleb from the ER and then fuse to form peroxisomes. Each type of preperoxisomal vesicle contains one-half of the peroxisomal translocon complex. Thus the protein-translocating complex only becomes fully functional when the two types of vesicles fuse. This prevents peroxisomal matrix proteins from being inserted into the ER lumen and thus allows the peroxisome to be biochemically distinct (van der Zand et al., 2012).

Thus, it seems that some of the cytosolically synthesized proteins required for peroxisome biogenesis enter the peroxisome directly, while others, including ascorbate peroxidase and a translocon (Hoepfner et al., 2005), enter the peroxisome indirectly by means of specialized ER-derived vesicles. More and more data are being amassed that show a precursor–product relationship for peroxisomal membrane proteins beginning in the ER, suggesting that the peroxisome, like the other endomembranes, is a derived organelle synthesized from the ER (Titorenko and Rachubinski, 1998; Hoepfner et al., 2005; Kragt et al., 2005; Kunau, 2005; Schekman, 2005; Kim et al., 2006; Titorenko and Mullen, 2006; Karnik and Trelease, 2007; Shen et al., 2010; Agrawal et al., 2011). The conclusion that peroxisomes are derived from the ER is supported by the discovery that yeast mutants that are defective in their ability to transport secretory proteins from the ER do not produce visible peroxisomes.

The peroxisomal membrane proteins seem to be derived from proteins that are synthesized on cytosolic free ribosomes that are posttranslationally inserted into the ER, and the peroxisomal matrix proteins seem to be derived from proteins that are synthesized on cytosolic free ribosomes but are posttranslationally targeted directly to the peroxisome. Blobel (1999) suggests that the peroxisomal protein translocators may not be able to open laterally to the lipid bilayer and thus may not be competent to integrate membrane proteins. Thus there is truth to both the ER-vesiculation model and the independent growth and division model. It is always wise to know the history of a subject because a synthesis of that subject requires putting together the thesis and the antithesis. Usually the accepted theory at any given time (the thesis) is only partly true, and the unaccepted theory at that time (the antithesis) is only partly false. Interestingly enough, the thesis of one generation of scientists often becomes the antithesis of the next. In all cases, the synthesis comes from combining the truth from both theories. At the risk of sounding too skeptical, a full resolution of the peroxisome controversy will have to await identification of the source or sources of all the peroxisomal proteins and a conformation or a refutation of the existence of a non-ER-derived, preperoxisomal element that can exist as a protoperoxisome (Lazarow, 2003).

Peroxisomes have a very limited capacity to synthesize lipids (Ballas et al., 1984; Chapman and Trelease, 1991a), so how do they get their lipids (Raychaudhuri and Prinz, 2008)? While some lipids may come directly from the ER membranes that vesiculate to form peroxisomes (Titorenko and Mullen, 2006), phospholipid exchange proteins (Dowhan, 1991; Cleves et al., 1991) may also be responsible for delivering lipids to the peroxisomal membrane. The exchange protein can extract a phospholipid from the ER membrane and then bury the phospholipid inside itself. Phospholipid-exchange proteins are water-soluble proteins that diffuse through the cytoplasm until they bump into another membrane. Then they release the phospholipids into the new bilayer. It is thought that exchange proteins randomly distribute phospholipids throughout the cytoplasmic membranes in a Robin Hood—like manner. Such a random exchange will lead to the net transfer of phospholipids from the phospholipid-rich ER (or lipid bodies) to the phospholipid-poor membranes (Ben Abdelkader et al., 1973; Tanaka and Yamada, 1979; Crain and Zilversmit, 1980; Yaffe and Kennedy, 1983; Dawidowicz, 1987; Bishop and Bell, 1988; Chapman and Trelease, 1991b). It is also possible that some of the lipids come from the ER already assembled into the peroxisomal membrane (Titorenko and Mullen, 2006).

Peroxisomes are born and peroxisomes die due to H_2O_2 damage and in times of nutrient limitation. They are digested in a bulk recycling process known as pexophagy (Till et al., 2012; Farmer et al., 2013; Kim et al., 2013; Shibata et al., 2013; Baker and Paudyal, 2014; Goto-Yamada et al., 2014; Yoshimoto et al., 2014), a type of autophagy (Chapter 7).

5.9 EVOLUTION OF PEROXISOMES

Peroxisomes are present in all eukaryotes except the *Archaezoa* (Cavalier-Smith, 1987a,b) and may have evolved endosymbiotically (see Chapter 15; de Duve, 1991). Alternatively, peroxisomes, like the ER, may have evolved from invaginations of regions of other membranes that contained the enzymes involved in what are now considered peroxisomal pathways (de Duve, 1969; Hoepfner et al., 2005). There are similarities between the functions of peroxisomes and mitochondria. Both organelles are capable of using O_2. In general, it seems that the mitochondria have taken over the functions of the peroxisomes. In some cases, however, it seems like the peroxisomes have taken over the functions of the mitochondria.

The peroxisomes in the cells of various phyla exhibit both morphological and biochemical diversity. Perhaps the range of innovations possible for peroxisomes is best seen in the green algae. In this class, which gave rise to the higher plants, there is evidence that generally throughout evolution, the peroxisome has become more important as an organelle by taking over some of the functions of the mitochondrion (Stabenau, 1992). Of course, evolution is not linear and there is also evidence that in some phyla the peroxisomes have become a more vestigial organelle, giving much of its biochemical capacity back to the mitochondria (de Duve, 1969; Stabenau, 1992). We must keep in mind that cells are dynamic not only over cellular time scales but also throughout geological time scales. Taking this into consideration, cytologically oriented systematists have used the presence of glycolate oxidase in the peroxisome or glycolate dehydrogenase in the mitochondria as characters used for classification purposes in the green algae (Stewart and Mattox, 1975; Betsche et al., 1992).

5.10 SUMMARY

We have learned about the structure and function of peroxisomes, and that these organelles are multifunctional. In some cases, their function varies from cell to cell within an organism or temporally within a cell. Peroxisomes are not static organelles but appear to have changed during the evolution of plants and animals. We have also learned about the concept of metabolite channeling, the importance of serial sectioning in interpreting electron micrographs, and the fact that a synthesis may require putting together the truths of the accepted and unaccepted theories.

5.11 QUESTIONS

5.1. How do the functions of peroxisomes change throughout the life of a cell?

5.2. What is metabolic channeling and what are its advantages?

5.3. What is the evidence that peroxisomes are autonomous organelles, and what is the evidence that they are derived from the ER?

The references for this chapter can be found in the references at the end of the book.

Chapter 6

Golgi Apparatus

Oh, to live on sugar mountain
With the barkers and the colored balloons
You can't be twenty on sugar mountain
Though you're thinking that you're leaving there too soon
You're leaving there too soon.

Neil Young (Sugar Mountain).

We are seeing the cells of plants and animals more and more clearly as chemical factories, where the various products are manufactured in separate workshops. The enzymes act as the overseers. Our acquaintance with these most important agents of living things is constantly increasing. Even though we may still be a long way from our goal, we are approaching it step by step. Everything is justifying our hopes. We must never, therefore, let ourselves fall into the way of thinking 'ignorabimus' ('We shall never know'), but must have every confidence that the day will dawn when even those processes of life which are still a puzzle today will cease to be inaccessible to us natural scientists.

Eduard Buchner (Nobel Lecture).

6.1 DISCOVERY AND STRUCTURE OF THE GOLGI APPARATUS

In the late 19th century, cytologists began to see that the cytoplasm was not homogeneous but contained previously unknown and invisible internal structures or "formed elements" of which the identity could be recognized by their characteristic staining patterns. In 1898, Camillo Golgi visualized a netlike reticulum of fibrils in the Purkinje cells of the owl, *Strix flammea* (Fig. 6.1). He could see this *internal reticular apparatus*, as he called it, because it reduced silver and thus became blackened and visible against the rest of the cytoplasm. The silver- or osmium-stained internal reticular apparatus was dubbed the *Golgi-Holmgren canals* by Santiago Ramón y Cajal (1937) and was later redubbed the *Golgi apparatus*.

In vertebrates, the Golgi apparatus usually appears morphologically as a fibrous network, whereas in invertebrates and plants, it appears as separate elements. According to Kirkman and Severinghaus (1938),

The Golgi apparatus appears to be the most protean of all cytoplasmic structures—it has been described as a fibrous reticulum, network, ring or cylinder, a very irregular fenestrated plate, a more or less incomplete hollow sphere, vesicle, or cup, a collection of small spheres, rodlets and platelets or discs, a series of anastomosing canals, a group of vacuoles, and a differentiated region of homogeneous cytoplasm crossed by irregular interfaces.

As a consequence of its polymorphous appearance as well as its change in chemical composition and stainability throughout the life cycle of a cell, the Golgi apparatus has been given many names, including the paraflagellar apparatus, the dictyosomes, the canaliculi of Holmgren, the fluid canaliculi, the trophospongium, and the osmiophilic platelets. However, Bowen (1926), Duboscq and Grassé (1933), Wilson and Pollister (1937), and many others showed the morphological and/or functional homology between these organelles and suggested that they all be called the Golgi apparatus.

In an attempt to determine the homologies between the newly discovered organelles of plant and animal cells, Robert Bowen (1928) began to characterize cytologically the osmiophilic platelets in a variety of plant cells, including the root-tip cells of barley and bean. He observed ring-like or disc-shaped structures that blackened selectively with osmic acid and tentatively concluded that these structures were homologous with the Golgi apparatus of animal cells (Fig. 6.2). However, he cautioned that this identity rested only on morphological grounds and staining characteristics, and that it would be important to test whether the osmiophilic discs had the same secretory function as the Golgi apparatus in animal cells. Because the staining reactions for the Golgi apparatus were selective, but not specific, interpretations on the reality of the Golgi apparatus based on staining remained controversial up through the 1960s (Weier, 1931, 1932a,b; Nahm, 1940; Guilliermond, 1941; Bourne, 1942, 1951, 1962, 1964; Worley, 1946; Palade and Claude, 1949; Bensley, 1951; Baker, 1957; Dalton, 1961; Beams and Kessel, 1968; Buvat, 1969).

In the mid-1950s, the Golgi apparatus in animal cells was shown with the electron microscope to be a real membranous structure with a distinct architecture and not just a chemical substance. This provided a means to distinguish unequivocally the Golgi apparatus from all the

Plant Cell Biology. https://doi.org/10.1016/B978-0-12-814371-1.00006-0

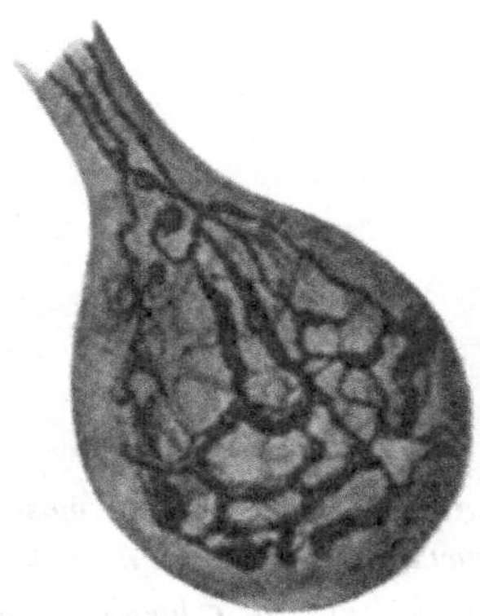

FIGURE 6.1 The internal reticular apparatus of Purkinje cells of an owl. *From Golgi, C., 1898. Sur la structure des cellules nerveuses. Arch. Ital. de Biologie 30, 60–71.*

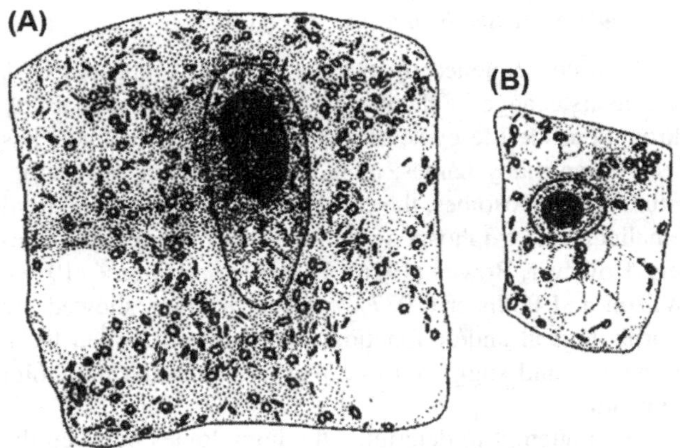

FIGURE 6.2 The osmiophilic platelets in (A) a large cell from the central core of a barley root tip and (B) a cell from the root tip of a kidney bean. *From Bowen, R.H., 1927. Golgi apparatus and vacuome. Anat. Rec. 35, 309–335.*

other cellular organelles (Sjöstrand and Hanzon, 1954; Palay and Palade, 1955; Dalton and Felix, 1956). The Golgi apparatus in plant cells was observed with the electron microscope by Hodge et al. in 1956, although they did not recognize it as such. They called it the *cytoplasmic lamellae* and proposed that the cytoplasmic lamellae participated in the formation of endoplasmic reticulum (ER) because the cytoplasmic lamellae looked like membrane factories where vesicle fusion was taking place.

The plant cytologist, Alexandre Guilliermond (1941), assumed that plants had enough organelles already with the plastids, mitochondria, and the vacuole, and any so-called Golgi apparatus stained with osmium or silver was actually one of the already-known organelles. Despite his the strong and persistent claims that the Golgi apparatus did not exist in plants (Porter and Moses, 1958), Keith Porter (1957), E. Perner (1957), Roger Buvat (1957) demonstrated the reality of the Golgi apparatus in plant cells with the electron microscope and concluded once and for all that plant cells, like animal cells, have a Golgi apparatus. Indeed, in optically favorable material such as the *Chara* rhizoid, it is possible to unequivocally identify the Golgi apparatus in living plant cells (Bartnik and Sievers, 1988) as it is in living animal cells (Brice et al., 1946; Oettlé, 1948). The Golgi apparatus can be easily visualized in living plant cells that have been transformed with fluorescent proteins fused to a resident Golgi apparatus protein (Boevink et al., 1998; Nebenführ et al., 1999, 2000; Brandizzi et al., 2002a,b; Saint-Jore et al., 2002; Neumann et al., 2003; daSilva et al., 2004; Zheng et al., 2004).

Perroncito (1910) noticed the Golgi apparatus split up into a number of elongated pieces during cell division and named each piece the *dictyosome*. Nowadays, each separate stack of Golgi membranes is called a Golgi stack or dictyosome (Mollenhauer and Morré, 1966). For the sake of uniformity among animal and plant cell biologists, I will refer to each separate stack as a Golgi stack as opposed to referring to it as a dictyosome. Although there can be between 0 and 25,000 Golgi stacks per cell (Rosen, 1968), there are typically hundreds (Satiat-Jeunemaitre and Hawes, 1994; Nebenführ et al., 1999). Golgi stacks are particularly abundant in secretory cells in plants as they are in animals (Bowen, 1926, 1927, 1929) and may be absent in dry seeds (Fahn, 1979).

The architecture of the Golgi apparatus, which consists of all the Golgi stacks in the cell, varies from cell to cell and throughout the life of the cell (Whaley et al., 1959, 1960; Manton, 1960; Bonneville and Voeller, 1963; Noguchi and Kakami, 1999; Noguchi et al., 2014). While the Golgi stacks may seem to be separate, electron microscopy of thick (1 μm) sections shows that the stacks may also be connected together in a three-dimensional Golgi reticulum (Rambourg and Clermont, 1990; Fig. 6.3). The Golgi apparatus is thus differentiated into a compact zone, called the *traditional Golgi stack*, and a noncompact zone that connects the stacks. Such a noncompact zone can be called the *cis-Golgi network* (CGN) when it is associated with the forming face of the Golgi stack and the *trans-Golgi network* (TGN) when it is associated with the maturing face. The Golgi apparatus may contain more or less differentiated cis- or trans-Golgi networks (Fig. 6.4).

In systematics and evolution, we learn over and over that nature mocks human categories (Bergson, 1911). In the systematics of organelles, we learn the same lesson again. That is, while the Golgi stack can be unambiguously identified, the identities of the membranes associated with the outskirts of the Golgi stacks are less certain. The associated membranes have been given a variety of names, including the CGN, TGN, the *Golgi-ER-lysosomal* continuum, the prevacuolar compartment, the *ER-Golgi intermediate compartment*, the *partially coated reticulum*, and the *vesicular-tubular cluster*. The degree of membrane elaboration probably reflects the developmental and/or

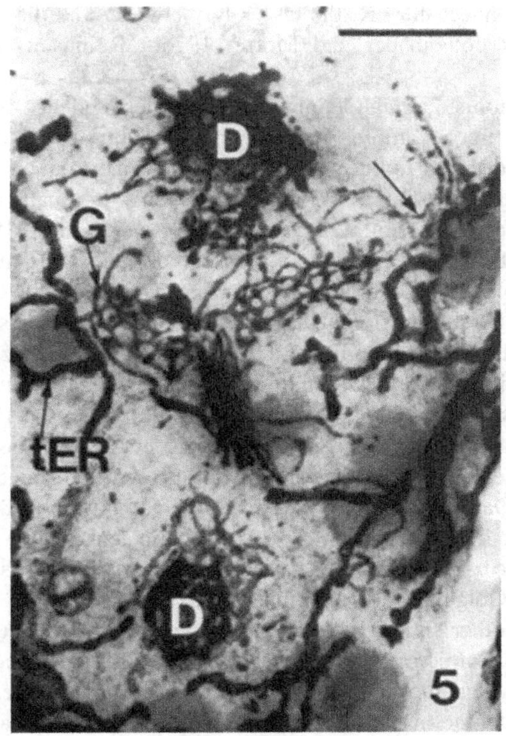

FIGURE 6.3 Electron micrograph of a vascular parenchyma cell of mung bean stained with zinc iodide—osmium tetroxide showing the tubular connections between the cisterna (D) of a Golgi stack and the tubular endoplasmic reticulum (tER). The fine tubules nearest the Golgi stack are approximately 10- to 20-nm thick. Bar, 500 nm. *From Harris, N., Oparka, K.J., 1983. Connections between dictyosomes, ER and GERL in cotyledons of mung bean (*Vigna radiata L.*). Protoplasma 114, 93–102.*

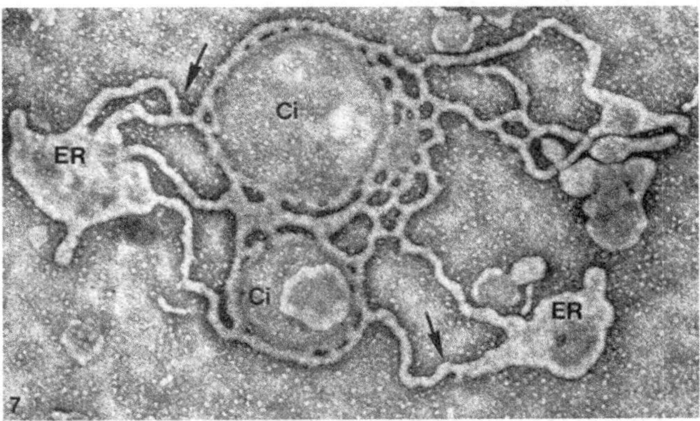

FIGURE 6.4 Electron micrograph of a negatively stained Golgi stack isolated from root-tip cells. The cisternae (Ci) are interconnected with each other and with the ER. ×55,000. *From Mollenhauer, H.H., Morré, D.J., 1976b. Transition elements between endoplasmic reticulum and Golgi apparatus in plant cells. Cytobiologie 13, 297–306.*

functional state of the cell, and perhaps even the taxon to which the organism belongs (Robinson, 2003). In any case, the reification of the organellar status of these membranes is at the same stage that the reality of the currently accepted organelles was in the past. After techniques are developed to unambiguously identify these membranes morphologically and cytochemically, and their specific functions are shown in cell-free systems following cell fractionation, a consensus on their individuality and uniqueness will be reached.

Each Golgi stack is a flattened disc about 1 μm in diameter and 0.25 μm long. A Golgi stack typically consists of a stack of 4–7 flattened membranes or cisternae, although more than 20 cisternae may be present (Mollenhauer et al., 1983; Kiss et al., 1990; Staehelin et al., 1990; Mollenhauer et al., 1991; Zhang and Staehelin, 1992; Hillmer et al., 2011). Each cisterna is separated from the others in the stack by a minimal space of 10–15 nm. Parallel fibers, called *intercisternal elements*, about 3–6 nm in diameter, exist between the cisternae (Mollenhauer, 1965; Turner and Whaley, 1965). The cytosolic region that surrounds the Golgi stack and the trans-Golgi network is devoid of ribosomes. Staehelin and Moore (1995) call this structurally specialized region the *Golgi matrix*. Application of brefeldin A to cells causes the resorption of the Golgi stacks into the ER, and removal of brefeldin A results in the regeneration of well-defined Golgi stacks within 90 minutes (Langhans et al., 2007). In *Gloeomonas kupfferi*, brefeldin A causes the medial cisternae to form a tubular network. Some Golgi stacks in BY-2 cells respond differently to brefeldin A than others (Ritzenthaler et al., 2002)—a possible indicator of differentiation between Golgi stacks in the same cell.

Not all Golgi apparati have their cisternae in stacks (Mowbrey and Dacks, 2009). The Golgi apparatus in the red alga, *Cyanidioschyzon merolae*, consists of only one or two cisternae (Okuwaki et al., 1996). Likewise, in the cells of many filamentous fungi, the Golgi apparatus appears as a single tubule known as a *Golgi equivalent* (Hoch and Staples, 1983; Roberson and Fuller, 1988; Bourett and Howard, 1996). While the Golgi apparatus in *Pichia pastoris* and *Schizosaccharomyces pombe* are organized into stacked cisternae, in the budding yeast, *Saccharomyces cerevisae*, the Golgi apparatus is composed of a cluster of randomly oriented cisternae (Preuss et al., 1992; Suda and Nakano, 2012). The Golgi equivalents are clearly homologous with the Golgi apparatus because they are associated with secretory vesicles. The Golgi apparatus of the budding yeast *Pichia* consists of four cisternae (Mogelsvang et al., 2003).

6.2 POLARITY OF THE GOLGI STACK

The Golgi stack is an organelle that exhibits polarity (Friend and Farquhar, 1967; Holtzman et al., 1967; Shannon et al., 1982; Staehelin and Kang, 2008; Day et al., 2013). It has two distinct faces: the forming or cis-face and the maturing or trans-face (Figs. 6.5 and 6.6). The cis-face is often, but not always, associated with the transition elements of the ER, or in many algae, the nuclear envelope (Massalski and Leedale, 1969; Mollenhauer and Morré, 1976b;

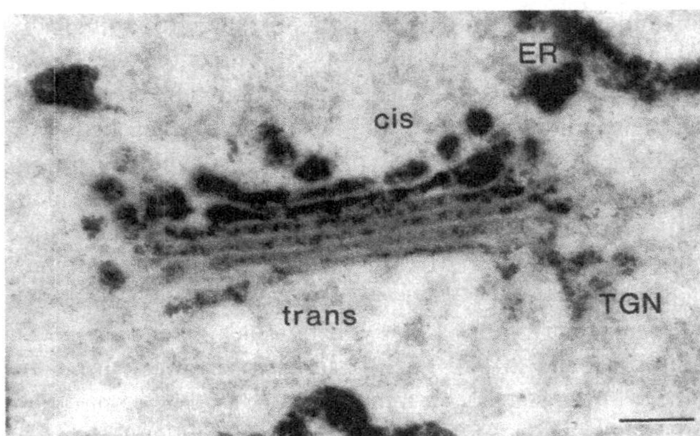

FIGURE 6.5 Electron micrograph of a Golgi stack in a cortical cell of a clover root stained with zinc iodide–osmium. *TGN*, trans-Golgi network. Bar, 100 nm. *From Moore, P.J., Sivords, K.M.M., Lynch, M.A., Staehelin, L.A., 1991. Spatial reorganization of the assembly pathways of glycoproteins and complex polysaccharides in the Golgi apparatus of plants. J. Cell Biol. 112, 589–602.*

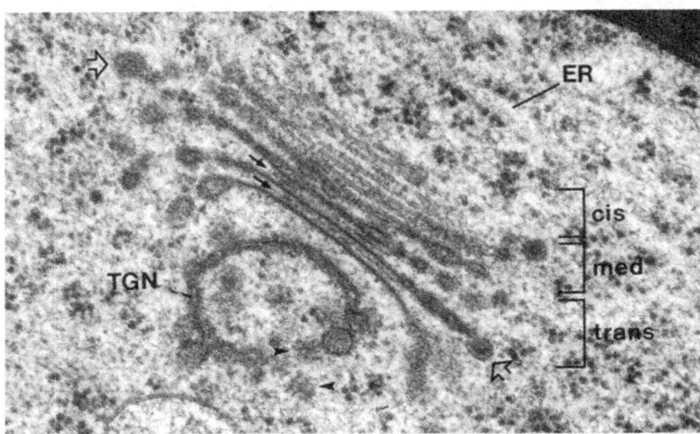

FIGURE 6.6 Electron micrograph of a Golgi stack in a columella cell of tobacco. Note the lightly stained *cis*-cisternae with wide lumena; the medial-cisternae contain darkly staining contents; and the *trans*-cisternae have tightly appressed membranes and very thin lumena. ×65,000. *From Staehelin Jr., L.A., Giddings, T.H., Kiss, J.Z., Sack, F., 1990. Macromolecular differentiation of Golgi stacks in root tips of* Arabidopsis *and* Nicotiana *seedings as visualized in high pressure frozen and freeze substituted samples. Protoplasma 157, 75–91.*

Robinson, 1980; Shannon et al., 1982). The cis-face is composed of many fenestrations that can be seen clearly in zinc iodide–osmium tetroxide–fixed cells (Dauwalder and Whaley, 1973). The fenestrations are about 50 nm in diameter. Small vesicles, 50 nm in diameter, appear between the transition elements of the ER and the cis-face of the Golgi stack. It appears as if these vesicles bleb off from the transition ER and fuse with the Golgi apparatus. The fenestrations in the cis-face of the Golgi apparatus may represent the fusion of these vesicles with the cisterna on the cis-face. Plants transformed with green fluorescent protein (GFP) reveal the close and dynamic association between the ER and the cis-face of the Golgi apparatus (Boevink et al., 1998, 1999; Batoko et al., 2000). Tubular connections may also connect the ER, the cis-Golgi network, and the Golgi stack (Mollenhauer and Morré, 1976b; Robinson et al., 2015).

Cytohistological staining gives further evidence of polarity. For example, under certain conditions, osmium tetroxide–zinc iodide precipitates heavily in the lumen of the ER and the cis-face of a Golgi stack but only minimally in the trans-face of the Golgi stack (Dauwalder and Whaley, 1973). Thus, Bowen (1928) probably observed the forming face of the Golgi in the light microscope. The cisternae of the Golgi stack also show different degrees of staining from the cis-side to the trans-side in terms of the cytohistological localization of various enzymes. Thiamine pyrophosphatase, inositol triphosphatase, and ATPase are localized on the trans-face, whereas CMPase, NADPase, and β-glycerolphosphatase are localized in vesicles emerging from the center of the trans-face (Dauwalder and Whaley, 1973; Domozych, 1989a,b; Rambourg and Clermont, 1990; Staehelin et al., 1990). GFP fused to α-1,2 mannosidase typically shows up in the cis-face of the Golgi stack (Nebenführ et al., 1999).

Another sign of polarity that is visible in conventional electron micrographs is that the cisternae become flatter from the cis-face to the trans-face. The flatness of the cisternae may depend on the intercisternal fibers as the number of these fibers increases from the cis-face to the trans-face (Turner and Whaley, 1965; Mollenhauer, 1965; Hawkins, 1974; Alley and Scott, 1977; Kristen, 1978). The membranes of the cisternae also increase in thickness from 5.6–6.4 nm at the cis-face to 6.4–9.1 nm at the trans-face (Morré and Mollenhauer, 1976a,b).

The Golgi stack maintains its polarity while membranes are continually flowing through it (Pelham, 2001; Pelham and Rothman, 2000; Beznoussenko and Mironov, 2002; Marsh and Howell, 2002; Storrie and Nilsson, 2002; Nebenführ, 2003). In some cases, discussed below, it is thought that whole cisternae move through the Golgi stack, and thus the membranous and lumenal components of each cisterna show a temporal pattern of polarity (Mogelsvang et al., 2003). However, there is also evidence that each cisterna maintains its position in the stack and transfers membranes and contents either by making direct membranous contacts between adjacent cisternae via membrane tubulization or by a process that involves vesicle blebbing and fusing (Morré and Keenan, 1994, 1997; Staehelin and Moore, 1995). In the latter two cases, the polarity would be exclusively spatial. It is not currently possible to unequivocally determine which model is correct without doing time-resolved, three-dimensional studies.

Once the membranous or lumenal components reach the trans-face of the Golgi stack, they become part of a tubular reticulum called the *TGN*, where they eventually bleb off for the last time and go to various destinations, including the plasma membrane and the vacuolar compartment (Grove et al., 1970; Dauwalder et al., 1969). There is evidence that not all traffic through a Golgi stack is anterograde from the cis-face to the trans-face. Consistent with the original proposal by Hodge et al. (1956), there is also retrograde movement from the trans-face to the cis-face (see discussion in Chapter 8). While in growing plant cells, it is likely that there is a net movement of membranous and lumenal components from the cis-face through the central or medial region to the trans-face. In stationary state cells, the movement in opposite directions throughout the endomembrane system must be balanced.

Four kinds of proteinaceous coats are associated with the Golgi apparatus of plants, and it is thought that these coat proteins facilitate the blebbing and fusion of membranes. Coat proteins, or Coatomers (COP I and COP II), appear predominantly on the transition elements of the ER, the transition vesicles, and on the cis- and medial-cisternae of the Golgi stacks. COP Ia is involved in the transfer of vesicles from the cis-Golgi and/or the cis-Golgi complex to the ER, and COP Ib is involved in the transfer of vesicles from one Golgi cisterna to another in either direction. COP II is involved in the transfer of vesicles from the ER to the cis-Golgi complex and/or the cis-Golgi (Pelham, 1994; Bednarek et al., 1995; Schekman and Orci, 1996; Pimpl et al., 2000; Phillipson et al., 2001; Robinson et al., 2007: Donohoe et al., 2007; Kang and Staehelin, 2008; Lee and Goldberg, 2010; Rothman, 2010; Miller and Schekman, 2013; Gomez-Navarro and Miller, 2016).

On the trans-Golgi network, buds are coated with either a clathrin coat or a lacelike coat. It is thought that the clathrin-coated vesicles originating from the trans-Golgi network are targeted to become part of the vacuolar compartment, whereas the lacelike-coated vesicles from the trans-Golgi network are destined to go to the plasma membrane.

The cis- and trans-side of the Golgi can be differentially labeled with fluorescent proteins and observed in living cells (Schoberer et al., 2010; Madison and Nebenführ, 2011; Schoberer and Strasser, 2011; Ito et al., 2012, 2014).

6.3 ISOLATION OF THE GOLGI APPARATUS

Intact Golgi stacks can be isolated from plant cells by homogenizing the tissue in the presence of a stabilizing agent such as 0.3% glutaraldehyde. The homogenate is filtered to remove the extracellular matrix and then centrifuged at low speed (100 g for 10 minutes) to remove the nuclei and plastids. The supernatant is recentrifuged at high speed (100,000 g for 60 minutes) to concentrate the microsomal membranes, which are then resuspended and layered on a discontinuous sucrose density gradient. After centrifugation at 100,000 g for 60 minutes, the Golgi are found at the 1.03−1.077 M sucrose interface, which is equivalent to a density between 1.132 and 1.138 g/mL (Green, 1983). Thus, the density of the Golgi membranes is intermediate between the density of the ribosomeless membranes of the ER and the density of the plasma membrane. D. James Morré (1987) developed a technique that uses free-flow electrophoresis to separate the isolated Golgi stacks into cis-, medial-, and trans-fractions (Parsons et al., 2012). The Golgi-derived vesicles can also be isolated (van Der Woude et al., 1971; Hasegawa et al., 1998).

6.4 COMPOSITION OF THE GOLGI APPARATUS

The lipids of the Golgi apparati isolated from soybean stems and rat liver cells are similar. Interestingly, the relative lipid composition of the Golgi membranes is intermediate between that of the ER and that of the plasma membrane (Keenan and Morré, 1970; Morré and Ovtracht, 1977). Soybean Golgi contains approximately (percent of total membrane phosphorous) 30% phosphatidylcholine, 20% phosphatidylethanolamine, 10% phosphatidylinositol, and 1% phosphatidylserine.

The marker enzymes for the Golgi apparatus include latent inosine diphosphatase and a number of glycosyl transferases, including glucan synthase I, α-1,4-galacturonosyltransferase, and UDPG:sterol glucosyl transferase (Green, 1983; Sterling et al., 2001). Nucleotide monophosphatase and diphosphatase have also been isolated from Golgi membranes (Staehelin and Moore, 1995; Gupta and Sharma, 1996). David Gibeaut and Nicholas Carpita (1993, 1994) have succeeded in getting isolated Golgi apparati to synthesize natural polysaccharides. Proteomic studies show that the Golgi apparatus is composed of about 500 proteins (Parsons et al., 2012, 2013; Nikolovski et al., 2012, 2014).

6.5 FUNCTION OF THE GOLGI APPARATUS

While the debate on the reality of the Golgi apparatus was going on, it was surmised by believers that the Golgi apparatus was involved in secretion because, of all the cells studied, it was most highly developed in gland cells (Bowen, 1926, 1927, 1929). In plant and animal cells, the Golgi apparatus is involved in the processing of secretory as well as other glycoproteins. In plant cells, the Golgi apparatus also participates in the secretion of a variety of extracellular materials, including fucose-rich mucilage by root cap cells, wall-degrading enzymes during abscission, hydrolases that degrade the food reserves during germination, and digestive enzymes in insectivorous plants (Northcote and Pickett-Heaps, 1966; Dauwalder and Whaley, 1973; Sexton and Hall, 1974; Sexton et al., 1977; Robinson, 1980; Cornejo et al., 1988; Jones and Robinson, 1989; Roy and Vian, 1991). The Golgi apparatus is also involved in an internal secretory pathway that ends in the vacuolar compartment (see Chapter 8).

6.5.1 Processing of Glycoproteins

Northcote and Pickett-Heaps (1966) discovered, in radioautographic studies, that the Golgi apparatus is the principal site of glucose incorporation in root cells. In plants, 80% of the glycosylation reactions result in the biosynthesis of complex polysaccharides and 20% are involved with processing glycoproteins (Driouich et al., 1993a). The glycoproteins that begin as polypeptides are synthesized on the rough ER. There, oligosaccharides, containing 14 sugar residues, are attached to the amino groups of certain asparagine residues. Subsequently, one mannose and three glucose residues are removed by glucosidases I and II and mannosidase, leaving a high-mannose glycoprotein (see Fig. 4.17 in Chapter 4). Once the glycoprotein reaches the Golgi apparatus, the oligosaccharides may be further processed by various glycosylases to form complex glycoproteins. Glycosylations that take place in the lumen of the ER or in the Golgi apparatus result in soluble glycosylated proteins that reside in the E-space or membrane proteins that are glycosylated on the regions exposed to the E-side.

A mutant of *Arabidopsis* (*cge*), which lacks N-acetyl glucosaminyl transferase I in the Golgi apparatus, is unable to process the N-linked glycoproteins (von Schaewen et al., 1993). When this mutant is transformed with a human cDNA that encodes this enzyme, the transformed mutant plant is capable of processing the N-linked glycoproteins (Gomez and Chrispeels, 1994). Interestingly, the mutants are identical to the wild-type plants, indicating that glycosylation of plant proteins may not be functional but fortuitous just because they pass through the Golgi apparatus (von Schaewen et al., 1993). While the function of some glycosylation reactions are understood (Höftberger et al., 1995), the functions of many remain a mystery!

In the Golgi apparatus, some sugars are also attached to the hydroxyl (OH) groups of serine or threonine. This is called O-linked glycosylation and is also catalyzed by glycosyl transferases. The O-linked glycosylation reactions necessary to form the arabinogalactan-rich proteins found in the extracellular matrix take place in the Golgi apparatus (Showalter, 1993).

6.5.2 Synthesis of Carbohydrates

In plants, the major secretory products of almost all cells are complex polysaccharides, including pectins and hemicelluloses that are secreted into the extracellular matrix (Conrad et al., 1982; Crosthwaite et al., 1994; Held et al., 2011; Atmodjo et al., 2013; Kim and Brandizzi, 2016). The Golgi apparati of plant cells secrete complex polysaccharides during cell plate formation (Hepler and Newcomb, 1967; Northcote et al., 1989; Rancour et al., 2002), and once the cell plate is completed, the Golgi-derived vesicles continue supplying material to the extracellular matrix of the primary and secondary cell walls. Not only do the vesicles supply complex polysaccharides but they may also contain glycoproteins as well as enzymes involved in wall formation and/or loosening (Schnepf, 1969a; Ray et al., 1976; Robinson et al., 1976a; Fry, 1986; Haigler and Brown, 1986; Moore and Staehelin, 1988; Moore et al., 1991; Paredez et al., 2006; Cai et al., 2011). Likewise, proteoglycans that are destined to reside in the extracellular matrix of animal cells are synthesized in the Golgi apparatus.

The complex polysaccharides found in the extracellular matrix of plant cells are often branched and are composed of more than 12 different monosaccharides, indicating that many enzymes, perhaps several hundred glycosyl transferases, may be necessary for their synthesis (Tezuka et al., 1992; Gibeaut and Carpita, 1993). Each enzyme may be capable of attaching a specific sugar at a specific position to make a given bond type. These enzymes are differentially localized within the Golgi apparatus and the Golgi-derived vesicles as shown by immunocytohistochemistry at the electron microscopic level (Brummell et al., 1990; Moore et al., 1991; Zhang and Staehelin, 1992; Fitchette-Lainé et al., 1994). Some xyloglucan-synthesizing enzymes may exist in the Golgi apparatus as multiprotein complexes (Chou et al., 2015).

Why are the glycosylation enzymes differentially localized within a Golgi stack? Moore et al. (1991) have surmised that there are two ways of having error-free synthesis of complex carbohydrates in the Golgi apparatus. That is, if the enzymes involved in the synthesis of each complex polysaccharide were extremely specific, then there would not be a need for these enzymes to be localized in any special manner. Alternatively, if the enzymes involved in the synthesis of complex carbohydrates were less specific, then specific complex carbohydrates could be synthesized by segregating the glycosylation enzymes into separate cisternae of the Golgi apparatus or into separate Golgi stacks. Moore et al. (1991), using immunocytochemistry, found that the synthesis of different complex carbohydrates occurs in different cisternae of a Golgi stack. Using antibodies that recognize various epitopes of certain polysaccharides, Zhang and Staehelin (1992) have shown which cisternae are involved in putting specific sugars on glycoproteins, xyloglucans, and rhamnogalacturonans in suspension cells (see Chapter 20).

The synthesis of hemicelluloses, including xyloglucans, takes place in the Golgi apparatus (Zhang and Staehelin, 1992; Lynch and Staehelin, 1992; Staehelin et al., 1992). Many of the enzymes required for hemicellulose synthesis, including glucosyl, xylosyl, fucosyl, and arabinosyl transferases, have been localized in the Golgi apparatus (Ray et al., 1969; Gardiner and Chrispeels, 1975; Green and Northcote, 1978; James and Jones, 1979; Ray, 1980; Hayashi and Matsuda, 1981; Camirand et al., 1987). Using a battery of monoclonal antibodies, including Anti-XG, which recognizes the β-1,4-linked glucosyl backbone of xyloglucan, and CCRC-M1, which recognizes the terminal fucosyl residue of the trisaccharide side chain of xyloglucan, it was determined that the synthesis and modification of xyloglucans take place exclusively in the trans-Golgi cisternae and the TGN (Staehelin et al., 1992; Zhang and Staehelin, 1992; Viotti et al., 2010).

By contrast, the backbone of pectins is initiated in the cis-Golgi cisternae and extended, and methyl esterified in the medial-Golgi cisternae, and the side chains are added in the trans-Golgi cisternae. This spatial localization is inferred from observations that monoclonal antibodies such as PGA/RG-I, which recognizes the esterified PGA/RG-I transition region, stain the cis-Golgi cisternae; JIM 7, which recognizes methyl esterified PGA, stains the medial- and trans-Golgi cisternae; and CCRC-M2 and CCRC-M7, which recognize the side chains of RG-I, stain the trans-Golgi cisternae and the trans-Golgi network (Zhang and Staehelin, 1992; Staehelin et al., 1992; Sherrier and VandenBosch, 1994; Toyooka et al., 2009; Driouich et al., 2012).

The pathway of movement through a Golgi stack associated with the synthesis and packaging of various complex polysaccharides appears to differ depending on the polysaccharide. Consistent with this observation, monensin, an Na^+/H^+ ionophore that inhibits Golgi sorting at or near the trans-face (Boss et al., 1984), inhibits the movement of xyloglucan but not pectins through the Golgi apparatus.

6.5.3 Transport of Sugars

All sugars must be activated before they are reactive enough to participate in the glycosylation reactions. They become activated after becoming bound to nucleotide diphosphates in the cytosol. Because these are relatively large polar molecules, the membranes of the Golgi apparatus must contain the transporters (dolichols or proteins) to transfer the nucleotide-activated sugars into the lumen (Chanson et al., 1984; Ali and Akazawa, 1985; Ali et al., 1985; Chanson and Taiz, 1985; Gogarten-Boekels et al., 1988). It is not known if the cisternae are differentiated in terms of which sugars they are competent to transport.

6.6 THE MECHANISM OF MOVEMENT FROM CISTERNA TO CISTERNA

No matter how much one would like to know that there is just one mechanism for movement through the Golgi, there are likely to be more than one mechanism (Emr et al., 2009; Glick and Nakano, 2009; Nakano and Luini, 2010; Glick and Luini, 2011). Perhaps the production of scales in the alga *Pleurochrysis* is an ideal system to visualize a specific type of cisterna-to-cisterna movement through the Golgi apparatus (see Figs. 6.7–6.11; Brown, 1969; Brown et al., 1970; McFadden and Melkonian, 1986; McFadden et al., 1986; Melkonian et al., 1991; Becker et al., 1995;

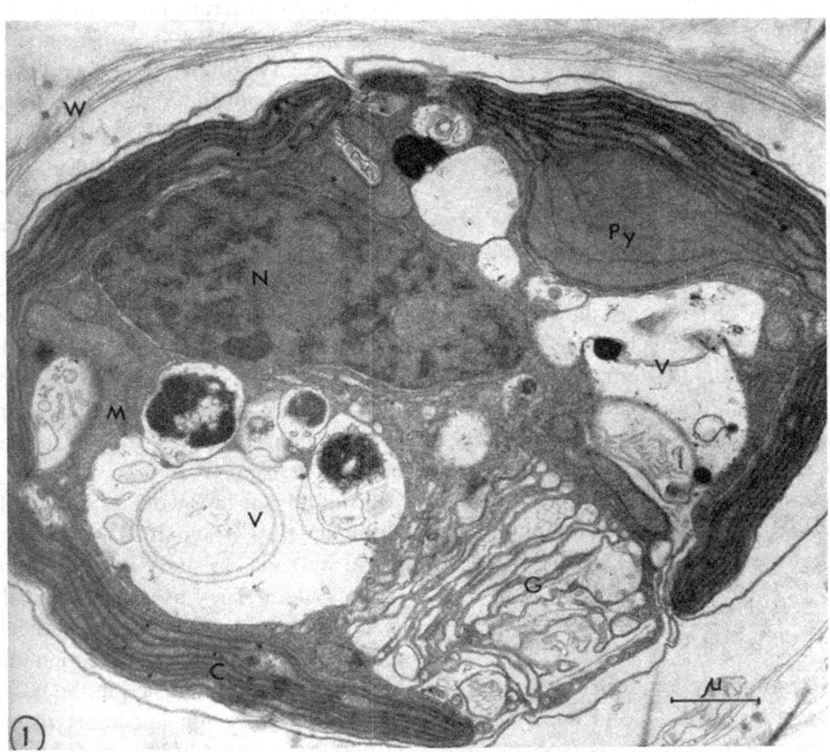

FIGURE 6.7 Electron micrograph of *Pleurochrysis scherffelii* showing a prominent Golgi stack, G. *N*, nucleus; *Py*, pyrenoid; *V*, vacuole; *W*, extracellular matrix. *From Brown Jr., R.M., Franke, W.W., Kleinig, H., Falk, H., Sitte, P., 1970. Scale formation in chrysophycean algae. I. Cellulosic and noncellulosic wall components made by the Golgi apparatus. J. Cell Biol. 45, 246–271.*

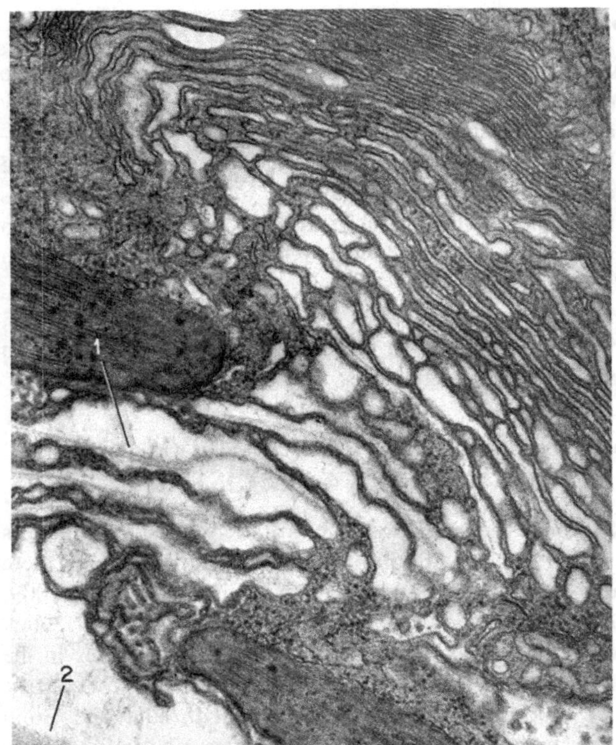

FIGURE 6.8 Medial section through a Golgi stack of *Pleurochrysis scherffelii* that contains a scale (1) destined to end up in the extracellular matrix (2). The cis-face of the Golgi stack is compact and the trans-face is inflated. ×64,800. *From Brown Jr., R.M., 1969. Observations on the relationship of the Golgi apparatus to wall formation in the marine chrysophycean alga,* Pleurochrysis scherffelii *Pringsheim. J. Cell Biol. 41, 109–123.*

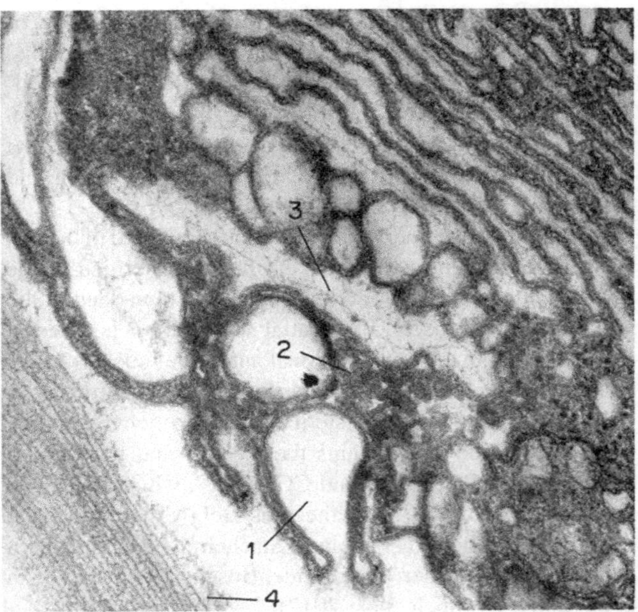

FIGURE 6.9 A Golgi stack of *Pleurochrysis scherffelii*. One cisterna (3) has a scale inside its lumen. Another cisterna (1) has probably just released a scale. Other cisternae (2) have previously deposited their scales. *Arrow* 4 points to the assembled laminate wall. ×89,600. *From Brown Jr., R.M., 1969. Observations on the relationship of the Golgi apparatus to wall formation in the marine chrysophycean alga,* Pleurochrysis scherffelii *Pringsheim. J. Cell Biol. 41, 109–123.*

Donohoe et al., 2013). This alga is covered by distinctive scales that are composed of cellulose microfibrils and pectins. The formation and secretion of these scales can be followed through the maturing Golgi cisternae, where they begin formation in a dilated polymerization center and end as a secreted exocytotic vesicle. The scales have an identical structure within the last cisternae as they do in the extracellular matrix. These observations indicate that in the case of *Pleurochrysis*, the individual cisterna move through the stack from the cis-face to the trans-face.

The secretion of scales can be viewed with the light microscope; one scale is secreted every minute. Because there are about 30 cisternae per Golgi stack, each Golgi stack must turn over every 30 minutes according to the following calculation:

$$\frac{1\text{ min}}{1\text{ scale}} \times \frac{1\text{ scale}}{1\text{ cisternae}} \times \frac{30\text{ cisternae}}{\text{Golgi stack}} = \frac{30\text{ min}}{\text{Golgi stack}} \quad (6.1)$$

Three-dimensional tomography of serial freeze-fixed electron micrographic sections of the Golgi apparatus indicates that the Golgi cisternae form at the cis-face from the fusion of COP II-coated vesicles derived from the transition ER. The cis-cisternae progressively mature into the medial-cisternae, the trans-cisternae, and trans-Golgi network cisternae. The trans-Golgi network cisterna eventually dissociates from the Golgi apparatus, becomes free in the cytoplasm, and gives rise to or receives clathrin-coated vesicles. Because Mogelsvang et al. (2003) and Donohoe et al. (2013) see no connections between the cisternae, they believe that each cisterna matures and moves through the stack as it does in *Pleurochrysis*. The cisternal progression model of intra-Golgi transport also appears to be sufficient to explain the movement of aggregates of procollagen in animal cells (Bonfanti et al., 1998; Nebenführ, 2003). However, COP Ib vesicles exist in the medial and trans-Golgi cisternae suggesting that intercisternal movements of membrane vesicles may also be important in intercisternal transport.

Building on the work of Günter Blobel (1999), James Rothman (2013) and Randy Schekman (2013) developed novel strategies to study movement through the endomembrane system. They used biochemical and genetic complementation approaches to demonstrate the precursor–product relationships between various membranes on a cell biological level. While interested in the transport of proteins from the ER to the Golgi apparatus in animal cells, Rothman (1992, 2013) serendipitously discovered that movement of membrane proteins between the cisternae of different Golgi

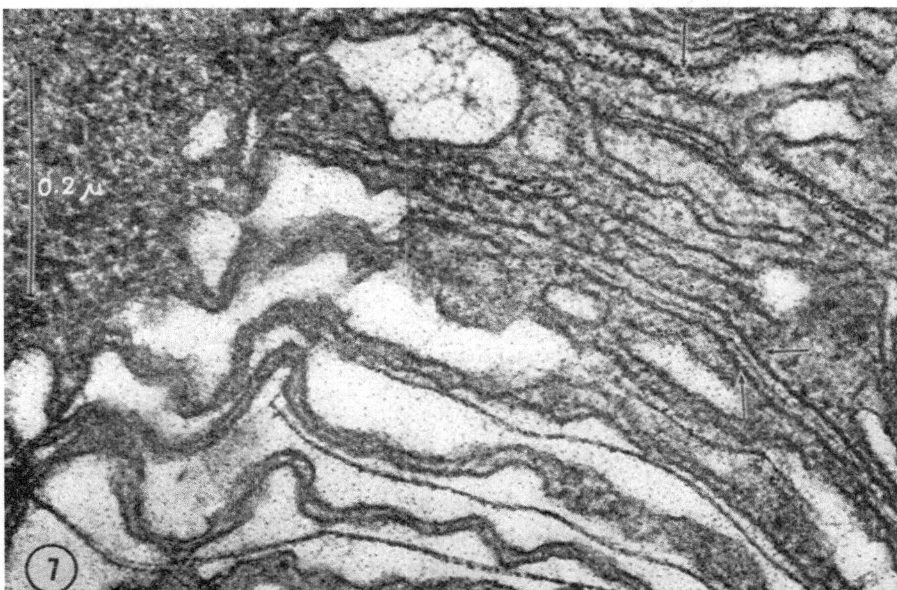

FIGURE 6.10 Electron micrograph showing the progressive maturation of scales in the Golgi stack of *Pleurochrysis scherffelii. From Brown Jr., R.M., Franke, W.W., Kleinig, H., Falk, H., Sitte, P., 1970. Scale formation in chrysophycean algae. I. Cellulosic and noncellulosic wall components made by the Golgi apparatus. J. Cell Biol. 45, 246–271.*

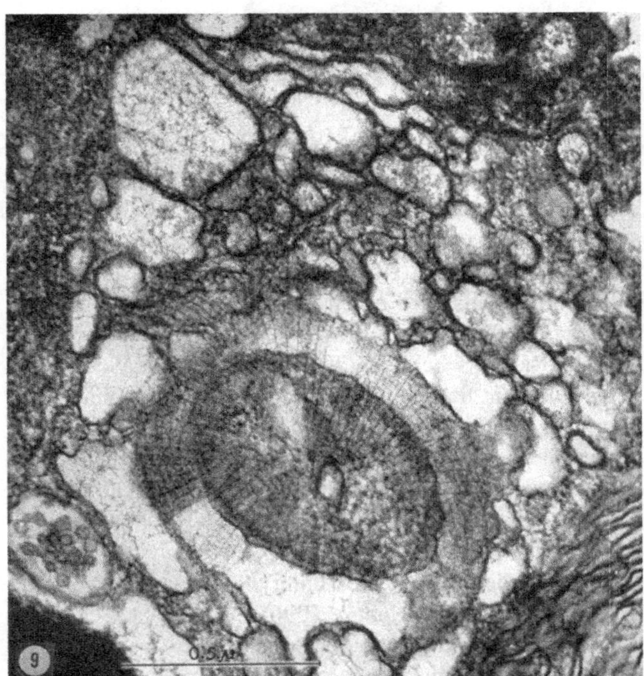

FIGURE 6.11 Electron micrograph showing a grazing section of a scale in the distal-most cisterna of the Golgi stack of *Pleurochrysis scherffelii. From Brown Jr., R.M., Franke, W.W., Kleinig, H., Falk, H., Sitte, P., 1970. Scale formation in chrysophycean algae. I. Cellulosic and noncellulosic wall components made by the Golgi apparatus. J. Cell Biol. 45, 246–271.*

stacks is also possible. Fascinated by George Palade's (1975) work on proteins that were synthesized on the ER and transported through the secretory pathway (see Chapter 8), Rothman (1992) wanted to know, "How could a membrane deform itself to pop out a vesicle? How could such a vesicle choose to fuse with the correct membrane? ... And, how can all of this be organized in time and in space so as to allow the cytoplasm to maintain and propagate its membrane compartments?" Working in an environment that encouraged risk-taking and believing that cell-free biochemistry would provide the only sure route to the underlying molecular mechanisms, Rothman (2013) decided to find the answer to the above questions by developing a cell-free, membrane-transfer system that has now become the model for developing all cell-free systems for the study of membrane transfer.

Fries and Rothman (1980) developed a cell-free system using Chinese hamster ovary cells (Fig. 6.12). A batch of mutant cells was infected with the vesicular stomatitis virus, which produces an abundant membrane protein called *viral G protein*. The mutants were missing an N-acetylglucosamine transferase, and thus their Golgi apparati were incapable of transferring N-acetylglucosamine to the viral G protein. Fries and Rothman mixed a homogenate of these cells with one from an uninfected wild type to see if the viral G protein would pinch off of the ER of the infected

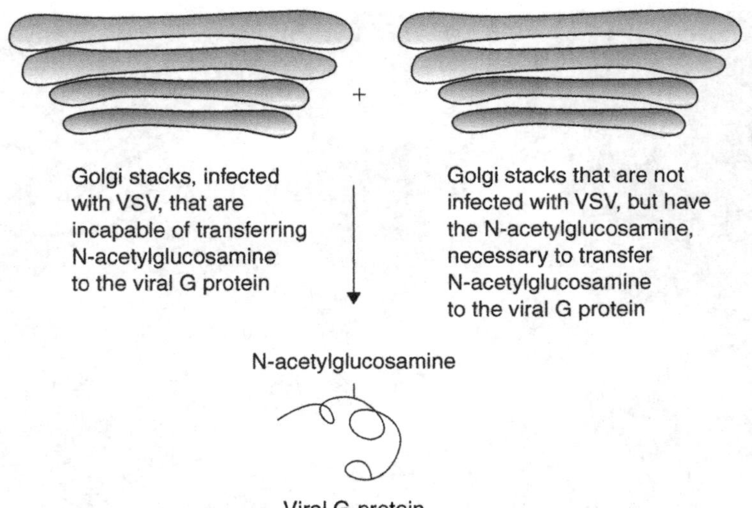

FIGURE 6.12 Mutant Chinese hamster ovary cells were infected with vesicular stomatitis virus (VSV), which produces a membrane protein known as viral G protein. The mutant cells were missing an N-acetylglucosamine transferase and were incapable of transferring N-acetylglucosamine to the viral G protein. When Golgi stacks isolated from these cells were mixed with Golgi stacks isolated from uninfected wild-type cells, the viral G protein was glycosylated, indicating that their transport between the Golgi cisternae can occur.

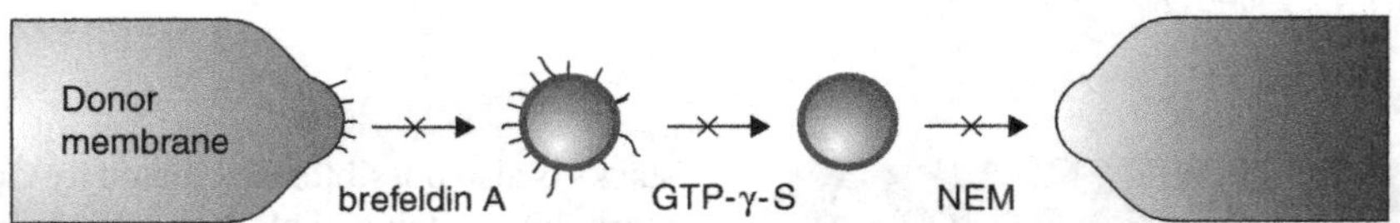

FIGURE 6.13 Diagram of GTP-dependent vesicular transport between two membranes, indicating possible sites of inhibitor action.

cells and move to the Golgi apparatus of the uninfected cells and become glycosylated with N-acetylglucosamine. It did! However, to make the experiments more exacting, Fries and Rothman reduced the time in which they pulse-labeled the proteins in the infected cell to determine the time it took for the viral G protein to move from the ER to the Golgi apparatus. In this way, they could make sure that they were obtaining homogenates from the infected mutant cells while the viral G protein was still in the ER.

Unhappily, when they redid the membrane-transfer experiment under conditions when they were sure that the viral G protein was starting in the ER, they found that the viral G protein was not processed by the wild-type Golgi apparatus, indicating that they had not yet developed the conditions necessary for ER-to-Golgi transfer. However, if they waited 10 minutes and used homogenates from the infected mutant cells in which the labeled protein had already moved to the Golgi apparatus, they found glycosylation of the protein by the wild-type Golgi apparatus! This indicated that there must be vesicular and/or tubular pathways to move material from cisterna to cisterna. While Rothman favors the view that vesicles are involved, the published electron micrographs are consistent with both the vesicular and tubular hypotheses (Morré and Keenan, 1997). Glycosylation of the viral G protein also occurs in a similar manner in vivo when wild-type cells are fused with infected mutant cells (Rothman et al., 1984b).

Rothman's in vitro cell-free work on mammalian systems has been complemented by in vivo genetic studies done by Randy Schekman (2013) on yeast cells. Schekman "was confident that the combined genetic and biochemical approach could prove crucial in the elucidation of …complex cellular processes." He treated yeast cells with the mutagen ethyl methanesulfonate and found a number of temperature-sensitive mutants known as *sec* mutants that had normal phenotypes at 24 C but at high temperature (37 C) were defective in their ability to *sec*rete (Novick and Schekman, 1979; Schekman, 1996).

To detect the mutants, Schekman used yeast cells that secrete a number of enzymes, including invertase, into the wall and also have a sulfate transporter on the plasma membrane. The transporter is incapable of distinguishing between sulfate and chromate. Because chromate is a toxin, the yeast cells would be killed when the cells were placed in chromate. Realizing that the sulfate transporter gets to the plasma membrane through the secretory pathway, Schekman searched for yeast cells that survived chromate treatment. In doing so he found many secretory mutants, where at 37 C, each mutant was blocked in a specific part of the secretory pathway, including the ER, the Golgi apparatus, and a variety of morphologically distinguishable vesicles between the ER and Golgi apparatus and between the Golgi apparatus and the plasma membrane (Novick et al., 1980; Esmon et al., 1981). Because the mutants retained the secretory proteins within the cell, they were denser than the wild type cells and could be readily separated by centrifugation through Ludox (Juhos, 1966), a commercial floor polish. This was an efficient way of selecting and capturing secretion mutants. Schekman used electron microscopy to see where in the secretory pathway the protein accumulated when the mutant was at 37 C but not at 24 C.

Using the classic double-mutation technique developed by Mitchell and Houlahan (1946) to determine the sequence of gene products, Novick et al. (1981) and Kaiser and Schekman (1990) determined the order of gene products necessary to identify the precursor–product relationship for secretion. In double mutants, the secretory protein accumulated at the earliest step blocked by the two mutations. To further characterize the genes that regulated secretion, Schekman used the complementation method developed by Hinnen et al. (1978) where they transfected the temperature-sensitive mutants with wild-type DNA to find the genes that restored the mutations. When they sequenced the genes that encoded the proteins involved in the secretory pathway, they found that many of the *sec* genes coded for guanosine triphosphate (GTP)–binding proteins (Salminen and Novick, 1987; Goud and McCaffrey, 1991; Schekman, 2013).

The in vitro and in vivo work on the mammalian and yeast systems came together when Vivek Malhotra added GTP-γ-S, a nonhydrolyzable analogue of GTP, to the mammalian cell-free system and found that GTP-γ-S inhibited membrane exchange between the cisternae of the Golgi apparatus (see Fig. 6.13; Malhotra et al., 1988, 1989; Wilson et al., 1989; Rothman, 1992). Moreover, this treatment caused the accumulation of vesicles the coats of which could be extracted by washing the isolated vesicles with 0.25 M KCl. In this way, many proteins could be isolated and their functions characterized. The proteins isolated from the coated vesicles included the ADP ribosylation factor (ARF), which is a GTP-binding protein, and the NEM-sensitive fusion protein (NSF), which is an ATPase.

These biochemical studies were combined with morphological and pharmacological studies to determine the precursor–product relationship in cisterna to cisterna transport (Balch et al., 1984a,b; Braell et al., 1984; Orci et al., 1986, 1989; Melançon et al., 1987; Block et al., 1988; Clary et al., 1990; Serafini et al., 1991; Waters et al., 1991; Ostermann et al., 1993; Söllner et al., 1993; Tanigawa et al., 1993; Søgaard et al., 1994; Rothman, 2010, 2013). For example, treatment with GTP-γ-S caused the accumulation of coated vesicles, whereas treatment with NEM caused the accumulation of naked vesicles. When the two drugs were added together, the coated vesicles accumulated, indicating that the coated vesicles give rise to the naked vesicles. These experiments suggest that GTP hydrolysis by ARF is necessary for the removal of the coat prior to vesicle fusion, and the NSF is required for the fusion of the naked vesicles to the target membrane. Other proteins isolated and characterized from the target membrane included the soluble NSF attachment proteins (SNAPs; Stenbeck, 1998), and the SNAP receptor proteins (SNAREs). Additional research showed that specific targeting occurs because each vesicle has a specific SNARE (v-SNARE) and each target membrane has a specific SNARE (t-SNARE) that allows for delivery at the right place and the right time (Rothman, 2013; Baker and Hughson, 2016).

Generally speaking, a region of the donor membrane, which contains the protein to be secreted, also has a protein known as v-SNARE that identifies the vesicle. The v-SNARE will eventually bind with the t-SNARE of the target membrane. The donor membrane then recruits proteins that coat the membrane and will become the coat of the vesicle that blebs off from the donor membrane. The specific proteins that form the coat determine the appearance of the vesicles and COP I-, COP II-, and clathrin-coated vesicles are made up of different coat proteins. The proteins in other coats, such as the lace-like coat, have not been characterized. The donor membrane also recruits a GTP-binding protein, known as ARF when a COP I- or clathrin-coated vesicle is going to form or SAR when a COP II-coated vesicle is going to form. As the coats polymerize on the bud of the donor membrane, the curvature of the coated vesicle increases until it pinches off. Once the vesicles are released from the donor membrane, the GTP-binding protein hydrolyzes the GTP to guanosine diphosphate (GDP) and the coats disperse thus exposing the v-SNARE on the naked vesicles. Brefeldin A binds to ARF and prevents vesicle formation. GTP-γ-S prevents the GTPase activity of ARF and prevents coat removal, and NEM binds to the NEM-sensitive fusion protein, which prevents the fusion of the vesicle with the target membrane.

Various proteins, including ARF, RAB, NSF, SNAP and SNARE, which are necessary for cisterna-to-cisterna movement, were purified based on their ability to restore intercisternal Golgi transport (Fries and Rothman, 1980; Balch et al., 1984a,b; Rothman et al., 1984a; Söllner et al., 1993; Warren, 1993; Whiteheart et al., 1993). RAB, is a GTP-binding protein that has an isoprenoid anchor that inserts into a vesicle membrane with a given v-SNARE and then binds to a specific receptor in the target membrane with a given t-SNARE (RAB stands for RAs-related in Brain, where RAS stands for a protein found in a RAt Sacrcoma). On bringing together the v-SNARE and t-SNARE, the vesicle fuses with the target membrane, and the GTP is hydrolyzed by the RAB-GTPase to GDP. This results in the release of RAB. The NSF and SNAPs are soluble proteins that first bind to the t-SNARE complex in the target membrane and then to the v-SNARE of the vesicle. In the presence of ATP-γ-S, a nonhydrolyzable form of adenosine triphosphate (ATP), the fusion complex is stable. However, in the presence of ATP, the ATPase activity of NSF is activated, which results in the dispersal of NSF, SNAP and the SNAREs. Thus the intercisternal transfer of membranes requires both ATP and GTP. Specific forms of ARF, which act in the budding process, and RAB and SNARE, which are involved in the targeting and tethering process, ensure the specificity of vesicle trafficking between donor and target membranes throughout the cell (Segev, 2001; Saito and Ueda, 2009; Bhuin and Roy, 2014).

The Rab and Arf genes are members of the Ras monomeric GTPase superfamily, which also includes Ran, Ras and Rop (Rho-related GTPase from Plants). These proteins contribute to many aspects of cell physiology (Brembu et al., 2006). In plants, Rab and Arf participate in the secretory pathway (se Chapter 8), Rop participates in the regulation of the actin cytoskeleton (see Chapter 10), and Ran participates in the translocation of proteins through the nuclear envelope (see Chapter 16).

In plants, genes that code for SAR, ARF, SNAP, RAB and SNARE, which are homologous to the *sec* and *rab* genes found in yeast and mammalian cells, have been identified. The identification of the genes and their proteins involved in the flow of membranes from cisterna to cisterna has been facilitated by finding yeast mutants that are defective in secretion. Many of these *sec* mutations act directly on the Golgi apparatus. The mutants defective in a given aspect of Golgi-mediated secretion are then transformed with plant DNA that putatively encodes a protein that acts in Golgi transport. If the mutant yeast strain is rescued by transfection with a given plant DNA sequence, then that sequence is assumed to code for a protein involved in the Golgi-mediated secretory system of that plant. The functions of many plant proteins have been discovered in this manner (Bednarek et al., 1994; Bassham and Raikhel, 2000; Bassham et al., 1995; Conceicao et al., 1997; d'Enfert et al., 1992; Zheng et al., 1999, 2004; Neumann et al., 2003; daSilva et al., 2004).

The plant genes involved in the secretory pathway have then been mutated and/or coupled to the sequence that codes for fluorescent proteins using genetic engineering techniques. In this way, the mechanism of action and intracellular localization of the gene products have been studied in plants (Staehelin and Moore, 1995; Sanderfoot et al., 2000; Pereira-Leal and Seabra, 2001; Rutherford and Moore, 2002; Vernoud et al., 2003; Pratelli et al., 2004; Uemura et al., 2004; Sutter et al., 2006; Lipka et al., 2007; Matheson et al., 2007; Min et al., 2007; Robinson et al., 2007; Sanderfoot, 2007; Zhang et al., 2007; Bassham and Blatt, 2008; Groen et al., 2008; Moshkov and Novikova, 2008; Nielsen et al., 2008; Rojo and Denecke, 2008; Woollard and Moore, 2008; Kang et al., 2011; Cevher-Keskin, 2013; Du et al., 2013; Yorimitsu et al., 2014).

Plants have many more isoforms of some of the secretory gene products than yeast. There is an enormous diversity in the genes that facilitate movement between membranous compartments. The function of a given homolog is determined by visualizing secretion microscopically in a given plant cell that has been transiently transformed using the gene gun with an engineered copy of a gene that encodes a protein that functions in the secretory system (Dairman et al., 1995). Such experiments show that specific homologs of a secretory protein function in the transport of vesicles between given organelles, in a given direction, in a given cell type, during a given stage of development, or in response to a given environmental stimulus (Cheung et al., 2002; Goncalves et al., 2007). In this regard, the genetic architecture of the genes involved in secretion in unicellular organisms such as yeast provides only a first-order approximation for the more complicated genetic architecture involved in the secretory processes that take place in the cells that make up multicellular plants.

6.7 POSITIONING OF THE GOLGI APPARATUS

In many plant taxa, the Golgi bodies are remarkably mobile organelles that can utilize the actomyosin system for their movement (Boevink et al., 1998; Nebenführ et al., 1999; Nebenführ and Staehelin, 2001; Avisar et al., 2008, 2009; Peremyslov et al., 2008) and can be closely associated with specific locations on the ER known as endoplasmic reticulum export sites (ERESs; daSilva et al., 2004; Yang et al., 2005; Donohoe et al., 2007, 2013; Staehelin and Kang, 2008; Osterrieder et al., 2010; Schoberer et al., 2010; Ito et al., 2012a,b; Brandizzi, 2017), which may be synonymous with the transition ER. In growing pollen tubes, Golgi bodies are at the base of the vesicular region and along the flank of the tube, whereas the Golgi-derived and endocytotic vesicles are at the growing tip (Lancelle and Hepler, 1992; Hepler and Winship, 2015). Somewhat similarly, in *Penium*, the 100–125 Golgi bodies are stationary and far from the site of wall deposition and the vesicles move to the site of wall growth (Domozych, 2014).

The cause of the geographical position of the Golgi apparatus and the Golgi-derived vesicles in the cell is a wonderful puzzle. As a consequence of the importance of cell polarity in many fascinating processes on plant and cell growth and development, plants serve as ideal organisms for studying the position of the Golgi apparatus and the Golgi-derived vesicles. Many plant and fungal cells have a specialized type of polarized growth called *tip growth* where Golgi-derived vesicles are targeted to a single locus in the growing cell (Sievers, 1963; Rosen et al., 1964; Rosen, 1968; Grove et al., 1970; Franke et al., 1972; Bartnik and Sievers, 1988; Steer and Steer, 1989; Hepler et al., 2013b; Hepler and Winship, 2015). Moreover, plant organs and presumably the cells within them also show differential growth in response to gravity and light. Thus, it is possible that differential growth results from the differential distribution of the Golgi stacks or Golgi-derived vesicles on opposite sides of the cell (Shen-Miller and Miller, 1972; Shen-Miller and Hinchman, 1974). Differential positioning of the Golgi apparatus or Golgi-derived vesicles may also result in the remarkable annular, spiral, scalariform, and reticulate secondary wall thickenings that occur in tracheary elements (Bierhorst, 1971).

6.8 SUMMARY

The Golgi apparatus plays a central role in the flow of membranes, proteins, and carbohydrates through the cell. It is a factory with many loading docks involved in bringing in raw materials, sending back defective parts, delivering processed materials to the plasma membrane and vacuolar compartments, and retrieving recycled merchandise. Perhaps its protean morphology results from the fact that it is the obligate intermediate between the ER on the one side and the cell surface and the vacuolar compartment on the other side and may constantly adjust to the demands of the rest of the secretory pathway. Indeed, determining the precise boundaries of this pivotal organelle with respect to the rest of the organelles involved in the secretory pathway is reminiscent of the recurrent problem of determining the relationships of the parts to the whole.

6.9 QUESTIONS

6.1. What is the function of the Golgi apparatus?
6.2. How does its structure reflect its function?
6.3. How may its polarity affect its function?
6.4. How might the structure and function of the Golgi apparati differ in a growing cell, in a cell in steady state, and in a senescing cell?

The references for this chapter can be found in the references at the end of the book.

Chapter 7

The Vacuole

Today we have gathered for the awarding of this year's Nobel Prizes in the Stockholm Concert Hall, one of Sweden's architectural masterpieces, which was inaugurated in 1926. The concert hall is in good shape despite intensive use over many decades, thanks to repairs and maintenance. Today it is also newly cleaned and spotless. The consumables that previous guests have left behind have been collected and hopefully recycled. We view this daily maintenance as something self-evident, but if the concert hall had not been repaired or cleaned in 90 years this would of course have been obvious to us. Similarly, our body's cells are constantly replacing defective components and removing waste materials. This year's Nobel Prize in Physiology or Medicine is about how the cell maintains its interior and recycles its own components in order to function optimally and adapt to new circumstances.

Nils Larsson (Presentation Speech for the 2016 Nobel Prize in Physiology or Medicine).

7.1 DISCOVERY OF THE VACUOLE

Most plant cells contain a conspicuous central region that appears empty in the light microscope (von Mohl, 1852). This region, which includes a transparent, or rarely colored, watery substance known as the cell sap, is called the *vacuole*, a term that comes from the Latin word for "empty." The large central vacuole can take up approximately 95% of the protoplast volume, although in typical higher plant cells, it takes up approximately 60% or more (Winter et al., 1994; Noguchi and Hayashi, 2014; see Fig. 7.1). A large central vacuole is not limited to plant cells. According to Robert Russell "R.R." Bensley (1951), a large central vacuole is also found in the cells of the flagellate *Noctiluca*, the ciliate *Trachelius*, the ectoderm and endoderm of the Coelenterates, and in the Heliozoa.

Vacuoles were first observed in protozoa. The contractile vacuoles or "stars" of many protozoa were seen by Lazzaro Spallanzani (1776), although he mistook them for respiratory organs (see Zirkle, 1937). These "stars" were named *vacuoles* by Félix Dujardin (1841). Although the optically structureless cell sap had been observed by botanists for years, the term *vacuole* was first applied to plant cells by Matthias Schleiden in 1842 when he distinguished the vacuole from the rest of the protoplasm (Zirkle, 1937).

The cell sap is surrounded by a differentially permeable membrane as determined from osmotic studies done by Hugo de Vries on *Tradescantia* epidermal cells and many other cell types (1884, 1885, 1888a,b). In these studies, he noticed that the cell walls bulged when the cells were placed in pure water. As he increased the concentration of solutes in the external solution, the walls relaxed, and at higher concentrations of solutes, he observed that the violet-colored vacuole shrank. de Vries concluded that a membrane must surround the cell sap for the vacuole to behave as an osmometer. He coined the term *tonoplast* to designate the membrane that surrounded the cell sap. The tonoplast was so named because he thought that it was the regulator of turgor, also known as tonicity, in the cell (de Vries, 1910). He mistakenly believed that the tonoplast was differentially permeable but the plasma membrane was not, and consequently, only the tonoplast regulated turgor.

de Vries (1885) also thought that the tonoplast was an autonomous self-replicating particle in the cell. However, Wilhelm Pfeffer (1900–06) showed that vacuoles are not autonomous but form de novo from the plasma membrane during phagocytosis. Nowadays it is possible to observe vacuoles develop in evacuolated (Hörtensteiner et al., 1992) or vacuoleless protoplasts (Davies et al., 1996). Because the tonoplast is neither self-replicating nor the primary site of turgor regulation, I will use the term *vacuolar membrane* to denote the differentially permeable membrane that surrounds the cell sap, as suggested by Pfeffer (1886).

7.2 STRUCTURE, BIOGENESIS, AND DYNAMIC ASPECTS OF VACUOLES

Although a few meristematic cells, including the apical cell in *Osmunda* and *Lunularia*, and the cambial initials in higher plants have prominent vacuoles (Bailey, 1930; Sharp, 1934), the vacuole is inconspicuous in most meristematic cells (Porter and Machado, 1960). The development of the vacuole can be followed in the light microscope. For example, Pensa (see Guilliermond, 1941) looked at the development of the vacuolar system in the cells of the teeth of young, living rose leaflets (Fig. 7.2).

Plant Cell Biology. https://doi.org/10.1016/B978-0-12-814371-1.00007-2

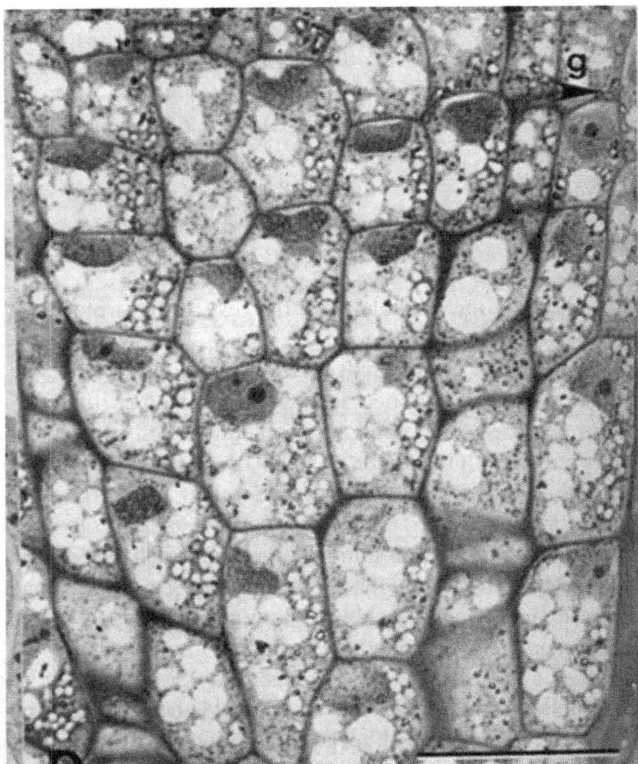

FIGURE 7.1 The vacuoles are a prominent component in the columella cells of this agravitropic mutant of barley. Bar, 50 μm. *From Moore, R., 1985. A morphometric analysis of the redistribution of organelles in columella cells in primary roots of normal seedlings and agravitropic mutants of Hordeum vulgare. J. Exp. Bot. 36, 1275–1286.*

The vacuolar systems in these cells are easy to observe because the vacuoles are filled with anthocyanin. In the youngest cells at the tip, the vacuoles appear as numerous, tiny filamentous elements. In slightly older cells, these filamentous elements appear to swell. Eventually, in the mature cells at the base, the swollen elements fuse into larger vacuoles and eventually form a large central vacuole. Dangeard (1919) gave the name *vacuome* or *vacuolar system* to all the vacuoles contained in the cell during all its phases of development.

A similar vacuolar development can be seen in maturing barley root cells stained with neutral red, a vital stain that is preferentially taken up into acidic compartments (Fig. 7.3). Other good examples of vacuolar development include the epidermal cells of young leaves and the hairs on the sepals of *Iris germanica*, the glandular hairs on the leaflets of walnut, and the leaves of *Anagallis arvensis*. By contrast, the vacuolar system of *Elodea canadensis* never goes through a filamentous stage but starts as small spherical vacuoles that later fuse into a large central vacuole (Guilliermond, 1941).

A more recent study of vacuolar development has been done by Palevitz and O'Kane (1981) and Palevitz et al. (1981) using the autofluorescent vacuole found in the developing guard cells of *Allium cepa* (Zeiger and Hepler, 1976, 1977, 1979). They find that the vacuoles of young guard mother cells are globular. As the guard mother cells develop, the vacuole is transformed into a reticulum of interlinked tubules and small chambers. The tubules appear

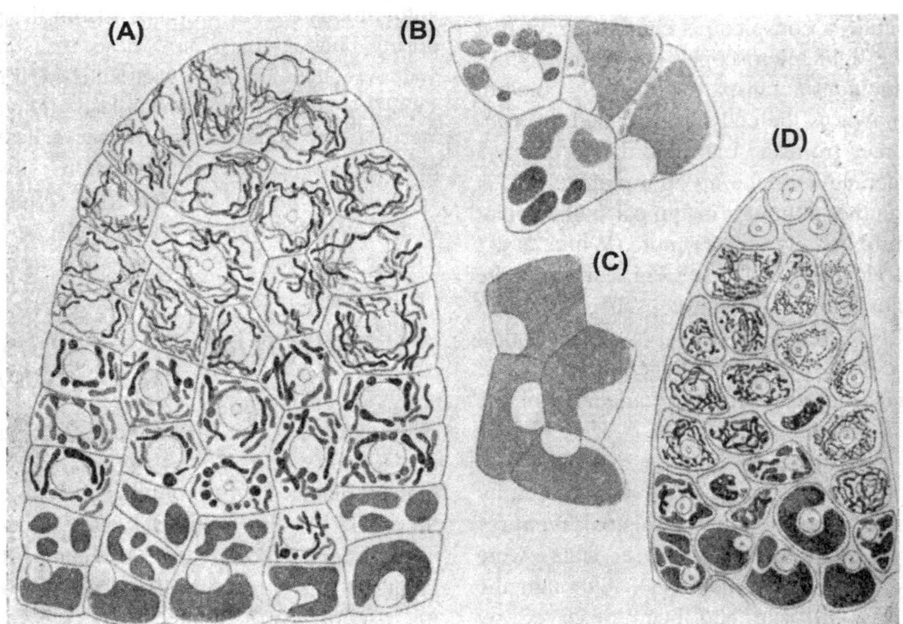

FIGURE 7.2 The anthocyanin-containing vacuolar system in the cells of the teeth of young, living rose leaflets. (A and D) Cells at tip; (Band C) older cells. *From Guilliermond, A., 1941. The Cytoplasm of the Plant Cell. Chronica Botanica Co., Waltham, MA.*

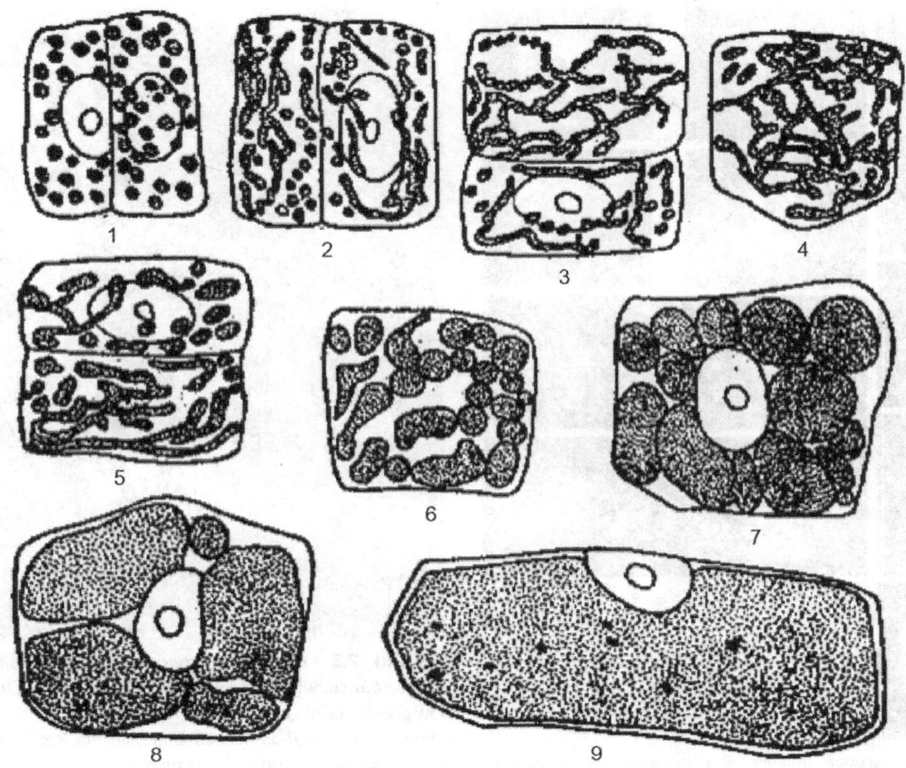

FIGURE 7.3 The vacuolar system in cells of a barley root vitally stained with neutral red. 1–5 are meristem cells, 6–8 are in the region of differentiation, and 9 is a mature cortical parenchyma cell. *From Guilliermond, A., 1941. The Cytoplasm of the Plant Cell. Chronica Botanica Co., Waltham, MA.*

to be approximately 100–500 nm in diameter. In the guard mother cell, the network continually undergoes changes in shape and remains reticulate during the division that gives rise to the two guard cells. The reticulate networks persist through the early stages of guard cell differentiation, and then they are transformed into two large globular vacuoles, one in each guard cell (Fig. 7.4). The dynamics of the vacuolar compartment have also been confirmed using various fluorescent vital stains and vacuolar proteins fused to green fluorescent protein (Flückiger et al., 2003; Hicks et al., 2004; Zouhar et al., 2004; Viotti et al., 2013; Hepler and Winship, 2015; Krüger and Schumacher, 2017). A vacuolar reticulum has also been observed with electron microscopy (Limbach et al., 2008).

The developmental pattern seen in vacuoles can be reversed under certain physiological conditions. For example, Charles Darwin (1897) noticed that the vacuole of the tentacle of the carnivorous plant *Drosera rotundifolia*, which is filled with anthocyanin, appears to breakup after the leaf is stimulated by an insect. (Actually, Darwin misidentified the vacuole as protoplasm.) The cells of the tentacles contain a single central anthocyanin-filled vacuole. At the moment of stimulation, the vacuole fragments into filamentous vacuoles. Immediately after stimulation, the filamentous vacuoles fuse to form a large central vacuole and the cell returns to its initial state (de Vries, 1886; Guilliermond, 1941; Lloyd, 1942; Juniper et al., 1989).

The vacuole can be defined operationally as a swollen terminally differentiated intracellular membrane–bound compartment of the secretory pathway (Marty, 1978). The minimal requirement for the formation of vacuoles is the synthesis of a vacuolar membrane that contains the transporters necessary to increase the osmotic pressure of the lumen. The increase in the osmotic pressure will allow the newly formed vacuoles to swell until the water potential of the vacuole is in equilibrium with the water potential of the cytosol. Moreover, continued membrane synthesis and/or delivery must occur if the vacuolar membrane thickness is to remain constant. The biogenesis of vacuoles does not occur by a single pathway (Robinson and Hinz, 1997; Marty, 1999), and at the electron microscopic level, we can see that vacuoles can form in a variety of ways (Marinos, 1963; Ueda, 1966; Matile and Moor, 1968).

Using electron microscopy, Francis Marty (1978, 1997, 1999) studied vacuole formation in meristematic cells of roots. The cells adjacent to the quiescent cells of the root are the most undifferentiated and do not have any vacuoles. In slightly older cells, primordial vacuole precursors, or provacuoles, arise from the trans-Golgi network

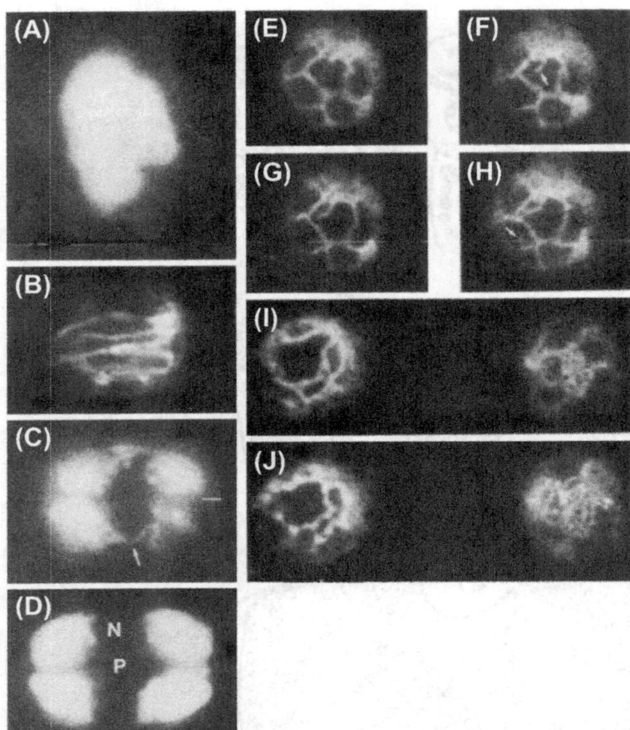

FIGURE 7.4 Autofluorescence images of various stages in vacuole development during stomatal differentiation in onion seedlings. *From Palevitz, B.A., O'Kane, D.J., 1981. Epifluorescence and video analysis of vacuole motility and development in stomatal cells of Allium. Science 214, 443–445.*

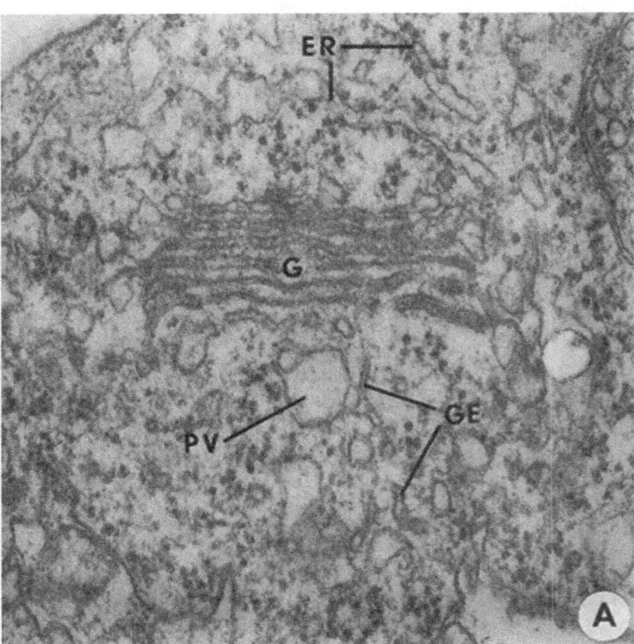

FIGURE 7.5 Electron micrograph of a meristematic root tip cell of *Euphorbia* showing the relationship between the Golgi stack (G) and the Golgi-associated endoplasmic reticulum (ER) from which lysosomes appear. *GE*, Golgi-associated Endoplasmic Reticulum from which Lysosomes apparently form; *PV*, provacuole.

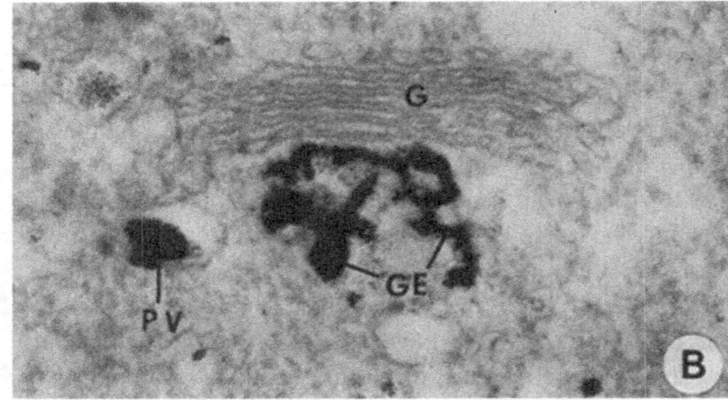

FIGURE 7.6 Electron micrograph of a meristematic root tip cell of *Euphorbia* stained with zinc iodide and osmium tetroxide showing the relationship between the Golgi stack (G), the Golgi-associated endoplasmic reticulum, from which lysosomes apparently form (GE), and a provacuole (PV). ×64,000. *From Marty, F., 1978. Cytochemical studies on GERL, provacuoles and vacuoles in root meristematic cells of Euphorbia. Proc. Natl. Acad. Sci. U.S.A. 75, 852–856.*

(Figs. 7.5–7.7). The provacuoles eventually form an anastomozing network of tubules, which then wrap themselves around portions of the cytoplasm like bars of a birdcage (Fig. 7.8). Subsequently, the tubules fuse, thus entrapping the enclosed cytoplasm in a double membrane. At this point, various hydrolases are probably released from the E-space between the two vacuolar membranes. This leads to autophagy of the enclosed cytoplasm, and the inner membrane of the provacuole becomes totally degraded (Fig. 7.9). The provacuole or phagophore, as it is called in mature cells, can selectively target protein aggregates as well as specific organelles such as peroxisomes, mitochondria, and chloroplasts. In these cases of selective autophagy, the processes are known as pexophagy, mitophagy, and chlorophagy, respectively (Wada et al., 2009; Mizushima et al., 2011; Floyd et al., 2012; Li and Vierstra, 2012; Shibata et al., 2013; Avin-Wittenberg and Fernie, 2014; Michaeli and Galili, 2014; Anding and Baehrecke, 2017; Izumi et al., 2017). Eventually the newly formed vacuoles fuse to form larger vacuoles (Fig. 7.10). *Arabidopsis* mutants have been found that do not form a large central vacuole (Isono et al., 2010; Kolb et al., 2015).

Vacuoles can also form directly from the endoplasmic reticulum (ER; see Fig. 7.11; Herman, 2008; Sabelli and Larkins, 2009; Viotti, 2014). Electron microscopy of cereal (e.g., wheat, maize, and rice) endosperm cells show that there is a population of ER, known as protein-body ER, which gives rise directly to protein storage vacuoles that store prolamins. Prolamins are proteins that are rich in sulfur-containing amino acids, soluble in 50%–95%

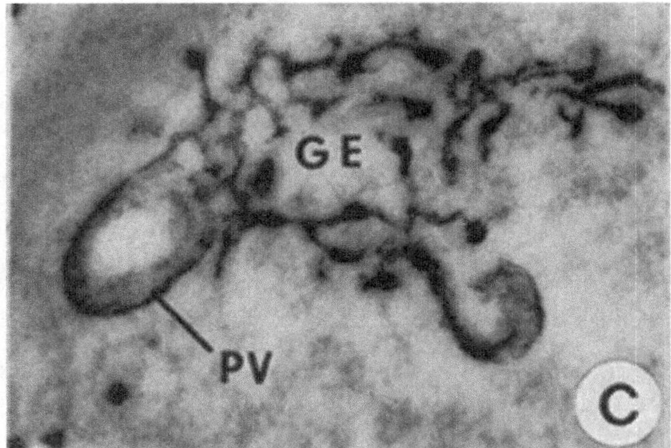

FIGURE 7.7 Electron micrograph of a meristematic root tip cell of *Euphorbia* stained with zinc iodide and osmium tetroxide showing the relationship between the Golgi-associated endoplasmic reticulum from which lysosomes apparently form (GE) and a provacuole (PV). ×51,500. *From Marty, F., 1978. Cytochemical studies on GERL, provacuoles and vacuoles in root meristematic cells of Euphorbia. Proc. Natl. Acad. Sci. U.S.A. 75, 852–856.*

aqueous ethanol, and as a result of forming disulphide bonds, they are insoluble in water. By contrast, protein storage vacuoles that store other storage proteins known as *glutelins* (which are soluble in dilute acid or base) form from clathrin-coated vesicles that bud from the trans-Golgi network and mature into protein bodies after the coats are shed (Nishimura and Beevers, 1978, 1979; Parker and Hawes, 1982; Nieden et al., 1984; Herman and Shannon, 1984a,b, 1985; Greenwood and Chrispeels, 1985a,b; Harris, 1986; Faye et al., 1988; Robinson et al., 1989; Hoh et al., 1991; Levanony et al., 1992; Li et al., 1993a). Replacing the cysteines of prolamins with serines using genetic engineering results in the engineered prolines becoming water soluble and moving through the secretory pathway (Mainieri et al., 2014). Specifics of the biogenesis of protein bodies vary in a species-specific manner and during different stages of seed development within a given species (Marty, 1997). Further evidence that the ER is involved in vacuole formation comes from studies in *Arabidopsis* that show that at least some of the anthocyanin that is ultimately found in the vacuole is transported initially into the ER lumen (Poustka et al., 2007).

The protein bodies contain the hydrolytic enzymes necessary for the breakdown of the resident food storage proteins (Van der Wilden et al., 1980; Herman et al., 1981). There is evidence that mature vacuoles can form from the merger of vacuoles produced by two independent pathways; one that produces the storage proteins and one that produces the hydrolases (Paris et al., 1996). There is also evidence that the protein bodies are directly transformed into lytic vacuoles (Zheng and Staehelin, 2011).

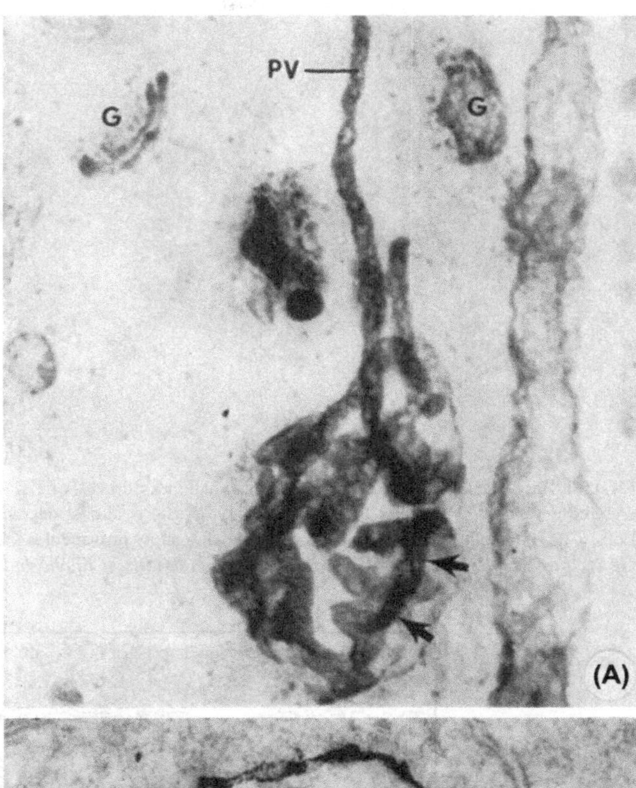

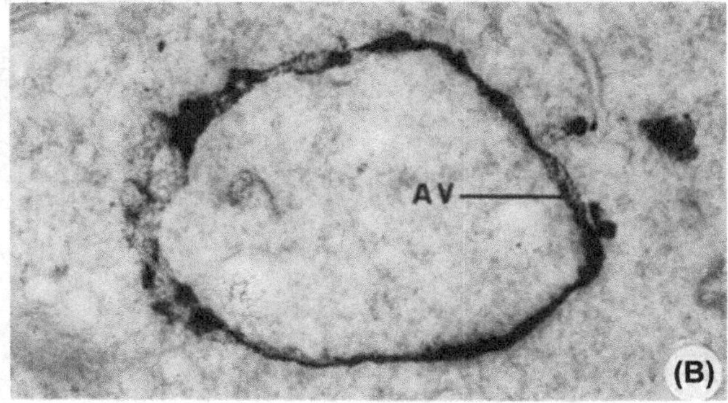

FIGURE 7.8 Electron micrographs of a meristematic root tip cell of *Euphorbia* stained with zinc iodide and osmium tetroxide showing the sequestration cage structure of a provacuole (PV) (A) and a transverse section of the same structure showing the typical ringlike structure of an autophagic vacuole (AV) (B). *G*, Golgi stack. *From Marty, F., 1978. Cytochemical studies on GERL, provacuoles and vacuoles in root meristematic cells of Euphorbia. Proc. Natl. Acad. Sci. U.S.A. 75, 852–856.*

When it comes to the interpretation of electron microscopic images, there is some contention concerning how vacuoles arise within the endomembrane system. In part, the disagreement comes from the lack of temporal resolution that is necessary to observe the dynamic three-dimensional behavior of the vacuolar compartment that is apparent in the light microscope. These dynamics are lost in static thin sections that have high two-dimensional spatial resolution, but low temporal and three-dimensional spatial resolution (Manton, 1962). Moreover, chemical fixation causes the transformation of a reticulate motile vacuolar

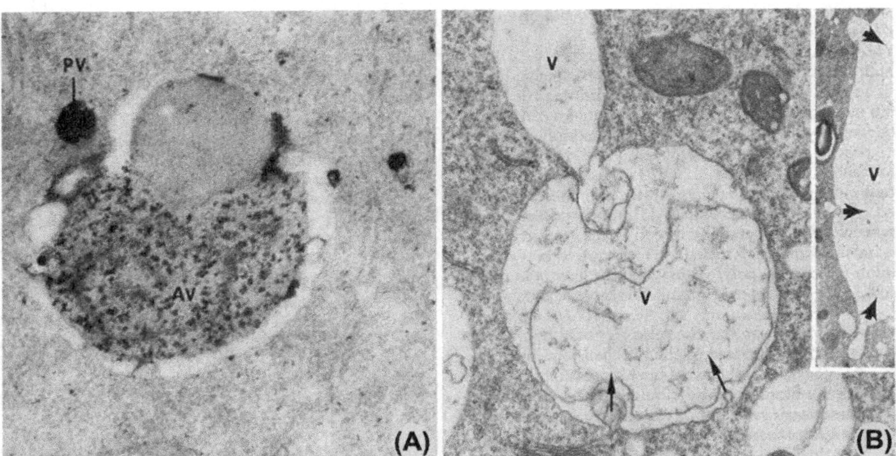

FIGURE 7.9 Electron micrograph of a meristematic root tip cell of *Euphorbia* showing the sequestered cytoplasm within an autophagic vacuole (AV) (A) and a vacuole (V) where the sequestered cytoplasm is almost degraded (B). The inset shows provacuoles (PVs) merging with the large vacuole. Left, ×42,100; right, ×20,600; inset, ×8000. *Arrows* show provacuoles merging with the large vacuole. *From Marty, F., 1978. Cytochemical studies on GERL, provacuoles and vacuoles in root meristematic cells of Euphorbia. Proc. Natl. Acad. Sci. U.S.A. 75, 852–856.*

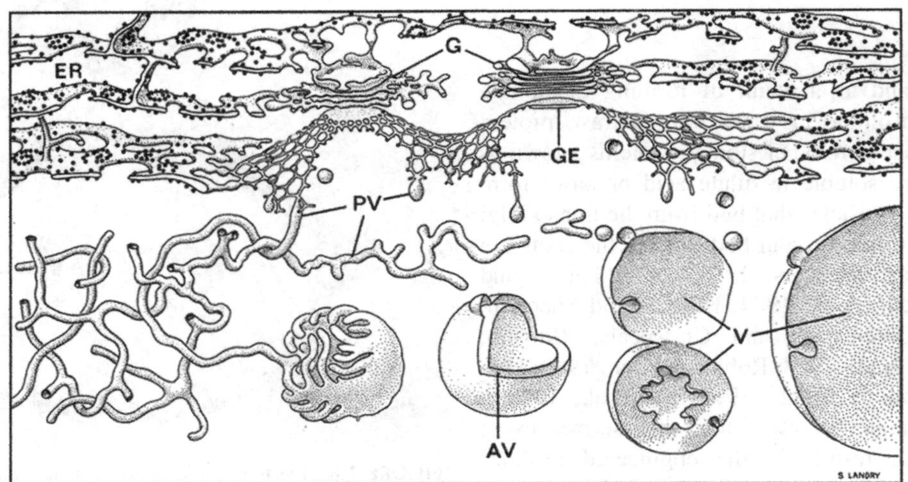

FIGURE 7.10 Various stages in the formation of vacuoles from the Golgi apparatus. *AV*, autophagic vacuole, *ER*, endoplasmic reticulum; *G*, Golgi stack; *GE*, Golgi-associated Endoplasmic Reticulum from which Lysosomes apparently form; *PV*, provacuole; *V*, vacuole. *From Marty, F., 1978. Cytochemical studies on GERL, provacuoles and vacuoles in root meristematic cells of Euphorbia. Proc. Natl. Acad. Sci. U.S.A. 75, 852–856.*

system into spherical vesicles, causing the connections between various compartments to become obscured (Wilson et al., 1990). Thus, we must be cautious of interpretations of structure based on chemically fixed specimens and single sections.

Building on a strong biochemical, biophysical, morphological, and physiological background, Yasuhiro Anraku and his colleagues have taken a genetic approach to understand vacuole biogenesis in yeast (Nishikawa et al., 1990; Wada et al., 1990, 1992; Wada and Anraku, 1992). They have made a series of mutants that block vacuole biogenesis at a variety of points and even inhibit vacuole formation altogether. Yoshihisa and Anraku (1990) have found that while most proteins enter the vacuole through the ER-Golgi pathway, and that the majority of the vacuole forms from the budding of membranes from the trans-Golgi network, α-mannosidase, a marker enzyme for the vacuolar membrane, does not have a signal peptide, is not glycosylated at its N-X-S/T site, and contains no complex carbohydrates, indicating that it enters the vacuole in a manner that does not involve the ER-Golgi pathway. Other proteins also enter the vacuole directly from the cytosol (Klionsky and Ohsumi, 1999). Thus, even a single vacuole may result from the work of many subcellular units (Pedrazzini et al., 2013).

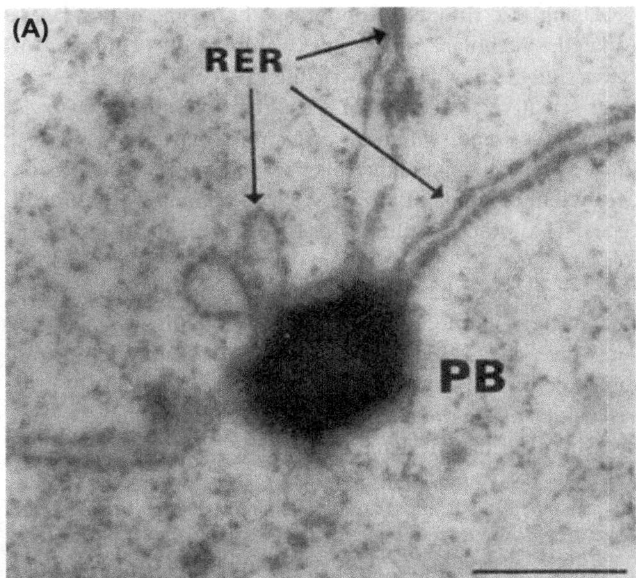

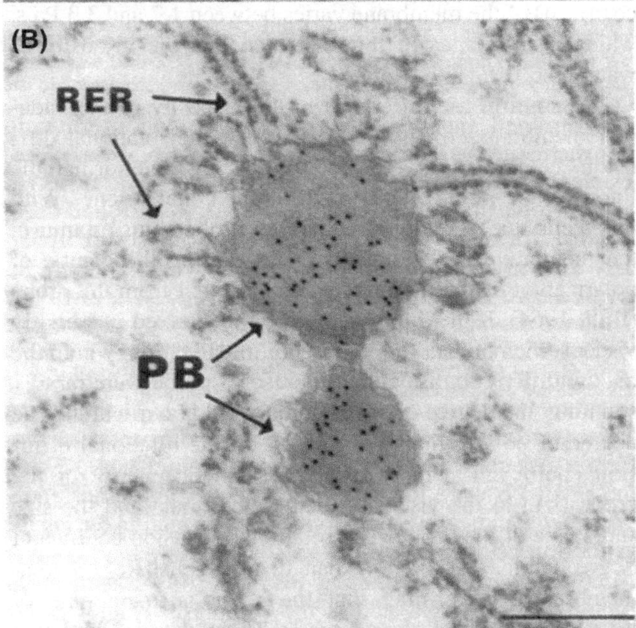

FIGURE 7.11 (A) A protein body (PB) forming directly from the rough endoplasmic reticulum (RER) in endosperm cells of wheat. (B) The PB is immunolabeled with gold particles that are indirectly attached to an antibody directed against prolamins. Bars, 500 nm. *From Levanony, H., Rubin, R., Altschuler, Y., Galili, G., 1992. Evidence for a novel route of wheat storage proteins to vacuoles. J. Cell Biol. 119, 1117–1128.*

7.3 ISOLATION OF VACUOLES

While studying the enzymology of carbohydrates and the mechanism of action of insulin, Christian de Duve assayed the activity of acid phosphatase, which had optimal activity at pH 5 and was soluble after the tissue was homogenized in a Waring blender. When de Duve started using the centrifugation method developed by Albert Claude, who separated cells into the nuclear, mitochondrial, microsomal, and supernatant fractions, he found that the acid phosphatase activity showed up in the mitochondrial fraction. And moreover, the activity of the enzyme was so much lower than it would be if the whole tissue had been homogenized in a Waring blender. de Duve obtained these puzzling results just before a weekend. They put the samples in the refrigerator. Five days later they assayed the enzyme again and found that the enzyme activity in the mitochondrial fraction increased dramatically. de Duve then showed that the effect of "aging" could be mimicked by homogenizing the fraction with the blender, subjecting the fraction to freeze-thaw cycles or by treating the fraction with detergent—all treatments that disrupted membranes. de Duve deduced that the "latency" was due to the enzyme being sequestered within a membrane, which made it inaccessible to the substrates. de Duve then used his centrifuge to separate the membrane fraction that contained latent acid phosphatase from the mitochondria. This was the first successful isolation of vacuoles (de Duve et al., 1955; de Duve, 1975, 2012; Bowers, 1998; Opperdoes, 2013; Sabatini and Adesnik, 2013). de Duve called the vacuoles lysosomes because they had a number of nonspecific hydrolytic enzymes, which can cause the lysis (dissolution) of the soma (body). This is why lysosomes are sometimes called *suicide sacs*. He defined the lysosome as an organelle surrounded by a single membrane that contains a number of nonspecific hydrolases, including acid phosphatase, an enzyme that can be easily localized cytohistochemically. The nonspecific hydrolases exhibit latency. That is, they do not show any activity when assayed with exogenous substrates. However, once the lysosomal membrane is permeabilized with a detergent such as Triton X-100, the enzymes are very active toward their substrates. de Duve coined the terms phagocytosis and autophagy to describe two pathways through which intercellular components and foreign components travel to be digested in the vacuole. Peroxisomes in rat liver cells were discovered as a contaminant of the lysosomal fraction (Chapter 5; Sabatini and Adesnik, 2013).

Vacuoles can be isolated from protoplasts. However, a large quantity of vacuoles can be isolated rapidly by mechanically slicing fresh or plasmolyzed tissue with a razor blade in a medium containing an osmoticum. The homogenate is filtered and centrifuged at 1300–3500 g to recover the vacuoles. The pellet is resuspended in 15% metrizamide. This suspension is overlayered with a layer of 10% metrizamide and an uppermost layer of 0% metrizamide. During centrifugation (500 g, 10 min), the vacuoles float and collect at the 10%–0% interface. The yield is low, but a large quantity of tissue can be processed in this manner (Cocking, 1960; Wagner and Siegelman, 1975; Leigh and Branton, 1976; Kringstad et al., 1980; Marty and Branton, 1980; Wagner, 1983; Trentmann and Haferkamp, 2013).

Isolated vacuoles can be further purified to obtain only tightly sealed vacuolar membrane vesicles. This can be done by centrifuging the membrane fraction in a density gradient made with a high–molecular mass polymer that is unable to penetrate intact vesicles. The density of intact vesicles will depend on the density of the cell sap (c. 1.01 g/mL; Kamiya and Kuroda, 1957), whereas the density of leaky vesicles will depend on the density of the membrane (c. 1.1–1.2 g/mL; Sze, 1985). Thus, the leaky vesicles will move down further into the gradient away from the intact vacuoles. Highly purified vacuolar membranes have also been isolated by using aqueous two-phase partitioning followed by using free-flow electrophoresis (Scherer et al., 1992).

7.4 COMPOSITION OF VACUOLES

There are over 100 proteins in the vacuole according to a biochemical analysis (Kenyon and Black, 1986), or as much as 650 proteins according to proteomic analyses (Carter et al., 2004; Sazuka et al., 2004, Shimaoka et al., 2004; Szponarski et al., 2004; Endler et al., 2006; Jaquinod et al., 2007; Schmidt et al., 2007; Whiteman et al., 2008a,b; Trentmann and Haferkamp, 2013). The cell sap contains a number of nonspecific hydrolytic enzymes that typically have acidic pH optima. These include proteases that split polypeptides into fragments by digesting internal peptide bonds (endopeptidases) and proteases that digest the terminal amino acids (exopeptidases) from the amino-terminus (aminopeptidases) or the carboxy-terminus (carboxypeptidases). The cell sap can also contain esterases (e.g., acid phosphatase), phosphodiesterases (e.g., RNase and DNase), and acyl-esterases (e.g., lipases). The cell is doubly protected from these nonspecific hydrolases because they are sequestered in the acidic vacuole, and if they were to be released, they would not function in the neutral cytosol due to their acidic pH optima (Matile, 1975).

α-Mannosidase is usually used as a cell sap marker in plant cells, although it is a vacuolar membrane marker in animal and yeast cells. In plant cells, the nitrate-sensitive, vanadate-insensitive H^+-ATPase is often used as a vacuolar membrane marker (Sze, 1985), although it also occurs on the membranes that give rise to the vacuole (Herman et al., 1994).

The lipids of the vacuolar membrane have been characterized, and they are similar to, but not identical with, the other membranes (see Table 7.1; Marty and Branton, 1980; Yoshida and Uemura, 1986). The similarities may be due to their similar function as a barrier and their differences may have a functional basis. For example, the activity of the vacuolar membrane H^+-ATPase is affected by its lipid environment (Yamanishi and Kasamo, 1993, 1994).

Using free-flow electrophoresis, Leborgne et al. (1992) isolated vacuolar membranes from cell cultures of *Eucalyptus*. They tested two lines of cells, one that was frost sensitive and one that was frost tolerant. They used fluorescence redistribution after photobleaching with a fluorescent phosphatidylcholine to determine the diffusion coefficient of lipids in the vacuolar membranes of both cell types. They find that the diffusion coefficients of the frost-tolerant type are greater than those of the frost-sensitive type, indicating that the vacuolar membrane of the frost-tolerant type is more fluid or less viscous than that of the sensitive type (Table 7.2). Given that the radius of a lipid molecule is approximately 0.4 nm (see Chapter 1), the viscosities of the vacuolar membrane can be obtained from the Stokes–Einstein equation:

$$D = \frac{kT}{6\pi r_H \eta} \tag{7.1}$$

The data of Leborgne et al. (1992) show that the viscosity (η) of the membrane varies between 1.7 and 3.3 Pa s, which is thousands of times greater than the viscosity of water (0.001 Pa s).

To confirm the difference in the viscosity of the vacuolar membranes of the two cell types, the rotational diffusion coefficients were determined by measuring the fluorescence polarization of 1,6-diphenylhexatriene. With this technique, the rate at which the probe spins around in the membrane is determined by measuring the degree of polarization of the fluorescent light emitted from the probe (Bull, 1964). If the fluorescent probe were fixed in a highly viscous membrane, the degree of polarization would be maximal. By contrast, if the fluorescent probe were rapidly spinning, the degree of polarization would be minimal. The degree of polarization is a measure of the rotational diffusion coefficient, which relates the kinetic energy of the probe (kT) to the viscosity of the membrane and the size and shape of the probe. The rotational diffusion coefficient is greater in frost-tolerant than in frost-sensitive vacuolar membranes, confirming that the vacuolar membrane of frost-tolerant cells is less viscous. These data indicate that through their effect on membrane viscosity, variations in lipid composition may be responsible for differences in frost tolerance or sensitivity.

7.5 TRANSPORT ACROSS THE VACUOLAR MEMBRANE

The lumen of the vacuole is an E-space and is topologically equivalent to the external space that surrounds the cell. The vacuolar membrane complements the plasma membrane and ER in its ability to transport many molecules and help maintain a cellular homeostasis (Fig. 7.12). The vacuolar membrane is also capable of generating an

TABLE 7.1 Lipid Composition of Vacuolar Membrane of Mung Beans

Lipid Phospholipids	Mol%						
Phosphatidylcholine (PC)	23.7						
Phosphatidylethanolamine (PE)	6.0						
Phosphatidylinositol (PI)	5.7						
Phosphatidylglycerol (PG)	2.3						
Phosphatidylserine (PS)	2.2						
Phosphatidic acid (PA)	1.1						
Subtotal	51.0						
Sterols							
Free sterols	18.2						
Acylated steryl glycoside	7.4						
Steryl glycoside	2.3						
Subtotal	27.9						
Other							
Ceramide monohexoside	16.6						
Monogalactosyl diglyceride	1.0						
Digalactosyldiglyceride	3.4						
Total	99.9						
Hydrocarbon Tails	PI	PS	PC	PE	PG	PA	Total
16:0	50.6	24.0	31.5	43.3	79.1	34.0	39.4
18:0	4.8	6.6	8.5	4.3	3.6	5.0	6.2
18:1	6.6	8.0	11.9	7.7	3.3	8.5	9.1
18:2	15.0	22.1	24.3	23.4	5.8	16.5	22.2
18:3	20.5	27.7	21.3	18.2	6.1	15.4	19.8
20:1	0.6	9.2	0.9	1.1	0.6	5.0	1.5
20:2	1.7	3.0	0.7	0.7	1.7	11.2	1.2
20:3		0.1	0.1	1.8			0.8
22:1		6.2	2.2	2.6			2.1
Unsaturated/saturated	0.81	2.27	1.50	1.10	0.21	1.56	1.19

From Yoshida, S., Uemura, M., 1986. Lipid composition of plasma membranes and tonoplasts isolated from eholated seedlings of mung bean (*Vigna radiata*). Plant Physiol. 82, 807–812.

action potential (Kikuyama and Shimmen, 1997). Various vacuolar membrane channels, carriers, and pumps have been characterized (Bennett and Spanswick, 1983, 1984a,b; O'Neill et al., 1983; Kaiser and Heber, 1984; Bennett et al., 1985; Blumwald and Poole, 1985, 1986; Lew et al., 1985; Rea and Poole, 1986; Bush and Sze, 1986; Hedrich et al., 1986; Blumwald et al., 1987; Hedrich and Kurkdjian, 1988; Hedrich et al., 1989; Johannes and Felle, 1989; Blackford et al., 1990; Schumaker and Sze, 1990; Maathuis and Sanders, 1992; Müller et al., 1996, 1997, 1999; Hirschi, 1999, 2001; Yamaguchi et al., 2001, 2003, 2005; Cheng et al., 2002; Müller and Taiz, 2002;

TABLE 7.2 Diffusion Coefficients for Phosphatidylcholine in Vacuolar Membrane

	D (in m^2/s)	
Cell line	**280 K**	**296 K**
Frost-tolerant	2.14×10^{-13}	3.22×10^{-13}
Frost-sensitive	1.65×10^{-13}	2.37×10^{-13}

From LeBorgne, N., Dupou-Cézanne, L., Teuliéres, C., Canut, H., Tocanne, J.-F., Boudet, A.M., 1992. Lateral and rotational mobilities of lipids in specific cellular membranes of Eucalyptus Gunnii cultivars exhibiting different freezing tolerance. Plant Physiol. 100, 246–254.

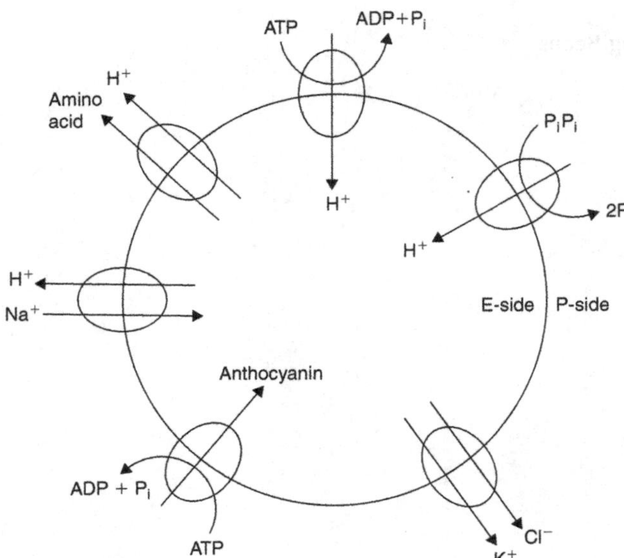

FIGURE 7.12 Diagram of the vacuolar membrane with a variety of transport proteins.

Pittman et al., 2002; Gaxiola et al., 2002; Sottosanto et al., 2004; Epimashko et al., 2006; Pottosin and Schönknecht, 2007; Schmidt et al., 2007; Shiratake and Martinoia, 2007; Schneider et al., 2008; Isayenkov et al., 2010; Hedrich and Marten, 2011; Chanroj et al., 2012; Emery et al., 2012; Beilby, 2016; Patel et al., 2016; Zhang et al., 2017). Some of the carriers and pumps are involved in Ca^{2+} homeostasis. At least two classes of primary proton pumps are involved in building up a proton difference across the vacuolar membrane that is utilized by secondary transporters to facilitate the transport of ions, sugars, amino acids, and other small molecules (Blumwald and Gelli, 1999). Other carriers, which are known as *ATP-binding cassette* (ABC) *transporters* or *traffic ATPases*, transport relatively large organic solutes (Rea et al., 1998; Klein et al., 1996, 1998, 2000, 2001). Work on the receptors that recognize cytosolically synthesized proteins and translocate them through the membrane is lagging behind the studies aimed at elucidating how proteins from within the secretory pathway enter the vacuole (Ahmed et al., 1997, 2000; Sanderfoot et al., 1998).

7.5.1 Proton-Translocating Pumps

Traditionally, studies on the vacuolar proton-translocating pumps have begun with biochemical studies (Churchill and Sze, 1983, 1984; Randall et al., 1985; Sze, 1985; Randall and Sze, 1986; Lai et al., 1988; Anraku et al, 1991; Nelson, 1992; Ward and Sze, 1995). The vacuolar H^+-pumping ATPase is known as the V-type ATPase and has been purified from isolated vacuolar membranes (see Fig. 7.12). It accounts for 6.5%–35% of the total vacuolar membrane protein and has a density of 970–3380 molecules per μm^2 (Ratajczak, 2000). The V-type ATPase is a 500-kDa protein complex (Rea et al., 1987a,b; Bowman et al., 1989; Nelson, 1989; Taiz et al., 1990; Ward et al., 1992; Ward and Sze, 1992b) that consists of two multipolypeptide components. One component (V_1) is a peripheral membrane complex, which contains the catalytic adenosine triphosphate (ATP)–hydrolyzing site. The other component (V_0) is an integral membrane complex that makes the proton channel. The ATPase can be dissociated from isolated vacuolar membranes by solubilizing it with Triton X-100. The solubilized protein is then separated from many of the other proteins by gel-filtration chromatography followed by anion-exchange chromatography. The canonical V-type H^+-ATPase is inhibited by NO_3, concanamycin A, bafilomycin, and dicyclohexylcarbodimide, but not by (VO_4) or azide.

The purified V-type ATPase is composed of 10–13 polypeptides. Gene sequence information has revealed that there are several isoforms of these polypeptides. A V-type ATPase has been reconstituted into proteoliposomes to determine its transport characteristics (Kasamo et al., 1991; Ward and Sze, 1992a). The ability of the reconstituted protein to form an ATP-dependent pH difference across the membrane is inhibited by gramicidin. Because gramicidin (Dubos, 1939) is an antibiotic that forms monovalent cation–conducting pores in membranes, the pH difference must be due to proton pumping as opposed to the transport of organic acids.

Moreover, H^+ pumping is stimulated by valinomycin, an antibiotic that functions as a K^+-selective ionophore. In the presence of valinomycin, one K^+ leaves the vesicle for every H^+ that is pumped in, and consequently, an electrical potential does not build up across the membrane in response to proton pumping. If the ATPase were electroneutral, that is, if it transported an anion in the same direction as the proton or a cation in the opposite direction, then valinomycin would have no effect on proton pumping. Because valinomycin stimulates proton pumping, the proton pump must be electrogenic. That is, the V-type ATPase generates an electrical potential across the vacuolar membrane. In the cell, the greater the electrical potential difference that the vacuolar H^+-pumping ATPase generates across the membrane, the more energy it will take to transport a proton from the more negative side to the more positive side.

The K_m for the Mg^{2+}-ATP complex ranges between 0.2 and 0.81 mol/m^3. The ATPase hydrolyzes about 30–50 ATP/s and pumps between 60 and 90 H^+/s. The H^+/ATP stoichiometry varies between 1.75 and 3.28 (Ratajczak, 2000). Müller et al. (1996) discovered an unusual vacuolar ATPase from the fruits of lemons that differs from the canonical V-type ATPase, which is found in lemon epicotyls. Unlike the canonical V-type ATPase, the vacuolar

ATPase in citrus fruits that are hyperacidic is sensitive to vanadate and insensitive to nitrate and bafilomycin. Interestingly, the unusual type of V-ATPase is found in the vacuolar membranes of acidic limes, but not in sweet limes, indicating that this protein is genetically adapted for vacuolar hyperacidity (Brune et al., 2002). By promiscuously exchanging DNA sequences, domains in a polypeptide can be swapped between the canonical ATPase polypeptides to form chimerical polypeptides with differing function, binding characteristics, localization, etc. This blurs the distinctions between the categories of ATPases one creates to pigeonhole transport proteins into a convenient system of classification before the diversity of proteins is established. So again, we learn that nature mocks human categories, and the vacuolar ATPases have evolved to ensure that lemons and acid limes are sour.

The vacuolar H^+-ATPase is a large protein and its structure can be seen in the electron microscope by negatively staining vesicles with phosphotungstic acid (Klink and Lüttge, 1991; Taiz and Taiz, 1991). The V_1 complex appears as an H- or V-shaped particle on a stalk with small projections emerging from the base. It is whimsical that the V-type ATPase looks like a V! Treatment with NO_3 inhibits the V-type ATPase because it causes the hydrophilic catalytic subunit to dissociate from the hydrophobic membrane channel complex (Adachi et al., 1990; Bowman et al., 1989). In electron micrographs of freeze-fractured preparations, the V-type ATPase appears as a particle 9.1 nm in diameter (Ratajczak, 2000). The vacuolar H^+-ATPase also occurs in other membranes in the cell and functions to acidify the compartments enclosed by these membranes (Maeshima et al., 1996). Grabe et al. (2000) suggest that the V-type ATPase is a mechanochemical enzyme, and ATP hydrolysis by the V_1 complex causes a rotary torque on the V_0 complex that results in the translocation of a proton across the vacuolar membrane.

The vacuolar membrane contains another major H^+-translocating pump (Martinoia et al., 2007). The second one is composed of a single polypeptide and is fueled by the hydrolysis of pyrophosphate (P_iP_i) (see Fig. 7.12; Rea and Sanders, 1987; Britten et al., 1992; Rea and Poole, 1993; Maeshima et al., 1996; Maeshima and Nakanishi, 2002). The H^+-$P_iP_i$ase is specifically inhibited by aminomethyldiphosphonate, a structural analogue of pyrophosphate (Zhen et al., 1994).

H^+-pumping by both the V-type ATPase and the pyrophosphatase is stimulated by Cl^-. Cl^- stimulates the electrogenic H^+ pumps because it enters the vacuole though a Cl^- transporter and reduces, without eliminating, the net positive charge in the lumen. Thus, the vacuoles are able to generate a substantial pH difference. In essence, Cl^- permits the conversion of an electrical potential difference into a chemical difference. Thus, the lumen of the vacuole, like the stomach, is acidified by HCl (Wada and Anraku, 1994).

Why does the vacuolar membrane have two different H^+ translocators, an ATPase and a pyrophosphatase? Perhaps, at different times in a cell's life, the two substrates are more or less prevalent. For example, meristematic cells that are synthesizing DNA and RNA produce a lot of pyrophosphate. The pyrophosphatase can use the free energy of this "waste product" to build a proton-motive force across the vacuolar membrane while helping to drive the reactions involving the synthesis of DNA and RNA. By contrast, in older cells, where biosynthetic reactions that generate pyrophosphate have slowed down or stopped, ATP is by far the most available substrate, so the ATPase should be more prevalent. This trend has been observed by Maeshima et al. (1996).

7.5.2 ATP-Binding Cassette Transporters or Traffic ATPases

After intensive work on the vacuolar ATPase and pyrophosphatase, there was still room for the discovery of new carriers that are involved in the active transport of organic solutes. These carriers, known as ABC transporters or traffic ATPases, directly bind Mg-ATP and transport organic solutes, including alkaloids, endogenous toxins, xenobiotic toxins, and anthocyanins, into the vacuole (see Fig. 7.12; Martinoia et al., 1993, 2000; Li et al., 1995, 1997; Bartholomew et al., 2002; Goodman et al., 2004; Marinova et al., 2007). The ABC transporters are recognized by their requirement for Mg-ATP, insensitivity to the electrochemical potential of protons across the membrane, and inhibition by vanadate (Rea et al., 1998). There are large families of ABC transporters, each with a unique cellular or subcellular localization and function (Martinoia et al., 2002; Rea, 2007; Hwang et al., 2016).

7.5.3 Slowly Activated Vacuolar Channels

Just as there are many channel types on the plasma membrane, there are also many channel types on the vacuolar membrane. Many of these kinds of channels may be related. However, each class of channels differs in terms of its organ, tissue, and cellular and intracellular localization. Moreover, each type of channel has a certain conductance, ion selectivity, and type of regulation. Some kinds of channels are rapidly activated, whereas others are slowly activated; some types of channels are voltage dependent or modified by pH or Ca^{2+}. It may be that there are sequences of amino acids that are capable of determining a given characteristic of a given class of channels. Combining such sequences over evolutionary time as a result of mixing and matching the gene sequences that encode the amino acid sequences may have created a single polypeptide or group of polypeptides that function as a channel for a specific ion with a certain type of regulation (Gilbert, 1978; Patel et al., 2016).

Using the patch-clamp technique invented by Bert Sakmann and Erwin Neher (1983), Rainer Hedrich and Erwin Neher discovered a channel in the vacuolar membrane of sugar beet vacuoles (Hedrich and Neher, 1987) that also exists in *Vicia faba* guard cells. This channel is neither a cation channel nor an anion channel but passes both K^+ and Cl^- with a $P_K/P_{Cl} \cong 3.5$. Because the permeability to K^+ is greater than the permeability to Cl^-, more positive charges pass through the channel than negative charges, and thus the net current is positive. Given the sign convention for patch clamping endomembranes, where the lumen of the vacuole is considered topologically equivalent to the external region surrounding a cell, a positive ionic current that passes from the vacuole to the cytosol is considered to be an inward current, and a positive current that passes from the cytosol to the vacuole is considered to be an outward current. The channel discovered by Hedrich and Neher (1987) is an inwardly rectifying channel, meaning that the net positive current, in the form of K^+ and Cl^-, passes from the lumen of the vacuole (E-space) to the cytosol (P-space). The single-channel conductance is approximately 280 pS, which is large for a single channel. The conductance is not constant, but varies with the KCl concentration. The channel is also activated by voltage when the potential on the E-side of the vacuolar membrane potential is approximately 0.06 V more positive than the potential on the P-side. Currently, it is not known whether the vacuolar membrane potential reaches this value in vivo. If it does, this class of channels, which is distributed on the vacuolar membrane with a density of $0.37/\mu m^2$, is extremely sensitive to the cytoplasmic concentrations of Ca^{2+} and H^+ and may function in the release of KCl from the vacuole during guard cell closure (Schulz-Lessdorf and Hedrich, 1995). When activated, this channel, which passes other cations, including Ca^{2+}, may function in releasing Ca^{2+} from the vacuole and into the cytosol (Hedrich and Marten, 2011).

7.5.4 Water Channels

An abundant protein in the vacuolar membrane is called γ-TIP, which stands for *tonoplast intrinsic protein*. The vacuolar membrane intrinsic protein (γ-TIP) can act as a water channel (Maurel et al., 1993; Maurel, 1997; Niemietz and Tyerman, 1997; Tyerman et al., 1999, 2002; Baiges et al., 2002) and has been given the name *aquaporin*, suggesting that this is its function in vivo. Aquaporins are also found in the other organelle membranes in the cell (Katsuhara et al., 2008; Maurel et al., 2008). Given the fact that water can permeate the lipid bilayer (with its low-specific hydraulic conductance but large area), as well as many proteins with aqueous channels (with their high-specific hydraulic conductance but small area), I feel that it is unlikely that the selective advantage of aquaporins in plant cell membranes is to facilitate the permeation of water. It is possible that the physiological function of aquaporins is to pass small, nonionic molecules, including carbon dioxide, that are similar in chemical structure to water (Wayne and Tazawa, 1990; Wayne et al., 1994; Ishibashi et al., 1994; Nakhoul et al., 1998; Terashima and Ono, 2002; Katsuhara et al., 2003; Uehlein et al., 2003, 2008, 2012, 2017; Hanba et al., 2004; Flexas et al., 2006; Kaldenhoff, 2006; Maurel et al., 2008; Warren, 2008; Otto et al., 2010; Kaldenhoff et al., 2014).

7.6 FUNCTIONS OF THE VACUOLE

The five-kingdom classification system of Robert Whittaker (1969) separates organisms, in part, based on their mode of nutrition. The vacuolar compartment may have evolved in the various kingdoms to reflect these differences. Animals typically acquire food as either organisms or macromolecules and must digest them. This has led to the evolution of a vacuolar compartment that is primarily involved in digestion and is thus usually termed the *lysosomal compartment* (de Duve and Wattiaux, 1966). Plants, on the other hand, make their food out of small inorganic molecules such as carbon dioxide, water, and nitrate, using the radiant energy of sunlight. To capture these molecules and energy, plants typically evolved an arborescent form and the vacuoles have evolved to take up space, which we will see allows the building of a structurally economical arborescent form. As a consequence of the multiplicity of functions of plant and fungal vacuoles, I will retain the name *vacuole* (Klionsky et al., 1990) to reflect its many functions (Marty, 1999; De, 2000; Robinson and Rogers, 2000). I will consider that one of the functions of the vacuole is to act as a lysosome.

7.6.1 Proteolysis and Recycling

In the plant, animal, and fungal kingdoms, cells must recycle their own protoplasm in times of starvation. They may recycle either within a cell or within the organism. During starvation conditions, the vacuolar compartment performs this function in a process known as autophagy (de Duve and Wattiaux, 1966; Dunn, 1990; Chen et al., 1994; Aubert et al., 1996; Niwa et al., 2004; Yano et al., 2004, 2007; Takatsuka et al., 2011; Umekawa and Klionsky, 2012; Li and Vierstra, 2012; Liu and Bassham, 2012; Shemi et al., 2015; Yano et al., 2015; Michaeli et al., 2016; Harnett et al., 2017; Reggiori and Ungermann, 2017). Life, like Ouroboros, the self-sustaining, tail-eating snake, is a balance between the synthesis and degradation of molecules (Bernard, 1865). Death is associated with a change in the balance, which leads to the destruction of biomolecules. Thus, within the cell, the basic unit of life, lies the very mechanism that can result in death. In fact, many diseases of humans result from malfunctions in the balance of synthesis and degradation, and even a decrease in degradation in the vacuole can be fatal (de Duve, 1981). Proteolysis, however, does not only occur in the vacuole but also in every cellular compartment (Vierstra, 1993, Chapter 17).

When Rudolf Schoenheimer (1942) introduced stable isotope tracers into the study of metabolism, he was surprised to find that almost all the macromolecules in mature bodies undergo turnover, and thus, even the material of which living organisms are made is in constant flux, and we are not composed of the same atoms and molecules for our whole life. According to de Duve (1981), the average liver cell lives for many years, yet it destroys and rebuilds its protoplasm approximately every week. The potatoes we eat today become our brain tomorrow (Feynman, 1955). Some cells, like those in senescing leaves or those that will give rise to laticifers or to conducting elements of the phloem and xylem, undergo almost total proteolysis (Wodzicki and Brown, 1973; Matile, 1975). Such programmed cell deaths are known as *apoptosis* (Fukuda, 1996; Groover et al., 1997). Partial proteolysis may be important for dedifferentiation and redifferentiation. It is the vacuolar compartment that specializes in cellular recycling, and in the vacuole, organelles can be seen in the process of being degraded (Sievers, 1966; Villiers, 1967).

Many nonspecific hydrolytic enzymes with acidic pH optima occur in plant vacuoles (Matile, 1975; Nishimura and Beevers, 1978, 1979; Moriyasu, 1995; Muntz, 2007). Moriyasu and Tazawa (1988) tested the proteolytic capability of the vacuole by introducing an exogenous protein such as bovine serum albumin (BSA) into the vacuole of giant algal cells. In this study, both ends of the cell were removed and about 10 μL of BSA were added to the vacuole, which contained approximately 50 μL of endogenous cell sap. The cell ends were then ligated, and the cells were allowed to sit for various times. Then the proteins in the vacuole were collected, run on SDS polyacrylamide gels, transferred to nitrocellulose paper, and immunoblotted with antibodies directed against BSA. Indeed, the BSA was hydrolyzed, indicating that the vacuole is capable of proteolysis. Moriyasu et al. (1987) have also purified and characterized vacuolar proteases from *Chara*.

Recognizing the importance of the fact that all living things are in a state of relentless and ephemeral flux, and influenced by the success of taking the genetic approach in yeast to understand the cell cycle (*cdc* mutants; Hartwell, 2001) and secretion (*sec* mutants; Schekman, 2013) in yeast and all eukaryotic organisms, Yoshinori Ohsumi (2016) combined genetics with biochemistry, pharmacology, and morphology to determine the temporal sequence of genes, proteins, supramolecular assemblies, and morphological structures involved in the precursor–product relationship that takes during autophagy in yeast and plants (Takeshige et al., 1992; Tsukada and Ohsumi, 1993; Baba et al., 1994, 1995; Moriyasu and Ohsumi, 1996; Matsuura et al., 1997; Mizushima et al., 1999; Kirisako et al., 2000; Klionsky and Emr, 2000; Kihara et al., 2001; Ohsumi, 2001, 2006, 2014; Shintani et al., 2001; Suzuki et al., 2001; Hanaoka et al., 2002; Kuma et al., 2002; Yoshimoto et al., 2004, 2009, 2014; Suzuki et al., 2005; Inoue et al., 2006; Ishida et al., 2008; Noda and Klionsky, 2008; Nakatogawa et al., 2009; Wada et al., 2009; Mizushima et al., 2011; Shibata et al., 2013). Pharmacologically, autophagy is inhibited by kinase inhibitors such as rapamycin and 3-methyladenine, phosphatase inhibitors such as okadaic acid, protease inhibitors such as leupeptin, and vacuolar H^+-ATPase inhibitors, such as concanamycin A and bafilomycin, which act by inhibiting the acidification of the vacuole. As in the case of studying cisternae-to-cisternae transport, precursor–product relationships can be determined by treating cells with two drugs at a time and seeing which drug has no effect because the cell is prevented by the first drug to progress to the stage where the second drug acts.

The first autophagy mutant was found using light microscopic selection to obtain mutants that fail to accumulate autophagic bodies under nitrogen-starvation conditions, but that grew well in a nutrient-rich medium. The use of double mutants allowed him to determine the temporal sequence of events encoded by the two genes. By screening a genomic library for DNA fragments that complemented the mutant phenotypes, Ohsumi could identify the genes involved in autophagy and subsequently sequence those genes—using the sequences to identify their functions. The functions of some of the genes deduced from sequence analysis confirmed the functions deduced from inhibitor studies; however, the functions of most of the genes were a mystery. Ohsumi utilized morphological, biochemical, and enzymological techniques to determine the roles of these genes in the precursor–product relationship that leads to autophagy in vitro and mutational analysis to show that these activities are required for autophagosome formation in vivo.

Autophagy is not only invoked in times of starvation. Selective autophagy allows dysfunctional proteins and organelles, such as chloroplasts that are damaged by UV light or peroxisomes that are damaged by H_2O_2, to be recycled throughout the life of a cell (Wada et al., 2009; Mizushima et al., 2011; Floyd et al., 2012; Li and Vierstra, 2012; Shibata et al., 2013; Michaeli and Galili, 2014; Farré and Subramani, 2016; Anding and Baehrecke, 2017; Izumi et al., 2017). Selective autophagy not only depends on the core autophagy machinery but also requires specific autophagy receptors that recognize the organelle to be degraded and to engage it with the autophagy machinery.

In the vacuole, the proteins are hydrolyzed into their constituent amino acids, and the amino acids are recycled back to the cytoplasm by way of an amino acid carrier on the vacuolar membrane. When the cell sap of *Chara* is replaced with artificial cell sap containing various amino acids, the amino acids leave the vacuole and enter the cytoplasm via an H^+/amino acid symporter (see Fig. 7.12; Sakano and Tazawa, 1985; Amino and Tazawa, 1989).

Why don't the hydrolytic enzymes in the vacuole destroy the vacuolar membrane itself? According to Christian de Duve (1981), we could reply in the manner of the "medical student" in the last act of Molière's (1673) *Le Malade Imaginaire*. He answered the question, "Why does opium put you to sleep?" with the answer, "Opium puts you to sleep because it is a soporific." That is, we can say that the proteins in the vacuolar membrane have a conformation that makes them resistant to the vacuolar proteases. To paraphrase de Duve, as well as Bacon (1620), Locke (1824), and Hayakawa (1941), it would be just as well to say, "We do not know" than to worship the "Idols of the Marketplace."

7.6.2 Taking up Space

Unlike animals, which can gather food, plants are usually sessile and have a dendritic form that helps them acquire light and the necessary nutrients that are dilute in the environment. The vacuole is essential for plant survival in that it allows the plant to attain a large open dendritic structure with a minimum investment in energy-intensive compounds such as cellulose or protein. Instead, the plant cell vacuole is filled with water, which is generally abundant and energetically cheap to obtain (Dixon and Joly, 1895; Dixon, 1938; Dainty, 1968; Wiebe, 1978; Taiz, 1992).

As a consequence of the large central vacuole, the cytoplasm is pushed to a parietal position, where the distance from the atmosphere to a chloroplast or mitochondrion is kept to a minimum. This can greatly enhance photosynthesis and respiration, as the diffusion rates of O_2 or CO_2 in air is approximately 10,000 times greater than they are in water (Table 7.3). The vacuole then ensures that resistance to diffusion of CO_2 and O_2 is kept to a minimum (Wiebe, 1978). According to Fick's Law, the flux of O_2 and CO_2 to the center of the cell will be proportional to the diffusion coefficient and inversely related to the distance it must travel:

$$J = -\left(\frac{DK}{dx}\right)dC \qquad (7.2)$$

If CO_2 and O_2 are transported through the plasma membrane and utilized by the chloroplasts and mitochondria faster than they are transported through the cytosol, then photosynthesis and respiration, respectively, will be limited by diffusion through the cytosol. Although the permeability coefficient of the plasma membrane to CO_2 is between 2×10^{-6} and 3.5×10^{-3} m/s (Gutknecht et al., 1977; Gimmler et al., 1990; Wayne et al., 1994), with a CO_2 difference of approximately 0.05 mol/m^3 and a partition coefficient of 1, the flux across the plasma membrane would be at least 10^{-7} mol/m^2 s. The flux of CO_2 through the aqueous cytosol to the center of a 2×10^{-4} m in diameter mesophyll cell would be 9.7×10^{-8} mol/m^2 s. This flux may be limiting to photosynthesis and respiration. However, if the chloroplasts were pushed within 10^{-6} m of the plasma membrane, the flux would increase 100 times to 9.7×10^{-6} mol/m^2 s, and then photosynthesis and respiration would most likely be limited by their enzymes and not by the length of the diffusion pathway. Of course, the light intensity at the chloroplast is greater when the chloroplast is at the periphery of the cell compared with when it is in the center of the cell, and this too may enhance photosynthesis.

TABLE 7.3 Diffusion Coefficients for Oxygen and Carbon Dioxide in Air and Water

		D (in m^2/s)	
Molecule	Temperature (K)	CO_2	O_2
Air	273K	1.04×10^{-5}	1.89×10^{-5}
Water	298K	1.94×10^{-9}	1.77×10^{-9}

From Weast, R. C. (Ed.), 1973–1974. The Handbook of Chemistry and Physics. 54th Edition. Cleveland, OH: CRC Press.

7.6.3 Storage and Homeostasis

Because vacuoles take up the better part of a cell, they contain the volume necessary to store levels of organic and inorganic molecules that would be toxic to the cytosol and the other organelles. In this way, vacuoles contribute to the protection of the cell and the maintenance of a cellular homeostasis in terms of ions, water, and amino acids (Matile, 1987; De, 2000). All vacuoles store water. This water is in equilibrium with the protoplasm and keeps the protoplasm hydrated so that enzymatic reactions can take place. This is a vital function in plants, which have undergone an evolutionary process from living in water to living on arid land. Water storage is particularly important to desert plants, which is one reason that they are so succulent (Walter and Stadelmann, 1968).

The acidic nature of most vacuoles has been known for a long time from looking at the color of natural or introduced dyes. Vacuoles are typically acidic (pH ~ 5) and thus act as a store of H^+. As a consequence of the large capacity of the vacuole to store H^+, it can function in pH regulation (Moriyasu et al., 1984; Takeshige et al., 1988; Takeshige and Tazawa, 1989b; Grabe and Oster, 2001). In fact, the vacuole is involved in the pH regulation necessary to protect plants from acid rain (Heber et al., 1994).

The pH of the vacuole of the brown alga *Desmarestia* is less than 1 (Wirth and Rigg, 1937; McClintock et al., 1982). The pH of the vacuoles of the juice cells of lemons and acid limes is also hyperacidic compared with most vacuoles, reaching values as low as 2–2.2 (Echeverria and Burns, 1989; Echeverria et al., 1992). Two factors may allow these hyper-acidic vacuoles to store so many protons. First, the V-ATPase in these cells is atypical and may transport only one H^+ per ATP hydrolyzed (Fig. 7.13), which would allow the vacuole to be more acidic, and second, the vacuolar membrane of the acidic juice cell vacuoles has less permeability to the passive movement of H^+ than vacuolar membranes in typical cells, so once H^+ are pumped into the vacuole, they will tend to stay there (Müller et al., 1996, 1997, 1999; Brune et al., 2002; Müller and Taiz, 2002).

The hydrolysis of a single molecule of ATP provides approximately 8×10^{-20} J of molecular free energy. According to the following equation, this is a sufficient quantity of energy to pump two protons from the cytosol to the vacuole when the pH of the cytoplasm ($-\log H_c^+$) is about 7, the pH of the vacuole ($-\log H_v^+$) is about 3, and the electrical potential (ψ_c) across the vacuolar membrane is about −0.02 V, but only enough to pump 1 H^+ from the cytosol to the vacuole when the pH of the vacuole is as low as 2.1. The energy needed to pump *n* protons from the cytosol into the vacuole is given by the following equation (Smith et al., 1982; Bennett and Spanswick, 1984b):

$$E = n\left(ze(\Psi_v - \Psi_c) + kT\ln\frac{H_v^+}{H_c^+}\right)$$

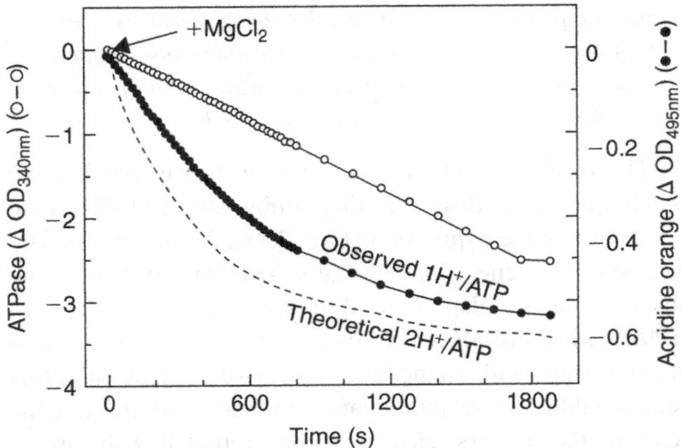

FIGURE 7.13 Simultaneous measurement of ATPase activity (*open circles*) and proton pumping (*closed circles*) from acid lime. *From Brune, A., Müller, M., Taiz, L., Gonzalez, P., Etxeberria, E., 2002. Vacuolar acidification in citrus fruit: Comparison between acid lime* (Citrus aurantifolia) *and sweet lime* (Citrus limmentiodes) *juice cells. J. Amer. Hort. Sci. 127, 171–177.*

where n is also known as the *coupling ratio* (Läuger, 1991; Schmidt and Briskin, 1993; Davies et al., 1994; Davies, 1999), which relates the number of protons transported per ATP molecule hydrolyzed at equilibrium:

$$n = \frac{E_{ATP}}{\left(ze(\Psi_v - \Psi_c) + kT\ln\frac{H_v^+}{H_c^+}\right)}$$

Remembering that death, not life, is characterized by the equilibrium state, we can nevertheless use equilibrium thermodynamics as a first approximation. However, once we are able to determine concentrations of metabolites and ions under nonequilibrium conditions, we can use irreversible or nonequilibrium thermodynamics to provide a more realistic description of life.

The electrochemical energy that is inherent in the extreme difference in the pH between the vacuole and the cytosol of the juice cells of lemons and acid limes provides the molecular free energy to transport high quantities of citric acid from the cytosol, where its concentration is less than 10 nM, to the vacuole, where its concentration is 325 mM (Brune et al., 1998; Ratajczak et al., 2003). The large pH gradient in juice cells is enhanced by the nonenzymatic hydrolysis of sucrose into organic acids in the vacuole (Echeverria and Burns, 1989; Echeverria et al., 1992).

While vacuoles are typically acidic, not all vacuoles are acidic, the blue color of the epidermal cells of heavenly blue morning glories is due to the alkaline nature of the vacuole as a consequence of Na^+/H^+ exchange (Yoshida et al., 1995, 2005). Interestingly, the vacuoles of the epidermal cells are basic when the flowers are ripe for pollination but become acidic before the flowers are ripe for pollination and after the flowers have been pollinated. When the epidermal cell vacuoles are acidic, the anthocyanins are purple and the purple flowers do not compete for the attention of the pollinator bees.

Vacuoles also store nutrients, such as PO_4 (Mimura et al., 1990a) and other ions (Leigh, 1997), as well as sugars (Fisher and Outlaw, 1979; Kaiser et al., 1982; Gerhardt and Heldt, 1984; Keller and Matile, 1985; Matile, 1987; Keller, 1992; Martinoia and Ratajczak, 1999) and amino acids (Wagner, 1979; Wayne and Staves, 1991; Riens et al., 1991). Many plants that are salt tolerant store Na^+ in the vacuole, thanks to a Na^+/H^+ antiporter (Staal et al., 1991; Epimashko et al., 2004).

Vacuoles can act as an intracellular toxic waste site and store substances that would be harmful if kept in the cytoplasm. Some of these compounds, including nicotine (Saunders, 1979; Steppuhn et al., 2004; Howe and Jander, 2008), protect the plant from would-be predators. Many secondary substances, particularly alkaloids that are useful to cell biologists and other human beings, are stored in the vacuole (Hobhouse, 1986; Ziegler and Facchini, 2008). These include trypsin inhibitors, antifungal phytoalexins, vinblastine, vincristine, colchicine, rubber, morphine, serpentine, caffeine, etc. (Sequeira, 1983; Blom et al., 1991; Sottomayor et al., 1996; Costa et al., 2008; Hagel et al., 2008). Some of these substances, which are membrane permeant, are trapped in the vacuole and do not leak into the cytoplasm because they form complexes with other molecules, including polyphenols and tannins, which increase their apparent size and polarity (Mösli Waldhauser and Baumann, 1996). High concentrations of heavy metals, found in abandoned industrial and mining sites, would be toxic if they were in the cytosol. These are also sequestered in the vacuole.

The beautiful reds, blues, and purples of autumn leaves as well as fruits and flowers are a consequence of the anthocyanins that are stored in the vacuole (Overton, 1899a; Matzke, 1942; Thimann, 1950; Moskowitz and Hrazdina, 1981; Andersen and Markham, 2006). The term *anthocyanin* was coined by L. Marguartin in 1835, and over the next century, the anthocyanins were isolated and characterized and the structures were deduced and confirmed by synthesis due to a large extent to the work of chemists Richard Willstätter and Sir Robert and Lady Robinson (Onslow, 1916; Robinson, 1955; Willstätter, 1965).

Not only have chemists been instrumental in understanding anthocyanins but anthocyanins have also been instrumental for the progress of chemists. Robert Boyle (1664) used the anthocyanins of violets as a pH indicator, and Jeremias Richter (1792–94) used the pH-indicating ability of the anthocyanins of violets to determine the quantity of acid needed to neutralize a quantity of base, and in doing so, he came to us with the concept of stoichiometry. The discovery of fixed stoichiometries formed the foundation necessary for the introduction of the mole concept in chemistry (Kieffer, 1963). The colors of the anthocyanins in the flower are determined by the pH of the vacuoles in the epidermal (Asen et al., 1975; Stewart et al., 1975; Kondo et al., 1992; Yoshida et al., 1995, 2005) or subepidermal (Yoshida et al., 2003) cells.

While studying the unstable inheritance of the mosaic pattern of blue, brown, and red spots that result from the differential production of vacuolar anthocyanins in the triploid aleurone cells of a single maize kernel, Barbara McClintock (1950) discovered transposable elements. Transposable elements are genes that are not immobilized in the chromosome but can move around landing in various regions of the genome and control the expression of the genes they jump out of and jump into. McClintock found that the transposable elements regulated the color of the kernels. When a transposable element moved into a gene-controlling anthocyanin synthesis (Grotewold, 2006; Lepiniec et al., 2006; Cone, 2007; Federoff, 2012; Ravindran, 2012), anthocyanin synthesis was suppressed in the aleurone cells of the kernel, and when it moved out of a gene-controlling anthocyanin synthesis, the pigment was produced. The randomness of the color mosaicism in the kernel reflects the randomness of the spatial insertion of the transposable element in the genome of the aleurone cells. The size of the colored spot on the kernel depends on the randomness of the timing of the insertion of the transposable element into a gene that leads to anthocyanin synthesis. By studying the unstable inheritance of the distribution of vacuolar coloration, McClintock realized that functionally differentiated cells in a multicellular organism must be a result of the differential expression of an identical genome. Moreover, McClintock (1983) realized that the genome itself was not static but could respond rapidly to challenges.

The dynamic nature of the genome as a result of "jumping genes" was not readily accepted by geneticists until they realized that the rapid evolution of antibiotic resistance in bacteria could be explained by a similar mechanism. McClintock's observations on the mosaic pattern of vacuolar anthocyanins in the aleurone cells of maize also lead to the understanding of how the movement of oncogenes from one position to another can result in cancer in humans.

Barbara McClintock was awarded the 1983 Nobel Prize in Physiology or Medicine. The presentation speech on behalf of the Nobel Assembly of the Karolinska Institute ended with the following words:

> *I have tried to summarize to this audience your work on mobile genetic elements in maize and to show how basic research in plant genetics can lead to new perspectives in medicine. Your work also demonstrates to scientists, politicians and university administrators how important it is that scientists are given the freedom to pursue promising lines of research without having to worry about their immediate practical applications. To young scientists, living at a time of economic recession and university cutbacks, your work is encouraging because it shows that great discoveries can still be made with simple tools.*

The readily visible anthocyanins in flower petals have also helped in the discovery that double-stranded RNAs are involved in gene expression (Fire, 2006; Mello, 2006). The discovery of gene silencing came unexpectedly from the observations of Napoli et al. (1990) and van der Krol (1990) who introduced a chalcone synthase gene into petunia plants with the hopes that the overexpression of this gene would increase anthocyanin synthesis and improve the color of the flowers. However, they found that the transformed plants lost the ability to produce anthocyanins in the vacuole and only produced white flowers. This led to a rethinking of what happens in the generation of transgenic plants and during normal gene expression (see Chapter 16; Dougherty and Parks, 1995).

Vacuoles also store flavones, and occasionally (e.g., in snapdragon flowers) the yellow color of petals are due to the presence of flavones, although typically, the yellow color of flowers comes from pigments in the plastids.

Desert plants and many other plants that have crassulacean acid metabolism (CAM) utilize the vacuole as a storage site for organic acids, including malic acid (Kenyon et al., 1978, 1985; Winter and Smith, 1996; Black and Osmond, 2003). Plants exhibiting CAM open their stomata at night to minimize transpirational water loss. These plants are able to use phosphoenolpyruvate carboxylase to fix CO_2 at night and store the fixed CO_2 as malic acid (Pucher et al., 1947; Vickery, 1953; Bandurski and Greiner, 1953; Bandurski, 1955; Epimashko et al., 2004). During the day, the stomata close to conserve transpirational water loss. In the presence of light, electron transport occurs and ATP and reduced nicotinamide adenine dinucleotide phosphate are formed by the light reactions of photosynthesis. Simultaneously, the CO_2 is released from the malic acid by the $NADP^+$-malic enzyme to become refixed by RuBP carboxylase (see Chapter 13).

In developing seeds, proteins are stored in the vacuole (Levanony et al., 1992; Li et al., 1993a,b; Jiang et al., 2000, 2001; Kumamaru et al., 2007; Ibl and Stoger, 2014). These protein-storing vacuoles are usually called *protein bodies*. During germination, the protein-storing vacuole acidifies (Swanson and Jones, 1996; Hwang et al., 2003). Subsequently, the proteins are hydrolyzed and the amino acids are mobilized to nourish the growing embryo (Filner and Varner, 1967; Graham and Gunning, 1970).

7.6.4 Role in Turgor Generation

The studies of osmotic and turgor pressure in plants done by Pfeffer (1877) and de Vries (1884) provided the experimental basis necessary for Jacobus van't Hoff (1888, 1901) to apply the gas laws to molecules in solution (Wald, 1986). Van't Hoff learned of Pfeffer's experimental results from his friend de Vries who asked van't Hoff to come up with a theoretical explanation of the results. Pfeffer had measured the osmotic pressure (P_π) at a given temperature of solutions made up of various concentrations of nonelectrolyes (C). Van't Hoff took Pfeffer's results and merely divided P_π by C and saw that this quotient was a constant at constant temperature (Fig. 7.14A). Pfeffer also determined the effect of various temperatures on the osmotic pressure of a given solution, and again van't Hoff noticed that, for a given concentration of solute, P_π/T was a constant (see Fig. 7.14B). By hypothesizing that liquids behave in an analogous manner to gases, he framed these two results in terms of Boyle's Law (PV = constant at constant temperature) and Gay-Lussac's Law (P/T = constant at constant volume). Van't Hoff combined these two equations to deduce that

$$P_\pi = RTC \tag{7.3}$$

where R, which is known as the universal gas constant, is a combination of the two constants mentioned above. As concentration (C) = amount in moles (s) divided by volume (V), then

$$PV = sRT \tag{7.4}$$

which is a restatement of the gas law.

The introduction of the gas constant gave the equation immediate significance, since now the colligative properties of solutions could be examined from a thermodynamic perspective. Van't Hoff used his new equation, which was backed up by thermodynamic theory, to deduce the laws of many diverse phenomena, including the effects of solutes on the vapor pressure and freezing point of a solution, as well as Guldberg and Waage's law of chemical equilibrium. Here is a case where a little mathematics applied to empirical physiological observations helped to not only generalize the observations within the field of plant cell biology but also to open up the field of physical chemistry.

One far-reaching effect of the van't Hoff equation is that it provided the theoretical and experimental basis for

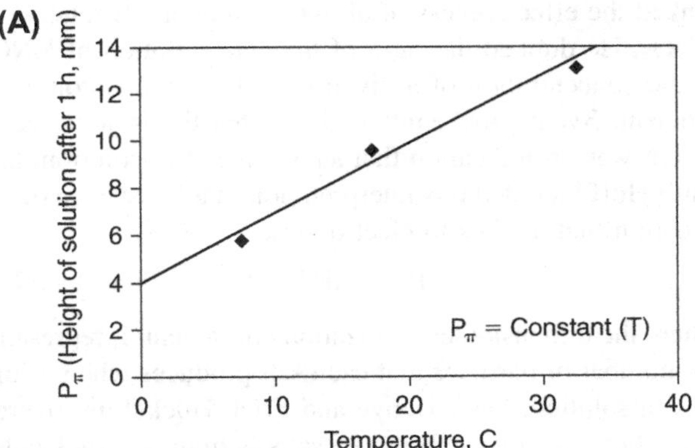

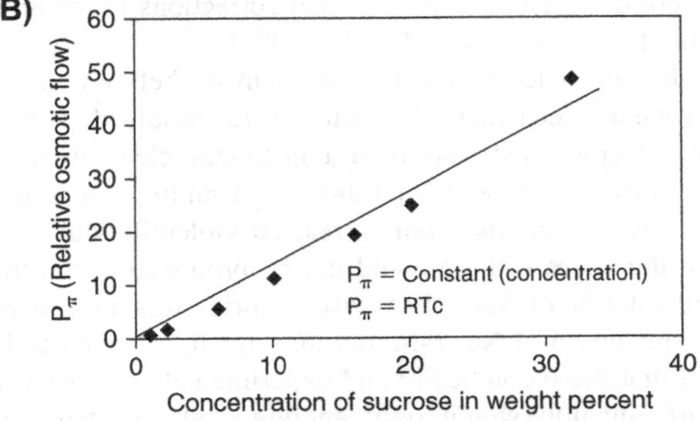

FIGURE 7.14 Two graphs of Pfeffer's (1877) tabular data. (A) The height of a sucrose solution in an osmometer held at various temperatures. (B) The osmotic flow of sucrose solutions of various concentrations at a given temperature.

calculating the molecular mass of nonvolatile substances, an important procedure that was uncertain until this time (Ostwald, 1891; Pattison Muir, 1909). Because $P_\pi = RTC$ and C is equal to the number of moles of a substance (s) divided by the volume of the solution (V), then

$$P_\pi = RT\frac{s}{V} \tag{7.5}$$

Furthermore, the molecular mass of a given solute could be determined because s is equal to the number of grams added to the volume divided by the molecular mass of the substance (M_r):

$$M_r = (\text{grams added})\frac{RT}{VP_\pi} \tag{7.6}$$

That is, the molecular mass of a solute could be determined by measuring the osmotic pressure of a known mass in a given volume at a given temperature.

Eq. (7.6) only holds for nonionized substances. However, van't Hoff also incorporated the observations of de Vries (1884, 1888a) on plasmolysis. de Vries noticed that it takes a lower concentration of KNO_3 compared with sucrose to plasmolyze various cells. de Vries determined the concentration of various compounds that were required to cause incipient plasmolysis. He then found the concentration of each chemical that was as effective as KNO_3 and ranked the effectiveness of all these compounds relative to KNO_3. He dubbed the ratio of the concentration of KNO_3 to the concentration of a given substance, the isotonic coefficient. Svante Arrhenius realized that the isotonic coefficient was an indication that salts ionized in solution, and van't Hoff included this interpretation in a latter form of his equation that applies to electrolytes:

$$P_\pi = iRTC \tag{7.7}$$

where the dimensionless ionization coefficient, i, represents the number of particles that each salt produces when it ionizes in solution. Peter Debye and Erich Hückel discovered that when the concentration of salts is high, the molecules do not dissociate completely, and corrections to the above formula must be made (Laidler, 1993).

It took a long time before chemists believed Svante Arrhenius' idea that when salts were dissolved in water they decomposed into their constituent charged atoms. The difficulty in believing this arose from the observations that pure metals like sodium reacted violently with water, and thus pure sodium could not be produced in the tranquil solution of NaCl. Likewise, chlorine was a green gas, yet a solution of NaCl did not turn green and bubble. The fact that the theoretically and experimentally robust van't Hoff equation would only apply to salts if they were considered to be ionized helped convince Arrhenius' contemporaries of the reality of ionization (Arrhenius, 1903, 1912).

Now back to plant cell biology! As a consequence of the presence of solutes in the cell, the differential permeability of the plasma membrane, and the rigidity of the extracellular matrix, water enters the cell and generates a turgor pressure of several 100,000 Pa. The potential energy of a volume of water in the cell is known as its *water potential* and it is given in J/m^3 or Pa. Because membranes that are not protected by the extracellular matrix typically lyse when the hydrostatic pressure difference across a membrane exceeds approximately 100 Pa (Wolfe et al., 1986), the water potential of every organelle must be the same as the cytosol or the organellar membranes will break. Therefore, the total concentration of solutes is essentially equal in all of the compartments of the cell, and turgor pressure is only generated across the plasma membrane.

Plants growing in environments with high osmotic pressure, due particularly to the presence of salts, must produce sufficient osmotic pressure within their cells to allow the uptake of water necessary for life and growth. Under these conditions, the high concentrations of osmoticum in the cytosol and the matrix of each organelle must allow the normal enzymatic reactions to take place, and thus the osmoticum must not adversely affect the enzymes. Depending on the cell, glycerol, trehalose, betaines, and proline act as "compatible solutes" that can be used to increase the osmotic pressure of the cytosol or an organelle without adversely affecting its enzymes. The compatible solutes often, but not always, mimic water by having many OH groups.

By contrast, the enzymes in the vacuole are not so particular and it appears that any osmoticum can be used to generate osmotic pressure. The osmotic pressure (P_π) of the vacuole and the cytoplasm must be equal. While the plasma membrane, and not the vacuolar membrane, may be the primary mediator of turgor regulation, the large size of the vacuole means that the number of moles of osmoticum is greater in the vacuole than in the rest of the cell. The components that usually contribute the most to the osmotic pressure of the vacuole are Na^+, K^+, and Cl^- (Bisson and Kirst, 1980; Okazaki, 1996). These are relatively "energetically cheap" turgor-generating substances. However, they are incompatible solutes and must be kept away from the enzymes localized in the cytosol.

Turgor pressure inside the cell (P_{ti}) results from the pressure that is exerted by the protoplast against the extracellular matrix when water moves down its water potential difference from outside the cell where $P_{to} = 0$ and $P_{\pi o}$ is small to inside the cell where $P_{\pi i}$ is large.

The water potential (P_w) is equal to the difference between the hydrostatic or turgor pressure (P_t) and the osmotic pressure (P_π). All quantities are given in Pascals.

$$P_w = P_t - P_\pi \tag{7.8}$$

Water passively moves from regions of high water potential (high energy/volume) to regions of low water potential (low energy/volume). Consequently, for passive flow, $P_{w\ final} - P_{w\ initial}$ is negative. Specifically, the water potential inside the cell (P_{wi}) and outside the cell (P_{wo}) is given by the following equations:

$$P_{wi} = P_{ti} - P_{\pi i} \quad (7.9)$$

$$P_{wo} = P_{to} - P_{\pi o} \quad (7.10)$$

At equilibrium, where there is no net water movement,

$$P_{wo} = P_{wi} \quad (7.11)$$

Therefore,

$$P_{to} - P_{\pi o} = P_{ti} - P_{\pi i} \quad (7.12)$$

As $P_{to} = 0$ (by definition in most cases),

$$P_{ti} = P_{\pi i} - P_{\pi o} \quad (7.13)$$

A change in $P_{\pi i}$ will cause the turgor to increase or decrease. As $P_{\pi i}$ depends mostly on Na^+, K^+, and Cl^-, an increase in these ions will result in an increase in turgor, whereas a decrease in these ions will result in a decrease in turgor (see Chapter 12). The turgor pressure of plant cells is typically between 10^5 and 10^6 Pa, although higher and lower values exist.

The turgor pressure that is generated by the cell is responsible for providing the motive force for continued cell expansion and shape generation (see Chapter 20; Harold, 1990). It also provides the motive force for leaflet movements, tendril curling, stomatal movements, and fungal invasions.

7.6.5 Other Functions

The small vacuoles in the tip of a *Chara* rhizoid are filled with barium sulfate crystals. These crystals act as statoliths and fall in a gravitational field. As they settle, they displace the Golgi-derived vesicles from the lowermost side. Growth that is dependent on the deposition of the Golgi-derived vesicles is restricted to the upper surface, and the rhizoid bends toward the earth in a positive gravitropic manner (Schröter et al., 1975).

The low density of the cell sap compared with the rest of the cytoplasm is important for understanding buoyancy regulation (Raven, 1984; Walsby, 1975, 1982, 1994), dispersal mechanisms (Gregory, 1961), and gravity sensing (Wayne and Staves, 1991) in single cells.

7.7 BIOTECHNOLOGY

In a series of studies that range from biophysics to biotechnology, Eduardo Blumwald has been able to overexpress in tomato and canola plants a vacuolar Na^+/H^+ antiporter, which allows the plants to grow in 200 mM NaCl. The salt that enters the xylem is transported to the leaves and is sequestered in the vacuoles of the cells of the leaf. Because water is supplied to the fruits from the phloem and not the xylem, and as salt in general does not enter the phloem, the fruits from the plants growing in high salt are not salty (Epstein, 1983; Apse et al., 1999; Shi et al., 2000; Zhang and Blumwald, 2001; Zhang et al., 2001; Lv et al., 2008; Uddin et al., 2008). Work is currently focused on making rice that can grow in high salt (Zeng et al., 2017).

7.8 SUMMARY

The vacuole is the most conspicuous organelle in plant cells. Because of its large volume, it is involved in storing many inorganic and organic molecules, and in so doing, it functions in homeostasis. The vacuole is also important in storing the molecules necessary for attracting pollinators and fruit dispersers, for plant defense, and for storing the osmoticum necessary for generating turgor pressure. The vacuole is also the recycling center of the cell and has the mechanisms necessary for collecting organelles and macromolecules, degrading them, and returning their constituent parts to the cytosol.

In this chapter, I discussed the interplay between plant cell biology and physical chemistry and the importance of the vacuole and its contents in developing the foundations of physical chemistry.

7.9 QUESTIONS

7.1. What are the many functions of the vacuole?

7.2. Why is the vacuolar compartment so well developed in plants compared with animals?

7.3. How have the readily visible anthocyanins in the vacuole been important for making discoveries in genetics?

The references for this chapter can be found in the references at the end of the book.

Chapter 8

Movement Within the Endomembrane System

Acutely aware that the division of science into disciplinary pigeonholes was an arbitrary administrative artifice, Frank was convinced that it did great harm by creating barriers for heuristically fertile communication.

George Adelman and Barry Smith on Francis Otto Schmitt.

Imagine this Nobel Prize Award Ceremony without any of the beautiful flowers that you can see here around me. These flowers are transported to Stockholm each year from Sanremo in Italy. But imagine if they were missorted and ended up in Copenhagen. Without a functioning transport system, this could easily be a reality. To avoid chaos, we are totally dependent on fine-tuned transport systems, where cargo is loaded into the right vehicle and transported to the right destination at the right time. The cell, with its different compartments, faces a similar transport challenge.

Juleen Zierath (Presentation Speech for the 2013 Nobel Prize in Physiology or Medicine).

Applying the mindset of a physicist to the complexities and mysteries of cell biology afforded me such a perspective and the approach to productively tackle the problem. Physicists seek universal laws to explain all related processes on a common basis, and achieve this by formulating the simplest hypothesis to explain the facts.

James Rothman (Nobel Lecture).

I get my best ideas while taking a bath. But with biology I have problems; I always have to jump out and look up a fact.

Leo Szilard (quoted in George Feher, 2002).

8.1 DISCOVERY OF THE SECRETORY PATHWAY

We have seen that cells are composed of a multitude of membranous motifs, including the plasma membrane, the endoplasmic reticulum (ER), the Golgi apparatus, the vacuole, and the peroxisome. These five organelles make up the non–self-replicating membranous organelles in the cell and serve as a framework to discuss the elaborations of the membranes in between these organelles, the dozens of proteins involved in the primary function of each organelle, the dozens of proteins involved in the transfer of cargo between organelles, and the hundreds of genes that regulate the endomembrane system in each and every cell type. In this chapter, I will discuss how the various membranes are generated, as well as the relationships between the various membrane systems (Claude, 1970; Griffiths, 1996; Robinson et al., 2007). The relationships between the various membranes were revealed to a large extent by studying the secretory process in pancreatic exocrine cells.

Henri Dutrochet (1824, quoted in Schwartz and Bishop, 1958) postulated that

> *it is within the cell that the secretion of the fluid peculiar to each organ is effected. … The cell is the secreting organ par excellence. It secretes, inside itself, substances which are, in some cases, destined to be transported to the outside of the body by way of the excretory ducts, and, in other cases, destined to remain within the cell which has produced them.*

In the 1870s and 1880s, Rudolf Heidenhain (1878) studied the exocrine cells of the pancreas of mammals. He noticed that shortly after an animal ate, microscopic granules disappeared from the apical part of their pancreatic exocrine cells and reappeared a few hours later. He correlated the disappearance of the apical granules with the appearance of digestive enzymes in the pancreatic juices that he measured biochemically and concluded that the granules, which he dubbed *zymogen granules*, contained the precursors of the digestive enzymes. The zymogen granules, he supposed, represented an available store of digestive enzymes that could be released on eating.

Limited by technology, Heidenhain was unable to elucidate the intracellular pathways involved in the secretion of the chymotrypsinogen, trypsinogen, and α-amylase that are stored in the zymogen granules. However, impressed with Heidenhain's work that combined morphology with biochemistry, George Palade (1959) set out to understand the intracellular part of the secretory process using and, more importantly, integrating the newly

Plant Cell Biology. https://doi.org/10.1016/B978-0-12-814371-1.00008-4

developed techniques of electron microscopy and cell fractionation. Palade (1959) considered these studies to be "a collaboration over almost a century between Rudolf Heidenhain, Philip Siekevitz, and myself." The integrated studies by Palade and his colleagues have become a watershed in the study of the intracellular secretory pathway (Moberg, 2012; Siekevitz and Jamieson, 2012). These papers constitute a good pedagogical example of the interplay between theory and experiment, inductive and deductive reasoning, and technique and interpretation.

While all cells secrete one thing or another, Palade chose to study secretion in cells that specialized in secretion. Palade and his colleagues injected ^{3}H-leucine into guinea pigs to radiolabel the newly synthesized proteins and then rapidly isolated the pancreas to follow the intracellular movement of the nascent proteins. Siekevitz and Palade (1958a,b,c, 1959, 1960a,b), using subcellular fractionation techniques, and Caro and Palade (1964), using radioautography at the electron microscope (EM) level, extended Heidenhain's conclusion by showing that the digestive enzymes, stored in the smooth membrane–enclosed zymogen granules at the apical end of pancreatic exocrine cells, were synthesized on the rough ER at the basal region of the cell (Figs. 8.1–8.3). However, their ability to resolve the pathway followed by the digestive enzymes as they moved from the rough membranes at the basal portion of the cell to the smooth membrane–enclosed vesicles at the apical region of the cell was compromised by the fact that it took too much time to label the newly synthesized proteins by intravenously supplying the pancreas with radioactive amino acids.

The lack of temporal resolution was overcome by Jamieson and Palade (1966, 1967a,b, 1968a,b, 1971a,b), who switched from whole guinea pigs to pancreatic tissue slices. This minimized the time it took for the tracer to travel to, and to diffuse into, the site of incorporation because the time it takes to diffuse into the exocrine cells is minimized when the surface-to-volume ratio is maximized. Moreover, thin sections maximized the uniformity of labeling in each cell as the tracer enters all the cells at approximately the same time. Now Jamieson and Palade were able to see a precursor–product relationship between organelles just as biochemists had seen in chemical reactions. Tissue slices had previously been used successfully by many biochemists, including Otto Warburg, Albert Szent-Györgyi, and Hans Krebs, to maximize the temporal resolution necessary to determine the sequences in a pathway (see Chapter 14).

Jamieson and Palade labeled the cells for only 3 min with radioactive leucine and then replaced the radioactive leucine with an excess concentration of unlabeled leucine. They followed the movement of the label with both electron microscopic radioautography and by cell fractionation. They discovered that the label moved like a wave through the cell (Fig. 8.4). It started at the ribosomes on the ER, which represent the site of protein synthesis. The protein then entered the lumen of the ER and was only released by treatments that break the membranes. After 7 min, the label appeared in the peripheral vesicles of the Golgi apparatus. The labeled protein probably took a membranous route from the ER to the Golgi apparatus as the label never increased in the cytosolic fraction. Moreover, the labeled protein traveled from lumen to lumen because it could only be released from the Golgi membranes by high-pH treatments that destroyed membrane integrity. Thirty-seven minutes following the pulse label, the protein appeared in the condensing vacuoles at the trans-Golgi network (TGN). Approximately 2 h after the pulse label, essentially all of the labeled proteins were in the zymogen granules at the apical end of the cell where they are stored. Thus, the fact that they selected favorable material for studying secretion,

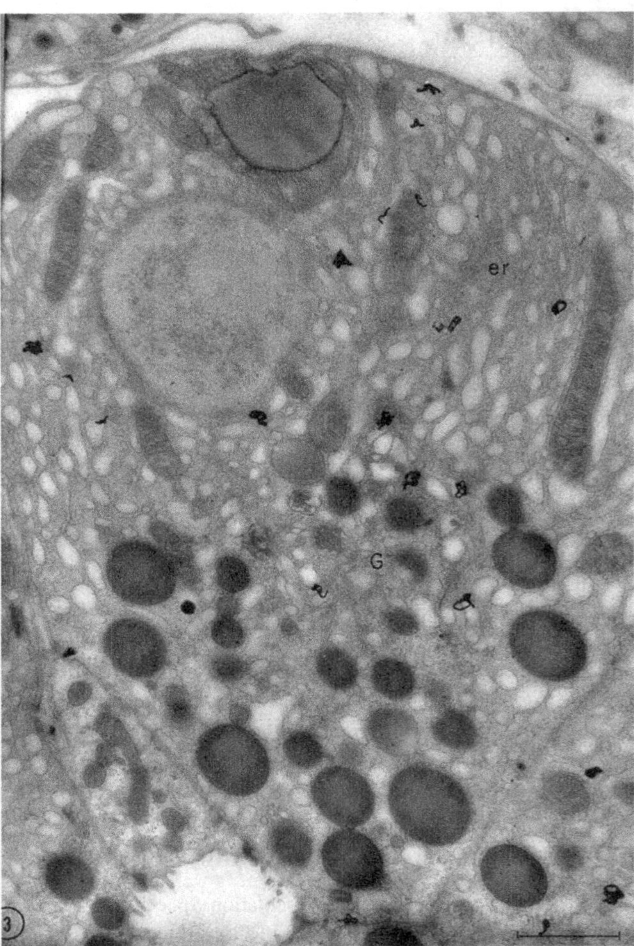

FIGURE 8.1 Electron microscopic autoradiograph of an exocrine cell 5 min after a pulse injection of ^{3}H-leucine into the guinea pig. Most of the grains are over the rough endoplasmic reticulum (ER). ×21,000. *G*, Golgi stack. *From Caro, L.G., Palade, G.E., 1964. Protein synthesis, storage, and discharge in the pancreatic exocrine cell. J. Cell Biol. 20, 473–495.*

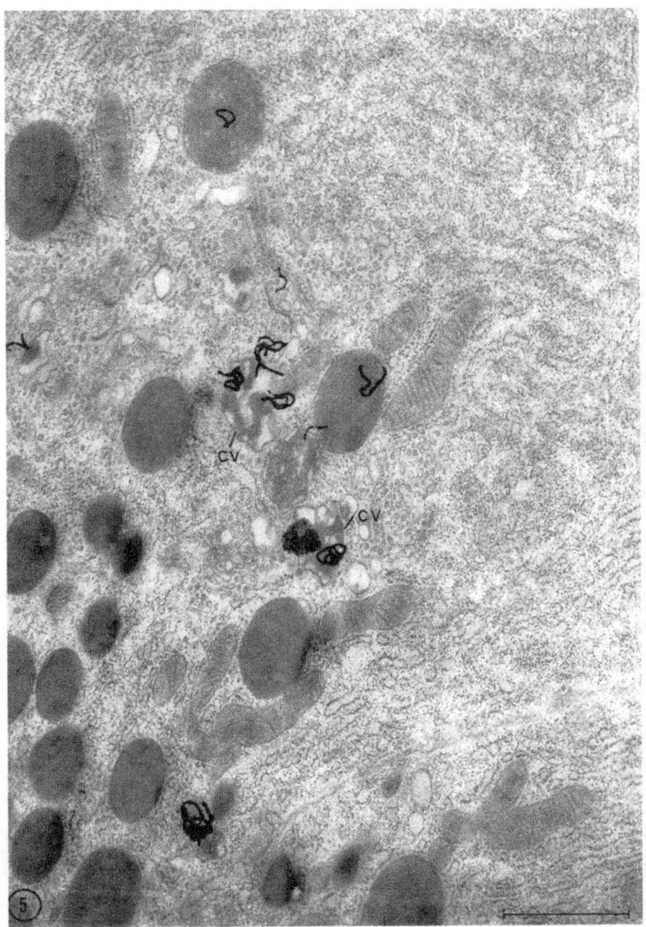

FIGURE 8.2 Electron microscopic autoradiograph of an exocrine cell 20 min after a pulse injection of ^{3}H-leucine into the guinea pig. The radioactivity has left the endoplasmic reticulum and most of the grains are over the Golgi stacks and condensing vacuoles (CVs). ×26,000. *From Caro, L.G., Palade, G.E., 1964. Protein synthesis, storage, and discharge in the pancreatic exocrine cell. J. Cell Biol. 20, 473–495.*

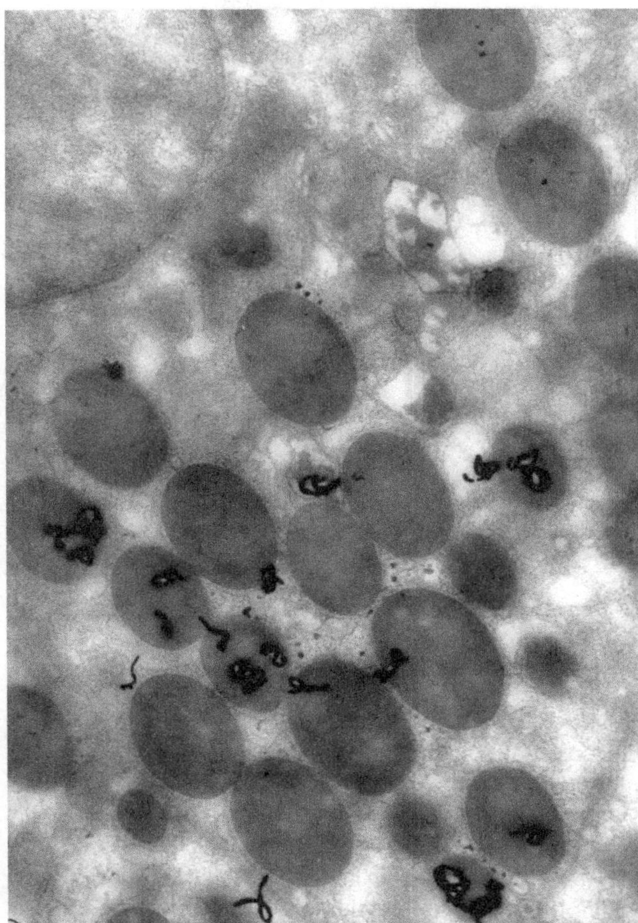

FIGURE 8.3 Electron microscopic autoradiograph of an exocrine cell 4 h after a pulse injection of ^{3}H-leucine into the guinea pig. The label has left the endoplasmic reticulum and Golgi stacks and most of the grains are over the mature zymogen granules. ×24,000. (Source: From Caro and Palade, 1964).

combined with their biophysical insight in deciding to use tissue slices to obtain a uniform and rapid uptake of labeled amino acids, allowed Jamieson and Palade to see the movement of proteins from the ER to the Golgi apparatus to the condensing vacuoles and then to the zymogen granules.

Interestingly, treating the tissue slices with chemicals that stimulate secretion caused an increase in secretory flow by increasing the amount of protein secreted, not by increasing the velocity of movement of a given protein through the secretory pathway. Thus, there must be an increase in the rate of protein synthesis and/or the area of the pathway. Indeed, the secretion-stimulating agents cause an elaboration of the Golgi apparatus!

While Jamieson and Palade followed intracellular transport through the secretory pathway with ^{14}C-leucine-labeled proteins, Northcote and Pickett-Heaps (1966) and Neutra and Leblond (1966) followed the movement of labeled glucose and discovered that the Golgi apparatus is a major site of glycosylation. Neutra and Leblond (1966) followed the movement of ^{3}H-glucose containing mucus glycoproteins through the secretory pathway of rat goblet cells with electron microscopic radioautography. They found that by 5 min after injecting the ^{3}H-glucose into a rat, the sugar was incorporated into glycoproteins in the Golgi apparatus. After 20 min of continuous labeling, the label is found in both the Golgi apparatus and the mucigen granules, and after 40 min, the label is in the mucigen granules. Unfortunately, these experiments were not done with tissue sections and with a short pulse, so the time resolution is marginal. However, at the same time, pulse-chase experiments with wheat root tips were done by Donald Northcote and Jeremy Pickett-Heaps although the extracellular matrix components they were interested in studying were not proteinaceous but were exclusively

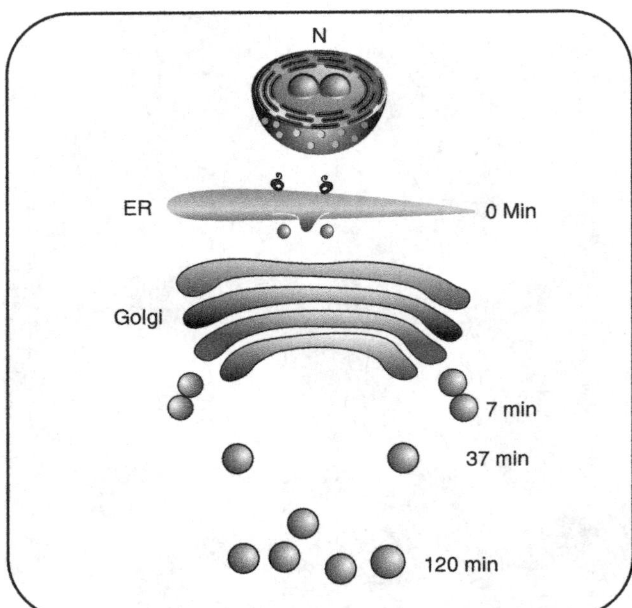

FIGURE 8.4 Spatiotemporal map of the secretory process discovered by Palade, Caro, Siekevitz, and Jamieson in pancreatic exocrine cells.

composed of polysaccharides. They found that after a 5-min pulse with radioactive glucose, the label showed up in the Golgi apparatus (Fig. 8.5). Following a 10-min pulse, the label was found in the Golgi apparatus and its associated vesicles. When the 10-min labeling period was followed with a chase period (for 10–60 min), the label declined in the Golgi apparatus and its associated vesicles and appeared in the extracellular matrix (Fig. 8.6). These data indicate that the Golgi apparatus in plant and animal cells serves as a site of protein glycosylation and polysaccharide synthesis in the intracellular secretory pathway.

Plant cells can have highly developed secretory systems, although the intracellular pathways have not been worked out with the high temporal resolution attained by Jamieson and Palade. While all cells in plants are involved in secretion to some extent, secretion often takes place in highly developed, taxonomically important, and truly gorgeous glandular trichomes. Many halophytes have glands that secrete salt to maintain a livable internal salt balance. Other plants secrete sweet nectar to attract pollinating insects, whereas insect-eating plants such as *Drosera* secrete slime that includes coniine, an alkaloid that paralyzes insects. Insectivorous plants also secrete the enzymes necessary to digest their prey. Cells in the abscission zone secrete wall-digesting enzymes that cause the leaves to fall in the autumn. Stigmas also secrete proteins that are involved in determining whether or not a given pollen grain will germinate or have an incompatible reaction. Plant cells secrete essential oils that we associate with herbs and resins that we associate with antimicrobial activity. Stinging trichomes secrete histamine, serotonin, and acetylcholine, molecules that we usually associate with nervous activity (Roshchina, 2001). Cereal grains have a layer of glandular cells known as the *aleurone layer* that surrounds the endosperm and secretes hydrolases that digest the stored food. The tapetal layer that surrounds the developing pollen

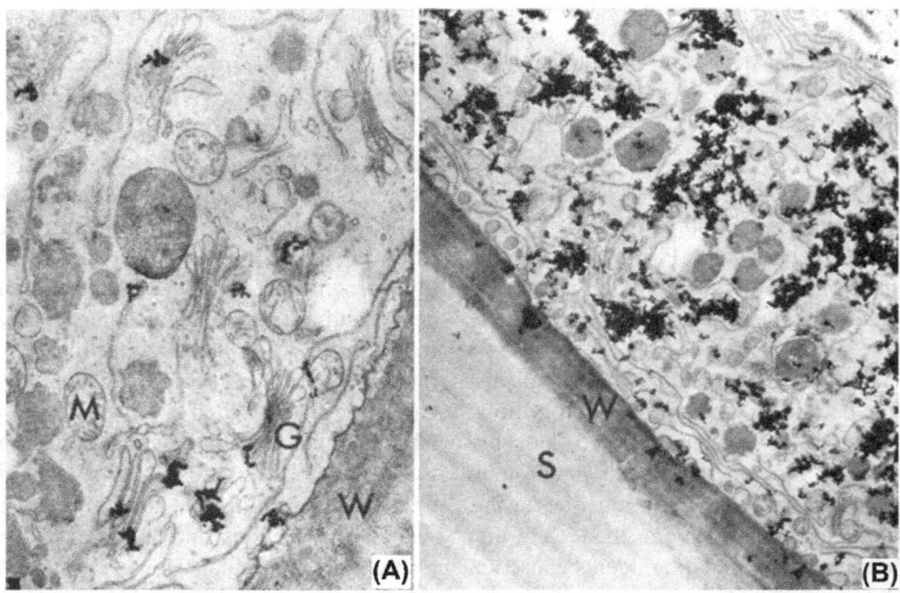

FIGURE 8.5 Root cap cells of wheat exposed to D-[6-^{3}H] glucose for 5 min (A) and 10 min (B). The majority of the labeled compounds are located in the Golgi apparatus after 5 min. By 10 min, some of the label shows up in the extracellular matrix. Left, ×11,625; right, ×14,435. *G*, Golgi stack; *M*, mitochondria; *W*, extracellular matrix. *From Northcote, D.H., Pickett-Heaps, J.D., 1966. A function of the Golgi apparatus in polysaccharide synthesis and transport in the root-cap cells of wheat. Biochem. J. 98, 159–167.*

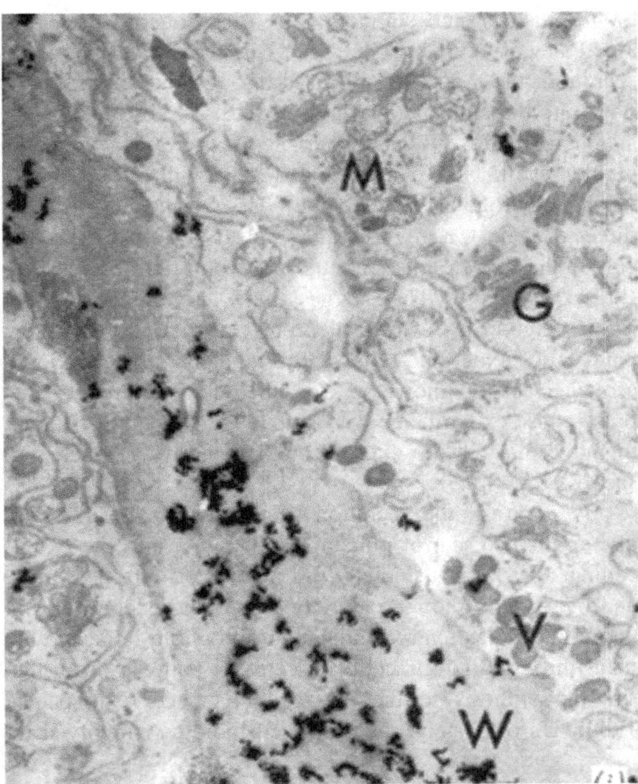

FIGURE 8.6 Root cap cells of wheat exposed to D-[6-^{3}H] glucose for 10 min followed by a 60-min chase period when the cells were treated with unlabeled glucose. The majority of the labeled compounds are now located in the extracellular matrix (W) and have left the Golgi apparatus. ×5625. *G*, Golgi stack; *M*, mitochondria; *V*, vacuole. *From Northcote, D.H., Pickett-Heaps, J.D., 1966. A function of the Golgi apparatus in polysaccharide synthesis and transport in the root-cap cells of wheat. Biochem. J. 98, 159–167.*

grains secretes the proteins involved in sporophytic incompatibility that discourages self-pollination (Hesse et al., 1993), and the pollen tubes secrete the wall material consisting of proteins and carbohydrates that are necessary for tip growth (Wang and Jiang, 2017). Many potentially fascinating aspects of the cell biology of these secretory pathways remain a mystery (Haberlandt, 1914; Fahn, 1979; Roshchina and Roshchina, 1993; Nicolson et al., 2007; Roshchina, 2008; Huchelmann et al., 2017) as well as a source of new discovery and biotechnological innovation.

How does a protein end up in a given organelle? In the past, when cell biologists thought about the biogenesis of each organelle in the endomembrane system, it was assumed that each and every component of the organelle is made at the same time and in the same way. Now the focus is on how each individual component gets to its targeted organelle. This approach allows for the possibility of multiple pathways—and which pathways are used may be cell-type or species specific and even developmentally or physiologically determined. Every protein has information encoded in its structure that determines where it will go, and as I have already discussed, proteins take different pathways to enter the ER, the peroxisome, and the vacuole. The information encoded in the sequence of the protein is affectionately known as its *molecular zip code* (see Chapter 17).

When I discuss the intracellular pathway for the movement of a protein, I follow the lead of Adolf Fick (1855), a physiologist who, under the influence of Hermann von Helmholtz, pioneered the application of physical thought to biological transport processes. Fick decided not to look at diffusion in isolation, but within the context of transport of heat and electricity. I will treat protein transport like any transport phenomenon and consider the structure of the molecule being transported, including its signal sequences, the receptor for this protein, the affinity of the protein for the receptor, the effect of the pH of the compartment on protein-receptor binding, and the proportion of this protein compared with other proteins that may compete with it for the receptor. Any of these factors can vary from cell to cell, during the life of a cell, and as a result of the overexpression of a given gene. Consequently, many factors can affect the transport of the protein at any stage in the pathway. Thus, as a result of differential concentrations of the translocated protein and its receptors as well as differences in the environment of an organelle which may affect protein folding and receptor binding, it is conceivable that proteins with identical amino acid sequences may be targeted to one place in one cell, but another location in a different cell.

Indeed, the uniformity found in targeting sequences in a single-celled organism such as yeast may not exist in multicellular organisms. An indication of this comes from the observations that phytohemagglutinin, a seed protein that is normally targeted to the vacuole of beans, is secreted to the external solution in transgenic monkey COS cells but transported to the vacuole in transgenic yeast (The acronym COS comes from the cells being **C**V-1 (simian) in **O**rigin, and having **S**V40 (viral) DNA; Voelker et al., 1986; Tague and Chrispeels, 1987). Moreover, inhibiting an H^+-ATPase in the Golgi-enriched fraction, which changes the pH of the organelle, results in the secretion of proteins to the extracellular matrix that are usually retained in the cell and the retention of proteins that are usually secreted (Matsuoka et al., 1997). It is also possible that two proteins that only differ by their signal sequence may be targeted to the same class of organelles in neighboring cells due to differences in the receptor proteins or the concentration of proteins competing for the same receptor (Frigerio et al., 1998). Elucidating the cellular and molecular components of the secretory system has been driven, in part, by the hope of using plants as light-powered bioreactors that are capable of secreting or storing high concentrations of economically valuable proteins that can be used as pharmaceuticals (Vitale and Pedrazzini, 2005; Sharma and Sharma, 2009; Ganapathy, 2016; Saveleva et al., 2016).

Proteins that lack a signal peptide are secreted from cells in response to pathogen attack or other forms of stress though unconventional protein secretory pathways, including the direct vesicleless translocation of the protein itself across the plasma membrane through a translocon (La Venuta et al., 2015; Zacherl et al., 2015), and the multivesicular body (MVB)—exosome pathway described anon (Ding et al., 2012, 2014; Drakakaki and Dandekar, 2013; Davis et al., 2016; Robinson et al., 2016; Chung and Zeng, 2017; Rabouille, 2017).

8.2 MOVEMENT TO THE PLASMA MEMBRANE AND THE EXTRACELLULAR MATRIX

In prokaryotes, the synthesis of new plasma membrane lipids (Dowhan, 2013) and proteins takes place in the existing plasma membrane. The insertion of proteins occurs both cotranslationally, with the help of the signal recognition particle, and posttranslationally, with the help of chaperones (Saraogi and Shan, 2013). In eukaryotes, the synthesis of new plasma membrane occurs on the endomembrane system. According to the endomembrane concept, the pathway for creating new plasma membrane results from the synthesis of proteins and lipids on the ER. As a result of ER membrane growth, transition vesicles, covered with a COP II coat, bleb off from the transition elements of the ER and fuse with the *cis*-face of the Golgi apparatus or its associated membranes. The membrane coats are composed of proteins that allow the blebbing off and fusion of vesicles with specific membranes (The acronym COP stands for **CO**at **P**rotein; Schekman, 1996; van de Meene et al., 2017). The new membrane and its luminal contents move through the Golgi stack where further processing of carbohydrates takes place. Once the membrane with its contents arrives at the trans-face of the Golgi apparatus, it becomes covered with another coat and moves to the plasma membrane. As it moves to the plasma membrane, it loses its coat. It then fuses with the plasma membrane in a process that involves bilayer adherence and bilayer joining. At this point, if the lumen of the vesicle contains something, that substance is secreted simultaneously in a process known as *exocytosis*. When the cell is no longer growing, plasma membrane replacement must be balanced by plasma membrane retrieval (Palade, 1959).

Plants secrete a variety of proteins into the extracellular space (Rose and Lee, 2010; Ruiz-May et al., 2012a,b; van de Meene et al., 2017). The secretion of α-amylase and the hydroxyproline-rich glycoprotein has been studied most extensively (Akazawa and Hara-Nishimura, 1985; Jones and Robinson, 1989; Jones and Jacobsen, 1991; Chrispeels, 1991). Paleg and Yomo independently discovered in 1960 that gibberellin, produced by the embryo of cereal grains, diffuses to a specialized secretory tissue known as the *aleurone layer* and stimulates the secretion of α-amylase (see Paleg, 1965; Chrispeels and Varner, 1967). Following secretion, the α-amylase diffuses through the extracellular matrix to the endosperm where it causes the breakdown of starch to maltose, which nourishes the growing embryo. Gubler et al. (1986) and Zingen-Sell et al. (1990) have labeled aleurone cells with a colloidal gold-tagged antibody directed against α-amylase and showed that α-amylase occurs in the ER, Golgi apparatus, and Golgi-derived vesicles (Figs. 8.7 and 8.8). They suggest that α-amylase follows the same secretory pathway in aleurone cells that it does in the pancreatic exocrine cells. Unfortunately, the temporal sequence is not known as the labeling time used in the pulse-chase experiments was as long as it takes the protein to move through the entire secretory pathway (Jones and Jacobsen, 1982).

By contrast, the secretion of the hydroxyproline-rich glycoprotein of carrot phloem parenchyma into the extracellular matrix was studied with a pulse of only 2–3 min, and by 30 min; the radioactivity was found exclusively in the extracellular matrix, indicating a possible rapid movement through the intracellular secretory pathway from the site of synthesis to the extracellular space (Chrispeels, 1969). Unfortunately, the plant cell biologists involved in

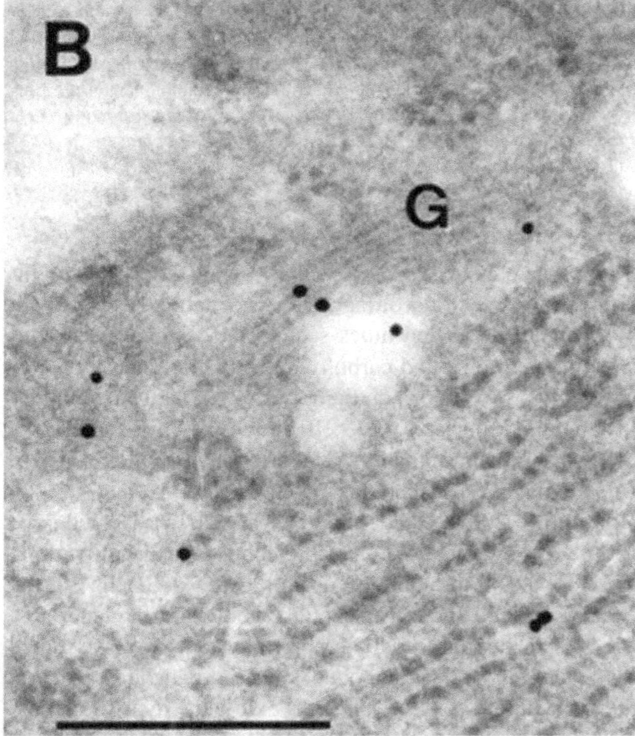

FIGURE 8.7 Immunolabeling of the endoplasmic reticulum and a Golgi stack in a cell of the aleurone of barley with an antibody to α-amylase. Bar, 300 nm. *G*, Golgi stack. *From Gubler, F., Jacobson, J.V., Ashford, A.E., 1986. Involvement of the Golgi apparatus in the secretion of a-amylase from gibberellin treated barley aleurone cells. Planta 168, 447–452.*

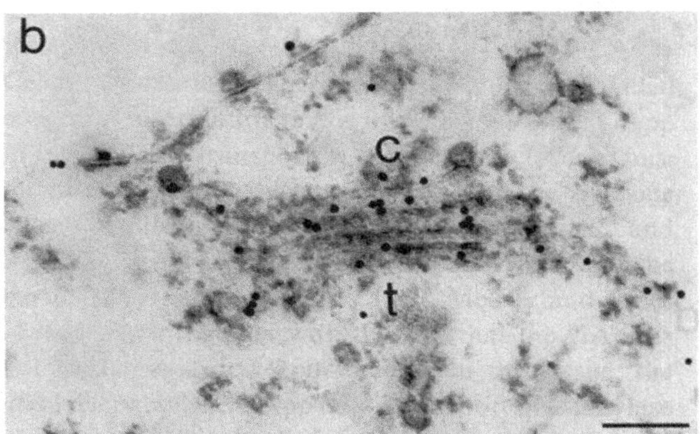

FIGURE 8.8 Immunolabeling of a Golgi stack in a cell of the aleurone of barley with an antibody to α-amylase. *c*, cis-face; *t*, trans-face. Bar, 500 nm. *From Zingen-Sell, I., Hillmer, S., Robinson, D.G., Jones, R.L., 1990. Localization of α-amylase isozymes within the endomembrane system of barley aleurone. Protoplasma 154, 16–24.*

studying secretion found it "impossible to obtain even moderately pure particulate fractions from plant tissue homogenates," and thus did not fractionate the material into ER, Golgi, and vesicular fractions but only into a membrane fraction and a supernatant fraction (Chrispeels, 1969). Apparently it was easier for Neil Armstrong to go to the moon in 1969 (Kennedy, 1962) than it was to fractionate plant organelles.

The hydroxyproline-rich glycoprotein must move through the Golgi apparatus because it is a complex arabinose-containing glycoprotein, and the enzymes necessary for the addition of arabinose in these cells are found in the Golgi apparatus (Chrispeels, 1970; Gardiner and Chrispeels, 1975). By 1982, David Robinson and his colleagues overcame the problems associated with cell fractionation, and determined, with a 5-min labeling period in a pulse-chase experiment, that, following the synthesis of the hydroxyproline-rich glycoprotein in the ER, the protein moves to the Golgi apparatus within 15 min and is secreted to the extracellular matrix within 30 min (Robinson and Glas, 1982; Wienecke et al., 1982).

The most common secretory products of all plant cells are hemicelluloses and pectins that will end up in the extracellular matrix (Ray, 1967). Jeremy Pickett-Heaps (1967b) has demonstrated in wheat root tips, with pulse-chase EM radioautography, that ^{3}H-glucose is incorporated into the Golgi apparatus within 10 min and transferred to the extracellular matrix after another 10 min (see Fig. 8.5). By 30 min, the entire label is in the extracellular matrix (see Fig. 8.6). Moore and Staehelin (1988) have shown that antibodies to the two polysaccharides are found localized in the Golgi apparatus and Golgi-derived vesicles but not in the ER, indicating that the Golgi apparatus is crucial for the secretion of the matrix of the cell wall.

Using the pulse-chase labeling method along with cell fractionation, Moreau et al. (1998) have shown that sterols are synthesized in the ER, transported through the Golgi apparatus, and transferred to the plasma membrane with a half time of about 30 min. This movement is inhibited by brefeldin A and monensin.

8.2.1 Movement Between the Endoplasmic Reticulum and the Golgi Apparatus

The ER not only produces and organizes membrane proteins but also folds them and performs a quality control function to ensure that they are folded correctly. The newly synthesized proteins in the ER are folded by members of a class of proteins known as *chaperonins*. Incorrectly folded proteins, proteins with hydrophobic surfaces, free sulfhydryl groups, or incomplete glycosylation are usually captured in the lumen of the ER by special chaperonins such as calreticulin and calrexin, which recognize incorrectly folded proteins. The chaperonins refold them to their correct configuration so they can leave the ER. This is an example of cellular quality control (Sonnichsen et al., 1994; Hammond and Helenius, 1995; Pelham, 1995; Opas et al., 1996).

There are two proposed mechanisms for the export of correctly folded proteins from the ER (Pimpl and Denecke, 2000). One proposal is that there is a constitutive, nonselective anterograde bulk flow of proteins into COP II–coated transition vesicles, which pinch off from the transition ER (TER) and move to the Golgi apparatus (Pelham, 1989a,b; Phillipson et al., 2001). Thus, the membrane proteins of the ER involved in lipid synthesis, ribosome docking, protein translocation, etc., as well as the chaperonins in the lumen of the ER, leave the TER by bulk flow in or on transition vesicles that are destined to arrive at the cis-Golgi network (CGN). These proteins must be reclaimed and returned to the ER by a retrograde transport mechanism (Sabatini et al., 1991; Pelham, 1991). COP I–coated membranes may be responsible for recycling membrane proteins back to the ER (Pimpl et al., 2000). The alternative proposal is that correctly folded proteins are actively selected and enriched in or on the COP II–coated vesicles destined to arrive at the Golgi apparatus. This is supported by the observation that these vesicles are enriched in some ER proteins but lacking in others. There may be truth in both proposals.

Some of the resident proteins in the lumen of the ER contain the amino acid sequence K/HDEL, whereas the resident membrane-bound proteins have the sequence KK or KXK on the cytoplasmically exposed carboxy-terminal region. These three sequences are recognized by receptor proteins in the CGN. In search of such receptors, Vaux et al. (1990) made an antibody to a KDEL-containing protein and then made an antibody to that antibody. They assumed that the second antibody had a shape similar to that of the original protein and would thus bind to the receptor. In this way, they found a receptor protein that is localized on the CGN. Somehow, binding to this receptor permits retrograde transport back to the ER. Thus, the vesicles or tubules seen in electron micrographs do not only move substances from the ER to the Golgi but also in the other direction.

Saint-Jore et al. (2002) and Brandizzi et al. (2002b) have shown, using green fluorescent protein (GFP) fused to the HDEL receptor, that the movement of this membrane protein from the ER to the Golgi apparatus does occur. Moreover, the ER-to-Golgi movement of this protein requires adenosine triphosphate (ATP), is inhibited by brefeldin A, and is independent of the actin and microtubular cytoskeletons. They also determined, using fluorescence redistribution after photobleaching, that the movement of this protein from the ER to the Golgi apparatus occurs within 5 min. There is currently a lot of work that is going on to understand the processes that take place in this "no organelles' land" between the ER and the Golgi apparatus, which is sometimes called the *ER–Golgi intermediate compartment*, the CGN, or the *vesicular-tubular cluster* (Hammond and Helenius, 1995; Pelham, 1995; Robinson, 2003). Many of the genes and proteins involved in ER-to-Golgi transport are being identified (Bassham and Raikhel, 2000; Batoko et al., 2000).

The majority of proteins that are synthesized in the ER continue moving to the Golgi apparatus. These proteins bleb off from the TER as transition vesicles and fuse with the Golgi apparatus and/or its cis-associated membranes (Fig. 8.9). Morré et al. (1989a) have shown that (50–70 nm) transition vesicles from the TER are able to fuse with the Golgi apparatus by using a reconstituted cell-free system. They isolated transition elements, and labeled them with ^{125}I. They added ATP (with an ATP regenerating system) to the TER that caused vesicles to bleb off. Then they isolated the ^{125}I-labeled transition vesicles. Concurrently, the Golgi apparati were isolated and adsorbed to nitrocellulose strips.

When the labeled transition vesicles were mixed with the isolated Golgi apparati, the Golgi apparati became labeled with ^{125}I, indicating that the transition vesicles fuse with the Golgi apparatus. Interestingly, transition vesicles isolated from rat liver could fuse with the Golgi apparatus isolated from soybeans and vice versa, indicating that a common "receptor protein" may exist. However, the specificity of this fusion reaction is not known because other subcellular fractions, such as the plasma membrane, mitochondria, etc., were not tested as acceptors.

Sturbois-Balcerzak et al. (1999) have shown that the transition vesicles that bleb off from the ER are enriched in phosphatidylserine compared with the ER membrane itself

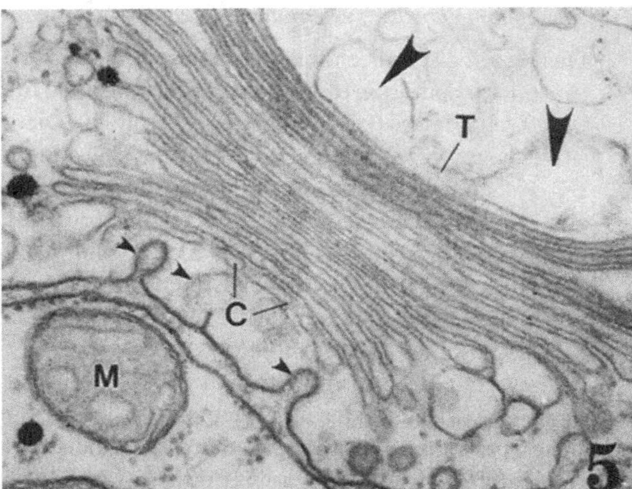

FIGURE 8.9 Small transition vesicles appear to bleb off from the ER that is associated with the cis-face (C) of the Golgi stack. *M*, mitochondrion; *T*, trans-face. ×78,200. *From Domozych, D.S., 1989a. The endomembrane system and mechanism of membrane flow in the green alga,* Gloeomonas kupfferi *(Volvocales, Chlorophyta). I. An ultrastructural analysis. Protoplasma 149, 95–107.*

(Table 8.1). Thus, in the process of transition vesicle formation, sorting of phospholipids and proteins occurs.

There are some proteins that seem to never leave the Golgi apparatus. It is not known how these proteins stay in the Golgi apparatus without continuing through the endomembrane system. In the case of membrane proteins, perhaps the membrane-spanning regions of the Golgi-localized proteins are shorter than those of the plasma membrane–localized proteins, and as Bretscher and Munro (1993) suggest, the Golgi-localized membrane proteins get stuck when they reach the region of the Golgi stack that has a certain membrane thickness. Brandizzi et al. (2002c) have shown, using genetically engineered GFP-containing fusion proteins with variable numbers of amino acids in the membrane-spanning region, that proteins with membrane-spanning regions that contain 17, 20, and 23 amino acids end up in the ER, Golgi apparatus, and plasma membrane, respectively.

8.2.2 Movement From the Golgi Apparatus to the Plasma Membrane

The vesicular movement of polysaccharides from the Golgi apparatus to the extracellular matrix was discussed in Chapter 6. During such movements, the lipids that compose the membranes of the Golgi-derived vesicles must fuse with the plasma membrane. Wait et al. (1990) have investigated the transfer of sterols from isolated Golgi stacks to the plasma membrane using a reconstituted cell-free system where the plasma membrane is adsorbed onto nitrocellulose strips and dipped in solutions containing radiolabeled

TABLE 8.1 Phospholipid Composition of Endoplasmic Reticulum and Endoplasmic Reticulum–Derived Transition Vesicles (TV)

	Phospholipid Composition (% of Total)			
Membrane Fraction	PC	PS	PI	PE
ER	75.9 +/− 5.8	1.7 +/− 1.2	3.2 +/− 1.3	19.2 +/− 4
TV(−ATP)	67.1 +/− 2.9	2.9 +/− 0.5	3.7 +/− 1.8	26.3 +/− 4.5
TV(+ATP)	69.1 +/− 3.6	6.9 +/− 2.6	3.5 +/− 1.9	20.5 +/− 3.2

ER, endoplasmic reticulum; *PC*, phosphatidylcholine; *PE*, phosphatidylethanolamine; *PI*, phosphatidylinositol; *PS*, phosphatidylserine. The PS increase in the vesicles TV + (ATP) compared with the ER was significant ($P < 0.01$).
Modified from Sturbois-Balcerzak, B., Vincent, P., Maneta-Peyret, L., Duvert, M., Satiat-Jeunemaitre, B., Cassagne, C., Moreau, P., 1999. ATP- dependent formation of phosphatidylserine-rich vesicles from the endoplasmic reticulum of leek cells. Plant Physiol. 120, 245–256.

donor membranes. They found that the Golgi stacks were more effective as donors than other membrane fractions. The transfer required ATP. The cell-free transfer system is not yet completely efficient; that is, less than 1% of the label is transferred, yet it will be a very powerful system to understand the cellular components that regulate membrane trafficking.

In plants, SAR, RAB (ROP), SNAP and SNARE genes that are homologous to secretory genes found in yeast and mammalian cells have been identified, mutated, and coupled to the sequence that codes for fluorescent proteins. In this way, the intracellular localization of the gene products and the effects of mutations have been studied in genetically engineered plants (Marsh and Goode, 1993; Denesvre and Malhotra, 1996; Seaman et al., 1996; Pimpl et al., 2000, 2003; Phillipson et al., 2001; Gu et al., 2003; Sohn et al., 2003; Happel et al., 2004; Pratelli et al., 2004; Surpin and Raikhel, 2004; Uemura et al., 2004; de Graaf et al., 2005; Hwang et al., 2005; Sutter et al., 2006; Lipka et al., 2007; Matheson et al., 2007; Min et al., 2007; Robinson et al., 2007; Sanderfoot, 2007; Zhang et al., 2007; Groen et al., 2008; Moshkov and Novikova, 2008; Nielsen et al., 2008; Rojo and Denecke, 2008; Woollard and Moore, 2008; Kang et al., 2011). The regulatory processes that determine whether a vesicle is secreted as soon as it is produced, stored in the cytoplasm until a stimulus induces its secretion, or stored in the cell indefinitely as a vacuole have yet to be elucidated.

8.3 MOVEMENT FROM THE ENDOPLASMIC RETICULUM TO THE GOLGI APPARATUS TO THE VACUOLE

The movement of proteins from the ER to the vacuole has been studied most extensively in seeds (Chrispeels, 1984, 1985). Protein bodies contain hydrolytic enzymes (and their inhibitors), indicating that they are part of the vacuolar compartment (see Chapter 7; Van der Wilden et al., 1980; Herman et al., 1981; Rasmussen et al., 1990). The best estimate of the temporal sequence of the intracellular secretory pathway to the vacuole comes from experiments done by Maarten Chrispeels (1983) who labeled bean cotyledons with ^{3}H-fucose and found, using cell fractionation, that the Golgi apparatus is labeled after approximately 45 min and the protein storage vacuoles are labeled after 60 min. The temporal resolution is poorer in all other vacuole-targeting studies because the labeling times used exceed the time it takes for the protein to move through the entire pathway (Chrispeels and Bollini, 1982; Vitale and Chrispeels, 1984; Lord, 1985).

Even though the temporal resolution of the pulse-chase experiments is not sufficient to determine the intracellular pathway, the Golgi apparatus must be involved in the trafficking of some storage proteins to the vacuole because some of the storage proteins contain complex carbohydrates (Vitale and Chrispeels, 1984), and immunocytohistochemistry at the EM level shows that storage proteins can be detected in the ER, Golgi apparatus, coated vesicles, and protein storage vacuoles of developing cotyledons (Nieden et al., 1982, 1984; Parker and Hawes, 1982; Herman and Shannon, 1984a,b, 1985; Greenwood and Chrispeels, 1985b; Boller and Wiemken, 1986; Harris, 1986; Faye et al., 1988; Kim et al., 1988; Robinson et al., 1989; Hoh et al., 1991). There is a gradient in the pH of the organelles in the pathway from the ER to the vacuole—decreasing from pH 7.5 in the ER to pH 6 in the vacuole (Martinière et al., 2013). Specialized forms of RAB and SNARE proteins that are related to the proteins that direct cisternae-to-cisternae transport in the Golgi provide the molecular basis for specifically delivering vesicles to the vacuole (Saito and Ueda, 2009; Wolfenstetter et al., 2012; Pedrazzini et al., 2013; Uemura and Ueda, 2014).

8.4 MOVEMENT FROM THE ENDOPLASMIC RETICULUM TO THE VACUOLE

After studying their electron micrographs of corn endosperm, Khoo and Wolf (1970) and Larkins and Hurkman (1978) concluded that protein storage vacuoles containing water-insoluble prolamins form directly from the ER. The results of these and other studies that indicated that protein-containing bodies may arise directly from the ER (Bonnett and Newcomb, 1965) were ignored, in part, because it was believed by many plant cell biologists that there was only one pathway of vacuole formation—and that one pathway involved the Golgi apparatus. Thus, it was generally thought that any electron micrographs that showed vacuole formation directly from the ER must be riddled with artifacts. However, there has been a paradigm shift, and now it is believed that there are many pathways involved in vacuole formation, and that some protein storage vacuoles do in fact form directly from the ER (see Chapter 7; Galili et al., 1996; Robinson and Hinz, 1996; Robinson et al., 1996; Herman, 2008).

At least two of these pathways exist in rice endosperm cells, and in these cells, there are two different populations of ER known as *cisternal ER* and *protein-body ER*. The cisternal ER is enriched in glutelin mRNA as evidenced from in situ hybridization with the cDNA that codes for the glutelins, which are proteins that are soluble in dilute acids or bases. The glutelins that are translated on the ER probably go through the Golgi to form protein storage vacuoles. The prolamins in contrast are translated on the ER that is connected to protein bodies as evidenced by in situ hybridization with the cDNA that codes for prolamin (Levanony et al., 1992; Li et al., 1993a,b). The water-insoluble prolamins that are retained in the ER lumen do not have the typical ER lumen-retention sequence KDEL or HDEL (Masumura et al., 1990). Perhaps the prolamins are retained in the ER as a consequence of their solubility properties. In rice endosperm cells, the prolamin-containing ER is sometimes engulfed by autophagosomes to form yet another kind of protein storage vacuole.

8.5 MOVEMENT FROM THE PLASMA MEMBRANE TO THE ENDOMEMBRANES

Christian de Duve (1963) coined the term *endocytosis* to name all the processes (e.g., phagocytosis, pinocytosis, micropinocytosis, etc.) whereby cells engulf small volumes of the external medium and pari passu internalize the plasma membrane. The movement of macromolecules into plant cells can occur by various mechanisms, including fluid-phase endocytosis and receptor-mediated endocytosis, both of which bring the macromolecules into the E-space. Many polypeptide toxins, including ricin and diphtheria, enter the cell through the endocytotic system (Lord and Roberts, 1998). Endocytosis occurs during plasmolysis, and it is has been suggested that the plasma membrane may be stored in the endocytotic vesicles readying the cell for deplasmolysis, although the endocytotic membranes have not yet been shown to be utilized during deplasmolysis (Oparka et al., 1990; Oparka, 1994; Lang-Pauluzzi, 2000). Another inwardly directed macromolecular transport system, sometimes called *piggyback endocytosis*, has been discovered, which brings macromolecules into the P-space of the cell. Specialized forms of SNAREs and proteins that affect the RAB GTPase provide the molecular basis for specifically delivering vesicles to the vacuole (Ebine et al., 2014; Singh et al., 2014). Using immunolocalization of proteins thought to represent the secretory pathway or the endocytotic pathway, Kang et al. (2011) show that these two pathways converge at the TGN.

8.5.1 Fluid-Phase Endocytosis

When wall products or enzymes are rapidly secreted, new plasma membrane may be added at a rate that is greater than that necessary to keep up with growth. Therefore, a mechanism is needed to retrieve excess or old membrane from the plasma membrane and either recycle it by sending it back to the Golgi apparatus or degrade it by sending it to the vacuole. This is accomplished by a process known as *endocytosis* (de Duve, 1963). Endocytosis occurs incessantly in the wall-less alga *Dunaliella* (Ginzburg et al., 1999). Although initially deemed impossible in walled plant cells due to the presence of turgor pressure (Cram, 1980), endocytosis does occur in such cells, as evidenced by the uptake of the relatively impermeant La^{3+} ion into root cells (Samuels and Bisalputra, 1990); the uptake of anionic and cationic colloidal gold particles into pollen tubes (Moscatelli et al., 2007; Moscatelli, 2008; Onelli and Moscatelli, 2013); the uptake of Lucifer Yellow into the inner cortical cells of roots (Baluska et al., 2004); the uptake of large, polar FITC-dextrans into suspension culture cells (Cole et al., 1990) and pollen tubes (O'Driscoll et al., 1993); the uptake of plasma membrane labeled with FM4-64 or FM1-43 (Vida and Emr, 1995; Bolte et al., 2004; Bove et al., 2008; Jelínková et al., 2010; Malínská et al., 2014; Rigal et al., 2015) in turgid guard cells (Meckel et al., 2004, 2005) and at the apex of growing pollen tubes (Zonia and Munnik, 2008); the uptake of CdSe/ZeS quantum dots into suspension culture cells (Etxeberria et al., 2006); and by using rapid-freezing techniques combined with electron microscopy (Ketelaar et al., 2008). Indeed, endocytosis may be important in recapturing molecules from the extracellular matrix that undergo turnover in the natural cycle of synthesis and degradation (Labavitch, 1981;

Herman and Lamb, 1992; Baluska et al., 2005), for recycling the plasma membrane (Parton et al., 2001; Meckel et al., 2004; Karahara et al., 2009; Bashline et al., 2013; Johnson and Vert, 2017), and in nutrient-starved cells undergoing autophagy (Yano et al., 2004).

As a first approximation, vesicle formation is a mechanical process, in which the free energy necessary to push a vesicle into a protoplast will be equal to the product of the volume of the vesicle (V) and the turgor pressure of the cell (P_t). Thus, for a given availability of free energy, cells with higher turgor pressures would have to have smaller endocytotic vesicles. The force needed to invaginate the membrane is supplied, in part, by a mechanochemical protein such as dynamin that is energized by the hydrolysis of guanosine triphosphate (GTP) (Collings et al., 2008; Taylor, 2011). The dynamin forms helical spirals around the neck of the forming vesicle. In vitro models of this mechanochemical process can be viewed under the microscope, and the change in the conductance of the membrane that occurs during the budding process can be followed with electrophysiology techniques (Bashkirov et al., 2008; Pucadyil and Schmid, 2008). The fusion of secretory vesicles with the plasma membrane can also be monitored with electrophysiological assays (Thiel and Battey, 1998). Dynamin is a mechanochemical GTPase that is also involved in pinching off the clathrin-coated vesicles from the TGN (Fujimoto and Tsutumi, 2014).

Let us study the parts of the endocytotic pathway individually. Plasma membrane vesicle formation begins where clathrin (see Chapter 6) binds to the plasma membrane. In the early stages of vesicle formation, the plasma membrane contains clathrin-coated pits that may have a surface density of about $0.1-4.5\ \mu m^{-2}$ (Roth and Porter, 1964; Pearse, 1976; Emons and Traas, 1986; Low and Chandra, 1994). The clathrin acts as a magnet for plasma membrane receptors, as coated pits in some animal cells contain 70% of certain plasma membrane receptors, while accounting for only 3% of the surface area (Pearse and Robinson, 1990; Robinson and Hillmer, 1990). As the membrane blebs in, the coated pits develop into clathrin-coated vesicles (Balusek et al., 1988; Robinson and Depta, 1988; Harley and Beevers, 1989; Coleman et al., 1987, 1991; Robinson et al., 1991; Beevers, 1996; Blackbourn and Jackson, 1996; Ito et al., 2012). Clathrin forms a polygonal network around the vesicle, which gives it a honeycomb-like appearance (Wiedenhoeft et al., 1988). Clathrin cages assemble into a honeycomb-like structure spontaneously in vitro when the proteins are above a critical concentration. Assembly does not require metabolic energy but depends on [Ca^{2+}], [H^+], and ionic strength. Clathrin does not bind directly to membranes. Consequently, the binding of clathrin to receptor proteins is mediated by adaptor proteins (Pearse and Robinson, 1990; Holstein et al., 1994; Drucker et al., 1995; Chen et al., 2011). Adaptor proteins are also required for the formation of clathrin-coated vesicles at the TGN (Fujimoto and Tsutsumi, 2014).

The coated vesicles originating from the plasma membrane lose their coats and then fuse with the partially coated reticulum (PCR) (see Chapter 6). The PCR was first observed in plants by Tom Pesacreta and Bill Lucas (1984, 1985). The PCR can be sparsely branched or extensively anastomosed. The membranes are coated along various regions throughout the reticulum. It seems to have a variable relationship with the Golgi apparatus (Hillmer et al., 1988; Mollenhauer et al., 1991). Some people believe that the PCR is an independent structure (Pesacreta and Lucas, 1984, 1985); others, after viewing three-dimensional reconstructions of serial sections, believe that it is interconnected with the trans-face of the Golgi apparatus and is thus equivalent to the TGN (Sluiman and Lokhorst, 1988; Hillmer et al., 1988). The PCR, at least in some soybean cells, may be synonymous with the early endosome in animal cells (Dettmer et al., 2006; Lam et al., 2007). The early endosome is defined as the first internal membranous body in the peripheral cytoplasm to which the endocytotic vesicles fuse (Brown et al., 1986; Mellman, 1996; Robinson et al., 2008).

As endocytosis proceeds, the endocytotic organelles change their appearance. As a result of maturation or of vesicle exchange, they become late endosomes (Tse et al., 2004; Ueda et al., 2004). The late endosomes in soybean cells look like multivesicular bodies (MVBs). At one time MVBs were considered to be fixation artifacts; however, Tanchak and Fowke (1987) demonstrated their importance in endocytosis. MVBs are usually 250−500 nm in diameter and contain a number of smaller vesicles, usually 40−100 nm in diameter. The MVBs may be specialized lysosomes that help degrade plasma membrane proteins. Some MVBs have been seen attached to tubules extending from the PCR or TGN (Noguchi and Kakami, 1999). The MVBs leave the vicinity of the TGN and then fuse with the central vacuole, or vesicles may bleb off the MVBs and fuse with the central vacuole, where final degradation takes place and the degraded components can be recycled. I would like to stress that, while fluid phase endocytosis may be a common process in all cell types, the actual intracellular pathway followed by the endocytotic vesicles is cell-type specific. For example, in many mammalian cells, the early endosomes appear as multivesicular vesicles. The membranous organelles that participate in endocytosis may also form an endosomal reticulum (Hopkins et al., 1990; Mironov et al., 1997). The pleomorphic nature of the early and late endosomes may reflect the particular balance of multidirectional transport processes that occur in these multifunctional organelles.

To determine the pathway and kinetics of endocytosis in soybean cells, Tanchak et al. (1984, 1988) treated

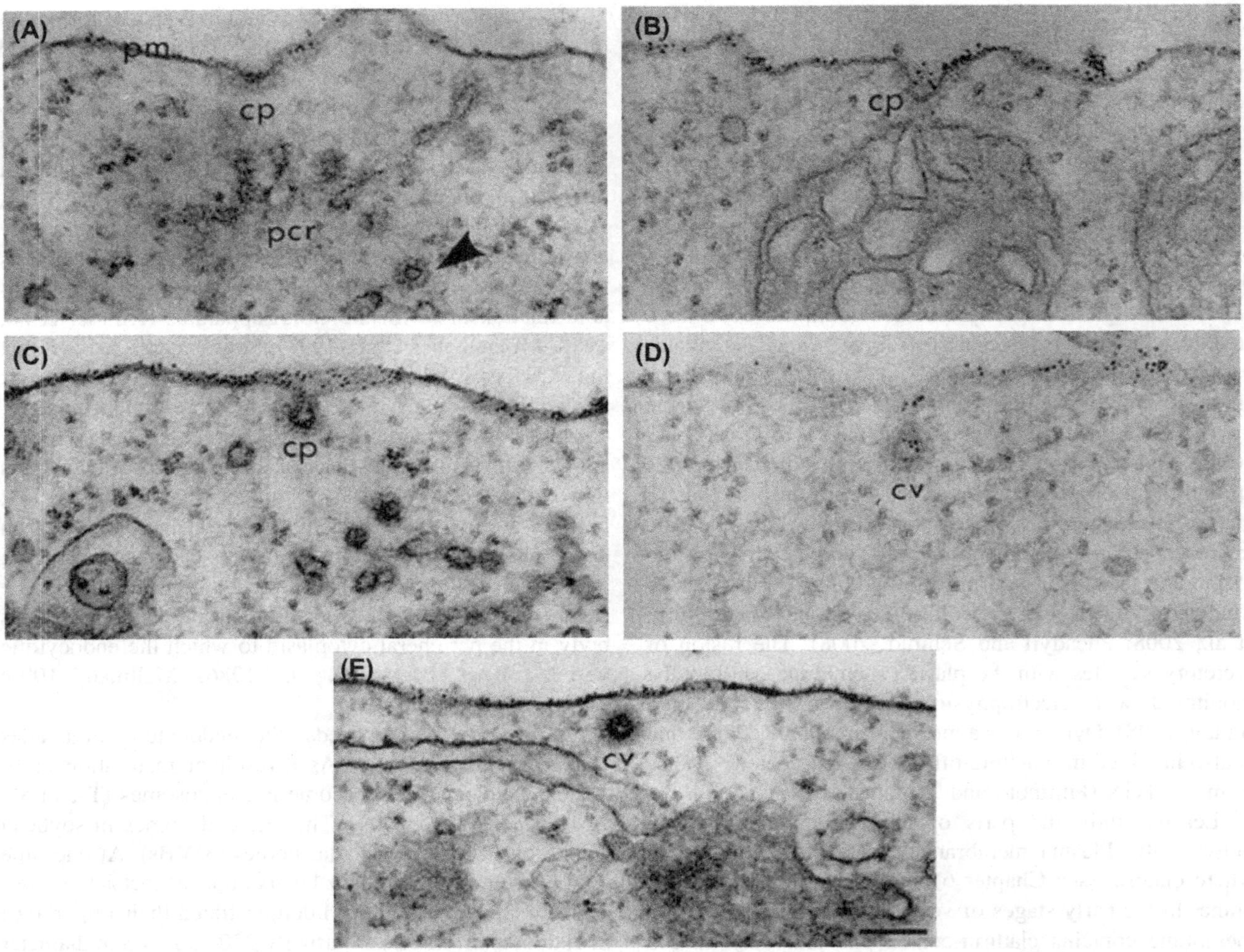

FIGURE 8.10 Endocytosis in plant cells. Cationized ferritin is seen as electron-dense dots in a coated pit (A). Cationized ferritin is seen as electron-dense *dots* in a deep, coated pit (B). Cationized ferritin is seen as electron-dense *dots* in a coated pit with a narrow neck (C). Cationized ferritin is seen as electron-dense *dots* in a coated vesicle (D). Cationized ferritin is seen as electron-dense *dots* in a coated vesicle (E). The soybean protoplast was treated with cationized ferritin for 10 s prior to fixation. *cp*, coated pit; *pcr*, unlabeled partially coated reticulum; *pm*, plasma membrane. *Arrow* points to an unlabeled coated vesicle. Bar, 100 nm. *From Tanchak, M.A., Griffing, L.R., Mersey, B.G., Fowke, L.C., 1984. Endocytosis of cationized ferritin by coated vesicles of soybean protoplasts. Planta 162, 481–486.*

protoplasts with cationized ferritin. Within 10 s the cationized ferritin is found evenly labeling the plasma membrane and coated pits (Fig. 8.10). After 30 s, the cationized ferritin is found in coated vesicles in the vicinity of the Golgi apparatus and in smooth vesicles (Fig. 8.11). After 30–120 s, the cationized ferritin is found in partially coated vesicles (Fig. 8.12). After 12 min, the cationized ferritin is found in the PCR and the Golgi stacks.

These results are consistent with those of Hübner et al. (1985), who looked at the uptake of heavy metals in intact root cap cells. They find that lead is localized in coated pits, coated vesicles, and the membranes near the trans-face of Golgi apparatus, which these authors believe to be the PCR. This could mean that vesicles can move from the plasma membrane to the PCR to the Golgi apparatus.

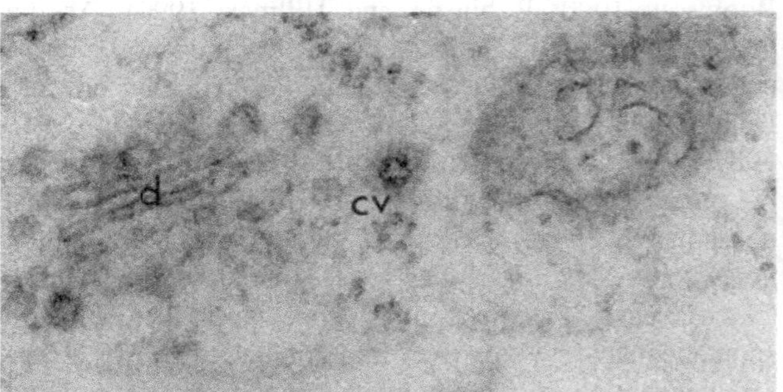

FIGURE 8.11 Endocytosis in plant cells. Cationized ferritin is seen as electron-dense *dots* in a coated vesicle. The soybean protoplast was treated with cationized ferritin for 30 s prior to fixation. *CV*, condensing vacuole; *d*, Golgi stack. *From Tanchak, M.A., Griffing, L.R., Mersey, B.G., Fowke, L.C., 1984. Endocytosis of cationized ferritin by coated vesicles of soybean protoplasts. Planta 162, 481–486.*

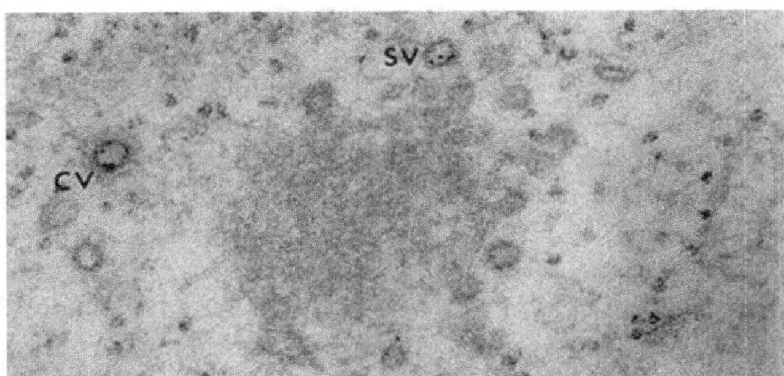

FIGURE 8.12 Endocytosis in plant cells. Cationized ferritin is seen as electron-dense dots in a smooth vesicle. The soybean protoplast was treated with cationized ferritin for 2 min prior to fixation. *From Tanchak, M.A., Griffing, L.R., Mersey, B.G., Fowke, L.C., 1984. Endocytosis of cationized ferritin by coated vesicles of soybean protoplasts. Planta 162, 481–486.*

TABLE 8.2 Time Course of Endocytosis in Protoplasts

Time	Structure Labeled
10 s	Coated pits, coated vesicles
30 s	Coated vesicles near Golgi apparatus and smooth vesicles
30–120 s	Partially coated vesicles
12 min	Partially coated reticulum and Golgi stacks
30 min	Golgi and multivesicular bodies
3 h	Large central vacuole

A more complete pathway of endocytosis was found by Tanchak and Fowke (1987) and Record and Griffing (1988; see Table 8.2). Soybean protoplasts were exposed to cationized ferritin for 5, 30, or 180 min and then the cells were fixed to localize the cationized ferritin and stained with acid phosphatase to visualize the vacuolar compartment. After 5 min, the cationized ferritin is found in the coated pits, coated vesicles, smooth vesicles, and PCR. After 30 min, the cationized ferritin is found in the Golgi complex and the MVBs (Fig. 8.13). After 3 h, the cationized ferritin is found in the large central vacuole. Acid phosphatase occurs in the smooth vesicles, Golgi apparatus, MVBs, and the vacuole (Fowke et al., 1991). A similar sequence is observed with the uptake of bovine serum albumin-gold, except that this probe, unlike the cationized ferritin, does not end up in the vacuole (Villanueva et al., 1993; Griffing et al., 1995). Recent work comparing the uptake of CdSe/ZeS quantum dots with soluble dextrans shows that the quantum dots also do not move all the way to the vacuole in suspension culture cells (Etxeberria et al., 2006). Thus, just as there are variations in the secretory pathway that depend on the substance secreted, there also seem to be variations in the endocytotic pathway that depend on the substance taken up (Onelli et al., 2008).

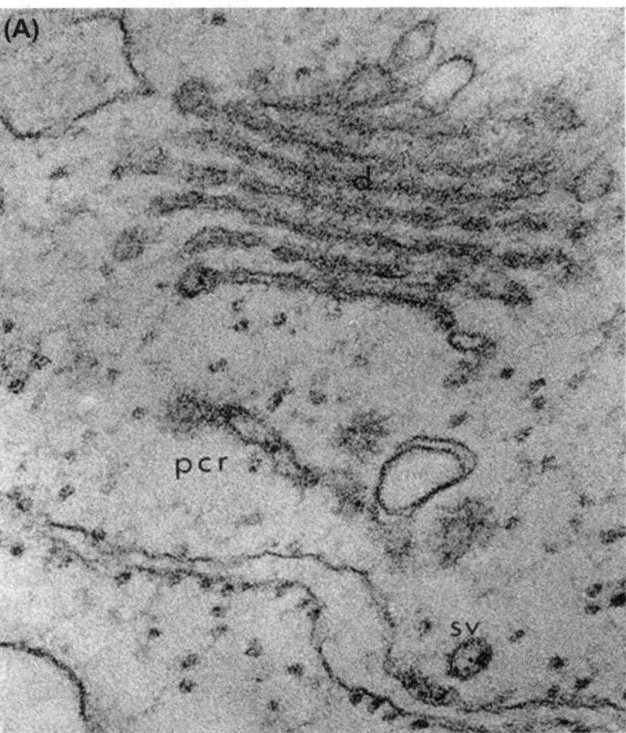

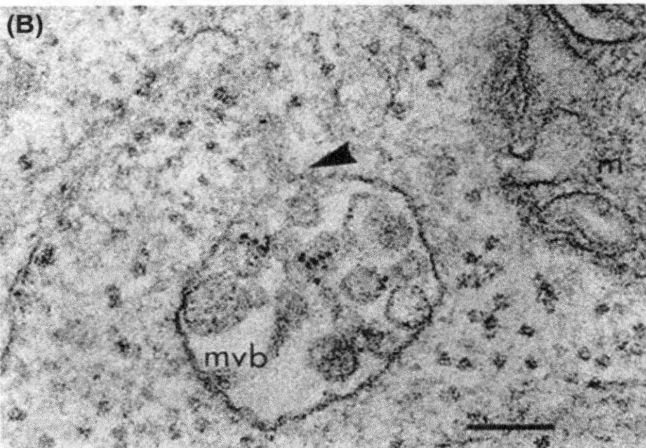

FIGURE 8.13 Endocytosis in plant cells. In (A), cationized ferritin is seen as electron-dense *dots* in the trans-cisternae of a Golgi stack (d for dictyosome), the partially coated reticulum (pcr), and a smooth vesicle (sv). In (B), cationized ferritin is seen as electron-dense *dots* in a multivesicular body (mvb). The *arrow* points to a tubular extension of the mvb. The soybean protoplast was treated with cationized ferritin for 12 min prior to fixation. Bar, 100 nm. *From Tanchak, M.A., Griffing, L.R., Mersey, B.G., Fowke, L.C., 1984. Endocytosis of cationized ferritin by coated vesicles of soybean protoplasts. Planta 162, 481–486.*

Endocytosis does not always begin with the formation of clathrin-coated vesicles (Moscatelli et al., 2007; Moscatelli and Idilli, 2009; Bandmann et al., 2012; Onelli and Moscatelli, 2013). There can be clathrin-dependent and clathrin-independent endocytotic pathways in the same cell.

In pollen tubes, the clathrin-independent pathways involve naked vesicles. An additional kind of membrane coat called *caveolin* has been found in animal cells and may be involved in capturing glycosyl-phosphatidylinositol–anchored proteins and low–molecular weight substances in a vesiculating process termed *potocytosis* (Anderson, 1993). Potocytosis and caveolin have not yet been found in plants.

In some cases when a plant cell is under attack, the MVBs fuse with the plasma membrane and release their contents into the periplasmic space between the plasma membrane and the extracellular matrix (Halperin and Jensen, 1967). The released vesicles, known as *exosomes*, are 30–150 nm and can be seen in transmission electron micrographs (Raposo and Stoorvogel, 2013). Because the occurrence of exosomes is enhanced by pathogen attack, they may play a role in the defense against pathogens (An et al., 2007; Wei et al., 2009; Wang et al., 2014; Samuel et al., 2015; Wang et al., 2017). Proteomic analysis shows that isolated exosomes contain over 100 proteins and are highly enriched in biotic and abiotic stress-response proteins. Eighty-four percent of the proteins found in the exosomes do not have a signal peptide that would first target the secreted protein to the ER for transport through the traditional secretory pathway (Rutter and Innes, 2017). Artificial exosomes are being developed as drug delivery systems for therapeutic purposes (Syn et al., 2017; García-Manrique et al., 2018).

Many symbionts and pathogens enter the cell by endocytosis (Son et al., 2003). The pathogens can be digested using the autophagy machinery in a process known as xenophagy (Jebanathirajah et al., 2002; Wileman, 2013). However, in some cases, the pathogen lives within the cell. After endocytotic uptake of *Listeria* and *Shigella*, the endocytotic membrane dissolves, and the pathogens reproduce in the cytoplasm (see Chapter 10). In other cases (e.g., *Chlamydia* and *Toxoplasma*), the endocytotic membrane is resistant to lysis and the pathogens reproduce within the endocytotic vesicle. In still others (e.g., *Coxiella* and *Leishmania*), the endocytotic vesicle fuses with the lysosome and the pathogen divides in the lysosomal compartment. In the most unusual case (e.g., *Legionella*), the endocytotic vesicle transforms into rough ER (Tilney et al., 2001). Is it possible for the endocytotic membranes in uninfected cells to transform into rough ER, providing yet another pathway for membrane traffic?

In some respects, the endosomal compartment is similar to the Golgi body in that they are both composed of multiple distinct compartments that are linked by anterograde and retrograde transport (Rothman, 2010). Why then are the Golgi cisternae stacked in most cell types, while the compartments of the endosomal pathway are not stacked? Does the stacking indicate one predominant pathway is being used in intracellular transport while nonstacked compartments indicate that more than one pathway is being used in intracellular transport?

8.5.2 Receptor-Mediated Endocytosis

Following the binding of extracellular ligands, including hormones, lectins, or antibodies to receptors in microdomains of the plasma membrane, the ligand and its receptor are typically taken up into the cell (Pastan and Willingham, 1985). This process is known as *receptor-mediated endocytosis* or *microdomain-associated endocytosis* (Geldner et al., 2003; Fan et al., 2015) and, like fluid-phase endocytosis, also involves clathrin-coated pits, clathrin-coated vesicles, and ADP ribosylation factor (ARF) GTPases (Chen et al., 2011). The removal of receptors from the plasma membrane is one way to terminate a given response and there is evidence that a part of the signal transduction chain occurs in the endosomal compartment (Geldner and Robatzek, 2008). The receptor proteins that are internalized by endocytosis have specific amino acid sequences in their cytoplasmic domains (Trowbridge, 1991; Geldner and Robatzek, 2008).

Horn et al. (1989, 1992) have developed a fascinating system to study receptor-mediated endocytosis in plants. Soybean suspension culture cells are induced to make fungal defense molecules in response to fungal attack. When the fungus begins to degrade the plant extracellular matrix it releases large, polar oligosaccharide molecules (M_r. 30,000 Da) that are known as *elicitors* (Sequeira, 1983). Elicitors are too large and polar to passively diffuse through the membrane; however, they bind to the plasma membrane (Schmidt and Ebel, 1987) and cause the cell to produce antifungal defense molecules, including glyceollin, pisatin, phaseolin, and H_2O_2 (Low and Heinstein, 1986; Apostol et al., 1987, 1989).

When cells are challenged with polygalacturonic acid elicitors, which are fluorescently labeled, the elicitors are first observed to bind to the plasma membrane. They enter the cytoplasm, and after approximately 2 h, they end up in the large central vacuole, where the β-glucanases, which digest the elicitors, are concentrated (van den Bulcke et al., 1989).

To test whether the elicitor entered the cell through nonspecific fluid-phase endocytosis or by receptor-mediated endocytosis, Horn et al. (1989) labeled bovine serum albumin and inulin to see if these large molecules, which presumably do not have receptors on the plasma membrane, are taken up by the plant cell. Neither molecule is taken up into the plant cell, indicating that the elicitor is taken up specifically by receptor-mediated endocytosis. Horn et al. (1989) studied the uptake of ^{125}I-labeled elicitors and found that 10^6 molecules are taken up per cell per minute (which is approximately 4.6×10^{-11} mol/m^2 s). They also found that 1 mM KCN and low temperatures (4°C), two treatments that inhibit energy-dependent processes, inhibit elicitor uptake. Furthermore, they found that uptake of the elicitor does not change its molecular size, eliminating the possibility that only small molecules or breakdown products are taken up. Robatzek et al. (2006) have shown that receptor-mediated endocytosis is also involved in the defense response stimulated by a bacterial flagellin peptide (Robatzek, 2007).

The discovery of hundreds of genes for receptor-like tyrosine kinases in plants stimulated research into receptor-mediated endocytosis in plants (Irani and Russinova, 2009; Di Rubbo and Russinova, 2012). The genes for the receptor-like kinases were fused to the gene for the GFP, and the fate of the receptor was followed in the cell. The receptors travel from the plasma membrane to the TGN to the MVBs and to the vacuole (Viotti et al., 2010; Nimchuk et al., 2011; Irani et al., 2012). Receptor-mediated endocytosis may also be important in understanding the mechanism of auxin and brassinosteroid action in plants (Geldner et al., 2001, 2003; Russinova et al., 2004; Geldner and Jürgens, 2006; Dhonukshe et al., 2007; Richter et al., 2010; Irani et al., 2012), flagellin sensing (Robatzek et al., 2006), cell-to-cell communication that occurs during pollination (Lind et al., 1996; Luu et al., 2000; Sanchez et al., 2004), boron transport (Takano et al., 2010), as well as other developmental and physiological responses (Geldner and Robatzek, 2008; Moriwaki et al., 2014).

8.5.3 Piggyback Endocytosis

Horn et al. (1990) reasoned that water-soluble vitamins such as folate, vitamin B_{12}, and biotin, which are too large to passively diffuse across the plasma membrane, may have receptors in the plasma membrane of plant cell membranes like those in animal cells. They also hypothesized that when compounds like biotin are attached to other large molecules that are normally impermeant, it may facilitate the transport of the large molecule across the plasma membrane and into the cytoplasm in a "piggyback" manner. In this way, large molecules could be introduced into the P-space of the cell. Indeed, Horn et al. found that fluorescently labeled molecules, including hemoglobin, RNAse, and bovine serum albumin, which normally do not pass the plant plasma membrane, can permeate the plasma membrane of soybean suspension cells when the macromolecules are tagged with biotin. This may be a great way to introduce antisense RNA, toxins (e.g., phalloidin, cholera toxin, pertussis toxin), antibodies, and individual genes into cells, especially as biotinylation of macromolecules is easy and the reagents are commercially available.

8.6 DISRUPTION OF INTRACELLULAR SECRETORY AND ENDOCYTOTIC PATHWAYS

While mutations have easily led to the discovery of hundreds or thousands of genes and gene products involved in the secretory and endocytotic pathways of various cells in various organisms from various kingdoms, finding specific inhibitors of the proteins involved in specific stages of the secretory pathway has been difficult. It is difficult to find a specific inhibitor of a specific stage in the secretory process because the proteins involved in vesicle blebbing, fusion, and other aspects of transport through the endomembrane system have many similarities with each other, as well as differences. Moreover, the identification of homologous stages of the secretory process in two cell types that have very protean secretory systems is difficult. Consequently, chemicals that influence the structure, function, or distribution of the membranes that comprise the endomembrane system have had only modest success in helping to elucidate the intracellular pathway taken by a given protein. Nevertheless, there are many chemicals that can be used to probe the endomembrane system (Irani and Russinova, 2009; Hicks and Raikhel, 2012; Mishev et al., 2013).

Brefeldin A is one such chemical that is used to test the importance of the Golgi apparatus in a given secretory pathway in plants (Klausner et al., 1992; Bauerfeind and Huttner, 1993; Driouich et al., 1993b, 1994; Satiat-Jeunemaitre et al., 1996; Kaneko et al., 1996). In animal cells, brefeldin A inhibits the ARF that results in the formation of anterograde membrane blebs without affecting retrograde membrane tubularization. This causes the Golgi stacks to be reabsorbed by the ER, and consequently, any trafficking that normally occurs through the Golgi apparatus is inhibited. In plant cells, brefeldin A does not always result in the reabsorption of the Golgi stacks (Langhans et al., 2007) but only induces a redistribution of the Golgi stacks (Satiat-Jeunemaitre and Hawes, 1994); thus, inhibition of secretion in plants by brefeldin A indicates either the Golgi stacks themselves or the arrangement of the Golgi stacks is important for a given secretory pathway.

Secretion can also be influenced by the monovalent cationophore, monensin, an antibiotic isolated from *Streptomyces cinnamonensis* that cause structural changes in the Golgi apparatus (Morré et al., 1983; Mollenhauer et al., 1983; Cornejo et al., 1988; Simon et al., 1990). Monensin has been shown to inhibit the secretion of α-amylase (Melroy and Jones, 1986). Monensin also inhibits the secretion of xyloglucans, but not pectins, in carrot and sycamore culture cells (Moore et al., 1991; Zhang et al., 1993), indicating that these two polysaccharides follow different secretory pathways in the cell. Interestingly, monensin redirects the movement of vicilin, legumin, and concanavalin A so that they are secreted extracellularly instead of being deposited in the protein bodies (Craig and Goodchild, 1984; Bowles et al., 1986).

Treating cells with 2-deoxyglucose to inhibit glycolysis inhibits ER to Golgi traffic, indicating that this transport process is not passive, but requires ATP (Brandizzi et al., 2002b). Other drugs that disrupt the secretory pathway include cyclopiazonic acid and tunicamycin. These agents inhibit the Ca^{2+}-ATPase in the ER and N-linked glycosylation in the ER, respectively (Höftberger et al., 1995).

Inhibitors of the V-ATPase, including bafilomycin and concanamycin A, influence the transport of proteins through the TGN (Matsuoka et al., 1997).

Inhibitor studies have also indicated that the cytoskeleton is involved in the transfer of vesicles from the Golgi apparatus to the plasma membrane. In plant cells, movement may be mediated by actin microfilaments (see Chapter 10) and not microtubules (see Chapter 11), because microfilament antagonists, but not microtubule disrupting agents, inhibit vesicle migration away from the Golgi stacks and the subsequent secretion (Franke et al., 1972; Mollenhauer and Morré, 1976a).

Endocytosis is inhibited by Wortmannin, LY294002, ikarugamycin, endosidin 1, dynasore, and tyrosine kinase inhibitors, such as tyrphostin (Aniento and Robinson, 2005; Moscatelli et al., 2007; Müller et al., 2007; Onelli et al., 2008; Robinson et al., 2008; Sharfman et al., 2011; Qi and Zheng, 2013; Foissner et al., 2016). It is also inhibited by latrunculin B in pollen tubes (Moscatelli et al., 2012).

The secretory pathway can be specifically disrupted by producing genes with altered sequences. Disruption of the intracellular secretory machinery as a result of producing mutants of RAB proteins causes the inhibition of tip growth (Cheung et al., 2002; de Graaf et al., 2005).

8.7 SUMMARY

Movement is the natural condition of living cells, and at the macromolecular and ultrastructural levels, we see that movement is incessantly occurring throughout the endomembrane system. Proteins and polysaccharides move throughout the endomembrane system of the cell in 30–60 min. It is still not known how each membrane maintains its unique mixture of proteins and lipids in the face of intense transfer between compartments (van Meer, 1993; Harryson et al., 1996). We can consider the plasma membrane, ER, Golgi apparatus, vacuole and peroxisome to be the framework in which to understand the morphology and physiology of the non–self-replicating membranous parts of the cell. These organelles are discrete and stable—yet they are also continuous. They exhibit the cellular version of the wave–particle duality, or of complementarity—the tension between continuous and discrete, of position and momentum. Indeed the measurement of the position and momentum of an organelle, like anything else, can never be done at the same time as its position has to be determined at a single time point and its momentum has to be determined at two time points. Depending on the specific physiological state of a given cell, there is an elaboration or depletion of the membranes that exist between these five relatively stable organelles. The elaborations may be controlled by hundreds of genes.

We have also seen how the ER and the Golgi apparatus cooperate in the synthesis and secretion of substances that are destined to go to either the vacuolar compartment or the plasma membrane. We have also seen that the plasma membrane and the vacuolar compartment are also in communication with each other through the endocytotic pathway, which includes the early and late endosomes (PCR and MVBs). It will be fascinating to determine how these organelles initiate, maintain, or change their spatial localization relative to each other. In Chapter 9, I will discuss the structure of the cytoplasm through which the vesicles, membrane tubules, and organelles move.

A deep and broad understanding of the secretory pathway was discovered in the "virtual century-long collaboration" between Dutrochet, Heidenhain, Palade, Blobel, Rothman, and Schekman. Since then, there have been an enormous number of papers published on the secretory pathways in plant, fungal, and animal cells, and many of these papers have provided deep insight into the importance of specific aspects of the secretory pathway in the life of a cell as it grows, develops, and responds in an adaptive manner to the environment. On the other hand, in elucidating the layers of a never-ending complexity in systems where the secretory system may play a number of roles at a single point in time and change those roles over time, orders of magnitude of more papers have provided little more than impressive-sounding buzzwords and acronyms that include three letters and a number followed by more letters, which can be used in grant proposals, reviews, and at cocktail parties.

In 1939, Winston Churchill (1939) described Russia as "a riddle wrapped in a mystery inside an enigma." I am confident that the plasma membrane, the ER, the peroxisome, the Golgi apparatus, and the vacuole are all well-characterized functional and structural entities that have both an independent and interdependent existence in the cell—just like the cell has an independent and interdependent existence in the plant and countries have an independent and interdependent existence on earth. However, the borderlands between the organelles are less well known and to this day remain a riddle wrapped in a mystery inside an enigma. I believe that to understand the membranes that represent the borderlands, one should first search the treasure house of nature to find the cells that function as a "secreting organ par excellence" as described by Dutrochet (1824) and then use these cells to investigate the secretory pathway. In addition to using cells because they can be easily visualized or plants because they can be easily transformed, Juniper et al. (1982) showed us that there is sufficient diversity within the endomembrane system of plant cells to provide systems that specialize in movement within the endomembrane system. Given the success in elucidating the secretory pathway in budding yeast cells, and that "*the budding process could be conceived as a special case of tip growth* (Matile et al., 1971)," tip growing cells that have not yet developed vigorous streaming, or vacuoles may serve as an exemplary experimental material to work out the molecular components of the secretory pathway. One could also investigate the differences that exist in the secretory pathway in single-celled organisms such as *Penium* (Domozych, 2014) as they progress from the growing phase to the stationary phase. Choosing the right experimental system allows one to use the functions of the better known organelles as a toehold to understand the functions of the lesser understood membranes between the better known organelles. In this way, the lesser known membranes could soon be characterized by meaningful biochemical markers in cell-free systems, in electron micrographs using immunolocalization techniques, and in live cells using fluorescence localization. Each of these techniques could be used on wild-type cells, cells expressing temperature-sensitive alleles (Vidali et al., 2009a), cells expressing a conditional RNAi, and cells whose genes have been altered by CRISPER/Cas9 (see Chapter 21). Once we know the markers for stable secretory structures and functions, we can identify the markers for specialized, transitory and ephemeral structures and functions. At this stage, we do not know which proteins define the borderland membranes as permanent markers and which proteins define the borderland membranes as temporary markers that are important at a given time in development or in response to stimulation. It is not impossible that we will find that the borderlands themselves will differ in space and time depending on their function, and consequently, there will be room to make sense of the fact that the experts find "themselves unable to reach a consensus about the modality of membrane traffic" (Robinson et al., 2015). Indeed, cells have no need to reach consensus as long as each mechanism does not go against the law of cause and effect, the conservation laws, and the laws of thermodynamics. As in natural law, *Nulla poena sine lege* or everything which is not forbidden is allowed.

I think that the situation can be compared with the discovery of the structure of the atom. The virtual collaboration between Dutrochet, Heidenhain, Palade, Blobel, Rothman, and Schekman is one biological equivalent to the "virtual collaboration" in physics between J. J. Thomson, Ernest Rutherford, and James Chadwick, who discovered the electron, proton, and neutron, respectively, and elucidated the structure of the atom. However, in the years following these pivotal discoveries, physicists imitating these discoverers searched for more elementary particles and found so many that the collection of them became known as the "particle zoo." There were so many new "elementary particles" discovered that they began to look like epicycles on epicycles in older versions of astronomy. Willis Lamb (1955) said in his Nobel Prize acceptance speech, "I have heard it said that 'the finder of a new elementary particle used to be rewarded by a Nobel Prize, but such a discovery now ought to be punished by a $10,000 fine.'"

Eventually, Murray Gell-Mann (1969) joined the "virtual collaboration" when he found a new way of looking at the growing list of new particles and developed a theory of elementary particles that simplified the "particle zoo" and provided a new foundation on which to build new physics. Likewise, while so many cell biologists work on discovering the growing list of ARF, RAB, SNAP SNARE, etc., genes and gene products involved in secretion, I hope that one of them will join the "virtual collaboration of secretory biologists" and find a new way of looking at the secretory pathways to find the fundamental laws that unify the "secretory gene zoo."

8.8 QUESTIONS

8.1. How may the flux of proteins influence the structure of the membranes in the endomembrane system of cells undergoing a development change or a physiological change in response to an environmental stimulus?

8.2. How can you explain the protean nature of the membranes of the endomembrane system?

8.3. How may the architecture of the membranes in the endomembrane system differ in growing cells, stationary-state cells, and senescing cells?

The references for this chapter can be found in the references at the end of the book.

Chapter 9

Cytoplasmic Structure

What community of form, or structure, is there between animalicule and the whale, or between the fungus and the fig-tree? And a fortiori, between all four?

Thomas H. Huxley (1890).

When Heisenberg tells us, as he did tell us, that the uncertainty principle presents an insurmountable barrier, a wall over which we shall never see, he is throwing physics back into the mysticism and vitalism of the distant past. That is the tragedy. Physics can claim its place of honor in the sun of knowledge only so long as it adheres to the principles of causality functioning in a realistic world. These principles were the emancipators of thought in the last century, and served as ports of refuge from the bigotry of intolerance and mysticism. I do not like DuBois-Reymond's 'Ignorabimus, we shall never know,' but I like it better than Heisenberg's 'We now know that we shall never know.'

William Seifriz (1943a).

He shook the bottle of catchup over his hamburger, violently. The bottle was almost full, but nothing came out. 'There —you see?' he said. 'When you shake catchup one way, it behaves like a solid. You shake it another way, and it behaves like a liquid.' He shook the bottle gently, and catchup poured over his hamburger. 'Know what that's called?'

No, I said.

'Thixotropy,' said Harry. He hit me playfully on the upper arm. 'There—you learned something today.'

Kurt Vonnegut (Unpaid Consultant).

9.1 HISTORICAL SURVEY OF THE STUDY OF CYTOPLASMIC STRUCTURE

In the previous chapters, I discussed the evidence that proteins, vesicles, membranous tubules, and organelles move throughout the cytoplasm. In this chapter, I will discuss the structure of the cytoplasm through which they move. Remember that when the cell was discovered by Robert Hooke (1665), he could only imagine that there was a possibility of an internal structure within the walls composed of passages, valves, instruments, and contrivances, which would be discovered by "some diligent observer, if helped by better microscopes."

In the 17th and 18th centuries, the lenses in light microscopes had various spherical and chromatic aberrations that made it difficult to see minutely detailed structures in nearly transparent objects (Wayne, 2009; Schickore, 2018). By the 19th century, the optics of microscopes were improved, thanks to the invention of the achromatic doublet by Chester Moor Hall, John Dollond, and/or James Ramsden, and the introduction of achromatic lenses into microscopes in the 1820s and 1830s by scientists and inventors, including Giovanni Battista Amici (1818) and Joseph Jackson Lister (1830), the father of the surgeon who pioneered the use of antiseptics. The newly developed lenses were corrected for spherical and chromatic aberrations and allowed light microscopists such as Félix Dujardin (1835, 1841) to resolve objects that were less than 1 μm, about 100 times smaller than that resolvable by the naked eye (Claude, 1948; Bradbury, 1967). The new microscopes with achromatic lenses provided the means to explore the structure of living beings at the subcellular level.

Robert Brown (1828, 1829) could see that cells consisted of spherical particles and molecules about 1/20,000 of an inch in diameter, and moreover, it was easy to see that these particles and molecules moved independently and incessantly when squeezed out into water. Dujardin (1835, 1841) could see and study the nature of the transparent, water-insoluble, glutinous, contractile substance that held together the food vacuoles of ciliates and gave it the name *sarcode*, from the Greek word for "flesh." In 1840, Jan Purkyně used the term *protoplasm*, a term long used in religious contexts to mean the first created thing (protoplast = Adam and protoplasmator = God), to designate the living substance of animal embryos. And in 1846, Hugo von Mohl independently applied the term *protoplasm* to the living substance of plant cells because he believed that the protoplasm was capable of giving rise to all other parts of the cell. By 1848, Alexander Ecker suggested that the sarcode is a fundamental substance of all animal life, from the cells of *Hydra* to those of muscles in higher animals, and Ferdinand Cohn (1853) further

Plant Cell Biology. https://doi.org/10.1016/B978-0-12-814371-1.00009-6

emphasized the ubiquity, constancy, and importance of protoplasm when he wrote

> *All these properties, however, are possessed by that substance in the plant-cell, which must be regarded as the prime seat of almost all vital activity, but especially of all the motile phenomena in its interior—the protoplasm. Not only do its optical, chemical and physical relations coincide with those of the 'Sarcode' or contractile substance, but it also possesses the faculty of forming 'vacuoles' … From these considerations it would therefore appear … that the protoplasm of the Botanists, and the contractile substance and sarcode of the Zoologists, if not identical, are at all events in the highest degree analogous formations.*

Calling attention to the similarity of the living substance of all cells and giving it a common name—protoplasm—propelled the search to find a definite structure within the protoplasm that would prove to be the essence of life itself (Beale, 1872; Drysdale, 1874; Brücke, 1898b; Reynolds, 2018). All the solutions to the problems of life were to be found in the identification of this one structure. The search ensued and dualistic theories, which distinguished between the living part of the cell and the lifeless part, were all the rage. Realizing that animal cells lacked the thick extracellular matrix typically found on the exterior of plant cells, Max Schultze (1863) decided that the extracellular matrix could be eliminated as a possible candidate (Kutschera, 2011). This left the naked protoplasm as the part of the cell that was endowed with all the attributes of life.

The 19th century biologists were interested in dissecting the protoplasm down to the ultimate constituent of life. Schleiden was enamored with the idea that the nucleus, or the *cytoblast* as he called it, was the elementary particle of life. This was because he could see that cells that had a nucleus were able to reproduce, whereas those without one could not. As the cytoblast was not always easy to see, Schleiden later believed that the cytoblast was an elaboration of the invisible cytoblastema, the true elementary substance (Schleiden, 1853). Later work showed that the nucleus existed in all living cells, divided prior to cell division, and as a consequence of its continuity, must house the living substance (von Mohl, 1852; Wilson, 1925; Goebel, 1926). Rudolf von Kölliker coined the term *cytoplasm* in 1862 to distinguish the nucleus from everything else in the protoplasm. The nucleus, like the protoplasm, showed substructure, and, of course, one part was thought to be more vital than its counterpart was. For example, the idiochromatin was considered to be the portion of the nucleus that contained the hereditary material and was thus more vital than the trophochromatin, which served merely to nourish the idiochromatin (Wilson, 1925).

While one school believed the nucleus or some of its contents were the true living substance, others felt that the surrounding elements in the cytoplasm were more vital. Thus, the cytoplasm was differentiated into various parts to distinguish the most vital part. For example, the cytoplasm was divided into the inner region of granular matter known as the *endoplasm* (Pringsheim, 1854; Hofmeister, 1867) and the outer border of a clearer substance called the *ectoplasm*. Johannes von Hanstein (1868) distinguished the protoplasm from the metaplasm, where the metaplasm performed certain duties necessary for life, but the protoplasm was the true living substance and retained all the properties of life. The metaplasm, which later became known as the ergastic substances, included the cell sap, starch grains, crystals, and the extracellular matrix.

Under the bright-field microscope, the cytoplasm appears as a fine dispersion of particles of different sizes (1−10 μm) freely suspended in a liquid medium. This led Hanstein to propose that the granules form the fundamental nature of cytoplasm (Fig. 9.1), that is, the fundamental nature of life. Hanstein (1882) named the granules *microsomes*—a term later used by Albert Claude to designate a membrane fraction isolated from rat liver cells (see Chapter 4). After Hanstein christened these granules, which were previously known as small bodies (i.e., *kleinkörperchen*), with the name *microsomes*, which comes from Greek for "small bodies," Otto Bütschli (1892, 1894) sarcastically wrote that microsomes had now "obtained the right of entry among the privileged and recognized units of cytoplasmic structure, for anything that is called by a Greek name at once seems to many people to be much better known, and as something which must be definitely reckoned with." Today acronyms fulfill this dysfunction. Richard Altmann suggested that the granules, which he called *bioblasts*, were equivalent to living bacteria, and the cell was really a colony of minute organisms, each of which

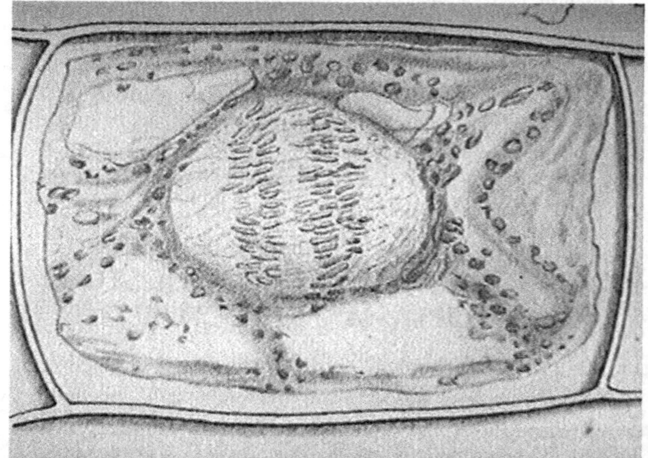

FIGURE 9.1 A dividing cell of *Equisetum* showing the granular nature of cytoplasm. *From von Hanstein, J., 1882. Einige Züge aus der Biologie des Protoplasmas. Botanische Abh. Herausgeg. v. Hanstein. Bd. 4./ Heft 2. Bonn.*

was the true vital agent in the cell (Altmann, 1890; see Chapters 14 and 15).

Others turned their attention to the elements that surrounded the granules. They felt that undue attention was being given to the motley collection of granules, which included vacuoles, crystals, oil droplets, etc., for surely not all the granules were important in understanding the vital nature of cytoplasm; some must only serve as food or contain wastes. The framework that surrounded the granules was studied by two groups of biologists, the histologists and the physical chemists, who did not see eye to eye.

Between 1870 and 1890, the techniques involved in the cytological staining of cells were being developed, and fixed and stained sections revealed an apparently three-dimensional meshwork or entanglement of fibers (see Fig. 9.2; Flemming, 1882; Wilson, 1895; Strasburger, 1897; Lee, 1893; Heidenhain, 1907, 1911; Mazia, 1975). The fibers were associated with the parts of the cytoplasm that moved, and Eduard Strasburger called the active, moving parts of the cytoplasm, which appeared fibrous in stained material, the *kinoplasm*. He gave the name *trophoplasm* to the substance that surrounded and supposedly nourished the kinoplasm (Strasburger et al., 1912). The kinoplasm included the plasma membrane, spindle fibers, centrosome, and cilia. Walther Flemming felt that the fibrils were the "seat of the energies on which life depends," whereas others felt that the hyaloplasm or substance that bathed the fibrillar framework was the real living substance and not just there to feed the kinoplasm (Seifriz, 1936).

While cytologists were discovering the unexpected and astonishing behavior of chromosomes and bringing to light new aspects of cell structure, as they observed the spindle fibers and cilia, they worked under the assumption that their techniques disclosed real, preexisting structures that were not visible in the optically transparent living cells. However, the physicochemically oriented cell biologists, influenced by the tenets of colloidal chemistry, argued bitterly that cytological techniques involving killing, fixing, staining, dehydrating, embedding, and sectioning material caused an artefactual phase separation of the hydrophilic and the hydrophobic substances, which resulted in the production of structures that do not exist in the living cell (Fischer, 1899; Hardy, 1899; Ostwald, 1922; Mazia and Dan, 1952). W. B. Hardy (1899) wrote

> *It is, I think, one of the most remarkable facts in the history of biological science that the urgency and priority of this question should have appealed to so few minds. ... It is notorious that the various fixing reagents are coagulants of organic colloids, and that they produce precipitates which have a certain figure or structure. It can also readily be shown ... that the figure varies ... according to the reagent used. It is therefore cause for suspicion when one finds that particular structures which are indubitably present in preparations are only found in cells fixed with certain reagents.*

Bütschli (1894) further admonished that many of the fibrous elements were probably diffraction artifacts because they could be seen best when using the poorest microscope illumination (Hacking, 1981).

Berthold (1886) and Bütschli (1894) among others looked at living cells and treated the cytoplasm as a semisolid/semiliquid or gel/sol colloidal system. Colloids, as we know them now, are macromolecules that are permanently dispersed in solution (Graham, 1842, 1843, 1850–57; Wilson, 1899; Zsigmondy, 1909, 1926; Zsigmondy and Spear, 1917; Staudinger, 1961). Colloids are approximately 1–1000 nm larger than low–molecular mass molecules but smaller than bacteria. Wolfgang Ostwald (1922) called this the domain of neglected dimensions (Frey-Wyssling, 1957). Colloids remain suspended because the electromagnetic force that results from their surface properties dominates over the gravitational force that results from their density.

To understand the structure of cytoplasm, the physicochemically oriented biologists followed the teachings of Lord Kelvin and made physical models that looked like protoplasm and imitated some of its properties, including movement (Seifriz, 1936). Bütschli (1894) considered the cytoplasm to be an emulsion of alveolae or vesicles that contained cell sap dispersed in a continuous phase that consisted of the vital element (Fig. 9.3). Ahead of his time, feeling like an alchemist and unappreciated (Goldschmidt, 1956), using simple ingredients and techniques originally developed by Quincke (1888) such as shaking up salt, water, and oil to create a foam, Bütschli created artificial physicochemical models of amoeboid motion to understand protoplasmic structure and its role in living processes.

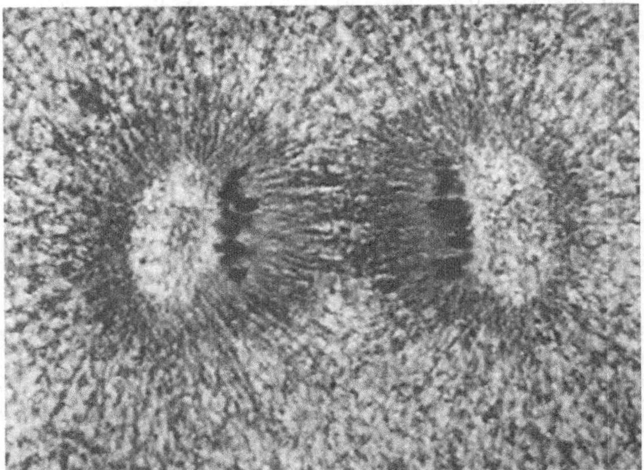

FIGURE 9.2 The fibrous nature of the cytoplasm can be seen in this zygote of *Toxopneustes* photographed in late anaphase. *From Wilson, E.B., 1895. An Atlas of the Fertilization and Karyokinesis of the Ovum. MacMillan, New York.*

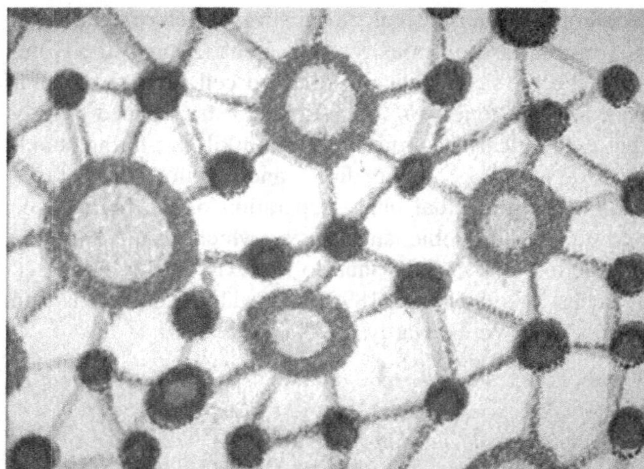

FIGURE 9.3 The aveolar protoplasm in the ovum of *Hydatina senta*. *From Bütschli, O., 1894. Investigations on Microscopic Foams and on Protoplasm. Adam and Charles Black, London.*

Others considered the cytoplasm to be a complex emulsion containing a mixture of oil in water and one of water in oil (Fig. 9.4). The dispersed phase was given the name *phaneroplasm* and the invisible, continuous phase was called the *cryptoplasm*. The cryptoplasm was considered to be vitally more important than the phaneroplasm.

As soon as any visible component of the cytoplasm seemed to lack the fundamental properties of life, the next most elusive and smaller component was considered to be the vital part of the cell. In fact, Gwendolyn Andrews (1897) wrote, "Nature might be well liked to a great spider, spinning and spinning the living stuff and weaving it into tapestries; and still hiding herself and the ever-lengthening thread of vital phenomena behind the web already spun."

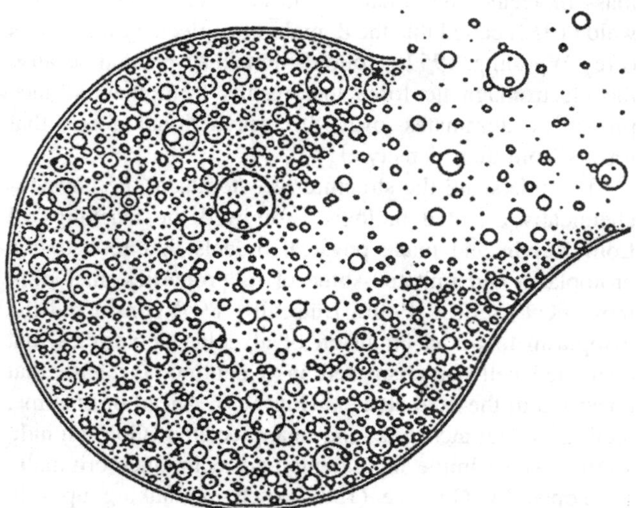

FIGURE 9.4 The protoplasm of a torn *Fucus* egg looks like an emulsion. *From Seifriz, W., 1938b. The Physiology of Plants. John Wiley and Sons, New York.*

Throughout history, monks, scientists, and philosophers have searched for essences and elixirs that they hoped would be the most fundamental unit of life (Berthelot, 1885; Taylor, 1953; Forbes, 1970). For example, Thales believed that water was the essence of all matter, and Jean Baptiste van Helmont provided evidence for this theory by showing that a tree grew and flourished when apparently all he provided it with was water (see Chapter 13). Alcohol was thought to be the general essence of all plants and was called aqua vitae, in Latin, which means water of life. The word "whiskey" comes from the Irish Gaelic "uisce beatha" and Scottish Gaelic "uisge beatha" which mean water of life. The French called brandy "eau de vie," which also means water of life. More recently, the putative vital elixir, or essence of life, has been given many names, including biogen, physiological units, bioblasts, micelles, plastidules, plasomes, ideoblasts, biophores, gemmules, pangenes, genes, genomes, transcriptomes, proteomes, metabolomes, ionomes, signalomes, etc. (Harper, 1919; Wilson, 1925; Seifriz, 1936; Conklin, 1940). Even more recently (in Lewontin, 2000), Sidney Brenner has said, "if he had the complete sequence of DNA of an organism and a large enough computer then he could compute the organism." With a like mind, Walter Gilbert has claimed, "that when we have the complete sequence of the human genome we will know what it is to be human." Currently, individuals and families are getting to know themselves through DNA testing provided by companies including Ancestry DNA, 23andMe, MyHeritageDNA, LivingDNA, GPS Origins, Vitagene and Futura Genomics. We know what Walter Gilbert said. I wonder what Socrates would say.

The search for the essence of life is based on the assumption that the characteristics of the whole can be found in the constitutive parts. So far the search to find a single structure or compound, smaller than that of a whole cell, that has all the properties of life, including the ability to take up molecules, generate electricity, grow, reproduce, and respond to external stimuli, has failed. According to Frederick Gowland Hopkins (1913),

> *it is clear that the living cell … is … a highly differentiated system; … a system of coexisting phases of different constitutions. Corresponding to the differences in their constitution, different chemical events may go on contemporaneously in the different phases, though every change in any phase affects the chemical and physico-chemical equilibrium of the whole system. … It is important to remember that change in any one of these constituent phases … must affect the equilibrium of the whole cell system, and because of this necessary equilibrium-relation it is difficult to say that any one of the constituents phases … is less essential than any other to the 'life' of the cell.*

The reductionist approach has led to the discovery of much cellular structure and function, though ironically in the quest to find the singular secret of life, these discoveries have only emphasized the intricate organization that is necessary for life, and indeed, that the whole is greater than the sum of its parts. None of the isolated microscopic parts exhibit all the properties of life, including assimilation, growth, reproduction, ability to respond to external stimuli, and adaptability (Blackman, 1906). To maintain the living condition for an extended period, all the parts of the protoplasm are necessary and together form the basis of life. "Life is not found in atoms or molecules or genes as such, but in organization" (Conklin, 1940).

Systems biology today is based on the idea that we will understand the organization of life once we know where and when *every* gene is expressed, *every* transcript is transcribed, *every* protein is translated, and *every* metabolite is measured (see Chapter 21).

In several notable articles, all entitled, "The Physical Basis of Life," T. H. Huxley (1890), W. B. Hardy (1906), E. B. Wilson (1923), and J. D. Bernal (1951) came to the conclusion that protoplasm is "the physical basis of life" rather than its essence. That is, protoplasm is necessary for the existence of life but is not necessarily the essence of life. I will discuss the structure of cytoplasm, with a view to understanding the physicochemical milieu in which organelles move and function, vesicles and membranous tubules move, and chemical reactions necessary for life take place.

9.2 CHEMICAL COMPOSITION OF PROTOPLASM

Protoplasm is not a chemical, but an elaborate organization of some of the most complex chemical substances known. Moreover, the chemical composition differs in every species and in every cell of the same organism. As studies on the transcriptome, posttranscriptome, proteome, lipidome, metabolome, signalome, ionome, and all other "-omes" intimate, the chemical composition also varies during the lifetime of a single cell. As a first approximation, however, protoplasm contains proteins, lipids, carbohydrates, nucleic acids, and their constituents (Table 9.1).

Henry Lardy (1965, 1995) defined the cytosol as the portion of the cell that is found in the supernatant fraction after centrifugation at 105,000 g for 1 h. At that time, it referred specifically to the cytoplasm minus the mitochondria and the endoplasmic reticulum (ER). In the cytosol of Escherichia *coli*, the protein concentration is about 200−320 mg/mL, the RNA concentration is about 75−120 mg/mL, and the DNA concentration is about 11−18 mg/mL (see Elowitz et al., 1999; Zimmerman and Trach, 1991).

The substances that make up the cytosol are dissolved in an aqueous salt solution that contains about 75% water and 100 mol/m^3 K^+, tens of mol/m^3 Cl^-, 1 mol/m^3 Mg^{2+}, 10^{-4} mol/m^3 each of H^+ and Ca^{2+}, and trace quantities of other ions. Most of the water may be free, forming an aqueous phase through which ions can freely diffuse; however, a portion of the water is bound to proteins, forming a glasslike phase (Garlid, 2000). Some *Halobacteria* contain up to 3000 mol/m^3 K (Ginzburg and Ginzburg, 1975). The concentrations of ions, as measured with fluorescent dyes, vary spatially and temporally throughout the protoplasm (Rathore et al., 1991; Pierson et al., 1994, 1996; Kropf et al., 1995). The redox potential (c. −0.309 to −0.325 V, depending on cell type) and pH of the cytosol (6.5−7.6, depending on cell type and gravistimulation) have been observed in transformed cells using redox- and pH-dependent forms of green fluorescent protein (GFP; Fasano et al., 2001; Moseyko and Feldman, 2001; Jiang et al., 2006a,b; Martinière et al., 2013; Shibata et al., 2013). The concentration of various small organic molecules, including adenosine triphosphate (ATP), amino acids, and sugars, usually falls between 0.1 and 10 mol/m^3 (Mimura et al., 1990b; Scott et al., 1995; Haritatos et al., 1996). The various ions and molecules are not necessarily uniformly distributed throughout the cell (Aw, 2000). Genetically engineered sensors based on fluorescence resonance energy transfer (FRET) have been developed creatively, and can be used to determine the concentration of almost any molecule of interest with high spatial and temporal resolution (Chudakov et al., 2010; Swanson et al., 2011; Haugh, 2012; Krebs et al., 2012; Okumoto et al., 2012; Fritz et al., 2013; Jones et al., 2013, 2014; Hamers et al., 2014).

Knowing the chemical composition at this level gives us little knowledge of the structure of the cytoplasm. While we might be tempted to conclude that the cytoplasm could behave like a viscous protein solution, we will find that it does not (Luby-Phelps and Weisiger, 1996).

TABLE 9.1 Composition of Dehydrated Protoplasm of the Slime Mold *Reticularia*

Substance	Percent Dry Weight
Protein	28
Nucleic acids	4
Other nitrogen-containing compounds	12
Fat	18
Lecithin	5
Cholesterin	1
Carbohydrates	23
Unknown	9

From Kiesel, A., 1930. Chemie des Protoplasmas. Protoplasmatologia Monograph 4, Berlin.

9.3 PHYSICAL PROPERTIES OF CYTOPLASM

Because of the primacy of manual labor in doing work, the ancient Greeks realized the importance of quantifying resistance to movement to optimize the number of people necessary for moving a given object (Cohen and Drabkin, 1958; Franklin, 1976). Friction is another word for mechanical resistance. As the internal resistance or friction of solutions influences the mobility of particles contained in it and the movement of the solution itself, the internal resistance must also be accounted for if we wish to understand the relationship between the motive force and the velocity of movement (Maxwell, 1873, 1878; Garber et al., 1986). The internal resistance of solutions is known as viscosity. Accounting for the resistance quantitatively and from first principles is very difficult and tedious (Tait and Steele, 1878; Synge and Griffith, 1949). Consequently, resistance is often ignored in mechanics, which starts with the ansatz, "*assume there is no friction* (Wayne, 2012a)." However, resistance is taken into consideration by athletes (Goff, 2010; Denny, 2011), by dancers (Laws, 2002), and by those in the fields of rheology and biorheology—fields that study the flow of matter (see Appendix 2). In general, resistance of gases and liquids results in viscous flow (Fig. 9.5), whereas resistance in solids results in the reversible (elastic) and irreversible (plastic) deformation of matter (see Fig. 9.6; Blair and Spanner, 1974).

The viscosity, internal resistance, or friction of the cytoplasm affects all aspects of cellular motion, including the transit of water on or off of ions, the translational diffusion of enzymes and the translational diffusion of substrates to enzymes, the rotational diffusion of substrates so that they can properly bind to an enzyme, and the movement of membrane vesicles, tubules, and organelles (Kramers, 1940; Fulton, 1982; Pocker and Janjić, 1987; Goodsell, 1991; Karplus and Weaver, 1994; Welch and Easterby, 1994; Jacob and Schmid, 1999; Luby-Phelps, 2000, 2013; Zhu et al., 2000; Forgacs and Newman, 2005; Weihs et al., 2006: Guigas et al., 2007; Jonas et al., 2008). The resistance to movement in the cytoplasm is quantified as the cytoplasmic viscosity (Fig. 9.7), which is the primary physical factor that influences the flow of material through the cell (Heilbrunn, 1958; Bereiter-Hahn, 1987; Hiramoto, 1987; Hiramoto and Kamitsubo, 1995; Tachikawa and Mochizuki, 2015). I will present viscosity explicitly as a physically real frictional and dissipative component that resists the movement of something with a constant mass in response to a constant applied force. If one did not account for the viscosity, it would appear that the motive force was weaker or the mass that is moving was greater. Studies on the viscosity and elasticity of cytoplasm have contributed greatly to our understanding of the structure of this living milieu. Rheological studies of the cell wall have also contributed to our understanding of cell growth (see Chapter 20).

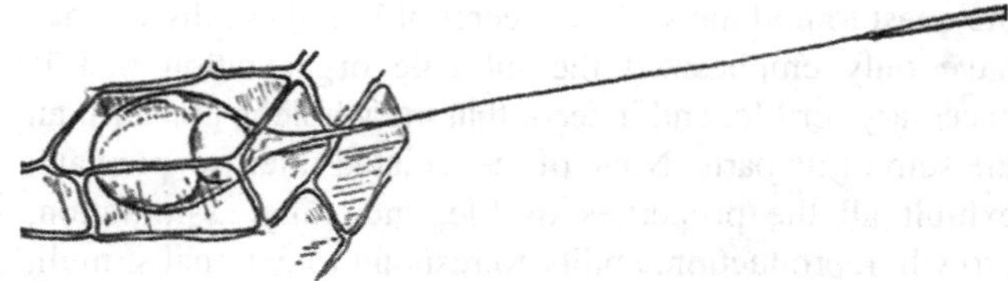

FIGURE 9.6 Stretching the protoplasm with the aid of a microneedle. *From Seifriz, W., 1936. Protoplasm. McGraw-Hill, New York.*

FIGURE 9.5 The behavior of the hanging strands of *Physarum polycephalum* indicates that the protoplasm is highly viscous. *From Seifriz, W., 1938b. The Physiology of Plants. John Wiley and Sons, New York.*

FIGURE 9.7 The viscous cytoplasm of a plant cell. *From Kahn, F., 1919. Die Zelle. Kosmos, Stuttgart, Germany.*

9.3.1 Viscosity of the Cytoplasm

Cytoplasm is a viscous fluid, and thus it resists flow. Viscosity is a measure of the resistance to flow and is given in units of Pa s, N/m^2 s, or kg/m^2 s. To get a feel for how the viscosity of bulk fluids is measured, imagine placing a fluid between two glass plates that are 1 m apart, and each plate has an area of 1 m^2 (Fig. 9.8). Imagine pushing the upper plate to the right with a force of 1 N. (The force exerted by the falling of a 100-g apple is approximately 1 N.) The more viscous the fluid, the longer it will take for the top plate to slide completely past the stationary lower plate. If the fluid (e.g., water) has a viscosity of 0.001 Pa s, it will take 0.001 s for the top plate to slide past the bottom plate. By contrast, if the fluid has a viscosity of about 0.1 Pa s (e.g., olive oil), about 1 Pa s (e.g., glycerol), or about 10 Pa s (e.g., honey), it will take 0.1, 1, or 10 s, respectively, for the top plate to slide past the bottom plate. In the examples given, the top plate travels at a velocity of 1000, 10, or 1 m/s, respectively, relative to the stationary plate.

9.3.1.1 Newtonian Fluids

A Newtonian fluid is one that obeys the law of fluid flow found in Isaac Newton's *Principia* (1729; Chandrasekhar, 1995). The law states that the velocity of flow (v) of a liquid is proportional to its fluidity (f, in Pa^{-1} s^{-1}) and the shearing stress (σ, in Pa) on either of two plates separated by a distance x (in m). The fluidity of a Newtonian fluid is independent of its velocity. The ratio of the velocity to the distance (v/x) is known as either the velocity gradient or the rate of shear (γ, in s^{-1}). The relationship between velocity and fluidity is given by the following formulae:

$$v = f\sigma x \tag{9.1}$$

and

$$v/x = f\sigma \tag{9.2}$$

Newton did not define *viscosity*, the term we use today. This was done by James Clerk Maxwell (1891), who stated that "the viscosity of a substance is measured by the tangential force on the unit area of either of two horizontal planes at the unit of distance apart, one of which is fixed, while the other moves with the unit of velocity, the space being filled with the viscous substance." The viscosity (η) is the reciprocal of fluidity, and it is given by the following formula:

$$\eta = 1/f = (\sigma x)/v = \sigma/(v/x) \tag{9.3}$$

Maxwell (1891) defined a shearing stress as one that moves tangentially along a fluid and induces movement within the fluid. The shearing stress exerted on a plane causes the adjacent substance to move with a rate of shear that depends on the viscosity of the fluid. The following formula gives the relationship between shearing stress, the rate of shear, and the viscosity:

$$\sigma = \eta(v/x) \tag{9.4}$$

To get a feel for the rate of shear, imagine the spinning of a compact disc or DVD. The center point does not move and the edge moves the fastest. Thus, there is a velocity gradient from outside to inside. The difference in the maximal and minimal velocity divided by the distance between them gives the rate of shear. The velocity (v_i) at any distance (x_i) from the nonmoving plate is given by:

$$v_i = (\sigma x_i)/\eta \tag{9.5}$$

and at $x_i = 0$, $v_i = 0$.

Newton's law of fluid flow was finally tested in the 1840s by Jean Poiseuille who came up with a formula that related the velocity of flow of a liquid in a capillary to the pressure difference across the capillary, the radius and length of the capillary, and the viscosity of the solution (Poiseuille, 1940). Poiseuille, who studied under the mathematicians and physicists Augustin-Louis Cauchy, André-Marie Ampere, François Arago and Alexis Petit (Sutera and Skalak, 1993), was a physiologist who was interested in studying the flow of blood in vertebrates. Specifically, he wanted to know why some organs receive more blood than others do. To this end, he studied how fluids moved through glass tubes the size of capillaries. He varied the pressure difference across the tube, the length of the tube, and the diameter of the tube and measured the time that it took a given volume of liquid to move through the tube.

When Poiseuille varied the pressure difference (dP), he found that the volume flow (Q, in m^3/s) was proportional to the pressure difference. However, as I will discuss below, the coefficient of proportionality (K) also depended on the geometry of the tube and the consistency of the fluid:

$$Q = K\,dP \tag{9.6}$$

When he varied the length of the tube (x), he found that the volume flow was inversely proportional to the length. However, the proportionality coefficient (K') still depended on the tube and the fluid:

$$Q = K'(dP/x) \tag{9.7}$$

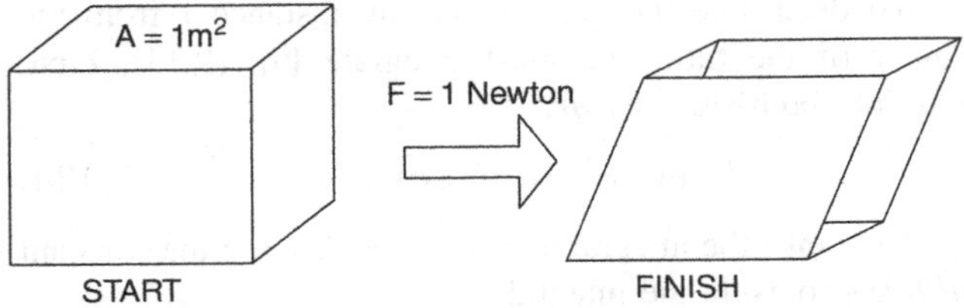

FIGURE 9.8 The deformation of a viscous liquid when it is exposed to a shearing stress.

When he varied the diameter ($2r$), he found that the volume flow was proportional to the fourth power of the diameter. Now all the variations due to the tube were accounted for, and the coefficient of proportionality (K'') was only a function of the fluid:

$$Q = K''(2r)^4(dP/x) \qquad (9.8)$$

Poiseuille noticed that his law did not hold when either the diameter or the pressure difference was too large or the length was too short. The law no longer holds under these conditions because the flow is no longer laminar but becomes turbulent (Niklas and Spatz, 2012).

Poiseuille also noticed that K'' was dependent on temperature and increased as the temperature increased. Thus, he reasonably assumed that K'' was inversely related to the density of the solution. He thus made mixtures of alcohol and water to make solutions of different densities. However, he found that as he increased the density of the solution by increasing the water content, the flow increased and then decreased at even higher densities. This indicated to him that K'' does not depend exclusively on the density but is inversely proportional to another property of the solution. In fact, the flow can be different in two solutions with the same density and the same in two solutions with different densities. We now know that the property of a solution that influences flow through pipes is not density, but viscosity (Khattab et al., 2012). Viscosity was defined by Maxwell years after Poiseuille's experiments.

Maxwell integrated Poiseuille's equation and showed its equivalence with Newton's equation that relates the rate of shear to the shear stress. Integrating Poiseuille's equation allows us to determine the shape of the velocity profile, the maximal velocity, the half-maximal velocity, and the average velocity.

To integrate the equation, we must define our parameters and assumptions. Let us assume that a fluid is moving in a tube at a constant velocity, which according to Newton indicates that there is no acceleration and thus no net force. Therefore, the sum of all the forces acting on the liquid must be zero. We will assume that there are two forces: one due to the pressure difference that pushes the liquid through the tube and one due to the friction that resists movement through the tube. Given the following tube, let us consider a section of length x and radius R. r_r is any distance starting from the center of the tube, parallel to a radius, and varies from 0 to R (Fig. 9.9).

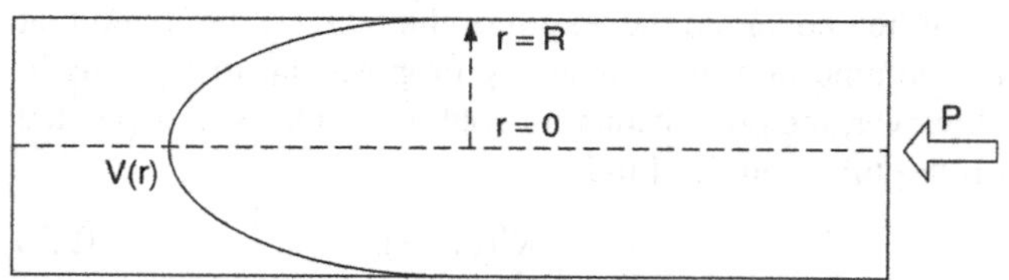

FIGURE 9.9 Parabolic flow of a Newtonian fluid through a tube.

The pressure difference along the tube is dP (in Pa or N/m^2) and it induces an inertial force (F_i, in N) that is equal to the product of the pressure difference and the area of a cylinder on which the pressure is exerted (πr_r^2). The inertial force is the force that tends to cause liquids to accelerate:

$$F_i = dP\pi r_r^2 \qquad (9.9)$$

The acceleration of a liquid is resisted by the viscous force (F_v). The flow is resisted because each molecule in a liquid is attracted to its neighbors. This attraction, which is due to electrostatic and/or steric interactions between molecules, causes friction between each layer of the fluid. The electrostatic and/or steric interactions that cause friction are quantified by measuring the viscosity of the fluid.

The viscosity of a fluid in a capillary can be measured by determining the relationship between the shear stress on the fluid and the rate of shear, just as it was done using two plates.

In a set of planes, $F_v = \eta(dv/dx)A$, where x is the distance between the planes and A is the product of the length and width of a plane. In a cylinder, $F_v = \eta(dv/dr)A$, where r is the radius of the cylinder and A is the area of the tube ($2\pi rx$). That is, the viscous force is equal to the product of the viscosity, the rate of shear, and the area of the outside skin of the cylinder fluid:

$$F_v = \eta(dv/dr)(2\pi rx) \qquad (9.10)$$

Consider a fluid that is flowing at a constant velocity; that is, it is exhibiting laminar flow and is not accelerating. In this case, the inertial force is balanced by the viscous force so that the net force on the liquid is zero:

$$F_i + F_v = 0 \qquad (9.11)$$

Thus, $F_i = -F_v$ and

$$dP\pi r^2 = -\eta(dv/dr)(2\pi rx) \qquad (9.12)$$

Cancel like terms and solve for dv/dr:

$$dv/dr = -dPr/(2x\eta) \qquad (9.13)$$

According to the rules of calculus, we can find the maximum or minimum velocity by finding where $dv/dr = 0$. Since dP, x and η all have non vanishing values, the maximum occurs in the center of the cylinder where $r = 0$.

To determine the velocity at any distance r from the center of the tube, we must integrate Eq. (9.13). First, multiply both sides by dr:

$$dv = -(dP/(2x\eta))r\,dr \qquad (9.14)$$

Now take the integral of both sides, leaving the constant $dP/(2x\eta)$ outside the integral:

$$\int dv = -(dP/(2x\eta))\int r\,dr \qquad (9.15)$$

After taking the antiderivative of each side, we get

$$v(r) + A = -(dP/(2x\eta))(r^2/2) + B \quad (9.16)$$

where A and B are constants of integration and can be combined into C as follows:

$$v(r) = -(dP/(2x\eta))(r^2/2) + C \quad (9.17)$$

To solve for C we use the "no slip" condition—that is, the velocity at the boundary of the cylinder and the wall of the tube is zero. So $v = 0$ when $r = R$:

$$v(R) = 0 = -(dP/(2x\eta))(R^2/2) + C \quad (9.18)$$

and

$$C = (dP/(2x\eta))(R^2/2) \quad (9.19)$$

Thus,

$$v(r) = -(dP/(2x\eta))(r^2/2) + (dP/(2x\eta))(R^2/2) \quad (9.20)$$

which can be simplified to

$$v(r) = (R^2 - r^2)(dP/(4x\eta)) \quad (9.21)$$

Solving this equation for various values of r shows that when a pressure is applied across a tube containing a solution with a constant viscosity, the velocity is maximal in the center of the tube and declines with the square of the distance as we move toward the edge of the cylinder where $v = 0$. Thus, there is parabolic flow. In 1860, Jacob Eduard Hagenbach named this equation after Poiseuille. Gotthilf Hagen, an engineer, independently discovered the law of parabolic flow, and in 1925, Wilhelm Ostwald, who invented an instrument to measure viscosity and who also defined happiness in terms of one's ability to overcome resistance (Boltzmann, 1904; Gadre, 2003), renamed the law of parabolic flow the Hagen–Poiseuille Law. This law has been useful for describing the flow of water through the xylem and the flow of sugar through the phloem (Zimmermann and Brown, 1971).

It is common, although not entirely correct, to refer to the half-maximal velocity as the average velocity. The average and half-maximal velocity are only equal when the rate of shear is linear. The half-maximal velocity is given by

$$v_{0.5\ \max} = dPR^2/(8x\eta) \quad (9.22)$$

The average velocity of parabolic flow, which I will not derive, is given by

$$v_{\text{ave}} = dPR^2/(6x\eta) \quad (9.23)$$

In honor of Poiseuille, viscosity can be measured in Poise (P). With this convention, water has a viscosity of 1 centiPoise (cP) at 20 C. In the SI system, the units of viscosity are Pa s, where 1 P = 0.1 Pa s and 1 cP = 0.001 Pa s.

TABLE 9.2 Viscosity of Various Newtonian Fluids

Substance (°C)	Viscosity (Pa s)
Glycerin (25)	0.954
Castor oil (20)	0.986
Heavy machine oil (15.6)	0.6606
Light machine oil (15.6)	0.113
Olive oil (20)	0.084
Water (20, 25, 37)	0.001, 0.0009, 0.0007
Air (18)	0.0000002

From Weast, R.C. (Ed.), 1973–74. The Handbook of Physics and Chemistry, 54th ed. CRC Press, Cleveland, OH.

The viscosity of a number of fluids is given in Table 9.2. The viscosity of a fluid can be measured in the manner described previously, or, as I will discuss later, it can also be calculated by measuring the velocity of a falling ball through the liquid (Stokes, 1922). Viscosity is important at the cellular level but also affects many organismal behaviors. The morphologies of plants and animals have evolved in part due to the influences of viscosity (Vogel, 1981; Niklas, 1992; Denny, 1993).

The cytoplasm of prokaryotic cells that have been transformed so as to express GFP appears to be a Newtonian fluid with a viscosity of about 0.010–0.015 Pa s, about 10–15 times the viscosity of water, as determined by measuring the diffusion coefficient of GFP using fluorescence redistribution after photobleaching (FRAP) and calculating the viscosity from the Einstein–Stokes equation (see Chapter 2; Elowitz et al., 1999; Mullineaux et al., 2006; Slade et al., 2009; Nenninger et al., 2010; Mika et al., 2011; Puchkov, 2013).

9.3.1.2 Non-Newtonian Fluids

Many solutions do not show parabolic flow and thus do not obey Poiseuille's Law. These are called *non-Newtonian fluids* (Seifriz, 1920, 1921, 1929, 1931, 1935, 1955a; Kamiya, 1956; Allen and Roslansky, 1959). Newtonian solutions obey Poiseuille's Law because they have a single viscosity. However, non-Newtonian solutions possess an infinite number of viscosity values, where the viscosity of the solution depends on the rate of shear, otherwise known as the *velocity gradient*. Noburô Kamiya (1950) used Poiseuille's Law to study the flow of the endoplasm of *Physarum* using the ectoplasm as the tube. He noticed that the velocity of the flowing endoplasm did not show a parabolic profile, where the rate of shear would be proportional to the shearing stress. Kamiya found that most of the particles in the endoplasm travel at the same speed (Fig. 9.10). More recent

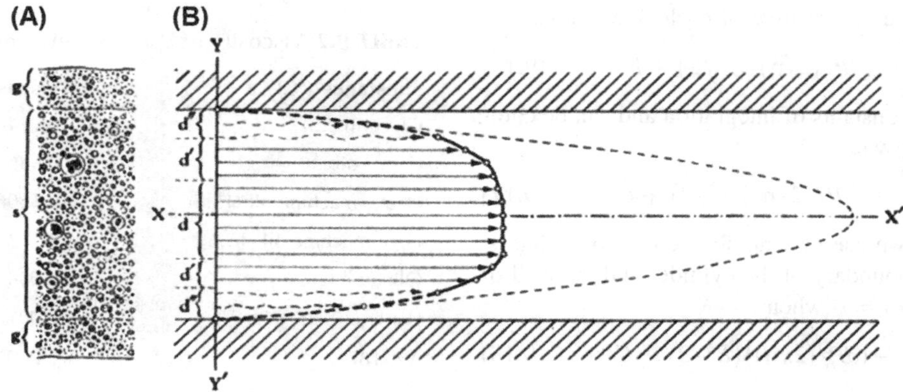

FIGURE 9.10 (A) A drawing of the sol-like endoplasm (s) and gellike ectoplasm (g) of a plasmodial strand of *Physarum*. (B) The velocity distribution of endoplasmic flow. The *small circles* represent experimental values, and the *broken line* represents what the velocity distribution would look like if the endoplasm were Newtonian and the flow were parabolic. *From Kamiya, N., 1950. The rate of protoplasmic flow in the myxomycete plasmodium. Cytologia 15, 183–193, 194–204.*

measurements using laser Doppler velocimetry show that there is more variation in the velocities than Kamiya observed, but still less than would be predicted if the cytoplasm moved by parabolic flow (Mustacich and Ware, 1977b; Earnshaw and Steer, 1979). Kamiya concluded that the nonparabolic flow indicated that the viscosity of the endoplasm depended on the rate of shear. That is, close to the ectoplasm, where the rate of shear was highest, the viscosity of the endoplasm was low, whereas in the center of the cell, where the rate of shear was lowest, the viscosity was highest, and the endoplasm there moved as a block. The flowing endoplasm of *Amoeba proteus* also shows shear thinning. However, at fast flow rates, the streaming endoplasm has a velocity profile close to Poiseuille flow and thus behaves as a Newtonian fluid (Rogers et al., 2008).

The dependence of the viscosity on the rate of shear depends on the molecular structure of the fluid. When the viscosity of the solution is independent of the rate of shear, the solution is probably composed of noninteracting spherical molecules. In non-Newtonian fluids, the electrostatic attraction between the molecules is not symmetrical and depends on the position of the molecules relative to each other. The relative position depends on the flow, and thus, the non-Newtonian properties depend on the relationship between the electrostatic energy between the molecules and the mechanical energy that can change their position.

If we know the viscous properties of a solution, we can make estimates of its molecular structure. For example, if the viscosity decreases as the rate of shear increases, we can infer that the shear-thinning solution is composed of asymmetrical molecules that may have the appearance of linear fibers. Solutions with these properties are called *thixotropic* (from the Greek words for "change by touch") solutions. When the viscosity increases as the rate of shear increases, the shear-thickening solution is called dilatant (Reynolds, 1885, 1886) or negatively thixotropic (Eliassaf et al., 1955). We can surmise that such a solution is compressible, and in response to a force, the particles come together to form a "tighter" solution. A dilatant solution may also be composed of highly branched, knotted, or hooked molecules that get entangled when exposed to a force. Fig. 9.11 shows the relationship between the shearing stress and rate of shear, and Fig. 9.12 shows the relationship between the rate of shear and viscosity for Newtonian, thixotropic, and dilatant solutions. In thixotropic solutions, the viscosity decreases as the rate of shear increases, whereas in dilatant solutions, the viscosity increases as the rate of shear increases. Catsup, an extract of plant cells, is thixotropic (Vonnegut, 1999). Cornstarch, another plant extract, is dilatant (Seifriz, 1936; Heilbrunn, 1958).

Because the viscosity of thixotropic solutions is highest when they are not disturbed or nothing is moving through them, and decrease as something moves through them, they possess what is called a *yield value*. The yield value is the minimum shearing stress required to produce a flow. In the case of cytoplasm, the yield value is a measure of the minimum force per unit area necessary to move a vesicle, chromosome, or organelle through the cytoplasm. The cytoplasmic motors I will discuss in Chapters 10 and 11 are capable of providing the force per unit area necessary to overcome the yield value.

9.3.1.3 Experimental Approaches to Measuring Cytoplasmic Viscosity

Cell biologists have come up with a variety of ingenious methods for measuring the viscosity of cytoplasm. For example, cytoplasmic viscosity can be studied with a centrifuge microscope, which is essentially a microscope with a rapidly rotating stage (Hiramoto and Kamitsubo, 1995).

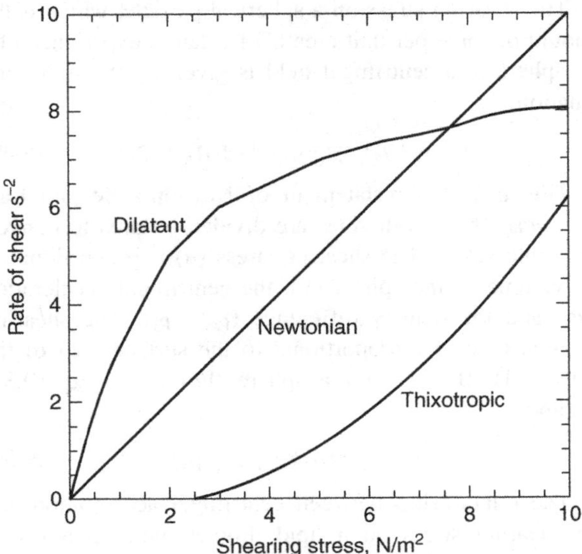

FIGURE 9.11 Graph of *curves* relating the rate of shear to the shearing stress. The viscosity of the solution is obtained from the reciprocal of the slope (or the tangent to the line at any point). The relationship between shearing stress and rate of shearing is linear for Newtonian fluids and nonlinear for non-Newtonian fluids. The non-Newtonian fluids are either dilatant or thixotropic. The viscosity of dilatant fluids increases as the shearing stress increases. Consequently, the rate of shear does not rise as fast at higher shearing stresses. The viscosity of thixotropic fluids decreases as the shearing stress increases. Consequently, the rate of shear rises faster at higher shearing stresses.

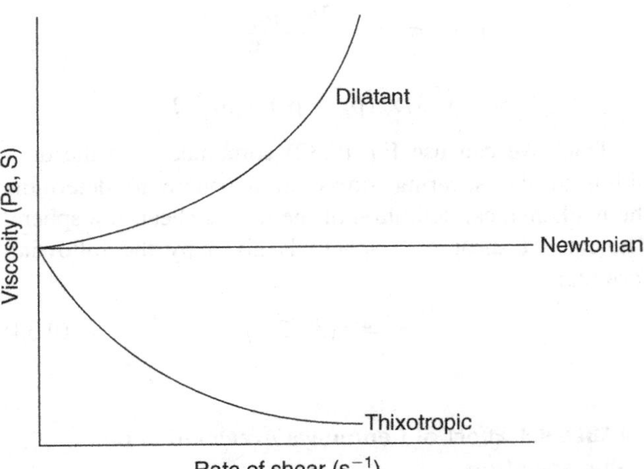

FIGURE 9.12 Newtonian and non-Newtonian (dilatant and thixotropic) viscosity. The viscosity of a dilatant solution can be proportional to the rate of shear, the rate of shear squared, or by a more complicated function of the rate of shear (see Appendix 2). The viscosity of a thixotropic solution can be inversely proportional to the rate of shear, the the rate of shear squared, or by a more complicated function of the rate of shear.

Kamitsubo et al. (1988) determined the viscosity of cytoplasm by observing the velocity of lipid droplets moving through the cytoplasm when the cell was exposed to a centrifugal force, which put a shearing stress on the lipid droplets. Eiji Kamitsubo and his colleagues used Stokes' Law to calculate the viscosity from the applied shearing stress and the observed rate of shear.

Stokes' Law was derived from empirical observations that determined the relationship between the forces that resist the movement of a sphere of a given radius through a viscous medium of a certain viscosity, with the velocity of the sphere. Stokes' Law is

$$F_v = 6\pi r_H \eta v \quad (9.24)$$

We can determine the velocity of a sphere, if it falls under the influence of gravity, because the inertial force exerted on the sphere due to gravity is given by Newton's Second Law:

$$F_i = mg \quad (9.25)$$

where m is the mass of the sphere and g is the acceleration due to gravity (9.8 m/s^2). In the absence of friction, the sphere will continue to travel due to its inertia, which is a function of its mass. In the absence of friction, the velocity will also increase over time as a result of the acceleration due to gravity.

As the mass of a sphere acted on by gravity depends on the density difference ($\rho_s - \rho_m$) between the sphere and the medium it passes through, the mass must be calculated from the following formula:

$$m = (\rho_s - \rho_m)(4/3)\pi r^3 \quad (9.26)$$

Thus, Newton's Second Law can be rewritten as

$$F_i = g(\rho_s - \rho_m)(4/3)\pi r^3 \quad (9.27)$$

Newton's First Law states that a body remains at rest or in uniform motion unless a force acts on it. According to this law, if a sphere falls at a constant velocity and thus without accelerating, there must be no net force exerted on it. In other words, the inertial force is opposed by the force resisting the travel; that is, the inertial force is opposed by a frictional force. Thus, the velocity of a sphere falling in a viscous fluid due to the force of gravity can be found by setting $F_i + F_v = 0$, where there is no acceleration, or simply, $F_i = -F_v$. Thus,

$$-6\pi r_H \eta v = g(\rho_s - \rho_m)(4/3)\pi r^3 \quad (9.28)$$

Assuming that the gravitational force is constant, we can solve for v. After rearranging terms, canceling, and simplifying, we get

$$v = \frac{2g(\rho_m - \rho_s)r^2}{9\eta} \quad (9.29)$$

Thus, the velocity of a falling sphere is directly proportional to the acceleration due to gravity and the difference in density between the medium and the sphere. The velocity is proportional to the square of the radius of the sphere and inversely proportional to the viscosity of the medium.

Moreover, in a viscous solution, the velocity is constant with respect to time. (On the other hand, in cases where the viscosity is nil, the change in velocity with respect to time will be constant. That is, $F = mg = mdv/dt$. Therefore, $g(dt) = dv$, $gdt = \int dv$, and $v(t) = gt$). It is important to realize that this case is only an idealization presented by Galileo and Newton and can never be realized because if η were equal to zero, the terminal velocity of any particle or projectile, which is the velocity at $t = \infty$, would be infinite.

To calculate the viscosity of the cytoplasm, we have to know the density of the lipid droplets and the cytoplasm through which they move, the radius of the lipid droplets, and their velocity.

The density of the endoplasm of characean cells was determined by cutting the cell and allowing the endoplasm to fall in a density gradient of dextran dissolved in artificial cell sap, which had an osmotic pressure equal to that of the cell. Due to its high molecular mass, the dextran exerts a negligible osmotic pressure so that the vesicles maintain their original density without swelling or shrinking. The endoplasmic drops are then left to fall into the gradient. Eventually, $= 0$ and the drop stops when it reaches its own density, that is, when $\rho_m = \rho_s$. The average density found by this method is 1.015×10^3 kg/m^3 (Kamiya and Kuroda, 1957). Have you noticed that I just explained the principle behind density gradient centrifugation?

It is also possible to measure the time it takes for a drop of endoplasm to fall a certain distance in a medium of known density and viscosity. Using this method, Kamiya and Kuroda (1957) found that the density of the endoplasm is 1.0145×10^3 kg/m^3. These values are very close to those found recently using optical methods (Wayne and Staves, 1991).

Once the density of the endoplasm is known, it is possible to determine the viscosity with a centrifuge microscope using Stokes' Law. Prior to the measurement of cytoplasmic viscosity, the endoplasm of a characean cell is centrifuged down to one end of the cell. This takes about 5–10 min at 1000 g. Then the viscosity of the endoplasm is determined by measuring the movement of oil droplets, which have a density of approximately 960 kg/m^3, through the endoplasm under various amounts of centrifugal acceleration (Kamitsubo et al., 1989).

If the cytoplasm were a Newtonian fluid, the viscosity would be constant at all rates of shear. Furthermore, in a Newtonian fluid, the rate of shear would be linearly related to the shear stress and the relationship, plotted on a graph, would go through the origin. To test whether or not the cytoplasm is a Newtonian or non-Newtonian fluid, Kamitsubo and his colleagues carried this experiment out at different shearing stresses and different rates of shear. At various amounts of centrifugal acceleration (100–500 g), the interface between the oil droplet and the endoplasm experiences various shearing stresses and rates of shear (Kamitsubo et al., 1988).

The shearing stress on a spherical particle, which is the amount of force per unit area (F/A), that is experienced by the sphere in a centrifugal field is given by the following equation:

$$\sigma = F/A = (V/A)\alpha g(\rho_m - \rho_s) \tag{9.30}$$

This is just a restatement of Newton's Second Law ($F = ma$) where both sides are divided by area to convert force into stress. The shearing stress (σ) is proportional to the volume of the sphere (V), the centrifugal acceleration (αg), and the density difference ($\rho_m - \rho_s$). The shearing stress is inversely proportional to the surface area of the sphere (A). Because for a sphere, $V/A = r/3$, Eq. (9.30) becomes

$$\sigma = (r/3)\alpha g(\rho_m - \rho_s) \tag{9.31}$$

The relationship between centrifugal acceleration and the shearing stress on a lipid droplet, with a radius of 10^{-5} m and a density of 960 kg/m^3, moving through a cytoplasm with a density of 1014.5 kg/m^3 is given in Table 9.3.

The rate of shear is the velocity gradient across a solution in which a body may or may not be immersed. The rate of shear is zero when the solution and/or object is at rest. The viscosity of a solution measured by the falling ball method, in which a spherical body falls, is given by Eq. (9.32), which I already derived:

$$\eta = \sigma/\gamma = \frac{2\alpha g(\rho_m - \rho_s)r^2}{9v} = ((r/3)\alpha g(\rho_m - \rho_s))/(3v/(2r)) \tag{9.32}$$

Thus, we can use Eq. (9.32) combined with the definition of the shearing stress on a sphere to determine the mathematical definition of the rate of shear of a sphere. The rate of shear of a sphere is given by the following formula:

$$\gamma = v(3/(2r)) \tag{9.33}$$

TABLE 9.3 Effect of Centrifugal Acceleration on Shearing Stress

Acceleration (g)	σ (N/m^2)
100	0.178
200	0.356
300	0.534
400	0.712
500	0.890

From Kamitsubo, E., Kikuyama, M., Kaneda, I., 1988. Apparent viscosity of the endoplasm of characean internodal cells measured by the centrifuge method. Protoplasma 5 (Suppl. 1), 10–14.

TABLE 9.4 Relationship Between Rate of Shear, Shearing Stress, and Cytoplasmic Viscosity

Lipid Droplet No.	γ (s^{-1})	σ (Pa)	η (Pa s)
1	0.69	0.545	0.790
2	0.79	0.673	0.852
3	1.88	0.609	0.324
4	2.56	0.660	0.258
5	3.85	0.644	0.167
6	5.18	0.805	0.155
7	8.09	0.877	0.108
8	11.00	1.45	0.132
9	11.90	0.887	0.075

From Kamitsubo, E., Kikuyama, M., Kaneda, I., 1988. Apparent viscosity of the endoplasm of characean internodal cells measured by the centrifuge method. Protoplasma 5 (Suppl. 1), 10–14.

The rate of shear (γ) of a sphere is proportional to the velocity of the sphere, and a geometric factor that is given by ($3/2r$).

Kamitsubo et al. (1988) measured the velocities of various lipid bodies moving in a centripetal direction under a variety of centrifugal forces and calculated the rate of shear of each particle and the shearing stress that caused that rate of shear. These data are presented in Table 9.4.

Plotting the rate of shear versus the shearing stress (Fig. 9.13), Kamitsubo et al. (1988) got a straight line that intercepts the x-axis at about 0.5 Pa. This means that at shearing stresses greater than 0.5 Pa, the lipid droplets move at a velocity that is proportional to the shearing stress. However, at shearing stresses less than 0.5 Pa, the cytoplasm has a very high resistance and the oil droplets are unable to move in the cytoplasm. The point where the relationship of rate of shear versus shearing stress crosses the x-axis is called the yield value. The possession of a *yield value* is characteristic of shear-thinning, thixotropic, non-Newtonian fluids.

We can obtain the cytoplasmic viscosity by dividing the shear stress by the rate of shear (Table 9.4):

$$\eta = \sigma/\gamma = \frac{2\alpha g(\rho_m - \rho_s)r^2}{9v} \quad (9.34)$$

Now we can plot the viscosity against the rate of shear (Fig. 9.14). We can see that the viscosity is very high (c. 10 Poise = 1 Pa s) at low rates of shear and very low (c. 1 Poise = 0.1 Pa s) at high rates of shear. That is, the cytoplasm is thixotropic.

These measurements come from characean cells that have very active streaming. As the endoplasm travels with a velocity of about 100 μm/s at the outside of the endoplasm and about 90 μm/s approximately 10 μm toward the center, the endoplasm travels with a rate of shear of (10 μm/s)/(10 μm) or about 1 s^{-1}. With this rate of shear, the viscosity of the streaming endoplasm must be very high (0.8 Pa s), about 800 times the viscosity of water (0.001 Pa s). The narrow range of velocities in the streaming cytoplasm observed with laser Doppler velocimetry is consistent with the non-Newtonian nature of the cytoplasm of characean cells (Mustacich and Ware, 1974, 1976, 1977a,b; Langley et al., 1976; Sattelle and Buchan, 1976).

The viscosity of neutrophils is approximately 0.131 Pa s, as measured by pulling the cells into a pipette and measuring the deformation. The viscosity decreases as the rate of shear increases, indicating that the cytoplasm of

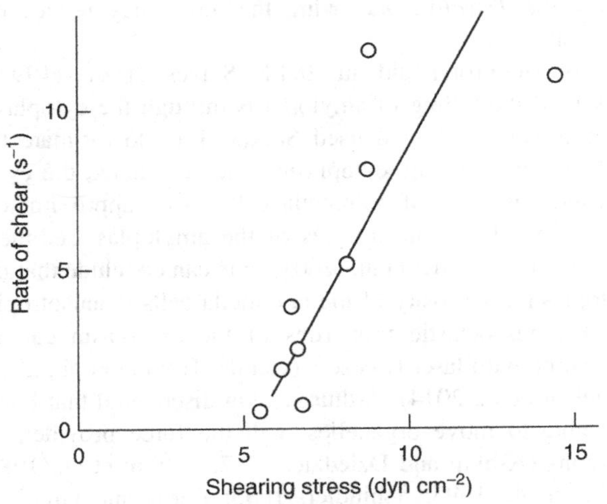

FIGURE 9.13 The relationship between shearing stress and rate of shear for the endoplasm of *Nitella axilliformis* determined with a centrifuge microscope. *From Kamitsubo, E., Kikuyama, M., Kaneda, I., 1988. Apparent viscosity of the endoplasm of characean internodal cells measured by the centrifuge method. Protoplasma 5 (Suppl. 1), 10–14.*

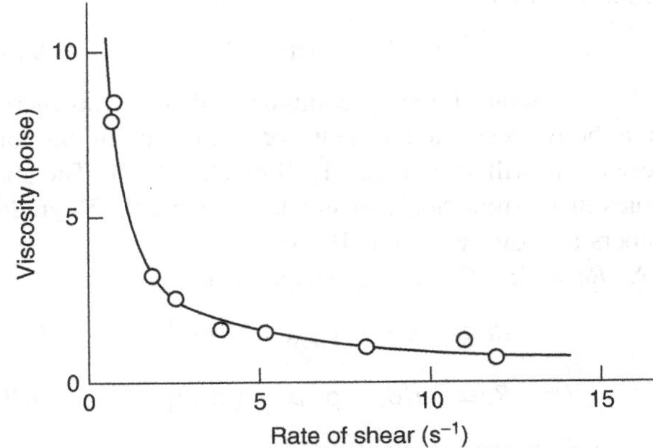

FIGURE 9.14 The relationship between rate of shear and viscosity for the endoplasm of *Nitella axilliformis* determined with a centrifuge microscope. *From Kamitsubo, E., Kikuyama, M., Kaneda, I., 1988. Apparent viscosity of the endoplasm of characean internodal cells measured by the centrifuge method. Protoplasma 5 (Suppl. 1), 10–14.*

neutrophils is also thixotropic (Tsai et al., 1993, 1994). As a rule, we can consider the bulk viscosity of cytoplasm to be between 0.1 and 0.8 Pa s, which is 100−800 times the viscosity of water, and between the viscosity of olive oil and glycerol.

According to Newton's Second Law, the acceleration is proportional to the force. By contrast, according to Stokes' Law, the velocity is proportional to the force. Which law is a better predictor of what happens in the cytoplasm? Newton's Second Law is not valid when friction is not negligible, and Stokes' Law typically applies when the movement of the sphere is slow and its size is small (Rayleigh, 1893). But how slow is slow, and how small is small? Quantitatively, this is measured with the Reynolds number (*Re*) (Reynolds, 1901). The *Re* (dimensionless) is the ratio of the inertial force to the viscous force. While the *Re* for moving solids is an approximation, which cannot be rigorously derived, it is valuable for approximations (Dodge and Thompson, 1937).

When the *Re* is greater than 1, friction is negligible, inertial forces dominate, and a body in motion will tend to stay in motion for a long time. Newton's Second Law best describes the motion of a spherical particle in an inertial system:

$$F_i = a(\rho_s - \rho_m)(4/3)\pi r^3 \quad (9.35)$$

For any shaped particle with a "characteristic length" of x, Newton's Second Law is

$$F_i = a(\rho_s - \rho_m)x^3 \quad (9.36)$$

When the *Re* is less than 1, viscous forces dominate and Stokes' Law best describes the motion of a spherical particle:

$$F_v = 6\pi r_H \eta v \quad (9.37)$$

For any shaped particle with a "characteristic length" of x, Stokes' Law is

$$F_v = x\eta v \quad (9.38)$$

When viscous forces predominate, there is no inertia and a body needs a constant force to stay in motion; otherwise it will stop instantly (Purcell, 1977). The mechanics of Aristotle apply to situations where the Reynolds numbers are low (Franklin, 1976).

As $Re = |F_i/F_v|$ for any shaped particle

$$Re \cong \left|\left[a(\rho_s - \rho_m)x^3\right]/(x\eta v)\right| \quad (9.39)$$

$$Re \cong \left|\left[a(\rho_s - \rho_m)x^3\right]/(x\eta v)\right| \quad (9.40)$$

As $a = v/t$, we can simplify

$$Re \cong \left|\left[(v/t)(\rho_s - \rho_m)(x^2)\right]/(\eta v)\right| = \left|\left[(\rho_s - \rho_m)x^2\right]/(\eta t)\right| \quad (9.41)$$

Because dimensionally, $t = x/v$, we can simplify again:

$$Re \cong \left|\left[v(\rho_s - \rho_m)x\right]/\eta\right| \quad (9.42)$$

Thus, the *Re* is proportional to the velocity, the density difference, and the characteristic length. It is inversely proportional to the viscosity.

In the cytoplasm, where the viscosity is typically between 0.1 and 0.8 Pa s, the density differences between a moving body and the cytoplasm are less than 250 kg/m^3, the velocity of movement is between 0.1 and 100 μm/s, the characteristic length of a moving body is between 1 and 10 μm, the *Re* is typically much less than 1, and viscous forces predominate over inertial forces by approximately one million times. Consequently, in the cytoplasm, Newton's Second Law is not valid. When viscous forces predominate, the inertial forces are negligible, and movement stops the instant the applied force is removed. A cell biologist cannot assume that there is no friction (Wayne, 2012a).

Kikuyama and Tazawa (1972) measured the viscosity of the cytoplasm of characean cells using a really clever method. They introduced *Tetrahymena* into the vacuolar space of *Nitella* by means of vacuolar perfusion. After closing both ends by cellular ligation, the *Tetrahymena* was forced into the endoplasm by centrifugal force. After centrifugation, the cell was ligated near the centrifugal end to obtain an endoplasm-rich cell fragment containing *Tetrahymena*. The viscosity of the cytoplasm was estimated by measuring the swimming speed of the *Tetrahymena* in the endoplasm. This was compared to a standard curve where the swimming speed of *Tetrahymena* was measured in media of various viscosities. Using this method, they estimated that the bulk viscosity was between 0.04 and 1 Pa s. We do not know whether the variation is due to the non-Newtonian properties of the cytoplasm, where the faster the *Tetrahymena* swim, the more they reduce the viscosity.

As Heilbronn did in 1914, Staves et al. (1997b) observed the falling of amyloplasts through the cytoplasm of columella cells and used Stokes' Law to estimate the bulk viscosity of the cytoplasm. They estimated the cytoplasmic viscosity of columella cells to be approximately 0.268 Pa s. From an analysis of the amyloplast sedimentation data of Yoder et al. (2001), one can conclude that the cytoplasmic viscosity of the columella cells is anisotropic.

The viscoelastic properties of the cytoplasm can be measured with laser tweezers (van der Honing et al., 2010; Ketelaar et al., 2014). Arthur Ashkin discovered that lasers are able to move organelles with the force provided by photons (Ashkin and Dziedzic, 1987; Ashkin et al., 1987; Leitz et al., 1995; Schindler, 1995; Berns and Greulich, 2007). The force per unit area provided by the laser is calculated from the following equation:

$$F/A = \mathrm{EFR}/(nc) \quad (9.43)$$

where EFR is the energy fluency rate of the laser (in J/m^2 s), n is the refractive index of the cytoplasm (dimensionless), and $c =$ speed of light (3×10^8 m/s). Because the energy fluency rate of the laser is approximately 10^{11} J/m^2 s, it is capable of producing a shearing stress of hundreds of N/m^2 (Ashkin et al., 1987).

Using laser tweezers to move particles through the cytoplasm, Ashkin and Dziedzic (1989) find that cytoplasm has a yield value and is thus a non-Newtonian fluid. They find that the yield value of the moving endoplasm of scallion epidermal cells is approximately 0.1 N/m^2, and the yield value of the stationary ectoplasm is between 10 and 1000 N/m^2. The yield value for the endoplasm is similar to those found previously in other cells. The yield value of the endoplasm is 0.5 Pa in *Nitella*, and in *Physarum* it is between 0.06 and 0.11 Pa, depending on the direction of movement (Sato et al., 1989). The presence of a yield value means that the cytoplasm in these cells is also non-Newtonian. Cytoplasmic viscosity has also been measured by injecting magnetic particles into cells or allowing them to be taken up by phagocytosis and then determining the ability of magnetic tweezers to move the particles (Seifriz, 1928; Bausch et al., 1999; Scherp and Hasenstein, 2007).

As I have already stated, the viscosity of the cytoplasm is, in part, the relationship between the velocity with which a particle moves and the force to which it is subjected. However, in a non-Newtonian cytoplasm, the viscosity that is measured also depends on the size of the particle. Thus, at a given shearing stress, the cytoplasm may have differing viscosities that depend on the size of the moving particle. The viscosity of the cytoplasm toward large particles such as organelles (0.100 nm) is known as the bulk viscosity. The viscosity experienced by metabolite-sized molecules (1 nm) is known as the *microviscosity*, and the viscosity experienced by macromolecules (1−100 nm) is called the *intermediate viscosity*. The studies mentioned above measured the bulk viscosity of cytoplasm using organelles as moving bodies.

Kate Luby-Phelps et al. (1985, 1986, 1988) have estimated the intermediate- and microviscosity of the cytoplasm by measuring the diffusion coefficient of a number of different molecules in the cytoplasm with the aid of the fluorescence redistribution after photobleaching (FRAP) technique (see Chapter 2). As I stated in Chapter 2, the diffusion coefficient of a spherical particle depends on the viscosity, according to the Stokes−Einstein equation: $D = kT/(6\pi r_H \eta)$.

Luby-Phelps et al. fluorescently labeled dextrans of various sizes and microinjected them into the cytoplasm. Then they measured the diffusion coefficient of the dextran in the cytoplasm by bleaching a region of the cytoplasm and watching the recovery of fluorescence over time. They also measured the diffusion coefficient of the dextran in water by bleaching a region of the water and watching the recovery of fluorescence over time. Then they plotted the ratio of the two diffusion coefficients, which is an estimate of the relative viscosity of the cytoplasm ($D_c/D_w = \eta_w/\eta_c$) versus the hydrodynamic radii of the molecules tested (Fig. 9.15). They estimate, by extrapolation, that the viscosity of the cytoplasm for infinitesimally small molecules is about four times the viscosity of water. They also find that the viscosity of the cytoplasm is not a constant but is proportional to the radius of the molecule for molecules with hydrodynamic radii from 2 to 15 nm. Thus, the larger the molecule, the greater the viscosity it experiences in the cytoplasm. This means that diffusion in the cytoplasm is hindered by some kind of network. For molecules with radii between 15 and 60 nm, the viscosity remains constant at a high value, approximately 12.5 times the viscosity of water. The constant viscosity for molecules with different hydrodynamic radii indicates that the elongated molecules may reptate through the netlike cytoplasm like worms. If we extrapolate the slope to the x-axis, where the diffusion in the cytoplasm becomes zero, we find that the radius is 20.7 nm. This may indicate that the average diameter of the holes in the mesh of the cytoplasm is about 41.4 nm (Provance et al., 1993). The ability to observe the relationship between intermediate cytoplasmic viscosity and the hydrodynamic radius of the diffusing molecule depends on the distance the probe is allowed to diffuse. It can be observed when the diameter of the bleached spot is 50 μm (Luby-Phelps et al., 1986) but not when it is 5 μm (Seksek et al., 1997; Swaminathan et al., 1997). When the FRAP

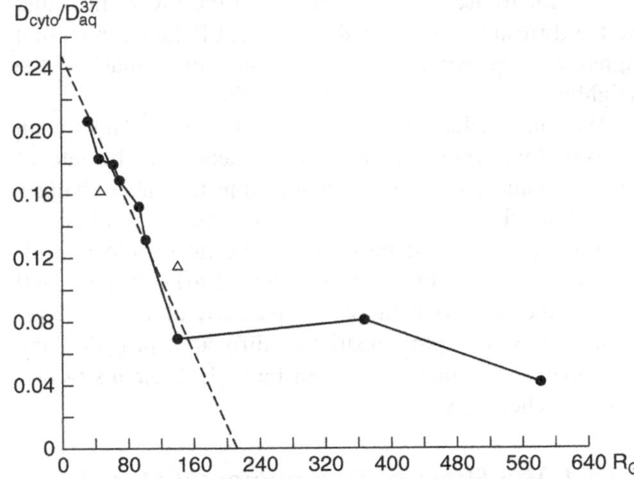

FIGURE 9.15 The effective viscosity experienced by dextrans in the cytoplasm depends on the size of the dextran (R_G, which is the radius of gyration, in Å). D_{cyto}/D_{aq} is the ratio of the diffusion coefficient of the dye in the cytoplasm relative to the diffusion coefficient of the dye in water. D_{cyto}/D_{aq} is related to the reciprocal of the effective viscosity. *From Luby-Phelps, K., Taylor, D.L., Lanni, F., 1986. Probing the structure of cytoplasm. J. Cell Biol. 102, 2015−2022.*

bleached area is too small, the probe cannot diffuse far enough to experience the difference between the intermediate viscosity and microviscosity of the cytoplasm (Kalwarczyk et al., 2011).

The intermediate viscosity of *Physcomitrella patens* as determined by measuring the diffusion coefficient of GFP with FRAP, is between 0.0048 and 0.0226 Pa s (Kingsley et al., 2016). The intermediate viscosity of HeLa cells as measured by the rotational frequency of bovine serum albumin-coated quantum dots is about 0.0047 Pa s (Nakane et al., 2012).

To visualize the microviscosity of the cytoplasm, Luby-Phelps et al. (1993) have used a very small fluorescent probe (Cy 3.18), the quantum yield of which varies with viscosity, and (Cy 5.18) another small fluorescent probe, the quantum yield of which is independent of viscosity. They then visualized the viscosity of the fluid phase of the cytoplasm by ratio imaging the two dyes. They find that the viscosity of the fluid phase is not significantly different from that of water (Luby-Phelps, 1994). Similar results have been obtained by Bicknese et al. (1993), Kao et al. (1993), and Parker et al. (2010).

Srivastava and Krishnamoorthy (1997) have shown that the microviscosity varies with location in the cell and Dijksterhuis et al. (2007) have shown that it varies over time during the germination of dormant spores. By contrast, Keith and Snipes (1974) found that the viscosity of this aqueous space in human cells, bean cells, and *Chlamydomonas* is about 100 times greater than the viscosity of water. Although the cells they studied may have been dead, Kuimova et al. (2009) found that the viscosity of cancer cells increases on death.

The microviscosity of the lumen of the ER as measured by the diffusion of GFP with a KDEL ER lumen retention signal is approximately 3–6 greater than that of the neighboring cytosol (Dayel et al., 1999).

We can conclude that the cytoplasm is a non-Newtonian viscous fluid where the viscosity depends on the rate of shear and the size of the moving object. It also changes throughout the life of the cell. As viscosity can affect all transport processes, it must always be taken into consideration. According to R.J.P. Williams (1961), "It may well be that the achievement of a separation of activated reagents in space plus restricted diffusion provides the fundamental distinction between biological chemistry and test-tube chemistry."

9.3.1.4 The Effect of Environmental Stimuli on Cytoplasmic Viscosity

Virgin (1954) observed that the chloroplasts in the leaf cells of *Elodea* were more easily displaced by centrifugation when they were treated with blue light. Similarly, Seitz (1967, 1979) observed the same phenomenon in the leaf cells of *Vallisneria spiralis*. Shingo Takagi has shown that blue light has the same effect in *Vallisneria gigantea*, whereas red light has the opposite effect (Takagi et al., 1989, 1991, 1992; Dong et al., 1998; Sakai and Takagi, 2005; Sakai et al., 2015). These results and others indicate that the viscosity of the cytoplasm is not constant but can be influenced by environmental factors, and perhaps these factors influence some cellular processes, in part through their effect on viscosity (Stafelt, 1955; Seitz, 1987; Virgin, 1987; Mansour et al., 1993). Thus, the translational diffusion of substrates to enzymes, the rotational diffusion of substrates so that they can properly bind to an enzyme, and the movement of membrane vesicles, tubules, and organelles may be affected by environmental factors. Moreover, as velocity in a non-Newtonian world depends on the difference between the inertial force and the viscous force, an increase in velocity may result from an increase in the inertial force and/or a decrease in the viscosity.

9.3.2 Elasticity of the Cytoplasm

The living cell is highly elastic and it loses its elasticity when it dies. The conclusion that the cytoplasm is constructed of fibrous elements is supported by studies on the elasticity of cytoplasm. Elasticity is the property of a body to resist deformation and to reversibly recover from deformation produced by force. The elasticity of a material is defined by Young's modulus of elasticity, which can be determined by stretching the body in question. In the case of a wire, the elastic modulus, M (in Pa or N/m^2), is given by determining the elongation (dx) of a wire of length x and radius r that is produced by a given force (mg). The relation between stress ($mg/\pi r^2$) and strain (dx/x) is given by the following formula:

$$M = \text{stress/strain} = \left((mg)/\left(\pi r^2\right)\right)/(dx/x) \quad (9.44)$$

The elastic modulus, also known as Young's modulus, is equal to the stress ($mg/\pi r^2$) needed to produce a doubling of the length (when $dx = x$), which is a unit strain (dx/x). Young's modulus is usually determined by extrapolation because many substances break before they double in length.

In William Seifriz's time (1924, 1928), elasticity was the best indicator of the structure of living cytoplasm. As elasticity depends on the presence of linear molecules, he assumed that the cytoplasm was composed of a network of linear molecules. This is also consistent with the thixotropic behavior of cytoplasm. He demonstrated that protoplasm is elastic by stretching it between microneedles. He found that live protoplasm is very elastic, whereas dead protoplasm is not elastic. Seifriz looked at plasmolyzed cells of the onion epidermis. After plasmolysis, the tissue is cut across to expose some cells without touching the protoplasm within. The cells were entered with a microneedle and the naked protoplasm touched. The protoplasm stuck to the needle and could be drawn out to great lengths. When the thread snapped, its elastic limit had been passed. Once the elastic limit was passed, the protoplasmic thread snapped back, becoming reincorporated into the protoplasm, usually without even disturbing the continuous cytoplasmic streaming (Fig. 9.6).

Crick and Hughes (1950) and Crick (1950) estimated that the elastic modulus of cytoplasm is about 10 N/m^2 using a magnetic particle technique. With this technique, magnetic particles were introduced into the cytoplasm of chick fibroblasts by phagocytosis and then "pulled" with a magnetic field to determine the relationship between the stress applied in the form of a magnetic field and the strain observed. When the magnetic field was shut off, the particles recoiled, providing further evidence for the elastic nature of cytoplasm. As a comparison, rubber has an elastic modulus of 5×10^8 N/m^2. Francis Crick and Arthur Hughes (1950) described the cytoplasm like so "If we were compelled to suggest a model we would propose Mother's Work Basket—a jumble of beads and buttons of all shapes and sizes, with pins and threads for good measure, all jostling about and held together by 'colloidal forces.'"

The elastic modulus can also be determined by coating magnetic beads with a peptide containing RGD (Arg-Gly-Asp), and subjecting the bead to a magnetic field. The RGD binds to the integrins in the plasma membrane and the magnetic field causes the bead to rotate. The degree of bead displacement is a function of the elastic modulus of the cell (Fabry et al., 2001).

9.4 MICROTRABECULAR LATTICE

The cytoplasm acts like a poroelastic material consisting of a porous elastic solid meshwork composed of cytoskeleton, organelles, and macromolecules bathed in an interstitial fluid known as the cytosol (Porter and Moses, 1958; Moeendarbary et al., 2013). Can we see the linear structures in the cytoplasm that lead to its thixotropic behavior and elastic properties? Keith Porter and others have observed a microtrabecular lattice in animal (see Fig. 9.16; Wolosewick and Porter, 1979; Porter and Tucker, 1981; Porter and Anderson, 1982; Porter et al., 1983; Porter, 1984) and plant cells (see Fig. 9.17; Hawes et al., 1983; Wardrop, 1983) using the million volt electron microscope that allows one to observe thick sections. Schliwa et al. (1981) observed the microtrabecular lattice in cells treated with Brij 58, a detergent that "opens" cells but does not interfere with chromosome motion. While there are arguments whether the microtrabecular lattice is fact or artifact (Heuser, 2002; McNiven, 2002; Sardet, 2002; Wolosewick, 2002), it may represent the fibrous network of the cytoplasm as envisioned by Rudolph Peters (1929, 1937) and Joseph Needham (1936).

Albert Szent-Györgyi (1941a,b) proposed that enzymes may form a solid-state structure in the cell, and such a

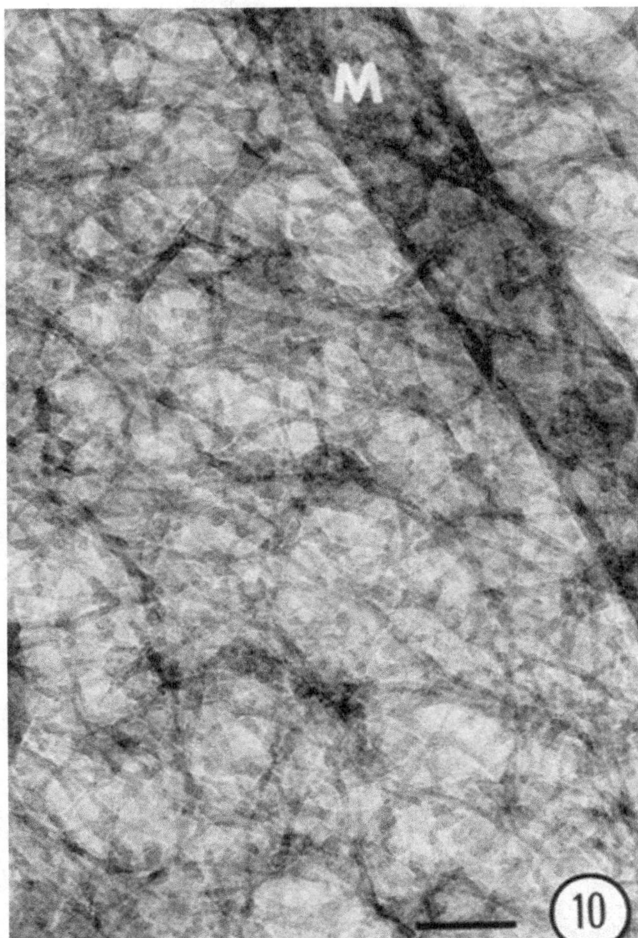

FIGURE 9.16 The microtrabecular lattice of a WI-38 cultured cell prepared by the freeze-substitution method. *M*, mitochondrion. ×100,000. *From Wolosewick, J.J., Porter, K.R., 1979. Microtrabecular lattice of the cytoplasmic ground substance; artifact or reality. J. Cell Biol. 82, 114–139.*

structure may be necessary for their action in vivo. The limitations of diffusion may have given rise to microscopic structures in the cell. That is, metabolite channeling may have resulted in the evolution of the microtrabecular lattice, which is actually composed of all the proteins in the cell that form a quintinary structure, including intrinsically disordered proteins (Tompa, 2013) and the enzymes of the glycolytic pathway (see Chapter 14; Clarke and Masters, 1975; Knull et al., 1980; McConkey, 1982; Srere, 1985; Svivastava and Bernhard, 1986; Schliwa et al., 1987; Pagliaro, 1993, 2000; Ovádi and Srere, 2000).

9.4.1 Function of the Microtrabecular Lattice in Polarity

Cells have the ability to sense and respond to mechanical signals (Hoffman and Crocker, 2009), and a deformation of the microtrabecular lattice may influence differentiation

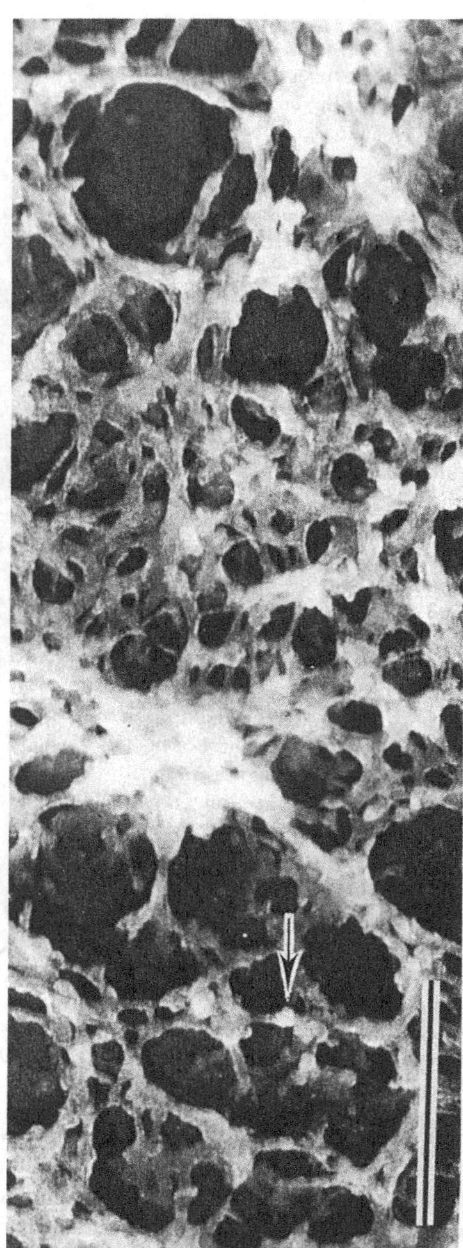

FIGURE 9.17 The microtrabecular lattice in a parenchyma cell of *Zea mays*. Bar, 500 nm. *From Wardrop, A.B., 1983. Evidence for the possible presence of a microtrabecular lattice in plant cells. Protoplasma 115, 81–87.*

of cell types. Any strains in the cytoplasm will be transmitted to the nucleus (Ofek et al., 2009; Wang et al., 2009; Isermann and Lammerding, 2013; Davidson and Lammerding, 2014), which may have an effect on gene expression (McBeath et al., 2004; Endler et al., 2006; Lee et al., 2013; Tatout et al., 2014; Ivanovska et al., 2015; Tamura et al., 2015; Zhou et al., 2015a,b). Moreover, cell differentiation may be brought about by the differential localization of various determinants in embryos (Wilson, 1925; Davison, 1976). These cytoplasmic determinants appear to be, in part, maternal messenger RNA molecules. The cytoplasmic determinants are localized in various regions of the egg cell and are selectively distributed to particular embryonic cell lineages where they may initiate specific developmental programs (Jeffrey, 1982, 1984a,b, 1985; Jeffrey and Wilson, 1983; Jeffrey et al., 1983; Jeffrey and Meier, 1984; Swalla et al., 1985). The eggs have three distinct regions with specific morphogenetic fates: the ectoplasm, endoplasm, and myoplasm. I will only discuss the myoplasm, which gives rise to muscle cells. When radiolabeled poly-U is used as an in situ hybridization probe for poly-A RNA, a general feature of all messenger RNAs (see Chapter 16), it is found that 45% of the poly-A RNA is in the ectoplasm, 50% is in the endoplasm, and 5% is in the myoplasm. However, when a radiolabeled actin cDNA is used, it is found that 40% of the actin mRNA is present in the ectoplasm, 15% is in the endoplasm, and 45% is in the myoplasm. Thus, the myoplasm, which is destined to be muscle, is specifically enriched in actin mRNA.

When the eggs are extensively extracted with Triton X-100 so that almost everything in the cell is washed away, a detergent-resistant lattice remains. This lattice includes actin, tubulin, and intermediate filaments. When this lattice is probed with radiolabeled cDNA using in situ hybridization, it is found that the actin mRNA is bound to the lattice and remains in the same position it was in before detergent extraction. It appears that the cytoplasmic lattice can help maintain a polarity in cells so that the daughter cells of a division get unequal components, which may determine their developmental fate.

9.5 SUMMARY

We have seen from cytological evidence provided by Walther Flemming and Eduard Strasburger; biophysical evidence provided by William Siefriz, Noburô Kamiya, Eiji Kamitsubo, Kate Luby-Phelps, Arthur Hughes, and Francis Crick; and electron microscopic evidence provided by Keith Porter and Alan Wardrop that the cytoplasm consists of a three-dimensional network of fibrous elements of unknown composition. This anatomizing reticulum may form a structure for the enzymes of various biochemical pathways and pari passu provides a resistance to the movement of intracellular macromolecules, vesicles, membranous tubules, and organelles. We see that the cytoplasm behaves as a non-Newtonian fluid. When the rate of shear is low, it behaves as a gel (gelatin) and has a high viscosity. When the rate of shear is high, it behaves as a sol (solution) and has a low viscosity. We see that the cytoplasm has a yield value, and at shearing stresses below the yield value, the cytoplasm provides such a high resistance to movement that increasing the shearing stress up to the yield value does not induce movement. Movement only occurs when the shearing stress is greater than the yield value. The low *Re* tells us that inertial forces are essentially nonexistent and viscous forces predominate. Thus, there is no inertia in the cell. Movement requires the application of a constant force, and it will stop instantly on the removal of the force.

In order to induce a vesicle to move in the cytoplasm, we have to apply a force per unit area on the vesicle that is greater than the yield value of the cytoplasm (0.5 Pa) and we have to continue to apply this force for as long as we want the vesicle to move. Thus, we need intracellular motors to move the vesicles. We have at least two classes of motors available: one that uses microfilaments as a track and one that uses microtubules as a track. These motors are discussed in Chapters 10 and 11.

The concentration of Ca^{2+} in the cytosol is approximately 10^{-7} M, whereas outside it is much higher. The concentration of H^+ in the cytosol is approximately 10^{-7} M, whereas outside it is probably much higher. The concentration of K^+ in the cytosol is approximately 10^{-1} M, whereas outside it is probably much lower. The concentration of ATP, amino acids, and sugars in the cytosol is between 10^{-4} M and 10^{-2} M, whereas outside their concentrations are much lower. The plasma membrane, in part energized by the H^+-pumping ATPase, sustains these great differences between the cytosol in the tiny living intracellular space and the enormous nonliving extracellular space. A puncture of the plasma membrane will result in an equalization of each chemical between the living and liveless spaces. Consequently, as Ca^{2+} increases in the living space, the cytoplasm will gelate, lose its elasticity, and will become lifeless.

9.6 QUESTIONS

9.1. Would it be productive today to try to discover the part of the protoplasm that is essential for life? Why or why not?

9.2. How would cellular processes differ if the cytoplasm were not viscous?

9.3. How would cellular processes differ if the cytoplasm were not thixotropic?

The references for this chapter can be found in the references at the end of the book.

Chapter 10

Actin- and Microfilament-Mediated Processes

You got to move
You got to move
You got to move, child
You got to move
But when the Lord
Gets ready
You got to move.

Mississippi Fred McDowell

Some fifteen years ago, I was convinced that there could be net principal difference between 'cabbages and kings'. Cabbages, then, being cheaper, I began to study cabbages. Biochemistry teaches us that many constituents of our body are found with equal frequency in plants and animals, fulfilling analogous functions in each. These substances of plants, just as they are, or with little alteration, fit into the machinery of our cells. Two machines, the parts of which are interchangeable, cannot be very different and so anything we learn about the plant will lead us closer to the understanding of ourselves. The plant, as an object of the study of life, has, compared to man, some very great advantages. The plant can dispense with many of the unessential complications found in our body which enable us to walk, hear, see, smell and think. Life in the plant will present itself in much simpler forms, and thus allow the great fundamental principles to come to the fore…

Albert Szent-Györgyi (Whatever a cell does…)

10.1 DISCOVERY OF ACTOMYOSIN AND THE MECHANISM OF MUSCLE MOVEMENT

Movement is one of the most easily distinguished characteristics of life. Theodor Engelmann (1879) noticed all kinds of motion in plants and protozoa, including amoeboid movement and cytoplasmic streaming (Fig. 10.1). He suggested that these activities might be a primitive version of the specialized movements that occur in muscle, and indeed, the same molecular mechanisms may be involved in them all. Seventy years later, Albert Szent-Györgyi (1949b) put it this way:

> *All living organisms are but leaves on the same tree of life. The various functions of plants and animals and their specialized organs are manifestations of the same living matter. This adapts itself to different jobs and circumstances, but operates on the same basic principles. Muscle contraction is only one of these adaptations.*

If all life shows motion, which cell, tissue, organ, or organism shall we choose to study to unravel the mysteries that underlie the vital process of movement in living organisms and to give us the clearest and most profound answers? Szent-Györgyi (1948) suggests that we use the cells that are most specialized for movement: skeletal muscle. The excitement of some of the pioneers in muscle research has been captured in their published lectures and monographs (Szent-Györgyi, 1947, 1948, 1953a,b; Mommaerts, 1950b; Weber, 1958; Huxley, 1966, 1969, 1996; Needham, 1971; Kaminer, 1977; Huxley, 1980; Straub, 1981; Engelhardt, 1982; Weber, 1988; Szent-Györgyi, 2004; Szent-Györgyi and Bagshaw, 2012; Franzini-Armstrong, 2018).

While most biochemists in the 1930s were studying water-soluble enzymes, the husband-and-wife team of Vladimir Engelhardt and Militza Ljubimowa violated one of the canons of biochemistry and studied the "residue instead of the extract" (Engelhardt, 1946, 1982). In those days, following the acceptance of Sumner's (1926, 1937) work, the residue was thought to be composed of mundane structural proteins and not exciting enzymes. However, while studying muscle, Engelhardt and Ljubimowa (1939) found that myosin, a "structural" protein that had previously been isolated from muscle by Wilhelm Kühne (1864), was also an enzyme capable of hydrolyzing adenosine triphosphate (ATP). Thus myosin had the properties to convert he chemical energy of ATP into mechanical work. Engelhardt (1946) compared myosin with a piston of a combustion engine and ATP with the explosive mixture.

Plant Cell Biology. https://doi.org/10.1016/B978-0-12-814371-1.00010-2

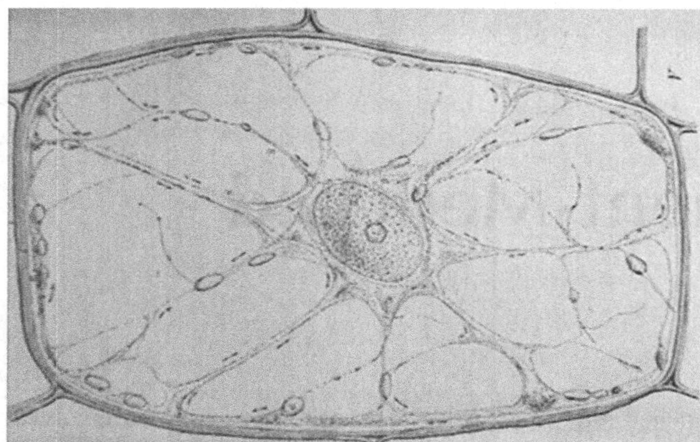

FIGURE 10.1 Cytoplasmic streaming in a parenchyma cell. *From von Hanstein, J., 1880. In: Frommel, W., Pfaff, F. (Eds.), Das Protoplasma als Träger der pflanzlichen und thierischen Lebensverrichtungen. Für Laien und Sachgenossen dargestellt. From Sammlung von Vorträgen für das deutsche Volk. Winter, Heidelberg.*

Szent-Györgyi became interested in muscle after he read about the ATPase activity of myosin. He thought that myosin might be the mechanochemical transducer that coupled the chemical energy of ATP to the mechanical energy of contraction, and he set out to test his hypothesis. Realizing that he was standing on the shoulders of giants, Szent-Györgyi repeated the work of the "old masters" and isolated myosin using the method of Engelhardt and Ljubimowa (Szent-Györgyi and Banga, 1941), He extracted the muscle for an hour with an alkaline 0.6 M KCl solution to get the typical syrupy myosin preparation. He precipitated the myosin by diluting the solution to 0.1 M KCl. Szent-Györgyi then put the precipitated threads of myosin on a slide and watched them under a microscope. Then he added ATP to the slide and, mirabile dictu, they contracted! It was as if he had seen life itself!

Ilona Banga continued to isolate myosin in Szent-Györgyi's laboratory, but she had to go home early one day and left the minced muscle in KCl all night. The next morning they realized that the extract was thicker than the usual extract and it also contracted more vigorously on the addition of ATP. They called the original extract myosin A and the thick extract myosin B. It turned out that the difference between the two extracts was that myosin A was extracted while the muscle still contained ATP, and myosin B was isolated after all the ATP had been hydrolyzed.

Szent-Györgyi suggested that Ferenc Brunó Straub investigate the difference between the weakly contracting myosin A and the forceful myosin B (Straub, 1981). Straub postulated that myosin B was enriched in a protein that was a contaminant in myosin A. Unbeknownst to Szent-Györgyi and Straub, the protein contaminant had been isolated by Halliburton in 1887 under the name *myosin-ferment* (see Finck, 1968). Straub extracted an ATP-containing muscle with 0.6 M KCl and then washed and dried the remaining muscle with acetone. The acetone powder, which was more or less the residue of the residue, was then extracted with water and a protein went into solution. This protein solution, when added to myosin A in the presence of ATP, caused the myosin to contract. Straub named this protein *actin* because it caused myosin to go into action (Bendiner, 1982; Moss, 1988; Rall, 2018), and then he and Szent-Györgyi renamed myosin B *actomyosin*. Actin had the ability to activate the ATPase activity of myosin by about 10-fold, in addition to being able to cause the actomyosin mixture to contract.

Szent-Györgyi resurrected an earlier proposal by Karl Lohmann (Meyerhof, 1944), the discoverer of ATP, that the chemical energy of ATP provided the energy for muscle contraction, and moreover, that muscle contraction was essentially due to the interaction of actomyosin and ATP. However, this conclusion was not widely accepted for a number of reasons, one of which was that the magnitude of the free energy released by the measured amount of ATP hydrolyzed was insufficient to account for the work performed by the contracting muscle (Mommaerts and Seraidarian, 1947; Perry et al., 1948; Hill, 1949; Mommaerts, 1950a; Szent-Györgyi, 1963a; Gergely, 1964).

Szent-Györgyi decided to demonstrate beyond a shadow of a doubt that ATP provides the chemical energy for contraction. Szent-Györgyi (1949a) and Varga (1950) developed a glycerinated muscle preparation. They extracted the muscle with 50% glycerol at low temperatures to make a permeabilized cell model (Arronet, 1973). Then, on addition of ATP, the model contracted and developed the same tension as if it were an intact muscle. Contraction is thus due to the conversion of the chemical energy of ATP to the mechanical energy of muscle contraction. The inability to detect the relationship between free energy release from ATP and work was due to the fact that the magnitude of ATP hydrolyzed by a contracting muscle was underestimated because in muscle, ATP is constantly being regenerated through a creatine phosphate system.

From the first observation of the contraction of actomyosin threads under the microscope, Szent-Györgyi (1948) believed that the proteins themselves contracted. Physical- and bio-chemists accepted his conjecture (Katchalsky and Lifson, 1954). However, structural data, which included X-ray diffraction images, as well as polarization, interference, and electron microscopic images obtained by Jean Hanson, Hugh Huxley, and Andrew Fielding Huxley (unrelated to Hugh, but related to T.H., Julian and Aldous), indicated that the contractile proteins were not contractile at all but slide past each other when they effected the shortening of a muscle (Hanson and Huxley, 1953, 1955; Huxley and Hanson, 1954, 1957; Huxley and Niedergerke, 1954; Maruyama, 1995).

Hugh Huxley was one of the nuclear physicists who left physics and entered biology in the late 1940s after the mass killing of Japanese by the atomic bomb. After all, the bomb was, and still is, the most visible by-product of nuclear physics. Will a similar migration occur from biology into fields concerned with human understanding if we allow genetically engineered diseases to be released accidentally or in an act of war? At present, there is a paucity of discussion on the ethical concerns of basic biological research (Bush, 1967; Chargaff, 1976), even though scientists should continually examine and reexamine the fruits of their labors and take responsibility for them (Williams, 1993a). Anyhow, Hugh Huxley decided to enter biology and figure out how muscles worked by combining the power of William Astbury's (1947a,b) X-ray diffraction technique, which Huxley believed provided true data in an enigmatic form, with the power of electron microscopy, a technique that provided tangible images even though, at that time, the images were laden with artifacts. Huxley decided to take a multidisciplinary approach, where he himself became well versed in many aspects of science. He already knew X-ray diffraction, and he went to the Massachusetts Institute of Technology to learn electron microscopy from Frank Schmitt (1990; Adelman and Smith, 1998). With his multidisciplinary approach, where he himself understood and combined many techniques, as opposed to an interdisciplinary approach, where each member of a team is an expert in a given technique, Hugh Huxley now saw in electron micrographs double hexagonal arrays of thick and thin filaments that he had deduced from the X-ray diffraction patterns of living and rigor muscles (Holmes, 2013a,b; Pollard and Goldman, 2013; Spudich, 2013; Weeds, 2013).

Although structural studies on muscle began in the 19th century when histologists observed that muscle cells contained repeating units called *sarcomeres* (Schäfer, 1902; Schmidt, 1924, 1937; Frey-Wyssling, 1975), they were forgotten or unknown to the muscle biochemists (Huxley, 1980). The observations by Theodor Engelmann that the sarcomeres, which were separated by the "in between bands" or Z-bands (zwichen-bands), were birefringent under a polarizing microscope were repeated by the structural biologists in the 1950s (Fig. 10.2). The birefringent area was given the English name *A-band* (meaning anisotropic) and the two nonbirefringent areas between the Z-bands and the A-band were given the name *I-bands* (meaning isotropic). X-ray diffraction data confirmed that the A-band had a repeating structure and provided data on the size and distribution of the repeating units.

Electron microscopy showed that the A-band was composed of thick filaments approximately 16 nm in diameter and 1.6 μm in length, and the I-bands were composed of thin filaments that were 5−6 nm in diameter and about 1 μm in length (Fig. 10.3). The thick filaments also contained globular regions, some of which made cross-bridges with the thin filaments. Studies utilizing phase and interference microscopy showed that treating the muscle fibers with high salt, which caused the extraction of myosin, simultaneously resulted in the disappearance of the A-bands! Longer extractions, which resulted in the subsequent loss of actin, also caused the I-bands to disappear. These results indicated that the thick filaments were made out of myosin, and the thin filaments were composed of actin. These results were later confirmed in situ using immunolocalization techniques.

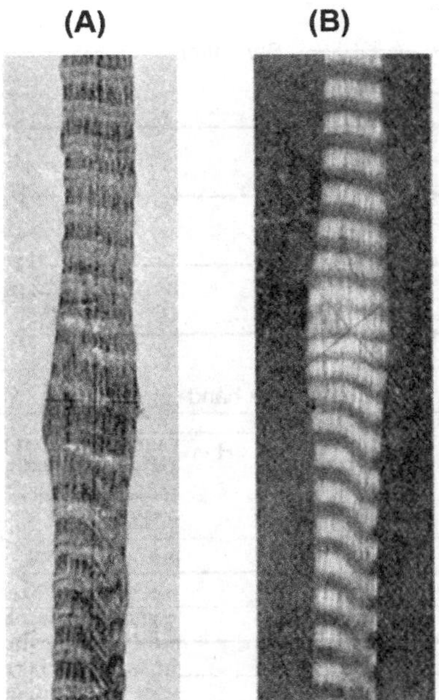

FIGURE 10.2 Photomicrographs taken by Professor Engelmann of a leg muscle fiber from *Chrysomela coerulea* observed with a polarized light microscope with (A) parallel and (B) crossed polars. *From Schäfer, E.A., 1902. Essentials of Histology. Sixth edition. Longmans, Green, and Co., New York.*

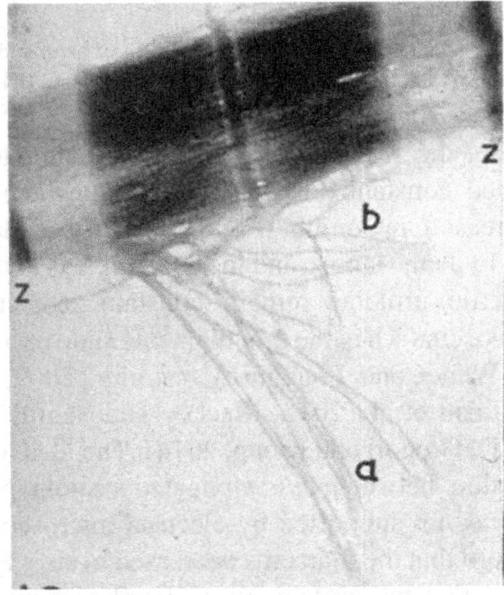

FIGURE 10.3 A myofibril from a toad muscle showing one sacromere. This sacromere has frayed, showing (a) the filamentous nature of its components. The Z-band, z; the A-band, b; actin microfilaments. ×28,000. *From Hodge, A.J., 1956. The fine structure of striated muscle. J. Biophys. Biochem. Cytol. 2 (4 Suppl.), 131−142.*

Phase and interference microscopy showed that the length of the A-bands and the distance between the Z-band and the edge of the H-band (an area of variable width in

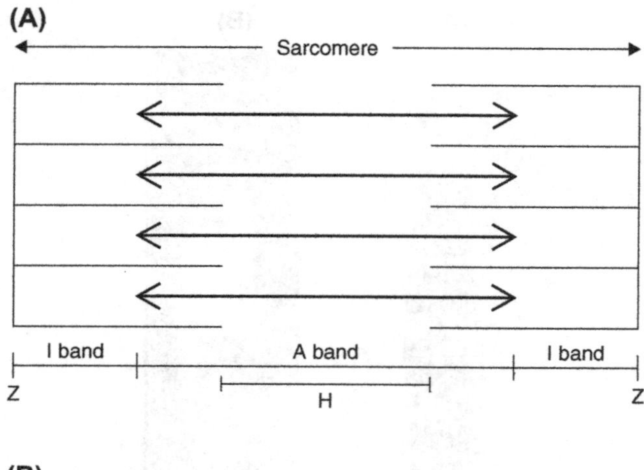

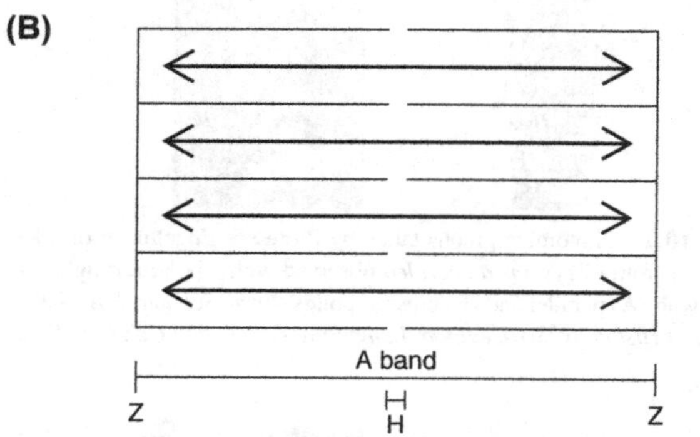

FIGURE 10.4 Diagram of the relative movement of actin and myosin as described by the sliding filament hypothesis: (A) relaxed and (B) contracted.

the middle of the A-band where the actin filaments do not reach) stayed constant during contraction, whereas the I-bands decreased in length (Fig. 10.4). These data were interpreted by Jean Hanson and the two Huxleys to mean that the contractile proteins remain constant in length, but contraction occurs when the thin filaments slide past the thick filaments (Weber and Franzini-Armstrong, 2002; Huxley, 2004; Goldman et al., 2012; Mackey and Santillán, 2013; Hitchcock-DeGregori and Irving, 2014). The idea of relative sliding motion between two elongated protein polymers, however, was not supported by electron microscopic data, which showed that the filaments decreased in size. However, it turned out that the proteins were depolymerizing during fixation, and later, good fixation procedures revealed that the filaments do not change size during contraction.

The filamentous nature of purified actin and myosin can be observed in the electron microscope using negatively stained preparations. Under high-salt conditions, actin forms filaments known as *F-actin*, and under low-salt conditions, the filaments depolymerize into globular subunits known as *G-actin*. An individual myosin molecule is a polar filamentous structure with two globular heads on necks and a long tail. Under physiological conditions, the myosin molecules join together to form a bipolar thick filament where the head groups are at the ends of the filament.

Further support for the sliding filament model came from experiments that showed that the actin filaments have a polarity, and moreover, the actin filaments on each side of the sarcomere are antiparallel. This was discovered as a result of Andrew Szent-Györgyi's (1953a,b) research on the proteolytic cleavage products of purified myosin. He found that the treatment of myosin with trypsin yields a rodlike segment known as *light meromyosin* and a head region known as *heavy meromyosin*. Huxley (1963) treated isolated Z-bands with the heavy meromyosin fragment and observed them in the electron microscope. He noticed that the heavy meromyosin bound to the actin filaments and decorated them with an arrowhead-like arrangement where the arrowheads pointed away from the Z-bands. This meant that the actin filaments had a polarity. During contraction, the myosin moves along the actin filament from the end farther away from the Z-band (the pointed end or minus end) toward the end nearer the Z-band (the barbed end or plus end).

Following the work of Jean Hanson, Hugh Huxley, Andrew Huxley, and Rolf Neidergerkie, the sliding filament model became universally accepted. Subsequently, biophysicists and biochemists have worked to understand how the chemical energy of ATP is converted into the mechanical energy of actomyosin by studying actomyosin kinetically (Lymm and Taylor, 1971) and structurally (Rayment et al., 1993). For the myosin molecule to generate movement along an actin microfilament, it must bind ATP. This causes the myosin to break its tight binding to the actin filament. Subsequently, myosin hydrolyzes the ATP and undergoes a conformational change so that the head is adjacent to the next actin monomer. Then the myosin releases the terminal phosphate of the ATP and binds actin tightly. This tight binding initiates a ratcheting of the myosin molecule that results in the power stroke and the release of adenosine diphosphate (ADP). The myosin head continues to bind tightly to the actin filament until it binds to another molecule of ATP, and the rowing motion continues as the myosin moves from the minus end of an actin filament to the plus end. When a cell dies and no longer produces ATP, the myosin head can no longer dissociate from the actin filament and the cell becomes nonelastic, a state known as rigor mortis. Andrew Huxley (1980) considers myosin to be a step-down transformer that converts the very strong chemical forces (involved in the hydrolysis of ATP) that act over a short distance (0.1 nm) into a much weaker mechanical force that acts over a greater distance (5 nm).

10.2 ACTIN IN NONMUSCLE CELLS

Actin is not only found in muscle cells but also occurs in all eukaryotic cells. Actin is one of the most abundant proteins in the world, second only to RuBP carboxylase (see Chapter 13). Actin has been purified from a number of cells, including pollen, root cells, protozoa, and slime molds, and it can make up as much as 5% of the cellular protein (Loewy, 1954; Ts'o et al., 1957; Nakajima, 1960; Hatano and Tazawa, 1968; Adelman and Taylor, 1969; Vahey and Scordilis, 1980; Vahey et al., 1982; Ma and Yen, 1989; Liu and Yen, 1992; Andersland and Parthasarathy, 1992, 1993; Andersland et al., 1992; Igarashi et al., 1999).

Actin filaments, or *microfilaments* as they are called, can be observed in nonmuscle cells at the electron microscopic level (Wohlfarth-Bottermann, 1962; Porter et al., 1965; Rhea, 1966; O'Brien and Thimann, 1966; Nagai and Rebhun, 1966; Parthasarathy and Mühlethaler, 1972; Lancelle et al., 1986, 1987, 1989; Ding et al., 1991a,b). The actin filaments in nonmuscle cells, like their muscle counterparts, have the ability to bind heavy meromyosin (Ishikawa et al., 1969; Nachmias et al., 1970; Condeelis, 1974) and the S-1 subfragment of myosin (Igarashi et al., 1999; Tominaga et al., 2000b; Lenartowska and Michalska, 2008). The arrowhead decorations indicate that nonmuscle actin filaments are also polar.

10.2.1 Temporal and Spatial Localization of Actin in Plant Cells

Actin filaments in nonmuscle cells form the actin cytoskeleton. However, the actin cytoskeleton in nonmuscle cells is a dynamic structure. The actin filaments polymerize and depolymerize and consequently appear in various places around the cell in a cell cycle–dependent manner. The three-dimensional architecture of the microfilament-based cytoskeleton can be visualized by fixing plant cells and immunolabeling the actin microfilaments (Lovy-Wheeler et al., 2005) or by treating fixed cells with fluorescently labeled phalloidin, a fungal toxin from *Amanita* that specifically binds to filamentous actin (see Fig. 10.5; Barak et al., 1980; Nothnagel et al., 1981, 1982; Parthasarathy, 1985; Parthasarathy et al., 1985). The introduction of this technique into plant biology resulted in an explosion of papers where the architecture of the actin cytoskeleton has been demonstrated in hundreds of cell types (Lloyd, 1987, 1988, 1989). Peter Hepler and his colleagues then developed the technology to microinject fluorescently labeled phalloidin or other actin-binding fluorescent molecules so that the dynamic aspects of actin filaments can be observed in living cells (Zhang et al., 1992, 1993; Meindl et al., 1994; Wasteneys et al., 1996; Vidali et al., 2009c). Actin can also be visualized in live cells that have been transformed with genes coding for fusion proteins composed of actin-binding proteins and green fluorescent protein (Kost et al., 1998; Kost et al., 1998; Cheung and Wu, 2004; Yamashita et al., 2011; Rosero et al., 2014). Seeds for plants expressing the fluorescent fusion proteins are available from stock collections (http://www.arabidopsis.org/ and http://arabidopsis.info/)

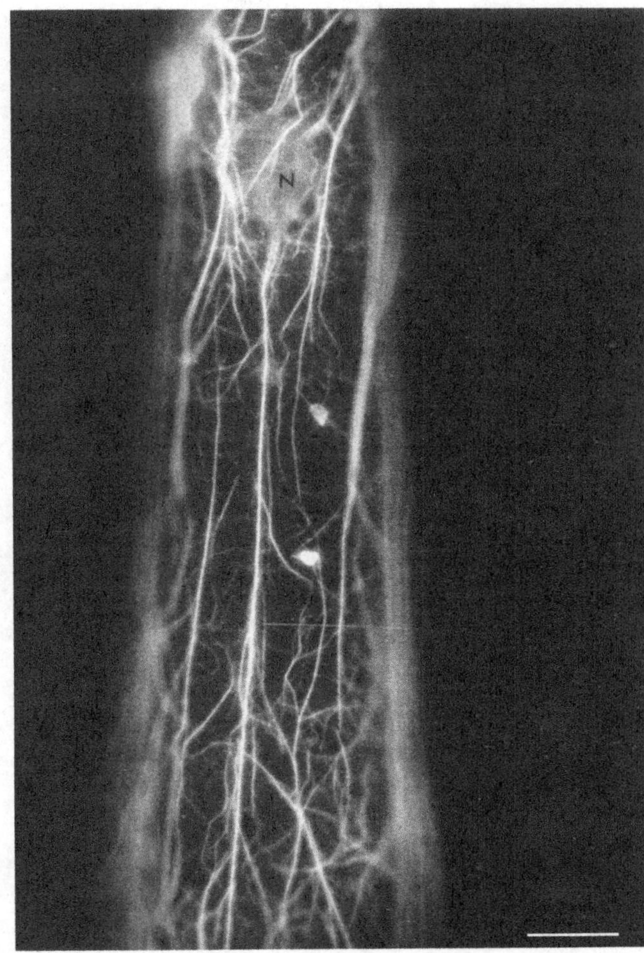

FIGURE 10.5 F-actin in a stem hair cell of a tomato. F-actin is stained with rhodamine phalloidin and viewed with a fluorescence microscope. Nucleus, N. Bar, 20 μm. *From Parthasarathy, M.V., Perdue, T.D., Witztum, A., Alvernaz, J., 1985. Actin network as a normal component of the cytoskeleton in many vascular plant cells. Am. J. Bot. 72, 1318–1323.*

While the actin cytoskeleton has a unique arrangement in each cell type, there are some basic similarities in all cells. For example, in the interphase cells of higher plants, the actin microfilaments typically appear either transverse to the long axis of the cell or as a random mesh in the cortical cytoplasm. Before the cell enters prophase, the actin forms a band that predicts the future site of cell division, and then the actin microfilaments disappear from the middle of the band (Karahara, 2014). During metaphase, there are actin filaments near the spindle poles and parallel to the spindle fibers. Some actin filaments are in the spindle, particularly associated with kinetochore fibers. During anaphase, the actin filaments become more and more aligned with the spindle fibers. In telophase, the actin filaments become part of the phragmoplast. These actin filaments are parallel to the spindle and connected to the reforming cortical actin network. Double labeling suggests that the actin in the phragmoplast is newly assembled during cell plate formation and does not come from the spindle-associated microfilaments (Gunning and Wick, 1985; Kakimoto and Shibaoka, 1987, 1988; Seagull et al., 1987; Traas et al., 1987; Clayton and Lloyd, 1985; Palevitz, 1988a; Schmidt and Lambert, 1990; Zhang et al., 1992). As cells elongate and stop dividing, the actin filaments are typically longitudinally oriented throughout the ectoplasm and traverse the transvacuolar strands (Thimann et al., 1992; Shimmen et al., 1995). There are reports that actin also occurs in mitochondria (Lo et al., 2003) and nuclei (Paves and Truve, 2004).

The arrangement of the actin cytoskeleton in epidermal cells is correlated with the ability of the cells to elongate. In cells that elongate slowly, the actin microfilaments are organized in dense bundles. On activation of elongation by continuous far-red light, which is mediated by phytochrome, the bundles split into fine strands (Waller and Nick, 1997). A similar response occurs after the addition of the elongation-inducing hormone auxin (Wang and Nick, 1998).

10.2.2 Biochemistry of Actin

Structural studies with fluorescently labeled actin filaments indicate that the actin cytoskeleton is an extremely dynamic structure. The biochemistry of actin and its associated proteins provide a molecular mechanism for the dynamic behavior of actin filaments. The behaviors of the replicate actin solutions that Straub (1981) isolated were erratic. Sometimes the solution was highly viscous and sometimes it was not. To increase the reproducibility of the extraction procedure, Straub varied the salt concentration and noticed that he obtained a highly viscous extract when the salt concentration was near 0.1 M NaCl or KCl and a less viscous extract when the salt concentration was lower. The highly viscous extract, but not the fluid one, had fibers that were visible in the electron microscope. The two extracts could be interconverted by adding or dialyzing away the salt. Later Straub discovered that concentrations of NaCl and KCl greater than 0.1 M caused a decrease in the viscosity of the extract and that the salts had an optimal concentration in which they promoted polymerization. The optimal concentration was around the point at which the salts were isoosmotic with intact muscle cells. Straub and Szent-Györgyi postulated that the actin was a fibrous polymer made out of globular subunits and named them *F-actin* and *G-actin*, respectively.

Actin can be purified by allowing it to go through several polymerization–depolymerization cycles using 0.1 and 0.6 M KCl, respectively, and centrifuging the filamentous actin away from the other cellular proteins. The physicochemical properties of purified actin have been studied (Janmey et al., 1990; Janmey, 1991). G-actin has now been studied by X-ray diffraction and it is a bilobed, pear-shaped, 42-kDa globular protein (Kabsch et al., 1990; Holmes et al., 1990; Otterbein et al., 2001; De La Cruz and Pollard, 2001). There are probably many isoforms of actin and actin-related proteins because there are many more or less related genes in a single organism (Meagher, 1991; Frankel and Mooseker, 1996; Meagher et al., 2011).

Each actin monomer is associated with one molecule of ATP. The polymerization of actin is accompanied by the hydrolysis of the terminal phosphate of the bound ATP. However, the energy released by the hydrolysis of ATP is not required for polymerization because either ADP or adenyl-imidodiphosphate, a nonhydrolyzable analogue of ATP, can substitute for ATP in the polymerization reaction.

Actin microfilaments can form and grow in vitro. When the ionic strength of the actin-ATP solution is increased from a nonphysiologically low level to a level closer to the physiological condition, there is a lag phase that reflects the initial step in polymerization. The slow step in the growth of polymeric actin involves the formation of a nucleating site or primer (Cori and Cori, 1939). Once nucleation occurs, polymerization takes place rapidly from the primer by the addition of monomers. The assembly reaction is reversible and eventually the monomer concentration decreases until disassembly proceeds at the same rate as assembly. This monomer concentration is known as the critical concentration (Korn et al., 1987).

The rate of polymerization of actin depends on the concentration of monomers ($[C]$, in M) and the on-rate constant (k_{on}, in $M^{-1}\,s^{-1}$) according to the following equation (Fig. 10.6):

$$\text{rate of polymerization} = k_{on}[C] \quad (10.1)$$

The on-rate constant is a measure of the rate of diffusion of the monomers to the site of polymerization. The rate that actin monomers dissociate from the filament is determined by the off-rate constant (k_{off}, in s^{-1}), which is independent of concentration according to the following equation:

$$\text{rate of depolymerization} = k_{off} \quad (10.2)$$

There is a critical concentration (C_c) where the rate of polymerization equals the rate of depolymerization:

$$k_{on}[C_c] = k_{off} \quad \text{and} \quad [C_c] = k_{off}/k_{on} \quad (10.3)$$

When the monomer concentration is greater than the critical concentration, polymerization continues. When the monomer concentration is less than the critical concentration, depolymerization occurs. At the critical concentration, there is no net filament growth. While polymerization does not require the hydrolysis of ATP, when hydrolyzable forms of ATP are present, as they are in the cell, a new property of actin known as *treadmilling* is exposed. In the presence of ATP, growth takes place at one end, whereas shrinkage occurs at exactly the same rate at the other end. Thus, even though the filament maintains a constant length,

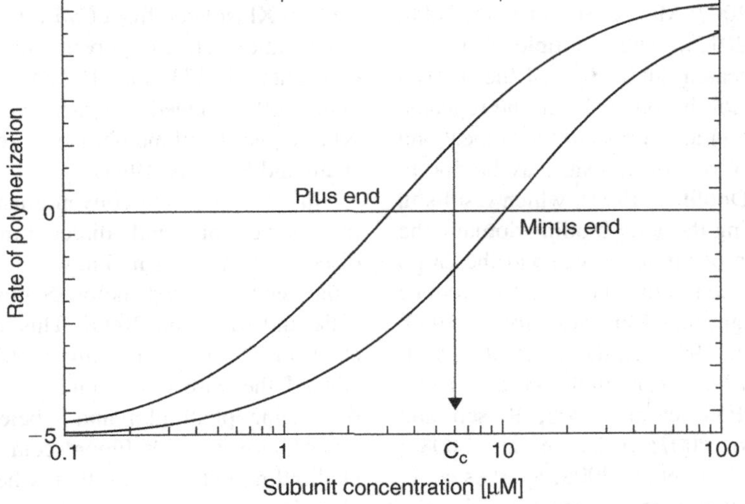

FIGURE 10.6 Graph of rate of polymerization versus concentration of actin subunits.

the individual actin monomers are constantly being transferred from one end to the other. This can be demonstrated by decorating short filaments of actin with heavy meromyosin, which will mark their polarity. The decorated actin filaments are then put back into the polymerizing solution. The end marked by the barb of the heavy meromyosin arrows grows 5–10 times faster than the end marked by the arrowheads (Bonder et al., 1983; Estes et al., 1992). The fast-growing end is known as the *plus end* and the slow-growing end is known as the *minus end*.

Cytochalasins, a family of fungal metabolites that bind to the plus end of actin filaments, prevent their further growth (Cooper, 1987). Mycalolide B or latrunculin, a toxin produced by a sponge, has a similar effect (Shimmen et al., 1995; Saito and Karaki, 1996). Phalloidin, a toxin produced by *Amanita*, on the other hand, stabilizes actin filaments so they cannot depolymerize. Treating motile processes with these pharmacological agents is a good way to test whether or not actin is involved in a given process.

The dynamic behavior of the actin filaments is an intrinsic property of the actin filament itself. However, the behavior of the actin filaments can be further modified by other cellular proteins (Hussey et al., 2002; Staiger and Hussey, 2004; Huang et al., 2011). For example, there are a number of proteins in plant and animal cells, including profilin, which interact with G-actin and prevent it from polymerizing (Giehl et al., 1994; Darnowski et al., 1996; Vidali and Hepler, 1997). Other proteins, including severin, fragmin, and gelsolin, bind to actin filaments and either cap the plus end and prevent polymerization or bind to the middle of the filament and cut it (Weeds and Maciver, 1993). A third class of proteins interacts with actin filaments and induces gel formation. This class of cross-linking proteins includes spectrin (de Ruijter and Emons, 1993; Faraday and Spanswick, 1993). A fourth class of actin-binding proteins, including villin, causes bundling (Yokota and Shimmen, 1999, 2000; Vidali et al., 1999; Yokota et al., 2000, 2003; Tominaga et al., 2000b; Huang et al., 2015). A fifth class of actin-binding proteins, including formin, acts as a primer in that it promotes nucleation of actin microfilaments (Banno and Chua, 2000; Cvrčková, 2000; Ingouff et al., 2005; Vidali et al., 2009b; Ye et al., 2009; Cheung and Wu, 2004; Blanchoin and Staiger, 2010; Cheung et al., 2010; Wang et al., 2012; van Gisbergen and Bezanilla, 2013). Other proteins are involved in the interaction between actin microfilaments and microtubules (Igarashi et al., 2000).

Formin is an integral plasma membrane protein, expressed in the tip of growing pollen tubes that causes the nucleation of actin microfilaments (Cheung et al., 2010). Because formins influence pollen tube growth, Cheung and Wu (2004) suggest that they may recognize extracellular signals from the female tissues to regulate pollen tube growth in the pistil.

10.2.3 Biochemistry of Myosins

Myosins are actin-binding mechanochemical transducer proteins, which are capable of generating force along actin filaments as a result of their ability to hydrolyze ATP (Yokota and Shimmen, 2011). There are at least 24 classes of myosin and probably one or more types occur in all plant and animal cells (Kato and Tonomura, 1977; Ohsuka and Inoue, 1979; Vahey and Scordilis, 1980; Vahey et al., 1982; Parker et al., 1986; Qiao et al., 1989; Kohno et al., 1991; Higashi-Fujime, 1991; Yokota and Shimmen, 1994; Yokota et al., 1995a,b, 1999a,b; Miller et al., 1995; Cope et al., 1996; Plazinski et al., 1997; Kashiyama et al., 2000;

Shimmen et al., 2000; Reddy and Day, 2001b; Lee and Liu, 2004; Li and Nebenführ, 2007; Mooseker and Foth, 2008; Hartman and Spudich, 2012). The multiple forms of myosin may be due to repeated duplication of the myosin gene combined with variation introduced into the repeated gene through mutation. As an alternative to the "repeat and vary" theme, the multiple forms of myosin may be due to the promiscuity of DNA (Doolittle, 1995), which results in the DNA sequences encoding the actin-binding domain, the ATPase activity, the length of the lever arm and the cargo binding regions being mixed and matched to produce each type of myosin (Knight and Kendrick–Jones, 1993; Kinkema and Schiefelbein, 1994; Kinkema et al., 1994; Hasson and Mooseker, 1995; Yamamoto et al., 1995; Yokota et al., 1999a,b; Foth et al., 2006; Reisen and Hanson, 2007; Yamamoto, 2007; Avisar et al., 2008b; Golomb et al., 2008; Hashimoto et al., 2008; Sparkes et al., 2008; Yokota et al., 2009) that is specialized to produce or maintain tension and elasticity in the cell or to pull a specific cargo at a given rate and a given direction. The rate of ATP hydrolysis of the head, the length of the lever arm at the neck, and the resistance produced by the cargo bound to the tail determine the velocity of the myosin. The activity of myosins can be regulated by calcium (Coluccio and Bretscher, 1987; Collins et al., 1990; Szent-Györgi, 1996; Szent-Györgi et al., 1999; Tominaga et al., 2012; see Chapter 12) and through phosphorylation (Karcher et al., 2001).

Myosin II is the conventional myosin found in skeletal muscle. The heavy chain of myosin II is a dimer, each monomer composed of one actin-binding, ATP-hydrolyzing head, one lever-like neck, and a tail. The tails are specialized to form bipolar filaments, which are necessary for skeletal muscle contraction. Myosin II is a plus end–directed motor that moves along an actin filament from the minus end or pointed end when decorated with heavy meromyosin to the plus end or barbed end when decorated with heavy meromyosin. Microscopic assays have been developed to image the hydrolysis of an individual ATP molecule by a single myosin II molecule (Funatsu et al., 1995).

Plants tend to have myosin VIII (Knight and Kendrick–Jones, 1993) and XI (Kinkema and Schiefelbein, 1994; Tominaga and Nakano, 2012), which are known as slow myosin and fast myosin, respectively (Buchnik et al., 2015). Myosin XIII has been found in *Acetabularia* (Cope et al., 1996; Yamamoto, 2008). Plant myosins are unconventional myosins that differ from myosin II in that they do not form bipolar filaments. Plant myosins are like myosin II in that they are plus end–directed motors. Plant myosins are dimers composed of monomers with an actin-binding, ATP-hydrolyzing head domain attached to a neck domain. The neck is attached to a tail domain. The heads and necks serve as the motors and levers, respectively, whereas the tails bind cargo such as membranes and liposomes (Adams and Pollard, 1989; Titus et al., 1989; Miyata et al., 1989; Haydon et al., 1990; Schroer, 1991; Zot et al., 1992). The tail of myosin XI isolated from *Chara* binds to vesicles made from phosphatidlyserine or phosphatidylinositol with dissociation constants of 273 and 157 nM, respectively (Nunokawa et al., 2007). Indeed, Golgi-derived vesicles contain myosin XI as a peripheral membrane protein on the cytosolic leaflet (Fath and Burgess, 1994).

Just as it became conventional wisdom that all myosin motors were plus end–directed motors, Wells et al. (1999) discovered a myosin, known as myosin VI, which is a minus end–directed motor (Schliwa, 1999; Cramer, 2000; Vale and Milligan, 2000). This was discovered by Wells et al. (1999) in an in vitro motility assay in which they labeled the barbed (+) end of actin microfilaments with rhodamine phalloidin and labeled the remainder of the microfilaments with Fluorescein IsoThioCyanate (FITC)-phalloidin and noticed that when these filaments were placed on a slide containing myosin V, a "typical myosin," the pointed (−) end moved first, that is, the myosin walked toward the plus end. However, when they placed the actin microfilaments on a slide coated with myosin VI, the barbed (+) end moved first, indicating that the myosin was walking toward the minus end.

Myosin VIII has been localized at the plasma membrane, in plasmodesmata, in endosomes, and in the developing cell plate (Reichelt et al., 1999; Baluska et al., 2001; Sattarzadeh et al., 2008; White and Barton, 2011; Madison and Nebenführ, 2013; Haraguchi et al., 2014). Myosin XI has been localized to the mitochondria, plastids (Wang and Pesacreta, 2004; Sattarzadeh et al., 2009, 2011, 2013), and the endoplasmic reticulum (ER) and provides the motive force to elongate ER tubules (Yokota et al., 2009, 2011). In addition, knockout mutants indicate that myosin XI is involved in the motility of the Golgi apparatus, peroxisomes, mitochondria, plastids, and the nucleus (Ueda et al., 2015).

10.3 FORCE-GENERATING REACTIONS INVOLVING ACTIN

10.3.1 Actomyosin

I have discussed the force-generating reactions that take place in muscle. Could actomyosin also be involved in moving vesicles through the cytoplasm? Vesicle movement is often inhibited by cytochalasin and latrunculin, as well as the sulfhydryl-binding agent N-ethylmaleimide. These agents are inhibitors of actin and myosin, respectively. Let us see if the interaction of actin and myosin provides enough force to overcome the yield value of the cytoplasm (0.5 N/m^2 or 0.5 Pa; see Chapter 9). Imagine a typical vesicle, with a diameter of 10^{-6} m and a surface area ($4\pi r^2$) of approximately 3.14×10^{-12} m^2, moving through the cytoplasm. It would need a force of (0.5 N/m^2) (3.14×10^{-12} m^2) = 1.6×10^{-12} N or 1.6 pN to overcome the viscous resistance of the cytoplasm (yield value) and move through it. Is this the ballpark for the forces exerted by myosin? How much force can each myosin molecule exert? It is possible to measure the force exerted by a single myosin molecule in a variety of ways.

Studies aimed at measuring the force of a single myosin molecule began when Shimmen and Tazawa (1982a,b) discovered that they could reconstitute cytoplasmic streaming in *Nitella* internodal cells using endoplasm from another source. Then Sheetz and Spudich (1983a,b), Sheetz et al. (1984), and Shimmen and Yano (1984) found that the characean actin bundles would support active streaming of myosin-coated latex beads. Thus, myosin molecules are capable of exerting force when they move along actin bundles in plant cells. Unfortunately, the researchers did not know how many myosin molecules on each bead were in contact with the actin bundles and thus could not determine the force exerted by a single myosin molecule. Chaen et al. (1989) opened up a characean cell to expose the actin cables. Then they placed a myosin-coated microneedle against the actin bundles, and the needle began to bend. The glass needle had an elastic coefficient of 40 pN/μm. If the researchers had known the number of myosin molecules attached to the glass, it would have been possible to calculate the force due to one myosin molecule after measuring how far the needle bent.

It is also possible to cover a glass slide with myosin so that the myosin molecules become attached to the glass slide, and then put fluorescently labeled actin filaments on top of them (Yanagida et al., 1984; Kron and Spudich, 1986). On the addition of ATP, the actin filaments move over the myosin. We can attach a small glass rod to the actin filament and measure how much it bends. Using the elastic coefficient of the glass and the number of myosin molecules touching the actin filament, it is possible to calculate the force due to one myosin molecule. According to Kishino and Yanagida (1988) and Ishijima et al. (1991, 1996), the minimum force that one myosin exerts is 0.2 pN. The movement of actin bundles over immobilized myosin molecules is a good functional assay that can be used for the purification of actin.

Finer et al. (1994) have used laser tweezers to stop actin filaments from moving across a slide sparsely coated with myosin so that only one myosin molecule will attach to an actin filament at a time. In this way, they determined that a single myosin molecule can exert a force of 3—4 pN. Optical trapping measurements of the force exerted by myosin XI show that the maximal force of this myosin is approximately 0.5 pN (Tominaga et al., 2003). Given that the average measured force of a single myosin molecule is 1.8 pN, and the yield value of the cytoplasm is approximately 0.5 Pa, a single myosin molecule would be capable of moving a typical 1 μm in diameter vesicle through the cytoplasm. Because of the low Reynolds numbers, which indicate that the viscous forces in the cytoplasm are greater than the inertial force exerted by myosin, myosin molecules must continually exert a force or the vesicles will stop.

Based on the results of in vitro motility assays, Leibler and Huse (1991, 1993) have come up with a theory of motor proteins that provides an understanding of the kinetics of the mechanochemical cycle that goes beyond that deduced by Lymm and Taylor (1971) for myosin in solution. Leibler and Huse conclude that when a motor protein, such as a single molecule of myosin, pulls a vesicle or organelle through the cytoplasm, it must remain attached to the actin microfilament for the majority of the mechanochemical cycle or the load will diffuse away. By contrast, a motor such as a myosin II molecule, which in skeletal muscle works in concert, yet asynchronously, with other myosin II molecules, must detach from the actin filament for a considerable portion of its mechanochemical cycle to not increase the friction against which the other myosin molecules must work. The details of the mechanochemical cycle of *Chara* myosin, which is approximately 20 times faster than skeletal muscle myosin, are still unknown (Higashi-Fujime et al., 1995; Uyeda, 1996; Kashiyama et al., 2000).

Szent-Györgyi (1947) wrote "Like most children, the biochemist, when he finds a toy, usually pulls it to pieces, and he can seldom keep his promise to put it together again." However, we can see that very definite progress is being made when it comes to reconstituting actin-based motility systems!

10.3.2 Polymerization of Actin Filaments

The mere polymerization of actin can provide a force (Tilney, 1983; Mahadevan and Matsudaira, 2000). This is well documented in the acrosomal reaction of some invertebrate sperm (Tilney, 1976). The acrosomal region of the sperm of *Thyone* is packed with monomeric actin. This actin stays as a monomer because it is bound to profilin, a protein that prevents the polymerization of actin (May, 2008). When the sperm touches the egg, the pH of the sperm cytoplasm rises, and the actin dissociates from the profilin and rapidly polymerizes at a rate of approximately 9 μm/s into a long thin acrosomal process "which punctures the egg coat like a harpoon." This allows the membranes of the egg and sperm to fuse.

The polymerization of actin in the host cell is responsible for providing the motive force for the movement of *Listeria* and *Shigella*, bacterial pathogens (Tilney and Portnoy, 1989; Tilney et al., 1990a, 1992a,b; Theriot et al., 1992; Welch and Way, 2013). The bacterium is taken up into a macrophage by phagocytosis. Subsequently, the phagosomal membrane dissolves and the bacterium causes the nucleation of actin filaments. The polymerization of actin provides the force necessary to propel the bacterium into an extended region of the cell. A neighboring macrophage then takes up the cell extension that contains the bacterium by phagocytosis, and the cycle continues. Other bacteria also harness the power of actin polymerization to propel them from cell to cell (Laine et al., 1997). It appears that viruses may also take advantage of the cytoskeleton to move around a cell (McLean et al., 1995).

10.4 ACTIN-BASED MOTILITY

Many motile processes in plant cells depend on the actin cytoskeleton (Grolig, 2004). Surveying cell motility as a whole, one cannot but be impressed by the relative simplicity in the arrangement, in as much as the total number of proteins necessary to accomplish a wide variety of motile processes throughout living cells is unexpectedly small (see Krebs and Kornberg, 1957). Actin can be definitively considered to be involved in a given process in a cell based on the following criteria:

- Descriptive analysis of the location and pattern of microfilaments
- Demonstration that actin antagonists inhibit the observed response
- Demonstrate that motility is impaired in vivo under nonpermissive conditions in conditional actin or myosin mutants, when cells express RNAi for actin or myosin, or that motility is impaired when the actin or myosin gene is edited with the Crisper/Cas 9 system (see Chapter 21).
- Test the characteristics of the system in a cell model
- Isolate the proteins involved in the process
- Reconstitute of a functional system

Given these criteria, cytoplasmic streaming is the best-characterized actin-based motile process in plants (Shimmen and Yokota, 2004; see Fig. 10.7).

10.4.1 Cytoplasmic Streaming

Cytoplasmic streaming is one of the most unforgettable processes that can be seen under the microscope, and a sensational and must-read account of it has been written by T.H. Huxley (1890). Cytoplasmic streaming, which occurs in almost all plant cells, facilitates the transport and mixing of substances in large cells by causing convection, which is much faster than diffusion (Darwin and Acton, 1894; Pickard, 1974; Hochachka, 1999; Goldstein et al., 2008; Verchot-Lubicz and Goldstein, 2010; Goldstein and van de Meent, 2015). Even the smell of smoke or perfume would take hours to cross a room if its movement depended on diffusion alone (Clausius, 1859, 1860, 1879; Maxwell, 1873, 1878; Garber et al., 1986). Cytoplasmic streaming also occurs in the large embryonic cells of animals, where diffusion would be rate limiting (Hird and White, 1993; Cramer et al., 1994; Niwayama et al., 2011, 2016).

The time it takes for a substance to diffuse a given distance can be calculated by Einstein's (1906) random walk equation:

$$t = x^2/(2D) \tag{10.4}$$

Given that $D = kT/(6\pi r_H \eta)$, T is usually close to 300 K, the microviscosity of the cytoplasm is approximately 0.004 Pa s, the radius of typical atoms and molecules falls between 10^{-10} and 10^{-9} m, and the diffusion coefficient of low–molecular mass molecules in the cell is typically between 0.5 and 5×10^{-10} m^2/s. If $D = 10^{-10}$ m^2/s, it would take 0.5 s to diffuse across a 10-μm long cell and 5×10^7 s (≈ 1.5 years) to diffuse from one end to the other in a 10-cm long characean cell.

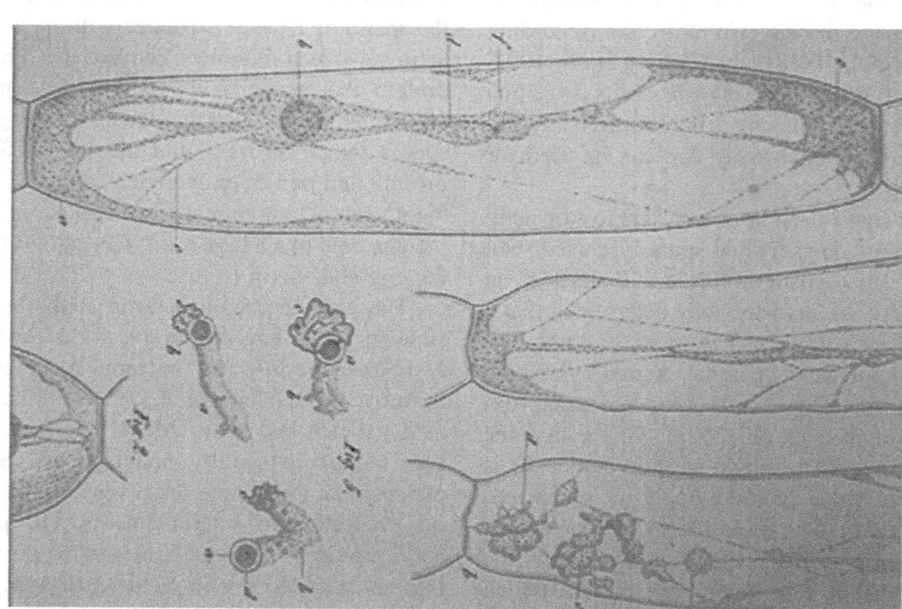

FIGURE 10.7 Cytoplasmic streaming in *Tradescantia virginica*: (A) plasma membrane, (B) nucleus, (C) protoplasm, (D) contracted area of protoplasm, (E and F) areas of flowing protoplasm. *From Kühne, W., 1864. Untersuchungen Über Das Protoplasma Und Die Kontraktilität. W. Engelmann, Leipzig.*

It can be seen that the time increases with the square of the distance, so movement of a given substance across a cell will be 100 times slower for a 10-μm cell than for a 1-μm cell and 100 times slower for a 100-μm cell than for a 10-μm cell. It will be really slow in a 100,000-μm long characean cell. Thus, it is understandable that the rate and organization of cytoplasmic streaming, as I will discuss in the following, are related to cell size.

The fascinating movements of the cytoplasm were first observed by Bonaventura Corti in 1774 and later by Giovanni Amici (1818) after he invented the achromatic lens. There are many manifestations of cytoplasmic streaming from the slow saltatory movement found in the small cells of *Spirogyra* to the rapid rotational streaming found in the giant *Chara* cells (Hofmeister, 1867; Berthold, 1886; Hörmann, 1898; Kamiya, 1959; Kuroda, 1990).

The slowest cytoplasmic movements, which occur in small cells like those of *Spirogyra*, are called *agitation* or *saltatory motion*. In this class of streaming, the motion of vesicles is erratic and haphazard, but statistically speaking, not devoid of directional movement. More organized and faster movements, known as circulation streaming, are characteristic of moderate-sized cells having transvacuolar strands such as the hair cells of *Tradescantia*. In these cells, vesicles of various sizes differ in speed and direction as they pass through the cytoplasm, indicating that there is a very complicated network of roads. The long root hair cells of *Limnobium* and the long pollen tubes of *Plantago major* have a type of streaming known as *fountain streaming*, which is highly organized and fast. Fountain streaming results when the stream of cytoplasm flows up the middle of the cell toward the tip and then flows back along the sides, looking much like a fountain in a town square. Reverse fountain streaming, where the flow reaches the apex from the sides, takes place in the long pollen tubes of *Camellia japonica*.

An extremely fast type of streaming, found in the giant internodal cells of characean algae, is called *rotational streaming*, as the protoplasm is limited to the periphery of the cell and it streams like a rotating conveyor belt. The fastest type of streaming is found in the slime mold *Physarum*. In the plasmodial stage of this giant single-celled organism, the cytoplasm moves back and forth in a rhythmic fashion with a velocity as great as 2 mm/s. The stream is reminiscent of a shuttle used in weaving and is thus called *shuttle streaming* (Kamiya, 1940, 1942, 1950a,b,c; Seifriz, 1943b, 1953; Kishimoto, 1958; Mustacich and Ware, 1977b; Newton et al., 1977; Dietrich, 2015). In addition and unrelated to shuttle streaming, a suspended plasmodium will rotate alternately clockwise and counterclockwise in a period of about 2 min (Kamiya and Seifriz, 1954).

Cytoplasmic streaming is affected by plant hormones (Sweeney and Thimann, 1942; Sweeney, 1944; Kelso and Turner, 1955; Ayling et al., 1990; Ayling and Butler, 1993), light (Nagai, 1993), electricity (Tazawa and Kishimoto, 1968), and gravity (Wayne et al., 1990; Staves et al., 1992) and can thus be used as an indicator of how cells respond to these stimuli (see Chapter 12). Moreover, cytoplasmic streaming is an excellent indicator of cell viability and can be used to determine whether or not a given treatment is lethal to a cell.

The velocity of cytoplasmic streaming depends on the magnitude of the inertial motive force and the viscous force that provides the resistance to flow. The motive force results from the conversion of chemical energy into mechanical energy by actomyosin, and the resistance to flow depends on the viscosity of the cytoplasm.

To determine the site where the motive force for streaming is generated, Kamiya and Kuroda (1956) measured the velocity gradient in a single characean internodal cell. They found that the velocity of the ectoplasm is zero and increases from 0 μm/s to about 100 μm/s, depending on temperature, approximately 1 μm into the interior of the endoplasm. Thus, in this region there is a large velocity gradient and a large rate of shear. As the cytoplasm of characean cells is non-Newtonian (see Chapter 9), the viscosity depends on the rate of shear. If the velocity gradient is 100 μm/s per μm, then the rate of shear at the ectoplasmic/endoplasmic interface is 100 s^{-1}, and consequently, the viscosity at the interface is approximately 0.01 Pa s. The velocity of endoplasm itself decreases from 100 to 90 μm/s over 10 μm. Thus, the rate of shear is only about 1 s^{-1}, and its viscosity is high at about 0.8 Pa s. Thus, the internal friction of the endoplasm resists the flow induced by the shearing stress.

The flowing endoplasm ruffles the vacuolar membrane and this transmits a force into the vacuole that causes streaming in the cell sap (Staves et al., 1995). The velocities of the cell sap particles are as fast next to the vacuolar membrane as they are in the endoplasm. The velocities of the cell sap inclusions decrease to zero near the middle of the vacuole and then they slowly increase in a symmetrical way, albeit in the opposite direction. A similar velocity gradient can be seen in cytoplasm-rich cells, which have had their vacuole removed by centrifugation, indicating that neither the vacuole nor the vacuolar membrane is a *sine qua non* for cytoplasmic streaming.

From the velocity profiles, Kamiya and Kuroda hypothesized that the ectoplasm/endoplasm interface is the site of the generation of the shear stress. Can you imagine the excitement when Eiji Kamitsubo (1966), using a phase-contrast microscope, first saw linear fibrillar structures at this interface, or when Reiko Nagai and Lionel Rehbun (1966), using electron microscopy, first observed the bundles of 5-nm diameter microfilaments, which were oriented parallel to the direction of flow, at this interface (Kamitsubo, 1972a,b,c, 1980)?

Barry Palevitz and Peter Hepler (1975) decorated the bundles at the ectoplasmic/endoplasmic interface with heavy meromyosin and confirmed that they were actin. Yolanda Kersey et al. (1976) then showed that the pointed ends (the minus ends) are directed away from the direction of streaming. So, characean myosin must move from the pointed (minus) end toward the barbed (plus) end, just as it does in skeletal muscle. Therefore, the microfilaments have the right polarity to act in concert with a plus end–directed myosin to provide the motive force for cytoplasmic streaming according to the sliding filament theory and not the contractile fiber theory (Frey-Wyssling, 1949; Seifriz, 1953). The fact that these bundles bind fluorescently labeled phallotoxins (Barak et al., 1980; Nothnagel et al., 1981, 1982; Foissner and Wasteneys, 2014) and antiactin antibodies (Grolig et al., 1988; Williamson et al., 1986, 1987) provides further evidence that these microfilaments are composed of actin.

Myosin is attached to the actin bundles (Yamamoto, 2008) and is attached to the organelles in the flowing endoplasm and may form a network in the endoplasm (Nagai and Hayama, 1979; Grolig et al., 1988). A network of some kind of filament seems to be important in coupling the motive force to the endoplasmic flow (Nothnagel and Webb, 1982). Katcher and Reese (1988) present beautiful pictures showing that the ER may be responsible for coupling the moving endoplasm so that it moves as a whole along the actin bundle (Figs. 10.8 and 10.9). However, I find that cytochalasin causes the separation of water and solids (syneresis) in the endoplasm, indicating that actin filaments may also provide the framework that gives the flowing cytoplasm a high viscosity and couples the bulk of the endoplasm to the moving vesicles.

The evidence that actin and myosin provide the motive force for streaming comes from experiments that show that treatments with cytochalasins, or cytochalasin given with latrunculin, which inhibit actin function, or N-ethylmaleimide, 2,3-butanedione monoxime, and heat, which inhibit myosin function (Chen and Kamiya, 1975, 1981; Kamitsubo, 1981; Kuroda, 1983; Tominaga et al., 2000a; Seki et al., 2003; Funaki et al., 2004; Foissner and Wasteneys, 2007), inhibit streaming. Evidence that a specific myosin is involved in the movement of a particular organelle comes from studies in which a given myosin gene was knocked out, overexpressed, or modified using genetic techniques (Peremyslov et al., 2008), including RNA interference (Avisar et al., 2008b).

The velocity distribution indicates that the endoplasmic layer moves passively as a unit within the ectoplasm. It also indicates that the motive force responsible for this streaming is provided by a shearing stress, generated at the boundary between the cortical gel and the streaming endoplasm (Kamiya and Kuroda, 1956, 1965; Tazawa, 1968, 2011; Donaldson, 1972; Pickard, 1972). The inertial motive force

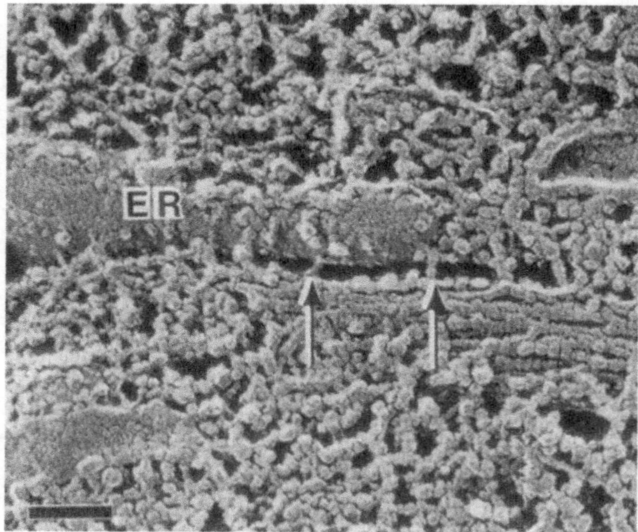

FIGURE 10.9 Freeze-etch micrograph showing the actin bundles and areas where they seem to be attached to the endoplasmic reticulum (ER) (*arrows*). Bar, 100 nm. *From Katcher, B., Reese, T.S., 1988. The mechanism of cytoplasmic streaming in characean algal cells: sliding of endoplasmic reticulum along actin filaments. J. Cell Biol. 106, 1545–1552.*

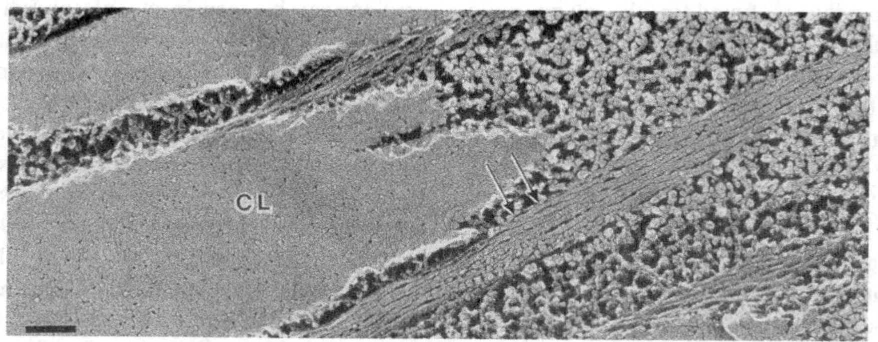

FIGURE 10.8 Freeze-etch micrograph showing the actin bundles (*arrows*). The flat region (CL) is part of the chloroplast envelope. Bar, 120 nm. *From Katcher, B., Reese, T.S., 1988. The mechanism of cytoplasmic streaming in characean algal cells: sliding of endoplasmic reticulum along actin filaments. J. Cell Biol. 106, 1545–1552.*

(F_m) exerted by the shearing stress (σ, in N/m^2) is equal to σA, where A is the area of the endoplasm acted on by the shearing stress on one side of the internodal cell.

How can we measure the magnitude of the inertial motive force per unit area responsible for cytoplasmic streaming? We can put the characean cell in a centrifuge microscope and determine the inertial force per unit area due to centrifugal acceleration (αg, in m/s^2) that is required to stop the cytoplasmic streaming in the centripetal direction. The acceleration needed to stop cytoplasmic streaming is also known as the *balance acceleration.* In a centrifugal field, the net shearing stress is given by the following equation:

$$\sigma = F_m/A - F_i/A \qquad (10.5)$$

The inertial force (F_i, in N) supplied by the centrifugal force and applied to the streaming endoplasm is given by Newton's Second Law:

$$F_i = ma = \alpha g(\rho_e - \rho_v)Ax \qquad (10.6)$$

where ($\rho_e - \rho_v$) is the density difference between the endoplasm and the vacuolar sap, A is the area against which the shearing stress exerts itself, and x is the thickness of the endoplasm under centrifugal accelerations. The volume of the endoplasm flowing in the centripetal direction is thus Ax. The motive force per unit area that powers cytoplasmic streaming can be determined at the balance acceleration, where the velocity gradient = 0. According to Eq. (9.4) (in Chapter 9)

$$\eta(v/x) = \sigma \qquad (10.7)$$

Thus, at the balance acceleration, the shearing stress also vanishes as shown in the following equation:

$$\eta(v/x) = \sigma = 0 \qquad (10.8)$$

Substituting Eq. (10.5) into Eq. (10.8), we get

$$\eta(v/x) = (F_m/A - F_i/A) = 0 \qquad (10.9)$$

As, according to Eq. (10.6), $F_i = \alpha g(\rho_e - \rho_v)Ax$, then

$$(F_m/A - (\alpha g(\rho_e - \rho_v)Ax)/A) = 0 \qquad (10.10)$$

Thus,

$$(F_m/A) = (\alpha g(\rho_e - \rho_v)Ax)/A \qquad (10.11)$$

and after canceling like terms

$$F_m/A = \alpha g x(\rho_e - \rho_v) \qquad (10.12)$$

Thus, the force per unit area that powers cytoplasmic streaming can be calculated from Eq. (10.12) as long as αg, x, and ($\rho_e - \rho_v$) are known. Assuming that g, x, and ($\rho_e - \rho_v$) were constants, Kamiya and Kuroda (1958) calculated the shearing stress to be about 0.16 N/m^2. Kamitsubo and Kikuyama (1994) calculated it to be higher. However, because the thickness of the endoplasm decreases as the centrifugal acceleration increases, due to pooling of the endoplasm at the centrifugal end of the cell, the actual centrifugal force applied to the cell decreases over time. Thus, Staves et al. (1995) propose that the shearing stress is overestimated by the balance acceleration and have determined it to be about 0.1 N/m^2 by extrapolation from the linear portion of a streaming velocity versus centrifugal force graph. If the motive force generated by a single myosin molecule is about 10^{-12} N, and the shearing stress powering cytoplasmic streaming is about 0.1 N/m^2, then there should be approximately 10^{11} myosin molecules per meter squared at the interface of the endoplasm/ectoplasm or 0.1 myosin molecule/μm^2. The concentration of myosin found in *Chara* cells is approximately 200 nM, which is equal to about 10 myosin molecules/μm^2—more than enough to account for the observed motive force. If the total population of myosin was attached to the actin cables, this myosin concentration, with its actin-activated ATPase activity, would hydrolyze ATP faster than respiration could produce it. Consequently, Yamamoto et al. (2006) suggest that the majority of myosin must not be attached to the actin cables at the same time.

Cell models have been important in the study of cytoplasmic streaming. Permeabilized and vacuolar membrane–free cell models have been used to show that ATP provides the energy for cytoplasmic streaming (Williamson, 1975; Shimmen, 1988b; Shimmen and Tazawa, 1982a,b, 1983) and that streaming is regulated through phosphorylation reactions (Tominaga et al., 1987; Awata et al., 2001, 2003; Morimatsu et al., 2002; see Chapter 12).

Characean actin bundles can be used as a tool for studying various aspects of actomyosin-based motility (Shimmen, 1988a; Shimmen and Tazawa, 1982a; Shimmen and Yano, 1984; Sheetz and Spudich, 1983a,b; Sheetz et al., 1984; Kohno and Shimmen, 1988; Katcher, 1985; Kohno et al., 1990; Rogers et al., 1999). For example, the actin bundles from characean internodal cells have been used in situ as a "common garden" to test the ability of various myosins to move just as yeast cells are currently being used as a common garden to test the function of a given DNA sequence. Interestingly, and perhaps unexpectedly, it turns out that myosin from plants exerts approximately 20 times more force than does skeletal muscle myosin (Shimmen, 1988a). Characean myosin is being studied to understand its interesting properties as the "fastest motor protein in the world" (Yamamoto et al., 1994, 1995; Kashiyama et al., 2000, 2001; Morimatsu et al., 2000; Awata et al., 2001, 2003; Ito et al., 2003, 2007; Kimura et al., 2003). Characean myosin hydrolyzes ATP faster, binds tightly to actin longer, and has a longer lever arm than other myosins (Ito et al., 2007). By genetically engineering the length of the lever arm of myosin, Schott et al. (2002) have been able to show that the transport velocity of exocytotic vesicles in living yeast cells is linearly related to the length of the lever arm.

Tominaga et al. (2013) transformed *Arabidopsis* plants with DNA that codes for the motor domain of high-speed *Chara* myosin or low-speed human myosin. They found that the plants that expressed the high-speed myosin motor domain were larger than the wild-type plants, whereas the plants that expressed the low-speed myosin motor domain were smaller than the wild-type plants. They suggest that cytoplasmic streaming can be a determinant of plant size (Tominaga and Ito, 2015).

10.4.2 Chloroplast Movements

There are many beautiful and fascinating light-stimulated, actomyosin-mediated motile responses in plant cells, including the chloroplast accumulation and avoidance response in *Arabidopsis* (Kong and Wada, 2011; Wada, 2013; Higa and Wada, 2015; Fujii and Kodama, 2018), the light-induced chloroplast turning response in *Mougeotia* (see Fig. 10.10; Haupt, 1965, 1983; Wagner et al., 1972; Wagner, 1979; Klein et al., 1980; Wagner and Klein, 1981); the light-stimulated chloroplast aggregation response in *Vaucheria* (Blatt and Briggs, 1980; Blatt et al., 1981; Blatt, 1983, 1987); the light-stimulated aggregation and dispersal response of chloroplasts in the protonema of the fern *Adiantum* (Yatsuhashi et al., 1985, 1987a,b; Wada and Kadota, 1989) and the moss *Physcomitrella* (Kadota et al., 2000; Yamashita et al., 2011; Shen et al., 2015); and the light-stimulated chloroplast orientation and induction of cytoplasmic streaming response in *Elodea*, *Egeria*, and *Vallisneria* (see Figs. 10.11 and 10.12; Ishigami and Nagai, 1980; Yamaguchi and Nagai, 1981; Takagi and Nagai, 1985, 1986; Dong et al., 1995, 1996, 1998; Ryu et al., 1995, 1997; Takagi, 1997, 2003; Takagi et al., 2003; Sakai and Takagi, 2017). Presumably, all these light-stimulated, actomyosin-mediated chloroplast movements optimize photosynthesis, and they are discussed further in Chapter 13 (Senn, 1908; Davis and Hangarter, 2012; Kataoka, 2015).

In *Physcomitrella*, the microtubule/kinesin system is important for the light avoidance response (MacVeigh-Fierro et al., 2017).

In C4 plants, the positioning of the chloroplasts in the mesophyll cells and the bundle sheath cells is an actomyosin-dependent process (Maai et al., 2011). The dependence on light for this response is species specific.

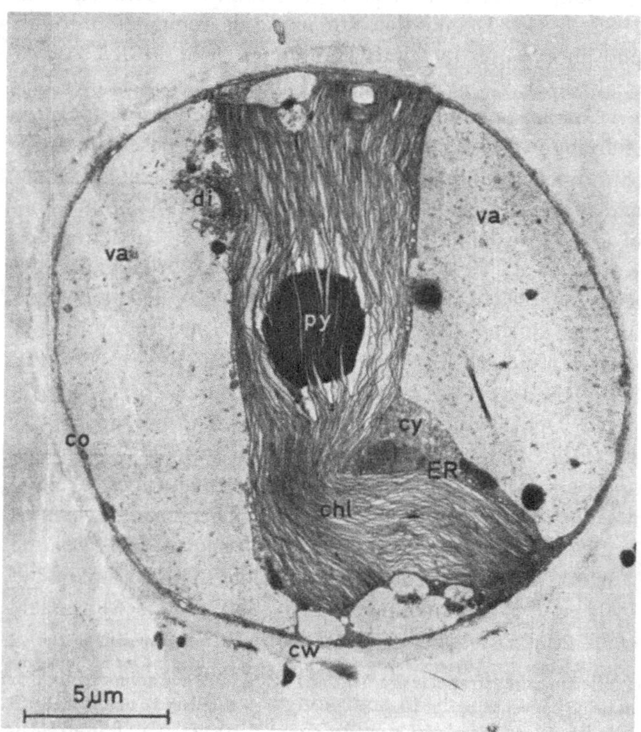

FIGURE 10.10 A cross section of a *Mougeotia* cell showing the shape and orientation of the chloroplast. *chl*, chloroplast; *co, cortical cytoplasm; cw, cell wall; cy*, cytoplasm; *di*, Golgi stack; *ER*, endoplasmic reticulum; *py*, pyrenoid; *va*, vacuole. *From Wagner, G., Klein, K., 1981. Mechanism of chloroplast movement in* Mougeotia. *Protoplasma 109, 169–185.*

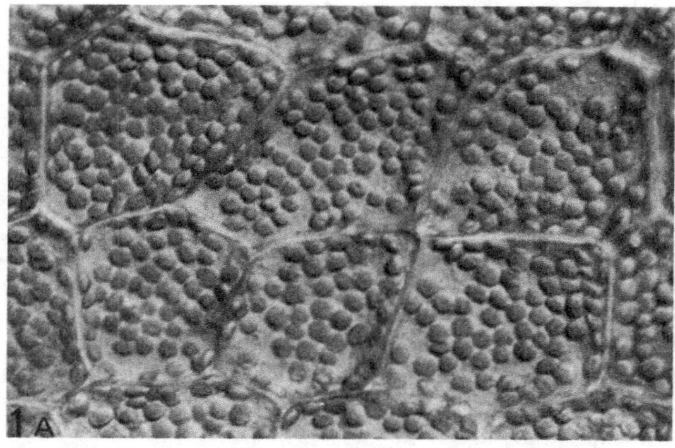

FIGURE 10.11 Epidermal cells of *Vallisneria gigantea* kept under low-intensity light. The chloroplasts are along the periclinal walls. *From Yamaguchi, Y., Nagai, R., 1981. Motile apparatus in* Vallisneria *leaf cells. I. Organization of microfilaments. J. Cell Sci. 48, 193–205.*

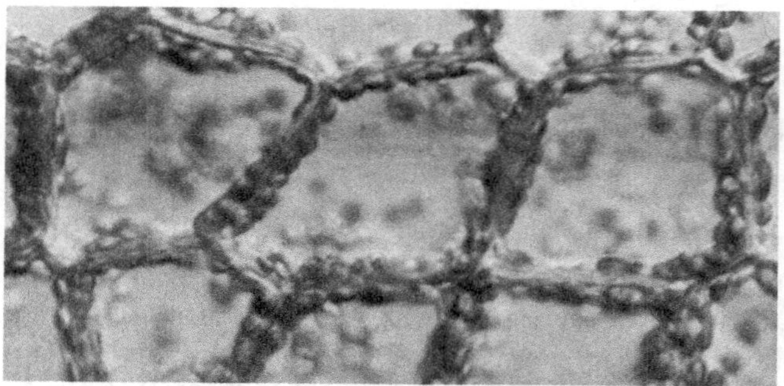

FIGURE 10.12 Epidermal cells of *Vallisneria gigantea* in which the chloroplasts have been induced to move to the anticlinal walls. *From Yamaguchi, Y., Nagai, R., 1981. Motile apparatus in* Vallisneria *leaf cells. I. Organization of microfilaments. J. Cell Sci. 48, 193–205.*

10.4.3 Cell Plate Reorientation in *Allium*

At the end of guard cell differentiation in *Allium*, the spindle in the guard mother cell ultimately lies along the longitudinal axis of the cotyledon, in contrast to the spindles of the epidermal cells proper. Initially, the spindle in the guard mother cell is oriented transversely to the long axis. However, during anaphase and telophase, the spindle reorients until it is aligned with the long axis of the cotyledon (Fig. 10.13). This process is inhibited by cytochalasin, indicating that actin is involved in the reorientation mechanism (Palevitz and Hepler, 1974a,b). Actin is involved in a number of other movement and morphogenetic processes (Mineyuki and Palevitz, 1990; Menzel, 1996; Kennard and Cleary, 1997).

10.4.4 Secretion of Vesicles Involved in Tip Growth and Auxin-Induced Growth

Many plant and fungal cells as well as neuronal cells grow predominantly at the tip. In all these cases of tip growth, actin filaments are involved. In pollen tubes and other tip-growing cells, actomyosin is involved in the delivery of Golgi-derived vesicles to the growing point (Picton and Steer, 1982; Kohno and Shimmen, 1988; Kohno et al., 1991, 1992; Lancelle and Hepler, 1988; Steer and Steer, 1989; Heath, 1990; Braun and Sievers, 1994; Yokota and Shimmen, 1994; Yokota et al., 1995a; Vidali et al., 2001; Lovy-Wheeler et al., 2005; Chebli et al., 2013; Rounds et al., 2014; Madison et al., 2015). Waller et al. (2002) suggest that an auxin-induced reconfiguration of the actin cytoskeleton induces growth in non–tip-growing cells by transporting vesicles containing cell wall components in the lumen and auxin efflux carriers in the membrane to the preferred region of the cells. The monomeric Rop GTPase may be involved in organizing the actin cytoskeleton during tip growth (Holdaway-Clarke and Hepler, 2003; Drøbak et al., 2004; Hwang et al., 2005).

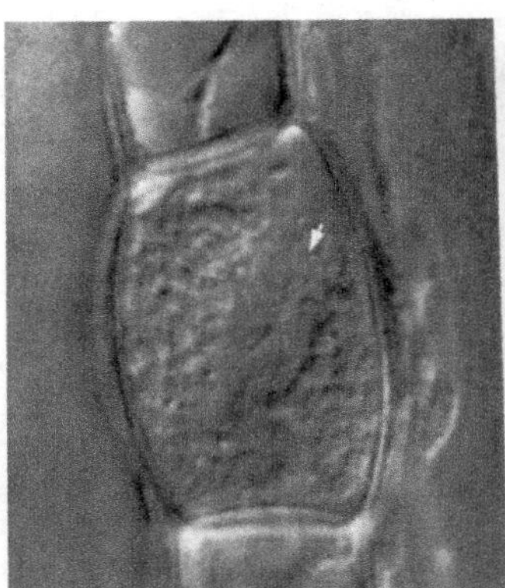

FIGURE 10.13 Nomarski differential interference contrast micrographs of an *Allium* guard mother cell in which the cell plate was caught in the process of reorienting. *From Palevitz, B.A., Hepler, P.K., 1974a. The control of the plane of cell division during stomatal differentiation in* Allium. *I. Spindle reorientation. Chromosoma 46, 297–341.*

10.4.5 Contractile Vacuoles

Contractile vacuoles were the first organelle seen in cells and they are just as exciting to see today (Allen and Naitoh, 2002; Allen et al., 2009) as they were over 200 years ago when Lazzaro Spallanzani (1776) observed the "stars he thought were respiratory organs." The contraction of these osmoregulatory organelles, which expel water and allow wall-less protozoan and algal cells to live in dilute solutions without bursting, is powered by actin and myosin (Zhu and Clarke, 1992; Domozych and Nimmons, 1992; Doberstein et al., 1993; Domozych and Dairman, 1993; Ishida et al., 1993; Nolta and Steck, 1993; Heuser et al., 1993; Nolta and Steck, 1994). Interestingly, the osmotic pressure of the cytosol of cells can be determined by increasing the osmotic pressure of the medium to the point where the contractile vacuoles disappear.

10.5 ROLE OF ACTIN IN MEMBRANE TRANSPORT

In animal cells, the actin cytoskeleton is intimately connected to the plasma membrane through such proteins as talin, vinculin, α-actinin, spectrin, and ankyrin (Luna, 1991; Luna and Hitt, 1992; Ervasti and Campbell, 1993; Hitt and Luna, 1994; Paller, 1994; Calderwood et al., 2000; Kawakatsu et al., 2000). Actin and spectrin are also associated with the plasma membrane of plant cells (de Ruijter and Emons, 1993; Faraday and Spanswick, 1993; Sonesson and Widell, 1993; Kobayashi, 1996). There is evidence in plant cells that the actin cytoskeleton may influence membrane permeability (Wayne and Tazawa, 1988; Tazawa and Wayne, 1989; Hwang et al., 1997; Khurana et al., 1997), as well as the position of Ca^{2+} channels (Brawley and Robinson, 1985; Saunders, 1986a,b).

The fact that the depolymerization of actin microfilaments causes plasmodesmata to dilate (Ding et al., 1996) and the inhibition of myosin ATPase activity constricts the neck of plasmodesmata (Radford and White, 1998) indicates that actin and myosin are involved in plasmodesmatal transport. Moreover, class VIII myosins are required to target proteins to the plasmodesmata for intercellular transport (Avisar et al., 2008a; White and Barton, 2011).

10.6 SUMMARY

Movement is one of the basic characteristics of life. In this chapter, I have provided evidence for the basic unity of nature that Theodor Engelmann and Thomas Huxley believed existed when they compared cytoplasmic streaming in the hair of a stinging nettle with muscle movements that allow a human mouth to recite poetry. However, among this unity we also found that there is diversity: some actin-mediated processes are driven by actomyosin, whereas others are driven by actin polymerization. We have also learned that there are many kinds of myosins. According to

Rafael Demos (1946) "The world is something that can be described, discussed, and argued about, a world admitting of' trenchant distinctions, about which sweeping generalizations are useless, a world that can be measured and whose parts can be compared and contrasted-in brief, a world which is rich in texture and meaning, and yet not so rich that it may not be rationally surveyed and brought together." In Chapter 11, we will see that there is even greater diversity among the various motile systems, and that many motile processes do not depend on actin at all.

10.7 QUESTIONS

10.1. How does actin participate in motile processes?

10.2. Why are there so many types of myosin?

10.3. What is the function of cytoplasmic streaming, and what is the relationship between the pattern and velocity of cytoplasmic streaming and cell size?

The references for this chapter can be found in the references at the end of the book.

Chapter 11

Tubulin and Microtubule-Mediated Processes

This is a marine biological station with her history of over sixty years. If you are from the Eastern Coast, some of you might know Woods Hole or Mt Desert or Tortugas. If you are from the West Coast, you may know Pacific Grove or Puget Sound Biological Station. This place is a place like one of these. Take care of this place and protect the possibility for the continuation of our peaceful research. You can destroy the weapons and the war instruments But save the civil equipments for Japanese students When you are through with your job here Notify to the University and let us come back to our scientific home.

The last one to go (*Katsuma Dan*, Time December 10, 1945; Inoué, 2016).

11.1 DISCOVERY OF MICROTUBULES IN CILIA AND FLAGELLA AND THE MECHANISM OF MOVEMENT

Imagine the excitement Antony van Leeuwenhoek (1677, 1678) felt when he first looked at a drop of water through the microscope he made with his own hands and saw a whole new world of little playful swimming creatures. Leeuwenhoek (1677) saw that "when these animalcula or living atoms did move, they put forth two little horns, continually moving themselves" and he noticed that others were "furnished with diverse incredibly thin feet, which moved very nimbly." 200 years later, with the advantage of better microscopes, cytologists could see that the flagella and cilia that powered the little protozoa were composed of fibers (see Fig. 11.1; Ballowitz, 1888), and Prénant (1913) suggested that these little fibers were contractile. The fibrous nature of the filaments within a cilium was confirmed using an electron microscope (see Fig. 11.2; Jakus and Hall, 1946; Grigg and Hodge, 1949). With the resolution attainable at the time, the filaments seemed similar to those found in muscle (Hall et al., 1946; Draper and Hodge, 1949). However, the introduction of the ultramicrotome allowed Fawcett and Porter (1954) to section cilia transversely and thin enough to reveal a structure different from that of muscle, a structure that has come to be known as the *9 + 2 arrangement* of tubules with which we are familiar today (see Fig. 11.3; Satir, 1974; Berger et al., 1975).

I will use the terms *cilia* and *flagella* interchangeably to describe the whiplike structures of eukaryotic cells. At one time, A.P. Shmagina suggested that the appendages be called *undulipodia*, but that term never caught on (Margulis, 1980; Corliss, 1980). Others have suggested that *flagella* be used to describe the whiplike appendages of prokaryotes, and *cilia* be used to describe those of eukaryotes. This suggestion, which also did not catch on, was based on the facts that prokaryotic and eukaryotic appendages are composed of different proteins and have different structures (Kobayashi et al., 1959; Asakura et al., 1964; Mohri et al., 1967; Mohri, 1968; Renaud et al., 1968). Thus, in eukaryotes, we are stuck with two terms for organelles with identical internal structure and composition. Many people consider flagella to be longer than cilia, more sparsely arranged on a cell, and to have a symmetrical beating motion compared with the asymmetrical beat of cilia. However, as I will discuss anon, the flagellar and ciliary beat can occur at different times in the same structure. Because of this, combined with the fact that there are intermediates in all the characteristics and that there is no universally accepted distinction, I will use the terms interchangeably.

While ciliary motion is widespread in unicellular plants, fungi, and animals, it also occurs in multicellular organisms, although it is restricted to specialized cells (Gray, 1928; Sleigh, 1962, 1974). In the plant kingdom, the sperm of some embryophytic taxa, including mosses, fern allies, ferns (Fig. 11.4), cycads, and *Ginkgo*, is ciliated (Manton, 1950, 1959; Hepler, 1976; Wolniak and Cande, 1980; Paolillo, 1981; Li et al., 1989; Sakai, 2014). In animals, sperms are powered by flagella; and cilia line the respiratory tract where they sweep mucus, dead cells, and dust up toward the mouth, and the oviduct, where they move the oocyte, egg, zygotes and blastocyst toward the uterus. The structure of cilia and flagella in all the above examples are extremely similar, which led Peter Satir (1961) to state that "cellular structure, down to its minute details, remains constant as long as function is constant." The exceptions often prove the rule, and some cilia, including the rods and

Plant Cell Biology. https://doi.org/10.1016/B978-0-12-814371-1.00011-4

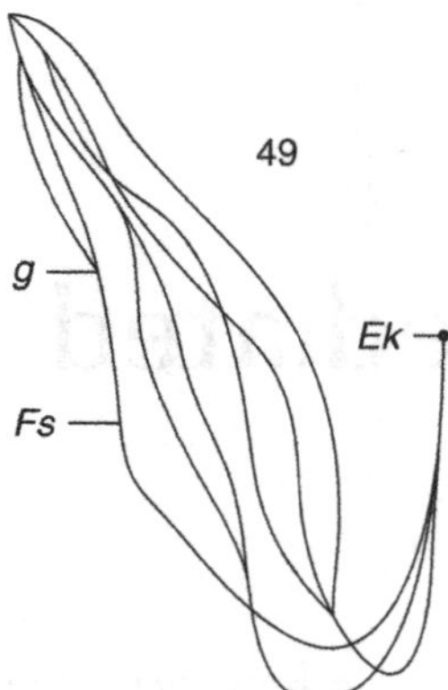

FIGURE 11.1 The sperm of *Passer domesticus* in which filamentous components are visible in the splayed flagella. *From* Ballowitz, E. 1888. Untersuchungen über die Struktur der Spermatozoën, zugleich ein Beitrag zur Lehre vom feineren Bau der contraktilen Elemente. Arck. Mikro. Anat. 32, 401–473.

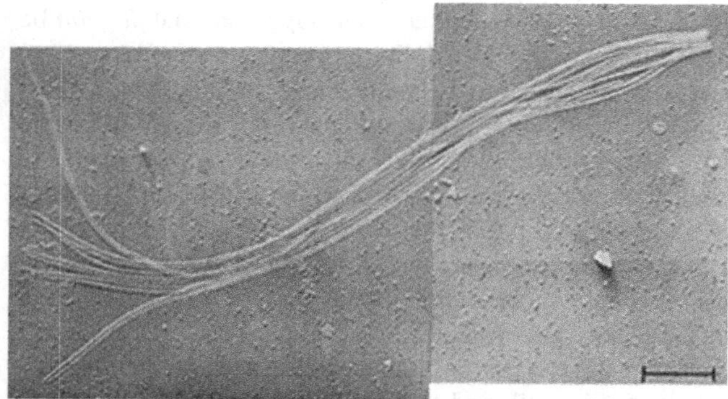

FIGURE 11.2 Electron micrograph of splayed filaments of a cilium from *Paramecium* that has been shadow cast with chromium. ×11,000. *From Jakus, M.A., Hall, C.E., 1946. Electron microscope observations of the trichocysts and cilia of* Paramecium. *Biol. Bull. 91, 141–144.*

cones in our retina, as well as our olfactory and auditory cells, have highly modified structures and consequently have lost their motile abilities in exchange for an enhancement of their sensory functions (Porter, 1957; Pazour and Witman, 2003).

Cilia, like muscle (Szent-Györgyi, 1949a; Varga, 1950), require adenosine triphosphate (ATP) for movement as was shown by Lardy et al. (1945) and Ivanov et al. (1946), who used respiratory inhibitors (see Chapter 14), and by Hartmut Hoffmann-Berling (1955), who used glycerinated cell models. According to Mazia et al. (1951), "The idea that 'working' molecules have the properties of ATPases has been a central theme in the study of the biochemistry of motility, even though no consensus exists even in the field of muscle biochemistry as to the role of ATP or of the splitting of ATP." Isolated cilia contain everything they need to beat except Mg^{2+}-ATP. When isolated cilia are given Mg^{2+}-ATP, they beat by themselves exactly as if they were intact and attached to a cell. To understand how cilia convert the chemical energy of ATP into the mechanical energy that causes the cilia to beat, Tibbs (1957) and Child (1959) took a biochemical approach and discovered that a protein isolated from the cilia of algae and protozoa has ATPase activity. Because nobody succeeded in isolating actin or myosin from cilia, it seemed that Engelmann's (1879) generalization that all motile processes were primitive version of the specialized movements that occur in muscle was limited (Engelhardt, 1946; Mohri and Shimomura, 1973; Mohri et al., 2012). There was diversity in the diversity.

Given the success of the structural approach in elucidating the mechanism of muscle contraction, Gibbons and Grimstone (1960) and Satir (1961) used electron microscopy to understand ciliary motion. Cilia are structurally

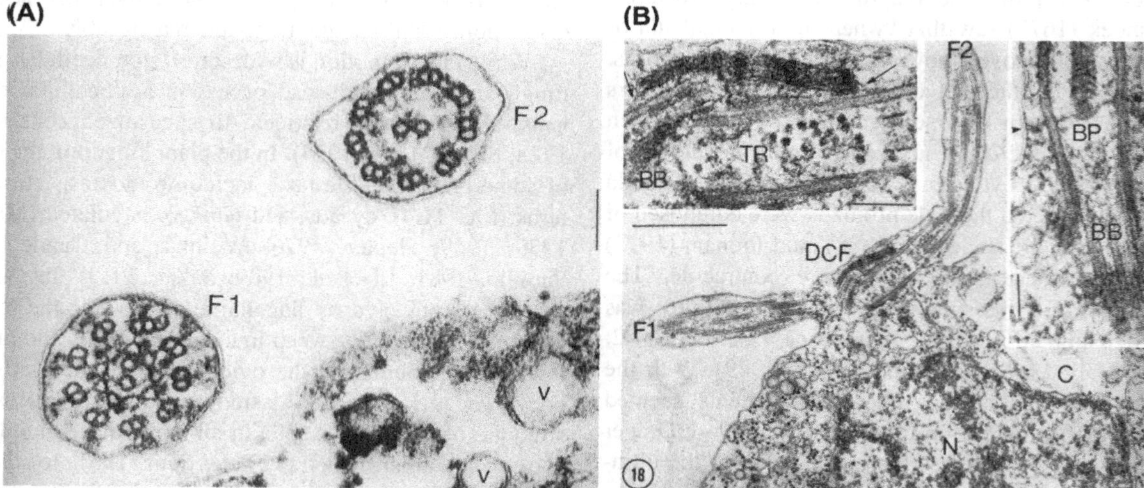

FIGURE 11.3 Electron micrograph of the cilia of the motile gametes of *Acetabularia:* (A) transverse view, bar, 100 nm, and (B) longitudinal view, bar, 500 nm. Inset bar, 100 nm. *From Berger, S., Herth, W., Franke, W.W., Falk, H., Spring, H., Schweiger, H.G., 1975. Morphology of the nucleo-cytoplasmic interactions during the development of Acetabularia cells. Protoplasma 84, 223–256.*

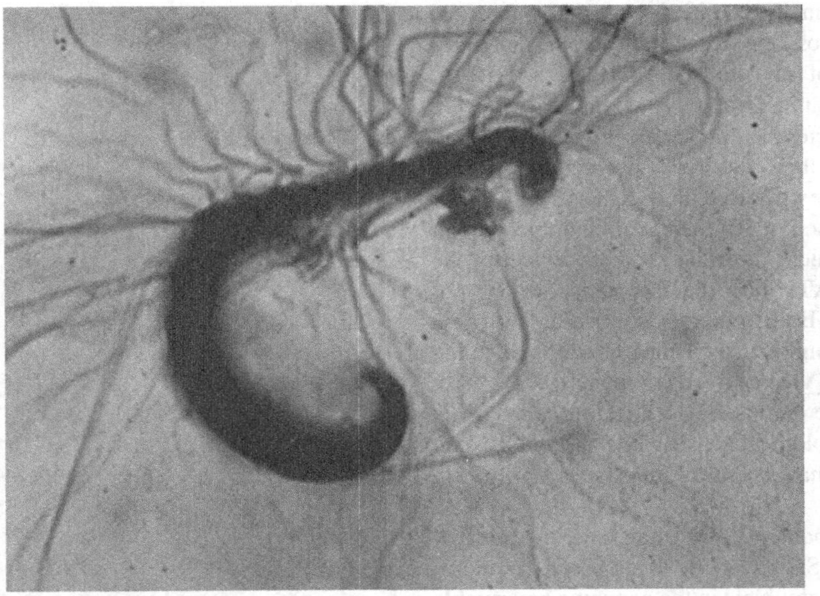

FIGURE 11.4 The sperm of *Dryopteris villarsii* taken with an ultraviolet microscope. The long and numerous cilia are clearly visible. *From Manton, I., 1950. Problems of Cytology and Evolution in the Pteridophyta. Cambridge University Press, Cambridge.*

complex, membrane-enclosed organelles approximately 0.2 μm in diameter and 10 μm long. Cilia can be as short as 5 μm and as long as 150 μm. The internal structure is known as the *axoneme* and is mainly composed of nine doublets of tubules that surround a central pair of tubules. The central pair is composed of two complete tubules, whereas each doublet is composed of one complete and one partial tubule called the *A tubule* and the *B tubule*, respectively (Pease, 1963; Andre and Thiery, 1963). Each tubule in the axoneme is approximately 24 nm in diameter and as long as the cilium. The cilia are asymmetric in every way, and thus the individuality of each doublet can be unambiguously recognized.

With the introduction of better fixation procedures, new structures appeared in the electron micrographs that hinted at how the cilia and flagella may produce force to generate movement. For example, Björn Afzelius (1959) discovered radial spokes that extended from the A tubule toward a central sheath. He also found arms along the length of the A tubule that form cross-bridges with the adjacent B tubule. Afzelius suggested that these arms, like the heads of myosin, might generate force by inducing sliding between adjacent doublets in a mechanism analogous to that found in muscle cells (see Chapter 10).

Gibbons and Grimstone (1960) suggested that the current microscopic data could not distinguish between the possibilities that the bending movement was due to a localized shortening of longitudinal contractile elements or to sliding of the tubules in a manner similar to that described by the sliding filament model of muscle contraction. By looking at cilia at different stages in their beat cycle with an electron microscope, Peter Satir (1965) concluded that there was no change in the tubule length during ciliary motion, as would be predicted if the tubules were contractile proteins. He also noticed that some tubules, which by this time were universally called *microtubules*, extended further at the tip of the cilia than other microtubules did. The time of their extension correlated perfectly with the position of the cilium during the beat cycle when it was fixed. That is, the microtubules on the concave side of the cilium were extended farther into the tip than the microtubules on the convex side. Because the concave side of the cilium was shorter than the convex side, these data were inconsistent with the microtubule contraction hypothesis but supported the sliding filament model for ciliary beating.

Biochemical, genetic, and proteomic data indicate that there are over 200 polypeptides in axonemes (Warner et al., 1989; Dutcher, 1995; Li et al., 2004; Pazour et al., 2005; Stolc et al., 2005). Presumably, most of the proteins are involved in the production of force and the regulation of motility. Gibbons (1963) isolated one protein from axonemes that had ATPase activity. He called the ATPase *dynein*, from the Greek for "force protein," and suggested that it may be important for many aspects of cell motility, including ciliary motion (Gibbons and Rowe, 1965). It is possible that dynein is the same protein identified by V.A. Engelhardt (1946) in sperm. Ian Gibbons localized the dynein in the axoneme by extracting the ATPase activity from the axonemes and then seeing which structure disappeared. When the dynein was extracted, the arms disappeared. The arms could be reconstituted by adding back the purified dynein. The purified dynein was observed with the electron microscope. It had a head and tail structure similar to that of myosin. The ATPase activity of dynein and the swimming activity of the sperm were simultaneously characterized in sperm by Gibbons and Gibbons (1972) after they dissolved the membrane from sea urchin sperm. They showed that ATPase activity of dynein paralleled the activity of the sperm and thus dynein is the mechanochemical transducer that converts the chemical energy of ATP into the mechanical energy of mechanical motion. Both ciliary motion and dynein ATPase activity are inhibited by 1 mM N-ethylmaleimide and 25 μM vanadate (Vale and Toyoshima, 1988, 1989). Thus, dynein has the structure, localization, enzymatic activity, and pharmacological sensitivity that is consistent with its being the mechanochemical transducer involved in microtubule sliding.

Summers and Gibbons (1971), using dark-field microscopy, and Sale and Satir (1977), using electron microscopy, showed that the doublets in the axoneme are capable of sliding past each other. They observed axonemes that had been treated with trypsin, a protease that disrupts the radial spokes, and a protein known as *nexin* that links the outer doublets but leaves the dynein and microtubules intact. When the trypsin-treated axonemes were treated with ATP, the axoneme elongated up to five times its original length before the cilia disintegrated, indicating that the microtubule doublets are capable of sliding. This makes it likely that the cilia beat as a result of the dynein arms walking along the adjacent doublets. The sliding of microtubules on one side of a cilium at a time results in the generation of a shearing stress that bends the cilium and propels the cell. While it seems certain that the sliding filament model explains how force is generated in cilia and flagella, we still do not know the role of most of the over 200 axonemal proteins. Some of them may provide elastic or rigid structures that help in the generation of shearing stresses. Others may be involved in regulating the activity of the motor and structural proteins to generate the three-dimensional beat that propels a cell through a viscous medium or a viscous medium over stationary cilia (Brokaw, 1972; Satir, 1974, 1975).

A cilium originates from a structure known as the *basal body*, which is a small organelle at the base of the cilium that is about 0.2 μm wide and 0.4 μm long. It is composed of nine groups of three tubules (Fig. 11.5). The basal bodies are in turn connected to filamentous structures known as *roots* that permeate the inner surface of the cell body (Pickett-Heaps, 1975; Lee, 1995; Yubuki and Leander, 2013). The outer tubules of the axoneme are continuous with the inner two tubules of each triplet of the basal body, while the central pair grows from an amorphous area at the distal end of the basal body. Basal bodies are formed either from self-duplicating centrioles or from blepharoplasts, which are essentially centrioles that form in the cell de novo (Sharp, 1914; Mizutani and Gall, 1966; Hepler, 1976; Duckett and Renzaglia, 1986; Kalnins, 1992; Keller et al., 2005; Nick, 2008a; Vaughn and Bowling, 2008). The cilia can be cut off from the basal bodies, and if they are, they regrow from the basal bodies synchronously in a couple of hours (Rosenbaum and Carlson, 1969). Thus, the assembly of all the proteins of the axoneme can be studied with extreme precision (Lefebvre and Rosenbaum, 1986).

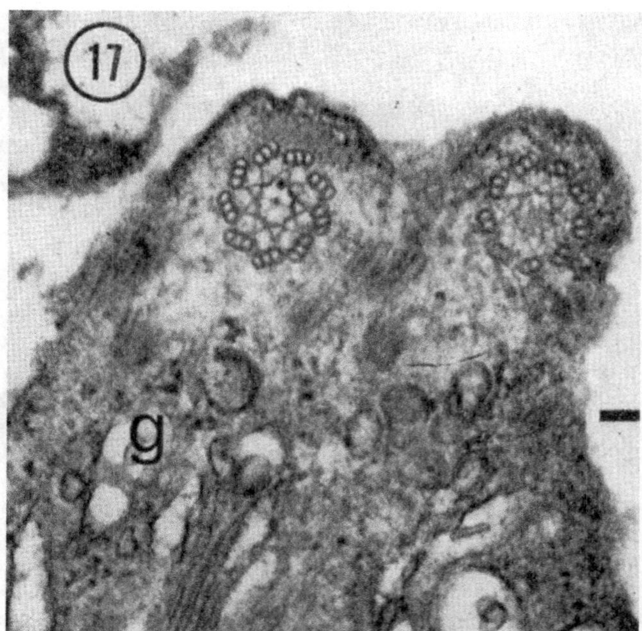

FIGURE 11.5 An electron micrograph of basal bodies in *Acetabularia*. Bar, 100 nm. *From Woodcock, C.L.F., Miller, G.J., 1973. Ultrastructural features of the life cycle of* Acetabularia mediterranea. *I. Gametog. Protoplasma 77, 313–329.*

While I am using the terms *cilia* and *flagella* to mean the same structure, I will use the term *flagellar motion* to mean a symmetrical snakelike motion that pushes the organism through a medium and the term *ciliary motion* to characterize an asymmetrical whiplike motion that is reminiscent of the breaststroke. Ciliary motion either pulls the organism through a medium or, if it is anchored to a stable structure, pulls medium over it.

The same axoneme is capable of both ciliary and flagellar motion. Indeed, an increase in the ciliary Ca^{2+} concentration from 0.1 to 1 μM causes a change from an asymmetrical ciliary beat to a symmetrical flagellar beat. This can be demonstrated by observing the shape of the ciliary beat in isolated axonemes placed in solutions containing different concentrations of Ca^{2+} (Hyams and Borisy, 1978). Environmental cues can stimulate the increase in intraciliary Ca^{2+}. For example, when cells of *Chlamydomonas* are subjected to an increase in light intensity they undergo a change from a ciliary beat shape to a flagellar beat shape. This is called the step-up photophobic response. Channelrhodopsin, a protein related to the photoreceptor for human vision, is the photoreceptor for this response (Foster, 2001; Foster and Smyth, 1980; Foster et al., 1984, 1988, 1989, 1991; Smyth et al., 1988; Saranak and Foster, 1994, 1997, 2000; Sineshchekov et al., 2002; Govorunova et al., 2004; Berthold et al., 2008; Hegemann, 2008). In intact cells, the light-stimulated reversal requires at least 1 μM extracellular Ca^{2+} and is inhibited by Ca^{2+} channel blockers (Schmidt and Eckert, 1976). Mechanical stimulation causes a similar reversal in *Paramecium* (Eckert, 1988). The contribution of Ca^{2+} in the coupling of a stimulus to a response is discussed in Chapter 12.

Due to the similarity between the cilia of *Paramecium* and the cilia of the respiratory organs, which are subjected to the nicotine from tobacco smoke, toxicity tests on *Paramecium* were routinely run by cigarette companies and other scientists (Kensler and Battista, 1963; Wang, 1966; Weiss and Weiss, 1964, 1966, 1967; Weiss, 1967). One test, referred to as the "hanging drop *Paramecium* test," exposes a hanging culture of *Paramecium* to puffs of cigarette smoke to determine the number of puffs required to stop all ciliary movement. Another test exposes *Paramecium* overnight to a homogenate of smoke collected in water to determine the concentration of smoke required to kill a standard volume of *Paramecium*.

11.2 MICROTUBULES IN NONFLAGELLATED OR NONCILIATED CELLS AND THE DISCOVERY OF TUBULIN

In the 19th century, Walther Flemming and Eduard Strasburger saw filamentous structures in the cytoplasm that were associated with movement. As I discussed in Chapter 9, many physicochemically oriented cytologists believed that the fibers, filaments, or kinoplasm were artifacts of the fixation process and could only be seen in fixed cells or as diffraction artifacts in living cells when the optics were poorly adjusted. However, with the introduction of polarizing microscopes, it became apparent that these fibers did exist in living cells and were very dynamic (Kuwada and Nakamura, 1934; Schmidt, 1936, 1937, 1939; Inoué, 1951, 1952, 1959; Inoué and Dan, 1951; Mazia and Dan, 1952; Steinbach Ulbrich, 2016). Many of the filamentous structures or kinoplasm seen in the 19th century (Ranvier, 1875; Flemming, 1880, 1882; Strasburger, 1897, 1898; Strasburger and Hillhouse, 1911) turned out to be tubules when they were visualized with the electron microscope using glutaraldehyde as a fixative (Sabatini et al., 1963) rather than the heavy metal–containing fixatives osmium tetroxide (Porter et al., 1945; Palade, 1952a) and potassium permanganate (Luft, 1956; Bradbury and Meek, 1960) (see Figs. 11.6 and 11.7; Bernhard and de Haven, 1956; Roth and Daniels, 1962; Ledbetter and Porter, 1963, 1964; Slautterback, 1963; Hepler and Newcomb, 1964; Esau and Grill, 1965; Hepler et al., 2013).

Tubules were seen by electron microscopists in the cytoplasm and spindle of a variety of nonciliated cells (Hepler et al., 2013). The diameters of the various tubules varied and the tubules were often misidentified as tubules of the endoplasmic reticulum. Nevertheless, David Slautterback (1963) reasoned that the variation may be artifactual and may result from the shrinkage and swelling that can occur during fixation, dehydration, and staining. Slautterback (1963), Ledbetter and Porter (1963), and Hepler and Newcomb (1964) independently proposed to call all the tubules in the cell body and cilia that were approximately 24 nm in diameter *microtubules*. Slautterback thought the interior of the microtubules might function in the transport

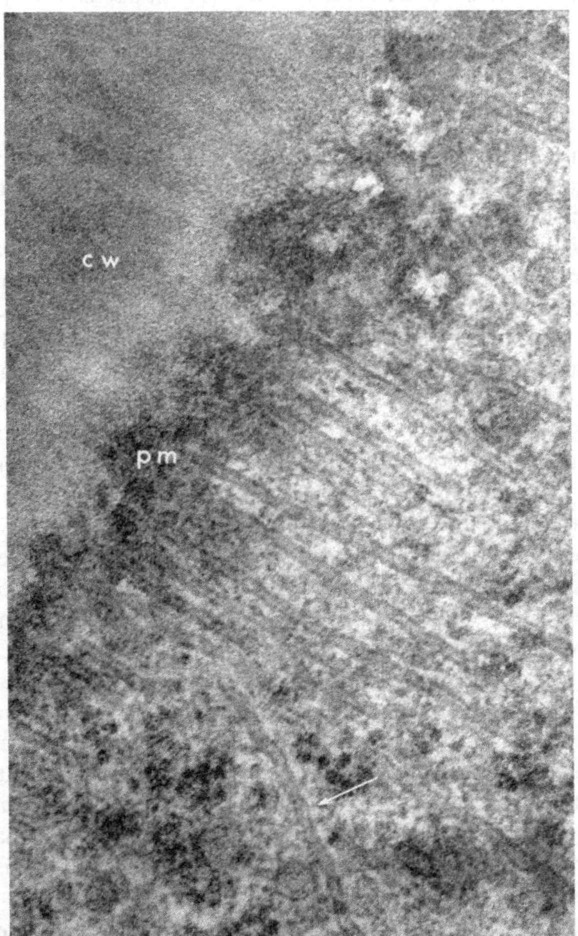

FIGURE 11.6 A grazing section of the extracellular matrix (CW), plasma membrane (pm), and cortex of a *Phleum* root cell showing microtubules that are coparallel and oriented circumferentially around the cell. *From Ledbetter, M.C., Porter, K.R., 1963. A "microtubule" in plant cell fine structure. J. Cell Biol. 19, 239–250.*

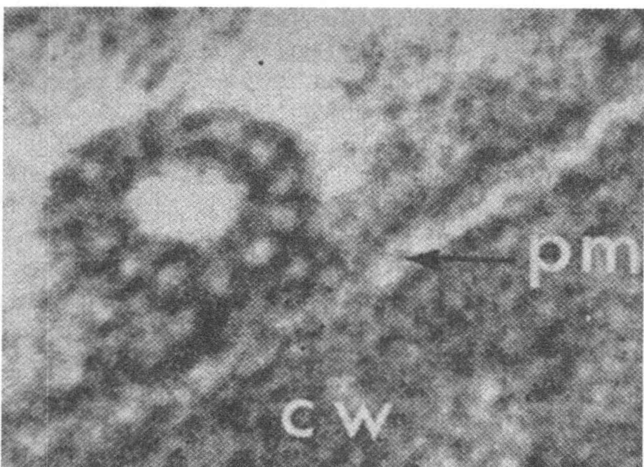

FIGURE 11.7 Transverse section of a microtubule from the cortex of the tannin-rich cells of *Juniperus chinensis*. The microtubule is composed of 13 protofilaments. The natural tannins in *Juniper* provide a natural contrasting agent for microtubules. ×740,000. *cw*, extracellular matrix; *pm*, plasma membrane. *From Ledbetter, M.C., Porter, K.R., 1964. Morphology of microtubules of plant cells. Science 144, 872–874.*

of water, ions, and small molecules, Ledbetter and Porter considered that the microtubules might be involved in regulating cell shape, and Hepler and Newcomb proposed that the microtubules might be involved in the deposition and alignment of cellulose microfibrils. Later, Ledbetter and Porter (1964) and Porter and Tilney (1965) proposed that microtubules were involved in intracellular motility based on their intracellular distribution, their relationship to the kinoplasm, and the effects of colchicine on morphogenesis.

Colchicine is a fascinating drug that was used by the Egyptians around 1550 BCE for treating gout and rheumatism (Eigsti and Dustin, 1955). It is extracted from the autumn crocus (*Colchicum autumnale*), which is a native plant of Colchis, a region on the Black Sea. Colchis is where Jason and the Argonauts went to capture the Golden Fleece, a symbol of authority. There, Medea helped Jason capture the Golden Fleece by giving him various life-saving brews. Perhaps one of the brews contained colchicine!

Colchicine inhibits mitosis (Pernice, 1889; Eigsti et al., 1949). However, Pernice initially believed that colchicine stimulated mitosis, as he found that after a dog had been treated with colchicine, there would be an inordinate number of mitotic figures in the cells of its stomach and intestine. However, by the 1930s, it was established that colchicine did not stimulate mitosis but prolonged mitosis, thus increasing the chance of catching dividing cells in a given section (Wellensiek, 1939). Dustin (1947) hypothesized that colchicine directly caused the spindle fibers to break down. However, he could not make a very good argument because at that time many people believed that the spindle was only an artifact of fixation (Mazia, 1975). Later, electron micrographs taken by Seder and Wilson (1951) showed that colchicine broke down the spindle fibers; however, the quality of the fixation was too poor to put much stock in their interpretation.

Acceptance of the reality of spindle fibers and the direct effect of colchicine on them arose from the polarized light microscopy studies of Shinya Inoué (1951, 1952), who showed that colchicine caused a decrease in the birefringence of the spindle fibers. Later, Harris and Bajer (1965) combined polarization microscopy with electron microscopy and showed that spindle fibers were indeed microtubules. Pickett-Heaps (1967a) and Shelanski and Taylor (1967) showed that colchicine caused microtubules to disappear from plant cells and isolated mitotic apparati, respectively. Taylor et al., using ^{3}H-colchicine, purified the colchicine-binding protein (Borisy and Taylor, 1967; Shelanski and Taylor, 1967; Adelman et al., 1968; Weisenberg et al., 1968; Wells, 2005), which Hideo Mohri (1968; Yanagisawa et al., 1968; Mohri and Hosoya, 1988) named *tubulin*. Plant tubulin differs from animal tubulin in terms of its drug sensitivity. For example, animal tubulin is depolymerized by 5×10^{-8} M colchicine and is insensitive to the herbicide oryzalin, whereas plant tubulin is depolymerized by millimolar concentrations of colchicine and is very sensitive to oryzalin (Morejohn and Fosket, 1984a,b, 1992).

Tubulin is a 110-kDa heterodimer that is composed of two globular subunits—α-tubulin and β-tubulin (Dustin, 1978)—encoded by two gene families (Breviario, 2008). Each subunit has a mass of approximately 55 kDa and is associated with one molecule of guanosine triphosphate (GTP; Yanagisawa et al., 1968). Initially, it was difficult to get tubulin to polymerize in vitro or to isolate microtubules. Then Richard Weisenberg (1972) noticed that the ability of tubulin to polymerize depended on which pH buffer he used. He discovered serendipitously that the ability of microtubules to polymerize in vitro was correlated with the Ca^{2+}-binding ability of the buffer. Subsequently, he routinely added EGTA (ethylene glycol-bis(β-aminoethyl ether)), a Ca^{2+} chelator, to the extraction buffer and polymerization solutions to obtain microtubules.

Ceteris paribus (all other factors being the same) isolated tubulin polymerizes into microtubules in vitro as long as the concentration of tubulin is greater than the critical concentration. Like actin filament growth, microtubule growth shows a lag phase, indicative of the need for a primer. Nucleation in vivo may require a third form of tubulin known as *γ-tubulin* (Oakley and Oakley, 1989; Oakley et al., 1990; Oakley, 1992; Ludueña et al., 1992; Liu et al., 1993; McDonald et al., 1993; Joshi, 1994; Hoffman et al., 1994; Marc, 1997; Binarová et al., 2006; Pastuglia et al., 2006; Murata et al., 2005, 2013; Fishel and Dixit, 2013), which, along with a protein known as *augmentin*, may also initiate microtubule branches along existing microtubules (Murata et al., 2005;

Wasteneys and Collings, 2007; Goshima et al., 2008; Yamada and Goshima, 2017).

The tubulin dimers are arranged in a specific orientation in microtubules, and consequently, microtubules, like microfilaments, are polar structures, and the two ends of microtubules are different. If purified tubulin is allowed to polymerize on fragments of a ciliary axoneme and the products are observed in the electron microscope, it can be seen that the microtubules elongate three times faster on one end than the other end. Growth at each end occurs when the polymerization reactions take place faster than the depolymerization reactions. As is the case with microfilaments, the rapidly polymerizing end is called the *plus end* and the slowly polymerizing end is called the *minus end*.

While GTP is necessary for polymerization, the hydrolysis of GTP is not, as nonhydrolyzable analogues of GTP support polymerization (Kirschner, 1978). Microtubules in the presence of GTP exhibit treadmilling just like actin filaments do in the presence of ATP. At the critical concentration, tubulin polymerization occurs at the plus end at the same rate that depolymerization occurs at the minus end, and, while the tubulin dimers are translocated along the microtubule from the plus end to the minus end, there is no net change in microtubule length (Kirschner, 1980; Bergen and Borisy, 1980; see Fig. 10.6 in Chapter 10). While treadmilling could presumably cause the movement of a vesicle bound to a tubulin dimer from one end of a microtubule to the other, it is probably not involved in intracellular motility as the flux of subunits occurs at an excruciatingly slow rate of about 0.5 μm/h, far slower than the slowest known microtubule-mediated motile process (1 μm/min = 60 μm/h).

In the dark-field microscope, microtubules reveal another fascinating behavior known as *dynamic instability* (Bayley, 1990). When microtubules assemble from pure tubulin, we can see the microtubules shrink and grow rapidly, alternating between the two states in a seemingly random manner. The growing end, or plus end, appears to switch between a slowly growing to a rapidly shrinking state. When the hydrolysis of GTP is slower than the rate of GTP-tubulin addition, GTP-tubulin dimers accumulate at the plus end and form a GTP-tubulin cap. GTP-tubulin dissociates 100 times less readily than guanosine diphosphate (GDP)-tubulin, and thus, when there is a GTP-tubulin cap, the microtubule is stable. However, when the rate of hydrolysis of GTP is greater than the rate of polymerization, the microtubule becomes capped with GDP-tubulin, and the microtubules rapidly depolymerize. Once the rapid depolymerization begins, the GTP-tubulin cap is hard to regain and the shrinking microtubule usually completely depolymerizes (Carlier, 1991). Shaw et al. (2003) characterized the dynamic behavior of cortical microtubules in living epidermal cells, and they found that the plus end grows and shrinks at a rate of 3.69 and 5.80 μm/min, respectively, wherease the minus end grows and shrinks at a rate of 1.98 and 2.78 μm/min, respectively. The rates, however, are cell-type specific (Shelden and Wadsworth, 1992b).

Microtubule polymerization and depolymerization can be affected by many natural products besides colchicine (Florian and Mitchison, 2016). These include mescaline, an abbreviated form of colchicine (Paulson and McClure, 1973; Harrisson et al., 1976), vinblastine and vincristine (isolated from *Catharanthus roseus*), podophyllotoxin (isolated from *Podophyllum peltatum*), taxol (isolated from *Taxus brevifolia*), and griseofulvin (isolated from *Penicillium griseofulvum*). Furthermore, many herbicides, including IPC (isopropyl-N-phenylcarbamate), CIPC (N-[3-chlorophenyl] carbamate) and trifluralin (trifluoro-2,6-dinitro-N,N-dipropyl-*p*-toluidine), APM (aminoprophos-methyl), and oryzalin affect microtubule polymerization or organization.

11.2.1 Temporal and Spatial Localization of Microtubules in Animal and Plant Cells

In vivo, microtubules always originate from regions of the cell known as *microtubule-organizing centers* (MTOCs; Fig. 11.8). In many mammalian cells, the microtubules typically radiate from a single region known as the *cell center* or *centrosome* (Mazia, 1987), whereas in plant cells, microtubules originate from many spatially separated MTOCs that contain γ-tubulin (Marc, 1997; Murata and Hasebe, 2011; Fishel and Dixit, 2013). While basal bodies and centrioles are MTOCs with an easily discernable structure, most MTOCs and the pericentriolar regions of the centrosome are amorphous areas. Studies in which the readily visible structures in the MTOCs have been removed genetically or mechanically show that the readily visible structures are more efficient in generating microtubule arrays than the amorphous regions alone (Ambrose and Cyr, 2007). MTOCs can be identified by depolymerizing microtubules with various microtubule-depolymerizing agents, and after removing the agent, watching where the microtubules reform (Falconer et al., 1988; Wacker et al., 1988; Cleary and Hardham, 1988; Wasteneys and Williamson, 1989). The MTOCs themselves are dynamic and move throughout the cell (Chan et al., 2003).

We can detect the polarity of microtubules in the cell by adding free tubulin molecules to the existing microtubules. Under special conditions, the tubulin does not add to the ends but forms curved protofilament sheets on the sides. In cross section, the sheets resemble hooks, and depending on the polarity of the microtubule, the hooks appear either clockwise or counterclockwise. When we look at a microtubule from the plus end, the hooks are oriented in the clockwise direction (Euteneuer and McIntosh, 1981a,b; Schliwa, 1984). Using this method, we can see that typically the plus ends are distal to the MTOCs and the minus

FIGURE 11.8 Immunofluorescence micrograph showing microtubules in an internodal cell of *Nitella tasmanica* regrowing from microtubule-organizing centers following the removal of oryzalin, a microtubule-depolymerizing agent. Bar, 20 μm. *From Wasteneys, G.O., Williamson, R.E., 1989. Reassembly of microtubules in Nitella tasmanica: Assembly of cortical microtubules in branching clusters and its relevance to steady-state microtubule assembly. J. Cell Sci. 93, 705–714.*

ends are embedded in the MTOCs. The polarity of microtubules can also be determined by decorating them with dynein (Telzer and Haimo, 1981) or various proteins that bind to either the plus end or to the minus end (Chan et al., 2003, 2005; Mathur et al., 2003; Komaki et al., 2010; Young and Bisgrove, 2011).

In some, but not all animal cells in interphase, the microtubules radiate from the centrosome in the center of the cell. In most cases, the end of the microtubules attached to the centrosome is the minus end, and the distal end is the plus end. This is known as the *plus end–distal arrangement*. As we will see in the following, the organization of the microtubules is important for moving organelles to the correct location. Roger Penrose (1994) has suggested that the microtubule cytoskeleton in brain cells has an even bigger role. He thinks that it is the material basis of the mind!

In higher plant cells, during interphase, microtubules occur in the cortical cytoplasm, and have a transverse orientation relative to the long axis of the cell (see Fig. 11.9; Vesk et al., 1996; Kumagai et al., 2001; Karahara, 2014). The cortical microtubules in the end wall are oriented randomly. As the cell elongates, the cortical microtubules along the cell flanks become oriented longitudinally (see Fig. 11.10; Lloyd, 1987). In naturally wall-less plant cells, the microtubules typically are oriented parallel to the long axis of the cell (Pickett-Heaps, 1975). Tobacco BY-2 cells are particularly good material in which to study microtubules in vitro and in vivo (Sonobe et al., 2001; Nagata et al., 2004; Dhonukshe et al., 2005).

Just before prophase, in late G2, a unique arrangement of microtubules known as the *preprophase band* forms in the cortex of most plant cells (see Figs. 11.11 and 11.12; Northcote, 1967). Pickett-Heaps and Northcote (1966b)

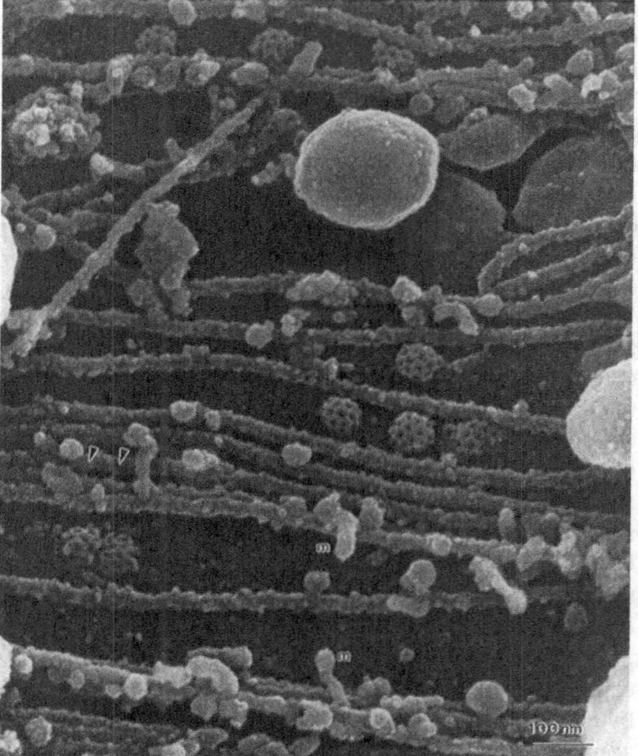

FIGURE 11.9 Cortical microtubules in an onion root cell that have been freeze-fixed, freeze-fractured, dried, and observed with a field-emission scanning electron microscope. There are many cross-bridges (*arrowheads*) and appendages (m) on the microtubules. The plasma membrane is in the background and coated vesicles can be seen. Bar, 100 nm. *From Vesk, P.A., Vesk, M., Gunning, B.E.S., 1996. Field emission scanning electron microscopy of microtubule arrays in higher plant cells. Protoplasma 195, 168–182.*

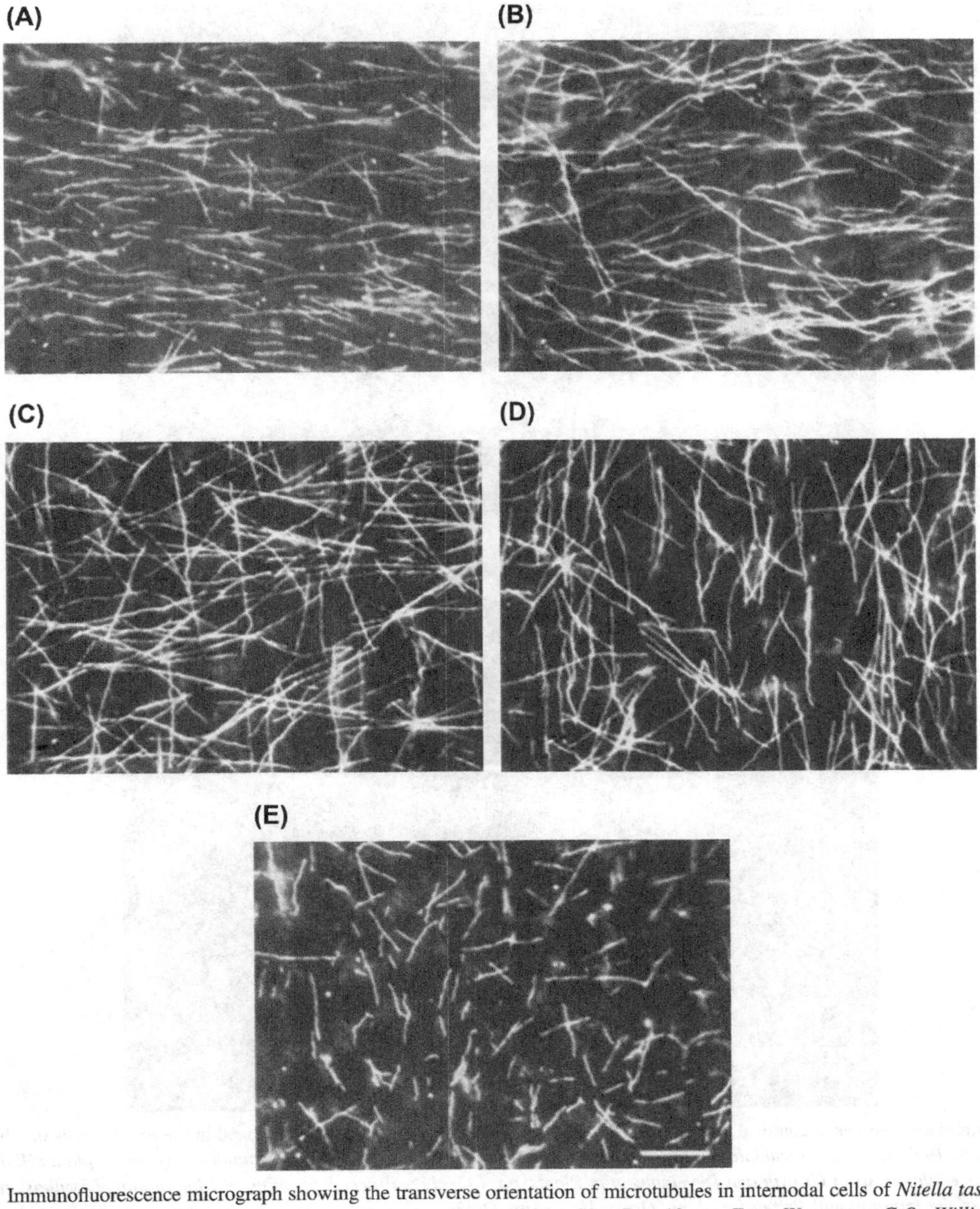

FIGURE 11.10 Immunofluorescence micrograph showing the transverse orientation of microtubules in internodal cells of *Nitella tasmanica*: (A) taken from the youngest internodal cell, and (B–E) taken from cells that are successively older. Bar, 10 μm. *From Wasteneys, G.O., Williamson, R.E., 1987. Microtubule orientation in developing internodal cells of Nitella: A qualitative analysis. Eur. J. Cell Biol. 43, 14–22.*

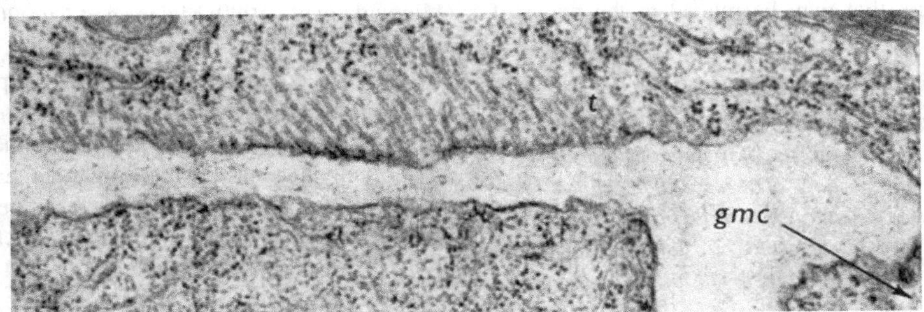

FIGURE 11.11 Electron micrograph of preprophase band in an epidermal cell of wheat that will be cut off a subsidiary cell. *gmc*, guard mother cell. × 36,000. *From Pickett-Heaps, J.D., Northcote, D.H., 1966a. Cell division in the formation of the stomatal complex of the young leaves of wheat. J. Cell Sci. 1, 121–128.*

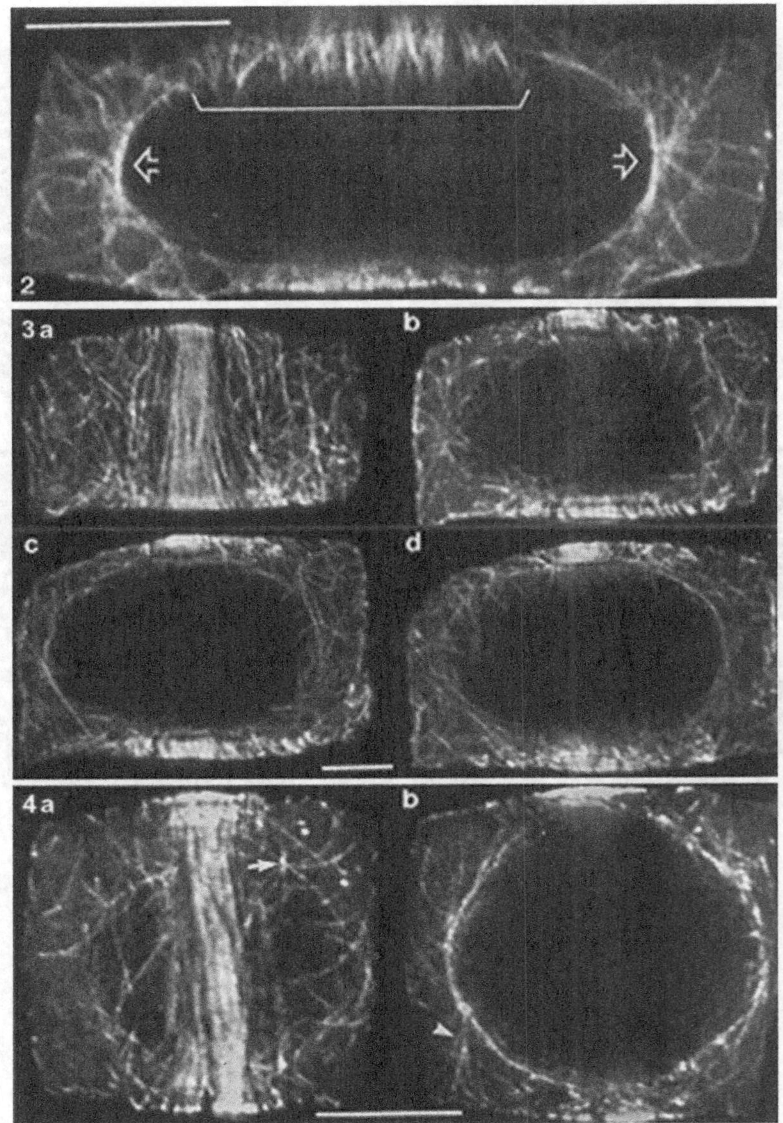

FIGURE 11.12 Immunofluorescence confocal micrographs showing the development of a preprophase band in the root tip cells of wheat. Bar, 10 μm. *From Gunning, B.E.S., 1992. Use of confocal microscopy to examine transitions between successive microtubule arrays in the plant cell division cycle. In: Shibaoka, H. (Ed.), Cellular Basis of Growth and Development in Plants, pp. 145–155. Proceedings of the VII International Symposium in Conjunction with the Awarding of the International Prize of Biology. Osaka University, Osaka.*

found the preprophase band while they were looking for a cytoplasmic structure that may be related to the plane of cell division:

> *It seemed most likely that spindle organization, and in particular microtubule synthesis, might be observable in what was going to be the future polar zone of these cells, but long and careful scrutiny of these regions failed to reveal any changes or structures that could be implicated in mitosis. However, a band consisting of a large number of microtubules was found near the wall of the cell, far removed from the polar zone.*

The preprophase band is formed from the gradual rearrangement of the randomly or transversely oriented cortical microtubules into a tightly packed transverse band (Mineyuki et al., 1989; Malcos and Cyr, 2011) that contains a number of other proteins (Duroc et al., 2011). The nucleus may control the position of the preprophase band because when the nucleus of an *Adiantum* protonema is displaced by centrifugation, the preprophase band appears at the new nuclear position (Murata and Wada, 1991; Wada, 1992). The nucleus may also be responsible for the disintegration of this transient structure because the preprophase band does not break down at metaphase if the nucleus is centrifuged away from it. In the case of cells that grow in three dimensions, the position of the preprophase band predicts the site of cell plate formation (Pickett-Heaps, 1969b; Cleary and Smith, 1998; Mineyuki, 1999).

The preprophase band disappears at prophase, (Hush et al., 1996) and in late prophase, the site where the preprophase band was localized becomes a localized site of clathrin-mediated endocytosis (Karahara et al., 2009, 2010; Cyr and Fisher, 2012). During prophase, the nuclear envelope serves as an MTOC for the polymerization of microtubules, forming one or two polar caps (Porter and Moses, 1958; Kosetsu et al., 2017), which give rise to the prophase spindle (Mizuno, 1993; Shibaoka and Nagai, 1994; Masoud et al., 2013). The nuclear envelope breaks down at prometaphase and the microtubules permeate the nucleus and form the spindle fibers (Zhang et al., 1990b). The shape of the spindle is determined by the compactness of the MTOCs, and consequently, animal spindles are more pointed at the poles than plant spindles are. Basically, there are two groups of microtubules in the spindle, ones that terminate at the kinetochores of chromosomes and ones that do not, although the two groups are interconnected (Euteneuer and McIntosh, 1980; McIntosh and Euteneuer, 1984; Bajer and Mole-Bajer, 1986; Kubiak et al., 1986; Schibler and Pickett-Heaps, 1987; Palevitz, 1988b; Fuge and Falke, 1991; Zhang and Dawe, 2011; Wadsworth et al., 2011). As a rule, the majority of microtubules in each half spindle are oriented so that their minus ends are embedded in the poles, and their plus ends are near the chromosomes (Euteneuer and McIntosh, 1981a,b). Kinesins localize with the microtubules in the spindle, (Miki et al., 2014) and RanGTP may nucleate microtubules from the kinetochore (Cavazza and Vernos, 2016).

Following nuclear division, a group of microtubules known as the *phragmoplast* organizes the developing cell plate (Nemec, 1899; Inoué, 1964; Esau and Grill, 1965; Hepler and Newcomb, 1967; Hepler and Jackson, 1968, 1969; Palevitz, 1987a,b; Shibaoka, 1992; Müller and Jürgens, 2016). The phragmoplast begins to form at the center of the future cell plate and moves in a centrifugal manner in most cases. However, in *Haemanthus* endosperm cells, the phragmoplast begins as a ring at some distance from the center and then moves both centrifugally and centripetally. The phragmoplast moves slightly ahead of the developing cell plate. The plus ends of the microtubules are embedded near the forming cell plate, and the minus ends stick out (Euteneuer and McIntosh, 1980). Kinesins localize with the microtubules in the phragmoplast (Miki et al., 2014). Microtubules can also be visualized in transgenic plant cells transformed with the gene for either tubulin or microtubule-associated proteins fused to the green fluorescent protein (GFP; Marc et al., 1998; Ueda et al., 1999; Granger and Cyr, 2000a,b; Dixit and Cyr, 2002; Chan et al., 2003; Shaw et al., 2003; Dhonukshe et al., 2005).

When microtubule organization is observed with fluorescence microscopy, we must remember that fluorescence-microscopic images may be misleading because microtubules are only 24 nm in diameter but appear to be about 240 nm in diameter in the light microscope due to diffraction (Williamson, 1990, 1991; Wayne, 2009). Therefore, microtubules near each other, but not touching, may appear to be connected to each other or grouped in bundles. Thus, while wide-field light microscopy gives a good impression of the three-dimensional architecture of the microtubule cytoskeleton, electron microscopy of serial sections and perhaps superresolution microscopy are essential for visualizing the true spatial relationship of microtubules.

The dynamic nature of the microtubule arrangements that occur throughout the cell cycle may result from a change in the distribution of MTOCs or of capping proteins that stabilize the plus ends of microtubules so that they do not depolymerize as a result of dynamic instability. The polymerization and depolymerization of microtubules and their organization can be modified by microtubule-associated proteins (Hotani and Horio, 1988; Cyr and Palevitz, 1989; Cyr, 1991a,b; Yasuhara et al., 1992; Chang-Jie and Sonobe, 1993; Chan et al., 2003; Shaw et al., 2003; Dixit and Petry, 2018; Fan et al., 2018; Nebenführ and Dixit, 2018).

11.2.2 Characterization of Microtubule-Associated Motor Proteins

When Melanie Pratt (1980) discovered dynein in nonciliated cells, the exciting implication was that the molecule could potentially participate in intracellular motility (Asai and Wilson, 1985; Vallee et al., 1988). Cytoplasmic dynein, like ciliary dynein, is a polymeric high–molecular mass protein where each monomer consists of a head, neck, and tail. When cytoplasmic dynein is immobilized on a glass cover slip, it is capable of translocating a microtubule in the direction of its plus end. That is, if the microtubule were immobilized, the dynein would walk to the minus end of the microtubule (Paschal et al., 1987; Lye et al., 1989). Dynein induces the movement of vesicles from the plus end to the minus end of microtubules at a rate of 1.25 μm/s. Both motility and the dynein ATPase activity are inhibited by 1 mM N-ethylmaleimide (NEM), 25 μM vanadate, and the adenine derivative erythro-9-(2-hydroxy-3-nonyl)adenine (EHNA).

Because vesicles can move both directions on a single microtubule, it seemed likely that another motor existed in cells that can walk to the plus end of microtubules (Svoboda et al., 1993; Vale and Milligan, 2000; Asbury et al., 2003; Yildiz et al., 2004; Mickolajczyk et al., 2015; Isojima et al., 2016; Peterman, 2016). Kinesin is just such a motor (Sheetz, 1989). Like dynein, it is a mechanochemical enzyme that converts the chemical energy of ATP into mechanical energy. Kinesin was first found in squid axons but probably occurs in most cells (Sheetz, 1989; Mitsui et al., 1993; Hoyt, 1994; Reddy and Day, 2011). It was discovered by squeezing out the axoplasm from the axon of a squid and watching organelles move along single microtubules with the aid of video-enhanced light microscopy. Adenylyl-imidodiphosphate (AMP-PNP), a nonhydrolyzable analogue of ATP, stopped the movement of organelles, and they become tightly bound to the microtubules (Brady, 1985; Lasek and Brady, 1985). This property helped in the identification and purification of the motor protein. Vale et al. (1985a,b,c) identified kinesin as a protein that bound to microtubules in the presence of AMP-PNP but was released on the addition of ATP. Like dynein and myosin, kinesin is a large elongated protein, where each monomer contains a head, neck, and tail (Vale, 1987; Sheetz, 1989). Kinesin transports vesicles from the minus end to the plus end at a rate of approximately 0.5 μm/s. In contrast to dynein, kinesin is relatively insensitive to 1 mM N-ethylmaleimide and 25 μM vanadate.

While rice contain the genes for the heavy chain of dynein (King, 2002), *Arabidopsis* lacks the genes for the heavy chain of dynein (Lawrence et al., 2001; Wickstead and Gull, 2007) but contains the genes for the light chain of dynein (Cao et al., 2017). Do the plants that lack the dynein heavy chain have another functional minus end–directed motor protein that uses microtubules as a track? Yes, they do. They have a minus end–directed kinesin (Jonsson et al., 2015; Yamada et al., 2017). This is an example of convergent evolution and analogy rather than divergent evolution and homology. When we speak of proteins, we should remember Richard Owen, a brilliant but forgotten scientist (Rupke, 1994, 2009), and clearly differentiate between the structure and function of a protein, which is important from a physiological point of view, from the shared ancestry of proteins, which is important from a phylogenetic point of view.

To make sense of the basic unity and diversity of organs, Richard Owen (1848, 1849) distinguished homology from analogy. Two or more parts that are related by common descent are considered to be homologous. The wing of a bat and the forearm of a rat is an example of a homology. If two or more parts have similar form and function, but are not related by common descent, they are considered to be analogous.

There is a family of kinesin proteins, all with similar gene sequences and with similar pharmacological properties (Lawrence et al., 2004; Lee and Liu, 2004, 2007; Richardson et al., 2006; Ambrose and Cyr, 2007; Lee et al., 2015). Most of the kinesins function as plus end–directed motors along microtubule tracks (Zhu and Dixit, 2011). However, one of them, known as *Ncd* in *Drosophila*, is a minus end–directed motor and not a plus end–directed motor as is the original kinesin (Sharp et al., 1997; Sablin et al., 1998; Liu and Lee, 2001; Ambrose et al., 2005). A single amino acid substitution from asparagine to lysine in the neck region of the protein is sufficient to transform *Ncd* from a minus end–directed motor to a plus end–directed motor (Endow and Higuchi, 2000). *Ncd* is a member of the kinesin-14 family. Kinesin-14 is a minus end–directed motor that occurs in plants (Reddy and Day, 2001a; Zhu and Dixit, 2012; Shen et al., 2012). Moreover, purified *Physcomitrella* kinesin-14 is able to move along microtubules in an in vitro assay toward the minus-end. It is also able to pull liposomes toward the minus end of the microtubule (Jonsson et al., 2015). Kinesin-14 is also able to move actin microfilaments along microtubules (Dixit, 2015; Walter et al., 2015).

Another translocator protein has been isolated from *Reticulomyxa* (Euteneuer et al., 1988). It has a high molecular mass and binds to microtubules in the absence of ATP and is released in the presence of ATP. This protein causes bidirectional movement at a rate of 3.6 μm/s. Phosphorylation of this protein by cyclic AMP-dependent protein kinase converts this bidirectional motor to a unidirectional motor.

Molecular biology has taught us that there are many genes that encode dynein-like proteins, kinesin-like proteins, and proteins that have some properties of each (Reddy and Day, 2001a; Shen et al., 2012). In general, each motor protein has a similar structure, even when the amino acid sequence differs greatly (Kull et al., 1996; Sablin et al., 1996). It is likely that nature is just as promiscuous when it comes to motor proteins as it is with ion channels, and sequences of DNA that code for functional domains in proteins were mixed and matched to form chimeric motor proteins that will transport a given cargo in the desired direction along a microtubule track (Gilbert, 1978; Doolittle, 1995). It even seems that some kinesin proteins act as transcription factors (Li et al., 2012).

11.3 FORCE-GENERATING REACTIONS INVOLVING TUBULIN

11.3.1 Sliding

The movement of dynein and kinesin along a microtubule is thought to occur in a similar manner to the way myosin slides along an actin microfilament (Warner et al., 1989). Kamimura and Takahashi (1981) and Oiwa and Takahashi (1988) have measured the force generated by the microtubule/dynein interaction using the glass microneedle method (see Chapter 10). They held a single demembranated sea urchin sperm flagellum between two microneedles and measured the amount the needle bent when the flagella were reactivated with Mg^{2+}-ATP. They found that a single microtubule/dynein association could produce a force of about 3 pN (assuming that there are 83 dynein molecules per μm flagellum). Using laser tweezers, Ashkin et al. (1990) estimate that a single dynein molecule exerts a force of approximately 2.6 pN. This is in the same ballpark as the force resulting from an actin/myosin association, and consequently, a single dynein molecule can move a 1 μm-diameter vesicle through a non-Newtonian cytoplasm that has a yield value of 0.5 Pa.

The force exerted by a single kinesin molecule has also been measured with laser tweezers. Block et al. (1990) estimated the force exerted by a single kinesin molecule to be between 0.5 and 5 pN. Moreover, a special interferometric version of the laser tweezers suggests that the kinesin molecule moves 8 nm with every stroke (Svoboda et al., 1993; Svoboda and Block, 1994). Thus, a single kinesin molecule, like dynein and myosin, is capable of moving a vesicle that is 1 μm in diameter through a non-Newtonian cytoplasm that has a yield value of 0.5 Pa. By contrast, it appears that more than one kinesin molecule is required to move a vesicle though the cytoplasm of some cells, including kidney epithelial cells, as the measured maximum force, steplike movement, and rate of ATP hydrolysis for a single kinesin molecule are not great enough (Holzwarth et al., 2002).

11.3.2 Polymerization/Depolymerization

The polymerization/depolymerization of microtubules can provide a motive force for cytoplasmic movement. It is possible that the depolymerization of microtubules plays a role in the movement of chromosomes in mitosis (see Chapter 19).

11.4 TUBULIN-BASED MOTILITY

Microtubules are involved in moving organelles around the cell. In Chapter 19, I will discuss the role of microtubules in moving chromosomes during mitosis. The involvement of microtubules in positioning the Golgi apparatus has been studied in Chinese hamster ovary (CHO) cells. The microtubules in these interphase cells have a plus end–distal arrangement typically found in centrosome-containing cells, and the minus ends are embedded in the centrosome adjacent to the central nucleus. When isolated Golgi stacks are added to semi-intact, permeabilized CHO cells, they are captured and transported to the nuclear periphery. Golgi capture and translocation is inhibited by nocodazole, a microtubule inhibitor, indicating that microtubules are necessary for the capture and translocation of the Golgi apparati. Capture and translocation also requires ATP. The capture and translocation do not occur in CHO cells that have been immunodepleted of dynein. Moreover, adding back dynein to the dynein-depleted CHO cell models yields a functional system. Thus, dynein is the translocator that moves the Golgi apparatus from the plus ends of the microtubules to the minus ends so that the Golgi apparatus can go to its typical position in the nuclear periphery of nonmitotic mammalian cells (Corthesy-Theulaz et al., 1992).

Nuclear migration is a prerequisite for asymmetric cell division, and in many plant cells, microtubules are associated with the migrating nucleus. Microtubule depolymerizing agents prevent nuclear migration, indicating that microtubules are involved in providing the tracks for nuclear migration (Kiermayer and Hepler, 1970; Kiermayer, 1972; Schnepf et al., 1982; Mineyuki and Furuya, 1985).

While the majority of movement in pollen tubes can be attributed to actin and myosin, the movement of some organelles is driven by dynein-like and kinesin-like motors, which are differentially localized along microtubules in various regions of the cell (Tiezzi et al., 1992; Cai et al., 1993, 2001, 2000; Moscatelli et al., 1995, 1998; Romagnoli et al., 2003).

Actin and myosin are typically responsible for cytoplasmic streaming. However, microtubules occasionally play a role too. Microtubules are involved in organizing the actin microfilaments, which provide the tracks for cytoplasmic streaming in *Hydrocharis* (Tominaga et al., 1997). Moreover, microtubules are directly involved in powering cytoplasmic streaming in the oocytes of *Drosophila* (Theurkauf, 1994) and the rhizoids of the alga *Caulerpa* (Manabe and Kuroda, 1984; Kuroda and Manabe, 1983). In *Caulerpa*, the cytoplasm streams at a rate of about 3 μm/s and is inhibited in this cell by colchicine, but not by cytochalasin. Microtubules also provide the tracks for cytoplasmic streaming in a *Chlorella*-containing autotrophic species of *Paramecium* (Sikora and Wasik, 1978; Wasik and Sikora, 1980; Cohen et al., 1984; Nishihara et al., 1999). Microtubules are also involved in the intracellular transport of viral movement proteins (Laporte et al., 2003; Heinlein, 2008).

11.5 MICROTUBULES AND CELL SHAPE

It has been known since the 1930s that colchicine causes plant organs and cells to lose their cylindrical form and swell isodiametrically. These early studies showed that colchicine has an effect on the deposition of the extracellular matrix (Eigsti and Dustin, 1955). These results were extended in 1962, when Paul Green showed that colchicine caused the cylindrical cells of *Nitella* to become isodiametric and proposed that a spindle fiber-like element may be responsible for ordering the wall microfibrils. Soon after, Hepler and Newcomb (1964) and Ledbetter and Porter (1963) independently visualized microtubules approximately 0.0175–1 μm from the plasma membrane—a position from where the microtubules may be able to affect cellulose microfibril orientation (VandenBosch et al., 1996; Hepler et al., 2013a; Wasteneys and Brandizzi, 2013). The cortical microtubules were parallel to each other, and as Ledbetter and Porter (1963) wrote, "They are … like hundreds of hoops around the cell."

The orientation of cellulose microfibrils is thought to regulate the direction of cell growth (see Chapter 20; Green, 1969, 1988; Laskowski, 1990; Harold, 1990; Williamson, 1990, 1991; Kropf et al., 1997; Nick, 2000, 2008; Inada and Shimmen, 2001; Inada et al., 2002; Hashimoto, 2011). Randomly arranged microfibrils will give rise to a roughly spherical cell, such as a cortical parenchyma cell. Cells with transversely oriented microfibrils will elongate in a direction perpendicular to the long axis of the microfibrils and will give rise to cylindrical cells like those of the procambium. Microtubules probably direct the orientation of cellulose microfibril deposition. This hypothesis is supported by the following observations:

1. Microtubules are parallel to cellulose microfibrils and predict the orientation of cellulose microfibrils.
2. Agents that inhibit microtubule polymerization or organization affect cellulose microfibril orientation.
3. Extracellular stimuli (e.g., light, gravity, hormones) that affect cellulose microfibril orientation also affect microtubule orientation.

While many studies support the relationship between microtubule orientation and growth, it must be remembered that growth requires the coordination of many cellular processes, and consequently, the transverse orientation of microtubules is not sufficient in itself for elongation (Kropf et al., 1997).

11.5.1 Apical Meristems

Let us look at the organization of microtubules and microfibrils in the cells of the apical meristem (Hanstein, 1870; Tsukaya, 2014). Schmidt (1924) proposed that the apical meristem of angiosperms was organized into two main layers. The surface layer or layers, which he called the *tunica*, divided anticlinally and gave rise to the epidermis. The internal layer, which he called the *corpus*, gave rise to the rest of the plant body. This theory was later incorporated into the cytohistological zonation theory (Foster, 1938), which describes many types of meristems, and is based on differences in cell staining. According to the cytohistological theory, the apical meristem is divided into the distal axial zone, the proximal axial zone, and the peripheral zone. The leaf primordia and procambium arise from the peripheral zone. The proximal axial zone becomes a rib meristem and gives rise to the pith. Genes involved in the structure and function of the meristematic cells have been identified (Ohtsu et al., 2007; Barton, 2010; Yadav et al., 2009, 2114; Takacs et al., 2012; Thompson et al., 2014; Gaillochet et al., 2015; Leiboff et al., 2015; Soyars et al., 2016). One gene in the apical meristem, WUSCHEL, promotes stem cell activity while another, CLAVATA3, suppresses proliferation, while promoting differentiation.

Sakaguchi et al. (1988a,b, 1990; Karahara, 2014) were interested in determining what causes the specific division and expansion patterns seen in the apex. They found that there is a relationship between the orientation of the microtubules in the tunica, corpus, and rib meristem cells as determined with immunofluorescence microscopy and the orientation of the cellulose microfibrils in these cells as determined by polarization microscopy. In the meristem region, the microtubules and microfibrils are coparallel. Briefly, the orientations of the microtubules and

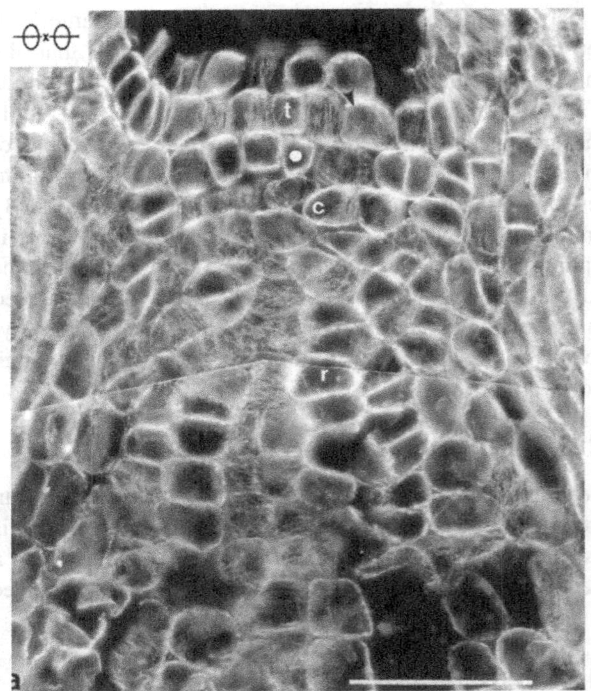

FIGURE 11.13 Immunofluorescence image of microtubules in the apical meristem of *Vinca major* showing the orientation of microtubules in the tunica (t), corpus (c), and rib meristem (r). Bar, 50 μm. *From Sakaguchi, S., Hogetsu, T., Hara, N., 1988a. Arrangement of cortical microtubules in the shoot apex of* Vinca major. *L. Planta 175, 403–411.*

microfibrils in the tunica cells are anticlinal (perpendicular to the surface); in the corpus cells, the orientations are random; and in the rib meristem, the orientations are transverse to the axis of the plant (Fig. 11.13).

These data support the hypothesis that cortical microtubules determine the alignment of adjacent cellulose microfibrils. The reinforcement exerted by the cellulose microfibrils causes the cell to expand at right angles to the long axis of the cellulose microfibrils, and the orientation of the cortical microtubules determines the structure of the stem apex.

11.5.2 Tracheary Elements

Tracheary elements are the cells that comprise the water-conducting system of the plant, which is what Brisseau-Mirbel (1808) referred to as *tubes and spirals*. Tracheary elements are dead at maturity, but of course, they are alive during their development. They have very elaborate and taxonomically distinctive cell wall patterns (Bierhorst, 1971). These include annular, spiral, scalariform, and reticulate thickenings (Fig. 11.14). Much elegant work has been done on determining the contribution of microtubules to cellulose microfibril orientation in these cell types.

In the root apex of *Azolla*, the fate of every cell is known (Gunning et al., 1978a,b,c). Using this material, Hardham and Gunning (1979, 1980) determined the orientation of microtubules in the cells that would give rise

FIGURE 11.14 Nomarski differential interference contrast micrograph of a wound xylem element of *Coleus* showing the secondary wall structure. ×1400. *From Hepler, P.K., Fosket, D.E., 1971. The role of microtubules in vessel member differentiation in* Coleus. *Protoplasma 72, 213–236.*

to the tracheary elements. In this case, the microtubules were coparallel with the microfibrils in the developing tracheids, and moreover, they predicted the site and orientation of microfibril deposition. Colchicine also prevented the normal-ordered deposition of microfibrils, and following colchicine treatment, the normal annular rings were not formed, and the secondary wall material was deposited in irregular masses. This is good evidence that microtubules determine the orientation of cellulose microfibrils in the tracheids of *Azolla*.

Microtubules are oriented parallel to the orientation of cellulose microfibrils in the tracheary elements of many species (Fig. 11.15), and microtubule depolymerizing or disorganizing agents disrupt normal wall deposition in these cells (see Fig. 11.16; Pickett-Heaps, 1967a; Roberts and Baba, 1968; Hepler and Fosket, 1971; Robinson and Quader, 1982; Kobayashi et al., 1988; Falconer and Seagull, 1988; Pesquet and Lloyd, 2011). The concentric orientation of the cellulose microfibrils in bordered pits of conifer tracheids is correlated with the concentric orientation of the microtubules beneath (Uehara and Hogetsu, 1993).

Interestingly, the cell wall deposition patterns and the formation of pits in one tracheary element are coordinated with those in the adjacent cells, indicating the possibility of transcellular or tissue level communication (Sinnott and Bloch, 1944, 1945; Witztum, 1978).

11.5.3 Guard Cells

Guard cells are another specialized cell type that have an unusual but very characteristic cell wall morphology. The cellulose microfibrils are arranged radially around the cell, and this arrangement, known as *radial micellation*, is important for proper stomatal function. Using polarization light microscopy combined with electron microscopy, Palevitz and Hepler (1976) showed that microtubules are coparallel with cellulose microfibrils and they are both

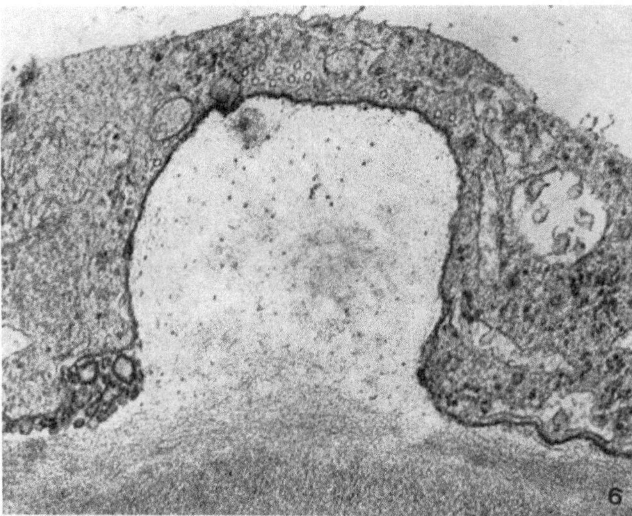

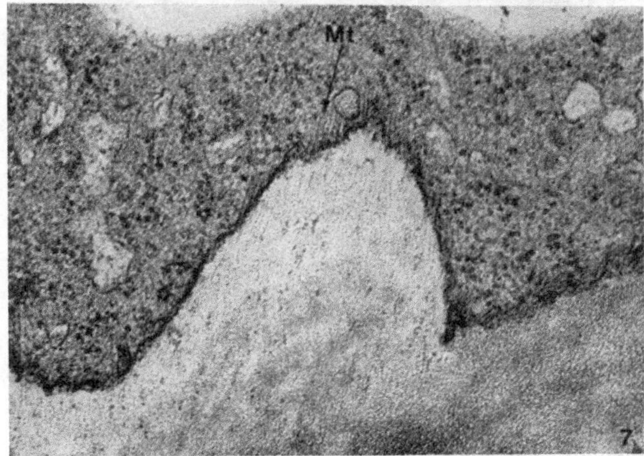

FIGURE 11.15 Transverse (top) and oblique (bottom) sections through a secondary wall thickening of a wound tracheary element of *Coleus*. In the transverse section, transverse sections of microtubules are evident. In the oblique section, coparallel microtubules and cellulose microfibrils are evident. ×60,000. *From Hepler, P.K., Fosket, D.E., 1971. The role of microtubules in vessel member differentiation in Coleus. Protoplasma 72, 213–236.*

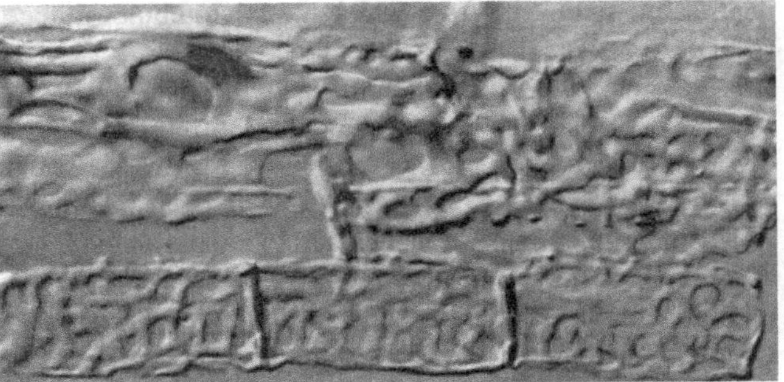

FIGURE 11.16 Nomarski differential interference contrast micrograph of a wound tracheary element of *Coleus* treated with colchicine. The secondary wall structure is unorganized. ×1400. *From Hepler, P.K., Fosket, D.E., 1971. The role of microtubules in vessel member differentiation in* Coleus. *Protoplasma 72, 213–236.*

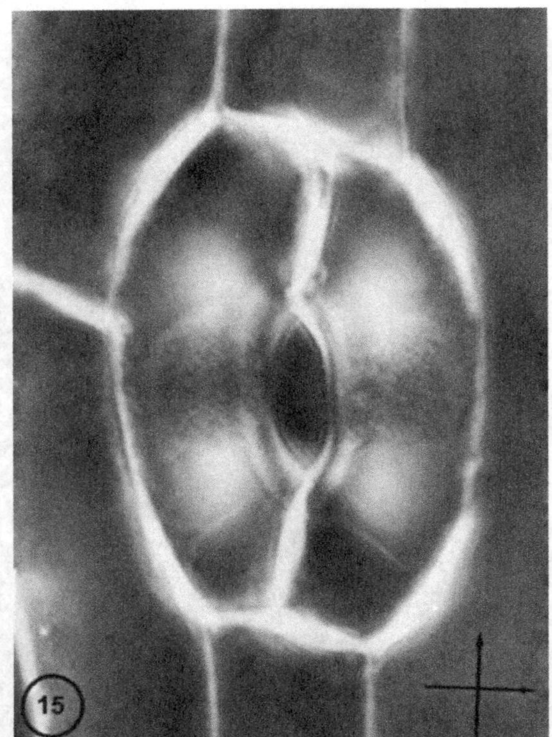

FIGURE 11.17 The guard cells of *Allium* viewed with polarization microscopy. The pattern of birefringence indicates the microfibrils in the extracellular matrix are arranged radially. ×1260. *From Palevitz, B.A., Hepler, P.K., 1976. Cellulose microfibril orientation and cell shaping in developing guard cells of* Allium*: the role of microtubules and ion accumulation. Planta 132, 71–93.*

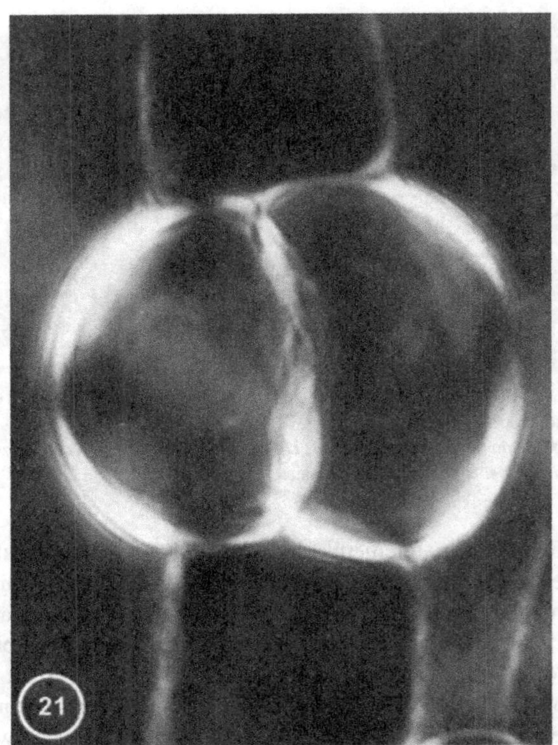

FIGURE 11.18 The guard cells of *Allium* that had been treated with colchicine during development. The pattern of birefringence indicates the microfibrils in the extracellular matrix are arranged randomly. ×1503. *From Palevitz, B.A., Hepler, P.K., 1976. Cellulose microfibril orientation and cell shaping in developing guard cells of* Allium*: the role of microtubules and ion accumulation. Planta 132, 71–93.*

arranged radially (Fig. 11.17). Furthermore, treatment of the cells with microtubule antagonists prevents the normal development of radial micellation and induces a random arrangement of microfibrils in the guard cells and a nonfunctional stomate that lacks a stoma (Fig. 11.18). The radial micellation is required for normal stomatal function (McCormick, 2017; Woolfenden et al., 2017). Interestingly, γ-tubulin, which is associated with sites of microtubule initiation, is present at the ventral side of the cell. Thus, the ventral side is the site from which the microtubules radiate (McDonald et al., 1993).

Microtubules probably influence the shape of all nonspherical plant cells. For example, the morphogenesis of puzzle piece-like epidermal cells proper (or pavement cells; Akita et al., 2015; Higaki et al., 2016; Sapala et al., 2018) and of the highly branched and lobed mesophyll cells also depends on microtubules (Lin et al., 2015). The microtubules are coparallel with the microfibrils that surround and "reinforce" the lobes (Jung and Wernicke, 1990; Panteris et al., 1993; Wernicke et al., 1993).

11.5.4 Extracellular Matrix of *Oocystis*

Perhaps the most dramatic example of a microtubule–microfibril relationship occurs in the alga *Oocystis* (Sachs et al., 1976; Quader et al., 1978). In this alga, the extracellular matrix is polylaminate and layers of microfibrils regularly change their orientation by 90 degrees (Fig. 11.19). This is correlated with a 90-degree change in microtubule orientation. Colchicine inhibits the 90-degree change in microfibril orientation. However, the newly formed microfibrils are not random but are deposited in the same alignment that they were before colchicine treatment (see Fig. 11.20; Robinson et al., 1976b; Grimm et al., 1976). Many other microtubule depolymerizing or disorganizing agents have similar effects (Robinson and Herzog, 1977; Quader, 1986). Perhaps we can conclude that microtubules are responsible for initiating microfibril orientation or changing a given pattern but are not required for the maintenance of a given orientation.

11.5.5 Mechanism of Microtubule-Mediated Cellulose Orientation

How do microtubules influence the orientation of cellulose microfibrils? Cellulose-synthesizing complexes can be readily visualized as rosettes on the P-leaflet of freeze-fractured plasma membranes (Brown, 1985). What is the spatial relationship between these rosettes and the cytoplasmic microtubules? Do the rosettes ride on the microtubules, or do the microtubules form membrane

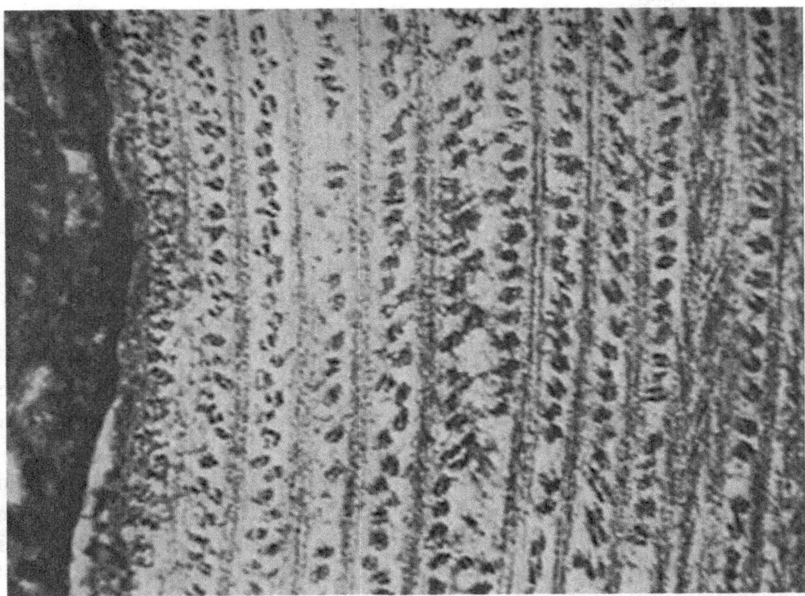

FIGURE 11.19 The polylaminate extracellular matrix of *Oocystis solitaria* showing the alternating perpendicular layers of cellulose microfibrils. Bar, 500 nm. *From Sachs, H., Grimm, I., Robinson, D.G., 1976. Structure, synthesis and orientation of microfibrils. I. Architecture and development of the wall of* Oocystis solitaria. *Cytobiologie 14, 49–60.*

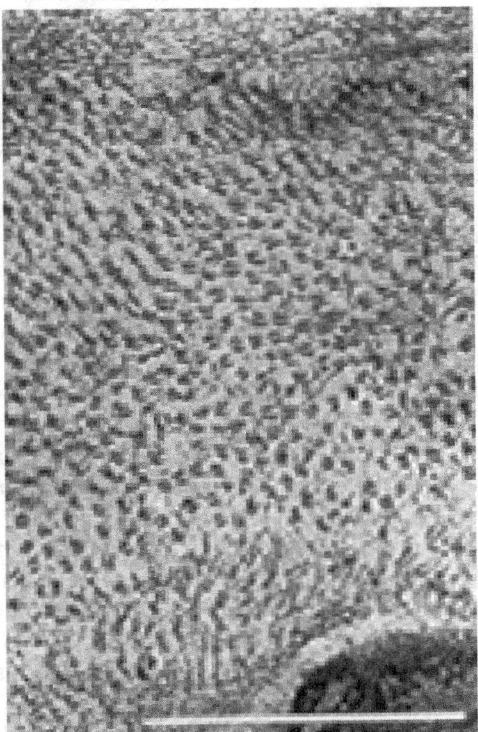

FIGURE 11.20 The extracellular matrix of *Oocystis solitaria* following the application of colchicine. The microfibrils are all coparallel and the extracellular matrix is no longer polylaminate. Bar, 500 nm. *From Grimm, I., Sachs, H., Robinson, D.G., 1976. Structure, synthesis and orientation of microfibrils. II. The effect of colchicine on the wall of Oocystis solitaria. Cytobiologie 14, 61–74.*

channels or riverbeds through which the rosettes ride? There is currently no simple answer to these questions, and the final answers may depend on the cell type and whether the cell is making a primary or secondary wall. Support for the direct involvement of microtubules comes from fluorescence studies in which the cellulose synthesizing complexes seem to ride on top of the microtubules (Lloyd, 2006; Paredez et al., 2006), although the limit of resolution of the light microscope does not allow one to strongly make this case. The direct involvement of microtubules is also supported by the fact that a microtubule-associated protein is one target of an inhibitor of cellulose synthesis (Rajangam et al., 2008). On the other hand, Giddings and Staehelin (1988), using rapid freeze-fixation combined with freeze-fracture and freeze-etching, show that two to seven rosettes, spaced at a constant interval of 30 nm, appear in a row in *Closterium*. Then they subjected the plasma membrane to prolonged etching, which caused the plasma membrane to collapse—except in the areas supported by microtubules. They find that the rosettes are always found adjacent to or between microtubules, but never directly over them. Occasionally they have observed filaments extending between the microtubule and the plasma membrane. They interpret these data to mean that the microtubules make canals or domains in the membrane through which cellulose-synthesizing centers move.

Once the cellulose begins polymerization, it continues in the same direction, thus becoming independent of the microtubule orientation. This may be why microtubules seem to be responsible for the initiation or reorientation of microfibrils, but not the maintenance of a specific orientation. Once initiated, the rigid microfibrils keep growing in the same direction, and the polymerization of cellulose provides the motive force for the movement of the cellulose-synthesizing centers through the canals (Diotallevi and Mulder, 2007).

11.5.6 Tip-Growing Cells

The correlation between microtubules and cellulose microfibrils is not clear in the tip-growing root hairs of *Equisetum hyemale*. Perhaps in this and other tip-growing cells, other mechanisms may influence the orientation of cellulose microfibrils (Emons and Wolters-Art, 1983; Traas et al., 1985; Emons, 1989). The causal relationship between the orientation of microtubules and the orientation of cellulose microfibrils in cortical cells is being questioned (Baskin, 2001). Himmelspach et al. (2003) have shown that the cellulose microfibrils orient transversely in cortical cells in which the cell wall had been disrupted previously with a cellulose synthesis inhibitor and in which the microtubules were disrupted as a result of a temperature-sensitive mutation. These data indicate that there are still yet to be discovered mechanisms that regulate the orientation of cellulose microfibrils, which results in the shapes of plant cells and the forms of plants (Wasteneys and Collings, 2006).

11.6 VARIOUS STIMULI AFFECT THAT MICROTUBULE ORIENTATION

Hormones, light, gravity, fungi, and other stimuli influence the orientation of microtubules in plant cells (Fischer and Schopfer, 1997; Genre and Bonfante, 1997). Hiroh Shibaoka has pioneered the study of the effect of hormones on microtubule orientation. He and his colleagues have shown that cells treated with hormones that cause elongation (e.g., auxin and gibberellic acid) have transverse microtubules, whereas cells treated with hormones that cause isodiametric growth (e.g., ethylene and cytokinins) have randomly arranged microtubules (Shibaoka, 1972, 1974; Shibaoka and Hogetsu, 1977; Takeda and Shibaoka, 1981a,b; Mita and Shibaoka, 1984a,b; Mita and Katsumi, 1986; Akashi and Shibaoka, 1987; Ishida and Katsumi, 1991; Hamada et al., 1994). These data support the hypothesis that microtubules control the orientation of microfibrils.

Light also affects the orientation of microtubules (Wada et al., 1981, 1983, 1990; Kadota et al., 1982, 1985; Murata et al., 1987; Murata and Wada, 1989a,b; Iino et al., 1990). The protonema of the fern *Adiantum* are phototropic toward red light. If the protonema are irradiated on one side of the cell with a microbeam, they bend toward the light. Preceding the bend, the microtubular band remains on the shaded side but disappears on the lighted side, perhaps allowing a randomization of the microfibrils on the lighted side and the formation of a new growing tip (Wada et al., 1990a,b).

Irradiation of the cells with polarized red light oriented 45 degrees relative to the long axis of the cell causes bending. This bending, which is known as *polarotropism*, is preceded by a shift in the angle of the microtubule band, so it predicts and surrounds the new growing tip (Wada et al., 1990a,b). Likewise, the microfibrils change their orientation in parallel with the microtubules (Wada et al., 1990a,b).

Hush and Overall (1991) find that both electrical and mechanical fields are capable of orienting cortical microtubules in the cells of pea roots. Electric fields of only 0.36 V/cm (<0.001 V/cell), oriented perpendicular to the long axis of the root, cause a change in the microtubule arrangement from transverse (relative to the axis of the root) to longitudinal (which is perpendicular to the applied electric field). Similar results were found by White et al. (1990) in *Mougeotia* protoplasts. Because the resistance of the plasma membrane is so much greater than the resistance of the cytoplasm, the voltage drop is almost exclusively across the plasma membrane. Therefore, it is possible that the electric field causes a reorientation of a dipole in a plasma membrane protein, which produces a conformational change in the rest of the protein that affects its ability to bind and/or orient microtubules or microtubule-associated proteins (Hush and Overall, 1991).

Hush and Overall (1991) have also applied mechanical forces of 0.12 N perpendicular to the longitudinal axis of the roots and observed that the microtubules reorient from transverse to longitudinal—that is, perpendicular to the direction of applied force. Williamson (1990) proposes that the mechanical stress applied to the extracellular matrix causes a strain in the cellulose microfibrils. This strain in turn causes a conformational change in transmembrane proteins that bind the cellulose microfibrils in the E-space and cortical microtubules in the P-space. The conformational change in these proteins can then cause a change in the orientation of cortical microtubules. Thus, microtubules can cause the orientation of cellulose microfibrils, and the microfibril orientation may also influence the orientation of microtubules in a feedback loop (Williamson, 1990, 1991). As I will discuss in Chapter 20, microtubules may be part of a structural continuum that includes the cytoskeleton, the plasma membrane, and the extracellular matrix (Akashi et al., 1990; Laporte et al., 1993; Joos et al., 1994; Gardiner et al., 2001; Verger et al., 2018).

11.7 MICROTUBULES AND CYTOPLASMIC STRUCTURE

Isolated microtubules behave as a non-Newtonian fluid consistent with the postulate that microtubules are one of the protein polymers that are responsible for the viscoelastic, thixotropic, and non-Newtonian properties of the cytoplasm (Sato et al., 1988). Similar to cytoplasm, the elastic modulus of microtubules is $<4\ N/m^2$ and the viscosity varies from 0.01 to 10 Pa s, as the rate of shear changes from 10^2 to $10^{-2}\ s^{-1}$ (Buxbaum et al., 1987; Sato et al., 1988).

11.8 INTERMEDIATE FILAMENTS

A third major cytoskeletal system exists that is composed of intermediate filaments (Fig. 11.21). These filaments are approximately 10 nm in diameter and therefore intermediate between actin microfilaments (5 nm) and microtubules (24 nm). There are a variety of proteins that make up the various types of intermediate filaments (Goodbody et al., 1989; Hargreaves et al., 1989a,b; Su et al., 1990a,b; Russ et al., 1991; Staiger and Lloyd, 1991; Li and Roux, 1992; McNulty and Saunders, 1992; Mizuno, 1995; Blumenthal et al., 2004). The immunological studies of Yang et al. (1992) suggest that the intermediate filament protein keratin may form the microtrabecular lattice of plant cells.

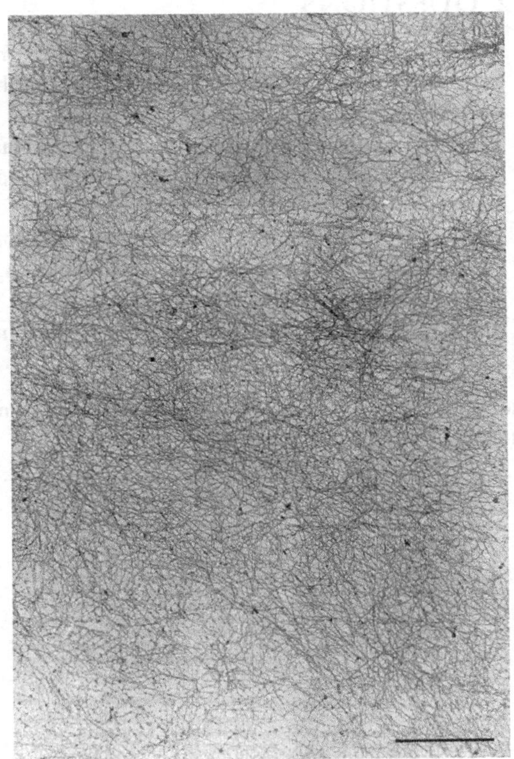

FIGURE 11.21 Intermediate filaments in a whole mount of a carrot suspension culture cell that has been extracted with detergent. Bar, 1 μm. *From Yang, C., Xing, L., Zhai, Z., 1992. Intermediate filaments in higher plant cells and their assembly in a cell-free system. Protoplasma 171, 44–54.*

Beven et al. (1991) suggest that in plants the intermediate filament protein lamin may form the network of 10-nm filaments that are found just inside the inner membrane of the nuclear envelope as it does in animal cells (Pappas, 1956; Fawcett, 1966; Gerace and Blobel, 1980; Aebi et al., 1986; Franke, 1987; Masuda et al., 1997; Fiserova et al., 2009; Fiserova and Goldberg, 2010; Wilson and Foisner, 2010). The lamins are involved in regulating the size and shape of the nucleus, chromatin organization, positioning the nuclear pore complex, and are part of the linker between the nucleoskeleton and the cytoskeleton that supports force transmission from the cytoplasm to the nucleus across the nuclear envelope (Dittmer et al., 2007; Dittmer and Richards, 2008; Dittmer and Misteli, 2011; Goto et al., 2014; Zhou et al., 2012, 2015a,b; Ciska and Moreno Díaz de la Espina, 2013, 2014; Sakamoto and Takagi, 2013; Tatout et al., 2014; Wang et al., 2013; Graumann, 2014; Tamura et al., 2015; Guo et al., 2017). However, while higher plants contain proteins that may have analogous functions to lamins, they lack the homologous DNA sequences for lamins. Thus, it is important that we take into consideration the contrasting ideas of analogy (form and function) and homology (shared ancestry), and convergent and divergent evolution, in understanding the life of a cell. That is, we must remember the thinking of Richard Owen (1848, 1849) and Charles Darwin (1859).

11.9 CENTRIN-BASED MOTILITY

Some motile proteins are actually contractile elements themselves (Stebbings and Hyams, 1979). These include the contractile protein in the spasmoneme of *Vorticella* and centrin, which was discovered in *Chlamydomonas* (Salisbury et al., 1984, 1988; Wright et al., 1985; McFadden et al., 1987; Melkonian, 1989). Centrin is a component of the MTOCs or pericentriolar material in algae, protozoa, mammals, and higher plants (Baron and Salisbury, 1988; Wick and Cho, 1988; Hiraoka et al., 1989; Katsaros et al., 1991). Centrin is also found in the multilayered structure in motile plant sperm cells (Vaughn et al., 1993; Hoffman et al., 1994).

11.10 TENSEGRITY IN CELLS

As a rule, the human-made structures around us are built "brick on brick" and are thus under compression. An exception to this rule is a suspension bridge, which is under tension. It turns out that nature, unlike humans, builds structures, like spider webs, that depend on tension to maintain their integrity. Buckminster Fuller coined the term *tensegrity* to describe these structures, the integrity of which is tension based, and Don Ingber (1993) has applied this concept to cellular structures. There are many proteins in the cell that are capable of making filamentous structures that are either elastic or rigid. There are also motor proteins that are able to produce shearing forces that can cause either tension or compression. Laser microbeam experiments have shown that the cytoplasm is under tension because irradiation causes a rapid retraction of the cytoplasm (Goodbody et al., 1991; Schmid, 1996).

The architectural entity formed from the tensile elements is capable of transmitting mechanical information throughout the cell. Ingber and Jamieson (1985) and Ingber and Folkman (1989a,b) suggest that the intermediate filaments, actin filaments, microtubules, and their associated motor proteins form a tension-like scaffold in the cell and have shown that the potential energy stored in this tension is capable of doing work and directing cell differentiation. A dynamic, information-bearing, and transmitting cellular structure was envisioned by Rudolph Peters as early as 1929 and was called the *cytoskeleton* by Joseph Needham in 1936.

11.11 SUMMARY

Mechanical movement within the cell or by the cell is a fundamental characteristic of life. These mechanical processes include cell locomotion, cytoplasmic streaming, chromosome segregation, vesicle movement, DNA replication, transcription, translation, protein translocation through a membrane, protein folding, and enzyme activity (Bustamante et al., 2004). The movements are characterized by their low Reynolds numbers. In Chapter 10, I discussed the involvement of actin in cell motility, and in this chapter, I have discussed another motility-generating system based on microtubules. Ciliary motion is the best characterized microtubule-based system. The shearing stress that powers the cilia is generated by the interaction of dynein with microtubules. We have learned that dynein is also present in nonciliated cells where it acts as a minus end–directed motor that walks down a microtubule to move vesicles through the cytoplasm. We also learned about kinesin, which is a mechanochemical transducer that moves vesicles along a microtubule or can move microtubules relative to a place where kinesin is anchored. The canonical kinesin is a plus end–directed motor, whereas some kinesin-related proteins, and especially those found in higher plants, are minus end–directed motors. A single amino acid substitution can change a minus end–directed motor to a plus end–directed motor. This indicates to me that assigning a function to a protein that has 99% homology with another protein of known function may give misleading results in cases where the identity of a single amino acid confers on the protein a specific and crucial function. It is as important to classify the components of life in terms of their form and function as it is to classify them by their shared ancestry. It is possible that there are two methods of deriving variation in protein sequences—one method may result from the promiscuous concatenation of discrete DNA sequences, and the other method may result from a gradual and continuous change in a given DNA sequence.

We have also learned that microtubules are involved in orienting cellulose microfibrils and thus determine the shape of the cell. In general then, microtubules are involved in cell motility and cell shape.

11.12 QUESTIONS

11.1. Why do you think there are two classes of motile systems in cells: one based on actin and one based on tubulin?

11.2. Why do you think there is more than one class of mechanochemical ATPases that move along microtubules?

11.3. How does the inclusion of analogy with homology, convergent evolution with divergent evolution, and discrete variation with continuous variation help one to understand the processes that contribute to the life of a cell?

The references for this chapter can be found in the references at the end of the book.

Chapter 12

Cell Signaling

Between stimulus and response there is a space. In that space is our power to choose our response. In our response lies our growth and our freedom.

Viktor E. Frankl (1985).

In the early 1960s, the idea that a simple inorganic ion such as calcium controls contraction was not popular among most biochemists — the prevailing belief was that such an important biological phenomenon as contraction should be regulated by sophisticated organic molecules. So Ebashi had a hard time getting his ideas taken seriously, despite his clear evidence. Only after the discovery of troponin and his elucidation of the mechanism did everybody accept the regulatory role of calcium ions.

Makoto Endo (2006).

12.1 THE SCOPE OF CELL REGULATION

Life, according to Herbert Spencer (1864), is "a continuous adjustment of internal relations to external relations." Morley Roberts (1938) wrote, "The irritability, or excitability, of the cell, or of the minutest possible portion of protoplasm, is a *sine qua non* of its existence and powers of reaction, and the sole source of feeling and life in all animals." What happens in the cell when it is faced with a stochastic or planned change in its environment? The cell undergoes adjustments that result in a variety of changes that range from the maintenance of homeostasis to a change in the developmental program. The regulatory mechanisms discussed in this chapter are relatively rapid processes that take place in the time scale of tens of milliseconds to several minutes. How much does the quality and quantity of our own life and the lives of others around us depend on the biophysical and biochemical events involved in cell signaling?

The vitality of plants is often underappreciated (Hall, 2002; Baluska et al., 2006). However, if we were to walk quietly and observantly through a garden, it would become increasingly clear that it is a normal and ubiquitous property of plants to sense and respond to their environment (Pfeffer, 1875; Darwin, 1881, 1897; Bose, 1906, 1913, 1926, 1985; Haberlandt, 1906, 1914; Goebel, 1920; Bünning, 1953, 1989; Jaffe and Galston, 1968; Sibaoka, 1969; Jaffe, 1980a,b; Bradbeer, 1988; Simons, 1992; Sopory et al., 2001; Darnowski, 2002; Trewavas, 2002, 2006, 2014, 2016; Ueda and Nakamura, 2007; Volkov et al., 2008; Trewavas and Baluska, 2011; Leopold, 2014; Karban, 2015). The sensing behavior of plants becomes most obvious when we watch their movements. The leaves of *Mimosa* and *Albizia*, for example, open during the day and show sleep movements at night (de Mairan, 1729); the leaves of *Mimosa* fall rapidly in response to the touch of an animal; the flowers of the daylily open at dawn and close at dusk. Even the seeds in the ground beneath our feet are able to exquisitely sense the temperature and light conditions, and then break dormancy so that the radical emerges in the appropriate season. If we were lucky enough to walk around a bog, we might even see the leaves of the Venus flytrap or the hairs of a sundew capture its meaty meal! We do not even have to leave our houses to see plants in action—houseplants in the window bend toward the light, and in every pot, the roots grow down, and the shoots grow up in response to gravity.

Alexander Pope (1745) wrote, "Know then thyself, presume not God to scan. The proper study of Mankind is Man." Perhaps we should ask: Could the study of plant behavior help us to understand man and the evolution of consciousness? Raoul Francé (1905) wrote, "What grander lesson could the speechless plants give than that which they have taught us: *that their sense life is a primitive form, the beginning of the human mind* ... it tells us that after all the living world is but mankind in the making, and that we are but a part of all." According to Lynn Margulis (2001), consciousness, as defined as an awareness of the external environment, began with life itself.

Herbert Jennings (1906) concluded that there is a biological basis for the distinctions between right and wrong that is based on the processes of cell signaling. Jennings (1933) wrote:

To determine what is to be done, what not to be done; in other words, to determine right and wrong, is an insistent problem for all organisms ... The daily, the hourly, occupation of most organisms—high or low—is the seeking of conditions that are favorable for life and the avoiding of conditions that are unfavorable ... With all organisms, life is a continuous process of selecting one line of action and

Plant Cell Biology. https://doi.org/10.1016/B978-0-12-814371-1.00012-6

rejecting another, of determining whether certain actions are right or wrong. The life of the single-celled organism is such a continual process of trial ... it has its dramatic crises as has the life of higher creatures.

Are the adaptive responses of an organism to the environment deterministic or subject to physical processes that are fundamentally probabilistic (Lillie, 1927)? According to Arthur Compton (1931), "the actions of the organism depend upon events on so small a scale that they are appreciably subject to Heisenberg uncertainty. This implies that the actions of a living organism cannot be predicted definitively on the basis of its physical condition." Indeed, Pascual Jordan (1938, 1939, 1942a,b, 1948; Jordan and Kronig, 1927) began a program to eliminate any mechanistic and deterministic basis for biology and tried to start "quantum biology" (Heilbron, 1986; Beyler, 1996, 2007; Popp and Beloussov, 2003).

Is a highly individualistic and courageous response determined by an individual's cellular balance of stochastic and deterministic material elements, or are one's actions dependent on nonmaterial elements, including free will? What causes a man like Martin Niemöller (1941) to shun cowardly, sheeplike, faddist behavior and stand up to an authority like Adolf Hitler (1943)? Niemöller initially supported Hitler, but by 1937, he was arrested by the Gestapo for his open opposition to Hitler and incarcerated in the Sachsenhausen and Dachau concentration camps. Nevertheless, he still berated himself for not doing more to fight the tyranny, and he was paraphrased in the *Congressional Record* (October 14, 1968, page 31,636) as having said:

When Hitler attacked the Jews I was not a Jew, therefore I was not concerned. And when Hitler attacked the Catholics, I was not a Catholic, and therefore, I was not concerned. And when Hitler attacked the unions and the industrialists, I was not a member of the unions and I was not concerned. Then Hitler attacked me and the Protestant church—and there was nobody left to be concerned.[1]

Does this kind of behavior depend on chemistry (Cashmore, 2010) or quantum uncertainty and statistical variation? Is free will a type of usable energy that, according to John Eccles (1979), is capable of inducing physicochemical reactions such as ion channel gating or the secretion of neurotransmitters (Popper and Eccles, 1977)?

12.2 WHAT IS STIMULUS-RESPONSE COUPLING?

Until recently, stimulus-response coupling was studied using the "black-box approach." The black-box approach is described by Ashby (1958):

The Problem of the Black Box arose in electrical engineering. The engineer is given a sealed box that has terminals for input, to which he may bring any voltages, shocks, or other disturbances he pleases, and terminals for output, from which he may observe what he can. He is to deduce what he can of its contents.

With the help of biophysical, biochemical, genetic, and various "-omic" tools, cell biologists are now cracking open the black box and understanding each step in the signal transduction chain.

We will consider a stimulus to be any environmental, physiological, or biological signal that induces a change in a biophysical, biochemical, physiological, morphological, or developmental process in the cell. Common stimuli include light (Bünning and Tazawa, 1957), hormones (Leopold, 1964; Leopold and Kriedemann, 1975; Thimann, 1977; Strader and Bartel, 2008), florigen (Ayre and Turgeon, 2004), neurotransmitters (Roshchina, 2001), touch (Jaffe and Galston, 1968; Jaffe, 1980a,b; Jaffe et al., 2002; McCormack et al., 2006), time (Cole, 1957; Hastings and Sweeney, 1957; Sweeney and Hastings, 1958; Sweeney and Haxo, 1961; Broda and Schweiger, 1981; Sweeney, 1987; Berger et al., 1992; Chandrashekaran, 1998; Suzuki and Johnson, 2001; Mittag et al., 2005), gravity (Nemec, 1899; Wayne and Staves, 1996a; Sack, 1997), and biological interactions (e.g., pollen and stigma or fungus and plant).

A stimulus must be able to impart a certain amount of free energy to the cell that is greater than the energy of thermal noise ($E < kT$) or a receptor will not be able to perceive the stimulus (Bialek, 1987; Block, 1992). A stimulus will have no effect on a cell unless that cell has the appropriate receptor, and if the cell has an appropriate receptor, the stimulus will provide the cell with information. The presence or absence of a certain constellation of receptor proteins will provide a certain degree of selectivity in terms of which cells respond to a stimulus. This competence of a cell to perceive a stimulus is predetermined by the genetic system. Substantial progress is going on in identifying hormone (auxin, gibberellin, cytokinin, ethylene) and other chemical (salicylic acid and nitric oxide) receptors, light receptors (phytochrome, cryptochrome, and phototropin), and gravity receptors

1. Others quote him to have said, "In Germany they came first for the Communists and I didn't speak up because I wasn't a Communist. Then they came for the Jews and I didn't speak up because I wasn't a Jew. Then they came for the trade unionists and I didn't speak up because I wasn't a trade unionist. Then they came for the Catholics and I didn't speak up because I was a Protestant. Then they came for me—and by that time no one was left to speak up." Sibylle Sarah Niemöller von Sell, in response to a student's question, "How could it happen?," quoted her husband, saying: "First they came for the Communists, but I was not a Communist so I did not speak out. Then they came for the Socialists and the Trade Unionists, but I was neither, so I did not speak out. Then they came for the Jews, but I was not a Jew, so I did not speak out. And when they came for me, there was no one left to speak out for me."

(Jones and Venis, 1989; Hooley et al., 1991; Wayne et al., 1992; Furuya, 1993, 2005; Ahmad and Cashmore, 1996; Lin et al., 1996a,b; Cashmore, 1997, 1998; Briggs, 1998, 2005; Christie et al., 1998, 1999; Trewavas, 2000; Salomon et al., 2000; Briggs et al., 2001a,b; Christie and Briggs, 2001; Briggs and Christie, 2002; Quail, 2005; Gilliham et al., 2006; Hegemann, 2008).

We will consider a response to be any biophysical, biochemical, physiological, morphological, or developmental process that changes after the cell receives the stimulus. Common responses include germination, flowering, osmoregulation, turgor regulation, chloroplast movement, phototaxis, leaf movements, gravitropism, senescence, abscission, and the processes involved in plant defense. The presence or absence of a constellation of response elements that are required for each of these responses will also provide a certain degree of selectivity. The competence of a cell to respond to a stimulus in a given manner is predetermined by the genetic system.

A signal transduction chain comprises all the biophysical and/or biochemical steps that occur between the perception of the stimulus and the response. In the simplest case, the signal transduction chain acts like a switch. It initiates the response when the cell is presented with a stimulus. As an analogy, think of the electric system in your house. All the switches are very similar, yet when you turn on the television set you see a television show; when you turn on the toaster, your bread gets brown; when you turn on your radio, you hear music. Your appliances are preprogrammed to respond to stimuli in a special way, so when you turn on a switch on the television, you see a show and your bread does not get brown. When a fern spore is given red light, it germinates and it does not make a nitrogen-fixing nodule; or when a characean cell gets an electrical stimulus, it stops streaming and it does not flower. In the simplest cases, the cell is preprogrammed to respond to one stimulus with one given response. A part of the response may be to reprogram the cell with new receptors and/or response elements so that it continues its developmental fate.

Considering the signal transduction chain as a switch has widespread appeal because of its elegant simplicity. Yet there is a growing awareness that the mechanisms that underlie cellular signaling are more complex and intricate. We already know that more than one switch exists: one switch is turned on by Ca^{2+} and others by cyclic adenosine monophosphate (cAMP) or cyclic guanosine monophosphate. Moreover, any two switches can be redundant, antagonistic, hierarchical, or sequential (Rasmussen, 1981; Barritt, 1992). The common intracellular switches such as Ca^{2+} and cyclic nucleotides are known as *second messengers*. A second messenger is the first relatively stable intracellular chemical that increases its concentration in response to the stimulus and can influence the response elements in the cell. cAMP at first was proposed to be the second messenger for all hormone responses, and Ca^{2+} as the second messenger in muscle contraction, secretion, and egg activation. However, through continued experimentation, it became apparent that the participation of both of these signaling systems was more widespread than had been thought originally. The totality of the components involved in cellular signaling has been dubbed the *signalome* (Reddy, 2001).

It is becoming clear that interacting and independent pathways are involved in coupling stimuli with responses when a number of responses can be identified in a single plant cell (Bowler et al., 1994; Wu et al., 1996; Neuhaus et al., 1997; Iseki et al., 2002). For example, *Dunaliella* cells have three independent responses to light: the phototactic response, the step-up photophobic response, and the step-down photophobic response. All these responses involve a rapid and subtle regulation of the ciliary beat (Wayne et al., 1991). The phototactic response involves a rapid turn toward blue light; the step-up photophobic response involves a change from a forward-swimming ciliary beat to a backward-swimming flagellar beat when green light is turned on; and the step-down photophobic response involves a turn of 90°degrees or more immediately after the green light is shut off. Therefore, different signal transduction chains must exist in the single *Dunaliella* cell or else all three responses would be turned on simultaneously. The three different chains are not completely independent but have common components. For example, they all require external Ca^{2+} (Noe and Wayne, 1990).

It is important to understand at the outset that we are discussing the effect of a stimulus on the response of single cells, whether they be individual organisms or part of a multicellular organism. This is important because neighboring cells in a multicellular organism can have quite different responses to the same stimulus. Hans Mohr (1972) has illustrated this elegantly in mustard seedlings. Red light causes the epidermal cells of the hypocotyl to differentiate hairs and the hypodermal cells to synthesize anthocyanins. Furthermore, the competence of the hypodermal cells to produce anthocyanins depends on the duration of time the seedling spent in the dark before it was irradiated. Thus, Mohr has shown that competence is a dynamic spatiotemporal phenomenon that depends on time, as well as on the position of the cell in the organism. Consequently, when investigating stimulus-response coupling, it is essential to isolate a given cell type at a given time to characterize how it responds to a stimulus, if we want to use established cellular paradigms to describe the signal transduction chain. These paradigms explain what happens in a single cell—not a whole cow or a whole plant.

Plant physiological experiments are often interpreted as if the plant were a single cell, or as if all the cells in the plant body were identical. If in fact each cell in a multicellular organism has a different constellation of receptors and response elements, as well as differences in the kinetics or types of signal transduction chains, then it would be ludicrous to provide a stimulus to the whole seedling and then grind it all up to find a change in the concentration of a second messenger. First of all, the temporal resolution would be low, and second, the response of all the cells would be averaged. Thus, if only the epidermal or hypocotyl cells responded to a stimulus, but the cortical or pith cells did not, a real change in the concentration of a second messenger in the responding cells could go undetected.

Of course, higher levels of regulation exist in plants where communication mediated by plasmodesmata, as well as chemical, electrical, and mechanical gradients, between cells in different tissues or tissue systems occurs (Chapter 3; von Sachs, 1887; Osterhout, 1906; Leopold and Kriedemann, 1975; Thimann, 1977; Sachs, 2006; Taiz and Zeiger, 2006). There is also integration at the organ level where two or more cells in an organ may communicate (e.g., abscission), or at the whole plant level where one or more cells in an organ may compare their states with one or more cells in another organ (e.g., apical dominance). An understanding of these higher-level processes requires experimental designs that take into consideration both the cellular and organismal levels of organization.

A stimulus contains energy, and a signal transduction chain involves the conversion of the energy of the primary stimulus, which may be light, gravity, chemical, heat, or electrical energy, to the energy of an intracellular molecule that can be coupled to the biochemical machinery in the cell. Indeed, as James Clerk Maxwell (1877) wrote, "The transactions of the material universe appear to be conducted, as it were, on a system of credit (except perhaps that credit can be artificially increased, or inflated). Each transaction consists of the transfer of so much credit or energy from one body to another." The energy of the stimulus is transferred to a receptor. The receptor often takes advantage of the potential energy already present, typically in the form of Ca^{2+} difference across the plasma membrane to activate the cell.

12.3 RECEPTORS

There are a variety of types of receptors in cells. The receptor proteins that act on ion channels are called *channel-linked receptors*, and the acetylcholine receptor is the canonical example of this class. Channel-linked receptors in plants (Demidchik, 2006) include channelrhodopsin, which mediates phototaxis and photophobic responses in *Chlamydomonas*. Channelrhodopsin is an example of a channel-linked receptor that passes cations, including H^+, K^+, Na^+, and Ca^{2+}, in response to light. The activation of this channel results in a depolarization of the plasma membrane (Sineshchekov et al., 2002; Govorunova et al., 2004; Berthold et al., 2008; Hegemann, 2008). By transforming cells with the gene for channelrhodopsin, the channelrhodopsin can be used as a nanoswitch to rapidly and noninvasively activate action potentials with light in normally light-insensitive neurons (Li et al., 2005; Nagel et al., 2005; Miller, 2006; Zhang et al., 2006a,b; Hegemann and Tsunoda, 2007; Zhang and Oertner, 2007).

Other receptors operate directly as enzymes after binding a ligand or after being activated by a physical stimulus. The insulin receptor is the canonical example of the catalytic receptor class. The insulin receptor is an integral, transmembrane protein with a cytoplasmic domain that functions as a protein kinase. In plants, the receptor involved with pollen-stigma incompatibility reactions (Stein and Nasrallah, 1993; Stein et al., 1996) and a blue-light photoreceptor, phototropin, which is a plasma membrane–localized photoreceptor involved in phototropism, stomatal opening, chloroplast movement, leaf expansion, and hypocotyl growth inhibition (Huala et al., 1997; Christie et al., 1998; Kagawa et al., 2001; Kinoshita et al., 2001; Sakai et al., 2001; Sakamoto and Briggs, 2002; Briggs, 2005), are integral membrane proteins with protein kinase activity. The blue-light photoreceptor, cryptochrome (Cashmore, 2005), and the red-light photoreceptor, phytochrome (Yeh and Lagarias, 1998; Suetsuga et al., 2005; Rockwell et al., 2006), both show protein kinase activity. However, the contribution of this activity to signal transduction is still being investigated and alternative hypotheses are being advanced (Quail, 2005). Indeed, just as related members of the rhodopsin family can initiate signal transduction chains through differing mechanisms (Hegemann, 2008); allied members of the phytochrome family may also initiate responses through different mechanisms. Interestingly, plants have made use of the promiscuity of DNA to create chimeric phytochrome-like photoreceptors from the canonical red-light and blue-light photoreceptors. For example, neochromes found in the alga *Mougeotia* and the fern *Adiantum* are red- and blue-light photoreceptors, the genes of which are composed of phytochrome and phototropin nucleotide sequences (Nozue et al., 1998; Suetsuga et al., 2005; Suetsuga and Wada, 2007).

Other types of catalytic receptors have been discovered in plants. These include the blue-light receptor for the step-up photophobic response in *Euglena* and the blue-light receptor for the branching response in golden algae. The photoreceptor in *Euglena*, which is not an integral membrane protein, is a flavoprotein that is found in the quasicrystalline paraflagellar body in *Euglena* and along the whole length of the flagellum in related genera. The photoreceptor protein catalyzes the conversion of ATP to cAMP in a light-dependent manner (Iseki et al., 2002; Häder et al., 2005). By transforming cells with the gene for this receptor, Schröder-Lang et al. (2007) have used this photoactivated adenylate cyclase as a tool to use light to increase the concentration of cAMP in transformed cells that are normally light insensitive.

A third class of receptors is known as G-protein–linked receptors (Ma et al., 1990, 1991; Coughlin, 1994; Ma, 1994, 2001; Mu et al., 1997; Assmann, 2002; Perfus-Barbeoch et al., 2004; Assmann, 2005; Pandey et al., 2006; Grill and Christmann, 2007; Liu et al., 2007a; Hegemann, 2008; Martinac et al., 2008). These receptors are mainly integral plasma membrane proteins with seven transmembrane domains that indirectly activate or inactivate plasma membrane–bound enzymes or ion channels through the activation of a G-protein. Vertebrate rhodopsin and the norepinephrine receptor are examples of G-protein–linked receptors, and similar G-protein–linked receptors can be found in plants. The best characterized G-protein–linked receptor in higher plants is the abscisic acid receptor (Liu et al., 2007a), although there is some controversy surrounding this research (Johnston et al., 2007; Liu et al., 2007b). The direct action of the AbA receptor is still unknown, although it appears to inhibit the activity of a protein phosphatase (Ma et al., 2009; Melcher et al., 2009; Park et al., 2009; Pennisi, 2009; Yin et al., 2009; Kline et al., 2010; Guo et al., 2011; Zhang et al., 2015; Guo et al., 2017a,b). This teaches us that there is more than one way that receptors can stimulate the phosphorylation of proteins—by having protein kinase activity itself, by activating the activity of a protein kinase, or by inhibiting the activity of a protein phosphatase. Other G-protein–linked AbA receptors have been found (Pandey et al., 2009). In almost all cases, activated G-protein–linked receptors bring about an increase in the concentration of Ca^{2+} or cyclic nucleotides.

G-proteins involved in cell signaling are composed of three subunits (α, β, and γ) and consequently are known as *heterotrimeric* G-proteins. The α subunit hydrolyzes guanosine triphosphate (GTP), and the β and γ subunits form a dimer that anchors the G-protein to the cytoplasmic side of the plasma membrane. In the inactive form, the G-protein exists as a trimer with guanosine diphosphate (GDP) bound to the α subunit. When a G-protein becomes activated, the α subunit binds a molecule of GTP in exchange for its bound GDP, and then dissociates from the $\beta\gamma$ dimer and diffuses in the plane of the membrane until it encounters a protein to which it can bind. The α subunit then binds tightly to the protein and either activates or inactivates it. The α subunit is active for only as long as the GTP molecule remains intact, which is typically 10–15 seconds. Once the GTP is hydrolyzed, the α subunit becomes inactive and dissociates from the protein to which it had been bound, and that protein is no longer activated or inactivated.

The activation of G-proteins provides an amplification step in cell signaling because the α subunit can remain activated for 10–15 seconds; long after the primary stimulus has dissociated. G-proteins can remain artificially activated for a long time by introducing GTP-γ-S into the cell. This molecule cannot be hydrolyzed, and thus the G-protein remains activated for a long time. This is an experimentally good way to determine whether G-proteins are involved in cell signaling. Stimulatory G-proteins can also be permanently activated by cholera toxin while inhibitory G-proteins can be inhibited by pertussis toxin.

A fourth class of receptor is a zinc-containing transcription factor that acts directly on gene expression (see Chapter 16). The steroid hormone receptor is the canonical example. The hormone binds to receptors in the cytosol, and the receptor–ligand complexes dimerize and move into the nucleus, where the dimer binds to a specific DNA sequence. Like the steroid hormone receptor, aureochrome, the blue-light photoreceptor that is involved in stimulating branching and initiating sex organs in gold-colored stramenopile algae, including *Vaucheria* and *Fucus*, is also a transcription factor (Takahashi et al., 2001, 2007a,b; Hisatomi et al., 2014; Kerruth et al., 2014; Nakatani and Hisatomi, 2015; Akiyama et al., 2016; Banerjee et al., 2016; Takahashi, 2016; Matiiv and Chekunova, 2018). Hironao Kataoka hopes to support science in fields other than plant photobiology by promoting aureochrome as a light-activated transcription factor that can be used to activate specific genes in the nucleus of cells that typically do not respond to light.

12.4 CARDIAC MUSCLE AS A PARADIGM FOR UNDERSTANDING THE BASICS OF STIMULUS-RESPONSE COUPLING

Movement is a characteristic of life, and muscle cells are specialized for movement. To be functional, the movement must be regulated. The regulation takes place in the form of cycles—cycles of contraction and relaxation. Setsuro Ebashi (Kumagai et al., 1955; Ebashi, 1960, 1961a,b, 1968; Ebashi et al., 1960; Ebashi and Ebashi, 1962; Ebashi and Lipmann, 1962; Endo, 2006, 2008, 2011; Godfraind, 2007; Gergely, 2008; Muscatello, 2008; Perry, 2008) discovered that actomyosin prepared from pure fractions of actin and myosin contracts when ATP is added, whereas actomyosin prepared directly from muscle does not contract unless Ca^{2+} is added along with the ATP. Subsequently, Ebashi discovered troponin, the factor in skeletal and cardiac muscle that conferred Ca^{2+} sensitivity to actomyosin.

Ebashi also found that a glycerinated muscle does not relax after being stimulated by ATP. So he homogenized the muscle, centrifuged it at 6000 × g, and collected the supernatant. He then added the supernatant along with ATP to a glycerinated muscle and the muscle relaxed. The relaxing factor turned out to be the Ca^{2+}-sequestering ability of the sarcoplasmic reticulum (endoplasmic reticulum [ER]; Bennett and Porter, 1953; Porter and Bonneville, 1968). In this chapter, I will discuss the role of Ca^{2+} in the contraction of cardiac muscle cells or cardio myocytes because cardiac muscle exhibits many ways in which Ca^{2+} acts as a second messenger, and thus serves as a good example (Fig. 12.1). Normally, our hearts beat with a rhythm set by the pacemaker cells. At rest, the cardiac muscle cells are hyperpolarized by a Na^+/K^+-ATPase or sodium pump that is similar to the H^+ pump in plant cells. The −0.08 V membrane potential across the 8 nm sarcolemma produces an electric field of 10,000,000 V/m. The pacemaker cells send a periodic electrical stimulus to the cardiac muscle cells that depolarizes the plasma membrane

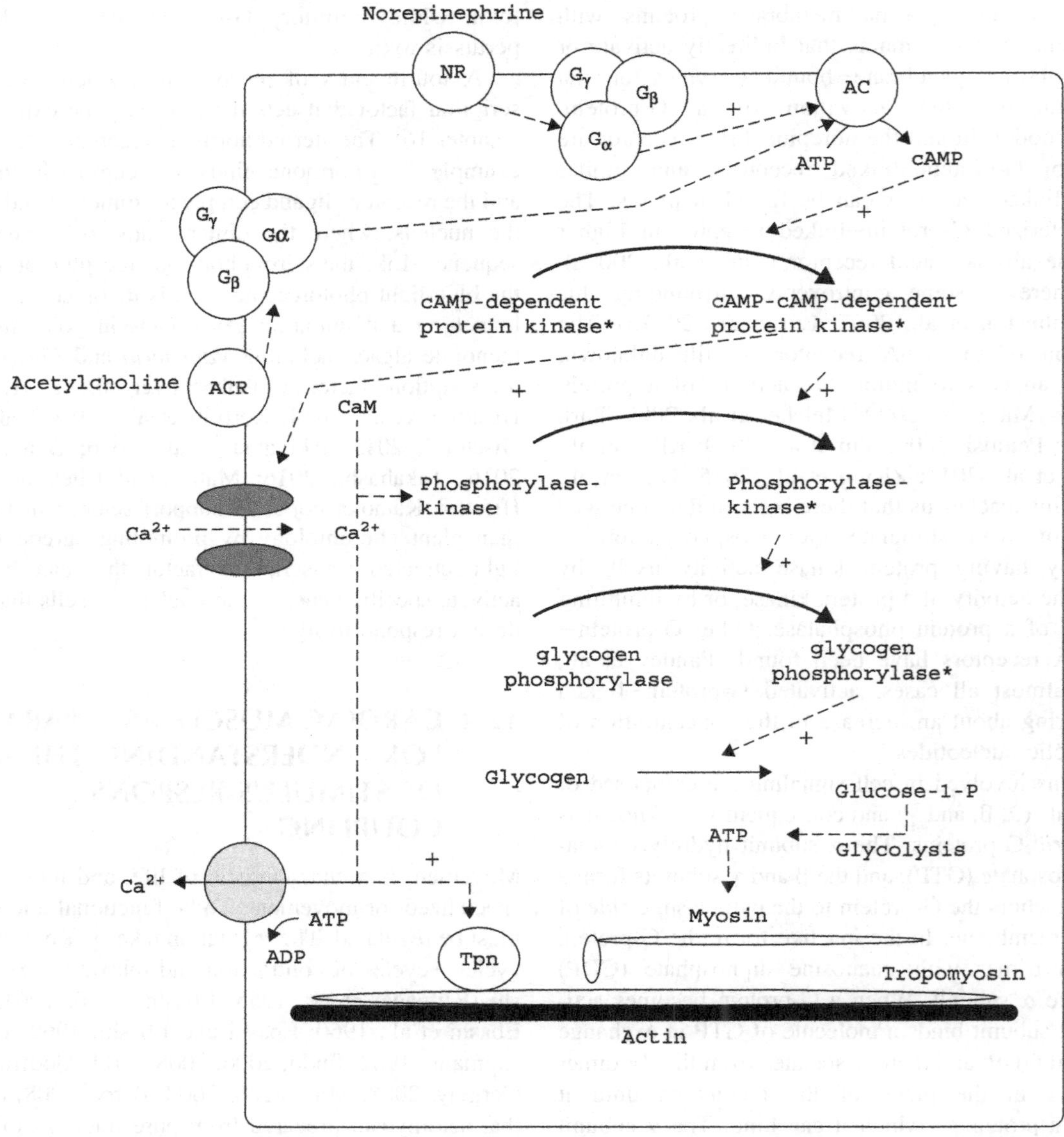

FIGURE 12.1 Scheme of a signal transduction network in a generalized cardiac muscle cell. Contraction is initiated when an electrical stimulus causes the plasma membrane of the muscle cell to depolarize. The depolarization activates the plasma membrane localized, membrane potential—dependent Ca^{2+} channels. The activated muscle cell then contracts due to the interaction of actin and myosin. Simultaneous activation of the receptors for norepinephrine results in the breakdown of glycogen and the formation of ATP necessary for contraction. Alternatively, activation of the acetylcholine receptor results in a diminished formation of ATP. An asterisk indicates the active form of an enzyme. *AC*, adenylate cyclase; *ACR*, acetylcholine receptor; *ADP, adenosine diphosphate; CaM*, calmodulin; *cAMP*, cyclic adenosine monophosphate; *NR*, norepinephrine receptor; *Tpn*, troponin.

or sarcolemma and induces the beat of the cardiac cells (Hoffman and Cranefield, 1960). As a consequence of the decrease in the electric field across the sarcolemma, the voltage-dependent or voltage-gated Na^+ and Ca^{2+} channels in the sarcolemma open and the membrane depolarizes and Na^+ and Ca^{2+} enter the cell along their electrochemical difference. The Na^+ channel closes spontaneously, whereas the Ca^{2+} channel remains open, which results in an increase in the cytosolic Ca^{2+} concentration. The increase of Ca^{2+} in the cytosolic P-space is augmented because the Ca^{2+} that enters the cell through the plasma membrane binds to the Ca^{2+}-release channel in the sarcoplasmic reticulum and causes a large release of Ca^{2+} as a result of a Ca^{2+}-induced Ca^{2+} release (Berridge, 2002, 2003). The membrane then repolarizes as a result of K^+ efflux through a K^+ channel.

As a result of this cascade effect, involving the sarcolemma and the sarcoplasmic reticulum, the Ca^{2+} concentration in the cytosol rises from 0.1 to 1–10 μM. At this elevated concentration, Ca^{2+} binds to one of the subunits of an intracellular Ca^{2+}-binding protein called *troponin* (Godfraind, 2007; Otsuka, 2007). Each troponin molecule binds four Ca^{2+} ions. Once troponin binds Ca^{2+}, the complex displaces a filamentous protein known as *tropomyosin* from actin so that myosin can interact with actin and cause contraction. Other proteins that bind Ca^{2+} include the Ca^{2+}-ATPase or calcium pump of the sarcolemma and the sarcoplasmic reticulum. The increase in cytosolic calcium during the heartbeat is transient, in part, because these proteins pump Ca^{2+} from the cytosol (P-space) into the extracellular space (E-space) and the lumen of the sarcoplasmic reticulum (E-space), which causes the myocyte to relax. In cardiac muscle cells, the increase in the concentration of intracellular-free Ca^{2+} lasts about 30 milliseconds. The activation of the actin-activated myosin ATPase depends on the increase in the concentration of intracellular Ca^{2+} and is thus an example of amplitude modulation.

When we are excited, our hearts beat faster. This is due to norepinephrine (=noradrenaline), an adrenaline-like neurotransmitter that is released by the sympathetic nervous system. Norepinephrine initiates a rise in cAMP. Norepinephrine does this by binding to a membrane receptor (β-adrenergic receptor). Subsequently, the receptor activates trimeric G-proteins by causing them to bind GTP in exchange for GDP. The GTP-binding proteins then activate adenylate cyclase molecules. Each adenylate cyclase molecule converts many ATP molecules to cAMP. The many cAMP molecules then bind to many cAMP-dependent protein kinases. The cAMP increase is only transient, and soon after the cAMP increases, it is converted to the inactive 3',5'-cyclic adenosine monophosphate by phosphodiesterase. Thus, like the Ca^{2+} signal, the cAMP signal is also transient.

One of the substrates activated by the cAMP-dependent protein kinase is phosphorylase kinase. Phosphorylase kinase is a regulatory protein that activates glycogen phosphorylase, the enzyme responsible for the breakdown of glycogen. Phosphorylase kinase is a calcium-binding protein. However, phosphorylase kinase does not bind calcium directly but binds a calcium-binding protein known as *calmodulin*. Calmodulin is a ubiquitous protein and is related to the Ca^{2+}-binding subunit of troponin.

The phosphorylation of phosphorylase kinase increases the affinity of phosphorylase kinase calmodulin for Ca^{2+} and lowers the concentration required for half-maximal activation from 3 to 0.3 μM Ca^{2+}. The activated phosphorylase kinase then phosphorylates many glycogen phosphorylase molecules. The phosphorylase catalyzes the breakdown of glycogen to glucose-1-phosphate. The sugar phosphate then goes through glycolysis and respiration to provide the chemical energy, in the form of ATP (see Chapter 14), needed for muscle contraction. As the activation of phosphorylase kinase depends on the change in its sensitivity to Ca^{2+}, not on an increase in the intracellular Ca^{2+} concentration, this type of regulation is known as *sensitivity modulation*.

Our heartbeat can also slow down when we are relaxed. This occurs as a consequence of the release of acetylcholine by the parasympathetic nervous system (Dale, 1936; Loewi, 1936; Valenstein, 2005). In cardiac muscle, acetylcholine binds to the inhibitory G-protein–linked muscarinic acetylcholine receptors. The inhibitory G-protein that is activated by this receptor inhibits the activity of adenylate cyclase and reduces the concentration of cAMP. This causes a slowdown of the heartbeat by reducing the production of ATP. At the same time, the inhibitory G-protein activates a K^+ channel in the sarcolemma. This leads to a slowdown of the heartbeat by causing a hyperpolarization of the plasma membrane that desensitizes the cell to the depolarization induced by the pacemaker cells.

In a cardiac muscle cell, a variety of stimuli act on a single cell in a coordinate manner to regulate muscle contraction. Cardiac muscle cells serve as a paradigm of the variety of signaling phenomena. There are examples of amplification, where a single molecule can activate many other molecules; amplitude modulation, where a change in the concentration of a chemical leads to a response; sensitivity modulation, where an increase in the affinity of a molecule for a second messenger leads to a response; covalent modifications, where the formation of an ester bond upon the addition of a phosphate group to a protein initiates a response; ionic regulation, where the electrostatic binding of an ion leads to a response; negative feedback, where the induction of a response leads to its termination; positive feedback, where the induction of a response leads to its amplification; and cross talk or interactions between two signal transduction chains (Rasmussen, 1981). Actually, the interactions are even more complex than I have already described. For example, the cAMP-dependent protein kinase phosphorylates Ca^{2+} channels in the plasma membrane and increases their opening probability. When the Ca^{2+} current increases, so does the force of contraction. The Ca^{2+} signaling system integrates many different regulatory elements in a cell, and consequently, a variety of pharmacological agents can have a similar effect. For example, the Ca^{2+} current is increased by isoproterenol, a β-adrenergic receptor activator; cholera toxin, an activator of G proteins; forskolin, an activator of adenylate cyclase; methylxanthines, inhibitors of phosphodiesterase; and okadaic acid, an inhibitor of protein phosphatases (Hille, 1992). Elucidating any physiological process depends on being able to reconstruct the relationship between the parts and the whole.

Why is Ca^{2+} such a good second messenger? Perhaps its fitness as a second messenger comes from the fact that it is abundant in the environment and thus there is always a reliable source for it to act as a regulatory chemical (Jaiswal, 2001). On the other hand, Ca^{2+} is a cytotoxin and at elevated cytoplasmic levels, it will bind to inorganic phosphate and form an insoluble precipitate known as *hydroxyapatite* (Weber, 1976). Thus, phosphate-based energy metabolism would be severely inhibited if the intracellular Ca^{2+} concentration approached the millimolar quantities found outside the cell. Rather than change energy metabolism, cells seemed to deal with this crisis by evolving an efficient method for removing Ca^{2+} from the cytosol, lowering its concentration to ~ 0.1 μM, at which point the reaction between Ca^{2+} and inorganic phosphate is insignificant (Kretsinger, 1977). The concentration of free Ca^{2+} in the cell is therefore 10,000 times lower than the concentration in the environment.

As the Ca^{2+} concentration is typically low on the P-sides of membranes and high on the E-sides of membranes, the entropy of the cell in terms of Ca^{2+} is low. According to Leo Szilard (1964), the information content of a system is proportional to the negative of the entropy. Thus, if a stimulus were to open Ca^{2+} channels in the membranes and the cytosolic concentration of Ca^{2+} were to rise as the concentration outside and inside equalized *pari passu*, the entropy would increase and information could be imparted to the cell. The increase of entropy (ΔS) results in a release of molecular free energy (ΔE) that can be harnessed to perform work, given that $\Delta E = [(\Delta H - T\Delta S)/N_A]$. An ion is thus able to do work on an intracellular receptor. The magnitude of the work depends on the magnitude of the change in entropy and the magnitude of heat loss accompanying the change in entropy. The ability of an ion to transmit information to the receptor depends, in part, on a change in the concentration of the ion, not on the absolute concentration. Let us look at enzyme kinetics to get a better feel for this.

12.5 A KINETIC DESCRIPTION OF REGULATION

For a cell to undergo a physiological or developmental change in response to its environment, the biophysical changes in the Ca^{2+} concentration must be converted into biochemical changes that involve proteins. Thus, we must investigate the affinity between Ca^{2+} ions and the proteins they bind. Either of the binding partners in a reaction can be called a *ligand*. The binding of Ca^{2+} to a ligand puts a process in motion. Kinetics comes from the Greek word *kinetikes*, which means "putting in motion." *Kinetos* is the verbal adjective of *kinein*, which means "to move." I will begin the discussion of kinetics from a historical point of view.

12.5.1 Early History of Kinetic Studies

The study of affinity began with Empedocles (<450 BCE), who thought that chemicals had the qualities of love and hate. To him, chemical combination and decomposition was analogous to marriage and divorce, respectively. Hippocrates generalized this idea somewhat and concluded that only chemicals that shared a kinship with each other combined to form compounds. This thinking has been captured in terms such as *hydrophilic* and *hydrophobic*! By contrast, Heraclitus argued that chemicals with opposite properties attract and thus form compounds. While Hippocrates was correct for the interactions between polar and nonpolar molecules, Heraclitus was right when it came to the interactions between charged chemicals. Neither theory was all-encompassing (Clark, 1952; Kaufmann, 1961).

Throughout the 18th century, chemists, including Torbern Bergman and Georg Stahl, set up affinity tables for various chemicals. They concluded that the order of affinity is $A > B > C$, if A displaced B and B displaced C from a given chemical. It was as if A bound with more force than B, and B bound with more force than C. At the end of the 18th century, Karl Wenzel applied Newtonian mechanics to the study of affinity. He proposed that chemical affinity is a force, and since a force causes a change in the velocity of a particle, an increase in the chemical force should cause an increase in the velocity of a chemical reaction. Thus, he studied the rate of decomposition of metals in various acids and concluded that the rate of the reaction, which was measured by the time it took the metal to dissolve, depended on both the affinity of the acid for the metal and the quantity of the acid. Unfortunately, Wenzel's work was unappreciated and forgotten (Ostwald, 1900, 1906; Winderlich, 1950).

The fact that the nature and quantity of a chemical are important in predicting the reactions in which it will participate came to light again in 1799 when Claude Berthollet, one of the scientists who accompanied Napoleon Bonaparte to Egypt, came up with a theory that explained how enormous quantities of sodium carbonate appeared on the shores of the salt lakes in Egypt. While it was already known that calcium chloride combined with sodium carbonate to make sodium chloride and the insoluble calcium carbonate according to the following reaction:

$$CaCl_2 + Na_2CO_3 \Leftrightarrow CaCO_3 + 2NaCl$$

the reverse reaction was not known. Berthollet postulated that the reverse reaction occurred when the amount of NaCl was very high, like it was in the salt lakes. He concluded that sodium carbonate was formed instead of calcium carbonate, even though calcium had a greater affinity than sodium for carbonate because the chemical force depended on both the concentration and the affinity (Mellor, 1914). Berthollet's work was ignored by fellow scientists because they were afraid that it insinuated that chemicals could combine in any proportion and this inference might undermine the atomic theory. In the 1860s, Cato Guldberg and Peter Waage built a theory that has been called the Law of Mass Action that incorporated all the known observations at the time and allowed the transformation of chemical reactions into mathematical equations (Guldberg and Waage, 1899; Bastiansen, 1964).

The Law of Mass Action may be limited when applied at the cell or organismal level. In biological systems composed of many enzymes, pathways, compartments, cells, tissues, and organs, a single chemical (e.g., drug, hormone, toxin, etc.) may have opposing effects at low and high concentrations. The slight stimulation caused by low concentrations of inhibitors, known as *hormesis*, may cause the organisms to "prepare" for larger doses by turning on pathways necessary to deal with higher concentrations of the chemical in question (Southam and Erhlich, 1943; Davis and Svendsgaard, 1990; Calabrese and Baldwin, 2000a,b; Kaiser, 2003; Calabrese, 2003, 2004). So, we must be careful in applying single-enzyme models to whole cells and organisms. Even so, the single-enzyme models have been very productive in understanding how cells respond to stimuli. I will now describe the importance of concentration and affinity in understanding enzyme reactions.

12.5.2 Kinetics of Enzyme Reactions

Invertase is an extracellular enzyme , first discovered in yeast (MacFadyen, 1908), that converts sucrose into glucose and fructose. Because sucrose rotates polarized light to the right while the mixture of glucose and fructose rotates polarized light to the left, this reaction inverts the sense of optical rotation measured with a polarimeter. This is how invertase got its name. While studying invertase, Leonor Michaelis and Maud Menten found that the rate of the reaction increased as the concentration of sucrose increased, but in a highly nonlinear manner (Michaelis and Menten, 1913, 2013; Briggs and Haldane, 1925; Haldane, 1930). They proposed a theory of enzyme action to explain the results that assumed that the enzyme interacted with the substrate in a particular manner. The mathematical model in the form of an equation fitted the experimental data, gave the affinity of the enzyme for the substrate, and gave the velocity of the reaction at any substrate concentration. The great success of their equation over the past century is evidence of the validity of their assumptions, and the limitations of their equation are evidence of additional regulatory effects a substrate has on an enzyme that were not taken into consideration by Michaelis and Menten (Gunawardena, 2012, 2013, 2014; Cherayil, 2013; Cornish-Bowden, 2013, 2015; Deichmann et al., 2014). Let us consider a generalized enzyme reaction that can be described by Michaelis–Menten kinetics. In this case, the substrate (S) binds to an enzyme (E) to make an enzyme–substrate complex (ES) that decomposes to form a product (P) and the regenerated enzyme (E).

$$S + E \underset{k_2}{\overset{k_1}{\Leftrightarrow}} ES \underset{k_4}{\overset{k_3}{\Leftrightarrow}} P + E$$

k_1 and k_4 (in $M^{-1}\,s^{-1}$) represent the rate constants that describe the formation of the ES complex, whereas k_2 and k_3 (in s^{-1}) represent the rate constants that describe the decomposition of the ES complex.

In analogy with the Law of Mass Action formulated by Guldberg and Waage, this equation tells us that the rate of formation of ES equals $k_1[E][S] + k_4[P][E]$ and the rate of decomposition of ES equals $k_2[ES] + k_3[ES]$. At steady state, the rate of formation of ES equals its rate of decomposition, and the concentration of ES does not change. Thus:

$$k_1[E][S] + k_4[P][E] = k_2[ES] + k_3[ES] \quad (12.1)$$

In spite of the current ubiquity of string theorists in physics, who call unknowns *free parameters*, there has been a long tradition among mathematically minded scientists to create equations where there are no more unknowns than can be measured. Among these traditional mathematically minded scientists, there is a saying, "1, 2, 3, infinity." That is, any equation that has more than three unknown variables is as useless as an equation with an infinite number of variables. Eq. (12.1) has too many unknown quantities. Thus, Leonor Michaelis and Maud Menten used a little algebra to combine all the unknown quantities into a single measurable quantity, known as the *Michaelis–Menten constant*, which turns out to be easy to determine experimentally, and very useful for understanding regulatory proteins and/or enzymes. I will derive the Michaelis–Menten equation by first rearranging the terms in Eq. (12.1):

$$[\mathrm{E}](k_1[\mathrm{S}] + k_4[\mathrm{P}]) = [\mathrm{ES}](k_2 + k_3) \quad (12.2)$$

and solve for [ES]/[E]:

$$[\mathrm{ES}]/[\mathrm{E}] = \frac{k_1[\mathrm{S}] + k_4[\mathrm{P}]}{(k_2 + k_3)} = \frac{k_1[\mathrm{S}]}{(k_2 + k_3)} + \frac{k_4[\mathrm{P}]}{(k_2 + k_3)} \quad (12.3)$$

Since the concentration of the product [P] in the initial stages of the reaction is very small and does not influence the initial velocity, we can simplify the model by studying the initial reaction where $[\mathrm{P}] = 0$. Thus:

$$[\mathrm{ES}]/[\mathrm{E}] = \frac{k_1[\mathrm{S}]}{(k_2 + k_3)} \text{ and } [\mathrm{E}]/[\mathrm{ES}] = \frac{k_2 + k_3}{k_1[\mathrm{S}]} \quad (12.4)$$

Since the total enzyme concentration $[\mathrm{E}]_T$ is equal to the concentration of free enzyme [E] plus the concentration of bound enzyme [ES], then $[\mathrm{E}] = [\mathrm{E}]_T - [\mathrm{ES}]$. Thus:

$$\begin{aligned} [\mathrm{E}]/[\mathrm{ES}] &= \frac{[\mathrm{E}]_T - [\mathrm{ES}]}{[\mathrm{ES}]} \\ &= ([\mathrm{E}]_T/[\mathrm{ES}]) - ([\mathrm{ES}]/[\mathrm{ES}]) \\ &= ([\mathrm{E}]_T/[\mathrm{ES}]) - 1 \end{aligned} \quad (12.5)$$

and

$$([\mathrm{E}]_T/[\mathrm{ES}]) - 1 = \frac{k_2 + k_3}{k_1[\mathrm{S}]} \quad (12.6)$$

If we define the Michaelis–Menten constant K_m as $(k_2 + k_3)/k_1$, then:

$$[\mathrm{E}]_T/[\mathrm{ES}] = (K_m/[\mathrm{S}]) + 1 \quad (12.7)$$

Unfortunately, $[\mathrm{E}]_T$ and [ES] cannot be readily determined. However, they can be expressed in terms of the initial velocities of the reaction (v) at any given substrate concentration [S] and the maximum initial velocity ($v_{\max}$) at saturating concentrations of [S].

The initial velocity (v) at any given substrate concentration is proportional to the concentration of the enzyme–substrate complex [ES]. Thus, v is proportional to [ES], and consequently v is an estimate of [ES]. Likewise, the maximal initial velocity ($v_{\max}$) is proportional to the total enzyme present when $[\mathrm{ES}] = [\mathrm{E}]_T$. That is, when all the enzyme is in the ES complex. Thus, $v_{\max}$ is proportional to $[\mathrm{E}]_T$, and consequently, $v_{\max}$ is an estimate of $[\mathrm{E}]_T$. Assuming that the proportionality constants are equal, $[\mathrm{E}]_T/[\mathrm{ES}] = v_{\max}/v$, and Eq. (12.7) becomes

$$v_{\max}/v = (K_m/[\mathrm{S}]) + 1 \quad (12.8)$$

After solving for v, we get the typical form of the Michaelis–Menten equation:

$$v = \frac{v_{\max}}{((K_m/[\mathrm{S}]) + 1)} \quad (12.9)$$

From Eq. (12.9), we see that the relative velocity of a reaction depends on the substrate concentration. When $[\mathrm{S}] = K_m$, $v = v_{\max}/2$. Thus, K_m is also defined as the concentration of S that supports a reaction that proceeds at the velocity of $v_{\max}/2$. When $[\mathrm{S}] = 10\ K_m$, $v = v_{\max}/1.1$ and the reaction proceeds at approximately 90% of its maximal velocity and we say the enzyme is activated. When $[\mathrm{S}] = 0.1\ K_m$, $v_{\max}/11$ and the reaction proceeds at approximately 10% of its maximal velocity and we say the enzyme is inactive. A reaction that obeys Michaelis–Menten kinetics is activated from $0.1\ v_{\max}$ to $0.9\ v_{\max}$ by an 81-fold change in the substrate concentration. When an increase in the substrate concentration causes an increase in the velocity of the reaction, the regulation is known as *amplitude modulation*.

The Michaelis–Menten constant is a steady-state constant attained under initial conditions, and not an equilibrium constant, so it cannot be analyzed with equilibrium thermodynamics. However, $K_m = (k_2 + k_3)/k_1$ is equivalent to the dissociation constant K_d ($=k_2/k_1$) when $k_2 >> k_3$. The dissociation constant (in M) is a measure of the affinity of two chemicals for each other. The dissociation constant is an equilibrium constant and consequently, can be treated thermodynamically. The dissociation constant of E for S in the reaction shown below is equal to k_2/k_1 where k_1 is called the *on-rate constant* (in $\mathrm{M}^{-1}\ \mathrm{s}^{-1}$) and k_2 is called the *off-rate constant* (in s^{-1}):

$$\mathrm{S} + \mathrm{E} \underset{k_2}{\overset{k_1}{\Leftrightarrow}} \mathrm{ES}$$

Let us consider the equilibrium state where the rate of formation of the active complex [ES] equals the rate of dissociation of the active complex [ES]. Thus:

$$k_1[\mathrm{S}][\mathrm{E}] = k_2[\mathrm{ES}] \quad (12.10)$$

Solving for [E]/[ES] and defining K_d as k_2/k_1 we get:

$$[\mathrm{E}]/[\mathrm{ES}] = (k_2/k_1)/[\mathrm{S}] = K_d/[\mathrm{S}] \quad (12.11)$$

Remember that $[\mathrm{E}]/[\mathrm{ES}] = ([\mathrm{E}]_T/[\mathrm{ES}]) - 1$. Thus:

$$([\mathrm{E}]_T/[\mathrm{ES}]) - 1 = K_d/\mathrm{S} \quad (12.12)$$

Since $[\mathrm{E}]_T/[\mathrm{ES}] = v_{\max}/v$, then

$$v_{\max}/v = \{K_d/[\mathrm{S}] + 1\} \quad (12.13)$$

and

$$v = v_{\max}/\{(K_d/[\mathrm{S}]) + 1\} \quad (12.14)$$

Thus, like a reaction that follows Michaelis–Menten kinetics, the velocity of a reaction activated by the ES complex depends on the concentration of the activator relative to the dissociation constant. The true dissociation constant is determined by incubating a known amount of enzyme in various concentrations of substrate and measuring how much substrate is bound at equilibrium (time $= \infty$). The results are plotted, and the K_d is equal to the substrate concentration at the inflection point. The association constant (K_a, in M^{-1}) is the reciprocal of K_d.

Most enzymes involved in cell regulation do not show typical Michaelis–Menten kinetics, where a plot of the reaction velocity versus the substrate concentration is a rectangular hyperbola. In the case of regulatory enzymes, a plot of velocity versus substrate concentration is typically sigmoidal. A sigmoidal response is indicative of multiple or cooperative binding sites for the substrate on the enzyme. If the binding of the substrate shows positive cooperativity, then the binding of the first substrate causes a conformational change in the enzyme so that the affinity of the next binding site is increased, etc. Thus, the substrate also acts as an activator. Multiple, positively cooperative binding sites allow more sensitive control of a reaction by a substrate/activator. Let us look at the following reaction where an enzyme binds four substrate/activator molecules.

$$[E] + [S] + [S] + [S] + [S] \underset{k_2'}{\overset{k_1'}{\Leftrightarrow}} [ES_4]$$

This equation is really a short hand version of four equations, including $[E] + [S] \Leftrightarrow [ES]$, $[ES] + [S] \Leftrightarrow [ES_2]$, $[ES_2] + [S] \Leftrightarrow [ES_3]$, and $[ES_3] + [S] \Leftrightarrow [ES_4]$, and thus k_1' is the product of four on-rate constants and is given in units of $M^{-n}\ s^{-n}$ where n is the number of substrate molecules that bind to the receptor. k_2' is the product of four off-rate constants and is given in units of s^{-n}. Thus:

$$k_1'[E][S][S][S][S] = k_2'[ES_4] \quad (12.15)$$

and if we define K' (in M^n) as k_2'/k_1', we get

$$[E]/[ES_4] = (k_2'/k_1')/[S]^4 = K'/[S]^4 \quad (12.16)$$

For simplicity, we assume that ES_4 is the only active form of the enzyme. Then we can use the following equation: $[E]/[ES_4] = ([E]_T/[ES_4]) - 1$. If we again assume that $v_{max}/v = ([E]_T/[ES_4])$, then

$$v = v_{max}/\{(K'/[S]^4) + 1\} \quad (12.17)$$

Eq. (12.17) can be written in a more generalized form:

$$v = v_{max}/\{(K'/[S]^n) + 1\} = v_{max}\Big/\left\{(K''/[S])^2 + 1\right\} \quad (12.18)$$

where K″ is the nth root of K', and K' is the concentration of substrate to the nth power that activates the enzyme to a level of $v_{max}/2$. This equation, which was originally produced to describe the binding of oxygen to hemoglobin, is known as the *Hill equation*; n is called the *Hill coefficient*, named after Archibald V. Hill (1962), and it represents the number of binding sites on the enzyme for the substrate (Hill, 1965). When $n = 1$, the Hill equation is identical to the equation that relates the velocity to the dissociation constant.

By solving the Hill Eq. (12.18), we see how the velocity of a reaction depends on the substrate concentration. For example, when $n = 4$ and $[S] = K''$, $v = v_{max}/2$, and the reaction will proceed at its half-maximal velocity. When $n = 4$ and $[S] = 2\ K''$, $v = v_{max}/1.06$, and the reaction will proceed at approximately its maximal velocity. When $n = 4$ and $[S] = 0.2\ K''$, $v = v_{max}/626$, and the reaction will proceed at an infinitesimally slow velocity, and we say that the enzyme is inactive. In general, the change in the concentration of a substrate required to activate an enzyme with n binding sites from 10% to 90% of its maximal activity is equal to $\sqrt[n]{81}$ (Segel, 1968). We can see this easily by solving Eq. (12.18) for $[S]^n$. I will solve for $[S]^n$ using a few algebraic steps:

$$v\{(K'/[S]^n) + 1\} = v_{max} \quad (12.19)$$

$$vK'/[S]^n + v = v_{max} \quad (12.20)$$

$$vK'/[S]^n = v_{max} - v \quad (12.21)$$

$$vK'/(v_{max} - v) = [S]^n \quad (12.22)$$

Now I will select two concentrations of $[S]^n$, which will result in two velocities, and set up Eq. (12.22) to solve for the ratio of $[S_1]^n/[S_2]^n$ that leads to a desired ratio of velocities.

$$[S_1]^n/[S_2]^n = \{v_1K'(v_{max} - v_2)\}/\{v_2K'(v_{max} - v_1)\} \quad (12.23)$$

Cancel like terms:

$$[S_1]^n/[S_2]^n = \{v_1(v_{max} - v_2)\}/\{v_2(v_{max} - v_1)\} \quad (12.24)$$

Let us find the ratio of $[S_1]^n/[S_2]^n$ that will lead to a reaction where $v_1 = 0.9\ v_{max}$ and $v_2 = 0.1\ v_{max}$:

$$[S_1]^n/[S_2]^n = \{0.9\ v_{max}(v_{max} - 0.1\ v_{max})\}/\{0.1\ v_{max}(v_{max} - 0.9\ v_{max})\} \quad (12.25)$$

$$[S_1]^n/[S_2]^n = \{0.9\ v_{max}(0.9\ v_{max})\}/\{0.1\ v_{max}(0.1\ v_{max})\} \quad (12.26)$$

$$[S_1]^n/[S_2]^n = \{0.81\ {v_{max}}^2\}/\{0.01\ {v_{max}}^2\} = 81 \quad (12.27)$$

$$S_1/S_2 = \sqrt[n]{([S_1]^n/[S_2]^n)} = \sqrt[n]{81} \quad (12.28)$$

When an enzyme has a Hill coefficient of 1, large changes in the concentration of the substrate cause small

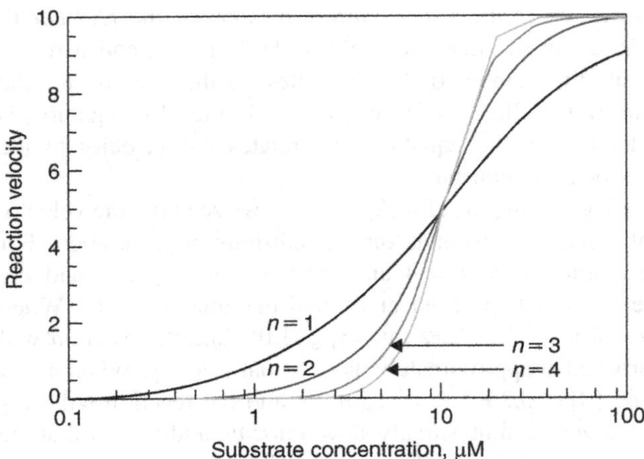

FIGURE 12.2 The relationship between reaction velocity and substrate concentration for enzymes with various Hill coefficients (n).

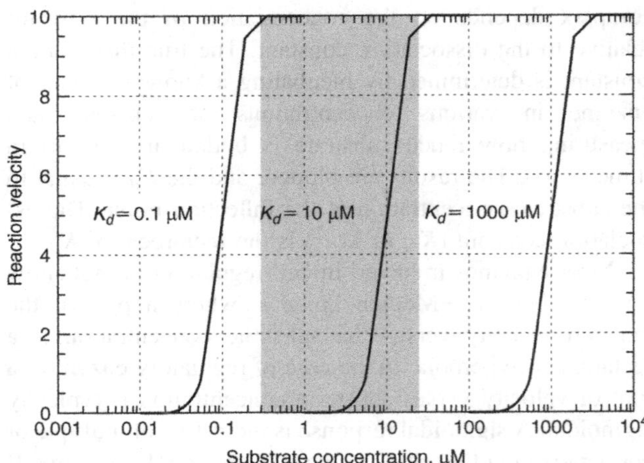

FIGURE 12.3 The relationship between reaction velocity and substrate concentration for proteins that have various affinities for the substrate. The smaller the dissociation constant (K_d), the greater the affinity. The *gray region* indicates the range in which the concentration of the substrate.

changes in the activity. Thus, the enzyme activity is stable with respect to that substrate. When the Hill coefficient is 4, small changes in substrate concentration around the K' lead to large changes in enzyme activity.

Thus, an enzyme that shows positive cooperativity is able to recognize small changes in the concentration (i.e., amplitude) of a substrate. Consequently, such an enzyme is exquisitely designed to act as a sensitive switch in a signal transduction chain (Fig. 12.2). A change in the activity of a protein implies a change in its shape, folding, or conformation. The greater the conformational change the longer the enzyme switching reaction takes. Fast switching requires that the response is limited by diffusion and not by the time it takes to fold the protein. Michaelis–Menten kinetics and their extensions described here can be used to describe and predict various characteristics of receptor proteins (Gunawardena, 2013), and a receptor protein can be considered to be a ligand-activated enzymatic switch that puts the regulatory machinery of the cell into motion.

I will now discuss the fundamental significance of the relationship between the dissociation constant of a regulatory protein and the cellular concentration of an activator (Fig. 12.3). Consider a receptor that has a $K_d = 10^{-3}$ M (=1000 μM) and a substrate of which the intracellular concentration varies from 3×10^{-7} to 3×10^{-5} M. This receptor does not have a high enough affinity for the substrate, so it will never bind it. If the substrate is needed to activate the receptor, this receptor will never be activated.

Consider a receptor that has a $K_d = 10^{-7}$ M (=0.1 μM) and a substrate of which the intracellular concentration varies from 3×10^{-7} to 3×10^{-5} M. This receptor has too high an affinity for the substrate, and even when the substrate is at its lowest concentration, the receptor will bind it. If the substrate is needed to activate the receptor, this receptor will always be activated.

Consider a receptor that has a $K_d = 10^{-6}$ M (=1 μM) and a substrate of which the intracellular concentration varies from 10^{-7} to 10^{-5} M. This receptor has too low of an affinity to bind the substrate when the substrate is at its lowest concentration (3×10^{-7} M) and will not be activated, but when the substrate reaches its highest concentration (3×10^{-5} M) the receptor will be activated. The substrate can act as a switch because the dissociation constant of the receptor is matched with the intracellular concentrations of the substrate in the resting and activated states. That is, the K_d is between the resting level and the activated level of the substrate. For example, when Ca^{2+} is the substrate, the K_d of the receptor should be about 10^{-6} M because the resting concentration is 10^{-7} M and the activated concentration is about 10^{-5} M. Of course, a cell may have a variety of intracellular Ca^{2+} receptors with K_ds near 10^{-6} M, but with different Hill coefficients. Thus, at a given rate of increase in the concentration of intracellular Ca^{2+}, each one binds Ca^{2+} at a different rate, and thus they become activated at different times. Differential activation can also occur because the intracellular Ca^{2+} receptors may be closer or farther from the site of Ca^{2+} entry. Let us look at the chemistry of Ca^{2+} and ask why Ca^{2+} is a ubiquitous second messenger while Mg^{2+}, a similar bivalent ion, is not.

12.5.3 Kinetics of Diffusion and Dehydration

Calcium ions must diffuse through the cell to a receptor before they are able to activate the receptor. In solution, water molecules are bound to ions as a consequence of the polar or electrical dipolar nature of water and the electrical charge of the ions. Thus, ions structure the water in the cytoplasm to some degree and the water of hydration surrounding the ions hinders the ions from binding to other ligands. The water that immediately surrounds an ion may thus have a different viscosity than the bulk water in a cell. According to R. J. P. Williams (1974, 1976, 1980), Ca^{2+} is more fit than Mg^{2+} to act as a second messenger because Ca^{2+} sheds its water of hydration more quickly than Mg^{2+}. But is the dehydration step the limiting step in ionic reactions in cells? I will discuss the factors that limit the overall rate of a reaction.

For Ca^{2+} and Mg^{2+} to enter a reaction, they must be dehydrated. The dehydration reactions are as follows:

$$Ca^{2+} + H_2O \underset{k_{off}}{\overset{k_{on}}{\Leftrightarrow}} Ca^{2+}H_2O$$

$$Mg^{2+} + H_2O \underset{k_{off}}{\overset{k_{on}}{\Leftrightarrow}} Mg^{2+}H_2O$$

In water, the concentration of free cations is less than the concentration of total cations because a majority of the ions is bound to molecules of water. These water molecules make concentric shells of more and more loosely bound water around the cation. The binding energies between water and a given ion are calculated using electrostatic models (Phillips and Williams, 1965, 1966). Because Mg^{2+} and Ca^{2+} have the same charge, but Ca^{2+} has a larger radius than Mg^{2+}, Ca^{2+} does not hold on to the negatively charged oxygen atoms in water as tightly as Mg^{2+} does. Due to its smaller charge density, Ca^{2+} sheds its water of hydration more than 1000 times more quickly than Mg^{2+}, and the off-rate constants (k_{off}) are $5 \times 10^8\ s^{-1}$ for Ca^{2+} and $10^5\ s^{-1}$ for Mg^{2+} (Eigen and Kruse, 1962). Because of this, dehydration limits the rate in which magnesium can bind to a ligand.

To convert these off-rate constants of dehydration into potential on-rate constants for the ion to bind with an intracellular ligand, the off-rate constants must be divided by the intracellular concentration of the ion that is able to activate the receptor (e.g., the K_d; see Table 12.1A). Likely

TABLE 12.1A On-Rate Constants (k_{on}) for Ca^{2+} and Mg^{2+} When Dehydration Is Limiting

	Ca^{2+}	Mg^{2+}
K_d	$k_{on}\ (M^{-1}\ s^{-1})$	$k_{on}\ (M^{-1}\ s^{-1})$
10^{-7}	5×10^{15}	10^{12}
10^{-6}	5×10^{14}	10^{11}
10^{-5}	5×10^{13}	10^{10}
10^{-4}	5×10^{12}	10^{9}
10^{-3}	5×10^{11}	10^{8}
10^{-2}	5×10^{10}	10^{7}

K_d is the ratio of the off-rate constant to the on-rate constant, and values of the on-rate constant are calculated by dividing the off-rate constants for dehydration by the K_d of the receptor ligand.

physiological on-rate constants are $12.5 \times 10^8\ M^{-1}\ s^{-1}$ and $10^8\ M^{-1}\ s^{-1}$ for Ca^{2+} and Mg^{2+}, respectively.

While the speed in which an ion sheds its water of hydration will influence its ability to bind a ligand, the dehydration step is not always the step that limits the rate of a reaction. The rate of a reaction can depend on the speed in which an ion diffuses to the ligand. While the rate of dehydration for similarly charged ions is proportional to the radius (r_H) of the ion, the rate of diffusion is inversely proportional to the radius of the ion. In addition, the rate of binding of an ion to a stationary ligand is proportional to the radius of the ligand (r_b). The on-rate constant for a nonelectrolyte to enter a diffusion-limited reaction is given by the equation derived by Marian von Smoluchowski (1917):

$$\begin{aligned} k_{on} &= [4\pi r_b RT/(6\pi r_H \eta)](10^3 L/m^3) \\ &= 4\pi r_b D N_A (10^3 L/m^3) \end{aligned} \tag{12.29}$$

where $4\pi r_b$ represents the size of the receptor, $RT/(6\pi r_H \eta)$ represents the molar diffusion coefficient of the ion (in $m^2/[s\ mol]$), and ($10^3\ L/m^3$) is the factor that converts cubic meters into liters. The on-rate constant for ions must take into consideration the charge of the ion and the strength of the electric field. Peter Debye (1942) has derived more sophisticated equations for describing the diffusion of ions to a receptor, and Glasstone et al. (1941) have formulated the Smoluchowski equation in quantum mechanical terms. For simplicity, we will estimate the on-rate constant for ions using Smoluchowski's classic equation. In general, the on-rate constant of diffusion-limited reactions can be considered to be approximately 10^8–$10^9\ M^{-1}\ s^{-1}$ given the viscosities, temperatures, and hydrodynamic radii frequently encountered in a cell. Consistent with the inverse relationship between on-rate constants and hydrodynamic radii of the diffusing ions, the diffusion-limited on-rate constants for Ca^{2+} are approximately 30% smaller than those for Mg^{2+} (Table 12.1B). Typically, Ca^{2+}-mediated reactions are limited by diffusion, whereas Mg^{2+}-mediated reactions are limited by dehydration.

As I discussed in the previous section, kinetics dictates that the K_d of a ligand for an ion must be close to the intracellular concentration of that ion if that ligand is to act as a signaling molecule. Thus, at a Ca^{2+} concentration of 0.1–10 μM and a Mg^{2+} concentration of 1–10 mM, the K_d for a Ca^{2+}-binding or an Mg^{2+}-binding ligand must be between 0.1–10 μM and 1–10 mM, respectively. Given these K_ds and the calculated maximum on-rate constants, for a given concentration of the ion, the off-rate constants of Ca^{2+}-activated and Mg^{2+}-activated reactions would be

- For Ca^{2+}: $(10^{-5} M)(12.5 \times 10^8 M^{-1}/s^{-1}) = 1.25 \times 10^4\ s^{-1}$
- For Ca^{2+}: $(10^{-6} M)(12.5 \times 10^8 M^{-1} s^{-1}) = 1.25 \times 10^3\ s^{-1}$

TABLE 12.1B On-Rate Constants (k_{on}) for Ca^{2+} and Mg^{2+} When Diffusion Is Limiting

Radius of Binding Site	Ca^{2+} k_{on} ($M^{-1}\,s^{-1}$)	Mg^{2+} k_{on} ($M^{-1}\,s^{-1}$)
1×10^{-10} m	4.2×10^8	6.3×10^8
2×10^{-10} m	8.3×10^8	12.7×10^8
3×10^{-10} m	12.5×10^8	19.0×10^8

Values are calculated from Smoluchowski's equation assuming $\eta = 0.004$ Pa s, T = 298 K, and the radii of Ca^{2+} and Mg^{2+} ions are 99 and 65 p.m., respectively. $k_{on} = [4\pi r_b RT/(6\pi r_H \eta)](10^3\ L/m^3)$. To convert on-rate constants into time, we must combine Einstein's random-walk equation ($t = x^2/(2D)$) with Smoluchowski's equation ($k_{on} = [4\pi r_b RT/(6\pi r_H \eta)](10^3\ L/m^3) = 4\pi r_b D N_A\ (10^3\ L/m^3)$). Since $D = k_{on}/(4\pi r_b N_A(10^3\ L/m^3))$, then $t = x^2(4\pi r_b N_A(10^3\ L/m^3))/(2k_{on})$ or $t = x^2(2\pi r_b N_A(10^3\ L/m^3))/(k_{on})$ or $t \alpha 1/(k_{on})$.

- For Ca^{2+}: $(10^{-7}\,M)\,(12.5 \times 10^8\,M^{-1}s^{-1}) = 1.25 \times 10^2\,s^{-1}$
- For Mg^{2+}: $(10^{-3}\,M)\,(10^8\,M^{-1}/s^{-1}) = 10^5\,s^{-1}$
- For Mg^{2+}: $(10^{-2}\,M)\,(10^7\,M^{-1}\,s^{-1}) = 10^5\,s^{-1}$

Wilhelm Ostwald defined the half-time of a reaction to be (ln 2)/k_{off}. The half-time of the reaction indicates how long an ion and a ligand will stay together before they dissociate. The shorter the half-time, the more rapidly the molecules involved can pass through the cycle of movements necessary to complete an elementary reaction and be ready for the next one. Regulatory proteins with short half-times can regulate rapid reactions with sharp and keen precision. Regulatory proteins with long half-times can regulate reactions with lower temporal resolution. If we consider diffusion to be limiting, the half-times for Ca^{2+}-mediated reactions will fall between 55 μs and 5.5 ms. If we consider diffusion to be limiting for Mg^{2+}-mediated reactions, the half-times are approximately 6.9 μs. As I discussed previously, when dehydration is limiting, the half-times for Ca^{2+}-activated and Mg^{2+}-activated reactions are 1.39 ns and 6.93 μs, respectively. Thus, Ca^{2+}-mediated reactions are typically limited by diffusion, whereas Mg^{2+}-mediated reactions are limited by dehydration. Thus, by considering diffusion, affinity, and dehydration, we find that, under typical cellular conditions, Mg^{2+} could actually be a faster signaling agent than Ca^{2+}.

Thus, it is incorrect to assume that Mg^{2+} is an inferior ion when it comes to cell signaling compared with Ca^{2+} because it is small and slow, whereas "Ca^{2+} is fat and fast." I do not know why Ca^{2+} is a ubiquitous second messenger while Mg^{2+} is not. Perhaps one reason is the fact that fewer Ca^{2+} ions have to enter the cell to raise the concentration 3- to 81-fold compared with Mg^{2+}. A 10-fold increase in the intracellular Mg^{2+} concentration may be an ionic and osmotic burden. Moreover, since the Mg^{2+} concentration outside the cell is likely to be equal to or less than the intracellular concentration, the electrochemical difference will be smaller for Mg^{2+} than it would be for Ca^{2+}, and an influx of Mg^{2+} may actually be the rate-limiting step.

It is possible that Ca^{2+} is a better signaling ion than Mg^{2+} because it is more flexible in forming coordinate bonds with a ligand. Mg^{2+} is a rigid ion that forms exactly six coordinate bonds with lengths that vary a little from 200 to 212 p.m., whereas Ca^{2+} can form six to eight coordinate bonds where the lengths of the various bonds are between 206 and 282 nm (Martell and Calvin, 1952). Perhaps the flexible Ca^{2+} ion can resonate and harmonize with conformational changes that occur in the protein to which it binds, whereas Mg^{2+} would be stiff and better suited to act as a bridging ion that can bring together a protein with ATP. Nevertheless, Mg^{2+} signaling is currently being investigated (Li et al., 2011; Wu and Veillette, 2011; Newton et al., 2016).

We can use thermodynamics to help us understand the relationship between affinity and the change in free energy of a receptor (Lewis and Randall, 1923). Given the likely on-rate constants, the off-rate constant varies inversely with the dissociation constant. The molecular free energy released on binding is related to K_d and K_a. The relationship is shown in Eq. (12.30):

$$E^{eq} - E^{Real} = kT \ln\{K_a/K\} \qquad (12.30)$$

$$E^{eq} - E^{Real} = kT \ln\{1/(KK_d)\} = -kT \ln\{KK_d\} \qquad (12.31)$$

where K is the ratio of products to reactants under real conditions, K_a represents the ratio of products to reactants at equilibrium, and $K_d = 1/K_a$. The product of K and K_d is always dimensionless no matter what the order of the reaction. As E^{eq} is defined to be zero, Eq. (12.31) becomes

$$E^{Real} = kT \ln\{KK_d\} \qquad (12.32)$$

The smaller the K_d, the greater the decrease in the molecular free energy, and consequently, the more stable the binding. Therefore, the high-affinity receptors regulate reactions with a low temporal precision compared with low-affinity receptors. Thus, for a given ion, a low-affinity receptor that becomes activated by an increase in the ion concentration (amplitude modulation) will always be able to regulate faster processes than the same receptor that increases its affinity for the ion (sensitivity modulation).

Calcium ions and H^+ are found in the cell at concentrations around 10^{-7} M, whereas other cations are either much more abundant (K^+, Na^+, Mg^{2+}, Mn^{2+}) or much less abundant (Cu^{2+}, Co^{2+}, Fe^{2+}, Zn^{2+}). Thus, the binding of Ca^{2+} or H^+ to a ligand will be intermediate between a stable binding and a loose binding. Perhaps this is important in its fitness as the universal second messenger.

12.5.4 A Thermodynamic Analysis of the Signal-To-Noise Problem

For a primary or secondary stimulus to be perceived, the energy of the signal must be greater than the ambient energy or noise. Thus, we must be concerned with the signal-to-noise ratio of reactions. The energy of a stimulus has to be greater than $\cong kT$ because each and every molecule in the cell has a certain amount of energy that results from the thermal energy of the cell. The thermal energy of a molecule at any temperature is approximately equal to kT. Thus, the minimum energy needed to activate a receptor is $\cong kT$, and using the tenets of quantum mechanics, the receptor will become active when it absorbs an amount of energy, in the form of gravitational energy, radiant energy, chemical energy, etc., which is equal to the difference in energy between that of the active and inactive states. I will use enzyme kinetics and thermodynamics to show the relationship between the energy input and the probability (K_s) that a receptor will become activated.

Let us assume that we have a receptor protein that becomes activated according to the following reaction:

$$\text{inactive} \Leftrightarrow \text{active}$$

Let us assume that to trigger a response, there has to be a probability of 100:1 that the receptor will become activated by the stimulus. Put another way, following stimulation, there must be a ratio of 100 active receptors to 1 inactive one. The probability of reaching this activation level depends on the energy of the stimulus. This is equal to the energy difference between the active and inactive states:

$$E - E^{\circ} = kT \ln(K_s) \tag{12.33}$$

where E and E° are the molecular free energies of the receptor in the active state and the inactive state, respectively, relative to a standard energy. In this case, we will take the standard energy to be kT, the approximate energy of thermal noise. K_s is the ratio of active to inactive receptors and is equal to [active]/[inactive]. Thus:

$$[\text{active}]/[\text{inactive}] = K_s = e^{((E-E^{\circ})/kT)} \tag{12.34}$$

The energy input necessary to induce a ratio of 100 active receptors to 1 inactive receptor is obtained by putting Eq. (12.34) in the following form:

$$\begin{aligned} \ln([\text{active}]/[\text{inactive}]) &= \ln(100) = 4.6 \\ &= (E - E^{\circ})/kT \end{aligned} \tag{12.35}$$

which simplifies to

$$(E - E^{\circ}) = 4.6\,kT = 1.89 \times 10^{-20}\text{J} \tag{12.36}$$

Thus, the difference between E and E° necessary to create a ratio of active to inactive receptors of 100:1 is equal to 4.6 kT. Thus, if $E^{\circ} = 1\,kT$, the energy of the activated receptor must be 5.6 kT. We can also look at this energy as the amount of energy needed to give a probability of 100:1 that a single receptor will become activated after an energy input of 4.6 kT. The difference in energy has to be 2.3 kT, 1.6 kT, and 0.69 kT to create a 10:1 probability, a 5:1 probability, and a 2:1 probability, respectively. In these cases, the energies of the activated receptors will be 3.3 kT, 2.6 kT, and 1.69 kT, respectively. The ratio of active to inactive receptors is a function of the energy difference between the inactive and active receptor. From a thermodynamic point of view, I cannot imagine cellular receptors that can be activated by energies that are less than that of thermal noise. Now that I have discussed the theoretical aspects of cell signaling, I will discuss the components of the cell that participate in cell signaling.

12.6 CA^{2+} SIGNALING SYSTEM

Sydney Ringer (1883) serendipitously discovered that Ca^{2+} is necessary to keep the heart beating when he used distilled water rather than water supplied by the New River Water Company to prepare a sodium chloride solution and noticed that the heart stopped beating. It turned out that the tap water supplied by the New River Water Company contained calcium. When Ringer added calcium to the saline solution made with distilled water, the heart did beat, indicating that calcium was necessary for contraction (Zimmer, 2005). Later L. V. Heilbrunn (1937, 1943, 1952) established the importance of Ca^{2+} in coupling a stimulus to a response (Campbell, 1986; Wiercinski, 1989). According to Heilbrunn (1952), "*The sensitivity of protoplasm and its response to stimulation are believed to be due to a sensitivity to free calcium ion and it is believed that the freeing of calcium and the reaction of this calcium with the protoplasm inside the cell is the most basic of all protoplasmic reactions.*" Heilbrunn's view of the role of Ca^{2+} in stimulus-response coupling was not appreciated during his lifetime (Mazia, 1975). Paul Gross (1986), the director of the Marine Biological Laboratory at Woods Hole wrote, "[Heilbrunn] *was a productive investigator, a notably successful teacher of young scientists, an MBL Trustee; certainly one of the influential American biologists of his generation. Yet not many of his peers paid attention to the arguments about calcium, not even at the MBL…Heilbrunn's work and its profound implications might be characterized, now, as 'rediscovered;' but in a fundamental sense they have been forgotten. The modern biology of calcium has grown up almost independently of the original advocate.*" Setsuro Ebashi, Lionel Jaffe, and Peter Hepler were leaders in the modern calcium renaissance.

As Ca^{2+} is a cytotoxin, cells have evolved an efficient method for lowering the Ca^{2+} concentration in the cytosolic P-space to 0.1 μM, approximately 10,000 times lower than the concentration in the environment. The low Ca^{2+} concentration in the cell is maintained by a plasma membrane–bound Ca^{2+}-ATPase, an ER-bound Ca^{2+}-ATPase, a vacuolar membrane–bound Ca^{2+}/H^{+} antiport system (Schumaker and Sze, 1985, 1986, 1987) and Ca^{2+}-ATPase (Berkelman and Lagarias, 1990), and a mitochondrial uptake system (Dieter and Marmé, 1980).

The challenge the cells faced to lower their intracellular-free Ca^{2+} concentration provided an opportunity for the cells to use this 10,000-fold gradient as a cellular switch to couple an extracellular stimulus with a cellular response. All that is needed is a control mechanism in the cell to induce a transient 10- to 100-fold increase in the Ca^{2+} concentration, which is still below the toxic level. This is accomplished by the opening of Ca^{2+} channels.

A Ca^{2+} channel is a protein made up of one or more polypeptides that form a hydrophilic pore in a membrane (Jammes et al., 2011; Spalding and Harper, 2011). The channel allows Ca^{2+} ions to pass relatively unimpeded at a rate of about $10^6\ s^{-1}$ or more. Ca^{2+}-permeable channels may be either nonselective or highly selective for Ca^{2+}. The selectivity depends on the pore size and the charge density of the binding sites at the mouth of the pore. These sites directly affect the ability of Ca^{2+} to shed its outer shells of water so that it can pass as a dehydrated ion (<0.2 nm in diameter).

All Ca^{2+} channels mediate the transfer of Ca^{2+} from E-spaces to P-spaces. Ca^{2+} channels occur in the plasma membrane and in the internal membranes. There can be many types of Ca^{2+} channels on the same membrane. The structure of channels is often deduced from molecular studies where a computer program calculates a hydropathy plot of the encoded amino acids (Kyte and Doolittle, 1982). The program is used to assess where the membrane-spanning regions of a protein may be, by equating a series of hydrophobic amino acids with a membrane-spanning region. The structure of channels can also be deduced from electrophysiological studies and biochemical studies. In the biochemical studies, radiolabeled channel blockers are used to follow the purification of the channel (Graziana et al., 1988; Harvey et al., 1989a,b; Thuleau et al., 1990).

The polypeptides of a Ca^{2+} channel complex isolated from muscle cells have been resolved by SDS polyacrylamide gel electrophoresis (Catterall et al., 1989; Catterall, 1995). The channel consists of five polypeptides. Their designations and relative molecular masses are α_1 (175 kDa), α_2 (143 kDa), β (54 kDa), γ (30 kDa), and δ (20 kDa). The α_1 and α_2 polypeptides run together in the absence of dithiothreitol (DTT), an agent that cleaves disulfide bonds, indicating that they are attached through disulfide bonds. The α_1 polypeptide probably forms the pore because it is the polypeptide that binds the Ca^{2+} channel blocker ^{3}H-azidopine.

The ability to determine which polypeptides have transmembrane-spanning segments comes from experiments where the polypeptides are challenged with the photoaffinity-labeling hydrophobic probe [^{125}I]3-(trifluoromethyl)-3-(m-iodo-phenyl)diazirine (TID). The α_1 and γ polypeptides are predominantly labeled, and the α_2 and δ polypeptides are slightly labeled. Thus, these polypeptides may have membrane-spanning regions. The α_1 polypeptide is the most abundantly labeled, indicating that it has the most membrane-spanning domains. The β polypeptide is not labeled at all. Therefore, it must be a peripheral protein. The α_2, γ, and δ polypeptides bind ^{125}I-wheat germ agglutinin, a sugar-binding agent, indicating that they are glycoproteins, and thus have domains on the E-side of the membrane.

Ca^{2+} channels are regulated through phosphorylation (Shiina et al., 1988) and it appears that the Ca^{2+} channel itself may be the substrate because the receptor complex becomes phosphorylated in the presence of the catalytic subunit of cAMP-dependent protein kinase and [^{32}P]ATP. Only the α_1 and β subunits are phosphorylated, indicating that they may contain the sites that are regulated by phosphorylation in vivo, and thus have domains on the P-side of the membrane.

A Ca^{2+} channel has been reconstituted in vitro, and its transport activity has been tested in a functional assay by inserting it into liposomes and determining their ability to take up $^{45}Ca^{2+}$, or by inserting it in a lipid bilayer and assaying its Ca^{2+}-transport activity with the patch clamp technique (Pelzer et al., 1989).

Once Ca^{2+} enters the cell, and its concentration rises from the resting level of 30–200 nM to the activated level of 0.3–40 μM (Bush, 1993a,b), it binds to specialized proteins, such as calmodulin, which influence many aspects of cell metabolism (Kawasaki and Kretsinger, 1995; Zeng et al., 2015). Activated calmodulin activates many enzymes, including -Ca^{2+}dependent protein kinases. A number of Ca^{2+} or Ca^{2+}-calmodulin-activated protein kinases have been found in plant cells. These include soluble, membrane-bound, and cytoskeleton-associated proteins (Polya et al., 1983, 1990; Harmon et al., 1987; Roberts, 1989a,b,c; Xie et al., 2014a,b). Many genes for Ca^{2+}-dependent protein kinases (CDPK) have also been discovered (Ranty et al.,

2016; Wang et al., 2016; Yang et al., 2017a,b; Zhang et al., 2017a,b,c,d). One of the actions of calmodulin is to activate the plasma membrane–bound Ca^{2+}-ATPase that pumps Ca^{2+} out of the cell and subsequently restores the cell to its resting state.

While the concentration of free Ca^{2+} is in the nanomolar range in the cytosol of resting cells, the total concentration of calcium is in the millimolar range (Wayne and Hepler, 1985b; Tazawa et al., 2001). This is because most of the calcium is sequestered in the E-space of the ER, which contains various high-capacity, low-affinity calcium-binding proteins, including calsequestrin and calreticulin, with millimolar dissociation constants (see Chapter 4).

12.7 MECHANICS OF DOING EXPERIMENTS TO TEST THE IMPORTANCE OF CA^{2+} AS A SECOND MESSENGER

To establish whether Ca^{2+} is involved in a given response, it is helpful to keep three rules, which have been named Jaffe's Rules in honor of Lionel Jaffe, in mind (Hepler and Wayne, 1985):

1. The response should be preceded or accompanied by an increase in intracellular Ca^{2+} concentration.
2. Blockage of the natural increase in the intracellular Ca^{2+} concentration should inhibit the response.
3. The experimental generation of an increase in the intracellular Ca^{2+} concentration should mimic the stimulus and stimulate the response.

Based on Jaffe's Rules, Ca^{2+} has been shown to be a ubiquitous regulator of plant growth and development (Hepler, 2005). Laboratory manuals that are helpful for planning and performing experiments to test the role of Ca^{2+} in stimulus-response coupling are available (Lambert and Rainbow, 2013; Parys et al., 2014). To determine whether or not a response is preceded or accompanied by an increase in the intracellular $[Ca^{2+}]$, the intracellular $[Ca^{2+}]$ must be measured. This can be done by microinjecting a luminescent protein from the jellyfish *Aequorea* into the cytoplasm (Shimomura, 2008). This protein, called *aequorin*, luminesces in the presence of Ca^{2+}. The amount of luminescence is a function of the Ca^{2+} concentration; however, the relationship is not linear, which makes the calibration somewhat difficult (Blinks et al., 1982). This protein is large enough (<20,000 Da) that once it is injected into the cytosol it stays there and reports on cytosolic Ca^{2+} concentrations, and not that of the ER, Golgi apparatus, etc., which could be measured by a small permeant Ca^{2+} indicator. Knight et al. (1991, 1992) have transformed plants with the aequorin gene and infiltrated the plant with the luminophore coelenterazine so that the luminescence of the protein will indicate the cytosolic Ca^{2+} concentrations without the difficulty of microinjection.

Let me tell you how Osamu Shimomura discovered that aequorin luminescence was Ca^{2+}-dependent. After Shimomura et al. (1957) crystallized the luciferin from *Cypridina*, Frank Johnson invited Shimomura to Princeton University to work on the luminescence of *Aequorea* thinking that it would be similar to the luminescence of *Cypridina*. Shimomura and Johnson tried every method they could think of to extract the luciferin and luciferase from *Aequorea* but nothing worked. Shimomura began thinking that maybe not all organisms use the luciferin–luciferase system to luminesce and maybe they should just try to isolate the luminescent substance whatever it might be. Johnson did not like his idea. In an awkward situation, Johnson worked on one side of the bench trying to isolate luciferin and luciferase, whereas Shimomura worked on the other side trying to isolate the luminescent substance.

Because the luminescing substance is consumed in the luminescent reaction, Shimomura tried to isolate the substance in the presence of anything he thought would inhibit the reaction. Nothing worked. To do some soul searching, Shimomura went out in a rowboat to get away from people. Then he thought, what if the luminescing substance was a protein? If it was, the luminescence may be inhibited by varying the pH. Sure enough, the luminescence was inhibited by a pH 4 buffer and returned on treatment with a pH 7 buffer. According to Shimomura (2008), "But a big surprise came the next moment. When I threw the extract into a sink, the inside of the sink lit up with a bright blue flash. The overflow of an aquarium was flowing into the sink, so I figured out that seawater had caused the luminescence. Because the composition of seawater is known, I easily found out that Ca^{2+} activated the luminescence. The discovery of Ca^{2+} as the activator suggested that the luminescence material could be extracted utilizing the Ca-chelator EDTA, and we devised an extraction method of the luminescent substance."

Shimomura et al. (1962, 1963, 1974; Shimomura and Johnson, 1969, 1972) isolated the protein and named it aequorin. Ridgway and Ashley (1967) realized that the aequorin could be used as a Ca^{2+} indictor and showed that in muscle fibers, a Ca^{2+} spike preceded contraction. Eventually, the gene for aequorin was cloned by Inouye et al. (1985, 1986) and Prasher et al. (1985), which made it possible to measure changes in the Ca^{2+} concentration in transformed organisms.

Johnson et al. (1963) isolated another protein from the jellyfish extract that fluoresced green and suggested that it fluoresced green as a result of Förster (1946, 1959) resonance energy transfer from aequorin when the aequorin was activated by Ca^{2+}. The energy transfer from aequorin to the green protein was confirmed by Morise et al. (1974). Morin and Hastings (1971) named the green fluorescing protein, green fluorescent protein (GFP). The chromophore of the wild-type GFP from *Aequorea* is composed of three amino acids (Ser65-Tyr66-Gly67; Shimomura, 1979; Ward et al., 1989; Cody et al., 1993). The gene for the GFP was cloned (Prasher et al., 1992) based on knowing the amino acid sequence of fragments, and it has been expressed in various organisms (Chalfie et al., 1994) or fused to genes that encode other proteins of interest to localize them in transformed organisms (Chalfie et al., 1993; Wang and Hazelrigg, 1994; Chalfie, 2008).

Roger Tsien (2008; Miyawaki et al., 1997) has designed a genetically engineered calcium-sensitive fluorescent protein by combining the gene sequences for the GFP (Chalfie, 2008; Shimomura, 2008) and calmodulin. Intracellular calcium concentrations can be measured in cells that have been transformed with the gene that encodes this protein (Palmer et al., 2004). Other genetically engineered fluorescent calcium-sensitive probes have been designed using the calcium-binding segment of troponin (Hein and Griesbeck, 2004). By adding certain promoters to the engineered DNA, the fluorescent calcium-sensitive probe can be expressed exclusively in a given cell type (Denninger et al., 2014; Hamamura et al., 2014). By adding targeting sequences to the expressed fluorescent calcium-sensitive protein, it can report on the calcium concentrations in the ER or mitochondria, organelles that store Ca^{2+} (Osibow et al., 2006; Tang et al., 2011; Wu et al., 2014).

Ca^{2+} concentrations can be measured with fluorescent dyes, including calcium green and Fura-2. Calcium green is an indicator whose emission increases on binding Ca^{2+} throughout the spectrum. Because calcium green is not ratiometric, the amount of fluorescence depends on the distribution and concentration of the dye, as well as the Ca^{2+} concentration. Consequently, the determination is only qualitative. Fura-2 is a ratiometric indicator, whose excitation spectrum changes on binding Ca^{2+}. Consequently, the ratio of fluorescence with 340 nm excitation to the fluorescence with 380 nm excitation is proportional to the cytosolic-free Ca^{2+} concentration, independent of dye concentration and distribution. Fura-2 can be loaded into the cells by diffusion if the cell is put in an acid environment. Acetoxymethyl ester derivatives of the dye readily penetrate the plasma membrane, and once inside, the intracellular esterases cleave off the hydrophobic ester making the dye hydrophilic so it stays in the cytosol. Unfortunately, over time the low−molecular mass dye diffuses into the membranous compartments, thus giving a misleading estimate of the cytosolic Ca^{2+} concentration (Bush and Jones, 1990). However, these dyes can be conjugated to high−molecular mass dextrans and injected into the cell so that they will stay in the cytosol (Miller et al., 1992; Gilroy and Jones, 1992; Ehrhardt et al., 1996; Plieth and Hansen, 1996; Felle and Hepler, 1997; Holdaway-Clarke et al., 1997, 2000; Plieth et al., 1997; Hepler et al., 2005; Tucker et al., 2005). This makes Fura-2-dextran the gold standard for quantitative Ca^{2+} measurement. Even though the genetically engineered probes are very clever and can be expressed in specific cells and cells that would be difficult to microinject Fura-2 dextran, none of the genetically engineered probes give quantitative measurements of the Ca^{2+} concentration the way Fura-2 dextran does.

Ca^{2+} concentrations can also be measured with Ca^{2+}-selective microelectrodes. These are glass microcapillary electrodes that have a synthetic resin membrane in them that has a high permeability to Ca^{2+} but not to other ions (Felle, 1989a). The difference in the concentration of Ca^{2+} in the cytosol relative to the inside of the Ca^{2+}-selective electrode generates an electrical potential. The Ca^{2+} concentration in the cytosol is determined using the Nernst equation. Each technique described previously has advantages and limitations and they really should be used in combination.

To test if a blockage of the stimulus-induced increase in the intracellular Ca^{2+} concentration inhibits the response, we must either lower the external Ca^{2+} concentration with Ca^{2+} chelators (e.g., ethylene glycol-bis(β-aminoethyl ether)-N,N,N′,N′-tetraacetic acid (EGTA)) or lower the intracellular Ca^{2+} concentration with the chelator BAPTA. Ca^{2+} entry can be prevented with inorganic blockers, including Nd^{3+}, La^{3+}, and Gd^{3+}, and organic blockers, including nifedipine and verapamil. Because each class of channels is only affected by some of the inhibitors, the identity of the channels involved in a given response can be deduced from their inhibitor specificity.

Lastly, according to Jaffe's Rules, it should be possible to generate an increase in the intracellular Ca^{2+} concentration in the absence of a stimulus and consequently induce the response. This can be done by microinjecting Ca^{2+} into the cell; microinjecting into the cell-caged Ca^{2+}, which can be released on demand by irradiating the cell with light (Gilroy et al., 1990) or by treating cells with Ca^{2+}-selective ionophores, including A23187 or ionomycin. An ionophore is an artificial carrier that diffuses through the membrane-carrying Ca^{2+} and thus collapses the Ca^{2+} difference, thereby increasing the intracellular Ca^{2+} concentration.

Currently, many scientists are taking an omic approach to establish the role of Ca^{2+} in various responses. This entails cataloging and mapping the genes that code for Ca^{2+}-dependent protein kinases and calmodulin-related proteins, subjecting them to phylogenetic analysis, and then localizing the transcripts in various tissues subject to biotic, chemical, and physical stimuli (McCormack and Braam, 2003; McCormack et al., 2005; Boonburapong and Buaboocha, 2007; Chinpongpanich et al., 2012; Zuo et al., 2013;

Hamel et al., 2014; Hu et al., 2016; Munir et al., 2016; Wang et al., 2016; Yang et al., 2017a,b; Zhang et al., 2017a,b,c,d). In his book, "La Biologie Cellulaire," J. B. Carnoy (1884) clearly defined cells, tissues, and organs as different levels of organization. Unfortunately, all too often the omic scientists define organs, such as roots, stems, and leaves, as tissues and are thus unaware that an organ is composed of many tissues, each composed of a distinct cell type, and each perhaps of different ages. In spite of the enormous amount of labor and funding that goes into this work, it does little to help us understand signaling in a single cell. As Alberts et al. (2014) lament in a paper entitled, "Rescuing US biomedical research from its systematic flaws," "The development of original ideas that lead to important scientific discoveries takes time for thinking, reading, and talking with peers. Today time for reflection is a disappearing luxury for the scientific community." If only all scientists had the time to read the experimental philosophy described in Hans Mohr's (1972) work discussed above.

12.8 SPECIFIC SIGNALING SYSTEMS IN PLANTS INVOLVING CA^{2+}

12.8.1 Double Fertilization

In flowering plants, the pollen tube migrates through the cells of the pistil transporting two sperm that may fertilize the egg and the central cell (Strasburger, 1884; Volkmann et al., 2012). Ca^{2+} plays the role of a second messenger in the activation of the egg by the sperm that occurs during fertilization in animal cells (Heilbrunn and Young, 1930; Steinhardt and Epel, 1974; Ridgway et al., 1977; Gilkey et al., 1978; Eisen et al., 1984; Miyazaki, 2006). More recently, Hamamura et al. (2014) and Denninger et al. (2014), using the ovules dissected from plants that were genetically engineered to express a Ca^{2+} sensor in the egg, synergids, or central cells, have shown that following the discharge of the two sperm from a pollen tube, a transient Ca^{2+} increase occurs in the egg and the central cells. On fusion of the sperm with the egg, another transient Ca^{2+} increase occurs in the egg cell.

12.8.2 Ca^{2+}-Induced Secretion in Barley Aleurone Cells

The secretion of α-amylase by aleurone cells is perhaps the best studied secretory system in plants and has also been used to study stimulus-response coupling (Bethke et al., 2006; Sreenivasulu et al., 2008). Secretion of α-amylase is stimulated by gibberellic acid (GA), and the GA-induced stimulation is inhibited by abscisic acid (AbA; Chrispeels, and Varner, 1967). The secretion of α-amylase is dependent on external Ca^{2+} (Moll and Jones, 1982; Jones and Jacobsen, 1983; Jones and Carbonell, 1984). GA-induced secretion of α-amylase is blocked by the Ca^{2+} channel blocker La^{3+} and by calmodulin inhibitors. Moreover, the release of α-amylase is stimulated in the absence of GA by the Ca^{2+} ionophore A23187 (Mitsui et al., 1984). These data suggest that Ca^{2+} acts as a second messenger in GA-induced α-amylase secretion.

Gilroy and Jones (1992) have shown that GA induces an increase in the intracellular Ca^{2+} concentration after 4 hours, which is 2 hours before it induces α-amylase secretion. The cytosolic Ca^{2+} concentration increases from 50 to 200 nM and is dependent on external Ca^{2+}. AbA reverses the GA-induced increase. Confocal microscopy shows that the GA-induced increase in intracellular Ca^{2+} is localized just inside the plasma membrane.

Bush et al. (1993) show that much of the GA-induced Ca^{2+} uptake is taken up into the ER by a Ca^{2+}-ATPase. They suggest that elevated levels of Ca^{2+} may be necessary in part for the correct folding of α-amylase in the ER. α-amylase is a Ca^{2+}-requiring metalloprotein and its correct folding may depend on the Ca^{2+}-binding chaperonin BiP (Jones and Bush, 1991). Heterotrimeric G-proteins are also involved in the GA-induced stimulation of α-amylase secretion (Ueguchi-Tanaka et al., 2000).

12.8.3 Excitation–Cessation of Streaming Coupling in Characean Internodal Cells

While secretion of α-amylase is a relatively slow response and has a protracted signal transduction chain, the cessation of cytoplasmic streaming is rapid and dramatic. Once the mechanism of cytoplasmic streaming was elucidated, Masashi Tazawa, Richard Williamson, and their colleagues turned their attention to how it is regulated and influenced by external stimuli (Kikuyama et al., 1996; Kikuyama, 2001). The cessation of streaming can be induced either electrically or mechanically (Kishimoto, 1968; Staves and Wayne, 1993; Wayne, 1994; Shimmen, 1996, 1997, 2001a,b, 2003, 2008; Iwabuchi et al., 2005; Kaneko et al., 2005; Tazawa, 2011). These stimuli transiently eliminate the motive force that drives cytoplasmic streaming. Cytoplasmic streaming stops instantly in response to these stimuli as would be expected in a cell in which viscous forces predominate over inertial forces (Tazawa and Kishimoto, 1968). While the cessation of streaming is rapid and occurs in less than a second, the recovery is slow and takes a few minutes. The slow recovery may be due, in part, to restarting movement in a thixotropic solution where the viscosity at the motor region increases 100-fold on stopping. It may also be a result, in part, from the slowness of the biochemical reactions needed to reactivate streaming.

Williamson and Ashley (1982) injected aequorin into cells to monitor the free calcium concentration and found that an action potential caused a transient increase in the cytoplasmic Ca^{2+} concentration from approximately 0.1 μM to tens of micromolar. Williamson (1975) removed the vacuolar membrane from cells so that he could artificially increase the Ca^{2+} concentration in the cytosol. He found that when the Ca^{2+} concentration was greater than 10 μM, streaming stopped. It follows that an action potential may induce an increase in the intracellular Ca^{2+} concentration to 10 μM and consequently stop streaming.

At that time, vacuolar membrane–free cells did not show much sensitivity to Ca^{2+} (Tominaga and Tazawa, 1981). This was probably due to a washing out, during the preparation of vacuolar membrane–free cells, of a Ca^{2+}-binding protein or other factor needed for Ca^{2+} sensitivity. This led to the development of a plasma membrane–permeabilized cell model. The permeabilized cell models are sensitive to Ca^{2+}, and increasing the cytoplasmic concentration above 1 μM causes the cessation of streaming (Tominaga et al., 1983). Streaming can be inhibited in intact cells by injecting the cells with Ca^{2+} (Kikuyama and Tazawa, 1982), indicating that artificially increasing the intracellular Ca^{2+} concentration mimics the stimulus. Restricting the influx of Ca^{2+} with the lanthanides or by placing the cells in media with submicromolar concentrations of Ca^{2+} prevents the stimulus-induced cessation of streaming (Staves and Wayne, 1993).

What does the Ca^{2+} bind to once it enters the cell? It is not currently known, but perhaps the Ca^{2+} activates a Ca^{2+}-dependent protein kinase, which can covalently modify one or more of the response elements crucial for streaming. This is likely because the ATP analogue ATP-γ-S, which acts as a substrate of protein kinases but cannot be removed by protein phosphatases, prevents the recovery of streaming that occurs following the Ca^{2+}-induced cessation of streaming. Thus, the phosphorylation of a protein may cause the cessation of streaming, and the dephosphorylation of this protein may be required for the recovery of streaming. This suggestion is supported by the fact that protein phosphatase, isolated from rabbit skeletal muscle and introduced into internodal cells, prevents or reverses the inhibition of streaming caused by Ca^{2+} (Tominaga et al., 1987). It is likely that this phosphatase also exists in characean cells because a very specific protein inhibitor of this phosphatase, also isolated from rabbit skeletal muscle, brings streaming to a standstill in the absence of Ca^{2+}. The phosphorylated protein may be myosin itself.

12.8.4 Regulation of Turgor in Cells

Before I discuss turgor regulation in characean cells, I will discuss the thermodynamic basis of the van't Hoff equation, which is fundamental for understanding turgor regulation. The van't Hoff equation relates the osmotic pressure of a solution to the concentration of solutes in it (see Chapter 7).

12.8.4.1 Thermodynamic Basis of the van't Hoff Equation

Depending on the field, the tendency of water to move is described in many ways. It is discussed in terms of energy, pressure, potential, etc. To put water movement on par with every other transport process I have discussed, I will begin by describing its movement in terms of free energy. The molecular free energy of water relative to pure water at atmospheric pressure is described, in part, by Eq. (12.37):

$$\Delta E = (\overline{V}_w/N_A)\Delta P + kT \ln a_w \tag{12.37}$$

where ΔE is the free energy (in J) of a single water molecule in the solution in question relative to a single molecule of pure water, $\overline{V}_w$ is the partial molar volume of water (in m^3/mol), N_A is Avogadro's number, k is Boltzmann's constant, T is the absolute temperature, ΔP is the hydrostatic pressure exerted on the water molecule relative to atmospheric pressure, and a_w is the activity of water in the solution in question compared with pure water (i.e., the relative activity). There are other terms that can be added to Eq. (12.37) to account for such things as the effect of gravity and electroosmosis (House, 1974; Spanner, 1979). However, for most cellular phenomena, Eq. (12.37) is sufficient.

Water will move from one place to another if there is a difference in the molecular free energy of water in the two regions. For example, water will spontaneously move into a cell if the molecular free energy of a water molecule inside the cell is smaller than the molecular free energy of a water molecule outside the cell—that is, if $\Delta E_i - \Delta E_o$ is negative. Thus, the movement of water into the cell depends on the following relationship:

$$\begin{aligned}\Delta E_i - \Delta E_o = {} & \{(\overline{V}_w/N_A)\Delta P + kT \ln a_w\}_i \\ & - \{(\overline{V}_w/N_A)\Delta P + kT \ln a_w\}_o\end{aligned} \tag{12.38}$$

Eq. (12.38) is not an easy equation to work with because the activity of water is not a readily measurable quantity. Therefore, I will describe how we can replace this term with one that describes directly measurable quantities.

The relative activity of water is defined as the product of the mole fraction of water (N_w) and its activity coefficient (γ_w) relative to pure water, where $a_w = \gamma_w N_w = 1$. The mole fraction of water is the ratio of moles of water to the total number of moles in the solution. That is,

$$N_w = n_w/(n_w + \Sigma n_j) \tag{12.39}$$

where n_w is the number of moles of water and n_j is the number of moles of solute j. To simplify our final equation, we will use the mathematical trick discussed in Chapter 2 and add zero to the numerator in the form of $(\Sigma n_j - \Sigma n_j)$.

$$N_w = (n_w + \Sigma n_j - \Sigma n_j)/(n_w + \Sigma n_j) \\ = 1 - \{\Sigma n_j/(n_w + \Sigma n_j)\} \tag{12.40}$$

Thus,

$$\ln a_w = \ln\{\gamma_w(1 - (\Sigma n_j/(n_w + \Sigma n_j)))\} \tag{12.41}$$

Remember that the concentration of pure water is 55.5 M. If we assume that we have an ideal solution, the activity coefficient of water is 1, and all the water is unbound or "free." A dilute solution where $n_w >> \Sigma n_j$ can be considered to be an ideal solution. In an ideal solution, where $\gamma_w = 1$, $a_w = N_w$, and

$$\ln a_w = \ln(1 - (\Sigma n_j/(n_w + \Sigma n_j))) \tag{12.42}$$

We can use the series $\{-x - x^2/2 - x^3/3 - \ldots - x^m/m\}$ to approximate $\ln(1 - x)$. When x is not small, the higher-order terms, known as the *virials*, must be used. When x is small, $\ln(1 - x) = -x$. Thus:

$$\ln a_w = \ln[1 - [\Sigma n_j/(n_w + \Sigma n_j)]] \cong - [\Sigma n_j/(n_w + \Sigma n_j)] \tag{12.43}$$

Since for a dilute solution, $n_w >> \sum n_j$, then

$$\ln a_w \cong - [\Sigma n_j/n_w] \tag{12.44}$$

Divide both sides by $\overline{V}_w$:

$$(\ln a_w)/\overline{V}_w \cong - [\Sigma n_j/n_w]/\overline{V}_w \tag{12.45}$$

Since $(\Sigma n_j/n_w)/\overline{V}_w$ is equal to the concentration of solute j, then

$$\ln a_w \cong - C\overline{V}_w \tag{12.46}$$

where C is the concentration of solute j (in mol/m^3).

As a first approximation, the osmotic pressure is a measure of the decrease in the concentration of water due to the displacement of water from a solution by a solute (Nobel, 1991). We can rewrite Eq. (12.38) like so:

$$\Delta E_i - \Delta E_o = \{(\overline{V}_w/N_A)\Delta P - kTC\overline{V}_w\}_i \\ - \{(\overline{V}_w/N_A)\Delta P - kTC\overline{V}_w\}_o \tag{12.47}$$

Water movement is typically discussed by plant biologists in terms of water potential (P_w; Dainty, 1998), which is given in units of pressure (Pa) and not energy. $\Delta E_i - \Delta E_o$, in units of Joules, can be converted into water potential, in units of Pascals, according to Eq. (12.48):

$$(P_w)_i - (P_w)_o = (\Delta E_i - \Delta E_o)(N_A/\overline{V}_w) \tag{12.48}$$

Thus, Eq. (12.47) becomes

$$(P_w)_i - (P_w)_o = (N_A/\overline{V}_w) \\ \left[\{(\overline{V}_w/N_A)\Delta P - kTC\overline{V}_w\}_i - \{(\overline{V}_w/N_A)\Delta P - kTC\overline{V}_w\}_o\right] \tag{12.49}$$

After canceling like terms, and substituting R for $N_A k$, Eq. (12.49) becomes

$$(P_w)_i - (P_w)_o = \{\Delta P - RTC\}_i - \{\Delta P - RTC\}_o \tag{12.50}$$

van't Hoff (1898–99) defined the osmotic pressure as RTC. I will denote it by P_π. Thus, Eq. (12.50) becomes

$$(P_w)_i - (P_w)_o = \{\Delta P - P_\pi\}_i - \{\Delta P - P_\pi\}_o \tag{12.51}$$

Assuming that the hydrostatic pressure outside the cell is equal to the atmospheric pressure, then $\Delta P_o = 0$ (by convention), and Eq. (12.51) becomes

$$(P_w)_i - (P_w)_o = \Delta P_i + P_{\pi o} - P_{\pi i} \tag{12.52}$$

Water will passively move into a cell when $(P_w)_i - (P_w)_o$ is negative. Decreasing ΔP_i and $P_{\pi o}$, or increasing $P_{\pi i}$, will favor the movement of water into a cell. Likewise, increasing ΔP_i and $P_{\pi o}$, or decreasing $P_{\pi i}$, will favor the movement of water out of the cell.

At equilibrium, when $(P_w)_i - (P_w)_o = 0$, there is no net movement of water into or out of the cell, and the hydrostatic pressure in the cell, which is referred to as the *turgor pressure*, is equal to the osmotic pressure difference:

$$P_{ti} = P_{\pi i} - P_{\pi o} \tag{12.53}$$

While the turgor pressure is a measure of the hydrostatic pressure experienced by the water molecules in the protoplast, turgor pressure typically considered to be the force per unit area that the protoplast exerts against the wall. However, consistent with its parity with hydrostatic pressure, as defined by Pfeffer (1877), it is equal to the pressure or the force per unit area that the wall exerts as it pushes back against the protoplast and forces water to leave the cell. A turgor pressure develops because the osmotic pressure is typically greater inside the cell than outside the cell, and thus water moves into the cell along its activity difference. However, the protoplast cannot expand indefinitely because of the presence of an extracellular matrix with a high tensile strength. Thus, at equilibrium, the wall exerts a pressure that is equal in magnitude but opposite in sign to the osmotic pressure difference. By contrast, animal cells and wall-less plant cells cannot develop a significant hydrostatic pressure due to the absence of an extracellular matrix with high tensile strength. These wall-less cells must live in isotonic conditions or possess a contractile vacuole to pump out the excess water. Turgor pressure is an important property of plant cells and may affect a number of responses, including sugar uptake (Diettrich and Keller, 1991; Keller, 1991), gravity sensing (Staves et al., 1992), wall extensibility and yield threshold (Pritchard et al., 1991), plasmodesmata conductance (Ding and Tazawa, 1989), AbA accumulation (Creelman and Mullet, 1991), wall deposition (Proseus and Boyer, 2006c), and tip growth (Hill et al., 2012; Hepler and Winship, 2015; Obermeyer, 2017). Consequently, many methods have been developed to measure turgor pressure (Tazawa, 1957; Boyer, 1995; Tomos and Leigh, 1999; Lintilhac et al., 2000; Wei et al., 2001; Geitmann, 2006).

Inside the cell, at equilibrium, the water potential of all the compartments must be equal or the various organelles will shrink or swell. The hydrostatic pressure in all the organelles must be similar because the tensile strength of a typical membrane is so low that the maximum tensile stress the membrane can withstand before breaking is less than 0.1 MPa. The relations between the tensile stress a membrane experiences (σ) and the difference in the hydrostatic pressure across that membrane (dP) is given by the following formula:

$$\sigma = rdP/(2w) \qquad (12.54)$$

where r is the radius of a spherical membrane–enclosed cell or organelle, and w is the thickness of the membrane (Nobel, 1983). Given that $\sigma = 0.1$ MPa, $w = 7 \times 10^{-9}$ m, and $r = 8 \times 10^{-6}$ m, the pressure difference across the membrane will be 175 Pa. Because this value is small compared to the osmotic pressure of each compartment (<700,000 Pa), we can assume that the hydrostatic pressure is essentially equal in all compartments. Since at equilibrium the water potential is equal in all compartments, the osmotic pressure must also be equal in all compartments.

12.8.4.2 Turgor Pressure in Characean Cells

Lamprothamnium is a characean alga that lives in brackish water (Okazaki, 1996; Beilby et al., 1999; Shepherd and Beilby, 1999; Shepherd et al., 1999; Beilby, 2015). As in other characean cells, the vacuole takes up 90%–95% of the cell volume, and thus accounts for most of the solutes that make up the osmotic pressure of the cell. For example, the volume of a 2×10^{-2} m long (x) and 0.5×10^{-3} m diameter ($2r$) cylindrical cell is $x\pi r^2 = 3.9 \times 10^{-9}$ m^3. If the vacuole takes up 95% of the cell volume, the volume of the vacuole is 3.7×10^{-9} m^3 and the volume of the cytoplasm is 0.2×10^{-9} m^3. If the osmotic pressure of the vacuole is 1.77 MPa, then it contains 715 mol/m^3 of osmotically active solutes. In the cytoplasm, the osmotic pressure must also be 1.77 MPa, and there must also be 715 mol/m^3 of osmotically active solutes in it. We can calculate how many solutes are in each compartment. In the vacuole, there are (715 mol/m^3) (3.7×10^{-9} m^3) = 2.6×10^{-6} mol of solute. In the cytoplasm, there are (715 mol/m^3) (3.7×10^{-9} m^3) = 1.43×10^{-7} mol of osmotically active solute. Therefore, the osmotic pressure of the cell sap usually gives us a good estimate of the osmotic pressure of the whole cell. *Lamprothamnium* internodal cells normally have an osmotic pressure of 1.77 MPa when they grow in brackish water with an osmotic pressure of 0.89 MPa. Thus, at equilibrium, their turgor pressure is 0.88 MPa.

12.8.4.3 Turgor Regulation in Characean Cells

Cells may regulate either their turgor pressure or their osmotic pressure (Bisson and Kirst, 1980). The internodal cell of *Lamprothamnium* is a particularly good example of a cell that regulates its turgor pressure (Okazaki, 1996). When the cells are transferred from their normal medium, which has an osmotic pressure of 0.89 MPa, to a hypotonic medium, which has an osmotic pressure of 0.51 MPa, the turgor pressure of the cells increases to 1.26 MPa. The cells then must lose solutes to decrease their internal osmotic pressure and to bring their turgor pressure back to 0.88 MPa (Fig. 12.4).

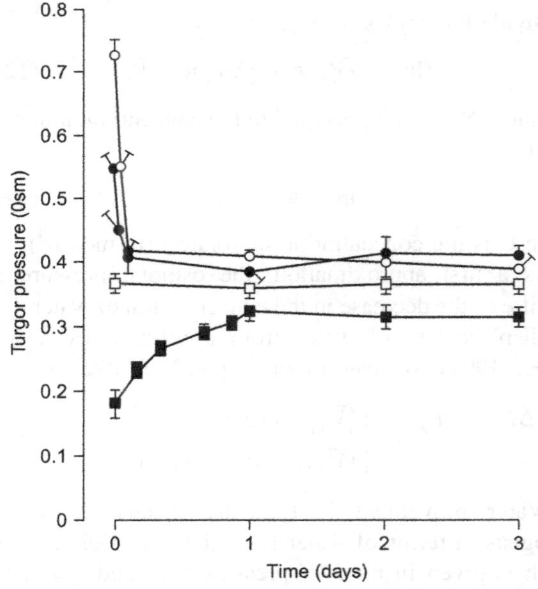

FIGURE 12.4 Time course of turgor regulation in *Lamprothamnium*. Cells were preconditioned in one-third–strength seawater and then placed in zero-strength seawater (*open circles*), one-sixth–strength seawater (*filled circles*), one-third–strength seawater (*open squares*), and one-half–strength seawater (*filled squares*). In all cases, the turgor pressure returned approximately to the control pressure. *From Okazaki, Y., Shimmen, T., Tazawa, M., 1984a. Turgor regulation in a brackish water charophyte,* Lamprothamnium succunctum. *I. Artificial modification of intracellular osmotic pressure. Plant Cell Physiol. 25, 565–571.*

How much must the osmotic pressure of the vacuole decrease to regain a turgor pressure of 0.88 MPa? It must decrease by 0.38 MPa. This represents a decrease in the concentration of osmotically active solutes of 153.4 mol/m^3. If the volume of the vacuole is 3.7×10^{-9} m^3, then 5.6×10^{-7} mol of osmotically active solutes must be lost. This is equivalent to 3.4×10^{17} solutes. Most of the solutes lost are ions, and thus to maintain electroneutrality, one-half of the solutes must be negatively charged, whereas the other half must be positively charged. Assuming that K^+ and Cl^- make up the greatest part of the osmoticum lost, 1.67×10^{17} Cl^- ions and 1.67×10^{17} K^+ ions must leave the cell.

Yoshiji Okazaki et al. (Okazaki and Tazawa, 1987b; Okazaki et al., 1984a,b; Okazaki and Iwasaki, 1991) have shown that Ca^{2+} acts as a second messenger in the turgor regulation response in *Lamprothamnium*. External Ca^{2+} is required for the transient increase in membrane conductance (Fig. 12.5) and the efflux of osmoticum from the cells (Fig. 12.6). The threshold concentration of external Ca^{2+} is 10 μM, and the optimum is approximately 1 mM (Okazaki and Tazawa, 1986a,b,c). The hypotonic medium–induced change in cellular osmotic pressure is inhibited by nifedipine, a Ca^{2+} channel blocker (Fig. 12.7).

The plasma membrane has a memory. If the hypotonic treatment is given in the absence of Ca^{2+}, the cell does not undergo an increase in the electrical conductance; however, if the Ca^{2+} concentration is increased about 30 minutes later, the membrane conductance rapidly increases following the addition of Ca^{2+} (Fig. 12.8).

The rate of cytoplasmic streaming in *Lamprothamnium* stops instantly and transiently on hypotonic treatment, but only if Ca^{2+} is in the external medium (see Fig. 12.9; Okazaki and Tazawa, 1986b). This is an indication that hypotonic treatment causes an increase in the intracellular Ca^{2+} concentration. An increase in the intracellular Ca^{2+} concentration can be measured with an atomic absorption spectrophotometer. Measurements with atomic absorption spectrophotometry show that the total cytoplasmic calcium concentration rises from 2 to 3.7 mM (Okazaki and Tazawa, 1987a). This means that there is an increase in the total cytoplasmic concentration of calcium of 1.7 mM. Using aequorin that had been microinjected into the cell, Okazaki et al. (1987) showed that hypotonic treatment causes an immediate transient increase in the intracellular-free Ca^{2+} concentration (Fig. 12.10). This increase requires external Ca^{2+} (Fig. 12.11). The intracellular-free calcium

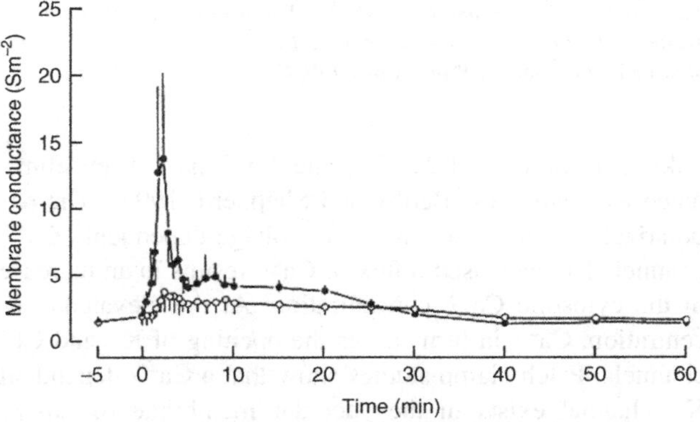

FIGURE 12.5 Time course of the change in membrane conductance and its dependence on external Ca^{2+}. *Open and closed circles* represent 0.01 and 3.9 mol/m^3 external Ca^{2+}, respectively. *From Okazaki, Y., Tazawa, M., 1986c. Involvement of calcium ion in turgor regulation upon hypotonic treatment in* Lamprothamnium succinctum. *Plant Cell Environ. 9, 185–190.*

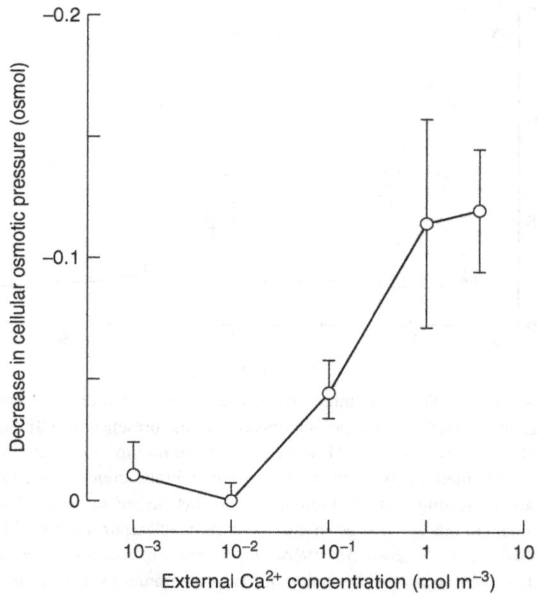

FIGURE 12.6 The dependence of the change in osmotic pressure on the external Ca^{2+} concentration. *From Okazaki and Tazawa, 1986c. Involvement of calcium ion in turgor regulation upon hypotonic treatment in Lamprothamnium succinctum. Plant Cell Environment 9, 185–190.*

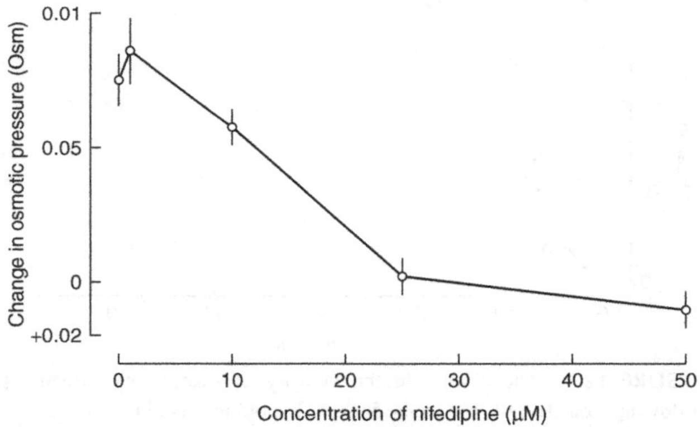

FIGURE 12.7 The effect of nifedipine, a Ca^{2+} channel blocker, on turgor regulation. *From Okazaki, Y., Tazawa, M., 1986a. Ca^{2+} antagonist nifedipine inhibits turgor regulation upon hypotonic treatment in internodal cells of* Lamprothamnium. *Protoplasma 134, 65–66.*

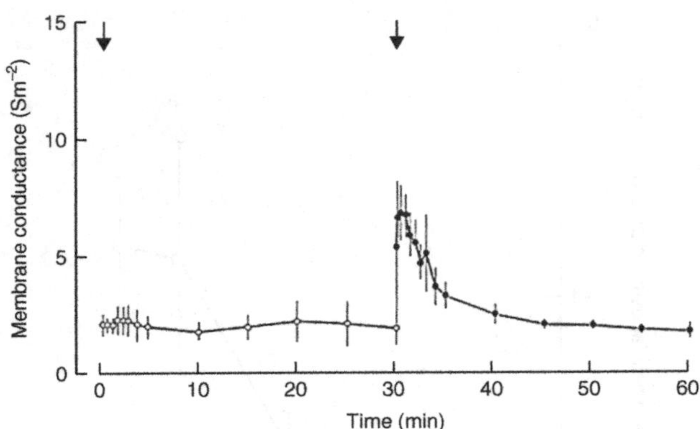

FIGURE 12.8 The membrane conductance does not increase when the cells are transferred into a hypotonic medium in the presence of 0.01 mol/m^3 external Ca^{2+} (*first arrow*). However, when 3.9 mol/m^3 external Ca^{2+} is added to the medium (*second arrow*), the membrane conductance rapidly increases, indicating that the membrane remains poised to respond to the turgor change, but the change in conductance itself requires external Ca^{2+}. *From Okazaki, Y., Tazawa, M., 1986c. Involvement of calcium ion in turgor regulation upon hypotonic treatment in* Lamprothamnium succinctum. *Plant Cell Environ. 9, 185–190.*

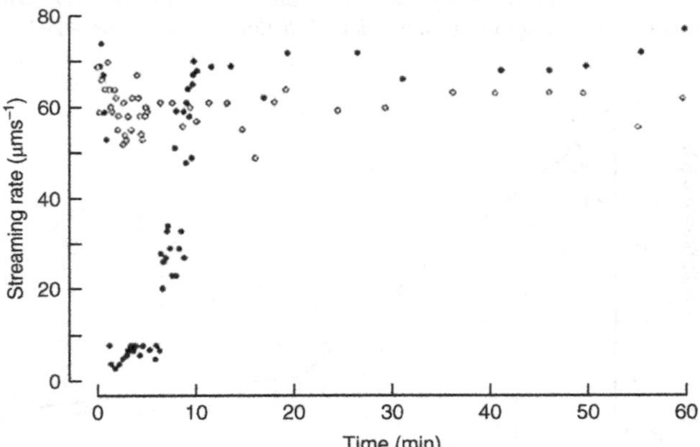

FIGURE 12.9 The change in the velocity of cytoplasmic streaming following transfer of the cells into a hypotonic medium containing 0.01 mol/m^3 external Ca^{2+} (*open circles*) or 3.9 mol/m^3 external Ca^{2+} (*filled circles*). *From Okazaki, Y., Tazawa, M., 1986b. Effect of calcium ion on cytoplasmic streaming during turgor regulation in a brackish water charophyte* Lamprothamnium. *Plant Cell Environ. 9, 491–494.*

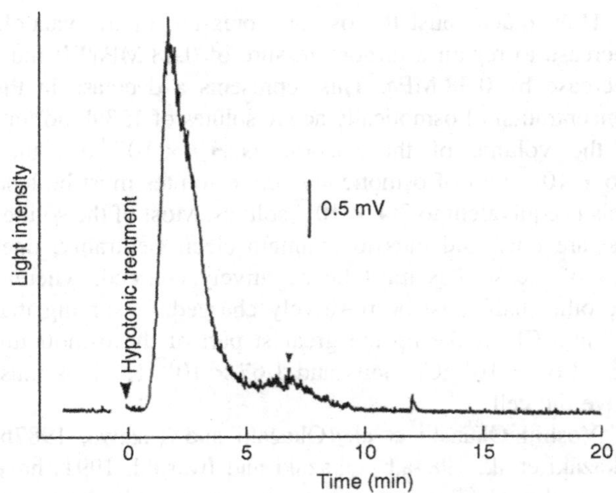

FIGURE 12.10 Time course of light emission of microinjected aequorin following hypotonic treatment. The light intensity is related to the intracellular-free [Ca^{2+}]. *From Okazaki, Y., Yoshimoto, Y., Hiramoto, Y., Tazawa, M., 1987. Turgor regulation and cytoplasmic free Ca^{2+} in the alga* Lamprothamnium. *Protoplasma 140, 67–71.*

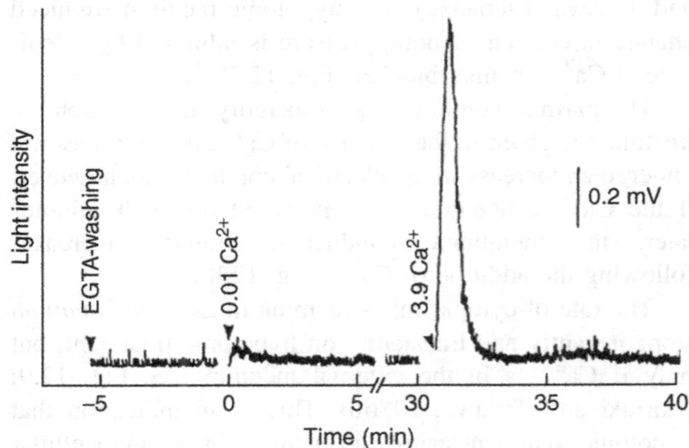

FIGURE 12.11 Time course of light emission of microinjected aequorin following hypotonic treatment in the presence of varying concentrations of extracellular Ca^{2+}. *From Okazaki, Y., Yoshimoto, Y., Hiramoto, Y., Tazawa, M., 1987. Turgor regulation and cytoplasmic free Ca^{2+} in the alga* Lamprothamnium. *Protoplasma 140, 67–71.*

concentration rises to an estimated 10–50 μM, which is approximately 30–100 times smaller than the rise in total cytoplasmic calcium. This means that there is a buffering capacity in the cytoplasm for Ca^{2+}. Some of the buffering may be due to intracellular Ca^{2+}-binding proteins and active sequestering by organelles.

According to Yoshiji Okazaki and Masashi Tazawa (1990), hypotonic stress causes an increase in the tension of the plasma membrane. This tension is sensed by a mechanoreceptor (Morris, 1990; Messerli and Robinson, 2007; Pickard, 2007), which causes a membrane depolarization. The ion(s) that carry the initial depolarizing current are unknown, although K^+, Cl^-, and Ca^{2+} have been eliminated as candidates (Beilby and Shepherd, 1996). The depolarization then activates a voltage-dependent Ca^{2+} channel. The increased influx of Ca^{2+} results in an increase in the cytosolic Ca^{2+} concentration. At the elevated concentration, Ca^{2+} in turn causes the opening of K^+ and Cl^- channels. Patch clamp studies show that a Ca^{2+}-dependent K^+ channel exists in the vacuolar membrane of *Lamprothamnium* (see Fig. 12.12; Katsuhara et al., 1989). Water follows the K^+ and Cl^- out of the cell and the turgor pressure returns to its normal value.

Calcium may activate the monovalent ion channels involved in turgor regulation through the mediation of a

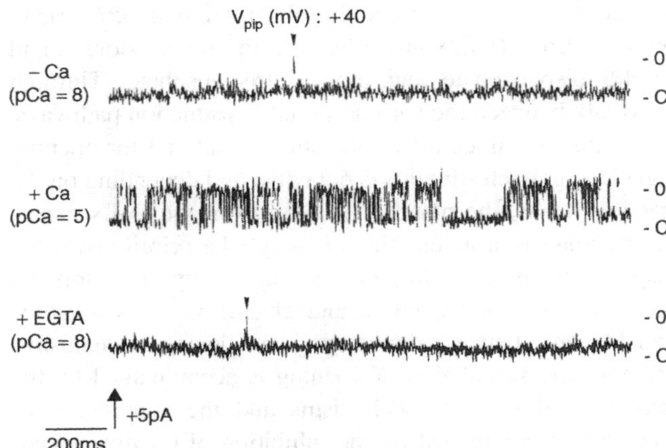

FIGURE 12.12 Recordings of K^+ channels from a cytoplasmic-side-out patch from the vacuolar membrane of *Lamprothamnium* at two different Ca^{2+} concentrations. The arrows indicate occasional channel openings at 10 nM Ca^{2+} (pCa 8). *From Katsuhara, M., Mimura, T., Tazawa, M., 1989. Patch clamp study on a Ca^{2+}-regulated K^+ channel in the tonoplast of the brackish characeae* Lamprothamnium succinctum. *Plant Cell Physiol. 30, 549–555.*

protein kinase. Support for this idea comes from the fact that a Ca^{2+}-dependent protein kinase has been isolated from *Lamprothamnium*; microinjection of antibodies directed against a Ca^{2+}-dependent protein kinase inhibits turgor regulation; and last, K-252a, an inhibitor of protein kinases, inhibits turgor regulation in *Lamprothamnium* cells (Yuasa et al., 1997). Older cells respond more slowly than young cells (Beilby et al., 1999).

Turgor is an important indicator of the status of plant cells. Wounding results in the loss of turgor pressure in a cell. In *Chara*, the change in pressure is converted into electrical signals that pass the "death message" from the wounded cell to the neighboring cells (Shimmen, 2001a, 2002, 2003).

12.9 PHOSPHATIDYLINOSITOL SIGNALING SYSTEM

The Ca^{2+} signaling system interacts with other signaling systems in plant cells, including one that is based on phosphatidylinositol (PI), which makes up less than 1% of membrane lipids (Boss and Massel, 1985; Berridge, 1987; Coté et al., 1987; Morse et al., 1987b, 1989; Pfaffmann et al., 1987; Sandelius and Morré, 1987; Wheeler and Boss, 1987; Drøbak et al., 1988; Ettlinger and Lehle, 1988; Sommarin and Sandelius, 1988; Murthy et al., 1989; Peeler et al., 1989; Einspahr and Thompson, 1990; Drøbak, 1991; Coté and Crain, 1993; Yang et al., 1993a,b; Munnik and Nielsen, 2011). The PI pathway is involved in stomatal movements, the regulation of mitosis (Chen and Wolniak, 1987; Wolniak, 1987; Larsen et al., 1991), plasmodesmata conductance (Tucker, 1988; Tucker and Boss, 1996), leaflet movements (Morse et al., 1987a, 1989, 1990; Kim et al., 1996a,b), osmotic adaptation in *Dunaliella* (Einspahr et al., 1988, 1989), the deflagellation response of *Chlamydomonas* (Yueh and Crain, 1993), pollen tube growth (Franklin-Tong et al., 1996), α-amylase secretion in rice (Kashem et al., 2000), water stress (Munnik and Meijer, 2001; Munnik and Vermeer, 2010), and chloroplast movements (Aggarwal et al., 2013). The inositol signaling system can be inhibited by lithium (Perdue et al., 1988; Berridge et al., 1989; Gillaspy et al., 1995; Wolpert and Richards, 1997) and neomycin (Chen and Boss, 1991).

12.9.1 Components of the System

PI is a lipid that is formed in the ER by the CDP-dependent PI exchange protein that catalyzes the attachment of inositol phosphate onto diacylglycerol (DAG) (see Chapter 4; Sandelius and Morré, 1987). Inositol lipids are enriched in the cytosolic leaflets of membranes. The PI gets to the plasma membrane through the mechanisms involved in membrane flow (Wheeler and Boss, 1987; Peeler et al., 1989; see Chapter 8). Phosphatidylinositol-4-phosphate (PIP) is formed from the PI by PI kinase, and phosphatidylinositol-4,5-bisphosphate (PIP_2) is formed from phosphatidylinositol-4-phosphate by PIP kinase according to the following reaction:

$$\text{PI kinase} \quad \text{PIP kinase}$$

$$PI \Rightarrow PI - 4 - P \Rightarrow PI - 4, 5 - P$$

Activation of the PI signaling pathway begins with the activation of a PIP_2-specific phospholipase C. This enzyme splits PIP_2 into DAG and the water-soluble inositol trisphosphate. There are many classes of phospholipase C; some are activated by G-proteins, others by Ca^{2+}, and some by both (Dillenschneider et al., 1986; Melin et al., 1987; Einspahr et al., 1989; Tate et al., 1989; Buffen and Hanke, 1990).

The water-soluble inositol phosphate IP_3 causes the release of Ca^{2+} from microsomal and vacuolar membrane fractions (Drøbak and Ferguson, 1985; Rincon and Boss, 1987; Schumaker and Sze, 1987; Ranjeva et al., 1988; Reddy et al., 1988; Alexandre et al., 1990; Brosnan and Sanders, 1990; Taylor and Marshall, 1992). This is one way the components of the PI pathway interact with Ca^{2+}-mediated signal transduction chains. In plants IP_6, which is also known as phytate, may be more active than IP_3 (Munnik and Nielsen, 2011).

The IP_3-stimulated Ca^{2+} release is transient because the IP_3 is rapidly dephosphorylated by a specific phosphatase to IP_2. Eventually the IP_2 is dephosphorylated to inositol where it will be a substrate of PI. IP_3 can also undergo phosphorylation. More and more isomers of inositol phosphates are being found and some of these may play a role in signaling (Berridge and Irvine, 1990).

Phospholipase C also causes the release of DAG from PIP_2 (Morse et al., 1989; Morré et al., 1989b). DAG is capable of stimulating the activity of a Ca^{2+} phospholipid-dependent protein kinase, called *protein kinase C* (Elliott and Kokke, 1987a,b). DAG increases the affinity of protein kinase C for Ca^{2+}. The activation by DAG is also transient because DAG is either cleaved to form arachidonic acid or phosphorylated to form phosphatidic acid within seconds. In plants phosphatidic acid, the phosphorylated form of DAG may be more active than DAG (Munnik, 2001; Munnik and Nielsen, 2011).

12.9.2 Phosphatidylinositol Signaling in Guard Cell Movement

In the epidermal layer of the portions of the plant that interface with a gaseous environment are the stomatal pores, or pass-ports as Nehemiah Grew (1682) called them, which are surrounded by guard cells that regulate the permeability of the plant at the organ level just as the plasma membrane regulates the permeability of the cell at the cellular level. The open pores facilitate the uptake of CO_2 for photosynthesis. Plants have the ability to close the stomatal pores in times of water stress to minimize wilting.

Stomatal opening requires the osmotic movement of water into the guard cells, whereas stomatal closing requires the osmotic movement of water out of the guard cells. The movement of water into the guard cells occurs when there is an increase in the osmotic pressure of the guard cells of approximately 300–400 mol/m^3 due primarily to influxes of K^+ and Cl^-. These ions flow into the guard cells across the plasma membrane through ion channels. As the hydrostatic pressure across the vacuolar membrane must be nil or it will burst, the ions must cross both the plasma membrane and the vacuolar membrane (Eisenach and De Angeli, 2017). The movement of water out of the guard cells occurs when there is a decrease in the osmotic pressure of approximately 300–400 mol/m^3, which is primarily due to a decrease in K^+ and Cl^-, which flows out of the guard cells across the plasma membrane through ion channels. Again, as the hydrostatic pressure across the vacuolar membrane must be nil, the ions must cross both the vacuolar membrane and the plasma membrane.

The opening and closing of the stomatal pore allows the plant to find a balance between the photosynthetic fixation of CO_2, which requires open stomatal pores and the prevention of wilting during water stress, which requires closed stomatal pores. The opening and closing of the stomatal pores is primarily affected by light and hormones. Each stimulus is recognized by one or more receptor proteins. The activated receptor proteins influence the change in the osmotic pressure of the guard cells through various signaling pathways that involve phosphorylation (protein kinases and protein phosphatases), heterotrimeric G proteins, Ca^{2+} (influx and release from internal stores), and PI 4,5 bisphosphate and inositol trisphosphate. Through cross talk between the various signal transduction pathways, the guard cells integrate all the stimuli that call for opening and closing. Each stimulus is not equal and depending on the species' needs, the plant's needs, or even the leaf's needs, the response to a given stimulus may take priority over the response to another stimulus, giving the optimal stomatal pore size. Stomatal opening and closing requires a differential balance between the opening and closing mechanisms, and thus the stimulation of opening is accompanied by the inhibition of closing mechanisms and the stimulation of closing is accompanied by the inhibition of opening mechanisms. While we understand the ionic basis of stomatal opening and closing quantitatively (see Chapter 2), our understanding of the factors that provide specificity to the regulatory mechanisms is still somewhat qualitative (Jezek and Blatt, 2017). Pathogens can enter the plant through open stoma, and some pathogens are able to co-opt the regulatory systems involved in opening and the closing of the stomatal pore to gain entrance. The stimulation of opening or the preventing of closing gives the pathogens access to the plant (McLachlan et al., 2014) and causes wilting. The opening and closing of guard cells has been studied from a systems biology perspective, where the interactions between 70 components implicated in stomatal opening have been modeled giving 10^{31} distinct states of the system (Sun et al., 2014). We will only discuss a few.

In general, the stomatal pore opens each morning in response to light (see Chapter 2). There are many receptors for light, but the main receptor is phototropin, which is a blue-light receptor with protein kinase activity (Inoue and Kinoshita, 2017). Active phototropin causes the phosphorylation of the plasma membrane H^+-ATPase through a phosphorylation cascade (Takemiya et al., 2006, 2013a,b; Takemiya and Shimazaki, 2010, 2016), which is followed by the binding of a 14-3-3 protein that activates the H^+-ATPase and hyperpolarizes the plasma membrane (Yamauchi et al., 2016). The hyperpolarized membrane activates the inward-rectifying K^+ channel, which increases the osmotic pressure of the guard cell. The increased concentration of this cation is matched by an increase in the concentration of anions, primarily malate and chloride. The Cl^- enters through a H^+-anion symporter. Malate is produced from the degradation of starch stored in the chloroplasts, and chloride enters the cell across the plasma membrane through anion channels. The increased osmotic pressure in the guard cells causes water to move into them. The increase in the water content enlarges the guard cells and the pore opens thanks to the radial micellation of the cellulose microfibrils in the walls.

The stomatal pore also opens in response to auxin (Fig. 12.13A; Marten et al., 1991), which is an indicator of

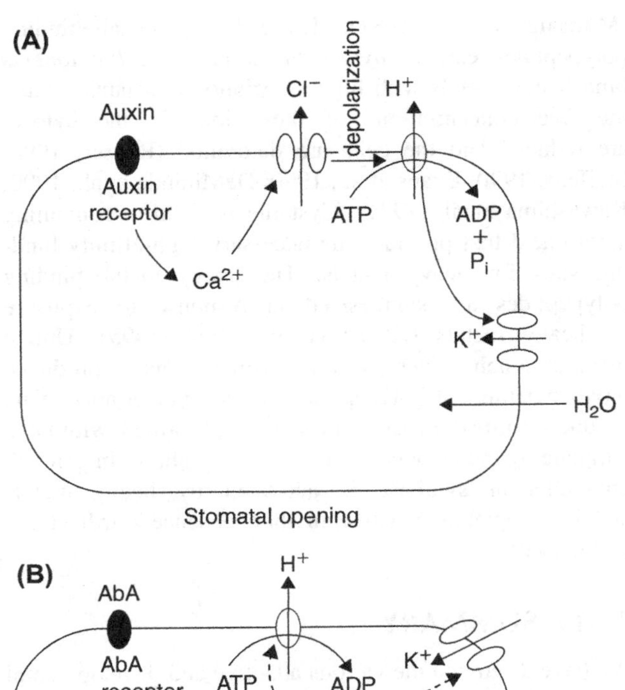

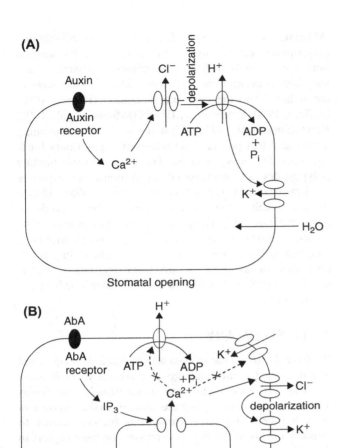

FIGURE 12.13 Regulation of stomatal closing due to abscisic acid and opening due to auxin (see text for details).

the younger tissues in a plant. Auxin binds to a receptor that causes an increase in the cytosolic Ca^{2+} concentration. The increased Ca^{2+} activates an anion channel, which causes a transient membrane depolarization. Subsequently, the H^+-pumping ATPase is activated and the membrane hyperpolarizes. This activates the inwardly rectifying K^+ channel. Thus, K^+ accumulates in the guard cell, water enters the guard cell, and the stoma opens.

The stomatal pore closes in response to water stress (Fig. 12.13B; Yamazaki et al., 2003). AbA is produced as a response to water stress. The AbA binds to one or more types of receptors in the guard cell. One of the receptors may activate a heterotrimeric G protein that activates phospholipase C, which results in the production of IP_3 or IP_6 (Lee et al., 1996a,b,c, 2007; Staxen et al., 1999; Hunt et al., 2003; Liu et al., 2007a). Such an increase in IP_3 or IP_6 may result in the release of Ca^{2+} from the ER and vacuole that results in an increase in the cytoplasmic-free $[Ca^{2+}]$ (McAinsh et al., 1990, 1992; Schroeder and Hagiwara, 1990; Gilroy et al., 1990, 1991; Schroeder and Thuleau, 1991). The increase in the cytoplasmic-free calcium concentration is sufficient to inhibit the plasma membrane H^+-ATPase (Kinoshita et al., 1995), inhibit the inwardly rectifying K^+ channels (Pei et al., 1998a), and activate a Ca^{2+}-activated anion channel in the plasma membrane (Keller et al., 1989). The inwardly rectifying K^+ channel can be inhibited experimentally by releasing caged IP_3 into the guard cell (Blatt et al., 1990). In fact, releasing either caged IP_3 or caged Ca^{2+} into the cytosol induces stomatal closure even in the presence of 1 mM La^{3+}. This indicates that AbA initiates stomatal closure by stimulating the formation of IP_3, which releases Ca^{2+} from internal stores (Mansfield et al., 1990; Gilroy et al., 1991). This Ca^{2+} release channel is also regulated through phosphorylation (Mori et al., 2000).

When another potential AbA receptor binds AbA, it impounds a protein phosphatase, which results in an enhanced Ca^{2+}-dependent phosphorylation of the slow anion channel by a Ca^{2+}-dependent protein kinase. The phosphorylation activates the slow anion channel (Pei et al., 1997; 1998b), which results in the efflux of Cl^- across the plasma membrane. The resulting depolarization activates the outwardly rectifying K^+ channel (see Chapter 2; Schroeder, 1988; Assmann and Shimazaki, 1999). This results in the efflux of K^+. The loss of K^+ and Cl^- from the guard cell is followed by a loss in water, a loss of turgor, and the closing of the stomatal pore.

In elegantly designed and performed reconstitution experiments in a heterologous system, the sequence of steps involved in AbA-induced stomatal closure has been determined by injecting into a frog oocyte with RNA that encoded the various components of the signal transduction chain in guard cells, such as the AbA receptor, the anion channel, various Ca^{2+}-dependent protein kinases, and a protein phosphatase. After the RNAs were translated and inserted in the plasma membrane of the frog oocyte, Brandt et al. (2012) measured the characteristics of the current passing through the anion channel using a two-electrode voltage clamp. They found that AbA stimulated an anion current. They also found that AbA caused the phosphorylation of a serine in the anion channel that is necessary for its conductance. The conductance of the anion channel was lost when they mutated the serine that was phosphorylated by the Ca^{2+}-dependent protein kinase to an alanine that could not be phosphorylated. By expressing RNA that codes for physically linked proteins or not, they could discern whether the expressed proteins have to be in physical contact or not in the signal transduction chain.

Irving et al. (1992) show that an increase in the intracellular Ca^{2+} concentration precedes both stomatal opening and closure. Thus, while Ca^{2+} may be necessary for the individual signal transduction chains, the specificity in each chain comes from other elements and/or differences in the spatial and/or temporal change in Ca^{2+} (Leckie et al., 1998; Blatt, 2000; McAinsh et al., 2000; Assmann, 2002; Luan, 2002; MacRobbie, 1997; 2000, 2002). One of the unknown elements may be H^+ because an initial acidification of the cytosol is correlated with stomatal opening, whereas an initial alkalinization of the cytoplasm is correlated with stomatal closure (Blatt, 1992; Blatt and Armstrong, 1993; Grabov and Blatt, 1997). Other regulatory elements include protein kinases and phosphatases (Thiel and Blatt, 1994; Armstrong et al., 1995; Li and Assmann, 1996; Pei et al., 1997; Mori and Muto, 1997; Kinoshita and Shimazaki, 1999; Allen et al., 1999a,b; Li et al., 2000, 2002; Hwang and Lee, 2001; Eun et al., 2001). Lastly, the channels involved in guard cell swelling and shrinking may be directly affected by turgor (Cosgrove and Hedrich, 1991).

Guard cell swelling may be accompanied by new membrane arriving at the plasma membrane through the endomembrane system (Duque et al., 2004). Likewise, shrinking may be accompanied by endocytosis. Vesicle movement may be mediated through the actin and/or the microtubular cytoskeleton. Presumably, these processes would also be regulated by the various signal transduction chains.

12.10 THE ROLE OF IONS IN CELLS

Calcium has been widely known to be a necessary element for plant growth ever since Benjamin Franklin wrote in large letters, formed by using ground plaster of Paris ($CaSO_4$), "This has been plastered" in a clover field along the road to Washington, DC, and every passerby noticed that the clover grew lushly in the plastered portion (Chaptal, 1836). I have discussed the function of Ca^{2+} in cell signaling. Ever since the early chemist's work on the inorganic constituents of plants, we have known that other ions, including K^+, Fe^{2+}, Cu^{2+}, Mo^{2+}, Mn^{2+}, Mg^{2+}, Co^{2+}, and Zn^{2+}, are important macro- and micronutrients for the normal functioning of cells (Davy, 1821, 1827; Chaptal, 1836; Liebig, 1841; Hoagland, 1948; Epstein, 1972; Clarkson and Hanson, 1980; Ochiai, 1987; Fraústo da Silva and Williams, 1991; Williams and Fraústo da Silva, 1996; Glusker et al., 1999). Each ion has particular properties that make each one unique for fulfilling its function.

Many ions, however, can be toxic, particularly those of heavy metals. Plant and animal cells contain polypeptides that bind toxic heavy metals, including Pb^{2+}, Cd^{2+}, Ni^{2+}, and Hg^{2+}, as well as micronutrients such as Cu^{2+} and Zn^{2+}, which become toxic at high concentrations (Murasugi et al., 1981). These heavy metal–binding polypeptides called *phytochelatins* and *metallothioneins* bind heavy metals with high-association constants so that the free concentrations of these ions in the cytosol are reduced and the cells are detoxified (Rauser, 1990; Steffens, 1990; Evans et al., 1990; DeMiranda et al., 1990; Kawashima et al., 1991). Cysteine is a sulfur-containing amino acid that provides the necessary high-affinity binding sites for heavy metals. The heavy metal–binding polypeptides are synthesized in response to exposure by heavy metals (Ahner et al., 1994, 1995). Unlike metallothioneins, which are primary gene products, phytochelatins are polypeptides that are not primary gene products formed in ribosomes but are formed without a template by the enzyme phytochelatin synthase. In general, phytochelatin synthase is activated by heavy metals and is completely inactive in their absence (Grill et al., 1985, 1989).

12.11 SUMMARY

We have discussed the various adaptive and developmental responses that plants undergo in response to their environment. We have studied the signal transduction chains that are involved in coupling the stimulus to the response in individual cells. We have learned that the competence to respond to a given stimulus is dependent on the receptors in the cell, and the response that the cell undergoes depends on its response elements. We have seen that Ca^{2+}, inositol, and phosphorylation are essential elements in signal transduction chains, and moreover that the specificity of a given response results from the fact that there are independent and interacting elements in the signal transduction chains that truly form a signal transduction network. We have also spent a good deal of time discussing kinetics to truly understand the regulatory aspects of cell signaling.

12.12 QUESTIONS

12.1. Why do various cell types have very similar signal transduction chains to couple the stimulus to the response?

12.2. Why do various cell types have very different signal transduction chains to couple a stimulus to a response?

12.3. Why is there more than one kind of signal transduction chain within a cell?

The references for this chapter can be found in the references at the end of the book.

Chapter 13

Chloroplasts

The MOUSE'S PETITION, Found in the TRAP where he had been confin'd all Night.*

Parcere subjectis, & debellare superbos. —VIRGIL.

OH! hear a pensive prisoner's prayer,
For liberty that sighs;
And never let thine heart be shut
Against the wretch's cries.

For here forlorn and sad I sit,
Within the wiry grate;
And tremble at th' approaching morn,
Which brings impending fate.

If e'er thy breast with freedom glow'd,
And spurn'd a tyrant's chain,
Let not thy strong oppressive force
A free-born mouse detain.

Oh! do not stain with guiltless blood
Thy hospitable hearth;
Nor triumph that thy wiles betray'd
A prize so little worth.

The scatter'd gleanings of a feast
My frugal meals supply;
But if thine unrelenting heart
That slender boon deny,

The cheerful light, the vital air,
Are blessings widely given;
Let nature's commoners enjoy
The common gifts of heaven.

The well-taught philosophic mind
To all compassion gives;
Casts round the world an equal eye,
And feels for all that lives.

If mind, as ancient sages taught,
A never dying flame,
Still shifts through matter's varying forms,
In every form the same,

Beware, lest in the worm you crush
A brother's soul you find;
And tremble lest thy luckless hand
Dislodge a kindred mind.

Or, if this transient gleam of day
Be all of life we share,
Let pity plead within thy breast
That little all to spare.

So may thy hospitable board
With health and peace be crown'd;
And every charm of heartfelt ease
Beneath thy roof be found.

So, when destruction lurks unseen,
Which men like mice may share,
May some kind angel clear thy path,
And break the hidden snare.

**To Doctor PRIESTLEY.*

The Author is concerned to find, that what was intended as the petition of mercy against justice, has been construed as the plea of humanity against cruelty. She is certain that cruelty could never be apprehended from the Gentleman to whom this is addressed; and the poor animal would have suffered more as the victim of domestic economy, than of philosophical curiosity.

From Anna Lætitia Aikin. *Poems*. London: Printed for Joseph Johnson, in St. Paul's Churchyard, 1773, pp. 37–40.

13.1 DISCOVERY OF CHLOROPLASTS AND PHOTOSYNTHESIS

Up until now, I have been discussing cells that are surrounded by a plasma membrane, contain endoplasmic reticulum (ER), Golgi stacks and their associated membranes, coated and naked vesicles, vacuoles, autophagosomes, endosomes, and peroxisomes. Such cells could have been from either a plant or an animal. In this chapter, I will discuss an organelle that is found exclusively in plant cells: the chloroplast. The chloroplast is involved in photosynthesis and consequently cells that contain chloroplasts are autotrophic—that is, able to make their own food from inorganic molecules using the radiant energy of sunlight (Lockhart, 1959b; Blankenship, 2002). It must be remembered that most plants contain both photosynthetic and colorless cells. In most plants, the majority of cells,

Plant Cell Biology. https://doi.org/10.1016/B978-0-12-814371-1.00013-8

including the cells in the root, the pith of the stem, and the epidermis of the leaves, do not perform photosynthesis, and thus, they must be considered to be heterotrophic. Even so, each year, all the chloroplasts in the world fix about 10–100 billion tons of carbon dioxide, which is approximately equal to the mass of metropolitan New York City (Kamen, 1963) and significant in mitigating the effect of this greenhouse gas on global warming (Arrhenius, 1896; Callendar, 1938, 1949; Gore, 2006). This chapter also begins the discussion of the organelles whose P-spaces are separated from the P-space of the cytosol by a double membrane. Consequently, unlike the organelles we have discussed previously, these organelles are compartmentalized into E- and P-spaces. The double-membraned organelles include the plastid, the mitochondrion, and the nucleus.

13.1.1 Discovery of Photosynthesis

Experimental studies of plant assimilation began when the alchemist Jean Baptiste van Helmont (1683) used his balance, one of the only tools available to scientists at the time, to study plant growth. He believed that all matter built was up from the single essence of water, an idea that can be traced back to Thales, who lived in the sixth century BC. van Helmont used the growth of plants to test this thesis. According to van Helmont,

> *I have learned from the following clear experiments that all plants make up their matter completely from the element of water. I conclude this because, I have taken an earthenware container and placed in it 200 pounds of earth, that I have placed in a baking oven and allowed to dry. I moistened this earth with rain-water and placed in it a willow stem which weighed five pounds. To ensure that the dust from the air didn't add to the weight, I threw a screen over the soil. I watered it when necessary with only rain-water or distilled water. The tree grew and set into the ground. After five years, the willow became a tree that weighed 169 pounds and about 3 ounces and I did not even take into consideration the weight of the leaves, which fell off every autumn for four years. Eventually I took the earth out of the container and found that it weighed only two ounces less than the original two hundred pounds. Thus the 164 pounds of wood, bark and knots had grown alone from the water.*

In performing experiments in other realms of alchemy, van Helmont found that when he burned 62 pounds of coal he was left with only 1 pound of ash. He called the material that made up the escaping 61 pounds the spiritis sylvestres, wild spirit or "gas," and although van Helmont was the discoverer of gases, he did not realize that the tree that grew in the pot was created from gas and from water. The idea that the leaves of plants assimilated air was originally proposed by Empedocles (Lambridis, 1976) but dismissed by Aristotle (Barnes, 1984) and his disciple Theophrastus (1916), who believed that all the nourishment came in through the roots. The proposal that plants assimilated air was not taken seriously until 1727. At this time, Stephen Hales (1727) burned plants and quantified the amount of gas given off and the amount of ash that remained. From the results of these experiments, Hales suggested that plants might assimilate air, just as animals do (Boyle, 1662; see Chapter 14). Experiments on the role of gases in plant growth were renewed at the end of the 18th century when Joseph Priestley decided to take up where Stephan Hales left off (Birch and Lee, 2007).

Like a burning candle, living beings require "clean air." A test for the presence of clean (or dephlogisticated) air is to see if a mouse can live, or a candle can burn, when placed in a container of the sample air (Faraday, 1860). Using such a test, Joseph Priestley (1774; Partington, 1933) accidently found that a sprig of mint could purify the air that had been previously fouled by the breathing of an animal or the burning of a candle. Perhaps Priestley was lucky to find the oxygenic (or dephlogisticated) property of plants and just chose mint to purify the air because of its refreshing smell. However, he also found that groundsel, a bad-smelling weed, and spinach also had the ability to purify the air, and this ability was a general property of plants.

Priestley later found his experiments to be irreproducible, perhaps because he did not control the light conditions in his laboratory. However, it was difficult for the politically incorrect Priestley to continue his experiments as his home and laboratory were burned down as a result of his antiauthoritarian views, including his support of the French Revolution and the Unitarian church (Priestley, 1809). John Ingen-Housz (1796) continued experiments on the purification of air by plants, and he solved the problem of irreproducibility by showing that oxygen evolution required light. Moreover, Ingen-Housz showed that only the green parts of the plant evolved oxygen.

Joseph Priestley had discovered how to make soda water by impregnating water with fixed air derived from acidified sodium carbonate. By varying the amount of carbonic acid in the soda water surrounding immersed leaves, Jean Senebier (1788) was then able to show that oxygen evolution was dependent on CO_2. Théodore de Saussure (1804) quantified the relationship between CO_2 fixation and O_2 evolution and established that the amount of O_2 given off was equivalent to the amount of O_2 taken up in the form of CO_2. However, the mass gained by the plant was greater than the mass contributed by the C and O in CO_2. Thus, he concluded that water must also be taken up and the C combined with H_2O.

While Johann Wolfgang von Goethe (1952), Alphonse de Candolle, and Henri Dutrochet accepted the results of

Ingen-Housz, Senebier, and de Saussure, it seemed ludicrous to many plant physiologists, including Franz Meyen (1837–39; Sachs, 1906; Werner and Holmes, 2002), to think that plants assimilated carbon from the tiny amounts found in air when they were surrounded by vast quantities in the form of humus or humic acid. They thought that it was more likely that plants absorbed their carbon from humus or humic acid, which surrounded the plants as was proposed by Aristotle (Sachs, 1906).

Justus von Liebig (1840) railed against his fellow German botanists, insisting that

> *... in botany the talent and labour of inquirers has been wholly spent in the examination of form and structure: chemistry and physics have not been allowed to sit in council upon the explanation of the most simple processes; their experience and their laws have not been employed, though the most powerful means of help in the acquirement of true knowledge.*

Liebig went on to say that "the art of experimenting is not known in physiology." The whole matter as to the origin of carbon in plant nutrition was finally settled in favor of the atmospheric source when Jean Boussingault succeeded in growing plants in a totally inorganic soil and proved once and for all that plants do not obtain their carbon from humus or humic acid, but from the carbon dioxide in the air (Dumas and Boussingault, 1844; Sachs, 1906).

In *The Vegetable Cell*, Hugo von Mohl (1852) offered two possible reaction mechanisms for the first step in carbon assimilation. First, one proposed by Humphry Davy (1827) that the C combined with H_2O to make $C(H_2O)$ or carbohydrate as the first product. Second, one proposed by Justus von Liebig that an organic acid was the first product as "it was far more probable that it was not the difficultly decomposable carbonic acid [bond energy $= 13.2 \times 10^{-19}$ J], but the readily decomposable water [bond energy $= 7.65 \times 10^{-19}$ J] which was separated into its elements." Liebig's idea, although correct, was not so simple and was forgotten for almost 100 years and independently rediscovered by van Niel (1941) using the perspective of comparative physiology.

The biotic carbon cycle was understood by chemists in the mid-19th century. J.B. Dumas (quoted in Fruton, 1972) wrote,

> *... green plants constitute the great laboratory of organic chemistry. It is they which, with carbon, hydrogen, nitrogen, and ammonium oxide, slowly build the most complex organic materials. They received from the solar rays, in the form of heat or chemical radiation, the power needed for this work.*
>
> *Animals assimilate or absorb the organic materials made by plants. They change them bit by bit. ... They therefore decompose bit by bit these organic materials created by plants; they bring them back bit by bit toward the state of carbonic acid, of water, of nitrogen, of ammonia, the state that permits them to be restored in the air.*

Thomas Huxley (1893) described the cycle by means of a sociological metaphor: "Thus the plant is the ideal *prolétaire* of the living world, the worker who produces; the animal, the ideal aristocrat, who mostly occupies himself in consuming."

While the identity of the first product of carbon assimilation would not be known for almost 100 years and would have to wait for the introduction of ^{14}C and paper chromatography, Julius von Sachs (1887) stained leaves with iodine and established that the starch grains contained in the chlorophyll granules were the end result of carbon assimilation. The starch grains only appeared when the plant was exposed to light and disappeared after the plant was put into darkness. Hans Molisch (1916) capitalized on this fact to use leaves as photographic paper. He put a negative image over a dark-adapted, starch-depleted leaf and exposed the leaf to bright sunlight. The light passed through the transparent parts of the negative and initiated starch formation. Molisch then stained the leaf with iodide and the image developed right on the leaf, just as it would have on photographic paper (Fig. 13.1)!

At the end of the 19th century, plant physiologists were realizing that the term *assimilation*, a term that had long been used by animal physiologists to designate the appropriation of digested food and its subsequent conversion into other substances, did not describe the unique aspects of the light-mediated synthesis of carbohydrates that occurred in plants. Plant physiologists began to look for a new term. In a published version of an address given before the American Association for the Advancement of Science, Charles Barnes (1893) wrote,

> *For the process of formation of complex carbon compounds out of simple ones under the influence of light, I propose that the term photosyntax be used. ... I have carefully considered the etymology and adaptation, as well as the expressiveness, of the word proposed, and consider it preferable to photosynthesis which naturally occurs as a substitute.*

The term *photosynthesis*, which was coined by Conway MacMillan, in a discussion following the presentation of Barnes's paper, gained universal acceptance (Oels, 1894; Barnes, 1898; Gest, 2002).

Isaac Newton (Qu.30, 1730) had suggested that light and matter may be convertible, and Stephen Hales (1727) added that this convertibility might play a role in plant nutrition. Eventually, chemists began searching for the pigment that aided in the conversion of sunlight into carbohydrate. The green pigment, which was soluble in an alcohol extract of leaves, was dubbed *chlorophyle*, from the

FIGURE 13.1 A leaf of *Tropaeolum majus* on which a photographic negative of a man's face had been placed. Then the leaf was placed in the light and allowed to photosynthesize. Starch was consequently formed under the clear areas of the negative. Subsequently, the leaf was stained with iodide and photographed. See Hangarter and Gest (2004) for more fascinating pictorial examples of photosynthesis. Instructions for this fascinating process can be found in a book by David Walker (1992) that covers photosynthesis from the quantum processes to its impact on ethical decisions. *From Molisch, H., 1916. Pflanzenphysiologie als Theorie der Gärtnerei. Gustav Fischer, Jena.*

Greek for "green leaf," by Joseph Pelletier and Joseph Caventou in 1818. Work by Jöns Jacob Berzelius, David Brewster, George Gabriel Stokes, Richard Willstätter (1920), Hans Fischer (1930), Harold Strain, James Conant, and others led to the elucidation of the structure of chlorophyll (Willstätter and Stoll, 1928; Strain, 1958; Willstätter, 1965). Paul Castelfranco and Sam Beale (1983), among others, elucidated how chlorophyll is synthesized in plants. In 1960, it was synthesized for the first time by humans at Harvard University (Woodward et al., 1960; Woodward, 1965). Interestingly, unlike the rhodopsin in our eyes, chlorophyll is terribly inefficient, as it does not absorb green light, which is the most prominent color of the sunlight that passes through the clouds to Earth (Barbour et al., 1980).

For a pigment to do photochemistry, it must absorb photons. Einstein's photochemical equivalence law states that each molecule in a linear photochemical process absorbs one photon of radiation, and the law predicts that the pigment involved in a photochemical process must have an absorption spectrum that closely follows the action spectrum for the process (Gerlach, 1921)—that is, in a linear photoreaction, the response at each wavelength will be proportional to the number of photons absorbed at that wavelength. Indeed, the action spectrum of carbon assimilation, originally obtained by Theodor Engelmann (1881, 1882) by observing the movement of oxytactic bacteria to various portions of photosynthetic cells that were irradiated with different-colored light, almost exactly mimics the absorption spectrum of chlorophyll—almost.

Robert Emerson and Charlton Lewis (1943) noticed that while red light between 700 and 730 nm was absorbed by chlorophyll, it was not very effective in stimulating photosynthesis. This was known as the *red-drop effect*. However, when the ineffective wavelengths were supplemented with short-wavelength light, the photosynthetic rate was greater than the rate induced by the sum of the two wavelengths alone (Emerson et al., 1957). In fact, the two wavelengths could still enhance photosynthesis when they were given sequentially (Myers and French, 1960). This enhancement, which is known as the *Emerson effect*, gave the first indication that there may be two sequential photosystems involved in photosynthesis: one that absorbs 700-nm light and one that does not (Govindjee and Rabinowitch, 1960; Rabinowitch and Govindjee, 1969). Hill and Bendall (1960) proposed that these results could be explained if there were two photosystems, and Duysens et al. (1961) named them *Photosystem I* (PS I) and *Photosystem II* (PS II).

The reactions involved in the conversion of light energy into carbohydrate have become crystal clear over the past 270 years, in part due to the introduction of favorable material by Otto Warburg and Cornelius van Niel, including single-celled algae and bacteria for the study of the rapid reactions of photosynthesis (Kluyver and van Niel, 1956; Myers, 1974; Wayne and Staves, 1996b; Feher, 2002). The success obtained in photosynthetic research is perhaps also a result of the excellence of the individual scientists involved, who as a rule, have acted on the admonishments of Liebig, and combined an interest in biology with chemistry and physics (Rabinowitch, 1945, 1955; Franck and Loomis, 1949; Hill and Whittingham, 1955; Bassham and Calvin, 1957; Rabinowitch and Govindjee, 1969; Clayton, 1980, 1988, 2002; Duysens, 1989, 1996; Calvin, 1992; Walker, 1997; Feher, 1998, 2002; Jagendorf, 1998, 2002; Fuller, 1999; Krogmann, 2000; Govindjee and Krogmann, 2004; Govindjee et al., 2004; Rosenberg, 2004; Chrispeels and Wraight, 2014; Wraight, 2014; Govindjee, 2017).

The individual reactions of photosynthesis span times from femtoseconds to hours, which provide work for scientists that range from physicists to ecologists. To help us comprehend the vast range in times, as well as making it easier to speak and write the various times, Martin Kamen (1963) introduced the symbol pt_s, which stands for "the minus log of the time in seconds." The symbol p was first employed by Sørensen (1909) when he used pH to stand for the minus log of the $[H^+]$.

The reactions of photosynthesis are divided into two major groups: the reactions that require light directly (i.e., the light reactions) and the reactions that do not require light directly (i.e., the dark reactions). The light reactions are involved in the conversion of light energy to chemical energy in the form of ATP and NADPH and take place on the time scale of pt_s15-pt_s3. The dark reactions, which take place on a time scale of pt_s4-pt_s0 or longer, are involved in the conversion of chemical energy in the form of ATP and NADPH to a more stable form of energy in the form of carbohydrate ($C[H_2O]$).

13.1.2 Discovery and Structure of Chloroplasts

Nehemiah Grew (1682) developed an interest in the theory of colors along with his contemporaries, Boyle, Descartes, Hooke, and Newton. Combining this interest with his interest in the anatomy of plants, he investigated the colors of plants and noticed green precipitates in leaves. With the introduction of achromatic lenses in the first part of the 19th century, it became apparent to plant cell biologists that the green precipitates, which came to be known as the *chlorophyll granules*, had fascinating shapes, including stars, plates, and spirals (Fig. 13.2), which were diagnostic of a given taxon, especially in the green algal order Zygnematales (Mohl, 1852). In the majority of plants, however, the chlorophyll granules are just ellipsoidal and about 2.5-μm long (Fig. 13.3). According to Haberlandt, "The immediate duty of every active chloroplast is the absorption of the carbon-dioxide that diffuses into the cell. Other things being equal, the rapidity and efficiency with which this task can be performed depends upon the surface available for purposes of absorption, and obviously the maximum exposure surface will be obtained when the functions of the chloroplasts are distributed among a number of small corpuscles." Indeed, a century later, Xiong et al. (2017) confirmed that a few enlarged chloroplasts are less efficient in photosynthesis than a large population of small chloroplasts.

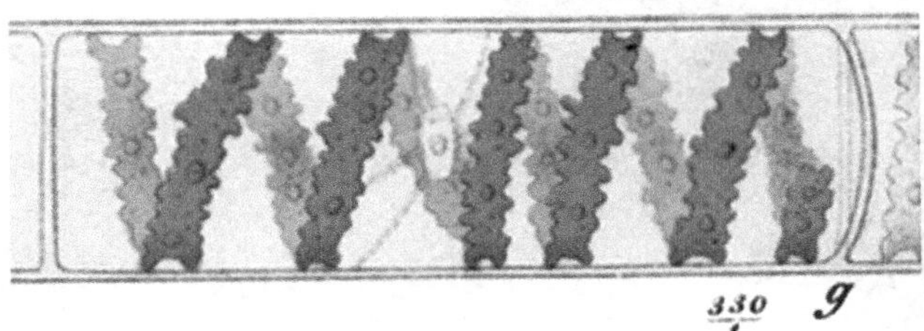

FIGURE 13.2 Chloroplast of *Spirogyra quinina*. ×330. *From Pringsheim, N., 1879–1881. Ueber Lichtwirkung und Chlorophyllfunction in der Pflanzen. Jahrbuch. Wiss. Bot. 12, 288–437.*

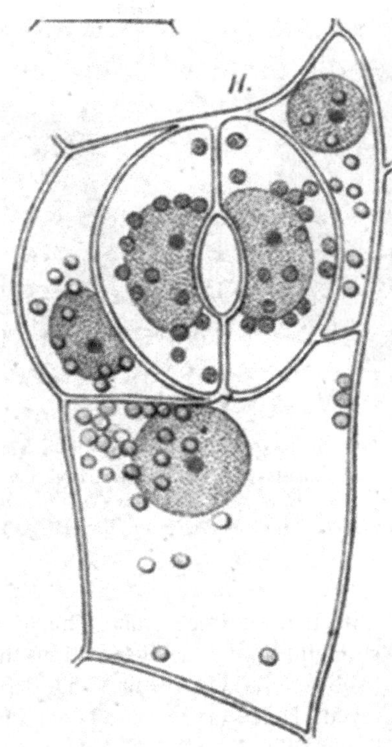

FIGURE 13.3 Chloroplasts in the guard cells of *Tradescantia subaspera. From Schimper, A.F.W., 1883. Ueber die Entwickelung der Chlorophyllkörner und Farbkörper. Bot. Z. 41,105,121, 137,153.*

Other granules, approximately the same size, were also observed using microscopes with achromatic objectives. These included the orange-yellow carotenoid-containing granules and clear granules, which due to the differences in color and shape were given various names. Schimper (1883, 1885), fascinated by the fact that these organelles could metamorphose into the other and were thus different facets of the same organelle, named the whole group *plastids* to emphasize their plasticity. He called the green plastids *chloroplasts*, the orange-yellow plastids *chromoplasts*, and the clear plastids *leucoplasts*.

Arthur Meyer (1883) noticed that the chlorophyll granules contained spherical grains that he called *grana* (Fig. 13.4). However, as the application of colloid chemistry to biological problems became popular, the grana were interpreted as an artifact of fixation, and the internal organization of the chloroplast in the living cell was considered to be a homogeneous emulsion of a very fine chlorophyll-containing lipoidal phase distributed throughout a hydrophilic phase (Guilliermond, 1941). Frederick Czapek (1911) believed that the fine emulsion separated during

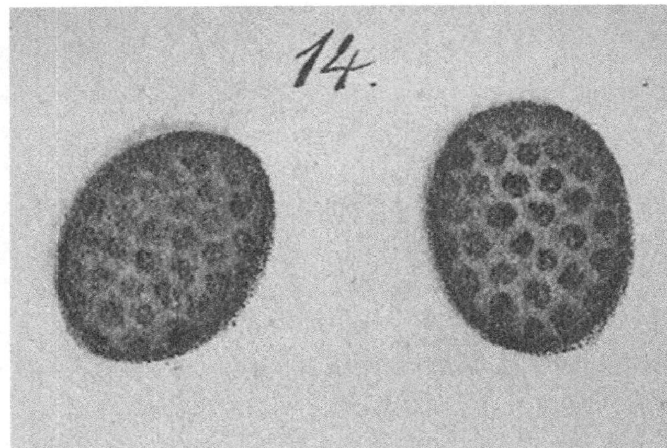

FIGURE 13.4 Chloroplasts with visible grana from the mesophyll of *Acanthephippium. From Meyer, A., 1883. Das Chlorophyllkorn in chemischer, morphologischer und biologischer Beziehung. Arthur Felix, Leipzig.*

fixation and artifactually formed grana. This argument was considerably strengthened by the observations that grana in fixed material showed varied structures that depend on the fixative employed. Heitz (1936, 1937) clearly resolved grana in the living cells of 180 different species of plants using a high numerical aperture apochromatic objective and blue light illumination. He described the 40–60 grana that were in each chloroplast as discs that varied in size from 0.3 to 2 μm, depending on the species. Heitz also observed the grana with polarized light microscopy and concluded that the chloroplast was not homogeneous. Even so, the reality of the grana continued to be debated until the 1960s (Küster, 1935; Strugger, 1935; Weier, 1938a,b, 1961; Mühlethaler, 1955).

The newly invented technique of electron microscopy was applied to understand the nature of the grana in disrupted chloroplasts. The ultramicrotome had not yet been invented, so the only way to determine whether or not the grana existed was to shadow cast the material (Algera et al., 1947; Granick and Porter, 1947; Frey-Wyssling and Mülethaler, 1949). However, these images of grana were not of the quality we are used to today and were not very convincing to the nonbelievers (Fig. 13.5). Following the invention of the ultramicrotome, however, Wolken and Palade (1952, 1953) obtained mediocre images while Steinmann and Sjöstrand (1955) were able to obtain excellent images of the grana that revealed that they were composed of 7- to 9-nm thick membranes (Fig. 13.6), which were later given the name *thylakoids* by Menke (1962). Thylakoids comes from a Greek word meaning "sacklike." The membranous structure of the plastid membranes of many species of plants was confirmed (Hodge et al., 1955; Heslop-Harrison, 1962; Paolillo, 1962; Weier and Thomson, 1962; Paolillo and Falk, 1966; Paolillo and Reighard, 1967; Paolillo et al., 1969; Shimoni et al., 2005; Brumfeld et al., 2008; Garab and Mannella, 2008; Mustárdy et al., 2008).

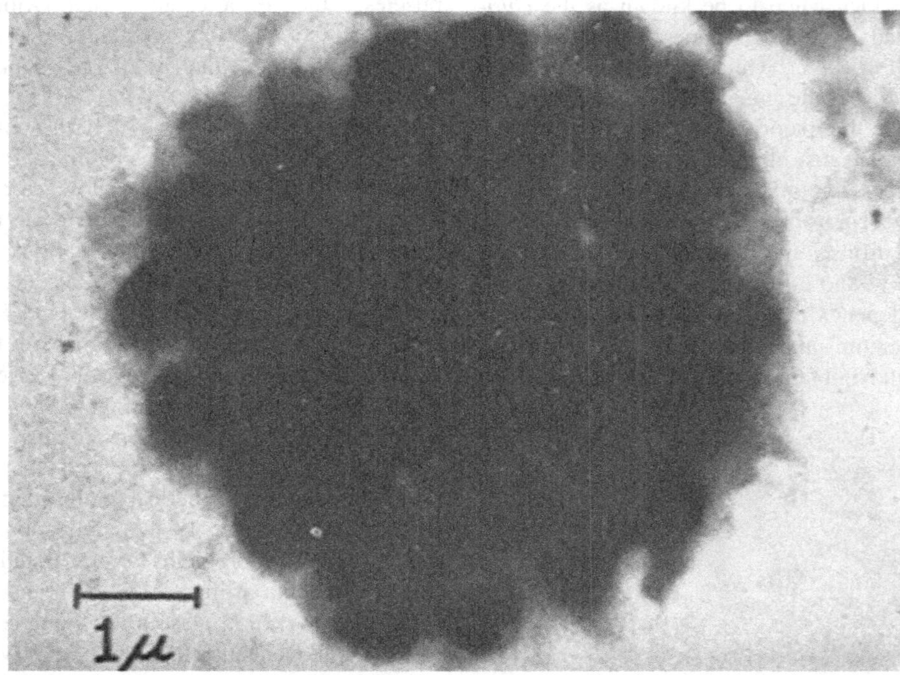

FIGURE 13.5 Electron micrograph of an isolated spinach chloroplast showing dense grana embedded in a paler matrix. The chloroplast preparation was dried directly on a wire mesh and observed with the electron microscope. *From Granick, S., Porter, K.R., 1947. The structure of the spinach chloroplast as interpreted with the electron microscope. Am. J. Bot. 34, 545–550.*

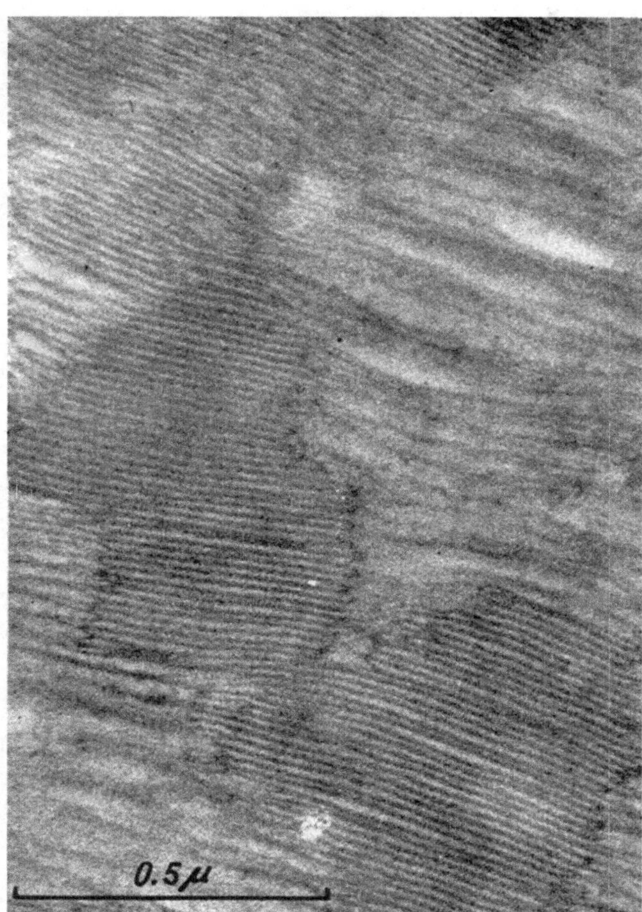

FIGURE 13.6 Electron micrograph of a thin section of the chloroplast of *Aspidistra elatior* showing grana. *From Steinmann, E., Sjöstrand, F.S., 1955. The ultrastructure of chloroplasts. Exp. Cell Res. 8, 15–23.*

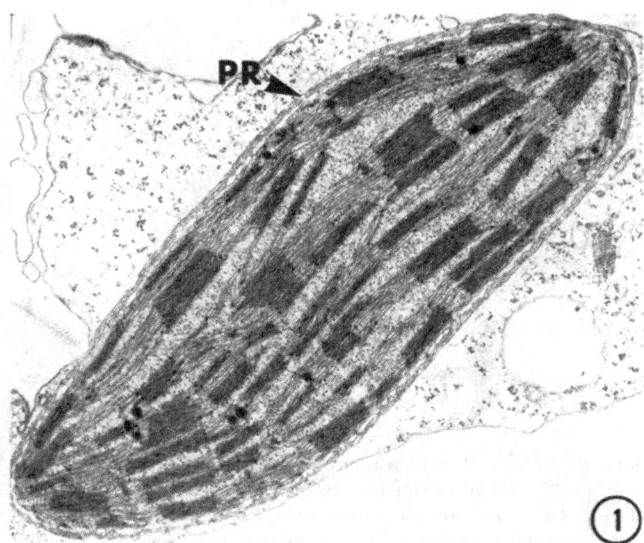

FIGURE 13.7 An electron micrograph of a chloroplast of *Zea mays* showing grana and a peripheral reticulum (PR) of membranes. *From Chollet, R., Paolillo Jr., D.J., 1972. Greening in a virescent mutant of maize. Z. Pfanzenphysiol. 68, 30–34.*

The chloroplasts are delineated by a double membrane called the *envelope*. The two 6- to 8-nm thick membranes are separated by a distance of approximately 10–20 nm. The membranes inside the chloroplast envelope are known as the thylakoids. A granum is formed when approximately 10–20 thylakoids, separated from each other by 3–4 nm, form a cylindrical stack, 300–600 nm in diameter and 200–600 nm in height (Shimoni et al., 2005). There are typically many grana per chloroplast (Fig. 13.7). The space between the thylakoid membrane and the inner membrane of the chloroplast is called the *stroma*. Topologically, the space within the two membranes of the envelope, as well as the lumen of the thylakoids, is the E-space; the stroma is a P-space.

Dominick Paolillo (1970) proposed a model for grana structure based on serial sections of known orientation (Fig. 13.8). He described the grana as cylinders that are surrounded by stromal lamellae that are arranged in right-handed helical frets (Fig. 13.9). The stromal lamellae interconnect the lumen of the thylakoids below and above each other within a grana stack and they also interconnect the grana within the chloroplast. The stromal lamellae rise progressively at an angle to the grana stack like a spiral staircase where each lamella is turned about 30 degrees from the one below. The interconnection between all the grana in a chloroplast means that the interior of the chloroplast is divided into only two topological compartments, not hundreds of compartments as we might think if we look

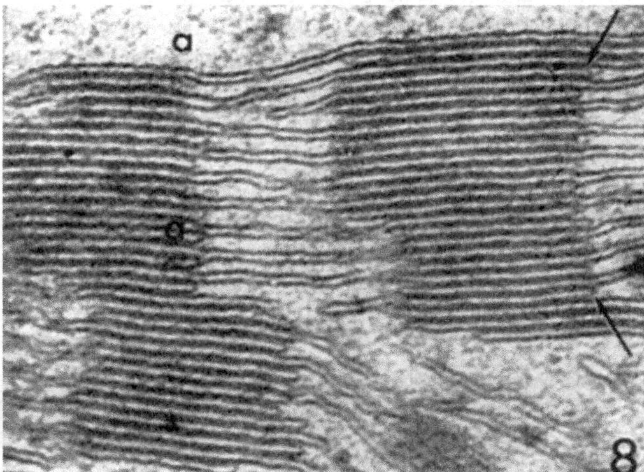

FIGURE 13.8 Thin-section electron micrograph of the grana and frets that connect the compartments in a chloroplast of *Zea mays*. Between a-a and the two *arrows*, there are 2:1 relationships between the number of thylakoids in a granum and the number of membranous frets running between grana. ×83,000. *From Paolillo Jr., D.J., Falk, R.H., 1966. The ultrastructure of grana in mesophyll plastids of* Zea mays. *Am. J. Bot. 53, 173–180.*

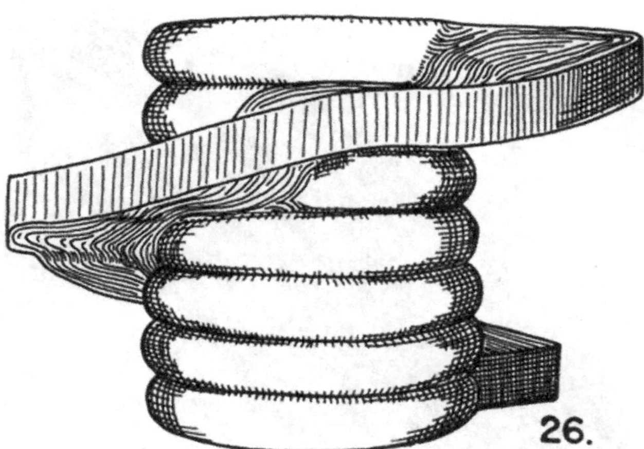

FIGURE 13.9 A three-dimensional model of a single granum showing only one fret. The two additional frets that would normally be attached to the granum are not shown for clarity. *From Paolillo Jr., D.J., Falk, R.H., 1966. The ultrastructure of grana in mesophyll plastids of* Zea mays. *Am. J. Bot. 53, 173–180.*

at only a single thin section. It is absolutely fascinating to consider how this structure is created. Do the thylakoid membranes fuse, bleb, twist, and wind? However they do it, the arrangement of the thylakoids is regulated both environmentally and genetically (Klekowski et al., 1994; Chuartzman et al., 2008).

In a continuation of the battle between chemists and cytologists, this model of chloroplast structure was challenged by Sam Wildman and his associates based on the assumption that cytologists were looking at small samples of dead cells (Spencer and Wildman, 1962; Wildman et al., 1962, 1974, 1980; Honda et al., 1971; Jope et al., 1980). Wildman and his colleagues described a new model of chloroplast structure that contradicted all the previous electron microscopic evidence. They proposed that in living cells, the grana were not arranged in overlapping rows interconnected in all directions but were arranged in a single plane like a spiral "string of beads." Paolillo and Rubin (1980) observed chloroplasts with the light microscope and concluded that the discrepancy between the light and electron microscopic images was only apparent and that the data used by Wildman et al. were based on a misunderstanding of the workings of a light microscope. van Spronsen et al. (1989) and Austin and Staehelin (2011) have confirmed Paolillo's model using confocal microscopy (Fig. 13.10) and electron tomography, respectively. More recently, Mehta et al. (1999) and Wildman et al. (2004) showed that in some chloroplasts in some cells of some taxa, the thylakoids are indeed arranged like strings of beads, although this is not a common arrangement.

Providing direct evidence that chloroplasts are the site of photosynthesis, Robert Hill (1937) showed that isolated chloroplasts were able to evolve oxygen when an artificial electron acceptor was added to the isolated chloroplast suspension. The reduction of the electron acceptor is easy to follow spectrophotometrically when the electron acceptor is a dye. This light-stimulated reduction is known as the *Hill reaction*. Given that the oxidation–reduction reactions that took place in the mitochondria depended on cytochromes, it seemed likely that cytochromes would also be involved in these reactions in the chloroplast, and soon such cytochromes were discovered in the chloroplast (Hill and Scarisbrick, 1951; Hill, 1954). Later, Sakae Katoh (1960, 1995) discovered a copper-containing protein in *Chlorella* and postulated that it too may be involved in photosynthetic electron transport. Katoh and Takamiya (1961) discovered that this protein could be reduced in the presence of light and grana and named it plastocyanin because it was a blue-colored protein that was localized in the chloroplast.

After it was discovered that NADP was the source of reducing power for malic enzyme in pigeon liver, the first enzyme discovered that could fix CO_2 (Ochoa et al., 1948), it seemed possible that NADP might be the natural electron acceptor for the light reactions of photosynthesis. Later, Vishniac and Ochoa (1951) found that it was. André Jagendorf (1956) and San Pietro and Lang (1956) showed that the reduction took place in the thylakoids. The NADP reduced by the light reactions serves as a source of reducing power for the so-called dark reactions. Daniel Arnon et al. (1954) discovered that isolated chloroplasts were capable of producing ATP, which is also necessary for the dark reactions. The various membrane-localized protein complexes involved in the light-dependent production of NADP and ATP can be visualized with freeze-fracture electron microscopy (Staehelin, 2003).

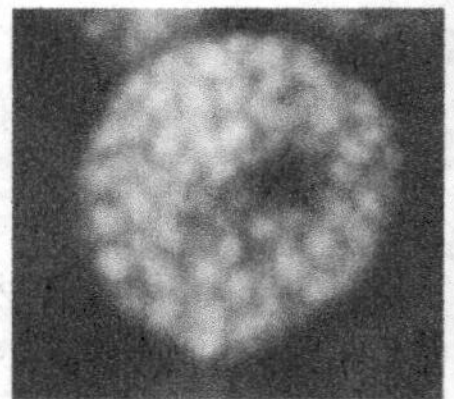
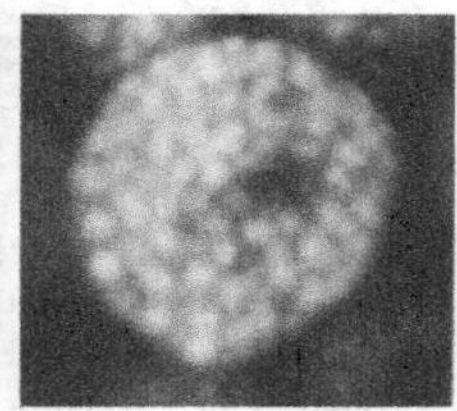
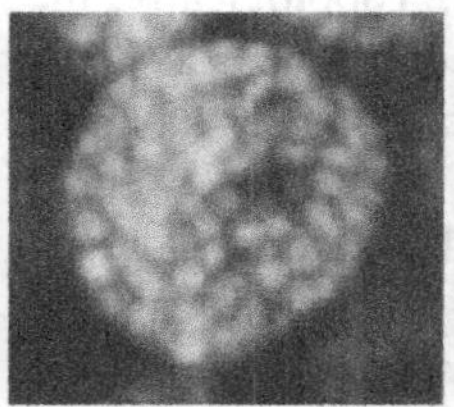
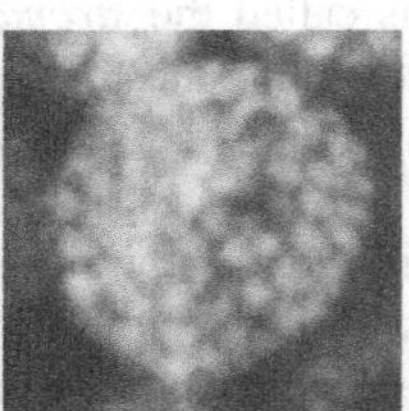

FIGURE 13.10 Successive optical section through a fluorescent chloroplast in a living cell of *Synnema triflorum*. *From van Spronsen, E.A., Sarafis, V., Brakenhoff, G.J., van der Voort, H.T.M., Nanninga, N., 1989. Three-dimensional structure of living chloroplasts as visualized by confocal scanning laser microscopy. Protoplasma 148, 8–14.*

13.2 ISOLATION OF CHLOROPLASTS

Haberlandt, Ewart, Mölisch, Hill, and Arnon all attempted to isolate chloroplasts that could perform all the aspects of photosynthesis in vitro with varying amounts of success (Hill and Scarisbrick, 1940; Hill and Whittingham, 1955). In fact, the amount of oxygen given off by the chloroplasts isolated from a moss or *Selaginella* was so small that Haberlandt and Ewart had to use luminescent bacteria in order to detect it. By adding exogenous electron acceptors, Hill was able to get chloroplast preparations that were able to evolve oxygen at a high rate, and eventually, Arnon et al. (1954) were able to get chloroplasts that were capable of making ATP in the presence of light. Through the work of many people, the conditions necessary for getting relatively high yields of intact and active chloroplasts were finally found.

To isolate chloroplasts, the tissue is homogenized, filtered, and centrifuged at 6000–8000 g for less than 90 s to separate the massive chloroplasts from the rest of the cellular components. As this chloroplast fraction may be contaminated by mitochondria and peroxisomes, it can then be further purified by centrifuging the chloroplasts in a Percoll density gradient at 500 g for 10 min. The intact chloroplasts are found as a green band near the bottom of the tube. The most common marker for the chloroplast is chlorophyll and the most common enzyme markers are $NADP^+$-dependent glyceraldehyde phosphate dehydrogenase and ribulose bisphosphate carboxylase.

To obtain isolated envelopes, the chloroplasts are gently lysed and layered on a discontinuous sucrose density gradient and centrifuged at 72,000 g for 60 min. The deep-yellow carotenoid-containing envelopes collect at the 0.6/0.93 M interface. The envelopes are removed from the interface, resuspended in a 0.3 M sucrose-containing buffer, and centrifuged at 113,000 g for 45 min to yield a pellet of envelopes. The marker enzyme for the envelope is a Mg^{2+}-dependent ATPase that is insensitive to N,N′-dicyclohexylcarbodiimide (DCCD).

The outer and inner membranes of the chloroplast envelope can be separated from each other by putting the chloroplast in a hypertonic medium, which causes the stroma volume to decrease so that the inner and outer envelope membranes separate as a function of the relative impermeability of the inner membrane compared with the outer membrane. The membranes are then ruptured by freeze-thawing, and then the two types of membranes are separated by density-gradient centrifugation because their densities are so different (Douce and Joyard, 1990).

To obtain thylakoids, the chloroplasts are lysed and then centrifuged at 6000 g for 5 min. The thylakoids, which are in the pellet, can be identified by the presence of ferredoxin: $NADP^+$ oxidoreductase.

13.3 COMPOSITION OF THE CHLOROPLASTS

Lipids make up 58% of the dry weight of the chloroplast envelope. The only other membranes that are as rich in lipids are the plasma membranes of Schwann cells and oligodendrocytes that form the myelin sheaths that surround the axonal processes of neurons. Plant lipids are used for cooking and for nutrition as well as for the production of varnish, soap and lubricants (Horn and Benning, 2016). Table 13.1 shows the lipid composition of the envelope. Treatment of the intact chloroplast with phospholipase C removes almost all the phosphatidylcholine (PC) in the chloroplast, indicating that this lipid, at least, is asymmetrically localized in the P-leaflet of the outer membrane of the envelope.

TABLE 13.1 Lipid Composition of the Envelope of a Chloroplast

Polar Lipid	Percent Dry Weight of Total Lipids
MGDG	20
DGDG	30
TGDG	4
TTGDG	1
SL	6
PC	20
PG	8
PI	1
PE	tr
Fatty Acid Precursor	Percent by Dry Weight
16:0	15
16:1	3
16:3	9
18:0	tr
18:1	6
18:2	10
18:3	57
Unsat/sat	5.7

DGDG, digalactosyldiglyceride; *MGDG*, monogalactosyldiglyceride; *PC*, phosphatidylcholine; *PE*, phosphatidylethanolamine; *PG*, phosphatidylglycerol; *PI*, phosphatidylinositol; *SL*, sulfoquinovosyldiacylglycerol; *TGDG*, trigalactosyldiglyceride; *tr*, trace; *TTGDG*, tetragalactosyldiglyceride.
From Douce, R., Holtz, R.B., Benson, A.A., 1973. Isolation and properties of the envelope of spinach chloroplast. J. Biol. Chem. 248, 7215–7222; Douce, R., Joyard, J., 1979. Structure and function of the plastid envelope. Adv. Bot. Res. 7, 1–116.

The outer envelope membrane is very permeable and contains a number of proteins (Ferro et al., 2003; Froehlich et al., 2003; Peltier et al., 2004; Bräutigam et al., 2008), including a pore protein that has an equivalent diameter of 3 nm that allows the passage of molecules up to 9–10 kDa (Flügge and Benz, 1984). It also contains a permease that facilitates the transport of lipids that are synthesized by the ER into the chloroplast (Xu et al., 2003; Wang and Benning, 2012; Hurlock et al., 2014). The inner membrane, by contrast, is a typical differentially permeable membrane and contains a number of proteins involved with metabolite transport. These transporters facilitate the flow of the substrates and products of photosynthesis, as well as the components needed for the synthesis of galactolipids, fatty acids, prenylquinones, and carotenoids (Heber and Heldt, 1981; Douce and Joyard, 1979; Flügge and Weber, 1994; Facchinelli and Weber, 2011; Kelly et al., 2016). The inner membrane also contains aquaporins that function to increase the permeability of the membrane to CO_2 (Uehlein et al., 2003, 2008; Evans et al., 2009).

The lipid composition of the thylakoid membranes has been determined by Douce et al. (1973; Table 13.2). The thylakoids are poor in phospholipids but rich in galactolipids, particularly digalactosyldiglyceride (DGDG) and monogalactosyldiglyceride (MGDG). In general, the glycerolipids make up the majority of the plant lipids (Horn and Benning, 2016). The glycerolipids with an 18 carbon fatty acyl group at the *sn-2* position are synthesized in the ER, whereas the glycerolipids with a 16 carbon fatty acyl group at the *sn-2* position are synthesized in the plastid itself (Hurlock et al., 2014). Different taxa utilize the two pathways differently (Heinz and Roughan, 1983). The chloroplast lipids that are synthesized in the chloroplast are synthesized on the membranes of the envelope, and it is not clear how they get to the thylakoids. The thylakoids also contain sulfur-containing lipids, including sulfoquinovosyldiglyceride (SL). MGDG, DGDG, and SL are endemic to the plastids and do not occur in any other organelle type. They do, however, also occur in the membranes of cyanobacteria, providing evidence that these organisms may share a common ancestor with chloroplasts (Douce and Joyard, 1990; see Chapter 15). The distribution of lipids in the thylakoid membranes is asymmetrical. Antibody studies show that phosphatidylglycerol is abundant on the stromal thylakoid leaflet, and MGDG and SL are abundant on the lumenal thylakoid leaflet (Murphy, 1986a,b). The thylakoid membranes contain many proteins (Peltier et al., 2002; Majeran et al., 2008), including the protein complexes that are involved with the light reactions of photosynthesis. All the enzymes necessary for the dark reactions of photosynthesis are found in the stroma.

TABLE 13.2 Lipid Composition of the Thylakoids

Polar Lipid	Percent Dry Weight of Total Lipids
MGDG	51
DGDG	26
SL	7
PG	9
PI	1
Fatty Acid Precursor	Percent by Dry Weight
16:0	8
16:1	5
16:3	13
18:0	tr
18:1	2
18:2	2
18:3	70
Unsaturated/saturated	12.2

DGDG, digalactosyldiglyceride; *MGDG*, monogalactosyldiglyceride; *PG*, phosphatidylglycerol; *PI*, phosphatidylinositol; *SL*, sulfoquinovosyldiacylglycerol.
From Douce, R., Holtz, R.B., Benson, A.A., 1973. Isolation and properties of the envelope of spinach chloroplast. J. Biol. Chem. 248, 7215–7222.

13.4 THERMODYNAMICS AND BIOENERGETICS IN PHOTOSYNTHESIS

Beacause the chloroplasts are organelles that specialize in the conversion of radiant energy to chemical energy, it is an appropriate time to review the laws of thermodynamics (Fenn, 1982; Dill and Bromberg, 2003). Heraclitus (540 BCE), who was interested in finding universal laws, proposed that fire was the basic unit of exchange for all things—just like gold was the basic unit of exchange in the marketplace (Hussey, 1995). This concept was further refined into the laws of thermodynamics in the mid-19th century, a revolutionary time when scientists were looking for unifying principles. Schleiden and Schwann came up with the idea that no matter how diverse living organisms are, they are made up of the same basic unit: cells. Darwin was also conceptualizing his theory of evolution by natural selection, indicating that all living organisms, no matter how different they look, are related. Thus, it seemed likely that the various forms of energy could also be related. These scientific discoveries that emphasized the unity of nature also had an influence on economists, including Frederick Engels (1934).

13.4.1 Laws of Thermodynamics

The First Law of Thermodynamics formulated independently by Hermann von Helmholtz, James Joule, and Robert Mayer states that energy cannot be created or destroyed, although it can be converted between different forms (Grove et al., 1867; Lenard, 1933). The Second Law of Thermodynamics, which was formulated by Sadi Carnot, Rudolf Clausius and Lord Kelvin, states that in the process of interconversion, there is a decrease in the ability of the energy to perform work. The decrease in the ability to do work is proportional to the increase in entropy. The Third Law of Thermodynamics, formulated by Walther Nernst, states that the entropy is zero at absolute zero. And the fourth law …? There isn't a fourth law, and there will never be one according to Nernst, who figured that it took three people to come up with the first law, two to come up with the second, and he was obliged to come up with the third by himself, and thus it follows by extrapolation that nobody could come up with a fourth law (Laidler, 1993).

Michael Faraday (1861) gave a very understandable and clear nonmathematical lecture on the convertibility of energy (then known as forces) to children. Likewise, Hermann von Helmholtz and Robert Mayer wrote clear accounts for the layperson about the "conservation of forces" (Grove et al., 1867; Spencer, 1880). William Thomson and Peter Tait (1862) wrote a very readable account for the layperson on the conservation of energy. In this account, they used the word *energy*, a term introduced into science by Thomas Young (1807). The concept of energy, instead of force, was applied by Rankine. Thomson and Tait then differentiated "kinetic energy" from "potential energy" and coined the term *conservation of energy*. Willard Gibbs (1875–78) wrote an important, but nearly incomprehensible, treatise on thermodynamics that gave a mathematical framework to relate all the forms of energy to each other and to chemical processes. Luckily, Lewis and Randall (1923) wrote a lively and readable account of their interpretation of Gibbs's idea! To get a solid understanding of the biological process of energy conversion, I will use equilibrium thermodynamics to relate the molecular free energies of each step in the conversion process to the one before it and the one after it. According to a fundamental theorem of thermodynamics, at constant temperature and pressure, the change in molecular free energy (ΔE) is composed of two parts, the change in molecular enthalpy ($\Delta H/N_A$) and the change in molecular entropy ($\Delta S/N_A$):

$$\Delta E = \frac{\Delta H}{N_A} - \frac{T\Delta S}{N_A} \tag{13.1}$$

The enthalpy of a reaction is determined by burning each reactant and product separately and determining how much the burning causes the temperature of water to rise. The enthalpy of a chemical is a property of the bonds that make up the chemical itself. Entropy, on the other hand, is a measure of the number of possible states in which a molecule can exist. The number of possible states is, in part, dependent on the possible number of resonant structures that can describe a given molecule (see Chapter 14). It is also dependent on the potential places a molecule can be localized in space. That is, the entropy depends, in part, on the relative disorder of molecules in a given volume, where the state with the maximum entropy is defined by a completely random distribution. Thus, *entropy* is a term that relates a molecule to its environment.

While energy is never created or destroyed, it can be redistributed in space, and this is what happens during every transformation. To understand what happens when free energy is transformed from one form of energy to another, we must take into consideration a property known as *entropy* (Clausius, 1879). Entropy, a term that comes from the Greek words for "in transformation," is a measure of the decrease in the amount of free energy available to perform work that occurs during a transformation. For this reason, entropy is a measure of the irreversibility of natural processes (Edsall and Gutfreund, 1983). To get a feel for entropy, consider a given amount of free energy in a system. If the entropy of this system were low, the energy would be concentrated among a few states (or atoms) or in a limited space and can thus be "rounded up" and utilized to do work. By contrast, if the entropy of this system were high, the energy would be partitioned among many states (or atoms) or in a large space and thus can be considered unavailable to do work. Given that the total amount of energy is conserved, when there is an increase in the amount of entropy, there is a decrease or dissipation in the amount of useful energy in a system.

While the Second Law of Thermodynamics allows one to predict the direction in which a reaction will proceed spontaneously (Fenn, 1982), many physicists believe that the world is fundamentally reversible and thus do not consider the Second Law of Thermodynamics to be a fundamental law. Whether the world is fundamentally reversible or fundamentally irreversible is a recurring topic of debate (Ehrenfest and Ehrenfest, 1912; Steckline, 1983), although an important one because the two views lead to contradictory physical descriptions of the world. Those who prefer an irreversible worldview see the Second Law of Thermodynamics, which states that a system evolves irreversibly toward a state of greater entropy, as being a fundamental law of physics, whereas those who prefer a reversible worldview see the Second Law of Thermodynamics as an approximate law that is only useful when the system under study is not observed for a long enough period of time for the initial state to recur. Scientists with a reversible worldview give priority to the recurrence theorem of statistical mechanics that states that any process is reversible in theory and can return to its initial condition with a finite probability after a long enough period. From the point of view of a cell biologist, the world is fundamentally irreversible as the physicochemical processes we see in biological systems are always irreversible (Wayne, 2012a). For example, cells always get older and not younger, nucleic acid and protein synthesis require the use of a template while degradation does not, and it is impossible to reverse mitosis so that cancer cells disappear.

13.4.2 Molecular Free Energy of Some Photosynthetic Processes

Because photosynthesis involves the absorption of light, I will introduce the equation that relates the energy in a photon of light (ΔE, in J) to its frequency (ν, in s^{-1}) or wavelength (λ, in m). h and c are Planck's constant and the speed of light, respectively (Clayton, 1970):

$$\Delta E = h\nu = \frac{hc}{\lambda} \qquad (13.2)$$

This equation was proposed by Max Planck (1920, 1949a) in 1900 while he was investigating black-body radiation, extended by Albert Einstein (1905) when he studied the photoelectric effect, used by Niels Bohr (1913) to describe the structure of the hydrogen atom, and directly tested by James Franck (1926) and Gustav Hertz (1926) by bombarding mercury with electrons that had various kinetic energies and observing the color of the light that was emitted. Franck continued his work on the quantum structure of atoms in Germany until Adolph Hitler came to power. Although Franck was a Jew, he was allowed to remain in his position as the Director of the Institute for Physical Chemistry in Göttingen, but only if he dismissed the other non-Aryan students and workers. Not willing to do this, Franck resigned and immigrated to the United States where he began work on the physical processes involved in the light reactions of photosynthesis (Rosenberg, 2004).

To quantify the energy transformations that occur during chemical reactions, we need to know the concentrations of each reactant and product under real conditions. However, the energetics of a reaction does not only depend on the absolute amount of the reactants and products but also on their distribution relative to the equilibrium distribution. We often do not know the real concentrations, and thus, to compare the molecular free energy of one reaction with that of others, we relate the free energies of a standard state, where the concentrations of all reactants and products are at any given concentration that we can choose. Let us choose 1 M for everything except water, which has a concentration of approximately 55.5 M. The constant K_{std} represents the ratio of products to reactants in the standard state. Consider the following reaction that describes the reversible synthesis of ATP:

$$ADP + P_i \leftrightarrow ATP + H_2O$$

$K_{std} = [ATP][H_2O]/[ADP][P_i]$ for the synthesis of ATP, and under the conditions that we have conveniently chosen to be standard, $K_{std} = [1][55.5]/[1][1] = 55.5$.

Let K_{eq} represent the equilibrium constant for the synthesis of ATP. K_{eq} describes the concentrations of ATP, ADP, P_i, and H_2O at equilibrium when they started out at the standard. Because the water concentration will always be approximately 55.5 M, it will cancel out when we take the ratio of K_{eq} to K_{std}. Taking water into consideration, the value of K_{eq} is 1.96×10^{-5}. The difference in the free energy between the initial and the equilibrium state is given by the following equation, which is always put in the form of the final state minus the initial state.

$$E_{eq} - E_{std} = kT\ln\frac{K_{eq}}{K_{std}} \qquad (13.3)$$

Equilibrium thermodynamic measurements depend on the fact that the system is reversible. The molecular free energy released as a reaction goes to equilibrium is equal, but opposite in sign, to the molecular free energy that must be added to push the reaction from equilibrium to the standard state. The amount of energy that must be added to drive the reaction from equilibrium to the standard state is known as the *standard molecular free energy*. It is given by the following equation:

$$E_{std} - E_{eq} = kT\ln\frac{K_{std}}{K_{eq}} = -kT\ln\frac{K_{eq}}{K_{std}} \qquad (13.4)$$

If $E_{std} - E_{eq}$ were negative, then the product ATP will be formed spontaneously as the reaction goes to equilibrium, and if $E_{std} - E_{eq}$ were positive, the reactants will be made spontaneously as equilibrium is reached. K_{eq}/K_{std} is equal to the equilibrium concentrations of all the products, reactants, and water divided by (1 M) (55.5 M)/(1 M) (1 M). K_{eq}/K_{std} is equal to approximately 3.53×10^{-7} for the above reaction involving ATP. The standard molecular free energy of ATP formation is positive. Consequently, the synthesis of ATP requires molecular free energy (6.1×10^{-20} J), and its hydrolysis yields molecular free energy (-6.1×10^{-20} J).

The actual molecular free energy available to do work by a given reaction depends on how far the products and reactants are from equilibrium in the cell. It is given by the following reaction (Davies et al., 1993):

$$E_{Real} - E_{eq} = kT\ln\frac{K_{Real}}{K_{eq}} \qquad (13.5)$$

where $K_{Real} = [H_2O][ATP]/[ADP][P_i]$ under real or cellular conditions. If K_{Real} is equal to K_{eq}, it will neither take energy to synthesize ATP nor will energy be made available by the hydrolysis of ATP. Because the concentrations of H_2O, ATP, ADP, and P_i in the cell are approximately 55.5 M,

10^{-3} M, 10^{-3} M, and 10^{-2} M, respectively (Mimura et al., 1984; Mimura and Kirino, 1984), then

$$K_{Real} = \frac{[55.5\ M][10^{-3}\ M]}{[10^{-3}\ M][10^{-2}\ M]} = 55.5 \times 10^2 \quad (13.6)$$

and

$$E_{Real} - E_{eq} = kT\ln\frac{55.5 \times 10^2}{1.96 \times 10^{-5}} = 8 \times 10^{-20}\ J \quad (13.7)$$

Thus, under typical cellular conditions it will take 8×10^{-20} J of free energy to synthesize a molecule of ATP, and 8×10^{-20} J will be made available to do work on its hydrolysis. As I will discuss in Chapter 14, ATP is used as an available source of free energy in the cell, in part, because it has very low entropy compared with ADP and P_i, and the concentrations of these molecules are far from equilibrium.

When using Eq. (13.5), the units of K_{Real} and K_{eq} must be identical or else the natural logarithm will be a dimensional number, and the result of the calculation will be dependent on the units used, which is not an attribute of a fundamental mathematical equation. Moreover, eliminating the contribution of water in either the equilibrium reaction or the real reaction could quantitatively change the calculated molecular free energy by a factor of 4 (i.e., ln 55.5). If water is not included in the fundamental equations, then the molecular free energies that are determined for the reactions that depend on the addition or elimination of water may not be comparable to those that do not involve water (Oesper, 1950). In fact, the molecular free energy required for ATP synthesis can be looked at as the free energy needed to pull a water molecule from ADP and P_i in an aqueous environment. The molecular free energy made available by the hydrolysis of ATP can be viewed as the free energy released when a water molecule is added back.

13.4.3 Molecular Free Energy of Oxidation—Reduction Reactions

Energy can also be made available to do work from oxidation—reduction reactions where an electron is passed from one molecule to another (Michaelis, 1930; Johnson, 1949; Cramer and Knaff, 1991; Deichmann et al., 2014). The loss of electrons is known as *oxidation*, and the gain of electrons is known as *reduction*. Remember LEO the lion goes GER? To quantify the energy state of the components involved in oxidation and reduction we cannot use Eq. (13.5) because the electrons do not have an independent existence in solution, and thus their concentration cannot be measured. Therefore, we use a technique that was developed to measure the free energy of a reaction that involves the ionization of a metal (Arrhenius, 1902). Consider the following two reactions:

$$Zn^{2+} + 2e^- \leftrightarrow Zn$$

$$2H^+ + 2e^- \leftrightarrow H_2$$

In these reactions, Zn^{2+} and $2H^+$ are being reduced to form Zn and H_2, respectively. To initiate these reactions, a piece of zinc metal is immersed in a 1 M Zn^{2+} solution; and in a separate vessel, a piece of platinum is immersed in a 1 M H^+ solution under 0.1 MPa of H_2 gas. The two solutions are connected with an agar tube, so that the two solutions can remain unmixed, yet electricity can pass between them. The zinc and platinum electrodes are connected by a wire. Because the solution in one vessel will have a different affinity for the electrons than the solution in the other vessel, the electrons will flow from the solution with low affinity to the solution with high affinity. A compound with a high affinity for electrons is known as an *oxidizing agent*, and a compound with a low affinity for electrons is known as a *reducing agent*. The affinity of the components of the solution for electrons can be quantified by inserting a voltmeter in place of the wire between the two electrodes. By definition the vessel that contains 1 M H^+, 1 atm H_2, and a platinum electrode has a standard reduction-oxidation potential ($z\Psi^m$) or standard redox potential of 0 V. If the test solution tends to donate electrons to the H^+, the test solution will have a negative standard redox potential ($z\Psi^m < 0$), and if it accepts electrons from H_2, the test solution will have a positive standard redox potential ($z\Psi^m > 0$). The two electrodes can be placed in a common ionic solution to form an electrical cell. The voltage generated across the two electrodes is equal to the sum of the two voltages generated by the half-cells. The positive pole represents the electron donor and the negative pole represents the electron acceptor.

To measure the redox potential of a biological molecule like cytochrome *b*, *a* platinum electrode and a reference electrode are placed in a solution of reduced and oxidized cyt b and small redox couples known as *mediators* are added to the solution (Caswell, 1968; Caswell and Pressman, 1968). The mediators transport electrons from the cytochrome to the electrodes. As it is not possible to make a cytochrome electrode and use solid cytochrome to define the standard state (as it is for zinc), the standard state is defined as the state where the concentration of the oxidized form and that of the reduced form are equal. Thus, the standard redox potential is called the *midpoint potential*, and together, the oxidized and reduced forms are known as a *redox couple*. The molecular free energy of our test reaction (relative to the molecular free energy of the hydrogen reaction) is given by the following equation:

$$E^m - E^H = \left(kT\ln\frac{[red]}{[ox]} + nez\Psi^m\right) - 0 \quad (13.8)$$

where $E^m - E^H$represents the difference in the molecular free energy between the test solution in the standard state (E^m) and the molecular free energy of the hydrogen solution (E^H). $z\Psi^m$ is the midpoint potential of the test solution, and n and e represent the number of electrons transported and the elementary charge, respectively.

Any standard state can be chosen—for example, H^+ can be 10^{-7} M (as the biologists prefer it) or 10^0 M (which is the chemists' choice). As kT ln([red]/[ox]) = 0 at the standard state, then the molecular free energy of the standard state of the test solution ($E^m - E^H$) is given by the following equation:

$$E^m - E^H = nez\Psi^m \quad (13.9)$$

The standard state is defined to be mathematically convenient but may not reflect the conditions found in the cell. As the ability of a redox couple to donate electrons to another redox couple will be greater when the concentration of the reduced form of the donor couple is greater than the concentration of the oxidized form of the acceptor couple, we must take the real concentration into account to quantify the energetics of redox couples in the cell (Mitchell, 1966; Prebble and Weber, 2003). Thus,

$$E^{Real} - E^m = \left(kT\ln\frac{[red]}{[ox]} + nez\Psi^{Real}\right) - nez\Psi^m \quad (13.10)$$

where $E^{Real} - E^m$ represents the difference in molecular free energy between a redox couple in the real state and the molecular free energy of the redox couple in the standard state. This is just another form of the Nernst equation. At equilibrium, $E^{Real} - E^m = 0$. The effect of [red]/[ox] on the oxidizing or reducing power of a couple can be quantified by observing the redox couple under equilibrium conditions where $z\Psi^{Real} - z\Psi^m = -\frac{kT}{ne}\ln([red]/[ox])$ and measuring the voltage necessary to induce each [red]/[ox]. When the electronic forces ($z\Psi^{Real} - z\Psi^m$) and the chemical forces (−kT/ne)ln([red]/[ox]) are equal, the couple is in equilibrium with the voltage:

$$-\frac{kT}{ne}\ln\frac{[red]}{[ox]} = z\Psi^{Real} - z\Psi^m \quad (13.11)$$

and

$$z\Psi^{Real} = z\Psi^m - \frac{kT}{ne}\ln\frac{[red]}{[ox]} \quad (13.12)$$

Thus, when [red]/[ox] = 1, $z\Psi^{Real} = z\Psi^m$. When [red]/[ox] > 1, $z\Psi^{Real}$ is more negative than $z\Psi^m$ and is a stronger reducing agent than it would be at the standard state. When [red]/[ox] < 1, $z\Psi^{Real}$ is more positive than $z\Psi^m$ and thus is a stronger oxidizing agent than it would be at the standard state.

When H^+ are also transferred in the reduction–oxidation reaction, then the complete dependency of the redox potential on concentration, number of electrons, and protons is given by the following equation:

$$z\Psi^{Real} = z\Psi^m - \frac{kT}{ne}\ln\frac{[red]}{[ox]} - m\frac{kT}{e}\ln\frac{[AH_m]}{[A^-]} \quad (13.13)$$

where m is the number of H^+ transferred to the proton acceptor A^-.

The maximal molecular free energy required for, or released by, the transfer of an electron between two carriers is given by the following equation:

$$\Delta E = nze\left(z\Psi^{final} - z\Psi^{initial}\right) \quad (13.14)$$

where $z\Psi^{final}$ is the real redox potential of the final carrier, and $z\Psi^{initial}$ is the real redox potential of the initial carrier. The real redox potentials depend on the relative concentrations of the reduced and oxidized forms. They are also affected by environmental factors, including ionic strength (Cramer and Knaff, 1991). Because all the relevant facts necessary to determine the real redox potentials are not known, we must be satisfied for the moment with using the midpoint potentials (Table 13.3). I have included the valence of the electron z outside the parentheses in Eq. (13.15) so that this equation is valid for the movement of positive charges (z > 0) as well as the movement of negative charges (z < 0).

TABLE 13.3 Approximate Midpoint Potentials ($z\Psi^m$, at pH 7) of Photosynthetic Electron Carriers

Redox Couple	$z\Psi^m$, at pH 7
Phaeophytin	−0.61
Ferridoxin	−0.43
(½)$NADP^+$/(½)NADPH	−0.32
Q_A	−0.30
Q_B	−0.15
cyt_{b6}	−0.14
cyt_f	0.36
Plastocyanin	0.37
P700 (reduced)	0.35
P700 (oxidized)	0.45
P700 (excited)	−1.20
P680 (reduced)	0.80
P680 (oxidized)	0.90
P680 (excited)	−0.80
Oxygen	0.81

As electrons move from reducing agents to oxidizing agents, free energy is made available; this free energy can be used to translocate protons against their electrochemical difference across a membrane. The energy needed to translocate one H^+ can be calculated from the differences in electrical and chemical potential of H^+ on both sides of the membrane according to the following equation (Nernst, 1923; see Chapter 2):

$$\Delta E_{(p-e)} = kT \ln\frac{[H^+]_p}{[H^+]_e} + ze(\Psi_p - \Psi_e) \quad (13.15)$$

where the subscripts p and e refer to the P- and E-spaces, and the form of the equation depends on how we integrate to get this equation (see Chapter 2). If we integrate from E- to P-space, we get Eq. (13.15) that tells us the magnitude of the energy gained or lost as a proton moves from the E- to the P-space. If we want to know the magnitude of the energy gained or lost as a proton moves from the P-space to the E-space, we must use the following equation[1]:

$$\Delta E_{(e-p)} = kT \ln\frac{[H^+]_e}{[H^+]_p} + ze(\Psi_e - \Psi_p) \quad (13.16)$$

As we will see below, the movement of an H^+ from the stroma to the lumen requires energy, which is provided by the electron transport chain and the movement of an H^+ from the lumen to the stroma releases energy that is captured in the chemical energy of ATP.

13.4.3.1 Measurement of the Electrical Potential Across a Membrane

The electrical potential across a membrane can be determined by measuring the distribution of a lipophilic anion, like tetraphenyl boron (TPB^-) or a lipophilic cation, such as tetraphenylphosphonium (TPP^+), both of which permeate the membrane passively. A lipophilic cation is used for membranes with a negatively charged interior, such as the inner membrane of the mitochondrion, and a lipophilic anion is used for membranes with a positively charged interior, such as the thylakoid. In doing these experiments, the thylakoid membranes are incubated with tracer amounts of TPB^-. Because TPB^- is negatively charged, it will distribute across the membrane in a manner related to the electrical potential of the membrane. For example, if the membrane is uncharged, the TPB^- will be equally distributed on both sides. If the membrane potential is inside-positive, the TPB^- will accumulate inside. The more positive the internal membrane potential, the more TPB^- will accumulate.

At equilibrium, the concentration of TPB^- on both sides of the membrane will be determined precisely by the membrane potential as described by the Nernst equation. The membrane potential (Ψ_p) across the thylakoids only amounts to about -0.01 to -0.04 V (lumen $= 0$ V). The membrane potential of the thylakoid membranes is low as a consequence of the presence of ion transporters, which are capable of neutralizing the charge caused by H^+ accumulation. For example, every time an H^+ enters the thylakoid lumen, a Cl^- also passes through the membrane. The Cl^- neutralizes the charge of the proton, and at the same time, it allows the buildup of a large pH difference.

13.4.3.2 Measurement of the pH Difference Across a Membrane

The distribution of a weak acid or weak base between two compartments is a function of the pK_{acid} or pK_{base} of the tracer and the pH of the two compartments. Thus, we can quantify the chemical potential for protons, which is related to the pH difference across a membrane, by measuring the distribution of a permeant weak base such as ^{14}C-methylamine or weak acid such as ^{14}C-dimethyloxazolidinedione, and calculating the internal pH of the membrane from the Henderson–Hasselbalch equation, which was originally derived by Lawrence Henderson (1908) and put in logarithmic form by Karl Hasselbalch. A weak acid is used for membranes with a basic interior, and a weak base is used for membranes with an acidic interior.

A weak base will only pass a membrane when it is uncharged. It will be uncharged in basic compartments and charged in acidic compartments. Thus, there will be a net movement across the membrane from the basic to the acidic compartment. By measuring the amount of tracer in the two compartments, it is possible to determine the pH in both compartments. The Henderson–Hasselbalch equation, used to determine the pOH with a weak base (BOH) could be derived from simple kinetics (Michaelis, 1926):

$$[OH^-] + [B^+] \underset{k_2}{\overset{k_1}{\leftrightarrow}} [BOH]$$

where k_1 is the on-rate constant for the association of the weak base, and k_2 is the off-rate constant for the dissociation of the base into a cation and a hydroxyl ion. At equilibrium, the rate of dissociation of BOH equals the rate of formation of BOH and

$$k_1[OH^-][B^+] = k_2[BOH] \quad (13.17)$$

1. Peter Mitchell defined the electrochemical proton gradient or the proton motive force (pmf, in V) as the molecular free energy (in J) divided by the elementary charge (e, in Coulombs). While the proton motive force has the advantage of being a vector, it has the disadvantage of having the wrong units for a force, which should be given in Newtons. For consistency, I will use molecular free energy as opposed to the proton motive force, or what Efraim Racker (1975b) called the Peter Mitchell Force.

If we define K_b, which is a measure of the ability of a weak base to produce OH^-, as k_2/k_1, then

$$K_{base} = \frac{[OH^-][B^+]}{[BOH]} \tag{13.18}$$

Here we can see that K_{base} is really a dissociation constant, and K_{acid}, which is a measure of the ability of a weak acid to produce H^+, is equal to 14 K_{base}. Putting the equation in logarithmic form, we get

$$\log K_{base} - \log [OH^-] = \log \frac{[B^+]}{[BOH]} \tag{13.19}$$

After multiplying both sides by −1, we get:

$$-\log K_{base} - (-\log[OH^-]) = -\log \frac{[B^+]}{[BOH]} \tag{13.20}$$

Because $pX = -\log X$

$$pK_{base} - p[OH^-] = -\log \frac{[B^+]}{[BOH]} \tag{13.21}$$

or

$$pK_{base} - p[OH^-] = \log \frac{[BOH]}{[B^+]} \tag{13.22}$$

If $pK_{base} = p[OH^-]$, then $[BOH] = [B^+]$. If $pK_{base} \gg p[OH^-]$, then $[BOH] \gg [B^+]$. If $pK_{base} \ll p[OH^-]$, then $[BOH] \ll [B^+]$.

The weak base (BOH) can pass across the thylakoid membrane. If the pK_{base} of the weak base is the same in both compartments, then

$$K_{base} = \frac{[OH^-]_i[B^+]_i}{[BOH]_i} = \frac{[OH^-]_o[B^+]_o}{[BOH]_o} \tag{13.23}$$

The log of the relative proportion of B^+ to BOH will be determined by the difference between pK_{base} and $p[OH^-]$. At equilibrium, $[BOH]_i = [BOH]_o$, and thus

$$[OH^-]_i[B^+]_i = [OH^-]_o[B^+]_o \tag{13.24}$$

After rearranging terms,

$$\frac{[OH^-]_o}{[OH^-]_i} = \frac{[B^+]_i}{[B^+]_o} \tag{13.25}$$

and we can calculate the pH, as $pH = 14 - pOH$. If the weak base has a pK_{base} that is one unit less than the expected pOH of the compartments, then, according to Eq. (13.22), $[B^+] > [BOH]$, and as a first approximation, the total concentration of radioactive substances measured is approximately equal to $[B^+]$. Using such techniques, the pH of the lumen was found to be about 5 and the pH of the stroma was found to be about 8 (Nicholls and Ferguson, 1992).

Now that we are able to quantify the energetics of photons, chemical reactions, reduction–oxidation reactions, and electrochemical differences, we will use the First Law of Thermodynamics to estimate the work that can be done as one form of energy is converted into another. In only using the First Law of Thermodynamics, we will be assuming that the cells are at equilibrium. I realize that this assumption is ridiculous and that while nonlife may be an equilibrium situation, life is certainly not, and needs a constant input of energy. Albert Claude (1975) wrote, "Life, this anti-entropy, ceaselessly reloaded with energy, is a climbing force, toward order amidst chaos, toward light among the darkness of the indefinite, toward the mystic dream of love, between the fire which devours itself and the silence of the cold." While nonequilibrium thermodynamics is important for a complete description of photosynthesis in living cells, equilibrium thermodynamics, as a first approximation, is still a powerful tool.

13.5 ORGANIZATION OF THE THYLAKOID MEMBRANE AND THE LIGHT REACTIONS OF PHOTOSYNTHESIS

Peter Mitchell (1977, 1980) realized that the structure of the cell with its membranes meant that using scalar quantities to describe bioenergetic processes left out half of the information about a molecule and vectors should be used instead of scalars to describe both the magnitude of a component and its directionality in space. Mitchell and Moyle (1960) wrote, "the studies of the autocatalytic processes of growth and the steady-state processes of cell maintenance at the molecular level of dimensions have been pursued without reference to spatial considerations—giving rise to the well-developed science of metabolism. The science was largely nourished from the domain of chemistry, and being mainly dependent on the language and symbolism of 'homogeneous' chemical reactions it was not adapted to take account of, let alone describe, the organisation of the chemical processes in space. Conversely...our conception of the processes by which solutes pass across the cell surface stems, to a great extent, from studies of the physiology of excitable cells...special consideration being given to the movement of stable ions and of water, but not to the metabolic processes accompanying these movements at the molecular level...perhaps the belief in the necessity for coupling separate membrane-transport systems and metabolic systems has partly been a reflection of the necessity for coupling physiologists to biochemists." Consequently, we must consider the specific location of a protein in space and its substrate and binding specificity.

The light reactions of photosynthesis take place on the thylakoids, which are a specialized membrane system that contains four kinds of protein complexes. These complexes function in the capture of light and the synthesis of ATP and NADPH (Murphy, 1986a,b; Mattoo et al., 1989; Camm and Green, 2004). The energy needed for the synthesis of ATP and NADPH comes from light (i.e., radiant energy). The light is absorbed by two protein-pigment complexes known as PS I and PS II. The radiant energy is converted to electrical energy, which flows through molecules instead of wire. The electrons flow from the PS II complex to the PS I complex, through the $cyt_{b6\text{-}f}$ complex. As electrons pass through the $cyt_{b6\text{-}f}$ complex, protons are translocated from the stroma to the lumen. Once a proton difference is established, the fourth complex, an ATP synthase, uses the energy of the proton difference to make ATP. Furthermore, the electrons that result from the photochemical process are used to reduce $NADP^+$ to make NADPH.

The protein complexes can be differentially extracted with detergents from the thylakoid. They can also be visualized on the thylakoid membranes using freeze-fracture electron microscopy (see Fig. 13.11; Anderson, 1975). The particles have been identified by correlating their presence with the presence or absence of biochemical activities that change with development or differ between mutants and the wild type. Their presence on either the granal or stromal thylakoids can also be correlated with the biochemical activities of each fraction, and last, the isolated protein complexes can be put into liposomes and freeze-fractured to see the size of the particles.

PS II particles, which are 8–11 nm in diameter, occur mostly (85%) on the lumenal leaflet of the membrane in the

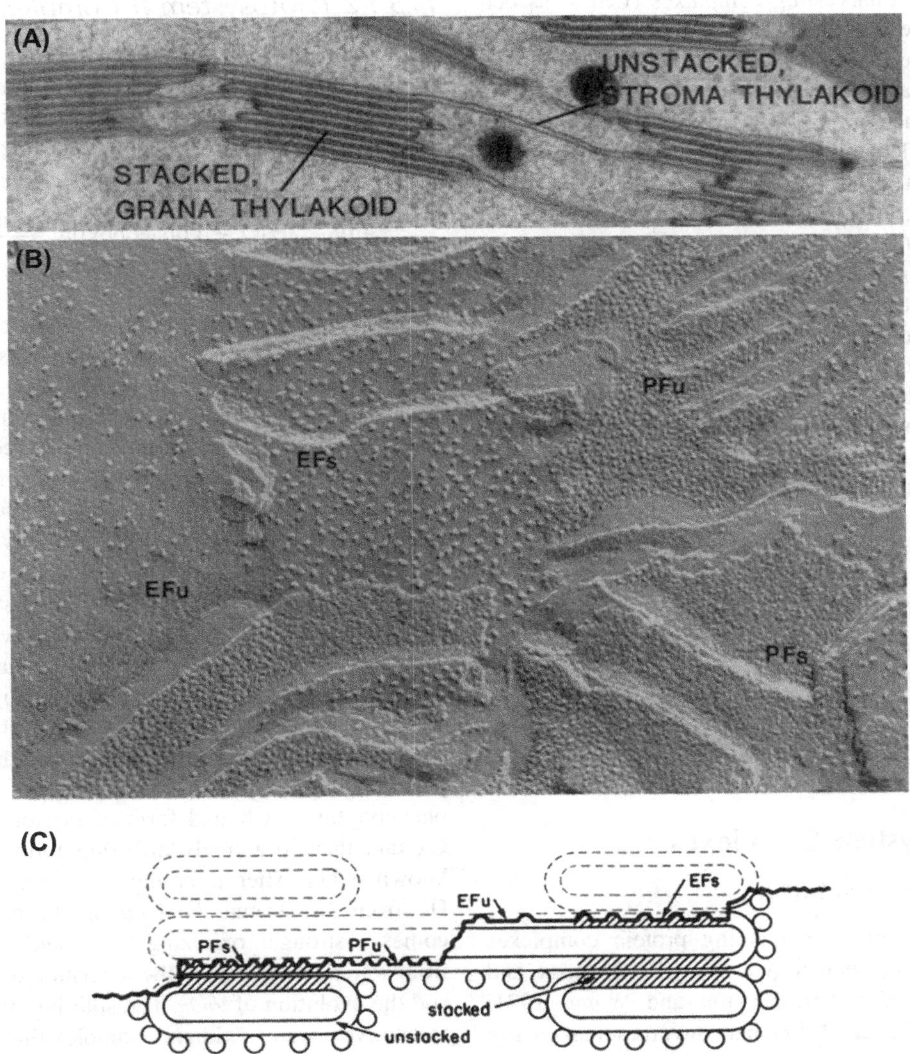

FIGURE 13.11 Thin-section electron micrograph (A), freeze-fracture micrograph (B), and model (C) of the thylakoid membranes of a pea. The membranes in the freeze-fracture micrograph were phosphorylated for 20 min in vitro before freezing. EFs and PFs are the E-face (lumen) and P-face (stroma), respectively, of the membranes from the stacked regions. EFu and PFu are the E-face and P-face, respectively, of the membranes from the unstacked regions. (A), ×100,000. (B) ×90,000. *From Staehelin, L.A., Arntzen, C.J., 1983. Regulation of chloroplast membrane function: protein phosphorylation changes the spatial organization of membrane components. J. Cell Biol. 97, 1327–1337.*

stacked regions of the thylakoids. There are 1200–1700 PS II complex particles per μm^2 in the stacked regions and 300–700 PS II complex particles per μm^2 in the unstacked regions (Staehelin, 1986). PS II mutants do not have this particle. Moreover, the chloroplasts of bundle sheath cells in plants exhibiting C_4 photosynthesis lack both PS II and these particles (Miller et al., 1977; Wrischer, 1989).

PS I complexes, which range in diameter from 10 to 13 nm, appear mostly (85%) on the stromal leaflet of the membrane in the unstacked regions of the thylakoids. There are about 2100–3300 PS I complex particles per μm^2 in the unstacked regions and 250–400 PS I complex particles per μm^2 in the stacked regions. PS I mutants lack this particle (Miller, 1980; Olive et al., 1983; Staehelin, 1983, 1986).

The size of PS I and PS II particles seem to be related to the number of light-harvesting complexes (LHCs) associated with the core complexes (Staehelin, 1986; Sprague et al., 1985). There is also a pool of mobile LHCs that are unbound and appear as 8-nm particles on the stromal leaflet of the thylakoid membrane (Simpson, 1979; Knoetzel and Simpson, 1991). The mobile particles may serve as the adhesion sites that allow the stacking of thylakoid membranes into grana. In the stacked thylakoid membranes, there are about 3600–5100 LHCs per μm^2.

The $cyt_{b6\text{-}f}$ complex is an 8.3-nm particle that is distributed in the stromal leaflet of the thylakoid membrane in both the stacked and unstacked regions. They range in density from 850 to 1300 $cyt_{b6\text{-}f}$ complex particles per μm^2.

The ATP synthase was first imaged as a lollipop-like structure using negative staining (Howell and Moudrianakis, 1967). It appears as a 10-nm particle on the stromal surface of the unstacked region of thylakoids and as a 6.5-nm particle when the membrane is fractured open. The peripheral protein complex, represented by the 10-nm particle, is known as CF_1 and is the hydrolytic portion of the complex. The integral protein complex, represented by the 6.5-nm particle, is known as CF_0 and is the proton-translocating portion (Miller and Staehelin, 1976; Mörschel and Staehelin, 1983). There are approximately 1000–1600 ATP synthase particles per μm^2 in the unstacked regions.

13.5.1 Photosystem Complexes

13.5.1.1 Charge Separation

Photosystems are pigment-containing protein complexes that contain reaction centers (Clayton, 1962; Reed and Clayton, 1968; Reed, 1969; Clayton and Wang, 1971; Feher, 1971; Gisriel et al., 2017) that convert radiant energy (hν) into chemical energy. On excitation, the pigment (P) becomes a strong reducing agent (P^+) that allows it to pass an electron to a primary acceptor (A), which then becomes reduced (A^-). This process, which takes place within the reaction center, is known as *charge separation* (Kluyver and van Niel, 1956) and is represented in the following reaction:

$$P + A + h\nu \rightarrow P^+ + A^-$$

This reaction is irreversible as a consequence of the rapid rereduction of P^+, which occurs as a result of the acquisition of an electron from an electron donor, as well as the rapid reoxidation of A^- that results from the reduction of the next electron acceptor. Both PS I and PS II are oriented in the thylakoid membrane such that the excited electron in the reaction center moves from the lumen side of the membrane to the stromal side of the membrane in an electrogenic manner.

13.5.1.2 Photosystem II Complex

In the case of PS II, phaeophytin is the primary electron acceptor, and water is the electron donor. The PS II complex can be said to function as a light-driven phaeophytin reductase and a water oxidase according to the following reaction:

$$4h\nu + 2H_2O + 4\ \text{phaeophytin}_{ox} \xrightarrow{\text{PS II}} O_2 + 4H^+ + 4\ \text{phaeophytin}_{red}$$

The PS II complex can be subdivided into three major functional units: the PS II core complex, surrounded by the PS II antenna complex (Emerson and Arnold, 1932a,b; Clayton, 2002), and the water-splitting complex (Ghanotakis and Yocum, 1990). The core complex contains about six polypeptides that bind four chlorophyll *a* molecules, two of which form the reaction center that has an absorption maximum of 680 nm. These chlorophyll molecules are known as P680. The core also contains two pheophytin, one β-carotene, and one nonheme iron. The β carotenes function as antioxidants that protect the photosystems from damage that may be caused by the free radicals that are formed as a natural result of being subjected to high light irradiances.

Once light is absorbed by P680, the energy of an electron is raised to an excited state and that energy is then passed to phaeophytin. The electron is passed from phaeophytin to a bound form of plastoquinone known as Q_A and then to a freely diffusing form of plastoquinone known as Q_B. After accepting two electrons, Q_B binds two H^+ from the stroma. *Pari passu*, the oxidized P680 becomes a stronger oxidizing agent and strips an electron from water, which causes the formation of H^+ in the lumen and the evolution of $\frac{1}{2}O_2$. The splitting of water is thought to involve a water-oxidizing complex that is situated in the PS II complex on the lumenal side of the thylakoid (Mayfield, 1991; Ettinger and Theg, 1991).

How much free energy is required to transfer one electron from the reduced form of P680 (0.8 V) to phaeophytin (−0.61 V)? According to Eq. (13.15), it will take 2.3×10^{-19} J, and it is thus an endergonic reaction. Is the energy in a photon of 680-nm light sufficient to raise the energy of the electron so that it can reduce phaeophytin? According to Eq. (13.2), there are 2.9×10^{-19} J associated with a photon of 680-nm light so that there is sufficient energy to reduce phaeophytin.

13.5.1.3 Photosystem I Complex

The PS I complex functions as a light-driven plastocyanin oxidase and a ferredoxin reductase (Golbeck, 1992; Ikeuchi, 1992). PS I is a pigment-containing protein complex that can be subdivided into a core complex and antenna complexes. The core complex, which contains the reaction center, contains seven polypeptides. The reaction center contains two chlorophyll molecules, known as *P700*, which have an absorption maximum at 700 nm. There are approximately 100 chlorophyll *a*, one β-carotene, two phylloquinone, and three 4Fe−4S centers per P700. The PS I complex passes a single electron from plastocyanin$_{red}$ to ferredoxin$_{ox}$ with the help of radiant energy according to the following formula:

$$4h\nu + 4Fd_{ox} + 4PC_{red} \xrightarrow{PS\ I} 4Fd_{red} + 4PC_{ox}$$

By inserting the appropriate values in Eqs. (13.14) and (13.2), we find that one photon of 700-nm light has sufficient energy to excite an electron from P700 to ferredoxin, an iron-sulfur protein. Once ferredoxin gets reduced by the electron from P700, it reduces $NADP^+$. It takes two electrons to reduce $NADP^+$ to NADPH:

$$2H^+ + 4Fd_{red} + 2NADP^+ \leftrightarrow 4Fd_{ox} + 2NADPH$$

In contrast to PS I, which is an integral membrane complex, ferredoxin and plastocyanin are water-soluble proteins. Ferredoxin is in the stroma, and plastocyanin is in the lumen. The ferredoxin-$NADP^+$ oxidoreductase (Avron and Jagendorf, 1956) is anchored to the stromal side of the thylakoid (Forti, 1999).

13.5.2 Cytochrome$_{b6\text{-}f}$ Complex

Once the cytochromes and plastocyanin were discovered in the chloroplast, their redox potentials and location in the thylakoids were established. The cytochrome$_{b6\text{-}f}$ complex has the appropriate midpoint redox potential to function as a plastoquinone−plastocyanin oxidoreductase according to the following reaction:

$$2PQ_B^{2-} + 4PC_{ox} + 4H^+_{stroma} \rightarrow 2PQ_B + 4PC_{red} + 4H^+_{lumen}$$

This complex is homologous to the cytochrome b-c complex of mitochondria (see Chapter 14). While the first cytochromes found were given the names a, b, and c, cytochrome f got its name because it is found in fronds (Hill and Scarisbrick, 1951; Davenport, 1952; Nelson and Racker, 1972; Bendall, 2005)! The electron transport chain between plastoquinone and plastocyanin transports at least one H^+ from the stroma to the lumen for every electron that moves from plastoquinone to plastocyanin. The transmembrane transfer of H^+ is effected by plastoquinone (Mitchell, 1975; Foyer et al., 2012).

Light induces the uptake of protons into the lumen, as determined by measuring the effect of light on the pH of the medium that surrounds isolated chloroplasts (Jagendorf and Hind, 1963; Hinkle and McCarty, 1978). Is the molecular free energy made available by the redox reaction sufficient to translocate a proton across the thylakoid membrane from the stroma to the lumen? Because we do not know the real redox potentials of the various components, we must settle with knowing the midpoint potentials ($z\Psi^m$). Given that the midpoint redox potential for Q_B is approximately −0.15 V and the midpoint redox potential of plastocyanin is approximately 0.37 V, then for every electron transported, 8.32×10^{-20} J of energy is made available to do work according to Eq. (13.14).

Given that $[H^+]_e = 10^{-5}$ M, $[H^+]_p = 10^{-8}$ M, $\Psi_e = 0$ V, and $\Psi_p = -0.02$ V, according to Eq. (13.16), it will take 3.2×10^{-20} J to transfer an H^+ from the stroma to the lumen. Thus, there is enough energy made available in the transfer of an electron from Q_B to plastocyanin to transfer two protons from the stroma into the lumen. Experiments show that one to two H^+ is transferred across the thylakoid membrane for every electron passed through the cytochrome$_{b6\text{-}f}$ complex (Harold, 1986).

The figures used in the calculations of the molecular free energy of proton transport come from measurements of the bulk pH of the lumen and stroma. It is possible that the local concentration of H^+ around the cytochrome$_{b6\text{-}f}$ complex may vary drastically from the average value, and thus the distribution of protons may be important in understanding the energetics of coupling proton translocation to electron transport (Junge, 2004).

13.5.3 ATP Synthase

The fourth complex in the thylakoid membrane is the ATP synthase, which transduces or couples the potential energy inherent in the transthylakoid proton difference into the chemical energy of ATP (Avron and Jagendorf, 1957; Avron et al., 1958; Krogmann et al., 1959; Good, 1961; Jagendorf, 1961; Avron, 1963; Good et al., 1966; McCarty, 1992). The chloroplast ATPase, also known as the coupling factor, is part of the F_0F_1-type (see Chapter 14) ATPase family and differs from the V- (see Chapter 7) and P-type (see Chapter 2) ATPases. For over 20 years, it was believed that a high-energy phosphorylated common intermediate known as "the squiggle" was responsible for coupling the energy of redox reactions into the chemical energy of ATP. There were, however, some skeptics (see Pasternak, 1993; Williams, 1993a,b).

Peter Mitchell (1961, 1966, 1978), a student of James Danielli, proposed an alternative explanation for energy coupling based on vectorial metabolism and osmochemistry (Orgel, 1999; Malmström, 2000). He proposed that the energy inherent in the electrochemical difference of protons across a membrane could provide the energy for ATP synthesis. That is, Mitchell proposed that biochemistry did not only depend on the concentration of substrates and the presence of enzymes but also on whether the substrates were on the p-side or the e-side of a membrane and the orientation of the enzymes in a membrane relative to the p-side and e-side. Mitchell, also realized that the algebra of thermodynamics would have to be augmented with vectors that were used to describe the directionality of driving forces in space and time. Mitchell's hypothesis of the importance of membranes and topology in bioenergetics was based on the need for intact membranes to couple oxidation and phosphorylation (see Chapters 1 and 2). Thus, it was more theoretical than factual, but a new paradigm has to start somewhere. According to Mitchell (see Wolpert and Richards, 1997), "Nature itself doesn't have physiologists, on the one hand, who look at the action, and biochemists, on the other, who look at the chemistry. Nature itself does it as one thing. So what you need to do is to try to bring together the notions of chemistry and the notions of physiology...." Mitchell (see Wolpert and Richards, 1997) went on to say, "On it was a very slow and painful process. Most people, who try to be creative, I think, have found that they've got to become craftspeople as well as art people. You have to go through the, dare I say, dreary business of school, and in my case, of learning chemistry out of the textbook. There is a huge amount of information you have to absorb before you can start walking about in it, as it were. I've often thought that the human mind is a bit of a garden. You prepare this garden, and you plant things there, and it's a sort of garden partly of facts, and partly of ideas, and you keep re-arranging it, and that's really quite hard work, but at the same time, it's well worth it because you can go for a lot of walks, especially if you don't sleep very well."

Mitchell's chemiosmotic hypothesis stated that the enzymes of the electron transport chain, which conducts electricity, and the ATP synthase were localized in a membrane with a well-defined orientation. The ATP synthase and the electron transport chain were functionally linked to a vectorial transfer of protons across a membrane or proticity—the positively charged version of electricity. Thus, the electron transport chain gave rise to an electrochemical proton difference across the membrane which served as a driving force for ATP synthesis. A requisite for the establishment of a proton gradient is that the membrane itself is impermeable to protons (Ernster, 1978). This explained the need for an intact membrane in oxidative and photosynthetic phosphorylation.

According to Prebble and Weber (2003), "Mitchell was a controversial figure in many ways. His unusual methods of approaching his science clearly contributed to his success in pursuing his ideas, but these approaches were often seen as weaknesses by his peers. His preferences for working things out from first principles was certainly a strength when a shift in paradigm thinking was required, but it was often perceived as a failure to review the literature carefully as a first step." Undaunted by Mitchell's hypothesis, the leaders in the field kept searching for the high-energy molecule known as the squiggle that would transfer the energy from the electron transport chain to ADP and Pi to form ATP (Mitchell, 1978).

The squiggle reaction can be represented by the following formula (Slater, 1953):

$$AH_2 + B + C \leftrightarrow A \sim C + BH_2$$

$$A \sim C + ADP + P_i \leftrightarrow A + C + ATP$$

The vectorial reaction proposed by Mitchell can be represented by the following formula:

$$ADP + P_i + 3H^+_{lumen} \overset{ATP\ synthase}{\leftrightarrow} ATP + H_2O + 3H^+_{stroma}$$

Work begun by Geoffrey Hind and André Jagendorf (Hind and Jagendorf, 1963), who showed that ATP formation can be temporally separated from the light requirement, was continued by Jagendorf and Neumann (1965), who showed that light causes a reversible rise in the pH of the medium that surrounded the chloroplasts. This work culminated in a publication by Jagendorf and Uribe (1966), in which they revealed that a proton difference across the thylakoid membrane was the high-energy intermediate that could be used to generate ATP, thus providing the first direct evidence for Peter Mitchell's chemosmotic hypothesis. In order to show that a pH difference across the thylakoid membranes could power ATP synthesis, Jagendorf and Uribe incubated chloroplasts in a permeant acid (i.e., weak acid) to acidify the inside of the chloroplast (pH 3.8) and then transferred the chloroplast to a neutral pH medium (pH 7.0) that contained ADP and P_i. On transfer, ATP was synthesized from the energy inherent in the proton difference (McCarty and Racker, 1966; Prebble and Weber, 2003; Govindjee, 2017).

Richard McCarty found that the rate of ATP synthesis is related to the third power of the pH difference across the thylakoid membranes. When viewed in light of kinetic theory (see Chapter 12), this result indicates that the generation of a single ATP probably results from the translocation of three H^+ from the lumen to the stroma through the ATP synthase (Hinkle and McCarty, 1978). Kaim and Dimroth (1999), and Junge (2004) believe that ATP synthesis depends on the energy inherent in the membrane potential and not the pH difference. Resolution of this problem will depend on better measurements of the steady-state $\Delta\Psi$ and ΔpH in living chloroplasts.

Given that the energy needed to synthesize a single ATP molecule is approximately 8.0×10^{-20} J under physiological conditions (see Eq. 13.6), and the energy made available by an H^+ moving from the lumen to the stroma is 3.2×10^{-20} J (see Eq. 13.16), then it will take the electrochemical potential energy of three H^+ to synthesize one ATP molecule. Thus, McCarty's finding is consistent with the Laws of Thermodynamics (Hinkle and McCarty, 1978).

13.5.4 Light Reactions of Photosynthesis

Light energy is absorbed by the antenna complex of PS II and is passed in a somewhat random fashion to P680 by resonance transfer (Laible et al., 1994). Once the energy reaches P680, an electron is raised to a higher energy state where it can reduce phaeophytin. Phaeophytin passes the electron to the plastoquinones, which in turn passes it to the cyt_{b6-f} complex. $P680^+$, because it has a more positive redox potential than P680, is able to take an electron from water to refill the hole left by the electron that was ejected by light. One H^+ is produced in the lumen from a water molecule for every photon absorbed by P680. The cyt_{b6-f} complex oxidizes PQH_2 and reduces PC_{ox}. As the electrons are passed through the cytochrome chain, protons are passed from the stroma to the lumen. Because the electrons take a *zigzag* course across the membrane, the biochemical pathway of the light reactions was dubbed Z-scheme of photosynthesis (see Fig. 13.12; Trebst, 1974), and it is reminiscent of the N-shaped diagram used to depict the redox potentials of the components of the light reactions, when it is turned on its side (Fig. 13.13).

When a great-enough proton concentration difference is developed across the thylakoid membrane, the protons move through the CF_0 portion of the ATP synthase, and the energy made available by the collapse of the pH difference is used for the synthesis of ATP from ADP and P_i. In this way, ATP is formed in the light reactions of photosynthesis.

Light is also absorbed by the antenna complex of PS I where the energy is passed by resonance transfer in a random manner to P700. Once it reaches P700, an electron is raised to a higher energy level where it is able to reduce ferredoxin. The $P700^+$ can then take an electron from PC to refill the hole left by the reduction of ferredoxin. The ferredoxin-$NADP^+$ oxidoreductase is then able to reduce $NADP^+$ to NADPH at the expense of the oxidation of two molecules of ferredoxin and a proton from the stroma. Thus, the formation of NADPH causes an alkalinization of the stroma, and this contributes to the pH difference across the thylakoid membrane. The ATP and the NADPH generated in the light reactions are then used to fix CO_2 into glucose in the dark reactions that take place in the stroma. As we will see, it takes two NADPH and three ATP molecules to fix a single CO_2 molecule.

It takes four electrons from PS I to reduce two $NADP^+$ molecules and form two NADPH. It takes four photons from PS II to induce the transfer of four electrons from water to PS I so that the four-electron deficit of PS I can be eliminated.

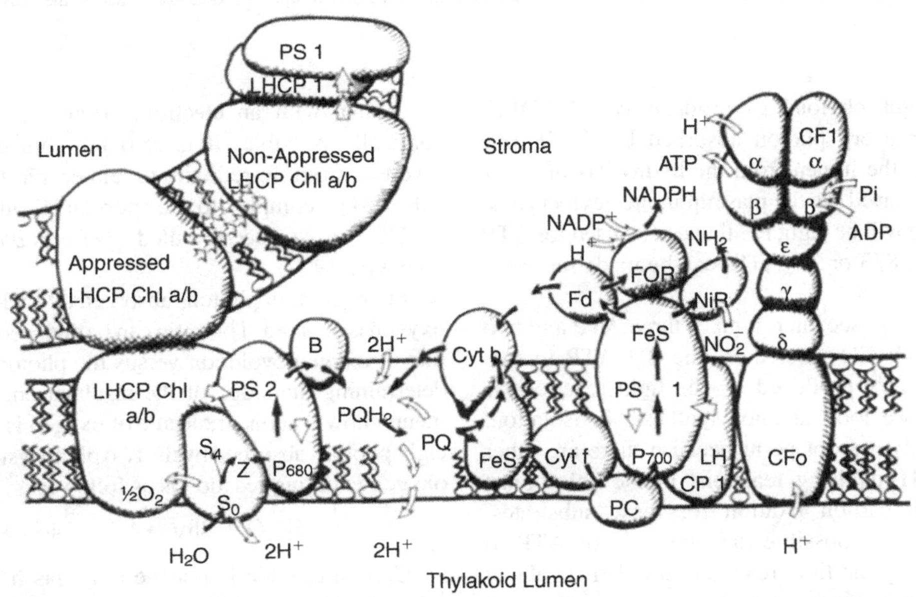

FIGURE 13.12 Diagram of the thylakoid membrane showing the topology of the complexes involved in the light reactions of photosynthesis. *From Wellburn, A.R., 1987. Plastids. Int. Rev. Cytol. Suppl. 17, 149–210.*

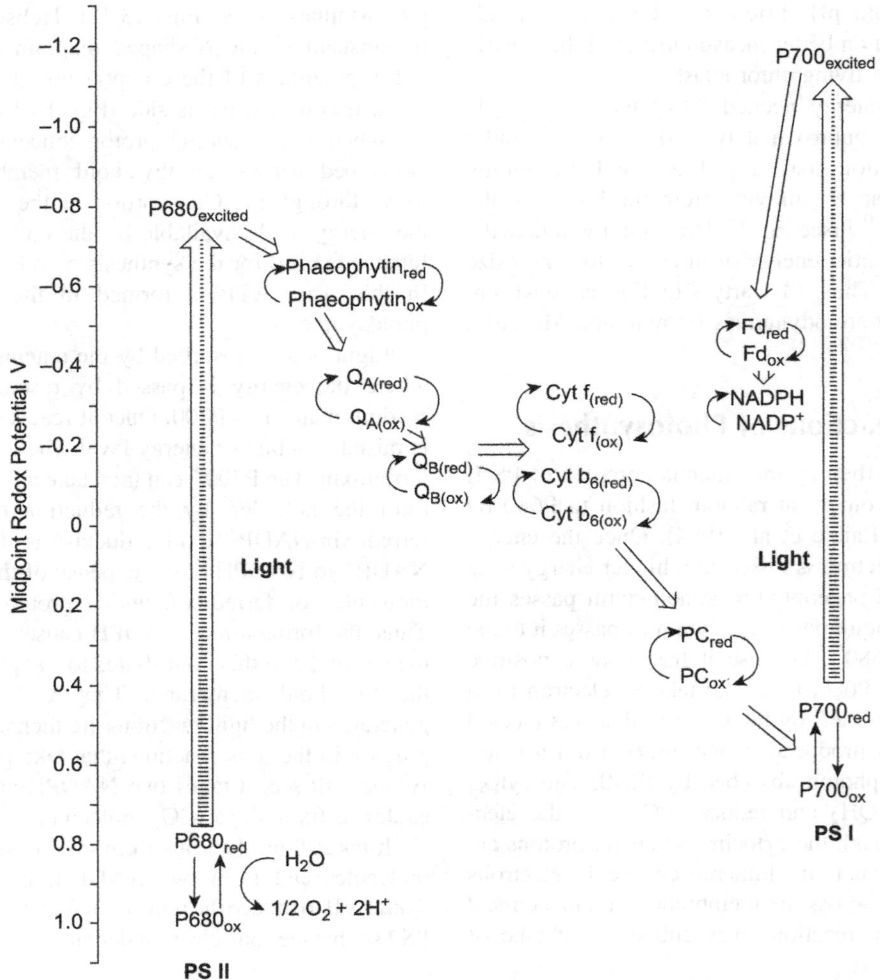

FIGURE 13.13 Diagram of the light reactions of photosynthesis showing the estimated midpoint redox potentials of the various electron transport components.

Thus, it takes eight photons to produce two NADPH. Assuming that for every photon absorbed by PS II, one H^+ is released into the lumen from the hydrolysis of water and one H^+ is transferred via the plastoquinone–cytochrome pathway, then there will be eight H^+ that can be used for ATP synthesis, and thus, 8/3 or 2.7 ATPs can be made for every eight photons.

Because, as we will see later, it takes three ATP and two NADPH to fix each CO_2 molecule, the 2.7 ATP to two NADPH stoichiometry produced by the light reactions is not quite right if we look at photosynthesis in isolation. Indeed, this is really a moot point because there are other ATP- and NADPH-requiring reactions in the chloroplast that compete with carbon fixation for these substrates. Nevertheless, it is also possible that the ratio of ATP to NADPH produced by the light reactions is either regulated or variable. A greater amount of ATP could be synthesized if occasionally two H^+ are transferred from the stroma to the lumen with an electron. After all, this is thermodynamically possible. It is also possible that electrons are diverted from ferredoxin to either plastoquinone or the cytochrome complex to transport an H^+ instead of reducing $NADP^+$ in a process called *cyclic photophosphorylation* (Havaux, 1992).

For every four photons absorbed by PS II, one molecule of oxygen is evolved. This ratio can be deduced by measuring the rate of oxygen evolution versus the photon fluence rate and determining the slope of the resultant curve. Many experiments show that one molecule of oxygen is evolved for every eight photons absorbed by the two photosystems together. The oxygen-evolving reaction is as follows:

$$2H_2O + 8h\nu \rightarrow 4H^+ + 4e^- + O_2$$

Carbon dioxide is not the only gas fixed by photosynthetic reactions in the biological world. Some bacteria use the products of the light reactions of photosynthesis to fix

N_2 (Gest and Kamen, 1949; Kamen and Gest, 1949; Kamen, 1955; Gest, 1994, 1999). Interestingly, nitrogenase, the enzyme that fixes N_2, produces H_2 in the absence of N_2. Imagine running your car on bacterial waste products by combining two volumes of the waste product of H_2-producing bacteria with one volume of the waste product of O_2-producing bacteria to form two volumes of H_2O and 4.75×10^{-19} J of heat per liquid water molecule formed.

Metabolic engineers are working on improving photosynthesis by transforming plants with new versions of genes or by overexpressing the genes that are normally present (Peterhänsel et al., 2008; Lin et al., 2014a,b; Hanson et al., 2016; Occhialini et al., 2016). On the other hand, nature is also the mother of invention, and photosynthesis has served as a model for engineers interested in converting solar energy into chemical energy. They have designed n-type semiconductors made out of titanium dioxide that use the energy of visible light to split water into oxygen and hydrogen with an efficiency of approximately 10% (Khan et al., 2002). The hydrogen can be recombined with the oxygen to generate heat and run an engine in which water is the only "waste product."

13.6 PHYSIOLOGICAL, BIOCHEMICAL, AND STRUCTURAL ADAPTATIONS OF THE LIGHT REACTIONS

The organization of the light-harvesting pigments undergoes dynamic adaptations in response to changes in the light quality to balance the energy distribution between PS I and PS II. These changes, which are referred to as *State I/State II transitions*, take place when leaves go between bright light and the limiting light conditions that occur when leaves shade each other (Bonaventura and Myers, 1969; Murata, 1969; Glazer and Melis, 1987; Allen, 2003; Chuartzman et al., 2008; Tikkanen and Aro, 2011; Rochaix, 2013, 2014; Wientjes et al., 2013; Crepin and Caffarri, 2015; Grabsztunowicz et al., 2017). Such shade contains a greater proportion of far-red light that is preferentially absorbed by PS I (P700) than does direct sunlight. Shady light has less blue and red because these colors have been absorbed by the chlorophyll in the leaves above it.

State I is the situation when a photosynthetic organism is exposed to the far-red light preferentially absorbed by PS I. A few minutes after being placed in this condition, a lateral movement of LHCs takes place that allows more of the absorbed light energy to be directed to PS II, so that PS I and PS II will absorb equal quanta to maximize the current flow in the series circuit between PS II and PS I and thus maximize photosynthetic efficiency.

State II is the condition that develops when a photosynthetic organism is exposed to light preferentially absorbed by PS II. Several minutes after a plant is put in this condition, a lateral movement of LHCs takes place that allows more of the absorbed light energy to be directed to PS I so that PS I and PS II will absorb equal quanta in order to maximize photosynthetic efficiency.

The lateral movement of the mobile LHCs between the stacked regions that are rich in PS II and the unstacked regions that are rich in PS I is the physical basis for the ability to undergo State I/State II transitions (Staehelin et al., 1982). The ability of the LHCs to pass light energy to the reaction centers by resonance energy transfer decreases with the sixth power of the distance, and consequently the spatial distribution of these complexes with respect to the reaction centers determines which reaction center gets the radiant energy absorbed by the LHCs. Quantitative freeze-fracture microscopy and immunological studies show that 20%–25% of the mobile LHC particles participate in the reversible migration between stacked and unstacked regions (Kyle et al., 1983; D'Paolo et al., 1990).

The position of the mobile LHCs is determined by the phosphorylation state of the complex (Schuster et al., 1986). The portion of the LHC that is exposed to the stroma contains a lot of positively charged amino acids, including lysine and arginine. These positive charges attract the negatively charged lipids and proteins of the thylakoid membranes and cause the thylakoid membranes to stack and form grana. However, when the LHC becomes phosphorylated at the threonine residues near the lysine-/arginine-rich area, this region of the protein becomes negatively charged, and it can no longer support stacking of the thylakoids into grana (Staehelin and Arntzen, 1983; Canaani et al., 1984). Consequently, the mobile LHCs are released from the grana stacks and are free to diffuse within the membrane. When they enter the adjacent unstacked regions, they associate with PS I. The diffusion coefficient of an LHC is approximately 2.5×10^{-16} m^2/s (Bennett, 1991), depending on the viscosity of the membrane (Haworth, 1983). According to Einstein's (1906) random walk equation, ($t = x^2/2D$), it will take approximately 0.2, 20, or 2000 s to diffuse 10, 100, or 1000 nm, respectively. The movement of the LHCs may also be due, in part, to blebbing and fusion events (Chuartzman et al., 2008).

The phosphorylated LHCs are found preferentially in the unstacked regions of thylakoids (Larsson et al., 1983). The protein kinase (LHC II protein kinase) that phosphorylates the mobile LHCs is activated when plastoquinone is reduced (PQ_BH_2) (Bennett, 1991). The State I/State II transition is absent in $cyt_{b6\text{-}f}$ mutants. Thus, it is thought that the LHC II protein kinase may be associated with the $cyt_{b6\text{-}f}$ complex, and that the kinase is directly regulated by the redox state of the $cyt_{b6\text{-}f}$ complex, which is regulated by the redox state of PQ (Havaux, 1992). The $cyt_{b6\text{-}f}$ complex may also migrate with the LHC during the state transitions (Vallon et al., 1991). The level of phosphorylation of the LHC depends on the relative activity of the protein kinase and a protein phosphatase (Sun et al., 1989).

In full sun conditions, when the activity of PS II is high relative to the activity of PS I, there will be a lot of PQH_2 and reduced $cyt_{b6\text{-}f}$, and the kinase will be active. Consequently, the LHC will become phosphorylated and negatively charged and it will migrate to the unstacked regions where it will be near PS I, and it will thus transfer its energy to PS I.

In the shade, PS I will be preferentially activated and the pool of PQH_2 will be relatively small, and thus the kinase will not be activated. Under this condition, the LHC will be dephosphorylated and will thus be positively charged. It will not be able to leave the stacked areas and will thus preferentially pass on its energy to PS II. Here is a case where the real redox state of components, in contrast to the midpoint potential, is important in understanding the regulation of photosynthesis.

The description/explanation of the State I/State II transitions is a beautiful example where the combination of structural, biochemical, biophysical, and physiological experiments have been used hand in hand and in just the right proportion to give a clear understanding of this very important and now extremely interesting biological process. It is not known if the differential stacking that results from the movement of the mobile LHCs explains the differences in chloroplast structure often seen in sun-versus shade-tolerant plants.

13.7 FIXATION OF CARBON

Adolf von Baeyer (1870) proposed that the carbohydrate formed by plants resulted from the direct light-mediated formation of formaldehyde (HCHO), followed by the condensation of six formaldehydes to make sugar. Other chemists and physiologists proffered additional one-step theories (Spoehr, 1926). However, we now know that carbohydrate formation is the result of a cyclic process that takes place in the stroma. The light-mediated synthesis of carbohydrate involves the addition of a one-carbon carboxyl group to a five-carbon receptor molecule. Because the end product of photosynthesis is a carbohydrate, then the carboxylate group must be reduced from the oxidized state (COO^-) to the state of carbohydrate (CH_2O). The reduction requires two NADPH molecules. In theory, one would be used to reduce COO^- to CO and H_2O and the other would be used to reduce the CO to CH_2O. In practice, many steps and intermediates are involved (Anderson and Beardall, 1991).

Along with Samuel Ruben, Martin Kamen developed a method for producing ^{14}C with the cyclotron in Ernest O. Lawrence's laboratory and began to work on finding the first products of photosynthesis (Ruben and Kamen, 1940b, 1941; Ruben, 1943; Gest, 2004). However, because of Kamen's liberal leanings, he was dismissed from the laboratory in 1944 (Davis, 1968; Kamen, 1985). Lawrence asked Melvin Calvin, who apparently had the "correct" political views, to continue the project. Calvin et al. combined the investigative power of ^{14}C with the newly developed technique of paper chromatography (Consden et al., 1944) to identify the first products of photosynthesis (Benson and Calvin, 1947; Calvin and Benson, 1948; Bassham and Calvin, 1957; Calvin, 1961; Calvin and Bassham, 1962; Bassham, 2003). Richard Goldsby (2017) described Calvin like so: "Calvin was a polymath. He was a biologist, he was a chemist, he was a physical chemist, he was something of a physicist, and he wasn't bad as a mathematician, and, of course, he worked on photosynthesis, which dropped you right in the midst of biology. Of course, Melvin's approach to biology was very quantitative, was very chemical, and he was one of the creators of the field of using radioisotopes to trace biological reactions, and so when you were in Melvin's lab, you were right at the interface in so many ways of what was most exciting in biological science, and he was a very exciting, kind of mercurial guy."

Calvin chose to work on *Chlorella*, a single-celled alga that could be easily cultivated (Bold, 1942; Pringsheim, 1972) and rapidly labeled with $^{14}CO_2$. The reaction could also be stopped rapidly by dropping the cells in boiling methanol, which both killed the cells and extracted the products of photosynthesis. The extracts were loaded onto a paper chromatogram, and each compound moved in a given solvent in a manner that was related to the affinity of the molecule for the polar water absorbed to the cellulose in the paper compared to the nonpolar solvent. In this way, the products could be compared to authentic standards. It could also be determined if the products were phosphorylated by treating the extract chemically or enzymatically to remove the phosphate, and seeing how they move on the chromatogram. In this way, Calvin et al. discovered that within 5 s, 3-phosphoglyceric acid was labeled and suggested that there was a two-carbon acceptor for CO_2 (Calvin, 1952). However, as soon as they increased the temporal resolution of their assay, they found that the first product labeled was a six-carbon molecule formed from the joining of CO_2 to a five-carbon molecule known as ribulose bisphosphate (RuBP; Krebs and Kornberg, 1957). A point of fact is that Calvin never saw the radioactive six-carbon doubly phosphorylated molecule on the paper chromatograms. The existence of this essential molecule was only inferred (Hans Kornberg, personal communication).

By following the position of the ^{14}C in the various products over time and analyzing the energetics of each step, Calvin et al. came up with the complete pathway of the photosynthetic carbon reduction cycle, which is known universally as the *Calvin cycle* (Fig. 13.14). According to the Calvin cycle, RuBP is carboxylated to form an unstable six-carbon intermediate that splits to form two molecules of 3-phosphoglyceric acid. Each 3-phosphoglyceric acid

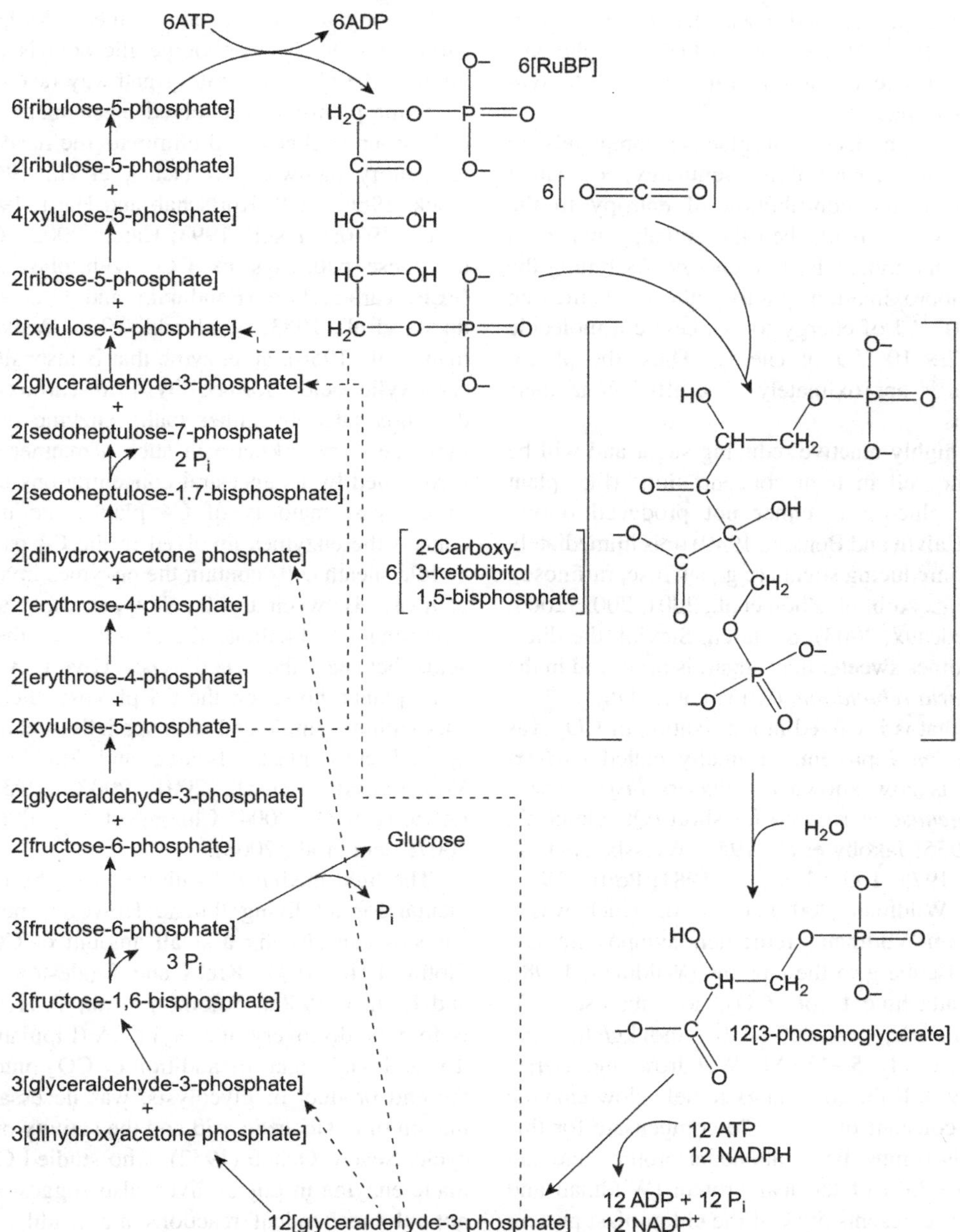

FIGURE 13.14 The Calvin cycle showing the intermediates involved in carbon fixation.

molecule is phosphorylated at the expense of ATP to form 1,3-phosphoglyceric acid. Each of these molecules is reduced by a molecule of NADPH to form glyceraldehyde-3-P and P_i. Some of the glyceraldehyde-3-P continues on to form starch or a translocatable sugar, while others are involved in the regeneration of RuBP. The regeneration of RuBP involves the hydrolysis of one ATP for every CO_2 fixed. Thus, two NADPHs and three ATPs are necessary for the fixation of a single CO_2 and the regeneration of the acceptor.

The chemical energy transformed from radiant energy during the light reactions is used for the fixation of CO_2 by rubisco in the stroma. The fixation of CO_2 into a sugar requires three ATP and two NADPH according to the following reaction scheme:

$$2H^+ + CO_2 + 3ATP + 3H_2O + 2NADPH \rightarrow CH_2O + 3P_i + 3ADP + 2NADP^+ + H_2O$$

Notice that the fixation of CO_2 also results in the alkalinization of the stroma and thus also contributes to the pH difference across the thylakoid membrane. The production of three ATP and two NADPH required eight photons.

Assuming that each photon had a wavelength of 680 nm, it would take 2.3×10^{-18} J to fix a single CO_2 and 48 photons, the energy of which is equivalent to 1.4×10^{-17} J, to synthesize a six-carbon sugar.

Burning a single molecule of glucose completely to CO_2 and H_2O yields a release of heat (enthalpy) equivalent to 4.7×10^{-18} J. If the contribution of entropy to the synthesis of glucose is small, then the enthalpy will be a good indicator of the molecular free energy. Assuming the validity of this approximation, photosynthesis requires an input of 1.4×10^{-17} J of energy to synthesize a molecule that contains 4.7×10^{-18} J of energy. Thus, the photosynthetic process is approximately 34% efficient in thermodynamic terms.

Glucose is a highly reactive reducing sugar and will be dangerous for the cell in high concentrations (like plant diabetes). Thus, glucose is either not produced during photosynthesis (Calvin and Benson, 1949) or is immediately converted into nonreducing sugars (e.g., sucrose, raffinose), sugar alcohols (e.g., sorbitol; Zhou et al., 2001, 2002, 2003; Zhou and Quebedeaux, 2003), or starch. Steviol, the diterpene that is 300 times sweeter than sugar, is produced in the chloroplast of *Stevia rebaudiana* (Kim et al., 1996).

The enzyme that is involved in the fixation of CO_2 was isolated as *Fraction I* protein, originally called *carboxydismutase*, and is now known as *ribulose bisphosphate carboxylase/oxygenase* or *rubisco* for short (Quayle et al., 1954; Racker, 1955; Jakoby et al., 1956; Weissbach et al., 1956; Wildman, 1979, 1998; Lorimer, 1981; Portis, 1992; Spreitzer, 1993; Wildman, 2002). Rubisco, which was a name given at Sam Wildman's retirement symposium as a joke by David Eisenberg to the enzyme (Wildman, 1998), catalyzes the attachment of CO_2 to ribulose 1,5-bisphosphate with a V_{max} of 200×10^{-3} mol kg/chl s and a K_m of approximately 5−15 μM (Woodrow and Berry, 1988; Nobel, 1991). Rubisco is an extremely slow enzyme with an off-rate constant of 3 s^{-1}. To compensate for this inefficiency, rubisco must be an abundant protein, and can account for 25%−75% of the leaf protein (Wildman and Bonner, 1947). It represents 50% of the chloroplast protein and is the most abundant protein on Earth. In fact, the total protein concentration of the stroma is 225 mg/mL (22.5% protein), which makes the stroma an extremely viscous gel! The protein concentration of the lumen is approximately 225 mg/mL (Kieselbach et al., 1998).

Perhaps one reason for the inefficiency of rubisco is that it reached its present state of evolution before plants poured O_2 into the environment. However, given the present atmospheric concentrations of oxygen and carbon dioxide, oxygen is a very effective competitor for the CO_2 binding site (Forrester et al., 1966; Tregunna et al., 1966; Andrews et al., 1971; Bowes et al., 1971; Ogren and Bowes, 1971; Bowes and Ogren, 1972). When rubisco oxidizes RuBP instead of carboxylating it, 3-phosphoglyceric acid and 2-phosphoglycollic acid are formed (Walker, 1992). The carbon lost to 2-phosphoglycollic acid is partially recaptured in the photorespiratory pathway (see Chapter 5).

Some plants have evolved mechanisms to concentrate CO_2 around rubisco and eliminate the need for the photorespiratory pathway (Kortschak et al., 1965; Hatch and Slack, 1966, 1998; Kortschak and Hartt, 1966; Burris and Black, 1976; Nickell, 1993; Hatch, 2002). Organisms that use these pathways fix CO_2 with phosphoenolpyruvate (PEP) carboxylase (Bandurski and Greiner, 1953; Bandurski et al., 1953; Bandurski, 1955; Walker, 1956; Saltman et al., 1956), an enzyme that is insensitive to O_2. The carboxylic acids formed by PEP carboxylase are then decarboxylated by either malic enzyme or phosphoenolpyruvate carboxykinase in such a manner that rubisco is surrounded by its ancestral concentrations of CO_2 and O_2. In the vast majority of C4 plants, the mesophyll cells contain the enzymes involved in the C4 pathway, and the bundle sheath cells contain the enzymes involved in the C3 pathway. Between the 2 cell types there are many plasmodesmata to facilitate the shuttling of the 3-C and 4-C acids between the 2 cell types (Evert et al., 1977). In some plants, however, the C4 photosynthetic pathway occurs within a single cell with spatially separated granal and agranal chloroplasts (Freitag and Stichler, 2000, 2002; Voznesenskaya et al., 2001, 2002, 2003, 2004, 2005; Edwards et al., 2004; Chuong et al., 2006; Boyd et al., 2007; Lara et al., 2008).

The light-mediated fixation of CO_2 by plants is fundamental for all living things. However, heterotrophic organisms can also fix a small amount of CO_2 (Evans and Slotin, 1940, 1941; Krebs and Eggleston, 1940a; Ruben and Kamen, 1940a; Solomon et al., 1941; Ochoa, 1980). Before the discovery of acetyl CoA (Lipmann et al., 1947), it was thought that the addition of CO_2 onto pyruvic acid, the end product of glycolysis, was necessary for the formation of oxaloacetic acid and the activity of the citric acid cycle. Severo Ochoa (1952), who studied CO_2 fixation by malic enzyme in pigeon liver, also suggested that "most if not all biochemical reactions are readily reversible and indicate that photosynthetic fixation of carbon dioxide may operate along the same basic patterns established for its fixation in animal tissues."

The fixation of CO_2 by single-celled algae is being used as an environmentally friendly way of removing CO_2 from the exhaust gas produced by industry and preventing it from entering the atmosphere (Ghosh and Kiran, 2017).

13.8 REDUCTION OF NITRATE AND THE ACTIVATION OF SULFATE

The readily available energy in the chloroplast makes it an ideal site for the reduction of other molecules besides CO_2 (Anderson and Beardall, 1991). Plants take up sulfate from the soil, and this relatively inert form of sulfur must be activated by the cell before it can be metabolized further (Schmidt and Jäger, 1992). The sulfate is activated by ATP sulfurylases in the chloroplast (Leustek et al., 1994) to form adenosine phosphosulfate (APS) and pyrophosphate. Once activated to form APS, the sulfur can be transferred to many different molecules. Methionine synthesis occurs in the chloroplasts of plants (Ravanel et al., 2004).

The nitrate taken up by plants must be reduced before it can be metabolized. The reduction of nitrate to ammonia takes place in a two-step process: the reduction of nitrate to nitrite and subsequently the reduction of nitrite to ammonia. While the first step takes place in the cytosol (Fedorova et al., 1994), the second reaction takes place in plastids (Guerrero et al., 1981; Solomonson and Barber, 1990; Ullich et al., 1990). To reduce nitrite to ammonia, six electrons are needed. These six electrons come from six ferredoxin molecules reduced in the light reactions. Thus, nitrogen is reduced at the expense of reducing $NADP^+$ that could be used for fixing carbon.

13.9 CHLOROPLAST MOVEMENTS AND PHOTOSYNTHESIS

Plants must balance the harmful effects of light that result in photodamage with the beneficial effects of light that result in photosynthesis (Long et al., 1994; Kasahara et al., 2002; Wada, 2005; Davis and Hangarter, 2012). Physiological adaptations to these needs include changes in the position of chloroplasts and leaf movements. Often the thick cell walls of epidermal cells act like lenses to focus the light (Gausman et al., 1974; Dennison and Vogelmann, 1989; Vogelmann, 1993), and under limiting light conditions, the chloroplasts move to the position of focused light (Stiles, 1925). Haberlandt (1914) also noticed that "chloroplasts adhere …to those walls which abut upon airspaces; by this means they evidently obtain the most favourable conditions for the absorption of carbondioxide."

The chloroplasts of many plants do not sit passively in the cell but constantly shift their position to optimize photosynthesis (Wada, 2013). Nowhere is this more spectacular than in the cells of *Mougeotia* (see Fig. 10.10 in Chapter 10). The ribbonlike chloroplast in this cell turns broadside to face light of moderate intensity. However, if the intensity of the light is too high, the chloroplast turns so that its thin edge faces the light, presumably preventing damage through photobleaching (Haupt, 1982, 1983a,b).

This photophysiological response is not mediated by the photosynthetic pigments themselves, but by a red light–absorbing pigment called *phytochrome* (or perhaps *neochrome*, which is a phytochrome–phototropin chimera; Suetsuga et al., 2005). Thus red light, absorbed through the photoreceptor pigment phytochrome, is most active in inducing the moderate fluence-turning response. Microbeam irradiation studies indicate that phytochrome is localized on or near the plasma membrane, not on the chloroplast (Bock and Haupt, 1961). In fact, the chloroplast-turning response shows an action dichroism (Haupt, 1960, 1968). That is, the effectiveness of red light depends on the azimuth of polarization of the red light, and only the cells that have their long axis normal to the electrical vector of the light show a response (Fig. 13.15). This action dichroism indicates that the photoreceptor pigments are oriented and therefore must be associated with a stable structure such as the plasma membrane. The chromophore

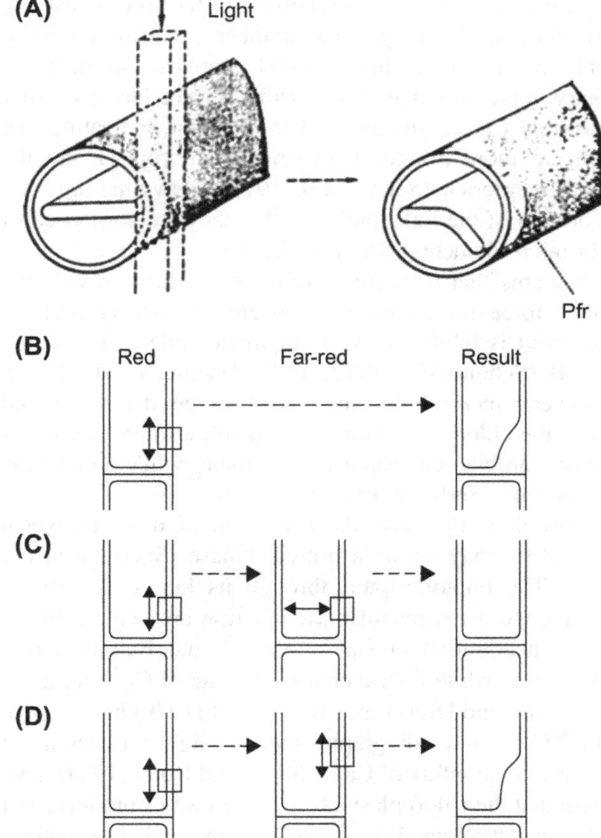

FIGURE 13.15 Microbeam irradiation of a *Mougeotia* cell. Overall view of experiment (A), and individual irradiation protocols (B–D). (B) The cell was irradiated with a red microbeam, the light of which vibrated parallel to the long axis of the cell. (C) The cell was first irradiated with a red microbeam vibrating parallel to the long axis of the cell and subsequently irradiated with a far-red microbeam vibrating perpendicular to the long axis of the cell. (dD The cell was first irradiated with a red microbeam vibrating parallel to the long axis of the cell and subsequently irradiated with a far-red microbeam vibrating parallel to the long axis of the cell. The chloroplast moved away from the microbeam in protocols (B and D), but not (C). *From Haupt, W., 1983a. Movement of chloroplasts under the control of light. Prog. Phycol. Res. 2, 227–281; Haupt, W., 1983b. The perception of light direction and orientation responses in chloroplasts. Soc. Exp. Biol. Symp. 36, 423–442.*

of the red light–absorbing form of phytochrome is oriented parallel to the cell surface. Interestingly, on the absorption of red light, the red light–absorbing form of phytochrome, P_r, is converted into the far–red light–absorbing form, P_{fr}, and the chromophore of P_{fr} is oriented normal to the surface of the cell (Haupt, 1970).

Beecause the flanks of the cell normal to the light rays will absorb more unpolarized light than the flanks parallel with the light rays, there will be more P_{fr} created on the membranes normal to the light rays compared with those parallel with the light rays. Under moderate light conditions, the chloroplast edges tend to move away from the highest concentration of P_{fr} and thus the face of the chloroplast ends up normal to the incident light.

How does P_{fr} cause the repulsion of the chloroplast? P_{fr} may increase the length and number of actin microfilaments. In fact, the ability of the chloroplast to remain stationary in the presence of a centrifugal force increases after exposure of a dark-treated cell to red light, indicating that there are more bridges between the chloroplast and the plasma membrane (Schönbohm, 1972). Heavy meromyosin decoration of the links indicates that they are composed of actin microfilaments (Klein et al., 1980).

It seems that the actomyosin system does provide the motive force for chloroplast movement since chloroplast movement is inhibited by N-ethylmaleimide and cytochalasin B (Schönbohm, 1972, 1975; Wagner et al., 1972). However, microtubules may need to be depolymerized before the chloroplast turns since taxol, a microtubule stabilizer, inhibits chloroplast movement while colchicine speeds it up (Serlin and Ferrell, 1989).

How does P_{fr} cause the activation of the actomyosin system? P_{fr} may act as a protein kinase (Suetsuga et al., 2005). The photoreceptor, through its kinase activity or through another mechanism, may cause an increase in the local concentration of Ca^{2+}. Wagner and Bellini (1976) showed that white light increases the rate of Ca^{2+} influx in *Mougeotia*, and Dreyer and Weisenseel (1979) have shown with $^{45}Ca^{2+}$ autoradiography that red light induces a net increase in the influx of Ca^{2+}. Serlin and Roux (1984) have shown that the chloroplast edges move away from localized high concentrations of Ca^{2+}, which are applied by putting the bivalent ion-selective ionophore A23187 on two microneedles on opposite sides of the cell. Ca^{2+} may induce its effect by binding to calmodulin, and the calcium-calmodulin complex may then bind to a protein kinase, like myosin light chain kinase, and activate the actomyosin system (Jacobshagen et al., 1986; Roberts, 1989). Calmodulin antagonists inhibit red light–induced chloroplast movement (Wagner et al., 1984; Serlin and Roux, 1984).

There is a splendid array of organisms that show chloroplast movements (Haupt and Scheuerlein, 1990; Haupt, 1999; Wada et al., 2003). The chloroplast-turning response in the single-celled alga *Mesotaenium* is similar to that in *Mougeotia* (Kraml et al., 1988). The streaming chloroplasts of *Vaucheria* aggregate within a microbeam of light (Fischer-Arnold, 1963; Zurzycki and Lelatko, 1969; Blatt and Briggs, 1980; Blatt et al., 1981; Blatt, 1983, 1987). Light-induced chloroplast movements are also seen in the leaves of mosses and *Selaginella* (Haupt, 1982). In these cases, mathematical modeling shows that under various light intensities, the chloroplasts are where they are expected to be to maximize photosynthesis (Haupt, 1982).

As we move up the phylogenetic tree, we also find that ellipsoidal chloroplasts in the protonema of *Adiantum* also move toward moderate light and avoid high light (Yatsuhashi et al., 1985, 1987; Wada and Kadota, 1989; Kadota et al., 1989) as well as mechanical stimulation (Sato et al., 1999, 2001a,b). Likewise, the chloroplasts in the angiosperm species of *Vallisneria*, *Lemna*, *Elodea* and *Arabidopsis* show a light-dependent change in position (Seitz, 1964, 1967, 1971; Ishigami and Nagai, 1980; Takagi and Nagai, 1985, 1988; Malec, 1994; Aggarwal et al., 2013). Under low-light conditions, the chloroplasts are immobilized on the periclinal wall, and they expose the greatest amount of surface area to the light. High-intensity light causes them to move to the anticlinal walls, where they participate in cytoplasmic streaming and minimize their exposure to the light. Phototropins are the photoreceptor for blue light–induced chloroplast movements (Kagawa et al., 2001; Sakai et al., 2001; Kasahara et al., 2002; Briggs, 2005), and the phosphatidylinositol and Ca^{2+} signaling systems are involved in the responses (Aggarwal et al., 2013). While much work has gone into studying the mechanisms involved in these intracellular phototactic movements, it remains a fascinating field of research for anyone interested in photobiology, cell signaling, and cell motility (Wada, 2005, 2013, 2016; Higa and Wada, 2015).

13.10 GENETIC SYSTEM OF PLASTIDS

Prokaryotes and eukaryotes both contain DNA; however, the DNA of eukaryotes shows higher levels of organization than that of prokaryotes. This difference served as the initial basis of the distinction between prokaryotes and eukaryotes (Dougherty, 1957). Hans Ris (1961) discovered DNA in the stroma of plastids, and this DNA looked like the naked DNA found in prokaryotes (Ris and Plaut, 1962), although in all cases, the DNA is associated with protein. Chloroplast DNA is typically circular, has a molecular mass of $85–95 \times 10^6$ Da, and is composed of $12–18 \times 10^5$ base pairs that make up approximately 50–200 genes (Ohyama et al., 1986; Shinozaki et al., 1986; Bogorad, 1998; Sugiura, 2003). The complete sequences of the chloroplast genomes from *Arabidopsis thaliana*, *Astasia longa*, *Atropa belladonna*, *Chaetosphaeridium globosum*, *Chlorella vulgaris*, *Cyanidium caldarium*,

Cyanophora paradoxa, *Epifagus virginiana*, *Euglena gracilis*, *Guillardia theta*, *Lotus japonicus*, *Marchantia polymorpha*, *Medicago truncatula*, *Mesostigma viride*, *Nephroselmis olivacea*, *Nicotiana tabacum*, *Odontella sinensis*, *Oenothera elata*, *Oryza sativa*, *Pinus thunbergii*, *Porphyra purpurea*, *Psilotum nudum*, *Spinacia oleracea*, *Toxoplasma gondii*, *Triticum aestivum*, and *Zea mays* are known. The DNA may be attached to the inner membrane of the envelope (Whitfield and Bottomley, 1983; Zurawski and Clegg, 1987; Sato et al., 1993). The chloroplast contains its own DNA gyrase (Wall et al., 2004), DNA polymerase (Tewari and Wildman, 1967; Spencer and Whitfield, 1969; Heinhorst et al., 1990; Moriyama and Sato, 2014), RNA polymerase (Polya and Jagendorf, 1971), DNA repair mechanisms (Cerutti and Jagendorf, 1993; Cerutti et al., 1992, 1993, 1995), and histone-like proteins (Kobayashi et al., 2002). The replication origins of the chloroplast DNA can be visualized with an electron microscope (Wu, 1998). Chloroplast genes contain promotor sequences (Kung, 1998) and they can be edited (Sun et al., 2013, 2015). There has been and continues to be a transfer of genes from the plastids to the nucleus (Huang et al., 2003; Timmis et al., 2004).

The chloroplast genome can be transformed with a gene that encodes a viral antigen that can be used to produce vaccines for animals and humans using solar energy. In this way, plants act as bioreactors to produce vaccines and other therapeutic proteins that can improve human health (Daniell et al., 2001, 2004, 2005; Walmsley and Arntzen, 2003; Rybicki, 2010; Takeyama et al., 2015; Stoger et al., 2014; Kashima et al., 2016; Laere et al., 2016; Marsian et al., 2017).

The plastid DNA encodes genes that are segregated in a non-Mendelian manner. The non-Mendelian inheritance is typically known as *maternal inheritance* (Anderson, 1923; Darlington, 1944). There are a variety of physical mechanisms that lead to maternal inheritance (Vaughn et al., 1980; Vaughn, 1981, 1985; Hagemann and Schröder, 1989; Mogensen, 1996). For example, in some plants, maternal inheritance of plastids results from the exclusion of plastids from the generative cell, which divides to form the two sperm cells. In these plants, the exclusive presence of plastids in the vegetative cell and the lack of plastids in the generative cell results from the exclusion of plastids from the portion of the microspore that gives rise to the generative cell. In cases where the plastids enter the generative cell, maternal inheritance can still occur if the plastids or the DNA within them is degraded.

When maternal inheritance depends on the degradation of DNA, the organellar DNA in the female gamete remains intact throughout syngamy, whereas the paternal organellar DNA is degraded either before or after fertilization. In isogamous plants, the male organellar DNA is degraded after syngamy (Kuroiwa et al., 1982; Tsubo and Matsuda, 1984; Coleman and Maguire, 1983; Kuroiwa, 1991), whereas in anisogamous or oogamous plants, the male organellar DNA is degraded before syngamy (Sun et al., 1988; Kuroiwa and Hori, 1986; Kuroiwa et al., 1988; Miyamura et al., 1987; Corriveau and Coleman, 1988).

More recently, it has been found in conifers and some species of angiosperms that paternal inheritance and biparental inheritance of chloroplast DNA occur (Szmidt et al., 1987; Chat et al., 1999; Hansen et al., 2007; Hu et al., 2008). Interestingly, in *Wisteria*, the mitochondrial genome can be inherited from the maternal parent, while the plastid genome is inherited from the paternal parent, indicating that the mitochondrial and plastid genomes are independently segregated or degraded during the pollination/fertilization process (Trusty et al., 2007).

Both the stroma and the thylakoids contain 70S ribosomes (Chua et al., 1976; Staehelin, 1986; Jagendorf and Michaels, 1990). In the plastids, the ribosomes are directed to the thylakoids by signal peptides (Chua et al., 1973) and by a protein homolog of the signal-recognition particle (Franklin and Hoffman, 1993).

The plastid genome does not code for all the proteins necessary for the plastid, and thus nuclear-encoded proteins are necessary for chloroplast function. Given that plant nuclei can assimilate exogenous DNA (Hemleben et al., 1975; Blascheck, 1979; Leber and Hemleben, 1979), the nuclear genes that encode the chloroplast proteins may have been transferred from the original endosymbiont that gave rise to the chloroplast during repeated exposures (Weeden, 1981; see Chapter 15). The nuclear and chloroplast genomes are in a dynamic equilibrium, in terms of evolutionary time scales, and genes continue to move from the chloroplast genome to the nuclear genome (Bubunenko et al., 1994; Kalanon and McFadden, 2008). The coordination of protein, cofactor, pigment, and enzyme production between the chloroplast and the rest of the cell occurs, and it seems likely that some kind of communication takes place between the chloroplast and the nucleus (Ellis, 1977). For example, the nuclear-encoded chlorophyll *a*/*b* binding protein is unstable in the absence of chlorophyll (Apel, 1979; Bennett, 1981). In the presence of chlorophyll, the protein concentration increases, without any changes in its mRNA concentration, indicating that there is coordination between the translation and/or stabilization of this protein and chlorophyll synthesis (Klein and Mullet, 1986). The chloroplast may have given rise to the apicoplast in *Toxoplasma* (Köhler et al., 1997; McFadden, 2000, 2011, 2014).

Given that many of the proteins that function in the chloroplast are encoded by the nucleus, there must be a mechanism to import proteins translated on cytosolic ribosomes into the chloroplast. Blair and Ellis (1973) proposed that there must be a protein carrier on the envelope. Chua and Schmidt (1979) proposed that carrier is likely to be found in contact sites where the inner membrane and outer membrane are in contact. The density of contact sites is greatest in developing plastids and decreases in mature chloroplasts (Pfisterer et al., 1982).

The imported polypeptides seemed to require a transit peptide on the N-terminus to recognize the chloroplast and to be imported into it (Dobberstein et al., 1977). When the transit peptide was cleaved from the polypeptide, the polypeptide could not enter the chloroplast (Mishkind et al., 1985). Moreover, if the transit peptide was added to polypeptides that usually did not enter the chloroplast, the latter then did (Lubben et al., 1988). It does not seem as if the primary structure of the transit peptide, once postulated to be the key to chloroplast recognition, is as important as once thought because as more transit peptides are sequenced, the sequences are more and more divergent from the canonical sequence. It may be that the secondary structure is important (Douce and Joyard, 1990; Flügge, 1990; Willey et al., 1991; Dreses-Werringloer et al., 1991; Cline and Henry, 1996; Bédard and Jarvis, 2005). In fact the targeting sequence does not have to be part of the N-terminus but can be on the C-terminus or in the middle of the protein. Variation around a theme is commonplace in biology. The fact that the cell is compartmentalized and that a specific protein is targeted to one or more compartments is the rule. However, there is variation in the targeting signal of the protein, its phosphorylation status, the receptor and translocator on the targeted plastid, the nature of the plastid, and the intracellular pathway taken to get to the organelle (Nakrieko et al., 2004; Inoue et al., 2010; Kessler, 2012; Shi and Theg, 2013; Kim and Hwang, 2013; Bölter and Soll, 2016). Indeed, some N-linked glycosylated proteins enter the chloroplast through the endomembrane pathway (Chen et al., 2004; Villarejo et al., 2005; Faye and Daniell, 2006; Nanjo et al., 2006; Radhamony and Theg, 2006; Agne and Kessler, 2007).

Given the variations in the signals used to transport proteins to the chloroplast, there may be a variety of protein translocator complexes. After discovering the translocon of the ER, Günter Blobel turned his attention to investigating how nuclear-encoded proteins that where synthesized on cytosolic ribosomes entered the chloroplast. Pain et al. (1988) initially identified a protein that is localized in the contact sites, although this protein turned out to be the phosphate transporter and not the protein transporter (Gray and Row, 1995). A protein translocator complex has been identified through molecular and biochemical studies (e.g., chemical cross-linking to translocated protein, immunoaffinity chromatography using protein A–translocated protein), genetic knockout studies as well as through reconstitution studies in planar lipid bilayers using detergent-solubilized translocator proteins (Schnell et al., 1994; Caliebe and Soll, 1999; May and Soll, 1999; Bédard and Jarvis, 2005; Shi and Theg, 2013; Köhler et al., 2015; Nakai, 2015; Bölter and Soll, 2016; Paila et al., 2016; O'Neil et al., 2017). The translocator complexes on the outer and inner chloroplast membranes are known as the Toc apparatus and Tic apparatus, respectively. The Tic-Toc translocons, with a conductance of 266 pS (Kikuchi et al., 2013) are localized at the contact sites (Schnell and Blobel, 1993; Schnell, 1995; Cline and Henry, 1996; Chen et al., 2000a,b; Lopez-Juez and Pyke, 2005) and are regulated through phosphorylation (Zufferey et al., 2017).

The first step in protein import is binding to the Toc component of the Tic-Toc translocon. This can be detected in vitro in the absence of nucleoside triphosphates and in the presence of cross-linking agents (Perry and Keegstra, 1994; Ma et al., 1996). Both ATP and GTP are required for the import of proteins into the chloroplast, which is reminiscent of the nucleotide requirements for cotranslational and posttranslation polypeptide import into the ER (Kim and Hwang, 2013; see Chapter 4). Similar to post-translational import into the ER, ATP is required for the action of chaperonins on the cytosolic side of the Toc complex that unfold the protein to be translocated. The exit of the protein on the stromal side of the Tic complex requires chaperonins and ATP (Gatenby and Ellis, 1990; Dessauer and Bartlett, 1994; Nielsen et al., 1997; Akita et al., 1997; Kouranov et al., 1998). The chaperonins provide the motive force for protein translocation into the stroma (Su and Li, 2010; Shi and Theg, 2010; Flores-Pérez and Jarvis, 2013). While GTP hydrolysis is required for the translocation of polypeptides across the envelope (Paila et al., 2015) as it is in cotranslational import into the ER, the function of GTP hydrolysis in the chloroplast translocon still remains unknown (Bölter and Soll, 2016). Once a polypeptide enters the stroma, the transit peptide is cleaved off by a stromal processing protease (Keegstra et al., 1989; Bassham et al., 1994; Su et al., 1999; Nishimura et al., 2017). Because the number of genes transferred from the chloroplast to the nucleus has evolved over time, the protein translocation complex itself must have evolved simultaneously to import the proteins necessary for chloroplast function (Kalanon and McFadden, 2008). Research over the past decade has revealed that many polypeptides are involved in the translocation of polypeptides across the chloroplast envelope. Some of the diversity found in the polypeptides that make up the translocon may depend on the species from which the chloroplasts are isolated, some of the diversity may reflect the changing roles of the chloroplast during development and under various environmental conditions, and some of the diversity may be the result of changes in the function of the chloroplast. Like always, we should look for the unity in the diversity to understand the general nature of protein translocation into the chloroplast. We should also look at the diversity in translocon polypeptides to understand the cause of the diversity and the reason why not all chloroplasts are identical. Richardson et al. (2014) are investigating how the components of the translocon are targeted to the chloroplast envelope and assembled.

Polypeptides targeted to the lumen of the thylakoids must cross the envelope and the thylakoid membrane and have two signal sequences (Keegstra et al., 1989). There is more than one protein transport pathway across the thylakoid membranes because a mutant form of a thylakoid membrane protein inhibits the translocation of thylakoid membrane proteins but not the lumenal proteins (Smith and Kohorn, 1994; Bédard and Jarvis, 2005).

13.11 BIOGENESIS OF PLASTIDS

All plastids in a plant are formed from proplastids, and they usually contain the same genome (Randolph, 1922; Sirks, 1938; Knudson, 1940; Kuroiwa, 1991). Proplastids are relatively small, spherical organelles (c. 0.5–1 $\times$ 10^{-6} m in diameter) that are present in meristematic cells. In their earliest stages they are called *eoplasts* (see Fig. 13.16; Whatley, 1977b, 1978). The proplastids contain nearly no internal membranes and only small amounts of DNA, RNA, ribosomes, and soluble proteins. The proplastid or eoplast then goes through an amyloplast stage where it remains spherical, yet accumulates starch. Then it goes through an amoeboid stage where it takes on a number of different amoeboid-like shapes in the cytoplasm (Senn, 1908; Weier, 1938a,b; Esau, 1944). Eldon Newcomb (1967) has suggested that the amoeboid stage is a "feeding stage" in the development of protein-containing plastids. The amoeboid plastids then enter the pregranal phase. The amoeboid stage can persist in chloroplasts formed in fern gametophytes following X-ray irradiation of the spores (Knudson, 1940; Howard and Haigh, 1968).

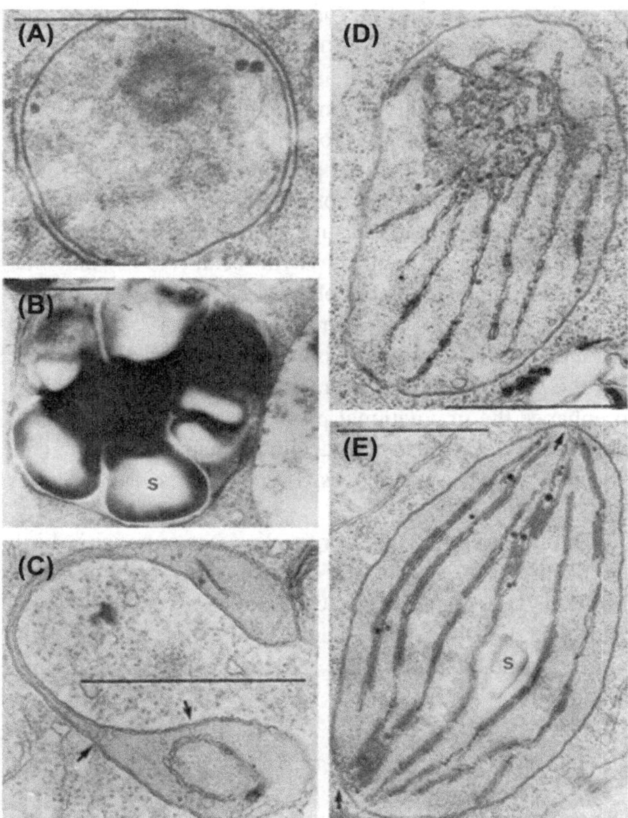

FIGURE 13.16 The development of plastids in light-grown *Phaseolus*: (A) eoplast, (B) amyloplast filled with starch (s), (C) amoeboid plastid, (D) immature chloroplast, and (E) mature chloroplast. Bars, 1 μm. *From Wellburn, A.R., 1982. Bioenergetic and ultrastructural changes associated with chloroplast development. Int. Rev. Cytol. 80, 133–191.*

When a plant develops in the light, the pregranal stage proplastids usually develop into chloroplasts. However, if the plant is kept in the dark, the proplastids develop into etioplasts, which are smaller than chloroplasts. Their largest diameter is ca. 1–2 μm. The protochorophyll in the etioplasts of dark-grown plants is confined to small optically dense 0.7–1.3 μm bodies originally known as 1 μ *centers*, and now known as *prolamellar bodies* (Boardman and Anderson, 1964; Virgin, 1981). The prolamellar bodies undergo extensive structural changes as they convert to stromal lamellae and granal stacks (Gunning, 1965). The crystalline membrane region appears to become more and more disorganized as the membrane protrusions appear to become longer and longer. Within 2 min after the plastid is exposed to light, the membrane becomes perforated and these membranes persist for about 2 h, which corresponds to the lag phase of chlorophyll synthesis (Simpson, 1978). Then more and more membranes flow over each other in the miraculous process of forming grana (Wellburn, 1987; Gunning and Steer, 1996). Phytochrome, Ca^{2+}, and cyclic GMP participate in the signal transduction chains that are involved in the greening response (Bowler et al., 1994; Reiss and Beale, 1995; Shiina et al., 1997).

Carl von Nägeli (1844) and Julius Sachs (1882a,b) observed that chloroplasts divide to give rise to other chloroplasts, and Tsuneyoshi Kuroiwa observed that contractile rings occur at the isthmus in the dividing plastid (Mita et al., 1986; Mita and Kuroiwa, 1988; Kuroiwa, 1991, 1998; Kuroiwa et al., 1998). The division rings in chloroplasts, like the division rings in the bacteria they evolved from, are composed of a tubulin-like protein known as Filamenting temperature-sensitive mutant Z (FtsZ). In bacteria, FtsZ forms a filamentous contractile ring on the cytosolic side of the plasma membrane at the site of division (Okazaki et al., 2010). FtsZ was discovered when a giant bacterial FtsZ was found that did not form a constricting ring around the midpoint (Lutkenhaus et al., 1980; Bi and Lutkenhaus, 1991; Bermudes et al., 1994; Erickson, 1995; Erickson et al., 1996; Faguy and Doolittle, 1998; Lu et al., 2000: Addinall and Holland, 2002).

Osteryoung and Vierling (1995) decided to see if the FtsZ protein is responsible, in part, for chloroplast division.

Both the overexpression or silencing of FtsZ genes result in the inhibition of plastid division and the formation of cells with only one to a few large chloroplasts (Pyke et al., 1994; Osteryoung et al., 1998; Strepp et al., 1998; Kiessling et al., 2000; Stokes et al., 2000; McAndrew et al., 2001; Osteryoung and McAndrew, 2001; Vitha et al., 2001; Jeong et al., 2002; Miyagishima et al., 2003). FtsZ can be visualized as a ring on the stromal side of the chloroplast envelope around the isthmus of the dividing plastids (Mori et al., 2001). The presence and function of FtsZ in the chloroplast are consistent with the endosymbiotic origin of chloroplasts (Osteryoung and Nunnari, 2003; Miyagishima et al., 2004). Xiong et al. (2017) demonstrated that the enlarged chloroplasts, which have a smaller surface-to-volume ratio, are less efficient in photosynthesis than a large population of small chloroplasts.

Other classes of mutants also show a reduction in the number of chloroplasts in cells. These mutants include mutations in proteins that are involved in the positioning of FtsZ on the stromal side of the envelope (e.g., Min, ARC and PDV proteins; Colletti et al., 2000; Kanamura et al., 2000; Dinkins et al., 2001; Itoh et al., 2001; Maple et al., 2002; Reddy et al., 2002; Glynn et al., 2007; Fujiwara et al., 2008; Miyagishima, 2011).

Dynamin is a protein with GTPase activity that is involved in membrane scission during clathrin-mediated endosytosis (see Chapter 9) and cell plate formation (see Chapter 19; Bednarek and Backues, 2010; Taylor, 2011; Fujimoto and Tsutsumi, 2014; Huang et al., 2015). Mutations in proteins with dynamin-like activity also show a reduction in the number of chloroplasts in cells, which indicates that a dynamin also drives the membrane scission necessary for chloroplast division. The dynamin forms a ring in the isthmus of the chloroplast where the FtsZ ring is except that the dynamin ring is on the cytosolic side of the envelope (Pyke and Leech, 1992, 1994; Robertson et al., 1995, 1996; Gao et al., 2003; McFadden and Ralph, 2003; Miyagishima et al., 2003, 2006).

Some of these cytoskeletal-like or rather plastoskeletal proteins may be involved in determining the structure and shape of the chloroplasts (Kiessling et al., 2000; McFadden, 2001). The shape and size of chloroplasts are also regulated by proteins (Maple et al., 2004), including mechanosensitive channels (Haswell and Meyerowitz, 2006).

Thin tubular connections between the plastids and thin tubular extensions of the plastids are called *stromatubules* (Köhler et al., 1997a,b; Bourett et al., 1999; Tirlapur et al., 1999; Shiina et al., 2000) or *stromules* (Köhler and Hanson, 2000; Gray et al., 2001; Pyke and Howells, 2002; Gunning, 2005; Hanson and Sattarzadeh, 2008, 2011). Their shape is determined by microfilaments and microtubules (Kwok and Hanson, 2003), and they can be induced by metabolic inhibitors (Itoh et al., 2010). Proteins may move from plastid to plastid in a plastid reticulum through the stromules (Gray et al., 1999, 2001; Köhler et al., 2000; Kwok and Hanson, 2004a,b; Hanson and Sattarzadeh, 2013; Hanson and Hines, 2018).

The plastids in many flowers and fruits lack chlorophyll but accumulate tremendous amounts of carotenoids. These chromoplasts attract insects and other animals that may help pollinate or disperse fruits and seeds. Chromoplasts are also responsible for the yellows of the fall foliage (Straus, 1953; Frey-Wyssling and Schwegler, 1965; Israel and Steward, 1967).

Some plastids in nonphotosynthetic tissues remain clear and function in the storage of macromolecules, including proteins, lipids, or starch. These plastids are known as *proteinoplasts* (Newcomb, 1967), *elaidoplasts* (Wellburn, 1987) and *amyloplasts*. The starch grains in the amyloplasts can be used for taxonomic characters (Reichert, 1919). Many people believe that amyloplast sedimentation is responsible for the perception of gravity (Kiss et al., 1989; Salisbury, 1993; Sack, 1997; Smith et al., 1997; Blancaflor et al., 1998; Fukaki et al., 1998; Kiss, 2000; Volkmann and Baluška, 2000; Fitzelle and Kiss, 2001; Tasaka et al., 2001; Yoder et al., 2001; Schwuchow et al., 2002; Blancaflor and Masson, 2003; Hou et al., 2003; Perbal and Driss-Ecole, 2003; Morita and Tasaka, 2004; Palmieri and Kiss, 2005; Perrin et al., 2005; Mano et al., 2006; Palmieri et al., 2007; Vitha et al., 2007; Shiva Kumar et al., 2008; Vandenbrink et al., 2014). Moulia and Fournier (2009) and Pouliquen et al. (2017) believe that the position of the amyloplasts is important for gravity perception. However, based on the facts that plant cells that do not contain sedimenting amyloplasts still sense gravity (Wayne et al., 1990) and that starchless mutants in higher plants are almost as sensitive to gravity as the wild-type plants (see Fig. 13.17; Caspar and Pickard, 1989; Weise and Kiss, 1999), others do not think that the amyloplasts act as gravity sensors, but as a ballast to enhance the gravitational pressure sensed by proteins at the plasma membrane–extracellular matrix junction (see Chapter 2; Wayne and Staves, 1996a; Staves, 1997; Staves et al., 1997a,b).

Chloroplasts, like other organelles, can experience birth and death. As a result of UV damage, or in times of stress, they can be degraded and recycled in a process known as chlorophagy (Dong and Chen, 2013; Ishida et al., 2014; Xie et al., 2015; Izumi et al., 2017; Lei, 2017; Williams, 2017a,b).

13.12 SUMMARY

The radiant energy of the sun is converted into chemical energy in the chloroplast that supports the life of every animal on the face of the Earth, including humans. The chloroplast converts light into life, both in terms of producing organic matter from carbon dioxide and water and

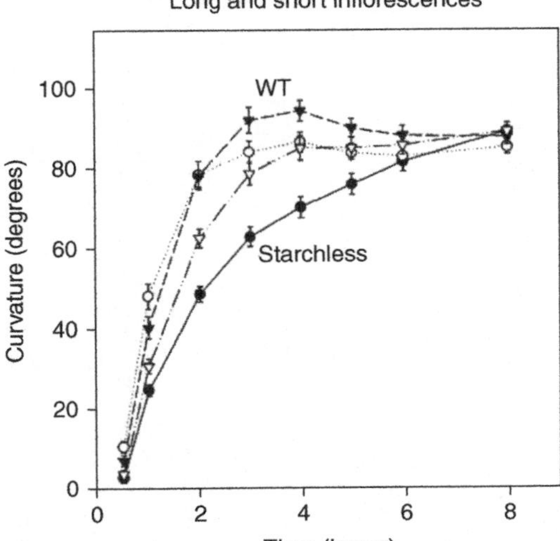

FIGURE 13.17 Gravitropic curvature of the inflorescences of wild-type (*filled triangles*) and two reduced-starch mutants (*open symbols*) and a starchless mutant (*filled circles*) of *Arabidopsis. From Weise, S.E., Kiss, J.Z., 1999. Gravitropism of inflorescence stems in starch-deficient mutants of Arabidopsis. Int. J. Plant Sci. 160, 521–527.*

in terms of providing the constant input of free energy necessary to prevent the dissipation of life expected from the Second Law of Thermodynamics. In this chapter, we have discussed the process of photosynthesis from a structural and functional perspective and a theoretical and experimental point of view.

The chemiosmotic theory of Peter Mitchell's, which describes how the proton-impermeant membranous structure of the cell powers the chemical synthesis of ATP, was a brilliant hypothesis that combined cell structure with cell chemistry in a way that revolutionized our understanding of the cell. Prebble and Weber (2003) described the human characteristics that made it possible for Mitchell to revolutionize cell biology. "One of the major characteristics of Mitchell's scientific work was his intuitive thinking. Thus, he was able to speculate about biological systems in a rational and creative way, which produced an understanding that often turned out to be close to reality. This gave him confidence to use his imagination, to build on hints from empirical evidence, to develop general explanations, and to generate a theory-driven research program. As he deepened his ideas about the fundamental relationship of metabolism and transport, he increasingly became convinced that he had an insight into basic principles that were congruent with the way nature was. Consequently, the chemiosmotic hypothesis was propounded with virtually no experimental basis and the Q cycle with only slight empirical support. He felt sufficiently sure of the basic correctness of his approach that when theory and experimental data were in conflict, he immediately suspected problems with the experiment or its interpretation. Further, he felt that his ideas, as embodies in his chemiosmotic theory, had been vindicated by the unsuccessful efforts of his many critics to falsify them."

The photosynthetic process reminds us that natural selection makes do with the components with which it has to work and selection is based on the organism as a whole and not based on a single process. For example, while photosynthesis is critical for all life on Earth, chlorophyll and rubisco are not particularly adapted to collect light and fix CO_2 under the present atmospheric conditions. How does this square with the idea that "nature does nothing in vain," an idea that really goes back to Aristotle's idea of final causes (teleology)? Perhaps a change in a double bond of chlorophyll that could change its absorption spectrum or an amino acid in rubisco that would decrease its affinity for oxygen would harm other processes irrevocably. Thus, in evolution there seems to be a collective bargaining that leads to a unity of the organism.

According to Charles Darwin (1859), natural selection is also relativistic in that "it adapts and improves the inhabitants of each country only in relation to their coinhabitants." In *The Origin of Species*, Darwin (1859) suggested that absolute perfection is a property of individual creation and not of natural selection, and that "the wonder indeed is, on the theory of natural selection, that more cases of the want of absolute perfection have not been detected."

13.13 QUESTIONS

13.1. What is the evidence that the radiant energy from the sun is converted into chemical energy in the chloroplast?

13.2. What is the mechanism by which the chloroplast converts radiant energy into chemical energy?

13.3. What are the limitations of thinking that the only function of the chloroplast is the conversion of radiant energy into chemical energy?

The references for this chapter can be found in the references at the end of the book.

Chapter 14

Mitochondria

If fruit juices or sugar solutions are left to stand in the open air, they show after a few days the processes which are covered by the name of fermentation phenomena. Gas is seen to develop, the clear solution becomes cloudy and a deposit appears which is called yeast. At the same time the sweet taste disappears and the liquid acquires an intoxicating effect. These observations are as old as the hills; at any rate, these processes have been used since the most ancient times of the human race for the production of fermented liquors. It is, however, only since the end of the eighteenth century - i.e. since Lavoisier - that we have known that during such a process sugar decomposes into carbon dioxide and ethyl alcohol.

Eduard Buchner (Nobel Lecture).

…in chemistry as in moral philosophy, it is extremely difficult to overcome prejudices imbibed in early education and to search for truth in any other road than the one we have been accustomed to follow.

Antoine Lavoisier (1790).

As each step of intermediary metabolism requires a specific enzyme, it is evident that a common pathway of oxidation results in an economy of chemical tools. Surveying the pathway of the degradation of foodstuffs as a whole, one cannot but be impressed by the relative simplicity of the arrangement, in as much as the total number of steps required to release the available energy from a multitude of different substrates is unexpectedly small.

Hans Krebs and Hans Kornberg (Energy Transformations in Living Matter: A Survey).

We cannot determine the truth of a hypothesis by counting the number of people who believe it.

Boris Ephrussi (1953).

In Chapter 13, I discussed how chloroplasts convert radiant energy to chemical energy in the form of carbohydrates. Carbohydrates are high-energy molecules in thermodynamic terms; however, they are relatively inert and cannot directly provide the energy necessary to fuel cellular processes. Thus, something must act as an intermediary between the carbohydrates and the energy-requiring cellular processes. Throughout the 1940s and 1950s, biologists came to the exciting conclusion that adenosine triphosphate (ATP) is the chemical compound that universally and directly acts as this intermediate (Krebs and Kornberg, 1957). ATP is involved in many cellular responses, including the synthesis of macromolecules (Cori, 1946) and coenzymes (Kornberg, 1950a,b; Schrecker and Kornberg, 1950); the activation of fatty acids (Kornberg and Pricer, 1953), amino acids (Hoagland et al., 1956), and sulfate (Bandurski et al., 1956); cell motility (Szent-Györgyi, 1949b; Hoffman-Berling, 1955; Weber, 1955); bioluminescence in the firefly (McElroy, 1951; McElroy and Strehler, 1954); the generation of electric currents in electric organs and nerves (Nachmansohn, 1955; Nachmansohn et al., 1943, 1946); the maintenance of the differential permeability of membranes (Ginzburg, 1959); and the transport of ions (Caldwell et al., 1960). The conversion of the chemical energy of carbohydrate into the chemical energy of ATP takes place in the mitochondrion, which Albert Claude (1948, 1975) called "the real power plants of the cell," and Philip Siekevitz (1957) called the "powerhouse of the cell."

14.1 DISCOVERY OF THE MITOCHONDRIA AND THEIR FUNCTION

This history of mitochondrial studies follows two quite separate paths: one physiological and the other structural. The unification of these two tracks in the 1940s led to a thrilling and satisfying understanding of the energetics of the cell (Claude, 1943, 1975; Lehninger, 1964). First, I will discuss the history of the physiological studies, and then the history of the structural studies.

14.1.1 History of the Study of Respiration

That most organisms require air to survive is common knowledge today, but this need was only discovered serendipitously. Robert Boyle (1662) was conducting experiments to see whether flies and butterflies could fly in air made thin by his newly invented vacuum pump. When he pulled a vacuum in the bell jar in which they were flying, the flies and butterflies fell down and died. He wondered

Plant Cell Biology. https://doi.org/10.1016/B978-0-12-814371-1.00014-X

whether they died as a consequence of the fall, or whether the animals themselves became weak because of their need for air. He realized that the air was necessary for life when he repeated his experiments using a lark with a broken wing. Because the lark could not fly, any adverse effects of the lack of air on the lark in the bell jar would not be caused by falling but would indicate the necessity of air for life. The lark died in the vacuum, and later experiments showed that a mouse died too. Boyle concluded from these experiments that air was necessary for respiration.

His fellow scientists did not believe his conclusions because he did not do the proper control experiment. That is, he did not test whether the mouse could live in the confined space in the bell jar in the absence of "creature comforts," even when there was plenty of air. In response to his critics, Boyle put a mouse in the bell jar overnight, gave it a paper bed and plenty of cheese, and then placed the bell jar by the fireside to keep the mouse warm during the night. In the morning, Boyle observed that the mouse was very much alive and ate the cheese like a normal mouse. However, as soon as he evacuated the air, the mouse started to die, showing that lack of air, not the lack of "creature comforts," was necessary for life.

Robert Hooke (1726), an assistant of Boyle, showed that the air was taken up by the lungs by demonstrating that a dog could be kept alive if air was continuously blown through the lungs, but not through other parts of the body. Stephen Hales (1727) suggested that air might also be important for plants after performing experiments in which he incinerated plants and found that they were composed of air, which at the time was considered to be an element. Hales also suggested that the leaves, which had passports or pores on their surface (Grew, 1682), might be analogous to the lungs of animals. Joseph Priestley (1774) and John Ingenhousz (1796) showed that oxygen (i.e., dephlogisticated air) was the vital part of the air that was taken up by the respiration of animals and plants, respectively.

Antoine Lavoisier, the founder of modern chemistry, determined that combustion results from the combination of oxygen with carbon and hydrogen. Lavoisier measured the amount of oxygen in a sample by treating metals with a given sample of air and measuring how much the mass of the metal increased as a result of oxidation. Lavoisier believed that respiration and combustion were analogous reactions, as he and Armand Séguin (Séguin and Lavoisier, 1789; translated in Fruton, 1972) observed:

> *In respiration, as in combustion, it is the atmospheric air which furnished oxygen and caloric; but since in respiration it is the substance itself of the animal, it is the blood, which furnishes the combustion matter, if animals did not regularly replace by means of food aliments that which they lose by respiration, the lamp would soon lack oil, and the animal would perish as a lamp is extinguished when it lacks nourishment. The proofs of this identity of effects in respiration and combustion are immediately deducible from experiment. Indeed, upon leaving the lung, the air that has been used for respiration no longer contains the same amount of oxygen; it contains not only carbonic acid gas but also much more water than it contained before it had been inspired.*

That is,

$$CH_2O + O_2 \rightarrow CO_2 + H_2O + \text{heat}$$

where Lavoisier believed that O_2 was a mixture of base of oxygen and caloric. Respiration was defined as a combustion process and measured by the uptake of O_2 and the expulsion of CO_2 and H_2O.

Séguin and Lavoisier (1789) believed that the function of respiration was to produce body heat. Lavoisier and the mathematician Pierre-Simon de Laplace collaborated on a set of experiments that showed that the amount of CO_2 produced by an animal is approximately equal to the amount of O_2 consumed. Using a calorimeter, they found that approximately the same amount of ice was melted by the respiration of a guinea pig and the burning of charcoal, which had equal outputs of CO_2. They concluded that the living body produces heat in a manner similar to inanimate objects. Lavoisier never finished his experiments on respiration because he was "politically incorrect" and lost his head in a guillotine during the French Revolution (Rolland, 1926). Following Lavoisier's beheading, Joseph LaGrange remarked (See Delambre, 1867), *"Il ne leur a fallu qu'un moment pour faire tomber cette tête, et cent années peut-être ne suffiront pas pour en reproduire une semblable."* ("It took them only an instant to cut off this head, and one hundred years might not suffice to reproduce its like.")

Scientists disagreed as to where the conversion of O_2 to CO_2 took place. Was it the lungs as Lavoisier often thought or was it the blood as Joseph Lagrange proposed (Mendelsohn, 1964)? It turned out that the conversion occurred in both places, as well as in every other part of the organism, as Lazzaro Spallanzani (1803) showed when he isolated various tissues and demonstrated that all tissues were capable of consuming oxygen and giving off CO_2. In 1804, Théodore de Saussure showed that in the dark, plants also take up O_2 and give off CO_2 and thus must have a respiratory mechanism similar to that of animals.

Eduard Pflüger (1872) also believed that every cell respired and that this was true for both plants and animals. Pflüger revealed that when the blood of a frog was replaced by saline, the frog respired just like a normal frog. He also mentioned that insects and plants that have no blood respire too. Pflüger concluded, "Here lies, and I want to state this once and for all, the crucial secret of the regulation of the total oxygen consumption by the organism, a gravity which is entirely determined by the cell itself."

By the end of the 19th century, it was becoming clear that respiration took place in each and every cell and was thus a cellular process (Krogh, 1916; Haldane, 1922). In 1924, it seemed likely that the uptake of oxygen was associated with the particulate parts of the cytoplasm, which at the time seemed to be the nuclei (Meyerhof, 1924)! Subsequent structural studies showed that the mitochondrion was the respiratory organelle.

14.1.2 History of the Structural Studies in Mitochondria

Mitochondria were first seen in plant cells (*Equisetum*) by Wilhelm Hofmeister in 1851 (see Fig. 9.1 in Chapter 9); however, structural studies of mitochondria began in earnest in the late 19th century when mitochondria were seen in many cell types following the introduction of the apochromatic lens (see Fig. 14.1; Newcomer, 1940). Excellent apochromatic lenses owe their existence, in part, to the botanist Matthias Schleiden, a teacher of Carl Zeiss. Schleiden personally encouraged Zeiss to develop good microscopes to make it possible to see cellular structure. Zeiss began a microscope company and hired Ernst Abbe to develop a theory of image formation and to design apochromatic lenses from first principles.

The organelles that were later called mitochondria were initially given the following names by the early cytologists: bioblasts, granules, microsomes, plastosomes, eclectosomes, histomeres, chromidia, chondriokonts, chondriosomes, polioplasma, and vibrioden. Of almost 30 terms coined for this structure, only *mitochondria* survived (Cowdry, 1924).

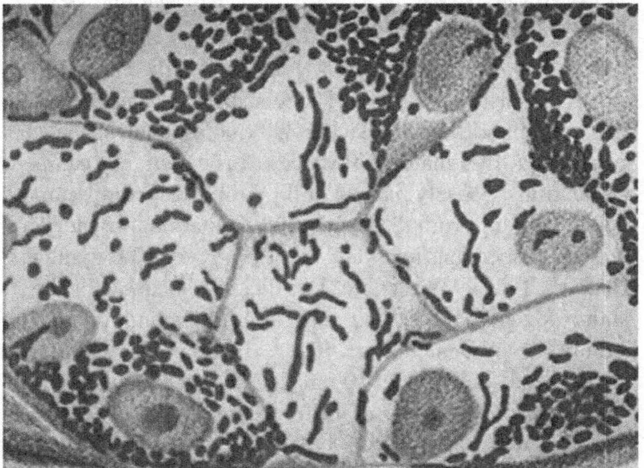

FIGURE 14.1 "Elementary organisms" in the mucilage-secreting cells of the stomach of a cat fixed with an osmium mixture. *From Altmann, R., 1890. Die Elementarorganismen und ihre Beziehung zu den Zellen. Veit & Co., Leipzig.*

Carl Benda (1899), while studying mouse sperm, gave the name "mitochondria" to the cytoplasmic threadlike granules. Benda defined mitochondrion as "granules, rods, or filaments in the cytoplasm of nearly all cells which are preserved by bichromates within a pH range approximately between 4.6 and 5.0, and which are destroyed by acids or fat solvents" (Newcomer, 1940; Goldschmidt, 1956; Sato, 1972). *Mitos* is Greek for thread, and *chondrin* is Greek for small grain. While Altmann, Flemming, Kölliker, Hanstein, Strasburger, and von Valette St. George are considered by various historians to have first discovered the mitochondrion, Richard Altmann made the most complete and accurate descriptions and drawings (see Fig. 14.1).

Such a large catalog of names followed from the limitation that the early cytologists could never reach a consensus about what was factual and what was artefactual when they and others observed mitochondria in fixed and stained sections (Fischer, 1899). This possible cause of error in describing facts should always be kept in mind, not only when doing microscopy but also when doing any experiments. Moreover, while the arguments were going on about what was fact and artifact, grandiose theories were being put forth to explain the function of these new organelles. For example, Altmann (1890) believed that the mitochondrion was the elementary particle of life and thus named it the bioblast. Others considered it to be the center of protein synthesis, fat synthesis, or the residence of genes. Last, but certainly not least, Kingbury (1912) proposed that the mitochondrion was the respiratory center of the cell.

A major breakthrough that proved that the mitochondrion was really a true organelle and not a preparation artifact came from the work of Leonor Michaelis (1900). At the suggestion of Paul Ehrlich, he tested the ability of the various newly invented fabric dyes to stain living tissue (Beer, 1959). Michaelis showed that the mitochondria in pancreatic exocrine cells were stained selectively and supravitally by a dilute solution of Janus Green. Four years later, Meves showed that the mitochondria of plant cells could be visualized with the same dye (Meves, 1904). The staining of mitochondria by Janus Green and other vital stains is only transient because the mitochondria reduce the dye and render it colorless (see Lazarow and Cooperstein, 1953). The ability of mitochondria to oxidize and reduce various dyes was what led Kingbury (1912) to propose that the mitochondria may be involved in cellular respiration.

Although Robert Bensley (1953), one of the originators of the technique of differential centrifugation, believed that the mitochondrion did not have the "dignity" of a real organelle and was at best "a journeyman worker who comes and goes," the question of the reality of mitochondria mostly ended in 1948 when Albert Claude used differential centrifugation to isolate the respiratory particles from rat liver cells. These particles stained with Janus Green, and thus he identified them as mitochondria and not

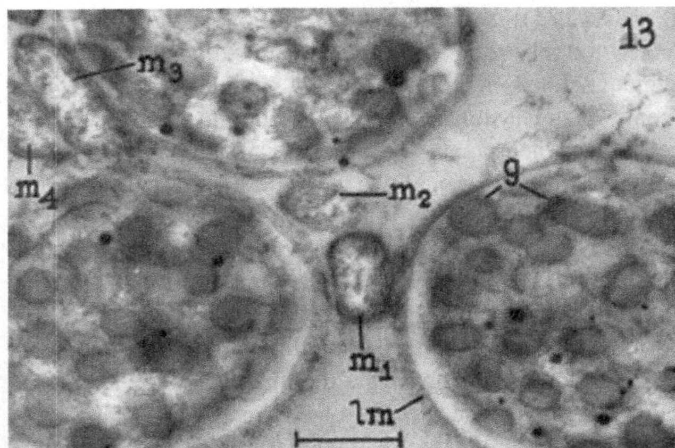

FIGURE 14.2 Thin section of mitochondria in the cytoplasm of a mesophyll cell of tobacco (m_1, m_2, m_3, m_4). The majority of the figure is taken up by chloroplasts that have limiting membranes (lm) and grana (g). *From Palade, G.E., 1953. An electron microscope study of the mitochondria structure. J. Histochem. Cytochem. 1, 188–211.*

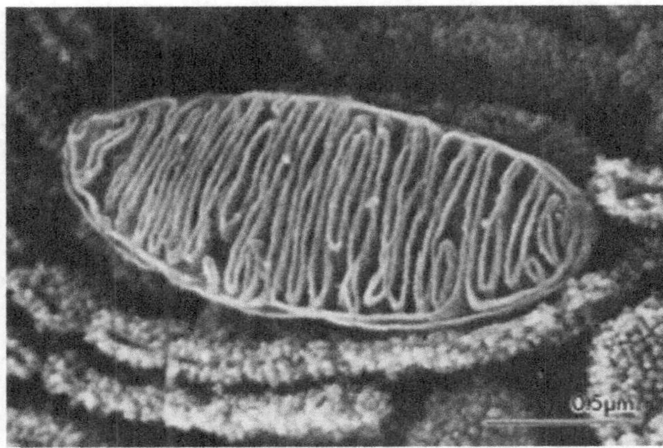

FIGURE 14.3 Scanning electron micrograph of a mitochondrion from a rat pancreatic acinar cell. Note the platelike cristae. *From Tanaka, K., 1987. Eukaryotes: scanning electron microscopy of intracellular structures. Int. Rev. Cytol. (Suppl. 17), 89–147.*

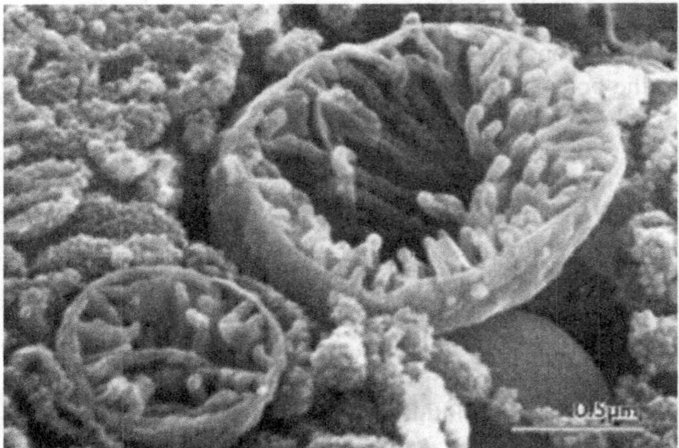

FIGURE 14.4 Scanning electron micrograph of a mitochondrion from a rat pancreatic acinar cell. Note the tubular cristae. *From Tanaka, K., 1987. Eukaryotes: scanning electron microscopy of intracellular structures. Int. Rev. Cytol. (Suppl. 17), 89–147.*

nuclei as was previously believed (Myerhof, 1924; Brachet, 1985). Further evidence that the particles were mitochondria came from electron microscopic studies, which showed that the particles were membrane-enclosed organelles, identical in appearance to mitochondria found in cells (see Fig. 14.2; Hogeboom et al., 1948; Palade, 1952b). Albert Claude (1974) and Kennedy and Lehninger (1949) then discovered that the isolated mitochondria contain all the enzymes involved in the Krebs cycle (Kennedy, 1992; Wickner, 2011), and Albert Claude (1974) and Chance and Williams (1955) found that mitochondria contain the cytochromes involved in electron transport. These were some of the first studies to combine biological chemistry and microscopy and give incontrovertible support to a hypothesis of function that originated from a morphological observation. At this point, the techniques and ways of thinking heretofore developed independently by biochemists and cytologists coalesced, and the new field of cell biology emerged. The great synthesizers of this time included Christian de Duve, George Palade, and Albert Claude.

The ultrastructure of mitochondria in plants (*Nicotiana*, *Lemna*, *Euglena*), animals, and protozoa was described by George Palade in 1952 and 1953. The mitochondria are cylindrical bodies, 1- to 4-μm long by 0.3- to 0.7-μm wide, which are physically separate from the other organelles (see Figs. 14.3 and 14.4; Douce, 1985; Rosamond, 1987). Palade (1953) described the mitochondria as "a system of internal ridges that protrude from the inner surface of the membrane toward the interior of the organelles. … In favorable electron micrographs the mitochondrial membrane appears to be double and the cristae appear to be folds of a second, internal mitochondrial membrane." The outer membrane is approximately 7-nm thick, and the inner membrane, originally called the *cristae mitochondriales*, is approximately 5-nm thick. The two membranes delineate two mitochondrial spaces: the intermembranal space (E-space) and the matrix (P-space). In negatively stained preparations, densely packed, lollipop-like structures can be seen emanating from the P-side of the inner membrane. The lollipop-shaped particles are the F_1F_0-type ATP synthase (Racker, 1965). While the F officially means "Factor," André Jagendorf told me that it might stand for "Ef," Efraim Racker's nickname. The F_1F_0-type ATP synthase has a head that is 9–10 nm in diameter and a stalk 3.5–4 nm in diameter and 4.5-nm long.

Today, the mitochondria in living cells can be easily observed with vital fluorescent dyes (e.g., Rhodamine 123 and MitoTracker) and green fluorescent protein (Partikian et al., 1998). Watching mitochondria that have been stained with these dyes show that they are amazingly plastic and

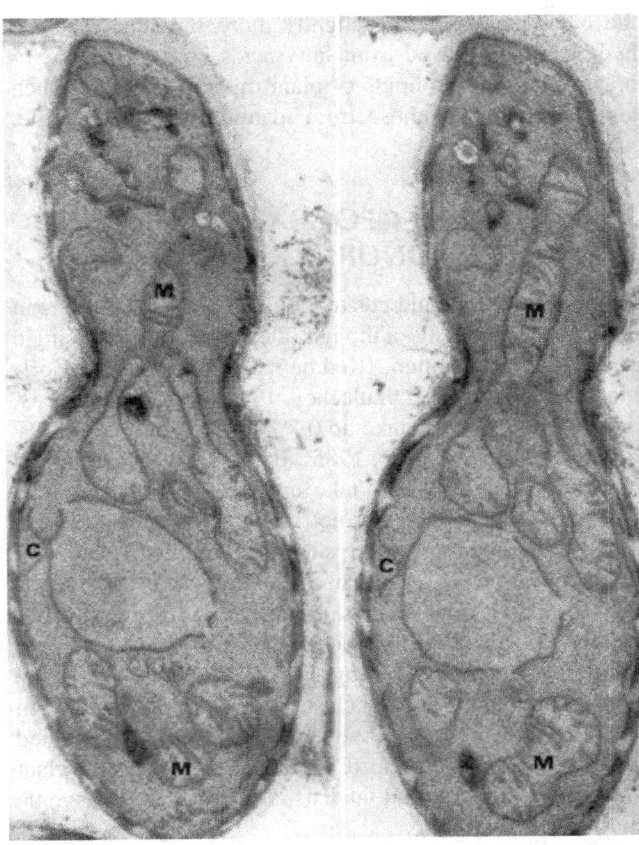

FIGURE 14.5 Two consecutive serial sections of *Pityrosporum* yeast cells that were used to reconstruct the mitochondrial reticulum. *C*, cytomembranes; *M*, mitochondrion. Similar mitochondrial reticula may be found in yeast (Hoffmann and Avers, 1973) and in some root tip cells of white lupine (Gunning and Steer, 1975). *From Keddie, F.M., Barajas, L., 1969. Three-dimensional reconstruction of* Pityosporum *yeast cells based on serial section electron microscopy. J. Ultrastruct. Res. 29, 260–275.*

mobile organelles (Rosamond, 1987). They appear to fuse, divide, and undergo amoeboid movement as they travel along actin filaments (Grolig, 1990). Vital staining, as well as serial sectioning, shows that mitochondria can even appear as a mitochondrial reticulum (see Fig. 14.5; Keddie and Barajas, 1969; Hoffmann and Avers, 1973; Skulachev, 1990; Kawano, 2014b). The lengths of mitochondria vary, and the frequency of fusion and division is under genetic control (Kawano et al., 1993). The proteins responsible for mitochondrial shape are being discovered (Youngman et al., 2004).

14.2 ISOLATION OF MITOCHONDRIA

Mitochondria were first isolated from animal cells by Bensley and Hoerr (1934) and from plant cells by Millerd et al. (1951). To isolate mitochondria, the tissues are homogenized, filtered, and centrifuged at 700–1000 g for 10 min. The pellet is discarded, and the supernatant is centrifuged at 10,000 g for 20 min. The pellet is then resuspended and centrifuged at 75,000–100,000 g for 2–4 h in a density gradient (Moore and Proudlove, 1988; Eubel et al., 2007; Millar et al., 2007; Taylor et al., 2014; Murcha and Whelan, 2015; Whelan and Murcha, 2015). Mitochondria have a density of 1.16–1.22 g/mL with a modal value of 1.18 g/mL. The intactness and purity of the isolated mitochondria can be determined with a light microscope after vital staining (Hogeboom et al., 1948) or with an electron microscope. A single mitochondrion can be isolated using laser tweezers (Kuroiwa et al., 1996).

The two mitochondrial membranes can be separated from each other after the mitochondria are lysed in a hypotonic medium. The lysed mitochondria are then layered on top of a discontinuous sucrose gradient and subjected to density gradient centrifugation. The outer membrane shows up in the 0.3/0.6 M (=1.04/1.08 g/mL) and 0.6/0.9-M (=1.08/1.12 g/mL) sucrose interfaces, and the mitoplast—that is, the inner membrane and matrix—is recovered as a pellet below the 1.25 M (=1.16 g/mL) sucrose layer (Douce, 1985). The matrix can be isolated by osmotically lysing the mitoplasts.

14.3 COMPOSITION OF MITOCHONDRIA

The outer membrane of plant mitochondria comprises only 6.8% of the mitochondrial protein, whereas the inner mitochondrial membrane accounts for 30% of the total protein. The inner membrane is composed of 60% protein and 40% lipid on a dry weight basis, whereas the outer membrane is composed of 20% protein and 80% lipid on a dry weight basis. The outer mitochondrial membrane contains a channel protein called porin that is permeable to molecules that are 10,000 Da or less. It is similar to the pore protein in the outer membrane of the chloroplast envelope.

14.3.1 Proteins

Proteomic analysis shows that the mitochondria are composed of about 3000 proteins (Bykova and Moller, 2006; Taylor and Millar, 2015). The proteins used to identify the mitochondria in vitro are those associated with respiration. They include succinate dehydrogenase, cytochrome c oxidase, and fumarase (Hogeboom and Schneider, 1955), as well as several cytochromes, including a, a_3, c, and b_5 (Chance and Williams, 1955). In animal cells, which lack plastids, the enzymes involved in lipid metabolism are present on the outer mitochondrial membrane.

The marker enzymes for the outer membrane are the succinate:cytochrome c oxidoreductase and the NADH: cytochrome c oxidoreductase. The marker enzymes for the inner mitochondrial membrane are malate dehydrogenase and succinate: $K_3Fe(CN)_6$ oxidoreductase.

14.3.2 Lipids

In contrast to the galactolipid-rich chloroplasts, the mitochondria are rich in phospholipids (23%–27% by weight). Phospholipids account for as much as 90% of the total mitochondrial lipids. Mitochondria contain high concentrations of phosphatidylcholine (PC) and phosphatidylethanolamine (PE) as well as diphosphatidylglycerol (DPG = cardiolipin), a lipid that is endemic to mitochondria (Table 14.1). Cardiolipin is also found in the plasma membrane of bacteria, giving support for the hypothesis that mitochondria evolved from bacteria (see Chapter 15).

The lipid compositions of the inner and outer membranes of the mitochondria are distinct. In particular, the inner membrane contains cardiolipin, which is absent from the outer membrane. The inner membrane is also richer in PE and poorer in PC relative to the outer membrane (Table 14.2). The hydrocarbons derived from fatty acids in the outer membrane are slightly more saturated than the hydrocarbons derived from fatty acids in the inner membrane. The phospholipids of plant mitochondria are much less saturated than those from mammalian mitochondria, which exist at 37°C.

TABLE 14.1 Phospholipid Composition (% Phospholipid by Weight) of Sycamore Cells and Mitochondria

Phospholipid	Cells	Mitochondria
Phosphatidylcholine	47	43
Phosphatidylethanolamine	29	35
Diphosphatidylglycerol[a]	1.8	13
Phosphatidylinositol	16	6
Phosphatidylglycerol	5	3

[a]*Also known as cardiolipin.*
From Douce, R., 1985. Mitochondria in Higher Plants. Structure, Function, and Biogenesis. Academic Press, Orlando.

TABLE 14.2 Phospholipid Composition (% Phospholipid by Weight) of Inner and Outer Membranes of Sycamore and Mung Bean Mitochondria

	Sycamore Cells		Mung Bean Cells	
Lipid	IM	OM	IM	OM
Phosphatidylcholine	41	54	29	68
Phosphatidylethanolamine	37	30	50	24
Diphosphatidylglycerol	14.5	0	17	0
Phosphatidylinositol	5	11	2	5
Phosphatidylglycerol	2.5	4.5	1	2

IM, inner membrane; *OM*, outer membrane.
From Douce, R., 1985. Mitochondria in Higher Plants. Structure, Function, and Biogenesis. Academic Press, Orlando.

14.4 CELLULAR GEOGRAPHY OF MITOCHONDRIA

There are usually hundreds to thousands of mitochondria in a typical plant cell, although they may form a single mitochondrial reticulum (Keddie and Barajas, 1969; Hoffman and Avers, 1973; Skulachev, 1990). In sycamore cells, the 250 mitochondria take up 0.7% of the protoplast or 7% of the cytoplasm (Douce, 1985). By contrast, mitochondria in mammalian liver cells take up to 20% of the volume. Mitochondria are typically not randomly arranged but are often positioned close to the site of carbohydrate storage or ATP utilization. In leaves, mitochondria are often situated next to peroxisomes and chloroplasts; the three organelles in this case function together in the process of photorespiration and carbon recycling. In onion epidermal cells, the mitochondria are typically clustered around the nucleus. In growing pollen tubes, the mitochondria are enriched 20–40 μm behind the tip (Cárdenas et al., 2006). In plant sperm, the spiral-shaped nucleus is surrounded by a single mitochondrion.

14.5 CHEMICAL FOUNDATION OF RESPIRATION

The lowest energy form of carbon and hydrogen are the most oxidized states CO_2 and H_2O, respectively. Reduced-carbon compounds, in the forms of carbohydrate, fats, and proteins, have a certain amount of energy that can be utilized for cellular energy following their oxidation. As Hans Weber (1958) wrote,

> *How is it possible that living creatures are able to utilize invariably the same amount of energy in performing vital work, when they derive this energy from such widely different foodstuffs as protein, carbohydrates, and fats? In order to realize how remarkable this is we need only imagine a diesel engine—a simple structure compared with the living cell—obtaining the necessary energy by consuming in turn gas, diesel oil, coal, or powdered milk.*

While the thermodynamists of the 19th century believed that foodstuffs energized organisms through the production of thermal energy (Grove et al., 1867; Guye, n.d.) or by creating an electrical intermediate (Thomson and Tait, 1862), the energy released in the oxidation of all these compounds is actually conserved in the synthesis of ATP, a molecule that is chemically unrelated to fats, carbohydrates, and lipids, which are the foodstuffs, but is related instead to the nucleic acids.

14.5.1 Fitness of Adenosine Triphosphate as a Chemical Energy Transducer

Why is ATP the universal energy source? It must have something to do with its chemical properties. Fritz Lipmann (1941) declared that ATP would turn out to be the universal energy currency by virtue of its large enthalpy and designated the bond that bound the terminal phosphate as the "high-energy bond," denoted by "the squiggle." By contrast, Herman Kalckar (1941, 1942) suggested that ATP could serve as the universal energy currency because of the many resonant structures in which a phosphate molecule could exist. That is, he suggested that entropy was the dominant factor in the selection of ATP as the universal energy currency. Lipmann's view gained widespread acceptance because of its simplicity, yet Kalckar's idea was closer to the truth. I will briefly discuss some of the concepts necessary to understand Kalckar's view.

The molecular free energy of a reaction is a function of the concentrations of the products and reactants. When water takes part in the reaction, the concentration of water must be factored into the equation. In almost all cases, water has been neglected, and thus the published values of dehydration and hydrolytic reactions may be incorrect by a factor of four (=ln 55.5) and should be redetermined (Oesper, 1950).

14.5.1.1 Free Energy, Enthalpy, and Entropy

The change in molecular free energy that occurs during a reaction (ΔE) is equal to $\Delta H/N_A - T\Delta S/N_A$, and thus the molecular free energy of a reaction depends on both the change in enthalpy and the change in entropy. At equilibrium, $\Delta H/N_A - T\Delta S/N_A = 0$. Reactions that have a negative change in molecular free energy ($\Delta H/N_A - T\Delta S/N_A < 0$) occur spontaneously and are called exergonic. Those that have a positive change in molecular free energy ($\Delta H/N_A - T\Delta S/N_A > 0$) do not occur spontaneously and are called endergonic (Coryell, 1940).

The enthalpy of a reaction is determined by measuring the difference between the heat generated by burning the products and the heat generated by burning the reactants at constant pressure. The enthalpy is greater for covalent bonds than for ionic bonds and is greater for double bonds than for single bonds. Misconceptions of molecular free energy come, in part, from the fact that enthalpy is often equated with free energy. When enthalpy is equated with free energy, it is assumed that reactions will proceed spontaneously if they give off heat and are exothermic and will require energy if they require heat and are endothermic. If this were true, ice could not spontaneously melt. Thus, we must take into consideration the entropy of the reaction. Then, if the enthalpy term is positive (endothermic) and the entropy term ($T\Delta S/N_A$) is both positive and greater in magnitude than the enthalpy term, the reaction will be exergonic and will proceed spontaneously. Likewise, if a reaction is exothermic, it may be endergonic if the entropy term is large and negative.

Typically, the combination of small molecules to make larger molecules requires the addition of free energy. This is because many biosynthetic reactions are dehydrations and consequently will not proceed spontaneously in the aqueous cellular environment (see Chapter 18). As I will discuss in the following, the molecular free energy made available by the hydrolysis of ATP provides the free energy necessary for these biosynthetic reactions. To appreciate this function of ATP, let us study the chemistry of a typical dehydration reaction. Consider the reaction where two carboxylic acids are joined to make an acid anhydride:

$$\text{RCOOH} + \text{HOOCR}' \leftrightarrow \text{RCO} - \text{OOCR}' + \text{H}_2\text{O}$$

Due to the high concentration of water in the cell, this reaction tends to go to the left. At equilibrium, the concentration of the carboxylic acids are high and the concentration of the anhydride is low. The reaction also proceeds to the left because the carboxylic acids each have two resonance structures, which tend to stabilize them (Lewis, 1923; Pauling, 1940). The electrons are not as localized in resonating structures as they are in a structure that only has one possible configuration, thus the probability of localizing an electron is low and the entropy of a resonating structure is high. The resonance energy of the carboxylic acid group reduces the overall energy of a molecule by approximately 10^{-19} J, and thus molecules with resonant structures are more stable than they would be if resonance did not occur (Oesper, 1950). In fact, the delocalized nature of the electrons makes it stable so that organic acids, including acetic acid and fatty acids, cannot enter directly into chemical reactions but must first be activated.

When two resonating structures are joined, the number of possible resonant structures is diminished, which results in a decrease in the entropy of the product. Thus, in the reaction just shown, the reactants have a lowered free energy due to resonance stabilization (Kalckar, 1942; Dugas, 1996). Given the chemical properties of molecules involved in dehydration reactions, an input of molecular free energy, in the form of ATP, is required to absorb the water molecule and drive these reactions in a cell.

14.5.1.2 Resonance Structures of Phosphates

Inorganic phosphate has 29 resonance structures (see Fig. 14.6; Oesper, 1950). Due to the stabilizing resonance energy of phosphate, the reaction of a phosphate and a carboxylic acid to make a mixed anhydride will tend to proceed to the left spontaneously when placed in water (Lehninger, 1959; Green and Baum, 1970; Becker, 1977):

$$RCOOH + PO_3OH \leftrightarrow RCOOPO_3 + H_2O$$

Reactions resulting in the formation of polyphosphoanhydrides, including adenosine diphosphate (ADP) and ATP, also tend to proceed to the left:

$$ADP - OH + PO_3OH \leftrightarrow ATP + H_2O$$

$$AMP - OH + PO_3OH \leftrightarrow ADP + H_2O$$

These reactions tend to go to the left because the reactants are stabilized by resonance stabilization, and in addition, there is an electrostatic repulsion between the phosphates in the polyphosphate in the compound with the greater number of phosphates. The resonance stabilization accounts for about 70% of the change in free energy, and the electrostatic repulsion accounts for about 30% (Hill and Morales, 1950, 1951).

It is possible to synthesize almost any molecule in vitro by taking advantage of the law of mass action and increasing the concentrations of the reactants. However, in the cell, this is usually not a viable option for highly endergonic (i.e., irreversible) reactions. In cells, highly endergonic reactions proceed because they are coupled to highly exergonic ones, with no net change in the water concentration.

$$RCOOH + PO_3OH \leftrightarrow RCOOPO_3 + H_2O \quad \text{Highly endergonic}$$

$$ATP + H_2O \leftrightarrow PO_3OH + ADP \quad \text{Highly exergonic}$$

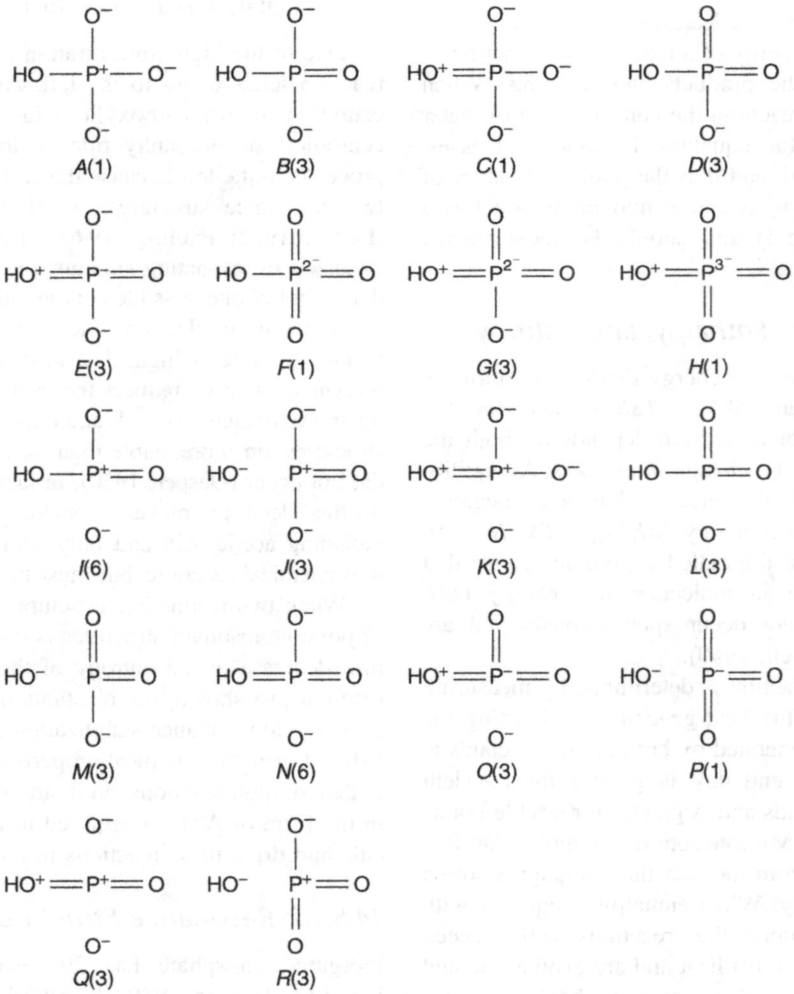

A(1) B(3) C(1) D(3)

E(3) F(1) G(3) H(1)

I(6) J(3) K(3) L(3)

M(3) N(6) O(3) P(1)

Q(3) R(3)

FIGURE 14.6 Resonant structures of phosphate. *From Oesper, P., 1950. Sources of the high energy content in energy-rich phosphates. Arch. Biochem. 27, 255–270.*

$$\text{RCOOH} + \text{ATP} \leftrightarrow \text{RCOOPO}_3 + \text{ADP}$$
Slightly exergonic

Universally, the exergonic reaction used for energetic coupling is the reaction involved in the hydrolysis of ATP. It is, in part, the chemical nature of phosphate that allows it to perform this function. The hydrolysis reaction is given in the following formula:

$$\text{ATP} + \text{H}_2\text{O} \leftrightarrow \text{ADP} + \text{P}_\text{i}$$

The constant of the reaction, which relates the concentrations of products to the concentrations of reactants, depends on the circumstances. At equilibrium,

$$K_{eq} = \frac{[\text{ADP}]_{eq}[\text{P}_\text{i}]_{eq}}{[\text{ATP}]_{eq}[\text{H}_2\text{O}]_{eq}} \tag{14.1}$$

In the standard state,

$$K_{std} = \frac{[\text{ADP}]_{std}[\text{P}_\text{i}]_{std}}{[\text{ATP}]_{std}[\text{H}_2\text{O}]_{std}} = \frac{1\text{ M } 1\text{ M}}{1\text{ M } 55.5\text{ M}} = 1.8 \times 10^{-2} \tag{14.2}$$

Under cellular conditions,

$$K_{Real} = \frac{[\text{ADP}]_{Real}[\text{P}_\text{i}]_{Real}}{[\text{ATP}]_{Real}[\text{H}_2\text{O}]_{Real}} = \frac{10^{-3}\text{ M } 10^{-2}\text{ M}}{10^{-3}\text{ M } 55.5\text{ M}} = 1.8 \times 10^{-4} \tag{14.3}$$

As a function of the chemistry of P_i, the equilibrium constant for the hydrolysis of ATP is large (Burton, 1958). That is, given the equilibrium and standard state concentrations of ATP, ADP, P_i, and H_2O, K_{eq} for the following reaction is approximately 5.11×10^4. K_{eq} can be calculated by finding values for ΔG° at 25 C, pH 7.4, and 1 mM Mg^{2+} in the *Handbook of Biochemistry and Molecular Biology*, third Edition (Boca Raton, FL: CRC Press, 1989), p. 302.

$$\Delta G^\circ = -8.8\frac{\text{kcal}}{\text{mol}} = -3.668 \times 10^4 \text{ J/mol}$$

Thus,

$$E_{std} = \frac{\Delta G^\circ}{N_A} = -6.1 \times 10^{-20} \text{ J} \tag{14.4}$$

K_{eq} can be calculated from the following equation:

$$E_{eq} - E_{std} = kT \ln \frac{K_{eq}}{K_{std}} \tag{14.5}$$

As, at equilibrium, $E_{eq} = 0$, then

$$E_{std} = -kT \ln \frac{K_{eq}}{K_{std}} \tag{14.6}$$

and

$$K_{eq} = 5.11 \times 10^4$$

The free energy that is made available by the hydrolysis of ATP in the cell depends on how far the real ratio of $[\text{ADP}][\text{P}_\text{i}]/[\text{ATP}][\text{H}_2\text{O}]$ is from the equilibrium ratio. The molecular free energy is given by the following equation:

$$E_{eq} - E_{Real} = kT \ln \frac{K_{eq}}{K_{Real}} \tag{14.7}$$

As shown by Tetsuro Mimura and his colleagues (Mimura et al., 1984; Mimura and Kirino, 1984), in the living cell, the concentrations of ADP, P_i, ATP, and H_2O are far from equilibrium. They are typically around 10^{-3} M, 10^{-2} M, 10^{-3} M, and 55.5 M, respectively. Thus, K_{Real} for the hydrolysis reaction is about 1.8×10^{-4}, and according to Eq. (14.7), the molecular free energy made available by the hydrolysis of ATP is about 8×10^{-20} J.

The large E_{Real} is a result of two things: the ATP concentration in the cell is much greater than it would be at equilibrium, and at equilibrium, the ATP concentration is very low due to the resonance stabilization of phosphate ions. The resonance stabilization accounts for 70% of the free energy of hydrolysis as ATP, ADP, P_i, and H_2O go from the standard state to equilibrium.

Consider the case where there is no resonance stabilization. That is, pretend that resonance and entropy are not important. Then,

$$E_{std}(w/o) = 0.3E_{std} \tag{14.8}$$

$$E_{std}(w/o) = 0.3(-6.11 \times 10^{-20}) \text{ J} = -1.83 \times 10^{-20} \text{ J} \tag{14.9}$$

At equilibrium, $E_{eq}(w/o) = 0$ and

$$E_{eq}(w/o) - E_{std}(w/o) = kT \ln \frac{K_{eq}(w/o)}{K_{std}(w/o)} \tag{14.10}$$

Thus,

$$E_{std}(w/o) = -kT \ln \frac{K_{eq}(w/o)}{K_{std}(w/o)} \tag{14.11}$$

Using E_{std} $(w/o) = -1.83 \times 10^{-20}$ J, and given that $K_{std}(w/o) = K_{std}$, we can solve Eq. (14.11) to get $K_{eq}(w/o) = 1.54$. With this value we can now calculate what the real molecular free energy made available by the hydrolysis of ATP would be if there were no resonance stabilization.

$$E_{eq}(w/o) - E_{Real}(w/o) = kT \ln \frac{K_{eq}(w/o)}{K_{Real}(w/o)} \tag{14.12}$$

At equilibrium, $E_{eq}(w/o) = 0$, and

$$E_{Real}(w/o) = -kT \ln \frac{K_{eq}(w/o)}{K_{Real}(w/o)} \tag{14.13}$$

Given that $K_{Real}(w/o) = K_{Real}$, and $K_{eq}(w/o) = 1.54$, we can use Eq. (14.13) to see that $E_{Real}(w/o) = -3.72 \times 10^{-20}$ J. Thus, in the absence of resonance energy, the energy made available by the hydrolysis of ATP would be significantly lower. Even if there were no resonance, the free energy of ATP hydrolysis could be increased to -8×10^{-20} J by raising the concentration of ATP in the cell or lowering the concentration of ADP and/or P_i.

What would $K_{Real}(w/o)$ have to become to increase the energy of hydrolysis of ATP to -8×10^{-20} J in the absence of resonance? Using Eq. (14.12), and assuming $K_{eq}(w/o) = 1.54$, then $K_{Real}(w/o)$ would have to be 5.5×10^{-9}. Using Eq. (14.3), we find that if [ADP] = 10^{-3} M, [P_i] = 10^{-2} M, and [H_2O] = 55.5 M, then the [ATP] would have to increase to 32.76 M. Likewise, if [ATP] remained at 10^{-3} M, the product of [ADP] and [P_i] would have to decrease to 3.05×10^{-10} M.

Thanks to the resonance stabilization of P_i, the hydrolysis of ATP can yield 8×10^{-20} J of energy when [ATP] = 10^{-3} M, [ADP] = 10^{-3} M, and [P_i] = 10^{-2} M.

The concept introduced by Fritz Lipmann (1941) that the phosphoanhydride bond of ATP is a high-energy bond is misleading because it does not take into consideration the resonance stabilization energy. When this is taken into consideration, the phosphoanhydride bond is actually a low-energy bond, and this is the reason it breaks easily (Gillespie et al., 1953).

Due to the dependence of the free energy on both K_{Real} and K_{eq}, other phosphate-containing molecules (e.g., glucose-6-P) would not be as good an energy-transducing molecule as ATP. This is because K_{eq} for glucose-6-P hydrolysis is about 2.6×10^2, approximately 0.1% of the equilibrium constant for the hydrolysis of ATP. Thus, if glucose-6-P served as the universal energy intermediate in the cell, the ratio of [glucose][P_i]/[glucose-6-P][H_2O] would have to be 1.3×10^{-7} for the hydrolysis of glucose-6-P to release as much energy as the hydrolysis of ATP. That means if the glucose-6-P concentration were 10^{-3} M, a value typical for substrates of energetic reactions, the product of the glucose and P_i concentrations would have to be 7.2×10^{-9} M, too low for glucose to act as an efficient foodstuff. If the product of the glucose and P_i concentrations were 10^{-4} M, the concentration of glucose-6-P would have to be 13.9 M to act as a chemical energy transducer. For this reason, it is best not to have one of the carbohydrate intermediates in the glycolytic pathway as the chemical energy transducer. Likewise, it is likely that if the chemical energy–transducing molecule were a component of lipids or protein, the other two most abundant classes of macromolecules in the cell, it too would wreak havoc with the metabolism of these foodstuffs. This may be the reason that a nucleotide became the chemical energy–transducing molecule. Why adenosine is the universal chemical energy–transducing molecule, but the other nucleotides are not, remains a mystery. Perhaps you will find out why.

14.5.1.3 The Principle of the Common Intermediate

The free energy released by the hydrolysis of ATP would be useless if it could not be coupled to endergonic chemical reactions. This energy cannot be used in the form of heat but can only be coupled if there is a transfer of mass between the ATP and a reactant of an endergonic reaction. This means that there must be a common intermediate in the endergonic and exergonic reaction. This is known as "the principle of the common intermediate." The reaction

$$\text{Glucose} + P_i \leftrightarrow \text{Glucose} - 6 - P + H_2O$$

is endergonic, and the molecular free energy required for the reaction to go to the right is 2.3×10^{-20} J. The hydrolysis of ATP, by contrast, is exergonic and proceeds with a release of free energy of 8.0×10^{-20} J. Thus, the following reactions will proceed simultaneously with a net release of molecular free energy of 5.7×10^{-20} J, if the ATP and glucose are in the same cellular compartment.

$$\text{Glucose} + P_i \leftrightarrow \text{Glucose} - 6 - P + H_2O$$

$$ATP + H_2O \leftrightarrow ADP + P_i$$

The terminal phosphate of ATP will be transferred to the glucose in a reaction catalyzed by hexokinase so that the net reaction is as follows:

$$\text{Glucose} + ATP \leftrightarrow \text{Glucose} - 6 - P + ADP$$

In this way, the molecular free energy released by the hydrolysis of ATP can be used to provide the molecular free energy to drive energetically unfavorable reactions.

14.5.2 Glycolysis

The majority of the ATP necessary to drive endergonic reactions is synthesized in the mitochondria. However, because energy cannot be created or destroyed, the synthesis of ATP in the mitochondria requires the input of material. In general, the input of material comes from the glycolytic reactions that occur in the cytosol. Louis Pasteur (1879), while studying wine and beer making, discovered that the breakdown of sugar, known as fermentation, was a general property of all living cells (Fruton, 2006). However, he could only detect fermentation in cells deprived of oxygen, as cells deprived of oxygen metabolize more sugar.

The hydrolysis of glucose is an extremely exergonic reaction. However, it does not occur spontaneously, even though carbon dioxide and water are thermodynamically more stable than glucose and O_2. The bonds are so stable that sugar dust requires temperatures of about 350 C to ignite. That is, glucose is kinetically stable even though it is thermodynamically unstable (Williams, 1961). Thus, while thermodynamics tells us about the possibility of a reaction, it does not tell us anything about its rate (Lewis and Randall, 1923; Glasstone et al., 1941). Thus, glucose will only break down at room temperature in the presence of glycolytic enzymes and/or activated oxygen. An enzyme decreases the activation energy of a reaction. That is, it increases the rate of a reaction by increasing the proportion of substrate molecules capable of entering it (Mehler, 1957).

14.5.2.1 The Discovery of the Glycolytic Pathway

While Justus von Liebig and Marcelin Berthelot were trying to eliminate the need for calling on "the vital force" for explaining biological reactions (anonymous, 1839), Louis Pasteur (1879) was unable to get ground-up yeast to ferment a sugar solution and declared that fermentation was a vital action that required living cells (Duclaux, 1920). However, Eduard Buchner (1897, 1907), using a German beer yeast instead of a French wine yeast, was able to obtain fermentation in vitro when he added sugar to a yeast extract. Actually, this was a lucky find because Buchner had no interest in glycolysis. He was making a health tonic, and only added the sugar as a preservative when the other antiseptics failed to keep the extract sterile (Harden, 1932). Buchner named the extract zymase, from *zyme*, the Greek word for yeast, and *diastasis*, the Greek word for making a breach. Willy Kühne named all biocatalysts, enzymes, from the Greek words *en zyme*, which mean in yeast.

Unlike the burning of wood, which only takes place at high temperatures[1], the oxidation of glucose at ambient temperatures depends on the intervention of enzymes and takes place in about a dozen steps. This leads to the production of many intermediates, and thus cellular respiration is also known as intermediary metabolism. It was astonishing for those biochemists who were working on either the study of fermentation in yeast and bacteria or the study of glycolysis in muscle to find that these two seemingly diverse processes taking place in two different kingdoms were, except for the last steps, identical (Pasteur, 1879; Kluyver, 1931; Plaxton, 1996; Plaxton and Podestá, 2006).

The studies of glycolysis and fermentation opened up an entirely new way of doing "biochemistry." Prior to these studies, most chemists felt that biochemistry should be a "super chemistry" and chemicals should be isolated, but attempts to define their relationship in the cell were rebuked with the words "More matter with less art." Feeling that it was time to say "Enough matter, more art," biochemists, including Arthur Harden (1932), Jan Kluyver (1931), Gustav Embden (Embden and Laquer, 1914), Otto Myerhof (1924), and Jacob Parnas (1910), began to make sense of the copious crowd of newly discovered chemicals that comprised "the chemical zoo" by elucidating the sequence of reactions that occur in living cells.

How could they determine the components and sequence of the pathway in the days before radioactive isotopes and genetic engineering? First, they found inhibitors of CO_2 evolution, including fluoride, and iodoacetate. Then, with classical organic chemical means, they determined which intermediates increased after treatment with the inhibitor. Iodoacetate causes the accumulation of fructose 1,6-bisphosphate, and fluoride causes the accumulation of 2-phosphoglycerate and 3-phosphoglycerate. The compounds were isolated and analyzed by conventional organic analysis (e.g., combustion and enzyme degradation) and volumetric analysis using the manometer developed by Warburg (Dixon, 1952). The chemical of which the concentration increased, after the cells or extract were treated with an inhibitor, was presumably the intermediate produced just prior to the step blocked by the inhibitor. The scientists working on glycolysis also added putative intermediates to the cells or extracts to see if they stimulated CO_2 evolution. If they did, they were presumably intermediates in the glycolytic pathway. However, not all presumptive intermediates of glucose oxidation turned out to be part of the glycolytic pathway, but part of a parallel pathway, known as the oxidative pentose phosphate pathway (Krebs and Kornberg, 1957; Racker, 1957; Horecker, 1982). This pathway is involved in the synthesis of ribose, an important substrate in the formation of nucleotides (Warburg and Christian, 1931, 1936; Hutchinson, 1964). Once the intermediates were discovered, it became possible to identify the enzymes that utilized and produced them.

The evolution of CO_2 from yeast extracts was not constant, but slowed down over time. Harden and Young (1906) found that the addition of phosphate maintained a high rate of CO_2 evolution. Thus, they guessed that phosphate must combine with glucose, and soon after fructose 1,6-bisphosphate, fructose-6-phosphate, and glucose-6-phosphate were isolated. It was also established that creatine was phosphorylated to creatine phosphate, and ADP was phosphorylated to ATP during glycolysis. From these data, the concept of coupled reactions began to emerge (Myerhof and Lohmann, 1928; Fiske and Subbarow, 1929; Lohmann, 1929; Lehmann, 1935; Myerhof, 1944).

Around the same time, Lundsgaard (1938) discovered that iodoacetic acid, an agent that inhibits glycolysis, has no qualitative effect on muscle contraction. This indicated that a store of energy exists in muscle cells. Lundsgaard noticed that in the poisoned cells, creatine phosphate is broken down at a faster rate than in the normal cells and proposed that the energy necessary for muscle contraction resulted from the hydrolysis of creatine phosphate. However, with the realization that creatine phosphate does not occur in all

1. Fahrenheit 451 is the temperature at which book paper catches fire, and burns (Ray Bradbury, 1953. Fahrenheit 451. Ballantine Books, New York).

cells, while another phosphoanhydride, ATP, does, it seemed likely that ATP may be the universal energy source (Kalckar, 1941, 1942, 1944, 1966; Lipmann, 1941). Kalckar and Lipmann independently concluded that the hydrolysis of ATP provided the molecular free energy necessary to perform all kinds of cellular work, including biosynthetic work, osmotic work, and mechanical work.

14.5.2.2 Metabolic Channeling in Glycolysis

Much of what we know about enzymes, coenzymes, and biochemical pathways came from the study of glycolysis by luminaries such as the Büchners, Otto Warburg, and Otto Meyerhof. Although the glycolytic enzymes can be isolated, purified, and even crystallized, they may not be in soluble form in the cell. In fact, these enzymes are part of the cytomatrix, and consequently even detergent-extracted cells are able to oxidize glucose to pyruvic acid (Schliwa et al., 1987). The fact that some of the glycolytic enzymes are immobilized on the cytoskeleton and thus have low diffusion coefficients has been established by fluorescence redistribution after photobleaching (FRAP studies) (Walsh and Knull, 1987; Pagliaro and Taylor, 1988, 1992; Wang et al., 1996; Azama et al., 2003). Giegé et al. (2003) have shown that in *Arabidopsis*, 3%−12% of the activities of the glycolytic enzymes are associated with the outer membrane of mitochondria. Clegg (1991) and Clegg and Jackson (1990) have shown that ^{12}C-glycolytic intermediates do not reduce the amount of $^{14}CO_2$ given off by cells supplied with ^{14}C-glucose, indicating that there is a diffusional barrier between the glycolytic intermediates produced endogenously by the glycolytic pathway and the intermediates supplied exogenously. All these data are consistent with the concept of metabolic channeling (Graham et al., 2007; Williams et al., 2011), although this interpretation is not widely held (Ovádi, 1991; Kühn-Velten, 1993; Mathews, 1993; Cascante et al., 1994).

14.5.3 Cellular Respiration

The breakdown of glucose and the formation of two pyruvic acid molecules during glycolysis require 9 phosphorylated intermediates, 10 enzymes, and 2 nucleotide coenzymes: NAD$^+$ and ATP (see Fig. 14.7; Cori, 1942; Myerhof, 1942; Michal, 1999). The net result of these reactions is the production of two molecules of ATP and two molecules of NADH for every glucose molecule oxidized. The production of two ATPs results in the conversion of only a few percent of the chemical energy available from the combustion of a glucose molecule to the chemical

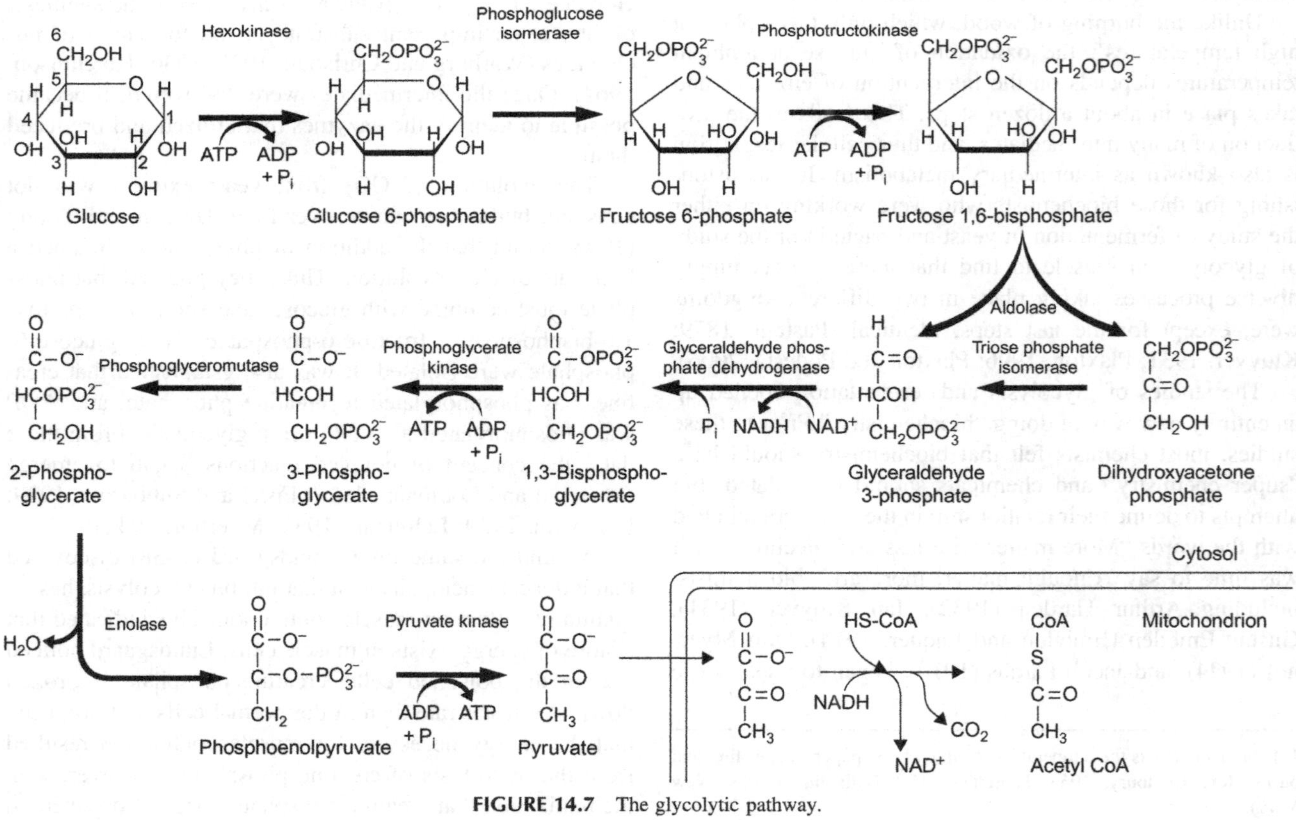

FIGURE 14.7 The glycolytic pathway.

energy available from the hydrolysis of ATP. Most of the energy of glucose is still contained in the two pyruvic acid molecules.

The pyruvic acid must leave the cytosol and enter the matrix of the mitochondria to generate more ATP (Ochoa, 1943; Bonner and Millerd, 1953; Millerd and Bonner, 1953). Here, the pyruvic acid encounters the pyruvate dehydrogenase complex, which is 20–25 nm in diameter. This complex oxidizes pyruvic acid to acetyl-CoA and produces one NADH and one CO_2 from each pyruvic acid molecule. The pyruvate dehydrogenase complex requires a variety of "vitamins," including FAD, NAD^+, CoA, thiamine diphosphate, and lipoic acid (1,2-dithiolane-3-valeric acid; Reed, 2001; Hackert et al., 2015). Acetyl CoA is the "activated" form of acetic acid (Lipmann, 1971).

Other pathways besides glycolysis enter and leave the Krebs cycle. One such pathway is involved in the breakdown of fats, which results in the formation of acetyl CoA, as shown by Feodor Lynen (1964). Moreover, amino acids, the breakdown products of proteins, also enter the Krebs cycle following their deamination.

14.5.3.1 Biochemistry of the Krebs Cycle

In the matrix of the mitochondrion, acetyl-CoA combines with oxaloacetic acid to form citric acid (Fig. 14.8). This step is catalyzed by citrate synthase. The citric acid is then oxidized to isocitric acid in a reaction catalyzed by aconitase. The isocitric acid is further oxidized to α-ketoglutaric acid by isocitrate dehydrogenase. This reaction releases CO_2 and forms one molecule of NADH per acetyl-CoA (Krebs, 1950).

The α-ketoglutaric acid then forms succinyl-CoA in a reaction catalyzed by α-ketoglutarate dehydrogenase. This reaction also releases a CO_2 and forms one molecule of NADH per acetyl-CoA. The succinyl-CoA then forms succinic acid in the presence of succinyl-CoA synthetase. This reaction causes the formation of one molecule of ATP per acetyl-CoA. The succinic acid then is converted to fumaric acid by succinic dehydrogenase, a membrane-bound enzyme (Borsook and Schott, 1931; Hogeboom, 1946; Glock and Jensen, 1953; Neufeld et al., 1954; Kearney and Singer, 1956; Avron and Biale, 1957). This

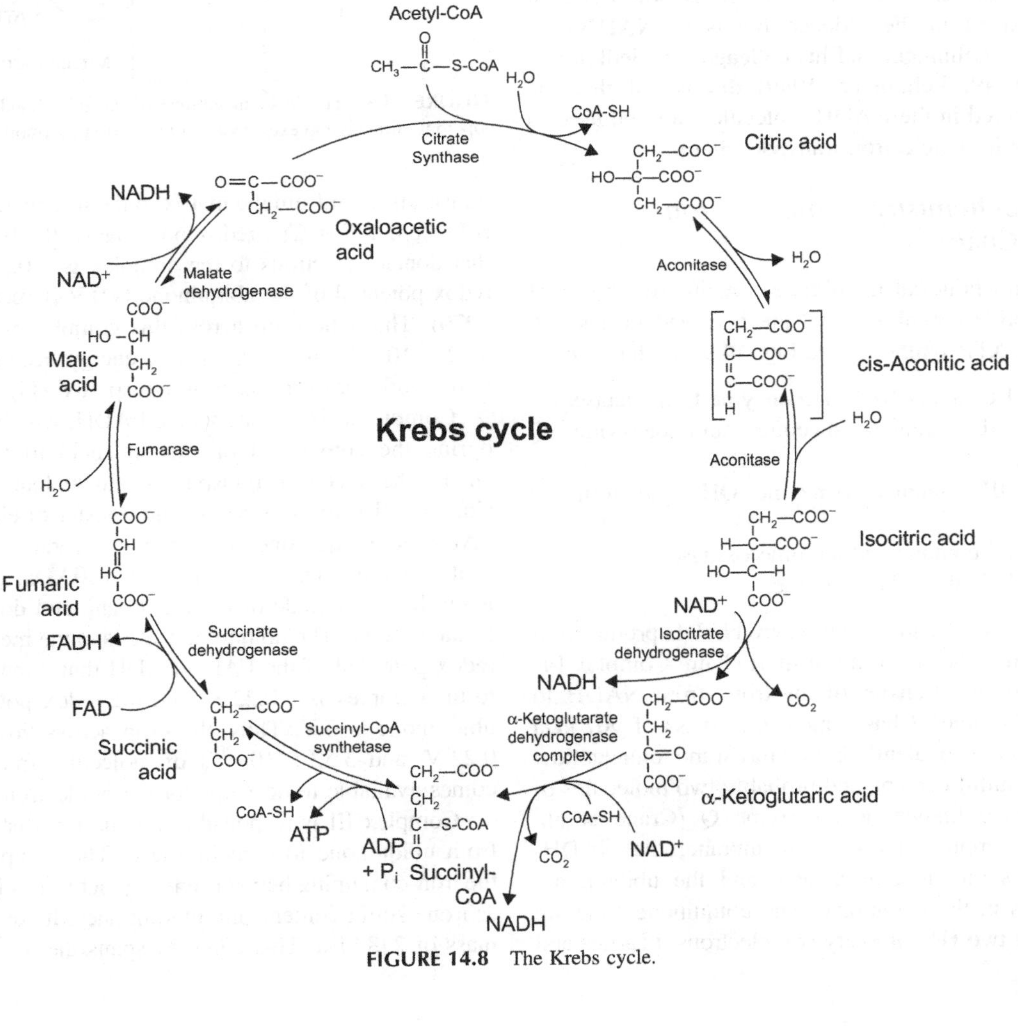

FIGURE 14.8 The Krebs cycle.

reaction causes the formation of one FADH from each molecule of acetyl-CoA. The fumaric acid is then converted into malic acid by the enzyme fumarase. The malic acid is then converted to oxaloacetic acid by the enzyme malate dehydrogenase. This reaction yields one NADH per acetyl-CoA. The oxaloacetic acid is then ready to combine with acetyl-CoA to undergo another round of the cycle. With each round of the cycle, acetyl-CoA is broken down into two CO_2 molecules, three NADH molecules, one FADH molecule, and one ATP. Because two acetyl molecules are derived from each glucose, four CO_2, six NADH, two FADH, and two ATP are produced in the citric acid cycle for each glucose molecule oxidized (Michal, 1999).

Including the one NADH and one CO_2 formed per pyruvate by the pyruvate dehydrogenase complex, a total of six CO_2, eight NADH, two FADH, and two ATP are produced per glucose molecule in the mitochondrion. If we add the two NADH and two ATP formed from each glucose molecule in the cytosol during glycolysis, there is a net yield of 6 CO_2, 10 NADH, 2 FADH, and 4 ATP from each glucose molecule. Only a small percentage of the free energy made available by the oxidation of glucose is conserved in the form of ATP. There is still a lot of energy contained in the reduced bonds of NADH and FADH. Albert Lehninger and his colleagues (Friedkin and Lehninger, 1949; Lehninger, 1949) discovered that the energy conserved in the NADH molecule could be used to produce ATP in the electron transport chain.

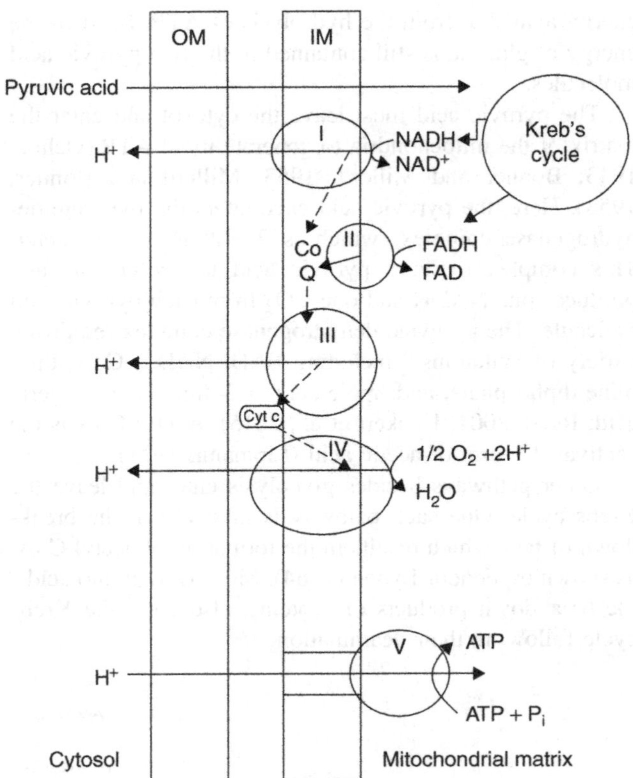

FIGURE 14.9 The inner membrane of the mitochondria showing the topology of the complexes involved in electron transport and ATP.

14.5.3.2 Biochemistry of the Electron Transport Chain

The inner mitochondrial membrane contains five kinds of detergent-soluble complexes that are involved in electron transport and ATP synthesis (see Fig. 14.9; Michal, 1999):

- Complex I contains NADH-coenzyme Q reductase.
- Complex II contains succinic acid-coenzyme Q reductase.
- Complex III contains coenzyme QH_2-cytochrome c reductase.
- Complex IV contains cytochrome oxidase.
- Complex V is the ATP synthase.

Complex I is the entry point where NADH produced in the matrix enters the electron transport chain. Complex I is responsible for the transfer of electrons from NADH to ubiquinone. Complex I has a molecular mass of 907 kDa and contains a noncovalently bound flavin mononucleotide, several iron−sulfur centers, and probably two molecules of ubiquinone, also known as coenzyme Q (Crane et al., 1957). This complex spans the membrane; the NADH-binding site is on the matrix side, and the ubiquinone-binding site is in the membrane. The ubiquinone picks up approximately two H^+ for every two electrons it carries and transports them from the matrix to the intermembranal space (Mitchell, 1975). The redox potential of the NAD^+/NADH that donates electrons to the complex is −0.32 V, and the redox potential of the ubiquinone is 0 V (Green and Baum, 1970). Thus, the drop across this complex is 0.32 V, and 5.12×10^{-20} J of molecular free energy becomes available to do work for every electron transported (Fig. 14.10).

Complex II is the site where FADH, which is produced during the conversion of succinic acid to fumaric acid, enters the electron transport chain (Chance, 1952a,b). Complex II is responsible for the transfer of electrons from FADH to ubiquinone. It contains several nonheme iron centers and has a molecular mass of 130 kDa. This complex is on the matrix side of the membrane and does not translocate protons. The ubiquinone site is in the membrane. The redox potential of the FAD^+/FADH that donates electrons to the complex is −0.22 V, and the redox potential of the ubiquinone is 0 V. Thus, the drop across this complex is 0.22 V, and 3.52×10^{-20} J of molecular free energy becomes available to do work for every electron transported.

Complex III is responsible for the transfer of electrons from ubiquinone to cytochrome c. The complex contains the iron-containing hemeproteins cytochromes b, c_1, as well as iron−sulfur centers, and ubiquinones. It has a molecular mass of 248 kDa. This complex spans the membrane. The

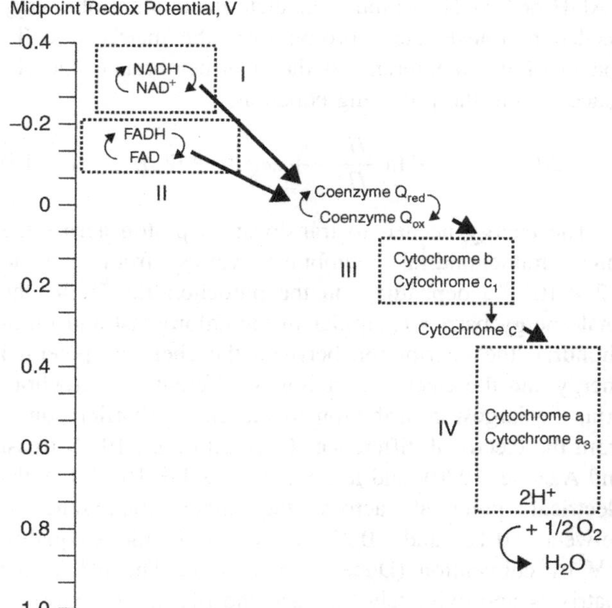

FIGURE 14.10 The estimated midpoint redox potentials of the electron carriers involved in respiration.

cytochrome c_1 and Fe—S protein are on the intermembranal side. Cytochrome b is in the membrane. This complex translocates 2 H^+ per $2e^-$. Cytochrome c is a water-soluble iron-containing hemeprotein that rapidly transfers electrons from complex III to complex IV, as the iron in the cytochrome is oxidized from the reduced ferric ($Fe^{2+)}$ ion to the oxidized ferrous (Fe^{3+}) ion. The redox potential of the ubiquinone that donates electrons to the complex is 0 V, and the redox potential of cytochrome c is 0.26 V. Thus, the drop across this complex is 0.26 V, and 4.16×10^{-20} J of molecular free energy becomes available to do work for every electron transported.

Plant mitochondria are able to oxidize cytosolic NADH directly, unlike animal mitochondria. There is an NADH dehydrogenase on the outer surface of the inner membrane that is specific for the β-hydrogen of NADH and feeds electrons directly to complex III. As I will discuss in the following, the oxidation of one molecule of cytosolic NADH then leads to the production of two molecules of ATP. By contrast, each molecule of NADH produced in the matrix of the mitochondria is capable of producing three molecules of ATP.

Complex IV is the terminal complex of the electron transport chain and is often called cytochrome c oxidase. It is the enzyme inhibited by cyanide and carbon monoxide (Commoner, 1940). This membrane-spanning complex transfers electrons from cytochrome c to O_2. These electrons bind to protons in the matrix and consequently water is formed along with the slight alkalinization of the matrix. Complex IV contains the iron-containing hemeprotein cytochrome a on the intermembranal side and the iron-containing hemeprotein cytochrome a_3 on the matrix side. It has a molecular mass of 210 kDa. This complex translocates protons with a stoichiometry of 4 $H^+/2e^-$ (Wikstrom, 1977). The protons are translocated through this complex in the opposite direction than the electrons are. The redox potential of the cytochrome c that donates electrons to the complex is 0.26 V, and the redox potential of the water is 0.82 V. Thus, the drop across this complex is 0.56 V, and 8.96×10^{-20} J of molecular free energy becomes available to do work for every electron transported.

These complexes do not exist in a linear chain in the inner membrane of the mitochondria, and electron transfers are mediated by random collisions between electron donors and electron acceptors. The small lipid-soluble ubiquinone diffuses rapidly in the plane of the membrane and transfer electrons from complex I and II to complex III. Electrons are transferred between complexes at a rate of one electron transferred per 5—20 ms, the rate of "typical" enzyme reactions. The nonlinear nature is deduced from the observation that the complexes are present in variable ratios. The four electron transport complexes are inserted into the membrane in a fixed orientation so that the protons are all pumped from the matrix (P-space) to the intermembrane space (E-space). This vectorial organization can be demonstrated by using membrane-impermeable probes to specifically label proteins.

The ordered transfer of electrons depends only on the redox potential and the concentrations of the various components. Using Einstein's random walk equation, Ogston and Smithies (1948) calculated that if the enzymes of the electron transport chain were not confined to the inner membrane of the mitochondria, but were free to diffuse throughout the cell, the rate of respiration would be miniscule, thus the "solid-state packing" of the enzymes makes it possible for the cell to produce ATP at a high rate (Ball and Barrnett, 1957; Ziegler et al., 1958; Lehninger, 1959; Smithies, 2007; Gitschier, 2015; Koller, 2017; Kucherlapati, 2017).

14.5.3.3 Oxidative Phosphorylation

In the 1950s, as a result of osmotic studies (Harman and Feigelson, 1952a,b), George Laties (1953) realized that "the remarkable biochemical integration of the mitochondria is closely related to their precise physical organization." The function of the electron transport chain on the inner membrane is to transport H^+ from the matrix side (P-space) to the intermembranal side (E-space). The proton translocation is electrogenic, thus the matrix side becomes negative relative to the intermembranal side. The electrical difference across the inner mitochondrial membrane can be monitored spectroscopically with safranine, and the pH difference across the inner mitochondrial membrane can be monitored spectroscopically with neutral red (Wikstrom, 1977). Alternatively, the electrical difference across the inner mitochondrial membrane can be monitored by measuring the distribution of lipophilic cations such as triphenylmethylphosphonium, and the pH difference across the inner mitochondrial membrane can be monitored by measuring the distribution of radioactive weak bases such as methylamine (Hoek et al., 1980). The free energy inherent in the electrical potential difference across the membrane can be harnessed to make ATP by complex V, the F_1F_0-ATP synthase (Mitchell, 1966; Boyer, 1997; Walker, 1997; Prebble and Weber, 2003). The ATP synthase synthesizes ATP when protons move passively through its proton channel from the E-space to the P-space. Thus the topology of the ATP synthase is the same for the mitochondrion and the chloroplast (Poyton, 1983). Reconstitution experiments pioneered by Ephraim Racker, in which he inserted purified proteins into liposomes, clearly demonstrated the components involved in oxidative phosphorylation and their individual functions (Penefsky et al., 1960; Pullman et al., 1960; Fessenden et al., 1966; Kagawa and Racker, 1966a,b, 1971, 1976; Fessenden-Raden et al., 1969; Racker and Stoeckenius, 1974; Jagendorf et al., 1991; Schatz, 1996; Kresge et al., 2006; Allchin, 2007). When the ATP synthase is stripped from isolated, inside-out, inner mitochondrial membranes, the inner membranes still oxidize NADH, although they lose their capacity to synthesize ATP (Racker, 1975a,b, 1976). When the ATP synthase is added back to the membranes, they synthesize ATP on the addition of NADH. However, ATP is only synthesized at the expense of NADH oxidation, when the membrane is intact and impermeable to H^+. Thus, in oxidative ATP synthesis, as in photophosphorylation, the free energy inherent in a proton, by virtue of its position, is used to make ATP. This is why proton ionophores, including dinitrophenol (Loomis and Lipmann, 1948), uncouple ATP synthesis from electron transport in mitochondria and chloroplasts.

14.5.3.4 Thermodynamics of Oxidative Phosphorylation

The amount of molecular free energy made available by each electron transport complex can be calculated from Eq. (14.14) (Ball, 1944). However, this only gives us an estimate, as the real value will depend on the real concentrations of the reduced and oxidized forms of the couple, and there will be some loss of molecular free energy due to friction.

$$\Delta E = nze\left(z\Psi^{\text{final}} - z\Psi^{\text{initial}}\right) \qquad (14.14)$$

Only the molecular free energy that can be coupled to proton transport will be conserved, and the rest will be dissipated as heat. To determine how many protons can be translocated for each pair of electrons that come from NADH or FADH, we must calculate the amount of energy needed to translocate a proton from the matrix side (P-space) of the membrane to the intermembranal side (E-space), using the following equation:

$$\Delta E_{(e-p)} = kT \ln \frac{[H^+]_e}{[H^+]_p} + ze(\Psi_e - \Psi_p) \qquad (14.15)$$

The energy needed to translocate a proton across the inner mitochondrial membrane varies from 2.5 to 4.2×10^{-20} J depending on the mitochondria. While the total energy barrier is similar in the chloroplast and mitochondria, the distribution between the chemical potential energy and the electrical energy is different. In mitochondria, the largest contribution to the energy barrier comes from the electrical difference (Cockrell et al., 1967; Rossi and Azzone, 1970), and it amounts to 2.4×10^{-20} J as the electrical potential across the inner membrane is between −0.126 and −0.250 V, and the E-side is equal to 0 V by convention (Ducet et al., 1983). The pH of the matrix is approximately 7.5, and the pH of the cytosolic side is approximately 7 (Nicholls and Ferguson, 1992), which only accounts for about 5×10^{-21} J, thus it takes approximately 2.4×10^{-20} J to pump a proton. Consequently, complexes I, III, and IV each provide enough molecular free energy to pump at least one proton per electron. This value may vary within a mitochondrion due to shape changes, as the energetics depend on the pH of the mitochondrion, and the pH depends on the volume (Lehninger, 1959). Sensors are currently being developed to determine the redox state, the pH, and the Ca^{2+} concentration in the mitochondrial matrix in plants (Wagner et al., 2015).

The energy inherent in the electrical potential difference generated by proton pumping can then be used to synthesize ATP and water from ADP and P_i. We use Eq. (14.16) to calculate the energy made available by a proton moving down its electrochemical potential difference from the intermembranal side (E-space) to the matrix side (P-space). Thus, each proton makes available approximately 3×10^{-20} J of free energy, which is able to do work.

$$\Delta E_{(p-e)} = kT \ln \frac{[H^+]_p}{[H^+]_e} + ze(\Psi_p - \Psi_e) \qquad (14.16)$$

The free energy of protons can be used to synthesize ATP from ADP and P_i. According to Eq. (14.17), the energy required for the synthesis of ATP will depend on K_{Real}, which depends on the relative concentrations of ATP, ADP, P_i, and water. Assuming that [ATP], [ADP], [P_i], and [H_2O] equal 10^{-3}, 10^{-3}, 10^{-3}, and 55.5 M, respectively, then the molecular free energy required for the synthesis of ATP is 8×10^{-20} J. However, just as the molecular free energy made available by the hydrolysis of ATP depends on the relative concentrations of ATP, ADP, and P_i, so does the molecular free energy required for ATP synthesis.

$$E_{Real} = -kT \ln \frac{K_{eq}}{K_{Real}} = kT \ln \frac{K_{Real}}{K_{eq}} \qquad (14.17)$$

These thermodynamic calculations reveal that it is possible to make one ATP with approximately three protons. They also tell us that complex I can pump 2 $H^+/2e^-$, complex II can pump 2 $H^+/2e^-$, and complex IV can pump 4 $H^+/2e^-$. Thus, for every NADH oxidized to water, eight protons are transported across the inner membrane. When these protons pass back into the matrix through the ATP synthase, they provide the energy for the synthesis of two to three ATPs. These theoretical determinations are corroborated by experiments performed by Peter Hinkle and his colleagues (Thayer and Hinkle, 1975a,b; Hinkle and McCarty, 1978; Hinkle and Yu, 1979; Berry and Hinkle, 1983; Scholes and Hinkle, 1984; Krishnamoorthy and Hinkle, 1984, 1988; Hinkle et al., 1991).

Why is the stoichiometry of H^+ translocation and ATP synthesis variable or uncertain? Stoichiometric calculations must take into account that mitochondria do not exist in isolation but must take up substrates, including inorganic phosphate and ADP from the cytosol, and export products, including ATP. Each of these processes requires energy and makes use of the proton difference. Another factor that affects the stoichiometry is that the volume of the mitochondria may not be constant. Consequently, the movement of molecules in and out of the mitochondria may result in a swelling or shrinking, which will affect the concentrations of all substances and the concentrations will affect the rates of the reactions.

How many ATPs are generated from each glucose oxidized? Because there are 6 NADHs coming from the Krebs cycle for each glucose, there will be between 12 and 18 ATPs formed from them. Two NADHs also come from the pyruvate dehydrogenase complex, which will give us four to six more ATPs. The two FADHs formed during the Krebs cycle will yield two ATPs each for a total of four ATPs from FADH. Thus, the mitochondria will produce 20–28 ATPs via the electron transport chain. Including the four ATPs formed by substrate-level phosphorylation during glycolysis and in the Krebs cycle, 24–32 ATPs are made. Lastly, if the two NADHs formed during glycolysis enter the electron transport chain at complex III, then four more ATPs will be made for each glucose. This yields a grand total or 28–36 ATPs per glucose.

Six molecules of molecular oxygen (O_2) are consumed in the oxidation of a six carbon sugar. The overall reaction of respiration is thus

$$C_6H_{12}O_6 + 6O_2 + nADP + n\,P_i \leftrightarrow 6CO_2 + (n+6)H_2O + nATP$$

14.5.3.5 The Alternative Oxidase

In addition to the cytochrome oxidase chain, plants have a cyanide-resistant alternative oxidase that is capable of pulling electrons from the reduced quinone pool (Henry and Nyns, 1975; Laties, 1998) and reducing oxidative stress (Møller, 2001). The alternative oxidase has been purified (Moore and Rich, 1985). The alternative pathway is usually thought to be uncoupled from ATP synthesis so that the electrons can move through the alternate oxidase pathway quickly, and most of the energy of the NADH is given off as heat rather than being conserved in the bonds of ATP. This explains the observations of Lamarck and Senebier that plants in the Araceae family, including skunk cabbage (*Symplocarpus foetidus*), the corpse plant (*Amorphophallus titanium*), and *Arum*, generate heat to melt the snow around their flowers in the late winter or use the heat to volatilize substances that attract flies (Lamarck and de Candolle, 1815; Church, 1908; James and Beevers, 1950; Knutson, 1974, 1979; Meeuse, 1975; Meeuse and Raskin, 1988; Seymour et al., 1983; Laties, 1998; Seymour, 2004; Gibernau et al., 2005). The male flowers of *Arum* produce salicylic acid, and the generation of heat by the alternative pathway in *Arum* and other plants can be induced by exogenous salicylic acid (Raskin et al., 1987, 1989, 1990; Kapulnik et al., 1992; Raskin, 1992a,b; Minorsky, 2003). It is also possible that in some cases the alternative pathway does produce ATP (Wilson, 1980). The alternative pathway in *Chara*, like that in protozoa, is capable of synthesizing ATP (Takano and Wayne, 2000).

The alternative pathway is expressed in some cells but not in others (Moore and Rich, 1985; Elthon et al., 1989; Guy et al., 1989; Ordentlich et al., 1991). I think that it is fascinating that this pathway is induced by chilling (Vanlerberghe, 2013) and as such may function in cellular temperature regulation (Moynihan et al., 1995). Indeed, as the generation of ATP is an enzymatic process, the diversion of electrons to the alternative pathway may actually maximize ATP formation by raising the temperature closer to the optimum for maximal enzymatic activity.

14.5.3.6 Historical Aspects of Cellular Respiration

While most biologists have memorized the Krebs cycle many times, few know how it was discovered. Hans Krebs had just elucidated the ornithine cycle, in which urea is synthesized from carbon dioxide and ammonia (Krebs, 1970, 1981b), and in the 1930s, he was ready to tackle the problem of how pyruvate is oxidized to carbon dioxide and water. Krebs was prepared to see everything as a cycle, and the burning of foodstuffs was no exception. Compared with a linear pathway, such a cycle would allow better regulation and integration with the whole cell (Krebs, 1971).

Krebs knew that foodstuffs undergo combustion to release energy, as Lavoisier pointed out, and that enzymes typically react with only one or two substrates at a time. Each step releases molecular free energy in the ballpark of 10^{-20} J. Thus, Krebs assumed that the combustion of pyruvic acid to CO_2 and H_2O probably involved many steps, and he decided to search for the intermediates.

As a prelude to Krebs' work on oxidation, Albert Szent-Györgyi (1935, 1937) developed a minced pigeon flight muscle system to study the oxidations that may be involved in cellular energetics. Szent-Györgyi added various compounds to the minced muscle to test if any of them were capable of stimulating pyruvate-dependent oxygen uptake. He assumed that the ones that stimulated oxygen uptake would turn out to be the natural intermediates between pyruvic acid and oxygen. Szent-Györgyi tested a number of dicarboxylic acids that were typically found in high concentrations as secondary products in the plants they are named after (e.g., fumaric acid from the fumatory; malic acid from *Malus*; Davy, 1827). Of the chemicals he tried, only succinic acid, fumaric acid, malic acid, and oxaloacetic acid stimulated oxygen uptake. Moreover, the amount of oxidation caused by each dicarboxylic acid was greater than that necessary to oxidize the added compound. Thus, he concluded that the dicarboxylic acids acted as catalysts in the oxidation of pyruvic acid. He also showed that added dicarboxylic acids could reduce methylene blue, an artificial hydrogen acceptor that mimicked oxygen. Szent-Györgyi used the reduction of methylene blue as a convenient way to assay the transfer of electrons because methylene blue changes from blue to colorless when it is reduced. Szent-Györgyi interpreted the observed reductions to mean that the electrons and protons that made up the hydrogens of pyruvic acid were passed in order from pyruvic acid to oxaloacetic acid to malic acid to fumaric acid to succinic acid in vivo. At this point, the electrons associated with the hydrogens would enter the electron transport chain.

Meanwhile Martius and Knoop (1937) were studying the oxidation of citric acid in both the test tube and in liver tissue using the techniques of organic chemistry (Martius, 1982). They showed that citric acid is converted to cis-aconitic acid, then to isocitric acid, and then to α-ketoglutaric acid.

It was already known that α-ketoglutaric acid could be converted to succinic acid. Thus, Krebs realized that if Szent-Györgyi made a mistake in the order of the pathway, then during pyruvic acid oxidation, citric acid could be converted all the way to oxaloacetic acid (Krebs and Johnson, 1937; Krebs and Eggleston, 1940a,b; Krebs, 1940a,b, 1943; Burton and Krebs, 1953; Krebs et al., 1953). Krebs and his colleagues showed that pyruvic acid stimulated both O_2 uptake and citric acid synthesis. They found that under anaerobic conditions, the addition of pyruvic and oxaloacetic acids to muscle suspensions led to the formation of citric acid, which they measured colorimetrically. Thus, Krebs postulated that a cycle was involved in intermediary metabolism, and he called it "the citric acid cycle."

The citric acid cycle was accepted by most, but not all, biochemists (Breusch, 1939; Thomas, 1939). The evidence for the cycle included the observations that

1. Succinic, fumaric, malic, oxaloacetic, α-ketoglutaric, citric, isocitric, and cis-aconitic acids stimulate O_2 uptake in pigeon muscle suspensions.
2. Catalytic amounts of these acids stimulate the oxidation of pyruvic acid.
3. The stimulations by catalytic amounts of these acids are prevented by malonic acid, an inhibitor of succinic dehydrogenase.

Malonic acid is an analogue of succinic acid and was known to inhibit O_2 uptake in whole cells (Thunberg, 1910; Quastel and Wheatley, 1930). Krebs treated cells with malonic acid and measured how much succinic acid was formed on addition of the intermediates. The amount of succinic acid was determined by measuring the volume of oxygen taken up in order of oxidize succinic acid to fumaric acid in the presence of added succinic dehydrogenase and an electron acceptor (Krebs and Eggleston, 1940b). The amount of succinic acid formed was equal to the amount of added citric acid, isocitric acid, cis-aconitic acid, or α-ketoglutaric acid. Moreover, in the presence of pyruvic acid, succinic acid was quantitatively formed in malonic acid–treated cells on the addition of fumaric acid, malic acid, or oxaloacetic acid.

Lastly, Krebs and his colleagues found that pyruvic acid oxidation was blocked in malonic acid–treated suspensions, but this inhibition was overcome by adding oxaloacetic acid in stoichiometric amounts (1:1) but not in catalytic amounts. This occurs because oxaloacetic acid is not regenerated in malonic acid–poisoned preparations as it is in untreated tissues. Because pyruvic acid is oxidized by a cycle, an infinite number of pyruvic acid molecules can be oxidized by just one molecule of oxaloacetic acid.

With the introduction of isotope-labeling techniques, it was found that the ^{14}carbon from the carboxyl group of pyruvate appeared only in the γ-carboxyl group of α-ketoglutaric acid and not in both the α- and the γ-carboxyl group. This asymmetrical labeling made it seem unlikely that citric acid was the first product of the cycle because citric acid is a symmetrical molecule, and consequently the labeled carbon in its terminal carboxyl group should appear in either of the two carboxyl carbons of α-ketoglutaric acid. Thus, it was proposed that either cis-aconitic acid or isocitric acid was the first condensation product of acetate and oxaloacetic acid, and the cycle was renamed the tricarboxylic acid cycle. The citric acid was thought to be a branch off of the asymmetrical cis-aconitic acid.

However, Sandy Ogston (1948) suggested that a symmetrically labeled molecule such as citric acid could give rise to an asymmetrically labeled molecule if the symmetrical molecule binds to an asymmetrical binding site in the aconitase (Bentley, 1978). This model has been elaborated by Mesecar and Koshland (2000) and Koshland (2002). That is, the substrate does not bind to the enzyme in a random position, but in a preferred manner. This seemed to solve the problem (Krebs, 1950), and thus the citric acid cycle, which was renamed the tricarboxylic acid cycle, came to be known as the *Krebs cycle*. According to Lehninger (1975), the asymmetrical action of aconitase comes from a stereochemical attribute (prochirality) of the citric acid molecule itself.

In 1951, coenzyme A was discovered by Fritz Lipmann (1951). He determined that pyruvic acid is decarboxylated to acetyl-CoA, and it is acetyl-CoA that enters the Krebs cycle. Thus, acetyl-CoA combines with oxaloacetic acid to form citric acid. This is the Krebs cycle as we know it today. Proteins and lipids are also oxidized for energy in the Krebs cycle by way of acetyl-CoA. Many reminiscences have been written about the discovery of the pathways involved in carbohydrate metabolism (Kennedy, 2001; Kornberg, 2001; Buchanan, 2002; Berg, 2003; Kornberg, 2003).

In the 1950s, it was shown that the Krebs cycle also occurs in plants (Bonner and Millerd, 1953; Millerd, 1953). Harry Beevers and David Walker (1956) showed that all the intermediates of the Krebs cycle stimulated O_2 uptake catalytically in the presence of pyruvic acid in castor bean mitochondria. Raymond and Burris (1953) showed that pyruvic acid-2-C^{14} was incorporated into the tri- and dicarboxylic acids of lupine mitochondria. And Avron and Biale (1957a) showed that malonic acid inhibited O_2 uptake in avocado mitochondria when succinic acid was used as the substrate. Thus, plant and animal mitochondria are similar in the manner that they combust foodstuffs (James, 1933, 1957; Beevers, 1961).

By the way, when Krebs first submitted his paper to *Nature*, he got the following rejection letter: "The editor of *Nature* presents his compliments to Dr. H. A. Krebs and regrets that as he has already sufficient letters to fill the correspondence column of *Nature* for seven or eight weeks, it is undesirable to accept further letters at the present time." Krebs and Johnson (1937) then submitted their paper to *Enzymologia* where it was accepted.

Now I will discuss how the electron transport chain was discovered. At the relatively low temperature of our body or in plant cells, oxygen does not directly attack carbohydrates. Thus, investigations ensued as to how oxygen was activated. It seemed likely that the catalyst would contain iron, which has a high affinity for oxygen and is able to bind it reversibly. In the mid-19th century it was widely held that the blood was involved in respiration, and indeed, an iron-containing compound was present in the blood. In fact, the concentration of iron in the blood is so high that the French nobility wore rings made from the iron extracted from the blood of their friends as a keepsake, much as people wear a lock of a loved one's hair in a ring around their finger or in a locket around their neck (anonymous, 1848). In the 1850s, Claude Bernard realized that the color of the blood depended on its state of oxygenation, and he and Felix Hoppe-Seyler independently found that when CO displaced O_2 in arterial blood, the color became even more vivid red. George Gabriel Stokes (1864) concluded that hemoglobin exists in oxidized and reduced states, and these can be distinguished by their absorption spectra.

After the general acceptance that respiration was a cellular phenomenon, respiratory pigments were searched for in many animal tissues. Charles MacMumm, in 1884–86, found a pigment of which the absorption spectrum varied with its oxidation state, just like hemoglobin. David Keilin, using yeast, showed that this pigment consisted of three pigments that he named cytochromes a, b, and c (see Keilin, 1966). Cytochrome c, the soluble cytochrome, was purified by Hugo Theorell in 1936. Oddly enough, none of the cytochromes a, b, or c interacted directly with O_2. Otto Warburg discovered another pigment, which he called the respiratory enzyme because it directly binds O_2 and oxidizes cytochrome. Warburg was a brilliant biochemist, and he was made an honorary Aryan by Hermann Goering so that he could continue his research on cancer in Germany. In contrast to the decision made by James Franck, Warburg decided to remain in Germany while other non-Aryan biochemists, including Carl Neuberg, Otto Meyerhof, and Hans Krebs were dismissed from their positions.

Warburg was able to isolate this important enzyme using a clever assay. He found that CO inhibited respiration in sea urchins and light reverses the inhibition. Moreover, light caused the release of CO from the respiratory enzyme, and consequently changed its absorption spectrum. In this way, he could assay each fraction to find the one of which the absorption spectrum in the presence of CO varied in the light compared with the dark. Later, Keilin showed that the respiratory enzyme, which had previously been isolated as indophenol oxidase, contains cytochromes a and a_3.

The notion that ATP synthesis is coupled to electron transport reactions comes from the observation that tissue extracts with high rates of O_2 uptake produced many phosphorylated products, and cyanide inhibited both O_2 uptake and phosphorylation (Kalckar, 1937). One of the phosphate acceptors was ADP, the phosphorylation of which resulted in ATP (Needham and Pillai, 1937; Kalckar, 1939). Kalckar suggested that the energy of respiration was conserved in phosphorylated products, and Albert Lehninger and his colleagues (Lehninger, 1949; Friedkin and Lehninger, 1949) showed that ATP is formed when NADH is given to isolated mitochondria.

14.6 OTHER FUNCTIONS OF THE MITOCHONDRIA

The Krebs cycle functions not only in energy conversion but also in the interconversion of carbon skeletons (Bandurski and Lipmann, 1956). The Krebs cycle has many sites where intermediates can enter and leave and as such, acts as a distribution center where the supplies and demands for carbon skeletons are analyzed and the carbon skeletons are converted to the required product and sent off (Krebs, 1954). Given the fact that intermediates enter and leave at all points of the cycle, it is unlikely that the enzymes involved in this cycle exhibit any form of metabolite channeling. Trefil et al., (2009) consider the Krebs cycle to be the most likely candidate that marks the conversion between geochemistry and biochemistry in the origin of life.

All mitochondria participate in respiration. However, mitochondria have evolved to perform various functions in a given cell type. There are differences between animal and plant mitochondria, and the DNA of plant mitochondria is much larger than the DNA of animal mitochondria (Douce and Neuburger, 1989). Some of the observed differences may reflect the autotrophic versus heterotrophic lifestyle of the mitochondrial host. However, remember that some plant cells are heterotrophic, whereas others are autotrophic, and consequently, the mitochondria of each tissue may be specialized. For example, Valerio et al. (1993) found that mitochondria isolated from autotrophic pea leaf cells are capable of hydrolyzing ATP and in doing so, generate a membrane potential and acidify the medium. Thus, the ATP synthase can "run backward" in these mitochondria. However, mitochondria isolated from heterotrophic potato tuber cells are unable to "run backward" (Valerio et al., 1993).

In plant mitochondria, the conversion of glycine to serine in the matrix of the mitochondria is an important aspect in photorespiration, which involves chloroplasts, peroxisomes, and mitochondria (see Chapter 5). Furthermore, the conversion of succinic acid to fumaric acid and fumaric acid to malic acid in the matrix of plant mitochondria is an important aspect of the conversion of fats to sucrose, which involves the lipid bodies, peroxisomes, mitochondria, and cytosol (see Chapter 5). The matrix of animal mitochondria contains the enzymes responsible for the β-oxidation of fatty acids, a function that takes place in the peroxisomes of plant cells.

The matrix of plant mitochondria contains the enzymes involved in proline oxidation. These enzymes include pyrroline-5-carboxylic acid dehydrogenase, which catalyzes the conversion of glutamic acid to pyrroline-5-carboxylic acid, and proline dehydrogenase, which catalyzes the conversion of pyrroline-5-carboxylic acid into proline. In several plants, proline is a compatible solute that accumulates in response to salt or water stress. Proline accumulation results from the stimulation of proline synthesis and the inhibition of proline oxidation (Steward et al., 1977). In this way, the mitochondria of plants are involved in the osmotic homeostasis of plant cells.

The inner membranes of plant and animal mitochondria contain the transport proteins that regulate the influx and efflux of substrates and products involved in energy metabolism. They also contain a transport protein involved in Ca^{2+} uptake, and thus the mitochondria participate in cellular calcium homeostasis. The mitochondria are high-capacity, low-affinity calcium stores.

14.7 GENETIC SYSTEM IN MITOCHONDRIA

Mitochondria, like chloroplasts, contain a complete genetic system, including DNA, ribosomes (70S), and the enzymes necessary to synthesize DNA (Christophe et al., 1981; Heinhorst et al., 1990; Chase, 1998; Moriyama and Sato, 2014), rRNA, tRNA, and proteins. There are typically 5–10 DNA molecules in the matrix of the mitochondrion. The DNA of liverwort mitochondria is circular or linear and contains 150–2500 kilobase pairs. This is much larger (10–100×) than the circular mitochondrial DNA of animals, which contains 16–19 kilobase pairs. By contrast, the mitochondrial DNA of *Chlamydomonas* is also small, containing 16 kilobase pairs; however, it is linear (Schuster and Brennicke, 1994). In some plants, the mitochondrial genome is organized into three circular "chromosomes" (Palmer and Shields, 1984). The mitochondrial DNA is associated with histone-like protein, as evidenced by cytological staining and protein purification (Kuroiwa, 1982). Some mitochondrial DNAs have introns (Fox and Leaver, 1981; Conklin et al., 1991; Pruitt and Hanson, 1991).

There is a tendency for mitochondrial genes to move to the nucleus through time. Within genera of the legume family, there are examples where the cox II gene, which codes for a polypeptide of cytochrome oxidase, has been transferred completely, partially, and not at all to the nucleus (Nugent and Palmer, 1991; Covello and Gray, 1992; Schuster et al., 1993; Schuster and Brennicke, 1994). Genetic transfer occurs between all the genomes of a cell: the mitochondrial DNA even has some promiscuous chloroplast sequences (Stern and Lonsdale, 1982; Stern et al., 1983; Koulintchenko et al., 2003).

The nucleotide sequence of the RNA transcribed in the mitochondria may differ from the DNA template due to editing (Araya et al., 1994; Bentolila et al., 2012, 2013; Diaz et al., 2017; Shi et al., 2017a,b), where CGG codons are converted to UGG. Editing was discovered by comparing the sequence of RNA transcripts with the sequence of the gene in question. This further strengthens the notion that life and information transfer is based on integration, and a single sequence, taken in isolation, has little meaning as far as regulation is concerned.

The genetic system of mitochondria is inhibited by the same antibiotics that inhibit bacterial growth. The transcription of mitochondrial RNA is inhibited by acridines, whereas the transcription of nuclear RNA is inhibited by α-amanitin (Bushnell et al., 2002). Likewise, translation on mitochondrial ribosomes is inhibited by chloramphenicol, tetracycline, and erythromycin, while cytoplasmic ribosomes are inhibited by cycloheximide.

The relative amounts of proteins, as well as the types of proteins synthesized by the mitochondrial genetic system, depend on the organ within which the mitochondria reside (Newton and Walbot, 1985; Rhoades and McIntosh, 1993; Conley and Hanson, 1994; Monéger et al., 1994). Therefore, there must be interesting regulatory mechanisms that control mitochondrial gene expression in a developmentally regulated manner. The mitochondria, like the plastids, do not have the capacity to code all their own proteins. The majority of the mitochondrial proteins coded by nuclear DNA, synthesized on cytoplasmic ribosomes, are imported through translocons from the cytosol into the mitochondria.

The presence of a mitochondrial genome allows for cytoplasmic inheritance (see Fig. 14.11; Sun et al., 1988) of maternal or paternal mitochondrial DNA. Maternal or paternal inheritance of mitochondrial DNA, like the maternal or paternal inheritance of chloroplast DNA, results from a variety of mechanisms, including the preferential distribution or destruction of mitochondria and its DNA (see Chapter 13).

While accepted today, the concept of cytoplasmic inheritance was frowned on amid the excitement of the rediscovery of Mendelian inheritance. The concept of an extranuclear genome led to a "cold war in biology" between those who believed that the Mendelian inheritance of nuclear genes can account for all aspects of heredity, and those who believed that there were not only genes in the nucleus but also independent self-replicating and developmentally influential genes in the cytoplasm as well (Darlington, 1944; Srb and Owens, 1952; Waddington, 1956; Lindegren, 1966). The first evidence for cytoplasmic inheritance came from the work of Carl Correns (1909), who showed that the color of the progeny from variegated plants depended on the color of the female and not of the male. This indicated that there were extra chromosomal factors involved in inheritance. Later, Boris Ephrussi (1953) also obtained results from genetic crosses that could not be easily explained by assuming that the nucleus was the sole regulator of heredity. Encouraged by the words of Joseph Henry Woodger (1948), who wrote "Admittedly, some hypotheses have become so well established that no one doubts them. But this does not mean that they are known to be true. We cannot determine the truth of a hypothesis by counting the number of people who believe it, and a hypothesis does not cease to be a hypothesis when a lot of people believe it," Ephrussi postulated that there were genetic factors in the mitochondria.

Ephrussi came to this conclusion after observing yeast that formed "petite plaques." It appeared that the gene that caused the petite plaques had an apparently high mutation rate, and the mutation was irreversible. As the "petite plaques" were smaller than the big plaques only in the presence of oxygen, yet grew at a similar rate in the absence of oxygen, Ephrussi deduced that the factor responsible for plaque size might be in the mitochondria. He suggested that the apparent mutation rate was high due to the unequal segregation of mitochondria that contained the "petite" gene or the "big" gene to the daughter cells. If a daughter cell only got the mitochondria with the "petite" gene, the so-called mutation would show up and be irreversible. Today, we know that these petite mutants lack large sections of DNA in their mitochondria. We also have many examples of cytoplasmic inheritance (Sager and Ryan, 1961).

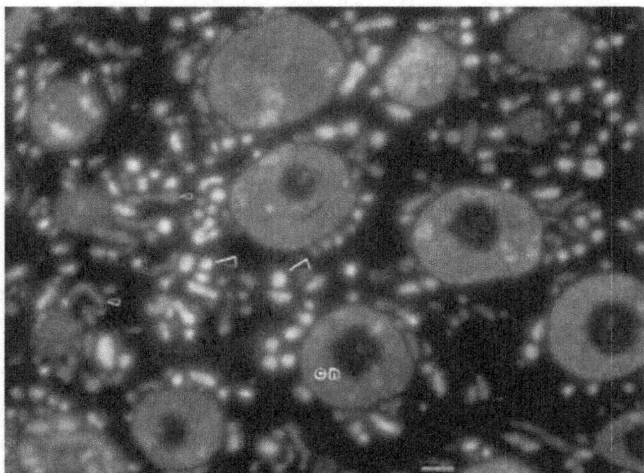

FIGURE 14.11 Cells in the quiescent center of the root of *Pelargonium zonale* stained with 4′-6-diamidino-2-phenylindole (DAPI), a DNA-specific stain. The large arrowheads point to mitochondria, and the small arrowheads point to plastids. Bar, 2 μm. *From Kuroiwa, T., Fujie, M., Kuroiwa, H., 1992. Studies on the behavior of mitochondrial DNA. J. Cell Sci. 101, 483–493.*

14.8 BIOGENESIS OF MITOCHONDRIA

Mitochondria divide to give rise to other mitochondria. Prior to division, a spherical mitochondrion elongates, becomes ovoid, and then becomes dumbbell shaped as the center begins to constrict and the two halves eventually separate (Fig. 14.12). Mitochondrial division, like plastid division, also requires filamentous GTP-binding FtsZ proteins (Osteryoung and Nunnari, 2003; Miyagishima et al., 2004) and/or the filamentous GTP-binding dynamin (Arimura et al., 2004b; Mano et al., 2004), where dynamin may have replaced FtsZ as an effecter of mitochondrial

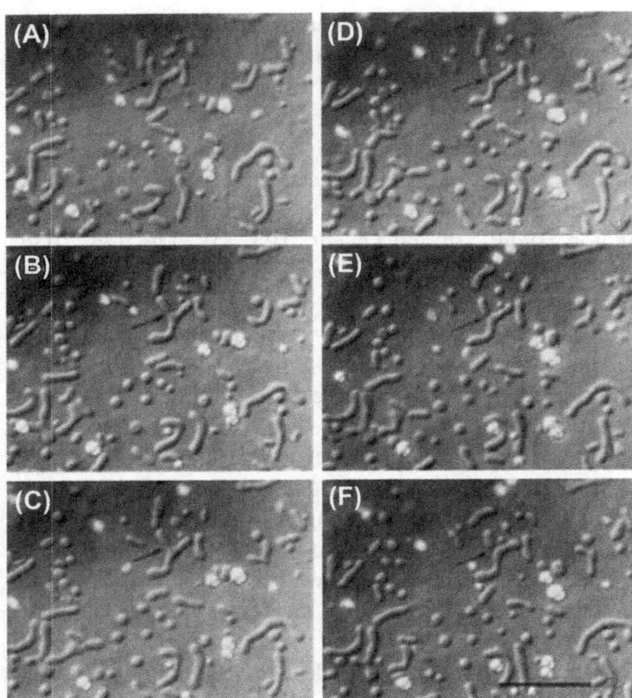

FIGURE 14.12 Differential interference photomicrographs taken 2 min apart of dividing mitochondria (arrow) in living cells of *Nitella flexilis*. Bar, 10 μm. *From Kuroiwa, T., 1982. Mitochondrial nuclei. Int. Rev. Cytol. 75, 1–59.*

division in the evolution of higher plants (Arimura and Tsutsumi, 2002; Logan, 2006). Studies in which mitochondria are expressing a fluorescent protein that can fluoresce either red or green show that mitochondria do not only divide but also fuse (Arimura et al., 2004a; Meeusen et al., 2004; Sheahan et al., 2005; Foisnner and Wasteneys, 2014; Kawano, 2014b). The ability to balance the fusion and division processes allows for a temporally and spatially distributable mitochondrial compartment.

While mitochondria are semiautonomous, their activities and numbers have to be coordinated with the activities of the rest of the cell. The continued growth and function of mitochondria require nuclear gene products. If isolated mitochondria are incubated in vitro with radiolabeled mitochondrial proteins, the proteins are taken up into the mitochondria (Kiebler et al., 1993; Moore et al., 1994). These proteins enter at special regions, called contact sites, where the inner and outer membranes touch. Like in the chloroplast, these regions are where the protein translocation complexes exist. A sequence of 12 amino acids on the amino-terminus of a protein is sufficient for the translocation of a protein into the mitochondria. These 12 amino acids can be attached to any protein, and they will direct its import into the mitochondria. The mitochondrial transit peptide forms an amphipathic α-helical structure where the positive charges are lined up on one side of the peptide. This sequence binds with the receptor at the contact site and subsequently the sequence is cleaved by a mitochondrial signal peptidase (Pfanner et al., 1994). A membrane potential across the inner membrane is required for the penetration of the mitochondrial signal peptide, and ATP hydrolysis is required for the import of the rest of the polypeptide. Chaperonins on both sides of the membrane participate in protein import (Moore et al., 1994; Burt and Leaver, 1994).

The canonical mitochondrial targeting sequence proved to be just one possible example of the many ways that a nuclear-encoded mitochondrial protein can enter the mitochondrion (Bédard and Jarvis, 2005). Some nuclear-encoded proteins involved in transcription and translation in the mitochondria are also involved in these processes in the chloroplast, and the signal sequence on these proteins targets these proteins to both organelles (Millar et al., 2006).

14.9 SUMMARY

In Chapter 13, I discussed how plants transform oxidized carbon and hydrogen into carbohydrate with the input of energy from the sun. In this chapter, I discussed how carbohydrate is converted to a readily usable form of energy in the form of ATP, and also how the carbon skeletons can be used in other biosynthetic reactions. I discussed the details of the grand cycle that is responsible for the conversion of energy and includes the autotrophic organisms and the heterotrophic organisms. This cycle was dubbed the Priestley cycle by Efraim Racker (1957). According to E.J. Conway (1953),

> *In the photosynthetic process ... the primary reaction consists in a splitting of water into hydrogen and oxygen atoms. The terminal stages of biological oxidation may be represented as the reforming of water from hydrogen and oxygen ions. The over-all process is a transference of electrons from the hydrogen atoms to oxygen atoms, with the absorption of free energy of the solar radiation bringing about the reverse change. Biological energetics, in all their episodic variety, may then be regarded as the life history of the electrons from hydrogen atoms to water, and all the directed energy of aerobic organisms lies between the splitting and the re-forming of* H_2O.

We must remember that heterotrophic cells do not convert everything into CO_2 and H_2O. This is the role of microorganisms, which are part of the great cycles involved in carbon, oxygen, sulfur, phosphorous, and nitrogen cycles. The great coal deposits have resulted from the fact that at certain times in Earth's history, microorganisms have not kept pace with the production of reduced carbon compounds. If microorganisms were not around to decompose organic matter into CO_2 and H_2O, eventually all the CO_2 in the atmosphere would be depleted, and photosynthesis and all life that depends on it would cease. Indeed, along with plants and animals, microorganisms are part of the great cycle of life.

Otto Rahn (1945) wrote in his book *Microbes of Merit*,

> *It may cause some readers a peculiar feeling to realize that the carbon, nitrogen and sulfur atoms which make up the entire living world of today are the same identical atoms which formed the living world of a million years ago, and that our own body may consist of some of the identical atoms which once were part of a dinosaur, or a tree fern of the coal age, or of one of our own ancestors. Only the pattern has changed, but the material from which the organic world has been formed, as long as there has been an organic world, has always remained the same. From dust to dust, as the Bible says; the same clay cast in ever-changing molds.*

14.10 QUESTIONS

14.1. What is the evidence that the mitochondria are the "power plants" of the cell?

14.2. What is the mechanism by which the mitochondria and their components provide energy in the form of ATP to the cell?

14.3. What are the limitations of thinking about the mitochondria exclusively as the power plants of the cell or as the only source of useful energy?

The references for this chapter can be found in the references at the end of the book.

Chapter 15

Origin of Organelles

Harvey's celebrated formula, ex ovo omnia-or, as usually quoted, omne vivum ex ovo- took with Redi the far more philosophical form, omne vivum e vivo, thus expressing a truth which forms the very foundation of all biological teaching at the present day. The development of the cell-theory, long afterwards, enabled Virchow to pronounce the more specific aphorism, omnis cellula e cellula (1855), a statement involving the highly interesting conclusion that protoplasm is never formed de novo, but always arises from or through the activity of preexisting protoplasm differentiated into the form of a cell....Altmann,- however, has sought to identify the elementary units, or 'bioblasts', with the visible protoplasmic granules; and, in his writings, the series-of Latin aphorisms initiated by Redi culminates in the saying, oinne granulum e granulo(!)....

E.B. Wilson (1899).

It is generally considered that eukaryotic cells evolved from prokaryotic cells. The progenitor of the eukaryotic cell may have consisted of a simple genome enclosed by a plasma membrane, and it may have looked much like the prokaryotic cells seen today (Stanier et al., 1963). The genome of the progenitor may have been attached to the plasma membrane, as it seems to be anchored to proteins in the plasma membrane of some species of prokaryotes (Jacob et al., 1963; Ryter, 1968; Fielding and Fox, 1970; Kavenoff and Ryder, 1976; Hendrickson et al., 1982; Funnell, 1993; Toro and Shapiro, 2010; Badrinarayanan et al., 2015). The plasma membrane of the progenitor probably functioned primarily to maintain the distinction between the inside of the cell and the environment. Over time, however, portions of the plasma membrane may have evolved enzymatic activities, including the ability to synthesize lipids, which would be necessary for the maintenance of the membrane. Such a plasma membrane may have also developed the ability to bind ribosomes, which would facilitate the insertion of nascent proteins into itself. The ancestral cell probably divided by a process involving the invagination of the plasma membrane, followed by its simultaneous breakage and fusion, in a manner that would prevent leakage.

15.1 AUTOGENOUS ORIGIN OF ORGANELLES

As time passed, the progenitor cell may have grown, although doing so would have caused the surface-to-volume ratio to decrease, thus limiting the supply of sufficient material to the interior of the cell. To maintain a sufficient surface-to-volume ratio, the plasma membrane may have invaginated to form a series of internal membranes, which led to the endoplasmic reticulum (ER). Such an intermediate can be seen in *Epulopiscium fishelsoni*, an organism with both eukaryotic and prokaryotic characteristics (see Fig. 15.1; Montgomery and Pollak, 1988; Angert et al., 1993, 1996; Robinow and Angert, 1998; Schultz and Jørgensen, 2001). The ER probably evolved further, differentiating into tubular and cisternal elements, as well as rough, smooth, and transitional regions. Further elaboration of the ER may have given rise to the Golgi apparatus and the peroxisomal reticulum. Alternatively or simultaneously, the Golgi apparatus may have arisen through invaginations of the plasma membrane by way of an endosomal intermediate. The DNA-binding region of the plasma membrane may have also invaginated in such a way as to sequester the hereditary material, and in so doing, it formed a nucleus (Cavalier-Smith, 1975), although it has also been suggested that the nucleus arose from an endosymbiotic event (López-García and Moreira, 2006).

The vacuolar membrane probably also evolved from invaginations of the plasma membrane, where the endosomal compartment forms the intermediate. Perhaps the plasma membrane developed the mechanism of endocytosis to take in food in the form of molecules that were too large and polar to cross the plasma membrane. Exocytosis may have evolved along with endocytosis so that a cell could continually take up food without decreasing in size. Perhaps the progenitor cell also developed an endocytotic mechanism that allowed it to take up other cells for food and to pass them to the vacuole where they would be digested into elementary molecules, including fatty acids, nucleotides, amino acids, and sugars.

The autogenous view of cell evolution suggests that each individual cell is genetically a single individual. What if some

Plant Cell Biology. https://doi.org/10.1016/B978-0-12-814371-1.00015-1

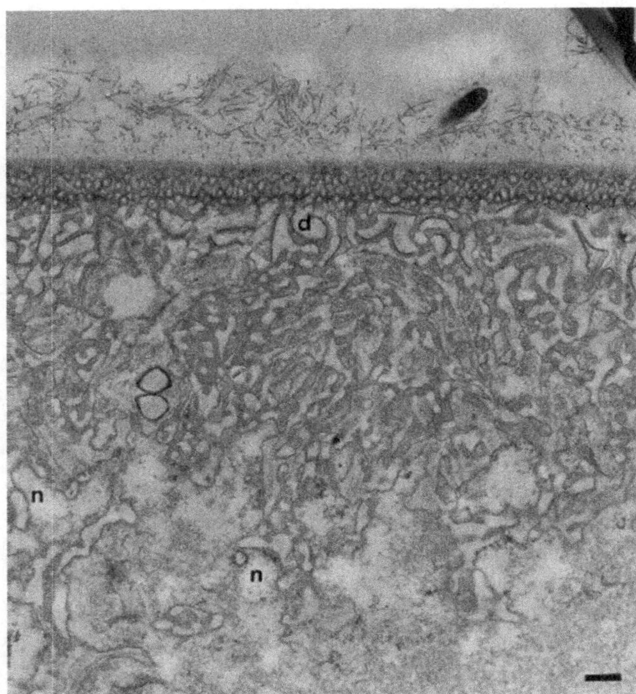

FIGURE 15.1 In the large prokaryotic cells of *Epulopiscium fishelsoni*, endoplasmic reticulum–like structures arise from invaginations of the plasma membrane. *n*, nucleoid; *d*, denser membranes than the ones below. *From Robinow, C., Angert, E.R., 1998. Nucleoids and coated vesicles of* "Epulopiscium" *spp. Arch. Microbiol. 170, 227–235.*

of the phagocytosed cells were not digested but developed a symbiotic relationship with the host cell instead? Then each cell would be a colony of evolutionarily distinct individuals. This possibility, which best describes the evolution of what we now call the *chloroplast* and *mitochondrion*, was explicitly proposed by Andreas Schimper (1883; Kutschera and Niklas, 2005). Schimper proposed that plastids might have evolved from cyanobacteria based on his observations that plastids and cyanobacteria have similar morphologies, and the fact that plastids divide to give rise to other plastids.

15.2 ENDOSYMBIOTIC ORIGIN OF CHLOROPLASTS AND MITOCHONDRIA

Schimper's proposal that chloroplasts evolved from cyanobacteria gained support from Constantin Mereschkowsky (1905, 1910, 1920) in the early part of the 20th century. Mereschowsky confirmed that chloroplasts share similar physical, metabolic, and reproductive properties with cyanobacteria, and further showed that chloroplasts can survive briefly in cytoplasm from which the nucleus had been removed.

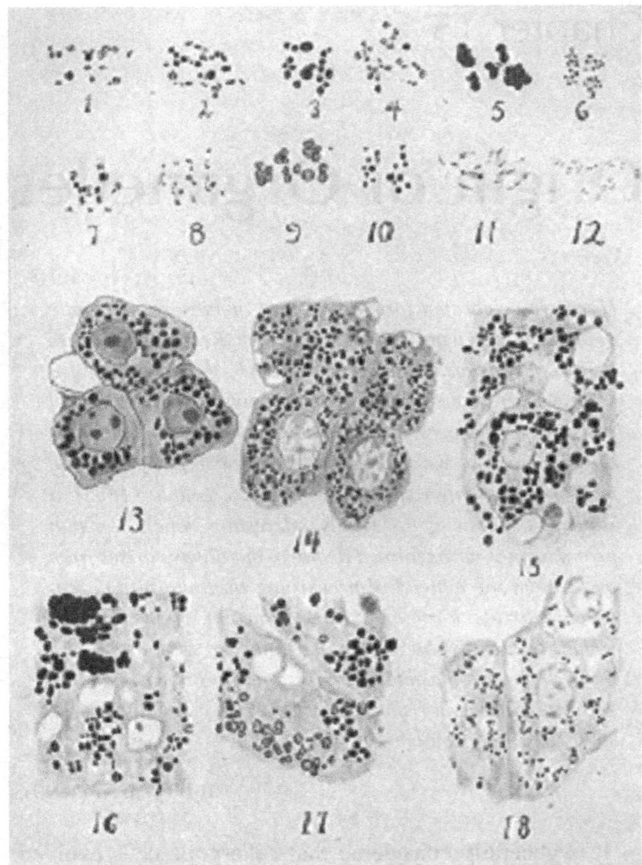

FIGURE 15.2 Camera lucida drawings of mitochondria supposedly cultured from newborn rabbit liver cultured in vitro (1-12) and similar-looking organisms in situ (13-18). *From Wallin, I.E., 1927. Symbionticism and The Origin of Species. Williams & Wilkins Co., Baltimore.*

Following the discovery of mitochondria, it was natural to propose that they too arose endosymbiotically. The way for this idea had already been prepared by Richard Altmann (1890), who believed that the mitochondria, or *bioblasts* as he called them, were not only free-living but also the fundamental unit of life, living within the lifeless milieu of the cell. Later, Paul Portier (1918) proposed that the mitochondria were not actually a normal component of the cell but bacteria that had been engulfed by the cell and lived within the cell symbiotically. Portier even thought that he had cultured them (Fig. 15.2), although his claim was not widely accepted (Lumiére, 1919). By 1922, Ivan Wallin (1922, 1927) proposed that mitochondria, like chloroplasts, evolved from symbiotic bacteria. Moreover, he proposed that the endosymbiotic relationship was important for speciation. He wrote, "Symbionticism insures the origin of new species."

By 1925, the endosymbiotic theory had entered mainstream cell biology textbooks. Edmund B. Wilson (1925) wrote about the endosymbiotic theory:

> *Such an hypothesis is of course unverifiable, and for this reason will to many appear worthless. As a purely speculative construction, however, it seems to the writer to offer possibilities concerning the early evolution of the cell that are worth considering, even though it brings us no nearer to a conception of the origin of life or a comprehension of organic individuality. ... To many, no doubt, such speculations may appear too fantastic for present mention in polite biological society; nevertheless it is within the range of possibility that they may some day call for more serious consideration.*

Further support for the endosymbiotic hypothesis came in the 1940s from studies using the newly discovered antibiotics. Luigi Provasoli and his colleagues treated *Euglena* cells with the antibiotic streptomycin (Lederberg, 1952) and found that this antibacterial agent "cured" the cells of their chloroplasts, providing further evidence for the bacterial origin of chloroplasts (Provasoli et al., 1948). Likewise, the antibiotics, including chloramphenicol, which function as inhibitors of protein synthesis in bacteria, also prevent protein synthesis in chloroplasts and mitochondria, supporting the homology of bacteria, chloroplasts, and mitochondria.

The hypothesis of the endosymbiotic origin of mitochondria and chloroplasts was bolstered by the observations that these organelles contain DNA. The first indication of extranuclear DNA came from the observation by Jean Brachet (1959) that tritiated thymidine, a component of DNA, was incorporated into the cytoplasm and the nucleus. Stocking and Gifford (1959) then showed that tritiated thymidine was taken up into the chloroplast. The DNA in chloroplasts was then visualized by Ris and Plaut (1962) using the electron microscope. They revealed that the chloroplasts contained fibrils of DNA that were similar in appearance to the naked DNA found in prokaryotic cells (Ris and Plaut, 1962). Nass and Nass (1963) later showed that DNA was also present in mitochondria. Ris (1962) suggested that these observations called for a serious reconsideration of the symbiotic hypothesis. Lynn Margulis (1970, 1981), a graduate student at the University of Wisconsin at the time when Hans Ris discovered chloroplast DNA, became very excited about this result and published a synthesis of the endosymbiotic theory and the evidence supporting it. She not only published the idea but also carried on what amounted to a personal crusade for the acceptance of the theory.

There are two polar ways to explain the genomic partitioning between organelles. The autogenous origin hypothesis asserts that the genome of a single individual became distributed between the nucleus and the mitochondria and chloroplasts. By contrast, the endosymbiotic hypothesis professes that the nuclear, mitochondrial, and chloroplastic genomes arose from different cells. Pigott and Carr (1972) performed a series of experiments aimed at distinguishing between these two hypotheses. They carried out a DNA–RNA hybridization experiment using nucleic acids isolated from the nucleus and chloroplast of the same cell and from a cyanobacterium. They found that the chloroplast DNA hybridized considerably more with the rRNA of free-living cyanobacteria than with the rRNA of the surrounding cytoplasm. This is not what would be expected if the plastid genome differentiated from the nuclear genome. It is, however, what would be expected if chloroplasts had an endosymbiotic origin. A host of protein and nucleic acid sequencing studies support the endosymbiotic origin of chloroplasts and mitochondria (Taylor, 1987; Martin and Müller, 2007; Sapp, 2007). Given the strength of the evidence for the endosymbiotic theory, Weeden (1981) suggests that the reason that the chloroplast and nucleus share responsibility for encoding chloroplast proteins is that many of the genes that encode these proteins have moved from the chloroplast to the nucleus throughout evolution (see Chapter 13).

Additional biochemical evidence lends support to the endosymbiotic origin of chloroplasts and mitochondria. The inner envelope membrane of chloroplasts and the plasma membrane of cyanobacteria both contain monogalactosyldiglyceride and sulfolipid (Douce and Joyard, 1990). Moreover, the inner membrane of mitochondria and the plasma membrane of bacteria uniquely contain cardiolipin (John and Whatley, 1977). Interestingly, chloroplasts and mitochondria have double membranes, and perhaps the inner one represents that of the endosymbiont and the outer one represents that of the host endosome.

The division of chloroplasts and some mitochondria depend on the morphogenetic motor based on dynamin and the GTP-binding filamentous protein FtsZ, which also mediates the division of eubacteria and some archaebacteria (McIntosh et al., 2010; Okazaki et al., 2010). While the endosymbiont contributed FtsZ to the contractile rings, the host contributed the dynamin (Miyagishima, 2011).

There are many candidates for the bacterium that is most closely related to the mitochondrion. John and Whatley (1975, 1977) proposed that mitochondria arose from an aerobic bacterium similar to *Paracoccus denitrificans*, based on the similarity of their respiratory chains. Woese (1977) suggested that mitochondria may have evolved from a bacterium related to purple nonsulfur bacteria (e.g., *Rhodospirillum*) based on the fact that these bacteria have infoldings of the plasma membrane that are similar in appearance to mitochondrial cristae (Fig. 15.3). Sequencing data, based on proteins and nucleic acids, support this alliance (Schwartz and Dayhoff, 1978). Like many purple nonsulfur bacteria, the original promitochondrion may have been a nonoxygenic photosynthetic anaerobe. These bacteria are facultative aerobes and shut down their photosynthetic ability in the presence of O_2.

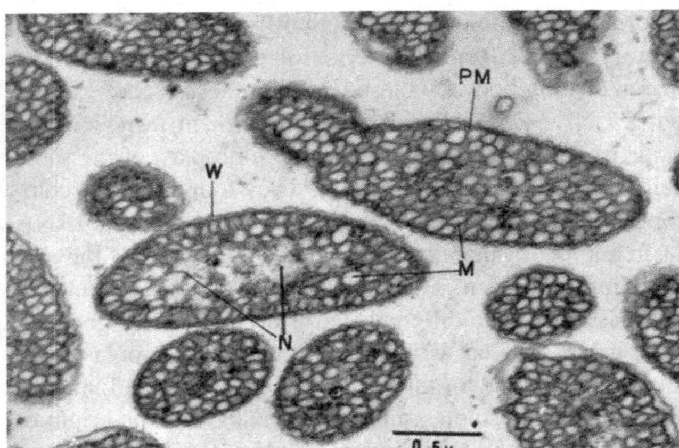

FIGURE 15.3 An electron micrograph of *Rhodospirillum rubrum* showing cells loaded with membrane-bound vesicles (M). *N*, nucleoplasm; *W*, extracellular matrix; *PM*, Peripheral membrane. *From Holt, S.C., Marr, A.G., 1965. Location of chlorophyll in* Rhodospirillum rubrum. *J. Bacteriol. 89, 1402–1412.*

Thus, if they became trapped inside a host cell, they would revert to respiring organelles (Taylor, 1987). The purple nonsulfur bacteria, like mitochondria, have an F_0F_1-type H^+-ATPase.

The splendid variety in the morphologies and pigments of the chloroplasts in the various groups of algae point to the likelihood that chloroplasts arose independently in each major taxon (Raven, 1970). However, it is also possible that the original endosymbiont already contained all the possible pigments, and the chloroplasts in each taxon evolved by the loss of pigments (Löffelhardt et al., 1997). A tremendous effort is now going into sequencing the chloroplast genomes of a variety of algae to distinguish between these hypotheses.

Prochlorococcus marinus, a prokaryotic cell that contains chlorophyll *a* and chlorophyll *b* as well as phycobillins, has also been nominated as the candidate that is most closely related to the cyanobacterium that gave rise to the chloroplasts (Hess et al., 1996). *Prochloron*, a prokaryotic cell that contains both chlorophyll *a* and chlorophyll *b* but no phycobillins, has been nominated as another possible candidate that is related to the cyanobacterium that gave rise to the chloroplasts of green algae and higher plants (see Fig. 15.4; Margulis, 1981). Margulis and Ober (1985) suggest that *Heliobacterium* is a contemporary descendant of the cyanobacterium that gave rise to the chloroplast of the golden-brown algae.

When did these endosymbiotic events take place? Fossil evidence indicates that eukaryotic cells originated 1.4 billion years ago, and the endosymbiotic events would have taken place about then. Because all eukaryotic cells except those of the Archezoa (Cavalier-Smith, 1989a) have mitochondria, the mitochondria probably entered the cells first. Plastids then entered the cells that were to give rise to plants.

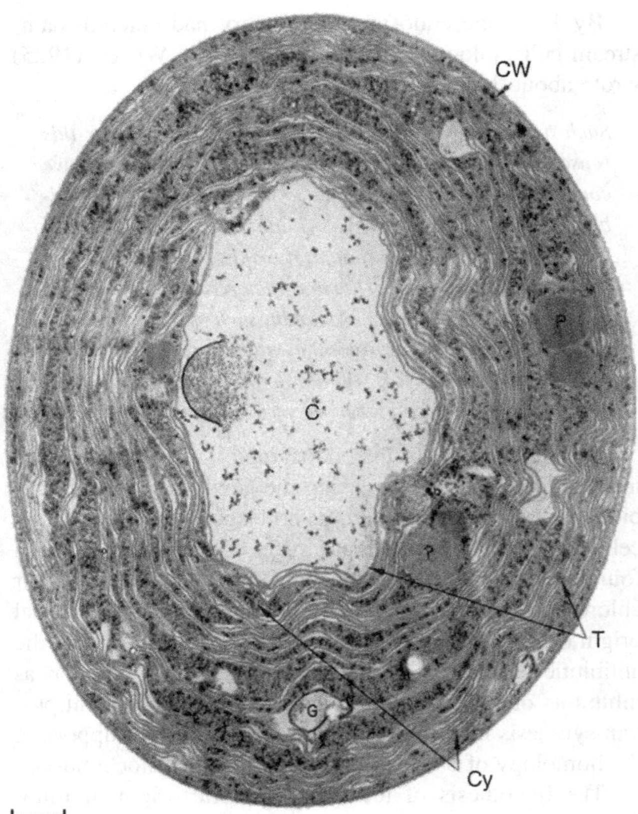

FIGURE 15.4 An electron micrograph of *Prochloron* showing the thylakoid-like membranes. *C*, central zone; *CW*, cell wall; *Cy*, cytoplasm; *G*, polyphosphate-like granule; *T*, thylakoids. *From Whatley, J.M., 1977a. The fine structure of prochloron. N. Phytol. 79, 309–313.*

15.3 ORIGIN OF PEROXISOMES, CENTRIOLES, AND CILIA

While Margulis (1970, 1981) has been the single most important advocate for the endosymbiotic theory in general, her personal research focuses on the proposal that cilia evolved from the endosymbiotic association of a spirochete-like bacteria (Sagan, 1967). The evidence in support of this hypothesis is the superficial morphological similarities between spirochetes and cilia and the similarities between the proteins FtsZ and tubulin (van Iterson et al., 1967; Margulis et al., 1978; Bermudes et al., 1994; Erickson, 1995). Tubulin genes have been identified in a new division of bacteria known as *Verrucomicrobia*, and Li and Wu (2005) have suggested that cilia and flagella may have originated from endosymbiotic events involving bacteria from this division.

There is also some evidence that centrioles have evolved from endosymbiotic organisms in that they both propagate by division, although a counterargument is that blepharoplasts, which are identical in structure to centrioles, do not propagate by division, but arise de novo in some cells that will give rise to sperm (Sharp, 1914; Hepler,

1976). Initial support for the endosymbiotic origin of centrioles came from the claim that centrioles contained DNA. However, this claim turned out to be irreproducible, and centrioles do not have DNA (Johnson and Rosenbaum, 1990; Kuriowa et al., 1990; Johnson and Rosenbaum, 1992; Johnson and Dutcher, 1991). However, centrioles do contain RNA (To, 1987), leaving the question of their origin still open.

Unlike chloroplasts, mitochondria, and perhaps centrioles, neither peroxisomes nor cilia contain DNA, RNA, or ribosomes. Thus, if peroxisomes or cilia are the descendents of free-living cells, their entire genetic apparatus has been transferred to the host or was lost. Evidence for the endosymbiotic origin of peroxisomes was supported by their apparent independent growth and division within the cell (Cavalier-Smith, 1989b; Dinis and Mesquita, 1994), but evidence that many of the peroxisomal membrane proteins are assembled in the ER argues in favor of an autogenous origin of peroxisomes (Gabaldón et al., 2006; see Chapter 5).

There is evidence that most of the genes that might have been present in the original endosymbiotic promitochondria and prochloroplasts have already been transferred to the nucleus (Timmis et al., 2004), and the complete transfer of DNA from an endosymbiont to the host nucleus may be the inevitable fate of all endosymbionts. Such a transfer effectively erases the evidence of the endosymbiotic event. However, the evidence supporting the endosymbiotic origin of peroxisomes, centrioles, and cilia is weak compared with the evidence supporting the endosymbiotic origin of chloroplasts and mitochondria. When the data at hand do not support a complicated hypothesis, it is more prudent to invoke Occam's razor, which states, "What can be accounted for by fewer assumptions is explained in vain by more" (Gillispie, 1974), and, for the time being, accept the simplest hypothesis (Newton, 1729; Watts, 1801, 1811; Arrhenius, 1915; Northrop, 1961; Galilei, 1962; Hoffman et al., 1997). Because the hypothesis of the autogenous origin of the peroxisome, cilia, and centrioles is simpler and requires fewer assumptions than the endosymbiotic origin hypothesis, I accept it for describing the origin of these organelles. On the other hand, I also believe that everything that is not physically impossible may be possible and that just because a decision is prudent does not mean it is correct.

15.4 ONGOING PROCESS OF ENDOSYMBIOSIS

Symbioses have been going on for the past 1.4 billion years and are going on all around us. Lichens are symbiotic relationships between algae and fungi. Symbiotic relationships also occur between *Hydra* and *Chlorella*, legumes and *Rhizobium*, and *Paramecium* and *Chlorella* (see Fig. 15.5; Pringsheim, 1928; Wichterman, 1986; Lewin, 1992; Reisser, 1992; Tanaka et al., 2002; Kadono et al., 2004). Perhaps even the Apicomplexan parasites, which include *Plasmodium*, the human malarial parasite, resulted from an endosymbiotic relationship with an organism that provided them with a plastid, which is called the *apicoplast* (Wilson et al., 1996; McFadden and Waller, 1997; Dzierszinski et al., 1999; McFadden and Roos, 1999; He et al., 2001; Maréchal and Cesbron-Delauw, 2001; Parsons et al., 2007; Tonkin et al., 2008). The plant partner in this endosymbiosis provides the possibility of preventing or treating malaria with herbicides (Maréchal and Cesbron-Delauw, 2001; Ralph et al., 2004). Symbiotic relationships that give rise to organelles are also still occurring. One example of two organisms caught in the process of one of them becoming a plastid is *Cyanophora paradoxa*. *Cyanophora* is a photoautotrophic protist that contains an obligate symbiotic cyanobacterium known as a *cyanelle* (Löffelhardt et al., 1997; Pfanzagl and Löffelhardt, 1999; Price et al., 2012; Kawano, 2014a). Cyanelles function as chloroplasts. However, unlike chloroplasts, they are each surrounded by a vestigial 7-nm thick peptidoglycan wall. The cyanelles have a reduced circular genome like plastids and import approximately 80% of their proteins from the cytosol (Schenk et al., 1987; Trench, 1991; Steiner et al., 2002).

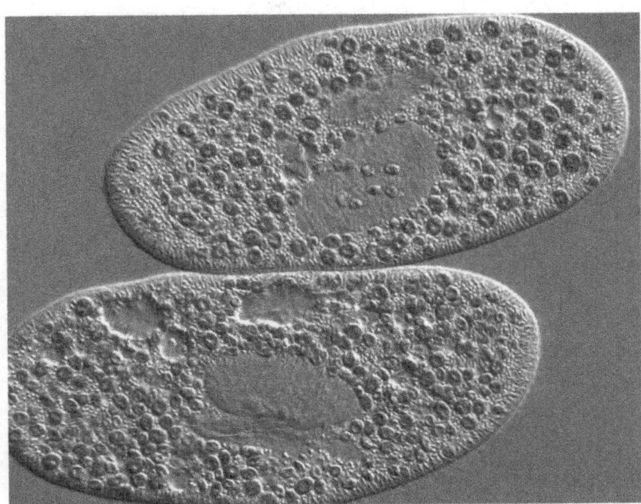

FIGURE 15.5 *Paramecium bursaria*, a symbiotic organism that contains *Chlorella vulgaris*. This symbiotic association has been growing on strictly inorganic media since 1999.

There are other examples of obligate intracellular symbioses among protists in which the endosymbiont has become an integral part of the host, yet still retains more of its cellular character than do typical chloroplasts. For example, there are small structures in the chloroplasts of *Cryptomonas* and *Chlorachnion* known as *nucleomorphs*

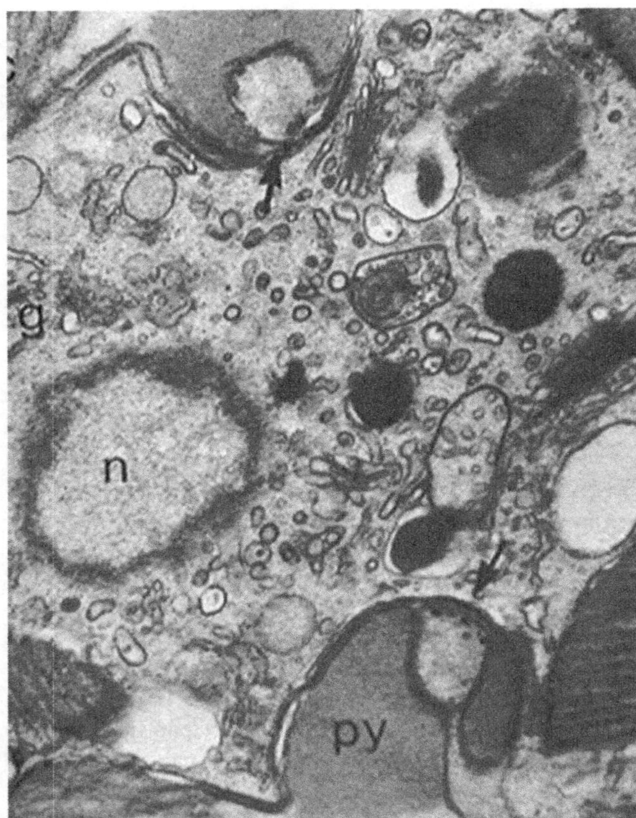

FIGURE 15.6 Nucleomorphs in *Chlorachnion reptans.* In a single cell, there is one nucleus (n) and two nucleomorphs (*arrows*) that are located along the periphery of the cell and adjacent to the pyrenoids (py). *c*, chloroplasts; *g*, Golgi; *m*, mitochondria. ×18,200. *From Ludwig, M., Gibbs, S., 1987. Are the nucleomorphs of cryptomonads and* Chlorachnion *vestigial nuclei of eukaryotic endosymbionts? Ann. N.Y. Acad. Sci. 503, 198–211.*

that contain DNA (see Fig. 15.6; Ludwig and Gibbs, 1987). As long as the nucleus of the endosymbiont continues to control its function, the endosymbiont is considered to be a foreign entity. There is a trend among such symbioses for a mutual dependence between the symbiont and the host and a concurrent loss of genetic material from both of them (Taylor, 1987; Kuroiwa and Uchida, 1996).

15.5 PRIMORDIAL HOST CELL

While the identity of the primordial host cell remains a mystery, a few candidates for this honor have been proffered. One family of hypotheses state that the *Archaebacteria* are the most likely candidates for the host organism, and this organism engulfed *Eubacteria*, which later evolved into mitochondria or chloroplasts (Cavalier-Smith, 1989a). The *Archaebacteria* include the sulfobacteria, halobacteria, and methanobacteria. (The *Eubacteria*, by contrast, include the Gram-positive bacteria, *Escherichia coli*, and the cyanobacteria.) The *Archaebacteria* hypothesis is based on molecular phylogenetic data, which show that the *Archaebacteria* are more closely related to the eukaryotes than they are to the *Eubacteria* (Balch et al., 1977; Huet et al., 1983; Iwabe et al., 1989; Gogarten et al., 1989; Brown and Doolittle, 1995; Keeling and Doolittle, 1995). Sequence data showing similarities between the H^+-ATPase of *Sulfolobus* and the V-type ATPase of eukaryotic vacuoles indicate that the host cell may be a relative of a contemporary archaebacterium like *Sulfolobus* (Zillig, 1987; Nelson, 1988). The *Archaebacteria* hypothesis is also supported by comparative biochemical research (Martin and Müller, 1998), including Dennis Searcy's (1987) research, which demonstrates that the thermophilic *Archaebacteria* possess actin-like proteins as well as histone-like proteins that form nucleosomes that protect the DNA from high temperatures (Searcy, 1975; Stein and Searcy, 1978; DeLange et al., 1981a,b; Green et al., 1983; Drlica and Rouviere-Yaniv, 1987; Mukherjee et al., 2008).

An alternative theory, based on comparative cytology, is that a member of the *Archezoa*, a eukaryotic taxon that is characterized by the lack of mitochondria, plastids, peroxisomes, and the Golgi apparatus, was the primordial host cell (Cavalier-Smith, 1987a, 1991; de Duve, 1991). However, others suggest that the *Archezoa* are not primitively "organelle-less," but lost their organelles through evolution (Keeling, 1998; Biagini and Bernard, 2000; Martin, 2000; Seravin, 2001). This interpretation is supported by the finding that nuclear genes, which produce proteins used in the mitochondria, are found in the nucleus of *Archezoa*. However, we do not know whether the genes present in bacteria that had been digested for food were transferred to the nucleus though a horizontal gene transfer mechanism before the evolution of mitochondria (Doolittle, 1998) or if those genes were transferred from mitochondria to the nucleus prior to the loss of the mitochondria (Palmer, 1997). Additional support for the theory that the *Archezoa* "lost" their mitochondria comes from the observation that remnants (50 nm × 90 nm) of mitochondria, with double membranes and that stain with antibodies directed against mitochondrial heat shock protein (Hsp70), can be detected in *Trachipleistophora hominis* (Williams et al., 2002).

According to Lynn Margulis (2001), the original host cell was already a merger between two unrelated prokaryotic organisms. She believes that a merger between an archaebacterium, such as *Thermoplasma*, and a eubacterium, such as *Spirochete*, gave rise to the emergent properties of eukaryotic cells.

We do not know the identity of the primordial host cell. Nevertheless, we can take comfort in knowing that *Archaebacteria* are eukaryote-like prokaryotes, the *Archezoa* are prokaryote-like eukaryotes, and that we have narrowed the apparent gap between prokaryotes and eukaryotes. There is no law that gives directionality to evolution, and thus evolution can be parallel, divergent, or convergent, and "progressive gradation" or elaboration as well as "progressive degradation" or reduction can take place (Seravin, 2001). Thus, we have a feeling for the missing link that gave rise to the host cell, even if we do not know its identity.

15.6 SYMBIOTIC DNA

E.B. Wilson (1925) considered the possibility that the genetic system of the cell, and consequently its development and evolution, could be affected by microorganisms, which we now call *viruses*. The viruses can cause "horizontal" gene transfer between taxa, making us all part of a genetic web of life (Doolittle, 1998; Doolittle et al., 2003; Won and Renner, 2003; Bergthorsson et al., 2004; Mower et al., 2004; Davis et al., 2005; Richardson and Palmer, 2007; Gao et al., 2014) rather than a tree of life. The prevalence of organelle-to-organelle gene transfer increases the possibility of evolutionary variation (Timmis et al., 2004). In the book *The Selfish Gene*, Richard Dawkins (1976) considered the possibility that the majority of nuclear DNA may be nonfunctional and "parasitic," using the organism as a means to reproduce the DNA. However, as time goes on, functions for seemingly nonfunctional DNA are being discovered (Fire, 2006; Mello, 2006). Will the artificial plasmids produced by genetic engineering techniques have any unforeseen evolutionary effects on cells?

There may also be symbiotic self-replicating proteins such as the prions, which are involved in mad cow disease (Pruisner, 1982) that can be transferred from organism to organism.

15.7 SUMMARY

The eukaryotic cell contains a number of organelles, some of which (nucleus, ER, Golgi apparatus, endosomes, and vacuoles) may have originated autogenously from the infolding of the plasma membrane. The origins of peroxisomes and centrioles are not clear, although peroxisomes may have also originated from the ER. Centrioles may also have an autogenous origin, but from nonmembranous components, such as those that make up the blepharoplast. By contrast, the chloroplasts and mitochondria appear to have arisen when eubacteria were taken up by phagocytosis but were not sent to the lysosomal compartment. Instead, they were retained by their archaebacterial or archaezoal host. Thus, throughout evolution these strangers developed a dependence on each other, cooperating in biosynthetic pathways and eventually forming a single organism.

Lewis Thomas (1974) wrote in his delightful book *The Lives of a Cell*,

> *Finally, there is the whole question of my identity, and, more than that, my human dignity. I did not mind it when I first learned of my descent from lower forms of life. I had in mind … ape-men. … I had never bargained on descent from single cells without nuclei. I could even make my peace with that, if it were all, but there is the additional humiliation that I have not, in a real sense, descended at all. I have brought them all along with me, or perhaps they have brought me. … If I concentrate, I can imagine that I feel them; they do not quite squirm, but there is, from time to time, a kind of tingle. I cannot help thinking that if only I knew more about them, and how they maintain our synchrony, I would have a new way to explain music.*
>
> *There is something intrinsically good-natured about all symbiotic relations, necessarily, but this one, which is probably the most ancient and most firmly established of all, seems especially equable. … If you are looking for something like natural law to take the place of the 'social Darwinism' of a century ago, you would have a hard time drawing lessons from the sense of life alluded to by my chloroplasts and mitochondria, but there it is.*

We have come a long way from the time when speculations on the endosymbiotic origin of the mitochondria and chloroplasts were "too fantastic for present mention in polite biological society." Indeed, Cohen (1971) wrote,

> *The problem [of endosymbiosis] may now be posed as a serious scientific challenge, warranting the most systematic and penetrating exploration. After all, if in the history of science we honor the Copernican revolution as the demonstration that man is not the center of the Universe, what effort should be accorded the proposition that man (and indeed all higher organisms) may be merely a social entity, combining within his cells the shared genetic equipment and cooperative metabolic systems of several evolutionary paths. We suspect that governments should be interested in such a possibility, but they may not yet have heard about it, nor might their responses be readily predictable.*

15.8 QUESTIONS

15.1. Is it reasonable to describe eukaryotic cells as a community of genetically distinct individuals that have intermarried throughout time?

15.2. What are the benefits derived by such a multicultural cell?

15.3. What problems must be overcome by such a multicultural cell?

The references for this chapter can be found in the references at the end of the book.

Chapter 16

The Nucleus

Molecular biology, as I envisaged it, "implies not so much a technique as an approach, an approach from the viewpoint of the so-called basic sciences with the leading idea of searching below the large-scale manifestations of classical biology for the corresponding molecular plan. It is concerned particularly with the forms of biological molecules, and with the evolution, exploitation and ramification of those forms in the ascent to higher and higher levels of organisation. Molecular biology is predominantly three-dimensional and structural—which does not mean, however, that it is merely a refinement of morphology. It must at the same time inquire into genesis and function.

William T. Astbury (1961).

For the biologist, Biology is a noun and molecular an adjective! Few molecular biologists would accept good heartedly this preeminence of Biology on molecules: they are former physicists or chemists and are more interested in the fine structure of macromolecules constituting living organisms than in their biological role.

Jean Brachet (1989).

All new news is old news happening to new people.

Attributed to Malcolm Muggeridge.

16.1 THE DISCOVERY OF THE NUCLEUS AND ITS ROLE IN HEREDITY, SYSTEMATICS, AND DEVELOPMENT

Robert Brown (1831) was the first to describe the structure that we call the *nucleus*. In an aside that was part of a paper devoted to pollination in orchids, Brown described the prominent nucleus as a "single circular areola, generally somewhat more opaque than the membrane of the cell. … This areola, which is more or less distinctly granular, is slightly convex, and although it seems to be on the surface is in reality covered by the outer lamina of the cell. There is no regularity as to its place in the cell; it is not infrequently however, central or nearly so." Brown did not mention the nucleus again in the paper.

It is conceivable that others, including Leeuwenhoek, may have seen the nucleus before Brown, and it may have been as unremarkable to them as it was to him. Matthias Schleiden (1849), on the other hand, speculated that the nucleus, which he called the *cytoblast*, was important for the formation of cells. He wrote, "I refer to Robert Brown … who has here, as in so many other instances, opened up a new path of inquiry. He first observed the cytoblast, as a body, frequently present in plants: he did not, however, know its significance in relation to the life of the cell; he called it 'nucleus of the cell.'" According to Schleiden, the cytoblastema was the essence of life, and the cytoblast was the first visible manifestation of the cytoblastema (Fig. 16.1). The cytoblast then gave rise to a cell (Fig. 16.2). Schleiden's (1838) idea that the nucleus was important in cell formation attracted the attention of a zoologist named Theodor Schwann. Schwann (quoted in Baker, 1988) recounted the surroundings in which he first realized that the cell and its constituents were important for development:

> *One day, when I was dining with Mr. Schleiden, this illustrious botanist pointed out to me the important role that the nucleus plays in the development of plant cells. I at once recalled having seen a similar organ in the cells of the notochord, and in the same instant I grasped the extreme importance that my discovery would have if I succeeded in showing that this nucleus plays the same role in the cells of the notochord as does the nucleus of plants in the development of plant cells.*

Schwann (1838, 1839) promoted the idea that cells, and their structural components (Fig. 16.3), were as important in determining the development and morphology of animals as they were in plants. Schleiden (1845) wrote about the importance of the nucleus in cell formation, "If this law is found essential to some plants and animals, this analogy forms a basis for enunciating this mode of formation as a universal law for both kingdoms." On the basis of studies made by Schleiden, a botanist, and Schwann, a zoologist, these two scientists are typically credited as the cofounders of the cell theory. The real founder was Henri Dutrochet (1824), a man who was ahead of his time.

As every introductory biology student knows, the nucleus can be difficult to see in a living cell. In fact, Schleiden (1845), von Nägeli (1844), Hofmeister (1849), von Mohl (1852) and Remak (1855) believed that nuclei

Plant Cell Biology. https://doi.org/10.1016/B978-0-12-814371-1.00016-3

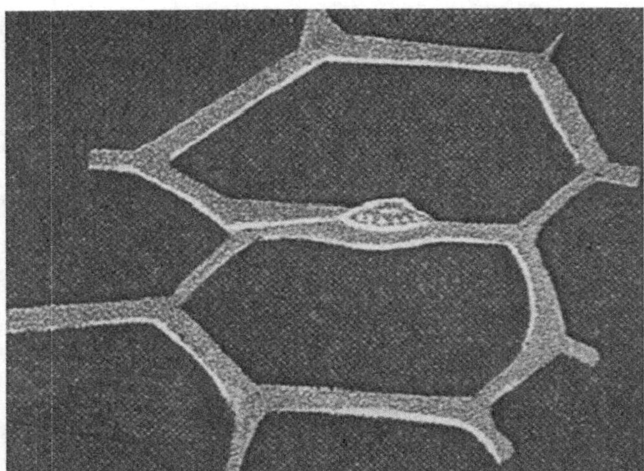

FIGURE 16.1 The cytoblast in cells of *Chamaedorea schiedeana. From Schleiden, M.J., 1838. 1867. Contribution to Phytogenesis. Sydenhan Soc., London.*

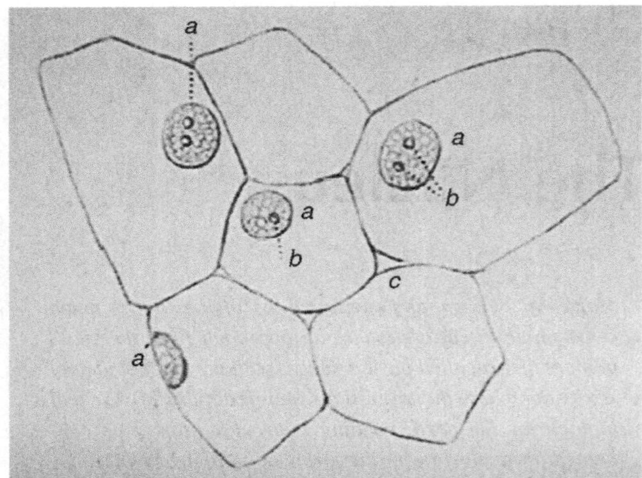

FIGURE 16.3 The nucleus (a) and nucleolus (b) of onion parenchyma cells. *From Schwann, 1847.*

were absent in the majority of cells and formed de novo just before the cells divided. They also believed, however, that in some cells, a nucleus arose from a preexisting nucleus. According to Hugo von Mohl (1852),

> *The ... origin of a nucleus, by division of a nucleus already existing in the parent-cell, seems to be much rarer than the new production of them, for as yet it has been observed only in a few cases, in the parent-cells of the spores of Anthoceros, in the formation of the stomates, in the hairs of the filaments of Tradescantia, &c., by myself, Nägeli, and Hofmeister; but it is possible that this process prevails very widely, since, as the preceding statements show, we know very little yet respecting the origin of nuclei.*

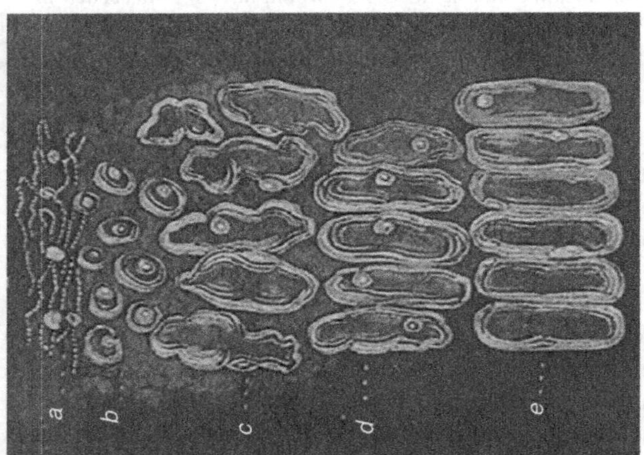

FIGURE 16.2 Free cell formation according to Schleiden. (a) The innermost mass, consisting of gum with intermingled mucous granules and cytoblasts. (b) Newly formed cells, still soluble in distilled water. (c–e) Further development of the cells, which, with the exception of the cytoblasts, may still coalesce, under slight pressure, into an amorphous mass. *From Schleiden, M.J. 1838. 1867. Contribution to Phytogenesis. Sydenhan Soc., London.*

The fact that the nucleus is present in all eukaryotic cells at all times was not obvious until Theodor Hartig (1854) and Lord Sidney Godolphin Osborne (1857) noticed that the nuclei in onion and wheat roots were stained by alkaline carmine solutions. Carmine is a dye made from the scale insect *Dactylopius coccus* (Von Georgievics and Grandmougin, 1920; Sandberg, 1994; Greenfield, 2005). Following the observation of carmine-stained nuclei in plants, it was discovered that carmine-stained nuclei were also common in animal cells (Gerlach, 1858; Clark and Kasten, 1983), and it became apparent that a nucleus was in each and every animal and plant cell (Virchow, 1860).

As a pioneer of the method of differential staining, Lionel Beale found that resting muscles stain with acid dyes, whereas active muscles, due to the accumulation of lactic acid, stain with basic dyes. Then it was discovered that the nucleus of active, living organisms also stained with basic dyes such as carmine, whereas those of nonliving organisms were apt to stain with acid dyes. Given these data and the theoretical paradigms of the day, it was reasonable to conclude that the more acid an organelle appeared to be, the more vital its activities were. Thus, Lionel Beale (1878) and others (Drysdale, 1874) came to believe that the nucleus, with all its acids, must contain the essence of life or *bioplasm*. Ranke (in Drysdale, 1874) wrote that,

> *... the chief activity of cell chemistry seems to originate in the nucleus. We see the vital activities if the organs run their course with the formation of organic acids, e.g., lactic acid, the production of which is copious in proportion to the heightened activity of the organs. Hence, we see the neutral or slightly alkaline reaction of the muscular or nerve tissues give place, under strongly exerted activity, to an acid one. These chemical transformations of the cell contents*

originate, as it appears, for the most part, in the nuclei, which in the living cell, exhibit constantly an acid reaction in contradistinction to their alkaline environment. The acid reaction is made known by the property of the nucleus to colour itself red quickly and permanently.

Prior to the introduction of fixatives and dyes, it seemed that the nucleus appeared and disappeared at various times during the life of a cell. However, the new cytological techniques revealed that the nucleus only seemed to appear and disappear in unstained material because of the lack of contrast, and in actuality, it fragmented into pieces. The fragments were equally distributed to the daughter cells during cell division. Moreover, Eduard Zacharias (1881) showed that the nuclei and their fragments were colored by the stains that he developed to test for the presence of nuclein (see Section 16.4). Because the nucleus was readily stained by colored dyes, its contents were given the name *chromatin* (Flemming, 1882), which means "colored substance." The fragments into which the chromatin divided were given the name chromatic elements by Theodor Boveri and *chromosomes*, which means "colored bodies," in a review by Wilhelm Waldeyer (1888, 1890; Goldschmidt, 1956; Zacharias, 2001).

Eduard Strasburger (1875) visualized the process of nuclear division while he was studying fertilization in *Picea*. He went on to show that nuclear division was not an artifact of fixation because he could also observe it in the living cells of *Spirogyra*. Strasburger searched for more favorable material than *Spirogyra* to see the process of nuclear division in living cells. He rediscovered the *Tradescantia* stamen hair, which had been previously used by Hofmeister (1849), and found that the process of nuclear division could be readily seen in vivo in these cells because of their leviathan-like chromosomes.

Walther Flemming (1880, 1882) began his cytological studies on salamander larval epithelial cells. He chose salamander larval epithelial cells because they were large and had exceptionally huge nuclei. The chromosomes in salamander cells are all long and thin. That is, the chromosomes are threadlike, and consequently he gave the name *mitosis* to the division of these threadlike chromosomes. As a consequence of the threadlike nature of the salamander chromosomes, Flemming could clearly see that the chromosomes divided longitudinally and not just by a nonquantitative fission process as was suspected by Strasburger, who studied cells that contained mostly globular chromosomes (Hughes, 1959; Baker, 1988).

The process of mitosis so clearly presented by Flemming indicated that the chromatin was not just a homogeneous material as might have been inferred from the results of the chemists Friedrich Miescher and Felix Hoppe-Seyler (see Section 16.4). Indeed, cytologists discovered that chromatin was not just equally distributed to each daughter cell, but the nucleus went through a complicated process to divide the chromosomes in such a way that each somatic cell had not only the same number of chromosomes but also identical copies of the chromosomes (Wilson, 1895). This led to the idea that the chromosomes contained the self-perpetuating information that was necessary to build an organism (Roux, 1883; de Vries, 1889; Weismann, 1893; Nägeli, 1914). E.B. Wilson (1923) described this remarkable finding like so:

The cytologist is first struck by the extraordinary pains that nature seems to take to ensure the perpetuation and accurate distribution of the components of the system in cell division, and hence heredity. Nothing is more impressive than the demonstration of this offered by the nucleus of the cell. … To our limited intelligence, it would seem a simple task to divide a nucleus into equal parts. The cell, manifestly, entertains a very different opinion. Nothing could be more unlike our expectation than the astonishing sight that is step by step unfolded to our view by the actual performance. The nucleus is cut in two in such a manner that every portion of its net-like inner structure is divided with exact equality between the two daughter-nuclei, and the cell performs this spectacular feat with an air of complete and intelligent assurance. The net-like framework is spun out into long threads or chromosomes; these are divided lengthwise into exactly similar halves; they shorten, thicken, separate and pass to opposite poles; and from the two groups formed, are built up two daughter-nuclei, while the cell-body divides between them. Such a process seems in some respects to contradict all physical principles; but its meaning has now become perfectly plain. In a general way it means, as Roux pointed out forty years ago, that the nucleus is not composed of a single homogeneous substance, but is made up of different and self-perpetuating components; and it means that these components are strung out in linear alignment in the threads so that they may be divided, or distributed in particular manner, by doubling of the thread.

The position of genes on chromosomes can be elegantly visualized today using fluorescence in situ hybridization (Hizume et al., 2002).

While the process of nuclear division was being studied in somatic cells, Strasburger observed that the nucleus of the sperm and egg fused during the process of fertilization in algae, conifers, and flowering plants (Ducker and Knox, 1985). Using the worm *Ascaris megalocephala* var. *univalens*, which only has two chromosomes in the diploid cell, Edouard Van Beneden (1883, in Hughes, 1959) observed that the sperm and the egg each contribute exactly one chromosome to the zygote. Thus, irrespective of the discrepancy in the size of the male and female gametes, each parent contributes the identical number of chromosomes. This observation was consistent with the observations by Aristotle in the fourth century BCE. Pierre-Louis Moreau de Maupertuis (1753) and Joseph Kölreuter discovered that each parent contributes equally to the inheritance of the offspring (see Voeller, 1968). Van Beneden also observed a reduction division that resulted in the gametes having half the number of the chromosomes as the parents to ensure that the diploid chromosome number remains constant over the generations from parent to offspring. Theodor Boveri and Oscar Hertwig independently discovered the mechanism of how the diploid number of chromosomes gave rise to the haploid number of chromosomes in reproductive cells (Hughes, 1959; Baltzer, 1967; Voeller, 1968).

This was a special kind of process that involved two cell divisions. The first cell division differed from the typical mitotic division because the chromosomes in a cell, which were originally derived from each parent, paired in such a way that the daughter cells received half the number of chromosomes that the mother cell contained. The daughter cells were thus haploid, although they carried two nearly identical copies of each chromosome. The second division did not involve the pairing of chromosomes, which occurred in the previous division, and was thus more similar to the typical mitotic division. Consequently, in 1887, Flemming named the first division *heterotype mitosis* and the second division *homotype mitosis*. The entire process, including the two divisions, was originally called *maiosis* from the Greek word μειωσιζ for reduction by Farmer and Moore (1905). The term was used in a few publications (Farmer et al., 1905a,b; Moore and Arnold, 1905; Moore and Embleton, 1905; Moore and Walker, 1905). Later, Charles Walker (1907) proposed the term *meiotic phase* as a synonym for heterotype mitosis, and *meiotic division* to describe anaphase of the heterotype mitosis. Misspellings and confusion occurred in the literature, and *meiosis*, a word already in use in the English language to mean understatement (Hughes, 1959), soon replaced the term *maiosis* (Fraser, 1908; Bateson, 1909; Walker, 1910; Farmer, 1913; Sharp, 1921).

Thus, the manner in which the chromatin was distributed from cell to cell in the processes of mitosis in somatic cells, meiosis in reproductive cells, and fertilization of gametes was understood by the end of the 19th century (Weismann, 1891, 1892, 1893; Overton, 1893; Wilson, 1895). For over a century, hereditary factors that could give and shape life were postulated by Gottfried Leibniz, Pierre-Louis de Maupertuis (1753), George-Louis Leclerc Compte de Buffon, John Needham, Herbert Spencer (1864), Charles Darwin (1868), August Weismann (1893), Carl von Nägeli (1914), and others, and given names such as *monads, particles, organic particles, elementary molecules, the vegetative force, physiological units, minute granules, gemmules, the idoplasm,* and *biophors*. Following the discovery of chromatin, it seemed self-evident to the 19th-century cytologists that chromatin must be the material basis of the theoretical hereditary devices (Baltzer, 1967).

Hugo de Vries (1889) believed that the chromosomes themselves carried the hereditary factors, which he called *pangens*. In his historical as well as prescient and prophetic work, *Intracellular Pangenesis*, de Vries suggested that a pangen, which was larger than a chemical, but still invisibly small, represented one hereditary character. According to de Vries, each zygote contains a complete set of pangens, and during cell division, each pangen is transmitted to both daughter cells so that each daughter cell contains a complete set. During the formation of gametes, there is a reduction division so that each gamete has half the copies of pangens relative to the somatic cells, and the zygote, which results from the fusion of two gametes, has the same number of copies as the somatic cells. In this book, de Vries urged the emergence of the field of molecular systematics when he wrote,

> *Systematic relationship is based on the possession of like pangens. The number of identical pangens in two species is a true measure of their relationship. The work of the systematist should be to make the application of this measure possible experimentally, by finding the limits of the individual hereditary characters. Systematic difference is due to the possession of unlike pangens.*

Twenty-first century systematics or molecular phylogenetics has been in essence a realization of de Vries's proposal (Davis et al., 2004; Davis and Soreng, 2010; Doyle and Egan, 2010; Egan and Doyle, 2010; Coate and Doyle, 2011;Coate et al., 2011a, 2012; Doyle, 2011; Havananda et al., 2011; Tel-Zur et al., 2011; Seberg et al., 2012; Sherman-Broyles et al., 2017).

Hugo de Vries (1889) also postulated that cell differentiation resulted from the secretion of a specific constellation of pangens from the nucleus to the cytoplasm, where they directed the synthesis of proteins. He wrote,

> *As soon as the moment arrived for certain pangens, which until then had been inactive, to be set into activity, they would obviously pass from the nucleus into the cytoplasm. However, in so doing they would retain their characters, and especially their power to grow and multiply. Only a few like pangens would therefore have to leave the nucleus every time in order, by further multiplication, to impress the characters of which they are the bearers, on a given part of the cytoplasm. This process would repeat itself at every change of function of a protoplast; every time new pangens would leave the nucleus in order to become active.*

De Vries's theory of intracellular pangenesis predicted in essence what was to be discovered 60 years later by Caspersson (1950) and Brachet (1957, 1985) and today using the platform of transcriptomics, which again has been a twenty-first century realization of de Vries's proposal (Dembinsky et al., 2007; Frank and Scanlon, 2015a,b; Harrison, 2015; Ranjan et al., 2015; see Chapter 21).

Soon after writing *Intracellular Pangenesis*, de Vries studied hybrids of *Oenothera* and developed a theory that mutations were a cause of speciation. While working on hybridization, de Vries rediscovered Gregor Mendel's work. Gregor Mendel introduced mathematical analysis, a technique he learned from Christian Doppler, into biology and specifically into the study of inheritance. Unfortunately, his mathematically rigorous, experimentally sound, botanically insightful, and biologically significant work received negative criticism from Carl Nägeli who responded by saying "Your results are only empirical data; nothing in them is rational" (Bateson, 1902; Rhoades, 1984; Chandrashekaran, 1998). The neglect by Nägeli and others may not be so unimaginable because the majority of traits in a species are continuous and not binary like the traits investigated by Mendel, and it may have been unimaginable to see clearly continuous traits as being constructed from many binary traits (Weldon, 1901, 1902). Such tensions are common between scientists who see the fundamentals of nature reduced to quantitative models and scientists who see the complexity of nature as irreducible.

The onset of the twentieth century was a time for appreciating the possibility that continuous appearances can result from summing many small discrete events. This was already well-known to chemists who build up molecules, both on paper and in test tubes from discrete atomic parts. Chemists were the rock stars of the time as a result of the dye and drug industry that could transform black coal tar to mauve, perfume, aspirin, and heroin (Beer, 1959).

In 1900, the atomistic, discrete, quantum of energy, which would later be known as the photon, was discovered by Max Planck (1900). This coincided with the rediscovery of Mendel's work by Hugo de Vries (1900a,b), Carl Correns (1900), and Erich von Tschermak (1900), botanists who were studying quantized, binary traits, such as sugary versus starchy endosperm and yellow versus white endosperm, reintroduced binary quantitative analysis as opposed to continuous statistical analysis into the study of heredity and energized the search for the mechanism of inheritance. Ironically, when de Vries, along with Carl Correns and Erich Tschermak, rediscovered the work of Gregor Mendel, de Vries's theory of intracellular pangenesis, which postulated how the genetic factors influenced morphology, fell into obscurity. de Vries (1907) also suggested that a quantitative approach to genetics due to the postulates of pangens and mutations would have a salutary effect on genetics like the postulate of atoms and molecules had on chemistry.

Perhaps de Vries's prophetic work may have fallen into obscurity as a consequence of the appearance of a new and comprehensive textbook by Edmund B. Wilson (1896, 1925). Throughout history, the appearance of comprehensive textbooks, such as Euclid's *Elements* and Ptolemy's *Almagest*, have had the effect of making it less necessary for scientists to consult the original literature, and consequently many ideas that are ahead of their time get forgotten. It is also possible that de Vries's ideas were lost as scientists began to specialize, becoming cytologists, geneticists, and developmental biologists.

Interested in unifying the tenets of cytology with those of developmental biology and evolution to find the cellular basis for development and heredity, E.B. Wilson marshalled all the known facts and interpretations in these fields into his great book, *The Cell in Development and Inheritance*. He wrote about de Vries's theory of intracellular pangenesis, and posed questions about the relationship between chromosomes, development, and heredity. Wilson's approach had a profound impact on generations of cytologists, embryologists, and cytogeneticists (Agar, 1920; McClung, 1924; Morgan, 1924; Darlington, 1932; Sharp, 1934; Brachet, 1957; Swanson, 1957; Swanson et al., 1967; Uhl, 1996), yet the theory of intracellular pangenesis of de Vries was all but forgotten.

By the close of the 19th century, it was already established that the nucleus contains the hereditary substance that allows life to maintain the appearance of being immortal through the constant succession of individuals, even when the individual organic bodies pass away. It was also postulated that the same substance provided the instructions that were necessary for the formation of the proteins in the cytoplasm that perform the functions that allow the individual cell to be considered alive. After learning about chromosomes and heredity from J.B.S. Haldane and C.D. Darlington, Erwin Schrödinger (1946) summarized the situation in his popular book, What is Life?, "It is these chromosomes…that contain in some kind of code-script the entire pattern of the individual's future development and of its functioning in the mature state. Every complete set of chromosomes contains the full code…In calling the structure of the chromosome fibres a code-script we mean that the all-penetrating mind, once conceived by Laplace, to which every causal connection lay immediately open, could tell from their structure whether the egg would develop…into a black cock or a speckled hen, into a fly or a maize plant…We believe a gene—or perhaps the whole chromosome fibre—to be an aperiodic solid. And the gene …is probably a large protein molecule (see Walsby and Hodge, 2017)." Schrödinger's book influenced many biologists, including Irene Manton (1945; Williams, 2016), who wrote, "When a great physicist takes the trouble to explain some of his matured thoughts on topics of general interest outside his own subject, it is an event for which one cannot be too grateful."

Hämmerling (1953) performed nuclear transplant experiments in *Acetabularia* that clearly demonstrated the importance of the nucleus and nuclear-derived instructions in development (see Chapter 17).

The chromosomal theory of inheritance was not universally accepted around the world. In the 20th century, it was rejected on ideological grounds by Vladimir Ilyich Ulyanov (a.k.a. V.I. Lenin) and Josif Vissarionovich Dzhugashvili (a.k.a. Joseph Stalin), who appointed Trofim Lysenko to come up with a theory of heredity that was more in keeping with their Bolshevik philosophy (Lysenko, 1946, 1948, 1954). Genetics and economics are still often interconnected, and experiments that show that the genomes of plants change when they are grown in different environments (Durrant, 1958, 1962; Evans et al., 1966; Timmis and Ingle, 1973; Nagl and Rucker, 1976; Cullis, 1977, 1981, 1990; Natali et al., 1986) are not well received by scientists working in an economy that patents genes as if they were patent medicines.

Yet there is more to development than genes. According to Conrad Waddington (1942), "Genetics ... has been the most successful in finding a way of analysing an animal into representative units, so that its nature can be indicated by a formula as we represent a chemical compound by its appropriate symbols." However, he also realized that the genetic landscape in which a given gene existed affected its expression. We now know that there are many opportunities during the development of the organism to influence the final outcome of a locus on a chromosome without changing its underlying sequence of DNA. He called the collection of modified loci the *epigenotype* of the organism and named the study of the epigenotype *epigenetics* (Goldberg et al., 2007), which is now known as epigenomics (see Chapter 21).

16.2 ISOLATION OF NUCLEI

Nuclei were first isolated from pus cells and sperm in the late 19th century. Currently, nuclei are isolated by homogenizing a tissue and then filtering and centrifuging the homogenate at 1900 g for 10 min to pellet the nuclei. The pellet is then resuspended and layered on top of a discontinuous Percoll gradient and subjected to density-gradient centrifugation at 7800 g for 30 min. The nuclei are removed from the 25%/50% interface, and then centrifuged at 1900 g to remove the Percoll (Datta et al., 1985). The purity and intactness are estimated by staining the nuclei with the Feulgen reagent (Ruzin, 1999) and assaying for DNA polymerase or RNA polymerase activity (Dunham and Bryant, 1988). Proteomic and bioinformatic studies have characterized the proteins that are in the nucleus (Petrovská et al., 2015; Tamura et al., 2015) and a proteomic study using green fluorescent protein has characterized the proteins in the nuclear pore complex (Tamura et al., 2010).

16.3 STRUCTURE OF THE NUCLEAR ENVELOPE AND MATRIX

Much of what we know about the ultrastructure of the nucleus comes from the work of Werner Franke, a cell biologist who also courageously worked to expose the East German program to dope athletes with anabolic steroids (Franke and Berendonk, 1997; Ungerleider, 2001; Gall, 2005; Sitte, 2005).

The nucleus is a large organelle about 10 μm in diameter. In meristematic cells, it is typically spherical, and it may become flattened and ellipsoidal in differentiated cells—where the precise morphology and lobulation depends on the activity of the cell (see Fig. 16.4; Mathews, 1899; Franke and Schinko, 1969; Collings et al., 2000). The nucleus is surrounded by an envelope composed of two concentric membranes (Dessev, 1992; Hurt et al., 1992) composed of lipids and proteins (Hendrix et al., 1989). The outer nuclear membrane is contiguous with the membranes of the endoplasmic reticulum (ER) and is sometimes regarded as a specialized form of the ER (Porter and Moses, 1958). Perhaps the nuclear envelope is the semiautonomous source of the endomembranes (see Chapters 5 and 8). The outer nuclear membrane is covered with ribosomes that are engaged in protein synthesis, and the proteins that are made on these ribosomes are transported to the space between the inner and outer nuclear membranes (Porter and Bonneville, 1968). This space, which is an E-space, is contiguous with the lumen of the ER. Like the lumen of the ER, it contains

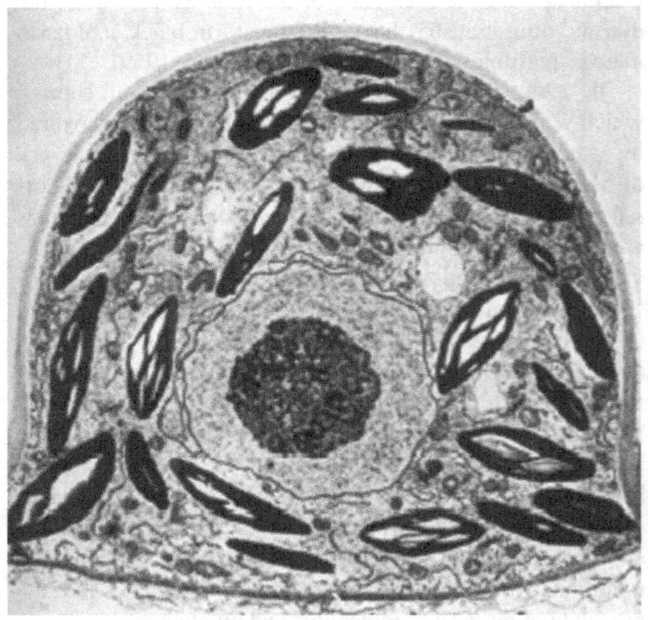

FIGURE 16.4 Electron micrograph of a nucleus from a bud of *Funaria hygrometrica*. ×3900. *From Conrad, P.A., Steucek, G.L., Hepler, P.K., 1986. Bud formation in Funaria: organelle redistribution following cytokinin treatment. Protoplasma 131, 211–223.*

proteins with the carboxy-terminal sequence KDEL as determined by immunolocalization (see Chapter 4; Herman et al., 1990). The outer membrane of the nuclear envelope can be embraced by a network of actin filaments (Seagull et al., 1987; Traas et al., 1987; Schmit and Lambert, 1987; Villanueva et al., 2005), and it can act as a microtubule-organizing center (MTOC; see Chapter 11; Mizuno, 1993).

The inner membrane of the nuclear envelope is associated with a network of filaments known as the *nuclear matrix*. The chromatin is embedded in this matrix, which E.B. Wilson (1895) called the *linin*. There are specific regions of the DNA that attach to the matrix (Hall et al., 1991; Breyne et al., 1992; Spiker and Thompson, 1996; Nomura et al., 1997). The matrix is defined operationally as the remains of the nucleus after salt and detergent extraction. Many of the protein components of the nuclear matrix, one of which is similar to the intermediate protein lamin (see Chapter 11), have been characterized (Fawcett, 1966; Gosh and Dey, 1986; Nigg, 1988; McNulty and Saunders, 1992; Frederick et al., 1992; Tong et al., 1993; Berezney et al., 1995; Bode et al., 1995; Moreno Díaz de la Espina, 1995; Masuda et al., 1997; Fiserova and Goldberg, 2010; Starr and Fridolfsson, 2010; Zhou et al., 2012, 2015a,b,c; Sakamoto and Takagi, 2013; Tapley and Starr, 2013; Wang et al., 2013; Zhou and Meier, 2013; Ciska and Moreno Díaz de la Espina, 2014; Evans et al., 2014; Goto et al., 2014; Guo and Fang, 2014; Tatout et al., 2014; Guo et al., 2017; Meier et al., 2017; Poulet et al., 2017).

The outer membrane of the nuclear envelope contains an integral membrane protein known as KASH, and the inner membrane of the nuclear envelope contains an integral membrane protein known as SUN (Matsunaga et al., 2013). The SUN–KASH complex is considered to be a linker of the nucleoskeleton to the cytoskeleton. The KASH protein in conjunction with other proteins interacts with the cytoskeleton on the cytosolic side of the envelope, and the SUN protein interacts with the lamin-like proteins of the nuclear matrix or nucleoskeleton on the nucleoplasmic side of the envelope. Because KASH and SUN may interact with each other, there may be direct mechanical communication between the cytoskeleton and the nuclear matrix, which influences the shape of the nucleus and may influence gene expression. There may be a connection between DNA and/or DNA-binding proteins such as transcription factors (Guo et al., 2017a,b) and the nuclear envelope in eukaryotes, as there may be a connection between replicating DNA and/or DNA-binding proteins and the plasma membrane in prokaryotes (Jacob et al., 1963; Funnell, 1993; Toro and Shapiro, 2010; Badrinarayanan et al., 2015).

Because of the large amount of chromatin, the interior of the nucleus is probably too viscous to allow the free diffusion of macromolecules within reasonable time frames (Agutter, 1991, 1995). Thus, the nuclear matrix may function in organizing the interphase nucleus (Heslop-Harrison and Bennett, 1990) and/or transporting RNA to the nuclear pore complex (Xing and Lawrence, 1991). The nuclear matrix may also be important in regulating both DNA replication and RNA transcription. The effect of nuclear structure and the position of genes within the nucleus on gene expression is a particularly fascinating and exciting area of research (Nagl, 1985; Heslop-Harrison et al., 1993; Shaw et al., 1993; Heslop-Harrison, 2000; Lengerova and Vyskot, 2001; Yano and Sato, 2002). The positions of many genes in the interphase nucleus can be found simultaneously using the FISH (fluorescent in situ hybridization) and GISH (genomic in situ hybridization; Heslop-Harrison et al., 1999) techniques.

Many enzymes and proteins that are synthesized in the cytoplasm, but function in the nucleus, including DNA and RNA polymerases and histones, must pass through the nuclear envelope from the cytoplasm into the nucleus. By contrast, messenger RNA (mRNA), transfer RNA (tRNA), and ribosomal RNA (rRNA) associated with protein synthesis must leave the nucleus and enter the cytoplasm. This bidirectional nucleocytoplasmic transport takes place through the pores in the nuclear envelope (see Fig. 16.5; Anderson and Beams, 1956; Hinshaw et al., 1992; Newmeyer, 1993). Many of the proteins of the nuclear pore complex involved in protein recognition and translocation have been identified (Gruber et al., 1988; Yamasaki et al., 1989; Burke, 1990; Carmo-Fonseca and Hurt, 1991; Scofield et al., 1992).

There is a family of glycoproteins in the nuclear pore complex known as *nucleoporins*. These glycoproteins contain *O*-linked *N*-acetylglucosamine (Snow et al., 1987; Holt et al., 1987), a linkage unlike those made in the ER

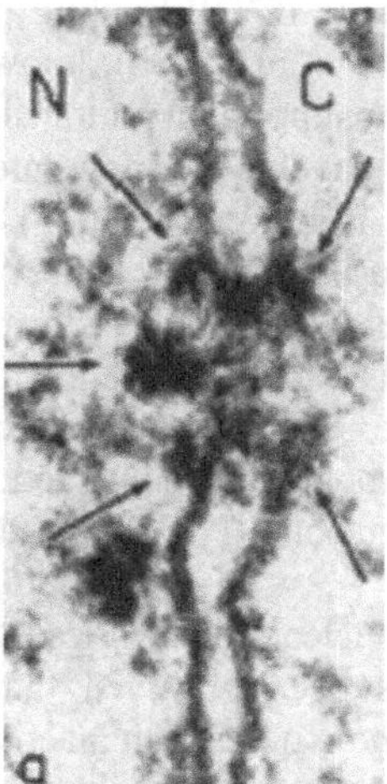

FIGURE 16.5 Nuclear pore of a *Canna generalis* pollen mother cell. ×160,000. *C*, cytoplasm; *N*, nucleus. *From Scheer, U., Franke, W.W., 1972. Annulate lamellae in plant cells: formation during microsporogenesis and pollen development in* Canna generalis *Bailey. Planta 107, 145–159.*

and Golgi apparatus. Davis and Blobel (1986, 1987) have shown that these proteins, which can be recognized by an antibody and localized to the nuclear pore, are made on cytoplasmic ribosomes and are glycosylated in the cytoplasm. There are approximately 30 different nucleoporins that make up each pore (Raices and D'Angelo, 2012). Given the eightfold symmetry in the plane of the nuclear envelope and the twofold symmetry perpendicular to the nuclear envelope, each pore consists of more than 480 proteins. The composition of the pores may vary from tissue to tissue (Raices and D'Angelo, 2012; Floch et al., 2014) and the variation in pore protein composition may reflect differences in pore function and transport specificity.

The importance of the glycoproteins in translocation through the nuclear pore is underscored by the observation that isolated nuclei can transport proteins into them, but not if the nuclei have been pretreated with wheat germ agglutinin, the glycoprotein-binding lectin (Finlay and Forbes, 1990).

Soluble proteins known as *importins* that are involved in facilitating translocation through the nuclear pore complex have been identified in in vitro assays that characterize factors that increase the efficiency of protein translocation into isolated nuclei or into nuclei in digitonin-permeabilized cells (Merkle et al., 1996; Merkle and Nagy, 1997; Han et al., 2011; Floch et al., 2014). Genetic screens have been used in yeast to identify the proteins known as *exportins* that facilitate export from the nucleus by finding mutants that are unable to export RNA from the nucleus (Haasen et al., 2000; Blanvillain et al., 2008; Floch et al., 2014). Evidence that importins and exportins are important in plants comes from the study of mutants. Mutations in the genes that code for importins and exportins result in developmental defects (Tamura and Hara-Nishimura, 2014). A small momomeric guanosine triphosphate (GTP) binding protein of the Ras superfamily (Brembu et al., 2006), known as *Ran*, and its associated proteins are important for translocation of proteins across the nuclear envelope during interphase (Rodrigo-Peiris et al., 2011; Boruc et al., 2015). Because the Ran GTPase is in the cytosol, RanGDP is the form typically found in the cytosol, whereas RanGTP is the form typically found in the nucleus. RanGTP stabilizes the exportin-cargo interaction so that the cargo is translocated out of the nucleus, and it destabilizes the importin-cargo interaction so that the cargo is released in the nucleus (Pay et al., 2002; Cavazza and Vernos, 2016).

The number of nuclear pores varies over time in a single nucleus and it has been observed that transcriptionally active nuclei have more pores than inactive ones. Typically, there are about 11 pores per μm^2, although the nuclear pore density ranges from 5 to 60 pores per μm^2. The nuclear pores can be distributed regularly or irregularly in the nuclear envelope (Wacker and Schnepf, 1990), and it has been postulated that the distribution of the pores may affect both gene expression and the position of the newly transcribed mRNA (Blobel, 1985).

Each pore is embedded in a large disklike structure that is composed of eight granules and has an eightfold symmetry (see Fig. 16.6; Franke, 1966; Gall, 1967; Unwin and Milligan, 1982; Tamura and Hara-Nishimura, 2013, 2014). The pore complex is about 100−150 nm in diameter and about 50−70 nm in thickness (Akey, 1989; Floch et al., 2014). Filaments extend 60 nm from the nuclear pore complex into the nucleoplasm, forming a basketlike structure (Goldberg and Allen, 1992). There is an aqueous pore in the center of the nuclear pore complex that has an opening of about 9 nm, as determined by measuring the rate of diffusion of fluorescent dyes of different molecular mass from the cytoplasm into the nucleus. A fluorescently labeled protein of 17 kDa equilibrates between the cytoplasm and the nucleus in about 2 min. A 44-kDa protein takes about 30 min, and a 60-kDa protein is hardly able to enter (Lang et al., 1986). The permeability of the nuclear pore depends on the cell type because a 9-kDa dextran, but not a 39-kDa dextran, can pass through the nuclear envelope of starfish oocytes, whereas in mouse oocytes, a 64.4-kDa dextran can pass through the nuclear envelope, but a 156.9-kDa dextran cannot (Hiramoto and Kaneda, 1988).

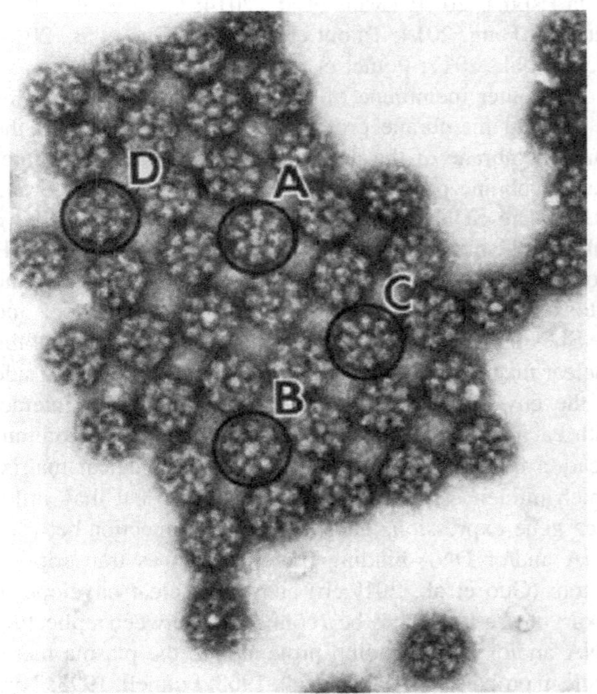

FIGURE 16.6 Nuclear pores on an isolated nuclear envelope from a *Xenopus* oocyte. The pores have been negatively stained and viewed with an electron microscope. (A−D) only refer to pore complexes that were analyzed by image-processing techniques to determine the three-dimensional structure of a pore complex. ×75,000. *From Unwin, P.N.T., Milligan, R.A., 1982. A large particle associated with the perimeter of the nuclear pore complex. J. Cell Biol. 93, 63−75.*

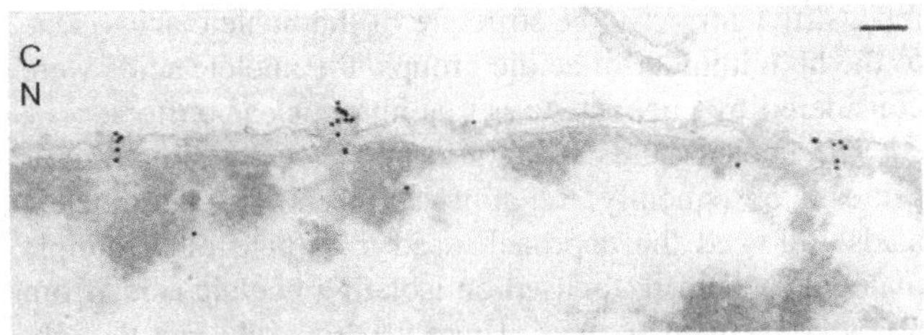

FIGURE 16.7 Colloidal gold particles (10 nm) coated with nucleoplasmin are seen being translocated through the nuclear pore complexes of an isolated rat liver nucleus. *C*, cytoplasm; *N*, nucleus. Bar, 100 nm. *From Newmeyer, D.D., Forbes, D.J., 1988. Nuclear import can be separated into distinct steps in vitro: nuclear pore binding and translocation. Cell 52, 641–653.*

Nucleoplasmin is one protein that must enter the nucleus. Nucleoplasmin is a chaperonin that is involved with the assembly of nucleosomes (see Section 16.5). Nuclear proteins, including nucleoplasmin, have been extracted from animal cells and microinjected back into the cytoplasm (Garcia-Bustos et al., 1991; Silver, 1991; Agutter, 1991; Davis, 1992). These proteins return to and accumulate in the nucleus. In fact, if nucleoplasmin is cleaved into a head and a tail region, and the tail region, which can enter the nucleus, is attached to a 20-nm colloidal gold sphere, the colloidal gold is transported into the nucleus, but only when attached to the tail (Fig. 16.7). The gold particles are not transported alone. The nuclear pore must recognize the tail protein and open up the pore in a specific manner (Dingwall and Laskey, 1986; Feldherr et al., 1984; Silver and Goodson, 1989). The translocation is coupled to the energy released by the hydrolysis of ATP (Newmeyer and Forbes, 1988).

Proteins that are translocated into the nucleus have a special nuclear import signal or nuclear localization signal. This nuclear localization sequence was first identified in the SV40 virus-encoded protein T-antigen. T-antigen is a 90-kDa protein that must enter the nucleus to facilitate viral DNA replication. The T-antigen normally accumulates in the nucleus after it is synthesized in the cytoplasm. However, a single mutation that introduces a threonine instead of a lysine into the protein prevents the import of this polypeptide into the nucleus. A short sequence around this lysine was considered to be the nuclear signal sequence. When a short piece of DNA that codes this region is attached to a gene that codes for a protein that is normally localized in the cytoplasm, the fusion protein is translocated into the nucleus. In this manner, it was determined that the shortest sequence that functioned in nuclear targeting was an eight–amino acid sequence in the middle of the polypeptide (Goldfarb et al., 1986; Kalderon et al., 1984; Lanford and Butel, 1984). Subsequent work showed that over half of the sequenced nuclear proteins have two regions interspersed in each polypeptide that act in concert to target the polypeptide to the nucleus (Restrepo et al., 1990; Carrington et al., 1991; Robbins et al., 1991; Varagona et al., 1991, 1992).

The import of proteins into the nucleus can be regulated through the phosphorylation of amino acids in the targeting sequence. The phosphorylation of an amino acid may affect the ability of the nuclear localization sequence to recognize its receptor. If the protein is one of the transcription factors, this can have a direct effect on gene expression (Moll et al., 1991; Davis, 1992; Hunter and Karin, 1992).

Not only must proteins enter the nucleus but also tRNA, mRNA, and ribosomal subunits containing rRNA must leave the nucleus (Mattaj, 1990). When 20-nm colloidal gold particles are complexed to poly-A RNA, rRNA, or tRNA and injected into the nucleus of a frog oocyte, they are transported through the pores into the cytoplasm. However, if they are injected into the cytoplasm, they remain there, indicating that the nuclear pore acts as a rectifier (Clawson et al., 1985; Dworetzky and Feldherr, 1988). The translocation of mRNA has been shown to be dependent on the association of the mRNA with snRNPs (small nuclear ribonucleoproteins, pronounced "snurps"; Davis, 1992; Legrain and Rosbash, 1989).

16.4 CHEMISTRY OF CHROMATIN

Friedrich Miescher was a physician who was interested in investigating the chemical composition of nuclei (Davidson and Chargaff, 1955; Mirsky, 1955; Fruton, 1972). In 1868, he isolated a compound from the nuclei of pus cells that had unique chemical properties. Unlike the well-characterized albuminous proteins, the new compound was insoluble in water, acetic acid, dilute HCl, and NaCl. When he treated the cells with alcohol to remove the fat, and the pepsin-containing pig gastric mucosa to remove the proteins, something remained in the nuclei. He called the new compound nuclein, which had the chemical formula $C_{29}H_{49}N_9P_3O_{22}$. In 1871, Hoppe-Seyler and his students showed that *nuclein* was ubiquitous, as it was present in yeast and the nuclei of snake and bird erythrocytes. Miescher separated from the nuclein a basic proteinaceous substance, which he named *protamine*, because it was less complex than most proteins. Albrecht Kossel (1884) showed that the protamine was actually a protein called *histone*, which he had already isolated from nuclei (see Kossel, 1928). Richard Altmann (1889) gave the name *nucleic acid* to the protein-free portion of nuclein. In a speculative note submitted along with his other work in 1869, Miescher suggested that the study of individual nucleins might reveal something about the differences between members of a group. However, this note was rejected (Miescher, 1897; Levene and Bass, 1931).

E.B. Wilson (1895) wrote about the possible function of nuclein:

The precise equivalence of the chromosomes contributed by the two sexes is a physical correlative of the fact that the two sexes play, on the whole, equal parts in hereditary transmission, and it seems to show that the chromosomal substance, the chromatin, is to be regarded as the physical basis of inheritance. Now, chromatin is known to be closely similar to, if not identical with, a substance known as nuclein ($C_{29}H_{49}N_9P_3O_{22}$, according to Miescher), which analysis shows to be a tolerably definite chemical composed of nucleic acid (a complex organic acid rich in phosphorous) and albumin. And thus we reach the remarkable conclusion that inheritance may, perhaps, be effected by the physical transmission of a particular chemical compound from parent to offspring.

Indeed, the possibility of testing the hypothesis that nucleic acids were the genetic material seemed possible to Emil Fischer (quoted in McCarty, 1995) in 1914 when he wrote,

With the synthetic approaches to this group we are now capable of obtaining numerous compounds that resemble, more or less, natural nucleic acids. How will they affect living organisms? Will they be rejected or metabolized or will they participate in the construction of the cell nucleus? Only the experiment will give us the answer. I am bold enough to hope that, given the right conditions, the latter may happen and that artificial nucleic acids may be assimilated without degradation of the molecule. Such incorporation should lead to profound changes of the organism, resembling perhaps permanent changes or mutations as they have been observed before in nature.

At the turn of the 20th century, nucleic acids were isolated from wheat germ, yeast, and thymus to study their structures. However, these studies were based on the perspectives of organic chemists with a reductionist approach (Jones, 1920; Feulgen, 1923; Levene and Bass, 1931). Based on degradation studies, the organic chemists considered that nucleic acids were simple macromolecules. When nucleic acids were hydrolyzed, they were broken down into their component parts, which turned out to be the purines (Fischer, 1902), adenine, and guanine, as well as a new class of compounds known as *pyrimidines*. This class included thymine (5-methyl uracil), cytosine, and uracil. Although the organic chemists made outstanding contributions to understanding the components of nucleic acids, they lost sight of the fact that the parent molecule was a macromolecule and may have a level of complexity that was greater than the complexity of its component parts (Leathes, 1926). Thus, they developed extraction procedures that were suitable for degradation studies, but would not allow them to isolate the native molecule, and to reconstruct correctly the structure of the nucleic acids. Due to the high number of acidic groups, the nucleic acids were considered by chemists to act as intranuclear buffers.

The nucleic acids seemed to be relatively simple molecules. Consequently, the approach used to isolate nucleic acids followed the approach used to isolate other simple materials and work focused on isolating nucleic acids from a few convenient sources. There was one odd fact though: the nucleic acids of plants seemed to be composed of ribonucleosides, whereas those of animals were composed of deoxyribonucleosides. Walter Jones (1920) concluded in his monograph on nucleic acids that "there are but two nucleic acids in nature, one obtainable from the nuclei of animal cells, and the other from the nuclei of plant cells." Thus, plant nucleic acids were known as either *yeast nucleic acid* or *ribonucleic acid*, and animal nucleic acids were known as *thymonucleic acid* or *desoxyribonucleic acid*. I would like to stress that the chemists' interpretations of the simplicity of nucleic acids resulted from their preconceived notions of the simplicity of nucleic acids, and thus they developed inadequate extraction procedures.

It appeared to the chemists that nucleic acids were composed of equal quantities of nucleotides, and Phoebus Levene (1921) concluded that nucleic acids were composed of repeating units of four nucleotides. According to this "tetranucleotide hypothesis," one proposed structure for thymonucleic acid was ATCGATCGATCG. The tetranucleotide hypothesis, introduced in 1921, dominated the thinking of scientists for more than 20 years. X-ray diffraction data, based on the X-ray diffraction technique introduced by Max von Laue (1915), William and Lawrence Bragg (1922, 1924), and William Astbury (Hall, 2011, 2014), were used to elucidate the structure of DNA. The data obtained by William Astbury and Florence Bell indicated regular repeats every 0.33−0.34 nm, which they believed were due to the nucleotides, and a 2.7-nm repeat, which they believed indicated that the DNA molecule was composed of a repeating sequence of 8 or 16 nucleotides (Astbury and Bell, 1938b; Astbury, 1939). William Astbury (1947b) concluded that "It seems improbable, too, to judge by the degree of perfection of the X-ray fibre diagram that these four different kinds of nucleotides are distributed simply at random." A nonuniform arrangement of nucleotides would be necessary if the nucleic acids were to function as the genetic material, whereas "the existence of tetranucleotide units, repeated throughout the molecule, would limit the potential number of isomers and hence diminish the possibilities of biological specificity" (Gulland, 1947). Elwyn Beighton took much better X-ray diffraction photographs of DNA on May 28 and June 1, 1951 in Astbury's laboratory. However, they had a much bigger effect on James Watson than they did on William Astbury (https://www.leeds.ac.uk/heritage/Astbury/Beighton_photo/index.html; Hall, 2011).

Ideas about the structure of thymonucleic acid began to change when Schmidt (1928, 1932, 1937) and Signer, Caspersson, and Hammarsten (1938), using physical techniques, studied the negative birefringence, viscosity, rate of diffusion, and rate of sedimentation of thymonucleic acid and concluded that nucleic acids behaved like thin rods, 300 times longer than they were wide, with molecular masses of between 500 and 1000 kDa. However, due to the strong influence of colloidal chemistry, it was assumed that these large fibers were actually aggregates of thymonucleic acid, and these physical studies, like the chemical studies, gave no impetus for considering DNA to be the genetic material.

Cytologists were obtaining data that conflicted with those of the organic chemists. Using a stain now known as *Feulgen's reagent*, Feulgen and Rossenbeck (1924; Brachet, 1989) discovered that DNA occurred in the nuclei of plants and animals, and not just in those of animal cells. They (quoted in Hughes, 1952) wrote,

> *This gave us great surprise, for the nuclei of the wheat embryo gave the nucleal reaction more intensively than we have ever seen in any animal tissues. This was confirmed in other plants, and so it was demonstrated that the plant nucleus contains nucleal bodies. The old dualism of yeast and thymonucleic acids is thus set aside.*

Conclusions about the importance of DNA based on Feulgen staining were not looked on favorably because the staining of various cells in the same organism seemed to vary in an unreasonable manner. Approximately two decades later, Mirsky and Ris (1948, 1949; Ris and Mirsky, 1950) determined that the DNA content of nuclei from somatic cells of the same organism is constant, and variations in the ratio of DNA to protein in the nucleus accounted for the variations in the staining.

Prior to the definitive work done by Erwin Chargaff in 1946–1950, there was little reason to even think that nucleic acid was the genetic material based on studies by organic chemists. Indeed, Wendell Stanley (1935) thought that he crystallized the hereditary component of the tobacco mosaic virus when he crystallized its protein. The chemistry of nucleic acids went through a revolution when Erwin Chargaff, "a licensed biochemist," became aware of the genetic evidence that DNA was the hereditary material (Chargaff, 1975). Chargaff was instantly captivated by the experiments performed by Oswald Avery, Colin MacLeod, and Maclyn McCarty in 1944 that indicated that DNA was the genetic material (McCarty, 1985). These experiments were based on the observations of Fred Griffith who discovered in 1928 that when mice were injected with virulent pneumococci that had been heat-killed, along with live avirulent pneumococci, the mice died and their blood contained living virulent pneumococci. Dawson and Sia (1931) found that the transformation of avirulent to virulent pneumatococci could also take place in a test tube. Avery et al. (1944) interpreted all of these results to mean that there was a transfer of genetic material between the bacteria during transformation. Avery et al. (1944) wrote,

> *Biologists have long attempted by chemical means to induce in higher organisms predictable and specific changes which thereafter could be transmitted in series as hereditary characters. Among microorganisms the most striking example of inheritable and specific alterations in cell structure and function that can be experimentally induced and are reproducible under well defined and adequately controlled conditions is the transformation of specific types of Pneumatococcus.*

Avery et al. (1944) fractionated the pneumatocci and meticulously isolated the various fractions and found that the transforming principle was none other than the protein-free DNA fraction. The experiments that pointed to DNA as the genetic material were continued by Rollin Hotchkiss (1995), although not everyone agreed that DNA was the genetic material (Mirsky and Pollister, 1946).

Hearing of Avery's results and believing that DNA was the genetic material, Chargaff began isolating DNA in earnest. He developed the techniques necessary to isolate intact macromolecules of DNA and used the newly introduced technique of paper chromatography (Consden et al., 1944) and a Beckman photoelectric quartz UV spectrophotometer (Cary and Beckman, 1941; Gibson and Balcom, 1947) to do quantitative work on nucleic acids (Vischer and Chargaff, 1948). Chargaff (1950a),

> *... started in our work from the assumption that nucleic acids were complicated and intricate high-polymers, comparable in this respect to the proteins, and that the determination of their structures and their structural differences would require the development of methods suitable for the precise analysis of all constituents of nucleic acids prepared from a large number of different cell types.*

Chargaff demonstrated that carefully isolated and analyzed molecules of DNA were not composed of equal quantities of the four deoxynucleotides as the tetranucleotide hypothesis predicted, but that DNA was composed of equal quantities of guanine and cytosine and equal quantities of adenine and thymine and that the ratio of the nucleotide quantities depended on the species from which the DNA was isolated. Moreover, Chargaff found that the ratio was constant for each species and was thus correlated with the genetic inheritance of the species. I cannot stress enough how important it was to approach DNA from the mindset that it may be a complicated, information-containing molecule (Butler, 1952). In this respect, Chargaff (1963b) was a real pioneer. Moreover, Chargaff confirmed the conclusion of Feulgen and Rossenbeck (1924) based on cytochemical data that even plants have

DNA (Mirsky, 1953; Chargaff, 1958). Chargaff (1950a,b) further observed "It is, however, noteworthy … that in all the deoxypentose nucleic acids examined thus far the molar ratios of total purines to total pyrimidines, and also of adenine to thymine and of guanine to cytosine, were not far from 1." Still, many scientists, including William Astbury, who coined the term molecular biology, still thought that protein was involved with inheritance—perhaps in the form of a nucleoprotein. In 1952, Alfred Hershey and Martha Chase (1952) performed an experiment that confirmed that DNA was the genetic material. They labeled the DNA of the T2 bacteriophage with ^{32}P, and in a separate experiment, they labeled its protein with ^{35}S. After the labeled bacteriophage had enough time to inject its genetic material into the bacteria, Hershey and Chase separated the bacteria from the phage by throwing them in a blender. They found that the T2 progeny was labeled with ^{32}P but not with ^{35}S, indicating that DNA, not protein, was the genetic material. DNA was finally accepted as the genetic material, and by 1952, research on nucleic acids had progressed so much that Arthur Hughes (1952), a mentor of Francis Crick's (Olby, 1972), wrote, "In recent years, there has been a tendency to explain every cellular change in terms of nucleic acids."

While Hotchkiss was working on the genetic nature of DNA, and Chargaff was working on its chemical nature and finding that each individual has different DNA molecules (Manchester, 2007), Rosalind Franklin was working in Maurice Wilkins's laboratory to determine the physical structure of DNA (Franklin and Gosling, 1953; Wilkins et al., 1953; Sayre, 1975). At the same time, Linus Pauling and Robert Corey (1953) were working on the structure of DNA from a theoretical perspective. They had already been successful in applying the concepts of quantum chemistry and the importance of hydrogen bonding to determine the structure of proteins, and they were utilizing this same approach to deduce the structure of DNA from first principles (Pauling, 1972). Pauling and Corey came up with a model of DNA that could be described as a triple helix, with the bases pointing out. Linus Pauling had planned to visit the X-ray crystallographers in Wilkins's laboratory to see the data that would confirm or refute his model, but he was not issued a passport by the State Department due to his liberal politics and concern about the ill effects of nuclear fallout (Pauling, 1963; Watson, 1968; Goertzel and Goertzel, 1995). Because Pauling was not politically correct, he was not privy to the X-ray data obtained by Wilkins et al. (1953) and by Franklin and Gosling (1953).

Unbeknownst to Rosalind Franklin, James Watson (1968) and Francis Crick (1988) did see a report of her unpublished results and rushed to publish a posteriori model of DNA that synthesized Avery et al.'s genetic data, Chargaff's chemical data, and the physical data obtained by Wilkins et al. (1953) and Franklin and Gosling (1953). The X-ray diffraction images taken by Franklin were excellent, and, when the DNA was observed in certain conditions of hydration, the images could only be interpreted to result from a structure that has a double helix (Crick, 1954). The double nature was beautiful from a biological and genetic perspective (Watson and Crick, 1953b), and Watson and Crick (1953a) wrote, "It has not escaped our notice that the specific pairing we have postulated immediately suggests a possible copying mechanism for the genetic material." Indeed, here was the chemical equivalent of Wilson's (1923) cytological description of the replication and division of the genetic material. It was also an excellent example of Pauling and Delbrück's (1940) hypothesis that complementary structures, as opposed to identical structures (Jordan, 1938, 1939, 1940), will be shown to be involved in autocatalytic processes.

From reading their 1953 paper, it would seem that the double helix sprang out of Watson and Crick's heads just like Athena sprang from the head of Zeus (Watson and Crick, 1953a; Crick, 1974b). Indeed, it even seems like the whole field of molecular biology, a term coined by William Astbury (1950, 1961; Waddington, 1961; Morange, 2008; Hall, 2014), started with the publication of their paper (Lamanna, 1968; Stent, 1968; Waddington, 1969; Hess, 1970; Weaver, 1970; Wyatt, 1972; Fruton, 1992). Nevertheless, I hope that I have shown that this discovery, like all other discoveries in cell biology, involve the work of many people. Unlike many other discoveries, however, this one has stimulated many people to write about the ethics of James Watson (Chargaff, 1968, 1974, 1978; Pauling, 1973; Sayre, 1975; Donohue, 1976; Stent, 1980; Crick, 1988, 1995).

Watson, like the eugenicists before him, is a man who contemplates ethical issues himself (Galton, 1869, 1892; Pearson, 1905; Davenport, 1910, 1912; Ellis, 1911; Castle, 1916; Popenoe and Johnson, 1918; East and Jones, 1919; East, 1920, 1927, 1931; Jennings, 1925; Pearson and Moul, 1925; Huxley, 1926; Popenoe, 1934; Huxley, 1946, 1962.). He helped launch the Human Genome Project, a large-scale initiative involved in sequencing the entire human genome, to better the human condition (Watson and Berry, 2003; McElheny, 2003; Lindee, 2003). In an autobiographical article entitled "Values from a Chicago Upbringing," Watson (1995) wrote of the importance of the human genome project and eugenics:

> *But diabolical as Hitler was, and I don't want to minimize the evil he perpetuated using false genetic arguments, we should not be held in hostage to his awful past. For the genetic dice will continue to inflict cruel fates on all too many individuals and their families who do not deserve this damnation. Decency demands that someone must rescue them from genetic hells. If we don't play God, who will?*

The cell contains a memory of its past and a potential of its future, and this special aspect of life is inseparable from the DNA. Each gene is on the order of about 4−8 nm, as determined by the efficacy of radiation in causing a mutation (Lea, 1947), and a typical diploid nucleus contains about 1−10 pg of DNA. Each gene present in the nucleus is related to the first ancestral gene (see Chapter 18) and has survived in some form throughout the evolution of life. To paraphrase Boris Ephrussi (1953), the nucleus, like Noah's ark, contains two of each gene, one from each parent. Each DNA molecule consists of two long antiparallel chains of deoxyribose phosphates. The deoxyribophosphates are held together with covalent phosphodiester bonds that link the 5′ carbon of one sugar to the 3′ carbon of the next. Each deoxyribose is connected to a nitrogenous base, adenine, cytosine, guanine, or thymine. The adenines pair with the thymines and the cytosines pair with the guanines (Watson and Crick, 1953a; Chargaff, 1955; Crick, 1955; Davidson, 1960; Jordan, 1960). This complementary base pairing results in the formation of a double helix (Herskowitz et al., 1993).

Francis Crick wrote to Erwin Schrödinger on August 12, 1953 (see Walsby and Hodge, 2017) "Watson and I were once discussing how we came to enter the field of molecular biology, and we discovered that we had both been influenced by your little book, 'What is Life?'. We thought you might be interested in the enclosed reprints—you will see that it looks as though your term 'aperiodic crystal' is going to be a very apt one. Yours sincerely, Francis Crick"

As a result of intrinsic genetic factors, including transposons (McClintock, 1983) and environmental mutagens (Klekowski and Berger, 1976; Klekowski and Klekowski, 1982), the DNA in the nuclei can mutate (Sanford, 2008). As is a consequence of these somatic mutations that accumulate in a nucleus over time, the DNA in each nucleus of a long-lived plant may not be necessarily identical (Klekowski, 1971, 1976; Natali et al., 1995; Hallé, 2002). The ability of the plant to keep or lose somatic mutations that take place in the growing tip that gives rise to the rest of the plant depends on the structure of the apical meristem (Klekowski and Kazarinova-Fukshansky, 1984a,b; Klekowski et al., 1989). It is important to remember that this genetic divergence, also known as *genomic plasticity*, is one strategy that long-lived plants use to adapt to a changing environment.

If we assume that a cell needs about 100 proteins to perform the basic functions of life (see Chapter 1), then the cell would need 100 genes. If each gene coded for a 100,000-Da protein (<1000 amino acids each), then the DNA would contain about 300,000 base pairs. The smallest genome is in *Mycoplasma genitalium*. This genome consists of 580,000 base pairs and encodes about 470 proteins (Fraser et al., 1995). Even plants with small genomes such as *Arabidopsis*, *Quercus*, and *Aesculus* have about 80×10^6 base pairs, which is enough to code for about 27,000 genes (Heslop-Harrison, 1991). Some plants have even more DNA. For example, *Fritillaria* contains $86{,}000 \times 10^6$ base pairs. As a comparison, humans have 3000×10^6 bp of DNA and Escherichia *coli* has 4.7×10^6 bp. It seems paradoxical that there is not any correlation between the amount of DNA (Gall, 1981) and the complexity of the organism. There are many kinds of DNA sequences in a genome. They range from those found in a single copy to those that exist in highly repetitive sequences (Britten and Kohne, 1968; Davidson et al., 1975; Goldberg, 1978, 2001; Peterson et al., 2002).

In Chapter 17, I discuss that the sequence of the base pairs of DNA determines the sequence of nucleotides in microRNA (miRNA), mRNA, tRNA, and rRNA, as well as the sequence of amino acids in proteins. Consequently, clever techniques have been devised to determine the sequence of bases in DNA (Gilbert, 1981; Sanger, 1981; see Chapter 21). Given the ease of obtaining sequences and the power of computers, the most prominent goal of informatically oriented theoretical biology has become to "compute the organism" (Brenner, 1999; Segal, 2001; see Chapter 21). Echoing the empty claim of LaPlace, Sidney Brenner declared that "if he had the complete sequence of DNA of an organism and a large enough computer, then he could compute the organism." Likewise, Walter Gilbert, a cofounder of the biotech firms Biogen and Myriad Genetics, claimed "that when we have the complete sequence of the human genome, we will know what it is to be human (see Lewontin, 2001).

A large proportion of biologists are now involved in sequencing the genes of model organisms, including *E. coli*, Saccharomyces *cerevisiae*, Caenorhabditis *elegans*, *Arabidopsis* (The Arabidopsis Genome Initiative, 2000), *Populus* (Tuskan et al., 2006), *Chlamydomonas* (Merchant et al., 2007), *Physcomitrella* (Rensing et al., 2008), *Zea* (Schnable et al., 2009), *Selaginella* (Banks et al., 2011), *Solanum* (The Tomato Genome Consortium, 2012; Bolger et al., 2014), *Amborella* (*Amborella* Genome Project, 2013), *Drosophila*, and humans. Many more plant genomes have been sequenced (Michael and Jackson, 2013). A quick look at the list of authors on the papers that present the genome sequences will show that there are more scientists involved in genome sequencing than were involved in discovering everything else I have presented in the preceding chapters. Indeed, biology is evolving from the era of craftsman science done in small laboratories to the era of big science performed in factories.

Most gene products are not studied directly, but their function is assigned by comparing the sequence of a gene of interest with the sequence of an *E. coli* or *S. cerevisiae* gene. The activity of sequencing and mapping genes has been given the intentionally nonacademic and magical-sounding name of *genomics* by Thomas H. Roderick (see McKusick and Ruddle, 1987; McKusick et al., 1993). It must be remembered that while some traits depend exclusively on the sequence of a single gene, the majority of traits depends on many factors, including the genetic background in which the gene is expressed (Grant, 1963; Dobzhansky et al., 1977) and the environment (Lewontin, 2001; see Wolpert and Richards, 1997). Thus, there is not a one-to-one correspondence between a "gene" and a trait, and when we speak of genes we must specify whether we are speaking about sequences of chemicals or of traits (Stadler, 1954). Manton (1945) suggested that it might be best to delete the word gene from the vocabulary.

In the so-called postgenomic period, people are beginning to realize once again that the genome, like the many others things that were considered to be the most fundamental aspect of life, is in itself not the fundamental unit of life (see Chapter 9). Sidney Brenner (2003) writes,

> *... I believe very strongly that the fundamental unit, the correct level of abstraction, is the cell and not the genome. In other words, I've been quoted as saying "forget the genome," you know we don't want to forget it, we'd like to thank all those people for their sterling work and give them all a gold watch and send them home, or better still send them back to the factory to sequence more genomes. But what we've got to do now is to get away from that and look at how we're going to give the true biological picture of it.*

16.5 MORPHOLOGY OF CHROMATIN

To evolve, the genome must be able to introduce variability into itself (de Vries, 1905, 1906, 1907). One source of this variability comes from single nucleotide changes in the DNA known as *point mutations*. Point mutations may lead to a protein with altered structures as evidenced by a change in the electrophoretic mobility of the protein (Pauling et al., 1949). The genome can also evolve through duplication of all the chromosomes (polyploidy). Having two sets of each gene, either from the same parent (autopolyploidy) or from divergent parents (allopolyploidy), makes it possible for one gene to accumulate mutations while the other one produces a functional protein. This "repeat and vary" theme can occur for individual genes and for entire genomes. Polyploidy is a common occurrence in plants—even so-called model plants (Blanc and Wolfe, 2004)—being the rule rather than the exception (Tate et al., 2005). This makes it difficult to do gene identification experiments on plants (see Chapter 21). Another source of variability comes from the rearrangement of chromosomes that takes place during the first prophase of meiosis. During this crossing-over process, variation can also result from the mixing of one gene with another. Such mixing may have given rise to the diversity in ion channels and motor proteins. The variation in chromosome structure correlated with the variation seen in the appearance of a species led to the conclusion that chromosomes provide a material basis for the mechanism of heredity and evolution (Morgan et al., 1915; Creighton and McClintock, 1931; Shine and Wrobel, 1976; Wayne, 2012b,c).

Flemming called the tangle of filamentous stands of chromatin visible in the light microscope the spireme (Reed, 1914). Electron micrographs of isolated chromatin show that chromatin looks like "beads on a string" (Fig. 16.8). The first electron micrographs that showed a regular repeating structure of chromatin were published by Olins and Olins (1974) and not by Christopher Woodcock (1973), due to the luck of the draw when it comes to reviewers. An anonymous reviewer (1973, quoted in van Holde, 1989) of a paper submitted by Woodcock wrote,

> *A eukaryotic chromosome made out of self-assembling 70 Å units, which could perhaps be made to crystallize, would necessitate rewriting our basic textbooks on cytology and genetics! I have never read such a naive paper purporting to be of such fundamental significance. Definitely it should not be published anywhere!*

The repeating units, which are 6–8 nm in diameter when viewed by negative staining and 10 nm in diameter when stained positively, are called *nucleosomes*. The nucleosome structure of chromatin had not been seen before 1973 because most protocols used for preparing samples for electron microscopy up until this time involve dehydrating the sample with ethanol. However, ethanol destroys the nucleosomes, and it was not until this step was removed or the chromatin was subsequently rehydrated after ethanol dehydration that nucleosomes could be readily observed (Gigot et al., 1976; Nicolaieff et al., 1976; Woodcock et al., 1976; Frado et al., 1977).

The nucleosomes are composed of some of the histones. The histones are relatively small proteins with a high proportion of positively charged amino acids such as lysine and arginine. The positive charges are responsible for the histones binding so tightly to the negatively charged phosphates of DNA. There are five different histones, named, unimaginatively, *H1*, *H2A*, *H2B*, *H3*, and *H4* (Fambrough and Bonner, 1966; Spiker, 1985; Luger et al., 1997; Gardiner et al., 2008). Nucleosomes are assembled

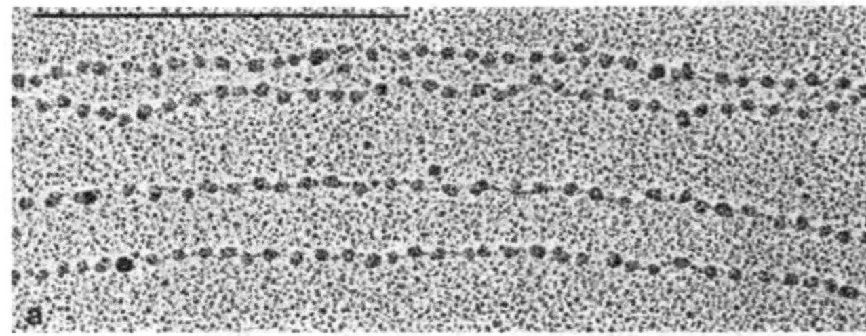

FIGURE 16.8 "Beads on a string" appearance of nucleosomes in *Zea mays* root cells. Bar, 500 nm. *From Greimers, R., Deltour, R., 1981. Organization of transcribed and nontranscribed chromatin in isolated nuclei of Zea mays root cells. Eur. J. Cell Biol. 23, 303–311.*

with the help of nucleoplasmin, a chaperonin (Laskey et al., 1978; Gatenby and Ellis, 1990).

The repeating structure of chromatin can be confirmed by treating plant or animal chromatin with micrococcal nuclease and running the digested chromatin through gels. The micrococcal nuclease digests the linker DNA, and consequently bands of integral lengths can be seen when the chromatin is run out on gels (Noll, 1974; Spiker et al., 1983; McGhee and Engel, 1975). An isolated nucleosome consists of a globular core of eight histone molecules (two H2A, two H2B, two H3, and two H4) that is surrounded by 146 base pairs of DNA wrapped in 1.75 turns around the nucleosome. Digestion of chromatin with DNAase I yields DNA fragments, indicating that the DNA is on the outside of the nucleosome. Histone H1 interacts with an additional 20–base pair linker that completes two turns of DNA around the histone core. The amount of DNA around a nucleosome varies. There are 160 base pairs in some fungi and 240 base pairs in some sea urchin sperm (Spiker, 1985). The amount of DNA within the linker varies from organism to organism, from cell type to cell type, and even within different regions of the nucleus. The degree of DNA bending depends on its position within the nucleosome (Hayes and Wolffe, 1992).

The nucleosomes with their associated DNA molecule compose the 10-nm fibers seen in nuclei in electron micrographs. The 10-nm fibers supercoil into a double helix or dinucleosomal ribbon to form the 30-nm fibers seen in interphase nuclei (Woodcock et al., 1984, 1991a,b; Hayes and Wolffe, 1992; Swedlow et al., 1993). The chromatin can be visualized as 10-nm fibers when the ionic strength is lower than about 60 mM and as 30-nm fibers when the ionic strength is greater than about 60 mM (Pederson et al., 1986). In general, chromatin will only form 30-nm fibers if histone H1 is present (Pederson et al., 1986; Yanagida, 1990). Acetylation of the histones causes the neutralization of the positive charge of lysines in the histones so that the positively charged histones do not bind as tightly to the negatively charged DNA. When the chromatin decondenses, it is more susceptible to degradation by DNase. The histones associated with transcribed genes are more acetylated than those associated with nontranscribed genes are. The acetylation of histone H4 is correlated with the formation of heterochromatin (Jasencakova et al., 2000). Measurements made with laser tweezers show that it takes approximately 15 pN to peel reversibly the DNA from the nucleosomes (Hayes and Hansen, 2002; Brower-Toland et al., 2002). Presumably, the DNA-processing enzymes in eukaryotic cells (e.g., helicase, DNA polymerase, and RNA polymerase) provide the motive force necessary to displace the DNA from the nucleosomes as they perform their better known functions (Bustamante et al., 2004).

Each DNA molecule contains between 50 and 250 million base pairs and is about 1.7–8.5 cm long. The entire human genome is about 2 m long. If these DNA molecules were only folded into 30-nm fibers, the chromosomes would be about 100 μm long, which is about 10 times longer than the nucleus! Scanning electron micrographs indicate that the 30-nm fibers are folded into looped domains. It appears that the loops contain about 20,000–100,000 base pairs. The long strands of chromatin are not randomly arranged in the nucleus. This was first indicated by the observations of Theodor Boveri (1888), who noticed that the individual chromosome positions in mitotic cells were correlated with the position of the same chromosomes during the preceding cell division. Using biotin-labeled probes, Lichter et al. (1988) showed that each chromosome was restricted to a specific portion of the interphase nucleus (Croft et al., 1999). Recent experiments using FISH and GISH are testing the hypothesis that the chromatin that makes up individual chromosomes exist in independent domains within the interphase nucleus (Matsunaga et al., 2013). Techniques are being developed that look at the localization of specific genes (Matzke et al., 2010; Jovtchev et al., 2011), and I look forward to work aimed at the localization of specific genes in cells in which they are expressed versus cells in which they are not.

The individual genes are not randomly arranged in the interphase nucleus but appear in reproducible positions from one cell division to another. This can be demonstrated using in situ hybridization, where a probe for a given gene hybridizes with the gene in the genome and thus marks the position of that gene (Hillier and Appels, 1989). In some higher eukaryotes, the regions of the chromosomes, called *centromeres* or *kinetochores*, which bind to the spindle fibers, are organized in regions near the poles (Hilliker and Appels, 1989). The nucleolar-organizing regions are located near the nuclear envelope (Hilliker and Appels, 1989; Hernandez-Verdun, 1991). The ends of the mitotic chromosomes, which are known as *telomeres*, can also be located predominantly at the nuclear periphery (Heslop-Harrison, 1991; Rawlins et al., 1991; Abranches et al., 1998; Cowan et al., 2001). It would be awesome to visualize the position of developmentally regulated or functionally associated genes in the nuclei of the cells in which they are expressed and those in which they are not. It would also be wonderful to localize developmentally regulated or functionally associated genes with respect to nuclear pore complexes in specific cells.

The morphology of the chromatin changes throughout the cell cycle (McLeish and Snoad, 1958). In the interphase nucleus, the chromatin is either highly condensed and it is called *heterochromatin* or it is much less condensed and it is called *euchromatin* (Figs. 16.9 and 16.10). Autoradiographic studies indicate that euchromatin is transcriptionally active, whereas heterochromatin is either less active or inactive (Lafontaine and Lord, 1974; Patient and Allan, 1989).

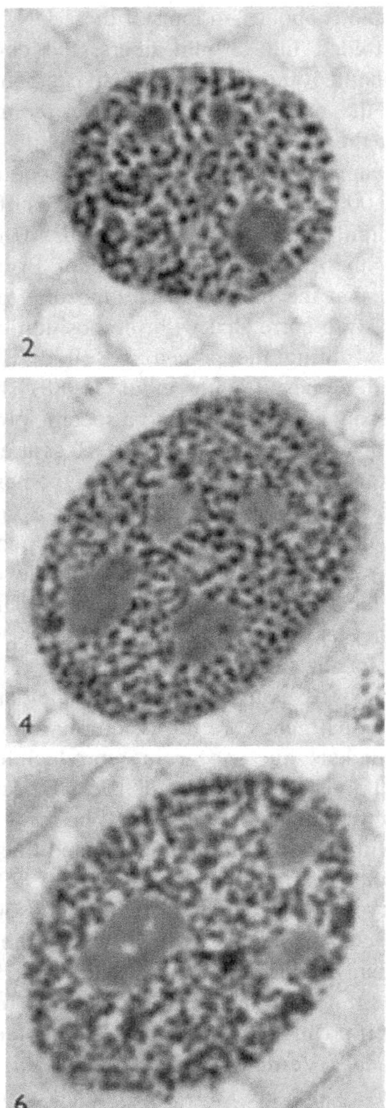

FIGURE 16.9 Phase-contrast micrograph of interphase nuclei of *Allium porrum* in late G1 (2), mid S-phase (4), and G2 (6). ×3800. *From Lafontaine, J.G., Lord, A., 1974. An ultrastructural and radioautographic study of the evolution of the interphase nucleus in plant cells (*Allium porrum*). J. Cell Sci. 14, 263–287.*

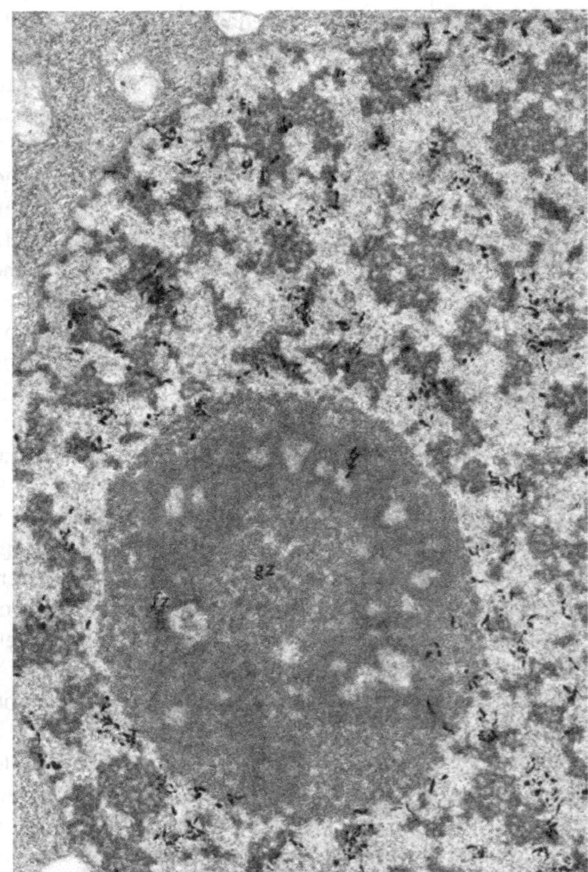

FIGURE 16.10 A radioautograph of the chromatin in a nucleus of *Allium porrum* in early S-phase. Notice that the ^{3}H-thymidine uptake during early S-phase occurs primarily in the lighter areas (euchromatin). ×31,000. *From Lafontaine, J.G., Lord, A., 1974. An ultrastructural and radioautographic study of the evolution of the interphase nucleus in plant cells (*Allium porrum*). J. Cell Sci. 14, 263–287.*

During prophase, the chromatin begins to condense. The condensation requires two ATP-hydrolyzing protein complexes known as *condensins*, which were discovered by characterizing the proteins necessary to reconstitute sperm chromatin into metaphase chromosomes in cell-free extracts of *Xenopus* eggs (Hirano and Mitchison, 1994; Hirano et al., 1997; Cobbe and Heck, 2000; Hirano, 2000, 2002, 2012; Jessberger, 2002; Hagstrom and Meyer, 2003). By metaphase, chromatin has condensed approximately 10,000 times; each DNA molecule condenses from about 5 cm long to about 5 μm long. The chromosomes not only shorten but also form helical structures with a given periodicity as they enter prophase of mitosis and meiosis. The distance between the gyres of the helical coils decreases as the chromosomes shorten in prometaphase and increase as the chromosomes uncoil in late anaphase (Baranetzky, 1880; Sharp, 1929; Sax and Sax, 1935; Darlington, 1940; Manton, 1943, 1945; Harman, 2004; Williams, 2016). The mitotic chromosomes have very characteristic forms that are invariant for each species (Baum and Appels, 1991). They also can be isolated in large quantities (Dolezel et al., 1992). The lining up and display of the mitotic chromosomes in a species is known as the *karyotype* of that species (Boyle, 1953; Haskell and Wills, 1968; LaCour and Darlington, 1975). The condensed chromosomes observed in mitotic cells represent chromatin that is in a package that is suitable for transport (see Chapter 19).

16.6 CELL CYCLE

The *cell cycle* is a term that describes the changes that occur through the life cycle of a cell, and, like most cellular processes, it is influenced by environmental conditions (Uchida and Furuya, 1997). The cell cycle is also influenced by circadian rhythms, which ensure that the ultraviolet-sensitive phases of the cell cycle occur at night (Nikaido and Johnson, 2000). The cell cycle is divided into four phases known as G1, S, G2, and M. The nucleus is typically in interphase and the chromatin is decondensed. Interphase is composed of the G1-, S-, and G2-phases. DNA synthesis takes place during the S-phase, and S stands for synthesis. S-phase is preceded by G1-phase and followed by G2-phase, where G stands for gap, because G1 and G2 were named at times when it appeared that nothing was happening during these phases. The G2-phase is followed by mitosis (or meiosis), which is known as M-phase. The newly formed nuclei soon return to the G1-phase. In some cells, notably the antipodal and synergid cells of the female gametophyte, the tapetal cells surrounding the male gametophyte, and the endosperm cells surrounding the developing embryo, multiple rounds of DNA synthesis occur without an intervening M-phase. Endoreduplication results in cells with large polytenic chromosomes with DNA contents around 24,576 times greater than the DNA content of normal haploid chromosomes (Nagl, 1978; Knowles et al., 1990; Polizzi et al., 1998; Sabelli and Larkins, 2007). The genes and proteins involved in regulating the phases and transitions of the cell cycle are particularly well known in yeast (Futcher, 1990), although new ones are being discovered in both yeast and other organisms.

The duration of each phase of the cell cycle can be measured by pulsing the cells with [^{3}H]-thymidine and preparing them for autoradiography (Swanson and Webster, 1985). The durations of the various phases of the cell cycle vary from species to species and from cell type to cell type (Webster, 1979). As an example, in *Vicia faba* meristematic cells, G1 takes about 2.5 h, S-phase takes about 6 h, G2 takes about 5 h, and M takes about 0.5 h. Most differentiated cells spend the majority of their life in G1 but are still capable of dividing as evidenced by the ability of plants and some animals to regenerate lost parts after wounding (Morgan, 1901; Loeb, 1924). The cell cycle can be best studied in cells, including tobacco BY-2 suspension cultured cells, in which the cell cycle can be readily synchronized (Joubès et al., 2004). While in other cell lines the mitotic index, which is the percentage of cells undergoing mitosis at the same time, is about 10%, in tobacco BY-2 cells, the mitotic index is greater than 70% (Nagata et al., 1992, 2006; Nagata and Kumagai, 1999; Nagata, 2004; Inzé, 2007).

What regulates the cell cycle? In 1970, it was found that a protein extract stimulated frog oocytes to divide when it was microinjected into the eggs (Dorée, 1990). The protein in the active extract was called *maturation-promoting factor*. Later it was found that a mutant of yeast, called *CDC* (for cell-division cycle), was unable to produce a protein necessary for cell division, and consequently, the mutant stops dead in its tracks during cell division. The CDC protein turned out to be the same as the protein that stimulates frog oocytes to divide. This protein is a protein kinase and occurs in all cell types (John et al., 1989, 1990, 1991; Feiler and Jacobs, 1990; Gorst et al., 1991; Ferreira et al., 1991). In plants, the level of the CDC protein kinase is enriched in meristems and is thus correlated with the ability of the cell to divide (John et al., 1990; Gorst et al., 1991; John and Wu, 1992). Intracellularly, the CDC protein kinase is enriched in the preprophase band (Mineyuki et al., 1991; Colasanti et al., 1993).

Surprisingly, the CDC protein kinase is always present in dividing cells (John and Wu, 1992), so how does it regulate the cell cycle? The CDC protein kinase is activated by another class of proteins called *cyclins*, and the concentration and type of cyclin change throughout the cell cycle. Each cyclin causes the CDC protein kinase to phosphorylate a different set of proteins necessary for a given phase. Interestingly, the injection of p34cdc2/cyclin B-like kinase into living cells of *Tradescantia* caused the preprophase band to disappear without affecting the interphase cortical, spindle, or phragmoplast microtubules. The p34cdc2/cyclin B-like kinase also accelerated chromatin condensation and nuclear envelope breakdown (Hush et al., 1996). Currently, more and more regulatory proteins are being discovered that are involved in regulating the cell cycle. The synthesis, availability, degradation, and phosphorylation states of these proteins are important in initiating each phase of the cell cycle (Nurse, 1990, 2001; John and Wu, 1992; Murray, 1992; Norbury and Nurse, 1992; Kirschner, 1992; Reeves, 1992; Hunt, 2001; Criqui and Genschik, 2002; Doerner, 2007; Shen, 2007). The genes involved in switching a plant cell over from a mitotic to a meiotic cell cycle, as well as the genes involved in all stages of meiosis including bivalent formation and recombination, are being discovered (Li et al., 2004; Pawlowski et al., 2004, 2007; Pawlowski and Cande, 2005; Ronceret et al., 2007; Bozza and Pawlowski, 2008).

Cells mature and may even senesce in time. Some of the developmental events that occur as cells age are regulated by extrinsic factors, whereas others are regulated by intrinsic factors. For example, cells may keep track of, in some way, the number of cell cycles and then initiate differentiation after a particular number of divisions. One way to register the number of cell divisions involves dilution. An initial cell may have a given amount of a particular molecule. Each time the cell divides, the compound will be diluted until there is a subthreshold level, and the cell can no longer divide. This compound or compounds, at least in part, are known to occur at the ends of chromosomes called *telomeres*. After each division, the telomere gets shorter and shorter. When it is no longer there, the cell stops dividing. By contrast, cancer cells produce a protein called *telomerase*. Telomerase is an enzyme that rebuilds the telomeres (Greider and Blackburn, 1985).

Theodor Boveri (1929) wrote that "the somatic cells cease to multiply. From an egoist being, the cell becomes an altruist, in the sense that it proceeds to new division only when the needs of the whole demand it." While no one knows how cells tell time, we know that some gene on the human chromosome #1 causes cells to lose their immortality. It is known that when human chromosome #1 is transferred into an immortal line of Syrian hamster cells, the cells acquire a limited life span (Suguwara et al., 1990; Pereira-Smith and Smith, 1988). On the other hand, oncogenes make products that cause the cell to be immortal (Groves et al., 1991). Likewise, *Agrobacterium* causes a cancer-like condition in plants (Smith and Townsend, 1907; Schilperoort, 1970).

The senescence of organisms may be a manifestation of the senescence of cells (McCormick and Campisi, 1991). This idea comes from studies in mammals, including humans, where the maximum number of divisions in culture cells is directly proportional to the life span of the species and inversely proportional to the age of the donor. Moreover, cells derived from donors with heritable premature aging syndromes usually senesce after fewer divisions than cells derived from people the same age who do not have the syndrome in question. The causes of programmed cell death, known as *apoptosis*, are being vigorously studied (Ellis et al., 1991; Groover et al., 1997).

16.7 CHROMOSOMAL REPLICATION

Jacques Loeb (1909; Brachet, 1975, 1989; Deichmann, 2009) suggested that before a cell divided, there was a total de novo synthesis of nucleic acids in the nucleus from phosphate, sugars, purines, and pyrimidines. William Astbury (1941) thought of the chromosome as an organism whose reproduction required the duplication of DNA and protein. Indeed, chromosomal replication includes the replication of the double-stranded DNA molecule that resides in the chromosome and the assembly of a new set of chromosomal proteins (Bloch and Godman, 1955; Stillman, 1989; Virshup, 1990; Hagstrom and Meyer, 2003; Groth et al., 2007). DNA replication depends on the ability of bases to undergo complementary base pairing in a process known as *DNA templating*. DNA synthesis is "semiconservative," and each strand acts as a template for a new strand (Watson and Crick, 1953c; Meselson and Stahl, 1958; Delbrück and Stent, 1957; Taylor et al., 1957; Sueoka, 1960). During DNA templating, adenine pairs with thymine and cytosine pairs with guanine. For this process to occur, the two strands of the DNA helix must be separated. This separation or melting, which is enhanced at higher temperatures, was once thought to be a spontaneous process in the cell. However, the separation process is catalyzed by mechanochemical enzymes (Kornberg, 2000). The DNA helix is unwound by an enzyme called *DNA helicase*. DNA helicase unwinds the DNA helix at the expense of ATP hydrolysis, so DNA helicase is a mechanochemical enzyme (Tuteja, 2003; Bustamante et al., 2004).

Once the DNA helicase unwinds the double helix, other proteins known as *helix-destabilizing proteins* or *single-strand DNA-binding proteins* bind to the single-stranded DNA and stabilize the opened strands without covering up the bases. This allows time for complementary base pairing. Once the DNA helix is unwound, the replication fork is created. The new DNA is synthesized semiconservatively in the tines of the fork. Once base pairing is accomplished to form a new $5'-3'$ strand on the template $3'-5'$ strand, an enzyme known as *DNA polymerase* connects the internal phosphate of the nucleotide at the $5'$ end of the nucleotide to the OH group of the $3'$ end of the last nucleotide in the polymer at the expense of the hydrolysis of the pyrophosphate from the newly added nucleotide (Kornberg, 1960, 1961, 1989a,b). The pyrophosphate made in dividing cells can be used by the vacuolar H^+-translocating pyrophosphatase. Moreover, it is possible that this pyrophosphatase can drive the biosynthetic reactions involved in DNA synthesis by using up the end product. Eukaryotes contain DNA polymerases in the nucleus, the mitochondria, and the plastids (Bollum and Potter, 1957; Kalf and Ch'ih, 1968; Hübscher et al., 2002; Moriyama and Sato, 2014; see Chapters 13 and 14).

It is important to know that the DNA polymerase, which was first isolated by Arthur Kornberg (1957, 1959, 1960) using a functional cell-free in vitro assay that measured the ability of the isolated enzyme to incorporate radioactive deoxyribonucleotides into acid-precipitable material, turned out not to be the main DNA polymerase involved in the semiconservative replication of DNA. This was a surprising revelation because Kornberg (1989a) had believed "that a biochemist devoted to enzymes could if persistent, reconstitute any metabolic event as well as the cell does it. In fact, better!" The surprising revelation came from experiments done by Paula De Lucia and John Cairns (1969; Wolpert and Richards, 1997), who isolated a temperature-sensitive mutant of *E. coli* that was defective in DNA polymerase activity yet capable of multiplying. The mutant was named *polA1* after Paula, and, relative to the wild type, it had an increased sensitivity to UV light (Gross and Gross, 1969). Other DNA polymerases were then isolated from *E. coli* by Arthur Kornberg's son Tom (Kornberg and Gefter, 1970, 1971), and the third type isolated, which is known as Pol III, is the polymerase involved in DNA replication (Friedberg, 2006). The original DNA polymerase isolated by Arthur Kornberg is now known as Pol I, and it is involved in DNA repair rather than replication.

While reconstitution studies using purified components are the gold standard to demonstrate the action and regulation of enzymatic mechanisms to a biochemist, extrapolation of the biochemical mechanisms to the living cell involves a jump that is based on interpretation. The geneticist can help with this jump by performing genetic studies using conditional mutants that demonstrate the function of an enzyme in vivo (see Chapter 21).

DNA polymerase only works in the 5′–3′ direction at a rate of about 50 nucleotides per second. So, how does the newly formed DNA on the other strand, which would appear to need to be synthesized in the 3′–5′ direction, get polymerized? The DNA in the "lagging strand" is synthesized in the 5′–3′ direction in a complicated manner that involves RNA to some extent. After DNA templating, DNA polymerase moves from the 5′ to the 3′ end, creating replication intermediates that are about 100–200 nucleotides long in eukaryotic cells. These fragments are called *Okazaki fragments* and are joined together by an enzyme known as *DNA ligase* to form a continuous chain of DNA.

Interestingly, autogradiographic studies show that, even though replication results in identical DNA molecules in each chromatid, during mitotic division (see Chapter 19), chromatids containing DNA strands of identical age segregate as a single unit in mammalian (Lark et al., 1966), plant (Lark, 1967), and fungal cells (Rosenberger and Kessel, 1968). This indicates that the template strands of DNA in all the chromatids are somehow connected with each other. In fact, in filamentous fungal cells, where it is possible to follow the movement and positioning of individual nuclei, Rosenberger and Kessel (1968) have shown that the chromatids that contain the template DNA are preferably positioned at the tip of the hypha.

Replication begins at specific sites known as the *replication origins* (Cairns, 1963a,b, 1966, 1980; Van't Hof, 1975, 1985, 1988; Van't Hof and Bjeeknes, 1977, 1981; Van't Hof et al., 1978, 1987a,b; Van't Hof and Lamm, 1992; Haaf, 1996; Bryant et al., 2001; Quélo and Verbelen, 2004). Replication forks originate at replication origins (Virshup, 1990). Replication origins, which are about 300 nucleotides long, form a structure known as the *replication bubble*. Many copies of an initiator protein bind to the replication origin to form a complex that then binds the DNA helicase and positions it correctly. Subsequently, the replication enzymes, including DNA polymerase, form complexes known as *primosomes*, and eventually two complete primosomes move in opposite directions along the two forks that are formed in one replication origin. The nuclear matrix may provide the necessary framework for replication to take place in eukaryotic cells. In fact, it is possible that the replication sites are fixed on the nuclear matrix and the unreplicated DNA is spooled through these sites (Pardee, 1989).

The existence of multiple replication origins and their bidirectional movement can be visualized by pulsing cells with [^{3}H]-thymidine, extracting the DNA in a relatively unfragmented state, and processing it for radioautography. Radioactive portions of DNA alternate with nonradioactive portions, indicating multiple sites of origin. Longer pulses result in longer-labeled DNA strands, indicating that the replicated segments of DNA are gradually being extended. Differential time pulses or pulses with various specific activities indicate that the DNA replicates in both directions, and eventually the whole strand of chromatin appears labeled (Swanson and Webster, 1985). Genetic screens in *Drosophila* and yeast aimed at discovering the glue that held together sister chromatids revealed mutants in which the sister chromatids separated precociously before anaphase. The genes responsible for the mutant phenotype coded for an ATP-hydrolyzing, mechanoenzymatic protein complex known as *cohesion*, which holds the two semiconservatively replicated strands of chromatin together from S-phase to anaphase (Uhlmann and Nasmyth, 1998; Cobbe and Heck, 2000; Hirano, 2000, 2002; Jessberger, 2002; Hagstrom and Meyer, 2003; Peters et al., 2008; Yuan et al., 2011; Bolaños-Villegas et al., 2017).

How does DNA replication take place in the presence of nucleosomes? Structural changes in chromatin may take place at the sites that are undergoing replication (Nagl et al., 1983; Nagl, 1982, 1985). There is a transient disruption of the nucleosomes located ahead of the replication fork and a transfer of the parental histones from these nucleosomes to the newly formed DNA strands. The nucleosomes on the nascent strand are then completed by the addition of newly synthesized histones (Groth et al., 2007). While the two DNA strands must inherit the nucleosomes in an organized manner to pass on the posttranslational modifications of the histones that define inherited epigenetic states, it is still unknown how the two strands inherit the nucleosomes (Pardee, 1989).

If all the replication origins were activated simultaneously, it would take on the order of 1 h to complete DNA replication. However, S-phase in Chinese hamster cells takes 6–8 h. This is because the 20–80 replication origins in a chromosome are not activated simultaneously. There seems to be a definite hierarchy in which different regions of the genome are replicated (Stubblefield, 1975). This was determined by pulsing synchronized cells with a thymidine analogue at different times during S-phase and then observing its distribution. The chromatin that corresponds to the G-C–rich bands of mitotic chromosomes replicates during the first half of S-phase, whereas the chromatin that corresponds to the A-T–rich bands in mitotic chromosomes replicates during the last half of S-phase. The heterochromatin that resides near the centromere replicates very late during S-phase (Lima de Faria and Jaworska, 1968).

Using DNA probes for specific genes it is possible to determine the time of replication of any gene. In these experiments, nonsynchronous cells are pulsed with a thymidine analogue, and the cells are separated by their size, which corresponds to their age. The younger cell's DNA would have been labeled early in S-phase, and the older cell's DNA would have been labeled late in S-phase. The thymidine analogue containing DNA is then separated from the rest of the DNA. Brown et al. (1987) showed that the housekeeping genes that are active in all cells are replicated very early in S-phase, whereas genes that are only active in a few cell types are replicated early in the cells in which they are expressed and later in other cell types.

This seems to be a very reasonable method of replicating DNA. First, the vital DNA is replicated, then the DNA that will be expressed in the cell in its specialized state is replicated, and finally the DNA that will not ever be expressed in that cell is replicated. So, why is S-phase so complicated? Perhaps the orderly turning on and off of replication units influence chromatin structure. The various resulting chromatin structures may then determine, in part, which genes are expressed in various cell types.

Leroy Hood (2008) developed an automated cell-free DNA synthesizer (see Chapter 21).

16.8 TRANSCRIPTION

Some cells isolated from plants are capable of giving rise to complete plants, and thus maintain complete copies of the genome. These cells are called *totipotent* (Steward et al., 1958; Thomas and Davey, 1975). Subsequently, a nucleus isolated from an intestinal cell of a frog and implanted into an enucleated egg cell of a frog was shown to be capable of directing the complete development of a frog (Gurdon[1], 1962, 2012; Gurdon and Uehlinger, 1966; Davidson, 1968). Today, clones of dogs and cats as well as sheep and cows are manufactured by injecting a totipotent somatic nucleus into an enucleated cell. There have also been claims that human beings have also been cloned in the same manner. I assume that all cells are totipotent, although for technical reasons this has not always been realized. If all cells are totipotent and contain a full complement of genes, then cell differentiation must result from the differential expression of those genes.

The chromatin of actively transcribing DNA is different from that of quiescent DNA, as evidenced from experiments in which the DNA in vertebrate cells was treated with DNAse I. In each cell, about 10% of the genome is digested, and the pattern of digestion varies in different cell types. The DNA that is degraded in each cell type corresponds to the genes transcribed in that cell type. Thus, transcribed genes are in a conformation susceptible to nuclease digestion (Vega-Palas and Ferl, 1995; Li et al., 1998). Long stretches of plant and animal DNA lack nucleosomes, and these regions contain nuclease-hypersensitive sites (Elgin, 1990; Grunstein, 1990). These stretches may represent locations where a sequence-specific protein displaced a nucleosome (Murray and Kennard, 1984; Spiker et al., 1983), or perhaps the nucleosomes may not form in these regions as a result of the DNA sequence itself (Elgin, 1990). Sawyer et al. (1987) found a greater DNase sensitivity of the legumin gene in actively transcribing cotyledons compared with leaves where the gene is not expressed. The structure of the chromatin in a cell type expressing a given gene differs from the chromatin structure of the gene in nonexpressing cell types (Vega-Palas and Ferl, 1995; Li et al., 1998a,b).

The type of attachment of the nucleosome to the DNA can be modified as a result of the chemical modification of the histones. The histones can be chemically modified in a number of ways, including methylation, acetylation, and phosphorylation (Hayes and Wolffe, 1992; Hansen and Ausio, 1992; Morse, 1992; Prymakowska-Bosak et al., 1999), and these posttranslational modifications may influence the ability of the transcribing RNA polymerase to access the associated DNA template (Clark and Felsenfeld, 1992), perhaps by influencing the ability of a group of nuclear proteins known as the *nonhistone proteins, acidic proteins, high-mobility group proteins, or transcription factors* (Liu et al., 1999; Chen et al., 2002) to bind to the DNA and initiate transcription. The addition of nucleosomes has been shown to inhibit transcription in vitro, and the inhibition is reversed when transcription factors are added (Jackson, 1991).

The physical position of a gene may also affect its expression. While the experiments performed by Gregor Mendel (1865; Meissner, 2012) indicated that genes assorted independently, further work on mice by Darbishire (1904) and sweet peas by Bateson et al. (1906) indicated that not all genes assorted independently, that is, some were linked. Genetic studies on *Drosophila* performed by members of Thomas Morgan's laboratory indicated that some genes were more closely linked to each other than to other genes. Morgan postulated that genes were linearly arranged on the chromosome, and the degree of genetic recombination between chromosomes is a reflection of the distance between the genes (Morgan, 1911, 1924; Morgan

1. John Gurdon (https://www.youtube.com/watch?v=c3RuJMTOnE4&feature=youtu.be&t=343) received the following report from his teacher Mr Gaddum when he was 15 years old and in the bottom of his class in biology at Eton College: "he will not listen, but will insist on doing his work in his own way. I believe he has ideas about becoming a scientist; on his present showing this is quite ridiculous. If he can't learn simple Biological facts he would have no chance of doing the work of a Specialist, and it would be sheer waste of time, both on his part, and of those who have to teach him." Gurdon has also pioneered the use of Xenopus oocytes for expressing microinjected mRNAs (see Gurdon, J.B., Lane, C.D., Woodland, H.R., Marbaix, G. 1971. Use of frog Eggs and oocytes for the study of messenger RNA and its translation in living cells. Nature 233, 177–182).

et al., 1915; Castle, 1919). Harriet Creighton and Barbara McClintock working with maize, and Curt Stern working with *Drosophila*, provided cytological evidence that showed that during meiosis, chromosomes do exchange material and this crossing over is the cytological basis of genetic recombination (McClintock, 1930; Creighton and McClintock, 1931; Stern, 1931; Kass, 2003).

Alfred Sturtevant (1913, 1951) performed genetic crosses between different mutants of *Drosophila* to construct maps of the chromosomes that showed the position of the known genes and soon he began to find that gene expression depended on the position of the gene (Sturtevant, 1925; Painter, 1939). Carl Swanson (1957) wrote in *Cytology and Cytogenetics*,

> *Early concepts of the gene, especially those relating to its particulate nature, were derived from crossing over data and from radiation studies. But a greater appreciation of the gene as a functional unit of inheritance and of the chromosome as an organized structure, has come from an analysis of position effects. Discovery of this phenomenon by Sturtevant [1925] has been followed by numerous analyses in Drosophila [Green and Green, 1949; Lewis, 1950, 1951, 1952, 1955; Green, 1954, 1955a,b], Oenothera [Catchside, 1947], and maize [McClintock, 1951, 1953], and it is becoming increasingly evident that the genetically determined phenotype is not only dependent upon the gene itself but upon the nature of the chromatin adjacent to it. The gene, therefore, even if it is of a particulate nature, is not an isolated unit operating simply in conjuction with other genes; it is actively influenced by them.*

The fact that gene expression depends on the position of the gene may in part relate to the physical arrangement of chromatin throughout the nucleus (Nagl, 1985; Bloom and Green, 1992; Reyes et al., 2002), and this may depend in part on the attachment of chromatin to the nuclear matrix (Bode et al., 1995). Genomes have more DNA than is necessary to code the mRNAs they produce. The exons make up a small percentage of the genome. We are just beginning to find that the large amount of non—protein-coding sequences plays a regulatory role in coding small regulatory RNAs. Perhaps, it also plays a role in the three-dimensional structure of the informational macromolecules. The fact that an enhancer or silencer sequences thousands of base pairs away from a gene can regulate the expression of gene that supports the idea that the physical conformation of chromatin is important for the regulation of gene expression (Tariq et al., 2002). Some regions of the genome are gene rich, whereas others are gene poor. We do not know if the gene density of a region has any meaning in the life of the organism. Moreover, some transformed plants express a given gene, whereas others do not, suggesting that the position of the inserted gene may be as important as its sequence in determining a phenotype.

The three-dimensional structure of chromatin may be important if stresses can cause the activation of a gene. When RNA polymerase opens a DNA helix that is anchored at both ends, tension will develop and the DNA molecule will tend to form supercoiled loops to relieve the tension. A moving RNA polymerase will tend to create positive superhelical tension in front of it and negative superhelical tension behind it. The positive tension may facilitate the opening of nucleosomes that may be required for transcription. If this occurs, the position of a gene will be important for its expression. We must begin to appreciate DNA as a three-dimensional structure instead of a linear sequence, just as we have already begun to appreciate the three-dimensional structure of small organic molecules and organelles (Wollaston, 1808; LeBel, 1874; Van't Hoff, 1874, 1967).

William Astbury (1941) asked, "*What then is the gene? Is it a structure that is handed on? Or is it a set of physico-chemical conditions?*" Richard Goldschmidt (1951) realized that if we define a gene as the hereditary material that determines a character of the organism, it is impossible to look at a gene in isolation of where it is in a chromosome. It is possible that every gene affects every other gene, and DNA sequences may not only code for proteins and regulatory RNAs but may also influence the physical properties of the chromatin that allow certain genes to be either expressed or silenced. Concerned with the relationship of genes to the whole organism, Barbara McClintock wondered, "Should one gene mutate by what mechanisms are adjustments made in every part of the complex whole?" (see Fedoroff, 1992).

Indeed this idea of looking at a gene in context goes back to Thomas Hunt Morgan (1909) who wrote the following admonition, "In the modern interpretation of Mendelism, facts are being transformed into factors at a rapid rate. If one factor will not explain the facts, then two are invoked; if two prove insufficient, three will sometimes work out. The superior jugglery sometimes necessary to account for the result, may blind us, if taken too naïvely, to the common-place that the results are often so excellently 'explained' because the explanation was invented to explain them. We work backwards from the facts to the factors, and then, presto! explain the facts by the very factors that we invented to account for them. I am not unappreciative of the distinct advantages that this method has in handling the facts. I realize how valuable it has been to us to be able to marshal our results under a few simple assumptions, yet I cannot but fear that we are rapidly developing a sort of Mendelian ritual by which to explain the extraordinary facts of alternative inheritance. So long as we do not lose sight of the purely arbitrary and formal nature of our formulae, little harm will be done; and it is only fair to state that those who are doing the actual work of progress along Mendelian lines are aware of the hypothetical nature of the factor-assumption. But those who know the results at second hand and hear the explanations given, almost invariably in terms of factors, are likely to exaggerate the importance of the interpretations and to minimize the importance of facts."

In any case, the expression of a gene that codes for a protein requires the synthesis of RNA through the action of RNA polymerase (Weiss and Gladstone, 1959). Eukaryotes have three different RNA polymerases called *RNA polymerase I, II,* and *III*. They each synthesize different kinds of RNAs. Polymerase I synthesizes the large ribosomal RNAs, polymerase II synthesizes mRNAs and miRNAs, and polymerase III synthesizes tRNAs and the 5S ribosomal RNA. Unlike bacterial RNA polymerase, eukaryotic RNA polymerases do not bind directly to DNA but must first bind to a transcription factor that recognizes a given promoter sequence (Pabo and Sauer, 1992). Each of the RNA polymerases recognizes a different set of transcription factors. The bacterial RNA polymerase is a mechanochemical enzyme that is capable of exerting a force of about 14 pN (Yin et al., 1995), almost 10 times greater than the force exerted by the cytosolic motor proteins. Thus, if the eukaryotic polymerases are similar, we can assume that the nucleus is extremely viscous and the physical stresses that may be important for gene expression exist in the nucleus.

Promoters are sequences of DNA, upstream from the transcribed region of the gene, which regulate the binding of RNA polymerase necessary for transcription. Some promoters cause the frequent initiation of transcription, whereas others are more pokey. As long as the position of the gene allows its transcription, the efficiency of transcription depends on the sequence of the promoter. RNA synthesis involves the opening of the two DNA strands to form an open complex. The RNA polymerase works by extending the newly formed polymer in the 5′−3′ direction. The RNA polymerase continues to add the ribonucleotides at a rate of about 30 nucleotides per second until it reaches a termination or stop signal (Cook, 1990). Once it reaches the stop signal, the newly synthesized RNA and the DNA separate from the RNA polymerase. While genes are always transcribed by RNA polymerase from the 5′ end to the 3′ end, transcription can occur from either strand and adjacent genes can be transcribed in opposite directions. It seems that the orientation of genes in the genome is random, and we do not know if the orientation of a gene has meaning in the life of the organism. In addition, genes can have alternative promoters that give rise to different isoforms of a protein, which in many cases give rise to membrane-associated and soluble forms of a protein (Ayoubi and van de Ven, 2017)

Transcription can be visualized in detergent-treated nuclei with the electron microscope. Among the normal nucleosome configuration of the chromatin, other globular particles are observed. These particles are RNA polymerase complexes that are attached to a trailing RNA molecule. Usually, RNA polymerase particles are seen as single units, indicating that most genes are transcribed infrequently into RNA so that one polymerase finishes before another one begins. However, sometimes many RNA polymerase particles can be seen attached to a gene, indicating that it is transcribed at a high frequency.

The majority of the RNA transcripts synthesized by RNA polymerase II are called *heterogenous nuclear RNA* (hnRNA; Birk, 2016). The 5′ end of an hnRNA transcript is immediately capped by the addition of a methylated guanine. Capping occurs after approximately 30 nucleotides have been polymerized. The methylguanosine cap may facilitate RNA export from the nucleus (Hamm and Mattaj, 1990; Hamm et al., 1990). The 3′ end of the hnRNAs is characterized by a poly-A tail. There are two sequences that signal the cleavage of the 3′ end. One sequence, AAUAAA, is on the RNA and signals that the RNA should be cleaved at a site 10−30 nucleotides downstream from this sequence, and there is another sequence that is downstream of the cleavage site. The poly-A tail, which includes 100−200 adenylic acid residues, is added by a separate poly-A polymerase to the 3′ end after cleavage. The poly-A tail regulates translation of the RNA transcript (Jackson and Standart, 1990). The newly made hnRNA associates with proteins to form spherical particles about 20 nm in diameter. These heterogenous nuclear ribonucleoprotein particles can be isolated and treated with ribonucleases at a concentration that just breaks the linker RNA.

The processing of hnRNA into mRNA requires the removal of long sequences, known as *introns*, from the middle of the RNA molecule (Gilbert, 1978). The introns can be visualized in the electron microscope using the R-loop technique. With this technique, mRNA is isolated and purified from a cell. This RNA is then mixed with a cloned double-stranded DNA from a genomic library. The mRNA displaces one of the strands of DNA where their sequences match and thus forms a DNA−RNA duplex. In the regions of the DNA where no matches with the mRNA occur, loops of double-stranded DNA form. The loops represent introns and the bound regions represent exons, sequences present in both the DNA and mRNA.

Processing involves the cutting out of introns and the annealing of exons in a process known as *RNA splicing*. Splicing occurs in spherical particles approximately 40−60 nm in diameter known as *spliceosomes* (Reed et al., 1988). A spliceosome contains several small nuclear ribonucleic acids (snRNAs), which have sequences complementary to those near the splice site, as well as a host of snRNPs, which facilitate the removal of introns. In some cases, the splicing may be accomplished by the nucleophilic attack by the 2′OH group of an adenosine in the intron on the 5′-phosphate of an exon. This step breaks the RNA molecule. The 3′OH produced by the first step then performs a nucleophilic attack on the 3′ phosphate of the adjacent exon. This step results in the release of the intron and the ligation of the two exons. The functions of many of the spliceosome proteins and snRNAs have been tested by reconstituting spliceosomes in vitro (Cech, 1986a,b; Gall, 1991; Guthrie, 1996). Splicing, known as trans-splicing, can even occur between two separate RNA molecules. Trans-splicing occurs in 10%–15% of the mRNAs of *C. elegans* and further supports the contention that a gene is not necessarily a single linear sequence of DNA.

Contrary to the central dogma, which states that information flow moves linearly from DNA to RNA to protein, RNA can also influence the expression of DNA. The realization that RNA can influence the expression of DNA came from deeply analyzing the unexpected results that plant biologists were obtaining when they transformed plants with gene sequences with apparently known functions (Dougherty and Parks, 1995). While overexpression of sense mRNA was supposed to increase the expression of a particular gene product and the expression of antisense RNA was supposed to decrease its expression, in many cases, both strategies seemed to downregulate the expression of the gene. This was particularly evident in the attempts to enhance the colors of flowers that resulted in colorless flowers (see Chapter 7; Napoli et al., 1990; van der Krol et al., 1990). de Carvalho et al. (1992) suggested that the suppression of gene expression caused by overexpression of sense RNA may occur at the post-transcriptional level.

The fact that plants transformed with sense RNA for a viral protein became resistant to the virus suggested that the plant cell could recognize too much RNA, incapacitate the viral RNA, and prevent it from replicating in the cell (Lindbo et al., 1993). But how?

Plant cells have a mechanism to eliminate RNA viruses using an RNA-dependent RNA polymerase (Schiebel et al., 1993a,b). This polymerase randomly copies RNA, making 10–75 nucleotide copies. These small RNAs bind to other RNAs in the cell and target their demise (Dougherty and Parks, 1995; Baulcombe, 2008).

The fact that miRNAs have the ability to regulate development and physiology naturally was discovered in *C. elegans* (Lee et al., 1993; Reinhart et al., 2000). MicroRNA is formed when RNA is transcribed in the nucleus by polymerase II from regions of DNA that do not code for proteins—the so-called junk DNA. These RNAs are then processed by a ribonuclease III–like nuclease to form miRNAs, which are approximately 18–25 nucleotides long (Kurihara and Watanabe, 2004). In the cytosol, these miRNAs pair with complementary regions of mRNA to inhibit their translation or target their degradation by an enzymatic process that utilizes enzymes known as *dicer* and *slicer* (Fire, 2006; Mello, 2006). Natural miRNAs that are capable of regulating development were discovered in plants by noticing that mutants that show an overproliferation of meristems, loss of polarity, or conversion of determinant to indeterminate structures had reduced levels of mRNA (Carrington and Ambros, 2003; Bonnet et al., 2004; Yoo et al., 2004; Sunkar and Zhu, 2004). MicroRNAs are also important in the regulation of nutrient uptake (Grennan, 2008).

Using a technique known as *posttranscriptional gene silencing, quelling*, or *RNA interference* (RNAi), double-stranded miRNAs can be used as a tool to test the function of a given gene product (Sørensen et al., 2014). This is done by treating a cell with a double-stranded miRNA with a given sequence so that it will silence or interfere with any gene that contains the complementary sequence to either strand. In this way, artificial miRNAs have been very successful in testing the function of a given gene product in a given response (Iseki et al., 2002). RNA interference techniques are a good way to test the importance of a particular class of gene products that are encoded by a family of genes, and high-throughput methods are under way to test the function of many classes of genes in many responses (Waterhouse and Helliwell, 2002).

The fate of the primary RNA transcripts can be followed by treating the cells with a pulse of tritiated uridine and following its fate. While the primary RNA transcripts are about 6000 nucleotides long, the mature mRNA molecules are about 1500 nucleotides long. The posttranscriptional processing of the hnRNA takes about 30 min. After 30 min, the transcripts begin to leave the nucleus, although only about 5% of these newly synthesized mRNA ever leaves the nucleus and the rest are degraded. Let us not always assume that natural selection acts to make every process in the cell efficient. Natural selection, like collective bargaining, must depend on compromise.

The mature mRNA is recognized by the nuclear pore complex and transported to the cytoplasm. The spliceosome proteins are stripped off in the nucleus and are never found in abundance in the cytoplasm. The spliceosome may contain special targeting sequences that directly target it to the nuclear pore. The mature mRNA is transferred from the spliceosome to the nuclear pore from which it is transported out of the nucleus to the cytoplasm where it is translated (Goldfarb and Michaud, 1991). The mRNA may move along the cytoskeleton to specific locations in the cytoplasm with the aid of cellular motors (Sundell and Singer, 1991).

Differentiated cells contain only a portion of the proteins the genome is capable of producing. Therefore, gene expression is selective. In a given cell, only a portion of the DNA is transcribed into functional mRNA (transcriptional control). Even if the gene is transcribed, only a small proportion transcribed into RNAs survives the RNA-processing steps (RNA-processing control). Perhaps only some of these mRNA are transported out of the nucleus (RNA-transport control), and once these mRNA get to the cytoplasm, they must be translated. This step too is regulated (translational control). Even if the protein is synthesized, it may need to be covalently modified or activated by ions (posttranslational control) to act, and if it is an enzyme, its substrates must be available. Thus, many steps are involved in regulating the expression of a gene and its product.

Selectivity requires the differential presence of only a few regulatory proteins (Singh, 1998). If one initial cell gives rise to two cells that differ in only one gene regulatory protein, and these cells divide and give rise to daughter cells, which differ in another protein, then the four cells will each have a different complement of proteins. According to this scheme, 10 different regulatory proteins can determine the differentiation of more than 1000 cell types, including the parental cells. Of course, as long as cells have the appropriate receptors (see Chapter 12), variations in extrinsic factors, including light, hormones, oxygen, gravity, and tissue stress, can result in differential gene expression.

By using techniques developed to introduce transgenes into plants stably or transiently (Wu, 1998; Liu and Vidali, 2011; Sørensen et al., 2014), we can take advantage of the genetic machinery in the cell to learn a lot about the cell itself, the proteins expressed in the cell, and the intracellular environment. Plants can be genetically engineered to express a particular *transgene* in a particular cell at a particular time. This is accomplished by fusing the gene of interest to a particular *promoter* that is known to be associated with genes that are expressed in a given cell at a given time in response to a given stimulus. The necessity of a particular cell type in the plant for a particular response can be tested by making a chimeric transgene with the DNA sequence that codes for diphtheria toxin fused to the desired promoter. The expression of such a DNA sequence in a given cell results in the ablation of the cell (Mariani et al., 1990, 1992; Thorsness et al., 1991, 1993; Guerineau et al., 2003) and a loss of its function. On the other hand, fusing a unique fluorescent protein to promoters unique to each cell type gives cell-specific markers (Ckurshumova et al., 2011).

The DNA sequence that encodes a particular protein can be fused to the DNA sequence that codes for a fluorescent protein such as green fluorescent protein (GFP) so that the expressed chimeric protein can be localized in living cells in a given organelle in both time and space and its movement, if any, can be followed (Chalfie et al., 1994; Hu and Cheng, 1995; Sheen et al., 1995; Chiu et al., 1996; Köhler et al., 1997; Leffel et al., 1997; Marc et al., 1998; Kleiner et al., 1999; Nebenführ et al., 1999; Collings et al., 2000; Cheung et al., 2002; Ritzenthaler et al., 2002; Pay et al., 2002; Sheahan et al., 2004; Bosch et al., 2005; Dettmer et al., 2006; Dixit et al., 2006; Berg and Beachy, 2008; Certal et al., 2008; Chalfie, 2008; Millwood et al., 2008; Thompson and Wolniak, 2008; McKenna et al., 2009; Mohanty et al., 2009; Vidali et al., 2009b; Rasmussen et al., 2011b; Furt et al., 2012; Ganguly and Cho, 2012; Viotti et al., 2013; Nebenführ, 2014; Sørensen et al., 2014; Zhang et al., 2016; Zhou et al., 2016; Krüger and Schumacher, 2017; Montes-Rodriguez and Kost, 2017; Weller et al., 2017). Any number of transgenes expressing fluorescent proteins that emit different colors can be introduced into an organism by performing genetic crosses (Méchali and Lutzmann, 2008; Newman and Zhang, 2008; Sakaue-Sawano et al., 2008; Ckurshumova et al., 2011; Furt et al., 2013). For example, two plants can be transformed with a different transgenes and crossed to produce plants that express both transgenes. In this way, two or more proteins can be localized in the cell at the same time, including cellulose synthase and tubulin (Paredez et al., 2006), actin and tubulin (Sampathkumar et al., 2011).

A genetically encoded biosensor can be constructed from a chimeric protein that unites the DNA sequences from two fluorescent proteins (e.g., blue fluorescent protein, cyan fluorescent protein, green fluorescent protein, yellow fluorescent protein, cherry, etc.) separated by the DNA sequence from another protein whose conformation changes in response to some factor (Tsien, 2008). The fusion protein is then illuminated with a wavelength that is absorbed by one of the fluorescent proteins, and the energy is transferred to the other fluorescent protein, which emits the energy as fluorescence. This is known as Fluorescence (or Förster) Resonance Energy Transfer or FRET. The intensity of the emitted light depends on the distance between the two fluorescent proteins, and the distance between the two fluorescent proteins depends on the conformation of the intervening protein. By including a DNA sequence that encodes for a Ca^{2+}-binding protein between the two fluorescent proteins, the fusion protein can be used to report on intracellular Ca^{2+}. By including a DNA sequence that encodes for pH-sensitive protein, the fusion protein can be used to report on intracellular pH. By including a DNA sequence that encodes for a voltage-dependent K^+-channel, a transgene that includes GFP can be used to report on membrane potential. By including a DNA sequence that changes its conformation in response to pressure, a transgene that includes GFP can be used to report on turgor pressure. By including a DNA sequence that encodes an ABA receptor and a PP2C-type phosphatase, a transgene that includes GFP can be used to report on the abscisic acid concentration (Hanson and Köhler, 2001; Lalonde et al., 2005; Swanson et al., 2011; Dang et al., 2012; Choi and Gilroy, 2014; Denninger et al., 2014; Jones et al., 2014a,b; Waadt et al., 2014). Such DNA sequences can also include a targeting sequence to report on the status of a given organelle or subcellular region.

16.9 NUCLEOLUS AND RIBOSOME FORMATION

A eukaryotic cell must synthesize about 10 million copies of each rRNA molecule to make the 10 million ribosomes it will need throughout its life. This is facilitated by the fact that a cell contains multiple copies of ribosomal genes that reside in a cluster known as the *nucleolar organizer region*. The nucleoli, an RNA-rich region (Brachet, 1940; Caspersson and Schultz, 1940) in which the ribosomes are made (Perry, 1969), form in this region at the end of mitosis and disappear again at prophase (Matsunaga and Fukui, 2010; Matsunaga et al., 2013; Matsunaga, 2014). The transcription of rRNA genes can be easily visualized in spread chromatin using the electron microscope (Fig. 16.11). The transcription of these genes by RNA polymerase I looks like a Christmas tree. The tip of the tree represents the point where transcription is initiated. The base of the tree is created by the detachment of the RNA polymerase and its transcript.

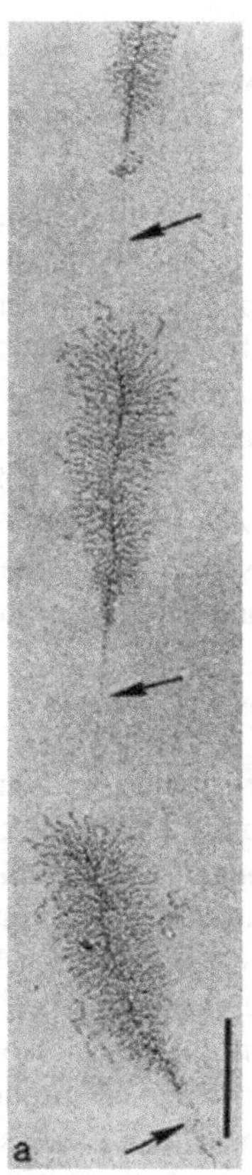

FIGURE 16.11 Transcriptional unit of rDNA of a nucleolus isolated from *Acetabularia cliftonii*. *Arrows* indicate the spacer regions. Bar, 1 μm. *From Franke, W.W., Scheer, U., Trendelenburg, M.F., Spring, H., Zentgraf, H., 1976. Absence of nucleosomes in transcriptionally active chromatin. Cytobiologie 13, 401–434.*

Each gene produces the same primary transcript, known as *45S rRNA*, which is about 13,000 nucleotides long. The 45S transcript is cleaved to yield one copy of 28S rRNA, one copy of 18S rRNA, and one copy of 5.8S rRNA. The remaining part of the primary transcript is degraded in the nucleus. The splicing of rRNA involves a nucleolus-specific small nucleoprotein particle (Fisher et al., 1991).

Ribosome formation requires a fourth rRNA, the 5S rRNA. It is transcribed from another set of tandemly arranged genes. This rRNA is synthesized by RNA polymerase III. It is not known why this rRNA is in a separate cluster and transcribed separately. The 5S rRNAs are made outside the nucleolus and imported into the nucleolus for ribosome formation.

The 28S, 5.8S, and 5S transcripts combine with about 50 polypeptides to form the large 60S ribosomal subunit. The 18S transcript combines with about 30 polypeptides to make the 40S ribosomal subunit. The large 60S and the small 40S subunits form the complete 80S ribosome. The packaging of the rRNA with the polypeptides takes place in the nucleus, in a specialized structure called the *nucleolus* (Warner, 1990; Hernandez-Verdun, 1991). The nucleolus is not surrounded by a membrane and is topologically part of the nuclear P-space. However, it appears to contract, expand, and undergo amoeboid movement, indicating it has different matrix properties than the nucleus (Fig. 16.12).

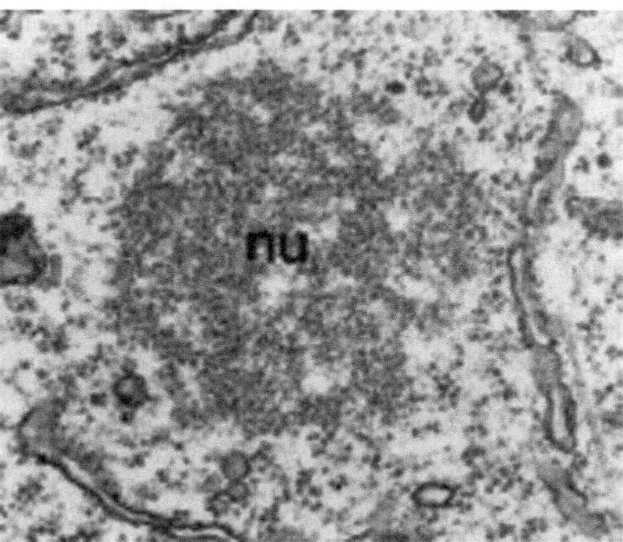

FIGURE 16.12 Electron micrograph of a nucleolus (nu) of *Funaria hygrometrica* that has been extruded intact into the cytoplasm. ×35,300. *From Conrad, P.A., Steucek, G.L., Hepler, P.K., 1986. Bud formation in* Funaria*: organelle redistribution following cytokinin treatment. Protoplasma 131, 211–223.*

The nucleolus, which is formed by the rRNA genes, contains large loops of DNA from several chromosomes that contain the 45S rRNA genes (Brown and Shaw, 1998). If an rRNA gene is added to a genome, a new nucleolus forms (Karpen et al., 1988). If a nucleolar organizing center is ablated with a laser, the remaining nucleolar organizing centers tend to exhibit higher activity, indicating that a nucleolus is capable of sensing the absence of another one and compensating for the loss (Berns et al., 1981). The increased activity of the remaining nucleolar organizing region may also be the result of an increased concentration of substrate due to the decreased competition for nucleotides.

Ribosomes are also composed of proteins (Liljas, 1991; Pendle et al., 2005), which enter the nucleolus from the cytoplasm. These proteins have a specialized signal (Underwood and Fried, 1990) in addition to the usual nuclear import signal (Hernandez-Verdun, 1991). The joining of the ribosomal polypeptides with the primary rRNA transcript can be seen in the electron microscope. The protein–rRNA combination appears as a large particle at the 5′ end of rRNA transcript.

Currently, the issue of exactly where rRNA transcription occurs is unresolved. Scheer and Benavente (1990) think that transcription takes place in the pale-staining fibrillar region of the nucleolus because this region is stained by anti-polymerase I antibodies (Scheer and Benavente, 1990). However, Hernandez-Verdun (1991) believe that the polymerases that are detected by the antibodies can only be in the free form and are not attached to DNA and are not actively involved in transcription. Thus, Hernandez-Verdun (1991) and Jordan (1991) believe that the dense fibrillar regions are involved in transcription. This is a fascinating argument because we do not know whether or not the antibody competes with the substrate, and so either interpretation is valid (Fisher et al., 1991; Jordan et al., 1992). Work that is more recent suggests that the lightly staining fibrillar region may be the site of transcription of rRNA. Using in situ hybridization techniques, Shaw et al. (1993) performed experiments that show that there is an inverse relationship between transcriptional activity (determined by the amount of antisense rRNA bound) and the amount of dense fibrillar regions in the nucleolus. Thus, they believe that transcription does not take place in the densely staining fibrillar regions. Also using in situ hybridization, Olmedilla et al. (1993), using tritiated uridine, found that rDNA was synthesized in the dense fibrillar regions. By looking at the incorporation of bromo-UTP and observing it with fluorescence microscopy, Dundr and Raska (1993), Hozak et al. (1994), Melcak et al. (1996), and Thompson et al. (1997) have shown that rRNA synthesis takes place in the dense fibrillar region. On the whole, it seems that rRNA is synthesized by RNA polymerase I in the dense fibrillar regions.

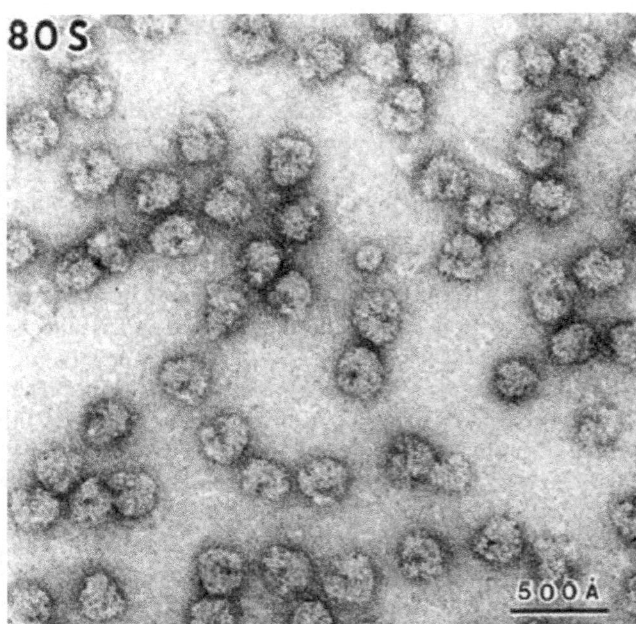

FIGURE 16.13 Electron micrograph of negatively stained 80S ribosomes from *Dictyostelium discoideum. From Boublik, M., 1987. Structural aspects of ribosomes. Int. Rev. Cytol. (Suppl. 17), 357–389.*

The accumulation of the two subunits of the ribosome gives certain regions of the nucleolus a granular appearance. Pulse chase experiments with tritiated uridine show that the process of ribosome formation and export to the cytoplasm takes about 60 min.

The last step in ribosome maturation takes place in the cytoplasm, making it impossible for functional ribosomes to come in contact with the hnRNA in the nucleus. In Chapter 17, I discuss the involvement of the ribosomes (Fig. 16.13), tRNA, and mRNA in protein synthesis.

16.10 SUMMARY

The nucleus is the organelle that contains the DNA. The DNA molecule, which is the only molecule that is unique in every living being, contains the hereditary information that is passed from generation to generation and the information necessary for the development of each living being. Each cell has a certain potential that is determined by the genes that are expressed in that cell. However, there is not necessarily a direct precursor–product relationship between a gene, as defined as a DNA sequence, and a trait. There are many factors that determine which genes are ultimately expressed. These factors may include the three-dimensional position of the gene in the genome and in the nuclear matrix, the presence of regulatory factors including protein transcription factors and miRNAs, the activity of the spliceosomes, the ability of the mRNA to recognize and be transported through the nuclear pores, the movement of the mRNA through the cytosol to specific sites, the translation of the mRNA at these sites, and the pH, Ca^{2+}, protein kinase, and phosphatase activities as well as protease activity at these sites that may modify the activity of the gene product. Thus, there are more than a few processes to elucidate before we are able to understand the role DNA plays in determining the individuality of a cell or an organism (Chargaff, 1997).

16.11 QUESTIONS

16.1. What is the evidence that the nucleus is the organelle involved in heredity and development?

16.2. What are the mechanisms by which the nucleus and its constituents participate in heredity and development?

16.3. What are the limitations of thinking about the nucleus as the sole regulator of heredity and development?

16.4. What are the limitations of thinking about the sequence of DNA as the sole means of directing heredity and development?

The references for this chapter can be found in the references at the end of the book.

Chapter 17

Ribosomes and Proteins

I remember Staudinger's lecture to the Zürich Chemical Society in 1925 on his high polymer thread molecules with a long series of Kekulé valence bonds. It was impossible to accommodate his view in the unit cell as established by X-ray analysis. All the great men present: the organic chemist, Karrer, the mineralogist, Niggli, the colloidal chemist, Wiegner, the physicist, Scherrer, and the X-ray crystallographer (subsequently cellulose chemist), Ott, tried in vain to convince Staudinger of the impossibility of his idea because it conflicted with exact scientific data. The stormy meeting ended with Staudinger shouting 'Hier stehe ich, ich kann nicht anders' [Here I stand, I cannot do otherwise] in defiance of his critics.

Albert Frey-Wyssling (1964; Klug, 2002).

17.1 NUCLEIC ACIDS AND PROTEIN SYNTHESIS

With the realization that there was a division of labor in cells, and that the chromatin in the nucleus carried the instructions for many vital processes that took place in the cytoplasm, the question arose of how the genetic material coded for the proteins (de Vries, 1889; Beadle and Tatum, 1941a,b; Tatum and Beadle, 1942; Dounce, 1952; Beadle, 1945, 1955, 1958; Tatum, 1958; Beadle and Beadle, 1966; Horowitz, 1985; Perkins, 1992). Immediately following the unveiling of the double-helix model for the structure of DNA by Watson and Crick, George Gamow (1954; Nanjundiah, 2004) realized that the sequence of amino acids in a protein could be deduced from the sequence of nucleotides in the gene that encoded it, and he suggested that amino acids might bind directly to the DNA double helix in a lock-and-key manner. Gamow (1954) envisioned that the DNA strands would directly act as a template for protein synthesis. While the proposal that a sequence of three out of four bases would be required to code for each of the 20 amino acids was correct; the idea that DNA acted directly as the template had already been shown to be untenable by cytological data.

In the 1930s and 1940s, Jean Brachet (1960) and Torbjörn Caspersson (1950) established that RNA is correlated with protein synthesis in both animal and plant cells. This was at the time when the organic chemists assumed that DNA was present only in animals, and RNA was present exclusively in plants (Jones, 1920; Levene and Bass, 1931). Caspersson and Schultz (1938) used ultraviolet (UV) microspectrophotometry combined with ribonuclease digestion to localize the RNA in the cell. In complementary experiments, Brachet (1957, 1960, 1985; Jeener and Brachet, 1944; Burny, 1988; Alexandre, 1992; Thomas, 1992) observed the distribution of RNA by staining cells that had been treated with or without ribonuclease with the basic dye methyl green-pyronine. They both found that RNA was most abundant in cells that were rapidly growing, such as onion meristematic cells, the imaginal discs of *Drosophila* larvae, and the protein secreting cells of the pancreas, the gastric mucosa, and the silk worm. By contrast, cells that were metabolically active, but did not secrete proteins such as kidney, heart, and skeletal muscle cells, had a low RNA content. Moreover, Brachet and Jeener (1944) also found that hemoglobin was specifically associated with the RNA-containing particles in red blood cells, amylase was specifically associated with the RNA-containing particles in salivary glands, and trypsin was specifically associated with the RNA-containing particles in the pancreas. While RNA was abundant in cells that synthesized large quantities of protein and was scarce in cells that did not, the DNA content, by contrast, was the same in all of the cells of the organism. Moreover, when cells were challenged with various physiological conditions that affected their rate of protein synthesis, the quantity of RNA changed in parallel with the rate of protein synthesis. By contrast, the DNA quantity remained constant. In the words of Brachet (1987), "After several not very encouraging attempts, I found in 1939 a simple cytochemical technique for detecting RNA in cells. To my great joy, I found RNA, like DNA, to be a universal constituent of cells-bacterial, vegetal, and animal. The intracellular localization of these two types of nucleic acid is, however, quite different: whereas DNA is found in chromatin and chromosomes, RNA accumulates in cytoplasm and nucleoli. In addition, whereas the amount of DNA per nucleus remains constant in a particular species (allowing for its doubling when cells prepare for division), the amount of RNA varies considerably from one tissue to another; I saw a completely unexpected correlation between the quantity of RNA in a cell and its capacity to synthesize proteins. This led me to another iconoclastic proposition: proteins are not

Plant Cell Biology. https://doi.org/10.1016/B978-0-12-814371-1.00017-5

synthesized by proteolytic enzymes operating backwards, as was generally thought, but by an unknown mechanism implicating RNA. The same conclusion was arrived at simultaneously (1941) by T. Caspersson in Stockholm, who was using a completely different technique for the cytochemical detection of nucleic acids."

The lack of a direct connection between DNA and protein synthesis was underscored by microsurgical experiments in which the nucleus was removed from the cell. The DNA-containing nucleus can be readily removed from a variety of cell types, including the giant alga *Acetabularia* (Hämmerling, 1953; Brachet, 1957; Harris, 1980; Mandoli, 1998). The enucleated stalks of *Acetabularia* live for several weeks and are capable of synthesizing proteins. They even undergo partial regeneration and differentiation of the cap. When a nucleus of another species of *Acetabularia* is inserted in the enucleated cell, the cap is completely regenerated, although it has characteristics of the two species. Thus, substances remained in the enucleate cells that were products of the nucleus, but stand between gene and character. Some evidence, although it was not compelling, suggested that this substance was RNA (Brachet, 1957, 1985).

Caldwell and Hinshelwood (1950) realized that the synthesis of a linear polymer of amino acids would require a template. The autosynthesis of a virus, which could be described as a nucleoprotein, provided the model for a hypothesis of protein synthesis. They proposed that the autosynthesis of the nucleoprotein required the coordination of the ribonucleic acid and the protein, so that the protein acted as the template for the synthesis of the ribonucleic acid and the ribonucleic acid acted as the template for the synthesis of the protein. According to Caldwell and Hinshelwood (1950), "This suggests some sort of correspondence between the units in the two kinds of polymer. In a protein, about 23 different amino-acids may occur, whereas in a nucleic acid only 5 basic units are found-two pyrimidine nucleotides, two purine nucleotides, and ribose phosphate. Clearly there cannot be a one-to-one correspondence between the position of an individual amino-acid in the protein part of a nucleoprotein and the position of an individual nucleotide in the nucleic acid part. If, however, it is assumed that, in the synthesis of a protein at the surface of a nucleic acid polymer, the amino-acid side-chain which is guided into a particular place depends on the nature and relative position of two adjacent nucleotide units, the difficulty can be overcome. Twenty-five different internucleotide arrangements are possible, and this is of the right order to give correspondence with the number of different possibilities in a protein chain."

In 1952, Alexander Dounce (1952) proposed a mechanism whereby the bases of RNA functioned as a template for protein synthesis. In his model, three adjacent nucleotides encoded for each amino acid. Dounce went on to say that a three-letter code could account for all the amino acids known to occur in proteins, and moreover, if there is "complete freedom of choice in arranging the order of four nucleotides … a sufficient number of nucleic acids could theoretically exist to account for the large variety of proteins in nature." It was unlikely that protein was the template for making more protein according to Dounce (1956), who wrote,

> *My interest in templates, and the conviction of their necessity, originated from a question asked me on my Ph.D. oral examination by Professor J.B. Sumner. He enquired how I thought proteins might be synthesized. I gave what seemed the obvious answer, namely, that enzymes must be responsible. Professor Sumner then asked me the chemical nature of enzymes, and when I answered that enzymes were proteins or contained proteins as essential components, he asked whether these enzyme proteins were synthesized by other enzymes and so on ad infinitum.*

As I previously described, DNA did not seem to be the direct template, and further experiments supported this conclusion. When cells were treated with X-ray or UV irradiation, DNA synthesis was inhibited, but these treatments had little effect on RNA or protein synthesis. On the other hand, treatments that interfered with RNA, including the addition of analogues of uracil, inhibit protein synthesis in vitro (Spiegelman, 1956). However, these in vitro experiments, where DNA or RNA were either selectively added or removed from broken cells, had ambiguous effects on protein synthesis (Spiegelman, 1956; Gale, 1957).

As I previously mentioned, Watson and Crick's (1953a) paper on the structure of DNA stimulated George Gamow to wonder how the information coded in DNA in the form of a sequence of four different bases could be translated into a sequence of 20 amino acids that make up a protein (see Gamow and Ycas, 1967; Gamow, 1970). The word information was first applied to genetics in 1953 (Ephrussi et al., 1953). While Gamow (1954) proposed a triplet code, it was not clear if the code was overlapping or not. Taking a theoretical approach to determine whether or not the code was overlapping required the best computers of the day (Gamow and Metropolis, 1954). Many mathematically inclined scientists from the Manhattan Project, including Richard Feynman, Nicholas Metropolis, Stanislaw Ulam, Edward Teller, and John von Neumann, contributed to cracking the code (Gamow et al., 1956; Rich, 1997). Gamow and Ycas (1955) used the computer at Los Alamos to perform a statistical analysis that compared the relative abundance of nucleotides in RNA with the relative abundance of amino acids in proteins, much as one would compare the relative abundance of letters in a code to determine which ones were vowels. They concluded from this analysis that there was a triplet code, and it was not overlapping.

Gamow and his colleagues provided further support for a nonoverlapping code by studying the sequence of amino acids in proteins and concluded that there were more combinations of adjacent amino acids possible than could be accounted for by an overlapping code (Gamow et al., 1956). After rejecting all possible overlapping codes, Gamow and Ycas (1955) wrote, "Thus it appears more probable that the number of determining nucleotides exceeds by a factor of 3 the number of amino acid residues in the synthesized protein, so that neighboring residues do not share determining nucleotides." Brenner (1957) and Crick et al. (1957, 1961) concurred with the concept of a three-base, nonoverlapping code. Experimental support for a nonoverlapping code came from studies that showed that mutagenesis of tobacco mosaic viruses usually causes a single amino acid substitution in any given part of the protein (Tsugita and Fraenkel-Conrat, 1960). If the code were overlapping, it would be likely that two or more adjacent amino acids would be substituted in each mutant. The fact that hemoglobin from people with sickle cell anemia only differed by one amino acid from the wild-type hemoglobin as determined with paper electrophoresis (Pauling et al., 1949; Hunt and Ingram, 1958; Ingram, 1958, 1959, 2004; Baglioni and Ingram, 1961) supported the idea that the code was not overlapping.

Progress in understanding the relationship between nucleic acids and proteins advanced along another tact when Albert Claude (1943a,b) used differential centrifugation to isolate the microsome fraction from cells (see Chapter 4). He found that this fraction contained approximately 40%–50% of the total cellular RNA and also stained with basic dyes. Claude concluded that the ribonucleic acid was responsible for binding the basic dyes, and thus the microsomes were the basophilic substance seen by cytologists. The importance of the microsome fraction in protein synthesis was demonstrated by Borsook et al. (1950) and Elizabeth Keller (1951), who injected rats with radioactive leucine and found that the microsomal fraction contained the highest specific activity. Palade (1955, 1958) suggested that the small particulate components on the endoplasmic reticulum (ER), and not the ER itself, were the cause of the basophilic staining observed by the cytologists and consequently may be involved in protein synthesis.

These particles that were known as the small particulate component, Palade's granules, opaque particles, ergastoplasmic particles, and ribonucleoprotein particles were given the "pleasant-sounding name," ribosome, at the first symposium of the Biophysical Society (Roberts, 1958). Ribosomes can be found attached to the ER, nuclear envelope, actin cytoskeleton (Yang et al., 1990; Davies et al., 1991; You et al., 1992; Durso et al., 1996), or free in the cytosol.

17.2 PROTEIN SYNTHESIS

Many reactions in the cell are reversible, and, in their search for simplicity, Max Bergmann and Joseph Fruton postulated that the well-characterized enzymes involved in protein degradation would also prove to be involved in protein synthesis, although during protein synthesis the reaction would run backward (Van't Hoff, 1903; Bergmann and Fruton, 1944; Linderstrøm-Lang, 1952; Brachet, 1957; Fruton, 1955, 1957; Fruton and Simmonds, 1958; Zamecnik, 1960). These experiments did not involve the use of a linear template, and consequently highly branched protein-like molecules were synthesized.

With the discovery of ^{14}carbon, and a means to produce it in copious amounts (Ruben and Kamen, 1940a,b), biochemists had a way of studying protein synthesis in preparations that contained high background levels of protein. Unfortunately, the early in vitro studies using radioactive alanine and glycine gave ambiguous results, as the amino acids were not only incorporated into protein but also participated in intermediary metabolism, including the glucose-alanine cycle, gluconeogenesis, purine synthesis, and porphyrin synthesis (Siekevitz, 1952; Zamecnik, 1969; Elliott and Elliott, 1997). As unreliable as these experiments turned out to be, they suggested that adenosine triphosphate (ATP) may be important for protein synthesis because amino acid incorporation required oxygen and was inhibited by dinitrophenol.

In vitro experiments on protein synthesis became more productive after Elizabeth Keller introduced radioactive leucine because leucine is not involved in as many nonprotein-synthetic pathways as alanine and glycine. Moreover, leucine is abundant in proteins (Keller et al., 1954). The incorporation of radioactive leucine into protein was inhibited by ribonuclease, indicating that RNA is important for protein synthesis. Leucine incorporation also required ATP (Zamecnik and Keller, 1954). Fritz Lipmann (1941) had already suggested that amino acids would not readily react with each other to form peptide bonds unless they were activated and suggested that ATP may be involved in the activation of amino acids. Indeed, Mahlon Hoagland, Elizabeth Keller, and Paul Zamecnik (1956) showed that when $^{32}P_iP_i$ was added to the soluble fraction of liver, ^{32}P-ATP was formed, and this exchange reaction was stimulated by adding amino acids. The active component in the soluble fraction was probably a protein because it was heat-labile, nondialyzable, and could be precipitated at pH 5. While Hoagland et al. were running the reaction backward for the convenience of assaying the enzyme, they concluded that the enzyme in the pH 5 fraction actually activates an amino acid by hydrolyzing ATP and forming an enzyme aminoacyl–adenosine monophosphate (AMP) complex and P_iP_i (Hoagland, 1990). Competition experiments showed that there was a different enzyme to activate each amino acid. These enzymes are now known as aminoacyl-tRNA synthetases. Each aminoacyl-tRNA synthetase is specific for a particular amino acid (Moras, 1992).

Thus, Zamecnik's group discovered that protein synthesis could take place in vitro, as long as ATP and the pH

5 fraction were added to the microsomes. Or could it? It turned out that when purified ATP was used in place of the crude source, amino acids were not incorporated into protein (Zamecnik, 1969). Keller and Zamecnik (1956) discovered that guanosine triphosphate (GTP) or guanosine diphosphate (GDP), in addition to pure ATP, was necessary for protein synthesis, and thus the crude source of ATP must have contained guanylate nucleotides. I will discuss the function of GTP below.

Littlefield et al. (1955) and Littlefield and Keller (1957) discovered that a ribonucleoprotein particle can be separated from the rest of the microsomal fraction by treating the microsomes with sodium deoxycholate and centrifuging the mixture at about 100,000 g. The ribonucleoprotein particles that were in the precipitate were capable of performing protein synthesis in the presence of ATP and GTP. The fact that a large volume of unlabeled amino acids added at the end of the assay did not reduce the amount of label incorporated into protein indicated that ATP- and GTP-dependent protein synthesis in the ribonucleoprotein particles was irreversible.

Robert Holley (1957, 1968) wondered: If activation of the amino acids was the first step in protein synthesis, what was the second? He surmised that the enzyme–aminoacyl–AMP complex might bind to a molecule X forming an aminoacyl–X complex and releasing the enzyme and AMP. Holley assumed that the two reactions could be described like the following:

$$\text{enzyme} + \text{amino acid} + \text{ATP}$$

$$\Leftrightarrow \text{enzyme} - \text{aminoacyl} - \text{AMP} + P_iP_i$$

$$\text{enzyme} - \text{aminoacyl} - \text{AMP} + \text{X}$$

$$\Leftrightarrow \text{aminoacyl} - \text{X} + \text{enzyme} + \text{AMP}$$

Holley conjectured that these two reactions were reversible and coupled, and thus he could discover X by adding various fractions to the assay system. The assay system included the pH 5 precipitate (Hoagland et al., 1956), alanine as the amino acid, and radioactive AMP. Holley fractionated the cytosol using a column as tall as the USDA building on the Cornell University campus, in which he was working. Holley assayed various soluble fractions for X by running the coupled reactions in the presence of radioactive AMP. He found a soluble fraction that catalyzed the alanine-dependent formation of radioactive ATP. ATP formation was inhibited by RNAse, indicating that the intermediate X was a soluble ribonucleic acid. These results were extended by Hoagland et al. (1957, 1958) and Ogata and Nohara (1957) for the activation of leucine and alanine, respectively. The soluble RNA was eventually given the name transfer RNA, which is abbreviated to tRNA.

Apparently unaware of much of the data described above obtained with classical biochemical techniques, Francis Crick (1958) came up with the "sequence hypothesis" and the "central dogma." In the hypothesis and the dogma, Crick said that the sequence of nucleic acids codes for the sequence of amino acids in a protein, and information is passed from nucleic acid to nucleic acid or from nucleic acid to protein, but never from protein to nucleic acid. The ribosomes contain an RNA that acts like a template for protein synthesis. The amino acids are carried to the template by small RNA molecules, which he called adapters. Crick (1958) went on to say,

> *It will be seen that we have arrived at the idea of common intermediates without using the direct experimental evidence in their favour; but there is one important qualification, namely that the nucleotide part of the intermediates must be specific for each amino acid, at least to some extent. It is not sufficient, from this point of view, merely to join adenylic acid to each of the twenty amino acids. Thus one is led to suppose that after the activating step, discovered by Hoagland ..., some other more specific step is needed before the amino acid can reach the template.*

Crick finished this paper with the following: "I shall be surprised if the main features of protein synthesis are not discovered within the next ten years." When I first read Crick's 1958 paper, I was under the mistaken impression that this paper formed the foundation of the field of protein synthesis. Now I realize that it, like Watson and Crick's (1953a) paper, in which they wrote that their theoretical model of DNA "must be regarded as unproved until it has been checked against more exact results" and that they "were not aware" of Franklin's more exact results, did not present an accurate version of the history of science.

It was initially assumed that the RNA in the ribosomes provided the template for protein synthesis. This came to be known as the "one gene–one ribosome–one protein hypothesis." However, evidence accumulated that indicated that the ribosomal RNA (rRNA) did not function as a template, and there may be another kind of RNA that acts as an intermediate between the DNA and proteins. The first evidence came when Volkin and Astrachan (1956) discovered that when *Escherichia coli* were infected with the T2 phage in the presence of ^{32}P, the radioactive phosphorous was immediately incorporated into RNA. Although they did not know the function of this RNA, they postulated that it may be a new species as its base composition was different than that of the average RNA in the cell. By 1958, Astrachan and Volkin tentatively suggested that this RNA may be involved in protein synthesis. Using yeast, Ycas and Vincent (1960) found that the RNA that was newly labeled had a base composition that was very similar to yeast DNA, except that in the RNA, uridylic acid replaced the thymidylic acid in DNA.

Further evidence hinted at the possibility that there was a transient RNA intermediate involved in gene expression. In a series of experiments, Riley et al. (1960) allowed bacteria to mate and then disrupted the mating process by vigorous shaking the conjugants at various times after the start of conjugation (Wollman et al., 1956; Jacob, 1965). They found that enzyme formation began in the recipient bacterium within 2 min of the time of injection of the β-galactosidase gene from the donor bacterium. Riley et al. felt that this was too short of a time to build a ribosome, which is what would be expected according to the one gene–one ribosome–one protein hypothesis. They suggested that the gene produced an unstable RNA intermediate in the recipient bacterium that moved to the preformed ribosomes and acted as a template for protein synthesis. Jacob and Monod (1961) dubbed this unstable RNA, which accounts for about 1–5 percent of the total cellular RNA, messenger RNA, or mRNA. rRNA makes up the majority of the cellular RNA. Subsequent studies showed that rapidly labeled mRNA molecules could be found in the ribosomes (Brenner et al., 1961; Gros et al., 1961; Loening, 1962; Risebrough et al., 1962; Watson, 1963).

Meanwhile, Marshall Nirenberg and Heinrich Matthaei found that protein synthesis only occurred when they combined the ribosomal fraction with the RNA fraction that was separated from the ribosomes by centrifugation (Matthaei and Nirenberg, 1961; Nirenberg and Matthaei, 1961; Matthaei et al., 1962; Portugal, 2015). The possibility that the RNA acted as mRNA lead them to wonder if they could induce protein synthesis in the cell-free system by adding synthetic enzymatically synthesized RNA of known composition as a template. If this worked, they could crack the genetic code by comparing the sequence of polynucleotide phosphorylase–synthesized RNA with the sequence of amino acids in the protein synthesized. When they added poly-U RNA synthesized by Leon Heppel and Maxine Singer to the ribosome fraction, they found that polyphenylalanine was synthesized! That is, when they added a complete mixture of 20 amino acids with only one being radioactive, they found that only radioactive phenylalanine was incorporated into acid-insoluble protein. The synthesis of polyphenylalanine required GTP, the tRNA for phenyalanine, and aminoacyl-tRNA synthetase that was specific for phenylalanine. The use of polynucleotides containing random arrangements of nucleotides promoted the incorporation of additional amino acids. Poly UA stimulated the incorporation of tyrosine, isoleucine, and lysine; poly UC stimulated the incorporation of serine, proline, and leucine; poly UG stimulated the incorporation of leucine, valine, cysteine, tryptophan, methionine, and glycine; and poly UGC stimulated the incorporation of arginine, serine, glutamic acid, and alanine (Martin et al., 1962). Because both UC and UG coded for leucine, Nirenberg realized that the code was degenerate. Crick (1966) described the degeneracy in his *wobble hypothesis*, which stated that "while the standard base pairs may be used rather strictly in the first two positions of the triplet, there may be some wobble in the pairing of the third base." The synthetic mRNA also proved helpful in determining that aggregates of ribosomes or polyribosomes typically bound to the mRNA (Barondes and Nirenberg, 1962; Spyrides and Lipmann, 1962). However, the big breakthrough with synthetic RNA came when Nirenberg and his colleagues (Nirenberg and Leder, 1964; Bernfield and Nirenberg, 1965) used synthetic mRNAs to crack the genetic code. Nirenberg's group added 64 synthetic RNAs that were three-nucleotides long to the ribosome fraction and analyzed which of the ^{14}C-aminoacyl-tRNAs bound to the ribosomes. When they used a synthetic UUU message, the phenylalanine-tRNA bound. Likewise, lysine-tRNA and proline-tRNA bound when AAA and CCC, respectively, were used. In this way, they could determine which amino acid was coded for by the 64 possible codons. They found that some of the codons represented an initiation codon (AUG) and others represented termination codons (UGA, UAA, and UAG; Caskey et al., 1968). When Nirenberg won the Nobel Prize, someone put a sign in his lab that said (Portugal, 2015), "UUU are great."

Gobind Khorana (1961, 1968, 1976), who shared the Nobel Prize with Nirenberg, provided complementary evidence for the genetic code, as it was not clear whether a three nucleotide message was equivalent to a polymeric template and whether the attachment of an amino acid to a ribosome was the same as the synthesis of a protein. Khorana and his associates used organic chemical methods to synthesize polymers of DNA with known sequences of repeating dinucleotides, trinucleotides, and tetranucleotides. They used organic chemistry to synthesize DNA rather than RNA because the organic chemical methods for DNA had advanced more than the methods for RNA. Then they used RNA polymerase to transcribe the DNA into RNA. They added the RNA to the cell-free ribosome preparation and analyzed the resultant polypeptides that were formed from these templates until they figured out the codes for each amino acid. Thus, the genetic code that related the sequence of RNA, which depended on the sequence of DNA in a gene to the sequence of amino acids in a protein, was cracked.

Work on the structure of tRNA continued during this miraculous decade of research in molecular biology, and by 1965, Robert Holley and his associates developed techniques of nucleic acid sequencing and sequenced the entire alanine tRNA from yeast. One end of the tRNA contains the terminal adenine and binds to the carboxyl group of the amino acid in an exchange reaction that released AMP and the aminoacyl-tRNA synthetase. The OH group of the

COOH of the amino acid binds to the 3′-OH group of the ribose. Another region of the folded tRNA molecule contains a sequence of three nucleotides, known as an anticodon, which is complementary to and thus can bind with the three-nucleotide codon of an mRNA. The gene that encodes the alanine tRNA of yeast was the first gene synthesized in vitro (Agarwal et al., 1970; Khorana et al., 1972; Caruthers, 2012).

During gene expression, the DNA coding for a gene is transcribed into RNA, and after this RNA is processed, the subsequent mRNA leaves the nucleus and attaches to a ribosome in the cytosol. The small subunit of the ribosome contains one binding site for mRNA and two for tRNA. One tRNA binding site is called the peptidyl-tRNA binding site (P-site). This site holds the tRNA that is linked to the growing end of the polypeptide chain. The other tRNA binding site is called the aminoacyl-tRNA binding site (A-site), and it holds the incoming aminoacyl-tRNA. The anticodons of the tRNAs pair with the codons of the mRNA in these sites. The P- and A-sites are so close that the two tRNAs form base pairs with adjacent codons. The pairing of codons and anticodons allows the formation of a polypeptide chain according to the sequence of the mRNA molecule. The tRNAs then act as translators of the genetic code from the nucleic acid sequence to the amino acid sequence from gene to polypeptide.

The initiation of protein synthesis is mediated in the small subunit by a group of polypeptides known as initiation factors. A special aminoacyl-tRNA is required for this step, and in eukaryotic ribosomes, it is always methionine-tRNA. The methionine-tRNA first binds to the small subunit of the ribosome where it is put in the correct position in the P-site by a protein known as a eukaryotic initiation factor. *Pari passu*, the small subunit binds an mRNA that it recognizes by the 7-methylguanosine residue at the 5′ end. The small subunit of the ribosome, containing the methionine-tRNA, then moves down the mRNA in search of the AUG codon (start codon). Then several polypeptides are released from the small subunit and the small and large subunit bind together, thus completing the formation of the ribosome. Protein synthesis continues as the next aminoacyl-tRNA binds to the A-site.

The polypeptide chain then elongates in a cycle that consists of three steps. Firstly, an aminoacyl-tRNA, which is bound to an elongation factor–GTP complex, binds to the vacant A-site. For elongation to proceed, the GTP must be hydrolyzed. This will only happen if the GTP is positioned correctly in the A-site for a sufficient amount of time. The GTP will only remain in the A-site for a long enough time if there is perfect matching between the codon and anticodon (at least for the first two nucleotides). Remember from Chapter 12 that

$$k_{\text{off}} = K_d(k_{\text{on}}) \tag{17.1}$$

and

$$\text{the half time of a reaction} = (\ln 2)/k_{\text{off}} \tag{17.2}$$

If the match is not perfect (the K_d is too large), and thus the binding strength is not sufficiently great, then the aminoacyl-tRNA complex is released from the ribosome before the GTP is hydrolyzed. If the codon–anticodon matching is correct, the tRNA molecule remains bound to the mRNA and the GDP-elongation factor complex is released from the ribosome.

The requirement for a correct binding before the GTP is hydrolyzed results in a proofreading mechanism and ensures the correct synthesis of a protein (Crick et al., 1957). In the next step of protein elongation, the carboxyl end of the polypeptide chain is uncoupled from the tRNA molecule in the P-site and linked to the amino group of the aminoacyl-tRNA in the A-site to form a peptide bond. This step is catalyzed by peptidyl transferase, which is a ribozyme, an enzyme made of RNA (Noller et al., 1992).

Subsequently, the free tRNA that was in the P-site is released from the ribosome as the peptidyl-tRNA that was in the A-site is translocated to the P-site as the ribosome moves three nucleotides down the mRNA (or the mRNA moves along a stationary ribosome). This translocation is also coupled to the hydrolysis of GTP. Thus, the addition of a single amino acid to a protein requires the hydrolysis of two GTP molecules. Counting the ATP that is necessary to activate the carboxylic acid group of the amino acid before it is attached to the tRNA molecule, the hydrolysis of three nucleoside triphosphates is necessary for the addition of each amino acid to a protein, and consequently, protein synthesis is an energy-intensive process. Soon after the process of protein synthesis was elucidated, the dynamic process was demonstrated in a dance that was filmed in 1971. The film entitled "Protein Synthesis: An Epic on the Cellular Level" can be seen at http://www.youtube.com/watch?v=u9dhO0iCLww.

The ribosome moves along the mRNA in the 5′–3′ direction. A protein is synthesized stepwise from its amino-terminal end to its carboxy-terminal end. The growing peptide moves through a tunnel in the large subunit of the ribosome. This tunnel accommodates approximately 35–39 amino acids as determined by protease-protection assays (Blobel and Sabatini, 1970; Bernabeu and Lake, 1982; Yonath et al., 1987). Approximately 10 amino acids are incorporated into a protein per second, and thus an average protein is synthesized in about 20–60 s.

Because the "average" amino acid has a molecular mass of 110 Da, on the average 1.1 kDa of protein is synthesized per ribosome per second. The time it takes to synthesize one protein with a molecular mass, M_r (in Da), is given by the following equation:

$$\text{Time} = M_r/(1100\ \text{Da/s}) \tag{17.3}$$

Highly expressed proteins are not synthesized by a single ribosome but by many ribosomes simultaneously (Galau et al., 1977). They appear as polyribosomes or "polysomes" in electron micrographs and density gradients.

Protein synthesis is terminated when the ribosome reaches a stop codon (UAA, UAG, or UGA). Cytosolic proteins known as release factors bind to the stop codon when it is in the A-site. This binding causes the peptidyl transferase to add a water molecule instead of an amino acid to the peptidyl-tRNA, and the carboxyl group of the polypeptide chain is freed from the tRNA and released from the ribosome. The ribosome then releases the mRNA and splits into its two subunits. Protein synthesis is inhibited by cycloheximide and puromycin.

If the amino-terminus of the nascent protein has a signal sequence, the ribosome that contains the nascent protein will be translocated to the surface of the ER; otherwise, it will remain either free in the cytosol or bound to the cytoskeleton. The mitochondria and chloroplasts also have their own ribosomes (Tao and Jagendorf, 1973; Chua et al., 1973, 1976; Bhaya and Jagendorf, 1985; Staehelin, 1986; Friemann and Hachtel, 1988). The ribosomes in the chloroplast and mitochondrion are smaller than the ribosomes in the cytosol, and like the ribosomes in prokaryotes, they are inhibited by chloramphenicol and not by cycloheximide.

The central dogma asserted that all the information necessary for protein folding must be in the amino acid sequence itself that is coded by the gene (Crick, 1958, 1974a). Indeed, this idea seemed likely at the time because, in vitro, many isolated proteins had been denatured, refolded, and renatured spontaneously due to the small difference in free energy between the native and denatured state. However, many and perhaps most proteins require chaperonins to fold correctly in vivo, where the concentration of a protein may be higher than it is in a test tube. Without chaperonins, proteins would bind to each other in a dysfunctional manner in the crowded conditions found in cells (Anfinsen, 1973; Hartl et al., 1992; Ellis, 1996).

Chaperonins have been identified biochemically by adding the putative chaperonin to a solution of denatured enzymes or structural proteins and testing whether or not there is an increase in the number of functional enzyme molecules or normal protein structures compared to the test tubes in which the proteins were able to self-assemble. By doing such experiments, it turned out that the assembly of significant protein complexes, such as nucleosomes and rubisco, required chaperonins. Chaperonins are involved in the folding of newly synthesized proteins as well as in the unfolding and refolding of proteins that are translocated through membranes. Consequently, chaperonins are found in most, if not every, compartment of the cell.

Leroy Hood (2008) developed the technology that made cell-free protein synthesis a reality (see Chapter 21). He synthesized the HIV-protease, crystalized it, performed X-ray diffraction on the crystal, and developed an HIV-protease drug that successfully fights AIDS.

17.3 PROTEIN ACTIVITY

Enzymes are catalysts that accelerate the rate of reactions as long as the free energy of the products is greater than the free energy of the reactants (Krebs and Kornberg, 1957). The free energy released depends on the concentrations of reactants and products, as well as the dissociation constant of the enzyme for the substrates (Michaelis and Menten, 1913; Haldane, 1930; Lineweaver and Burk, 1934). Enzymes catalyze chemical reactions by reducing the activation energy necessary for bringing the substrates together in the correct orientation so that a reaction can take place. Enzymes also reduce the activation energy by creating the right environment in the active site for the reaction to take place. For example, to dehydrate substrates, enzymes create a water-free, hydrophobic pocket out of amino acids, including tryptophan, phenylalanine, glycine, alanine, valine, leucine, isoleucine, proline, and methionine. About one-third of enzymes contain metal ions in their active sites that are able to orient substrates. To hold Ca^{2+} in the active site, the enzymes may contain negatively charged amino acids, including aspartic acid and glutamic acid. By contrast, Zn^{2+} and Mg^{2+} are bound by nitrogen-containing amino acids, including asparagine, histidine, and glutamine. Once in the active site, metal ions often act as electrophiles, accepting electrons so that two substrates can bond (Glusker et al., 1999). Enzymes are remarkable in that they work at ambient temperatures. Chemical reactions that take place in the vats of the chemical industry typically require high temperatures and pressures to occur at acceptable rates.

Enzymes can be extremely specific for a given substrate and even discriminate between optical isomers (Czapek, 1911; Euler, 1912; Pfeiffer, 1955). In 1894, Emil Fischer (quoted in Clark, 1952) wrote about enzymes and substrates: "The one may be said to fit into the other as a key fits into a lock." While today we take for granted that the majority of enzymes are proteins, this was a revolutionary idea when James Sumner (1926, 1927, 1937; Sumner et al., 1924; Sumner and Graham, 1925) announced that he crystallized urease from jack bean and found that it was a protein. Kirk and Sumner (1931) made an antibody to the enzyme that inhibited its activity in vitro and in vivo. Sumner chose to work on urease because research on it would be inexpensive. He used a very simple procedure to isolate the enzyme. He ground up dried jack beans in a coffee mill, dissolved the meal in 30% acetone at room temperature, poured the solution through a filter, and placed the filtrate on ice. The next morning he centrifuged the filtrate, observed the precipitate with a microscope, and saw that it contained crystals. The crystals tested positive for protein and could decompose their own weight in urea in 1.4 s. Richard Willstätter (1927) and J.B.S. Haldane (1930), who were giants in the field, did not believe that Sumner purified the enzyme and suggested that the real nonproteinaceous enzyme was probably just absorbed to a protein colloid in the crystal and would eventually be separated from it (Sumner, 1933, 1935, 1937, 1946; Waldschmidt-Leitz, 1934; Tiselius, 1946; Edsall, 1976; Simoni et al., 2002). Hans Pringsheim, a German biochemist and the Baker nonresident lecturer at Cornell University, said to Sumner (1937), "I have read about your isolation of urease, which is not true." Sumner asked, "But why is it not true?" Pringshein replied, "Oh, I know too little about it to tell you why, but if you come to Baker Laboratory some time and describe your work on urease before my class then I shall tell you where you are wrong."

Eventually all the enzymes that were crystallized turned out to be proteins, and every schoolchild knew that "all enzymes are proteins although not all proteins are enzymes." Revolutionary history repeated itself in the 1980s, when Sidney Altman and Thomas Cech independently proved that not all enzymes are proteins—some, in fact, were made out of RNA (Guerrier-Takada et al., 1983; Guerrier-Takada and Altman, 1984; Cech, 1985, 1988; Altmann, 1990; Noller et al., 1992). In whichever chemical class an enzyme belongs, I can empathize with Arthur Kornberg (1989a,b) who wrote, "I never met a dull enzyme."

Crick (1958) wrote, "Watson said to me, a few years ago, 'The most significant thing about the nucleic acids is that we don't know what they do.' By contrast the most significant thing about proteins is that they can do almost anything." Crick went on to say,

> *Biologists should not deceive themselves with the thought that some new class of biological molecules, of comparable importance to the proteins, remains to be discovered. This seems highly unlikely. In the protein molecule Nature has devised a unique instrument in which an underlying simplicity is used to express great subtlety and versatility; it is impossible to see molecular biology in proper perspective until this peculiar combination of virtues has been clearly grasped.*

Perhaps this is why Erwin Chargaff (1997) wrote, "I do not believe that there will be much objection to the claim that biology, in its struggle to become an exact science, has made enormous advances in the last 50 years. To make them, biology has had, however, to abjure its very name-that is, to be the science of life. The more exact it became, the more distant from the state of living-an appraisal that is certain to be rejected by all practitioners of what has become a new scientific discipline, molecular biology, whose very name strikes me as something like ambrosia with garlic."

Proteins are intimately involved in every aspect of cellular life. Proteins form the ion channels and pumps in the membranes; the cytoskeleton and the motors to move organelles through the cell; the enzymes involved in carbohydrate, fat, protein, and nucleic acid metabolism; and energy production. Protein activity is not invariant but can be regulated by varying the rate of synthesis, degradation, and/or substrate availability. Protein activity can also be regulated through the binding of ions or small molecules (e.g., Ca^{2+}, cAMP, GTP) or the covalent modification of the protein (e.g., phosphorylation/dephosphorylation, acylation, glycosylation, prenylation, and acetylation; Gruhler and Jensen, 2006; Angel et al., 2012). These posttranslational modifications cause a change in the charge or hydrophobicity/hydrophilicity of the protein that result in a change in the structure of the protein. These kinds of modifications can control protein activity because the structure of a protein determines its function. It is especially important to understand how the activity of an enzyme that controls the rate-limiting step in a biochemical pathway is controlled (Blackman, 1905; Krebs and Kornberg, 1957). Various types of posttranslational modifications are being studied with proteomic techniques (Chen et al., 2012a,b; Fíla et al., 2012, 2016; Mayank et al., 2012; Chao et al., 2016) (Fig. 17.1).

According to Arthur Kornberg (2000), "Based on the conviction that all reactions in the cell are catalysed and directed by enzymes, the first commandment commands that enzymology can be relied on to clarify a biologic question…The first and crucial step is to find a way to observe the phenomenon of interest in a cell-free system.

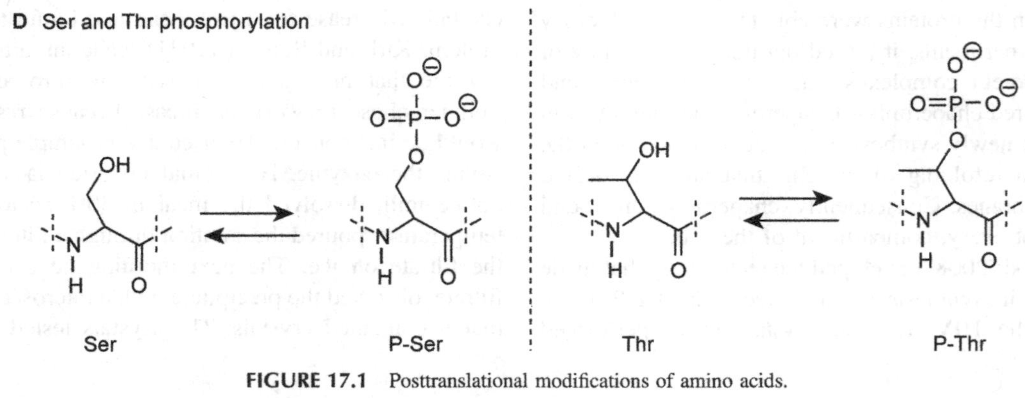

D Ser and Thr phosphorylation

Ser ⇌ P-Ser | Thr ⇌ P-Thr

FIGURE 17.1 Posttranslational modifications of amino acids.

Should that succeed, then one should be able to reduce the event to its molecular components by enzyme fractionation…With a cell-free system in hand that recreates a biologic event, the biochemist should be able to perform the process as well as the cell does it." Successful applications of this commandment in the development of functional assays can be seen in the Nobel Lectures of Eduard Buchner (1907), Otto Meyerhof, (1923), Arthur Harden (1929), Otto Warburg (1931a), Albert Szent-Györgyi (1937a), James Sumner (1946), Hans Krebs (1953), Fritz Lipmann (1953), Arthur Kornberg (1959), Severo Ochoa (1959), Marshall Nirenberg (1968), Erwin Neher (1991), Bert Sakmann (1991), James Rothman (2013), Randy Schekman (2013), Yoshinori Ohsumi (2016) and others. Kornberg (2000) laments that in the age of genomics "Striking phenotypes are produced by mutations and transfections, but the alternations in enzymes are only inferred. Rarely are they verified by the isolation of proteins with demonstrable functions. To many cell biologists and developmental biologists, the need to examine an event in a cell-free system does not come up on the radar screen." Erwin Chargaff (1997) described the situation like so: "Practitioners of this discipline [molecular biology] are forced to use many half-understood techniques, yielding many results they are not really competent to evaluate."

17.4 PROTEIN TARGETING

Once a protein is synthesized it must be targeted to its correct location in the cell (Blobel and Sabatini, 1971; Blobel and Dobberstein, 1975a,b; Schekman, 1985; Chrispeels and Staehelin, 1992; Schatz and Dobberstein, 1996). This intracellular sorting is accomplished by the presence of a signal peptide or a signal patch (Blobel, 1980). A signal peptide is a linear stretch of 15–60 amino acid residues and is often, but not always, removed from the mature protein once translocation into the targeted organelle is completed. Signal peptides direct proteins from the cytosol to the ER, to the vacuole, to the peroxisomes, to the mitochondria, to the chloroplasts, and to the nucleus. A signal patch is a three-dimensional arrangement of atoms on the protein surface that once forms the protein folds up. The amino acids that form this patch may be widely separated from one another in the linear sequence, and they generally remain in the mature protein.

The importance of the individual signal peptides for protein targeting has been demonstrated by putting the signal sequence on another peptide using genetic engineering techniques (Chua and Schmidt, 1978, 1979; Huisman et al., 1978; Garoff, 1985; White and Scandalios, 1988). The targeting of a protein can then be controlled artificially. The import of the protein into the targeted organelle is then assayed using immunoprecipitation, immunocytochemistry, or reporter molecules such as green fluorescent protein, which can be placed into the transported protein using genetic engineering techniques that allow the protein of interest to be autofluorescent (Cutler et al., 2000; Tian et al., 2004; Koroleva et al., 2005; Goodin et al., 2007; Shemer et al., 2008; Thompson and Wolniak, 2008). Such experiments clearly show that the peptide is necessary and sufficient for correct targeting. However, they do not discern whether the primary sequence, the secondary structure, or the chemical or physical properties are important for specificity. A variety of signals may target different proteins to the same compartment (Hunter et al., 2007). Each organelle has its own complex that can recognize and translocate the targeted protein (Soll and Schleiff, 2004; Bédard and Jarvis, 2005; Kalanon and McFadden, 2008; Koenig et al., 2008).

It has become clear that each organelle has a diversity of translocons that facilitate the transport of proteins translated on cytosolic ribosomes into the organelle. What is not clear is how much this diversity depends on the developmental and physiological state of the organelle, and how much it depends on the techniques used for the purification and characterization of the translocon.

17.5 PROTEIN–PROTEIN INTERACTIONS

The binding of or interaction between two polypeptides or two domains of a single protein can be visualized by engineering a fluorescent peptide into each of the two polypeptides of interest. To characterize the proximity or change in proximity of the two polypeptides, which may give an indication of the protein activity, the ability of the second fluorescent peptide to fluorescence in response to excitation of the first fluorescent peptide is assayed (Dixit et al., 2006; Shaw, 2006; Goodin et al., 2007). The transfer of energy between the first and second fluorescent peptide is dependent on the sixth power of the distance between the two fluorescent peptides. The transfer is known as Förster (or fluorescence) resonance energy transfer (FRET). The interaction between two peptides can also be monitored using the bimolecular fluorescence complementation technique, whereby one engineers one part of a bioluminescent protein such as luciferase into one polypeptide and another part of the same bioluminescent protein into another polypeptide and sees if the two regions are close enough to each other to luminescence under the appropriate conditions (Chen et al., 2008).

Protein–protein interactions can also be analyzed by other techniques, including the yeast two-hybrid system and the split-ubiquitin system (Braun and Schmitz, 2006; Liu et al., 2007a,b; Rahim et al., 2008). In the split-ubiquitin system, two parts of ubiquitin are expressed by two different engineered fusion genes. Each gene is an engineered fusion of a part of the ubiquitin gene, a reporter gene, and a gene for a protein of which the interaction with another protein will be tested. If the two proteins interact, then the ubiquitin parts will be brought close enough together to allow the binding of a ubiquitin-specific protease and the release of the reporter protein. The field of computational cell biology is emerging to quantify and describe the complex interaction of proteins within the cell in space and time (Slepchenko et al., 2002).

17.6 PROTEIN DEGRADATION

The rapid turnover of molecules in biological tissues was inferred by Peter Mark Roget (1834) while he was studying regeneration. However, these observations were forgotten. Rudolf Schoenheimer (1942; Kohler, 1977; Clarke, 1941; Quastel, 1942; Kennedy, 2001) introduced stable and rare isotopes of H, O, N, and S into the study of metabolism. These isotopes were made by Harold Urey. Schoenheimer et al. (1939a,b) fed rats with ^{15}N-tyrosine and used a mass spectrometer to analyze where the heavy nitrogen went. Unexpectedly, they found that the heavy nitrogen appeared in every amino acid, except the essential ones, in every tissue that they examined. They concluded that "all constituents of living matter, whether functional or structural, of simple or of complex constitution, are in a steady state of flux." Prior to Schoenheimer's astonishing results, it seemed reasonable that mature living matter was like a machine that used nutritive substances as fuel to run the machine, and the excretions were the part of the food that were not used. After all, most cells and organisms spend most of their life at a constant size. Schoenheimer's work indicated that the living matter was constantly being built up and broken down, and each molecule had a lifetime that was less than that of the cell or organism itself (Cohn, 1989; Hogness et al., 1955; Schoenheimer and Rittenberg, 1938). Schoenheimer (1942) wrote, *"The simile of the combustion engine pictured the steady state flow of fuel into a fixed system, and the conversion of this fuel into waste products. The new results imply that not only the fuel, but the structural materials are in a steady state of flux. The classical picture must thus be replaced by one which takes account of the dynamic state of body structure."* After studying the induction of β-galatocidase in *E. coli*, Hogness et al. (1955; Cohn, 1989) observed that enzyme synthesis is stimulated by an inducer which causes an increase in the amount of the enzyme but any decrease in the amount of the enzyme on the removal of the inducer is only due to dilution by the increasing cell mass. That is, there is no protein turnover. They concluded, *"To sum up: there seems to be no conclusive evidence that the protein molecules within the cells of mammalian tissues are in a dynamic state. Moreover, our experiments have shown that the proteins of growing E. coli are static. Therefore it seems necessary to conclude that the synthesis and maintenance of proteins within growing cells is not necessarily or inherently associated with a 'dynamic state.'"* In the following decades encompassing the golden age of molecular biology, research on the synthesis of proteins made great strides while research on protein degradation was more or less neglected.

Recognizing that the concentration of proteins in a cell was dependent both on degradation and synthesis (Schimke and Doyle, 1970), Avram Hershko, Aaron Ciechanover and Irwin Rose began to study the mechanisms of protein degradation. Hershko (2004) explained, "It was clear to me that the only way to find out how a completely novel system works is that of classical biochemistry. This consists of using a cell-free system that faithfully reproduces the process in the test tube, fractionation to separate its different components, purification and characterization of each component and reconstitution of the system from isolated and purified components." Using this method Hershko (2004), Ciechanover (2004) and Rose (2004) discovered the role of ubiquitin, so named because it was found everywhere (Goldstein et al. 1975), in protein degradation.

All proteins are eventually degraded by proteases, and plants produce hundreds of proteases (van der Hoorn, 2008). The proteases polarize the carbonyl group of the peptide bond of the substrate by stabilizing the oxygen in an oxyanion hole. This makes the carbon atom of the carbonyl group vulnerable for attack by an activated nucleophile (Dunn, 2001; Goldstein et al., 1975; Schimke and Doyle, 1970). If the nucleophile is an amino acid in the proteases, then the protease is named after the nucleophilic amino acid: serine proteases, cysteine proteases, aspartic acid proteases, threonine proteases, and glutamate proteases. The action of proteases can be inhibited experimentally with protease inhibitors. Serine proteases are inhibited by phenylmethylsulfonyl fluoride, 4-(2-aminoethyl)benzenesulfonyl fluoride hydrochloride (AEBSF HCl), aprotinin, and leupeptin; aminopeptidases are inhibited by bestatin; cysteine proteases are inhibited by E−64 and leupeptin; aspartic acid proteases are inhibited by pepstatin; and metalloproteases are inhibited by ethylenediaminetetraacetic acid. Plant pathogens secrete proteases (Fedatto et al., 2006), and plants also produce protease inhibitors to protect themselves from insect attack (Ryan, 1990; Zhu-Salzman and Zeng, 2015). Proteases exist in many organelles, including the ER, nucleus, mitochondrion, and chloroplast, and in the cytosol as well as in the autophagosomal or lysosomal vacuolar compartment (Dice, 1990; Goldberg, 1990). The lysosomal vacuolar compartment is probably involved in the degradation of endocytosed extracellular proteins and intracellular proteins under times of stress (Takeshige et al., 1992; Tsukada and Ohsumi, 1993; Baba et al., 1994, 1995; Moriyasu and Ohsumi, 1996; Matsuura et al., 1997; Mizushima et al., 1999; Kirisako et al., 2000; Klionsky and Emr, 2000; Kihara et al., 2001; Ohsumi, 2001, 2006, 2014; Shintani et al., 2001; Suzuki et al., 2001; Hanaoka et al., 2002; Kuma et al., 2002; Yoshimoto et al., 2004, 2009, 2014; Suzuki et al., 2005;

Inoue et al., 2006; Ishida et al., 2008; Noda and Klionsky, 2008; Nakatogawa et al., 2009; Wada et al., 2009; Mizushima et al., 2011; Shibata et al., 2013).

Selective autophagy allows dysfunctional proteins and organelles, such as chloroplasts that are damaged by UV light or peroxisomes that are damaged by H_2O_2 to be recycled (Wada et al., 2009; Mizushima et al., 2011; Floyd et al., 2012; Li and Vierstra, 2012; Shibata et al., 2013; Michaeli and Galili, 2014; Farré and Subramani, 2016; Anding and Baehrecke, 2017; Izumi et al., 2017). Selective autophagy not only depends on the core autophagy machinery but also requires specific autophagy receptors that recognize the organelle to be degraded and to engage it with the autophagy machinery (see Chapter 7).

Otherwise, most selective degradation probably takes place in the other organelles of the cytoplasm and in the cytosol (Ciechanover and Gonen, 1990; Ciechanover, 2004, Hershko, 2004; Rose, 2004). The various proteases hydrolyze the substrate proteins into free amino acids. As recycled amino acids, they can participate again in protein synthesis. The activities of some proteases are activated by calcium ions (Moriyasu and Tazawa, 1987; Reddy et al., 1994; Lid et al., 2002; Moriyasu and Wayne, 2004).

All polypeptides in eukaryotes are synthesized with a methionine at the amino-terminal end. The methionine is often cleaved by a specific aminopeptidase. The amino-terminal amino acid determines the stability of the cytosolic proteins. If the amino-terminal amino acid is Met, Ser, Thr, Ala, Val, Cys, Gly, or Pro, the cytosolic polypeptide will be stable and long-lived. However, if the amino-terminal amino acid is not one of the above amino acids, the cytosolic protein is targeted to be degraded in the ubiquitin-dependent pathway (Rechsteiner, 1987; Ciechanover and Gonen, 1990). In this pathway, many molecules of the small protein ubiquitin are covalently attached to the target protein. The ubiquitin-linked proteins are then degraded by an ATP-dependent protease that exists in a multienzyme complex known as the proteasome (Arrigo et al., 1988; Goldberg, 1990; Schliephacke et al., 1991; Vierstra, 1993, 2003; Moriyasu and Malek, 2004; Pickart, 2004). Phytochrome is degraded by this pathway (Shanklin et al., 1987). Proteasomes can be experimentally inhibited by bortezomib, lactacystin, and ritonavir.

17.7 STRUCTURE OF PROTEINS

The primary sequence of a protein is determined by a number of constraints. First and foremost, the catalytic portion must bind the substrate and release the product with appropriate binding constants and reaction rates. Secondly, other sites on the protein must either bind or not bind the regulatory molecules that are present in the cell so that the protein is activated or inactivated at the appropriate times. Thirdly, the protein must be soluble in the compartment that it is active in—for example, membrane (lipid), stroma (pH 8), vacuole (pH 5). Lastly, a protein must have the correct targeting sequences to get it to the correct compartment. Any of these constraints may compete with the other constraints, and consequently cells and organisms may have many variations of a given enzyme. Such varieties of enzymes are known as isozymes. Proteins function in the milieu of a given compartment. Consequently, an enzyme that exists in two very different compartments may need different amino acid sequences to perform identical functions, whereas enzymes in the same compartment may perform very different functions, yet have similar amino acid sequences. As far as I can tell, this set of fundamental constraints have not been considered when deducing gene ontologies. Gene ontology (GO) is a major initiative that uses bioinformatics to unify the representation of attributes of genes and gene products across all species (The Gene Ontology Consortium, 2008).

The structure of proteins can be deduced from the X-ray diffraction pattern. Soon after Sirs William and Lawrence Bragg (1915) deduced the structure of sodium chloride and other simple substances, R.O. Herzog and W. Jancke (1921) took the first X-ray diffraction photographs of the protein-rich samples of hair, horn, muscle, silk, and tendon. This work was followed up by W.T. Astbury (1933, 1939, 1941, 1947a), C.H. Bamford, W.E. Hanby, F. Happey, and Max Perutz (Pauling et al., 1955). The structures of many proteins and protein complexes, including the H^+-ATPase (Kühlbrandt et al., 2002), a K^+ channel (MacKinnon, 2003), the reaction center of a photosystem (Deisenhofer and Michel, 1988; Huber, 1988), and a protein-translocating channel (Van den Berg et al., 2004), are being determined to relate their structure to their function and regulation. The 2017 Nobel Prize in Chemistry was awarded to Jacques Dubochet (2017), Joachim Frank (2017) and Richard Henderson (2017) for developing cryo-electron microscopy for the high-resolution structure determination of biomolecules, including proteins, in solution. With cryoEM, it is possible to freeze proteins in mid-movement in order to visualize the sequence of structural changes that occur in the proteins that make life possible.

Some proteins that were once known as "dancing protein clouds" cannot be crystallized because they are intrinsically disordered and lack one inherently stable three-dimensional structure (Dunker et al., 2001, 2008, 2014; Dunker and Uversky, 2010; Sun et al., 2012; Niklas et al., 2015, 2018; Niklas and Dunker, 2016; Uversky, 2016; Oldfield et al., 2017; Yruela et al., 2017, 2018). Because transcription factors and signaling proteins, including many calcium-binding proteins, are *intrinsically disordered proteins* (IDPs), it is likely that it is the temporal dynamics of the three-dimensional structures and not a single three-dimensional structure that is important for the function of these regulatory proteins (See Chapter 12). Of particular interest, the number of IDPs in a taxon is correlated with the number of cell types.

17.8 FUNCTIONS OF PROTEINS

There are typically tens of thousands of genes in an organism that encode proteins with various functions. Currently, there is an emphasis on discovering a vast array of mutants that are unable to undergo a given process and guess the function of the protein encoded by the gene of interest using computer databases. Indeed, as technology advances, the rate of discovery of genes postulated to be directly involved in a given process coming from experiments done in silico seems to double each year or two following a biotechnological analogue of Moore's Law for integrated circuits, which states that the number of components in an integrated circuit will double every year or two (Moore, 1965). However, conclusions about the function of a protein drawn from in silico work must be confirmed with conclusions drawn from functional assays done in vitro.

Throughout this book, I have emphasized the importance of well-defined in vitro functional assays in creating the understanding that we have concerning the activity of a given protein in cellular processes, including membrane transport, plasmodesmatal selectivity, secretion, motility, signaling, photosynthesis, photorespiration, β-oxidation, glycolysis, respiration, replication, transcription, translation, mitosis, cell plate formation, and growth. I look forward to the day when the identification of mutants is routinely complemented with a well-defined in vitro functional assay of the wild-type and mutated-gene products.

Enzymes, such as DNA polymerase, DNA ligase, RNase, and proteases, are produced by nature to perform the functions of life. Such enzymes have been crystalized (Northrop, 1929, 1930; Northrop and Kunitz, 1931; Kunitz and Northrop, 1934, 1936; Dubos, 1937; Dubos and Thompson, 1938; Kunitz, 1939, 1940) and used by biochemists, cytologists, and molecular biologists in many clever ways to probe the nature of life (see Chapter 21), including the determination of the sequences of DNA (Wu and Kaiser, 1967, 1968; Sanger, 1980, 1981), RNA (Holley et al., 1965a,b; Holley, 1968), and proteins (Sanger, 1958).

17.9 TECHNIQUES OF PROTEIN PURIFICATION

The word protein was first used by Jöns Jacob Berzelius and first published by Gerardus Johannes Mulder (1838) to describe the constituents of the first rank (πρώτειος) found in animals and plants (Vickery, 1950; Hartley, 1951). Plant cells, particularly those involved in protein storage, have provided a rich source for purified proteins (Osborne, 1924; Sørensen, 1925; Chibnall, 1939; Synge and Williams, 1990). These studies demonstrated the variety and individuality of proteins. The proteins, which reside in a primarily water-rich protoplasm, were originally classified by their solubilities. The albumins are soluble in water; the globulins are insoluble in water but soluble in salt solutions; the prolamins are insoluble in water, but soluble in 50%–90% ethanol; and the glutelins are insoluble in neutral solutions, but soluble in dilute acid or base (Chibnall, 1939; Bidwell, 1974).

To study proteins, they should be purified to homogeneity. To purify a protein, we must first decide on the source of the protein. Do we want the protein from a particular cell type, or do we want large amounts of protein but do not care which cell type it comes from? Do we think that there may be isozymic differences in the proteins in different cell types, and consequently we must separate the cells before we isolate the protein? Do we want to isolate the protein from a given organelle and therefore must first isolate the organelle? We may also opt to isolate a gene expressed in a given cell type and then express the protein in bacteria using molecular-cloning techniques.

Once we decide on a source, we must solubilize the protein. We must homogenize the organ, tissue, or cell. If we want to isolate a protein from a given organelle, we must use centrifugation and/or aqueous two-phase partitioning to isolate the given organelle or membrane first. While we homogenize the starting material, we must make sure that the protein is stable, so we must include in our homogenization medium a buffer with the correct pH, ion activity, ionic strength, and protease inhibitors. We must perform the purification at temperatures that minimize damage due to proteolysis.

To differentially purify a protein to homogeneity, we can precipitate it, chromatograph it, or ultracentrifuge it (Svedberg, 1937; Sumner and Somers, 1943). Starting in 1926, proteins were purified to homogeneity by finding, with great patience, the conditions that allowed them to crystallize out of a solution (Northrop et al., 1948). It is again becoming popular to crystallize proteins and apply biophysical techniques to determine the structure of the protein at atomic resolution. Remember, to ensure that we are isolating the intact protein, we must perform a functional assay at each stage of purification (Racker, 1985; Li et al., 2005).

The field of proteomics strives to identify all the proteins in a cell, tissue, or organ at a given point in development or following a treatment (Patterson, 2004a,b; Heazlewood and Miller, 2006; Isaacson and Rose, 2006; Rajjou et al., 2006; Sun et al., 2009; Zou et al., 2009; Li et al., 2012, 2014, 2016; Chaturvedi et al., 2013). Complex mixtures of hundreds to thousands of proteins can be analyzed by mass spectrometry and their identity determined from the protein sequence predicted by data sets produced as a result of genome sequencing. The proteomic approach has the advantage of listing all the proteins in a given compartment that may interact with each other to accomplish a given function. While the various types of mass spectrometers used for proteomic studies are excellent in identifying proteins, particularly in plants of which the genome has been sequenced, it is a fundamental truism that to get meaningful data from the analysis, it is necessary to solubilize all the proteins of interest, to protect them from degradation, and to prevent proteins present in unwanted compartments from contaminating the sample.

17.10 PLANTS AS BIOREACTORS TO PRODUCE PROTEINS FOR VACCINES

Certain plants can annoy various people as a result of the production of proteins that function well for the plant but act as allergens that induce an immune response in people (Cosgrove et al., 1997; Yennawar et al., 2006; Valdivia et al., 2007a,b). Charles Arntzen has long thought of taking advantage of the ability of plants to synthesize immunogenic proteins by using plants as bioreactors to produce vaccines against viruses that plague humans and animals. Arntzen and others have transformed plants and plant cells with genes that encode proteinaceous antigens against a number of viruses, including the Newcastle virus and the hepatitis B virus. The proteinaceous antigens produced by the "plant cell factories" have been purified from whole plants and cultured plants cells and used to immunize humans and animals (Mason et al., 1992, 2002; Raskin et al., 2002; Ma et al., 2003; Walmsley and Arntzen, 2003; Goldstein and Thomas, 2004; Arntzen et al., 2005; Sunil Kumar et al., 2005; Twyman et al., 2005; Vitale and Pedrazzini, 2005; Kim et al., 2007; Pascual, 2007; Santi et al., 2008).

17.11 SUMMARY

Richard Vierstra (2003) describes the life of a protein in terms of birth, taxes, and death. This chapter follows the life history of proteins from their synthesis (birth), through the regulation of their activity (taxation), to their degradation (death), and even to their isolation and characterization. I have described the experiments that showed that nucleic acids act as a template for protein synthesis; the experiments showing that RNA may be an intermediate between DNA and protein synthesis; the experiments done to determine the mechanism of protein synthesis; and the experiments done to show that protein synthesis takes place in ribosomes. I have also shown that the expression of a single gene product, that is, a single polypeptide, requires the integrated action of many proteins, and by extension, the coordinated action of many genes. Thus, a change in the sequence or expression of a single gene could influence the correct expression of many other genes. These experiments, which were aimed at understanding the "language of life," involved the work of many people.

The physicists who entered genetics, including Francis Crick and George Gamow, like the colloidal chemists before them, did not take the work and perspective of the cytologists and licensed biochemists seriously. This resulted in the creation of untenable hypotheses and many missed opportunities for advancement on previous data (Cohn, 1989). It seems to me that to understand the "language of life," people trained in many disciplines must respect each other and speak a common language.

17.12 QUESTIONS

17.1. What is the evidence that proteins are involved in the functions of the cell mandated by the genes?

17.2. What are the limitations about thinking of proteins as the only catalysts in the cell?

17.3. Why are there so many amino acids, and what are the properties of the amino acids that are important for protein function, targeting, interaction, etc.?

The references for this chapter can be found in the references at the end of the book.

Chapter 18

The Origin of Life

I am inclined to look at everything as resulting from designed laws, with the details, whether good or bad, left to the working out of what we may call chance. Not that this notion at all satisfies me. I feel most deeply that the whole subject is too profound for the human intellect. A dog might as well speculate on the mind of Newton.— Let each man hope & believe what he can.—

Charles Darwin (1860a).

When I was going into science, people were concerned with questions of where we came from. Some people gave mystical answers—for example, 'the truth came from revelation.' But as a college kid I was influenced by Linus Pauling, who said, 'We came from chemistry.'

James D. Watson (1992).

We dance round in a ring and suppose,
But the Secret sits in the middle and knows.

Robert Frost, "The Secret Sits."

Time goes, you say? Ah no!
Alas, Time stays, we go;

Henry Dobson, "The Paradox of Time."

One of the attributes of life is the ability to store and express hereditary information. In Chapters 16 and 17, I discussed DNA, RNA, and proteins—three classes of molecules that are necessary for the storage and expression of genetic information. In this chapter, I discuss the origin of life primarily in terms of the origin of these three classes of molecules, which make up the genetic apparatus. The overriding philosophical question is: does order come from disorder, an idea that is familiar to physicists, or does order come from order, an idea that is familiar to cell biologists (Manton, 1945; Schrödinger, 1946; Harold, 2014) and theologists (Ravi Zacharias, *Let My People Think*; https://rzim.org/)?

18.1 SPONTANEOUS GENERATION

Where did life come from, and how can we answer this question? The obvious answer is to look in the Bible, although we would not get a clear answer as two alternative creation stories are given in Chapters 1 and 2 of *Genesis* (Airy, 1876; Paine, 1880). We could also turn to the writings of the great philosophers; for example, Aristotle, who synthesized the teachings of the times into a theory of life that envisioned that living beings can either come from other living beings or be formed spontaneously. Aristotle believed that plants originated spontaneously from the earth: Frogs and mice sprang up from mud; fireflies came from the morning dew; and mosquitoes, maggots, flies, fleas, bed bugs, and lice came from manure, the slime of wells, human excrement, decaying meat, and other filth. Actually, this conclusion is supported by casual observations of the world (Virgil, 1952). Even van Helmont had a recipe for producing mice by combining human sweat with wheat germ and leaving them alone in a jar for 21 days (Oparin, 1938; Fothergill, 1958; Horowitz, 1958). However, in 1668, Francesco Redi saw things differently. Redi showed that maggots did not appear in meat when he placed the meat in a jar, and carefully covered it with muslin. In fact, he noticed that maggots did not arise spontaneously but only developed when flies were allowed to lay their eggs on the meat.

The belief in spontaneous generation of large plants and animals began to wane throughout the 17th and 18th centuries, due in part to observations on sperm by Antony van Leeuwenhoek (1678, 1699a,b, 1700, 1701; Cole, 1930; Castellani, 1973; Ruestow, 1983, 1996) and on embryo development by William Harvey, Marcello Malpighi, and Pierre Louis Moreau de Maupertuis (Maupertuis, 1753; Needham and Hughes, 1959). However, with the discovery of microbes by Leeuwenhoek (1677), the belief in the spontaneous generation of microorganisms became the standard belief because the microbes seemed to appear out of nothing (Farley, 1977). The apparent spontaneous generation of microorganisms was confirmed experimentally when John Needham (1749) boiled mutton gravy, stoppered it, and found that microbes grew in the boiled broth.

Lazzaro Spallanzani (1769, 1784) repeated Needham's experiment and showed that if you boiled chicken broth *extensively* before you stoppered it tightly, microbes would not appear in the broth. They only appeared after the stopper was opened. Thus, it appeared that microbes only seemed to arise spontaneously because they were ubiquitous. They were either already in any preparation that had not been

Plant Cell Biology. https://doi.org/10.1016/B978-0-12-814371-1.00018-7

properly sterilized or were capable of contaminating any preparation that they could enter. Spallanzani's supporters believed that he had shown that spontaneous generation was impossible, whereas Needham's supporters believed that Spallanzani had only shown that microbes need air.

In the middle of the 19th century, Louis Pasteur performed the critical experiment. With his now famous swan-shaped flasks that allowed air, but not microbes, to pass, Pasteur showed that as long as a solution is properly sterilized (e.g., pasteurized) and airborne contaminants excluded, no microbes were generated in the broth, even when air was able to freely pass through the long neck. He concluded that there is no such thing as spontaneous generation of microbes (Dubos, 1960). John Tyndall (1898) showed that filtration also prevented the "spontaneous" appearance of organisms (Arrhenius, 1908).

If living organisms cannot originate spontaneously, and the early Earth was a molten ball incapable of supporting life, then how did they originate on Earth? Some scientists, including H. E. Richter, Lord Kelvin (King, 1925), Hermann von Helmholtz (1881), Svante Arrhenius (1908), Fred Hoyle (1983; Hoyle and Wickramasinghe, 1981; Steele et al., 2018), and Francis Crick (1981), realizing that no one has yet created life in the laboratory, suggested that life cannot be created but must come from existing life. If life can only originate from life, then life on Earth must have originated in outer space and come to Earth on meteorites in the form of cosmozoa, microbes, spores, or seeds. This theory is called *panspermia*, which means seeds everywhere. Arrhenius (1908) wrote, "The Universe in its essence has always been what it is now. Matter, energy, and life have only varied as to shape and position in space." This sounds very much like conservation of life! After an extensive review of the observational and experimental data, Steele et al. (2018) conclude that "*life may have been seeded here on Earth by life-bearing comets as soon as conditions on Earth allowed it to flourish (about or just before 4.1 Billion years ago); and living organisms such as space-resistant and space-hardy bacteria, viruses, more complex eukaryotic cells, fertilised ova and seeds have been continuously delivered ever since to Earth so being one important driver of further terrestrial evolution which has resulted in considerable genetic diversity and which has led to the emergence of mankind.*"

Even if the panspermia theory is true, we are still faced with the question of how living organisms originated in the universe. So although it is possible that life on Earth originated on another planet in another solar system or another galaxy, I will use Occam's razor and assume that life on Earth originated from lifeless matter on Earth itself. This does not mean that life did not also arise from lifeless matter elsewhere in the universe, and the arguments I make apply to the origin of life anywhere (Muller, 1973a; Levy et al., 2000).

18.2 CONCEPT OF VITALISM

The notion that life arose from lifeless matter will seem more outrageous to many readers than the idea of special creation, as there seems to be an enormous gap between living and nonliving matter. However, as I discussed in Chapter 1, it is impossible to devise a classification system to separate the living from the lifeless. In this book, I present the materialist view that there is a continuum between lifeless matter and living beings but personally subscribe to the dualist view that recognizes that all effects require a cause, and as far as scientists have observed, lifeless matter has never been able to give rise to living matter. I do recognize that artificial classification systems designed to separate animate from inanimate molecules generally have been found wanting. For example, in the past century, chemists thought that the organic molecules present in living beings were different from the inorganic molecules present in rocks. The term *organic* was originally coined by Torbern Bergman in 1780 to indicate one of the three main classes of organizations based on complexity:

- The first class was composed of minerals, a relatively simple group.
- The second class was composed of the pure chemicals extracted from living organisms. These chemicals were considered more complex than minerals, but not living.
- The third class, designated organic, included the constituents of the fluids and tissues of plants and animals.

Jöns Jacob Berzelius, a leading authority in chemistry at the beginning of the 19th century, believed that organic compounds had properties that were determined by the *vital force*, a term coined by Friedrich Medicus in 1774. The concept of a vital force has arisen many times throughout history to explain how organic molecules were made. Galen called it the *pneuma* or "breath of life." Aureolus Philippus Theophrastus Bombastus von Hohenheim, who called himself Paracelsus, called it the "vital spirit" or *Archeus*. To distinguish chemical reactions that take place in solution or in minerals (and depend solely on electrical forces) from those reactions that take place in living organisms (and depend on the vital force), Berzelius divided chemistry into two branches: inorganic and organic.

On account of the requirement for a vital force for the synthesis of organic compounds, Berzelius believed that inorganic compounds could be prepared in the laboratory, but organic compounds could only be formed in living organisms. Berzelius wrote (quoted in Needham, 1971),

> *In living, as compared with inanimate nature elements appear to obey quite different laws. The products of their mutual reactions are quite different from those in the sphere of inorganic nature. The discovery of the cause of this*

difference of the behaviour of the elements in living and non-living nature would furnish the key to the theory of organic chemistry.

In 1828, Friedrich Wöhler synthesized urea from ammonium cyanate. He was clearly excited about this result and recognized the importance of this synthesis in relation to vitalistic thought when he wrote to his colleague and former teacher, Berzelius "I live in the daily, nay hourly, intense hope of receiving a letter from you, however I will no longer wait but instead write you again because I cannot, as it were, hold my chemical water and must tell you that I can make urea, without a kidney and without any animal be it human or dog" (see Wallach, 1901). Of course, the urea was actually synthesized by a living organism, Wöhler himself! Almost 50 years later, August Hofmann (1876) wrote of this work

Wöhler had demonstrated the possibility of building up from its elements this very urea, the formation of which, up to that period, had been supposed to take place exclusively under the influence of vitality—an experiment very memorable, since it removed at a single blow the artificial barrier which had been raised between organic and inorganic chemistry.

Louis Pasteur realized that many natural organic molecules, including sugars, starch, quinine, morphine, and tartaric acid, were optically active, whereas many inorganic molecules and organic molecules synthesized in the laboratory were not (Findlay, 1948). Inspired by this realization, Pasteur proposed a new distinction between living and nonliving matter (Meierhenrich, 2008). He proposed that only living organisms could synthesize asymmetric molecules and that optically active molecules could never be synthesized in the laboratory. Pasteur said, "This important criterion [molecular asymmetry] constitutes perhaps the only sharply defined difference which can be drawn at the present time between the chemistry of dead and of living matter" (quoted in Dubos, 1960). However, this idea was dropped when, in 1860, William Henry Perkin and Baldwin Francis Duppa synthesized racemic acid, a mixture of optically active forms of tartaric acid from optically inactive succinic acid (Hudson, 1992). Later, Vladimir Vernadsky (1944) proposed that living organisms could be distinguished from nonliving matter by their ability to discriminate among different isotopes. This classification too was shown to be artificial when scientists found that they could separate the isotopes using physical techniques, including diffusion, centrifugation, and mass spectrometry (Urey, 1939). Assuming that there are no clear distinctions between living and lifeless matter, I will discuss how living matter may have arisen from nonliving matter on Earth. I will start at the very beginning, with the origin of the universe.

18.3 THE ORIGIN OF THE UNIVERSE

Our present idea of the origin of the universe is intimately connected with our concept of its size. In Aristotle's time, it was believed that Earth was the center of a spherical universe (White, 1913; Thiel, 1957). In the third century BCE, Eratosthenes measured the diameter of Earth by noticing that during the summer solstice, the rays from the sun went straight down a well in Syene, whereas at the same time, about 500 miles away in Alexandria, the sun's rays struck the side of a well, making a 7-degree angle with the vertical. This angle was too large to be explained by a flat Earth, in which case the sun's rays would hit Alexandria at approximately the same angle that they hit Syene. Assuming then that Earth was spherical, and the circumference is equal to both πd and 360 degrees, Eratosthenes set up the relationship: 7 degrees/360 degrees $= 500/\pi d$ to determine the circumference and diameter of Earth. He found that the circumference was approximately 25,000 miles, and the diameter was approximately 8000 miles (Asimov, 1971). Except for the writings of a few minor clergymen and the members of the Flat Earth Society, knowledge that Earth was round was commonplace from ancient times through the present (White, 1877, 1913; Draper, 1898; Russell, 1991; Principe, 2006; Garwood, 2008).

Using the diameter of Earth and the rules of trigonometry, Hipparchus, in the second century BCE, estimated the distances of the sun and moon from Earth to be 1210 and 59 times the radius of Earth, respectively (Hirshfeld, 2001). However, without sensitive screw micrometers (which were invented by Joseph Fraunhöfer in 1820) to measure the tiny angles that occur when measuring large distances, it was not possible to measure the distance to the stars, which all seemed to be equally far from Earth. Thus, the ancient Greeks believed that the universe was a finite sphere where the edge consisted of a layer that contained the fixed stars (see Ptolemy, 1984).

Nevertheless, in the 17th century, Isaac Newton (1728) and Christiaan Huygens (1762, in Munitz, 1957) independently made attempts to measure the distance to the stars by using the Principle of Uniformity of Nature and provisionally assuming that the sun was a star and the distant stars had the same intrinsic brightness as the sun. They assumed that the stars only appeared dimmer than the sun due to their distance. Using the inverse square law, which states that the apparent brightness decreases with the square of the distance, Huygens estimated that Sirius, the brightest star, is 28,000 times farther from Earth than the sun is. (This was an overestimate because Sirius has an intrinsic brightness that is about 25 times as bright as that of the sun.) Throughout the 18th century, telescope designs improved, and astronomers, including William Herschel, began to see more distant stars, which had not been seen with the older telescopes. With each improvement of the telescope, the known universe became larger and larger

(Shapley, 1926; de Sitter, 1932). By the 19th century, technology had advanced sufficiently to allow Friedrich Bessel, Thomas Henderson, and Otto Struve to measure the distances to the closest stars using triangulation techniques by relating the apparent positions of the stars to the diameter of Earth's orbit (Jeans, 1930).

Our present understanding of the size of the universe came unexpectedly from the study of Cepheids, which are stars of which the brightness varies in a regular and periodic manner. While Harlow Shapley was studying the spectra of these variable stars in the beginning of the 20th century, he noticed that the spectra were blue-shifted when the Cepheid was at its brightest and red-shifted when it was at its dimmest (Shapley, 1943). Normally the spectrum of stars is red-shifted when a star is moving away from us and blue-shifted when it is approaching. The Cepheids seemed to do both. Shapley thought that perhaps it was more likely that the Cepheids were expanding and contracting in a regular manner, a behavior that would give the observed patterns of spectral shifts. This is in fact what happens. As an aside, the variable light originating from pulsars is produced by a different mechanism involving a rapidly spinning neutron star that emits electromagnetic radiation in jets from its poles (Gold, 1968, 1969; Hewish et al., 1968; Sullivan, 1964; Harrison, 1981; Taylor, 1993; Hulse, 1993; McNamara, 2008).

In 1908, Henrietta Leavitt discovered a relationship between the average apparent brightness of a group of Cepheids in the Magellanic Clouds and the periodicities with which they winked (Leavitt, 1908; Pickering, 1912; Johnson, 2005). That is, the greater the average apparent brightness of the Cepheid, the longer the period. Because all the Cepheids in the Magellanic Clouds are approximately the same distance from Earth, Ejnar Hertzsprung assumed that the period of each one was probably related to its intrinsic brightness. Hertzsprung studied Cepheids that were closer to Earth as determined from published records of their apparent tangential movement. He compared the apparent brightness of the near and far Cepheids with similar periods. By assuming that the intrinsic brightnesses of Cepheids with the same period are the same, he could estimate the distance to the Cepheids in the Magellanic Clouds by assuming that the difference in the apparent brightness of Cepheids with similar periods was due to a difference in their distance from Earth (Leuschner, 1937). Using the same method, Shapley (1943) mapped the galaxies in the universe by using the period–distance relationship of Cepheid stars found in the various galaxies and concluded that the universe was very large.

Meanwhile, Vesto Slipher measured the spectral shifts in a number of galaxies. Using the Doppler Principle, he concluded that, as a rule, the galaxies were receding from Earth at tremendous velocities. Edwin Hubble (1937) noticed that the recession velocities of galaxies were proportional to their distance from Earth and concluded that the universe was expanding. This result was consistent with one formulation of the Theory of General Relativity (de Sitter, 1932; Einstein, 1961). Hubble determined the constant of proportionality between the recession velocity and the distance from Earth. The proportionality constant, which is known as *Hubble's constant*, is 1.58×10^{-18} km s^{-1}/km. If we assume that prior to the large-scale expansion of the universe all the galaxies were clumped together, we can estimate the age of the universe from the reciprocal of Hubble's constant. Using this estimate, the universe began 20×10^9 years ago. Actually, the relationship between recession velocity and distance is not linear and is influenced by the mass-density of the universe, a number that is not known with certainty (Guth, 1997). In addition, it is difficult to measure both the recession velocity and the distance of a given galaxy, and consequently, the Hubble constant is continually being refined (Ferris, 1997). Taking everything into consideration, the best guess for the age of the universe is 13.8 billion years (Planck Collaboration, 2015). This is far older than the estimates made by biblical scholars including Moses Maimonides, and Archbishop Ussher, who determined, by tracing back the lineages given in various ancient translations of the Bible(s), that the universe was created between 6000 and 4000 BCE (many authors, 1747; Lamarck, 1802; Wallace, 1844; Ussher, 1658, 1864). According to John Lightfoot, another biblical scholar, "heaven and earth, center and circumference, were created together" and "this work took place and man was created by the Trinity on the 23rd of October, 4004 BC, at nine o'clock in the morning" (see White, 1913).

How was the universe formed? The current consensus among cosmologists is that 13.8 billion years ago, space and time, as well as all the matter and energy contained in the universe, came into being in one gigantic explosion. This theory, proffered by George Gamow, Ralph Alpher, and Robert Herman, is called the *Big Bang Theory* (Alpher et al., 1948; Gamow, 1952; Marateck, 2008), a moniker given by Fred Hoyle to mock this cosmological creation theory. Hoyle believed that the universe was eternal and thus could not have had a beginning. Although I will only discuss the Big Bang Theory, which is strongly supported by the discovery of the cosmic microwave background radiation (Penzias and Wilson, 1965; Peebles, 1993), there is an alternative cosmological theory, known as the *Steady-State Theory*, based on the idea of continuous creation (Hoyle, 1953; Bondi, 1960; Alfvén, 1966; Bondi et al., 1995).

According to the modern version of the Big Bang Theory (Lederman and Teresi, 1993; Guth, 1997), at time zero, the universe exploded from an infinitesimally tiny and infinitely hot point. There was only one thing, and this force/particle traveled in this infinitesimally small space with an energy of kT, where T equals infinity. Of course, this infinitesimally small point could not exist inside anything because that would be something. So according to the Big Bang Theory, and *Genesis* for that matter, in the beginning there was a unity, a singularity, a primeval atom (Lemaître, 1931, 1950). Some people may call it *God*, others *love, intelligence, the spirit of life*, or *consciousness* (Spinoza, 1492; Ray, 1691; Grew, 1701; Sartres, 1946, 2007; Kafatos and Nadeau, 1990; Zacharias, 2008; Marks et al., 2013). It is also called the *unified field*, for which Albert Einstein spent his life searching. By redefining "nothing" in terms of special relativity and quantum mechanics, Lawrence Krauss (2012) considers that the singularity, which is the product of chance and quantum uncertainty, comes from nothing. If essence is prior to existence (Plato, 1892; Gilson, 1956), then according to Krauss, nothing would be the essence of existence. This view is consistent with the uncertainty principle but contrary to the first law of thermodynamics. Whatever we call the first cause, the violent explosion caused the universe to expand, and as a consequence of the expansion, the universe began to cool. Approximately 10^{-43} seconds after the Big Bang, the universe had already cooled to 10^{32} K. At this temperature, the single particle in the universe no longer had enough energy to prevent its splitting into two particles, and when it split, there was not enough energy (kT) to fuse the two split particles together (Weinberg, 1979). Thus, the particles that carry the gravitational force separated from the particles that carry the grand unified (GUT) force. This "phase change" is not so different from the changes that occur when water separates from steam or ice separates from water. The energy of a given particle is typically expressed in electron volts (eV). The energy of a particle can be related to temperature with the following identity:

$$eV = kT \tag{18.1}$$

The universe continued to expand, and by 10^{-35} seconds after the creation of the universe, the temperature cooled to 10^{26} K, which is low enough to allow the separation of the GUT particle into particles that carry the electroweak and the strong force. Ten nanoseconds after the creation of the expanding universe, the temperature cooled to 10^{16} K, and the particles that carry the electroweak force separated into particles that carry the weak force and the electromagnetic force. Between 1 μs and 1 ms after the creation of the universe, the temperature cooled to 10^{11} K, and at this lowered temperature, it became possible for protons, neutrons, electrons, and other elementary particles and their antiparticles to persist. Determining the behavior of particles at various temperatures after the Big Bang comes from extrapolating from the results of experiments done in particle accelerators that mimic these early conditions. The times are obtained from estimations of the rate of expansion of the universe. If we assume that the first law of thermodynamics was true at the Big Bang, then we can see clearly how gravitational, electrical, and nuclear energy are convertible.

Three minutes after the Big Bang, the universe cooled to 10^9 K, which allowed the formation of hydrogen and helium nuclei (Alpher et al., 1948; Schramm and Turner, 1998). Even at this temperature, the nuclei in this plasma collided with each other with energies (kT) that were too high to allow the formation of atoms. Ten thousand years after the creation of the universe, the universe cooled to 10^4 K, which allowed the association of electrons with nucleons, and hydrogen and helium atoms were formed. Thus, in accordance with Einstein's equation (Rothman, 2003)

$$E = mc^2 \tag{18.2}$$

energy was converted into mass as we know it (Einstein, 1946). Currently, the temperature of the universe is about 3 K (Penzias and Wilson, 1965; Peebles, 1993; Smoot and Davidson, 1993; Mather and Boslough, 1996). The current laws of physics suggest that universe is a four-dimensional space—time continuum in which matter is a minor component and dark matter and dark energy preponderate. By taking friction into consideration and by assuming that the photon is not a mathematical point but has extension, I have identified dark matter as antimatter (Wayne, 2015d, 2016c), dark energy as a decrease in light energy (Wayne, 2016c), and gravity as a force (Wayne, 2012d,e, 2015b, 2017). In this way, the universe can be known in terms of better known entities instead of lesser known entities.

About 10.5—12 billion years ago, the atoms began to coalesce into dense areas as a result of gravitational attraction. The aggregation of these atoms gave rise to stars and galaxies. As the atoms in the stars were pulled together as a result of gravitational attraction, the gravitational energy was transformed into radiant energy, and the masses of helium and hydrogen ignited to become glowing stars. The high temperatures and pressures developed inside the stars provided the energy necessary to fuse the hydrogen into helium and other light elements, including carbon, nitrogen, oxygen, sulfur, and phosphorous—the elements so important for life (Hoyle, 1979). The fusion of hydrogen into helium, one of the reactions that causes sunlight, takes place in a cyclic reaction involving the catalytic participation of carbon, nitrogen, and oxygen (Prout, 1815, 1816; Bethe and Critchfield, 1938; Bethe, 1939, 1968, 2003; Weizsäcker, 1949; Brock, 1985). Eventually these first-generation stars exploded, sending fragments of dust into the universe. The energy of the explosion formed the heavier elements, including Na, Mg, Ca, Fe, and Co, which were spread over the universe in the form of cosmic dust (Oparin, 1964; Salpeter, 1974, 2002; Mason, 1992). Very heavy elements like platinum and gold might have been made by the collision of neutron stars (Barish, 2017)

Approximately 4.6 billion years ago, on the edge of a spiral galaxy known as the Milky Way, a rotating cloud of gas and dust known as a *nebula* collapsed and began to spin faster and faster, just like a figure skater does, according to the law of conservation of angular momentum. The center of the cloud became so massive and dense; it collapsed under gravitational pressure and ignited the gasses within it to form a star, which we call our sun. Around the sun, other dust particles clumped together into what we now call the *planets*. One of these clumps formed our home planet, Earth (Urey, 1979; Struve, 1962; Cameron, 1979; Kelley, 1988).

The age of Earth was estimated in the past few centuries by estimating how long it would take a fresh water ocean to become salty, given the present salt concentration and the rate of salt transport to the oceans by estimating how long a molten body would take to cool given the diameter of Earth and the temperature difference between the Earth and sky, or by estimating the time it would take for the sediments carried by rivers to reach their current heights, given the rate of sediment flow (Lyell, 1872; Kelvin, 1897; Rutherford, 1909; Jeans, 1930; Lindley, 2004). Currently, the age of Earth is determined by comparing, in ancient rocks, the quantity of radioactive isotopes with the quantity of their decay product (Jeans, 1930; Libby, 1952; Dalrymple, 1991; Lewis, 2000). For example, potassium-40 decays to argon-40 by electron capture with a half-life of 1.3×10^9 years, and an analysis of the ratios of ^{40}K to ^{40}Ar in meteorites found on Earth indicates that their age is between 4.5 and 4.6 billion years old. Analysis of ^{87}Rb and ^{87}Sr ratios from lunar rocks gives approximately the same age. Analysis of $^{142}Nd/^{144}Nd$ ratios that result from the radioactive decay of ^{146}Sm to ^{142}Nd indicates that the oldest known rocks on Earth are 4.28 billion years old (O'Neil et al., 2008). Hence, the entire solar system is approximately 4.3–4.6 billion years old.

18.4 GEOCHEMISTRY OF THE EARLY EARTH

Four and a half billion years ago Earth was becoming fully formed, although it was extremely hot and essentially oceanless and atmosphereless. Heat was primarily generated by radioactive decay, although some heat may have been due to gravity, pulling Earth's components together. As a result of gravity, the dense nickel and iron sank to the core, whereeas the lighter rocky materials containing aluminum, silicon, calcium, magnesium, sodium, and potassium floated to the surface (Siever, 1979). Earthquakes, volcanism, and impacts caused gasses in rocks to be released, probably producing an atmosphere of H_2O, CO_2, N_2, as well as CO, CH_4, NH_3, and H_2S. The gravitational attraction of Earth was not great enough to hold onto the lightest elements, including H_2 and He_2, and thus most of the original atmosphere of hydrogen and helium was lost (Mendeleev, in Oparin, 1924). The loss of hydrogen does not mean that the atmosphere became oxidizing because there was no molecular oxygen in the atmosphere yet. The accumulation of molecular oxygen (O_2) only occurred after the origin of life and the evolution of the photosynthetic mechanisms. It is still a mystery whether or not the early atmosphere was oxidizing, reducing, or something in between.

Water from outgassing reacted with CO_2 in the air to produce carbonic acid. Returning to Earth as rain, the carbonic acid probably leached Ca^{2+} and Mg^{2+} from rocks and formed limestone and dolomite. In this way, the CO_2 was removed from the atmosphere and precipitated in sediments. Atmospheric CO_2 would have acted as a greenhouse gas to keep the early Earth warm; thus, knowledge of the CO_2 concentration would be useful in determining the climate (Arrhenius, 1908). While the actual concentration during the formation of Earth is not known, the amount of CO_2 in the atmosphere was determined by the balance between outgassing and precipitation (Walker, 1983; Kasting and Ackerman, 1986).

From the formation of Earth 4.6 billion years ago until approximately 3.8 billion years ago, Earth may have been bombarded with meteorites or fragments of rocks that were not included in the initial process of planet formation. Any one of these impactors may have hit with so much energy that it would have vaporized any organic molecules or living organism that may have already formed (Sleep et al., 1989). The idea that Earth was hot at this time is currently being challenged by investigators looking at zircon formed during this period that would not have been able to form if Earth were hot (Hopkins et al., 2008). It is not clear whether the zircons formed during this time represent microenvironments on Earth or generalized conditions. Thus, if Earth were hot from 4.6 to 3.8 billion years ago, attempts at the creation of life, at least in locations far from the zircons, would have been frustrated by the enormous energy provided by the impactors, and life neither could have formed nor continued (Sleep et al., 1989). According to Tommy Gold (1992, 1999), thermophilic microbes may have originated in the igneous rocks deep within a hot planet using the abundant quantities of primordial methane (natural gas) as an energy source. Gold also contends that the CO_2 present in the early earth atmosphere was more likely a product of oxidation by these thermophilic microbes using methane as a food source than the abiotic source of carbon for the bottom of the food chain. In fact, Gold suggests that the remains of these methane-fueled chemotrophic microbes or the molecules they secreted may be the source of deep oil deposits.

Some of the oldest known rocks, which are 3.5 billion years old, formed on Earth contain fossils that resemble cyanobacteria and stromatolites (Schopf, 1992; Tice and Lowe, 2004). Thus, prokaryotic-like cells evolved between 3.8 and 3.5 billion years ago, only 300 million years after what may have been repeated sterilizations of the planet by impactors from space. Eukaryotic cells originated approximately 1.4 billion years ago.

18.5 PREBIOTIC EVOLUTION

Life, as we know it, requires carbon-containing compounds, and we must ask: what was the source of the organic compounds that made up the first life on Earth? It is possible that organic compounds, including urea, formaldehyde, amino acids, purines, sugars, etc., came from asteroids, comets, or meteorites, and these compounds have been found in meteorites (Chamberlin and Chamberlin, 1908; Kvenvolden et al., 1970, 1971; Cronin et al., 1988; Chyba, 1990a,b; Chyba et al., 1990; Chyba and Sagan, 1987, 1988, 1992; McKee, 2005), and cyanide and acetylene have been found by NASA's Spitzer Space Telescope surrounding a star named IRS 46 in a galaxy 3 billion light years away from Earth (*www.nasa.gov/vision/universe/starsgalaxies/spitzer-20051220.html*). However, according to Matthias Schleiden (1853), Charles Darwin, Eduard Pflüger (see Arrhenius, 1908), and Frederick Engels (1934), it is likely that prebiotic chemical evolution took place on Earth. The idea of prebiotic chemical evolution was taken up by Jacques Loeb, who in 1916 proposed that proteins could be produced from CO_2 and nitrogen.

Alexander Oparin (1924) strongly pushed the idea of prebiotic evolution, perhaps due to its consistency with the socialist philosophy of dialectic materialism that has no place for God. John Burdon Sanderson Haldane, another socialist, felt prebiotic evolution was also the mechanism that best explains the origin of life. Haldane (1929) wrote,

> *Now, when ultra-violet light acts on a mixture of water, carbon dioxide, and ammonia, a vast variety of organic substances, including sugars and … proteins are built up. … In this present world, such substances, if left about, decay—that is to say, they are destroyed by micro-organisms. But before the origin of life they must have accumulated till the primitive oceans had reached the consistency of hot dilute soup.*

"Impressed by the precise way in which all biological phenomena, when carefully investigated, turned out to be in accordance with physical laws, including those of chemistry, and not to involve any special vital principles," John Bernal (1951, 1962, 1967), who was also a socialist interested in the material basis of life, turned his interest in using X-rays to study crystal structure to the idea that in prebiotic times, a self-replicating inorganic crystal may have given rise to life itself.

In 1951, experiments on prebiotic evolution began when Melvin Calvin and his associates succeeded in fixing carbon dioxide into a more reduced, organic form. They irradiated a mixture of water and carbon dioxide in a closed chamber with a helium ion beam from a cyclotron. This resulted in the formation of formic acid and formaldehyde (Fig. 18.1). Formic acid, like all organic acids, has the structure RCOOH, where R = H, and formic acid is the first acid in a homologous series. Formaldehyde, like all aldehydes, has the structure RCHO, where R = H, and formaldehyde is the first aldehyde in the homologous series. This experiment represented the first-accepted production of organic molecules under presumed prebiotic conditions (Garrison et al., 1951; Fox, 1959; Miller and Urey, 1959b; Calvin and Calvin, 1964; Calvin, 1969; Fox and Dose, 1972; Miller and Orgel, 1974).

At about this same time, Harold Urey (1952a,b), who had been studying the atmosphere of Jupiter, wrote that the atmosphere of the early Earth, like that of Jupiter's, may have been reducing and thus may have consisted largely of hydrogen, methane, ammonia, and water. Indeed, a preponderance of the reduced forms of molecules would have more chemical energy than a preponderance of their oxidized forms, and the additional chemical energy would promote synthetic reactions. Urey suggested that Calvin's experiment be repeated using a reducing, not an oxidizing, atmosphere. Stanley Miller, a graduate student of Urey's, created an apparatus designed to mimic this presumed early-Earth condition (Bada and Lazcano, 2012). A gaseous mixture of methane, ammonia, hydrogen, and water was connected to a flask of boiling water. The steam created by the boiling water caused the gasses to move past electrodes, the electrical discharges of which simulated lightning in the atmosphere. A cold-water jacket caused molecules to condense and fall out of the "atmosphere." The reaction was allowed to run for a week, after which the solution, which had become deep red, was analyzed. Miller had succeeded in producing not only the formic acid and formaldehyde formed in the experiment of Garrison et al. (1951), but as he included nitrogen, he could also form

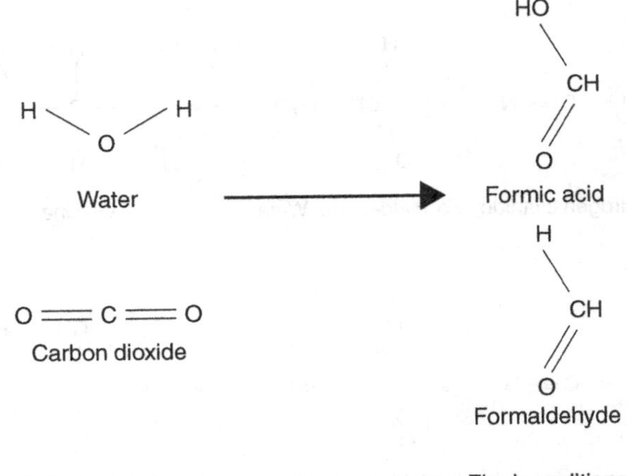

FIGURE 18.1 The synthesis of formic acid and formaldehyde.

hydrogen cyanide, which can combine with water and aldehydes to form amino acids (Miller, 1953, 1955; Miller and Urey, 1959a,b; Bada and Lazcano, 2003; Lazcano and Bada, 2004).

In fact, glycine and alanine were created in these experiments. This exciting result indicated that amino acids may have been present on the early Earth before the advent of life. How was glycine made in Miller's experiment? It is likely that hydrogen cyanide and formaldehyde combined with water to form glycine via an amino nitrile intermediate. A more complex aldehyde yields more complex amino acids. For example, substituting acetaldehyde (CH_3CHO) for formaldehyde ($HCHO$) in the reaction results in the production of alanine (Fig. 18.2).

Under prebiotic conditions, amino acids can polymerize into polypeptides without the aid of enzymes or a template (Flores and Ponnamperuma, 1972). The peptide bonds between the amino acids occur as a result of dehydration reactions. Even more complex structures like proteinoid microspheres can form under prebiotic conditions. Proteinoids are large, branched molecules produced when amino acid mixtures containing large amounts of aspartic acid, glutamic acid, or lysine are heated without water. When these dry proteinoids are placed in warm water and allowed to cool, microspheres are produced (Fox and Dose, 1972; Fox, 1988b). Such proteinoid microspheres may have joined together with phospholipids, which can also be synthesized under prebiotic conditions (Hargreaves et al., 1977), to form the first plasma membranes in a process of self-assembly (Deamer and Fleischaker, 1994). The proteinoid microspheres look similar to the microspheres found in rocks that are 3.8 billion years old (Strother and Barghoorn, 1980). Ernest Just (1939) wrote in his book *The Biology of the Cell Surface*, "In the differentiation of ectoplasm [plasma membrane] from the ground-substance we thus must seek the cause of evolution." And E. Newton Harvey (1952) wrote that "by their membrane ye shall know them."

Nucleic acids can also be synthesized under early-Earth conditions (Orgel, 1973; Miller and Orgel, 1974; Powner et al., 2009, 2011; Stairs et al., 2017). The purines turned out to be relatively easy to synthesize, and Juan Oró (1960, 1976) and Sanchez et al. (1968) have shown that under prebiotic conditions, adenine can be formed from hydrogen cyanide. Adenine is made according to the simple overall reaction: 5 HCN $\Leftrightarrow$ adenine (Fig. 18.3). Thus, HCN, a poison to aerobic organisms today, may have been the giver of life billions of years ago. Ponnamperuma et al. (1963a) showed that adenine can be formed by electron irradiation of methane, ammonia, and water. Ribose and other sugars can also be made under similar conditions by the overall reaction: 5 $\times$ formaldehyde (CH_2O) $\Leftrightarrow$ ribose ($C_5H_{10}O_5$). The adenine and ribose can lose a single water molecule and form adenosine (see Fig. 18.4; Ponnamperuma et al., 1963b; Fuller et al., 1972). When inorganic phosphate is added to the prebiotic mix, adenosine triphosphate (ATP) is formed (Ponnamperuma et al., 1963c).

The pyrimidines uracil and cytosine have been formed from cyanoacetaldehyde and urea by Robertson and Miller (1995). Shapiro (1999) worries that the small amount of any cytosine formed under prebiotic conditions would break down faster than it was formed. As a consequence, before he died, Stanley Miller found a way to increase the yield of cytosine and uracil formed under presumed prebiotic conditions (Cleaves et al., 2006). He succeeded by concentrating the precursors by eutectic freezing (Sanchez et al., 1966; Orgel, 2004). That is, Miller and his colleagues froze a dilute solution of precursors, and as the water in the solution crystallized, the concentrations of the precursors in the solution between the ice crystals rose to concentrations high enough to form pyrimidines. By including phosphate in the presumed early-Earth conditions, nucleoside monophosphates, including adenosine monophosphate, guanosine monophosphate, cytidine monophosphate, thymidine

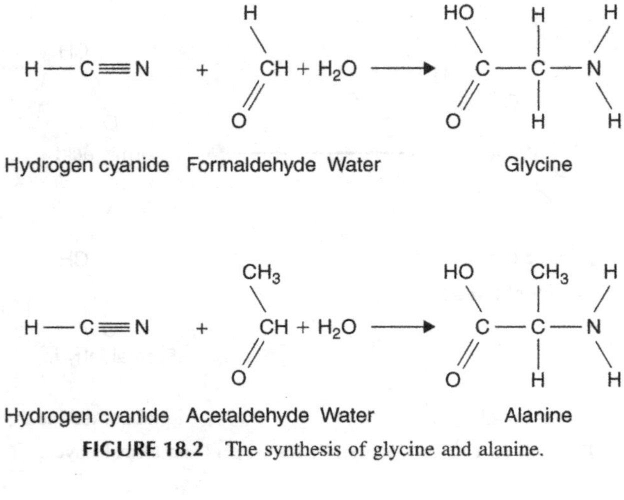

FIGURE 18.2 The synthesis of glycine and alanine.

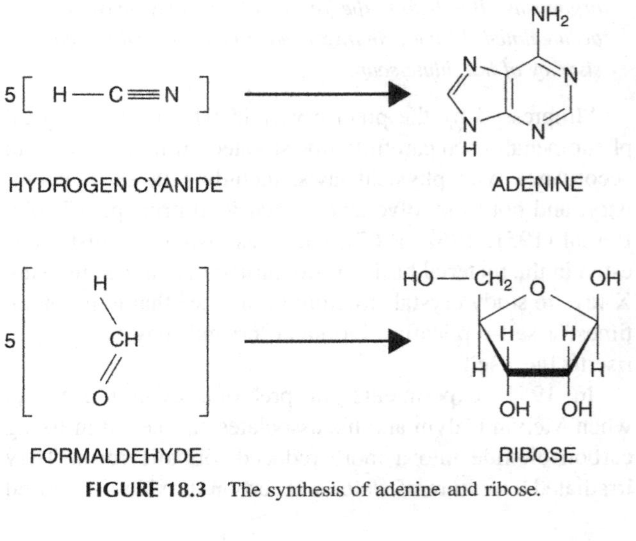

FIGURE 18.3 The synthesis of adenine and ribose.

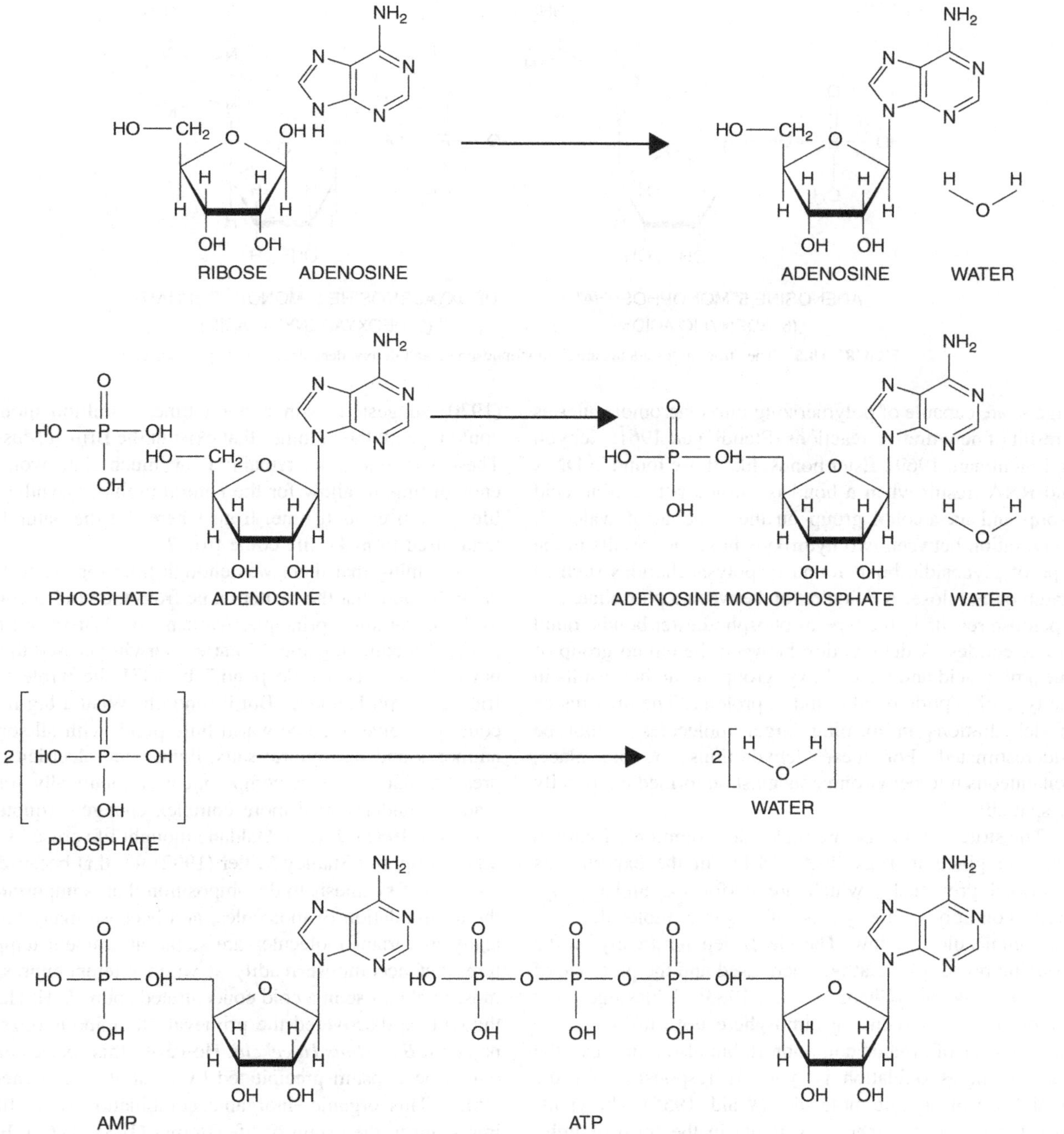

FIGURE 18.4 The synthesis of adenosine, adenosine monophosphate, and ATP.

monophosphate, and uridine monophosphate, can also be formed (Ponnamperuma and Mack, 1965). Nicotinamide, a coenzyme that contains adenine, is also formed under presumed prebiotic conditions (Cleaves and Miller, 2001).

Urea is formed under prebiotic conditions (Lohrmann, 1972). This is important as urea catalyzes many reactions. For example, adenosine 2′,3′-phosphate forms in a urea-catalyzed reaction (Lohrmann and Orgel, 1971). The adenosine 2′,3′-phosphate will polymerize via a dehydration reaction to form ribonucleic acids with a 3′,5′ linkage (i.e., RNA; Verlander et al., 1973). A reduction of the 2′ hydroxy group of ribose would result in deoxyribose, and the polymerization of deoxyribonucleotides would result in DNA (Fig. 18.5). Peptide nucleic acids, which are more stable than RNA alone, are also formed under prebiotic conditions (Nelson et al., 2000).

Only molecules with more than one functional group are capable of polymerizing. Monomers with at least two functional groups, including amino acids, nucleotides, and

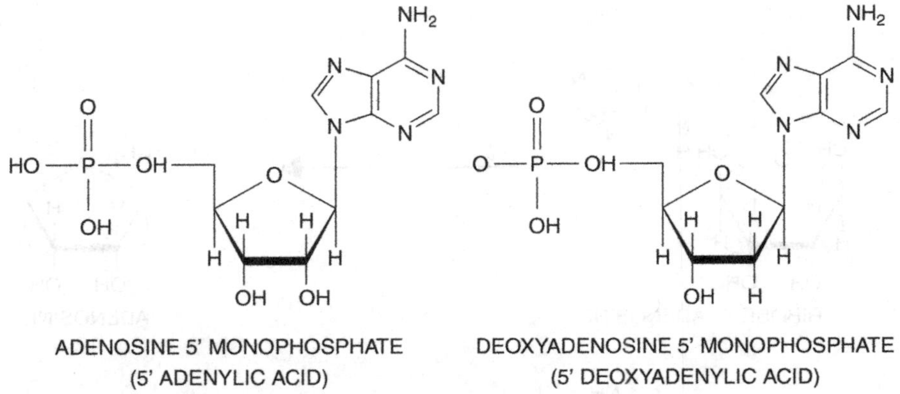

ADENOSINE 5′ MONOPHOSPHATE (5′ ADENYLIC ACID)

DEOXYADENOSINE 5′ MONOPHOSPHATE (5′ DEOXYADENYLIC ACID)

FIGURE 18.5 The structure of adenosine 5′ monophosphate and deoxyadenosine 5′ monophosphate.

sugars, are capable of polymerizing into macromolecules as a result of dehydration reactions (Staudinger, 1961; Kenyon and Steinman, 1969). Ester bonds, like those found in DNA and RNA, result when a bond is formed between an acid group and an alcohol group on the removal of water. A dehydration between two hydroxyls in sugars results in the type of glycosidic bond found in polysaccharides such as starch or cellulose. A dehydration between a phosphate and a pentose results in the type of phosphodiester bonds found in nucleotides. A dehydration between the amino group of one amino acid and the carboxyl group of another results in the type of peptide bond found in proteins. The importance of dehydrations in forming larger molecules cannot be underestimated. For these dehydrations to take place, nonaqueous microenvironments must be formed temporally or spatially.

The structures of organic molecules commonly found in cells are given in Figs. 18.6–18.11. In the experiments described previously, which are performed under early-Earth conditions, the yields of organic molecules and macromolecules are low. The yields depend greatly on the reducing power of the atmosphere used and the presence of activating agents (Chang et al., 1983; Schlesinger and Miller, 1983). A reducing atmosphere not only enhances the synthesis of organic compounds, but also minimizes the breakdown, as oxidation is typically responsible for the breakdown of organic molecules (Wald, 1955). The yields also depend on the energy available in the form of light, heat, lightning, cosmic rays, etc., and the availability of dehydrating conditions (Ponnamperuma, 1972; Fox, 1988a; Miller, 1992). While the probability of various molecules coming together to form a living organism is low, it only had to happen once. During a long enough time and with a large enough number of mixtures, every possible combination will eventually occur and improbable combinations eventually occur (Guye, no date; Bohm, 1961). As Herodotus (c.450 BCE) said, "If one is sufficiently lavish with time, everything possible happens." And as Émile Borel (1913), Arthur Eddington (1929), and Sir James Jeans (1930) suggested, with enough time, a million monkeys could type all the volumes that exist in the British Museum. These two questions remain: How much time would be enough time to allow for the natural prebiotic synthesis of life? Given enough time, from where did the natural material used to make life come from?

Assuming that there was enough time for life to being naturally and that the matter came from the void according to the uncertainty principle, we can ask: Where did these molecules come together? Charles Darwin guessed that life began in a "warm little pond." In 1871, he wrote to his friend, Joseph Hooker, "But if (and oh! What a big if!) we could conceive in some warm little pond, with all sorts of ammonia and phosphoric salts, light, heat, electricity, etc. present, that a protein compound was chemically formed ready to undergo still more complex changes" (quoted in Harrison, 1981). J. B. S. Haldane thought life arose in a hot dilute soup, but Stanley Miller (1992) felt that because it is the ratio of synthesis to decomposition that is important for the accumulation of molecules, and because many biologically important molecules are stable at ambient temperatures but decompose readily at warmer temperatures, life most likely arose in a cold concentrated soup. T. H. Huxley thought he discovered the primeval albuminous ooze and named it *Bathybius Haeckelii*. However, this species turned out to be gypsum precipitated by alcohol (see Arrhenius, 1908). This organic–inorganic combination may still be important in the origin of life. Bernal (1951, 1967), Broda (1975), and Arrhenius et al. (1997) think it is more likely that submerged claylike rocks facilitated the formation of living molecules due to their ability to adsorb and concentrate molecules. The clay would promote dehydration reactions by bringing molecules together close enough to from a bond.

There seems to be a thin line that separates chemistry from biology. Jacobus van't Hoff (quoted in de Vries, 1889) wrote,

> *From the chemical properties of carbon it appears that this element is able, with the help of two or three others, to form*

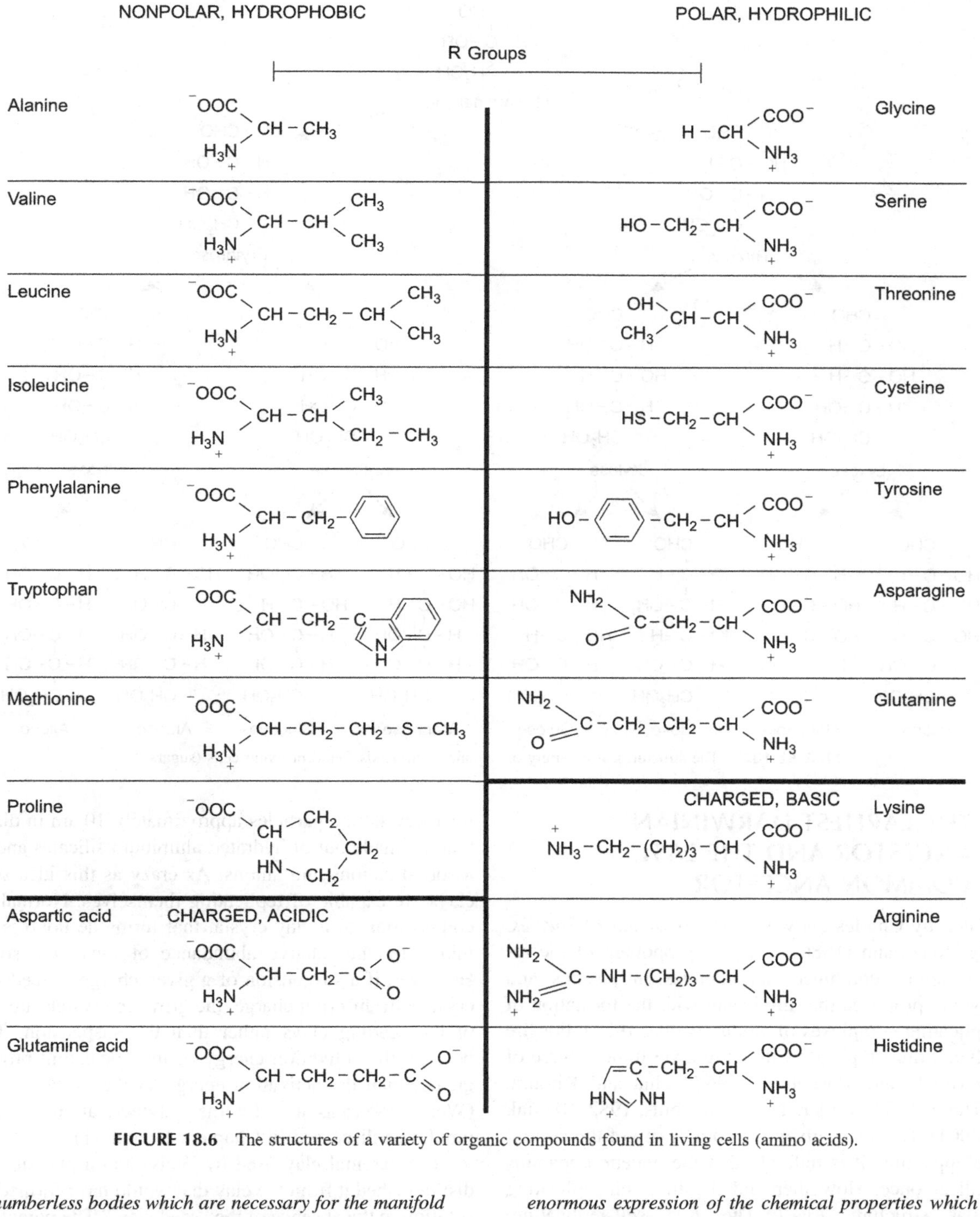

NONPOLAR, HYDROPHOBIC	POLAR, HYDROPHILIC
Alanine	Glycine
Valine	Serine
Leucine	Threonine
Isoleucine	Cysteine
Phenylalanine	Tyrosine
Tryptophan	Asparagine
Methionine	Glutamine
Proline	CHARGED, BASIC: Lysine
CHARGED, ACIDIC: Aspartic acid	Arginine
Glutamine acid	Histidine

R Groups

FIGURE 18.6 The structures of a variety of organic compounds found in living cells (amino acids).

the numberless bodies which are necessary for the manifold needs of a living being; from their almost equal tendency to combine with hydrogen and oxygen, follows the capacity of the carbon-compounds to be adapted alternately for processes of reduction and oxydation as the simultaneous existence of a vegetable and an animal kingdom requires. … Therefore, one does not go too far in assuming that the existence of the vegetable and animal world is the enormous expression of the chemical properties which the carbon-atom has at the temperature of our earth.

The importance of the chemical properties of carbon for producing life, as we know it, is further explained and extolled by Ernst Haeckel (1866), Lawrence Henderson (1913), John Edsall and Jeffries Wyman (1958), and George Wald (1963).

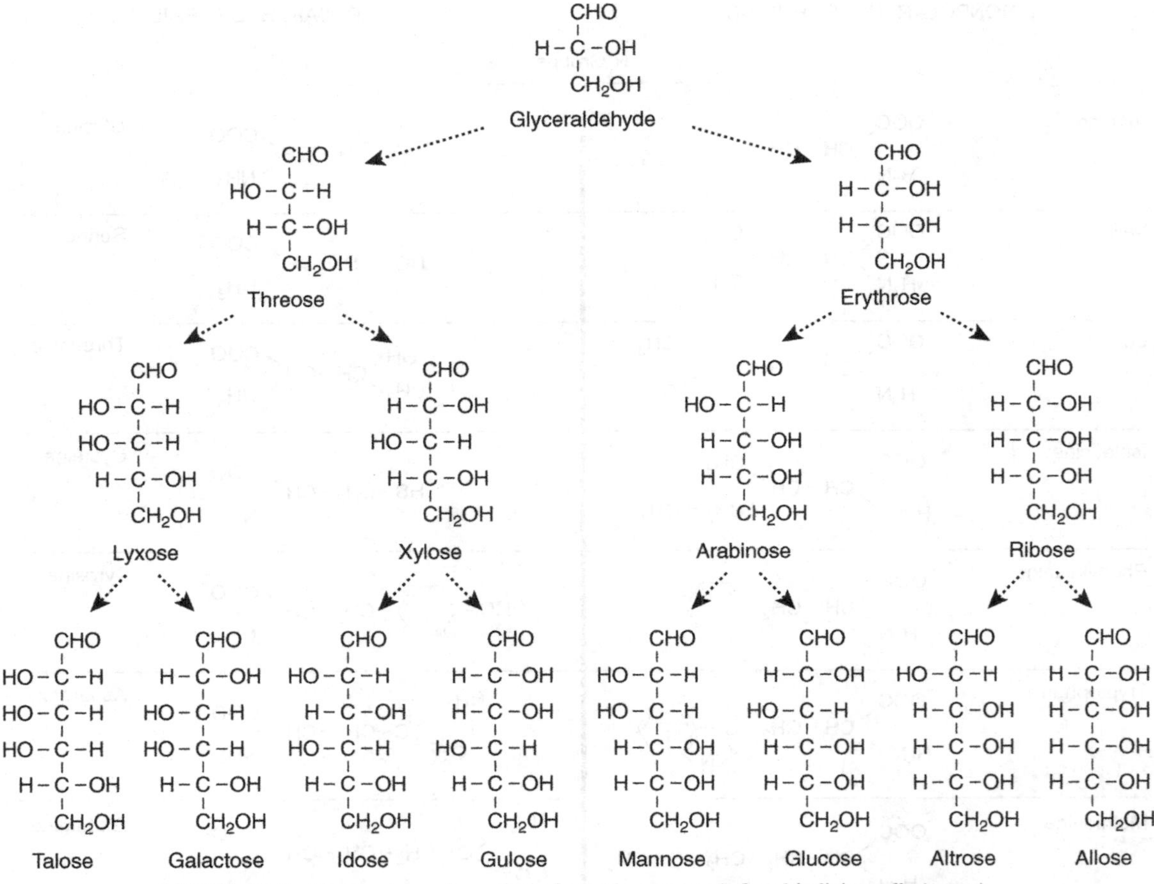

FIGURE 18.7 The structures of a variety of organic compounds found in living cells (sugars).

18.6 THE EARLIEST DARWINIAN ANCESTOR AND THE LAST COMMON ANCESTOR

Stimulated by Charles Darwin's (1859) *Origin of Species*, Ludwig Boltzmann (1886), a strong proponent of the reality of atoms, combined his interests in physics and biology and proposed that life began with the formation of self-replicating complexes of atoms (Broda, 1983). For life to evolve, it must replicate with a high, yet finite, degree of fidelity (Fitch and Margoliash, 1967; Ohta and Kimura, 1971; Ulam, 1972; Muller, 1973b; Stebbins, 1982; Szostak et al., 2001). However, given the complexity of the current genetic apparatus, it is unlikely that the genetic apparatus arose all at once. How then did the first self-replicating molecular structure arise? The first self-replicating, information-bearing structure of which the replication was not quite perfect, and was thus capable of evolving through natural selection, is known as the earliest Darwinian ancestor (Bloch and Staves, 1986).

One candidate for the earliest Darwinian ancestor, alluded to in *Genesis* (2:7), is clay (Cairns-Smith, 1982; Cairns-Smith and Hartman, 1986). Clays are inorganic microcrystalline particles approximately 10 μm in diameter that are made out of hydrated aluminum silicates and other assorted cations and anions. As crazy as this idea sounds, clays are capable of replicating themselves. Normally, the composition of a clay crystal that forms de novo is determined by the relative abundance of ions in a solution. However, if a suspension of a given charge is seeded with crystals of differing charge, the growing crystals are typical of the seeding clays rather than the suspension. This is because the activation energy of the nucleating process is greater than the activation energy of the growing process (Weiss, 1981; as it is for the polymerization of actin or tubulin as discussed in Chapters 10 and 11).

The parental clay used by Weiss had a phenotype that distinguished it from the clay that would have formed in the solution in the absence of the "seed clay." The parental clay had the proper redox potential to reduce Co^{3+} to Co^{2+}. In Weiss's experiments, this phenotype was transferred unchanged to the descendants for 22 generations.

Clays can have many other catalytic properties that are important for life (Pinnavaia, 1983). For example, clays can facilitate important biochemical processes, including the

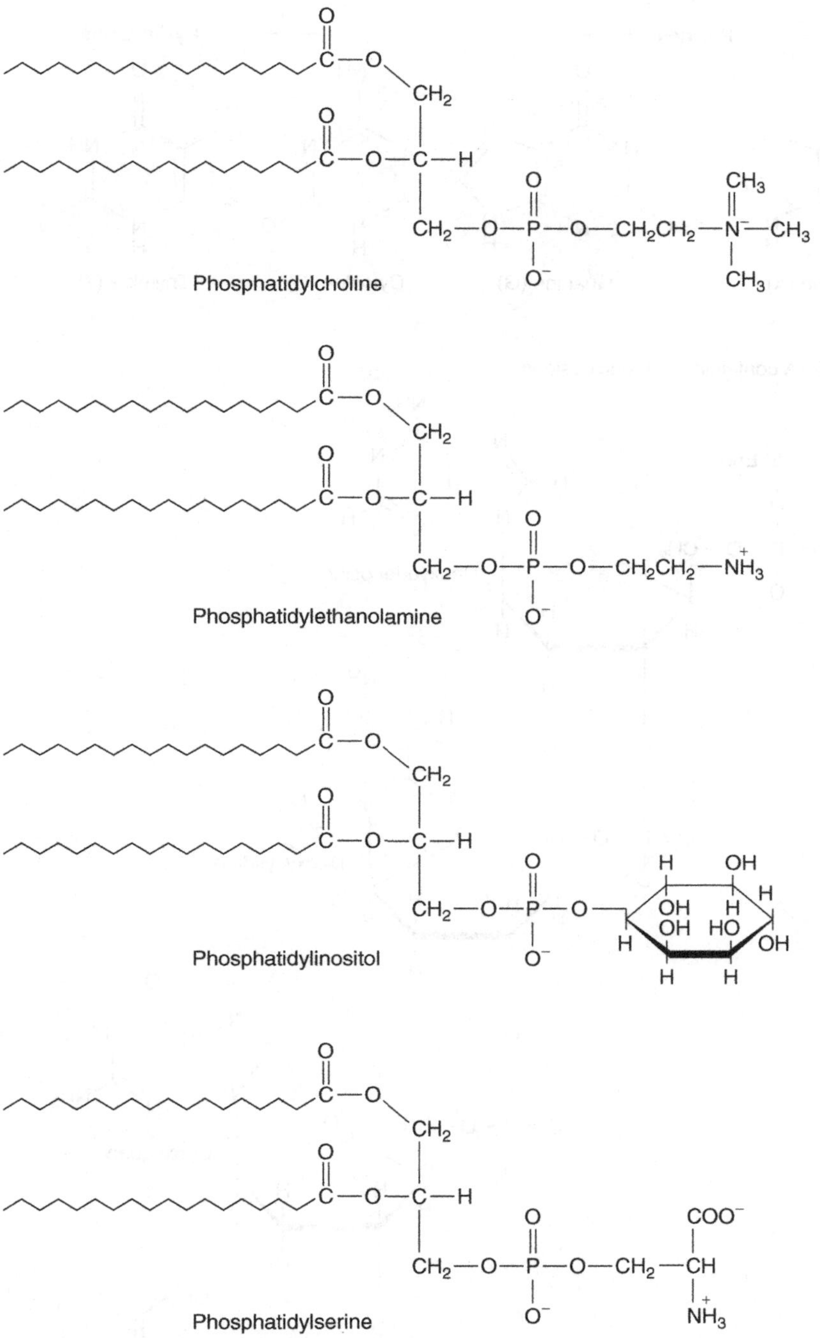

FIGURE 18.8 The structures of a variety of organic compounds found in living cells (lipids).

polymerization of activated amino acids. The presence of clays also increases the length of polypeptides formed under presumed prebiotic conditions (Paecht-Horowitz et al., 1970; Paecht-Horowitz, 1976; Paecht-Horowitz and Eirich, 1988; Böhler et al., 1996; Ferris et al., 1996). Clays containing ZnS or TiO_2 act like semiconductors that can use ultraviolet light energy to reduce organic acids and even run the Krebs in the reverse direction to fix CO_2 into organic acids (Zhang and Martin, 2006; Guzman and Martin, 2008, 2009, 2010; Saladino et al., 2011; Habisreutinger et al., 2013; Zhou and Guzman, 2014, 2016; Zhou et al., 2017). Clays are currently used to facilitate a number of industrial syntheses.

The clays may have facilitated the formation of organic molecules in prebiotic conditions. These organic molecules, in turn, might have facilitated the growth of the clays by

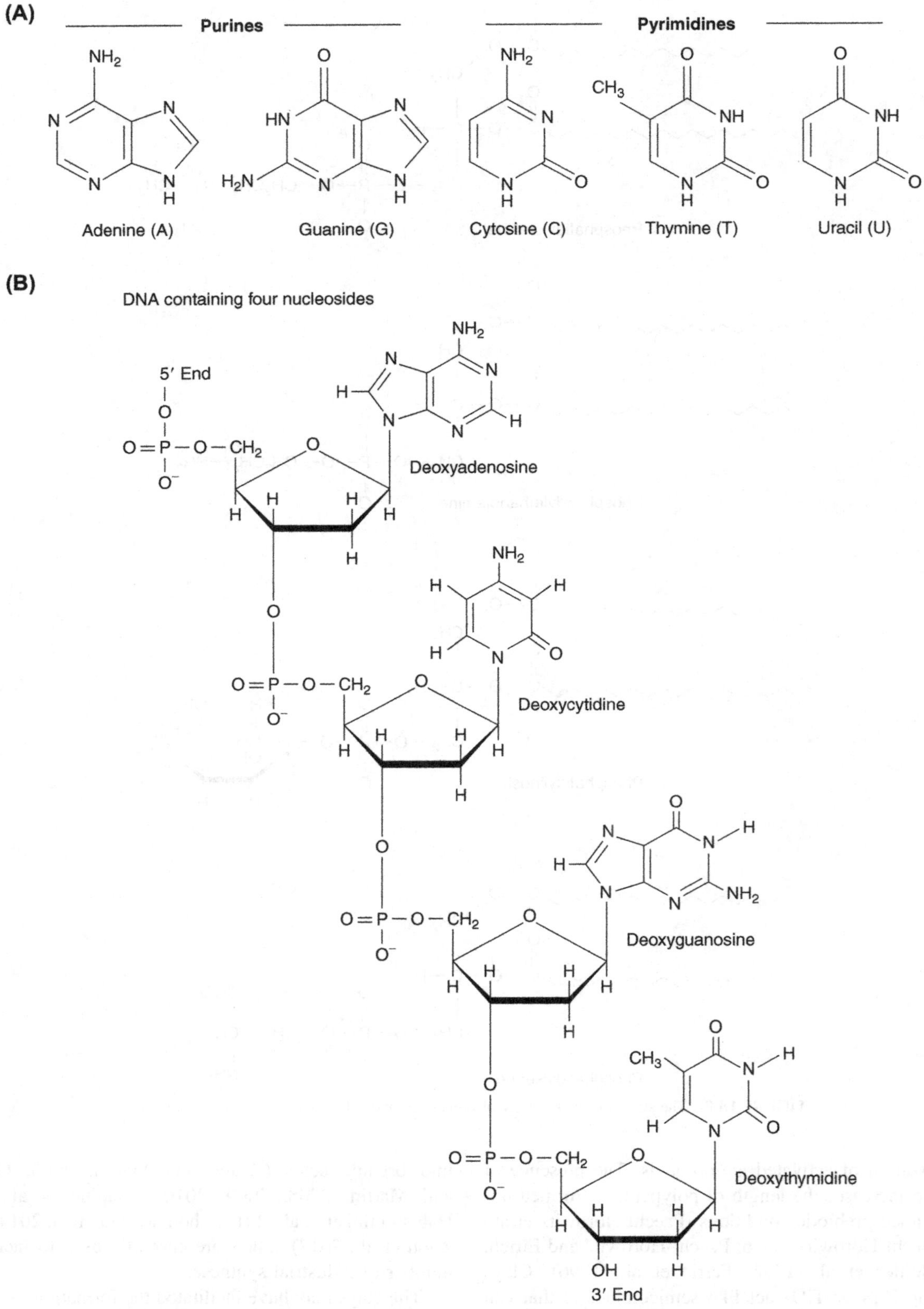

(A)

(B)

FIGURE 18.9 The structures of a variety of organic compounds found in living cells (purines, pyrimidines, nucleotides).

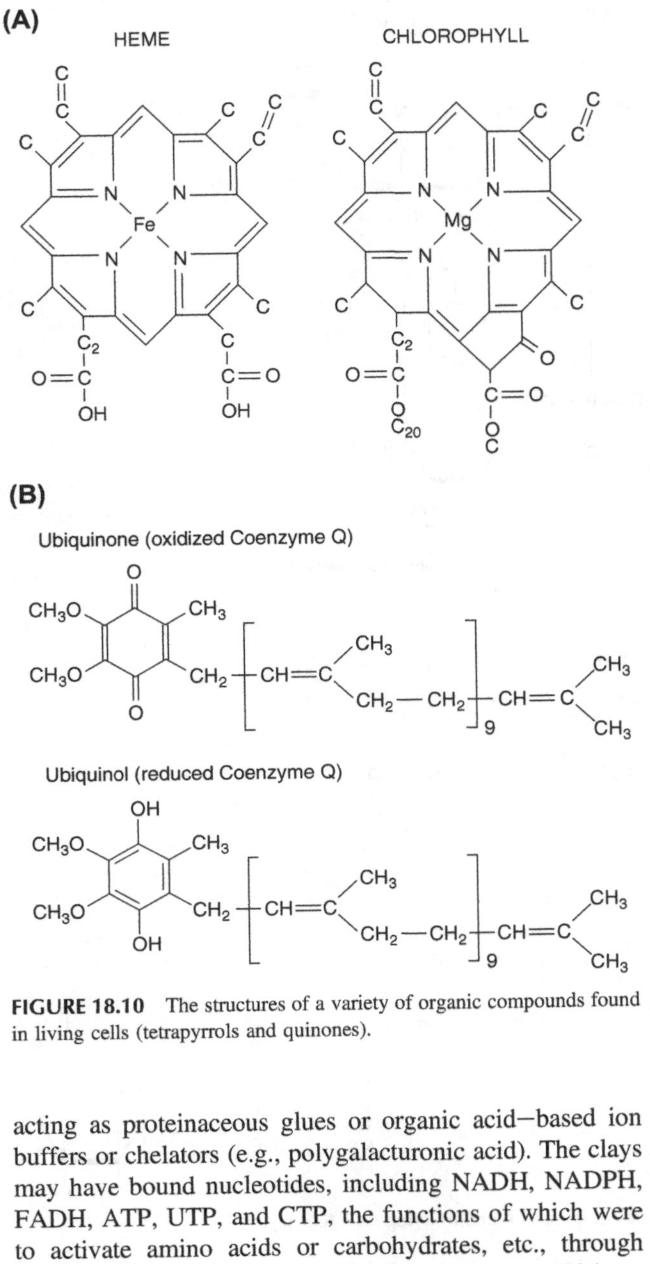

FIGURE 18.10 The structures of a variety of organic compounds found in living cells (tetrapyrrols and quinones).

acting as proteinaceous glues or organic acid–based ion buffers or chelators (e.g., polygalacturonic acid). The clays may have bound nucleotides, including NADH, NADPH, FADH, ATP, UTP, and CTP, the functions of which were to activate amino acids or carbohydrates, etc., through group transfer so that the activated molecules could have participated in complex synthetic reactions. A given sequence of charge density on a clay might have resulted in the binding and ordering of a particular linear sequence of nucleotide coenzymes that might have resulted in the performance of sequential reactions. The sugar phosphates of closely bound nucleotides might have esterified to form a backbone so that the macromolecular complex could have performed sequential reactions free in solution.

Clays have the additional life-generating ability to accelerate vesicle (0.5–30 μm in diameter) formation from fatty acid containing micelles by about 100-fold (Hanczyc et al., 2003). The clay acts catalytically, as the generated membrane surface area is 50-fold greater than the maximum possible surface area of the clay. The vesicles have the ability to retain fluorescently labeled RNA associated with the clay.

A sequence of clay-bound nucleotides might have contained the information necessary to form a polymer and to allow a sequence of reactions. As an added bonus, however, the nucleic acid polymer would have the ability to bind with a "complementary nucleotide" through the formation of hydrogen bonds and form an intermediate template so that it could reproduce itself. If it could reproduce faster than the clays, the nucleic acids would outcompete the clays for the replicating function. This is what Graham Cairns-Smith (1982) calls *genetic takeover*. Thus, the first Darwinian ancestor may have gone from a clay-based replicating system to a nucleotide-based replicating system. Eventually, the nucleotides left the evolutionarily challenged clays behind, and the nucleotide-based genetic code went through its own evolutionary development.

Manfred Eigen and his colleagues suggest that RNA, and not clay, is the most likely candidate for the earliest Darwinian ancestor (Eigen et al., 1981). RNA is a good candidate because it is both an information-bearing molecule, which has the ability to replicate itself due to its endogenous polymerase and ligase activities (Guerrier-Takada et al., 1983; Inoue and Orgel, 1983; Cech, 1986a,b; Doudna and Szostak, 1989), and a molecule with other enzymatic activities, for example, the ability to polymerize amino acids (Noller et al., 1992; Piccirilli et al., 1992). Peptide nucleic acids may have served as a precursor to RNA (Nelson et al., 2000).

David Bloch, Mark Staves and their associates figured that they might be able to find the primeval RNA sequence if they scanned the sequences of the ribosomal and transfer RNAs from phylogenetically diverse organisms, including archaebacteria, eubacteria, yeast, and bovine mitochondria, to find an area rich in nucleotide matches. Using this molecular archeological approach, they found such sequences (Bloch et al., 1983, 1985). Because there were no differences in the matches found among intraspecific or interspecific searches, the matches probably reflect ancient similarities derived from common origins rather than a more recent convergence of the molecules. They believe the length of the ancestral RNA was nine bases long (Nazarea et al., 1985; Bloch, 1986, 1988). Eigen et al. (1981) have shown that short sequences of RNA that are allowed to replicate together in a test tube compete for nucleotides and are thus subject to Darwinian evolution (Eigen and Schuster, 1979; Dyson, 1985; Eigen, 1992, 1993).

RNA may have also provided the mechanism for polypeptide formation (Dounce, 1952; Lacey and Mullins, 1983; Lacey et al., 1988, 1990a,b, 1991, 1992, 1993; Wickramasinghe et al., 1991; Wickramasinghe and Lacey, 1994; Bailey, 1998). In fact, a single ribonucleoside,

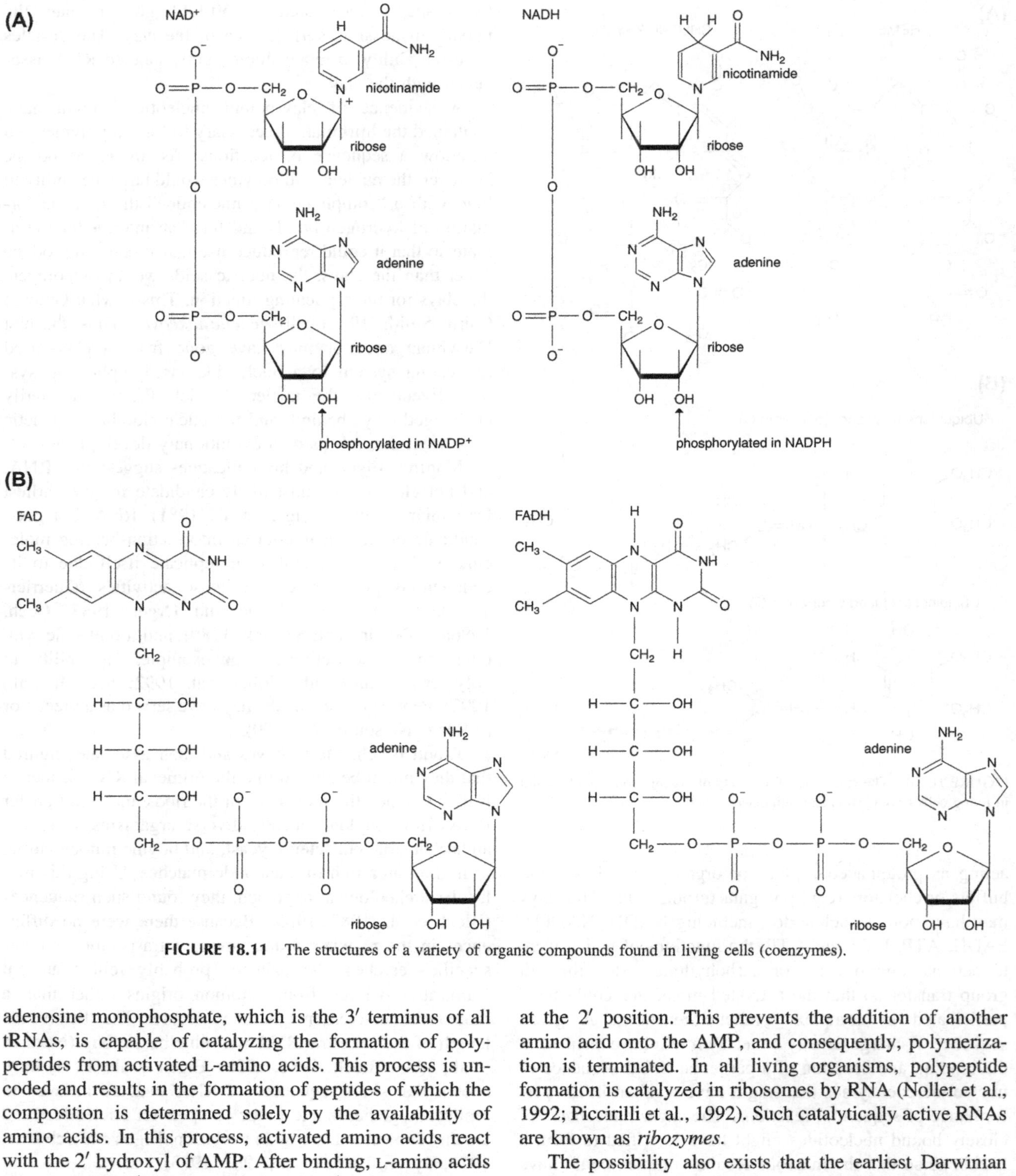

(A) NAD+ NADH

(B) FAD FADH

FIGURE 18.11 The structures of a variety of organic compounds found in living cells (coenzymes).

adenosine monophosphate, which is the 3′ terminus of all tRNAs, is capable of catalyzing the formation of polypeptides from activated L-amino acids. This process is uncoded and results in the formation of peptides of which the composition is determined solely by the availability of amino acids. In this process, activated amino acids react with the 2′ hydroxyl of AMP. After binding, L-amino acids migrate to the 3′ position, leaving the 2′ position open for further attack by another amino acid. Because of their proximity, a peptide bond forms between the adjacent amino acids and this process occurs indefinitely. However, D-amino acids do not migrate to the 3′ hydroxyl but remain at the 2′ position. This prevents the addition of another amino acid onto the AMP, and consequently, polymerization is terminated. In all living organisms, polypeptide formation is catalyzed in ribosomes by RNA (Noller et al., 1992; Piccirilli et al., 1992). Such catalytically active RNAs are known as *ribozymes*.

The possibility also exists that the earliest Darwinian ancestor was a protein. Lee et al. (1996) have shown that a 32–amino acid–polypeptide sequence, typically found in a nuclear transcription factor, is capable of replicating itself when supplied with amino acids. Moreover, a prion, which is a modified cellular protein (e.g., PrP) that causes infectious neurodegenerative diseases such as mad cow disease, is able to replicate in the absence of nucleic acids. Each strain-specific phenotype is encoded by the tertiary structure of the protein (Telling et al., 1996; Prusiner, 1997a,b; Hegde et al., 1998; Legname et al., 2004). The replication of the yeast prion requires a heat-shock protein (Serio et al., 2000; Shorter and Lindquist, 2004).

Whether clay, proteins, or RNA was the earliest Darwinian ancestor, the genetic apparatus probably evolved from RNA alone, into the trinity of molecules that carry the information of life: DNA, RNA, and protein (Woese, 1967; Crick, 1968). DNA would be selected for over RNA as an informational molecule, in part, because its stability is greater than that of RNA due to the reduction of the 2′OH to 2′H. The 2′OH is able to perform a nucleophilic attack on the phosphorous atom participating in the phosphodiester bond with the 3′ carbon (Lilley, 2011). This results in the breakage of the phosphodiester bond that links the 3′ carbon of one ribose with the 5′ carbon of the next and causes the formation of a 2′3′ cyclic phosphate on the 3′ ribose.

Proteins, on the other hand, out perform RNA in enzymatic functions and thus would be selected through natural selection, due perhaps to the variety of functional groups found in the amino acid monomers compared with the nucleotides. These functional groups are capable of interacting with many different atoms, bonds, and molecules. Eventually, RNA not only provided the link between the coding function of DNA, and the catalytic function of proteins, but also became intimately involved in the regulation of gene expression through small nonprotein-coding RNAs (Fire et al., 1998; Ambros, 2008; Baulcombe, 2008).

While the earliest Darwinian ancestor may have evolved into the genetic apparatus of cells, other processes are also necessary for life. Throughout this book, I have discussed how various organelles make life possible. In terms of making environmental energy available for cellular work, the chloroplast converts radiant energy into the chemical energy of carbohydrate, as well as the chemical energy of readily usable activated nucleotides, including ATP and NADPH; the mitochondria convert the chemical energy of organic macromolecules into the chemical energy of the activated nucleotide ATP. The plasma membrane, due to its capacitance, converts differences in ionic concentrations on both sides of the membrane into electrical energy that can be used for secondary transport and for signaling in response to changes in environmental conditions. The membranes, as well as the solutions they surround, provide the surfaces and volumes necessary for catalysts to perform chemical reactions. To visualize the original cell, we must imagine the minimum components necessary to perform the functions necessary for life. Albert Libchaber and his associates (Noireaux et al., 2003, 2005, 2011; Noireaux and Libchaber, 2004) are trying to create artificial life by adding a given mRNA or DNA to cytosolic homogenates enclosed within an artificial lipid bilayer and expressing one gene at a time, whereas Craig Venter and his associates at Synthetic Genomics (http://www.syntheticgenomics.com/about-us/) are mutating the individual genes in the genome of *Mycoplasma genitalium*, which has 517 genes, to discover the minimal genome necessary for life (Hutchison et al., 1999). Because genes that are individually dispensable may not be simultaneously dispensable, the scientists at Synthetic Genomics would like to create an artificial chromosome to find the minimal genome. Synthetic Genomics is also adding laboratory-made genomes to cells to produce novel organisms capable of synthesizing metabolites that can be used for biofuels or pharmaceuticals.

The earliest Darwinian ancestor may have formed symbioses with other prebiotic entities that may have been good at performing both endergonic and exergonic chemical reactions. The endergonic reactions may have been performed by prebiotic chemicals (e.g., porphorins) that were capable of capturing radiant energy and reducing sulfur with the aid of mineral catalysts. Perhaps the molecules involved in these bioenergetic processes passed the excited electrons from one molecule to another by making use of an iron intermediate. Perhaps these entities associated with others that were able to make lipids. Perhaps a lipid bilayer formed on the surface of a puddle, and wave action or the falling of a rock induced the formation of a liposome that encapsulated the other entities to form a protocell. The membrane surrounding this protocell may have been able to allow the passage of necessary nutrients into the cell, while still being capable of maintaining an ionic difference. Perhaps the polypeptides created by clays or from a nucleic acid template could have facilitated the transport functions. At some point, the protocell would have to be able to divide and reproduce. At this stage it probably formed the last common ancestor of all organisms (Deamer, 1997; Woese, 1998). William Kirby (1853), describing Lamarck's view of the origin of the first cell, wrote,

> *We know, by observation, that the most simple organizations, whether vegetable or animal, are never met with but in minute gelatinous bodies, very simple and delicate; in a word, only in frail bodies almost without consistence and mostly transparent. These minute bodies he supposes nature forms, in the waters, by the power of attraction; and that next, subtile and expansive fluids, such as caloric and electricity, penetrate these bodies, and enlarge the interstices of their agglutinated molecules, so as to form utricular cavities and so produce irritability and life, followed by a power of absorption, by which they derive from without.*

The similarities in molecules, mechanisms, metabolic pathways, and structures in living organisms point to a single common ancestor (Horowitz, 1945; Granick, 1957; Gaffron, 1960; Blum, 1962; Lipmann, 1965; Ycas, 1974; Jensen, 1976; Morowitz, 1992; Dyer and Obar, 1994). In his Nobel Lecture, Albert Claude (1974) said, "*In the long course of cell life on this earth it remained, for our age for our generation, to receive the full ownership of our inheritance. We have entered the cell, the Mansion of our birth, and started the inventory of our acquired wealth. For over two billion years, through the apparent fancy of her endless differentiations and metamorphosis the Cell, as regards its basic physiological mechanisms, has remained one and the same. It is life itself, and our true and distant ancestor...In the course of the past 30 or 40 years, we have learned to appreciate the complexity and perfection of the cellular mechanisms, miniaturized to the utmost at the molecular level, which reveal within the cell an unparalleled knowledge of the laws of physics and chemistry. If we examine the accomplishments of man in his most advanced endeavors, in theory and in practice, we find that the cell has done all this long before him, with greater resourcefulness and much greater efficiency...Man, like other organisms, is so perfectly coordinated that he may easily forget, whether awake or asleep, that he is a colony of cells in action, and that it is the cells which achieve, through him, what he has the illusion of accomplishing himself. It is the cells which create and maintain in us, during the span of our lives, our will to live and survive, to search and experiment, and to struggle... I am afraid that in this description of the cell, based on experimental facts, I may be accused of reintroducing a vitalistic and teleological concept which the rationalism and the scientific materialism of the 19th and early 20th centuries had banished from our literature and from our scientific thinking. Of course, we know the laws of trial and error, of large numbers and probabilities. We know that these laws are part of the mathematical and mechanical fabric of the universe, and that they are also at play in biological processes. But, in the name of the experimental method and out of our poor knowledge, are we really entitled to claim that everything happens by chance, to the exclusion of all other possibilities?*"

Throughout history, the idea of common descent was espoused in one form or another by Empedocles (see Lambridis, 1976), Pierre Louis Moreau de Maupertuis (1753), Jean Lamarck (1809), Robert Chambers (n.d.), Patrick Matthew, W. C. Wells, Herbert Spencer, and others (see Darwin, 1859−1882). Matthias Schleiden (1853) wrote,

> *This view, that the whole fullness of the vegetable world has been gradually developed out of a single cell and its descendants, by gradual formation of varieties, which became stereotyped into species, and then, in like manner, became the producers of new forms, is at least quite as possible as any other, and is perhaps more probable and correspondent than any other, since it carries back the Absolutely Inexplicable, namely the production of Organic Being, into the very narrowest limits which can be imagined.*

Charles Darwin (1859) presented evidence that because variation could be acted on by artificial selection, evolution must take place as a gradual result of natural selection. Relying on anatomical evidence, Richard Owen (1848, 1849) thought it was more likely that evolution was a result of congenital changes or jumps (Rupke, 1994, 2009). With a like mind, Richard Goldschmidt (1933, 1940, 1955) offered an alternative to Darwin's theory of the gradual origin of species by natural selection, which was based on comparative morphology and genetics. Goldschmidt proposed that new species evolve through drastic changes that result from a mutation in a gene that influences the relative rates of various developmental processes. Such a change would create "hopeful monsters which would start a new evolutionary line if fitting into some empty environmental niche" (Gould, 1977). Irene Manton (1950; Swanson, 1951; Williams, 2016) also believed that "to understand evolution in general terms we need to look not outside but inside the organism and in particular to study, with all the new tools...not merely the external attributes of chromosomes...but rather their intimate molecular structure." A minute change in the DNA that encodes a transcription factor, an element in a signal transduction cascade, or a regulatory RNA, may provide the mechanism that leads to such a drastic change and a new evolutionary line (Kirschner and Gerhart, 2005).

18.7 DIVERSITY IN THE BIOLOGICAL WORLD

Charles Darwin (1859−1882) believed that "All the members of whole classes are connected together by as chain of affinities." Darwin (1860b) wrote in *The Origin of Species*,

> *It is interesting to contemplate a tangled bank, clothed with many plants of many kinds, with birds singing on the bushes, with various insects flitting about, and with worms crawling through the damp earth, and to reflect that these elaborately constructed forms, so different from each other, and dependent upon each other ... have all been produced by laws acting around us. ... There is grandeur in this view of life, with its several powers, having been originally breathed by the Creator into a few forms or into one; and that whilst this planet has gone cycling on according to the fixed law of gravity, from so simple a beginning endless forms most beautiful and most wonderful have been, and are being evolved.*

Alfred Russel Wallace (1869), often credited as the cofounder of the theory of evolution by natural selection

(Wallace, 1858, 1870, 1889, 1905, 1908, 1910; Lloyd et al., 2010; Hossfeld and Olsson, 2013; Kutschera and Hossfeld, 2013), wrote "*This subject is a vast one, and would require volumes for its proper elucidation, but enough, we think, has now been said, to indicate the possibility of a new stand-point for those who cannot accept the theory of evolution as expressing the whole truth in regard to the origin of man. While admitting to the full extent the agency of the same great laws of organic development in the origin of the human race as in the origin of all organized beings, there yet seems to be evidence of a Power which has guided the action of those laws in definite directions and for special ends. And so far from this view being out of harmony with the teachings of science, it has a striking analogy with what is now taking place in the world, and is thus strictly uniformitarian in character. Man himself guides and modifies nature for special ends. The laws of evolution alone would perhaps never have produced a grain so well adapted to his uses as wheat; such fruits as the seedless banana, and the bread-fruit; such animals as the Guernsey milch-cow, or the London dray-horse. Yet these so closely resemble the unaided productions of nature, that we may well imagine a being who had mastered the laws of development of organic forms through past ages, refusing to believe that any new power had been concerned in their production, and scornfully rejecting the theory that in these few cases a distinct intelligence had directed the action of the laws of variation, multiplication, and survival, for his own purposes. We know, however, that this has been done; and we must therefore admit the possibility, that in the development of the human race, a Higher Intelligence has guided the same laws for nobler ends.*

Such, we believe, is the direction in which we shall find the true reconciliation of Science with Theology on this most momentous problem. Let us fearlessly admit that the mind of man (itself the living proof of a supreme mind) is able to trace, and to a considerable extent has traced, the laws by means of which the organic no less than the inorganic world has been developed. But let us not shut our eyes to the evidence that an Overruling Intelligence has watched over the action of those laws, so directing variations and so determining their accumulation, as finally to produce an organization sufficiently perfect to admit of, and even to aid in, the indefinite advancement of our mental and moral nature."

This is a good time to put this book down, walk outside, and marvel at the beauty and wonder of nature (Agassiz, 1863; Burroughs, 1908; Muir, 1911). Aldo Leopold (1949) wrote in *A Sand County Almanac*,

> *It is a century now since Darwin gave us the first glimpse of the origin of species. We know now what was unknown to all the preceding caravan of generations: that men are only fellow-voyagers with other creatures in the odyssey of evolution. This new knowledge should have given us, by this time, a sense of kinship with fellow creatures; a wish to live and let live; a sense of wonder over the magnitude and duration of the biotic enterprise.*

Think about the biochemical, genetic, structural, and physiological unity that underlies the apparent diversity. All living things are related, yet unique. Each living being contains unique molecules of deoxyribonucleic acid that probably never existed before or will ever exist again. I marvel in the richness of the tapestry, the tightness of the weave, the beauty with which each thread complements every other one. It is amazing that even as we look at smaller and smaller pieces of the tapestry of nature, the weave still looks just as fine, elaborate, harmonious, and beautiful. However, even if we did not have an eye for the beauty and poetry of nature, or the innocence of children, nature's treasure house is important to cell biologists because it provides a vast array of organisms, each one specialized to fit in its niche, and because of this, each species has a distinct set of traits that make it amenable to the study of a given biological process (Krogh, 1929; Krebs, 1975; Wayne and Staves, 1996b). Many of these organisms hold the key to help us understand biology in a whole new light or to cure a disease of mankind.

Liberty Hyde Bailey (1915), Aldo Leopold (1949), and others developed a system of ethics based on the interrelationships between organisms and the land (Leopold, 2004). Leopold (1949) wrote, "A thing is right when it tends to preserve the integrity, stability, and beauty of the biotic community. It is wrong when it tends otherwise." Pierre Teilhard de Chardin (1966) felt that the connections were even greater in that the whole universe was living. Do we have any indication that life may actually be supra-organismal? According to Albert Szent-Györgyi (1948),

> *In our discussion we have passed several levels of organization from electrons ... [to] the whole rabbit, but I doubt whether the list is herewith complete. Everyone knows this much of biology—that one rabbit could never reproduce itself, and if life is characterized by self-reproduction, one rabbit could not be called alive at all, and one rabbit is no rabbit, and only two rabbits are one rabbit, and so we may go on calling in the end only the whole of living nature alive.*

18.8 THE ORIGIN OF CONSCIOUSNESS

In this chapter, I have discussed how the original quantum particle evolved into atoms, how atoms gave rise to molecules, how molecules gave rise to self-replicating systems, and how self-replicating systems gave rise to cellular life. In each stage of the evolution of life in the universe, new and surprising properties emerged from the combination of previous entities (Bergson, 1911; Morgan, 1923, 1926, 1933). Louis de Broglie (1946) maintains that thought is an essential condition for the progressive evolution of the human race. Some cells may specialize in higher functions of thought and self-identity.

A small group of large spindle-shaped cells has been discovered in the brains of humans and primates (Nimchinsky et al., 1999; Allman et al., 2001; Hof et al., 2001). These cells may be involved in self-identity and self-awareness. When these cells are damaged, people become "vegetables." These cells are less active in depressed people, disappear in people afflicted with Alzheimer's disease, and are more active in people with manic disorders. These cells alone are probably not sufficient to make us human (Bullock, 2003).

What is the relationship between the origin of consciousness and the origin of life (Drysdale, 1874; Wythe, 1880; Troglodyte, 1891; Shapley, 1958, 1967; Sinnott, 1961; Luria, 1973; Broda, 1983; Crick, 1994; Cairns-Smith, 1996; Margulis, 2001)? I discussed in Chapter 12 how the processes involved in cell signaling in single cells could provide a material basis for the concepts of consciousness and morality. I discussed that plants as well as animals can have consciousness (Gardiner, 2012; Leopold, 2014; Trewavas, 2014) and that action potentials and electrical signaling provide a means of communication in plant and animal cells. In Chapter 3, I discussed how plasmodesmata in plants and gap junctions in animal cells provide another means to integrate the activities between cells. Although truly amazing, is it not possible that when you put together billions of cells that are specialized for communication that consciousness is a natural outcome?

George Wald (1963) captured this awe and rational thinking when he spoke in front of the president of the United States and said

> *We have been told so often and on such tremendous authority as to seem to put it beyond question, that the essence of things must remain forever hidden from us; that we must stand forever outside nature, like children with their noses pressed against the glass, able to look in, but unable to enter. This concept of our origins encourages another view of matter. We are not looking into the universe from outside. We are looking at it from inside. Its history is our history; its stuff, our stuff. From that realization we can take some assurance that what we see is real.*
>
> *Judging from our experience upon this planet, such a history that begins with elementary particles, leads perhaps inevitably toward a strange and moving end; a creature that knows, a science-making animal that turns back upon the process that generated him and attempts to understand it. Without his like, the universe could be, but not be known, and that is a poor thing.*
>
> *Surely this is a great part of our dignity as men, that we can know, and that through us matter can know itself; that beginning with protons and electrons, out of the womb of time and the vastness of space, we can begin to understand; that organized as in us, the hydrogen, the carbon, the nitrogen, the oxygen, those 16 to 20 elements, the water, the sunlight—all, having become us, can begin to understand what they are, and how they came to be.*

The existence of creative creatures that can begin to understand what we are and how we came to be is a miracle, statistical or otherwise. A visualization of the first wonderer was created by Arthur Putnam. It is a sculpture created for the Bohemian Club of San Francisco, entitled, "The First Wonderer" (see Beebe, 1934).

Throughout this book, I have treated cells as if they act as mechanical machines that obey the laws of physics and chemistry. Alan Turing (1992a) and John von Neumann (1958; Neumann and Burks, 1966) predicted that computers would be able to think faster than humans and even replicate. In the future, will silicon-based machines become better thinkers than carbon-based machines and accomplish all the requirements of life?

18.9 CONCEPT OF TIME

Any discussion of evolution is predicated on the assumption that time flows in a linear manner as Isaac Newton assumed in *Principia* (Wayne, 2012a, 2016b). Indeed, time is probably the most important independent variable in all biological research, and biological organisms have an endogenous biological clock to tell time (de Mairan, 1729; Vernon, 1960; Konopka and Benzer, 1971; Bargiello et al., 1984; Zehring et al., 1984; Siwicki et al., 1988; Hardin et al., 1990; Renvoize, 1991; Liu et al., 1992; Vosshall et al., 1994; Millar et al., 1995a,b; Emery et al., 1998; Price et al., 1998; Salisbury, 1998; Somers et al., 1998; van der Hoorst et al., 1999; McClung, 2001, 2006, 2008, 2011, 2013, 2015; Nakajima et al., 2005; Xu et al., 2010; Nair, 2011; Chow and Kay, 2013; Abe et al., 2015; Beilby et al., 2015; Chang et al., 2015; Mori et al., 2015; Nohales and Kay, 2016; Ibáñez, 2017). Thus it is important to ask the question, what is the true nature of time? Throughout human existence, our concept of time has wavered between the idea of absolute and real linear flowing time and the idea that time, like space, is nothing more than an arbitrary axis in the four-dimensional space–time continuum we call our universe (Newton, 1729; Minkowski, 1909; Whitrow, 1961, 1972).

Aristotle even questioned the reality of time, realizing that the past and the future have duration but no existence, whereas the present has existence but no duration (Physics Book IV in Barnes, 1984). According to many religious philosophies, time had a dual nature, and by achieving enlightenment, people could transmigrate from the world of flowing time to the eternal timeless world of the spirit (Macey, 1987). In ancient times, philosophers generally considered time to be a cyclic process as opposed to a linear flow, and in this scenario, the world would periodically undergo a recreation. While the Hindus and Buddhists believed that the universe evolved while undergoing a cyclic process of dissolution and recreation, the Stoics believed that in each recreation, the world and all its events would be identical. Saint Augustine, however, reasoned that time must flow in a linear fashion, for if it did not; the crucifixion of Christ would not be a unique event. However, Saint Augustine also proposed that time was a mental construct.

In constructing his system of the world, Isaac Newton (1729) postulated that time is absolute, and it flows linearly and independently of both mind and matter. He wrote,

> *Absolute, true, and mathematical time, of itself, and from its own nature, flows equably without relation to anything external, and by another name is called duration: relative, apparent, and common time, is some sensible and external (whether accurate or unequable) measure of duration by the means of motion, which is commonly used instead of true time; such as an hour, a day, a month, a year.*

Disagreeing with Newton, Immanuel Kant proposed that time is not a characteristic of physical reality but is only an outcome of the way humans make the world intelligible. He came to this conclusion based on the idea that he could imagine that there were other species that could experience space and time differently from the way human beings do. According to Kant, human beings structure perceptions in such a way that time appears to flow along a mathematical line. Kant's belief in the illusion of linear time was dismissed in the 19th century when geologists, including Charles Lyell (1872) and James Croll (1885), and biologists, including Jean Lamarck (1802) and Charles Darwin (1859), used the concept of absolute linear time to make sense of the evolutionary processes they were proposing. In a recent move that deemphasizes the putative linear nature of time, biologists interested in the relationships between taxa have minimized the concept of the linear flow of time in their conclusions and have developed cladograms to represent the similarities between taxa without committing to any evolutionary processes that are coexistent with a given direction or duration of time.

Henri Poincaré reintroduced Kant's rejection of absolute time based on the fact that it takes a finite time for information from an event to be perceived by an observer, and, as a result, two observers will perceive the order of events depending on their spatial position relative to the events. Albert Einstein (1905, 1926, 1961) expanded this view in his theories of special and general relativity. In his Theory of Special Relativity, Einstein concluded that the measurement of time depends on the velocity of the observer, and the order of events seen by one observer may not be the order of events seen by another. Einstein came to this conclusion by assuming that the speed of light was constant, and any visual information that came to the observer would take a finite time. Thus, an event that took place at a given instant near the observer would seem to take place simultaneously with another event that occurred at a distant place much earlier, if the light rays that formed the images of the two events arrived at the observer at the same time (Bondi, 1964).

According to Ilya Prigogine (Tucker, 1983), "*The attitude of Einstein toward science, for example, was to go beyond the reality of the moment. He wanted to transcend time. But this was the classical view: Time was an imperfection, and science, a way to get beyond this imperfection to eternity. Einstein wanted to travel away from the turmoil, from the wars. He wanted to find some kind of safe harbor in eternity. For him science was an introduction to a timeless reality behind the illusion of becoming. My own attitude is very different because, to some extent, I want to feel the evolution of things. I don't believe in transcending, but in being embedded in a reality that is temporal.*"

Given the strength of relativity theory in describing the world, a reasonable person could conclude that there is no absolute, linear flow of time, and everything that we consider to be past, present, and future was created at once and coexists as coordinates in the four-dimensional space−time continuum. Given this scenario, the flow of time and thus evolution is an illusion. On the other hand, if one accepts the evolution of life as an objective fact, a reasonable person could conclude that there is reason to look for an alternative to the Theory of Special Relativity (see Appendix 2; Hupfeld, 1931).

18.10 SUMMARY

In this chapter, I discussed the role of astronomy and astrophysics in determining that the universe had a beginning. This means that life too must have had a beginning, and consequently, as biologists, we must consider the origin of life. Astronomy also teaches us how to make a series of approximations in a complicated system that is not so easily isolated and subject to experimental measurements and tests. Thus, there are many similarities in the methods used to study the heavens (astronomy) and life (biology).

In this chapter, I have also stressed the chemical nature of life as I was discussing the origin of life. I would like to relate an anecdote told by the biochemist Arthur Kornberg (1989a,b) in his book *For the Love of Enzymes*. In it, he tells of the importance of chemistry to medical science. It goes like this

Physicians are inclined to action. There is the story, often retold, of a surgeon who, while jogging around a lake, spotted a man drowning. He pulled off his clothes, dove in, dragged the victim ashore, and resuscitated him. He resumed his jogging, only to see another man drowning. After he dragged the second one out and got him breathing again, he wearily resumed his jogging. Soon he saw several more drowning. He also saw a professor of biochemistry nearby, absorbed in thought. He called to the biochemist to go after one while he went after another. When the biochemist was slow to respond, he asked him why he wasn't doing something. The biochemist said: 'I am doing something. I'm desperately trying to figure out who's throwing all these people in the lake.'

Kornberg goes on to say,

This parable is not intended to convey a disregard of fundamental issues among physicians nor a callousness among scientists. Rather, it portrays the reality that, in the war on disease, some must contribute their special skills to the distressed individual while others must try to gain the broad knowledge base necessary to outwit both present and future enemies.

In this chapter, I discussed the prebiotic origin of organic chemicals. This can be compared with the mechanism used by autotrophic plant cells to produce the carbon skeletons necessary for life of all other cells. Because autotrophic plant cells directly or indirectly provide the carbon skeletons to all heterotrophic cells, the biblical conclusion that "All flesh is grass" is not far from correct. Now we also know that all life is also made up of a little clay and a smattering of stardust. This scenario does not negate the words of Thomas Jefferson written in the Declaration of Independence, "*We hold these truths to be self-evident, that all men are created equal, that they are endowed by their Creator with certain unalienable Rights, that among these are Life, Liberty and the pursuit of Happiness.*" Is it possible that carbon-based life is making it possible for silicon-based life to evolve on Earth, just like clay-based life may have made it possible for carbon-based life to evolve? Currently, there are discussion on whether robots who look and act like humans have rights (http://blogs.discovermagazine.com/crux/2017/12/05/human-rights-robots/#.W4w-ZLpFzVI; https://money.cnn.com/2018/04/12/technology/robots-rights-experts-warn-europe/index.html; https://www.newyorker.com/magazine/2016/11/28/if-animals-have-rights-should-robots; http://www.euronews.com/2018/04/13/robot-rights-violate-human-rights-experts-warn-eu).

In this chapter, I have considered the universe and the life in it to arise spontaneously by chance or accident. Alfred Russel Wallace disagreed with this conclusion. In his book entitled "The World of Life," Alfred Russel Wallace (1911) wrote about his observations on evolution: "*I argue, that they necessarily imply first, a Creative .Power, which so constituted matter as to render these marvels possible; next, a directive Mind which is demanded at every step of what we term growth, and often look upon as so simple and natural a process as to require no explanation; and, lastly, an ultimate Purpose, in the very existence of the whole vast life-world in all its long course of evolution throughout the eons of geological time. This Purpose, which alone throws light on many of the mysteries of its mode of evolution, I hold to be the development of Man, the one crowning product of the whole cosmic process of life-development; the only being which can to some extent comprehend nature; which can perceive and trace out her modes of action; which can appreciate the hidden forces and motions everywhere at work, and can deduce from them a supreme and overruling Mind as their necessary cause. For those who accept some such view as I have indicated, I show…how strongly it is supported and enforced by a long series of facts and co-relations which we can hardly look upon as all purely accidental co-incidences. Such are the infinitely varied products of living things which serve man's purposes and man's alone not only by supplying his material wants, and by gratifying his higher tastes and emotions, but as rendering possible many of those advances in the arts and in science which we claim to be the highest proofs of his superiority to the brutes, as well as of his advancing civilisation.*"

18.11 QUESTIONS

18.1. What is the strongest evidence for the origin of life from nonliving molecules?

18.2. What is the weakest part of the evidence for the origin of life from nonliving molecules?

The references for this chapter can be found in the references at the end of the book.

Chapter 19

Cell Division

More and more so-called "artifacts" have been shown to be normal 'realfacts'

Edwin G. Conklin, 1950.

The gifts of the microscopes to our understanding of cells and organisms are so profound that one has to ask: What are the gifts of the microscopist? Here is my opinion. The gift of the great microscopist is the ability to THINK WITH THE EYES AND SEE WITH THE BRAIN.

Dan Mazia, 1996.

Cells multiply by dividing.

Dan Mazia (in Epel and Schatten, 1998).

19.1 MITOSIS

In Chapters 16 and 17, I discussed the events that take place during interphase, which includes the S, G1, and G2 phases, when the decondensed chromatin is in a metabolically active condition consistent with its template functions and is undergoing replication and/or transcription (Mazia and Prescott, 1954). The DNA content is 2C during G1, 4C during G2, and intermediate during S. In this chapter, I concentrate on the events that take place during the mitotic phase (M), where RNA synthesis stops and protein synthesis slows down to about 25% of its normal rate. In fact, many cell activities, including cytoplasmic streaming and pinocytosis, stop.

Mitosis was first defined by Walther Flemming (1882) as a mode of indirect nuclear division that provides the daughter nuclei with identical chromosomes (Fig. 19.1). Mitosis is part of the process where the cell changes from oneness to twoness. While mitosis involves the equal division of chromosomes between two daughter cells, I will not discuss the chromosomes themselves as much as I will discuss the structural and motile elements that may be involved in bringing them together to the metaphase plate and then distributing them to the daughter cells (Dumont and Mitchison, 2012). Daniel Mazia (1961b; Earnshaw and Carmena, 2003) considered the chromosomes to be like the corpse at the funeral where everybody expects it to be there, but nobody expects it to play an active role. When we study the movements of chromosomes, we must remember the admonitions of Aristotle in *The Movement of Animals* (Barnes, 1984). That is, the movement of one part depends on the nonmovement of an adjoining part. Thus, to have motion, there must be an anchor that is capable of providing resistance to motion. Archimedes realized the importance of an anchor when using a lever when he said, "Give me a place to stand—and I shall move the world" (see Heath, 1897; John Kennedy, 1960, 1963; Robert Kennedy, 1966; Edward Kennedy, 2000). In mitosis, it is possible that separate proteins act as the lever and the anchor, although a given protein may have dual functions. That is, a cross-bridging protein may act as a motor to induce motion at one time, and as an anchor, against which another motor may exert a force at another time.

The segregation of the genome occurs in all dividing cells; however, mitosis only occurs in eukaryotes. In prokaryotes, genomes may be segregated by the growth of the plasma membrane between the attachment points of the two replicated DNA molecules (Jacob et al., 1963; Funnell, 1993; Toro and Shapiro, 2010; Badrinarayanan et al., 2015). The process of mitosis is far more complex and it has gone through evolutionary change in each eukaryotic kingdom. In general, the segregation of the genome depends less and less on membranes and more and more on polymeric spindle fibers as one progresses up the evolutionary ladder (Kubai, 1975, 1978). The dinoflagellates provide examples of intermediates between typical prokaryotic and eukaryotic genome division (Kubai and Ris, 1969). Knowing that there are both similarities and differences in the structure of the mitotic apparati in various organisms, we might expect that there also will be similarities and differences in the mechanisms involved in mitosis.

Mitosis is a continuous process (Figs. 19.2 and 19.3). For convenience, however, it is divided into six phases: prophase, prometaphase, metaphase, anaphase, telophase, and cytokinesis. Karyokinesis, or the division of the nucleus, is typically, but not always, followed by cytokinesis, where the cytoplasm is divided into two. It should always be kept in mind that mitosis is a three-dimensional process. Charles Walker (1907) wrote in *The Essentials of Cytology*, "It is of the greatest importance to the right understanding of the phenomenon of mitosis that all mental concepts of the various phases of the process should be definitely three

Plant Cell Biology. https://doi.org/10.1016/B978-0-12-814371-1.00019-9

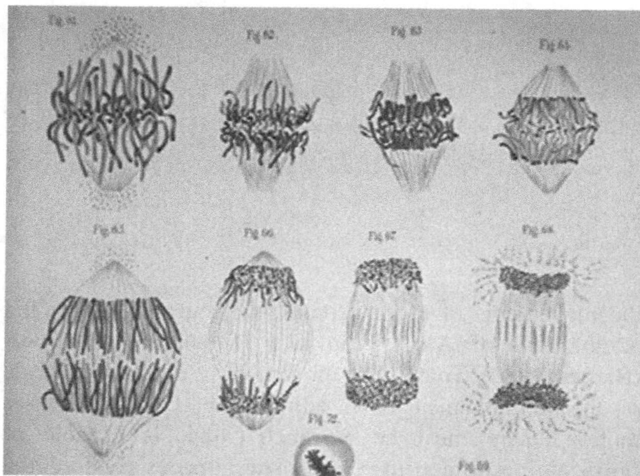

FIGURE 19.1 Mitosis in *Lilium croceum*. *From Flemming, W., 1882. Zellsubstanz. Kern und Zeiltheilung. Vogel, Leipzig. The word mitosis comes from the Greek word mitos (μίτος), which means thread. It describes the thread-like chromosomes observed by Flemming.*

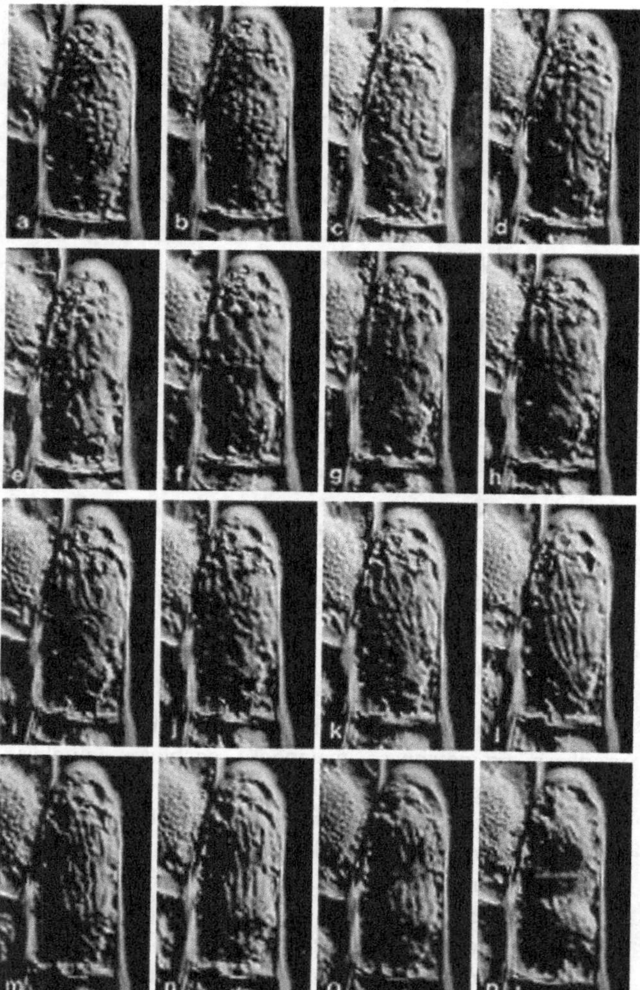

FIGURE 19.2 Time-lapse photographs of mitosis in a stamen hair cell of *Tradescantia*. The entire sequence was shot in 52 minutes. ×680. *From Hepler, P.K., 1985. Calcium restriction prolongs metaphase in dividing* Tradescantia *stamen hair cells. J. Cell. Biol. 100, 1363–1368.*

dimensional." Consequently, Walker included in the book a series of stereopticon cards, which gives one three-dimensional views of mitosis.

Mitosis has been traditionally studied in two types of cells. One type, which is good for morphological studies, is optically clear and has few large chromosomes (Bajer, 1955, 1957, 1958a,b; Bajer and Molé-Bajer, 1956, 1972; Rieder and Khodjakov, 2003; Wakefield et al., 2011). The other type, which is good for biochemical studies, can be easily synchronized and grown in large quantities (Zachleder and van den Ende, 1992). Since the first edition of this book was published (Wakefield et al., 2011), "hypothesis-driven experimentation ...is being replaced by cook-book and largely automated isolations, purifications and rote characterizations of genes and their function." Over the past decade, much work on mitosis has involved using model organisms with small genomes such as yeast, *Drosophila*, and *Arabidopsis*, to discover the genes that code for the proteins that are important for mitosis (Endow et al., 1994; Burgos-Rivera and Dawe, 2012; Ding et al., 2012; Miki et al., 2014; Sedwick, 2014; Petry, 2016; Yamada and Goshima, 2017). Because the genomes of the model organisms have been sequenced and stock centers maintain and make available a wide spectrum of mutants that were produced by a variety of mutagenic techniques, the genes involved in mitosis can be readily identified with standard techniques (see Chapter 21). These *functional genomic* techniques provide a parts list for the protein components essential for chromosome condensation and segregation. Proteomic studies show a minimum of 795 different proteins associated with the mitotic and meiotic spindles (Collado-Romero et al., 2014; Sauer et al., 2005; Petry, 2016), over 14,000 unique sites of phosphorylation on >1000 different proteins (Dephoure et al., 2008), and over 4000 proteins in isolated mitotic chromosomes (Ohta et al., 2010). Moreover, because a polypeptide that is found in a protein complex essential for mitosis may also be involved in protein complexes required for other processes, the mutation may be pleiotropic, and it may be more difficult than expected to resolve the precursor–product relationship between the gene and the trait.

One of the functional genomic techniques used for gene identification is *forward genetics*, which asks which genes are necessary for the completion of mitosis. With this technique, a population of plants is mutagenized chemically or with radiation. The resulting plants are then screened for a phenotype indicative of a mitotic defect. The screen should not be too mild as to produce false positives that will waste resources investigating genes unrelated to mitosis or too stringent as to produce false negatives and miss genes essential for mitosis. It is also possible to screen for a temperature-sensitive or any other conditional mutant so that it is possible to turn a functional protein on or off in a temperature or other conditional way. Once the mutants have been identified, the mutants are crossed with the wild type and the locus of a gene involved in mitosis can be

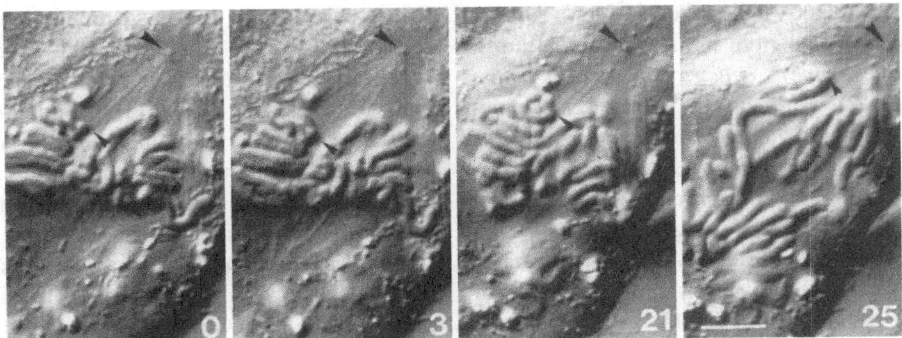

FIGURE 19.3 Mitosis in cultured newt lung cells visualized by Nomarski differential interference contrast microscopy. Metaphase (0 and 3 minutes); anaphase (21 and 25 minutes). The *arrow heads* point to the centrosome and one kinetochore. Bar, 10 μm. The time in minutes as metaphase is given by the numbers. *From Mitchison, T.J., Salmon, E.D., 1992. Poleward kinetochore fiber movement occurs during both metaphase and anaphase-A in newt lung cell mitosis. J. Cell. Biol. 119, 569–582.*

genetically mapped using recombination frequencies based on its linkage to other genes whose positions are already known. If the mapping is fine enough, the segment of the DNA that codes for a polypeptide involved in mitosis can be isolated, cloned, sequenced, and subjected to a bioinformatic analysis to characterize the nature of the polypeptide.

Another functional genomic technique used for gene identification is *reverse genetics*, which asks which aspects of mitosis are affected by a given gene. Using recombinant DNA techniques, a gene product can be eliminated by making a knockout mutant, decreased by making a knockdown mutant, or introduced by making a knockin mutant. A given gene can also be knocked down by transforming the plant with antisense RNA, which binds to the normal mRNA and neutralizes it so that it does not participate in translation. It is also possible to knockout genes with RNA interference (RNAi), which results in the degradation of the transcripts in a particular cell type produced by the gene in question (Bezanilla et al., 2003, 2005; McGinnis et al., 2005; Vidali et al., 2009b; Burgos-Rivera and Dawe, 2012; Nakaoka et al., 2012; Miki et al., 2015; Naito and Goshima, 2015; Shen et al., 2015; MacVeigh-Fierro et al., 2017). The results from reverse genetics are not always easy to interpret because transgenes integrate into the genome at unpredictable sites, and it is possible that the position of the transgene will affect its expression. Moreover, the transgene may be inserted into another gene thus generating a new mutation.

The unintended and complicating consequences of reverse genetics described above can be overcome using *homologous recombination* techniques. To achieve homologous recombination, a piece of DNA that codes for the gene of interest is engineered to code for a nonfunctional protein. The DNA is then ligated to two flanking sequences that will recognize the position of the gene of interest in the genome. The DNA repair machinery in the cell will then replace the functional gene in its natural location with the engineered gene. Gene editing using homologous recombination or CRISPR/Cas9 to identify genes involved in mitosis are particularly effective in the moss *Physcomitrella* (Schaefer and Zrÿd, 1997; Reski, 1998a,b; Collonnier et al., 2016; Lopez-Obando et al., 2016; Yamada and Goshima, 2017).

While it is possible to use the above techniques to identify novel sequences of DNA involved in mitosis, most genetic studies in plants have documented the presence of sequences of DNA in plants that are already known to be involved in mitosis in yeast or other model organisms. In principle, novel sequences of DNA can also be discovered doing the unbiased omic approach discussed in Chapter 21. To establish a context in which the genetic and omic studies can be understood, I will discuss the mechanics, morphology and movements that occur during each phase of mitosis that can be observed with light and electron microscopy. If we apply synthesis with analysis and integrate the molecular with the structural and theoretical work, progress in understanding cell division will not be described as it was by Edwin Conklin (1950) as like "that of a squid, which moves rapidly backward, at the same time emitting large quantities of ink (Mazia, 1960)."

19.1.1 Prophase

Prophase begins as the chromatin starts to condense from its decondensed metabolically active state into well-defined chromosomes that are more easily moved through the cytoplasm (Belmont, 2006). Each chromosome is composed of two sister chromatids. Each sister chromatid contains one of the two strands of DNA that result from the semiconservative DNA replication process that took place during the previous S phase (see Chapter 16). The sister chromatids are linked by a protein complex known as *cohesin*, which is incorporated during DNA replication (Kerrebrock et al., 1995; Mitchison and Salmon, 2001; Liu et al., 2002; Kitajima et al., 2004; Marston et al., 2004; Salic et al., 2004;

McIntosh et al., 2012; Bolaños-Villegas et al., 2017). During prophase, cohesins dissociate from the chromosomal arms, remaining only at the *primary constriction* or *centromere* (McIntosh et al., 2012). At the same time, a mechanochemical enzyme known as *condensin* is added to the chromosome, causing it to condense (Hirano, 2000, 2002, 2016; Liu et al., 2002; Shintomi and Hirano, 2011; Shintomi et al., 2015, 2017). Typically, each somatic cell has an even number of homologous chromosomes, one set from each parent. (Unfortunately, textbooks often show diagrams of mitotic cells with an odd number of chromosomes, as if each parent contributed unequally to the offspring.)

During prophase, the nuclear envelope, which contains the SUN complex (see Chapter 16: Oda and Fukuda, 2011), is intact and the mitotic spindle begins to form outside the nuclear envelope from microtubule organizing centers (MTOCs), which are sometimes called *centrosomes* or *centrospheres* (De Mey et al., 1982; McIntosh, 1983; Smirnova and Bajer, 1994, 1998; see Chapter 11). The centrosome, which contains a pair of perpendicular centrioles, is surrounded by pericentriolar material consisting of *γ-tubulin*, microtubule-associated proteins, protein kinases and protein phosphatases (Scheibel, 2000; McIntosh et al., 2012; O'Toole et al., 2012). In the fungi, the centrosomes are known as *spindle pole bodies* (Aist and Williams, 1972). The centriole has been often, although erroneously, considered to be the organizer of the spindle during prophase for over 100 years. In 1895, E. B. Wilson wrote, "It seems certain that the centriole is in many cases a definite morphological body lying within the reticulated centrosphere. ... There are, however, many grounds for accepting the view that the centriole, though a frequent, is not a necessary element of the centrosphere."

Evidence that the centriole is not required for spindle formation or mitosis comes from the observation that most plant cells and some animal cells do not have centrioles, although they still form mitotic spindles and undergo mitosis (Mazia, 1961b; Inoué, 1964; Pickett-Heaps, 1969a, 1971; Duncan and Wakefield, 2011; Malcos and Cyr, 2011; Zhang and Dawe, 2011; Meunier and Vernos, 2016). Moreover, when the centriole of animal cells is destroyed by laser irradiation, the spindle still forms, and the cell undergoes mitosis (Khodjakov et al., 2000; Magidson et al., 2007). By contrast, irradiation of the pericentriolar material prevents spindle formation, indicating that it is something in the centrosome besides the centrioles that is required for spindle formation (Berns et al., 1981). The pericentriolar material is rich in γ-tubulin. Experiments aimed at forming spindles in vitro indicate that the factors necessary for spindle formation include dynein, dynactin, a dynein-activating protein, and kinesin (Merdes and Cleveland, 1997).

According to Mazia (1960), "One is inclined to conclude that if plant cells do not have centrioles they do possess equivalents that we may call whatever we please. Out of deference to those who are impressed by the absence of morphologically distinct centrioles, let us call the polar organizers in plant mitosis 'euphemisms'." We will call them MTOCs. Thus, the MTOC contains the fundamental substances necessary to initiate the prophase spindle. These substances include γ-tubulin and additional proteins, which provide the nucleation site for microtubule growth (Liu et al., 1993; Murata et al., 2005; Wiese and Zheng, 2006; Goshima et al., 2008; Uehara and Goshima, 2010). The MTOC can have a diverse array of shapes, and in many plant cells, the MTOC may be dispersed along the nuclear envelope (Wick et al., 1981; Mazia, 1984; Mizuno, 1993). The microtubules generated from this MTOC form an optically clear zone around the nucleus. The clear zone may be thicker at the polar regions, forming polar caps (Bajer, 1953; Bajer and Molé-Bajer, 1972; Vandré and Borisy, 1989; Marc, 1997; Yamada and Goshima, 2017). The division of the MTOC is another sign that a cell is going from a state of oneness to a state of twoness.

The formation of the mitotic apparatus differs in liverworts, hornworts, and mosses compared with higher plants (Yamada and Goshima, 2017). In addition, the mitotic apparatus of seed plants can exist in a multitude of forms. It can appear spindle shaped, barrel shaped, or almost amorphous (Vejdovský, 1926–1927; Belar, 1929; Mazia, 1961a,b; Ambrose and Cyr, 2007). Forms that are even more diverse can appear in hybrids (Darlington and Thomas, 1937; Walters, 1958). Ivor Cornman (1944) suggested that each chromosome has an independent spindle, and Mottier (1903) and Bungo Wada (1966) suggested that each more or less independent spindle joins together to form a bipolar mitotic apparatus of which the shape depends on the shape of the cell. Thus, the concept of independent spindles explains the lack of bipolar organization that exists in the mitotic apparatus of long, thin pollen tubes (Sax and O'Mara, 1941; Palevitz and Cresti, 1989; Liu and Palevitz, 1991), the oblique metaphase plate that occurs in polyploid wheat (Wada, 1966), and other short squat cells with large chromosomes (Palevitz and Hepler, 1974a). Smirnova and Bajer (1993) think that multiple, short-lived nucleating sites may regulate spindle formation. Perhaps the diverse organizations of spindles identifiable in the biological world also represent varying degrees of unity of chromosomes that came together into the same genome by various mechanisms, including duplication and hybridization. Interestingly, when two independent spindles form in a cell created by cell fusion techniques, the two spindles show some degree of coordination (Rieder et al., 1997).

During spindle formation in prophase, the poles move away from the nuclear envelope as the microtubules grow. The separation of the poles, which results in the formation of the bipolar spindle, requires a motive force. The pushing force may be provided by a kinesin-related protein because it has been shown that immunodepletion of a kinesin-related protein prevents bipolar spindle formation (Sawin et al., 1992a,b). A kinesin-related protein may not be the only motor protein involved in spindle formation because injection of dynein antibodies also inhibits spindle formation during prophase (Vaisberg et al., 1993). Of course, it is possible that kinesin or dynein act as an anchor and not a motor and that the antibody interferes with its anchoring function. Spindle formation may be a result of the actions and reactions of mediated by a variety of microtubule-associated proteins. Spindle orientation also depends on kinesin and dynein in some cells; in others, it depends on actin-related proteins (Stearns, 1997).

During prophase, the viscosity of the nucleus in the regions surrounding the chromosomes is still fairly low, and small particles move in all directions. Just before prometaphase, the spindle undergoes a marked contraction and the oscillations of the particles become linear in the direction of the poles (Bajer, 1958a; Molé-Bajer, 1958). The viscosity of the nucleus at this stage is about 0.282 Pa s (Alexander and Rieder, 1991). The nucleolus usually disappears during prophase, although it may persist throughout mitosis in some cases (Bajer, 1953; Pickett-Heaps, 1970; Vaughan and Braselton, 1985).

19.1.2 Prometaphase

The onset of prometaphase is marked by the breakdown of the nuclear envelope, although the nucleoplasm and the cytoplasm remain visually distinct (Yasui, 1939; Wada, 1950, 1966, 1970, 1972). While the nuclear envelope typically breaks down and becomes indistinguishable from the endoplasmic reticulum (ER), the integrity of the nuclear envelope is not completely lost as the nuclear envelope–ER complex forms a loose sheath that encircles and isolates the mitotic apparatus from the rest of the cell (Hepler, 1980, 1989b). Nuclear pores are no longer seen at this stage, but fenestrated lamellae, which look like nuclear pores without the pore complex, are prevalent (see Fig. 4.6 in Chapter 4). The SUN complex is probably associated with the ER in the spindle (Oda and Fukuda, 2011). A lamin-like protein associates with the chromosomes from prometaphase through anaphase (Sakamoto and Takagi, 2013). It is not known whether the proteins of the pore complexes associate with specific parts of the chromosomes during mitosis. If this happens, the nuclear pores may be positioned next to specific genes during a given stage in development in each cell type. The nuclear envelope never breaks down in many lower plants, animals, and fungi, and consequently, the spindle forms inside the nucleus (Wise, 1988; Rose, 2007).

If and when the nuclear envelope breaks down, the spindle microtubules permeate the nuclear region and attach to the kinetochores (Walczak and Heald, 2008; Gatlin and Bloom, 2010). The nuclear envelope–ER complex also permeates the spindle and runs coparallel with the microtubules that attach to the kinetochores (Hepler and Wolniak, 1984). When the nuclear envelope breaks down, cytosolic *condensin* enters the nucleoplasm and its mechanochemical action condenses the chromosomes further (Kimura and Hirano, 1997; Sakamoto et al., 2011).

The spindle fibers typically attach to a visibly constricted region of the chromosome known as the *primary constriction*, *kinetochore*, or *centromere* (Sharp, 1934). The terms *kinetochore* and *centromere*, along with 25 other terms, define the part of the chromosome that is attached to a spindle fiber (Schrader, 1944). The terms were defined to mean the same structure, and consequently, I will use them interchangeably. However, there is a tendency in the literature to refer to the DNA and protein in the primary constriction as the centromere and kinetochore, respectively. In some monocots, as well as in the arthropods, the chromosomes do not have a visible primary constriction, and the microtubules attach along the whole length of the chromosome. In this case, each of the chromatids has a diffuse kinetochore.

Localized kinetochores appear to be crescent-shaped, trilaminar plates or diffuse balls in the primary constriction, which have similar staining properties as centrioles (Schrader, 1936). Kinetochores can be isolated from mitotic chromosomes, and the proteins that make up the kinetochore are being identified. They include a centromere-specific form of histone H3 (CENP-A), microtubule-associated proteins, plus end– and minus end–directed motors, as well as protein kinases and protein phosphatases (Brinkley et al., 1989; Brinkley, 1990; Molé-Bajer et al., 1990; Compton et al., 1991; Liu and Palevitz, 1992; McEwen et al., 1993; Starr et al., 1997; Yu et al., 2000; Welburn and Cheeseman, 2008; Zhang and Dawe, 2012; McIntosh et al., 2012; Ishii et al., 2015). When a localized kinetochore is in the middle of a chromosome, the chromosome is said to be metacentric; when it occurs between a short arm and a long arm, the chromosome is said to be *acrocentric*; and when it occurs at the very end, the chromosome is said to be *telocentric*.

Van Beneden and Neyt (Haydon et al., 1990) first proposed in the late 1880s that the chromosomes attach to fibers generated by the spindle poles. The attachment process results in rapid, irregular oscillations of the chromosomes. This frenzied movement, which is called *metakinesis*, takes place at a rate of 2.5 μm/s. As a result of this movement, the chromosomes become drawn out to fine points at their kinetochores and then relax as they continue to interact with the spindle. Initially, the kinetochore of each chromosome moves toward one or the other pole. By the end of four to six of these rapid oscillations, the arms of the chromosomes have been moved to the metaphase plate where they extend perpendicular to the central spindle (Pickett-Heaps et al., 1980; Pickett-Heaps, 1991; Nagele et al., 1995; Magidson et al., 2011).

Tippit et al. (1980) interpret the frenzied movement of chromosomes to indicate that microtubules radiate from the pole and capture the kinetochore. Usually one kinetochore of a pair is captured by a microtubule before the other. The attachment of a microtubule from a pole leads to the movement of the chromosome toward that pole. Eventually the other kinetochore of the pair interacts with a microtubule from the other pole, and the chromosome undergoes a series of linear oscillations between the two poles until the chromosomes align along the metaphase plate, equidistant between the two spindle poles. If both kinetochores of the same chromosome interact with two microtubules from the same pole, the chromosomes undergo unstable oscillations with a maximal velocity of 2.5 μm/s. During these unstable oscillations, one of the kinetochores is able to interact with microtubules from the other pole, and the chromosome then undergoes stable oscillations (0.5 μm/s). During the stable oscillations, additional microtubules attach to the kinetochore, and a discrete kinetochore spindle fiber is built. If chromatids are separated by laser microsurgery so that each "chromosome" has only one kinetochore, the chromosome still congresses to the metaphase plate as long as it is captured by microtubules from both poles (Khodjakov et al., 1997). This means that a single kinetochore may have two independent domains, and the motile function of each domain acts independently.

Using newt lung cells, which have a particularly optically clear nuclear region, Haydon et al. (1990) have shown that the microtubules do radiate from the centrosome during prometaphase and then capture kinetochores. These microtubules exhibit dynamic instability (see Chapter 11). They elongate at a rate of 0.24 μm/s and shorten at a rate of 0.26 μm/s. Each microtubule repeatedly elongates toward the chromosome and catastrophically shortens until it finds a kinetochore within its "casting range." When a microtubule attaches to the kinetochore, the chromosome moves poleward (Figs. 19.4 and 19.5). According to Rieder and Alexander (1990),

> *The MTs [microtubules] can therefore be envisioned to be analogous to a stationary 'fisherman' who casts radially about in search of fish in the surrounding water. As with chromosomes in NPs [Newt pnematocytes] hungry fish not within the range of casts will not be caught until either the fisherman moves closer to the fish or the fish wanders closer to the fisherman. Our data indicates that the 'casting' range of the NP centrosome is seldom > 50 μm.*

After the microtubule captures the kinetochore, the kinetochore moves along its surface (Rieder and Alexander, 1990). The microtubule does not terminate at the kinetochore, but some distance past it, indicating that poleward movement of the chromosome does not have to be coupled to microtubule disassembly. Moreover, this observation indicates that the motor is on the surface of the microtubule, on the corona of the kinetochore, or both. Rieder and Alexander suggest that dynein provides the force to move the chromosomes poleward at the observed speeds of 0.4–0.9 μm/s. Dynein has been localized in the kinetochores (Pfarr et al., 1990; Steuer et al., 1990), and antibodies directed against dynein inhibit prometaphase movements (Wise and Bhattacharjee, 1992). Other kinetochore proteins may also be involved in prometaphase movement (Bernat et al., 1990; Cooke et al., 1990; Simerly et al., 1990; Lombillo et al., 1995b). Laser microsurgery experiments show that the kinetochore that follows the leading kinetochore to the pole does not exert any pushing force (Khodjakov and Rieder, 1996). These data indicate that, in these cells, each kinetochore can exist in two states: the pulling state and the neutral state.

Gorbsky and Ricketts (1993) found an antibody that recognizes a phosphoprotein in kinetochores. In mid-prometaphase, when the chromosomes are oscillating

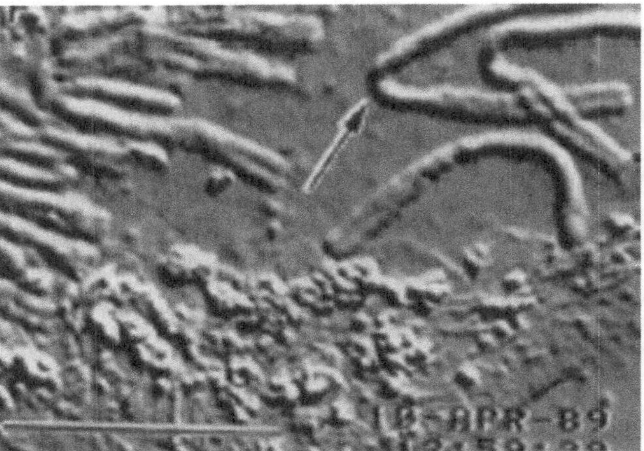

FIGURE 19.4 Nomarski differential interference micrograph of a newt lung cell undergoing prometaphase movement. Bar, 20 μm. *arrow*, kinetochore. *From Rieder, C.L., Alexander, S.P., 1990. Kinetochores are transported poleward along single astral microtubule during chromosome attachment to the spindle in newt lung cells. J. Cell Biol. 110, 81–95.*

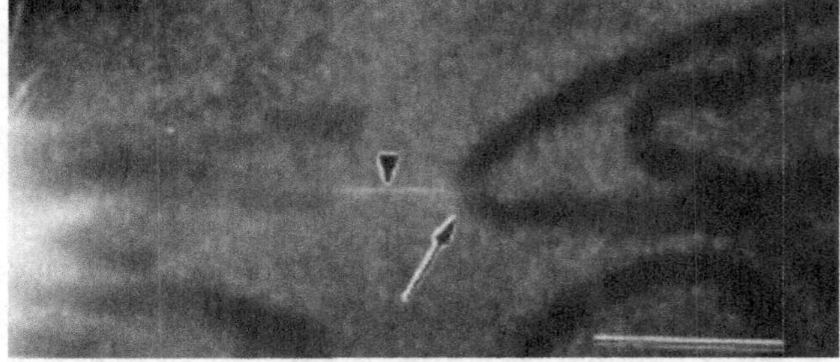

FIGURE 19.5 Immunofluorescence micrograph showing a single microtubule that attached to a kinetochore during prometaphase movement in newt lung cells. Bar, 10 μm. *arrow*, kinetochore; *arrowhead*, single microtubule. *From Rieder, C.L., Alexander, S.P., 1990. Kinetochores are transported poleward along single astral microtubule during chromosome attachment to the spindle in newt lung cells. J. Cell Biol. 110, 81–95.*

rapidly, some chromosomes show staining on both sister kinetochores, and others show staining on only one sister kinetochore. It turns out that in the case where one kinetochore is stained, only the leading kinetochore is stained. Gorbsky and Ricketts conclude that dephosphorylation prevents the activity of a minus end–directed motor such as dynein or a minus end–directed kinesin.

Currently there are two hypotheses to explain how the chromosomes move away from the pole during prometaphase. One is that the polymerization of microtubules from the poles push the chromosomes away as the microtubules polymerize (Haydon et al., 1990). The alternative hypothesis is that kinesin-related proteins provide the plus end–directed force. These hypotheses are not mutually exclusive. The first hypothesis is supported by the observation that chromosome oscillations are inhibited in vivo by the microtubule-stabilizing agent taxol, and that chromosome fragments are pushed away from the closest pole (Ault et al., 1991; Rieder, 1991). The polar ejection forces do not occur in all cells. In fact, in *Haemanthus* endosperm cells, chromosome fragments are pushed toward the closest pole during prometaphase. The poleward force is inhibited by colchicine. Thus, in contrast to vertebrate cells, where polar ejection forces are typical, in *Haemanthus*, polar ejection forces are absent, and microtubule-dependent, kinetochore-independent poleward forces exist (Khodjakov et al., 1996).

The hypothesis that chromosomes move away from the poles due to the action of a kinesin-related protein is supported by the observations that kinesin-related proteins are present in kinetochores of mitotic cells (Sawin et al., 1992b), and that kinesin-related genes appear to regulate mitosis and meiosis (Rieder, 1991; Roof et al., 1992; Hoyt et al., 1992). However, if in fact the kinesin only acted as a bridge or anchor between the kinetochore and a kinetochore microtubule (MT) (Desai and Mitchison, 1995), then the polar ejection force could be the only pushing force, and the prometaphase motor in the kinetochore would have only a pulling and a neutral state (Khodjakov and Rieder, 1996; McEwen et al., 1997).

During prometaphase, the microtubules that become attached to kinetochores are more stable than nonkinetochore microtubules (Cassimeris et al., 1988; Falconer et al., 1988; Huitorel and Kirschner, 1988; Salmon, 1989). Cooling, heating, cell lysis, high pressure, and treatment with colchicine and other microtubule inhibitors cause the nonkinetochore microtubules to depolymerize before the kinetochore microtubules. If biotinylated tubulin is microinjected into the cell at prometaphase, nonkinetochore microtubules become fully labeled within 1 minute, whereas kinetochore microtubules take 10 minutes to become fully labeled (Mitchison et al., 1986). These data indicate that the tubulin dimers in kinetochore microtubules turn over more slowly than the tubulin dimers in nonkinetochore microtubules.

Current data indicate that the kinetochores themselves do not initiate the majority of the microtubules under normal conditions in vivo (Summers and Kirschner, 1979), although the possibility remains that they initiate one or two. At one time it appeared that the kinetochore was the primary site of microtubule initiation based on the observations that microtubules are attached to the kinetochores in normal cells (Harris and Bajer, 1965); the kinetochore has similar staining properties as the centriole (Schrader, 1936); microtubules can grow from the kinetochore in vivo (Carothers, 1936; Pollister and Pollister, 1943) and in vitro (McGill and Brinkley, 1975; Snyder and McIntosh, 1975; Telzer et al., 1975; Gould and Borisy, 1978); kinetochores nucleate microtubules in vivo in cells that had been previously treated with microtubule-depolymerizing agents and later released from inhibition (Ris and Witt, 1981), and in *Xenopus* egg extracts, beads coated with mitotic chromatin containing RanGTP nucleate microtubules and form spindles (Heald et al., 1996; Dinarina et al., 2009; Duncan and Wakefield, 2011; Cavazza and Vernos, 2016; Meunier and Vernos, 2016). However, when the concentration of tubulin in the various experiments is taken into consideration, it seems that the kinetochores are only capable of initiating microtubules under conditions where the concentration of free tubulin dimers is greater than would be expected under normal conditions in vivo. In normal mitotic cells, the poles and nuclear envelope are the main, if not the only, sites of mitotic microtubule initiation or stabilization (Salmon, 1989). However, chromosomes are necessary for spindle formation because the removal of the chromosomes with micromanipulators prevents the formation of the spindle (Zhang and Nicklas, 1995).

The kinetochores are capable of organizing microtubules that have been initiated elsewhere, and therefore, Euteneuer and McIntosh (1981) suggest that the kinetochores should still be considered MTOCs, but the concept of MTOCs should be broken up into two parts: microtubule-initiating sites and microtubule-positioning sites. Sometime after prometaphase, centrosomes become dispensable because they can be removed by micromanipulation and mitosis still proceeds normally (Hiramoto and Nakano, 1988).

19.1.3 Metaphase

Metaphase begins when the chromosomes line up in a plane midway between the two poles (Walker, 1907; Bajer, 1958b; see Figs. 19.6 and 19.7). The formation of a metaphase plate is not essential for mitosis, and in cells, particularly those of the fungi, the chromosomes do not line up on a metaphase plate (Aist and Williams, 1972). The mitotic apparatus is fully developed during metaphase (Petry, 2016). While the fibers in the mitotic apparatus were readily observed in fixed and stained cells, they were not

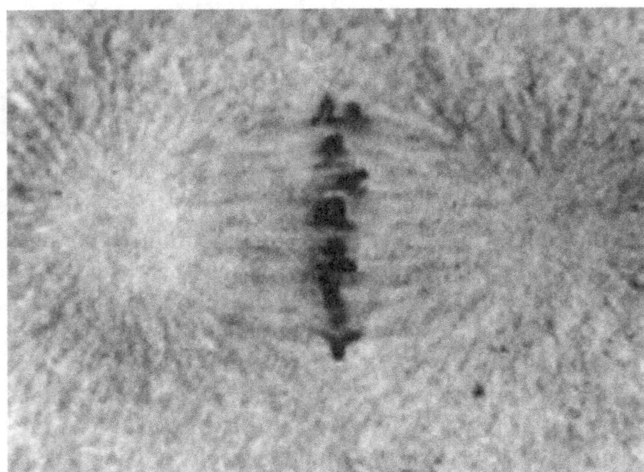

FIGURE 19.6 Photomicrograph of metaphase in the zygote of *Toxopneustes. From Wilson, E.B., 1895. An Atlas of the Fertilization and Karyokinesis of the Ovum. MacMillan, New York.*

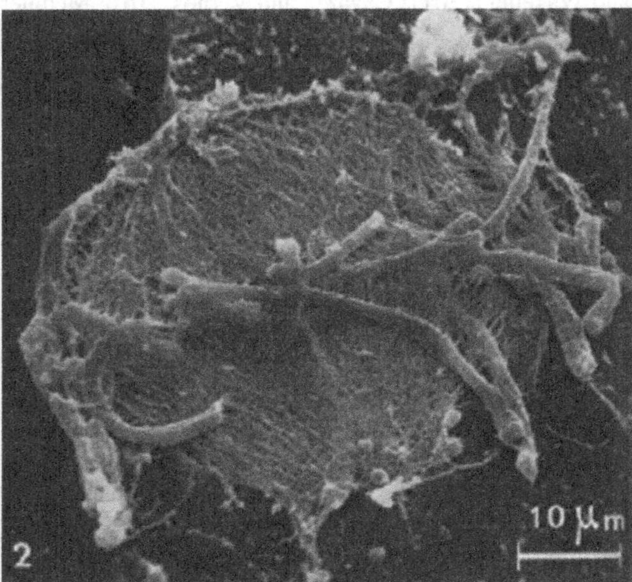

FIGURE 19.7 Scanning electron micrograph of metaphase in the endosperm cells of *Haemanthus katherinae. From Heneen, W.K., Czajkowski, J., 1980. Scanning electron microscopy of the intact mitotic apparatus. Biol. Cell. 37, 13–22.*

readily observed in living cells and were considered a fixation artifact (Schrader, 1934). The reality of the spindle fibers began to be appreciated when they could be seen in living cells with polarized light microscopy (Kuwada and Nakamura, 1934; Schmidt, 1936, 1937, 1939; Schmitt, 1939; Swann, 1951a,b, 1952; Inoué and Dan, 1951; Fuseler, 1975; Dell and Vale, 2004; Gouveia, 2010; Inoué, 2011, 2016; Tani et al., 2016), and the spindle could be isolated as a semi-functional unit (Mazia et al., 1951, 1961; Mazia and Dan, 1952; Mazia, 1953, 1956, 1961a, 1972, 1975; Paweletz, 1996; Schatten and Epel., 1997; Epel and Schatten, 1998). It was not until the introduction of electron microscopy, however, that we could really get an appreciation of how the fibers were arranged in the mitotic apparatus. We know that in general, the spindle is mainly composed of two kinds of microtubules—those that run from the pole to the kinetochore, which are called *kinetochore microtubules*, and the others that do not attach to kinetochores (Heath, 1981; McDonald, 1989). This distinction is less than clear because the microtubules join together and appear as "fir trees" (Bajer and Molé-Bajer, 1986), the formation of which may require the *augmin* protein complex (Meunier and Vernos, 2016; Yamada and Goshima, 2017).

The arrangement of microtubules varies from spindle to spindle (Pickett-Heaps, 1969a; Kubai, 1975; Fuller, 1976). Paweletz (1967) finds that the nonkinetochore microtubules do not run continuously from pole to pole, but different sets of microtubules are formed from each pole and their plus ends overlap in the middle. On the other hand, Ding et al. (1993) show that some microtubules are continuous from pole to pole. Likewise, three-dimensional reconstructions of serial-sectioned spindles indicate that the kinetochore microtubules are continuous between the kinetochore and the spindle pole in some cells (McDonald et al., 1992; Ding et al., 1993), but not in others (Schibler and Pickett-Heaps, 1987).

The orientation of the microtubules can be inferred from the observations by Ritter et al. (1978) who observed microtubules grow from the poles to the kinetochores in *Barbulanympha*. This would suggest that the minus end is situated at the centrosome and the plus end extends toward the kinetochore. Euteneuer and McIntosh (1980, 1981a,b) and McIntosh et al. (1980) have confirmed this polarity of microtubules in PtK_1 and *Haemanthus* cells using neurotubulin to decorate the microtubules. When they add neurotubulin to fixed material, it forms hooklike projections from the microtubules that are clockwise when we are looking from the plus end. Euteneuer and McIntosh (1981a,b) find that 90%–95% of the microtubules between the kinetochores and the pole are oriented with their plus end toward the kinetochore and their minus end toward the pole. The polarities of the microtubules in each half-spindle are opposing. Using dynein to decorate the microtubules of the meiotic spindle of the clam *Spisula*, Telzer and Haimo (1981) also show that the majority of the microtubules in the half-spindle have the same polarity; that is, the minus end is oriented toward the pole and the plus end is oriented toward the kinetochore.

During metaphase, the microtubules of the spindle are not static but continuously exchange subunits with the pool of free tubulin dimers. This can be seen by microinjecting fluorescently labeled tubulin into the cell and observing the redistribution of fluorescence after photobleaching

(Salmon et al., 1984a,b,c; Wadsworth and Salmon, 1986; Hush et al., 1994). These experiments show that the turnover time for the microtubules in the spindle is about 30 seconds. These experiments confirm that the kinetochore microtubules are more stable than the nonkinetochore microtubules during metaphase as they are during prometaphase (Mitchison et al., 1986).

During metaphase, as in prometaphase, the attachment of microtubules to the kinetochore does not prevent tubulin exchange at the plus end of the kinetochore microtubule. Mitchison et al. (1986) microinjected biotinylated tubulin during metaphase and fixed the cells at various times after microinjection. They found that within 1 minute, biotinylated tubulin was incorporated into the kinetochore microtubules next to the kinetochore. Wise et al. (1986) found the same thing at the light microscope level using fluorescently labeled tubulin. Thus, tubulin exchange occurs at the plus ends of microtubules stabilized by the kinetochore.

The haphazard metakinetic movements of the chromosomes that occur during prometaphase give rise to the apparently ordered positioning of chromosomes midway between the two poles. This position is known as the *metaphase plate* (see Figs. 19.6 and 19.7). How do the chromosomes congress exactly to this position? van Beneden (1883), Cornman (1944), and Östergren (1949, 1951; Östergren and Prakken, 1946; Heneen, 2014) proposed that the fibers, which appear between the kinetochores and the poles, exert force on the chromosomes. The tensile force (F, in N) produced by elastic fibers, according to Hooke's Law (Hooke, 1678; Moyer, 1977), depends on the spring constant (K, in N/m) and length (x, in m) of the contractile fiber:

$$F = -Kx \qquad (19.1)$$

As the tensile force is proportional to the length, the longer the fiber, the greater the force available to pull the chromosome to the pole. If a chromosome was not on the metaphase plate, the traction fiber on one side would be longer than the traction fiber on the other side, and the longer fiber would pull with greater force until the chromosome lined up midway between the poles. Currently, there is a consensus that the microtubules make up the traction fibers. However, this conclusion may be provisional, and microtubules may only play a part in the tensile system (Pickett-Heaps et al., 1997).

Gunnar Östergren (1949, 1951; Heneen, 2014) believed that the traction fibers were made out of protein subunits that were themselves contractile. However, it now seems more likely that the tension may be due to the presence of plus end–directed motors that are anchored in the spindle matrix near the pole. If the motors bound to microtubules in a spatially periodic manner, then the force they generated would be proportional to the length of the spindle fiber (Mitchison, 1989; Salmon, 1989).

The tension that holds the chromosomes at the metaphase plate has been demonstrated in experiments in which the kinetochore microtubules are irradiated with an ultraviolet or laser microbeam that severs the microtubule–kinetochore connection. On microbeam irradiation, the distance between the nonirradiated kinetochore of the pair and the spindle pole becomes shortened, indicating that the chromosome was under tension (McNeill and Berns, 1981; Leslie and Pickett-Heaps, 1983; Hays and Salmon, 1990; Skibbins et al., 1995). The proteins involved in sister chromatid cohesion are also necessary for the full development of tension (Salic et al., 2004).

Arthur Forer (1965) helped pioneer the technique of optically identifying and analyzing the movements of microtubules in the mitotic apparatus of living cells. Forer's technique, which is presented in a paper that uniquely uses peace signs for symbols in some of the figures, has undergone many modifications to ensure that we can visualize the kinetochore microtubules and not the other elements of the spindle. Mitchison (1989), who has helped pioneer the modifications, injected a type of tubulin into the cell that becomes fluorescent on relatively low-light irradiation. He then activates the fluorescence in a small area of the spindle with a microbeam. Mitchison found that the fluorescent bar moves poleward during metaphase at a velocity of 0.4–0.6 μm/minute. He concluded that the forces that cause the poleward flux might also cause the tension in the metaphase spindle (Fig. 19.8).

The fact that the tension is necessary to provide stability to the metaphase plate can be demonstrated by micromanipulation of the meiotic chromosomes of grasshopper spermatocytes (Nicklas and Koch, 1969; Henderson and Koch, 1970). When both kinetochores of a chromosome are experimentally attached with a microneedle to one pole, the chromosome oscillates rapidly. However, when the chromosome fibers are restrained with a microneedle, the unstable oscillations stop, and the chromosome becomes stable. Thus, the kinetochores cannot intrinsically distinguish between the two poles, but respond to the tension normally imposed on them (Nicklas and Staehly, 1967; Nicklas, 1967). The bipolar orientation of kinetochores at the metaphase plate, which ensures the equal distribution of the hereditary material, is the most stable configuration.

R. Bruce Nicklas (1997) and his colleagues have shown that not only do the cells sense tension but also the tension directly affects the phosphorylation state of a protein in the kinetochore. Li and Nicklas (1995) have shown that when a univalent chromosome is present in a meiotic cell, the cell does not enter anaphase for 5–6 hours. However, if the chromosome is artificially held under tension with microneedles, the cell enters anaphase in less than an hour. When the chromosome is not under tension, a protein in the kinetochore is phosphorylated by a protein kinase known as *aurora b* (Terada et al., 1998; Adams et al., 2001; Zeitlin

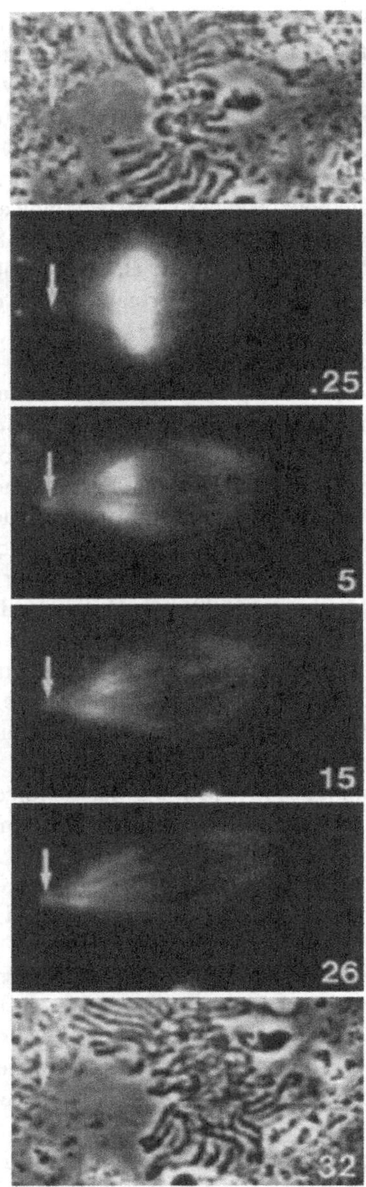

FIGURE 19.8 The visualization of tubulin in newt lung cells by fluorescence microscopy following a photoactivation pulse. The tubulin that has incorporated into the spindle fibers is photoactivated and can then be visualized to move toward the centrosome (*arrows*) during metaphase. The numbers in the corner indicate time following photoactivation. The top and bottom phase-contrast micrographs show the cell.

et al., 2001; Kallio et al., 2002; Demidov et al., 2005, 2014; Welburn et al., 2010). However, holding the univalent under tension with a microneedle causes a dephosphorylation of the kinetochore protein (Nicklas et al., 1995). Tension may result in an intrakinetochore stress that pulls the substrate from the kinase and increases the access of the substrate of a protein phosphatase that dephosphorylates the protein (Khodjakov and Pines, 2010; Maresca and Salmon, 2010; Lampson and Cheeseman, 2011). The greater the tension to which a kinetochore is exposed, the greater the number of microtubules that are attached to that kinetochore. It is likely that once a sufficient number of microtubules attach to the kinetochore, the attachment-sensitive proteins become dephosphorylated (Nicklas et al., 2001). Here is an excellent demonstration of how physical energy can be converted into chemical changes in biological systems.

Errors in the attachment of chromosomes to the spindle could lead to cells with missing or extra chromosomes. If this were to occur during crucial stages of development, birth defects could occur (Nicklas et al., 2001). The attachment-sensitive proteins act as a checkpoint during mitosis and meiosis. Unattached kinetochores have a phosphoprotein known as *MAD2* (mitotic arrest deficient) that binds to the attachment-sensitive protein (Yu et al., 1999; Kimbara et al., 2004). On attachment of the kinetochore to the spindle, the phosphoprotein is dephosphorylated and MAD2 detaches from the dephosphorylated protein. Anaphase will not commence until all the MAD2 proteins have been lost from the kinetochores. It appears that the loss of the MAD2 protein acts as a "proceed to anaphase" signal that prevents errors in attachment from occurring during mitosis and meiosis (Nicklas et al., 2001).

19.1.4 Anaphase

Anaphase begins abruptly as the paired kinetochores of each chromosome split (Pauliulis and Nicklas, 2000). However, microtubule-mediated tension is not required for chromosome separation because the chromatids separate synchronously from each other even when the microtubules in the spindle have been depolymerized by colchicine (Beams and King, 1938; Levan, 1938; Nebel and Ruttle, 1938; Walker, 1938; Shimamura, 1939; Berger and Witkus, 1943; Molé-Bajer, 1958). The separation of sister chromatids seems to be due to the activation of a protease known as *separase* that degrades the *cohesin* complex at the centromeres that hold together sister chromatids (Earnshaw and Bernat, 1991; Cooke et al., 1987; Uhlmann et al., 1999, 2000; Peters, 2002; Onn et al., 2008; Bolaños-Villegas et al., 2017). Chromosomes separate at a permissive temperature and fail to separate at a restrictive temperature in a temperature-sensitive separase mutant in *Arabidopsis* (Wu et al., 2010). Moreover, RNAi experiments show that separase is required for bivalent separation in meiosis (Liu and Makaroff, 2006; Yang et al., 2009). Separation may also result from the activation of a topoisomerase II (Shamu and Murray, 1992) or a change in the charge of the chromosomes (Kamiya, cited in Molé-Bajer, 1958). Following the longitudinal splitting of the chromosomes, they continue to separate at a rate of 0.2–9 μm/minute, which is one of the slower manifestations of cell motility (Barber, 1939; Ris, 1943, 1949; Hughes and Swann, 1948; Aist and Williams, 1972), and on the order of the speed of continental drift (Wegener, 1966; Forer, 1978) and fingernail growth (Yaemsiri et al., 2010).

Chromosome segregation can be effected by two kinds of movements that distinguish two substages of anaphase: anaphase A and anaphase B. During anaphase A, the chromosomes move to the poles, and during anaphase B, the two poles move apart. The two forms of anaphase occur to different extents in various cell types, and in plants, anaphase B is rarely seen. Sometimes anaphase A precedes anaphase B, sometimes anaphase B precedes anaphase A, and sometimes they occur simultaneously (Fuseler, 1975; Hayashi et al., 2007). Intact microtubules are required for both types of anaphase movements (Inoué, 1952; Inoué and Sato, 1967; Sato et al., 1975). Interestingly, when the chromosomes are removed with micromanipulators from dividing cells, "anaphase" still continues, indicating that the chromosomes themselves are not necessary for anaphase movement (Zhang and Nicklas, 1996).

To determine the forces that are necessary to move chromosomes to the pole at the observed velocity, we must determine whether inertial forces or viscous forces predominate. To do this, we must estimate the Reynolds number (*Re*; see Chapter 9):

$$Re = \frac{wv\rho}{\eta} \tag{19.2}$$

Given that the observed velocity (v) of chromosome movement is 1.6×10^{-8} m/s, the characteristic length (w) of the chromosome is approximately 10^{-5} m, the density (ρ) of the cytoplasm is 1.0145×10^3 kg/m^3, and the viscosity (η) of the cytoplasm for chromosomes is about 1 Pa s, therefore, the Reynold's number is about 3×10^{-10}. Thus, any inertial forces will be minuscule, and only viscous forces need to be considered.

The force necessary to move the chromosomes through the cytoplasm can thus be determined with Stokes' Law (see Chapter 9):

$$F_v = wv\eta \tag{19.3}$$

where w is the size factor for nonspherical objects and is considered to be $25-59 \times 10^{-6}$ m for chromosomes (Nicklas, 1965). Thus, when w is 59×10^{-6} m, then $F_v = 9.4 \times 10^{-13}$ N, which is approximately 1 pN.

This is approximately the force produced by a single molecule of dynein, kinesin, or myosin, and thus a single motor protein would be sufficient to move a chromosome to the pole (Forer, 1969). Indeed, some chromosomes are attached to only one microtubule (Church and Lin, 1985; Nicklas and Kubai, 1985).

While each chromosome only requires about 1 pN to move to the pole, the force that is available to pull the chromosomes to the pole may be much larger. The fact that chromosomes, which have been experimentally shortened by X-ray irradiation, move to the pole at the same rate as normal chromosomes can be interpreted to mean that the motor that pulls the chromosomes is capable of sensing the load and adjusting its force (Nicklas, 1965). Indeed, during anaphase, meiotic chromosomes with two centromeres, known as *bridge chromosomes*, are pulled to the two poles at a rate that is identical to the rate that the other chromosomes are pulled, indicating that the motive force is capable of pulling chromosomes at a constant velocity, even when the load is greater (Cornman, 1944; Östergren et al., 1960). The force that is exerted on a kinetochore by a spindle fiber has been measured with the aid of glass microneedles (Nicklas, 1983). These experiments show that chromosome motion is unaffected by pulling on a chromosome with a force of 100 pN. The velocity decreases linearly as the force is increased and falls to zero at 700 pN. Above this level of force, the chromosomes will unravel (Houchmandzadeh et al., 1997). Thus, the actual force is about 100 times greater than was expected. How can we explain this?

It is possible that the actual force produced by the motor is regulated by the load that the motor has to pull. Thus, as the load is increased with a microneedle, the force increases up to a maximum of 700 pN. Thus, the motor may be able to sense the load and try to maintain a constant velocity. It could do this by regulating the number of motor proteins involved or regulating such properties as the length of the displacement, the rate of ATP hydrolysis, or other things that affect the duration of the mechanochemical cycle.

It is also possible that the discrepancy between the actual force and the calculated force results from the presence of a governor (Nicklas, 1983). The governor effectively increases the viscous force. The kinetochore microtubules may be the governor, and their slow rate of depolymerization may limit the movement of the chromatids, no matter how hard the motor works.

David Begg and Gordon Ellis (1979) used micromanipulators to push a chromosome toward the pole in anaphase. When they did this, the chromosome waited until the other chromosomes caught up to it before it continued to travel to the pole. According to Inoué (1981), "The chromosomes behave as though they were all being reeled in to the pole by individual fishing lines each attached to the kinetochore, but all sharing a common reel." Thus, any theory of anaphase motion must be able to account for the possibility that the chromosomes can move to the pole in an interdependent fashion.

There has been no dearth of theories introduced to explain anaphase motion. Wilson (1925) summarizes the rise and fall of many of the early theories, including the magnetic and electrical theories based on the fact that the spindle "vividly recalls the arrangement of iron filings about the two poles of a magnet." I will discuss a few hypotheses. Any theory must be consistent with the cytological observations that (1) chromosomes move together to the poles as an "anaphase plate" at a constant velocity, and thus the motive force equals the viscous force, resulting in no net force, or else the chromosomes would accelerate; and (2) the kinetochores lead the way to the poles. Like every generalization, however, there are exceptions (Darlington, 1932; Metz, 1936), and until the problem is solved we never know whether the exceptions to the generalizations are just red herrings or whether they provide the insight necessary to discover the underlying mechanism.

Although currently, the mechanism(s) responsible for anaphase motion during mitosis and meiosis are not known, cell biologists continue to experiment and piece together information to fully understand the mechanisms, just as Walther Flemming did when he wrote the following words in 1880: "If, for the time being, we assemble and compare observations on the life processes of the cell, we do so with the hope that they will and must aid in the ultimate insights and physical explanation of the same phenomena. Without this conviction I would have no basis for sitting at my microscope any longer." The mitotic puzzle, the process that ironically is described both as cell *multiplication* and cell *division*, is one of the most intriguing aspects of cell biology and will never lose its appeal. In a 2010 interview with Conly Rieder, Herbert Macgregor and James Wakefield stated and asked him, "There are now a bewildering number of molecules, all with their own special acronym that seem to be involved in regulating the various integrated processes involved in the mitotic cell cycle. Do you think the day will come when we will be able to put the jigsaw together and present a comprehensive description of the entire process, G1 through M, with a full understanding of the roles of each of the molecules involved?" To which Conly Rieder answered, "No!" I too believe that we must learn the biological equivalent of musical scales and understand the relationships of the notes in an octave before we can play all the parts in a symphony. If we do not, the sounds are not musical with a rich rhythm, harmony, and melody but cacophonous and tuneless.

19.1.4.1 The Sliding Filament Hypothesis

After the mechanism of ciliary beating (see Chapter 11) was described it was natural to look for cross bridges that may represent the dynein-like mitotic motor responsible for anaphase motion (Hepler et al., 1970). Once such cross bridges were found, McIntosh et al. (1969) proposed that sliding occurs between antiparallel kinetochore and nonkinetochore microtubules driven by dynein. This model was supported by experiments that showed that vanadate and other inhibitors of dynein inhibit anaphase motion in permeabilized PtK_1 cell models (Cande and Wolniak, 1978; Cande, 1982a,b). It is not supported by experiments that show that antibodies directed against dynein do not inhibit anaphase motion (Vaisberg et al., 1993).

To describe anaphase motion, the sliding filament hypothesis requires that the nonkinetochore and kinetochore microtubules be antiparallel. When Euteneuer and McIntosh (1980, 1981a,b) and Telzer and Haimo (1981) discovered that 90%–95% of the microtubules in each half-spindle have identical polarities, this model lost favor. However, because one pair of antiparallel microtubules would be sufficient to pull a chromosome to the pole, and it is possible that the kinetochore initiates the formation of one or two microtubules, for some cells, this model is still a contender. The sliding filament hypothesis has found support in explaining anaphase B movement because the microtubules are often antiparallel at the zone of microtubule overlap midway between the two poles (Cande and McDonald, 1985, 1986; Masuda and Cande, 1987; Wordeman and Cande, 1987; McIntosh and Koonce, 1989). However, to push the poles apart, the motor would have to be a plus end–directed motor such as kinesin. Laser microbeam studies in fungal and mammalian cells indicate that anaphase B is predominantly powered by dynein-mediated pulling forces on the astral microtubules that radiate from the poles, and the overlap region actually slows down movement (Aist and Berns, 1981; Aist et al., 1991, 1993; Waters et al., 1993).

19.1.4.2 Microtubule Depolymerization and Poleward Flux Theories

The mitotic spindle is a dynamic structure that reversibly breaks down when exposed to elevated hydrostatic pressures (Pease, 1941, 1946; Salmon, 1975a,b; Salmon and Ellis, 1975) and microtubule-depolymerizing drugs, including colchicine (Inoué, 1952). While observing the mitotic spindle of *Chaetopterus* with polarization microscopy, Shinya Inoué (1953, 1959, 1964) noticed a remarkable and completely unexpected phenomenon. He observed that as he increased the temperature of the stage, the amount of birefringence of the mitotic spindle increased. Because a completely random specimen is isotropic, and birefringence typically indicates that the specimen is ordered, an increase in the birefringence of the specimen means an increase in order. This was surprising because typically increasing the temperature increases the disorder in a system.

This unusual effect of temperature on protein had been observed before with muscle protein and the tobacco mosaic virus protein. It turns out that protein monomers have a shell of water around them, just as ions do (see Chapter 12), and this bound water prevents the protein monomers from interacting among themselves and forming a polymer. The removal of the bound water requires an input of energy, which comes from raising the temperature. As the temperature increases, the shell of water surrounding the protein monomers is removed, and hydrophobic interactions between the protein monomers themselves can take place. Inoué proposed that the fibers in the spindle were made out of proteins, and the fibers formed when the protein subunits polymerized.

Inoué quantified his work in the following way: He assumed that the maximum birefringence occurred when all the subunits were polymerized. The birefringence of a specimen is measured with a polarizing microscope by determining the phase of the light passing through the specimen in one direction relative to the phase of the light passing through the specimen in an orthogonal direction. Let's call the total concentration of the protein subunits A_o and the concentration of polymerized subunits B. B is proportional to the amount of retardation (Γ, in nm) measured with the polarizing microscope. Retardation is the product of the birefringence and thickness. When all the subunits are polymerized $A_o = B$, and when all the subunits are unpolymerized $B = 0$, the free subunits can be calculated as $A_o - B$. Inoué assumed that polymerization occurred according to the following reaction:

$$A_o - B \underset{k_{off}}{\overset{k_{on}}{\leftrightarrow}} B$$

where

$$K_{eq} = \frac{[B]}{[A_o - B]} \tag{19.4}$$

Inoué determined the concentration of polymerized subunits at a given temperature by measuring the amount of retardation of the spindle at that temperature. The maximal amount of retardation he obtained was an estimate of the total concentration of subunits, and he calculated the concentration of free subunits at a given temperature by subtraction.

The degree of polymerization is determined by the equilibrium constant (K_{eq}), which represents the ratio of polymerized to free subunits $\left(\frac{[B]}{[A_o - B]}\right)$. The standard molecular free energy (E_{std}) relative to the energy that occurs at equilibrium ($E_{eq} = 0$) when a subunit is added to the polymer can be calculated from the following equations (see also Chapter 13):

$$E_{eq} - E_{std} = kT \ln \frac{K_{eq}}{K_{std}} = kT \ln \frac{\frac{[B]}{[A_o - B]}}{1/1} \tag{19.5}$$

$$E_{std} - E_{eq} = -kT \ln \frac{K_{eq}}{K_{std}} = -kT \ln \frac{\frac{[B]}{[A_o - B]}}{1/1} \tag{19.6}$$

At standard pressure, the change in standard molecular free energy (relative to the equilibrium state) can be related to the change in standard molecular enthalpy ($\Delta H_{std}/N_A$, in J/molecule) and to the change in standard molecular entropy ($\Delta S_{std}/N_A$, in J/molecule K) according to the following equation:

$$\Delta E_{std} = \frac{\Delta H_{std}}{N_A} - \frac{T\Delta S_{std}}{N_A} \tag{19.7}$$

Therefore,

$$\Delta E_{std} = \frac{\Delta H_{std}}{N_A} - \frac{T\Delta S_{std}}{N_A} = -kT \ln \frac{[B]}{[A_o - B]} \tag{19.8}$$

which can be rewritten as a linear equation:

$$-\ln \frac{[B]}{[A_o - B]} = \frac{\Delta H_{std}}{kN_A}\left(\frac{1}{T}\right) - \frac{\Delta S_{std}}{kN_A} \tag{19.9}$$

If we plot $\ln \frac{[B]}{[A_o - B]}$ versus $(1/T)$, we get a van't Hoff plot, where the slope is equal to $(\Delta H_{std}/kN_A)$ and the y-intercept is equal to $\Delta S_{std}/kN_A$. Thus, at a given temperature we can estimate $\Delta H_{std}/N_A$ and $(T/N_A)\Delta S_{std}$. Inoué determined the standard molecular enthalpy of polymerization at room temperature to be 1.9×10^{-19} J/subunit and the product of the standard molecular entropy and the temperature to be 2.1×10^{-19} J/subunit. Thus, the standard molecular free energy made available by the polymerization reaction is 2×10^{-20} J. As $(T/N_A)\Delta S_{std}$ is greater than $\Delta H_{std}/N_A$, polymerization of the subunits is an entropy-driven process. Thus, Inoué concluded that the subunits of the spindle fibers were surrounded by bound water, and the displacement of the bound water caused an increase in entropy. This increase in entropy provided the molecular free energy necessary to drive the polymerization reaction. Moreover, he proposed that the controlled depolymerization of the spindle fibers provided the motive force that pulled the chromosomes to the poles during anaphase. These experiments are remarkable in light of the fact that they were done approximately 20 years before the discovery of tubulin.

After the discovery of microtubules with the electron microscope, Inoué and Sato (1967) recast the dynamic equilibrium hypothesis in terms of microtubules (Sato et al., 1975; Inoué, 1981). The force (F) exerted by a microtubule that is either polymerizing or depolymerizing is given by the following equation:

$$\left|\frac{E_{Real}}{x}\right| = |F| = \left|\frac{kT}{x} \ln [C_T] \frac{k_{on}}{k_{off}}\right| \tag{19.10}$$

where x is the length that the microtubule changes on the addition or loss of a dimer, E_{Real} is the molecular free energy of the polymerization reaction under cellular conditions, E_{Real}/x is the force exerted by the polymerization reaction under cellular conditions, $[C_T]$ is the concentration of tubulin, k_{on} is the on-rate constant for the addition of tubulin, k_{off} is the off-rate constant for the loss of tubulin, and $[C_T]k_{on}/k_{off}$ is equal to K_{eq}. Assuming $x = 0.61$ nm, approximately the length of a tubulin dimer, Inoué and

Salmon (1995) estimated that the pushing force ($|F|$) is 0, 16, 31, or 46 pN when $[C_T]k_{on}/k_{off} = 1$, 10, 100, and 1000, respectively, and the pulling force ($|F|$) is 0, 16, 31, or 46 pN when $[C_T]k_{on}/k_{off} = 1$, 0.1, 0.01, and 0.001, respectively. Thus, a change in $[C_T]$, k_{on}, and/or k_{off} will result in a force either toward or away from the poles (Cohn et al., 1986).

Koshland et al. (1988) tested the ability of microtubule depolymerization to generate force in vitro. They have shown that chromosomes can move from the plus end to the minus end of isolated microtubules in the absence of ATP. Coue et al. (1991) estimated the force generated by microtubule depolymerization using *Tetrahymena* cells that had been lysed and extracted such that the radial arrays of microtubules remained. They then perfused a solution containing isolated chromosomes by the microtubules until some bound to the microtubules. They then decreased the tubulin concentration so that the microtubules would depolymerize. The depolymerizing microtubules pulled the chromosomes. This movement took place in the absence of ATP and in the presence of vanadate, indicating that motor proteins like kinesin or dynein were not responsible for the movement. Coue et al. (1991) calculated the force exerted by depolymerization to be greater than 1 pN by measuring the depolymerization-mediated movement against a buffer that flowed in the opposite direction.

When microtubules depolymerize in vitro, the typical motor proteins, including kinesin, kinesin-related protein, and dynein, facilitate the minus end—directed movement of microspheres in the absence of ATP, indicating that they may not act as a mechanochemical ATPase, but can act as a bridge or anchor between the kinetochore and the depolymerizing microtubules (Desai and Mitchison, 1995; Lombillo et al., 1995b). McIntosh et al. (2008, 2010) suggest that the microtubules pull the chromosome to the pole as they shorten because the protofilaments that make up the microtubules bend, flare, and change their curvature as they depolymerize. They remind us that the force-generating system powered by FtsZ—the tubulin like protein that effects the division of bacteria, chloroplasts, and mitochondria (see Chapters 13—15)—is driven by the depolymerization associated bending or contraction of the ring.

The dynamic equilibrium model has been extended to include the properties of microtubules that lead to dynamic instability (see Chapter 11). Microtubules show catastrophic changes in length that are a result of the differing stability of microtubules capped with either guanosine triphosphate (GTP)-tubulin or guanosine diphosphate (GDP)-tubulin. The GTP is spontaneously hydrolyzed over time so that the minus end, the end closest to the spindle pole, is composed entirely of GDP-tubulin, whereas the plus end, the end attached to the kinetochore, is composed of GTP-tubulin. The GTP-tubulin readily binds new monomers of GTP-tubulin, and the microtubule grows from the plus end. However, if the rate of GTP hydrolysis exceeds the rate of polymerization, then the plus end will be capped with a GDP-tubulin, which does not bind GTP-tubulin well, and a catastrophic depolymerization at the plus end will ensue. The model then states that the microtubule depolymerizes at the kinetochore in a catastrophic manner and pulls the chromosome along with it.

Biotinylated tubulin is incorporated into the kinetochore microtubules during early, but not late, anaphase (Wadsworth et al., 1989). Shelden and Wadsworth (1992a) injected biotinated tubulin into the spindle of PtK_1 cells and observed with the electron microscope that it was incorporated into the kinetochore microtubules at the kinetochore. When biotin-labeled tubulin was injected into living cells during anaphase, the chromosomes changed their movement from going toward the poles at 1 μm/minute to moving away from the poles at 0.5 μm/minute. Moreover, they observed that the kinetochore was compressed during the reversal, indicating that the reversal force was applied at or near the kinetochore. These data support the original hypothesis of Inoué and Sato (1967) that the force for chromosome motion is generated by kinetochore microtubule disassembly. Shelden and Wadsworth (1992a) propose that the rate and direction of chromosome-to-pole motion are regulated by the concentration of tubulin subunits at the kinetochore, but it is unclear whether the microtubule depolymerization acts as the motor or the governor (Nicklas, 1965, 1975). During polymerization and depolymerization, the kinetochores remain attached to the plus ends of the microtubules (Hyman and Mitchison, 1990; Coue et al., 1991) because the electrostatic or hydrophobic interaction energy between the kinetochore and the microtubule is greater than the thermal energy necessary for the diffusion of the chromosome away from the microtubule.

To determine whether the depolymerization of microtubules at the kinetochore is limiting for anaphase motion, Sawin and Mitchison (1991a,b) labeled cell extracts from *Xenopus* eggs with photoactivatable tubulin. During anaphase, the distance between the bright band arising from the photoactivated fluorescent tubulin and the chromosomes does not change while the bright band of tubulin moves poleward at a rate of 2.9 μm/minute, similar to the rate of chromosome movement to the pole. This poleward flux was inhibited by AMP-PNP and unaffected by vanadate. Because AMP-PNP inhibits both poleward flux and kinesin-related proteins, and because both poleward flux and kinesin-related proteins are insensitive to vanadate, Sawin and Mitchison propose that the poleward flux is driven by a kinesin-related protein. Kinesin-related proteins move along microtubules from the minus end to the plus end in an ATP-dependent manner. Thus, if kinesin-related proteins were immobilized in the spindle matrix, they would pull the microtubule toward the pole (Mitchison and

Sawin, 1990). Thus, it seems that depolymerization at the minus end combined with a kinesin-related motor is responsible for anaphase motion in *Xenopus* eggs. Higher-resolution studies show that depolymerization of MTs at the plus end contributes a little to the movement of chromosomes to the pole (Maddox et al., 2003).

Mitchison and Salmon (1992) used the same technique to study newt lung cells. They find that in these cells, the chromosomes move to the pole at a rate of 1.7 μm/minute, while poleward flux occurs at a rate of only 0.44 μm/minute, indicating that in these cells, poleward flux can only account for about one-quarter of the motive force, and the majority of the force necessary for anaphase motion must be provided at the kinetochore. Likewise, Zhai et al. (1995) find that more than two-thirds of the force necessary to move chromosomes during anaphase in PtK_1 cells must be provided at the kinetochore. This force could be provided by depolymerizing microtubules or by an MT-associated motor. By contrast, poleward flux occurs at the same rate as chromosome movement in tobacco BY-2 cells, indicating that poleward flux provides the lion's share of the force necessary for anaphase movement in these cells (Dhonukshe et al., 2006).

19.1.4.3 The Motile Kinetochore Model

Cytoplasmic dynein has been localized in kinetochores (Pfarr et al., 1990; Steuer et al., 1990; Lombillo et al., 1995a,b; Wordeman and Mitchison, 1995) and is capable of moving a chromosome from the plus end to the minus end of a microtubule. The observation that anaphase can be inhibited by inhibitors of dynein in PtK_1 cells (Cande and Wolniak, 1978; Cande, 1982a,b) indicates that dynein is important for anaphase motion. Dynein is capable of producing velocities around 1 μm/second. However, anaphase motion is 1—3 μm/minute. Perhaps, the slow velocity at anaphase results from the possibility that microtubule depolymerization at the kinetochore limits the velocity that dynein is capable of producing. According to this model, the kinetochore microtubules, due to their slow rate of depolymerization, act as a governor and the dynein molecules are the motor.

The motile kinetochore model is supported by observations by Hiramoto and Nakano (1988) on sand dollar eggs and Nicklas (1989) on spermatocytes. They have come up with a novel approach to test whether the kinetochores are passively pulled to the poles by traction forces or whether they actively move down the microtubules. They cut the spindle with a microneedle by pressing the spindle between the microneedle and a coverslip. When they separate the chromosomes from the pole in this way, the chromosome still moves to within 0.5 μm of the ends of the microtubules during anaphase, indicating that the motor must be in or near the kinetochore.

19.1.4.4 Actin

Arthur Forer (1969, 1978, 1988) has long held the view that, until proved otherwise, actin is the universal cellular motile protein and most likely participates in any given motile process, just as ATP is the universal energy currency and DNA is the universal genetic molecule (see Chapter 10). This has not been a very popular view as cytochalasins (Schmit and Lambert, 1988), antiactin antibodies (Molé-Bajer et al., 1988), or phalloidin (Schmit and Lambert, 1990a,b) do not typically inhibit chromosome motion during anaphase. However, cytochalasin B inhibits mitosis in mycoplasma (Ghosh et al., 1978), and Pickett-Heaps et al. (1996) and Sampson et al. (1996) have shown that in the cells of *Oedogonium*, actin microfilaments are coaligned with kinetochore microtubules and cytochalasin inhibits mitosis in these cells. Cytochalasin also inhibits meiosis in crane-fly spermatocytes (Forer and Pickett-Heaps, 1998). In these cells, antiactin and antimyosin (2,3 butanedione 2-monoxine) drugs inhibit poleward flux during metaphase (Silverman-Gavrila and Forer, 2000, 2001). Thus, the actomyosin system may provide the force for anaphase movement in some cells.

19.1.4.5 Double Insurance and Functional Redundancy

There is no "one spindle" (Mogilner and Craig, 2010), and there are probably multiple mechanisms that account for the coordinated movement of chromosomes during anaphase in all cells. Charles Metz (1936) wrote,

> *Considering the fundamental nature of the process of mitosis and the essential uniformity of its results throughout both plants and animals it is only to be expected that a mechanism would exist which would insure its correct operation under widely varying conditions. A single agency could hardly be effective to this extent, and it seems only natural that in most organisms definite insurance should be provided by the presence of additional agencies. One is reminded here of Spemann's principle of 'double insurance' in the regulation of development of the embryo. I suspect that in mitosis we have not only double, but probably multiple insurance and that ultimately several distinct agencies will be identified here, any one of which could perhaps bring about the necessary chromosomes movements alone if the others failed to act.*

Arthur Hughes (1952) wrote, "Mitosis seems to me to be the supreme example in science of diversity within a unity." The study of mitosis is truly the study of beauty and harmony. In fact, according to Samuel Taylor Coleridge (1835, 1884), in Roman times, the definition of beauty was "multitude in unity." To Gottfried Leibniz (Carlin, 2000), harmony was "unity in variety." How do we study the beauty and harmony of biological processes? Kornberg and Krebs (1957) note that "some authors prefer to lay stress on endless variety and complexity, and others to trace the common ancestry from which the variety has sprung, but most will agree on the need fro both approaches."

Chromosome segregation is an extremely reliable process, where the chance of losing a chromosome is about one in 10,000 divisions (Hartwell and Smith, 1985). I have described many ways that chromosomes could be pulled to the poles to ensure reliable chromosome segregation. They could be pulled by sliding microtubules or by kinesin-related proteins pulling the microtubules to the pole while they were depolymerizing at the minus end. They could be pulled by depolymerization of microtubules at the plus end, with or without an active mechanochemical ATPase. It is also possible that the microtubules act only as a governor and the actomyosin system may provide the motive force. Perhaps each force-generating reaction takes part to a greater or lesser extent in the mitotic or meiotic apparatus of each cell type, and their relative roles represent the balance reached in each cell to have equal division in a nucleus composed of chromosomes, which may have come together by various mechanisms, including duplication and hybridization.

Pickett-Heaps et al. (1996, 1997) have stressed that cell biologists have concentrated too long on the microtubules in the spindle and should pay more attention to the other proteins in the spindle matrix. Forer (1969) argues that the microtubules make up only a few percent of the volume of the spindle. Perhaps the nuclear matrix continues to contribute to chromatin structure and the position of chromosomes during mitosis, and the nuclear membrane–matrix–chromatin association never completely breaks down but reforms to make the spindle matrix. This spindle matrix may reform into the nuclear matrix during telophase.

19.1.5 Telophase

Telophase begins when the decondensing daughter chromatids arrive at the poles, the kinetochore microtubules disappear, and the nuclear envelope reforms around the decondensing chromosomes to form the two daughter nuclei (Benavente, 1991). Nuclear lamins may specifically interact with chromatin to promote nuclear envelope reassembly (Glass and Gerace, 1990). The chromatin begins to decondense and the nucleoli reappear. During telophase, all the nuclear proteins, including RanGTP (Pay et al., 2002), must be rounded up and brought back to the nucleus (see Chapter 16).

19.2 REGULATION OF MITOSIS

Lewis V. Heilbrunn (1956) postulated that Ca^{2+} regulated mitosis after seeing the gelation or contraction that occurs during mitosis and equated it to the gelation seen in Ca^{2+}-dependent blood clotting. The Ca^{2+} hypothesis seemed likely when Weisenberg (1972) showed that Ca^{2+} caused the depolymerization of microtubules. This observation has been repeated in a variety of systems, both in vivo and in vitro (Kiehart, 1981; Keith, 1987; Cyr, 1991a,b). Petzelt (1972) identified a Ca^{2+}-ATPase in the mitotic apparatus, and Hepler and his colleagues (Wick and Hepler, 1980, 1982; Wolniak et al., 1980, 1981; Hepler and Wolniak, 1983, 1984), as well as Silver et al. (1980), showed that Ca^{2+}-sequestering membrane systems exist in the mitotic apparatus, and mitosis is inhibited by antibodies and inhibitors directed against the Ca^{2+}-sequestering system (Silver, 1986). Molecular genetic studies have shown that calmodulin is necessary for mitosis (Ohya and Anraku, 1989; Sun et al., 1991).

Ca^{2+} is required for mitosis because calcium channel blockers inhibit mitosis, injection of Ca^{2+} into the cell affects mitosis, and the cellular concentration of Ca^{2+} varies during mitosis. Thus, according to Jaffe's Rules (see Chapter 12), Ca^{2+} is involved in the mitotic regulatory system, but we are still trying to find out how (Kao et al., 1990; Hepler, 1992, 1994; Whitaker, 1997; Machaca, 2010; Martin-Romero et al., 2012).

19.3 ENERGETICS OF MITOSIS

All movement requires energy, and chromosome motion is no different—in theory. However, the energetics of mitosis has been very difficult to sort out. Indeed, the first experiments on the energetics of mitosis done by Otto Warburg were harbingers of the difficulties that have been encountered. Otto Warburg (1931a,b) showed that normal cells required respiration for mitosis, whereas rapidly dividing cancerous cells got their energy from the energetically inefficient glycolytic reactions and not the efficient respiratory reactions (see Krebs, 1981a). While Amoore (1962, 1963), Hepler and Palevitz (1986), Spurck and Pickett-Heaps (1987), and Armstrong and Snyder (1987) have tested the effect of metabolic inhibitors on mitosis, the data are conflicting, and there is no general conclusion.

To see how difficult this problem is, I will only consider the molecular entropy contribution to the free energy and not even discuss ATP. Let's assume that the concentration of tubulin increases so that microtubules spontaneously form on MTOCs. Let's assume the microtubules eventually attach to kinetochores. Those microtubules will be present as long as the tubulin concentration remains high. Imagine that a tubulin-binding protein appears that begins to sequester the tubulin subunits. As the free concentration of tubulin falls, the microtubules will depolymerize at a rate that depends on the concentration of tubulin. The depolymerization is capable of pulling the chromosomes to the pole. So while the molecular free energy is important for any motile process, we have to seriously consider how the increase in molecular entropy generated by a variety of molecules may influence molecular free energy used to move the chromosomes to the poles.

19.4 DIVISION OF ORGANELLES

For cells to pass on their organelles, the organelles must grow by taking up newly synthesized proteins, enlarge their membrane surface area, divide, and be distributed between the daughter cells (Hoepfner et al., 2005). Interestingly, as a reflection of their evolutionary past, these requirements are of differential importance for the various organelles in the cell. For example, the plastids, mitochondria, and presumably the ER are semiautonomous organelles, of which the loss during division would be fatal. By contrast, the Golgi apparatus, its associated membranes, the plasma membrane, the endosomal compartment, and the vacuolar compartment are all derived organelles and can be regenerated from the ER. While the preponderance of current data indicates that the peroxisomes should be included in the class of organelles that do not have to divide prior to cell division, but can be derived de novo in the daughter cells, the jury is still out (see Chapter 5; Lazarow, 2003; Kunau, 2005; Schekman, 2005).

During cell division, all the organelles are typically distributed to the daughter cells (Wilson, 1925; Cleary et al., 1992; Langhans et al., 2007; Nebenführ, 2007; Sano et al., 2007; Jongsma et al., 2014). While it is unknown how most of the organelles divide, actin and other motile proteins, including FtsZ and dynamin, mediate the division of the plastids and mitochondria, which make up the autonomous organelles (see Chapters 13 and 14; Sheahan et al., 2004).

19.5 CYTOKINESIS

Cytokinesis, or division of the cytoplasm, takes place in microorganisms, animals, and plants (Pollard, 2010). Typically, but not always, karyokinesis is followed by cytokinesis. As they do not always occur together, I will consider them to be related but independent processes. I would like to mention that mitosis and cytokinesis are a *sine qua non* for differentiation of some cell types because gamma-irradiated wheat embryos that are incapable of dividing are unable to differentiate guard cells, subsidiary cells, and trichomes—cells that typically differentiate following an asymmetrical division (Foard and Haber, 1961; Haber et al., 1961).

19.5.1 Cell Plate Formation

Melchior Treub (1878) first described cytokinesis in the fixed and stained epidermal cells of orchid ovules. Eduard Strasburger, in his *Textbook of Botany* (Strasburger et al., 1912), called the newly formed cell wall the *cell plate*. It appeared that the cell plate arose from a system of fibrils that was called the *phragmoplast* by Leo Errera[1] (1888). Becker (1932, 1938) saw that the vesicles at the cell plate accumulated stain like vacuoles do and concluded that the cell plate formed from the fusion of petite vacuoles. Yasui (1939) observed the phragmoplast and the granules fusing to make the cell plate in living cells. Using polarization microscopy, Becker (1938), Sato (see Wada, 1972), and Inoué and Bajer (1961) showed that the phragmoplast existed in living cells.

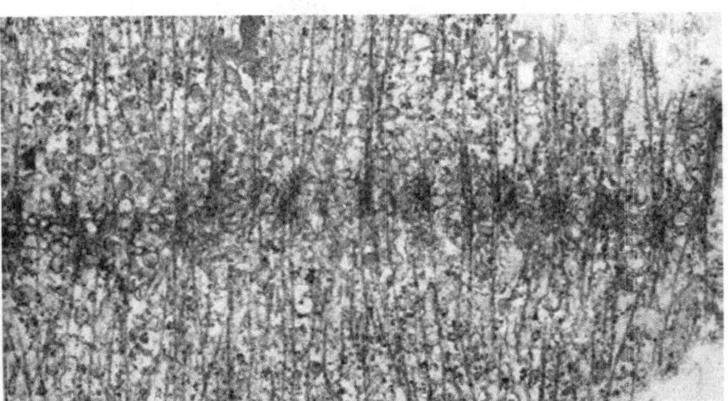

FIGURE 19.9 Microtubules at an early stage of cell plate formation in *Haemanthus katherinae*. ×29,000. *From Hepler, P.K., Jackson, W.T., 1968. Microtubules and early stages of cellplate formation in the endosperm of Haemanthus Katherinae Baker. J. Cell Biol. 38, 437–446.*

With the advent of electron microscopy, Buvat and Puissant (1958), Porter and Caulfield (1958), and Porter and Machado (1960) observed cells in the process of cell plate formation and concluded that the phragmoplast fibers are real, the membranes of the vesicles become the new plasma membrane, and the contents of the vesicles form the new extracellular matrix. Later, Whaley and Mollenhauer (1963) and Frey-Wyssling et al. (1964) concluded that the vesicles involved in cell plate formation were derived from the Golgi apparatus. With the advent of glutaraldehyde fixations, it became clear that the phragmoplast was composed of microtubules (see Fig. 19.9; Hepler and Newcomb, 1967; Hepler and Jackson, 1968; Bajer and Jensen, 1969). The proteins associated with the phragmoplast have also been visualized using immunofluorescence (Pay et al., 2002) and green fluorescent protein (GFP) fused to microtubule-associated proteins (Granger and Cyr, 2000b; Hasezawa et al., 2000; Smertenko et al., 2000; Twell et al., 2002).

Typically, the microtubules of the phragmoplast begin as the remaining nonkinetochore microtubules at the end of anaphase (Yasui, 1939), but this is not necessarily always

1. Leo Errera (1894) wrote The Russian Jews; Extermination or Emancipation? David Nutt, London.

the case (Baskin and Cande, 1990). For example, in the algae that are classified in the lines that did not give rise to higher plants, a phragmoplast does not exist; rather, a phycoplast exists. A phycoplast is a group of microtubules that run parallel, instead of perpendicular, to the developing cell plate between the two daughter nuclei in telophase (Pickett-Heaps, 1972, 1975). The independence of the spindle microtubules and the phragmoplast microtubules is underscored by the observation that irradiating the spindle microtubules with an ultraviolet (UV) microbeam prevents spindle formation but has no effect on phragmoplast formation (Wada and Izutsu, 1961). In addition, chloral hydrate or ethidium bromide inhibits spindle formation and function but has no effect on the phragmoplast (González-Fernández, 1970; Zachariadis et al., 2000; Gunning, 1982), and last, the phragmoplast can form in cytoplasmic drops of endosperm that lack nuclei (Bajer and Allen, 1966).

Bohumil Nemec (1899) noted that the fibers that are perpendicular to the cell plate seem to have free ends, indicating that they may originate at the cell plate. In modern terms, this would make the cell plate area an MTOC. In 1964, Inoué irradiated either the distal ends of the microtubules or the ends that were embedded in the cell plate. Using polarization microscopy to assay the integrity of the microtubules, he found that when he irradiated the distal ends, the proximal ends survived, whereas when he irradiated the proximal ends, the whole microtubule disappeared. This indicated that the end embedded in the cell plate was the growing or plus end. Euteneuer and McIntosh (1980) confirmed this interpretation of the polarity of the phragmoplast microtubules using neurotubulin to decorate the phragmoplast microtubules, and GFP fusions of plus end–tracking proteins are also found in the phragmoplast (Liu et al., 2011). Thus, the phragmoplast is composed of two sets of interdigitated microtubules with opposite polarities. Fluorescence microscopy shows that the addition of fluorescent tubulin to the plus ends of microtubules occurs at the ends that are embedded in the cell plate (Asada et al., 1991, 1997), while γ-tubulin, which is involved in the nucleation of microtubules at the minus ends, is absent in the cell plate but present in the distal ends of the microtubules (Liu et al., 1993; Marc, 1997). These data allow us to conclude that the phragmoplast is an MTOC, but the microtubules grow from the center, pushing out the minus ends of the growing microtubules. This contrasts with the microtubule–MTOC relationship in the centrosome and basal body, where the minus ends of the microtubules are embedded in the recognizable microtubule organizing structure.

Typically, the phragmoplast begins in the center of the cell and vesicles move toward the cell plate parallel to the microtubules. Kinesin-like proteins provide the motive force (Lee and Liu, 2000; Liu and Lee, 2001; Lee et al., 2001). Just as the 64-nm diameter vesicles fuse to form a vesicular-tubular structure, the microtubules in that area begin to shorten and appear C-shaped in cross section (Lambert and Bajer, 1972; Samuels et al., 1995; Otegui et al., 2001). New microtubules are added to the centrifugal side of the cell plate at a rate of about 8000 $minute^{-1}$. The phragmoplast appears lens shaped as it moves centrifugally because microtubules are polymerizing on the outside edge of it, are maximally polymerized in the center of the moving phragmoplast, and are depolymerizing at the inner edge. Actin microfilaments are also present in the developing cell plate (Kakimoto and Shibaoka, 1987; Staiger and Schliwa, 1987; Palevitz, 1987a; Molé-Bajer and Bajer, 1988; van Lammeren et al., 1989; Schopfer and Hepler, 1991).

As the Golgi vesicles move toward the cell plate, their stainability changes, indicating that chemical changes are taking place as they "mature." The Golgi-derived vesicles form tubular, irregularly branched, or star-shaped bodies with fuzzy-coated arms in the plane of the cell plate as they fuse (see Figs. 19.10–19.13; Jones and Payne, 1978; Samuels et al., 1995; Staehelin and Hepler, 1996; Seguí-Simarro et al., 2004, 2007). Observations of the developing cell plate in cells that have been fixed by freeze fixation show details of the variety of membrane morphologies that occur in the initial and later stages of cell plate formation (Figs. 19.12 and 19.13). The Golgi-derived vesicles have the endoxyloglucan transferase necessary for

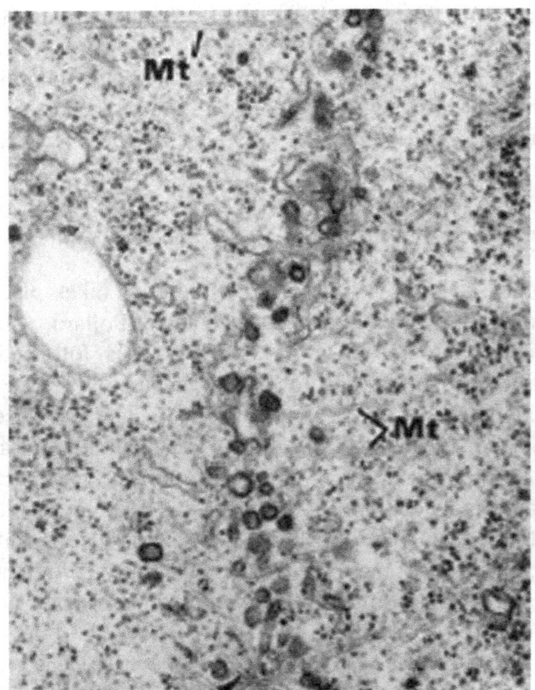

FIGURE 19.10 Early stage of cell plate formation in *Phaseolus vulgaris*. *Mt*, microtubule. × 38,000. *From Hepler, P.K., Newcomb, E.H., 1967. Fine structure of cell plate formation in the apical meristem of* Phaseolus *roots. J. Ultrastruct. Res. 19, 498–513.*

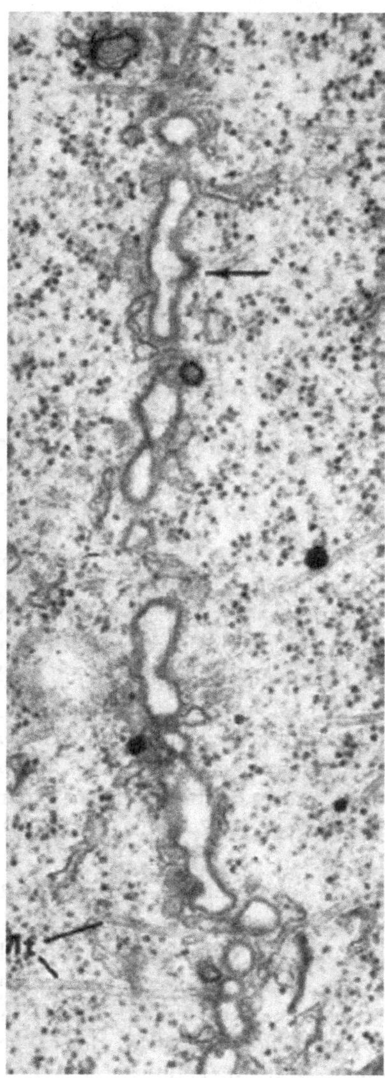

FIGURE 19.11 Mid-stage of cell plate formation in *Phaseolus vulgaris*. ×62,000. *From Hepler, P.K., Newcomb, E.H., 1967. Fine structure of cell plate formation in the apical meristem of* Phaseolus *roots. J. Ultrastruct. Res. 19, 498–513.*

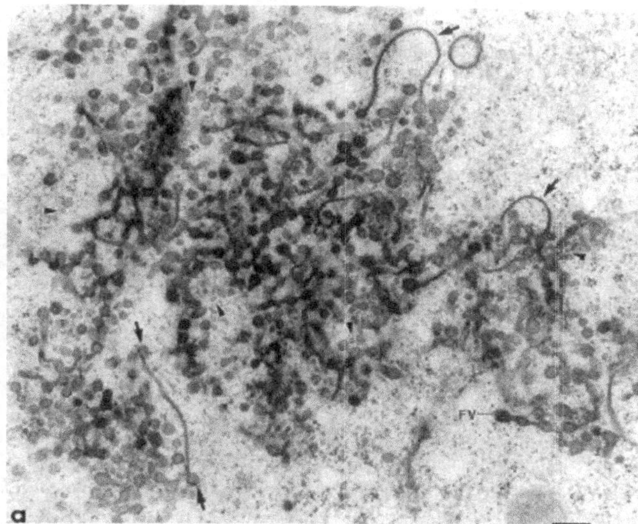

FIGURE 19.12 *En face* view of fusing cell plate vesicles in a tobacco root tip cell. The *arrows* show fuzzy-coated tubular membrane structures that are continuous with vesicle-like structures. *FV*, fusing vesicle. Bar, 250 nm. *From Samuels Jr., A.L., Giddings, T.H., Staehelin, L.A., 1995. Cytokinesis in tobacco BY-2 and root tip cells: a new model of cell plate formation in higher plants. J. Cell Biol. 130, 1345–1357.*

the formation of the cell plate (Yokoyama and Nishitani, 2001). Phragmoplastin, a member of the dynamin GTPase family of proteins involved in membrane tubule formation and the pinching off of vesicles, is specifically localized in the cell plate (Dombrowski and Raikhel, 1995; Gu and Verma, 1996; Kang et al., 2003; Hong and Verma, 2007; Ito et al., 2012).

The centrifugal microtubules of the phragmoplast form as the centripetal microtubules depolymerize. The depolymerization of the centripetal microtubules is inhibited by brefeldin A, which indicates that a mechanism exists that ties together the formation of vesicles with the depolymerization of the microtubules that bring those vesicles to the cell plate (Yasuhara and Shibaoka, 2000). That is, the microtubules bring the vesicles to the cell plate, and the vesicles keep the microtubules there. The membranes of the ER are also evident in the developing cell plate (Porter and Machado, 1960; Hepler, 1982; Gupton et al., 2006) where they give rise to the desmotubule of the plasmodesmata (see Figs. 3.3 and 3.5 in Chapter 3). Immediately after division, the membranes of the ER are tightly appressed to the nascent plasma membrane in the plasmodesmata. The cytoplasmic annulus develops subsequently during cell expansion (Nicolas et al., 2017).

The first wall products detectable in the cell plate are callose (Yasui, 1939; Fulcher et al., 1976; Northcote et al., 1989; Jones and Payne, 1978) and hemicelluloses (Moore and Staehelin, 1988), although this generalization may turn out to be an oversimplification, and the first products may be cell-type specific. The callose remains only until the cell plate fuses with the parental wall and then disappears. The polymerization of callose may provide the motive force for cell plate growth (Samuels et al., 1995), or it may not be functional, but only a consequence of the high [Ca^{2+}] that may exist in the developing cell plate. The hemicellulose is persistent. Pectins do not seem to be incorporated into the wall at cell plate formation, but only later in development (Yasui, 1939; Moore and Staehelin, 1988). This, however, may turn out to depend on the cell type.

When the cells are challenged with caffeine, an agent that interferes with Ca^{2+} homeostasis, the vesicles move to the cell plate as they do in untreated cells (see Fig. 19.14); however, remarkably, after the vesicles come together to

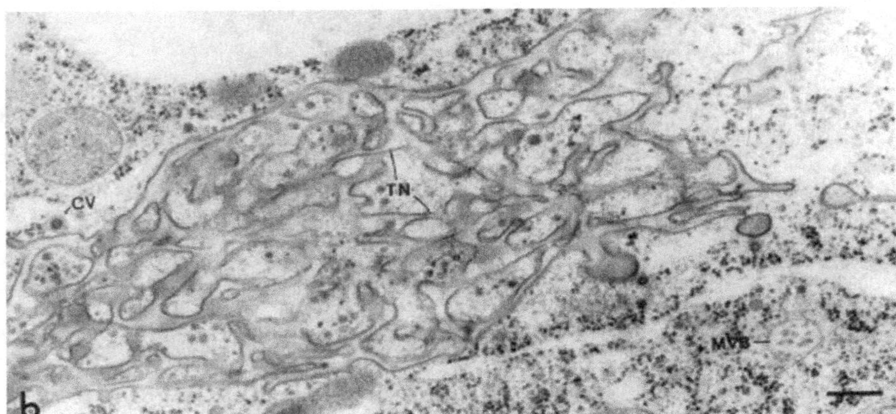

FIGURE 19.13 A tangential section through a tobacco BY-2 cell plate showing the extensiveness of the tubular network of membranes. *CV*, coated vesicle; *TN*, tubular network; *MVB*, multivesicular body. Bar, 250 nm. *From Samuels Jr., A.L., Giddings, T.H., Staehelin, L.A., 1995. Cytokinesis in tobacco BY-2 and root tip cells: A new model of cell plate formation in higher plants. J. Cell Biol. 130, 1345–1357.*

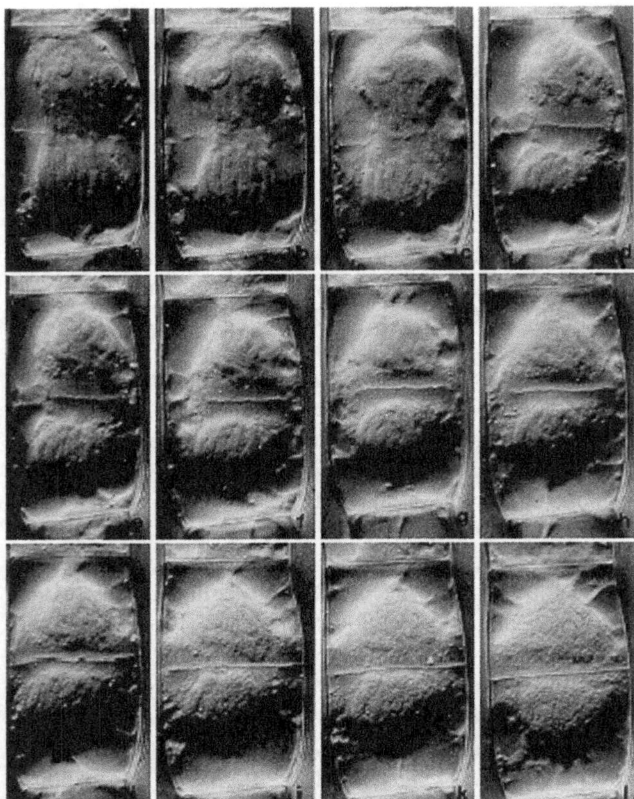

FIGURE 19.14 Time-lapse photographs of cell plate formation in a stamen hair cell of *Tradescantia*. The entire sequence was shot in 27 minutes. The vesicles begin to fuse in the midzone of the cell (c), and thereafter, the cell plate grows centrifugally. (*Source:* From Bonsignore, C.A., Hepler, P.K., 1985. Caffeine inhibition of cytokinesis: Dynamics of cell plate formation-deformation *in vivo*. Protoplasma 129, 28–35.).

form the cell plate, they do not fuse to form the extensive tubulovesicular network seen in normal cells (see Fig. 19.13) but scatter in the cytoplasm. Consequently, the cell becomes binucleate (see Fig. 19.15; López-Sáez et al., 1966b; Paul and Goff, 1973; Bonsignore and Hepler, 1985; Hepler and Bonsignore, 1990; Samuels and Staehelin, 1996). While caffeine may affect vesicle fusion and the formation of the vesicotubular network, it has no effect on the cytoskeletal elements of the phragmoplast (Valster and Hepler, 1997).

Cell plate formation depends, in part, on actomyosin because inhibitors of myosin and agents that affect actin polymerization retard or inhibit the lateral expansion of the cell plate (Valster et al., 1997; Hepler et al., 2002; Molchan et al., 2002).

Since the first edition of this book was published, the study of cytokinesis mutants in *Arabidopsis* and *maize* have revealed scores of genes that code for proteins that serve as molecular motors or are involved in membrane tethering and fusion (Rasmussen et al., 2011a; McMichael and Bednarek, 2013; Lipka et al., 2015). It is a good time to look for the unity in the diversity.

19.5.2 Isolation of Cell Plates

Hiroh Shibaoka (1992) developed a system to isolate cell plates for biochemical work. He isolated spindles from synchronized tobacco BY-2 suspension cells. After the cells were synchronized, cells were isolated and treated with cellulase and pectinase to make protoplasts. Then, after a 90-minute incubation with the enzymes, most of the cells were in anaphase or telophase. The protoplasts were then washed free of enzymes and resuspended in a buffer in which they were lysed by gently pushing them through a 16 μm nylon mesh. The cell plates were then concentrated by centrifugation at 170 g for 2 minutes and resuspended in the same buffer. Using isolated cell plates, Kakimoto and Shibaoka (1992) studied enzymes involved in carbohydrate synthesis and microtubule motility. A protein kinase was found in the cell plate (Nishihama and Machida, 2001; Nishihama et al., 2001).

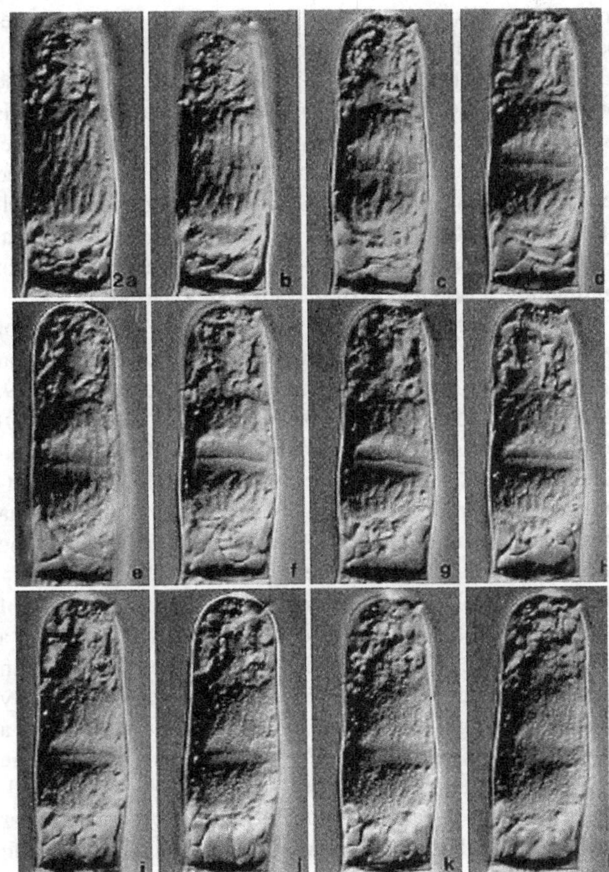

FIGURE 19.15 Time-lapse photographs of cell plate formation in a caffeine-treated stamen hair cell of *Tradescantia*. The entire sequence was shot in 32 minutes. The cell plate begins to form just as it does in the control cells. However, the vesicles never fuse and the cell plate dissolves, leaving a binucleate cell. *From Bonsignore, C.A., Hepler, P.K., 1985. Caffeine inhibition of cytokinesis: dynamics of cell plate formation-deformation* in vivo. *Protoplasma 129, 28–35.*

19.5.3 Orientation of the Cell Plate

Cell division in plants plays an important role in cell differentiation, and often a change in the plane of cell division or a change in the symmetry of cell division is one of the first signs that a cell will differentiate (Lindsey, 2004; Schmit and Nick, 2008). The genetically determined orientation or symmetry of cell division may also be important in the development of form in plants and plant organs. For example, the fruits of cucurbids and tomatoes that come from plants that differ by a single gene have tremendous variations in shape (Sinnott et al., 1950). Sinnott (1960) proposed that there are "shape genes" that control the orientation of the cell plate and/or the polarity of cell growth, and such genes have been found (Liu et al., 2002). What determines the orientation of the cell plate?

In general, the preprophase band, which is formed late in the G2 phase and persists until prometaphase, predicts the position where the developing cell plate will fuse with the parent cell wall (Pickett-Heaps and Northcote, 1966a,b; Hardham and Gunning, 1980; Gunning, 1982, 1992; Duroc et al., 2011). The analysis of mutants that exhibit defects in the plane of cell division has revealed a number of proteins that are localized in the preprophase band or the cortical division zone that underlies the preprophase band and persists and maintains a distinct protein composition until the new cross wall is formed (Rasmussen et al., 2011a,b, 2013; McMichael and Bednarek, 2013; Lipka et al., 2015; Smertenko et al., 2017). It is thought that microfilament arrays radiate from the preprophase band to the nucleus, which provides a mechanism for the orientation of the cell plate (Valster and Hepler, 1997).

However, the preprophase band does not occur in all cell types that have predetermined planes of cell division, and prior to an asymmetric division, the nucleus may migrate to the site of cell division hours before the formation of the preprophase band (Pickett-Heaps, 1969b; Gunning, 1982). Thus, the preprophase band may not directly be involved in determining where the cell plate fuses with the parental wall but may be a consequence of an earlier polarizing event.

If not the preprophase band, what determines where the cell plate forms and the orientation of the developing cell plate? Sinnott and Bloch (1940, 1941) coined the term *phragmosome* to describe the cytoplasmic structure providing the earliest possible indication of the plane of cell division. They found that in vacuolate cells, the nucleus moves from a parietal position to a central position and strands radiate in a three-dimensional network from the nucleus, but eventually, the strands move into a two-dimensional arrangement, into a plane that predicts the future plane of the cell plate. They emphasized that the cytoplasm, not the nucleus, is involved in determining the plane of cell division.

When cells are wounded, the phragmosome aligns perpendicular to the wound (Lloyd, 1991a,b). Venverloo et al. (1980) studied phragmosome formation after stimulating epidermal cells to divide in the periclinal direction by excising a part of the leaf. They found that the nucleus moves from a peripheral part of the cell to the center about 14 hours before the onset of prophase. Then the number of transvacuolar strands increases, generating a three-dimensional system of transvacuolar strands that radiates from the nucleus. A few hours prior to prophase, the strands become two-dimensional in the periclinal plane. Again, the formation of the phragmosome is a gradual process that precedes karyokinesis. Moreover, the phragmosome precedes the preprophase band by several hours (Venverloo et al., 1980). Panteris et al. (2004) have found that the earliest sign indicating the position of the future cell plate in vacuolate cells is an actin- and ER-rich region that rings the cortical cytoplasm. This cortical cytoplasmic ring persists through the cell cycle, and its formation is inhibited by actin and myosin antagonists.

The cytoplasmic strands that radiate to the nuclei contain microtubules and are under tension (Hahne and Hoffman, 1984; Goodbody et al., 1991). D'Arcy Wentworth Thompson (1959) predicted that the cell plate formed under conditions of minimal tension. Now we know that each strand of the phragmosome is indeed under tension. It appears that the strands may initiate anywhere but then move along the cortex until they reach the position of minimum tension. As the strands are contractile and under tension, according to Hooke's Law ($F = -Kx$), the position that would lead to minimum tension is where the distance from the nucleus to the plasma membrane is the shortest. As the corners of a cubic cell are farther away from the nucleus than the sides, a new cell plate does not form in the vertex of three cells, and consequently, each vertex has a maximum of three, but not four, walls. Thus, cells are packed in a tissue the way that bricks are placed in a wall—alternating to give maximum strength. It will be important to determine whether the cytoplasmic strands themselves, the microtubules, the actin filaments, asnd/or the plasma membrane are sensing and responding directly to the various tensions, stresses, and strains (Pickett-Heaps et al., 1999).

For more than 100 years, it has been observed that the cell plate usually forms so as to assume an orientation that results in its occupying the minimum area (Gunning, 1982; Furuya, 1984). Can we be more mechanistic about this observation? There is evidence that the plane of cell division is regulated by the mechanical forces to which the dividing cell is subjected. French and Paolillo (1975) observed that the guard cells that developed on the capsule, under the calyptra of *Funaria* and *Physcomitrium*, always divide longitudinally in the anticlinal plane. However, if the calyptra is removed, thus relieving the capsule of compressive forces, the guard cell mother cells divide in any anticlinal plane, from transverse to longitudinal, and the stomatal complexes appear random.

Philip Lintilhac (1974) and Lintilhac and Jensen (1974) investigated the effects of mechanical stress in developing embryos by combining theoretical models with classic anatomical techniques. They conclude that the cell plate forms in the only plane that is completely free from shearing stresses, perpendicular to the direction of the maximal principal stress (either compression or tension). Indeed, Lintilhac and Vesecky (1980, 1981, 1984) show that when a mechanical stress is applied to callus tissue the new cell plate forms perpendicular to the source of the compressive stress. Thus, within an organ, the orientation of the division plane depends in part on the mechanical forces set up by the entire organ. By contrast, Lynch and Lintilhac (1997) find that in protoplasts, the new cell plate forms primarily parallel to the principal compressive stress. While the callus cells tend to divide perpendicular to the compressive stress and the protoplasts divide parallel to the compressive stress, in both situations, the cell plate is free from shearing stresses. Indeed, the behavior of the part depends on the physical influence of the whole (Heitler, 1963).

How may the mechanical forces actually control the orientation of the cell plate? Perhaps the mechanical energy is converted into electrical energy that may orient the cell plate. Mary Jane Saunders (1986a,b) finds that a cell-generated ionic current enters the site that predicts the position of cell plate fusion with the parental cell in *Funaria* protonemata. There is more evidence that ionic currents are involved in regulating the orientation of the cell plate (Harold, 1990).

Light and other factors, including gravity and the site of sperm penetration, can determine the plane of cell division in the zygotes of the Fucaceae. Lionel Jaffe (1977, 1979, 1980, 1981, 1990, 2004, 2005) and Franklin Harold (1990) proposed that all of these stimuli act by inducing a transcellular current. The photoreceptor responsible for the light-induced polarization is probably localized in the plasma membrane as the direction of the cell plate depends on the azimuth of polarization of the light (Jaffe, 1958).

Jaffe (1966) was interested to test if the zygotes of *Fucus* were polarized by an electrical event but needed a way to measure the tiny currents and voltages that exists in cells. To measure whether or not the zygotes are electrically polarized by light, he placed lots of zygotes in series in a tube. They then secreted a jelly that held them firmly in the tube. He then exposed them to unilateral light. Each cell, like a cell in a battery, contributed a tiny transcellular electrical potential that summed up to something that Jaffe could measure. As the cells were electrically polarized, he assumed that the zygotes passed a current through themselves. The rhizoidal and thalloidal ends acted like negative and positive poles, respectively. After measuring the voltage generated by the eggs and the resistances of the eggs, Jaffe calculated that each egg generates a current of approximately 2 pA through itself.

Jaffe and his associates then developed the vibrating electrode to measure the magnitude and direction of the current. Concurrently, they also determined the ionic composition of the current and found that the electrical current consists mainly of K^+ and Cl^-, but it appears to be activated by an initial influx of Ca^{2+} (Robinson and Jaffe, 1973, 1975; Chen and Jaffe, 1978, 1979; Nuccitelli, 1978; Speksnijder et al., 1989; Kühtreiber and Jaffe, 1990) and by cyclic guanosine monophosphate (Robinson and Muller, 1997). This current, as well as the rotation of the cell plate, is disrupted by cytochalasins (Brawley and Robinson, 1985; Allen and Kropf, 1992), indicating that actin, which is only observed in the cortex (Kropf et al., 1989), organizes the current. In the zygotes of the Fucaceae and the protonemata of *Funaria*, an electrical current is associated with cell plate formation. However, an important difference exists in these two systems: In the Fucaceae, the current is perpendicular to the plane of the forming cell plate; in *Funaria*, it is parallel.

What causes the cell plate to become oriented in a certain position is still one of the most fascinating mysteries in plant cell biology. It appears to involve actin, mechanical stress, and ionic currents. We could tentatively assume that stress due to growth causes a strain in the plasma membrane, which activates mechanosensitive Ca^{2+} channels, which in turn may activate mechanochemical ATPases that rotate the spindle poles so that they are perpendicular to the direction of growth and so that there is equal tension from both sides. As maximal growth causes maximal strain in the membrane, the new cell plate will form perpendicular to the direction of growth. In isodiametric cells, the influx of Ca^{2+} would be almost the same everywhere, and the cell plate is random.

The developing cell plate does not always form perpendicular to the parental cell walls. Sometimes it is oblique (Saunders and Hepler, 1981), sometimes lens shaped, sometimes funnel shaped, and sometimes it can even take on a cylindrical shape so that the daughter cells form concentric cylinders (Bierhorst, 1971; Gunning, 1982). These uncommon shapes are common in guard cell mother cells and antheridia. In *Allium* guard cell mother cells, the cell plate begins to form in a transverse orientation but rotates to take its final place in a longitudinal orientation (see Fig. 10.11 in Chapter 10; Palevitz and Hepler, 1974a). The reorientation is inhibited by cytochalasins (Palevitz and Hepler, 1974b; Palevitz, 1980), although cytochalasin has no effect on the formation of a normal cell plate (Molé-Bajer and Bajer, 1988). Cytochalasin also prevents the normal orientation of the cell plate that gives rise to the subsidiary cells (Cho and Wick, 1990, 1991). An antimyosin drug inhibits the normal positioning of the cell plate in *Tradescantia* (Molchan et al., 2002). In general, the actomyosin system affects the orientation, but not the formation, of the cell plate. Some of the genes involved in cell plate orientation and asymmetric cell divisions are being identified (Smith, 2001; Falbel et al., 2003; Sack, 2004).

Plants are exceptional organisms for studying the orientation of the division plane (Wright and Smith, 2007). Particularly impressive are the divisions that give rise to two-dimensional growth from one-dimensional growth in moss and fern gametophytes (Miller, 1968, 1980b; Stetler and DeMaggio, 1972; Davis et al., 1974; Stockwell and Miller, 1974; Dyer, 1979; Raghavan, 1979; Cooke and Paolillo, 1980; Miller, 1980b; Saunders and Hepler, 1981, 1982; Cooke and Racusen, 1982; Racusen and Cooke, 1982; Racusen et al., 1988; Racusen, 2002; Whitewoods et al., 2018). The plane of cell division is also important in animals. Early studies in experimental embryology showed that the cleavage plane of a singled-celled embryo is correlated with the future development of the two protoplasmic parts (Roux, 1888), and one way to ensure that the two daughter cells formed by a dividing cell are different is to produce an asymmetrical division (Bisgrove and Kropf, 2007). An asymmetrical division precedes the formation of many differentiated cell types, including rhizoids in ferns (Murata and Sugai, 2000) and trichomes and guard cells in the epidermis of angiosperms (Nadreau and Sack, 2003; Lucas et al., 2006). The placement of the division plane is just beginning to be studied in animal cells (White and Borisy, 1983; Canman et al., 2003).

19.6 SUMMARY

S phase is the time when the genome in a given cell doubles. Mitosis is the process in which one cell divides into two. Typically, nuclear division (karyokinesis) is followed by the division of the whole cell (cytokinesis). To ensure an equal distribution of the hereditary material during nuclear division, numerous motile processes occur, including the condensation of chromatin during prophase, metakinesis, congression to the metaphase plate, chromatid separation, and anaphase motion. Each motile process involves force-generating reactions, and the establishment of an anchor, against which force can be generated. In general, it is believed that tubulin-, actin-, dynein-, and kinesin-related proteins are involved in force generation and anchoring. Progress is being made on understanding how these and other proteins induce the various mitotic movements in individual cell types. However, a unified description of mitosis is not yet at hand.

I also described the process of cytokinesis and how it involves the fusion of Golgi-derived vesicles to form the cell plate. The formation of the cell plate is influenced by a portion of the cytoskeleton known as the phragmoplast. The orientation of the cell plate is also dependent on the cytoskeleton. It is also influenced by ionic currents passing through the plasma membrane.

We are still searching for a unified theory of mitosis that takes into consideration all the experimental and observational data and can describe and explain the great diversity observed in different organisms undergoing mitosis. Such a theory will take into consideration the various physicochemical mechanisms or "means" (which must obey the laws of physics) to attain the same "end" (equal division of hereditary material). Aldous Huxley (1937) wrote in Ends and Means, "The human mind has an invincible tendency to reduce the diverse to the identical. That which is given us, immediately, by our senses, is multitudinous and diverse. Our intellect, which hungers and thirsts after explanation, attempts to reduce this diversity to identity. Any proposition stipulating the existence of an identity underlying diverse phenomena, or persisting through time and change, seems to us intrinsically plausible. We derive a deep satisfaction from any doctrine which reduces irrational multiplicity to rational and comprehensible unity. To this fundamental psychological fact is due the existence of science, of philosophy, of theology. If we were not always trying to reduce diversity to identity, we should find it almost impossible to think at all. The world would be a mere chaos, an unconnected series of mutually irrelevant phenomena. The effort to reduce diversity to identity can be, and generally is, carried too far. This is particularly true in regard to thinkers who are working in fields not subjected to the discipline of one of the well-organized natural sciences. Natural science recognizes the fact that there is a residue of irrational diversity which cannot be reduced to the identical and the rational. There are two tendencies in science; the tendency towards identification and generalization and the tendency towards the exploration of brute reality, accompanied by a recognition of the specificity of phenomena."

19.7 QUESTIONS

19.1. Do you think that the concept of double insurance and functional redundancy is valuable in understanding biological systems and planning and interpreting experiments? Why or why not?

19.2. What limits the various means a cell can use to accomplish a specific end, such as the duplication and division of the hereditary material?

The references for this chapter can be found in the references at the end of the book.

Chapter 20

Extracellular Matrix

Something there is that doesn't love a wall,
That sends the frozen-ground-swell under it,
And spills the upper boulders in the sun,
And makes gaps even two can pass abreast....

There where it is we do not need the wall:
He is all pine and I am apple orchard.
My apple trees will never get across
And eat the cones under his pines, I tell him.
He only says, 'Good fences make good neighbors'.

Robert Frost, "Mending Wall."

20.1 RELATIONSHIP OF THE EXTRACELLULAR MATRIX OF PLANT AND ANIMAL CELLS

In the 17th century, the presence of thick walls in wood and cork made it possible for cells to be readily identified in plants (Hooke, 1665). The presence of a thick wall was the original *sine qua non* for defining a cell, and cells were defined and characterized by their walls and not their contents. The fact that the extracellular matrix of most animal cells is thin or nonexistent was an impediment to the realization that animals were also composed of cells.

Most cells, to a greater or lesser degree, secrete macromolecules into the surrounding medium. These macromolecules amalgamate into an organized structure, which has been called a cell wall, an extracellular matrix, an apoplast, a periplast, slime, a glycocalyx, or a cell covering (O'Brien and McCully, 1969; Okuda, 2002; Niklas, 2004). Although plant cells generally have a thicker extracellular matrix than animal cells, there is a continuum between the two extremes. Some plant cells, like those of *Dunaliella*, have a thin or nonexistent extracellular matrix, and some animal cells, such as those of the tunicates, have an extracellular matrix composed of cellulose (Huxley, 1853a; Hall and Saxl, 1960, 1961). Among plant biologists, the terms *cell wall* and *extracellular matrix* are used interchangeably, and the continuum of cell walls is usually referred to as the *apoplast*. Each term carries with it a slightly different shade of meaning. The term *cell wall* emphasizes the supporting role of the structure and accentuates the apparent gulf between plant and animal cells. The term *extracellular matrix* emphasizes the primary importance of the plasma membrane and not the cell wall as an active barrier to diffusion in separating the protoplasm from the external environment. The term *extracellular matrix* also emphasizes the lively and dynamic aspects of the region external to the plasma membrane and has never carried the connotation of being a dead part of cells, as does the term *apoplast*, which comes from the Greek for "without form." The use of the term *extracellular matrix* has been productive in understanding the unity of nature while still allowing an appreciation of its diversity. This is not a new view. Thomas Huxley (1853b), in his essay "The Cell-Theory," considered the extracellular matrix of plant and animal cells to be homologous and gave them a common name: *periplast*. There are similarities between cabbages and kings.

Young plant cells are surrounded by a thin extracellular matrix that prevents them from lysing when placed in dilute solutions typical of lakes, streams, and soils. By contrast, animal cells with a thin extracellular matrix will lyse unless they are surrounded with a solution that is isotonic with the cell. In fact, the differences in the ionic basis of action potentials in plant and animal cells can be traced to the fact that plant cells, unlike animal cells, are not typically bathed in solutions containing high concentrations of NaCl (Wayne, 1994; Johnson et al., 2002). To live on land or in dilute aqueous environments, plant cells have evolved specializations in the extracellular matrix (including the hollow tracheids and vessel elements), whereas animal cells have evolved specializations in the circulatory systems. While the extracellular matrix allows plant cells to live in dilute environments, it contains numerous fixed charges that accumulate considerable amounts of essential nutrients in a nonosmotic form (Grignon and Sentenac, 1991; Gabriel and Kesselmeier, 1999). Thus, the cell carries a suitcase around itself that contains the necessary nutrients found in its ancestral seas.

In the meristems of plants, the extracellular matrix begins to form during cytokinesis (see Chapter 19). The primary cell wall is approximately 100 nm thick (Roberts, 1989, 1990, 1994) and is only slightly rigid so that the cell can still expand. The primary cell wall contains even thinner regions, known as *primary pit fields*, through which plasmodesmata pass (see Chapter 3; Fig. 20.1). Correlated

Plant Cell Biology. https://doi.org/10.1016/B978-0-12-814371-1.00020-5

FIGURE 20.1 The primary wall of *Zea mays* showing a pit field through which the plasmodesmata traverse. ×15,000. *From Mühlenthaler, K., 1950. Electron microscopy of developing plant cell walls. Biochim. Biophys. Acta 5, 1–9.*

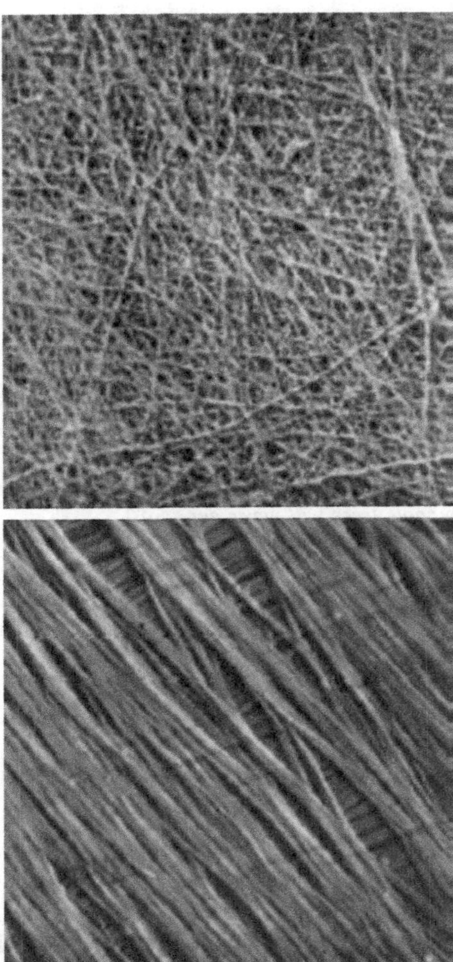

FIGURE 20.2 Surface view of an early stage (top) and a late stage (bottom) of development of the extracellular matrix of *Valonia ocellata*.

with the cessation of expansion, the secondary cell wall begins to be deposited either by thickening the primary cell wall (intussusception) and/or by apposition, which is the deposit of new layers of wall material approximately 10 μm thick between the plasma membrane and the primary cell wall (Fig. 20.2). The primary and secondary cell walls are defined based on their order of formation—the primary wall is formed while a cell is still expanding, whereas the secondary wall is formed as and after expansion ceases (Wardrop, 1962; Esau, 1965; Bowes, 1996). However, wall growth is a continuous process so there is no hard and fast separation between the primary and secondary walls (Keegstra, 2010). The secondary wall is specialized for mechanical support, and often, cells surrounded by a secondary wall are dead at maturity. The thick-walled support tissue, including sclerenchyma tissue, composed of fiber cells with thick secondary walls, and collenchyma tissue, composed of cells with thick primary walls, is distributed within the plant body so as to minimize the effect of mechanical stress on the plant due to wind and other factors (Bower, 1925). In fact, Venning (1949) and Walker (1960) have shown that exposure to wind causes an increase in the amount of collenchyma tissue in a plant. The correlation between thick cell walls and dead cells has historically caused plant biologists to view the cell wall as dead wood instead of the dynamic organelle it is (Bolwell, 1993; Kieliszewski and Lamport, 1994; Carpita et al., 1996; He et al., 1996; Somerville et al., 2004; Hepler and Winship, 2010).

The extracellular matrix of plants thickens during development. This allows each cell to generate an internal hydrostatic pressure or turgor pressure in the order of 0.1–1 MPa, which provides the cell with a certain amount of rigidity and mechanical strength. Consequently, plants can grow tall and wide to maximize their ability to intercept the sun's rays for photosynthesis. While a thick extracellular matrix is extremely useful for a phototrophic organism, it prevents plants from moving, thereby making them susceptible to attack by predators.

It is a general biological principle that crises lead to opportunities, and consequently species evolve by natural selection (Darwin, 1859). The very extracellular matrix that prevents plants from moving has evolved a number of properties that allow it to function in the defense of plant cells (Lionetti and Métraux, 2014; Ebine and Ueda, 2015; Lionetti et al., 2017; Stavolone and Lionetti, 2017). In some cases, the extracellular matrix becomes extremely thick to prevent the entrance of pathogens (Aist, 1977; Israel et al., 1980; Kobayashi et al., 1997). In other cases, it acts as a lookout at the frontier of the cell and signals the appearance of a pathogen to the rest of the cell (Bergey et al., 1996; Brady and Fry, 1997; Gelli et al., 1997; Stratmann and Ryan, 1997; Xing et al., 1997; Seifert and Blaukopf, 2010).

In this role, fragments of the extracellular matrix known as *elicitors* are released by the hydrolytic enzymes of pathogens and signal the rest of the cell to form antiseptic compounds known as *phytoalexins* (Darvill and Albersheim, 1984; Ryan and Farmer, 1991; Zhao and Last, 1996). The extracellular matrix of epidermal cells in the aerial portion of the plant are clad with a cuticle, which contains cutin and wax that prevents pathogen attack (Cutler et al., 1982; Kerstiens, 1996; Riederer and Muller, 2006). Agrobacterium-mediated transformation also depends on a functional extracellular matrix (Zhu et al., 2003). The cuticle also waterproofs the outer epidermal cell walls (Schreiber, 2010). Sporopollenin is a hydrocarbon that waterproofs the wall of spores and pollen and protects them from microbial attack.

The extracellular matrix that makes up the secondary wall of tracheids, vessel elements, and fibers contains lignins (Higuchi, 1998), which are cross-linked polymers composed of derivatives from the polypropanoid pathway. Cellulosic walls absorb water easily and swell, as evidenced by what happens to paper when it gets wet. Lignin, however, waterproofs the cellulose so that it maintains its rigidity and allows for the conduction of water (Niklas, 1992; Niklas and Spatz, 2012; Witztum and Wayne, 2016b; Niklas et al., 2017). The new paper towels in our building are not very absorbent, and a test for lignin using the phloroglucinol reaction shows that they are highly lignified.

Suberin, another waxy substance, which along with lignin occurs in a specialized thickening of the anticlinal walls of the endodermis in the root known as the Casparian strip (Ursache et al., 2018). The waterproofing functions to ensure that water and solutes enter the stele through the symplast—specifically through the periclinal surfaces of the plasma membrane of the endodermal cells, which are specialized for ion transport (see Chapter 2; Van Fleet, 1961; Bonnett, 1968; Karahara and Shibaoka, 1988, 1992; Ma et al., 2006, 2007; Alassimone et al., 2010; Takano et al., 2010; Naseer et al., 2012; Roppolo and Geldner, 2012; Geldner, 2013a; Andersen et al., 2015; Kamiya et al., 2015; Barberon et al., 2016; Pauluzzi and Bailey-Serres, 2016; Barberon, 2017; Foster and Miklavcic, 2017).

The extracellular matrix of plant cells can be considered a keeper of positional information and as such can be involved in many aspects of cell physiology and development in addition to those mentioned above. The influence of the extracellular matrix can be tested with cell wall mutants and/or by treating cells with cell wall enzymes and their inhibitors (Hofmannová et al., 2008; Paredez et al., 2008). Such studies have shown that the extracellular matrix is involved in organizing the actin and microtubular cytoskeletons (Akashi et al., 1990; Ryu et al., 1997; Fisher and Cyr, 1998; Paredez et al., 2008). It is also required for positioning the nucleus (Katsuta and Shibaoka, 1988), for cell division (Meyer and Herth, 1978; Schindler et al., 1989), for adhesion (Kaminskyj and Heath, 1995; Henry et al., 1996), for appressorium formation (Corrêa et al., 1996), for fixation of the embryonic axis (Kropf et al., 1988), for gravity sensing (Wayne et al., 1992; Staves, 1997), and for embryo formation (Barthou et al., 1999). Likewise, the extracellular matrix of animal cells is involved in parasite interactions (Finley, 1990), signaling during fertilization (Snell, 1990), adhesion (O'Rourke and Mescher, 1990), motility (Zetter and Brightman, 1990), differentiation (Ginsberg et al., 1992; Damsky and Werb, 1992), and gene expression (Slavkin, 1972; Slavkin and Greulich, 1975; Juliano and Haskill, 1993; Sastry and Horwitz, 1993). The thick extracellular matrix of endosperm cells, in which plasmodesmata were first discovered by Tangl (1879), serves as a source of carbon for developing embryo (Otegui, 2007).

There are many fascinating structural specializations of the extracellular matrix that result in extraordinary colors in plants (Fox and Wells, 1971; Lee, 2007). The iridescent blue colors of the leaves of *Selaginella willdenowii* and of the fern *Danaea nodosa* and the "brilliant blue fruits" of rudraksha (*Elaeocarpus angustifolius*) are caused by the interference of light resulting from the multilayered nature of the extracellular matrix in their outer epidermal walls. The blue color of the leaves of the blue spruce results from the selective scattering of short-wavelength light from the small waxy particles that cover the leaves. While the function of these extracellular matrix specializations in the epidermal cells of leaves is not well understood, the striated cuticle in petals that occurs in over 50% of angiosperm taxa acts as a diffraction grating and produces colors that may be important as a visual cue for bumblebees during pollination (Whitney et al., 2009).

20.2 ISOLATION OF THE EXTRACELLULAR MATRIX OF PLANTS

To isolate the extracellular matrix, tissues are homogenized (Fry, 1988; Harris, 1988), and either centrifuged at 250–1000 g for 5–20 min to pellet the extracellular matrix or filtered through a mesh with pores 5–10 μm in diameter, and the material that is pelleted in the centrifuge tube or trapped in the filter is washed and saved. The purity of the extracellular matrix fraction is assayed by observing it with a light microscope. The composition of the extracellular matrix can be characterized by mass spectrometry (Günl et al., 2011), electrophoresis (Fry, 2011; Goubet et al., 2011; Mort and Wu, 2011), microarrays (Sørensen and Willats, 2011), histochemically (Soukup, 2014), and immunologically (Hervé et al., 2011; Popper, 2011).

20.3 CHEMICAL COMPOSITION AND ARCHITECTURE OF THE EXTRACELLULAR MATRIX

The composition and architecture of the extracellular matrix of plants has been of interest to humanity for millennia due to its importance in the clothing, paper, food, musical instruments, and building industries (Kidd, 1852; Cross and Bevan, 1895; Heuser, 1944; Percival, 1950; Wegst, 2006). Plant cell walls are used in dendrochronology to build tree ring chronologies that are used to date archaeological finds and reconstruct the natural hydroclimate (Douglass, 1929; Manning and Bruce, 2009; Malevich et al., 2013). Plant cell walls were also used in the manufacture of celluloid and rayon (Roberts, 1989c), and currently plant cell walls and the enzymes that degrade them are important for understanding what causes rot and for the production of cellulosic biofuels (Mandels and Reese, 1965; Reese, 1976; Mousdale, 2008; Rubin, 2008; Somerville, 2008). Consequently, the extracellular matrix of the plants as a whole, and not the extracellular matrix of individual cells, has been studied (Keegstra, 2010). Thus, we have a pretty good idea about the "average extracellular matrix," but we are just beginning to understand how unique the extracellular matrices of individual cells are (McCann and Rose, 2010). In general, the plant extracellular matrix is approximately 60% water and the dry matter is composed primarily of polysaccharides, including cellulose, hemicelluloses, and pectins (Bauer et al., 1973; Keegstra et al., 1973; Talmadge et al., 1973; Carpita and McCann, 2002; Noguchi, 2014; Wang et al., 2012, 2015). There are similarities and differences in the composition of the extracellular matrix between the various taxa of plants and algae (Popper and Fry, 2003; Popper and Tuohy, 2010; Domozych et al., 2012; Domozych and Domozych, 2014).

According to Malcolm Brown (2004), more than 10^{11} tons of cellulose is produced each year, making it the most abundant macromolecule on Earth. Cellulose is a long linear chain of at least 500 glucose molecules linked together with β-1,4 glycosidic bonds. Adjacent cellulose molecules are held together by hydrogen bonds, and approximately 18–24 cellulose molecules, all with the same polarity, are held together to form a cellulose microfibril 3 nm in diameter (Preston, 1939; Burgess, 1985; McCann et al., 1990; Wolters-Arts et al., 1993; Cosgrove, 2014). The cellulose microfibrils are crystalline and are birefringent when viewed under polarized light (see Fig. 11.17 in Chapter 11 and Fig. 20.3; Herzog and Janke, 1920). Cellulose accounts for approximately 25%–30% of the dry weight of the extracellular matrix (Burgess, 1985).

Hemicelluloses are a heterogeneous group of branched polysaccharides composed of a linear backbone of a β-1,4-linked homopolymer of a sugar (e.g., glucose), from which short side chains of other sugars (e.g., xylose, galactose,

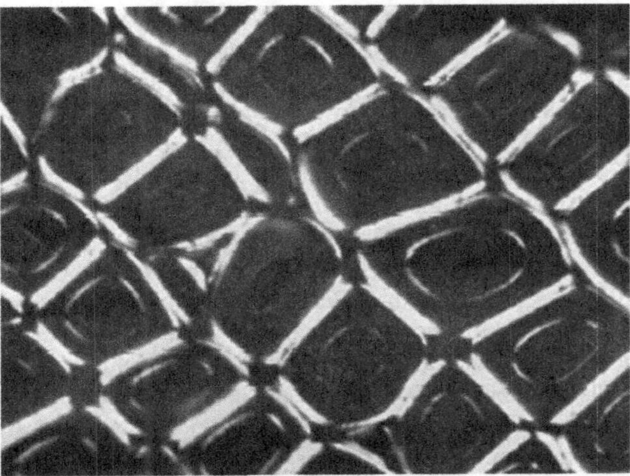

FIGURE 20.3 Transverse section of the tracheids of *Pinus radiate* photographed with a polarizing microscope. Due to the alignment of microfibrils, the extracellular matrix is birefringent and thus appears bright on a dark background. *From Preston, R.D., 1952. The Molecular Architecture of Plant Cell Walls. John Wiley & Sons, New York.*

fucose) protrude (Scheller and Ulvskov, 2010). The hemicelluloses include xyloglucans and arabinoxylans. The backbone of the hemicelluloses binds tightly but noncovalently to the surface of each microfibril, thus cross-linking them via hydrogen bonds (Hayashi, 1989, 1991; McCann et al., 1990). The arabinoxylans bind to each other by means of coupled phenols. Hemicelluloses account for about 15%–20% of the dry weight of the extracellular matrix (Burgess, 1985).

Pectins are a heterogeneous group of molecules that contain a backbone of many α-1,4-linked negatively charged galacturonic acid residues. Unbranched homopolymers of galacturonic acid are anionic and readily bind Ca^{2+} to form a stiff gel with the texture of jelly (Geitmann and Parre, 2004; Proseus and Boyer, 2006a; Carpita, 2007). Interestingly, the binding affinity of pectins for calcium ions may be modified by changing the tension on the pectins (Proseus and Boyer, 2007). When the carboxylic acid groups of the galacturonic acids are methylesterified, the anions are neutralized and the pectin methylester is unable to bind Ca^{2+}; the pectin methylester forms a sol or solution that is less viscous than the gel formed by the anionic homopolymer in the presence of Ca^{2+}. Rhamnogalacturonan I is a pectin that has a backbone of alternating rhamnosyl-galacturonosyl dimers, where the rhamnosyl residues may have side chains consisting of fucosyl, galacturonosyl, or O-methyl galacturonosyl residues. Rhamnogalacturonan II is a pectin that has a backbone of galacturonosyl residues and four different side chains containing peculiar sugar residues, such as apiose, aceric acid, 3-deoxy-lyxo-2-heptulosaric acid, and 3-deoxy-manno-2-octulosonic acid (Hayashi, 1989; Varner and Taylor, 1989; Shea et al., 1989; Vreeland et al., 1989; Knox et al., 1990; Voragen et al., 2009). A borate ester is necessary for cross-linking rhamnogalacturonan II so that it forms dimers (O'Neill et al., 1996, 2001; Ishii et al., 2001). The pectins bind to the cellulose microfibrils via ester bonds and to each other by means of calcium bridges or glycosidic bonds. Pectins make up about 35% of the dry weight of the extracellular matrix (Burgess, 1985).

Proteins account for 5%−10% of the dry weight of the extracellular matrix of plants. Many of the major proteins are glycoproteins and have a high proportion of attached sugars. The most well-known extracellular matrix protein is a protein named *extensin*, which was prematurely named because it has no relationship to cell extension (Lamport, 1965; Cleland and Karlsnes, 1967; Sadava and Chrispeels, 1973), although more recent work suggests that extensin may be required as a scaffold for pectins in the developing cell plate of some cells (Cannon et al., 2008). About 30% of the amino acids of extensin and related proteins are hydroxyproline (Cassab and Varner, 1988; Adair and Appel, 1989; Langan and Nothnagel, 1997; Ellis et al., 2010). When proline is hydroxylated, its hydroxyl group, like those in serine and threonine, can be glycosylated in the Golgi apparatus to form glycoproteins (Mohnen and Tierney, 2011; Velasquez et al., 2011). Collagen, a major component of the animal extracellular matrix, is also rich in hydroxyproline. The plant extracellular matrix also contains many nonstructural proteins, particularly glycosidases, that are involved in the building and restructuring of its architecture (Cassab and Varner, 1988; Fischer and Bennett, 1991; del Campillo and Lewis, 1992), pectin methylesterases that are involved with the transformation of methylesterified pectins into anionic pectins (Bosch and Hepler, 2006; Kohli et al., 2015), expansins (Cosgrove, 2016b), yieldins (Okamoto-Nakazato, 2002), and proteins involved in interacting with the plasma membrane (Pennell et al., 1989; Schindler et al., 1989; Sanders et al., 1991; Quatrano, 1990; Wagner et al., 1992; Pickard and Ding, 1992, 1993; Lord and Sanders, 1992; He et al., 1998, 1999; Caderas et al., 2000; Kim et al., 2000; Rose et al., 2000; Catalá et al., 2001; Madson et al., 2003; Yokoyama et al., 2004, 2007). There are also hundreds of cell wall−related genes (Schindelman et al., 2001; Vanzin et al., 2002; Lao et al., 2003; Yokoyama and Nishitani, 2004) and the proteins they encode (Robertson et al., 1997; Bayer et al., 2006; Isaacson and Rose, 2006; Jarnet et al., 2008).

Why is the sugar coating surrounding each cell so diverse in terms of its composition of sugars and glycosidic linkages? To be in a position to answer these questions, the structure of the plant extracellular matrix surrounding each cell should be thought of in the same way that Hermann Staudinger (1961) viewed the structure of macromolecules; Emil Fischer, James Sumner, and Frederick Sanger viewed the structure of proteins (Fruton, 1972); and Erwin Chargaff (1950a, 1978) viewed the structure of nucleic acids. That is, begin the isolation procedure with the notion that the extracellular matrix is an extremely complex macromolecule, involved in positional information, of which the activity varies with the arrangement of monomers.

Keith Roberts, Bruce Knox, and Andrew Staehelin have changed our ways of looking at the extracellular matrix. They have taught us that the extracellular matrix has both a history and a geography by demonstrating immunocytochemically that the extracellular matrix varies from cell to cell, from one side of the cell to the other, and during the development of the cell (see Fig. 20.4; Moore and Staehelin, 1988; Stafstrom and Staehelin, 1988; Moore

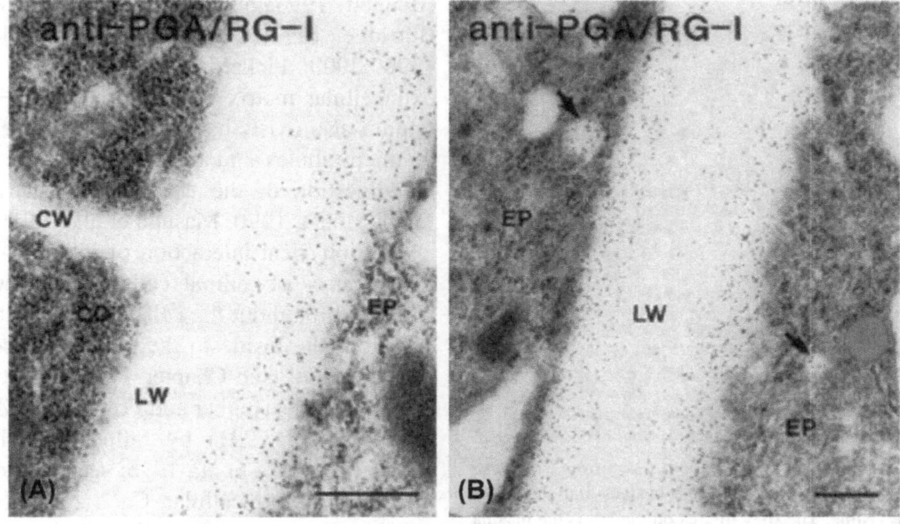

FIGURE 20.4 (A) An antibody to pectins labels the portion of the extracellular matrix next to the epidermal cell (EP) but not next to the cortical cell (CO). (B) An antibody to pectins labels the portion of the extracellular matrix next to the EP but not the portion midway between the two epidermal cells. Bars, 500 nm. *CW*, cross wall; *LW*, longitudinal wall. *From Lynch, M.A., Staehelin, L.A., 1992. Domain-specific and cell type-specific localization of two types of cell wall matrix polysaccharides in the clover root tip. J. Cell Biol. 118, 467−479.*

et al., 1991; Knox, 1992a,b; Li et al., 1992, 1994; Lynch and Staehelin, 1992; Levy and Staehelin, 1992; Ferguson et al., 1998; Parre and Geitmann, 2005; Chebli et al., 2012). There are now over 100 different antibodies available to localize specific carbohydrates in the extracellular matrix (Pattathil et al., 2010). The chemical composition of individual cells can now be studied with infrared microscopes, which can identify the various chemical bonds found in carbohydrates (McCann et al., 1997; Himmelsbach et al., 1999; Carpita et al., 2001a,b). Many studies are showing how dynamic the extracellular matrix is and how the chemical composition of the extracellular matrix changes with the environment (Iraki et al., 1989) and during development (Gorshkova et al., 1997; Peña and Carpita, 2004; Peña et al., 2004; Dunn et al., 2007). Thus, the models of wall architecture derived from grinding up the whole plant serve as a first approximation of the architecture of the extracellular matrix, but an understanding of the role of the extracellular matrix in growth and development will depend on an understanding of the extracellular matrix of the cell in question.

As a first approximation, the extracellular matrix of plants can be considered to be a gellike matrix of hemicelluloses, pectins, and proteins surrounding cellulose microfibrils just like reinforced concrete is composed of steel reinforcing bars (rebar) embedded in cement (see Fig. 20.5; Varner and Lin, 1989; McCann et al., 1990; Geitmann, 2010; Kirby, 2011). However, in reality, the gellike matrix is also made out of approximately 20- to 40-nm long macromolecules that can cross-link the cellulose microfibrils (McCann et al., 1990). This means that a 100-nm thick extracellular matrix has about three layers of cellulose microfibrils.

Like the extracellular matrix of plant cells, the extracellular matrix of animal cells is also composed of high−molecular mass polysaccharides, including hyaluronan and cellulose (Huxley, 1898; van Daele et al., 1991; Kimura and Itoh, 1995), and proteins such as collagen, elastin, fibronectin, laminin, and vitronectin (Sanders et al., 1991). However, in the extracellular matrices of animal cells, there are typically a greater proportion of proteins than polysaccharides.

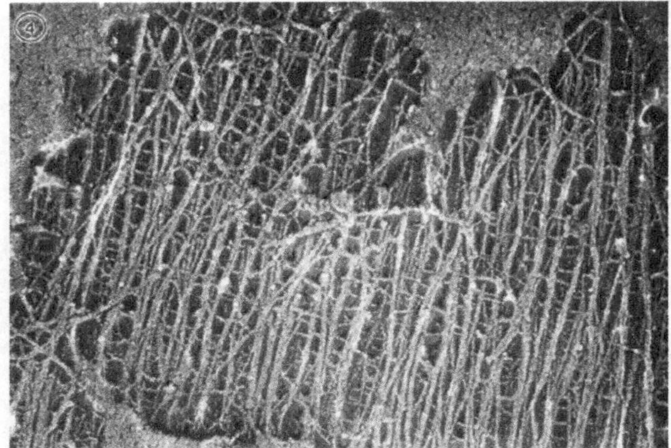

FIGURE 20.5 The extracellular matrix of a root tip cell from *Zea mays* revealed by deep etching. Regularly spaced cross-bridges that attach the cellulose microfibrils are visible. The smooth area on the left is the plasma membrane. ×213,000. *From Satiat-Jeunemaitre, B., Martin, B., Hawes, C., 1992. Plant cell wall architecture is revealed by rapid-freezing and deep-etching. Protoplasma 167, 33−42.*

20.4 EXTRACELLULAR MATRIX−PLASMA MEMBRANE−CYTOSKELETAL CONTINUUM

While the extracellular matrix is on the E-side of the plasma membrane, it is not totally isolated from the internal contents of the cell. The plasma membranes of fungal, plant, and animal cells are attached to the proteins in the extracellular matrix by protein−protein interactions (Ruoslahti, 1988). The integral plasma membrane proteins that recognize and bind various amino acid sequences in the proteins of the extracellular matrix are known as *integrins* (Schindler et al., 1989; Haas and Plow, 1994; Pickard, 2007; Knepper et al., 2011). Arg-Gly-Asp (RGD) is one of the sequences in the extracellular matrix proteins, which is recognized by integrins, and the binding of integrins with these extracellular matrix proteins can be inhibited by adding oligo-peptides containing the sequence RGD (Humphries, 1990).

The cytoplasmic domain of integrins can interact with the peripheral membrane proteins on the cytosolic side of the membrane known as the membrane skeleton (Steck, 1974; Luna and Hitt, 1992; Gens et al., 2000; Pickard and Fujiki, 2005). The *membrane skeleton* is composed of proteins, including spectrin and ankyrin, which can link directly and/or indirectly to cytoskeletal elements, including actin, tubulin, and intermediate filament proteins (Bennett, 1990a,b; Magee and Buxton, 1991; Turner and Burridge, 1991; Quaranto and Jones, 1991; Gens et al., 1996, 2000; Pickard, 2008). There is evidence that the extracellular matrix−plasma membrane−cytoskeletal continuum also exists in plant cells because the organizations of microtubules and microfilaments are altered following the digestion of the extracellular matrix (Akashi et al., 1990; Melan, 1990; Masuda et al., 1991; Ryu et al., 1997).

The physical interaction of integrins with the extracellular matrix of animal cells allows the transmission of stresses throughout the cell, from the outside to the inside, and from the inside to the outside. According to the tensegrity model (see Chapter 11), these stresses can lead to morphogenesis (Ingber and Folkman, 1989a,b; Williamson, 1990; Ingber, 1991) by influencing signal transduction chains (Schwartz et al., 1995; see Chapter 12) that involve increases in intracellular Ca^{2+} (Wacholtz et al., 1989), changes in intracellular pH (Schwartz et al., 1991), and increases in tyrosine phosphorylation (Juliano and Haskill, 1993; Kornberg et al., 1991, 1992).

20.5 BIOGENESIS OF THE EXTRACELLULAR MATRIX

20.5.1 Plasma Membrane

I will use the biogenesis of the extracellular matrix as a pedagogical tool to briefly review most of the concepts covered in the previous chapters. If one were to go through this book backward, this chapter could serve as an introduction to the rest of the cell because there is a relationship between the extracellular matrix and the organelles that make up the protoplast (Newcomb, 1963). The cellulose microfibrils in the extracellular matrix are synthesized from uridine diphosphate (UDP) glucose by the cellulose-synthesizing complexes on the plasma membrane. The existence of an enzymatic cellulose-synthesizing complex was first postulated by Roelofsen (1958) and was later discovered by Brown and Montezinos (1976), Mueller et al. (1976), and others (Robenek and Peveling, 1977; Mueller and Brown, 1980; Wada and Staehelin, 1981; Hogetsu, 1983; Herth and Weber, 1984; Reiss et al., 1984; Emons, 1985; Schmid and Meindl, 1992) using freeze-fracture electron microscopy. In freeze-fracture micrographs, it appears as a single globule on the E-face of the plasma membrane and as a rosette of six particles, 8 nm in diameter, arranged in a hexagon on the complementary P-face. The cytoplasmic domain of the cellulose-synthesizing complex has been observed in plasma membrane ghosts isolated from protoplasts. The cellulose-synthesizing complexes are associated with the ends of cellulose microfibrils and appear as hexagonal structures 45–50 nm in diameter and extending 30–35 nm into the cytoplasm (Bowling and Brown, 2008).

The organization of the cellulose-synthesizing complex determines the size and shape of the resulting cellulose microfibrils (Okuda et al., 2004; Saxena and Brown, 2005). The complex occurs with a density of approximately 0.5–5/μm^2 and binds antibodies to a cellulose synthase polypeptide (Kimura et al., 1999), confirming that it is indeed the cellulose-synthesizing complex. Fluorescence studies in living cells show that the cellulose-synthesizing complex fused to green fluorescent protein moves along the plasma membrane at a rate of 330 nm/min—a rate that is equivalent to the polymerization of 300–1000 glucose monomers per glucan chain per minute (Lloyd, 2006; Paredez et al., 2006; Nair and DeBolt, 2011).

The rosette arrangement of particles occurs on the plasma membrane of higher plants and the algae in the evolutionary lines that may have given rise to higher plants; for example, *Nitella* (Hotchkiss and Brown, 1987), *Chara* (McLean and Juniper, 1986), and *Coleochaete* (Okuda and Brown, 1992). In the other algae, as well as in bacteria and *Dictyostelium*, the cellulose-synthesizing complex on the E-face is typically linear, and the length of the complex seems to be correlated with the number of cellulose molecules that make up the cellulose microfibril (Robinson and Preston, 1972; Brown and Montezinos, 1976; Brown et al., 1976; Zaar, 1977; Peng and Jaffe, 1976; Willison and Brown, 1978; Quader and Robinson, 1981; Itoh et al., 1984; Brown, 1985; Mizuta, 1985; Itoh, 1992). The structure of the cellulose-synthesizing complex in tunicates is also linear (Kimura and Itoh, 1996).

Evidence that the linear terminal complexes are involved in cellulose synthesis comes from the observations that the addition of ethylenediaminetetraacetic acid (EDTA), Congo red, or Tinopal (Calcofluor) causes the inhibition of cellulose synthesis and the perturbation or disappearance of the linear terminal complexes (Montezinos and Brown, 1979; Robinson and Quader, 1981; Itoh et al., 1984).

The integrity of the plasma membrane is important for cellulose synthesis. The plasma membrane participates in maintaining a submicromolar concentration of intracellular Ca^{2+} in the cytosol that is necessary for cellulose synthesis. Calcium ionophores, which increase the concentration of intracellular-free Ca^{2+}, inhibit the synthesis of cellulose, and stimulate the synthesis of callose, a β-1,3 glucan (Quader and Robinson, 1979). Heinrich Kauss (1987) proposed that the same enzyme complex is involved in the synthesis of both polymers, but it synthesizes cellulose under submicromolar Ca^{2+} conditions and callose under micromolar Ca^{2+} conditions. The transient increase in the intracellular Ca^{2+} concentration may function as a component of the signal transduction chain that signals an attack so that the cell can seal itself off from the rest of the plant with wound callose (Kauss, 1987; Delmer, 1991). Callose deposition around plasmodesmata may also be important for sealing off one cell from another (Sager and Lee, 2018).

The wound callose response is inhibited by the Ca^{2+} channel blockers, nifedipine, and La^{3+}. Furthermore, treatment of the cell with fungal wall products that induce the callose response causes the total Ca^{2+} concentration of the cell to rise by 100 μM. Thus, the wound callose response satisfies all three of Jaffe's Rules for implicating Ca^{2+} as a second messenger in the elicitor-induced formation of callose (see Chapter 12).

Another fascinating observation concerning callose synthesis is that a membrane potential is required for the polymerization of UDP-glucose into callose by a crude membrane fraction from bacteria (Bacic and Delmer, 1981; Delmer, 1987). The membrane potential was established across the membrane vesicles by adding K^+ and valinomycin, a potassium ionophore, which allowed K^+ to diffuse across the membrane and establish a diffusion potential across the membrane in response to the imposed concentration difference (see Chapter 2). If these experiments that include a membrane potential were to be repeated in the presence of Ca^{2+}/EGTA buffers that maintain a submicromolar concentration of Ca^{2+}, it is possible that cellulose would be formed in vitro instead of callose in the crude membrane fraction. While genetic evidence suggests that callose synthase and cellulose synthase come from independent genes (Jacobs et al., 2003; Nishimura et al., 2003; Somerville, 2006), the idea of assaying the function of membrane proteins in the presence of a membrane potential and physiological ion concentrations and even asymmetries is still valuable when it comes to looking at enzymatic processes from a cellular perspective.

Nevertheless, hard work on isolating a cellulose synthase from higher plants has come to fruition and the various genes that encode cellulose-synthesizing enzymes and the cellulose synthase itself have been isolated and characterized (Delmer, 1991, 1999; Mayer et al., 1991; Amor et al., 1991; Delmer et al., 1991; Kudlicka et al., 1995, 1996; Pear et al., 1996; Kudlicka and Brown, 1997; Arioli et al., 1998; Carpita and Veraga, 1998; Blanton et al., 2000; Richmond and Somerville, 2000; Holland et al., 2000; Carpita et al., 2001a,b; Nobles et al., 2001; Peng et al., 2001; Vergara and Carpita, 2001; Wu et al., 2001; Gillmor et al., 2002; Pagant et al., 2002; Read and Bacic, 2002; Roberts et al., 2002; Kurek et al., 2003; Tanaka et al., 2003; Roberts and Roberts, 2004; Saxena and Brown, 2005; Somerville, 2006; Lu et al., 2008; Taylor, 2008; Wang et al., 2008). Karl Niklas (2004) and Malcolm Brown (2004) emphasize the importance of horizontal transfer of the cellulose-synthesizing genes, from eubacteria to the nucleus of eukaryotic plant cells, in the evolution of land plants.

20.5.2 Cytoskeleton

The orientation of the cellulose microfibrils is thought to be regulated, in part, by microtubules, as microtubule orientation is parallel to, and correlated with, the orientation of cellulose microfibrils (Hepler and Newcomb, 1964). Moreover, microtubule orientation predicts the orientation of newly deposited cellulose microfibrils. Last, agents that inhibit microtubule polymerization or organization affect cellulose microfibril orientation (Hepler et al., 2013). Remember that the majority of the data favor the hypothesis that microtubules control the cellulose-synthesis complex and orient microfibrils (Baskin, 2001; Gardiner et al., 2003; Wasteneys and Collings, 2006; Rajangam et al., 2008), but there is strong evidence for alternative hypotheses (see Chapter 11).

20.5.3 Endomembrane System

The endomembrane system is fundamental in the synthesis and delivery of the components of the extracellular matrix (Kim and Brandizzi, 2014, 2016; Lütz-Meindl, 2016; van de Meene et al., 2017). The synthesis of hemicellulose takes place predominantly or even exclusively in the Golgi apparatus (Zhang and Staehelin, 1992; Lynch and Staehelin, 1992; Staehelin et al., 1992). Many of the (100−1000) enzymes required for hemicellulose synthesis, including glucosyl, xylosyl, fucosyl, and arabinosyl transferases, have been localized in the Golgi apparatus (Ray et al., 1969; Gardiner and Chrispeels, 1975; Green and Northcote, 1978; James and Jones, 1979; Ray, 1980; Hayashi and Matsuda, 1981; Urbanowicz et al., 2004). The resulting hemicelluloses are transported to the plasma membrane through the secretory pathway.

The synthesis of pectins, including homogalacturonan, rhamnogalacturonan I, and rhamnogalacturonan II, also takes place in the Golgi apparatus (Zhang and Staehelin, 1992; Staehelin et al., 1992; Sherrier and VandenBosch, 1994; Staehelin and Moore, 1995; Sterling et al., 2001; Bosch and Hepler, 2005; Harholt et al., 2010; Anderson, 2016). The resulting pectins are transported to the plasma membrane through the secretory pathway (Toyooka et al., 2009).

The proteins that make up the cellulose-synthesizing centers pass through the Golgi apparatus (Haigler and Brown, 1986; Wightman and Turner, 2010a,b; Bashline et al., 2011, 2014a,b; Cai et al., 2011; Oikawa et al., 2013; McFarlane et al., 2014). Perhaps, the same vesicles that contain the cellulose-synthesizing complexes also contain the hemicelluloses and pectins. Because the hemicelluloses and pectins in each cell type may be different, the enzymatic composition of the Golgi apparati in these cells must also be different. Moreover, the position of the Golgi body or the delivery of Golgi-derived vesicles to various sides of the cell must be regulated in cells that are surrounded by an extracellular matrix that is not uniform all the way around the cell. We do not know what controls the position of the Golgi apparatus and its enzymatic composition in these cases.

The synthesis of extracellular matrix proteins such as extensin begins when the gene becomes transcriptionally active. Then RNA polymerase II transcribes the DNA into hnRNA that subsequently binds to RNA-binding proteins and Small NUclear RibonucleoProteins (SNURPs) to form spliceosomes. The introns are then removed from the hnRNA to make mRNA and the 5′ end is capped with methylguanosine and the 3′ end is polyadenylated. The protein−mRNA complex then interacts with the nuclear pores and moves into the cytoplasm (see Chapter 16).

The mRNA for extensin must bind to the ribosomes, and as translation begins, the nascent polypeptide must bind the signal recognition particle and move to the endoplasmic reticulum (ER) where the signal recognition particle binds to its receptor (see Chapter 4). Subsequently, the amino-terminus containing the signal peptide is inserted through a protein-translocating channel of the ER and enters the secretory pathway, where it moves to the Golgi apparatus. The prolines participating in the O-linked glycosylations are hydroxylated in the ER (Hieta and Myllyharju, 2002; Tiainen et al., 2005). Extensin is glycosylated, and the *O*-linked arabinosylation of extensin begins in *cis*-Golgi cisternae (Moore et al., 1991). Double immunolabeling experiments with colloidal gold show that extensin moves through the entire Golgi apparatus and can be processed in the same Golgi stack as xyloglucan (Moore et al., 1991). Eventually, the extracellular matrix proteins end up in the lumen of a secretory vesicle (see Chapter 8).

Enzymes that affect the physicochemical properties of the extracellular matrix, including expansin (Vannerum et al., 2011; Cosgrove, 2016a,b), yieldin (Okamoto-Nakazato, 2002), xyloglucan endotransglycosylase/hydrolase, polygalacturonase, pectate lyase, pectin methylesterase (Bosch et al., 2005; Tian et al., 2006), and pectin methylesterase inhibitor (Giovane et al., 2004), follow the same pathway as extensin, as they travel from the rough ER to the extracellular matrix.

The vesicles that bud off the Golgi cisternae or the *trans*-Golgi network move through the viscous cytosol (see Chapter 9) with the aid of a mechanochemical ATPase (see Chapters 10 and 11) to the plasma membrane. The vesicles fuse with the plasma membrane, and contents are secreted into the extracellular matrix. Of course, if too much plasma membrane is added relative to the wall material needed, the membrane is recycled by endocytosis (see Chapter 8; Moscatelli, 2008; Zonia and Munnik, 2008; Bashline et al., 2013). Moreover, the extracellular matrix is dynamic and is also undergoing turnover (Labavitch, 1981; Gorshkova et al., 1997). We do not know how the exocytotic and endocytotic vesicles that contain the cellulose-synthesizing complexes, hemicelluloses, pectins, and proteins are regulated, related, or coordinated.

20.5.4 Self-Assembly of the Extracellular Matrix

The final assembly of the extracellular matrix takes place in the extracellular matrix itself and there are enzymes endemic to the extracellular matrix that can facilitate the assembly (Fry, 2004). The extracellular matrix varies in space, and thus the endemic enzymes of the extracellular matrix must be localized in various regions of the extracellular matrix. Moreover, the extracellular matrix changes during cell development, and thus the enzymatic components from the cell plate to the mature extracellular matrix must also vary temporally. This kind of spatiotemporal complexity has been elegantly demonstrated in developing transfer cells (Talbot et al., 2007; Vaughn et al., 2007). The ubiquity and importance of the extracellular matrix along with its structural complexity and unique and dynamic constellation of enzymes really give it organellar status.

20.6 PERMEABILITY OF THE EXTRACELLULAR MATRIX

The extracellular matrix is not solid and amorphous but contains numerous aqueous pores, and like other polysaccharide networks (Laurent, 1995; Ogston, 1995), it acts as a sieve through which water and polar molecules can easily diffuse when the extracellular matrix is hydrated (Kamiya et al., 1962, 1963). The extracellular matrix becomes limiting to water movement when it is dehydrated, and evaporation is the driving force for water movement (Tazawa and Okazaki, 1997). The average sizes of pores in semidehydrated extracellular matrices were originally estimated to be approximately 3–5 nm in diameter (Carpita et al., 1979; Miller, 1980a; Carpita, 1982), whereas those in hydrated extracellular matrices were estimated to be approximately twice as large (Tepfer and Taylor, 1981; Baron-Epel et al., 1988a; Shepherd and Goodwin, 1989; Meiners et al., 1991a; McCann and Roberts, 1991). Therefore, when the extracellular matrix is hydrated, its permeability is typically greater than that of the plasma membrane and can be ignored as a first approximation when measuring the permeation of solutes into or out of the cell (see Chapter 2).

The original experiments to determine pore size of the extracellular matrix were done in 100–300 mol/m^3 solutions that contained solutes that were either small enough to pass through the extracellular matrix and plasmolyze the cell or were too large to pass through the extracellular matrix and caused it to crinkle (cytorhysis). Either way, the 0.1- to 0.3-M solutions dehydrated the extracellular matrix during the experiment and thus provided a minimum estimate of the average pore size (Carpita et al., 1979; Miller, 1980a; Carpita, 1982).

Melvin Schindler and his colleagues performed experiments involving fluorescence redistribution after photobleaching to study the permeability of the extracellular matrix of soybean cells in a more sensitive manner (Meiners et al., 1991a). Although these cells were also plasmolyzed, and thus the extracellular matrix was dehydrated, they find that the average pore size is between 8.6 and 8.8 nm. The minimum estimate of the pore size of the extracellular matrix may not only depend on the technique used to measure it but it may also vary from cell type to cell type and depend on the physiological state and age of the cell (Read and Bacic, 1996; Titel et al., 1997; Berestovsky et al., 2001; Proseus and Boyer, 2005; Kramer et al., 2007).

Tepfer and Taylor (1981) used a clever technique to estimate the porosity of the hydrated extracellular matrix. They ground it into a powder, loaded it into a typical chromatography column tube, and used the powder as the matrix of a gel exclusion column. Then proteins of various sizes were added to the top of the column. If a protein was too large to enter the "pore space," it eluted in the void volume—that is, it was not retarded by the column. Proteins that were retarded must have traveled through the tortuous porous pathways. Using this technique, Tepfer and Taylor found that proteins less than 7 nm in diameter were retarded by the column, and thus most of the pores must have an average diameter of less than 7 nm. While the pore size is likely to vary from cell to cell and during the development of a single cell, the extracellular matrix of living cells is porous enough to allow the movement of proteins (Stafstram and Staehelin, 1988; Kandasamy et al., 1989, 1991; Hoson et al., 1991; Inouhe and Nevins, 1991). Such a porocity is necessary for cell-to-cell communication between pollen and stigma (Dzelzkalns et al., 1992).

The permeability of the plasma membrane can be tested directly using patch-clamping techniques after removing the extracellular matrix enzymatically, mechanically, or by using lasers and forming protoplasts or spheroplasts (Martinac et al., 1990; Kaiser et al., 1998; Roberts et al., 1999). The pressure difference across the membranes of wall-less cells is different from the pressure across the membranes of walled cells. In addition, the extracellular matrix proteins and carbohydrates may directly influence the ion channels, as would the ionic and osmotic (matrix potential) properties of the extracellular matrix. We always must consider the relationship of the part to the whole.

20.7 MECHANICAL PROPERTIES OF THE EXTRACELLULAR MATRIX

Most studies on the mechanical properties of the extracellular matrix have been carried out on plant tissues and organs (Geitmann, 2006), and consequently, average mechanical properties of the extracellular matrix have typically been measured. This limits the ability to resolve the subtle and transient mechanical properties that control the growth of a given cell. Progress is being made, however, in measuring the mechanical properties of the extracellular matrix of single cells (Wang et al., 2004; Wei et al., 2006; Zerzour et al., 2009).

The extracellular matrix of plant cells is almost as strong as steel. The strength of a material can be quantified by its elastic modulus and its tensile strength. The elastic modulus (M, in N/m^2) is a measure of the amount of stress (in N/m^2) required to produce a doubling of the length of a substance. The greater the elastic modulus, the greater the amount of stress necessary to obtain a given strain. A doubling of the length is equivalent to unit strain, where *strain* is defined as the change in length divided by the initial length.

$$\text{Elastic modulus} = \text{stress/strain, when strain} = 1 \quad (20.1)$$

The tensile strength is the amount of stress needed to break a material. The tensile strength is generally less than the elastic modulus of a given material, and consequently, the elastic modulus is obtained by extrapolation from a plot of strain versus stress and thus assumes that the ratio of strain to stress is proportional. The elastic modulus is typically greater than the tensile strength as most materials break before they double in length. The elastic modulus and tensile strength of a cotton fiber are $6-12 \times 10^9$ and $0.25-0.8 \times 10^9$ N/m^2, which is not that much less than the elastic modulus and tensile strength of steel, which are 200×10^9 and 10^9 N/m^2, respectively (Siegel, 1962; Mark, 1967; Harris, 1980; Niklas, 1992; Niklas and Spatz, 2012).

Building a plant out of materials with high elastic moduli and tensile strengths allows plants to withstand many mechanical stresses. It is not only the mechanical properties of the walls but also the spatial arrangement of cells with an extra thick cell wall within the plant that allows the organ to withstand and/or recover from such forces as those experienced by plants on a windy day or the force on a branch caused by the weight of the fruits it bears (Roget, 1834; Farmer, 1913; Niklas, 1992; Witztum and Wayne, 2014, 2015, 2016a,b). The mechanical properties of plants, their organs, and their cells depend on the mechanical properties of the materials and the arrangement and geometry of the materials. Moreover, the tensile strength of tissues and organs depend on the strength and arrangement of molecules that cause cells to adhere to one another (Verger et al., 2018).

The high tensile strength of the extracellular matrix protects the protoplast from lysing and allows the plant to survive in a dilute aqueous environment. When the extracellular matrix is removed, as in the case of protoplasts, the protoplast will burst unless placed in an isoosmotic medium. This is because the tensile strength of a plasma membrane is approximately 175 N/m^2 (Nobel, 1983). When the extracellular matrix is present, the cell takes up water but can only swell to a limited extent. It stops taking up water when the difference in the osmotic pressure becomes balanced by the hydrostatic or turgor pressure that develops within the cell (see Chapters 7 and 12).

The high turgor pressure that results from the osmotic difference between the outside of the cell and the inside, combined with the high tensile strength of the extracellular matrix, provides the plant with mechanical strength and rigidity. This strength allows plants to stand tall and wide, spreading their leaves to maximize solar light capture and thus photosynthetic processes that take place within the chloroplast (see Chapter 13). Thus, in general, the exceedingly strong extracellular matrix is very useful for an autotrophic organism, but the resultant immobility would be a hindrance to a heterotrophic organism.

On the other hand, the high tensile strength of the extracellular matrix will impede growth, which can be looked at as a type of movement. Therefore, mechanisms must exist to loosen the extracellular matrix to permit cell expansion. Cell expansion results when the pressure exerted by the protoplast causes a stress in the extracellular matrix that is great enough to overcome the strength of the load-bearing bonds between the polymers in the extracellular matrix. That is, sufficient stress is needed to shear the elements of the extracellular matrix past each other. The extracellular matrix acts as if it were composed both of elastic and viscous components, both of which may be important for growth. How do we measure the viscoelastic properties of the extracellular matrix that limit cell expansion? We would like to measure the parameter, known as the *viscoelastic extensibility* (Φ, in $Pa^{-1}\ s^{-1}$), which relates the volumetric relative rate of strain (dV/Vdt, in s^{-1}) to the stress (σ, in N/m^2) according to the following equation:

$$\Phi = \frac{\frac{dV}{Vdt}}{\sigma} \tag{20.2}$$

The viscoelastic extensibility is the reciprocal of the viscosity of the wall. It is possible to determine it by measuring the volumetric relative rate of strain after imposing multiaxial stresses of various magnitudes (Kamiya et al., 1963; Green, 1968; Richmond et al., 1980) or by measuring the stress at various volumetric relative rates of strain. However, there are no adequate techniques to measure the viscoelastic extensibility in many cell types, and consequently, people somehow arbitrarily measure the relationship between stress and strain in just one dimension (Masuda, 1978a,b; Gertel and Green, 1977; Probine and Preston, 1961, 1962; Probine and Barber, 1966; Métraux and Taiz, 1977, 1979; Métraux et al., 1980; Taiz, 1984; Cosgrove, 1993b; Burgert and Fratzl, 2006; Geitmann and Ortega, 2009; Milani et al., 2011; Fernandes et al., 2012; Braybrook, 2015; Kim et al., 2015; Vogler et al., 2015, 2017; Weber et al., 2015; Xi et al., 2015; Geitmann, 2017). To estimate the mechanical properties of the wall, one can measure strain under constant stress, the stress relaxation properties, or stress under various strains. Induced changes in strain or stress should be gradual as opposed to sudden, to best mimic physiologically relevant conditions (Wei and Lintilhac, 2003). Each of these techniques measures unidirectional properties of the extracellular matrix, when in reality, the three-dimensional viscoelastic extensibility, which relates the volumetric rate of strain to various stresses, is needed (Baskin, 2005; Cosgrove, 2016b).

Even so, potentially important mechanical properties can be determined by measuring strain under constant stress. With this technique, a weight is attached to a dead strip of extracellular matrix, and the strain or percent change in length is measured over time (Fig. 20.6). Initially there is an instantaneous percent change in length, followed by a time-dependent percent change in length known as *creep*. Creep is constant with the logarithm of time. After a period, the weight is removed and the percent change in length decreases almost instantaneously to a constant level. The irreversible percent change in length is known as the *plastic deformation*, and the reversible percent change in length is called the *elastic deformation*. The initial rate of creep is correlated with growth conditions of the cell, and thus growth is considered by some to be a metabolically sustained creep process.

Métraux and Taiz (1978) and Richmond et al. (1980) have measured the strain under constant stress in the walls from growing cylindrical cells of *Nitella*. They find that, *Ceteris paribus*, at a given stress, the amount of creep as well as the amounts of plastic and elastic deformations are greater in the longitudinal direction than in the transverse direction. They also find that the relationship between stress and strain is nonlinear, and the extracellular matrix has a yield value (Fig. 20.7). The yield value in the transverse direction is about twice as large as the yield value in the longitudinal direction. As I will discuss later, the stress in the transverse direction is twice as great as the stress in the longitudinal direction. As long as the deformation at a physiologically relevant stress is greater in the longitudinal direction than in the transverse direction, the cell will elongate. Under physiological pressures, the elastic modulus and the viscoelastic extensibility in the longitudinal direction are 4–7 times greater than they are in the transverse direction.

A second and complementary way to measure the potentially important mechanical properties of the wall is to stretch rapidly a dead piece of wall by applying a weight. The wall is then held at that constant length (isometric) and the stress in the wall is determined by dividing the force measured with a force transducer by the cross-sectional area of the wall. Over a short period of time, the stress decreases while the length stays the same (Fig. 20.8).

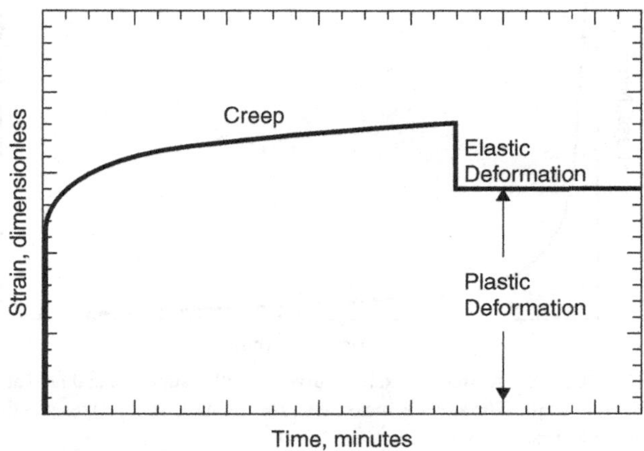

FIGURE 20.6 A representative time course of strain of a dead extracellular matrix. A constant stress was applied to the cell wall and it extended. The log-linear portion is known as *creep*. When the stress is removed, the wall shrinks again. The irreversible extension is known as the *plastic deformation* and the reversible extension is known as the *elastic deformation*.

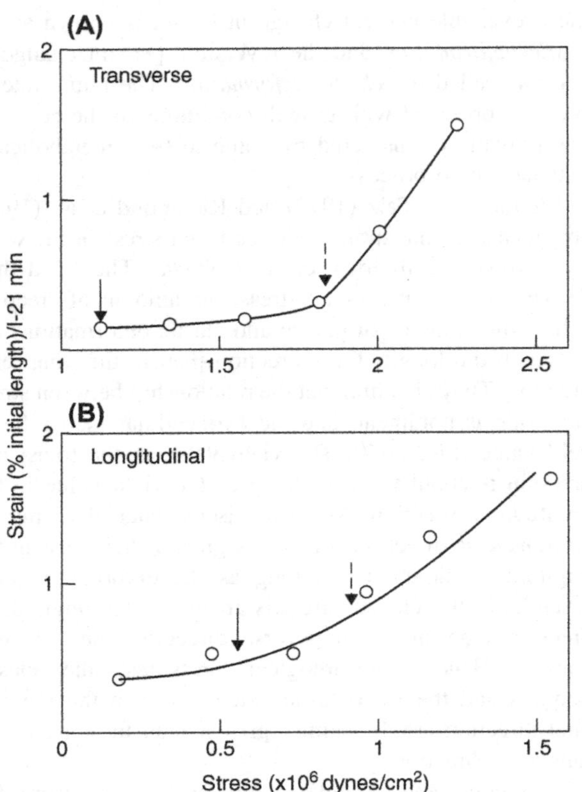

FIGURE 20.7 A rate of strain versus stress curve for transversely extended (A) and longitudinally extended (B) extracellular matrices of *Nitella axillaris*. The *solid and broken arrows* indicate the in vivo stresses when the turgor pressure equals 0.5 and 0.8 MPa, respectively. *From Métraux, J.P., Taiz, L., 1978. Transverse viscoelastic extension in Nitella. I. Relationship to growth rate. Plant Physiol. 61, 135–138.*

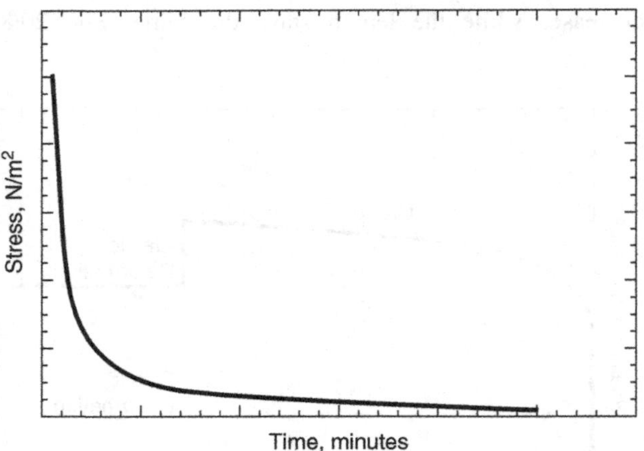

FIGURE 20.8 A stress-relaxation curve. A wall is stretched and held at a constant length. The stress is measured. Over time the wall relaxes and the stress decreases.

The stress relaxation time is different in tissues with different growth rates and is inversely correlated with the rate of growth when measured before or after the application of auxin (Hoson, 1991).

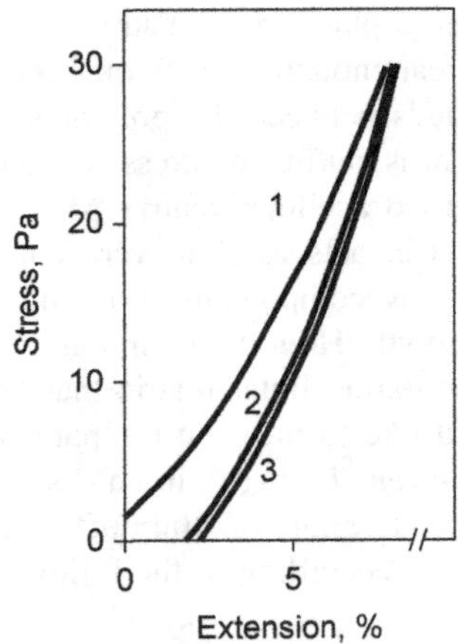

FIGURE 20.9 A curve that represents the results from an experiment where a wall is lengthened at a constant rate and the force in the wall is measured. The stress is determined by dividing the force by the cross-sectional area of the wall. The slope of the first pull (1) gives the total compliance of extensibility (in Pa^{-1}), and the slope of the second pull (2) gives the elastic compliance of extensibility. The plastic compliance of extensibility is obtained by subtraction.

Other potentially important mechanical properties of the extracellular matrix can be determined by measuring stress under various strains. This is done by attaching a dead segment to an instrument known as an Instron (Cleland, 1971a). This instrument has a force transducer, and the stress can be calculated by dividing the cross-sectional area of the wall by the measured force. The force is measured as the extracellular matrix is lengthened at a constant rate. Once a predetermined force is reached, the extension is stopped and the clamps are returned to their original distance. Then the wall is lengthened again at a constant rate (Fig. 20.9). The first lengthening includes both elastic and plastic extension, whereas the second lengthening includes only elastic extension. The total compliance extensibility (strain/stress, in Pa^{-1}) is obtained from the initial slope of the curve from the first extension. The elastic component of the total compliance of extensibility is obtained from the initial slope of the curve from the second extension. Lastly, the plastic component of the total compliance of extensibility is obtained by subtracting the elastic component from the total extensibility. The plastic component of a wall preparation is correlated with the growth rates of the tissues from which the wall came. The fact that the slope of the second extension is different from that of the first indicates that the measurements of extensibility depend on the previous history of the wall—that is, how much it has been extended already. Thus, measurements of mechanical properties of the cell must be interpreted carefully. Each one of the three techniques just mentioned gives information about some of the mechanical properties of the polymeric architectural system found in the wall. In theory, each one of the mechanical properties allows one to know something about the molecular arrangements and degree of cross-linking of polymers with known mechanical properties in the wall.

20.8 CELL EXPANSION

Long-term growth is a complex process that depends on a number of factors in addition to those involved in cell wall loosening. Continual cell expansion is a hormone-modulated, energy-dependent process that requires the synthesis of RNA, protein, and cell wall–directed carbohydrates; the delivery of proteins and carbohydrates to the extracellular matrix; the uptake of water across the plasma membrane and vacuolar membrane into the vacuole; and the maintenance of the turgor pressure of the cell and tension in the extracellular matrix above the yield threshold (Sachs, 1882a,b, 1887; Jost, 1907; Davenport, 1908; Vanderhoef and Stahl, 1975; Vanderhoef, 1985; Cleland, 1971a; Wei and Lintilhac, 2003, 2007; Wei et al., 2006; Fricke and Chaumont, 2006; Proseus and Boyer, 2006c, 2007; Verbelen and Vissenberg, 2006; Chebli and Geitmann, 2017). Growth can be defined many ways, but I will define it simply as a change in the volume of the cell in a given time. A change in volume per unit time is a rate of change in position (in three dimensions), and thus it follows from Newton's Second Law that a force is needed to change the rate of movement.

20.8.1 Forces, Pressures, and Stresses and Their Relationship to Strain

At a constant growth rate, the inertial force exactly balances the viscous force, and the net force is zero. When the acceleration is zero and the cell expands at constant velocity, the inertial force due to turgor is exactly balanced by the viscous force that results from the viscosity of the extracellular matrix. The inertial force results from the hydrostatic pressure in the cell acting against the wall. The force exerted by the hydrostatic pressure is three-dimensional but for convenience can be resolved into two components: one that pushes on the end walls and thus acts parallel to the long axis of the cell and one that pushes perpendicular to the long axis of the cell. The force against the end walls provides a shearing force on the extracellular matrix along the length of the cell. The viscous force results from the viscosity of the extracellular matrix. As a consequence of the low Reynolds number, growth stops immediately when the inertial force stops (see Chapter 9).

If we initially assume that the extracellular matrix is a Newtonian fluid, the velocity of movement closest to the site of generation of the motive force will be the highest, and it will decrease farther and farther away from the protoplast. The higher the viscosity of the extracellular matrix, the smaller the velocity gradient in the extracellular matrix will be. The higher the viscosity, the more uniformly the extracellular matrix will extend. If the extracellular matrix were thixotropic, then there will be no movement until the shearing stress is great enough to overcome the yield value (see Chapter 9). While the micrographs of cell walls are consistent with their being non-Newtonian fluids, there is currently a paucity of rheological data to determine directly if many cell walls are non-Newtonian (Métraux and Taiz, 1978; Richmond et al., 1980; Taguchi et al., 1999; Okamoto-Nakazato, 2002; Proseus and Boyer, 2006b). To determine the magnitude of the shearing stress in the extracellular matrix we must consider the forces that cause it.

The forces involved in growth depend on the geometry of the cell. For simplicity, I will consider a cylindrical cell of radius (r) and length (l). This cell has a wall with a thickness (x). The inertial force against the end wall of the cell is equal to the hydrostatic pressure, (P_t), times the area of the end wall, (πr^2), and is given by the following equation:

$$F_i = P_t \pi r^2 \tag{20.3}$$

That is, the inertial force against the end wall equals (P_t) (πr^2), and in the absence of acceleration, the inertial force is exactly balanced by the viscous force. If we want to calculate the shearing stress (σ_L) induced by the viscous force, we must determine the area over which the viscous force acts. This area is equal to the cross-sectional area of the cylindrical wall ($2\pi rx$), and consequently,

$$F_v = 2\pi rx\sigma_L \tag{20.4}$$

Many people confuse the quantities of mass (kg), force (N, kg m/s^2), and pressure or stress (Pa, N/m^2, kg(/m s^2)). To differentiate between force and shearing stress, or force and pressure, consider two wires of identical lengths made of the same material, one with a radius of 1 mm and the other with a radius of 3 mm. Experiments show that we must add a force (e.g., hang a weight) to the second wire that is nine times as heavy as the weight that we must add to the first to get identical changes in length. As the two wires are made of the same material and have the same elastic modulus, we must have added equal stresses to get equal strains. This could only be true if the stress is equal to the force divided by the area on which the force acts.

Thus, the shear stress parallel to the long axis of the cell (σ_L) is given by the following equation:

$$\sigma_L = \frac{F_v}{2\pi rx} \tag{20.5}$$

and because, at constant growth rate, $|F_i| = |F_v|$

$$2\pi rx\sigma_L = P_t \pi r^2 \tag{20.6}$$

and

$$\sigma_L = \frac{P_t \pi r^2}{2\pi rx} = \frac{P_t r}{2x} \tag{20.7}$$

Thus, the shearing stress that acts on the extracellular matrix is proportional to the turgor pressure and the radius of the cell. It is inversely proportional to twice the thickness of the extracellular matrix.

Let us also consider the hydrostatic pressure that pushes perpendicularly to the extracellular matrix along the long axis of the cell. Again, let us assume that the extracellular matrix moves outward at a constant velocity with respect to time. Again, the inertial force due to turgor equals the viscous force that is a result of the viscosity of the extracellular matrix.

Consider that the inertial force perpendicular to the long axis of the cell acts to split the cell in two longitudinally, and thus the inertial force is equal to the turgor pressure times the area against which it pushes. While we really need calculus to show this (see Appendix 3), for now assume that the total inertial force is equal to $P_t(2rl)$. At constant velocity, the viscous force that resists the inertial force must be equal to $P_t 2rl$. The stress ($\sigma_\perp$) in the extracellular matrix in the direction perpendicular to the long axis of the cell is equal to this force divided by the area over which it acts, which is approximately equal to $2xl$, where x is the thickness of the extracellular matrix on each side of the hemicylinder and l is the length of the cell (Castle, 1937; Preston, 1952; Nobel, 1974). Thus, $F_v = (\sigma_\perp)2xl$, and as, at constant velocity, $|F_i| = |F_v|$, then

$$\sigma_\perp 2xl = P_t 2rl \tag{20.8}$$

and

$$\sigma_\perp = \frac{P_t 2rl}{2xl} = \frac{P_t r}{x} \tag{20.9}$$

Thus, the perpendicular stress that acts on the extracellular matrix is proportional to the turgor pressure and the radius of the cell. In contrast to the shearing stress, it is inversely proportional the thickness of the extracellular matrix and not to twice the thickness.

I have shown that $\sigma_L = \frac{P_t r}{2x}$ and $\sigma_\perp = \frac{P_t r}{x}$, and consequently,

$$\sigma_\perp = 2\sigma_L \tag{20.10}$$

That is, for a given turgor pressure, the tangential stress is twice as great as the longitudinal stress.

The strain is proportional to the stress, and according to Eqs. (20.11) and (20.12), the elastic modulus (M) is the proportionality constant that relates the strain to the stress. Consequently, the linear extension rates parallel and perpendicular to the long axis of the cell will be given by the following equations:

$$\text{strain}_\perp = \frac{\sigma_\perp}{M} \tag{20.11}$$

$$\text{strain}_L = \frac{\sigma_L}{M} \tag{20.12}$$

and because $\sigma_\perp = 2\sigma_L$,

$$\text{strain}_\perp = 2\ \text{strain}_L \tag{20.13}$$

Likewise, the rate of strain in the transverse and longitudinal directions are given by the following equations:

$$\text{Rate of strain}_\perp = \sigma_\perp \Phi \tag{20.14}$$

$$\text{Rate of strain}_L = \sigma_L \Phi \tag{20.15}$$

and because $\sigma_\perp = 2\sigma_L$,

$$\text{Rate of strain}_\perp = 2\ \text{Rate of strain}_L \tag{20.16}$$

Thus, if the elastic moduli or viscoelastic extensibilities are the same in the tangential and longitudinal directions, then the cell should grow twice as wide as it grows long. Because cells typically elongate, the elastic modulus perpendicular to the long axis of the cell must be greater than the elastic modulus parallel to the long axis of the cell. Likewise the viscoelastic extensibility parallel to the long axis of the cell must be greater than the viscoelastic extensibility perpendicular to the long axis of the cell.

In fact, this is true, and the reason the elastic modulus and viscoelastic extensibility are unequal in the two directions is that the cellulose microfibrils in the wall are not arranged randomly but transversely around a growing cell. This can be observed with both polarization and electron microscopy (Frey-Wyssling, 1959; Green, 1954, 1958a,b, 1961, 1962, 1963).

In a young cell, the cellulose microfibrils are deposited in a transverse orientation perpendicular to the long axis of the cell. Consequently, close to the plasma membrane, the microfibrils are transverse to the direction of cell expansion. As we move out into the extracellular matrix, we encounter older microfibrils, and these have become obliquely oriented as a result of growth. Still farther out, the oldest microfibrils are passively reoriented longitudinally as a result of growth. Thus, the elastic modulus in each layer of the extracellular matrix could theoretically be different (Green, 1960; Frey-Wyssling and Mühlethaler, 1965; Frey-Wyssling, 1953, 1959). Consequently, due to the differences in orientation, as well as potential differences in the numbers and types of cross-bridges between the microfibrils, each layer could influence growth differently (Baskin, 2005).

The orientation of cellulose microfibrils, which may be dependent on the orientation of microtubules (see Chapter 11), is thought to regulate the direction of cell growth in response to the motive force provided by turgor pressure. Thus, randomly arranged microfibrils will give rise to a roughly spherical cell such as a cortical parenchyma cell. Cells that have transversely arranged microfibrils will elongate in a direction perpendicular to the direction of the microfibrils and will give rise to elongate cells like those in the procambium. The orientation of cellulose microfibrils thus determines the direction of growth and consequently, the shape of the cell. While the polarity of growth depends on the orientation of cellulose microfibrils, the rate of elongation is not invariantly related to the orientation of cellulose microfibrils (Paolillo, 2000; Baskin, 2005). This is reasonable when we take into consideration the fact that cellulose microfibrils are often involved in the two, sometimes opposing, functions of determining the polarity of growth and of mechanical support.

The mechanical properties of the wall are only one of the physical factors involved in cell expansion. The irreversible expansion of cells depends on two interdependent physical processes: the uptake of water and the yielding of the extracellular matrix (Lockhart, 1965; Lockhart et al., 1967; Cleland, 1958, 1971a,b, 1981; Yamamoto et al., 1970; Masuda, 1978a,b; Cosgrove, 1986, 1987; Hayashi, 1991; Kutschera, 1991; Hoson, 1992; Ortega, 1985, 1990; Proseus et al., 1999). According to one model of cell expansion, growth occurs when the extracellular matrix is elastically stretched by the protoplast as a result of hydrostatic pressure. This gives rise to a stress in the load-bearing bonds. Irreversible growth results when the load-bearing bonds in the extracellular matrix break and allow the elastic components in the wall to relax. This relaxation allows a reduction in the stress in the extracellular matrix. Because the stress in the wall and the hydrostatic pressure constitute equal and opposite forces, the hydrostatic pressure and the wall stress decrease simultaneously. The decrease in the hydrostatic pressure or turgor pressure reduces the water potential of the cell. This gives rise to a water influx, which increases the cell volume and extends the extracellular matrix, and the load-bearing bonds form in a new place. The cycle repeats itself until the extracellular matrix is no longer extensible. Thus, like the Krebs, glyoxylate and Calvin cycles, growth is also cyclic, and there is a growth cycle.

This growth cycle only describes short-term growth that takes place in time scales less than an hour. Long-term growth requires the expression of genes and the synthesis on new wall material to make new load-bearing bonds. Without these processes, growth would stop (Masuda, 1990). There may be feedback loops between the extracellular matrix and the cytoplasmic components involved in the synthesis and positioning of the polymers in the extracellular matrix. Thus, there may be a growth cycle that involves the whole cell.

James Lockhart (1965) made a mathematical model to describe the dynamic balance between wall yielding, which tends to dissipate turgor, and the uptake of water, which tends to restore turgor. I will present the derivation of the Lockhart equation. This equation describes how the growth rate depends on the viscoelastic extensibility of the wall (Φ, in $Pa^{-1}\,s^{-1}$), the difference in the osmotic pressure inside and outside the cell (π), the yield value of the wall (Y, in N/m^2), and the hydraulic conductivity of the plasma membrane (L_p, in m/s Pa). L_p is equal to the osmotic permeability coefficient P_{os} (in m/s; see Chapter 2) times $(-\overline{V}_w/RT)$, where $-\overline{V}_w$ is the partial molar volume of water (see Chapter 12). Growth is positively correlated with Φ, Π, and L_p and negatively correlated with Y. It is possible that one or another parameter becomes dominant in controlling growth at different times in the growth process, in different cells in an organ or in different plants. Therefore, in each tissue type it is necessary to find out which cells limit growth, and in that cell type, it is necessary to find out which parameter is limiting. This is another specific example of Blackman's (1905; Kornberg and Krebs, 1957) law of limiting factors, which states that "When a process is conditioned as to its rapidity by a number of separate factors, the rate of the process is limited by the pace of the 'slowest' factor."

The rate in which a cell grows depends on how big it is at the start of the measurement. Thus, I will use the elemental rate of growth, which accounts for the initial size of a cell (Lockhart, 1971). The elemental rate of growth is a "specific growth rate" or "relative growth rate" and is defined to be $(1/V)\,(dV/dt)$. The elemental rate of volume growth (in s^{-1}) is a function of the viscoelastic extensibility of the extracellular matrix (Φ) and the difference between the turgor pressure (P_{ti}) and the yield value (Y) of the extracellular matrix as defined in the following equation:

$$(1/V)(dV/dt) = \Phi(P_{ti} - Y) \tag{20.17}$$

The elemental rate of volume growth is also dependent on hydraulic factors according to the following equation:

$$(1/V)(dV/dt) = AL_p(P_{wo} - P_{wi})/V \tag{20.18}$$

where V is the cell volume (in m^3), t is the time (in s), A is the surface area of the cell (in m^2), L_p is the hydraulic conductivity (in m/s Pa), and $(P_{wo} - P_{wi})$ is the water potential difference between the outside and the inside of the cell. As

$$\begin{aligned} P_{wo} - P_{wi} &= (P_{to} - sP_{\pi o}) - (P_{ti} - sP_{\pi i}) \\ &= sP_{\pi i} - sP_{\pi o} - P_{ti} = s\Pi - P_{ti} \end{aligned} \tag{20.19}$$

where P_{to} and P_{ti} are the hydrostatic pressures outside and inside the cell, respectively (P_{to} is defined as 0 by definition); $P_{\pi o}$ and $P_{\pi i}$ are the osmotic pressures outside and inside the cell, respectively; $\Pi = (P_{\pi i} - P_{\pi o})$; and s is the reflection coefficient for the solutes. Thus,

$$(1/V)(dV/dt) = AL_p(s\Pi - P_{ti})/V \tag{20.20}$$

and

$$P_{ti} = s\Pi - (dV/dt)(1/AL_p) \tag{20.21}$$

Substituting $P_{ti} = s\Pi - (dV/dt)\,(1/AL_p)$ into

$$(1/V)(dV/dt) = \Phi(P_{ti} - Y)$$

we get

$$(1/V)(dV/dt) = \Phi(s\Pi - Y - (dV/dt)(1/AL_p)) \quad (20.22)$$

which is equivalent to

$$\begin{aligned} dV/dt &= V\Phi(s\Pi - Y - (dV/dt)(1/AL_p)) \\ &= V\Phi(s\Pi - Y) - V\Phi(dV/dt)(1/AL_p) \end{aligned} \quad (20.23)$$

Bring all the terms with dV/dt to the left side of the equation:

$$dV/dt + (dV/dt)(V\Phi/AL_p) = V\Phi(s\Pi - Y) \quad (20.24)$$

Using the distributive rule,

$$(dV/dt)(1 + (V\Phi/AL_p)) = V\Phi(s\Pi - Y) \quad (20.25)$$

After solving for dV/dt, we get

$$dV/dt = V\Phi(s\Pi - Y)/(1 + (V\Phi/AL_p)) \quad (20.26)$$

and the elemental growth rate is given by the following equation:

$$dV/(Vdt) = \Phi(s\Pi - Y)/(1 + (V\Phi/AL_p)) \quad (20.27)$$

The foundation of the Lockhart equation is the assumption that the growth rate results from a difference between the force exerted on the wall due to turgor and its restraint due to the mechanical properties of the wall. It is clear from Eqs. (20.26) and (20.27) that a change in any number of factors that influence turgor and the mechanical properties of a cell has the potential of influencing the growth rate and elemental growth rate. Because V/A in the denominator will typically be a very small number, the denominator will most likely reduce to unity. Generally, the growth rate will increase when Φ, s, Π , and/or L_p increases and when Y decreases. Potentially, each of these variables can be affected by a stimulus, including hormones and light. Thus, experiments performed to elucidate the mechanism of how stimuli influence growth must not only show which factors are modulated but also the ones which remain constant.

The growth equation and the others we have used in this book were derived with a few assumptions from a limited number of examples, and any of the equations given in this book may have to be modified or expanded to efficiently describe a given situation and to have predictive value. These equations should not be used as laws but as approximations that simplify the information that we know and allow us to plan experiments and consider the effect of more than one factor at a time. Most scientific arguments start because one group may have tested one parameter and assumed the others were constant, whereas another group tested a different parameter assuming that another set of parameters were constant. These equations help us to see that more than one group may be partially right, and that there may be more than one factor that influences a process. Indeed, the synthesis comes from combining the thesis with the antithesis. Deriving an equation by ourselves reminds us of the assumptions that we have included in the equation and the information that we have left out. The experiments, however, can guide experiments, and they can help refine the equation (Einstein, 1950).

The words of T.H. Huxley (1890) give us perspective on the value of a quantitative approach and its limits. He wrote in an essay, "On the Physical Basis of Life"

> *But the man of science, who, forgetting the limits of philosophical inquiry, slides from these formulae and symbols into what is commonly understood by materialism, seems to me to place himself on a level with the mathematician, who should mistake the x's and y's with which he works his problems, for real entities—and with this further disadvantage, as compared with the mathematician, that the blunders of the latter are of no practical consequence, while the errors of systematic materialism may paralyse the energies and destroy the beauty of a life.*

The growth equation derived above describes the growth of cylindrical single cells, such as the internodal cells of *Nitella* or *Chara*. It has also been used to describe the growth of cylindrical organs. However, the application of this equation to multicellular organs depends on the assumption that to some degree, all the cells of the organ are identical or the internal cells provide the pressure for growth, whereas the outer epidermal cell wall resists growth. The growth of a multicellular organ is a complicated problem. Throughout the organ there are a number of cell types in each tissue (epidermal, cortical, pith, phloem, or xylem) that have different wall architectures, thicknesses, extensibilities, etc. In sunflower hypocotyls, for example, the pith and cortical cells have a very thin extracellular matrix, and thus their turgor pressure will keep them extended. This may cause the thick-walled epidermal cells to be under tension. Then the tension in the epidermal cells may keep the cortical and pith cells under compression and prevent water uptake. Stéphane Verger et al. (2018) has shown using certain mutants of *Arabidopsis*, the the outer wall of an epidermal cell is under tension and the inner wall is under compression.

The tissue stresses in an organ can be readily observed by peeling the epidermis and seeing that the isolated epidermis contracts approximately 20%, whereas the inner tissues expand immediately due to the water uptake, which was prevented by the tissue compression. In sunflower hypocotyls, growth appears to be driven by water uptake into the compressed parenchyma tissues and limited by the extensibility of the epidermal cells (Kutschera, 1991). Thus, there is a supracellular level of control that is not included in the growth equation.

Auxin is a hormone that promotes growth in most plants, and the possibility remains that auxin may affect one or all of the parameters in the growth equation and to a different degree in each cell type or in each plant. Robert Cleland and David Rayle have provided a wealth of evidence that auxin acts on the plasma membrane proton-pumping ATPase that acidifies the extracellular matrix, thus activating enzymes that break hydrogen bonds between polymers (Cleland and Rayle, 1977; Rayle and Cleland, 1977). The breakage of the load-bearing bonds increases extensibility and/or decreases the yield value (Taguchi et al., 1999). Auxin may act directly on the Golgi apparatus or other parts of the secretory pathway to cause the secretion of new extracellular matrix material (Ray and Baker, 1965; Ray, 1977), including pectins (Albersheim and Bonner, 1959). Auxin may stimulate elongation growth by acting directly on microtubule orientation and keeping microtubules in a transverse orientation for a longer period of time, particularly in the outer epidermal wall (Bergfeld et al., 1988; Nick et al., 1990; Mayumi and Shibaoka, 1996). Auxin may also increase the hydraulic conductivity of the plasma membrane (Kang and Burg, 1971; Boyer and Wu, 1978; Loros and Taiz, 1982). And lastly, auxin may act directly on gene expression by activating transcription of growth-limiting mRNAs and/or translation of growth-limiting proteins (Key, 1969; Ulmasov et al., 1997). These effects are not mutually exclusive, and auxin may influence one or more processes in a given cell at the same time (Vanderhoef et al., 1977).

Dan Cosgrove (1993; 2015, 2016a,b) and his colleagues (McQueen-Mason et al., 1992, 1993, 2006; Li et al., 1993a,b,c; Cosgrove et al., 2002) discovered some enzymes that may be activated by an auxin-induced acidification of the extracellular matrix. Cosgrove and his associates developed a functional assay to look for the enzymes necessary for extension. First, they isolated extracellular matrices from either cucumber hypocotyls or oat coleoptiles. The native walls extend in response to low pH (4.5) under constant stress. However, when the walls are heat-treated, they fail to extend. Proteins were then isolated from other walls, fractionated, and tested to see if they cause extension of the heat-treated walls. In this way, Cosgrove and his associates isolated expansins, which are proteins that induce extension, with a pH optimum of 4.5. Expansins may bind to xylans and induce extension by transiently breaking hydrogen bonds between xyloglucans and cellulose (Whitney et al., 2000). There is a web site devoted to these interesting proteins (www.bio.psu.edu/expansins/), and it includes many references. It covers the evolutionary origin, mode of action, and the role of expansins in various developmental responses, including rachis elongation in semiaquatic ferns (Kim et al., 2000a,b), elongation of wheat coleoptiles (Gao et al., 2008), and cell wall disassembly during fruit ripening (Rose and Bennett, 1999; Rose et al., 2000).

Interestingly, localized application or expression of expansins induces the generation of leaf primordia in apical meristems, indicating that localized expansion of the cell wall is capable of triggering morphogenesis (Fleming et al., 1997; Pien et al., 2001). In addition, leaf primordia also seem to form at the meristematic sites with lower elastic moduli due to the demethylation of pectin (Peaucelle et al., 2008, 2011, 2012; Fleming, 2011; Palin and Geitmann, 2012; Wolf and Greiner, 2012; Braybrook and Peaucelle, 2013; Milani et al., 2013). These results, which still require more analysis and understanding, would have pleased Paul Green (1996, 1997), who pioneered the role of physical mechanisms in initiating morphogenesis at the apical meristem (Silk, 2000).

In some cases, auxin increases the extensibility of the wall and expansins may be the enzymes involved in increasing the extensibility in response to auxin. Other studies show that auxin also decreases the yield threshold of the wall (Okamoto et al., 1989, 1990, 1995; Mizuno et al., 1993). Okamoto and Okamoto (1994, 1995), Okamoto et al. (1997), and Okamoto-Nakazato et al. (2000a,b, 2001) have isolated two proteins from the extracellular matrix of *Vigna*: one that is involved in regulating extensibility and one that is involved in regulating the yield threshold (Fig. 20.10). She calls the extracellular matrix–localized protein that regulates the yield threshold of the wall *yieldin* (Okamoto-Nakazato, 2002; 2018).

Gibberellin is another hormone that affects plant growth (Brian and Hemming, 1955; Phinney, 1956; Lockhart, 1959a; Stowe and Yamaki, 1959), and a lack of gibberellin results in the dwarf phenotype observed in garden peas by Gregor Mendel (Lester et al., 1997). Gibberellin and auxin do not act independently in inducing growth in plants and organs (Yang et al., 1996), and consequently it is difficult

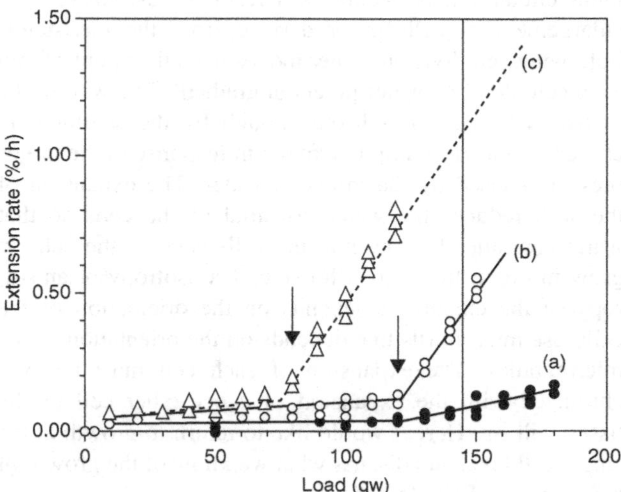

FIGURE 20.10 The demonstration of the function of yieldin. (a) extension at pH 6.2; (b), extension at pH 4; and (c), extension at pH 4 plus yieldin. *From Okamoto-Nakazato, A., 2002. A brief note on the study of yieldin, a wall-bound protein that regulates the yield threshold of the cell wall. J. Plant Res. 115, 309–313.*

to separate the cellular effects of each hormone. Hiroh Shibaoka has pioneered studies aimed at elucidating the cellular mechanisms of gibberellin-induced growth. Shibaoka (1972) first noticed that gibberellin promoted elongation growth at the expense of stem thickening and the lateral expansion of cells and concluded that gibberellin controls the direction of expansion of plant cells.

Shibaoka (1972) studied the effect of gibberellin in stem segments of the adzuki bean. The control segments were incubated in auxin and the treated segments were incubated in auxin plus gibberellin. The gibberellin-treated segments elongated over the controls in the absence of cell division. Shibaoka postulated that gibberellin caused the microtubules to orient transversely to the cell axis. As a result, the newly synthesized cellulose microfibrils were also laid down transversely to the cell axis. Consequently, the cell expanded longitudinally but not transversely. Shibaoka showed that colchicine inhibited gibberellin-induced growth. An inhibitor of gibberellin biosynthesis causes swelling of cells and a randomization of microtubules (Mita and Shibaoka, 1983, 1984a,b). Shibaoka and his colleagues observed that the microtubules and the cellulose microfibrils were transversely oriented in gibberellin-treated cells (Shibaoka, 1974; Takeda and Shibaoka, 1981a,b). Although Shibaoka (1994) originally believed that auxin did not affect microtubule orientation in adzuki beans like it does in other plants, Mayumi and Shibaoka (1996) found that auxin is required for the gibberellin-induced suppression of the reorientation of microtubules from a transverse to a longitudinal arrangement.

The expansion of the cell wall and the growth of an organism and its organs have many levels of control from the subcellular level to the organismal level. The growth of multicellular plants occurs as a result of the irreversible enlargement of cells in or derived from the meristems. Following cell division, water moves from the apoplast into each cell along its water potential gradient. The wall in the meristematic cells is relaxed enough by the acidification caused by the H^+ pump to expand in response to the turgor pressure caused by the influx of water. The expansion of the wall reduces the water potential of the cells so that water continues to flow into the cells causing the cells to grow in this cellular growth cycle. The isotropy or anisotropy of the expansion depends on the orientation of the cellulose microfibrils that depends on the orientation of the microtubules. The expansion of each cell must be synchronized with the expansion of every other cell or the tissue will rip. Here, I would like to return to growth at the single cell level and discuss what we know of the growth of a few types of single cells.

The intermodal cells of *Chara* do not elongate in response to acid as do coleoptile, hypocotyl, and epicotyl segments of higher plants, but they may have a lot to teach us about cell elongation. Proseus and Boyer (2006b) wanted to know the cause of elongation in *Chara* intermodal cells. They pressure-injected oil into cell walls that had been isolated from the internodal cells of *Chara* that had been boiled to inactivate any enzymes. The pressure of the oil caused the walls to grow irreversibly for over an hour at the same rate that they would have grown in the intact plant. However, the isolated walls stopped growing after 1−2 h while they would have continued to grow in the live cells. Proseus and Boyer (2006b) asked, "What specific molecules allowed growth for 1 to 2 h, then were depleted or missing when the walls were isolated from the cytoplasm?"

In live *Chara* cells, like live higher plant cells, cellulose microfibrils are synthesized by rosettes on the plasma membrane, and hemicelluloses and pectins synthesized in the Golgi apparatus are added by the exocytosis of secretory vesicles (Boyer, 2009, 2016). Proseus and Boyer (2005) showed that oil pressure caused fluorescent dextran polymers to move into the existing wall, reptate through it, and bind to the polymers already there in a process known as *intussusception*. Proseus and Boyer (2006c) suggested that, in a similar manner, in the living cell, turgor pressure drives the snake-like movement of the pectins released by exocytosis from the periplasmic space into the existing wall. The pectins were the molecules depleted from boiled walls after 1−2 h.

In the periplasmic space of living cells, the carboxylic acid groups of the homogalacturonans form coordinate bonds with Ca^{2+} to form a stiff gel, which can be pushed into the existing wall by the turgor pressure. Any pushing apart of the existing wall polymers results in growth. However, as the wall expands as a consequence of turgor pressure, the homogalacturonans will be put under tension causing their carboxylic acids groups to be pulled apart so that they will no longer be capable of binding Ca^{2+}. Consequently, Ca^{2+} will be released from the older homogalacturonans that are under tension and the wall will relax. The released Ca^{2+} will bind to the newly secreted homogalacturonans in the periplasm that are not under tension, and thus they are capable of forming coordinate bonds with Ca^{2+}. In this way, the extracellular matrix goes through a cycle of relaxing and stiffening, which allows sustained growth in response to a relatively constant turgor pressure of about 0.5 MPa. Here is another case where tension influences chemistry as it does during mitosis and meiosis (see Chapter 19). Growth slows when turgor is decreased more than 0.1 MPa because the turgor pressure can no longer distort the load-bearing calcium−pectate bonds, causing them to release Ca^{2+} and the calcium pectate cycle stops (Boyer, 2016).

The calcium pectate cycle in *Chara* involved in cell expansion does not require enzymes in the cell wall (Proseus and Boyer, 2012a,b). However, in the Desmid, *Penium Margaritaceum*, the calcium pectate cycle requires the enzyme pectin methylesterase, which demethylesterifies the methylesterified homogalacturonans that are secreted into the periplasmic space. Growth, in *Penium*, occurs at the isthmus, where a highly methylesterified homogalacturonan is deposited into the existing wall. Antibody studies show that the intussuscepted methylesterified homogalacturonan becomes demethylesterified and cross-linked by Ca^{2+} to form a gel as it is displaced from the growing zone (Domozych et al., 2009; Domozych, 2014; Domozych et al., 2014a,b).

There are many other specialized forms of growth, including the polar growth that occurs in rhizoids, trichomes, root hairs, pollen tubes, protonematal cells, and fungal hyphae, that allow us to dissect further the calcium pectate cycle, as in these cell types, the growth rate is not constant but oscillates five-to six-fold over with a period of 20–50 s (Pierson et al., 1996; Holdaway-Clarke et al., 1997; Holdaway-Clarke and Hepler, 2003; Bosch and Hepler, 2005; Cárdenas et al., 2006; Hepler et al., 2013b; van Hemelryck et al., 2017). Clearly the inertial force due to turgor pressure and the viscous force due to the extracellular matrix are not equal and opposite, as they would be if the growth rate were constant (Money and Harold, 1992; Money, 1997; Money and Hill, 1997; Bosch and Hepler, 2005; Obermeyer, 2017). For the growth rate to oscillate, one force must be periodically greater than the other. It is currently being tested if the oscillatory growth results because the resistive force provided by the calcium pectate cycle involving demethylesterification oscillates while the turgor pressure stays nearly constant (Chebli and Geitmann, 2007; Rojas et al., 2011; Rounds et al., 2011a; Winship et al., 2010, 2011) or if the turgor pressure oscillates and the resistive force stays nearly constant (Messerli et al., 2000; Messerli and Robinson, 2003; Zonia et al., 2006; Zonia and Munnik, 2007, 2009, 2011; Zonia, 2010). Because each component of tip growth can affect the other components, investigations are currently trying to resolve the cause and effect or precursor–product relationship between the components that involve almost every aspect of cell biology by performing cross-correlation analysis to determine the phase relations between the components with the same period as the growth rate (Messerli and Robinson, 1997; Messerli et al., 1999; Zonia et al., 2002; Monshausen et al., 2008; McKenna et al., 2009; Rounds et al., 2011a; Furt et al., 2013; Hepler et al., 2013b; Damineli et al., 2017).

Observation that the turgor pressure of growing pollen tubes measured by a pressure probe appears to be a constant and independent of growth rate (Benkert et al., 1997) supports the calcium pectate cycle hypothesis for tip-growing pollen tubes. The intracellular components that contribute to pollen tube growth include the energy-dependent (Rounds et al., 2010, 2011b), actin-mediated transfer of Golgi-derived secretory vesicles filled with methylesterified homogalacturonans and pectin methylesterase synthesized on the ER to the growing tip (Bosch and Hepler, 2005; Rounds et al., 2014). The secretion of the vesicles at the growing tip anticipates the increase in growth rate, indicating that the turgor pressure–driven intussusception of the methylesterified pectin into the extracellular matrix and its subsequent demethylesterification may relax the extracellular matrix by robbing the load-bearing calcium–pectate bonds of its Ca^{2+} resulting in a slightly delayed yet increased growth rate (McKenna et al., 2009; Hepler et al., 2013b). Endocytosis retrieves the excess of plasma membrane formed during the secretion of wall material (Moscatelli et al., 2007; Moscatelli, 2008; Onelli and Moscatelli, 2013). As in *Penium*, the methylesterified pectins secreted into the extracellular matrix at the tip become demethylesterified by pectin methylesterase. The removal of the methoxy groups in the pectins at the flanks of the apical dome unmasks their negatively charged carboxylate groups. The anionic homogalacturonans then bind Ca^{2+} and become stiffer as the new apical dome, which will incorporate more methylesterified pectins and pectin methylesterase, grows away from the stiffened flanks composed of calcium pectate (Bosch et al., 2005; Bosch and Hepler, 2006; Rounds et al., 2011a; Chebli et al., 2012). The external Ca^{2+} concentration is critical. When the external Ca^{2+} concentration is below 10 μM, the amount of calcium pectate is so low that the extracellular matrix is too weak and the pollen tube bursts. When the external Ca^{2+} concentration is above 10 mM, the amount of calcium pectate is so high that the extracellular matrix is too stiff and the pollen tube will not grow (Hepler and Winship, 2010; Hill et al., 2012; Hepler et al., 2013b).

Tip growth allows these cells to "search out" the nutrients or cell types necessary for growth and/or reproduction. To orient themselves, tip-growing cells are capable of sensing tiny gradients in chemical, electrical, gravitational, and mechanical stimuli. Tip growth is dependent on an influx of Ca^{2+} into the tip (Schnepf, 1986; Miller et al., 1992; Kühtreiber and Jaffe, 1990; Franklin-Tong et al., 1993, 1996; Malhó et al., 1994; Pierson et al., 1994, 1996; Malhó and Trewavas, 1996; Lancelle et al., 1997; Messerli and Robinson, 1997; Messerli et al., 2000). Let's not forget that there is even wall-to-wall communication between a tip-growing pollen tube and the pistil. In this case, the pistil produces an extracellular matrix glycoprotein that stimulates the growth of pollen tubes and attracts them to the ovules (Cheung, 1995; Cheung et al., 1995).

Growth of a single cell is not always straight. One side of a cell can grow faster than the other side. This can be observed clearly in the filaments of *Vaucheria* (Kataoka, 1975a,b), the large internodal cells of characean algae, or the sporangiophores of *Phycomyces* when these cells bend in response to light and gravity (Bergman et al., 1969; Staves et al., 2005). In these cases, either the viscoelastic extensibility or the yield value for the wall on each side must be different as the turgor pressure that drives growth is the same on both sides. Of course, differential growth also occurs in the phototropic and gravitropic responses of the organs of higher plants, but in these cases, it is not clear whether the cells on two sides of an organ grow differentially or if the bending of an organ is due to the summation of the differential growth of each cell (Tomos et al., 1989; Cosgrove, 1997; Evans and Ishikawa, 1997).

We can use the Lockhart equation, repeated below, to get a glimpse of the difficulty facing people using a reductionist approach to cell biology based on the study of mutants:

$$dV/(Vdt) = \Phi(s\pi - Y)/(1 + (V\Phi/AL_p)) \qquad (20.28)$$

When one sees a mutation that affects growth, one could imagine that the mutation could affect the extensibility, the turgor pressure, the reflection coefficient, the osmotic pressure, the yield value, or the hydraulic conductivity. That is not all: Imagine that the yield threshold of the wall is directly affected by the enzyme yieldin. How would a mutation in transcription, RNA processing, translation, translocation at the ER, membrane trafficking, vesicle transport, synthesis of the substrate for yieldin, etc., affect the yield value of the cell wall? Likewise, imagine that the extensibility of the wall is directly affected by the enzyme expansin. Because there are numerous isozymes of each cell wall protein, one must know if the mutation affects the expression in the cell in question. Cell biologists must be able to recognize the relationships between the part and the whole in all aspects of "informatic" and "-omic" analysis.

20.9 SUMMARY

The extracellular matrix is a dynamic organelle that surrounds the cell. It is a matrix of polysaccharide and protein polymers that prevents cell lysis and allows the development of a hydrostatic pressure within the cell that provides mechanical support for the plant. The extracellular matrix contains its own cadre of enzymes to assemble and rearranges itself to allow for growth. The extracellular matrix has a history and a geography and may act as an informational macromolecule that encodes three-dimensional positional information.

20.10 EPILOG

Now that I have discussed the components of wood, and by extension, the properties of logs, I will end this chapter with the "epilog." I am awed by how much we know about the cell, and how much is understood in terms of the cellular mechanisms that make life possible. When I realize how much there still is for me to learn about the cell, it is hard to stop writing this book. But I must publish what I know and what I think I know or else I will have created the cellular version of "The Madonna of the Future" (James, 1962).

I hope that this book has fulfilled James Clerk Maxwell's goal as a teacher. He wrote (quoted in Mahon, 2003),

> *In this class, I hope you will learn not merely results, or formulae applicable to cases that may possibly occur in our practice afterwards, but the principles on which those formulae depend, and without which the formulae are mere mental rubbish. I know the tendency of the human mind is to do anything rather than think. But mental labour is not thought, and those who have with labour acquired the habit of application often find it much easier to get up a formula than to master a principle.*

Maxwell also wrote,

> *My duty is to give you the requisite foundation and to allow your thoughts to arrange themselves freely. It is best that every man should be settled in his own mind, and not be led into other men's ways of thinking under the pretence of studying science. By a careful and diligent study of natural laws I trust that we shall at least escape the dangers of vague and desultory modes of thought and acquire a habit of healthy and vigorous thinking which will enable us to recognise error in all the popular forms in which it appears and to seize and hold fast truth whether it be old or new.*

Throughout this book, I have emphasized the unity of nature and knowledge (Kidd, 1852; Butcher, 1891; Bohr, 1963; Schrödinger, 1996). I have used the teachings of mathematicians, physicists, and chemists as well as those of poets to help us understand the nature of life and the mechanisms cells use to transform matter and energy into life. I have often emphasized the relationship between the part and the whole, and how the combination of parts leads to a whole with new emergent properties, which are based on, but were not found in, the properties of the component parts. In this respect, I have discussed the formation of atoms from elementary particles, the formation of molecules from atoms, the formation of macromolecules from monomers, the formation of cells from macromolecules, the formation of multicellular organisms from cells, and their continual development of multicellular organisms into rational, feeling organisms that are capable of understanding and appreciating the wonder, beauty, and spirit of the universe (Bergson, 1911). One theme of *Zorba the Greek*, written by Bergson's PhD student, Nikos Kazantzakis (1953), is that every day each of us repeats this evolutionary process as we convert food into fat, work, or spirit. The proportion we convert into each product describes to some extent who we are. Physiology and behavior recapitulate phylogeny.

If there truly is a unity of knowledge and nature, an understanding of the nature of life from the biological point of view should be able to help us appreciate, value, and understand higher levels of the organizations of life—where the entire organism is a part of a family, a town, a state, a nation, and a world (Priestley, 1771; Boveri, 1929; Browne, 1944; Emerson, 1947; Thoreau, 1947; Jefferson, 1955; Picken, 1960; Hamilton et al., 1961; Olson, 1965; Claude, 1975; Cohen, 1997; Elwick, 2003). Albert Szent-Györgyi (1957a, 1963a, 1970) defined ethics as the set of rules that makes living together possible in these higher levels of organization of life.

At every level I have studied in biology, there is a relationship between the part and the whole. Every process can be seen as a compromise between what is optimal for the part and what is optimal for the whole. Society is the same way, albeit more complicated than a cell. Yet, we must look at every situation in terms of the complementary relationship between the part and the whole. As citizens of the same planet, we must not only live for ourselves but also ask what we can do for the freedom of humankind (Kennedy, 1961), dedicate ourselves to the struggle for human rights (Carter, 1979), be tolerant of people with different values, and occasionally rise above principle (Martin, 1988)—that is, we must live and let live. As President Kennedy said at the Commencement Address at American University in Washington on June 10, 1963,

> *So, let us not be blind to our differences—but let us also direct attention to our common interests and to the means by which those differences can be resolved. And if we cannot end now our differences, at least we can help make the world safe for diversity. For, in the final analysis, our most basic common link is that we all inhabit this small planet. We all breathe the same air. We all cherish our children's future. And we are all mortal.*

Now I have shared will you my personal outlook on cell biology. I have discussed my views on the origin of cells, the mechanisms of how cells transform matter and energy into life, and the relationship of cells to organisms and even higher levels of organization. I have presented cell biology to the best of my ability. I still have much to learn, so I encourage you to question all my conclusions. After all, this book is only a snapshot of my current understanding of plant cell biology—an understanding that continually grows, develops, and evolves much like the plants from which the cells come. According to Szent-Györgyi (1964), "Books are there to keep the knowledge in while we use our heads for something better. ... So I leave knowledge, for safe-keeping, to books and libraries and go fishing, sometimes for fish, sometimes for new knowledge."

In this book, I have presented cell biology to you from my own point. I encourage you to find your own point of view. I am but one of the blind men described by the poet John Godfrey Saxe, who studied elephants. You can be another, and all together, with our diverse viewpoints, we may understand the cell of life. God save us from scientists who lack the courage to find their own viewpoints. For if Richard Feynman (1955) is correct in believing that the greatest value of science is the "freedom to doubt," then a healthy science depends on the courage and convictions of the individuals (Conklin, 1923; Blackett, 1935; Einstein et al., 1936; Rutherford, 1936; Brady, 1937; Stark, 1938; Bernal, 1939, 1949; Szent-Györgyi, 1943, 1972; Sax, 1944; Baker, 1945; Lysenko, 1946, 1948, 1954; Zirkle, 1949; Kennedy, 1964; Fermi, 1971; Marshall, 1982a,b; Kamen, 1985; King, 1992; Williams, 1993a; Krogmann, 2000; Levi-Montalcini, 1988; Deichmann and Müller-Hill, 1998). Today, people are discussing the socio-politico-economic influence of scientists on each other and the need for a Declaration of Academic Freedom (Rabounski, 2006).

I have tried to present to you the fashionable (Stevens, 1932; Chargaff, 1976; Bauer, 1994) and the unfashionable theories and experiments in cell biology. Science requires both (Wangensteen, 1947). According to Richard Feynman (1965a),

> *... possibly the chance is high that the truth lies in the fashionable direction. But, on the off-chance that it is in another direction—a direction obvious from an unfashionable view ...—who will find it? Only someone who has sacrificed himself by teaching himself ... from a peculiar and unusual point of view; one that he may have to invent for himself. I say sacrificed himself because he most likely will get nothing from it, because the truth may lie in another direction, perhaps even the fashionable one.*

So be yourself, trust yourself, be an explorer or a tourist, but either way, take responsibility for your assumptions, your conclusions, and your actions. I look forward to hearing your views and your criticisms of mine. For now, dear reader, we must end our journey together through the cell, and I will close the chapters that describe pre-omics biology with the words Matthias Schleiden (1853) used to conclude his beautiful book, *Poetry of the Vegetable World*:

> *A gentle sound trembles through the fragrant evening air. The bell of the native village calls him home, returned after*

restless travel over the great God's World, after rich impressions, exciting adventures, pressing hardships, and strange delights, back to rest, to that, which, in spite of all intervening things, he never does nor can forget, the paradise of childhood, the house of his parents, his mother's arms.

20.11 QUESTIONS

20.1. What is the relationship between the extracellular matrix and the plasma membrane?

20.2. What is the relationship between the extracellular matrix and the nucleus?

20.3. What is the relationship between the extracellular matrix and the endoplasmic reticulum?

20.4. What is the relationship between the extracellular matrix and the Golgi apparatus?

20.5. What is the relationship between the extracellular matrix and the endosomal compartment?

20.6. What is the relationship between the extracellular matrix and the cytoskeleton?

20.7. What is the relationship of the extracellular matrix, the plasma membrane, nucleus, endoplasmic reticulum, Golgi apparatus, vacuole, and cytoskeleton to growth?

The references for this chapter can be found in the references at the end of the book.

Chapter 21

Omic Science: Platforms and Pipelines

An intellect which at a certain moment would know all forces that set nature in motion, and all positions of all items of which nature is composed, if this intellect were also vast enough to submit these data to analysis, it would embrace in a single formula the movements of the greatest bodies of the universe and those of the tiniest atom; for such an intellect nothing would be uncertain and the future just like the past would be present before its eyes.

Pierre-Simon Laplace (1814).

'The time has come,' the Walrus said,
'To talk of many things'

Lewis Carroll (1872).

Doubtless other glories lie ahead. Bigger and better capsules carried to the moon; down in the test-tube something stirs; 'I think, therefore you're not,' says the computer.

Malcolm Muggeridge (2006).

There is also an effort to sequence a plant genome, arabadopsis [sic], which we hope will be led by the National Science Foundation with help from other agencies, including ourselves. This is roughly seventy megabases, and the project should be a real boon to botany.

James D. Watson (1992).

21.1 ONE, TWO, THREE, INFINITY

Technology is being developed with the goal of fulfilling system biology's version of Laplace's (1814) vision of an intellect who acquired all data and was vast enough to submit these data to analysis. Aebersold et al. (2000) define systems biology as the study "*of all the elements in a biological system both before and after chemical or genetic perturbation. Ultimately, systems biology aims to establish computational models that are predictive of the behavior of the system or its emergent properties in response to any given perturbation.*" The ambitious goals of systems biology are being accomplished through the *omic* sciences that use high-throughput technologies to generate the panoply of data required for a systems-level understanding of the organism (Hood, 1992; Blanchard and Hood, 1996; Sheth and Thaker, 2014). To know an organism, in a systems biology way, requires a variety of technologies or *platforms*, a term borrowed from computer science, to gain knowledge of the organism's genome—the complete sequence of nucleotides in its DNA, knowledge of its phenome—every morphological and physiological feature, as well as knowledge of everything in between (transcriptome, proteome, metabolome, etc.; Agrawal et al., 2015; Davies and Barh, 2015). To ensure that the data obtained and analyzed in each platform are comprehensive, accurate, and precise, there are a number of techniques, tools, methods, or *pipelines* available for the acquisition and analysis of data (Fondi and Liò, 2015). Obtaining such comprehensive data sets necessitates the invocation of A.J.P. Martin's (1952) principle of scientific research: "*Nothing is too much trouble if somebody else does it.*"

Hood et al. (2012) make the analogy between a biological process and a machine designed by Rube Goldberg that performs a simple chore in a complex manner such as cooling soup in a process involving 14 steps. To understand the Rube Goldberg machine "*one would have to have a parts list of all the components, know how the parts are connected together and understand the dynamics of how the parts move with respect to one another to cool the soup (the dynamics of the soup-cooling machine). These are three of the major requirements for a systems approach to understanding biological systems — the parts list, their interconnections, and the dynamics of the parts — interactions — to determine how the system functions or exhibits dysfunction.*"

To manage the nearly unbounded quantity of data required to have a systems-level knowledge of all the taxa in the treasure house of nature, *omic* analysis starts with the study of human beings and model organisms with small genomes. Informational and computational sciences are at the heart of omic data gathering, management, and analysis. The analyses performed by the omic equivalent of Laplace's intellect goes by the name of computational biology or bioinformatics (Schneider and Orchard, 2011; Kristensen et al., 2014).

Omic science is described as hypothesis-generating or discovery-based science as opposed to the more focused, hypothesis-based science that took advantage of the economy of nature (Krebs and Kornberg, 1957) that was

Plant Cell Biology. https://doi.org/10.1016/B978-0-12-814371-1.00021-7

presented in the previous chapters (Aebersold et al., 2000; Wiley, 2008; Horgan and Kenny, 2011). The use of the word discovery in discovery-based science indicates using the newest technology to see what nobody has seen before—at least nobody in your immediate group, similar to the way that Columbus used the newly invented carracks and caravels, ships that were large and stable enough to sail across the ocean, to discover America—something that few Europeans had ever seen before. This contrasts with Albert Szent-Gyorgyi's definition of discovery as recalled by J. D. Bernal (1962): "*Discovery consists of seeing what everybody has seen and thinking what nobody has thought.*" Science requires both kinds of discovery. Now that we have performed a limited exegesis on the definition of the word discovery, in terms of methods and thinking, we will examine the word science, which comes from *scientia*, the Latin word for knowledge.

The goal of science, stripped down to its essence, is the demonstration of what is necessarily true about the natural world. Theoretical, experimental, and observational science is based on inference, metaphysics, and technique. Inference requires the enunciation of foundational principles that serve as a conception of truth; metaphysics requires an answer to the question, what is reality; and technique requires the development and intelligent use of technology to peer deeply into the nature of life and the processes by which life is defined (see Chapter 1). An attempt to discern truth, to test our model of reality, and to explore the uses and limitations of technology undergirds all scientific research. Coming full circle, scientific research helps us to understand the nature of truth and reality, and to develop technology (Bacon, 1928a,b). Moreover, the critical thinking skills necessary to become an intelligent cell biologist develop while practicing this reiterative process based on checks and balances—a process that can be called ground truthing, which in the remote sensing field describes the process of verifying a satellite image with what is known about the location on the ground (Rao and Ulaby, 1977; Story and Congalton, 1986; Kloser et al., 2001).

Truth, that is, what is true for all time, can be expressed according to Gottfried Leibniz; in the general form, subject is predicate (Burnham, no date). For example, the cell is the basic unit of life. Once we make a truth statement, we use the *principle of sufficient reason* to explain why things are as they are and the *principle of noncontradiction* to explain why things are not otherwise. That is, we find a set of true predicates that provide sufficient reason without contradiction for our truth statements such as

- The cell is the basic unit of life because it has a differentially permeable plasma membrane that regulates the passage of matter, energy, and information between the living space and the rest of the universe. Without a differentially permeable plasma membrane, the molecules necessary for life would be diluted, and life would cease (see Chapter 2).
- The cell is the basic unit of life because it contains information in the form of a sequence of nucleic acids that can replicate with less than perfect fidelity to allow for inheritance and evolution, that is, stability and change. Without nucleic acids that can replicate, there would be no reproduction, and life would cease (see Chapter 16).

Additional, although more limited, truth statements include

- The plant cell is the basic unit of life because it can contain chloroplasts that convert radiant energy into the kinetically stable chemical energy that is necessary for all living organisms. (This is a limited truth statement because it does not take prokaryotic cells into consideration and most eukaryotic plant cells do not contain chloroplasts; see Chapter 13.) Nevertheless, without chloroplasts, there would be little to no transformation of radiant energy to chemical energy, and life on earth as we know it would cease.
- The cell is the basic unit of life because it can contain mitochondria that convert kinetically stable energy in the form of carbohydrate and lipid into kinetically available energy in the form of adenosine triphosphate (ATP) (This is a limited truth statement because it does not take prokaryotic cells into consideration; see Chapter 14.) Nevertheless, without mitochondria, there would be little to no transformation of kinetically stable energy to kinetically available energy, and life on earth as we know it would cease.

The cell is a *contingent entity*, which cannot account for the existence of itself, and consequently, a complete concept of the cell cannot explain the existence of cells. The principle of sufficient reason must extend to why the cell exists at all, why is it that the cell, and not something else, the basic unit of life, and even more basically, why does the cell or life itself exist at all? Simply put, all cells require an energy source—photoautotrophic cells require sunlight and heterotrophic cells require food. But beyond the totality of contingent things, there must be something that requires no explanation other than itself but is the cause necessary to explain the existence and intelligibility of all contingent things. That something, depending on your worldview, could be called anything from a creator (Bacon, 1928a; Lack, 1957; Gingerich, 2006; Polkinghorne and Beale, 2009) to chaos (Prigogine and Stengers, 1984; Dawkins, 1995).

The universal principle of inference, which detects and eliminates false conclusions and establishes true conclusions, is the basis of scientific knowledge. A systematic presentation of what constitutes scientific knowledge can be found in Aristotle's *Organon*, which means an instrument of thought, in the chapter on *Posterior Analytics* (Owen, 1853):

- All demonstration must be founded on known principles, which themselves are demonstrable, or on a priori or first principles that are considered self-evident—that is, necessary, general, and eternal truths (Watts, 1806; Spencer, 1880).
- There cannot be an infinite number of middle terms between the first principle and the conclusion.
- The demonstrations cannot be circular, where the conclusion is supported by the premises and the premises by the conclusion.

In all demonstrations, the conclusion, and all the intermediate propositions, must be necessary, general, and eternal truths. There is no demonstration when things happen by chance or contingently.

Through experiment and observation, based on Aristotle's *Posterior Analytics*, updated as "*surest rules and demonstrations*" in Francis Bacon's (2007) *Novum Organon*, and in William Whewell's (1858) *Novum Organon Renovatum*, a cell biologist strives to demonstrate the connection between the subject and the predicate, the cause and the effect, and the precursor and the product, to explain the material basis of the fundamental processes of life. That is, a cell biologist tries to explain the phenomena of life in terms of better known processes that occur in the nonliving world (Mathews, 1916). Discovery has consisted of finding the most essential elements of the fundamental processes of life that not only describe the physicochemical processes that make life possible but also provide substantive models about the nature of truth and reality.

Before the omics approach was introduced, a cell biologist typically (1) chose an important problem concerning the cellular basis of life; (2) devised alternative hypotheses that could explain the phenomenon; (3) devised crucial experiments or modes of observation that would support or exclude one or more of the hypotheses; (4) developed the technical skill or technology necessary to carry out a small and elegant experiment or observation to get a clean result; (5) analyzed the experimental results and observations; (6) refined the hypothesis, experiments, and observations and then repeated the procedure at a more refined level or on a different system.

The first 20 chapters in this book document how productive hypothesis-based science has been in understanding the cellular basis of life. It is, however, possible that the observations made, the experiments performed, and the explanations given in the search for truth were not based on multiple working hypotheses but merely confirmed an underlying and unconscious bias in favor of the ruling theory (Chamberlain, 1890, 1897, 1965; Platt, 1964; Jewett, 2005). To avoid any possibility of a pathological science based on confirmation bias (Langmuir, 1989a,b) due to careerism, grant seeking, or political views, omic science pushes the method of multiple working hypotheses to the limit of infinity which ironically at the level of reductio ad absurdum is indistinguishable from Newton's phrase *Hypotheses non fingo*. While omic science aims to attack problems in a comprehensive and unbiased fashion (International Human Genome Sequencing Consortium, 2001), it may be wise to remember what the pointed man said in Harry Nilsson's (1971) film, *The Point*: "*A point in every direction is the same as no point at all.*"

There is a tacit assumption in omic research that the research is bias free. As we will see below, bias can arise from technical issues because it is never possible to isolate completely, amplify equally, and analyze all the components in a mixture without bias when the components exist in vivo with different volatilities, solubilities, polarities, and in vastly different quantities.

Using the omics approach, a demonstration is typically made in the following manner: (1) choose an important problem concerning the cellular basis of life; (2) find a mutant that cannot perform the phenomenon; (3) chemically analyze any differences in the genome, transcriptome, proteome, metabolome, etc., between the mutant and the wild type using high-throughput technology and imaging; (4) analyze the results using bioinformatics and computational biology to tie together all the data in a way that maximizes "*the distinction between a pile of bricks and a true edifice*" (Forscher, 1963). Richard Dawkins (1995) saw the revolutionary ideas of Watson and Crick as being on par with Aristotle and Plato, adding "*What is truly revolutionary about molecular biology in the post-Watson-Crick era is the is has become digital.*" Indeed digital technology can be considered to be at the heart, if not the heart itself, of omics.

To relate the observed patterns of molecular components with the phenotype or traits of the whole plant, the field of phenomics or the high-throughput analysis of form and function has developed. The phenome is then correlated with molecular components analyzed by high-throughput analysis: genomics, transcriptomics, proteomics, metabolomics, etc. When pronouncing these words, follow the rule for words ending in "*ics*" by putting the stress on the penultimate syllable, which is "*om*." Indeed, Lederberg and McCray (2001) describe the philosophy of omics as being best described by the sacred Sanskrit sound, "*Om*," which signifies fullness and completeness, encompassing the entire universe in its divine unlimitedness. The word "*Om*" appears in the story of *Kãthaka Upaniṣad* (1.2.15–16), where Naciketas talks with Yama about the nature of knowledge and how it is obtained (Deussen, 1920):

The word which all the Vedas proclaim,
That which is expressed in every self-mortification (penance),
That for which they live the life of a Brahmacārin,
understand that word in its essence,
Om! that is the word.
Yes, this syllable is Brahman,
This syllable is the highest;
To him who possesses knowledge of this syllable,
whatever he may wish, is allotted.

In the words of Lederberg and McCray (2001), "*What could resonate more with today's –ome terms*!" Building on the work of the microbe hunters, the vitamin hunters, the enzyme hunters, and the gene hunters who discovered the biological role of microbes, vitamins, enzymes, and genes, respectively, the omic hunters are discovering the biological role of everything.

As the suffix implies, omics, which comes from the Greek word "nomos" (νόμος), which means management, is not far removed from econ*omics* when it comes to posing or not posing questions in science. The market approach to science permeates the Academy (Gardner et al., 2001; Anonymous, 2010) as "*exceptionally bright science PhD holders from elite academic institutions are slogging through five or ten years of poorly paid postdoctoral studies, slowly becoming disillusioned by the ruthless and often fruitless fight for a permanent academic position*" (Anonymous, 2011). Cyranoski et al. (2011) note that supply has exceeded demand and there are not enough good jobs for graduate students who have trained at great length and expense to be researchers. They conclude that "*it is not clear that spending years securing this high-level qualification is worth it for a job as, for example, a high-school teacher*." On the other hand, Daniel Lametti (2012) writes that "*it's worth it*," and Alan Jones (2014) concludes that the planet needs more plant scientists.

While I will describe the current state of the applications of omic science, I will emphasize the creative development of omic techniques from physical, chemical, and biological principles that have stood the test of time. This is because the applications of omic science are so far from reaching the promised goal of gathering and analyzing a nearly infinite quantity of unbiased data despite serious technical challenges in discriminating between bona fide low-abundance molecules and contaminants and despite of the inconsistent predictions that arise from using different bioinformatic databases and programs to relate each sequence to something that is directly observable (The Gene Ontology Consortium, 2008, 2013, 2015; Wei et al., 2017). Indeed, up until the present, the results obtained from omic studies in plants can be characterized in general as results that verify, corroborate, and confirm rather than improve on or contradict the results obtained from using preomic techniques. This is probably why so many people read papers or go to seminars primarily to see which techniques are being used. In all honesty, the history of omic studies could be characterized as a history of promises—perhaps even recycled promises (Comfort, 2012)—deferred. To paraphrase Langston Hughes (1999), what happens to a promise deferred? Does is dry up like a raisin in the sun, or does it get even more funding from the taxpayers?

Systems biology makes use of two basic high-throughput approaches to understand the structure and function of living organisms. The *genomic* or *structural approach* uses mapping and sequencing to determine the structure and organization of the genes that make up the genome and the genomes between and among taxa. The *postgenomic* or *functional genomic* approach analyzes gene expression in terms of the transcriptome, the proteome, the metabolome, and the phenome. A definition of a gene is fundamental to understanding structural and functional genomics. In the past, genes were formally identified by observing mutations in the phenotype that results in a distinct trait. With omic technologies, a gene is now identified as a DNA sequence, and mutations are identified as differences in the DNA sequence. This is a clear but limited and limiting definition, and the full definition of a gene is still murky.

21.2 GENOMICS

Genomics is the study of the structure of the genome. The goal of genomics is to determine the complete sequences of deoxyribonucleotides in a given organism, in the populations of organisms that make up a given species, and in the various species that make up the web of life. Additionally, the goal of genomics is to understand the meaning of the nucleotide sequence. The sequence of nucleotides in each three-dimensional genomic DNA molecule represents the linear sequence of nucleotides in the protein-encoding genes, including the sequences that make up the *promoters* (Lifton et al., 1978; Kiran et al., 2006), the *introns*, and the protein-encoding *exons* (Gilbert, 1978; Jones et al., 2013). An open reading frame (ORF) is the linear sequence of deoxyribonucleotides in the introns and exons. Some of the ORFs contain overlapping genes (Barrell et al., 1976; Berget et al., 1977; Smith et al., 1977; Berk, 2016). The promoters influence the transcription of the adjacent introns and exons (Ayoubi and van de Ven, 2017). Some of the nucleotide sequences in the genomic DNA represent regulatory elements, including *enhancers* and *silencers* that influence the transcription of distant genes (Timko et al., 1985; Ott and Chua, 1990; Biłas et al., 2016; Weber et al., 2016). Additional nucleotide sequences encode microRNAs that function in the posttranscriptional regulation of gene expression. Together, these nucleotide sequences, the interaction between them, and their response to the environment influence the traits of an individual organism. Genomics aims to understand all of these interactions. Variation in the nucleotide sequence between individuals in a species are correlated with traits that influence health and disease, resistance or susceptibility to drugs, including herbicides or pesticides, biosynthetic growth and development, energy metabolism, chemical composition, and the ability to respond to environmental signals. Variations in the nucleotide sequence between populations are correlated with speciation. The variations may be due to point mutations or larger-scale rearrangements of the DNA, including chromosome translocations.

21.2.1 The Relationship Between a Trait and the Sequence of DNA

Gregor Mendel (1865), a Gregorian monk who studied under Christian Doppler, set out to understand the number of different forms a trait could exhibit and to analyze those traits statistically. By studying discontinuous binary traits that could be analyzed with the binomial theorem ($(p+q)^n$), Mendel laid the foundations of our understanding of inheritance. The binary traits that Mendel studied could be simply interpreted in terms of dominant or recessive alleles. Mendel's research was rejected by his contemporaries (East, 1923), probably because in nature, most observed traits are not discrete but continuous, due to the contribution of many genes each of which has a quantitative and limited effect. Compared with the inheritance of discrete traits ($n=1$, alleles $=2$), the inheritance of quantitative traits ($n \to \infty$) is far less tractable in terms of analysis using the binomial theorem. To Mendel's contemporaries, the inheritance of continuous traits rather than discrete traits seemed to provide a more realistic picture of nature. Of course the continuous traits could be subjected to analysis, although it would require large-scale hybridizations that would merely give a statistical result (Bateson, 1899).

Following the success of the nineteenth century chemists, who rearranged discrete atoms to make valuable dyes and medicines (Beer, 1959), Max Planck (1900) reassessed the continuous nature of energy and provisionally accepted the discrete nature of energy that led to the principles of quantum theory. Likewise, Hugo de Vries (1900a,b), Carl Correns (1900) and Erich Tschermak (1900) reassessed the continuous nature of variation and accepted the discrete nature of traits to provide the underlying natural principles necessary for plant breeding (Bateson, 1902; Roberts, 1929). Following the rediscovery of Mendelian traits in plants, other breeders identified Mendelian traits in animals and eugenic and dysgenic traits in humans (Davenport, 1910, 1921). Archibald Garrod (1902) demonstrated that alkaptonuria was a disease in humans caused by a variant of a single gene that caused an alternative course of metabolism. Likewise, Beet (1949) and Neel (1949) demonstrated that sickle cell anemia was also a single gene disease complicated by the fact that the heterozygote, which did not have the disease, was resistant to malaria (Allison, 1954; Lederberg, 1999). Other single-gene genetic disorders in humans include hemophilia A, Duchenne's muscular dystrophy, thalassemia, and Huntington's chorea.

In the plant world, Harold Flor (1942, 1947, 1955, 1971; Sequeira, 2000) demonstrated that a single gene provided resistance in a flax plant against a pathogen, and more recently, Martin et al. (1993) were able to transform tomato with a single avirulence gene that conferred resistance to *Pseudomonas syringe* pv. *tomato*. Extrapolating from a limited number of quantified cases that show discrete traits and discounting the large number of qualitative observations of continuous traits led to the idea that traits, including genetic diseases, were in general unit characters controlled by single genes with two competing alleles.

Sewell Wright (1931), who served as a voice in the wilderness, warned that "*As genetic studies continued, ever smaller differences were found to mendelize, and any character, sufficiently investigated, turned out to be affected by many factors.*" Through mapping techniques, the polygenic effects were attributed to quantitative trait loci (QTL; East, 1910a,b, 1916; Grant, 1975; Paterson et al., 1988; Vicedo, 1991; Altmüller et al., 2001; Wilson et al., 2001; Wolyn et al., 2004; Niklas, 2016) that could be mapped onto chromosomes (Botstein et al., 1980; Tanksley et al., 1989, 1992; Tanksley, 1993; Kole and Abbott, 2008). According to Karl Sax (1923), "*The multiple-factor hypothesis…does not necessarily mean that the various factors are equal in effect or that they are inter-dependent for their expression.*" The traits involving multiple factors must be analyzed statistically, and the assumptions concerning the differential effectiveness of each sequence, the additivity or multiplicity of their effects, and the different amounts of linkage between the sequences must be factored into the statistical model.

Early on it was understood that genes exerted their influence on normal development, physiology, and disease by producing enzymes (Bateson, 1909; Loeb and Chamberlain, 1915; Goldschmidt, 1916). Searching for even greater reduction, Leonard Troland (1917) in his "*intentionally polemical*" paper entitled *Biological Enigmas and the Theory of Enzyme Action* suggested that the nucleic acids that made up "*the actual Mendelian factors are enzymes.*"

The transformative work of Avery et al. (1944) demonstrated beyond a reasonable doubt that DNA was the chemical that held the hereditary information in the transformation of nonencapsulated avirulent pneumococci into capsulated virulent pneumococci (McCarty and Avery, 1946a,b; McCarty, 1946); Avery's work was a turning point (or turning back point if one takes Miescher's (1897) views into consideration; see Chapter 16) in the way that chemists approached the study of DNA. Prior to 1944, chemists used not only the tools of the chemist but also the thinking of chemists to analyze DNA (Levene and Bass, 1931). With their reductionist worldview, the chemists concluded that DNA was too simple a molecule to carry hereditary information; Erwin Chargaff (1950a), on the other hand, realized that one must take the biological function of DNA (teleology) into consideration in the development of chemical tools to analyze DNA. Realizing that the chemical identity and biological significance of a molecule of DNA was based on "*the sequence on which these* [nucleotide] *constituents are arranged in the molecule*," Chargaff worked against the current of his time and used chemical methods that would reveal the information-carrying function of the DNA. As a consequence of his revolutionary worldview, Chargaff, who understood the importance of one's worldview in the activity of science, took the chemistry of DNA as an informational molecule seriously. Consequently, he developed paper chromatographic methods to separate the nitrogenous bases (adenine, guanine, the purines; and cytosine, thymine, the pyrimidines) and ultraviolet absorption spectroscopy to quantify the amount of each base (Vischer and Chargaff, 1948).

Chargaff (1955) wrote, "*An account—not practical in the present framework—of the evolutionary sequence of our understanding of the nucleic acids would not be uninstructive: it would show how often in the natural sciences, even behind seemingly trivial experiments, there lies and entire vision of nature; it would demonstrate that the answers which the investigator receives often are contained in the questions he asks. As long as, in the field under consideration here, feeble and faltering questions were asked, the answers were vague; but when the direction of the quest became reasonably well defined, the results sharpened in definition. Though we still are very far from a solution of the many problems that keep multiplying as our knowledge progresses, it would be hazardous to erect a sign saying* 'Ignorabimus.' *But much is left to future generations*."

Chargaff (1950a) asked, "*How different must complicated substances be, before we can recognize their difference?*" He realized, that "*the disappearance of one guanine molecule out of a hundred, could produce far-reaching changes*." As a start, he characterized the ratio of nucleotides in DNA from human beings, animals, plants, and microorganisms with decimal precision (Chargaff, 1947; Chargaff et al., 1949, 1950, 1951; Chargaff, 1950b, 1951a,b; Zamenhof and Chargaff, 1949, 1950a,b; Zamenhof et al., 1950). Chargaff (1950a) concluded that "*Generalizations in science are both necessary and hazardous; they carry a semblance of finality which conceals their essentially provisional character; they drive forward, as they retard; they add, but they also take away. Keeping in mind all these reservations, we arrive at the following conclusions. The desoxypentose nucleic acids* [DNA] *from animal and microbial cells contain varying proportions of the same four nitrogenous constituents, namely, adenine, guanine, cytosine, thymine. Their composition appears characteristic of the species, but not of the tissue, from which they are derived. The presumption, therefore, is that there exists an enormous number of structurally different nucleic acids; a number, certainly much larger than the analytical methods available to us at present can reveal. It cannot yet be decided, whether what we call the desoxypentose nucleic acid of a given species is one chemical individual, representative of the species as a whole, or whether it consists of a mixture of closely related substances, in which case the constancy of its composition merely is a statistical expression of the unchanged state of the cell. The latter may be the case if, as appears probable, the highly polymerized desoxypentose nucleic acids form an essential part of the hereditary processes…It would be gratifying if one could say—but this is for the moment no more than an unfounded speculation—that just as the desoxypentose nucleic acids of the nucleus are species-specific and concerned with the maintenance of the species, the pentose nucleic acids* [RNA] *of the cytoplasm are organ-specific and involved in the important task of differentiation*." Chargaff (1951b) left it to future generations to determine the sequence of nucleotides in the DNA of a species and the individuals that compose the species, stating "*The elaboration of methods for sequence analysis is, perhaps, one of the most urgent problems in nucleic acid chemistry, since differences in nucleotide sequence may very well be among the determinants of chemical and biological specificity*."

The study of mutants has clearly shown that the sequence of nucleotides in genes contribute to nearly every aspect of the life of an organism and the lives of all the cells that compose the organism. As the physical basis of the cell's memory of the past and its potential for the future, DNA is one of the most interesting and complex of the molecules found in the cell. Paradoxically, the complexity arises from the simplicity of its basic subunit structure composed of four different nucleotides. To understand the complexity within the simplicity, it became imperative to fulfill Chargaff's (1947, 1951b, 1963, 1965) vision and develop techniques to sequence DNA. The genome sequencing projects, which purportedly enumerate the elements of a system irrespective of any hypotheses on how the system functions, are the archetype of discovery science (Aebersold et al., 2000). The success of the projects, however, depends on understanding traits that are controlled by a limited number of genes ($n \rightarrow 1$), as well as understanding what is a gene.

DNA was the last class of macromolecules comprising the trinity of the central dogma to be sequenced (Crick, 1958, 1970, 1974a; Thieffry and Sarkar, 1998; Cobb, 2017). Due to the existence of specific proteases that cleaved proteins at a single amino acid and specific ribonucleases that cleaved RNA at a single nucleotide (Gilham, 1970), it was possible to sequence proteins (Sanger, 1958) and RNA (Holley et al., 1965a,b) by sequential chemical degradation. On the other hand, because there are no single nucleotide-specific DNA hydrolases, DNA had to be sequenced by synthesis. Sequencing by synthesis, which is the biochemical basis of omic technology, is based on the discoveries made by heroic researchers whose goal was to understand the chemical basis of heredity in the decade before and the three decades following the publication of the double helix by Watson and Crick (1953a,b,c). I call this period the golden age of molecular biology.

21.2.2 Sequencing DNA by Synthesis

In presenting the 1959 Nobel Prize in Physiology or Medicine to Severo Ochoa and Arthur Kornberg, Hugo Theorell (1959) acknowledged, "*For proteins it has been proved, and for the nucleic acids it is highly probable, that the order of the different building blocks in the chains is by no means left to chance, but on the contrary is planned in detail for each kind of molecule and for each kind of living organism.*"

21.2.2.1 Sequencing by Measuring the Incorporation of Radioactive Deoxyribonucleotides Into a Primer-Template

Given the importance of the sequence of nucleotides in DNA to the central dogma and to understanding inheritance, Ray Wu (1993, 1994) decided to develop a method, using nature's toolbox, to reproduce a biological process that would allow him to sequence DNA by synthesis. Wu's sequencing method built on Arthur Kornberg's discovery of DNA polymerase, which polymerized nucleotides into acid-insoluble DNA in the presence of DNA. Interestingly, although Kornberg performed the experiment in 1954, a year after the publication of Watson and Crick's (1953a,b,c) paper, Kornberg did not add the DNA to the reaction mixture to act as a template for the DNA polymerase (Kornberg and Baker, 1992) but to serve as a primer for the nascent DNA chain just as glycogen served as a primer for glycogen phosphorylase to reduce the lag phase in the synthesis of a growing glycogen chain (Cori and Cori, 1939). Kornberg also added DNA to protect the nascent chains from being digested by nucleases in the extract.

Kornberg (1989a,b) demonstrated that, using DNA isolated from the bacteriophage ΦX174 (Goulian et al., 1967) and DNA polymerase and DNA ligase (Weiss and Richardson, 1967; Howell and Hecht, 1971; Howell and Stern, 1971; Lehman, 1974; Elder et al., 1987; Daniel and Bryant, 1988; Babiychuk et al., 1998; Shuman, 2009; Waterworth et al., 2009, 2010) isolated from *Escherichia coli*, he could "*assemble a 5000-nucleotide DNA chain with identical form, composition, and genetic activity of DNA from a natural virus.*"

The time had come to develop a method to determine the sequence of DNA to understand its many-fold roles in development, physiology, and inheritance. Wu's pioneering but primitive DNA sequencing method allowed him to determine the sequence of DNA at a rate of about 12 nucleotides per year. Given the extremely long chain length of naturally occurring DNA, the complete sequence of the molecule was a daunting task. Wu reduced the task to bite-sized portions by choosing to sequence the short single-stranded cohesive ends of the lysogenic bacteriophage lambda DNA (Lederberg and Lederberg, 1953; Hershey et al., 1963; Ris and Chandler, 1963). When inside a bacterium, the DNA of bacteriophage lambda forms a ring and either incorporates into the bacterial DNA where it remains latent or becomes free from the bacterial DNA and replicates producing 100–10,000 identical molecules of bacteriophage DNA. The bacteriophage DNA codes for proteins that are translated on the bacterial ribosomes. The translated proteins encapsulate the nascent DNA as the DNA changes form from circular to linear. The bacteriophage proteins also lyse the bacterial cell wall, which results in the release of the bacteriophage to the surrounding medium.

The transformation between linear and circular DNA is effected by the cohesive ends of the DNA where the 5′ end of each strand was assumed to be mutually complementary (Fig. 21.1). Because single-stranded DNA can act as a template and DNA polymerase can only bind to double-stranded DNA, Wu and Kaiser (1967, 1968) surmised that if the 3′ ends of each strand of the DNA molecule were shorter than the 5′ ends, then the shorter 3′ end would act as a primer, whereas the longer 5′ end would act as a template. Thus, the cohesive ends but not the rest of the DNA might serve as natural primer-templates for the DNA polymerase. The short labeled ends could then be sequenced without the complication of simultaneous DNA synthesis from the rest of the long DNA molecule.

Using DNA polymerase isolated from *E. coli*, the short 3′-OH ends of the lambda DNA were extended stepwise in the 5′ → 3′ direction by incorporating deoxyribonucleoside triphosphates (dNTPs) using the complementary 5′ end as a template. On incorporation, two phosphates are cleaved from each deoxyribonucleoside triphosphate forming a deoxyribonucleoside monophosphate (dNMP) and pyrophosphate (P_iP_i). Each reaction mixture contained one of

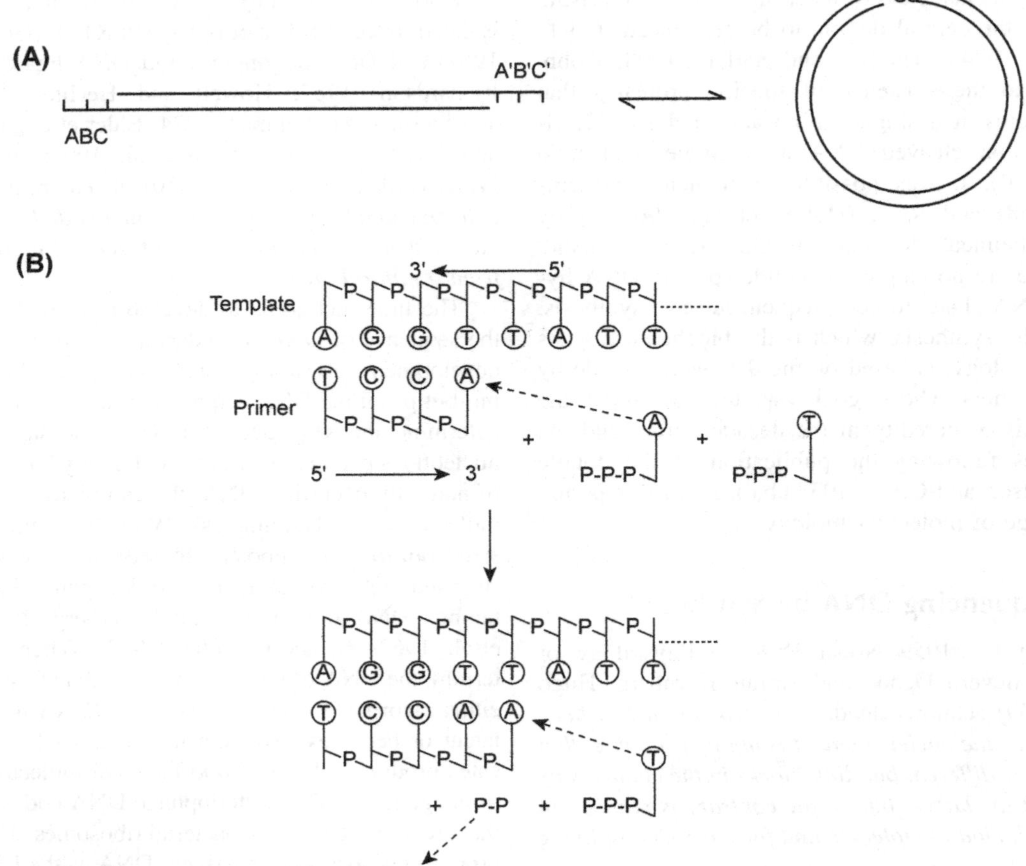

FIGURE 21.1 (A) Diagram of the linear and circular structures of lambda bacteriophage DNA. When the DNA circularizes, the cohesive ends are held together by complementary base pairing. (B) Polymerization occurs by the stepwise addition of deoxyribonucleoside triphosphates to the 3′-OH end of the primer. DNA polymerase adds the deoxyribonucleoside triphosphate that matches the first available deoxynucleotide monophosphate in the single-stranded template. *(A) From Wu, R., Kaiser, A.D., 1968. Structure and base sequence in the cohesive ends of bacteriophage lambda DNA. J. Mol. Biol. 35, 523–537. (B)Modified from Sanger, F., December 8, 1980. Determination of Nucleotide Sequences in DNA. Nobel Lect.*

the four necessary dNTP labeled with ^{32}P on the 5′ end. The label was incorporated into the newly synthesized DNA when the dNTP coded by the template was present. In the case of lambda DNA, deoxyguanosine triphosphate (dGTP) was the only dNTP that matched the first dNMP in the template. Quantifying the amount of label per DNA molecule indicated that the initial sequence appeared to be 5′-GGGG-3′-OH (Table 21.1).

The next deoxyribonucleotide in the sequence was determined by finding out which ^{3}H-dNTP was incorporated into the lambda DNA in the presence of ^{32}P-dGTP. ^{3}H-dCTP was the only dNTP whose incorporation was dGTP dependent. Therefore the sequence appeared to be 5′-GGGGC-3′-OH. In the presence of dGTP and dCTP, the ratio of dGTP:dCTP incorporated into the lambda DNA was 7.2:3.2. Subtracting 4 dGTP and 1 dCTP for the first five nucleotides means that there is a dGTP:dCTP ratio of 3.2:2.2 in the next part of the sequence (Table 21.2). This meant that the sequence might be 5′-GGGGCCC-3′-OH, but there were other possibilities. Wu and Kaiser (1968) performed a nearest neighbor analysis, a technique developed by Robert Sinsheimer (1954, 1955; Josse et al., 1961) to determine the order of these nucleotides.

In the nearest neighbor analysis experiment, the phosphate adjacent to the 5′ carbon of dCTP was labeled with ^{32}P and ^{32}P–dCTP was incorporated in the presence of unlabeled dGTP.Then the DNA polymerase "repaired" the single-stranded region by polymerizing nucleotides in the 5′ → 3′ direction to extend the 3′-OH end of the lambda DNA using the protruding 5′-terminated single strand as a template until it reached the nucleotide on the template that coded for the omitted nucleotides (dATP, dTTP).

The newly synthesized DNA fragments were then partially digested with micrococcal DNase and fractionated

TABLE 21.1 The Incorporation of Single ^{32}P-Deoxyribonucleotides Into Lambda DNA

Labeled Nucleotide Added (No Unlabeled Nucleoside Triphosphates)	Residues Incorporated/Molecule of DNA
$dAT^{32}P$	1.2
$dCT^{32}P$	0.3
$dTT^{33}P$	0.1
$dGT^{32}P$	4.1[a]
$dCT^{32}P$ no polymerase	0.2
$dGT^{33}P$ no polymerase	0.1

Incorporation was measured after 3 hours under conditions described in Materials and Methods. Note that 4 dGTPs are incorporated per molecule of DNA in a DNA polymerase dependent manner.
[a]*Average of six independent determinations. The individual values were 3.8, 4.0, 4.0, 4.1, 4.2, and 4.4.*
From Wu, R., Kaiser, A.D., 1968. Structure and base sequence in the cohesive ends of bacteriophage lambda DNA. J. Mol. Biol. 35, 523–537.

TABLE 21.2 The dGTP-Dependent Incorporation of Single ^{3}H-Deoxyribonucleotides Into Lambda DNA

Labeled Nucleoside Triphosphates Added (No Unlabeled Ones Present)	Residues Incorporated/ Molecule of λ DNA
$dGT^{32}P + [^{3}H]dATP$	4.0 dG; 1.4 dA
$dGT^{32}P + [^{3}H]dTTP$	4.1 dG; 0.2 dT
$dGT^{32}P + [^{3}H]dCTP$	7.3 dG; 3.2 dC

Note that dCTP is the only deoxyribonucleotide significantly incorporated in a dGTP-dependent manner.
From Wu, R., Kaiser, A.D., 1968. Structure and base sequence in the cohesive ends of bacteriophage lambda DNA. J. Mol. Biol. 35, 523–537.

on a diethylaminoethyl—cellulose (DEAE)—cellulose) column. The oligonucleotide fractions were then completely digested into 3′-monodeoxyribonucleotides with phosphodiesterase isolated from spleen (Fig. 21.2). Spleen phosphodiesterase is an exonuclease that digests single-stranded DNA from the 5′ end to yield dNMPs that retain the phosphate bound to the 3′ carbon of deoxyribose. Thus, the 3′ monodeoxyribonucleoside monophosphate that is just before the ^{32}P—dCMP labeled at the 5′ position gains the labeled phosphorus on hydrolysis by spleen phosphodiesterase. The deoxyribomononucleotides were then separated by paper chromatography to determine the dNMP composition of each fragment. Because the number of dCTP incorporated is approximately equal to the number of dGMP that becomes radioactive, then all three residues of dCMP have dGMP as their nearest neighbor. The sequence could be 5′-GGGGCGCGCG-3′-OH (Table 21.3). Because the number of labeled 3′dGMP was close to, but not equal to, the number of 5′ dCTP incorporated, the determination of the exact sequence required the experiment to be refined (Wu, 1970; Wu and Taylor, 1971).

When the reaction mixture was scaled up approximately 100-fold from 0.15 to 15 mL and higher specific activity nucleotides were available, Wu found that when dGTP was the only nucleotide, three not four dGTP were incorporated into the newly synthesized DNA. Moreover, the sequence became clearer when ^{32}P—dGTP and ^{32}P—dCTP was used for nearest neighbor analysis.

Wu et al. (1973) adopted a separation system (Brownlee and Sanger, 1969) involving two-dimensional paper chromatography (Consden et al., 1944) and paper electrophoresis (Michl, 1958) that was able to separate partially digested fragments of DNA up to 25 nucleotides long that differed by a single nucleotide. This allowed Wu et al. (1973) to read out the sequence of nucleotides without having to isolate each oligonucleotide and sequence it separately (Fig. 21.3). The sequence of the cohesive end was found to be 5′-GGGCGGCGACCT-3′-OH. Nevertheless, there is a limit on how useful this method was for a long and complex sequence of DNA.

The cohesive end of the lambda DNA was a naturally occurring template-primer system. Wu (1970) realized that once the sequence of the cohesive ends were known, he could determine the sequence of the adjacent region, which coded for lysozyme, by using an exonuclease to lengthen the template at the 5′ end, which would expose the gene coding for lysozyme. This could be done by performing a limited digestion of the double-stranded DNA with exonuclease III purified from *E. coli*. This exonuclease nicks and then removes nucleotides from the 3′ end of double-stranded DNA. By creating a new primer, DNA polymerase would bind to this locus and the DNA coding for lysozyme could be sequenced (Padmanabhan and Wu, 1972a,b).

As we discussed in the case of microfilaments (see Chapter 10) and microtubules (see Chapter 11), primers are important for polymerization. Carl and Gert Cori discovered the importance of a primer in eliminating the lag phase for glycogen synthesis when they began using purified glycogen phosphorylase (Cori and Cori, 1939). Likewise, a primer in the form of oligonucleotides was important for eliminating the lag phase in the enzymatic synthesis of high—molecular mass polyribonucleotides by polynucleotide phosphorylase (Mii and Ochoa, 1957; Singer et al., 1957, 1960; Heppel, 2004) and by DNA polymerase (Kornberg, 1957). Knowing that DNA synthesis required a

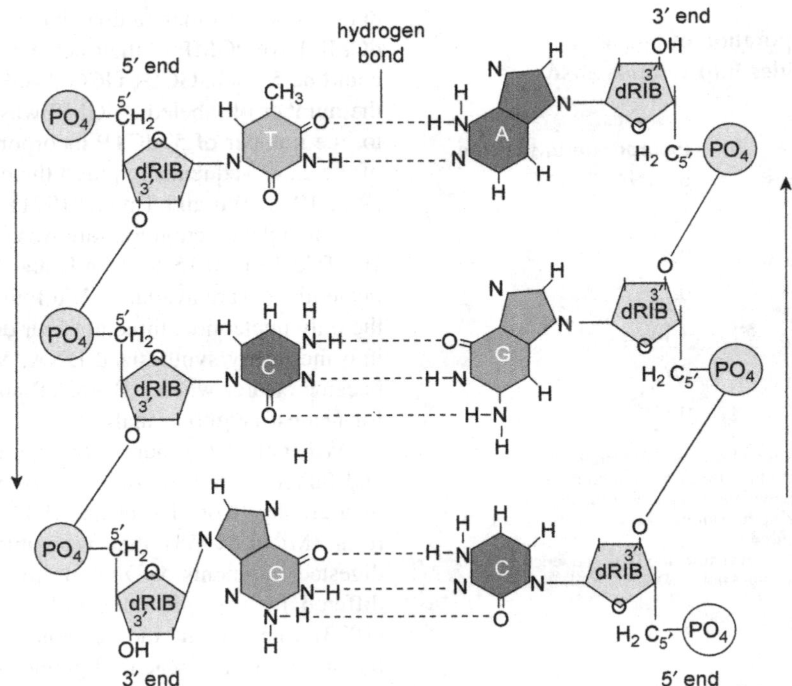

FIGURE 21.2 The two single strands that make up double-stranded DNA are antiparallel. The phosphates in the backbone of DNA are bound to the 3′ and 5′ carbons of deoxyribose. Spleen phosphodiesterase is an exonuclease that digests single-stranded DNA from the 5′ end to yield deoxyribonucleoside monophosphates that retain the phosphate bound to the 3′ carbon of deoxyribose. Snake venom phosphodiesterase differs from spleen phosphodiesterase in that it is an exonuclease that digests single-stranded DNA from the 3′ end to yield deoxyribonucleoside monophosphates that retain the phosphate bound to the 5′ carbon of deoxyribose. In addition, there are three hydrogen bonds between cytosine (C) and guanine (G) and two hydrogen bonds between adenine (A) and thymine (T). Consequently, the binding between cytosine and guanine is more stable than the binding between adenine and thymine.

TABLE 21.3 Nearest Neighbor Analysis of DNA After Incorporation of dCT^{32}P and dGTP

		Molecules of Labeled 3′-Nsucleotide Found After Digestion			
Nucleotides Added	**Residues of dC Incorporated/DNA Molecule**	**dC**	**dG**	**dA**	**dT**
dCT^{32}P + dGTP	2.9	0.2	2.4	0.1	0.2

λ DNA was allowed to incorporate dGTP and [α-^{32}P]dCTP, 2 × 10^{9} counts/min/μmole, under conditions described in Materials and Methods for 3 hours. The product was washed and digested with micrococcal DNase and spleen phosphodiesterase to 3′ mononucleotides. The mononucleotides were separated by paper chromatography. Note that while most of the^{32}P incorporated was found as dGT^{32}P after digestion with spleen phosphodiesterase, 2.4 is not equal to 2.9, and neither 2.9 nor 2.4 indicate an integral number of deoxyribonucleotides.
From Wu, R., Kaiser, A.D., 1968. Structure and base sequence in the cohesive ends of bacteriophage lambda DNA. J. Mol. Biol. 35, 523–537.

single strand template, Wu sought to create specific primers that would recognize and bind to unique sequences on the single strand of DNA and thus target the DNA of interest without sequencing the remainder of the DNA. That is, the DNA primer did not only act as a nucleation site for polymerization but it also recognized a specific sequence on the template.

DNA polymerase can only add nucleotides to the 3′-end of the primer, and accordingly, the copy of the template DNA is made in the 5′ → 3′ direction. By choosing an oligonucleotide as a primer that was complementary to the desired region of the template and long enough to uniquely bind to a region of single-stranded DNA, the starting point of the newly synthesized DNA would be certain. The nucleotide sequence of the primer that Wu synthesized reflected the predicted sequence of codons that coded for the known sequence of amino acids in lysozyme. It took Wu (1972) four months to synthesize a primer composed of 12 nucleotides, even under the guidance of H. G. Khorana. It took Sanger (1988) more than a year to synthesize a

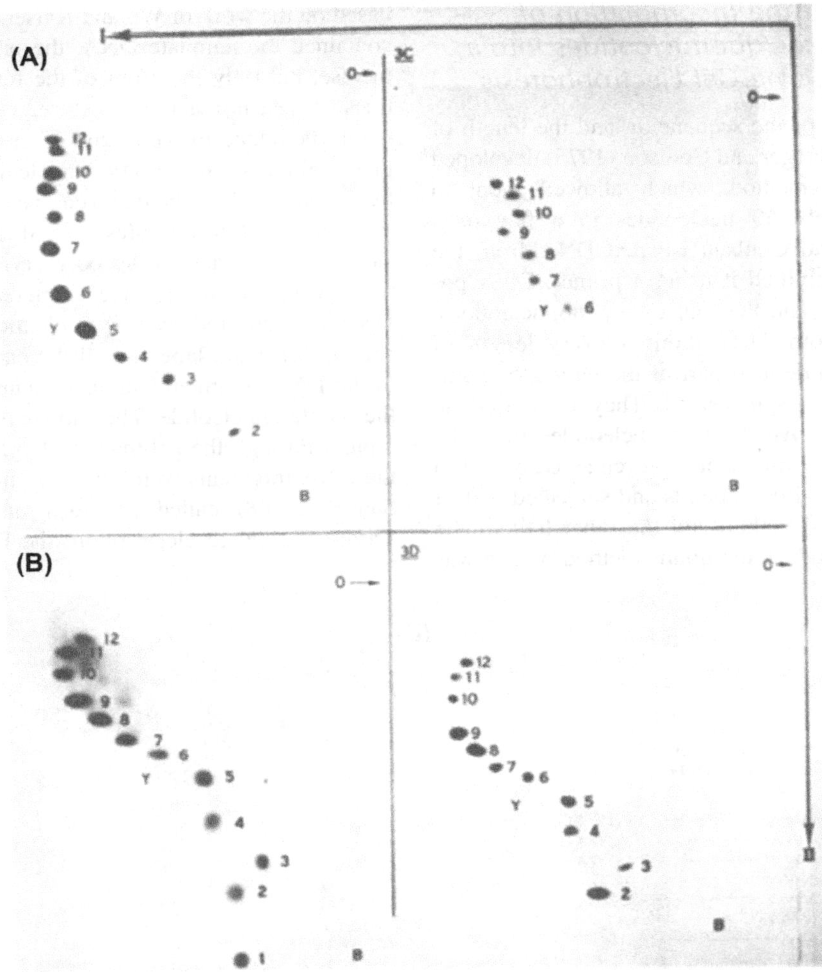

FIGURE 21.3 Determination of the sequence of synthetic oligonucleotides by partially digesting the radioactive oligonucleotides and observing the mobility of the products on two-dimensional cellulose thin-layer chromatography. (A) Limited snake venom phosphodiesterase digestion of ^{32}P-oligonucleotides. The complete sequence was deduced from the mobility of the oligonucleotide fragments and knowing that spot 1 was dCM^{32}P; (B) Limited spleen phosphodiesterase digestion of ^{32}P-oligonucleotides. The complete sequence was deduced from the mobility of the oligonucleotide fragments; (C) Limited snake venom phosphodiesterase digestion of ^{32}P-oligonucleotides followed by labeling at the 3′ ends with deoxyribonucleotidyl terminal transferase. Each spot was eluted, completely digested with spleen phosphodiesterase, and the mononucleotide composition was determined; (D) Limited spleen phosphodiesterase digestion of ^{32}P-oligonucleotides followed by labeling at the 3′ ends with polynucleotide kinase. Each spot was eluted, completely digested with snake venom phosphodiesterase, and the mononucleotide composition was determined. *From Wu, R., Tu, C.D., Padmanabhan, R., 1973. Nucleotide sequence analysis of DNA. XII. The chemical synthesis and sequence analysis of a dodecadeoxynucleotide which binds to the endolysin gene of bacteriophage lambda. Biochem. Biophys. Res. Commun. 55, 1092–1098.*

primer composed of eight nucleotides. Using synthetic primers, Sanger et al. (1973) sequenced a portion of the DNA that coded for the major coat protein in phage f1, and Padmanabhan et al. (1974) sequenced the lysozyme gene from bacteriophage T4.

By combining biological, physical, and chemical knowledge, it became possible to determine the sequence of DNA (Wu, 1994). It then became clear that knowledge of the sequence of DNA would provide an important avenue to understand the biology of organisms, their development, pathology, and evolution. Subsequently, technology developed to increase the speed and lower the cost of DNA sequencing. The intelligent use of primers is the foundation of all sequencing techniques and all enzymatic sequencing techniques are based on the use of a primer-template, DNA polymerase, and deoxyribonucleotides.

21.2.2.2 Analysis of the Incorporation of Radioactive Dideoxy-Ribonucleotides into a Primer-Template Using Gel Electrophoresis

To increase the speed of the sequencing and the length of the DNA sequenced, Sanger and Coulson (1975) developed the plus and minus method, which allowed them to sequence approximately 50 nucleotides in a few days (Fig. 21.4). Sanger and Coulson isolated DNA from the ΦX174 phage and amplified it using a primer, DNA polymerase from *E. coli*, and the four deoxyribonucleotides. Consequently, a random distribution of every length of DNA fragment from a defined part of the template identified by the primer was synthesized. They then used an agarose column to remove the free nucleotides from the newly synthesized DNA fragments that remained bound to the template through hydrogen bonds and subjected half of the DNA to the minus method and the other half of the DNA to the plus method. In the minus method, which was based on the work of Wu and Kaiser, each reaction mixture contained the template DNA, the primer, and DNA polymerase, but only three out of the four radioactive nucleotides. In each reaction mixture, radioactive nucleotides would be added to the 3′ end of the primer until the template called for the omitted nucleotide. There were four treatments, each of them lacked one of the four nucleotides. The four reaction mixtures were then fractionated side by side using thin denaturing polyacrylamide gel electrophoresis, which could resolve fragments of single-stranded DNA that differed by one nucleotide (Sanger and Coulson, 1978). Each lane had all the bands that corresponded to the DNA fragment lengths containing nucleotides up to the omitted nucleotide. The shorter fragments moved more rapidly through the gel than the larger fragments, and thus the DNA fragments were separated in the thin gels by size. Sanger (1988) called the idea of using a thin polyacrylamide gel to electrophore the DNA "*the best idea I*

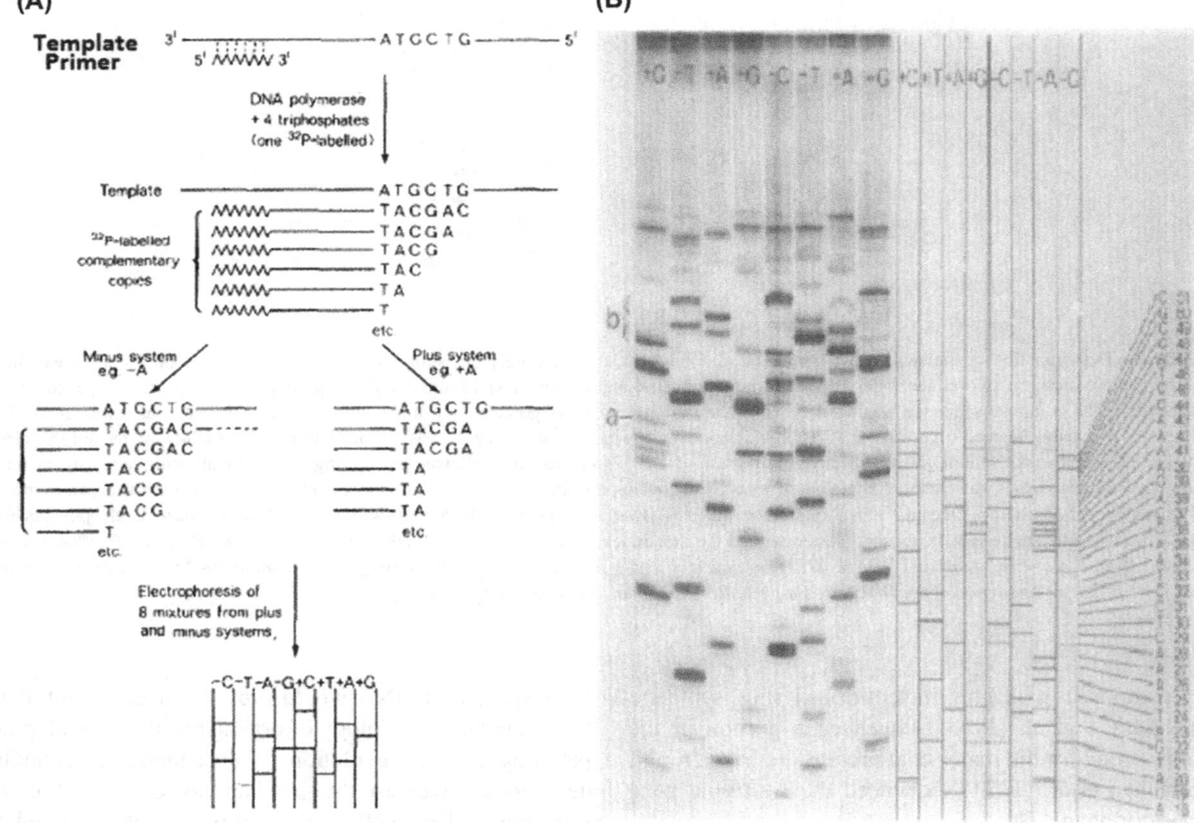

FIGURE 21.4 The principle (A) of the plus and minus method, and an autoradiograph (B) of a plus and minus sequencing experiment with the interpretation of the experiment, and the deduced sequence. In the minus method, each reaction mixture contained the template DNA, the primer, DNA polymerase, but only three out of the four radioactive nucleotides. Each lane had all the bands that corresponded to the DNA fragment lengths containing nucleotides up to the omitted nucleotide (−C, −T, −A, and −G). In the plus method, each reaction mixture contained the primer-template, T4 DNA polymerase, and only one nucleotide. The bands in the +C, +T, +A, and +G lanes represent the sum of the preceding nucleotides and the included nucleotide. *From Sanger, F. and A. R. Coulson. 1975. A rapid method for determining sequences in DNA by primed synthesis with DNA polymerase. J. Mol. Biol. 94, 441–448.*

have ever had." The thin polyacrylamide gel made it possible to read the sequence of nucleotides directly from the developed X-ray film. Using this method, the time needed for DNA sequencing was reduced to days.

The bands observed in the minus lanes represented the lengths of the fragments up until the template called for the omitted nucleotide. Thus, the bands in the −C, −T, −A, and −G lanes represented the sum of the nucleotides preceding the missing nucleotide. Reading the opaque bands in the minus lanes in the autoradiogram starting with the shortest fragment and moving up, we see that the sequence of nucleotides gives 5′-CATGATTAA…3′-OH. This sequence of bands, can be compared with the sequence of bands obtained using the plus method when only one nucleotide is in the reaction mixture.

In the plus method, each reaction mixture contained the primer-template, T4 DNA polymerase, and only one nucleotide. The T4 DNA polymerase differs from the *E. coli* DNA polymerase in that the T4 DNA polymerase has exonuclease activity that degrades double-stranded DNA from its 3′ end, but the exonuclease activity stops when the enzyme reaches a nucleotide that is the same as the one in the reaction mixture (Englund, 1971). This is because the polymerase reaction balances the exonuclease reaction for the one included nucleotide. Digests made in the presence of each nucleotide are then fractionated side by side using polyacrylamide gel electrophoresis so that each lane had all the bands that terminate with the added nucleotide. The sequence of nucleotides in the DNA is read directly from the developed X-ray film.

The bands observed in the plus lanes represented the lengths of the fragments that terminated with the included nucleotide. Thus, the bands in the +C, +T, +A, and +G lanes represented the sum of the preceding nucleotides and the included nucleotide. Reading the opaque bands in the plus lanes in the autoradiogram starting with the shortest fragment and moving up, we see that the sequence of nucleotides is 5′-ACATGATTAA…3′-OH.

Compared with Wu's method, the plus and minus method was a rapid and simple technique for determining the sequence of nucleotides in DNA. The fact that there were bands that occurred in either the plus or minus lanes but not both was a problem. It was probably due to the formation of based-paired loops of DNA that resulted from incomplete denaturation of the DNA (Sanger and Coulson, 1978). Sanger and Coulson (1975) did not regard the plus and minus method as a completely reliable method and believed that it was necessary to confirm the sequence of deoxyribonucleotides with the sequence of ribonucleotides in the transcript or the sequence of amino acids in the protein coded by the DNA.

Sanger et al. (1977a) figured out another trick to speed up the sequencing process. Instead of chemically synthesizing primers, they made primers by digesting DNA with *restriction enzymes*, which were new components of nature's toolbox. Restriction enzymes were originally discovered to be part of a defense mechanism that restricts bacteriophage from attacking the bacteria (Luria and Human, 1952; Bertani and Weigle, 1953; Hartwell et al., 2015). In certain combinations of bacteriophage and bacteria, in which the growth of the bacteriophage is poor or restricted, the DNA of the bacteriophage is digested within the bacterium. Arber (1965) proposed that the restricted growth may be the result of yet-to-be discovered restriction enzymes that recognize specific base sequences in the invading bacteriophage DNA. Indeed, restriction endonucleases that recognized specific sequences of nucleotides 4−6 base pairs were discovered in bacteria (Linn and Arber, 1968; Meselson and Yuan, 1968; Kelly and Smith, 1970; Smith and Wilcox, 1970; Hedgpeth et al., 1972; Arber, 1978; Nathans, 1978; Smith, 1978). The bacteria also produce enzymes that methylate their own DNA so that the sequences that normally would be susceptible to digestion by the restriction endonucleases are protected. Sometimes the methylating enzymes methylate the viral DNA, which allow the viruses to replicate and kill the bacteria.

Richard Roberts (1978) realized that restriction endonucleases could be useful for sequencing DNA and began systematically organizing the restriction endonucleases that had been discovered by sampling the bacteria from the American Type Culture Collection. Many different restriction enzymes were discovered as a consequence of being able to readily observe and differentiate the restriction fragments produced by various enzymes after gel electrophoresis and ethidium bromide staining of DNA (Sharp et al., 1973).

The sequence of nucleotides recognized by a restriction enzyme was originally determined by Kelley and Smith (1970) using a sequential degradation method. Kelley and Smith labeled the 5′ termini at the cleavage site of the restriction fragments with ^{32}P−dATP using T4 polynucleotide kinase. Then the labeled restriction fragments were digested with pancreatic DNase for various amounts of time to produce double-stranded DNA with one, two, three, or four base pairs with the same labeled 5′ end. The fragments were then separated by cellulose thin-layer electrophoresis. Each fraction was eluted from the thin-layer plate and digested with snake venom phosphodiesterase. Snake venom phosphodiesterase differs from spleen phosphodiesterase in that it is an exonuclease that digests single-stranded DNA from the 3′ end to yield dNMPs that retain the phosphate bound to the 5′ carbon of deoxyribose. The resultant mononucleotides produced by snake venom phosphodiesterase were separated by cellulose thin-layer electrophoresis, and the nucleotide composition of each fraction was determined by comparison with known standards. The sequence of the nucleotides that made up the cleavage site at the 5′ termini was deduced by comparing the composition of the nucleotides, dinucleotides, trinucleotides and tetranucleotides that made up the 5′ termini at the cleavage site and seeing which nucleotide was lost in the successively smaller fragments (Kelley and Smith, 1970; Fig. 21.5).

The lengths of the fragments produced by restriction enzymes depend in general on the number of base pairs that are recognized by the restriction enzyme. Assume that the base pairs are equally represented and randomly arranged in the DNA, then if the restriction enzyme recognizes four base pairs (e.g., TaqI or RsaI), the average length of the fragment will be about $4^4 = 256$ base pairs; if the restriction enzyme recognizes six base pairs (e.g., EcoRI, BamHI, or HindIII), the average length of the fragment will be about $4^6 = 4096$ base pairs; if the restriction enzyme recognizes eight base pairs (e.g., NotI), the average length of the fragment will be about $4^8 = 65{,}536$ base pairs; the reciprocal of these lengths being equal to the probability that a restriction site of the given size will be present in the DNA. The number of unique fragments produced will be approximately equal to the ratio of the number of base pairs in the whole genome to the number of base pairs in the average fragment.

By digesting the ΦX174 DNA with various restriction enzymes to make overlapping templates and by using primers that recognize the ends of the fragments where the restriction sites are, Sanger et al. (1977a) used the plus and minus method to sequence the entire 5375 nucleotide long genome of the bacteriophage ΦX174. They noted that "*the plus and minus technique used by itself cannot be regarded as a completely reliable system and occasional errors may occur.*" The number of errors increased as the number of nucleotides exceeded 50, where it became more and more difficult to resolve consecutive sequences involving the same nucleotide. At the time, it seemed that such errors could only be eliminated by "*more laborious experiments.*" Sanger et al. (1977b) developed the dideoxy method, which was a faster and more reliable way of sequencing DNA.

Dideoxy-ribonucleoside triphosphates lack the 3′-OH group. When a dideoxy-ribonucleosidetriphosphate without the 3′-OH residue is incorporated into a strand, the polymerization is terminated because without the terminal 3′-OH group, a 5′-deoxyribonucleotide cannot be added (Atkinson et al., 1969).

The Sanger (1980) technique used a single-stranded DNA as a template. The stand is incubated with a DNA primer made from a denatured restriction enzyme fragment to form a segment of double-stranded DNA that acts as an initiation site for the DNA polymerase. In the presence of all of the deoxyribonucleotides and one radioactive dideoxyribonucleotide added at a concentration equal to about 1/100 of the concentrations of deoxyribonucleotides, the DNA polymerase initiates DNA synthesis at the primer and synthesis

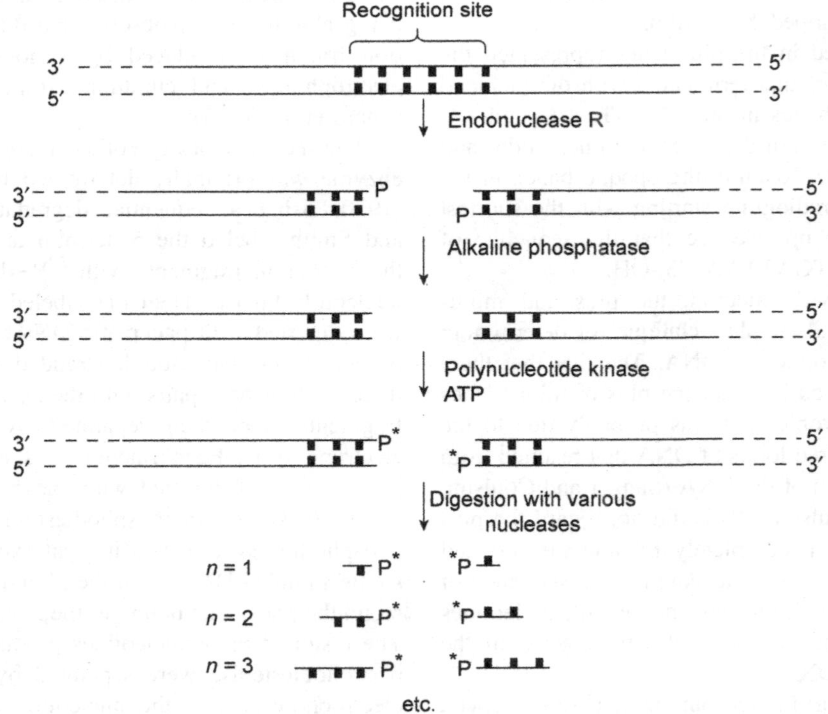

FIGURE 21.5 The method to analyze the deoxyribonucleotide sequence at the 5′ ends on a restriction fragment. *From Kelley, T.J., Smith, H.O., 1970. A restriction enzyme from* Hemophilus influenzae. *II. Base sequence of the recognition site. J. Mol. Biol. 51, 393–409.*

continues until the dideoxy-ribonucleosidetriphosphate is added instead of the related deoxyribonucleotide. Four preparations are made, one with each of the radioactive dideoxy-ribonucleotides. The newly synthesized DNA all have the same 5′ ends but different lengths depending on when the dideoxy-ribonucleotide was added instead of the deoxyribonucleotide (Fig. 21.6A).

Each of the preparations are placed side by side in wells in the electrophoresis unit so that each of four lanes in the acrylamide gel contain the newly synthesized DNA fragments that are terminated by only one of radioactive dideoxy-ribonucleotides. The newly synthesized DNA fragments are run through an acrylamide gel that separates the fragments by size. The small fragments migrate faster than the large fragments. When the fragments have separated, the gel is subject to autoradiography, and the radioactive dideoxy-ribonucleotides expose the film. In each lane, a sequence of opaque bands is resolved, where each band in the lane is terminated by the same dideoxy-ribonucleotide. The dideoxy-ribonucleotide technique gives beautiful autoradiograms with sharp bands of equal intensity over a long sequence (Sanger, 1988). The sequence of the DNA is read directly from opaque bands in the autoradiogram (Fig. 21.6B). Starting from the smallest fragment, you read up. Further sequencing work using the dideoxy method revealed that there were overlapping genes in the same DNA fragment, and a given deoxyribonucleotide sequence could be transcribed in more than one reading frame to produce different proteins as a result of alternative splicing (Barrell et al., 1976; Smith et al., 1977). That is, the theory of the one-to-one relationship between a gene and a protein (Beadle and Tatum, 1941a; Horowitz, 1995, 1996) was no longer a valid theory. The genetic definition of a gene as a single locus on the chromosome that regulates a single trait also had to be modified. A single locus could control many traits.

Maxam and Gilbert (1977) also developed a procedure for sequencing DNA using polyacrylamide gel electrophoresis. The Maxam–Gilbert technique was not a synthetic technique based on natural workings of the cell but a degradative technique based on the traditions of chemical analysis. With this technique, one could sequence at a rate of about 5000 bases per year (Gilbert, 1992). As speed became a virtue, Wu's contribution in developing the primer extension method to sequence DNA became forgotten (Onaga, 2014; Xue et al., 2016; Heather and Chain, 2017), and Walter Gilbert (1980) and Fred Sanger (1980; García-Sancho, 2010; Jeffers, 2017) shared the Nobel Prize for their first-generation DNA sequencing technologies.

DNA sequencing depended on purifying large quantities of DNA. Because DNA is so long and thin, it breaks when exposed to the shearing forces needed to isolate DNA. Thus, DNA isolated from 10,000 cells will have 10,000 copies of the same gene, but they will be on DNA fragments of different lengths. Restriction enzymes made it possible to cut DNA into reproducible fragments so that the gene of interest would be enriched in a fragment of known

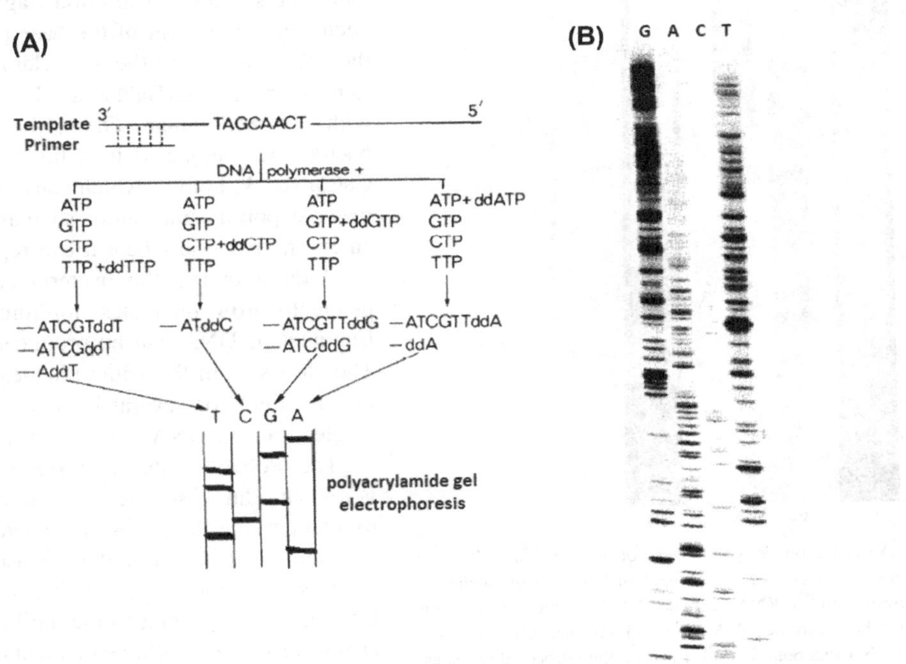

FIGURE 21.6 The principle of the dideoxy method (A) and an autoradiograph of a DNA sequencing gel (B). *Modified from Sanger, F., December 8, 1980. Determination of Nucleotide Sequences in DNA. Nobel Lect.*

length. Edwin Southern (1975), whose "*desk would be a pile of glass and plastic and glue and wires and batteries and agarose and he would sort of not be visible for long periods of time and then suddenly he would pop his head up with some invention*," (Harding, 2005) was a problem solver who discovered a new and simple way to identify which fragment of DNA contained a given gene.

Southern developed an apparatus that would transfer the DNA fragments from an agarose gel to nitrocellulose paper. The nitrocellulose paper was sandwiched between the hydrated agarose gel and blotting paper so that the DNA would move from the agarose gel to the nitrocellulose paper along the water potential gradient. The pores in the nitrocellulose paper were large enough for water to pass but too small for the DNA to pass, so the DNA stuck to the nitrocellulose paper. The DNA on the nitrocellulose paper was then allowed to hybridize with a manually synthesized radioactive oligonucleotide of known sequence. After hybridization, the nitrocellulose paper was placed on X-ray film. The exposed autoradiogram allowed one to identify the fragment of DNA that contained the DNA sequence or gene of interest (Southern, 1975; Fig. 21.7). Many of the genes involved in plant development, including *Deficiens*, *AGAMOUS*, *SCARECROW*, and *SHORT-ROOT*, were originally identified as restriction fragments whose mobility

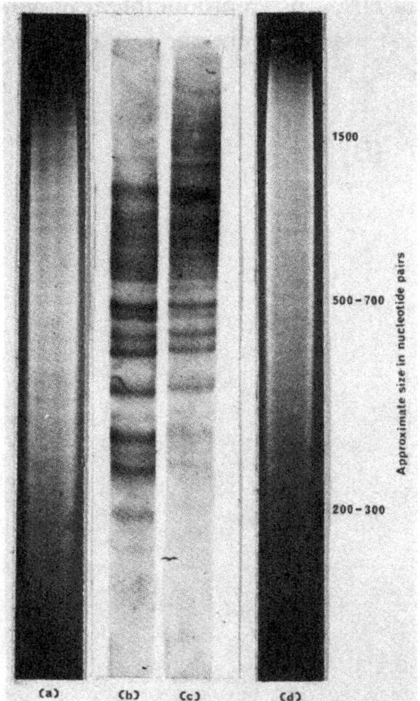

FIGURE 21.7 A restriction enzyme digest of Escherichia *coli* DNA analyzed by agarose gel electrophoresis and transferred to nitrocellulose paper and hybridized with ^{32}P-RNA. (A) and (D) show ethidium bromide fluorescence after it bound to the DNA in the gel; (B) and (C) show autoradiograms of the DNA that bound to ^{32}P-RNA on the nitrocellulose paper. *From Southern, E.M. 1975. Detection of specific sequences among DNA fragments separated by gel electrophoresis. J. Mol. Biol. 98, 503–517.*

on Southern blots differed between the wild-type and the insertion mutant. The DNA fragment from the wild type was then sequenced using the Maxam–Gilbert chemical or Sanger dideoxyribonucleotide enzymatic method (Sommer et al., 1990; Yanofsky et al., 1990; Di Laurenzio et al., 1996; Helariutta et al., 2000).

The fact that one needed to start with sufficient quantities of purified DNA to fractionate it (Work and Burdon, 1983), placed a limitation on the types of organisms that one could analyze, making it nearly impossible to study the nuclear DNA of eukaryotic organisms, or even the DNA in prokaryotes or eukaryotic organelles. If one could insert restriction fragments in a vector using recombinant DNA techniques, transform the bacteria, let them divide, and purify the recombinant DNA from them, large genomes could be sequenced by putting together the overlapping deoxyribonucleotide sequences from different clones.

Gronenborn and Messing (1978) discovered a method to purify DNA by amplifying or *cloning* it rather than *fractionating* it from the organism that contained the DNA of interest. They generated a bacteriophage (M13mp2) that is unique in that the DNA while in the phage particle is single stranded. It becomes double stranded in the replicative form within the bacterium. The replicative form has a restriction enzyme site in the middle of the β-galactosidase gene. The functioning β-galactosidase produces a blue plaque on agar-containing substrate for the β-galactosidase. If a fragment of double-stranded DNA is inserted into same restriction enzyme site, the β-galactosidase gene is no longer functional and the plaque is white. The base pairing that occurs when a restriction fragment is inserted into the vector is reminiscent of the base pairing that occurs when the cohesive ends of the linear lambda bacteriophage DNA join to form a ring (Bahl et al., 1976). To transform bacteria with the recombinant DNA, the recombinant DNA and bacteria are subjected to a high-voltage shock or a cold $CaCl_2$ shock, both of which cause the plasma membrane to become porous transiently so that the recombinant DNA enters the bacteria where it can replicate.

After allowing the bacteria containing the M13mp2 phage to grow on plates, millions of copies of a single fragment of DNA can be produced through reproduction. The clones from the white plaques are then grown in culture medium for several hours, and the phage containing single-stranded DNA is secreted into the medium.

The probability that even one vector enters a bacterium is low and thus a stringent selective procedure is necessary to capture the 0.1% of the bacterial cells that contain recombinant DNA. For these reason a selectable marker, such as antibiotic resistance, is also cloned into the vector. Cells that are resistant to the antibiotic form colonies consisting of tens of millions of identical cells or clones, each containing hundreds of molecules of the same recombinant DNA (Cohen et al., 1973).

Sanger et al. (1980) isolated the cloned copies of single-stranded recombinant DNA and sequenced them using the four dideoxy-ribonucleoside triphosphates and a primer that matched the template just upstream from the inserted DNA. Because the primer is complementary to a part of the bacteriophage DNA, the same synthetic primer can be conveniently used to prime the reaction to sequence any random piece of inserted DNA. Large fragments of DNA are typically lost from the clones so that most of the clones only contain small fragments of DNA.

Clones that contain the recombinant DNA are organized into libraries, where each clone contains a different fragment of DNA. The fragments should cover the entire genome and overlap with each other. The size of the recombinant DNA fragment in each clone depends on the vector used. Plasmids can hold 4 kb of DNA, bacteriophage lambda (Cosmids) can contain 40 kb of DNA, bacterial or yeast artificial chromosomes (BAC, YAC) can contain 100–200 kb of DNA, P1-derived artificial chromosomes (PAC) or bacterial genomes can contain 1–2 Mb of DNA. The genomic libraries of *Arabidopsis* that are readily available for purchase contain approximately 10,000 clones with an average insert of 80 kb or 1000 clones with an average insert of 420 kb (https://www.arabidopsis.org/abrc/catalog/genomic_library_1.html).

Sanger et al. (1982) used bacteriophage as a vector. Sanger originally believed that developing methods to sequence-specific regions of the genome using restriction fragments would be the best way to determine the complete sequence; however, experience showed that this *directed*, *hierarchical*, or *divide and conquer* approach did not yield as many useful clones as the random shotgun approach (Hunkapiller et al., 1991). The random shotgun approach involved breaking apart a large DNA fragment into small fragments using nucleases (Anderson, 1981) or sonication (Fuhrman et al., 1981; Deininger, 1983), fractionating the fragments using electrophoresis, chromatography, or centrifugation through a sucrose gradient, cloning the fragments of DNA with approximately 200–300 nucleotides into a single-stranded bacteriophage vector, and sequencing randomly chosen clones. The random shotgun approach also had the advantage of producing fragments with overlapping sequences that would be helpful in reconstructing the sequence of the DNA molecule in question. The sequence of a fragment is now known as a *read*.

Rodger Staden (1983) developed algorithms to reconstruct the sequence of a gene by analyzing the sequences attained from different clones that had overlapping sequences. The algorithms then assembled the reads into one continuous sequence known as a *contig* (Staden, 1980) using a computer program known as SEQFIT (Staden, 1977). The quality of the sequence was assessed at the level of the individual nucleotide as read from each autoradiogram. It was also assessed by the number of times each position was given a particular assignment from different gels. Then, according to Staden (1979), "*if the 5′ end of the sequence from one gel reading is the same as the 3′ end of the sequence from another the data is said to overlap. If the overlap is of sufficient length to distinguish it from being a repeat in the sequence the two sequences must be contiguous. The data from the two gel readings can then be joined to form one longer continuous sequence.*"

Using the shotgun approach, the dideoxy method, and the computer program, Sanger sequenced a 187 kilobase (kb) human mitochondrial DNA (Sanger et al., 1980) and a 48.5 kb bacteriophage lambda DNA (Sanger et al., 1982). They sequenced at a rate of more than 1000 nucleotides per day—the limiting factor being the computer analysis. Enough sequence data had accumulated by 1982 (Bilofsky et al., 1986) so that GenBank (https://www.ncbi.nlm.nih.gov/sra) was created to provide a "*timely, centralized, accessible repository for genetic sequences.*"

There is a fundamental barrier to the read length achieved using electrophoresis (Chan, 2005). Because a strand that is $n + 1$ nucleotides long migrates slower than a strand that is n nucleotides long, the DNA fragments fractionate by size during polyacrylamide gel electrophoresis. However, as n gets large, the ratio between $n + 1$ and n asymptotically approaches 1 and the gel becomes unable to resolve large fragments of DNA. This asymptotic effect due to large numbers is an important touchstone to remember when scaling up physicochemical procedures.

DNA fragments can be separated by size using acrylamide or agarose gel electrophoresis. Acrylamide is better at separating smaller fragments, and agarose is better at separating larger fragments. During the electrophoresis run, the DNA migrates through the gel towards the positive pole as a result of the negatively charged phosphate groups in the backbone. The larger the DNA fragment, the greater the friction it experiences in moving through the pores in the gel. Consequently, the mobilities of the fragments in the gel are inversely related to the sizes of the fragments. At the end of the run, the DNA can be visualized in the gel with a fluorescent dye, such as ethidium bromide.

Sanger (1988) shared his thoughts on the dideoxy method in a review article entitled, Sequences, sequences, and sequences: "*I suppose the dideoxy method could be regarded as the climax of my research career and the fulfilment of an ambition that had gradually been forming as I became more and more involved in sequencing. It is of course very exciting and gratifying to read of the method now being used in many laboratories and of vast regions of the genomes of both simple and higher organisms becoming exposed in the form of sequences, and these are helping in the understanding of some of the fundamental problems of life. But I think I have derived even more pleasure from the development of the work—seeing the method gradually improving until we were able to read a sequence straight from an autoradiograph. Before this reading became automated, people complained that it was a tedious process, but to me it was always a delight, having in the back of my mind the way we used to do sequences one residue at a time by painstaking partial hydrolysis, fractionation, and analysis. At one stage in the work I would take the autoradiographs home with me and look forward to the pleasure of reading them in the peace of the evening.*"

According to Erika Check Hayden (2014), "*for all of Sanger sequencing's high cost, it still remains the benchmark for accuracy. And sequencing costs are no longer dropping as quickly as they were a few years ago.*" Sanger's appreciation of hard work reminds me of John F. Kennedy's Address at Rice University on September 12, 1962: "*we choose to go to the moon in this decade and do the other things, not because they are easy, but because they are hard.*"

The first 5 years of the human genome project, which has been compared with the Apollo program by the National Institutes of Health, focused on mapping the genome genetically and physically. The genetic map represented the loci of genes along the chromosomes based on the analysis of recombinants in genetic crosses. The physical map represented the positions of overlapping fragments of DNA produced by various restriction enzymes (Roach et al., 1995). Venter et al. (1996) challenged the assumption that it was necessary to map a DNA fragment before it was sequenced and suggested that they just sequence the genome and figure out how to assemble it afterward. J. Craig Venter developed the fastest, if not the most accurate, way to sequence and assemble the genome and consequently led the field of genomics, in which speed was of the essence.

21.2.2.3 Analysis of the Incorporation of Fluorescent Dideoxy-Ribonucleotides Into a Primer-Template Using Gel Electrophoresis and Automated Detection

Because the Sanger method of DNA sequencing was "*labour-intensive, fairly expensive and involve*[s] *the use of radioisotopes*," Smith et al. (1986) developed an automated DNA sequencer that became commercially available in 1987 (Connell, 1987; Smith, 1988; Halloran et al., 1993; Burzik, 2006; Springer, 2006). In the automated sequencer, the sequence is detected by a photomultiplier tube that records the fluorescent signal given off by either fluorescent primers whose 3′-OH ends were extended by DNA polymerase and terminated by dideoxy-ribonucleotides or by nonfluorescent primers whose 3′-OH ends were extended by DNA polymerase and terminated by fluorescent dideoxy-ribonucleotides. The fluorescent primer approach with a single label required four separate reaction vessels and four electrophoretic lanes, and the four fluorescent terminators approach required only a single reaction vessel and a single electrophoretic lane. The fluorescence was recorded in real time as the bands ran past the detector. The signal from the photomultiplier tube was digitized and used as the input to a computer that had software that could analyze the sequence. The first genes sequenced with this technique included the genes for the rat cardiac β-adrenergic and muscarinic cholinergic receptors (Gocayne et al., 1987). The sequencing rate was greater than 30,000 bases per week. Other automated DNA synthesizers were also designed and built (Edwards et al., 1990).

By 1991, six automatic sequencers had been set up in a core facility at the National Institutes of Health, and the small pox virus genome, six cosmids containing human DNA, and 11,000 expressed sequence tags (ESTs, which is a segment of a sequence from a cDNA clone that corresponds to a mRNA) from humans, and *Caenorhabditis elegans* were sequenced from 1990 to 1992 (Adams et al., 1991, 1992, 1993a,b; Martin-Gallardo et al., 1992; McCombie et al., 1992a,b; Massung et al., 1993; Sikela and Auffray, 1993). J. Craig Venter had a falling out with his colleagues at NIH and created The Institute for Genomic Research (TIGR; Adams et al., 1994), a private institute that is now the J. Craig Venter Institute, a nonprofit genomics institute founded in 2006 (https://www.jcvi.org/about).

By 1992, 30 automated DNA sequencers were running simultaneously at The Institute for Genome Research. Two principles guided the design of TIGR: flexibility to incorporate improvements at each stage and the integration of data analysis with all phases of the sequencing process. According to Adams et al. (1994), "*The key to successful large-scale sequencing, regardless of the particular application, is balance. If data analysis is held until the end, systematic problems with the data will not be discovered until considerable time and money have been spent…There is no such thing as a static production line for sequencing. Every step must be continually monitored and analyzed, both for consistency of data and opportunities for improvements, by knowledgeable, trained scientists.*"

By using Sanger's dideoxy-ribonucleotide method with four fluorescent dyes—one for each dideoxy-ribonucleotide, instead of four radioactive dideoxy-ribonucleotides—the data could be obtained quicker because there was no need to spend time developing X-ray film. Moreover, the reactions with the four fluorescent dideoxy-ribonucleotides, unlike the reactions with four ^{32}P-labeled dideoxy-ribonucleotides, could take place simultaneously in a single reaction vessel before being subjected to electrophoresis. The fluorescently labeled DNA fragments could be separated in a single lane using capillary gel electrophoresis instead of four lanes using slab gel electrophoresis. The DNA fragments with a terminal fluorescent nucleotide move down the gel in temporal order from shortest to longest. As they move past the detector, the sequence of colors is captured in real time by the automated image processing system that has a different channel for each of the four colors, where each color represents a different base.

If the raw data were readily interpretable, there would be no need for image analysis. However, because the four fluorescent dyes have significant spectral overlap, the actual base that is being measured must be resolved from the four channels by multicomponent analysis. Moreover, the unequal mobility shifts of the dideoxyribonucleotides in the gel caused by the different dyes must also be taken into consideration. The error rate in determining the sequence was about 1%.

In general, improvements in automatic sequencing required the further development of four different fluorescently labeled dideoxy-ribonucleotide terminators, where each of the four bases was conjugated to a different fluorophore that, on excitation by a laser, fluoresced a unique and identifying color (Prober et al., 1987), the engineering of DNA polymerases that worked efficiently with fluorescently labeled dideoxy-ribonucleotides (Wilson et al., 1990b; Reeve and Fuller, 1995; Tabor and Richardson, 1995), the change from tube gel electrophoresis to slab gel electrophoresis to capillary electrophoresis that allowed the gels to run faster because the electric field could be increased in the capillary without causing a proportional increase in heat (Luckey et al., 1990), scaling up the number of capillaries that can be sequenced simultaneously to 96 or even 384 (Shendure and Ji, 2008), synthesizing oligonucleotide primers with an automated DNA synthesizer (Strauss et al., 1986; Horvath et al., 1987; Boysen et al., 1997) rather than manually, and increasing the degree of automation through robotics (Wilson et al., 1988, 1990a; Hunkapiller et al., 1991). The 3 GB sequence of the human genome was completed using several hundred sequencers based on capillary gel electrophoresis (International Human Genome Sequencing Consortium, 2001, 2004). According to Wheeler et al. (2008), using this technique, it cost $100,000,000 to sequence J. Craig Venter's DNA (Levy et al., 2008). Capillary gel electrophoresis was used to sequence the 125 Mb genome of *Arabidopsis* (Lin et al., 1999; Mayer et al., 1999; Pennisi, 2000; Theologis et al., 2000; Salanoubat et al., 2000; Tabata et al., 2000; The *Arabidopsis* Initiative, 2000) and 466 Mb genome of rice (Yu et al., 2002).

21.2.2.4 Using Polymerase Chain Reaction to Clone DNA Outside of Microorganisms

Soon automated oligonucleotide synthesizers replaced the nucleotide chemists who synthesized the oligonucleotides manually. Kary Mullis, who became an underemployed nucleotide chemist, had time to think and putter. He developed a new method, known as the *polymerase chain reaction* (PCR), that uses the primer extension method to amplify DNA automatically without the need of a bacterial intermediate. According to Saiki et al. (1988), "*By virtue of the exponential accumulation of literally billions of copies derived from a single progenitor sequence, PCR based on Taq DNA polymerase represents a form of* 'cell-free molecular cloning' *that can accomplish in an automated 3- to 4-h* in vitro *reaction what might otherwise take days or weeks of biological growth and biochemical purification.*" In finding a way that DNA polymerase could be utilized to make large quantities of DNA, Mullis (1990) had fulfilled the heretofore unrealized hope of Arthur Kornberg and Joshua Lederberg. Kary Mullis (1993) won the Nobel Prize for inventing PCR—a technique that not only permits the amplification of DNA but also literally, if one has the right primers, allows one to find a needle in the haystack (Varki and Altheide, 2005; Starr, 2016). PCR has been an invaluable tool to detect genetic diseases, to provide evidence of paternity (Burzik, 2006), to provide evidence of guilt and innocence (https://www.innocenceproject.org/), and to identify pathogens in food (Acuri et al., 1999), water (Jones et al., 2014), soil, people, animals, and plants.

PCR is a cyclic process involving the heating and cooling of DNA. During the first heating phase, the hydrogen bonds between complementary bases break and the double-stranded DNA melts into single-stranded DNA, which can serve as a template. This occurs in the presence of excess deoxyribonucleotides and primers. During the cooling phase, the forward primer that recognizes the 3′ end of the sense strand and the reverse primer that recognizes the 3′ end of the antisense strand of a particular sequence bind to the DNA, forming primer-templates for the DNA polymerase that will be extended in the 5′ → 3′ direction from the 3′ ends of the two primers to create a double-stranded copy of the DNA approximately 25 kb long that is situated between the two primers. If the temperature is too cool, the primers will bind to any sequence that is close to the complementary sequence, and if the temperature is too high, the primer will not bind to any sequence. The correct hybridization temperature depends in part on the nucleotide composition of the primer as cytosine–guanine pairs have three hydrogen bonds and are more stable than adenine–thymine pairs, which have only two (Fig. 21.2). The correct hybridization temperature also depends on the length of the oligonucleotide primer because the greater the length, the more stable is the binding. During the second heating phase, the DNA polymerase adds deoxyribonucleotides to the primers in the 5′ → 3′ direction that are complementary to the template. The replicated DNA then acts as a template for replication in the following cycle so that after *n* cycles the particular sequence is amplified exponentially (2^n) in a chain reaction (Saiki et al., 1985; Mullis and Faloona, 1987). By realizing that nature's toolbox contains a variety of polymerases, Mullis used *Taq* polymerase, a heat-stable DNA polymerase isolated from a thermophilic bacterium *Thermus aquaticus* (Chien et al., 1976; Saiki et al., 1988), instead of the DNA polymerase isolated from *E. coli*, in the PCR because the *Taq* polymerase survives the heating cycle, and fresh DNA polymerase does not have to be added after each heating cycle. The amplified DNA fragments produced by PCR can be directly sequenced using the Sanger dideoxy (Engelke et al., 1988) or any other method.

21.2.2.5 Massively Parallel Analysis of the Incorporation of Deoxyribonucleotides Into a Primer-Template Using Pyrosequencing

A completely different method to sequence DNA by synthesis was developed by Edward Hyman (1988; Rothberg and Leamon, 2008). This method, known as *pyrosequencing*, does not use electrophoresis, dideoxy-ribonucleotides, radioactivity, or fluorescence. The pyrosequencing method works by measuring the P_iP_i formed by the DNA polymerase as it converts a free deoxyribonucleoside triphosphate into a free P_iP_i and a dNMP bound to the 3′ end of the primer. In the original method, the template-primer and the DNA polymerase were bound to a DEAE-Sepharose column. When a solution that contained the deoxyribonucleoside triphosphate that matched the next available base in the template was added to the column, P_iP_i was released. During pyrosequencing, the released P_iP_i was converted into ATP by ATP sulfurylase. The ATP was then used as a substrate for the luminescent firefly luciferin–luciferase reaction, where the amount of light given off was proportional to the P_iP_i concentration (Nyrén and Lundin, 1985; Nyrén, 1987). The sequence of DNA being produced on the primer-template was determined by measuring the amount of light given off as each of the four dNTPs was pumped through the column.

Nyrén et al. (1993) miniaturized Hyman's (1988) method by immobilizing the template-primers on paramagnetic beads using a streptavidin–biotin reaction (Fig. 21.8). The streptavidin-coated beads were bound to biotin containing primers, which then hybridized with the template DNA. The beads containing the primer-templates were then divided among four 50 μL reaction vessels each of which contained DNA polymerase and one of the four possible single dideoxy-ribonucleotides. Each vessel contained a different dideoxy-ribonucleotide. The bead bound template-primers were then washed, and all four deoxyribonucleotides were added. Three of the reaction vessels gave off light as a result of the P_iP_i released from the added deoxyribonucleotide and one did not because that one had the dideoxy-ribonucleotide bound to it that matched the next available site on the template. In this way the first base after the primer was deduced.

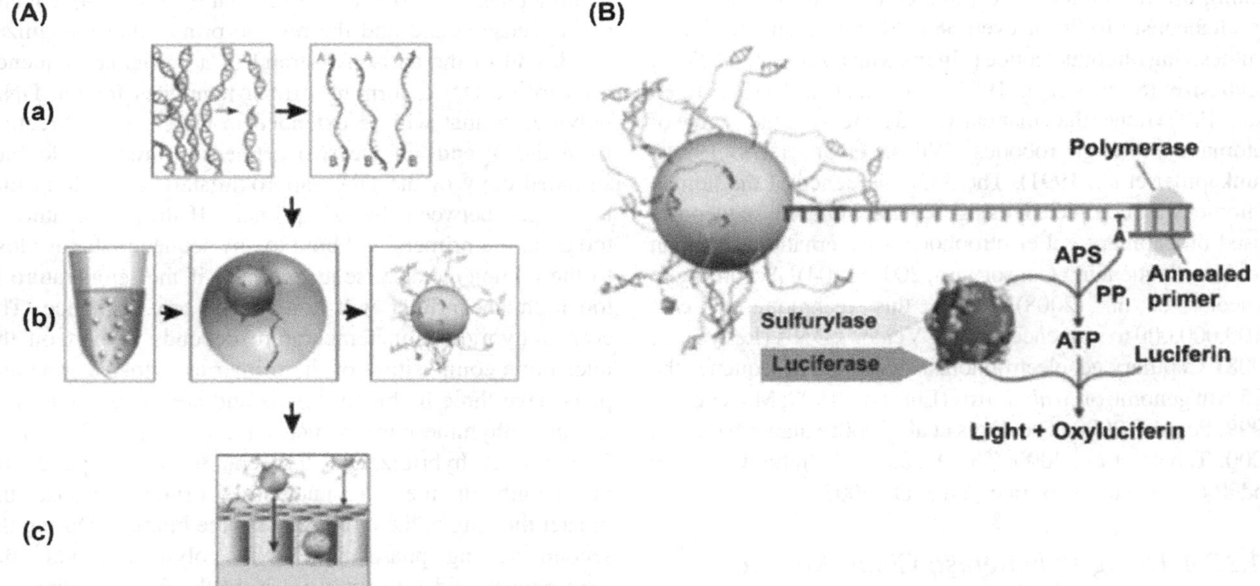

FIGURE 21.8 (A) Outline of workflow for pyrosequencing. To construct a library, (a) ligate adapters to DNA fragments with DNA ligase and then (b) couple magnetic beads to DNA in an emulsion where PCR amplifies the fragments. (c) Then load the beads into the picotiter plate and add the appropriate primer. (B) When a solution that contains the deoxyribonucleoside triphosphate that matches the next available base in the template DNA is added to the flow cell, pyrophosphate is released. The released pyrophosphate is converted into ATP by ATP sulfurylase. The ATP is then used as a substrate for the luciferin–luciferase reaction, where the amount of light given off is proportional to the pyrophosphate concentration. The sequence of DNA being produced on the primer-template is determined by measuring the amount of light given off as each of the four deoxyribonucleoside triphosphates is pumped through the flow cell. *From Ansorge, W.J., 2009. Next-generation DNA sequencing techniques. New Biotechnol. 25, 195–203.*

In a single reaction vessel, Ronaghi et al. (1996) subjected 291 deoxyribonucleotide long template-primers immobilized on the streptavidin-coated beads to each deoxyribonucleotide and observed the luminescence. If they added a deoxyribonucleotide that did not match the next available base on the template, there was no signal. If they added a deoxyribonucleotide that matched, there was a signal. If the deoxyribonucleotide matched the next two bases, the signal would be doubled. By performing a wash step after every luminescence reading, they could sequence the entire template. Ronaghi et al. (1998) added apyrase, a nucleotide-degrading enzyme, to the wash to decrease the wash time and increase the signal-to-noise ratio.

The pyrosequencing technique made it possible to perform massively parallel sequencing reactions in picoliter volumes. In so doing, the sequencing speed increased and the cost decreased dramatically (Margulies et al., 2005). DNA isolated from a Neanderthal and a woolly mammoth were sequenced by pyrosequencing (Noonan et al., 2006; Poinar et al., 2006). James D. Watson's DNA was sequenced on a 454 Inc. Genome Sequencer FLX instrument using pyrosequencing (http://sequencing.roche.com/en.html) for less than $1,000,000 (Wheeler et al., 2008). Plant genomes have also been sequenced with this technology (Brenchley et al., 2012; Garcia-Mas et al., 2012; Hernandez et al., 2012). The downside of this automated pyrosequencing technique is that the templates cannot be longer than 100–700 bases, depending on the system, or the accuracy decreases (Siqueira et al., 2012). Thus, the genome has to be assembled from the short reads using bioinformatics programs.

When performing the automated cyclic array pyrosequencing technique, DNA from an entire genome is mechanically sheared and separated by electrophoresis into fragments approximately 100–700 bases long. Single molecules of DNA, which will serve as templates, are obtained by limiting dilution. Common oligonucleotide adapters are ligated to the DNA templates and primers complementary to the common oligonucleotide adapter are attached to each bead. Due to hybridization at limited dilution, each bead contains a single DNA template. The other PCR primer is in the reaction mixture so that the individual templates are clonally amplified by the PCR reaction inside a reaction mixture-in-oil emulsion (see Chapter 9). After 10 million copies of each DNA template are made, the droplets are broken, the DNA strands are denatured, and the beads containing single-stranded DNA clones are placed in 1.6 million wells. Small beads carrying immobilized enzymes necessary for P_iP_i sequencing are added to each well. Solutions that contain one of the dNTPs at a time or apyrase in the wash flow over the wells and the production of light from any well is captured by a pixel in the charge-coupled device (CCD) sensor mounted directly to the wells. Each cycle includes an enzyme-driven biochemistry step and an image-based data acquisition step (Shendure and Ji, 2008; Voelkerding et al., 2009; Shendure et al., 2017). The sequence of the DNA template in each well is deduced from the pattern of luminescence caused by the enzymatic reaction with the flowing deoxyribonucleoside triphosphates.

21.2.2.6 Massively Parallel Sequencing by Hybridization

Shendure et al. (2005) developed a technique to sequence DNA that shares similarities with pyrosequencing in that the DNA fragments are ligated to common oligonucleotide adapters, attached to paramagnetic beads by the 5′ end of the adapter-template, and amplified by PCR in a reaction mixture-in-oil emulsion (Fig. 21.9). Approximately 100 million beads randomly attach to the flow cell to give 100 million reads in a cyclic process involving a reaction and imaging. Unlike pyrosequencing, it is not a sequencing by synthesis method using DNA polymerase but a sequencing by hybridization method using DNA ligase. Sequencing by hybridization is based on the specificity of DNA ligases to ligate fluorescent oligonucleotides to the template DNA in a sequence-specific manner. The method was commercialized by Applied Biosystems and was known as SOLiD (Supported Oligotide Ligation and Detection; Shendure et al., 2017).

The beads containing clonally amplified DNA molecules are attached to the glass surface of a flow cell. Then oligonucleotide primers that are complementary to the common adapter are added to the flow cell and the primers hybridize with the adapter at the adapter-template junction. Because the 5′ of the adapter-template DNA is attached to the bead, the 3′ OH end of the primer is adjacent to the bead and the 5′-phosphate is available to be ligated to an oligonucleotide probe. There are 16 probes, where each one represents a unique two-base sequence. Each probe contains two probe specific bases at the 3′ end, trailed by six degenerate bases and one of four fluorophores bound to the 5′ end. In the presence of DNA ligase, the probe that hybridizes with the first two bases of the template DNA is ligated to the primer. Then the unbound probes are washed away and the color of the fluorescence of the probe that hybridized is captured. Then the degenerate sequence and the fluorophore are cleaved from the probe and the 5′ phosphate is regenerated. Subsequently, the 5′ P_i is challenged with the set of oligonucleotide probes and the cycle continues. After seven hybridization cycles are finished, the nascent strand is removed and a new primer ($n - 1$) that is one base farther from the template is hybridized to the adapter until another round consisting of seven cycles is finished. After completing five rounds using five different primers

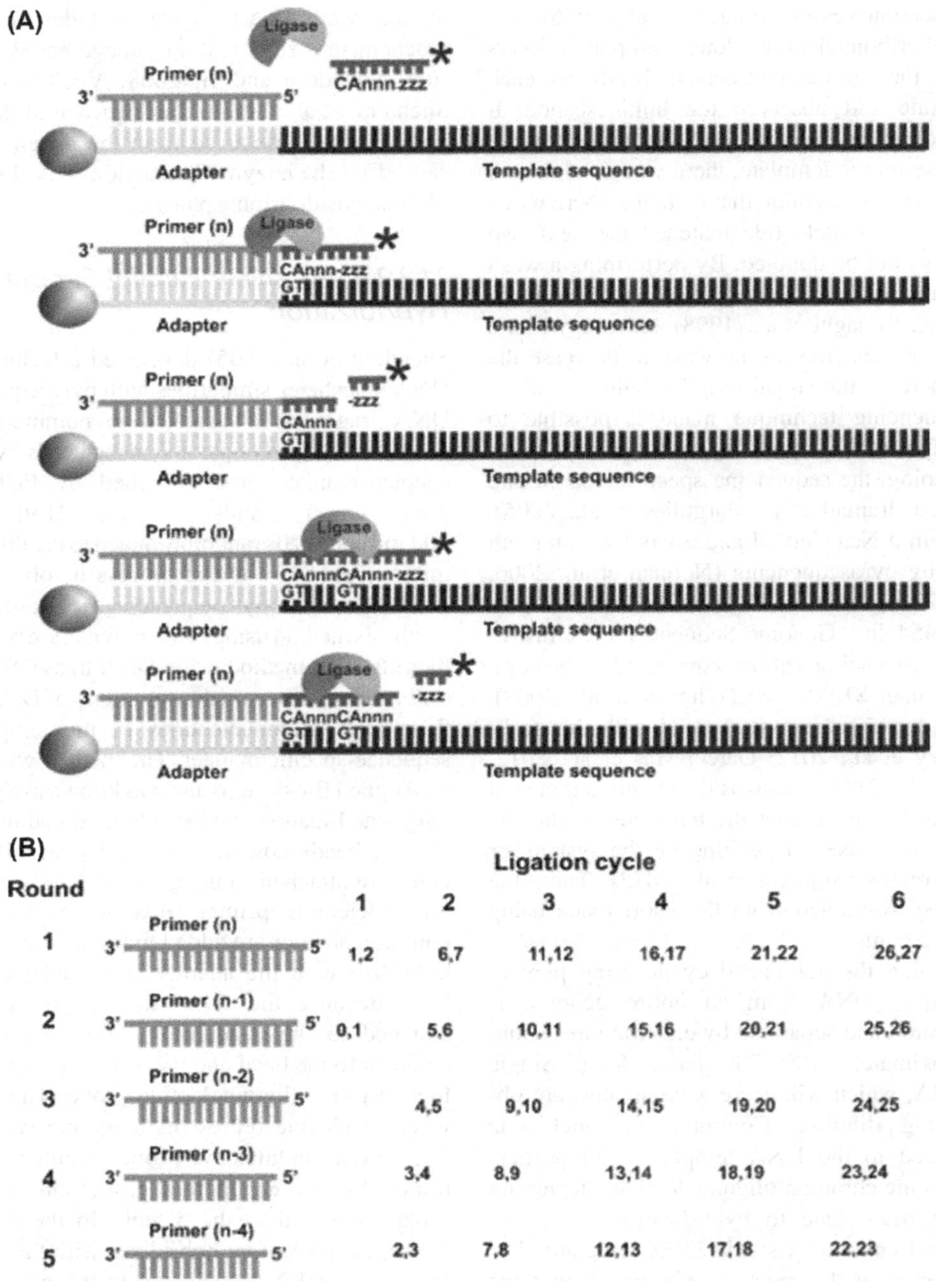

FIGURE 21.9 Sequence by ligation. (A) To construct a library, ligate adapters to DNA fragments using DNA ligase. Then add primers that hybridize to the adapter. A set of four fluorescent oligonucleotide probes compete for ligation to the primer. Following ligation of the matching oligonucleotide probe, the fluorescent label emits light that identifies which oligonucleotide was ligated, and then the degenerate deoxyribonucletides of the probe are enzymatically removed. (B) The sequence determination is performed in multiple cycles using primers where each one is shorter than the previous one by a single base. *From Ansorge, W.J., 2009. Next-generation DNA sequencing techniques. New Biotechnol. 25, 195–203.*

(n, $n-1$, $n-2$, $n-3$, $n-4$), the sequence is completed. Because each base is tested sequentially with two probes, once when the base in question is the base closer to the 5′ end of the template DNA and once when it is the base in question is the base closer to the 3′ end of the template DNA, the dinucleotide sequence is called by the light emitted by two fluorophores.

21.2.2.7 Massively Parallel Analysis of the Incorporation of Reversible Fluorescent Dideoxy-Ribonucleotides Into a Primer-Template

Bentley et al. (2008; Balasubramanian, 2013) and Turcatti et al. (2008) discovered a way to speed up the sequencing process and reduce the cost of sequencing the entire human genome to \$250,000 by using terminator deoxyribonucleotides that were both fluorescent and reversible (Ruparel et al., 2005; Seo et al., 2005; Ansorge, 2009). Using reversible terminators, the Illumina HiSeqX System can sequence more than 45 human genomes per day at a cost of approximately \$1000/genome. According to Shendure et al. (2017), the Illumina platform "*quickly became, and remains today, the most widely used sequencing instrument.*" It is often used to sequence plant genomes (The International Brachypodium Initiative, 2010; Argout et al., 2011; Dassanayake et al., 2011; Potato Genome Sequencing Consortium, 2011; The International Barley Genome Sequencing Consortium, 2012; The Tomato Genome Consortium, 2012; Wang et al., 2012; Birol et al., 2013; Chen et al., 2013; Bolger et al., 2014; Dohm et al., 2014; Kim et al., 2014; Zimin et al., 2014; Türktas et al., 2015; Bertioli et al., 2016; Clouse et al., 2016; Bauer et al., 2017). Currently, DNA sequencers available from Illumina can analyze 6000 Tb (6×10^{15} bases) in a single run.

In the Illumina massively parallel, high-throughput method of sequence by synthesis, a complete set of four reversible terminator deoxyribonucleotides flows over an immobilized primer-template in the presence of genetically engineered DNA polymerase (https://www.illumina.com/documents/products/techspotlights/techspotlight_sequencing.pdf). The terminator that matches the template is added to the primer, and a camera captures the color of the fluorescence in four sequential images, which indicates which deoxyribonucleotide was added. Red indicates dCTP was added, green indicates dTTP was added, yellow indicates that dATP was added, and blue indicates that dGTP was added. Following the capture of the four images, another solution flows over the primer-template that removes the fluorophore and the terminator and restores the OH group to the 3′ carbon, so another reversible terminator deoxyribonucleotide can bind to the nascent DNA strand. This process takes place in a cyclic manner until the complete sequence of the DNA template is obtained. The sequencing process has been sped up using the same four fluorophores but recording only two images using only two colors, red and green, where red represents the addition of dCTP, green represents the addition of dTTP, yellow, which is a mixture of red and green, represents the addition of dATP, and darkness (neither red nor green) represents the addition of dGTP. This process is known as two-channel sequencing.

The amount of light emitted by each base is not great enough to be captured by a CCD camera, and thus the DNA template has to be amplified. Unlike the massively parallel pyrosequencing technique, where the template DNA is immobilized on beads, in Illumina's massively parallel sequencing technique, the DNA template is immobilized on the surface of one or two flow cells, each with eight lanes (Fig. 21.10; Adessi et al., 2000; Fedurco et al., 2006). This is done by ligating adapters to each end of the fragmented template DNA. The adapters are oligonucleotides composed of an identifier region or index and a sequence that will recognize a forward or reverse primer. To speed up the process of making a shotgun library to about 15 minutes, the fragmentation and ligation of adapters can be done in a single step called tagmentation (Tatsumi et al., 2015).

Under limiting dilution, the resulting DNA is then hybridized to forward and reverse PCR primers that are covalently bound to the substrate by their 5′ ends and are, depending on the orientation, complementary to one adapter or the other. The adapters hybridize with their respective primer.

The single-stranded DNA is then amplified by an isothermal amplification process, in which the primers are extended in a template-dependent manner from their 3′ end with DNA polymerase and the four reversible, fluorescent deoxyribonucleotides. Then the DNA is denatured into single-stranded DNA so that the original DNA fragments, which hybridized to the bound primers and were never covalently bound to the flow cell, are washed away. While the 5′ end of the nascent strands are covalently bound to the substrate, the 3′ end of the nascent DNA strands are able to bend over and hybridize with the primer complementary to the adapter at the 5′ end. These strands serve as templates for the new nascent strands that are extended from the primers that are covalently bound to the flow cell. After 10 cycles or so, the solid-phase PCR amplification creates approximately 1000 cloned copies of each template DNA in a cluster. This is known as *bridge* or *cluster amplification*. There are approximately 10 million clusters in each flow cell, each one representing a fragment of DNA.

The reverse strands that are covalently bound to the flow cell are cleaved and washed away leaving only the forward strands for sequencing. The 3′−OH ends of the reverse primers are blocked, and the forward primer is added to the reaction mixture along with DNA polymerase and the four fluorescent reversible terminators, and the sequence along with the index are read as described above. The color of each location captured by the CCD camera records the base added to the DNA template that makes up the cluster in each location. The number of cycles determines the length of the read of the forward strand. The accuracy of sequencing a strand of template DNA decreases with the length of the template (Dohm et al., 2008; Erlich et al., 2008; Quail et al., 2008; Shendure and Ji, 2008). However, it is possible to increase one's confidence in the sequence by sequencing the forward and reverse strands of double-stranded template DNA using the same library preparation (Korbel et al., 2007; Campbell et al., 2008). To sequence the reverse stand, the 3′−OH ends of the reverse

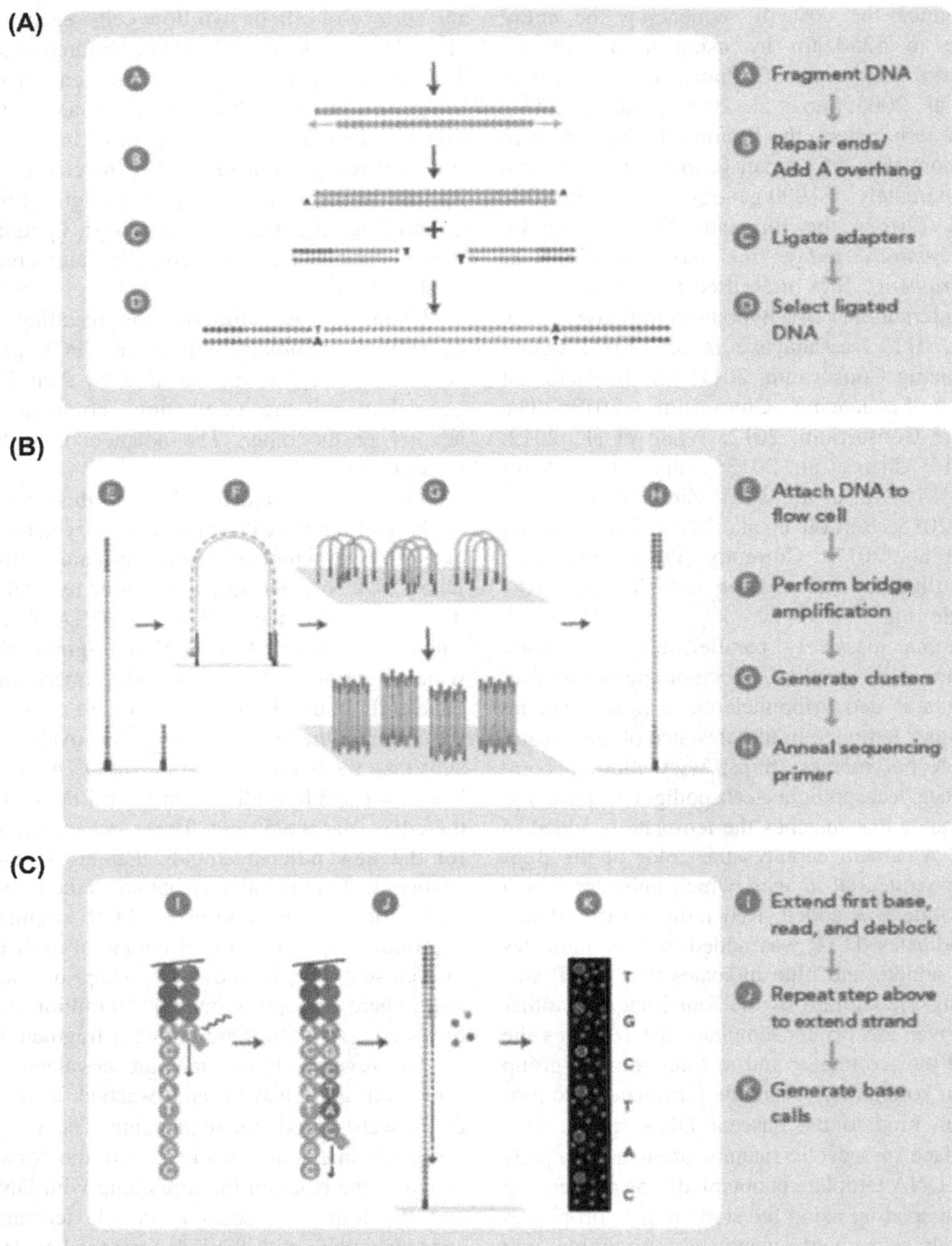

FIGURE 21.10 Outline of the Illumina sequence by synthesis workflow. To construct a library in a single step (A) the DNA is fragmented and adapters are ligated to the fragments with DNA ligase. (B) The adapters hybridize with the primers that are covalently bound to the surface of the flow cell. (C) the base sequences are read. The DNA fragments form bridge molecules that are amplified by an isothermal amplification process to produce a cluster of identical fragments. The fragments are subsequently denatured, and the sequencing primer is added. The amplified DNA fragments are sequenced by synthesis using reversible fluorescent dideoxy-ribonucleotides. The identity of the deoxyribonucleotide incorporated is given by the color emitted on incorporation. *From Ansorge, W.J., 2009. Next-generation DNA sequencing techniques. New Biotechnol. 25, 195–203.*

primer are unblocked and the nascent forward strands serve as templates for the reverse nascent strands. The reverse nascent strands are covalently bound to the flow cell. The isothermal amplification process is repeated and then the forward strands are cleaved and washed away leaving only the reverse strands. The 3′–OH ends of the forward primers are blocked and the reverse primer is added along with the DNA polymerase and the four fluorescent terminators. The sequence along with the index is read as described above. The number of cycles determines the length of the read.

By using forward and reverse primers and sequencing both ends of the DNA fragments in an indexed fashion, the forward and reverse reads of the same fragment of double-stranded DNA can be aligned as read pairs to give more accurate read alignment and identification of insertions and deletions (indels) (Kircher et al., 2012).

The error rates of raw sequence data produced by massively parallel sequencing platforms are greater than those obtained with Sanger sequencing. However, the overall error rate of massively parallel sequencing platforms is less than the error rate of Sanger sequencing when the sequence is oversampled approximately 40-fold (Tucker et al., 2009). The amount of oversampling is known as the *sequencing depth*. The reads are analyzed bioinformatically to produce a consensus sequence of the gene and its location in the genome.

The computational tools necessary to assemble the sequence of each chromosome in a genome were originally developed at The Institute for Genomic Research and Celera Genomics (Fleischmann et al., 1995; Myers et al., 2000). Assembly of reads into contigs requires overlapping sequences (Tiwary, 2015). As the lengths of the reads decrease, the more discerning algorithms must be to accurately assemble the reads into the DNA molecules that contain the hereditary information for an individual, a population, or a species. If the software is robust, the deoxyribonucleotide-by-deoxyribonucleotide view of the genomes of individuals possessing various traits provides a clear view of all genomic alterations, including single nucleotide polymorphisms (SNPs), indel, copy number variants, and structural variations that give rise to various phenotypes.

It is important to remember that any software has its limitations (Staden, 1983), and the end user is responsible to know them. The end user must be aware of using models that may overfit the data (Prohaska and Stadler, 2011). Overfitting occurs when an algorithm or model has too many parameters relative to the number of observations. While such an excessively complex model may perform well on the samples used, its usefulness may not extend beyond the sample that the model was contingent on. When thinking about how many free parameters are used to determine the sequence, it is good to remember the words of Erico Fermi (Dyson, 2004): "*I remember my friend Johnny von Neumann used to say, with four parameters I can fit an elephant, and with five I can make him wiggle his trunk.*" That is, as the number of free parameters (n) increases, and $n \to \infty$, the algorithm is no longer dependent on evidence.

Elizabeth Pennisi (2017b) notes that, "*among existing eukaryotic genomes, only the relatively small ones of the much-studied brewer's yeast and the nematode* Caenorhabditis elegans, *sequenced by early, slow, and costly techniques, have all the bases correctly positioned and located to specific chromosomes with no gaps.*" Despite the hype, promotion, marketing, overwhelming adoption, and use of next-generation massively parallel sequencing technologies (Goodwin et al., 2016), next-generation, also known as second-generation, massively parallel sequencing technologies have produced incomplete and erroneous assemblies (International Human Genome Sequencing Consortium, 2001; Salzberg and Yorke, 2005; Schatz et al., 2010; Eisenstein, 2015). This is a result of the large number of repetitive sequences that occur in most genomes (Schnable et al., 2009; Yuan et al., 2017).

According to Phillippy (2017), "*the goal of genome assembly is a gapless, haplotype-resolved reconstruction, but these genomic jigsaw puzzles are so difficult that we have not yet finished the human genome…which still contains more than 800 gaps after decades of work and billions of dollars spent.*" The genomes of plants, which are typically polyploid, are even more difficult to assemble from short reads because the genome contains a large number of repetitive sequences (Gregory, 2005; Gonzalo Claros et al., 2012; Bolger et al., 2017a,b; Schmutzer et al., 2017). As an alternative to creating more powerful bioinformatic genome assembly algorithms and software, which may be even less understandable from an end user's perspective, third-generation sequencing technology is being developed to produce longer (10,000 base) reads by analyzing the sequence of a single long DNA fragment at a time (Pareek et al., 2011; English et al., 2012; Berlin et al., 2015; Sakai et al., 2015; Gordon et al., 2016; Jain et al., 2017; Jiao and Schneeberger, 2017; Pennisi, 2017a).

21.2.2.8 Single Molecule Real Time Sequencing Analysis of the Incorporation of Fluorescent Dideoxy-Ribonucleotides Into a Primer-Template

Given the difficulty in assembling short reads into a contig that represents a whole molecule of DNA, methods to sequence a long single molecule of DNA are being explored. The concept of sequencing a single molecule of DNA has been around for almost as long as multimolecular DNA synthesizing techniques. Jett et al. (1989) and Werner et al. (2003) at the Los Alamos National Laboratories proposed a method to sequence DNA based on a technique they developed to detect a single fluorescent molecule flowing in a hydrodynamic system (Nguyen et al., 1987). A single-stranded template DNA composed of bases conjugated to distinguishing fluorochromes would be immobilized and suspended in a flow system where it would be subjected to the activity of an exonuclease that successively removed bases from either the 5′ end or the 3′ end of the template DNA. The released nucleotides would flow through the system to a detector that would identify the base by its color. The analysis of the sequence of colors would give the sequence of the DNA template.

The development of microfluids technology that reduces the volume necessary for the polymerization reaction to zeptoliters (Craighead, 2000; Foquet et al., 2002; Reccius et al., 2005) and zero mode waveguides that have a diameter less than the wavelength of light (Levene et al., 2003), not only made pyrosequencing possible but it also made it possible to observe efficiently the polymerization of a single strand of DNA and to sequence it. The sequence by synthesis method, developed by Eid et al. (2009), uses four different fluorescently labeled deoxyribonucleotides, in which the fluorophores are linked to the terminal phosphate of the deoxyribonucleotides (Fig. 21.11). The bases are identified by the color of the fluorescent light emitted. The template-primer and a DNA polymerase that has been genetically engineered to be unaffected by the fluorophore on the deoxyribonucleotide are immobilized together on the bottom of a 20 zL reaction volume, just under the zero mode waveguide. There are tens of thousands of reaction cells on a glass chip.

As the DNA polymerase extends the primer with a new fluorophore-deoxyribonucleotide, the fluorophore-conjugated deoxyribonucleotide fluoresces and its fluorescence vanishes when the terminal phosphate is cleaved as the next deoxyribonucleotide is added by the DNA polymerase. Then the newly added fluorophore-deoxyribonucleotide fluoresces. The sensitivity of the system is so high that the sequential addition of nucleotides can be observed in a single molecule of DNA. While the accuracy of sequencing a single-stranded molecule of DNA is less than 90% due to deletions, insertions, and mismatches, Eid et al. (2009) can achieve 99.3% accuracy by sequencing 15 single-stranded molecules of DNA.

This third-generation single molecule–sequencing technique has been used to sequence the genomes of plants (Badouin et al., 2017; Brozynska et al., 2017; Byrne et al., 2017; Clavijo et al., 2017; Du et al., 2017; Jarvis et al., 2017; Lan et al., 2017; Mascher et al., 2017; Minio et al., 2017; Teh et al., 2017; Zimin et al., 2017).

21.2.2.9 Single Molecule Real Time DNA Sequencing Using Nanopore Technology

Up until now, I have been describing biochemical means to sequence DNA using enzymes. It is also possible to take a physical approach to sequence DNA using nanopores (Deamer and Akeson, 2000: Branton et al., 2008; Zwolak and Di Ventra, 2008). We have discussed the movement of ions (see Chapter 1) and proteins (see Chapter 4) through pores. Kasianowicz et al. (1996) showed that individual single-stranded polynucleotide molecules would move through a 2.6 nm wide pore composed of α-hemolysin, a pore-forming toxin isolated from the pneumonia- and sepsis-causing bacterium, *Staphylococcus aureus*. After a single pore created by six α-hemolysin molecules is added to a black lipid membrane, nucleic acids are added to one side, and a voltage is placed across the membrane. The voltage drives an ionic current through the pore. Because the phosphates of nucleic acids are negatively charged at any pH greater than 1, the applied voltage will pull a single stand of DNA through the pore. Because of the vastly slower velocity of the DNA through the nanopore compared with ions, there is a transient decrease in the current caused by the movement of ions when the nucleic acid is moving through the nanopore. The duration of the

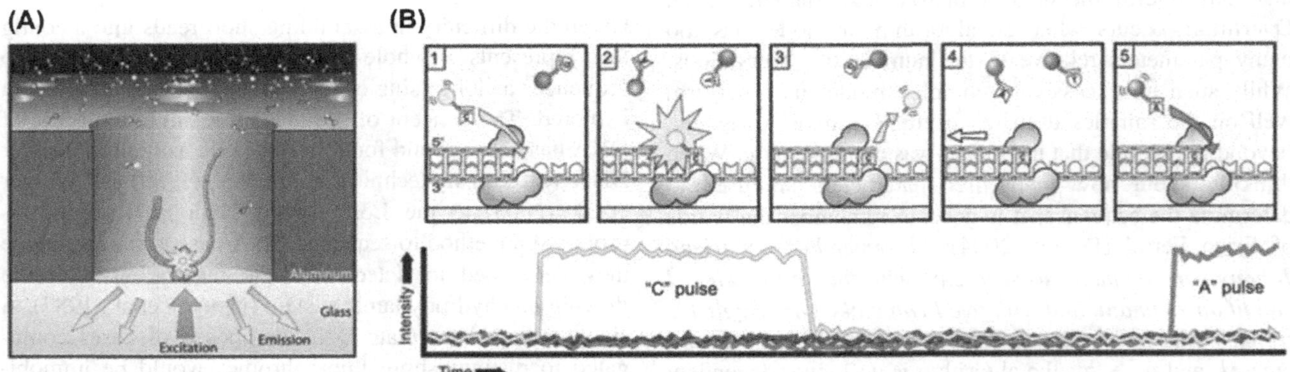

FIGURE 21.11 Single molecule real time sequencing. (A) The DNA polymerase and primer-template are bound to the bottom of a well adjacent to the zero-mode waveguide. (B) Each of the four deoxyribonucleotides is labeled with a different fluorescent dye so that they have distinct emission spectrums. As a fluorescent deoxyribonucleotide is held next to the zero-mode waveguide by the polymerase, light that identifies the deoxyribonucleotide is emitted. (1) A fluorescently labeled deoxyribonucleotide associates with the template in the active site of the polymerase. (2) Fluorescence emission of the color corresponding to the incorporated deoxyribonucleotide increases. (3) The dye–pyrophosphate product is cleaved from the deoxyribonucleotide and diffuses away from the waveguide. Consequently, the fluorescence emission decreases. (4) The polymerase translocates to the next position. (5) The next deoxyribonucleotide associates with the template in the active site of the polymerase, and the cycle repeats. *From Rhoads, A., Au, K.F., 2015. PacBio sequencing and its applications. Genom. Proteom. Bioinform. 13, 278–289.*

transient decrease in current is correlated with the length of the nucleic acid. This method has been improved so that nanopore sequencing can directly and rapidly detect the sequence of nucleotides in individual single molecules of DNA (Branton et al., 2008). By using synthetic nanopores made from a 1 nm think graphene, it is possible to control the characteristics of a nanopore, including its size and shape (Garaj et al., 2010; Merchant et al., 2010; Schneider et al., 2010; Ansorge, 2016).

The MinION manufactured by Oxford Nanopore Technologies identifies individual DNA bases that pass through a proteinaceous nanopore—produced by protein engineering techniques (Fig. 21.12). The MinION has been used to sequence DNA from plants (Michael et al., 2017; Mondal et al., 2017; Schmidt et al., 2017). There are no real limits on the length of the DNA sequence that it can read (Lu et al., 2016). The MinION has also been used on the International Space Station (https://www.nasa.gov/mission_pages/station/research/news/biomolecule_sequencer; https://nanoporetech.com/resource-centre/posters/dna-sequencing-microgravity-international-space-station-iss-using-minion).

Massively parallel sequencing technology continues to be developed that can produce longer reads (McCoy et al., 2014; Li et al., 2015), although it is important to note the observations of Jiao and Schneeberger (2017) that "*despite all of the impressive recent progress in long-read DNA sequencing, it was so far not possible to assemble a complete plant genome from sequence reads alone.*" Jiao and Schneeberger (2017) hope that sequences of long reads can be easily aligned with a map of restriction fragments obtained by optical mapping (Nagarajan et al., 2008; Tang et al., 2015). Optical mapping produces a map of restriction sites on each chromosome. To perform optical mapping, the chromosomes are stretched in agarose, fluorescently stained, and then digested by various restriction enzymes (Schwartz et al., 1993). The cleavage sites are visualized by the appearance of gaps in the fluorescence of the chromosome. The restriction fragments observed by fluorescence are in the same order as they were in the parent DNA molecule that makes up the chromosome.

21.2.2.10 Sequenced Genomes, Phylogenomics, and Resequencing

Scores of plant genomes have been sequenced and the up-to-date reference genomes can be found at the National Center for Biotechnology Information Website (ftp://ftp.ncbi.nlm.nih.gov/genomes/refseq/plant/). Hugo de Vries (1889; see Chapter 16) suggested that, "*systematic relationship is based on the possession of like pangens. The number of identical pangens in two species is a true measure of their relationship. The work of the systematist should be to make the application of this measure possible experimentally, by finding the limits of the individual hereditary characters. Systematic difference is due to the possession of unlike pangens.*" Today, the deoxyribonucleotide sequences of the genomes of various taxa are being compared to make phylogenetic inferences (Tateno et al., 1982; Eisen, 1998; Eisen and Hanawalt, 1999; Eisen and Fraser, 2003; Delsuc et al., 2005; Ciccarelli et al., 2006; Rannala and Yang, 2008; Wu and Eisen, 2008; Thomas, 2010; Chan and Ragan, 2013).

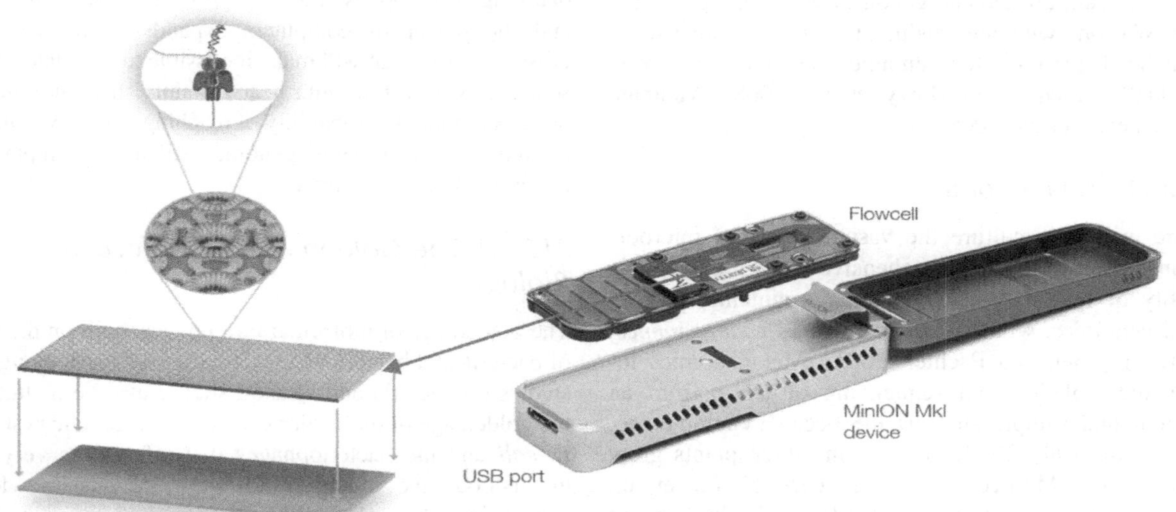

FIGURE 21.12 MinION single molecule real time sequencing using nanopore technology. The double-stranded DNA is denatured and a single strand is pulled through the nanopore as a result of the electric potential difference. As the single-stranded DNA passes through the nanopore, the current changes in a way that identifies the deoxyribonucleotide in the nanopore. The deoxyribonucleotide sequence of the single-stranded DNA is deduced by monitoring the current. *From Lu, H., Giordano, F., Ning, Z., 2016. Oxford nanopore MinIon sequencing and genome assembly. Genom. Proteom. Bioinform. 14, 265–279.*

The study of evolutionary relationships based on the comparative analysis of genome-scale data from multiple taxa is known as *phylogenomics* (Eisen, 2012). In phylogenomics, each nucleotide or each codon is considered a character state. The evolutionary relationships are obtained from algorithms that align the DNA sequences of each taxa and determine the similarity of the nucleotide or codon sequences. The algorithms then group, order, and organize the sequences into a tree that quantifies the differences between the aligned DNA sequences and gives the genetic distance between the taxa. The algorithms produce a gene tree, using algorithms based on parsimony, maximum likelihood (Whelan, 2008), or Bayesian inference (Huelsenbeck et al., 2001). By analyzing the genomes of all taxa, one can fulfill Darwin's (1857) dream of producing *"fairly true genealogical trees for each great kingdom of Nature."*

It is wise to remember that the reference genome sequence for a given species is an abstraction that represents the average sequence of several individuals and the actual sequence of no one. In reality, the sequences of individuals all differ slightly, but these differences have profound biological effects that represent individuality (Kaiser, 2008). A base substitution, deletion, or insertion that results in difference between individuals in the deoxyribonucleotide at a specific locus is known as a SNP. There are more than one million SNPs in the human genome. For some biological studies the reference sequence is sufficient, but if one wants to know why an outstanding individual differs from the group, either a candidate gene or the entire genome must be individually sequenced. This is known as *resequencing* (Huang et al., 2009; Xu et al., 2012). The genomes of Craig Venter and James Watson, who are leading the way in the use of personalized genomes for genomic medicine, have been individually resequenced (Levy et al., 2008; Wadman, 2008; Wheeler et al., 2008).

21.2.2.11 Metagenomics

We are unable to culture the vast majority of microorganisms. Thus, to get a comprehensive view of the genetic diversity of microbes in a given environment, environmental genomics, which is also known as *metagenomics*, has arisen (Chen and Pachter, 2005). Metagenomics involves the isolation and sequencing of DNA from an environmental sample such as the ocean (Venter et al., 2004; Yutin et al., 2007), the soil in which plants grow (Daniel, 2005; Montecchia et al., 2015; Nesme et al., 2016; Castañeda and Barbosa, 2017), or the human gut (Gill et al., 2006; Grice and Segre, 2012) and comparing the sequences with those in a metagenomics database to determine the degree of microbial diversity (Alvarenga et al., 2017). The reads are then assembled into a genome that represents the microbial community. The metagenomic approach has identified millions of previously unknown DNA sequences that will not only help us to understand the biological diversity of our environment but will also serve as a resource of new genes for the genetic engineering of organisms (Hutchinson, 2007; Gutleben et al., 2017).

Ray Wu, Fred Sanger, and Leroy Hood were model scientists in the way they developed the art and science of DNA sequencing (Wayne and Staves, 2008). Following in their footsteps, corporations such as Applied Biosystems (https://www.thermofisher.com/us/en/home/life-science/sequencing/sanger-sequencing.html), PACBIO (http://www.pacb.com/products-and-services/pacbio-systems/), Oxford Nanopore Technologies (https://nanoporetech.com/), and Illumina (https://www.illumina.com/systems.html) have also creatively contributed innovations that led to the development of even higher throughput and lower cost sequencing platforms that should make the $1000 genome, which was temporarily incentivized by the Archon Genomics X Prize, a reality (Hayden, 2014; https://genomics.xprize.org/).

21.2.3 Gene Discovery and Annotation

While there is no comprehensive definition of a gene, in theory, the assembled reference genomes will be important in providing insights into the gene content and location, the variation in the genes and the genome, the genetic and genomic basis of human health and disease, and the agronomic traits necessary to feed the growing population in a changing climate. In practice, the probability of finding a DNA sequence that gives rise to a trait of interest is daunting. The hope is, that as the price of sequencing drops, and the power of computers expands, there will be a crossover point that will make it possible to correlate a DNA sequence with a trait with near certainty. It is important to remember that the probability of reaching certainty is neither limited by the fact that no genome sequence is complete nor is any completely accurate.

21.2.3.1 The Golden Age of Molecular Biology

The golden age of molecular biology was built on the study of bacteria and bacteriophages. Studies on yeast, along with studies on bacteria and viruses, should also be included in the golden age of molecular genetics. The genetic system of *E. coli* and the bacteriophages that infect it is very well understood, and much, if not most, of our knowledge, concerning the relationship between a gene, a DNA sequence, and a trait, has come from the use of mutants produced spontaneously or with mutagens such as gamma rays, X-rays, UV light, and chemicals that interact with DNA (Watson et al., 1987).

The genetic system of plants, by contrast, is complex and not as well understood. Therefore an understanding of the function of an organ, cell, organelle, or pathway in plants was determined by plant biologists with various expertises and techniques, including physical and biophysical, chemical and biochemical, physiological, anatomical and morphological, and genetic. In the omics era, where much of this kind of evidence is considered indirect and correlative, the understanding of the function in question now comes from using specific mutants to decipher the genetic network that controls the function (Barberon, 2017). This is a shame because there was never a golden age of molecular biology in plants to build on. Indeed a Google search for "golden age of plant molecular biology," the "golden age of molecular biology of plants," and the "golden age of the molecular biology of plants" yielded no results. I think that there was no golden age of plant molecular biology because too many scientists followed the technology instead of where the biology took them. Perhaps if plant molecular biologists began studying the relationship between DNA sequences and traits in single-celled organisms (Hippler, 2017) or single cells in organisms (Dixit and Nasrallah, 2001; Nasrallah, 2002, 2005; Cheung et al., 2010), and/or depended more on functional assays to investigate the gene and gene product (Munemasa et al., 2016), gene annotation in plants would have been more conclusive and foundational, and there might have been a golden age of plant molecular biology to build on.

Let's take a look back on the golden age of molecular biology, where biochemists isolated from nature's toolkit, enzymes such as DNA polymerase and DNA ligase and determined their function in vitro using functional assays (Gefter et al., 1967; Olivera and Lehman, 1967; Lehman, 1974; Friedberg, 2006). DNA polymerase was assayed by its ability to incorporate radioactive deoxyribonucleotides into acid-precipitable material (Kornberg, 1959), and DNA ligase was assayed by its ability to covalently bind the two cohesive ends of lambda phage to form circular DNA molecules. This activity results in a change in the chromatographic and sedimentation properties of the DNA and a change in its morphology (Cozzarelli et al., 1967; Gellert, 1967; Fig. 21.13). The roles of these enzymes in vivo were determined when geneticists analyzed the phenotype of nonlethal mutants that produced a defective enzyme or underexpressed or overexpressed a functional enzyme. Geneticists also found conditional mutants in which the thermolabile enzymes were active at a permissive temperature but lethal at a higher temperature (Fareed and Richardson, 1967; de Lucia and Cairns, 1969; Gottesman et al., 1973; Konrad et al., 1973). These techniques furnished a direct demonstration of the functional relationship between a gene, a DNA sequence, and a trait. These studies provide an excellent study of the relationship between

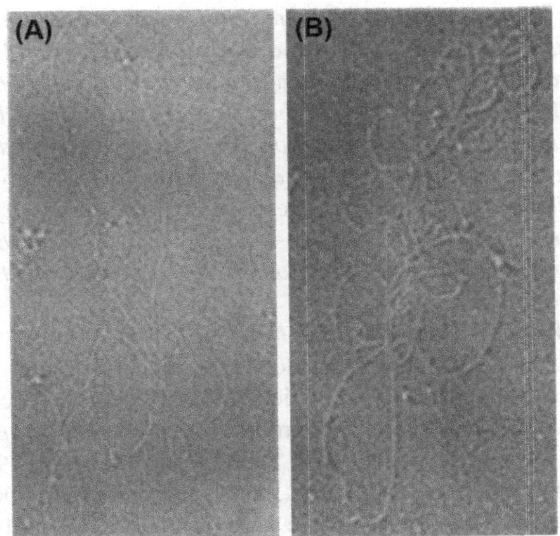

FIGURE 21.13 Electron micrographs of covalent circles of lambda DNA formed in *Escherichia coli* extracts. Magnification: (A) ×30,000; (B) ×43,000. *From Gellert, M., 1967. Formation of covalent circles of lambda DNA by* E. coli *extracts. Proc. Natl. Acad. Sci. USA 57, 148–155.*

ontology, which is the study of the nature of reality, and epistemology, which is the study of how we determine what is real.

21.2.3.2 The Golden Age of Biotechnology

Near the end of the golden age of molecular biology and the beginning of the golden age of biotechnology, recombinant DNA technology allowed molecular biologists to isolate a gene from a given organism or chemically synthesize it. Molecular biologists could ligate the gene into an autonomously replicating plasmid and then introduce the recombinant plasmid into bacteria. The bacteria would produce the gene product, which could then be isolated in vast quantities. This allowed one to demonstrate the relationship between a gene, a DNA sequence, and a trait while harnessing the power of the gene to produce pharmaceuticals such as insulin and human growth hormone or to modify the bacterium to produce enzymes for doing laundry (Sinsheimer, 1977; Villa-Komaroff et al., 1978; Goeddel et al., 1979; Cohen, 2013; Gurung et al., 2013).

Keeping in mind the science of eugenics, the science used for genetic engineering must be used within an ethical framework, (Nirenberg, 1967; Chargaff, 1977, 1978; Pollack, 1999; Allen, 2001; Gardner et al., 2001) and a prepared mind must be engaged to minimize unintended consequences. On September 12, 1963, *The New York Times* published (See Anonymous, 1963), "*Ultimately we may be able to fashion living species to order, to* 'manufacture' *living organisms with specific properties just as we now produce machines or instruments. It is no longer fantastic to contemplate the time when the stuff of heredity may be manipulated to produce babies with predetermined traits and features. Is mankind ready for such powers? The moral, economic and political implications of these possibilities are staggering, yet they have as yet received little organized public consideration. The danger exists that the scientists will make at least some of these God-like powers available to us in the next few years, well before society—on present evidence—is likely to be even remotely prepared for the ethical and other dilemmas with which we shall be faced.*"

Two decades after *The New York Times* article was published, technology developed so that eukaryotes too could be genetically modified (Wu, 1998). Genetically modified plants have been produced that are able to resist viral attack (Sequeira, 1984; Sanford and Johnston, 1985; Abel et al., 1986; Fitch et al., 1992; Ferreira et al., 2002), which incidentally saved the papaya industry (Gonsalves, 2016); to resist insect attack (Vaeck et al., 1987; Charles, 2001); to be herbicide resistant (Padgette et al., 1995; Funke et al., 2006; Pollegioni et al., 2011); to produce vaccines for animals and humans (Richter et al., 1996; Zhang et al., 2006), and to glow like a firefly (Ow et al., 1986). Currently, it is possible for biotechnologists to engineer prokaryotic genes into eukaryotes by introducing DNA modifying machinery from a prokaryote into a eukaryote to modify a DNA sequence in its normal context within the organism of interest to elucidate its relationship to a given trait. Joshua Lederberg (1963) wrote, "*The ultimate application of molecular biology would be the direct control of nucleotide sequences in human chromosomes, coupled with recognition, selection and integration of the desired genes, of which the existing population furnishes a considerable variety.*" Indeed, two decades later, it became possible to modify human DNA to treat genetic diseases such as severe combined immunodeficiency, which is the result of a defective adenosine deaminase gene (Blaese et al., 1995). By 2015, scientists modified the β-globin gene of human zygotes using CRISPR/Cas9, although the efficiency was low (Liang et al., 2015).

There are many methods to introduce DNA into yeast (Begg, 1978; Hinnen et al., 1978), plants (Fraley et al., 1983; Klein et al., 1987, 1988a,b; Sanford, 1990; Chilton and Que, 2003), and animals (Capecchi, 1980, 2007; Smithies et al., 1985; Smithies, 2007a,b). Typically, the sequence one wants to transform an organism with is engineered into a viral or plasmid DNA vector to increase the efficiency of transformation because viral and plasmid DNA can incorporate efficiently into the genome of the host. The vectors are deactivated in a way that they cannot transmit a disease. The insertion of DNA into the eukaryotic genome can be either random or targeted. As I will discuss below, nonrandom gene targeting involves including the nucleotide sequences in the vector that are necessary to make use of the enzymatic machinery involved in the native homologous recombination process (Struhl et al., 1979; Szostak and Wu, 1979; Orr-Weaver et al., 1981; Orr-Weaver and Szostak, 1983; Szostak et al., 1983, Figs. 21.14 and 21.15). Adding enhancer sequences to the vector promotes the expression of the DNA sequence incorporated into the genome, and including

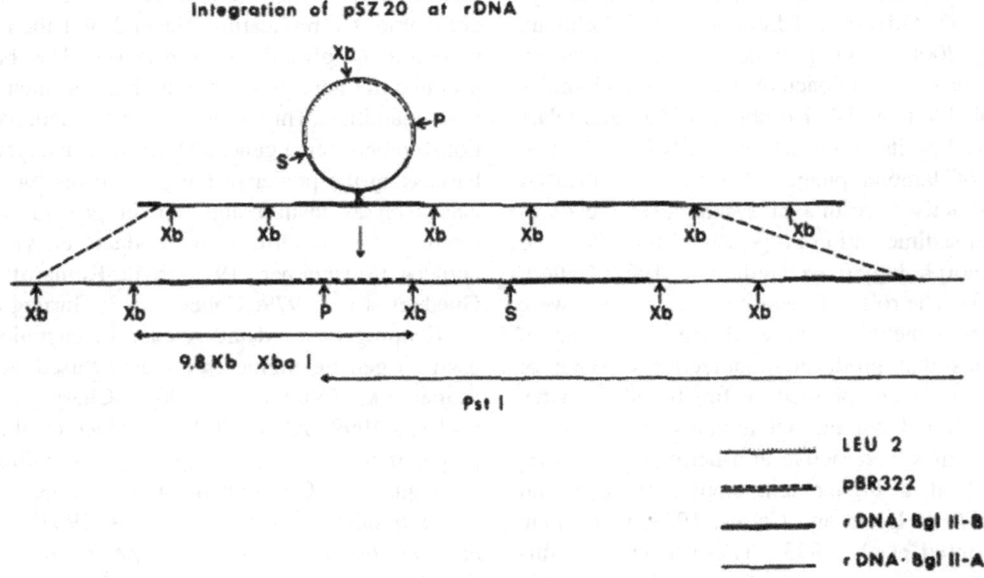

FIGURE 21.14 Homologous recombination during the integration of plasmid sZ20 DNA into the rDNA region of chromosome XII in yeast. Homologous recombination depends on sequence homology between the regions adjacent to the double-strand break and the ends of a donor DNA molecule. *From Szostak, J., Wu, R., 1979. Insertion of a genetic marker into the ribosomal DNA of yeast. Plasmid 2, 536–554.*

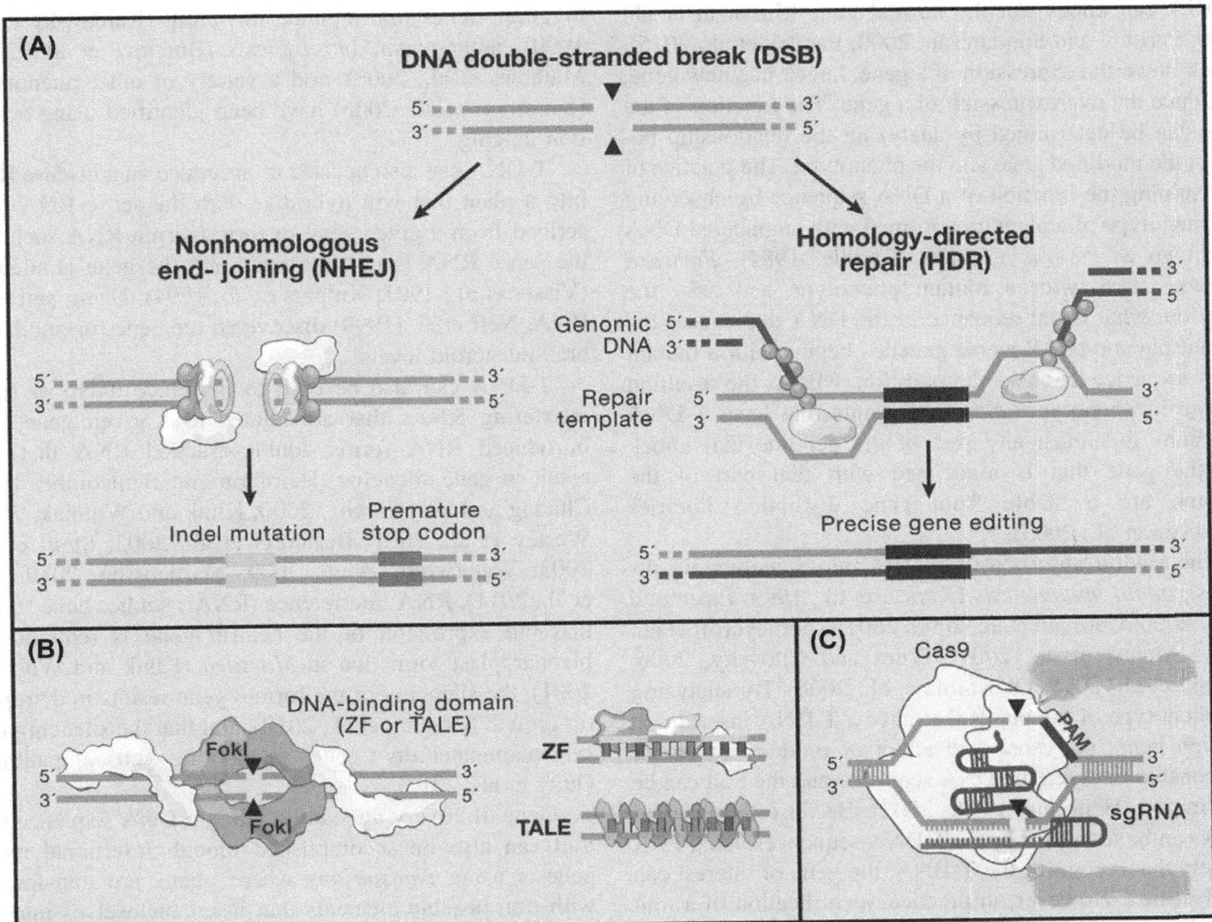

FIGURE 21.15 (A) DNA double-strand breaks are repaired by nonhomologous end-joining (NHEJ) or homology-directed repair. In the NHEJ pathway, repair proteins bind to the double-strand breaks, the breaks are repaired but insertions and deletions (indels) or substitutions are introduced. In homology-directed repair, other repair proteins bind to the double-strand breaks that direct recombination with homology arms on a donor DNA. (B) Zinc finger proteins and transcription activator–like effectors are naturally occurring DNA-binding domains that can be fused with an endonuclease to target specific sequences and introduce double-strand breaks that result in indel or base-substitution mutants. (C) The Cas9 nuclease is targeted to specific DNA sequences via the single guide RNA (sgRNA). Arrowheads indicate site of double-stranded break. The sgRNA directly pairs with the target DNA, and the Cas9 protein introduces a double-stand break between the recognized sequence and the protospacer adjacent motif. The Cas9 nuclease then introduces a double-strand break. In the presence of donor DNA with homologous arms, a portion of the donor DNA will be inserted into the DNA with the double-strand break through homologous recombination. In the presence of donor DNA, the result can be a knock-in or gene replacement mutation. In the absence of donor DNA, the result can be an indel or nucleotide replacement mutation. *From Hsu, P.D., Lander, E.S., Zhang, F., 2014. Development and applications of CRISPR-Cas9 for genome engineering. Cell 157, 1262–1278.*

certain promoter sequences in the vector allows one to control the expression of the incorporated DNA sequence. Because the efficiency of transformation of eukaryotes is in general low, a stringent selection procedure is necessary to isolate transformants.

21.2.3.3 Methods Used for Gene Discovery

While we still lack a clear definition of the gene, it is not impossible that an operational or working definition of a gene will emerge from attempts at gene discovery using molecular approaches. In general the molecular approaches make use of the genetic engineering techniques developed during the golden age of biotechnology to transform the genome to understand the relationship between a gene, as defined by a nucleotide sequence, and a trait. The initial techniques took advantage of the ability of the tumor-inducing plasmid of *Agrobacterium* to transform plants.

Agrobacterium, the cause of crown gall disease in plants, has been genetically engineered to deliver DNA to plants without causing crown gall disease. The DNA, which is randomly incorporated into the plant genome, is known as transfer DNA (T-DNA). Depending on the composition of the T-DNA and the site of insertion, the

T-DNA can knock out the normal gene (Sussman et al., 2000; Parinov and Sundaresan, 2000; Pandey et al., 2015), knock down the expression of a gene, knock in a new gene, or induce the overexpression of a gene. The function of the gene can be determined by analyzing the relationship between the modified gene and the phenotype. The practice of determining the function of a DNA sequence by observing the phenotype of a plant transformed with engineered DNA is known as *reverse genetics* (Ruddle, 1984). *Forward genetics* starts with a mutant phenotype and asks the question, what is the sequence of the DNA that causes the mutant phenotype? Reverse genetics begins with a mutant DNA sequence and asks the question, what is the resulting change in phenotype? Seeds of plants that have T-DNA insertions in almost any part of the genome that knock out the gene that is associated with that part of the genome are available from gene disruption libraries (Schween et al., 2005).

The random insertion of T-DNA into a genome results in *insertional mutagenesis* (Krysan et al., 1999; Eastmond et al., 2000; Germain et al., 2000, 2001; Thorneycroft et al., 2001; Alonso et al., 2003; Tzfira and Citovsky, 2006; Inagaki et al., 2015; Okushima et al., 2005). By analyzing the phenotype of the plants that have a T-DNA insertion in a given gene, the cause and effect or product–precursor relationship between the DNA sequence and the trait can be determined (Henriques et al., 2002). Because the inserted DNA can be identified by its T-DNA sequences using PCR primers that recognize the T-DNA, the gene of interest can be identified. However, in practice, identification of a trait related to the mutated DNA sequence is hampered because of off targeting, where the insertion occurs in more than one locus, because transformed plants die as a result of the lethal effects of the mutated gene in the embryonic stage, or because there is no visible phenotype as a result of the action of redundant genes (Weigel et al., 2000; Bouche and Bouchez, 2001; Jander et al., 2002; Cutler and McCourt, 2005; Mathieu et al., 2009; Waterworth et al., 2009). Because knockouts were and continue to be effective in demonstrating the connection between a gene and a trait in bacteria, yeast, *Drosophila*, and mice, the difference in effectiveness between microorganisms, animals, and plants may be due to the omnipresent nature of polyploidy in plants (Stebbins, 1985; Meyers and Levin, 2006).

The function of a DNA sequence can also be determined by using insertional mutagenesis to overexpress a gene. Overexpression involves transforming plants with T-DNA vectors that contain several transcriptional enhancers from the cauliflower mosaic virus 35S gene. The enhancers are randomly inserted into the genome, and if there are enough plants, an insertion is bound to occur adjacent to the gene of interest (Koncz et al., 1989; Weigel et al., 2000; Cowperthwaite et al., 2002; Fladung et al., 2004; Guo and Li, 2012; Cui et al., 2013). This is known as *activation tagging*. Genes that regulate flowering (Kardailsky et al., 1999), anthocyanin biosynthesis (Borevitz et al., 2000; Mathews et al., 2003), and a variety of other phenotypes (Kuromori et al., 2006) have been identified using activation tagging.

T-DNA can also be used to introduce an antisense RNA into a plant that will hybridize with the sense RNA transcribed from a given gene to form hairpin RNA such that the sense RNA is not translated and the gene is silenced (Visser et al., 1991; Kuipers et al., 1994). Using antisense RNA, Neff et al. (1999) discovered the gene for regulating brassinosteroid levels.

T-DNA can also be used to introduce micro or small interfering RNAs that are related to a given gene. The introduced RNA forms double-stranded RNA that will result in gene silencing (Hamilton and Baulcombe, 1999; Chuang and Meyerowitz, 2000; Klink and Wolniak, 2000; Wesley et al., 2001; Bezanilla et al., 2003; Stout et al., 2003; Waterworth et al., 2009; Martienssen, 2010; Lee et al., 2011). RNA interference (RNAi) studies have shown that the expression of the centrin gene is required for blepharoplast formation in *Marsilea* (Klink and Wolniak, 2001), the silencing of the formin gene results in disrupted tip growth (Cheung et al., 2010), and that the silencing of a self-incompatability gene results in self-compatibility (Jung et al., 2012).

Gene discovery aimed at relating a DNA sequence to a trait can also be accomplished though insertional mutagenesis using *transposons* where plants are transformed with transposable elements that insert themselves into the genome and disrupt the gene into which it inserts (McClintock, 1953). Because the sequence of the transposon is known, the gene in question can be identified using PCR and transposon-specific primers that hybridize to the sequence of deoxyribonucleotides that contains the transposon (Martienssen, 1998; Cheng et al., 2011; Cui et al., 2013). Seeds of plants that have a transposon in nearly any gene of interest are available (Tadege et al., 2008). This high-throughput knockout technique has been used successfully in many plant taxa (Cui et al., 2013), although like T-DNA, the usefulness of transposon insertion is also limited by polyploidy, where other genes may compensate for the targeted gene (Hua and Meyerowitz, 1998), or pleiotropy, where one gene may have multiple effects and be required during certain phases of the life cycle so that a gene knockout results in early embryonic or gametophytic lethality.

Another road to gene discovery is to introduce nucleases that are capable of directly targeting and mutagenizing a sequence in the genome (Fig. 21.15). Genetic engineers are making use of nature's toolbox by fusing the DNA sequences that code for restriction endonucleases with the DNA sequences that code DNA binding proteins made from either transcription factors or transcription activator–like effectors to create sequence-specific nucleases that are able to modify a single gene (Baltes and Voytas, 2015; Schiml and Puchta, 2016). These engineered sequence-specific nucleases, which are targeted to the DNA by proteins, include zinc finger nucleases (Wright et al., 2005; Townsend et al., 2009; Osakabe et al., 2010; Zhang et al., 2013a,b; http://www.zincfingers.org/), meganucleases (Gao et al., 2010; Watanabe et al., 2016), and transcription activator–like effector nucleases (Zhang et al., 2013). These nucleases are capable of directly targeting and mutagenizing a sequence in the genome in which they are expressed. They introduce a targeted DNA double-strand break that, in eukaryotic cells, is repaired by either nonhomologous end-joining (NHEJ) or by homologous recombination (HR). In NHEJ repair, which is predominant in somatic cells, the break is rapidly rejoined in an error-prone process that results in single nucleotide indel that result in frameshift mutations that effectively knocks out the gene. In homologous recombination repair, which is a high fidelity process that is rare in somatic cells and prevalent in cells undergoing meiosis, repair depends on sequence homology between the regions adjacent to the double-strand break and the ends of a donor DNA molecule that abut a modified version of the targeted gene (Puchta and Fauser, 2013).

None of the methods of gene discovery described have proven to be magic bullets for gene discovery. Currently, there is a fair amount of hope that the CRISPR/Cas9 system (CRISPR [clustered regularly interspaced short palindromic repeat]-Cas9 [CRISPR-associated nuclease 9]) will fulfill the role of a magic bullet in gene discovery (Songstad et al., 2017). The CRISPR/Cas9 system is a sequence-specific nuclease that can be used for site-directed mutagenesis (Gaj et al., 2013; Malzahn et al., 2017). Unlike the sequence-specific nucleases described above where the nucleases are targeted to the DNA by proteins, in the CRISPR/Cas9 system, the nuclease is targeted to the DNA by RNA.

In nature, CRISPR/Cas9 and other related systems are part of an adaptive immune system found in many bacteria and archaea. It protects them from invading bacteriophages and plasmids much like the RNAi system protects eukaryotic cells from invading viruses (Mojica et al., 2005; Horvath and Barrangou, 2010; Lander, 2016). The CRISPR/Cas9 system was discovered serendipitously and independently by people working to understand microbes growing in saltmarshes, microbes involved in biological warfare, and microbes involved in yogurt production. Their submitted manuscripts, all of which showed a novel DNA sequence, were rejected from the top tier journals. Then a bioinformatic approach showed the relationship of these studies to each other by showing that the spacer DNA in the cluster of short palindromic repeated sequences (CRISPR) was homologous to viral DNA. This suggested that CRISPR may function in an adaptive immunity response (Lander, 2016).

The adaptive immunity response requires the Cas family of proteins, which include nucleases, helicases, polymerases, and polynucleotide-binding proteins. For the adaptive immunity response, the bacterium acquires a record of the invader by using Cas proteins to cleave the invading DNA and insert it into its genome as a spacer within a cluster of short palindromic repeated sequences (CRISPR). The inserted DNA is transcribed into RNA that is then processed into small RNAs, each of which represents one of the original invaders. The small RNAs are complementary to a region of a transactivating CRISPR RNA (tracrRNA). The tracrRNA is transcribed from a region of the genome immediately adjacent to CRISPR. The complementary sequences hybridize and bind to the Cas9 nuclease. On infection by DNA with a complementary sequence, the small RNA-tracrRNA-Cas9 ribonucleoprotein binds to the complementary sequence in the newly invading DNA. Once bound, the Cas9 nuclease inactivates the DNA of the invader by making a double-strand break in the sequence between the Protospacer adjacent motif (PAM) and the targeted sequence. The double-stranded break prevents the invading DNA from replicating. Because the DNA sequence in the bacterial genome does not include the PAM, the Cas9 nuclease is able to recognize the difference between self-DNA and foreign non-self DNA. As a result, the Cas9 nuclease does not introduce a replication-inhibiting double-strand break into its own genome. The adaptive immunity system has been reconstituted in vitro (Gasiunas et al., 2012; Jinek et al., 2012; Bortesi and Fischer, 2015; Raitskin and Patron, 2016).

Virginijus Siksyns (Gasiunas et al., 2012), Jennifer Doudna and Emmanuelle Charpentier (Jinek et al., 2012; Wright et al., 2016), Feng Zhang (Ran et al., 2013; Cong et al., 2013; Hsu et al., 2014), and George Church (Mali et al., 2013) independently proposed and developed ways to harness the RNA-guided CRISPR/Cas9 mechanism in nature's toolbox to perform "*RNA-directed DNA surgery*" that would target and edit genes in any cell (Gasiunas et al., 2012; Fig. 21.15). The CRISPR/Cas9 system utilizes an engineered single guide RNA (sgRNA) that binds the Cas9 nuclease and brings it to the genome where a portion of the sgRNA binds to the complementary DNA sequence. The Cas9 nuclease then cuts the two strands of DNA between the PAM and the sequence associated with the sgRNA. The sgRNA can be multiplexed to contain various sequences that recognize a number of different genes. In eukaryotic cells, the double-strand breaks are repaired by the endogenous DNA repair systems. Repair by nonhomologous end joining results in the repaired DNA having a deletion, insertion, or nucleotide substitution, while homology-directed repair results in the gene being edited in a manner specified by the homologous yet altered DNA sequence.

The CRISPR/Cas9 system has been engineered into T-DNA, geminivirus, or other vectors so that specific sequences can be targeted for editing through nonhomologous end joining, resulting in indel or nucleotide substitution mutants or through homology-directed repair, resulting in knock in or gene replacement mutants (Jiang et al., 2013; Li et al., 2013; Mao et al., 2013; Nekrasov et al., 2013; Upadhyay et al., 2013; Feng et al., 2014; Jia and Wang, 2014; Čermák et al., 2015; Lowder et al., 2016; Ma et al., 2016; Paul and Qi, 2016; Schiml and Puchta, 2016; Song et al., 2016; Baltes et al., 2017; Curtin et al., 2017a; Gil-Humanes et al., 2017; Kui et al., 2017; Liu et al., 2017; Sánchez-León et al., 2017; Songstad et al., 2017; Wang et al., 2017; Watanabe et al., 2017). sgRNAs, based on known genome sequences (Xie et al., 2014), can be designed and constructed using the CRISPR-PLANT platform (http://genome.arizona.edu/crispr/). The RNA guide sequence can be engineered to target transcription factors that activate or silence genes on a genome-wide scale (Larson et al., 2014).

The use of CRISPR/Cas9 may make it feasible to provide demonstrations of cause and effect or precursor–product relations in plants, even in polyploid or paleopolyploid plants where the CRISPR/Cas9 system can target multiple genes or multiple homologous copies of a gene with no detectable off-targeting (Wang et al., 2014a,b,c; Svitashev et al., 2015; Lopez-Obando et al., 2016; Nomura et al., 2016; Yan et al., 2016; Andersson et al., 2017; Braatz et al., 2017). An excellent example of the use of CRISPR/Cas9 is the simultaneous elimination of the genes for three plasma membrane receptors that make wheat susceptible to powdering mildew (Wang et al., 2014). Being able to modify several genes at a time not only expedites multiple gene modifications or trait stacking for plant breeding (Bortesi and Fischer, 2015) but can also expedite the understanding of traits that are products of many genes.

Targeted mutagenesis using CRISPR/Cas9 and related systems is not only an effective way of relating the DNA sequence to the trait using reverse genetics in plants with polyploid genomes but is also being touted as a way of editing genes in a way that would be useful for metabolic engineering and molecular farming, where plant cells or plants are used to produce valuable metabolites or proteins (Baltes and Voytas, 2015). The CRISPR/Cas9 system is also marketed as a way of increasing crop yields in response to a growing population and shrinking amount of arable land. Because the edited gene has no relationship on the chromosome to the CRISPR/Cas9 DNA, the CRISPR/Cas9 DNA can be bred out of the transformed plant which heretofore has allowed CRISPR/Cas9 edited plants to be produced unhindered by government regulation (Jones, 2015; Woo et al., 2015; Waltz, 2016; Wolt et al., 2016; Malzahn et al., 2017).

The CRISPR craze (Pennisi, 2013) has an incredible potential to improve human health and agriculture that is based on one, two, three, and even many genes. According to Songstad et al. (2017), "*The traits necessary to support modern agriculture, such as tolerance to environmental and biotic stresses and efficient use of nutrients, are typically encoded by combinations of multiple, interacting genes. The challenges created by the complexity of these traits cannot be overemphasized and will continue to tax current genome modification technology. For example, the unpredictable outcomes of random DNA integration and the difficulty of simultaneously handling multiple genes are just some of the key contributors to the significant challenges ahead. It is anticipated that genome editing technology and the ability to create targeted KI* [Knock-in] *mutations via site-specific DNA integration will play a major role in addressing these challenges.*" Thus, the realization of the full potential of the CRISPR/Cas9 system in enhancing human health and agriculture depends on the interplay between the hypothesis-driven approach where $n \rightarrow 1, 2, 3$ and the omics approach where $n \rightarrow \infty$.

Heretofore, none of the reverse-genetic approaches to gene discovery has been a magic bullet to demonstrate the cause and effect or precursor–product relationship between a specific deoxyribonucleotide sequence and a trait, although each of them have been effective in demonstrating a cause and effect or precursor–product relationship between some gene and some trait. Forward-genetic approaches using chemically induced mutants are also an option for gene discovery. Map-based or *positional cloning* allows one to identify the DNA sequence that gives rise to a mutant phenotype by determining the linkage of the trait in question to DNA markers whose physical position in the genome is known. This technique was used to successfully find the gene for Huntington's disease (Gusella et al., 1983). According to Jander et al. (2002), "*The big advantage to map-based cloning is that it is a process without prior assumptions. Essentially, one is looking at all of the genes in the genome at the same time to find the ones that affect the phenotype of interest. It is a process of discovery that makes it possible to find mutations anywhere in the genome....*"

Genome-wide association (GWA) studies have been used to identify DNA sequences that are correlated with a trait. To do a GWA study, a particular trait is selected and then the SNPs that exist in various individuals are correlated with the variation in that trait (Hirschhorn and Daly, 2005; Welcome Trust Case Control Consortium, 2007; Atwell et al., 2010; Han and Huang, 2013; Korte and Farlow, 2013; Seren et al., 2017; Thoen et al., 2017). The regions of the genome where the correlated SNPs occur are candidate regions where a gene that controls the trait may occur.

While DNA can be isolated from populations that contain different expression of that trait, GWA studies are typically performed in silico using public databases

(Suwabe and Yano, 2008) that contain SNPs from various populations of humans and model taxa. In plants, GWA studies have identified the sequences and the complex genetic architecture that underlie flowering time (Aranzana et al., 2005), the shade avoidance response (Filiault and Maloof, 2012), root development (Meijón et al., 2014), seed oil composition (Branham et al., 2016a,b), tolerance to fungicides (Atanasov et al., 2016), nodulation (Curtin et al., 2017b), flowering times (The 1001 Genomes Consortium, 2016), growth (Fusari et al., 2017; Luo et al., 2017), resistance to biotic and abiotic stresses (Bartoli and Roux, 2017; Thoen et al., 2017), and other agronomic traits (Fang et al., 2017; Huang et al., 2010; Zheng et al., 2017).

A trait that is controlled by a small number of sequences, each one with a large effect, is amenable to GWA studies, whereas a trait controlled by a plethora of sequences, each with a small effect, is not. Moreover, the effect of a given sequence might depend on the genetic background and the environmental conditions. It turns out that most human diseases (Chial, 2008) and many of the traits of interest in plants are highly quantitative and involve complex genetic architectures made of many sequences with small effects (Morrell et al., 2011). Thus, the results of GWA studies typically report correlation rather than causes, implication rather than demonstration. Ioannidis et al. (2009) caution us that GWA studies "*will often fail to identify the causal variants with certainty. Both the genetic and phenotypic architecture could have dense correlation patterns that are difficult to decipher. Functional insights can also often be tenuous. Although the discovery of GWA signals is exciting, the amount of work required to achieve and confirm causal variants should not be underestimated.*"

Gene ontology (GO) is a bioinformatic method with a set of rules (http://www.geneontology.org; https://www.arabidopsis.org/portals/nomenclature/guidelines.jsp) on how to annotate DNA sequences and give them a name that describes their cellular localization (e.g., nucleus or plastid), the biological process in which they participate (e.g., signal transduction or photosynthesis), and their molecular function (e.g., adenylate cyclase activity, calcium transport activity). The annotations also contain *evidence codes* that state whether the annotation is based on experimental evidence, computational analysis, or author statements (Bolger et al., 2017a,b). The robustness of homology-based functional annotation depends on the assumptions that the original annotation is accurate and that sequence similarity is equivalent to functional homology. In building an Aristotelian demonstration, it is important to know that "*evidence codes cannot be used as a measure of the quality of the annotation*" (http://geneontology.org/page/guide-go-evidence-codes).

The in silico bioinformatics approach known as homology-based functional annotation is the most common way to determine the function of a given sequence of deoxyribonucleotides. Homology-based annotation compares the sequence of interest to sequences already annotated in other genomes (Pearson, 1991). When the number of nucleotides in the sequence (*n*) is greater than 10, the probability that the sequence of interest and any other sequence that is *n* nucleotides long would be identical by chance is $(1/4)^n$, which is less than one in a million. For these reason, sequences with sufficient similarity are considered to be homologous and derived from a common ancestor in a process of descent with modification as a result of natural selection. If the sequence of interest is found in many different species, the sequence is considered *conserved*. It is assumed that the greater the conservation, the more indispensable is the gene with this particular sequence of nucleotides. FASTA (Pearson and Lipman, 1988) and BLAST (Basic Local Alignment Search Tool; Altschul et al., 1990) were the original programs that allowed for rapid sequence comparisons between a DNA sequence of interest and DNA sequences annotated and housed in various databases. The Gene Ontology Consortium (2000) contends that "*there is likely to be a single limited universe of genes and proteins, many of which are conserved in most or all living cells. This recognition has fuelled a grand unification of biology; the information about the shared genes and proteins contributes to our understanding of all the diverse organisms that share them. Knowledge of the biological role of such a shared protein in one organism can certainly illuminate, and often provide strong inference of, its role in other organisms.*"

Not all the sequences that are known have known functions. There are sequences that are annotated: "molecular function unknown", "cellular component unknown", and/or "biological process unknown" (Berardini et al., 2004). According to Furbank (2009), a significant proportion of the *Arabidopsis* genome is annotated as "gene of unknown function" or annotated using sequences with limited if any homology.

The gold standard for annotating the function of a given sequence comes from doing biochemical and/or biophysical functional assays. One way to determine the function of a DNA sequence is to compare the DNA sequence with the sequence predicted from the amino acid sequence obtained by Edman digestion from a conserved region of a purified protein whose function has already been well characterized. This technique was instrumental in discovering the gene sequences of the plasma membrane–localized H^+-ATPase (Harper et al., 1989; Pardo and Serrano, 1989; see Chapter 2). Another way to determine the function of a DNA sequence is to see if transforming yeast cells with the sequence of interest can complement mutant yeast cells that do not express that sequence. This technique was instrumental in discovering the gene sequences of the plasma membrane–localized K^+ channel (Anderson et al., 1992) and H^+-ATPase (Palmgren and Christensen, 1993; de Kerchove d'Exaerde et al., 1995). Once the DNA sequence for the H^+-ATPase was known, it was discovered that multiple bands with the sequence in question showed up on Southern blots (Boutry et al., 1989). This indicated that the H^+-ATPase gene was part of a gene family. The different members of a gene family expressed in a taxon are considered *paralogous* genes, whereas the same gene expressed in different taxa are considered *orthologous* genes.

A quarter of a century ago, James Watson (1992) wrote, "*I can imagine that typical work for undergraduates will be to find the gene once all the sequence has been obtained. Professors could tell their students: If you identify a gene, we will let you go to graduate school and do real science.*" The difficulty in identifying genes was underestimated. Often, the annotations given in papers written by gene hunters are best described as demonstrating technical progress toward the goal of identifying the nucleotide sequence that controls a given trait, rather than an actual demonstration of the biological relation between a nucleotide sequence and a trait. The annotations adopted from papers written to demonstrate technical progress should be viewed differently from the annotations that come from direct functional biochemical and/or biophysical demonstrations.

Building a functional ontology for complex developmental or physiological processes that is based on poor annotations results in an unprecedented level of hand waving and an unsustainable line of reasoning. Hand waving (Houk, 1977) in the analog age depended on order-of-magnitude estimates and dimensional analysis (see Preface to the first edition), and it was limited to ten fingers. Interestingly, the word digital comes from the Latin word *digitus*, which means finger. In the omics age where knowledge is often based on bioinformatics, the number of hands waving while making an argument is nearly unlimited. As Henry David Thoreau (1854) wrote, "*Our life is frittered away by detail. An honest man has hardly need to count more than his ten fingers, or in extreme cases he may add his ten toes, and lump the rest. Simplicity, simplicity, simplicity! I say, let your affairs be as two or three, and not a hundred or a thousand; instead of a million count half a dozen....*"

The process of gene discovery that leads to the accurate annotation of a DNA sequence is difficult—especially in an era when use of new technologies takes priority over the demonstration of cause and effect in a DNA sequence-trait relationship. Typically the techniques of gene discovery become obsolete quickly, and the use of the new method becomes essential in a scientific technopoly. In the study of development and evolution in plants, annotations that demonstrate the direct relationship between a DNA sequence and a trait are rare, yet the annotations provide the ground truth of all omic technologies. By contrast, in the study of biochemical pathways where the enzymology is established, annotations that demonstrate the relationship between DNA sequence and trait are providing new ways, understanding the regulation of the biosynthetic pathways and their relationship to other pathways (Chapple et al., 1992; Ruegger et al., 1999; Ruegger and Chapple, 2001; Bonawitz et al., 2014).

Without reliable annotations, the tie between sequence acquisition and the biological purpose of the sequence becomes severed. According to Neil Postman (1992) when "*information appears indiscriminately, directed at no one in particular, in enormous volume and at high speeds, and disconnected from theory, meaning, or purpose,*" *technopoly* thrives. We are currently in what I call the golden age of biotechnopoly, where the research priorities focus on the demonstrated use of omic technology that produces data with a high throughput rather than the demonstration of a single biological truth. Ernest Rutherford (1908), who discovered the nucleus of the atom, was known to say (Andrade, 1964), "*We've got no money, so we've got to think.*" In the age of biotechnopoly, it seems that "*We've got so much money to generate data, we have to hire computer scientists and bioinformaticists to tell us what to think.*" I cannot emphasize enough that gene annotation is the quicksand in the foundation of all omic science.

21.2.4 Targeted Characterization of the Genome

Instead of sequencing the entire genome in an untargeted and unbiased manner, it is possible to target certain aspects of the genome. One could target the nucleotide sequences that encode sequences of amino acids that make up polypeptides. This portion of the genome, which includes the exons, is known as the *exome*. One could also target the DNA sequences that make up the promoters on a genome-wide scale. This portion of the genome is known as the *cistrome* (Liu et al., 2011; Yu et al., 2016). One could target the DNA sequences that bind to a particular transcription factor. One could also target the portion of the genome that contains methylated cytosines or posttranslationally modified histones. The set of methylated cytosines is known as the *methylome* (Kim et al., 2014), and the set of methylated cytosines and posttranslationally modified histones is known as the *epigenome*. One could also target the *introme*, which consists of introns that may have regulatory activity. For example, some introns give rise to circular RNAs when the introns are spliced out of the hnRNA in the process of making mRNA. The circular RNA may have a function in regulating translation (Servick, 2017). To investigate any regulatory functions that might be ascribed to the repetitive sequences, one could target the *repeatome*. Lastly, one could target the unannotated portions of the genome that make up the dark matter of the genome (Santuari and Hardtke, 2010; Maumus and Quesneville, 2014; Ellens et al., 2017).

21.2.4.1 Exomics: Exome Enrichment Using Hybridization

A very small portion of the genome appears to directly code for proteins. To enrich the genomic DNA for the portions that encode proteins, the genomic DNA that hybridizes to ESTs or cDNAs is selected and then sequenced using any one of the methods described above in a procedure known as *whole-exome sequencing* (Neves et al., 2013). While the exome contains the elements of the gene that are transcribed into mRNA and are necessary to understand the function of the expressed protein, it lacks the regulatory elements that are important for the expression of the gene in time and space. Because the exome represents less than 2% of the human genome, yet contains approximately 85% of the sequences correlated with disease, whole-exome sequencing provides a cost-effective alternative to whole-genome sequencing and has the further advantage of producing smaller more manageable data sets. After all, how does one handle, store, and analyze terabytes (10^7), petabytes (10^8), exabytes (10^9), zettabytes (10^{10}), or yottabytes (10^{11}) of sequence data?

21.2.4.2 Epigenomics: Detecting Cytosine Modification Using Bisulfite Conversion

Chromatin is composed of DNA and histones (see Chapter 16). Chargaff (1951b) discovered that a proportion of cytosines isolated from DNA were methylated at the 5′-position. The histones also experience covalent modifications due to acetylation, phosphorylation, methylation, ubiquitylation, ADP-ribosylation, and sumoylation (Tariq and Paszkowski, 2004; Pfluger and Wagner, 2007; Saleh et al., 2008). These modifications, which are brought about by developmental and environmental signals, inhibit the binding of transcription factors and/or change the three-dimensional spatial organization of the genome (Wang et al., 2015). Gene silencing is the result of methylation of cytosines in the promoter region of the gene (Das and Messing, 1994; Richards, 2006; Lister et al., 2008; Lister and Ecker, 2009; Martin et al., 2009; Zhu et al., 2013; Giovannoni et al., 2017; Raza et al., 2017). The study of the heritable changes in the genome that occur as a result of covalent modification without any alteration in the sequence of deoxyribonucleotides in DNA is known as *epigenomics*.

The field of epigenomics began by taking a phenotype-first approach, which aimed to find the chemical cause of heritable polymorphisms that did not result from differences in the sequence of DNA. However, this work was considered to be time- and labor-intensive. Consequently, according to Seymour and Becker (2017), "*the advent of new sequencing technologies, coupled to steady decreases in the cost of sequencing, led to a flurry of genome-wide surveys of DNA methylation in both model-plant and crop species*."

DNA methylation is analyzed using the bisulfite conversion method, which converts unmethylated cytosine to uracil in a deamination process that leaves 5-methylcytosine intact. The treated DNA is then subjected to any of the sequencing methods described above. As in other omic techniques, the bottleneck in epigenomics is rapidly shifting from data acquisition to data analysis (Laird, 2010).

The methylomes of over 1001 naturally inbred lines of *Arabidopsis* have been characterized and their methylation patterns are correlated with the geography and climate in which the inbred lines originated (Kawakatsu et al., 2016; Lang et al., 2016). However, according to Seymour and Becker (2017), "*the relationship between DNA methylation and gene expression is still surprisingly muddled*."

21.2.4.3 Finding Transcription-Factor Specific Promoters Using ChiP-Seq

Regions of the genome that bind a specific transcription factor can be identified by immunoprecipitating chromatin fragments with an antibody that binds to the particular transcription factor (Ren et al., 2000). The DNA fragments enriched in the transcription factor binding site are then separated from the proteins in the immunoprecipitate and sequenced using any of the techniques described above. This technique where DNA that has been isolated from chromatin that has been immunoprecipitated is sequenced is known as ChIP-Seq.

ChIP-Seq makes it possible to identify simultaneously regions of the genome that serve as binding sites for transcription factors and other DNA-associated proteins that activate or repress gene expression (Fields, 2007; Johnson et al., 2007; Kaufmann et al., 2010a; Muiño et al., 2011; Heyndrickx et al., 2014; Yamaguchi et al., 2014). This makes it possible to characterize all the genes, activated or repressed by a specific transcription factor, that contribute to any given process. ChiP-Seq can also be used to identify the regions of the genome that are associated with the epigenomic covalent modifications of histones.

When doing ChIP-Seq, the cells are treated with a crosslinking agent so that the proteins associated with the DNA remain attached. The nuclei are then isolated, lysed, and sonicated into chromatin fragments composed of approximately 500 bp of DNA. These fragments are then immunoprecipitated by beads containing an antibody that recognizes the DNA-binding protein of interest. The chromatin is eluted from the beads, the proteins are removed from the chromatin, and the DNA is sequenced.

By determining the DNA sequence that interacts with the particular DNA-binding protein, the binding sites for the protein of interest can be mapped throughout the genome. In this way, binding sites for a number of transcription factors and the genes that they regulate have been discovered in plants (Kaufmann et al., 2009, 2010b; Mathieu et al., 2009a,b; Morohashi and Grotewold, 2009; Zheng et al., 2009; Ricardi et al., 2014; Sun et al., 2015a,b).

21.2.5 The Postgenomic Era: Functional Genomics

Walter Gilbert (Lewontin, 2001) claimed "*...that when we have the complete sequence of the human genome, we will know what it is to be human.*" Gilbert (1992) elaborated, "*Three billion bases of sequence can be put on a single compact disk (CD), and one will be able to pull a CD out of one's pocket and say,* 'Here is a human being; it's me!' But this will be difficult for humans. Not only do we look upon the human race as having tremendous variation; we look upon ourselves as having an infinite potential. To recognize that we are determined, in a certain sense, by a finite collection of information that is knowable will change our views of ourselves. It is the closing of an intellectual frontier, with which we will have to come to terms."

Structural genomics aimed to understand life through the sequencing of genomes, and big data sets of deoxyribonucleotide sequences have been produced by big science initiatives and consortiums (The *Arabidopsis* Genome Initiative, 2000; International Human Genome Sequencing Consortium, 2001, 2004). Whether the big data sets obtained with high-throughput technology, but without focused questions, have helped us to understand the fundamental nature of life any better is a real question.

Multicellular organisms are composed of a number of cell types, each with the same genome, and those cell types change during development and in response to the environment. In theory, bioinformaticists can produce algorithms that rely on many free parameters that can explain and predict all cell behavior from genome sequences. But is the theory good enough to produce algorithms that do not overfit the data?

Following the sequencing of various genomes, the relationship between a DNA sequence and a trait remained, as Winston Churchill (1939) described Russia, "*a riddle, wrapped in a mystery, inside an enigma; but perhaps there is a key.*" The *functional genomic* approach was proposed to be the key to understand the relationship between the gene, the DNA sequence, and a trait. Functional genomics investigates the function of the gene at every other level, including the transcriptional level, the posttranscriptional level, the translational level, the posttranslational level, the metabolic level, and the phenotypic level (Anonymous, 1997; Rastan and Beeley, 1997; Bunnik and Le Roch, 2013; Jha et al., 2015). Each aspect of functional genomics and the techniques developed to probe them will be discussed below.

According to the International Human Genome Sequencing Consortium (2001), "*The human genome project is but the latest increment in a remarkable scientific program whose origins stretch back a hundred years to the rediscovery of Mendel's laws and whose end is nowhere in sight. In a sense, it provides a capstone for efforts in the past century to discover genetic information and a foundation for efforts in the coming century to understand it.*

We find it humbling to gaze upon the human sequence now coming into focus. In principle, the string of genetic bits holds long-sought secrets of human development, physiology and medicine. In practice, our ability to transform such information into understanding remains woefully inadequate. This paper simply records some initial observations and attempts to frame issues for future study. Fulfilling the true promise of the Human Genome Project will be the work of tens of thousands of scientists around the world, in both academia and industry."

Before tens of thousands of scientists commit to fulfilling the true promise of the human genome project, it is worth assessing the value of what we have learned from sequencing the human genome and the genomes of crop plants and model organisms such as *Arabidopsis*, *Zea*, *Solanum*, *Physcomitrella*, and *Selaginella*? Does the value of the true promise of the big science of omics overwhelmingly tip the balance relative to the value of small science and non−omic research? Is it wise to collect massive amounts of data without a focused question? Is it wise to collect massive amounts of data without an intelligent way to analyze it? Perhaps, the shift in emphasis from focused research at the single gene level to unbiased research at the whole-genome level can be compared with a shift in focusing on one needle at a time to focusing on the haystack and cataloging all the pieces of hay in hopes of finding the needles (Cervantes, 1956; Hilgartner, 2017). Remember the words of Erwin Chargaff (1980), who said, "*We all know that what is cannot be otherwise. The existence of anything weights the scales most unfairly against everything else that could have been in its place but is not.*" Are we making the right decisions as to the future direction of science?

21.3 TRANSCRIPTOMICS

After the human genome was sequenced, James Watson (2003) wrote, "*With a fully formed dynamic understanding of when and where each of our 35,000—plus genes functions during normal development from fertilized egg to functioning adult, we would have a basis of comparison by which to understand every affliction: what we need is the complete human* 'transcriptome.' *This is the next holy grail of genetics, the next big quest in need of superfunding.*"

Likewise, Borevitz and Ecker (2004) wrote, "*The genome sequence of* Arabidopsis thaliana *has not been fully characterized, nor will such a large task be completed for some time. We have only touched the tip of the iceberg in our understanding of the information in the raw plant genome sequences. Deep comprehension of how the complex instructions of plant life are written in the linear code will be an enormous challenge for the future. Such an understanding will require the development of a third wave of new technologies to glean the next level of genomic complexity. Fortunately, new tools such as whole genome arrays (WGAs) can integrate genomic data on a common platform. Global genome data, including transcriptome atlases with alternatively spliced messages, DNA binding site profiles, and chromatin state surveys, will give a more holistic picture of the cells' activity.*"

The goals of transcriptomics are to catalog all species of RNA, including mRNAs, noncoding RNAs, and small RNAs; to determine the structure of genes, in terms of their start sites, their 5′ and 3′ ends, their splicing patterns, and other posttranscriptional modifications; and to quantify the changing expression levels of each transcript in space and time.

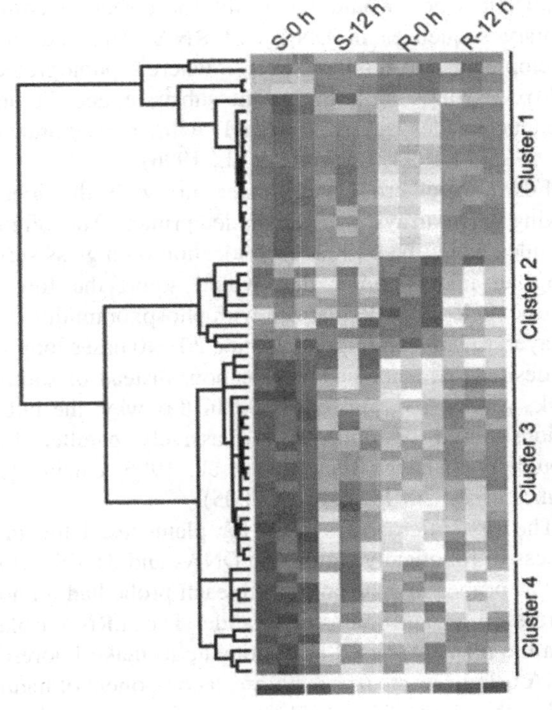

FIGURE 21.16 Results of a microarray analysis of 60 transcripts that are differentially expressed in resistant (R) and susceptible (S) wheat leaves in response to infection by powdery mildew 0 and 12 hours after infection. The 60 transcripts are separated into four clusters. Within a cluster, the transcripts show similar expression patterns; between clusters, the transcripts show distinct expression patterns. Each horizontal line represents one of the 60 transcripts. The vertical lines represent the treatment with three replicates. *From Xin, M., Wang, X., Peng, H., Yao, Y., Xie, C., Han, Y., Ni, Z., Sun, Q., 2012. Transcriptome comparison of susceptible and resistant wheat in response to powdery mildew infection. Genom. Proteom. Bioinform. 10, 94–106.*

21.3.1 Microarrays

Transcriptomics involves the characterization of the RNA transcripts produced in each and every cell at a given time. The sum total of the transcripts is known as the *transcriptome* (Velculescu et al., 1997). Originally the transcripts were characterized by their ability to hybridize to fragments of DNA on northern blots, which in principle are similar to Southern blots. In northern blots, mRNA with its poly A tail is isolated from various samples using a polyT column, separated by agarose gel electrophoresis, and transferred to a diazobenzyloxymethyl-paper membrane, where it is allowed to hybridize with radioactive cDNA probes. The patterns of hybridization in the samples are visualized on X-ray film, and the relative amounts of the probed mRNAs in the samples are determined by the relative opacity of the bands (Alwine et al., 1977; Evans et al., 1990; Memelink et al., 1994; Baldwin et al., 1999).

The desire for higher throughputs led to the development and use of microarrays to measure the differential expression of thousands of mRNAs (Fig. 21.16). Microarrays facilitate the hybridization between mRNA and probes in a massively parallel manner. Microarrays composed of various oligonucleotide probes are made using photolithography, similar to how integrated circuits are made (Drmanac et al., 1989; Khrapko et al., 1989; Kehoe et al., 1999). Indeed the microarrays produced by one brand are known as GeneChips. Fodor et al. (1991; 1993) initially used photolithography to synthesize different polypeptides, amino acid by amino acid, on a glass slide and used the synthetic polypeptides to discover the nature of the ligands that bound to biological receptors. They simultaneously synthesized a variety of oligopeptides from amino acids derivatized with nitroveratryloxycaronyl, a photolabile protecting group, on a glass surface using laser light to control the synthesis. The light removed the photolabile protecting group, allowing the introduced amino acids to bind to the amino functional group bound to the surface or to the free amino functional group of the last amino acid in the oligopeptide. By using a mask containing opaque and transparent areas to vary the exposure of light at each location on the glass surface, they could control the chemical synthesis of polypeptides. The polypeptide sequence at each location was determined by the sequence of reactants and the patterns of the masks used to irradiate each reactant. To explore the generality of combining photolithography with solid-phase chemistry, they also synthesized dinucleotides and suggested that the photolithography technique which allowed light-directed synthesis of nucleic acids would be useful for detecting complementary sequences in DNA and RNA. This led to the development of microarrays, where photoprotected hydroxyl groups are bound to the substrate, and the deoxyribonucleosides are derivatized with phosphoramidite (Pease et al., 1994; Lockhart et al., 1996).

Edwin Southern (1996) came up with the idea of making microarrays with an inkjet printer. The oligonucleotides would be built at each location on a glass surface using an inkjet printer that would squirt the four deoxyribonucleotides derivatized with phosphoramidite, layer by layer, to build an oligonucleotide 20–40 bases long with the desired sequence at each location, instead of squirting black, yellow, red, and blue ink. In this way, the hybridizations could be done in a massively parallel, high-throughput manner (Blanchard et al., 1996; Cuzin, 2001; Hughes et al., 2001; Harding, 2005).

The first microarrays used for plants used the inkjet process to robotically squirt 45 cDNAs and 31 ESTs to be used as probes in an array, where each probe had a known address on a glass microscope slide. The mRNA isolated from the plant was used as a template to make fluorescent cDNA using *reverse transcriptase*, a component of nature's toolkit that was discovered in retroviruses, which have RNA genomes. The virus uses reverse transcriptase to make a DNA copy of its genome that can be inserted into the host's DNA (Anonymous, 1970; Baltimore, 1970, 1975; Temin and Mizutani, 1970; Temin, 1975). The mRNA from roots and leaves were labeled with different fluorophores. The cDNAs from the two treatments were then hybridized with the probes on the slide, and the difference in the intensity of color emitted from each location indicated the difference in the transcripts expressed in the two organs (Schena et al., 1995). In time, the probes in the microarrays for plant became based on synthetic oligonucleotides and the number of probes in the microarray increased by orders of magnitude (Zhu and Wang, 2000; Alba et al., 2004).

Microarrays soon became commercialized by Affymetrix, Agilent, and other companies. Microarrays now hold millions of probes (O'Donnell-Maloney et al., 1996; Southern et al., 1992; Southern, 1996; Cuzin, 2001; Redman et al., 2004; Rensink and Buell, 2005; Galbraith, 2006). Microarrays contain specific oligonucleotides, typically 25 bases long, that are synthesized at known locations on an 18 μm × 18 μm glass wafer. Each location contains oligonucleotide sequences that represent perfect matches to different target sequences of a gene. Adjacent to each of those locations is a location that contains oligonucleotides that differ from a perfect match by a single nucleotide. These mismatch probes serve as a control for nonspecific hybridization. The different oligonucleotide sequences, which are known as probes, are chosen to represent the whole genome.

RNA can be isolated with high temporal resolution from flash-frozen tissues by immunopurifying the mRNA that is bound to ribosomes (Mustroph and Bailey-Serres, 2010). This population of RNA is known as the *translatome*. RNA can be isolated with high spatial resolution from a single cell type in multicellular plants by using microcapillary pipettes (Karrer et al., 1995; Brandt et al., 1999, 2002), laser capture microdissection (Asano et al., 2002; Kerk et al., 2003; Day et al., 2005, 2007a,b; Ohtsu et al., 2007; Yu et al., 2007; Scanlon et al., 2009; Yeats et al., 2010; Martin et al., 2016a), or a laser cell sorter (Brady et al., 2007). The small quantities of RNA must then be separated from the other components of the cell, reverse transcribed to make cDNA, and amplified (van Gelder et al., 1990).

To measure the expression of mRNA using a microarray, mRNA is isolated from the sample, and the amplification and biotinylation is done using nature's toolkit. A single-stranded copy of cDNA is made on the mRNA template using reverse transcriptase and a poly dT primer. The RNA–DNA hybrid is separated by heating the reaction mixture, and the RNA is digested with RNase. The remaining single-stranded cDNA is used as a template for DNA polymerase to form double-standed cDNA. In a reaction mixture that includes T7 RNA polymerase, a primer with the T7 promoter sequence, an oligo dT primer, and biotin-labeled ribonucleotides, the double-stranded cDNA serves as a template for the transcription of multiple copies of biotinylated cRNA. The biotinylated RNA fragments each represent a different mRNA from the sample and the quantity of biotinylated cDNA formed is related to the quantity of the original mRNA transcript.

The biotinylated cRNAs are then allowed to hybridize with the oligonucleotide probes on the microarray. The strength of the binding between the cRNA and the bound oligonucleotides depends on the degree in which the two sequences match. The greater the match, the better the two fragments hold together at high temperatures. By performing the hybridization at the appropriate temperature, a perfectly matched fragment will bind at a given temperature, whereas with a fragment with a single nucleotide difference, the hybridization will fall apart. Following hybridization, the microarray is washed, treated with a primary antibody that recognizes biotin, and then treated with streptavidin-phycoerythrin, which is a fluorescent molecule that binds to the primary antibody. The microarray is scanned with a laser and the relative light output from each spot on the chip is recorded. Because the oligonucleotide probes are synthesized in known locations on the array, the hybridization patterns and signal intensities can be interpreted in terms of gene identity and relative expression levels (Beier et al., 2004; Brownstein, 2006; Dalma-Weiszhausz et al., 2006).

The resolution of the microarrays depend on the choice of oligonucleotide probes. While microarrays contain oligonucleotides that cover the whole genome, some regions of the genome are better represented than others. In *tiling arrays*, the entire nonrepetitive genome sequence is synthesized as probes that are either overlapping, end-to-end, or equally spaced. The average distance between each pair of neighboring segments gives the resolution of the tiled path. An overlapping tiled path can give a resolution of one nucleotide (Yamada et al., 2003; Mockler and Ecker, 2005; Li et al., 2006; Yazaki et al., 2007).

The analysis of the RNA transcripts produced in a given cell type at a given time is known as *expression profiling* (Fig. 21.17). The goal of transcriptomics is to characterize the ribonucleotide sequence of the transcripts produced in every cell type at every time point as a function of normal development, environmental stimuli, mutation, or disease. To this end, the transcripts involved in cell differentiation and specific cell functions have been characterized from cells of the phloem (Asano et al., 2002; Hannapel et al., 2013), the root (Birnbaum et al., 2003; Lee et al., 2006; Brady et al., 2007), the epidermal and vascular tissues (Nakazono et al., 2003), pollen (Becker et al., 2003; Pina et al., 2005), the embryo (Casson et al., 2005; Spencer et al., 2007), the root cap (Jiang et al., 2006), the endosperm (Day et al., 2007a,b), the pericycle (Woll et al., 2005; Dembinsky et al., 2007), hairs (Wu et al., 2007; Deal and Henikoff, 2010; Chen et al., 2014), the root apical meristem (Brady et al., 2007), the abscission zone of the stamen (Cai and Lashbrook, 2008), the shoot apical meristem (Yadav et al., 2009; Brooks et al., 2009), in response to symbionts and pathogens (Klink et al., 2005; Wise et al., 2007; Balestrini et al., 2009), UV light (Wolf et al., 2010), and from stages throughout the life cycle of *Physcomitrella* (Hiss et al., 2014; Ortiz-Ramírez et al., 2016). The expression profiling procured from hybridization on thousands of transcripts can be compared and validated, in part, with in situ hybridization studies done with fluorescent oligonucleotide probes (Birnbaum et al., 2003; Yadav et al., 2009; Li et al., 2014; Frank et al., 2015).

In general, the microarray results show that each cell type contains thousands of transcripts represented by the gene ontology (GO) categories, a portion of which are specific to the cell type and a portion of which change in time or in response to a stimulus. The relative numbers of transcripts in each GO category implicates the relative importance of that particular GO category to the cell under investigation. Comparing mutant and wild-type cells sharpens the results by reducing the number of hypothetically relevant transcripts from thousands to tens while providing a workable resource for gene discovery. Again, the microarray studies are still cataloging every piece of hay in order to find the needle. Results from microarray studies can be said to provide an implication rather than a demonstration, as defined by Aristotle.

The goal in using microarrays for expression analysis is to provide a precise, accurate, quantitative, and unbiased estimate of the abundance of all transcripts in the sample of interest. The analysis of the transcriptome by microarrays has its limitations in that the technique has a limited dynamic range such that only the transcripts of average abundance can be quantified, whereas the transcripts with the least and most abundant transcripts escape accurate quantification. Thus, the importance of lowly or highly expressed RNAs in a given cell type or response cannot be excluded. Moreover, as probes with different ratios of CG to AT have different optimal temperatures for binding to

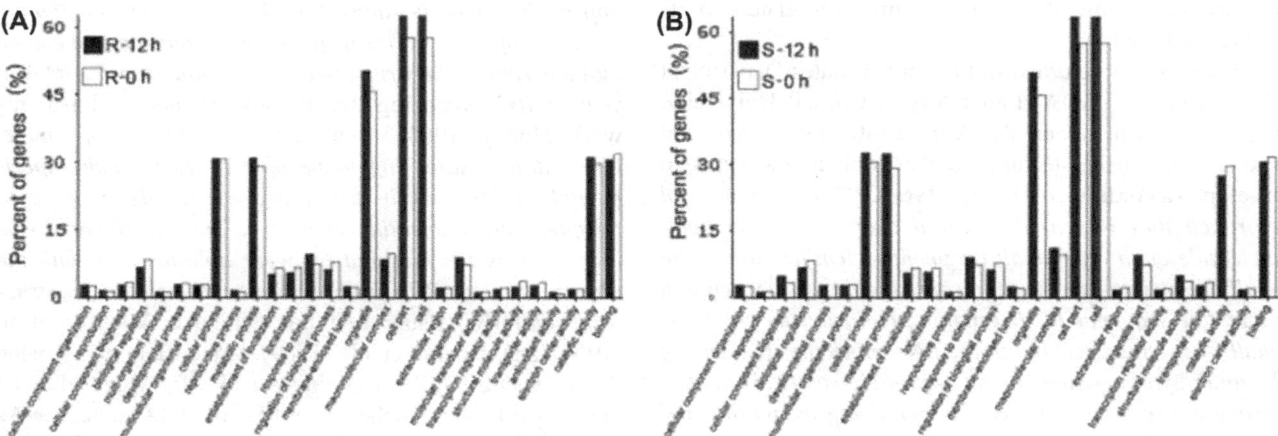

FIGURE 21.17 Analysis of the transcriptome of resistant (A) and susceptible (B) wheat leaves in response to infection by powdery mildew 0 and 12 hours after infection. The Affymetrix Wheat Genome Array containing 61,127 probe sets was used to quantify the transcriptome. The transcripts are categorized by gene ontology terms. The quality of the analysis depends on the accuracy of the gene annotation. *From Xin, M., Wang, X., Peng, H., Yao, Y., Xie, C., Han, Y., Ni, Z., Sun, Q., 2012. Transcriptome comparison of susceptible and resistant wheat in response to powdery mildew infection. Genom. Proteom. Bioinform. 10, 94–106.*

their complementary strands, more technical biases are introduced. This might reflect the fact that microarrays tend to favor transcripts related to membrane, cell surface, and secreted proteins (Zhang et al., 2016a,b). There is systematic bias introduced because hybridization tends to be greater at the boundary locations in the microarray (Steger et al., 2011). Another demerit of microarrays is that the microarrays are only available for taxa whose genome has been sequenced. RNA-Seq was developed as a possible method to decrease the bias and increase the accuracy of the data (Wang et al., 2009a,b).

21.3.2 RNA-Seq

RNA-Seq is an alternative to microarrays for characterizing the transcriptome. RNA-Seq makes use of massively parallel DNA sequencing technology instead of massively parallel hybridization technology (Lister et al., 2008; Wilhelm et al., 2008; Wang et al., 2009; Yassour et al., 2009; McGettigan, 2013). According to Kumar et al. (2012), "*Compared to other technologies such as hybridization-based microarrays and Sanger sequencing-based methods, RNA-Seq provides a more comprehensive understanding of transcriptome complexity and the ability to detect a dynamic range of expression levels....*"

Historically, ribonucleic acid was the first nucleic acid to be sequenced. While doing a sabbatic in James Bonner's laboratory at Cal Tech in 1956, Robert Holley (1968) became interested in protein synthesis and began his search to characterize the acceptor of activated amino acids, which at the time was known as soluble RNA and is now known as transfer RNA (see Chapter 17). As a structural biochemist, Holley was intrigued by the existence of a low−molecular mass, amino acid−specific RNA and set out to isolate the alanine tRNA, purify it, and sequence it, to understand how its chemical structure related to its biological function.

When Holley returned to the United States Department of Agriculture (USDA) Laboratory at Cornell University, he purified 1 g of alanine tRNA from 140 kg of commercial baker's yeast using techniques that took a few years to develop. According to Holley (1968), "*The experimental approach that was used involved cleavage of the polynucleotide chain into small fragments, identification of the small fragments, and then reconstruction of the original nucleotide sequence by determining the order in which the small fragments occurred in the RNA molecule. In terms of the analogy of a sentence, the approach was equivalent to breaking a sentence into words, identifying the words, and reconstructing the sequence of the letters in the sentence by determining the order of the words.*"

To determine the sequence of the alanine tRNA, Holley digested the purified tRNA with pancreatic ribonuclease, which cleaved the tRNA in such a way that each fragment ended in a pyrimidine ribonucleotide, or he digested it with Taka-Diastase ribonuclease T1, which cleaved the tRNA in such a way that each fragment ended with a guanine ribonucleotide (Holley et al., 1961, 1965a,b; Apgar et al., 1965; Penswick and Holley, 1965; Zamir et al., 1965). Holley then separated the fragments produced by each digestion into mononucleotides, dinucleotides, and higher oligonucleotides using DEAE-Sephadex chromatography. Holley hydrolyzed each of the separated fragments with alkali, and identified the composition of the component mononucleotides using UV spectroscopy and paper electrophoresis or two-dimensional paper chromatography. To determine the sequence of the ribonucleotides in each oligonucleotide, he performed a limited digestion using snake venom phosphodiesterase. Snake venom phosphodiesterase is an exonuclease that digests RNA from the 3′ end to yield ribonucleoside monophosphates that retain the phosphate bound to the 5′ carbon of ribose. The snake venom phosphodiesterase gave a series of stepwise degradation products with decreasing sizes. The successive peaks in the chromatogram represented the successive stepwise degradation products. By identifying the ribonucleosides obtained from the successive peaks, Holley got the nucleotide sequence of the fragment.

By matching up the sequences of fragments created by long or short digestions with the two ribonucleases, Holley and his colleagues discovered the complete sequence of the alanine tRNA. It took two and a half years to get the sequence of each fragment. The most time-consuming parts were the sequencing of the long fragments and the identification of the unusual ribonucleotides we now know are present in tRNA such as 1-methylinosinic acid and 5,6-dihydrouridylic acid. Holley (1968) realized that he revealed, "*the first nucleotide sequence known for a nucleic acid. Also, it can be said that the sequence gives, with appropriate modifications for DNA, the first nucleotide sequence of a gene. This would be the sequence of the gene that determines the structure of the alanine transfer RNA in yeast cells.*" Realizing that it took 9 years to finish this work, Holley (1968) when on to say, "*It was, of course, tremendously satisfying to be able to solve each experimental problem as it arose, and eventually be able to complete the nucleotide sequence. The satisfaction was increased by the fact that we were able to work with the alanine transfer RNA from discovery to isolation to structural analysis.*" Using a similar technique, Madison et al. (1966) and Doctor et al. (1969) sequenced the tyrosine tRNA. Using similar techniques and ^{32}P, which allowed one to work with smaller quantities of RNA, other RNAs were also sequenced (Brownlee and Sanger, 1967; Cory et al., 1968; Dube et al., 1968; Goodman et al., 1968; Adams et al., 1969; Min-Jou et al., 1972; Fiers et al., 1976).

Holley purified 1 g of RNA to sequence the alanine tRNA. Today, orders of magnitude less RNA has to be purified to sequence the entire transcriptome. In one of the original RNA-Seq papers, Nagalakshmi et al. (2008) isolated 200 ng of mRNA in the following manner: "*Total RNA was extracted from Ribopure Yeast kit (Ambion, Austin, TX) and treated for 30 min at 370C with RNAse free DNase I (Ambion). Poly(A) RNA was purified with the Micro Poly(A) purist Kit according to the manufacturer's instructions (Ambion).*"

The mRNA serves as a template for the sense strand of cDNA. Using Nature's toolkit 200 ng of poly(A) RNA is mixed with 50 ng of random primers or 50 ng of oligo (dT) primer, incubated at 650°C for 5 minutes, and then transferred to ice. The first strand cDNA synthesis is performed in reaction mixture containing reverse transcriptase and 0.25 mM deoriboxyribonucleotides in a total volume of 20 μL. The reaction mixture is incubated at 420°C for 50 minutes, and then subjected to RNase treatment for 20 minutes at 370°C. The resulting first strand cDNA is used to make second strand cDNA in a reaction mixture containing 10 mM deoxyribonucleotides, DNA ligase, DNA polymerase, and RNase in a total volume of 150 μL. This reaction mixture is incubated at 160°C for 2 hours. The resulting double-stranded cDNA is purified using the QIAquick PCR purification kit (Qiagen, Valencia, CA). Samples are then fragmented with DNase I at 370°C to yield fragment sizes of 100−300 bps.

The cDNAs that represent each fragment of mRNA are ligated to adapters that include identifying barcodes. If one is using single molecule sequencing methods, there is no need to amplify the cDNA before sequencing, which ensures that the relative abundance of the cDNAs to be sequenced is equal to the relative abundance of the mRNA transcripts. If one is using one of the massively parallel sequencing methods, then, the cDNA must be amplified with the PCR using adapter-specific primers. It is important to ascertain that the relative abundance of amplified cDNA is equal to the relative abundance of the mRNA transcripts. It is possible to retain the information as to which cDNA strand represents the sense strand and which one represents the antisense strand and then use adapters and adapter-specific primers that preserve that information (Levin et al., 2010; Tafolla-Arellano et al., 2017). The resulting cDNAs are sequenced using any of the technologies described above, and the reads are assembled de novo or aligned with a reference genome with software (Grabherr et al., 2011; Kukurba and Montgomery, 2015). An analysis pipeline calculates the significant genes that define the different samples.

RNA-Seq has been used to characterize the transcriptome of flowers (Leydon et al., 2017; Zhang et al., 2017), the transcriptome in epidermal cells (Tafolla-Arellano et al., 2017), the transcriptome in leaves in response to a pathogen (Zuluaga et al., 2016a,b; Frenandez-Pozo et al., 2017), and the transcriptomes in leaves of 120 cultivated and wild accessions of spinach (Xu et al., 2017). However, isolating transcripts from a whole organ or tissue system represents the average transcriptome of all cells and is not necessarily representative of any cell type (Brandt, 2005). This makes it imperative to isolate mRNA from a tissue with identical cells or single cells. RNA-seq is beginning to be used to study how plant evolution will keep pace with climate change (Voelckel et al., 2017).

A single cell has approximately 1−60 pg of total RNA (Brennecke et al., 2013; Zheng et al., 2017a,b), and it is possible to do expression profiling of single cells using RNA-Seq (Hashimshony et al., 2012; Efroni and Birnbaum, 2016; Ziegenhain et al., 2017). Teichert et al. (2012) have characterized the transcriptome during fungal development; Li et al. (2014a,b) and Thakare et al. (2014) have characterized the transcriptome of endosperm cells; Frank and Scanlon (2015a,b) and Frank et al. (2015) have characterized the transcriptome of apical meristematic cells in evolutionarily diverse species; and Efroni et al. (2015) have characterized the transcriptome in root cells.

It is important to be aware when analyzing the results of global transcript profiling that technical biases can be introduced by the methods. For example, different amplification methods preferentially amplify different transcripts resulting in the possibility that the relative abundance of the amplified transcripts is not necessarily equivalent to the relative abundances of transcripts in the starting mRNA (Meyers et al., 2004; Day et al., 2007a; Scanlon et al., 2009; Martin et al., 2013; Coenen et al., 2015). To ensure that transcriptome analyses are quantitative, especially when starting with picogram or nanogram quantities of RNA, it is important to know that the relative abundance of transcripts in the amplified product is equal to the relative abundance of transcripts in the cell (Iscove et al., 2002).

Each high-throughput transcriptomic study provides a list, collection, or catalog of transcripts, the first of "*three of the major requirements for a systems approach to understanding biological systems — the parts list, their interconnections, and the dynamics of the parts*" as outlined by Leroy Hood (Hood et al., 2012). According to Aristotle, a systematic presentation of what constitutes scientific knowledge demands that a demonstration must be founded on known principles, which themselves are demonstrable or on first principles that are considered self-evident; there cannot be an infinite number of middle terms between the first principle and the conclusion; and the demonstrations cannot be circular, where the conclusion is supported by the premises and the premises by the conclusion. Primarily, as a result of the weakness in gene annotation (Bolger et al., 2017a,b), but also due to the uncertainty caused by contamination and technical bias when trying to isolate quantitatively every transcript (Scanlon et al., 2009), I am not aware of any published high-throughput transcriptomic research that has produced more than lists and that goes beyond that which is already known to explain an unknown process in terms of a better known process.

According to Meyers et al. (2004), "*The ability to simultaneously measure the expression of thousands of genes is a powerful analytical system, and the availability of technologies for this has presented scientists with many new opportunities....The massive datasets generated by gene expression technologies present novel statistical and analytical problems, resulting in a convergence of biology, mathematics, and computer science.*" As the movie *Moneyball* revealed, data scientists can make sense of big data to identify young baseball players among thousands who will be successful once they reached the majors. In a like manner, bioinformatics gurus are being called on to find algorithms to make sense of the transcriptomics data and to extract meaningful insights, particularly if it leads to the uncovering of transcripts that serve as biomarkers of human disease. A true biological understanding of the number and identity of transcripts that contribute primarily to a given developmental process or physiological effect, in contrast to a statistical understanding of the massive data sets produced by transcriptomics, has not been attained. While each new technology gives us a new way of cataloging all the pieces of hay, we are still looking for the needle in a haystack by cataloging all the pieces of hay.

21.4 PROTEOMICS

As described in the previous 20 chapters, proteins such as the plasma membrane and vacuolar H^+-ATPases, the K^+ channels, the viral movement protein, the signal recognition particle, the protein translocator, actin, myosin, tubulin, kinesin, dynein, calmodulin, cohesion, condensin, and DNA polymerase were isolated one or two at a time, and their purpose and regulation were revealed through functional assays. Since the last edition of this book, there has been a great effort to increase the breadth and depth of protein characterization through high-throughput technological pipelines. The complete catalog of all the proteins in an organism is known as the *proteome,* and the investigation of the proteome is known as *proteomics*. Characterizing the proteome in space and time is known as *protein profiling*.

The genome is relatively stable from cell to cell and during the life of a plant. The genome can be considered digital information composed of two copies of linear sequences of four different bases composing tens of thousands of genes, which are primarily localized in the nucleus. The proteome, such as the transcriptome, differs from cell to cell and from time to time. The proteome, unlike the genome or transcriptome, is analog in nature, the proteins being dynamic three-dimensional structures composed of a linear sequence of 20 different amino acids that may be folded in various ways (Kim et al., 2015), posttranslationally modified and/or proteolytically processed, and that are present in every compartment in the cell in varying quantities from one to more than tens of million of copies (Picotti et al., 2009; Hood et al., 2012). Alternative splicing results in even more isoforms. Thus, it is a Herculean challenge to cover the proteome comprehensively and discriminate between low-abundance proteins and contaminants (Horgan and Kenny, 2011; Angel et al., 2012). Consequently, international collaborations have been proposed to fulfill the promise of proteomics (Tyers and Mann, 2003).

21.4.1 Characterizing the Proteome Using Polyacrylamide Gel Electrophoresis

The introduction of one-dimensional (Raymond and Weintraub, 1959; Raymond et al., 1962; Raymond, 1964; Shapiro et al., 1966, 1967; Laemmli, 1970; Maizel, 2000) polyacrylamide gel electrophoresis made it possible to separate many proteins simultaneously, even when starting with small quantities (see Chapter 2). This made it possible to characterize the protein composition of different cells and organelles and to observe how it changed during development or in response to various environmental signals. By itself, polyacrylamide gel electrophoresis did not yield enough chemical information to identify the proteins separated in the gel. However, the electrophoretic transfer of the proteins to nitrocellulose paper, so that they could be stained with antibodies, made it possible to identify one or two proteins in the gel using a technique known as western blotting (Towbin, 1979, 1988; Burnette, 1981, 1991). However, most of the proteins were left unidentified.

The introduction of two-dimensional (Klose, 1975; O'Farrell, 1975; Berth et al., 2007) polyacrylamide gel electrophoresis made it possible to resolve many more proteins. In two-dimensional gel electrophoresis, the polypeptides are first separated in a tube in one dimension by their isoelectric points (pI). This process is known as *isoelectric focusing*. When doing isoelectric focusing, the charged polypeptides migrate through an immobilized pH gradient along an electric field. As long as the pI of the polypeptide is not equal to the pH of the gel, the polypeptide will be charged and will move along the applied electric field. However, when the polypeptide moves to a region where its pI is equal to the pH, the protein will lose its charge and will no longer migrate along the electric field.

The gel containing the polypeptides separated by isoelectric focusing is then treated with a detergent to denature the polypeptides. Then the cylindrical gel containing the isoelectrically focused polypeptides is placed along the top of a denaturing slab gel made of polyacrylamide. The denatured polypeptides move from the isoelectric focusing gel into the slab gel along an electric field and are separated by their molecular masses. Two-dimensional polyacrylamide gel electrophoresis resolves most proteins because most proteins have a unique combination of pI and molecular mass. Two protein samples, each one labeled with a different fluorescent probe, can be run at the same time on the same two-dimensional polyacrylamide gel, allowing one to visualize differential protein expression on the same gel (Unlü et al., 1997). Polypeptides isolated from two-dimensional polyacrylamide gel electrophoresis can be sequenced by mass spectrometry (Yang et al., 2007).

The polypeptides can also be digested by proteases in the gel and then isolated. In this case, the sequence of the protein is determined from the proteolytic fragments.

21.4.2 Sequencing of a Polypeptide by Stepwise Degradation

To sequence a protein, it must be purified using methods such as differential solubilization, chromatography, or ultracentrifugation (see Chapter 17). The sequence of amino acids in a polypeptide was originally determined by partially digesting the polypeptide, separating the fragments with paper chromatography, and performing a stepwise chemical degradation on the N-terminal amino acid of each fragment using a method initially developed to specifically determine the N-terminal amino acid. Each fragment was treated with or without nitrous acid and then completely hydrolyzed, and the amino acid content of the hydrolyzates treated with or without nitrous acid was analyzed and compared. The amino acid missing in the nitrous acid–treated preparation was the N-terminal amino acid. The procedure was repeated until the whole polypeptide was degraded amino acid by amino acid. This method was sufficient to sequence short polypeptides such as gramicidin S, an antibiotic isolated from the soil bacterium *Bacillus brevis*, but gave ambiguous results for longer polypeptides due to the lack of specificity of the nitrous acid reaction (Consden et al., 1947; Wieland and Bodanszky, 1991). When n represents the number of amino acids in a polypeptide, the accuracy of the nitrous acid sequencing method declines as $n \to \infty$.

Bergmann et al. (1927), Abderhalden and Brockmann (1930), and Bergmann and Zervas (1937) developed a better method to identify the N-terminal amino acid by using phenylisocyanate to make a phenylureido derivative of the polypeptide and then treating the polypeptide with methanolic HCl at high temperature. With this treatment, the polypeptide splits into the phenylhydantoin derivative of the N-terminal amino acid with the free amino group and a polypeptide that was one amino acid shorter than the original polypeptide. The procedure is repeated until the whole polypeptide is degraded amino acid by amino acid. Pehr Edman (1949, 1950) improved the procedure by using phenylisothiocyanate, a sulfur-containing analogue of phenylisocyanate, to form a phenylthioureido derivative of the polypeptide in the presence of an anhydrous medium—thus, preventing any nonspecific cleavage of the polypeptide and ensuring that only the amino acid at the N-terminus was cleaved from the polypeptide in each round. He was able to determine the sequence of short synthetic polypeptides.

Frederick Sanger began working on the nitrogenous compounds in the potato (Neuberger and Sanger, 1942) and decided to work on understanding proteins. Sanger (1958) replaced the Edman's phenylisocyanate with fluorodinitrobenzene to determine the N-terminal amino acid of insulin. It seemed that insulin had more than one N-terminal amino acid and was composed of more than one polypeptide chain. Because insulin contains a number of cystines, Sanger guessed that the polypeptide chains were held together by disulfate bonds. Sanger reduced the cystines to cysteic acids with performic acid and then separated a polypeptide with glycine in the N-terminal position from another polypeptide with phenylalanine in the N-terminal position by their differential solubility. Sanger partially digested each polypetide chain separately with acid and proteolytic enzymes, separated the fragments by paper chromatography, determined the N-terminal amino acid of each fragment, and then subjected each fragment to complete hydrolysis to determine the amino acid composition of the fragment. By comparing the amino acid sequences of the various fragments, he could reconstruct the amino acid sequence of both polypeptides of insulin.

Sanger (1958) observed, "*Examination of the sequences of the two chains reveals neither evidence of periodicity of any kind, nor does there seem to be any basic principle which determines the arrangement of the residues. They seem to be put together in an order that is random, but nevertheless unique and most significant, since on it must depend the important physiological action of the hormone.*" Sanger won the 1958 Nobel Prize in Chemistry for sequencing insulin (Sansom, 2006). Soon many other proteins were sequenced using his methods. As larger and larger proteins were sequenced, we learned the importance of the sequence in terms of the biological activity of the protein. We also learned the biological principle known as the signal hypothesis, which states that the targeting of a polypeptide to a given organelle depends roughly on its sequence (Blobel, 1999; see Chapter 4).

The degradation techniques used for protein sequencing could only sequence one polypeptide at a time. The protein sequencing process was sped up when Edman and Begg (1967) automated the Edman reaction. Decades later, Leroy Hood and his colleagues developed a protein sequencer that was so sensitive when it performed the cyclic Edman chemistry that cleaved the amino acids one at a time from the N-terminus of a protein that only one hundredth of the protein was necessary to sequence the protein compared with conventional chemistry (Hunkapiller and Hood, 1980; Hewick et al., 1981). Hood (2008) used the protein sequencer to sequence erythropoietin, which led to the identification and cloning of the erythropoietin gene and the development of the first billion dollar drug produced using biotechnology. Hood (2008) also developed an automated peptide synthesizer that was able to synthesize the 99 amino acid long HIV protease (Kent et al., 1984). The chemically synthesized protein was crystallized, its crystalline structure was resolved, and an antiprotease drug to treat AIDS was the result (Miller et al., 1989). The protein sequencer and peptide synthesizer were commercialized by Applied Biosystems (Burzik, 2006; Springer, 2006).

21.4.3 Sequencing of a Polypeptide by Mass Spectrometry

The successful application of the comparative genomic approach in showing that membrane receptors and G-proteins have similar sequences and the realization that the sequences of the genes gave information concerning the function of the proteins seemed to make protein sequencing unnecessary, given that the sequencing of genes was so much easier (James, 1997). This trend changed when biochemistry became proteomics and proteomics became an application of the genome projects. According to Domon and Aebersold (2006), "*The trend toward mass spectrometry as the technique of choice for identifying and probing the covalent structure of proteins was accelerated by the genome project. Genomics demonstrated the power of high-throughput, comprehensive analyses of biological systems. Genomics also provides complete genomic sequences, which are a critical resource for identifying proteins quickly and robustly by the correlation of mass spectrometric measurements of peptides with sequence databases. The systematic analysis of all the proteins in a tissue or cell was popularized under the name proteomics, with mass spectrometry central to most proteomic strategies.*"

Protein analysis by mass spectrometry was made possible by Klaus Biemann (2002; Biemann et al., 1959), who is known as the father of organic mass spectrometry of terrestrial and extraterrestrial samples (Chung, 2006). Biemann used a mass spectrometer (MS), which had previously been used to identify atoms, isotopes, and organic molecules by their mass-to-charge ratio, to identify the sequence of amino acids in polypeptides. Biemann reduced the polypeptides to polyamines because polyamines could be volatilized, which is a necessary prerequisite for mass spectrometry.

In a mass spectrometer, a volatilized sample moves along an electric potential (V) in a vacuum at a velocity (v) that depends inversely on its mass-to-charge ratio (m/z). The mass-to-charge ratio is determined from the time (t) it takes for a molecule to travel a distance (L) along the electric potential. The mass-to-charge ratio is obtained from the following equations that make use of the first law of thermodynamics that states that the kinetic energy $\left(\frac{1}{2}mv^2\right)$ is equal to the electrical potential energy (zV) and the relationship between velocity, time, and distance $\left(v = \frac{L}{t}\right)$.

$$zV = \frac{1}{2}mv^2 \tag{21.1}$$

$$v = \sqrt{\frac{2Vz}{m}} \tag{21.2}$$

$$t = \frac{L}{v} = L\sqrt{\frac{m}{2Vz}} \tag{21.3}$$

$$t^2 = \frac{L^2}{2V}(m/z) \tag{21.4}$$

Initially the time of flight was determined by the time it took for a molecule to pass a UV or IR detector. Subsequently, molecules could be identified by their mass-to-charge ratio derived from the time of flight. The UV and IR spectroscopic detectors were later replaced by electrical detectors that produced an electrical signal in response to being bombarded by a charged molecule. The greater the number of charges that hit and deposited charge on the detector at a given time, the greater the electrical signal at that time.

A mass spectrometer also includes a magnetic field that accelerates the molecules according to their mass-to-charge ratio in a direction perpendicular to the acceleration by the electric potential. The molecules are thus deflected according to their mass-to-charge ratio so that different electrical detectors arranged in an array report on molecules with different mass-to-charge ratios, and the magnitude of the electrical signal from a molecule with a given mass-to-charge ratio is presented graphically. There are currently a variety of detectors used in mass spectrometers, each with a different resolution and sensitivity (Zhang et al., 2010). R. Graham Cooks (2004) reminisces that "*the only thing that has changed since then is that the human mind has been removed from the loop.*"

As the number of monomers (n) in a molecule increases and $n \rightarrow \infty$, it gets harder and harder to volatilize the molecule. Moreover, as the molecules must be charged to move along the electric potential in the mass spectrometer and as $n \rightarrow \infty$, the molecules become more fragile. This means that the violent photoionization, electron ionization, and chemical ionization techniques used to ionize atoms and small molecules with low mass-to-charge ratios for mass spectrometry cause extensive decomposition of polypeptides with a high mass-to-charge ratio (McLafferty, 2011).

John Fenn, who had been interested in jet propulsion to power antiaircraft missiles (Kolb and Herschbach, 1984; Robinson, 2011), was going to solve this problem so that he could perform mass spectrometry on polypeptides. Fenn began his career creating supersonic molecular beams containing volatile low–molecular mass molecules with high translational energies under steady state conditions that were far from equilibrium. His purpose was to study molecule–molecule and molecule–surface interactions, where the proportion of translational, vibrational, and rotational energies could be controlled (Abuaf et al., 1967).

Fenn adopted the technique of *electrospray ionization* (ESI) to volatilize the polypeptides after learning about it from Malcolm Dole (1989). Malcolm Dole was a chemist who worked as a consultant for a paint company. Electrospraying was used by the automobile industry to paint cars. By establishing an electric potential between the sprayer and the surface to be sprayed, the nebulized paint droplets obtained a great enough surface charge to direct them toward the surface being painted thus minimizing the paint lost to the surroundings. Dole realized that electrospraying could be used in mass spectrometry to volatilize droplets of high-polymeric molecules without degrading them. Moreover, the solution in which the protein was dissolved could be adjusted to give the protein any given charge. By electrospraying the high polymers into dry nitrogen gas, Dole hoped that the solution surrounding them would evaporate as the polymers with the charge headed toward the detector. However, Dole et al. (1968) were unable to find a relationship between the mass-to-charge ratio of the polymers and their time of flight in the mass spectrometer.

The lack of relation between the mass-to-charge ratio and the mobility was due to the fact that the desolvated ions acted as condensation nuclei, and the charged polymers reacquired the water from the nitrogen gas and became resolvated (Fenn, 2002a,b). Fenn overcame the condensation problem that Dole experienced by replacing the stagnant volume of dry nitrogen gas with a countercurrent of dry nitrogen gas. Consequently, when the fine spray of charged droplets, each containing a single charged polypeptide, was injected into the mass spectrometer, the water evaporated and traveled with the dry nitrogen gas perpendicular to the detector, while the dry polypeptide with its given mass-to-charge ratio was accelerated toward the detector. Fenn (2002a) had given electrospray wings to molecular elephants. This led to more sophisticated studies on larger and larger molecules worthy of many Nobel Prizes in Chemistry (Herschbach, 1986; Lee, 1986; Polanyi, 1986; Curl, 1996; Kroto, 1996; Smalley, 1996; Zewail, 1999). By the mid 1990's, mass spectrometry replaced Edman degradation for the sequencing of polypeptides (Domon and Aebersold, 2006).

The mass spectrum of a polypeptide exhibits a number of peaks (Fig. 21.18). Each peak differs from the adjacent peaks by having one more or one fewer proton adduct, which provides the charge. There is a coherence in the spectrum in that adjacent peaks differ by one charge. Thus, it is relatively easy to calculate the mass of the charge adducts in each peak, making it possible to calculate the mass of the polypeptide independently from each peak. The first proteins to be characterized by mass spectrometry using ESI were gramicidin S, cyclosporine, bradykinin, insulin, lysozyme, α-amylase, cytochrome *c*, myoglobin, α-chymotrypsin, alcohol dehydrogenase, and conalbumin (Fenn et al., 1989).

As an alternative to volatilizing polypeptides by electrospray, one can use a laser pulse to ionize proteins from a matrix on which a sample has been deposited (Tanaka et al., 1988; Kara and Hillenkamp, 1988). Again it became important that the radiant energy of the laser does not break the bonds of the molecule, an unwanted phenomenon that became more and more difficult to avoid as $n \rightarrow \infty$ and the molecular mass increased. This problem was solved by Koichi Tanaka by combining the protein sample with

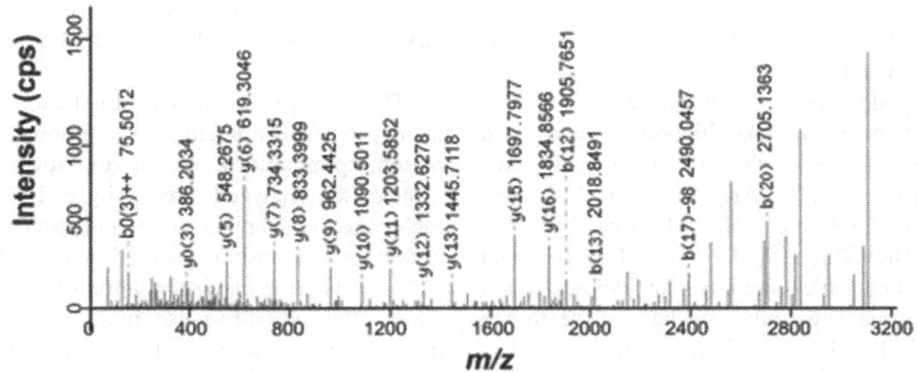

FIGURE 21.18 Example of a mass spectrum showing the intensity versus mass-to-charge ratio. *From Quintas-Granados, L.I., Lopez-Camarillo, C., Fandiño Armas, J., Mendoza Hernandez, G., Alvarez-Sánchez, M.E., 2013. Identification of the phosphorylated residues in TvelF5A by mass spectrometry. Genom. Proteom. Bioinform. 11, 378–384.*

ultrafine metal powder, a few tens of nanometers in diameter, suspended in glycerin. The small size of the ultrafine metal powder ensured that the laser light would be absorbed more efficiently by the matrix thus maximizing the heating of the sample and minimizing the heat loss to the environment. The desorbed proteins form a cloud of freely hovering, electrically charged polypeptide–ion combinations that can be accelerated in a vacuum chamber so that the time it takes to travel a given distance can be measured. The first proteins to be characterized by mass spectrometry using matrix-assisted laser desorption ionization (MALDI) were lysozyme, chymotrypsinogen, trypsin, β-lactoglobulin A, and bovine albumin.

The development of soft ionization techniques such as ESI, in which the sample is sprayed by a strong electric field (Yamashita and Fenn, 1984; Fenn et al., 1989; Fenn, 2002b), and MALDI, in which the sample is released by a laser from a solid or viscous matrix, has made it possible to produce steady state molecular beams containing intact ions from complex and nonvolatile molecules, including polypeptides. These discoveries also were worthy of Nobel Prizes in Chemistry (Fenn, 2002a; Tanaka, 2002) and opened the possibility for proteomics.

It is possible to couple two or more mass spectrometers (MS; McLafferty et al., 1980). In the MS/MS combination, the first mass spectrometer fractionates the sample by mass-to-charge ratio, and then the fraction with the desired mass-to-charge ratio is selected by electromagnetic focusing and passes into the second mass spectrometer where each polypeptide is fragmented when it collides with an argon atom or a nitrogen molecule, and the kinetic energy is converted into vibrational energy. The collisions that occur in the second mass spectrometer produces a spectrum of fragments, and the difference in the mass between adjacent fragments with the same charge gives the mass of the amino acid that is present in the more massive fragment and absent in the lighter one. The daughter fragments can be isolated and passed to a third mass spectrometer in which granddaughter fragments are produced and analyzed. In some cases, 10 generations of fragments have been characterized (James, 1997). By analyzing the amino acid difference in each fragment pair, the amino acid sequence of the entire polypeptide can be deduced. However, this method does not give an unambiguous amino acid sequence because only 18 of the 20 amino acids have unique masses (in Da; Ala, 71; Arg, 156; Asn, 114; Asp, 115, Cys, 103; Glu, 129; Gly, 57; His, 137; Met, 131; Phe, 147; Pro, 97; Ser, 87; Thr, 101; Trp, 186; Tyr, 163, and Val, 99). Leucine and isoleucine have an identical mass (113.1594), and lysine and glutamine have close to the same mass (128.1741 and 128.1307, respectively).

The sequence of a polypeptide can also be deduced by treating the polypeptide with an exopeptidase and measuring the mass-to-charge ratio of the fragments produced by limited digestion (James, 1997). This is known as ladder sequencing.

A mass spectrometer or tandem MS/MS can be directly or indirectly coupled to a high-performance liquid chromatograph (HPLC) so that the polypeptides can be fractionated in the liquid chromatograph by polarity prior to their mass-to-charge ratios being determined in the first mass spectrograph and their sequences being determined in the second mass spectrometer.

The sequence of a given polypetide can be deduced by peptide mass fingerprinting, in which the molecular masses of the fragments generated by a specific protease, typically trypsin, are compared to the masses of the fragments already in protein databases (http://www.expasy.org/proteomics; Elias et al., 2004; Beausoleil et al., 2006). The masses of the fragments can also be analyzed by web-based search engines, such as Protein Prospector (http://prospector.ucsf.edu/prospector/mshome.htm) and Profound (http://prowl.rockefeller.edu/prowl-cgi/profound.exe), both of which predict the proteolytic digestion pattern of a protein encoded by a given nucleotide sequence (Oliver et al., 1983). Comparisons made with scrambled databases are used to estimate the false discovery rate.

21.4.4 Characterizing the Proteome by Mass Spectrometry

To get the protein fraction from a given organelle, cells must be homogenized, the homogenate must be filtered, and the filtrate must be centrifuged in a manner that selects for the proper organelle or suborganellar compartment. The fraction may also be subjected to aqueous two-phase partitioning or free-flow electrophoresis. The proteins are then solubilized, and the solubilized proteins are either partially digested with a protease and fractionated by liquid chromatography or fractionated by one- or two-dimensional polyacrylamide gel electrophoresis where the separated proteins are cut out of the gel and partially digested with a protease (Friso et al., 2011). Each spot containing proteolytic fragments is then analyzed by mass spectrometry (Sabidó et al., 2012; Gemperline et al., 2016; Feng et al., 2017).

There are two approaches to characterize the proteome using mass spectrometry: the top-down approach and the bottom-up approach. In the top-down approach, intact proteins are analyzed directly by ESI-MS/MS to give the molecular mass, composition, and sequence of each protein as well as any posttranslational modifications (McLafferty, 2011). While the characterization of a given protein is accurate, the signal-to-noise ratio, which depends on the mass-to-charge ratio, is low (Valaskovic et al., 1996; Compton et al., 2011; Tran et al., 2011). Therefore, the top-down approach should be used for purified proteins and not used for high-throughput studies. Moreover, it is difficult to ionize intact proteins with molecular masses greater than 30 kD. Therefore, the top-down approach introduces a bias based on molecular mass when trying to profile the entire proteome.

In the bottom-up approach (Domon and Aebersold, 2010), it is not the intact molecule that is directly analyzed by the ESI-MS/MS but either a shotgun mix of proteins or the proteolytic fragments of a protein that has been separated by polyacrylamide gel electrophoresis. The MS gives a list of masses corresponding to the fragments produced by the original protein. Proteolytic cleavage of the intact protein erases the complete sequence information from the protein, and it has to be resurrected bioinformatically by comparing the fragments observed in the mass spectrograph with the proteolytic fragments predicted from knowing the sequence of the gene that encodes the protein.

Comparing the fragmentation pattern observed when using the bottom-up approach with the fragmentation pattern predicted from the gene sequence is complicated due to any posttranslational modifications that may be part of the fragment (Kelleher, 2004). Zabrouskov et al. (2003) compared the top-down and bottom-up methodologies and concluded that the bottom-up method is preferable for the identification of proteins in organisms whose genome has been sequenced as it is capable of identifying more proteins (97) than the top-down method (22). However, the top-down method was capable of showing that the majority of proteins characterized by the top-down method had molecular masses that differed from the mass predicted by gene sequence data, which is indicative of posttranslational modifications. Top-down analysis gives in-depth coverage of each protein, whereas the bottom-up approach gives high-throughput coverage of the entire proteome without giving in-depth coverage of any polypetide. Because bottom-up proteomic studies use peptides as a proxy for proteins, the nature of the intact protein is inferred and not determined directly by the high-throughput, bottom-up approach (Toby et al., 2016).

When the goal of proteomics is to maximize the sequence coverage, then it is sufficient to perform bottom-up proteomics that provides an accurate measurement of the masses of thousands of proteins in minute amounts of sample that are or are not consistent with the predicted gene sequences. The masses that are inconsistent with the predicted gene sequence may represent proteins formed by alternative splicing or proteins that have been modified posttranslationally. Because these modifications are particularly important for signal transduction and regulatory control, a comprehensive description of the proteome, which depends on detecting the varied and changing forms of protein molecules, is quite a challenge.

One can reduce the challenge of characterizing the complete proteome by using targeted proteomics (Picotti et al., 2009). Targeted proteomics involves making hypotheses concerning which proteins are important biomarkers of normal development or disease and then tailoring the extraction procedure by including immunoaffinity depletion, cation exchange, and other types of chromatography to ensure their inclusion. Polypeptides that are posttranslationally modified through phosphorylation, which makes up the phosphoproteome (Reiland et al., 2009; Schulze, 2010; Fíla et al., 2012, 2016; Dephoure et al., 2013; Facette et al., 2013; P. Wang et al., 2013b; X. Wang et al., 2013c; Zhang et al., 2013a,b; Stecker et al., 2014; van Wijk et al., 2014; Friso and van Wijk, 2015; Li et al., 2015a,b; Marcon et al., 2015; Minkoff et al., 2015; Silva-Sanchez et al., 2015; Lohscheider et al., 2016), or by glycosylation, which makes up the glycoproteome (Catalá et al., 2011; Ruiz-May et al., 2012a,b; Yang et al., 2017), can be affinity purified and then analyzed by mass spectrometry. The affinity purified postranslationally modified proteins are then separated into two samples. One sample is directly analyzed by mass spectrometry, and the other sample is treated enzymatically or chemically to remove the posttranslational modification and then analyzed by mass spectrometry. The patterns of posttranslational modification are deduced with algorithms that analyze and compare the mass spectra from the two samples.

Plant biologists working on crop plants (e.g., Tan et al., 2017) and model systems (e.g., Wienkoop et al., 2010) are using proteomic techniques. Proteomic studies of all subcellular compartments (Dunkley et al., 2006; Wang et al., 2016), including the plasma membrane (Natera et al., 2008; de Michele et al., 2016), plasmodesmata (Fernandez-Calvino et al., 2011), endoplasmic reticulum (Maltman et al., 2007; Komatsu et al., 2012), lipid droplets (Davidi et al., 2015); peroxisome (Arai et al., 2008a,b; Eubel et al., 2007, 2008; Reumann et al., 2007, 2009; Reumann, 2011), Golgi (Asakura et al., 2006; Nikolovski et al., 2012; Parsons et al., 2012, 2013; Ford et al., 2016), trans-Golgi network (Drakakaki et al., 2012), endosomes (Heard et al., 2015), the vacuole (Carter et al., 2004) and vacuolar membrane (Shimaoka et al., 2004; Ohnishi et al., 2018), plastids (Peltier et al., 2000, 2002, 2004; van Wijk, 2000; Schubert et al., 2002; van Wijk and Baginsky, 2011; Barsan et al., 2012; Lundquist et al., 2012, 2013; Offermann et al., 2015; van Wijk and Kessler, 2017), mitochondria (Salvato et al., 2014; Taylor and Millar, 2015), the nucleus (Yin and Komatsu, 2016), the nucleolus (Pendle et al., 2005), and the extracellular matrix (Isaacson and Rose, 2006; Rose and Lee, 2010; Yeats et al., 2010; Lee and Rose, 2011; Lee and Rose, 2011; Ruiz-May and Rose, 2013) have been done. Seaton et al. (2018) have done a proteomic study to see how the proteins in many of these organelles change with the photoperiod. The proteome of the pericarp of tomato in response to a cutin mutation (Martin et al., 2016b) and the proteome of plant in response to pathogens (Kaffarnik et al., 2009; Shah et al., 2012) have been studied. According to Fíla et al. (2017), more than 100 omic studies have already been performed on the male gametophyte. Have those studies made a single addition to the fundamental knowledge we have about pollen? Wei et al. (2017) studied the response of the single-celled alga *Dunaliella* to a salinity shock that induced the immotile palmella stage. They report changes in everything...but what are the important changes? An effort into increasing confidence in the identification of proteins in a single cell is being made (Angel et al., 2012).

Each of the proteomic studies provide a list, collection, or catalog of proteins, the first of "*three of the major requirements for a systems approach to understanding biological systems – the parts list, their interconnections, and the dynamics of the parts*" as outlined by Leroy Hood (Hood et al., 2012). The long lists of proteins are usually reduced by presenting the proteins in terms of their GO classification (Fig. 21.19).

Making a decision about the number and identity of polypeptides in a list that exists in a given organelle or that contribute primarily to a given developmental process or physiological effect involves navigating between the Scylla of false negatives and the Charybdis of false positives. Megan Scudellari (2011) reported, "Proteomics still awaits its own version of RNA-Seq, some fast technology to boost the output of protein data. Technology in the field is improving, says Eugene Kolker, chief data officer at Seattle Children's Hospital, who is involved in numerous proteomics studies. Still, for proteomics (and metabolomics, the profiling of metabolites to fully characterize a cell's metabolic pathways and processes) to be truly quantitative, he says, scientists must be able to accurately measure concentrations of proteins and other molecules in a sample, not just determine whether they are present or not—a refinement that is still in the works."

The rapid generation of vast quantities of data has outpaced the analysis of data being done by specialists in computational biology or bioinformatics—specialists who should also understand the biological complexity of the sample and the extraction techniques to ensure that the data

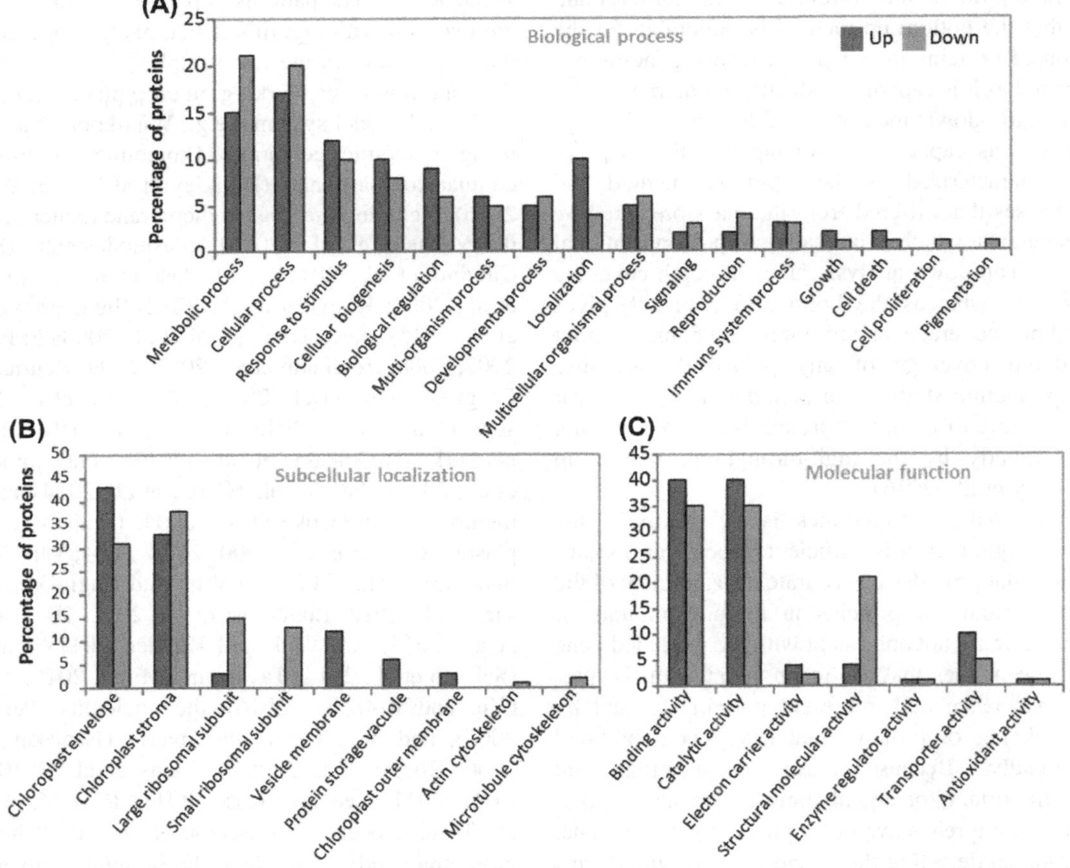

FIGURE 21.19 A functional analysis of the proteins that are up regulated or down regulated in a protein phosphatase inhibitor mutant of *Arabidopsis* compared with wild type. The proteins are catagorized by gene ontology terms. (A) biological processes; (B) subcellular localization; (C) molecular functions. The quality of the analysis depends on the accuracy of the gene annotation. *From Ahsan, N., Chen, M., Salvato, F., Wilson, R.S., Prasad Rao, R.S., Thelen, J.J., 2017. Comparative proteomic analysis provides insight into the biological role of protein phosphatase inhibitor-2 from Arabidopsis. J. Proteom. 165, 51–60.*

obtained are as meaningful as they are measureable. Ralph Waldo Emerson worried about such fragmentation and wrote a jeremiad stating, "*The state of society is one in which the members have suffered amputation from the trunk, and strut about so many walking monsters, - a good finger, a neck, a stomach, an elbow, but never a man.*" Indeed the imbalance between specialists and generalists is affecting many aspects of science (Sequeira, 1988; Niklas et al., 2013; Ledford, 2018).

21.5 INTERACTOMICS

As we discussed in Chapter 9, to maintain the living condition for an extended period of time, all the parts of the protoplasm are necessary and together form the basis of life. As Edwin Conklin (1940) wrote, "*Life is not found in atoms or molecules or genes as such, but in organization.*" There are interactions at every level in a plant. Understanding the relationship of the cell to the organism is fundamental in understanding life in multicellular organisms and the interactions within cells and among cells (see Chapter 3). Moreover, there are proteins in the cell, including photoreceptors, protein kinases, and protein phosphatases, that are involved in cell signaling and regulate many other proteins (see Chapter 12), just as transcription factors are proteins that regulate the synthesis of mRNA that encodes other proteins (see Chapter 16). *Interactomics* is the study of all the interactions between all the proteins in an organism (Braun et al., 2013). The protein−protein interactome, or simply the *interactome*, should be determined by systematic experiments performed with proteins under physiological conditions (Kiemer and Cesareni, 2007). The next best way to study interactions between two proteins at a time is to use the various bimolecular complementation systems such as the yeast two-hybrid system. In the yeast two-hybrid system, one of the proteins is genetically fused to the DNA binding domain of a transcription factor, whereas the other protein is genetically fused to the activation domain of the transcription factor. If the two proteins interact when they are expressed in yeast, the transcription factor is reconstituted and the protein−protein interaction activates a reporter gene (Fields and Song, 1989; Yazaki et al., 2016). The newer bimolecular complementation systems release a transcription factor or emit light as a result of Förster resonance energy transfer or by bringing two halves of GTP or similar fluorescent proteins together when the two proteins interact. The bimolecular complementation systems are amenable to high-throughput studies in yeast, where interactions between every pair of proteins in the proteome can be studied (Uetz et al., 2000; Miller et al., 2005). False positives and false negatives occur, and just because two proteins interact in yeast does not mean that they interact in the plant cell. Jones et al. (2014a,b,c) developed a low-throughput technique that ensures the interaction takes place in a given plant cell. The high-throughput interactome is a description of every possible interacting pair of polypeptides encoded by the genome. For example, a genome with 25,000 genes encodes $25{,}000^2 = 625$ million possible pairwise interactions. In the past, proteins were considered a priori to act independently (Ovádi, 1991; Kühn-Velten, 1993; Mathews, 1993; Cascante et al., 1994; see Chapter 5), today protein−protein interactions are considered the norm, even if they are not in the same place at the same time.

One can use the algorithms of "network science" to produce a wiring diagram of the cell by drawing the interconnections that describe the interactions between different proteins (Barabási and Oltvai, 2004; Braun, 2012). For example, if protein A interacts with proteins B, C, and F, then you draw links from A to B, A to C, and A to F. If protein C interacts with proteins D, E, and F, then you draw links from C to D, from C to E, and from C to F. The proteins are considered nodes, and the combination of nodes and links make a network of networks that describes the protein−protein interactions that are predicted to take place. The wiring diagram of wheat leaves, or dust bunny diagram, as I call it, is shown in Fig. 21.20. The nodes are characterized by their connectivity, which is the number of links they have to other nodes. The nodes with the most connectivity are called hubs and are considered particularly important in the regulation of cell development, metabolism, and physiology. According to Barabási and Oltvai (2004), the application of network science to omic science "*has led to the realization that the architectural features of*

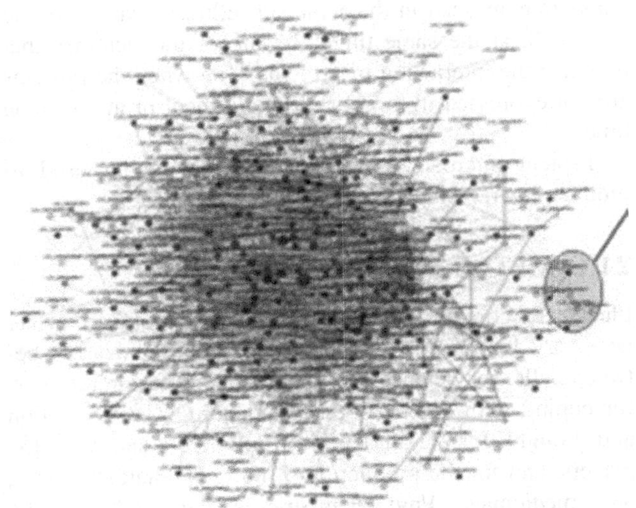

FIGURE 21.20 The predicted interactome of rice that shows the 76,585 protein−protein interactions that are predicted to occur among 5049 proteins. *From Zhu, P., Gu, H., Jiao, Y., Huang, D., Chen, M., 2011. Computational identification of protein-protein interactions in rice based on the predicted rice interactome network. Genom. Proteom. Bioinform. 9, 128−137.*

molecular interaction networks within a cell are shared to a large degree by other complex systems, such as the Internet, computer chips and society. This unexpected universality indicates that similar laws may govern most complex networks in nature, which allows the expertise from large and well-mapped non-biological systems to be used to characterize the intricate interwoven relationships that govern cellular functions... It is impossible to ignore the apparent universality we have witnessed by delving into the totality of pairwise interactions among the various molecules of a cell. Instead of chance and randomness, we have found a high degree of internal order that governs the cell's molecular organization." I think that it is fair to compare the intelligibility and usefulness of the science that yields dust bunny diagrams to the science that gave us intelligible diagrams that emphasized the economy of nature (Krebs and Kornberg, 1957), including diagrams of glycolysis, the ornithine cycle (Kornberg, 2000; Wilson et al., 2010), the Krebs cycle, the Calvin cycle, and the glyoxylate cycle, which, if we were to give it an eponymous name, would be the Kornberg cycle.

Geisler-Lee et al. (2007) initially deduced the interactome of *Arabidopsis* from the experimental interactomes obtained in other species. They built an interactome by assuming that a pair of interacting orthologs or interologs in the reference species predicted an interaction in *Arabidopsis*. Since then, genes from plant taxa have been used in the bimolecular complementation assay to produce interactomes (Arabidopsis Interactome Mapping Consortium, 2011; Gu et al., 2011; Mochida and Shinozaki, 2011; Yang et al., 2012; Jones et al., 2014a,b,c; Yue et al., 2016). However, the interactome is an abstraction that only has meaning if the two genes that encode the interacting proteins are expressed in the same subcellular location, in the same cell, at the same time. Beware of interactomes that illustrate the interaction of proteins even when the proteins are not expressed in the same compartment or at the same time.

Protein microarray methods are now being used to produce interactomes (Yazaki et al., 2016).

21.6 METABOLOMICS

Plant cells contain a diverse array of metabolites that are necessary for the life of the cell, for communication between cells, for defense against pathogens and herbivores, for coping with abiotic stresses, including UV irradiation and drought, for the attraction of pollinators and seed dispersers, and for the production of nutrients, hallucinogens, and medicinals. Phytochemistry is the study of the chemicals in plants, particularly alkaloids, glycosides, polyphenols and terpenes, and includes the isolation of the chemicals, the structural analysis of the chemicals, the biosynthesis of the chemicals, and the function of the chemicals in the plant, in animals, and in humans (Lind, 1757; Eijkman, 1929; Szent-Györgyi, 1939a,b; Terris, 1964; Pauling, 1976; Schultes, 1976, 1988; Schultes and Hofmann, 1979, 1980; Hofmann, 1980; Rodriguez et al., 1985; Wragham and Rodriguez, 1989; Schultes and Raffauf, 1990, 1992; Robinson, 1991a,b; Arnason et al., 1995; Tang and Han, 1999; Sneader, 2000; Kandell, 2001; DeJoseph et al., 2002; Meskin et al., 2002; Osbourn and Lanzotti, 2009; Tu, 2011, 2015; Weng et al., 2012; Li and Weng, 2017). Plant *metabolomics* differs from phytochemistry in being the large-scale and unbiased analysis of the relative changes or differences in the concentrations of metabolites that occur (Fiehn, 2002; Fernie, 2003; Dixon and Strack, 2003; Goodacre et al., 2004) with reference to the genome of the organism in question (Hirai et al., 2004; Sharma and Shrivastava, 2016; Moghe and Kruse, 2018).

Truly unbiased metabolomics is inherently challenging as a consequence of the broad differences in the volatility, solubility, polarity, and concentration of the metabolites. In addition, chemicals that do not exist naturally in the plant form when some metabolites react with other metabolites in an extraction mixture (Lu et al., 2017). As the number of metabolites (n) one tries to collect in an unbiased manner increases, $n \rightarrow \infty$, and misinformation may result when one tries to collect everything without quantifying anything well.

The *metabolome*, which is defined as the total complement of metabolites in a cell (Ward et al., 2003), represents the end products of gene expression and the biochemical phenotype of a cell (Sumner et al., 2003). Traditionally, a gene responsible for the production of a given metabolite was discovered using a cell-free lysate that contains the enzymes that produce a product from a precursor. The enzymes were characterized by their binding constants to various precursors, their rate of product production (see Chapter 12), and their sensitivity to inhibitors. The enzyme, with known characteristics, was then isolated and purified to homogeneity. The protein was then digested by various proteases so that the fragments would be sequenced by Edman degradation. The likely nucleotide sequence was determined from the amino acid sequence, and then an oligonucleotide was chemically synthesized to fish out the strand of DNA from the organism that produces the metabolite using hybridization in Southern blots or the PCR.

As a given metabolite may be produced in the wild type but not in a mutant, or it may only be produced in a given cell type at a given stage of development or in response to a given stimulus, differential genomic and transcriptomic techniques can be used to identify the genes responsible for the enzymes that produce the metabolite in question under one condition but not in the other (Fig. 21.21; Torrens-Spence et al., 2016; Wisecaver et al., 2017). Once the DNA sequence has been identified, the enzymes necessary

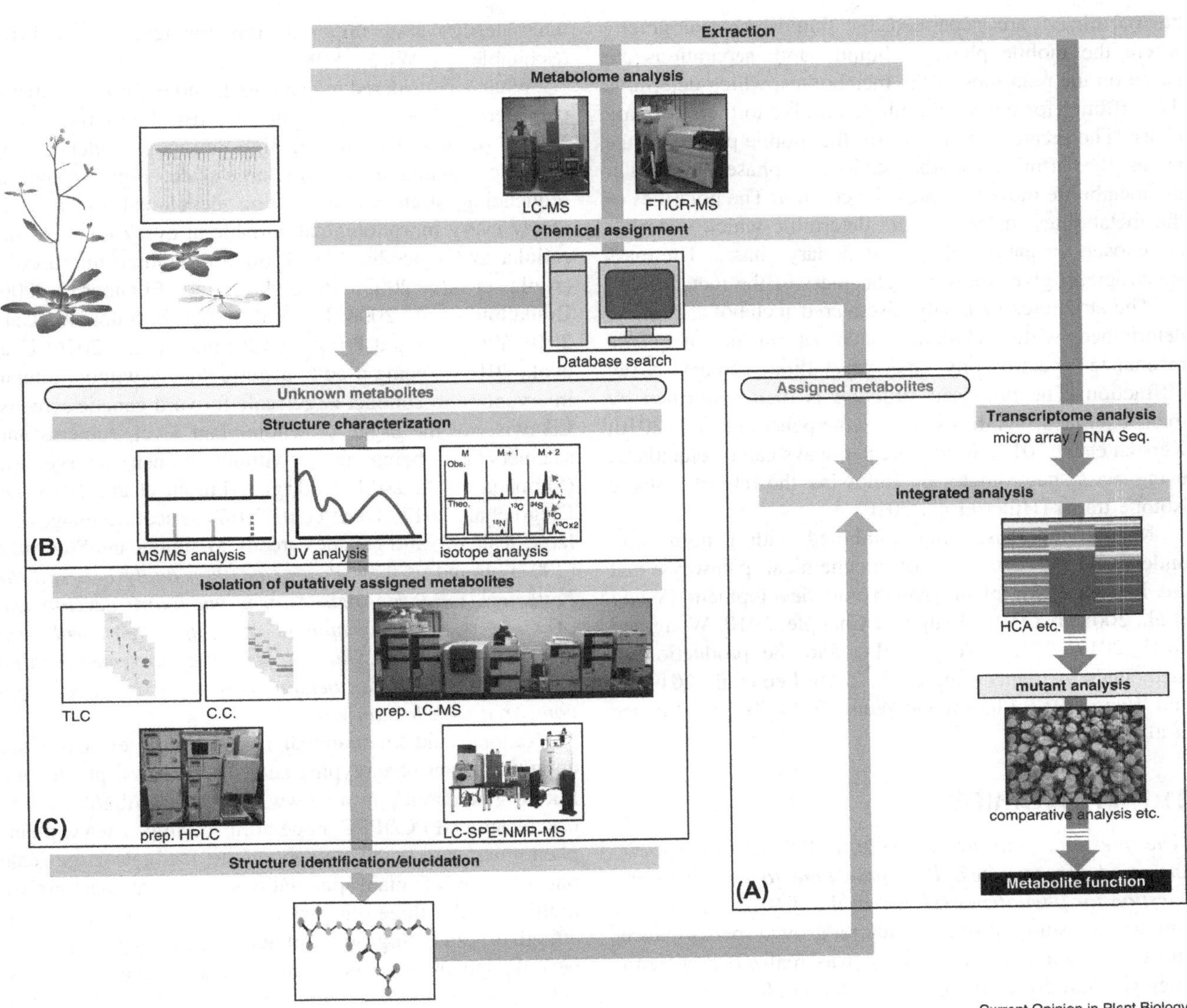

FIGURE 21.21 Metabolomic workflow. (A) The function of the metabolite is determined in a mutant whose transcriptome has been profiled to identify the genes responsible for the metabolites; (B) the chemical features of the metabolites are characterized; and (C) the structures of the metabolites are elucidated. *From Nakabayashi, R., Saito, K., 2015. Integrated metabolomics for abiotic stress response in plants. Curr. Opin. Plant Biol. 24, 10–16.*

to produce the metabolite in question can be produced using recombinant DNA techniques. Compared with purifying the enzyme from the original organism where it is likely to be of low abundance, purification of the overexpressed protein produced by bacteria or yeast and secreted into the medium is relatively easy—especially in strains of bacteria or yeast that have been selected for this purpose (Racker, 1976; Torrens-Spence et al., 2016).

Metabolic pathways have been totally reconstituted in a heterologous host system such as *E. coli*, *Saccharomyces cerevisiae*, and *Nicotiana benthamiana*. The reconstitution of pathways in an organism that has been transformed with the genes that encode the enzymes in the metabolic pathway in question not only provides proof of the pathway but also provides a way to produce valuable natural products, including artemisinin and opioids, using *metabolic engineering* and *synthetic biology* on an industrial scale (Paddon et al., 2013; Galanie et al., 2015; Nielsen, 2015; Torrens-Spence et al., 2016).

Once solubilized and extracted, the metabolites in a cell from a wild type or mutant, or in response to a given stimulus or developmental stage, can be isolated and characterized using gas chromatography—mass spectrometry (Broeckling et al., 2005; Chmielewska et al., 2016; Fiehn, 2016) or HPLC—mass spectrometry (Farag et al., 2008; Sadeghnezhad et al., 2016). Metabolites that can be volatilized are separated by gas chromatography where the mobile phase is gaseous, and the metabolites that cannot be volatilized are separated by liquid chromatography where the mobile phase is liquid. Both separations are based on the polarities of the metabolites, which determine their affinity for the mobile phase relative to the stationary phase. The greater the affinity for the mobile phase and the lesser the affinity for the stationary phase, the faster the metabolite moves through the column. The polarities of the metabolites in the mixture determine which materials are chosen for the mobile and stationary phases. The mass spectrograph gives the molecular mass of the metabolite.

The structures of newly discovered metabolites can be determined with techniques such as nuclear magnetic resonance spectroscopy and crystalline sponge X-ray diffraction. The proposed structure is then confirmed by total chemical synthesis (Torrens-Spence et al., 2016; Kersten et al., 2017). Metabolic pathways can be elucidated using mass spectrometry by following the fate of a stable isotope tracer (Hiller et al., 2010).

Metabolomics is being combined with genomics to understand the evolution of biochemical pathways that are important for plant growth and development (Weng et al., 2008a,b; 2010; Weng and Chapple, 2010; Weng and Noel, 2012, 2013; Weng, 2014) and the production of medicinals by plants (Kim et al., 2016; Lee et al., 2017; Li and Weng, 2017; Pluskal and Weng, 2017; Torrens-Spence et al., 2017).

21.7 PHENOMICS

"*The question,* 'why not measure it all?' *was fortunately, answered for genomes; it is now time to ask the same question for phenotypes* (Houle et al., 2010)." *Phenomics*, which is the study of plant structure, growth, performance, and composition, began when it was realized that genomics generated massive data sets that could not be correlated to detailed phenotypic data sets (Furbank, 2009; Furbank and Tester, 2011). That is, the relationship between the sequence of nucleotides and the traits they encoded was questionable. The solution was to begin industrial-scale, high-throughput phenotyping (Lu et al., 2008). Heretofore, plant biologists predominantly studied the morphological and physiological variables that seemed most relevant, and plant breeders kept a keen eye out for sports that would provide the genetic material for understanding the material basis of heredity (Wayne, 2012c) and for breeding better plants (Darwin, 1868; Bailey, 1901; Harlan, 1992). For example, in the summer of 1953, while Henry Munger was picking blueberries in his mother-in-law's backyard in Cape Cod, he noticed a wild carrot plant that had pink petals rather than white petals. There was no pollen in the pink flowers. That Fall, Munger collected seeds from the sport and then bred the cytoplasmic male sterility trait, which the sport had, into cultivated carrots (Mutschler, 1986). Today, cytoplasmic male sterility is an important trait for generating hybrids (Schnable and Wise, 1998).

What is considered relevant to the non−omic scientist is considered biased by the omic scientist. To minimize any chance of bias, large-scale, high-throughput phenotyping strategies analogous to the large-scale, high-throughput sequencing strategies are being developed to measure nearly every morphological variable at every level. At the cellular and subcellular level, high-throughput microscopy (Avila et al., 2003; Pepperkok and Ellenberg, 2006; Boulaflous et al., 2008; Lu et al., 2008a,b; Salomon et al., 2010; Zwiewka and Friml, 2012; Styles et al., 2016; Usaj et al., 2016) is being used to acquire images without human intervention to conduct large-scale forward genetic screens. Likewise, at the organ or whole-plant level, cameras and scanners are being used without human intervention (Shimada et al., 2011; Rodriguez-Furlán et al., 2016; De Diego et al., 2017; Tomé et al., 2017) to acquire images for large-scale forward genetic screens. Wollman and Stuurman (2007) remind us that "*Post-acquisition analysis is still the bottleneck for large-scale screens but future development of more powerful computers and algorithms will help mitigate this further. Still, it is important to remember that computer vision is far behind in some of its capabilities compared with human vision.*"

National and international facilities now exist (https://eppn2020.plant-phenotyping.eu/; http://www.plantphenomics.org.au/about/; http://www.fz-juelich.de/ibg/ibg-2/EN/organisation/JPPC/JPPC_node.html; https://www.plant-phenomics.ac.uk/index.php/about/) to facilitate large-scale phenotyping of plant populations using nondestructive, multispectral, three-dimensional, high-resolution, cost-effective, and high-throughput technology that can be mounted on tractors, cranes, drones, helium balloons, helicopters, and manned aircraft (Fig. 21.22; Finkel, 2009; Kelley, 2009; Berger et al., 2012; Brown et al., 2012; Vankudavath et al., 2012; White et al., 2012; Yazdanbakhsh and Fisahn, 2012; Yang et al., 2013; Andrade-Sanchez et al., 2014; Cozzolino et al., 2015; Rahman et al., 2015; Haghighattalab et al., 2016; Zhang et al., 2016; Salas Fernandez et al., 2017). Laser radar or lidar is being used to measure growth rates (Hosoi and Omasa, 2009; Konishi et al., 2009), infrared cameras are being used to measure temperature profiles, photosynthesis, and transpiration rates (Qiu et al., 2009; Sirault et al., 2009), fluorescence is being used to screen for mutants in photorespiration and disease resistance (Badger et al., 2009; Scholes and Rolfe, 2009), X-ray and magnetic resonance imaging are being used to measure root growth, and positron emission tomography is being used to measure translocation.

Once the data are collected, bioinformatics is used to extract the phenotypic information from the data and to create a searchable and meaningful phenotypic databases that will link gene sequence to the phenotype. According to

FIGURE 21.22 Sensors attached to a high-clearance tractor simultaneously characterize plant height, temperature, and spectral reflectance of cotton plants. *From White, J.W., Andrade-Sanchez, P., Gore, M.A., Bronson, K. F., Coffelt, T.A., Conley, M.M., Feldmann, K.A., Heun, J.T., Hunsaker, D.J., Kimball, B.A., Roth, R.L., Strand, R.J., Thorp, K.R., Wall, G.W., Wang, G., 2012. Field-based phenomics for plant genetics research. Field Crops Research 133, 101–112.*

Jennifer Normanly (2012), the premise of phenomics is that "*the higher the resolution of the phenotype analysis the more likely that new genes and complex interactions will be revealed...As robotics, computing, and imaging technologies all continue to advance at a rapid rate, the list of quantifiable assays that can be carried out in high-throughput and at high resolution will continue to expand, providing more tools to understand plant growth and development.*"

When doing large-scale, high-throughput remote sensing studies in the field, one must remember that when comparing the phenotypes of various genotypes in the field, bias can be introduced serendipitously, for example, if the soil in which one genotype is growing happens to be different from the soil in which the other genotypes are growing. Sometimes it is the little things that count, and an experienced farmer or breeder walking through the field just might recognize something that the computers were not programmed to analyze.

Furbank and Tester (2011) refer to plant phenomics "*as simply plant physiology in* 'new clothes.'" While the phenomic scientist wants to ensure that every important variable is measured and analyzed, the phenotypes vary from cell to cell and from moment to moment, which led Houle et al. (2010) to acknowledge that "*phenomics will always involve prioritizing what to measure and a balance between exploratory and explanatory goals.*" That is, the scientist must use his/her brain as well as technical tools when doing science. After all, "*Science,*" according to Erwin Chargaff (1980), "*is the application of reason, and mainly of logic, to the study of the phenomena of nature. Therefore, the most important scientific tool is the human brain. Each brain sits in its own head. Hence, the all-important unit in research is the individual scientist.*" Otherwise plant phenomics will become plant physiology with no clothes.

21.8 PAN-OMICS

The limited success of any one omic technique in fulling the promise of systems biology has led to *pan-omics* or the simultaneous and comprehensive measurement of the genome, transcriptome, proteome, metabolome, etc. (Voros, 2014). Pan-omics, also known as multiomics (Balcke et al., 2017; Haas et al., 2017), is the Holy Grail of omic science. According to Angel et al. (2012), "*the power of pan-omics will be increasingly realized by the integration of information from a range of measurements, enabling modeling and predicting biological processes and response to external stimuli, which collectively constitutes a systems biology approach to biological sciences.*" Laplace would smile.

21.9 SINGLE-CELL OMICS

A. J. P. Martin (1964), a chemist who was once a biochemist, realized that nature was clever in the use of small quantities. Consequently he realized that the *"appetite of the chemist to work on a small scale will grow as it becomes more possible. He will be able to analyze and experiment on single cells. There is obviously an almost limitless field in making and using apparatus for measuring various physical properties on small objects."* As technology allows the movement of chemists toward single cells, we will no longer be held hostage to knowing the chemical composition that represents an average of all cells that may or may not be representative of any cell (Kennedy et al., 1989; Badiei et al., 2002; Toriello et al., 2008; Spiller et al., 2010; Liang et al., 2014; de Vargas Roditi and Claassen, 2015). Indeed, Leroy Hood (Hood et al., 2012), one of the founders of the human genome project, realized that *"in order to understand fundamental biological or disease mechanisms −single cell analyses will be critical."* Will we be doing pan-omics at the single cell level?

21.10 ONE, TWO, THREE...INFINITY REVISITED

A year after the publication of the first edition of this book, plant biologists met to determine a common research goal for the following decade. Chory et al. (2000) decided that, *"In order to most efficiently and safely manipulate plants to meet growing societal needs, we must create a wiring diagram of a plant through its entire life cycle: from germinating seed to production of the next generation of seeds in mature flowers. These processes are guided by genes and the proteins they encode. They are directed by both intrinsic developmental cues and environmental signals. The long-term goal for plant biology following complete sequencing of the Arabidopsis genome is to understand every molecular interaction in every cell throughout a plant lifecycle. In essence, to understand the function of every gene by the year 2010. The ultimate expression of our goal is nothing short of a virtual plant which one could observe growing on a computer screen, stopping this process at any point in that development, and with the click of a computer mouse, accessing all the genetic information expressed in any organ or cell under a variety of environmental conditions."* To accomplish this goal, *"New experimental tools that investigate gene function at the subcellular, cellular, organ, organismal, and ecosystem level need to be developed. New bioinformatics tools to analyze and extract meaning from increasingly systems-based datasets will need to be developed. These will require, in part, creation of entirely new tools. An important and revolutionary aspect of The (2010) Project is that it implicitly endorses the allocation of resources to attempts to assign function to genes that have no known function. This represents a significant departure from the common practice of defining and justifying a scientific goal based on the biological phenomena. The rationale for endorsing this radical change is that for the first time it is feasible to envision a whole-systems approach to gene and protein function. This whole-systems approach promises to be orders of magnitude more efficient than the conventional approach."* I argue that the goal was not met for two reasons: the focus was on every gene and every cell, rather on some essential genes and particular cells, and the focus was on technology and not on biology. This reminds me of Muggeridge's (2006) quote, "I think, therefore you're not says the computer."

Neil Postman (1992) defines a technopoly as a society in which technology is deified, meaning *"the culture seeks its authorization in technology, finds its satisfactions in technology, and takes its orders from technology."* A technopoly is characterized by a new kind of social order and a surplus of technology-generated information, which is analyzed by technological tools to provide direction and purpose for society and individuals. *"Those who feel most comfortable in Technopoly are those who are convinced that technical progress is humanity's supreme achievement and the instrument by which our most profound dilemmas may be solved."* Systems biology may be a technopolistic science:

- We put our faith in technology to generate data with high throughput, even when we know that the reliability of many techniques, including nucleotide sequencing, gel electrophoresis, and mass spectrometry, decreases as the number of monomers $n \rightarrow \infty$.
- We put our faith in a discovery-based science that is unbiased, even when we know that the experimental techniques themselves introduce bias as the number of things we want to assay $n \rightarrow \infty$.
- We put our faith in the computer software that analyzes the high-throughput data, even though we are unaware of the number of free parameters the algorithms use to fit the data, realizing that as the number of free parameters $n \rightarrow \infty$, the algorithms may be overfitting the data.

If these three points are true, systems biologists no longer rely on inference to make a demonstration but have surrendered their reasoning to an elite group of technopolists who are given *"authority and prestige by those who have no such competence."* Indeed, inferentially speaking, omic science has failed to demonstrate its initial proposition that it is an unbiased way of doing science. Indeed, in statistical science, limitations, and thus bias, must be introduced to eliminate false positives because at any given level of significance, the number of false positives increases as $n \rightarrow \infty$. *"As with all measurements, but in particular large-scale omics-type studies, false discoveries are made"* (Friso and van Wijk, 2015).

The analysis of high-throughput data is organically understood by few if any of the people who generate it. Eugene Kolker, the editor in chief of OMICS: A Journal of Integrative Biology says (Scudellari, 2011), "*We produce so much data, we're not even always sure what we produce.*" This reminds me of a quote from Erwin Chargaff (1980): "*It is not recognized sufficiently that there can be an inflation of scientific facts: the more are being produced, the less the value of each. The knowledge industry is no less absurd than other industries appeared to Charlie Chaplin in the film Modern Times.*" According to T. C. Chamberlain (1897), "*the vitality of study quickly disappears when the object sought is a mere collocation of unmeaning facts.*"

When the sequences and lists generated by omic scientists, who are cataloging all the pieces of hay in hopes of finding the needle, contain so many entries (n) and $n \to \infty$, how is one to know which entries are true demonstrations that have been validated and verified to be the ground truth (Garrity, 2009)? When the sequences and lists generated by omic science contains so many entries (n) and $n \to \infty$, how is one to know which sequences and entries on the lists will stand the test of time and continue to be of value into the distant future and is thus future-proofed (Garrity and Lyons, 2003)? The material presented in the first 20 chapters of this book serve as examples of the future-proofed work of gathering data, making interpretations, and providing biological insights that have stood the test to time.

It is worth remembering that before the age of high-throughput technologies, scientists had to purify enzymes, synthesize their own radioactive nucleotides and dideoxyribonucleotides, and develop their own fractionation techniques before they could intelligently employ them to do an experiment that made a demonstration by giving direct and unambiguous evidence (Gilbert, 1991; Dodson, 2005; Szostak, 2009). Important components such as primers were discovered serendipitously as the enzymes became purer and purer. The conversation with nature that took place involved the interaction of brains and brawn, of tactile and mental processes. Something is gained and something is lost in today's age of high-throughput technology. About a decade ago, Clyde Hutchison III (2007) wrote, "*It appears possible that methods for collecting sequence data could soon outstrip our capacity to adequately analyze that data...The obvious importance of computational analysis of sequence data has led to a greater overall appreciation of the role of theory in biology. A relationship between theory and experiment, not unlike that found in 20th century physics, seems to be taking shape.*" As $n \to \infty$, where n stands for the number of genes, transcripts, proteins, metabolites, dollars, or dimensions, we should ask, are we getting a more global and universal view of reality and the natural world based on the central limit theorem, or are we just getting a better understanding of the law of diminishing returns?

The scientists described in the first 20 chapters and Frederick Sanger, Robert Holley, Ray Wu, and Leroy Hood, described in this chapter, devised simple systems so that they were able to listen to one, two, or three processes, organelles, proteins, or genes at a time—at the level which best suited the scientist and the question being asked. They were then able to reconstruct a cellular symphony by orchestrating those voices. How can the omic scientists of today learn to listen to the voice of a given molecule and reconstruct the cellular symphony when everyone's talking at once? Will signal processing algorithms that separate the signals from the noise produce the promised cellular symphony? Is it possible that Laplace's intellect is really Laplace's fool?

21.11 SUMMARY

Biology was defined by the physician-scientist Thomas Beddoes (1799) as "*the doctrine of the living system in all its states.*" Biology is a combination of the Greek word "bios" (βίος), which means life, course, or way of living; and the suffix "ology," which is related to the Greek word logos (λόγος), which means word, speech, discourse, or reason. To be a biologist is to be able to speak and reason about living systems in all their states. According to James Watson (2003), "*We are now in the new era of comprehensiveness in biology ushered in by the once-unimaginable feat of the Human Genome Project.*" The comprehensive plan was extended to include the study of plant genomes, which has been deemed necessary to feed and fuel the growing population as the climate is changing (Chory et al., 2000; Türktas et al., 2015; Van Emon, 2016; Wang et al., 2017a,b,c,d).

It is not unimportant to ask if the comprehensiveness is well thought out. Erwin Chargaff (1986) had a parable that I think is not inappropriate: "*Once upon a time, the proverbial Man from Mars was, for bad behavior, exiled to our globe, with the injunction to investigate the structure and function of the first thing he came across. Encountering a standing automobile, he took it apart, inspected the different organs of the contraption, put the whole thing together again, and came to the conclusion that it consisted essentially of a combustion engine that was driving the wheels. Since he had no intention of constructing himself another car, he declared his task as done: he had investigated the structure and function of a motorcar. He was even proud of his truly extraterrestrial intelligence. But when he reported to his Martian probation officer, there came the order* 'Go deeper.' *That command was repeated after each interim report. And so the Man from Mars is still at it, analyzing the exact composition of each part, rubber and glass, plastic steel, and alloys, glue and paint. He had to build a huge laboratory and all sorts of apparatus for the quantitative estimation of many minute components. Structure and function? Do not ask him about it, for he does not know anymore that he is studying an automobile; he is in the middle of working out a new method for the microdetermination of manganese. He is correct in declaring that the vistas are endless*."

It is also not unimportant to ask if the systems biology approach is delivering on its promises, and what, if any, are the unintended consequences of the high-technology omics approach (Gilbert, 1991, 1992; Ankeny, 2003; Alberts et al., 2014; Cyranoski et al., 2011a,b; Scudellari, 2011; Vastag, 2012; Sauermann and Roach, 2016; Roach and Sauermann, 2017; Wallis, 2017). According the Stephen Hilgartner (2017), from the onset, scientists critical of omics contended that the sequencing procedures would "*tie up talent in uninteresting and repetitive work*." In a conversation between Lewis Wolpert and Leroy Hood (see Wolpert and Richards, 1997) Wolpert said, "*I'm sorry to pursue this point about technique, but there are people who say that the trouble with the field—not necessarily your field, but certainly part of my field, developmental biology—is that people don't think any more. There are all these enormously powerful tools, some of which you've developed, and they simply apply what's there. So it's almost become very high-class cookery, just following recipes. As a result, a lot of the intellectual excitement, they argue, has gone out of the field. Do you think that's true*?" Hood answered, "*I would agree with that point of view completely*."

Words that contain the suffix "omics" instantly gain buzzword status and attention (Van Emon, 2016), especially among provosts and deans, but do the promises of omics outweigh the educational consequences? Are graduate students and postdocs becoming "*tools of our tools*," as Henry David Thoreau (1854) described in *Walden*? Aldous Huxley (1960) predicted in *Brave New World Revisited* the kind of dehumanization of scientists that is being caused by the massive advances in organization that have accompanied the advances in omic technology spurred by the human genome project, "*During the past century the successive advances in technology have been accompanied by corresponding advances in organization. Complicated machinery has had to be matched by complicated social arrangements, designed to work as smoothly and efficiently as the new instruments of production. In order to fit into these organizations, individuals have had to deindividualize themselves, have had to deny their native diversity and conform to a standard pattern, have had to do their best to become automata*." James Watson (1992) told a story of a mutiny that broke out in a team of researchers in Japan who were sequencing the chloroplast genome. According to Watson, "*It is imaginable that an American graduate student might tell his supervisor to go to hell; it is unimaginable that a Japanese graduate student might do the same. In the face of the extraordinary mutiny, the Japanese supervisors decided that forced-labor sequencing was too inhumane and resolved to change the system*." Now robots have replaced humans in doing the sequencing. Ironically, as "*the training pipeline produces more scientists than relevant positions in academia, government, and the private sector are capable of absorbing* (Alberts et al., 2014), robots are gaining status in society (White, 2015; Prodhan, 2016; Satell, 2016; Eidenmueller, 2017). This reminds me of the end of Neil Postman's (1985) book, Amusing Ourselves to Death, where he wrote, *For in the end, he* [Aldous Huxley] *was trying to tell us what afflicted the people in* Brave New World *was not that they were laughing instead of thinking, but they did not know what they were laughing about and why they had stopped thinking.*

Science is not immune from deindividualization. A decade after the publication of the double helix, Chargaff (1965, 1978) wrote, "*The fashion of our times favors dogmas. Since a dogma is something that everybody is expected to accept, this has led to the incredible monotony of our journals. Very often it is sufficient for me to read the title of a paper in order to reconstruct its summary and even some of the graphs. Most of these papers are very competent; they use the same techniques and arrive at the same results. This is often called the confirmation of a scientific fact. Every few years the techniques change; and then everybody will use the new techniques and confirm a new set of facts. This is called the progress of science. Whatever originality there may be, must be hidden in the crevices of an all-embracing conventional makeshift: a huge kitchen midden in which the successive layers of scientific habitation will be dated easily through the various apparatuses and devices and tricks, and even more through the several concepts and terms and slogans, that were fashionable at a given moment*."

Stanley Fields (2014), who called Fred Sanger the most important biologist in the latter half of the twentieth century, wondered "*whether his* [Sanger's] *style of science could survive in today's environment*." He goes on to say, "*In an era of large collaborations, multi-authored papers, and enormous datasets, is there still room for the single creative idea that proves to be a game-changer? I for one surely want to believe there is still room*." According to Sidney Brenner (2014), "*A Fred Sanger would not survive today's world of science. With continuous reporting and appraisals, some committee would note that he published little of import between insulin in 1952 and his first paper on RNA sequencing in 1967 with another long gap until DNA sequencing in 1977. He would be labeled as unproductive, and his modest personal support would be denied. We no longer have a culture that allows individuals to embark on long-term—and what would be considered today extremely risky—projects.*" Recently other Nobel Laureates, including Jeffrey Hall (2008, 2017; https://www.jacobinmag.com/2018/07/capitalism-science-research-academia-funding-publishing) and Peter Higgs (https://www.theguardian.com/science/2013/dec/06/peter-higgs-boson-academic-system; https://www.jacobinmag.com/2018/07/capitalism-science-research-academia-funding-publishing) echoed similar worries.

While science, on the average has a positive effect on society, we cannot turn our eyes from the negative consequences. Aldous Huxley (1938), the grandson of T. H. Huxley, brother of Julian Huxley, and half brother of Andrew Fielding Huxley, knew both the value and limitations of science when he noted, "*We are living now, not in the delicious intoxication induced by the early successes of science, but in a rather grisly morning-after, when it has become apparent that what triumphant science has done hitherto is to improve the means for achieving unimproved or actually deteriorated ends*." I ask you dear reader, to add up the costs and benefits of omic technologies and ask if they have been worth it, or if they just provide, what Thoreau (1854) called "*improved means to an unimproved end*." I look forward to the intelligent use of technology to make demonstrations that are grounded in truth (Garrity, 2009) and future-proofed (Garrity and Lyons, 2003). Demonstrations that put Laplace's intellect in its place on the back burner and are made according to the logic of Aristotle.

The shift in scale from personal science performed by individuals or small groups to technology-based science performed by consortiums of scientists mirrors the shift in art and architecture that accompanied the industrial revolution (Jones, 1856). The Arts and Crafts Movement emerged as a reaction against the decline in standards that the reformers associated with factory production. Owen Jones (1853) observed that as a result of the industrial revolution, artists "*have no principles*" and were producing "*novelty without beauty, or beauty without intelligence*." The pre-Raphaelite Brotherhood reacted to the consequences of the industrial revolution by rejecting creations based on technical skill without inspiration. William Michael Rossetti expressed the doctrines of the pre-Raphaelite Brotherhood in four declarations (Latham, 2003):

- have genuine ideas to express;
- study nature attentively, so as to know how to express them;
- sympathize with what is direct and serious and heartfelt in previous art, to the exclusion of what is conventional and self-parading and learned by rote;
- and most indispensable of all, to produce thoroughly good pictures and statues (or in the case of plant cell biology observations, experiments, and theories).

The Arts and Crafts Movement influenced many other fields of endeavor (Gropius, 1962; Michel et al., 2010).

This chapter has focused on the scientists and their discoveries that made the omic revolution possible. The promises of the application of these technologies and the superfunding granted to use them have been promises deferred. When the architects of the Human Genome Project asked the American people for $3 billion dollars, they were signing a promissory note to which every American was to fall heir (see King, 1963). Now is the time to make real the promises of omics or to divest the superfunding into other investments, scientific or otherwise. Now is the time to lift our nation from the quick sands of the promises of using omic technology in the search for everything omic in an unbiased manner to the solid rock of posing limited scientific questions and using technology intelligently to answer the questions asked. Now is the time for students and postdocs in STEM to be treated as creative individuals. Now is the time to teach STEM students how to think critically and question nature rather than memorize answers. Now is the time to analyze data using intelligence and programs that are understood by the user rather than robotically using artificial unintelligence that is mascarading as artificial intelligence. Now is the time to make science a reality for people not for robots.

In some respects, this chapter was an attempt at *omology*—the systematic study of −omes (Prohaska and Stadler, 2011). I have concluded that omic science is being admired in the way that the emperor's new clothes were, and this is, in part, the reason I wrote this chapter. Eve Marder (2003) wrote in a paper entitled, *The emperor's new clothes*, "As a young child I also read science books and autobiographies of the great men and women of science. In those tales scientists forged forward in search of truth despite great adversity. Today the practice of good science rarely involves great acts of courage, but instead requires multiple small acts of personal courage, among them the strength to speak one's mind even when tact suggests silence. It requires the willingness to be wrong publicly. Above all it demands that we remember that the purpose of science is the pursuit of truth, even if that pursuit becomes uncomfortable for ourselves or others. Sometimes I find myself wishing there were a wise gnome or gentle wizard living under the fume hood to give me words of wisdom!" I will sum up my findings with a few lines from *Choruses from the Rock*, written by T. S. Eliot (1936), in response to the Modernism of the nineteen-thirties soon after the origin of quantum mechanics:

Where is the life we have lost in the living?
Where is the wisdom we have lost in knowledge?
Where is the knowledge we have lost in information?

21.12 QUESTIONS

21.1. Compare and contrast the approach and fundamental findings of big science and craftsman science.

21.2. Compare and contrast the creativity that went into the development of the techniques discussed in this chapter and the creativity that went into the use of the techniques in probing the mysteries of the cell.

21.3. Kahvejian et al. (2008) wrote a paper entitled, "What would you do if you could sequence everything?" They concluded the paper by saying, "*Sequencing everything is only the first step, as we then need to process that data into information that can be used broadly to benefit human health and productivity.*" Argue for or against the thesis that sequencing everything is a necessary first step.

21.4. Can you foresee robots replacing the developers of the techniques? Can you foresee robots replacing the users of the techniques?

21.5. The omics approach has been prominent in plant biology for two decades. Describe one example where genomics, epigenomics, transcriptomics, proteomics, phosphoproteomics, metabolomics, lipidomics, or phenomics have contributed something new to the fundamental knowledge of plant biology. Hint: When I asked this question to Deans, Directors, Department Chairs, Professors and Postdocs, I got the same answer: "That's a good question."

21.6. The ability to ask a question of nature has been the heart of the natural sciences since the Renaissance. In a book entitled, The Improvement of the Mind, Isaac Watts (1801) wrote about how to consider a question. He wrote, *When a subject is proposed to your thoughts, consider whether it be knowable at all, or no; and then whether in the present state; and remember that it is great waste of time to busy yourselves too much amongst unsearchables... Consider again whether the matter be worthy of your enquiry at all; and then how far it may be worthy of your present search and labour according to your age, your time of life, your station in the world, your capacity, your profession, your chief design and end...Consider whether the subject of your enquiry be easy or difficult; whether you have sufficient foundation or skill...Consider whether the subject be in any ways useful or no, before you engage in the study of it. Consider what tendency it has to make you wiser and better...If the question appear to be well worth your diligent application, and you are furnished with the necessary requisites to pursue it, then consider whether it be dressed up and entangled in more words than is necessary; and if so, endeavour to reduce it to a greater simplicity and plainness which will make the enquiry and argument easier and plainer all the way.* Which platform and pipeline do you consider the best to answer a question worthy of yourself?

21.7. Erwin Chargaff (1978), to whom this book is dedicated, wrote, *What science has done to the universities is that it has inflated and disfigured them; it has left them more bankrupt than they were before. The large private universities have been turned into huge corporations whose only business is to lose money. There are exceptions, but, in general, power-hungry, empty-headed money grabbers have taken over. The true and only function of a university, namely to help young people find themselves by bringing them to the accumulated memory of mankind, has been swept aside. By misunderstanding, through overemphasis, of the old adage of the unity of research and learning, research has been made into a teaching tool, into a most expensive and stultifying one, forcing every student to become a researcher and trivializing the purpose of scientific research.* From your own observation, do you agree or disagree with Chargaff, and why?

21.8. Erwin Chargaff (1978) also wrote, *Now to my second question: What have universities done to science? They have bled it for overhead; they have cheapened and vulgarized it to the point of nonrecognition; they have made it into a public-relations "gimmick." If the products of this kind of education often still are so good, it testifies only to the resilience of young minds, But many are damaged irreversibly.* From your own observation, do you agree or disagree with Chargaff, and why?

21.9. The average number of authors on a given paper seems to have increased as we moved from the second to the third millennium. What are the advantages and disadvantages of multiauthored papers? What is correlated with the increase in authors? An increase in rigor? An increase in knowledge? An increase in academic positions? An increase in grant funds from government agencies given to universities for Facilities and Administration (F&A)?

21.10. Science like any other human endeavor moves in a given direction, and sometimes it is hard to know whether it is more profitable to stay the course or to change directions. Alberts et al. (2014) and Richard Harris (2017) have recently questioned the direction of science. Such questioning occurred in the 1960s and early 1970s when it came to the Vietnam War. Pete Seeger (https://www.youtube.com/watch?v=uXnJVkEX8O4) wrote "Waist Deep in the Big Muddy" to remind us that we should evaluate the direction we are marching:

Now I'm not going to point any moral –
I'll leave that for yourself.
Maybe you're still walking, you're still talking,
You'd like to keep your health.
But every time I read the papers, that old feeling comes on,
We're waist deep in the Big Muddy
And the big fool says to push on.

How would you evaluate the direction science is going?

The references for this chapter can be found in the references at the end of the book.

Appendix 1

SI Units, Constants, Variables, and Geometric Formulae

1. SI UNITS

Concept	Unit Name	Units	Symbol
Basic Units			
Length	Meter		m
Mass	Kilogram		kg
Time	Second		s
Electric Current	Ampere		A
Temperature	Kelvin		K
Amount of substance	Mole		mol
Derived Units			
Electrical potential	Volt	$kg\ m^2\ s^{-3}\ A^{-1}$	V
Force	Newton	$kg\ m\ s^{-2}$	N
Energy, work	Joule	N m	J
Pressure, stress	Pascal	$N\ m^{-2}$	Pa
Electric charge	Coulomb	A s	C
Capacitance	Farad	C/V	F
Resistance	Ohm	V/A	Ω
Conductance	Siemens	A/V	S

2. CONSTANTS

Symbol	Name	Approximate Value
π	Pi	3.14
N_A	Avogadro's number	$6.02 \times 10^{23}\ mol^{-1}$
μ_o	Magnetic permeability of vacuum	$4\pi \times 10^7\ N/A^2$
ε_o	Electrical permittivity of vacuum	8.85×10^{-12} F/m
$c = (\varepsilon_o \mu_o)^{21/2}$	Speed of light	3×10^8 m/s
Da	Dalton, atomic mass unit (amu)	1.66×10^{-27} kg
e	Elementary charge	1.60×10^{-19} C
$F = eN_A$	Faraday constant	9.65×10^4 C/mol
g	Gravitational acceleration	$9.8\ m/s^2$
h	Planck's constant	6.63×10^{-34} J s
$\hbar = h/2\pi$	Reduced Planck's constant	1.05×10^{-34} J s
k	Boltzmann's constant	1.38×10^{-23} J/K
$R = kN_A$	Universal gas constant	$8.31\ J\ mol^{-1}\ K^{-1}$
w	Wien coefficient	2.89784×10^{-3} m K
σ_B	Stefan–Boltzmann's constant	$5.67 \times 10^{-8}\ J\ K^{-4}\ m^{-2}\ s^{-1}$

3. VARIABLES

Symbol	Name	Units
a	Acceleration	m/s^2
a_w	Relative activity of water	Dimensionless
A	Area	m^2
c'	Local speed of light	Depends on relative velocity
C	Concentration	mol/m^3
C_{sp}	Specific capacitance	F/m^2
D	Diffusion coefficient	m^2/s
E	Molecular free energy	J
EFR	Energy fluence rate	$J\ m^{-2}\ s^{-1}$
f	Fluidity	$Pa^{-1}\ s^{-1}$
F	Force	N
G	Conductance	S
H	Molar enthalpy	J/mol
i	Ionization coefficient	Dimensionless
I	Current	A, C/s
J	Flux	$mol\ m^{-2}\ s^{-1}$
k	Rate constants	Defined for each equation
k	Angular wave number	m^{-1}
K	Constants	Defined for each equation
K	Partition coefficient	Dimensionless
L_p	Hydraulic conductivity	$m\ s^{-1}\ Pa^{-1}$
m	Mass	kg
M	Elastic modulus	N/m^2
M_r	Molecular mass	Daltons
n	Refractive index	Dimensionless
n	Quantity of matter	Dimensionless
P	Permeability coefficient	m/s
P	Pressure	Pa
P_t	Hydrostatic pressure	Pa
P_w	Water potential	Pa
P_π	Osmotic pressure	Pa
q	Charge	C
Q	Volume flow	m^3/s
r	Radius	m
R	Maximal radius	m
R	Resistance	Ω
S	Molar entropy	$J\ mol^{-1}\ K^{-1}$
t	Time	s
T	Absolute temperature	K
u	Mobility coefficient	$m^2\ J^{-1}\ s^{-1}$
u'	Electrical mobility coefficient	$m^2\ V^{-1}\ s^{-1}$
u	Energy density per unit wavelength	J/m^4
U	Energy density	J/m^3
v	Velocity	m/s; other units are possible when talking about reaction velocities
V	Volume	m^3
$\overline{V}_w$	Partial molar volume of water	m^3/mol
w	Length	m
x	Length	m
Y	Yield value	N/m^2
z	Valence	Dimensionless
α	P_{na}/P_K	Dimensionless
β	P_{cl}/P_K	Dimensionless
ε	Relative permittivity	Dimensionless
ρ	Density	kg/m^3
ρ	Photon density	$photons/m^3$
σ	Shearing stress	N/m^2
σ	Cross section	m^2
η	Viscosity	Pa s
λ	Wavelength	m
ν	Frequency	s^{-1}
γ	Rate of shear	s^{-1}
Π	Osmotic pressure difference	Pa
Φ	Viscoelastic extensibility	$Pa^{-1}\ s^{-1}$
ψ	Electrical potential	V
Ψ	Wave function	Depends on type of wave
Δ	Indicates a finite difference between two quantities	
d, δ	When used in calculus indicates an infinitesimally small difference between two quantities	

4. GEOMETRIC FORMULAE

Circle
Radius = r
Diameter = $2r$
Circumference = $2\pi r$
Area = πr^2

Sphere
Radius = r
Diameter = $2r$
Area = $4\pi r^2$
Volume = $(4/3)\pi r^3$

Right Circular Cylinder
Length = x
Radius = r
Area of curved portion = $2\pi rx$
Area of two ends = $2\pi r^2$
Total area = $2\pi rx + 2\pi r^2$
Volume = $\pi r^2 x$

Square
Length = x
Perimeter = $4x$
Area = x^2

Cube
Area = $6x^2$
Volume = x^3

Appendix 2

A Cell Biologist's View of Non-Newtonian Physics

Ask not what physics can do for biology, ask what biology can do for physics.

Stanislaw Ulam (quoted in Knight, 2002).

Cells live in the world of neglected dimensions between the world of macroscopic physics and the world of microscopic physics. Studying physicochemical processes in such a world has its advantages and disadvantages. One disadvantage of working in this world of neglected dimensions is that it is not easy to assume that a given subset of physical laws can be neglected to model biological processes and solve the equations easily. One advantage of working in the world of neglected dimensions is that a cell biologist has the opportunity to look for fundamental laws that are applicable to microscopic systems and macroscopic systems and thus help to unify macrophysics and microphysics. Such laws could provide a parsimonious toolbox for modeling and solving a wide range of physicochemical problems.

Although cells and the particles within them do not travel anywhere near the speed of light, as a cell biologist, I have gained a perspective to suggest why charged particles do not travel faster than the speed of light. In this appendix, I present a hypothesis that light itself in the form of a dilatant photon gas prevents charged particles from moving faster than the speed of light. It may be right, it may be wrong, but it is definitely thought provoking.

Newton's Second Law ($F = m\, dv/dt$) implies that any particle with mass (m) can be accelerated to any velocity (v) in time (t) by the application of a large enough constant force (F). However, experience shows that particles do not accelerate to infinite velocity and the amount of force required to accelerate a charged particle increases nonlinearly as its velocity asymptotically approaches the speed of light (c). This indicates that there must be a non-Newtonian process that decreases the effectiveness of the constant force (F). The non-Newtonian process may increase the particle's mass (m) nonlinearly, nonlinearly influence its reckoning of the duration of time the force is administered, or add a nonlinear velocity-dependent resistance to the acceleration as a particle's velocity approaches the speed of light.

According to Einstein's Special Theory Of Relativity, the duration of time (dt) is relative. Consequently, according to an observer in the inertial frame of a moving particle, the force applied to the particle is attenuated because a moving particle experiences a constant force for a shorter duration of time (dt_{proper}) than does the experimenter who is at rest with respect to the force ($dt_{improper}$). According to the Special Theory of Relativity, the duration of time differs in a velocity-dependent manner according to the following equation:

$$dt_{proper}/dt_{improper} = \sqrt{(1 - v^2/c^2)} \qquad \text{(A2.1)}$$

The question for the physicochemically minded cell biologist is, Is the non-Newtonian behavior observed for particles moving at speeds approaching the speed of light best explained by invoking the relativity of space and time given by the Special Theory of Relativity, or by invoking the presence of a resisting force that causes the moving particle to experience a nonlinear relationship between force and acceleration? A cell biologist is at an advantage in discovering a resisting force that causes a moving particle to respond to a constant force in a nonlinear manner as a result of working in the world of neglected dimensions.

Ever since Antony van Leeuwenhoek (1677) observed the incessant movement of animalcules and Bonaventura

Corti (1774) and Giovanni Amici (1818) observed rotational cytoplasmic streaming in giant algal cells, cell biologists have realized that movement is one of the fundamental attributes of life (Huxley, 1890). To understand the physicochemical mechanisms that make life itself possible, cell biologists have studied the movement of ions through channels (see Chapter 2); the movement of small molecules through plasmodesmata (see Chapter 3); the movement of proteins from their site of synthesis to their site of action (see Chapters 4 and 17); the movement of membrane vesicles, tubules (see Chapters 6, 7, and 8), and organelles (see Chapters 13 and 14) throughout the viscous cytoplasm (Chapter 9); the movement of polymerases along the DNA that comprises the hereditary material (see Chapter 16); the movement of the hereditary material during mitosis (see Chapter 19) and meiosis when it is packaged in the bodies suitable for transport; the movement of the polymers of the extracellular matrix that allow cell growth (see Chapter 20); and the movement of proteins, including myosin, dynein, and kinesin, which convert chemical energy into mechanical energy to facilitate the movement of organelles and membrane vesicles along the cytoplasmic tracks known as microfilaments (see Chapter 10) and microtubules (see Chapter 11). Each one of these cellular movements, as well as others not listed here, involves a motive force that overcomes a nonnegligible resistive force. The study of these movements gives a cell biologist a special talent in identifying resistances. Moreover, because the resistance of the cytoplasm and the extracellular matrix is nonlinear, a cell biologist has profound experience in understanding non-Newtonian physics (Seifriz, 1936, 1938b; Kamiya, 1950; Métraux and Taiz, 1978; Kamitsubo et al., 1988; Okamoto-Nakazato et al., 2000a,b, 2001; Okamoto-Nakazato, 2002).

When one studies the movement of a vesicle or a chromosome through the cytoplasm or nucleoplasm, one must ask, "What is the nature of the cytoplasmic or nucleoplasmic space through which the vesicle or chromosome moves?" Likewise, when one studies the movement of a particle through a space, one must also ask, "What are the properties of the space through which the particles move?" Let us consider the movement of an electron through space where the motive force is provided by an electric field. At any temperature above 0K, the space consists of a radiation field composed of photons. The photons can be considered to have a black-body distribution (Fig. A2.1).

An electron moving through a photon gas, in which the radiation relative to the center of momentum of the radiation field is distributed according to Planck's radiation law, experiences the photons as being Doppler shifted. The photons that collide with the front of the moving electron will be blue-shifted and the photons that collide with the back of the moving electron will be red-shifted (Fig. A2.2).

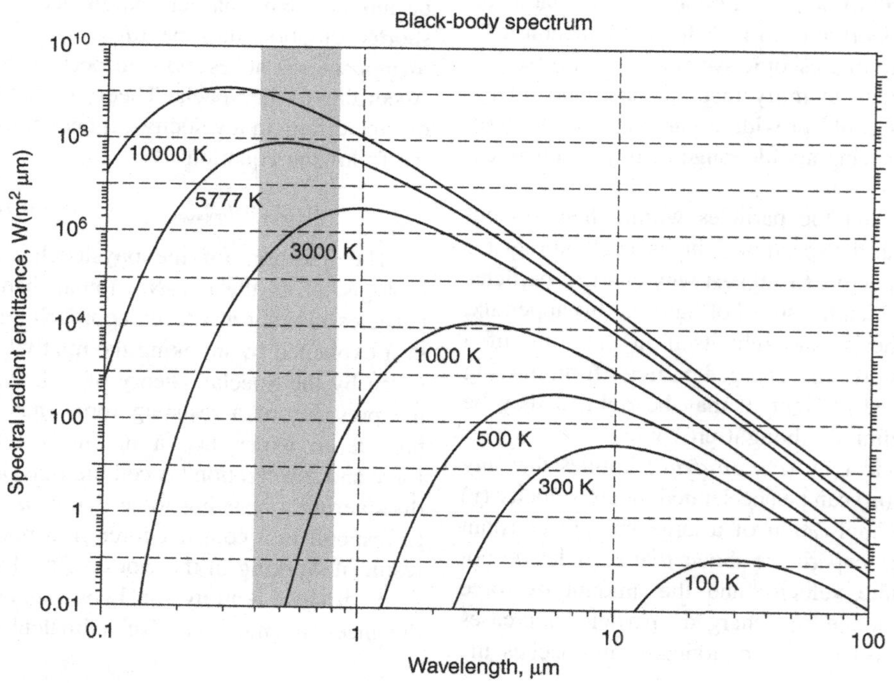

FIGURE A2.1 Black-body distribution of radiation at various temperatures. The gray band indicates the visible wavelengths.

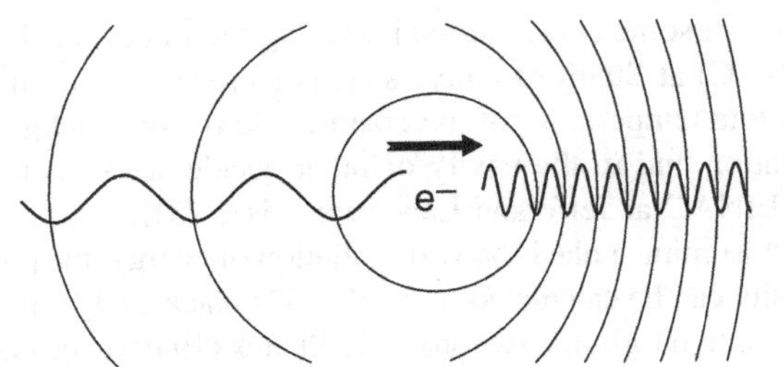

FIGURE A2.2 An electron moving through a photon gas consisting of a black-body distribution of radiation experiences the photons as being Doppler shifted. The photons striking the front of an electron are blue-shifted and the photons striking the rear of the electron are red-shifted.

Because the electron is moving at a velocity (v) relative to the center of momentum of the radiation field, I will describe the radiation experienced by the moving particle with an original relativistic wave equation. It describes the propagation of light waves between inertial frames moving relative to each other at velocity v. This new relativistic wave equation is given by:

$$\partial^2\Psi/\partial t^2 = cc'\left(\sqrt{(1-v/c)}\Big/\sqrt{(1+v/c)}\right)\partial^2\Psi/\partial x^2 \tag{A2.2}$$

where $v > 0$ when the source and observer move away from each other, and $v < 0$ when the source and observer move toward each other. Different aspects of the speed of a light wave are represented by c and c'. The parameter c, which is absolute and independent of the velocity of the source or the observer, gives the speed of the wave through space and is equal to the square root of the reciprocal of the product of the electric permittivity («$_o$) and the magnetic permeability (μ_o) of the vacuum. By contrast, c' is local and depends on the relative velocity of the source and observer. c' gives the ratio of the angular frequency (Ω) of the source in its inertial frame to the angular wave number (k) observed in any inertial frame ($c' = \Omega_{source}/k_{observer}$). The ratio of c' to c is equal to the ratio of the angular frequencies in the inertial frame of the source and the inertial frame of the observer ($c'/c = \Omega_{source}/\Omega_{observer}$). When there is no relative motion between the source and the observer, $v = 0$, $cc'\left(\sqrt{(1-v/c)}/\sqrt{(1+v/c)}\right) = c^2$, and the new relativistic wave equation reduces to d'Alembert's or Maxwell's wave equations.

By introducing the perspicuous correction factor $\left(\sqrt{(1-v/c)}/\sqrt{(1+v/c)}\right)$ to ensure the invariance of this new relativistic wave equation, I obtain the relativistic Doppler equation naturally as the dispersion relation:

$$\begin{aligned} k_{observer} &= \pm k_{source}\left[\sqrt{(1-v/c)}\Big/\sqrt{(1+v/c)}\right] \\ &= \pm k_{source}\left[1\Big/\sqrt{(1-v^2/c^2)}\right] \end{aligned} \tag{A2.3}$$

The experimental observations of Ives and Stillwell (1938) on the displacement of the spectral lines of hydrogen with velocity confirm the utility and validity of the new relativistic wave equation.

The linear momentum of a photon is given by $\hbar\boldsymbol{k}$, where $\hbar$ is Planck's constant (h) divided by 2π. As a consequence of the relativistic Doppler effect, the linear momentum of the radiation perceived by an electron moving with speed v is velocity dependent and is given by:

$$\begin{aligned} \hbar\boldsymbol{k}_{observer} &= \hbar\boldsymbol{k}_{source}\left[\sqrt{(c-v)}\Big/\sqrt{(c+v)}\right] \\ &= \hbar\boldsymbol{k}_{source}\left[(1-v/c)\Big/\sqrt{(1-v^2/c^2)}\right] \end{aligned} \tag{A2.4}$$

For convenience, I will use the absolute value of the velocity and split Eq. (A2.4) into two equations—one for an electron moving parallel relative to the waves propagating from the source and one for an electron moving antiparallel relative to the waves propagating from the source. The momentum of the light experienced by an electron traveling parallel to light propagating from the source is

$$\hbar\boldsymbol{k}_{electron} = \hbar\boldsymbol{k}_{source}(1-v/c)\big/\left(1-v^2/c^2\right)^{1/2} \tag{A2.4a}$$

Eq. (A2.4a) is identical to the equation that describes the Compton effect, where $\hbar\boldsymbol{k}_{source}$ is the momentum of a photon before a collision, and $\hbar\boldsymbol{k}_{electron}$, the momentum transferred to the electron, is equivalent to the momentum lost by the photon after a collision (Compton, 1924). The momentum of the light experienced by a particle traveling antiparallel to light propagating from the source is

$$\hbar\boldsymbol{k}_{electron} = \hbar\boldsymbol{k}_{source}(1+v/c)\big/\left(1-v^2/c^2\right)^{1/2} \tag{A2.4b}$$

Eq. (A2.4b) is identical to the equation that describes the inverse Compton effect, where $\hbar\boldsymbol{k}_{source}$ is the momentum of a photon before a collision, and $\hbar\boldsymbol{k}_{electron}$, the momentum transferred from the electron, is equivalent to the momentum gained by the photon after a collision (Feenberg and Primakoff, 1948).

Assume that the moving electron interacts with one photon from the front and one photon from the back. In this case, there is a net momentum transferred from the electron to the radiation field, and the vector of the momentum transferred to the radiation field is antiparallel to the velocity vector of the electron and is given by

$$\begin{aligned} \hbar\boldsymbol{k}_{electron} &= \hbar\boldsymbol{k}_{source}[(1-v/c)-(1+v/c)]\big/\left(1-v^2/c^2\right)^{1/2} \\ &= \hbar\boldsymbol{k}_{source}(-2v/c)\big/\left(1-v^2/c^2\right)^{1/2} \end{aligned} \tag{A2.5}$$

The average decrease of momentum ($m_o v$) experienced by a moving particle on colliding with one photon in an isotropic radiation field would be

$$\begin{aligned} \hbar \boldsymbol{k}_{\text{electron}} &= -(1/2)\hbar \boldsymbol{k}_{\text{source}}(2v/c)/\left(1 - v^2/c^2\right)^{1/2} \\ &= -\hbar \boldsymbol{k}_{\text{source}}(v/c)/\left(1 - v^2/c^2\right)^{1/2} \end{aligned} \tag{A2.6}$$

where the negative sign indicates that the momentum of the electron moving through the radiation field decreases.

As the "average photon" can strike the moving particle at any angle from 0 to $\pm\pi/2$ with differing effectiveness, the average transfer of momentum[1] from the radiation field to the particle is $\hbar \boldsymbol{k}_{\text{source}}(1/4)(v/c)/(1 - v^2/c^2)^{½}$ for a single collision. Consequently, at all temperatures greater than absolute zero, contrary to Newton's First Law, all bodies will slow down, in principle. In an adiabatic radiation field, this will result in an increase in temperature and an increase in the peak angular wave number of the photons in the radiation field. In an isothermal radiation field, this will result in an expansion of the field. Because an isotropic radiation field exists at all temperatures above absolute zero, Newton's First Law is only valid for charged particles at absolute zero. By taking the thermodynamics of the radiation field into consideration, I will recast Newton's Second Law in an alternative form that applies to charged particles[2] moving at velocities close to the speed of light and that parallels Einstein's Special Theory of Relativity.

The velocity-dependent relativistic Doppler-shifted momentum of the radiation field provides the basis for a velocity-dependent counterforce ($\boldsymbol{F}_{\text{Dopp}}$) on a particle accelerated by a constant applied force ($\boldsymbol{F}_{\text{app}}$).

$$\boldsymbol{F}_{\text{app}} + \boldsymbol{F}_{\text{Dopp}} = m\,dv/dt \tag{A2.7}$$

where $\boldsymbol{F}_{\text{app}}$ and $\boldsymbol{F}_{\text{Dopp}}$ are antiparallel.

The force exerted by the radiation field on a moving particle is a function of the collision rate between the moving particle and the photons in the field. The collision rate (dn/dt) depends on the photon density (ρ), the speed of the particle (v), and the cross section of the photon (σ) according to the following equation:

$$dn/dt = \rho v \sigma \tag{A2.8}$$

The photon density is a function of the absolute temperature. The absolute temperature on Earth and in the cavity of some accelerators, including the linear accelerator (LINAC) at Stanford University, is close to 300K, and the absolute temperature of the cosmic microwave background radiation and in the cavity of other accelerators, including the LINAC at Jefferson Laboratory, is 2.73K.

Assuming a black-body distribution of energy, the photon density can be calculated from Planck's black-body radiation distribution formula. According to Planck (1949a), the energy density per unit wavelength interval (u) is given by

$$u = \left(8\pi hc/\lambda^5\right)\left(1/(\exp[hc/\lambda kT] - 1)\right) \tag{A2.9}$$

The peak wavelength can be obtained by differentiating Eq. (A2.9) with respect to wavelength or by simply using Wien's distribution law:

$$\lambda_{\text{peak}} = 2.89784 \times 10^{-3}\ \text{mK}/T = w/T \tag{A2.10}$$

where w is called the *Wien coefficient* and is equal to $2.89{,}784 \times 10^{-3}$ mK. The peak wavelengths (λ_{peak}) in 300 and 2.73K radiation fields are 9.66×10^{-6} m and 1.87×10^{-3} m, respectively. The energies of photons with these wavelengths are given by Planck's equation:

$$E = hc/\lambda_{\text{peak}} \tag{A2.11}$$

and are 2.06×10^{-20} and 1.06×10^{-22} J/photon for the peak photons in a 300 and 2.73K radiation field, respectively. The total energy density (U) of a radiation field can be determined by integrating Eq. (A2.9) over wavelengths from zero to infinity.[3]

$$\begin{aligned} \text{U} &= \int u\,d\lambda \\ &= \int \left(8\pi hc/\lambda^5\right)\left(1/(\exp[hc/\lambda kT] - 1)\right) d\lambda \\ &= \left(8\pi k^4T^4/c^3h^3\right)\int x^3[\exp(x) - 1]^{-1}\,dx \\ &= \left(8\pi k^4T^4/c^3h^3\right)\left(\pi^4/15\right) \\ &= \left(8\pi^5k^4/15c^3h^3\right)\left(T^4\right) \\ &= 7.57 \times 10^{-16}\left(T^4\right) \end{aligned} \tag{A2.12}$$

The quantity 7.57×10^{-16}, known as the radiation constant, is equal to ($4\sigma_B/c$), where σ_B represents the Stefan–Boltzmann constant. The total energy densities of radiation fields with temperatures of 300 and 2.73K radiation are 6.13×10^{-6} and 4.02×10^{-14} J/m^3, respectively. The photon densities (ρ) in 300 and 2.73K radiation fields, which are obtained by dividing Eq. (A2.12) by Eq. (A2.11), are 2.98×10^{14} photons/m^3 and 3.79×10^{8} photons/m^3, respectively.

Although light is often modeled as an infinite plane wave or a mathematical point, the phenomena of diffraction and interference indicate that a photon has neither an

---

1. Assuming isotropy, where the linear momentum coming from any direction is $\hbar\boldsymbol{k}(\theta, \phi) = \hbar\boldsymbol{k}(0, 0)$, the total linear momentum coming from all directions is $\hbar\boldsymbol{k} = \int_0^{2\pi}\int_0^{\pi} \hbar\boldsymbol{k}(\theta, \phi)\sin\theta\,d\theta\,d\phi = \hbar\boldsymbol{k}\int_0^{2\pi}\int_0^{\pi/2}$ $\cos\theta\sin\theta d\theta\,d\phi$. The total linear momentum coming from all directions per unit area per unit time is $\hbar\boldsymbol{k}e_0^2\pi e_0^{\pi/2}\cos\theta\sin\theta d\theta d\phi = \hbar k/4$.
2. This way of thinking also occurs to neutral particles, including neutrons and neutrinos, with a magnetic moment that may form an electrical dipole that can couple to the radiation field.
3. Let $x = hc/\lambda kT$; $\int x^3[\exp(x) - 1]^{-1}dx = \pi^4/15$.

TABLE A2.1 Counterforce Produced by Thermal Background Radiation (300K)

Velocity (3×10^{-8} m/second)	$\frac{\rho\sigma h}{4\lambda_o}$ ($3 \times 10_{26}$ Ns/m)	$(v^2/c)/(1 - v^2/c^2)^{1/2}$ (3×10^{29} m/second)	F_{Dopp} (N)
2.9	3.80	1.11	4.22×10^{-17}
2.99	3.80	3.98	1.51×10^{-16}
2.997	3.80	12.1	4.60×10^{-16}
2.9979	3.80	74.0	2.81×10^{-15}
2.99792	3.80	172.0	6.54×10^{-15}

infinite nor a vanishing width (Lorentz, 1924; Wayne, 2009). I assume that a photon has a finite wave width as well as a wavelength and that the geometrical cross section[4] (σ) of a photon is given by:

$$\sigma = \pi r^2 \tag{A2.13}$$

where r is the radius of the photon. The radius of a photon can be estimated using a mixture of classical and quantum reasoning following the example of Niels Bohr—by making use of the fact that all photons have the same quantized angular momentum ($L = \hbar$), independent of their wavelength, and that classically, angular momentum is equal to $m\omega r^2$.

By using $E = mc^2 = \hbar\omega$ and assuming the equivalent mass (m) of a photon is given by $\hbar\omega/c^2$, then its radius (r) will be equal to $\sqrt{(\hbar/m\omega)} = \sqrt{(\hbar c^2/\hbar\omega^2)} = \sqrt{(c^2/\omega^2)} = c/\omega = 1/k = \lambda/2\pi$, which is the reciprocal of the angular wave number.[5] Thus, the geometrical cross section, which is related to its angular wave number, is given by

$$\sigma = \pi(1/k)^2 = \pi(\lambda/2\pi)^2 = \lambda^2/4\pi \tag{A2.14}$$

According to this reasoning, the cross sections of thermal (300K) photons and microwave (2.73K) photons are 7.43×10^{-12} and 2.78×10^{-7} m^2, respectively.

According to Eq. (A2.8), the collision rate (dn/dt) between a moving electron and photons in an isotropic thermal or microwave radiation field is dependent on the velocity of the electron. After factoring in the photon densities and the cross section of the peak photons, I find that the collision rate is equal to (2214.14 collisions/m)v and (105.36 collisions/m)v for 300 and 2.73K radiation fields, respectively, where v is the velocity of the electron relative to the observer. For a given velocity, the collision rate increases with the temperature of the radiation field. In a 300K radiation field, at speeds approaching the speed of light, the collision rate will be about 6.64×10^{11} per second, while it will be about 3.16×10^{10} per second for a 2.73K microwave radiation field.

The velocity-dependent counterforce ($\boldsymbol{F}_{\mathbf{Dopp}}$) exerted by the radiation field is given by the product of the collision rate and the average velocity-dependent momentum of a photon in the radiation field:

$$\begin{aligned} \boldsymbol{F}_{\mathrm{FDopp}} &= -(\rho\sigma v)(\hbar k_{\mathrm{source}})(1/4)(v/c)/\left(1 - v^2/c^2\right)^{1/2} \\ &= -(\rho\sigma h/4\lambda_{\mathrm{source}})\left(v^2/c\right)/\left(1 - v^2/c^2\right)^{1/2} \end{aligned} \tag{A2.15}$$

At $v = 0$, there is no net counterforce and the average momenta ($h/4\lambda_o$) of photons in a thermal radiation field and a microwave radiation field are 1.72×10^{-29} and 8.86×10^{-32} kg m/second, respectively, and an electron will exhibit Brownian motion in the photon gas. Tables A2.1 and A2.2 give the velocity-dependent counterforce exerted by black-body distributions of thermal radiation (300K) and microwave radiation (2.73K). The influence of temperature and velocity on the counterforce is presented in Fig. A2.3. The nonlinear temperature-dependent coefficient of friction (r) of the radiation field

4. While these geometrical cross sections appear large, using Eq. (A2.13), the geometrical cross section calculated for a 10-MeV photon is 1. 23×10^{-27} m^2 or 0.123 barn, within the range of the experimentally determined photon cross sections. The cross section is typically a measure of the probability that any given reaction will occur, and the total cross section is a measure of the probability that all possible reactions will occur. The cross sections for individual processes that make up the total cross sections vary by many orders of magnitude and may be less than, equal to, or greater than the geometrical crosssection. Here, I assume that a charged particle in thermal equilibrium with the black-body radiation field has a resonance for photons in the radiation field with every possible angular wave number, and thus the probability of an electron interacting with the radiation field is unity. Consequently, the effective cross section equals the geometrical cross section.

5. This calculation can also be based on the fact that light radiated from an object provides that object with linear momentum parallel to the direction of radiation (Lewis, 1908). Semiclassically, the equivalent momentum of the photon of light is equal to mv. Because the photon travels at the speed of light (c), its equivalent momentum is given by mc. According to quantum theory, the momentum of the photon is given by $\eta\boldsymbol{k}$. By equating the classical and quantum descriptions of momentum, the equivalent mass of a photon is given by the absolute value of $\eta k/c$, which is equal to $\eta\Omega/c^2$. Friedrich Hasenöhl derived the relationship $E \simeq mc^2$, entirely based on classical reasoning making use of Maxwell's light pressure and equating the Poynting vector to the momentum vector multiplied by c^2 (Lenard, 1933; Pauli, 1958).

TABLE A2.2 Counterforce Produced by Cosmic Microwave Background Radiation (2.7K)

Velocity (3×10^{-8} m/second)	$\frac{\rho\sigma h}{4\lambda_o}$ (3 × 10_{30} Ns/m)	$(v^2/c)/(1 - v^2/c^2)^{½}$ (3×10^{-9} m/second)	F_{Dopp} (N)
2.9	9.33	1.11	1.04×10^{-20}
2.99	9.33	3.98	3.71×10^{-20}
2.997	9.33	12.1	1.13×10^{-19}
2.9979	9.33	74.0	6.90×10^{-19}
2.99792	9.33	172.0	1.60×10^{-18}

is given by $\boldsymbol{F}_{\mathbf{Dopp}}/v = (\rho\sigma h/4\lambda_o)(v/c)/(1 - v^2/c^2)^{½}$, and the power dissipated by the radiation field is given by $v\boldsymbol{F}_{\mathbf{Dopp}}$ (Fig. A2.4). The dissipated power will increase the temperature and/or increase the volume of the radiation field. Such an effect may have been important in the expansion of the universe.

We can define the product of ρ and σ as the linear photon density (ρ_L): Replace h with $e^2/«_oc\alpha$, using the definition of the fine-structure constant (α), substitute μ_o for $1/«_oc^2$, and replace λ_{source} with w/T to get the following:

$$F_{FDopp} = -\left[\rho_L T e^2 \mu_o/4w\alpha\right]\left(v^2\right)/\left(1 - v^2/c^2\right)^{1/2} \tag{A2.16}$$

This form of the counterforce shows explicitly that the counterforce depends on the temperature, the square of the charge of the moving particle, and the fine-structure constant, which quantifies the strength of the interaction between a charged particle and the radiation field. The counterforce vanishes as either the charge of the moving particle or the temperature goes to zero.

The equation of motion that accounts for the temperature and velocity-dependent resistance due to the optomechanical Doppler effect is

$$F_{app} - \left[\rho_L T e^2 \mu_o/4w\alpha\right]\left(v^2\right)/\left(1 - v^2/c^2\right)^{1/2} = m_o dv/dt \tag{A2.17}$$

Eq. (A2.17), which is based on thermodynamics, contrasts with Planck's (1906) relativistic version of Newton's Second Law:

$$F_{app} = (d/dt)\left[m_o v\left(1 - v^2/c^2\right)^{-1/2}\right] \tag{A2.18}$$

Depending on the assumptions, Eq. (A2.18) can take the following forms:

$$F_{app} = m_o\left[\left(1 - v^2/c^2\right)^{-1/2}\right](dv/dt) \tag{A2.19}$$

$$F_{app} = m_o\left[\left(1 - v^2/c^2\right)^{-3/2}\right](dv/dt) \tag{A2.20}$$

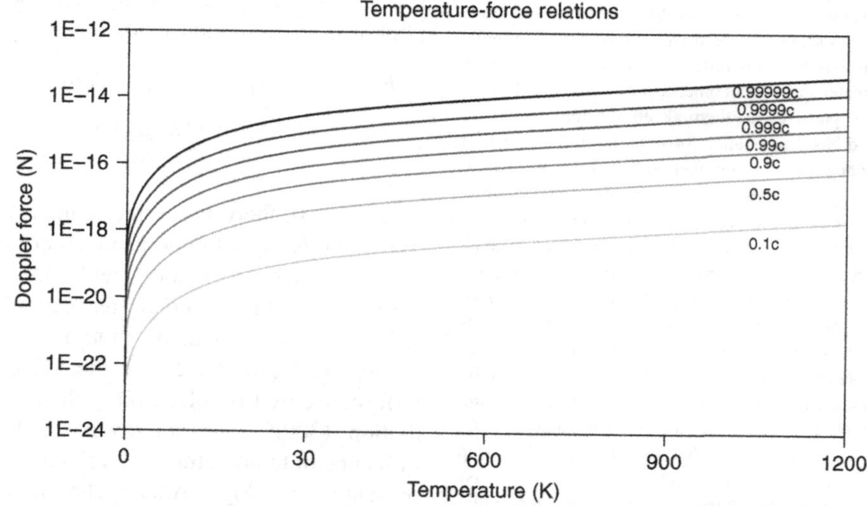

FIGURE A2.3 The effect of temperature on the magnitude of the Doppler force acting on an electron moving at various velocities relative to the speed of light.

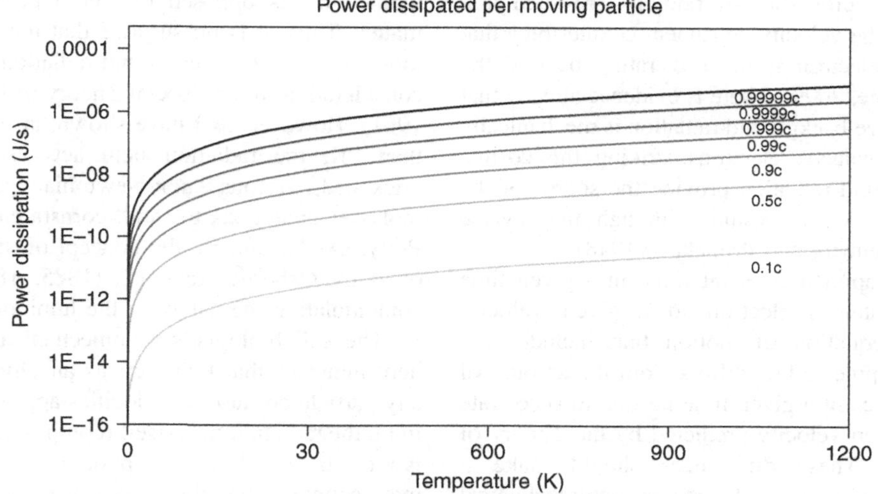

FIGURE A2.4 The effect of temperature on the power dissipated by an electron moving at various velocities.

Eqs. (A2.17), (A2.19), and (A2.20) all predict that the force–velocity relationship for a given interval of time will be nonlinear and thus non-Newtonian. That is, the optomechanical Doppler effect model and the Special Theory of Relativity both predict that the force necessary to accelerate a constant mass electron from rest to velocity v in 1 second will be velocity dependent. However, the optomechanical Doppler effect model, in contrast to the Special Theory of Relativity, further predicts that the force necessary to overcome the friction resulting from the optomechanical Doppler effect will be temperature dependent. The predictions given by the optomechanical Doppler effect equation solved for 2.73 and 300K and both equations of Special Relativity are given in Fig. A2.5. Notice that the temperature-independent relativistic Eq. (A2.20) approximates Eq. (A2.17) for temperatures between 2.73 and 300K.

Is the counterforce exerted by the thermal radiation field in an accelerator on Earth sufficient to limit the speed of an electron to the speed of light? Yes, if one takes the approach of a cell biologist and *assumes* that time and/or mass are not local quantities, and that a nonlinear relationship between a constant force exerted for a given time and velocity reveals that the electron with a constant mass must be traveling through a space that contains something that can resist the movement of the electron in a velocity-dependent manner.

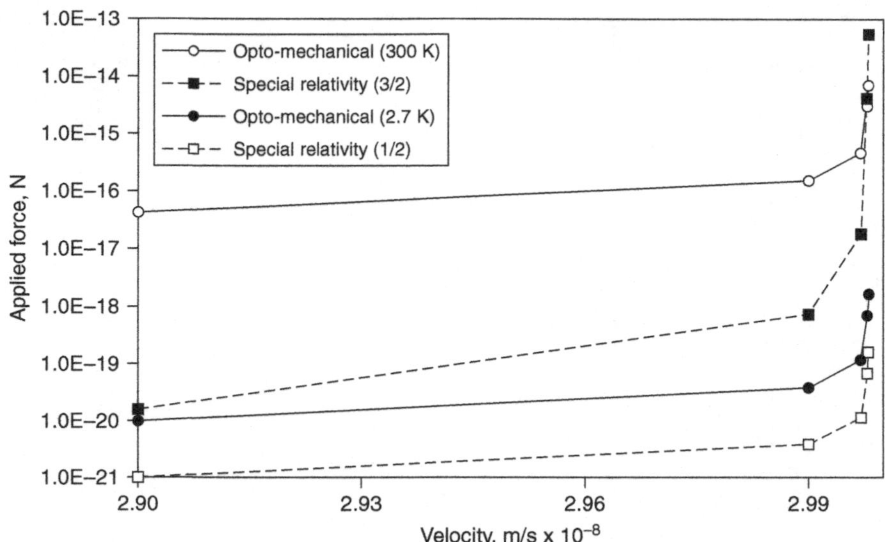

FIGURE A2.5 Prediction of how much force will be necessary to accelerate an electron to a given velocity at a given temperature according to the temperature-dependent optomechanical Doppler theory and the temperature-independent Theory of Special Relativity.

The optomechanical properties of photons are such that they could provide the velocity-dependent counterforce that would prevent an electron from accelerating beyond the speed of light (Wayne, 2009). There is evidence already that the cosmic microwave background radiation is mechanically active in that intergalactic electrons striking the cosmic microwave background radiation provide the source of the high-energy cosmic X-rays resulting though the inverse Compton effect (Feenberg and Primakoff, 1948).

The amount of applied constant force in a given time needed to accelerate an electron to a given velocity predicted by the equation of motion that includes the optomechanical Doppler effect differs from the amount of applied constant force in a given time needed to accelerate an electron to a given velocity predicted by the Theory of Special Relativity. These differences should make it possible to test the temperature-dependent, optomechanical resistance of the radiation field by performing experiments similar to those done by Kaufmann (1902), Bucherer (1904, 1908 and 1909), Neumann, Jones (1922), Guye (1915), Rogers (1940), and Bertozzi (1962), to determine the effect of velocity on mass under defined temperatures. To this end, it would be interesting to compare the dynamics of electron acceleration in the linear accelerators at Jefferson Laboratory that runs at 2K and the Stanford Linear Accelerator that runs at 300K, both of which reach energies in the GeV range.

In 1905, when the Special Theory of Relativity was published, Einstein (1905) had just conceived the idea of the quantum of radiation. It would be another 12 years before he would publish his ideas on the transfer of linear momentum as opposed to inertia between radiation and matter. Thus, it is no surprise that the effect of the linear momentum of the background radiation was not taken into consideration in the Special Theory of Relativity (Einstein, 1961). However, as I have shown, at temperatures greater than 0K, the radiation field acts as a dilatant, shear-thickened, viscous "non-Newtonian solution" or optical molasses analogous to sand, cornstarch, Kevlar, and Silly Putty. Oddly enough, the concept of dilatancy was developed by Osborne Reynolds (1885, 1886) when he was contemplating the nature of the luminous ether.

The cell biologist's optomechanical picture presented here indicates that light acts as an ultimate speed limit to any particle because, at velocities approaching the speed of light, the radiation field itself (composed of light *sensu latu*) is more than a Newtonian photon gas but is a dilatant gas that becomes infinitely viscous as the velocity of a moving particle approaches the speed of light. That is, the radiation field provides a velocity-dependent optomechanical counterforce that prevents charged particles from accelerating as they reach the speed of light. This cell biologist's explanation contrasts with that given by the Special Theory of Relativity, which asserts that either time or mass is relative and local quantities that depend on velocity, and consequently, the force acting on an accelerating electron, are not as efficient as the same force acting on a stationary electron, and the force becomes less and less efficient as the velocity of the electron increases. The optomechanical version of Newton's Second Law, developed from the perspective of a plant cell biologist and given in this appendix, applies to microscopic and macroscopic physical processes.

Appendix 3

Calculation of the Total Transverse Force and Its Relation to Stress

To show that the transverse stress (σ_∞) is equal to $(rP_t)/x$ and that the magnitude of the tangential force is equal to $(P_t)2rl$, where P_t is the turgor pressure of the cell, I need to use calculus. Consider a cylindrical cell to be composed of two hemicylinders. In spite of the turgor pressure inside acting to push them apart, each hemicylinder is held together by two tiny pieces of the extracellular matrix with a combined area of $2xl$ (Fig. A3.1). The total force that pushes the hemicylinders apart at any one point is the product of the area of the flat face of each hemicylinder ($2rl$) and the effective pressure that is pushing against this surface area. That is,

$$F = (2rl)\ \text{effective pressure} \tag{A3.1}$$

The turgor pressure acting on the wall can be resolved into two components: one that is parallel to the axis of the cylinder ($P_{\text{longitudinal}}$) and one that is perpendicular to the axis of the cylinder ($P_{\text{transverse}}$). Only the perpendicular component will be effective in pushing the two hemicylinders apart. Given the definition of sine, $P_{\text{transverse}}$ is related to P_t by the following equation:

$$P_{\text{transverse}} = 1/2P_t \sin\theta \tag{A3.2}$$

where θ is the angle between the turgor pressure vector and the flat portion of the hemisphere. At θ equal to either 0 or π, $P_{\text{transverse}}$ will vanish; at ½, it will be equal to ½ P_t and will be maximally effective in pushing the two hemicylinders apart. At all other angles, $P_{\text{transverse}}$ will have intermediate values according to Eq. (A3.3). The ½ is included in Eq. (A3.2) because we are only considering the portion of the turgor pressure exerted by one-half of the cylinder on the other:

$$F_i = (2rl)\ 1/2P_t \sin\theta = (rl)\ P_t \sin\theta \tag{A3.3}$$

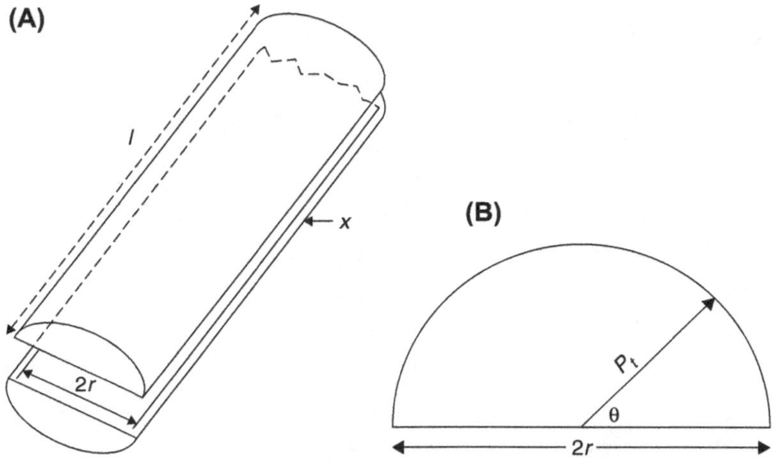

FIGURE A3.1 (A) The geometry of a cell presented as two hemicells. l, length; r, radius; x, thickness of wall. (B) The cross-section of a hemicell showing the position of θ, the angle between the flat portion of the hemicell, and the radial vector.

To get the total inertial force exerted from $\theta = 0$ to $\theta = \pi$, we must integrate Eq. (A3.3) with respect to θ from 0 to π. The total force pushing the hemispheres apart is thus

$$F = rl\,P_t \int_o^{\pi} \sin\theta\, d\theta \quad \text{(A3.4)}$$

As e $\sin\theta = 2\cos\theta$, Eq. (A3.4) becomes

$$F = -rl\,P_t\,[\cos(\pi) - \cos(0)] \quad \text{(A3.5)}$$

As $\cos(\pi) = -1$ and $\cos(0) = 1$, Eq. (A3.5) becomes

$$F = 2rl\,P_t \quad \text{(A3.6)}$$

When the velocity of movement is constant, the inertial force must be equal to the viscous force. As described in Chapter 20, the viscous force that resists the inertial force is given by

$$F_v = \sigma 2xl \quad \text{(A3.7)}$$

where x is the thickness of the extracellular matrix and l is the length of the hemicylinder. Thus, at constant velocity,

$$F_i = 2rl\,P_t = F_v = \sigma 2xl \quad \text{(A3.8)}$$

and after solving for σ_∞, we get

$$\sigma = rP_t/x \quad \text{(A3.9)}$$

As shown in Chapter 20, the transverse stress is twice the longitudinal stress, yet cells tend to elongate instead of bulge as a result of the transverse orientation of cellulose microfibrils in the extracellular matrix.

Appendix 4

Laboratory Exercises

1. LABORATORY 1: INTRODUCTION TO THE LIGHT MICROSCOPE

The purpose of this and the next two laboratory exercises is to introduce you to the light microscope so that you will feel empowered to make detailed and accurate observations of plant cells. You will learn various types of microscopy that will help you to see different aspects of cell structure and cellular processes. See Wayne (2009) for detailed explanations of the microscopy methods presented here. We will begin with a historical approach that will provide you with the opportunity to be a virtual witness to the observations made by the founders of cell biology.

1.1 Bright-Field Microscopy: Establish Köhler Illumination on the Olympus BH-2

Get a prepared slide and put it on the microscope stage. Focus on the specimen with the coarse and fine focus knobs. Close down the field diaphragm and focus an image of the field diaphragm on the specimen plane by raising or lowering the substage condenser. Adjust the interpupillary distance by moving the oculars together or apart until you find the position that gives you the most comfortable binocular vision. Adjust the diopter by looking at the specimen with your right eye through the right ocular and focusing using the fine focus control. Then look at the same image point with your left eye through the left ocular, and without using the coarse and fine focus knobs, rotate the diopter adjustment ring until the image point is in clear focus for your left eye.

The interpupillary distance adjustment and the diopter adjustment allow you to set up the microscope for the distance between your eyes and the difference in focal length of your two eyes, respectively. Open the field diaphragm so that it almost fills the field and center it with the centering screws on the substage condenser. Open and close the aperture diaphragm so that you optimize resolution and contrast. Remove the ocular and look down the microscope tube. At the position where the resolution and contrast are optimal, the light will fill about 80% of the width of the optical tube. All microscope work from here on will be done with the room lights off. You can use a flashlight to find things, prepare specimens, and write in your notebook. Set the light path selector one stop so that 80% of the light goes to the camera or two stops so that 100% of the light goes to the camera. Take a picture of your specimen.

1.2 Measurements With a Microscope

To measure the length of an object, insert an ocular micrometer in the ocular and place the stage micrometer on the stage. Carefully focus the stage micrometer. Both the stage micrometer and the ocular micrometer will be in focus and sharply defined. Turn the ocular so that the lines of the eyepiece are parallel to each other.

Determine how many intervals on the ocular micrometer correspond to a certain distance on the stage micrometer and then calculate the length that corresponds to one interval of the ocular micrometer. For example, if 35 intervals correspond to 200 μm (0.2 mm), then one interval equals (200/35) = 5.7 μm. This value is specific for each objective. Calibrate the ocular micrometer for each objective on your microscope. Put the calibration in a convenient place in your laboratory notebook.

Find the field-of-view number on the ocular. Divide the field-of-view number by the magnification of the objective to get the diameter (in mm) of the field of view. Remember to divide this distance again by the magnification introduced by any additional intermediate pieces. List the diameter of the field for each objective next to its calibration.

Take a picture of the stage micrometer using each objective. The photographs of the stage micrometers can be used as convenient rulers to measure the size of objects in photomicrographs taken under the same conditions as the stage micrometer.

1.3 Observations of Cells and Organelles Using Bright-Field Microscopy

Cells were first observed and described by Robert Hooke in his book *Micrographia* in 1665. Make a thin hand section of a piece of cork, as Robert Hooke did. Mount it in a drop of lens cleaner on a Gold Seal microscope slide and cover it with a Gold Seal #1½ cover glass (www.tedpella.com/). Keep the lids closed so the slides and cover glasses remain dry and dust free. The lens cleaner helps eliminate air bubbles. What is the most prominent structure of the cell? Can you see why Hooke chose to name the building blocks of plants by using the word to describe the austere dwelling of a monk? Adjust the aperture diaphragm to optimize resolution and contrast. Take a picture of the cork.

When you take photomicrographs, make a note of the specimen identity, the type of objective you use, its numerical aperture, its magnification, the total magnification, and the type of microscopy (e.g., bright-field). Also, note the camera and/or film type and the exposure adjustment. How do you set the exposure adjustment? If the specimen is completely distributed throughout the bright background, set the exposure adjustment knob to 1×. If the specimen covers approximately 25% of the bright background, set the exposure adjust to 0.5× to make the exposure longer. If the specimen covers less than 25% of the bright background, set the exposure adjust to 0.25× to make the exposure even longer.

Nehemiah Grew compared the cells of the root of *Asparagus* in his book *The Anatomy of Plants* in 1682 to the froth of beer. Repeat Grew's observations on the root of *Asparagus* by making transverse hand sections with a sharp razor blade. What is the most prominent structure of the cell? Can you see why cell biology was known as cytology, which means the study of hollow places, for such a long time? Adjust the aperture diaphragm to optimize resolution and contrast and take a picture of a cross section of an *Asparagus* root.

Robert Brown discovered the nucleus, an organelle he serendipitously found in the seemingly hollow cavities of the cell, while he was studying pollination. Brown described the nucleus in the epidermal cells of orchids in 1831 in a paper entitled "On the Organs and Mode of Fecundation in Orchideae and Asclepiadeae." Repeat Brown's observations of the nucleus by putting an epidermal peel from the top of a vanilla orchid leaf on a drop of water on a microscope slide and observing it with the Olympus BH-2 microscope set up for Köhler illumination. Can you see the nucleus?

Now, I will teach you a trick. Instead of using Köhler illumination, we will use oblique illumination, which was readily available to microscopists in Brown's time. To obtain oblique illumination with your microscope, slightly rotate the substage condenser turret until a pseudorelief image appears. The nucleus in each cell will appear as prominent as they did to Robert Brown. Do you remember how difficult it was to see the nucleus in high school, freshmen biology, and just a minute ago, before you knew about oblique illumination? Document your observations with photographs.

The bright-field microscope is optimal for visualizing colored objects. Pull off a young leaf of a dark-adapted *Elodea* with forceps, and place it on a drop of water on a slide. Carefully lower a coverslip on top of the specimen and observe the epidermal cells with a microscope. Unlike most epidermal cells, the epidermal cells of *Elodea* contain chloroplasts that are easy to see because of their natural pigmentation. Notice the positions of the chloroplasts and take a photograph. What happens to the chloroplasts over time as they become irradiated by the microscope light? Photographically document the changes that occur in response to light.

To get a feeling for the diversity in the form of chloroplasts, make slides of *Spirogyra* (ER-15-2525), *Zygnema* (ER-15-2695), and *Micrasterias* (ER-15-2345), all of which are green algae available from Carolina Biological Supply (www.carolina.com). Document your observations with photographs.

To visualize the vacuole, peel off the epidermis from the lower side of a *Rhoeo discolor* leaf by snapping the leaf and pulling it so that epidermal peels stick out of the broken ends of the leaf. The vacuole of *Rhoeo* contains anthocyanins, which makes it a convenient specimen for observation. Mount one epidermal peel in water and another one in 0.5-M sucrose (17.1 g/100 mL H_2O). Observe the epidermal cells that were mounted in water. Estimate the percentage of an epidermal cell that is taken up by the vacuole. Observe the protoplasts in the plasmolyzed cells in the sucrose solution. How can you explain the various shapes of the protoplasts in these cells? This plasmolysis technique can be used quantitatively to determine the osmotic pressure of the cell (Bennet-Clark et al., 1936) and the osmotic permeability coefficient for water (Stadelmann and Lee, 1974). Can you figure out how?

1.4 Observations of Cells Using Dark-Field Microscopy

Focus on the diatoms on a diatom exhibition slide or a diatom test plate. These slides are available from Klaus D. Kemp, Microlife Services (www.diatoms.co.uk/pg.htm). Raise the dark-field condenser on the Olympus BH-2 microscope so that it almost touches the bottom of the slide. Turn the substage condenser turret to DF. Open the field diaphragm only until all the specimens in the field are evenly illuminated, and observe the diatoms. Do some of the diatoms appear colored? Are they still colored when viewed with bright-field microscopy? Can you guess the cause of the colors? Document your observations with photographs.

Antony van Leeuwenhoek (1679), a linen draper, made microscopes in his spare time and discovered animalcules or "living atoms" with his microscopes. Leeuwenhoek was able to see the single-celled organisms with his simple microscope because he used dark-field illumination. Repeat Leeuwenhoek's observations by putting a drop of pond water, a drop from a soil–water mixture, a drop of pepper water, or a drop of hay infusion on a microscope slide and cover it with a cover glass. Observe the animalcules that Leeuwenhoek saw 300 years ago. If you have a slowly moving organism in the preparation, document your observations with photographs.

2. LABORATORY 2: PHASE-CONTRAST, POLARIZATION, AND DIFFERENTIAL INTERFERENCE CONTRAST MICROSCOPY

While the bright-field microscope is optimal for colored objects, transparent objects appear almost invisible when observed under conditions that maximize the resolving power of the microscope. That is when the aperture formed by the diaphragm is large. The phase-contrast microscope, the polarizing microscope, and the differential interference contrast microscope can be used to introduce contrast in transparent living specimens. The polarizing microscope has the added advantage in that it can allow the user to differentiate between macromolecules with different spatial orientations. The purpose of this laboratory is to learn how to set up a microscope for phase-contrast, polarization, and differential interference microscopy and to gain experience in observing the nature of transparent cells with these contrast-generating techniques.

2.1 Observation of Cells with Phase-Contrast Microscopy

Before you obtain cells to view with phase-contrast microscopy, put a drop of water on a microscope slide so that the cells you will obtain will remain hydrated. You can make a peel of the epidermis from the convex side of the bulb scale by cutting out a 1 cm × 1 cm piece of a bulb scale that is four or five bulb scales deep into the onion. Snap the bulb scale so that most of it breaks in two. The epidermal layer will not break clean. Pull this layer back with forceps, and place it on a drop of water on a microscope slide. To make a peel of the epidermis on the concave side of the bulb scale, remove a bulb scale that is four or five layers deep within the onion. Make a checkerboard pattern of cuts with a razor blade on the concave side of the bulb scale. Pick up several 3 mm × 3 mm squares of epidermal tissue with forceps, and place the epidermal tissue sections on the drop of water.

Set up Köhler illumination in the Olympus BH-2 microscope. Observe an onion epidermal cell with bright-field optics using the 10× phase-contrast objective lens. Make sure that the condenser ring is in the 0 position. The cells are virtually invisible and the contrast is best when the specimen is slightly defocused. Contrast can also be enhanced by closing down the aperture diaphragm. Although the contrast increases, diffraction rings and lines are produced that obscure the round and linear details, respectively, that make up the specimen.

Phase-contrast microscopy can be used to observe transparent cells with both high resolution and high contrast. Turn the substage condenser turret to the 10 position to observe the cells with phase-contrast microscopy. In general, the number on the turret matches the magnification of the objective lens. Center the phase ring in the substage condenser turret by removing one ocular and inserting the centering telescope. Focus the centering telescope so that the phase ring and the phase plate are in focus. Use the phase annulus-centering screws to center the phase ring, and align it with the phase plate. Remove the centering telescope and replace the ocular. Repeat for the 40× phase objective.

Observe the onion epidermal cells with the 40× PL phase-contrast objective. Are the epidermal cells on the convex side different from the epidermal cells from the concave side? The cells certainly do not appear hollow. Do you see the cytoplasm circulating throughout the cell? Which organelles move and which stay still? Can you see structure within the nucleus? What is the three-dimensional shape of the nucleus? Can you see mitochondria? Can you see the peripheral endoplasmic reticulum? Document your observations with photographs.

Look at the pollen tubes of periwinkle (*Catharanthus roseus*) that have been growing on a modified Brewbaker–Kwack medium for about 2 h. Notice the fountain, reverse fountain, or rotational streaming of the cytoplasm in the pollen tube. Notice the muscular nature of the cytoplasm and the variety of organelles that are visible within it. Document your observations with photographs.

To make 10 mL of modified Brewbaker–Kwack medium, mix together 3 g sucrose, 1 mg boric acid, 2 mg $MgSO_4 \times 7H_2O$, 1 mg KNO_3, 3 mg $Ca(NO_3)_2 \times 4H_2O$, and 58 mg 2-(*N*-morpholino)ethanesulfonic acid (MES). Add 7 mL distilled water and titrate to pH 6.5 with 1 N NaOH. Bring up to 10 mL with distilled water.

Place a segment of a root of *Hydrocharis* or *Limnobium* in a drop of water on a slide and look at the root hairs. Notice the fountain streaming of the cytoplasm in the root hair, the muscular nature of the cytoplasm, and the variety of organelles within it. Document your observations with photographs.

2.2 Observation of Cells with Polarization Microscopy

The polarizing microscope allows one to see anisotropic objects with high contrast. Set up the Olympus BH-2 microscope for bright-field microscopy and then insert the polarizer and the analyzer. Leave the first-order wave plate in the out position.

Obtain a lightly stained prepared slide of bordered pits or make your own specimen out of a pine board. Observe the bordered pits under crossed polars with the polarized light microscope with and without the first-order wave plate. Document your observations with photographs, making sure to document the orientation of the polarizer, the analyzer, and the slow axis (z') of the first-order wave plate. The cellulose microfibrils that make up the bordered pit are positively birefringent. How are they oriented in the bordered pit? Estimate the retardation of the bordered pits using the Michel-Lévy color chart.

Grind a cube of potato in water with a mortar and pestle. Pipette a drop of the starch grain solution on a microscope slide and cover with a cover glass. Observe the starch grains under crossed polars with the polarized light microscope with and without the first-order wave plate. Document your observations with photographs, making sure to document the orientation of the polarizer, the analyzer, and the slow axis (z') of the first-order wave plate. The starch molecules that make up the starch grains are positively birefringent. How are the starch molecules oriented in the grain? Using the Michel-Lévy color chart, estimate the retardation of the starch grain. If you like, prepare a thin section of the potato to see the starch grains in situ.

Using forceps, mount thin strands of herring sperm DNA in a drop of lens cleaner on a microscope slide. Observe with a polarized light microscope under crossed polars with and without the first-order wave plate. Document your observations with photographs, making sure to document the orientation of the polarizer, the analyzer, and the slow axis (z') of the first-order wave plate. The strands of DNA are negatively birefringent. Do the strands, the physical axes of which are parallel to the slow axis of the first-order wave plate, show additive colors (blue) or subtraction colors (yellow-orange)? Put a lot of DNA on the slide and observe retardation-dependent colors with and without the first-order wave plate. Herring sperm deoxyribonucleic acid is available from Sigma Aldrich (www.sigmaaldrich.com/).

Make a thin transverse hand section of an *Asparagus* root using a razor blade. Mount it in distilled water. Observe it with a polarized light microscope with and without the first-order wave plate. Document your observations with photographs. What can you say about the orientation of the positively birefringent cellulose microfibrils? Using the Michel-Lévy color chart, estimate the retardation of the walls.

Make an epidermal peel from the bottom of a vanilla leaf. Mount it in distilled water. Observe the stomata with a polarized light microscope with and without the first-order wave plate. How are the positively birefringent cellulose microfibrils arranged in the guard cells of the stomatal complex? Using the Michel-Lévy color chart, estimate the retardation of the walls. Document your observations with photographs.

2.3 Differential Interference Contrast Microscopy (Demonstration)

Mount a *Tradescantia* stamen hair from the second-youngest bud in water on a slide and focus on the hair cells using bright-field optics and Köhler illumination. If you are lucky you will be able to see mitosis and/or cytokinesis in the cells at or close to the tip of the hair. Use the SPLAN objectives. Slide in the polarizer that is attached to the substage condenser. Slide in the differential interference contrast beam-recombining prism with the built-in analyzer and first-order wave plate. Rotate the substage condenser turret to the red number that matches the magnification of the objective lens. Vary the contrast by turning the knob on the beam-recombining prism. Which color gives the best contrast? How does the direction of shear influence the image? Optically section the cells. Document your observations with photographs.

3. LABORATORY 3: FLUORESCENCE MICROSCOPY

The purpose of this laboratory is to learn how to set up a microscope for fluorescence microscopy to visualize transparent cells with optimal resolution and contrast. You should begin to get a sense of the absolute and relative size of each organelle, their shapes, their number, and their distribution.

3.1 Setting Up Köhler Illumination With Incident Light

Before you begin, set up Köhler illumination for both the transmitted light and the incident light on the Olympus

BH-2. To set up Köhler illumination for the incident mercury light

1. Turn on the Hg lamp.
2. Place a slide on the stage.
3. Open the shutter slider all the way.
4. Rotate the iris diaphragm (A) and the field diaphragm (F) counterclockwise so they are open maximally.
5. Close down the field diaphragm so that you can just see the edges. Center the field diaphragm by adjusting the centering screws. Open the field diaphragm.
6. Center the lamp carefully and gently with the two lamp-centering screws, until the center of the field is maximally bright.
7. Adjust the collector lens with the focusing handle until the field is maximally bright and evenly illuminated.
8. You may find that you get better contrast with some specimens and even-enough illumination by focusing the lamp on the image plane.

3.2 Observe Organelles (e.g., Mitochondria and/or Peroxisomes) in Tobacco Cells Transformed With Organelle-Targeted GFP

Tobacco plants that have been transformed with the gene for green fluorescent protein (GFP) combined with sequences that code for the targeting of the GFP to the various organelles have been prepared. For each plant, make several hand sections of a piece of tobacco leaf in water or make epidermal peels of the leaf. Place the sections or peels in a drop of water on a microscope slide. Using the blue excitation cube, look for the organelles in the epidermal hairs. For each organelle, estimate its size and describe its distribution in the cell. Estimate how many of each organelle exists in the cell. Do the proportion and distribution vary in the various cells that make up the hair? Document your observations with photographs.

3.3 Visualizing Organelles With Fluorescent Organelle-Selective Stains

To visualize the mitochondria of onion epidermal cells, place epidermal peels in a droplet of 1–10 μg/mL rhodamine 123 on Parafilm in a covered petri dish in the dark for 30–60 min. Transfer epidermal peels to distilled water in a petri dish and let them sit overnight in the dark. Place the epidermal peels on 0.05% n-propyl gallate and observe the mitochondria using the green excitation cube. What can you say about the size, distribution, and motility of the mitochondria? Document your observations with photographs.

Place an unstained onion epidermal peel on a drop of water and observe it with fluorescence microscopy. What do you see? As you did not stain the cells, anything you see is due to autofluorescence. Are the mitochondria autofluorescent? If so, then, how do you know that the rhodamine 123 selectively stained the mitochondria? If not, you can be more certain that the rhodamine 123 selectively stained the mitochondria.

To take micrographs of bright fluorescent specimens scattered over a dark field, you will typically have to set the exposure meter so that the camera takes shorter exposures. If the staining is too intense and there is too much background staining, dilute the working solution down to 1:50 with distilled water. As a rule, as one gets to know the specimen better, one requires less stain and consequently achieves better selectivity and contrast.

To observe the endoplasmic reticulum of onion epidermal cells, make a stock solution of $DiOC_6(3)$ by dissolving 1 mg of $DiOC_6(3)$ in 1 mL ethanol and dilute the stock solution with water (1:1000) to make the working solution. Then prepare several pieces of onion epidermis and mount them on a drop of staining solution. Wait 5–10 min. Remove the staining solution with a pipette and immediately replace it with 0.05% n-propyl gallate. Observe the endoplasmic reticulum using the blue excitation cube. Describe its structure. Document your observations with photographs.

To observe the lipid bodies or spherosomes in onion epidermal cells, make a 1-mg/mL stock solution of Nile Red in ethanol and dilute the stock solution to make a 1 μg/mL working solution. Then prepare several pieces of onion epidermis and mount them on a drop of staining solution. Wait 5–10 min. Remove the staining solution with a pipette and immediately replace it with 0.05% *n*-propyl gallate. Observe the spherosomes using the blue excitation cube. Document your observations with photographs.

It is important to shut off the mercury lamp when you finish with your observations because the lamp has a limited lifetime and may explode if left on for extended periods when you leave the laboratory. One should never turn a hot mercury lamp back on until it has cooled.

4. LABORATORY 4: PHOTOMICROGRAPHY

The purpose of this laboratory is to learn to take photomicrographs that capture the content of the specimens in the most artistic way possible. It also gives you a chance to observe new kinds of cells and reobserve the cells you have already seen now that you have experience with many types of light microscopy.

The procedures are as follows:

1. Go to the conservatory and choose plants that interest you to use as specimens, bring in your own research materials, or use the materials listed below:

a. Cork
b. *Asparagus* root
c. Orchid epidermis
d. Beetroot cells
e. *Elodea* leaves
f. Onion epidermis
g. Diatoms (from a drainage ditch and prepared slides)
h. Fern sporangia
i. *Tradescantia* or *Setcreasea* stamens
j. *Catharanthus* pollen
k. Cotton hairs
l. Herring sperm DNA
m. Potato starch grains
n. Circular-bordered pits (*Chamycyperis* or *Pinus*)
o. Iris leaf
p. *Tradescantia* stem
q. Impatiens stem
r. African violet leaves
s. Tobacco plants transformed with GFP targeted to the nucleus, endoplasmic reticulum, mitochondria, or peroxisomes.

2. Prepare each specimen and set up each type of microscope optimally. Photograph each specimen, taking into consideration the composition, exposure, resolution, and contrast.
3. Take your time. You have 2 weeks to make your own *Atlas of the Plant Cell.* Your photographs may also be included in making this year's Plant Cell Biology class calendar.

5. LABORATORY 5: MEMBRANE PERMEABILITY

The purpose of this laboratory is to develop the technical and mathematical skills necessary to describe membrane permeability, a biophysical aspect of cells.

5.1 Determination of the Osmotic Permeability Coefficient of the Plasma Membrane of *Chara corallina*

In this lab, we will measure the osmotic permeability coefficient of the plasma membrane of *Chara corallina* using the transcellular osmosis (TCO) technique.

Movement of water across a membrane occurs when there is a difference in the water potential (Ψ, in Pa) on the two sides of the membrane. When the difference in water potential is due to a difference in the osmotic pressure of the solutions on both sides of the membrane, the movement of water is known as *osmosis*. The rate of osmotic water movement across a membrane (J_v, in m^3/s) is proportional to the difference in the water potential outside the cell (Ψ_o) and inside the cell (Ψ_i). The rate of osmotic water movement through a membrane is also proportional to the surface area of the membrane (A, in m^2).

$$J_v \propto A(\Psi_o - \Psi_i)$$

The coefficient that relates the osmotic water flow to the difference in the water potential is the hydraulic conductivity (L_p, in m s^{-1} Pa^{-1}):

$$J_v = L_p A(\Psi_o - \Psi_i)$$

The water potential can be represented as the difference between the hydrostatic pressure (P) and the osmotic pressure (π). Consequently,

$$\Psi_o = P_o - \pi_o$$

and

$$\Psi_i = P_i - \pi_i$$

The hydrostatic pressure inside the cell is commonly called the *turgor pressure* of the cell. The hydrostatic pressure outside the cell is equal to zero by convention, and consequently,

$$\Psi_o = -\pi_o$$

The difference in water potential between the outside of the cell and the inside of the cell is

$$\Psi_o - \Psi_i = -\pi_o - (P_i - \pi_i) = \pi_i - \pi_o - P_i$$

If we define $\pi_i - \pi_o$ as $\Delta\pi$, then the above equation becomes

$$\Psi_o - \Psi_i = \Delta\pi - P_i$$

The most general way to measure the hydraulic conductivity coefficient is to subject the cell to plasmolysis and to measure the change in volume of the protoplast induced by an osmotic pressure difference across the plasma membrane. However, as the shrinking of the plasma membrane that occurs during plasmolysis may affect the water-transporting properties of the plasma membrane itself, it is desirable to find a method to measure water movement without plasmolyzing the cell. TCO is such a method (Osterhout, 1949; Kamiya and Tazawa, 1956; Kiyosawa and Ogata, 1987; Wayne and Tazawa, 1988; Wayne and Tazawa, 1990). To perform TCO, a cell is partitioned into two hydraulically isolated compartments A and B (Fig. A4.1). The difference in water potential across the plasma membrane in chamber A is

$$\Psi_A = \pi_a - \pi_A - P_i$$

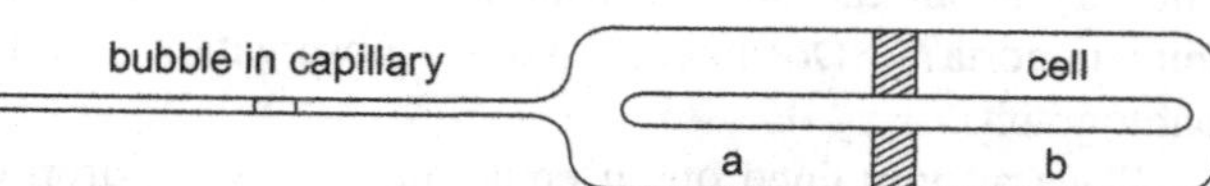

FIGURE A4.1 A chamber for transcellular osmosis.

where π_a is the osmotic pressure of the cell part in chamber A and π_A is the osmotic pressure of the solution in chamber A. P_i is the turgor pressure of the cell. The turgor pressure, which moves at the speed of sound, equalizes throughout the cell rapidly and is the same in the cell part in chamber A and the cell part in chamber B. The difference in water potential across the plasma membrane in chamber B is:

$$\Psi_B = \pi_b - \pi_B - P_i$$

where π_b is the osmotic pressure of the cell part in chamber B and π_B is the osmotic pressure of the solution in chamber B. If π_A is set to zero,

$$\Psi_A = \pi_a - P_i$$

The water potential difference between the outside of the plasma membrane in chamber A and the outside of the plasma membrane in chamber B is

$$\Psi_A - \Psi_B = (\pi_a - P_i) - (\pi_b - \pi_B - P_i)$$

$$\Psi_A - \Psi_B = \pi_a - P_i - \pi_b + \pi_B + P_i$$

$$\Psi_A - \Psi_B = \pi_a - \pi_b + \pi_B$$

and if we assume that initially $\pi_a - \pi_b = 0$, then

$$\Psi_A - \Psi_B = \pi_B$$

At the start of TCO, the difference in the water potential across the plasma membrane is equal to the osmotic potential of the solution placed in chamber B. Thus,

$$J_v = K\pi_B$$

where K is a coefficient that takes into consideration the area of the cell and the hydraulic conductivity coefficient of the plasma membrane. The osmotic pressure (π_B) is given by the following equation:

$$\pi_B = RTc$$

where R is the universal gas constant 8.31 J/mol K, T is the absolute temperature in K, and c is the concentration of the osmoticum in mol/m^3. As J = Nm and Pa = N/m^2, the universal gas constant can be given in pressure units instead of energy units and is equal to 8.31 Pa m^3/mol K. Calculate the osmotic pressure of a 100-mM sorbitol solution.

If we assume that the cell is partitioned into equal halves in the two chambers and that the L_p for water movement into the cell (endosmosis in chamber A) is equal to the L_p for water movement out of the cell (exosmosis in chamber B), then

$$K = AL_p/2$$

where A (in m^2) is the surface area of the cell in either chamber. Therefore,

$$L_p = 2J_v/A\pi_B$$

To measure the hydraulic conductivity coefficient of the plasma membrane of *C. corallina*, measure the length (l) of an isolated light-green internodal cell, approximately 4 cm long, with a ruler. Measure the diameter (d) of the cell under the microscope using a calibrated ocular micrometer. Calculate the surface area of the cell in either chamber by assuming the cell is a cylinder and using the equation:

$$A = \pi dl$$

Fill chamber A of the TCO apparatus with distilled water, put grease in the area between chambers A and B, and grease the block that will separate the two chambers. Blot the surface of the internodal cell with toilet paper until it is almost dry and place it in the silicone grease-filled groove of the TCO chamber. Place the cell so that equal lengths are in chambers A and B. Cover the cell part in the groove and apply the chamber cover in such a way that there are no air bubbles in the chamber, and the water meniscus will be visible in the calibrated capillary tube.

Add distilled water to chamber B, and let the cell equilibrate. Put the chamber under the Olympus CH microscope so that the meniscus in the calibrated capillary tube is visible with a 10× objective lens. For a clear image of the meniscus, put a drop of water on the capillary above the meniscus and put a tiny piece of a coverslip on top of the water. When the meniscus is stable, replace the water in chamber B with 100 mM sorbitol. With the help of your laboratory partner and a stopwatch, measure the rate that the meniscus moves across a calibrated ocular micrometer from 6 to 18 s after the addition of 100 mol/m^3 sorbitol.

The rate that the water moves in this time can be calculated. First, calibrate the ocular micrometer with the stage micrometer using the 10× objective to determine the number of meters per unit.

Next, calibrate the volume of water moved per unit distance that the meniscus moves with the ocular micrometer. This is done my measuring with a ruler the length between the calibrated volume marks (e.g., 0.01 mL = 15 mm). Determine the number of cubic meters per meter along the calibrated capillary tube.

To determine the volume flow (J_v) of the water (in m^3/s), we use the following equation:

$$J_v = (\text{units/s})(\text{meters/unit})(\text{meters cubed/meter length})$$

where units/s is measured between 6 and 18 s.

The hydraulic permeability coefficient can be calculated from the volume flow, the surface area of the cell part in one chamber, and the osmotic pressure used to instantiate water flow according to the following equation:

$$L_p = 2J_v/A\pi_B$$

By expressing the difference in the osmotic pressure across the membrane as a difference in the water concentration, the osmotic permeability coefficient can be expressed as a permeability coefficient known as the *osmotic permeability coefficient* (P_{os}, in m/s).

$$P_{os} = (RT/\overline{V}_w)L_p$$

where $\overline{V}_w$ represents the partial molar volume of water and is equal to 1.818×10^{-5} m^3/mol. How does the osmotic permeability coefficient compare with other permeability coefficients you can find in the literature?

5.2 Measurement of the Reflection Coefficient of Nonelectrolytes

To measure the hydraulic conductivity coefficient in *C. corallina*, we need to use a nonpermeating solute like sorbitol to induce the water movement. In essence, the sorbitol is not transmitted through the membrane but is reflected by it. The ability of a membrane to reflect a solute is known as the membrane's *reflection coefficient* (σ, dimensionless) for that solute (Staverman, 1951; Owen and Eyring, 1975; Steudle and Tyerman, 1983). By comparing the reflection coefficients of a membrane, one gets a feeling for the permeability of that membrane. The reflection coefficient for a solute can be determined by taking the ratio of the volume flow induced by the solute to the volume flow induced by the same concentration of sorbitol according to the following equation:

$$\sigma_{\text{solute}} = J_{v\text{-solute}}/J_{v\text{-sorbitol}}$$

Determine the volume flow of water induced by 100-mol/m^3 solutions of the solutes given in the following table:

Solute	Molecular Mass (g/mol)	Density (g/mL)	Radius (nm)
Diethylamine	73.14	0.707	—
Butyl alcohol	74.12	0.803	—
Glucose	180.16	—	0.35
Glycerol	92.09	1.261	—
Ethanol	46.07	0.785	—
Propanol	60.10	0.785	—
Methanol	32.04	0.791	—
Ethylene glycol	62.07	1.113	—
Urea	60.06	1.335	—
Thiourea	76.12	—	—

How do your results compare with Overton's Rules? Can the reflection coefficients be explained in terms of the lipophilicity of the molecules or the relative size of the molecules compared with the size of aqueous pores in the membrane?

The volume of a molecule (in mL) can be approximated with the following formula:

$$\text{Volume} = (\text{molecular mass})/(\text{density} \times \text{Avogadro's number})$$

The volume in meters cubed can be determined by dividing the volume in mL by 10^6. The radius of the molecule can be determined by assuming that the molecule is spherical and that the radius cubed = $(3/4\pi)$ (volume of the molecule). Estimate the radii of the molecules used to measure the reflection coefficient. What is the relationship between the reflection coefficient and the radius of each molecule?

5.3 Determination of the Permeability of Living Cells to Acids and Bases

The anthocyanins in the vacuole of the lower epidermal cells of *Rhoeo* can be used as an intrinsic pH indicator to report on the pH of the vacuole. When we observe a color change in the vacuole, we can infer that the H^+ concentration in the vacuole has changed. When these changes occur in response to the application of acids and bases to the solution surrounding the cell, we can infer that the acids or bases must have crossed the plasma and vacuolar membranes. We can get a sense of the permeability of membranes to acids and bases by following the rate in which the color changes in response to the addition of a given acid or base to the external solution.

Prepare a number of strips of the lower epidermis of *Rhoeo* and float them on distilled water. The vacuole is reddish purple when anthocyanin is protonated and blue when it is unprotonated. It is protonated below its pK_a and unprotonated above it. Is the vacuole basic or acidic?

Place two strips into 0.25 N KOH and six into 0.025 N NH_4OH. What happens? Record the time it takes for the anthocyanin to turn blue.

After the strips have turned blue in the 0.025 N NH_4OH solution, transfer three strips from the 0.025 N NH_4OH solution to water and then transfer them to the 0.025 N acetic acid solution. At the same time, transfer three strips from 0.025 N NH_4OH to 0.025 N HCl. What happens? Record the time it takes for a color change to take place.

After the color change is completed, transfer the strips from the acid solutions into the water and then transfer them back to the 0.025 N NH_4OH solution. Describe and interpret your results. What can you say about the permeability of the membrane to acids, bases, and H^+? How do your results relate to the chemiosmotic theory?

5.4 Observation of Spatial Heterogeneity of the Proton Pump in the Plasma Membrane of *Chara*

The plasma membrane is not homogeneous throughout but is differentiated into regions with different transporter protein activities. The plasma membrane of *C. corallina* cells is differentiated into regions that have a net proton efflux and regions that have a net proton influx (Spear et al., 1969; Lucas and Smith, 1973). The acid and alkaline regions of *C. corallina* can be visualized by placing cells on 1% agar containing the following components: 0.1 mM KCl, 0.1 mM $NaHCO_3$, 0.1 mM NaCl, 0.1 mM mM $CaCl_2$, 0.1 mM phenol red, and titrated to pH 7.2 with NaOH. The phenol red is a pH indicator that turns yellow where the pH is acidic and red where the pH is basic. Describe the possible distribution of transporters on the plasma membrane that would give the observed color pattern.

How could you use regions of these agar plates far from the cells to measure the diffusion coefficient of protons through the agar?

6. LABORATORY 6: CELL DIVERSITY AND CELL MOTILITY

The purpose of this laboratory is to observe a variety of single-celled organisms to get a feel for the diversity among cells. Each of these cells is at once a single cell and a complete organism and in these capacities performs all the functions of life.

One attribute of life is spontaneous movement. We will see many motile processes, including swimming powered by cilia and flagella, cytoplasmic streaming based predominantly on the microfilament and/or microtubule system, as well as endocytosis, exocytosis, and vesicle movements that take place as a part of feeding.

Use the following materials:

- *Paramecium bursaria*
- *Amoeba proteus*
- *Chaos carolinensis*
- *Volvox globator*
- *Green Hydra*
- *Vorticella*
- *Stentor*
- *Acetabularia*
- *Caulerpa*
- *Bursaria truncatella*
- *Micrasterias*
- *Rhodospirillum rubrum*
- *Dictyostelium discoideum*
- *Spirogyra grevilleana*
- *Alternaria alternata*
- *Coleochaete scutata*
- *Caulerpa*

These organisms can be obtained from Carolina Biological Supply (www.carolina.com/).

6.1 Observation of Cells with Your Microscope

Using the hanging drop method with depression slides, or flat slides with two strips of fishing line supporting the coverslip, observe the variety of single-celled organisms provided. You may slow down their movements by adding a drop of Protoslo to your preparation.

Look at *Vorticella* and watch it move. See its spasmoneme? Can you see its cilia? Can you see the organelles move inside? Look at *Stentor* and *Bursaria*. Look at how *Amoeba proteus* and *Chaos carolinensis* move across the slide. Look at the pseudopods. Can you see their organelles move inside? Look at *Volvox*. How does it move? Can you see the protoplasmic strands that connect all the cells together? Look at *Dictyostelium discoideum*. How does it move?

Look at *Micrasterias*. Imagine the cellular processes that must take place to generate the elaborate shape of each semicell!

Look at *Spirogyra grevilleana*. Isn't the chloroplast beautiful? Can you see saltational movements in the cytoplasm? Can you find the suspended nucleus?

Look at *Coleochaete scutata*. Notice how the cells form a tissue.

Look at *Rhodospirillum rubrum*, a photosynthetic, purple, nonsulfur bacteria. This bacterium has infoldings of the membrane that are similar in appearance to the cristae of mitochondria and may be related to the bacteria that gave rise to mitochondria.

Look at *Alternaria alternata*. Can you see the spores?

Look at *P. bursaria*. Can you see the *Chlorella* cells? Can you see the contractile vacuole? Can you see the cytoplasmic streaming? Can you see them conjugating?

Look at a drop of pond water. What other fascinating organisms and types of movements can you see?

6.2 Observation of Giant Cells Using the Dissecting Microscope

Look at *Acetabularia*, which is a large single-celled and uninucleate green alga. The cells can be up to 20 cm long and have beautiful cytoplasmic streaming. *Acetabularia* has a rapid wound-healing response that allows one to remove the nucleus by cutting off the rhizoid with a pair of scissors. A nucleus from another species can then be implanted in the enucleated species to test the role of nuclear determinants in the generation of form. From such studies done with *Acetabularia* in the 1930s, Hämmerling demonstrated that molecules involved in the genetic determination of the phenotype of the cap are produced in the nucleus and transported to the cytoplasm.

Look at *Hydra*. We have two forms of *Hydra*: one is transparent and the other is green. The green one is an endosymbiotic association between *Hydra* and *Chlorella*. Notice how the translucent *Hydra* buds. Notice how both types of *Hydra* respond to mechanical stimulation. Try making a hanging drop mount of the *Hydra* so that you can look at it with your microscope.

Look at *Caulerpa*. You are looking at a single cell that has differentiated into leaflike, stemlike, and rootlike structures, showing that, at least in this case, the differentiation of the organism into cells is not necessary for complex development.

6.3 Using Cytoskeletal Inhibitors to Probe Cell Motility

Use inhibitors of the cytoskeleton to test whether microtubules or microfilaments are involved in swimming, cytoplasmic streaming, and/or vacuole contraction in *P. bursaria*. The contribution of actin microfilaments can be tested by treating the cells with cytochalasin D (10 μM) or latrunculin B (30 nM). The contribution of microtubules can be tested by treating the cells with colchicine (5 μM or 5 mM) or oryzalin (10 μM).

6.4 Observing Actin Filaments in Onion Epidermal Cells

To observe actin microfilaments in onion epidermal cells, prepare 2 mL of the staining solution by mixing 1.8 mL of Part A with 0.2 mL of Part B. To make 10 mL of Part A, add 5.5 mL of a 100 mM stock solution of piperazine-1,4-bis(2-ethanesulfonic acid) (PIPES) buffer (pH 7.0), 0.055 mL of a 10% stock solution of Triton X-100 (to permeabilize the cells), 0.55 mL of a 100 mM stock solution of $MgCl_2$, 0.275 mL of a stock solution of ethylene glycol tetraacetic acid (pH 7), 0.165 mL of a 100 mM stock solution of dithiothreitol, 0.165 mL of a 100 mM stock solution of phenylmethylsulfonyl fluoride, 0.275 mL of 200 mM Na^+ phosphate buffer (pH 7.3), and 0.44 g NaCl. To make 10 mL of 200 mM Na^+ phosphate buffer (pH 7.3), mix together 2.3 mL of 200 mM monobasic sodium phosphate and 7.7 mL of 200 mM dibasic sodium phosphate. Part B consists of a 3.3 mM stock solution of rhodamine-labeled phalloidin dissolved in methanol.

Place several peels of the onion epidermis in the staining solution for 10 min in a warm place (35°C). Mount the epidermal peels in phosphate-buffered saline (PBS) that contains 0.05% (w/v) n-propyl gallate. To make PBS, add 0.85 g NaCl, 0.039 g $NaH_2PO_4 \times H_2O$, 0.0193 g $Na_2HPO_4 \times 7H_2O$, and enough distilled water to bring the solution up to 100 mL. Observe the actin microfilaments using the green excitation cube. Document your observations with photographs.

Bibliography

Ever since the publication of Euclid's Elements and Ptolemy's Almagest, textbooks have had the unfortunate effect of stifling the search for the original papers cited within. In order to prevent this, I have provided a somewhat comprehensive bibliography to facilitate the readers research into the original literature. I am particularly indebted to the staff of Mann Library, Kroch Library and the University Annex, for finding, storing, and making it possible for me to read these references.

Abderhalden, E., Brockmann, H., 1930. Beitrag zur Konstitutionsermittlung von Proteinen bzw. Polypeptiden. Biochem. Z. 225, 386–425.

Abe, J., Hiyama, T.B., Mukaiyama, A., Son, S., Mori, T., Saito, S., Osako, M., Wolanin, J., Yamashita, E., Kondo, T., Akiyama, S., 2015. Circadian rhythms. Atomic-scale origins of slowness in the cyanobacterial circadian clock. Science 349, 312–316.

Abel, P.P., Nelson, R.S., De, B., Hoffmann, N., Rogers, G.G., Fraley, R.T., Beachy, R.N., 1986. Delay of disease development in transgenic plants that express the tobacco mosaic virus coat protein gene. Science 232, 738–743.

Abel, S., Theologis, A., 2010. Odyssey of auxin. Cold Spring Harb. Persepct. Biol. 2, a004572.

Abell, G.O., 1976. Realm of the Universe. Holt, Rinehart and Winston, New York.

Abranches, R., Beven, A.F., Aragón-Alcaide, L., Shaw, P.J., 1998. Transcription sites are not correlated with chromosome territories in wheat nuclei. J. Cell Biol. 143, 5–12.

Abuaf, N., Anderson, J.B., Andres, R.P., Fenn, J.B., Marsden, D.G.H., 1967. Molecular beams with energies above one electron volt. Science 155, 997–999.

Acharya, B.R., Jeon, B.W., Zhang, W., Assmann, S.M., 2013. Open stomata 1 (OST1) is limiting in abscisic acid responses of *Arabidopsis* guard cells. New Phytol. 200, 1049–1063.

Acuri, E.F., Wiedmann, M., Boor, K.J., 1999. Development of a PCR assay for etection of spore-forming bacteria. J. Rapid Methods Autom.Microbiol. 7, 251–262.

Adachi, Y., Luke, M., Laemmli, U.K., 1991. Assembly in vitro: topoisomerase II is required for condensation. Cell 64, 137–148.

Adachi, P., Puopolo, K., Marquez-Sterling, N., Arav, H., Forgar, M., 1990. Dissolution, cross-linking and glycosylation of the coated-vesicle proton pump. J. Biol. Chem. 265, 967–973.

Adair, G.S., 1937. The chemistry of the proteins and amino acids. Ann. Rev. Biochem. 6, 163–192.

Adair, W.S., Appel, H., 1989. Identification of a highly conserved hydroxyproline–rich glycoprotein in the cell walls of *Chlamydomonas reinhardtii* and two other Volvocales. Planta 179, 381–386.

Adam, N.K., 1941. The Physics and Chemistry of Surfaces, third ed. Oxford University Press, Oxford.

Adamowski, M., Friml, J., 2015. PIN-dependent auxin transport: action, regulation, and evolution. Plant Cell 27, 20–32.

Adams, M.D., Dubnick, M., Kerlavage, A.R., Moreno, R., Kelley, J.M., Utterback, T.R., Nagle, J.W., Fields, C., Venter, J.C., 1992. Sequence identification of 2,375 human brain genes. Nature 355, 632–634.

Adams, J., Jeppesen, P., Sanger, F., Barrell, B., 1969. Nucleotide sequence from the coat protein cistron of R17 bacteriophage RNA. Nature 228, 1009–1014.

Adams, M.D., Kelley, J.M., Gocayne, J.D., Dubnick, M., Polymeropoulos, M.H., Xiao, H., Merril, C.R., Wu, A., Olde, B., Moreno, R.F., Kerlavage, A.R., McCombie, W.R., Venter, J.C., 1991. Complementary DNA sequencing: expressed eequence tags and human genome project. Science 252, 1651–1656.

Adams, M.D., Kerlavage, A.R., Fields, C., Venter, J.C., 1993a. 3,400 new expressed sequence tags identify diversity of transcripts in human brain. Nat. Genet. 4, 256–267.

Adams, M.D., Kerlavage, A.R., Kelley, J.M., Gocayne, J.D., Fields, C., Fraser, C.M., Venter, J.C., 1994. A model for high-throughput automated DNA sequencing and analysis core facilities. Nature 368, 474–475.

Adams, R.R., Maiato, H., Earnshaw, W.C., Carmena, M., 2001. Essential roles of *Drosophila* inner centromere protein (INCENP) and aurora B in histone H3 phosphorylation, metaphase chromosome alignment, kinetochore disjunction, and chromosome segregation. J. Cell Biol. 153, 865–880.

Adams, R.J., Pollard, T.D., 1989. Binding of myosin I to membrane lipids. Nature 340, 565–568.

Adams, M.D., Soares, M.B., Kerlavage, A.R., Fields, C., Venter, J.C., 1993b. Rapid cDNA sequencing (expressed sequence tags) from directionally cloned human infant brain cDNA library. Nat. Genet. 4, 373–380.

Adams, G., Whicher, O., 1949. The Living Plant and the Science of Physical and Ethereal Spaces. Goethean Science Foundation, Clent, England.

Addinall, S.G., Holland, B., 2002. The tubulin ancestor, FtsZ, draughtsman, designer and driving force for bacterial cytokinesis. J. Mol. Biol. 318, 219–236.

Adelman, M.R., Borisy, G.G., Shelanski, M.L., Wisenberg, R.C., Taylor, E.W., 1968. Cytoplasmic filaments and tubules. Fed. Proc. 27, 1186–1193.

Adelman, G., Smith, B., 1998. Francis Otto Schmitt: 1903-1995. Biographical Memoir. National Academies Press, Washington, DC.

Adelman, M.R., Taylor, E.W., 1969. Further purification and characterization of slime mold myosin and slime mold actin. Biochemistry 8, 4976–4988.

Adessi, C., Matton, G., Ayala, G., Turcatti, G., Mermod, J.-J., Maye, P., Kawashima, E., 2000. Solid phase DNA amplification: characterisation of primer attachment and amplification mechanisms. Nucleic Acids Res. 28, e87.

Adler, M.J., van Doren, C., 1967. How to Read a Book. The Classic Guide to Intelligent Reading. Simon and Schuster, New York.

Aebersold, R., Hood, L.E., Watts, J.D., 2000. Equipping scientists for the new biology. Nature Biotechnol. 18, 359.

Aebi, U., Cohn, J., Buhle, L., Gerace, L., 1986. The nuclear lamina is a meshwork of intermediate-type filaments. Nature 323, 560–564.

Afzelius, B., 1959. Electron microscopy of the sperm tail. Results obtained with a new fixative. J. Biophys. Biochem. Cytol. 5, 269–278.

Agar, W.E., 1920. Cytology with Special Reference to the Metazoan Nucleus. MacMillan and Co., London.

Agarwal, K.L., Buchi, H., Caruthers, M.H., Gupta, N., Khorana, H.G., Kleppe, K., Kumar, E., Ohtsuka, E., Rajbhandary, U.L., van de Sande, J.H., Sgaramella, V., Weber, H., Yamada, T., 1970. Total synthesis of the gene for an alanine transfer ribonucleic acid from yeast. Nature 227, 27–34.

Agassiz, L., 1863. Methods of Study in Natural History. Ticknor and Fields, Boston.

Aggarwal, C., Łabuz, J., Gabrys, H., 2013. Phosphoinositides play differential roles in regulating phototropin1- and phototropin2-mediated chloroplast movements in *Arabidopsis*. PLoS One 8, e55393.

Agne, B., Kessler, F., 2007. Protein import into plastids. Top. Curr. Genet. 19, 339–370.

Agrawal, P.K., Babu, B.K., Saini, N., 2015. Omics of model plants. In: Barh, D., Sarwar Khan, M., Davies, E. (Eds.), PlantOmics: The Omics of Plant Science. Springer India, pp. 1–32.

Agrawal, G., Joshi, S., Subramani, S., 2011. Cell-free sorting of peroxisomal membrane proteins from the endoplasmic reticulum. Proc. Natl. Acad. Sci. U.S.A. 108, 9113–9118.

Agre, P., Brown, D., Nielsen, S., 1995. Aquaporin water channels: unanswered questions and unresolved controversiea. Curr. Opin. Cell Biol. 7, 472–483.

Agutter, P.S., 1991. Between Nucleus and Cytoplasm. Chapman and Hall, London.

Agutter, P.S., 1995. Intracellular structure and nucleocytoplasmic transport. Int. Rev. Cytol. 162B, 183–224.

Ahmad, M., Cashmore, A.R., 1996. Seeing blue: the discovery of cryptochrome. Plant Mol. Biol. 30, 851–861.

Ahmed, S.U., Bar-Peled, M., Raikhel, N.V., 1997. Cloning and subcellular location of an *Arabidopsis* receptor-like protein that shares common features with protein-sorting receptors of eukaryotic cells. Plant Physiol. 114, 325–336.

Ahmed, S.U., Rojo, E., Kovaleva, V., Venkataraman, S., Dombrowski, J.E., Matsuoka, K., Raikhel, N.V., 2000. The plant vacuolar sorting receptor AtELP is involved in transport of NH2-terminal propeptide-containing vacuolar proteins in *Arabidopsis thaliana*. J. Cell Biol 149, 1335–1344.

Ahner, B.A., Kong, S., Morel, F.M.M., 1995. Phytochelatin production in marine algae. Limnol. Oceanogr. 40, 649–657, 658–665.

Ahner, B.A., Price, N.M., Morel, F.M.M., 1994. Phytochelatin production by marine phytoplankton at low free metal ion concentrations: laboratory studies and field data from Massachusetts Bay. Proc. Natl. Acad. Sci. U.S.A. 91, 8433–8436.

Ahsan, N., Chen, M., Salvato, F., Wilson, R.S., Prasad Rao, R.S., Thelen, J.J., 2017. Comparative proteomic analysis provides insight into the biological role of protein phosphatase inhibitor-2 from *Arabidopsis*. J. Proteom. 165, 51–60.

Airy, G.B., 1876. Notes on the Early Hebrew Scriptures. Longmans, Green, and Co., London.

Aist, J.R., 1977. Mechanically induced wall appositions of plant cells can prevent penetration by a parasitic fungus. Science 197, 568–571.

Aist, J.R., Bayles, C.J., Tao, W., Berns, M.W., 1991. Direct experimental evidence for the existence, structural basis and function of astral forces during anaphase B in vivo. J. Cell Sci. 100, 279–288.

Aist, J.R., Berns, M.W., 1981. Mechanisms of chromosome separation during mitosis in *Fusarium* (fungi imperfecti): new evidence from ultrastructural and laser microbeam experiments. J. Cell Biol. 91, 446–458.

Aist, J.R., Liang, H., Berns, M.W., 1993. Astral and spindle forces in mitotic PtK2 cells. J. Cell Sci. 104, 1207–1216.

Aist, J.R., Williams, P.H., 1972. Ultrastructure and time course of mitosis in the fungus *Fusarium oxysporum*. J. Cell Biol. 55, 368–389.

Akashi, T., Kawasaki, S., Shibaoka, H., 1990. Stabilization of cortical microtubules by the cell wall in cultured tobacco cells effects of extension on the cold stability of cortical microtubules. Planta 182, 363–369.

Akashi, T., Shibaoka, H., 1987. Effects of gibberellin on the arrangement and the cold stability of cortical microtubules in epidermal cells of pea internodes. Plant Cell Physiol. 28, 334–348.

Akashi, T., Shibaoka, H., 1991. Involvement of transmembrane proteins in the association of cortical microtubules with the plasma membrane in tobacco BH-2 cells. J. Cell Sci. 98, 169–172.

Akazawa, T., Hara-Nishimura, I., 1985. Topographic aspects of biosynthesis, extracellular secretion, and intracellular storage of proteins in plant cells. Ann. Rev. Plant Physiol. 36, 441–472.

Akey, C.W., 1989. Interactions and structure of the nuclear pore complex revealed by cryo-electron microscopy. J. Cell Biol. 109, 955–970.

Akita, K., Higaki, T., Kutsuna, N., Hasezawa, S., 2015. Quantitative analysis of microtubule orientation in interdigitated leaf pavement cells. Plant Signal. Behav. 10, e1024396.

Akita, M., Nielsen, E., Keegstra, K., 1997. Identification of protein transport complexes in the chloroplastic envelope membranes via chemical cross-linking. J. Cell Biol. 136, 983–994.

Akiyama, Y., Nakasone, Y., Nakatani, Y., Hisatomi, O., Terazima, M., 2016. Time-resolved detection of light-induced dimerization of monomeric Aureochrome-1 and change in affinity for DNA. J. Phys. Chem. B 120, 7360–7370.

Alassimone, J., Naseer, S., Geldner, N., 2010. A developmental framework for endodermal differentiation and polarity. Proc. Natl. Acad. Sci. U.S.A. 107, 5220–5225.

Alassimone, J., Roppolo, D., Geldner, N., Vermeer, J.E.M., 2012. The endodermis—a development and differentiation of the plant's inner skin. Protoplasma 249, 433–443.

Alba, R., Fei, Z., Liu, Y., Debbie, P., Gordon, J., Rose, J.K.C., Martin, G., Tanksley, S., Bouzayen, M., Jahn, M.J., Giovannoni, J.J., 2004. ESTs, cDNA microarrays, and gene expression profiling: tools for dissecting plant physiology and development. Plant J. 39, 697–714.

Albersheim, P., Bonner, J., 1959. Metabolism and hormonal control of pectic substances. J. Biol. Chem. 234, 3105–3108.

Alberts, B., Kirschner, M.W., Tilghman, S., Varmus, H., 2014. Rescuing US biomedical research from its systemic flaws. Proc. Natl. Acad. Sci. U.S.A. 111, 5773–5777.

Aldini, J., 1803. An Account of the Late Improvements in Galvanism: With a Series of Curious and Interesting Experiments Performed before the Commissioners of the French National Institute, and Repeated Lately in the Anatomical Theatres of London. Wilks and Taylor, London.

Alexander, D., 1972. Beyond Science. A. J. Holman, Philadelphia, PA.

Alexander, D., 2001. Rebuilding the Matrix. Science and Faith in the 21st Century. Lion Publishing, Oxford.

Alexander, S.P., Rieder, C.L., 1991. Chromosome motion during attachment to the vertebrate spindle: initial saltatory-like behavior of chromosomes and quantitative analysis of force production by nascent kinetochore fibers. J. Cell Biol. 113, 805–815.

Alexander, D., White, R.S., 2004a. Beyond Belief. Science, Faith and Ethical Challenges. Lion Publishing, Oxford.

Alexander, D., White, R.S., 2004b. Science, Faith, and Ethics: Grid or Gridlock? Henrickson, Peabody, MA.

Alexandersson, E., Saalbach, G., Larsson, C., Kjellbom, P., 2004. *Arabidopsis* plasma membrane proteomics identifies components of transport, signal transduction and membrane trafficking. Plant Cell Physiol. 45, 1543–1556.

Alexandre, H., 1992. Jean Brachet and his school. Int. J. Dev. Biol. 36, 29–41.

Alexandre, J., Lassalles, J.P., Kado, R.T., 1990. Opening of Ca^{2+} channels in isolated red beet vacuole membrane by inositol (1,4,5) triphosphate. Nature 343, 567–570.

Alfvén, H., 1966. Worlds-Antiworlds. Antimatter in Cosmology. W.H. Freeman and Co., San Fransisco.

Algera, L., Beijer, J.J., van Iterson, W., Karsten, W.K.H., Thung, T.H., 1947. Some data on the structure of the chloroplast, obtained by electron microscopy. Biochem. Biophys. Acta 1, 517–526.

Ali, M.S., Akazawa, T., 1985. Association of H^+-translocating ATPase in the Golgi membrane system from suspension- cultured cells of sycamore (*Acer pseudoplatanus* L.). Plant Physiol. 81, 222–227.

Ali, M.S., Nishimura, M., Mitsui, T., Akazawa, T., Kojima, K., 1985. Isolation and characterization of Golgi membranes from suspension-cultured cells of sycamore (*Acer pseudoplatanus* L.). Plant Cell Physiol. 26, 1119–1133.

Allaway, W.G., Carpenter, J.L., Ashford, A.E., 1985. Amplification of inter-symbiont surface by root epidermal transfer cells in the *Pisonia* mycorrhiza. Protoplasma 128, 227–231.

Allchin, D., 2007. Racker, Efraim. New Dictionary of Scientific Biography. http://douglasallchin.net/papers/racker.pdf.

Allen, R.D., 1969. Mechanism of the seismonastic reaction in *Mimosa pudica*. Plant Physiol. 44, 1101–1107.

Allen, G.E., 2001. Is a new eugenics afoot? Science 294, 59–60.

Allen, J.F., 2003. State transitions—A question of balance. Science 299, 1530–1532.

Allen, N.S., Brown, D.T., 1988. Dynamics of the endoplasmic reticulum in living onion epidermal cells in relation to movement. Cell Motil. Cytoskelet. 10, 153–163.

Allen, R.D., Kamiya, N. (Eds.), 1964. Primative Motile Systems in Cell Biology. Academic Press, New York.

Allen, V.W., Kropf, D.L., 1992. Nuclear rotation and lineage specification in *Pelveta* embryos. Development 115, 873–883.

Allen, G.J., Kuchitsu, K., Chu, S.P., Murata, Y., Schroeder, J.I., 1999a. *Arabidopsis* abi1-1 and abi2-1 phosphatase mutations reduce abscisic acid-induced cytoplasmic calcium rises in guard cells. Plant Cell 11, 1785–1798.

Allen, G.J., Kwak, J.M., Chu, S.P., Llopis, J., Tsien, R.Y., Harper, J.F., Schroeder, J.I., 1999b. Cameleon calcium indicator reports cytoplasmic calcium dynamics in *Arabidopsis* guard cells. Plant J. 19, 735–747.

Allen, R.D., Naitoh, Y., 2002. Osmoregulation and contractile vacuoles of protozoa. Int. Rev. Cytol. 215, 351–394.

Allen, R.D., Roslansky, J.D., 1959. The consistency of Ameba cytoplasm and its bearning on the mechanism of ameboid movement. I. An analysis of endoplasmic velocity profiles of *Chaos chaos*. J. Biophys. Biochem. Cytol. 6, 437–446.

Allen, R.D., Tominaga, T., Naitoh, Y., 2009. The contractile vacuole complex and cell volume control in protozoa. In: Evans, D.H. (Ed.), Osmotic and Ionic Regulation. Cells and Animals. CRC Press, Boca Ration, FL, pp. 69–105.

Alley, C.D., Scott, J.L., 1977. Unusual dictyosome morphology and vesicle formation in tetrasporangia of the marine red alga *Polysiphonia denudada*. J. Ultrastruct. Res. 58, 289–298.

Allison, A.C., 1954. Protection afforded by sickle-cell trait against subtertian malareal infection. Br. Med. J. 1, 290–294.

Allman, J.M., Hakeem, A., Erwin, J.M., Nimchinsky, E., Hof, P., 2001. The anterior cingulate cortex: the evolution of an interface between emotion and cognition. Proc. N.Y. Acad. Sci. 935, 107–117.

Allman, A.H., Watson, K., 2002. Two phylogenetic specializations in the human brain. Neuroscientist 8, 335–346.

Alonso, J.M., Stepanova, A.N., Leisse, T.J., Kim, C.J., Chen, H., Shinn, P., Stevenson, D.K., Zimmerman, J., Barajas, P., Cheuk, R., Gadrinab, C., Heller, C., Jeske, A., Koesema, E., Meyers, C.C., Parker, H., Prednis, L., Ansari, Y., Choy, N., Deen, H., Geralt, M., Hazari, N., Hom, E., Karnes, M., Mulholland, C., Ndubaku, R., Schmidt, I., Guzman, P., Aguilar-Henonin, L., Schmid, M., Weigel, D., Carter, D.E., Marchand, T., Risseeuw, E., Brogden, D., Zeko, A., Crosby, W.L., Berry, C.C., Ecker, J.R., 2003. Genome-wide insertional mutagenesis of *Arabidopsis thaliana*. Science 301, 653–657.

Alpher, R.A., Bethe, H., Gamow, G., 1948. The origin of chemical elements. Phys. Rev. 73, 803–804.

Altmann, R., 1889. Ueber nucleinsäuren. Arch. Anat. Physiol. Physiol. Abt. 524–536.

Altmann, R., 1890. Die Elementarorganismen und ihre Beziehung zu den Zellen. Veit & Co., Leipzig.

Altmann, S., 1990. Enzymatic cleavage of RNA by RNA. Angew. Chem. (Int. Ed.) 29, 749–758.

Altmüller, J., Palmer, L.J., Fischer, G., Scherb, H., Wjst, M., 2001. Genomewide scans of complex human diseases: true linkage is hard to find. Am. J. Hum. Genet. 69, 936–950.

Altschul, S.F., Gish, W., Miller, W., Myers, E.W., Lipman, D.J., 1990. Basic local alignment search tool. J. Mol. Biol. 215, 403–410.

Alvarenga, D.O., Fiore, M.F., Varani, A.M., 2017. A metagenomics approach to cyanobacterial genomics. Front. Microbiol. 8, 809.

Alwine, J.C., Kemp, D.J., Stark, G.R., 1977. Method for detection of specific RNAs in agarose gels by transfer to diazobenzyloxymethyl-paper and hybridization with DNA probes. Proc. Natl. Acad. Sci. U.S.A. 74, 5350–5354.

Amborella Genome Project, 2013. The *Amborella* genome and the evolution of flowering plants. Science 342, 1241089.

Ambros, V., 2008. The evolution of our thinking about micro- RNAs. Nat. Medods 14, 10–14.

Ambrose, J.C., Cyr, R., 2007. Mitotic spindle assembly and function. Plant Cell Monogr. 9, 141–167.

Ambrose, J.C., Li, W., Marcus, A., Ma, H., Cyr, R., 2005. A minus-end-directed kinesin with plus-end tracking protein activity is involved in spindle morphogenesis. Mol. Biol. Cell 16, 1584–1592.

Amici, G.B., 1818. Osservazioni sulla circolazione del succhio nella *Chara*. Memorie della Società Italiana delle Scienze 18, 183–198.

Amino, S., Tazawa, M., 1989. Dependence of tonoplast transport of amino acids on vacuolar pH in *Chara* cells. Proc. Jpn. Acad. Sci. 65 (B), 34–37.

Amoore, J.E., 1962. Participation of a non-respiratory ferrous complex during mitosis in roots. J. Cell Biol. 13, 373–381.

Amoore, J.A., 1963. Non-identical mechanisms of mitotic arrest by respiratory inhibitors in pea root tips and sea urchin eggs. J. Cell Biol. 18, 555–567.

Amor, Y., Mayer, R., Benziman, M., Delmer, D., 1991. Evidence for a cyclic diguanylic acid dependent cellulose synthase in plants. Plant Cell 3, 989–995.

Amsbury, S., Kirk, P., Benitez-Alfonso, Y., 2018. Emerging models on the regulation of intercellular transport by plasmodesmata-associated callose. J. Exp. Bot. 69, 105–115.

An, Q., van Bel, A., Hückelhoven, R., 2007. Do plant cells secrete exosomes derived from multivesicular bodies? Plant Signal. Behav. 2, 4–7.

Anderberg, H.I., Kjellbom, P., Johanson, U., 2012. Annotation of *Selaginella moellendorffii* major intrinsic proteins and the evolution of the protein family in terrestrial plants. Front. Plant Sci. 3, 33.

Andersen, T.G., Barberon, M., Geldner, N., 2015. Suberization—the second life of an endodermal cell. Curr. Opin. Plant Biol. 28, 9–15.

Andersen, O.M., Markham, K.R., 2006. Flavonoids: Chemistry, Biochemistry and Applications. CRC Press, Boca Raton, FL.

Andersland, J.M., Jagendorf, A.T., Parthasarathy, M.V., 1992. The isolation of actin from pea roots by DNAse I affinity chromatography. Plant Physiol. 100, 1716–1723.

Andersland, J.M., Parthasarathy, M.V., 1992. Phalloidin binds and stabilizes pea root actin. Cell Motil. Cytoskelet. 22, 245–249.

Andersland, J.M., Parthasarathy, M.V., 1993. Conditions affecting depolymerization of actin in plant homogenate. J. Cell Sci. 104, 1273–1279.

Anderson, E.G., 1923. Maternal inheritance of chlorophyll in maize. Bot. Gaz. 76, 411–418.

Anderson, J.M., 1975. The molecular organization of chloroplast thylakoids. Biochim. Biophys. Acta 416, 191–235.

Anderson, S., 1981. Shotgun DNA sequencing using clones DNase I-generated fragments. Nucleic Acids. Res. 9, 3015–3027.

Anderson, R.G.W., 1993. Plasmalemmal caveolae and GPI-anchored membrane proteins. Curr. Opin. Cell Biol. 5, 647–652.

Anderson, C.T., 2016. We be jammin': an update on pectin biosynthesis, trafficking and dynamics. J. Exp. Bot. 67, 495–502.

Anderson, J.T., Beams, H.W., 1956. Evidence from electron micrographs for the passage of material through pores of the nuclear membrane. J. Biophys. Biochem. Cytol. 2 (Suppl.), 439–444.

Anderson, J.W., Beardall, J., 1991. Molecular Activities of Plant Cells. An Introduction to Plant Biochemistry. Blackwell Scientific Publications, Oxford.

Anderson, J.A., Huprikar, S.S., Kochian, L.V., Lucas, W.J., Gaber, R.F., 1992. Functional expression of a probable *Arabidopsis thaliana* potassium channel in *Saccharomyces cerevisiae*. Proc. Natl. Acad. Sci. U.S.A. 89, 3736–3740.

Anderson, J.V., Li, Q.-B., Haskell, D.W., Guy, C.L., 1994. Structural organization of the spinach endoplasmic reticulumluminal 70-kilodalton heat-shock cognate gene and expression of 70-kilodalton heat-shock genes during cold acclimation. Plant Physiol. 104, 1359–1370.

Andersson, M.X., Goksor, M., Sandelius, A.S., 2007a. Membrane contact sites: physical attachment between chloroplasts and endoplasmic reticulum revealed by optical manipulation. Plant Signal. Behav. 2, 185–187.

Andersson, M.X., Goksor, M., Sandelius, A.S., 2007b. Optical manipulation reveals strong attracting forces at membrane contact sites between endoplasmic reticulum and chloroplasts. J. Biol. Chem. 282, 1170–1174.

Andersson, M., Turesson, H., Nicolia, A., Fäit, A.S., Samuelsson, M., Hofvander, P., 2017. Efficient targeted multiallelic mutagenesis in tetraploid potato (*Solanum tuberosum*) by transient CRISPR-Cas9 expression in protoplasts. Plant Cell Rep. 36, 117–128.

Anding, A.L., Baehrecke, E.H., 2017. Cleaning house: selective autophagy of organelles. Dev. Cell 41, 10–22.

Andrade, E.N., 1964. Rutherford and the Nature of the Atom. Doubleday & Co., Garden City, NY.

Andrade-Sanchez, P., Gore, M., Heun, J.T., Thorp, K.R., Carmo-Silva, A.E., French, A.N., Salvucci, M.E., White, J.W., 2014. Development and evaluation of a field-based high-throughput phenotyping platform. Funct. Plant Biol. 41, 68–79.

Andre, J., Thiery, J.P., 1963. Mise evidence d'une sous-structure fibrillaire dans les filaments axonmatiques des flagelles. J. Microsc. 2, 71–80.

Andrews, G.F., 1897. The Living Substance as Such: And as Organism. Athenaeum Press, Boston.

Andrews, E.R., 2002. Multicellular Organs and Organisms. Encyclopedia of Life Sciences. John Wiley & Sons, Ltd. www.els.net.

Andrews, T.J., Lorimer, G.H., Tolbert, N.E., 1971. Incorporation of molecular oxygen into glycine and serine during photorespiration in spinach leaves. Biochemistry 10, 4777–4782.

Anfinsen, C.B., 1973. Principles that govern the folding of protein chains. Science 181, 223–230.

Angel, T.E., Aryal, U.K., Hengel, S.M., Baker, E.S., Kelly, R.T., Robinson, E.W., Smith, R.D., 2012. Mass spectrometry-based proteomics: existing capabilities and future directions. Chem. Soc. Rev. 41, 3912–3928.

Angert, E.R., Brooks, A.E., Pace, N.R., 1996. Phylogenetic analysis of *Metabacterium polyspora*: clues to the evolutionary origin of *Epulopiscium* spp., the largest bacteria. J. Bacteriol. 178, 1451–1456.

Angert, E.R., Clements, K.D., Pace, N.R., 1993. The largest bacterium. Nature 362, 239–241.

Aniento, F., Robinson, D.G., 2005. Testing for endocytosis in plants. Protoplasma 226, 3–11.

Ankeny, R.A., 2003. Sequencing the genome from nematode to human: changing methods, changing science. Endeavour 27, 87–92.

Anonymous, 1839. Das enträthselte Geheimniss der geistigen Gährung. perhaps F. Wöhler and J. Liebig Ann. Pharm. (Heidelberg) 29, 100–104.

Anonymous, 1848. The best heat generator in the world. Am. Phrenol. J. Misc. 10, 61–67.

Anonymous, 1970. Central dogma reversed. Nature 226, 1198–1199.

Anonymous, 1997. Genomics: structural and functional studies of genomes. Genomics 45, 244–249.

Anonymous, 1999. A Nobel Prize for cell biology. Nat. Cell Biol. 1, E169.

Anonymous, 2011. Fix the PhD. Nature 472, 259–260.

Anonymous, December 16, 2010. The Disposable Academic. Why Doing a PhD is Often a Waste of Time. The Economist. http://www.economist.com/node/17723223.

Anonymous, March 2, 1905. Dr. Loeb's incredible "discovery." New York Times.

Anonymous, September 12, 1963. Probing Heredity's secrets. New York Times.

Anraku, Y., Hirata, R., Umemoto, N., Ohya, Y., 1991. Molecular aspects of the yeast vacuolar membrane ATPase. In: Mukohata, Y. (Ed.), New Era in Bioenergetics. Academic Press, Tokyo, pp. 133–168.

Ansorge, W.J., 2009. Next-generation DNA sequencing techniques. New Biotechnol. 25, 195–203.

Ansorge, W.J., 2016. Next generation DNA sequencing (II): techniquest applications. Next Gener. Seq. Appl. S1, 005.

Anthon, G.E., Spanswick, R.M., 1986. Purification and properties of the H^+-translocating ATPase from the plasma membrane of tomato roots. Plant Physiol. 81, 1080–1085.

Apel, K., 1979. Phytochromeinduced appearance of mRNA activity for the apoprotein of the light-harvesting chlorophyll a/b protein of barley (*Hordeum vulgare*). Eur. J. Biochem. 97, 183–188.

Apgar, J., Everett, G.A., Holley, R.W., 1965. Isolation of large oligonucleotide fragments from the alanine RNA. Proc. Natl. Acad. Sci. U.S.A. 53, 546–548.

Apostol, M., 2007. Comment on the "Declaration of the Academic Freedom" by D. Rabounski. Prog. Phys. 3, 81–84.

Apostol, I., Heinstein, P.F., Low, P.S., 1989. Rapid stimulation of an oxidative burst during elicitation of cultured plant cells: role in defense and signal transduction. Plant Physiol. 90, 109–116.

Apostol, I., Low, P.S., Heinstein, P.F., Stipamovic, D., Altonan, D.W., 1987. Inhibition of elicitor-induced phytoalexin formation in cotton and soybean cells by citrate. Plant Physiol. 84, 1276–1280.

Apse, M.P., Aharon, G.S., Snedden, W.A., Blumwald, E., 1999. Salt tolerance conferred by overexpression of a vacuolar Na^+/H^+ antiport in *Arabidopsis*. Science 285, 1256–1258.

Arabidopsis Interactome Mapping Consortium, 2011. Evidence for network evolution in an *Arabidopsis* interactome map. Science 333, 601–607.

Arai, Y., Hayashi, M., Nishimura, M., 2008a. Proteomic analysis of highly purified peroxisomes from etiolated soybean cotyledons. Plant Cell Physiol. 49, 526–539.

Arai, Y., Hayashi, M., Nishimura, M., 2008b. Proteomic identification and characterization of a novel peroxisomal adenine nucleotide transporter supplying ATP for fatty acid beta-oxidation in soybean and *Arabidopsis*. Plant Cell 20, 3227–3240.

Arai, S., Lee, S.-C., Zhai, D., Suzuki, M., Chang, Y.T., 2014. A molecular fluorescent probe for targeted visualization of temperature at the endoplasmic reticulum. Sci. Rep. 4, 6701.

Aranzana, M.J., Kim, S., Zhao, K., Bakker, E., Horton, M., Jakob, K., Lister, C., Molitor, J., Shindo, C., Tang, C., et al., 2005. Genome-wide association mapping in *Arabidopsis* identifies previously known flowering time and pathogen resistance genes. PLoS Genet. 1, e60.

Araya, A., Bégu, D., Litvak, S., 1994. RNA editing in plants. Physiol. Plant 91, 543–550.

Arber, W., 1965. Host-controlled modification of bacteriophage. Annu. Rev. Microbiol. 19, 365–378.

Arber, W., December 8, 1978. Promotion and limitation of genetic exchange. Nobel Lecture.

Argout, X., Salse, J., Aury, J.M., Guiltinan, M.J., Droc, G., Gouzy, J., Allegre, M., Chaparro, C., Legavre, T., Maximova, S.N., et al., 2011. The genome of *Theobroma cacao*. Nat. Genet. 43, 101–108.

Arimura, S., Aida, G.P., Fujimoto, M., Nakazono, M., Tsutsumi, N., 2004b. *Arabidopsis* dynamin-like protein 2a (ADL2a), like ADL2b, is involved in plant mitochondrial division. Plant Cell Physiol. 45, 236–242.

Arimura, S., Tsutsumi, N., 2002. A dynamin-like protein (ADL2b), rather than FtsZ, is involved in *Arabidopsis* mitochondrial division. Proc. Natl. Acad. Sci. U.S.A. 99, 5727–5731.

Arimura, S.-iI., Yamamoto, J., Aida, G.P., Nakazono, M., Tsutsumi, N., 2004a. Frequent fusion and fission of plant mitochondria with unequal nucleoid distribution. Proc. Natl. Acad. Sci. U.S.A. 101, 7805–7808.

Arioli, T., Peng, L., Betzner, A.S., Burn, J., Wittke, W., Herth, W., Camilleri, C., Hofte, H., Plazinski, J., Birch, R., et al., 1998. Molecular analysis of cellulose biosynthesis in *Arabidopsis*. Science 279, 717–720.

Armstrong, F., Leung, J., Grabov, A., Brearley, J., Giraudat, J., Blatt, M.R., 1995. Sensitivity to abscisic acid of guard cell K^+ channels is suppressed by abl-1, a mutant Arabidopsis gene encoding a putative protein phosphatase. Proc. Natl. Acad. Sci. U.S.A. 92, 9520–9524.

Armstrong, L., Snyder, J.A., 1987. Quinicrine-induced changes in mitotic PtK1 spindle microtubule organization. Cell Motil. Cytoskelet. 7, 10–19.

Arnason, J.T., Mata, R., Romero, J.T. (Eds.), 1995. Phytochemistry of Medicinal Plants. Recent Advances in Phytochemistry, vol. 29. Springer Science+Business Media, New York.

Arnon, D.I., 1951. Extracellular photosynthetic reactions. Nature 167, 1008–1010.

Arnon, D.I., Allen, M.B., Whatley, F.R., 1954. Photosynthesis by isolated chloroplasts. Nature 174, 394–396.

Arntzen, C., Plotkin, S., Dodet, B., 2005. Plant-derived vaccines and antibodies: potential and limitations. Vaccine 23, 1753–1756.

Arrhenius, S., 1896. On the influence of carbonic acid in the air upon the temperature of the ground. Philos. Mag. 41, 237–276.

Arrhenius, S., 1902. Textbook of Electrochemistry. Longmans, Green, and Co., London.

Arrhenius, S., 1908. Worlds in the Making. Harper & Row, New York.

Arrhenius, S., 1912. Theories of Solutions. Yale University Press, New Haven, CT.

Arrhenius, S., 1915. Quantitative Laws in Biological Chemistry. G. Bell and Sons, Ltd., London.

Arrhenius, S., December 11, 1903. Development of the Theory of Electrolyte Dis-Sociation. Nobel Lecture.

Arrhenius, G., Sales, B., Mojzsis, S., Lee, T., 1997. Entropy and charge in molecular evolution—the case of phosphate. J. Theor. Biol. 187, 503–522.

Arrigo, A.P., Tanaka, K., Goldberg, A.L., Welch, W.J., 1988. Identity of the 19S "prosome" particle with the large multifunctional protease complex of mammalian cells (the proteosome). Nature 331, 192–194.

Arronet, N.I., 1973. Motile Muscle and Cell Models. Translated from the Russian by Basil Haigh. Consultants Bureau, New York.

Asada, T., Kuriyama, R., Shibaoka, H., 1997. TKRP125, a kinesin- related protein involved in the centrosome-independent organization of the cytokinetic apparatus in tobacco BY-2 cells. J. Cell Sci. 110, 179–189.

Asada, T., Sonobe, S., Shibaoka, H., 1991. Microtubule translocation in the cytokinetic apparatus of cultured tobacco cells. Nature 350, 238–241.

Asai, D.J., Wilson, L.S., 1985. A latent activity dynein-like cytoplasmic magnesium adenosine triphosphatase. J. Biol. Chem. 260, 699–702.

Asakura, S., Eguchi, G., Iino, T., 1964. Reconstitution of bacterial flagella *in vitro*. J. Mol. Biol. 10, 42–56.

Asakura, T., Hirose, S., Katamine, H., Kitajima, A., Hori, H., Sato, M.H., Fujiwara, M., Shimamoto, K., Mitsui, T., 2006. Isolation and proteomic analysis of rice Golgi membranes: cis-Golgi membranes labeled with GFP-SYP31. Plant Biotechnol. 23, 475–485.

Asano, T., Maumura, T., Kusano, H., Kikuchi, S., Kurita, A., Shimada, H., Kadowaki, K., 2002. Construction of a specialized cDNA library from plant cells isolated by laser capture microdissection: toward comprehensive analysis of the genes expressed in the rice phloem. Plant J. 32, 401–408.

Asbury, C.L., Fehr, A.N., Bloch, S.M., 2003. Kinesin moves by an asymmetric hand-over-hand-mechanism. Science 302, 2130–2134.

Asen, S., Stewart, R.N., Norris, K.H., 1975. Anthocyanin, flavonol copigments and pH responsible for larkspur flower color. Phytochemistry 14, 2677–2682.

Asenjo, J.A., Andrews, B.A., 2011. Aqueous two-phase systems for protein separation: a perspective. J. Chromatogr. A 1218, 8826–8835.

Ashby, W.R., 1958. An Introduction to Cybernetics. John Wiley & Sons, New York.

Ashford, A.E., Allaway, W.E., 1985. Transfer cells and hartig net in the root epidermis of the sheathing mycorrhiza of *Pisonia grandis* R.Br. from Seychelles. New Phytol. 100, 595–612.

Ashkin, A., Dziedzic, J.M., 1987. Optical trapping and manipulation of viruses and bacteria. Science 235, 1517–1520.

Ashkin, A., Dziedzic, J.M., 1989. Internal cell manipulation using infrared laser traps. Proc. Natl. Acad. Sci. U.S.A. 86, 7914–7918.

Ashkin, A., Dziedzic, J.M., Yamaore, T., 1987. Optical trapping and manipulation of single cells using infrared laser beams. Nature 330, 769–771.

Ashkin, A., Schuetze, K., Dziedzic, J.M., Euteneuer, U., Schliwa, M., 1990. Force generation of organelle transport measured in-vivo by an IR laser trap. Nature 348, 346–348.

Asimov, I., 1954. The Chemicals of Life: Enzymes, Vitamins and Hormones. Abelard-Schuman, New York.

Asimov, I., 1971. The Universe. From Flat Earth to Quasar, Revised Ed. Walker and Co., New York.

Assmann, S.M., 1993. Signal transduction in guard cells. Ann. Rev. Cell Biol. 9, 345–375.

Assmann, S.M., 2002. Heterotrimeric and unconventional GTP binding proteins in plant cell signaling. Plant Cell 14, S355–S373.

Assmann, S.M., 2005. G proteins go green: a plant G protein signaling FAQ sheet. Science 310, 71–73.

Assmann, S.M., Shimazaki, K., 1999. The multisensory guard cell. Stomatal responses to blue light and abscisic acid. Plant Physiol. 119, 809–816.

Astbury, W.T., 1933. Some problems in the X-ray analysis of the structure of animal hairs and other protein fibers. Trans. Faraday Soc. 29, 193–211.

Astbury, W.T., 1939. X-ray studies of the structure of compounds of biological interest. Annu. Rev. Biochem. 8, 113–133.

Astbury, W.T., 1941. Protein and virus studies in relation to the problem of the gene. In: Punnett, R.C. (Ed.), Proceedings of the Seventh International Genetical Congress, Edinburgh, Scotland, 23-30 August 1939. Cambridge University Press, Cambridge, pp. 49–51.

Astbury, W.T., 1947a. Croonian lecture: on the structure of biological fibres and the problem of muscle. Proc. R. Soc. Lond. B 134, 303–328.

Astbury, W.T., 1947b. X-ray studies of nucleic acids. Symp. Soc. Exp. Biol. 1, 66–67.

Astbury, W.T., 1950/51. Adventures in molecular biology. Harvey Lect. 46, 3–44.

Astbury, W.T., 1961. Molecular biology or ultrastructural biology? Nature 190, 1124.

Astbury, W.T., Bell, F.O., 1938a. Some recent developments in the X-ray study of proteins and related structures. Cold Spring Harb. Symp. Quant. Biol. 6, 109–118.

Astbury, W.T., Bell, F.O., 1938b. X-ray studies of thymonucleic acid. Nature 141, 747–748.

Astbury, W.T., Sisson, W.A., 1935. X-ray studies of the structure of hair, wool, and related fibres. III. The configuration of the keratin molecule and its orientation in the biological cell. Proc. R. Soc. A 150, 533–551.

Astbury, W.T., Street, A., 1932. X-ray studies of the structure of hair, wool, and related fibres. I. General. Philos. Trans. R. Soc. Lond. A 230, 681–693.

Astbury, W.T., Woods, H.J., 1934. X-ray studies of the structure of hair, wool, and related fibres. II. The molecular structure and elastic properties of hair keratin. Trans. R. Soc. Lond. A 232, 333–394.

Aston, F.W., December 12, 1922. Mass spectra and isotopes. Nobel Lecture.

Astrachan, L., Volkin, E., 1958. Properties of ribonucleic acid turnover in T2-infected *Escherichia coli*. Biochim. Biophys. Acta 29, 536–544.

Atanasov, K.E., Barboza-Barquero, L., Tiburcio, A.F., Alcázar, R., 2016. Genome wide association mapping for the tolerance to the polyamine oxidase inhibitor guazatine in *Arabidopsis thaliana*. Front. Plant Sci. 7, 401.

Atkinson, M.R., Deutscher, M.P., Kornberg, A., Russell, A.F., Moffatt, J.G., 1969. Enzymatic synthesis of deoxyribonucleic acid. XXXIV. Termination of chain growth by a 2′,3′-dideoxyribonucleotide. Biochemistry 8, 4897–4904.

Atmodjo, M.A., Hao, Z., Mohnen, D., 2013. Evolving views of pectin biosynthesis. Annu Rev. Plant Biol. 64, 747–779.

Atwell, S., Huang, Y.S., Vihjálmsson, B.J., Willems, G., Horton, M., Li, Y., Meng, D., Platt, A., Tarone, A.M., Hu, T.T., et al., 2010. Genome-wide association study of 107 phenotypes in *Arabidopsis thaliana* inbred lines. Nature 465, 627–631.

Aubert, S., Gout, E., Bligny, R., Marty-Mazars, D., Barrieu, F., Alabouvette, J., Marty, F., Douce, R., 1996. Ultrastructural and biochemical characterization of autophagy in higher plant cells subjected to carbon deprivation: control by the supply of mitochondria with respiratory substrates. J. Cell Biol. 133, 1251–1263.

Ault, J.G., Demarco, A.J., Salmon, E.D., Rieder, C.L., 1991. Studies on the ejection properties of asters; astral microtubule turnover influences the oscillatory behavior and positioning of non-oriented chromosomes. J. Cell Sci. 99, 701–710.

Austin II, J.R., Staehelin, L.A., 2011. Three-dimensional architecture of grana and stroma thylakoids of higher plants as determined by electron tomography. Plant Physiol. 155, 1601–1611.

Avery, O.T., MacLeod, C.M., McCarty, M., 1944. Studies on the chemical nature of the substance inducing transformation of pneumococcal types. J. Exp. Med. 79, 137–158.

Avila, E.L., Zouhar, J., Agee, A.E., Carter, D.G., Chary, S.N., Raikhel, N.V., 2003. Tools to study plant organelle biogenesis: point mutation lines with disrupted vacuoles and high-speed confocal screening of green fluorescent protein-tagged organelles. Plant Physiol. 133, 1673–1676.

Avin-Wittenberg, T., Fernie, A.R., 2014. At long last: evidence for pexophagy in plants. Mol. Plant 1257–1260.

Avisar, D., Abu-Abied, M., Belausov, E., Sadot, E., Hawes, C., Sparkes, I.A., 2009. A comparative study if the involvement of 17 *Arabidopsis* myosin family members on the motility of Golgi an dother organelles. Plant Physiol. 150, 700–709.

Avisar, D., Prokhnevsky, A.I., Dolja, V.V., 2008a. Class VIII myosins are required for plasmodesmatal localization of a *Closterovirus* Hsp70 homolog. J. Virol. 82, 2836–2843.

Avisar, D., Prokhnevsky, A.I., Makarova, K.S., Koonin, E.V., Dolja, V.V., 2008b. Myosin XI-K is required for rapid trafficking of Golgi stacks, peroxisomes and mitochondria in leaf cells of *Nicotiana benthamiana*. Plant Physiol. 146, 1098–1108.

Avogadro, A., 1837. Fisica de' corpi ponderabili; ossia Trattato della costituzione generale de' corpi, del cavaliere Amedeo Avogadro.

Avron (Abramsky), M., Biale, J.B., 1957a. Metabolic processes in cytoplasmic particles of the avocado fruit. Plant Physiol. 32, 100–105.

Avron (Abramsky), M., Biale, J.B., 1957b. Metabolic processes in cytoplasmic particles of the avocado fruit. J. Biol. Chem. 225, 699–708.

Avron (Abramsky), M., Jagendorf, A.T., 1956. A TPNH diaphorase from chloroplasts. Arch. Biochem. Biophys. 65, 475–490.

Avron (Abramsky), M., Jagendorf, A.T., 1957. An extractable factor in photosynthetic phosphorylation. Nature 179, 428–429.

Avron, M., 1963. A coupling factor in photophosphorylation. Biochim. Biophys. Acta 77, 699–702.

Avron, M., Krogmann, D.W., Jagendorf, A.T., 1958. The relation of photosynthetic phosphorylation to the Hill reaction. Biochim. Biophys. Acta 30, 144–153.

Aw, T.Y., 2000. Intracellular compartmentation of organelles and gradients of low molecular weight species. Int. Rev. Cytol. 192, 223–253.

Awata, J., Kashiyama, T., Ito, K., Yamamoto, K., 2003. Some motile properties of fast characean myosin. J. Mol. Biol. 326, 659–663.

Awata, J., Saitoh, K., Shimada, K., Kashiyama, T., Yamamoto, K., 2001. Effects of Ca^{2+} and calmodulin on the motile activity of characean myosin *in vitro*. Plant Cell Physiol. 42, 828–834.

Axelrod, D., Koppel, D.E., Schlessinger, J., Elson, E., Webb, W.W., 1976. Mobility measurements by analysis of fluorescence photobleaching recovery kinetics. Biophys. J. 16, 1055–1069.

Axelsen, K.B., Palmgren, M.G., 2001. Inventory of the superfamily of P-type ion pumps in *Arabidopsis*. Plant Physiol. 126, 696–706.

Ayling, S.M., Butler, R.C., 1993. Time-series analysis of measurements on living cells illustrated by analysis of particle movement in the cytoplasm of tomato root hairs. Protoplasma 172, 124–131.

Ayling, S.M., Clarkson, D.T., Brownlee, C., 1990. Cytoplasmic free calcium levels in tomato root hairs are altered by indole-3-acetic acid. Proc. R. Micro. Soc. 25, 275.

Ayoubi, T.A.Y., van de Ven, W.J.M., 2017. Regulation of gene expression by alternative promotors. FASEB J. 10, 453–460.

Ayre, B.G., Turgeon, R., 2004. Graft transmission of a floral stimulant derived from CONSTANS. Plant Physiol. 135, 2271–2278.

Baas-Becking, I.G.M., Parks, G.S., 1927. Energy relations in the metabolism of autotrophic bacteria. Physiol. Rev. 7, 85–106.

Baba, M., Osumi, M., Ohsumi, Y., 1995. Analysis of the membrane structures involved in autophagy in yeast by freeze-replica method. Cell Struct. Funct. 20, 465–471.

Baba, M., Takeshige, K., Baba, N., Ohsumi, Y., 1994. Ultrastructural analysis of the autophagic process in yeast: detection of autophagosomes and their characterization. J. Cell Biol. 124, 903–913.

Babiychuk, E., Cottrill, P.B., Storozhenko, S., Fuangthong, M., Chen, Y., O'Farrell, M.K., Van Montagu, M., Inzé, D., Kushnir, S., 1998. Higher plants possess two structurally different poly(ADP-ribose) polymerases. Plant J. 15, 635–645.

Bacic, A., Delmer, D.P., 1981. Stimulation of membrane-associated polysaccharide synthetases by a membrane potential in developing cotton fibers. Planta 152, 346–351.

Bacon, F., 1620. Novum Organum. The philosophers of science, 1947. In: Commins, S., Linscott, R.N. (Eds.), Man and the Universe. Random House, New York. Book 1. Reprinted in The World's Greatest Thinkers.

Bacon, R., 1928a. The Opus Majus of Roger Bacon. Volume I. University of Pennsylvania Press, Philadelphia, PA. Translated by Robert Belle Burke.

Bacon, R., 1928b. The Opus Majus of Roger Bacon. Volume II. University of Pennsylvania Press, Philadelphia, PA. Translated by Robert Belle Burke.

Bacon, J.S.D., 1970. Life outside the cell (or, what the biologist saw). In: Bartley, W., Kornberg, H.L., Quayle, J.R. (Eds.), Essays in Cell Metabolism. Hans Krebs Dedicatory Volume. Wiley-Interscience, London, pp. 45–66.

Bacon, F., 2007. The New Organon or: True Directions Concerning the Interpretation of Nature. http://www.earlymoderntexts.com/assets/pdfs/bacon1620.pdf.

Bada, J.L., Lazcano, A., 2003. Prebiotic soup—revisiting the Miller experiment. Science 300, 745–746.

Bada, J.L., Lazcano, A., 2012. Stanley L. Miller: March 7, 1930–May 20, 2007. Biogr. Mem. http://www.nasonline.org/publications/biographical-memoirs/memoir-pdfs/miller-stanley.pdf.

Badelt, K., White, R.G., Overall, R.L., Vesk, M., 2016. Ultrastructural specializations of the cell wall sleeve around plasmodesmata. Am. J. Bot. 81, 1422–1427.

Badger, M.R., Fallahi, H., Kaines, S., Takahashi, S., 2009. Chlorophyll fluorescence screening of *Arabidopsis thaliana* for CO_2 sensitive photorespiration and photoinhibition mutants. Funct. Plant Biol. 36, 867–873.

Badiei, H.R., Rutzke, M., Karanassios, V., 2002. Calcium content of individual, microscopic, (sub) nanoliter volume *Paramecium* sp. cells using rhenium-cup in-torch vaporization (ITV) sample introduction and axially viewed ICP-AES. J. Anal. At. Spectrom. 17, 1007–1010.

Badouin, H., Gouzy, J., Grassa, C.J., Murat, F., Staton, S.E., Cottret, L., et al., 2017. The sunflower genome provides insights into oil metabolism, flowering and Asterid evolution. Nature 546, 148–152.

Badrinarayanan, A., Le, T.B.K., Laub, M.T., 2015. Bacterial chromosome organization and segregation. Annu. Rev. Cell Dev. Biol. 31, 171–199.

Baglioni, C., Ingram, V.M., 1961. Abnormal human haemoglobins. V. Chemical investigation of haemoglobins A, G,C, X from one individual. Biochim. Biophys. Acta 48, 253–265.

Bahl, C.P., Marians, R.J., Wu, R., Stawinski, J., Narang, S.A., 1976. A general method for inserting specific DNA sequences into cloning vehicles. Gene 1, 81–92.

Baiges, I., Schäffner, A.R., Affenzeller, M.J., Mas, A., 2002. Plant aquaporins. Physiol. Plant 115, 175–182.

Bailer, S.M., Eppenberger, H.M., Griffiths, G., Nigg, E.A., 1991. Characterization of a 54-kD protein of the inner nuclear membrane: evidence for cell cycle dependent interaction with the nuclear lamina. J. Cell Biol. 114, 398–400.

Bailey, L.H., 1901. A maker of new fruits and flowers. In: The World's Work. Volume II. Doubleday, Page and Co., NY, pp. 1209–1214. May 1901 to October 1901.

Bailey, L.H., 1915. The Holy Earth. Charles Schribner's Sons, New York.

Bailey, I.W., 1930. The cambium and its derivative tissues. V. A reconnaissance of the vacuome in living cells. Zeit. Zellf. Mikro. Anat. 10, 651–682.

Bailey, J., 1997. Building a plasmodium: development in the acellular slime mould *Physarum polycephalum*. BioEssays 19, 985–992.

Bailey, J.M., 1998. RNA-directed amino acid homochirality. FASEB J. 12, 503–507.

Bailey, K., Sanger, F., 1951. The chemistry of amino acids and proteins. Ann. Rev. Biochem. 20, 103–130.

Baitsell, G.A., 1940. The cell theory II. A modern concept of the cell as a structural unit. Am. Nat. 74, 5–24.

Bajer, A., 1953. Observations on spindle structure and persisting nucleoli. Acta Soc. Bot. Pol. 22, 653–666.

Bajer, A., 1955. Living smears from endosperm. Experientia 11, 221–222.

Bajer, A., 1957. Ciné-micrographic studies on mitosis in endosperm. III. The origin of the mitotic spindle. Exp. Cell Res. 13, 493–502.

Bajer, A., 1958a. Ciné-micrographic studies on mitosis in endosperm. IV. The mitotic contraction stage. Exp. Cell Res. 14, 245–256.

Bajer, A., 1958b. Ciné-micrographic studies on mitosis in endosperm. V. Formation of the metaphase plate. Exp. Cell Res. 15, 370–383.

Bajer, A.S., Allen, R.E., 1966. Role of phragmoplast filaments in cell plate formation. J. Cell Sci. 1, 455–462.

Bajer, A.S., Jensen, C., 1969. Detectability of mitotic spindle microtubules with the light and electron microscopes. J. Microsc. 8, 343–354.

Bajer, A., Mole-Bajer, J., 1956. Ciné-micrographic studies on mitosis in endosperm. II. Chromosome, cytoplasmic and Brownian movements. Chromosoma 7, 558–607.

Bajer, A., Molé-Bajer, J., 1956. Ciné-micrographic studies on mitosis in endosperm. II. Chromosome, cytoplasmic and Brownian movements. Chromosoma 7, 558–607.

Bajer, A.S., Molè-Bajer, J., 1972. Spindle dynamics and chromosome movements. Int. Rev. Cytol. (Suppl. 3), 1–273.

Bajer, A.S., Molé-Bajer, J., 1972. Spindle dynamics and chromosome movements. Int. Rev. Cytol. (Suppl. 3), 1–273.

Bajer, A.S., Molè-Bajer, J., 1986. Reorganization of microtubules in endosperm cells and cell fragments of the higher plant *Haemanthus* in vivo. J. Cell Biol. 102, 263–281.

Bajer, A.S., Molé-Bajer, J., 1986. Reorganization of microtubules in endosperm cells and cell fragments of the higher plant Haemanthus in vivo. J. Cell Biol. 102, 263–281.

Baker, J.R., 1945. Science and the Planned State. MacMillian, New York.

Baker, J.R., 1957. The Golgi controversy. Symp. Soc. Exp. Biol. 10, 1–10.

Baker, J.R., 1988. The Cell Theory. A Restatement, History and Critique. Garland Publishing, New York.

Baker, A., Graham, I. (Eds.), 2002. Plant Peroxisomes. Biochemistry, Cell Biology and Biotechnological Implications. Kluwer Academic Publishers, Dordrecht.

Baker, R.W., Hughson, F.M., 2016. Chaperoning SNARE assembly and disassembly. Nat. Rev. Mol. Cell Biol. 17, 465–479.

Baker, A., Paudyal, R., 2014. The life of the peroxisome: from birth to death. Curr. Opin. Plant Biol. 22, 39–47.

Balasubramanian, S., 2013. An interview with Shankar Balasubramanian. Trends Biochem. Sci. 38, 170–171.

Balch, W.E., Dunphy, W.G., Braell, W.A., Rothman, J.E., 1984a. Reconstitution of the transport of protein between successive compartments of the Golgi measured by the coupled incorporation of N-acetylglucosamine. Cell 39, 405–416.

Balch, W.E., Glick, B.S., Rothman, J.E., 1984b. Sequential intermediates in the pathway of intercompartmental transport in a cell-free system. Cell 39, 525–536.

Balch, W.E., Magrum, L.J., Fox, G.E., Wolfe, R.S., Woese, C.R., 1977. An ancient divergence among the bacteria. J. Mol. Evol. 9, 305–311.

Balcke, G.U., Bennewitz, S., Bergau, N., Athmer, B., Henning, A., Majovsky, P., Jiménez-Gómez, J.M., Hoehenwarter, W., Tissier, A., 2017. Multi-omics of tomato glandular trichomes reveals distinct features of central carbon metabolism supporting high productivity of specialized metabolites. Plant Cell 29, 960–983.

Baldwin, D., Crane, V., Rice, D., 1999. A comparison of gel-based, nylon filter and microarray techniques to detect differential RNA expression in plants. Curr. Opin. Plant Biol. 2, 96–103.

Balestrini, R., Gómez-Ariza, J., Klink, V.P., Bonfante, P., 2009. Application of laser microdissection to plant pathogenic and symbiotic interactions. J. Plant Interact. 4, 81–92.

Balke, N.E., Hodges, T.K., 1975. Plasma membrane adenosine triphosphatase of oat roots. Activation and inhibition by Mg^{2+} and ATP. Plant Physiol. 55, 83–86.

Ball, E.G., 1944. Energy relationships of the oxidative enzymes. Ann. N.Y. Acad. Sci. 45, 363–375.

Ball, P., 2000. Life's Matrix: A Biography of Water. Farrar, Straus, and Giroux, New York.

Ball, E.G., Barrnett, R.J., 1957. An integrated morphological and biochemical study of a purified preparation of the succinate and DPNH oxidase system. J. Biochem. Biophys. Cytol. 3, 1023–1036.

Ballas, L.M., Lazarow, P.B., Bell, R.M., 1984. Glycerolipid synthetic capacity of rat liver peroxisomes. Biochim. Biophys. Acta 795, 297–300.

Ballowitz, E., 1888. Untersuchungen über die Struktur der Spermatozoën, zugleich ein Beitrag zur Lehre vom feineren Bau der contraktilen Elemente. Arck. Mikro. Anat. 32, 401–473.

Baltes, N.J., Gil-Humanes, J., Voytas, D.F., 2017. Genome engineering and agriculture: opportunities and challenges. Prog. Mol. Biol. Transl. Sci. 149, 1–26.

Baltes, N.J., Voytas, D.F., 2015. Enabling plant synthetic biology through genome engineering. Trends Biotechnol. 33, 120–131.

Baltimore, D., 1970. Viral RNA-dependent DNA polymerase. Nature 226, 1209–1211.

Baltimore, D., December 12, 1975. Viruses, polymerases and cancer. Nobel Lecture.

Baltzer, F., 1967. Theodor Boveri. Life and Work of a Great Biologist 1862–1915. University of California Press, Berkeley.

Balusek, K., Depta, H., Robinson, D.G., 1988. Two polypeptides (30 and 38 kDa) in plant coated vesicles with clathrin light chains properties. Protoplasma 146, 174–176.

Baluska, F., Baroja-Fernandez, E., Pozueta-Romero, J., Hlavcka, A., Etxeberria, E., Samaj, J., 2005. Endocytic uptake of nutrients, cell wall molecules, and fluidized cell wall portions into heterotrophic plant cells. Plant Cell Monogr. 1, 19–35.

Baluska, F., Cvčkova, F., Kendrick-Jones, J., Volkmann, D., 2001. Sink plasmodesmata as gateway for phloem unloading. Myosin VIII and calreticulin as molecular determinants of sink strength? Plant Physiol. 126, 39–46.

Baluska, F., Samaj, J., Hlavacka, A., Kendrick-Jones, J., Volkmann, D., 2004a. Actin-dependent fluid-phase endocytosis in inner cortex cells of maize roots. J. Exp. Bot. 55, 463–473.

Baluska, F., Samaj, J., Napier, R., Volkmann, D., 1999. Maize calreticulin localizes preferentially to plasmodesmata in root apex. Plant J. 19, 481–488.

Baluska, F., Volkmann, D., Barlow, P.W., 2004b. Eukaryotic cells and their cell bodies: cell theory revisited. Ann. Bot. 94, 9–32.

Baluska, F., Volkmann, D., Hlavacka, A., Mancuso, S., Barlow, P.W., 2006. Neurobiological view of plants and their body plan. In: Baluska, F., Mancuso, S., Volkmann, D. (Eds.), Communication in Plants. Springer, Berlin, pp. 19–35.

Bandmann, V., Homann, U., 2012. Clathrin-independent endocytosis contributes to uptake of glucose into BY-2 protoplasts. Plant J. 70, 578–584.

Bandmann, V., Müller, J.D., Köhler, T., Homann, U., 2012. Uptake of fluorescent nano beads into BY2-cells involves clathrin-dependent and clathrin-independent endocytosis. FEBS Lett. 586, 3626–3632.

Bandurski, R.S., 1955. Further studies on the enzymatic synthesis of oxalacetate from phosphorylenolpyruvate and carbon dioxide. J. Biol. Chem. 217, 137–150.

Bandurski, R.S., Greiner, C.M., 1953. The enzymatic synthesis of oxalacetate from phosphoryl-enolpyruvate and carbon dioxide. J. Biol. Chem. 204, 781–786.

Bandurski, R.S., Greiner, C.M., Bonner, J., 1953. Enzymatic carboxylation of phosphoenolpyruvate to oxalacetate. Fed. Proc. 204, 781–786.

Bandurski, R.S., Lippmann, F., 1956. Studies on an oxaloacetic carboxylase from liver mitochondria. J. Biol. Chem. 219, 741–752.

Bandurski, R.S., Wilson, L.G., Squires, C.L., 1956. The mechanism of active sulfate formation. J. Am. Chem. Soc. 78, 6408–6409.

Banerjee, A., Herman, E., Kottke, T., Essen, L.-O., 2016. Structure of a native-like Aureochrome 1a LOV domain dimer from Phaeodactylum tricornutum. Structure 24, 171–178.

Banks, J.A., Nishiyama, T., Hasebe, M., Bowman, J.L., Gribskov, M., dePamphilis, C., Albert, V.A., Aono, N., Aoyama, T., Ambrose, B.A., Ashton, N.W., Axtell, M.J., Barker, E., Barker, M.S., Bennetzen, J.L., Bonawitz, N.D., Chapple, C., Cheng, C., Correa, L.G., Dacre, M., DeBarry, J., Dreyer, I., Elias, M., Engstrom, E.M., Estelle, M., Feng, L., Finet, C., Floyd, S.K., Frommer, W.B., Fujita, T., Gramzow, L., Gutensohn, M., Harholt, J., Hattori, M., Heyl, A., Hirai, T., Hiwatashi, Y., Ishikawa, M., Iwata, M., Karol, K.G., Koehler, B., Kolukisaoglu, U., Kubo, M., Kurata, T., Lalonde, S., Li, K., Li, Y., Litt, A., Lyons, E., Manning, G., Maruyama, T., Michael, T.P., Mikami, K., Miyazaki, S., Morinaga, S., Murata, T., Mueller-Roeber, B., Nelson, D.R., Obara, M., Oguri, Y., Olmstead, R.G., Onodera, N., Petersen, B.L., Pils, B., Prigge, M., Rensing, S.A., Riano-Pachon, D.M., Roberts, A.W., Sato, Y., Scheller, H.V., Schulz, B., Schulz, C., Shakirov, E.V., Shibagaki, N., Shinohara, N., Shippen, D.E., Sorensen, I., Sotooka, R., Sugimoto, N., Sugita, M., Sumikawa, N., Tanurdzic, M., Theissen, G., Ulvskov, P., Wakazuki, S., Weng, J.K., Willats, W.W., Wipf, D., Wolf, P.G., Yang, L., Zimmer, A.D., Zhu, Q., Mitros, T., Hellsten, U., Loque, D., Otillar, R., Salamov, A., Schmutz, J., Shapiro, H., Lindquist, E., Lucas, S., Rokhsar, D., Grigoriev, I.V., 2011. The *Selaginella* genome identifies genetic changes associated with the evolution of vascular plants. Science 332, 960–963.

Banno, H., Chua, N.-H., 2000. Characterization of the *Arabidopsis* formin-like protein AFH1 and its interacting protein. Plant Cell Physiol. 41, 617–626.

Bar, R.S., Deamer, D.W., Cornwell, D.G., 1966. Surface area of human erythrocyte lipids: reinvestigation of experiments on plasma membrane. Science 153, 1010–1012.

Barabási, A.-L., Oltvai, Z.N., 2004. Network biology: understanding the cell's functional organization. Nat. Rev. Genet. 5, 101–113.

Barak, L.S., Yocum, R.R., Nothnagel, E.A., Webb, W.W., 1980. Fluorescence staining of the actin cytoskeleton in living cells with 7-nitrobenz-2-oxa-1,3-diazole phallacidin. Proc. Natl. Acad. Sci. U.S.A. 77, 980–984.

Baranetzky, J., 1880. Kerntheilung in den Pollenmutterzellen einiger *Tradescantien*. Bot. Z. 38, 241–248, 265–274, 281–296.

Barber, H.N., 1939. The rate of movement of chromosomes on the spindle. Chromosoma 1, 33–50.

Barberon, M., 2017. The endodermis as a checkpoint for nutrients. New Phytol. 213, 1604–1610.

Barberon, M., Vermeer, J.E.M., De Bellis, D., Wang, P., Naseer, S., Ander Diesen, T.G., Humbel, B.M., Nawrath, C., Takano, J., Salt, D.E., Geldner, N., 2016. Adaptation of root function by nutrient-induced plasticity of endodermal differentiation. Cell 164, 447–459.

Barbour, M.G., Burk, J.H., Pitts, W.D., 1980. Terrestrial Plant Ecology. Benjamin/Cummings Publishing, Menlo Park, CA.

Bargiello, T.A., Jackson, F.R., Young, M.W., 1984. Restoration of circadian behavioural rhythms by gene transfer in *Drosophila*. Nature 312, 752–754.

Barish, B.B., 2017. Nobel Lecture. December 8, 2017. https://www.nobelprize.org/nobel_organizations/nobelmedia/channels/widget/live.html.

Barlow, P.W., 1982. "The plant forms cells, not cells the plant": the origintof De Bary's aphorism. Ann. Bot. 49, 269–271.

Barnes, C.R., 1893. On the food of green plants. Bot. Gaz. 18, 403–411.

Barnes, C.R., 1898. So-called "Assimilation." Bot. Centralbl. 76, 257–259.

Barnes, J. (Ed.), 1984. The Complete Works of Aristotle. Princeton University Press, Princeton, NJ.

Baroja-Fernandez, E., Etxeberria, E., Muñoz, J.F., Gonzalez, P., Pozueta-Romero, J., 2006. An important pool of sucrose linked to starch biosynthesis is taken up by endocytosis in heterotrophic cells. Plant Cell Physiol. 47, 447–456.

Baron, A.T., Salisbury, J.L., 1988. Identification and localization of a novel, cytoskeletal, centrosome-associated protein in PtK2 cells. J. Cell Biol. 107, 2669–2678.

Baron-Epel, O., Gharayal, P.K., Schindler, M., 1988a. Pectin as mediators of wall porosity in soybean cells. Planta 175, 389–395.

Baron-Epel, O., Hernandez, D., Jiang, L.-W., Meiners, S., Schindler, M., 1988b. Dynamic continuity of cytoplasmic and membrane compartments between plant cells. J. Cell Biol. 106, 715–721.

Barondes, S.H., Nirenberg, M.W., 1962. Fate of a synthetic polynucleotide directing cell-free protein synthesis. Science 138, 810–817.

Barrell, B.G., Air, G.M., Hutchison III, C.A., 1976. Overlapping genes in bacteriophage ΦX174. Nature 264, 34–41.

Barritt, G.J., 1992. Communication Within Animal Cells. Oxford University Press, Oxford.

Barroso, J.B., Corpas, F.J., Carreras, A., Sandalio, L.M., Valderrama, R., Palma, J.M., Lupiáñez, J.A., del Río, L.A., 1999. Localization of nitric-oxide synthase in plant peroxisomes. J. Biol. Chem. 274, 36729–36733.

Barsan, C., Zouine, M., Maza, E., Bian, W., Egea, I., Rossignol, M., Bouyssie, D., Picheraux, C., Purgatto, E., Bouzayen, M., Latché, A., Pech, J.-C., 2012. Proteomic analysis of chloroplast-to-chromoplast transition in tomato reveals metabolic shifts coupled with disrupted thylakoid biogenesis machinery and elevated energy-production components. Plant Physiol. 160, 708–725.

Bartholomew, D.M., Van Dyk, D.E., Lau, S.-M.C., O'Keefe, D.P., Rea, P.A., Viitanen, P.V., 2002. Alternate energy-dependent pathways for the vacuolar uptake of glucose and glutathione conjugates. Plant Physiol. 130, 1562–1572.

Barthou, H., Petitprez, M., Brière, C., Souvré, A., Alibert, G., 1999. RGD-mediated membrane-matrix adhesion triggers agarose-induced embryoid formation in sunflower protoplasts. Protoplasma 206, 143–151.

Bartnik, E., Sievers, A., 1988. *In-vivo* observations of a spherical aggregate of endoplasmic reticulum and of Golgi vesicles in the tip of fast-growing *Chara* rhizoids. Planta 176, 1–9.

Bartoli, C., Roux, F., 2017. Genome-wide association studies in plant pathosystems: toward an ecological approach. Front. Plant Sci. 8, 763.

Barton, M.K., 2010. Twenty-years on: the inner torkings of the shoot apical meristem, a developmental dynamo. Dev. Biol. 341, 95–113.

Bashkirov, P.V., Akimov, S.A., Evseev, A.E., Schmid, S.L., Zimmerberg, J., Frolov, V.A., 2008. GTPase cycle of dynamin is coupled to membrane squeeze and release, leading to spontaneous fission. Cell 135, 1–11.

Bashline, L., Du, J., Gu, Y., 2011. The trafficking and behavior of cellulose synthase and a glimpse of potenatial cellulose synthesis regulators. Front. Biol. 6, 377–383.

Bashline, L., Lei, L., Li, S., Gu, Y., 2014a. Cell wall, cytoskeleton, and cell expansion in higher plants. Mol. Plant 7, 586–600.

Bashline, L., Li, S., Anderson, C.T., Lei, L., Gu, Y., 2013. The endocytosis of cellulose synthase in *Arabidopsis* is dependent on μ2, a clathrin-mediated endocytosis adaptin. Plant Physiol. 163, 150–160.

Bashline, L., Li, S., Gu, Y., 2014b. The trafficking of the cellulose synthase complex in higher plants. Ann. Bot. 114, 1059–1067.

Basir, S.N., Yussof, H., Zahari, N.I., 2015. Simulation analysis of *Mimosa pudica* main pulvinus towards biological tactile sensing modelling. Proced. Comput. Sci. 76, 425–429.

Baskin, T.I., 2001. On the alignment of cellulose microfibrils by cortical microtubules: a review and a model. Protoplasma 215, 150–171.

Baskin, T.I., 2005. Anisotropic expansion of the plant cell wall. Annu. Rev. Cell Dev. Biol. 21, 203–222.

Baskin, T.I., Cande, W.Z., 1990. The structure and function of the mitotic spindle in flowering plants. Annu. Rev. Plant Physiol. Plant Mol. Biol. 41, 277–315.

Bassham, J.A., 2003. Mapping the carbon reduction cycle: a personal retrospective. Photosynth. Res. 76, 35–52.

Bassham, D.C., Blatt, M.R., 2008. SNAREs: cogs and coordinators in signaling and development. Plant Physiol. 147, 1504–1515.

Bassham, J.A., Calvin, M., 1957. The Path of Carbon in Photosynthesis. Prentice-Hall, Englewood Cliffs, NJ.

Bassham, D.C., Creighton, A.M., Arretz, M., Brunner, M., Robinson, C., 1994. Efficient but aberrant cleavage of mitochondrial precursor proteins by the chloroplast stromal processing peptidase. FEBS J. 221, 523–528.

Bassham, D.C., Gal, S., Conceicao, A.dD.S., Raikhel, N.V., 1995. An *Arabidopsis* syntaxin homologue isolated by functional complementation of a yeast pep12 mutant. Proc. Natl. Acad. Sci. U.S.A. 92, 7262–7266.

Bassham, D.C., Raikhel, N.V., 2000. Plant cells are not just green yeast. Plant Physiol. 122, 999–1001.

Bastiansen, O., 1964. The Law of Mass Action. A Centenary Volume 1864–1964. Det Norske Videnskaps-Akademi I, Oslo.

Bateson, W., 1899. Hybridisation and cross-breeding as a method of scientific investigation. J. R. Hort. Soc. 24, 59–66.

Bateson, W., 1902. Mendel's Principles of Heredity. A Defence. Cambridge University Press, Cambridge.

Bateson, W., 1909. Mendel's Principles of Heredity. Cambridge University Press, Cambridge.

Bateson, W., Saunders, E.R., Punnett, R.C., 1906. Report III. Reports to the Evolution Committee of the Royal Society. Harrison and Sons, London.

Batoko, H., Zheng, H.-Q., Hawes, C., Moore, I., 2000. A Rab1 GTPase is required for transport between the endoplasmic reticulum and Golgi apparatus and for normal Golgi movement in plants. Plant Cell 12, 2201–2218.

Baudenbacher, F., Fong, L.E., Thiel, G., Wacke, M., Jazbinsek, V., Holzer, J.R., Stampfl, A., Trontelj, Z., 2005. Intracellular axial current in Chara coralline reflects the altered kinetics of ions in cytoplasm under influence of light. Biophys. J. 88, 690–697.

Baudhuin, P., Beaufay, H., de Duve, C., 1965. Combined biochemical and morphological study of particulate fractions from rat liver. J. Cell Biol. 26, 219–243.

Bauer, H.H., 1994. Scientific Literacy and the Myth of the Scientific Method. University of Illinois Press, Champaign, IL.

Bauer, E., Schmutzer, T., Barilar, I., Mascher, M., Gundlach, H., Martis, M.M., Twardziok, S.O., Hackauf, B., Gordillo, A., Wilde, P., Schmidt, M., Korzun, V., Mayer, K.F.X., Schmid, K., Schön, C.-C., Scholz, U., 2017. Towars a whole-genome sequence for rye (*Secale cereale* L.). Plant J. 89, 853–869.

Bauer, W.D., Talmadge, K.W., Keegstra, K., Albersheim, P., 1973. The structure of plant cell walls. II. The hemicellulose of the walls of suspension cultured sycamore cells. Plant Physiol. 51, 174–187.

Bauerfeind, R., Huttner, W.B., 1993. Biogenesis of constitutive secretory vesicles, secretory granules and synaptic vesules. Curr. Opin. Cell Biol. 5, 628–635.

Baulcombe, D., 2008. Of maize and men, or peas and people: case histories to justify plants and other model systems. Nat. Medods 14, 20–23.

Baum, M., Appels, R., 1991. The cytogenetic and molecular architecture of chromosome 1R-one of the most widely utilized sources of alien chromation in wheat varieties. Chromosoma 101, 1–10.

Bausch, A.R., Moeller, W., Sackmann, E., 1999. Measurements of local viscoelasticity and forces in living cells by magnetic tweezers. Biophys. J. 76, 573–579.

Bauwe, H., 2010. Recent developments in photorespiration research. Biochem. Soc. Trans. 38, 677–682.

Bayer, E.M., Bottrill, A.R., Walshaw, J., Vigouroux, M., Naldrett, M.J., Thomas, C.L., Maule, A.J., 2006. Arabidopsis cell wall proteome using multidimensional protein identification technology. Proteomics 6, 301–311.

Bayer, E., Thomas, C.L., Maule, A.J., 2004. Plasmodesmata in *Arabidopsis thaliana* suspension cells. Protoplasma 223, 93–102.

Bayley, P.M., 1990. What makes microtubules dynamic? J. Cell Sci. 95, 329–334.

Beadle, G.W., 1945. Genetics and metabolism in *Neurospora*. Physiol. Rev. 25, 643–663.

Beadle, G.W., 1955. The genes of men and molds. In: Scientific American (Ed.), The Physics and Chemistry of Life. Simon and Schuster, New York, pp. 151–162.

Beadle, G., Beadle, M., 1966. The Language of Life. An Introduction to the Science of Genetics. Doubleday & Co., Garden City, New York.

Beadle, G.W., December 11, 1958. Genes and chemical reactigns in Neurospora. Nobel Lecture.

Beadle, G.W., Tatum, E.L., 1941a. Genetic control of biochemical reactions in *Neurospora*. Proc. Natl. Acad. Sci. U.S.A. 27, 499–506.

Beadle, G.W., Tatum, E.L., 1941b. Experimental control of development and differentiation. Genetic control of developmental reactions. Am. Nat. 75, 107–116.

Beale, L.S., 1872. Bioplasm: An Introduction to the Study of Physiology and Medicine. J. & A. Churchill, London.

Beale, L.S., 1878. The Microscope in Medicine, fourth ed. Much Enlarges. J. & A. Churchill, London.

Beale, L.S., 1892. Protoplasm, fourth ed. Harrison & Sons, London.

Beams, H.W., Kessel, R.G., 1968. The Golgi apparatus: structure and function. Int. Rev. Cytol. 23, 209–276.

Beams, H.W., King, R.L., 1938. An experimental study on mitosis in the somatic cells of wheat. Biol. Bull. 75, 189–207.

Beaudoin, F., Wikinson, B.M., Striring, C.J., Napier, J.A., 2000. In vivo targeting of a sunflower oil body protein in yeast secretory (sec) mutants. Plant J. 23, 159–170.

Beausoleil, S.A., Villén, J., Gerber, S.A., Rush, J., Gygi, S.P., 2006. A probability-based approach for high-throughput protein phosphorylation analysi sand site localization. Nat. Biotechnol. 24, 1285–1292.

Becher, P., 1965. Emulsions. Theory and Practice, second ed. Reinhold, New York.

Becker, W.A., 1932. Recherches expérimentales sur la cytocinèse et la formation de la plaque cellulaire dans la cellule vivante. C. R. Acad. Sci. Paris 194, 1850–1852.

Becker, W.A., 1938. Recent investigations in vivo on the division of plant cells. Bot. Rev. 4, 446–472.

Becker, W.M., 1977. Energy and the Living Cell. An Introduction to Bioenergetics. J. B. Lippincott Co., Philadelphis, PA.

Becker, J.D., Boavida, L.C., Carneiro, J., Haury, M., Feijó, J.A., 2003. Transcriptional profiling of *Arabidopsis* tissues reveals the unique characteristics of the pollen transcriptome. Plant Physiol. 133, 713–725.

Becker, B., Bölinger, B., Melkonian, M., 1995. Anterograde transport of algal scales through the Golgi complex is not mediated by vesicles. Trends Cell Biol. 5, 305–307.

Beckett, A., Heath, I.B., McLaughlin, D.J., 1974. An Atlas of Fungal Ultrastructure. Longman, London.

Bédard, J., Jarvis, P., 2005. Recognition and envelope translocation of chloroplast proteins. J. Exp. Bot. 56, 2287–2320.

Beddoes, T., 1799. Contributions to Physical and Medical Knowledge, Principally from the West of England. Biggs & Cottle, Bristol.

Bednarek, S.Y., Backues, S.K., 2010. Plant dynamin-related protein families DRP1 and DRP2 in plant development. Biochem. Soc. Trans. 38, 797–806.

Bednarek, S.Y., Ravazzola, M., Hosobuchi, M., Amherdt, M., Perrelet, A., Schekman, R., 1995. COPI- and COPII-coated vesicles bud directly from the endoplasmic reticulum in yeast. Cell 83, 1183–1196.

Bednarek, S.Y., Reynolds, T.L., Schroeder, M., Grabowski, R., Hengst, L., Gallwitz, D., Raikhel, N.V., 1994. A small GTP binding protein from *Arabidopsis thaliana* functionally complements the yeast YPT6 null mutant. Plant Physiol. 104, 591–596.

Beebe, W., 1934. Half Mile Down. Harcourt, Brace and Co., New York.

Beebe, D.U., Evert, R.F., 1992. Photoassimilate pathway(s) and phloem loading in the leaf of *Moricandia arvensis* (L.) DC (Brassicaceae). Int. J. Plant Sci. 153, 61–77.

Beebe, D.U., Turgeon, R., 1991. Current perspectives on plasmodesmata: structure and function. Physiol. Plant 83, 194–199.

Beer, J.J., 1959. The Emergence of the German Dye Industry. University of Illinois Press, Urbana.

Beet, E.A., 1949. The genetics of the sickle-cell trait in a Bantu tribe. Ann. Eugen. 14, 279–284.

Beevers, H., 1961. Respiratory Metabolism in Plants. Harper & Row, New York.

Beevers, H., 1979. Microbodies in higher plants. Ann. Rev. Plant Physiol. 30, 159–193.

Beevers, L., 1996. Clathrin-coated vesicles in plants. Int. Rev. Cytol. 167, 1–35.

Beevers, H., Walker, D.A., 1956. The oxidative acivity of particulate fractions from germinating castor beans. Biochem. J. 114–120.

Begg, J.D., 1978. Transformation of yeast by a replicating hybrid plasmid. Nature 275, 104–109.

Begg, D.A., Ellis, G.W., 1979. Micromanipulation studies of chromosome movement. J. Cell Biol. 82, 528–541.

Beier, V., Bauer, A., Baum, M., Hoheisel, J.D., 2004. Fluorescent sample labeling for DNA microarray analysis. Methods Mol. Biol. 283, 127–135.

Beilby, M.J., 2015. Salt tolerance at a single cell level in giant-celled Characeae. Front. Plant Sci. 6, 226.

Beilby, M.J., 2016. Multi-scale characean experimental system: from electrophysiology of membrane transporters to cell-to-cell connectivity, cytoplasmic streaming and auxin metabolism. Front. Plant Sci. 7, 1052.

Beilby, M.J., Cherry, C.A., Shepherd, V.A., 1999. Dual turgor regulation response to hypotonic stress in *Lamprothamnium papulosum*. Plant Cell Environ. 22, 347–360.

Beilby, M.J., Shepherd, V.A., 1996. Turgor regulation in *Lamprothamnium papulosum*. I. I/V analysis and pharmacological dissection of the hypotonic effect. Plant Cell Environ. 19, 837–847.

Beilby, M.J., Turi, C.E., Baker, T.C., Tymm, F.J.M., Murch, S.J., 2015. Circadian changes in endogenous concentrations of indole-3-acetic acid, melatonin, serotonin, abscisic acid and josmonic acid in Characeae (*Chara australis* Brown). Plant Signal. Behav. 10, e108.2697.

Belar, K., 1929. Der Formwechsel der Protistenkerne. Ergesbn. u. Fortschr. Zool 6, 235–654.

Belmont, A.S., 2006. Mitotic chromosome structure and condensation. Curr. Opin. Cell Biol. 18, 632–638.

Ben Abdelkader, A., Cherif, A., Demandre, C., Mazliak, P., 1973. The oleyl-coenzyme-A desaturase of potato tubers: enzymatic properties, intracellular localization and induction during "aging" of tuber slices. Eur. J. Biochem. 32, 155–165.

Benavente, R., 1991. Postmitotic nuclear reorganization events analyzed in living cells. Chromosoma 100, 215–220.

Benda, C., 1899. Weitere Mitteilungen über die Mitochondria. Arch. Anat. Physiol. (Physiol. Abt.) 376.

Bendall, D.S., 2005. The unfinished story of cytochrome f. In: Govindjee, Beatty, J.T., Gest, H., Allen, J.F. (Eds.), Discoveries in Photosynthesis. Springer, pp. 531–542.

Bendiner, E., May 1982. Albert Szent-Gyorgyi: the art of being wrong. Hosp. Pract. 17, 179–192.

Benitez-Alfonso, Y., Faulkner, C., Pendle, A., Miyashima, S., Helariutta, A., Maule, A., 2013. Symplastic intercellular connectivity regulates lateral root patterning. Dev. Cell. 26, 136–147.

Benito, A., Rodriguez-Navarro, B., 2003. Molecular cloning and characterization of a sodium-pump ATPase of the moss Physcomitrella patens. Plant J. 36, 382–389.

Benkert, R., Obermeyer, G., Bentrup, F.-W., 1997. The turgor pressure of growing lily pollen tubes. Protoplasma 198, 1–8.

Bennet-Clark, T.A., Greenwood, A.D., Barker, J.W., 1936. Water relations and osmotic pressure of plant cells. New Phytol. 35, 277–291.

Bennett, C.W., 1956. Biological relations of plant viruses. Ann. Rev. Plant Physiol. 7, 143–170.

Bennett, J., 1981. Biosynthesis of the light-harvesting chlorophyll a/b protein. Eur. J. Biochem. 118, 61–70.

Bennett, V., 1990a. Spectrin: a structural mediation between diverse plasma membrane proteins and the cytoplasm. Curr. Opin. Cell Biol. 2, 51–56.

Bennett, V., 1990b. Spectrin-based membrane skeleton: a multipotential adaptor between plasma membrane and cytoplasm. Physiol. Rev. 70, 1029–1065.

Bennett, J., 1991. Protein phosphorylation in green plant chloroplasts. Annu. Rev. Plant Physiol. Plant Mol. Biol. 42, 281–311.

Bennett, V., Gilligan, D.M., 1993. The spectrin-based membrane skeleton and micron-scale organization of the plasma membrane. Annu. Rev. Cell Biol. 9, 27–66.

Bennett, A.B., O'Neill, S.D., Eilmann, N., Spanswick, R.M., 1985. H^+-ATPase activity from storage tissue of Beta vulgaris. III. Modulation of ATPase activity by reaction substrates and products. Plant Physiol. 78, 495–499.

Bennett, H.S., Porter, K.R., 1953. An electron microscope study of sectioned breast muscle of the domestic fowl. Am. J. Anat. 93, 61–105.

Bennett, A.B., Spanswick, R.M., 1983. Optical measurements of ΔpH and $\Delta\Psi$ in corn root membranes vesicles: kinetic analysis of Cl^- effects on a proton-translocating ATPase. J. Membr. Biol. 71, 95–107.

Bennett, A.B., Spanswick, R.M., 1984a. H^+-ATPase activity from storage tissue of Beta vulgaris. I. Identification and characterization of an anion-sensitive H^+-ATPase. Plant Physiol. 74, 538–544.

Bennett, A.B., Spanswick, R.M., 1984b. H^+-ATPase activity from storage tissue of Beta vulgaris. II. H^+/ATP stoichiometry of an anion sensitive H^+-ATPase. Plant Physiol. 74, 545–548.

Bensley, R.R., 1951. Facts versus artefacts in cytology: the Golgi apparatus. Exp. Cell Res. 2, 1–9.

Bensley, R.R., 1953. Introduction and greetings. Symposium: the structure and biochemistry of mitochondria. J. Histochem. Cytochem. 1, 179–182.

Bensley, R.R., Hoerr, N.L., 1934. Studies on cell structure by the freezing-drying method. VI. The preparation and properties of mitochondria. Anat. Rec. 60, 449–455.

Benson, A.A., 2002. Paving the path. Annu. Rev. Plant Biol. 53, 1–25.

Benson, A.A., Calvin, M., 1947. The dark reductions of photosynthesis. Science 106, 648–652.

Bent, H.A., 1965. The Second Law. Oxford University Press, New York.

Bentley, R., 1978. Ogston and the development of the prochirality theory. Nature 276, 673–676.

Bentley, D.R., Balasubramanian, S., Swerdlow, H.P., Smith, G.P., Milton, J., Brown, C.G., et al., 2008. Accurate whole genome sequencing using reversible terminator chemistry. Nature 456, 53–59.

Bentolila, S., Heller, W.P., Sun, T., Babina, A.M., Friso, G., van Wijk, K.J., Hanson, M.R., 2012. RIP1, a member of an *Arabidopsis* protein family, interacts with the protein RARE1 and broadly affects RNA editing. Proc. Natl. Acad. Sci. U.S.A. 109, E1453–E1461.

Bentolila, S., Oh, J., Hanson, M.R., Bukowski, R., 2013. Comprehensive high-resolution analysis of the role of an *Arabidopsis* gene family in RNA editing. PLoS Genet. 9, e1003584.

Berardini, T.Z., Mundodi, S., Reiser, L., Huala, E., Garcia-Hernandez, M., Zhang, P., Mueller, L.A., Yoon, Y., Doyle, A., Lander, G.,

Moseyko, N., Yoo, D., Xu, I., Zoeckler, B., Montoya, M., Miller, N., Weems, D., Rhee, S.Y., 2004. Functional annotation of the *Arabidopsis* genome using controlled vocabularies. Plant Physiol. 135, 745–755.

Bereiter-Hahn, J., 1987. Mechanical principles of architecture of Eurkaryotic cells. In: Bereiter-Hahn, J., Anderson, O.R., Reif, W.E. (Eds.), Cytomechanics. Springer Verlag, Berlin, pp. 3–30.

Berestovsky, G.N., Ternovsky, V.I., Kataev, A.A., 2001. Through pore diameter in the cell wall of Chara. J. Exp. Bot. 52, 1173–1177.

Berezney, R., Mortillaro, M.J., Ma, H., Wei, X., Samarabandu, J., 1995. The nuclear matrix: a structural milieu for genomic function. Int. Rev. Cytol. 162A, 1–65.

Berg, P., 2003. Moments of discovery: my favorite experiment. J. Biol. Chem. 278, 40417–40424.

Berg, R.H., Beachy, R.N., 2008. Fluorescent protein applications in plants. Methods Cell Biol. 85, 153–177.

Bergen, L.G., Borisy, G.G., 1980. Head-to-tail polymerization of microtubules in vitro. J. Cell Biol. 84, 141–150.

Berger, B., de Regt, B., Tester, M., 2012. High-throughput phenotyping of plant shoots. In: Normanly, J. (Ed.), High-Throughput Phenotyping in Plants. Methods and Protocols. Springer, New York, pp. 9–20.

Berger, S., Dirk, J., von Lindern, L., Wolff, D., Mergenhagen, D., 1992. Temperature dependency of circadian periodvin a single cell (Acetabularia). Bot. Acta 105, 382–386.

Berger, S., Herth, W., Franke, W.W., Falk, H., Spring, H., Schweiger, H.G., 1975. Morphology of the nucleo-cytoplasmic interactions during the development of Acetabularia cells. Protoplasma 84, 223–256.

Berger, C., Witkus, E., 1943. A cytological study of c-mitosis in the polysomatic plant *Spinacea oleracea*, with comparative observation on *Allium cepa*. Bull. Torrey Bot. Club 70, 457–467.

Berget, S.M., Moore, C., Sharp, P.A., 1977. Spliced segments at the 5′ terminus of adenovirus 2 late mRNA. Proc. Natl. Acad. Sci. U.S.A. 74, 3171–3175.

Bergey, D.R., Howe, G.A., Ryan, C.A., 1996. Polypeptide signaling for plant defensive genes exhibits analogies to defense signaling in animals. Proc. Natl. Acad. Sci. U.S.A. 93, 12053–12058.

Bergfeld, R., Speth, V., Schopfer, P., 1988. Reorientation of microfibrils and microtubules at the outer epidermal wall of maize coleoptiles during auxin-mediated growth. Bot. Acta 101, 57–67.

Bergman, K., Burke, P.V., Cerdá-Olmedo, E., David, C.N., Delbrück, M., Foster, K.W., Goodell, E.W., Heisenberg, M., Meissner, G., Zalokar, M., Dennison, D.S., Shropshire Jr., W., 1969. Phycomyces. Bacteriol. Rev. 33, 99–157.

Bergmann, M., Fruton, J.S., 1944. The significance of coupled reactions for the enzymatic hydrolysis and synthesis of proteins. Ann. N.Y. Acad. Sci. 45, 409–423.

Bergmann, M., Kann, E., Miekeley, A., 1927. Verfahren zur Strukturbestimmung von Dipeptiden. Liebig's Ann. Chem. 458, 56–59.

Bergmann, M., Niemann, C., 1938. The chamistry of amino acids and proteins. Ann. Rev. Biochem. 7, 99–124.

Bergmann, M., Zervas, L., 1937. A method for the stepwise degradation of polypeptides. J. Biol. Chem. 113, 341–357.

Bergmans, A.C.J., De Boer, A.D., Derksen, J.W.M., van der Schoot, C., 1997. The symplasmic coupling of L-2-cells diminishes in early floral development of Iris. Planta 203, 245–252.

Bergson, H., 1911. Creative Evolution. Henry Holt and Co., New York. Translated by A. Mitchell.

Bergthorsson, U., Richardson, A.O., Young, G.J., Goertzen, L.R., Palmer, J.D., 2004. Massive horizontal transfer of mitochondrial genes from diverse land plant donors to the basal angiosperm *Amborella*. Proc. Natl. Acad. Sci. U.S.A. 101, 17747–17752.

Berk, A.J., 2016. Discovery of RNA splicing and genes in pieces. Proc. Natl. Acad. Sci. U.S.A. 113, 801–805.

Berkelman, T., Lagarias, J.C., 1990. Calcium transport in the green alga Mesotaenium caldariorum. Plant Physiol. 93, 748–757.

Berlin, K., Koren, S., Chin, C.-S., Drake, J.P., Landolin, J.M., Phillippy, A.M., 2015. Assembling large genomes with single-molecule sequencing and locality-sensitive hashing. Nat. Biotechnol. 33, 623–630.

Berlinski, D., 2008. The Devil's Delusion. Atheism and Its Scientific Pretensions. Crown Forum, New York.

Bermudes, D., Hinkle, G., Margulis, L., 1994. Do prokaryotes contain microtubules? Microbiol. Rev. 58, 387–400.

Bernabeu, C., Lake, J.A., 1982. Nascent polypeptide chains emerge from the exit domain of the large ribosomal subunit. Proc. Natl. Acad. Sci. U.S.A. 79, 311–3115.

Bernal, J.D., 1939. The Social Function of Science. George Routledge & Sons, London.

Bernal, J.D., 1949. The Freedom of Necessity. Routledge & Kegan Paul, London.

Bernal, J.D., 1951. The Physical Basis of Life. Routledge & Kegan Paul, London.

Bernal, J.D., 1962. Biochemical evolution. In: Kasha, M., Pullman, B. (Eds.), Horizons in Biochemistry. Albert Szent-Györgyi Dedicatory Volume. Academic Press, New York, pp. 11–22.

Bernal, J.D., 1962a. Biochemical evolution. In: Kasha, M., Pullman, B. (Eds.), Horizons in Biochemistry. Albert Szent-Györgyi Dedicatory Volume. Academic Press, New York, pp. 11–22.

Bernal, J.D., 1962b. The place of speculation in modern technology and science. In: Good, I.J., Mayne, A.J., Smith, J.M. (Eds.), The Scientist Speaks. Basic Books, New York, pp. 11–28.

Bernal, J.D., 1967. The Origin of Life. Weidenfeld and Nicolson, London.

Bernard, C., 1865. An Introduction to the Study of Experimental Medicine, 1957. Dover Publications, New York.

Bernard, C., 1965. An Introduction to the Study of Experimental Medicine, 1957. Dover Publications, New York.

Bernat, R.L., Borisy, G.G., Rothfield, N.F., Earnshaw, W.C., 1990. Injection of anticentromere antibodies in interphase disrupts events required for chromosome movement at mitosis. J. Cell Biol. 111, 1519–1533.

Bernfield, M.R., Nirenberg, M.W., 1965. RNA codewords and protein synthesis. The nucleotide sequences of multiple codewords for phenylalanine, serine, leucine, and proline. Science 147, 479–484.

Bernhard, W., de Haven, E., 1956. Sur la présence dans certaines cellulles de mammifères d'un organite de nature probablement centriolaire. Étude au microscope électronique. C. R. Acad. Sci. Paris 242, 288–290.

Berns, M.W., Aist, J., Edwards, J., Strahs, K., Girton, J., McNeill, P., Ratner, J.B., Kitzes, M., Hammer-Wilson, M., Liaw, L.-H., Siemens, A., Koonce, M., Peterson, S., Brenner, S., Burt, J., Walter, R., Bryant, P.J., van Dyk, D., Coulombe, J., Cahill, T., Berns, G.S., 1981. Laser microsurgery in cell and developmental biology. Science 213, 505–513.

Laser Maniputation of Cells and Tissues. In: Berns, M.W., Greulich, K.O. (Eds.), 2007. Methods in Cell Biology, vol. 82. Elsevier, Amsterdam.

Berridge, M.J., 1987. Inositol triphosphate and discylglycerol: two interacting second messengers. Ann. Rev. Biochem. 56, 159–193.

Berridge, M.J., 2002. The endoplasmic reticulum: a multifunctional signaling organelle. Cell Calcium 32, 235–249.

Berridge, M.J., 2003. Cardiac calcium signalling. Biochem. Soc. Trans. 31, 930–933.

Berridge, M.J., Downs, C.P., Hanley, M.R., 1989. Neural and developmental action sof lithium: a unifying hypothesis. Cell 59, 411–419.

Berridge, M.J., Irvine, R.F., 1990. Inositol phosphates and cell signaling. Nature 341, 197–205.

Berry, W., 2000. Life is a Miracle. An Essay Against Modern Superstition. Counterpoint, Washington, DC.

Berry, E.A., Hinkle, P.C., 1983. Measurements of the electrochemical proton gradient in submitochondrial particles. J. Biol. Chem. 258, 1474–1486.

Berry, J.A., Osmond, C.B., Lorimer, G.H., 1978. Fixation of $^{18}O_2$ during photorespiration. Kinetic and steady state studies of the photorespiratory carbon oxidation cycle with intact leaves and isolated cloroplasts of C3 plants. Plant Physiol. 62, 954–967.

Bertani, G., Weigle, J.J., 1953. Host controlled variation in bacterial viruses. J. Bacteriol. 65, 113–121.

Berth, M., Moser, F.M., Kolbe, M., Bernhardt, J., 2007. The state of the art in the analysis of two-dimensional gel electrophoresis images. Appl. Microbiol. Biotechnol. 76, 1223–1243.

Berthelot, M., 1885. Les Origines de l'Alchime. Georges Steinheil, Paris.

Berthold, G., 1886. Studien über Protoplasmamechanik. Verlag Arthur Felix, Leipzig.

Berthold, P., Tsunoda, S.P., Ernst, O.P., Mages, W., Gradmann, D., Hegemann, P., 2008. Channelrhodopsin-1 initiates phototaxis and photophobic responses in *Chlamydomonas* by immediate light-induced depolarization. Plant Cell 20, 1665–1677.

Bertioli, D.J., Cannon, S.B., Froenicke, L., Huang, G., Farmer, A.D., Cannon, E.K.S., Liu, X., Gao, D., Clevenger, J., 2016. The genome sequences of *Arachis duranensis* and *Arachis ipaensis*, the diploid ancestors of cultivated peanut. Nat. Genet. 48, 438–446.

Bertozzi, W., 1962. The Ultimate Speed - An Exploration with High Energy Electrons. https://www.youtube.com/watch?v=B0BOpiMQXQA.

Bertozzi, W., 1964. Speed and kinetic energy of relativistic electrons. Am. J. Phys. 32, 551–555.

Bethe, A., 1930. The permeability of the surface of marine animals. J. Gen. Physiol. 13, 437–444.

Bethe, H.A., 1939. Energy production in stars. Phys. Rev. 55, 434–456.

Bethe, H.A., 1968. Les Prix Nobel en 1967. Energy Production in Stars. Nobel Foundation, Stokholm.

Bethe, H.A., 2003. My life in astrophysics. Annu. Rev. Astron. Astrophys. 41, 1–14.

Bethe, H.A., Critchfield, C.L., 1938. The formation of deuterons by proton combination. Phys. Rev. 54, 248–254.

Bethke, P.C., Hwang, Y.-S., Zhu, T., Jones, R.L., 2006. Global patterns of gene expression in the aleurone of wild-type and dwarf1 mutant rice. Plant Physiol. 140, 484–498.

Betsche, T., Schaller, D., Melkonian, M., 1992. Identification and characterization of glycolate oxidase and related enzymes from the endocyanotic alga Cyanophora paradoxa and from pea leaves. Plant Physiol. 98, 887–893.

Beven, A., Quan, Y., Pearl, J., Cooper, C., Shaw, P., 1991. Monoclonal antibodies to plant nuclear matrix reveal intermediate filament-related components within the nucleus. J. Cell Sci. 98, 293–302.

Bevoda, R., Pleskot, R., Žárský, V., Potocký, M., 2014. Antisense oligodeoxynucleotide-mediated gene knockdown in pollen tubes. Methods Mol. Biol. 1080, 231–236.

Beyler, R.H., 1996. Targeting the organism. The scientific and cultural context of Pascual Jordan's quantum biology, 1932–1947. Isis 87, 248–273.

Beyler, R.H., 2007. Exporting the quantum revolution: pascual Jordan's biophysical initiatives. In: Pascual Jordan (1902–1980) Mainzer Symposium zum 100. Geburtstag, pp. 69–81. http://www.mpiwg-berlin.mpg.de/Preprints/P329.PDF.

Bezanilla, M., Pan, A., Quatrano, R.S., 2003. RNA interference in the moss *Physcomitrella patens*. Plant Physiol. 133, 470–474.

Bezanilla, M., Perroud, P.-F., Pan, A., Klueh, P., Quatrano, R.S., 2005. An RNAi system in *Physcomitrella patens* with an internal marker for silencing allows for rapid identification of loss of function phenotypes. Plant Biol. 7, 251–257.

Beznoussenko, G.V., Mironov, A.A., 2002. Models of intracellular transport and evolution of the Golgi complex. Anat. Rec. 268, 226–238.

Bhaskar, V., 2003. Root Hairs. The "Gills" of Roots. Science Publishers, Enfield, NH.

Bhat, R.A., Panstruga, R., 2005. Lipid rafts in plants. Planta 223, 1–15.

Bhaya, D., Jagendorf, A.T., 1985. Synthesis of the a and ß subunits of coupling factor 1 by polysomes from pea chloroplasts. Arch. Biochem. Biophys. 237, 217–223.

Bhuin, T., Roy, J.K., 2014. Rab proteins: the key regulators of intracellular vesicle transport. Exp. Cell Res. 328, 1–19.

Bi, E.F., Lutkenhaus, J., 1991. FtsZ ring structure associated with division in *Escherichia coli*. Nature 354, 161–164.

Biagini, G.A., Bernard, C., 2000. Primitive anaerobic protozoa: a false concept? Microbiology 146, 1019–1020.

Bialek, W., 1987. Physical limits to sensation and perception. Ann. Rev. Biophys. Biophys. Chem. 16, 455–478.

Bicknese, S., Periasamy, N., Shohet, S.B., Verkman, A.S., 1993. Cytoplasmic viscosity near the cell plasma membrane: measurement by evanescent field frequency-domain microfluorimetry. Biophys. J. 65, 1272–1282.

Bidwell, R.G.S., 1974. Plant Physiology. MacMillan, New York.

Biemann, K., 2001. Four decades of structure determination by mass spectrometry: from alkaloids to heparin. J. Am. Soc. Mass Spectrom. 13, 1254–1272.

Biemann, K., 2002. Four decades of structure determination by mass spectrometry: from alkaloids to heparin. J. Am. Soc. Mass Spectrom. 13, 1254–1272.

Biemann, K., Gapp, F., Seibl, J., 1959. Application of mass spectrometry to structure problems. I. Amino acid sequence in peptides. J. Am. Chem. Soc. 81, 2274–2275.

Bierhorst, D.W., 1971. Morphology of Vascular Plants. Macmillan, New York.

Biłas, R., Szafran, K., Hnatuszko-Konka, K., Kononowicz, A.K., 2016. *Cis*-regulatory elements used to control gene expression in plants. Plant Cell Tiss. Cult. 127, 269–287.

Bilofsky, H.S., Burks, C., Fickett, J.W., Goad, W.B., Lewitter, F.I., Rindone, W.P., Swindell, C.D., Tung, C.S., 1986. The GenBank genetic sequence databank. Nucleic Acids Res. 14, 1–4.

Binarová, P., Cenklová, V., Procházková, J., Doskočilová, A., Volc, J., Vrlik, M., Bögre, L., 2006. γ-tubulin is essential for acentrosomal microtubule nucleation and coordination of late mitotic events in *Arabidopsis*. Plant Cell 18, 1199–1212.

Binns, D., Januszewski, T., Chen, Y., Hill, J., Markin, V.S., Zhao, Y., Gilpin, C., Chapman, K.D., Anderson, R.G.W., Goodman, J.M., 2006. An intimate collaboration between peroxisomes and lipid bodies. J. Cell Biol. 173, 719–731.

Biological Sciences Curriculum Study, 1963. Biological Science: Molecules to Man. Blue Version. Houghton Mifflin, Boston, MA.

Birch, J.S., Lee, J., 2007. Joseph Priestley. A Celebration of His Life and Legacy. Priestley Society, Lancaster, England.

Birmingham, K., 1999. A Nobel for Blobel. Nat. Med. 5, 1230.

Birnbaum, K., Shasha, D.E., Wang, J.Y., Jung, J.W., Lambert, G.M., Galbraith, D.W., Benfey, P.N., 2003. A gene expression map of the *Arabidopsis* root. Science 302, 1956–1960.

Birol, I., Raymond, A., Jackman, S.D., Pleasance, S., Coope, R., Taylor, G.A., Yuen, M.M., Keeling, C.I., Brand, D., Vandervalk, B.P., Kirk, H., Pandoh, P., Moore, R.A., Zhao, Y., Mungall, A.J., Jaquish, B., Yanchuk, A., Ritland, C., Boyle, B., Bousquet, J., Ritland, K., Mackay, J., Bohlmann, J., Jones, S.I., 2013. Assembling the 20 Gb white spruce (*Picea glauca*) genome from whole-genome shotgun sequencing data. Bioinformatics 29, 1492–1497.

Bischof, M., 1996. Some remarks on the history of biophysics (and its future) paper delivered at the 1st Hombroich symposium on biophysics, Neuss, Germany, October 3–6, 1995. In: Zhang, C.L., Popp, F.A., Bischof, M. (Eds.), Current Development of Biophysics. Hangzhou University Press, Hangzhou, China.

Bisgrove, S.R., Kropf, D.L., 2007. Asymmetric cell divisions: zygotes of fucoid algae as a model system. Plant Cell Monogr. 9, 323–341.

Bishop, W.R., Bell, R.M., 1988. Assembly of phospholipids into cellular membranes: biosynthesis, transmembrane movement and intracellular translocation. Ann. Rev. Cell Biol. 4, 579–610.

Bisson, M.A., Kirst, G.O., 1980. Lamprothamnium, a euryhaline charophyte. I. Osmotic relations and membrane potential at steady state. J. Exp. Bot. 31, 1223–1235.

Black, C.C., Osmond, C.B., 2003. Crassulacean acid metabolism photosynthesis: "Working the night shift." Photosynth. Res. 76, 329–341.

Blackbourn, H.D., Jackson, A.P., 1996. Plant clathrin heavy chain: sequence analysis and restricted localization in growing pollen tubes. J. Cell Sci. 109, 777–787.

Blackett, P.M.S., 1935. The frustration of science. Forward by F. Soddy. In: The Frustration of Science. George Allen & Unwin, London, pp. 129–144.

Blackford, S., Rea, P.A., Sanders, D., 1990. Voltage sensitivity of H^+/Ca^{2+} antiport in higher plant tonoplast suggests a role in vacuolar calcium accumulation. J. Biol. Chem. 265, 9617–9620.

Blackman, F.F., 1905. Optima and limiting factors. Ann. Bot. 19, 281–295.

Blackman, F.F., 1906. Incipient vitality: an account of some recent work throwing light upon the chemical mechanism of the cell. New Phytol. 5, 22–34.

Blackman, L.M., Harper, J.D.I., Overall, R.L., 1999. Localization of a centrin-like protein to higher plant plasmodesmata. Eur. J. Cell Biol. 78, 297–304.

Blackman, L.M., Overall, R.L., 1998. Immunolocalisation of the cytoskeleton to plasmodesmata in *Chara corallina*. Plant J. 14, 733–741.

Blaese, R.M., Culver, K.W., Miller, A.D., Carter, C.S., Fleisher, T., Clerici, M., Shearer, G., Chang, L., Chiang, Y., Tolstoshev, P., Greenblatt, J.J., Rosenberg, S.A., Klein, H., Berger, M., Mullen, C.A., Ramsey, W.J., Mullen, C.L., Morgan, R.A., Anderson, W.F., 1995. T-lymphocyte-directed gene therapy for ADA-SCID: initial trial results after 4 years. Science 270, 475–480.

Blair, G.E., Ellis, R.J., 1973. Protein synthesis in chloroplasts. I. Light-driven synthesis of the large subunit of fraction I protein by isolated pea chloroplasts. Biochim. Biophys. Acta 319, 223–234.

Blair, G.W.S., Spanner, D.C., 1974. An Introduction to Biorheology. Elsevier, Amsterdam.

Blanc, G., Wolfe, K.H., 2004. Widespread paleopolyploidy in model plant species inferred from age distribution of duplicate genes. Plant Cell 16, 1667–1678.

Blancaflor, E.B., Fasano, J.M., Gilroy, S., 1998. Mapping the functional roles of cap cells in the response of *Arabidopsis* primary roots to gravity. Plant Physiol. 16, 213–222.

Blancaflor, E.B., Masson, P.H., 2003. Plant gravitropism. Unraveling the ups and downs of a complex process. Plant Physiol. 133, 1677–1690.

Blanchard, A.P., Hood, L., 1996. Sequence to array: probing the genome's secrets. Nat. Biotechnol. 14, 1649.

Blanchard, A.P., Kaiser, R.J., Hood, L.E., 1996. High-density oligonucleotide arrays. Biosens. Bioelectron. 11, 687–690.

Blanchoin, L., Staiger, C.J., 2010. Plant Formins: diverse isoforms and unique molecular mechanism. Biochim. Biophys. Acta 1803, 201–206.

Blankenship, R.E., 2002. Molecular Mechanisms of Photosynthesis. Blackwell Science, Oxford.

Blanton, R.L., Fuller, D., Iranfar, N., Grimson, M.J., Loomis, W.F., 2000. The cellulose synthase gene of Dictyostelium. Proc. Natl. Acad. Sci. U.S.A. 97, 2391–2396.

Blanvillain, R., Boavida, L.C., McCormick, S., Ow, D.W., 2008. Exportin1 genes are essential for development and function of the gametophytes in *Arabidopsis thaliana*. Genetics 180, 1493–1500.

Blascheck, W., 1979. Binding of exogenous DNA by chromatin from isolated plant cell nuclei. Plant Sci. Lett. 15, 139–149.

Blatt, M.R., 1983. The action spectrum for chloroplast movements and evidence for blue-light photoreceptor cycling in the alga *Vaucheria*. Planta 159, 267–276.

Blatt, M.R., 1987. Toward the link between membrane transport and photoreception in plants. Photochem. Photobiol. 45, 933–938.

Blatt, M.R., 1991. Ion channel gating in plants: physiological implications and integration for stomatal function. J. Membr. Biol. 124, 95–112.

Blatt, M.R., 1992. K^+ channels of stomatal guard cells. Characteristics of the inward rectifier and its control by pH. J. Gen. Physiol. 99, 615–644.

Blatt, M.R., 2000. Cellular signaling and volume control in stomatal movements in plants. Ann. Rev. Cell Dev. Biol. 16, 221–224.

Blatt, M.R., Armstrong, F., 1993. K^+ channels of stomatal guard cells: abscisic acid-evoked control of the outward rectifier mediated by cytoplasmic pH. Planta 191, 330–341.

Blatt, M.R., Briggs, W.R., 1980. Blue-light-induced cortical fiber reticulation concomitant with chloroplast aggregation in the alga *Vaucheria sessilis*. Planta 147, 355–362.

Blatt, M.R., Theil, G., Trentham, D.R., 1990. Reversible activation of K^+ channels of *Vicia* stomatal guard cells following photolysis of caged inositol 1,4,5-triphosphate. Nature 346, 766–769.

Blatt, M.R., Weisenseel, M.H., Haupt, W., 1981. A light-dependent current associated with chloroplast aggregation in the alga *Vaucheria sessilis*. Planta 152, 513–526.

Blinks, L.R., 1930. The direct current resistance of *Nitella*. J. Gen. Physiol. 13, 495–508.

Blinks, L.R., 1939. The effects of current flow on bioelectric potential. III. *Nitella*. J. Gen. Physiol. 20, 229–265.

Blinks, J.R., Wier, W.G., Hess, P., Predergast, F.G., 1982. Measurement of Ca^{2+} concentrations in living cells. Prog. Biophys. Mol. Biol. 40, 1–114.

Blobel, G., 1980. Intracellular protein topogenesis. Proc. Natl. Acad. Sci. U.S.A. 77, 1496–1500.

Blobel, G., 1985. Gene gating: a hypothesis. Proc. Natl. Acad. Sci. U.S.A. 82, 8527–8529.

Blobel, G., 2012. Protein targeting. In: Moberg, C.L. (Ed.), Entering an Unseen World. Rockefeller University Press, New York, pp. 177–191.

Blobel, G., December 8, 1999. Protein Targeting. Nobel Lecture.

Blobel, G., Dobberstein, B., 1975a. Transfer of proteins across membranes. I. Presence of proteolytically processed and unprocessed nascent immunoglobulin light chains on membrane-bound ribosomes of murine myeloma. J. Cell Biol. 67, 835–851.

Blobel, G., Dobberstein, B., 1975b. Transfer of proteins across membranes. II. Reconstitution of functional rough microsomes from heterologous components. J. Cell Biol. 67, 852–862.

Blobel, G., Potter, V.R., 1967a. Studies on free and membrane- bound ribosomes in rat liver. I. Distribution as related to total cellular RNA. J. Mol. Biol. 26, 279–292.

Blobel, G., Potter, V.R., 1967b. Studies on free and membrane- bound ribosomes in rat liver. II. Interaction of ribosomes and membranes. J. Mol. Biol. 26, 293–301.

Blobel, G., Sabatini, D.D., 1970. Controlled proteolysis of nascent polypeptides in rat liver cell fractions. I. Location of the polypeptides within ribosomes. J. Cell Biol. 45, 130–145.

Blobel, G., Sabatini, D.D., 1971. Ribosome-membrane interaction in eukaryotic cells. In: Manson, L.A. (Ed.), Biomembranes, vol. 2. Plenum Press, New York, pp. 193–195.

Bloch, D.P., 1986. Stages in the development of an evolutionary process. In: Cairns-Smith, A.G., Harman, H. (Eds.), Clay Minerals and the Origin in Life. Cambridge University Press, London, pp. 153–158.

Bloch, D.P., 1988. Cybernetic origins of replication. Orig. Life 18, 87–96.

Bloch, D.P., Godman, G.C., 1955. A microphotometric study of the syntheses of desoxyribonucleic acid and nuclear histone. J. Biophys. Biochem. Cytol. 1, 17–28.

Bloch, D.P., McArthur, B., Mirrop, S., 1985. tRNA-rRNA sequence homologies: evidence for an ancient modular format shared by tRNAs and rRNAs. BioSystems 17, 209–225.

Bloch, D.P., McArthur, B., Widdowson, R.B., Spector, D., Guimaraes, R.C., Smith, J., 1983. tRNA-rRNA sequence homologies: evidence for a common evolutionary origin? J. Mol. Evol. 19, 420–428.

Bloch, D.P., Staves, M.P., 1986. Phylogenetic pedigrees marking events lying between the first Darwinian ancestors and the last common ancestor. Orig. Life 16, 309–310.

Block, S.M., 1992. Biophysical principles of sensory transduction. In: Corey, D.P., Roper, S.D. (Eds.), Sensory Transduction. The Rockefeller Press, New York, pp. 1–17.

Block, M.R., Glick, B.S., Wilcox, C.A., Wieland, F.T., Rothman, J.E., 1988. Purification of an N-ethylmaleimide-sensitive protein catalyzing vesicular transport. Proc. Natl. Acad. Sci. U.S.A. 85, 7852–7856.

Block, S.M., Goldstein, L.S.B., Schnapp, B.J., 1990. Bead movement by single kinesin molecules studied with optical tweezers. Nature 348, 348–352.

Blom, T.J.M., Sierra, M., van Vliet, T.B., Franke-van Dijk, M.E., De Koning, P., van Iren, F., Verpoorte, R., Libbenga, K.R., 1991. Uptake and accumulation of ajmalicine into isolated vacuoles of cultured cells of *Catharanthus roseus* (L.) G. Don. and its conversion into serpentine. Planta 183, 170–177.

Bloom, A.J., Burger, M., Rubio Asenio, J.S., Cousins, A.B., 2010. Carbon dioxide enrichment inhibits nitrate assimilation in wheat and *Arabidopsis*. Science 328, 899–903.

Bloom, K., Green, M., 1992. Nucleus and gene expression. Editorial overview. Curr. Opin. Cell Biol. 4, 377–378.

Blow, J.J., Nurse, P., 1990. A cdc2-like protein is involved in the initiation of DNA replication in *Xenopus* egg extracts. Cell 62, 855–862.

Blum, H.F., 1962. Time's Arrow and Evolution. Harper & Brothers, New York.

Blumenthal, S.S., Clark, G.B., Roux, S.J., 2004. Biochemical and immunological characterization of pea nuclear intermediate filament proteins. Planta 218, 965–975.

Blumwald, E., Cragoe, E.J., Poole, R.J., 1987. Inhibition of Na^+/H^+ antiport activity in sugar been tonoplast by analogs of amiloride. Plant Physiol. 85, 30–33.

Blumwald, E., Gelli, A., 1999. Secondary inorganic ion transport at the tonoplast. Adv. Bot. Res. 25, 401–417.

Blumwald, E., Poole, R.J., 1985. Na^+/H^+ antiport in isolated tonoplast vesicles from storage tissues of *Beta vulgaris*. L. Plant Physiol. 78, 163–167.

Blumwald, E., Poole, R.J., 1986. Kinetics of Ca^{2+}/H^+ antiport in isolated tonoplast vesicles form storage tissue of Beta vulgaris. L. Plant Physiol. 870, 727–731.

Boardman, N.K., Anderson, J.M., 1964. Studies on the greening of dark-grown bean: formation of chloroplasts from proplastids. Aust. J. Biol. Sci. 17, 86–92.

Bock, G., Haupt, W., 1961. Die Chloroplastendrehung bei *Mougeotia*. III. Die Frage der Lokalisierung des Hellrot- Dunkelrot-Pigmentsystems in der Zelle. Planta 57, 518–530.

Bode, J., Schlake, T., Rios-Ramírez, M., Mielke, C., Stengert, M., Kay, V., Klehr-Wirth, D., 1995. Scaffold/matrix-attached regions: structural properties creating transcriptionally active loci. Int. Rev. Cytol. 162A, 389–454.

Boevink, P., Cruz, S.S., Hawes, C., Harris, N., Oparka, K.J., 1996. Virus-mediated delivery of the green fluorescent protein to the endoplasmic reticulum of plant cells. Plant J. 10, 1179–1201.

Boevink, P., Martin, B., Oparka, K., Santa Cruz, S., Hawes, C.R., 1999. Transport of virally expressed green fluorescent protein through the secretory pathway in tobacco leaves is inhibited by cold shock and brefeldin A. Planta 208, 392–400.

Boevink, P., Oparka, K., Santa Cruz, S., Martin, B., Betteridge, A., Hawes, C.R., 1998. Stacks on tracks: the plant Golgi apparatus traffics on an actin/ER network. Plant J. 15, 441–447.

Bogorad, L., 1998. Discovery of chloroplasr DNA, genomes and genes. Discov. Plant Biol. 2, 15–43.

Böhler, C., Hill Jr., A.R., Orgel, L.E., 1996. Catalysis of the oligomerization of o-phospho-serine, aspartic acid, or glutamic acid by cationic micelles. Orig. Life 26, 1–5.

Bohm, D., 1961. Causality and Chance in Modern Physics. Harper Torchbooks, New York.

Bohn, M., Heinz, E., Lüthje, S., 2001. Lipid composition and fluidity of plasma membranes isolated fromscorn (*Zea mays* L.) roots. Arch. Biochem. Biophys. 387, 35–40.

Bohr, N., 1913. On the constitution of atoms and molecules. Phil. Mag. 26, 1–24, 476–502, 857–875.

Bohr, N., 1933. Light and life. Nature 131, 421–423, 457–459.

Bohr, N., 1950. On the notions of causality and complementarity. Science 111, 51–54.

Bohr, N., 1963. Essays 1958–1962 on Atomic Physics and Human Knowledge. Richard Clay and Co., Bungay, Suffolk.

Bolaños-Villegas, P., De, K., Pradillo, M., Liu, D., Makaroff, C.A., 2017. In favor of establishment: regulation of chromatid cohesion in plants. Front. Plant Sci. 8, 846.

Bold, H.C., 1942. The cultivation of algae. Bot. Rev. 8, 69–138.

Bold, H.C., Wynne, M.J., 1976. Introduction to the Algae. Structure and Reproduction. Prentice-Hall, Inc., Engelwood Cliffs, NJ.

Bold, H.C., Wynne, M.J., 1978. Introduction to the Algae. Structure and Reproduction. Prentice-Hall, Inc., Engelwood Cliffs, NJ.

Bolger, M.E., Arsova, B., Usadel, B., 2017b. Plant genome and transcriptome annotations: from misconceptions to simple solutions. Brief. Bioinform. 2017, 1–13.

Bolger, M., Schwacke, R., Gundlach, H., Schmutzer, T., Chen, J., Arend, D., Oppermann, M., Weise, S., Lange, M., Fiorani, F., Spannagl, M., Scholz, U., Mayer, K., Usadel, B., 2017a. From plant genomes to phenotypes. J. Biotechnol. 261, 46–52.

Bolger, A., Scossa, F., Bolger, M.E., Lanz, C., Maumus, F., Tohge, T., Quesneville, H., Alseekh, S., Sørensen, I., Lichtenstein, G., Fich, E.A., Conte, M., Keller, H., Schneeberger, K., Schwacke, R., Ofner, I., Vrebalov, J., Xu, Y., Osorio, S., Aflitos, S.A., Schijlen, E., Jiménez-Goméz, J.M., Ryngajllo, M., Kimura, S., Kumar, R., Koenig, D., Headland, L.R., Maloof, J.N., Sinha, N., van Ham, R.C., Lankhorst, R.K., Mao, L., Vogel, A., Arsova, B., Panstruga, R., Fei, Z., Rose, J.K., Zamir, D., Carrari, F., Giovannoni, J.J., Weigel, D., Usadel, B., Fernie, A.R., 2014. The genome of the stress-tolerant wild tomato species *Solanum pennellii*. Nat Genet. 46, 1034–1038.

Boller, T., Wiemken, A., 1986. Dynamics of vacuolar compartmentation. Ann. Rev. Plant Physiol. 37, 137–164.

Bollum, F.J., Potter, V.R., 1957. Thymidine incorporation into deoxyribonucleic acid of rat liver homogenates. J. Am. Chem. Soc. 79, 3603–3604.

Bolte, S., Talbot, C., Boutte, Y., Catrics, O., Read, N.D., Satiat-Jeunemaitre, B., 2004. FM-dyes as experimental probes for dissecting vesicle trafficking in living plant cells. J. Microsc. 214 (2), 159–173.

Bölter, B., Soll, J., 2016. Once upon a time—chloroplast protein import research from infancy to future challenges. Mol. Plant 9, 798–812.

Boltzmann, L., 1886. The second law of thermodynamics. In: McGuinness, B. (Ed.), Ludwig Boltzmann: Theoretical Physics and Philosophical Problems. D. Reidel Publishing Co., Dordrecht, Holland, pp. 13–32. Selected Writings.

Boltzmann, L., 1904. Reply to a lecture on happiness given by Professor Ostwald. In: McGuinness, M. (Ed.), Ludwig Boltzmann: Theoretical Physics and Philosophical Problems. D. Reidel Publishing Co., Dordrecht, Holland, pp. 173–184.

Boltzmann, L., 1964. Lectures on Gas Theory. Translated by S. G. Brush. University of California Press, Berkeley.

Bolwell, G.P., 1993. Dynamic aspects of the plant extracellular matrix. Int. Rev. Cytol. 146, 261–324.

Bonaventura, C., Myers, J., 1969. Fluorescence and oxygen evolution from *Chlorella pyrenoidosa*. Biochim. Biophys. Acta 189, 366–383.

Bonawitz, N.D., Kim, J.I., Tobimatsu, Y., Ciesielski, P.N., Anderson, N.A., Ximenes, E., Maeda, J., Ralph, J., Donohoe, B.S., Ladisch, M., Chapple, C., 2014. Distruption of Mediator rescues the stunted growth of a lignin-deficient *Arabidopsis* mutant. Nature 509, 376–380.

Bonder, E.M., Fishkind, D.J., Mooseker, M.S., 1983. Direct measurement of critical concentrations and assembly rate constants at the two ends of an actin filament. Cell 34, 491–501.

Bondi, H., 1960. The Universe at Large. Doubleday & Co., Garden City, NJ.

Bondi, H., 1964. Relativity and Common Sense. Doubleday & Co., Garden City, NJ.

Bondi, H., Gold, T., Hoyle, F., 1995. Origins of steady-state theory. Nature 373, 10.

Bonfanti, L., Martella Jr., A.A., Fusella, A., Baldassarre, M., Buccione, R., Geuze, H.J., Mironov, A.A., Luini, A., 1998. Procollagen traverses the Golgi stack without leaving the lumen of cisternae: evidence for cisternal maturation. Cell 95, 993–1003.

Bonner, J.T., 1962. The Ideas of Biology. Harper & Row, New York.

Bonner, J., Millerd, A., 1953. Oxidative phosphorylation by plant mitochondria. Arch. Biochem. Biophys. 42, 135–148.

Bonnet, E., Wuyts, J., Rouze, P., Van de Peer, Y., 2004. Detection of 91 potential conserved plant microRNAs in *Arabidopsis thaliana* and *Oryza sativa* identifies important target genes. Proc. Natl. Acad. Sci. U.S.A. 101, 11511–11516.

Bonnett Jr., H.T., 1968. The root endodermis: fine structure and function. J. Cell Biol. 37, 199–205.

Bonnett Jr., H.T., Newcomb, E.H., 1965. Polyribosomes and cisternal accumulations in root cells of radish. J. Cell Biol. 27, 423–432.

Bonneville, M.A., Voeller, B.R., 1963. A new cytoplasmic component of plant cells. J. Cell Biol. 18, 703–708.

Bonneville, M.A., Voeller, B.R., 1993. A new cytoplasmic component of plant cells. J. Cell Biol. 18, 703–708.

Bonsignore, C.A., Hepler, P.K., 1985. Caffeine inhibition of cytokinesis: dynamics of cell plate formation-deformation *in vivo*. Protoplasma 129, 28–35.

Boonburapong, B., Buaboocha, T., 2007. Genome-wide identification and analysis of the rice calmodulin and related potential calcium sensors. BMC Plant Biol. 7, 4.

Boorstin, D.J., 1948. The Lost World of Thomas Jefferson. H. Holt, New York.

Borel, E., 1913. Mécanique Statistique et Irréversibilité. J. Phys. 5th Ser. 3, 189–196.

Borevitz, J.O., Ecker, J.R., 2004. Plant genomics: the thord wave. Annu. Rev. Genom. Hum. Genet. 5, 443–477.

Borevitz, J.O., Xia, Y., Blount, J., Dixon, R.A., Lamb, C., 2000. Activation tagging identifies a conserved MYB regulator of phenylpropanoid biosynthesis. Plant Cell 12, 2383–2393.

Borisy, G.G., Taylor, E.W., 1967. The mechanism of action of colchicine. Binding of colchicine-^{3}H to cellular protein. J. Cell Biol. 34, 525–548. https://www.ibiology.org/ibiomagazine/issue-6/gary-borisy-edwin-taylor-the-discovery-of-tubulin.html.

Borner, G.H.H., Sherrier, D.J., Weimar, T., Michaelson, L.V., Hawkins, N.D., MacAskill, A., Napier, J.A., Beale, M.H., Lilley, K.S., Dupree, P., 2005. Analysis of detergent-resistant membranes in *Arabidopsis*. Evidence for plasma membrane lipid rafts. Plant Physiol. 137, 104–116.

Borsook, H., Deasy, C.L., Haagen-Smit, A.J., Keighley, G., Lowy, P.H., 1950. Metabolism of C14-labeled glycine, L-histidine, L-leucine, and L-lysine. J. Biol. Chem. 187, 839–848.

Borsook, H., Schott, H.F., 1931. The role of the enzyme in the succinate-enzyme-fumarate equilibrium. J. Biol. Chem. 92, 535–557.

Borstlap, C., Schuurmans, J.A.M., 2001. Proton-symport of L-valine in plasma membrane vesicles isolated from leaves of the wild-type and the Valr-2 mutant of *Nicotiana tabacum* L. Plant Cell Physiol. 41, 1210–1217.

Bortesi, L., Fischer, R., 2015. The CRISPR/Cas9 system for plant genome editing and beyond. Biotechnol. Adv. 33, 41–52.

Boruc, J., Griffis, A.H., Rodrigo-Peiris, T., Tilford, B., van Damme, D., Meier, I., 2015. GAP activity, but not subcellular targeting, is required for *Arabidopsis* RanGAP cellular and developmental functions. Plant Cell 27, 1985–1998.

Bosch, M., Cheung, A.Y., Hepler, P.K., 2005. Pectin methylesterase, a regulator of pollen tube growth. Plant Physiol. 138, 1334–1346.

Bosch, M., Hepler, P.K., 2005. Pectin methylesterases and pectin dynamics in pollen tubes. Plant Cell 17, 3219–3226.

Bosch, M., Hepler, P.K., 2006. Silencing of the tobacco pollen pectin methylesterase NtPPME1 results in retarded in vivo pollen tube growth. Planta 223, 736–745.

Bose, J.C., 1906. Plant Response as a Means of Physiological Investigation. Longmans, Green, and Co., London.

Bose, J.C., 1913. Researches on Irritability in Plants. Longmans. Green, and Co., London.

Bose, J.C., 1926. The Nervous Mechanism of Plants. Longmans. Green, and Co., London.

Bose, J.C., 1985. Life Movements in Plants. B. R. Publishing Co., New Delhi.

Boss, W.F., Massel, M.O., 1985. Polyphosphoinositides are present in plant tissue culture cells. Biochem. Biophys. Res. Commun. 132, 1018–1023.

Boss, W.F., Morré, D.J., Mollenhauer, H.H., 1984. Monenson-induced swelling of Golgi apparatus cisternae mediated by a proton gradient. Eur. J. Cell Biol. 34, 1–8.

Botha, C.E.J., Cross, R.H.M., Liu, L., 2005. Comparative structures of specialized monocotyledonous leaf blade plasmodesmata. In: Oparka, K.J. (Ed.), Plasmodesmata. Blackwell Publishing, Oxford, pp. 73–89.

Botstein, D., White, R.L., Skolnick, M., Davis, R.W., 1980. Construction of a genetic linkage map in man using restriction fragment length polymorphisms. Am. J. Hum. Genet. 32, 314–331.

Boublik, M., 1987. Structural aspects of ribosomes. Int. Rev. Cytol. (Suppl. 17), 357–389.

Bouche, N., Bouchez, D., 2001. *Arabidopsis* gene knockout, phenotypes wanted. Curr. Opin. Plant Biol. 4, 111–117.

Bouchè-Pillon, S., Fleurat-Lessard, P., Serrano, R., Bonnemain, J.L., 1994. Asymmetric distribution of plasma membrane H^+-ATPase in embryos of *Vicia faba* L. with special reference to transfer cells. Planta 193, 392–397.

Boudaoud, A., 2003. Growth of walled cells: from shells to vesicles. Phys. Rev. Lett. 91, 0181041–0181044.

Boulaflous, A., Faso, C., Brandizzi, F., 2008. Deciphering the Golgi apparatus: from imaging to gene. Traffic 9, 1613–1617.

Bourett, T.M., Czymmek, K.J., Howard, R.J., 1999. Ultrastructure of chloroplast protuberences in rice leaves preserved by high-pressure freezing. Planta 208, 472–479.

Bourett, T.M., Howard, R.J., 1996. Brefeldin A-induced structural changes in the endomembrane system of a filamentous fungus, *Magnaporthe grisea*. Protoplasma 190, 151–163.

Bourne, G.H. (Ed.), 1942, 1951, 1964. Cytology and Cell Physiology, first ed. Clarendon Press, Oxford, second ed. Clarendon Press, Oxford, third ed. Academic Press, New York.

Bourne, G.H., 1962. Division of Labor in Cells. Academic Press, New York.

Bourne, H.R., 2009. Ambition and Delight: A Life in Experimental Biology. Xlibris.

Boutry, M., Michelet, B., Goffeau, A., 1989. Molecular cloning of a family of plant genes encoding a protein homologous to plasma membrane H^+-translocating ATPase. Biochem. Biophys. Res. Commun. 162, 567–574.

Bouvier, F., Camera, B., 2007. The role of plastids in ripening fruits. In: Wise, R.R., Hoober, J.K. (Eds.), The Structure and Function of Plastids. Springer, Dordrecht, pp. 419–432.

Bovarnick, M., 1955. Rickettsiae. In: Scientific American (Ed.), The Physics and Chemistry of Life. Simon and Schuster, New York, pp. 143–149.

Bove, J., Vaillancourt, B., Kroeger, J., Hepler, P.K., Wiseman, P.W., Geitmann, A., 2008. Magnitude and direction of vesicle dynamics in growing pollen tubes using spatiotemporal image correlation spectroscopy and fluorescence recovery after photobleaching. Plant Physiol. 147, 1646–1658.

Boveri, T., 1888. Die Befruchtung und Teilung des Eles von Ascaris megalocephala. In: Zellen Studien H.2. G. Fisher, Jena, pp. 1–189.

Boveri, T., 1929. The Origin of Malignant Tumors. Williams & Wilkins, Baltimore.

Bowen, R.H., 1926. The Golgi-apparatus—Its structure and functional significance. Anat. Rec. 32, 151–193.

Bowen, R.H., 1927. Golgi apparatus and vacuome. Anat. Rec. 35, 309–335.

Bowen, R.H., 1928. Studies on the structure of plant protoplasm. I. The osmiophilic platelets. Zeitschrift für Zellforschung und Mikroskopische Anatomie 6, 689–725.

Bowen, R.H., 1929. The cytology of glandular secretion. Q. Rev. Biol. 4, 299–324, 484–519.

Bower, F.O., 1883. On plasmolysis and its bearning upon the reduction between cell wall and protoplasm. Q. J. Microsc. Sci. 23, 151–168.

Bower, F.O., 1925. Plants and Man. MacMillan, London.

Bowers, W.E., 1998. Christian de Duve and the discovery of lysosomes and peroxisomes. Trends Cell Biol. 8, 330–333.

Bowes, B.G., 1996. A Color Atlas of Plant Structure. Iowa State University Press, Ames, IA.

Bowes, G., Ogren, W.L., 1972. Oxygen inhibition and other properties of soybean ribulose 1,5-diphosphate carboxylase. J. Biol. Chem. 247, 2171–2176.

Bowes, G., Ogren, W.L., Hageman, R.H., 1971. Phosphoglycolate production catalyzed by ribulose diphosphate carboxylase. Biochem. Biophys. Res. Commun. 45, 716–722.

Bowler, C., Neuhaus, G., Yamagata, H., Chua, N.-H., 1994. Cyclic GMP and calcium mediate phytochrome phototransduction. Cell 77, 73–81.

Bowles, D.J., Marcus, S.E., Pappin, D.J.C., Findlay, J.B.C., Maycox, P.R., Burgess, J., 1986. Post-translational processing of concanavalin A precursors in jackbean cotyledons. J. Cell Biol. 102, 1284–1297.

Bowling, A.J., Brown Jr., R.M., 2008. The cytoplasmic domain of the cellulose-synthesizing complex in vascular plants. Protoplasma 233, 115–127.

Bowman, B.J., Uschida, W.J., Harris, T., Bowman, E.J., 1989. The vacuolar ATPase of *Neurospora crassa* contains an F1-like structure. J. Biol. Chem. 264, 15606–15612.

Boyd, C.N., Franceschi, V.R., Chuong, S.D.X., Akhani, H., Kiirats, O., Smith, M., Edwards, G.E., 2007. Flowers of *Bienertia cycloptera* and *Suaeda aralocaspica* (Chenopodiaceae) complete the life cycle performing single-cell C4 photosynthesis. Funct. Plant Biol. 34, 268–281. Special Issue in Memory of Vincent R. Franceschi.

Boyer, J.S., 1995. Measuring the Water Status of Plants and Soils. Academic Press, San Diego.

Boyer, J.S., 2009. Cell wall biosynthesis and the molecular mechanism of plant enlargement. Funct. Plant Biol. 36, 383–394.

Boyer, J.S., 2015a. Impact of cuticle on calculations of the CO_2 concentration inside leaves. Planta 242, 1405–1412.

Boyer, J.S., 2015b. Turgor and the transport of CO_2 and water across the cuticle (epidermis) of leaves. J. Exp. Bot. 66, 2625–2633.

Boyer, J.S., 2016. Enzyme-less growth in *Chara* and terrestrial plants. Front. Plant Sci. 7, 866.

Boyer, P.D., December 8, 1997. Energy, life, and ATP. Nobel Lecture.

Boyer, J.S., Wu, G., 1978. Auxin increases the hydraulic conductivity of auxin-sensitive hypocotyl tissue. Planta 139, 227–237.

Boyle, R., 1662. New Experiments Physico-Mechanical, Touching the Spring of the Air and its Effects. H. Hall, Oxford.

Boyle, R., 1664. Experiments and Considerations Touching Colours. Henry Herringman, London.

Boyle, R., 1665. Some Motives and Incentives to the Love of God: Pathetically Discours'd of in a Letter to a Friend, fourth ed. Henry Herringman, London.

Boyle, W.S., 1953. Principles and Practice in Plant Cytology. Burgess Publishing Co., Minneapolis.

Boysen, C., Simon, M.I., Hood, L., 1997. Fluorescence-based sequencing directly from bacterial and P1-derived artificial chromosomes. BioTechniques 23, 978–982.

Bozza, C.G., Pawlowski, W.P., 2008. The cytogenetics of homologous chromosome pairing in meiosis in plants. Cytogenet. Genome. Res. 120, 313–319.

Bozzone, D., Martin, D.A., 1998. Chemotaxis in the plasmodial slime mold, *Physarum polycephalum*: an experimental system for student exploration and investigation. Am. Biol. Teach. 60, 59–62.

Braatz, J., Harloff, H.-J., Masher, M., Stein, N., Himmelbach, A., Jung, C., 2017. CRISPR-Cas9 targeted mutagenesis leads to simultaneous modification of different homeologous gene copies in polyploid oilseed rape (*Brassica napus*). Plant Physiol. 174, 935–942.

Brachet, J., 1940. La détection histochimique desacides pentosenucléiques. Compt. Rend. Soc. Biol. 133, 88–90.

Brachet, J., 1957. Biochemical Cytology. Academic Press Inc., New York.

Brachet, J., 1959. New observations on biochemical interactions between nucleus and cytoplasm in *Amoeba* and *Acetabularia*. Exp. Cell Res. (Suppl. 6), 78–96.

Brachet, J., 1960. Ribonucleic acids and the synthesis of cellular proteins. Nature 186, 194–199.

Brachet, J., 1975. From chemical to molecular sea urchin embryology. Am. Zool. 15, 485–491.

Brachet, J., 1985. Molecular Cytology. Academic Press, Orlando.

Brachet, J., 1989. Recollections on the origin of molecular biology. Biochim. Biophys. Acta 1000, 1–5.

Brachet, J., Jeener, R., 1944. Recherches sur des particules cytoplasmiques de dimensions macromoléculaires riches en acide pentosenucléique. I. Propriétes generals, relations avec les hydrolases. Les hormones, les proteins de structure. Enzymologia 11, 196–212.

Bradbeer, J.W., 1988. Seed Dormancy and Germination. Chapman & Hall, New York.

Bradbury, S., 1967. The Evolution of the Microscope. Pergamon Press, Oxford.

Bradbury, S., Meek, G.A., 1960. A study of potassium permanganate 'fixation' for electron microscopy. Q. J. Microsc. Sci. 101, 241–250.

Brady, R.A., 1937. The Spirit and Structure of German Fascism, 1971. Citadel Press, New York.

Brady, S.T., 1985. A novel brain ATPase with properties expected for fast axonal transport motor. Nature 317, 73–75.

Brady, J.D., Fry, S.C., 1997. Formation of di-isodityrosine and loss of isoditryrosine in the cell walls of tomato cell-suspension cultures treated with fungal elicitors or H_2O_2. Plant Physiol. 115, 87–92.

Brady, S.T., Lasek, R.J., Allen, R.D., 1985. Video microscopy of fast axonal transport in extruded axoplasm: a new model for study of molecular mechanisms. Cell Motil. 5, 81–101.

Brady, S.M., Orlando, D.A., Lee, J.-Y., Wang, J.Y., Koch, J., Dinneny, J.R., Mace, D., Ohler, U., Benfey, P.N., 2007. A high-resolution root spatiotemporal map reveals dominant expression patterns. Science 318, 801–806.

Braell, W.A., Balch, W.E., Dobbertin, D.C., Rothman, J.E., 1984. The glycoprotein that is transported between successive compartments of the Golgi in a cell-free system resides in stacks of cisternae. Cell 39, 511–524.

Bragg, W.H., Bragg, W.L., 1915. X Rays and Crystal Structure. G. Bell and Sons, London.

Bragg, W.H., Bragg, W.L., 1924. X Rays and Crystal Structure, fourth ed. G. Bell and Sons, London.

Bragg, L., September 6, 1922. The diffraction of X-rays by crystals. Nobel Lecture.

Brand, E., Edsall, J.T., 1947. The chemistry of the proteins and amino acids. Ann. Rev. Biochem. 16, 223–272.

Brandizzi, F., 2017. Transport from the endoplasmic reticulum to the Golgi in plants: where are we now? Semin. Cell Dev. Biol. 80, 94–105.

Brandizzi, F., 2018. Transport from the endoplasmic reticulum to the Golgi in plants: where are we now? Semin. Cell Dev. Biol. (in press).

Brandizzi, F., Frangne, N., Marc-Martin, S., Hawes, C., Neuhaus, J.-M., Paris, N., 2002c. The destination for single-pass membrane proteins is influenced markedly by the length of the hydrophobic domain. Plant Cell 14, 1–17.

Brandizzi, F., Fricker, M., Hawes, C., 2002a. A greener world: the revolution in plant bioimaging. Nat. Rev. Mol. Cell Biol. 3, 520–530.

Brandizzi, F., Snapp, E.L., Roberts, A.G., Lippincott-Schwartz, J., Hawes, C., 2002b. Membrane protein transport between the endoplasmic reticulum and the Golgi in tobacco leaves is energy dependent but cytoskeleton independent: evidence from selective photobleaching. Plant Cell 14, 1293–1309.

Brandt, S.P., 2005. Microgenomics: gene expression analysis at the tissue-specific and single-cell levels. J. Exp. Bot. 56, 495–505.

Brandt, B., Brodsky, D.E., Xue, S., Negi, J., Iba, K., Kangasjärvi, J., Ghassemian, M., Stephan, A.B., Hu, H., Schroeder, J.I., 2012. Reconstitution of abscisic acid activation of SLAC1 anion channel by CPK6 and OST1 kinases and branched ABI1 PP2C phosphatase action. Proc. Natl. Acad. Sci. U.S.A. 109, 10593–10598.

Brandt, S.P., Kehr, J., Walz, C., Imlau, A., Willmitzer, L., Fisahn, J., 1999. A rapid method for detection of plant gene transcripts from epidermal, mesophyll and companion cells of intact leaves. Plant J. 20, 245–250.

Brandt, S., Kloska, S., Altmann, T., Kehr, J., 2002. Using array hybridization to monitor gene expression at the single cell level. J. Exp. Bot. 53, 2315–2323.

Branham, S.E., Wright, S.J., Reba, A., Morrison, G.D., Linder, C.R., 2016a. Genome-wide association study in *Arabidopsis thaliana* of natural variation in seed oil melting point: a widespread adaptive trait in plants. J. Hered. 107, 257–265.

Branham, S.E., Wright, S.J., Reba, A., Morrison, G.D., Linder, C.R., 2016b. Genome-wide association study of *Arabidopsis thaliana* identifies determinants of natural variation in seed oil composition. J. Hered. 107, 248–256.

Branton, D., 1966. Fracture faces of frozen membranes. Proc. Natl. Acad. Sci. U.S.A. 55, 1048–1056.

Branton, D., Deamer, D.W., Marziali, A., Bayley, H., Benner, S.A., et al., 2008. The potential and challenges of nanopore sequencing. Nat. Biotechnol. 26, 1146–1153.

Braun, P., 2012. Interactome mapping for analysis of complex phenotypes: insights from benchmarking binary interaction assays. Proteomics 12, 1499–1518.

Braun, P., Aubourg, S., Van Leene, J., De Jaeger, G., Lurin, C., 2013. Plant protein interactomes. Annu. Rev. Plant Biol. 64, 161–187.

Braun, H.-P., Schmitz, U.K., 2006. Strategies for the investigation of protein–protein interactions in plants. Annu. Plant Rev. 28, 55–70.

Braun, M., Sievers, A., 1993. Centrifugation causes adaptation of microfilaments. Studies on the transport of statoliths in gravity sensing *Chara* rhizoids. Protoplasma 174, 50–61.

Braun, M., Sievers, A., 1994. Role of the microtubule cytoskeleton in gravisensing Chara rhizoids. Eur. J. Cell Biol. 63, 289–298.

Bräutigam, A., Hoffmann-Benning, S., Weber, A.P.M., 2008. Comparative proteomics of chloroplast envelopes from C3 and C4 plants reveals specific adaptations of the plastid envelope to C4 photosynthesis and candidate proteins required for maintaining C4 metabolite fluxes. Plant Physiol. 148, 568–579.

Brawley, S.H., Robinson, K.R., 1985. Cytochalasin treatment disrupts the endogenous currents associated with cell polarization in fucoid zygotes: studies on the role of F-actin in embryogenesis. J. Cell Biol. 100, 1173–1184.

Bray, D., 1992. Cell Movements. Garland Publishing Inc., New York.

Braybrook, S., 2015. Measuring the elasticity of plant cells with atomic force microscopy. Methods Cell Biol. 125, 237–254.

Braybrook, S.A., Peaucelle, A., 2013. Mechano-chemical aspects of organ formation in *Arabidopsis thaliana*: the relationship between auxin and pectin. PLoS One 8, e57813.

Brecknock, S., Dibbayawan, T.P., Vesk, M., Vesk, P.A., Faulkner, C., Barton, D.A., Overall, R.L., 2011. High resolution scanning electron microscopy of plasmodesmata. Planta 234, 749–758.

Breidenbach, R.W., Beevers, H., 1967. Association of the glyoxylate cycle enzymes in a novel subcellular particle from castor bean endosperm. Biochem. Biophys. Res. Commun. 27, 462–469.

Brembu, T., Winge, P., Bones, A.M., Yang, Z., 2006. A RHOse by any other name: a comparative analysis of animal and plant Rho GTPases. Cell Res. 16, 435–445.

Brenchley, R., Spannagl, M., Pfeifer, M., Barker, G.L., D'Amore, R., Allen, A.M., McKenzie, N., Kramer, M., Kerhornou, A., Bolser, D., et al., 2012. Analysis of the bread wheat genome using whole-genome shotgun sequencing. Nature 491, 705–710.

Brennecke, P., Anders, S., Kim, J.K., Kołodziejczyk, A.A., Zhang, X., Proserpio, V., Baying, B., Benes, V., Teichmann, S.A., Marioni, J.C., Heisler, M.G., 2013. Accounting for technical noise ib single-cell RNA-seq experiments. Nat. Methods 10, 1093–1095.

Brenner, S., 1957. On the impossibility of all overlapping triplet codes in information transfer from nucleic acid to proteins. Proc. Natl. Acad. Sci. U.S.A. 43, 687–694.

Brenner, S., 1999. Theoretical biology in the third millenium. Philos. Trans. R. Soc. 354B, 1963–1965.

Brenner, S., 2014. Retrospective. Frederick Sanger (1918–2013). Science 343, 262.

Brenner, D.M., Carroll, G.C., 1968. Fine-structural correlate of growth in hyphae of *Ascodesmis sphaerospora*. J. Bacteriol. 95, 658–671.

Brenner, S., Jacob, F., Meselson, M., 1961. An unstable intermediate carrying information from genes to ribosomes fro protein synthesis. Nature 190, 576–581.

Brenner, S., October 16, 2003. From genes to organisms. In: Genes and Genomes: Impacts on Medicine and Society. Genes, Genomes, and Evolution. Conference at Columbia University.

Brennicke, A., Kück, U. (Eds.), 1993. Plant Mitochondria. VCH, Weinheim.

Bresler, V., Montgomery, W.L., Fishelson, L., Pollak, P.E., 1998. Gigantism in a bacterium, *Epulopiscium fishelsoni*, correlates with complex patterns in arrangement, quantity, and segregation of DNA. J. Bacteriol. 180, 5601–5611.

Bret-Harte, M.S., 1993. Total epidermal cell walls of pea stems respond differently to auxin than does the outer epidermal wall alone. Planta 190, 368–379.

Bret-Harte, M.S., Talbott, L.D., 1993. Changes in composition of the outer epidermal wall of pea stems during auxin-induced growth. Planta 190, 369–378.

Bretscher, M.S., 1971a. A major protein which spans the human erythrocyte membrane. J. Mol. Biol. 59, 351–357.

Bretscher, M.S., 1971b. Major human erythrocyte glycoprotein spans the cell membrane. Nat. New Biol. 231, 229–232.

Bretscher, M.S., 1973. Membrane structure: some general principles. Science 181, 622–629.

Bretscher, M.S., Munro, S., 1993. Cholesterol and the Golgi apparatus. Science 261, 1280–1281.

Bretscher, M.S., Raff, M.C., 1975. Mammalian plasma membranes. Nature 258, 43–49.

Breusch, F.L., 1939. The fate of oxaloacetic acid in different organs. Biochem. J. 33, 1757–1770.

Breviario, D., 2008. Plant tubulin genes: regulatory and evolutionary aspects. Plant Cell Monogr. 11, 207–232.

Brewster, D., 1831. The Life of Sir Isaac Newton. J. & J. Harper, New York.

Brewster, D., 1854. More Worlds Than One. Robert Carter & Bros, New York.

Breyne, P., van Montagu, M., Depicker, A., Gheysen, G., 1992. Characterization of plant scaffold attachment region in a DNA fragment that normalizes transgene expression in tobacco. Plant Cell 4, 463–471.

Brian, P.W., Hemming, H.G., 1955. The effect of gibberellic acid on shoot growth of pea seedlings. Physiol. Plant 8, 669–681.

Brice, A.T., Jones, R.P., Smyth, J.D., 1946. Golgi apparatus by phase contrast microscopy. Nature 157, 553–554.

Briggs, W.R., 1998. Discovery of phytochrome. Discov. Plant Biol. 2, 115–135.

Briggs, W.R., 2005. Phototropin overview. In: Wada, M., Shimazaki, K., Iino, M. (Eds.), Light Sensing in Plants. Springer, Tokyo, pp. 139–146.

Briggs, W.R., Beck, C.F., Cashmore, A.R., Christie, J.M., Hughes, J., Jarillo, J.A., Kagawa, T., Kanegae, H., Liscum, E., Nagatani, A., Okada, K., Salomon, M., Rüdiger, W., Sakai, T., Takano, M., Wada, M., Watson, J.C., 2001a. The phototropin family of photoreceptors. Plant Cell 13, 993–997.

Briggs, W.R., Christie, J.M., 2002. Phototropin 1 and phototropin 2: two versatile plant blue-light receptors. Trends Plant Sci. 7, 204–210.

Briggs, W.R., Christie, J.M., Salomon, M., 2001b. Phototropins: a new family of flavin-binding blue light receptors in plants. Antiox. Redox Sig. 3, 775–788.

Briggs, G.E., Haldane, J.B.S., 1925. A note on the kinetics of enzyme action. Biochem. J. 19, 338–339.

Briggs, G.E., Petrie, A.H.K., 1931. Respiration as a factor in the ionic equilibria between plant tissues and external solutions. Proc. R. Soc. 108B, 317–326.

Briggs, G.E., Robertson, R.N., 1957. Apparent free space. Annu. Rev. Plant Physiol. 8, 11–30.

Brinkley, B.R., 1990. Toward a structural and molecular definition of a kinetochore. Cell Motil. Cytoskelet. 16, 104–106.

Brinkley, B.R., Valdivia, M.M., Tousson, A., Balczon, R.D., 1989. The kinetochore: structure and molecular organization. In: Hyams, J.S., Brinkley, B.R. (Eds.), Mitosis: Molecules and Mechanisms. Academic Press, London, pp. 77–118.

Briskin, D.P., Basu, S., Assmann, S.M., 1995. Characterization of the red beet plasma membrane H+-ATPase reconstituted in a planar bilayer system. Plant Physiol. 108, 393–398.

Briskin, D.P., Leonard, R.T., Hodges, T.K., 1987. Isolation of plasma membrane: membrane markers and general principles. Methods Enzymol. 48, 548–568.

Brisseau-Mirbel, C.F., 1808. Exposition et Défence de Ma Théorie de L'organisation Végétale. Chez Les Frères van Cleef, La Haye.

Britten, R.J., Kohne, D.E., 1968. Repeated sequences in DNA. Science 161, 529–540.

Britten, C.J., Zhen, R.-G., Kim, E.J., Rea, P.A., 1992. Reconsitution of transport function of vacuolar H^+-translocating inorganic pyrophosphatase. J. Biol. Chem. 267, 21850–21855.

Brock, W.H., 1985. From Protyle to Proton, William Prout and the Nature of Matter. 1785–1985. Adam Hilger, Bristol.

Broda, E., 1975. The Evolution of Bioenergetic Processes. Pergamon Press, Oxford.

Broda, E., 1983. Ludwig Boltzmann. Man—Physicist—Philosopher. Ox Bow Press, Woodbridge, CT.

Broda, H., Schweiger, H.G., 1981. Long-term measurement of endeogenous diurnal oscillations of electrical potential in an individual *Acetabularia* cell. Eur. J. Cell Biol. 26, 1–4.

Broeckling, C.D., Huhman, D.V., Faraq, M.A., Smith, J.T., May, G.D., Mendes, P., Dixon, R.A., Sumner, L.W., 2005. Metabolic profiling of *Medicago truncatula* cell cultures reveals the effects of biotic and abiotic elicitors on metabolism. J. Exp. Bot. 56, 323–336.

Brokaw, C.J., 1972. Flagellar movement: a sliding filament model. Science 178, 455–462.

Brooks, C. Mc.C., Cranefield, P.F. (Eds.), 1959. The Historical Development of Physiological Thought. Hafner Publishing Co., New York.

Brooks III, L., Strable, J., Zhang, X., Ohtsu, K., Zhou, R., Sarkar, A., et al., 2009. Microdissection of shoot meristem functional domains. PLoS Genet. 5, e1000476.

Brosnan, J.M., Sanders, D., 1990. Inositol triphosphate-medicated Ca^{2+} release in beet microsomes is inhibited by heparin. FEBS Lett. 260, 70–72.

Brower, D.L., Hepler, P.K., 1976. Microtubules and secondary wall deposition in xylem: the effects of isopropyl *N*-phenylcarbamate. Protoplasma 87, 91–111.

Brower-Toland, B.D., Smith, C.L., Yeh, R.C., Lis, J.T., Peterson, C.L., Wang, M.D., 2002. Mechanical disruption of individual nucleosomes reveals a reversible multistage release of DNA. Proc. Natl. Acad. Sci. U.S.A. 99, 1960–1965.

Brown, R., 1829. Additional Remarks on Active Molecules, pp. 479–486. Richard Taylor, London 17–20. Republished by the Ray Society 1866.

Brown, R., 1831. On the organs and mode of fecundation in Orchideae and Asclediadeae. New Series 4, 161–179. Republished by the Philosophical Magazine and Annals of Philosophy.

Brown, R., 1831. On the organs and mode of fecundation in Orchideae and Asclediadeae. Trans. Linn. Soc. Lond. 16, 685–745.

Brown Jr., R.M., 1969. Observations on the relationship of the Golgi apparatus to wall formation in the marine chrysophycean alga, Pleurochrysis scherffelii Pringsheim. J. Cell Biol. 41, 109–123.

Brown Jr., R.M., 1985. Cellulose microfibril assembly and orientation: recent developments. J. Cell Sci. (Suppl. 2), 13–32.

Brown Jr., R.M., 2004. Cellulose structure and biosynthesis: what is in store for the 21st century? J. Polym. Sci. A1 42, 487–495.

Brown, J.R., Doolittle, W.F., 1995. Root of the universal tree of life based on ancient aminoacyl-tRNA synthetase gene duplications. Proc. Natl. Acad. Sci. U.S.A. 92, 2441–2445.

Brown, D.J., DuPont, F.M., 1989. Lipid composition of plasma membranes and endomembranes prepared from roots of barley (*Hordeum vulgare* L.). Plant Physiol. 90, 955–961.

Brown Jr., R.M., Franke, W.W., Kleinig, H., Falk, H., Sitte, P., 1970. Scale formation in chrysophycean algae. I. Cellulosic and noncellulosic wall components made by the Golgi apparatus. J. Cell Biol. 45, 246–271.

Brown, W.J., Goodhouse, J., Farquhar, M.G., 1986. Mannose 6- phosphate receptors for lysosomal enzymes cycle between the Golgi complex and endosomes. J. Cell Biol. 103, 1235–1247.

Brown, E.H., Iqbal, M.A., Stuart, S., Hatton, K.S., Valinsky, J., Schildkraut, C.L., 1987. Rate of replication of the murine immunoglobin heavy-chain locus: evidence that the region is part of a single replicon. Mol. Cell. Biol. 7, 450–457.

Brown Jr., R.M., Montezinos, D., 1976. Cellulose microfibrils: visualization of biosynthetic and orienting complexes in association with the plasma membrane. Proc. Natl. Acad. Sci. U.S.A. 73, 143–147.

Brown, J.W.S., Shaw, P.J., 1998. Small nucleolar RNAs and pre- RNA processing in plants. Plant Cell 10, 649–657.

Brown Jr., R.M., Willison, J.H.M., Richardson, C.L., 1976. Cellulose biosynthesis in Acetobacter xylinum: visualization of the site of synthesis and direct measurement of the in vivo process. Proc. Natl. Acad. Sci. U.S.A. 73, 4565–4569.

Brown, T., Zimmermann, C., Panneton, W., Noah, N., Borevitz, J., 2012. High-resolution, time-lapse imaging for ecosystem-scale phenotyping in the field. In: Normanly, J. (Ed.), High-Throughput Phenotyping in Plants. Methods and Protocols. Springer, New York, pp. 71–96.

Browne, C.A., 1944. Thomas Jefferson and the scientific trends of his time. Chronica. Bot. 8, 361–424.

Brownlee, G.G., Sanger, F., 1967. Nucleotide sequences from the low molecular weight ribosomal RNA of *Escherichia coli*. J. Mol. Biol. 23, 337–353.

Brownlee, G.G., Sanger, F., 1969. Chromatography of ^{32}P-labelled oligonucleotides on thin layers of DEAE-cellulose. Eur. J. Biochem. 239, 251–258.

Brownstein, M., 2006. Sample labeling: an overview. Methods Enzymol. 410, 222–237.

Brozynska, M., Copetti, D., Furtado, A., Wing, R.A., Crayn, D., Fox, G., Ishikawa, R., Henry, R.J., 2017. Sequencing of Australian wild rice genomes reveals ancestral relationships with domesticated rice. Plant Biotechnol. 15, 765–774.

Brücke, E. von., 1898a. Ueber die Bewegungen der Mimosa pudica, 1848. In: Pflanzenphysiologische Abhangdlungen. Ostwald's Klassiker der Exakten Wissenschaften. Nr. 95. Wilhelm Engelmann, Leipzig, pp. 34–53.

Brücke, E. von, 1898b. Die elementarorganismen, 1861. In: Pflanzenphysiologische Abhangdlungen. Ostwald's Klassiker der Exakten Wissenschaften. Nr. 95. Wilhelm Engelmann, Leipzig, pp. 54–79.

Brüke, E., 1898. Pflanzenphysiologische Abhangdlungen Astwald's Klassiker der Exakten Wissenschaften. Wilhelm Engelmann, Leipzig. Nr. 95.

Brumfeld, V., Charuvi, D., Nevo, R., Chuartzman, S., Tsabari, O., Ohad, I., Shimoni, E., Reich, Z., 2008. A note on three-dimensional models of higher-plant thylakoid networks. Plant Cell 20, 2546–2549.

Brummell, D.A., Camirand, A., MacLachlan, G.A., 1990. Differential distribution of xyloglucan glycosyl transferases in pea Golgi dictyosomes and secretory vesicles. J. Cell Sci. 96, 705–710.

Brummond, D.O., Burris, R.H., 1953. Transfer of C14 by lupine mitochondria through reactions of the tricarboxylic acid cycle. Proc. Natl. Acad. Sci. U.S.A. 39, 754–759.

Brune, A., Gonzalez, P., Goren, R., Zehavi, U., Echeverria, E., 1998. Citrate uptake into tonoplast vesicles from acid lime (*Citrus aurantifolia*) juice cells. J. Membr. Biol. 166, 197–203.

Brune, A., Müller, M., Taiz, L., Gonzalez, P., Etxeberria, E., 2002. Vacuolar acidification in citrus fruit: comparison between acid lime (*Citrus aurantifolia*) and sweet lime (*Citrus limmentiodes*) juice cells. J. Am. Hort. Sci. 127, 171–177.

Brunkard, J.O., Runkel, A.M., Zambryski, P.C., 2013. Plasmodesmata dynamics are coordingated by intracellular signaling pathways. Curr. Opin. Plant Biol. 16, 614–620.

Brunkard, J.O., Runkel, A.M., Zambryski, P.C., 2015. The cytosol must flow: intercellular transport through plasmodesmata. Cur. Opin. Cell Biol. 35, 13–20.

Brunkard, J.O., Zambryski, P.C., 2017. Plasmodesmata enable multicellularity: new insights into their evolution, biogenesis, and functions in development and immunity. Curr. Opin. Plant Biol. 35, 76–83.

Brush, S.G., 1983. Statistical Physics and the Atomic Theory of Matter from Boyle and Newton to Landau and Onsager. Princeton University Press, Princeton, NJ.

Bryant, J.A., Moore, K., Aves, S.J., 2001. Origins and complexes: the initiation of DNA replication. J. Exp. Bot. 52, 193–202.

Bubunenko, M.G., Schmidt, J., Subramanian, A.R., 1994. Protein substitution in chloroplast ribosome evolution. A eukaryotic cytosolic protein has replaced its organelle homologue (L23) in spinach. J. Mol. Biol. 240, 28–41.

Buch-Pedersen, M.J., Palmgren, M.G., 2003. Mechanism of proton transport by plant plasma membrane proton ATPases. J. Plant Res. 116, 507–515.

Buchanan, J.M., 2002. Biochemistry during the life and times of Hans Krebs and Fritz Lipmann. J. Biol. Chem. 277, 33531–33536.

Bucherer, A.H., 1904. Mathematische Einführung in die Elektronentheorie. Teubner, Leipzig.

Bucherer, A.H., 1908. Messungen an Becquerelstrahlen. Die experimentelle Bestätigung der Lorentz-Einsteinschen Theorie. Phys. Z. 9 (22), 755–762.

Bucherer, A.H., 1909. Die experimentelle Bestätigung des Relativitätsprinzips. Ann. Phys. 333 (3), 513–536.

Buchner, E., 1897. Alkoholische Gärung ohne Hefezellen. Ber. Dtsch. Chem. Ges. 30, 117–124.

Buchner, E., December 11, 1907. Cell-free fermentation. Nobel Lecture.

Buchnik, L., Abu-Abied, M., Sadot, E., 2015. Role of plant myosins in motile organelles: is a direct interaction required? J. Integr. Plant Biol. 57, 23–30.

Buckhout, T.J., 1984. Characterization of Ca^{2+} transport in endoplasmic reticulum membrane vesicles from *Lepidium sativum* L. roots. Plant Physiol. 76, 962–967.

Buckle, H.T., 1872. The influence of women on the progress of knowledge. A Discourse delivered at the Royal Institution, on Friday, the 19th of March, 1858. In: Taylor, H. (Ed.), Miscellaneous and Posthumous Works of Henry Thomas Buckle. Volume I. Longmans, Green, and Co., London, pp. 1–19.

Buer, C.S., Weathers, P.J., Swartzlander Jr., G.A., 2000. Changes in Hechtian strands in cold-hardened cells measured by optical microsurgery. Plant Physiol. 122, 1365–1378.

Buffen, M., Hanke, D.E., 1990. Polyphosphoinositidase activity in soybean membranes is Ca^{2+}-dependent and shows no requirement for cyclic-nucleotides. Plant Sci. 69, 147–155.

Bukrinsky, J.T., Buch-Pedersen, M.J., Larsen, S., Palmgren, M.G., 2001. A putative proton binding site of plasma membrane H^+-ATPase identified through homology modeling. FEBS Lett. 494, 6–10.

Bull, H.B., 1952. The chemistry of amino acids and proteins. Ann. Rev. Biochem. 21, 179–208.

Bull, H.B., 1964. An Introduction to Physical Biochemistry. F. A. Davis Co., Philadelphia.

Bullock, T.H., 2003. Have brain dynamics evolved? Should we look for unique dynamics in sapient species? Neural Comput. 15, 2013–2027.

Bunnik, E.M., Le Roch, K.G., 2013. An introduction to functional genomics and systems biology. Adv. Wound Care 2, 490–498.

Bünning, E., 1953. Entwicklungs—Und Bewegungsphysiologggie der Pflanzen. Springer, Berlin.

Bünning, E., 1988. Wilhelm Pfeffer (1845–1920). Botanica Acta 101, 91–92.

Bünning, E., 1989. Ahead of His Time: Wilhelm Pfeffer. Early Advances in Plant Biology. Carleton University Press, Ottawa, Canada. Translated by H. W. Pfeffer.

Bünning, E., Tazawa, M., 1957. Uber die negativ-phototaktische Reaktion von Euglena. Arch. Mikrobiol. 27, 306–310.

Burbulis, I.A., Winkel-Shirley, B., 1999. Interactions among enzymes of the Arabidopsis flavanoid biosynthetic pathway. Proc. Natl. Acad. Sci. U.S.A. 96, 12929–12934.

Burch-Smith, T.M., Brunkard, J.O., Choi, Y.G., Zambryski, P.C., 2011. Organelle-nucleus cross-talk regulates plant intercellular communication via plasmodesmata. Proc. Natl. Acad. Sci. U.S.A. 108, E1451–E1460.

Burch-Smith, T.M., Zambryski, P.C., 2010. Loss of increased size exclusion limit (ise)1 or ise2 increases the formation of secondary plasmodesmata. Curr. Biol. 20, 989–993.

Bures, J., Petran, M., Zachar, J., 1967. Electrophysiological Methods in Biological Research. Academia Publishing House of the Czechoslovak Academy of Sciences, Prague.

Burgert, I., Fratzl, P., 2006. Mechanics of the expanding cell wall. Plant Cell Monogr. 5, 191–215.

Burgess, J., 1985. An Introduction to Plant Cell Development. Cambridge University Press, London.

Burgos-Rivera, B., Dawe, R.K., 2012. An *Arabidopsis* tissue-specific RNAi method for studying genes essential to mitosis. PLoS One 7, e51388.

Burke, B., 1990. The nuclear envelope and nuclear transport. Curr. Opin. Cell Biol. 2, 514–520.

Burnette, W.N., 1981. "Western Blotting": electrophoretic transfer of proteins from sodium dodecyl sulfate-polyacrylamide gels to unmodified nitrocellulose and radiographic detection with antibody and radioiodinated protein A. Anal. Biochem. 112, 195–203.

Burnette, W.N., 1991. Citation classic: Western Blotting. http://bit.ly/Nzqyml.

Burnham, D., no date. Gottfried Leibniz: Metaphysics. Internet Encyclopedia of Philosophy. http://www.iep.utm.edu/leib-met/.

Burny, A., 1988. Jean Brachet (1909-1988). Nature 335, 768.

Burridge, K., Fath, K., Kelly, T., Nuckolls, G., Turner, C., 1988. Focal adhesions: transmembrane junctions between the extracellular matrix and the cytoskeleton. Annu. Rev. Cell Biol. 4, 487–525.

Burris, R.H., Black, C.C., 1976. CO_2 Metabolism and Plant Productivity. University Park Press, Baltimore.

Burroughs, J., 1908. Leaf and Tendril. Houghton Miflin, Boston.

Burt, W.J.E., Leaver, C.J., 1994. Identification of a chaperonin-10 homologue in plant mitochondria. FEBS Lett. 339, 139–141.

Burton, K., 1958. Energy of adenosine triphosphate. Nature 181, 1594–1595.

Burton, M., Moore, J., 1974. The mitochondrion of the flagellate, *Polytomella agilis*. J. Ultrastruct. Res. 48, 414–419.

Burzik, C.M., 2006. Twenty-Five Years of Advancing Science. http://home.appliedbiosystems.com/about/presskit/pdfs/25_years_advancing_science.pdf.

Busch, M.B., Sievers, A., 1993. Membrane traffic from the endoplasmic reticulum to the Golgi apparatus is disturbed by an inhibitor of the Ca^{2+}-ATPase in the ER. Protoplasma 177, 23–31.

Bush, V., 1967. Science is Not Enough. William Morrow & Co., New York.

Bush, D., 1993a. Inhibitors of proton-sucrose symport. Arch. Biochem. Biophys. 307, 355–360.

Bush, D.S., 1993b. Regulation of cytosolic calcium in plants. Plant Physiol. 103, 7–13.

Bush, D.S., Biswas, A.K., Jones, R.L., 1989a. Gibberellic-acid-stimulated Ca^{2+} accumulation in endoplasmic reticulum of barley aleurone: Ca^{2+} transport and steady-state levels. Planta 178, 411–420.

Bush, D.S., Biswas, A.K., Jones, R.L., 1993. Hormonal regulation of Ca^{2+} transport in the endomembrane system of the barley aleurone. Planta 189, 507–515.

Bush, D.S., Jones, R.L., 1990. Measuring intracellular Ca^{2+} levels in plant cells using the fluorescent probes, Indo-1 and Fura-2. Progress and prospects. Plant Physiol. 93, 841–845.

Bush, D.S., Sticher, L., Huystee, R.V., Wagner, D., Jones, R.L., 1989b. The calcium requirement for stability and enzymatic activity of two isoforms of barley aleurone α-amylase. J. Biol. Chem. 264, 19392–19398.

Bush, D.S., Sze, H., 1986. Calcium transport in tonoplast and endoplasmic reticulum vesicles isolated from cultured carrot cells. Plant Physiol. 80, 549–555.

Bushnell, D.A., Cramer, P., Kornberg, R.D., 2002. Structural basis of transcription: alpha-amanitin-RNA polymerase II cocrystal at 2.8 A resolution. Proc. Natl. Acad. Sci. U.S.A. 99, 1218–1222.

Bustamante, C., Chemla, Y.R., Forde, N.R., Izhaky, D., 2004. Mechanical processes in biochemistry. Annu. Rev. Biochem. 73, 705–748.

Butcher, S.H., 1891. Some Aspects of the Greek Genius. MacMillan, London.

Butler, J.A.V., 1952. The nucleic acid of the chromosomes. Endeavour 11, 154–158.

Bütschli, O., 1892. Untersuchungen über mikroskopische Schäume und das Protoplasma. W. Engelmann, Leipzig.

Bütschli, O., 1894. Investigations on Microscopic Foams and on Protoplasm. Adam and Charles Black, London.

Buvat, R., 1957. Formations de Golgi dans les cellules radiculaires d'Allium cepa. L. Compte Rendus des Séances 244, 1401–1403.

Buvat, R., 1969. Plant Cells. An Introduction to Plant Protoplasm. McGraw-Hill, New York.

Buvat, R., Carasso, N., 1957. Mise en évidence de l'ergastoplasme (réticulum endoplasmique) dans les cellules méristématiques de la racine d'Allium cepa L. Compte Rendus des Séances 244, 1532–1534.

Buvat, R., Puissant, A., 1958. Observations sur la cytodierese et l'origine des plasmodesmes. Comptes Rendus Acad. Sci. Paris 247, 23–236.

Buxbaum, R.E., Dennerll, T., Weiss, S., Heidemann, S.R., 1987. F-actin and microtubule suspensions as indeterminate fluids. Science 235, 1511–1514.

Bykova, N.V., Moller, I.M., 2006. Proteomics of plant mitochondria. Annu. Plant Rev. 28, 211–243.

Byrne, S.L., Erthmann, P.O., Agerbirk, N., Bak, S., Hauser, T.P., Nagy, I., Paina, C., Asp, T., 2017. The genome sequence of *Barbarea vulgaris* facilitates the study if ecological biochemistry. Sci. Rep. 7, 40728.

Bywater, J., 1817. Observations on the nature of animalcules and principles of vegetable physiology. In: Travels in the Interior of America by John Bradbury. Sherwood, Neely, and Jones, London.

Bywater, J., 1824. Physiological Fragments. R. Hunter, London.

Cacas, J.L., Buré, C., Grosje, K., Gerbeau-Pissot, P., Lherminier, J., Rombouts, Y., Maes, E., Bossard, C., Gronnier, J., Furt, F., Fouillen, L., Germain, V., Bayer, E., Cluzet, S., Robert, F., Schmitter, J.-M., Magali Deleu, M., Lins, L., Simon-Plas, F., Mongrand, S., 2016. Revisiting plant plasma membrane lipids in tobacco: a focus on sphingolipids. Plant Physiol. 170, 367–384.

Caderas, D., Muster, M., Vogler, H., Mandel, T., Rose, J.K.C., McQueen-Mason, S., Kuhlemeier, C., 2000. Limited correlation between expansin gene expression and elongation growth rate. Plant Physiol. 123, 1399–1413.

Cahoon, E.B., Lynch, D.V., 1991. Analysis of glucocerebrosides of Rye (*Secale cereale* L. cv. Puma). Leaf and plasma membrane. Plant Physiol. 95, 58–68.

Cai, G., Bartalesi, A., Del Casino, C., Moscatelli, A., Tiezzi, A., Cresti, M., 1993. The kinesin-immunoreactive homologue from *Nicotiana tabaccum* pollen tube: biochemical properties and subcellular localization. Planta 191, 496–506.

Cai, G., Faleri, C., Del Casino, C., Emons, A.M.C., Cresti, M., 2011. Distribution of callose synthase, cellulose synthase, and sucrose synthase in tobacco pollen tube is controlled in dissimilar ways by actin filaments and microtubules. Plant Physiol. 155, 1169–1190.

Cai, S., Lashbrook, C.C., 2008. Stamen abscission zone transcriptome profiling reveals new candidates for abscission control: enhanced retention of floral organs in transgenic plants overexpressing *Arabidopsis* zinc finger protein 2. Plant Physiol. 146, 1305–1321.

Cai, G., Romagnoli, S., Cresti, M., 2001. Microtubule motor proteins and the organization of the pollen tube cytoplasm. Sex. Plant Reprod. 14, 27–34.

Cai, G., Romagnoli, S., Moscatelli, A., Olvidi, E., Gambellini, G., Tezzi, A., Cresti, M., 2000. Identification and characterization of a novel microtubule-based motor associated with membranous organelles in tobacco pollen tubes. Plant Cell 12, 1719–1736.

Cairns, J., 1963a. The bacterial chromosome and its manner of replication as seen by autoradiography. J. Mol. Biol. 6, 208–213.

Cairns, J., 1963b. The chromosome of *Escherichia coli*. Cold Spring Harb. Symp. 28, 43–46.

Cairns, J., 1966. Autoradiography of HeLa cell DNA. J. Mol. Biol. 15, 372–373.

Cairns, J., July 21, 1980. This week's citation classic. Cit. Class. 29, 34.

Cairns-Smith, A.G., 1982. Genetic Takeover and the Mineral Origins of Life. Cambridge University Press, Cambridge.

Cairns-Smith, A.G., 1996. Evolving the Mind: On the Nature of Matter and the Origin of Conscienceness. Cambridge University Press, Cambridge.

Cairns-Smith, A.G., Hartman, H., 1986. Clay Minerals and the Origin of Life. Cambridge University Press, Cambridge.

Calabrese, E.J., 2002. Hormesis: changing view of the dose-response, a personal account of the history and current status. Mutat. Res. 511, 181–189.

Calabrese, E.J., 2004. Hormesis: a revolution in toxicology, risk assessment and medicine. EMBO Rep. 5 (Suppl. 1), S37–S40.

Calabrese, E.J., Baldwin, L.A., 2000a. Chemical hormesis: its historical foundations as a biological hypothesis. Hum. Exp. Toxicol. 19, 2–31.

Calabrese, E.J., Baldwin, L.A., 2000b. The marginalization of hormesis. Hum. Exp. Toxicol. 19, 32–40.

Calderwood, D.A., Shattil, S.J., Ginsberg, M.H., 2000. Integrins and actin filaments: reciprocal regulation of cell adhesion and signaling. J. Biol. Chem. 275, 22607–22610.

Caldwell, P.C., Hinshelwood, C., 1950. Some considerations on autosynthesis in bacteria. J. Chem. Soc. 3156–3159.

Caldwell, P.C., Hodgkin, A.L., Keynes, R.D., Shaw, T.I., 1960. The effects of injecting "energy-rich" phosphorous compounds on the active transport of ions in the giant axons of *Loligo*. J. Physiol. 152, 561–590.

Calebrese, E.J., 2003. Welcome to Nonlinearity in Biology, Toxicology and Medicine. Nonlinearity in Biology, Toxicology, and Medicine, vol. 1, p. 1.

Caliebe, A., Soll, J., 1999. News in chloroplast protein import. Plant Mol. Biol. 39, 641–645.

Callendar, G.S., 1938. The artificial production of carbon dioxide and its influence on climate. Q. J. R. Meteorol. Soc. 64, 223–240.

Callendar, G.S., 1949. Can carbon dioxide influence climate? Weather 4, 310–314.

Calvin, M., 1952. The path of carbon in photosynthesis. Harvey Lect. 46, 218–251.

Calvin, M., 1969. Chemical Evolution. Molecular Evolution Towards the Origin of Living Systems on the Earth and Elsewhere. Clarendon Press, Oxford.

Calvin, M., 1992. Following the Light. A Scientific Odyssey. American Chemical Society, Washington, DC.

Calvin, M., Bassham, J.A., 1962. The Photosynthesis of Carbon Compounds. W. A. Benjamin, Inc., New York.

Calvin, M., Benson, A.A., 1948. The path of carbon in photosynthesis. Science 107, 476–480.

Calvin, M., Benson, A.A., 1949. The path of carbon in photosynthesis IV. The identity and sequence of the intermediates in sucrose synthesis. Science 109, 140–142.

Calvin, M., Calvin, G.J., 1964. Atom to Adam. Am. Sci. 52, 163–186.

Calvin, M., December 11, 1961. The path of carbon in photosynthesis. Nobel Lecture.

Cameron, A.G.W., 1979. The origin and evolution of the solar system. In: Life. Origin and Evolution. Readings From Scientific American. W. H. Freeman and Co., San Francisco, pp. 9–17.

Camirand, A., Brummell, D., MacLachlan, G., 1987. Fucosylation of xyloglucan: localization of the transferase in dictyosomes of pea stem cells. Plant Physiol. 84, 753–756.

Camm, E.L., Green, B.R., 2004. How the chlorophyll-proteins got their names. Photosynth. Res. 80, 189–196.

Campbell, A.K., 1986. Lewis Victor Heilbrunn: pioneer of calcium as an intracellular regulator. Cell Calcium 7, 287–296.

Campbell, C., 2014. Whole: Rethinking the Science of Nutrition. BenBella Books, Dallas, TX.

Campbell, P.J., Stephens, P.J., Pleasance, E.D., O'Meara, S., Li, H., Santarius, T., et al., 2008. Identification of somatically acquired rearrangements in cancer using genome-wide massively parallel paired end sequencing. Nat. Genet. 40, 722–729.

Campbell, N.A., Stika, K.M., Morrison, G.M., 1979. Calcium and potassium in the motor organ of the sensitive plant: localization by ion microscopy. Science 204, 185–187.

Campbell, N.A., Thomson, W.W., 1977. Effects of lanthanum and ethylenediaminetetraacetate on leaf movements of *Mimosa*. Plant Physiol. 60, 635–639.

Canaani, O., Barber, J., Malkin, S., 1984. Evidence for phosphorylation and dephosphorylation regulated the distribution of excitation energy between the two photosystems of photosynthesis in vivo: photoacoustic and fluorimetric study of an intact leaf. Proc. Natl. Acad. Sci. U.S.A. 81, 1614–1618.

Cande, W.Z., 1982a. Inhibition of spindle elongation in permeabilized mitotic cells by erythron-9-[3-(hydroxynonyl)] adenine. Nature 295, 700–701.

Cande, W.Z., 1982b. Nucleotide requirements for anaphase chromosome movements in permeabilized mitotic cells: anaphase B but not Anaphase A requires ATP. Cell 28, 15–22.

Cande, W.Z., McDonald, K.L., 1985. *In vitro* reactivation of anaphase spindle elongation using isolated diatom spindles. Nature 316, 168–170.

Cande, W.Z., McDonald, K.L., 1986. Physiological and ultrastructural analysis of elongating spindles reactivated *in vitro*. J. Cell Biol. 103, 593–604.

Cande, W.Z., Wolniak, S.M., 1978. Chromosome movement in lysed mitotic cells is inhibited by vanadate. J. Cell Biol. 79, 573–580.

Canguilhem, G., 1969. La Connaissance de la Vie, déuxième edition revue et augmentée. Churchhill, Paris.

Canman, J., Cameron, L., Maddox, P., Straight, A., Tirnauer, J., Mitchison, T., Fang, G., Kapoor, T., Salmon, E.D., 2003. Determining the position of the cell division plane. Nature 424, 1074–1078.

Cannan, R.K., Levy, M., 1950. Chemistry of amino acids and proteins. Ann. Rev. Biochem. 19, 125–148.

Cannon, W.B., 1932. The Wisdom of the Body. W. W. Norton & Sons, New York.

Cannon, W.B., 1941. The body physiologic and the body politic. Science 93, 1–11.

Cannon, M.C., Terneus, K., Hall, Q., Tan, L., Wang, Y., Wegenhart, B.L., Chen, L., Lamport, D.T.A., Chen, Y., Kieliszewski, M.J., 2008. Self-assembly of the plant cell wall requires an extensin scaffold. Proc. Natl. Acad. Sci. U.S.A. 105, 2226–2231.

Canut, H., Brightman, A.O., Boudet, A.M., Morré, D.J., 1987. Determination of sidedness of plasma membrane and tonoplast vesicles isolated from plant stems. In: Leaver, C., Sze, H. (Eds.), Plant Membranes: Structure, Function, Biogenesis. Alan R. Liss, New York, pp. 141–159.

Canut, H., Carrasco, A., Galaud, J.P., Cassan, C., Bouyssou, H., Vita, N., Ferrara, P., Pont-Lezica, R., 1998. High affinity RGD-binding sites at the plasma membrane of Arabidopsis thaliana links the cell wall. Plant J. 16, 63–71.

Cao, Y., Crawford, N.M., Schroeder, J.I., 1995. Amino terminus and the first four membrane-spanning segments of the *Arabidopsis* K^+ channel KAT1 confer inward-rectification property of plant-animal chimeric channels. J. Biol. Chem. 270, 17697–17701.

Cao, J., Li, X., Lv, Y., 2017. Dynein light chain family gees in 15 plant species: identification, evolution and expression profiles. Plant Sci. 254, 70–81.

Capecchi, M.R., 1980. High efficiency transformation by direct microinjection of DNA into cultured mammalian cells. Cell 22, 479–488.

Capecchi, M.R., December 7, 2007. Gene targeting 1977—present. Nobel Lecture.

Cárdenas, L., McKenna, S.T., Kunkel, J.G., Hepler, P.K., 2006. NAD(P)H oscillates in pollen tubes and is correlated with tip growth. Plant Physiol. 142, 1460–1468.

Carlier, M.F., 1991. Nucleotide hydrolysis in cytosketetal assembly. Curr. Opin. Cell Biol. 3, 12–17.

Carlin, L., 2000. On the very concept of harmony in leibniz. Rev. Metaphys. 54, 99–125.

Carlisle, M.J., 1970. Nutrition and chemotaxis in th emyxomycete *Physarum polycephalum*: the effect of carbohydrates on the plasmodium. J. Gen. Microbiol. 63, 221–226.

Carmo-Fonseca, M., Hurt, E.C., 1991. Across the nuclear pores with the help of nucleoporins. Chromosoma 101, 199–205.

Carnoy, J.B., 1884. La Biologie Cellulaire. Joseph Van In & CieLierre.

Caro, L.G., Palade, G.E., 1964. Protein synthesis, storage, and discharge in the pancreatic exocrine cell. J. Cell Biol. 20, 473–495.

Carothers, E.E., 1936. Components of the mitotic spindle with especial reference to the chromosomal and interzonal fibers in the Acrididae. Biol. Bull. 71, 469–491.

Carpita, N.C., 1982. Limiting diameter of pores and the surface structure of plant cell walls. Science 218, 813–814.

Carpita, N., 2007. Cell wall-ion interactions. In: Sattelmacher, B., Horst, W.J. (Eds.), The Apoplast of Higher Plants; Compartments of Storage, Transport and Reactions. Springer, Dordrecht, The Netherlands, pp. 15–18.

Carpita, N.C., Defernez, M., Findlay, K., Wells, B., Shoue, D.A., Catchpole, G., Wilson, R.H., McCann, M.C., 2001b. Cell wall architecture of the elongating maize coleoptile. Plant Physiol. 127, 551–565.

Carpita, N.C., McCann, M.C., 2002. The functions of cell wall polysaccharides in composition and architecture revealed through mutations. Plant Soil 274, 71–80.

Carpita, N.C., McCann, M., Griffing, L.R., 1996. The plant extracellular matrix: news from the cell's frontier. Plant Cell 8, 1451–1463.

Carpita, N., Sabularse, D., Montezinos, D., Delmer, D.P., 1979. Determination of the pore size of cell walls of living plants cells. Science 205, 1144–1147.

Carpita, N., Tierney, M., Campbell, M., 2001a. Molecular biology of the plant cell wall: searching for the genes that define structure, architecture and wall dynamics. Plant Mol. Biol. 47, 1–5.

Carpita, N., Vergara, C., 1998. A recipe for cellulose. Science 279, 672–673.

Carraro, N., Tisdale-Orr, T.E., Clouse, R.M., Knöller, A.S., Spicer, R., 2012. Diversification and expression of the PIN, AUX/LAX, and ABCB families of putative auxin transporters in *Populus*. Front. Plant Sci. 3, 17.

Carrington, J.C., Ambros, V., 2003. Role of microRNAs in plant and animal development. Science 301, 336–338.

Carrington, J.C., Freed, D.D., Leinicke, A.G., 1991. Bipartate signal sequence mediates nuclear translocation of the plant potyviral Nla protein. Plant Cell 3, 953–962.

Carroll, L., 1872. The walrus and the carpenter. In: Through the Looking-Glass and What Alice Found There. Macmillan and Co., London.

Carroll, G., 1967. The fine structure of the ascus septum in *Ascodesmis sphaerospora* and *Saccobolus kerverni*. Mycologia 59, 527–532.

Carroll, F.E., 1972. A fine-structural study of conidium initiation in *Stemphylium botryosum* Wollroth. J. Cell Sci. 11, 33–47.

Carter, J., 1979. In: State of the Union, 45. Vital Speeches of the Day, pp. 226–229.

Carter, C., Pan, S., Zouhar, J., Avila, E.L., Girke, T., Raikhel, N.V., 2004. The vegetative vacuole proteome of *Arabidopsis thaliana* reveals predicted and unexpected proteins. Plant Cell 16, 3285–3303.

Caruthers, M.H., 2012. Gene synthesis with H.G. Khorana. Resonance 1143–1146.

Carvalho, J.De F., Madgwick, P.J., Powers, S.J., Keys, A.J., Lea, P.J., Parry, M.A., 2011. An engineered pathway for glyoxylate metabolism in tobacco plants aimed to avoid the release of ammonia in photorespiration. BMC Technol. 11, 111.

Cary, H.H., Beckman, A.O., 1941. A quartz photoelectric spectrophotometer. J. Opt. Soc. Am. 31, 682–689.

Cascante, M., Sorribas, A., Canela, E.I., 1994. Enzyme-enzyme interactions and metabolite channelling: alternative mechanisms and their evolutionary significance. Biochem. J. 298, 313–320.

Cashmore, A.R., 1997. The cryptochrome family of photoreceptors. Plant Cell Environ. 20, 764–767.

Cashmore, A.R., 1998. The cryptochrome family of blue/UV-A photoreceptors. J. Plant Res. 111, 267–270.

Cashmore, A.R., 2005. Cryptochrome overview. In: Wada, M., Shimazaki, M.K., Iino, M. (Eds.), Light Sensing in Plants. Springer, Tokyo, pp. 121–130.

Cashmore, A.R., 2010. The Lucretian swerve: the biological basis of human behavior and the criminal justice system. Proc. Natl. Acad. Sci. U.S.A. 107, 4499–4504.

Caskey, C.T., Tompkins, R., Scolnick, E., Caryk, T., Nirenberg, M., 1968. Sequential translation of trinucleotide codons for the initiation and termination of protein synthesis. Science 162, 135–138.

Caspar, T., Pickard, B.G., 1989. Gravitropism in a starchless mutant of *Arabidopsis*. Planta 177, 185–197.

Caspersson, T.O., 1950. Cell Growth and Cell Function. A Cytochemical Study. Norton, New York.

Caspersson, T., Schultz, J., 1938. Pentose nucleotides in the cytoplasm of growing tissues. Nature 143, 602–603.

Caspersson, T., Schultz, J., 1940. Ribonucleic acids in both nucleus and cytoplasm and the function of the nucleolus. Proc. Natl. Acad. Sci. U.S.A. 26, 507–515.

Caspi, A., McClay, J., Moffitt, A.T.E., Mill, J., Martin, J., Craig, I.W., Taylor, A., Poulton, R., 2002. Role of genotype in the cycle of violence in maltreated children. Science 297, 851–854.

Cassab, G.I., Varner, J.E., 1988. Cell wall proteins. Ann. Rev. Plant Physiol. Plant Mol. Biol. 39, 321–353.

Cassimeris, L., Pryer, N.K., Salmon, E.D., 1988. Real-time observations of microtubule dynamic stability in living cells. J. Cell Biol. 107, 2223–2231.

Casson, S., Spencer, M., Walker, K., Lindsey, K., 2005. Laser capture microdissection for the analysis of gene expression during embryogenesis of *Arabidopsis*. Plant J. 42, 111–123.

Castañeda, L.E., Barbosa, O., 2017. Metagenomic analysis exploring taxonomic and functional diversity of soil microbial communities in Chilean vinyards and surrounding native forests. Peerj 5, e3098.

Castelfranco, P.A., Beale, S.I., 1983. Chlorophyll biosynthesis: recent advances and areas of current interest. Ann. Rev. Plant Physiol. 34, 241–278.

Castellani, C., 1973. Spermatozoan biology from Leeuwenhoek to Spallanzani. J. Hist. Biol. 6, 37–68.

Castle, W.E., 1916. Genetics and Eugenics. Harvard University Press, Cambridge, MA.

Castle, W.E., 1919. Is the arrangement of the genes in the chromosome linear? Proc. Natl. Acad. Sci. U.S.A. 5, 25–32.

Castle, E.S., 1937. Membrane tension and orientation of structure in the plant cell wall. J. Cell. Comp. Physiol. 10, 113–121.

Caswell, A.H., 1968. Potentiometric determination of interrelationships of energy conservation and ion gradients in mitochondria. J. Biol. Chem. 243, 5827–5836.

Caswell, A.H., Pressman, B.C., 1968. Electromeric analysis of cytochromes in mitochondria. Arch. Biochem. Biophys. 125, 318–325.

Catalá, C., Howe, K.J., Hucko, S., Rose, J.K.C., Thannhauser, T.W., 2011. Towards characterization of the glycoproteome of tomato (*Solanum lycopersicum*) fruit using Concanavalin A lectin affinity chromatography and LC-MALDI-MS/MS analysis. Proteomics 11, 1530–1544.

Catalá, C., Rose, J.K.C., York, W.S., Albersheim, P., Darvill, A.G., Bennett, A.B., 2001. Characterization of a tomato xyloglucan endotransglycosylase gene that is down-regulated by auxin in etiolated hypocotyls. Plant Physiol. 127, 1180–1192.

Catchside, D.C., 1947. Th eP-locus position effect in Oenothera. J. Genet. 48, 31–42.

Cattell, J. (Ed.), 1940. The Cell Theory, Biological Symposia. The Jaques Cattell Press, Lancaster, PA.

Catterall, W.A., 1994. Molecular properties of a super family of plasma-membrane cation channels. Curr. Opin. Cell Biol. 6, 607–615.

Catterall, W.A., 1995. Structure and function of voltage-gated ion channels. Ann. Rev. Biochem. 64, 493–531.

Catterall, W.A., Seagar, M.J., Takahasi, M., Nunoki, K., 1989. Molecular properties of dihydropyridine-sensitive calcium channels. Ann. N.Y. Acad. Sci. 560, 1–14.

Cavalier-Smith, T., 1975. The origin of nuclei and of eukaryotic cells. Nature 256, 463–468.

Cavalier-Smith, T., 1987a. Eukaryotes with no mitochondria. Nature 326, 332–333.

Cavalier-Smith, T., 1987b. The simultaneous symbiotic origin of mitochondria, chloroplasts and microbodies. Ann. N.Y. Acad. Sci. 503, 55–71.

Cavalier-Smith, T., 1989a. Archaebacteria and Archezoa. Nature 339, 100–101.

Cavalier-Smith, T., 1989b. Symbiotic origin of peroxisomes. Endocytobiology IV, 515–521.

Cavalier-Smith, T., 1991. Archamoebae: the ancestral eukaryote? BioSystems 25, 25–38.

Cavazza, T., Vernos, I., 2016. The RanGTP pathway: from nucleo-cytoplasmic transport to spindle assembly and beyond. Front. Cell Dev. Biol. 3, 82.

Cech, T.R., 1985. Self-splicing RNA. Implications for evolution. Int. Rev. Cytol. 93, 3–22.

Cech, T.R., 1986a. A model for the RNA-catalyzed replication of RNA. Proc. Natl. Acad. Sci. U.S.A. 83, 4360–4363.

Cech, T.R., 1986b. The generality of self-splicing RNA: relationship to nuclear mRNA splicing. Cell 44, 207–210.

Cech, T.R., 1988. Biological catalysis by RNA. Harvey Lect. 82, 123–144.

Čermák, T., Baltes, N.J., Čegan, R., Zhang, Y., Voytas, D.F., 2015. High-frequency, precise modification of the tomato genome. Genome Biol. 16, 232.

Certal, A.C., Almeida, R.B., Cavalho, L.M., Wong, E., Moreno, N., Michard, E., Carneiro, J., Rodriguéz-Léon, J., Wu, H.-M., Cheung, A.Y., Feijó, J.A., 2008. Exclusion of a proton ATPase from the apical membrane is associated with cell polarity and tip growth in *Nicotiana tabacum* tubes. Plant Cell 20, 614–634.

Cerutti, H., Ibrahim, H.Z., Jagendorf, A.T., 1993. Treatment of pea (*Pisum sativum* L.) protoplasts with DNA-damaging agents induces a 39-kilodalton chloroplast protein immunologically related to *Escherichia coli* RecA. Plant Physiol. 102, 155–163.

Cerutti, H., Jagendorf, A.T., 1993. DNA strand-transfer activity in pea (*Pisum sativum* L.) chloroplasts. Plant Physiol. 102, 145–153.

Cerutti, H., Johnson, A.M., Boynton, J.E., Gillham, N.W., 1995. Inhibition of chloroplast DNA recombination and repair by dominant negative mutants of *Escherichia coli* RecA. Mol. Cell. Biol. 15, 3003–3011.

Cerutti, H., Osman, M., Grandoni, P., Jagendorf, A.T., 1992. A homology of *Escherichia coli* RecA protein in plastids of higher plants. Proc. Natl. Acad. Sci. U.S.A. 89, 8068–8072.

Cevher-Keskin, B., 2013. ARF1 and SAR1 GTPases in endomembrane trafficking in plants. Int. J. Mol. Sci. 14, 18181–18199.

Chaen, S., Oiwa, K., Shimmen, T., Iwamoto, H., Sugi, H., 1989. Simultaneous recordings of force and sliding movement between a myosin – coated glass microneedle and actin cables *in vitro*. Proc. Natl. Acad. Sci. U.S.A. 86, 1510–1514.

Chalfie, M., December 8, 2008. Lighting up life. Nobel Lecture.

Chalfie, M., Tu, Y., Euskirchen, G., Ward, W.W., Prasher, D.C., 1994. Green fluorescent protein as a marker for gene expression. Science 263, 802–805.

Chalfie, M., Tu, Y., Prasher, D.C., 1993. Glow Worms – A New Method of Looking at *C. elegans* Gene Expression. Worm Breeder's Gaz. 13, 19. www.wormbase.org/db/misc/paper?name=WBPaper00014747; class=Paper.

Chamberlain, T.C., 1890. The method of multiple working hypotheses. Science 15, 92–96.

Chamberlain, T.C., 1897. Studies for Students: the method of multiple working hypotheses. J. Geol. 5, 837–848.

Chamberlain, T.C., 1965. The method of multiple working hypotheses. Science 148, 754–759.

Chamberlin, T.C., Chamberlin, R.T., 1908. Early terrestrial conditions that may have favored organic synthesis. Science 28, 897–911.

Chambers, R., 1922. A microinjection study on the permeability of the starfish egg. J. Gen. Physiol. 5, 189–194.

Chambers, R., n.d. Vestiges of the Natural Histroy of Creation. With a Sequel, Harper & Brothers, New York.

Chan, E.Y., 2005. Advances in sequencing technology. Mutat. Res. 573, 13–40.

Chan, J., Calder, G.M., Doonan, J.H., Lloyd, C.W., 2003. EB1 reveals mobile microtubule nucleation sites in *Arabidopsis*. Nat. Cell Biol. 5, 967–971.

Chan, J., Calder, G., Fox, S., Lloyd, C., 2005. Localization of the microtubule end binding protein EB1 reveals alternative pathways of spindle development in *Arabidopsis* suspension cells. Plant Cell 17, 1737–1748.

Chan, C.X., Ragan, M.A., 2013. Next-generation phylogenomics. Biol. Direct 8, 3.

Chance, B., 1952a. The kinetics and inhibition of cytochrome components of succinic oxidase system. I. Activity determination and purity criteria. J. Biol. Chem. 197, 557–565.

Chance, B., 1952b. The kinetics and inhibition of cytochrome components of succinic oxidase system. II. Steady state properties and difference spectra. J. Biol. Chem. 197, 567–576.

Chance, B., 1967. Is there a proton noise in mitochondrial membranes leading to chance fluctuations in pH? Nature 214, 399–400.

Chance, B., Williams, G.R., 1955. Respiratory enzymes in oxidative phosphorylation. J. Biol. Chem. 217, 383–393, 395–407, 409–427, 429–438.

Chandrasekhar, S., 1995. Newton's Principia for the Common Reader. Clarendon Press, Oxford.

Chandrashekaran, M.K., 1998. Biological rhythms research: a personal account. J. Biosci. 23, 545–555.

Chang, Y.-G., Cohen, S.E., Phong, C., Myers, W.K., Kim, Y.-I., Tseng, R., Lin, J., Zhang, L., Boyd, J.S., Lee, Y., Kang, S., Lee, D., Li, S., Britt, R.D., Rust, M.J., Golden, S.S., LiWang, A., 2015. Circadian rhythms. A protein fold switch joins the circadian oscillator to clock output in cyanobacteria. Science 349, 324–328.

Chang, S., DesMarais, D., Mack, R., Miller, S.L., Strathearn, G.E., 1983. Prebiotic organic syntheses and the origin of life. In: Schopf, J.W. (Ed.), Earth's Earliest Biosphere. Princeton University Press, Princeton, NJ, pp. 53–92.

Chang-Jie, J., Sonobe, S., 1993. Identification and preliminary characterization of a 65 kDa higher-plant microtubule-associated protein. J. Cell Sci. 105, 891–901.

Chanroj, S., Wang, G., Venema, K., Zhang, M.W., Delwiche, Sze, H., 2012.

Chanson, A., McNaughton, E., Taiz, L., 1984. Evidence for a KCl-stimulated Mg^{2+}-ATPase on the Golgi of corn coleoptiles. Plant Physiol. 76, 498–507.

Chanson, A., Taiz, L., 1985. Evidence for an ATP-dependent proton pump on the Golgi of corn coleoptiles. Plant Physiol. 78, 232–240.

Chantrenne, H., 1958. Newer developments in relation to protein biosynthesis. Ann. Rev. Biochem. 27, 35–56.

Chao, Q., Gao, Z.F., Wang, Y.F., Li, Z., Huang, X.H., Wang, Y.C., Mei, Y.C., Zhao, B.G., Li, L., Jiang, Y.B., Wang, B.C., 2016. The proteome and phosphoproteome of maize pollen uncovers fertility candidate proteins. Plant Mol. Biol. 91, 287–304.

Chapman, K.D., Trelease, R.N., 1991a. Acquisition of membrane lipids by differentiating glyoxysomes: role of lipid bodies. J. Cell Biol. 115, 995–1007.

Chapman, K.D., Trelease, R.N., 1991b. Intracellular localization of phosphatidylcholine and phosphatidyl-ethanolamine synthesis in cotyledons of cotton seedlings. Plant Physiol. 95, 69–76.

Chapple, C.C.S., Vogt, T., Ellis, B.E., Somerville, C.R., 1992. An *Arabidopsis* mutant defective in the general phenylpropanoid pathway. Plant Cell 4, 1413–1424.

Chaptal, J.A., 1836. Chymistry Applied to Agriculture. Hilliard, Gray Co., Boston.

Chargaff, E., 1947. On the nucleoproteins and nucleic acids of microorganisms. Cold Spring Harb. Symp. Quant. Biol. 12, 28–34.

Chargaff, E., 1950a. Chemical specificity of nucleic acids and mechanism of their enzymatic degradation. Experientia 6, 201–210.

Chargaff, E., 1950b. Composition of human desoxypentose nucleic acid. Nature 165, 756–757.

Chargaff, E., 1951a. Some recent studies on the composition and structure of nucleic acids. J. Cell. Comp. Physiol. 38 (Suppl. 1), 41–59.

Chargaff, E., 1951b. Structure and function of nucleic acids as cell constituents. Fed. Proc. 10, 654–659.

Chargaff, E., 1955. Isolation and composition of the deoxypentose nucleic acids and the corresponding nucleoproteins. In: Chargaff, E., Davidson, J.N. (Eds.), The Nucleic Acids, vol. 1. Academic Press, New York, pp. 307–371.

Chargaff, E., 1958. Of nucleic acids and nucleoproteins. Harvey Lect. 52, 57–73.

Chargaff, E., 1963. Essays on Nucleic Acids. Elsevier, Amsterdam.

Chargaff, E., 1963a. On some of the biological consequences of base-pairing in the nucleic acids. In: Developmental and Metabolic Control Mechanisms and Neoplasia. A Collection of Papers Presented at the Nineteenth Annual Symposium on Fundamental Cancer Research, 1965. Williams and Wilkins, Baltimore, MD, pp. 7–25, 1965.

Chargaff, E., 1963b. Essays on Nucleic Acids. Elsevier, Amsterdam.

Chargaff, E., 1965. On some of the biological consequences of base-pairing in the nucleic acids. In: Developmental and Metabolic Control Mechanisms and Neoplasia. A Collection of Papers Presented at the Nineteenth Annual Symposium on Fundamental Cancer Research, 1965. Williams and Wilkins, Baltimore, MD, pp. 7–25.

Chargaff, E., 1968. A quick climb up Mount Olympus. Science 159, 1448–1449.

Chargaff, E., 1974. Building the tower of Babble. Nature 248, 776–779.

Chargaff, E., 1975. Voices in the labyrinth: dialogues around the study of nature. Perspect. Biol. Med. 18, 251–285, 313–330.

Chargaff, E., 1976. Triviality in science: a brief meditation on fashions. Perspect. Biol. Med. 19 (Spring), 324–333.

Chargaff, E., 1977. Voices in the Labyrinth. Nature, Man and Science. Seabury Press, New York.

Chargaff, E., 1978. Heraclitian Fire: Sketches from a Life Before Nature. Rockefeller University Press, New York.

Chargaff, E., 1980. In praise of smallness—How can we return to small science? Perspect. Biol. Med. 23 (3), 370–385.

Chargaff, E., 1986. Serious Questions. An ABC of Skeptical Reflections. Birkäuser, Boston.

Chargaff, E., 1997. In dispraise of reductionism. BioScience 47, 795–797.

Chargaff, E., Lipshitz, R., Green, C., Hodes, M.E., 1951. The composition of the desoxyribonucleic acid of salmon sperm. J. Biol. Chem. 192, 223–230.

Chargaff, E., Vicher, E., Doniger, R., Green, C., Misani, F., 1949. The composition of the desoxypentose nucleic acids of thymus and spleen. J. Biol. Chem. 177, 405–416.

Chargaff, E., Vischer, E., 1948. Nucleoproteins, nucleic acids, and related substances. Annu. Rev. Biochem. 17, 201–226.

Chargaff, E., Zamenhof, S., Green, C., 1950. Composition of human desoxypentose nucleic acid. Nature 165, 756–757.

Charles, D., 2001. Lords of the Harvest: Biotech, Big Money, and the Future of Food. Perseus, Cambridge, MA.

Chase, C., 1998. The discovery of plant mitochondrial genomes. Discov. Plant Biol. 2, 85–103.

Chat, J., Chalak, L., Petit, R.J., 1999. Strict paternal inheritance of chloroplast DNA and maternal inheritance of mitochondrial DNA in intraspecific crosses of kiwi fruit. Theor. Appl. Genet. 99, 314–322.

Chaturvedi, P., Ischebeck, T., Egelhofer, V., Lichtscheidl, I., Weckwerth, W., 2013. Cell-specific analysis of the tomato pollen proteome from pollen mother cell to mature pollen provides evidence for developmental priming. J. Proteome Res. 12, 4892–4903.

Chau, R., Bisson, M.A., Siegel, A., Elkin, G., Klim, P., Straubinger, R.M., 1994. Distribution of charasomes in *Chara*: re-establishment and loss in darkness and correlation with banding and inorganic carbon uptake. Aust. J. Plant Physiol. 21, 113–123.

Chebli, Y., Geitmann, A., 2007. Mechanical principles governing pollen tube growth. Funct. Plant Sci. Biotechnol. 1, 232–245.

Chebli, Y., Geitmann, A., 2017. Cellular growth in plants requires regulation of cell wall biochemistry. Curr. Opin. Cell Biol. 44, 28–35.

Chebli, Y., Kanead, M., Zerzour, R., Geitmann, A., 2012. The cell wall of the *Arabidopsis* pollen tube-spatial distribution, recycling, and network formation of polysaccharides. Plant Physiol. 160, 1940–1955.

Chebli, Y., Kroeger, J., Geitmann, A., 2013. Transport logistics in pollen tubes. Mol. Plant 20, 614–634.

Chen, Q., Boss, W.F., 1991. Neomycin inhibits the phosphatidylinositol monophosphate and phosphatidylinositol bisphosphate stimulation of plasma membrane ATPase activity. Plant Physiol. 96, 340–343.

Chen, K., Chen, X., Schnell, D.J., 2000a. Initial binding of preproteins involving the Toc159 receptor can be bypassed during protein import into chloroplasts. Plant Physiol. 122, 813–822.

Chen, K., Chen, X., Schnell, D.J., 2000b. Mechanism of protein import across the chloroplast envelope. Biochem. Soc. Trans. 28, 485–491.

Chen, J., Doyle, C., Qi, X., Zheng, H., 2012a. The endoplasmic reticulum: a social network in plant cells. J. Integr. Plant Biol. 4, 840–850.

Chen, R., Hilson, P., Sedbrook, J., Rosen, E., Caspar, T., Masson, P.H., 1998. The *Arabidopsis thaliana* AGRAVITROPIC 1 gene encodes a component of the polar-auxin-transport efflux carrier. Proc. Natl. Aad. Sci. U.S.A. 95, 14112–15117.

Chen, J., Huang, Q., Gao, D., Wang, J., Lang, Y., Liu, T., Li, B., Bai, Z., Goicoechea, J.L., Liang, C., 2013. Whole-genome sequencing of *Oryza brachyantha* reveals mechanisms underlying *Oryza* genome evolution. Nat. Commun. 4, 1595.

Chen, M.-H., Huang, L.-F., Li, H.-M., Chen, Y.-R., Yu, S.M., 2004. Signal peptide-dependent targeting of a rice α-amylase and cargo proteins to plastids and extracellular compartments of plant cells. Plant Physiol. 135, 1367–1377.

Chen, X., Irani, N.G., Friml, J., 2011. Clathrin-mediated endocytosis: the gateway into plant cells. Curr. Opin. Plant Biol. 14, 674–682.

Chen, T.-H., Jaffe, L.F., 1978. Effects of membrane potential on calcium fluxes of *Pelvetia* eggs. Planta 140, 63–67.

Chen, T.-H., Jaffe, L.F., 1979. Forced calcium entry and polarized growth of *Funaria* spores. Planta 144, 401–406.

Chen, J.C.W., Kamiya, N., 1975. Localization of myosin in the internodal cell of *Nitella* as suggested by differential treatment with N-ethylmaleimide. Cell Struct. Funct. 1, 1–9.

Chen, J.C.W., Kamiya, N., 1981. Differential heat treatment of the *Nitella* internodal cell and its relation to cytoplasmic streaming. Cell Struct. Funct. 6, 201–207.

Chen, M.-H., Liu, L.-F., Chen, Y.-R., Wu, H.-K., Yu, S.-M., 1994. Expression of α-amylases, carbohydrate metabolism, and autophagy in cultured rice cells is coordinately regulated by sugar nutrient. Plant J. 6, 625–636.

Chen, Y., Liu, P., Hoehenwarter, W., Lin, J., 2012b. Proteomic and phosphoproteomic analysis of *Picea* wilsonii pollen development under nutrient limitation. J. Proteome Res. 11, 4180–4190.

Chen, C., Liu, M., Jiang, L., Liu, X., Zhao, J., Yan, S., Yang, S., Ren, H., Liu, R., Zhang, X., 2014. Transcriptome profiling reveals roles of meristem regulators and polarity genes during fruit trichome development in cucumber (Cucumis sativus L.). J. Exp. Bot. 65, 4943–4958.

Chen, K., Pachter, L., 2005. PLoS Bioinformatics for whole-genome shotgun sequencing of microbial communities. Comput. Biol. 1, e24.

Chen, W., Provart, N.J., Glazebrook, J., Katagiri, F., Chang, H.-U., Eulgen, T., Mauch, F., Luan, S., Zou, G., Whitham, S.A., Budworth, P.R., Tao, Y., Xie, Z., Chen, X., Lam, S., Kreps, J.A., Harper, J.F., Si-Ammour, Z., Mauch-Mani, B., Heinlein, M., Kobayashi, K., Hohn, T., Dangl, J.L., Wang, X., Zhu, T., 2002. Expression profile matrix of *Arabidopsis* transcription factor genes suggests their putative functions in response to environmental stresses. Plant Cell 14, 559–574.

Chen, T.L., Wolniak, S.M., 1987. Lithium induces cell plate dispersion during cytokinesis in *Tradescantia*. Protoplasma 141, 56–63.

Chen, H., Zou, Y., Shang, Y., Lin, H., Wang, Y., Cai, R., Tang, X., Zhou, J.-M., 2008. Firefly luciferase complementation imaging assay for protein-protein interactions in plants. Plant Physiol. 146, 368–376.

Cheng, N.-H., Pittman, J.K., Shigaki, T., Hirschi, K.D., 2002. Characterization of CAX4, an *Arabidopsis* H^+/cation antiporter. Plant Physiol. 128, 1245–1254.

Cheng, X., Wen, J., Tadge, M., Ratet, P., Mysore, K.S., 2011. Reverse genetics in Medicago truncatula using Tnt1 insertion mutants. Methods Mol. Biol. 678, 179–190.

Cheng, X., Wen, J., Tadge, M., Ratet, P., Mysore, K.S., 2013. Reverse genetics in *Medicago truncatula* using *Tnt1* insertion mutants. Methods Mol. Biol. 678, 179–190.

Cherayil, B.J., 2013. Michaelis and Menten at 100:still going strong. Resonance 18, 969–995.

Chet, I., Naveh, A., Henis, Y., 1977. Chemotaxis of *Physarum polycephalum* towards carbohydrates, amino acids and nucleotides. J. Gen. Microbiol. 102, 145–148.

Cheung, A.Y., 1995. Pollen-pistil interactions in compatible pollination. Proc. Natl. Acad. Sci. U.S.A. 92, 3077–3080.

Cheung, A.Y., Chen, C.Y., Glaven, R.H., de Graaf, B.H., Vidali, L., Hepler, P.K., Wu, H.M., 2002. Rab2 GTPase regulates vesicle trafficking between the endoplasmic reticulum and the Golgi bodies and is important to pollen tube growth. Plant Cell 14, 945–962.

Cheung, A.Y., Niroomand, S., Zou, Y., Wu, H.M., 2010. A transmembrane formin nucleates subapical actin assembly and controls tip-focused growth in pollen tubes. Proc. Natl. Acad. Sci. U.S.A. 107, 16390–16395.

Cheung, A.Y., Wang, H., Wu, H.-M., 1995. A floral transmitting tissue-specific glycoproteïn attracts pollen tubes and stimulates their growth. Cell 82, 383–393.

Cheung, A.Y., Wu, H.M., 2004. Overexpression of an *Arabidopsis* formin stimulates supernumerary actin calble formation from pollen tube cell membrane. Plant Cell 16, 257–269.

Chial, H., 2008. Rare genetic disorders: learning about genetic disease through gene mapping, SNPs, and microarray data. Nat. Educ. 1, 192.

Chibnall, A.C., 1939. Protein Metabolism in the Plant. Yale University Press, New Haven, CT.

Chien, A., Edgar, D.B., Trela, J.M., 1976. Deoxyribonucleic acid polymerase from the extreme thermophile *Thermus aquaticus*. J. Bacteriol. 127, 1550–1557.

Child, F.M., 1959. The characterization of the cilia of *Tetrahymena pyriformis*. Exp. Cell Res. 18, 258–267.

Chilton, M.-D.M., Que, Q., 2003. Targeted integration of T-DNA into the tobacco genome at double-strand breaks: new insights on the mechanism of T-DNA integration. Plant Physiol. 133, 956–965.

Chinpongpanich, A., Limruengroj, K., Phean-o-pas, S., Limpaseni, Buaboocha, T., 2012. Expression analysis of calmodulin and calmodulin-like genes from rice, *Oryza sativa* L. BMC Res. Notes 5, 625.

Chisti, Y., 2007. Biodiesel from microalgae. Biotechnol. Adv. 25, 294–306.

Chisti, Y., 2008. Biodiesel from microalgae beats bioethanol. Trends Biotechnol. 26, 126–131.

Chiu, W., Niwa, Y., Zeng, W., Hirano, T., Kobayashi, H., Sheen, J., 1996. Engineered GFP as a vital reporter in plants. Curr. Biol. 6, 325–330.

Chmielewska, K., Rodziewicz, P., Swarcewicz, B., Sawikowska, A., Kriewski, P., Marczak, Ł., Ciesiołka, D., Kuczyńska, A., Mikołajczak, K., Ogrodowicz, P., Krystkowiak, K., Surma, M., Adamski, T., Bednarek, P., Stobiecki, M., 2016. Analysis of drought-induced proteomic and metabolomics changes in barley (*Hordeum vulgare* L.) leaves and roots unravels some aspects of biochemical mechanisms involved in drought tolerance. Front. Plant Sci. 7, 1108.

Cho, S.-O., Wick, S.M., 1990. Distribution and function of actin in the developing stomatal complex of winter rye (*Secale cereale* cv. Puma). Protoplasma 157, 154–164.

Cho, S.-O., Wick, S.M., 1991. Actin in the developing stomatal complex of winter rye: a comparision of actin antibodies and Rh-phalloidin labelling of control and CB-treated tissues. Cell Motil. Cytoskelet. 19, 25–36.

Choi, W.-G., Gilroy, S., 2014. Plant biologists FRET over stress. eLife 3, e02763.

Chollet, R., Ogren, W.L., 1975. Regulation of photorespiration in C3 and C4 species. Bot. Rev. 41, 137–179.

Chollet, R., Paolillo Jr., D.J., 1972. Greening in a virescent mutant of maize. Z. Pfanzenphysiol. 68, 30–34.

Cholodny, N., Sankewitsch, E., 1933. Plasmolyseform und Ionenwirkung. Protoplasma 20, 57–72.

Chory, J., Ecker, J.R., Briggs, S., Caboche, M., Coruzzi, G.M., Cook, D., Dangl, J., Grant, S., Guerinot, M.L., Henikoff, S., Martienssen, R., Okada, K., Raikhel, N.V., Somerville, C.R., Weigel, D., 2000. National Science Foundation-Sponsored Workshop Report: "The 2010 Project," functional genomics and the virtual plant: a blueprint for understanding how plants are built and how to improve them. Plant Physiol. 123, 423–426.

Chou, M., Krause, K.-H., Campbell, K.P., Jensen, K.G., Sjolund, R.D., 1989. Antibodies against the calcium-binding protein calsequestrin from *Strepthanthus tortuosus* Brassicaceae. Plant Physiol. 91, 1259–1261.

Chou, Y.-H., Pogorelko, G., Young, Z.T., Zabotina, O.A., 2015. Protein-protein interactions among xyloglucan-synthesizing enzymes and formation of Golgi-localized multiprotein complexes. Plant Cell Physiol. 56, 255–267.

Chow, B.Y., Kay, S.A., 2013. Global approaches for telling time: omics and the arabidopsis circadian clock. Semin. Cell Dev. Biol. 24, 383–392.

Chrick, F.H.C., 1966. Codon-Anticodon pairing: the wobble hypothesis. J. Mol. Biol. 19, 548–555.

Chrispeels, M.J., 1969. Synthesis and secretion of hydroxyproline containing macromolecules in carrots. I. Kinetic analysis. Plant Physiol. 44, 1187–1193.

Chrispeels, M.J., 1970. Synthesis and secretion of hydroxyproline containing macromolecules in carrots. Plant Physiol. 45, 223–227.

Chrispeels, M.J., 1983. The Golgi apparatus mediates the transport of phytohemagglutinin to the protein bodies in bean cotyledons. Planta 158, 140–151.

Chrispeels, M.J., 1984. Biosynthesis, processing and transport of storage proteins and lectins in cotyledons of developing legume seeds. Philos. Trans. R. Soc. Lond. B. 304, 309–322.

Chrispeels, M.J., 1985. The role of the Golgi apparatus in the transport and post-translational modification of vacuolar (protein body) proteins. Oxf. Surv. Plant Mol. Cell Biol. 2, 43–68.

Chrispeels, M.J., 1991. Sorting of proteins in the secretory system. Annu. Rev. Plant Physiol. Plant Mol. Biol. 42, 21–53.

Chrispeels, M.J., Bollini, R., 1982. Characteristics of membrane-bound lectin in developing *Phaseolus vulgaris* cotyledons. Plant Physiol. 70, 1425–1428.

Chrispeels, M.J., Staehelin, L.A., 1992. Budding, fission, transport, targeting, fusion—Frontiers in secretion research. Plant Cell 4, 1008–1015.

Chrispeels, M.J., Varner, J.E., 1967. Gibberellic acid-enhanced synthesis and release of α-amylase and ribonuclease by isolated barley aleurone layers. Plant Physiol. 42, 398–406.

Chrispeels, M.J., Wraight, C.A., 2014. Roderick K. Clayton: 1922-2011. Biographical Memoirs. National Academy of Sciences, Washington, DC. http://www.nasonline.org/publications/biographical-memoirs/memoir-pdfs/clayton-roderick.pdf.

Christie, J.M., Briggs, W.R., 2001. Blue light sensing in higher plants. J. Biol. Chem. 276, 11457–11460.

Christie, A.E., Leopold, A.C., 1965a. On the manner of triiodobenzoic acid inhibition of auxin transport. Plant Cell Physiol. 6, 337–345.

Christie, A.E., Leopold, A.C., 1965b. Entry and exit of indoleacetic acid in corn coleoptiles. Plant Cell Physiol. 6, 453–465.

Christie, J.M., Reymond, P., Powell, G.K., Bernasconi, P., Raibekas, A.A., Liscum, E., Briggs, W.R., 1998. *Arabidopsis* NPH1: a flavoprotein with the properties of a photoreceptor for phototropism. Science 282, 1698–1701.

Christie, J.M., Salomon, M., Nozue, K., Wada, M., Briggs, W.R., 1999. LOV (light, oxygen, or voltage) domains of the blue-light photoreceptor phototropin (nph1): binding sites for the chromophore flavin mononucleotide. Proc. Natl. Acad. Sci. U.S.A. 96, 8779–8783.

Christophe, L., Tarrago-Litvak, L., Castroviejo, M., Litvak, S., 1981. Mitochondrial DNA polymerase from wheat embryos. Plant Sci. Lett. 21, 181–192.

Chu, Z., Jin, X., Yang, B., Zeng, Q., 2007. Buoyancy regulation of *Microcystis flos-aquae* during phosphorous-linited and nitrogen-limited growth. J. Plankton Res. 29, 739–745.

Chua, N.-H., 1973. Attachment of chloroplast polysomes to thylakoid membranes in *Chlamydomonas reinhardtii*. Proc. Natl. Acad. Sci. U.S.A. 70, 1554–1558.

Chua, N.-H., Blobel, G., Siekevitz, P., Palade, G.E., 1973. Attachment of chloroplast polysomes to thylakoid membranes in *Chlamydomonas reinhardtii*. Proc. Natl. Acad. Sci. U.S.A. 70, 1554–1558.

Chua, N.-H., Blobel, G., Siekevitz, P., Palade, G.E., 1976. Periodic variations in the ratio of free to thylakoid–bound chloroplast ribosomes during the cell cycle of *Chlamydomonas reinhardtii*. J. Cell Biol. 71, 497–514.

Chua, N.-H., Schmidt, G.W., 1978. Post-translational transport into intact chloroplasts of a precursor to the small subunit of ribulose-1,5-bisphosphate carboxylase. Proc. Natl. Acad. Sci. U.S.A. 75, 6110–6114.

Chua, N.-H., Schmidt, G.W., 1979. Transport of proteins into mitochondria and chloroplasts. J. Cell Biol. 81, 461–483.

Chuang, C.-F., Meyerowitz, E.M., 2000. Specific and heritable genetic interference by double stranded RNA in *Arabidopsis thaliana*. Proc. Natl. Acad. Sci. U.S.A. 97, 4985–4990.

Chuartzman, S.G., Nevo, R., Shimoni, E., Charuvi, D., Kiss, V., Ohad, I., Brumfeld, V., Reich, Z., 2008. Thylakoid membrane remodeling during state transitions in *Arabidopsis*. Plant Cell 20, 1029–1039.

Chudakov, D.M., Matz, M.V., Lukyanov, S., Lukyanov, K.A., 2010. Fluorescent proteins and their applications in imaging living cells and tissues. Physiol. Rev. 90, 1103–1163.

Klaus Biemann: the father of organic mass spectrometry. In: Chung, D.D.L. (Ed.), 2006. The Road to Scientific Success. Inspiring Life Stories of Prominent Researchers, vol. 1. World Scientific, Hackensack, NJ, pp. 143–180.

Chung, K.P., Zeng, Y., 2017. An overview of protein secretion in plant cells. Methods Mol. Biol. 1662, 19–32.

Chuong, S.D.X., Franceschi, V.R., Edwards, G.E., 2006. The cytoskeleton maintains organelle partitioning required for single-cell C4 photosynthesis in chenopodiaceae species. Plant Cell 18, 2207–2223.

Church, A.H., 1908. Types of Floral Mechanism. Oxford University Press, Oxford.

Church, K., Lin, H.-P.P., 1985. Kinetochore microtubules and chromosome movement during, prometaphase in *Drosophila melanogaster* spermatocytes studied in life and with the electron microscope. Chromosoma 92, 273–282.

Churchill, W., October 1, 1939. The Russian Enigma. Broadcast. http://www.churchill-society-london.org.uk/RusnEnig.html.

Churchill, K.A., Sze, H., 1983. Anion-sensitive, H^+-pumping ATPase in membrane vesicles from oat roots. Plant Physiol. 71, 610–617.

Churchill, K.A., Sze, H., 1984. Anion-sensitive, H^+-pumping ATPase of oat roots: direct effects of Cl^-, NO_3^-, and a disulfonic stilbene. Plant Physiol. 76, 490–497.

Chyba, C.F., 1990a. Impact delivery and erosion of planetary oceans in the early inner solar system. Nature 343, 129–133.

Chyba, C.F., 1990b. Extraterrestrial amino acids and terrestrial life. Nature 348, 113–114.

Chyba, C.F., Sagan, C., 1987. Cometary organics but no evidence for bacteria. Nature 329, 208.

Chyba, C.F., Sagan, C., 1988. Cometary organic matter, still a contentious issue. Nature 332, 592.

Chyba, C., Sagan, C., 1992. Endogenous production, exogenous delivery and impact-shock synthesis of organic molecules: an inventory for the origins of life. Nature 255, 125–132.

Chyba, C.F., Thomas, P.J., Brookshawl, L., Sagan, C., 1990. Cometary delivery of organic molecules to the early earth. Science 249, 366–373.

Ciccarelli, F.D., Doerks, T., von Mering, C., Creevey, C.J., Snel, B., Bork, P., 2006. Toward automatic reconstruction of a highly resolved tree of life. Science 311, 1283–1287.

Ciechanover, A., December 8, 2004. Intracellular protein degradation: from a vague idea thru the lysosome and the ubiquitin-proteasome system and onto human diseases and drug targeting. In: Nobel Lecture.

Ciechanover, A., Gonen, H., 1990. The ubiquitin-mediated proteolytic pathway: enzymology and mechanisms of recognition of the proteolytic substrates. Semin. Cell Biol. 1, 415–422.

Cilia, M.L., Jackson, D., 2004. Plasmodesmata form and function. Curr. Opin. Cell Biol. 16, 500–506.

Ciska, M., Moreno Díaz de la Espina, S., 2013. NMCP/LINC proteins. Putative lamin analogs in plants? Plant Signal. Behav. 8, e2666.9.

Ciska, M., Moreno Díaz de la Espina, S., 2014. The intriguing plant nuclear lamina. Front. Plant Sci. 5, 166.

Ckurshumova, W., Caragea, A.E., Goldstein, R.S., Berleth, T., 2011. Glow in the dark: fluorescent proteins as cell and tissue-specific markers in plants. Mol. Plant 4, 794–804.

Clark, F.L.G., 1890. Paley's Natural Theology. E. & J. B. Young & Co., New York.

Clark, W.M., 1952. Topics in Physical Chemistry, second ed. Williams & Wilkins Co., Baltimore.

Clark, D.J., Felsenfeld, G., 1992. A nucleosome core is transferred out of the path of a transcribing polymerase. Cell 71, 11–22.

Clark, G., Kasten, F.H., 1983. History of Staining, third ed. Williams & Wilkins, Baltimore.

Clark, G.B., Thompson Jr., G., Roux, S.J., 2001. Signal transduction mechanisms in plants: an overview. Curr. Sci. 80, 170–177.

Clarke, H.T., 1941. Rudolf Schoenheimer, 1898-1941. Science 94, 553–554.

Clarke, F.M., Masters, C.J., 1975. On the association between glycolytic enzymes with structural proteins of skeletal muscle. Biochim. Biophys. Acta 381, 37–46.

Clarkson, D.T., Hanson, J.B., 1980. The mineral nutrition of higher plants. Annu. Rev. Plant Physiol. 31, 239–298.

Claros, M.G., Bautista, R., Guerrero-Fernández, Benzerki, H., Seoane, P., Fernández-Pozo, N., 2012. Why assembling plant genome sequences is so challenging. Biology 1, 439–459.

Clary, D.O., Griff, I.C., Rothman, J.E., 1990. SNAPs, a family of NSF attachment proteins involved in intracellular membrane fusion in animals and yeast. Cell 61, 709–721.

Claude, A., 1938. Concentration and purification of chicken tumor I agent. Science 87, 467–468.

Claude, A., 1939. Chemical composition of the tumor-producing fraction of chicken tumor I. Science 90, 213–214.

Claude, A., 1940. Particulate components of normal and tumor cells. Science 91, 77–78.

Claude, A., 1941. Cold Spring Harb. Symposia Quant. Biol. 9, 263–271.

Claude, A., 1943a. Distribution of nucleic acids in the cell and the morphological constitution of cytoplasm. In: Hoerr, N.L. (Ed.), Frontiers in Cytochemistry. The Jaques Cattell Press, Lancaster, PA, pp. 111–129.

Claude, A., 1943b. The constitution of protoplasm. Science 97, 451–456.

Claude, A., 1946a. Fractionation of mammalian liver cells by differential centrifugation. I. Problems, methods, and preparation of extract. J. Exp. Med. 84, 51–60.

Claude, A., 1946b. Fractionation of mammalian liver cells by differential centrifugation. II. Experimental procedures and results. J. Exp. Med. 84, 61–89.

Claude, A., 1948. Studies on cells: morphological, chemical constitutution, and distribution of biochemical functions. Harvey Lect. 43, 121–164.

Claude, A., 1970. Growth and differentiation of cytoplasmic membranes in the course of lipoprotein granule synthesis in the hepatic cell. J. Cell. Biol. 47, 745–766.

Claude, A., 1975. The coming of age of the cell. Science 189, 433–435.

Claude, A., December 12, 1974. The Coming of Age of the Cell. Nobel Lecture.

Clausius, R., 1859. On the mean length of the paths described by the separate molecules of gaeous bodies on the occurrence of molecular motion: together with some other remarks upon the mechanical theory of heat. Philos. Mag. Fourth Ser. 17, 81–91.

Clausius, R., 1860. On the dynamical theory of gases. Philos. Mag. Fourth Ser. 19, 484–486.

Clausius, R., 1879. The Mechanical Theory of Heat. MacMillan, London.

Clavijo, B.J., Venturini, L., Schudoma, C., Garcia Accinelli, G., Kaithakottil, G., et al., 2017. AN improved assembly and annotation of the allohexaploid wheat genome identifies complete families of agronomic genes and provides genomic evidence for chromosomal translocations. Genome Res. 27, 885–896.

Clawson, G.A., Feldherr, C.M., Smuckler, E.A., 1985. Nucleocytoplasmic RNA transport. Mol. Cell. Biochem. 67, 87–100.

Clayton, R.K., 1962. Primary reactions in bacterial photosynthesis. III. Reactions of carotenoids and cytochromes in illuminated bacterial chromatophore. Photochem. Photobiol. 1, 313–323.

Clayton, R.K., 1970. Light and Living Matter: Volume 1-The Physical Part; Volume 2-The Biological Part. Chemistry-Biology Interface Series. McGraw Hill Book Company, New York.

Clayton, R.K., 1980. Photosynthesis. Physical Mechanisms and Chemical Patterns. Cambridge University Press, Cambridge.

Clayton, R.K., 1988. Personal perspectives—memories of many lives. Photosynth. Res. 19, 207–224.

Clayton, R.K., 2002. Research on photosynthetic reaction centers from 1932 to 1987. Photosynth. Res. 73, 63–71.

Clayton, L., Lloyd, C.W., 1985. Actin organization during the cell cycle in meristematic plant cells. Exp. Cell Res. 156, 231–238.

Clayton, R.K., Wang, R.T., 1971. Photochemical reaction centers from *Rhodopseudomonas spheroides*. Methods Enzymol. 69, 696–704.

Cleary, A.L., Brown, R.C., Lemmon, B.E., 1992. Establishment of division plane and mitosis in monoplastidic guard mother cells of *Selaginella*. Cell Motil. Cytoskelet. 23, 89–101.

Cleary, A.L., Hardham, A.R., 1988. Depolymerization of microtubule arrays in root tip cells by oryzalin and the recovery by modified nucleation patterns. Can. J. Bot. 66, 2353–2366.

Cleary, A.L., Smith, L.G., 1998. The tangled1 gene is required for spatial control of cytoskeletal arrays associated with cell division during maize leaf development. Plant Cell 10, 1875–1888.

Cleaves, H.J., Miller, S.L., 2001. The nicotinamide biosynthetic pathway is a by-product of the RNA world. J. Mol. Evol. 52, 73–77.

Cleaves II, H.J., Nelson, K., Miller, S.L., 2006. The prebiotic synthesis of pyrimidines in frozen solution. Naturwissenschaften 93, 228–231.

Clegg, J.S., 1991. The organization of aqueous compartments in cultured animal cells. In: Tigyi, J., Kellermayer, M., Hazlewood, C.F. (Eds.), The Physical Aspect of the Living Cell. Akadémiai Kiadó, Budapest, pp. 129–143.

Clegg, J.S., Jackson, S.A., 1990. Glucose metabolism and the channeling of glycolytic intermediates in permeabilized L-929 cells. Arch. Biochem. Biophys. 278, 452–460.

Cleland, R.E., 1958. A separation of auxin-induced cell wall loosening into its plastic and elastic components. Physiol. Plant 11, 599–609.

Cleland, R.E., 1971a. Cell wall extension. Annu. Rev. of Plant Physiol. 22, 197–222.

Cleland, R.E., 1971b. The mechanical behavior of isolated *Avena* coleoptile walls subjected to constant stress. Properties and relation to cell elongation. Plant Physiol. 47, 805–811.

Cleland, R.E., 1981. Wall extensibility. In: Tanner, W., Loewus, F.A. (Eds.), Plant carbohydrates II. Extracellular carbohydrates, 13B, pp. 255–273. Encycl. Plant Physiol.

Cleland, R.E., 2002. The role of th eapoplastic pH in cell wall extension and cell enlargement. In: Handbook of Plant Growth. pH as the Master Variable. Marcel Dekker, New York, pp. 131–148.

Cleland, R.E., Karlsnes, A.M., 1967. A possible role of hydroxyproline-containing proteins in the cessation of cell elongation. Plant Physiol. 42, 669–671.

Cleland, R.E., Rayle, D.L., 1977. Reevaluation of the effect of calcium ions on auxin-induced elongation. Plant Physiol. 60, 709–712.

Cleves, A., McGee, T., Bankaitis, V.A., 1991. Phospholipid transfer proteins: A biological debut. Trends in Cell Biol. 1, 30–34.

Cline, K., Henry, R., 1996. Import and routing of nucleus-encoded chloroplast proteins. Annu. Rev. Cell Dev. Biol. 12, 1–26.

Clinton, H., Kaine, T., 2016. Stronger Together: A Blueprint for America's Future. Simon & Schuster, New York.

Clouse, J.W., Adhikary, D., Page, J.T., Ramaraj, T., Deyholos, M.K., Udall, J.A., Fairbanks, D.J., Jellen, E.N., Maughan, P.J., 2016. The amaranth genome: genome, transcriptome, and physical map assembly. Plant Genome 9. https://doi.org/10.3835/plantgenome 2015.07.0062.

Coate, J.E., Doyle, J.J., 2011. Divergent evolutionary fates of major photosynthetic gene networks following gene and whole genome duplications. Plant Signal. Behav. 6, 594–597.

Coate, J.E., Luciano, A.K., Seralathan, V., Minchew, K.J., Owens, T.G., Doyle, J.J., 2012. Anatomical, biochemical and photosynthetic responses to recent allopolyploidy in Glycine dolichocarpa. Am. J. Bot. 99, 55–67.

Coate, J.E., Schlueter, J.A., Whaley, A., Doyle, J.J., 2011. Comparative evolution of photosynthetic genes in response to polyploid and non-polyploid duplication. Plant Physiol. 155, 2081–2095.

Cobb, M., 2017. 60 years ago, Francis Crick changed the logic of biology. PLoS One 15, e2003243.

Cobbe, N., Heck, M.M.S., 2000. SMCs in the world of chromosome biology—from prokaryotes to higher eukaryotes. J. Struct. Biol. 129, 123–143.

Cocking, E.C., 1960. A method for the isolation of plant protoplasts and vacuoles. Nature 187, 962–963.

Cockrell, R.S., Harris, E.J., Pressman, B.C., 1967. Synthesis of ATP driven by a potassium gradient in mitochondria. Nature 215, 1487–1488.

Cody, C.W., Prasher, D.C., Westler, W.M., Prendergast, F.G., Ward, W.W., 1993. Chemical structure of the hexapeptide chromophore of the *Aequorea* green-fluorescent protein. Biochemistry 32, 1212–1218.

Coenen, C., Liedtke, S., Kogler, G., 2015. RNA amplification protocol leads to biased polymerase chain reaction results especially for low-copy transcripts of human bone marrow-derived stromal cells. PLoS One 10, e0141070.

Cohen, S.S., 1971. Are/were mitochondria and chloroplasts microorganisms. In: Monod, J., Borek, E. (Eds.), Of Microbes and Life. Columbia University Press, New York, pp. 129–149.

Cohen, I.B., 1997. Science and the Founding Fathers. W. W. Norton & Co., New York.

Cohen, S.N., 2013. DNA cloning: a personal view after 40 years. Proc. Natl. Acad. Sci. U.S.A. 110, 15521–15529.

Cohen, S.N., Chang, A.C.Y., Boyer, H.W., Helling, R.B., 1973. Construction of biologically functional bacterial plasmids in vitro. Proc. Natl. Acad. Sci. U.S.A. 70, 3240–3244.

Cohen, M.R., Drabkin, I.E., 1958. A Source Book in Greek Science. Harvard University Press, Cambridge, MA.

Cohen, J., Garreau de Loubresse, N., Beisson, J., 1984. Actin microfilaments in *Paramecium*: localization and role in intracellular movements. Cell Motil. 4, 443–468.

Cohn, F., 1853. On the natural history of Protococcus pluvialis. In: Henfrey, A. (Ed.), Botanical and Physiological Memoirs. Ray Society, London, pp. 515–564.

Cohn, E.J., 1935. The chemistry of the proteins and amino acids. Ann. Rev. Biochem. 4, 93–148.

Cohn, M., 1989. The way it was: a commentary by Melvin Cohn. Biochim. Biophys. Acta 1000, 109–112.

Cohn, S.A., Tippit, D.H., Spurck, T.P., 1986. Microtubule dynamics in the spindle. II. A thermodynamic and kinetic description. J. Theor. Biol. 122, 277–301.

Colasanti, J., Cho, S.-O., Wick, S., Sundaresan, V., 1993. Localization of the functional p34cdc2 homology of maize in root tip and stomatal complex cells: association with predicted division sites. Plant Cell 5, 1101–1111.

Cole, F.J., 1930. Early Theories of Sexual Generation. Clarendon Press, Oxford.

Cole, L.C., 1957. Biological clock in the unicorn. Science 125, 874–876.

Cole, K.S., 1968. Membranes, Ions and Impulses. University of California Press, Berkeley.

Cole, K.S., 1979. Mostly membranes. Annu. Rev. Physiol. 41, 1–24.

Cole, L., Coleman, J., Evans, D., Hawes, C., 1990. Internalization of fluorescein isothiocyanate and fluorescein isothiocyanate-dextran by suspension-cultured plant cells. J. Cell Sci. 96, 721–730.

Cole, K.S., Curtis, H.J., 1938. Electrical impedance of *Nitella* during activity. J. Gen. Physiol. 22, 37–64.

Cole, K.S., Curtis, H.J., 1939. Electrical impedance of the squid axon during activity. J. Gen. Physiol. 22, 649–670.

Coleman, J.O.D., Evans, E.D., Hawes, C., Horsley, D., Cole, L., 1987. Structure and molecular organization of higher plant coated vesicles. J. Cell Sci. 88, 35–45.

Coleman, J.O.D., Evans, E.D., Horsley, D., Hawes, C.R., 1991. The molecular structure of plant clathrin and coated vesicles. In: Hawes, C.R., Coleman, J.O.D., Evans, D.E. (Eds.), Endocytosis, Exocytosis and Vesicle Traffic in Plants. Cambridge University Press, Cambridge, pp. 41–63.

Coleman, A.W., Maguire, M.J., 1983. Cytological detection of the basis of uniparental inheritance of plastid DNA in *Chlamydomonas moewusii*. Curr. Genet. 7, 211–218.

Coleridge, S.T., 1835. In: Specimens of the Table Talk of the Late Samuel Taylor Coleridge, vol. 2. John Murray, London.

Coleridge, S.T., 1884. In: Shedd, W.G.T. (Ed.), The Complete Works of Samuel Taylor Coleridge, vol. VI. Harper & Brothers, New York.

Collado-Romero, M., Alós, E., Prieto, P., 2014. Unraveling the proteomic profile of rice meiocytes during early meiosis. Front. Plant Sci. 5, 356.

Collander, R., 1924. Über die Durchlässigkeit der Kupferferrozyanidniederschlagsmembran für Nichtelektrolyte. Kolloidchem. Beihefte 19, 72–105.

Collander, R., 1925. Über die Durchlässigkeit der Kupferferrocyanidmembran für Säuren, nebst Bemerkungen zur Ultrafilterfunktion des Protoplasmas. Kolloidchem. Beihefte 20, 273–287.

Collander, R., 1937. The permeability of plant protoplasts to non-electrolytes. Trans. Faraday Soc. 33, 985–990.

Collander, R., 1959. Cell membranes: their resistance to penetration and their capacity for transport. In: Stewart, F.C. (Ed.), Plant Physiology. A treatise. Plants in relation to water and solutes, 2. Academic Press, New York, London, pp. 3–102.

Collander, P.R., 1965. Ernest Overton (1865-1933): a pioneer to remember. Leopoldina 3 (8/9), 242–254.

Collander, R., Bärlund, H., 1933. Permeabilitätsstudien an *Chara ceratophylla*. Acta Botanica Fennica 11, 1–114.

Colletti, K.S., Tattersall, E.A., Pyke, K.A., Froelich, J.E., Stokes, K.D., Osteryoung, K.W., 2000. A homologue of the bacterial cell division site-determining factor MinD mediates placement of the chloroplast division apparatus. Curr. Biol. 10, 507–516.

Collings, D.A., Carter, C.N., Rink, J.C., Scott, A.C., Wtatt, S.E., 2000. Plant nuclei can contain extensive grooves and invaginations. Plant Cell 12, 2425–2439.

Collings, D.A., Gebbie, L.K., Howles, P.A., Hurley, U.A., Birch, R.J., Cork, A.H., Hocart, C.H., Arioli, T., Williamson, R.E., 2008. *Arabidopsis* dynamin-like protein DRP1A: a null mutant with widespread defects in endocytosis, cellulose synthesis, cytokinesis, and cell expansion. J. Exp. Bot. 59, 361–376.

Collings, D.A., Harper, J.D.I., Marc, J., Overall, R.L., Mullen, R.T., 2002. Life in the fast lane: actin-based motility of plant peroxisomes. Can. J. Bot. 80, 430–441.

Collings, D.A., Zsuppan, G., Allen, N.S., Blancaflor, E.B., 2001. Demonstration of prominent actin filaments in the root columella. Planta 212, 392–403.

Collins, F.S., 1992. Positional cloning: let's not call it reverse anymore. Nat. Genet. 1, 3–6.

Collins, K., Sellers, J.R., Matsudaira, P., 1990. Calmodulin dissociation regulates brush border myosin I 110K calmodulin mechanochemical activity in vitro. J. Cell Biol. 120, 1137–1148.

Collonnier, C., Epert, A., Mara, K., Maclot, F., Guyon-Debast, A., Charlot, F., White, C., Schaefer, D.G., Nogué, F., 2016. CRISPR-Cas9-mediated efficient directed mutagenesis and RAD51-dependent and RAD51-independent gene targeting in the moss *Physcomitrella patens*. Plant Biotechnol. J. 15, 122–131.

Coluccio, L.M., Bretscher, A., 1987. Calcium-regulated cooperative binding of the microvillar 110K-calmodulin complex to F actin. Formation of decorated filaments. J. Cell Biol. 105, 325–334.

Coman, D.R., 1940. Additional observations on positive and negative chemotaxis: experiments with a myxomycete. Arch. Pathol. 29, 220–228.

Comfort, N., 2012. The Science of Human Perfection. How Genes Became the Heart of American Medicine. Yale University Press, New Haven, CT.

Commoner, B., 1940. Cyanide inhibition as a means of elucidating the mechanisms of cellular respiration. Biol. Rev. 15, 168–201.

Compton, A.H., 1921a. Possible magnetic polarity of free electrons. Philos. Mag. 16, 279–281.

Compton, A.H., 1921b. The magnetic electron. J. Franklin Inst. 192, 145–155.

Compton, A.H., 1924. A general quantum theory of the wavelength of scattered x-rays. Phys. Rev. 24, 168–176.

Compton, A.H., 1931. The uncertainty principle and free will. Science 74, 172.

Compton, D.A., Yen, T.J., Cleveland, D.W., 1991. Identification of novel centromere/kinetochore-associated proteins using monoclonal antibodies generated against human mitotic chromosome scaffolds. J. Cell. Biol. 112, 1083–1097.

Compton, P.D., Zamdborg, L., Thomas, P.M., Kelleher, N.L., 2011. On the scalability and requirements of whole protein mass spectrometry. Anal. Chem. 83, 6868–6874.

Comtet, J., Jensen, K.H., Turgeon, R., Stroock, A.D., Hosoi, A.E., 2017a. Passive phloem loading and long-distance transport in a synthetic tree-on-a-chip. Nat. Plants 3, 17032.

Comtet, J., Turgeon, R., Stroock, A.D., 2017b. Phloem leading through plasmodesmata: a biophysical analysis. Plant Physiol. https://doi.org/10.1104/pp.16.01041.

Conant, J.B., 1947. On Understanding Science. An Historical Approach. Yale University Press, New Haven, CT.

Conant, J.B., 1970. My Several Lives. Memoirs of a Social Inventor. Harper & Row, New York.

Conceicao, A.dD.S., Marty-Mazars, D., Bassham, D.C., Sanderfoot, A.A., Marty, F., Raikhel, N.V., 1997. The syntaxin homolog AtPEP12p resides on a late post-Golgi compartment in plants. Plant Cell 9, 571–582.

Condeelis, J.S., 1974. The identification of F actin in the pollen tube and protoplast of *Amaryllis belladonna*. Exp. Cell Res. 88, 435–439.

Cone, K.C., 2007. Anthocyanin synthesis in maize aleurone tissue. Plant Cell Monogr. 8, 121–139.

Coneva, V., Chitwood, D.H., 2015. Plant architecture without multicellularity: quandries over patterning and the soma-germline divide in siphonous algae. Front. Plant Sci. 6, 287.

Cong, L., Ran, A.A., Cox, D., Lin, S., Barretto, R., Habib, N., Hsu, P.S., Wu, X., Jiang, W., Marraffini, L.A., Zhang, F., 2013. Multiplex genome engineering using CRISPR/Cas systems. Science 339, 819–823.

Conklin, E.G., 1923. Heredity and Environment in the Development of Man, fifth ed. Princeton University Press, Princeton, NJ.

Conklin, E.G., 1940. Cell and protoplasm concepts: historical account. In: Moulton, F.R. (Ed.), The Cell and Protoplasm. The Science Press, Washington, DC, pp. 6–19.

Conklin, E.G., 1950. Cleavage and differentiation in marine eggs. Ann. N.Y. Acad. Sci. 51 (8), 1279.

Conklin, P.L., Wilson, R.K., Hanson, M.R., 1991. Multiple transsplicing events are required to produce a mature nad1 transcript in a plant mitochondrion. Genes Dev. 5, 1407–1415.

Conley, C.A., Hanson, M.R., 1994. Tissue-specific protein expression in plant mitochdondria. Plant Cell 6, 85–91.

Connell, C., Fung, S., Heiner, C., Bridgham, J., Chakerian, V., Heron, E., Jones, B., Menchen, S., Mordan, W., Raff, M., Recknor, M., Smith, L.M., Springer, J., Woo, S., Hunkapiller, M.W., 1987. Automated DNA sequence analysis. BioTechniques 5, 342–348.

Connolly, J.H., Berlyn, G., 1996. The plant extracellular matrix. Can. J. Bot. 74, 1545–1546.

Conrad, P.A., Binari, L.L.W., Racusen, R.H., 1982. Rapidly-secreting, cultured oat cells serve as a model system for the study of cellular exocytosis. Characterization of cells and isolated secretory vesicles. Protoplasma 112, 196–204.

Conrad, P.A., Steucek, G.L., Hepler, P.K., 1986. Bud formation in *Funaria*: organelle redistribution following cytokinin treatment. Protoplasma 131, 211–223.

Consden, R., Gordon, A.H., Martin, A.J.P., 1944. Qualitative analysis of proteins: a partition chromatographic method using paper. Biochem. J. 38, 224–232.

Consden, R., Gordon, A.H., Martin, A.J.P., Synge, R.L.M., 1947. Gramicidin S: the sequence of the amino acid residues. Biochem. J. 41, 596–602.

Contreras, F.-X., Sanchez-Magraner, L., Alonso, A., Goñi, F.M., 2010. Transbilayer (flip-flop) lipid motion and lipid scrambling in membranes. FEBS Lett. 584, 1779–1786.

Conway, E.J., 1953. A redox pump for the biological performance of osmotic work, and its relation to the kinetics of free ion diffusion across membranes. Int. Rev. Cytol. 2, 419–445.

Cook, P.R., 1990. How mobile are active RNA polymerases? J.Cell Sci. 96, 189–192.

Cooke, C.A., Bernat, R.L., Earnshaw, W.C., 1990. CENP-B: a major human centromere protein located beneath the kinetochore. J. Cell Biol. 110, 1475–1488.

Cooke, M.E., Graham, L.E., 1999. Evolution of plasmodesmata. In: van Bel, A.J.E., van Kesteren, W.J.P. (Eds.), Plasmodesmata. Structure, Function, Role in Cell Communication. Springer, Berlin, pp. 101–117.

Cooke, M.E., Graham, L.E., Botha, C.E.J., Lavin, C.A., 1997. Comparative ultrastructure of plasmodesmata of *Chara* and selected bryophytes: toward an elucidation of the evolutionary origin of plant plasmodesmata. Am. J. Bot. 84, 1169–1178.

Cooke, C.A., Heck, M.M.S., Earnshaw, W.C., 1987. The inner centromere protein (INCENP) antigens: movement from inner centromere to midbody during mitosis. J. Cell Biol. 105, 203–2067.

Cooke, T.J., Paolillo, D.J., 1980. The control of the orientation of cell divisions in fern gametophytes. Am. J. Bot. 67, 1320–1333.

Cooke, T.J., Racusen, R.H., 1982. Cell expansion in the filamentous gametophyte of the fern *Onoclea sensibilis* L. Planta 155, 449–458.

Cooks, R.G., 2004. John Beynon at Purdue. Rapid Commun. Mass Spectrom. 18, 7–10.

Cooper, J.A., 1987. Effects of cytochalasins and phalloidin on actin. J. Cell Biol. 105, 1473–1478.

Cooper, T.G., Beevers, H., 1969. ß–oxidation in glyoxysomes from castor bean endosperm. J. Biol. Chem. 244, 3514–3520.

Cooper, M.S., Keller, R.E., 1984. Perpendicular orientation and directional migration of amphibian neural crest cells in DC fields. Proc. Natl. Acad. Sci. U.S.A. 81, 160–164.

Cooper, M.S., Schliwa, M., 1985. Electrical and ionic control of tissue cell locomotion in DC electric fields. J. Neurosci. Res. 13, 223–244.

Cope, M.J.T.V., Whisstock, J., Rayment, I., Kendrick-Jones, J., 1996. Conservation within the myosin motor domain—implication for structure and function. Structure 4, 969–987.

Cori, C.F., 1942. Phosphorylation of carbohydrates. In: A Symposium on Respiratory Enzymes. The University of Wisconsin Press, Madison, pp. 175–189.

Cori, C.F., 1946. Enzymatic reactions in carbohydrate metabolism. Harvey Lect. 41, 253–272.

Cori, G.T., Cori, C.F., 1939. The activating effect of glycogen on the enzymatic synthesis of glycogen from glucose-1 phosphate, 131, 397–398.

Corliss, J.O., 1980. Objection to "undulipodium" as an inappropriate and unnecessry term. BioSystems 12, 109–110.

Cornejo, M.J., Platt–Aloia, K.A., Thomson, W.W., Jones, R.L., 1988. Effects of GA_3 and Ca^{2+} on barley aleurone protoplasts: a freeze-fracture study. Protoplasma 146, 157–165.

Cornish-Bowden, A., 2013. The origina of enzyme kinetics. FEBS J. 587, 2725–2730.

Cornish-Bowden, A., 2015. One hundred years of Michaelis-Menten kinetics. Perspect. Sci. 4, 3–9.

Cornman, I., 1944. A summary of evidence in favor of the traction fiber in mitosis. Am. Nat. 78, 410–422.

Corpas, F.J., Barroso, J.B., Carreras, A., Quiros, M., Leon, A.M., Romero-Puertas, M.C., Esteban, F.J., Valderrama, R., Palma, J.M., Sandalio, L.M., Gomez, M., del Rio, L.A., 2004. Cellular and subcellular localization of endogenous nitric oxide in young and senescent pea plants. Plant Physiol. 136, 2722–2733.

Corpas, F.J., Barroso, J.B., del Rio, L.A., 2001. Peroxisomes as a source of reactive oxygen species and nitric oxide signal molecules in plant cells. Trends Plant. Sci. 6, 145–150.

Corpas, F.J., Sandalio, L.M., Brown, M.J., del Río, L.A., Trelease, R.N., 2000. Identification of porin-like polypeptide(s) in the boundary membrane of oilseed glyoxysomes. Plant Cell Physiol. 41, 1218–1228.

Corrêa Jr., A., Staples, R.C., Hoch, H.C., 1996. Inhibition of thigmostimulated cell differentiation with RGD-peptides in *Uromyces* germlings. Protoplasma 194, 91–102.

Correns, C., 1900. G. Mendel's law concerning the behavior of progeny of varietal hybrids, 1950 Genetics 35, 33–41.

Correns, C., 1909. Vererbungsversuche mit blas (gleb) grunen und bunt blattigen sippen bei *Mirabilis, Urtica* and *Lunaria*. Z. Vererb. 1, 291–329.

Corriveau, J.L., Coleman, A.W., 1988. Rapid screening method to detect potential biparental inheritance of plastid DNA and results for over 200 angiosperm species. Am. J. Bot. 75, 1443–1458.

Corthesy-Theulaz, I., Pauloin, A., Pfeffer, S.R., 1992. Cytoplasmic dynein participates in the centrosomal localization of the Golgi complex. J. Cell Biol. 118, 1333–1345.

Corti, B., 1774. Osservazioni Microscopiche Sulla Tramella e Sulla Circulazione del Fluido in una Pianta Aquajuola. Lucca.

Cory, S., Marcker, K.A., Dube, S.K., Clark, B.F., 1968. Primary structure of a methionine transfer RNA of *Escherichia coli*. Nature 220, 1039–1040.

Coryell, C.D., 1940. The proposed terms "exergonic" and "endergonic" for thermodynamics. Science 92, 380.

Cosgrove, D.J., 1986. Biophysical control of plant growth. Ann. Rev. Plant Physiol. 37, 377–405.

Cosgrove, D.J., 1987. Wall relaxation and the driving forces for cell expansive growth. Plant Physiol. 84, 561–564.

Cosgrove, D.J., 1993a. How do plant cell walls extend? Plant Physiol. 102, 1–6.

Cosgrove, D.J., 1993b. Wall extensibility: its nature measurement and relationship to plant cell growth. New Phytol. 124, 1–23.

Cosgrove, D.J., 1993c. Water uptake by growing cells: an assessment of the controlling roles of wall relaxation, solute uptake, and hydraulic conductance. Int. J. Plant Sci. 154, 10–21.

Cosgrove, D.J., 1997. Cellular mechanisms underlying growth asymmetry during stem gravitropism. Planta 203, S130–S135.

Cosgrove, D.J., 1999. Enzymes and other agents that enhance cell wall extensibility. Annu. Rev. Plant Physiol. Plant Mol. Biol. 50, 391–417.

Cosgrove, D.J., 2014. Re-constructing our models of cellulose and primary cell wall assembly. Curr. Opin. Cell Biol. 22, 122–131.

Cosgrove, D.J., 2015. Plant expansins: diversity and interactions with plant cell walls. Curr. Opin. Plant Biol. 25, 162–172.

Cosgrove, D.J., 2016a. Catalysts of plant cell wall loosening. F1000 Research 5, 119.

Cosgrove, D.J., 2016b. Plant cell wall extensibility: connecting plant cell growth with cell wall structure, mechanics, and the action of wall-modifying enzymes. J. Exp. Bot. 67, 463–476.

Cosgrove, D.J., Bedinger, P.A., Durachko, D.M., 1997. Group I allergens of grass pollen as cell wall loosening agents. Proc. Natl. Acad. Sci. U.S.A. 94, 6559–6564.

Cosgrove, D.J., Gilroy, S., Kao, T.-H., Ma, H., Schultz, J.C., 2000. Plant signaling 2000. Cross talk among geneticists, physiologists, and ecologists. Plant Physiol. 124, 499–506.

Cosgrove, D.J., Hedrich, R., 1991. Stretch activated chloride, potassium and calcium channels coexisting in plasma membranes of guard cells of *Vicia faba*. L. Planta 186, 143–153.

Cosgrove, D.J., Li, L.C., Cho, H.-T., Hoffmann-Benning, S., Moore, R.C., Blecker, D., 2002. The growing world of expansins. Plant Cell Physiol. 43, 1436–1444.

Costa, M., Hilliou, F., Duarte, P., Pereira, L., Almeida, I., Leech, M., Memelink, J., Bartceló, A., Sottomayor, M., 2008. Molecular cloning and characterization of a vacuolar class III peroxidase involved in the metabolism of anticancer alkaloids in *Catharanthus roseus*. Plant Physiol. 146, 403–417.

Coté, G.G., Crain, R.C., 1993. Biochemistry of phosphoinositides. Ann. Rev. Plant Physiol. Plant Mol. Biol. 44, 333–356.

Coté, G.G., Morse, M.J., Crain, R.C., Satter, R.L., 1987. Isolation of soluble metabolites of the phosphatidlyinositol cycle from *Samanea saman*. Plant Cell Rep. 6, 352–355.

Coue, M., Lombillo, V.A., McIntosh, J.R., 1991. Microtubule depolymerization promotes particle and chromosome movement in vitro. J. Cell Biol. 112, 1165–1175.

Coughlan, S.J., Hastings, C., Winfrey Jr., R.J., 1996. Molecular characterization of plant endoplasmic reticulum. Identification of protein disulfide-isomerase as the major reticuloplasmin. Eur. J. Biochem. 235, 215–224.

Coughlin, S.R., 1994. Expanding horizons for receptors coupled to G-proteins: diversity and disease. Curr. Opin. Cell Biol. 6, 191–197.

Covello, P.S., Gray, M.W., 1992. Silent mitochondria and active nuclear genes for subunit 2 of cytochrome c oxidase (cox 2) in soybean: evidence for RNA-mediated gene transfer. EMBO J. 11, 3815–3820.

Cowan, C.R., Carlton, P.M., Cande, W.Z., 2001. The polar arrangement of telomeres in interphase and meiosis. Rab1 organization and the bouquet. Plant Physiol. 125, 532–538.

Cowdry, E.V., 1924. Cytological constituents—mitochondria, Golgi apparatus, and chromidial substance. In: Cowdry, E.V. (Ed.), General Cytology. University of Chicago Press, Chicago, pp. 313–382.

Cowperthwaite, M., Park, W., Xu, Z., Yan, X., Maurais, S.C., Dooner, H.K., 2002. Use of the transposon *Ac* as a gene-searching engine in the maize genome. Plant Cell 14, 713–726.

Cox, R.P. (Ed.), 1974. Cell Communication. John Wiley & Sons, New York.

Cozzarelli, N.R., Melechen, N.E., Jovin, T.M., Kornberg, A., 1967. Polynucleotide cellulose as a substrate for a polynucleotide ligase induced by phage T4. Biochem. Biophys. Res. Commun. 28, 578–586.

Cozzolino, D., Fassio, A., Restaino, E., Vicente, E., 2015. Instrumental techniques and methods: their role in plant omics. In: Barh, D., Sarwar Khan, M., Davies, E. (Eds.), PlantOmics: The Omics of Plant Science. Springer India, pp. 33–52.

Craig, S., Goodchild, D.J., 1984. Golgi–mediated vicilin accumulation in pea cotyledon cells is redirected by monensin and nigericin. Protoplasma 122, 91–97.

Craighead, H.C., 2000. Nanoelectromechanical systems. Science 290, 1532–1535.

Crain, R.C., Zilversmit, D.B., 1980. Two nonspecific exchange proteins from beef liver. Biochemistry 19, 1440–1447.

Cram, W.J., 1980. Pinocytosis in plants. New Phytol. 84, 1–17.

Cramer, L.P., 2000. Myosin VI: roles for a minus end-directed actin motor in cells. J. Cell Biol. 150, F121–F126.

Cramer, W.A., Knaff, D.B., 1991. Energy Transduction in Biological Membranes. A Textbok of Bioenergetics. Springer-Verlag, New York.

Cramer, L.P., Mitchison, T.J., Theriot, J.A., 1994. Actin-dependent motile forces and cell motility. Curr. Opin. Cell Biol. 6, 82–86.

Crane, F., Hatefi, Y., Lester, R., Widmer, C., 1957. Isolation of a quinone from beef heart mitochondria. Biochim. Biophys. Acta 25, 220–221.

Crawford, K.M., Zambryski, P.C., 2000. Subcellular localization determined the availability of non-targeted proteins to plasmodesmal transport. Curr. Biol. 10, 1032–1040.

Creed, R.S., Denny-Brown, D., Eccles, J.C., Liddell, E.G.T., Sherrington, C.S., 1932. Reflex Activity of the Spinal Cord. Clarendon Press, Oxford.

Creelman, R.A., Mullet, J.E., 1991. Abscisic acid accumulates at positive turgor potential in excised soybean seedling growing zones. Plant Physiol. 95, 1209–1213.

Creighton, H.B., McClintock, B., 1931. A correlation of cytological and genetical crossing-over in *Zea mays*. Proc. Natl. Acad. Sci. U.S.A. 17, 492–497.

Crepin, A., Caffarri, S., 2015. The specific localizations of phosphorylated Lhcb1 and Lhcb2 isoforms reveal the role of Lhcb2 in the formation of the PSI-LHCII supercomplex in *Arabidopsis* during state transitions. Biochim. Biophys. Acta 1847, 1539–1548.

Crick, F.H.C., 1950. The physical properties of cytoplasm. A study by means of the magnetic particle method. Part II. Theoretical treatment. Exp. Cell Res. 1, 505–533.

Crick, F.H.C., 1954. The structure of the hereditary material. Sci. Am. 191, 54–61.

Crick, F.H.C., 1955. The structure of the hereditary material. In: Scientific American (Ed.), The Physics and Chemistry of Life. Simon and Schuster, New York, pp. 118–142.

Crick, F.H.C., 1958. On protein synthesis. Soc. Exp. Biol. 12, 138–163.

Crick, F.H.C., 1966. Codon—anticodon pairing: the wobble hypothesis. J. Mol. Biol. 19, 548–555.

Crick, F.H.C., 1968. The origin of the genetic code. J. Mol. Biol. 38, 367–379.

Crick, F., 1970. Central dogma of molecular biology. Nature 227, 561–563.

Crick, F., 1974a. Central dogma of molecular biology. Nature 227, 561–563.

Crick, F., 1974b. The double helix: a personal view. Nature 248, 66–769.

Crick, F., 1981. Life Itself. Simon and Schuster, New York.

Crick, F., 1988. What Mad Pursuit. A Personal View of Scientific Discovery. Basic Books, New York.

Crick, F., 1994. The Astonishing Hypothesis. Charles Scribner's Sons, New York.

Crick, F.H.C., 1995. DNA: a cooperative discovery. Ann. N.Y. Acad. Sci. 758, 198–199.

Crick, F.H.C., Barnett, L., Brenner, S., Watts-Tobin, R.J., 1961. General nature of the genetic code for proteins. Nature 192, 1227–1232.

Crick, F.H.C., December 11, 1962. On the genetic code. Nobel Lecture.

Crick, F.H.C., Griffith, J.S., Orgel, L.E., 1957. Codes without commas. Proc. Natl. Acad. Sci. U.S.A. 43, 416–421.

Crick, F.H.C., Hughes, A.F.W., 1950. The physical properties of the cytoplasm. A study by means of the magnetic particle method. Exp. Cell Res. 1, 37–80.

Criqui, M.C., Genschik, P., 2002. Mitosis in plants: how far we have come at the molecular level? Curr. Opin. Plant Biol. 5, 487–493.

Croft, J.A., Bridger, J.M., Boyle, S., Perry, P., Teague, P., Bickmore, W.A., 1999. Differences in the localization and morphology of chromosomes in the human nucleus. J. Cell. Biol. 145, 1119–1131.

Croll, J., 1885. Climate and Time in Their Geological Relations: A Theory of Secular Changes of the Earth's Climate. D. Appleton and Co., New York.

Cronin, J.R., Pizzarello, S., Cruikshank, D.P., 1988. Organic matter in carbonaceous chondrites, planetary satellites, asteroids, and comets, 819–857.

Cross, C.F., Bevan, E.J., 1895. Cellulose. An Outline of the Chemistry of the Structural Elements of Plants with Reference to their Natural History and Industrial Uses. Longmans, Green, and Co., London.

Cross, L.L., Ebeed, H.T., Baker, A., 2016. Peroxisome biogenesis, protein targeting mechanisms and PEX gene functions in plants. Biochim. Biophys. Acta 1863, 850–862.

Crosthwaite, S.C., MacDonald, F.M., Baydoun, E.A.-H., Bret, C.T., 1994. Properties of a protein-linked glucuronoxylan formed in the plant Golgi apparatus. J. Exp. Bot. 45, 471–475.

Crowley, K.S., Liao, S., Worrell, V.E., Reinhard, G.D., Johnson, A.E., 1994. Secretory proteins move through the endoplasmic reticulum membrane via an aqueous pore. Cell 78, 461–471.

Crowley, K.S., Reinhard, G.D., Johnson, A.E., 1993. The signal sequence moves through a ribosomal tunnel into a noncytoplasmic aqueous environment at the ER membrane early in translocation. Cell 73, 1101–1115.

Cruz Castillo, M., Martínez, C., Buchala, A., Métraux, J.-P., León, J., 2004. Gene-specific involvement of β-oxidation in wound-activated responses in *Arabidopsis*. Plant Physiol. 135, 85–94.

Cui, Y., Barampuram, S., Stacey, M.G., Hancock, C.N., Findley, S., Mathieu, M., Zhang, Z., Parrott, W.A., Stacey, G., 2013. Tnt1 retrotransposon mutagenesis: a tool for soybean functional genomics. Plant Physiol. 161, 36–47.

Cui, L.L., Lu, Y.-S., Li, Y., Yang, C., Peng, X.-X., 2016. Overexpression of glycolate oxidase confers improved photosynthesis under high light and high temperature in rice. Front. Plant Sci. 7, 1165.

Cullis, C.A., 1977. Molecular aspects of the environmental induction of heritable changes in flax. Heredity 38, 129–154.

Cullis, C.A., 1981. Environmental induction of heritable changes in flax: defined environments inducing changes in rDNA and peroxidase isozyme band pattern. Heredity 47, 87–94.

Cullis, C.A., 1990. DNA rearrangements in response to environmental stress. Adv. Genet. 28, 73–97.

Cullis, C.A., 2008. AN overview on plant genome initiatives. In: Kole, C., Abbott, A.G. (Eds.), Principles and Practices of Plant Genomics. Volume 1: Genome Mapping. Science Publishers, Enfield, NH, pp. 303–328.

Curl Jr., R.F., December 7, 1996. Dawn of the fullerenes: Experiment and conjecture. Nobel Lecture.

Curtin, S.J., Tiffin, P., Guhlin, J., Trujillo, D.I., Burghardt, L.T., Atkins, P., Baltes, N.J., Denny, R., Voytas, D.F., Stupar, R.M., Young, N.D., 2017b. Validating genome-wide association candidates controlling quantitative variation in nodulation. Plant Physiol. 173, 921–931.

Curtin, S.J., Xiong, Y., Michno, J.M., Campbell, B.W., Stec, A.O., Čermák, T., Starker, C., Voytas, D.F., Eamens, A.L., Stupar, R.M., 2017a. CRISPR/Cas9 and TALENs generate heritable mutations for genes involved in small RNA processing of Glycine max and *Medicago truncatula*. Plant Biotechnol. J. https://doi.org/10.1111/pbi.12857.

Cutler, D.F., Alvin, K.L., Price, C.E. (Eds.), 1982. The Plant Cuticle. Academic Press, New York.

Cutler, S.R., Ehrhardt, D.W., Griffitts, J.S., Somerville, C.R., 2000. Random GFP: cDNA fusions enable visualization of subcellular structures in cells of *Arabidopsis* at a high frequency. Proc. Natl. Acad. Sci. U.S.A. 97, 3718–3723.

Cutler, S., McCourt, P., 2005. Dude, where's my phenotype? Dealing with redundancy in signaling networks. Plant Physiol. 138, 558–559.

Cuzin, M., 2001. DNA chips: a new tool for genetic analysis and diagnostics. Transfus. Clin. Biol. 8, 291–296.

Cvrčková, F., 2000. Are plant formins integral membrane proteins? Genome Biol. 1, 0011.

Cyr, R.J., 1991a. Calcium/calmodulin affects microtubule stability in lysed protoplasts. J. Cell Sci. 100, 311–317.

Cyr, R.J., 1991b. Microtubule-associated proteins in higher plants. In: Lloyd, C. (Ed.), The Cytoskeletal Basis of Plant Growth and Form. Academic Press, London, pp. 57–67.

Cyr, R.J., Fisher, D., 2012. Plant Cell Division and Its Unique Features. eLS.

Cyr, R.J., Palevitz, B.A., 1989. Microtubule–binding proteins from carrot. I. Initial characterization and microtubule bundling. Planta 177, 245–260.

Cyranoski, D., Gilbert, N., Ledford, H., Nayar, A., Yahia, M., 2011. The PhD factory: the world is producing more PhDs than ever before. Is it time to stop? Nature 472, 276–279.

Cyranoski, D., Gilbert, N., Ledford, H., Nayar, A., Yahia, M., 2011a. Education: the PhD factory. Nature 472, 276–279.

Cyranoski, D., Gilbert, N., Ledford, H., Nayar, A., Yahia, M., 2011b. The PhD factory: the world is producing more PhDs than ever before. Is it time to stop? Nature 472, 276–279.

Czapek, F., 1911. Chemical Phenomena in Life. Harper & Brothers, London.

Dainty, J., 1960. Ion transport across plant cell membranes. Proc. R. Phys. Soc. Edin. 28, 3–14.

Dainty, J., 1962. Ion transport and electrical potentials in plant cells. Annu. Rev. Plant Physiol. 13, 379–402.

Dainty, J., 1968. The structure and possible function of the vacuole. In: Pridheim, J.B. (Ed.), Plant Cell Organelles. Academic Press, London, pp. 40–46.

Dainty, J., 1990. Prefatory chapter. Annu. Rev. Plant Physiol. Plant Mol. Biol. 41, 1–20.

Dainty, J., 1998. The discovery of water potential. Discov. Plant Biol. 2, 271–285.

Dairman, M., Donofrio, N., Domozych, D.S., 1995. The effects of brefeldin A upon the Golgi apparatus of the green algal flagellate, Gloeomonas kupffer 283, 181–186.

Dale, H., December 12, 1936. Some Recent Extensions of the Chemical Transmission of the Effects of Nerve Impulses. Nobel Lecture.

Dalma-Weiszhausz, D.D., Warrington, J., Tanimoto, E.Y., Miyada, C.G., 2006. The Affymetrix GeneChip® Platform: an overview. Methods Enzymol. 410, 3–28.

Dalrymple, G.B., 1991. The Age of the Earth. Stanford University Press, Stanford, CA.

Dalton, A.J., 1961. Golgi apparatus and secretion granules. In: Brachet, J., Mirsky, A.E. (Eds.), The Cell. Biochemistry. Physiology, Morphology. Cells and their Component Parts, vol. 2. Academic Press, New York, pp. 603–619.

Dalton, A.J., Felix, M.D., 1956. A comparative study of the Golgi complex. J. Biophys. Biochem. Cytol. 2 (Suppl.), 79–84.

Damineli, D.S.C., Portes, M.T., Feijó, J.A., 2017. One thousand and one oscillators at he pollen tube tip: the quest for a central pacemaker. In: Obermeyer, G., Feijó, J. (Eds.), Pollen Tip Growth. Springer International, pp. 391–413.

Damsky, C.H., Werb, Z., 1992. Signal transduction by integrin receptors for extracellular matrix: cooperative processing of extracellular information. Curr. Opin. Cell Biol. 4, 772–781.

Dang, Y.-Q., Li, Q., Wang, K., Wu, Y., Lian, L., Zou, B., 2012. Hydrostatic pressure effects on the fluorescence and FRET behavior of Cy3-labeled phycocyanin system. J. Phys. Chem. B 116, 11010–11016.

Dangeard, P.A., 1919. Sur la distinction du chondriome des auteurs en vacuome, plastidome et sphérome. C. R. Acad. Sci. 169, 1005–1010.

Daniel, R., 2005. The metagenomics of soil. Nat. Rev. Microbiol. 3, 470–478.

Daniel, P.P., Bryant, J.A., 1988. DNA ligase in pea (*Pisum sativum* L.) seedlings: changes in activity during germination and effects of deoxyribonucleotides. J. Exp. Bot. 39, 481–486.

Daniell, H., Carmona-Sanchez, O., Burns, B.B., 2004. Chloroplast-derived vaccine antibodies, biopharmaceuticals, and edible vaccines in transgenic plants engineered via the chloroplast genome. In: Schillberg, S. (Ed.), Molecular Farming. Wiley– VCH Verlag, Berlin, pp. 113–133.

Daniell, H., Chebolu, S., Kumar, S., Singleton, M., Falconer, R., 2005. Chloroplast-derived vaccine antigens and other therapeutic proteins. Vaccine 23, 1779–1783.

Daniell, H., Lee, S.B., Panchal, T., Wiebe, P.O., 2001. Expression of cholera toxin B subunit gene and assembly as functional oligomers in transgenic tobacco chloroplasts. J. Mol. Biol. 311, 1001–1009.

Danielli, J.F., 1975. The bilayer hypothesis of membrane structure. In: Weissmann, G., Claiborne, R. (Eds.), Cell Membranes. Biochemistry, Cell Biology & Pathology. HP Publishing Co., New York, pp. 3–11.

Danielli, J.F., Harvey, E.N., 1935. The tension at the surface of mackerel egg oil with remarks on the nature of the cell surface. J. Cell. Comp. Physiol. 5, 483–494.

Darbishire, A.D., 1904. On the result of crossing Japanese waltzing with albino mice. Biometrika 3, 1–51.

Darlington, C.D., 1932. Recent Advances in Cytology. P. Blakiston's Son & Co., Philadelphia.

Darlington, C.D., 1940. The prime variables of meiosis. Biol. Rev. 15, 307–322.

Darlington, C.D., 1944. Heredity, development and infection. Nature 154, 164–169.

Darlington, C.D., Thomas, P.T., 1937. The breakdown of cell division in a Festuca-Lolium derivative. Ann. Bot. 1, 747–761.

Darnowski, D.W., 2002. Triggerplants. Rosenberg Publications, New South Wales, Australia.

Darnowski, D.W., Valenta, R., Parthasarathy, M.V., 1996. Identification and distribution of profilin in tomato (*Lycopersicon esculentum* Mill.). Planta 198, 158–161.

Darnsky, C.H., Werb, Z., 1992. Signal transduction by integrin receptors for extracellular matrix: cooperative processing of extracellular information. Curr. Opin. Cell Biol. 4, 772–781.

Darvill, A.G., Albersheim, P., 1984. Phytoalexins and their elicitors - A defense against microbial infection in plants. Ann. Rev. Plant Physiol. 35, 243–275.

Darwin, C., 1859–1882. The Origin of Species, 1939, sixth ed. J.M. Dent & Sons, London.

Darwin, C., 1859. The Origin of Species. John Murray, London.

Darwin, C., 1860b. The Origin of Species, second ed. John Murray, London.

Darwin, C., 1868. The Variation of Animals and Plants Under Domestication. J. Murray, London.

Darwin, C., 1897. Insectivorous Plants. D. Appleton and Co., New York.

Darwin, C., 1966. The Power of Movement in Plants, 1881. DaCapo Press, New York.

Darwin, F., Acton, E.H., 1894. Practical Physiology of Plants. Cambridge University Press, Cambridge.

Darwin, C., May 22, 1860a. Letter to Asa Gray. https://www.darwinproject.ac.uk/letter/DCP-LETT-2814.xml.

Darwin, C., September 18, 1861. Letter to Henry Fawcett. https://www.darwinproject.ac.uk/letter/DCP-LETT-3257.xml.

Darwin, C., September 26, 1857. Letter to T. H. Huxley. https://www.darwinproject.ac.uk/letter/DCP-LETT-2143.xml.

Das, O.P., Messing, J., 1994. Variegated phenotype and developmentsl methylation changes of a maize allele originating from epimutation. Genetics 136, 1121–1141.

daSilva, L.L.P., Snapp, E.L., Denecke, J., Lippincott-Schwartz, J., Hawes, C., Brandizzi, F., 2004. Endoplasmic reticulum export sites and Golgi bodies behave as single mobile secretory units in plant cells. Plant Cell 16, 1753–1771.

Dassanayake, M., Oh, D.H., Haas, J.S., Hernandez, A., Hong, H., Ali, S., Yun, D.J., Bressan, R.A., Zhu, J.K., Bohnert, H.J., et al., 2011. The genome of the extremophile crucifer *Thellungiella parvula*. Nat. Genet. 43, 913–918.

Datta, M., 1957. Vacuoles and movement in the pulvinus of Mimosa pudica. Nature 179, 253–254.

Datta, N., Chen, Y.-R., Roux, S.J., 1985. Phytochrome and calcium stimulation of protein phosphorylation in isolated pea nuclei. Biochem. Biophys. Res. Commun. 128, 1403–1408.

Dauwalder, M., Whaley, W.G., 1973. Staining of cells of Zea mays root apices with the osmium–zinc iodide and osmium impregnation technique. J. Ultrastruc. Res. 45, 279–296.

Dauwalder, M., Whaley, W.G., 1974. Patterns of incorporation of [3H] galactose of Zea mays root tips. J. Cell Sci. 14, 11–27.

Dauwalder, M., Whaley, W.G., Kephart, J.E., 1969. Phosphatases and differentiation of the Golgi apparatus. J. Cell Sci. 4, 455–497.

Davenport, C.B., 1908. Experimental Morphology. Macmillan, New York.

Davenport, C.B., 1910. Eugenics. The Science of Human Improvement by Better Breeding. Henry Holt and Co., New York.

Davenport, C.B., 1912. The Trait Book. Eugenics Record Office. Cold Spring Harbor, New York. Bulletin No. 6.

Davenport, C.B., 1921. Research in eugenics. Science 54, 391–397.

Davenport, H.E., 1952. Cytochrome components in chloroplasts. Nature 170, 1112–1114.

Davidi, L., Levin, Y., Ben-Dor, S., Pick, U., 2015. Proteome analysis of cytoplasmic and plastidic β-carotene lipid droplet in *Dunaliella bardawil*. Plant Physiol. 167, 60–79.

Davidson, J.N., 1960. The Biochemistry of the Nucleic Acids. Methuen & Co., London.

Davidson, E.H., 1968. Gene Activity in Early Development. Academic Press, New York.

Davidson, J.N., Chargaff, E., 1955. Introduction. In: Chargaff, E., Davidson, J.N. (Eds.), The Nucleic Acids, vol. 1. Academic Press, New York, pp. 1–8.

Davidson, E.H., Galau, G.A., Angerer, R.C., Britten, R.J., 1975. Comparative aspects of DNA organization in metazoa. Chromosoma 51, 253–259.

Davidson, P.M., Lammerding, J., 2014. Broken nuclei—lamins, nuclear mechanics, and disease. Trends Cell Biol. 24, 247–256.

Davies, D.D., 1953. The Krebs cycle enzyme system of pea seedlings. J. Exp. Bot. 4, 173–183.

Davies, J.M., 1999. The bioenergetics of vacuolar H^+ pumps. Adv. Bot. Res. 25, 339–363.

Davies, E., Barh, D., 2015. Plantomics and Futuromics. In: Barh, D., Sarwar Khan, M., Davies, E. (Eds.), PlantOmics: The Omics of Plant Science. Springer India, pp. 821–825.

Davies, E., Fillingham, B.D., Oto, Y., Abe, Y., 1991. Evidence for the existence of cytoskeleton-bound polysomes in plants. Cell Biol. Int. Rep. 15, 973–981.

Davies, J.M., Hunt, I., Sanders, D., 1994. Vacuolar H^+-pumping ATPase variable transport coupling ratio controlled by pH. Proc. Natl. Acad. Sci. U.S.A. 91, 8547–8551.

Davies, J.M., Poole, R.J., Sanders, D., 1993. The computed free energy change of hydrolysis of inorganic pyrophosphate and ATP: apparent significance for inorganic-pyrophosphate-driven reactions of intermediary metabolism. Biochim. Biophys. Acta 1141, 29–36.

Davies, T.G.E., Steele, S.H., Walker, D.J., Leigh, R.A., 1996. An analysis of vacuole development in oat aleurone protoplasts. Planta 198, 356–364.

Davis, N.P., 1968. Lawrence and Oppenheimer. Simon and Schuster, New York.

Davis, B.K., 1971. Genetic analysis of a meiotic mutant resulting in precocious sister centromere separation. Mol. Gen. Genet. 113, 251–272.

Davis, L.I., 1992. Control of nucleocytoplasmic transport. Curr. Opin. Cell Biol. 4, 424–429.

Davis, C.C., Anderson, W.R., Wurdack, K.J., 2005. Gene transfer from a parasitic flowering plant to a fern. Proc. R. Soc. B 272, 2237–2242.

Davis, L.I., Blobel, G., 1986. Identification and characterization of a nuclear pore complex protein. Cell 45, 699–709.

Davis, L.I., Blobel, G., 1987. Nuclear pore complex contains a family of glycoproteins that includes p62: glycosylation through a previously unidentified cellular pathway. Proc. Natl. Acad. Sci. U.S.A. 84, 7552–7556.

Davis, B.D., Chen, J.C.W., Philpott, M., 1974. The transition from filamentous to two-dimensional growth in fern gametophytes IV. Initial events. Am. J. Bot. 61, 722–729.

Davis, P.A., Hangarter, R.P., 2012. Chloroplast movement provides photoprotection to plants by redistributing PSII damage within leaves. Photosynth. Res. 112, 153–161.

Davis, D.J., Kang, B.-H., Heringer, A.S., Wilkop, T.E., Drakakaki, G., 2016. Unconventional protein secretion in plants. Methods Mol. Biol. 1459, 47–63.

Davis, J.I., Soreng, R.J., 2010. Migration of endpoints of two genes relative to boundaries between regions of the plastid genome in the grass family (Poaceae). Am. J. Bot. 97, 874–892.

Davis, J.I., Stevenson, D.W., Petersen, G., Seberg, O., Campbell, L.M., Freudenstein, J.V., Goldman, D.H., Hardy, C.R., Michelangeli, F.A., Simmons, M.P., Specht, C.D., Vergara-Silva, F., Gandolfo, M., 2004. A phylogeny of the monocots, as inferred from rbcL and atpA sequence variation, and a comparison of methods for calculating jackknife and bootstrap values. Syst. Bot. 29, 467–510.

Davis, J.M., Svendsgaard, D.J., 1990. U shaped dose–response curves: their occurrence and implications for risk assessment. J. Toxicol. Environ. Health 30, 71–83.

Davison, E.H., 1976. Gene Activity in Early Development. Academic Press, New York.

Davson, H., Danielli, J.F., 1943. The Permeability of Natural Membranes. Cambridge University Press, Cambridge.

Davson, H., Danielli, J.F., 1952. The Permeability of Natural Membranes, second ed. Cambridge University Press, Cambridge.

Davy, H., 1808. I. The Bakerian Lecture, on some new phenomena of chemical changes produced by electricity, particularly the decomposition of the fixed Alkalies, and the exhibition of the new substances which constitute their bases; and on the general nature of alkaline bodies. Phil. Trans. R. Soc. Lond. 98, 1–44.

Davy, H., 1821. Elements of Agricultural Chemistry. B. Warner, Philadelphia.

Davy, H., 1827. Elements of Agricultural Chemistry, fourth ed. Longman, Rees, Ormo, Brown and Green, London.

Dawidowicz, E.A., 1987. Lipid exchange: transmembrane movement, spontaneous movement, and proteinmediated transfer of lipids and cholesterol. Curr. Top. Membr. Trans. 29, 175–202.

Dawkins, R., 1976. The Selfish Gene. Oxford University Press, Oxford.

Dawkins, R., 1995. River Out of Eden: A Darwinian View of Life. Weidenfeld & Nicolson, London.

Dawkins, R., 2006. The God Delusion. Houghton Mifflin Co., Boston.

Dawson, M.H., Sia, R.H.P., 1931. In vitro transformations of pneumatococcal types. I. A techniques for inducing transformation of pneumatococcal types in vitro. J. Exp. Med. 54, 681–700.

Day, R.C., Grossniklaus, U., Macknight, R.C., 2005. Be more specific! Laser-assisted microdissection of plant cells. Trends in Plant Sci. 10, 397–406.

Day, R.C., McNoe, L., Macknight, R.C., 2007a. Transcript analysis of laser microdissected plant cells. Physiol. Plant 129, 267–282.

Day, R.C., McNoe, L., Macknight, R.C., 2007b. Evaluation of global RNA amplification and its use for high-throughput transcript analysis of laser-microdissected endosperm. Int. J. Plant Genom. 61028.

Day, K.J., Staehelin, L.A., Glick, B.S., 2013. A three-stage model of Golgi structure and function. Histochem. Cell Biol. 140, 239–249.

Dayel, M.J., Horn, E.F.Y., Verkman, A.S., 1999. Diffusion of green fluorescent protein in the aqueous-phase lumen of endoplasmic reticulum. Biophys. J. 76, 2843–2851.

de Broglie, L., 1946. Matter and Light. Dover Publications, New York.

de Carvalho, F., Gheysen, G., Kushnir, S., van Montagu, M., Inze, D., Castresana, C., 1992. Suppresion of β-1,2-glucanase transgene expression in homozygous plants. EMBO J. 11, 2595–2602.

de Cervantes, M., 1956. The History of the Ingenious Gentleman Don Quixote of La Mancha. Translated by P. A. Motteaux. Volume III, chapter 10 page 128. Harper & Brothers, New York.

De Diego, N., Fürst, T., Humplík, J.F., Ugena, L., Podlešáková, K., Spíchal, L., 2017. An automated method for high-throughput screening of Arabidopsis rosette growth in multi-well plates an dits validation in stress conditions. Front. Plant Sci. 8, 1702.

de Duve, C., 1963. Note. In: De Reuck, A.V.S., Cameron, M.P. (Eds.), Lysosomes. CIBA Foundation Synposium. Little, Brown, and Co., Boston, p. 126.

de Duve, C., 1969. Evolution of the peroxisome. Ann. N.Y. Acad. Sci. 168, 369–381.

de Duve, C., 1975. Exploring cells with a centrifuge. Science 189, 186–194.

de Duve, C., 1981. A Guided Tour of the Living Cell. Scientific American Books, New York.

de Duve, C., 1991. Blueprint for a Cell: The Nature and Origin of Life. Neil Patterson Publishers, Burlington, NC.

de Duve, C., 2012. Exploring cells with a centrifuge: discoveries of the lysosome and peroxisome. In: Moberg, C.L. (Ed.), Entering an Unseen World. A Founding Laboratory and Origins of Modern Cell Biology 1910–1974. Rockefeller University Press, New York, pp. 273–304.

de Duve, C., Baudhuin, P., 1966. Peroxisomes (microbodies and related particles). Physiol. Rev. 46, 323–357.

de Duve, C., Pressman, B.C., Gianetto, R., Wattiaux, R., Applemans, F., 1955. Tissue fractionation studies. 6. Intracellular distribution patterns of enzymes in rat-liver tissue. Biochem. J. 60, 604–617.

de Duve, C., Wattiaux, R., 1966. Functions of lysosomes. Ann. Rev. Physiol. 28, 435–492.

de Graaf, B.H., Cheung, A.Y., Andreyeva, T., Levasseur, K., Kieliszewski, M., Wu, H.M., 2005. Rab11 GTPase-regulated membrane trafficking is crucial for tip-focused pollen tube growth in tobacco. Plant Cell 17, 2564–2579.

de Kerchove d'Exaerde, A., Supply, P., Dufour, J.P., Thinés, D., Goffeau, A., Boutry, M., 1995. Functional complementation of a null mutation of the yeast *Saccharomyces cerevisiae* plasma membrane H^+-ATPase by a plant H^+-ATPase gene. J. Biol. Chem. 270, 23828–23837.

de Kruif, P., 1926. The Microbe Hunters. Harcourt, Brace and Co., New York.

De La Cruz, E.M., Pollard, T., 2001. Structural biology. Actin' up. Science 293, 616–618.

de LaFayette, M., Jefferson, T., 1789. Declaration of the Rights of Man and the Citizen. http://www.constitution.org/ fr/fr_drm.htm.

de Lucia, P., Cairns, J., 1969. Isolation of *an E. coli* strain with a mutation affecting DNA polymerase. Nature 224, 1164–1166.

de Mairan, J.-J., 1729. Observation botanique. Histoire de l'Académie royale des sciences avec les mémoires de mathématique et de physique tirés des registres de cette Académie, p. 35.

de Maupertuis, P.-L.M., 1753. The Earthly Venus, 1966. Johnson Reprint Corp., New York.

De Mey, J., Lambert, A.M., Bajer, A.S., Moeremans, M., De Brabander, M., 1982. Visualization of microtubules in interphase and mitotic plant cells of *Haemanthus* endosperm with immunogold staining method. Proc. Natl. Acad. Sci. U.S.A. 79, 1898–1902.

de Michele, R., McFarlane, H.E., Parsons, H.T., Meents, M.J., Lao, J., González Fernández-Niño, S.M., Petzold, C.J., Frommer, W.B., Samuels, A.L., Heazlewood, J.L., 2016. Free-flow electrophoresis of plasma membrane vesicles enriched by two-phase partitioning enhances the quality of proteome from *Arabidopsis* seedlings. J. Proteome Res. 15, 900–913.

de Ruijter, N., Emons, A., 1993. Immunodetection of spectrin antigens in plant cells. Cell Biol. Int. 17, 169–182.

de Saussure, T., 1804. Recherches Chimiques Sur LaVégétation, 1957. Gauthier-Villars, Paris.

de Sitter, W., 1932. Kosmos. Harvard University Press, Cambridge, MA.

de Vargas Roditi, L., Claassen, M., 2015. Computational and experimental single cell biology techniques for the definition of cell type heterogeneity, interplay and intracellular dynamics. Curr. Opin. Biotechnol. 2015 (34), 9–15.

de Vries, H., 1884. Eine Methode zur Analyse der Turgorkraft. Jahr. Wiss. Bot. 14, 427–601.

de Vries, H., 1885. Plasmolytische Studien über die Wand der Vacuolen. Jahr. Wiss. Bot. 16, 465–598.

de Vries, H., 1886. Ueber die Aggregation im Protoplasma von *Drosera rotundifolia*. Bot. Z. 44, 57–64.

de Vries, H., 1888a. Osmotische Versuche mit lebenden Membranen. Z. Physikal. Chem. 2, 415–432.

de Vries, H., 1888b. Über eine neue Anwendung der plasmolytischen Methode. Bot. Z. 46, 393–397.

de Vries, H., 1889. Intracellular Pangensis. Translated by C. Stuart Gager, 1910. The Open Court Publishing Co., Chicago.

de Vries, H., 1900a. Concerning the laws of segregation of hybrids, 1950 Genetics 35, 30–32.

de Vries, H., 1900b. Das Spaltungsgesetz der Bastarde. Berichte der Deutschen Botanische Gesellschaft 18, 83–90.

de Vries, H., 1905. Species and Varieties, their Origin by Mutation: Lectures Delivered at the University of California. In: MacDougal, Daniel T. (Ed.). The Open Court Publishing Company, Chicago.

de Vries, H., 1906. Species and Varieties, their Origin by Mutation: Lectures Delivered at the University of California. In: MacDougal, D.aniel T. (Ed.), second ed. The Open Court Publishing Company, Chicago.

de Vries, H., 1907. Plant-Breeding. Comments on the Experiments of Nilsson and Burbank. The Open Court Publishing Company, Chicago.

De, D.N., 2000. Plant Cell Vacuoles. CSIRO Publishing, Collingswood, Australia.

Deal, R.B., Henikoff, S., 2010. A simple method for gene expression and chromatin profiling of individual cell types within a tissue. Dev. Cell. 18, 1030–1040.

Deamer, D.W., 1997. The first living systems: a bioenergetic perspective. Microbiol. Mol. Biol. Rev. 61, 239–261.

Deamer, D.W., Akeson, M., 2000. Nanopores and nucleic acids: prospects for ultrarapid sequencing. Trends Biotechnol. 18, 147–151.

Deamer, D.W., Fleischaker, G.R., 1994. Origins of Life. The Central Concepts. Jones and Bartlett Publishers, Sudbury, MA.

Debye, P., 1942. Reaction rates in ionic solutions. Trans. Electrochem. Soc. 82, 265–325.

deCandolle, A.P., Sprengel, K., 1821. Elements of the Philosophy of Plants. William Blackwood, Edinburgh.

Decker, J.P., 1955. A rapid, postillumination deceleration of respiration in green leaves. Plant Physiol. 30, 82–84.

Decker, J.P., 1959. Comparative responses of carbohydrate outburst and uptake in tobacco. Plant Physiol. 34, 100–102.

DeFalco, T.A., Marshall, C.B., Munro, K., Kang, H.-G., Moeder, W., Ikura, M., Snedden, W.A., Yoshioka, K., 2016. Multiple calcmodulin-binding sites positively and negatively regulate *Arabidopsis* cyclic nucleotide-gated channel 12. Plant Cell 28, 1738–1751.

Deichmann, U., 2009. Chemistry and the engineering of life around 1900: research and reflections by Jacques Loeb. Biol. Theor. 4, 323–332.

Deichmann, U., Müller-Hill, B., 1998. The fraud of Abderhalden's enzymes. Nature 393, 109–111.

Deichmann, U., Schuster, S., Mazat, J.-P., Cornish-Bowden, A., 2014. Commemorating the 1913 Michaelis-Menten paper Die Kinetik der Invertinwirkung: three perspectives. FEBS J. 281, 435–463.

Deininger, P.L., 1983. Random subcloning of sonicated DNA: application to shotgun DNA sequence analysis. Anal. Biochem. 129, 216–223.

Deisenhofer, J., Michel, H., 1991. Structures of bacterial photosynthetic reaction centers. Ann. Rev. Cell Biol. 7, 1–23.

Deisenhofer, J., Michel, H., December 8, 1988. The photosynthetic reaction centre from the purple bacterium *Rhodopseudomonas viridis*. Nobel Lecture.

DeJoseph, M., Taylor, R.S.L., Baker, M., Aregullin, M., 2002. Fur-rubbing behavior of capuchin monkeys. J. Am. Acad. Dermatol. 46, 924–925.

del Campillo, E., Lewis, L.N., 1992. Occurrence of 9.5 cellulase and other hydrolases in flow reproductive organs under going major cell wall disruption. Plant Physiol. 99, 1015–1020.

del Rio, L.A., 2011. Peroxisomes as a cellular source of reactive nitrogen species signal molecules. Arch. Biochem. Biophys. 506, 1–11.

del Rio, L.A., Corpa, F.J., Sandalio, L.M., Palma, J.M., Gomez, M., Barroso, J.B., 2002. Reactive oxygen species, antioxidant systems and nitric oxide in peroxisomes. J. Exp. Bot. 53, 1255–1272.

Delambre, J.-B., Delambre, J., 1867. Œuvres de Lagrange. Gauthier-Villars, Paris.

DeLange, R.J., Green, G.R., Searcy, D.G., 1981a. A histone-like protein (HTa) from *Thermoplasma acidophilum*. I. Purification and properties. J. Biol. Chem. 256, 900–904.

DeLange, R.J., Green, G.R., Searcy, D.G., 1981b. A histone-like protein (HTa) from *Thermoplasma acidophilum*. II. Complete amino acid sequence. J. Biol. Chem. 256, 905–911.

Delbrück, M., December 10, 1969. A physicist's renewed look at biology— Twenty years later. Nobel Lecture.

Delbrück, M., Stent, G.S., 1957. On the mechanism of DNA replication. In: McElroy, W.D., Glass, B. (Eds.), The Chemical Basis of Heredity. Johns Hopkins Press, Baltimore, pp. 699–736.

Dell, K.R., Vale, R.D., 2004. A tribute to Shinya Inoue and innovation in light microscopy. J. Cell Biol. 165, 21–26.

Delmer, D.P., 1987. Cellulose biosynthesis. Ann. Rev. Plant Physiol. 38, 259–290.

Delmer, D.P., 1991. The biochemistry of cellulose synthesis. In: Lloyd, C.W. (Ed.), The Cytoskeletal Basis of Plant Growth and Form. Academic Press, London, pp. 101–107.

Delmer, D.P., 1999. Cellulose biosynthesis: exciting times for a difficult field. Ann. Rev. Plant Physiol. Plant Mol. Biol. 50, 245–276.

Delmer, D.P., Solomon, M., Read, S.M., 1991. Direct photolabeling with [32P] UDP-glucose for identification of a subunit of cotton fiber cellose synthase. Plant Physiol. 95, 556–563.

Delsuc, F., Brinkmann, H., Philippe, H., 2005. Phylogenomics and the reconstruction of the tree of life. Nat. Rev. Genet. 6, 361–375.

DeMaggio, A.E., Greene, C., Stetler, D., 1980. Biochemistry of fern spore germination. Glyoxylate and glycolate cycle activity in *Onoclea sensibilis* L. Plant Physiol. 66, 922–924.

Dembinsky, D., Woll, K., Saleem, M., Liu, Y., Fu, Y., Borsuk, L.A., Lamkemeyer, T., Fladerer, C., Madlung, J., Barbazuk, B., Nordheim, A., Nettleton, D., Schnable, P.S., Hochholdinger, F., 2007. Transcriptomic and proteomic analysis of pericycle cells of the maize primary root. Plant Physiol. 145, 575–588.

Demidchik, V., 2006. Physiological roles of nonselective cation channels in the plasma membrane of higher plants. In: Baluska, F., Mancuso, S., Volkmann, D. (Eds.), Communication in Plants. Springer, Berlin, pp. 235–248.

Demidov, D., Lermontova, I., Weiss, O., Fuchs, J., Rutten, T., Kumke, K., Sharbel, T.F., van Damme, D., de Storme, N., Geelen, D., Houben, A., 2014. Altered expression of aurora kinases in *Arabidopsis* results in aneu- and polyploidization. Plant J. 80, 449–461.

Demidov, D., van Damme, D., Geelen, D., Blattner, F.R., Houben, A., 2005. Identification and dynamics of two classes of aurora-like kinases in *Arabidopsis* and other plants. Plant Cell 17, 836–848.

DeMiranda, J.R., Thomas, M.A., Thurman, D.A., Tomset, A.B., 1990. Metallothionein genes from the flowering plant Mimulus guttathus. FEBS Lett. 260, 277–280.

Demos, R., 1946. Types of unity according to Plato and Aristotle. Philos. Phenomenol. Res. 6, 534–546.

Denesvre, C., Malhotra, V., 1996. Membrane fusion in organelle biogenesis. Curr. Opin. Cell Biol. 8, 519–523.

Dennett, D.C., 1978. Where am I?. In: Brainstorms. Philosophical Essays on Mind and Psychology. Brandford Books, Montgomery, VT, pp. 310–323.

Denninger, P., Bleckmann, A., Lausser, A., Vogler, F., Ott, T., Ehrhardt, D.W., Frommer, W.B., Sprunck, S., Dresselhaus, T., Grossmann, G., 2014. Male-female communication triggers calcium signatures during fertilization in *Arabidopsis*. Nat. Commun. 5, 4645.

Dennison, D.S., Vogelmann, T.C., 1989. The *Phycomyces* lens: measurement of the sporangiophore intensity profile using a fiber optic microprobe. Planta 179, 1–10.

Denny, M.W., 1993. Air and Water. The Biology and Physics of Life's Media. Princeton University Press, Princeton, NJ.

Denny, M., 2011. Gliding for Gold. The Physics of Winter Sports. The Johns Hopkins University Press, Baltimore, MD.

Deom, C.M., Oliver, M.J., Beachy, R.N., 1987. The 30-kilodalton gene product of tobacco mosaic virus potential virus movement. Science 337, 389–394.

Deom, C.M., Schubert, K.I., Wolf, S., Holt, C., Lucas, W.J., Beachy, R.N., 1990. Molecular characterization and biological function of the movement protein of tobacco mosaic virus in transgenic plants. Proc. Natl. Acad. Sci. U.S.A. 87, 3284–3288.

Deom, C.M., Wolf, S., Holt, C.A., Lucas, W.J., Beachy, R.N., 1991. Altered function of the tobacco mosaic virus movement protein in a hypersensitive host. Virology 180, 251–256.

Dephoure, N., Gould, K.L., Gygi, S.P., Kellogg, D.R., 2013. Mapping and analysis of phosphorylation sites: a quick guide for cell biologists. Mol. Biol. Cell 24, 535–542.

Dephoure, N., Zhou, C., Villén, J., Beauoleil, S.A., Bakalarski, C.E., Elledge, S.J., Gygi, S.P., 2008. A quantitative atlas of mitotic phosphorylation. Proc. Natl. Acad. Sci. U.S.A. 105, 10762–10767.

Desai, A., Mitchison, T.J., 1995. A new role for motor proteins as couplers to depolymerizing microtubules. J. Cell Biol. 128, 1–4.

Deslattes, R.D., 1980. The avogadro constant. Ann. Rev. Phys. Chem. 31, 435–461.

Desnuelle, P., 1953. The general chemistry of amino acids and peptides. In: Neurath, H., Bailey, K. (Eds.), The Proteins. Academic Press, New York, pp. 87–180.

Desrosiers, M.F., Bandurski, R.S., 1988. Effect of a longitudinally applied voltage upon the growth of *Zea mays* seedlings. Plant Physiol. 87, 874–877.

Dessauer, C.W., Bartlett, S.G., 1994. Idenification of a chaperonin binding site in a chloroplast precursor protein. J. Biol. Chem. 269, 19766–19776.

Dessev, G.N., 1992. Nuclear envelope structure. Curr. Opin. Cell Biol. 4, 430–435.

Dettmer, J., Hong-Hermesdorf, A., Stierhof, Y.D., Schumacher, K., 2006. Vacuolar H^+-ATPase activity is required for endocytic and secretory trafficking in *Arabidopsis*. Plant Cell 18, 715–730.

Deussen, P., 1920. Sixty Upanisads of the Veda. Part One. Motilal Banarsidass, Delhi.

Dewey, J., 1910. How we think. D.C. Heath, Lexington, MA.

Dhonukshe, P., Aniento, F., Hwang, I., Robinson, D.G., Mravec, J., Stierhof, Y.D., Friml, J., 2007. Clathrin-mediated constitutive endocytosis of PIN auxin efflux carriers in *Arabidopsis*. Curr. Biol. 17, 520–527.

Dhonukshe, P., Mathus, J., Hülskamp, M., Gadella, T.W.J., 2005. Microtubule plus-ends reveal essential links between intracellular polarization and localized modulation of endocytosis during division-plane establishment in plant cells. BMC Biol. 3, 11–26.

Dhonukshe, P., Vischer, N., Gadella Jr., T.W.J., 2006. Contribution of microtubule growth polarity and flux to spindle assembly and functioning in plant cells. J. Cell Sci. 119, 3193–3205.

Di Laurenzio, L., Wysocka-Diller, J., Malamy, J.E., Pysh, L., Helariutta, Y., Freshour, G., Hahn, M.G., Feldmann, K.A., Benfey, P.N., 1996. The *SCARECROW* gene regulates an asymmetric cell division that is essential for generating the radial organization of the *Arabidopsis* root. Cell 86, 423–433.

Di Palma, J.R., Mohl, R., Best Jr., W., 1961. Action potential and contraction of *Dioaea muscipula* (Venus flytrap). Science 133, 878–879.

Di Rubbo, S., Russinova, E., 2012. Receptor-mediated endocytosis in plants. In: Šamaj, J. (Ed.), Endocytosis in Plants. Springer-Verlag, Berlin, pp. 151–164.

Dias, A.F., Francisco, T., Rodrigues, Grou, C.P., Azevedo, J.E., 2016. The first minutes of the life of a peroxisomeal matrix protein. Biochim. Biophys. Acta 1863, 814–820.

Diaz, M.F., Bentolila, S., Hayes, M.L., Hanson, M.R., Mulligan, R.M., 2017. A protein with an unusually short PPR domain, MEF8, affects editing at over 60 Arabidopsis mitochondrial C targets of RNA editing. Plant J. 92, 638–649.

Dice, J.F., 1990. Introduction: pathways of intracellular proteolysis. Semin. Cell Biol. 1, 411–413.

Diechmann, U., 2009. Chemistry and the engineering of life around 1900: research and reflections by Jacques Loeb. Biol. Theory 4, 323–332.

Dieter, P., Marmé, D., 1980. Ca^{2+} transport in mitochondrial and microsomal fractions from higher plants. Planta 150, 1–8.

Dietrich, M.R., 2015. Explaining the "pulse of protoplasm": the search for molecular mechanisms of protoplasmic streaming. J. Integr. Plant Biol. 57, 14–22.

Dietrich, P., Anschütz, U., Kugler, A., Becker, D., 2010. Physiology and biophysics of plant ligand-gated ion channels. Plant Biol. 12 (Suppl. 1), 80–93.

Diettrich, B., Keller, F., 1991. Carbohydrate transport in discs of storage parenchyma of celery petioles. 1. Uptake of glucose and fructose. New Phytol. 117, 413–422.

Dijksterhuis, J., Nijsse, J., Hoekstra, F.A., Golovina, E.A., 2007. High viscosity and anisotropy characterize the cytoplasm of fungal dormant stress-resistant spores. Eukaryot. Cell 6, 157–170.

Dill, K.A., Bromberg, S., 2003. Molecular Driving Forces. Statistical Thermodynamics in Chemistry and Biology. Garland Science, New York.

Dillenschneider, M., Hetherington, A., Graziana, A., Alibert, G., Berta, P., Haiech, J., Ranjeva, R., 1986. The formation of inositol phosphate derivatives by isolated membranes from *Acer pseudoplatanus* is stimulated by guanine nucleotides. FEBS Lett. 208, 413–417.

Dinarina, A., Pugieux, C., Corral, M.M., Loose, M., SPatz, J., Karsenti, E., Nedelec, F., 2009. Chromatin shapes the mitotic spindle. Cell 17, 836–848.

Ding, B., 1998. Intercellular protein trafficking through plasmodesmata. Plant Mol. Biol. 38, 279–310.

Ding, B., Haudenshield, J.S., Hull, R.J., Wold, S., Beachy, R.N., Lucas, W.J., 1992a. Secondary plasmodesmata are specific sites of localization of the tobacco mosaic virus movement protein in transgenic tobacco plants. Plant Cell 4, 915–928.

Ding, B., Itaya, A., 2007. Control of macromolecular trafficking across specific cellular boundaries: a key to integrative plant biology. J. Integr. Plant Biol. 49, 1227–1234.

Ding, B., Itaya, A., Woo, Y., 1999. Plasmodesmata and cell-to-cell communication in plants. Int. Rev. Cytol. 190, 251–316.

Ding, B., Kwon, Mo.O., Warnberg, I., 1996. Evidence that actin filaments are involved in controlling the permeability of plasmodesmata in tobacco mesophyll. Plant J. 10, 157–164.

Ding, R., McDonald, K.L., McIntosh, J.R., 1993. Three-dimensional reconstruction and analysis of mitotic spinkles from the yeast, *Schizosaccharomyces pombe*. J. Cell Biol. 120, 141–151.

Ding, D., Muthuswamy, S., Meier, I., 2012. Functional interaction between the *Arabidopsis* orthologs of spindle assembly checkpoint proteins MAD1 and MAD2 and the nucleoporin NUA. Plant Mol. Biol. 79, 203–216.

Ding, Y., Robinson, D.G., Jiang, L., 2014. Unconventional protein secretion (UPS) pathways in plants. Curr. Opin. Cell Biol. 29, 107–115.

Ding, D.-Q., Tazawa, M., 1989. Influence of cytoplasmic streaming and turgor pressure gradient on the transnodal transport of rubidium and electrical conductance in *Chara corallina*. Plant Cell Physiol. 30, 739–748.

Ding, B., Turgeon, R., Parthasarathy, M.V., 1991a. Microfilament organization and distribution in freeze-substituted tobacco plant tissue. Protoplasma 165, 96–105.

Ding, B., Turgeon, R., Parthasarathy, M.V., 1991b. Microfilaments in preprophase band of freeze-substituted tobacco root cells. Protoplasma 165, 209–211.

Ding, B., Turgeon, R., Parthasarathy, M.V., 1992b. Substructure of freeze-substituted plasmodesmata. Protoplasma 169, 28–41.

Ding, Y., Wang, J., Wang, J., Stierhof, Y.-D., Robinson, D.G., Jiang, L., 2012b. Unconventional protein secretion. Trens. Plant Sci. 17, 606–615.

Dingwall, C., Laskey, R.A., 1986. Protein import into the cell nucleus. Ann. Rev. Cell Biol. 2, 367–390.

Dinis, A.M., Mesquita, J.F., 1994. Ultrastructural and cytochemical evidence for the presence of peroxisomes in the generative cell of *Magnolia x Soulageana* pollen grain. Ann. Bot. 73, 83–90.

Dinkins, R., Reddy, M.S., Leng, M., Collins, G.B., 2001. Overexpression of the *Arabidopsis thaliana* MinD1 gene alters chloroplast size and number in transgenic tobacco plants. Planta 214, 180–188.

Diotallevi, F., Mulder, B., 2007. The cellulose synthase complex: a polymerization driven supramolecular motor. Biophys. J. 92, 2666–2673.

Dittmer, T.A., Misteli, T., 2011. The lamin protein family. Genome Biol. 12, 222.

Dittmer, T.A., Richards, E.J., 2008. Role of LINC proteins in plant nuclear morphology. Plant Signal. Behav. 3, 485–487.

Dittmer, T.A., Stacey, N.J., Sugimoto-Shirasu, K., Richards, E.J., 2007. Little nuclei genes affecting nuclear morphology in Arabidopsis thaliana. Plant Cell 19, 2793–2803.

Dixit, R., 2015. Teamster's union. Nat. Plants 1, 15126.

Dixit, R., Cyr, R., 2002. Golgi secretion is not required for marking the preprophase band site in cultured tobacco cells. Plant J. 29, 99–108.

Dixit, R., Cyr, R., Gilroy, S., 2006. Using intrinsically fluorescent proteins for plant cell imaging. Plant J. 45, 599–615.

Dixit, R., Nasrallah, J.B., 2001. Recognizing self-incompatibility response. Plant Physiol. 125, 105–108.

Dixit, R., Petry, S., 2018. The life of a microtubule. Mol. Biol. Cell 29, 689.

Dixon, H.H., 1938. The Croonian lecture: transport of substances in plants. Proc. R. Soc. Lond. 125B, 1–25.

Dixon, M., 1952. Manometric Mehods as Applied to the Measurement of Cell Respiration and other Processes, third ed. Cambridge University Press, Cambridge.

Dixon, H.H., Joly, J., 1895. On the ascent of sap. Philos. Trans. R. Soc. Lond. 186B, 563–576.

Dixon, R.A., Strack, D., 2003. Phytochemistry meets genome analysis, and beyond. Phytochemistry 62, 815–816.

Djerassi, C., Hoffmann, R., 2001. Oxygen. Wiley-VCH, Weinheim, Germany.

Dobberstein, B., Blobel, G., Chua, N.-H., 1977. In vitro synthesis an dprocessing of a putative precursor for the small subunit of ribuose-1,5-bisphosphate carboxylase of *Chlamydomonas reinhardtii*. Proc. Natl. Acad. Sci. U.S.A. 74, 1082–1085.

Doberstein, S.K., Baines, I.C., Wiegland, G., Korn, E.D., Pollard, T.D., 1993. Inhibition of contractile vacuole function in vivo by antibodies against myosin-I. Nature 365, 841–843.

Dobson, A., 1893. Old-World Idylls and other Verses. Kegan Paul, Trench, Trüber & Co., London.

Dobzhansky, T., Ayala, F.J., Stebbins, G.L., Valentine, J.W., 1977. Evolution. W. H. Freeman and Co., San Francisco.

Doctor, B.P., Loebel, J.E., Sodd, M.A., Winter, D.B., 1969. Nucleotide sequence of *Escherichia coli* tyrosine transfer ribonucleic acid. Science 163, 693–695.

Dodge, R.A., Thompson, M.J., 1937. Fluid Mechanics. McGraw-Hill, New York.

Dodson, G., 2005. Fred Sanger: sequencing pioneer. Biochemist 27 (6), 31–35.

Doerner, P., 2007. Transcriptional control of the plant cell cycle. Plant Cell Monogr. 9, 13–32.

Dohm, J.C., Lottaz, C., Borodina, T., Himmelbauer, H., 2008. Substantial biases in ultra-short read data sets from high-throughput DNA sequencing. Nucleic Acids Res. 36, e105.

Dohm, J.C., Minoche, A.E., Holtgrawe, D., Capella-Gutierrez, S., Zakrzewski, F., Tafer, H., Rupp, O., Sorensen, T.R., Stracke, R., Reinhardt, R., et al., 2014. The genome of the recently domesticated crop plant sugar beet (*Beta vulgaris*). Nature 505, 546–549.

Dole, M., 1989. My Life in the Golden Age of America. Vantage Press, NY.

Dole, M., Mach, L.L., Hines, R.L., Mobley, R.C., Ferguson, L.P., Alice, A.B., 1968. Molecular beams of macroions. J. Chem. Phys. 49, 2240–2249.

Dolezel, J., Cihalikova, J., Lucretti, S., 1992. A high-yield procedure for isolation of metaphase chromosomes from root tips of *Vicia faba* L. Planta 188, 93–98.

Dombrowski, J.E., Raikhel, N.V., 1995. Isolation of cDNA encoding a novel GTP-binding protein of *Arabidopsis thaliana*. Plant Mol. Biol. 28, 1121–1126.

Domon, B., Aebersold, R., 2006. Mass spectrometry and protein analysis. Science 312, 212–217.

Domon, B., Aebersold, R., 2010. Options and considerations when selecting a quantitative proteomics strategy. Nature Biotechnol. 28, 710–721.

Domozych, D.S., 1989a. The endomembrane system and mechanism of membrane flow in the green alga, *Gloeomonas kupfferi* (Volvocales, Chlorophyta). I. An ultrastructural analysis. Protoplasma 149, 95–107.

Domozych, D.S., 1989b. The endomembrane system and mechanism of membrane flow in the green alga, *Gloeomonas kupfferi* (Volvocales, Chlorophyta). II. A cytochemical analysis. Protoplasma 149, 108–119.

Domozych, D.S., 2014. *Penium margaritaceum*: a unicellular model organism for studying plant cell wall architecture and dynamics. Plants 3, 543–558.

Domozych, D.S., Ciancia, M., Fangel, J.U., Mikkelsen, M.D., Ulvskov, P., Willats, W.G.T., 2012. The cell walls of green algae: a journey through evolution and diversity. Front. Plant Sci. 3, 82.

Domozych, D.S., Dairman, M., 1993. Synthesis of the inner cell wall layer of the chlamydomonad flagellate, Gloeomonas kupfferi. Protoplasma 176, 1–13.

Domozych, D.S., Domozych, C.E., 2014. Multicellularity in green algae: upsizing in a walled complex. Front. Plant Sci. 5, 649.

Domozych, D.S., Lambiasse, L., Kiemle, S.N., Gretz, M.R., 2009. Cell-wall development and bipolar growth in the desmid *Penium Margaritaceum* (Zygnematophyceae, Streptpphyta). Asymmetry in a symmetrical world. J. Phycol. 45, 879–893.

Domozych, D.S., Nimmons, T.T., 1992. The contractile vacuole as an endocytic organelle of the chlamydomonal flagellate *Gloeomonas kupfferi*. Volvocales Chlorophyta. J. Phycol. 28, 809–816.

Domozych, D.S., Sørensen, I., Popper, Z.A., Ochs, J., Andreas, A., Fangel, J.U., Pielach, A., Sacks, C., Brechka, H., Willats, W.G.T., Rose, J.K.C., 2014a. Pectin metabolism and assembly in the cell gall of the Charophyte Green alga *Penium margaritaceum*. Plant Physiol. 165, 105–118.

Domozych, D.S., Sørensen, I., Sacks, C., Brechka, H., Andreas, A., Fangel, J.U., Rose, J.K.C., Willats, W.G.T., Popper, Z.A., 2014b. Disruption of the microtubule network alters cellulose deposition and causes major changes in pectin distribution in the cell wall of the green alga, *Penium margaritaceum*. J. Exp. Bot. 65, 465–479.

Donaldson, I.G., 1972. The estimation of the motive force for protoplasmic streaming in *Nitella*. Protoplasma 74, 329–344.

Donaldson, R.P., Beevers, H., 1977. Lipid composition of organelles from germinating castor bean endosperm. Plant Physiol. 59, 259–263.

Donaldson, R.P., Tolbert, N.E., Schnarrenberger, C., 1972. A comparison of microbody membranes with microsomes and mitochondria from plant and animal tissue. Arch. Biochem. Biophys. 152, 199–215.

Donaldson, R.P., Tully, R.E., Young, O.A., Beevers, H., 1981. Organelle membranes from germinating castor bean endosperm. II. Enzymes, cytochromes, and permeability of the glyoxysome membrane. Plant Physiol. 67, 21–25.

Donelson, J.E., Wu, R., 1972. Nucleotide sequence analysis of deoxyribonucleic acid. VI. Determination of 3'-terminal dinucleotide sequences of several species of duplex deoxyribonucleic acid using Escherichia coli deoxyribonucleic acid polymerase I. J. Biol. Chem. 247, 4654–4660.

Dong, J., Chen, W., 2013. The role of autophagy in chloroplast degradation and chlorophagy in immune defenses during *Pst* DC3000 (*AvrRps4*) infection. PLoS One 8, e73091.

Dong, X.-J., Nagai, R., Takagi, S., 1998. Microfilaments anchor chloroplasts along the outer periclinal wall in *Vallisneria* epidermal cells through cooperation of Pfr and photosynthesis. Plant Cell Physiol. 39, 1299–1306.

Dong, X.-J., Ryu, J.H., Takagi, S., Nagai, R., 1996. Dynamic changes in the organization of microfilaments associated with the photocontrolled motility of chloroplasts in epidermal cells of *Vallisneria*. Protoplasma 195, 18–24.

Dong, X.-J., Takagi, S., Nagai, R., 1995. Regulation of the orientation movement of chloroplasts in epidermal cells of *Vallisneria*: cooperation of phytochrome with photosynthetic pigment under low-fluence-rate light. Planta 197, 257–263.

Donnan, F.G., 1928. The mystery of life. Nature 122, 512–514.

Donnan, F.G., 1937. Concerning the applicability of thermodynamics to the phenomena of life. J. Gen. Physiol. 8, 685–688.

Donne, J., 1634. Devotions upon Emergent Occasions, fourth ed. A. Mathewes, London.

Donohoe, B.S., Kang, B.-H., Gerl, M.J., Gergely, Z.R., McMichael, C.M., Bednarek, S.Y., Staehelin, L.A., 2013. *Cis*-Golgi cisternal assembly and biosynthetic activation occur sequentially in plants and algae. Traffic 14, 551–567.

Donohoe, B.S., Kang, B.-H., Staehelin, L.A., 2007. Identification and characterization of COPIa and COPIb-type vescicle classes associated with plant and algal Golgi. Proc. Natl. Acad. Sci. U.S.A. 104, 163–168.

Donohue, J., 1976. Honest Jim? Q. Rev. Biol. 51, 285–289.

Doolittle, R.F., 1995. The multiplicity of domains in proteins. Ann. Rev. Biochem. 64, 287–314.

Doolittle, W.F., 1998. You are what you eat: a gene transfer ratchet could account for bacterial genes in eukaryotic nuclear genomes. Trends in Genet. 14, 307–311.

Doolittle, W.F., Boucher, Y., Nesbo, C.L., Douady, C.J., Andersson, J.O., Roger, A.J., 2003. How big is the iceberg of which organellar genes in nuclear genomes are but the tip? Philos. Trans. R. Soc. B 358, 39–57.

Dorée, M., 1990. Control of M-phase by maturation-promoting factor. Curr. Opin. Cell Biol. 2, 269–273.

Dörmann, P., 2007. Lipid synthesis, metabolism and transport. In: Wise, R.R., Hoober, J.K. (Eds.), The Structure and Function of Plastids. Springer, Dordrecht, pp. 335–353.

Dorn, A., Weisenseel, M.H., 1984. Growth and the current pattern around internodal cells of *Nitella flexilis* L. J. Exp. Bot. 35, 373–383.

Douce, R., 1985. Mitochondria in Higher Plants. Structure, Function, and Biogenesis. Academic Press, Orlando.

Douce, R., Holtz, R.B., Benson, A.A., 1973. Isolation and properties of the envelope of spinach chloroplast. J. Biol. Chem. 248, 7215–7222.

Douce, R., Joyard, J., 1979. Structure and function of the plastid envelope. Adv. Bot. Res. 7, 1–116.

Douce, R., Joyard, J., 1980. Plant galactolipids. In: Stumpf, P.K. (Ed.), The Biochemistry of Plants, vol. 4. Academic Press, New York, pp. 321–362.

Douce, R., Joyard, J., 1990. Biochemistry and function of the plastid envelope. Annu. Rev. Cell Biol. 6, 173–216.

Douce, R., Neuburger, M., 1989. The uniqueness of plant mitochondria. Ann. Rev. Plant Physiol. Plant Mol. Biol. 40, 371–414.

Doudna, J.A., Szostak, J.W., 1989. RNA-catalyzed synthesis of complementary-strand RNA. Nature 339, 519–522.

Dougherty, E.C., 1957. Neologisms needed for structures of primitive organisms. I. Types of nuclei. J. Protozool. 4 (Suppl.), 14.

Dougherty, W.G., Parks, T.D., 1995. Transgenes and gene suppression: telling us something new? Curr. Opin. Cell Biol. 7, 399–405.

Douglass, A.E., 1929. The secret of the Southwest solved by talkative tree rings. Natl. Geogr. Mag. 56 (6), 736–770.

Douglass, A.D., Vale, R.D., 2005. Single-molecule microscopy reveals plasma membrane microdomains created by protein-protein networks that exclude or trap signaling molecules in T cells. Cell 121, 937–950.

Dounce, A.L., 1952. Duplicating mechanism for peptide chain and nucleic acid synthesis. Enzymologia 15, 251–258.

Dounce, A.L., 1956. Nucleoproteins. Round table discussion. Symposium on structure of enzmes and proteins. J. Cell. Comp. Physiol. 47 (Suppl. 1), 103–112.

Dove, W.F., Dee, J., Hatano, S., Haugli, F.B., Wohlfath-Bottermann, K.E. (Eds.), 1986. The Molecular Biology of Physarum polycephalum. Plenum Press, New York.

Dove, W.F., Rusch, H.P., 1980. Growth and Differentiation in Physarum Polycephalum. Princeton University Press, Princeton, NJ.

Dowhan, W., 1991. Phospholipid-transfer proteins. Curr. Opin. Cell Biol. 3, 621–625.

Dowhan, W., 2013. A retrospective: use of *Escherichia coli* as a vehicle to study phospholipid synthesis and function. Biochim. Biophys. Acta 1831, 471–494.

Doyle, J.J., 2011. Phylogenetic perspectives on the origins of nodulation. Mol. Plant Microbe Interact. 24, 1289–1295.

Doyle, J.J., Egan, A.N., 2010. Dating the origins of polyploidy events. New Phytol. 186, 73–85.

Drakakaki, G., Dandekar, A., 2013. Protein secretion: how many secretory routes does a plant cell have? Plant Sci. 203-204, 74–78.

Drakakaki, G., van de Ven, W., Pan, S., Miao, Y., Wang, J., Keinath, N.F., Weatherly, B., Jiang, L., Schumacher, K., Hicks, G., Raikhel, N., 2012. Isolation and proteomic analysis of the syp61 compartment reveal its role in exocytic trafficking in *Arabidopsis*. Cell Res. 22, 413–424.

Draper, J.W., 1898. History of the Conflict between Religion and Science. D. Appleton and Co., New York.

Draper, M.H., Hodge, A.J., 1949. Studies on muscle with the electron microscope. I. The ultrastructure of toad striated muscle. Aust. J. Exp. Biol. Med. Sci. 27, 465–503.

d'Enfert, C., Gensse, M., Gaillardin, C., 1992. Fission yeast and a plant have functional homologues of the Sar1 and Sec12 proteins involved in ER to Golgi traffic in budding yeast. EMBO J. 11, 4205–4211.

Dreses-Werringloer, V., Fischer, K., Wachter, E., Link, T.A., Flügge, U.-I., 1991. cDNA sequence and deduced amino acid sequence of the precursor of the 37-kDa inner envelope membrane polypeptide from spinach chloroplasts. Its transit peptide contains an amphiphilic α-helix as the only detectable structural element. Eur. J. Biochem. 195, 361–368.

Dreyer, E.M., Weisenseel, M.H., 1979. Phytochrome-mediated uptake of calcium in *Mougeotia* cells. Planta 146, 31–39.

Driesch, H., 1914. The History and Theory of Vitalism. Translated by C. K. Ogden. MacMillan, London.

Driesch, H., 1929. The Science and Philosophy of the Organism, second ed. A. & C. Black, Ltd., London.

Driouich, A., Faye, L., Staehelin, L.A., 1993a. The plant Golgi apparatus: a factory for complex polysaccharides and glycoproteins. Trends Biochem. Sci. 18, 210–214.

Driouich, A., Follet-Gueye, M.-L., Bernard, S., Kousar, S., Chevalier, L., Vicré-Gibouin, M., Lerouxel, O., 2012. Golgi mediated synthesis and secretion of matrix polysaccharides of the primary cell wall of higher plants. Front. Plant Sci. 3, 79.

Driouich, A., Levy, S., Staehelin, L.A., Faye, L., 1994. Structural and functional organization of the Golgi apparatus in plant cells. Plant Physiol. Biochem. 32, 731–749.

Driouich, A., Zhang, G.F., Staehelin, L.A., 1993b. Effect of Brefelden A on the structure of the Golgi apparatus and on the synthesis and secretion of proteins and polysaccharides in sycamore maple (*Acer pseudoplatanus*) suspension-cultured cells. Plant Physiol. 101, 1363–1373.

Drlica, K., Rouviere-Yaniv, J., 1987. Histonelike proteins of bacteria. Microbiol. Rev. 51, 301–319.

Drmanac, R., Labat, I., Brukner, I., Crkvenjakov, R., 1989. Sequencing of megabase plus DNA by hybridization: theory of the method. Genomics 4, 114–128.

Drøbak, B.K., 1991. Plant signal perception and transduction: the role of the phosphoinositide system. Essays Biochem. 26, 27–37.

Drøbak, B.K., Ferguson, I.B., 1985. Release of Ca^{2+} from plant hypocotyl microsomes by inositol–1,4,5–triphosphate. Biochem. Biophys. Res. Commun. 130, 1241–1246.

Drøbak, B.K., Ferguson, I.B., Dawson, A.P., Irvine, R.F., 1988. Inositol–containing lipids in suspension-cultured plant cells. Plant Physiol. 87, 217–222.

Drøbak, B.K., Franklin-Tong, V.E., Staiger, C.J., 2004. The role of the actin cytoskeleton in plant cell signaling. New Phytol. 163, 13–30.

Drucker, M., Herkt, B., Robinson, D.G., 1995. Demonstration of a beta-type adaptin at the plant plasma membrane. Cell Biol. Int. 19, 191–201.

Drucker, S.C., Knox, R.B., 1985. Pollen and pollination: a historical review. Taxon 34, 401–419.

Drysdale, J., 1874. The Protoplasmic Theory of Life. Baillière, Tindall, and Cox, London.

D'Arcy Thompson, W., 1959. On Growth and Form. Cambridge University Press, Cambridge.

D'Paolo, M.L., Perutto, A.D.B., Bassi, R., 1990. Immunological studies on chlorophyll-a/b proteins and their distribution in thylakoid membrane domains. Planta 181, 275–286.

Du Bois-Reymond, E.H., 1848. Untersuchungen über thierische Elektricität, 84. G. Reimer, Berlin.

Du, W., Tamura, K., Stefano, G., Brandizzi, F., 2013. The integrity of the plant Golgi apparatus depends on cell growth-controlled activity of GNL1. Mol. Plant 6, 905–915.

Du, H., Yu, Y., Ma, Y., Gao, Q., Cao, Y., Chen, Z., Ma, B., Qi, M., Li, Y., Zhao, X., Wang, J., Liu, K., Qin, P., Yang, X., Zhu, L., Li, S., Liang, C., 2017. Sequencing and de novo assembly of a near complete *indica* rice genome. Nat. Commun. 8, 15324.

Dube, S.K., Marchker, K.A., Clark, B.F., Cory, S., 1968. Nucleotide sequence of N-formyl-methionine-transfer RNA. Nature 218, 232–233.

Dubochet, J., 2017. Early Cryo-Electron Microscopy. https://www.nobelprize.org/prizes/chemistry/2017/dubochet/lecture/.

DuBois-Reymond, E., 1872. Ueber die Grenzen der Naturerkenntniss. Leipzig.

Dubos, R.J., 1937. The decomposition of yeast nucleic acid by a heat resistant enzyme. Science 85, 549–550.

Dubos, R.J., 1939. Studies on a bacterial agent extracted from a soil bacillis. J. Exp. Med. 70, 1–10.

Dubos, R., 1960. Pasteur and Modern Science. Anchor Books, Garden City, New York.

Dubos, R.J., Thompson, R.H.S., 1938. The decomposition of yeast nucleic acid by a heat-resistant enzyme. J. Biol. Chem. 124, 501–510.

Duboscq, O., Grassé, P.P., 1933. L'appareil parabasal des flagellés Avec des remarques sur le trophosponge, l'appareil de Golgi, des mitochondries et le vacuome. Arch. Zool. Exp. Et. Gén 73, 381–621.

Ducet, G., Gidrol, X., Richaud, P., 1983. Membrane potential changes in coupled potato mitochondria. Physiol. Veg. 21, 385–394.

Duchenne, G.B., 1862. Méchanisme de la Physionomie Humaine. Jules Renouard, Paris.

Duchenne, G.B., 1871. A Treatise on Localized Electrization. Lindsay & Blakiston, Philadelphia, PA.

Duchenne, G.B., 1949. Physiology of Motion. J. B. Lippincott, Philadelphia.

Ducker, S.C., Knox, R.B., 1985. Pollen and pollination: a historical review. Taxon 34, 401–419.

Duckett, J.G., Renzaglia, K.S., 1986. The blepharoplast of *Hypnum*. J. Bryol. 14, 375–385.

Duclaux, E., 1920. Pasteur. The History of a Mind. W. B. Saunders, Philadelphia.

Duden, R., Allan, V., Kreis, T., 1991. Involvement of β-COP in membrane traffic through the Golgi complex. Trends Cell Biol. 1, 14–19.

Dugas, H., 1996. Bioorganic Chemistry. A Chemical Approach to Enzyme Action, third ed. Springer-Verlag, New York.

Dugas, C.M., Li, Q., Kham, I.A., Nothnagel, E.A., 1989. Lateral diffusion in the plasma membrane of maize protoplasts with implications for cell culture. Planta 179, 387–396.

Dujardin, F., 1835. Recherches sur les organismes inférieurs. Ann. Sc. Nat. Zool. Second Series 4, 343–377.

Dujardin, F., 1841. Histoire naturelle des zoophytes: Infusoires. Librairie Encyclopédique de Roret, Paris.

Dulbecco, R., 1986. A turning point in cancer research: sequencing the human genome. Science 231, 1055–1056.

Dumas, M.J., Boussingault, M.J.B., 1844. The Chemical and Physiological Balance of Organic Nature. An Essay. H. Bailliere Publisher, London.

Dumont, S., Mitchison, T., 2012. Mechanical forces in mitosis. Compr. Biophys. 4, 298–320.

Duncan, E.H., Eakin, W.R., 1981. Thomas Reid's Lectures on Natural Theology (1780). University Press of America, Washington, DC.

Duncan, T., Wakefield, J.G., 2011. 50 ways to build a spindle: the complexity of microtubule generation during mitosis. Chromosome Res. 19, 321–333.

Dundr, M., Raska, I., 1993. Nonisotopic ultrastructural mapping of transcription sites within the nucleolus. Exp. Cell Res. 208, 275–281.

Dunham, V.L., Bryant, J.A., 1988. Nuclei. In: Hall, J.L., Moore, A.L. (Eds.), Isolation of Membranes and Organelles from Plant Cells. Academic Press, London, pp. 237–275.

Dunker, A.K., Bondos, S.E., Huang, F., Oldfield, C.J., 2014. Intrinsically disordered proteins and multicellular organisms. Semin. Cell Dev. Biol. 37, 44–55.

Dunker, A.K., Lawson, J.D., Brown, C.J., Williams, R.M., Romero, P., Oh, J.S., Oldfield, C.J., Campen, A.M., Ratliff, C.M., 2001. Intrinsically disordered protein. J. Mol. Gr. Model. 19, 26–59.

Dunker, A.K., Silman, I., Uversky, V.N., Sussman, J.L., 2008. Function and structure of inherently disordered proteins. Curr. Opin. Struct. Biol. 18, 756–764.

Dunker, A.K., Uversky, V.N., 2010. Understanding protein non-folding. Biochim. Biophys. Acta 1804, 1231–1264.

Dunkley, T.P.J., Svenja, H., Shadforth, I.P., John, R., et al., 2006. Mapping the *Arabidopsis* organelle proteome. Proc. Natl. Acad. Sci. U.S.A. 103, 6518–6523.

Dunn, M.S., 1941. Chemistry of amino acids and proteins. Ann. Rev. Biochem. 10, 91–124.

Dunn Jr., W.A., 1990. Studies on the mechanisms of autophagy: formation of the autophagic vacuole. J. Cell Biol. 110, 1923–1933.

Dunn, B.M., 2001. Determination of protease mechanism. In: Beynon, R., Bond, J.S. (Eds.), Proteolytic Enzymes: A Practical Approach, second ed. Oxford Univiversity Press, Oxford, pp. 77–104.

Dunn, E.K., Shoue, D.A., Huang, X., Kline, R.E., MacKay, A.L., Carpita, N.C., Taylor, I.E.P., Mandoli, D.F., 2007. Spectroscopic and biochemical analysis of regions of the cell wall of the unicellular "mannan weed," *Acetabularia acetabulum*. Plant Cell Physiol. 48, 122–133.

Duque, P., Sánchez, J.-P., Chua, H.-N., 2004. Signalling and the cytoskeleton in guard cells. Annu. Plant Rev. 10, 290–317.

Durham, A.C.H., Ridgway, E.B., 1976. Control of chemotaxis in *Physarum polycephalum*. J. Cell Biol. 69, 218–223.

Duroc, Y., Bouchez, D., Pastuglia, M., 2011. The preprophase band and division site determination in land plants. The Plant Cytoskeleton. Adv. Plant Biol. 2, 145–185.

Durrant, A., 1958. Environmental conditioning of flax. Nature 181, 928–929.

Durrant, A., 1962. The environmental induction of heritable changes in *Linum*. Heredity 17, 27–61.

Durso, N.A., Leslie, J.D., Cyr, R.J., 1996. In situ immunocytochemical evidence that a homolog of protein translation elongation factor EF-1α is associated with microtubules in carrot cells. Protoplasma 190, 141–150.

Dustin, P., 1947. Some new aspects of mitotic poisoning. Nature 159, 794–797.

Dustin, P., 1978. Microtubules. Springer-Verlag, Berlin.

Dutcher, S.K., 1995. Flagellar assembly in two hundred and fifty easy-to-follow steps. Trends Genet. 11, 398–404.

Dutrochet, M.H., 1824. Recherches Anatomiques et Physiologiques sur La Structure Intime des Animaux et des Végétaux et sur Leur Motilité. Chez J. B. Baillière, Paris.

Dutt, A.K., 1957. Vacuoles of the pulvinus and the mechanism of movement. Nature 179, 254.

Dutta, R., Robinson, K.R., 2004. Identification and characterization of stretch-activated channels in pollen protoplasts. Plant Physiol. 135, 1398–1406.

Duysens, L.N.M., 1989. The discovery of the two photosystems: a personal account. Photosynth. Res. 21, 61–80.

Duysens, L.N.M., 1996. Arnold's inspiring experiments. Photosynth. Res. 48, 25–29.

Duysens, L.N.M., Amesz, J., Kamp, B.M., 1961. Two photochemical systems in photosynthesis. Nature 190, 510–511.

Dworetzky, S.I., Feldherr, C.M., 1988. Translocation of RNA- coated gold particles through the nuclear pores of oocytes. J. Cell Biol. 106, 575–584.

Dyer, A.F., 1979. The Experimental Biology of Ferns. Academic Press, London.

Dyer, B.D., Obar, R.A., 1994. Tracing the History of Eukaryotic Cells. The Enigmatic Smile. Columbia University Press, New York.

Dyson, F., 1985. Origins of Life. Cambridge University Press, Cambridge.

Dyson, F., 2004. A meeting with Enrico Fermi. Nature 427, 297.

Dyson, F., 2006. The Scientist as Rebel. New York Review Books, New York.

Dzelzkalns, J., Nasrallah, J.B., Nasrallah, M.E., 2002. Cell-cell communication in plants: self-incompatibility in flower development. Dev. Biol. 153, 70–82.

Dzierszinski, F., Popescu, O., Toursel, C., Slomianny, C., Yahiaoui, B., Tomavo, S., 1999. The protozoan parasite *Toxoplasma gondii* expresses two functional plant-like glycolytic enzymes. Implications for evolutionary origin of apicomplexans. J. Biol. Chem. 274, 24888–24895.

Earnshaw, W.C., Bernat, R.L., 1991. Chromosomal passengers: toward an integrated view of mitosis. Chromosoma 100, 139–146.

Earnshaw, W.C., Carmena, M., 2003. A perfect funeral with no corpse. J. Cell Biol. 160, 989–990.

Earnshaw, J.C., Steer, M.W., 1979. Studies of cellular dynamics by laser Doppler microscopy. Pestic. Sci. 10, 358–368.

East, E.M., 1910a. A Mendelian interpretation of variation that is apparently continuous. Am. Nat. 44, 65–82.

East, E.M., 1910b. Notes on an experiment concerning the nature of unit characters. Science 32, 93–95.

East, E.M., 1916. Studies on size inheritance in *Nicotiana*. Genetics 1, 164–176.

East, E.M., 1920. Population. Sci. Mon. 10 (6), 603–624.

East, E.M., 1923. Mendel and his contemporaries. Sci. Mon. 16, 225–236.

East, E.M., 1927. Heredity and Human Affairs. Charles Scribner's Sons, New York.

East, E.M., 1931. The future of man in the light of his past: the view-point of a geneticist. Sci. Mon. 32, 301–308.

East, E.M., Jones, D.F., 1919. Inbreeding and Outbreeding. Their Genetic and Sociological Significance. J. B. Lippincott Co., Philadelphia.

Eastmond, P., Germain, V., Lange, P., Bryce, J.H., Smith, S.M., Graham, I.A., 2000. Post-germinative growth and lipid catabolism in oilseeds lacking the glyoxylate cycle. Proc. Natl. Acad. Sci. U.S.A. 97, 5669–5674.

Ebashi, S., 1960. Calcium binding and relaxation in the actomyosin system. J. Biochem. 48, 150–151.

Ebashi, S., 1961a. Calcium binding activity of vesicular relaxing factor. J. Biochem. 50, 236–244.

Ebashi, S., 1961b. The role of relaxing factor in contraction-relaxation cycle of muscle. Prog. Theort. Phys. (Suppl. 17), 33–40.

Ebashi, S., 1985. Relaxing factor, sarcoplasmic reticulum and troponin-A historical survey. In: Fleisher, S., Tonomora, Y. (Eds.), Structure and Function of Sarcoplasmic reticulum. Academic Press, Orlando, pp. 1–18.

Ebashi, F., Ebashi, S., 1962. Removal of calcium and relaxation in actomyosin systems. Nature 194, 378–379.

Ebashi, S., Ebashi, F., Fujie, Y., 1960. The effect of EDTA and its analogues on glycerinated muscle fibers and myosin adenosinetriphosphatase. J. Biochem. 47, 54–59.

Ebashi, S., Lipmann, F., 1962. Adenosine triphosphate-linked concentration of calcium ions in a particulate fraction of rabbit muscle. J. Cell Biol. 14, 389–400.

Ebashi, S., October 29, 1968. The first correct prediction. Curr. Contents 44, 18.

Ebine, K., Inoue, T., Ito, E., Uemura, T., Goh, T., Abe, H., Sato, K., Nakano, A., Ueda, T., 2014. Plant vacuolar trafficking occurs through distinctly regulated pathways. Curr. Biol. 24, 1375–1382.

Ebine, K., Ueda, T., 2015. Roles of membrane trafficking in plant cell wall dynamics. Front. Plant Sci. 6, 878.

Eccles, J.C., 1964. The Physiology of Synapses. Springer-Verlag, Berlin.

Eccles, J.C., 1979. The Human Mystery. Springer-Verlag, Berlin.

Eccles, J., December 11, 1963. The Ionic Mechanism of Postsynaptic Inhibition. Nobel Lecture.

Echeverria, E., Burns, J.K., 1989. Vacuolar acid hydrolysis as a physiological mechganism for sucrose breakdown. Plant Physiol. 90, 530–533.

Echeverria, E., Burns, J.K., Felle, H., 1992. Compartmentalization and cellular conditions controlling sucrose breakdown in mature acid lime fruit. Phytochemistry 31, 4091–4095.

Ecker, A., 1848. Zur Lehre vom Bau und Leben der contractilen Substanz der niedesten Thiere. Z. Wiss. Zool. 1, 218–245.

Eckert, R., 1988. Animal Physiology, third ed. W. H. Freeman and Co., New York.

Eddington, A.S., 1929. The Nature of the Physical World: The Gifford Lectures, 1927. Macmillan, New York.

Eddington, A.S., 1992. The Nature of the Physical World: The Gifford Lectures, 1927. Macmillan, New York.

Edidin, M., 1992. Patches, posts and fences: proteins and plasma membrane domains. Trends Cell Biol. 2, 376–380.

Edidin, M., 2001. Shrinking patches and slippery rafts: scales of domains in the plasma membrane. Trends Cell Biol. 11, 492–496.

Edman, P., 1949. A method for the determination of the amino acid sequence of peptides. Arch. Biochem. Biophys. 22, 475–483.

Edman, P., 1950. Method for determination of the amino acid sequence in peptides. Acta Chem. Scand. 4, 283–293.

Edman, P., Begg, G., 1967. A protein sequenator. Eur. J. Biochem. 1, 80–91.

Edsall, J.T., 1942. The chemistry of the proteins and amino acids. Ann. Rev. Biochem. 11, 151–182.

Edsall, J.T., 1976. James Sumner and the crystallization of urease. Trends Biochem. Sci. 1, 21.

Edsall, J.T., Gutfreund, H., 1983. Biothermodynamics. The Study of Biochemical Processes at Equilibrium. John Wiley & Sons, New York.

Edsall, J.T., Wyman, J., 1958. Biophysical Chemistry. Academic Press, New York.

Edström, J.-E., December 12, 1974. Award ceremony speech.

Edwards, G.E., Franceschi, V.R., Voznesenskaya, E.V., 2004. Single cell C4 photosynthesis versus the dual-cell (Kranz) paradigm. Annu. Rev. Plant Physiol. Plant Mol. Biol. 55, 173–196.

Edwards, A., Voss, H., Rice, P., Civitello, A., Stegmann, J., et al., 1990. Automated DNA sequencing of the HPRT locus. Genomics 6, 593–608.

Efroni, I., Birnbaum, K.D., 2016. The potential of single-cell profiling in plants. Genome Biol. 17, 65.

Efroni, I., Ip, P.L., Nawy, T., Mello, A., Birnbaum, K.D., 2015. Quantification of cell identity from single-cell gene expression profiles. Genome Biol. 16, 9.

Egan, A.N., Doyle, J.J., 2010. A comparison of global, gene-specific, and relaxed clock methods in a comparative genomics framework: dating the polyploid history of soybean (Glycine max). Syst. Biol. 59, 534–547.

Ehlers, K., Kollmann, R., 2000. Synchronization of mitotic activity in protoplast-derived *Solanum nigrum* L. microcalluses is correlated with plasmodesmal connectivity. Planta 210, 269–278.

Ehlers, K., Kollmann, R., 2001. Primary and secondary plasmodesmata: structure, origin, and function. Protoplasma 216, 1–30.

Ehrenfest, P., Ehrenfest, T., 1912. The Conceptual Foundations of the Statistical Approach in Mechanics, 1959. Cornell University Press, Ithaca, New York.

Ehrhardt, D.W., Wais, R., Long, S.R., 1996. Calcium spiking in plant root hairs responding to Rhizobium nodulation signals. Cell 85, 673–681.

Eid, J., Fehr, A., Gray, J., Luong, K., Lyle, J., Otto, G., Peluso, P., Rank, D., Baybayan, P., Bettman, B., Bibillo, A., Bjornson, K., Chaudhuri, B., Christians, F., Cicero, R., Clark, S., Dalal, R., deWInter, A., Dixon, J., Foquet, M., Gaertner, A., Hardenbol, P., Heiner, C., Hester, K., Holden, D., Kearns, G., Kong, X., Kuse, R., Lacroix, Y., Lin, S., Lungquist, P., Ma, C., Marks, P., Maxham, M., Murphy, D., Park, I., Pham, T., Phillips, M., Roy, J., Sebra, R., Shen, G., Sorenson, J., Tomaney, A., Travers, K., Trulson, M., Vieceli, J., Wegener, J., Wu, D., Yang, A., Zaccarin, D., Zhao, P., Zhong, F., Korlach, J., Turner, S., 2009. Real-time DNA sequencing from single polymerase molecules. Science 323, 133–138.

Eidenmueller, H., 2017. The Rise of Robots and the Law of Humans. Oxford Legal Studies Research Paper No. 27/2017. https://papers.ssrn.com/sol3/papers.cfm?abstract_id=2941001.

Eigen, M., 1992. Steps Towards Life: A Perspective on Evolution. Translated by P. Woolley. Oxford University Press, Oxford.

Eigen, M., 1993. Viral quasispecies. Sci. Am. 269, 42–49.

Eigen, M., Gardiner, W., Schuster, P., Winkler-Oswatitsch, R., 1981. The origin of genetic information. Sci. Am. 244, 88–118.

Eigen, M., Kruse, W., 1962. Some fast protolytic reactions of complexes. In: Seventh International Conference on Coordination Chemistry. Butterworths, London.

Eigen, M., Schuster, P., 1979. The Hypercycle. A Principle of Natural Self-Organization. Springer-Verlag, Berlin.

Eigsti Jr., O.J., Dustin, P., 1955. Colchicine. In: Agriculture Medicine Biology and Chemistry. Iowa State College Press, Ames.

Eigsti Jr., O.J., Dustin, P., Gay-Winn, N., 1949. On the discovery of the action of colchicine on mitosis in 1889. Science 110, 692.

Eijkmann, C., December 10, 1929. Antineuritic Vitamin and Beriberi. Nobel Lecture.

Eilers, T., Schwartz, G., Brinkmann, H., Witt, C., Richter, T., Nieder, J., Koch, B., Hille, R., Hänsch, R., Mendel, R.R., 2001. Identification and biochemical characterization of *Arabidopsis thaliana* sulfite oxidase. J. Biol. Chem. 276, 46989–46999.

Einspahr Jr., K.J., Peeler, T.C., Thompson, G.A., 1988. Rapid changes in polyphosphoinositide metabolism associated with the response of *Dunaliella salina* to hypoosmotic shock. J. Biol. Chem. 263, 5775–5779.

Einspahr Jr., K.J., Peeler, T.C., Thompson, G.A., 1989. Phosphatidylinositol 4,5-bisphosphate phospholipase C and phosphomonoesterase in *Dunaliella salina* membranes. Plant Physiol. 90, 1115–1120 photon concept—A translation of the Annalen der Physik paper of 1905. Amer. J. Physics, 33, 367–374.

Einspahr Jr., K.J., Tompson, G.A., 1990. Transmembrane signaling via phosphatidylinositol 4 5-bisphosphate hydrolysis in plants. Plant Physiol. 93, 361–366.

Einstein, A., 1905. Concerning a heuristic point of view toward the emission and transformation of light, 1965 Annalen der Physik 17, 132. Translated by A. B. Arons and M. B. Peppard. in Einstein's proposal of the.

Einstein, A., 1905. Concerning an heuristic point of view toward the emission and transformation of light. Ann. Phys. 17, 132–148. http://einsteinpapers.press.princeton.edu/vol2-trans/100.

Einstein, A., 1906. Zur Theorie der Brownschen Bewegung. Annalen der Physik 19, 371–381.

Einstein, A., 1950. On the generalized theory of gravitation. Sci. Am. 182 (April), 13–17.

Einstein, A., 1956. Investigations on the Theory of the Brownian Movement. Dover Publications, New York.

Einstein, A., 1961. Relativity. The Special and the General Theory. Translated by R. W. Lawson. Bonanza Books, New York.

Einstein, A., April 1946. EMc2. Sci. Illus. 1, 16–17.

Einstein, A., Schrödinger, E., Tchernavin, V., 1936. The freedom of learning. Science 83, 372–373.

Eisen, J.A., 1998. Phylogenomics: improving functional predictions for uncharacterized genes by evolutionary analysis. Genome Res. 8, 163–167.

Eisen, J., 2012. Badomics words and the power and peril of the omememe. Gigascience 1, 6.

Eisen, J.A., Fraser, C.M., 2003. Phylogenomics: intersection of evolution and genomics. Science 300, 1706–1707.

Eisen, J.A., Hanawalt, P.C., 1999. A phylogenomic study of DNA repair genes, proteins, and processes. Mutat. Res. 435, 171–213.

Eisen, A., Kiehart, D.P., Wieland, S.J., Reynolds, G.T., 1984. Temporal sequence and spatial distribution of early events of fertilization in single sea urchin eggs. J. Cell Biol. 99, 1647–1654.

Eisenach, C., De Angeli, A., 2017. Ion transport at he vacuole during stomatal movements. Plant Physiol. 174, 520–530.

Eisenberg, R.S., 1990. Channels as enzymes. J. Membr. Biol. 115, 1–12.

Eisenhut, M., Bräutigam, A., Timm, S., Florian, A., Tohge, T., Fernie, A.R., Bauwe, H., Weber, A.P.M., 2017. Photorespiration is crucial for dynamic response of photosynthetic metabolism and stomatal movement to altered CO_2 availability. Mol. Plant 10, 47–61.

Eisenstein, M., 2015. Startups use short-read data to expand long-lead sequencing market. Nat. Biotechnol. 33, 433–435.

Elbein, A.D., 1979. The role of lipid-linked saccharides in the biosynthesis of complex carbohydrates. Ann. Rev. Plant Physiol. 30, 239–272.

Elder, R.H., Dell Aquila, A., Mezzina, M., Sarasin, A., Osborne, D.J., 1987. DNA ligase in repair and replication in the embryos of rye *Secale cereale*. Mutat. Res. 181, 61–71.

Elgin, S.C.R., 1990. Chromatin structure and gene activity. Curr. Opin. Cell Biol. 2, 437–445.

Elias, J.E., Gibbons, F.D., King, O.D., Roth, F.P., Gygi, S.P., 2004. Intensity-based protein identification by machine learning from a library of tandem mass spectra. Nat. Biotechnol. 22, 214–219.

Eliassaf, J., Silberberg, A., Katchalsky, A., 1955. Negative thixotropy of aqueous solutions of polymethacrylic acid. Nature 176, 1119.

Eliot, T.S., 1936. Choruses from the Rock. Harcourt. In: Collected Poems 1909-1962. Brace & World, New York, 1963.

Ellens, K.W., Christian, N., Singh, C., Satagopam, V.P., May, P., Linster, C.L., 2017. Confronting the catalytic dark matter encoded by sequenced genomes. Nucleic Acids Res. 45, 11495–11514.

Elliott, W.H., Elliott, D.C., 1997. Biochemistry and Molecular Biology. Oxford University Press, Oxford.

Elliott, D.C., Kokke, Y.S., 1987a. Cross-reaction of a plant protein kinase with antiserum raised against a sequence from bovine brain protein kinase C regulatory subunit. Biochem. Biophys. Res. Commun. 145, 1043–1047.

Elliott, D.C., Kokke, Y.S., 1987b. Partial purification and properties of a protein kinase C type enzyme from plants. Phytochemistry 26, 2929–2935.

Ellis, H., 1911. The Problem of Race-Regeneration. Moffat, Yard & Co, New York.

Ellis, R.J., 1977. Protein synthesis by isolated chloroplasts. Biochim. Biophys. Acta 463, 185–215.

Ellis, R.J. (Ed.), 1996. The Chaperonins. Academic Press, San Diego.

Ellis, M., Egelund, J., Schultz, C., Bacic, A., 2010. Arabinogalactan-proteins: key regulators at the cell surface. Plant Physiol. 153, 403–419.

Ellis, R., Yuan, J., Horvitz, R., 1991. Mechanisms and functions of cell death. Ann. Rev. Cell Biol. 7, 663–698.

Ellison, H., 1833. Mad Moments, or First Verse Attempts, by a Born Natural, addressed respectfully to the light-headed of Society at large, but intended more particularly for the use of that World's Madhouse. London.

Elowitz, M.B., Surette, M.G., Wolf, P.-E., Stock, J.B., Leibler, S., 1999. Protein mobility in the cytoplasm of *Escherichia coli*. J. Bacteriol. 181, 197–203.

Elsberg, L., 1883. Memoirs: plant Cells and Living Matter. Q. J. Microsc. Sci. 23, 87–98.

Elthon, T.E., Nickels, R.L., McIntosh, L., 1989. Mitochondrial events during development of thermogenesis in Sauromatum gultatum (Schott). Planta 180, 82–89.

Elwick, J., 2003. Herbert Spencer and the disunity of the social organism. Hist. Sci. 16, 35–72.

Emans, N., Zimmermann, S., Fischer, R., 2002. Uptake of a fluorescent marker in plant cells is sensitive to brefeldin A and wortmannin. Plant Cell 14, 71–86.

Emanuelsson, O., Elofsson, A., von Heijne, G., Cristobal, S., 2003. In silico prediction of the peroxisomal proteome in fungi, plants and animals. J. Mol. Biol. 330, 443–456.

Embden, G., Laquer, F., 1914. Über die Chemie des Lactacidogens. Z. Physiol. Chem. 94, 123.

Emerson, R.W., 1837. The American Scholar. Speech delivered on August 31, 1837 to the Phi Beta Kappa Society of Harvard College. http://digitalemerson.wsulibs.wsu.edu/exhibits/show/text/the-american-scholar.

Emerson, R.W., 1947. Compensation. In: Cummins, S., Linscott, R.N. (Eds.), Man and Man. The Social Philosophers. Random House, New York, pp. 427–446.

Emerson, A.E., 1954. Dynamic homeostasis: a unifying principle in organic, social, and ethical evolution. Sci. Mon. 78, 67–85.

Emerson, R.W., 1970. America the Beautiful. In the Words of Ralph Waldo Emerson. Country Beautiful Association, Waukesha, WI.

Emerson, R., Arnold, W.A., 1932a. A separation of the reactions in photosynthesis by means of intermittent light. J. Gen. Physiol. 15, 391–420.

Emerson, R., Arnold, W.A., 1932b. The photochemical reaction in photosynthesis. J. Gen. Physiol. 16, 191–205.

Emerson, R., Chalmers, R., Cederstand, C., 1957. Some factors influencing the long-wave limit of photosynthesis. Proc. Natl. Acad. Sci. U.S.A. 43, 133–143.

Emerson, R., Lewis, C.M., 1943. The dependence on the quantum yield of *Chlorella* photosynthesis on the wave length of light. Am. J. Bot. 30, 165–178.

Emery, P., So, W.V., Kaneko, M., Hall, J.C., Rosbash, M., 1998. CRY, a *Drosophila* clock and light-regulated cryptochrome, is a major contributor to circadian rhythm resetting and photosensitivity. Cell 95, 669–679.

Emery, L., Whelan, S., Hirschi, K.D., Pittman, J.K., 2012. Protein phylogenetic analysis of Ca^{2+}/cation antiporters and insights into the evolution of plants. Front. Plant Sci. 3, 1.

Emons, A.M.C., 1985. Plasma membrane rosettes in root hairs of *Equisetum hyemale*. Planta 163, 350–359.

Emons, A.M.C., 1989. Helicoidal microfibril deposition in a tip growing cell and microtubule alignment during tip morphogenesis: a dry-cleaving and freeze-substitution study. Can. J. Bot. 67, 2401–2408.

Emons, A.M.C., Traas, J.A., 1986. Coated pits and coated vesicles on the plasma membrane of plant cells. Eur. J. Cell Biol. 41, 57–64.

Emons, A.M.C., Wolters-Art, A.M.C., 1983. Cortical microtubules and microfibril deposition in the cell wall of root hairs of *Equisetum hyemale*. Protoplasma 117, 68–81.

Emr, S., Glick, B.S., Linstedt, A.D., Lippincott-Schwartz, J., Luini, A., Malhotra, V., Marsh, B.J., Nakano, A., Pfeffer, S.R., Rabouille, C., Rothman, J.E., Warren, G., Wieland, F.T., 2009. Journys through the Golgi—taking stock in a new era. J. Cell Biol. 187, 449–453.

Endler, A., Meyer, S., Schelbert, S., Schneider, T., Weschke, W., Peters, S.W., Keller, F., Baginsky, S., Martinoia, E., Schmidt, U.G., 2006. Identification of a vacuolar sucrose transporter in barley and *Arabidopsis* mesophyll cells by a tonoplast proteomic approach. Plant Physiol. 141, 196–207.

Endo, M., 2006. Setsuro Ebashi (1922-2006). Nature 442, 996.

Endo, M., 2008. Calcium ion and troponin: professor S. Ebashi's epoch-making achievement. Biochem. Biophys. Res. Commun. 369, 30–33.

Endo, M., 2011. Reiji Natori, Setsuro Ebashi, and excitation-contraction coupling. Prog. Biophys. Mol. Biol. 105, 129–133.

Endow, S.A., Higuchi, H., 2000. A mutant of the motor protein kinesin that moves in both directions on microtubules. Nature 406, 913–916.

Endow, S.A., Kang, S.J., Satterwhite, L.L., Rode, M.D., Skeen, V.P., Salmon, E.D., 1994. Yeast Kar3 is a minus-end microtubule motor protein that destabilizes microtubules preferentially at the minus end. EMBO J. 13, 2708–2713.

Engel, J., 1991. Common structural motifs in proteins of the extracellular matrix. Curr. Opin. Cell Biol. 3, 779–785.

Engelhardt, V.A., 1946. Adenosinetriphosphatase properties of myosin. Adv. Enzymol. Relat. Areas Mol. Biol. 6, 147−191.

Engelhardt, W.A., 1982. Life and science. Ann. Rev. Biochem. 51, 1−19.

Engelhardt, W.A., Ljubimowa, M.N., 1939. Myosine and Adenosinetriphosphate. Nature 144, 668−669.

Engelke, D.R., Hoener, Collins, F.S., 1988. Direct sequewncing of enzymatically amplified human genomic DNA. Proc. Natl. Acad. Sci. U.S.A. 85, 544−548.

Engelman, D.M., 2005. Membranes are more mosaic than fluid. Nature 438, 578−580.

Engelmann, Th.W., 1879. Physiologie der Protoplasma- und Flimmerbewegung. In: Hermann, L. (Ed.), Handbuch der Physiologie, Vol. I. Vogel, Leipzig, pp. 341−408.

Engelmann, Th.W., 1881. Neue Methode zur Untersuchung der Sauerstoffausscheidung pflanzlicher und thierischer Organismen. Bot. Z. 39, 441−448.

Engelmann, Th.W., 1882. On the production of oxygen by plant cells in a microspectrum, 1955. In: Gabriel, M.L., Fogel, S. (Eds.), Great Experiments in Biology. Prentice-Hall, Englewood Cliffs, NJ, pp. 166−170.

Engels, F., 1934. Dialectics of Nature. Progress Publishers, Moscow.

Engle, A.J., Sen, S., Sweeney, H.L., Discher, D.E., 2006. Matrix elasticity directs stem cell lineage specification. Cell 126, 677−689.

English, A.C., Richards, S., Han, Y., Wang, M., Vee, V., Qu, J., Qin, X., Muzny, D.M., Reid, J.G., Worley, K.C., et al., 2012. Mind the gap: upgrading genomes with Pacific Biosciences RS long-read sequencing technology. PLoS One 7, e47768.

English, A.R., Voeltz, G.K., 2013. Endoplasmic reticulum structure and interconnections with other organelles. Cold Spring Harb. Perspect. Biol. 5, a013227.

Englund, P.T., 1971. Analysis of nucleotide sequences at 3' termini of duplex deoxyribonucleic acid with the use of T4 deoxyribonucleic acid polymerase. J. Biol. Chem. 246, 3269−3276.

Enke, F., Erlangen, Nägeli, C. von, 1914. A Mechanico-Physiological Theory of Organic Evolution. Open Court Publishing Co., Chicago.

Epel, D., Schatten, G., 1998. Daniel Mazia: a passion for understanding how cells reproduce. Trends Cell Biol. 8, 416−418.

Epel, B.L., van Lent, J.W.M., Cohen, L., Kotlizsky, G., Katz, A., Yahalom, A., 1996. A 41 kDa protein from maize mesocotyl cell walls immunolocalizes to plasmodesmata. Protoplasma 191, 70−78.

Ephrussi, B., 1942. Chemistry of "eye color hormones" of *Drosophila*. Q. Rev. Biol. 17, 327−338.

Ephrussi, B., 1953. Nucleo-Cytoplasmic Relations in Microorganisms. Clarendon Press, Oxford.

Ephrussi, B., Leopold, U., Watson, J.D., Weigle, J.J., 1953. Terminologiy in bacterial genetics. Nature 171, 701.

Epimashko, S., Fischer-Schliebs, E., Christian, A.-L., Thiel, G., Lüttge, U., 2006. Na^+/H^+-transporter, H^+-pumps and an aquaporin in light and heavy tonoplast membranes from organic acid and NaCl accumulating vacuoles of the annual facultative CAM plant and halophyte Mesembryanthemum crystallinum L. Planta 224, 944−951.

Epimashko, S., Meckel, T., Fischer-Schliebs, E., Lüttge, U., Thiel, G., 2004. Two functionally different vacuoles for static and dynamic purposes in one plant mesophyll leaf cell. Plant J. 37, 294−300.

Epstein, E., 1972. Mineral Nutrition of Plants: Principles and Perspectives. John Wiley & Sons, New York.

Epstein, E., 1983. Crops tolerant to salinity and other mineral stresses. In: Nugent, J., O'Connor, M. (Eds.), Better Crops for Food, Ciba Foundation Symposium, vol. 97, pp. 61−82. Pitman, London.

Epstein, E., Bloom, A.J., 2005. Mineral Nutrition of Plants: Principles and Perspectives, second ed. Sinauer Associates, Sunderland, MA.

Epstein, E., Rains, D.W., Elzam, O.E., 1963. Resolution of dual mechanisms of potassium absorption by barley roots. Proc. Natl. Acad. Sci. U.S.A. 49, 684−692.

Erickson, H.P., 1995. FtsZ, a prokaryotic homolog of tubulin? Cell 80, 367−370.

Erickson, C.A., Nuccitelli, R., 1982. Embryonic cell motility can be guided by weak electric fields. J. Cell Biol. 95 (2. Pt. 2), 314a.

Erickson, H.P., Taylor, D.W., Taylor, K.A., Bramhill, D., 1996. Bacterial cell division protein FtsZ assembles into protofilament sheets and minirings, structural homologs of tubulin polymers. Proc. Natl. Acad. Sci. U.S.A. 93, 519−523.

Erlich, Y., Mitra, P.P., delaBastide, M., McCombie, W.R., Hannon, G.L., 2008. Alta-Cyclic: a self-optimizing base caller for next-generation sequencing. Nat. Methods 5, 679−682.

Ernster, L., 1978. Award Ceremony Speech. The Nobel Prize in Chemistry 1978.

Errera, L., 1888. Ueber Zellformen und seifenblasen. Bot. Centralbl. 34, 395−399.

Ervasti, J.M., Campbell, K.P., 1993. Dystrophin and the membrane skeleton. Curr. Opin. Cell Biol. 5, 82−87.

Erwee, M.G., Goodwin, P.B., 1985. Symplastic domains in extrastelar tissues of Egeria densa Planch. Planta 163, 9−19.

Erwee, M.G., Goodwin, P.B., van Bel, A.J.E., 1985. Cell-cell communication in the leaves of Commelina cyanea and other plants. Plant Cell Environ. 8, 173−178.

Esau, K., 1944. Anatomical and cytological studies on beet mosaic. J. Agric. Res. 69, 95−117.

Esau, K., 1956. An anatomist's view of virus disease. Am. J. Bot. 43, 739−748.

Esau, K., 1961. Plants, Viruses, and Insects. Harvard University Press, Cambridge, MA.

Esau, K., 1965. Plant Anatomy, second ed. John Wiley & Sons, New York.

Esau, K., Grill, R.H., 1965. Observations on cytokinesis. Planta 67, 168−181.

Esau, K., Thorsch, J., 1985. Sieve plate pores and plasmodesmata, the communication channels of the symplast: ultrastructural aspects and developmental relations. Am. J. Bot. 72, 1641−1653.

Escobar, N.M., Haupt, S., Thow, G., Boevink, P., Chapman, S., Oparka, K., 2003. High-throughput viral expression of cDNA-green fluorescent protein fusions reveals novel subcellular addresses and identifies unique proteins that interact with plasmodesmata. Plant Cell 15, 1507−1523.

Esmon, B., Novick, P., Schekman, R., 1981. Compartmentalized assembly of oligosaccharides on exported glycoproteins in yeast. Cell 25, 451−460.

Estes, J.E., Selden, L.A., Kinosian, H.J., Gershman, L.C., 1992. Tightly-bound divalent cation of actin. J. Muscle Res. Cell Motil. 13, 272−284.

Etherton, B., 1963. Relationship of cell transmembrane electropotential to potassium and sodium accumulation ratios in oat and pea seedlings. Plant Physiol. 38, 581−585.

Etherton, B., Higinbotham, N., 1960. Transmembrane potential. measurements of cells of higher plants as related to salt uptake. Science 131, 409−410.

Etherton, B., Rubinstein, B., 1978. Evidence for amino acid-H^+ co-transport in oat coleoptiles. Plant Physiol. 61, 933−937.

Ettinger, W.F., Theg, S.M., 1991. Physiologically active chloroplasts contain pools of unassembled extrinsic proteins of the photosynthetic oxygen-evolving enzyme complex in the thylakoid lumen. J. Cell. Biol. 115, 321–328.

Ettlinger, C., Lehle, L., 1988. Auxin induces rapid changes in phosphotidylinositol metabolites. Nature 331, 176–178.

Etxeberria, E., Baroja-Fernandez, E., Muñoz, F.J., Pozueta- Romero, J., 2005a. Sucrose inducible endocytosis as a mechanism for nutrient uptake in heterotrophic plant cells. Plant Cell Physiol. 46, 474–481.

Etxeberria, E., Gonzalez, P., Baroja-Fernandez, E., Pozueta Romero, J., 2006. Fluid phase endocytotic uptake of artificial nano-spheres and fluorescent quantum dots by sycamore cultured cells. Plant Signal. Behav. 1, 196–200.

Etxeberria, E., Gonzalez, P.C., Pozueta-Romero, J., 2005b. Sucrose transport into the vacuole of Citris juice cells: evidence for an endocytotic transport system. J. Am. Soc. Hort. Sci. 130, 269–274.

Etxeberria, E., Gonzalez, P.C., Pozueta-Romero, J., 2007. Mannitol-enhanced, fluid-phase endocytosis in storage parenchyma cells of celery (*Apium graveolens*; Apiaceae) petioles. Am. J. Bot. 94, 1041–1045.

Etxeberria, E., Gonzalez, P.C., Tomlinson, P., Pozueta-Romero, J., 2005c. Existence of two parallel mechanisms for glucose uptake in heterotrophic plant cells. J. Exp. Bot. 56, 1905–1912.

Eubel, H., Heazlewood, J.L., Millar, A.H., 2007. Isolation and subfractionation of plant mitochondria for proteomic analysis. Methods Mol. Biol. 355, 49–62.

Eubel, H., Meyer, E.H., Taylor, N.L., Bussell, J.D., O'Toole, N., Heazlewood, J.L., Castleden, I., Small, I.D., Smith, S.M., Millar, A.H., 2008. Novel proteins, putative membrane transporters, and an integrated metabolic network are revealed by quantitative proteomic analysis of *Arabidopsis* cell culture peroxisomes. Plant Physiol. 148, 1809–1829.

Euler, H., 1912. General Chemistry of the Enzymes. Translated by T. H. Pope. John Wiley & Sons, New York.

Eun, S.-O., Bae, S.-H., Lee, Y., 2001. Cortical actin filaments in guard cells respond differently to ABA in wild type and abi-1 mutant *Arabidopsis*. Planta 212, 466–469.

Eun, S.-O., Lee, Y., 1997. Actin filaments of guard cells are reorganized in response to light and abscisic acid. Plant Physiol. 115, 1491–1498.

Eun, S.-O., Lee, Y., 2000. Stomatal opening by fusicoccin is accompanied by depolymerization of actin filaments in guard cells. Planta 210, 1014–1017.

Euteneuer, U., Koonce, M.P., Pfister, K.K., Schliwa, M., 1988. An ATPase with properties expected for the organelle motor of the giant amoeba, *Reticulomyxa*. Nature 332, 176–178.

Euteneuer, U., McIntosh, J.R., 1980. Polarity of midbody and phragmoplast microtubules. J. Cell Biol. 87, 509–515.

Euteneuer, U., McIntosh, J.R., 1981. Polarity of some motility-related microtubules. Proc. Natl. Acad. Sci. U.S.A. 78, 372–376.

Euteneuer, U., McIntosh, J.R., 1981a. Polarity of some motility-related microtubules. Proc. Natl. Acad. Sci. U.S.A. 78, 372–376.

Euteneuer, U., McIntosh, J.R., 1981b. Structural polarity of kinetochore microtubules in PtK1 cells. J. Cell Biol. 89, 338–345.

Evans, G.M., Durrant, A., Rees, H., 1966. Associated nuclear changes in the induction of flax genotrophs. Nature 212, 697–699.

Evans, I.M., Gatehouse, L.N., Gatehouse, J.A., Robinson, N.J., Croy, R.R.D., 1990. A gene from pea (*Pisum sativum* L.) with homology to metallothionein genes. FEBS Lett. 262, 29–32.

Evans, M.L., Ishikawa, H., 1997. Computer based imaging and analysis of root gravitropism. Gravit. Space Biol. Bull. 10, 65–73.

Evans, J.R., Kaldenhoff, R., Genty, B., Terashima, I., 2009. Resistances along the CO_2 diffusion pathway inside leaves. J. Exp. Bot. 60, 2235–2248.

Evans, D.E., Pawa, V., Smith, S.J., Graumann, K., 2014. Protein interactions at he higher plant nuclear envelope: evidence for a linker of nucleoskeleton and cytoskeleton complex. Front. Plant Sci. 7, 5.

Evans Jr., E.A., Slotin, L., 1940. The utilization of carbon dioxide in the synthesis of α-ketoglutaric acid. J. Biol. Chem. 136, 301–302.

Evans Jr., E.A., Slotin, L., 1941. Carbon dioxide utilization by pigeon liver. J. Biol. Chem. 141, 439–450.

Evert, R.F., Eschrich, W., Heyser, W., 1977. Distribution and structure of the plasmodesmata in mesophyll and bundle-sheath cells of *Zea mays* L. Planta 136, 77–89.

Ewart, A.J., 1906. The ascent of water in trees. Philos. Trans. R. Soc. Lond. 198B, 41–85.

Eychmans, J., Boudou, T., Yu, X., Chen, C.S., 2011. A hitchhiker's guide to mechanobiology. Dev. Cell 21, 35–47.

Fabry, B., Maksym, G.N., Butler, J.P., Glogauer, M., Navajas, D., Fredberg, J.J., 2001. Scaling the microrheology of living cells. Phys. Rev. Lett. 87, 148102-1-4.

Facchinelli, F., Weber, A.P.M., 2011. The metabolite transporters of the plastid envelope. An update. Front. Plant Sci. 2, 50.

Facette, M.R., Shen, Z., Björnsdóttir, F.R., Briggs, S.P., Smith, L.G., 2013. Parallel proteomic and phosphoproteomic analyses of successive stages of maize leaf development. Plant Cell 25, 2798–2812.

Faguy, D.M., Doolittle, D.F., 1998. Cytoskeletal proteins: the evolution of cell division. Curr. Biol. 8, R338–R341.

Fahn, A., 1979. Secretory Tissues in Vascular Plants. Academic Press, London.

Falbel, T.G., Koch, L.M., Nadeau, J.A., Segui-Simarro, J.M., Sack, F.D., Bednarek, S.Y., 2003. SCD1 is required for cytokinesis and polarized cell expansion in *Arabidopsis*. Development 130, 4011–4024.

Falconer, M.M., Donaldson, G., Seagull, R.W., 1988. MTOCs in higher plant cells: an immunofluorescent study of microtubule assembly sites following depolymerization by APM. Protoplasma 144, 46–55.

Falconer, M.M., Seagull, R.W., 1988. Xylogenesis in tissue cul-ture III. Continuing wall deposition during tracheary element development. Protoplasma 144, 10–16.

Falhof, J., Pedersen, J.T., Fuglsang, A.T., Palmgren, M., 2016. Plasma membrane H^+-ATPase regulation in the center of plant physiology. Mol. Plant 9, 323–337.

Falke, L.C., Edwards, K.L., Pickard, B.G., Misler, S., 1988. A stretch-activated anion channel in tobacco protoplasts. FEBS Lett. 237, 141–144.

Fambrough, D.M., Bonner, J., 1966. On the similarity of plant and animal histones. Biochemistry 5, 2563–2570.

Fan, Y., Burkart, G.M., Dixit, R., 2018. The Arabidopsis SPIRAL2 protein targets and stabilizes microtubule minus ends. Curr. Biol. 28, 987–994.

Fan, L., Li, R., Pan, J., Ding, Z., Lin, J., 2015. Endocytosis and its regulation in plants. Trends Plant Sci. 20, 388–397.

Fang, C., Ma, Y., Wu, S., Liu, Z., Wang, Z., Yang, R., Hu, G., Zhou, Z., Yu, H., et al., 2017. Genome-wide association studies dissect the genetic networks underlying agronomical traits in soybean. Genome Biol. 18, 161.

Faraday, M., 1860. The chemical history of a candle, 1962. In: W. Crookes with the Original Illustrations and a New Introduction by L. P. Williams. Collier Books, New York.

Faraday, M., 1861. On the Forces of Nature and their Relations to Each Other, 1960. Viking Press, New York.

Faraday, C.D., Spanswick, R.M., 1992. Maize root plasma membranes isolated by aqueous polymer two-phase partitioning: assessment of residual tonoplast ATPase and pyrophosphatase activities. J. Exp. Bot. 43, 1583–1590.

Faraday, C.D., Spanswick, R.M., 1993. Evidence for a membrane skeleton in higher plants a spectrin-like polypeptide co-isolates with rice root plasma membranes. FEBS Lett. 318, 313–316.

Faraday, C.D., Spanswick, R.M., Bisson, M.A., 1996. Plasma membrane isolation from freshwater and salt-tolerant species of *Chara*: antibody cross-reactions and phosphohydrolase activities. J. Exp. Bot. 47, 589–594.

Faraday, C.D., Thomson, W.W., 1986a. Functional aspects of the salt glands of the Plumbagenaceae. J. Exp. Bot. 37, 1129–1135.

Faraday, C.D., Thomson, W.W., 1986b. Morphometric analysis of *Limonium* salt glands in relation to ion efflux. J. Exp. Bot. 37, 471–481.

Faraday, C.D., Thomson, W.W., 1986c. Structural aspects of the salt glands of the Plumbaginaceae. J. Exp. Bot. 37, 461–470.

Farag, M.A., Huhman, D.V., Dixon, R.A., Sumner, L.W., 2008. Metabolomics reveals novel pathways and differential mechanistic and elicitor-specific responses in phenylpropanoid and isoflavonoid biosynthesis in *Medicago truncatula* cell cultures. Plant Physiol. 146, 387–402.

Farber, J., 1969. he Student as Nigger. Essays and Stories. In: T. Contact Books, North Hollywood, CA.

Fareed, G.C., Richardson, C.C., 1967. Enzymatic breakage and joining of deoxyribonucleic acid, II. The structural gene for polynucleotide ligase in bacteriophage *T4*. Proc. Natl. Acad. Sci. U.S.A. 58, 665–672.

Farley, J., 1977. The Spontaneous Generation Controversy from Descartes to Oparin. The Johns Hopkins University Press, Baltimore.

Farmer, J.B., 1913. Plant Life. Wiliams & Norgate, London.

Farmer, J.B., Moore, J.E., 1905. On the maiotic phase (reduction divisions) in animals and plants. Q. J. Microsc. Sci. 48, 489–557.

Farmer, J.B., Moore, J.E.S., Walker, C.E., 1905a. On the cytology of malignant growths. Proc. R. Soc. B 77, 336–352.

Farmer, J.B., Moore, J.E.S., Walker, C.E., 1905b. On the resemblances existing between the "Plimmers Bodies" of malignant growths, and certain normal constituents of reproductive cells of animals. Proc. R. Soc. B 76, 230–234.

Farmer, L.M., Rinaldi, M.A., Young, P.G., Danan, C.H., Burkhart, S.E., Bartel, B., 2013. Distrupting autophagy restores peroxisome function to an *Arabidopsis* lon2 mutant and reveals a role for LON2 protease in peroxisomal protein degradation. Plant Cell 25, 4085–4100.

Farquharson, K.L., 2015. Sterols modulate cell-to-cell connectivity at plasmodesmata. Plant Cell 27, 948.

Farré, J.-C., Subramani, S., 2016. Mechanistic insights into selective autophagy pathways: lessons from yeast. Nat. Rev. Mol. Cell Biol. 17, 537–552.

Fasano, J.M., Swanson, S.J., Blancaflor, E.B., Dowd, P.E., Kao, T.-H., Gilroy, S., 2001. Changes in root cap pH are required for the gravity response of the arabidopsis root. Plant Cell 13, 907–921.

Fath, K.R., Burgess, D.R., 1994. Membrane motility mediated by unconventional myosin. Curr. Opin. Cell Biol. 6, 131–135.

Faulkner, C., Akman, O.E., Ball, K., Jeffree, C., Oparka, K., 2008. Peeking into pit fields: a multiple twinning model of secondary plasmodesmata formation in tobacco. Plant Cell 20, 1504–1518.

Faulkner, C.R., Blackman, L.M., Cordwell, S.J., Overall, R.L., 2005. Proteomic identification of putative plasmodesmatal proteins from *Chara corallina*. Proteomics 5, 2866–2875.

Faulkner, C.R., Maule, A., 2011. Opportunities and successes in the search for plasmodesmal proteins. Protoplasma 248, 27–38.

Fawcett, D.W., 1966. On the occurence of a fibrous lamina on the inner aspect of the nuclear envelope in certain cells of vertebrates. Am. J. Anat. 119, 129–145.

Fawcett, D.W., Porter, K.R., 1954. A study of the fine structure of ciliated epithelia. J. Morphol. 94, 282.

Faye, L., Daniell, H., 2006. Novel pathways for glycoprotein import into chloroplasts. Plant Biotechnol. J. 4, 275–279.

Faye, L., Greenwood, J.S., Herman, E.M., Sturm, A., Chrispeels, M.J., 1988. Transport and posttranslational processing of the vacuolar enzyme α-mannosidase in jack-bean cotyledons. Planta 174, 271–282.

Faye, L., Johnson, D.D., Stern, A., Crispeels, M.J., 1989. Structure, biosynthesis, and function of asparagine-linked glycans on plant glycoproteins. Physiol. Plant 75, 309–314.

Featherstone, C., Russell, P., 1991. Fission yeast p107weel mitotic inhibitor is a tyrosine/serine kinase. Nature 349, 808–811.

Fedatto, L.M., Silva-Stenico, M.E., Etchegaray, A., Pacheco, F.T.H., Rodrigues, J.L.M., Tsai, S.M., 2006. Detection and characterization of protease secreted by the plant pathogen *Xylella fastidiosa*. Microbiol. Res. 161, 263–272.

Fedoroff, N., 1992. Barbara McClintock: the Geneticist, the Genius, the Woman. Cell 71, 181–182.

Fedoroff, N., 2012. McClintock's challenge in the 21st century. Proc. Natl. Acad. Sci. U.S.A. 109, 20200–20203.

Fedorova, E., Greenwood, J.S., Oaks, A., 1994. In situ localization of nitrate reductase in maize roots. Planta 194, 279–286.

Fedurco, M., Romieu, A., Williams, S., Lawrence, I., Turcatti, G., 2006. BTA, a novel reagent for DNA attachment on glass and efficient generation of solid-phase amplified DNA colonies. Nucleic Acids Res. 36, e25.

Feenberg, E., Primakoff, H., 1948. Interaction of cosmic-ray primaries with sunlight and starlight. Phys. Rev. 73, 449–469.

Feher, G., 1971. Some chemical and physical properties of a bacterial reaction center particle and its primary photochemical reactants. Photochem. Photobiol. 14, 373–387.

Feher, G., 1998. Three decades of research in bacterial photosynthesis and the road leadin to it: a personal account. Photosynth. Res. 55, 1–40.

Feher, G., 2002. My road to biophysics: picking flowers on the way to photosynthesis. Annu. Rev. Biophys. Biomol. Struct. 31, 1–44.

Feijó, J.A., Malhó, R., Obermeyer, G., 1995. Ion dynamics and its possible role during *in vitro* pollen germination and tube growth. Protoplasma 187, 155–167.

Feijó, J.A., Sainhas, J., Hackett, G.R., Kunkel, J.G., Hepler, P.K., 1999. Growing pollen tubes possess a constitutive alkaline band in the clear zone and a growth-dependent acidic tip. J. Cell Biol. 144, 483–496.

Feijó, J.A., Sainhas, J., Holdaway-Clarke, T., Cordiero, S., Kunkel, J.K.G., Hepler, P.K., 2001. Cellular oscillations and the regulation of growth: the pollen tube paradigm. BioEssays 23, 86–94.

Feiler, H.S., Jacobs, T.W., 1990. Cell division in higher plants; a cdc2 gene, its 34kDa product, and histone H1 kinase activity in pea. Proc. Natl. Acad. Sci. U.S.A. 87, 5397–5401.

Feldherr, C.M., Kallenbach, E., Schultz, N., 1984. Movement of a karyophilic protein through the nuclear pores of oocytes. J. Cell Biol. 99, 2216–2222.

Felle, H.H., 1982. Effects of fusicoccin upon membrane potential, resistance and current-voltage characteristics in root hairs of *Sinapis alba*. Plant Sci. Lett. 25, 219–225.

Felle, H.H., 1989a. Ca^{2+} microelectrodes and their application to plant cells and tissues. Plant Physiol. 35, 397–439.

Felle, H.H., 1989b. pH as a second messenger in plants. In: Boss, W.F., Morré, D.J. (Eds.), Second Messengers in Plant Growth and Development. Alan R. Liss, Inc., New York, pp. 145–166.

Felle, H.H., 2002. pH as a signal and regulator of membrane transport. In: Rengel, Z. (Ed.), Handbook of Plant Growth. pH as the Master Variable. Marcel Dekker, New York, pp. 107–130.

Felle, H.H., Hepler, P.K., 1997. The cytosolic Ca^{2+} concentration gradient in *Sinapsis alba* root hairs as revealed by Ca^{2+}-selective microelectrode tests and fura-dextran ratio imaging. Plant Physiol. 114, 39–45.

Fellman, J.K., Loescher, W.H., 1987. Comparative studies of sucroose and mannitol utilization in celery (Apium graveolens). Physiol. Plant 69, 337–341.

Feng, Y., Cappelletti, V., Picotti, P., 2017. Quantitative proteomics of model organisms. Curr. Opin. Syst. Biol. 6, 58–66.

Feng, Z., Mao, Y., Xu, N., Zhang, B., Wei, P., Yang, D.-L., Wang, Z., Zhang, Z., Zheng, R., Yang, L., Zeng, L., Liu, X., Zhu, J.-K., 2014. Multigeneration analysis reveals the inheritance, specificity, and patterns of CRISPR/Cas-induced gene modifications in Arabidopsis. Proc. Natl. Acad. Sci. U.S.A. 111, 4632–4637.

Fenn, J.B., 1982. Engines, Energy, and Entropy. A Thermodynamics Primer. W. H. Freeman, San Francisco.

Fenn, J.B., 1989. Electrospray ionization for mass spectrometry of large biomolecules. Science 246, 64–71.

Fenn, J.B., December 8, 2002a. Electrospray Wings for Molecular Elephants. Nobel Lecture.

Fenn, J.B., 2002b. Electrospray ionization mass spectrometry: how it all began. J. Biomol. Tech. 13, 101–118.

Fenn, J.B., Mann, M., Meng, C.K., Wong, S.F., Whitehouse, C.M., 1989. Electrospray ionization for mass spectrometry of large biomolecules. Science 246, 64–71.

Ferguson, C., Teeri, T.T., Siika, A.M., Read, S.M., Bacic, A., 1998. Location of cellulose and callose in pollen tubes and grains of *Nicotiana tabacum*. Planta 206, 452–460.

Fermi, L., 1954. Atoms in the Family. My Life with Enrico Fermi. University of Chicago Press, Chicago.

Fermi, L., 1971. Illustrious Immigrants, second ed. The Univeristy of Chicago Press, Chicago.

Fernandes, A.N., Chen, X., Scotchford, C.A., Walker, J., Wells, D.M., Roberts, C.J., Everitt, N.M., 2012. Mechanical properties of epidermal cells of whole living roots of *Arabidopsis thaliana*: an atomic force microscopy study. Phys. Rev. E 85, 021916.

Fernandez-Calvino, L., Faulkner, C., Walshaw, J., Saalbach, G., Bayer, E., Benitz-Alfonso, Y., Maule, A., 2011. *Arabidopsis* plasmodesmal proteome. PLoS One 6, e18880.

Fernie, A.R., 2003. Metabolome characterization in plant system analysis. Funct. Plant Biol. 30, 111–120.

Ferreira, R.M.B., Bird, B., Davies, D.D., 1989. The effect of light on the structure and organization of *Lemna* peroxisomes. J. Exp. Bot. 40, 1029–1035.

Ferreira, P.C.G., Hemerly, A.S., Villarruel, R., van Montagu, M., Inzé, D., 1991. The *Arabidopsis* functional homolog of the p34cdc2 protein kinase. Plant Cell 3, 531–540.

Ferreira, S.A., Pitz, K.Y., Manshardt, R., Zee, F., Fitch, M., Gonsalves, D., 2002. Virus coat protein transgenic papaya provides practical control of *Papaya ringspot virus* in Hawaii. Plant Dis. 86, 101–105.

Ferris, T., 1997. The Whole Shebang. Simon & Schuster, New York.

Ferris Jr., J.P., Hill, A.R., Liu, R., Orgel, L.E., 1996. Synthesis of long prebiotic oligomers on mineral surfaces. Nature 381, 59–61.

Ferro, M., Salvi, D., Brugière, S., Miras, S., Kowalski, S., Louwagie, M., Garin, J., Joyard, J., Rolland, N., 2003. Proteomics of the chloroplast envelope membranes from *Arabidopsis thaliana*. Mol. Cell. Proteom. 2, 325–345.

Fessenden, J.M., Dennenberg, M.A., Racker, E., 1966. Effect of coupling factor 3 on oxidative phosphorylation. Biochem. Biophys. Res. Commun. 25, 54–59.

Fessenden-Raden, J.M., Lange, A.J., Dannenberg, M.A., Racker, E., 1969. Partial resolution of the enzymes catalyzing oxidative phosphorylation. XIX. Purification and characterization of a new coupling factor (F_5). J. Biol. Chem. 244, 6656–6661.

Feulgen, R., 1923. Chemie und Physiologie der Nukleinstoffe. Borntraeger, Berlin.

Feulgen, R., Rossenbeck, H., 1924. Mikroskopisch- chemischer Nachweis einer Nucleinsäure vom Typus der Thymonucleinsäure und die daraut beruhende elektive Färbung von Zellkernen in mickroskopischen Präparaten. Z. Physiol. Chem. 135, 203–248.

Feynman, R.P., 1955. The value of science. Eng. Sci. 19 (December), 13–15.

Feynman, R.P., 1965a. The Character of Physical Law. The MIT Press, Cambridge, MA.

Feynman, R.P., 1969. What is science? Phys. Teach. 313–320. September.

Feynman, R.P., 1996. Feynman Lectures on Computation. Addison-Wesley, Reading, MA.

Feynman, R.P., December 11, 1965b. The Development of the Space-Time View of Quantum Electrodynamics. Nobel Lecture.

Fichmann, J., Taiz, L., Gallagher, S., Leonard, R.T., Depta, H., Robinson, D.G., 1989. Immunological comparison of the coated vesicle H^+-ATPases of plants and animals. Protoplasma 153, 117–125.

Fick, A., 1855. On liquid diffusion. Philos. Mag. 10, 30–39.

Fiehn, O., 2002. Metabolomics: the link between genotypes and phenotypes. Plant Mol. Biol. 48, 155–171.

Fiehn, O., 2016. Metabolomics by gas chromatography–mass spectrometry: combined targeted and untargeted profiling. Curr. Protoc. Mol. Biol. Unit 30.4.1–30.4.32.

Fielding, P., Fox, C.F., 1970. Evidence for stable attachment of DNA to membrane at the replication origin of *Escherichia coli*. Biochem. Biophys. Res. Commun. 41, 157–162.

Fields, S., 2007. Site-seeing by sequencing. Science 316, 1441–1442.

Fields, S., 2014. Would Fred Sanger get funded today? Genetics 197, 435–439.

Fields, S., Song, O., 1989. A novel genetic system to detect protein–protein interactions. Nature 340, 245–246.

Fiers, W., Contreras, R., Duerinck, F., Haegeman, G., Iserentant, D., Merregaert, J., Min Jou, W., Molemans, F., Raeymaekers, A., van den Berghe, A., Volckaert, G., 1976. Complete nucleotide sequence of bacteriophage MS2 RNA: primary and secondary structure of the replicase gene. Nature 260, 500–507.

Fíla, J., Matros, A., Radau, S., Zahedi, R.P., Čapková, V., Mock, H.-P., Honys, D., 2012. Revealing phosphoproteins playing role in tobacco pollen activated in vitro. Proteomics 12, 3229–3250.

Fíla, J., Radau, S., Matros, A., Hartmann, A., Scholz, U., Feciková, J., Mock, H.P., Čapková, V., Zahedi, R.P., Honys, D., 2016. Phosphoproteomics profiling of tobacco mature pollen and pollen activated in vitro. Mol. Cell Proteom. 15, 1338–1350.

Fíla, J., Záveská Drábková, L., Gibalová, A., Honys, D., 2017. When simple meets complex: Pollen and the –Omics. In: Obermeyer, G., Feijó, J. (Eds.), Pollen Tube Growth. Springer International, pp. 247–292.

Filiault, D.L., Maloof, J.N., 2012. A Genome-wide association study identifies variants underlying the *Arabidop sis thaliana* shade avoidance response. PLoS Genet. 8 (3), e1002589.

Filner, P., Varner, J.E., 1967. A test of de novo synthesis of enzymes: density labelling with H_2O^{18} of barley α-amylase induced by gibberellic acid. Proc. Natl. Acad. Sci. U.S.A. 58, 1520–1526.

Finck, H., 1968. On the discovery of actin. Science 160, 332.

Findlay, A., 1948. A Hundred Years of Chemistry, second ed. Gerald Duckworth & Co., London.

Finer, J.T., Simmons, R.M., Spudich, J.A., 1994. Single myosin molecule mechanics: piconewton forces and nanometre steps. Nature 368, 113–119.

Finkel, E., 2009. With 'phenomics' plant scientists hope to shift breeding into overdrive. Science 325, 380–381.

Finlay, D.R., Forbes, D.J., 1990. Reconstitution of biochemically altered nuclear pores: transport can be eliminated and restored. Cell 60, 17–29.

Finley, B.B., 1990. Cell adhesion and invasion mechanisms in microbial pathogenesis. Curr. Opin. Cell Biol. 2, 815–820.

Fire, A.Z., December 8, 2006. Gene silencing by double stranded RNA. Nobel Lecture.

Fire, A., Xu, S.Q., Montgomery, M.K., Kostas, S.A., Driver, S.E., Mello, C.C., 1998. Potent and specific genetic interference by double-stranded RNA in Caenorhabditis elegans. Nature 391, 806–811.

Fischer, A., 1899. Fixierung, Färbung und Ban des Protoplasmas.

Fischer, E., 1923. Untersuchungen über Aminosäuren, Polypeptide und Proteine. II. 1907–1919. Springer, Berlin.

Fischer, R.L., Bennett, A.B., 1991. Role of cell wall hydrolases in fruit ripening. Annu. Rev. Plant Physiol. Plant Mol. Biol. 42, 675–703.

Fischer, E.H., December 8, 1992. Protein phosphorylation and cellular regulation II. Nobel Lecture.

Fischer, H., December 11, 1930. On haemin and the relationship between haemin and chlorophyll. Nobel Lecture.

Fischer, E., December 12, 1902. Syntheses in the purine and sugar group. In: Nobel Lect.

Fischer, Jena, Fischer, E., December 12, 1902. Syntheses in the Purine and Sugar Group. Nobel Lecture.

Fischer, K., Schopfer, P., 1997. Interaction of auxin, light, and mechanical stress in orienting microtubules in relation to tropic curvature in the epidermis of maize coleoptiles. Protoplasma 196, 108–116.

Fischer-Arnold, G., 1963. Untersuchungen über die Chloroplastenbewegung bei *Vaucheria sessili*. Protoplasma 56, 495–520.

Fiserova, J., Goldberg, M.W., 2010. Relationships at the nuclear envelope: lamins and nuclear pore complexes in animals and plants. Biochem. Soc. Trans. 38, 829–831.

Fiserova, J., Kiseleva, E., Goldberg, M.W., 2009. Nuclear envelope and nuclear pore complex structure and organization in tobacco BY-2 cells. Plant J. 59, 243–255.

Fishel, E.A., Dixit, R., 2013. Role of nucleation in cortical microtubule array organization: variations on a theme. Plant J. 75, 270–277.

Fisher, D.B., 1999. The estimated pore diamter for plasmodesmatal channels in the *Abutalon* nectory trichome should be about 4 nm, rather than 3 nm. Planta 208, 299–300.

Fisher, D.D., Cyr, R.J., 1998. Extending the microtubule/microfibril paradigm: cellulose synthesis is required for normal cortical microtubule alignment in elongating cells. Plant Physiol. 116, 1043–1051.

Fisher, J.D., Hansen, D., Hodges, T.K., 1970. Correlation between ion fluxes and ion-stimulated adenosine triphosphatase activity in plant roots. Plant Physiol. 46, 812–814.

Fisher, J.D., Hodges, T.K., 1969. Monovalent ion stimulated adenosine triphosphatase from oat roots. Plant Physiol. 44, 385–395.

Fisher Jr., D.B., Outlaw, W.H., 1979. Sucrose compartimentation in the palisade parenchyma of. Vicia Faba. Plant Physiol. 64, 481–483.

Fisher, D., Weisenberger, D., Scheer, U., 1991. Assigning functions to nucleolar structures. Chromosoma 101, 133–140.

Fisher, D.B., Wu, Y., Ku, M.S.B., 1992. Turnover of soluble proteins in the wheat sieve tube. Plant Physiol. 100, 1433–1441.

Fiske, C.H., Subbarow, Y., 1929. Phosphorous compounds of muscle and liver. Science 70, 381–382.

Fitch, M.M.M., Manshardt, R.M., Gonsalves, D., Slightom, J.L., Sanford, J.C., 1992. Virus resistant papaya derived from tissues bombarded with coat protein gene of papaya ringspot virus. Nat. Biotechnol. 10, 1466–1472.

Fitch, W.M., Margoliasch, E., 1967. Construction of phylogenetic trees. Science 155, 279–284.

Fitchette-Lainé, A.C., Gomord, V., Chekkafi, A., Faye, L., 1994. Distribution of xylosylation and fucosylation in the plant Golgi apparatus. Plant J. 5, 673–682.

Fitzelle, K.J., Kiss, J.Z., 2001. Restoration of gravitropic sensitivity in starch-deficient mutants of *Arabidopsis* by hypergravity. J. Exp. Bot. 52, 265–275.

Fladung, M., Deutsch, D., Hönick, H., Kumar, S., 2004. T-DNA Transp. Tagg. Aspen Plant Biol. 6, 5–11.

Fleischmann, R.D., Adams, M.D., White, O., Clayton, R.A., Kirkness, E.F., Kerlavage, A.R., Bult, C.J., et al., 1995. Whole-genome random sequencing and assembly of *Haemophilus influenzae* Rd. Science 269, 469–512.

Fleming, A., 2011. Morphogenesis: forcing the tissue. Curr. Biol. 21, R840–R841.

Fleming, A.J., McQueen-Mason, S., Mandel, T., Kuhlemeier, C., 1997. Induction of leaf primordial by the cell wall protein expansin. Science 276, 1415–1418.

Flemming, W., 1880. Contributions to the knowledge of the cell and its vital processes, 1965 J. Cell Biol. 25 (Suppl. on Mitosis), 1–69.

Flemming, W., 1882. Zellsubstanz. Kern und Zeiltheilung. Vogel, Leipzig.

Fleurat-Lessard, P., Bouche-Pillon, S., Leloup, C., Lucas, W.J., Serrano, R., Bonnemain, J.L., 1995. Absence of plasma membrane H^+-ATPase in plasmodesmata located in pit-fields of the young reactive pulvinus if *Mimosa pudica* L. Protoplasma 188, 180–185.

Flexas, J., Ribas-Carbó, M., Hanson, D.T., Bota, J., Otto, B., Cifre, J., McDowell, N., Medrano, H., Kaldenhoff, R., 2006. Tobacco aquaporin NtAQP1 is involved in mesophyll conductance to CO_2 in vivo. Plant J. 48, 427–439.

Floch, A.G., Palancade, B., Doye, V., 2014. Fifty years of nuclear pores and nucleocytoplasmic transport studies: multiple tools revealing complex rules. Methods Cell Biol. 122, 1–40.

Flor, H.H., 1942. Inheritance of pathogenicity in *Melampsora lini*. Phytopathology 32, 653–669.

Flor, H.H., 1947. Inheritance of reaction to rust in flax. J. Agric. Res. 74, 241–262.

Flor, H.H., 1955. Host-parasite interaction in flax rust - its genetics and other implications. Phytopathology 45, 680–685.

Flor, H.H., 1971. Current status of the gene-for-gene concept. Annu. Rev. Phytopathol. 9, 275–296.

Flores, J.J., Ponnamperuma, C., 1972. Polymerization of amino acids under primitive Earth conditions. J. Mol. Evol. 2, 1–9.

Flores-Pérez, Ú., Jarvis, P., 2013. Molecular chaperone involvement in chloroplast protein import. Biochim. Biophys. Acta 1833, 332–340.

Florian, S., Mitchison, T.J., 2016. Anti-microtubule drugs. Methods Mol. Biol. 1413, 403–421.

Floyd, B.E., Morriss, S.C., MacIntosh, Bassham, D.C., 2012. What to eat: evidence for selective autophagy in plants. J. Integr. Plant Biol. 54, 907–920.

Flückiger, R., De Caroli, M., Piro, G., Dalessandro, G., Neuhaus, J.-M., Di Sansebastiano, G.-P., 2003. Vacuolar system distribution in *Arabidopsis* tissues, visualized using GFP-fusion proteins. J. Exp. Bot. 54, 1577–1584.

Flügge, U.-I., 1990. Import of proteins into chloroplasts. J. Cell Sci. 96, 351–354.

Flügge, U.-I., Benz, R., 1984. Pore-forming activity in the outer membrane of the chloroplast envelope. FEBS Lett. 169, 85–89.

Flügge, U.-I., Weber, A., 1994. A rapid method for measuring organelle-specific substrate transport in homogenates from plant tissues. Planta 194, 181–185.

Flynn, C.R., Mullen, R.T., Trelease, R.N., 1998. Mutational analyses of a type 2 peroxisomal targeting signal that is capable of directing oligomeric import into tobacco glyoxysomes. Plant J. 16, 709–720.

Foard, D.E., Haber, A.H., 1961. Anatomic studies of gamma- irradiated wheat growin without cell division. Am. J. Bot. 48, 438–446.

Fodor, S.P.A., Rava, R.P., Huang, X.C., Pease, A.C., Holmes, C.P., Adams, C.L., 1993. Multiplexed biochemical assays with biological chips. Nature 364, 555–556.

Fodor, S.P.A., Read, J.L., Pirrung, M.C., Stryer, L., Tsai Lu, A., Solas, D., 1991. Light-directed, spatially addressable parallel chemical synthesis. Science 251, 767–773.

Foissner, I., Sommer, A., Hoeftberger, M., Hoepflinger, M.C., Absolonova, M., 2016. Is Wortmannin-induced reorganization of the trans-Golgi network the key to explain charasome formation? Front. Plant Sci. 7, 756.

Foissner, I., Wasteneys, G.O., 2007. Wide-ranging effects of eight cytochalasins and latrunculin A and B on intracellular motility and actin filament reorganization in characean internodal cells. Plant Cell Physiol. 48, 585–597.

Foissner, I., Wasteneys, G.O., 2014. Characean intermodal cells as a model system for the study of cell organization. Int. Rev. Cytol. Mol. Biol. 311, 307–364.

Fondi, M., Liò, P., 2015. Multi-omics and metabolic modelling pipelines: challenges and tools for systems microbiology. Microbiol. Res. 171, 52–64.

Fontes, E.B.P., Shank, B.B., Worbel, R.L., Moose, S.P., O'Brian, G.R., Wutzel, E.T., Boston, R.S., 1991. Characterization of an imnunoglobin binding protein homolog in the maize floury-2 endosperm mutant. Plant Cell 3, 483–496.

Footitt, S., Slocombe, S.P., Larner, V., Kurup, S., Wu, Y., Larson, T., Graham, I., Baker, A., Holdsworth, M., 2002. Control of germination and lipid mobilization by COMATOSE, the *Arabidopsis* homologue of human ALDP. EMBO J. 21, 2912–2922.

Foquet, M., Korlach, J., Zipfel, W., Webb, W.W., Craighead, H.G., 2002. DNA fragment sizing by single molecule detection in submicrometer-sized closed fluidic channels. Anal. Chem. 74, 1415–1422.

Forbes, R.J., 1970. A Short History of the Art of Distillation. E. J.Brill Leiden, The Netherlands.

Ford, B.J., 2001. The Royal Society and the microscope. Notes Rec. R. Soc. 55, 29–49.

Ford, K.L., Chin, T., Srivastava, V., Zeng, W., Doblin, M.S., Bulone, V., Bacic, A., 2016. CoGolgi" proteome study of *Lolium multiflorum* and *Populus trichocarpa*. Proteomes 4, 23.

Forer, A., 1965. Local reduction of spindle fiber birefringence in living Nephrotoma suturalis (Loew) spermatocytes induced by ultraviolet microbeam irradiation. J. Cell Biol. 25, 95–117.

Forer, A., 1969. Chromosome movements during cell-division. In: Lima-de-Faria, A. (Ed.), Handbook of Molecular Cytology. North Holland Publishing Co., Amsterdam, pp. 553–601.

Forer, A., 1978. Chromosome movements during cell-division: possible involvement of actin filaments. In: Heath, I.B. (Ed.), Nuclear Division in the Fungi. Academic Press, New York, pp. 21–88.

Forer, A., 1988. Do anaphase chromosomes chew their way to the pole or are they pulled by actin? J. Cell Sci. 91, 449–453.

Forer, A., Pickett-Heaps, J.D., 1998. Cytochalasin D and latrunculin affect chromosome behaviour during meiosis in crane- fly spermatocytes. Chromosome Res. 6, 533–549.

Forestan, C., Farinati, S., Vrotto, S., 2012. The maize PIN family of auxin transporters. Front. Plant Sci. 3, 16.

Forgacs, G., Newman, S.A., 2005. Biological Physics of the Developing Embryo. Cambridge University Press, Cambridge.

Forrester, M.L., Krotkov, G., Nelson, C.D., 1966. Effect of oxygen on photosynthesis, photorespiration, and respiration in detached leaves. II. Soybean. Plant Physiol. 41, 422–427.

Forscher, B.K., 1963. Chaos in the brickyard. Science 142, 339.

Förster, T., 1946. Energiewanderung und Fluoreszenz. Naturwissenshaften 33, 166–175.

Förster, T., 1959. 10th spiers memorial lecture: transfer mechanisms of electronic excitation. Discuss. Faraday Soc. 27, 7–17.

Forti, G., 1999. Personal recollections of 40 years in photosynthesis research. Photosynth. Res. 60, 99–110.

Fortman, J.L., Chhabra, S., Mukhopadhyay, A., Chou, H., Lee, T.S., Steen, E., Keasling, J.D., 2008. Biofuel alternatives to ethanol: pumping the microbial well. Trends Biotechnol. 26, 375–381.

Foster, A.S., 1938. Structure and growth of the shoot apex in *Ginkgo biloba*. Bull. Torrey Bot. Club 65, 531–556.

Foster, K.W., 2001. Action spectroscopy of photomovement. In: Hader, D.-P., Lebert, M. (Eds.), Photomovement. Comprehensive Series in the Photosciences, vol. 1, pp. 51–115.

Foster, K.J., Miklavcic, S.J., 2017. A comprehensive biophysical model of ion and water transport in plant roots. I. Clarifying the roles of endodermal barriers in the salt stress response. Front. Plant Sci. 8, 1326.

Foster, K.W., Saranak, J., 1989. The Chlamydomonas (Chlorophyceae) eye as a model of cellular structure, intracellular signaling and rhodopsin activation. In: Coleman, A., Goff, L., Stein-Taylor, J.R. (Eds.), Algae as Experimental Systems. Alan R. Liss, New Yorlk, pp. 215–230.

Foster, K.W., Saranak, J., Derguini, F., Zarrilli, G.R., Johnson, R., Okabe, M., Nakanishi, K., 1989. Activation of *Chlamydomonas* rhodopsin *in vivo* does not require isomerization of retinal. Biochemistry 28, 819–824.

Foster, K.W., Saranak, J., Dowben, P.A., 1991. Spectral sensitivity, structure, and activation of eukaryotic rhodopsins: activation spectroscopy of rhodopsin analogs in *Chlamydomonas*. J. Photochem. Photobiol. B Biol. 8, 385–408.

Foster, K.W., Saranak, J., Patel, N., Zarilli, G., Okabe, M., Kline, T., Nakanishi, K., 1984. Rhodopsin is the functional photoreceptor for phototaxis in the unicellular eukaryote *Chlamydomonas*. Nature 311, 756–759.

Foster, K.W., Saranak, J., Zarrilli, G., 1988. Autoregulation of rhodopsin synthesis in *Chlamydomonas reinhardtii*. Proc. Natl. Acad. Sci. U.S.A. 85, 6379–6383.

Foster, K.W., Smyth, R.D., 1980. Light antennas in phototactic algae. Microbiol. Rev. 44, 572–630.

Foth, B.J., Goedecke, M.C., Solsati, D., 2006. New insights into myosin evolution and classification. Proc. Natl. Acad. Sci. U.S.A. 103, 3681–3686.

Fothergill, P.G., 1958. Life and Its Origin. Sheed and Ward, London.

Fowke, L.C., Tanchak, M.A., Galway, M.E., 1991. Ultrastructural cytology of the endocytic pathway in plants. In: Hawes, C.R., Coleman, J.O.D., Evans, D.E. (Eds.), Endocytosis, Exocytosis and Vesicle Traffic in Plants. Cambridge University Press, Cambridge, pp. 15–40.

Fox, S.W., 1959. Origin of life. Science 130, 1622.

Fox, R.F., 1988a. Energy and the Evolution of Life. W. H. Freeman and Co., New York.

Fox, S., 1988b. The Emergence of Life. Basic Books, New York.

Fox, S.W., Dose, K., 1972. Molecular Evolution and the Origin of Life. W. H. Freeman, San Francisco.

Fox, D.L., Wells, J.R., 1971. Schemochromic blue lead-surfaces of *Selaginella*. Am. Fern. J. 61, 137–139.

Foyer, C.H., Neukermans, J., Queval, G., Noctor, G., Harbinson, J., 2012. Photosynthetic control of electron transport and the regulation of gene expression. J. Exp. Bot. 63, 1637–1661.

Frado, L.-L., Annunziato, A.T., Woodcock, C.L.F., 1977. Structural repeating units of chromatin. III. A comparison of subunits from vertebrate, ciliate and angiosperm species. Biochem. Biophys. Acta 475, 514–520.

Fraenkel-Conrat, H., 1956. The chemistry of proteins and peptides. Ann. Rev. Biochem. 25, 291–330.

Fraenkel-Conrat, H., 1962. Design and Function at the Threshold of Life: The Viruses. Academic Press, New York.

Fraenkel-Conrat, H. (Ed.), 1968. Molecular Basis of Virology. Reinhold, New York.

Fraenkel-Conrat, H., 1969. Chemistry and Biology of Viruses. Academic Press, New York.

Fraenkel-Conrat, H., 1985. The Viruses: Catalogue, Characterization, and Classification. Plenum Press, New York.

Fraenkel-Conrat, H., Kimball, P.C., Levy, J.A., 1988. Virology. Plenum Press, New York.

Fraley, R.T., Rogers, S.G., Horsch, R.B., Sanders, P.R., Flick, J.S., Adams, S.P., Bittner, M.L., Brand, L.A., Fink, C.L., Fry, J.S., Galluppi, G.R., Goldberg, S.B., Hoffmann, N.L., Woo, S.C., 1983. Expression of bacterial genes in plant cells. Proc. Natl. Acad. Sci. U.S.A. 80, 4803–4807.

Francé, R.H., 1905. Germs of Mind in Plants. Translated by A. M. Simons. Charles H. Kerr & Co., Chicago.

Francé, R.H., 1909. Pflanzenpsychologie als Arbeithypothese der Pfanzenphysiologie. Franck'sche Verlagshandlung, Stuttgart.

Franceschi, V.R., Ding, B., Lucas, W.J., 1994. Mechanism of plasmodesmata formation in characean algae in relation to evolution of intercellular communication in higher plants. Planta 192, 347–358.

Franck, J., December 11, 1926. Transformation of Kinetic Energy of Free Electrons into Excitation Energy of Atoms by Impacts. Nobel Lecture.

Franck, J., Loomis, W.E. (Eds.), 1949. Photosynthesis in Plants. Iowa State College Press, Ames.

Frank, J., 2017. Single-Particle Reconstruction – Story in a Sample. https://www.nobelprize.org/prizes/chemistry/2017/frank/lecture/.

Frank, M.H., Edwards, M.B., Schultz, E.R., McKain, M.R., Fei, Z., Sørensen, I., Rose, J.K.C., Scanlon, M.J., 2015. Dissecting the molecular signatures of apical cell-type shoot meristems from two ancient land plant lineages. New Phytol. 207, 893–904.

Frank, M.H., Scanlon, M.J., 2015a. Cell-specific transcriptomic analyses of three-dimensional shoot developmentin the moss, *Physcomitrella patens*. Plant J. 83, 743–751.

Frank, M.H., Scanlon, M.J., 2015b. Transcriptomic evidence for the evolution of shoot meristem function in sporophyte-dominant land plants through concerted selection of ancestral gametophytic and sporophytic genetic programs. Mol. Biol. Evol. 32, 355–367, 3033.

Franke, W.W., 1966. Isolated nuclear membranes. J. Cell Biol. 31, 619–623.

Franke, W.W., 1987. Nuclear lamins and cytoplasmic intermediate filament proteins: a growing multigene family. Cell 48, 3–4.

Franke, W.W., Berendonk, B., 1997. Hormonal doping and androgenization of athletes: a secret program of the German Democratic Republic government. Clin. Chem. 43, 1262–1279.

Franke, W.W., Herth, W., VanDerWoude, W.J., Morré, D.J., 1972. Tubular and filamentous structures in pollen tubes: possible involvement as guide elements in protoplasmic streaming and vectorial migration of secretory vesicles. Planta 105, 317–341.

Franke, W.W., Scheer, U., Trendelenburg, M.F., Spring, H., Zentgraf, H., 1976. Absence of nucleosomes in transcriptionally active chromatin. Cytobiologie 13, 401–434.

Franke, W.W., Schinko, W., 1969. Nuclear shape in muscle cells. J. Cell Biol. 42, 326–331.

Frankel, S., Mooseker, M.S., 1996. The actin-related proteins. Curr. Opin. Cell Biol. 8, 30–37.

Frankl, V., 1973. The Doctor and the Soul: From Psychotherapy to Logotherapy. Second Expanded Edition. Vintage Books, New York.

Frankl, V.E., 1985. Man's Search for Meaning. Washington Square Press, New York.

Franklin, A., 1976. Principle of inertia in the middle ages. Am. J. Phys. 44, 529–545.

Franklin, R.E., Gosling, R.G., 1953. Molecular configuration in sodium thymonucleate. Nature 171, 740–741.

Franklin, A.E., Hoffman, N.E., 1993. Characterization of a chloroplast homologue of the 54-kDa subunit of the signal recognition particle. J. Biol. Chem. 268, 22175–22180.

Franklin-Tong, V.E., 1999. Signalling and the modulation of pollen tube growth. Plant Cell 11, 727–738.

Franklin-Tong, V.E., Drøbak, B.K., Allan, A.C., Watkins, P.A.C., Trewavas, A.J., 1996. Growth of pollen tubes of Papaver rhoeas is regulated by a slow-moving calcium wave propagated by inositol 1,4,5-trisphosphate. Plant Cell 8, 1305–1321.

Franklin-Tong, V.E., Ride, J.P., Read, N.D., Trewavas, A.J., Franklin, F.C.H., 1993. The self-incompatibility response in *Papaver rhoeas* is mediated by cytosolic free calcium. Plant J. 4, 163–177.

Franks, F., 2000. Water: A Matrix of Life, second ed. Royal Society of Chemistry, Cambridge.

Franzini-Armstrong, C., 2018. The relationship between form and function throughout the history of excitation–contraction coupling. J. Gen. Physiol. https://doi.org/10.1085/jgp.201711889.

Fraser, H.C.I., 1908. Nuclear fusions and reductions in the ascomycetes. Brit. Assoc. Adv. Sci. 688–689. Report of the seventy-seventh meeting.

Fraser, C.M., Gocayne, J.D., White, O., Adams, M.D., et al., 1995. The minimal gene complement of Mycoplasma genitalium. Science 270, 397–404.

Fraústo da Silva, J.J.R., Williams, R.J.P., 1991. The Biological Chemistry of the Elements. The Inorganic Chemistry of Life. Clarendon Press, Oxford.

Frederick, S.E., Gruber, P.J., Newcomb, E.H., 1975. Plant microbodies. Protoplasma 84, 1–29.

Frederick, S.E., Mangan, M.E., Carey, J.B., Gruber, P.J., 1992. Intermediate filament antigens of 60 and 65 kDa in the nuclear matrix of plants: their detection and location and localization. Exp. Cell Res. 199, 213–222.

Frederick, S.E., Newcomb, E., 1969. Cytochemical localization of catalase in leaf microbodies (peroxisomes). J. Cell Biol. 43, 343–352.

Frederick, S.E., Newcomb, E.H., Virgil, E.L., Wergin, W.P., 1968. Fine structural characterization of plant microbodies. Planta 81, 229–252.

Freitag, H., Stichler, W., 2000. A remarkable new leaf type with unusual photosynthetic tissue in a Central Asiatic genus of Chenopodiaceae. Plant Biol. 2, 154–160.

Freitag, H., Stichler, W., 2002. Bienertia cycloptera Bunge ex Boiss., Chenopodiaceae, another C4 plant without Kranz tissues. Plant Biol. 4, 121–132.

Frenandez-Pozo, N., Zheng, Y., Snyder, S., Nicolas, P., Shinozaki, Y., Fei, Z., Catalá, C., Giovannoni, J.J., Rose, J.K.C., Mueller, L.A., 2017. The tomato expression atlas. Bioinformatics 33, 2397–2398.

French Jr., J.C., Paolillo, D.J., 1975. The effect of the calyptra on the plane of guard cell mother cell division in *Funaria* and *Physcomitrium* capsules. Ann. Bot. 39, 233–236.

Frevert, J., Koller, W., Kindl, H., 1980. Occurrence and biosynthesis of glyoxysomal enzymes in ripening cucumber seeds. Hoppe-Seyler's Z. Physiol. Chem. 361, 1557–1565.

Frey-Wyssling, A., 1949. Physiochemical behaviour of cytoplasm. Research 2, 300–307.

Frey-Wyssling, A., 1953. Submicroscopic Morphology of Protoplasm. Elsevier, Amsterdam.

Frey-Wyssling, A., 1957. Macromolecules in Cell Structure. Harvard University Press, Cambridge, MA.

Frey-Wyssling, A., 1959. Die Pflanzliche Zellwand. Springer- Verlag, Berlin.

Frey-Wyssling, A., 1964. Frühgeschichte und Ergebnisse der submikroskopischen Morphologie. Mikroskopie 19, 2–12.

Frey-Wyssling, A., 1975. The scientific work of W. J. Schmidt. Microsc. Acta 77 (2), 105–113.

Frey-Wyssling, A., 1991. The plant cell wall, 3rd rev. Barntrceger, Berlin.

Frey-Wyssling, A., Kreutzer, E., 1958. The submicroscopic development of chromoplasts in the fruit of Capsicum annuum L. J. Ultrastruct. Res. 1, 397–411.

Frey-Wyssling, A., Lopez-Saez, J.F., Mühlenthaler, K., 1964. Formation and development of the cell plate. J. Ultrastruct. Res. 10, 422–432.

Frey-Wyssling, A., Mühlethaler, K., 1965. Ultrastructural Plant Cytology. Elsevier, Amsterdam.

Frey-Wyssling, A., Mülethaler, K., 1949. Über den Feinbau der Chorophyllköner. Vierteljahrsschr. Naturf. Gesell (Zürich) 94, 179–183.

Frey-Wyssling, A., Schwegler, F., 1965. Ultrastructure of the chromoplasts in the carrot root. J. Ultrastruct. Res. 13, 543–559.

Frias, I., Caldeira, M.T., Perez-Castineira, J.R., Navarro- Avino, J.P., Culianez-Macia, F.A., Kuppinger, O., Stransky, H., Pages, M., Hager, A., Serrano, R., 1996. A major isoform of the maize plasma membrane H^+-ATPase: characterization and induction by auxin in coleoptiles. Plant Cell 8, 1533–1544.

Fricke, H., 1925. The electric capacity of suspensions with special reference to blood. J. Gen. Physiol. 9, 137–152.

Fricke, W., Chaumont, F., 2006. Solute and water relations of growing plant cells. Plant Cell Monogr. 5, 7–31.

Friedberg, E.C., 2006. The eureka enzyme: the discovery of DNA polymerase. Nat. Rev. Mol. Cell Biol. 7, 143–147.

Friedkin, M., Lehninger, A.L., 1949. Esterificataon of inorganic phosphate coupled to electron transport between dihydrodiphosphopyridine nucleotide and oxygen. I. J. Biol. Chem. 178, 611–623.

Friemann, A., Hachtel, W., 1988. Chloroplast messenger RNAs of free and thylakoid-bound polysomes from *Vicia faba* L. Planta 175, 50–59.

Friend, D.S., Farquhar, M.G., 1967. Functions of coated vesicles during protein absorption in the rat vas deferens. J. Cell Biol. 35, 357–376.

Fries, E., Rothman, J.E., 1980. Transient activity of Golgi-like membranes as donors of vesicular stomatitis viral glycoprotein in vitro. J. Cell Biol. 90, 697–704.

Frigerio, L., de Virgilio, M., Prada, A., iaoro, F., Vitale, A., 1998. Sorting of phaseolin to the vacuole is saturable and requires a short c-terminal peptide. Plant Cell 10, 1031–1042.

Friml, J., Jones, A.R., 2010. Endoplasmic reticulum: the rising compartment in auxin biology. Plant Physiol. 154, 458–462.

Friso, G., Olinares, P.D., van Wijk, K.J., 2011. The workfoow for quantitative proteome analysis of chloroplast development and differentiation, chloroplast mutants, and protein interactions by spectral counting. Methods Mol. Biol. 775, 265–282.

Friso, G., van Wijk, K.J., 2015. Posttranslational protein modifications in plant metabolism. Plant Physiol. 169, 1469–1487.

Fritz, R.D., Letzelter, M., Reimann, A., Martin, K., Fusco, L., Ritsma, L., Ponsioen, B., Fluri, E., Schulte-Merker, S., van Rheenen, J., Pertz, O., 2013. A versatile toolkit to produce sensitive FRET biosensors to visualize signaling in time and space. Sci. Signal. 6 (285), rs12.

Froehlich, J.E., Wilkerson, C.G., Ray, W.K., Mcondrew, R.S., Osteryoung, K.W., Gage, D.A., Phinney, B.S., 2003. Proteomic study of the *Arabidopsis thaliana* chloroplastic envelope membrane utilizing alternatives to traditional two-dimensional electrophoresis. J. Proteome Res. 2, 413–425.

Fromageot, L.L., Jutisz, M., 1953. Chemistry of amino acids, peptides, and proteins. Ann. Rev. Biochem 22, 629–678.

Fromm, J., Lautner, S., 2007. Electrical signals and their physiological significance in plants. Plant Cell Environ. 30, 249–257.

Fruton, J.S., 1955. Proteins. In: Scientific American (Ed.), The Physics and Chemistry of Life. Simon and Schuster, New York, pp. 58–73.

Fruton, J.S., 1957. Enzymic radrolysis and syntpesis of peptide bonds. Harvey Lect. 51, 64–87.

Fruton, J.S., 1972. Molecules and Life: Historical Essays on the Interplay of Chemistryuand Biology. Wiley, New York.

Fruton, J.S., 1992. A Skeptical Biochemist. Harvard University Press, Cambridge, MA.

Fruton, J.S., 2006. Fermentation: Vital or Chemical Process? Brill Academic Publishers, Leiden, The Netherlands.

Fruton, J.S., Simmonds, S., 1958. General Biochemistry, tecond ed. Wiley, New York.

Fry, S.C., 1986. Cross-linking of matrix polymers in the growing cell walls of angiosperms. Ann. Rev. Plant Physiol. 37, 165–186.

Fry, S.C., 1988. The Growing Plant Cell Wall: Chemical and Metabolic Analysis. Longman Scientific & Technical, Harlow Essex, UK.

Fry, S.C., 2004. Primary cell wall metabolism: tracking the careers of wall polymers in living plant cells. New Phytol. 161, 641–675.

Fry, S.C., 2011. High-voltage paper electrophoresis (HVPE) of cell-wall building blocks and their metabolic precursors. Methods Mol. Biol. 715, 55–80.

Frye, L.D., Edidin, M., 1970. The rapid intermixing of cell surface antigens after formation of mouse-human heterokaryons. J. Cell Sci. 7, 319–325.

Fuge, H., Falke, D., 1991. Morphological Aspects of chromosome spindle fibers in *Mesostoma*. "Microtubular fir-tree" structures and microtubule association with kinetochores and chromation. Protoplasma 160, 39–48.

Fuhrman, S.A., Deininger, P.L., La Porte, P., Friedmann, T., Geiduschek, E.P., 1981. Analysis of transcription of the human Alu family ubiquitous repeating element by eukaryotic RNA polymerase III. Nucleic Acids Res. 9, 6439–6456.

Fujii, H., Chinnusamy, V., Rodrigues, A., Rubio, S., Antoni, R., Park, S.-Y., Cutler, S.R., Sheen, J., Rodriguez, P.L., Zhu, J.-K., 2009. *In vitro* reconstitution of an abscisic acid signalling pathway. Nature 462, 660–664.

Fujii, Y., Kodama, Y., 2018. Refinements to light sources used to analyze the chloroplast cold-avoidance response over the past century. Plant Signal. Behav. 13, e1411452.

Fujimoto, M., Tsutsumi, N., 2014. Dynamin-related proteins in plant post-Golgi traffic. Front. Plant Sci. 5, 408.

Fujiwara, T., Giesman-Cookmeyer, D., Ding, B., Lommel, S.A., Lucas, W.J., 1993. Cell-to-cell trafficking of macromolecules through plasmodesmata potentiated by the red clover necrotic mosaic virus movement protein. Plant Cell 5, 1783–1794.

Fujiwara, M.T., Hashimoto, H., Kazama, Y., Abe, T., Yoshida, S., Sato, N., Itoh, R.D., 2008. The assembly of the FtsZ ring at the mid chloroplast division site depends on a balance between the activities of AtMinE1 and ARC11/AtMinD1. Plant Cell Physiol. 49, 345–361.

Fukaki, H., Wysocka-Diller, J., Kato, T., Fujisawa, H., Benfey, P.N., Tasaka, M., 1998. Genetic evidence that the endodermis is essential for shoot gravitropism in *Arabidopsis thaliana*. Plant J. 14, 425–430.

Fukuda, H., 1996. Xylogenesis: initiation, progression, and cell death. Ann. Rev. Plant Physiol. Plant Mol. Biol. 47, 299–325.

Fulcher, R.G., McCully, M.E., Setterfield, G., Sutherland, J., 1976. β-1,3-glucans may be associated with cell plate formation during cytokinesis. Can. J. Bot. 54, 539–542.

Fuller, M.S., 1976. Mitosis in fungi. Int. Rev. Cytol. 45, 113–153.

Fuller, R.C., 1999. Forty years of microbial photosynthesis research: where it came from and what it led to. Photosynth. Res. 62, 1–29.

Fuller, W.D., Sanchez, R.A., Orgel, L.E., 1972. Studies in prebiotic synthesis. VI. Synthesis of purine nucleosides. J. Mol. Biol. 67, 25–33.

Fuller, M.T., Wilson, P.G., 1992. Force and couterforce in the mitotic spindle. Cell 71, 547–550.

Fulsang, A.T., Guo, Y., Cuin, T.A., Qiu, Q.-S., Song, C., Kristiansen, K.A., Bych, K., Schulz, A., Shabala, S., Schumaker, K.S., Palmgren, M.G., Zhu, J.-K., 2007. *Arabidopsis* protein kinase PKS5 inhibits the plasma membrane H^+-ATPase by preventing interaction with 14-3-3 protein. Plant Cell 19, 1617–1634.

Fulton, A.B., 1982. How crowded is the cytoplasm? Cell 30, 345–347.

Funaki, K., Nagata, A., Akimoto, Y., Shimada, K., Ito, K., Yamamoto, K., 2004. The motility of Chara corallina myosin was inhibited reversibly by 2.3-butanedione monoxime (BDM). Plant Cell Physiol. 45, 1342–1345.

Funatsu, T., Harada, Y., Tokunaga, M., Saito, K., Yanagida, T., 1995. Imaging of single fluorescent molecules and individual ATP turnovers by single myosin molecules in aqueous solution. Nature 374, 555–559.

Funke, T., Han, H., Healy-Fried, M.L., Fischer, M., Schönbrunn, E., 2006. Molecular basis for the herbicide resistance of Roundup Ready crops. Proc. Natl. Acad. Sci. U.S.A. 103, 13010–13015.

Funnell, B.E., 1993. Participation of the bacterial membrane in DNA replication and chromosome partitioning. Trends Cell Biol. 3, 20–25.

Furbank, R.T., 2009. Plant phenomics: from gene to form and function. Funct. Plant Biol. 36, 5–6.

Furbank, R.T., Tester, M., 2011. Phenomics—technologies to relieve the phenotyping bottleneck. Trends Plant Sci. 16, 635–644.

Furt, F., Lemoi, K., Tüzel, E., Vidali, L., 2012. Quantitative analysis of organelle distribution and dynamics in Physcomitrella patens protonemal cells. BMC Plant Biol. 12, 70.

Furt, F., Liu, Y.-C., Bibeau, J.P., Tüzel, E., Vidali, L., 2013. Apical myosin XI anticipates F-actin during polarized growth of *Physcomitrella patens* cells. Plant J. 73, 417–428.

Furtado, J.S., 1971. The septal pore and other ultrastructural features of the pyrenomycete *Sordaria fimicola*. Mycologia 63, 104–113.

Furuya, M., 1984. Cell division patterns in multicellular plants. Ann. Rev. Plant Physiol. 35, 349–373.

Furuya, M., 1993. Phytochromes: their molecular species gene families and functions. Ann. Rev. Plant Physiol. Plant Mol. Biol. 44, 617–645.

Furuya, M., 2001. Differential perception of environmental light by phytochromes. In: Sopory, S.K., Oelmüller, R., Maheshwari, S.C. (Eds.), Signal Transduction in Plants. Kluwer Academic, New York, pp. 1–7.

Furuya, M., 2005. History and insights. In: Wada, M., Shimazaki, K., Iino, M. (Eds.), Light Sensing in Plants. Springer, Tokyo, pp. 3–18.

Fusari, C.M., Kooke, R., Lauxmann, M.A., Annunziata, M.G., Encke, B., Hoehne, M., Krohn, N., Becker, F.F.M., Schlereth, A., Sulpice, R., Stitt, M., Keurentjes, J.J.B., 2017. Genome-wide association mapping reveals that specific and pleiotropic regulatory mechanisms fine-tune central metabolism and growth in *Arabidopsis*. Plant Cell 29, 2349–2373.

Fuseler, J.W., 1975. Mitosis in Tilia americana endosperm. J. Cell Biol. 64, 159–171.

Futcher, A.B., 1990. Yeast cell cycle. Curr. Opin. Cell Biol. 2, 246–251.

Gabaldón, T., Snel, B., van Zimmeren, F., Hemrika, W., Tabak, H., Huynen, M.A., 2006. Origin and evolution of the peroxisomal proteome. Biol. Direct 1, 8.

Gabathuler, R., Cleland, R.E., 1985. Auxin regulation of a proton translocating ATPase in pea root plasma membrane vesicles. Plant Physiol. 79, 1080–1085.

Gabriel, R., Kesselmeier, J., 1999. Apoplastic solute concentrations of organic acids and mineral nutrient in the leaves of several fagaceae. Plant Cell Physiol. 40, 604–612.

Gachotte, D., Meens, R., Benveniste, P., 1995. An *Arabidopsis* mutant deficient in sterol biosynthesis: heterologous complementation by ERG3 encoding a Δ7-sterol-C5-desaturase from yeast. Plant J. 8, 407–416.

Gadella, T.W.J., Van der Krogt, G.N.M., Bisseling, T., 1999. GFP- based FRET microscopy in living plant cells. Trends Plant Sci. 4, 287–291.

Gadre, R.S., 2003. Century of nobel prizes: 1909 chemistry lecture. Resonance 8, 77–83.

Gaffron, H., 1960. The origin of life. In: Tax, S. (Ed.), The Evolution of Life. Its Origin, History and Future. University of Chicago Press, Chicago, pp. 39–84.

Gaillochet, C., Daum, G., Lohmann, J.U., 2015. O cell, where art thou? The mechanisms of shoot meristem patterning. Curr. Opin. Plant Biol. 23, 91–97.

Gaj, T., Gersbach, C.A., Barbas III, C.F., 2013. ZFN, TALEN and CRISPR/Cas-based methods for genome engineering. Tends Biotechnol. 31, 397–405.

Galanie, S., Thodey, K., Trenchard, I.J., Interrante, M.F., Smolke, C.D., 2015. Complete biosynthesis of opiods in yeast. Science 349, 1095–1100.

Galau, G.A., Lipson, E.D., Britten, R.J., Davidson, E.H., 1977. Synthesis and turnover of polysomal mRNAs in sea urchin embryos. Cell 10, 415–432.

Galbraith, D.W., 2006. DNA microarray analysis in higher plants. OMICS 10, 455–473.

Galcheva-Gargova, Z.I., Marinova, E.L., Koleva, S.T., 1988. Isolation of nuclear shells from plant cells. Plant Cell Environ. 11, 819–825.

Gale, E.F., 1957. Nucleic acids and protein synthesis. Harvey Lect. 51, 25–63.

Galilei, G., 1962. Dialogue Concerning the Two Chief World Systems, Ptolemaic and Copernican. Translated by S. Drake. University of Californis Press, Berkeley.

Galili, G., Shimoni, Y., Giorini-Silfen, S., Levanony, H., Altschuler, Y., Shani, N., 1996. Wheat storage proteins: assembly, transport and deposition in protein bodies. Plant Physiol. Biochem. 34, 245–252.

Gall, J.Z.G., 1967. Octagonal nuclear pores. J. Cell Biol. 32, 391–399.

Gall, J.G., 1981. Chromosome structure and the C-value paradox. J. Cell Biol. 91, 3s–14s.

Gall, J.G., 1991. Spliceosomes and snurposomes. Science 252, 1499–1500.

Gall, J.G., 2005. Werner Franke. Eur. J. Cell Biol. 84, 89–90.

Galton, F., 1869. Hereditary Genius: An Inquiry into its Laws and Consequences. MacMillan, London.

Galton, F., 1874. English Men of Science. MacMillan, London.

Galton, F., 1892. Hereditary Genius: An Inquiry into Its Laws and Consequences. MacMillan, London.

Galvani, L., 1953a. Commentary on the Effects of Electricity on Muscular Motion. Translated by M. G. Foley. Burndy Library, Norwalk, CT.

Galvani, L., 1953b. Commentary on the Effects of Electricity on Muscular Motion. Translated by R. M. Green. Elizabeth Licht, Cambridge, MA.

Gälweiler, L., Guan, C., Müller, A., Wisman, E., Mendgen, K., Yephremov, A., Palme, K., 1998. Regulation of polar auxin transport by AtPIN1 in *Arabidopsis* vascular tissue. Science 282, 2226–2230.

Gamow, G., 1952. The Creation of the Universe. The New American Library, New York.

Gamow, G., 1954. Possible relation between deoxyribonucleic acid and protein strutures. Nature 173, 318.

Gamow, G., 1970. My World Line. An Informal Autobiography. Viking Press, New York.

Gamow, G., 1988. The Great Physicists from Galileo to Einstein. Dover Publications, New York.

Gamow, G., Metropolis, N., 1954. Numerology of polypeptide chains. Science 120, 779–780.

Gamow, G., Rich, A., Ycas, M., 1956. The problem of information transfer from the nucleic acids to proteins. Adv. Biol. Med. Phys. 4, 23–68.

Gamow, G., Ycas, M., 1955. Statistical correlation of protein and ribonucleic acid compostion. Proc. Natl. Acad. Sci. U.S.A. 41, 1011–1019.

Gamow, G., Ycas, M., 1967. Mr. Tompkins Inside Himself. Adventures in the New Biology. Viking Press, New York.

Ganapathy, M., 2016. Plants as bioreactors- a review. Adv. Tech. Biol. Med. 4, 161.

Ganguly, A., Cho, H.-T., 2012. The phosphorylation code is implicated in cell-specific trafficking of PIN-FORMEDs. Plant Signal. Behav. 7, 1215–1218.

Gant, J.R., Hendershot, L.M., 1993. The modification and assembly of proteins in the endoplasmic reticulum. Curr. Opin. Cell Biol. 5, 589–595.

Gao, H.-B., 2014. Corrigendum: ER network dynamics are differentially controlled by myosins XI-K, XI-C, XI-E, XI-I, an dXI-2. Front. Plant Sci. 5, 637.

Gao, H.B., Kadirjan-Kalbach, D., Froehlich, J.E., Osteryoung, K.W., 2003. ARC5, a cytosolic dynamin-like protein from plants, is part of the chloroplast division machinery. Proc. Natl. Acad. Sci. U.S.A. 10 (0), 4328–4333.

Gao, C., Ren, X., Mason, A.S., Liu, H., Xiao, M., Li, J., Fu, D., 2014. Horizontal genh transfer in plants. Funct. Integr. Genom. 14, 23–29.

Gao, H., Smith, J., Yang, M., Jones, S., Djukanovic, V., Nicholson, M.G., West, A., Bidney, D., Falco, S.C., Jantz, D., Lyznik, L.A., 2010. Heritable targeted mutagenesis in maize using a designed endonuclease. Plant J. 61, 176–187.

Gao, Q., Zhao, M., Li, F., Guo, Q., Xing, S., Wang, W., 2008. Expansins and coleoptile elongation in wheat. Protoplasma 233, 73–81.

Garab, G., Mannella, C.A., 2008. Reply: on three-dimensional models of higher-plant thylakoid networks: elements of consensus, controversies, and future experiments. Plant Cell 20, 2549–2551.

Garaj, S., Hubbard, W., Reina, A., Kong, J., Branton, D., Golovchenko, J.A., 2010. Graphene as a subnanometre trans-electrode membrane. Nature 467, 190–193.

Garber, E., Brush, S.G., Everitt, C.W.F. (Eds.), 1986. Maxwell on Molecules and Gases. MIT Press, Cambridge, MA.

Garcia-Bustos, J., Hetman, J., Hall, M.N., 1991. Nuclear protein localization. Biochim. Biophys. Acta 1071, 83–101.

García-Manrique, P., Gutiérrez, G., Blanco-López, M.C., 2018. Exosomes: towards new theranostic biomaterials. Trends Biotechnol. 36, 10–14.

Garcia-Mas, J., Benjak, A., Sanseverino, W., Bourgeois, M., Mir, G., González, V.M., Hénaff, E., Câmara, F., Cozzuto, L., Lowy, E., et al., 2012. The genome of melon (*Cucumis melo* L.). Proc. Natl. Acad. Sci. U.S.A. 109, 11872–11877.

García-Sancho, M., 2010. A new insight into Sanger's development of sequencing: from proteins to DNA, 1943-1977. J. Hist. Biol. 43, 265–323.

Gardiner, J., 2012. Insights into plant consciousness from neuroscience, physics and mathematics: a role for quasicrystals? Plant Signal. Behav. 7, 1049–1055.

Gardiner, M., Chrispeels, M.J., 1975. Involvement of the Golgi apparatus in the synthesis and secretion of hydroxyproline- rich cell wall glycoproteins. Plant Physiol. 55, 536–541.

Gardiner, J.C., Harper, J.D.I., Weerakoon, N.D., Collings, D.A., Ritchie, S., Gilroy, S., Cyr, R.J., Marc, J., 2001. A 90-kD phospholipase D from tobacco binds to microtubules and the plasma membrane. Plant Cell 13, 2143–2158.

Gardiner, J., McGee, P., Overall, R., Marc, J., 2008. Are histones, tubulin, and actin derived from a common ancestral protein? Protoplasma 233, 1–5.

Gardiner, J.C., Taylor, N.G., Turner, S.R., 2003. Control of cellulose synthase complex localization in developing xylem. Plant Cell 15, 1740–1748.

Gardner, H., Csikszentmihalyi, M., Damon, W., 2001. Good Work: When Excellence and Ethics Meet. Basic Books, New York.

Garlid, K.D., 2000. The state of water in biological systems. Int. Rev. Cytol. 192, 281–302.

Garoff, H., 1985. Using recombinant DNA techniques to study protein targeting in the eukaryotic cell. Ann. Rev. Cell Biol. 1, 403–445.

Garrison, W.M., Morrison, D.H., Hamilton, J.G., Benson, A.A., Calvin, M., 1951. The reduction of carbon dioxide by ionizing radiation. Science 114, 416–418.

Garrity, G.M., 2009. Ground truth. Stand. Genom. Sci. 1, 91–92.

Garrity, G.M., Lyons, C., 2003. Future-proofing biological nomenclature. OMICS 7, 31–33.

Garrod, A.E., 1902. The incidence of alkaptonuria: a study in chemical individauality. Lancet 2, 1616–1620.

Garrod, A.E., 1923. Inborn Errors of Metabolism, second ed. Oxford Medical Publishers, Oxford.

Garvey, M., 1938. Marcus Garvey speaking in Menelik Hall, Sydney, Nova Scotia. The Work that has been done. The Black Man 3 (10), 7–11.

Garwood, C., 2008. Flat Earth. The History of an Infamous Idea. Thomas Dunne Books, New York.

Gasiunas, G., Barrangou, R., Horvath, P., Siksnys, V., 2012. Cas9-crRNA ribonucleoprotein complex mediates specific DNA cleavage for adaptive immunity in bacteria. Proc. Natl. Aad. Sci. U.S.A. 109, E2579–E2586.

Gatenby, A.A., Ellis, R.J., 1990. Chaperone function: the asmembly of ribulose bisphosphate carboxylase-oxygenase. Annu. Rev. Cell Biol. 6, 125–149.

Gatlin, J.C., Bloom, K., 2010. Microtubule motors in eukaryotic spindle assembly and maintenance. Semin. Cell Dev. Biol. 21, 248–254.

Gausman, H.W., Allen, W.A., Escobar, D.E., 1974. Refractive index of plant cell walls. Appl. Optics 13, 109–111.

Gaxiola, R.A., Fink, G.R., Hirschi, K.D., 2002. Genetic manipulation of vacuolar proton pumps and transporters. Plant Physiol. 129, 967–973.

Gefter, M.L., Becker, A., Hurwitz, J., 1967. The enzymatic repair of DNA, I. Formation of a circular λDNA. Proc. Natl. Acad. Sci. U.S.A. 58, 240–247.

Geiger, D.R., Giaquinta, R.T., Sovonick, S.A., Fellows, R.J., 1973. Solute distribution in sugar beet leaves in relation to phloem loading and translocation. Plant Physiol. 52, 585–589.

Geiger, D.R., Sovonick, S.A., Shock, T.L., Fellows, R.J., 1974. Role of free space in translocation in sugar beet. Plant Physiol. 54, 892–898.

Geisler-Lee, J., O'Toole, N., Ammar, R., Provart, N.J., Millar, A.H., Geisler, M., 2007. A predicted interactome for *Arabidopsis*. Plant Physiol. 145, 317–329.

Geitmann, A., 2006. Experimental approaches used to quantify physical parameters at cellular and subcellular levels. Am. J. Bot. 93, 1380–1390.

Geitmann, A., 2010. Mechanical modeling and structural analysis of the primary plant cell wall. Curr. Opin. Plant Biol. 13, 693–699.

Geitmann, A., 2017. System (MEMS)-based platforms for experimental analysis of pollen tube growth behavior and quantification of cell mechanical properties. In: Obermeyer, G., Feijó, J. (Eds.), Pollen Tip Growth. Springer International, pp. 86–103.

Geitmann, A., Li, Y.-Q., Cresti, M., 1995. Ultrastructural immunolocalization of periodic pectin depositions in the cell wall of *Nicotiana tabacum* pollen tubes. Protoplasma 187, 168–171.

Geitmann, A., Ortega, J.K.E., 2009. Mechanics and modeling of plant cell growth. Trends Plant Sci. 14, 467–478.

Geitmann, A., Parre, E., 2004. The local cytomechanical properties of growing pollen tubes correspond to the axial distribution of structural cellular elements. Sex. Plant Reprod. 17, 9–16.

Geldner, N., 2009. Cell polarity in plants—a PARspective on PINs. Curr. Opin. Plant Biol. 12, 42–48.

Geldner, N., 2013a. Casparian strips. Curr. Biol. 23, R1025–R1026.

Geldner, N., 2013b. The endodermis. Annu. Rev. Plant Biol. 64, 531–558.

Geldner, N., Anders, N., Wolters, H., Keicher, J., Kornberger, W., Muller, P., Delbarre, A., Ueda, T., Nakano, A., Jürgens, G., 2003. The Arabidopsis GNOM ARF-GEF mediates endosomal recycling, auxin transport, and auxin −dependent plant growth. Cell 112, 219–230.

Geldner, N., Friml, J., Stierhof, Y.D., Jürgens, G., Palme, K., 2001. Auxin transport inhibitors block PIN1 cycling and vesicle trafficking. Nature 413, 425–428.

Geldner, N., Jürgens, G., 2006. Endocytosis in signaling and development. Curr. Opin. Plant Biol. 9, 589–594.

Geldner, N., Robatzek, S., 2008. Plant receptors go endosomal: a moving view on signal transduction. Plant Physiol. 147, 1565–1574.

Gell-Mann, M., December 10, 1969. Nobel Banquet Speech.

Gellert, M., 1967. Formation of covalent circles of lambda DNA by *E. coli* extracts. Proc. Natl. Acad. Sci. U.S.A. 57, 148–155.

Gelli, A., Higgins, V.J., Blumwald, E., 1997. Activation of plant plasma membrane Ca^{2+}-permeable channels by race-specific fungal elicitors. Plant Physiol. 113, 269–279.

Gemperline, E., Keller, C., Li, L., 2016. Mass spectrometry in plant-omics. Anal. Chem. 88, 3422–3434.

Genre, A., Bonfante, P., 1997. A mycorrhizal fungus changes microtubule orientation in tobacco root cells. Protoplasma 199, 30–38.

Gens, J.S., Fujiki, M., Pickard, B.G., 2000. Arabinogalatan protein and wall-associated kinase in a plasmalemmal reticulum with specialized vertices. Protoplasma 212, 115–134.

Gens, J.S., Reuzeau, C., Doolittle, K.W., McNally, J.G., Pickard, B.G., 1996. Covisualization by computational optical- sectioning microscopy of integrin and associated proteins at the cell membrane of living onion protoplasts. Protoplasma 194, 215–230.

Gerace, L., Blobel, G., 1980. The nuclear envelope lamina is reversibly depolymerized during mitosis. Cell 19, 277–287.

Gergely, J. (Ed.), 1964. Biochemistry of Muscle Contraction. Little, Brown, and Co., Boston.

Gergely, J., 2008. Key events in the hisory of calcium regulation of striated muscle. Biochem. Biophys. Res. Commun. 369, 49–51.

Gerhardt, R., Heldt, H.W., 1984. Measurement of subcellular metabolite levels in leaves by fractionation of freeze stopped material in nonaequeous media. Plant Physiol. 75, 542–547.

Gerlach, J., 1858. Mikroskopische Studien aus dem Gebiete der menschlichen Morphologie.

Gerlach, W., 1921. Die experimenteller Grunlagen der Quantentheorie. Heft. 58. Vieweg, Summlung.

Germain, V., Footitt, S., Dieuaide-Noubhani, M., Raymond, P., Renaudin, J.-P., Bryce, J.H., Smith, S.M., 2000. Role of malate synthase and the glyoxylate cycle in oilseed plants. Plant Mol. Biol. Rep. 18, S20–S26.

Germain, V., Rylott, E.L., Larson, T.R., Sherson, S.M., Bechtold, N., Carde, J.-P., Bryce, J.H., Graham, I.A., Smith, S.M., 2001. Requirement for the 3-ketoacyl-CoA thiolase-2 in peroxisome development, fatty acid β-oxidation and breakdown of triacylglycerol in lipid bodies of *Arabidopsis* seedlings. Plant J. 28, 1–12.

Gertel, E.T., Green, P.B., 1977. Cell growth pattern and wall microfibrillar arrangement. Plant Physiol. 60, 247–254.

Gest, H., 1994. A microbiologist's odyssey: bacterial viruses to photosynthetic bacteria. Photosynth. Res. 40, 129–146.

Gest, H., 1999. Memoir of a 1949 railway journey with photosynthetic bacteria. Photosynth. Res. 61, 91–96.

Gest, H., 2002. History of the word photosynthesis and evolution of its definition. Photosynth. Res. 73, 7–10.

Gest, H., 2004. Samuel Ruben's contributions to research on photosynthesis and bacterial metabolism with radioactive carbon. Photosynth. Res. 80, 77–83.

Gest, H., Kamen, M.D., 1949. Photoproduction of molecular hydrogen by Rhodospirillum rubrum. Science 109, 558–559.

Ghanotakis, D.F., Yocum, C.F., 1990. Photosystem II and the oxygen-evolving complex. Annu. Rev. Plant Physiol. Plant Mol. Biol. 41, 255–276.

Ghosh, A., Kiran, B., 2017. Carbon concentration in algae: reducing CO_2 from exhaust gas. Trends Biotechnol. 35, 806–808.

Ghosh, A., Maniloff, J., Gerling, D.A., 1978. Inhibition of mycoplasma cell division by cytochalasin. B. Cell 13, 57–64.

Ghoshroy, S., Lartey, R., Sheng, J., Citovsky, V., 1997. Transport of proteins and nucleic acids through plasmodesmata. Ann. Rev. Plant Physiol. Plant Mol. Biol. 48, 27–50.

Giannini, J.L., Gildensoph, L.H., Reynolds-Niesman, I., Briskin, D.P., 1987a. Calcium transport in sealed vesicles from beet (*Beta vulgarus* L.) storage tissue. I. Plant Physiol. 85, 1129–1136.

Giannini, J.L., Ruiz-Critin, J., Briskin, D.P., 1987b. Calcium transport in sealed vesicles form red beet (*Beta vulgarus* L.) storage tissue. II. Plant Physiol. 85, 1137–1142.

Giaquinta, R., 1976. Evidence for phloem loading from the apoplast. Chemical modification of membrane sulfhydryl groups. Plant Physiol. 57, 872–875.

Gibbons, I.R., 1963. Studies on the protein components of cilia from *Tetrahymena pyriformis*. Proc. Natl. Acad. Sci. U.S.A. 50, 1002–1010. https://www.ibiology.org/ibiomagazine/the-discovery-of-dynein.html.

Gibbons, B.H., Gibbons, I.R., 1972. Flagellar movement and adenosine triphosphatase activity in sea urchin sperm extracted with Triton X-100. J. Cell Biol. 54, 75–97.

Gibbons, I.R., Grimstone, A.V., 1960. On flagellar structure in certain flagellates. J. Biophys. Biochem. Cytol. 7, 697–716.

Gibbons, I.R., Rowe, A.J., 1965. Dynein: a protein with adenosine triphosphatase activity from cilia. Science 149, 424–426.

Gibbs, J.W., 1875–1878. The Collected Works of J. Willard Gibbs. Volumes I and II, 1931. Longmans, Green and Co., New York.

Gibcus, J.H., Dekker, J., 2013. The hierarchy of the 3D genome. Mol. Cell 49, 773–782.

Gibeaut, D.M., Carpita, N.C., 1993. Synthesis of (1→3), (1→4)-β-D-glucan in the Golgi apparatus of maize coleoptiles. Proc. Natl. Acad. Sci. U.S.A. 90, 3850–3854.

Gibeaut, D.M., Carpita, N.C., 1994. Improved recovery of (1→3), (1→4)-β-D-glucan synthase (MG) activity from Golgi apparatus of *Zea mays* (L.) using differential flotation centrifugation. Protoplasma 180, 92–97.

Gibernau, M., Barabé, D., Moisson, M., Trombe, A., 2005. Physical constraints on temperature difference in some thermogenic aroid inflorescences. Ann. Bot. 96, 117–125.

Gibson, K.S., Balcom, M.M., 1947. Transmission measurements with the Beckman quartz spectrophotometer. J. Opt. Soc. Am 37, 593–608.

Giddings Jr., T.H., Brower, D.L., Staehelin, L.A., 1980. Visualization of particle complexes in the plasma membrane of Micrasterias denticulata associated with the formation of cellulose fibrils in primary and secondary cell walls. J. Cell Biol. 84, 327–339.

Giddings Jr., T.H., Staehelin, L.A., 1988. Spatial relationship between microtubules and plasma-membrane rosettes during the deposition of primary wall microfibrils in *Closterium*. Planta 173, 22–30.

Giehl, K., Valenta, R., Rothkegel, M., Ronsiek, M., Mannherz, H.B., Jockusch, B.M., 1994. Interaction of plant profilin with mammalian actin. Eur. J. Biochem. 226, 681–689.

Gierth, M., Mäser, P., 2007. Potassium transporters in plants— Involvement in K^+ acquisition, redistribution and homeostasis. FEBS Lett. 581, 2348–2356.

Gietl, C., 1990. Glyoxysomal malate dehydrogenase from watermelon is synthesized with an amino-terminal transit peptide. Proc. Natl. Acad. Sci. U.S.A. 87, 5773–5777.

Gigot, C., Phillips, G., Nicolaieff, A., Hirth, L., 1976. Some properties of tobacco protoplast chromatin. Nucleic Acids Res. 3, 2315–2329.

Gil-Humanes, J., Wang, Y., Liang, Z., Shan, Q., Ozuna, C.V., Sánchez-León, S., Baltes, N.J., Starker, C., Barro, F., Gao, C., Voytas, D.F., 2017. High-efficiency gene targeting in hexaploid wheat using DNA replicons and CRISPR/Cas9. Plant J. 89, 1251–1262.

Gilbert, W., 1958. On the Magnet, 1600. Basic Books, New York.

Gilbert, W., 1978. Why genes in pieces? Nature 271, 501.

Gilbert, W., 1981. DNA sequencing and gene structure. Science 214, 1305–1312.

Gilbert, W., 1991. Towards a paradigm shift in biology. Nature 349, 99.

Gilbert, W., 1992a. A vision of the grail. In: Kevles, D.J., Hood, L. (Eds.), The Code of Codes. Harvard University Press, Cambridge, MA, pp. 83–97.

Gilbert, W., 1992b. Towards a paradigm shift in biology. Nature 349, 99.

Gilbert, W., December 8, 1980. DNA Sequencing and Gene Structure. Nobel Lecture.

Gilham, P.T., 1970. RNA sequence analysis. Ann. Rev. Biochem 39, 227–250.

Gilkey, J.C., Jaffe, L.F., Ridgway, E.B., Reynolds, G.T., 1978. A free calcium wave traverses the activating eggs of the medaka. Orzias Latipes. J. Cell Biol. 76, 448–466.

Gill, S.R., Pop, M., Deboy, R.T., Eckburg, P.B., Turnbaugh, P.J., Samuel, B.S., Gordon, J.I., Relman, D.A., Fraser-Liggett, C.M., et al., 2006. Metagenomic analysis of the human distal gut microbiome. Science 312, 1355–1359.

Gillaspy, G.E., Keddie, J.S., Oda, K., Gruissem, W., 1995. Plant inositol monophosphatase is a lithium-sensitive enzyme encoded by a multigene family. Plant Cell 7, 2175–2185.

Gillespie, R.J., Maw, G.A., Vernon, C.A., 1953. The concept of phosphate bond-energy. Nature 171, 1147–1149.

Gilliham, M., Campbell, M., Dubos, C., Becker, D., Davenport, R., 2006. The *Arabidopsis thaliana* glutamate-like receptor family. In: Baluska, F., Mancuso, S., Volkmann, D. (Eds.), Communication in Plants. Springer, Berlin, pp. 187–204.

Gillispie, C.C., 1974. Dictionary of Scientific Biography. Volume X. Charles Scribner's Sons, New York.

Gillmor, A., Poindexter, C.S.P., Lorieau, J., Palcic, M.M., Somerville, C., 2002. α-Glucosidase I is required for cellulose biosynthesis and morphogenesis in Arabidopsis. J. Cell Biol. 156, 1003–1013.

Gilmore, R., Blobel, G., Walter, P., 1982a. Protein translocation across the endoplasmic reticulum. I. Detection in the microsomal membrane of a receptor for the signal recognition particle. J. Cell Biol. 95, 463–469.

Gilmore, R., Walter, P., Blobel, G., 1982b. Protein translocation across the endoplasmic reticulum. II. Isolation and characterization of the signal recognition particle receptor. J. Cell Biol. 95, 470–477.

Gilroy, S., Fricker, M.D., Read, N.D., Trewavas, A.J., 1991. Role of calcium in signal transduction of *Commelina* guard cells. Plant Cell 3, 333–344.

Gilroy, S., Jones, R.L., 1992. Gibberellic acid and abscisic acid coordinately regulate cytoplasmic calcium and secretory activity in barley aleurone protoplasts. Proc. Natl. Acad. Sci. U.S.A. 89, 3591–3595.

Gilroy, S., Jones, R.L., 1993. Calmodulin-stimulation of unidirectional calcium uptake by the endoplasmic reticulum of barley aleurone. Planta 190, 289–296.

Gilroy, S., Read, N.D., Trewavas, A.J., 1990. Elevation of cytosolic calcium by caged calcium or caged inositol triphosphate inhibits closure. Nature 346, 768–771.

Gilson, E., 1956. The Christian Philosophy of Thomas Aquinas. Random House, New York.

Gimmler, H., Weiss, C., Baier, M., Hartung, W., 1990. The conductance of the plasmalemma for carbon dioxide. J. Exp. Bot. 41, 785–794.

Gingerich, O., 2006. God's Universe. Harvard University Press, Cambridge.

Ginsberg, M.H., Du, X., Plow, E.F., 1992. Inside-out integrin signalling. Curr. Opin. Cell Biol. 4, 766–771.

Ginzburg, B., 1959. Maintenance of semipermeability of plant cell membranes in the absence of metabolic energy supply. Nature 184, 1073.

Ginzburg, M., Ginzburg, B.Z., 1975. Factors influencing the retention of K in a *Halobacterium*. Biomembranes 3, 219–251.

Ginzburg, M., Ginzburg, B., Wayne, R., 1999. Ultrarapid endocytotic uptake of large molecules in *Dunaliella* species. Protoplasma 206, 73–86.

Giovane, A., Servillo, L., Balestrieri, C., Raiola, A., D'Avino, R., Tamburrini, M., Ciardiello, M.A., Camardella, L., 2004. Pectin methylesterase inhibitor. Biochim. Biophys. Acta 1696, 245–252.

Giovannoni, J., Nguyen, C., Ampofo, B., Zhong, S., Fei, Z., 2017. The epigenome and transcriptional dynamics of fruit rpiening. Annu. Rev. Plant Biol. 68, 61–84.

Gisel, A., Barella, S., Hempel, F.D., Zambryski, P.C., 1999. Temporal and spatial regulation of symplastic trafficking during development in *Arabidopsis thaliana* apices. Development 126, 1879–1889.

Gisel, A., Hempel, F.D., Barella, A.S., Zambryski, P.C., 2001. Leaf-to-shoot apex movement of symplastic tracer is restricted coincident with flowering in *Arabidopsis*. Proc. Natl. Acad. Sci. U.S.A. 99, 1713–1717.

Gisriel, C., Sarrou, I., Ferlez, B., Golbeck, J.H., Redding, K.A.E., Fromme, R., 2017. Structure of a symmetric photosynthetic reaction center—photosystem. Science 357, 1021–1025.

Gitschier, J., 2015. The whole of a scientific career: an interview with Oliver Smithies. PLoS Genet. 11, e1005224.

Glass, J.R., Gerace, L., 1990. Lamins A and C bind and assembly at the surface of mitotic chromosomes. J. Cell Biol. 111, 1047–1057.

Glasstone, S., Laidler, K.J., Eyring, H., 1941. The Theory of Rate Processes. McGraw-Hill, New York.

Glazer, A.N., Melis, A., 1987. Photochemical reaction centers: structure, organization, and function. Ann. Rev. Plant Physiol. 38, 11–45.

Glick, B.S., Luini, A., 2011. Models for Golgi traffic: a critical assessment. Cold Spring Harb. Perspect. Biol. 3, a005215.

Glick, B.S., Nakano, A., 2009. Membrane traffic within the Golgi apparatus. Ann. Rev. Cell Dev. Biol. 25, 113–132.

Glick, B.S., Rothman, J.E., 1987. Possible role for fatty acyl-coenzyme A in intracellular protein transport. Nature 326, 309–312.

Glock, E., Jensen, C.O., 1932. The colorimetric determination of plant succinic dehydrogenase. J. Biol. Chem. 201, 271–278.

Glotzer, M., Murray, A.W., Kirschner, M., 1991. Cyclin is degraded by the ubiquitin pathway. Nature 349, 132–138.

Glusker, J.P., Katz, A.K., Bock, C.W., 1999. Metal ions in biological systems. Rigaku J. 16, 8–16.

Glynn, J.M., Miyagishima, S.-Y., Yoder, D.W., Osteryoung, K.W., Vitha, S., 2007. Chloroplast division. Traffic 8, 451–461.

Gocayne, J., Robinson, D.A., FitzGerald, M.G., Chung, F.-Z., Kerlavage, A.R., Lentes, K.-U., Lai, J., Wang, C.-D., Fraser, C.M., Venter, J.C., 1987. Primary structure of rat cardiac β-adrenergic and muscarinic cholinergic receptors obtained by automated DNA sequence analysis: further evidence for a multigene family. Proc. Natl. Acad. Sci. U.S.A. 84, 8296–8300.

Godfraind, T., 2007. Dr. Setsuro Ebashi. J. Pharmacol. Sci. 103, 1–3.

Goebel, K., 1920. Die Entfaltungsbewegungen der Pflanzen und deien teleologische Deutung. Gustav Fischer, Jena.

Goebel, K. von, 1926. Wilhelm Hofmeister. The Work and Life of a Nineteenth Century Botanist. Ray Society, London.

Goebl, M.G., Winey, M., 1991. The yeast cell cycle. Curr. Opin. Cell Biol. 3, 242–246.

Goeddel, D.V., Kleid, D.G., Bolivar, F., Heyneker, H.L., Yansura, D.G., Crea, R., Hirose, T., Kraszewski, A., Itakura, K., Riggs, A.D., 1979. Expression in *Escherichia coli* of chemically synthesized genes for human insulin. Proc. Natl. Acad. Sci. U.S.A. 76, 106–110.

Goertzel, T., Goertzel, B., 1995. Linus Pauling. A Life in Science and Politics. Basic Books, New York.

Goff, J.E., 2010. Gold Medal Physics. The Science of Sports. The Johns Hopkins University Press, Baltimore, MD.

Gogarten, J.P., Kiblak, H., Dittrich, P., Taiz, L., Bowman, E.J., Bowman, B.J., Manolson, N.F., Poole, R.J., Date, T., Oshima, T., Konishi, J., Denda, K., Yoshida, M., 1989. Evolution of the H^+- ATPase: implications for the origin of eukaryotes. Proc. Natl. Acad. Sci. U.S.A. 86, 6661–6665.

Gogarten-Boekels, M., Gogarten, J.P., Bentrup, F.-W., 1988. Sugar nucleotides dissipate ATP-generated transmembrane pH gradient in Golgi vesicles from suspension-cell protoplasts of *Chenopodium rubrum* L. Planta 174, 349–357.

Golbeck, J.H., 1992. Structure and function of photosystem I. Annu. Rev. Plant Physiol. Plant Mol. Biol. 43, 293–324.

Gold, T., 1968. Rotating neutron stars as the origin of the pulsating radio sources. Nature 218, 731–732.

Gold, T., 1969. Rotating neutron stars and the nature of pulsars. Nature 221, 25–27.

Gold, T., 1989. New ideas in Science. J. Sci. Exp. 3, 103–112.

Gold, T., 1992. The deep, hot biosphere. Proc. Natl. Acad. Sci. U.S.A. 89, 6045–6049.

Gold, T., 1999. The Deep Hot Biosphere. Springer, Berlin.

Goldberg, R.B., 1978. DNA sequence organization in the soybean plant. Biochem. Genet. 16, 45–68.

Goldberg, A.L., 1990. ATP-dependent proteases in prokaryotic and eukaryotic cells. Semin. Cell Biol. 1, 423–432.

Goldberg, R.B., 2001. From Cot curves to genomics. How gene cloning established new concepts in plant biology. Plant Physiol. 125, 4–8.

Goldberg, M.W., Allen, T.D., 1992. High resolution scanning electron microscopy of the nuclear envelope: demonstration of a new regular, fibrous lattice attached to the baskets of the nucleoplasmic face of the nuclear pores. J. Cell Biol. 119, 1429–1440.

Goldberg, A.D., Allis, C.D., Bernstein, E., 2007. Epigenetics: a landscape takes shape. Cell 128, 638–835.

Goldfarb, D.S., Gariépy, J., Schoolnik, G., Kornberg, R.D., 1986. Synthetic peptides as nuclear localization signals. Nature 322, 641–644.

Goldfarb, D.S., Michaud, N., 1991. Pathways for the nuclear transport of proteins and RNAs. Trends. Cell. Biol. 1, 20–24.

Goldman, Y.E., Franzini-Armstrong, C., Armstrong, C.M., 2012. Andrew Fielding Huxley (1917-2012). Nature 486, 474.

Goldsby, R.A., 2017. The American Association of Immunologists Oral History Project Transcript. Richard A. Goldsby, Ph.D. May 13, 2017. Interview conducted by Brien R. Williams, Ph.D. http://www.aai.org/AAISite/media/About/History/OHP/Transcripts/Trans-Inv-037_Goldsby_Richard_A-2017_Final.pdf.

Goldschmidt, R., 1916. Genetic factors and enzyme reaction. Science 43, 98–100.

Goldschmidt, R., 1933. Some aspects of evolution. Science 78, 539–547.

Goldschmidt, R., 1940. The Material Basis of Evolution. Yale University Press, New Haven, CT.

Goldschmidt, R.B., 1951. Understanding Heredity. An Introduction to Genetics. John Wiley & Sons, Inc., New York.

Goldschmidt, R.B., 1955. Theoretical Genetics. University of California Press, Berkeley.

Goldschmidt, R.B., 1956. The Golden Age of Zoology. University of Washington Press, Seattle, WA.

Goldsmith, M.H.M., 1966a. Maintenance of polarity of auxin movement by basipetal transport. Plant Physiol. 41, 749–754.

Goldsmith, M.H.M., 1966b. Movement of indoleacetic acid in coleoptiles of Arena sativa L. II. Suspension of polarity by total inhibition of the basipetal transport. Plant Physiol. 41, 15–27.

Goldsmith, M.H.M., 1977. The polar transport of auxin. Annu. Rev. Plant Physiol. 28, 439–478.

Goldsmith, M.H.M., Goldsmith, T.H., Martin, M.H., 1981. Mathematical analysis of the chemosmotic polar diffusion of auxin through plant tissues. Proc. Natl. Acad. Sci. U.S.A. 78, 976–980.

Goldsmith, M.H.M., Ray, P.M., 1973. Intracellular localization of the active process in polar transport of auxin. Planta 111, 297–314.

Goldstein, G., Scheid, M., Hammerling, U., Schlesinger, D.H., Niall, H.D., Boyse, E.A., 1975. Isolation of a polypeptide that has lymphocyte-differentiating properties and is probably represented universally in living cells. Proc. Natl. Acad. Sci. U.S.A. 72, 11–15.

Goldstein, D.A., Thomas, J.A., 2004. Biopharmaceuticals derived from genetically modified plants. QJM 97, 705–716.

Goldstein, R.E., Tuval, I., van de Meent, J.-W., 2008. Microfluidics of cytoplasmic streaming and its implications for intracellular transport. Proc. Natl. Acad. Sci. U.S.A. 105, 3663–3667.

Goldstein, R.E., van de Meent, J.-W., 2015. A physical perspective on cytoplasmic streaming. Interface Focus 5, 20150030.

Goldsworthy, A., 1976. Photorespiration. Carolina Biology Readers. Carolina Biological Supply, Burlington, NC.

Goldsworthy, A., 1983. The evolution of plant action potentials. J. Theor. Biol. 103, 645–648.

Goldsworthy, A., Rathore, K.S., 1985. The electrical control of growth in plant tissue cultures: the polar transport of auxin. J. Exp. Bot. 36, 1134–1141.

Golgi, C., 1898. Sur la structure des cellules nerveuses. Arch. Ital. de Biologie 30, 60–71.

Golgi, C., December 11, 1906. The neuron doctrine—theory and facts. Nobel Lecture.

Golomb, L., Abu-Abied, M., Belausov, E., Sadot, E., 2008. Different subcellular localizations and functions of *Arabidopsis* myosin VIII. BMC Plant Biol. 8, 3.

Gomez, L., Chrispeels, M.J., 1994. Complementation of an *Arabidopsis thaliana* mutant that lacks complex asparagine- linked glycans with the human cDNA encoding N-acetyl glucosaminyl transferase I. Proc. Natl. Acad. Sci. U.S.A. 91, 1829–1833.

Gomez-Navarro, N., Miller, E.A., 2016. COP—coated vesicles. Curr. Biol. 26, R47–R59.

Goncalves, S., Cairney, J., Rodriquez, M.P., Canovas, F., Oliveira, M., Miquel, C., 2007. PpRab1, a Rab GTPase from maritime pine is differentially expressed during embryogenesis. Mol. Genet. Genom. 278, 273–282.

Goñi, F.M., 2014. The basic structure and dynamics of cell membranes: an update of the Singer-Nicolson model. Biochim. Biophys. Acta 1838, 1467–1476.

Gonsalves, D., 2016. The Hawaiian Transgenic Papaya Story: "And the Beat goes on." ISHS Acta Horticulturae. https://doi.org/10.17660/ActaHortic.2016.1111.3, 1111.

Gonzalez, E., 1982. Aggregated forms of malate and citrate lyase are localized in endoplasmic reticulum of endosperm of germinating castor bean. Plant Physiol. 69, 83–87.

Gonzalez, E., Beevers, H., 1976. Role of the endoplasmic reticulum in glyoxysome formation in castor bean endosperm. Plant Physiol. 57, 406–409.

González-Fernández, A., Giménez-Martín, G., López-Sáez, J.E., 1970. Cytokinesis at prophase in plant cells treated with ethidium bromide. Exp. Cell Res. 62, 464–467.

Good, N., 1961. Photophosphorylation and the Hill reaction. Nature 191, 678–679.

Good, N., Izawa, S., Hind, G., 1966. Uncoupling and energy transfer inhibition in photophosphorylation. Curr. Topics Bioenerg. 1, 75–112.

Goodacre, R., Vaidyanathan, S., Dunn, W.B., Harrigan, G.G., Kell, D.B., 2004. Metabolomics by numbers: acquiring and understanding global metabolite data. Trends Biotechnol. 22, 245–252.

Goodbody, K.C., Hargreaves, A.J., Lloyd, C.W., 1989. On the distribution of microtubule-associated intermediate filament anti-gens in plant suspension cells. J. Cell Sci. 93, 427–438.

Goodbody, K.C., Venverloo, C.J., Lloyd, C.W., 1991. Laser microsurgery demonstrates that cytoplasmic strands anchoring the nucleus across the vacuole of premitotic plant cells are under tension. Implications for division plane alignment. Development 113, 931–939.

Goodin, M.M., Chakrabarty, R., Banerjee, R., Yelton, S., DeBolt, S., 2007. Update on live-cell imaging in plants. New gateways to discovery. Plant Physiol. 145, 1100–1109.

Goodman, H.M., Abelson, J., Landy, A., Brenner, S., Smith, J.D., 1968. Amber suppression: a nucleotide change in the anticodon of a tyrosine transfer RNA. Nature 217, 1019–1024.

Goodman, H.M., Abelson, J., Landy, A., Brenner, S., Smith, J.D., 1969. Amber suppression: a nucleotide change in the anticodon of a tyrosine transfer RNA. Nature 217, 1019–1024.

Goodman, C.D., Casati, P., Walbot, V., 2004. A multidrug resistance associated protein involved in anthocyanin transport in Zea mays. Plant Cell 16, 1812–1826.

Goodsell, D.S., 1991. Inside a living cell. TIBS 16, 203–206.

Goodwin, P.B., 1983. Molecular size limit for movement in the symplast of the *Elodea* leaf. Planta 157, 124–130.

Goodwin, S., McPherson, J.D., McCombie, W.R., 2016. Coming of age: ten years of next-generation sequencing technologies. Nat. Rev. Genet. 17, 333–351.

Goodwin, P.B., Shepherd, V., Erwee, M.G., 1990. Compartmentation of fluorescent tracers injected into the epidermal cells of *Egeria densa* leaves. Planta 181, 129–136.

Gorbsky, G.J., Ricketts, W.A., 1993. Differential expression of a phosphoepitope at the kinetochores of moving chromosomes. J. Cell Biol. 122, 1311–1321.

Gordon, D., Huddleston, J., Chaisson, M.J.P., Hill, C.M., Kronenberg, Z.N., Munson, K.M., Malig, M., Raja, A., Fiddes, I., Hillier, L.W., Dunn, C., Baker, C., Armstrong, J., Diekhans, M., Paten, B., Shendure, J., Wilson, R.K., Haussler, D., Chin, C.-S., Eichler, E.E., 2016. Long-read sequwence assembly of the gorilla genome. Science 352, aae0344.

Gore, A., 2006. An Inconvenient Truth. Rodale Press, NewYork.

Gorgas, K., 1984. Peroxisomes in sebaceous glands. V. Complex peroxisomes in the mouse preputial gland: serial sectioning and three-dimensional reconstruction studies. Anat. Embryol. 169, 261–270.

Görlich, D., Rapoport, T.A., 1993. Protein translocation into proteoliposomes reconstituted from purified components of the endoplasmic reticulum membrane. Cell 75, 615–630.

Gorshkova, T.A., Chemikosova, S.B., Lozovaya, V.V., Carpita, N.C., 1997. Turnover of galactans and other cell wall polysaccharides during development of flax plants. Plant Physiol. 114, 723–729.

Gorst, J.R., John, P.C.L., Sek, F.J., 1991. Levels of p34cdc2-like protein in dividing, differentiating and dedifferentiating cells of carrot. Planta 185, 304–310.

Gorter, E., Grendel, F., 1925. On bimolecular layers of lipoids on the chromocytes of the blood. J. Exp. Med. 41, 439–443.

Gosh, S., Dey, R., 1986. Nuclear matrix network in *Allium cepa*. Chromosoma 93, 429–434.

Goshima, G., Mayer, M., Zhang, N., Stuurman, N., Vale, R.D., 2008. Augmin: a protein complex required for centrosome-independent microtubule generation within the spindle. J. Cell Biol. 181, 421–429.

Goshima, G., Scholey, J.M., 2010. Control of mitotic spindle length. Annu. Rev. Cell Dev. Biol. 26, 21–57.

Goto, C., Tamura, K., Fukao, Y., Shimada, T., Hara-Nishimura, I., 2014. The novel nuclear envelope protein KAKU4 modulates nuclear morphology in *Arabidopsis*. Plant Cell 26, 2143–2155.

Goto-Yamada, S., Mano, S., Nakamori, C., Kondo, M., Yamawaki, R., Kato, A., Nishimura, M., 2014. Chaperone and protease functions of LON protease 2 modulate the peroxisomal transition and degradation with autophagy. Plant Cell Physiol. 55, 482–496.

Gottesman, M.M., Hicks, M., Gellert, J., 1973. Genetics and function of DNA ligase in *Escherichia coli*. J. Mol. Biol. 77, 531–547.

Gottschalk, M., Dolgener, E., Xoconostle-Cázares, B., Lucas, W.J., Komor, E., Schobert, C., 2008. *Ricinus communus* cyclophilin: functional characterization of a sieve tube protein involved in protein folding. Planta 228, 687–700.

Goubet, F., Dupree, P., Salomon Johansen, K., 2011. Carbohydrate gel electrophoresis. Methods Mol. Biol. 715, 81–92.

Goud, B., McCaffrey, M., 1991. Small GTP-binding proteins and their role in transport. Curr. Opin. Cell Biol. 3, 626–633.

Goudsmit, S.A., 1971. The discovery of the electron spin. http://www.lorentz.leidenuniv.nl/history/spin/goudsmit.html.

Gould, S.J., 1977. The return of hopeful monsters. Nat. Hist. 22–30 (June/July).

Gould, R.R., Borisy, G.G., 1978. Quantitative initiation of microtubule assembly by chromosomes from chinese hamster ovary cells. Exp. Cell. Res. 113, 369–374.

Gould, S.J., Keller, G.-A., Schneider, M., Howell, S.H., Garrard, L.J., Goodman, J.M., Distel, B., Tabak, H., Subramani, S., 1990. Peroxisomal protein import is conserved between yeast, plants, insects and mammals. EMBO J. 9, 85–90.

Gould, K.L., Moreno, S., Owen, D.J., Sazer, S., Nurse, P., 1991. Phosphorylation at Thr161 is required for Schizosaccharomyces pombe. p34cdc2 function. EMBO J. 10, 3297–3309.

Goulian, M., Kornberg, A., Sinsheimer, R.L., 1967. Enzymatic synthesis of DNA. XXIV. Synthesis of infectious phage phi-X174 DNA. Proc. Natl. Acad. Sci. U.S.A. 58, 1723–1730.

Gout, E., Rébeillé, F., Douce, R., Bligny, R., 2014. Interplay of Mg^{2+}, ADP, and ATP in the cytosol and mitochondria: unravelling the role of Mg^{2+} in cell respiration. Proc. Natl. Acad. Sci. U.S.A. 111, E4560–E4567.

Gouveia, A., June 11, 2010. Cape scientist earns top award. Cape Cod Times. http://www.capecodtimes.com/article/20100611/NEWS/6110316?start=2.

Govindjee, 2017. André Tridon Jagendorf (1926-2017): a personal tribute. Photosynth. Res. 132, 235–243.

Govindjee, Allen, J.F., Beatty, J.T., 2004. Celebrating the millennium: historical highlights of photosynthsis research, Part 3. Photosynth. Res. 80, 1–13.

Govindjee, F., Krogmann, D., 2004. Discoveries in oxygenic photosynthesis (1727–2003): a perspective. Photosynth. Res. 80, 15–57.

Govindjee, Rabinowitch, E., 1960. Two forms of chlorophyll a in vivo with distinct photochemical functions. Science 132, 355–356.

Govorunova, E.G., Jung, K.H., Sineshchekov, O.A., Spudich, J.L., 2004. Chlamydomonas sensory rhodopsin A and B: cellular content and role in photophobic response. Biophys. J. 86, 2342–2349.

Goyal, U., Blackstone, C., 2013. Untangling the web: mechaisms underlying ER network formation. Biochim. Biophys. Acta 1833, 2492–2498.

Grabe, M., Oster, G., 2001. Regulation of organelle acidity. J. Gen. Physiol. 117, 329–343.

Grabe, M., Wang, H., Oster, G., 2000. The mechanochemistry of V-ATPase proton pumps. Biophys. J. 78, 2796–2813.

Grabherr, M.G., Haas, B.J., Yassour, M., Levin, J.Z., Thompson, D.A., Amit, I., Adiconis, X., Fan, L., Raychowdhury, R., Zeng, Q., Chen, Z., Mauceli, E., Hacohen, N., Gnirke, A., Rhind, N., di Palma, F., Birren, B.W., Nusbaum, C., Lindblad-Toh, K., Friedman, N., Regev, A., 2011. Full-length transcriptome assembly from RNA-Seq data without a reference genome. Nat. Biotechnol. 29, 644–652.

Grabov, A., 2007. Plant KT/KUP/HAK potassium transporters: single family-multiple functions. Ann. Bot. 99, 1035–1041.

Grabski, S., deFeijter, A.W., Schindler, M., 1993. Endoplasmic reticulum forms a dynamic continuum for lipid diffusion between contiguous soybean root cells. Plant Cell 5, 25–38.

Grabsztunowicz, M., Koskela, M.M., Mulo, P., 2017. Post-translational modifications in regulation of chloroplast function: recent Advances. Front. Plant Sci. 8, 240.

Gradmann, D., Klempke, W., 1974. Current-voltage relationship of the electrogenic pump in *Acetabularia mediterranea*. In: Zimmermann, U., Dainty, J. (Eds.), Membrane Transport in Plants. Springer-Verlag, Berlin, pp. 131–138.

Graham, T., 1842. Elements of Chemistry, Including the Applications of the Science in the Arts. Hippolyte Bailliere, London.

Graham, T., 1843. Elements of Chemistry, Including the Applications of the Science in the Arts With Notes and Additions by Robert Bridges. M. D. Lea & Blanchard, Philadelphia.

Graham, T., 1850–1857. Elements of Chemistry, including the applications of the science in the arts. Baillière, New York.

Graham, T.A., Gunning, B.E.S., 1970. The localization of legumin and vicilin in bean cotyledon cells using fluorescent antibodies. Nature 228, 81–82.

Graham, J.W.A., Williams, T.C.R., Morgan, M., Fernie, A.R., Ratcliffe, R.G., Sweetlove, L.J., 2007. Glycolytic enzymes associate dynamically with mitochondria in response to respiratory demand and support substrate channeling. Plant Cell 19, 3723–3738.

Granger, C.L., Cyr, R.J., 2000a. Expression of GFP-MAP4 reporter gene in a stably transformed tobacco cell line reveals dynamics of microtubule reorganization. Planta 210, 502–509.

Granger, C.L., Cyr, R.J., 2000b. Microtubule reorganization in tobacco BY-2 cells stably expressing GFP-MBD. Planta 210, 502–509.

Granick, S., 1957. Speculations on the origins and evolution of photosynthesis. Ann. N.Y. Acad. Sci. 69, 292–308.

Granick, S., Porter, K.R., 1947. The structure of the spinach chloroplast as interpreted with the electron microscope. Am. J. Bot. 34, 545–550.

Grant, V., 1963. The Origin of Adaptations. Columbia University Press, New York.

Grant, V., 1975. Genetics of Flowering Plants. Columbia University Press, New York.

Graumann, K., 2014. Evidence for LINC1-SUN associations at the plant nuclear periphery. PLoS One 9, e93406.

Gray, J., 1928. Ciliary Movement. Cambridge University Press, Cambridge.

Gray, J.C., Hibberd, J.M., Linley, P.J., Uijtewaal, B., 1999. GFP movement between chloroplasts. Nat. Biotechnol. 17, 1146.

Gray, J.C., Row, P.E., 1995. Protein translocation across chloroplast envelope membranes. Trends Cell Biol. 5, 243–247.

Gray, J.C., Sullivan, J.A., Hibberd, J.M., Hansen, M.R., 2001. Stromules: mobile protrusions and interconnections between plastids. Plant Biol. 3, 223–233.

Graziana, A., Fosset, M., Ranjeva, R., Hetherington, A.M., Lazdunski, M., 1988. Ca^{2+} channel inhibitors that bind to plant cell membranes block Ca^{2+} entry into protoplasts. Biochemistry 27, 764–768.

Grebe, M., 2011. Unveiling the Casparian strip. Nature 473, 294–295.

Green, P.B., 1954. The spiral growth pattern in the cell wall in *Nitella axillaris*. Am. J. Bot. 41, 403–409.

Green, M.M., 1954. Pseudoallelism at the vermilion locus in Drosophila melanogaster. Proc. Natl. Acad. Sci. U.S.A. 40, 92–99.

Green, M.M., 1955a. Pseudoallelism and th egene concept. Am. Nat 89, 65–74.

Green, M.M., 1955b. Phenotypic variation and pseudoallelism in th eforked locus in Drosophila melanogaster. Proc. Natl. Acad. Sci. U.S.A. 41, 375–379.

Green, P.B., 1958a. Concerning the site of the addition of new wall substances to the elongating *Nitella* cell wall. Am. J. Bot. 45, 111–116.

Green, P.B., 1958b. Structural characteristics of developing *Nitella* internodal cell walls. J. Biophys. Biochem. Cytol. 1, 50–519.

Green, P.B., 1960. Multinet growth in the cell wall of *Nitella*. J. Biophys. Biochem. Cytol. 7, 289–297.

Green, P.B., 1961. Structural changes in growing cell walls. Recent Adv. Bot. 8, 746–749.

Green, P.B., 1962. Mechanism for plant cellular morphogenesis. Science 138, 1404–1405.

Green, P.B., 1963. On mechanisms of elongation. In: Locke, M. (Ed.), Cytodifferentiation and Macromolecular Synthesis; A Symposium. Academic Press, New York, pp. 203–234.

Green, P.B., 1968. Growth physics in *Nitella*: a method for continuous in vivo analysis of extensibility based in a micro-manometer technique for turgor pressure. Plant Physiol. 43, 1169–1184.

Green, P.B., 1969. Cell morphogenesis. Ann. Rev. Plant Physiol. 20, 365–394.

Green, J.R., 1983. The Golgi apparatus. In: Hall, J.L., Moore, A.L. (Eds.), Isolation of Membranes and Organelles from Plant Cells. Academic Press, London, pp. 135–152.

Green, P.B., 1988. A theory for inflorescence development and flower formation based on morphological and biophysical analysis in Echeveria. Planta 175, 153–169.

Green, P.B., 1996. Expression of form and pattern in plants—a role for biophysical fields. Semin. Cell Dev. Biol. 7, 903–911.

Green, P.B., 1997. Expansin and morphology: a role for biophysics. Trends Plant Sci. 2, 365–366.

Green, D.E., Baum, H., 1970. Energy and the Mitochondrion. Academic Press, New York.

Green, P.B., Chapman, G.B., 1955. On the development and structure of the cell wall in Nitella. Am. J. Bot. 42, 685–693.

Green, P.B., Erickson, R.O., Buggy, J., 1971. Metabolic and physical control of cell elongation rate. Plant Physiol. 47, 423–430.

Green, M.M., Green, K.C., 1949. Crossing over between alleles at the *lozenge* locus in *Drosophila melanogaster*. Proc. Natl. Acad. Sci. U.S.A. 35, 586–591.

Green, J.R., Northcote, D.H., 1978. The structure and function of glycoproteins synthesized during slime-polysaccharide production by membranes of the root-cap cells of maize (Zea mays). Biochem. J. 170, 599–608.

Green, G.R., Searcy, D.G., DeLange, R.J., 1983. Histonelike protein in the archaebacterium *Sulfolobis acidocaldariius*. Biochim. Biophys. Acta 741, 251–257.

Greenbaum, E., Lee, J.W., Tevault, C.V., Blankenship, S.L., Mets, L.J., 1995. CO_2 fixation and photoevolution of H_2 and O_2 in a mutant of Chlamydomonas lacking photosystem I. Nature 376, 438–441.

Greenfield, A.B., 2005. A Perect Red. HarperCollins, New York.

Greenwood, J.S., Chrispeels, M.J., 1985a. Correct targeting of the bean storage protein haseolin in the seeds of transformed tobacco. Plant Physiol. 79, 65–71.

Greenwood, J.S., Chrispeels, M.J., 1985b. Immunocytochemical localization of phaseolin and phytohemagglutin in the endoplasmic reticulum and Golgi complex of developing bean cotyledons. Planta 164, 295–302.

Gregory, P.H., 1961. The Microbiology of the Atmosphere. Interscience Publishers, New York.

Gregory, T.R., 2005. The C-value enigma in plants and animals: a review of parallels and an appeal for a partnership. Ann. Bot. 95, 133–146.

Greider, C.W., Blackburn, E.H., 1985. Identification of a specific telomere terminal transferase activity in *Tetrahymena* extracts. Cell 43, 405–413.

Greimers, R., Deltour, R., 1981. Organization of transcribed and nontranscribed chromatin in isolated nuclei of Zea mays root cells. Eur. J. Cell Biol. 23, 303–311.

Grennan, A.K., 2008. Arabidopsis microRNAs. Plant Physiol. 146, 3–4.

Grew, N., 1682. The Anatomy of Plants with an Idea of a Philosophical History of Plants and Several other Lectures Read before The Royal Society, 1965. Johnson Reprint Corp., New York.

Grew, N., 1701. Cosmologia Sacra: or A Discourse of the Universe as it is the Creature and Kingdon of God. Printed for W. Rogers, S. Smith and B. Walford, London.

Gribbin, J., 1984. Search of Schödinger's Cat. Bantam Books, Toronto.

Gribbin, J., 1995. Schrödinger's Kittens and the Search for Reality. Little, Brown, and Co., Boston.

Grice, E.A., Segre, J.A., 2012. The human microbiome: our second genome. Annu. Rev. Genom. Hum. Genet. 13, 151–170.

Griessner, M., Obermeyer, G., 2003. Characterization of whole- cell K^+ currents across the plasma membrane of pollen grain and tube protoplasts of Lillium longiflorum. J. Membr. Biol. 1193, 99–108.

Griffing, L.R., 2010. Networking in the endoplasmic reticulum. Biochem. Soc. Trans. 38, 747–753.

Griffing, L.R., Gao, H.T., Sparkes, I., 2014. ER network dynamics are differentially controlled by myosins XI-K, XI-C, XI-E, XI-I, an dXI-2. Front. Plant Sci. 5, 218.

Griffing, L.R., Villanueva, M.A., Taylor, J., Moon, S., 1995. Confocal epipolarization microscopy of gold probes in plant cells and protoplasts. Methods Cell Biol. 49, 109–121.

Griffiths, G., 1996. On vesicles and membrane compartments. Protoplasma 195, 37–58.

Griffiths, R.B., 2008. A review of: quantum physics and theology. An unexpected kinship. Phys. Today 61 (2), 65–66.

Grigg, G.W., Hodge, A.J., 1949. Electron microscopic studies of spermatozoa. I. The morphology of the spermatozoon of the common domestic fowl (Gallus domesticus). Aust. J. Sci. Res. Ser. B. 2, 271–286.

Grignon, C., Sentenac, H., 1991. pH and ionic conditions in the apoplast. Annu. Rev. Plant Physiol. Plant Mol. Biol. 42, 103–128.

Grill, E., Christmann, A., 2007. A plant receptor with a big family. Science 315, 1676–1677.

Grill, E., Loeffler, S., Winnacker, E.-L., Zenk, M.H., 1989. Phytochelatins, the heavy-metal-binding peptides of plants, are synthesized from glutatione by a specific γ-glutamylcysteine dipeptide transpeptidase (phytochelatin synthase). Proc. Natl. Acad. Sci. U.S.A. 86, 6838–6842.

Grill, E., Winnacker, E.-L., Zenk, M.H., 1985. Phytochelatins: the principle heavy-metal complexing peptides of higher plants. Science 230, 674–676.

Grimm, I., Sachs, H., Robinson, D.G., 1976. Structure, synthesis and orientation of microfibrils. II. The effect of colchicine on the wall of Oocystis solitaria. Cytobiologie 14, 61–74.

Grison, M.S., Brocard, K., Fouillen, L., Nicolas, W., Wewer, V., Dormann, P., Nacir, H., Benitez-Alfonso, Y., Claverol, S., Germain, V., Boutté, Y., Mongrand, S., Bayer, E.M., 2015a. Specific membrane lipid composition is important for plasmodesmata function in *Arabidopsis*. Plant Cell 27, 1228–1250.

Grison, M.S., Fernandez-Calvino, L., Mongrand, S., Bayer, E.M.F., 2015b. Isolation of plasmodesmata from *Arabidopsis* suspension cells. Methods Mol. Biol. 1217, 83–93.

Groen, A.J., de Vries, S.C., Lilley, K.S., 2008. A proteomics approach to membrane trafficking. Plant Physiol. 147, 1584–1589.

Grolig, F., 1990. Actin-based organelle movements in interphase Spirogyra. Protoplasma 155, 29–42.

Grolig, F., 2004. Organelle movements: transport and positioning. Annu. Plant Rev. 10, 148–175.

Grolig, F., Williamson, R.E., Parke, J., Miller, C., Anderson, B.H., 1988. Myosin and Ca^{2+}-sensitive streaming in the alga *Chara*: detection of two polypeptides reacting with a monoclonal antimyosin and their localization in the streaming endoplasm. Eur. J. Cell Biol. 47, 22–31.

Gronegress, P., 1974. The structure of chromoplasts and their conversion to chloroplasts. J. Microsc. 19, 183–192.

Gronenborn, B., Messing, J., 1978. Methylation of single-stranded DNA in vitro introduces new restriction endonuclease cleavage sites. Nature 272, 375–377.

Groover, A., deWitt, N., Heidel, P., Jones, A., 1997. Programmed cell death of plant tracheary elements differentiating in vitro. Protoplasma 196, 197–211.

Gropius, W., 1962. The Scope of Total Architecture. Collier Books, New York.

Gros, F., Hiatt, H., Gilbert, W., Kurland, C.G., Risebrough, R.W., Watson, J.D., 1961. Unstable ribonucleic acid revealed by pulse labelling of *Escherichia coli*. Nature 190, 581–585.

Gross, P.R., 1986. The truth about calcium. MBL Sci. 2 (1), 17–18.

Gross, J., Gross, M., 1969. Genetic analysis of an *E. coli* strain with a mutation affecting DNA polymerase. Nature 224, 1166–1168.

Grotewold, E., 2006. The genetics and biochemistry of floral pigments. Annu. Rev. Plant Biol. 57, 761–780.

Groth, A., Rocha, W., Almouzni, G., 2007. Chromatin challenges during DNA replication and repair. Cell 128, 721–733.

Grove, S.N., Bracker, C.E., Morré, D.J., 1970. An ultrastructural basis for hyphal tip growth in Pythium ultimum. Am. J. Bot. 57, 245–266.

Grove, W.R., Helmholtz, H., Mayer, J.R., Faraday, M., Liebig, J., Carpenter, W.B., Youmans, E.L., 1867. The Correlation and Conservation of Forces. A Series of Expositions. D. Appleton and Co., New York.

Groves, A.K., Bögler, O., Jat, P.S., Noble, M., 1991. The cellular measurement of time. Curr. Opin. Cell Biol. 3, 224–229.

Gruber, P.J., Becker, W.M., Newcomb, E.H., 1972. The occurrence of microbodies and peroxisomal enzymes in achlorophyllous leaves. Planta 105, 114–138.

Gruber, P.J., Sweeney, K.A., Frederick, S.E., 1988. The detection of fucose residues in plant nuclear envelopes. Planta 174, 298–304.

Gruber, P.J.R., Trelease, N., Becker, W.M., Newcomb, E.H., 1970. A correlative ultrastructural and enzymatic study of cotyledonary microbodies following germination of fat storing seeds. Planta 93, 269–288.

Gruhler, A., Jensen, O.N., 2006. Proteomic analysis of posttranslational modifications by mass spectrometry. Annu. Plant Rev. 28, 33–53.

Grunstein, M., 1990. Nucleosomes: regulators of transcription. Trends Genet. 6, 395–400.

Gu, X., Verma, D.P.S., 1996. Phragmoplastin, a dynamin-like protein associated with cell plate formation in plants. EMBO J. 15, 695–704.

Gu, Y., Vernoud, V., Fu, Y., Yang, Z., 2003. ROP GTPase regulation of pollen tube growth through the dynamics of tip-localized F-actin. J. Exp. Bot. 54, 93–101.

Gu, H., Zhu, P., Jiao, Y., Meng, Y., Chen, M., 2011. PRIN: a predicted rice interactome network. BMC Bioinform. 12, 161.

Gubler, F., Jacobson, J.V., Ashford, A.E., 1986. Involvement of the Golgi apparatus in the secretion of a-amylase from gibberellin treated barley aleurone cells. Planta 168, 447–452.

Guerineau, F., Sorensen, A.M., Fenby, N., Scott, R.J., 2003. Temperature sensitive diphtheria toxin confers conditional male-sterility in *Arabidopsis thaliana*. Plant Biotechnol. J. 1, 33–42.

Guerrero, M.G., Vega, J.M., Losada, M., 1981. The assimilatory nitrate-reducing system and its regulation. Ann. Rev. Plant Physiol. 32, 169–204.

Guerrier-Takada, C., Altman, S., 1984. Catalytic activity of an RNA molecule prepared by transcription in vitro. Science 223, 285–286.

Guerrier-Takada, C., Gardiner, K., Marsh, T., Pace, N., Altmann, S., 1983. The RNA moiety of ribonuclease P is the catalytic subunit of the enzyme. Cell 35, 849–857.

Guigas, G., Kalla, C., Weiss, M., 2007. Probing the nanoscale viscoelasticity of intracellular fluids of living cells. Biophys. J. 93, 316–323.

Guilliermond, A., 1941. The Cytoplasm of the Plant Cell. Chronica Botanica Co., Waltham, MA.

Guldberg, C.M., Waage, P., 1899. Untersuchungen über die Chemischen Affinitäten. W. Engelmann, Leipzig.

Gulland, J.M., 1947. The structures of nucleic acids. Symp. Soc. Exp. Biol. 1, 1–14.

Gunawardena, J., 2012. Some lessons about models from Michaelis and Menten. Mol. Biol. Cell 23, 517–519.

Gunawardena, J., 2013. Biology is more theoretical than physics. Mol. Biol. Cell 24, 1827–1929.

Gunawardena, J., 2014. Time-scale separation—Michaelis and Menten's old idea, still bearing fruit. FEBS J. 281, 473–488.

Günl, M., Kraemer, F., Pauly, M., 2011. Oligosaccharide mass profiling (OLIMP) of cell wall polysaccharides by MALDI-TOF/MS. Methods Mol. Biol. 715, 43–54.

Gunning, B.E.S., 1965. The greening process of plastids. I. The prolamellar body. Protoplasma 60, 111–130.

Gunning, B.E.S., 1982. The cytokinetic apparatus: its development and spatial regulation. In: Lloyd, C.W. (Ed.), The Cytoskeleton in Plant Growth and Development. Academic Press, London, pp. 229–292.

Gunning, B.E.S., 1992. Use of confocal microscopy to examine transitions between successive microtubule arrays in the plant cell division cycle. In: Shibaoka, H. (Ed.), Cellular Basis of Growth and Development in Plants, pp. 145–155. Proceedings of the VII International Symposium in Conjunction with the Awarding of the International Prize of Biology. Osaka University, Osaka.

Gunning, B.E.S., 1998. The identity of mystery organelles in Arabidopsis plants expressing GFP. Trends Plant Sci. 3, 417.

Gunning, B.E.S., 2005. Plastid Stromules: Video Microscopy of their Outgrowth, Retractions, Tensioning, Anchoring, Branching, Bridging, and Tip-Shedding.

Gunning, B.E.S., Hardham, A.R., Hughes, J.E., 1978b. Evidence for initiation of microtubules in discrete regions of the cell cortex in Azolla roottip cells, and an hypothesis on the development of cortical arrays of microtubules. Planta 143, 161–179.

Gunning, B.E.S., Hardham, A.R., Hughes, J.E., 1978c. Pre-prophase bands of microtubules in all categories of formative and proliferative cell division in *Azolla* roots. Planta 143, 145–160.

Gunning, B.E.S., Hughes, J.E., Hardham, A.R., 1978a. Formative and proliferative cell divisions, cell differentiation, and developmental changes in the meristem of Azolla roots. Planta 143, 121–144.

Gunning, B.E.S., Overall, R.L., 1983. Plasmodesmata and cell-to-cell transport in plants. BioScience 33, 260–265.

Gunning, B.E.S., Pate, J.S., 1974. Transfer cells. In: Robards, A.W. (Ed.), Dynamic Aspects of Plant Ultrastructure. McGraw-Hill, New York, pp. 441–480.

Gunning, B.E.S., Robards, A.W. (Eds.), 1976. Intercellular Communication in Plants: Studies on Plasmodesmata. Springer-Verlag, Berlin.

Gunning, B.E.S., Steer, M.W., 1996. Plant Cell Biology. Structure and Function. Jones and Bartlett Publishers, Sudbury, MA.

Gunning, B.E.S., Wick, S.M., 1985. Preprophase bands, phragmoplasts, and spatial control of cytokinesis. J. Cell Sci. (Suppl.) 2, 157–179.

Guo, T., Fang, Y., 2014. Functional organization and dynamics of the cell nucleus. Front. Plant Sci. 5, 378.

Guo, X., Li, J., 2012. Activation tagging. Methods Mol. Biol. 876, 117–133.

Guo, T., Mao, X., Zhang, H., Zhang, Y., Fu, M., Sun, Z., Kuai, P., Lou, Y., Fang, Y., 2017. Lamin-like proteins negatively regulate plant immunity through NAC WITH TRANSMEMBRANE MOTIF1-LIKE9 and NONEXPRESSOR OF PR GENS1 in Arabidopsis thaliana. Mol. Plant 10, 1334–1348.

Guo, T., Mao, X., Zhang, H., Zhang, Y., Fu, M., Sun, Z., Kuai, P., Lou, Y., Fang, Y., 2017b. Lamin-like proteins negatively regulate plant immunity through NAC with transmembrane MOTIF1-LIKE9 and nonexpressor of PR GENS1 in *Arabidopsis thaliana*. Mol. Plant 10, 1334–1348.

Guo, J., Yang, X., Weston, D.J., Chen, J.G., 2011. Abscisic acid receptors: past, present and future. J. Integr. Plant Biol. 53, 469–479.

Guo, D., Zhou, Y., Li, H.-L., Zhu, J.H., Wang, Y., Chen, X.T., Peng, S.-Q., 2017a. Identification and characterization of the abscisic acid (ABA) receptor gene family and its expression in response to hormones in the rubber tree. Sci. Rep. 7, 45157.

Gupta, A., Sharma, C.B., 1996. Purification to homogeneity and characterization of a plasma membrane and Golgi apparatus-specific 5′-adenosine monophosphatase from peanut cotyledons. Plant Sci. 117, 65–74.

Gupton, S.L., Collings, D.A., Allen, N.S., 2006. Endoplasmic reticulum targeted GFP reveals ER organization in tobacco NT-1 cells during cell division. Plant Physiol. Biochem. 44, 95–105.

Gurdon, J.B., 1962. Adult frogs derived from the nuclei of single somatic cells. Dev. Biol. 4, 256–273.

Gurdon, J., December 7, 2012. The Egg and the nucleus: a battle for supremacy. In: Nobel Lecture.

Gurdon, J.B., Uehlinger, V., 1966. "Fertile" intestine nuclei. Nature 210, 1240–1241.

Gurung, N., Ray, S., Bose, S., Rai, V., 2013. A broader view: microbial enzymes and their relevance in industries, medicine, and beyond. BioMedRes. Int. 2013, 329121.

Gusella, J.F., Wexler, N.S., Conneally, P.M., Naylor, S.L., Anderson, M.A., Tanzi, R.E., Watkins, P.C., Ottina, K., Wallace, M.R., Sakaguchi, A.Y., Young, A.B., Shoulson, I., Bonilla, E., Martin, J.B., 1983. A polymorphic DNA marker genetically linked to Huntingtons deisease. Nature 306, 234–238.

Guth, A., 1997. The Inflationary Universe. Addison-Wesley, Reading, MA.

Guthrie, C., 1996. The spliceosome is a dynamic ribonucleoprotein machine. Harvey Lect. 90, 59–80.

Gutknecht, J., Bisson, M.A., Tosteson, D.C., 1977. Diffusion of carbon dioxide through lipid bilayer membranes: effects of carbonic anhydrase, bicarbonate and unstirred layers. J. Gen. Physiol. 69, 779–794.

Gutleben, J., de Mares, M.C., van Elsas, J.D., Smidt, H., Overmann, J., Spikema, D., May 31, 2017. The muti-omics promise in context: from sequence to microbial isolate. Crit. Rev. Microbiol. 1–18.

Guy, R.D., Berry, J.A., Fogel, M.L., Hoering, T.C., 1989. Differential fractionation of oxygen isotypes by cyanide- resistant and cyanide-sensitive respiration in plants. Planta 177, 483–491.

Guye, C.E., Lavanchy, C., 1915. Vérification expérimentale de la formule de Lorentz–Einstein par les rayons cathodiques de grande vitesse. Comptes Rendus Acad. Sci. 161, 52–55.

Guye, C. E. n.d. Physico-Chemical Evolution. E. P. Dutton and Co., New York.

Guzman, M.I., Martin, S.T., 2008. Oxaloacetate-to-malate conversion by mineral photoelectrochemistry: implications for the viability of the reductive tricarboxylic acid cycle in prebiotic chemistry. Int. J. Astrobiol. 7, 271–278.

Guzman, M.I., Martin, S.T., 2009. Prebiotic metabolism: production by mineral photoelectrochemistry of a ketocarboxylic acids in the reductive tricarboxylic acid cycle. Astrobiology 9, 833–842.

Guzman, M.I., Martin, S.T., 2010. Photo-production of lactate from glyoxylate: how minerals can facilitate energy storage in a prebiotic world. Chem. Commun. 46, 2265–2267.

Haaf, T., 1996. High-resolution analysis of DNA replication in released chromatin fibers containing 5-bromodeoxyuridine. BioTechniques 21, 1050–1054.

Haas, T.A., Plow, E.F., 1994. Integrin-ligand interactions: a year in review. Curr. Opin. Cell Biol. 6, 656–662.

Haas, R., Zelezniak, A., Iacovacci, J., kamrad, S., Townsend, S., Ralser, M., 2017. Designing and interpreting 'multi-omic' experiments that may change our understanding og biology. Curr. Opin. Syst. Biol. 6, 37–45.

Haasen, D., Köhler, C., Neuhaus, G., Merkle, T., 2000. Nuclear export of proteins in plants. AtXPO1 is the export receptor for leucine-rich nuclear export signals in *Arabidopsis thaliana*. Plant J. 20, 695–705.

Haber, A.H., Carrier, W.L., Foard, D.E., 1961. Metabolic studies of gamma-irradiated wheat growing without cell divisions. Am. J. Bot. 48, 431–438.

Haberlandt, G., 1906. Sinnesorgane im Pflanzenreich, zur Perception mechanischer Reize. W. Engelmann, Leipzig.

Haberlandt, G., 1914. Physiological Plant Anatomy. MacMillan, London.

Habisreutinger, S.N., Schmidt-Mende, L., Stolarczyk, J.K., 2013. Photocatalytic reduction of CO_2 on TiO_2 and other semiconductors. Angew. Chem. Int. Ed. 52, 7372–7408.

Hackert, M.L., Appling, D.R., Lambowitz, A.M., 2015. Lester reed: a "complex" man who loved science. Proc. Natl. Acad. Sci. U.S.A. 112, 6247.

Hacking, I., 1981. Do we see through a microscope? Pac. Philos. Q. 62, 305–322.

Häder, D.-P., Ntefidou, M., Iseki, M., Watanabe, M., 2005. Phototaxis photoreceptor in *Euglena gracilis*. In: Wada, M., Shimazaki, K., Iino, M. (Eds.), Light Sensing in Plants. Springer, Tokyo, pp. 223–229.

Hadjantonakis, A.K., Nagy, A., 2001. The color of mice: in the light of GFP-variant reporters. Histochem. Cell. Biol. 115, 49–58.

Haeckel, E., 1866. Generelle Morphologie der Organismen. Georg Reimer, Berlin.

Haeckel, E., 1879a. Freedom in Science and Teaching. D. Appleton and Co., New York.

Haeckel, E., 1879b. The Evolution of Man. A popular Exposition of the Principle Points of Human Ontogeny and Phylogeny. D. Appleton and Co., New York.

Hagel, J.M., Yeung, E.C., Facchini, P.J., 2008. Got milk? The secret life of laticifers. Trends Plant Sci. 14, 631–639.

Hagemann, M., Bauwe, H., 2016. Photorespiration and the potential to improve photosynthesis. Curr. Opin. Chem. Biol. 35, 109–116.

Hagemann, R., Schröder, M.B., 1989. The cytological basis of the plastid inheritance in angiosperms. Protoplasma 152, 57–64.

Hagendoorn, M.J.M., Poorhnga, A.M., Wong Fong Song, H.W., Van der Plas, L.H.W., van Waltrave, H.S., 1991. Effect of electrons on the plasma membrane of *Petunia hybrida* cell suspensions. Role of pH in signal transduction. Plant Physiol. 96, 1261–1267.

Hager, A., Menzel, H., Krauss, A., 1971. Versuche und Hypothese zue Primärwirkung des Auxins beim Streckungswachstum. Planta 100, 47–75.

Haghighattalab, A., González Pérez, L., Mondal, S., Singh, D., Schinstock, D., Rutkoski, J., Ortiz-Monasterio, I., Prakash Singh, R., Goodin, D., Poland, J., 2016. Application of unmanned aerial systems for high throughput phenotyping of large wheat breeding nurseries. Plant Methods 12, 35.

Hagstrom, K.A., Meyer, B.J., 2003. Condensin and cohesion: more than chromosome compactor and glue. Nat. Rev. Genet. 4, 520–534.

Haguenau, F., 1958. The ergastoplasm: its history, ultrastructure, and biochemistry. Int. Rev. Cytol. 7, 425–483.

Hahne, G., Hoffman, F., 1984. The effect of laser microsurgery on cytoplasmic strands and cytoplasmic streaming in isolated plant protoplasts. Eur. J. Cell Biol. 33, 175–179.

Haigler Jr., C.H., Brown, R.M., 1986. Transport of rosettes from the Golgi apparatus to the plasma membrane in isolated mesophyll cells of Zinnia elegans during differentiation to tracheary elements in suspension culture. Protoplasma 134, 111–120.

Haldane, J.S., 1922. Respiration. Yale University Press, New Haven, CT.

Haldane, J.B.S., 1929. In: Maynard Smith, J. (Ed.), On Being the Right Size and other Essays. Oxford University Press, Oxford, 1985.

Haldane, J.B.S., 1930. Enzymes. Longmans, Green and Co., London.

Haldane, J.B.S., 1942. New Paths in Genetics. Harper, New York.

Hales, S., 1727. Vegetable Staticks, 1961. Oldbourne, London.

Hall, J.L., 1983. Plasma membranes. In: Hall, J.L., Moore, A.L. (Eds.), Isolation of Membranes and Organelles from Plant Cells. Academic Press, London, pp. 55–81.

Hall, J., 2008. Jeffrey C. Hall. Current Biology 18, R101-103.

Hall, K.T., 2011. William Astbury and the biological significance of nucleic acids, 1938-1951. Stud. Hist. Philos. Bio Biomed. Sci. 42, 119–128.

Hall, K.T., 2014. The Man in the Monkeynut Coat. Oxford University Press, Oxford.

Hall, J., 2017. A Nobel pursuit may not run like clockwork. Cell 171, 1246–1251.

Hall Jr., G., Allen, G.C., Loer, D.S., Thompson, W.F., Spiker, S., 1991. Nuclear scaffolds and scaffold-attachment regions in higher plants. Proc. Natl. Acad. Sci. U.S.A. 88, 9320–9324.

Hall, C.E., Jakus, M.A., Schmitt, F.O., 1946. An investigation of cross striations and myosin filaments in muscle. Biol. Bull. 90, 32–50.

Hall, D.A., Saxl, H., 1960. Human and other animal cellulose. Nature 187, 547–550.

Hall, D.A., Saxl, H., 1961. Studies on human and tunicate cellulose and of their relation to reticulin. Proc. R. Soc. Lond. 155B, 202–217.

Hallé, F., 2002. In: Praise of Plants. Timber Press, Portland, OR.

Halloran, N., Du, Z., Wilson, R.K., 1993. Sequencing reactions for the Applied Biosystems 373A automated DNA sequencer. Methods Mol. Biol. 23, 297–315.

Halperin, W., Jensen, W.A., 1967. Ultrastructural changes during growth and embryogenesis in carrot cell cultures. J. Ultrastruct. Res. 18, 428–443.

Hamada, H., Mita, T., Shibaoka, H., 1994. Stabilization of cortical microtubules in maize mesocotyl cells by gibberellin A3. Plant Cell Physiol. 35, 189–196.

Hamamura, Y., Nishimaki, M., Takeucki, H., Geitmann, A., Kurihara, D., Higashiya, T., 2014. Live imaging of calcium spikes during double fertilization in *Arabidopsis*. Nat. Commun. 5, 4722.

Hamel, L.P., Sheen, J., Séguin, A., 2014. Ancient signals: comparative genomics of green plant CDPKs. Trends Plant Sci. 19, 79–89.

Hamers, D., van Voorst, V.L., Borst, J.W., Goedhart, J., 2014. Development of FRET biosensors for mammalian and plant systems. Protoplasma 251, 333–347.

Hamill, O.P., Marty, A., Neher, E., Sakmann, B., Sigworth, F.J., 1981. Improved patch-clamp techniques for high-resolution current recording from cells and cell-free membrane patches. Pflügers Arch. 391, 85–100.

Hamilton, A.J., Baulcombe, D.C., 1999. A species of small antisense RNA in post-transcriptional gene silencing in plants. Science 286, 950–952.

Hamilton, A., Madison, J., Jay, J., 1961. The Federalist Papers. New American Library, New York.

Hamilton, E.S., Schlegel, A.M., Haswell, E.S., 2015. United in diversity: mechanosensitive ion channels in plants. Annu. Rev. Plant Biol. 66, 113–137.

Hamm, J., Darzynkiewicz, E., Tahara, S.L.M., Mattaj, I.W., 1990. The trimethylquanosine cap structure of U1 snRNA is a component of a bipartite nuclear targeting signal. Cell 62, 569–577.

Hamm, J., Mattaj, I.W., 1990. Monomethylated cap structures facilitate RNA export from the nucleus. Cell 63, 109–118.

Hamman, B.D., Chen, J.-C., Johnson, E.E., Johnson, A.E., 1997. The aqueous pore through the translocon has a diameter of 40-60 Å during cotranslational protein translocation at the ER membrane. Cell 89, 535–544.

Hämmerling, J., 1953. Nucleo-cytoplasmic relationships in the development of *Acetabularia*. Int. Rev. Cytol. 2, 475–498.

Hammond, C., Helenius, A., 1995. Quality control in the secretory pathway. Curr. Opin. Cell Biol. 7, 523–529.

Han, B., Chen, S., Dai, S., Yang, N., Wang, T., 2010. Isobaric tags for relative and absolute quantification-based comparative proteomics reveals the features of plasma membrane associated proteomes of pollen grains and pollen tubes from *Lilium davidii*. J. Integr. Plant Biol. 52, 1043–1058.

Han, B., Huang, X., 2013. Sequencing-bases genome-wide association study in rice. Curr. Opin. Plant Biol. 16, 133–138.

Han, X., Hyun, T.K., Zhang, M., Koh, E., Kang, B.H., Lucas, W.J., Kim, J.Y., 2014. Auxin-callose mediated plasmodesmal gating is essential for effective auxin gradient formation and signaling. Dev. Cell 28, 132–146.

Han, M.-., Jung, K.-., Yi, G., An, G., 2011. Rice *importin β1* gene affects pollen tube elongation. Mol. Cells 31, 523–530.

Hanaoka, H., Noda, T., Shirano, Y., Kato, T., Hayashi, H., Shibata, D., Tabata, S., Ohsumi, Y., 2002. Leaf senescence and starvation-induced chlorosis are accelerated by the disruption of an *Arabidopsis* autophagy gene. Plant Physiol. 129, 1181–1193.

Hanba, Y.T., Shibasaka, M., Hayashi, Y., Hayakawa, T., Kasamo, K., Terashima, I., Katsuhara, M., 2004. Overexpression of the barley aquaporin HvPIP2;1 increases internal CO_2 conductance and CO_2 assimilation in the leaves of transgenic rice plants. Plant Cell Physiol. 45, 521–529.

Hanczyc, M.M., Fujikawa, S.M., Szostak, J.W., 2003. Experimental model of primative cellular compartments: encapsulation, growth, and division. Science 302, 618–622.

Hanein, D., Matlack, K.E.S., Jungnickel, B., Plath, K., Kalies, K.-U., Miller, K.R., Rapoport, T.A., Akey, C.W., 1996. Oligomeric rings of the Sec61p complex induced by ligands required for protein translocation. Cell 87, 721–732.

Hangarter, R.P., Gest, H., 2004. Pictorial demonstrations of photosynthesis. Photosynth. Res. 80, 421–425.

Hannapel, D.J., Sharma, P., Lin, T., 2013. Phloem-mobile messenger RNAs and root development. Front. Plant Sci. 4, 257.

Hansen, J.C., Ausio, J., 1992. Chromatin dynamics and the modulation of genetic activity. Trends Biochem. Sci. 17, 187–191.

Hansen, A.K., Escobar, L.K., Gilbert, L.E., Janse, R.K., 2007. Paternal, maternal, and biparental inheritance of the chloroplast genome in *Passiflora* (Passifloraceae): implications for phylogenetic studies. Am. J. Bot. 94, 42–46.

Hanser, R., 1979. A noble treason: The revolt of the Munich students against Hitler. Putnam, New York.

Hanson, M.R., Hines, K.M., 2018. Stromules: probing formation and function. Plant Physiol. 176, 128–137.

Hanson, J., Huxley, H.E., 1953. Structural basis of the cross-striations in muscle. Nature 172, 530–532.

Hanson, J., Huxley, H.E., 1955. The structural basis of muscle contraction in striated muscle. Symp. Soc. Exp. Biol. 9, 228–264.

Hanson, M.R., Köhler, R.H., 2001. GFP imaging: methodology and application to investigate cellular compartmentation in plants. J. Exp. Bot. 52, 529–539.

Hanson, M.R., Lin, M.T., Carmo-Silva, A.E., Parry, M.A., 2016. Towards engineering carboxysomes into C3 plants. Plant J. 87, 38–50.

Hanson, M.R., Sattarzadeh, A., 2008. Dynamic morphology of plastids and stromules in angiosperm plants. Plant Cell Environ. 31, 646–657.

Hanson, M.R., Sattarzadeh, A., 2011. Stromules: recent insights into a long neglected feature of plastid morphology and function. Plant Physiol. 155, 1486–1492.

Hanson, M.R., Sattarzadeh, A., 2013. Trafficking of proteins through plastid stromules. Plant Cell 25, 2774–2782.

Hanstein, J., 1868. Ueber die Organe der Harz- und Schleim- Absonderung in den Laubknospen. Bot. Z. 26, 697, 721,745,769.

Hanstein, J. von, 1880. Das Protoplasma als Träger der pflanzlichen und thierischen Lebensverrichtungen. In: Frommel, W., Pfaff, F. (Eds.), Für Laien und Sachgenossen dargestellt. From Sammlung von Vorträgen für das deutsche Volk. Winter, Heidelberg.

Hanstein, J., von, 1870. Die Entwickelung des Keimes der Monokotylen und der Dikotylen. Bot. Abhandl. 1, 1–112.

Happel, N., Höning, S., Neuhaus, J.-M., Paris, N., Robinson, D.G., Holstein, S.E.H., 2004. Arabidopsis μA-adaptin interacts with the tyrosine motif of the vacuolar sorting receptor VSR-PS1. Plant J. 37, 678–693.

Haraguchi, T., Tominaga, M., Matsumoto, R., Sato, K., Nakano, A., Yamamoto, K., Ito, K., 2014. Molecular characterization and subcellular localization of *Arabidopsis* class VIII myosin, ATM1. J. Biol. Chem. 289, 12343–12355.

Harden, A., 1932. Alcoholic Fermentation. Longmans, Green and Co., London.

Harden, A., December 12, 1929. The function of phosphate in alcoholic fermentation. Nobel Lecture.

Harden, A., Young, W.J., 1906. The alcoholic ferment of yeast-juice. Proc. R. Soc. B77, 405–520.

Hardham, A.R., Gunning, B.E.S., 1979. Interpolation of microtubules into cortical arrays during cell elongation and differentiation in roots of *Azolla pinnata*. J. Cell Sci. 37, 411–442.

Hardham, A.R., Gunning, B.E.S., 1980. Some effects of colchicine on microtubules and cell division in roots of *Azolla pinnata*. Protoplasma 102, 31–51.

Hardin, P.E., Hall, J.C., Rosbash, M., 1990. Feedback of the *Drosophila* period gene product on circadian cycling of its messenger RNA levels. Nature 343, 536–540.

Harding, A., 2005. Sir Edwin Southern: scientist as problem solver. Lancet 366, 1919.

Hardy, W.B., 1899. On the structure of cell protoplasm. J. Physiol. 24, 158–210.

Hardy, W.B., 1906. The Physical Basis of Life. Sci. Prog. 1, 177–205.

Hardy, G.H., 1940. A Mathematician's Apology. Cambridge University Press, Cambridge.

Hardy, P.M., 1985. The protein amino acids. In: Barrett, G.C. (Ed.), Chemistry and Biochemistry of the Amino Acids. Chapman and Hall, London, pp. 6–24.

Hargreaves, A.J., Dawson, P.J., Butcher, G.W., Larkins, A., Goodbody, K.C., Lloyd, C.W., 1989a. A monoclonal antibody raised against cytoplasmic fibrillar bundles from carrot cells, and its cross-reaction with animal intermediate filaments. J. Cell Sci. 92, 371–378.

Hargreaves, A.J., Goodbody, K.C., Lloyd, C.W., 1989b. Reconstitution of intermediate filaments from a higher plant. Biochem. J. 261, 679–682.

Hargreaves, W.R., Mulvihill, S.J., Deamer, D.W., 1977. Synthesis of phospholipids and membranes in prebiotic conditions. Nature 266, 78–80.

Harholt, J., Suttangkakul, A., Scheller, H.V., 2010. Biosynthesis of pectin. Plant Physiol. 153, 384–395.

Haritatos, E., Keller, F., Turgeon, R., 1996. Raffinose oligosaccharide concentrations measured in individual cells and tissue types in *Cucumis melo* L. leaves: implications for phloem loading. Planta 198, 614–622.

Harlan, J.R., 1992. Crops and man, 2nsecond ed. American Society of Agronomy and Crop Science Society of America, Madison, WI.

Harley, S.M., Beevers, L., 1989. Isolation and partial characterization of clathrin-coated vesicles from pea (*Pisum sativum* L.) cotyledons. Protoplasma 150, 103–109.

Harman, O.S., 2004. The Man Who Invented the Chromosome. Harvard University Press, Cambridge, MA.

Harman, J.W., Feigelson, M., 1952a. Studies on mitochondria. III. The relationship of structure and function of mitochondria from heart muscle. Exp. Cell Res. 3, 47–58.

Harman, J.W., Feigelson, M., 1952b. Studies on mitochondria. V. The relationship of structure and oxidative phosphorylation in mitochnodra of heart muscle. Exp. Cell Res. 3, 509–525.

Harmon, A.C., Putnam-Evans, C., Cormier, M.J., 1987. A calcium-dependent but calmodulin-independent protein kinase from soybean. Plant Physiol. 83, 830–837.

Harnett, M.M., Pineda, M.A., Latré de Laté, P., Eason, R.J., Besteiro, S., Harnett, W., Langsley, G., 2017. From Christian de Duve to Yoshinori Ohsumi: more to autophagy than just dining at home. Biomed. J. 40, 9–22.

Harold, F.M., 1986. The Vital Force: A Study of Bioenergetics. W.H. Freeman and Co., New York.

Harold, F.M., 1990. To shape a cell: an inquiry into the causes of morphogenesis of microorganisms. Microbiol. Rev. 54, 381–431.

Harold, F.M., 2001. The Way of the Cell. Molecules. Organisms and the Order of Life. Oxford University Press, Oxford.

Harold, F.M., 2014. In: Search of Cell History. The Evolution of Life's Building Blocks. University of Chicago Press, Chicago.

Harper, R.A., 1919. The structure of protoplasm. Am. J. Bot. 6, 273–300.

Harper, J.F., Manney, L., DeWitt, N.D., Yoo, M.H., Sussman, M.R., 1990. The *Arabidopsis thaliana* plasma membrane H^+-ATPase multigene family. J. Biol. Chem. 265, 13601–13608.

Harper, J.F., Surowy, T.K., Sussman, M.R., 1989. Molecular cloning and sequence of cDNA encoding the plasma membrane proton pump (H^+-ATPase) of *Arabidopsis thaliana*. Proc. Natl. Acad. Sci. U.S.A. 86, 1234–1238.

Harris, B., 1980. The mechanical behavior of composite materials. Soc. Exp. Biol. 34, 37–74.

Harris, B., 1980a. The mechanical behavior of composite materials. Soc. Exp. Biol. 34, 37–74.

Harris, H., 1980b. Joachim Hämmerling. 9 March 1901-5 August 1980. Biogr. Mem. Fellows R. Soc. 28, 110–124.

Harris, N., 1986. Organization of the endomembrane system. Ann. Rev. Plant Physiol. 37, 73–92.

Harris, P.J., 1988. Cell walls. In: Hall, J.L., Moore, A.L. (Eds.), Isolation of Membranes and Organelles from Plant Cells. Academic Press, London, pp. 25–53.

Harris, R., 2017. Rigor Mortis: How Sloppy Science Creates Worthless Cures, Crushes Hope, and Wastes Billions. Basic Books, New York.

Harris, P., Bajer, A., 1965. Fine structure studies on mitosis in endosperm metaphase of Haemanthus katherinea Bak. Chromosoma 16, 624–636.

Harris, N., Oparka, K.J., 1983. Connections between dictyosomes, ER and GERL in cotyledons of mung bean (*Vigna radiata* L.). Protoplasma 114, 93–102.

Harrison, E.R., 1981. Cosmology. The Science of the Universe. Cambridge University Press, Cambridge.

Harrison, C.J., 2015. Shooting through time: new insights from transcriptomic data. Trends Plant Sci. 20, 468–470.

Harrisson, C.M.H., Page, B.M., Keir, H.M., 1976. Mescaline as a mitotic spindle inhibitor. Nature 260, 138–139.

Harrop, L.C., Kornberg, H.L., 1966. The role of isocitrate lyase in the metabolism of algae. Proc. R. Soc. Lond. 166 B, 11–29.

Harryson, P., Morré, D.J., Sandelius, A.S., 1996. Cell-free transfer of phosphatidylinositol between membrane fractions isolated from soybean. Plant Physiol. 110, 631–637.

Hartig, T., 1854. Über des verholten des Zellkerns bei der Zellentheilung. Bot. Z. 12, 893–902.

Hartl, F.U., Martin, J., Neupert, W., 1992. Protein folding in the cell: the role of molecular chaperones Hsp 70 and Hsp 60. Ann. Rev. Biophys. Biomol. Struct. 21, 293–322.

Hartley, H., 1951. Origin of the word 'protein'. Nature 168, 244.

Hartman, M.A., Spudich, J.A., 2012. The myosin superfamily at a glance. J. Cell Sci. 125, 1627–1632.

Hartmann, M.-A., 1998. Plant sterols and the membrane environment. Trends Plant Sci. 3, 170–175.

Hartmann, M.-A., Benveniste, P., 1987. Plant membrane sterols: isolation, identification and biosynthesis. Methods Enzymol. 148, 632–650.

Hartwell, L.H., December 9, 2001. Yeast and Cancer. Nobel Lecture.

Hartwell, L.H., Goldberg, M.L., Fischer, J.A., Hood, L., Aquadro, C.F., 2015. Genetics: From Genes to Genomes, Fififth ed. McGraw Hill, New York.

Hartwell, L.H., Hopfield, J.J., Leibler, S., Murray, A.W., 1999. From molecular to modular cell biology. Nature 402, C47–C52.

Hartwell, L.H., Smith, D., 1985. Altered fidelity of mitotic chromosome transmission in cell cycle mutants of *S. cerevisiae*. Genetics 110, 381–395.

Harvey, E.N., 1952. Forward. In: Davson, H., Danielli, J.F. (Eds.), The Permeability of Natural Membranes. Cambridge University Press, Cambridge, p. vii.

Harvey, H.J., Venis, M.A., Trewavas, A.J., 1989a. The detection and purification of a verapmil-binding protein from plant tissue. Ann. N.Y. Acad. Sci. 560, 56–58.

Harvey, H.J., Venis, M.A., Trewavas, A.J., 1989b. Partial purification of protein from maize (Zea mays) coleoptile membranes binding the Ca^{2+}-channel antagonist verapamil. Biochem. J. 257, 95–100.

Hasegawa, Y., Nakamura, S., Katizoe, S., Sato, M., Nakamura, N., 1998. Immunocytochemical and chemical analysis of Golgi vesicles isolated from germinated pollen of *Camellia japonica*. J. Plant Res. 111, 421–429.

Hasezawa, S., Ueda, K., Kumagai, F., 2000. Time-sequence observations of microtubule dynamics throughout mitosis in living cell suspensions of stable transgenic *Arabidopsis*: direct evidence for the origin of cortical microtubules at M/G(1) interface. Plant Cell Physiol. 41, 244–250.

Hashimoto, T., 2011. Microtubule and cell shape determination. The Plant Cytoskeleton. Adv. Plant Biol. 2, 245–257.

Hashimoto, K., Igarashi, H., Mano, S., Takenaka, C., Shiina, T., Yamaguchi, M., Demura, T., Nishimura, M., Shimmen, T., Yokota, E., 2008. An isoform of *Arabidopsis* myosin XI interacts with small GTPases in its C-terminal tail region. J. Exp. Bot. 59, 3523–3531.

Hashimoto, K., Igarashi, H., Nishimura, M., Shimmen, T., Yokota, E., 2005. Peroxisomal localization of a myosin XI isoform in *Arabidopsis thaliana*. Plant Cell Physiol. 46, 782–789.

Hashimshony, T., Wagner, F., Sher, N., Yanai, I., 2012. CEL-Seq: single-cell RNA-Seq by multiplexed linear amplification. Cell Rep. 2, 666–673.

Haskell, G., Wills, A.B., 1968. Primer of Chromosome Practice. Oliver and Boyd, Edinburgh.

Hasson, T., Mooseker, M.S., 1995. Molecular motors, membrane movements and physiology: emerging roles for myosins. Curr. Opin. Cell Biol. 7, 587–594.

Hastings, J.W., Sweeney, B.M., 1957. On the mechanism of temperature independence in a biological clock. Proc. Natl. Acad. Sci. U.S.A. 43, 804–811.

Haswell, E.S., 2007. MscS-like proteins in plants. Curr. Topics Membr. 58, 329–359.

Haswell, E.S., Meyerowitz, E.M., 2006. MscS-like proteins control plastid size and shape in *Arabidopsis thaliana*. Curr. Biol. 16 (16), 1–11.

Haswell, E.S., Peyronnet, R., Barbier-Brygoo, H., Meyerowitz, E.M., Frachisse, J.-M., 2008. Two MscS homologs provide mechanosensitive channel activites in the *Arabidopsis* root. Curr. Biol. 18, 730–734.

Hatano, S., Tazawa, M., 1968. Isolation, purification and characterization of myosin B from myxomycete plasmodium. Biochim. Biophys. Acta 154, 507–519.

Hatch, M.D., 2002. C4 photosynthesis: discovery and resolution. Photosynth. Res. 73, 251–256.

Hatch, M.D., Slack, C.R., 1966. Photosynthesis by sugar-cane leaves: a new carboxylation reaction and the pathway of sugar formation. Biochem. J. 101, 103–111.

Hatch, M.D., Slack, C.R., 1998. C4 photosynthesis: discovery, resoltuion, recognition, and significance. Discov. Plant Biol. 1, 175–196.

Hato, M., 1979. Membrane phenomena in chemotaxis in true slime mold *Physarum polycephalum* (II). Colloid Polymer Sci. 257, 745–747.

Haugh, J.M., 2012. Live-cell fluorescence microscopy with molecular biosensors: what are we really measuring? Biophys. J. 102, 2003–2011.

Haupt, W., 1960. Die Chloroplastendrehung bei *Mougeotia*. II. Die Induktion der Schwachlichtbewegung durch linear polarisiertes Licht. Planta 55, 465–479.

Haupt, W., 1965. Perception of environmental stimuli orienting growth and movement in lower plants. Ann. Rev. Plant Physiol. 16, 267–290.

Haupt, W., 1968. Die Orientierdur Phytochrom-Moleküle in der Mougeotia Zelle: Ein neues Modell zue Deutung der experimentellen Befunde. Z. Pflanzenphysiol. 58, 331–346.

Haupt, W., 1970. Über den Dichroismus von Phytochrom660 und Phytochrom730 bie Mougeotia. Z. Pflanzenphysiol. 62, 287–298.

Haupt, W., 1982. Light-mediated movement of chloroplasts. Ann. Rev. Plant Physiol. 33, 204–233.

Haupt, W., 1983a. Movement of chloroplasts under the control of light. Prog. Phycol. Res. 2, 227–281.

Haupt, W., 1983b. The perception of light direction and orientation responses in chloroplasts. Soc. Exp. Biol. Symp. 36, 423–442.

Haupt, W., 1999. Chloroplast movement: from phenomenology to molecular biology. Prog. Bot. 60, 3–36.

Haupt, W., Mörtel, G., Winkelnkemper, I., 1969. Demonstration of different dichroic orientation of phytochrome PR and PFR. Planta 88, 183–186.

Haupt, W., Scheuerlein, R., 1990. Chloroplast movement. Plant Cell Environ. 13, 595–614.

Hauri, H.P., Schweizer, A., 1992. The endoplasmic reticulum Golgi intermediate compartment. Curr. Opin. Cell Biol. 4, 600–608.

Havananda, T., Brummer, E.C., Doyle, J.J., 2011. Complex patterns of autopolyploid evolution in alfalfa and allies (Medicago sativa: Leguminosae). Am. J. Bot. 98, 1633–1646.

Havaux, M., 1992. Photoacoustic measurements of cyclic election flow around photosystem I in leaves adapted to light-states 1 and 2. Plant Cell Physiol. 33, 799–803.

Hawes, C.R., Juniper, B.E., Horne, J.C., 1983. Electron microscopy of resin-free sections of plant cells. Protoplasma 115, 88–93.

Hawes, C., Kiviniemi, P., Keichbaumer, V., 2015. The endoplasmic reticulum: a dynamic and well connected organelle. J. Integr. Biol. 57, 50–62.

Hawes, C., Saint-Jore, C., Martin, B., Zheng, H.-Q., 2001. ER confirmed as the location of mystery organelles in *Arabidopsis* plants expressing GFP! Trends Plant Sci. 6, 245–246.

Hawkins, E.K., 1974. Golgi vesicles of uncommon morphology and wall formation in the red alga, *Polysiphonia*. Protoplasma 80, 1–14.

Haworth, P., 1983. Protein phosphorylation-induced State I-State II transitions are dependent on thylakoid membrane microviscosity. Arch. Biochem. Biophys. 226, 145–154.

Hay, E.D., 1983. Cell and extracellular matrix: their organization and mutual dependence. Mod. Cell Biol. 2, 509–548.

Hayakawa, S.I., 1941. Language in Action. Harcourt, Brace and Co, New York.

Hayashi, T., 1989. Xyloglucans in the primary cell wall. Ann. Rev. Plant Physiol. Plant Mol. Biol. 40, 139–168.

Hayashi, T., 1991. Biochemistry of xyloglucans in regulating cell elongation and expansion. In: Lloyd, C.W. (Ed.), The Cytoskeletal Basis of Plant Growth and Form. Academic Press, London, pp. 131–144.

Hayashi, Y., Hayashi, M., Hayashi, H., Hara-Nishimura, I., Nishimura, M., 2001. Direct interaction between glyoxysomes and lipid bodies in cotyledons of the *Arabidopsis thaliana* ped1 mutant. Protoplasma 218, 83–94.

Hayashi, T., Matsuda, K., 1981. Biosynthesis of xyloglucan in suspension-cultured soybean cells. Evidence that the enzyme system of xyloglucan synthesis does not contain ß-1,4-glucan 4-ß-D-glucosyltransferase activity (EC 2.4.1.12). Plant Cell Physiol. 22, 1571–1584.

Hayashi, M., Nito, K., Takei-Hoshi, R., Yagi, M., Kondo, M., Suenaga, A., Yamaya, T., Nishimura, M., 2002. Ped3p is a peroxisomal ATP-binding cassette transporter that might supply substrates for fatty acid β-oxidation. Plant Cell Physiol. 43, 1–11.

Hayashi, T., Sano, T., Kutsuna, N., Kumagai-Sano, F., Hasezawa, S., 2007. Contribution of anaphase B to chromosome separation in higher plant cells estimated by image processing. Plant Cell Physiol. 48, 1509–1513.

Hayden, E.C., 2014. The $1,000 genome. Nature 507, 294–295.

Haydon, J.H., Bowser, S.S., Rieder, C.L., 1990a. Kinetochores capture astral microtubules during chromosome attachment to the mitotic spindle: direct visualization in live newt lung cells. J. Cell Biol. 111, 1039–1045.

Haydon, S.M., Wolenski, J.S., Mooseker, M.S., 1990b. Binding of brush border myosin I to phospholipid vesicles. J. Cell Biol. 111, 443–451.

Hayes, J.J., Hansen, J.C., 2002. Commentary. New insights into unwrapping DNA from the nucleosome from a single-molecule optical tweezers method. Proc. Natl. Acad. Sci. U.S.A. 99, 1752–1754.

Hayes, J.J., Wolffe, A.P., 1992. The interaction of transcription factors with nucleosomal DNA. BioEssays 14, 597–603.

Hays, T.S., Salmon, E.D., 1990. Poleward force at the kinetochore in metaphase depends on the number of kinetochore microtubules. J. Cell Biol. 110, 391–404.

Haywood, V., Kragler, F., Lucas, W.J., 2002. Plasmodesmata: pathways for protein and ribonucleoprotein signaling. Plant Cell 14 (Suppl), S303–S325.

He, Z.H., Cheeseman, I., He, D., Kohorn, B.D., 1999. A cluster of five cell wall associated receptor kinase genes, Wak1–5, are expressed in specific organs of Arabidopsis. Plant Mol. Biol. 39, 1189–1196.

He, Z.H., Fujiki, M., Kohorn, B.D., 1996. A cell wall-associated, receptor-like protein kinase. J. Biol. Chem. 271, 19789–19793.

He, Z.H., He, D., Kohorn, B.D., 1998. Requirement for the induced expression of a cell wall associated receptor kinase for survival during the pathogen response. Plant J. 14, 55–63.

He, C.Y., Shaw, M.K., Pletcher, C.H., Striepen, B., Tilney, L.G., Roos, D.S., 2001. A plastid segregation defect in the protozoan parasite *Toxoplasma gondii*. EMBO J. 20, 330–339.

Heald, R., Tournebize, R., Blank, T., Sandaltzopoulos, R., Becker, P., Hyman, A., Karsenti, E., 1996. Self-organization of microtubules into bipolar spindles around artificial chromosomes in *Xenopus* egg extracts. Nature 382, 420–425.

Heard, W., Sklenář, J., Tomé, D.F.A., Robatzek, S., Jones, A.M.E., 2015. Identification of regulatory and cargo proteins of endosomal and secretory pathways in *Arabidopsis thaliana* by proteomic dissection. Mol. Cell. Proteom. 14, 1796–1813.

Heath, T.L., 1897. The Works of Archimedes. Cambridge University Press, Cambridge.

Heath, I.B., 1981. Mitosis through the electron microscope. In: Zimmerman, A.M., Forer, A. (Eds.), Mitosis/Cytokinesis. Academic Press, New York, pp. 245–275.

Heath, I.B., 1990. The roles of actin in tip growth of fungi. Int. Rev. Cytol. 123, 95–127.

Heather, J.M., Chain, B., 2017. The sequence of sequencers: the history of sequencing DNA. Genomics 107, 1–8.

Heazlewood, J.L., Miller, A.H., 2006. Plant proteomics: challenges and resources. Annu. Plant Rev. 28, 1–31.

Heber, U., Heldt, H.W., 1981. The chloroplast envelope: structure, function, and role in leaf metabolism. Ann. Rev. Plant Physiol. 32, 139–168.

Heber, U., Wagner, V., Kaiser, W., Neimanis, S., Bailey, K., Walker, D., 1994. Fast cytoplasmic pH regulation in acid-stressed leaves. Plant Cell Physiol. 35, 479–488.

Hecht, K., 1912. Studien über den vorgang der plasmolyse. Beiträge zur Biologie der Pflanzen 11, 133–145.

Hedgpeth, J., Goodman, H.M., Boyer, H.W., 1972. DNA nucleotide sequence restricted by the RI endonuclease. Proc. Natl. Acad. Sci. U.S.A. 69, 3448–3452.

Hedrich, R., 1995. Technical approaches to studying specific properties of ion channels in plants. In: Sakmann, B., Neher, E. (Eds.), Single-Channel Recording, second ed. Plenum Press, New York, pp. 277–305.

Hedrich, R., Flügge, U.I., Fernandez, J.M., 1986. Patch-clamp studies of ion transport in isolated plant vacuoles. FEBS Lett. 204, 228–232.

Hedrich, R., Kurkdjian, A., 1988. Characterization of an anion-permeable channel from sugar beet vacuoles: effect of inhibitors. EMBO J. 7, 3661–3666.

Hedrich, R., Kurkdjian, A., Guern, J., Flügge, U.I., 1989. Comparative studies on the electrical properties of the H^+ translocating ATPase and phyrophosphatase of the vacuolar- lysosomal compartment. EMBO J. 8, 2835–2841.

Hedrich, R., Marten, I., 2006. 30-year progress of membrane transport in plants. Planta 224, 725–739.

Hedrich, R., Marten, I., 2011. TPC1-SV channels gain shape. Mol. Plant 4, 428–441.

Hedrich, R., Neher, E., 1987. Cytoplasmic calcium regulates voltage dependent ion channels in plant vacuoles. Nature 329, 833–835.

Hegde, R.S., Mastrianni, J.A., Scott, M.R., DeFea, K.A., Tremblay, P., Torchia, M., DeArmond, S.J., Prusiner, S.B., Lingappa, V.R., 1998. A transmembrane form of the prion protein in neurode- generative disease. Science 279, 827–834.

Hegemann, P., 2008. Algal sensory photoreceptors. Annu. Rev. Plant Biol. 59, 167–189.

Hegemann, P., Tsunoda, S., 2007. Light tools for neuroscience: channelrhodopsin and light-activated enzymes. Cell Sci. Rev. 3, 108–123.

Heidegger, M., 2010. Being and Time. Translated by Joan Stambaugh, revised by Dennis J. Schmidt. SUNY Press, Albany, NY.

Heidenhain, R., 1878. Ueber secretorische und trophische Drüsennerven. Pflüger's Archiv für die gesamte Physiologie des Menschen und der Tiere 17, 1–67.

Heidenhain, M., 1907. Plasma und Zelle. Bd. 1. Fischer, Jena.

Heidenhain, M., 1911. Plasma und Zelle. Bd. 2. Fischer, Jena.

Heilbron, J.L., 1986. The Dilemmas of an Upright Man. Max Planck as Spokesman for German Science. University of California Press, Berkeley, CA.

Heilbronn, A., 1914. Zustand des Plasmas und Reizbarkeit. Ein Beitrag zur Physiologie der lebenden Substanz. Jahr. Wiss. Bot. 54, 357–390.

Heilbrunn, L.V., 1937. An Outline of General Physiology. Saunders, Philadelphia, PA.

Heilbrunn, L.V., 1943. An Outline of General Physiology, Sesecond ed. Saunders, Philadelphia, PA.

Heilbrunn, L.V., 1952. An Outline of General Physiology, Ththird ed. Saunders, Philadelphia, PA.

Heilbrunn, L.V., 1956. The Dynamics of Living Protoplasm. Academic Press, New York.

Heilbrunn, L.V., 1958. The Viscosity of Protoplasm. Protoplasmatologia Bd II. Springer-Verlag, Wein.

Heilbrunn, L.V., Young, R.A., 1930. The activation of ultra-violet rays on *Arbacia* egg protoplasm. Physiol. Zool. 3, 330–341.

Hein, N., Griesbeck, O., 2004. Genetically encoded indicators of cellular calcium dynamics based on troponin C and green fluorescent protein. J. Biol. Chem. 279, 14280–14286.

Heinhorst, S., Cannon, G.C., Weissbach, A., 1990. Chloroplast and mitochondrial DNA polymerases from cultured coybean cells. Plant Physiol. 92, 939–945.

Heinlein, M., 2005. Systemic RNA silencing. In: Oparka, K.J. (Ed.), Plasmodesmata. Blackwell Publishing, Oxford, pp. 212–240.

Heinlein, M., 2008. Microtubules and viral movement. Plant Cell Monogr. 11, 141–173.

Heinlein, M., 2015. Plasmodesmata. Methods and Protocols. Methods in Plant Biology 1217. Springer, New York.

Heinlein, M., Epel, B.L., 2004. Macromolecular transport and signaling through plasmodesmata. Int. Rev. Cytol. 235, 93–164.

Heinz, E., Roughan, P.G., 1983. Similarities and differences in lipid metabolism of chloroplasts isolated from 18:3 and 16:3 plants. Plant Physiol. 72, 273–279.

Heitler, W., 1963. Man and Science. Oliver and Boyd, Edinburgh.

Heitz, E., 1936a. Untersuchung über den Bau der Plastiden. Planta 26, 134–163.

Heitz, E., 1936b. Gerichtete Chlorophyllscheiben als strukturelle Assimilationseinheiten der Chloroplasten. Ber. Deutsch. Bot. Ges. 54, 362–368.

Heitz, E., 1937. Untersuchungen uiber den Bau der Plas tiden. Planta 26, 134–163.

Helariutta, Y., Fukaki, H., Wysocka-Diller, J., Nakajima, K., Jung, J., Sena, G., Hauser, M.-T., Benfey, P.N., 2000. The *SHORT-ROOT* gene controls radial patterning of the *Arabidopsis* root through radial signaling. Cell 101, 555–567.

Held, M.A., Be, E., Zemelis, S., Withers, S., Wilkerson, C., Brandizzi, F., 2011. CGR3: a Golgi-localized protein influencing homogalacturonan methylesterification. Mol. Plant 5, 832–844.

Held, M.A., Boulaflous, A., Brandizzi, F., 2008. Advances in fluorescent protein-based imaging for the analysis of plant endomembranes. Plant Physiol. 147, 1469–1481.

Hemleben, V., Ermisch, N., Kimmich, D., Leber, B., Peter, G., 1975. Studies on the fate of homologous DNA applied to seedlings of *Matthiola incana*. Eur. J. Biochem. 56, 403–411.

Hemmersbch, R., Braun, M., 2006. Gravity-sensing and gravity- related signaling pathways in unicellular model systems of protests and plants. Signal Transduct. 6, 432–442.

Henderson, L.J., 1908. Concerning the relationship between the strength of acids and their capacity to preserve neutrality. Am. J. Physiol. 21, 173–179.

Henderson, L.J., 1913. The Fitness of the Environment. MacMillan, New York.

Henderson, L.J., 1917. The Order of Nature. An Essay. Harvard University Press, Cambridge, MA.

Henderson, R., 2017. From Electron Crystallography to Single Particle cryoEM. https://www.nobelprize.org/prizes/chemistry/2017/henderson/lecture/.

Henderson, S.A., Koch, C.A., 1970. Co-orientation stability by physical tension: a demonstration with experimentally interlocked bivalents. Chromosoma 29, 207–216.

Hendrickson, W.G., Kusano, T., Yamaki, H., Balakrishnan, R., King, M., Murchie, J., Schaechter, M., 1982. Binding of the origin of replication of *Escherichia coli* to the outer membrane. Cell 30, 915–923.

Hendrix, K.W., Assefa, H., Boss, W.F., 1989. The polyphosphoinositides, phosphatidylinositol monophosphate and phosphatidylinositol bisphosphate, are present in nuclei isolated from carrot protoplast. Protoplasma 151, 62–72.

Heneen, W.K., 2014. Kinetochore structure and chromosome orientation: a tribute to Gunner östergren. Hereditas 151, 115–118.

Heneen, W.K., Czajkowski, J., 1980. Scanning electron microscopy of the intact mitotic apparatus. Biol. Cell. 37, 13–22.

Henriques, R., Jásik, J., Klein, M., Martinoia, E., Feller, U., Schell, J., Pais, M.S., Koncz, C., 2002. Knock-out of Arabidopsis metal transporter gene *IRAT1* results in iron deficiency accompanied by cell differentiation effects. Plant Mol. Biol. 50, 587–597.

Henry, C.A., Jordan, J.R., Kropf, D.L., 1996. Localized membrane- wall adhesions in *Pelvetia* zygotes. Protoplasma 190, 39–52.

Henry, M.-F., Nyns, E.-J., 1975. Cyanide-insensitive respiration. An alternative mitochondrial pathway. Sub-Cell. Biochem. 4, 1–65.

Hensel, W., 1987. Cytodifferentiation of polar plant cells: formation and turnover of endoplasmic reticulum in root statocytes. Exp. Cell Res. 172, 377–384.

Hepler, P.K., 1976. The blepharoplast of *Marsilea*: its *de novo* formation and spindle association. J. Cell Sci. 21, 361–390.

Hepler, P.K., 1980. Membranes in the mitotic apparatus of barley cells. J. Cell Biol. 86, 490–499.

Hepler, P.K., 1981. The structure of the endoplasmic reticulum revealed by osmium tetroxide-potassium ferricyanide staining. Eur. J. Cell Biol. 26, 102–110.

Hepler, P.K., 1982. Endoplasmic reticulum in the formation of the cell plate and plasmodesmata. Protoplasma 111, 121–133.

Hepler, P.K., 1985. Calcium restriction prolongs metaphase in dividing *Tradescantia* stamen hair cells. J. Cell. Biol. 100, 1363–1368.

Hepler, P.K., 1989a. Calcium transients during mitosis: observations in flux. J. Cell Biol. 109, 2567–2573.

Hepler, P.K., 1989b. Membranes in the mitotic apparatus. In: Hyams, J.S., Brinkley, B.R. (Eds.), Mitosis. Molecules and Mechanisms. Academic Press, London, pp. 241–271.

Hepler, P.K., 1992. Calcium and mitosis. Int. Rev. Cytol. 138, 239–268.

Hepler, P.K., 1994. The role of calcium in cell division. Cell Calcium 16, 322–330.

Hepler, P.K., 2005. Calcium: a central regulator of plant growth and development. Plant Cell 17, 2142–2155.

Hepler, P.K., Bonsignore, L., 1990. Caffeine inhibition of cytokinesis: ultrastructure of cell plate formation/degradation. Protoplasma 157, 182–192.

Hepler, P.K., Bosch, M., Cardenas, L., Lovy-Wheeler, A., McKenna, S.T., Wilsen, K.L., Kunkel, J.G., 2005. Oscillatory pollen tube growth: imaging the underlying structures and physiological processes. Microsc. Microanal. 11 (Suppl. 2), 148–149.

Hepler, P.K., Fosket, D.E., 1971. The role of microtubules in vessel member differentiation in *Coleus*. Protoplasma 72, 213–236.

Hepler, P.K., Jackson, W.T., 1968. Microtubules and early stages of cellplate formation in the endosperm of *Haemanthus Katherinae* Baker. J. Cell Biol. 38, 437–446.

Hepler, P.K., Jackson, W.T., 1969. Isopropyl N-phenylcarbamate affects spindle microtubule orientation in dividing endosperm cells of Haemanthus Katherinae. Baker. J. Cell Sci. 5, 727–743.

Hepler, P.K., McIntosh, J.R., Cleland, S., 1970. Intermicrotubule bridges in mitotic spindle apparatus. J. Cell Biol. 45, 438–444.

Hepler, P.K., Newcomb, E.H., 1964. Microtubules and fibrils in the cytoplasm of *Coleus* cells undergoing secondary wall deposition. J. Cell Biol. 2, 529–533.

Hepler, P.K., Newcomb, E.H., 1967. Fine structure of cell plate formation in the apical meristem of *Phaseolus* roots. J. Ultrastruc. Res. 19, 498–513.

Hepler, P.K., Palevitz, B.A., 1986. Metabolic inhibitors block Anaphase A in vivo. J. Cell Biol. 102, 1995–2005.

Hepler, P.K., Palevitz, B.A., Lancelle, S.A., McCauley, M.M., Lichtschidl, I., 1990. Cortical endoplasmic reticulum in plants. J. Cell Sci. 96, 355–373.

Hepler, P.K., Pickett-Heaps, J.D., Gunning, B.E.S., 2013. Some retrospectives on early studies of plant microtubules. Plant J. 75 (2), 189–201.

Hepler, P.K., Rounds, C.M., Winship, L.J., 2013a. Control of cell wall extensibility during pollen tube growth. Mol. Plant 6, 998–1017.

Hepler, P.K., Valster, A., Molchan, T., Vos, J.W., 2002. Roles for kinesin and myosin during cytokinesis. Philos. Trans. R. Soc. Lond 357B, 761–766.

Hepler, P.K., Vidali, L., Cheung, A.Y., 2001. Polarized cell growth in plants. Annu. Rev. Cell Dev. Biol. 17, 159–187.

Hepler, P.K., Wayne, R.O., 1985. Calcium and plant development. Ann. Rev. Plant Physiol. 36, 397–439.

Hepler, P.K., Winship, L.J., 2010. Calcium at the cell wall-cytoplast interface. J. Integr. Plant Biol. 52, 147–160.

Hepler, P.K., Winship, L.J., 2015. The pollen tube clear zone: clues to the mechanism of polarized growth. J. Integr. Plant Biol. 57, 79–92.

Hepler, P.K., Wolniak, S.M., 1983. Membranous compartments and ionic transients in the mitotic apparatus. Mod. Cell Biol. 2, 93–112.

Hepler, P.K., Wolniak, S.M., 1984. Membranes in the mitotic apparatus: their structure and function. Int. Rev. Cytol. 90, 169–238.

Heppel, L.A., 2004. Reminiscences of Leon A. Heppel. J. Biol. Chem. 279, 52807–52811.

Herman, E.M., 1988. Immunocytochemical localization of macromolecules with the electron microscope. Ann. Rev. Plant Physiol. 39, 139–155.

Herman, E.M., 2008. Endoplasmic reticulum bodies: solving the insoluble. Curr. Opin. Plant Biol. 11, 672–679.

Herman, E.M., Baumgartner, B., Chrispeels, M.J., 1981. Uptake and apparent digestion of cytoplasmic organelles by protein bodies. Eur. J. Cell Biol. 24, 226–235.

Herman, E.M., Lamb, C.J., 1992. Arabinogalactan-rich glycoproteins are localized on the cell surface and in intravacuolar multivesicular bodies. Plant Physiol. 98, 264–272.

Herman, E.M., Li, X., Su, R.T., Larsen, P., Hsu, H.T., Sze, H., 1994. Vacuolar-type H^+-ATPases are associated with the endoplasmic reticulum and provacuoles of root tip cells. Plant Physiol. 106, 1313–1324.

Herman, E.M., Shannon, L.M., 1984a. Immunocytochemical evidence for the involvement of Golgi apparatus in the deposition of seed lectin of Bauhinia purpurea (Leguminosae). Protoplasma 121, 163–170.

Herman, E.M., Shannon, L.M., 1984b. Immunocytochemical localization of concanavalin A in developing jack bean cotyledons. Planta 161, 97–104.

Herman, E.M., Shannon, L.M., 1985. Accumulation and subcellular localization of α-galactosidase in developing soybean cotyledons. Plant Physiol. 77, 886–890.

Herman, E.M., Tague, B.W., Hoffman, L.M., Kjemtrup, S.E., Chrispeels, M.J., 1990. Retention of phytohemagglutinin with carboxyterminal tetrapeptide KDEL in the nuclear envelope and the endoplasmic reticulum. Planta 182, 305–312.

Hernandez, P., Martis, M., Dorado, G., Pfeifer, M., Galvez, S., Schaaf, S., Jouve, N., Simkova, H., Valarik, M., Dolezel, J., et al., 2012. Next-generation sequencing and syntenic integration of flow-sorted arms of wheat chromosome 4A exposes the chromosome structure and gene content. Plant J. 69, 377–386.

Hernandez-Verdun, D., 1991. The nucleolus today. J. Cell Sci. 99, 465–471.

Herschbach, D.R., December 8, 1986. Molecular Dynamics of Elementary Chemical Reactions. Nobel Lecture.

Hershey, A.D., Burgi, E., Ingraham, L., 1963. Cohesion of DNA molecules isolated from phage lambda. Proc. Natl. Acad. Sci. U.S.A. 49, 748–755.

Hershey, A.D., Chase, M., 1952. Independent functions of viral protein and nucleic acid in growth of bacteriophage. J. Gen. Physiol. 36, 39–56.

Hershey, A.D., December 12, 1969. Idiosyncrasies of DNA Structure. Nobel Lecture.

Hershko, A., December 8, 2004. The ubiquitin system for protein degradation and some of its roles in the control of the cell division cycle. In: Nobel Lecture.

Herskowitz, J., Herslowitz, I., Cuddihy, J.G., 1993. Double talking helix blues. In: Book and Tape Forms. Cold Spring Harbor Laboratory Press, Plainview, New York.

Hertel, R., Leopold, A.C., 1963. Versuche zur Analyse des Auxintransports in der Koleoptile von Zea mays L. Planta 59, 535–562.

Herth, W., 1983. Arrays of plasmamembrane "rosettes" involved in cellulose microfibril formation of Spirogyra. Planta 159, 347–356.

Herth, W., 1984. Oriented "rosette" alignment during cellulose formation in mung bean hypocotyl. Naturwissenschaften 71, 216–217.

Herth, W., 1985. Plasmamembrane rosettes involved in localized wall thickening during xylem vessel formation of Lepidium sativum. L. Planta 164, 12–21.

Herth, W., Weber, G., 1984. Occurrence of the putative cellulose synthesizing "rosettes" in the plasma membrane of Glycine max suspension culture cells. Naturwissenschaften 71, 153–154.

Hertwig, O., 1895. The Cell. Outlines of General Anatomy and Physiology. Translated by M. Campbell. Swan Sonnenscein & Co., London.

Hertz, G., December 11, 1926. The results of the electron-impact tests in the light of Bohr's theory of atoms. Nobel Lecture.

Hervé, C., Marcus, S.E., Knox, J.P., 2011. Monoclonal antibodies, carbohydrate-binding modules, and the detection of polysaccharides in plant cell walls. Methods Mol. Biol. 715, 103–113.

Herzog, R.O., Jancke, W., 1920. Über den physikalischen Aufbau einiger hochmolekularer organischer Verbindungen. Ber. Dtsch. Chem. Ges. 53, 2162–2164.

Herzog, R.O., Jancke, W., 1921. Verwendung von Röntgenstrahlen zur Untersuchung metamikroskopischer biologischer Strukturen. In: Neuberg, C. (Ed.), Festschrift der Kaiser Wilhelm Gesellschaft zur Förderung der Wissenschaften zu ihrem Zehnjährigen Jubiläum Dargebracht von ihren Instituten. Springer, Berlin, pp. 118–120.

Herzog, R.O., Janke, W., 1920. Über den physikalischen Aufbau einiger hochmolekularer organischer Verbindungen. Ber. Dtsch. Chem. Ges 53, 2162–2164.

Herzog, R.O., Janke, W., 1921. In: Neuberg, C. (Ed.), Verwendung von Röntgenstrahlen zur Untersuchung metamikroskopischer biologischer Strukturen. Festschrift der Kaiser Wilhelm Gesellschaft zur Förderung der Wissenschaften zu ihrem Zehnjährigen Jubiläum Dargebracht von ihren Instituten. Springer, Berlin, pp. 118–120.

Heslop-Harrison, J., 1962. Evanescent and persistent modifications of chloroplast ultrastructure induced by an unnatural pyrimidine. Planta 58, 237–256.

Heslop-Harrison, J.S., 1991. The molecular cytogenetics of plants. J. Cell. Sci. 100, 15–21.

Heslop-Harrison, J.S., 2000. Comparaytive genome organization in plants: from sequence and markers to chromatin and chromosomes. Plant Cell 12, 617–635.

Heslop-Harrison, J.S., Bennett, M.D., 1990. Nuclear architecture in plants. Trends Genet. 6, 401–405.

Heslop-Harrison, J.S., Leitch, A.R., Schwarzacher, T., 1993. The physical organization of interphase nuclei. In: Heslop-Harrison, J.S., Flavell, R.B. (Eds.), The Chromosome. BIOS Scientific Publishers, Oxford, pp. 221–232.

Heslop-Harrison, P., Osuji, J., Hull, R., Harper, G., D'Hort, A., Carreel, F., 1999. Fluorescent in situ Hybridization of Plant Chromosomes: Illuminating the Musa Genome INIBAP Annual Report 1998. INIBAP, Montpellier, France, pp. 26–29.

Hess, E.L., 1970. Origins of molecular biology. Science 168, 664–669.

Hess, W.R., Partensky, F., Van der Staay, G.W., Garcia–Fernandez, J.M., Börner, T., Vaulot, D., 1996. Coexistence of phycoerythrin and a chlorophyll a/b antenna in a marine prokaryote. Proc. Natl. Acad. Sci. U.S.A. 93, 11126–11130.

Hesse, M., 1993. The Tapetum. Cytology, Function, Biochemistry and Evolution. Springer-Verlag, Wien.

Hesse, T., Feldwisch, J., Balshüsemann, D., Puype, M., Vandekerckhove, J., Löbler, M., Klämbt, D., Schell, J., Palme, K., 1989. Molecular cloning and structural analysis of a gene from *Zea mays* (L.) coding for a putative receptor for the plant hormone auxin. EMBO J. 8, 2453–2461.

Hesse, M., Pacini, E., Willemse, M., 1993. The Tapetum. In: Cytology, Function, Biochemistry and Evolution. Springer-Verlag, Wien.

Heupel, R., Heldt, H.W., 1994. Protein organization in the matrix of leaf peroxisomes. A multienzyme complex involved in photo respiratory metabolism. Eur. J. Biochem. 220, 165–172.

Heupel, R., Markgraf, T., Robinson, D.G., Heldt, H.W., 1991. Compartmentation studies on spinach leaf peroxisomes. Evidence for channeling of photorespiratory metabolites in peroxisomes devoid of intact boundary membrane. Plant Physiol. 96, 971–979.

Heuser, E., 1944. The Chemistry of Cellulose. John Wiley & Sons, New York.

Heuser, J., 2002. Whatever happened to the 'microtrabecular concept'? Biol. Cell 94, 561–596.

Heuser, J., Zhu, Q., Clarke, M., 1993. Proton pumps populate the contractile vacuoles of Dictyostelium amoebae. J. Cell Biol. 121, 1311–1327.

Hewick, R.M., Hunkapiller, M.W., Hood, L.E., Dreyer, W.J., 1981. A gas-liquid-solid-phase peptide and protein sequenator. J. Biol. Chem. 256, 7990–7997.

Hewish, A., Bell, S.J., Pilkington, J.D.H., Scott, P.F., Collins, R.A., 1968. Observations of a rapidly pulsating radio source. Nature 217, 709–713.

Hewitt, L.F., Kekwick, R.A., McFarlane, A.S., 1943. The chemistry of the proteins and amino acids. Ann. Rev. Biochem. 12, 81–114.

Heyndrickx, K.S., Van de Velde, J., Wanf, C., Weigel, D., Vandepoele, K., 2014. A functional and evolutionary perspective on transcription factor binding in *Arabidopsis thaliana*. Plant Cell 26, 3894–3910.

Hicks, S.J., Drysdale, J.W., Munro, H.N., 1969. Preferential synthesis of ferritin and albumin by different populations of liver polysomes. Science 164, 584–585.

Hicks, G.R., Raikhel, N.V., 2012. Small molecules present large opportunities in plant biology. Annu. Rev. Plant Biol. 63, 261–282.

Hicks, G.R., Rojo, E., Hong, S., Carter, D.G., Raikhel, N.V., 2004. Geminating [sic] pollen has tubular vacuoles, displays highly dynamic vacuole biogenesis, and requires VACUOLESS1 for proper function. Plant Physiol. 134, 1227–1239.

Hieta, R., Myllyharju, J., 2002. Cloning and characterization of a low molecular weight prolyl 4-hydroxylase from *Arabidopsis thaliana*. J. Biol. Chem. 277, 23965–23971.

Higa, T., Wada, M., 2015. Clues to the signals for chloroplast photorelocation from the lifetimes of accumulation and avoidance responses. J. Integr. Plant Biol. 57, 120–126.

Higaki, T., Kutsuna, N., Akita, K., Takigawa-Imamura, H., Yoshimura, K., Miura, T., 2016. A theoretical model of jigsaw-puzzle pattern formation by plant leaf epidermal cells. PLoS Comput. Biol. 12, e1004833.

Higashi Jr., T., Peters, T., 1963. Studies on rat liver catalase. II. Incorporation of C14leucine into catalase of liver cell fractions *in vivo*. J. Biol. Chem. 238, 3952–3954.

Higashi-Fujime, S., 1991. Reconstitution of active movement in vitro based on the actin-myosin interaction. Int. Rev. Cytol. 125, 95–138.

Higashi-Fujime, S., Ishikawa, R., Ishikawa, H., Kagami, O., Kurimoto, E., Kohama, K., Hozumi, T., 1995. The fastest actin-based motor protein from the green algae, *Chara*, and its distinct mode of interaction with actin. FEBS Lett. 375, 151–154.

High, S., Dobberstein, B., 1991. The signal sequence interacts with the methionine-rich domain of the 54-kD protein of signal recognition particle. J. Cell Biol. 113, 229–233.

High, S., Dobberstein, B., 1992. Mechanisms that determine the transmembrane disposition of proteins. Curr. Opin. Cell Biol. 4, 581–586.

Higuchi, T., 1998. The discovery of lignin. Disoceries Plant Biol. 2, 233–269.

Hilgartner, S., 2017. Reordering Life. Knowledge and Controll in the Genomics Revolution. MIT Press, Cambridge, MA, pp. 50–86.

Hill, A.V., 1926. The physical environment of the living cell. In: Dale, H.H., Drummond, J.C., Henderson, L.J., Hill, A.V. (Eds.), Lectures on Certain Aspects of Biochemistry. University of London Press, Ltd., London, pp. 251–280.

Hill, A.V., 1928. The rôle of oxidation in maintaining the dynamic equilibrium of the muscle cell. Proc. R. Soc. Lond. 103B, 138–162.

Hill, A.V., 1929. The maintenance of life and irritability in isolated animal tissues. Nat. Suppl. 123, 723–730.

Hill, A.V., 1932. The revolution in muscle physiology. Physiol. Rev. 12, 56–67.

Hill, R., 1937. Oxygen evolved by isolated chloroplasts. Nature 139, 881–882.

Hill, A.V., 1949. Adenosine triphosphate and muscular contraction. Nature 163, 320.

Hill, R., 1954. The cytochrome b component of chloroplasts. Nature 174, 501–503.

Hill, A.V., 1962. The Ethical Dilemma of Science and Other Writings. Scientific Book Guild, London.

Hill, A.V., 1965. Trails and Trials in Physiology. Williams & Wilkins Co., Baltimore.

Hill, R., Bendall, F., 1960. Function of two cytochrome components in chloroplasts: a working hypothesis. Nature 186, 136–137.

Hill, A.V., December 12, 1923. The mechanism of muscular contraction. Nobel Lecture.

Hill, R.L., Kimmel, J.R., Smith, E.L., 1959. The structure of proteins. Ann. Rev. Biochem. 28, 97–144.

Hill, T.L., Morales, M.F., 1950. Sources of the high energy content in energy-rich phosphates. Arch. Biochem. 29, 450–451.

Hill, T.L., Morales, M.F., 1951. On "high energy phosphate bonds" of biochemical interest. J. Am. Chem. Soc. 73, 1656–1660.

Hill, R., Scarisbrick, R., 1940. Production of oxygen by illuminated chloroplasts. Nature 146, 61–62.

Hill, R., Scarisbrick, R., 1951. The haematin compounds of leaves. New Phytol. 50, 98–111.

Hill, A.E., Shar char-Hill, B., Skepper, J.N., Powell, J., Shachar-Hill, Y., 2012. An osmotic model of the growing pollen tube. PLoS One 7, e36585.

Hill, R., Whittingham, C.P., 1955. Photosynthesis. Methuen & Co., London.

Hille, B., 1992. Ionic Channels in Excitable Membranes, second ed. Sinauer Associates, Inc., Sunderland, MA.

Hiller, K., Metallo, C.M., Kelleher, J.K., Stephanopoulos, G., 2010. Nontargeted elucidation of metabolic pathways using stable-isotope tracers and mass spectrometry. Anal. Chem. 82, 6621–6628.

Hilliker, A.J., Appels, R., 1989. The arrangement of interphase chromosomes: structural and functional aspects. Exp. Cell. Res. 185, 297–318.

Hillman, H., 2001. Research practices in need of examination and improvement. Sci. Eng. Ethics 7, 7–14.

Hillmer, S., Freundt, H., Robinson, D.G., 1988. The partially coated reticulum and its relationship to the Golgi apparatus in higher plant cells. Eur. J. Cell Biol. 47, 206–212.

Hillmer, S., Viotti, C., Robinson, D.G., 2011. An improved procedure for low-temperature embedding of high-pressure frozen and freeze-substituted plant tissues resulting in excellent structural preservation and contrast. J. Microsc. 247 (Pt. 1), 43–47.

Himmelsbach, D.S., Khahili, S., Akin, D.E., 1999. Near-infrared-Fourier-transform Raman microspectroscopic imaging of flax stems. Vib. Spectrosc. 19, 361–367.

Himmelspach, R., Williamson, R.E., Wasteneys, G.O., 2003. Cellulose microfibril alignment recovers from DCB-induced disruption despite microtubule disorganization. Plant J. 36, 565–575.

Hind, G., Jagendorf, A.T., 1963. Separation of light and dark stages in photophosphorylation. Proc. Natl. Acad. Sci. U.S.A. 49, 715–722.

Hinkle, P.C., Kumar, M.A., Resetar, A., Harris, D.C., 1991. Mechanistic stoichiometry of mitochondrial oxidative phosphorylation. Biochemistry 30, 3576–3582.

Hinkle, L., McCaig, C.D., Robinson, K.R., 1981. The direction of growth of differentiating neurons and myoblasts from frog embryos in an applied electric field. J. Physiol. 314, 121–135.

Hinkle, P.C., McCarty, R.E., 1978. How cells make ATP. Sci. Am. 238, 104–123.

Hinkle, P.C., Yu, M.L., 1979. The phosphorous/oxygen ratio of mitochrondrial oxidative phosphorylation. J. Bio. Chem. 254, 2450–2455.

Hinnen, A., Hicks, J.B., Fink, G.R., 1978. Transformation of yeast. Proc. Natl. Acad. Sci. U.S.A. 75, 1929–1933.

Hinshaw, J.E., Carragher, B.O., Milligan, R.A., 1992. Architecture and design of the nuclear pore complex. Cell 69, 1133–1141.

Hippler, M. (Ed.), 2017. Chlamydomonas: Biotechnology and Biomedicine. Springer International, Cham, Switzerland.

Hirai, M.Y., Yano, M., Goodenowe, D.B., Kanaya, S., Kimura, T., Awazuhara, M., Arita, M., Fujiwara, T., Saito, K., 2004. Integration of transcriptomics and metabolomics for understanding of global responses to nutritional stresses in *Arabidopsis thaliana*. Proc. Natl. Acad. Sci. U.S.A. 101, 10205–10210.

Hiramoto, Y., 1987. Evaluation of cytomechanical properties. In: Bereiter-Hahn, J., Anderson, O.R., Reif, W.E. (Eds.), Cytomechanics. Springer Verlag, Berlin, pp. 31–46.

Hiramoto, Y., Kamitsubo, E., 1995. Centrifuge microscope as a tool in the study of cell motility. Int. Rev. Cytol. 157, 99–128.

Hiramoto, Y., Kaneda, I., 1988. Diffusion of substances in the cytoplasm and across the nuclear envelope in egg cells. Protoplasm (Suppl. 2), 88–94.

Hiramoto, Y., Nakano, Y., 1988. Micromanipulation studies of the mitotic apparatus in sand dollar eggs. Cell Motil. Cytoskelet. 10, 172–184.

Hirano, T., 2000. Chromosome cohesion, condensation, and separation. Annu. Rev. Biochem. 69, 673–745.

Hirano, T., 2002. The ABCs of SMC proteins: two-armed ATPases from chromosome condensation, cohesion, and repair. Genes Dev. 16, 399–414.

Hirano, T., 2012. Condensins: universal organizers of chromosomes with diverse functions. Genes Dev. 26, 1659–1678.

Hirano, T., 2016. Condensin-based chromosome organization from bacteria to vertebrates. Cell 164, 847–857.

Hirano, T., Kobayashi, R., Hirano, M., 1997. Condensins, chromosome condensation protein complexes containing XCAP-C, XCAP-E and a *Xenopus* homolog of the *Drosophila* Barren protein. Cell 89, 511–521.

Hirano, T., Mitchison, T.J., 1991. Cell cycle control of higher-order chromatin assembly around naked DNA in vitro. J. Cell Biol. 115, 1479–1489.

Hirano, T., Mitchison, T.J., 1994. A heterodimeric coiled-coil protein required for mitotic chromosome condensation in vitro. Cell 79, 449–458.

Hiraoka, L., Golden, W., Magnuson, T., 1987. Spindle pole organization during early mouse development. Dev. Biol. 133, 24–36.

Hiraoka, L., Golden, W., Magnuson, T., 1989. Spindle pole organization during early mouse development. Dev. Biol. 133, 24–36.

Hird, S.N., White, J.G., 1993. Cortical and cytoplasmic flow polarity in early embryonic cells of *Caenorhadbitis elegans*. J. Cell Biol. 121, 1343–1355.

Hirschhorn, J.N., Daly, M.J., 2005. Genome-wide association studies for common diseases and complex traits. Nat. Rev. Genet. 6, 95–108.

Hirschi, K.D., 1999. Expression of *Arabidopsis* CAX1 in tobacco: alterned calcium homeostasis and increased stress sensitivity. Plant Cell 11, 2113–2122.

Hirschi, K., 2001. Vacuolar H^+/Ca^{2+} transport: who's directing the traffic? Trends Plant Sci. 6, 100–104.

Hirshfeld, A.W., 2001. Parallax: The Race to Measure the Cosmos. W. H., Freeman, New York.

Hisatomi, O., Nakatani, Y., Takeuchi, K., Takahashi, F., Kataoka, H., 2014. Blue light induced dimerization of monomeric aureochrome-1 enhances its affinity for the target sequence. J. Biol. Chem. 289, 17379–17391.

Hiss, M., Laule, O., Meskauskiene, R.M., Arif, M.A., Decker, E.L., Erxleben, A., Frank, W., Hanke, S.T., Lang, D., Martin, A., et al., 2014. Large-scale gene expression profiling data for the model moss *Physcomitrella patens* aid understanding of developmental progression, culture and stress conditions. Plant J. 79, 530–539.

Hitchcock, D.I., 1940. Chemistry of amino acids and proteins. Ann. Rev. Biochem. 9, 173–198.

Hitchcock-DeGregori, S.E., Irving, T.C., 2014. Hugh E. Huxley: the compleat biophysicist. Biophys. J. 107, 1493–1501.

Hitler, A., 1943. Mein Kampf. Translated by Ralph Manheim. Houghton Mifflin, Boston.

Hitt, A.L., Luna, E.J., 1994. Membrane interactions with the actin cytoskeleton. Curr. Opin. Cell Biol. 6, 120–130.

Hizume, M., Shibata, F., Matsusaki, Y., Garajova, Z., 2002. Chromosome identification and comparative karyotypic analyses of four *Pinus* species. Theor. Appl. Genet. 105, 491–497.

Hoagland, D.R., 1948. Lectures on the Inorganic Nutrition of Plants. Chronica Botanica, Waltham, MA.

Hoagland, M., 1990. Toward the Habit of Truth. A Life in Science. W. W. Norton & Co., New York.

Hoagland, M.B., Keller, E.B., Zamecnik, P.C., 1956. Enzymatic carboxyl activation of amino acids. J. Biol. Chem. 218, 345–358.

Hoagland, M.B., Stephenson, M.L., Scott, J.F., Hecht, L.I., Zamecnik, P.C., 1958. A soluble ribonucleic acid intermediate in protein synthesis. J. Biol. Chem. 231, 241–257.

Hoagland, M.B., Zamecnik, P.C., Stephenson, M.L., 1957. Intermediate reactions in protein biosynthesis. Biochim. Biophys. Acta 24, 215–216.

Hobbes, T., 1651. Leviathan. http://www.constitution.org/th/leviatha.htm.

Höber, R., 1930. The present conception of the structure of the plasma membrane. Biol. Bull. 58, 1–17.

Höber, R., 1945. Physical Chemistry of Cells and Tissues. Blakiston Co., Philadelphia.

Hobhouse, H., 1986. Seeds of Change. Five Plants that Transformed Mankind. Harper & Row, New York.

Hobson, E.W., 1923. The Domain of Natural Science. MacMillan, New York.

Hoch, H.C., Staples, R.C., 1983. Ultrastructural organization of the non-differentiated uredospore germling of Uromyces phaseoli variety typica. Mycologia 75, 795–824.

Hoch, H.C., Staples, R.C., Whitehead, B., Comeau, J., Wolf, E.D., 1987. Signaling for growth orientation and cell differentiation by surface topography in *Uromyces*. Science 235, 1659–1662.

Hochachka, P.W., 1999. The metabolic implications of intracellular circulation. Proc. Natl. Acad. Sci. U.S.A. 96, 12233–12239.

Hodge, A.J., 1956. The fine structure of striated muscle. J. Biophys. Biochem. Cytol 2 (4 Suppl.), 131–142.

Hodge, A.J., McLean, J.D., Mercer, F.V., 1955. Ultrastructure of the lamellae and grana in the chloro plasts of Zea mays L. J. Biophys. Biochem. Cytol. 1, 605–613.

Hodge, A.J., McLean, J.D., Mercer, F.V., 1956. A possible mechanism for the morphogenesis of lamellar systems in plant cells. J. Biophys. Biochem. Cytol. 2, 597–608.

Hodges, T.K., Leonard, R.T., 1974. Purification of a plasma membrane-bound adenosine triphosphatase from plant roots. Methods Enzymol. 32, 392–406.

Hodges, T.K., Leonard, R.T., Bracker, C.E., Keenan, T.W., 1972. Purification of an ion-stimulated adenosine triphosphatase from plant roots: association with plasma membranes. Proc. Natl. Acad. Sci. U.S.A. 69, 3307–3311.

Hodgkin, A.L., December 11, 1963. The Ionic Basis of Nervous Conduction. Nobel Lecture.

Hodgkin, D.C., December 11, 1964. The X-ray Analysis of Complicated Molecules. Nobel Lecture.

Hoek, J.B., Nicholls, D.G., Williamson, J.R., 1980. Determination of the mitochondrial protonmotive force in isolated hepatocytes. J. Biol. Chem. 255, 1458–1464.

Hoepfner, D., Schildknegt, D., Braakman, I., Phillippsen, P., Tabak, H.F., 2005. Contribution of the endoplasmic reticulum to peroxisome formation. Cell 122, 85–95.

Hof, P.R., Nimchinsky, E.A., Perl, D.P., Erwin, J.M., 2001. An unusual cingulate cortex of hominids contains the calcium-binding protein calretinin. Neurossci. Lett. 307, 139–142.

Hoffman, B.F., Cranefield, P.F., 1960. Electrophysiology of the Heart. McGraw-Hill, New York.

Hoffman, B.D., Crocker, J.C., 2009. Cell mechanics: dissecting the physical responses of cells to force. Annu. Rev. Biomed. Eng. 11, 259–288.

Hoffman, R., Minkin, V.I., Carpenter, B.K., 1997. Ockham's razor and chemistry. Int. J. Philos. Chem. 3, 3–28.

Hoffman, J.C., Vaughn, K.C., Joshi, H.C., 1994. Structural and immunocytochemical characterization of microtubule organizing centers in pteridophyte spermatogenous cells. Protoplasma 179, 46–60.

Hoffmann, H.-P., Avers, C.J., 1973. Mitochondrion of yeast: ultrastructural evidence for one giant, branched organelle per cell. Science 181, 749–751.

Hoffmann-Berling, H., 1955. Geisselmodelle und Adenosintriphosphat (ATP). Biochim. Biophys. Acta 16, 146–154.

Hofmann, A., 1980. LSD: My Problem Child. McGraw Hill, New York.

Hofmannová, J., Schwarzerová, K., Havelková, L., Boriková, P., Petrásek, J., Opatrný, Z., 2008. A novel, cellulose synthesis inhibitory action of ancymidol impairs plant cell expansion. J. Exp. Bot. 59, 3963–3974.

Hofmeister, W., 1849. Die Entstehung des Embryos der Phanerogamen. Wilhelm Engelmann, Leipzig.

Hofmeister, W., 1867. Die Lehre von der Pflanzenzelle. Wilhelm Engelmann, Leipzig.

Höftberger, M., Url, T., Meindl, U., 1995. Disturbance of the secretory pathway in *Micrasterias denticulata* by tunicamycin and cyclopiazonic acid. Protoplasma 189, 173–179.

Hogeboom, G.H., 1946. Succinic dehydrogenase of mammalian liver. J. Biol. Chem. 162, 739–740.

Hogeboom, G.H., Schneider, W.C., 1955. The cytoplasm. In: Chargaff, E., Davidson, J.N. (Eds.), The Nucleic Acids. Chemistry and Biology. Academic Press, New York, pp. 199–246.

Hogeboom, G.H., Schneider, W.C., Pallade, G.E., 1948. Cytochemical studies of mammalian tissues. J. Biol. Chem. 172, 619–636.

Hogetsu, T., 1983. Distribution and local activity of particle complexes synthesizing cellulose microfibrils in the plasma membrane of *Closterium acerosum* (Schrank) Ehrenberg. Plant Cell Physiol. 24, 777–781.

Hogness, D.S., Cohn, M., Monod, J., 1955. Studies on the induced synthesis of β-galactosidase in Escherichia coli: the kinetics and mechanism of sulfur incorporation. Biochim. Biophys. Acta 16, 99–116.

Hoh, B., Schauermann, G., Robinson, D.G., 1991. Storage protein polypeptides in clathrin coated vesicle fractions from developing pea cotyledons are not due to endomembrane contamination. J. Plant Physiol. 138, 309–316.

Holdaway-Clarke, T.L., Feijó, J.A., Hackett, G.R., Kunkel, J.G., Hepler, P.K., 1997. Pollen tube growth and the intracellular cytosolic calcium gradient oscillate in phase while extracellular calcium influx is delayed. Plant Cell 9, 1999–2010.

Holdaway-Clarke, T.L., Hepler, P.K., 2003. Control of pollen tube growth: role of ion gradients and fluxes. New Phytol. 159, 539–563.

Holdaway-Clarke, T.L., Walker, N.A., Hepler, P.K., Overall, R.L., 2000. Physiological elevations in cytoplasmic free calcium by cold or ion injection result in transient closure of higher plant plasmodesmata. Planta 210, 329–335.

Holden, M.J., Norman, H.A., Britz, S.J., 1994. Spectral quality during pod development affects omega-6 desaturase activity in soybean seen endoplasmic reticulum. Physiol. Plant 91, 346–351.

Holland, N., Holland, D., Helentjaris, T., Dhugga, K.S., Xoconostle-Cazeres, B., Delmer, D.P., 2000. A comparative analysis of the plant cellulose synthase (CesA) gene family. Plant Physiol. 123, 1313–1323.

Holley, R.W., 1957. An alanine-dependent, ribonuclease-inhibited conversion of AMP to ATP, and its possible relationship to protein synthesis. J. Am. Chem. Soc. 79, 658–662.

Holley, R.W., Apgar, J., Everett, G.A., Madison, J.T., Marquisee, M., Merrill, S.H., Penswick, J.R., Zamir, A., 1965a. Structure of a ribonucleic acid. Science 147, 1462–1465.

Holley, R.W., Everett, G.A., Madison, J.T., Zamir, A., 1965b. Nucleotide sequences in the yeast alanine transfer ribonucleic acid. J. Biol. Chem. 240, 2122–2128.

Holley, R.W., Apgar, J., Merrill, S.H., Zubkoff, P.L., 1961. Nucleotide and oligonucleotide compostions of the alanine-, valine-, and tyrosine-acceptor "soluble" ribonucleic acids of yeast. J. Am. Chem. Soc. 83, 4861–4862.

Holley, R.W., December 12, 1968. Alanine transfer RNA. Nobel Lecture.

Hollins, D.L., Jaffe, M.J., 1997. On the role of tannin vacuoles in several nastic leaf responses. Protoplasma 199, 215–222.

Holmes, K.C., 2013a. Hugh Esmor Huxley (1924–2013). J. Muscle Res. Cell Motil. 34, 311–315.

Holmes, K.C., 2013b. Hugh Esmor Huxley (1924–2013). Proc. Natl. Acad. Sci. U.S.A. 110, 18344–18345.

Holmes, K.C., Popp, D., Gebhand, W., Kabasch, W., 1990. Atomic model of the actin filament. Nature 347, 44–49.

Holstein, S.E.H., Drucker, M., Robinson, D.G., 1994. Identification of a beta-type adaptin in plant clathrin-coated vesicles. J. Cell Sci. 107, 945–953.

Holt, S.C., Marr, A.G., 1965. Location of chlorophyll in *Rhodospirillum rubrum*. J. Bacteriol. 89, 1402–1412.

Holt, G.D., Snow, C.M., Senior, A., Haltiwanger, R.S., Gerace, L., Hart, G.W., 1987. Nuclear pore complex glycoproteins contain cytoplasmically disposed O-linked N-acetylglucosamine. J. Cell Biol. 104, 1157–1164.

Holton, G., 1978. The Scientific Imagination: Case Studies. Harvard University Press, Cambridge.

Holtzman, E., Novikoff, A.B., Villaverde, H., 1967. Lysosomes and GERL in normal and chromatolytic neurons of the rat ganglion nodosum. J. Cell Biol. 33, 419–435.

Holzwarth, G., Bonin, K., Hill, D.B., 2002. Forces required of kinesin during processive transport through cytoplasm. Biophys. J. 82, 1784–1790.

Honda, S.I., Hongladadarom-Honda, T., Kwanyuen, P., Wildman, S.G., 1971. Interpretations on chloroplast reproduction derived from correlations between cells and chloroplasts. Planta 97, 1–15.

Hong, Z., Verma, D.P.S., 2007. Molecular analysis of the cell plate forming machinery. Plant Cell Monogr. 9, 303–320.

Hood, L., 1992. Biology and medicine in the twenty-first century. In: Kevles, D.J., Hood, L. (Eds.), The Code of Codes: Scientific and Social Issues in the Human Genome Project. Harvard University Press, Cambridge, MA, pp. 136–163.

Hood, L., 2002. A personal view of molecular technology and how it changed biology. J. Proteome Res. 1, 399–409.

Hood, L., 2008. A personal journey of discovery: developing technology and changing biology. Annu. Rev. Anal. Chem. 1, 1–43.

Hood, L., 2011. Deciphering complexity: a personal view of systems biology and the coming of "Big" science. Genet. Eng. Biotechnol. News 31, 131.

Hood, L., Balling, R., Auffray, C., 2012. Revolutioning medicine in the 21st century through systems approaches. Biotechnol. J. 7, 992–1001.

Hood, L., Flores, M., 2012. A personal view on systems medicine and the emergence of proactive P4 medicine: predicitive, preventive, personalized and participatory. New Biotechnol. 29, 613–624.

Hooke, R., 1665. Micrographia or Some Physiological Descriptions of Minute Bodies Made by Manifying Glasses with Observations and Inquiries thereupon, 1961. Dover Publications, New York.

Hooke, R., 1678. Lectures de Potentia Restitutiva, Or of Spring Explaining the Power of Springing Bodies. Printed for John Martyn, London.

Hooke, R., 1726. Philosophical Experiments and Observations of the Late Eminent Dr. Robert Hooke. W. and J. Innys, London.

Hooley, R., 2001. Progress towards the identification of cytokinin receptors. In: Sopory, S.K., Oelmüller, R., Maheshwari, S.C. (Eds.), Signal Transduction in Plants. Kluwer Academic, New York, pp. 193–199.

Hooley, R., Beale, M.H., Smith, S.J., 1991. Gibberellin perception at the plasma membrane of Avena fatua aleurone protoplasts. Planta 183, 274–280.

Hope, A.B., Walker, N.A., 1975. The Physiology of Giant Algal Cells. Cambridge University Press, Cambridge.

Hopkins, F.G., 1913. The dynamic side of biochemistry. Nature 92, 213–223.

Hopkins, F.G., 1914. Presidential address: The dynamic side of biochemistry. Report of the Eighty-Third Meeting of the British Association for the Advancement of Science, pp. 652–668.

Hopkins, C.R., Gibson, A., Shipman, M., Miller, K., 1990. Movement of internalized ligand-receptor complexes along a continuous endosomal reticulum. Nature 346, 335–339.

Hopkins, M., Mark Harrison, T., Manning, C.E., 2008. Low heat flow inferred from >4 Gyr zircons suggests Hadean plate boundary interactions. Nature 456, 493–496.

Horecker, B.L., 1982. Cytochrome reductase, the pentose phosphate pathway, and Schiff base mechanisms. In: Semeza, G. (Ed.), Of Oxygen, Fuels, and Living Matter, Part 2. John Wilry & Sons, Chichester, UK, pp. 59–172.

Horgan, R.P., Kenny, L.C., 2011. 'Omic' technologies: genomics, transcriptomics, proteomics and metabolomics. Obstet. Gynacol. 13, 189–195.

Hörmann, G., 1898. Studien über die Protoplasmaströmung bein den Characeen. Sustav Fischer, Jena.

Horn, P.J., Benning, C., 2016. The plant lipidome in human health and environmental health. Science 353, 1228–1232.

Horn, M.A., Heinstein, P.F., Low, P.S., 1989. Receptor-mediated endocytosis in plant cells. Plant Cell 1, 1003–1009.

Horn, M.A., Heinstein, P.F., Low, P.S., 1990. Biotin-mediated delivery of exogenous macromolecules into soybean cells. Plant Physiol. 93, 1492–1496.

Horn, M.A., Heinstein, P.F., Low, P.S., 1992. Characterization of parameters influencing receptor-mediated endocytosis in cultured soybean cells. Plant Physiol. 98, 673–679.

Horowitz, N.H., 1945. On th eevolution of biochemical syntheses. Proc. Natl. Acad. Sci. U.S.A. 31, 153–157.

Horowitz, N.H., 1958. The origin of life. In: Hutchings Jr., E. (Ed.), Frontiers in Science. A Survey. Basic Books, New York, pp. 19–27.

Horowitz, N.H., 1985. The origins of molecular genetics: one gene, one enzyme. BioEssays 3, 37–39.

Horowitz, N.H., 1995. One-gene-one-enzyme: remembering biochemical genetics. Protein Sci. 4, 1017–1019.

Horowitz, N.H., 1996. The sixtieth anniversary of biochemical genetics. Genetics 143, 1–4.

Horowitz, N.H., Beadle, G.W., 1943. A microbiological method for the determination of choline by use of a mutant of Neurospora. J. Biol. Chem. 150, 325–333.

Horowitz, N.H., Bonner, D., Mitchell, H.K., Tatum, E.L., Beadle, G.W., 1945. Genic control of biochemical reactions in Neurospora. Am. Nat. 79, 304–317.

Horowitz, P., Hill, W., 1989. The Art of Electronics, second ed. Cambridge University Press, Cambridge.

Hörtensteiner, S., Martinoia, E., Amrhein, N., 1992. Reappearance of hydrolytic activities and tonoplast proteins in the regenerated vacuole of vacuolated protoplasts. Planta 187, 113–121.

Horvath, P., Barrangou, R., 2010. CRISPR/Cas, the immune system of bacteria and archaea. Science 327, 167–170.

Horvath, S.J., Firca, J.R., Hunkapiller, T., Hunkapiller, M.W., Hood, L., 1987. An automated DNA synthesizer employing deoxynucleoside 3′ phosphoramidites. Methods Enzymol. 154, 314–326.

Hosoi, F., Omasa, K., 2009. Detecting seasonal change of broad-leaved woody canopy leaf area density profile using 3D portable LIDAR imaging. Funct. Plant Biol. 36, 998–1005.

Hoson, T., 1991. Structure and function of plant cell walls: immunological approaches. Int. Rev. Cytol. 130, 233–268.

Hoson, T., 1992. Auxin-induced cell wall extension. In: Shibaoka, H. (Ed.), Cellular Basis of Growth and Development, Proc. of the VII International Symposium in Conjunction with the Awarding of the International Prize for Biology. November 26–28, 1992. Toyonaka, Osaka, pp. 27–34.

Hoson, T., Masuda, Y., Sone, Y., Misaki, A., 1991. Xyloglucan antibodies inhibit auxin-induced elongation and cell wall loosening of azuki bean epicotyls but not oat coleoptiles. Plant Physiol. 96, 551–557.

Hossfeld, U., Olsson, L., 2013. The prominent absence of Alfred Russel Wallace at the Darwin anniversaries in Germany in 1909, 1959 and 2009. Theory Biosci. 132, 251–257.

Hosy, E., Vavasseur, A., Mouline, K., Dreyer, I., Gaymard, F., Porée, F., Boucherez, J., Lebaudy, A., Bouchez, D., Véry, A.-A., Simonneau, T., Thibaud, J.-B., Sentenac, H., 2003. The *Arabidopsis* outward K^+ channel GORK is involved in regulation of stomatal movements and plant transpiration. Proc. Natl. Acad. Sci. U.S.A. 100, 5549–5554.

Hotani, H., Horio, T., 1988. Dynamics of microtubules visualized by darkfield microscopy: treadmilling and dynamic instability. Cell Motil. Cytoskelet. 10, 229–236.

Hotchkiss, R.D., 1995. DNA in the decade before the double helix. Ann. N.Y. Acad. Sci. 758, 55–73.

Hotchkiss Jr., A.T., Brown Jr., R.M., 1987. The association of rosette and globule terminal complexes with cellulose microfibril assembly in *Nitella translucens* var. axillaris (Charophyceae). J. Phycol. 23, 229–237.

Hoth, S., Geiger, D., Becker, D., Hedrich, R., 2001. The pore of plant K^+ channels is involved in voltage and pH sensing: domain swapping between different K^+ channel α-subunits. Plant Cell 13, 943–952.

Hou, G.C., Mohamalawari, D.R., Blancaflor, E.B., 2003. Enhanced gravitropism of roots with a disrupted cap actin cytoskeleton. Plant Physiol. 131, 1360–1373.

Houchmandzadeh, B., Marko, J.F., Chatenay, D., Libchaber, A., 1997. Elasticity and structure of eukaryote chromosomes studied by micromanipulation and micropipette aspiration. J. Cell Biol. 139, 1–12.

Houk, K.N., 1977. Frontier molecular orbital theory and organic reactions. Nature 266, 662.

Houle, D., Govindaraju, D.R., Omholt, S., 2010. Phenomics: the next challenge. Nat. Rev. Genet. 11, 855–866.

House, C.R., 1974. Water Transport in Cells and Tissues. Edward Arnold, London.

Howard, A., Haigh, M.V., 1968. Chloroplast aberrations in irradiated fern spores. Mutat. Res. 6, 263–280.

Howe, G.A., Jander, G., 2008. Plant immunity to insect herbivores. Annu. Rev. Plant Biol. 59, 41–66.

Howell, S.H., Hecht, N.B., 1971. The appearance of polynucleotide ligase and DNA polymerase during synchronous mitotic cycle in lilium microscopes. Biochim. Biophys. Acta 240, 343–352.

Howell, S.H., Moudrianakis, E.N., 1967. Hill reaction site in chloroplast membranes: nonparticipation of the quantasome particle in photoreduction. J. Mol. Biol. 27, 323–333.

Howell, S.H., Stern, H., 1971. The appearance of DNA breakage and repair activities in the synchronous meiotic cycle of *Lilium*. J. Mol. Biol. 55, 357–378.

Howey, W. (Ed.), 1948. The Faith of Great Scientists. A Collection of "My Faith" Articles from The American Weekly. Hearst, New York.

Hoyle, F., 1953. The Nature of the Universe. Basil Blackwell, Oxford.

Hoyle, F., 1979. An astronomer's view of the evolution of man. In: Huff, D., Prewett, O. (Eds.), The Nature of the Physical Universe. 1976 Nobel Conference. John Wiley & Sons, New York, pp. 63–82.

Hoyle, F., 1983. The Intelligent Universe. Holt, Reinhart and Winston, New York.

Hoyle, F., Wickramasinghe, C., 1981. Space Travellers. The Bringers of Life. University College Cardiff Press, Cardiff, UK.

Hoyt, M.A., 1994. Cellular roles of kinesin and related proteins. Curr. Opin. Cell Biol. 6, 63–68.

Hoyt, M.A., He, L., Loo, K.K., Saunders, W.S., 1992. Two Saccharomyces cerevisiae kinesin-related gene products required for mitotic spindle assembly. J. Cell Biol. 118, 109–120.

Hozak, P., Cook, P.R., Schofer, C., Mosgoller, W., Wachtler, F., 1994. Site of transcription of ribosomal RNA and intranucleolar structure in HeLa cells. J. Cell Sci. 107, 639–648.

Hrazdina, G., Jensen, R.A., 1992. Spatial organization of enzymes in plant metabolic pathways. Annu. Rev. Plant Physiol. Plant Mol. Biol. 43, 241–267.

Hrazdina, G., Wagner, G.J., 1985. Metabolic pathways as enzyme complexes: evidence for the synthesis of phenylpropanoids and flavonoids on membrane associated enzyme complexes. Arch. Biochem. Biophys. 237, 88–100.

Hrazdina, G., Zobel, A.M., Hoch, H.C., 1987. Biochemical, immunological and immunocytochemical evidence for the association of chalcone synthase with endoplasmic reticulum membranes. Proc. Natl. Acad. Sci. U.S.A. 84, 8966–8970.

Hsieh, T.-S., 1992. DNA topoisomerases. Curr. Opin. Cell Biol. 4, 396–400.

Hsieh, K., Huang, A.H.C., 2004. Endoplasmic reticulum, oleosins, and oils in seeds and tapetum cells. Plant Physiol. 136, 3427–3434.

Hsu, P.D., Lander, E.S., Zhang, F., 2014. Development and applications of CRISPR-Cas9 for genome engineering. Cell 157, 1262–1278.

Hu, J., Aguirre, M., Peto, C., Alonso, J., Ecker, J., Chory, J., 2002. A role for peroxisomes in photomorphogenesis and development of *Arabidopsis*. Science 297, 405–409.

Hu, J., Baker, A., Bartel, B., Linka, N., Mullen, R.T., Reumann, S., Zolman, B., 2012. Plant peroxisomes: biogenesis and function. Plant Cell 24, 2279–2303.

Hu, W., Cheng, C.L., 1995. Expression of Aequoria green fluorescent protein in plant cells. FEBS Lett. 369, 331–334.

Hu, Z., Lv, X., Xia, X., Zhou, J., Shi, K., Yu, J., Zhou, Y., 2016. Genome-wide identification and expression analysis of calcium-dependent protein kinase in tomato. Fron. Plant Sci. 7, 469.

Hu, Y., Zhang, Q., Rao, G., Sodmergen, 2008. Occurrence of plastids in the sperm cells of Caprifoliaceae: biparental plastid inheritance in angiosperms is unilaterally derived from maternal inheritance. Plant Cell Physiol. 49, 958–968.

Hua, J., Meyerowitz, E.M., 1998. Ethylene responses are negatively regulated by a receptor gene family in *Arabidopsis thaliana*. Cell 94, 261–271.

Huala, E., Oeller, P.W., Liscum, E., Han, I.S., Larsen, E., Briggs, W.R., 1997. *Arabidopsis* NPH1: a protein kinase with rutative redox-sensing domain. Science 278, 2120–2123.

Huang, C.Y., Ayliffe, M.A., Timmis, J.N., 2003. Direct measurement of the transfer rate of chloroplast DNA into the nucleus. Nature 422, 72–76.

Huang, X., Feng, Q., Qian, Q., Zhao, Q., Wang, L., Wang, A., Guan, J., Fan, D., Weng, Q., Huang, T., Dong, G., Sang, T., Han, B., 2009. High-throughput genotyping by whole-genome resequencing. Genome Res. 19, 1068–1076.

Huang, J., Fujimoto, M., Fujiwara, M., Fukao, Y., Arimura, S., Tsutsumi, N., 2015. Arabidopsis dynamin-related proteins, DRP2A and DRP2B, function coordinately in post-Golgi trafficking. Biochem. Biophys. Res. Commun. 456, 238–244.

Huang, S., Qu, X., Zhang, R., 2015. Plant villins: versitile actin regulatory proteins. J. Integr. Plant Biol. 57, 40–49.

Huang Jr., A.H.C., Trelease, R.N., Moore, T.S., 1983. Plant Peroxisomes. Academic Press, New York.

Huang, X., Wei, X., Sang, T., Zhao, Q., Feng, Q., Zhao, Y., Li, C., Zhu, C., Lu, T., Zhang, Z., Li, M., Fan, D., Guo, Y., Wang, A., Wang, L., Deng, L., Li, W., Lu, Y., Weng, Q., Liu, K., Huang, T., Zhou, T., Jing, Y., Li, W., Lin, Z., Buckler, E.S., Qian, Q., Zhang, Q.-., Li, J., Han, B., 2010. Genome-wide association studies of 14 agronomic traits in rice landraces. Nat. Genet. 42, 961–967.

Huang, S., Xiang, Y., Ren, H., 2011. Actin-binding proteins and actin dynamics in plant cells. Plant Cytoskelet. Adv. Plant Biol. 2, 57–80.

Hubble, E., 1937. The Obervational Approach to Cosmology. Clarendon Press, Oxford.

Hubble, E., 1954. The Nature of Science and Other Lectures. The Huntington Library, San Marino, CA.

Huber, R., December 8, 1988. A Structural Basis of Light Energy and Electron Transfer in Biology. Nobel Lecture.

Hübner, R., Depta, H., Robinson, D.G., 1985. Endocytosis in maize root cap cells. Evidence obtained using heavy metal salt solutions. Protoplasma 129, 214–222.

Hübscher, U., Maga, G., Spadari, S., 2002. Eukaryotic DNA polymerases. Annu. Rev. Biochem. 71, 133–163.

Huchelmann, A., Boutry, M., Hachez, C., 2017. Plant glandular trichomes: natural cell factories of high biotechnological interest. Plant Physiol. 175, 6–22.

Hudson, J., 1992. The History of Chemistry. Chapman & Hall, New York.

Huelsenbeck, J.P., Ronquist, F., Nielsen, R., Bollback, J.P., 2001. Bayesian inference of phylogeny and its impact on evolutionary biology. Science 294, 2310–2314.

Huet, J., Schnabel, R., Sentenac, A., Zillig, W., 1983. Archaebacteria and eukaryotes posses DNA-dependent RNA polymerases of a common type. EMBO J. 2, 1291–1294.

Hughes, A., 1952. The Mitotic Cycle. Academic Press, New York.

Hughes, A., 1959. A History of Cytology. Abelard-Schuman, London.

Hughes, L., 1999. Langston Hughes. Everyman's Library, Alfred A. Knopf, New York.

Hughes, T.R., Mao, M., Jones, A.R., Burchard, J., Marton, M.J., Shannon, K.W., et al., 2001. Expression profiling using microarrays fabricated by an ink-jet oligonucleotide synthesizer. Nat. Biotechnol. 19, 342–347.

Hughes, W.L., Sinex, F.M., 1954. Chemistry of the proteins, peptides, and amino acids. Ann. Rev. Biochem. 23, 177–214.

Hughes, A.F., Swann, M.M., 1948. Anaphase movement in the living cell. J. Exp. Biol. 25, 45–70.

Huisman, J.G., Moorman, A.F., Verkley, F.N., 1978. In vitro synthesis of chloroplast ferredoxin as a high molecular weight precursor in a cell-free protein synthesizing system from wheat germs. Biochem. Biophys. Res. Commun. 82, 1121–1131.

Huitorel, P., Kirschner, M.W., 1988. The polarity and stability of microtubule capture by the kinetochore. J. Cell Biol. 106, 151–159.

Hulse, R.A., December 8, 1993. The Discovery of the Binary Pulsar. Nobel Lecture.

Hume, D., 1748. Essays and Treastises on Several Subjects. http://www.constitution.org/dh/hume.htm.

Hume, D., 1952. An Enquiry Concerning Human Understanding. Great Books of the Western World. Volume 35, 1748. Encyclopedia Britannica. Inc., Chicago.

Humphries, M.J., 1990. The molecular basis and specificity of integrin-ligand interactions. J. Cell Sci. 97, 585–592.

Hunkapiller, M.W., Hood, L., 1980. New protein sequenator with increased sensitivity. Science 207, 523–525.

Hunkapiller, T., Kaiser, R., Koop, B., Hood, L., 1991. Large-scale and automated DNA sequence determination. Science 254, 59–67.

Hunt, T., December 9, 2001. Protein synthesis, proteolysis, and cell cycle transitions. Nobel Lecture.

Hunt, J.A., Ingram, V.M., 1958. Abnormal human haemoglobins. II. The chymotryptic digestion of the trypsin-resistant "core" of haemoglobins A and S. Biochim. Biophys. Acta 28, 546–549.

Hunt, L., Mills, L.N., Pical, C., Leckie, C.P., Aitken, F.L., Kopka, J., Mueller-Roeber, B., McAinsh, M.R., Hetherington, A.M., Gray, J.E., 2003. Phospholipase C is required for the control of stomatal aperture by ABA. Plant J. 34, 47–55.

Hunter, P.R., Craddock, C.P., Di Benedetto, S., Roberts, L.M., Frigerio, L., 2007. Fluorescent reporter proteins for the tonoplast and the vacuolar lumen identify a single vacuolar compartment in *Arabidopsis* cells. Plant Physiol. 145, 1371–1382.

Hunter, T., Karin, M., 1992. The regulation of transcription by phosphorylation. Cell 70, 375–387.

Hupfeld, H., 1931. As Time Goes By. Harms Incorporated, New York.

Hurlock, A.K., Roston, R.L., Wang, K., Benning, C., 2014. Lipid trafficking in plant cells. Traffick 15, 915–932.

Hurst, A.C., Mackel, T., Tayefeh, S., Thiel, G., Homann, U., 2004. Trafficking of the plant potassium inward rectifier KAT1 in guard cell protoplasts of *Vicia faba*. Plant J. 37, 391–397.

Hurt, E.D., Mutvei, A., Carmo-Fonseca, M., 1992. The nuclear envelope of the yeast *Saccharomyces* cerevisiae. Int. Rev. Cytol. 136, 145–184.

Hush, J.M., Overall, R.L., 1991. Electrical and mechanical fields orient cortical microtubules in higher plant tissues. Cell Biol. Int. Rep. 15, 551–560.

Hush, J.M., Wadsworth, P., Callaham, D.A., Hepler, P.K., 1994. Quantification of microtuble dynamics in living plant cells using fluorescence redistribution after photobleaching. J. Cell Sci. 107, 775–784.

Hush, J., Wu, L.P., John, P.C.L., Hepler, L.H., Hepler, P.K., 1996. Plant mitosis promoting factor disassembles the microtubule preprophase band and accelerates prophase progression in *Tradescantia*. Cell Biol. Int. 20, 275–287.

Hussey, E., 1995. The Presocratics. Hackett Publishing, Indianapolis.

Hussey, P.J., Allwood, E.G., Smertenko, A.P., 2002. Actin-binding proteins in the *Arabidopsis* genome database: properties of functionally distinct actin-depolymerizing factors/cofilins. Philos. Trans. R. Soc. Lond. B 357, 791–798.

Hutchinson, D.W., 1964. Nucleotides and Coenzymes. Methuen & Co., London.

Hutchinson III, C.A., 2007. DNA sequencing: bench to bedside and beyond. Nucleic Acids Res. 35, 6227–6237.

Hutchison III, C.A., Peterson, S.N., Gill, S.R., Cline, R.T., White, O., Fraser, C.M., Smith, H.O., Venter, J.C., 1999. Global transposon mutagenesis and a minimal *Mycoplasma* genome. Science 286, 2165–2169.

Huxley, J., 1June 5, 1946. UNESCO: Its Purpose and its Philosophy. Preparatory Commission of the United Nations Educational, Scientific and Cultural Organization.

Huxley, T.H., 1853. The cell-theory. Br. For. Medico Chir. Rev. 12, 285–314.

Huxley, T.H., 1853a. Observations on the existence of cellulose in the tunic of ascidians. Q. J. Microsc. Sci. 1, 22–24.

Huxley, T.H., 1853b. The cell-theory. Brit. For. Medico Chir. Rev. 12, 285–314.

Huxley, T.H., 1890. On the physical basis of life. In: Lay Sermons, Addresses, and Reviews. D. Appleton and Co., New York, pp. 120–146.

Huxley, T.H., 1893. Science and Culture and Other Essays. D. Appleton and Co., New York.

Huxley, T.H., 1898. On the identity of structure of plants and animals. Sci. Mem. 1, 216–220.

Huxley, J., 1912. The Individual in the Animal Kingdom. Cambridge University Press, Cambridge.

Huxley, L., 1926. Progress and the Unfit. Watts & Co., London.

Huxley, A., 1937. Ends and Means. Chatto & Windus, London.

Huxley, A., 1938. Ends and Means. Chatto & Windus, London.

Huxley, H.E., 1957. The double array of filaments in cross-striated muscle. Biophys. Biochem. Cytol. 3, 631–648.

Huxley, H.E., 1958. The contraction of muscle. Sci. Am. 199, 66–82.

Huxley, A., 1960. Brave New World & Brave New World Revisited. Harper & Row, New York.

Huxley, J., 1962. Eugenics in evolutionary perspective. Nature 195, 227–228.

Huxley, H.E., 1963. Electron microscope studies on the structure of natural and synthetic filaments from striated muscle. J. Mol. Biol. 7, 281–308.

Huxley, H.E., 1966. The fine structure of striated muscle and its functional significance. Harvey Lect. 60, 85–118 (1964/65).

Huxley, H.E., 1969. The mechanism of muscular contraction. Science 164, 1356–1366.

Huxley, A.F., 1980. Reflections on Muscle. Princeton University Press, Princeton, NJ.

Huxley, A.F., 1992. Kenneth Stewart Cole. Biogr. Mem. Fellows R. Soc. 38, 99–110.

Huxley, A.F., 1994. Kenneth Stewart Cole. Biogr. Mem. Natl. Acad. Sci. 70, 25–45.

Huxley, H.E., 1996. A personal view of muscle and motility mechanisms. Ann. Rev. Physiol. 58, 1–19.

Huxley, H.E., 2004. Fifty years of muscle and the sliding filament hypothesis. Eur. J. Biochem. 271, 1403–1415.

Huxley, A.F., December 11, 1963a. The Quantitative Analysis of Excitation and Conduction in Nerve. Nobel Lecture.

Huxley, H.E., Hanson, J., 1954. Changes in the cross-striations of muscle during contraction and stretch and their structural interpretation. Nature 173, 973–976.

Huxley, H.E., Hanson, J., 1957. Quantitative studies on the structure of cross-striated myofibrils. I. Investigations by interference microscopy. Biochim. Biophys. Acta 23, 229.

Huxley, A.F., Niedergerke, R., 1954. Structural changes in muscle during contraction. Nature 173, 971–973.

Huygens, C., 1762. Cosmotheros. Printed for Robert Urie, Glasgow.

Hwang, Y.-S., Bethke, P.C., Gubler, F., Jones, R.L., 2003. cPrG-HCl a potential H^+/Cl^- symporter prevents acidification of storage vacuoles in aleurone cells and inhibits GA-dependent hydrolysis of storage protein and phytate. Plant J. 35, 154–163.

Hwang, J.-U., Gu, Y., Lee, Y.-J., Yang, Z., 2005. Oscillatory ROP GTPase activation leads the oscillatory polarized growth of pollen tubes. Mol. Biol. Cell 16, 5385–5399.

Hwang, J.-U., Lee, Y., 2001. ABA-induced actin reorganization in guard cells of *Commelina communis* is mediated by cytosolic calcium levels and by protein kinase and protein phosphatase activities. Plant Physiol. 125, 2120–2128.

Hwang, J.-U., Song, W.-Y., Hong, D., Ko, D., Yamaoka, Y., Jang, S., Yim, S., Lee, E., Khare, D., Kim, K., Palmgren, H.S., Yoon, E., Martinoia, Lee, Y., 2016. Plant ABC transporters enable many uniqueaspects of a terrestrial plant's lifestyle. Mol. Plant 9, 338–355.

Hwang, J.-U., Suh, S., Yi, H., Kim, J., Lee, Y., 1997. Actin filaments modulate both stomatal opening and inward K^+-channel activities in guard cells of *Vicia faba*. L. Plant Physiol. 115, 335–342.

Hyams, J.S., Borisy, G.G., 1978. Isolated flagellar apparatus of *Chlamydomonas*: characterization of forward swimming and alteration of waveform and reversal of motion by calcium ions in vitro. J. Cell Sci. 33, 235–253.

Hyman, E.D., 1988. A new method for sequencing DNA. Anal. Biochem. 174, 423–436.

Hyman, A.A., Mitchison, T.J., 1990. Modulation of microtubule stability by kintochores in vitro. J. Cell Biol. 110, 1607–1616.

Ibáñez, C., 2017. Scientific Background. Discoveries of Molecular Mechanisms Controlling Circadian Rhythm. https://www.nobelprize.org/nobel_prizes/medicine/laureates/2017/advanced-medicineprize2017.pdf.

Ibl, V., Stoger, E., 2014. Live cell imaging during germination reveals dynamic tubular structures derived from protein storage vacuoles of barley aleurone cells. Plants 3, 442–457.

Igarashi, H., Orii, H., Mori, H., Shimmen, T., Sonobe, S., 2000. Isolation of a novel 190 kDa protein from tobacco BY-2 cells: possible involvement in the interaction between actin filaments and microtubules. Plant Cell Physiol. 41, 920–931.

Igarashi, H., Vidali, L., Yokota, E., Sonobe, S., Hepler, P.K., Shimmen, T., 1999. Actin filaments purified from tobacco cultured BY-2 cells can be translocated by plant myosin. Plant Cell Physiol. 40, 1167–1171.

Iino, M., Shitanishi, K., Kadota, A., Wada, M., 1990. Phytochrome-mediated phototropism in *Adiantum* protonemata. I. Phototropism as a function of the lateral Pfr gradient. Photochem. Photobiol. 51, 469–476.

Ikeda, S., Nasrallah, J.B., Dixit, R., Preiss, S., Nasrallah, M.E., 1997. An aquaporin-like gene required for the *Brassica* self-incompatibility response. Science 276, 1564–1566.

Ikeuchi, M., 1992. Subunit proteins of photosystem I. Plant Cell Physiol. 33, 669–676.

Imanaka, T., Small, G.M., Lazarow, P.B., 1987. Translocation of acylCoA oxidase into peroxisomes requires ATP hydrolysis but not a membrane potential. J. Cell Biol. 105, 897–905.

Imlau, A., Truernit, E., Sauer, N., 1999. Cell-to-cell and long-distance trafficking of the green fluorescent protein in the phloem and symplastic unloading of the protein into sink tissues. Plant Cell 11, 309–322.

Inada, S., Shimmen, T., 2001. Involvement of cortical microtubules in plastic extension regulated by gibberellin in *Lemna minor* root. Plant Cell Physiol. 42, 395–403.

Inada, S., Sonobe, S., Shimmen, T., 2002. Regulation of directional expansion by the cortical microtubule array in roots of *Lemna minor*. Funct. Plant Biol. 29, 1273–1278.

Inagaki, S., Henry, I.M., Lieberman, M.C., Comai, L., 2015. High-throughput analysis of T-DNA location and structure using sequence capture. PLoS One 10, e0139672.

Ingber, D.E., 1991. Integrins as mechanochemical transducers. Curr. Opin. Cell Biol. 3, 841–848.

Ingber, D.E., 1993. Cellular tensegrity: defining new rules of biological design that govern the cytoskeleton. J. Cell Sci. 104, 613–627.

Ingber, D.E., Folkman, J., 1989a. How does the extracellular matrix control capillary morphogenesis? Cell 58, 803–805.

Ingber, D.E., Folkman, J., 1989b. Tension and compression as basic determinants of cell form and function: utilization of a tensegrity mechanism. In: Stein, W.D., Bronner, F. (Eds.), Cell Shape: Determinants, Regulation, and Regulatory Role. Academic Press, San Diego, pp. 3–31.

Ingber, D.E., Jamieson, J.D., 1985. Cells as tensegrity structures: architectural regulation of histodifferentiation by physical forces transduced over basement membrane. In: Andersson, L.C., Gahmberg, C.G., Ekblom, P. (Eds.), Gene Expression during Normal and Malignant Differentiation. Academic Press, Orlando, pp. 13–32.

Ingen-Housz, J., 1796. An Essay on the Food of Plants and the Renovation of Soils, 1933. Printed for Private Distribution by J. Christian Bay.

Ingouff, M., Fitz Gerald, J.N., Guérin, C., Robert, H., Sørensen, M.B., Van Damme, D., Geelen, D., Blanchoin, L., Berger, F., 2005. Plant formin AtFH5 is an evolutionarily conserved actin nucleator involved in cytokinesis. Nat. Cell Biol. 7, 374–380.

Ingram, V.M., 1958. Abnormal human haemoglobins. I. The comparison of normal human and sickle-cell haemoglobins by "fingerprinting." Biochim. Biophys. Acta 28, 539–545.

Ingram, V.M., 1959. Abnormal human haemoglobins. III. The chemical difference between normal and sickle cell haemoglobins. Biochim. Biophys. Acta 36, 402–411.

Ingram, V.M., 2004. Sickle-cell anemia hemoglobin: the molecular biology of the first "molecular disease"—The crucial importance of serendipity. Genetics 167, 1–7.

Inohara, N., Shimomura, S., Fukui, T., Futai, M., 1989. Auxin binding protein located in the endoplasmic reticulum of maize shoots: molecular cloning and complete primary structure. Proc. Natl. Acad. Sci. U.S.A. 86, 3564–3568.

Inoué, S., 1951. A method for measuring small retardations of structure in living cells. Exp. Cell Res. 2, 513–517.

Inoué, S., 1952. The effect of colchicine on the microscopic and submicroscopic structure of the mitotic spindle. Exp. Cell Res. (Suppl. 2), 305–318.

Inoué, S., 1953. Polarization optical studies of the mitotic spindle. I. The demonstration of spindle fibers in living cells. Chromosoma 5, 487–500.

Inoué, S., 1959. Motility of cilia and the mechanism of mitosis. In: Oncley, J.L., Schmitt, F.O., Williams, R.C., Rosenberg, M.D., Holt, R.H. (Eds.), Biophysical Science—A Study Program. John Wiley & Sons, New York, pp. 402–408.

Inoué, S., 1964. Organization and function of the mitotic spindle. In: Allen, R.D., Kamiya, N. (Eds.), Primative Motile Systems in Cell Biology. Academic Press, New York, pp. 549–598.

Inoué, S., 1981. Cell division and the mitotic spindle. J. Cell Biol. 91, 131s–147s.

Inoué, S., 2011. Shinya Inoue: lighting the way in microscopy. Interviewed by Caitlin Sedwick. J. Cell Biol. 194, 810–811.

Inoué, S., 2016. Pathways of a Cell Biologist. Through Yet another Eye. Springer Biographies, Singapore.

Inoué, S., Bajer, A., 1961. Birefringence in endosperm mitosis. Chromosoma 12, 48–63.

Inoué, S., Dan, K., 1951. Birefringence of the dividing cell. J. Morphol. 89, 423–456. https://www.ibiology.org/ibiomagazine/issue-6/shinya-inoue-the-dynamic-mitotic-spindle.html.

Inoue, S.-I., Kinoshita, T., 2017. Blue light regulation of stomatal opening and the plasma membrane H^+-ATPase. Plant Physiol. 174, 531–538.

Inoue, S.-I., Kinoshita, T., Matsumoto, M., Nakayama, K.I., Doi, M., Shimazaki, K.-I., 2008. Blue light-induced autophosphorylation of phototropin is a primary step for signaling. Proc. Natl. Acad. Sci. U.S.A. 105, 5626–5631.

Inoue, T., Orgel, L.E., 1983. A nonenzymatic RNA polymerase model. Science 219, 859–862.

Inoue, H., Rounds, C., Schnell, D.J., 2010. The molecular basis for distinct pathways for protein import into *Arabidopsis* chloroplasts. Plant Cell 22, 1947–1960.

Inoué, S., Salmon, E.D., 1995. Force generation by microtubule assembly/disassembly in mitosis and related movements. Mol. Biol. Cell 6, 1619–1640.

Inoué, S., Sato, H., 1967. Cell Motility by labile association of molecules. J. Gen. Physiol. (Suppl. 50), 259–292.

Inoue, Y., Suzuki, T., Hattori, M., Yoshimoto, K., Ohsumi, Y., Moriyasu, Y., 2006. *AtATG* genes, homologs of yeast autophagy genes, are involved in constituitive autophagy in *Arabidopsis* root tip cells. Plant Cell Physiol. 47, 1641–1652.

Inouhe, M., Nevins, D.J., 1991. Inhibition of auxin-induced cell elongation of maize coleoptiles by antibodies specific for cell wall glucanases. Plant Physiol. 96, 426–431.

Inouye, S., Noguchi, M., Sakaki, Y., Takagi, Y., Miyata, T., Iwanaga, S., Miyata, T., Tsuji, F.I., 1985. Cloning and sequence analysis of cDNA for the luminescent protein aequorin. Proc. Natl. Acad. Sci. U.S.A. 82, 3154–3158.

Inouye, S., Sakaki, Y., Goto, T., Tsuji, F.I., 1986. Expression of apoaequorin complementary DNA in *Escherichia coli*. Biochemistry 25, 8425–8429.

International Human Genome Sequencing Consortium, 2001. Initial sequencing and analysis of the human genome. Nature 409, 860–921.

International Human Genome Sequencing Consortium, 2004. Finishing the euchromatic sequence of the human genome. Nature 431, 931–945.

Inzé, D. (Ed.), 2007. Cell Cycle Control and Plant Development. Annual Plant Reviews, vol. 32. Blackwell Publishing, Oxford.

Ioannidis, J.P., Thomas, G., Daly, M.J., 2009. Validating, augmenting and refining genome-wide association signals. Nat. Rev. Genet. 10, 318–329.

Iraki, N.M., Bressan, R.A., Hasegawa, P.M., Carpita, N.C., 1989. Alteration of the physical and chemical structure of the primary cell wall of growth-limited plant cells adapted to osmotic stress. Plant Physiol. 91, 39–47.

Irani, N.G., Di Rubbo, S., Mylle, E., Van Den Begin, J., Schneider-Pizon, J., Hniliková, J., Šíša, M., Buyst, D., Vilarrasa-Blasi, J., Szatmári, A.M., Van Damme, D., 2012. Fluorescent castasterone reveals BRI1 signaling from the plasma membrane. Nat. Chem. Biol. 8, 583–589.

Irani, N.G., Russinova, E., 2009. Receptor endocytosis and signaling in plants. Curr. Opin. Plant Biol.

Irving, H.R., Gehring, C.A., Parish, R.W., 1992. Changes in cytosolic pH and calcium of guard cells precede stomatal movements. Proc. Natl. Acad. Sci. U.S.A. 89, 1790–1794.

Isaacson, T., Rose, J.K.C., 2006. Surveying the plant cell wall proteome, or secretome. Annu. Plant Rev. 28, 185–209.

Isayenkov, S., Isner, J.C., Maathuis, F.J.M., 2010. Vacuolar ion channels: roles in plant nutrition and signalling. FEBS Lett. 584, 1982–1988.

Iscove, N.N., Barbara, M., Gu, M., Gibson, M., Modi, C., Winegarden, N., 2002. Representation is faithfully preserved in global cDNA amplified exponentially from sub-picogram quantities of mRNA. Nat. Biotechnol. 20, 940–943.

Iseki, M., Matsunaga, S., Murakami, A., Ohno, K., Shiga, K., Yoshida, K., Sugai, M., Takahashi, T., Hori, T., Watanabe, M., 2002. A blue-light-activated adenylyl cyclase mediates photoavoidance in *Euglena gracilis*. Nature 415, 1047–1051.

Isermann, P., Lammerding, J., 2013. Nuclear mechanics and mechanotransduction in health an ddisease. Curr. Biol. 23, R1113–R1121.

Ishibashi, K., Sasaki, S., Fushimi, K., Uchida, S., Kuwahara, M., Saito, H., Furukawa, T., Nakajima, K., Yamaguchi, Y., Gojobori, T., Marumo, F., 1994. Molecular cloning and expression of a member of the aquaporin family with permeability to glycerol and urea in addition to water expressed at the baseolateral membrane of kidney collection duct cells. Proc. Natl. Acad. Sci. U.S.A. 91, 6269–6273.

Ishida, M., Aihara, M.S., Allen, R.D., Fok, A.K., 1993. Osmoregulation in *Paramecium*: the locus of fluid sergregation in the contractile vacuole complex. J. Cell Sci. 106, 693–702.

Ishida, H., Izumi, M., Wada, S., Makino, A., 2014. Roles of autophagy in chloroplast recycling. Biochim. Biophys. Acta 1837, 512–521.

Ishida, K., Katsumi, M., 1991. Immunofluorescence microscopical observation of cortical microtubule arrangement as affected by gibberellin in d5 mutant of *Zea mays* L. Plant Cell Physiol. 32, 409–417.

Ishida, H., Yoshimoto, K., Izumi, M., Reisen, D., Yano, Y., Makino, A., Ohsumi, Y., Hanson, M.R., Mae, T., 2008. Mobilization of rubisco and stroma-localized fluorescent proteins of chloroplasts to the vacuole by an *ATG* gene-dependent autophagic process. Plant Physiol. 148, 142–155.

Ishigami, M., Nagai, R., 1980. Motile apparatus in *Vallisneria* leaf cells. II. Effects of cytochalasin B and lead acetate on the rate and direction of streaming. Cell Struct. Funct. 5, 13–20.

Ishii, T., Matsunaga, T., Hayashi, N., 2001. Formation of rhamnogalacturonan II-borate dimer in pectin determined cell wal thickness of pumpkin tissue. Plant Physiol. 126, 1698–1705.

Ishii, T., Sunamura, N., Matsumoto, A., Eltayeb, A.E., Tsujimoto, H., 2015. Preferential recruitment of the maternal centromere-specific H3 (CENH3) in oat (Avena sativa L.) × pearl millet (Pennisetum glaucum L.) hybrid embryos. Chromosome Res. 23, 709–718.

Ishijima, A., Doi, T., Sakurada, K., Yanagida, T., 1991. Sub-piconewton force fluctuations of actomyosin in vitro. Nature 352, 301–306.

Ishijima, A., Kojima, H., Higuchi, H., Harada, Y., Funatsu, T., Yanagida, T., 1996. Multiple-and single-molecule analysis of the actomyosin motor by nanometer-piconewton manipulation with a microneedle: unitary steps and forces. Biophys. J. 70, 383–400.

Ishikawa, H., Bishoff, R., Holtzer, H., 1969. Formation of arrowhead complexes with heavy meromyosin in a variety of cell types. J. Cell Biol. 43, 312–328.

Isojima, H., Iino, R., Niitani, Y., Noji, H., Tomishige, M., 2016. Direct observation of intermediate states during the stepping motion of kinesin-1. Nat. Chem. Biol. 12, 290–298.

Isono, E., Katsiarimpa, A., Müller, I.K., Anzenberger, F., Stierhof, Y.-D., Geldner, N., Chory, J., Schwechheimer, C., 2010. The deubiquitinating enzyme AMSH3 is required for intracellular trafficking and vacuole biogenesis in *Arabidopsis thaliana*. Plant Cell 22, 1826–1837.

Israel, H.W., Steward, F.C., 1967. The fine structure and development of plastids in cultured cells of *Daucus carota*. Ann. Bot. 31, 1–18.

Israel, H.W., Wilson, R.G., Aist, J.R., Kunoh, H., 1980. Cell wall appositions and plant disease resistance: acoustic microscopy of papillae that block fungal ingress. Proc. Natl. Acad. Sci. U.S.A. 77, 2046–2049.

Iswanto, A.B.B., Kim, J.-Y., 2017. Lipid raft, regulator of plasmodesmatal homeostasis. Plants 6, 15.

Itaya, A., Ma, F., Qi, Y., Matsuda, Y., Zhu, Y., Liang, G., Ding, B., 2002. Plasmodesma-mediated selective protein traffic between "symplasmically-isolated" cells probed by a viral movement protein. Plant Cell 14, 2071–2083.

Itaya, A., Woo, Y.M., Masuta, C., Bao, Y., Nelson, R.S., Ding, B., 1998. Developmental regulation of intercellular protein trafficking through plasmodesmata in tobacco leaf epidermis. Plant Physiol. 118, 373–385.

Ito, E., Fujimoto, M., Ebine, K., Uemura, T., Ueda, T., Nakano, A., 2012a. Dynamic behavior of clathrin in *Arabidopsis thaliana* unveiled by live imaging. Plant J. 69, 204–216.

Ito, K., Ikebe, M., Kashiyama, T., Mogami, T., Kon, T., Yashimoto, K., 2007. Kinetic mechanism of the fastest motor protein, *Chara* myosin. J. Biol. Chem. 282, 19534–19545.

Ito, K., Kashiyama, T., Shimada, K., Yamaguchi, A., Awata, J., Hachikubo, Y., Manstein, D.J., Yamamoto, K., 2003. Recombinant motor domain constructs of *Chara corallina* myosin display fast motility and high ATPase activity. Biochem. Biophys. Res. Commun. 312, 958–964.

Ito, Y., Uemura, T., Nakano, A., 2014. Formation and maintenance of the Golgi apparatus in plant cells. Int. Rev. Cell Mol. Biol. 310, 221–287.

Ito, Y., Uemura, T., Shoda, K., Fujimoto, M., Ueda, T., Nakano, A., 2012b. *cis*-Golgi proteins accumulate near the ER exit sites and act as the scaffold for Golgi regeneration after brefeldin A treatment in tobacco BY-2 cells. Mol. Biol. Cell 23, 3203–3214.

Itoh, T., 1990. Cellulose synthesizing complexes in some giant marine algae. J. Cell Sci. 95, 309–319.

Itoh, T., 1992. Assembly of cellulose microfibrils in giant marine algae. In: Shibaoka, H. (Ed.), Cellular Basis of Growth and Development in Plants, Proc. of the VII International Symposium in Conjunction with the Awarding of the International Prize for Biology November 26–28, 1991. Toyonaka, Osaka, pp. 35–43.

Itoh, T., Brown Jr., R.M., 1984. The assembly of cellulose microfibrils in *Valonia macrophysa* Kütz. Planta 160, 372–381.

Itoh, R., Fujiwara, M., Nagata, N., Yoshida, S., 2001. A chloroplast protein homologous to the eubacterial topological specificity factor MinE plays a role in chloroplast division. Plant Physiol. 127, 1644–1655.

Itoh, T., Kimura, S., 2001. Immunogold labeling of terminal cellulose-synthesizing complexes. J. Plant Res. 114, 483–489.

Itoh, T., O'Neil, R.M., Brown Jr., R.M., 1984. Interference of cell wall regeneration of *Boergesenia forbesii* protoplasts by Tinopal LPW, a fluorescent brightening agent. Protoplasma 123, 174–183.

Itoh, R.D., Yamasaki, H., Septiana, A., Yoshida, S., Fujiwara, M.T., 2010. Chemical induction of rapid and reversible plastic filamentation in *Arabidopsis thaliana* roots. Physiol. Plant 139, 144–158.

Ivanov, I.I., Kassavina, B.S., Fomenko, L.D., 1946. Adenosine triphosphate in mammalian spermatozoa. Nature 158, 624.

Ivanovska, I.L., Shin, J.-W., Swift, J., Discher, D.E., 2015. Stem cell mechanobiology: diverse lessons from bone marrow. Trends Cell Biol. 25, 523–532.

Ives, H.E., Stillwell, G.R., 1938. An experimental study of the rate of a moving atomic clock. J. Opt. Soc. Am. 28, 215–226.

Iwabe, N., Kuma, K.-I., Hasegawa, M., Osawa, S., Miyata, T., 1989. Evolutionary relationship of archaebacteria, eubacteria, and eukaryotes inferred from phylogenetic trees of duplicated genes. Proc. Natl. Acad. Sci. U.S.A. 86, 9355–9359.

Iwabuchi, K., Kaneko, T., Kikuyama, M., 2005. Ionic mechanism of mechano-perception in characeae. Plant Cell Physiol. 46, 1863–1871.

Iwabuchi, K., Kaneko, T., Kikuyama, M., 2008. Mechanosensitive ion channels in *Chara*: influence of water channel inhibitors, $HgCl_2$ and $ZnCl_2$ on generation of receptor potential. J. Membr. Biol. 221, 27–37.

Izumi, M., Ishida, H., Nakamura, S., Hidema, J., 2017. Entire photodamaged chloroplasts are transported to the central vacuole by autophagy. Plant Cell 29, 377–394.

Jackson, M.E., 1991. Negative regulation of eukaryotic transcription. J. Cell Sci. 100, 1–7.

Jackson, D., 2005. Transcription factor movement through plasmodesmata. In: Oparka, K.J. (Ed.), Plasmodesmata. Blackwell Publishing, Oxford, pp. 113–134.

Jackson, R.J., Standart, N., 1990. Do the poly(A) tail and 3-untranslated region control mRNA translation? Cell 62, 15–24.

Jacob, F., Brenner, S., Cuzin, F., 1963. Onthe regulation of DNA replication in bacteria. Cold Spring Harb. Symp. Quant. Biol. 28, 329–348.

Jacob, F., December 11, 1965. Genetics of the bacterial cell. Nobel Lecture.

Jacob, F., Monod, J., 1961. Genetic regulatory mechanisms in the synthesis of proteins. J. Mol. Biol. 3, 318–356.

Jacob, M., Schmid, F.X., 1999. Protein folding as a diffusion process. Biochemistry 38, 13773–13779.

Jacobs, A.K., Lipka, V., Burton, R.A., Panstruga, R., Strizhov, N., Schulze-Lefert, P., Fincher, G.B., 2003. An *Arabidopsis* callose synthase, GSL5, is required for wound and papillary formation. Plant Cell 15, 2503–2513.

Jacobshagen, S., Altmüller, D., Grolig, F., Wagner, G., 1986. Calcium pools, calmodulin and light-regulated chloroplast movements in *Mougeotia* and *Mesotaenium*. In: Trewavas, A.J. (Ed.), Molecular and Cellular Aspects of Calcium in Plant Development. Plenum Press, New York, pp. 201–209.

Jaffe, L.F., 1958. Tropistic responses of zygotes of the Fucaceae to polarized light. Exp. Cell Res. 15, 282–299.

Jaffe, L.F., 1966. Electrical currents through the developing *Fucus* egg. Proc. Natl. Acad. Sci. U.S.A. 56, 1102–1109.

Jaffe, L.F., 1977. Electrophoresis along cell membranes. Nature 265, 600–602.

Jaffe, L.F., 1979. Control of development by ionic currents. In: Cone, R.A., Dowling, J.E. (Eds.), Membrane Transduction Mechanisms. Raven Press, New York, pp. 199–231.

Jaffe, L.F., 1980a. Control of plant development by steady ionic currents. In: Spanswick, R.M., Lucas, W.J., Dainty, J. (Eds.), Plant Membrane Transport: Current Conceptual Issues. Elsevier/ North-Holland Biomedical Press, Amsterdam, pp. 381–388.

Jaffe, M.J., 1980b. Morphogenetic responses of plants to mechanical stimuli or stress. BioScience 30, 239–243.

Jaffe, L.F., 1981. The role of ionic currents in establishing developmental pattern. Philos. Trans. R. Soc. Lond. (Biol.) 295, 553–566.

Jaffe, L.F., 1990. Calcium ion currents and gradients in fucoid eggs. In: Leonard, R.T., Hepler, P.K. (Eds.), Calcium in Plant Growth and Development. Amer. Soc. Plant Physiol. Symp. Series Volume, 4, pp. 120–126.

Jaffe, L.F., 2004. Marine plants may polarize remote *Fucus* eggs via luminescence. Biol. Bull. 207, 160.

Jaffe, L.F., 2005. Marine plants may polarize remote *Fucus* eggs via luminescence. Luminescence 20, 414–418.

Jaffe, L.F., 2006. The discovery of calcium waves. Semin. Cell Dev. Biol. 17, 229.

Jaffe, L.F., 2007. Stretch-activated calcium channels relay fast calcium waves propagated by calcium induced calcium influx. Biol. Cell 99, 175–184.

Jaffe, M.J., Galston, A.W., 1967. Phytochrome control of rapid nyctinastic movements and membrane permeability in Albizzia julibrissin. Planta 77, 135–141.

Jaffe, M.J., Galston, A.W., 1968. The physiology of tendrils. Annu. Rev. Plant Physiol. 19, 417–434.

Jaffe, M.J., Leopold, A.C., Staples, R.C., 2002. Thigmo responses in plants and fungi. Am. J. Bot. 89, 375–382.

Jaffe, L.F., Poo, M.-, 1979. Neurites grow faster towards the cathode than the anode in a steady field. J. Exp. Zool. 209, 115–128.

Jagendorf, A.T., 1956. Oxidation and reduction of pyridine nucleotides by purified chloroplasts. Arch. Biochem. Biophys. 62, 141–150.

Jagendorf, A.T., 1961. Photophosphorylation and the Hill reaction. Nature 191, 679–680.

Jagendorf, A.T., 1998. Chance, luck and photosynthesis research: an inside story. Photosynth. Res. 57, 215–229.

Jagendorf, A.T., 2002. Photophosphorylation and the chemiosmotic perspective. Photosynth. Res. 73, 233–241.

Jagendorf, A.T., Hind, G., 1963. Studies on the mechanism of photophosphorylation. In: Photosynthetic Mechanisms of Green Plants. National Academy of Sciences-National Research Council., Washington, DC, pp. 599–610.

Jagendorf, A.T., MacDonald, J.F., Hinkle, P.C., 1991. Efraim Racker: June 28, 1913-September 9, 1991. https://ecommons.cornell.edu/bitstream/handle/1813/18718/Racker_Efraim_1991.pdf;jsessionid=1B86F3E06A6A99389CB9D242F7F8DF47?sequence=2.

Jagendorf, A.T., Michaels, A., 1990. Rough thylakoids: translation on photosynthetic membrane. Plant Sci. 71, 137–145.

Jagendorf, A.T., Neumann, J., 1965. Effect of uncouplers on the light-induced pH rise with spinach chloroplasts. J. Biol. Chem. 240, 3210–3214.

Jagendorf, A.T., Uribe, E., 1966. ATP formation caused by acid- base transition of spinach chloroplasts. Proc. Natl. Acad. Sci. U.S.A. 55, 170–177.

Jahn, T., Baluska, F., Michalke, W., Harper, J.F., Volkmann, D., 1998. Plasma membrane H^+-ATPAse in the maize root apex: evidence for strong expression in xylem parenchyma and asymmetric localization within cortical and epidermis cells. Physiol. Plant 104, 311–316.

Jahn, G., December 10, 1964. Presentation speech.

Jahn, T., Fuglsang, A.T., Olsson, A., Bruntrup, I.M., Collinge, D.B., Volkmann, D., Sommarin, M., Palmgren, M.G., Larrson, C., 1997. The 14-3-3 protein interacts directly with the C terminal region of the plant plasma membrane H^+-ATPase. Plant Cell 9, 1805–1814.

Jahn, T., Palmgren, M.G., 2002. H^+-ATPases in the plasma membrane: physiology and molecular biology. In: Rengel, Z. (Ed.), Handbook of Plant Growth. pH as the Master Variable. Marcel Dekker, New York, pp. 1–22.

Jain, M.K., 1972. The Bimolecular Lipid Membrane. Vam Nostrand Reinhold, New York.

Jain, M., Koren, S., Quick, J., Rand, A.C., Sasani, T.A., Tyson, J.R., Beggs, A.D., et al., 2017. Nonopore sequencing and assembly of a human genome with ultra-long reads. bioRxiv. https://www.biorxiv.org/content/biorxiv/early/2017/04/20/128835.full.pdf.

Jaiswal, J.K., 2001. Calcium-how and why? J. Biosci. 26, 357–363.

Jaki, S.L., 1978. The Road of Science and the Ways to God. University of Chicago Press, Chicago, IL.

Jakoby, W.B., Brummond, D.O., Ochoa, S., 1956. Formation of 3-phosphoglyceric acid by carbon dioxide fixation with spinach leaf enzymes. J. Biol. Chem. 218, 811–822.

Jakus, M.A., Hall, C.E., 1946. Electron microscope observations of the trichocysts and cilia of *Paramecium*. Biol. Bull. 91, 141–144.

James, W.O., 1953. The use of respiratory inhibitors. Annu. Rev. Plant Physiol. 4, 59–90.

James, W.O., 1957. Reaction paths in the respiration of higher plants. Adv. Enzymol. 18, 281–318.

James, H., 1962. The Madonna of the Future, and Other Early Stories. New American Library, New York.

James, P., 1997. Protein identification in the post-genome era: the rapid rise of proteomics. Q. Rev. Biophys. 30, 279–331.

James, W.O., Beevers, H., 1950. The respiration of the *Arum* spadix. A rapid respiration, resistant to cyanide. New Phytol. 49, 353–374.

James, D.W., Jones, R.L., 1979. Intracellular localization of GDP-fucose polysaccharide fucosyl transferase in corn roots. Plant Physiol. 64, 914–918.

Jamieson, J.D., Palade, G.E., 1966. Role of the Golgi complex in the intracellular transport of secretory proteins. Proc. Natl. Acad. Sci. U.S.A. 55, 424–431.

Jamieson, J.D., Palade, G.E., 1967a. Intracellular transport of secretory proteins in the pancreatic exocrine cell. I. J. Cell Biol. 34, 577–596.

Jamieson, J.D., Palade, G.E., 1967b. Intracellular transport of secretory proteins in the pancreatic exocrine cell. II. J. Cell Biol. 34, 597–615.

Jamieson, J.D., Palade, G.E., 1968a. Intracellular transport of secretory proteins in the pancreatic exocrine cell. III. J. Cell Biol. 39, 580–588.

Jamieson, J.D., Palade, G.E., 1968b. Intracellular transport of secretory proteins in the pancreatic exocrine cell. IV. J. Cell Biol. 39, 589–603.

Jamieson, J.D., Palade, G.E., 1971a. Condensing vacuole conversion and zymogen granule discharge in pancreatic exocrine cells: metabolic studies. J. Cell Biol. 48, 503–522.

Jamieson, J.D., Palade, G.E., 1971b. Synthesis, intracellular transport, and discharge of secretory proteins in stimulated pancreatic exocrine cells. J. Cell Biol. 50, 135–158.

Jammes, F., Hu, H.C., Villers, F., Bouten, R., Kwak, J.M., 2011. Calcium-permeable channels in plant cells. FEBS J. 278, 4262–4276.

Jan, L.Y., Jan, Y.N., 1994. Membrane permeability: editorial overview. Curr. Opin. Cell Biol. 6, 569–570.

Jander, G., Norris, S.R., Rounsley, S.D., Bush, D.F., Levin, I.M., Last, R.L., 2002. *Arabidopsis* map-based cloning in the post-genome era. Plant Physiol. 129, 440–450.

Jang, C.S., Lee, T.G., Kim, J.Y., Park, J.H., Kim, D.S., Park, J.H., Seo, Y.W., 2005. The molecular characterization of a cDNA encoding the putative integral membrane protein, HvSec61α, expressed during early stage of barley kernel development. Plant Sci. 168, 233–239.

Janmey, P.A., 1991. Mechanical properties of cytoskeletal polymers. Curr. Opin. Cell Biol. 3, 4–11.

Janmey, P.A., Hvidt, S., Oster, G.F., Lamb, J., Stossel, T.P., Hartwig, J.H., 1990. The effect of ATP on actin filament stiffness. Nature 347, 95–99.

Jaquinod, M., Villiers, F., Kieffer-Jaquinod, S., Hugouvieux, V., Bruley, C., Garin, J., Bourguignon, J., 2007. A proteomic dissection of *Arabidopsis thaliana* vacuoles isolated from cell culture. Mol. Cell Proteom. 6, 394–412.

Jamet, E., Albenne, C., Boudart, G., Irshad, M., Canut, H., Pont- Lezica, R., 2008. Recent advances in plant cell wall proteomics. Proteomics 8, 893–908.

Jarvis, D.E., Ho, Y.S., Lightfoot, D.J., Schmöckel, S.M., Li, B., Borm, T.J.A., et al., 2017. The genome of *Chenopodium quinoa*. Nature 542, 307–312.

Jasencakova, Z., Meister, A., Walter, J., Turner, B.M., Schubert, I., 2000. Histone H4 acetylation of euchromatin and heterochromatin is cell cycle dependent and correlated with replication rather than with transcription. Plant Cell 12, 2087–2100.

Jasinski, M., Stukkens, Y., Degand, H., Purnelle, B., Marchand-Brynaert, J., Boutry, M., 2001. A plant plasma membrane ATP binding cassette–type transporter is involved in antifungal terpenoid secretion. Plant Cell 13, 1095–1107.

Jeans, J., 1930. The Mysterious Universe. Cambridge University Press, Cambridge.

Jeans, J., 1962. An Introduction to the Kinetic Theory of Gases. Cambridge University Press, Cambridge.

Jebanathirajah, J.A., Peri, S., Pandey, A., 2002. Toll and interleukin-1 receptor (TIR) domain-containing proteins in plants: a genomic perspective. Trends Plant Sci. 7, 388–391.

Jedd, G., Chua, H.N., 2002. Visualization of peroxisomes in living plant cells reveals acto-myosin-dependent cytoplasmic streaming and peroxisome budding. Plant Cell Physiol. 43, 384–392.

Jeener, R., Brachet, J., 1944. Recherches sur l'acide ribonucléique des levures. Enzymologia 11, 222–234.

Jeffers, J.S., 2017. Frederick Sanger: Two-Time Nobel Laureate in Chemistry. Springer Briefs in Molecular Science. History of Chemistry, Cham, Switzerland.

Jefferson, T., 1955. Notes on the State of Virginia. University of North Carolina Press, Chapel Hill, NC.

Jefferson, T., 1984. Letter to Isaac McPherson, 1813. In: Jefferson Writings. The Library of America, New York. August 13, 1813.

Jeffrey, W.R., 1982. Calcium ionophore polarizes ooplasmic segregation in Ascidian eggs. Science 216, 545–547.

Jeffrey, W.R., 1984a. Pattern formation by ooplasmic segregation in Ascidian eggs. Biol. Bull. 166, 277–298.

Jeffrey, W.R., 1984b. Spatial distribution of messenger RNA in the cytoskeletal framework of Ascidian eggs. Dev. Biol. 103, 482–492.

Jeffrey, W.R., 1985. The spatial distribution of maternal mRNA is determined by a cortical cytoskeletal domain in *Chaetopterus* eggs. Dev. Biol. 110, 217–229.

Jeffrey, W.R., Meier, S., 1984. Ooplasmic segregation of the myoplasmic actin network in stratified ascidian eggs. Rouxs Arch. Dev. Biol. 193, 257–262.

Jeffrey, W.R., Tomlinson, C.R., Brodeur, R.D., 1983. Localization of actin messenger RNA during Ascidian development. Dev. Biol. 99, 408–417.

Jeffrey, W.R., Wilson, L.J., 1983. Localization of messenger RNA in the cortex of *Chaetopterus* eggs and early embryos. J. Embryol. Exp. Morph. 75, 225–239.

Jelínková, A., Malínská, K., Simon, S., Kleine-Vehn, J., Pařezova, M., Pejchar, P., Kubeš, M., Martinec, J., Friml, J., Zažímalová, E., Petrášek, J., 2010. Probing plant membranes with FM dyes: tracking, dragging or blocking? Plant J. 61, 883–892.

Jencks, W.P., 1975. Free energies of hydrolysis and decarboxylation. In: Fasman, G.D. (Ed.), Handbook of Biochemistry and Molecular Biology, third ed., vol. 1. CRC Press, Boca Raton, FL, pp. 296–304.

Jennings, H.S., 1906. Behavior of the Lower Organisms. Columbia University Press, New York.

Jennings, H.S., 1925. Prometheus or Biology and the Advancement of Man. E. P. Dutton & Co., New York.

Jennings, H.S., 1927. Diverse doctrines of evolution, their relation to the practice of science and of life. Science 65, 19–25.

Jennings, H.S., 1933. The Universe and Life. Yale University Press, New Haven, CT.

Jensen, R.A., 1976. Enzyme recruitment in evolution of new function. Annu. Rev. Microbiol. 30, 409–425.

Jeong, W.J., Jeong, S.K., Min, S.R., Yoo, O.J., Liu, J.R., 2002. Growth retardation of plants transformed by overexpression of NtFtsZ1-2 in tobacco. J. Plant Biol. 45, 107–111.

Jessberger, R., 2002. The many functions of SMC proteins in chromosome dynamics. Nat. Rev. Mol. Cell Biol. 3, 767–778.

Jett, J.H., Keller, R.A., Martin, J.C., Marrone, B.L., Moyzis, R.K., Ratliff, R.L., Seitzinger, N.K., Shera, E.B., Stewart, C.C., 1989. High-speed DNA sequencing: an approach based upon fluorescence detection of single molecules. J. Biomol. Struct. Dyn. 7, 301–309.

Jewett, D.L., 2005. What's wrong with a single hypothesis?: why it is time for Strong-Inference-PLUS. Scientist 19 (21), 10.

Jezek, M., Blatt, M.R., 2017. The membrane transport system of the guar cell and its integration for stomatal dynamics. Plant Physiol. 174, 487–519.

Jha, U.C., Bhat, J.S., Patil, B.S., Hossain, F., Bahr, D., 2015. Functional genomics: applications in plant science. In: Barh, D., Sarwar Khan, M., Davies, E. (Eds.), PlantOmics: The Omics of Plant Science. Springer India, pp. 65–111.

Jia, H., Wang, N., 2014. Targeted genome editing of sweet orange using Cas9/sgRNA. PLoS One 9, e93806.

Jiang, L., Phillips, R.C., Hamm, C.A., Drozdowicz, Y.M., Rea, P.A., Maeshima, M., Rogers, S.W., Rogers, J.C., 2001. The protein storage vacuole: a unique compound organelle. J. Cell Biol. 155, 991–1002.

Jiang, L., Phillips, T.E., Rogers, S.W., Rogers, J.C., 2000. Biogenesis of the protein storage vacuole crystalloid. J. Cell Biol. 150, 755–770.

Jiang, K., Schwarzer, C., Lally, E., Zhang, S., Ruzin, S., Machen, T., Remington, S.J., Feldman, L., 2006a. Green fluorescent protein (reduction-oxidation-sensitive green fluorescent protein) in *Arabidopsis*. Plant Physiol. 141, 397–403.

Jiang, K., Zhang, S., Lee, S., Tsai, G., Kim, K., Huang, H., Chilcott, C., Zhu, T., Feldman, L.J., 2006b. Transcription profile analyses identify genes and pathways central to root cap functions in maize. Plant Mol. Biol. 60, 343–363.

Jiang, W., Zhou, H., Bi, H., Fromm, M., Yang, B., Weeks, D.P., 2013. Demonstration of CRISPR/Cas9/sgRNA-mediated targeted gene modification in Arabidopsis, tobacco, sorghum and rice. Nucleic Acids Res. 41, e188.

Jiao, Y., Peluso, P., Shi, J., Liang, T., Stitzer, M.C., Wang, B., et al., 2017. Improved maize reference genome with single-molecule technologies. Nature 546, 524–527.

Jiao, W.-B., Schneeberger, K., 2017. The impact of thrird generation genomic technologists on plant genome assembly. Curr. Opin. Plant Biol. 36, 64–70.

Jinek, M., Chylinski, K., Fonfara, I., Hauer, M., Doudna, J.A., Charpentier, E., 2012. A programmable dual-RNA-guided DNA endonuclease in adaptive bacterial immunity. Science 337, 816–821.

Johannes, E., Felle, H., 1990. Proton gradient across the tonoplast of *Riccia fluitans* as a result of the joint action of two electroenzymes. Plant Physiol. 93, 412–417.

John, P.C.L., Sek, F.J., Carmichael, J.P., McCurdy, D.W., 1990. p34cdc2 homologue level, cell division, phytohormone responsiveness and cell differentiation in wheat leaves. J. Cell Sci. 97, 627–636.

John, P.C.L., Sek, F.J., Hayes, J., 1991. Association of the plant p34cdc2 like protein with p13suc1: implications for control of cell division cycles in plants. Protoplasma 161, 70–74.

John, P.C.L., Sek, F.J., Lee, M.G., 1989. A homologue of the cell cycle control protein p34cdc2 participates in the cell division cycle of *Chlamydomonas* and a similar protein is detectable in higher plants and remote taxa. Plant Cell 1, 1185–1193.

John, P., Whatley, F.R., 1975. Paracoccus denitrificans and the evolutionary origin of the mitochondrion. Nature 254, 495–498.

John, P., Whatley, F.R., 1977. Paracoccus denitrificans Davis (*Micrococcus denitrificans*) as a mitochondrion. Adv. Bot. Res. 4, 51–115.

John, P.C.L., Wu, L., 1992. Cell cycle control proteins in division and development. In: Shibaoka, H. (Ed.), Cellular Basis of Growth and Development. Proc. of the VII International Symposium in Conjunction with the Awarding of the International Prize of Biology. Toyonaka, Osaka, pp. 119–126.

Johnson, T., 1934. A tropic response in germ tubes of urediospores of *Puccinia graminis tritici*. Phytopathology 24, 80–82.

Johnson, M.J., 1949. Oxidation-reduction potentials. In: Lardy, H.A. (Ed.), Respiratory Enzymes. Burgess Publishing Co., Minneapolis, MN, pp. 58–70.

Johnson, G., 2005. Miss Leavitt's Stars: The Untold Story of the Woman Who Discovered How to Measure the Universe. W. W. Norton & Co., New York.

Johnson, D.E., Dutcher, S.K., 1991. Molecular studies of linkage group XIX of Chlamydomonas reinhardtii. Evidence against a basal body location. J. Cell Biol. 113, 339–346.

Johnson, F.H., Eyring, H., Stover, B.J., 1974. The Theory of Rate Processes in Biology and Medicine. John Wiley & Sons, New York.

Johnson, D.S., Mortazavi, A., Myers, R.M., Wold, B., 2007. Genome-wide mapping of in vivo protein-DNA interactions. Science 316, 1497–1502.

Johnson, N., Powis, K., High, S., 2013. Post-translational translocation into the endoplasmic reticulum. Biochim. Biophys. Acta 1833, 2403–2409.

Johnson, K.A., Rosenbaum, J.L., 1990. The basal bodies of Chlamydomonas reinhardtii do not contain immunologically detectable DNA. Cell 62, 615–619.

Johnson, K.A., Rosenbaum, J.L., 1992. Replication of basal bodies and centrioles. Curr. Opin. Cell Biol. 4, 80–85.

Johnson, F.H., Shimomura, O., Saiga, Y., Gershman, L.C., Reynolds, G.T., Waters, J.R., 1963. Quantum efficiency of *Cypridina* luminescence, with a note on that of *Aequorea*. J. Cell. Comp. Physiol. 60, 85–104.

Johnson, A., Vert, G., 2017. Single event resolution of plant plasma membrane protein endocytosis by TIRF microscopy. Front. Plant Sci. 8, 612.

Johnson, B.R., Wytenbach, R.R., Wayne, R., Hoy, R.R., 2002. Action potentials in a giant algal cell: a comparative approach to mechanisms and evolution of excitability. J. Undergrad. Neurosci. Educ. 1, A23–A27.

Johnston, C.A., Temple, B.R., Chen, J.-G., Gao, Y., Moriyama, E.N., Jones, A.M., Siderovski, D.P., Willard, F.S., 2007. Comment on "A G protein-coupled receptor is a plasma membrane receptor for the plant hormone abscisic acid." Science 318, 914c.

Jolly, J., 1920. Hamties des tylopodes. C. R. Soc. Biol. Paris 93, 125–127.

Jonas, M., Huang, H., Kamm, R.D., So, P.T.C., 2008. Fast fluorescence laser tracking microrheometry, II. Quantitative studies of cytoskeletal mechanotransduction. Biophys. J. 95, 895–909.

Jones, O., 1853. Colour in the decorative arts. In: Lectures on the Results of the Great Exhibition of 1851. Second Series. David Bogue, London, pp. 255–300.

Jones, O., 1856. The Grammar of Ornament. Day and Son, London.

Jones, W., 1920. Nucleic Acids, second ed. Longmans Green and Co, London.

Jones, R.L., 1969a. The fine structure of barley aleurone cells. Planta 85, 359–375.

Jones, R.L., 1969b. Gibberellic acid and the fine structure of barley aleurone cells. I. Changes during the lag-phase of α-amylase synthesis. Planta 87, 119–133.

Jones, R.L., 1969c. Gibberellic acid and the fine structure of barley aleurone cells. II. Changes during the synthesis and secretion of α-amylase. Planta 88, 73–86.

Jones, A.M., 2014. Opinion: The planet needs more plant scientists. The Scientist. https://www.the-scientist.com/?articles.view/articleNo/41133/title/Opinion–The-Planet-Needs-More-Plant-Scientists/.

Jones, H.D., 2015. Regulatory uncertainty over genome editing. Nat. Plants 1, 3.

Jones, R.L., Bush, D.S., 1991. Gibberellic acid and abscissic acid regulate the level of a BiP cognate in the endoplasmic reticulum of barley aleurone cells. Plant Physiol. 97, 456–459.

Jones, R.L., Carbonell, J., 1984. Regulation of the synthesis of barley aleurone α-amylase by gibberellic acid and calcium ions. Plant Physiol. 76, 218–231.

Jones, A.M., Danielson, J.Å.H., ManojKumar, S.N., Lanquar, V., Grossman, G., Frommer, W.B., 2014a. Abscisic acid dynamics in roots detected with genetically encoded FRET sensors. eLife 3, e01741.

Jones, A.M., Grossmann, G., Danielson, J.A., Sosso, D., Chen, L.Q., Ho, C.H., Frommer, W.B., 2013. In vivo biochemistry: applications for small molecule biosensors in plant biology. Curr. Opin. Plant Biol. 16, 389–395.

Jones, L.T., Holte, H.O., 1922. The mass of the electron at slow velocity. Science 55, 647.

Jones, R.L., Jacobsen, J.V., 1982. The role of the endoplasmic reticulum in the synthesis and transport of α-amylase in barley aleurone layers. Planta 156, 421–432.

Jones, R.L., Jacobsen, J.V., 1983. Calcium regulation of the secretion of α-amylase isozymes and other proteins form barley aleurone layers. Planta 158, 1–9.

Jones, R.L., Jacobsen, J.V., 1991. Regulation of synthesis and transport of secreted proteins in cereal aleurone. Int. Rev. Cytol. 125, 49–88.

Jones, R., Ougham, H., Thomas, H., Waaland, S., 2013b. The Molecular Life of Plants. John Wiley & Sons, Chichester, West Sussex, UK.

Jones, M.G.K., Payne, H.L., 1978. Cytokinesis in Impatiens balsamina and the effect of caffeine. Cytobios 20, 79–91.

Jones, R.L., Robinson, D.G., 1989. Protein secretion in plants. New Phytol. 111, 567–597.

Jones, A.M., Venis, M.A., 1989. Photoaffinity labeling of indole-3-acetic acid binding proteins in maize. Proc. Natl. Acad. Sci. U.S.A. 86, 6153–6156.

Jones, L.A., Worobo, R.W., Smart, C.D., 2014a. Plant-pathogenic oomycetes, *Escherichia coli* strains, and *Salmonella* spp. Frequently found in surface water used for irrigation of fruit and vegetable crops in Nrew York State. Appl. Environ. Microbiol. 80, 4814–4820.

Jones, A.M., Xuan, Y., Xu, M., Wang, R.-S., Ho, C.-H., Lalonde, S., You, C.H., Sardi, M.I., Parsa, S.A., Smith-Valle, E., Su, T., Frazer, K.A., Pilot, G., Pratelli, R., Grossmann, G., Acharya, B.R., Hu, H.-C., Engineer, C., Villiers, F., Ju, C., Takeda, K., Su, Z., Dong, Q., Assmann, S.M., Chen, J., Kwak, J.M., Schroeder, J.I., Albert, P., Rhee, S.Y., Frommer, W.B., 2014b. Border control – a membrane-linked interactome of *Arabidopsis*. Science 344, 711–716.

Jongsma, M.L.M., Berlin, I., Neefjes, J., 2014. On the move: organelle dynamics during mitosis. Trends Cell Biol. 25, 112–124.

Jonsson, E., Yamada, M., Vale, R.D., Goshima, G., 2015. Clustering of a kinesin-14 motor enables progressive retrograde microtubule-based transport in plants. Nat. Plants 1, 15087.

Joos, V., VanAken, J., Kristen, U., 1994. Microtubules are involved in maintaining the cellular polarity in pollen tubes of *Nicotiana sylvestris*. Protoplasma 179, 5–15.

Jope, C.A., Atchison, B.A., Pringle, R.C., 1980. A computer analysis of a spiral string-of-grana model of the three dimensional structure of chloroplasts. Bot. Gaz. 141, 37–47.

Jordan, P., 1938. Die Verstärkertheorie der Organismen in ihrem gegenwärtigen Stand. Naturwissenschaften 26, 537–545.

Jordan, P., 1939. Zur Quanten-Biologie. Biologisches Zentralblatt 59, 1–39.

Jordan, P., 1942a. Begriff und Umgrenzung der Quantenbiologie. Physis (Stuttgart) 1, 13–26.

Jordan, P., 1942b. Zukunftsaufgaben der quantenbiologischer Forschung. Physis (Stuttgart) 1, 64–79.

Jordan, P., 1948. Die Physik und das Geheimnis des organischen Lebens. Friedrich Vieweg & Sohn, Braunschweig.

Jordan, D.O., 1960. The Chemistry of Nucleic Acids. Butterworths, London.

Jordan, E.G., 1991. Interpreting nucleolar structure; where are the transcribing genes. J. Cell Sci. 98, 437–449.

Jordan, P., Kronig, R.de.L., 1927. Movements of the lower jaw of cattle during mastication. Science 120, 807.

Jordan, E.G., Zatsepina, O.V., Shaw, P.J., 1992. Widely dispersed DNA within plant and animal nucleoli visualized 3-D fluorescence microscopy. Chromosoma 101, 478–482.

Jørgensen, P.L., Pedersen, P.A., 2001. Structure-function relationships of Na^+, K^+, ATP, or Mg^{2+} binding and energy transduction in a Na,K-ATPase. Biochim. Biophys. Acta 1505, 57–74.

Joshi, H.C., 1994. Microtubule organizing centers and γ-tubulin. Curr. Opin. Cell Biol. 6, 55–62.

Josse, J., Kaiser, A.D., Kornberg, A., 1961. Enzymatic synthesis of deoxyribonucleic acid. VIII. Frequencies of nearest neighbor base sequences in deoxyribonucleic acid. J. Biol. Chem. 236, 864–875.

Jost, L., 1907. Lectures on Plant Physiology. Translated by J. H. Gibson. Clarendon Press, Oxford.

Joubès, J., Inzé, D., Geelen, D., 2004. Improvement of the molecular toolbox for cell cycle studies in tobacco BY-2 cells. Biotechnol. Agric. For. 53, 7–23.

Joule, J.P., 1843. On the calorific effects of magneto-electricity, and on the mechanical value of heat. Phil. Mag. 23, 263–276, 347–355, 435–443.

Joule, J.P., 1852. On the heat disengaged in chemical combinations. Philos. Mag. Fourth Ser. 3 (Suppl. 21), 481–504.

Joule, J.P., 1943. On the calorific effects of magneto-electricity, and on the mechanical value of heat. Philos. Mag. 23, 263–276, 347–355, 435–443.

Jovtchev, G., Borisova, B., Kuhlmann, M., Fuchs, J., Watanabe, K., Schubert, I., Mette, M., 2011. Pairing of lacO tandem repeats in *Arabidopsis thaliana* nuclei requires the presence of hypermethylated, large arrays at wo chromosomal positions, but does not depend on H3-lysine-dimethylation. Chromosoma 120, 609–619.

Joyce, G.F., 1991. The rise and fall of the RNA world. New Biol. 3, 399–407.

Juhos, E.T., 1966. Density gradient centrifugation of bacteria and nonspecific bacteriophage in silica sol. J. Bacteriol. 91, 1376–1377.

Juliano, R.L., Haskill, S., 1993. Signal transduction from the extracellular matrix. J. Cell Biol. 120, 577–585.

Jung, H.-J., Jung, H.-J., Ahmed, N.U., Park, J.-I., Kang, K.-K., et al., 2012. Development of self-compatible *B. rapa* by RNAi-mediated S locus gene silencing. PLoS One 7, e49497.

Jung, J.-Y., Kim, Y.-W., Kwak, J.M., Hwang, J.-D., Young, J., Schroeder, J.I., Hwang, I., Lee, Y., 2002. Phosphatidylinositol 3-and 4-phosphate are required for normal stomatal movements. Plant Cell 14, 2397–2412.

Jung, G., Wernicke, W., 1990. Cell shaping and microtubules in developing mesophyll of wheat (*Triticum aestivum* L.). Protoplasma 153, 141–148.

Junge, W., 2004. Protons, proteins and ATP. Photosynth. Res. 80, 197–221.

Juniper, B.E., Hawes, C.R., Horne, J.C., 1982. The relationships between the dictyosomes and the forms of endoplasmic reticulum in plant cells with different export programs. Bot. Gaz. 143, 135–145.

Juniper, B.E., Robins, R.J., Joel, D.M., 1989. The Carnivorous Plants. Academic Press, London.

Junker, S., 1977. Ultrastructure of tactile papillae on tendrils of *Eccremocarpus scaber* R. et P. New Phytol. 78, 607–610.

Just, E.E., 1939. The Biology of the Cell Surface. P. Blakiston's Son & Co., Philadelphia.

Kabsch, W., Mannherz, H.G., Sock, D., Pai, E.F., Holmes, K.C., 1990. Atomic structure of the actin: DNAse I complex. Nature 347, 37–43.

Kadono, T., Kawano, T., Hosoya, H., Kosaka, T., 2004. Flow cytometric studies of the host-regulated cell cycle in algae symbiotic with green *Paramecium*. Protoplasma 223, 133–141.

Kadota, A., Kohyama, I., Wada, M., 1989. Polarotropism and photomovement of chloroplasts in the protonemata of the ferns *Pteris* and *Adiantum*: evidence for the possible lack of dichroic phytochrome in *Pteris*. Plant Cell Physiol. 30, 523–531.

Kadota, A., Sato, Y., Wada, M., 2000. Intracellular chloroplast photorelocation in the moss *Physcomitrella patens* is mediated by phytochrome as well as a blue-light receptor. Planta 210, 932–937.

Kadota, A., Wada, M., Furuya, M., 1982. Phytochrome-mediated phototropism and different dichroic orientation of Pr and Pfr in protonemata of the fern *Adiantum capillus-veneris* L. Photochem. Photobiol. 35, 533–536.

Kadota, A., Wada, M., Furuya, M., 1985. Phytochrome-mediated polarotropism of *Adiantum capillus-veneris* L. protonemata as analyzed by microbeam irradiation with polarized light. Planta 165, 30–36.

Kafatos, M., Nadeau, R., 1990. The Conscious Universe. Part and Whole in Modern Physical Theory. Springer Verlag, New York.

Kaffarnik, F.A.R., Jones, A.M.E., Rathjen, J.P., Peck, S.C., 2009. Effector proteins of the bacterial pathogen *Pseudomonas syringae* alter the extracellular proteome of the host plant, *Arabidopsis thaliana*. Mol. Cell. Proteom. 8, 145–156.

Kagawa, T., Beevers, H., 1975. The development of microbodies (glyoxysomes and leaf peroxisomes) in cotyledons of germinating watermelon seedlings. Plant Physiol. 55, 258–264.

Kagawa, T., McGregor, D.I., Beevers, H., 1973. Development of enzymes in the cotyledons of watermelon seedlings. Plant Physiol. 51, 66–71.

Kagawa, Y., Racker, E., 1966a. Partial resolution of the enzymes catalyzing oxidative phosphorylation. IX. Reconstruction of oligomycin-sensitive adenosine triphosphatase. J. Biol. Chem. 241, 2467–2474.

Kagawa, Y., Racker, E., 1966b. Partial resolution of the enzymes catalyzing oxidative phosphorylation. X. Correlation of morphology and function in submitochondrial particles. J. Biol. Chem. 241, 2475–2482.

Kagawa, Y., Racker, E., 1971. Partial resolution of the enzymes catalyzing oxidative phosphorylation. XXV. Reconstitution of vesicles catalyzing 32Pi-adenosine triphosphate exchange. J. Biol. Chem. 246, 5477–5487.

Kagawa, T., Sakai, T., Suetsugu, N., Oikawa, K., Ishiguro, S., Kato, T., Tabata, S., Okada, K., Wada, M., 2001. *Arabidopsis* NPL1: a phototropin homolog controlling the chloroplast high-light avoidance response. Science 291, 2138–2141.

Kahn, F., 1919. Die Zelle. Kosmos, Stuttgart, Germany.

Kahvejian, A., Quackenbush, J., Thompson, J.F., 2008. What would you do if you could sequence everything? Nat. Biotechnol. 26, 1133–1135.

Kaim, G., Dimroth, P., 1999. ATP synthesis by F-type ATP synthase is obligatorily dependent on the transmembrane voltage. EMBO J. 18, 4118–4127.

Kaiser, J., 2003. Sipping from a posioned chalice. Science 302, 376–379.

Kaiser, J., 2008. A plan to capture human diversity in 1000 genomes. Science 319, 395.

Kaiser, B.N., Finnegan, P.M., Tyerman, S.D., Whitehead, L.F., Bergersen, F.J., Day, D.A., Udvardi, M.K., 1998. Characterization of an ammonium transport protein from the peribacteroid membrane of soybean nodules. Science 281, 1202–1206.

Kaiser, G., Heber, U., 1984. Sucrose transport into vacuoles isolated from barley mesophyll protoplasts. Planta 161, 562–568.

Kaiser, R.J., MacKellar, S.L., Vinayak, R.S., Sanders, J.Z., Saavedra, R.A., Hood, L.E., 1989. Specific-primer-directed DNA sequencing using automated fluorescence detection. Nucleic Acids Res. 17, 6087–6102.

Kaiser, G., Martinoia, E., Wiemken, A., 1982. Rapid appearance of photosynthetic products in the vacuoles isolated from barley mesophyll protoplasts by a new fast method. Z. Pflanzenphysiol. 107, 103–113.

Kaiser, C.A., Schekman, R., 1990. Distinct sets of *SEC* genes govern transport vesicle formation and fusion early in the secretory pathway. Cell 61, 723–733.

Kakimoto, T., Shibaoka, H., 1987. Actin filaments and microtubules in the preprophase band and phragmoplast of tobacco cells. Protoplasma 140, 151–156.

Kakimoto, T., Shibaoka, H., 1988. Cytoskeletal ultrastructure of phragmoplast-nuclei complexes isolated from cultured tobacco cells. Protoplasma (Suppl. 2), 95–103.

Kakimoto, T., Shibaoka, H., 1992. Synthesis of polysaccharides in phragmoplasts isolated from tobacco BY-2 cells. Plant Cell Physiol. 33, 353–361.

Kalanon, M., McFadden, G.I., 2008. The chloroplast protein translocation complexes of *Chlamydomonas reinhardtii*: a bioinformatic comparison of Toc and Tic components in plants, green algae and red algae. Genetics 179, 95–112.

Kalckar, H., 1937. Phosphorylation in kidney tissue. Enzymologia 2, 47–52.

Kalckar, H., 1939. Coupling between phosphorylations and oxidations in kidney extracts. Enzymologia 6, 209–212.

Kalckar, H.M., 1941. The nature of energetic coupling in biological syntheses. Chem. Rev. 28, 71–178.

Kalckar, H.M., 1942. The function of phosphate in cellular assimilations. Biol. Rev. 17, 28–45.

Kalckar, H.M., 1944. The fuction of phosphate in enzymatic syntheses. Ann. N.Y. Acad. Sci. 45, 395–408.

Kalckar, H.M., 1966. Lipmann and the "squiggle". In: Kaplan, N.O., Kennedy, E.P. (Eds.), Current Aspects of Biochemical Energetics. Fritz Lipmann Dedicatory Volume. Academic Press, New York, pp. 1–8.

Kaldenhoff, R., 2006. Besides water: functions of plant membrane intrinsic proteins and aquaporins. Prog. Bot. 67, 206–218.

Kaldenhoff, R., Kai, L., Uehlein, N., 2014. Aquaporins and membrane diffusion of CO_2 in living organism. Biochim. Biophys. Acta 1840, 1592–1595.

Kalderon, D., Roberts, B.L., Richardson, W.D., Smith, A.E., 1984. A short amino acid sequence able to specify nuclear location. Cell 39, 499–509.

Kalf, G.F., Ch'ih, J.J., 1968. Purification and properties of deoxyribonucleic acid polymerase from rat liver mitochondria. J. Biol. Chem. 243, 4904–4916.

Kallio, M.J., McCleland, M.L., Stukenberg, P.T., Gorbsky, G.J., 2002. Inhibition of aurora B kinase blocks chromosome segregation, overrides the spindle checkpoint, and perturbs microtubule dynamics in mitosis. Curr. Biol. 12, 900–905.

Kalnins, V.I. (Ed.), 1992. The Centrosome. Cell Biology: A Series of Monographs. Academic Press, San Diego.

Kalwarczyk, T., Ziebacz, N., Bielejewska, A., Zaboklicka, E., Koynov, K., Szymański, J., Wilk, A., Patkowski, A., Gapiński, J., Butt, H.-J., Hołyst, R., 2011. Comparative analysis of viscosity of complex liquids and cytoplasm of mammalian cells at the nanoscale. Nano Lett. 11, 2157–2163.

Kamen, M.D., 1955. Nitrogen fixation. In: Scientific American. The Physics and Chemistry of Life. Simon and Schuster, New York, pp. 48–55.

Kamen, M.D., 1963. Primary Processes in Photosynthesis. Academic Press, New York.

Kamen, M.D., 1985. Radiant Science, Dark Politics. A Memoir of the Nuclear Age. University of California Press, Berkeley.

Kamen, M.D., 1995. Nitrogen fixation. In: Scientific American (Ed.), The Physics and Chemistry of Life. Simon and Schuster, New York, pp. 48–55.

Kamen, M.D., Gest, H., 1949. Evidence for a nitrogenase system in a photosynthetic bacterium Rhodospirillum rubrum. Science 109, 560.

Kamimura, S., Takahashi, K., 1981. Direct measurement of the force of microtubule sliding in flagella. Nature 293, 566–568.

Kaminer, B., 1977. Search and Discovery: A Tribute to Albert Szent-Györgyi. Academic Press, New York.

Kaminskyj, S.G.W., Heath, I.B., 1995. Integrin and spectrin homologues, and cytoplasm-wall adhension in tip growth. J. Cell Sci. 108, 849–856.

Kamitsubo, E., 1966. Motile protoplasmic fibrils in cells of characeae. II. Proc. Jpn. Acad. Sci. 42, 640–643.

Kamitsubo, E., 1972a. Destruction and restoration of the protoplasmic fibrillar structure responsible for streaming in the *Nitella* cell. Symp. Int. Soc. Cell. Biol. 23, 123–130.

Kamitsubo, E., 1972b. Motile protoplasmic fibrils in cells of characeae. Protoplasma 74, 53–70.

Kamitsubo, E., 1972c. Protoplasmic fibrillar structure responsible for cytoplasmic streaming in the characean cells. Cell 4, 12–19.

Kamitsubo, E., 1980. Cytoplasmic streaming in characean cells: role of subcortical fibrils. Can. J. Bot. 58, 760–765.

Kamitsubo, E., 1981. Effect of supraoptimal temperatures on the function of the subcortical fibrils and an endoplasmic factor in *Nitella* internodes. Protoplasma 109, 3–12.

Kamitsubo, E., Kikuyama, M., 1994. Measurement of motive force for cytoplasmic streaming in characean internodes. Protoplasma 180, 153–157.

Kamitsubo, E., Kikuyama, M., Kaneda, I., 1988. Apparent viscosity of the endoplasm of characean internodal cells measured by the centrifuge method. Protoplasma 5 (Suppl. 1), 10–14.

Kamitsubo, E., Ohashi, Y., Kikuyama, M., 1989. Cytoplasmic streaming in internodal cells of *Nitella* under centrifugal acceleration: a study done with a newly constructed centrifuge microscope. Protoplasma 152, 148–155.

Kamiya, N., 1940. Control of protoplasmic streaming. Science 92, 462–463.

Kamiya, N., 1942. Physical aspects of protoplasmic streaming. The Structure of Protoplasm. In: Seifriz (Ed.), Monogr. Amer. Soc. Plant Physiol., Ames-Iowa, pp. 199–244.

Kamiya, N., 1950a. The rate of protoplasmic flow in the myxomycete plasmodium. Cytologia 15, 183–193, 194–204.

Kamiya, N., 1950b. The rate of the protoplasmic flow in the myxomycete plasmodium. II. Cytologia 15, 194–204.

Kamiya, N., 1950c. The protoplasmic flow in the myxomycete plasmodium as revealed by a volumetric analysis. Protoplasma 39, 344–357.

Kamiya, N., 1956. Memoriam William Seifriz. Protoplasma 45, 513–524.

Kamiya, N., 1959. Protoplasmic Streaming, Protoplasmatologia. Bd 8, 3a. Springer-Verlag, Vienna.

Kamiya, T., Borghi, M., Wang, P., Danku, J.M.C., Kalmbach, L., Hosmani, P.S., Naseer, S., Fujiwara, T., Geldner, N., Salt, D.E., 2015. The MYB36 transcription factor orchestrates Casparian strip formation. Proc. Natl. Acad. Sci. U.S.A. 112, 10533–10538.

Kamiya, N., Kuroda, K., 1956. Velocity distribution of the protoplasmic streaming in *Nitella* cells. Bot. Mag. (Tokyo) 69, 544–554.

Kamiya, N., Kuroda, K., 1957. Cell operation in *Nitella* III. Specific gravity of the cell sap and endoplasm. Proc. Jpn. Acad. Sci. 33, 403–406.

Kamiya, N., Kuroda, K., 1958. Measurement of the motive force of the protoplasmic rotation in *Nitella*. Protoplasma 50, 144–148.

Kamiya, N., Kuroda, K., 1965. In: Rotational protoplasmic streaming in Nitella and some physical properties of the endoplasm. Proc. of the Fourth International Congress on Rheology. Part 4. Symposium on Biorheology. Brown University, Aug 26–30, 1963. John Wiley and Sons, New York.

Kamiya, N., Seifriz, W., 1954. Torsion in a protoplasmic thread. Exp. Cell Res. 6, 1–16.

Kamiya, N., Tazawa, M., 1956. Studies on water permeability of a single plant cell by means of transcellular osmosis. Protoplasma 46, 394–422.

Kamiya, N., Tazawa, M., Takata, T., 1962. Water permeability of the cell wall in *Nitella*. Plant Cell Physiol. 3, 285–292.

Kamiya, N., Tazawa, M., Takata, T., 1963. The relation of turgor pressure to cell volume in Nitella with special reference to mechanical properties of the cell wall. Protoplasma 57, 501–521.

Kanamaru, K., Fujiwara, M., Kim, M., Nagashima, A., Nakazato, E., Tanaka, K., Takahashi, H., 2000. Chloroplast targeting, distribution and transcriptional fluctuation of AtMinD1, a eubacteria-type factor critical for chloroplast division. Plant Cell Physiol. 41, 1119–1128.

Kanczewska, J., Marco, S., Vandermeeren, C., Maudoux, O., Rigaud, J.L., Boutry, M., 2005. Activation of the plant plasma membrane H^+-ATPase by phosphorylation and binding of 14-3-3 proteins converts a dimer into a hexamer. Proc. Natl. Acad. Sci. U.S.A. 102, 11675–11680.

Kandasamy, M.K., Paolillo, D.J., Faraday, C.D., Nasrallah, J.B., Nasrallah, M.E., 1989. The S-locus specific glycoproteins of *Brassica* accumulate in the cell wall of developing stigma papillae. Dev. Biol. 134, 426–472.

Kandasamy, M.K., Parthasarathy, M.V., Nasrallah, M.E., 1991. High pressure freezing and freeze substitution improve immunolabeling of S-locus specific glycoproteins in the stigma papillae of *Brassica*. Protoplasma 162, 187–191.

Kandell, J., April 13, 2001. Richard E. Schultes, 86, dies; Trailblazing authority on hallucinogenic plants New York Times. http://www.nytimes.com/2001/04/13/us/richard-e-schultes-86-dies-trailblazing-authority-on-hallucinogenic-plants.html.

Kaneko, T., Saito, C., Shimmen, T., Kikuyama, M., 2005. Possible involvement of mechanosensitive Ca^{2+} channels of plasma membrane in mechanoperception in *Chara*. Plant Cell Physiol. 46, 130–135.

Kaneko, T.S., Sato, M., Osumi, M., Muroi, M., Takahashi, A., 1996. Two isoforms of acid phosphatase secreted by tobacco protoplasts: differential effect of brefeldin A on their secretion. Plant Cell Rep. 15, 409–413.

Kang, B.G., Burg, S.P., 1971. Rapid change in water flux induced by auxins. Proc. Natl. Acad. Sci. U.S.A. 68, 1730–1733.

Kang, B., Busse, J., Bednarek, S., 2003. Members of the *Arabidopsis* dynamin-like gene family, ADL1, are essential for plant cytokinesis and polarized cell growth. Plant Cell 15, 899–913.

Kang, B.-H., Nielsen, E., Preuss, M.L., Mastronarde, D., Staehelin, L.A., 2011. Electron tomography of RabA4b- and PI-4Kβ1-labeled *trans* Golgi network compartments in *Arabidopsis*. Traffic 12, 313–329.

Kang, B.-H., Staehelin, L.A., 2008. ER-to-Golgi transport by COPII vesicles in *Arabidopsis* involves a ribosome-excluding scaffold that is transferred with the vesicles to the Golgi matrix. Protoplasma 234, 51–64.

Kanigel, R., 1986. Apprentice to Genius: The Making of a Scientific Dynasty. Macmillan, New York.

Kao, H.P., Abney, J.R., Verkman, A.S., 1993. Translational mobility of a small solute in cell cytoplasm. J. Cell Biol. 120, 613–627.

Kao, J.P.Y., Alderton, J.M., Tsien, R.Y., Steinhardt, R.A., 1990. Active involvement of Ca^{2+} in mitotic progression of Swiss 3T3 fibroblasts. J. Cell Biol. 111, 183–196.

Kaplan, D.R., 1992. The relationship of cells to organisms in plants: problem and implications of an organismal perspective. Int. J. Plant Sci. 153, S28–S37.

Kaplan, D.R., Hagemann, W., 1991. The relationship of cell and organism in vaqcular plants: are cells the building blocks of plant form? BioScience 41, 693–703.

Kapulnik, Y., Yalpani, N., Raskin, I., 1992. Salicylic acid induces cyanide-resistant respiration in tobacco cell-suspension cultures. Plant Physiol. 100, 1921–1926.

Kara, M., Hillenkamp, F., 1988. Laser desorption ionization of proteins with molecular masses exceeding 10,000 daltons. Anal. Chem. 60, 2299–2301.

Karahara, I., 2014. Cytoskeletons. In: Noguchi, T., Kawano, S., Tsukaya, H., Matsunaga, S., Sakai, A., Karahara, I., Hayashi, Y. (Eds.), Atlas of Plant Cell Structure. Springer, Japan, pp. 107–136.

Karahara, I., Shibaoka, H., 1988. Effects of brefeldin A on the development of the Casparian strip in pea epicotyls. Protoplasma 77, 243–269.

Karahara, I., Shibaoka, H., 1992. Isolation of Casparian strips from pea roots. Plant Cell Physiol. 33, 555–561.

Karahara, I., Suda, J., Tahara, H., Yokota, E., Shimmen, T., Misaki, K., Yonemura, S., Staehelin, L.A., Mineyuki, Y., 2009. The preprophase band is a localized center of clathrin-mediated endocytosis in late prophase cells of the onion cotyledon epidermis. Plant J. 57, 819–831.

Karban, R., 2015. Plant Sensing & Communication. University of Chicago Press, Chicago.

Karcher, R.L., Roland, J.T., Zappacosta, F., Huddleston, M.J., Annan, R.S., Carr, S.A., Gelfand, V.I., 2001. Cell cycle regulation of myosin-V by calcium/calmodulin-dependent protein kinase II. Science 293, 1317–1320.

Kardailsky, I., Shukla, V.K., Ahn, J.H., Dagenais, N., Christensen, S.K., Nguyen, J.T., Chory, J., Harrison, M.J., Weigel, D., 1999. Activation tagging of the floral inducer FT. Science 286, 1962–1965.

Karling, J.S., 1939. Schleiden's contribution to the cell theory. Am. Nat. 73, 517–537.

Karnik, S.K., Trelease, R.N., 2007. *Arabidopsis* peroxin 16 trafficks through the ER and an intermediate compartment to pre-existing peroxisomes via overlapping molecular targeting signals. J. Exp. Bot. 58, 1677–1693.

Karpen, G.H., Schaefer, J., Laird, C.D., 1988. A *Drosophila* rRNA gene located in euchromatin is active in transcription and nucleolus formation. Genes Dev. 2, 1745–1763.

Karplus, M., Weaver, D.L., 1994. Protein folding dynamics: the diffusion-collision model and experimental data. Protein Sci. 3, 650–668.

Karrer, E.E., Lincoln, J.E., Hogenhout, S., Bennett, A.B., Bostock, R.M., Martineau, B., Lucas, W.J., Gilchrist, D.G., Alexander, D., 1995. In situ isolation of mRNA from individual plant cells: creation of cell-specific cDNA libraries. Proc. Natl. Acad. Sci. U.S.A. 92, 3814–3818.

Kasahara, M., Kagawa, T., Oikawa, K., Suetsugu, N., Miyao, M., Wada, M., 2002. Chloroplast avoidance movement reduces photodamage in plants. Nature 420, 829–832.

Kasamo, K., 2003. Regulation of plasma membrane H^+-ATPase activity by the membrane environment. J. Plant Res. 116, 517–523.

Kasamo, K., Sakakibara, Y., 1995. The plasma membrane H^+-ATPase from higher plants: functional reconstitution into liposomes and its regulation by phospholipids. Plant Sci. 111, 117–131.

Kasamo, K., Yamanishi, H., Kagita, F., Saji, H., 1991. Reconstruction of tonoplast H^+-ATPase from mung bean (Vigna radiata L.) hypocotyls in liposomes. Plant Cell Physiol. 32, 643–651.

Kashem, M.A., Itoh, K., Iwabuchi, S., Hori, H., Mitsui, T., 2000. Possible involvement of phosphoinositide-Ca^{2+} signaling in the regulation of α-amylase expression and germination of rice seed (*Oryza sativa* L. Plant Cell Physiol. 41, 399–407.

Kashima, K., Yuki, Y., Mejima, M., Kurokawa, S., Suzuki, Y., Minakawa, S., Takeyama, N., Fukuyama, Y., Azegami, T., Tanimoto, T., Kuroda, M., Tamura, M., Gomi, Y., Kiyono, H., 2016. Good manufacturing practices production of a purification-free oral cholera vaccine expressed in transgenic rice plants. Plant Cell Rep. 35, 667–679.

Kashiyama, T., Ito, K., Yamamoto, K., 2001. Functional expression of a chimeric myosin-containing motor domain of Chara myosin and neck and tail domains of *Dictyostelium* myosin II. J. Mol. Biol. 311, 461–466.

Kashiyama, T., Kimura, N., Mimura, T., Yamamoto, K., 2000. Cloning and characterization of a myosin from characean alga, the fastest motor protein in the world. Biochem. J. 127, 1065–1070.

Kasianowicz, J.J., Brandin, E., Branton, D., Deamer, D.W., 1996. Characterization of individual polynucleotide molecules using a membrane channel. Proc. Natl. Acad. Sci. U.S.A. 93, 13770–13773.

Kass, L.B., 2003. Anecdotal, historical and critical commentaries on genetics. Records and recollections: a new look at Barbara McClintock, Nobel-prize-winning geneticist. Genetics 164, 1251–1260.

Kasting, J.F., Ackerman, T.P., 1986. Climatic consequences of very high carbon dioxide levels in the early Earth's atmosphere. Science 234, 1383–1385.

Katahara, I., Staehelin, L.A., Mineyuki, Y., 2010. A role for endocytosis in plant cytokinesis. Commun. Integr. Biol. 3, 36–38.

Kataoka, H., 1975a. Phototropism in *Vaucheria geminata*. I. The action spectrum. Plant Cell Physiol. 16, 427–437.

Kataoka, H., 1975b. Phototropism in *Vaucheria geminata*. II. The mechanism of bending and branching. Plant Cell Physiol. 16, 439–448.

Kataoka, H., 2015. Gustav Senn (1875-1945): the pioneer of chloroplast movement research. J. Integr. Plant Biol. 57, 4–13.

Katchalsky, A., Curran, P.F., 1965. Nonequilibrium Thermodynamics in Biophysics. Harvard University Press, Cambridge, MA.

Katchalsky, A., Lifson, S., 1954. Muscle as a machine. Sci. Am. 190 (3), 72–76.

Katcher, B., 1985. Direct visualization of organelle movement along actin filaments dissociated from characean algae. Science 227, 1355–1357.

Katcher, B., Reese, T.S., 1988. The mechanism of cytoplasmic streaming in characean algal cells: sliding of endoplasmic reticulum along actin filaments. J. Cell Biol. 106, 1545–1552.

Kato, A., Hayashi, M., Nishimura, M., 1999. Oligomeric proteins containing N-terminal targeting signals are imported into peroxisomes in transgenic *Arabidopsis*. Plant Cell Physiol. 40, 586–591.

Kato, T., Tonomura, Y., 1977. Identification of myosin in *Nitella flexilis*. J. Biochem. 82, 777–782.

Katoh, S., 1960. A new copper protein from *Chlorella ellipsoidea*. Nature 186, 533–534.

Katoh, S., 1995. The discovery and function of plastocyanin: a personal account. Photosynth. Res. 43, 177–189.

Katoh, S., Takamiya, A., 1961. A new leaf copper protein "plastocyanin," a natural Hill oxidant. Nature 189, 665–666.

Katsaros, C., Kreimer, G., Melkonian, M., 1991. Localization of tubulin and a centrin-homologue in vegetative cells and developing gametangia of *Ectocarpus siliculosus* (Dillio) Lyngb. (Phaeophyceae Ectocarpales). A combined immunofluorescence and confocal laser scanning microscope study. Bot. Acta 104, 87–92.

Katsuhara, M., Hanba, Y., Shibasaka, M., Hayashi, Y., Hayakawa, T., Kasamo, K., 2003. Increase in CO_2 permeability (diffusion conductance) in leaves of transgenic rice plant over-expressing barley aquaporin. Plant Cell Physiol. 44, S86.

Katsuhara, M., Hanba, Y.T., Shiratake, K., Maeshima, M., 2008. Expanding roles of plant aquaporins in plasma membranes and cell organelles. Funct. Plant Biol. 35, 1–14.

Katsuhara, M., Mimura, T., Tazawa, M., 1989. Patch clamp study on a Ca^{2+}-regulated K^+ channel in the tonoplast of the brackish characeae *Lamprothamnium succinctum*. Plant Cell Physiol. 30, 549–555.

Katsuta, J., Shibaoka, H., 1988. The roles of the cytoskeleton and the cell wall in nuclear positioning in tobacco BY-2 cells. Plant Cell Physiol. 29, 403–413.

Kaufmann, W., 1902. Über die elektromagnetische Masse des Elektrons. Göttinger Nachrichten (5), 291–296.

Kaufmann, W., 1961. Philosophic Classics. Thales to St. Thomas. Prentice-Hall, Engelwood Cliffs, NJ.

Kaufmann, K., Muiño, J.M., Jauregui, R., Airoldi, C.A., Smaczniak, C., Krajewski, P., Angenent, G.C., 2009. Target genes of the MADS transcription factor SEPALLATA3: integration of developmental and hormonal pathways in the *Arabidopsis* flower. PLoS Biol. 7, e1000090.

Kaufmann, K., Muiño, J.M., Østerås, M., Farinelli, L., Krajewski, P., Angenent, G.C., 2010a. Chromatin immunoprecipitation (ChIP) of plant transcription factors followed by sequencing (ChIP-SEQ) or hybridization to whole genome arrays (ChIP-CHIP). Nat. Protoc. 5, 457–472.

Kaufmann, K., Wellmer, F., Muiño, J.M., Ferrier, T., Wuest, S.E., Kumar, V., Serrano-Mislata, A., Madueno, F., Krajewski, P., Meyerowitz, E.M., Angenent, G.C., Riechmann, J.L., 2010b. Orchestration of floral initiation by APETALA1. Science 328, 85–89.

Kaur, N., Hu, J., 2011. Defining the plant peroxisome proteome: from *Arabidopsis* to rice. Front. Plant Sci. 2, 103.

Kauss, H., 1987. Some aspects of calcium-dependent regulation in plant metabolism. Ann. Rev. Plant Physiol. 38, 47–72.

Kavenoff, R., Ryder, O.A., 1976. Electron microscopy of membrane-associated folded chromosomes of *Escherichia coli*. Chromosoma 55, 13–25.

Kawai-Toyooka, H., Kuramoto, C., Orui, K., Motoyama, K., Kikuchi, K., Kanegae, T., Wada, M., 2004. DNA interference: a simple and efficient gene-silencing system for high-throughput functional analysis in the fern *Adiantum*. Plant Cell Physiol. 45, 1648–1657.

Kawakatsu, T., Huang, S.S., Jupe, F., Sasaki, E., Schmitz, R.J., Urich, M.A., Castanon, R., Nery, J.R., Barragan, C., He, Y., et al., 2016. Epigenomic diversity in a global collection of *Arabidopsis thaliana* accessions. Cell 166, 492–505.

Kawakatsu, T., Kikuchi, A., Shimmen, T., Sonobe, S., 2000. Interaction of actin filaments with the plasma membrane in Amoeba proteus: studies using a cell model and isolated plasma membrane. Cell Struct. Funct. 25, 269–277.

Kawano, S., 2014a. Chloroplasts. In: Noguchi, T., Kawano, S., Tsukaya, H., Matsunaga, S., Sakai, A., Karahara, I., Hayashi, Y. (Eds.), Atlas of Plant Cell Structure. Springer, Japan, pp. 45–70.

Kawano, S., 2014b. Mitochondria. In: Noguchi, T., Kawano, S., Tsukaya, H., Matsunaga, S., Sakai, A., Karahara, I., Hayashi, Y. (Eds.), Atlas of Plant Cell Structure. Springer, Japan, pp. 25–44.

Kawano, S., Takano, H., Imai, J., Mori, K., Kuroiwa, T., 1993. A genetic system controlling mitochondrial fusion in the slime mould, *Physarum polycephalum*. Genetics 133, 213–224.

Kawasaki, H., Kretsinger, R.H., 1995. Protein Profile. Volume 2. Calcium-Binding Proteins 1: EF-hands. Academic Press, London.

Kawashima, I., Inokuchi, Y., Chino, M., Kimura, M., Shimizu, N., 1991. Isolation of a gene for metalothionein-like protein from soybean. Plant Cell Physiol. 32, 913–916.

Kayum, M.A., Park, J.I., Nath, U.K., Biswas, M.K., Kim, H.T., Nou, I.S., 2017. Genome-wide expression profiling of aquaporin genes confer responses to abiotic and biotic stresses in *Brassica rapa*. BMC Plant Biol. 17, 23.

Kazantzakis, A., 1953. Zorba the Greek. Simon and Schuster, New York.

Kearney, E.B., Singer, T.P., 1956. Studies on succinic dehydrogenase. Int. J. Biol. Chem. 219, 963–975.

Kebeish, R., Niessen, M., Thiruveedhi, K., Bari, R., Hirsch, H.J., Rosenkranz, R., Stäbler, N., Schönfeld, B., Kreuzaler, F., Peterhänsel, C., 2007. Chloroplastic photorespiratory bypass increases photosynthesis and biomass production in *Arabidopsis thaliana*. Nat. Biotechnol. 25, 539–540.

Keddie, F.M., Barajas, L., 1969. Three-dimensional reconstruction of *Pityosporum* yeast cells based on serial section electron microscopy. J. Ultrastruct. Res. 29, 260–275.

Keegstra, K., 2010. Plant cell walls. Plant Physiol. 154, 483–486.

Keegstra, K., Olsen, L.J., Theg, S.M., 1989. Chloroplastic precursors and their transport across the envelope membranes. Ann. Rev. Plant Physiol. Mol. Biol. 40, 471–501.

Keegstra, K., Talmadge, K.W., Bauer, W.D., Albersheim, P., 1973. The structure of plant cell walls. III. A model of the walls of suspension cultured sycamore cells based on the inter connections of the macromolecular components. Plant Physiol. 51, 188–196.

Keeling, P.J., 1998. A kingdom's progress: archezoa and the origin of eukaryotes. BioEssays 20, 87–95.

Keeling, P.J., Doolittle, W.F., 1995. Archaea: narrowing the gap between prokaryotes and eukaryotes. Proc. Natl. Acad. Sci. U.S.A. 92, 5761–5764.

Keenan, T.W., Morré, D.J., 1970. Phospholipid class and fatty acid composition of Golgi apparatus isolated from rat liver and comparison with other cell fractions. Biochemistry 9, 19–25.

Kehoe, D.M., Villand, P., Somerville, S., 1999. DNA microarrays for studies of higher plants and other photosynthetic organisms. Trends Plant Sci. 4, 38–41.

Keifer, D.W., Spanswick, R.M., 1978. Activity of the electrogenic pump in *Chara corallina* as inferred from measurements of the membrane potential, conductance, and potassium permeability. Plant Physiol. 62, 653–661.

Keifer, D.W., Spanswick, R.M., 1979. Correlation of adenosine triphosphatate levels in *Chara corallina* with the activity of the electrogenic pump. Plant Physiol. 64, 165–168.

Keilin, D., 1966. The History of Cell Respiration and Cytochrome. Cambridge University Press, Cambridge.

Keith, C.H., 1987. Effect of microinjected calcium calmodulin on mitosis in PtK2 cells. Cell Motil. Cytoskelet. 7, 1–9.

Keith, A.D., Snipes, W., 1974. Viscosity of cellular protoplasm. Science 183, 666–668.

Kelleher, N.L., 2004. Top-down proteomics. Anal. Chem. 76, 196A–203A.

Keller, E.B., 1951. Turnover of proteins of adult rat liver in vivo. Fed. Proc. 10, 206.

Keller, F., 1991. Carbohydrate transport in discs of storage parenchyma of celery petioles. 2. Uptake of glucose and fructose. New Phytol. 117, 413–422.

Keller, F., 1992. Transport of stachyose and sucrose by vacuoles of Japanese artichoke (*Stachys sieboldii*) tubers. Plant Physiol. 98, 442–445.

Keller, B.U., Hedrich, R., Raschke, K., 1989. Voltage-dependent anion channels in the plasma membrane of guard cells. Nature 341, 450–453.

Keller, G.-A., Krisans, S., Gould, S.J., Sommer, J.M., Wang, C.C., Schliebs, W., Kunau, W., Bordy, S., Subramini, S., 1991. Evolutionary conservation of microbody targeting signal that targets proteins to peroxisomes, glyoxysomes, and glycosomes. J. Cell Biol. 114, 893–904.

Keller, F., Matile, P., 1985. The role of the vacuole in storage and mobilization of stachyose in tubers of Stachys sieboldii. J. Plant Physiol. 119, 369–380.

Keller, L.C., Romijn, E.P., Zamora, I., Yates, J.R., Marshall, W.F., 2005. Proteomic analysis of isolated *Chlamydomonas* centrioles reveals orthologs of ciliary-disease genes. Curr. Biol. 15, 1090–1098.

Keller, E.B., Zamecnik, P.C., 1956. The effect of guanosine diphosphate and triphosphate on the incorporation of labeled amino acids into proteins. J. Biol. Chem. 221, 45–59.

Keller, E.B., Zamecnik, P.C., Loftfield, R.B., 1954. The role of microsomes in the incorporation of amino acids into proteins. J. Histochem. Cytochem. 2, 378–386.

Kelley, K.W. (Ed.), 1988. The Home Planet. Addison-Wesley Publishing Co., Reading, MA, and Mir Publishers, Moscow.

Kelley, B., 2009. Agri-photonics. SPIE 7, 14–17. https://spie.org/membership/spie-professional-magazine/spie-professional-archives-and-special-content/july2009-spie-professional/agri-photonics.

Kellner, R., De la Concepcion, J.C., Maqbool, A., Kamoun, S., Dagdas, Y.F., 2017. ATG8 expansion: a driver of selective autophagy diversification? Trends Plant Sci. 22, 204–214.

Kelly, A.A., Kalisch, B., Hölzl, G., Schulze, S., Thiele, J., Melzer, M., Roston, R.L., Benning, C., Dörmann, P., 2016. Synthesis and transfer of galactolipids in the chloroplast envelope membranes of *Arabidopsis thaliana*. Proc. Natl. Acad. Sci. U.S.A. 113, 10714–10719.

Kelly, T.J., Smith, H.O., 1970. A restriction enzyme from *Hemophilus influenzae*. II. Base sequence of the recognition site. J. Mol. Biol. 51, 393–409.

Kelso, J.M., Turner, J.S., 1955. Protoplasmic streaming in *Tradescantia* 1. Effects of indole acetic acid and other growth promoting substanes on streaming. Aust. J. Bio. Sci. 8, 19–35.

Kelvin, Lord (William Thomson). 1889–1894, Popular Lectures and Addresses. MacMillan, London.

Kelvin, L., 1897. The age of the Earth as an abode for life. Annu. Rep. Smithson. Inst. 337–357.

Kendrew, J.C., December 11, 1962. Myoglobin and the structure of proteins. Nobel Lecture.

Kennard, J.L., Cleary, A.L., 1997. Pre-mitotic nuclear migration in subsidiary mother cells of Tradescantia occurs in G1 of the cell cycle and required F-actin. Cell Motil. Cytoskelet. 36, 55–67.

Kennedy, J.F., 1961. For the freedom of man. We must all work together. Vital Speeches Day 27, 226–227.

Kennedy, J.F., 1962. Address at Rice University, Houston, Texas, 12 September 1962. https://er.jsc.nasa.gov/seh/ricetalk.htm.

Kennedy, J.F., 1963. Commencement Address at American University in Washington on June 10, 1963. In: Meyers, J. (Ed.), John Fitzgerald Kennedy… As We Remember Him. Athenium, New York, 1965.

Kennedy, J.F., 1963. Address before the 18th General Assembly of the United Nations. New York, New York, USA, September 20, 1963.

Kennedy, J.F., 1964. A Nation of Immigrants. Harper & Row, New York.

Kennedy, E.P., 1992. Sailing to Byzantium. Annu. Rev. Biochem. 61, 1–28.

Kennedy, E.M., 2000. Bentley College, Waltham, Massachusetts, USA. May 20, 2000.

Kennedy, E.P., 2001. Hitler's gift and the era of biosynthesis. J. Biol. Chem. 276, 42619–42631.

Kennedy, R.F., June 6, 1966. Day of Affirmation Address. University of Capetown, Capetown, South Africa.

Kennedy, E.P., Lehninger, A.L., 1949. Oxidation of fatty acids and tricarboxylic acid cycle intermediates by isolated rat liver mitochondria. J. Biol. Chem. 179, 957–972.

Kennedy, J.F., November 2, 1960. Presidential Campaign Speech Proposing the United States. Peace CorpsSan, Francisco, California, USA.

Kennedy, R.T., Oates, M.D., Cooper, B.R., Nickerson, B., Jorgenson, J.W., 1989. Microcolumn separations and the analysis of single cells. Science 246, 57–63.

Kensler, C.J., Battista, S.P., 1963. Components of cigarette smoke with ciliary-depressant activity—their selective removal by filters containing activated charcoal. N. Engl. J. Med. 269, 1161–1166.

Kent, S.B., Hood, L.E., Beilan, H., Marriot, M., Meister, S., Geiser, T., 1984. A novel approach to automated peptide synthesis based on new insights into solid phase chemistry. In: Isymiya, N. (Ed.), Proceedings of the Japanese Peptide Symposium. Protein Resaerch Foundation, Osaka, Japan, pp. 217–222.

Kenyon, W.H., Black Jr., C.C., 1986. Electrophoretic analysis of protoplast, vacuole, and tonoplast vesicle proteins in crassulacean acid metabolism plants. Plant Physiol. 82, 916–924.

Kenyon, W.H., Kringstad, R., Black, C.C., 1978. Diurnal changes in th emalic acid content of vacuoles isolated from leaves of the Crassulacean acid metabolism plant, *Sedum telephium*. FEBS Lett. 94, 282–283.

Kenyon, W.H., Severson, R.F., Black Jr., C.C., 1985. Maintenance carbon cycle in crassulacean acid metabolism plant leaves. Plant Physiol. 77, 183–189.

Kenyon, D.H., Steinman, G., 1969. Biochemical Predestination. McGraw-Hill, New York.

Kerk, M.M., Ceserani, T., Tausta, S.L., Sussex, I.M., Nelson, T.M., 2003. Laser capture microdissection of cells from plant tissues. Plant Physiol. 132, 27–35.

Kerkeb, L., Donaire, J.P., Venema, K., Rodriguez-Rosales, M.P., 2001. Tolerance to NaCl induces changes in plasma membrane lipid composition, fluidity and H^+-ATPase activity of tomato calli. Phys. Plant 113, 217–224.

Kerkut, G.A., York, B., 1971. The Electrogenic Sodium Pump. Scientechnica Ltd., Bristol.

Kerrebrock, A.W., Moore, D.P., Wu, J.S., Orr-Weaver, I.L., 1995. Mei-S332, a *Drosophila* protein required for sister-chromatid cohesion, can localize to meiotic centromere regions. Cell 83, 247–256.

Kerridge, J.F., Matthews, M.S. (Eds.), University of Arizona Press, Chicago, IL.

Kerruth, S., Ataka, K., Frey, D., Schlichting, I., Heberle, J., 2014. Aureochrome 1 illuminated: structural changes of a transcription factor probed by molecular spectroscopy. PLoS One 9, e103307.

Kersey, Y.M., Hepler, P.K., Palevitz, B.A., Wessels, N.K., 1976. Polarity of actin filaments in characean algae. Proc. Natl. Acad. Sci. U.S.A. 73, 165–167.

Kersten, R.D., Lee, S., Fujita, D., Pluskal, T., Kram, S., Smith, J.E., Iwai, T., Noel, J.P., Fujita, M., Weng, J.K., 2017. A red algal bourbonane sesquiterpene synthase defined by microgram-scale NMR-coupled crystalline sponge XRD analysis. J. Am. Chem. Soc. 139, 16838–16844.

Kerstiens, G. (Ed.), 1996. Plant Cuticles: An Integrated Functional Approach. BIOS Scientific, Oxford.

Kessler, F., 2012. Chloroplast delivery by UPS. Science 338, 622–623.

Ketelaar, T., de Ruiiter, N., Niehren, S., 2014. Optical trapping in plant cells. Methods Mol. Biol. 1080, 259–265.

Ketelaar, T., Galway, M.E., Mulder, B.M., Emons, A.M.C., 2008. Rates of exocytosis and endocytosis in *Arabidopsis* root hairs and pollen tubes. J. Microsc. 231, 265–273.

Key, J.L., 1969. Hormones and nucleic acid metabolism. Ann. Rev. Plant Physiol. 20, 449–474.

Khan, S.U.M., Al-Shahry, M., Ingler Jr., W.B., 2002. Efficient photochemical water splitting by a chemically modified n- TiO_2. Science 297, 2243–2245.

Khattab, I.S., Bandarkar, F., Fakhree, M.A.A., Jouyban, A., 2012. Density, viscosity, and surface tension of water+ethanol mixtures from 293 to 323 K. Korean J. Chem. Eng. 29, 812–817.

Khodjakov, A., Cole, R.W., Bajer, A.S., Rieder, C.L., 1996. The force for poleward chromosome motion in Haemanthus cells acts along the length of the chromosome during metaphase but only at the kinetochore during anaphase. J. Cell Biol. 132, 1093–1104.

Khodjakov, A., Cole, R.W., McEwen, B.F., Buttle, K.F., Rieder, C.L., 1997. Chromosome fragments possessing only one kinetochore can congress to the spindle equator. J. Cell Biol. 136, 229–240.

Khodjakov, A., Cole, R.W., Oakley, B.R., Rieder, C.L., 2000. Centrosome-independent mitotic spindle formation in vertebrates. Curr. Biol. 10, 59–67.

Khodjakov, A., Pines, J., 2010. Centromere tension: a divisive issue. Nat. Cell Biol. 12, 919–923.

Khodjakov, A., Rieder, C.L., 1996. Kinetochores moving away from their associated pole do not exert significant pushing force on their chromosome. J. Cell Biol. 135, 315–327.

Khoo, U., Wolf, M.J., 1970. Origin and development of protein granules in maize endosperm. Am. J. Bot. 57, 1042–1050.

Khorana, H.G., 1961. Some Recent Developments in the Chemistry of Phosphate Esters of Biological Interest. John Wiley & Sons, New York.

Khorana, H.G., December 12, 1968a. Nucleic acid synthesis in the study of the genetic code. Nobel Lecture.

Khorana, H.G., 1968b. Polynucleotide synthesis and the genetic code. Harvey Lect. 62, 79–105.

Khorana, H.G., 1976. Synthesis in the study of nucleic acids. In: Kornberg, A., Horecker, B.L., Cornudella, L., Oro, J. (Eds.), Reflections on Biochemistry in Honour of Severo Ochoa. Pergamon Press, Oxford, pp. 273–281.

Khorana, H.G., Agarwal, K.L., Buchi, H., Caruthers, M.H., Gupta, N.K., Kleppe, K., Kumar, A., Otsuka, E., RajBhandary, U.L., van de Sande, J.H., Sgaramella, V., Terao, T., Weber, H., Yamada, T., 1972. Studies on polynucleotides. CIII. Total synthesis of the structural gene for an alanine transfer ribonucleic acid from yeast. J. Mol. Biol. 72, 209–217 and the following twelve manuscripts in this volume.

Khrapko, K.R., Lysov, Y.P., Khorlyn, A.A., Shick, V.V., Florentiev, V.L., Mirzabekov, A.D., 1989. An oligonucleotide hybridization approach to DNA sequencing. FEBS Lett. 256, 118–122.

Khurana, S., Arpin, M., Patterson, R., Donowitz, M., 1997. Ileal microvillar protein villin is tyrosine-phosphorylated and associates with PLC-gamma1. 1997. Role of cytoskeletal rearrangement in the carbachol-induced inhibition of ileal NaCl absorption. J. Biol. Chem. 272, 30115–30121.

Kidd, J., 1852. On the Adaptation of External Nature to the Physical Condition of Man. H. G. Bohn, London.

Kiebler, M., Becker, K., Pfanner, N., Neupert, W., 1993. Mitochondrial protein import: specific recognition and membrane translocation of preproteins. J. Membr. Biol. 135, 191–207.

Kieffer, W.F., 1963. The Mole Concept in Chemistry. Reinhold Publishing, New York.

Kiehart, D.P., 1981. Studies on the in vivo sensitivity of spindle microtubules to calcium ions and evidence for a vesicular calcium-sequestering system. J. Cell Biol. 88, 604–617.

Kieliszewski, M.J., Lamport, D.T.A., 1994. Extensin: repetitive motifs, functional sites, post-translational codes, and phylogeny. Plant J. 5, 157–172.

Kiemer, L., Cesareni, G., 2007. Comparative interactomics: comparing apples and pears? Trends Biotechnol. 25, 448–454.

Kiermayer, O., 1972. Beeinflussung der postmititischen Kernmigration von Microsterias denticulata Bréb. durch das Herbizid Trifluralin. Protoplasma 75, 421–426.

Kiermayer, O., Hepler, P.K., 1970. Hemmung der Kernmigration von Jochalgen (Micrastrias) durch Isopropyl-N-phenylcarbamat. Die Naturwissenschaften 5, 252.

Kiesel, A., 1930. Chemie des Protoplasmas. Protoplasmatologia Monograph 4, Berlin.

Kieselbach, T., Hagman, Å., Andersson, B., Schröder, W.P., 1998. The thylakoid lumen of chloroplasts − isolation and characterization. J. Biol. Chem. 273, 6710−6716.

Kiessling, J., Kruse, S., Rensing, S.A., Harter, K., Decker, E.L., Reski, R., 2000. Visualization of a cytoskeleton-like FtsZ network in chloroplasts. J. Cell Biol. 151, 945−950.

Kihara, A., Noda, T., Ishihara, N., Ohsumi, Y., 2001. Two distinct Vps34 phosphatidylinositol 3-kinase complexes function in autophagy and carboxypeptidase Y sorting in *Saccharomyces cerevisiae*. J. Cell Biol. 152, 519−530.

Kikuchi, S., Bédard, J., Hirano, M., Hirabayashi, Y., Oishi, M., Imai, M., Takase, M., Ide, T., Nakai, M., 2013. Uncovering the protein translocon at the chloroplast inner envelope membrane. Science 339, 571−574.

Kikuyama, M., Hara, Y., Shimada, K., Yamamoto, K., Hiramoto, Y., 1992. Intercellular transport of macromolecules in *Nitella*. Plant Cell Physiol. 33, 413−417.

Kikuyama, M., Shimmen, T., 1997. Role of Ca^{2+} on triggering tonoplast action potential in intact *Nitella flexilis*. Plant Cell Physiol. 38, 941−944.

Kikuyama, M., Tazawa, M., 1972. An attempt to introduce various unicellular organisms into the cytoplasm of *Nitella flexilis*. In: Proc. 37th Annual Meeting of the Botanical Society of Japan, p. 161. In Japanese.

Kikuyama, M., Tazawa, M., 1982. Ca^{2+} ion reversibly inhibits the cytoplasmic streaming of *Nitella*. Protoplasma 113, 241−243.

Kikuyama, M., Tazawa, M., Tominaga, Y., Shimmen, T., 1996. Membrane control of cytoplasmic streaming in characean cells. J. Plant Res. 109, 113−118.

Kim, T.-H., Böhmer, M., Hu, H., Nishimura, N., Schroeder, J.I., 2010. Guard cell signal transduction network: advances in understanding abscisic acid, CO_2 and Ca^{2+} signaling. Annu. Rev. Plant Biol. 61, 561−591.

Kim, S.-J., Brandizzi, F., 2014. The plant secretory pathway: an essential factory for building the plant cell wall. Plant Cell Physiol. 55, 687−693.

Kim, S.-J., Brandizzi, F., 2016. The plant secretory pathway for the trafficking of cell wall polysaccharides and glycoproteins. Glycobiol. 26, 940−949.

Kim, M., Canio, W., Kessler, S., Sinha, N., 2001. Developmental changes due to long-distance movement of a homeobox fusion transcript in tomato. Science 293, 287−289.

Kim, I., Cho, E., Crawford, K.M., Hempel, F.D., Zambryski, P.C., 2005. Cell-to-cell movement of GFP during embryogenesis and early seedling development in *Arabidopsis*. Proc. Natl. Acad. Sci. U.S.A. 102, 2227−2231.

Kim, J.H., Cho, H.T., Kende, H., 2000b. α-expansins in the semi-aquatic ferns *Marsilea quadrifolia* and *Regnellidium diphyllum*: evolutionary aspects and physiological role in rachis elongation. Planta 212, 85−92.

Kim, H.Y., Coté, G.G., Crain, R.C., 1996. Inositol 1,4,5-trisphosphate may mediate closure of K^+ channels by light and darkness in *Samanea saman* motor cells. Planta 198, 279−287.

Kim, K.-D., El Baidouri, M., Jackson, S.A., 2014. Accessing epigenetic variation in the plant methylome. Brief. Funct. Genom. 13, 318−327.

Kim, W.T., Franceschi, V.R., Krishnan, H.B., Okita, T.W., 1988. Formation of wheat protein bodies: involvement of the Golgi apparatus in gliadin transport. Planta 176, 173−182.

Kim, S.W., Gupta, R., Lee, S.H., Min, C.W., Agrawal, G.K., Rakwal, R., Kim, J.B., Jo, I.H., Park, S.-Y., Kim, Y.-C., Bang, K.H, Kim, S.T., 2016. An integrated biochemical, proteomics, and metabolomics approach supporting medicinal value of *Panax ginsing* fruits. Front. Plant Sci. 7, 994.

Kim, I., Hempel, F.D., Sha, K., Pfluger, J., Zambryski, P.C., 2002a. Identification of a developmental transition in plasmodesmatal function during embryogenesis in *Arabidopsis thaliana*. Development 129, 1261−1272.

Kim, M., Hepler, P.K., Eun, S.-O., Ha, K.S., Lee, Y., 1995. Actin filaments in mature guard cells are radially distributed and involved in stomatal movement. Plant Physiol. 109, 1077−1084.

Kim, D.H., Hwang, I., 2013. Direct targeting of proteins from the cytosol to organelles: the ER versus endosymbiotic organelles. Traffic 14, 613−621.

Kim, T.G., Kim, M.Y., Kim, B.G., Kang, T.J., Kim, Y.S., Jang, Y.S., Arntzen, C.J., Yang, M.S., 2007. Synthesis and assembly of *Escherichia coli* heat-labile enterotoxin B subunit in transgenic lettuce (*Lactuca sativa*). Protein Expr. Purif. 51, 22−27.

Kim, I., Kobayashi, K., Cho, E., Zambryski, P.C., 2005. Subdomains for transport via plasmodesmata corresponding to the apical-basal axis are established during *Arabidopsis* embryogenesis. Proc. Natl. Acad. Sci. U.S.A. 102, 11945−11950.

Kim, J., Lee, H., Lee, H.N., Kim, S.H., Shin, K.D., Chung, T., 2013. Autophagy-related proteins are required for degradation of peroxisomes in *Arabidopsis* hypocotyls during seedling growth. Plant Cell 25, 4956−4966.

Kim, P.K., Mullen, R.T., Schumann, U., Lippincott-Schwartz, J., 2006. The origin and maintenance of mammalian peroxisomes involves a de novo PEX16-dependent pathway from the ER. J. Cell Biol. 173, 521−532.

Kim, J.-B., Olek, A.T., Carpita, N.C., 2000. Cell wall and membrane-associated exo-β-D-glucanases from developing maize seedlings. Plant Physiol. 123, 471−485.

Kim, S., Park, M., Yeom, S.-I., Kim, Y.-M., Lee, J.M., Lee, H.-A., Seo, E., Choi, J., Cheong, K., Kim, K.-T., Jung, K., Lee, G.-W., Oh, S.-K., Bae, C., Kim, S.-B., Lee, H.-Y., Kim, S.-Y., Kim, M.-S., Kang, B.-C., Jo, Y.D., Yang, H.B., Jeong, H.-J., Kang, W.-H., Kwon, J.-K., Shin, C., Lim, J.Y., Park, J.H., Huh, J.H., Kim, J.-S., Kim, B.-D., Cohen, O., Paran, I., Suh, M.C., Lee, S.B., Kim, Y.-K., Shin, Y., Noh, S.J., Park, J., Seo, Y.S., Kwon, S.-Y., Kim, H.A., Park, J.M., Kim, H.-J., Choi, S.-B., Bosland, P.W., Reeves, G., Jo, S.-H., Lee, B.-W., Cho, H.-T., Choi, H.-S., Lee, M.-S., Yu, Y., Choi, Y.D., Park, B.-S., van Deynze, A., Ashrafi, H., Hill, T., Kim, W.T., Pai, H.-S., Ahn, H.K., Yeam, I., Giovannoni, J.J., Rose, J.K.C., Sørensen, I., Lee, S.-J., Kim, R.W., Choi, I.-K., Choi, B.-S., Lim, J.-S., Lee, Y.-H., Choi, D., 2014b. Genome sequence of the hot pepper provides insights into the evolution of pungency in Capsicum species. Nat. Genet. 46, 270–278.

Kim, K.K., Sawa, Y., Shibata, H., 1996. Hydroxylation of ent-kaurenoic acid to steviol in Stevia rebaudiana Bertoni—purification and partial characterization of the enzyme. Arch. Biochem. Biophys. 332, 223–230.

Kim, K., Yi, H., Zamil, M.S., Haque, M.A., Puri, V.M., 2015a. Multiscale stress-strain characterization of onion outer epidermal tissue in wet and dry states. Am. J. Bot. 102, 12–20.

Kim, S.J., Yoon, J.S., Shishido, H., Yang, Z., Rooney, L.A., Barral, J.M., Skach, W.R., 2015b. Translational tuning optimizes nascent protein folding in cells. Science 348, 444–447.

Kim, J.-Y., Yuan, Z., Cilia, M., Khalfan-Jagani, Z., Jackson, D., 2002b. Intercellular trafficking of a KNOTTED1 green fluorescent protein fusion in the leaf and shoot meristem of *Arabidopsis*. Proc. Natl. Acad. Sci. U.S.A. 99, 4103–4108.

Kim, J.Y., Yuan, Z., Jackson, D., 2003. Developmental regulation and significance of KNOX protein trafficking in *Arabidopsis*. Development 130, 4351–4362.

Kimbara, J., Endo, T.R., Nasuda, S., 2004. Characterization of the genes encoding for MAD2 homologues in wheat. Chromosome Res. 12, 703–714.

Kimura, K., Hirano, T., 1997. ATP-dependent positive supercoiling of DNA by 13S condensin: a biochemical implication for chromosome condensation. Cell 90, 625–634.

Kimura, S., Itoh, T., 1995. Evidence for the role of the glomerulocyte in cellulose synthesis in the tunicate, *Metandrocarpa uedai*. Protoplasma 186, 24–33.

Kimura, S., Itoh, T., 1996. New cellulose synthesizing complexes (terminal complexes) involved in animal cellulose biosynthesis in the tunicate *Metandrocarpa uedai*. Protoplasma 194, 151–163.

Kimura, S., Kondo, T., 2002. Recent progress in cellulose biosynthesis. J. Plant Res. 115, 297–302.

Kimura, S., Laosinchai, W., Itoh, T., Cui, X., Linder, C.R., Brown Jr., R.M., 1999. Immunogold labeling of rosette terminal cellulose-synthesizing complexes in the vascular plant *Vigna angularis*. Plant Cell 11, 2075–2085.

Kimura, Y., Toyoshima, N., Hirakawa, N., Okamoto, K., Ishijima, A., 2003. A kinetic mechanism for the fast movement of *Chara myosin*. J. Mol. Biol. 328, 939–950.

Kincaid, R.L., Mansour, T.E., 1978a. Chemotaxis toward carbohydrates and amino acids in *Physarum polycephalum*. Exp. Cell Res. 116, 377–385.

Kincaid, R.L., Mansour, T.E., 1978b. Measurement of chemotaxis in the slime mold *Physarum polycephalum*. Exp. Cell Res. 116, 365–375.

Kindl, H., 1982a. Glyoxysome biogenesis via cytoplasmic pools in cucumber. Ann. N.Y. Acad. Sci. 386, 314–328.

Kindl, H., 1982b. The biosynthesis of microbodies (peroxisomes, glyoxysomes). Int. Rev. Cytol. 80, 193–229.

King, A.G., 1925. Kelvin the Man. A Biographical Sketch by his Niece, Agnes Gardner King. Hodder and Stoughton, London.

King Jr., M.L., 1963. I have a dream, 1999. In: Copeland, L., Lamm, L.W., McKenna, S.J. (Eds.), The World's Greatest Speeches. Dover, Mineola, NY, pp. 751–754.

King Jr., M.L., 1992. I Have a Dream: Writing and Speeches that Changed the World. Harper, San Fransisco.

King, S.M., 2002. Dyneins motor on in plants. Traffic 3, 930–931.

King Jr., M.L., December 11, 1964. The Quest for Peace and Justice. Nobel Lecture.

Kingsbury, B.F., 1912. Cytoplasmic fixation. Anat. Rec. 6, 39–52.

Kingsley, J.L., Bibeau, J.P., Chen, Z., Huang, X., Vidali, L., Tüzel, E., 2016. Probing Cytoplasmic Viscosity in the Confined Geometry of Tip-Growing Plant Cells via FRAP. https://www.biorxiv.org/content/early/2016/06/15/059220.

Kinkema, M., Schiefelbein, J., 1994. A myosin from a higher plant has structural similarities to Class V myosins. J. Mol. Biol. 239, 591–597.

Kinkema, M., Wang, H., Schielbein, J., 1994. Molecular analysis of the myosin gene family in *Arabidopsis* thaliana. Plant Mol. Biol. 26, 1139–1153.

Kinoshita, T., Doi, M., Suetsugu, N., Kagawa, T., Wada, M., Shimazaki, K.-I., 2001. phot1 and phot2 mediate blue light regulation of stomatal opening. Nature 414, 656–660.

Kinoshita, T., Emi, T., Tominaga, M., Sakamoto, K., Shigenaga, A., Doi, M., Shimazaki, K.-I., 2003. Blue-light-and phosphorylation-dependent binding of a 14-3-3 protein to phototropins in stomatal guard cells of broad bean. Plant Physiol. 133, 1453–1463.

Kinoshita, T., Nishimura, M., Shimazaki, K.-I., 1995. Cytosolic concentration of Ca^{2+} regulates the phosphorylation of the C-terminus in stomatal guard cells. EMBO J. 18, 5548–5558.

Kinoshita, T., Shimazaki, K.-I., 1999. Blue light activates the plasma membrane H^+-ATPase by phosphorylation of the C-terminus in stomatal guard cells. EMBO J. 18, 5548–5558.

Kinoshita, T., Shimazaki, K.-I., 2001. Analysis of the phosphorylation Level in guard-cell plasma membrane H^+-ATPase in response to Fusicoccin. Plant Cell Physiol. 42, 424–432.

Kinoshita, T., Shimazaki, K.-I., 2002. Biochemical evidence for the requirement of 14-3-3 protein binding in activation of the guard cell plasma membrane H^+-ATPase by blue light. Plant Cell Physiol. 43, 1359–1365.

Kinraide, T.B., Etherton, B., 1980. Electrical evidence for different mechanisms of uptake for basic, neutral, and acidic amino acids in oat coleoptiles. Plant Physiol. 65, 1085–1089.

Kiran, K., Ansari, S.A., Srivastava, R., Lodhi, N., Chaturvedi, C.R., Sawant, S.V., Tuli, R., 2006. The TATA-box sequence in the basal promoter contributes to determining light-dependent gene expression in plants. Plant Physiol. 142, 364–376.

Kirby, W., 1853. On the Power, Wisdom, and Goodness of God, as Manifested in the Creation of Animals. Henry Bohn, London.

Kirby, A.R., 2011. Atomic force microscopy of plant cell walls. Methods Mol. Biol. 715, 169–178.

Kircher, M., Sawyer, S., Meyer, M., 2012. Double indexing overcomes inaccuracies in multiplex sequencing on the Illumina platform. Nucleic Acids Res. 40, 2513–2524.

Kirisako, T., Ichimura, Y., Okada, H., Kabeya, Y., Mizushima, N., Yoshimori, T., Ohsumi, M., Takao, T., Noda, T., Ohsumi, Y., 2000. The reversible modification regulates the membrane-binding state of Apg8/Aut7 essential for autophagy and the cytoplasm to vacuole targeting pathway. J. Cell Biol. 151, 263–276.

Kirk, J.S., Sumner, J.B., 1931. Antiurease. J. Biol. Chem. 94, 21–28.

Kirkman, H., Severinghaus, A.E., 1938. A review of the Golgi apparatus. Anat. Rev. 70, 413–431, 557–573, 71, 79–103.

Kirschner, M.W., 1978. Microtubule assembly and nucleation. Int. Rev. Cytol. 54, 1–71.

Kirschner, M.W., 1980. Implications of treadmilling for the stability and polarity of actin and tubulin polymers *in vivo*. J. Cell Biol. 86, 330–334.

Kirschner, M., 1992. The cell cycle then and now. Trends Biochem. Sci. 17, 281–285.

Kirschner, M.W., Gerhart, J.C., 2005. The Plausibility of Life. Yale University Press, New Haven, CT.

Kishi-Kaboshi, M., Murata, T., Hasebe, M., Watanabe, Y., 2005. An extraction method for tobacco mosaic virus movement protein localizing in plasmodesmata. Protoplasma 225, 85–92.

Kishimoto, U., 1958. Rhythmicity in the protoplasmic streaming of a slime mold, *Physarum polycephalum*. II. Theoretical treatment of the electric potential rhythum. J. Gen. Physiol. 41, 1223–1244.

Kishimoto, U., 1968. Response of Chara internodes to mechanical stimulation. Ann. Rep. Biol. Works. Fac. Sci. Osaka Univ. 16, 61–66.

Kishino, A., Yanagida, T., 1988. Force measurement by micromanipulation of a single actin filament by glass needles. Nature 334, 74–76.

Kiss, J.Z., 2000. Mechanisms of the early phases of plant gravitropism. Crit. Rev. Plant Sci. 19, 551–573.

Kiss, J.Z., Giddings Jr., T.H., Staehehin, L.A., Sack, F., 1990. Comparison of the ultrastructure of conventionally fixed and high pressure frozen/freeze substituted root tips of Nicotiana and *Arabidopsis*. Protoplasma 157, 64–74.

Kiss, J.Z., Hertel, R., Sack, F.D., 1989. Amyloplasts are necessary for full gravitropic sensitivity in roots of *Arabidopsis thaliana*. Planta 177, 19198.

Kitagawa, M., Fujita, T., 2013. Quantitative imaging of directional transport through plasmodesmata in moss protonemata via single-cell photoconversion of Dendra26. J. Plant Res. 126, 577–585.

Kitagawa, M., Fujita, T., 2015. A model system for analyzing intercellular communication through plasmodesmata using moss protonemata and leaves. J. Plant Res. 128, 63–72.

Kitajima, T.S., Kanashima, S.A., Watanabe, Y., 2004. The conserved kinetochore protein shugushin protects centromeric cohesion during meiosis. Nature 427, 510–517.

Kitasato, H., 1968. The influence of H^+ on the membrane potential and ion fluxes of *Nitella*. J. Gen. Physiol. 52, 60–87.

Kitasato, H., 2003. Membrane potential genesis in *Nitella* cells, mitochondria, and thylakoids. J. Plant Res. 116, 401–418.

Kiyosawa, K., Ogata, K., 1987. Influence of external osmotic pressure on water permeability and electrical conductance of *Chara* cell membrane. Plant Cell Physiol. 28, 1013–1022.

Kjellbom, P., Larsson, C., 1984. Preparation and polypeptide composition of chlorophyll-free plasma membranes from leaves of light-grown spinach and barley. Physiol. Plant 62, 501–509.

Klausner, R.D., Donaldson, J.G., Lippincott-Schwartz, J., 1992. Brefeldin A: insights into the control of membrane traffic and organelle structure. J. Cell Biol. 116, 1071–1080.

Klein, T.M., Fromm, M.E., Gradziel, T., Sanford, J.C., 1988b. Factors influencing gene delivery into Zea mays cells by high-velocity microprojectiles. BioTechnology 6, 559–563.

Klein, T.M., Fromm, M.E., Weissinger, A., Tomes, D., Schaaf, S., Sleeten, M., Sanford, J.C., 1988a. Transfer of foreign genes into intact maize cells using high velocity microprojectiles. Proc. Natl. Acad. Sci. U.S.A. 85, 4305–4309.

Klein, M., Martinoia, E., Hoffmann-Thoma, G., Weissenböck, G., 2000. A membrane-potential dependent, ubiquitous ABC- like transporter mediates the vacuolar uptake of rye flavone glucuronides—regulation of glucuronide uptake by glutathione and its conjugates. Plant J. 21, 289–304.

Klein, M., Martinoia, E., Hoffmann-Thoma, G., Weissenböck, G., 2001. The ABC-like vacuolar transporter fro rye mesophyll flavone glucuronides is not species-specific. Phytochemistry 56, 153–159.

Klein, M., Martinoia, E., Weissenböck, G., 1998. Directly energized uptake of β-estradiol 17-(β-D-glucuronide) in plant vacuoles is strongly stimulated by glutathione conjugates. J. Biol. Chem. 273, 262–270.

Klein, R.R., Mullet, J.E., 1986. Regulation of chloroplast-encoded chlorophyll-binding protein translation during higher plant chloroplast biogenesis. J. Biol. Chem. 261, 11138–11145.

Klein, K., Wagner, G., Blatt, M.R., 1980. Heavy-meromyosin- decoration of microfilaments from *Mougeotia* protoplasts. Planta 150, 354–356.

Klein, M., Weissenböck, G., Dufaud, A., Gaillard, C., Kreuz, K., Martinoia, E., 1996. Different energization mechanisms drive the vacuolar uptake of a flavonoid glucoside and a herbicide glucoside. J. Biol. Chem. 271, 29666–29671.

Klein, T.M., Wolf, E.D., Wu, R., Sanford, J.C., 1987. High-velocity microprojectiles for delivering nucleic acids into living cells. Nature 327, 70–73.

Kleiner, O., Kircher, S., Harter, K., Batschauer, A., 1999. Nuclear localization of the Arabidopsis blue light receptor cryptochrome 2. Plant J. 19, 289–296.

Kleinig, H., 1989. The role of plastids in isoprenoid biosynthesis. Ann. Rev. Plant Physiol. Mol. Biol. 40, 39–59.

Klekowski Jr., E.J., 1971. Ferns and genetics. BioScience 21, 317–322.

Klekowski Jr., E.J., 1976. Mutational load in a fern population growing in a polluted environment. Am. J. Bot. 63, 1024–1030.

Klekowski Jr., E.J., Berger, B.B., 1976. Chromosome mutations in a fern population growing in a polluted environment: a bioassay for mutagens in aquatic environments. Am. J. Bot. 63, 239–246.

Klekowski Jr., E.J., Kazarinova-Fukshansky, N., 1984a. Shoot apical meristems and mutation: fixation of selectively neutral cell genotypes. Am. J. Bot. 71, 22–27.

Klekowski Jr., E.J., Kazarinova-Fukshansky, N., 1984b. Shoot apical meristems and mutation: selective loss of disadvantageous cell genotypes. Am. J. Bot. 71, 28–34.

Klekowski Jr., E.J., Kazarinova-Fukshansky, N., Fukshansky, L., 1989. Patterns of plant ontogeny that may influence genomic stasis. Am. J. Bot. 76, 189–195.

Klekowski Jr., E.J., Klekowski, E., 1982. Mutation in ferns growing in an environment contaminated with polychlorinated biphenyls. Am. J. Bot. 69, 721–727.

Klekowski Jr., E.J., Lovenfeld, R., Helper, P.K., 1994. Mangrove genetics. II. Outcrossing and lower spontaneous mutation rates in Puerto Rican rhizophora. Int. J. Plant Sci. 155, 373–381.

Klima, A., Foissner, I., 2008. FM dyes label sterol-rich plasma membrane domains and are internalized independently of the cytoskeleton in characean internodal cells. Plant Cell Physiol. 49, 1508–1521.

Kline, K.G., Sussman, M.R., Jones, A.M., 2010. Abscisic acid reseptors. Plant Physiol. 154, 479–482.

Klink, V.P., Alkharouf, N., MacDonald, M., Matthews, B., 2005. Laser capture microdissection (LCM) and expression analyses of *Glycine max* (soybean) syncytium containing root regions formed by the plant pathogen *Heterodera glycenes* (soybean cyst nematode). Plant Mol. Biol. 59, 965–979.

Klink, R., Lüttge, U., 1990. Electron microscope demonstration of a head and stalk structure of the leaf vacuolar ATPase in *Mesembryanthemum crystallinum* L. Bot. Acta 104, 122–131.

Klink, R., Lüttge, U., 1991. Electron microscope demonstration of a head and stalk structure of the leaf vacuolar ATPase in *Mesembryanthemum crystallinum* L. Bot. Acta 104, 122–131.

Klink, V.P., Wolniak, S.M., 2000. The efficacy of RNAi in the study of the plant cytoskeleton. J. Plant Growth Regul. 19, 371–384.

Klink, V.P., Wolniak, S.M., 2001. Centrin is necessary for the formation of the motile apparatus in spermatids of *Marsilea*. Mol. Biol. Cell 12, 761–776.

Klionsky, D.J., Emr, S.D., 2000. Autophagy as a regulated pathway of cellular degradation. Science 290, 1717–1721.

Klionsky, D.J., Herman, P.K., Emr, S.D., 1990. The fungal vacuole: composition, function, and biogenesis. Microbiol. Rev. 54, 266–292.

Klionsky, D.J., Ohsumi, Y., 1999. Cytoplasm to vacuole protein transport. Annu. Rev. Cell Dev. Biol. 15, 1–32.

Klose, J., 1975. Protein mapping by combined isoelectric focusing and electrophoresis of mouse tissues. A novel approach to testing for induced point mutations in mammals. Humangenetik 26, 231–243.

Kloser, R.J., Bax, N.J., Williams, R.A., Barker, B.A., 2001. Remote sensing of seabed types in the Australian South East Fishery: development and application of normal incident acoustic techniques and associated 'ground truthing.'. Mar. Freshwater Res. 52, 475–489.

Klug, J., 2002. Commentary on Gáspár Jékely's article in EMBO reports, July 2002. EMBO Rep. 3, 1003–1005.

Kluyver, A.J., 1931. The Chemical Activities of Micro-Organisms. University of London Press, London.

Kluyver, A.J., van Niel, C.B., 1956. The Microbe's Contribution to Biology. Harvard University Press, Cambridge, MA.

Knebel, W., Quader, H., Schnepf, E., 1990. Mobile and immobile endoplasmic reticulum in onion bulb epidermis cells: short and long-term observations with a confocal laser scanning microscope. Eur. J. Cell Biol. 52, 328–340.

Kneller, K.A., 1911. Christianity and the Leaders of Modern Science. B. Herder, London.

Knepper, C., Savory, E.A., Day, B., 2011. *Arabidopsis* NDR1 is an integrin-like protein with a role in fluid loss an dplasma membrane-cell wall adhesion. Plant Physiol. 156, 286–300.

Knight, J., 2002. Bridging the culture gap. Nature 419, 244–246.

Knight, M.R., Campbell, A.K., Smith, S.M., Trewavas, A.J., 1991. Transgenic plant aequorin reports the effect of touch and cold-shock and elicitors on cytoplasmic calcium. Nature 352, 524–526.

Knight, A.E., Kendrick-Jones, J., 1993. A myosin-like protein from a higher plant. J. Mol. Biol. 231, 148–154.

Knight, M.R., Smith, S.M., Trewavas, A.J., 1992. Wind-induced plant motion immediately increases cytosolic calcium. Proc. Natl. Acad. Sci. U.S.A. 89, 4967–4971.

Knoetzel, J., Simpson, D., 1991. Expression and organization of antenna proteins in the light- and temperature-sensitive barley mutant chlorina-104. Planta 185, 111–123.

Knowles, R.V., Srienc, F., Phillips, R.L., 1990. Endoreduplication of nuclear-DNA in the developing maize endosperm. Dev. Genet. 11, 125–132.

Knox, J.P., 1990. Emerging patterns of organization at the plant cell surface. J. Cell Sci. 96, 557–561.

Knox, J.P., 1992a. Cell adhesion, cell separation and plant morphogenesis. Plant J. 2, 137–141.

Knox, J.P., 1992b. Molecular probes for the plant cell surface. Protoplasma 167, 1–9.

Knox, J.P., Umstead, P.J., King, J., Cooper, C., Roberts, K., 1990. Pectin esterification is spatially regulated both within cell walls and between developing tissues of root apices. Planta 181, 512–521.

Knudson, L., 1940. Permanent changes of chloroplasts induced by X rays in the gametophyte of *Polypodium aureum*. Bot. Gaz. 101, 721–758.

Knull, H.R., Bronstein, W.W., des Jardins, P., Niehaus, W.G., 1980. Interaction of selected brain glycolytic enzymes with an F-actin-tropomyosin complex. J. Neurochem. 34, 222–225.

Knutson, R.M., 1974. Heat production and temperature regulation in eastern skunk cabbage. Science 186, 746–747.

Knutson, R.M., 1979. Plants in heat. Nat. Hist. 88, 42–47.

Kobae, Y., Sekino, T., Yoshioka, H., Nakagawa, T., Martinoia, E., Maeshima, M., 2006. Loss of AtPDR8, a plasma membrane ABC transporter of *Arabidopsis thaliana*, causes hypersensitive cell death upon pathogen infection. Plant Cell Physiol. 47, 309–318.

Kobayashi, H., 1996. Changes in the relationship between actin filaments and the plasma membrane in cultured *Zinnia* cells during tracheary element differentiation investigated by using plasma membrane ghosts. J. Plant Res. 109, 61–65.

Kobayashi, H., Fukuda, H., Shibaoka, H., 1988. Interrelation between the spatial disposition of actin filaments and microtubules during the differentiation of tracheary elements in cultured *Zinnia* cells. Protoplasma 143, 29–37.

Kobayashi, K., Kim, I., Cho, E., Zambryski, P., 2005. Plasmodesmata and plant morphogenesis. In: Oparka, K.J. (Ed.), Plasmodesmata. Blackwell Publishing, Oxford, pp. 90–112.

Kobayashi, T., Rinker, J.N., Koffler, H., 1959. Purification and chemical properties of flagellin. Arch. Biochem. Biophys. 64, 342–362.

Kobayashi, T., Takahara, M., Miyagishima, S.Y., Kuroiwa, H., Sasaki, N., Ohta, N., Matsuzaki, M., Kuroiwa, T., 2002. Detection and localization of a chloroplast-encoded HU-like protein that organizes chloroplast nucleoids. Plant Cell 14, 1579–1589.

Kobayashi, Y., Yamada, M., Kobayashi, I., Kunoh, H., 1997. Actin microfilaments are required for the expression of nonhost resistance in higher plants. Plant Cell Physiol. 38, 725–733.

Koch, A.L., 1996. What size should a bacterium be? A question of scale. Annu. Rev. Microbiol. 50, 317–348.

Koch, A., Thiemann, M., Grabenbauer, M., Yoon, Y., McNiven, M.A., Schrader, M., 2003. Dynamin-like protein 1 is involved in peroxisomal fission. J. Biol. Chem. 278, 8597–8605.

Koenig, F.O., 1959. On the history of science and of the second law of thermodynamics. In: Evans, H.M. (Ed.), Men and Moments in the History of Science. University of Washington Press, Seattle, pp. 57–111.

Koenig, P., Oreb, M., Muhle-Goll, C., Sinning, I., Schleiff, E., Tews, I., 2008. On the significance of the Toc-GTPase homodimers. J. Biol. Chem. 283, 23104–23112.

Koenigsberger, L., 1906. Hermann von Helmholtz. Clarendon Press, Oxford.

Kohler Jr., R.E., 1977. Rudolf Schoenheimer, isotopic tracers, and biochemistry in the 1930's. Hist. Studies Phys. Sci. 8, 257–298.

Köhler, R.H., Cao, J., Zipfel, W.R., Webb, W.W., Hanson, M.R., 1997b. Exchange of protein molecules through connections between higher plant plastids. Science 276, 2039–2042.

Köhler, P., Carr, D.J., 2006. A somewhat obscure discoverer of plasmodesmata: Eduard Tangl (1848–1905). In: Kokowski, M. (Ed.), The Global and the Local: The History of Science and the Cultural Integration of Europe. Proceedings of the 2nd ICESHS Cracow, Poland, September 6–9, 2006, pp. 208–211.

Köhler, S., Delwiche, C.F., Denny, P.W., Tilney, L.G., Webster, P., Wilson, R.J.M., Palmer, J.D., Roos, D.S., 1997. A plastid of probable green algal origin in apicomplexan parasites. Science 275, 1485–1488.

Köhler, R.H., Hanson, M.R., 2000. Plastid tubules of higher plants are tissue-specific and developmentally regulated. J. Cell Sci. 113, 81–89.

Köhler, D., Montandon, C., Hause, G., Majovsky, P., Kessler, F., Baginsky, S., Agne, B., 2015. Characterization of chloroplast protein import without Tic56, a component of the 1-megadalton translocon at the inner envelope membrane of chloroplasts. Plant Physiol. 167, 972–990.

Köhler, R.H., Schwillem, P., Webbm, W.W., Hansonm, M.R., 2000. Active protein transport through plastid tubules: velocity quantified by fluorescence correlation spectroscopy. J. Cell Sci. 113, 3921–3930.

Köhler, R.H., Zipfel, W.R., Webb, W.W., Hanson, M.R., 1997a. The green fluorescent protein as a marker to visualize plant mitochondria in vivo. Plant J. 11, 613–621.

Kohli, P., Kalia, M., Gupta, R., 2015. Pectin methylesterases: a review. J. Bioprocess. Biotech. 5, 5.

Kohli, P., Kalia, M., Gupta, R., 2015. Pectin methylesterases: a review. J. Bioprocess Biotech. 5, 1000227.

Kohno, T., Chaen, S., Shimmen, T., 1990. Characterization of the translocator associated with pollen tube organelles. Protoplasma 154, 179–183.

Kohno, T., Ishikawa, R., Nagata, T., Kohama, K., Shimmen, T., 1992. Partial purification of myosin from lily pollen tubes by monitoring with in vitro motility assay. Protoplasma 170, 77–85.

Kohno, T., Okagaki, T., Kohama, K., Shimmen, T., 1991. Pollen tube extract supports the movement of actin filaments in vitro. Protoplasma 161, 75–77.

Kohno, T., Shimmen, T., 1988. Accelerated sliding of pollen tube organelles along characeae actin bundles regulated by Ca^{2+}. J. Cell Biol. 196, 1539–1543.

Koketsu, K., Murata, T., Yamada, M., Nishina, M., Boruc, J., Hasebe, M., van Damme, D., Goshima, G., 2017. Cytoplasmic MTOCs control spindle orientation for asymmetric cell division in plants. Proc. Natl. Acad. Sci. U.S.A. http://www.pnas.org/content/early/2017/09/27/1713925114.abstract.

Kolb, C.E., Herschbach, D.R., 1984. John Bennett Fenn. J. Phys. Chem. 88, 4447–4448.

Kolb, C., Nagel, M.-K., Kalinowska, K., Hagmann, J., Ichikawa, M., Anzenberger, F., Alkofer, A., Sato, M.H., Braun, P., Isono, E., 2015. FYVE1 is essential for vacuole biogenesis and intracellular trafficking in *Arabidopsis*. Plant Physiol. 167, 1361–1373.

Kole, C., Abbott, A.G. (Eds.), 2008. Principles and Practices of Plant Genomics. Volume 1: Genome Mapping. Science Publishers, Enfield, NH.

Koller, B.H., 2017. Oliver Smithies (1925-2017). Cell 168, 743–744.

Kollist, H., Nuhkat, M., Roelfsema, M.R.G., 2014. Closing gaps: linking elements that control stomatal movement. New Phytol. 203, 44–62.

Kollmann, R., Glockmann, C., 1991. Studies on graft Unions. III. On the mechanism of secondary formation of plasmodesmata at the graft interface. Protoplasma 165, 71–85.

Komaki, S., Abe, T., Coutuer, S., Inze, D., Russinova, E., Hashimoto, T., 2010. Nuclear-localized subtype of end-binding 1 protein regulates spindle organization in Arabidopsis. J. Cell Sci. 123, 451–459.

Komatsu, S., Kuji, R., Nanjo, Y., Hiraga, S., Furukawa, K., 2012. Comprehensive analysis of endoplasmic reticulum-enriched fraction in root tips of soybean under flooding stress using proteomics techniques. J. Proteome 77, 531–560.

Konar, R.N., Linskins, H.F., 1966. The morphology and anatomy of the stigma of *Petunia hybrida*. Planta 71, 356–371.

Koncz, C., Martini, N., Mayerhofer, R., Koncz-Kalman, Z., Körber, H., Redei, G.P., Schell, J., 1989. High-frequency T-DNA-mediated gene tagging in plants. Proc. Natl. Acad. Sci. U.S.A. 86, 8467–8471.

Kondo, T., Yoshida, K., Nakagawa, A., Kawai, T., Tamura, H., Goto, T., 1992. Structure basis of blue-colour development in flower petals from *Commelina communis*. Nature 358, 515–518.

Kong, S.-G., Wada, M., 2011. New insights into dynamic actin-based chloroplast photorelocation movement. Mol. Plant 4, 771–781.

Konijn, T.M., Koevenig, J.L., 1971. Chemotaxis in myxomycetes or true slime molds. Mycologia 63, 901–906.

Konishi, A., Eguchi, A., Hosoi, F., Omasa, K., 2009. 3D monitoring spatio-temporal effects of herbicide on a whole plant using combined range and chlorophyll a fluorescence imaging. Funct. Plant Biol. 36, 874–879.

Konopka, R.J., Benzer, S., 1971. Clock mutants of *Drosophila melanogaster*. Proc. Natl. Acad. Sci. U.S.A. 68, 2112–2116.

Konrad, E.B., Modrich, P., Lehman, I.R., 1973. Genetic and enzymatic characterization of a conditional lethal mutant of *Eschericia coli* K12 with a temperature-sensitive DNA ligase. J. Mol. Biol. 77, 529–531.

Korbel, J.O., Urban, A.E., Affourtit, J.P., Godwin, B., Grubert, F., Simons, J.F., et al., 2007. Paired end mapping reveals extensive structural variation in the human genome. Science 318, 420–426.

Korn, E.D., 1966. Structure of biological membranes. Science 153, 1491–1498.

Korn, E.D., Carlier, M.-F., Pantaloni, D., 1987. Actin polymerization and ATP hydrolysis. Science 238, 638–644.

Kornberg, A., 1950a. Enzymatic synthesis of triphosphopyridine nucleotide. J. Biol. Chem. 182, 805–813.

Kornberg, A., 1950b. Reversible enzymatic synthesis of diphosphopyridine nucleotide and inorganic pyrophosphate. J. Biol. Chem. 182, 779–793.

Kornberg, A., 1957. Pathways of enzymatic synthesis of nucleotides and polynucleotides. In: D McElroy, W., Glass, B. (Eds.), The Chemical Basis of Heredity. Johns Hopkins Press, Baltimore, MD, pp. 579–590.

Kornberg, A., 1960. Biological synthesis of deoxyribonucleic acid. Science 131, 1503–1508.

Kornberg, A., 1961. Enzymatic Synthesis of DNA. John Wiley & Sons, New York.

Kornberg, A., 1989a. For the Love of Enzymes. The Odyssey of a Biochemist. Harvard University Press, Cambridge, MA.

Kornberg, A., 1989b. Never a dull enzyme. Annu. Rev. Biochem. 58, 1–30.

Kornberg, A., 2000. Ten Commandments: lessons from the enzymology of DNA replication. J. Bacteriol. 182, 3613–3618.

Kornberg, H., 2000. Krebs and his trinity of cycles. Nat. Rev. Mol. Cell Biol. 1, 225–228.

Kornberg, A., 2001. Remembering our teachers. J. Biol. Chem. 276, 3–11.

Kornberg, H.L., 2003. Memoirs of a biochemical Hod carrier. J. Biol. Chem. 278, 9993–10001.

Kornberg, A., Baker, T.A., 1992. DNA Replication, second ed. W. H. Freeman and Co., New York.

Kornberg, H.L., Beevers, H., 1957a. A mechanism of conversion of fat to carbohydrate in castor beans. Nature 180, 35–36.

Kornberg, H.L., Beevers, H., 1957b. The glyoxylate cycle as a stage in the conversion of fat to carbohydrate in castor beans. Biochim. Biophys. Acta 26, 531–537.

Kornberg, A., December 11, 1959. The biologic synthesis of deoxyribonucleic acid. Nobel Lecture.

Kornberg, L., Earp, H.S., Parsons, J.T., Shalla, M., Juliano, R.L., 1992. Cell adhesion of integrin clustering increases phosphorylation of a focal adhesion associated tyrosine kinase. J. Biol. Chem. 267, 23439–23442.

Kornberg, L., Earp, H.S., Turner, C., Prokop, C., Juliano, R.L., 1991. Signal transduction by integrins: increased protein tyrosine phosphorylation caused by clustering of beta 1 integrins. Proc. Natl. Acad. Sci. U.S.A. 88, 8392–8396.

Kornberg, T., Gefter, M.L., 1970. DNA synthesis in cell-free extracts of a DNA polymerase-defective mutant. Biochem. Biophys. Res. Commun. 40, 1348–1355.

Kornberg, T., Gefter, M.L., 1971. Purification and DNA synthesis in cell-free extracts: properties of DNA polymerase II. Proc. Natl. Acad. Sci. U.S.A. 68, 761–764.

Kornberg, H.L., Krebs, H.A., 1957. Synthesis of cell constituents from C_2-units by a modified tricarboxylic acid cycle. Nature 179, 988–991.

Kornberg, H.L., Madsen, N.B., 1957. Synthesis of C_4-dicarboxylic acids from acetate by a "glyoxylate bypass" of the tricarboxylic acid cycle. Biochim. Biophys. Acta 24, 651–653.

Kornberg, R.D., McConnell, H.M., 1971a. Inside-outside transitions of phospholipids in vesicle membranes. Biochemistry 10, 1111–1120.

Kornberg, R.D., McConnell, H.M., 1971b. Lateral diffusion of phospholipids in a vesicle membrane. Proc. Natl. Acad. Sci. U.S.A. 68, 2564–2568.

Kornberg, A., Pricer Jr., W.E., 1953. Enzymatic synthesis of the coenzyme A derivatives of long chain fatty acids. J. Biol. Chem. 204, 329–343.

Koroleva, O.A., Tomlinson, M.L., Leader, D., Shaw, P., Doonan, J.H., 2005. High-throughput protein localization in *Arabidopsis* using Agrobacterium-mediated transient expression GFP-ORF fusions. Plant J. 41, 162–174.

Korte, A., Farlow, A., 2013. The advantages and limitations of trait analysis with GWAS: a review. Plant Methods 9, 29.

Kortschak, H.P., Hartt, C.E., 1966. The effect of varied conditions on carbon dioxide fixation in sugar cane leaves. Naturwissenschaften 53, 253.

Kortschak, H.P., Hartt, C.E., Burr, C.O., 1965. Carbon dioxide fixation in sugar cane leaves. Plant Physiol. 40, 209–213.

Kosetsu, K., Murata, T., Yamada, M., Nishina, M., Boruc, J., Hasebe, M., van Damme, D., Goshima, G., 2017. Cytoplasmic MTOCs control spindle orientation for asymmetric cell division in plants. Proc. Natl. Acad. Sci. U.S.A. 114 (42), E8847–E8854.

Koshland, D.E., 2002. Case of hidden assumptions. Biochem. Mol. Biol. Educ. 30, 27–29.

Koshland, D.E., Mitchison, T.J., Kirschner, M.W., 1988. Polewards chromosome movement driven by microtubule depolymerization in vitro. Nature 331, 499–504.

Kossel, A., 1884. Über ein peptonartigen Bestandteil des Zellkerns. Z. Physiol. Chem. 8, 511–515.

Kossel, A., 1928. The Protamines and Histones. Translated by W. V. Thorpe. Longmans, Green and Co., London.

Kossel, A., December 12, 1910. The chemical composition of the cell nucleus. Nobel Lecture.

Kost, B., Spielhofer, P., Chua, N.-H., 1998. A GFP-mouse talin fusion protein labels plant actin filaments in vivo and visualizes the actin cytoskeleton in growing pollen tubes. Plant J. 16, 393–401.

Kotlizsky, G., Shurtz, S., Yahalom, A., Malik, Z., Traub, O., Epel, B.L., 1992. An improved procedure for the isolation of plasmodesmata embedded in clean maize cell walls. Plant J. 2, 623–630.

Koulintchenki, M., Konstantinov, Y., Dietrich, A., 2003. Plant mitochondria actively import DNA via the permeability transition pore complex. EMBO J. 22, 1245–1254.

Kouranov, A., Chen, X., Fuks, B., Schnell, D.J., 1998. Tic20 and Tic22 are new components of the protein import apparatus at the chloroplast inner envelope membrane. J. Cell Biol. 143, 991–1002.

Kozaki, A., Takeba, G., 1996. Photorespiration protects C3 plants from photooxidation. Nature 384, 557–560.

Kragler, F., 2005. Plasmodesmata: protein transport signals and receptors. In: Oparka, K.J. (Ed.), Plasmodesmata. Blackwell Publishing, Oxford, pp. 53–72.

Kragler, F., 2013. Plasmodesmata: intercellular tunnels facilitating transport of macromolecules in plants. Cell Tissue Res. 352, 49–58.

Kragler, F., Monzer, J., Xoconostle-Cázares, B., Lucas, W.J., 2000. Peptide antagonists of the plasmodesmatal macromolecular trafficking pathway. EMBO J. 19, 2856–2868.

Kragt, A., Voorn-Brouwer, T., van den Berg, M., Distel, B., 2005. Endoplasmic reticulum-directed Pex3p routes to peroxisomes and restores peroxisome formation in a *Saccharomyces cerevisiae* pex3Δ strain. J. Biol. Chem. 280, 34350–34357.

Kramer, E.M., Frazer, N.L., Baskin, T.I., 2007. Measurement of diffusion within the cell wall in living roots of *Arabidopsis thaliana*. J. Exp. Bot. 58, 3005–3015.

Kramers, H.A., 1940. Brownian motion in a field of force and the diffusion model of chemical reactions. Physica 7, 284–304.

Kraml, M., Büttner, G., Haupt, W., Herrmann, H., 1988. Chloroplast orientation of Mesotaenium: the phytochrome effect is strongly potentiated by interaction with blue light. Protoplasma (Suppl. 1), 172–179.

Krause, K.-H., Chou, M., Thomas, M.A., Sjolund, R.D., Campbell, K.P., 1989. Plant cells contain calsequestrin. J. Biol. Chem. 264, 4269–4272.

Krauss, L., 2012. A Universe from Nothing. Free Press, New York. https://www.youtube.com/watch?v=7ImvlS8PLIo.

Krebs, H.A., 1940a. The citric acid cycle. Biochem. J. 34, 460–463.

Krebs, H.A., 1940b. The citric acid cycle and the Szent-Györgyi cycle in pigeon breast muscle. Biochem. J. 34, 775–779.

Krebs, H.A., 1943. The intermediary stages in the biological oxidation of carbohydrate. Adv. Enzymol. 3, 191–252.

Krebs, H.A., 1950. The tricarboxylic acid cycle. Harvey Lect. 44, 165–199.

Krebs, H.A., 1954. Considerations concerning the pathways of syntheses in living matter. Bull. Johns Hopkins 95, 19–33.

Krebs, H.A., 1957. Control of metabolic processes. Endeavour 16, 125–132.

Krebs, H.A., 1967. The making of a scientist. Nature 215, 1441–1445.

Krebs, H.A., 1970. The history of the tricarboxylic acid cycle. Perspect. Biol. Med. 14, 154–170.

Krebs, H.A., 1971. How the whole becomes more than the sum of the parts. Perspect. Biol. Med. 14, 448–457.

Krebs, H.A., 1975. The August Krogh Principle: for many problems there is an animal on which it can be most conveniently studied. J. Exp. Zool. 194, 221–226.

Krebs, H.A., 1981a. Otto Warburg. Cell Physiologist Biochemist and Eccentric. Oxford University Press, New York.

Krebs, H.A., 1981b. Hans Krebs: Reminiscences and Reflections. Clarendon Press, Oxford.

Krebs, E.G., 1982. Protein phosphorylation and cellular regulation. I. Nobel Lecture. December 8, 1992.

Krebs, H.A., December 11, 1953. The citric acid cycle. Nobel Lecture.

Krebs, H.A., Eggleston, L.V., 1940a. Biological synthesis of oxaloacetic acid from pyruvic acid and carbon dioxide. Biochem. J. 34, 1383–1395.

Krebs, H.A., Eggleston, L.V., 1940b. The oxidation of pyruvate in pigeon breast muscle. Biochem. J. 34, 442–459.

Krebs, M., Held, K., Binder, A., Hashimoto, K., Den Herder, G., Parniske, M., Kudla, J., Schumacher, K., 2012. FRET-based genetically encoded sensors allow high-resolution live cell imaging of Ca^{2+} dynamics. Plant J. 69, 181–192.

Krebs, H.A., Johnson, W.A., 1937. The role of citric acid in intermediate metabolism in animal tissues. Enzymologia 4, 148–156.

Krebs, H.A., Kornberg, H.L., 1957. Energy Transformations in Living Matter: A Survey. Springer-Verlag, Berlin.

Křeček, P., Skůpa, P., Libus, J., Naramoto, S., Tejos, R., Friml, J., Zažimalová, E., 2009. The PIN-FORMED (PIN) protein family of auxin transporters. Genome Biol. 10, 249.

Kresge, N., Simoni, R.D., Hill, R.L., 2006. Unraveling the enzymology of oxidative phosphorylation: the work of Efraim Racker. J. Biol. Chem. 281, e4.

Kresge, N., Simoni, R.D., Hill, R.L., 2010. The discovery of glyoxysomes: the work of Harry Beevers. J. Biol. Chem. 285, e6–e7.

Kretsinger, R.H., 1977. Evolution of the informational role of calcium in eukaryotes. In: Wassermann, R.H., Corradino, R.A., Carafoli, E., Kretsinger, R.H., MacLennan, D.H., Siegel, F.L. (Eds.), Calcium-Binding Proteins and Calcium Function. North Holland, NY, pp. 63–72.

Kreuz, K., Tommasini, R., Martinoia, E., 1996. Old enzymes for a new job. Plant Physiol. 111, 349–353.

Krieg, U.C., Johnson, A.E., Walter, P., 1989. Protein translocation across the endoplasmic reticulum membrane: identification by photocross-linking of a 39-kD integral membrane glycoprotein as part of a putative translocation tunnel. J. Cell Biol. 109, 2033–2043.

Kringstad, R., Kenyon, W.H., Black, C.C., 1980. The rapid isolation of vacuoles from leaves of Crassulacean acid metabolism plants. Plant Physiol. 66, 379–382.

Krishnamoorthy, G., Hinkle, P.C., 1984. Non-ohmic proton conductance of mitochondria and liposomes. Biochemistry 23, 1640–1645.

Krishnamoorthy, G., Hinkle, P.C., 1988. Studies on the electron transfer pathway, topography of iron-sulfur centers, and site of coupling in NADH-Q oxidoreductase. J. Biol. Chem. 263, 17566–17575.

Kristen, U., 1978. Ultrastructure and a possible function of the intercisternal elements in dictyosomes. Planta 138, 29–33.

Kristensen, V.N., Lingjærde, O.C., Russnes, H.G., Vollan, H.K.M., Frigessi, A., Børresen-Dale, A.-L., 2014. Principles and methods of integrative genomic analyses in cancer. Nat. Rev. Cancer 14, 299–313.

Krogh, A., 1916. The Respiratory Exchange of Animals and Man. Longmans, Green and Co., London.

Krogh, A., 1929. Progress in physiology. Am. J. Physiol. 90, 243–251.

Krogmann, D.W., 2000. The golden age of biochemical research in photosynthesis. Photosynth. Res. 653, 109–121.

Krogmann, D.W., Jagendorf, A.T., Avron, M., 1959. Uncouplers of spinach chloroplast photosynthetic phosphorylation. Plant Physiol. 34, 272–277.

Kron, S.J., Spudich, J.A., 1986. Fluorescent actin filaments move on myosin fixed to a glass surface. Proc. Natl. Acad. Sci. U.S.A. 83, 6272–6276.

Kropf, D.L., Berge, S.K., Quatrano, R.S., 1989. Actin localization during Fucus embryogenesis. Plant Cell 1, 191–200.

Kropf, D.L., Henry, C.A., Gibbon, B.C., 1995. Measurement and manipulation of cytosolic pH in polarizing zygotes. Eur. J. Cell Biol. 68, 297–305.

Kropf, D.L., Kloareg, B., Quatrano, R.S., 1988. Cell wall is required for fixation of the embryonic axis in Fucus zygotes. Science 239, 187–190.

Kropf, D.L., Williamson, R.E., Wasteneys, G.O., 1997. Microtubule orientation and dynamics in elongating characean internodal cells following cytosolic acidification, induction of pH bands, or premature growth arrest. Protoplasma 197, 188–198.

Kroto, H.W., December 7, 1996. Symmetry, space, stars and C_{60}. Nobel Lecture.

Krüger, F., Schumacher, K., 2018. Pumping up the volume—vacuole biogenesis in Arabidopsis thaliana. Semin. Cell Dev. Biol. in press.

Krüger, F., Schumaker, K., 2017. Pumping up the volume—vacuole biogenesis in Arabidopsis thaliana. Semin. Cell Dev. Biol. 80, 106–112.

Kruse, C., Frevert, J., Kindl, H., 1981. Selective uptake by glyoxysomes of in vitro translated malate synthase. FEBS Lett. 129, 36–38.

Kruse, C., Kindl, H., 1983. Oligomerization of malate synthase during glyoxysome biosynthesis. Arch. Biochem. Biophys. 223, 629–638.

Krysan, P.J., Young, J.C., Sussman, M.R., 1999. T-DNA as an insertional mutagen in *Arabidopsis*. Plant Cell 11, 2283–2290.

Kubai, D.F., 1975. The evolution of the mitotic spindle. Int. Rev. Cytol. 43, 167–227.

Kubai, D.F., 1978. Mitosis and fungal phylogeny. In: Heath, I.B. (Ed.), Nuclear Division in the Fungi. Academic Press, New York, pp. 177–229.

Kubai, D.F., Ris, H., 1969. Division in the dinoflagellate Gyrodinium Cohnii (Schiller). J. Cell Biol. 40, 508–528.

Kubiak, J., De Brabander, M., De May, J., Tarkowska, J.A., 1986. Origin of the mitotic spindle in onion root cells. Protoplasma 130, 51–56.

Kucherlapati, R., 2017. Oliver Smithies (1925–2017). Nature 542, 166.

Kudlicka, K., Brown Jr., R.M., 1997. Cellulose and callose biosynthesis in higher plants. I. Solubilization and separation of $(1\rightarrow3)$- and $(1\rightarrow4)$-β-glucan synthase activities from mung bean. Plant Physiol. 115, 643–656.

Kudlicka, K., Brown Jr., R.M., Li, L., Lee, J.H., Shin, H., Kuga, S., 1995. β-Glucan synthesis in the cotton fiber. IV. In vitro assembly of the cellulose 1 allomorph. Plant Physiol. 107, 111–123.

Kudlicka, K., Lee, J.H., Brown Jr., R.M., 1996. A comparative analysis of in vitro cellulose synthesis from cell-free extracts of mung bean (*Vigna radiata*, Fabaceae) and cotton (*Gossypium hirsutum*, Malvaceae). Am. J. Bot. 83, 274–284.

Kühlbrandt, W., Zeelen, J., Dietrich, J., 2002. Structure, mechanism, and regulation of the Neurospora plasma membrane H^+-ATPase. Science 297, 1692–1696.

Kühn-Velten, W.N., 1993. Fusion of enzyme kinetics with quantitative cell biology constitutes a novel, synergistic approach to calculate local enzyme activities *in situ* from subcellular- site concentrations of enzymes and their substrates: an endocrinologist's contribution to the "metabolite channelling" debate. J. Theor. Biol. 165, 447–453.

Kühne, W., 1864. Untersuchungen Über Das Protoplasma Und Die Kontraktilität. W. Engelmann, Leipzig.

Kühtreiber, W.M., Jaffe, L.F., 1990. Detection of extracellular Ca^{2+} gradients with a calcium-specific vibrating electrode. J. Cell Biol. 110, 1565–1573.

Kui, L., Chen, H., Zhang, W., He, S., Xiong, Z., Zhang, Y., Yan, L., Zhong, C., He, F., Chen, J., Zeng, P., Zhang, G., Yang, S., Dong, Y., Wang, W., Cai, J., 2017. Building agenetic manipulation tool box for orchid biology: identification of constitutive promoters and application of CRISPR/Cas9 in the orchid, *Dendrobium officinale*. Front. Plant Sci. 7, 2036.

Kuimova, M.K., Botchway, S.W., Parker, A.W., Balaz, M., Collins, H.A., Anderson, H.L., Suhling, K., Ogilby, P.R., 2009. Imaging intracellular viscosity of a single cell during photoinduced cell death. Nat. Chem. 1, 69–73.

Kuipers, A.G.J., Soppe, W.J.J., Jacobsen, E., Visser, R.G.F., 1994. Field-evaluation of transgenic potato giants expressing an antisense granule-bound starch synthase gene-increase of the antisense effect during tuber growth. Plant Mol. Biol. 26, 1759–1773.

Kukurba, K.R., Montgomery, S.B., 2015. RNA sequencing and analysis. Cold Spring Harb. Protoc. 2015, 951–969.

Kull, F.J., Sablin, E.P., Lau, R., Fletterick, R.J., Vale, R.D., 1996. Crystal structure of the kinesin motor domain reveals structural similarity to myosin. Nature 380, 550–554.

Kuma, A., Mizushima, N., Ishihara, N., Ohsumi, Y., 2002. Formation of the approximately 350-kDa Apg12-Apg5.Apg16 multimeric complex, mediated by Apg16 oligomerization, is essential for autophagy in yeast. J. Biol. Chem. 277, 18619–18625.

Kumagai, K., Ebashi, S., Takeda, F., 1955. Essential relaxing factor in muscle other than myokinase and creatine phosphokinase. Nature 176, 166.

Kumagai, F., Yoneda, A., Tomida, T., Sano, T., Nagata, T., Hasegawa, S., 2001. Fate of nascent microtubules organized at the M/G interface, as visualized by synchronized tobacco BY-2 cells stably expressing GFP-tubulin: time-sequence observations of the reorganization of cortical microtubules in living plant cells. Plant Cell Physiol. 42, 723–732.

Kumamaru, T., Ogawa, M., Satoh, H., Okita, T.W., 2007. Protein body biogenesis in cereal endosperms. Plant Cell Monogr. 8, 141–158.

Kumar, R., Ichihashi, Y., Kimura, S., Chitwood, D.H., Headland, L.R., Peng, J., Maloof, J.N., Sinha, N.R., 2012. A high-throughput method for Illumina RNA-Seq library preparation. Front. Plant Sci. 3, 202.

Kunau, W.H., 2005. Peroxisome biogenesis: end of the debate. Curr. Biol. 15, R774–R776.

Kung, S.-D., 1998. Discovery of promotor sequewnces of chloroplast genes. Discov. Plant Biol. 2, 61–72.

Kunitz, M., 1939. Isolation from beef pancrease of a crystalline protein possessing ribonuclease activity. Science 90, 112–113.

Kunitz, M., 1940. Crystalline ribonuclease. J. Gen. Physiol. 24, 15–32.

Kunitz, M., Northrop, J.H., 1934. The isolation of crystalline trypsinogen and its conversion into crystalline trypsin. Science 80, 505–506.

Kunitz, M., Northrop, J.H., 1936. Isolation from beef pancreas of crystalline trypsinogen, trypsin, a trypsin inhibitor, and an inhibitor-trypsin compound. J. Gen. Physiol. 19, 991–1007.

Kurek, I., Kawagoe, Y., Jacob-Wilk, D., Dobin, M., Delmer, D., 2003. Dimerization of cotton fiber cellulose synthase catalytic subunits occurs via oxidation of the zinc-binding domains. Proc. Natl. Acad. Sci. U.S.A. 99, 11109–11114.

Kurihara, Y., Watanabe, Y., 2004. *Arabidopsis* micro-RNA biogenesis through Dicer-like 1 protein functions. Proc. Natl. Acad. Sci. U.S.A. 101, 12753–12758.

Kuroda, K., 1983. Cytoplasmic streaming in characean cells cut open by microsurgery. Proc. Jpn. Acad. Sci. 59, 126–130.

Kuroda, K., 1990. Cytoplasmic streaming in plant cells. Int. Rev. Cytol. 121, 267–307.

Kuroda, K., Manabe, E., 1983. Microtubule-associated cytoplasmic streaming in *Caulerpa*. Proc. Jpn. Acad. 59, 131–134.

Kuroiwa, T., 1982. Mitochondrial nuclei. Int. Rev. Cytol. 75, 1–59.

Kuroiwa, T., 1991. The replication, differentiation, and inheritance of plastids with emphasis on the concept of organelle nuclei. Int. Rev. Cytol. 128, 1–62.

Kuroiwa, T., 1998. The primitive red algae *Cyanidium caldarium* and *Cyanidioschyzon merolae* as model system for investigating the dividing apparatus of mitochondria and plastids. BioEssays 20, 344–354.

Kuroiwa, T., Fujie, M., Kuroiwa, H., 1992. Studies on the behavior of mitochondrial DNA. J. Cell Sci. 101, 483–493.

Kuroiwa, T., Hori, T., 1986. Preferential digestion of male chloroplast nuclei during gametogenesis of *Bryopsis maxima* Okamura. Protoplasma 133, 85–87.

Kuroiwa, T., Ishibashi, K., Takano, H., Higashiyama, T., Sasaki, N., Nishimura, Y., Matsunaga, S., 1996. Optical isolation of individual mitochondria of *Physarum polycephalum* for PCR analysis. Protoplasma 194, 274–279.

Kuroiwa, T., Kawano, S., Nishibayashi, S., Sato, C., 1982. Epifluorescence microscopic evidence for maternal inheritance of chloroplast DNA. Nature 298, 481–484.

Kuroiwa, T., Kuroiwa, H., Sakai, A., Takahashi, H., Toda, K., Itoh, R., 1998. The division apparatus of platids and mitochondria. Int. Rev. Cytol. 181, 1–41.

Kuroiwa, H., Sugai, M., Kuroiwa, T., 1988. Behavior of chloroplast and chloroplast nuclei during spermatogenesis in the fern *Pteris vittata*. Protoplasma 146, 89–100.

Kuroiwa, T., Uchida, H., 1996. Organelle division and cytoplasmic inheritance. BioScience 46, 827–835.

Kuroiwa, T., Yorihuzi, T., Yabe, N., Ohta, T., Uchida, H., 1990. Absence of DNA in the basal body of Chlamydomonas reinhardtii by fluorimetry using a video-intensified microscopy photon-counting system. Protoplasma 158, 155–164.

Kuromori, T., Wada, T., Kamiya, A., Yuguchi, M., Yokouchi, T., Imura, Y., Takabe, H., Sakurai, T., Akiyama, K., Hirayama, T., Okada, K., Shinozaki, K., 2006. A trial of phenome analysis using 4000 Ds-insertional mutants in gene-coding regions of *Arabidopsis*. Plant J. 47, 640–651.

Küster, E., 1935. Anisotropic elements of the plant cell. J. R. Microsc. Soc. 55, 99–101.

Küster, E., 1996. Pathology og the Plant Cell. Part 1. Pathology of Protoplasm, 1929, Translated by E. Stadelmann, W. R. Bushnell, and A. H. Bushnell. Saad Publications, Karachi, Pakistan.

Kusumi, A., Sako, Y., 1996. Cell surface organization by the membrane skeleton. Curr. Opin. Cell Biol. 8, 566–574.

Kutschera, U., 1991. Regulation of cell expansion. In: Lloyd, C.W. (Ed.), The Cytoskeletal Basis of Plant Growth and Form. Academic Press, London, pp. 149–158.

Kutschera, U., 2011. The cell was defined 150 years ago. Nature 480, 457.

Kutschera, U., Hossfeld, U., 2013. Alfred Russel Wallace (1823–1913): the forgotten co-founder of the Neo-Darwinian theory of biological evolution. Theory Biosci. 132, 207–214.

Kutschera, U., Nick, P., 2017. Peter Sitte (1929–2015): a theistic cell biologist. Protoplasma 254, 1821–1822.

Kutschera, U., Niklas, K., 2005. Endosymbiosis, cell evolution, and speciation. Theory Biosci. 124, 1–24.

Kuwada, Y., Nakamura, T., 1934. Behaviour of chromomemata in mitosis. IV. Double refraction of chromosomes in *Tradescantia reflexa*. Int. J. Cytol. 6, 78–86.

Kvenvolden, K.A., Lawless, J., Pering, K., Peterson, E., Flores, J., Ponnamperuma, C., Kaplan, I., Moore, C., 1970. Evidence for extraterrestrial amino acids and hydrocarbons in the Murchison meteorite. Nature 228, 923–926.

Kvenvolden, K.A., Lawless, J.G., Ponnamperuma, C., 1971. Non-protein amino acids in the Murchison meteorite. Proc. Natl. Acad. Sci. U.S.A. 68, 486–490.

Kwiatkowska, M., 2003. Plasmodesmal changes are related to different developmental stages of antheridia of *Chara* species. Protoplasma 222, 1–11.

Kwok, E.Y., Hanson, M.R., 2003. Microfilaments and microtubules control the morphology and movement of non-green plastids and stromules in *Nicotiana tabacum*. Plant J. 35, 16–26.

Kwok, E.Y., Hanson, M.R., 2004a. GFP-labeled Rubisco and aspartate aminotransferase accumulate in plastid stromules and traffic between plastids. J. Exp. Bot. 55, 595–604.

Kwok, E.Y., Hanson, M.R., 2004b. Stromules and the dynamic nature of plastid morphology. J. Microsc. 214, 124–137.

Kyle, D.J., Staehelin, L.A., Arntzen, C.J., 1983. Lateral mobility of the light harvesting complex in chloroplast membranes controls excitation energy distribution in higher plants. Arch. Biochem. Biophys. 222, 527–541.

Kyte, J., Doolittle, R.F., 1982. A simple method for displaying the hydropathic character of a protein. J. Mol. Biol. 157, 105–132.

La Venuta, G., Zeitler, M., Steringer, J.P., Müller, H.-M., Nickel, W., 2015. The starting proterties of fibroblast growth factor 2: how to exit mammalian cells without a signal peptide at hand. J. Biol. Chem. 290, 27015–27020.

Labavitch, J.M., 1981. Cell wall turnover in plant development. Ann. Rev. Plant Physiol. 32, 385–406.

Labonte, M.L., 2017. Blobel and Sabatini's "beautiful idea": visual representations of the conception and refinement of the signal hypothesis. J. Hist. Biol. https://link.springer.com/content/pdf/10.1007%2Fs10739-016-9462-7.pdf.

Lacey Jr., J.C., Hawkins, A.F., Thomas, R.D., Watkins, C.L., 1988. Differential distribution of D and L amino acids between the 2′ and 3′ positions of the AMP residue at the 3′ terminus of transfer ribonucleic acid. Proc. Natl. Acad. Sci. U.S.A. 85, 4996–5000.

Lacey Jr., J.C., Mullins Jr., D.W., 1983. Experimental studies related to the origin of the genetic code and the process of protein synthesis—A review. Orig. Life 13, 3–42.

Lacey Jr., J.C., Staves, M.P., Thomas, R.D., 1990a. Ribonucleic acids may be catalysts for the prepeptides: a minireview. J. Mol. Evol. 31, 244–248.

Lacey, J.C., Thomas, R.D., Staves, M.P., Watkins, C.L., 1991. Stereoselective formation of bis(alpha-aminoacyl) esters of 5′- AMP suggests a primative peptide synthesizing system with a preference for L-amino acids. Biochim. Biophys. Acta 1076, 395–400.

Lacey Jr., J.C., Thomas, R.D., Wickaramasinghe, N.S., Watkins, C.L., 1990b. Chemical esterification of 5′-AMP occurs predominantly at the 2′ position. J. Mol. Evol. 31, 251–256.

Lacey, J.C., Wickramasinghe, N.S., Cook, G.W., Anderson, G., 1993. Couplings of character and of chirality in the origin of the genetic system. J. Mol. Evol. 37, 233–239.

Lacey, J.C., Wickramasinghe, N.S., Sabatini, R.S., 1992. Preferential hydrophobic interactions are responsible for a preference of D-amino acids in the aminoacylation of 5′-AMP with hydrophobic amino acids. Experientia 48, 379–383.

Lachaud, S., Maurousset, L., 1996. Occurrence of plasmodesmata between differentiating vessels and other xylem cells in *Sorbus torminalis* L. Crantz and their fate during xylem maturation. Protoplasma 191, 220–226.

Lack, D., 1957. Evolutionary Theory and Christian Belief: The Unresolved Conflict. Methuen & Co., London.

LaCour, L.F., Darlington, C.D., 1975. The Handling of Chromosomes. John Wiley & Sons, New York.

Laemmli, U.K., 1970. Cleavage of structural proteins during assembly of the head of bacteriophage T4. Nature 227, 680–685.

Laere, E., Ling, A.P.K., Wong, Y.P., Koh, Y.P., Lia, M.A.M., Hussein, S., 2016. Plant-based vaccines: production and challenges. J. Bot. 2016, 4928637.

Laffe, L.F., 1983. Sources of calcium in egg activation: a review and hypothesis. Dev. Biol. 99, 265–276.

Lafontaine, J.G., Lord, A., 1974. An ultrastructural and radioautographic study of the evolution of the interphase nucleus in plant cells (*Allium porrum*). J. Cell Sci. 14, 263–287.

Lai, S.P., Randall, S.K., Sze, H., 1988. Peripheral and integral subunits of the tonoplast H^+-ATPase from oat roots. J. Biol. Chem. 263, 16731–16737.

Laible, P.D., Zipfel, W., Owens, T.G., 1994. Excited state dynamics in chlorophyll-based antennae: the role of transfer equilibrium. Biophys. J. 66, 844–860.

Laidler, K.J., 1993. The World of Physical Chemistry. Oxford University Press, Oxford.

Laine, R.O., Zeile, W., Kang, F., Purich, D.L., Southwick, F.S., 1997. Vinculin proteolysis unmasks an ActA homolog for actin-based Shigella motility. J. Cell Biol. 138, 1255–1267.

Laird, P.W., 2010. Principles and challenges of genomewide DNA methylation analysis. Nat. Rev. Genet. 11, 191–203.

Lakshminarayanaiah, N., 1965. Transport phenomena in artificial membranes. Chem. Rev. 65, 491–565.

Lakshminarayanaiah, N., 1969. Transport Phenomena in Membranes. Academic Press, New York.

Lalonde, S., Ehrhardt, D.W., Frommer, W.B., 2005. Shining light on signaling and metabolic networks by genetically encoded biosensors. Curr. Opin. Plant Biol. 8, 574–581.

Lam, S.K., Siu, C.L., Hillmer, S., Jang, S., An, G., Robinson, D.G., 2007. Rice SCAMP1 defines clathrin-coated, trans-Golgi-located tubular-vesicular structures as an early endosome in tobacco BY- 2 cells. Plant Cell 19, 296–319.

Lamanna, C., 1968. DNA discovery in perspective. Science 160, 1397–1398.

Lamarck, J.B., 1802. (An X). Hydrogéologie. Lamarck, Paris.

Lamarck, J.B., 1809. Zoological Philosophy, 1914, Translated by H. Elliot. MacMillan, London.

Lamarck, J.B., de Candolle, A.P., 1815. Flore Français 3, 151.

Lamb Jr., W.E., December 12, 1955. Fine structure of the hydrogen atom. Nobel Lecture.

Lambert, A.-M., Bajer, A.S., 1972. Dynamics of spindle fibers and microtubules during anaphase and phragmoplast formation. Chromosoma 39, 101–144.

Lambert, D.G., Rainbow, R.D. (Eds.), 2013. Calcium Signaling Protcols, third ed. Springer, New York.

Lambridis, H., 1976. Empedocles. A Philosophical Investigation. University of Alabama Press, Tuscaloosa, AL.

Lametti, D., 2012. Is a science Ph.D. a waste of time? Don't feel too sorry for graduate students. It's worth it. Slate.com. http://www.slate.com/articles/health_and_science/science/2012/08/what_is_the_value_of_a_science_phd_is_graduate_school_worth_the_effort_.html.

Lamont, L., 1965. Day of Trinity. Atheneum, New York.

Lamport, D.T.A., 1965. The protein component of primary cell walls. Adv. Bot. Res. 2, 151–218.

Lampson, M.A., Cheeseman, I.M., 2011. Sensing centromere tension: aurora B and the regulation of kinetochore function. Trends Cell Biol. 21, 133–140.

Lan, T., Renner, T., Ibarra-Laclette, E., Farr, K.M., Chang, T.-H., Cervantes-Pérez, S.A., Zheng, C., Sankoff, D., Tang, H., Purbojati, R.W., Putra, A., Drautz-Moses, D.I., Schuster, S.C., Herrera-Estrella, L., Albert, V.A., 2017. Long-read sequencing uncovers the adaptive topography of a carnivorous plant genome. Proc. Natl. Acad. Sci. U.S.A. 114, E4435–E4441.

Lancelle, S.A., Callaham, D.A., Hepler, P.K., 1986. A method for rapid freeze fixation of plant cells. Protoplasma 131, 153–165.

Lancelle, S.A., Cresti, M., Hepler, P.K., 1987. Ultrastructure of the cytoskeleton in freeze-substituted pollen tubes of *Nicotiana alata*. Protoplasma 140, 141–150.

Lancelle, S.A., Cresti, M., Hepler, P.K., 1997. Growth inhibition and recovery in freez-substituted *Lilium longiflorum* pollen tubes: structural effects of caffeine. Protoplasma 196, 21–33.

Lancelle, S.A., Hepler, P.K., 1988. Cytochalasin-induced ultrastructural alterations in *Nicotiana* pollen tubes. Protoplasma (Suppl. 2), 65–75.

Lancelle, S.A., Hepler, P.K., 1989. Immunogold labeling of actin on sections of freeze substituted plant cells. Protoplasma 150, 72–74.

Lancelle, S.A., Hepler, P.K., 1992. Ultrastructure of freeze-substituted pollen tubes of *Lilium longiflorum*. Protoplasma 167, 215–230.

Lander, E.S., 2016. The heroes of CRISPR. Cell 164, 18–28.

Lanford, R.E., Butel, J.S., 1984. Construction and characterization of an SV40 mutant defective in nuclear transport of T antigen. Cell 37, 801–813.

Lang, I., Scholz, M., Peters, R., 1986. Molecular mobility and nucleocytoplasmic flux in hepatoma cells. J. Cell Biol. 102, 1183–1190.

Lang, Z., Xie, S., Zhu, J.-K., 2016. The 1001 Arabidopsis DNA methylomes: an important resource for studying natural genetic, epigenetic, and phenotypic variation. Trends Plant Sci. 21, 906–908.

Lang-Pauluzzi, I., 2000. The behaviour of the plasma membrane during plasmolysis: a study by UV microscopy. J. Microsc. 198, 188–198.

Lang-Pauluzzi, I., Gunning, B.E.S., 2000. A plasmolytic cycle: the fate of cytoskeletal elements. Protoplasma 212, 174–185.

Langan, K.J., Nothnagel, E.A., 1997. Cell surface arabinogalactan-proteins and their relation to cell proliferation and viability. Protoplasma 196, 87–98.

Langhans, M., Hawes, C., Hillmer, S., Hummel, E., Robinson, D.G., 2007. Golgi regeneration after brefeldin A treatment in BY-2 cells entails stack enlargement and cisternal growth followed by division. Plant Physiol. 145, 527–538.

Langley, J.N., 1921. The Autonomic Nervous System. W. Heffer & Sons, Cambridge.

Langley, K.H., Piddington, R.W., Ross, D., Sattelle, D.B., 1976. Photon correlation analysis of cytoplasmic streaming. Biochem. Biophys. Acta 444, 893–898.

Langmuir, I., 1916. The constitution and fundamental properties of solids and liquids. Part 1. Solids. J. Am. Chem. Soc. 38, 2221–2295.

Langmuir, I., 1917. The constitution and fundamental properties of solids and liquids. Part 2. Liquids. J. Am. Chem. Soc. 39, 1848–1906.

Langmuir, I., 1989a. Pathological science. Phys. Today 42 (10), 36–48.

Langmuir, I., 1989b. Pathological science. Res. Technol. Manag. 32 (5), 11–17.

Lao, N.T., Long, D., Kiang, S., Coupland, G., Shoue, D.A., Carpita, N.C., Kavanagh, T.A., 2003. Mutation of a family 8 glycosyl- transferase gene alters cell wall carbohydrate composition and causes a humidity-sensitive semi-sterile dwarf phenotype in *Arabidopsis*. Plant Mol. Biol. 53, 687–701.

Laplace, P.-S., 1814. Essai Philosophique sur les Probabilités. Mme. Ve Courcie, Paris.

Laporte, K., Rossignol, M., Traas, J.A., 1993. Interaction of tubulin with the plasma membrane: tubulin is present in purified plasmalemma and behaves as an integral membrane protein. Planta 191, 413–416.

Laporte, C., Vetter, G., Loudes, A.-M., Robinson, D.G., Hillmer, S., Stussi-Garaud, C., Ritzenthaler, C., 2003. Involvement of the secretory pathway and the cytoskeleton in intracellular targeting and tubule assembly of Grapevine fanleaf virus movement protein in tobacco BY-2 cells. Plant Cell 15, 2058–2075.

Lara, M.V., Offermann, S., Smith, M., Okita, T.W., Andreo, C.S., Edwards, G.E., 2008. Leaf development in the single-cell C4 System in *Bienertia sinuspersici*: expression of genes and peptide levels for C4 metabolism in relation to chlorenchyma structure under different light conditions. Plant Physiol. 148, 593–610.

Lardy, H.A., 1965. On the direction of pyridine nucleotide oxidation-reduction reactions in gluconeogenesis and lipogenesis. In: Chance, B., Estabrook, R.W., Williamson, J.R. (Eds.), Control of Energy Metabolism. Academic Press, New York, pp. 245–248.

Lardy, H.A., 1995. So it's cytosol! Trends Cell Biol. 5, 416.

Lardy, H.A., Winchester, B., Phillips, P.H., 1945. The repiratory metabolism of ram spermatozoa. Arch. Biochem. 6, 33–40.

Lark, K.G., 1967. Nonrandom segregation of sister chromatids in *Vicia faba* and *Triticum boeoticum*. Proc. Natl. Acad. Sci. U.S.A. 58, 352–359.

Lark, K.G., Consigli, R.A., Minocha, H.C., 1966. Segregation of sister chromatids in mammalian cells. Science 154, 1202–1205.

Larkins, B.A., Hurkman, W.J., 1978. Synthesis and deposition of protein bodies of maize endosperm. Plant Physiol. 62, 256–263.

Larmor, J., 1905. Note on the mechanics of the ascent of sap in trees. Proc. R. Soc. Lond. 76B, 460–463.

Larsen, P.M., Chen, T.L.L., Wolniak, S.M., 1991. Neomycin reversibly disrupts mitotic progression in stamen hair cells of Tradescantia. J. Cell. Sci. 98, 159–168.

Larson, M.H., Gilbert, L.A., Wang, X., Lim, W.A., Weissman, J.S., Qi, L.S., 2014. CRISPR interference (CRISPRi) for sequence-specific control of gene expression. Nat. Protoc. 8, 2180–2196.

Larsson, N., December 10, 2016. Award Ceremony Speech. Nobel Prize in Physiology or Medicine.

Larsson, U., Jergil, B., Anderson, B., 1983. Changes in the lateral distribution of the light-harvesting chlorophyll a/b-protein complex induced by its phosphorylation. Eur. J. Biochem. 136, 25–29.

Lasek, R.J., Brady, S.T., 1985. Attachment of transported vesicles to microtubules in axoplasm is facilitated by AMP-PMP. Nature 316, 645–647.

Laskey, R.A., Honda, B.M., Mills, A.D., Finch, J.T., 1978. Nucleosomes are assembled by an acidic protein which binds histones and transfers then to DNA. Nature 275, 416–420.

Laskowski, M.J., 1990. Microtubule orientation in pea stem cells: a change in orientation follows the initiation of growth rate decline. Planta 181, 44–52.

Latham, D. (Ed.), 2003. Haunted Texts: Studies in Pre-Raphaelism in Honour of William E. Fredeman. University of Toronto Press, Toronto.

Laties, G.G., 1953. The physical environment and oxidative and phosphorylative capacities of higher plant mitochondria. Plant Physiol. 28, 557–575.

Laties, G.G., 1953. The physical environment and oxidative and phosphorylative capacities of higher plant mitochondria. Plant Physiol. 28, 557–575.

Laties, G.G., 1959. Active transport of salt into plant tissue. Ann. Rev. Plant Physiol. 10, 87–112.

Laties, G.G., 1964. Physiological aspects of membrane function in plant cells during development. In: Locke, M. (Ed.), Cellular Membranes in Development. Academic Press, New York, pp. 299–320.

Laties, G.G., 1998. The discovery of the cyanide-resistant alternative path: and its aftermath. Discov. Plant Biol. 1, 233–255.

Laue, M. von, December 12, 1915. Concerning the detection of X-ray interferences. In: Nobel Lect.

Läuger, P., 1991. Electrogenic Ion Pumps. Sinauer Associates Inc., Sunderland, MA.

Laurent, T.C., 1995. An early look at macromolecular crowding. Biophys. Chem. 57, 7–14.

Lauterborn, R., 1911. Die biologiche Selbstreinigung unserer Gewässer.

Laval, V., Chabannes, M., Carrié, R.E., Canut, H., Barre, A., Rougé, P., Pont-Lezica, R., Galaud, J., 1999. A family of *Arabidopsis* plasma membrane receptors presenting animal β-integrin domains. Biochim. Biophys. Acta 1435, 61–70.

Lavoisier, A., 1790. Elements of Chemistry. William Creech, Edinburgh.

Lavoisier, A., Laplace, P., 1780. Memoir on heat. Memoires de l'academie des sciences. In: Gabriel, M.L., Fogel, S. (Eds.), Translated in Great Experiments in Biology. Prentice-Hall, Engelwood Cliffs, NJ, p. 337, 1955.

Lawrence, C.J., Dawe, R.K., Christie, K.R., Cleveland, D.W., Dawson, S.C., Endow, S.A., Goldstein, L.S., Goodson, H.V., Hirokawa, N., Howard, J., Malmberg, R.L., McIntosh, J.R., Miki, H., Mitchison, T.J., Okada, Y., Reddy, A.S., Saxton, W.M., Schliwa, M., Scholey, J.M., Vale, R.D., Walczak, C.E., Wordeman, L., 2004. A standardized kinesin nomenclature. J. Cell Biol. 167, 19–22.

Lawrence, C.J., Morris, N.R., Meagher, R.B., Dawe, R.K., 2001. Dyneins have run their course in plant lineage. Traffic 2, 362–363.

Lawrence, W.J.C., Price, J.R., 1940. The genetics and chemistry of flower color variation. Biol. Rev. 15, 35–38.

Laws, K., 2002. Physics and the Art of Dance. Oxford University Press, Oxford.

Lazarow, P.B., 2003. Peroxisome biogenesis: advances and conundrums. Curr. Opin. Cell Biol. 15, 489–497.

Lazarow, A., Cooperstein, S.J., 1953. Studies on the enzymatic basis for the Janus Green B staining reaction. J. Histochem. Cytochem. 1, 234–241.

Lazarow, P.B., de Duve, C., 1976. A fatty acyl-CoA oxidizing system in rat liver peroxisomes; enhancement by clofibrate, a hypolipidemic drug. Proc. Natl. Acad. Sci. U.S.A. 73, 2043–2046.

Lazarow, P.B., Fujiki, Y., 1985. Biogenesis of peroxisomes. Ann. Rev. Cell Biol. 1, 489–530.

Lazarowitz, S.G., 1999. Probing plant cell structure and function with viral movement proteins. Curr. Opin. Plant Biol. 2,332–338.

Lazarowitz, S.G., Beachy, R.N., 1999. Viral movement proteins as probes for intracellular trafficking in plants. Plant Cell 11, 535–548.

Lazcano, A., Bada, J.L., 2004. The 1953 Stanley L. Miller Experiment: fifty years of prebiotic chemistry. Orig. Life Evol. Biosph. 33, 235–242.

Lea, D.E., 1947. Actions of Radiations on Living Cells. Cambridge University Press, Cambridge.

Leathes, J.B., 1926. Function and design. Science 64, 387–394.

Leavitt, H.S., 1908. 1777 variables in the Magellanic Clouds. Ann. Harv. Coll. Obs. 60, 87–108.

LeBel, J.A., 1874. On the relations which exist between the atomic formulas of organic compounds and the rotatory power of their solutions. Bull. Soc. Chim. 22, 337–347.

Leber, B., Hemleben, V., 1979. Uptake of homologous DNA into nuclei of seedlings and by isolated nuclei of a higher plant. Z. Pflanzenphysiol. 91, 305–316.

Leborgne, N., Dupou-Cézanne, L., Teuliéres, C., Canut, H., Tocanne, J.-F., Boudet, A.M., 1992. Lateral and rotational mobilities of lipids in specific cellular membranes of Eucalyptus Gunnii cultivars exhibiting different freezing tolerance. Plant Physiol. 100, 246–254.

Leckie, C.P., McAinsh, M.R., Montgomery, L., Priestley, A.J., Staxen, I., Webb, A.A.R., Hetherington, A.M., 1998. Second messengers in guard cells. J. Exp. Bot. 49 (Special Issue), 339–349.

Lecomte du Nouy, P., 1947. Human Destiny. Longmans, Green and Co., London.

Ledbetter, M.C., Porter, K.R., 1963. A "microtubule" in plant cell fine structure. J. Cell Biol. 19, 239–250.

Ledbetter, M.C., Porter, K.R., 1964. Morphology of microtubules of plant cells. Science 144, 872–874.

Lederberg, J., 1952. Cell genetics and hereditary symbiosis. Physiol. Rev. 32, 403–430.

Lederberg, J., 1963. Biological future of man. In: Wolstenholme, G. (Ed.), Man and His Future. A Ciba Foundation volume. Little, Brown and CO., Boston, pp. 263–273.

Lederberg, J., 1999. J. B. S. Haldane (1949) on infectious disease and evolution. Genetics 153, 1–3.

Lederberg, E.M., Lederberg, J., 1953. Genetic studied of lysogenicity in *Escherichia coli*. Genetics 38, 51–64.

Lederberg, J., May 29, 1959. A view of genetics. Nobel Lecture.

Lederberg, J., McCray, A.T., 2001. 'Ome Sweet 'Omics—A genealogical treasury of words. Scientist 15 (7), 8.

Lederman, L., Teresi, D., 1993. The God Particle. If the Universe is the Answer, What is the Question? Houghton Miflin, Boston.

Ledford, H., 2018. Botanical Renaissance. Nature 553, 396–398.

Lee, A.B., 1893. The Microtomist's Vade-mecum, third ed. J. & A. Churchill, London.

Lee, R.E., 1995. Phycology, second ed. Cambridge University Press, Cambridge.

Lee, D., 2007. Nature's Palette. The Science of Color. University of Chicago Press, Chicago.

Lee, J.-Y., 2015. Plasmodesmata: a signaling hub at he cellular boundary. Curr. Opin. Plant Biol. 27, 133–140.

Lee, J., Abdeen, A.A., Zhang, D., Kilian, K.A., 2013b. Directing stem cell fate on hydrogel substrates by controlling geometry, matrix mechanics and adhesion ligand composition. Biomaterials 34, 8140–8148.

Lee, Y., Choi, Y.B., Suh, S., Lee, J., Assmann, S.M., Joe, C.O., Kelleher, J.F., Crain, R.C., 1996c. Abscisic acid-induced phosphoinositide turnover in guard cell protoplasts of *Vicia faba*. Plant Physiol. 110, 987–996.

Lee, E.K., Cibrian-Jaramillo, A., Kolokotronis, S.O., Katari, M.S., Stamatakis, A., Ott, M., Chiu, J.C., Little, D.P., Stevenson, D.W., McCombie, W.R., Martienssen, R.A., Coruzzi, G., DeSalle, R., 2011a. A functional phylogenomics view of the seed plants. PLoS Genet. 7, e1002411.

Lee, J.-Y., Colinas, J., Wang, J.Y., Mace, D., Ohler, U., Benfey, P.N., 2006. Transcriptional and posttranscriptional regulation of transcription factor expression in *Arabidopsis* roots. Proc. Natl. Acad. Sci. U.S.A. 103, 6055–6060.

Lee, Y.T., December 8, 1986. Molecular beam studies of elementary chemical processes. Nobel Lecture.

Lee, R.C., Feinbaum, R.L., Ambros, V., 1993. The *C. elegans* heterochronic gene lin-4 encodes small RNAs with antisense complementarity to lin-14. Cell 75, 843–854.

Lee, C., Ferguson, M., Chen, L.B., 1989. Construction of the endoplasmic reticulum. J. Cell Biol. 109, 2045–2055.

Lee, Y.-R.J., Giang, H.M., Liu, B., 2001. A novel plant kinesin-related protein specifically associates with the phragmoplast organelles. Plant Cell 13, 2427–2440.

Lee, C., Goldberg, J., 2010. Structure of coatamer cage proteins and the relationship among COPI, COPII, and clathrin vesicle coats. Cell 142, 123–132.

Lee, D.H., Granja, J.R., Martinez, J.A., Severin, K., Ghadiri, M.R., 1996. A self-replicating peptide. Nature 382, 525–528.

Lee, Y., Kim, Y.W., Jeon, B.W., Park, K.Y., Suh, S.J., Seo, J., Kwak, J.M., Martinoia, E., Hwang, I., Lee, Y., 2007. Phosphatidylinositol 4,5-bisphosphate is important for stomatal opening. Plant J. 52, 803–816.

Lee, Y.-R.J., Liu, B., 2000. Identification of a phragmoplast- associated kinesin-related protein in higher plants. Curr. Biol. 10, 797–800.

Lee, Y.-R.J., Liu, B., 2004. Cytoskeletal motors in *Arabidopsis*. Sixty-one kinesins and seventeen myosins. Plant Physiol. 136, 3877–3883.

Lee, Y.-R.J., Liu, B., 2007. Cytoskeletal motor proteins in plant cell division. Plant Cell Monogr. 9, 169–193.

Lee, J.-Y., Lu, H., 2011. Plasmodesmata: the battleground against intruders. Trends Plant Sci. 16, 201–210.

Lee, M.S., Mullen, R.T., Trelease, R.N., 1997. Oil seed isocitrate lyases lacking their essential type 1 peroxisomal targeting signal are piggybacked to glyoxysomes. Plant Cell 9, 185–197.

Lee, Y.S., Park, H.-S., Lee, D.-K., Jayakodi, M., Kim, N.-H., Lee, S.-C., Kundu, A., Lee, D.-Y., Kim, Y.C., In, J.G., Kwon, S.W., Yang, T.-J., 2017. Comparative analysis of the transcriptomes and primary metabolite profiles of adventitious roots of five *Panax ginsing* cultivars. J. Ginsing Res. 41, 60–68.

Lee, Y.-R., Qiu, W., Liu, B., 2015. Kinesin motors in plants: from subcellular dynamics to motility regulation. Curr. Opin. Plant Biol. 28, 120–126.

Lee, S.J., Rose, J.K., 2011. Charaterization of the plant cell wall proteome using high-throughput screens. Methods Mol. Biol. 715, 255–272.

Lee, H., Sparkes, I., Gattolin, S., Dzimitrowicz, N., Roberts, L.M., Hawes, C., Frigerio, L., 2013a. An *Arabidopsis* reticulon and atlastin homologue RHD3-like2 act together in shaping the tubular endoplasmic reticulum. New Phytol. 197, 481–489.

Lee, J.-Y., Taoka, K.I., Yoo, B.-C., Ben-Nissan, G., Kim, D.-J., et al., 2005. Plasmodesmal-associated protein kinase in tobacco and *Arabidopsis* recognizes a subset of non-cell- autonomous proteins. Plant Cell 17, 2817–2831.

Lee, J.W., Tevault, C.V., Owens, T.G., Greenbaum, E., 1996b. Oxygenic photoautotrophic growth without photosystem I. Science 273, 364–367.

Lee, J.-Y., Wang, X., Cui, W., Sager, R., Modla, S., Czymmek, K., Zybaliov, B., van Wijk, K., Zhang, C., Lu, H., Lakshmanan, V., 2011b. A plasmodesmata-localized protein mediates crosstalk between cell-to-cell communication and innate immunity in *Arabidopsis*. Plant Cell 23, 3353–3373.

Leegood, R.C., 2007. A welcome diversion from photorespiration. Nat. Biotechnol. 25, 539–540.

Lefebvre, B., Furt, F., Hartmann, M.-A., Michaelson, L.V., Carde, J.-P., Sargueil-Boiron, F., Rossignol, M., Napier, J.A., Cullimore, J., Bessoule, J.-J., Mongrand, S., 2007. Characterization of lipid rafts from Medicago truncatula root plasma membranes: a proteomic study reveals the presence of a raft-associated redox system. Plant Physiol. 144, 402–418.

Lefebvre, P.A., Rosenbaum, J.L., 1986. Regulation of the synthesis and assembly of ciliary and flagellar proteins during regeneration. Ann. Rev. Cell Biol. 2, 517–546.

Leff, H.S., Rex, A.F., 1990. Maxwell's Demon: Entropy, Information, Computing. Princeton University Press, Princeton, NJ.

Leffel, S.M., Mabon, S.A., Stewart Jr., C.N., 1997. Applications of green fluorescent protein in plants. BioTechniques 23, 912–918.

Legname, I., Baskakov, V., Nguyen, H.-O.B., Riesner, D., Cohen, F.E., DeArmond, S.J., Prusiner, S.B., 2004. Synthetic mammalian prions. Science 305, 673–676.

Legrain, P., Rosbash, M., 1989. Some cis and transacting mutants for splicing target pre m-RNA to the cytoplasm. Cell 57, 573–583.

Lehman, I.R., 1974. DNA ligase Structure, mechanism, and function. Science 186, 790–797.

Lehmann, H., 1935. Über die enzymatische Synthese der Kreatinphosphorsäure durch Umesterung der Phosphobrenzt raubensäure. Biochem. Z. 281, 271–291.

Lehninger, A.L., 1949. Esterification of inorganic phosphate coupled to electron transport between dihydrodiphosphopyridine nucleotide and oxygen. J. Biol. Chem. 178, 625–644.

Lehninger, A.L., 1955. Oxidative phosphorylation. Harvey Lect. 49, 176–215.

Lehninger, A.L., 1959. Respiratory-energy transformation. In: Oncley, J.L., Schmitt, F.O., Williams, R.C., Rosenberg, M.D., Holt, R.H. (Eds.), Biophysical Science—A Study Program. John Wiley & Sons, New York, pp. 136–146.

Lehninger, A.L., 1964. The Mitochondrion. Molecular Basis of Structure and Function. W. A. Benjamin, New York.

Lehninger, A.L., 1965. Bioenergetics. W. A. Benjamin, New York.

Lehninger, A.L., 1975. Biochemistry, second ed. Worth Publishers, New York.

Lei, L., 2017. Chlorophagy: preventing sunburn. Nat. Plants 3, 17026.

Leibler, S., Huse, D.A., 1991. A physical model for motor proteins. C. R. Acad. Sci. Series III. 313, 27–36.

Leibler, S., Huse, D.A., 1993. Porters versus rowers. A unified stochastic model of motor proteins. J. Cell Biol. 121, 1357–1368.

Leiboff, S., Li, X., Hu, H.-C., Todt, N., Yang, J., Li, X., Yu, X., Muehlbauer, G.J., Timmermans, M.J.P., Yu, J., Schnable, P.S., Scanlon, M.J., 2015. Genetic control of morphometric diversity in the maize shoot apical meristem. Nat. Commun. 6, 8974.

Leigh, R.A., 1997. The solute composition of plant vacuoles. Adv. Bot. Res. 25, 171–194.

Leigh, R., Branton, D., 1976. Isolation of vacuoles from root storage tissue of *Beta vulgaris* L. Plant Physiol. 58, 656–662.

Leitz, G., Schnepf, E., Greulich, K.O., 1995. Micromanipulation of statoliths in gravi-sensing Chara rhizoids by optical tweezers. Planta 197, 278–288.

Lemaître, G., 1931. The beginning of the world from the point of view of quantum theory. Nature 127, 706.

Lemaître, G., The Primeval Atom: An Essay on Cosmogony. van Nostrand, New York.

Lenard, P., 1933. Great Men of Science. A History of Scientific Progress. G. Bell and Sons, London.

Lenartowska, M., Michalska, A., 2008. Actin filament organization and polarity in pollen tubes revealed by myosin II subfragment 1 decoration. Planta 228, 891–896.

Lengerova, M., Vyskot, B., 2001. Sex chromatin and nucleolar analyses in *Rumex acetosa* L. Protoplasma 217, 147–153.

Leonard, R.T., Hansen, D., Hodges, T.K., 1973. Membrane-bound adenosine triphosphatase activities of oat roots. Plant Physiol. 51, 749–754.

Leonard, R.T., Hodges, T.K., 2000. Discovery of plasma membrane proton pumping ATPase: our point of view. Discov. Plant Biol. 3, 291–304.

Leonard, D.A., Zaitlin, M., 1982. A temperature-sensitive strain of tobacco mosaic virus defective in cell-to-cell movement generates an altered viral-encoded protein. Virology 117, 416–424.

Leopold, A., 1949. A Sand County Almanac and Sketches Here and There. Oxford University Press, Oxford.

Leopold, A.C., 1964. Plant Growth and Development. McGraw-Hill, New York.

Leopold, A.C., 2004. Living with the land ethic. BioScience 54, 149–154.

Leopold, A.C., 2014. Smart plants: memory and communication without brains. Plant Signal. Behav. 9 (10), e972268.

Leopold, A.C., Hall, O.F., 1966. Mathematical model of polar auxin transport. Plant Physiol. 41, 1476–1480.

Leopold, A.C., Jaffe, M.J., Brokaw, C.J., Goebel, G., 2000. Many modes of movement. Science 288, 2131–2132.

Leopold, A.C., Kriedemann, P.E., 1975. Plant Growth and Development, second ed. McGraw-Hill, New York.

Lepiniec, L., Debeaujon, I., Routaboul, J.-M., Baudry, A., Pourcel, I., Nesi, N., Caboche, M., 2006. Genetics and biochemistry of seed flavonoids. Annu. Rev. Plant Biol. 57, 405–430.

Lerner, E.J., Griffiths, R.B., 2008. Mixing science and theology. Two Lett. Phys. Today 61 (9), 12.

Lersten, N.R., Czlapinski, A.R., Curtis, J.D., Freckmann, R., Horner, H.T., 2006. Oil bodies in leaf mesophyll cells of angiosperms: overview and a selected survey. Am. J. Bot. 93, 1731–1739.

Leslie, R.J., Pickett-Heaps, J.D., 1983. Ultraviolet microbeam irradiations of mitotic diatoms: investigation of spindle elongation. J. Cell Biol. 96, 548–561.

Lester, D.R., Ross, J.J., Davies, P.J., Reid, J.B., 1997. Mendel's stem length gene (Le) encodes a gibberellin 3-β-hydroxylase. Plant Cell 9, 1435–1443.

Leuschner, A.O., 1937. The award of the Bruce gold medal to professor Ejnar Hertzsprung. Publ. Astron. Soc. Pac. 49, 65–81.

Leustek, T., Murillo, M., Cervantes, M., 1994. Cloning of a cDNA encoding ATP sulfurylase from *Arabidopsis thaliana* by functional expression in Saccharomyces cerevisiae. Plant Physiol. 105, 897–902.

Levan, A., 1938. The effect of colchicine on root mitoses in *Allium*. Hereditas 24, 471–486.

Levanony, H., Rubin, R., Altschuler, Y., Galili, G., 1992. Evidence for a novel route of wheat storage proteins to vacuoles. J. Cell Biol. 119, 1117–1128.

Levene, P.A., 1921. On the structure of thymus nucleic acid and on its possible bearing on the structure of the plant nucleic acid. J. Biol. Chem. 48, 119–125.

Levene, P.A., Bass, L.W., 1931. Nucleic acids. American Chemical Society Monograph Series. The Chemical Catalog Co., New York.

Levene, M.J., Korlach, J., Turner, S.W., Foquet, M., Craighead, H.G., Webb, W.W., 2003. Zero-mode waveguides for single-molecule analysis at high concentrations. Science 299, 682–686.

Levi-Montalcini, R., 1988. In Praise of Imperfection. My Life and Work. Basic Books, New York.

Levin, P.A., Angert, E.R., 2015. Small but mighty: cell size and bacteria. Cold Spring Harb. Prospect. Biol. a019216.

Levin, J.Z., Yassour, M., Adiconis, X., Nusbaum, C., Thompson, D.A., Friedman, N., Gnirke, A., Regev, A., 2010. Comprehensive comparative analysis of strand-specific RNA sequencing methods. Nat. Methods 7, 709−715.

Levy, A., Erlanger, M., Rosenthal, M., Epel, B.L., 2007. A plasmodesmata-associated beta-1,3-glucanase in *Arabidopsis*. Plant J. 49, 669−682.

Levy, J.A., Fraenkel-Conrat, H., Owens, R.A., 1994. Virology. Prentice-Hall, Englewood Cliffs, N.J.

Levy, M., Miller, S.L., Brinton, K., Bada, J.L., 2000. Prebiotic synthesis of adenine and amino acids under Europa-like conditions. Icarus 145, 609−613.

Levy, S., Staehelin, L.A., 1992. Synthesis, assembly and function of plant cell wall macromolecules. Curr. Opin. Cell Biol. 4, 856−862.

Levy, S., Sutton, G., Ng, P.C., Feuk, L., Halpern, A.L., et al., 2008. The diploid genome sequence of an individual human. PLoS Biol. 5, e254.

Lew, R.R., Bushunov, N., Spanswick, R.M., 1985. ATP-dependent proton pumping activities of zucchini fruit microsomes. A study of tonoplast and plasma membrane activities. Biochem. Biophys. Acta 821, 341−347.

Lewenhoeck, A., 1678. Observations D. Anthonii Lewenhoeck de natis é semine genitali animalculis. Philos. Trans. R. Soc. Lond. 12, 1040−1046.

Lewin, R.A., 1992. Origins of Plastids. Symbiogenesis, Prochlorophytes, and the Origins of Chloroplasts. Chapman & Hall, New York.

Lewis, G.N., 1908. A revision of the fundamental laws of matter and energy. Philos. Mag. Ser. 6 (16), 705−717.

Lewis, G.N., 1926. The Anatomy of Science. Yale University Press, New Haven, CT.

Lewis, E.B., 1950. The phenomenon of position effect. Adv. Genet. 3, 73−116.

Lewis, E.B., 1951. Pseudoallelism and gene evolution. Cold Spring Harbor Symp. Quant. Biol. 16, 159−174.

Lewis, E.B., 1952. The pseudoallelism of white and apricot in Drosophila melanogaster. Proc. Natl. Acad. Sci. U.S.A. 39, 953−961.

Lewis, E.B., 1955. Some aspects of poosition pseudoallelism. Am. Nat. 89, 73−90.

Lewis, G.N., 1966. Valence and the Structure of Atoms and Molecules, 1923. Dover Publications, New York.

Lewis, C., 2000. The Dating Game: One Man's Search for the Age of the Earth. Cambridge University Press, Cambridge.

Lewis, M.J., Pelham, H.R., 1992. Sequence of a second human KDEL receptor. J. Mol. Biol. 226, 913−916.

Lewis, G.N., Randall, M., 1923. Thermodynamics and the Free Energy of Chemical Substances. McGraw-Hill, New York.

Lewontin, R.C., 2000. The Triple Helix: Gene, Organism, and Environment. Harvard University Press, Cambridge, MA.

Lewontin, R.C., 2001. The Triple Helix: Gene, Organism, and Environment. Harvard University Press, Cambridge, MA.

Leydig, F., 1857. Lehrbuch der Histologie des Menschen und der Thiere. Meidinger, Frankfurt.

Leydon, A.R., Weinreb, C., Venable, E., Reinders, A., Ward, J.M., Johnson, M.A., 2017. The molecular dialog between flowering plant reproductive partners defined by SNP-informed RNA-sequencing. Plant Cell 29, 984−1006.

Li, Z.-S., Alfenito, M., Rea, P.A., Walbot, V., Dixon, R.A., 1997. Vacuolar uptake of the phytoalexin medicarpin by the glutathione conjugate pump. Phytochemistry 45, 689−693.

Li, J., Assmann, S.M., 1996. An abscisic acid-activated and calcium-independent protein kinase from guard cells of fava bean. Plant Cell 8, 2359−2368.

Li, J., Assmann, S.M., 2000. Regulation of guard cell function by phosphorylation and dephosphorylation events. In: Walker, J.A. (Ed.), Phosphorylation in Plant Function, Annal. Bot., pp. 459−479.

Li, L., Brown Jr., R.M., 1993. β glucan synthesis in the cotton fiber II. Regulation and kinetic properties of β glucan synthases. Plant Physiol. 101, 1143−1148.

Li, Y.-Q., Bruun, L., Pierson, E.S., Cresti, M., 1992. Periodic deposition of arabinogalactan epitopes in the cell wall of pollen tubes of *Nicotiana tabacum* L. Planta 188, 532−538.

Li, F.-Y., Chaigne-Delalande, B., Kanellopoulou, C., Davis, J.C., Matthews, H.F., Douek, D.C., Cohen, J.I., Uzel, G., Su, H.C., Lenardo, M.J., 2011. Second messenger role for Mg^{2+} revealed by human T-cell immunodeficiency. Nature 475, 471−476.

Li, G., Chandler, S.P., Wolffe, A.P., Hall, T.C., 1998a. Architectural specificity in chromatin structure at the TATA box in vivo: nucleosome displacement upon β-phaseolin gene activation. Proc. Natl. Acad. Sci. U.S.A. 95, 4772−4777.

Li, Y.-G., Chen, F., Linskens, H.F., Cresti, M., 1994. Distribution of unesterified and esterified pectins in cell walls of pollen tubes of flowering plants. Sex. Plant Reprod. 7, 145−152.

Li, W., Chen, C., Markmann-Mulisch, U., Timofejeva, L., Schmelzer, E., Ma, H., Reiss, B., 2004b. The *Arabidopsis* AtRAD51 gene is dispensible for vegetative development but required for meiosis. Proc. Natl. Acad. Sci. U.S.A. 101, 10596−10601.

Li, Q., Du, W., Liu, D., 2008. Perspectives of microbial oils for biodiesel production. Appl. Microbiol. Biotechnol. 80, 749−756.

Li, Z.C., Durachko, D.M., Cosgrove, D.J., 1993. An oat coleoptile wall protein that induces wall extension in vitro and that is antigenically related to a similar protein from cucumber hypocotyls. Planta 191, 349−356.

Li, X., Franceschi, V.R., Okita, T.W., 1993a. Segregation of storage protein mRNAs on the rough endoplasmic reticulum membranes of rice endosperm cells. Cell 72, 869−879.

Li, J.B., Gerdes, J.M., Haycraft, C.J., Fan, Y., Teslovich, T.M., May-Simera, H., Li, H., Blacque, O.E., Li, L., Leitch, C.C., et al., 2004a. Comparative genomics identifies a flagellar and basal body proteome that includes the BBS5 human disease gene. Cell 117, 541−552.

Li, X., Gould, S.J., 2003. The dynamin-like GTPase DLP1 is essential for peroxisome division and is recruited to peroxisomes in part by PEX11. J. Biol. Chem. 278, 17012−17020.

Li, X., Gutierrez, D.V., Hanson, M.G., Han, J., Mark, M.D., Chiel, H., Hegemann, P., Landmesser, L.T., Herlitze, S., 2005. Fast noninvasive activation and inhibition of neural and network activity by vertebrate rhodopsin and green algae channelrhodopsin. Proc. Natl. Acad. Sci. U.S.A. 102, 17816–17821.

Li, R., Hsieh, C.-L., Young, A., Zhang, Z., Ren, X., Zhao, Z., 2015. Illumina synthetic long read sequencing allows recovery of missing sequences even in the "finished" *C. elegans* genome. Sci. Rep. 5, 10814.

Li, J., Kinoshita, T., Pandey, S., et al., 2002. Modulation of an RNA-binding protein by abscisic-acid-activated protein kinase. Nature 418, 793–797.

Li, J., Lee, Y.-R., Assmann, S.M., 1998b. Guard cells possess a CDPK that phosphorylates the KAT1 K^+ channel. Plant Physiol. 116, 785–795.

Li, J.-F., Nebenführ, A., 2007. Organelle targeting of myosin XI is mediated by two globular tail subdomains with separate cargo binding sites. J. Biol. Chem. 282, 20593–20602.

Li, X., Nicklas, R.B., 1995. Mitotic forces control a cell-cycle checkpoint. Nature 373, 630–632.

Li, J.F., Norville, J.E., Aach, J., McCormack, M., Zhang, D., Bush, J., Church, G.M., Sheen, J., 2013. Multiplex and homologous recombination-mediated genome editing in *Arabidopsis* and *Nicotiana benthamiana* using guide RNA and Cas9. Nat. Biotechnol. 31, 688–691.

Li, H., Roux, S.J., 1992. Casein kinase II protein kinase in bound to lamina-matrix and phosphorylates lamin-like protein in isolated pea nuclei. Proc. Natl. Acad. Sci. U.S.A. 89, 8434–8438.

Li, M., Sha, A., Zhou, X., Yang, P., 2012b. Comparative proteomic analyses reveal the changes of metabolic features in soybean (*Glycine max*) pistils upon pollination. Sex. Plant Reprod. 25, 281–291.

Li, J., Silva-Sanchez, C., Zhang, T., Chen, S., Li, H., 2015a. Phosphoproteomics technologies and applications in plant biology research. Front. Plant Sci. 6, 430.

Li, F., Vierstra, R.D., 2012. Autophagy: a multifaceted intracellular system for bulk and selective recycling. Trends Plant Sci. 17, 526–537.

Li, Y., Wang, F.H., Knox, R.B., 1989. Ultrastructural analysis of the flagellar apparatus in sperm cells of Ginkgo biloba. Protoplasma 149, 57–63.

Li, M., Wang, K., Li, S., Yang, P., 2016. Exploration of rice pistil responses during early postpollination through a combined proteomic and transcriptomic analysis. J. Proteom. 131, 214–226.

Li, L., Wang, X.F., Stolc, V., Li, X.Y., Zhang, D.F., Su, N., et al., 2006. Genome-wide transcription analyses in rice using tiling microarrays. Nat. Genet. 38, 124–129.

Li, M., Wang, K., Wang, X., Yang, P., 2014b. Morphological and proteomic analysis reveal the role of pistil under pollination in *Liriodendron chinense* (Hemsl.) Sarg. PLoS One 9, e99970.

Li, J., Wang, X.-Q., Watson, M.B., Assmann, S.M., 2000. Regulation of abscisic acid-induced stomatal closure and anion channels by guard cell AAPK kinase. Science 287, 300–303.

Li, G., Wang, D., Yang, R., Logan, K., Chen, H., Zhang, S., Skaggs, M.I., Lloyd, A., Burnett, W.J., Laurie, J.D., Hunter, B.G., Dannenhoffer, J.M., Larkins, B.A., Drews, G.N., Wang, X., Yadegari, R., 2014. Temporal patterns of gene expression in developing endosperm identified through transcriptome sequencing. Proc. Natl. Acad. Sci. U.S.A. 111, 7582–7587.

Li, F.-S., Weng, J.-K., 2017. Demystifying traditional herbal medicine with modern approaches. Nat. Plants 3, 17109.

Li, J.Y., Wu, C.F., 2005. New symbiotic hypothesis on the origin of eukaryotic flagella. Naturwissenschaften 92, 305–309.

Li, X., Wu, Y., Zhang, D.-Z., Gillikin, J.W., Boston, R.S., Franceschi, V.R., Okita, T.W., 1993b. Rice prolamine protein body biogenesis: a BiP-mediated process. Science 262, 1054–1056.

Li, J., Xu, Y., Chong, K., 2012. The novel functions of kinesin motor proteins in plants. Protoplasma 249 (Suppl 2), S95–S100.

Li, Z.-S., Zhao, Y., Rea, P.A., 1995. Magnesium adenosine 5′-trisphosphate-energized transport of glutathione *S*-conjugates by plant vacuolar membrane vesicles. Plant Physiol. 107, 1257–1268.

Liang, J., Cai, W., Sun, Z., 2014. Single-cell sequencing technologies: current and future. J Genet. Genom. 41, 513–528.

Liang, P., Xu, Y., Zhang, X., Ding, C., Huang, R., Zhang, Z., Lv, J., Xie, X., Chen, Y., Li, Y., Sun, Y., Fai, Y., Songyang, Z., Ma, W., Zhou, C., Huang, J., 2015. CRISPR/Cas9-mediated gene editing in human tripronuclear zygotes. Protein Cell 6, 363–372.

Liao, S., Lin, J., Do, H., Johnson, A.E., 1997. Both lumenal and cytosolic gating of the aqueous ER translocon pore are regulated from inside the ribosome during membrane protein integration. Cell 90, 31–41.

Liarzi, O., Epel, B.L., 2005. Development of a quantitative tool fo measuring changes in the coefficient of conductivity of plasmodesmata induced by developmental, biotic and abiotic signals. Protoplasma 225, 67–76.

Libby, W.F., 1952. Radiocarbon Dating. University of Chicago Press, Chicago.

Lichter, P., Cremer, T., Borden, J., Manuelidis, L., Ward, D.C., 1988. Delineation of individual human chromosomes in metaphase and interphase cells by in situ suppression hybridization using recombinant DNA libraries. Hum. Genet. 80, 224–234.

Lichtscheidel, I.K., Foissner, I., 1996. Video microscopy of dynamic plant cell organelles: principles of the technique and practical application. J. Microsc. 181, 117–128.

Lichtscheidl, I.K., Lancelle, S.A., Hepler, P.K., 1990. Actin- endoplasmic reticulum complexes in Drosera. Their structural relationship with the plasmalemma, nucleus, and organelles in cells prepared by high pressure freezing. Protoplasma 155, 116–126.

Lichtscheidl, I.K., Url, W.G., 1990. Organization and dynamics of cortical endoplasmic reticulum in inner epidermal cells of onion bulb scales. Protoplasma 157, 203–215.

Lid, S.E., Gruis, D., Jung, R., Lorentzen, J.A., Ananiev, E., Chamberlin, M.M., Niu, X., Meeley, R., Nichols, S., Olsen, O.A., 2002. The defective kernel 1 (dek1) gene required for aleurone cell development in the endosperm of maize grains encodes a membrane protein of the calpain gene superfamily. Proc. Natl. Acad. Sci. U.S.A. 99, 5460–5465.

Liebe, S., Quader, H., 1994. Myosin in onion (*Allium cepa*) bulb scale epidermal cells: involvement in dynamics of organelles and endoplasmic reticulum. Physiol. Plant 90, 114–124.

Liebig, J., 1840. Organic Chemistry in its Applications to Agriculture and Physiology. Taylor and Walton, London.

Liebig, J., 1841. Organic Chemistry in its Application to Agriculture and Physiology. John Owen, Cambridge.

Lifton, R.P., Goldberg, M.L., Karp, R.W., Hogness, D.S., 1978. The organization of the histone genes in *Drosophila melanogaster*: functional and evolutionary implications. Cold Spring Harb. Symp. Quant. Biol. 42 (Pt 2), 1047–1051.

Liljas, A., 1991. Comparative biochemistry and biophysics of ribosomal proteins. Int. Rev. Cytol. 124, 103–135.

Lilley, D.M.J., 2011. Mechanisms of RNA catalysis. Phil. Trans. R. Soc. Lond. B 366, 2910–2917.

Lillie, R., 1927. Physical indeterminism and vital action. Science 66, 139–144.

Lima de Faria, A., Jaworska, H., 1968. Late DNA synthesis in heterochromatin. Nature 217, 138–142.

Limbach, C., Staehelin, L.A., Sievers, A., Braun, M., 2008. Electron tomographic characterization of a vacuolar eticulum and of six vesicle types that occupy different cytoplasmic domains in the apex of tip-growing *Chara* rhizoids. Planta 227, 1101–1114.

Lin, C., Ahmad, M., Cashmore, A.R., 1996a. *Arabidopsis* cryptochrome 1 is a soluble protein mediating blue light-dependent regulation of plant growth and development. Plant J. 10, 893–902.

Lin, C., Ahmad, M., Chan, J., Cashmore, A.R., 1996b. CRY2: a second member of the *Arabidopsis* cryptochrome gene family. Plant Physiol. 110, 1047.

Lin, Y., Cluette-Brown, J.E., Goodman, H.M., 2004. The peroxisome deficient *Arabidopsis* mutant sse1 exhibits impaired fatty acid synthesis. Plant Physiol. 135, 814–827.

Lin, S., Fischl, A.S., Bi, X., Parce, W., 2003. Separation of phospholipids in microfluid chip device: application to high-throughput screening assays for lipid-modifying enzymes. Anal. Biochem. 314, 97–107.

Lin, X., Kaul, S., Rounsley, S., Shea, T.P., Benito, M.-I., et al., 1999. Sequence and analysis of chromosome 2 of the plant *Arabidopsis thaliana*. Nature 402, 761–768.

Lin, M.T., Occhialini, A., Andralojc, P.J., Devonshire, J., Hines, K.M., Parry, M.A., Hanson, M.R., 2014b. ß-carboxysomal proteins assemble into highly organized structures in *Nicotiana* chloroplasts. Plant J. 79, 1–12.

Lin, M.T., Occhialini, A., Andralojc, P.J., Parry, M.A., Hanson, M.R., 2014c. A faster rubisco with potential to increase photosynthesis in crops. Nature 513, 547–550.

Lin, D., Ren, H., Fu, Y., 2015. ROP-GTPase-mediated auxin signaling regulates pavement cell interdigitation. J. Integr. Plant Biol. 57, 93–105.

Lin, C., Zhang, Y., Sparkes, I., Ashwin, P., 2014. Structure and dynamics of ER: minimal networks and biophysical constraints. Biophys. J. 107, 763–772.

Lind, J., 1757. A Treatise on the Scurvy in Three Parts. A Millar, London.

Lind, J.L., Boenig, I., Clarke, A.E., Anderson, M.A., 1996. A style-specific 120-kDa glycoprotein enters pollen tubes of *Nicotiana alata in vivo*. Sex. Plant Reprod. 9, 75–86.

Lindbo, J.L., Silva-Rosales, L., Proebsting, W.M., Dougherty, W.G., 1993. Induction of a highly specific antiviral state in transgenic plants: implications for regulation of gene expression and virus resistance. Plant Cell 5, 1749–1759.

Lindee, M.S., 2003. Watson's world. Science 300, 432–434.

Lindegren, C.C., 1966. The Cold War in Biology. Planarian Press, Ann Arbor, MI.

Linderstrøm-Lang, K.U., 1952. Proteins and Enzymes. Stanford University Press, Stanford, CA.

Lindley, D., 2004. Degrees Kelvin. Joseph Henry Press, Washington, DC.

Lindsay, R.B., 1973. Julius Robery Mayer: Prophet of Energy. Pergamon Press, Oxford.

Lindsey, K. (Ed.), 2004. Polarity in Plants. Annual Plant Reviews, vol. 12. Blackwell Publishing, Oxford.

Lineweaver, H., Burk, D., 1934. The determination of enzyme dissociation constants. J. Am. Chem. Soc. 56, 658–666.

Ling, G.N., 1984. In Search of the Physical Basis of Life. Plenum, New York.

Ling, G.N., 2001. At the Cell and Below-Cell Level. The Hidden History of a Fundamental Revolution in Biology. Pacific Press, New York.

Linka, N., Esser, C., 2012. Transport proteins regulate the flux of metabolites and cofactors across the membrane of plant peroxisomes. Front. Plant Sci. 3, 3.

Linka, N., Weber, A.P.M., 2010. Intracellular metabolite transporters in plants. Mol. Plant 3, 21–53.

Linn, S., Arber, W., 1968. Host specificity of DNA produced by *Escherichia coli*, X. In vitro restriction of phage fd replicative form. Proc. Natl. Acad. Sci. U.S.A. 59, 1300–1306.

Lintilhac, P.M., 1974. Differentiation, organogenesis, and the tectonics of cell wall orientation. Am. J. Bot. 61 (135–140), 230–237.

Lintilhac, P.M., 1999. Toward a theory of cellularity-speculations on the nature of the living cell. BioScience 49, 59–68.

Lintilhac, P.M., Jensen, W.A., 1974. Differentiation, organogenesis, and the tectonics of cell wall orientation. I. Preliminary observations on the development of the ovule in cotton. Am. J. Bot. 61, 129–134.

Lintilhac, P.M., Vesecky, T.B., 1980. Mechanical stress and cell wall orientation in plants. I. Photoelastic derivation of principal stresses with a discussion of the concept of axillarity and the significance of the "accurate shell zone". Am. J. Bot. 67, 1477–1483.

Lintilhac, P.M., Vesecky, T.B., 1981. Mechanical stress and cell wall orientation in plants. III. The application of controlled directional stress to growing plants; with a discussion on the nature of the wound reaction. Am. J. Bot. 68, 222–1230.

Lintilhac, P.M., Vesecky, T.B., 1984. Stress-induced alignment of division plane in plant tissues grown in vitro. Nature 307, 363–364.

Lintilhac, P.M., Wei, C., Tanguay, J.J., Outwater, J.O., 2000. Ball tonometry: a rapid nondestructive method for measuring cell turgor pressure in thin-walled plant cells. J. Plant Growth Regul. 19, 90–97.

Lionetti, V., Fabri, E., De Caroli, M., Hansen, A.R., Willats, W.G., Piro, G., Bellincampi, D., 2017. Three pectin methylesterase inhibitors protect cell wall integrity for Arabidopsis immunity to Botrytis. Plant Physiol. 173, 1844–1863.

Lionetti, V., Métraux, J.P., 2014. Plant cell wall in pathogenesis, parasitism and symbiosis. Front. Plant Sci. 5, 612.

Lipka, E., Herrmann, A., Mueller, S., 2015. Mechanisms of plant cell division. WIREs Dev. Biol. 4, 391–405.

Lipka, V., Kwon, C., Panstruga, R., 2007. SNARE-ware: the role of SNARE-domain proteins in plant biology. Annu. Rev. Cell Dev. Biol. 23, 147–174.

Lipmann, F., 1941. Metabolic generation and utilization of phosphate bond energy. Adv. Enzymol. 1, 99–162.

Lipmann, F., 1965. Projecting backward from the present stage of evolution of biosynthesis. In: Fox, S.M. (Ed.), The Origins of Prebiological Systems. Academic Press, New York, pp. 259–280.

Lipmann, F., 1971. Wanderings of a Biochemist. Wiley-Interscience, New York.

Lipmann, F.A., December 11, 1953. Development of the acetylation problem: A personal account. Nobel Lecture.

Lipmann, F., Kaplan, N.O., Novelli, G.D., Tuttle, L.C., Guirard, B.M., 1947. Coenzyme for acetylation, a pantothenic acid derivative. J. Biol. Chem. 167, 869–870.

Lisenbee, C.S., Heinze, M., Trelease, R.N., 2003. Peroxisomal ascorbate peroxidase resides within a subdomain of rough endoplasmic reticulum in wild-type *Arabidopsis* cells. Plant Physiol. 132, 870–882.

Lister, J.J., 1830. On some properties in achromatic object-glasses applicable to the improvement of the microscope. Philos. Trans. R. Soc. 120, 187–200.

Lister, R., Ecker, J.R., 2009. Finding the fifth base: genome-wide sequencing of cytosine methylation. Genome Res. 19, 959–966.

Lister, R., O'Malley, R.C., Tonti-Filippini, J., Gregory, B.D., Berry, C.C., Millar, A.H., Ecker, J.R., 2008. Highly integrated single-base resolution maps of the epigenome in *Arabidopsis*. Cell 133, 523–536.

Littlefield, J.W., Keller, E.B., 1957. Incorporation of C14-amino acids into ribonucleoprotein particles from the Ehrlich mouse ascites tumor. J. Biol. Chem. 224, 13–30.

Littlefield, J.W., Keller, E.B., Gross, J., Zamecnik, P.C., 1955. Studies on cytoplasmic ribonucleoprotein particles from the liver of the rat. J. Biol. Chem. 217, 111–123.

Liu, D., 2018. Heads and tails: molecular imagination and the lipid bilayer, 1917-1941. In: Maienschein, K.S., Laubichler, M.D. (Eds.), Visions of Cell Biology. Reflections Inspired by Cowdry's General Cytology. University of Chicago Press, Chicago, pp. 209–245.

Liu, Y., Bassham, D.C., 2012. Autophagy: pathways for self-eating in plant cells. Annu. Rev. Plant Biol. 63, 215–237.

Liu, B., Hotta, T., Ho, C.-M.K., Lee, Y.-R.J., 2011a. Microtubule organization in the phragmoplast. The plant cytoskeleton. Adv. Plant Biol. 2, 207–225.

Liu, B., Hotta, T., Ho, C.-M.K., Lee, Y.-R.J., 2011a. Microtubule organization in the phragmoplast. The plant cytoskeleton. Adv. Plant Biol. 2, 207–225.

Liu, B., Lee, Y.-R.J., 2001. Kinesin-related proteins in plant cytokinesis. J. Plant Growth Regul. 20, 141–150.

Liu, Z., Makaroff, C.A., 2006. *Arabidopsis* separase AESP is essential for embryo development and the release of cohesion during meiosis. Plant Cell 18, 1213–1225.

Liu, B., Marc, J., Joshi, H.C., Palevitz, B.A., 1993. A γ-tubulin-related protein associated with the microtubule arrays of higher plants in a cell cycle-dependent manner. J. Cell Sci. 104, 1217–1228.

Liu, C.-M., McElver, J., Tzafrir, I., Joosen, R., Wittich, P., Patton, D., van Lammeren, A.A.M., Meinke, D., 2002. Condensin and cohesion knockouts in *Arabidopsis* exhibit a titan seed phenotype. Plant J. 29, 405–415.

Liu, T., Ortiz, J.A., Taing, L., Meyer, C.A., Lee, B., Zhang, Y., Shin, H., Wong, S.S., Ma, J., Lei, Y., Pape, U.J., Poidinger, M., Chen, Y., Yeung, K., Brown, M., Turpaz, Y., Liu, X.S., 2011b. Cistrome: an integrative platform for transcriptional regulation studies. Genome Biol. 12, R83.

Liu, B., Palevitz, B.A., 1991. Kinetochore fiber formation in dividing generative cells of Tradescantia. J. Cell Sci. 98, 475–482.

Liu, B., Palevitz, B.A., 1992. Organization of cortical microfilaments in dividing root cells. Cytoskeleton 23, 252–264.

Liu, Y.-C., Vidali, L., 2011. Efficient polyethylene glycol (PEG) mediated transformation in the moss *Physcomitrella patens*. J. Vis. Exp. 50, 2560.

Liu, L., White, M.J., MacRae, T.H., 1999. Transcription factors and their genes in higher plants. Eur. J. Biochem. 262, 247–257.

Liu, X., Wu, S., Xu, J., Sui, C., Wei, J., 2017. Application of CRISPR/Cas9 in plant biology. Acta Pharmaceutica Sinica B 7, 292–302.

Liu, X., Yen, L.F., 1992. Purification and characterization of actin from maize pollen. Plant Physiol. 99, 1151–1155.

Liu, X., Yue, Y., Li, W., Ma, L., 2007b. Response to comment on "A G protein-coupled receptor is a plasma membrane receptor for the plant hormone abscisic acid." Science 318, 914d.

Liu, X., Yue, Y., Li, B., Nie, Y., Li, W., Wu, W.-H., Ma, L., 2007a. A G protein-coupled receptor is a plasma membrane receptor for the plant hormone abscisic acid. Science 315, 1712–1716.

Liu, X., Zwiebel, L.J., Hinton, D., Benzer, S., Hall, J.C., Rosbash, M., 1992. The *period* gene encodes a predominantly nuclear protein in adult *Drosophila*. J. Neurosci. 12, 2735–2744.

Livingston, L.G., 1933. The nature and distribution of plasmodesmata in the tobacco plant. Am. J. Bot. 32, 75–87.

Lloyd, F.E., 1942. The Carnivorous Plants. Chronica Botanica, Waltham, MA.

Lloyd, C.W., 1987. The plant cytoskeleton: the impact of fluorescence microscopy. Ann. Rev. Plant Physiol. 38, 119–139.

Lloyd, C.W., 1988. Actin in plants. J. Cell Sci. 90, 185–188.

Lloyd, C.W., 1989. The plant cytoskeleton. Curr. Opin. Cell Biol. 1, 30–35.

Lloyd, C.W., 1991a. Cytoskeletal elements of the phragmosome establish the division plane in vacuolated higher plant cells. In: Lloyd, C.W. (Ed.), The Cytoskeletal Basis of Plant Growth and Form. Academic Press, London, pp. 245–257.

Lloyd, C.W., 1991b. How does the cytoskeleton read the laws of geometry in aligning the division plane of plant cells? Development (Suppl. 1), 55–66.

Lloyd, C., 2006. Microtubules make tracks for cellulose. Science 312, 1482–1483.

Lloyd, C.W., Traas, J.A., 1988. The role of F-actin in determining the division plane of carrot suspension cells: drug studies. Development 102, 211–222.

Lloyd, D., Wimpenny, J., Venables, A., 2010. Alfred Russel Wallace deserves better. J. Biosci. 35, 339–349.

Lo, Y.-S., Wang, W.-N., Jane, Y.-T., Hsaio, L.-J., Chen, L.F., Dai, H., 2003. The presence of actin-like protein filaments in higher plant mitochondria. Bot. Bull. Acad. Scn. 44, 19–24.

Locke, J., 1690. The Second Treatise of Civil Government. http://www.constitution.org/jl/2ndtreat.htm.

Locke, J., 1824. An Essay Concerning Human Understanding, twentyfourth ed. C. and J. Rivington, London.

Lockhart, J.A., 1959a. Gibberellin—a new plant hormone. Eng. Sci. 22 (Nov.), 15–19.

Lockhart, J.A., 1965. An analysis of irreversible plant cell elongation. J. Theor. Biol. 8, 264–275.

Lockhart, J.A., 1971. An interpretation of cell growth curves. Plant Physiol. 48, 245–248.

Lockhart, J.A., Bretz, C., Kenner, R., 1967. An analysis of cell- wall extension. Ann. N.Y. Acad. Sci. 144, 19–33.

Lockhart, D.J., Dong, H., Byrne, M.C., Follettie, M.T., Gallo, M.V., Chee, M.S., Mittmann, M., Wang, C., Kobayashi, M., Horton, H., Brown, E.L., 1996. Expression monitoring by hybridization to high-density oligonucleotide arrays. Nat. Biotechnol. 14, 1675–1680.

Lockhart, J.A., May 1959b. The care and feeding of spacemen. Eng. Sci. 22, 11–13.

Lockhart, J.A., November 1959a. Gibberellin—a new plant hormone. Eng. Sci. 22, 15–19.

Loeb, J., 1906. The Dynamics of Living Matter. Columbia University Press, New York.

Loeb, J., 1909. Die chemische Entwicklungserregung des tierischen Eies. Springer, Berlin.

Loeb, J., 1912. The Mechanistic Conception ot Life; Biological Essays. University of Chicago Press, Chicago.

Loeb, J., 1916. The Organism as a Whole. G. P. Putnam's Sons, New York.

Loeb, J., 1924. Regeneration. McGraw-Hill, New York.

Loeb, L.B., 1961. The Kinetic Theory of Gases. Dover Publications, New York.

Loeb, J., Chamberlain, M.M., 1915. An attempt at a physico-chemical explanation of certain groups of fluctuating variations. J. Exp. Zool. 19, 559–568.

Loening, U.E., 1962. Messenger ribonucleic acid in pea seedlings. Nature 195, 467–469.

Loew, O., 1896. The Energy of Living Protoplasm. K. Paul, Trench, Trübner & Co., London.

Loewi, O., December 12, 1936. The chemical transmission of nerve action. Nobel Lecture.

Loewy, A.G., 1954. An actomyosin-like substance from the plasmodium of a myxomycete. J. Cell. Comp. Physiol. 40, 127–156.

Löffelhardt, W., Bohnert, H.J., Bryant, D.A., 1997. The cyanelles of Cyanophora paradoxa. Crit. Rev. Plant Sci. 16, 393–413.

Logan, D.C., 2006. Plant mitochondrial dynamics. Biochim. Biophys. Acta 1763, 430–441.

Logan, D.C., 2010. The dynamic plant chondriome. Semin. Cell Dev. Biol. 21, 550–557.

Lohmann, K., 1929. Über die pyrophosphatfraktion im Muskel. Naturwissenschaften 17, 624–625.

Lohrmann, R., 1972. Formation of urea and guanidine by irradiation of ammonium cyanide. J. Mol. Evol. 1, 263–269.

Lohrmann, R., Orgel, L.E., 1971. Urea-inorganic phosphate mixtures as prebiotic phosphorylating agents. Science 171, 490–494.

Lohscheider, J.N., Friso, G., van Wijk, K.J., 2016. Phosphorylation of plastoglobular proteins in Arabidopsis thaliana. J. Exp. Bot. 67, 3975–3984.

Lombard, J., 2014. Once upon a time the cell membrane: 175 years of cell boundary research. Biol. Direct 9, 32.

Lombillo, V.A., Nislow, C., Yen, T.J., Gelfand, V.I., McIntosh, J.R., 1995a. Antibodies to the kinesin motor domain and CENP-E inhibit depolymerization-dependent motion of chromosomes in vitro. J. Cell Biol. 128, 107–115.

Lombillo, V.A., Stewart, R.J., McIntosh, J.R., 1995b. Minus- end-directed motion of kinesin-coated microspheres driven by microtubule depolymerization. Nature 373, 161–164.

Long, S.P., Humphries, S., Falkowski, P.G., 1994. Photoinhibition of photosynthesis in nature. Ann. Rev. Plant Physiol. Plant Mol. Biol. 45, 633–662.

Loomis, W.F., Lipmann, F., 1948. Reversible inhibition of the coupling between phosphorylation and oxidation. J. Biol. Chem. 173, 807–808.

López-García, P., Moreira, D., 2006. Selective forces for the origin of the eukaryotic nucleus. BioEssays 28, 525–533.

Lopez-Huertas, E., Oh, J., Baker, A., 1999. Antibodies against Pex14p block ATP-independent binding of matrix proteins to peroxisomes in vitro. FEBS Lett. 459, 227–229.

Lopez-Juez, E., Pyke, K.A., 2005. Plastids unleashed: their development and their integration in plant development. Int. J. Dev. Biol. 49, 557–577.

Lopez-Obando, M., Hoffmann, B., Géry, C., Guyon-Debast, A., Téoulé, E., Rameau, C., Bonhomme, S., Nogué, F., 2016. Simple and efficient targeting of multiple genes through CRISPR-Cas9 in *Physcomitrella patens*. G3 6, 3647–3653.

López-Sáez, J.F., Giménez-Martin, G., Risueno, M.C., 1966a. Fine structure of plasmodesma. Protoplasma 61, 81–84.

López-Sáez, J.F., Risueno, M.C., Gimenez-Martin, G., 1966b. Inhibition of cytokinesis in plant cells. J. Ultrastruct. Res. 14, 85–94.

Lord, J.M., 1983. Endoplasmic reticulum and ribosomes. In: Hall, J.L., Moore, A.L. (Eds.), Isolation of Membranes and Organelles from Plant Cells. Academic Press, London, pp. 119–134.

Lord, J.M., 1985. Precursors of ricin and Ricinus communis agglutinin. Glycosylation and processing during synthesis and intracellular transport. Eur. J. Biochem. 146, 411–416.

Lord, J.M., Roberts, L.M., 1982. Glyoxysome biogenesis via the endoplasmic reticulum in castor bean endosperm? Ann. N.Y. Acad. Sci. 386, 362–376.

Lord, J.M., Roberts, L.M., 1998. Toxin entry: retrograde transport through the secretory pathway. J. Cell Biol. 140, 733–736.

Lord, E.M., Sanders, L.C., 1992. Roles for the extracellular matrix in plant development and pollination: a special case for cell movements in plants. Dev. Biol. 153, 16–28.

Lorentz, H.A., 1924. The radiation of light. Nature 113, 608–611.

Lorimer, G.H., 1981. The carboxylation and oxygenation of ribulose 1,5-bisphosphate: the primary events in photosynthesis and photorespiration. Ann. Rev. Plant Physiol. 32, 349–383.

Loros, J., Taiz, L., 1982. Auxin increases the water permeability of Rhoeo and Allium epidermal cells. Plant Sci. Lett. 26, 93–102.

Lörz, H., Pazkowski, J., Dierks-Ventling, C., Potrykus, I., 1981. Isolation and characterization of cytoplasts and miniprotoplasts derived from protoplasts of cultured cells. Physiol. Plant 53, 385–391.

Loschmidt, J., 1865. On the size of the air molecules. Proc. Acad. of Sci. Vienna 52, 395–413. Translated in J. Chem. Educ. 72,870–875.

Lough, T.J., Lucas, W.J., 2006. Integrative plant biology: role of phloem long-distance macromolecular trafficking. Annu. Rev. Plant Biol. 57, 203–232.

Lovejoy, A.O., 1911. The meaning of vitalism. Science 33, 610–614.

Lovy-Wheeler, A., Wilsen, K.L., Baskin, T.I., Hepler, P.K., 2005. Enhanced fixation reveals the apical cortical fringe of actin filaments as a consistent feature of the pollen tube. Planta 221, 95–104.

Low, P.S., Chandra, S., 1994. Endocytosis in plants. Ann. Rev. Plant Physiol. Plant Mol. Biol. 45, 609–631.

Low, P.S., Heinstein, P.F., 1986. Elicitor stimulation of the defense response in cultured plant cells monitored by fluorescent dyes. Arch. Biochem. Biophys. 249, 472–479.

Lowder, L., Malzahn, A., Qi, Y., 2016. Rapid evolution of manifold CRISPR systems for plant genome editing. Front. Plant Sci. 7, 1683.

Lowey, A.G., 1952. An actomyosin-like substance from the plasmodium of a myxomycete. J. Cell Comp. Physiol. 40, 127–156.

Lu, H., Giordano, F., Ning, Z., 2016. Oxford Nanopore MinIon sequencing and genome assembly. Genom. Proteom. Bioinform. 14, 265–279.

Lu, S., Li, L., Yi, X., Joshi, C.P., Chiang, V., 2008a. Differential expression of three eucalyptus secondary cell wall-related cellulose synthase genes in response to tension stress. J. Exp. Bot. 59, 681–695.

Lu, C., Reedy, M., Erickson, H.P., 2000. Straight and curved conformations of FtsZ are regulated by GTP hydrolysis. J. Bacteriol. 182, 164–170.

Lu, Y., Savage, L.J., Ajjawi, I., Imre, K.M., Yoder, D.W., Benning, C., DellaPenna, D., Ohlrogge, J.B., Osteryoung, K.W., Weber, A.P., Wilkerson, C.G., Last, R.L., 2008b. New connections across pathways and cellular processes: industrialized mutant screening reveals novel associations between diverse phenotypes in *Arabidopsis*. Plant Physiol. 146, 1482–1500.

Lu, W., Su, X., Klein, M.S., Lewis, I.A., Fiehn, O., Rabinowitz, 2017. Metabolite measurements: pitfalls to avoid and practices to follow. Annu. Rev. Biochem. 86, 277–304.

Luan, S., 2002. Tyrosine phosphorylation in plant cell signaling. Proc. Natl. Acad. Sci. U.S.A. 99, 11567–11569.

Lubben, T.H., Theg, S.M., Keegstra, K., 1988. Transport of proteins into chloroplasts. Photosynth. Res. 17, 173–194.

Luby-Phelps, K., 1994. Physical properties of the cytoplasm. Curr. Opin. Cell Biol. 6, 3–9.

Luby-Phelps, K., 2000. Cytoarchitecture and physical properties of cytoplasm: volume, viscosity, diffusion, intracellular surface area. Int. Rev. Cytol. 192, 189–221.

Luby-Phelps, K., 2013. The physical chemistry of cytoplasm and its influence on cell function: an update. Mol. Biol. Cell 24, 2593–2596.

Luby-Phelps, K., Lanni, F., Taylor, D.L., 1985. Behavior of a fluorescent analogue of calmodulin in living 3T3 cells. J. Cell Biol. 101, 1245–1256.

Luby-Phelps, K., Lanni, F., Taylor, D.L., 1988. The submicroscopic properties of cytoplasm as a determinant of cellular function. Ann. Rev. Biophys. Biophys. Chem. 17, 369–396.

Luby-Phelps, K., Mujumdar, S., Mujumdar, R., Ernst, L., Galbraith, W., Waggoner, A., 1993. A novel fluorescence ratiometric method confirms the low solvent viscosity of the cytoplasm. J. Biophys. 65, 236–242.

Luby-Phelps, K., Taylor, D.L., Lanni, F., 1986. Probing the structure of cytoplasm. J. Cell Biol. 102, 2015–2022.

Luby-Phelps, K., Weisiger, R.A., 1996. Role of cytoarchitecture in cytoplasmic transport. Comp. Biochem. Physiol. B. 115, 295–306.

Luby–Phelps, K., Ning, G., Fogerty, J., Besharsea, J.C., 2003. Visualization of identified GFP-expressing cells by light and electron microscopy. J. Histochem. Cytochem. 51, 271–274.

Lucas, K., 1917. The Conduction of the Nervous Impulse. Longmans, Green and Co., London.

Lucas, W.J., 1999. Plasmodesmata and the cell-to-cell transport of proteins and nucleoprotein complexes. J. Exp. Bot. 50, 979–987.

Lucas, W.J., Bouché-Pillon, S., Jackson, D.P., Nguyen, L., Baker, L., Ding, B., Hake, S., 1995. Selective trafficking of KNOTTED1 homeodomain protein and its mRNA through plasmodesmata. Science 270, 1980–1983.

Lucas, J.R., Nadeau, J.A., Sack, F.D., 2006. Microtubule arrays and *Arabidopsis* stomatal development. J. Exp. Bot. 57, 71–79.

Lucas, W.J., Olesinski, A., Hull, R.J., Haudensheild, J.S., Deom, C.M., Beachy, R.N., Wolf, S., 1993. Influence of the tobacco mosaic virus 30-kDa movement protein on carbon metabolism and photosynthate partitioning in transgenic tobacco plants. Planta 190, 88–96.

Lucas, W.J., Smith, F.A., 1973. The formation of alkaline and acid regions at the surface of Chara corallina cells. J. Exp. Bot. 24, 1–14.

Luckey, J.A., Drossman, H., Kostichka, A.J., Mead, D.A., D'Cunha, J., Norris, T.B., Smith, L.M., 1990. High speed DNA sequencing by capillary electrophoresis. Nucleic Acids Res. 18, 4417–4421.

Ludueña, R.F., Banerjee, A., Khan, I.A., 1992. Tubulin structure and biochemistry. Curr. Topics Cell Biol. 4, 53–57.

Ludwig, J., Canvin, D.T., 1971. The rate of photorespiration during photosynthesis and the relationship of the substrate of light respiration to the products of photosynthesis in sunflower leaves. Plant Physiol. 48, 712–719.

Ludwig, M., Gibbs, S., 1987. Are the nucleomorphs of cryptomonads and *Chlorachnion* vestigial nuclei of eukaryotic endosymbionts? Ann. N.Y. Acad. Sci. 503, 198–211.

Luef, B., Frischkorn, K.R., Wrighton, K.C., Holman, H.-Y.N., Birarda, G., Thomas, B.C., Singh, A., Williams, K.H., Siergerist, C.E., Tringe, S.G., Downing, K.H., Comolli, L.R., Banfield, J.F., 2015. Diverse uncultivated ultra-small bacterial cells in groundwater. Nat. Commun. 6, 6372.

Luft, J.H., 1956. Permanganate—a new fixative for electron microscopy. J. Biophys. Biochem. Cytol. 2, 799–802.

Luger, K., Mäder, A.W., Richmond, R.K., Sargent, D.F., Richmond, T.J., 1997. Crystal structure of the nucleosome core particle at 2.8 Å resolution. Nature 389, 251–260.

Lumiére, A., 1919. Le Myth des Symbiotes. Mason et Cie, Paris.

Luna, E.J., 1991. Molecular links between the cytoskeleton and membranes. Curr. Opin. Cell Biol. 3, 120–126.

Luna, E.J., Hitt, A.L., 1992. Cytoskeleton-plasma membrane interactions. Science 258, 955–964.

Lund, E.J., 1947. Bioelectric Fields and Growth. University of Texas Press, Austin.

Lundegardh, H., 1955. Mechanisms of absorption, transport, accumulation and secretion of ions. Ann. Rev. Plant Physiol. 6, 1–24.

Lundgren, K., Walworth, N., Booher, R., Dembski, N., Kirschner, M., Beach, D., 1991. Mik1 and wee1 cooperate in the inhibitory tyrosine phosphorylation of cdc2. Cell 64, 1111–1122.

Lundgren, H.P., Ward, W.H., 1949. Chemistry of amino acids and proteins. Ann. Rev. Biochem. 18, 115–154.

Lundquist, P.K., Poliakov, A., Bhuiyan, N.H., Zybailov, B., Sun, Q., van Wijk, K.J., 2012. The functional network of the Arabidopsis thaliana plastoglobule proteome based on quantitative proteomics and genome-wide co-expression analysis. Plant Physiol. 58, 1172–1192.

Lundquist, P.K., Poliakov, A., Giacomelli, L., Friso, G., Appel, M., McQuinn, R.P., Krasnoff, S.B., Rowland, E., Ponnala, L., Sun, Q., vanWijk, K.J., 2013. Loss of plastoglobule-localized kinases ABC1K1 and ABC1K3 leads to a conditional degreening phenotype, a modified prenyl-lipid composition and recruitment of JA biosynthesis. Plant Cell 25, 1818–1839.

Lundsgaard, E., 1938. The Pasteur-Myerhof reaction in muscle metabolism. Harvey Lect. 33, 65–88.

Lunn, J.E., 2006. Compartmentation in plant metabolism. J. Exp. Bot. 58, 35−47.

Luo, X., Xue, Z., Ma, C., Hu, K., Zeng, Z., Dou, S., Tu, J., Shen, J., Yi, B., Fu, T., 2017. Joint genome-wide association and transcriptome sequencing reveals a complex polygenic network underlying hypocotyl elongation in rapeseed (*Brassica napus* L.). Sci. Rep. 7, 41561.

Luria, S.E., 1970. Molecular biology: past, present, future. BioScience 20, 1289−1296.

Luria, S.E., 1973. Life...The Unfinished Experiment. Charles Scribner's Sons, New York.

Luria, S.E., December 10, 1969. Phage, colicins and macroregulatory phenomena. Nobel Lecture.

Luria, S.E., Human, M.L., 1952. A nonhereditary, host-induced variation of bacterial viruses. J. Bacteriol. 64, 557−569.

Luschnig, C., Vert, G., 2014. The dynamics of plant plasma membrane proteins: PINs and beyond. Development 141, 2924−2938.

Lutkenhaus, J.F., Wolf-Watz, H., Donachie, W.D., 1980. Organization of genes in the ftsA-envA region of the Escherichia coli genetic map and identification of a new fts locus (ftsZ). J. Bacteriol. 142, 615−620.

Lüttge, U., 2016. Transport processes: the key integrators in plant biology. Prog. Bot. 77, 3−65.

Lütz-Meindl, U., 2016. *Micrasterias* as a model system in plant cell biology. Front. Plant Sci. 7, 999.

Luu, D.-T., Qin, X.K., Morse, D., Cappadocia, M., 2000. S-RNase uptake by compatible pollen tubes in gametophytic self-incompatibility. Nature 407, 649−651.

Lv, S., Zhang, K., Gao, Q., Lian, L., Song, Y., Zhang, J., 2008. Overexpression of an H^+-PPase gene from *Thellungiella halophila* in cotton enhances salt tolerance and improves growth and photosynthetic performance. Plant Cell Physiol. 49, 1150−1164.

Lwoff, A., December 11, 1965. Interaction among virus, cell, and organism. Nobel Lecture.

Lye, R.J., Pfarr, C.M., Porter, M.E., 1989. Cytoplasmic dynein and microtubule translocators. In: Warner, F.D., McIntosh, J.R. (Eds.), Cell Movement, Kinesin, Dynein and Microtubule Dynamics, vol. 2. Alan R. Liss, Inc., New York, pp. 141−154.

Lyell, C., 1872. Principles of Geology. D. Appleton & Co., New York.

Lymm, R.W., Taylor, E.W., 1971. Mechanism of adenosine triphosphate hydrolysis by actomyosin. Biochemistry 10, 4617−4624.

Lynch, T.M., Lintilhac, P.M., 1997. Mechanical signals in plant development: a new method for single cell studies. Dev. Biol. 181, 246−256.

Lynch, M.A., Staehelin, L.A., 1992. Domain-specific and cell type-specific localization of two types of cell wall matrix polysaccharides in the clover root tip. J. Cell Biol. 118, 467−479.

Lynch, D.V., Steponkus, P.L., 1987. Plasma membrane lipid alterations associated with cold acclimation of winter rye seedlings (*Secale cereale* L. cv Puma). Plant Physiol. 83, 761−767.

Lyne, F., December 11, 1964. The pathway from "activated acetic acid" to the terpenes and fatty acids. In: Nobel Lecture.

Lysenko, T.D., 1946. Heredity and its Variability (T. Dobzhansky, Trans.). King's Crown Press, Morningside Heights, New York.

Lysenko, T., 1948. The Science of Biology Today. International Publishers, New York.

Lysenko, T.D., 1954. Agrobiology. Foreign Languages Publishing House, Moscow.

Ma, H., 1994. GTP-binding proteins in plants: new members of an old family. Plant Mol. Biol. 26, 1611−1636.

Ma, H., 2001. Plant G proteins: the different faces of GPA1. Curr. Biol. 11, R869−R871.

Ma, J.K.-C., Drake, P.M., Christou, P., 2003. The production of recombinant pharmaceutical proteins in plants. Nat. Rev. Genet. 4, 794−805.

Ma, Y., Kouranov, A., LaSala, S.E., Schnell, D.J., 1996. Two components of the chloroplast protein import apparatus, IAP86 and IAP75, interact with the transit sequence during the recognition and translocation of precursor proteins at the outer envelope. J. Cell Biol. 134, 315−327.

Ma, Y., Szostkiewicz, I., Korte, A., Moes, D., Yang, Y., Christmann, A., Grill, E., 2009. Regulators of PP2C phosphatase activity function as abscisic acid sensors. Science 324, 1064−1068.

Ma, J.F., Tamai, K., Yamaji, N., Mitani, N., Konishi, S., Katsuhara, M., Ishiguro, M., Murata, Y., Yano, M., 2006. A silicon transporter in rice. Nature 440, 688−691.

Ma, J.F., Yamaji, N., Mitani, N., Tamai, K., Konishi, S., Fujiwara, T., Katsuhara, M., Yano, M., 2007. An efflux transporter of silicon in rice. Nature 448, 209−212.

Ma, H., Yanofsky, M.F., Huang, H., 1991. Isolation and sequence analysis of TGA1 cDNAs encoding tomato G protein α subunit. Gene 107, 189−195.

Ma, H., Yanofsky, M.F., Meyerowitz, E.M., 1990. Molecular cloning and characterization of GPA1, a G protein α subunit gene from *Arabidopsis thaliana*. Proc. Natl. Acad. Sci. U.S.A. 87, 3821−3825.

Ma, Y.Z., Yen, L.F., 1989. Actin and myosin in pea tendril. Plant Physiol. 89, 586−589.

Ma, X., Zhu, Q., Chen, Y., Liu, Y.-G., 2016. CRISPR/Cas9 platforms for genome editing in plants: developments and applications. Mol. Plant 9, 961−974.

Maai, E., Miyaki, H., Taniguchi, M., 2011. Differential positioning of chloroplasts in C_4 mesophyll and bundle sheath cells. Plant Signal. Behav. 6, 1111−1113.

Maathuis, F.J.M., Sanders, D., 1992. Plant membraneptransport. Curr. Opin. Cell Biol. 4, 661−669.

Macer, D.R.J., Koch, G.L.E., 1988. Identification of a set of calcium-binding proteins in reticuloplasm, the luminal content of the endoplasmic reticulum. J. Cell Sci. 91, 61−70.

Macey, S.L., 1987. Patriarchs of Time: Dualism in Saturn-Cronus, Father Time, the Watchmaker God, and Father Christmas. University of Georgia Press, Athens, GA.

MacFadyen, A., 1908. The Cell as the Unit of Life and Other Lectures. J. & A. Churchill, London.

Mach, E., 1986. Principles of the Theory of Heat, 1900. D. Reidel Publishing Co., Dordrecht.

Machaca, K., 2010. Ca^{2+} signaling, genes and the cell cycle. Cell Calcium 48, 243−250.

Machamer, C.E., 1993. Targeting and retention of Golgi membrane proteins. Curr. Opin. Cell Biol. 5, 606−612.

Mackey, M.C., Santillán, M., 2012. Andrew fielding Huxley (1917-2012). Not. Am. Math. Soc. 60, 576−584.

MacKinnon, R., December 8, 2003. Potassium Channels and the Atomic Basis of Selective Ion Conduction. Nobel Lecture.

MacRobbie, E., 1997. Signalling in guard cells and regulation of ion channel activity. J. Exp. Bot. 48, 515−528.

MacRobbie, E.A.C., 2000. ABA activates multiple Ca^{2+} fluxes in stomatal guard cells, triggering vacuolar K^+ (Rb^+) release. Proc. Natl. Acad. Sci. U.S.A. 97, 12361−12368.

MacRobbie, E.A.C., 2002. Evidence for a role for protein tyrosine phosphatase in the control of ion release from the guard cell vacuole in stomatal closure. Proc. Natl. Acad. Sci. U.S.A. 99, 11963–11968.

MacRobbie, E.A.C., Dainty, J., 1957. On the existence of a membrane at the outer surface of the protoplasm of vacuolated plant cells. Proc. R. Phys. Soc. Edin. 26, 37–41.

MacVeigh-Fierro, D., Tüzel, E., Vidali, L., 2017. The motor kinesin 4II is important for growth and chloroplast light avoidance in the moss *Physcomitrella patens*. Am. J. Plant Sci. 8, 791–809.

Maddox, P., Desai, A., Oegema, K., Micheson, T.J., Salmon, E.D., 2002. Poleward microtubule flux is a major component of spindle dynamics and anaphase a in mitotic *Drosophila* embryos. Curr. Biol. 12, 1670–1674.

Maddox, P., Straight, A., Coughlin, P., Mitchison, T.J., Salmon, E.D., 2003. Direct observation of microtubule dynamics at kinetochores in *Xenopus* extract spindles: implications for spindle mechanics. J. Cell Biol. 162, 377–382.

Madison, S.L., Buchanan, M.L., Glass, J.D., McClain, T.F., Park, E., Nebenführ, A., 2015. Class XI myosins move specific organells in pollen tubes and are required for normal fertility and pollen tube growth in *Arabidopsis*. Plant Physiol. 169, 1946–1960.

Madison, J.T., Everett, G.A., Kung, H., 1966. Nucleotide sequence of a yeast tyrosine transfer RNA. Science 153, 531–534.

Madison, S.L., Nebenführ, A., 2011. Live-cell imaging of dual labeled Golgi stacks in tobacco BY-2 cells reveals similar behaviors for different cisternae during movement and Brefeldin A treatment. Mol. Plant 4, 896–908.

Madison, S.L., Nebenführ, A., 2013. Understanding myosin functions in plants: are we there yet? Curr. Opin. Plant Biol. 16, 710–717.

Madson, M., Dunand, C., Verma, R., Vanzin, G.F., Caplan, J., Li, X., Shoue, D.A., Carpita, N.C., Reiter, W.-D., 2003. Xyloglucan galactosyltransferase, a plant enzyme in cell wall biogenesis homologous to animal exostosins. Plant Cell 15, 1662–1670.

Maeshima, M., Nakanishi, Y., 2002. H^+ -ATPase abd H^+ – PPase in the vacuolar membrane: physiology and molecular biology. In: Rengel, Z. (Ed.), Handbook of Plant Growth. pH as the Master Variable. Marcel Dekker, New York, pp. 23–47.

Maeshima, M., Nakanishi, Y., Matsuura-Endo, C., Tanaka, Y., 1996. Proton pumps of the vacuolar membrane in growing plant cells. J. Plant Res. 109, 119–125.

Magee, A.L., Buxton, R.S., 1991. Transmembrane molecular assemblies regulated by the greater cadherin family. Curr. Opin. Cell Biol. 3, 861–864.

Magidson, V., Lončarek, J., Hergert, P., Rieder, C.L., Khodjakov, A., 2007. Laser microsurgery in the GFP era: a cell Biologist's perspective. Methods Cell Biol. 82, 239–266.

Magidson, V., O'Connell, C.B., Lončarek, J., Paul, R., Mogilner, A., Khodjakov, A., 2011. The spatial arrangement of chromosomes during prometaphase facilitates spindle formation. Cell 146, 555–567.

Magie, W.F., 1899. The Second Law of Thermodynamics. Harper & Bros, New York.

Mahadevan, L., Matsudaira, P., 2000. Motility powered by supramolecular springs and ratchets. Science 288, 95–100.

Mahon, N., 2003. The Man Who Changed Everything. The Life of James Clerk Maxwell. Wiley & Sons, Chicester, West Sussex, England.

Mainieri, D., Morandini, F., Maîtrejean, M., Saccani, A., Pedrazzini, E., Vitale, A., 2014. Protein body formation in the endoplasmic reticulum as an evolution of storage sorting to vacuoles: insights from maize γ-zein. Front. Plant Sci. 5, 331.

Maizel Jr., J.V., 2000. SDS polyacrylamide gel electrophoresis. Trends Biochem. Sci. 25, 590–592.

Majeran, W., Zybailov, B., Ytterberg, A.J., Dunsmore, J., Sun, Q., van Wijk, K.J., 2008. Consequence of C4 differentiation for chloroplast membrane proteomes in maize mesophyll and bundle sheath cells. Mol. Cell. Proteom. 7, 1609–1638.

Malcos, J.L., Cyr, R., 2011. Acentrosomal spindle formation through the heroic age of microscopy: past techniques, present thoughts, and future directions. The Plant Cytoskeleton. Adv. Plant Biol. 2, 187–205.

Malec, P., 1994. Kinetic modelling of chloroplast phototranslocations in *Lemma trisulca* L: two rate limiting components? J. Theor. Biol. 169, 189–195.

Malevich, S.B., Woodhouse, C.A., Meko, D.M., 2013. Tree-ring reconstructed hydroclimate of the Upper Klamath basin. J. Hydrol. 495, 13–22.

Malhó, R., Read, N.D., Pais, M.S., Trewavas, T.J., 1994. Role of cytosolic free calcium in the reorientation of pollen tube growth. Plant J. 5, 331–341.

Malhó, R., Trewavas, A.J., 1996. Localized apical increases of cytosolic free calcium control pollen tube orientation. Plant Cell 8, 1935–1949.

Malhotra, V., Orci, L., Glick, B.S., Block, M.R., Rothman, J.E., 1988. Role of N-ethylmaleimide-sensitive transport component in promoting fusion of transport vesicles with cisternae of the Golgi stack. Cell 54, 221–227.

Malhotra, V., Serafini, T., Orci, L., Shepherd, J.C., Rothman, J.E., 1989. Purification of a novel class of coated vesicles mediating biosynthetic protein transport through the Golgi stack. Cell 58, 329–336.

Mali, P., Yang, L., Esvelt, K.M., Aach, J., Guell, M., DiCarlo, J.E., Norville, J.E., Church, G.M., 2013. RNA-guided human genome engineering via Cas9. Science 339, 823–826.

Malínská, K., Jelínová, A., Petrášek, J., 2014. The use of FM dyes to analyze plant endocytosis. Methods Mol. Biol. 1209, 1–11.

Malinsky, J., Opekarov á, M., Grossman, G., Tanner, W., 2013. Membrane microdomains, rafts, and detergent-resistant membranes in plants and fungi. Annu. Rev. Plant Biol. 64, 501–529.

Malmström, B.G., 2000. Mitchell saw the new vista, if not the details. Nature 403, 356.

Maltman, D.J., Gadd, S.M., Simon, W.J., Slabas, A.R., 2007. Differential proteomic analysis of the endoplasmic reticulum from developing and germinating seeds of castor (*Ricinus communis*) identifies seed protein precursors as significant components of the endoplasmic reticulum. Proteomics 7, 1513–1528.

Malzahn, A., Lowder, L., Qi, Y., 2017. Plant genome editing with TALEN and CRISPR. Cell Biosci. 7, 21.

Manabe, E., Kuroda, K., 1984. Ultrastructural basis of the microtubule-associated cytoplasmic streaming in Caulerpa. Proc. Jpn. Acad. 60, 118–121.

Manchester, K.L., 2007. Historical opinion: Erwin Chargaff and his 'rules' for the base composition of DNA: why did he fail to see the possibility of complementarity? Trends Biochem. Sci. 33, 65–70.

Mandels, M., Reese, E.T., 1965. Inhibition of cellulases. Annu. Rev. Phytopathol. 3, 85–102.

Mandoli, D.F., 1998. What ever happened to *Acetabularia*? Bringing a once-classical model system into the age of molecular genetics. Int. Rev. Cytol. 182, 1–67.

Mandon, E.C., Jiang, Y., Gilmore, R., 2003. Dual recognition of the ribosome and the signal recognition particle by the SRP receptor during protein targeting to the endoplasmic reticulum. J. Cell Biol. 162, 575–585.

Maniloff, J., Morowitz, H.J., 1972. Cell biology of the mycoplasmas. Bacteriol. Rev. 36, 263–290.

Manjoo, F., 2011. Will Robots Steal Your Iob? Scientists are Approaching "The End of Insight." Can Computers Replace them, too? Slate Sepetember 30. http://www.slate.com/articles/technology/robot_invasion/2011/09/robot_invasion_can_computers_replace_scientists_.html.

Manning, S.W., Bruce, M.J. (Eds.), 2009. Tree-Rings, Kings and Old World Archaeology and Environment: Papers Presented in Honor of Peter Ian Kuniholm. Oxbow Books, Oxford.

Mano, E., Horiguchi, G., Tsukaya, H., 2006. Gravitropism in leaves of Arabisopsis thaliana (L.) Heynh. Plant Cell Physiol. 47, 217–223.

Mano, S., Nakamori, C., Hayashi, M., Kato, A., Kondo, M., Nishimura, M., 2002. Distribution and characterization of peroxisomes in *Arabidopsis* by visualization with GFP: dynamic morphology and actin-dependent movement. Plant Cell Physiol. 43, 331–334.

Mano, S., Nakamori, C., Kondo, M., Hayashi, M., Nishimura, M., 2004. An *Arabidopsis* dynamin-related protein, DRP3A, controls both peroxisomal and mitochondrial division. Plant J. 38, 487–498.

Mansfield, T.A., Hetherington, A.M., Atkinson, C.J., 1990. Some current aspects of stomatal physiology. Annu. Rev. Plant Physiol. Plant Mol. Biol. 41, 55–75.

Mansour, M.M.F., Lee-Stadelmann, O.Y., Stadelmann, E.J., 1993. Solute potential and cytoplasmic viscosity in Triticum aestivum and Hordeum vulgare under salt stress. A comparison of salt resistant and salt sensitive lines and cultivars. J. Plant Physiol. 142, 623–628.

Manton, I., 1936. Spiral structure of chromosomes in *Osmunda*. Nature 138, 1058.

Manton, I., 1943. Observations on the spiral structure of somatic chromosomes in *Osmunda* with the aid of ultraviolet light. Ann. Bot. 7, 195–212.

Manton, I., 1945. Comments on chromosome structure. Nature 155, 471–473.

Manton, I., 1950. Problems of Cytology and Evolution in the Pteridophyta. Cambridge University Press, Cambridge.

Manton, I., 1957. Observations with the electron microscope on the cell structure of the antheridium and spermatozoid of *Sphagnum*. J. Exp. Bot. 8, 382–400.

Manton, I., 1959. Observations on the microanatomy of the spermazoid of the bracken fern (*Pteridium aquilinum*). J. Biophys. Biochem. Cytol. 6, 413–417.

Manton, I., 1960. On a reticular derivative from Golgi bodies in the meristem of *Anthoceros*. J. Biophys. Biochem. Cytol. 8, 221–231.

Manton, I., 1962. Observations on stellate vacuoles in the meristem of *Anthoceros*. J. Exp. Bot. 13, 161–167.

Many authors, 1747. In: An Universal History, from the Earliest Account of Time to the Present, vol. 1. T. Osborne, London.

Mao, Y., Zhang, H., Xu, N., Zhang, B., Gou, F., Zhu, J.-K., 2013. Application of the CRISPR-Cas system for efficient genome editing in plants. Mol. Plant 6, 2008–2011.

Maple, J., Chua, N.H., Moller, S.G., 2002. The topological specificity factor AtMinE1 is essential for correct plastid division site placement in *Arabidopsis*. Plant J. 31, 269–277.

Maple, J., Fujiwara, M.T., Kitahata, N., Lawson, T., Baker, N.R., Yoshida, S., Moller, S.G., 2004. GIANT CHLOROPLAST 1 is essential for correct plastid division in *Arabidopsis*. Curr. Biol. 14, 776–781.

Marateck, S.L., 2008. Alpher, Bethe, Gamow. Phys. Today 61 (9), 11–12.

Marc, J., 1997. Microtubule-organizing centres in plants. Trends Plant Sci. 2, 223–230.

Marc, J., Granger, C.L., Brincat, J., Fisher, D.D., Kao, T.-H., McCubbin, A.G., Cyr, R.J., 1998. Use of a GFP-MAP4 reporter gene for visualizing cortical microtubule rearrangements in living epidermal cells. Plant Cell 11, 1927–1939.

Marcon, C., Malik, W.A., Walley, J.W., Shen, Z., Paschold, A., Smith, L.G., Piepho, H.-P., Briggs, S.P., Hochholdinger, F., 2015. A high-resolution tissue-specific proteome nd phosphoproteome atlas of maize primary roots reveals functional gradients along the root axis. Plant Physiol. 168, 233–246.

Marder, E., 2003. The emperor's new clothes. Curr. Biol. 13, R166.

Maréchal, E., Cesbron-Delauw, M.-F., 2001. The apicoplast: a new member of the plastid family. Trends Plant Sci. 6, 200–205.

Maresca, T.J., Salmon, E.D., 2010. Welcome to a new kind of tension: translating kinetochore mechanics into a wait-anaphase signal. J. Cell Sci. 123, 825–835.

Margulies, M., Egholm, M., Altman, W., Attiya, S., et al., 2005. Genome sequencing in microfabricated high-density picolitre reactors. Nature 437, 376–380.

Margulis, L., 1970. Origin of Eukaryotic Cells. Yale University Press, New Haven, CT.

Margulis, L., 1980. Undulipodia, flagella and cilia. BioSystems 12, 105–108.

Margulis, L., 1981. Symbiosis in Cell Evolution. W. H. Freeman and Co., San Francisco.

Margulis, L., 2001. The conscious cell. Ann. N.Y. Acad. Sci. 929, 55–70.

Margulis, L., Ober, R., 1985. Heliobacterium and the origin of chloroplasts. BioSystems 17, 317–325.

Margulis, L., To, L., Chase, D., 1978. Microtubules in prokaryotes. Science 200, 1118–1124.

Mariani, C., De Beuckeleer, M., Truettner, J., Leemans, J., Goldberg, R.B., 1990. lnduction of male sterility in plants by a chimaeric ribonuclease gene. Nature 347, 737–741.

Mariani, C., Gossele, V., De Beuckeleer, M., De Block, M., Goldberg, R.B., De Greef, W., Leemans, J., 1992. A chimaeric ribonuclease-inhibitor gene restores fertility to male sterile plants. Nature 357, 384–387.

Marinos, N.G., 1963. Vacuolation in plant cells. J. Ultrastruc. Res. 9, 177–185.

Marinova, K., Pourcel, L., Weder, B., Schwarz, M., Barron, D., Routaboul, J.-M., Debeaujon, I., Klein, M., 2007. The *Arabidopsis* MATE transporter TT12 acts as a vacuolar flavonoid/H^+-antiporter active in proanthocyanidin-accumulating cells of the seed coat. Plant Cell 19, 2023–2038.

Mark, R.E., 1967. Cell Wall Mechanics of Tracheids. Yale University Press, New Haven, CT.

Marks, R.J.III., Behe, M.J., Demski, W.A., Gordon, B.L., Sanford, J.C., 2013. Biological Information New Perspectives. In: Proceedings of a Symposium held May 31 through June 3, 2011 at Cornell University. World Scientific, New Jersey.

Marmagne, A., Ferro, M., Meinnel, T., Bruley, C., Kuhn, L., Garin, J., Barbier-Brygoo, H., Ephritikhine, G., 2007. A high content in lipid-modified peripheral proteins and integral receptor kinases features in the *Arabidopsis* plasma membrane proteome. Mol. Cell. Proteom. 6, 1980–1996.

Marmagne, A., Rouet, M.-A., Ferro, M., Rolland, N., Alcon, C., Joyard, J., Garin, J., Barbier-Brygoo, H., Ephrithine, G., 2004. Identification of new intrinsic proteins in *Arabidopsis* plasma membrane proteome. Mol. Cell. Proteom. 3, 675–691.

Marom, Z., Shtein, I., Bar_on, B., 2017. Stomatal opening: the role of cell-wall mechanical anisotropy and its analytical relations to the bio-composite characteristics. Front. Plant Sci. 8, 2061.

Marsh, J., Goode, J. (Eds.), 1993. The GTPase Superfamily. John Wiley & Sons, Chicester.

Marsh, B.J., Howell, K.E., 2002. The manmmalian Golgi-complex debates. Nat. Rev. Mol. Cell Biol. 3, 789–795.

Marshall, E., 1982a. Security checks on USDA reviewers. Science 216, 600.

Marshall, E., 1982b. USDA official defends loyality checks. Science 216, 1391.

Marsian, J., Fox, H., Bahar, M.W., Kotecha, A., Fry, E.E., Stuart, D.I., Macadam, A.J., Rowlands, D.J., Lomonossoff, G.P., 2017. Plant-made polio type 3 stabilized VLPs—a candidate synthetic polio vaccine. Nat. Commun. 8, 245.

Marston, A.L., Tham, W.H., Shah, H., Amon, A., 2004. A genome- wide screen identifies genes required for centromeric cohesion. Science 303, 1367–1370.

Martell, A.E., Calvin, M., 1952. Chemistry of the Metal Chelate Compounds. Prentice-Hall, New York.

Marten, H., Hedrich, R., Roelfsema, M.R.G., 2007. Blue light inhibits guard cell plasma membrane anion channels in a phototropin-dependent manner. Plant J. 50, 29–39.

Marten, I., Hoshi, T., 1998. The N-terminus of the K^+ channel KAT1 controls its voltage-dependent gating by altering the membrane electric field. Biophys. J. 74, 2953–2962.

Marten, I., Lohse, G., Henrich, R., 1991. Plant growth hormones control voltage-dependent activity of anion channels in plasma membrane of guard cells. Nature 353, 758–762.

Martienssen, R., 1998. Functional genomics: probing plant gene function and expression with transposons. Proc. Natl. Acad. Sci. U.S.A. 95, 2021–2026.

Martienssen, R., 2010. Small RNA makes its move. Science 328, 834–835.

Martin, A.J.P., 1964. Future possibilities in micro-analysis. In: van Swaay, M. (Ed.), Gas Chromatography 1962. Butterworth, London, pp. xxvii–xxxiii.

Martin, J., 1988. To Rise Above Principle: The Memoirs of an Unreconstructed Dean. University of Illinois Press, Champaign, IL.

Martin, W., 2000. Primitive anerobic protozoa: the wrong host for mitochondria and hydrogenosomes? Microbiology 146, 1021–1022.

Martin, G.B., Brommonschenkel, S.H., Chunwongse, J., Frary, A., Ganal, M.W., Spivey, R., Wu, T., Earle, E.D., Tanksley, S.D., 1993. Map-based cloning of a protein kinase gene conferring disease resistance in tomato. Science 262, 1432–1436.

Martin, A.J.P., December 12, 1952. The Development of Partition Chromatography. Nobel Lecture.

Martin, L.B.B., Fei, Z., Giovannoni, J.J., Rose, J.K.C., 2013. Catalyzing plant science research with RNA-seq. Front. Plant Sci. 4, 66.

Martin, S.W., Glover, B.J., Davies, J.M., 2005. Lipid microdomains—plant membrane get organized. Trends Plant Sci. 10, 263–265.

Martin, R.G., Matthaei, J.H., Jones, O.W., Nirenberg, M.W., 1962. Ribonucleotide composition of the genetic code. Biochem. Biophys. Res. Commun. 6, 410–414.

Martin, E.M., Morton, R.K., 1956. Enzymic properties of microsomes and mitochondria from silver beet. Biochem. J. 62, 696–704.

Martin, W., Müller, M., 1998. The hydrogen hypothesis for the first eukaryote. Nature 392, 37–41.

Martin, W.F., Müller, M. (Eds.), 2007. Origin of Mitochondria and Hydrogenosomes. Springer, Berlin.

Martin, L.B.B., Nicolas, P., Matas, A.J., Shinozaki, Y., Catalá, C., Rose, J.K.C., 2016a. Laser microdissection of tomato fruit cell and tissue types for transcriptome profiling. Nat. Protoc. 11, 2376–2388.

Martin, L.B.B., Sherwood, R.W., Nicklay, J.J., Yang, Y., Muratore-Schroeder, T.L., Anderson, E.T., Thannhauser, T.W., Rose, J.K.C., Zhang, S., 2016b. Application of wide selected-ion monitoring (WiSIM)-data independent acquisition (DIA) to identify tomato fruit proteins regulated by the CUTIN DEFICIENT2 transcription factor. Proteomics 16, 2081–2094.

Martin, A., Troadec, C., Boualem, A., Rajab, M., Fernandez, R., Morin, H., Pitrat, M., Dogimont, C., Bendahmane, A., 2009. A transposon-induced epigenetic change leads to sex determination in melon. Nature 461, 1135–1138.

Martin-Gallardo, A., McCombie, W.R., Gocayne, J.D., FitzGerald, M.G., Wallace, S., Lee, B.M.B., Lamerdin, J., Trapp, S., Kelley, J.M., Liu, L.-I., Dubnick, M., Johnston-Dow, L.A., Kerlavage, A.R., de Jong, P., Carrano, A., Fields, C., Venter, J.C., 1992. Automated DNA sequencing and analysis of 106 kilobases from human chromosome 19q13.3. Nat. Genet. 1, 34–39.

Martin-Romero, F.J., López-Guerrero, A.M., Alvarez, I.S., Pozo-Guisado, E., 2012. Role of store-operated calcium entry during meiotic progression and fertilization of mammalian oocytes. Int. Rev. Cell. Mol. Biol. 295, 291–328.

Martinac, B., Saimi, Y., Kung, C., 2008. Ion channels in microbes. Physiol. Rev. 88, 1449–1490.

Martinac, B., Zhu, H., Kubalski, A., Zhou, X., Culbertson, M., Bussey, H., Kung, C., 1990. Yeast K1 killer toxin forms ion channels insensitive yeast spheroplasts and in artificial liposomes. Proc. Natl. Acad. Sci. U.S.A. 87, 6228–6232.

Martinière, A., Bassil, E., Jublanc, E., Alcon, C., Reguera, M., Sentenac, H., Blumwald, E., Paris, N., 2013. *In vivo* intracellular pH measurements in tobacco and *Arabidopsis* reveal an unexpected pH gradient in the endomembrane system. Plant Cell 25, 4028–4043.

Martinoia, E., Grill, E., Tommasini, R., Kreuz, K., Amrhein, N., 1993. ATP-dependent glutathione S-congugate "export" pumps in the vacuolar membrane of plants. Nature 364, 247–249.

Martinoia, E., Klein, M., Geisler, M., Bovet, L., Forestier, C., et al., 2002. Multifunctionality of plant ABC transporters: more than just detoxifiers. Planta 214, 345–355.

Martinoia, E., Klein, M., Geisler, M., Sánchez-Fernández, R., Rea, P.A., 2000. Vacuolar transport of secondary metabolites and xenobiotics. Annu. Rev. Plant Rev. 5, 221–253.

Martinoia, E., Maeshima, M., Neuhaus, H.E., 2007. Vacuolar transporters and their essential role in plant metabolism. J. Exp. Bot. 58, 83–102.

Martinoia, E., Ratajczak, R., 1999. Transport of organic molecules across the tonoplast. Adv. Bot. Res. 25, 365–400.

Martius, C., 1982. How I became a biochemist. In: Semenza, G. (Ed.), Of Oxygen, Fuels, and Living Matter, Part 2. John Wiley & Sons, Chichester, pp. 1–57.

Martius, C., Knoop, F., 1937. Der physiologische Abbau der Citronensäure. Z. Physiol. Chem. 246, I–II.

Martoglio, B., Hofmann, M.W., Brunner, J., Dobberstein, B., 1995. The protein-conducting channel in the membrane of the endoplasmic reticulum is open laterally toward the lipid bilayer. Cell 81, 207–214.

Marty, F., 1978. Cytochemical studies on GERL, provacuoles and vacuoles in root meristematic cells of Euphorbia. Proc. Natl. Acad. Sci. U.S.A. 75, 852–856.

Marty, F., 1997. The biogenesis of vacuoles: insights from microscopy. Adv. Bot. Res. 25, 1–42.

Marty, F., 1999. Plant vacuoles. Plant Cell 11, 587–599.

Marty, F., Branton, D., 1980. Analytical characterization of beetroot vacuole membrane. J. Cell Biol. 87, 72–83.

Maruyama, K., 1995. Birth of the sliding filament concept in muscle contraction. J. Biochem. 117, 1–6.

Mascher, M., Gundlach, H., Himmelbach, A., Beier, S., Twardziok, S.O., Wicker, T., et al., 2017. A chromosome conformation capture ordered sequence of the barley genome. Nature 544, 427–433.

Mäser, P., Hosoo, Y., Goshima, S., Horie, T., Eckelman, B., Yamada, K., Yoshida, K., Bakker, E.P., Shinmyo, A., Oiki, S., Schroeder, J.I., Uozumi, N., 2002. Glycine residues in potassium channel-like selectivity filters determine potassium selectivity in four-loop-per-subunit HKT transporters from plants. Proc. Natl. Acad. Sci. U.S.A. 99, 6428–6433.

Mäser, P., Thomine, S., Schroeder, J.I., Ward, J.M., Hirschi, K., Sze, H., Talke, I.N., Amtmann, A., Maathuis, F.J.M., Sanders, D., Harper, J.F., Tchieu, J., Gribskov, M., Persans, M.W., Salt, D.E., Kim, S.A., Guerinot, M.L., 2001. Phylogenetic relationships within cation transporter families of *Arabidopsis*. Plant Physiol. 126, 1646–1667.

Mason, S.F., 1992. Chemical Evolution. Origin of the Elements, Molecules, and Living Systems. Clarendon Press, Oxford.

Mason, H.S., Lam, D.M., Arntzen, C.J., 1992. Expression of hepatitis B surface antigen in transgenic plants. Proc. Natl. Acad. Sci. U.S.A. 89, 11745–11749.

Mason, H.S., Warzecha, H., Mor, T., Arntzen, C.J., 2002. Edible plant vaccines: applications for prophylactic and therapeutic molecular medicine. Trends Mol. Med. 8, 324–329.

Masoud, K., Herzog, E., Chaboute, M.-E., Schmidt, A.-C., 2013. Microtubule nucleation and establishment of the mitotic spindle in vascular plant cells. Plant J. 75, 245–257.

Massalski, A., Leedale, G.F., 1969. Cytology and ultrastructure of the Xanthophyceae. Br. Phycol. J. 4, 159–180.

Masson, C., Andre, C., Arnoult, J., Gerand, C., Hernandez-Verdun, D., 1990. A 116000 Mr nucleolar antigen specific for the dense fibrillar component for the nucleoli. J. Cell Sci. 95, 371–381.

Massung, R.F., Esposito, J.J., Liu, L,-i., Qi, J., Utterback, T.R., Knight, J.C., Aubin, L., Yuran, T.E., Parsons, J.M., Loparev, V.N., Selivnov, N.A., Cavallaro, K.F., Kerlavage, A.R., Mahy, B.W.J., Venter, J.C., 1993. Potential virulence determinants in terminal regions of variola smallpox virus genome. Nature 366, 748–751.

Massung, R.F., Esposito, J.J., Liu, L,-i., Qi, J., Utterback, T.R., Knight, J.C., Aubin, L., Yuran, T.E., Parsons, J.M., Loparev, V.N., Selivnov, N.A., Cavallaro, K.F., Kerlavage, A.R., Mahy, B.W.J., Venter, J.C., 2003. Potential virulence determinants in terminal regions of variola smallpox virus genome. Nature 366, 748–751.

Mast, S.O., 1924. Structure and locomotion of Amoeba proteus. Anat. Rec. 29, 88.

Masters, S.C., Crane, D., 1995. The Peroxisome: A Vital Organelle. Cambridge University Press, Cambridge.

Masuda, Y., 1978a. Auxin-induced cell wall loosening. Bot. Mag. (Tokyo) 1 (Special Issue), 103–123.

Masuda, Y., 1978b. Hormone interactions in regulation of cell wall loosening. In: Schütte, H.R., Gross, D. (Eds.), Regulation of Developmental Processes in Plants. Gustav Fischer Verlag, Jena, pp. 319–330.

Masuda, Y., 1990. Auxin-induced cell elongation and cell wall changes. Bot. Mag. 103, 345–370. Tokyo.

Masuda, H., Cande, W.Z., 1987. The role of tubulin polymerization during spindle elongation in vitro. Cell 49, 193–202.

Masuda, Y., Takagi, S., Nagai, R., 1991. Protease-sensitive anchoring of microfilament bundles provides tracks for cytoplasmic streaming in Vallisneria. Protoplasma 162, 151–159.

Masuda, K., Xu, Z.J., Takahashi, S., Ito, A., Ono, M., Nomura, K., Inoue, M., 1997. Peripheral framework of carrot cell nucleus contains a novel protein predicted to exhibit a long alpha-helical domain. Exp. Cell Res. 232, 173–181.

Masumura, T., Hibino, T., Kidzu, K., Mitsukawa, N., Tanaka, K., Fujii, S., 1990. Cloning and characterization of a complementary DNA encoding a rice 13-kda prolamin. Mol. Gen. Genet. 221, 1–7.

Mather, J.C., Boslough, J., 1996. The Very First Light. Basic Books, New York.

Matheson, L.A., Hanton, S.L., Rossi, M., Latijnhouwers, M., Stefano, G., Renna, L., Brandizzi, F., 2007. Multiple roles of ADP-ribosylation factor 1 in plant cells include spatially regulated recruitment of coatomer and elements of the Golgi matrix. Plant Physiol. 143, 1615–1627.

Mathews, A., 1899. The changes in structure of the pancreas cell. J. Morphol. 15, 171–216.

Mathews, A.P., 1916. Physiological Chemistry, second ed. William Wood and Co., New York.

Mathews, C.K., 1993. The cell—Bag of enzymes or network of channels? J. Bacteriol. 175, 6377–6381.

Mathews, H., Clendennen, S.K., Caldwell, C.G., Liu, X.L., Connors, K., Matheis, N., Schuster, D.K., Menasco, D.J., Wagoner, W., Lightner, J., Wagner, D.R., 2003. Activation tagging in tomato identifies a transcriptional regulator of anthocyanin biosynthesis, modification, and transport. Plant Cell 15, 1689–1703.

Mathieu, M., Winters, E.K., Kong, F., Wan, J., Wang, S., Eckert, H., Luth, D., Paz, M., Donovan, C., Zhang, Z., Somers, D., Wang, K., Nguyen, H., Shoemaker, R.C., Stacey, G., Clemente, T., 2009. Establishment of a soybean (Glycine max Merr. L) transposon-based mutagenesis repository. Planta 229, 279–289.

Mathieu, J., Yant, L.J., Murdter, F., Kuttner, F., Schmid, M., 2009a. Repression of flowering by the miR172 target SMZ. PLoS Biol. 7, e1000148.

Mathur, J., Mathur, N., Hulskamp, M., 2002. Simultaneous visualization of peroxisomes and cytoskeletal elements reveals actin and not microtubule-based peroxisome motility in plants. Plant Physiol. 128, 1031–1104.

Mathur, J., Mathur, N., Kernebeck, B., Srinivas, B.P., Hulskamp, M., 2003. A novel localization pattern for an EB1-like protein links microtubule dynamics to endomembrane organization. Curr. Biol. 13, 1991–1997.

Matiiv, A.B., Chekunova, E.M., 2018. Aureochromes—blue light photoreceptors. Biochemistry (Mosc.) 83, 662–673.

Matile, P., 1975. The Lytic Compartment of Plant Cells. SpringerVerlag, Wein.

Matile, P., 1987. The sap of plant cells. New Phytol. 105, 1–26.

Matile, P., Cortat, M., Wiemken, A., Frey-Wyssling, A., 1971. Isolation of glucanase-containing particles from budding *Saccharomyces cerevisiae*. Proc. Natl. Acad. Sci. U.S.A. 68, 636–640.

Matile, P., Moor, H., 1968. Vacuolation: origin and development of the lysosomal apparatus in root-tip cells. Planta 80, 159–175.

Matlin, K.S., 2013. In: eLS (Ed.), History of the signal hypothesis. John Wiley & Sons, Ltd, Chichester. https://doi.org/10.1002/9780470015902.a002508.

Matsunaga, S., 2014. Nuclei and chromosomes. In: Noguchi, T., Kawano, S., Tsukaya, H., Matsunaga, S., Sakai, A., Karahara, I., Hayashi, Y. (Eds.), Atlas of Plant Cell Structure. Springer, Japan, pp. 1–24.

Matsunaga, S., Fukui, K., 2010. The chromosome peripheral proteins play an active role in chromosome dynamics. Biomol. Concepts 1, 157–164.

Matsunaga, S., Katagiri, Y., Nagashima, Y., Sugiyama, T., Hasegawa, J., Hayashi, K., Sakamoto, T., 2013. New insights into the dynamics of plant cell nuclei and chromosomes. Int. Rev. Cell Mol. Biol. 305, 253–301.

Matsuoka, K., Higuchi, T., Maeshima, M., Nakamura, K., 1997. A vacuolar-type H^+-ATPase in a nonvacuolar organelle is required for the sorting of soluble vacuolar protein precursors in tobacco cells. Plant Cell 9, 533–546.

Matsushima, R., Hayashi, Y., Yamada, K., Shimada, T., Nishimura, M., Hara-Nishimura, I., 2003. The ER body, a novel endoplasmic reticulum-derived structure in Arabidopsis. Plant Cell Physiol. 44, 661–666.

Matsuura, A., Tsukada, M., Wada, Y., Ohsumi, Y., 1997. Apg1p, a novel protein kinase required for the autophagic process in *Saccharomyces cerevisiae*. Gene 192, 245–250.

Mattaj, I.W., 1990. Splicing stories and poly(A) tales: an update on RNA processing and transport. Curr. Opin. Cell Biol. 2, 528–538.

Matthaei, J.H., Jones, O.W., Martin, R.G., Nirenberg, M.W., 1962. Characteristics and composition of RNA coding units. Proc. Natl. Acad. Sci. U.S.A. 48, 666–677.

Matthaei, J.H., Nirenberg, M.W., 1961. Characteristics and stabilization of DNAase-sensitive protein synthesis in *E. coli* extracts. Proc. Natl. Acad. Sci. U.S.A. 47, 1580–1588.

Mattoo, A.K., Marder, J.B., Edelman, M., 1989. Dynamics of the photosystem II reaction center. Cell 56, 241–246.

Matzke, E., 1942. The finest show on earth. Sci. Mon. 55, 349–354.

Matzke, A.J., Watanabe, K., van der Winden, J., Naumann, U., Matzke, M., 2010. High frequency, cell type-specific visualization of fluorescent-tagges genomic sites in interphase and mitotic cells of *Arabidopsis* plants. Plant Methods 6, 2.

Maumus, F., Quesneville, H., 2014. Deep investigation of *Arabidopsis thaliana* junk DNA reveals a continuum between repetitive elements and genomic dark matter. PLoS One 9, e94101.

Maurel, C., 1997. Aquaporins and water permeability of plant membranes. Ann. Rev. Plant Physiol. Plant Mol. Biol. 48, 399–429.

Maurel, C., Reizer, J., Schroeder, J.I., Chrispeels, M.J., 1993. The vacuolar membrane protein γ-TIP creates water specific channels in Xenopus oocytes. EMBO J. 12, 2241–2247.

Maurel, C., Verdoucq, L., Luu, D.-T., Santoni, V., 2008. Plant aquaporins: membrane channels with multiple integrated functions. Annu. Rev. Plant Biol. 59, 595–624.

Maurino, V.G., Peterhansel, C., 2010. Photorespiration: current status and approaches for metabolic engineering. Curr. Opin. Plant Biol. 13, 249–256.

Mauseth, J.D., 2009. Botany. An Introduction to Plant Biology. Jones and Bartlett, Sudbury, MA.

Maxam, A.M., Gilbert, W., 1977. A new method for sequencing DNA. Proc. Natl. Acad. Sci. U.S.A. 74, 560–564.

Maxwell, J.C., 1860. Illustrations of the dynamical theory of gases. Part I. On the motions and collisions of perfectly elastic spheres. Philos. Mag. Fourth Ser. 19, 19–32.

Maxwell, J.C., 1873. Molecules. Nature 8, 437–441.

Maxwell, J.C., 1875. On the dynamical evidence of the constitution of bodies. J. Chem. Soc. (Lond.) 28, 493–508.

Maxwell, J.C., 1878. Diffusion, ninth ed. 7. Encyclopedia Britannica, pp. 214–221.

Maxwell, J.C., 1891. Theory of Heat Tenth Edition (with corrections and additions by Lord Rayleigh). Longmans, Green, and Co., London.

Maxwell, J.C., 1897. Theory of Heat, with corrections and additions by Lord Rayleigh. Longmans, Green, and Co., New York.

Maxwell, J.C., 1991. Matter and Motion [1877]. Dover Publications, New York.

May, P.W., 2008. Molecules with Silly or Unusual Names. Imperial College Press, London. http://www.chm.bris.ac.uk/sillymolecules/sillymols2.htm.

May, T., Soll, J., 1999. Chloroplast precursor protein translocon. FEBS Lett. 452, 52–56.

Mayank, P., Grossman, J., Wuest, S., Boisson-Dernier, A., Roschitzki, B., Nanni, P., Nuehse, T., Grossniklaus, U., 2012. Characterization of the phosphoproteome of mature *Arabidopsis* pollen. Plant.

Mayer, R., Ross, P., Weinhouse, H., Amicam, D., Volmna, G., Ohana, P., Calhoon, R.D., Wong, H.C., Emerick, A.W., Benziman, M., 1991. Polypeptide composition of bacterial cyclic diguanlic acid dependent cellulose synthase and the occurrence of immunologically cross reacting proteins in higher plants. Proc. Natl. Acad. Sci. U.S.A. 88, 5472–5476.

Mayer, K., Schüller, C., Wambutt, R., Murphy, G., Volckaert, G., et al., 1999. Sequence and analysis of chromosome 4 of the plant *Arabidopsis thaliana*. Nature 402, 769–777.

Mayfield, S.P., 1991. Over-expression of the oxygen-evolving enhancer 1 protein and its consequences on photosystem II accumulation. Planta 185, 105–110.

Maynard, J.W., Lucas, W.J., 1982. Sucrose and glucose uptake into Beta vulgaris leaf tissues. A case for general (apoplastic) retrieval systems. Plant Physiol. 70, 1436–1443.

Mayow, J., 1926. Tractatus Quinque Medico-Physici [1674], Translated as Medico-Physical Works. Oxford University Press, Oxford.

Mayumi, K., Shibaoka, H., 1996. The cyclic reorientation of cortical microtubules on walls with a crossed polylamellate structure: effects of plant hormones and an inhibitor of protein kinases on the progression of the cycle. Protoplasma 195, 112–122.

Mazia, D., 1953. Cell division. Sci. Am. 189 (2), 53–63.

Mazia, D., 1956. Materials for the biophysical and biochemical study of cell division. Adv. Biol. Med. Phys. 4, 69–118.

Mazia, D., 1960. The analysis of cell reproduction. Ann. N.Y. Acad. Sci. 90, 455–469.

Mazia, D., 1961a. How cells divide. Sci. Am. 205 (3), 101–120.

Mazia, D., 1961b. Mitosis and the physiology of cell division. In: Brachet, J., Mirsky, A. (Eds.), The Cell: Biochemistry, Physiology, Morphology. Academic Press, New York, pp. 77–394.

Mazia, D., 1975. Microtubule research in perspective. Ann. N.Y. Acad. Sci. 253, 7–13.

Mazia, D., 1984. Centrosomes and mitotic poles. Exp. Cell Res. 153, 1–15.

Mazia, D., 1987. The chromosome cycle and the centrosome cycle in the mitotic cycle. Int. Rev. Cytol. 100, 49–92.

Mazia, D., 1996. Dan Mazia (1912-1996). https://web.stanford.edu/group/Urchin/mazia.html.

Mazia, D., Chaffee, R.R., Iverson, R.M., 1951. Adenosine triphosphtase in the mitotic apparatus. Proc. Natl. Acad. Sci. U.S.A. 47, 788–790.

Mazia, D., Dan, K., 1952. The isolation and biochemical characterization of the mitotic apparatus of dividing cells. Proc. Natl. Acad. Sci. U.S.A. 38, 826–838.

Mazia, D., Mitchison, J.M., Medina, H., Harris, P., 1961. The direct isolation of the mitotic apparatus. J. Biophys. Biochem. Cytol. 10, 467–474.

Mazia, D., Petzelt, C., Williams, R.O., Meza, I., 1972. A Ca-activated ATPase in the mitotic apparatus of the sea urchin egg (isolated by a new method). Exp. Cell Res. 70, 325–332.

Mazia, D., Prescott, D.M., 1954. Nuclear function and mitosis. Science 120, 120–122.

McAinsh, M.R., Brownlee, C., Hetherington, A.M., 1990. Abscisic acid-induced elevation of guard cell cytosolic Ca^{2+} precedes stomatal closure. Nature 343, 186–188.

McAinsh, M.R., Brownlee, C., Hetherington, A.M., 1992. Visualizing changes in cytosolic-free Ca^{2+} during the response of stomatal guard cells to abscisic acid. Plant Cell 4, 1113–1122.

McAinsh, M.R., Gray, J.E., Hetherington, A.M., Leckie, C.P., Ng, C., 2000. Ca^{2+} signalling in stomatal guard cells. Biochem. Soc. Trans. 28 (4), 476–481.

McAndrew, R.S., Froehlich, J.E., Vitha, S., Stokes, K.D., Osteryoung, K.W., 2001. Colocalization of plastid division proteins in the chloroplast stromal compartment establishes a new functional relationship between FtsZ1 and FtsZ2 in higher plants. Plant Physiol. 127, 1656–1666.

McBeath, R., Pirone, D.M., Nelson, C.M., Bhadriraju, K., Chen, C.S., 2004. Cell shape, cytoskeletal tension, and RhoA regulate stem cell lineage commitment. Dev. Cell 6, 483–495.

McCabe, M., 1967. Mitochondria and pH. Nature 213, 280–281.

McCann, M.C., Chen, L., Roberts, K., Kemsley, E.K., Séné, C., Carpita, N.C., Stacey, N.J., Wilson, R.H., 1997. Infrared microspectroscopy: sampling heterogeneity in plant cell wall composition and architecture. Physiol. Plant 100, 729–738.

McCann, M.C., Roberts, K., 1991. Architecture of the primary cell wall. In: Lloyd, C.W. (Ed.), The Cytoskeletal Basis of Plant Growth and Form. Academic Press, London, pp. 109–129.

McCann, M., Rose, J., 2010. Blueprint for building plant cell walls. Plant Physiol. 153, 365.

McCann, M.C., Wells, B., Roberts, K., 1990. Direct visualization of cross-links in the primary plant cell wall. J. Cell Sci. 96,323–334.

McCarty, M., 1946. Chemical nature and biological specificity of the substance inducing transformation of pneumococcal types. Bact. Rev. 10, 63–71.

McCarty, M., 1985. The Transforming Principle. Discovering that Genes Are Made of DNA. W. W. Norton & Co., New York.

McCarty, R.E., 1992. A plant biochemist's view of H^+-ATPases and ATP synthases. J. Exp. Biol. 172, 431–441.

McCarty, M., 1995. A fifty-year perspective of the genetic role of DNA. Ann. N.Y. Acad. Sci. 758, 48–54.

McCarty, M., Avery, O.T., 1946a. Studies on the chemical nature of the substance inducing transformation of pneumococcal types II. J. Exp. Med. 83, 89–96.

McCarty, M., Avery, O.T., 1946b. Studies on the chemical nature of the substance inducing transformation of pneumococcal types III. J. Exp. Med. 83, 97–104.

McCarty, R.E., Racker, E., 1966. Effects of antisera to the chloroplast coupling factor on photophosphorylation and related processes. Abstract # 208. Fed. Proc. 25 (2), 226.

McClendon, J.F., 1927. The permeability and the thickness of the plasma membrane as determined by electric currents of high and low frequency. Protoplasma 3, 71–81.

McClintock, B., 1930. A cytological demonstration of the location of an interchange between two non-homologous chromosomes of Zea mays. Proc. Natl. Acad. Sci. U.S.A. 16, 791–796.

McClintock, B., 1950. The origin and behavior of mutable loci in maize. Proc. Natl. Acad. Sci. U.S.A. 36, 344–355.

McClintock, B., 1951. Chromosome organization and genic expression. Cold. Spring Harbor Symp. Quant. Biol. 16, 13–47.

McClintock, B., 1953. Induction of instability at selected loci in maize. Genetics 38, 579–599.

McClintock, B., December 8, 1983. The Significance of Responses of the Genome to Challenge. Nobel Lecture.

McClintock, M., Higinbotham, N., Uribe, E.G., Cleland, R., 1982. Active, irreversible accumulation of extreme levels of H_2SO_4 in the brown alga. Desmerestia Plant Physiol. 70,771–774.

McClung, C.E., 1924. The chromosome theory of heredity. In: Cowdry, E.V. (Ed.), General Cytology. University of Chicago Press, Chicago, pp. 609–689.

McClung, C.R., 2001. Circadian rhythms in plants. Annu. Rev. Plant Physiol. Plant Mol. Biol. 52, 139–162.

McClung, C.R., 2006. Plant circadian rhythms. Plant Cell 18, 792–803.

McClung, C.R., 2008. Comes a time. Curr. Opin. Plant Biol. 11, 514–520.

McClung, C.R., 2011. Circadian rhythms: lost in post-translation. Curr. Biol. 21, R400–R402.

McClung, C.R., 2013. Beyond arabidopsis: the circadian clock in non-model species. Semin. Cell Dev. Biol. 24, 430–436.

McClung, C.R., 2015. Circadian clocks: who knows where the time goes. Nat. Plants 1, 1–2.

McCombie, W.R., Martin-Gallado, A., Gocayne, J.D., FitzGerald, M., Dubnick, M., Kelley, J.M., Castilla, L., Liu, L.I., Wallace, S., Trapp, S., Tagle, D., Whaley, W.L., Cheng, S., Gusella, J., Frischauf, A.-M., Poustka, A., Lehrach, H., Collins, F.S., Kerlavage, A.R., Fields, C., Venter, J.C., 1992a. Expressed genes, *Alu* repeats and polymorphisms in cosmids sequenced from chromosome 4p16.3. Nat. Genet. 1, 348–353.

McConkey, E.H., 1982. Molecular evolution, intracellular organization, and the quinary structure of proteins. Proc. Natl. Acad. Sci. U.S.A. 79, 3236–3240.

McCormack, E., Braam, J., 2003. Calmodulins and related potential calcium sensors of *Arabidopsis*. New Phytol. 159, 585–598.

McCormack, E., Tsai, Y.-C., Braam, J., 2005. Handling calcium signaling: Arabidopsis CaMs and CMLs. Trends Plant Sci. 10, 383–389.

McCormack, E., Velasquez, L., Delk, N.A., Braam, J., 2006. Touch-responsive behaviors and gene expression in plants. In: Baluska, F., Mancuso, S., Volkmann, D. (Eds.), Communication in Plants. Springer, Berlin, pp. 248–260.

McCormick, S., 2017. A 3-dimensional biomechanical model of guard cell mechanics. Plant J. 92, 3–4.

McCormick, A., Campisi, J., 1991. Cellular aging and senescence. Curr. Opin. Cell Biol. 3, 230–234.

McCoy, R.C., Taylor, R.W., Blauwkamp, T.A., Kelley, J.L., Kertesz, M., Pushkarev, D., Petrov, D.A., Fiston-Vavier, A.-S., 2014. Illumina TruSeq synthetic long-reads empower de novo assembly and resolve complex, highly-repetitive transposable elements. PLoS One 9, e10668.

McCrombie, W.R., Adams, M.D., Kelley, J.M., FitzGerald, M.G., Utterback, T.R., Khan, M., Dubnick, M., Kerlavage, A.R., Venter, J.C., Fields, C., 1992b. Caenorhabditis elegans expressed sequence tags identify gene families and potential disease gene homologues. Nat. Genet. 1, 124–131.

McDonald, J.A., 1988. Extracellular matrix assembly. Annu. Rev. Cell Biol. 4, 183–208.

McDonald, K., 1989. Mitotic spindle ultrastructure and design. In: Hyams, J.S., Brinkley, B.R. (Eds.), Mitosis. Molecules and Mechanisms. Academic Press, London, pp. 1–38.

McDonald, A.R., Liu, B., Joshi, H.C., Palevitz, B.A., 1993. γ-tubulin is associated with a cortical-microtubule-organizing zone in the developing guard cells of *Allium cepa* L. Planta 191, 357–361.

McDonald, K.L., O'Toole, E.T., Mastronarde, D.N., McIntosh, J.R., 1992. Kinetochore microtubules in PtK1 cells. J. Cell Biol. 118, 369–383.

McDonough, F., 2009. Sophie Scholl: The Real Story of the Woman Who Defied Hitler. The History Press, Stroud, U.K.

McElheny, V.K., 2003. Watson and DNA. Making a Scientific Revolution. Perseus, Cambridge, MA.

McElroy, W.D., 1951. Properties of the reaction utilizing adenosine triphosphate for bioluminescence. J. Biol. Chem. 191, 547–557.

McElroy, W.D., Strehler, B.L., 1954. Bioluminescence. Bacteriol. Rev. 18, 177–194.

McEwen, B.F., Arena, J.T., Frank, J., Rieder, C.L., 1993. Structure of the colecemid-treated P & K, kinetodure outer plate as determined by high voltage electron microscopic tomography. J. Cell Biol. 120, 301–312.

McEwen, B.F., Heagle, A.B., Cassels, G.O., Buttle, K.F., Rieder, C.L., 1997. Kinetochore fiber maturation in PtK1 cells and its implications for the mechanisms of chromosome congression and anaphase onset. J. Cell Biol. 137, 1567–1580.

McFadden, G.I., 2000. Mergers and acquisitions: malaria and the great chloroplast heist. Genome Biol. 1 reviews1026.

McFadden, G.I., 2001a. Chloroplast origin and integration. Plant Physiol. 125, 50–53.

McFadden, G.I., 2011b. The apicoplast. Protoplasma 248, 641–650.

McFadden, G.I., 2014. Apicoplast. Curr. Biol. 24, R262–R263.

McFadden, G.I., Melkonian, M., 1986. Golgi-apparatus activity and membrane flow during scale biogenesis in the green flagellate *Scherffelia-Dubia* (Prasinophyceae). 1. Flagellar regeneration. Protoplasma 130, 186–198.

McFadden, G.I., Presig, H.R., Melkonian, M., 1986. Golgi-apparatus activity and membrane flow during scale biogenesis in the green flagellate *Scherffelia-Dubia* (Prasinophyceae). 2. Cell-wall secretion and assembly. Protoplasma 131, 174–184.

McFadden, G.I., Ralph, S.A., 2003. Dynamin: the endosymbiosis ring of power? Proc. Natl. Acad. Sci. U.S.A. 100, 3557–3559.

McFadden, G.I., Roos, D.S., 1999. Apicomplexan plastids as drug targets. Trends Microbiol. 7, 328–333.

McFadden, G.I., Schulze, D., Surek, B., Salisbury, J.L., Melkonian, M., 1987. Basal body reorientation mediated by a Ca^{2+}-modulated contractile protein. J. Cell Biol. 105,903–912.

McFadden, G.I., Waller, R.F., 1997. Plastids in parasites of humans. BioEssays 19, 1033–1040.

McFarlane, H.E., Döring, A., Persson, S., 2014. The cell biology of cellulose synthesis. Annu. Rev. Plant Biol. 65, 69–94.

McGettigan, P.A., 2013. Transcriptomics in the RNA-seq era. Curr. Opin. Chem. Biol. 17, 4–11.

McGhee, J.D., Engel, J.D., 1975. Subunit structure of chromatin is the same in plants and animals. Nature 254, 449–450.

McGill, M., Brinkley, B.R., 1975. Human chromosomes and centrioles as nucleating sites for the in vitro assembly of microtubules from bovine brain tubulin. J. Cell Biol. 67, 189–199.

McGinnis, K., Chandler, V., Cone, K., Kaeppler, H., Kaeppler, S., Kerschen, A., Pikaard, C., Richards, E., Sidorenko, L., Smith, T., Springer, N., Wulan, T., 2005. Transgene-induced RNA interference as a tool for plant functional genomics. Methods Enzymol. 392, 1–24.

McGrath, A., 2005. Dawkins' God: Genes, Memes, and the Meaning of Life. Blackwell, Oxford.

McGrath, A., McGrath, J.C., 2007. The Dawkins Delusion. Atheist Fundamentalism and the Denial of the Divine. InterVarsity Press, Downers Grove, IL.

McGregor, D.I., Beevers, H., 1969. Development of enzymes in watermelon seedlings. Plant Physiol. 44, S-33.

McIntosh, J.R., 1983. The centrosome as an organizer of the cytoskeleton. Mol. Cell Biol. 2, 115–142.

McIntosh, J.R., Euteneuer, U., 1984. Tubulin hooks as probes for microtubule polarity: an analysis of the method and an evaluation of data on microtubule polarity in the mitotic spindle. J. Cell Biol. 98, 525–533.

McIntosh, J.R., Euteneuer, U., Neighbors, B., 1980. Initial characterization of conditions for displaying polarity of microtubules. In: De Brabander, M., De May, J. (Eds.), 2nd International Symposium on Microtubules and Microtubule Inhibitors. Elsevier North-Holland, Amsterdam, pp. 357–371.

McIntosh, J.R., Grishchuk, E.L., Morphew, M.K., Efremov, A.K., Zhudenkov, K., Volkov, V.A., Cheeseman, I.M., Desai, A., Mastronarde, D.N., Ataullakhanov, F.I., 2008. Fibrils connect microtubule tips with kinetochores: a mechanism to couple tubulin dynamics to chromosome motion. Cell 135, 322–333.

McIntosh, J.R., Hepler, P.K., VanWie, D.G., 1969. Model for mitosis. Nature 244, 659–663.

McIntosh, J.R., Koonce, M.P., 1989. Mitosis. Science 246, 622–628.

McIntosh, J.R., Molodtsov, M.I., Ataullakhanov, F.I., 2012. Biophysics of mitosis. Q. Rev. Biophys. 45, 147–207.

McIntosh, J.R., Volkov, V., Ataullakhanov, F.I., Grishchuk, E.L., 2010. Tubulin depolymerization may be an ancient biological motor. J. Cell Sci. 123, 3425–3434.

McKee, M., August 24, 2005. Space radiation may select amino acids for life. New Sci.

McKeekin, T.L., Warner, R.C., 1946. The chemistry of the proteins and amino acids. Ann. Rev. Biochem. 15, 119–154.

McKenna, S.T., Kunkel, J.G., Bosch, M., Rounds, C.M., Vidali, L., Winship, L.J., Hepler, P.K., 2009. Exocytosis precedes and predicts the increase in growth in oscillating pollen tubes. Plant Cell 21, 3026–3040.

McKusick, V.A., Kucherlapati, R.S., Ruddle, F.H., 1993. Genomics: stock-taking after 5 years. Genomics 15, 1–2.

McKusick, V.A., Ruddle, F.H., 1987. A new discipline, a new name, a new journal. Genomics 1, 1–2.

McLachlan, D.H., Kopischke, M., Robatzek, S., 2014. Gate control: guard cell regulation by microbial stress. New Phytol. 203, 1049–1063.

McLafferty, F., 2011. A century of progress in molecular mass spectrometry. Annu. Rev. Anal. Chem. 4, 1–22.

McLafferty, F.W., Todd, P.J., McGilvery, D.C., Baldwin, M.A., 1980. High-resolution tandem mass spectrometry (MS/MS) of increased sensitivity and mass range. J. Am. Chem. Soc. 102, 3360–3363.

McLaren, A.D., Babcock, K.L., 1959. Some characteristics of enzyme reactions at surfaces. In: Hayashi, T. (Ed.), Subcellular Particles. Ronald Press Co., New York, pp. 23–36.

McLean, B., Juniper, B.E., 1986. The plasma membrane of young Chara internodal cells revealed by rapid freezing. Planta 169, 153–161.

McLean, B.G., Zupan, J., Zambryski, P.C., 1995. Tobacco mosaic virus movement protein associates with the cytoskeleton in tobacco cells. Plant Cell 7, 2101–2114.

McLeish, J., Snoad, B., 1958. Looking at Chromosomes. St. Martin's Press, London.

McMichael, C.M., Bednarek, S.Y., 2013. Cytoskeletal and membrane dynamics during higher plant cytokinesis. New Phytol. 197, 1039–1057.

McMillan, 1968. Electron Paramagnetism. Reinhold Book Co., New York.

McNamara, G., 2008. Clocks in the Sky. Springer, Berlin.

McNeill, P.A., Berns, M.W., 1981. Chromosome behavior after laser microirradiation of a single kinetochore in mitotic PtK2 cells. J. Cell Biol. 88, 543–553.

McNiven, M.A., 2002. The solid state cell. Biol. Cell 94, 555–556.

McNulty, A.K., Saunders, M.J., 1992. Purification and immunological detection of pea nuclear intermediate filaments: evidence for plant nuclear lamins. J. Cell Sci. 103, 407–414.

McNutt, N.S., Weinstein, R.S., 1973. Membrane ultrastructure at mammalian intercellular junctions. Prog. Biophys. Mol. Biol. 26, 45–101.

McQueen-Mason, S., Durachko, D.M., Cosgrove, D.J., 1992. Endogenous proteins that induce cell wall expansion in plants. Plant Cell 4, 1425–1433.

McQueen-Mason, S., Fry, S.C., Durachko, D.M., Cosgrove, D.J., 1993. The relationship between xyloglucan endotransglycosylase and in-vitro cell wall extension in cucumber hypocotyls. Planta 190, 327–331.

McQueen-Mason, S., Le, N.T., Brocklehurst, D., 2006. Expansins. Plant Cell Monogr. 5, 117–138.

Meagher, R.B., 1991. Divergence and differential expression of actin gene families in higher plants. Int. Rev. Cytol. 125, 139–163.

Meagher, R.B., Kandasamy, M.K., King, L., 2011. Actin functions in the cytoplasmic and nuclear compartments. The Plant Cytoskeleton. Adv. Plant Biol. 2, 3–32.

Méchali, M., Lutzmann, M., 2008. The cell cycle: now live and in color. Cell 132, 341–343.

Meckel, T., Hurst, A.C., Thiel, G., Homann, U., 2004. Endocytosis against high turgor: intact guard cells of Vicia faba constitutively endocytose fluorescently labeled plasma membrane and GFP-tagged K-channel KAT1. Plant J. 39, 182–193.

Meckel, T., Hurst, A.C., Thiel, G., Homann, U., 2005. Guard cells undergo constitutive and pressure-driven membrane turnover. Protoplasma 226, 23–29.

Medeiros, D.B., Daloso, D.M., Fernie, A.R., Nikoloski, Z., Araújo, W.L., 2015. Utilizing systems biology to unravel stomatal function and the hierarchies underpinning its control. Plant Cell Environ. 38, 1457–1470.

Meeuse, A.D.J., 1957. Plasmodesmata (Vegetable Kingdom). Protoplasmatologia Band II. Springer-Verlag, Wien.

Meeuse, B.J.D., 1975. Thermogenic respiration in aroids. Ann. Rev. Plant Physiol. 26, 117–126.

Meeuse, B.J.D., Raskin, I., 1988. Sexual reproduction in the arum lily family, with emphasis on thermogenicity. Sex. Plant Reprod. 1, 3–15.

Meeusen, S., McCaffery, J.M., Nunnari, J., 2004. Mitochondrial fusion intermediates revealed in vitro. Science 305, 1747–1752.

Mehler, A.H., 1957. Introduction to Enzymology. Academic Press, New York.

Mehrshahi, P., Johnny, C., DellaPenna, D., 2014. Redefining the metabolic continuity of chloroplasts and ER. Trends Plant Sci. 19, 501–507.

Mehrshahi, P., Stefano, G., Andaloro, J.M., Brandizzi, F., Froehlich, J.E., DellaPenna, D., 2013. Transorganellar complementation redefines the biochemical continuity of endoplasmic reticulum and chloroplasts. Proc. Natl. Acad. Sci. U.S.A. 110, 12131–12136.

Mehta, M., Sarafis, V., Critchley, C., 1999. Thylakoid membrane architecture. Aust. J. Plant Physiol. 26, 2039–2042.

Meier, I., 2007. Composition of the plant nuclear envelope: theme and variations. J. Exp. Bot. 58, 27–34.

Meier, I., Brkljacic, J., 2009. Adding pieces to the puzzling plant nuclear envelope. Curr. Opin. Plant Biol. 12, 752–759.

Meier, I., Richards, E.J., Evans, D.E., 2017. Cell biology of the plant nucleus. Annu. Rev. Plant Biol. 68, 139–172.

Meierhenrich, U., 2008. Amino Acids and the Asymmetry of Life. Springer, Berlin.

Meijón, M., Satbhai, S.B., Tsuchimatsu, T., Busch, W., 2014. Genome-wide association study using cellular traits identifies a new regulator of root development in *Arabidopsis*. Nat. Genet. 46, 77–81.

Meindl, U., Zhang, D., Hepler, P.K., 1994. Actin microfilaments are associated with the migrating nucleus and the cell cortex in the green alga Micrasterias. J. Cell Sci. 107, 1929–1934.

Meiners, S., Gharyal, P.K., Schindler, M., 1991a. Permeabilization of the plasmalemma and wall of soybean root cells to macromolecules. Planta 184, 443–447.

Meiners, S., Schindler, S., 1987. Immunological evidence for gap junction polypeptide in plant cells. J. Biol. Chem. 262, 951–953.

Meiners, S., Schindler, M., 1989. Characterization of a connexin homologue in cultured soybean cells and diverse plant organs. Planta 179, 148–155.

Meiners, S., Xu, A., Schindler, M., 1991b. Gap junctions protein homologue from *Arabidopsis thaliana*: evidence for connexins in plants. Proc. Natl. Acad. Sci. U.S.A. 88, 4119–4122.

Meissner, S.T., 2012. How classical segregation can fit within modern cell biological segregation. Open Sci. Repos. Biol. e70081908. Online.

Melan, M.A., 1990. Taxol maintains organized microtubule patterns in protoplasts which lead to a resynthesis of organized cell wall microfibrils. Protoplasma 153, 169–177.

Melançon, P., Glick, B.S., Malhotra, V., Weidman, P.J., Serafini, T., Gleason, M.L., Orci, L., Rothman, J.E., 1987. Involvement of GTP-binding "G" proteins in transport through the Golgi stack. Cell 51, 1053–1062.

Melcak, I., Risueno, M.C., Raska, I., 1996. Ultrastructural nonisotropic mapping of nucleolar transcription sites in onion protoplasts. J. Struct. Biol. 116, 253–263.

Melcher, K., Ng, L.-M., Zhou, X.E., Soon, F.-F., Xu, Y., Suino-Powell, K.M., Park, S.-Y., Weiner, J.J., Fujii, H., Chinnusamy, V., Kovach, A., Li, J., Wang, Y., Li, J., Peterson, F.C., Jensen, D.R., Yong, E.-L., Volkman, B.F., Cutler, S.R., Zhu, J.-K., Xu, H.E., 2009. A gate-latch-lock mechanism for hormone signalling by abscisic acid receptors. Nature 462, 602–608.

Melin, P.M., Sommarin, M., Sandelius, A.S., Jergil, B., 1987. Identification of Ca^{2+}-stimulated polyphosphoinositide phospholipase C in isolated plant plasma membranes. FEBS Lett. 223, 87–91.

Melkonian, M., 1989. Centrin-mediated motility. A novel cell motility mechanism in eukaryotic cells. Bot. Acta 102, 3–4.

Melkonian, M., Becker, B., Becker, D., 1991. Scale formation in algae. J. Electron Microsc. Tech. 17, 165–178.

Mellman, I., 1996. Endocytosis and molecular sorting. Ann. Rev. Cell Biol. 12, 575–625.

Mellman, I., Simons, K., 1992. The Golgi complex: in vitro veritas? Cell 68, 829–840.

Mello, C.C., December 8, 2006. Return to the RNAi World: Rethinking Gene Expression and Evolution. Nobel Lecture.

Mellor, J.W., 1914. Chemical Statics and Dynamics. Longmans, Green and Co., London.

Melroy, D., Jones, R.L., 1986. The effect of monensin on intracellular transport and secretion of α-amylase isozymes in barley aleurone. Planta 167, 252–259.

Memelink, J., Swords, K.M.M., Staehelin, L.A., Hoge, J.H.C., 1994. Southern, Northern and Western blot analysis. Plant Mol. Biol. Man. F1, 1–23.

Mendel, G., 1865. Experiments in Plant Hybridization, 1926. Harvard Univesity Press, Cambridge, MA.

Mendelsohn, E., 1964. Heat and Life. The Development of the Theory of Animal Heat. Harvard Univeristy Press, Cambridge, MA.

Menke, W., 1962. Structure and chemistry of plastids. Ann. Rev. Plant Physiol. 13, 27–44.

Menzel, D., 1996. The role of the cytoskeleton in polarity and morphogenesis of algal cells. Curr. Opin. Cell Biol. 8, 38–42.

Merchant, C.A., Healy, K., Wanunu, M., Ray, V., Peterman, N., Bartel, J., Fischbein, M.D., Venta, K., Luo, Z., Johnson, A.T.C., Drndic, M., 2010. DNA translocation through graphene nanopores. Nano Lett. 10, 2915–2921.

Merchant, S.S., Prochnik, S.E., Vallon, O., Harris, E.H., Karpowicz, S.J., Witman, G.B., Terry, A., Salamov, A., Fritz-Laylin, L.K., Maréchal-Drouard, L., Marshall, W.F., Qu, L.-H., Nelson, D.R., Sanderfoot, A.A., Spalding, M.H., Kapitonov, V.V., Ren, Q., Ferris, P., Lindquist, E., Shapiro, H., Lucas, S.M., Grimwood, J., Schmutz, J., Chlamydomonas Annotation Team, JGI AnnotationTeam, Grigoriev, I.V., Rokhsar, D.S., Grossman, A.R., 2007. The Chamydomonas genome reveals the evolution of key animal and plant functions. Science 318, 245–251.

Merdes, A., Cleveland, D.W., 1997. Pathways of spindle pole formation: different mechanisms; conserved components. J. Cell Biol. 138, 953–956.

Mereschkowsky, C., 1905. Über Natur und Ursprung der Chromatophoren im Pflanzenreiche. Biol. Cent. 25, 593–604.

Mereschkowsky, C., 1910. Theorie der zwei Plasmaarten als Grundlage der Symbiogenesis, einer neuen Lehre von der Entstehung der Organismen. Biol. Cent. 30, 278–303, 321–347, 353–367.

Mereschkowsky, C., 1920. La plante considérée comme une complex symbiotique. Bull. Soc. Nat. Sci. Ouest France 6, 17–21.

Merilo, E., Jõesaar, I., Brosché, M., Kollist, H., 2014. To open or to close: species-specific stomatal responses to simultaneously applied opposing environmental factors. New Phytol. 202, 499–508.

Merkle, T., Leclerc, D., Marshallsay, C., Nagy, F., 1996. A plant in vitro system for the nuclear import of proteins. Plant J. 10, 1177–1186.

Merkle, T., Nagy, F., 1997. Nuclear import of proteins: putative import factors and development of in vitro systems in higher plants. Trends Plant Sci. 2, 458–464.

Merlot, S., Leonhardt, N., Fenzi, F., Valon, C., Costa, M., Piette, L., Vavasseur, A., Genty, B., Boivin, K., Müller, A., Giraudat, J., Leung, J., 2007. Constitutive activation of a plasma membrane H^+-ATP prevents abscisic acid-mediated stomatal closure. EMBO J. 26, 3216–3226.

Mersey, B.G., McCully, M.E., 1978. Monitoring the course of fixation of plant cells. J. Microsc. 166, 43–56.

Mesecar, A.D., Koshland, D.E., 2000. A new model for protein stereospecificity. Nature 403, 614–615.

Meselson, M., Stahl, F.W., 1958. The replication of DNA in *Escherichia coli*. Proc. Natl. Acad. Sci. U.S.A. 44, 671–682.

Meselson, M., Stahl, F.W., Vinograd, J., 1957. Equilibrium sedimentation of macromolecules in density gradients. Proc. Natl. Acad. Sci. U.S.A. 43, 581–588.

Meselson, M., Yuan, R., 1968. DNA restriction enzyme from *E. coli*. Nature 217, 1110–1114.

Meskin, M.S., Bidlack, W.R., Davies, A.J., Omaye, S.T. (Eds.), 2002. Phytochemicals in Nutrition and Health. CRC Press, Boca Raton, FL.

Messerli, M.A., Creton, R., Jaffe, L.F., Robinson, K.R., 2000. Periodic increases in elongation rate precede increases in cytosolic Ca^{2+} during pollen tube growth. Dev. Biol. 222, 84–98.

Messerli, M., Danuser, G., Robinson, K.R., 1999. Pulsatile influxes of H^+, K^+ and Ca^{2+} lag growth pulses of *Lilium longiflorum* pollen tubes. J. Cell Sci. 112, 1497–1509.

Messerli, M., Robinson, K.R., 1997. Tip localized Ca^{2+} pulses are coincident with peak pulsatile growth rates in pollen tubes of Lilium longiflorum. J. Cell Sci. 110, 1269–1278.

Messerli, M.A., Robinson, K.R., 2003. Ionic and osmotic disruptions of the lily pollen tube oscillator: testing proposed models. Planta 217, 147–157.

Messerli, M.A., Robinson, K.R., 2007. MS channels in tip-growing systems: mechanosensitive ion channels. Curr. Top. Membr. 58, 393–412.

Metaxas, E., 2007. William Wilberforce and the Heroic Campaign to End Slavery. HarperCollins, New York.

Metaxas, E., 2010. Bonhoeffer: Pastor, Martyr, Prophet, Spy. Thomas Nelson, Nashville, TN.

Metcalf III, T.N., Villanueva, M.A., Schindler, M., Wang, J.L., 1986a. Monoclonal antibodies directed against protoplasts of soybean cells: analysis of lateral mobility of plasma membrane-bound antibody. J. Cell Biol. 102, 1350–1357.

Metcalf III, T.N., Wang, J.L., Schindler, M., 1986b. Lateral diffusion of phospholipids in the plasma membrane of soybean protoplasts: evidence for membrane lipid domains. Proc. Natl. Acad. Sci. U.S.A. 83, 95–99.

Métraux, J.P., Richmond, P.A., Taiz, L., 1980. Control of cell elongation in *Nitella* by endogenous cell wall pH gradients. Plant Physiol. 65, 204–210.

Métraux, J.P., Taiz, L., 1977. Cell wall extension in *Nitella* as influenced by acids and ions. Proc. Natl. Acad. Sci. U.S.A. 74, 1565–1569.

Métraux, J.P., Taiz, L., 1978. Transverse viscoelastic extension in *Nitella*. I. Relationship to growth rate. Plant Physiol. 61, 135–138.

Métraux, J.P., Taiz, L., 1979. Transverse viscoelastic extension in *Nitella*. II. Effects of acids and ions. Plant Physiol. 63, 657–659.

Metz, C.W., 1936. Factors influencing chromosome movements in mitosis. Cytologia 7, 219–231.

Meunier, S., Vernos, I., 2016. Acentrosomal microtubule assembly in mitosis: the where, when, and how. Trends Cell Biol. 26, 80–87.

Meves, F., 1904. Ueber das Vorkmmen von Mitochonrien bezw: Condriomitten in Pflanzenzellen. Ber. d. Deut. Bot. Gesells. 22, 284–286.

Meves, F., 1911. Gesammelte Studien an den roten Blutkrperchen der Amphibien. Arch. Mikrosc. Anat. 77, 465–540.

Meyen, F.J.F., 1837–1839. Neues System der Pflanzen-Physiologie. Haude und Spenersche Buchhandlung, Berlin.

Meyer, A., 1883. Das Chlorophyllkorn in chemischer, morphologischer und biologischer Beziehung. Arthur Felix, Leipzig.

Meyer, Y., Herth, W., 1978. Chemical inhibition of cell wall formation and cytokinesis, but not of nuclear division, in protoplasts of *Nicotiana tabacum* L. cultivated *in vitro*. Planta 142, 253–262.

Meyerhof, O., 1924. Chemical Dynamics of Life Phaenomena. J.B. Lippincott Co., Philadelphia.

Meyerhof, O., 1942. Intermediate carbohydrate metabolism. In: A Symposium on Respiratory Enzymes. The University of Wisconsin Press, Madison, pp. 3–15.

Meyerhof, O., 1944. Energy relationships on glycolysis and phosphorylation. Ann. N.Y. Acad. Sci. 44, 377–393.

Meyerhof, O.F., December 12, 1923. Energy Conversions in Muscle. Nobel Lecture.

Meyerhof, O., Lohmann, K., 1928. Über die natürlichen Guanidino phosphorsäuren (Phosphogene) in der quergestreiften Muskulatur. Biochem. Z. 196, 22–48, 49–72.

Meyers, B.C., Galbraith, D.W., Nelson, T., Agrawal, V., 2004. Methods for transcriptional profiling in plants. Be fruitful and replicate. Plant Physiol. 135, 637–652.

Meyers, L.A., Levin, D.A., 2006. On the abundance of polyploids in flowering plants. Evolution 60, 1198–1206.

Michael, T.P., Jackson, S., 2013. The first 50 plant genomes. Plant Genome 6. https://dl.sciencesocieties.org/publications/tpg/articles/6/2/plantgenome2013.03.0001in.

Michael, T.P., Jupe, F., Bemm, F., Motley, S.T., Sandoval, J.P., Loudet, O., Weigel, D., Ecker, J.R., 2017. High Contiguity *Arabidopsis thaliana* Genome Assembly with Single Nanopore Flow Cell. bioRxiv. https://www.biorxiv.org/content/early/2017/06/14/149997.

Michaeli, S., Galili, G., 2014. Degradation of organelles or specific organelle components via selective autophagy in plant cells. Int. J. Mol. Sci. 15, 7624–7638.

Michaeli, S., Galili, G., Genschik, P., Fernie, A.R., Avin-Wittenberg, T., 2016. Autophagy in plants—What's new on the menu? Trends Plant Sci. 21, 134–144.

Michaelis, L., 1900. Die vitale Färbung, eine Darstellungsmethode der Zellgranula. Arch. Mikr. Anat. 55, 558–575.

Michaelis, L., 1925. Contribution to the theory of permeability of membranes for electrolytes. J. Gen. Physiol. 8, 33–59.

Michaelis, L., 1926. Hydrogen Ion Concentration. Williams & Wilkins Co., Baltimore.

Michaelis, L., 1926a. Die Permeabililit: at von Membranen. Naturwissenschaften 3, 33–42.

Michaelis, L., 1926b. Hydrogen Ion Concentration. Williams & Wilkins Co., Baltimore.

Michaelis, L., 1930. Oxidation-Reduction Potentials. Translated by L. B. Flexner. J. B. Lippincott Co., Philadelphia.

Michaelis, L., 1958. Leonor Michaelis. An autobiography with additions by D. A. MacInnes and S. Granick. Bio. Mems. Proc. Natl. Acad. Sci. U.S.A. 31, 282–321.

Michaelis, L., Menten, M., 1913. Die Kinetik der Invertinwirkung. Biochem. Z. 49, 333–369.

Michaelis, L., Menten, M., 2013. The kinetics of invertin action. FEBS Lett. 587, 2712–2720.

Michal, G. (Ed.), 1999. Biochemical Pathways. John Wiley & Sons, New York.

Michalak, M., Milner, R.E., Burns, K., Opas, M., 1992. Calreticulin Biochem. J. 285, 681–692.

Michel, J.-B., Shen, Y.K., Aiden, A.P., Veres, A., Gray, M.K., The Google Books Team, Pickett, J.P., Hoiberg, D., Clancy, D., Norvig, P., Orwant, J., Pinker, S., Nowak, M.A., Aiden, E.L., 2010. Quantitative analysis of culture using millions of digitized books. Science 331, 176–182.

Michelet, B., Lukazewicz, M., Dupriez, V., Boutry, M., 1994. A plant plasma membrane proton-ATPase gene is regulated by development and environment and shows signs of translational regulation. Plant Cell 6, 1375–1389.

Michl, H., 1958. Hochvoltelektrophorese. J. Chromatogr. 1, 93–121.

Mickolajczyk, K.J., Deffenbaugh, N.C., Arroyo, J.O., Andrecka, J., Kukura, P., Hancock, W.O., 2015. Kinetics of nucleotide-dependent structural transitions in the kinesin-1 hydrolysis cycle. Proc. Natl. Acad. Sci. U.S.A. 112, E186–E193.

Mickolajczyk, K.J., Deffenbaugh, N.C., Arroyo, J.O., Andrecka, J., Kukura, P., Hancock, W.O., 2016. Kinetics of nucleotide-dependent structural transitions in the kinesin-1 hydrolysis cycle. Proc. Natl. Acad. Sci. U.S.A. 112, E186–E193.

Miescher, F., 1897. Die histochemischen und physiologischen Arbeiten von Friedrich Miescher, vols. I and II. F. C. W. Vogel, Leipzig.

Migliaccio, G., Nicchitta, C.V., Blobel, G., 1992. The signal sequence receptor, unlike the signal recognition particle receptor, is not essential for protein translocation. J. Cell Biol. 117, 15–25.

Mii, S., Ochoa, S., 1957. Polyribonucleotide synthesis with highly purified polynucleotide phosphorylase. Biochem. Biophys. Res. Commun. 26, 445–446.

Mika, J.T., Krasnikov, V., van den Bogaart, G., de Haan, F., Poolman, B., 2011. Evaluation of pulsed-FRAP and conventional-FRAP for determination of protein mobility in prokaryotic cells. PLoS One 6, e25664.

Miki, T., Naito, H., Nishina, M., Goshima, G., 2014. Endogenous localizome identifies 43 mitotic kinesins in a plant cell. Proc. Natl. Acad. Sci. U.S.A. 111, E1053−E1061.

Miki, T., Nishina, M., Goshima, G., 2015. RNAi screening identifies the armadillo repeat-containing kinesins responsible for microtubule-dependent nuclear positioning in *Physcomitrella patens*. Plant Cell Physiol. 56, 737−749.

Milani, P., Braybrook, S.A., Boudaoud, A., 2013. Striking the hammer: micromechanical approaches to morphogenesis. J. Exp. Bot. 64, 4651−4662.

Milani, P., Gholamirad, M., Traas, J., Arneodo, A., Boudaoud, A., Argoul, F., Hamant, O., 2011. In vivo analysis of local wall stiffness at he shoot apical meristem in *Arabidopsis* using atomic force microscopy. Plant J. 67, 1116−1123.

Millar, A.J., Carré, I.A., Stayer, C.A., Chua, N.H., Kay, S.A., 1995a. Circadian clock mutants in *Arabidopsis* identified by luciferase imaging. Science 267, 1161−1163.

Millar, A.H., Liddell, A., Leaver, C.J., 2007. Isolation and subfractionation of mitochondria from plants. Methods Cell. Biol. 80, 65−90.

Millar, A.J., Straume, M., Chory, J., Chua, N.-H., Kay, S.A., 1995b. The regulation of circadian period by phototransduction pathways in *Arabidopsis*. Science 267, 1163−1166.

Millar, A.H., Whelan, J., Small, I., 2006. Recent surprises in protein targeting to mitochondria and plastids. Curr. Opin. Plant Biol. 9, 610−615.

Miller, S.L., 1953. A production of amino acids under possible primitive Earth conditions. Science 117, 528−529.

Miller, S.L., 1955. Production of some organic compounds under possible primitive Earth conditions. J. Am. Chem. Soc. 77, 2351−2361.

Miller, J.H., 1968. An evaluation of specific and non-specific inhibition of 2-dimensional growth in fern gametophytes. Physiol. Plant 21, 699−710.

Miller, J.H., 1980a. Differences in the apparent permeability of spore walls and prothalial cell walls in *Onoclea sensibilis*. Am. Fern J. 70, 119−123.

Miller, K.R., 1980a. A chloroplast membrane lacking changes in unstacked membrane regions. Biochim. Biophys. Acta 592, 143−152.

Miller, J.H., 1980b. Orientation of the plane of cell division in fern gametophytes: the roles of cell shape and stress. Am. J. Bot. 67, 534−542.

Miller, J.H., 1980b. Differences in the apparent permeability of spore walls and prothalial cell walls in Onoclea sensibilis. Am. Fern J. 70, 119−123.

Miller, K.R., 1980c. A chloroplast membrane lacking changes in unstacked membrane regions. Biochim. Biophys. Acta 592, 143−152.

Miller, S.L., 1992. The prebiotic synthesis of organic compounds as a step toward the origin of life. In: Schopf, J.W. (Ed.), Major Events in the History of Life. Jones and Bartlett, Sudbury, MA, pp. 1−28.

Miller, G., 2006. Optogenetics: shining new light on neural circuits. Science 314, 1674−1676.

Miller, D.D., Callaham, D.A., Gross, D.J., Hepler, P.K., 1992. Free Ca gradient in growing pollen tubes of Lillium. J. Cell Sci. 101, 7−12.

Miller, J.P., Lo, R.S., Ben-Hur, A., Desmarais, C., Stagljar, I., Noble, W.S., Fields, S., 2005. Large-scale identification of yeast integral membrane protein interactions. Proc. Natl. Acad. Sci. U.S.A. 102, 12123−12128.

Miller, K.R., Miller, G.J., McIntyre, K.R., 1977. Organization of the photosynthetic membrane in maize mesophyll and bundle-sheath chloroplasts. Biochim. Biophys. Acta 459, 145−156.

Miller, S.L., Orgel, L.E., 1974. The Origins of Life on the Earth. Prentice-Hall, Engelwood Cliffs, NJ.

Miller, E.A., Schekman, R., 2013. COPII—a flexible vesicle formation system. Curr. Opin. Cell Biol. 25, 420−427.

Miller, M., Schneider, J., Sathyanarayana, B.K., Toth, M.V., Marshall, G.R., Clawson, L., Selk, L., Kent, S.B.H., Wlodawer, A., 1989. Structure of a complex of synthetic HIV-1 protease with a substrate-based inhibitor at 2.3 Å resolution. Science 246, 1149−1152.

Miller, D.D., Scordilis, S.P., Hepler, P.K., 1995. Identification and localization of three classes of myosins in pollen tubes of *Lilium longiflorum* and *Nicotiana alata*. J. Cell Sci. 108, 2549−2563.

Miller, K.R., Staehelin, L.A., 1976. Analysis of the thylakoid outer surface: coupling factor is limited to unstacked membrane regions. J. Cell Biol. 68, 30−47.

Miller, S.L., Urey, H.C., 1959a. Lett. Sci. 130, 1622−1624.

Miller, S.L., Urey, H.C., 1959b. Organic compound synthesis on the primitive Earth. Science 130, 245−251.

Millerd, A., 1953. Respiratory oxidation of pyruvate by plant mitochondria. Arch. Biochem. Biophys. 42, 149−163.

Millerd, A., Bonner, J., 1953. The biology of plant mitochondria. J. Histochem. Cytochem. 1, 254−264.

Millerd, A., Bonner, J., Axelrod, B., Bandurski, R., 1951. Oxidative and phosphorylative activity of plant mitochondria. Proc. Natl. Acad. Sci. U.S.A. 37, 855−862.

Millikan, R.A., 1935. Electrons (+ and −), Protons, Photons, Neutrons, and Cosmic Rays. University of Chicago Press, Chicago.

Millikan, R.A., May 23, 1924. The Electron and the Light-Quant from the Experimental Point of View. Nobel Lecture.

Millwood, R.J., Moon, H.S., Stewart Jr., C.N., 2008. Fluorescent proteins in transgenic plants. Rev. Fluoresc. 2008, 387−403.

Mimura, T., Dietz, K.-J., Kaiser, W., Schramm, M.J., Kaiser, G., Heber, U., 1990a. Phosphate transport across biomembranes and cytosolic phosphate homeostasis in barley leaves. Planta 180, 139−146.

Mimura, T., Kirino, Y., 1984. Changes in cytoplasmic pH measured by 31P-NMR in cells of *Nitellopsis obtusa*. Plant Cell Physiol. 25, 813−820.

Mimura, T., Sakano, K., Tazawa, M., 1990b. Changes in the subcellular distribution of free amino acids in relation to light conditions in cells of *Chara corallina*. Bot. Acta 103, 42−47.

Mimura, T., Shimmen, T., Tazawa, M., 1984. Adenine-nucleotide levels and metabolism-dependent membrane potential in cells of *Nitellopsis obtusa* Groves. Planta 162, 77−84.

Min Jou, W., Haegeman, G., Ysebaert, M., Fiers, W., 1972. Nucleotide sequence of the gene coding for the bacteriophage MS2 coat protein. Nature 237, 82−88.

Min, M.K., Kim, S.J., Miao, Y., Shin, J., Jiang, L., Hwang, I., 2007. Overexpression of *Arabidopsis* AGD7 causes relocation of golgi-localized proteins to the endoplasmic reticulum and inhibits protein trafficking in plant cells. Plant Physiol. 143, 1601−1614.

Mineyuki, Y., 1999. The preprophase band of microtubules: its function as a cytokinetic apparatus in higher plants. Int. Rev. Cytol. 187, 1−49.

Mineyuki, Y., Furuya, M., 1985. Involvement of microtubules on nuclear positioning during apical growth in *Adiantum* protonemata. Plant Cell Physiol. 26, 213−220.

Mineyuki, Y., Marc, J., Gunning, B.E.S., 1989. Development of the preprophase band from random cytoplasmic microtubules in guard mother cells of *Allium cepa* L. Planta 178, 291–296.

Mineyuki, Y., Palevitz, B.A., 1990. Relationship between the preprophase band organization, F-actin, and the division site in *Allium*. J. Cell Sci. 97, 283–295.

Mineyuki, Y., Yamashita, M., Nagahama, Y., 1991. P34cdc2 kinase homologue in the preprophase band. Protoplasma 162, 182–186.

Minio, A., Lin, J., Gaut, B.S., Cantu, D., 2017. How single molecule real-time sequencing and haplotype phasing have enabled reference-grade diploid genome assembly of wine grapes. Front. Plant Sci. 8, 826.

Minkoff, B.B., Stecker, K.E., Sussman, M.R., 2015. Rapid phosphoproteomic effects of Abscisic Acid (ABA) on wildtype and ABA receptor-deficient *A. thaliana* mutants. Mol. Cell. Proteom. 14, 1169–1182.

Minkowski, H., 1909. Raum und Zeit. Jahresber. Dtsch. Math. Ver. 75–88.

Minorsky, P.V., 2002. The hot and the classic. Peroxisomes: organelles of diverse function. Plant Physiol. 130, 517–518.

Minorsky, P.V., 2003. The hot and the classic. Plant thermogenesis and thermoregulation. Plant Physiol. 132, 25–26.

Mirchev, R., Golan, D.E., 2001. Single-particle tracking and laser optical tweezers studies of the dynamics of individual protein molecules in membranes of intact human and mouse red cells. Blood Cell 27, 143–147.

Mirmajlessi, S.M., Loit, E., Mänd, M., Mansouripour, S.M., 2015. Real-time PCR applied to study on plant pathogens: potential applications in diagnosis—A review. Plant Protect. Sci. 51, 177–190.

Mironov, A.A., Weidman, P., Luini, A., 1997. Variations on the intracellular transport theme: maturing cisternae and trafficking tubules. J. Cell Biol. 138, 481–484.

Mirsky, A.E., 1953. The chemistry of heredity. Sci. Am. 188, 47–57.

Mirsky, A.E., 1955. The chemistry of heredity. In: The Physics and Chemistry of Life. Simon and Schuster, New York, pp. 99–117.

Mirsky, A.E., Pollister, A.W., 1946. Chromosin, a desoxyribose nucleoprotein complex of the cell nucleus. J. Gen. Physiol. 30, 121–148.

Mirsky, A.E., Ris, H., 1948. The chemical composition of isolated chromosomes. J. Gen. Physiol. 31, 7–18.

Mirsky, A.E., Ris, H., 1949. Variable and constant components of chromosomes. Nature 163, 666–667.

Mishev, K., Dejonghe, W., Russinova, E., 2013. Small molecules for dissecting endomembrane trafficking: a cross-systems view. Chem. Biol. 20, 475–486.

Mishkind, M.L., Wessler, S.R., Schmidt, G.W., 1985. Functional determinants in transit sequences: import and partial maturation by vascular plant chloroplasts of the ribulose-1,5-bisphosphate carboxylase small subunit of *Chlamydomonas*. J. Cell Biol. 100, 226–234.

Misra, B.B., Assmann, S.M., Chen, S., 2014. Plant single-cell and single-cell-type metabolomics. Trends Plant Sci. 19, 637–646.

Mita, T., Kanbe, T., Tanaka, K., Kuroiwa, T., 1986. A ring structure around the dividing plane of the *Cyanidium caldarium* chloroplast. Protoplasma 130, 211–213.

Mita, T., Katsumi, M., 1986. Gibberellin control of microtubule arrangements in the mesocotyl epidermal cells of the d5 mutant of *Zea mays* L. Plant Cell Physiol. 27, 651–659.

Mita, T., Kuroiwa, T., 1988. Division of plastids by a plastid dividing ring in *Cyanidium caldarium*. Protoplasma (Suppl. 1), 133–152.

Mita, T., Shibaoka, H., 1983. Changes in microtubules in onion leaf sheath cells during bulb development. Plant Cell Physiol. 29, 109–117.

Mita, T., Shibaoka, H., 1984a. Effects of S-3307, an inhibitor of gibberellin biosynthesis on seedling of leaf sheaf cells and on the arrangement of cortical microtubules in onion seedlings. Plant Cell Physiol. 25, 1531–1539.

Mita, T., Shibaoka, H., 1984b. Gibberellin stabilizes microtubules in onion leaf sheath cells. Protoplasma 119, 100–109.

Mitchell, P., 1954a. Transport of phosphate across the osmotic barrier of *Micrococcus pyrogenes*: specificity and kinetics. J. Gen. Microbiol. 11, 73–82.

Mitchell, P., 1954b. Transport of phosphate through an osmotic barrier. Symp. Soc. Exp. Biol. 8, 254–261.

Mitchell, P., 1959. The origin of life and the formation and organizing functions of natural membranes. In: Proceedings of the First International Symposium in The Origin of Life on the Earth. Held at Moscow 19–24 August 1957. Pergamon Press, New York, pp. 437–443.

Mitchell, P., 1960. Introducing the cell surface. Proc. R. Phys. Soc. Edin. 28, 1–2.

Mitchell, P., 1961. Coupling of phosphorylation to electron and hydrogen transfer by a chemi-osmotic type of mechanism. Nature 191, 144–148.

Mitchell, P., 1966. Chemiosmotic coupling in oxidative and photosynthetic phosphorylation. Biol. Rev. 41, 445–502. An expanded version of this article was published in May 1966 by Glynn Research Ltd., Bodmin, Corwall.

Mitchell, P., 1967. Intramitochondrial pH: a statistical question of hydrogen ion concentration in a small element of space-time. Nature 214, 400.

Mitchell, P., 1975. The protonmotive Q cycle: a general formulation. FEBS Lett. 59, 137–139.

Mitchell, P., 1977. Vectorial chemiosmotic processes. Annu. Rev. Biochem. 46, 996–1005.

Mitchell, P., 1980. The culture of imagination. J. R. Inst. Cornwall New Series 8, 173–191.

Mitchell, P., December 8, 1978. David Keilin's Respiratory Chain Concept and its Chemiosmotic Consequences. Nobel Lecture.

Mitchell, H., Houlahan, M., 1946. Adenine-requiring mutants of *Neurospora crassa*. Fed. Proc. 5, 370–381.

Mitchell, P., Moyle, J., 1960. Coupling of metabolism and transport by enzymic translocation of substrates through membranes. Proc. R. Phys. Soc. Edin. 28, 19–27.

Mitchison, T.J., 1989. Polewards microtubule flux in the mitotic spindle: evidence from photoactivation of fluorescence. J. Cell Biol. 109, 637–652.

Mitchison, T.J., Evans, L., Schulze, E., Kirschner, M., 1986. Sites of microtubule assembly and disassembly in the mitotic spindle. Cell 45, 515–527.

Mitchison, T.J., Salmon, E.D., 1992. Poleward kinetochore fiber movement occurs during both metaphase and anaphase-A in newt lung cell mitosis. J. Cell Biol. 119, 569–582.

Mitchison, T.J., Salmon, E.D., 2001. Mitosis: a history of division. Nat. Cell Biol. 3, E17–E21.

Mitchison, T.J., Sawin, K.E., 1990. Tubulin flux in the mitotic spindle: where does it come from, where is it going? Cell Motil. Cytoskelet. 16, 93–98.

Mitsui, T., Christeller, J.T., Hara-Nishimura, I., Akazawa, T., 1984. Possible role of calcium and calmodulin in the biosyntheses and secretion of α-amylase in rice seed scutellar epithelium. Plant Physiol. 75, 21–25.

Mitsui, H., Yamaguchi-Shinozaki, K., Shinozaki, K., Nishikawa, K., Takahashi, H., 1993. Identification of a gene family (Kat) encoding kinesin-like proteins in *Arabidopsis thaliana* and characterization of secondary structure of KatA. Mol. Gen. Genet. 238, 362–368.

Mittag, M., Kiaulehn, S., Johnson, C.H., 2005. The circadian clock in *Chlamydomonas reinhardtii*. What is it for? What is it similar to? Plant Physiol. 137, 399–409.

Miyagishima, S.-Y., 2011. Mechanism of plastid division: from a bacterium to an organelle. Plant Physiol. 155, 1533–1544.

Miyagishima, S., Froehlich, J., Osteryoung, K., 2006. PDV1 and PDV2 mediate recruitment of the dynamin-related protein ARC5 to the plastid division site. Plant Cell 18, 2517–2530.

Miyagishima, S.Y., Nishida, K., Mori, T., Matsuzaki, M., Higashiyama, T., Kuroiwa, H., Kuroiwa, T., 2003. A plant- specific dynamin-related protein forms a ring at the chloroplast division site. Plant Cell 15, 655–665.

Miyagishima, S.Y., Nozaki, H., Nishida, K., Matzuzaki, M., Kuroiwa, T., 2004. Two types of FtsZ proteins in mitochondria and red-lineage chloroplasts: the duplication of FtsZ is implicated in endosymbiosis. J. Mol. Evol. 58, 291–303.

Miyamura, S., Kuroiwa, T., Nagata, T., 1987. Disappearance of plastid and mitochondria nucleoids during the formation of generative cells of higher plants revealed by fluorescence microscopy. Protoplasma 141, 149–159.

Miyata, H., Bowers, B., Korn, E.D., 1989. Plasma membrane association of *Acanthamoeba* myosin I. J. Cell Biol. 109, 1519–1528.

Miyawaki, A., Llopis, J., Heim, R., McCaffery, J.M., Adams, J.A., Ikura, M., Tsien, R.Y., 1997. Fluorescent indicators for Ca^{2+} based on green fluorescent proteins and calmodulin. Nature 388, 882–887.

Miyazaki, S., 2006. Thirty years of calcium signals at fertilization. Semin. Cell Dev. Biol. 17, 233–243.

Mizuno, K., 1993. Microtubule-nucleation sties on nuclei of higher plant cells. Protoplasma 173, 77–85.

Mizuno, K., 1995. A cytoskeletal 50 kDa protein in higher plants that forms intermediate-sized filaments and stabilizes microtubules. Protoplasma 186, 99–112.

Mizuno, A., Nakahori, K., Katou, K., 1993. Acid-induced changes in the in vivo wall-yielding properties of hypocotyl sections of Vigna unguiculata. Physiol. Plant 89, 693–698.

Mizushima, N., Noda, T., Ohsumi, Y., 1999. Apg16p is required for the function of the Apg12p-Apg5p conjugate in the yeast autophagy pathway. EMBO J. 18, 3888–3896.

Mizushima, N., Yoshimori, T., Ohsumi, Y., 2011. The role of Atg proteins in autophagosome formation. Annu. Rev. Cell Dev. Biol. 27, 107–132.

Mizuta, S., 1985. Assembly of cellulose synthesizing complexes on the plasma membrane of *Boodlea coacta*. Plant Cell Physiol. 26, 1443–1453.

Mizutani, I., Gall, J., 1966. Centriole replication II. Sperm formation in the fern, *Marsilea*, and the cycad, *Zamia*. J. Cell Biol. 29, 97–111.

Moberg, C.L., 2012. Entering an Unseen World. Rockefeller University Press, New York.

Moberg, C.L., 2012. The Palade model of combining structure with function. In: Moberg, C.L. (Ed.), Entering and Unseen World. Rockefeller University Press, New York, pp. 156–157.

Mochida, K., Shinozaki, K., 2011. Advances in omics and bioinformatics tools for systems analyses of plant functions. Plant Cell Physiol. 52, 2017–2038.

Mockler, T.C., Ecker, J.R., 2005. Applications of DNA tiling arrays for whole-genome analysis. Genomics 85, 1–15.

Moeendarbary, E., Valon, L., Fritzsche, M., Harris, A.R., Moulding, D.A., Thrasher, A.J., Stride, E., Mahadevan, L., Charras, G.T., 2013. The cytoplasm of living cells behaves as a poroelastic material. Nat. Mater. 12, 253–261.

Mogelsvang, S., Gomez-Ospina, N., Soderholm, J., Glick, B.S., Staehelin, L.A., 2003. Tomographic evidence for continuous turnover of Golgi cisternae in *Pichia pastoris*. Mol. Biol. Cell 14, 2277–2291.

Mogensen, H.L., 1996. The hows and whys of cytoplasmic inheritance in seed plants. Am. J. Bot. 83, 383–404.

Moghe, G.D., Kruse, L.H., 2018. The study of plant specialized metabolism: challenges andprospects in the genomics era. Am. J. Bot. 105, 1–4.

Mogilner, A., Craig, E., 2010. Towards a quantitative understanding of mitotic spindle assembly and mechanics. J. Cell Sci. 123, 3435–3445.

Mohanty, A., Luo, A., DeBlasio, S., Ling, X., Yang, Y., Tuthill, D.E., Williams, K.E., Hill, D., Zadrozny, T., Chan, A., Sylvester, A.W., Jackson, D., 2009. Advancing cell biology and functional genomics in maize using fluorescent protein-tagged lines. Plant Physiol. 149, 551–557.

Mohl, H. von, 1852. Principles of the Anatomy and Physiology of the Vegetable Cell. Translated by A. Henfrey. John van Voorst, London.

Mohnen, D., Tierney, M.L., 2011. Plants get Hyp to O-glycosylation. Science 332, 1393–1394.

Mohr, H., 1972. Lectures on Photomorphogenesis. Springer-Verlag, Berlin.

Mohri, H., 1968. Amino-acid composition of "tubulin" constituting microtubules of sperm flagella. Nature 217, 1053–1054.

Mohri, H., Hosoya, N., 1988. Two decades since the naming of tubulin—The multi-facets of tubulin—. Zool. Sci. 5, 1165–1185 (in Japanese).

Mohri, H., Inaba, K., Ishijima, S., Baba, S.A., 2012. Tubulin-dynein system in flagellar and ciliary movement. Proc. Jpn. Acad. Ser. B 88, 397–415.

Mohri, H., Murakami, S., Maruyama, K., 1967. On the protein constituting 9+2 fibers of th etil of sea urchin spermatozoa. J. Biochem. 61, 518–519.

Mohri, H., Shimomura, M., 1973. Comparison of tubulin and actin. J. Biochem. 74, 209–220.

Moir, R.D., Goldman, R.D., 1993. Lamin dynamics. Curr. Opin. Cell Biol. 5, 408–411.

Mojica, F.J.M., Díez-Villaseñor, C., García-Martínez, J., Soria, E., 2005. Intervening sequences of regularly spaced prokaryotic repeats derive from foreign genetic elements. J. Mol. Evol. 60, 174–182.

Molchan, T.M., Valster, A.H., Hepler, P.K., 2002. Actomyosin promotes cell plate alignment and late lateral expansion in *Tradescantia* stamen hair cells. Planta 214, 683–693.

Molé-Bajer, J., 1958. Cine-micrographic analysis of c-mitosis in endosperm. Chromosoma 9, 332–358.

Molé-Bajer, J., Bajer, A.S., 1988. Relation of F-actin organization to microtubules in drug treated Haemanthus mitosis. Protoplasma (Suppl. 1), 91–112.

Molè-Bajer, J., Bajer, A.S., Inoué, S., 1988. Three dimensional localization and redistribution of F-actin in higher plant mitosis and cell plate formation. Cell Motil. Cytoskelet. 10, 217–228.

Molé-Bajer, J., Bajer, A.S., Zenkowski, R.P., Balczon, R.D., Brinkley, B.R., 1990. Autoantibodies from a patient with scleroderma CREST recognize kinetochores of the higher plant Haemanthus. Proc. Natl. Acad. Sci. U.S.A. 87, 3599–3603.

[Jean-Baptiste Poquelin] Molfère, 1673. Le malade Imaginaire. The Would-Be Invalid, 1950. Appleton-Century-Crofts, New York. Translated and Edited by Morris Bishop.

Molfère, 1957. Le Bourgeois Gentilhomme. The Would-Be Gentleman [Jean-Baptiste Poquelin]. [1670]. The Modern Library, New York. Translated by Morris Bishop.

Molisch, H., 1916. Pflanzenphysiologie als Theorie der Gärtnerei. Gustav Fischer, Jena.

Moll, B.A., Jones, R.L., 1982. α-amylase secretion by single barley aleurone layers. Plant Physiol. 70, 1149–1155.

Moll, T., Tebb, G., Surana, U., Robitsch, H., Nasmyth, K., 1991. The role of phosphorylation and the cdc28 protein kinase in cell cycle regulated nuclear import of the *S. cerevisiae* transcription factor SW15. Cell 66, 743–758.

Mollenhauer, H.H., 1965. An intercisternal structure in the Golgi apparatus. J. Cell Biol. 24, 504–511.

Mollenhauer, H.H., Morré, D.J., 1966. Golgi apparatus and plant secretion. Ann. Rev. Plant Physiol. 17, 27–46.

Mollenhauer, H.H., Morré, D.J., 1976a. Cytochalasin B, but not colchicine, inhibits migration of secretory vesicles in root tips of maize. Protoplasma 87, 39–48.

Mollenhauer, H.H., Morré, D.J., 1976b. Transition elements between endoplasmic reticulum and Golgi apparatus in plant cells. Cytobiologie 13, 297–306.

Mollenhauer, H.H., Morré, D.J., 1987. Some unusual staining properties of tannic acid in plants. Histochemistry 88, 17–22.

Mollenhauer, H.H., Morré, D.J., Droleskey, R., 1983. Monensin affects the trans half of *Euglena* dictyosomes. Protoplasma 114, 119–124.

Mollenhauer, H.H., Morré, D.J., Griffing, L.R., 1991. Post Golgi apparatus structures and membrane removal in plants. Protoplasma 162, 55–60.

Møller, I.M., 2001. Plant mitochondria and oxidative stress: electron transport, NADPH turnover, and metabolism of reactive oxygen species. Annu. Rev. Plant Physiol. Plant Mol. Biol. 52, 561–591.

Møller, I.M., Lin, W., 1986. Membrane-bound NAD(P)H dehydrogenases in higher plant cells. Ann. Rev. Plant Physiol. 37, 309–334.

Mommaerts, W.F.H.M., 1950a. A consideration of experimental facts pertaining to the primary reaction in muscular activity. Biochim. Biophys. Acta 4, 50–57.

Mommaerts, W.F.H.M., 1950b. Muscular Contraction. A Topic in Molecular Physiology. Interscience Publishers, New York.

Mommaerts, W.F.H.M., Seraidarian, K., 1947. A study of the adenosine triphosphatase activity of myosin and actomyosin. J. Gen. Physiol. 30, 401–422.

Mondal, T.K., Rawal, H.C., Gaikwad, K., Sharma, T.R., Singh, N.K., 2017. First de novo genome sequence of *Oryza coarctata*, the only halophytic species in the genus *Oryza*. F1000 Research 6, 1750.

Monéger, F., Smart, C.J., Leaver, C.J., 1994. Nuclear restoration of cytoplasmic male sterility in sunflower is associated with the tissue-specific regulation of a novel mitochondrial gene. EMBO J. 13, 8–17.

Money, N.P., 1997. Wishful thinking of turgor revisited: the mechanics of fungal growth. Fungal Genet. Biol. 21, 173–187.

Money, N.P., Harold, F.M., 1992. Extension growth of the water mold *Achlya*: interplay of turgor and wall strength. Proc. Natl. Acad. Sci. U.S.A. 89, 4245–4249.

Money, N.P., Hill, T.W., 1997. Correlation between endoglucanase secretion and cell wall strength in oomycete hyphae: implications for growth and morphogenesis. Mycologia 89, 777–785.

Mongrand, S., Morel, J., Laroche, J., Claverol, S., Carde, J.P., Hartmann, M.A., Bonneu, M., Simon-Plas, F., Lessire, R., Bessoule, J.J., 2004. Lipid rafts in higher plant cells: purification and characterization of Triton X-100-insoluble microdomains from tobacco plasma membrane. J. Biol. Chem. 279, 36277–36286.

Monod, J., 1965. From Enzymatic Adaptation to Allosteric Transitions. Nobel Lecture. December 11, 1963.

Monshausen, G.B., Messerli, M.A., Gliroy, S., 2008. Imaging of the yellow cameleon 3.6 indicator reveals that elevations in cytosolic Ca^{2+} follow oscillating increases in growth in root hairs of *Arabidopsis*. Plant Physiol. 147, 1690–1698.

Montecchia, M.S., Tosi, M., Soria, M.A., Vogrig, J.A., Sydorenko, O., Correa, O.S., 2015. Pyrosequencing reveals changes in soil bacterial communities after conversion of Yungas forests to Agriculture. PLoS One 10, e0119426.

Montes-Rodriguez, A., Kost, B., 2017. Direct comparison of the performance of commonly employed in vivo F-actin markers (Lifeact-YFP, YFP-mTn and YFP-FABD2) in tobacco pollen tubes. Front. Plant Sci. 8, 1349.

Montezinos, D., Brown Jr., R.M., 1979. Cell wall biogenesis in Oocystis: experimental alteration of microfibril assembly and orientation. Cytobios 23, 119–139.

Montgomery, W.L., Pollak, P.E., 1988. *Epulopiscium fishelsoni* N. G., N. Sp., a protist of uncertain taxonomic affinities from the gut of an herbivorous reef fish. J. Protozool. 35, 565–569.

Mooney, B.P., Miernyk, J.A., Randall, D.D., 2002. The complex fate of α-ketoacids. Annu. Rev. Plant Biol. 53, 357–375.

Moor, H., Mühlethaler, K., 1963. Fine-structure in frozen-etched yeast cells. J. Cell Biol. 17, 60628.

Moor, H., Mühlethaler, K., Waldner, H., Frey-Wyssling, A., 1961. A new freezing ultramicrotome. J. Biophys. Biochem. Cytol. 10, 1–14.

Moore, G.E., 1965. Cramming more components onto integrated circuits. Electronics 38 (8), 114–117.

Moore, T.S., 1982. Phospholipid biosynthesis. Ann. Rev. Plant Physiol. 33, 235–259.

Moore, R., 1985. A morphometric analysis of the redistribution of organelles in columella cells in primary roots of normal seedlings and agravitropic mutants of Hordeum vulgare. J. Exp. Bot. 36, 1275–1286.

Moore, T.S., 1987. Phosphoglyceride synthesis in endoplasmic reticulum. Methods Enzymol. 148, 585–596.

Moore, J.E.S., Arnold, A.L., 1905. On the existence of permanent forms among the chromosomes of the first maiotic division in certain animals. Proc. R. Soc. B 77, 563–570.

Moore, J.E.S., Embleton, A.L., 1905. On the synapsis in amphibia. Proc. R. Soc. B 77, 555–562.

Moore, R.T., McAlean, J.H., 1961. Fine structure of mycota. V. Lomasomes—previously uncharacterized hyphal structures. Mycologia 53, 194–200.

Moore, A.L., Proudlove, M.O., 1988. Mitochondria and submitochondrial particles. In: Hall, J.L., Moore, A.L. (Eds.), Isolation of Membranes and Organelles from Plant Cells. Academic Press, London, pp. 153–184.

Moore, A.L., Rich, P.R., 1985. Organization of the respiratory chain and oxidative phosphorylation. In: Douce, R., Day, D.A. (Eds.), Higher Plant Cell Respiration. Springer Verlag, Berlin, pp. 134–171.

Moore, P.J., Sivords, K.M.M., Lynch, M.A., Staehelin, L.A., 1991. Spatial reorganization of the assembly pathways of glycoproteins and complex polysaccharides in the Golgi apparatus of plants. J. Cell Biol. 112, 589–602.

Moore, P.J., Staehelin, L.A., 1988. Immunogold localization of the cellwallmatrix polysaccharides rhamnogalacturonan I and xyloglucan during cell expansion and cytokinesis in *Trifolium pratense* L. Implications for secretory pathways. Planta 174, 433–445.

Moore, J.E.S., Walker, C.E., 1905. The Maiotic Process in Masmmalia. Thompson Yates and Johnston Laboratories Report. University of Liverpool, pp. 75–88.

Moore, A.L., Wood, C.K., Watts, F.Z., 1994. Protein import into plant mitochondria. Ann. Rev. Plant Physiol. Plant Mol. Biol. 45, 545–575.

Mooseker, M.S., Foth, B.J., 2008. The structural and functional diversity of the myosin family of actin-based molecular motors. In: Coluccio, L.M. (Ed.), Myosins. Springer, Berlin, pp. 1–34.

Moran, N., 2007. Osmoregulation of leaf motor cells. FEBS Lett. 581, 2337–2347.

Moran, N., Ehrenstein, G., Iwasa, K., Mischke, C., Bare, C., Satter, R.L., 1988. Potassium channels in motor cells of *Samanea saman*: a patch-clamp study. Plant Physiol. 88, 643–648.

Morange, M., 2008. The death of molecular biology? Hist. Philos. Life Sci. 30, 31–42.

Moras, D., 1992. Structural and functional relationships between aminoacyl-tRNA synthetases. Trends Biochem. Sci. 17, 159–164.

Moreau, P., Hartmann, M.-A., Perret, A.-M., Sturbois-Balcerzak, B., Cassagne, C., 1998. Transport of sterols to the plasma membrane of leek seedlings. Plant Physiol. 117, 931–937.

Morejohn, L.C., Bureau, T.E., Mole-Bajer, J., Bajer, A.S., Fosket, D.E., 1987. Oryzalin, a dinitroaniline herbicide, binds to plant tubulin and inhibits microtubule polymerization *in vitro*. Planta 172, 252–264.

Morejohn, L.C., Fosket, D.E., 1982. Higher plant tubulin identified by self-assembly into microtubules in vitro. Nature 297, 426–428.

Morejohn, L.C., Fosket, D.E., 1984a. Inhibition of plant microtubule polymerization in vitro by the phosphoric amide herbicide amiprophos-methyl. Science 224, 874–876.

Morejohn, L.C., Fosket, D.E., 1984b. Taxol-induced rose microtubule polymerization *in vitro* and its inhibition by colchicine. J. Cell Biol. 99, 141–147.

Moreno Díaz de la Espina, S., 1995. Nuclear matrix isolated from plant cells. Int. Rev. Cytol. 162B, 75–139.

Morgan, T.H., 1901. Regeneration. MacMillan, New York.

Morgan, T.H., 1909. What are "factors" in Mendelian explanations? Am. Breed. Assoc. Rep. 5, 365–368.

Morgan, T.H., 1911. Random segregation versus coupling in Mendelian inheritance. Science 34, 384.

Morgan, C.L., 1923. Emergent Evolution. Williams and Norgate, London.

Morgan, T.H., 1924. Mendelian heredity in relation to cytology. In: Cowdry, E.V. (Ed.), General Cytology. University of Chicago Press, Chicago, pp. 691–734.

Morgan, C.L., 1926. Life, Mind, and Spirit. Williams and Norgate, London.

Morgan, C.L., 1933. The Emergence of Novelty. Williams and Norgate, London.

Morgan, T.H., June 4, 1934. The Relation of Genetics to Physiology and Medicine. Nobel Lecture.

Morgan, T.H., Sturtevant, A.H., Muller, H.J., Bridges, C.B., 1915. The Mechanism of Mendelian Heredity. H. Holt and Co., New York.

Mori, T., Kuroiwa, H., Takahara, M., Miyagishima, S.-Y., Kuroiwa, T., 2001. Visualization of an FtsZ ring in chloroplasts of *Lilium longiflorum* leaves. Plant Cell Physiol. 42, 555–559.

Mori, T., Mchaourab, H., Johnson, C.H., 2015. Circadian clocks: unexpected biochemical cogs. Curr. Biol. 25, R827–R844.

Mori, I.C., Muto, S., 1997. Abscisic acid activates a 48-kilodalton protein kinase in guard cell protoplasts. Plant Physiol. 113, 295–303.

Mori, I.C., Uozumi, N., Muto, S., 2000. Phosphorylation of the inward-rectifying potassium channel KAT1 by ABR kinase in *Vicia* guard cells. Plant Cell Physiol. 41, 850–856.

Morimatsu, M., Hasegawa, S., Higashi-Fujimi, S., 2002. Protein phosphorylation regulates actomyosin-driven vesicle movement in cell extracts isolated from green algae, *Chara corallina*. Cell Motil. Cytoskelet. 53, 66–76.

Morimatsu, M., Nakamura, A., Sumiyoshi, H., Sakaba, N., Taniguchi, H., Kohama, K., Higashi-Fujime, S., 2000. The molecular structure of the fastest myosin from green algae, *Chara*. Biochem. Biophys. Res. Commun. 270, 147–152.

Morin, J.G., Hastings, J.W., 1971. Energy transfer in a bioluminescent system. J. Cell. Physiol. 77, 313–318.

Morise, H., Shimomura, O., Johnson, F.H., Winant, J., 1974. Intermolecular energy transfer in the bioluminescent system of *Aequorea*. Biochemistry 13, 2656–2662.

Morita, M.T., Tasaka, M., 2004. Gravity sensing and signaling. Curr. Opin. Plant Biol. 7, 712–718.

Moriwaki, T., Miyazawa, Y., Fujii, N., Takahashi, H., 2014. GNOM regulates root hydrotropism and phototropism independently of PIN-mediated auxin transport. Plant Sci. 215–216, 141–149.

Moriyama, T., Sato, N., 2014. Enzymes involved in organellar DNA replication in photosynthetic eukaryotes. Front. Plant Sci. 5, 480.

Moriyasu, Y., 1995. Examination of the contribution of vacuolar proteases to intracellular protein degradation in *Chara corallina*. Plant Physiol. 109, 1309–1315.

Moriyasu, Y., Malek, L., 2004. Purification and characterization of the 20S proteasome from the alga *Chara corallina* (Charophyceae). J. Phycol. 40, 333–340.

Moriyasu, Y., Ohsumi, Y., 1996. Autophagy in tobacco-suspension cultured cells in response to sucrose starvation. Plant Physiol. 111, 1233–1241.

Moriyasu, Y., Sakano, K., Tazawa, M., 1987. Vacuolar/extravacuolar distribution of aminopeptidases in giant alga *Chara australis* and partial purification of one such enzyme. Plant Physiol. 84, 720–725.

Moriyasu, Y., Shimmen, T., Tazawa, M., 1984. Vacuolar pH regulation in *Chara australis*. Cell Struct. Funct. 9, 225–234.

Moriyasu, Y., Tazawa, M., 1986. Distribution of several proteases inside and outside the central vacuole of *Chara australis*. Cell Struct. Funct. 11, 37–42.

Moriyasu, Y., Tazawa, M., 1987. Calcium-activated protease in the giant alga *Chara australis*. Protoplasma 140, 72–74.

Moriyasu, Y., Tazawa, M., 1988. Degradation of proteins artificially introduced into vacuoles of *Chara australis*. Plant Physiol. 88, 1092–1096.

Moriyasu, Y., Wayne, R., 2004. Presence of a novel calcium- activated protease in Chara corallina. Eur. J. Phycol. 39, 57–66.

Morohashi, K., Grotewold, E., 2009. A systems approach reveals regulatory circuitry for Arabidopsis trichome initiation by the GL3 and GL1 selectors. PLoS Genet. 5, e1000396.

Morowitz, H.J., 1992. Beginnings of Cellular Life. Metabolism Recapitulates Biogenesis. Yale University Press, New Haven, CT.

Morré, D.J., 1987. The Golgi apparatus. Int. Rev. Cytol. 17, 211–253.

Morré, D.J., Boss, W.J., Grimes, H., Mollenhauer, H.H., 1983. Kinetics of Golgi apparatus membrane flux following monensin treatment of embryogenic carrot cells. Eur. J. Cell Biol. 30, 25–32.

Morré, D.J., Keenan, T.W., 1994. Golgi apparatus buds—vesicles or coated ends of tubules? Protoplasma 179, 1–4.

Morré, D.J., Keenan, T.W., 1997. Membrane flow revisited. BioScience 47, 489–498.

Morré, D.J., Mollenhauer, H.H., 1976. Interactions among cytoplasm, endomembranes, and the cell surface. In: Stocking, C.R., Heber, U. (Eds.), Transport in Plants. III. Intracellular Interactions and Transport Processes. Springer-Verlag, Berlin, pp. 290–344.

Morré, D.J., Nowack, D.D., Paulik, M., Brightman, A.O., Thornborough, K., Yim, J., Auderset, G., 1989a. Transitional endoplasmic reticulum membranes and vesicles isolated from animals and plants. Homologous and heterologous cell-free membrane transfer to Golgi apparatus. Protoplasma 153, 1–13.

Morré, D.J., Ovtracht, L., 1977. Dynamics of the Golgi apparatus: membrane differentiation and membrane flow. Int. Rev. Cytol. (Suppl. 5), 61–188.

Morré, D.J., Pfaffman, H., Drøbak, B., Wilkinson, F.E., Hartmann, E., 1989b. Diacylglycerol levels unchanged during auxin- stimulated growth of excised hypocotyl segments of soybean. Plant Physiol. 90, 275–279.

Morrell, P.L., Buckler, E.S., Ross-Ibarra, J., 2011. Crop genomics: advances and applications. Nat. Rev. Genet. 13, 85–96.

Morris, C.E., 1990. Mechanosensitive ion channels. J. Membr. Biol. 113, 93–107.

Morris, B.M., Gow, N.A.R., 1993. Mechanism of electrotaxis of zoospores of phytopathogenic fungi. Phytopathology 8, 877–882.

Mörschel, E., Staehelin, L.A., 1983. Reconstitution of cytochromeb6-f and CF0/CF1 ATP synthase complexes into phospholipid and galactolipid liposomes. J. Cell Biol. 97, 301–310.

Morse, R.H., 1992. Transcribed chromatin. Trends Biochem. Sci. 17, 23–26.

Morse, M.J., Crain, R.C., Coté, G.G., Satter, R.L., 1989. Light stimulated inositol phospholipid turnover in *Samanea saman* pulvini. Increased levels of diacylglycerol. Plant Physiol. 89, 724–727.

Morse, M.J., Crain, R.C., Coté, G.G., Satter, R.L., 1990. Light- signal transduction via accelerated inositol phospholipid turn- over in Samanea pulvini. In: Morré, D.J., Boss, W.F., Loewus, F.A. (Eds.), Inositol Metabolism in Plants. Wiley-Liss, New York, pp. 201–215.

Morse, M.J., Crain, R.C., Satter, R.L., 1987a. Light-stimulated inositol phospholipid turnover in Samanea saman pulvini. Proc. Natl. Acad. Sci. U.S.A. 84, 7075–7078.

Morse, M.J., Crain, R.C., Satter, R.L., 1987b. Phosphatidyl inositol cycle metabolites in Samanea saman pulvini. Plant Physiol. 83, 640–644.

Mort, A., Wu, X., 2011. Capillary electrophoresis with detection by laser-induced fluorescence. Methods Mol. Biol. 715, 93–102.

Moscatelli, A., 2008. Endocytic pathways in pollen tube. Plant Signal. Behav. 3, 325–327.

Moscatelli, A., Cai, G., Ciampolini, F., Cresti, M., 1998. Dynein heavy chain-related polypeptides are associated with organelles in pollen tubes of *Nicotiana tabcacum*. Sex. Plant Reprod. 11, 31–40.

Moscatelli, A., Ciampolini, F., Rodighiero, S., Onelli, E., Cresti, M., Santo, N., Idilli, A., 2007. Distinct endocytic pathways identified in tobacco pollen tubes using charged nanogold. J. Cell Sci. 120, 3804–3819.

Moscatelli, A., Del Casino, C., Lozzi, L., Cai, G., Scali, M., Tiezzi, A., Cresti, M., 1995. High molecular weight polypeptides related to dynein heavy chains in *Nicotiana tabacum* pollen tubes. J. Cell Sci. 108, 1117–1125.

Moscatelli, A., Idilli, A.I., 2009. Pollen tube growth: a delicate equilibrium between secretory and endocytic pathways. J. Integr. Plant Biol. 51, 727–739.

Moscatelli, A., Idilli, A.I., Rodighero, S., Caccianga, M., 2012. Inhibition of actin polymerization by low concentration Latrunculin B affects endocytosis and alters exocytosis in shank and tip of tobacco pollen tubes. Plant Biol. 14, 770–782.

Moseyko, N., Feldman, L.J., 2001. Expression of pH-sensitive green fluorescent protein in *Arabidopsis thaliana*. Plant Cell Environ. 24, 557–563.

Moshelion, M., Becker, D., Czempinski, K., Mueller-Roeber, B., Hedrich, R., Attali, B., Moran, N., 2002a. Diurnal and circadian regulation of putative potassium channels in a leaf-moving organ. Plant Physiol. 128, 634–642.

Moshelion, M., Becker, D., Hedrich, R., Biela, A., Otto, B., Levi, H., Moran, N., Kaldenhoff, R., 2002b. PIP aquaporins in the motor cells of *Samanea saman*. Diurnal and circadian regulation. Plant Cell 14, 727–739.

Moshelion, M., Moran, N., 2000. K^+-efflux in extensor and flexor cells of Samanea saman are not identical. Effect sof cytosolic Ca^{2+}. Plant Physiol. 124, 911–919.

Moshkov, I.E., Novikova, G.V., 2008. Superfamily of plant monomeric GTP-binding proteins:2. Rab proteins are the regulators of vesicles trafficking and plant responses to stresses. Russ. J. Plant Physiol. 55, 119–129.

Moskowitz, A.H., Hrazdina, G., 1981. Vacuolar contents of fruit subepidermal cells from Vitis species. Plant Physiol. 68, 686–692.

Mösli Waldhauser, S.S., Baumann, T.W., 1996. Compartmentation of caffeine and related purine alkaloids depends exclusively on the physical chemistry of their vacuolar complex formation with chlorogenic acids. Phytochemistry 42, 985–996.

Moss, R.W., 1988. Free Radical. Albert Szent-Gyorgyi and the Battle over Vitamin C. Paragon House, New York.

Mottier, D.M., 1903. The behavior of the chromosomes in the spore mother cells of higher plants and the homology of the pollen and embryo sac mother cells. Bot. Gaz. 35, 250–282.

Moulia, B., Fournier, M., 2009. The power and control of gravitropic movements in plants: a Biomechanical and systems biology view. J. Exp. Bot. 60, 461–486.

Mousdale, D.M., 2008. Biofuels: Biotechnology, Chemistry, and Sustainable Development. CRC Press, Boca Raton, FL.

Mowbrey, K., Dacks, J.B., 2009. Evolution and diversity of the Gogi body. FEBS Lett. 583, 3738–3745.

Mower, J.P., Stefanovic, S., Young, G.J., Palmer, J.D., 2004. Gene transfer from parasitic to host plants. Nature 432, 165–166.

Moyer, A.E., 1977. Robert Hooke's ambiguous presentation of "Hooke's Law". Isis 68, 266–275.

Moynihan, M.R., Ordentlich, A., Raskin, I., 1995. Chilling-induced heat evolution in plants. Plant Physiol. 108, 995–999.

Mu, J.H., Chua, N.-H., Ross, E.M., 1997. Expression of human muscarinic cholinergic receptors in tobacco. Plant Mol. Biol. 34, 357–362.

Mueckler, M., Lodish, H.F., 1986. Post-translational insertion of the fragment of the glucose transporter into microsome requires phosphoanhydride bond cleavage. Nature 322, 549–552.

Mueller Jr., S.C., Brown, R.M., 1980. Evidence for an intramembrane component associated with a cellulose microfibril-synthesizing complex in higher plants. J. Cell Biol. 84, 315–346.

Mueller Jr., S.C., Brown, R.M., Scott, T.K., 1976. Cellulosic microfibrils: nasent stages of synthesis in higher plant cell. Science 194, 949–951.

Muggeridge, M., 2006. Chronicles of Wasted Time. Regent College, Vancouver, BC.

Mühlethaler, K., 1950. Electron microscopy of developing plant cell walls. Biochim. Biophys. Acta 5, 1–9.

Mühlethaler, K., 1955. The structure of chloroplast. Int. Rev. Cytol. 4, 197–220.

Muiño, J.M., Angenent, G.C., Kaufmann, K., 2011. Visualizing and characterizing in vivo DNA-binding events and direct target genes of plant transcription factors. Methods Mol. Biol. 754, 293–305.

Muir, J., 1911. My First Summer in the Sierra, 1987. Penguin Books, New York.

Mukherjee, A., Sokunbi, A.O., Grove, A., 2008. DNA protection by histone-like protein HU from the hyperthermophilic eubacterium *Thermotoga maritima*. Nucleic Acids Res. 36, 3956–3968.

Mulder, G.J., 1838. Sur la compostion de quelques substances animals. Bulletin des Sciences Physiques et Naturelles en Néerlande 104–119.

Mulder, G.J., 1849. The Chemistry of Vegetable & Animal Physiology. William Blackwood and Sons, Edinburgh.

Mullen, R.T., 2002. Targeting and import of matrix proteins into peroxisomes. In: Baker, A., Smith, I. (Eds.), Plant Peroxisomes: Biochemistry, Cell Biology and Biotechnological Applications. Kluwer Academic Publishers.

Mullen, R.T., Flynn, C.R., Trelease, R.N., 2001. How are peroxisomes formed? The role of the endoplasmic reticulum and peroxins. Trends Plant Sci. 6, 256–261.

Mullen, R.T., Lisenbee, C.S., Miernyk, J.A., Trelease, R.N., 1999. Peroxisomal membrane ascorbate peroxidase is sorted to a membranous network that resembles a subdomain of the endoplasmic reticulum. Plant Cell 11, 2167–2185.

Mullen, R.T., Trelease, R.N., 2000. The sorting signals for peroxisomal membrane-bound ascorbate peroxidase are within its C-terminal tail. J. Biol. Chem. 275, 16337–16344.

Muller, H.J., 1973a. Man's Future Birthright. Essays on Science and Humanism. State University of New York Press, Albany, New York.

Muller, H.J., 1973b. The Modern Concept of Nature. Essays on Theoretical Biology and Evolution. State University of New York Press, Albany, New York.

Müller, I., 2007. A History of Thermodynamics. The Doctrine of Energy and Entropy. Springer, Berlin.

Muller, H.J., December 12, 1946. The Production of Mutations. Nobel Lecture.

Müller, M.L., Irkens-Kiesecker, U., Kramer, D., Taiz, L., 1997. Purification and reconstitution of the vacuolar H^+-ATPases from lemon fruits and epicotyls. J. Biol. Chem. 272, 12762–12770.

Müller, M.L., Irkens-Kiesecker, U., Rubinstein, B., Taiz, L., 1996. On the mechanism of hyperacidification in lemon: comparison of the vacuolar H^+-ATPase activities of fruits and epicotyls. J. Biol. Chem. 271, 1916–1924.

Müller, M.L., Jensen, M., Taiz, L., 1999. The vacuolar H^+-ATPase of lemon fruits is regulated by variable H^+/ATP coupling and slip. J. Biol. Chem. 274, 10706–10716.

Müller, S., Jürgens, G., 2016. Plant cytokinesis—No ring, no constriction of the portioning membrane. Semin. Cell Dev. Biol. 53, 10–18.

Müller, J., Mettbach, U., Menzel, D., Samaj, J., 2007. Molecular dissection of endosomal compartments in plants. Plant Physiol. 145, 293–304.

Müller, M.L., Taiz, L., 2002. Regulation of the lemon-fruit V-ATPase by variable stoichiometry and organic acids. J. Membr. Biol. 185, 209–220.

Mullineaux, C.W., Nenninger, A., Ray, N., Robinson, C., 2006. Diffusion of green fluorescent protein in three cell environments in *Escherichia coli*. J. Bacteriol. 188, 3442–3448.

Mullis, K.B., 1990. The unusual origin of the polymerase chain reaction. Sci. Am. 262 (4), 56–65.

Mullis, K.B., December 8, 1993. The Polymerase Chain Reaction. Nobel Lecture.

Mullis, K.B., Faloona, F.A., 1987. Specific synthesis of DNA in vitro via a polymerase-catalyzed chain reaction. Methods Enzymol. 155, 335–350.

Mummert, H., Hansen, U.-P., Gradmann, D., 1981. Current voltage curve of electrogenic Cl^- pump predicts voltage-dependent Cl efflux in *Acetabularia*. J. Membr. Biol. 62, 139–148.

Munemasa, S., Hauser, F., Park, J., Waadt, R., Brant, B., Schroeder, J.I., 2016. Mechanisms of abscisic acid-mediated control of stomatal aperture. Curr. Opin. Plant Biol. 28, 154–162.

Munir, S., Khan, M.R.G., Song, J., Munir, S., Zhang, Y., Ye, Z., Wang, T., 2016. Genome-wide identification, characterization and expression analysis of calmodulin-like (CML) proteins in tomato (*Solanum lycopersicum*). Plant Physiol. Biochem. 102, 167–179.

Munitz, M.K. (Ed.), 1957. Theories of the Universe. The Free Press, Glencoe, IL.

Munnik, T., 2001. Phosphatidic acid – an emerging plant lipid second messenger. Trends Plant Sci. 6, 227–233.

Munnik, T., Meijer, H.J.G., 2001. Osmotic stress activates distinct lipid- and MAPK signalling pathways in plants. FEBS Lett. 498, 172–178.

Munnik, T., Nielsen, E., 2011. Green light for polyphosphoinositide signals in plants. Curr. Opin. Plant Biol. 14, 489–497.

Munnik, T., Vermeer, J.E.M., 2010. Osmotic stress-induced phosphoinositide and inositolphosphate signaling in plants. Plant Cell Environ. 33, 655–669.

Muntz, K., 2007. Protein dynamics and proteolysis in plant vacuoles. J. Exp. Bot. 58, 2391–2407.

Murasugi, A., Wada, C., Hayashi, Y., 1981. Cadmium-binding peptide induced in fission yeast, *Schizosaccharomyces pombe*. J. Biochem. 90, 1561–1564.

Murata, N., 1969. Control of excitation transfer in photosynthesis. I. Light-induced change of chlorophyll a fluorescence in *Porphyridium cruentum*. Biochim. Biophys. Acta 172, 242–251.

Murata, T., Hasebe, M., 2011. Microtubule nucleation and organization in plant cells. The Plant Cytoskeleton. Adv. Plant Biol. 2, 81–94.

Murata, T., Kadota, A., Hogetsu, T., Wada, M., 1987. Circular arrangement of cortical microtubules around the subapical part of a tipgrowing fern protonema. Protoplasma 141, 135–138.

Murata, Y., Mori, I.C., Munemasa, S., 2015. Diverse stomatal signaling and the signal integration mechanism. Annu. Rev. Plant Biol. 66, 369–392.

Murata, T., Sano, T., Sasabe, M., Nonaka, S., Higashiyama, T., Hasezawa, S., Machida, Y., Hasabe, M., 2013. Mechanism of microtubule array expansion in the cytokinetic phragmoplast. Nat. Commun. 4, 1967.

Murata, T., Sonobe, S., Baskin, T.I., Hyodo, S., Hasezawa, S., Nagata, T., Horio, T., Hasebe, M., 2005. Microtubule- dependent microtubule nucleation based on recruitment of γ-tubulin in higher plants. Nat. Cell Biol. 7, 961–968.

Murata, T., Sugai, M., 2000. Protoregulation of asymmetric cell division followed by rhizoid development in the fern *Ceratopteris prothalli*. Plant Cell Physiol. 41, 1313–1320.

Murata, T., Wada, M., 1989a. Effects of colchicine and amiprophosmethyl on microfibril arrangement and cell shape in *Adiantum* protonemal cells. Protoplasma 151, 81–87.

Murata, T., Wada, M., 1989b. Organization of cortical microtubules and microfilament deposition in response to blue-light- induced apical swelling in a tip-growing *Adiantum* protonema cell. Planta 178, 334–341.

Murata, T., Wada, M., 1991. Effects of centrifugation on preprophase-band formation in *Adiantum* protonemata. Planta 83, 391–398.

Murata, T., Wada, M., 1997. Formation of a phragmosome-like structure in centrifuged protonemal cells of *Adiantum capillus- veneris*. L. Planta 201, 273–280.

Murcha, M.W., Whelan, J., 2015. Isolation of intact mitochondria from the model plant species *Arabidopsis thaliana* and *Oryza thaliana*. Methods Mol. Biol. 1305, 1–12.

Murphy, D.J., 1986a. Structural properties and molecular organization of the acyl lipids of photosynthetic membranes. Encyl. Plant Physiol. Photosynth. III. 19, 713–725.

Murphy, D.J., 1986b. The molecular organization of the photosynthetic membrane in higher plants. Biochim. Biophys. Acta 864, 33–94.

Murphy, D.J., Vance, J., 1999. Mechanisms of lipid-body formation. Trends Biochem. Sci. 24, 109–115.

Murray, A.W., 1992. Creative blocks: cell cycle checkpoints and feedback controls. Nature 359, 599–604.

Murray, M.G., Kennard, W.C., 1984. Altered chromatin conformation of the higher plant gene phaseolin. Biochemistry 23, 4225–4232.

Murthy, P.P.N., Renders, J.M., Keranen, L.M., 1989. Phosphoinositides in barley aleurone layers and gibberellic acid-induced changes in metabolism. Plant Physiol. 91, 1266–1269.

Muscatello, U., 2008. A period of convergence in the studies on muscle contraction and relaxation: the Ebashi's contribution. Biochem. Biophys. Res. Commun. 369, 52–56.

Mustacich, R.V., Ware, B.R., 1974. Observation of protoplasmic streaming by laser-light scattering. Phys. Rev. Lett. 33, 617–620.

Mustacich, R.V., Ware, B.R., 1976. A study of protoplasmic streaming in *Nitella* by laser Doppler spectroscopy. Biophys. J. 16, 373–388.

Mustacich, R.V., Ware, B.R., 1977a. Velocity distributions of the streaming protoplasm in *Nitella flexilis*. Biophys. J. 17, 229–241.

Mustacich, R.V., Ware, B.R., 1977b. A study of protoplasmic streaming in *Physarum* by laser Doppler spectroscopy. Protoplasma 91, 351–367.

Mustárdy, L., Buttle, K., Steinbach, G., Garab, G., 2008. The three-dimensional network of the thylakoid membranes in plants: quasihelical model of the granum-stroma assembly. Plant Cell 20, 2552–2557.

Mustroph, A., Bailey-Serres, J., 2010. The *Arabidopsis* translatome cell-specific mRNA atlas. Mining suberin and cutin tipid monomer biosynthesis genes as an example for data application. Plant Signal. Behav. 5, 320–324.

Mutschler, M.A., 1986. Dedication: Henry M. Munger: vegetable breeder and educator. Plant Breed. Rev. 4, 1–8.

Myers, J., 1974. Conceptual developments in photosynthesis, 1924–1974. Plant Physiol. 54, 420–426.

Myers, J., French, C.S., 1960. Evidences from action spectra for specific participation of chlorophyll b in photosynthesis. J. Gen. Physiol. 43, 723–736.

Myers, E.W., Sutton, G.G., Delcher, A.L., Dew, I.M., Fasulo, D.P., Flanigan, M.J., Kravitz, S.A., Mobarry, C.M., et al., 2000. A whole-genome assembly of *Drosophila*. Science 287, 2196–2204.

Nachmansohn, D., 1955. Metabolism and function of the nerve cell. Harvey Lect. 49, 57–90.

Nachmansohn, D., Coates, C.W., Rothenberg, M.A., Brown, M.V., 1946. Action potential and enzyme activity in the electric organ of Electrophorus electricus. J. Biol. Chem. 165, 223–231.

Nachmansohn, D., Cox, R.T., Coates, C.W., Machado, A.L., 1943. Action potential and enzyme activity in the electric organ of Electrophorus electricus. II. Phospho-creatine as energy source of the action potential. J. Neurophysiol. 6, 383–396.

Nachmias, V.T., Huxley, H.E., Kessler, D., 1970. Electron microscope observations on actomyosin and actin preparations from *Physarum polycephalum*, and on their interaction with heavy meromyosin subfragment I from muscle myosin. J. Mol. Biol. 50, 83–90.

Nadreau, J.A., Sack, F.D., 2003. Stomatal development: cross talk puts mouths in place. Trends Plant Sci. 8, 294–299.

Naeve, G.S., Sharma, A., Lee, A.S., 1991. Temporal events regulating the early phases of the mammalian cell cycle. Curr. Opin. Cell Biol. 3, 261–268.

Nagai, R., 1993. Regulation of intracellular movements in plant cells by environmental stimuli. Int. Rev. Cytol. 145, 251–310.

Nagai, R., Hayama, T., 1979. Ultrastructure of the endoplasmic factor responsible for cytoplasmic streaming in *Chara* internodal cells. J. Cell Sci. 36, 121–136.

Nagai, R., Rebhun, L.I., 1966. Cytoplasmic microfilaments in streaming *Nitella* cells. J. Ultrastruct. Res. 14, 571–589.

Nagalakshmi, U., Wang, Z., Waern, K., Shou, C., Raha, D., Gerstein, M., Snyder, M., 2008. The transcriptional landscape of the yeast genome defined by RNA sequencing. Science 320, 1344–1349.

Nagarajan, N., Read, T.D., Pop, M., 2008. Scaffolding and validation of bacterial genome assemblies using optical restriction maps. Bioinformatics 24, 1229–1235.

Nagata, T., 2004. When I encountered tobacco BY-2 cells. Biotechnol. Agric. For. 53, 1–6.

Nagata, T., Hasezawa, S., Inzé, D. (Eds.), 2004. Tobacco BY-2 Cells. Biotechnology in Agriculture and Forestry 53. Springer-Verlag, Berlin.

Nagata, T., Kumagai, F., 1999. Plant cell biology through the window of the highly synchronized tobacco BY-2 cell line. Methods Cell Sci. 21, 123–127.

Nagata, T., Matsuoka, K., Inzé, D. (Eds.), 2006. Tobacco BY-2 Cells: From Cellular Dynamics to Omics. Biotechnology in Agriculture and Forestry 58. Springer-Verlag, Berlin.

Nagata, T., Nemoto, Y., Hasegawa, S., 1992. Tobacco BY-2 cell line as the "HeLa" cells in the cell biology of higher plants. Int. Rev. Cytol. 132, 1–30.

Nagel, G., Brauner, M., Liewald, J.F., Adeishvili, N., Bamberg, E., Gottschalk, A., 2005. Light activation of channelrhodopsin-2 in excitable cells of *Caenorhabditis elegans* triggers rapid behavioral responses. Curr. Biol. 15, 2279–2284.

Nagele, R., Freeman, T., McMorrow, L., Lee, H.-Y., 1995. Precise spatial positioning of chromosomes during prometaphase: evidence for chromosomal order. Science 270, 1831–1835.

Nagl, W., 1978. Endopolyploidy and Polyteny in Differentiation and Evolution. North-Holland, Amsterdam.

Nagl, W., 1982. Nuclear chromatin. In: Parthier, B., Boulter, D. (Eds.), Nucleic Acids and Proteins in Plants. II. Structure, Biochemistry and Physiology of Nucleic Acids. Encyl. Plant Physiol. Springer Verlag, Berlin, pp. 1–45.

Nagl, W., 1985. Chromatin organization and the control of gene activity. Int. Rev. Cytol. 94, 21–56.

Nagl, W., Jeanjour, M., Kling, H., Kühner, S., Michels, I., Müller, T., Stein, B., 1983. Genome and chromatin organization in higher plants. Biol. Zbl. 102, 129–148.

Nagl, W., Rucker, W., 1976. Effect of phytohormones on thermal denaturation profiles of *Cymbidium* DNA: indication of differential DNA replication. Nucleic Acid. Res. 3, 2033–2039.

Nagpal, P., Quatrano, R.S., 1999. Isolation and characterization of a cDNA clone from *Arabidopsis thaliana* with partial sequence similarity to integrins. Gene 230, 33–40.

Nahm, L.J., 1940. The problem of Golgi material in plant cells. Bot. Rev. 6, 49–72.

Nair, P., 2011. Profile of Anthony R. Cashmore. Proc. Natl. Acad. Sci. U.S.A. 108, 443–445.

Nair, M., DeBolt, S., 2011. Analysing cellulose biosynthesis with confocal microscopy. Methods Mol. Biol. 715, 141–152.

Naito, H., Goshima, G., 2015. NACK kinesin is required for metaphase chromosome alignment in the moss *Physcomitrella patens*. Cell Struct. Funct. 40, 31–41.

Nakabayashi, R., Saito, K., 2015. Integrated metabolomics for abiotic stress response in plants. Curr. Opin. Plant Biol. 24, 10–16.

Nakai, M., 2015. The TIC complex uncovered: the alternative view on the molecular mechanism of protein translocation across the inner envelope membrane of chloroplasts. Biochim. Biophys. Acta 1847, 957–967.

Nakajima, H., 1960. Some properties of a contractile protein in a myxomycete plasmodium. Protoplasma 52, 413–436.

Nakajima, M., Imai, K., Ito, H., Nishiwaki, T., Murayama, Y., Iwasaki, H., Oyama, T., Kondo, T., 2005. Reconstitution of circadian oscillation of cyanobacterial KaiC phosphorylation *in vitro*. Science 308, 414–415.

Nakajima, K., Sena, G., Nawy, T., Benfey, P.N., 2001. Intercellular movement of the putative transcription factor SHR in root patterning. Nature 413, 307–311.

Nakamura, T., Meyer, C., Sano, H., 2002. Molecular cloning and characterization of plant genes encoding novel peroxisomal molybdoenzymes of the sulfite oxidase family. J. Exp. Bot. 53, 1833–1836.

Nakane, Y., Sasaki, A., Kinjo, M., Jin, T., 2012. Bovine serum albumin-coated quantum dots as a cytoplasmic viscosity probe in a single living cell. Anal. Methods 4, 1903–1905.

Nakano, A., Luini, A., 2010. Passage through the Golgi. Curr. Opin. Cell Biol. 22, 471–478.

Nakano, K., Yamamoto, T., Kishimoto, T., Noji, T., Tanaka, K., 2008. Protein kinases Fpk1p and Fpk2p are novel regulators of phospholipids asymmetry. Mol. Biol. Cell 19, 1783–1797.

Nakaoka, Y., Miki, T., Fujioka, R., Uehara, R., Tomioka, A., Obuse, C., Kubo, M., Hiwatashi, Y., Goshima, G., 2012. An inducible RNA interference system in *Physcomitrella patens* reveals a dominant role of augmin in phragmoplast microtubule regeneration. Plant Cell 24, 1478–1493.

Nakatani, Y., Hisatomi, O., 2015. Molecular mechanism of photozipper, a light-regulated dimerizing module consisting if the bZIP and LOV domains of aureochrome-1. Biochemistry 54, 3302–3313.

Nakatogawa, H., Suzuki, K., Kamada, Y., Ohsumi, Y., 2009. Dynamics and diversity in autophagy mechanisms: lessons from yeast. Nat. Rev. Mol. Cell Biol. 10, 458–467.

Nakazono, M.F., Qiu, F., Borsuk, L.A., Schnable, P.S., 2003. Laser-capture microdissection, a tool for the global analysis of gene expression in specific plant cell types: identification of genes expressed differentially in epidermal cells or vascular tissues of maize. Plant Cell 15, 583–596.

Nakhoul, N.L., Davis, B.A., Romero, M.F., Boron, W.F., 1998. Effect of expressing the water channel aquaporin-1 on the CO_2 permeability of *Xenopus* oocytes. Am. J. Physiol. Cell Physiol. 43, C543–C548.

Nakrieko, K.A., Mould, R.M., Smith, A.G., 2004. Fidelity of targeting to chloroplasts is not affected by removal of the phosphorylation site from the transit peptide. Eur. J. Biochem. 271, 509–516.

Nanjo, Y., Oka, H., Ikarashi, N., Kaneko, K., Kitajima, A., Mitsui, T., Muñoz, F.J., Rodríguez-López, M., Baroja-Fernández, E., Pozueta-Romero, J., 2006. Rice plastidial N-glycosylated nucleotide pyrophosphatase/phosphodiesterase is transported from the ER-golgi to the chloroplast through the secretory pathway. Plant Cell 18, 2582–2592.

Nanjundiah, V., 2004. George Gamow and the genetic code. Resonance 9, 44–49.

Napoli, C., Lemieux, C., Jorgensen, R., 1990. Introduction of a chimeric chalcone synthase gene into petunia results in reversible co-suppression of homologous genes in trans. Plant Cell 2, 279–289.

Naseer, S., Lee, Y., Lapierre, C., Franke, R., Nawrath, C., Geldner, N., 2012. Casparian strip diffusion barrier in *Arabidopsis* is made of a lignin polymer without suberin. Proc. Natl. Acad. Sci. U.S.A. 109, 10101–10106.

Nasrallah, J.B., 2002. Recognition and rejection of self in plant reproduction. 296, 305–308.

Nasrallah, J.B., 2005. Recognition and rejection of self in plant self-incompatibility: comparisons to animal histocompatibility. Trends Immunol. 26, 412–418.

Nass, M.M.K., Nass, S., 1963. Intramitochondrial fibers with DNA characteristics. I. Fixation and electron staining reactions. J. Cell Biol. 19, 593–611.

Natali, L., Cavallini, A., Cremonini, R., Bassi, P., Cionini, P.G., 1986. Amplification of nuclear DNA sequences during induced plant cell dedifferentiation. Cell Differ. 18, 157–161.

Natali, L., Giordani, T., Cionini, G., Pugliesi, C., Fambrini, M., Cavallini, A., 1995. Heterochromatin and repetitive DNA frequency variation in regenerated plants of *Helianthus annus* L. Theor. Appl. Genet. 91, 395–400.

Natera, S.H.A., Ford, K.L., Cassin, A.M., Patterson, J.H., Newbigin, E.J., Bacic, A., 2008. Analysis of the *Oryza sativa* plasma membrane proteome using combined protein and peptide fractionation approaches in conjunction with mass spectrometry. J. Proteome Res. 7, 1159–1187.

Nathans, D., December 8, 1978. Restriction Endonucleases, Simian Virus 40, and the New Genetics. Nobel Lecture.

Naulin, P.A., Alveal, N.A., Barrera, N.P., 2014. Toward atomic force microscopy and mass spectrometry to visualize and identify lipid rafts in plasmodesmata. Front. Plant Sci. 5, 234.

Navari-Izzo, F., Quartacci, M.F., Izzo, R., 1989. Lipid changes in maize seedlings in response to field water deficits. J. Exp. Bot. 40, 675–680.

Nazarea, A., Bloch, D.P., Semrau, A., 1985. Detection of a principle modular format in tRNAs and rRNAs: second order spectral analysis. Proc. Natl. Acad. Sci. U.S.A. 82, 5337–5341.

Nebel, B.R., Ruttle, M.L., 1938. The cytological and genetical significance of colchicine. J. Hered. 29, 3–9.

Nebenführ, A., 2003. Intra-Golgi transport: escalator or bucket brigade? Annu. Plant Rev. 9, 76–89.

Nebenführ, A., 2007. Organelle dynamics during cell division. Plant Cell Monogr. 9, 195–206.

Nebenführ, A., 2014. Identifying subcellular protein localization with fluorescent protein fusions after transient expression in onion epidermal cells. Methods Mol. Biol. 1080, 77–85.

Nebenführ, A., Dixit, R., 2018. Kinesins and myosins: molecular motors that coordinate cellular functions in plants. Annu. Rev. Plant Biol. 69, 329–361.

Nebenführ, A., Frohlick, J.A., Staehelin, L.A., 2000. Redistribution of Golgi stacks and other organelles during mitosis and cytokinesis in plant cells. Plant Physiol. 124, 135–151.

Nebenführ, A., Gallagher, L., Dunahay, T.G., Frohlick, J.A., Mazurkiewicz, A.M., Meehl, J.B., Staehelin, L.A., 1999. Stop- and-go movements of plant Golgi stacks are mediated by the actomyosin system. Plant Physiol. 121, 1127–1141.

Nebenführ, A., Staehelin, L.A., 2001. Mobile factories: Golgi dynamics in plant cells. Trends Plant Sci. 6, 160–167.

Needham, J.T., 1749. Observations upon the Generation, Composition, and Decomposition of Animal and Vegetable Substances. London.

Needham, J., 1930. The Skeptical Biologist. Yale University Press, New Haven, CT.

Needham, J., 1936. Order and Life. Yale University Press, New Haven, CT.

Needham, D.M., 1971. Machina Carnis. The Biochemistry of Muscular Contraction in its Historical Development. Cambridge University Press, Cambridge.

Needham, J. (Ed.), 1971. The Chemistry of Life. Eight Lectures on the History of Biochemistry. Cambridge University Press, Cambridge.

Needham, J., Hughes, A., 1959. A History of Embryology, second ed. Abelard-Schuman, New York.

Needham, D.M., Pillai, R.K., 1937. The coupling of oxido-reductions and dismutations with esterification of phosphate in muscle. Biochem. J. 31, 1837–1851.

Neel, J.V., 1949. The inheritance of sickle cell anemia. Science 110, 64–66.

Neff, M.M., Nguyen, S.M., Malancharuvil, E.J., Fujioka, S., Noguchi, T., Seto, H., Tsubuki, M., Honda, T., Takatsuro, S., Yoshida, S., Chory, J., 1999. *BAS1*: a gene regulating brassinosteroid levels and light responsiveness in *Arabidopsis*. Proc. Natl. Acad. Sci. U.S.A. 96, 15316–15323.

Neher, E., December 9, 1991. Ion Channels for Communication Between and Within Cells. Nobel Lecture.

Nekrasov, V., Staskawicz, B., Weigel, D., Jones, J.D.G., Kamoun, S., 2013. Targeted mutagenesis in the model plant Nicotiana benthamiana using CAs9 RNA-guided endonuclease. Nat. Biotechnol. 31, 691–693.

Nelson, N., 1988. Structure, function, and evolution of proton ATPases. Plant Physiol. 86, 1–3.

Nelson, N., 1989. Structure, molecular genetics and evolution of vacuolar H^+ ATPases. J. Bioenerg. Biomembr. 21, 553–571.

Nelson, N., 1992. The vacuolar H^+-ATPase–-one of the most fundamental ion pumps in nature. J. Exp. Biol. 172, 19–27.

Nelson, K., Levy, M., Miller, S.L., 2000. Peptide nucleic acids rather than RNA may have been the first genetic molecule. Proc. Natl. Acad. Sci. U.S.A. 97, 3868–3871.

Nelson, N., Racker, E., 1972. Partial resolution of the enzymes catalyzing photophosphorylation. X. Purification of spinach cytochrome f and its photooxidation by resolved Photosystem I particles. J. Biol. Chem. 247, 3848–3853.

Nemec, B., 1899. Uebuer die Karyokinetische Kerntheilung in der Wurzelsspitz von Allium cepa. Jahr. Wiss. Bot. 33, 313–336.

Nenninger, A., Mastroianni, G., Mullineaux, C.W., 2010. Size dependence of protein diffusion in the cytoplasm of *Escherichia coli*. J. Bacteriol. 192, 4535–4540.

Nernst, W., 1888. On the kinetics of substances in solution, 1979. In: Kepner, G.R. (Ed.), Cell Membranes permeability and Transport. Dowden, Hutchinson & Ross, Inc., Stroudsburg, PA, pp. 174–183.

Nernst, W., 1923. Theoretical Chemistry from the Standpoint of Avogadro's Rule of Thermodynamics. Translated by L. W. Codd. MacMillan, London.

Nes, W.R., 1977. The biochemistry of plant sterols. Adv. Lipid Res. 15, 233–324.

Nesme, J., Achouak, W., Agathos, S.N., Bailey, M., Baldrian, P., Brunel, D., et al., 2016. Back to the future of soil metagenomics. Front. Microbiol. 7, 73.

Neuberger, A., Sanger, F., 1942. The nitrogen of the potato. Biochem. J. 36, 662–671.

Neufeld, H.A., Scott, C.R., Stotz, E., 1954. Purification of heart muscle succinic dehydrogenase. J. Biol. Chem. 210, 869–876.

Neuhaus, G., Bowler, C., Hiratsuka, K., Yamagata, H., Chua, N.-H., 1997. Phytochrome-regulated repression of gene expression requires calcium and cGMP. EMBO J. 16, 2554–2564.

Neumann, U., Brandizzi, F., Hawes, C., 2003. Protein transport in plant cells: in and out of the Golgi. Ann. Bot. 92, 167–180.

Part A. In: Neurath, H., Bailey, K. (Eds.), 1953. The Proteins, vol. 1. Academic Press, New York.

Neurath, H., Greenstein, J.P., 1944. The chemistry of the proteins and amino acids. Ann. Rev. Biochem. 13, 117–154.

Neutra, M., LeBlond, C.P., 1966. Synthesis of the carbohydrate of mucus in the Golgi complex as shown by electron microscope radioautography of goblet cells from rats injected with glucose-H3. J. Cell Biol. 30, 119–136.

Neves, L.G., Davis, J.M., Barbazuk, W.B., Kirst, M., 2013. Whole-exome targeted sequencing of the uncharacterized pine genome. Plant J. 75, 146–156.

Newcomb, E.H., 1963. Cytoplasm-cell wall relationships. Annu. Rev. Plant Physiol. 14, 43–64.

Newcomb, E.H., 1967. Fine structure of protein-storing plastids in bean root tips. J. Cell Biol. 33, 143–163.

Newcomer, E.H., 1940. Mitochondria in plants. Bot. Rev. 6, 85–147.

Newman, T., de Bruijn, F.J., Green, P., Keegstra, K., Kende, H., McIntosh, L., Ohlrogge, J., Raikhel, N., Somerville, S., Thomashow, M., Retzel, E., Somerville, C., 1994. Genes galore: a summary of methods for accessing results from large-scale partial sequencing of anonymous *Arabidopsis* cDNA clones. Plant Physiol. 106, 1241–1255.

Newman, R.H., Zhang, J., 2008. Fucci: street light on the road to mitosis. Chem. Biol. 15, 97–98.

Newmeyer, D.D., 1993. The nuclear pore complex and nucleocytoplasmic transport. Curr. Opin. Cell Biol. 5, 395–407.

Newmeyer, D.D., Forbes, D.J., 1988. Nuclear import can be separated into distinct steps in vitro: nuclear pore binding and translocation. Cell 52, 641–653.

Newton, I., 1728. A Treatise of the System of the World. Printed for F. Fayram, London.

Newton, I., 1729. Sir Isaac Newton's Mathematical Principles of Natural Philosophy and his System of the World. Fourth Printing, 1960. University of California Press, Berkeley, CA.

Newton, I., 1730. Opticks or A Treatise of the Reflections, Refractions, Inflections & Colours of Light, 1979, fourth ed. Dover Publications, New York.

Newton, A.C., Bootman, M.D., Scott, J.D., 2016. Second messengers. Cold Spring Harb. Perspect. Biol. https://doi.org/10.1101/cshperspect.a005926.

Newton, S.A., Ford, N.C., Langley, K.H., Sattelle, D.B., 1977. Laser light-scattering analysis of protoplasmic streaming in the slime mold *Physarum polycephalum*. Biochim. Biophys. Acta 496, 212–224.

Newton, K.J., Walbot, V., 1985. Maize mitochondria synthesize organ-specific polypeptides. Proc. Natl. Acad. Sci. U.S.A. 82, 6879–6883.

Ng, D.T.W., Walter, P., 1994. Protein translocation across the endoplasmic reticulum. Curr. Opin. Cell Biol. 6, 510–516.

Nguyen, D.C., Keller, R.A., Jett, J.H., Martin, J.C., 1987. Detection of single molecules of phycoerythrin in hydrodynamically focused flows by laser-induced fluorescence. Anal. Chem. 59, 2158–2161.

Nicchitta, C.V., Blobel, G., 1993. Lumenal proteins of the mammalian endoplasmic reticulum are required to complete protein translocation. Cell 73, 989–998.

Nicholls, D.G., Ferguson, S.J., 1992. Bioenergetics 2. Academic Press, London.

Nick, P., 2000. Plant Microtubules. Potential for Biotechnology. Springer-Verlag, Berlin.

Nick, P., 2008a. Control of cell axis. Plant Cell Monogr. 11, 3–46.

Nick, P., 2008b. Cellular footprints of plant evolution—news from the blepharoblast. Protoplasma 233, 175.

Nick, P., Bergfeld, R., Schäfer, E., Schopfer, P., 1990. Unilateral reorientation of microtubules at the outer epidermal wall during photo- and gravitropic curvature of maize coleoptiles and sunflower hypocotyls. Planta 181, 162–168.

Nickell, L.G., 1993. A tribute to Hugo Kortschak. Photosynth. Res. 35, 201–204.

Nicklas, R.B., 1965. Chromosome velocity during mitosis as a function of chromosome size and position. J. Cell Biol. 25, 119–135.

Nicklas, R.B., 1967. Chromosome micromanipulation. II. Induced reorientation and the experimental control of segregation in meiosis. Chromosoma 21, 17–50.

Nicklas, R.B., 1975. Chromosome movement: current models and experiments in living cells. In: Inoué, S., Stephens, R.E. (Eds.), Molecules and Cell Movement. Raven Press, New York, pp. 97–117.

Nicklas, R.B., 1983. Measurement of the force produced by the mitotic spindle in anaphase. J. Cell Biol. 97, 542–548.

Nicklas, R.B., 1989. The motor for poleward chromosome movement in anaphase is in or near the kinetochore. J. Cell Biol. 109, 2245–2255.

Nicklas, R.B., 1997. How cells get the right chromosomes. Science 275, 632–637.

Nicklas, R.B., Koch, C.A., 1969. Chromosome micromanipulation. III. Spindle fiber tension and reorientation of maloriented chromosomes. J. Cell Biol. 43, 40–50.

Nicklas, R.B., Kubai, D.F., 1985. Microtubules, chromosome movement, and reorientation after chromosomes are detached from the spindle by micromanipulation. Chromosoma 92, 313–324.

Nicklas, R.B., Staehly, C.A., 1967. Chromosome micromanipulation. I. The mechanics of chromosome attachment to the spindle. Chromosoma 21, 1–16.

Nicklas, R.B., Ward, S.C., 1994. Elements of error correction in mitosis: microtubule capture, release, and tension. J. Cell Biol. 126, 1241–1253.

Nicklas, R.B., Ward, S.C., Gorbsky, G.J., 1995. Kineotchore chemistry is sensitive to tension and may link mitotic forces to a cell cycle checkpoint. J. Cell Biol. 130, 929–939.

Nicolaieff, A., Phillips, G., Gigot, C., Hirth, L., 1976. Ring subunit associated with plant chromatin. J. Microsc. Biol. Cell 26, 1–4.

Nicolas, W.J., Grison, M.S., Trépout, S., Gaston, A., Fouché, M., Cordelières, F.P., Oparka, K., Tilsner, J., Brocard, L., Bayer, E.M., 2017. Architecture and permeability of post-cytokinesis plasmodesmata lacking cytoplasmic sleeves. Nat. Plants 3, 17082.

Nicolas, W.J., Grison, M.S., Trépout, S., Gaston, A., Fouché, M., Cordelières, F.P., Oparka, K., Tilsner, J., Brocard, L., Bayer, E.M., 2017. Architecture and permeability of post-cytokinesis plasmodesmata lacking cytoplasmic sleeves. Nat. Plants 3, 17082.

Nicolson, G.L., 2014. The fluid-mosaic model of membrane structure: still relevant to understanding the structure, function and dynamics of biological membranes after more than 40 years. Biochim. Biophys. Acta 1838, 1451–1466.

Nicolson, W., Carlisle, A., 1800. Account of the new electrical or galvanic apparatus of Sig. Alex. Volta, and experiments performed with the same. Nicolsons J. Nat. Philos. 4, 179–187.

Nicolson, S.W., Nepi, M., Pacini, E., 2007. Nectaries and Nectar. Springer, Dordrecht, The Netherlands.

Nieden, U.Z., Manteuffel, R., Weber, E., Neumann, D., 1984. Dictyosomes participate in the intracellular pathway of storage proteins in developing *Vicia faba* cotyledons. Eur. J. Cell Biol. 34, 9–17.

Nieden, U.Z., Neumann, D., Manteuffel, R., Weber, E., 1982. Electron microscopic immunocytochemical localization of storage proteins in *Vicia faba* seeds. Eur. J. Cell Biol. 26, 228–233.

Niehl, A., Heinlein, M., 2011. Cellular pathways for viral transport through plasmodesmata. Protoplasma 248, 75–99.

Nielsen, J., 2015. Yeast cell factories on the horizon. Science 349, 1050–1051.

Nielsen, E., Akita, M., Davila-Aponte, J., Keegstra, K., 1997. Stable association of chloroplastic precursors with protein translocation complexes that contain proteins from both envelope membranes and a stromal Hsp100 molecular chaperone. EMBO J. 16, 935–946.

Nielsen, E., Cheung, A.Y., Ueda, T., 2008. The regulatory RAB and ARF GTPases for vesicular trafficking. Plant Physiol. 147, 1516–1526.

Nielsen, S.O., Lehninger, A.L., 1954. Oxidative phosphorylation in the cytochrome system of mitochondria. J. Am. Chem. Soc. 76, 3860–3861.

Niemietz, C.M., Tyerman, S.D., 1997. Characterization of water channels in wheat root membrane vesicles. Plant Physiol. 115, 561–567.

Nigg, E.A., 1988. Nuclear function and organization: the potential of immunochemical approaches. Int. Rev. Cytol. 110, 27–92.

Nikaido, S.S., Johnson, C.H., 2000. Daily and circadian variation in survival from ultraviolet radiation in *Chlamydomonas rheinhardtii*. Photochem. Photobiol. 71, 758–765.

Niklas, K.J., 1992. Plant Biomechanics. University of Chicago Press, Chicago.

Niklas, K.J., 2004. The cell walls that bind the tree of life. BioScience 54, 831–841.

Niklas, K.J., 2014. The evolutionary-developmental origins of multicellularity. Am. J. Bot. 101, 6–25.

Niklas, K.J., 2016. Plant Evolution: An Introduction to the History of Life. Chicago University Press, Chicago, IL.

Niklas, K.J., Bondos, S.E., Dunker, A.K., Newman, S.A., 2015. Rethinking gene regulatory networks in light of alternative splicing, intrinsically disordered protein domains, and post-translational modifications. Front. Cell Dev. Biol. 3, 8.

Niklas, K.J., Cobb, E.D., Matas, A.J., 2017. The evolution of hydrophobic cell wall polymers: from algae to angiosperms. J. Exp. Bot. 68, 5261–5269.

Niklas, K.J., Dunker, A.K., 2016. Alternative splicing, intrinsically disordered. In: Niklas, K.J., Newman, S.A. (Eds.), Multicellularity: Origins and Evolution. MIT Press, Cambridge, MA, pp. 17–40.

Niklas, K.J., Dunker, A.K., Yruela, I., 2018. The evolutionary origins of cell type diversification and the role of intrinsically disordered proteins. J. Exp. Bot. 69, 1437–1446.

Niklas, K.J., Newman, S.A., 2013. The origins of multicellular organisms. Evol. Dev. 15, 41–52.

Niklas, K.J., Owens, T.G., Wayne, R.O., 2013. Unity and disunity in biology. BioScience 63, 811–816.

Niklas, K.J., Spatz, H.-C., 2012. Plant Physics. University of Chicago Press, Chicago.

Niko, N., Shliaha, P.V., Gatto, L., Dupree, P., Lilley, K.S., 2014. Label-free protein quantification for plant Golgi protein localization abundance. Plant Physiol. 166, 1033–1043.

Nikolovski, N., Rubtsov, D., Segura, M.P., Miles, G.P., Stevens, T.J., Dunkley, T.P.J., Munro, S., Lilley, K.S., Dupree, P., 2012. Putative glycosyltransferases and other plant Golgi apparatus proteins are revealed by LOPIT proteomics. Plant Physiol. 160, 1037–1051.

Nikolovski, N., Shliaha, P.V., Gatto, L., Dupree, P., Lilley, K.S., 2014. Label-free protein quantification for plant Golgi protein localization and abundance. Plant Physiol. 166, 1033–1043.

Nilsson, H., 1971. The Point. https://www.youtube.com/watch?v=hHgj1uQ5FH8&list=PL1F7FBBEDD6073213.

Nimchinsky, E.A., Gilissen, E., Allman, J.M., Perl, D.P., Erwin, J.M., Hof, P.R., 1999. A neuronal morphologic type unique to humans and great apes. Proc. Natl. Acad. Sci. U.S.A. 96, 5268–5273.

Nimchuk, Z.L., Tarr, P.T., Ohno, C., Qu, X., Meyerowitz, E.M., 2011. Plant Stem Cell Signaling Involves.

Nirenberg, M.W., 1967. Will society be prepared? Science 167, 633.

Nirenberg, M., 1997. Reminiscences of George Gamow. In: Harper, E., Parke, W.C., Anderson, D. (Eds.), George Gamow Symposium. ASP Conference Series, vol. 129, p. 130.

Nirenberg, M., December 12, 1968. The Genetic Code. Nobel Lecture.

Nirenberg, M., Leder, P., 1964. RNA codewords and protein synthesis. The effect of trinucleotides upon the binding of sRNA to ribosomes. Science 145, 1399–1407.

Nirenberg, M.W., Matthaei, J.H., 1961. The dependence of cell-free protein synthesis in *E. coli* upon naturally occuring or synthetic polyribonucleotides. Proc. Natl. Acad. Sci. U.S.A. 47, 1538–1602.

Nirenberg, M.W., Matthaei, J.H., Jones, O.W., 1962. An intermediate in the biosynthesis of polyphenylalanine directed by synthetic template RNA. Proc. Natl. Acad. Sci. U.S.A. 48, 104–109.

Nishiguchi, M., Motoyoshi, F., Oshima, N., 1980. Further investigation of a temperature-sensitive strain of tobacco mosaic virus: its behavior in tomato leaf epidermis. J. Gen. Virol. 45, 497–500.

Nishihama, R., Ishikawa, M., Araki, S., Soyano, T., Asada, T., Machida, Y., 2001. The NPK1 mitogen-activated protein kinase kinase kinase is a regulator of cell-plate formation in plant cytokinesis. Genes Dev. 15, 352–363.

Nishihama, R., Machida, Y., 2001. Expansion of the phragmoplast during plant cytokinesis: a MAPK pathway may not MAP out. Curr. Opin. Plant Biol. 4, 507–512.

Nishihama, R., Soyano, T., Ishikawa, M., Araki, S., Tanaka, H., Asada, T., Irie, K., Ito, M., Terada, M., Banno, H., 2002. Expansion of the cell plate in plant cytokinesis requires a kinesin-like protein/MAPKKK complex. Cell 109, 87–99.

Nishihara, N., Horiike, S., Oka, Y., Takahashi, T., Kosaka, T., Hosoya, H., 1999. Microtubule-dependent movement of symbiotic algae and granules in Paramecium bursaria. Cell Motil. Cytoskelet. 43, 85–98.

Nishikawa, S., Umemoto, N., Ohsumi, Y., Nakano, A., Anraku, Y., 1990. Biogenesis of vacuolar membrane glycoproteins of yeast *Saccharomyces cerevisiae*. J. Biol. Chem. 265, 7440–7448.

Nishimura, M., Beevers, H., 1978. Hydrolases in vacuoles from castor bean endosperm. Plant Physiol. 62, 44–48.

Nishimura, M., Beevers, H., 1979. Hydrolysis of protein in vacuoles isolated from higher plant tissue. Nature 277, 412–413.

Nishimura, K., Kato, Y., Sakamoto, W., 2017. Essential of proteolytic machineries in chloroplasts. Mol. Plant 10, 4–19.

Nishimura, M.T., Stein, M., Hou, B.-H., Vogel, J.P., Edwards, H., Somerville, S.C., 2003. Loss of a callose synthase results in salicylic acid-dependent disease resistance. Science 301, 969–972.

Nishimura, M., Takeuchi, Y., DeBellis, L., Hara-Nishimura, I., 1993. Leaf peroxisomes are directly transformed to glyoxysomes during senescence of pumpkin cotyledons. Protoplasma 175, 131–137.

Nishimura, M., Yamaguchi, J., Mori, H., Akazawa, T., Yokota, S., 1986. Immunocytochemical analysis shows that glyoxysomes are directly transformed to leaf peroxisomes during greening of pumpkin cotyledons. Plant Physiol. 81, 313–316.

Nito, K., Kamigaki, A., Kondo, M., Hayashi, M., Nishimura, M., 2007. Functional classification of *Arabidopsis* peroxisome biogenesis factors proposed from analyses of knockdown mutants. Plant Cell Physiol. 48, 763–774.

Niwa, Y., Kato, T., Tabata, S., Seki, M., Kobayashi, M., Shinozaki, K., Moriyasu, Y., 2004. Disposal of chloroplasts with abnormal function into the vacuole in *Arabidopsis thaliana* cotyledon cells. Protoplasma 223, 229–232.

Niwayama, R., Nagao, H., Kitajima, T.S., Hufnagel, L., Shinohara, K., Higuchi, T., Ishikawa, T., Kimura, A., 2016. Baysian inference of forces causing cytoplasmic streaming in *Caenorhabiditis elegans* embryos and mouse oocytes. PLoS One 11, e0159917.

Niwayama, R., Shinohara, K., Kimura, A., 2011. Hydrodynamic property of the cytoplasm is sufficient to mediate cytoplasmic streaming in the *Caenorhabiditis elegans* embryo. Proc. Natl. Acad. Sci. U.S.A. 108, 11900–11905.

Nobel, P.S., 1974. Introduction to Biophysical Plant Physiology. W. H. Freeman and Co., New York.

Nobel, P.S., 1983. Biophysical Plant Physiology and Ecology. W. H. Freeman and Co., New York.

Nobel, P.S., 1991. Physicochemical and Evironmental Plant Physiology. Academic Press, New York.

Nobel, P.S., 2005. Physicochemical and Evironmental Plant Physiology, third ed. Elsevier Academic Press, New York.

Nobles Jr., D.R., Romanovicz, D.K., Brown, R.M., 2001. Cellulose in cyanobacteria. Origin of vascular plant cellulose synthase? Plant Physiol. 127, 529–542.

Noda, T., Klionsky, D.J., 2008. The quantitative Pho8Delta60 assay of nonspecific autophagy. Methods Enzymol. 451, 33–42.

Noe, K., Wayne, R., 1990. The phototactic response of *Dunaliella* requires nanomolar concentrations of external calcium. Plant Physiol. 93, S-90.

Noguchi, T., 2014. Cell walls. In: Noguchi, T., Kawano, S., Tsukaya, H., Matsunaga, S., Sakai, A., Karahara, I., Hayashi, Y. (Eds.), Atlas of Plant Cell Structure. Springer, Japan, pp. 137–156.

Noguchi, T., Hayashi, Y., 2014. Vacuoles and storage organelles. In: Noguchi, T., Kawano, S., Tsukaya, H., Matsunaga, S., Sakai, A., Karahara, I., Hayashi, Y. (Eds.), Atlas of Plant Cell Structure. Springer, Japan, pp. 89–106.

Noguchi, T., Kakami, F., 1999. Transformation of trans-Golgi network during the cell cycle in a green alga, *Botryococcus braunii*. J. Plant Res. 112, 175–186.

Noguchi, T., Matsunaga, S., Hayashi, Y., 2014. The endoplasmic reticulum, Golgi apparatuses, and endocytic organelles. In: Noguchi, T., Kawano, S., Tsukaya, H., Matsunaga, S., Sakai, A., Karahara, I., Hayashi, Y. (Eds.), Atlas of Plant Cell Structure. Springer, Japan, pp. 71–88.

Nohales, M.A., Kay, S.A., 2016. Molecular mechanisms at he core of the plant circadian oscillator. Nat. Struct. Mol. Biol. 23, 1061–1069.

Noireaux, V., Bar-Ziv, R., Godefroy, J., Salman, H., Libchaber, A., 2005. Toward an artificial cell based on gene expression in vesicles. Phys. Biol. 2, P1–P8.

Noireaux, V., Bar-Ziv, R., Libchaber, A., 2003. Principles of cell-free genetic circuit assembly. Proc. Natl. Acad. Sci. U.S.A. 100, 12672–12677.

Noireaux, V., Libchaber, A., 2004. A vesicle bioreactor as a step toward an artificial cell assembly. Proc. Natl. Acad. Sci. U.S.A. 101, 17669–17674.

Noireaux, V., Maeda, Y.T., Libchaber, A., 2011. Development of an artificial cell, from self-organization to computation and self-reproduction. Proc. Natl. Acad. Sci. U.S.A. 108, 3473–3480.

Noll, M., 1974. Subunit structure of chromatin. Nature 251, 249–251.

Noller, H.F., Hoffarth, V., Zimniak, L., 1992. Unusual resistance of peptidyl transferase to protein extraction procedures. Science 256, 1416–1419.

Nolta, K.V., Steck, T.L., 1994. Isolation and initial characterization of the bipartite contractile vacuole complex from *Dictyostelium discoideum*. J. Biol. Chem. 269, 2225–2233.

Nomura, K., Saito, W., Ono, K., Moriyama, H., Takahashi, S., Inoue, M., Masuda, K., 1997. Isolation and characterization of matrix associated region DNA fragments in rice (*Oryza sativa* L.). Plant Cell Physiol. 38, 1060–1068.

Nomura, T., Sakurai, T., Osakabe, Y., Osakabe, K., Sakakibara, H., 2016. Efficient and heritable targeted mutagenesis in mosses using the CRISPR/Cas9 system. Plant Cell Physiol. 57, 2600–2610.

Noonan, J.P., Coop, G., Kudaravalli, S., Smith, D., Krause, J., Alessi, J., Chen, F., Platt, D., Pääbo, S., Pritchard, J.K., Rubin, E.M., 2006. Sequencing and analysis of Neanderthal DNA. Nature 444, 330–336.

Norbury, C., Nurse, P., 1992. Animal cell cycles and their control. Ann. Rev. Biochem. 61, 441–470.

Normanly, J., 2012. High-Throughput Phenotyping in Plants. Methods and Protocols. Springer, New York.

Northcote, D.H., 1967. The Living Cell. University of Hull, Hull Printers, Hull, Yorkshire.

Northcote, D.H., Davey, R., Lay, J., 1989. Use of antisera to localize callose, xylan and arabinogalactan in the cell-plate, primary and secondary walls of plant cells. Planta 178, 353–366.

Northcote, D.H., Pickett-Heaps, J.D., 1966. A function of the Golgi apparatus in polysaccharide synthesis and transport in the root-cap cells of wheat. Biochem. J. 98, 159–167.

Northrop, J.H., 1929. Crystalline pepsin. Science 69, 580.

Northrop, J.H., 1930. Crystalline pepsin. Isolation and tests of purity. J. Gen. Physiol. 13, 739–766.

Northrop, F.S.C., 1931. Science and First Principles. MacMillan, New York.

Northrop, J.H., 1961. Biochemists, biologists, and William of Occam. Ann. Rev. Biochem. 30, 1–10.

Northrop, J.H., Herriottt, R.M., 1938. Chemistry of the crytalline enzymes. Ann. Rev. Biochem. 7, 37–50.

Northrop, J.H., Kunitz, M., 1931. Isolation of protein crystals possessing tryptic activity. Science 73, 262–263.

Northrop, J.H., Kunitz, M., Herriott, R.M., 1948. Crystalline Enzymes. Columbia University Press, New York.

Northrup, J.H., Herriottt, R.M., 1938. Chemistry of the crystalline enzymes. Ann. Rev. Biochem. 7, 37–50.

Northrup, J.H., Kunitz, M., 1931. Isolation of protein crystals possessing tryptic activity. Science 73, 262–263.

Northrup, J.H., Kunitz, M., Herriott, R.M., 1948. Crystalline Enzymes. Columbia University Press, New York.

Nothnagel, E.A., Barak, L.S., Sanger, J.W., Webb, W.W., 1981. Fluorescence studies on modes of cytochalasin B and phallotoxin action on cytoplasmic streaming in *Chara*. J. Cell Biol. 88, 364–372.

Nothnagel, E.A., Sanger, J.W., Webb, W.W., 1982. Effects of exogenous proteins on cytoplasmic streaming in perfused *Chara* cells. J. Cell Biol. 93, 735–742.

Nothnagel, E.A., Webb, W.W., 1982. Hydrodynamic models of viscous coupling between motile myosin and endoplasm in characean cells. J. Cell Biol. 94, 444–454.

Noueiry, A.O., Lucas, W.J., Gilbertson, R.L., 1994. Two proteins of a plant DNA virus coordinate nuclear and plasmodesmatal transport. Cell 76, 925–932.

Novick, P., Ferro, S., Schekman, R., 1981. Order of events in the yeast secretory pathway. Cell 25, 461–469.

Novick, P., Field, C., Schekman, R., 1980. The identification of 23 complementation groups required for post-translational events in the yeast secretory pathway. Cell 21, 205–215.

Novick, P., Schekman, R., 1979. Secretion and cell surface growth are blocked in a temperature sensitive mutant of *Saccharomyces cerevisiae*. Proc. Natl. Acad. Sci. U.S.A. 76, 1858–1862.

Novikoff, A.B., Shin, W.-Y., 1964. The endoplasmic reticulum in the Golgi zone and its relations to microbodies, Golgi apparatus and autophagic vacuoles in rat liver cells. J. Microsc. 3, 187–206.

Nozue, K., Kanegae, T., Imaizumi, T., Fukuda, S., Okamoto, H., Yeh, K.-C., Lagarias, J.C., Wada, M., 1998. A phytochrome from the fern *Adiantum* with features of the putative photoreceptor NPH1. Proc. Natl. Acad. Sci. U.S.A. 95, 15826–15830.

Nuccitelli, R., 1978. Ooplasmic segregation and secretion in *Pelvetia* egg is accompanied by a membrane-generated electrical current. Dev. Biol. 62, 13–33.

Nuccitelli, R., 1983. Transcellular ion currents: signals and effectors of cell polarity. Mod. Cell Biol. 2, 451–481.

Nuccitelli, R., Erickson, C.A., 1983. Embryonic cell motility can be guided by physiological electric fields. Exp. Cell Res. 147, 195–201.

Nugent, J.M., Palmer, J.D., 1991. RNA-mediated transfer of the gene cox II from the mitochondrion to the nucleus during flowering plant evolution. Cell 66, 473–481.

Numbers, R.L., 2007. Science and Christianity in Pulpit and Pew. Oxford University Press, Oxford.

Nunokawa, S.-Y., Anan, H., Shimada, K., Hachikubo, Y., Kashiyama, T., Ito, K., Yamamoto, K., 2007. Binding of *Chara* myosin globular tail domain to phospholipids vesicles. Plant Cell Physiol. 48, 1558–1566.

Nurse, P., 1990. Universal control mechanism regulating onset of M-phase. Nature 344, 503–508.

Nurse, P., December 9, 2001. Cyclin Dependent Kinases and Cell Cycle Control. Nobel Lecture.

Nyathi, Y., Wilkinson, B.M., Pool, M.R., 2013. Co-translational targeting and translocation of proteins to the endoplasmic reticulum. Biochim. Biophys. Acta 1833, 2392–3402.

Nyquist, H., 1924. Certain factors affecting telegraph speed. Bell Syst. Tech. J. 3, 324–346.

Nyquist, H., 1928. Certain topics in telegraph transmission theory. Trans. AIEE 47, 617–644.

Nyrén, P., 1987. Enzymatic method for continuous monitoring of DNA polymerase activity. Anal. Biochem. 238, 235–238.

Nyrén, P., 1987. Enzymatic method for continuous monitoring of DNA polymerase activity. Anal. Biochem. 238, 235–238.

Nyrén, P., Lundin, A., 1985. Enzymantic method for continuous monitoring of inorganic pyrophosphate synthesis. Anal. Biochem. 509, 504–509.

Nyrén, P., Pettersson, B., Uhlén, M., 1993. Solid phase DNA minisequencing by an enzymatic luminometric inorganic pyrophosphate detection assay. Anal. Biochem. 208, 171–175.

Nziengui, H., Bouhidel, K., Pillon, D., Der, C., Marty, F., Schoefs, B., 2007. Reticulon-like proteins in *Arabidopsis thaliana*: structural organization and ER localization. FEBS Lett. 581, 3356–3362.

O'Brien, T.P., McCully, M.E., 1969. Plant Structure and Developmemnt. A Pictorial and Physiological Approach. Macmillan, London.

O'Brien, T.P., Thimann, K.V., 1966. Intracellular fibers in oat coleoptile cells and their possible significance in cytoplasmic streaming. Proc. Natl. Acad. Sci. U.S.A. 56, 888–894.

O'Donnell-Maloney, M.J., Smith, C.L., Cantor, C.R., 1996. The development of microfabricated arrays for DNA sequencing and analysis. Trends Biotechnol. 14, 401–407.

O'Driscoll, D., Hann, C., Read, S.M., Steer, M.W., 1993. Endocytotic uptake of fluorescent dextrans by pollen tubes grown in vitro. Protoplasma 175, 126–130.

O'Farrell, P.H., 1975. High resolution two-dimensional electrophoresis of proteins. J. Biol. Chem. 250, 4007–4021.

Oakley, B.R., 1992. γ-tubulin: the microtubule organizer? Trends Cell Biol. 2, 1–6.

Oakley, C.E., Oakley, B.R., 1989. Identification of gamma tubulin, a new member of the tubulin super family by mipA gene of *Aspergillus nidulans*. Nature 338, 662–664.

Oakley, B.R., Oakley, C.E., Yoon, Y., Jung, M.K., 1990. Gamma-tubulin is a component of the spindle pole body that is essential for microtubule function in *Aspergillus nidulans*. Cell 61, 1289–1301.

O'Neil, J., Carlson, R.W., Francis, D., Stevenson, R.K., 2008. Neodymium-142 evidence for hadean mafic crust. Science 321, 1828–1831.

O'Neil, P.K., Richardson, L.G.L., Paila, Y.D., Piszczek, G., Chakravarthy, S., Noinaj, N., Schnell, D., 2017. The POTRA domains of Toc75 exhibit chaperone-like function to facilitate import into chloroplasts. Proc. Natl. Acad. Sci. U.S.A. 114, E4868–E4876.

O'Neill, S.D., Bennett, A.B., Spanswick, R.M., 1983. Characterization of a NO3-sensitive H^+-ATPase from corn roots. Plant Physiol. 72, 837–846.

O'Neill, M.A., Eberhard, S., Albersheim, P., Darvill, A.G., 2001. Requirement of borate cross-linking of cell wall rhamnogalacturonan II. For *Arabidopsis* growth. Science 294, 846–849.

O'Neill, M.A., Warrenfetz, D., Kates, K., Pellerin, P., Doco, T., Darvill, A.G., Albersheim, P., 1996. Rhamnogalacturonan-II, a pectic polysaccharide in the walls of growing plant cells, forms a dimer that is covalently cross-linked by borate ester. In vitro conditions for the formation and hydrolysis of the dimer. J. Biol. Chem. 271, 22923–22930.

O'Rourke, A.M., Mescher, M.F., 1990. T-cell receptor-activated adhesion systems. Curr. Opin. Cell Biol. 2, 850–856.

O'Toole, E., Greenan, G., Lange, K.I., Srayko, M., Müller-Reichert, T., 2012. The role of γ-tubulin in centrosomal microtubule organization. PLoS One 7, e29795.

Obermeyer, G., 2017. Water transport in pollen. In: Obermeyer, G., Feijó, J. (Eds.), Pollen Tip Growth. Springer International, pp. 13–34.

Obermeyer, G., Lützelschwab, M., Heumann, H.-G., Weisenseel, M.H., 1992. Immunolocalisation of H^+ ATPases in the plasma membrane of pollen grains and pollen tubes of *Lilium longiflorum*. Protoplasma 171, 55–63.

Occhialini, A., Lin, M.T., Andralojc, P.J., Hanson, M.R., Parry, M.A., 2016. Transgenic tobacco plants with improved cyanobacterial rubisco expression but no extra assembly factors grow at near wild-type rates if provided with elevated CO_2. Plant J. 85, 148–160.

Ochiai, E.-I., 1987. General Principles of Biochemistry of the Elements. Plenum Press, New York.

Ochoa, S., 1943. Efficiency of aerobic phosphorylation in cell-free heart extracts. J. Biol. Chem. 151, 493–505.

Ochoa, S., 1952. Enzyme studies in biological oxidations and synthesis. Harvey Lect. 46, 153–180.

Ochoa, S., 1980. The thrill of discovery. Comp. Biochem. Physiol. 67B, 359–365.

Ochoa, S., December 11, 1959. Enzymatic Synthesis of Ribonucleic Acid. Nobel Lecture.

Ochoa, S., Mehler, A.H., Kornberg, A., 1948. Biosynthesis of dicarboxylic acids by carbon dioxide fixation. J. Biol. Chem. 174, 979–1000.

Oda, Y., Fukuda, H., 2011. Dynamics of *Arabidopsis* SUN proteins during mitosis and their involvement in nuclear shaping. Plant J. 66, 629–641.

Oels, W., 1894. Experimental Plant Physiology. Translated and edited by D. T. MacDougal. Morrris & Wilson, Minneapolis.

Oesper, P., 1950. Sources of the high energy content in energy-rich phosphates. Arch. Biochem. 27, 255–270.

Oettlé, A.G., 1948. Golgi apparatus of living human testicular cells seen with phase-contrast microscopy. Nature 162, 76–77.

Ofek, G., Natoli, R.M., Athanasiou, K.A., 2009. In situ mechanical properties of the chondrocyte cytoplasm and nucleus. J. Biomech. 42, 873–877.

Offermann, S., Friso, G., Doroshenk, K.A., Sun, Q., Sharpe, R.M., Okita, T.W., Wimmer, D., Edwards, G.E., van Wijk, K.J., 2015. Developmental and subcellular organization of single-cell C_4 photosynthesis in *Bienertia sinuspersici* determined by large-scale proteomics and cDNA assembly from 454 DNA sequencing. J. Proteome Res. 14, 2090–2108.

Offler, C.E., McCurdy, D.W., Patrick, J.W., Talbot, M.J., 2003. Transfer cells: cells specialized for a special purpose. Annu. Rev. Plant Biol. 54, 431–454.

Ogata, K., Nohara, H., 1957. The possible role of the ribonucleic acid (RNA) of the pH 5 enzyme in amino acid activation. Biochim. Biophys. Acta 25, 659–660.

Ogren, W.L., 2003. Affixing the O to rubisco: discovering the source of photorespiratory glycolate and its regulation. Photosynth. Res. 76, 53–63.

Ogren, W.L., Bowes, G., 1971. Ribulose diphosphate carboxylase regulates soybean photorespiration. Nat. New Biol. 230, 159–160.

Ogston, A.G., 1948. Interpretation of experiments on metabolic processes, using isotope tracer elements. Nature 162, 963.

Ogston, A.G., 1955. Chemistry of proteins, peptides, and amino acids. Ann. Rev. Biochem. 24, 191–206.

Ogston, A.G., 1995. Person reflections. Biophys. Chem. 57, 3–5.

Ogston, A.G., Smithies, O., 1948. Some thermodynamic and kinetic aspects of metabolic phosphorylation. Physiol. Rev. 28, 283–303.

Ohana, P., Delmer, D.P., Volman, G., Steffens, J.C., Matthews, D.E., Benziman, M., 1992. ß-furfuryl-ß glucoside: an endogenous activator of higher plant UDP-glucose:(1,3)-ß-glucan synthase. Plant Physiol. 98, 708–715.

Ohlrogge, J.B., Kuhn, D.N., Stumpf, P.K., 1979. Subcellular localization of acyl carrier protein in leaf protoplasts of *Spinacia oleracea*. Proc. Natl. Acad. Sci. U.S.A. 76, 1194–1198.

Ohm, G.S., 1827. Die galvanische Kette, mathematisch bearbeitet, 1966, Translated by W. Francis. In: Taylor, R. (Ed.), The Galvanic Current invesitgated Mathematically. Scientific Memoirs, Selected from the Transactions of Foreign Academies of Science and Learned Societies and from Foreign Journals, Printed by Richard and John E. Taylor, London, vol. 2. Reprinted by Johnson Reprint Corp., New York, pp. 401–506.

Ohnishi, M., Yoshida, K., Mimura, T., 2018. Analyzing the vacuolar (tonoplast) proteome. Methods Mol. Biol. 1696, 107–116.

Ohno, T., Takamatsu, N., Meshi, T., Okada, Y., Nishiguchi, M., Kiho, Y., 1983. Single amino acid substitution in 30 K protein of TMV defective in virus transport function. Virology 131, 255–258.

Ohsuka, K., Inoue, A., 1979. Identification of myosin in a flowering plant *Egeria densa*. J. Biochem. 85, 375–378.

Ohsumi, Y., 2001. Molecular dissection of autophagy: two ubiquitin-like systems. Nat. Rev. Mol. Cell Biol. 2, 211–216.

Ohsumi, Y., 2006. Protein turnover. IUBMB Life 58, 363–369.

Ohsumi, Y., 2014. Historical landmarks of autophagy research. Cell Res. 24, 9–23.

Ohsumi, Y., December 7, 2016. Autophagy: An Intracellular Recycling System. Nobel Lecture.

Ohta, S., Bukowski-Wills, J.-C., Sanchex-Pulido, L., de Lima Alves, F., Wood, L., Chen, Z.A., Plantani, M., Fischer, L., Hudson, D.F., Ponting, C.P., Fukagawa, T., Earnshaw, W.C., Rappsilber, J., 2010. The protein composition of mitotic chromosomes determined using multiclassifier combinatorial proteomics. Cell 142, 810–821.

Ohta, S., Bukowski-Wills, J.C., Sanchez-Pulido, L., Alves Fde, L., Wood, L., Chen, Z.A., Platani, M., Fischer, L., Hudson, D.F., Ponting, C.P., Fukagawa, T., Earnshaw, W.C., Rappsilber, J., 2010. The protein composition of mitotic chromosomes determined using multiclassifier combinatorial proteomics. Cell 142, 810–821.

Ohta, T., Kimura, M., 1971. Amino acid composition of proteins as a product of molecular evolution. Science 174, 150–153.

Ohtsu, K., Smith, M.B., Emrich, S.J., Borsuk, L.A., Zhou, R., Chen, T., Zhang, X., Timmermans, M.C.P., Beck, J., Buckner, B., Janick-Buckner, D., Nettleton, D., Scanlon, M.J., Schnable, P.S., 2007a. Global gene expression of the shoot apical meristem of maize (*Zea mays* L.). Plant J. 52, 391–404.

Ohtsu, K.H., Takahashi, H., Schnable, P.S., Nakazono, M., 2007b. Cell type-specific gene expression profiling in plants by using a combination of laser microdissection and high-throughput technologies. Plant Cell Physiol. 48, 3–7.

Ohya, Y., Anraku, Y., 1989. A galactose-dependent and I mutant of *Saccharomyces cerevisiae*: involvement of calmodulin in nuclear division. Am. Genet. 15, 113–120.

Ohyama, K., Fukuzawa, H., Kohchi, T., Shirai, H., Sano, T., Sano, S., Umesono, K., Shiki, Y., Takeuchi, M., Chang, Z., Aota, S.-I., Inokuchi, H., Ozeki, H., 1986. Chloroplast gene organization deduced from complete sequence of liverwort, *Marchantia polymorpha* chloroplast DNA. Nature 322, 572–574.

Oikawa, A., Lund, C.H., Sakuragi, Y., Scheller, H.V., 2013. Golgi-localized enzyme complexes for plant cell wall biosynthesis. Trends Plant Sci. 18, 49–58.

Oikawa, K., Matsunaga, S., Mano, S., Kondo, M., Yamada, K., Hayashi, M., Kagawa, T., Kadota, A., Sakamoto, W., Higashi, S., Watanabe, M., Mitsui, T., Shigemasa, A., Iino, T., Hosokawa, Y., Nishimura, M., 2015. Physical interaction between peroxisomes and chloroplasts elucidated by *in situ* laser analysis. Nat. Plants 1, 15035.

Oiwa, K., Takahashi, K., 1988. The force-velocity relationship for microtubule sliding in demembranated sperm flagella of the sea urchin. Cell Struct. Funct. 13, 193–205.

Okada, K., Ueda, J., Komaki, M.K., Bell, C.J., Shimura, Y., 1991. Requirement of the auxin polar transport system in early stages of *Arabidopsis* floral bud formation. Plant Cell 3, 677–684.

Okamoto, A., Katsumi, M., Okamoto, H., 1995. The effects of auxin on the mechanical properties in vivo of cell wall in hypocotyl segments from gibberellin-deficient cowpea seedlings. Plant Cell Physiol. 36, 645–651.

Okamoto, H., Liu, Q., Nakahori, K., Katou, K., 1989. A pressure-jump method as a new tool in growth physiology for monitoring physiological wall extensibility and effective turgor. Plant Cell Physiol. 30, 979–985.

Okamoto, H., Miwa, C., Masuda, T., Nakahori, K., Katou, K., 1990. Effects of auxin and anoxia on the cell wall yield threshold determined by negative pressure jumps in segments of cowpea hypocotyl. Plant Cell Physiol. 31, 783–788.

Okamoto, A., Nakamura, T., Nakamura, N., Okamoto, H., 1997. Two isolated wall proteins restore independently the pH-dependencies of the extensibility and the yield threshold tension of the glycerinated hollow cylinder of cowpea hypocotyl. Plant Physiol. 114 (Suppl.), 23.

Okamoto, H., Okamoto, A., 1994. The pH-dependent yield threshold of cell wall in a glycerinated hollow cylinder (in vitro system) of cowpea hypocotyl. Plant Cell Environ. 17, 979–983.

Okamoto, A., Okamoto, H., 1995. Two proteins regulate the cell wall extensibility and the yield threshold in glycerinated hollow cylinder of cowpea hypocotyl. Plant Cell Environ. 18, 827–830.

Okamoto-Nakazato, A., 2002. A brief note on the study of yieldin, a wall-bound protein that regulates the yield threshold of the cell wall. J. Plant Res. 115, 309–313.

Okamoto-Nakazato, A., 2002. A brief note on the study of yieldin, a wall-bound protein that regulates the yield threshold of the cell wall. J. Plant Res. 115, 309–313.

Okamoto-Nakazato, A., 2018. Implications of the galactosidase activity of yieldin in the regulatory mechanism of yield threshold that is fundamental to cell wall extension. Physiol. Plant 163, 259–266.

Okamoto-Nakazato, A., Nakamura, T., Okamoto, H., 2000a. The isolation of wall-bound proteins regulating yield threshold tension in glycerinated hollow cylinders of cowpea hypocotyl. Plant Cell Environ. 23, 145–154.

Okamoto-Nakazato, A., Takahashi, K., Katoh-Semba, R., Katou, K., 2001. Distribution of yieldin, a regulatory protein of the cell wall yield threshold, in etiolated cowpea seedlings. Plant Cell Physiol. 42, 952–958.

Okamoto-Nakazato, A., Takahashi, K., Kido, N., Owaribe, K., Katou, K., 2000. Molecular cloning of yieldins regulating the yield threshold of cowpea cell walls: cDNA cloning and characterization of recombinant yieldin. Plant Cell Environ. 23, 155–164.

Okazaki, Y., 1996. Turgor regulation in a brackish water charophyte, Lamprothamnium succinctum. J. Plant Res. 109, 107–112.

Okazaki, Y., 2002. Blue light inactivates plasma membrane H(+)-ATPase in pulvinar motor cells of *Phaseolus*. Plant Cell Physiol. 43, 860–868.

Okazaki, Y., Ishigami, M., Iwasaki, N., 2002. Temporal relationship between cytosolic free Ca^{2+} and membrane potential during hypotonic turgor regulation in a brackish water charophyte *Lamprothamnium succinctum*. Plant Cell Physiol. 43, 1027–1035.

Okazaki, Y., Iwasaki, N., 1991. Injection of a Ca^{2+} chelating agent into the cytoplasm retards the progress of turgor regulation upon hypotonic treatment in the alga, *Lamprothamnium*. Plant Cell Physiol. 32, 185–194.

Okazaki, K., Kabeya, Y., Miyagishima, S.-Y., 2010. The evolution of the regulatory mechanism of chloroplast division. Plant Signal. Behav. 5, 164–167.

Okazaki, Y., Shimmen, T., Tazawa, M., 1984a. Turgor regulation in a brackish water charophyte, *Lamprothamnium succunctum*. I. Artificial modification of intracellular osmotic pressure. Plant Cell Physiol. 25, 565–571.

Okazaki, Y., Shimmen, T., Tazawa, M., 1984b. Turgor regulation in a brackish water charophyte, *Lamprothamnium succunctum*. II. Changes in K^+, Na^+ and Cl^- concentrations, membrane potential and membrane resistance during turgor regulation. Plant Cell Physiol. 25, 573–581.

Okazaki, Y., Tazawa, M., 1986a. Ca^{2+} antagonist nifedipine inhibits turgor regulation upon hypotonic treatment in internodal cells of *Lamprothamnium*. Protoplasma 134, 65–66.

Okazaki, Y., Tazawa, M., 1986b. Effect of calcium ion on cytoplasmic streaming during turgor regulation in a brackish water charophyte *Lamprothamnium*. Plant Cell Environ. 9, 491–494.

Okazaki, Y., Tazawa, M., 1986c. Involvement of calcium ion in turgor regulation upon hypotonic treatment in *Lamprothamnium succinctum*. Plant Cell Environ. 9, 185–190.

Okazaki, Y., Tazawa, M., 1987a. Dependence of plasmalemma conductance and potential on intracellular free Ca^{2+} in tonoplast-removed cells of a brackish water characeae *Lamprothamnium*. Plant Cell Physiol. 28, 703–708.

Okazaki, Y., Tazawa, M., 1987b. Increase in cytoplasmic calcium content in internodal cells of *Lamprothamnium* upon hypotonic treatment. Plant Cell Environ. 10, 619–621.

Okazaki, Y., Tazawa, M., 1990. Calcium ion and turgor regulation in plant cells. J. Membr. Biol. 114, 189–194.

Okazaki, Y., Yoshimoto, Y., Hiramoto, Y., Tazawa, M., 1987. Turgor regulation and cytoplasmic free Ca^{2+} in the alga *Lamprothamnium*. Protoplasma 140, 67–71.

Okihara, K., Kiyosawa, K., 1988. Ion composition of the *Chara* internode. Plant Cell Physiol. 29, 21–25.

Okuda, K., 2002. Structure and phylogeny of cell coverings. J. Plant Res. 115, 283–288.

Okuda, K., Brown Jr., R.M., 1992. A new putative cellulose- synthesizing complex of Coleochaete scutata. Protoplasma 168, 51–63.

Okuda, K., Li, L., Kudlicka, K., Kuga, S., Brown Jr., R.M., 1993. β glucan synthesis in the cotton. Identification of β-1,4 and β-1,3 glucans synthesized in vitro. Plant Physiol. 101, 1131–1142.

Okuda, K., Sekida, S., Yoshinaga, S., Suetomo, Y., 2004. Cellulose-synthesizing complexes in some chromophyte algae. Cellulose 11, 365–376.

Okumoto, S., Jones, A., Frommer, W.B., 2012. Quantitative imaging with fluorescent biosensors. Annu. Rev. Plant Biol. 63, 663–706.

Okushima, Y., Overvoorde, P.J., Arima, K., Alonso, J.M., Chan, A., Chang, C., Ecker, J.R., Hughes, B., Lui, A., Nguyen, D., Onodera, C., Quach, H., Smith, A., Yu, G., Theologis, A., 2005. Functional genomic analysis of the AUXIN RESPONSE FACTOR gene family members in *Arabidopisis thaliana*: unique and overlapping functions of ARF7 and ARF19. Plant Cell 17, 444–463.

Okuwaki, T., Takahashi, H., Itoh, R., Toda, K., Kawazu, T., Kuroiwa, H., Kuroiwa, T., 1996. Ultrastructures of the Golgi body and cell surface in *Cyanidioschyzon merolae*. Cytologia 61, 69–74.

Olby, R., 1972. Francis Crick, DNA, and the central dogma. In: Holton, G. (Ed.), The Twentieth-Century Sciences. W. W. Norton & Co., New York, pp. 227–280.

Oldfield, Y.I., Oldfield, C.J., Niklas, K.J., Dunker, A.K., 2017. Evidence for a strong correlation between transcription factor protein disorder and organismic complexity. Genome Biol. Evol. 9, 1248–1265.

Olesen, P., 1979. The neck construction in plasmodesmata: evidence for a peripheral sphincter-like structure revealed by fixation with tannic acid. Planta 144, 349–358.

Olins, A.L., Olins, D.E., 1974. Spheroid chromatin units (v bodies). Science 183, 330–332.

Olive, J., Wollman, F.A., Bennoun, P., Recouvreur, M., 1983. Localization of the core and peripheral antennae of photosystem I in the thylakoid membranes of *Chlamydomonas reinhardtii*. Biol. Cell 48, 81–84.

Oliver, D.F., 1994. The glycine decarboxylase complex from plant mitochondria. Ann. Rev. Plant Physiol. Plant Mol. Biol. 45, 323–327.

Oliver, N.A., Greenberg, B.D., Wallace, D.C., 1983. Assignment of a polymorphic polypeptide to the human mitochondrial DNA unidentified reading frame 3 gene by a new peptide mapping strategy. J. Biol. Chem. 258, 5834–5839.

Olivera, B.M., Lehman, I.R., 1967. Linkage of polynucleotides through phosphodiester bonds by an enzyme from *Escherichia coli*. Proc. Natl. Acad. Sci. U.S.A. 57, 1426–1433.

Olmedilla, A., Testillano, P.S., Vincente, O., Delseny, M., Risueño, M.C., 1993. Ultrastructural rRNA localization in plant cell nuclei. J. Cell Sci. 106, 1333–1346.

Olsen, L.J., Ettinger, W.F., Damsz, B., Matsudaira, K., Webb, M.A., Harada, J.J., 1993. Targeting of glyoxysomal proteins to peroxisomes in leaves and roots of a higher plant. Plant Cell 5, 941–952.

Olsen, L.J., Harada, J.J., 1995. Peroxisomes and their assembly in higher plants. Ann. Rev. Plant Physiol. Plant Mol. Biol. 46, 123–146.

Olson, E.C., 1965. The Evolution of Life. Weidenfeld and Nicolson, London.

Onaga, L.A., 2014. Ray Wu as Fifth Business: deconstructing collective memory in the history of DNA sequencing. Stud. Hist. Philos. Sci. C Stud. Hist. Philos. Biol. Biomed. Sci. 46, 1–14.

Onelli, E., Moscatelli, A., 2013. Endocytic pathways and recycling in growing pollen tubes. Plants 2, 211–229.

Onelli, E., Prescianotto-Baschong, C., Caccianiga, M., Moscatelli, A., 2008. Clathrin-dependent and independent endocytotic pathways in tobacco protoplasts revealed by labeling with charged nanogold. J. Exp. Bot. 59, 3051–3068.

Onn, I., Heidinger-Pauli, J.M., Guacci, V., Unal, E., Koshland, D.E., 2008. Sister chromatid cohesion: a simple concept with a complex reality. Ann. Rev. Cell Dev. Biol. 24, 105–129.

Onslow, M.W., 1916. The Anthocyanin Pigments of Plants. Cambridge University Press, Cambridge.

Oparin, A.I., 1924. The Origin of Life. Reprinted in The Origin of Life by J. D. Bernal. 1967. The World Publishing Co., Cleveland, OH.

Oparin, A.I., 1938. Origin of Life, 1953. Dover Publications, New York.

Oparin, A.I., 1964. The Chemical Origin of Life. Charles E. Thomas Publisher, Springfield, IL.

Oparka, K.J., 1994. Plasmolysis: new insights into an old process. New Phytol. 126, 571–591.

Oparka, K., Boevink, P., 2005. Techniques for imaging intercellular transport. In: Oparka, K.J. (Ed.), Plasmodesmata. Blackwell Publishing, Oxford, pp. 241–262.

Oparka, K.J., Prior, D.A.M., 1992. Direct evidence for pressure- generated closure of plasmodesmata. Plant J. 2, 741–750.

Oparka, K.J., Prior, D.A.M., Harris, N., 1990. Osmotic induction of fluid phase endocytosis in onion epidermal cells. Planta 180, 555–561.

Oparka, K.J., Roberts, A.G., 2001. Plasmodesmata. A not so open and shut case. Plant Physiol. 125, 123–126.

Oparka, K.J., Roberts, A.G., Boevink, P., Santa Cruz, S., Roberts, I.M., Pradel, K.S., Imlau, A., Kotlizky, G., Sauer, N., Epel, B., 1999. Simple, but not branched, plasmodesmata allow the nonspecific trafficking of proteins in developing tobacco leaves. Cell 97, 743–754.

Oparka, K.J., Turgeon, R., 1999. Sieve elements and companion cells—traffic control centers of the phloem. Plant Cell 11, 739–750.

Opas, M., Tharin, S., Milner, R.E., Michalak, M., 1996. Identification and localization of calreticulin in plant cells. Protoplasma 191, 164–171.

Opperdoes, F., 2013. A feeling for the cell: Christian de Duve (1917-2013). PLoS Biol. 11, e1001671.

Orci, L., Glick, B.S., Rothman, J.E., 1986. A new type of coated vesicular carrier that appears not to contain clathrin: its possible role in protein transport within the Golgi stack. Cell 46, 171–184.

Orci, L., Malholtra, V., Amherdt, M., Serafini, T., Rothman, J.E., 1989. Dissection of a single round of vesicular transport: sequential intermediates for intercisternal movement in the Golgi stack. Cell 56, 357–368.

Ordentlich, A., Linzer, R.A., Raskin, I., 1991. Alternative respiration and heat evolution in plants. Plant Physiol. 97, 1545–1550.

Orgel, L.E., 1973. The Origins of Life: Molecules and Natural Selection. John Wiley & Sons, New York.

Orgel, L., 1999. Are you serious Dr. Mitchell? Nature 402, 17.

Orgel, L., 2004. Prebiotic adenine revisited: eutectics and photochemistry. Orig. Life Evol. Biosph. 34, 361–369.

Ormenese, S., Bernier, G., Périlleux, C., 2006. Cytokinin application to the shoot apical meristem of *Sinapsis alba* enhances secondary plasmodesmata formation. Planta 224, 1481–1484.

Oró, J., 1960. Synthesis of adenine from ammonium cyanide. Biochem. Biophys. Res. Commun. 2, 407–412.

Oró, J., 1976. Prebiological chemistry and the origin of life. A personal account. In: Kornberg, A., Horecker, B.L., Cornudella, L., Oró, J. (Eds.), Reflections on Biochemistry. In Honour of Severo Ochoa. Pergamon Press, Oxford, pp. 423–443.

Orr-Weaver, T.L., Szostak, J.W., 1983. Yeast recombination: the association between double-strand gap repair and crossing over. Proc. Natl. Acad. Sci. U.S.A. 80, 4417–4421.

Orr-Weaver, T.L., Szostak, J.W., Rothstein, R.J., 1981. Yeast transformation: a model system for the study of recombination. Proc. Natl. Acad. Sci. U.S.A. 78, 6354–6358.

Orr-Weaver, T.L., Szostak, J.W., Rothstein, R.J., 1983. Yeast transformation: a model system for the study of recombination. Proc. Natl. Acad. Sci. U.S.A. 78, 6354–6358.

Ortega, J.K.E., 1985. Augmented growth equation for cell wall expansion. Plant Physiol. 79, 318–320.

Ortega, J.K.E., 1990. Governing equations for plant cell growth. Physiol. Plant 79, 116–121.

Ortiz-Ramírez, C., Hernandez-Coronado, M., Thamm, A., Catarino, B., Wang, M., Dolan, L., Feijó, J.A., Becker, J.D., 2016. A transcriptome atlas of *Physcomitrella patens* provides insights into the evolution and development of land plants. Mol. Plant 9, 205–220.

Osakabe, Y., Osakabe, K., 2015. Genome editing with engineered nucleases in plants. Plant Cell Physiol. 56, 389–400.

Osakabe, K., Osakabe, Y., Toki, S., 2010. Site-directed mutagenesis in Arabidopsis using custom-designed zinc finger nucleases. Proc. Natl. Acad. Sci. U.S.A. 107, 12034–12039.

Osborne, S.G., 1857. Vegetable cell-structure and its formation, as seen in the early stages of the growth of the wheat plant. Q. J. Microsc. Sci. 5, 104–122.

Osborne, T.B., 1924. The Vegetable Proteins, second ed. Longmans, Green and Co., New York.

Osbourn, A.E., Lanzotti, V. (Eds.), 2009. Plant-Derived Natural Products. Synthesis, Function and Application. Springer Science+Businesss Media, New York.

Osibow, K., Malli, R., Kostner, G.M., Graier, W.F., 2006. A new type of non-Ca^{2+}-buffering apo(a)-based fluorescent indicator for intraluminal Ca^{2+} in the endoplasmic reticulum. J. Biol. Chem. 281, 5017–5025.

Östergren, G., 1949. *Luzula* and the mechanism of chromosome movements. Hereditas 35, 445–468.

Östergren, G., 1951. The mechanism of co-orientation in bivalents and multivalents. The theory of orientation by pulling. Hereditas 37, 85–156.

Östergren, G., Molè-Bajer, J., Bajer, A., 1960. An interpretation of transport phenomena at mitosis. Ann. N.Y. Acad. Sci. 90, 381–408.

Östergren, G., Prakken, R., 1946. Behavior on the spindle of the actively mobile chromosome ends of rye. Heriditas 32, 473–494.

Osterhout, W.J.V., 1924. The Nature of Life. The Colver Lectures in Brown University. 1922. Henry Holt, New York.

Osterhout, W.J.V., 1949. Movements of water in cells of *Nitella*. J. Gen. Physiol. 32, 553–557.

Ostermann, J., Orci, L., Tani, K., Amherdt, M., Ravazzola, M., Elazar, Z., Rothman, J.E., 1993. Stepwise assembly of functionally active transport vesicles. Cell 75, 1015–1025.

Osterrieder, A., Hummel, E., Carvalho, C.M., Hawes, C., 2010. Golgi membrane dynamics after induction of a dominant-negative mutant Sar1 GTPase in tobacco. J. Exp. Bot. 61, 405–422.

Osteryoung, K.W., McAndrew, R.S., 2001. The plastid division machine. Annu. Rev. Plant Physiol. Plant Mol. Biol. 52, 315–333.

Osteryoung, K.W., Nunnari, J., 2003. The division of endosymbiotic organelles. Science 302, 1698–1704.

Osteryoung, K.W., Stokes, K.D., Rutherford, S.M., Percival, A.L., Lee, W.Y., 1998. Chloroplast division in higher plants requires members of two functionally divergent gene families with homology to bacterial FtsZ. Plant Cell 10, 1991–2004.

Osteryoung, K.W., Vierling, E., 1995. Conserved cell and organelle division. Nature 376, 473–474.

Ostwald, W., 1891. Solutions. Translated by M. M. Pattison Muir. Longmans, Green, and Co., London.

Ostwald, W., 1900. The Scientific Foundations of Analytical Chemistry. MacMillan and Co., London.

Ostwald, W., 1906. The historical development of general chemistry. Sch. Mines Q. 27, 1–84.

Ostwald, W., 1910. Natural Philosophy. Translated by T. Seltzer. Henry Holt and Co., New York.

Ostwald, W., 1922. An Introduction to Theoretical and Applied Colloid Chemistry The World of Neglected Dimensions. John Wiley & Sons, New York.

Ostwald, W., 1980. Electrochemistry. History and Theory, vols. I and II. Amerind Publishing Co., New Delhi.

Otegui, M.S., 2007. Endosperm cell walls: formation, composition, and functions. Plant Cell Monogr. 8, 159–177.

Otegui, M.S., Mastronarde, D.N., Kang, B.-H., Bednarek, S.Y., Staehelin, L.A., 2001. Three-dimensional analysis of syncytial-type cell plates during endosperm cellularization visualized by high resolution electron tomography. Plant Cell 13, 2033–2051.

Otsuka, M., 2007. Setsuro Ebashi (1922-2006). Proc. Jpn. Acad. Ser. B 83, 179–180.

Ott, R.W., Chua, N.H., 1990. Enhancer sequences from *Arabidopsis thaliana* obtained by library transformation of *Nicotiana tabacum*. Mol. Gen. Genet. 223, 169–179.

Otterbein, L.R., Graceffa, P., Dominguez, R., 2001. The crystal structure of uncomplexed actin in the ADP state. Science 293, 708–711.

Otto, B., Uehlein, N., Sdorra, S., Fischer, M., Ayaz, M., Belastegui-Macadam, X., Heckwork, M., Lachnit, M., Pede, N., Priem, N., Reinhard, A., Siegfart, S., Urban, M., Kaldenhoff, R., 2010. Aquaporin tetramer composition modifies the function of tobacco aquaporins. J. Biol. Chem. 285, 31253–31260.

Ovádi, J., 1991. Physiological significance of metabolite channelling. J. Theor. Biol. 152, 1–22.

Ovádi, J., Srere, P.A., 2000. Macromolecular compartmentation and channeling. Int. Rev. Cytol. 192, 255–280.

Overall, R.L., Gunning, B.E.S., 1982. Intercellular communication in *Azolla* roots. II. Electrical coupling. Protoplasma 111, 151–160.

Overall, R.L., Wolfe, J., Gunning, B.E.S., 1982. Intercellular communication in *Azolla* roots I. Ultrastructure of plasmodesmata. Protoplasma 111, 134–150.

Overton, E., 1893. On the reduction of the chromosomes in the nuclei of plants. Ann. Bot. 7, 139–143.

Overton, C.E., 1899a. Beobachtungen und Versuche über das Auftreten von rothem Zellsaft bei Pflanzen. Jahrb. Fuer Wissenschaftliche Botanik 33, 171–233.

Overton, E., 1899b. Ueber die allgemeinen osmotischen Eigenschaften der Zelle, ihre vermutlichen Ursachen und ihre Bedeutung für die Physiologie. Vierteljahrschr. Naturf. Gesell (Zürich) 44, 88–135. Translated by R. B. Park In: Papers on Biological Membrane Structure. 1968. Selected by D. Branton and R. B. Park. Little, Brown, and Co., Boston.

Overton, E., 1900. Studien über die Aufnahme der Anilinfarben durch die lebende Zelle. Jahr. Wiss. Bot. 34, 669–701.

Ow, D.W., Wood, K.V., DeLuca, M., de Wet, J.R., Helsinki, D.R., Howell, S.H., 1986. Transient and stable expression of the firefly luciferase gene in plant cells and transgenic plants. Science 234, 856–859.

Owen, R., 1848. On the Achetype and Homologies of the Vertebrate Skeleton. Richard and John E. Taylor, London.

Owen, R., 1849. On the Nature of Limbs. John van Voorst, London.

Owen, O.F., 1853. Organon, or Logical Treatises, of Aristotle, vol. I. Henry G. Bohn, London.

Owen, J.D., Eyring, E.M., 1975. Reflection coefficients of permeant molecules in human red cell suspensions. J. Gen. Physiol. 66, 251–265.

Oxford English Dictionary, 1933. A New English Dictionary on Historical Principles. Clarendon Press, Oxford.

Pabo, C.O., Sauer, R.T., 1992. Transcription factors: structural families and principles of DNA recognition. Annu. Rev. Biochem. 61, 1053–1095.

Paddon, C.J., Westfall, P.J., Pitera, D.J., Benjamin, K., Fisher, K., McPhee, D., Leavell, M.D., Tai, A., Main, A., Eng, D., Polichuk, D.R.,

Teoh, K.H., Reed, D.W., Treynor, T., Lenihan, J., Fleck, M., Bajad, S., Dang, G., Dengrove, D., Diola, D., Dorin, G., Ellens, K.W., Fickes, S., Galazzo, J., Gaucher, S.P., Geistlinger, T., Henry, R., Hepp, M., Horning, T., Iqbal, T., Jiang, H., Kizer, L., Lieu, B., Melis, D., Moss, N., Regentin, R., Secrest, S., Tsuruta, H., Vazquez, R., Westblade, L.F., Xu, L., Yu, M., Zhang, Y., Zhao, L., Lievense, J., Covello, P.S., Keasling, J.D., Reiling, K.K., Renninger, N.S., Newman, J.D., 2013. High-level semi-synthetic production of the potent antimalarial artemisinin. Nature 496, 528–532.

Padgette, S.R., Kolacz, K.H., Delannay, X., Re, D.B., LaVallee, D.J., Tinius, C.N., Rhodes, W.K., Otero, Y.I., Barry, G.F., Eichholtz, D.A., Peschke, V.M., Nida, D.L., Taylor, N.B., Kishore, G.M., 1995. Development, identification, and characterization of a glyphosate-tolerant soybean line. Crop Sci. 35, 1451–1461.

Padmanabhan, R., Jay, E., Wu, R., 1974. Chemical synthesis of a primer and its use in the sequence analysis of the lysozyme gene of bacteriophage T4. Proc. Natl. Acad. Sci. U.S.A. 71, 2510–2514.

Padmanabhan, R., Wu, R., 1972a. Nucleotide sequence analysis of DNA: IV. Complete nucleotide sequence of the left-hand cohesive ends of coliphage 186 DNA. J. Mol. Biol. 65, 447–467.

Padmanabhan, R., Wu, R., 1972b. Nucleotide sequence analysis of DNA: IX. Use of oligonucleotides of defined sequence as primers in DNA sequence analysis. Biochem. Biophys. Res. Commun. 48, 1295–1302.

Paecht-Horowitz, M., 1976. Clays as possible catalysts for peptide formation in the prebiotic era. Orig. Life 7, 369–381.

Paecht-Horowitz, M., Berger, J., Katchalsky, A., 1970. Prebiotic synthesis of polypeptides by heterogenous polycondensation of amino-acid adenylates. Nature 228, 636–639.

Paecht-Horowitz, M., Eirich, F.R., 1988. The polymerization of amino acid adenylates on sodium montmorillonite with preadsorbed polypeptides. Orig. Life 18, 359–388.

Pagant, S., Bichet, A., Sugimoto, K., Lerouxel, O., Desprez, T., McCann, M., Lerouge, P., Vernhettes, S., Höfte, H., 2002. KOBITO1 encodes a novel plasma membrane protein necessary for normal synthesis of cellulose during cell expansion in arabidopsis. Plant Cell 14, 2001–2013.

Pagliaro, L., 1993. Glycolysis revisited—A funny thing happened on the way to the Krebs cycle. News Physiol. Sci. 8, 219–223.

Pagliaro, L., 2000. Mechanisms for cytoplasmic organization: an overview. Int. Rev. Cytol. 192, 303–318.

Pagliaro, L., Taylor, D.L., 1988. Aldolase exists in both the fluid and solid phases of cytoplasm. J. Cell Biol. 107, 981–992.

Pagliaro, L., Taylor, D.L., 1992. 2-deoxyglucose and cytochalasin D modulate aldolase mobility in living 3T3 cells. J. Cell Biol. 118, 859–863.

Paila, Y.D., Richardson, L.G.L., Inoue, H., Parks, E.S., McMahon, J., Inoue, K., Schnell, D.J., 2016. Multi-functional roles for the polypeptide transport associated domains of Toc75 in chloroplast protein import. eLife 5, e12631.

Paila, Y.D., Richardson, L.G.L., Schnell, D.J., 2015. New insights into the mechanism of chloroplast protein import and its integration with protein quality control, organelle biogenesis and development. J. Mol. Biol. 427, 1038–1060.

Pain, D., Kanwar, Y.S., Blobel, G., 1988. Identification of a receptor for protein import into chloroplasts and its localization to envelope contact zones. Nature 331, 232–237.

Paine, T., 1880. Letter to Mr. Erskine. In: Blanchard, C. (Ed.), Complete Works of Thomas Paine. Belford, Clarke & Co., Chicago, pp. 237–265.

Painter, T.S., 1939. The structure of salivary gland chromosomes. Am. Nat. 73, 31530.

Painter, T.S., 1940. Recent developments in our knowledge of chromosome structure and their application to genetics. In: Baitsell, G.A. (Ed.), Science in Progress, vol. 1. Yale University Press, New Haven, CT, pp. 210–232.

Palade, G.E., 1952a. A study of fixation for electron microscopy. J. Exp. Med. 95, 285–298.

Palade, G.E., 1952b. The fine structure of mitochondria. Anat. Rec. 114, 427–451.

Palade, G.E., 1953. An electron microscope study of the mitochondria structure. J. Histochem. Cytochem. 1, 188–211.

Palade, G.E., 1955. A small particulate component of the cytoplasm. J. Biophys. Biochem. Cytol. 1, 59–68.

Palade, G.E., 1956. The endoplasmic reticulum. J. Biophys. Biochem. Cytol. 2 (Suppl.), 85–97.

Palade, G.E., 1958. Microsomes and ribonuclear particles. In: Roberts, R.B. (Ed.), Microsomal Particles and Protein Synthesis. Pergamon Press, New York, pp. 36–61.

Palade, G.E., 1959. Functional changes in the structure of cell components. In: Hayashi, T. (Ed.), Subcellular Particles. Ronald Press Co., New York, pp. 64–83.

Palade, G.E., 1963. The organization of living matter. In: The Scientific Endeavor. The Rockefeller Press, New York, pp. 179–203.

Palade, G.E., 1975. Intracellular aspects of the processing of protein synthesis. Science 189, 347–358.

Palade, G.E., 1977. Keith Roberts Porter and the development of contemporary cell biology. J. Cell Biol. 75, D1–D19.

Palade, G.E., Claude, A., 1949. The nature of the Golgi apparatus. J. Morphol. 85, 35–70 71–112.

Palade, G.E., Porter, K.R., 1954. Studies on the endoplasmic reticulum. I. Its identification in cells in situ. J. Exp. Med. 100, 641–656.

Palade, G.E., Siekevitz, P., 1956. Liver microsomes. An integrated morphological and biochemical study. J. Biophys. Biochem. Cytol. 2, 171–199.

Palay, S.L., Palade, G.E., 1955. The fine structure of neurons. J. Biophys. Biochem. Cytol. 1, 69–88.

Paleg, L.G., 1965. Physiological effects of gibberellins. Ann. Rev. Plant Physiol. 16, 291–322.

Palevitz, B.A., 1980. Comparative effects of phalloidin and cytochalasin B on motility and morphogenesis in *Allium*. Can. J. Bot. 58, 773–785.

Palevitz, B.A., 1987a. Accumulation of F-actin during cytokinesis in Allium. Correlation with microtubule distribution and the effects of drugs. Protoplasma 141, 24–32.

Palevitz, B.A., 1987b. Actin in the preprophase band of *Allium cepa*. J. Cell Biol. 104, 1515–1519.

Palevitz, B.A., 1988a. Cytochalasin-induced reorganization of actin in *Allium* root cells. Cell Motil. Cytoskelet. 9, 283–298.

Palevitz, B.A., 1988b. Microtubular fir-trees in mitotic spindles of onion roots. Protoplasma 142, 74–78.

Palevitz, B.A., Cresti, M., 1989. Cytoskeletal changes during generative cell division and sperm formation in *Tradescantia virginiana*. Protoplasma 150, 54–71.

Palevitz, B.A., Hepler, P.K., 1974a. The control of the plane of cell division during stomatal differentiation in *Allium*. I. Spindle reorientation. Chromosoma 46, 297–341.

Palevitz, B.A., Hepler, P.K., 1974b. The control of the plane of division during stomatal differentiation in *Allium*. II. Drug studies. Chromosoma 46, 327–341.

Palevitz, B.A., Hepler, P.K., 1975. Identification of actin in situ at the ectoplasm-endoplasm interface of Nitella. J. Cell Biol. 65, 29–38.

Palevitz, B.A., Hepler, P.K., 1976. Cellulose microfibril orientation and cell shaping in developing guard cells of *Allium*: the role of microtubules and ion accumulation. Planta 132, 71–93.

Palevitz, B.A., Hepler, P.K., 1985. Changes in dye coupling of stomatal cells of *Allium* and *Commelina* demonstrated by microinjection of Lucifer yellow. Planta 164, 473–479.

Palevitz, B.A., O'Kane, D.J., 1981. Epifluorescence and video analysis of vacuole motility and development in stomatal cells of *Allium*. Science 214, 443–445.

Palevitz, B.A., O'Kane, D.J., Kobres, R.E., Railhel, N.V., 1981. The vacuole system in stomatal cells of *Allium*. Vacuole movements and changes in morphology in differentiating cells as revealed by epifluorescence, video and electron microscopy. Protoplasma 109, 23–55.

Palin, R., Geitmann, A., 2012. The role of pectin in plant morphogenesis. BioSystems 109, 397–402.

Paller, M.S., 1994. Lateral mobility of Na,K-ATPase and membrane lipids in renal cells. Importance of cytoskeletal integrity. J. Membr. Biol. 142, 127–135.

Palmer, J.D., 1997. Organelle genomes: going, going, gone! Science 275, 790–791.

Palmer, J.D., Deleiche, C.F., 1996. Second-hand chloroplasts and the case of the disappearing nucleus. Proc. Natl. Acad. Sci. U.S.A. 93, 7432–7435.

Palmer, A.E., Jin, C., Reed, J.C., Tsien, R.Y., 2004. Bcl-2-mediated alterations in endoplasmic reticulum Ca^{2+} analyzed with an improved genetically encoded fluorescent sensor. Proc. Natl. Acad. Sci. U.S.A. 101, 17404–17409.

Palmer, J.D., Shields, C.R., 1984. Tripartite structure of the Brassica campestria mitochondrial genome. Nature 307, 437–440.

Palmgren, M.G., Christensen, G., 1993. Complementation in situ of the yeast plasma membrane H^+-ATPase gene PMA1 by an H^+-ATPase gene from a heterologous species. FEBS Lett. 317, 216–222.

Palmgren, M.G., Sommarin, M., Serrano, R., Larsson, C., 1991. Identification of an autoinhibitory domain in the C-terminal region of the plant plasma membrane H^+-ATPase. J. Biol. Chem. 266, 20470–20475.

Palmieri, M., Kiss, J.Z., 2005. Disruption of the F-actin cytoskeleton limits statolith movement in *Arabidopsis* hypocotyls. J. Exp. Bot. 56, 2539–2550.

Palmieri, M., Schwind, M.A., Stevens, M.H.H., Edelmann, R.E., Kiss, J.Z., 2007. Effects of the myosin ATPase inhibitor 2,3-butanedione monoxime on amyloplast kinetics and gravitropism of *Arabidopsis* hypocotyls. Phys. Plant 130, 613–626.

Pandey, S., Chen, J.G., Jones, A.M., Assmann, S.M., 2006. G-protein complex mutants are hypersensitive to ABA-regulation of germination and post-germination development. Plant Physiol. 141, 243–256.

Pandey, S., Nelson, D.C., Assmann, S.M., 2009. Two novel GPCR-type G proteins are abscisic acid receptors in *Arabidopsis*. Cell 136, 136–148.

Pandey, P., Senthil-Kumar, M., Mysore, K.S., 2015. Advances in plant gene silencing methods. In: Mysore, K.S., Senthil-Kumar, M. (Eds.), Plant Gene Silencing: Methods and Protocols Methods in Molecular Biology, vol. 1287. Springer Science+Business Media, New York, pp. 3–23.

Panteris, E., Apostolakos, P., Galatis, B., 1993. Microtubule organization, mesophyll cell morphogenesis, and intercellular space formation in *Adiantum capillus* veneris leaflets. Protoplasma 172, 97–110.

Panteris, E., Apostolakos, P., Quader, H., Galatis, B., 2004. A cortical ring predicts the division plane in vacuolated cells of *Coleus*: the role of actomyosin and microtubules in the establishment and function of the division site. New Phytol. 163, 271–286.

Paolillo Jr., D.J., 1962. The plastids of *Isoetes Howellii*. Am. J. Bot. 49, 590–598.

Paolillo Jr., D.J., 1967. On the structure of the axoneme in flagella of *Polytrichum juniperinum*. Trans. Am. Microsc. Soc. 86, 428–433.

Paolillo Jr., D.J., 1970. The threedimensional arrangement of intergranal lamellae in chloroplasts. J. Cell Sci. 6, 243–255.

Paolillo Jr., D.J., 1981. The swimming sperms of land plants. BioScience 31, 367–373.

Paolillo Jr., D.J., 2000. Axis elongation can occur with net longitudinal orientation of wall microfibrils. New Phytol. 145, 449–455.

Paolillo Jr., D.J., Falk, R.H., 1966. The ultrastructure of grana in mesophyll plastids of *Zea mays*. Am. J. Bot. 53, 173–180.

Paolillo Jr., D.J., MacKay, N.C., Reighard, J.A., 1969. The structure of grana in flowering plants. Am. J. Bot. 56, 344–347.

Paolillo Jr., D.J., Reighard, J.A., 1967. On the relationship between mature structure and ontogeny in the grana of chloroplasts. Can. J. Bot. 45, 773–782.

Paolillo Jr., D.J., Rubin, G., 1980. Reconstruction of the grana-fretwork system of a chloroplast. Am. J. Bot. 67, 575–584.

Pappas, G.D., 1956. The fine structure of the nuclear envelope of *Amoeba proteus*. J. Biophys. Biochem. Cytol. 2, 431–444.

Parcharoenwattana, L., Smith, S.M., 2008. When is a peroxisome not a peroxisome? Trends Plant Sci. 13, 522–525.

Pardee, A.B., 1989. G1 events and regulation of cell proliferation. Science 246, 603–608.

Pardo, J.M., Serrano, R., 1989. Structure of a plasma membrane H^+-ATPase gene from the plant *Arabidopsis thaliana*. J. Biol. Chem. 264, 8557–8562.

Paredez, A.R., Persson, S., Ehrhardt, D.W., Somerville, C.R., 2008. Genetic evidence that cellulose synthase activity influences microtubule cortical array organization. Plant Physiol. 147, 1723–1734.

Paredez, A.R., Somerville, C.R., Erhardt, D.W., 2006. Visualization of cellulose synthase demonstrates functional association with microtubules. Science 312, 1491–1495.

Pareek, C.S., Smoczynski, R., Tretyn, A., 2011. Sequencing technologies and genome sequencing. J. Appl. Genet. 52, 413–435.

Parinov, S., Sundaresan, V., 2000. Functional genomics in *Arabidopsis*: large-scale insertional mutagenesis complements the genome sequencing project. Curr. Opin. Biotechnol. 11, 157–161.

Paris, N., Stanley, C.M., Jones, R.L., Rogers, J.C., 1996. Plant cells contain two functionally distinct vacuolar compartments. Cell 85, 563–572.

Park, S.Y., Fung, P., Nishimura, N., Jensen, D.R., Fujii, H., Zhao, Y., Lumba, S., Santiago, J., Rodrigues, A., Chow, T.F., Alfred, S.E., Bonetta, D., Finkelstein, R., Provart, N.J., Desveaux, D., Rodriguez, P.L., McCourt, P., Zhu, J.-K., Schroeder, J.I., Volkman, B.F., Cutler, S.R., 2009. Abscisic acid inhibits type 2C protein phosphatases via the PYR/PYL family of START proteins. Science 324, 1068−1107.

Parker, W.C., Chakraborty, N., Vrikkis, R., Elliot, G., Smith, S., Moyer, P.J., 2010. High-resolution intracellular viscosity measurement using time-dependent fluorescence anisotropy. Optics Express 18, 16607−16617.

Parker, M.L., Hawes, C.R., 1982. The Golgi apparatus in developing endosperm of wheat (*Triticum aestivum* L.). Planta 154, 277−283.

Parker, J., Miller, C., Anderson, B.H., 1986. Higher plant myosin heavy-chain identified using a monoclonal antibody. Eur. J. Cell Biol. 41, 9−13.

Parnas, J.K., 1910. Ueber fermentative Beschleunigung der Cannizzaroschen Aldehydumlagerung durch Gewebssäfte. Biochem. Z. 28, 274−294.

Parpart, A.K., Dziemian, A.J., 1940. The chemical composition of the red cell membrane. Cold Spring Harb. Symp. Quant. Biol. 8, 17−24.

Parre, E., Geitmann, A., 2005. Pectin and the role of the physical properties of the cell wall in pollen tube growth of *Solanum chacoense*. Planta 220, 582−592.

Parsons, H.T., Christiansen, K., Knierim, B., Carroll, A., Ito, J., Batth, T.S., Smith-Moritz, A.M., Morrison, S., McInerney, P., Hadi, M.Z., Auer, M., Mukhopadhyay, A., Petzold, C.J., Scheller, H.V., Loqué, D., Heazlewood, J.L., 2012. Isolation and proteomic characterization of the *Arabidopsis* Golgi defines functional and novel components involved in plant cell wall biosynthesis. Plant Physiol. 159, 12−16.

Parsons, H.T., Drakakaki, G., Heazlewood, J.L., 2013. Proteomic dissection of the *Arabidopsis* Golgi and trans-Golgi network. Front. Plant Sci. 3, 298.

Parsons, M., Karnataki, A., Feagin, J.E., DeRocher, A., 2007. Protein trafficking to the Apicoplast. Deciphering the apicomplexan solution to secondary endosymbiosis. Eukaryot. Cell 6, 1081−1088.

Parthasarathy, M.V., 1974. Ultrastructure of phloem in palms. III. Mature phloem. Protoplasma 79, 265−315.

Parthasarathy, M.V., 1985. F-actin architecture in coleoptile epidermal cells. Eur. J. Cell Biol. 39, 1−12.

Parthasarathy, M.V., Mühlethaler, K., 1972. Cytoplasmic microfilaments in plant cells. J. Ultrastruc. Res. 38, 46−62.

Parthasarathy, M.V., Perdue, T.D., Witztum, A., Alvernaz, J., 1985. Actin network as a normal component of the cytoskeleton in many vascular plant cells. Am. J. Bot. 72, 1318−1323.

Partikian, A., Ölveczky, B., Swaminathan, R., Li, Y., Verkman, A.S., 1998. Rapid diffusion of green fluorescent protein in the mitochondrial matrix. J. Cell Biol. 140, 821−829.

Partington, J.R., 1933. The scientific work of Joseph Priestley. Nature 131, 348−350.

Parton, R.M., Fischer-Parton, S., Watahiki, M.K., Trewavas, A.J., 2001. Dynamics of the apical vesicle accumulation and the rate of growth are related in individual pollen tubes. J. Cell Sci. 114, 2685−2695.

Parys, J.B., Bootman, M.D., Yule, D.I., Bultynck, G. (Eds.), 2014. Calcium Techniques. A Laboratory Manual. Cold Spring Harbor Laboratory Press, Cold Spring Harbor, New York.

Paschal, B.M., Shpetner, H.S., Vallee, R.B., 1987. MAP 1C is a microtubule-activated ATPase which translocates microtubules in vitro and has dynein-like properties. J. Cell Biol. 105, 1273−1282.

Pascual, D.W., 2007. Vaccines are for dinner. Proc. Natl. Acad. Sci. U.S.A. 104, 10757−10758.

Pastan, I., Willingham, M.C., 1985. Endocytosis. Plenum Press, New York.

Pasternak, C.A., 1993. A glance back over 30 years. Biosci. Rep. 13, 191−212.

Pasteur, L., 1910. The physiological theory of fermentation [1879]. In: Scientific Papers. The Harvard Classics, 38, pp. 275−363.

Pastuglia, M., Azimzadeh, J., Goussot, M., Camilleri, C., Belcram, K., Evard, J.L., Schmitt, A.-C., Guerche, P., Bouchez, D., 2006. γ-tubulin is essential for microtubule organization and development in *Arabidopsis*. Plant Cell 18, 1412−1425.

Pate, J.S., Gunning, B.E.S., 1972. Transfer cells. Annu. Rev. Plant Physiol. 23, 173−196.

Patel, S., Penny, C.J., Rahman, T., 2016. Two-pore channels enter the atomic era: structure of plant TPC revealed. Trends Biochem. Sci. 41, 475−477.

Patel, N.B., Poo, M.-M., 1984. Orientation of neurite growth by extracellular electric fields. J. Neurosci. 2, 483−496.

Paterson, A.H., Lander, E.S., Hewett, J.W., Peterson, S., Lincoln, S.E., Tanksley, S.D., 1988. Resolution of quantitative traits into Mendelian factors by using a complete linkage map of restriction fragment polymorphisms. Nature 335, 721−726.

Patient, R.K., Allan, J., 1989. Active chromatin. Curr. Opin. Cell Biol. 1, 454−459.

Pattathil, S., Avci, U., Baldwin, D., Swennes, A.G., McGill, J.A., Popper, Z., Bootten, T., Albert, A., Davis, R.H., Chennareddy, C., Dong, R., O'Shea, B., Rossi, R., Leoff, C., Freshour, G., Narra, R., O'Neil, M., Yorkm, W.S., Hahn, M.G., 2010. A comprehensive toolkit of plant cell wall glycan-directed monoclonal antibodies. Plant Physiol. 153, 514−525.

Patterson, S.D., 2004a. How much of the proteome do we see with discovery-based proteomics methods and how much do we need to see? Curr. Proteom. 1, 3−12.

Patterson, S.D., 2004b. Proteomics: beginning to realize its promise? Arthritis Rheum. 50, 3741−3744.

Pattison Muir, M.M., 1909. A History of Chemical Theories and Laws. John Wiley & Sons, New York.

Paul, P., Chaturvedi, P., Selymesi, M., Ghatak, A., Mesihovic, A., Scharf, K.D., Weckwerth, W., Simm, S., Schleiff, E., 2016. The membrane proteome of male gametophyte in *Solanum lycopersicum*. J. Proteom. 131, 48−60.

Paul, D.C., Goff, S.C., 1973. Comparative effects of caffeine, its analogues and calcium deficiency on cytokinesis. Exp. Cell Res. 78, 399−413.

Paul III, J.W., Qi, Y., 2016. CRISPR/Cas9 for plant genome editing: accomplishments, problems and prospects. Plant Cell Rep. 35, 1417−1427.

Pauli, W., 1907. Physical Chemistry in the Service of Medicine. Translated by Martin H. Fischer. John Wiley & Sons, New York.

Pauli, W., 1934. The chemistry of the amino acids and the proteins. Ann. Rev. Biochem. 3, 111−132.

Pauli, W., 1958. Theory of Relativity. Dover Publications, New York.

Paulik, M.A., Collignon, C.M., Keenan, T.W., Morré, D.J., 1987. Transition vesicle-enriched fractions isolated from rat liver exhibit a unique peptide pattern. Protoplasma 141, 180−184.

Pauling, L., 1932. The nature of the chemical bond. IV. The energy of single bonds and the relative electronegativity of atoms. J. Am. Chem. Soc. 54, 3570−3582.

Pauling, L., 1940. The Nature of the Chemical Bond, second ed. Cornell University Press, Ithaca, New York.

Pauling, L., 1954a. General Chemistry, second ed. W, H. Freeman and Co., San Francisco.

Pauling, L., 1970. General Chemsitry. Dover Publications, New York.

Pauling, L., 1972. Fifty years of progress in structural chemistry and molecular biology. In: Holton, G. (Ed.), The Twentieth- Century Sciences. W. W. Norton & Co., New York, pp. 281–307.

Pauling, P., 1973. DNA—The race that never was? New Sci. 58, 558–560.

Pauling, L., 1976. Vitamin C, The Common Cold, and the Flu. W.H. Freeman, San Francisco.

Pauling, L., Corey, R.B., 1953. A proposed structure for the nucleic acids. Proc. Natl. Acad. Sci. U.S.A. 39, 84–97.

Pauling, L., Corey, R.B., Hayward, R., 1955. The structure of proteins. In: The Physics and Chemistry of Life. Simon and Schuster, New York, pp. 74–86.

Pauling, L., December 11, 1954b. Modern Structural Chemistry. Nobel Lecture.

Pauling, L., December 11, 1963. Science and Peace. Nobel Lecture.

Pauling, L., Delbrück, M., 1940. The nature of the intermolecular forces operative in biological processes. Science 92, 77–79.

Pauling, C., Hamm, L., 1968. Properties of a temperature-sensitive radiation-sensitive mutant of *Escherichia coli*. Proc. Natl. Acad. Sci. U.S.A. 60, 1495–1502.

Pauling, L., Itano, H.A., Singer, S.J., Wells, I.C., 1949. Sickle cell anemia, a molecular disease. Science 110, 543–548.

Pauling, L., Wilson Jr., E.B., 1935. Introduction to Quantum Mechanics with Applications to Chemistry. McGraw-Hill, New York.

Pauliulis, L.V., Nicklas, R.B., 2000. The reduction of chromosome number in meiosis is determined by properties built into the chromosomes. J. Cell Biol. 150, 1223–1232.

Paulson, J.C., McClure, W.O., 1973. Inhibition of axoplasmic transport by mescaline and other trimethoxyphenylalkylamines. Mol. Pharmacol. 9, 41–50.

Pauluzzi, G., Bailey-Serres, J., 2016. Flexible ion barrier. Cell 164, 345–346.

Paves, H., Truve, E., 2004. Incorporation of mammalian actin into microfilaments in plant cell nucleus. BMC Plant Biol. 4, 7–15.

Paweletz, N., 1967. Zur Funktion des "Flemming-Körpers" bei der Teilung tirisher Zellen. Naturwissenschaften 54, 533–535.

Paweletz, N., 1996. A tribute to Daniel Mazia (1912–1996)—pioneer in the science of the mitotic apparatus. Chromosome Res. 4, 409–410.

Pawlowski, W.P., Cande, W.Z., 2005. Coordinating the events of the meiotic prophase. Trends Cell Biol. 15, 674–681.

Pawlowski, W.P., Harper, L.C., Cande, W.Z., 2004. Meiosis. Encylopedia of Plant and Crop Science. Marcel Dekker, New York, pp. 711–713.

Pawlowski, W.P., Sheehan, M.J., Ronceret, A., 2007. In the beginning: the initiation of meiosis. BioEssays 29, 511–514.

Pay, A., Resch, K., Frohnmeyer, H., Fejes, E., Nagy, F., Nick, P., 2002. Plant RanGAPs are localized at the nuclear envelope in interphase and associated with microtubules in mitotic cells. Plant J. 30, 699–709.

Pazour, G.J., Agrin, N., Lesyk, J., Whitman, G.B., 2005. Proteomic analysis of a eukaryotic cilium. J. Cell Biol. 170, 103–113.

Pazour, G.J., Witman, G.B., 2003. The vertebrate primary cilium is a sensory organelle. Curr. Opin. Cell Biol. 15, 105–110.

Peachey, L.D., 2013. Keith R. Porter. 1912–1997. Biographical Memoirs of the National Academy of Sciences. http://www.nasonline.org/publications/biographical-memoirs/memoir-pdfs/porter-keith.pdf.

Peachey, L.D., Brinkley, B.R., 1983. Scientific achievements and contributions of Keith R. Porter to modern cell biology. Mod. Cell Biol. 2, 1–12.

Pear, J.R., Kawagoe, Y., Schreckengost, W.E., Delmer, D.P., Stalker, D.M., 1996. Higher plants contain homologs of the bacterial celA genes encoding the catalytic subunit of cellulose synthase. Proc. Natl. Acad. Sci. U.S.A. 93, 12637–12642.

Pearse, B.M.F., 1976. Clathrin: a unique protein associated with intracellular transfer of membrane by coated vesicles. Proc. Natl. Acad. Sci. U.S.A. 73, 1255–1259.

Pearse, B.M.F., Robinson, M.S., 1990. Clathrin, adaptors and sorting. Annu. Rev. Cell Biol. 6, 151–171.

Pearson, K., 1905. National Life from the Standpoint of Science. Adam and Charles Black, London.

Pearson, W.R., 1991. Identifying distantly related protein sequences. Curr. Opin. Struct. Biol. 1, 321–326.

Pearson, W.R., Lipman, D.J., 1988. Improved tools for biological sequence comparison. Proc. Natl. Acad. Sci. U.S.A. 85, 2444–2448.

Pearson, K., Moul, M., 1925. The problem of alien immigration into Great Britain, illustrated by an examination of Russian and Polish Jewish children. Ann. Eugenics 1, 5–127.

Pease, D.C., 1941. Hydrostatic pressure effects upon the spindle figure and chromosome movement. I. Experiments on the first mitotic division of Urechis eggs. J. Morphol. 69, 405–441.

Pease, D.C., 1946. Hydrostatic pressure effects upon the spindle figure and chromosome movement. II. Experiments on the meiotic divisions of *Tradescantia* pollen mother cells. Biol. Bull. 91, 145–169.

Pease, D.C., 1963. The ultrastructure of flagellar fibrils. J. Cell Biol. 18, 313–326.

Pease, D.C., Porter, K.R., 1981. Electron microscopy and ultramicrotomy. J. Cell Biol. 91, 287s–292s.

Pease, A.C., Solas, D., Sullivan, E.J., Cronin, M.T., Holmes, C.P., Fodor, S.P.A., 1994. Light-directed oligonucleotide arrays for rapid DNA sequence analysis. Proc. Natl. Acad. Sci. U.S.A. 91, 5022–5026.

Peaucelle, A., Braybrook, S., Höfte, H., 2012. Cell wall mechanics and growth control in plants: the role of pectins revisited. Front. Plant Sci. 3, 121.

Peaucelle, A., Braybrook, S.A., Le Guillou, L., Bron, E., Kuhlemeier, C., Höfte, H., 2011. Pectin-induced changes in cell wall mechanics underlie organ initiation in *Arabidopsis*. Curr. Biol. 21, 1720–1726.

Peaucelle, A., Louvet, R., Johansen, J.N., Höfte, H., Laufs, P., Pelloux, J., Mouille, G., 2008. *Arabidopsis* phyllotaxis is controlled by the methyl-esterification status of cell-wall pectins. Curr. Biol. 18, 1943–1948.

Pedersen, K.O., 1948. The chemistry of the proteins and amino acids. Ann. Rev. Biochem. 17, 169–200.

Pedersen, C.N.S, Axelsen, K.B., Harper, J.F., Palmgren, M.G., 2012. Evolution of plant P-type ATPases. Front. Plant Sci. 3, 31.

Pederson, D.S., Thomas, F., Simpson, R.T., 1986. Core particle, fiber, and transcriptionally active chromatin structure. Ann. Rev. Cell Biol. 2, 117–147.

Pedrazzini, E., Komarova, N.Y., Rentsch, D., Vitale, A., 2013. Traffic routes and signals for the tonoplast. Traffic 14, 622–628.

Peebles, P.J.E., 1993. Principles of Physical Cosmology. Princeton University Press, Princeton, NJ.

Peeler, T.C., Stephenson, M.B., Einspar, K.J., Thompson Jr., G.A., 1989. Lipid characterization of an enriched plasma membrane fraction of *Dunaliella salina* grown in media of varying salinity. Plant Physiol. 89, 970–976.

Pei, Z.-M., Baizabal-Aguirre, V.W., Allen, G.J., Schroeder, J.I., 1998a. A transient outward-rectifying K^+ channel current down-regulated by cytosolic Ca^{2+} in *Arabidopsis thaliana* guard cells. Proc. Natl. Acad. Sci. U.S.A. 95, 6548–6553.

Pei, Z.-M., Ghassemian, M., Kwak, C.M., McCourt, P., Schroeder, J.I., 1998b. Role of farnesyltransferase in AbA regulation of guard cell anion channels and plant water loss. Science 282, 287–290.

Pei, Z.-M., Kuchitsu, K., Ward, J.M., Schwartz, M., Schroeder, J.I., 1997. Differential abscisic acid regulation of guard cell slow anion channels in *Arabidopsis* wild-type and abi1 and abi2 mutants. Plant Cell 9, 409–423.

Pelham, H.R.B., 1989a. Control of protein exit from the endoplasmic reticulum. Annu. Rev. Cell Biol. 5, 1–23.

Pelham, H.R.B., 1989b. Heat shock and the sorting of lumenal ER proteins. EMBO J. 8, 3171–3176.

Pelham, H.R.B., 1991. Recycling of proteins between the endoplasmic reticulum and Golgi complex. Curr. Opin. Cell Biol. 3, 585–591.

Pelham, H.R.B., 1994. About turn for the COPs? Cell 79, 1125–1127.

Pelham, H.R.B., 1995. Sorting and retrieval between the endoplasmic reticulum and Golgi apparatus. Curr. Opin. Cell Biol. 7, 530–535.

Pelham, H.R.B., 2001. Traffic through the Golgi apparatus. J. Cell Biol. 155, 1099.

Pelham, H.R.B., Rothman, J.E., 2001. The debate about transport in the Golgi—two sides of the same coin? Cell 102, 713–719.

Pelletier, F., Caventou, J.B., 1818. Sur la Matière verte des Feuilles. Ann. Chim. Phys. 9, 194–196.

Peltier, J.B., Emanuelsson, O., Kalume, D.E., Ytterberg, J., Friso, G., Rudella, A., Liberles, D.A., Soderberg, L., Roepstorff, P., von Heije, G., van Wijk, K.J., 2002. Central functions of the lumenal and peripheral thylakoid proteome of *Arabidopsis* determined by experimentation and genome-wide prediction. Plant Cell 14, 211–236.

Peltier, J.B., Friso, G., Kalume, D.E., Roepstorff, P., Nilsson, F., Adamska, I., van Wijk, K.J., 2000. Proteomics of the chloroplast: systematic identification and targeting analysis of lumenal and peripheral thylakoid proteins. Plant Cell 12, 319–342.

Peltier, J.B., Ytterberg, A.J., Sun, Q., van Wijk, K.J., 2004. New functions of the thylakoid membrane proteome of *Arabidopsis thaliana* revealed by a simple, fast and versatile fractionation strategy. J. Biol. Chem. 279, 49367–49383.

Pelzer, D., Grant, A.O., Cavalie, A., Sieber, M., Hofmann, F., Trautwein, W., 1989. Calcium channels reconstituted from the skeletal muscle dihydropyridine receptor protein complex and its α1 peptide subunit in lipid bilayers. Ann. N.Y. Acad. Aci 560, 138–154.

Peña, M.J., Carpita, N.C., 2004. Loss of highly branches arabinans and debranching of rhamnogalacturonan I accompany loss of firm texture and cell separation during proloinged storage of apple. Plant Physiol. 135, 1305–1313.

Peña, M.J., Ryden, P., Madson, M., Smith, A., Reiter, W.-D., Carpita, N.C., 2004. Galactosylation of xyloglucans is essential for maintenance of cell wall tensile strength during cell growth in plants. Plant Physiol. 134, 443–451.

Pendle, A.F., Clark, G.P., Boon, R., Lewandowska, D., Lam, Y.W., Andersen, J., Mann, M., Lamond, A.I., Brown, J.W.S., Shaw, P.J., 2005. Proteomic analysis of the *Arabidopsis* nucleolus suggests novel nucleolar functions. Mol. Biol. Cell 16, 260–269.

Penefsky, H.S., Pullman, M.E., Datta, A., Racker, E., 1960. Partial resolution of the enzymes catalyzing oxidative phosphorylation. II. Participation of a soluble adenosine triphosphatase in oxidative phosphorylation. J. Biol. Chem. 235, 3330–3336.

Peng, H.B., Jaffe, L.F., 1976. Cell-wall formation in Pelveta embryos. A freeze-fracture study. Planta 133, 57–71.

Peng, L., Xiang, F., Roberts, E., Kawagoe, Y., Greve, L.C., Kreuz, K., Delmer, D.P., 2001. The experimental herbicide CGA 325'615 inhibits synthesis of crystalline cellulose and causes accumulation of non-crystalline β-1,4-glucan associated with CesA protein. Plant Physiol. 126, 981–992.

Pennell, R.I., Knox, J.P., Scofield, G.N., Selvendra, R.R., Roberts, K., 1989. A family of abundant plasma membrane-associated glycoproteins related to the arabinogalactan proteins is unique to flowering plants. J. Cell Biol. 108, 1967–1977.

Pennisi, E., 2000. Plants join the genome sequencing bandwagon. Science 290, 2054–2055.

Pennisi, E., 2009. Stressed out over a stress hormone. Science 324, 1012–1013.

Pennisi, E., 2013. The CRISPR craze. Science 341, 833–836.

Pennisi, E., 2017a. Pocket-sized sequencers start to pay off big. Science 356, 572–573.

Pennisi, E., 2017b. New technologies boost genome quality. Earlier sequences are riddled with gaps, overlapping repeats, misplaces segments. Science 356, 10–11.

Penrose, R., 1994. Shadows of the Mind. A Search for the Missing Science of Consciousness. Oxford University Press, Oxford.

Penswick, J.R., Holley, R.W., 1965. Specific cleavage of the yeast alanine RNA into large fragments. Proc. Natl. Acad. Sci. U.S.A. 53, 543–546.

Penzias, A.A., Wilson, R.W., 1965. A measurement of excess antenna temperature at 4080 Mc/s. Astrophys. J. 142, 419–421.

Pepper, D.A., Brinkley, B.R., 1979. Microtubule initiation at kinetochores and centrosomes in lysed mitotic cells. J. Cell Biol. 82, 585–591.

Pepperkok, R., Ellenberg, J., 2006. High-throughput fluorescence microscopy for systems biology. Nat. Rev. Mol. Cell Biol. 7, 690–696.

Perbal, G., Driss-Ecole, D., 2003. Mechanotransduction in gravisensing cells. Trends Plant Sci. 8, 498–504.

Percival, E.G.V., 1950. Structural Carbohydrate Chemistry. Prentice-Hall, New York.

Perdue, D.O., LaFavre, A.K., Leopold, A.C., 1988. Calcium in the regulation of gravitropism by light. Plant Physiol. 86, 1276–1280.

Pereira-Smith, O.M., Fisher, S.F., Smith, J.R., 1985. Senescent and quiescent cell inhibitors of DNA synthesis: membrane-associated proteins. Exp. Cell Res. 160, 297–306.

Pereira-Smith, O.M., Smith, J.R., 1988. Genetic analysis of indefinite division in human cells: identification of four complementation groups. Proc. Natl. Acad. Sci. U.S.A. 85, 6042–6046.

Pereira-Smith, O.M., Smith, J.R., 1988. Genetic analysis of indefinite division in human cells: identification of four complementation groups. Proc. Natl. Acad. Sci. U.S.A. 85, 6042–6046.

Peremyslov, V.V., Prokhnevsky, A.I., Avisar, D., Dolja, V.V., 2008. Two class XI myosins function in organelle trafficking and root hair development in *Arabidopsis*. Plant Physiol. 146, 1109–1116.

Pererra-Leal, J.B., Seabra, M.C., 2001. Evolution of the Rab family of small GTP-binging proteins. J. Mol. Biol. 313, 889–901.

Perez-Prats, E., Narasimhan, M.L., Niu, X., Botella, M.A., Bressan, R.A., Valpuesta, V., Hasegawa, P.M., Binzel, M.L., 1994. Growth-cycle stage-dependent NaCl induction of plasma membrane H^+-ATPase messenger-RNA accumulation in deadapted tobacco cells. Plant Cell Environ. 17, 327–333.

Perfus-Barbeoch, L., Jones, A.M., Assmann, S.M., 2004. Plant heterotrimeric G protein function: insights from *Arabidopsis* and rice mutants. Curr. Opin. Plant Biol. 7, 719–731.

Perkins, D.D., 1992. *Neurospora*: the organism behind the molecular revolution. Genetics 130, 687–701.

Perlmann, G.E., Diringer, R., 1960. The structure of proteins. Ann. Rev. Biochem. 29, 151–182.

Perner, E.S., 1957. Zum elekromikroskopischen Nachweis des "Golgi apparates" in Zellen höherer Pflanzen. Naturwissenschaften 44, 336.

Pernice, B., 1889. Sulla cariocinesi della cellule epiteliali e dell'endotelio dei vasi della mucosa dello stomaco e dell'intestino, nelle studio della gastroenterite sperimentale (nell'avvelenamento per colchico). Sicilia Med. 1, 265–279.

Perouansky, M., 2015. The Overton in Meyer-Overton: a biographical sketch commemorating the 150th anniversary of Charles Ernest Overton's birth. Br. J. Anaesth. 144, 537–541.

Perrin, J.B., December 11, 1926. Discontinuous Structure of Matter. Nobel Lecture.

Perrin, R.M., Young, L.-S., Narayana Murthy, U.M., Harrison, B.R., Wang, Y., Will, J.L., Masson, P.H., 2005. Gravity signal transduction in primary roots. Ann. Bot. 96, 737–743.

Perroncito, A., 1910. Contribution à l'étude de la biologie cellulaire. Mitochondres, chromidies et appareil réticulaire interne dans les cellules spermatiques. Le phénomène de la dictyokinèse. Arch. Ital. Biol. 54, 307–345.

Perry, R.P., 1969. Nucleoli: the cellular sites of ribosome production. In: Lima-de-Faria, A. (Ed.), Handbook of Molecular Cytology. North Holland Publishing Co., Amsterdam, pp. 620–636.

Perry, J., 1978. A Dialogue on Personal Identity and Immortality. Hackett Publishing, Indianapolis.

Perry, S.V., 2008. Background to the discovery of troponin and Setsuro Ebashi's contribution to our knowledge of the mechanism of relaxation in striated muscle. Biochem. Biophys. Res. Commun. 369, 43–48.

Perry, S.E., Keegstra, K., 1994. Envelope membrane proteins that interact with chloroplastic precursor proteins. Plant Cell 6, 93–105.

Perry, S.V., Reed, R., Astbury, W.T., Spark, L.C., 1948. An electron microscope and X-ray diffraction study of the synaeresis of actomyosin. Biochim. Biophys. Acta 2, 674–694.

Pertl, H., Poeckl, M., Blaschke, C., Obermeyer, G., 2010. Osmoregulation in lilium pollen grains occurs via modulation of the plasma membrane H^+ ATPase activity by 14-3-3 proteins. Plant Physiol. 154, 1921–1928.

Pertl, H., Schulze, W.X., Obermeyer, G., 2009. The pollen organelle membrane proteome reveals highly spatial-temporal dynamics during germination and tube growth of lily pollen. J. Proteome Res. 8, 5142–5152.

Pertl-Obermeyer, H., 2017. The pollen membrane proteome. In: Obermeyer, G., Feijó, J. (Eds.), Pollen Tip Growth. Springer International, pp. 293–318.

Pertl-Obermeyer, H., Obermeyer, G., 2013. Pollen cultivation and preparation for proteome studies. Methods Mol. Biol. 1072, 435–449.

Pertl-Obermeyer, H., Schulze, W.X., Obermeyer, G., 2014. In vivo cross-linking combined with mass spectrometry analysis reveals receptor-like kinases and Ca^{2+} signalling proteins as putative interaction partners of pollen plasma membrane H(+) ATPases. J Proteom.

Perutz, M.F., 1962b. Proteins and Nuclei Acids. Structure and Function. Elsevier, Amsterdam.

Perutz, M.F., December 11, 1962a. X-ray Analysis of Haemoglobin. Nobel Lecture.

Pesacreta, T.C., Lucas, W.J., 1984. Plasma membrane coat and a coated vesicle-associated reticulum of membranes: their structure and possible interrelationship in *Chara corallina*. J. Cell Biol. 98, 1537–1545.

Pesacreta, T.C., Lucas, W.J., 1985. Presence of a partially-coated reticulum in angiosperms. Protoplasma 125, 173–184.

Pesquet, E., Lloyd, C., 2011. Microtubules, MAPs and xylem formation. The Plant Cytoskeleton. Adv. Plant Biol. 2, 277–306.

Peterhänsel, C., Maurino, V.G., 2011. Photorespiration redesigned. Plant Physiol. 155, 49–55.

Peterhänsel, C., Niessen, M., Kebeish, R.M., 2008. Metabolic engineering towards the enhancement of photosynthesis. Photochem. Photobiol. https://doi.org/10.1111/j.1751-1097.2008.00427.x.

Peterman, E.J.G., 2016. Kinesin's gait captured. Nat. Chem. Biol. 12, 206–207.

Peters, R.A., 1929. Co-ordinative bio-chemistry of the cell and tissues. J. State Med. 37, 683–709.

Peters, R.A., 1937. Biochemical conception of cell organization. Proc. R. Soc. 121B, 587–592.

Peters, J.M., 2002. The anaphase-promoting complex: proteolysis in mytosis and beyond. Mol. Cell 9, 931–943.

Peters, J.M., Tedeschi, A., Schmitz, J., 2008. The cohesin complex an dits roles in chromosome biology. Genes Dev. 22, 3089–3114.

Peterson, F.C., Burgie, E.S., Park, S.-Y., Jensen, D.R., Weiner, J.J., Bingman, C.A., Chang, C.-E.A., Cutler, S.R., Phillips Jr., G.N., Volkman, B.F., 2010. Structural basis for selective activation of ABA receptors. Nat. Struct. Mol. Biol. 17, 1109–1114.

Peterson, D.G., Schulze, S.R., Sciara, E.B., Lee, S.A., Bowers, J.E., Negel, A., Jiang, N., Tibbits, D.C., Wessler, S.R., Paterson, A.H., 2002. Integration of Cot analysis, DNA cloning, and high- throughput sequencing facilitates genome characterization and gene discovery. Genome Res. 12, 795–807.

Petrášek, J., Friml, J., 2009. Auxin transport routes in plant development. Development 136, 2675–2688.

Petrášek, J., Mravec, J., Bouchard, R., Blakeslee, J.J., Abas, M., Seifertová, D., Wiśniewska, J., Tadele, Z., Kubeš, M., Čovanová, M., Dhonukshe, P., Skůpa, P., Benková, E., Perry, L., Křeček, P., Lee, O.R., Fink, G.R., Geisler, M., Murphy, A.S., Luschnig, C., Zažímalová, E., Friml, J., 2006. PIN proteins perform a rate-limiting function in cellular auxin efflux. Science 312, 914–918.

Petrovská, B., Šebela, M., Doležel, J., 2015. Inside a plant nucleus: discovering the proteins. J. Exp. Bot. 66, 1627–1640.

Petry, S., 2016. Mechanisms of mitotic spindle assembly. Annu. Rev. Biochem. 85, 659–683.

Petterssen, R., December 10, 1999. Award Ceremony Speech.

Petzelt, C., 1972. Ca^{++}-activated ATPase during the cell cycle of the sea urchin *Strongylocentrotus purpuratus*. Exp. Cell Res. 70, 333–339.

Pfaffmann, H., Hartmann, E., Brightman, A.O., Morré, D.J., 1987. Phosphatidylinositol specific phospholipase C of plant stems. Plant Physiol. 85, 1151–1155.

Pfanner, N., Craig, E.A., Meyer, M., 1994. The protein import machinery of the mitochondrial inner membrane. TIBS 19, 368–372.

Pfanzagl, B., Löffelhardt, W., 1999. In vitro synthesis of peptidoglycan precursors modified with N-acetylputrescine by *Cyanophora paradoxa* cyanelle envelope membranes. J. Bacteriol. 181, 2643–2647.

Pfarr, C.M., Cove, M., Grissom, P.M., Hays, T.S., Porter, M.E., McIntosh, J.R., 1990. Cytoplasmic dynein is localized to kinetochores during mitosis. Nature 345, 263–265.

Pfarr, C.M., Cove, M., Grissom, P.M., Hays, T.S., Porter, M.E., McIntosh, J.R., 1990. Cytoplasmic dynein is localized to kinetochores during mitosis. Nature 345, 263–265.

Pfeffer, W., 1875. Die Periodischen Bewegungen der Blattorgane. Engelmann, Leipzig.

Pfeffer, W., 1877. Osmotische Untersuchungen: Studien zur Zell-Mechanik. Engelmann, Leipzig.

Pfeffer, W., 1886. Kritische Besprechung von de Vries: Plasmolytische Studien über Wand der Vakuolen. Bot. Z. 44, 114–125.

Pfeffer, W., 1900–1906. The Physiology of Plants. Translated by A. J. Ewart. Clarendon Press, Oxford.

Pfeiffer, J., 1955. Enzymes. In: Simon, M., Schuster, M. (Eds.), The Physics and Chemistry of Life. A Scientific American Book, New York, pp. 163–180.

Pfisterer, J., Lachmann, P., Kloppstech, K., 1982. Transport of proteins into chloroplasts. Binding of nuclear-coded chloroplast proteins to the chloroplast envelope. Eur. J. Biochem. 126, 143–148.

Pflüger, E., 1872. Ueber die Diffusion des Sauerstoffs, den Ort und die Gesetze der Oxydationsprocesse im thierischen Organismen. Pflügers Arch 6, 43–64.

Pfluger, J., Wagner, D., 2007. Histone modifications and dynamic regulation of genome accessibility in plants. Curr. Opin. Plant Biol. 10, 645–652.

Philipp, E.-I., Franke, W.W., Keenan, T.W., Stadler, J., Jarasch, E.-D., 1976. Characterization of nuclear membranes and endoplasmic reticulum isolated from plant tissue. J. Cell Biol. 68, 11–29.

Phillippy, A.M., 2017. New advances in sequence assembly. Genome Res. 27, xi–xiii.

Phillips, C.S.G., Williams, R.J.P., 1965. Inorganic Chemistry. I. Principles and Non-Metals. Oxford University Press, Oxford.

Phillips, C.S.G., Williams, R.J.P., 1966. Inorganic Chemistry. II. Metals. Oxford University Press, Oxford.

Phillipson, B.A., Pimpl, P., Lamberti Pinto daSilva, L., Crofts, A.J., Taylor, J.P., Movafeghi, A., Robinson, D.G., Denecke, J., 2001. Secretory bulk flow of soluble proteins is efficient and COP II dependent. Plant Cell 13, 2005–2020.

Phillipson, B.A., Pimpl, P., Lamberti Pinto daSilva, L., Crofts, A.J., Taylor, J.P., Movafeghi, A., Robinson, D.G., Denecke, J, 2001. Secretory bulk flow of soluble proteins is efficient and COP II dependent. Plant Cell 13, 2005–2020.

Phinney, B.O., 1956. Growth response of single-gene dwarf mutants in maize to gibberellic acid. Proc. Natl. Acad. Sci. U.S.A. 42, 185–189.

Piccirilli, J.A., McConnell, T.S., Zaug, A.J., Noller, H.F., Cech, T.R., 1992. Aminoacyl esterase activity of the Tetrahymena ribozyme. Science 256, 1420–1424.

Pickard, W.F., 1972. Further observations on cytoplasmic streaming in *Chara braunii*. Can. J. Bot. 50, 703–711.

Pickard, W.F., 1974. Hydrodynamic aspects of protoplasmic streaming in *Chara braunii*. Protoplasma 82, 321–339.

Pickard, B.G., 2007. Delivering force and amplifying signals in plant mechanosensing. Curr. Top. Membr. 58, 362–392.

Pickard, B.G., 2008. "Second extrinsic organizational mechanism" for orienting cellulose: modeling a role for the plasmalemmal reticulum. Protoplasma 233, 29–67.

Pickard, B.G., Beachy, R.N., 1999. Intercellular connections are developmentally controlled to help move molecules through the plant. Cell 98, 5–8.

Pickard, B.G., Ding, J.-P., 1992. Gravity sensing by higher plants. Adv. Comp. Environ. Physiol. 10, 81–110.

Pickard, B.G., Ding, J.-P., 1993. The mechanosensory calcium-selective ion channel: key component of a plasmalemmal control centre? Aust. J. Plant Physiol. 20, 439–459.

Pickard, B.G., Fujiki, M., 2005. Ca^{2+} pulsation in BY-2 cells and evidence for control of mechanosensory Ca^{2+}-selective channels by the plasmalemmal reticulum. Funct. Plant Biol. 2, 863–879.

Pickart, C.M., 2004. Back to the future with ubiquitin. Cell 116, 181–190.

Picken, L., 1960. The Organization of Cells. Clarendon Press, Oxford.

Pickering, E.C., March 3, 1912. Periods of 26 variable stars in the small Magellanic Clouds. Harvard College Observatory. Circular 173.

Pickett-Heaps, J.D., 1967a. The effects of colchicine on the ultrastructure of dividing plant cells, xylem wall differentiation and distribution of cytoplasmic microtubules. Dev. Biol. 15, 206–236.

Pickett-Heaps, J.D., 1967b. The use of radioautography for investigating wall secretion in plant cells. Protoplasma 64, 49–66.

Pickett-Heaps, J.D., 1969a. The evolution of the mitotic apparatus: an attempt at comparative ultrastructural cytology in dividing cells. Cytobios 1, 257–280.

Pickett-Heaps, J.D., 1969b. Preprophase microtubules and stomatal differentiation in *Commelina cyanea*. Aust. J. Biol. Sci. 22, 375–391.

Pickett-Heaps, J.D., 1970. The behavior of the nucleolus during mitosis in plants. Cytobios 6, 69–78.

Pickett-Heaps, J.D., 1971. The autonomy of the centriole: fact or fallacy. Cytobios 3, 205–214.

Pickett-Heaps, J.D., 1972. Variation in mitosis and cytokinesis in plant cells: its significance in the phylogeny and evolution of ultrastructural systems. Cytobios 5, 59–77.

Pickett-Heaps, J.D., 1975. Green Algae. Structure, Reproduction and Evolution in Selected Genera. Sinauer Associates, Inc., Sunderland, MA.

Pickett-Heaps, J.D., 1991. Cell division in Diatoms. Int. Rev. Cytol. 128, 63–108.

Pickett-Heaps, J.D., Forer, A., 2001. Pac-man does not resolve the enduring problem of anaphase chromosome movement. Protoplasma 215, 16–20.

Pickett-Heaps, J.D., Forer, A., Spurck, T., 1996. Rethinking anaphase: where "Pac-Man" fails and why a role for the spindle matrix is likely. Protoplasma 192, 1–10.

Pickett-Heaps, J.D., Forer, A., Spurck, T., 1997. Traction Fibre: toward a "Tensegral" model of the spindle. Cell Motil. Cytoskelet. 37, 1–6.

Pickett-Heaps, J.D., Gunning, B.E.S., Brown, R.C., Lemmon, B.E., Cleary, A.L., 1999. The cytoplast concept in dividing plant cells: cytoplasmic domains and the evolution of spatially organized cell division. Am. J. Bot. 86, 153–172.

Pickett-Heaps, J.D., Northcote, D.H., 1966a. Cell division in the formation of the stomatal complex of the young leaves of wheat. J. Cell Sci. 1, 121–128.

Pickett-Heaps, J.D., Northcote, D.H., 1966b. Organization of microtubules and endoplasmic reticulum during mitosis and cytokinesis in wheat meristems. J. Cell Sci. 1, 109–120.

Pickett-Heaps, J.D., Tippit, D.H., Leslie, R., 1980. Light and electron microscopic observations in two large pennate diatoms: *Hantzschia* and *Nitzschia*. Eur. J. Cell Biol. 21, 1–11, 12–27.

Picotti, P., Bodenmiller, B., Mueller, N., Domon, B., Aebersold, R., 2009. Full dynamic range proteome analysis of *S. cerevisiae* by targeted proteomics. Cell 138, 795–806.

Picton, J.M., Steer, M.W., 1982. A model for the mechanism of tip extension in pollen tubes. J. Theor. Biol. 98, 15–20.

Pien, S., Wyrzykowska, J., McQueen-Mason, S., Smart, C., Fleming, A., 2001. Local expression of expansin induces the entire process of leaf development and modifies leaf shape. Proc. Natl. Acad. Sci. U.S.A. 98, 11812–11817.

Pierson, E.S., Miller, D.D., Callaham, D.A., Shipley, A.M., Rivers, B.A., Cresti, M., Hepler, P.K., 1994. Pollen tube growth is coupled to the extracellular calcium ion flux and the intracellular calcium gradient: effect of BAPTA-type buffers and hypertonic media. Plant Cell 6, 1815–1828.

Pierson, E.S., Miller, D.D., Callaham, D.A., van Aken, J., Hackett, G., Hepler, P.K., 1996. Tip-localized calcium entry fluctuates during pollen tube growth. Dev. Biol. 174, 160–173.

Pigott, G.H., Carr, N., 1972. Homology between nucleic acids of blue0green algae and chloroplast of *Euglena gracilis*. Science 175, 1259–1261.

Pimpl, P., Denecke, J., 2000. ER retention of soluble proteins: retrieval, retention, or both? Plant Cell 12, 1517–1519.

Pimpl, P., Hanton, S.L., Taylor, J.P., Pinto-daSilv, L.L., Denecke, J., 2003. The GTPase ARF1p controls the sequence-specific vacuolar sorting route to the lytic vacuole. Plant Cell 15, 1242–1256.

Pimpl, P., Movafeghi, A., Coughlan, S., Denecke, J., Hillmer, S., Robinson, D.G., 2000. In situ localization and *in vitro* induction of plant COP I-coated vesicles. Plant Cell 12, 2219–2236.

Pina, C., Pinto, F., Feijó, J.A., Becker, J.D., 2005. Gene family analysis of the *Arabidopsis* pollen transcriptome reveals biological implications for cell growth, division control, and gene expression regulation. Plant Physiol. 138, 744–756.

Pinheiro, H., Samalova, M., Geldner, N., Chory, J., Martinez, A., Moore, I., 2009. Genetic evidence that the higher plant Rab-D1 and Rab-D2 GTPases exhibit distinct but overlapping interactions in the early secretory pathway. J. Cell Sci. 122, 3749–3758.

Pinnavaia, T.H., 1983. Intercalated clay catalysts. Science 220, 365–371.

Pinto da Silva, P., Branton, D., 1970. Membrane splitting in freeze-etching. J. Cell Biol. 45, 598–605.

Piperno, G., 1984. Monoclonal antibodies to dynein subunits reveal the existence of cytoplasmic antigens in sea urchin egg. J. Cell Biol. 98, 1842–1850.

Pirie, N.W., 1938. The meaninglessness of the terms life and living. In: Needham, J., Green, D.E. (Eds.), Perspectives in Biochemistry. Cambridge University Press, Cambridge, pp. 11–22.

Pittman, J.K., Shigaki, T., Cheng, N.-H., Hirschi, K.D., 2002. Mechanism of N-terminal autoinhibition in the *Arabidopsis* Ca^{2+}/H^{+} antiporter CAX1. J. Biol. Chem. 277, 26452–26459.

Planck Collaboration, 2015. Planck 2015 results. XIII. Cosmological parameters. Astron. Astrophys. 594, A13. See Pdf, Page 31, Table IV.

Planck, M., 1900. Zur Theorie des Gesetzes der Energieverteilung im Normalspektrum. Verh. Deutsch. Phys. Ges. 2, 237–245.

Planck, M., 1906. Das Prinzip der Relativität und die Grundgleichungen der Mechanik. Verh. der Deutschen Physikalische Gesellschaft 8, 136–141.

Planck, M., 1932. Where is Science Going? W. W. Norton & Co., New York.

Planck, M., 1949a. Scientific Autobiography, and Other Papers. Philosophical Library, New York.

Planck, M., 1949b. Theory of Heat. Introduction to Theoretical Physics. Macmillan, London.

Planck, M., June 2, 1920. The Genesis and Present State of Development of the Quantum Theory. Nobel Lecture.

Plato, 1892. Timaeus. In: The Dialogues of Plato. vol. III, third ed. Translated by B. Jowett. Oxford University Press, Oxford. pp. 339–515.

Plato, 1959. Timaeus. In: The Dialogues of Plato, third ed., vol. III. Macmillan, New York, pp. 339–515. Translated by B. Jowett.

Platt, J.R., 1964. Strong inference. Science 146, 347–353.

Plaxton, W.C., 1996. The organization and regulation of plant glycolysis. Ann. Rev. Plant Physiol. Plant Mol. Biol. 47, 185–214.

Plaxton, W.C., Podestá, F.E., 2006. The functional organization and control of plant respiration. Crit. Rev. Plant Sci. 25, 159–198.

Plazinski, J., Elliot, J., Hurley, U.A., Burch, J., Arioli, T., Williamson, R.E., 1997. Myosins from angiosperms, ferns, and algae. Protoplasma 196, 78–86.

Plieth, C., Hansen, U.-P., 1996. Methodological aspects of pressure leading of Fura-2 into characean cells. J. Exp. Bot. 47, 1601–1612.

Plieth, C., Sattelmacher, B., Hansen, U.-P., 1997. Cytoplasmic Ca^{2+}-H^{+}-exchange buffers in green algae. Protoplasma 198, 107–124.

Plowe, J.Q., 1931. Membranes in the plant cell. II. Localization of differential permeability in the plant protoplast. Protoplasma 12, 221–240.

Pluskal, T., Weng, J.K., 2017. Natural product modulators of human sensations and mood: molecular mechanisms and therapeutic potential. Chem. Soc. Rev. https://doi.org/10.1039/c7cs00411g.

Pocker, Y., Janjić, N., 1987. Enzyme kinetics in solvents of increased viscosity. Dynamic aspects of carbonic anhydrase catalysis. Biochemistry 26, 2597–2606.

Poinar, H.N., Schwarz, C., Qi, J., Shapiro, B., MacPhee, R.D.E., Buigues, B., Tikhonov, A., Huson, D.H., Tomsho, L.P., Auch, A., Rampp, M., Miller, W., Schuster, S.C., 2006. Metagenomics to paleogenomics: large-scale sequencing of mammoth DNA. Science 311, 392–394.

Poiseuille, J.L.M., 1940. Experimental Investigations upon the Flow of Liquids in Tubes of Very Small Diameter. Translated by W.H. Herschel. In: Bingham, E.C. (Ed.), Rheological Memoirs, vol. 1. Lancaster Press, Easton, PA. No. 1.

Polanyi, J.C., December 8, 1986. Some Concepts in Reaction Dynamics. Nobel Lecture.

Polizzi, E., Natali, L., Muscio, A.M., Giordani, T., Cionini, G., Cavallini, A., 1998. Analysis of chromatin and DNA during chromosome endoreduplication in the endosperm of *Triticum durum*. Desf. Protoplasma 203, 175–185.

Polkinghorne, J., Beale, N., 2009. Questions of Truth. Fifty-one Responses to Questions about God, Science, and Belief. Westminster John Knox Press, Louisville, KY.

Pollack, R., 1999. The Missing Moment. Houghton Mifflin, Boston.

Pollack, R., 2015. Eugenics lurk in the shadow of CRISPR. Science 348, 871.

Pollard, T.D., 2010. Mechanics of cytokinesis in eukaryotes. Curr. Opin. Cell Biol. 22, 50–56.

Pollard, T.D., Goldman, Y.E., 2013. Remembrance of Hugh E. Huxley, a founder of our field. Cytoskeleton 70, 471–475.

Pollegioni, L., Schonbrunn, E., Siehl, D., 2011. Molecular basis of glyphosate resistance: different approaches through protein engineering. FEBS J. 278, 2753–2766.

Pollister, A.W., Pollister, P.F., 1943. The relation between centriole and centromere in atypical spermatogenesis of viviparid snails. Ann. N.Y. Acad. Sci. 45, 1–48.

Polya, G.M., Davies, J.R., Micucci, V., 1983. Properties of a calmodulin-activated Ca^{2+}-dependent protein kinase from wheat germ. Biochem. Biophys. Acta 761, 1–12.

Polya, G.M., Jagendorf, A.T., 1971. Wheat leaf RNA polymerases. I. Partial purification and characterization of nuclear, chloroplast and soluble DNA-dependent enzymes. Arch. Biochem. Biophys. 146, 635–648.

Polya, G.M., Klucis, E., Haritou, M., 1987. Resolution and characterization of two soluble calcium-dependent protein kinases from silver beet leaves. Biochem. Biophys. Acta 931, 68–77.

Polya, G.M., Morrice, N., Wettenhall, R.E.H., 1989. Substrate specificity of wheat embryo calcium-dependent protein kinase. FEBS Lett. 253, 137–140.

Polya, G.M., Nott, R., Lucis, E., Minichiello, J., Chandra, S., 1990. Inhibition of plant calcium-dependent protein kinases by basic polypeptides. Biochem. Biophys. Acta 1037, 259–2623.

Ponnamperuma, C., 1972. The Origins of Life. E.P. Dutton, New York.

Ponnamperuma, C., Lemmon, R.M., Mariner, R., Calvin, M., 1963a. Formation of adenine by electron irradiation of methane, ammonia and water. Proc. Natl. Acad. Sci. U.S.A. 49, 737–740.

Ponnamperuma, C., Mack, R., 1965. Nucleotide synthesis under possible primative Earth Conditions. Science 148, 1221–1223.

Ponnamperuma, C., Mariner, R., Sagan, C., 1963b. Formation of adenosine by ultra-violet irradiation of a solution of adenine and ribose. Nature 198, 1199–1200.

Ponnamperuma, C., Sagan, C., Mariner, R., 1963c. Synthesis of adenosine triphosphate under possible primative Earth conditions. Nature 199, 222–226.

Poo, M.-M., 1981. In situ electrophoresis of membrane components. Annu. Rev. Biophys. Bioeng. 10, 245–276.

Poo, M.-M., Robinson, K.R., 1977. Electrophoresis of concanavalin A receptors along embryonic muscle cell membrane. Nature 297, 602–605.

Pool, M.R., Lopez-Huertas, E., Hong, J.-T., Baker, A., 1998a. NADPH ois a specific inhibitor of protein import into glyoxysomes. Plant J. 15, 1–14.

Pool, M.R., Lopez-Huertas, E., Baker, A., 1998b. Characterization of intermediates in the process of plant peroxisomal protein import. EMBO J. 17, 6854–6862.

Pool, M.R., Stumm, J., Fulga, T.A., Sinning, I., Dobberstein, B., 2002. Distinct modes of signal recognition particle interaction with the ribosome. Science 297, 1345–1348.

Pope, A., 1745. An Essay on Man. Enlarged and Improved by the Author with Notes by William Warburton. John and Paul Knapton, London.

Popenoe, P., 1934. The German sterilization law. J. Hered. 25, 257–260.

Popenoe, P., Johnson, R.H., 1918. Appled Eugenics. Macmillan, New York.

Popp, F.-A., Beloussov, V. (Eds.), 2003. Integrative Biophysics. Biophotonics. Kluwer Academic Publishers Group, Dordrecht.

Popper, Z.A. (Ed.), 2011. The Plant Cell Wall: Methods and Protocols. Humana Press. Springer, New York.

Popper, K.R., Eccles, J.C., 1977. The Self and Its Brain. Springer-Verlag, Berlin.

Popper, Z.A., Fry, S.C., 2003. Primary cell wall compostion of bryophytes and charophytes. Ann. Bot. 91.

Popper, Z.A., Tuohy, M.G., 2010. Beyond the green: understanding the evolutionary puzzle of plant and algal cell walls. Plant Physiol. 153, 373–383.

Porter, K.R., 1957. The submicroscopic morphology of protoplasm. Harvey Lect. 51, 175–228, 1955/1956.

Porter, K.R., 1984. The cytomatrix: a short history of its study. J. Cell Biol. 99, 3s–12s.

Porter, K.R., Anderson, K.L., 1982. The structure of the cytoplasmic matrix preserved by freeze-drying and freeze substitution. Eur. J. Cell Biol. 29, 83–96.

Porter, K.R., Beckerle, M., McNiven, M., 1983. The cytoplasmic matrix. Mod. Cell Biol. 2, 259–302.

Porter, K.R., Bonneville, M.A., 1968. Fine Structure of Cells and Tissues, third ed. Lea & Febiger, Philadelphia, PA.

Porter, K.R., Caulfield, J.B., 1958. The formation of the cell plate during cytokinesis in *Allium cepa* L. Proc. 4th Int. Cong. Electron Microcopy (Berlin) 2, 503–507.

Porter, K.R., Claude, A., Fullam, E., 1945. A study of tissue culture cells by electron microscopy. J. Exp. Med. 81, 233–241.

Porter, K.R., Kawakami, N., Ledbetter, M.C., 1965. Structural basis of streaming in *Physarum polycephalum*. J. Cell Biol. 27, 78A.

Porter, K.R., Machado, R.D., 1960. Studies on the endoplasmic reticulum IV. Its form and distribution during mitosis in cells of onion root tip. J. Biophys. Biochem. Cytol. 7, 167–180.

Porter, K., Moses, M., 1958. The Cell. A Scope Monograph on Cytology. The Upjohn Co., Kalamazoo, MI. http://www.upjohn.net/other/brain_cell/brain_cell.htm.

Porter, K.R., Thompson, H.P., 1948. A particulate body associated with epithelial cells cultured from mammary carcinomas of mice of a milk-factor strain. J. Exp. Med. 88, 15–24.

Porter, K.R., Tilney, L.G., 1965. Microtubules and intracellular motility. Science 150, 382.

Porter, K.R., Tucker, J.B., March 1981. The ground substance of the living cell. Sci. Am. 244 (3), 41–51.

Portier, P., 1918. Les Symbiotes. Masson et Cie, Paris.

Portillo, F., 2000. Regulation of plasma membrane H^+-ATPase in fungi and plants. Biochim. Biophys. Acta 1469, 31–42.

Portis Jr., A.R., 1992. Regulation of ribulose 1,5-bisphosphate carboxylase/oxygenase activity. Ann. Rev. Plant Physiol. Plant Mol. Biol. 43, 415–437.

Portugal, F.H., 2015. The Least Likely Man. Marshall Nirenberg and the Discovery of the Genetic Code. MIT Press, Cambridge, MA.

Postman, N., 1985. Amusing Ourselves to Death. Public Discourse in the Age of Show Business. Penguin Books, New York.

Postman, N., 1992. Technopoly: The Surrender of Culture to Technology. Vintage Books, New York.

Postman, N., 1993. Technopoly: The Surrender of Culture to Technology. Vintage Books, New York.

Potato Genome Sequencing Consortium, 2011. Genome sequence and analysis of the tuber crop potato. Nature 475, 189–195.

Pottosin, I.I., Schönknecht, G., 2007. Vacuolar calcium channels. J. Exp. Bot. 58, 1559–1569.

Poulet, A., Probst, A.V., Graumann, K., Tatout, C., Evans, D., 2017. Exploring the evolution of the proteins of the plant nuclear envelope. Nucleus 8, 46–59.

Pouliquen, O., Forterre, Y., Bérut, A., Chauvet, H., Bizet, F., Legué, V., Moulia, B., 2017. A new scenario for gravity detection in plants: the postion sensor hypothesis. Phys. Biol. 14, 035005.

Poustka, F., Irani, N.G., Feller, A., Lu, Y., Pourcel, L., Frame, K., Grotewold, E., 2007. A trafficking pathway for anthocyanins overlaps with the endoplasmic reticulum-to-vacuole protein-sorting route in *Arabidopsis* and contributes to the formation of vacuolar inclusions. Plant Physiol. 145, 1323–1335.

Power, H., 1664. Experimental Philosophy in Three Books, 1966. Johnson Reprint Corporation, New York.

Powner, M.W., Gerland, B., Sutherland, J.D., 2009. Synthesis of activated pyrimidine ribonucleotides in prebiotically plausible conditions. Nature 459, 239–242.

Powner, M.W., Sutherland, J.D., Szostak, J.W., 2011. The origins of nucleotides. Synlett 14, 1956–1964.

Poyton, R.O., 1983. Memory and membranes: the expression of genetic and spatial memory during the assembly of organelle macrocompartments. Mod. Cell Biol. 2, 15–72.

Pozueta-Romero, D., Gonzalez, P., Etxeberria, E., Pozueta-Romero, J., 2008. The hyperbolic and linear phases of the sucrose accumulation curve in turnip storage cells denote carrier-mediated and fluid phase endocytotic transport, respectively. J. Am. Soc. Hort. Sci. 133, 612–618.

Prasher, D.C., Eckenrode, V.K., Ward, W.W., Prendergast, F.G., Cormier, M.J., 1992. Primary structure of the *Aequorea victoria* green fluorescent protein. Gene 111, 229–233.

Prasher, D., McCann, R.O., Cormier, M.J., 1985. Cloning and expression of the cDNA coding for aequorin, a bioluminescent calcium-binding protein. Biochem. Biophys. Res. Commun. 126, 1259–1268.

Pratelli, R., Sutter, J.U., Blatt, M.R., 2004. A new catch in the SNARE. Trends Plant Sci. 9, 187–195.

Pratt, M.M., 1980. The identification of a dynein ATPase in unfertilized sea urchin eggs. Dev. Biol. 74, 364–378.

Prebble, J.N., 1981. Mitochondria Chloroplasts and Bacterial Membranes. Longmans, London.

Prebble, J., Weber, B., 2003. Peter Mitchell and the Making of Glynn. Oxford University Press, Oxford.

Prénant, A., 1913. Les appareils ciliés et leurs dérivés. J. de l'Anat. et Physiol. 49, 88–108 34–382, 506–553, 565–617.

Presig-Müller, R., Muster, G., Kindl, H., 1994. Heat shock enhances the amount of pienylated Dnaj protein at membranes of glyoxysomes. Eur. J. Biochem. 219, 57–63.

Preston, R.D., 1939. The molecular chain structure of cellulose and its botanical significance. Biol. Rev. 14, 281–313.

Preston, R.D., 1952. The Molecular Architecture of Plant Cell Walls. John Wiley & Sons, New York.

Preston, R.D., 1988. Cellulose-microfibril-orienting mechanisms in plant cell walls. Planta 174, 67–74.

Preuss, D., Mulholland, J., Franzusoff, A., Segev, N., Botstein, D., 1992. Characteriztion of the *Saccharomyces* Golgi complex through the cell cycle by immunoelectron microscopy. Mol. Biol. Cell 3, 789–803.

Price, J.L., Blau, J., Rothenfluh, A., Abodeely, M., Kloss, B., Young, M.W., 1998. *double-time* is a novel *Drosophila* clock gene that regulates PERIOD protein accumulation. Cell 94, 83–95.

Price, D.C., Chan, C.X., Yoon, H.S., Yang, E.C., Qiu, H., Weber, A.P.M., Schwacke, R., Gross, J., Blouin, N.A., Lane, C., Reyes-Prieto, A., Durnford, D.G., Neilson, J.A.D., Lang, B.F., Burger, G., Steiner, J.M., Löffelhardt, W., Meuser, J.E., Posewitz, M.C., Ball, S., Arias, M.C., Henrissat, B., Coutinho, P.M., Rensing, S.A., Symeonidi, A., Doddapaneni, H., Green, B.R., Rajah, V.D., Boore, J., Bhattacharya, D., 2012. *Cyanophora paradoxa* genome elucidates origin of photosynthesis in algae and plants. Science 335, 843–847.

Priestley, J., 1771. Essay on the First Principles of Government. J. Johnson, London.

Priestley, J., 1774. Experiments and Observations on Different Kinds of Air. J. Johnson, London.

Priestley, J., 1809. Memoirs of Dr. Joseph Priestley Written by Himself (to the Year 1795) with a Continuation to the Time of his Decease by His Son, Joseph Priestley, 1893. City Road, London.

Priestley, J., 1974. Experiments and Observations on Different Kinds of Air. J. Johnson, London.

Prigogine, I., Stengers, I., 1984. Order Out of Chaos: Man's New Dialogue with Nature. Bantam Books, Toronto.

Principe, L.M., 2006. Science and Religion. The Teaching Company, Chantilly, VA.

Pringsheim, N., 1854. Untersuchungen über den Bau und die Bildung der Pfanzenzelle. Hirschwald, Berlin.

Pringsheim, N., 1879–1881. Ueber Lichtwirkung und Chlorophyllfunction in der Pflanzen. Jahr. Wiss. Bot. 12, 288–437.

Pringsheim, E.G., 1928. Physiologische Untersuchungen an Paramaecium bursaria. Archiv f. Protistenkunde 64, 289–418.

Pringsheim, E.G., 1972. Pure Cultures of Algae. Hafner Publishing Co., New York.

Pritchard, J., Wyn-Jones, R.G., Tomos, A.D., 1991. Turgor, growth and rheological gradients of wheat roots following osmotic stress. J. Exp. Bot. 42, 1043–1049.

Prober, J.M., Trainor, G.L., Dam, R.J., Hobbs, F.W., Robertson, C.W., Zagursky, R.J., Cocuzza, A.J., Jensen, M.A., Baumeister, K., 1987. A system for rapid DNA sequencing with fluorescent chain-terminating dideoxynucleotides. Science 238, 336–341.

Probine, M.C., Barber, N.F., 1966. The structure and plastic properties of the cell wall of *Nitella* in relation to extension growth. Aust. J. Biol. Sci. 19, 439–457.

Probine, M.C., Preston, R.D., 1961. Cell growth and the structure and mechanical properties of the wall in internodal cells of *Nitella opaca*. I. Wall structure and growth. J. Exp. Bot. 121, 261–282.

Probine, M.C., Preston, R.D., 1962. Cell growth and the structure and mechanical properties of the wall in internodal cells of *Nitella opaca* II. Mechanical properties of the walls. J. Exp. Bot. 13, 111–127.

Proctor, R.A., 1896. Other Worlds than Ours. D. Appleton and Co., New York.

Prodhan, G., 2016. Europe's robots to become 'electronic persons' under draft plan. Reuters, #Science News, June 21. https://www.reuters.com/article/us-europe-robotics-lawmaking/europes-robots-to-become-electronic-persons-under-draft-plan-idUSKCN0Z72AY.

Prohaska, S.J., Stadler, P.F., 2011. The use and abuse of −omes. Methods Mol. Biol. 719, 173–196.

Proseus, T.E., Boyer, J.S., 2005. Turgor pressure move polysaccharides into growing cell walls of *Chara corallina*. Ann. Bot. 95, 967–979.

Proseus, T.E., Boyer, J.S., 2006a. Calcium pectate chemistry controls growth rate of *Chara corallina*. J. Exp. Bot. 57, 3989–4002.

Proseus, T.E., Boyer, J.S., 2006b. Identifying cytoplasmic input to the cell wall of growing *Chara corallina*. J. Exp. Bot. 57, 3231–3242.

Proseus, T.E., Boyer, J.S., 2006c. Periplasm turgor pressure controls wall deposition and assembly in growing *Chara corallina* cells. Ann. Bot. 98, 93–105.

Proseus, T.E., Boyer, J.S., 2007. Tension required for pectate chemistry to control growth in *Chara corallina*. J. Exp. Bot. 58, 4283–4292.

Proseus, T.E., Boyer, J.S., 2012a. Pectate chemistry links cell expansion to wall deposition in *Chara corallina*. Plant Signal. Behav. 7, 1490–1492.

Proseus, T.E., Boyer, J.S., 2012b. Calcium deprivation disrupts enlargement of *Chara corallina* cells. Further evidence for the calcium pectate cycle. J. Exp. Bot. 63, 3953–3958.

Proseus, T.E., Ortega, J.K.E., Boyer, J.S., 1999. Separating growth from elastic deformation during cell enlargement. Plant Physiol. 119, 775–784.

Prout, W., 1815. On the relation between the specific gravities of bodies in their gaseous state and the weights of their atoms. Ann. Philos. 6, 321–330.

Prout, W., 1816. Correction of a mistake in the essay on the relation between the specific gravities of bodies in their gaseous state and the weights of their atoms. Ann. Philos. 7, 111–113.

Provance, D.W., MacDowall, A., Marko, M., Luby-Phelps, K., 1993. Cytoarchitecture of size-excluding compartments in living cells. J. Cell Sci. 106, 565–578.

Provasoli, L., Hutner, S.H., Schatz, A., 1948. Streptomycin-induced chlorophyll-less races of *Euglena*. Proc. Soc. Exp. Biol. Med. 69, 279–282.

Pruitt, K.D., Hanson, M.R., 1991. Splicing of the petunia cytochrome oxidase subunit II intron. Curr. Genet. 19, 191–198.

Prusiner, S.B., 1982. Novel proteinaceous infectious particles cause scrapie. Science 216, 136–144.

Prusiner, S.B., December 8, 1997a. Prions. Nobel Lecture.

Prusiner, S.B., 1997b. Prion diseases and the BSE crisis. Science 278, 245–251.

Prymakowska-Bosak, M., Przewloka, M.R., Slusarczyk, J., Kuras, M., Lichota, J., Kilianczyk, B., Jerzmanowski, A., 1999. Linker histones play a role in male meiosis and the development of pollen grains in tobacco. Plant Cell 11, 2317–2329.

Ptolemy, 1984. The Almagest, 2nd Century. Springer-Verlag, New York. Translated by G.J. Toomer.

Pucadyil, T.J., Schmid, S.L., 2008. Real-time visualization of dynamin-catalyzed membrane fission and vesicle release. Cell 135, 1–13.

Pucher, G.W., Leavenworth, C.S., Ginter, W.D., Vickery, H.B., 1947. Studies in the metabolism of crassulacean plants: the diurnal variation in organic acid and starch content of *Bryophyllum calycinum*. Plant Physiol. 22, 360–376.

Puchkov, E.O., 2013. Intracellular viscosity: methods of measurement and role in metabolism. Biochem. (Moscow) Suppl. Series A Membr. Cell Biol. 7, 270–279.

Puchta, H., Fauser, F., 2013. Gene targeting in plants: 25 years later. Int. J. Dev. Biol. 57, 629–637.

Pullman, M.E., Penefsky, H.S., Datta, A., Racker, E., 1960. Partial resolution of the enzymes catalyzing oxidative phosphorylation. I. Purification and properties of soluble, Ddnitrophenol-stimulated adenosine triphosphatase. J. Biol. Chem. 235, 3322–3329.

Purcell, E.M., 1977. Life at low Reynolds number. Am. J. Phys. 45, 3–11.

Pyke, K.A., Howells, C.A., 2002. Plastid and stromule morphogenesis in tomato. Ann. Bot. 90, 559–566.

Pyke, K.A., Leech, R.M., 1992. Chloroplast division and expansion is radically altered by nuclear mutations in *Arabidopsis thaliana*. Plant Physiol. 99, 1005–1008.

Pyke, K.A., Leech, R.M., 1994. A genetic analysis of chloroplast division and expansion in *Arabidopsis thaliana*. Plant Physiol. 104, 201–207.

Pyke, K.A., Rutherford, S.M., Robertson, E.J., Leech, R.M., 1994. Arc6, a fertile *Arabidopsis* mutant with only two mesophyll cell chloroplasts. Plant Physiol. 106, 1169–1177.

Qi, Z., Kishigami, A., Nakagawa, Y., Iida, H., Sokabe, M., 2004b. A mechanosensitive anion channel in *Arabidopsis thaliana* mesophyll cells. Plant Cell Physiol. 45, 1704–1708.

Qi, Y., Pelissier, T., Itaya, A., Hunt, E., Wassenegger, M., Ding, B., 2004. Direct role of a viroid RNA motif in mediating directional RNA trafficking across a specific cellular boundary. Plant Cell 16, 1741–1752.

Qi, X., Zheng, H., 2013. Rab-A1c GTPase defines a population of the trans-Golgi network that is sensitive to endosidin1 during cytokinesis in Arabidopsis. Mol. Plant 6, 847–859.

Qiao, L., Grolig, F., Jablonsky, P.P., Williamson, R.E., 1989. Myosin heavy chains: detection by immunoblotting in higher plants and localization by immunofluorescence in the alga, *Chara*. Cell Biol. Int. Rep. 13, 107–117.

Qiu, G.Y., Omasa, K., Sase, S., 2009. An infrared-based coefficient to screen plant environmental stress: concept, test and applications. Funct. Plant Biol. 36, 990–997.

Quader, H., 1986. Cellulose microfibril orientation in *Oocystis solitaria*: proof that microtubules control the alignment of the terminal complexes. J. Cell Sci. 83, 223–234.

Quader, H., 1990. Formation and disintegration of cisternae of the endoplasmic reticulum visualized in live cells by conventional fluorescence and confocal laser scanning microscopy: evidence for the involvement of calcium and the cytoskeleton. Protoplasma 155, 166–175.

Quader, H., Fast, H., 1990. Influence of cytosolic pH changes on the organization of the endoplasmic reticulum in epidermal cells of onion bulb scales: acidification by loading with weak organic acids. Protoplasma 157, 216–224.

Quader, H., Hofmann, A., Schnepf, E., 1987. Shape and movement of the endoplasmic reticulum in onion bulb epidermis cells: possible involvement of actin. Eur. J. Cell Biol. 44, 17–26.

Quader, H., Robinson, D.G., 1979. Structure, synthesis, and orientation of microfibrils. VI. The role of ions in microfibril deposition in *Oocystis solitaria*. Eur. J. Cell Biol. 20, 51–56.

Quader, H., Robinson, D.G., 1981. *Oocystis solitaria*: a model organism for understanding the organization of cellulose synthesis. Ber. Dtsch. Bot. Ges. 94, 75–84.

Quader, H., Schnepf, E., 1986. Endoplasmic reticulum and cytoplasmic streaming: fluorescence microscopical observations in adaxial epidermis cells of onion bulb scales. Protoplasma 131, 250–252.

Quader, H., Wagenbreth, I., Robinson, D.G., 1978. Structure, synthesis and orientation of microfibrils. V. On the recovery of *Oocystis solitaria*. Cytobiologie 15, 463–474.

Quail, P.H., 1979. Plant cell fractionation. Ann. Rev. Plant Physiol. 30, 425–484.

Quail, P.H., 2005. Phytochrome overview. In: Wada, M., Shimazaki, K., Iino, M. (Eds.), Light Sensing in Plants. Springer, Tokyo, pp. 21–35.

Quail, M.A., Korarewa, I., Smith, F., Scally, A., Stephens, P.J., Durbin, R., Swerdlow, H., Turner, D.J., 2008. A large genome center's improvements to the Illumina sequencing system. Nat. Methods 5, 1005–1010.

Quaranto, V., Jones, J.C.R., 1991. The internal affairs of an integrin. Trends Cell Biol. 1, 2–4.

Quartacci, M.F., Glisic, O., Stevanovic, B., Navari-Izzo, F., 2002. Plasma membrane lipids in the resurrection plant *Ramonda serbica* following dehydration and rehydration. J. Exp. Bot. 53, 2159–2166.

Quastel, J.H., 1942. Prof. Rudolf Schenheimer. Nature 149, 15–16.

Quastel, J.H., Wheatley, A.H.M., 1930. Biological oxidations in the succinic acid series. Biochem. J. 25, 117–128.

Quatrano, R.S., 1990. Polar axis fixation and cytoplasmic localization in *Fucus*. In: Mahowald, A.P. (Ed.), Genetics of Pattern Formation and Growth control. Wiley-Liss, Inc., New York, pp. 31–46.

Quayle, J.R., Fuller, R.C., Benson, A.A., Calvin, M., 1954. Enzymatic carboxylation of ribulose diphosphate. J. Am. Chem. Soc. 76, 3610–3612.

Quekett, J., 1852–1854. In: Lectures on Histology, vols. I and II. Hippolyte Balliere, London.

Quélo, A.-H., Verbelen, J.-P., 2004. Bromodeoxyuridine DNA fiber technology in plants: replication origins and DNA synthesis in tobacco BY-2 cells under prolonged treatement with aphidocolin. Protoplasma 223, 197–202.

Quincke, G., 1888. Ueber periodische Ausbreitung von Flössigkeitsoberflächen und dadurch hervorgerufene Bewegungserscheinungen Sitzb, pp. 791–804. D. kgl. Preusi. Akad. D. Wiss. Zu. Berlin 34.

Quintas-Granados, L.I., Lopez-Camarillo, C., Fandiño Armas, J., Mendoza Hernandez, G., Alvarez-Sánchez, M.E., 2013. Identification of the phosphorylated residues in TveIF5A by mass spectrometry. Dev. Reprod. Biol. 11, 378–384.

Rabinowitch, E.I., 1945. Photosynthesis and Related Processes.

Rabinowitch, E., 1955. Photosynthesis. In: The Physics and Chemistry of Life. Simon and Schuster, New York, pp. 27–47.

Rabinowitch, E., Govindjee, 1969. Photosynthesis. John Wiley & Sons, New York.

Rabouille, C., 2017. Pathways of unconventional protein secretion. Trends Cell Biol. 27, 230–240.

Rabounski, D., 2006. Declaration of academic rights. (Scientific human rights). Prog. Phys. 1, 57–60.

Rachmilevitch, S., Cousins, A.B., Bloom, A.J., 2004. Nitrate assimilation in plant shoots depends on photorespiration. Proc. Natl. Acad. Sci. U.S.A. 101, 11506–11510.

Racker, E., 1955. Synthesis of carbohydrates from carbon dioxide and hydrogen in a cell-free system. Nature 175, 249–251.

Racker, E., 1957. Micro and macrocycles in carbohydrate metabolism. Harvey Lect. 51, 143–174.

Racker, E., 1965. Mechanisms in Bioenergetics. Academic Press, New York.

Racker, E., 1975a. Inner mitochondrial membranes: basic and applied aspects. In: Cell Membranes: Biochemistry, Cell Biology & Pathology. HP Publishing Co., New York, pp. 135–141.

Racker, 1975b. Reconstitution, mechanism of action and control of ion pumps. Biochem. Soc. Trans. 3, 785–802.

Racker, E., 1976. A New Look at Mechanisms in Bioenergetics. Academic Press, New York.

Racker, E., 1985. Reconstitution of Transporters, Receptors, and Pathological States. Academic Press, Orlando, FL.

Racker, E., Stoeckenius, W., 1974. Reconstitution of purple membrane vesicles catalyzing light-driven proton uptake and adenosine triphosphate formation. J. Biol. Chem. 249, 662–663.

Racusen, R.H., 1976. Phytochrome control of electrical potentials and intracellular coupling in oat-coleoptile tissue. Planta 132, 25–29.

Racusen, R.H., 2002. Early development in fern gametophytes: interpreting the transition to prothallial architecture in terms of coordinated photosynthate production and osmotic ion uptake. Ann. Bot. 89, 227–240.

Racusen, R.H., Cooke, T.J., 1982. Electrical changes in the apical cell of the fern gametophyte during irradiation with photomorphogenetically active light. Plant Physiol. 70, 331–334.

Racusen, R.H., Ketchum, K.A., Cooke, T.J., 1988. Modifications of extracellular electric and ionic gradients preceding the transition from tip growth to isodiametric expansion in the apical cell of the fern gametophyte. Plant Physiol. 87, 69–77.

Radford, J.E., Vesk, M., Overall, R.L., 1998. Callose deposition at plasmodesmata. Protoplasma 201, 30–37.

Radford, J.E., White, R.G., 1998. Localization of a myosin-like protein to plasmodesmata. Plant J. 14, 743–750.

Radford, J.E., White, R.G., 2001. Effect of tissue-preparation induced callose synthesis on estimates of plasmodesma size exclusion limits. Protoplasma 216, 47–55.

Radhamony, R.N., Theg, S.M., 2006. Evidence for an ER to Golgi to chloroplast protein transport pathway. Trends Cell Biol. 16, 385–387.

Raghavan, V., 1979. Developmental Biology of Fern Gametophytes. Cambridge University Press, Cambridge.

Raghavendra, A., Reumann, S., Heldt, H.W., 1998. Participation of mitochondrial metabolism in photorespiration. Plant Physiol. 116, 1333–1337.

Rahim, G., Bischof, S., Kessler, F., Agne, B., 2008. In vivo interaction between atToc33 and atToc159 GTP-binding domains demonstrated in a plant split-ubiquitin system. J. Exp. Bot. 60, 257–267.

Rahman, H., Ramanathan, V., Jagadeeshselvam, N., Ramasamy, S., Rajendran, S., Ramachandran, M., Sudheer, P.D.V.N., Chauhan, S., Natesan, S., Muthurajan, R., 2015. Phenomics: technologies and applications in plant and agriculture. In: Barh, D., Sarwar Khan, M., Davies, E. (Eds.), PlantOmics: The Omics of Plant Science. Springer India, pp. 385–411.

Rahn, O., 1945. Microbes of Merit. The Jaques Cattell Press, Lancaster, PA.

Raices, M., D'Angelo, M.A., 2012. Nuclear pore complex composition: a new regulator of tissue specific and developmental functions. Nat. Rev. Mol. Cell Biol. 13, 687–699.

Raitskin, O., Patron, N.J., 2016. Multi-gene engineering in plants with RNA-guided Cas9 nuclease. Curr. Opin. Biotechnol. 37, 69–75.

Rajangam, A.S., Kumar, M., Aspeborg, H., Guerriero, G., Arvestad, L., Pansri, P., Brown, C.J.-L., Hober, S., Blomqvist, K., Divne, C.,

Ezcurra, I., Mellerowicz, E., Sundberg, B., Bulone, V., Teeri, T.T., 2008. MAP20, a microtubule-associated protein in the secondary cell walls of hybrid aspen, is a target of the cellulose synthesis inhibitor 2,6-dichlorobenzonitrile. Plant Physiol. 148, 1283–1294.

Rajjou, L., Gallardo, K., Job, C., Job, D., 2006. Proteome analysis for the study of developmental processes in plants. Annu. Plant Rev. 28, 152–184.

Rall, J.A., 2018. Generation of life in a test tube: Albert Szent-Györgyi, Bruno Straub, and the discovery of actin. Adv. Physiol. Educ. 42, 277–288.

Ralph, S.A., Foth, B.J., Hall, N., McFadden, G.I., 2004. Evolutionary pressures on apicoplast transit peptides. Mol. Biol. Evol. 21, 2183–2194.

Rambourg, A., Clermont, Y., 1990. Three-dimensional electron microscopy: structure of the Golgi apparatus. Eur. J. Cell Biol. 51, 189–200.

Ramirez-Weber, F.A., Kornberg, T.B., 1999. Cytonemes: cellular processes that project to the principal signaling center in Drosophila imaginal discs. Cell 97, 599–607.

Ramón y Cajal, S., 1937. Recollections of My Life. Translated by E. Horne Craigie. MIT Press, Cambridge, MA.

Ramón y Cajal, S., 1999. Advice for a Young Investigator. Translated by N. Swanson and L. W. Swanson. MIT Press, Cambridge, MA.

Ramón y Cajal, S., December 12, 1906. The Structure and Connexions of Neurons. Nobel Lecture.

Ran, F.A., Hsu, P.D., Wright, J., Garwala, V.A., Scott, D.A., Zhang, F., 2013. Genome engineering using CRISPR-Cas9 system. Nat. Protoc. 8, 2281–2308.

Rancour, D.M., Dickey, C.E., Park, S., Bednarek, S.Y., 2002. Characterization of AtCDC48. Evidence for multiple membrane fusion mechanisms at the plane of cell division in plants. Plant Physiol. 130, 1241–1253.

Randall, S.K., Sze, H., 1986. Properties of the partially purified tonoplast H^+-pumping ATPase from oat roots. J. Biol. Chem. 261, 1364–1371.

Randall, S.K., Wang, Y., Sze, H., 1985. Purification and characterization of the soluble F(1)-ATPase of oat root mitochondria. Plant Physiol. 79, 957–962.

Randolph, L.F., 1922. Cytology of chlorophyll types of maize. Bot. Gaz. 73, 337–375.

Ranjan, A., Townsley, B.T., Ichihashi, Y., Sinha, N.R., Chitwood, D.H., 2015. An intracellular transcriptomic atlas of the giant coenocyte *Caulerpa taxifolia*. PLoS Genet. 11, e1004900.

Ranjeva, R., Carrasco, A., Boudet, A.M., 1988. Inositol triphosphate stimulates the release of calcium from intact vacuoles isolated from *Acer* cells. FEBS Lett. 230, 137–141.

Rannala, B., Yang, Z.H., 2008. Phylogenetic inference using whole genomes. Annu. Rev. Genom. Hum. Genet. 9, 217–231.

Ranty, B., Aldon, D., Catelle, V., Galaud, J.-P., Thuleau, P., Mazars, C., 2016. Calcium sensors as key hubs in plant responses to biotic and abiotic stresses. Front. Plant Sci. 7, 327.

Ranvier, L., 1875. Recherches sur les l'ments du sang. Arch. Physiol. 2, 1–15.

Rao, R.G.S., Ulaby, F.T., 1977. Optimal spatial sampling techniques for ground truth data in microwave remote sensing of soil moisture. Rem. Sens. Environ. 6, 289–301.

Rapoport, A., 1975. Semantics. Thomas Y. Crowell Co., New York.

Raposo, G., Stoorvogel, W., 2013. Extracellular vesicles: exosomes, microvesicles, and friends. J. Cell Biol. 200, 273–283.

Rapp, S., Saffrich, R., Anton, M., Jaekle, U., Ansorge, W., Gorgas, K., Just, W.-W., 1996. Microtubule-based peroxisomal movement. J. Cell Sci. 109, 837–849.

Raschke, K., Hedrich, R., Rechmann, U., Schroeder, J.I., 1988. Exploring biophysical and biochemical components of the osmotic motor that drives stomatal movements. Bot. Acta 101, 283–294.

Rasi-Caldogno, F., DeMichelis, M.I., Pugliarello, M.C., Marrè, E., 1986. H^+-pumping driven by the plasma membrane ATPase in membrane vesicles from radish: stimulation by fusicoccin. Plant Physiol. 82, 121–125.

Raskin, I., 1992a. Role of salicylic acid in plants. Annu. Rev. Plant Physiol. Mol. Biol. 43, 439–463.

Raskin, I., 1992b. Salicylate, a new plant hormone. Plant Physiol. 99, 799–803.

Raskin, I., Ehmann, A., Melander, W.R., Meeuse, B.J.D., 1987. Salicylic acid—a natural inducer of heat production in Arum lilies. Science 237, 1601–1602.

Raskin, I., Ribnicky, D.M., Komarnytsky, S., Ilic, N., Poulev, A., Borisjuk, N., Brinker, A., Moreno, D.A., Ripoll, C., Yakoby, N., et al., 2002. Plants and human health in the twenty-first century. Trends Biotechnol. 20, 522–531.

Raskin, I., Skubatz, H., Tang, W., Meeuse, B.J.D., 1990. Salicylic acid levels in thermogenic and non-thermogenic plants. Ann. Bot. 66, 373–376.

Raskin, I., Turner, I.M., Melander, W.R., 1989. Regulation of heat production in the inflorescences of an Arum lily by endogenous salicylic acid. Proc. Natl. Acad. Sci. U.S.A. 86, 2214–2218.

Rasmussen, H., 1981. Calcium and cAMP as Synarchic Messengers. John Wiley & Sons, New York.

Rasmussen, C.G., Humphries, J.A., Smith, L.G., 2011a. Determination of symmetric and asymmetric division planes in plant cells. Annu. Rev. Plant Biol. 62, 387–409.

Rasmussen, C.G., Sun, B., Smith, L.G., 2011b. TANGLED localization at the cortical division site of plant cells occurs by several mechanisms. J. Cell Sci. 124, 270–279.

Rasmussen, U., Munck, L., Ullrich, S.E., 1990. Immunogold localization of chymotrypsin inhibitor-2, a lysine-rich protein in developing barley endosperm. Planta 150, 272–277.

Rasmussen, C.G., Wright, A.J., Müller, S., 2013. The role of the cytoskeleton and associated proteins in determination of the plant cell division plane. Plant J. 75, 258–269.

Rastan, S., Beeley, L.J., 1997. Functional genomics: going forwards from the databases. Curr. Opin. Genet. Dev. 7, 777–783.

Ratajczak, R., 2000. Structure, function and regulation of the plant vacuolar H^+-translocating ATPase. Biochim. Biophys. Acta 1465, 17–36.

Ratajczak, R., Lüttge, U., Gonzalez, P., Etxeberria, E., 2003. Malate and malate-channel antibodies inhibit electrogenic and ATP-dependent citrate transport across the tonoplast of citrus juice cells. J. Plant Physiol. 160, 1313–1317.

Rathore, K.S., Cork, R.J., Robinson, K.R., 1991. A cytoplasmic gradient of calcium is correlated with the growth of lily pollen tubes. Dev. Biol. 148, 612–619.

Rauser, W.E., 1990. Phytochelatins. Annu. Rev. Biochem. 59, 61–86.

Ravanel, S., Block, M.A., Rippert, P., Jabrin, S., Curien, G., Rébeillé, F., Douce, R., 2004. Methionine metabolism in plants. J. Biol. Chem. 279, 22548–22557.

Raven, P.H., 1970. A multiple origin for plastids and mitochondria. Science 169, 641–646.

Raven, J.A., 1984. Energetics and Transport in Aquatic Plants. Alan R. Liss. Inc., New York.

Raven, J.A., 1997. Multiple origins of plasmodesmata. Eur. J. Phycol. 32, 95–101.

Raven, J.A., 2005. Evolution of plasmodesmata. In: Oparka, K.J. (Ed.), Plasmodesmata. Blackwell, Oxford, pp. 33–52.

Ravindran, S., 2012. Barbara McClintock and the discovery of jumping genes. Proc. Natl. Acad. Sci. U.S.A. 109, 20198–20199.

Rawlins, D.J., Highett, M.I., Shaw, P.J., 1991. Localization of telomeres in plant interphase nuclei by in situ hybridization and 3D confocal microscopy. Chromosoma 100, 424–431.

Ray, J., 1691. The Wisdom of God Manifested in the Works of the Creation. Printed for S. Smith, London.

Ray, J., 1962. Miscellaneous Discourses Concerning the Dissolution and Changes of the World Wherein the Primitive Chaos and Creation, the General Deluge, Fountains, Formed Stones, Sea Shells Found in the Earth, Subterraneous Trees, Mountains, Earthquakes, Vulcanoes, the Universal Conflagration and Future State, are Largely Discussed and Examined. Printed for S. Smith, London.

Ray, P.M., 1967. Radioautographic study of cell wall deposition in growing plant cells. J. Cell Biol. 35, 659–674.

Ray, P.M., 1977. Auxin binding sites of maize coleoptiles are localized on membranes of the endoplasmic reticulum. Plant Physiol. 59, 594–599.

Ray, P.M., 1980. Cooperative action of ß-glucan synthetase and UDPxylose xylosyl transferase of Golgi membranes in the synthesis of xyloglucan-like polysaccharide. Biochim. Biophys. Acta 629, 431–444.

Ray, P.M., Baker, D.B., 1965. The effect of auxin on synthesis of oat coleoptile cell wall constituents. Plant Physiol. 40, 353–360.

Ray, P.M., Eisinger, W.R., Robinson, D.G., 1976. Organelles involved in cell wall polysaccharide formation and transport in pea cells. Ber. Dtsch. Bot. Ges. 89, 121–146.

Ray, P.M., Shininger, T.L., Ray, M.M., 1969. Isolation of glucan synthetase particles from plant cells and identification with Golgi membranes. Proc. Natl. Acad. Sci. U.S.A. 64, 605–612.

Raychaudhuri, S., Prinz, W.A., 2008. Nonvesicular phospholipid transfer between peroxisomes and the endoplasmic reticulum. Proc. Natl. Acad. Sci. U.S.A. 105, 15785–15790.

Rayle, D.L., Cleland, R., 1977. Control of plant cell enlargement by hydrogen ions. Curr. Top. Dev. Biol. 11, 187–214.

Rayleigh, L., 1893. On the flow of viscous liquids, especially in two dimensions. Philos. Mag. 36, 354–372.

Rayment, I., Holden, H.M., Whittaker, M., Yohn, C.B., Lorenz, M., Holmes, K.C., Milligan, R.A., 1993. Structure of the actin-myosin complex and its implications for muscle contraction. Science 261, 58–65.

Raymond, S., 1964. Acrylamide gel electrophoresis. Ann. N.Y. Acad. Sci. 121, 350–365.

Raymond, S., Nakamichi, M., Aurell, B., 1962. Acrylamide gel as an electrophoresis medium. Nature 195, 697–698.

Raymond, S., Weintraub, L., 1959. Acrylamide gel as a supporting medium for zone electrophoresis. Science 130, 711.

Raza, M.A., Yu, N., Wang, D., Cao, L., Gan, S., Chen, L., 2017. Differential DNA methylation and gene expression in reciprocal hybrids between *Solanum lycopersicum* and *S. pimpinellifolium*. DNA Res. 24, 597–607.

Rea, P.A., 2007. Plant ATP-binding cassette transporters. Annu. Rev. Plant Biol. 58, 347–375.

Rea, P.A., Griffith, C.J., Manolson, M.F., Sanders, D., 1987a. Irreversible inhibition of H^+-ATPase of higher plant tonoplast by chaotropic anions: evidence for peripheral location of nucleotide-binding subunits. Biochem. Biophys. Acta 904, 1–12.

Rea, P.A., Griffith, C.J., Sanders, D., 1987b. Purification of the N,N′-dicyclohexylcarbodiimide-binding proteolipid of a higher plant tonoplast H^+-ATPase. J. Biol. Chem. 262, 14745–14752.

Rea, P.A., Li, Z.-S., Lu, Y.-P., Drozdowicz, Y.M., Martinoia, E., 1998. From vacuolar GS-X pumps to multispecific ABC transporters. Annu. Rev. Plant Physiol. Plant Mol. Biol. 49, 727–760.

Rea, P.A., Poole, R.J., 1986. Chromatographic resolution of H^+-translocating pyrophosphatase from H^+-translocating ATPase of higher plant tonoplast. Plant Physiol. 81, 126–129.

Rea, P.A., Poole, J., 1993. Vacuolar H^+-translocating pyrophosphatase. Ann. Rev. Plant Physiol. Plant Mol. Biol. 44, 157–180.

Rea, P.A., Sanders, D., 1987. Tonoplast energization: two H^+ pumps, one membrane. Physiol. Plant 71, 131–141.

Read, S.H., Bacic, A., 1996. Cell wall porosity and its determination. Mod. Methods Plant Anal. 17, 63–80.

Read, S.M., Bacic, T., 2002. Prime time for cellulose. Science 295, 59–60.

Reccius, C.H., Mannion, J.T., Cross, J.D., Craighead, H.G., 2005. Compression and free expansion of single DNA molecules in nanochannels. Phys. Rev. Lett. 95, 268101.

Rechsteiner, M., 1987. Ubiquitin-mediated pathways for intracellular proteolysis. Annu. Rev. Cell Biol. 3, 1–30.

Record, R.D., Griffing, L.R., 1988. Convergence of the endocytotic and lysosomal pathways in soybean protoplasts. Planta 176, 425–432.

Reddy, A.S., 2001. Calcium: silver bullet in signaling. Plant Sci. 160, 381–404.

Reddy, A.S.N., Carrasco, A., Boudet, A.M., 1988. Inositol 1,4,5-triphosphate induced calcium release from corn coleoptile microsomes. J. Biochem. 101, 569–573.

Reddy, A.S., Day, I.S., 2001a. Kinesins in the Arabidopsis genome: a comparative analysis among eukaryotes. BMC Genom. 2, 2.

Reddy, A.S., Day, I.S., 2001b. Analysis of myosins encoded in the recently completed *Arabidopsis thaliana* genome sequence. Genome Biol. 2. RESEARCH0024.

Reddy, A.S.N., Day, I.S., 2011. Microtubule motor proteins in the eukaryotic green lineage: functions and Regulation. The Plant Cytoskeleton. Adv. Plant Biol. 2, 119–141.

Reddy, M.S., Dinkins, R., Collins, G.B., 2002. Overexpression of the *Arabidopsis thaliana* MinE1 bacterial division inhibitor homologue gene alters chloroplast size and morphology in transgenic *Arabidopsis* and tobacco plants. Planta 215, 167–176.

Reddy, A.S.N., Safadi, F., Beyette, J.R., Mykles, D.L., 1994. Calcium-dependent proteinase activity in root cultures of *Arabidopsis*. Biochem. Biophys. Res. Commun. 199, 1089–1095.

Redman, C.M., 1969. Biosynthesis of serum proteins and ferritin by free and attached ribosomes of rat liver. J. Biol. Chem. 244, 4308–4315.

Redman, J.C., Haas, B.J., Tanimoto, G., Town, C.D., 2004. Development and evaluation of an *Arabidopsis* whole genome Affymetrix probe array. Plant J. 38, 545–561.

Reed, T., 1914. The nature of the double spireme in *Allium cepa*. Ann. Bot. 28, 271–281.

Reed, D.W., 1969. Isolation and composition of a photosynthetic reaction center complex from Rhodopseudomonas spheroids. J. Biol. Chem. 244, 4936–4941.

Reed, L.J., 2001. A trail of research from lipoic acid to α-keto acid dehydrogenase complexes. J. Biol. Chem. 276, 38329–38336.

Reed, D.W., Clayton, R.K., 1968. Isolation of a reaction center fraction from *Rhodopseudomonas spheroides*. Biochem. Biophys. Res. Commun. 30, 471–475.

Reed, R., Griffith, J., Maniatis, T., 1988. Purification and visualization of native spliceosomes. Cell 53, 949–961.

Reese, E.T., 1976. History of the cellulase program at the U.S. army Natick Development Center. Biotechnol. Bioeng. Symp. 6, 9–20.

Reeve, M.A., Fuller, C.W., 1995. A novel thermostable polymerase for DNA sequencing. Nature 376, 796–797.

Reeves, R., 1992. Chromatin changes during the cell cycle. Curr. Opin. Cell Biol. 4, 413–423.

Reggiori, F., Ungermann, C., 2017. Autophagosome maturation and fusion. J. Mol. Biol. 429, 486–496.

Reichelt, S., Knight, A.E., Hodge, T.P., Baluska, F., Samaj, J., Volkmann, D., Kendrick-Jones, J., 1999. Characterization of the unconventional myosin VIII in plant cells and its localization at the post-cytokinetic cell wall. Plant J. 19, 555–567.

Reichert, E.T., 1919. A Biochemic Basis for the Study of Problems of Taxonomy, Heredity, Evolution, etc., with Especial Reference to the Starches and Tissues of Parent-Stocks and Hybrid-Stocks and the Starches and Hemoglobins of Varieties, Species, and Genera. Parts I and II. Carnegie Institution of Washington, Washington, DC.

Reichle, R.E., Alexander, J.V., 1965. Multiporate septations, Woronin bodies, and septal plugs in Fusarium. J. Cell Biol. 24, 489–496.

Reil, J.C., 1796. On the vital force. Archiv für die Physiologie, 1992. In: A Documentary History of Biochemistry, 1770–1940, 1. Farleigh Dickenson University Press, Rutherford, NJ, p. 8. Translated by K. Teich.

Reiland, S., Messerle, G., Baerenfaller, K., Gerrits, B., Endler, A., Grossmann, J., Gruissem, W., Baginsky, 2009. Large-scale *Arabidopsis* phosphoproteome profiling reveals novel chloroplast kinase substrates and phosphorylation networks. Plant Physiol. 150, 889–903.

Reinhart, B.J., Slack, F.J., Basson, M., Pasquinelli, A.E., Bettinger, J.C., Rougvie, A.E., Horvitz, H.R., Ruvkun, G., 2000. The 21-nucleotide let-7 RNA regulates developmental timing in *Caenorhabditis elegans*. Nature 403, 901–906.

Reisen, D., Hanson, M.R., 2007. Association of six YFP-myosin XI-tail fusions with mobile plant cell organelles. BMC Plant Biol. 7, 6.

Reiss, C., Beale, S.I., 1995. External calcium requirements for light induction of chlorophyll accumulation and its enhancement by red light and cytokinin pretreatments in excised etiolated cucumber cotyledons. Planta 196, 635–641.

Reiss, H.D., Schnepf, E., Herth, W., 1984. The plasma membrane of the *Funaria* caulonema tip cell: morphology and distribution of particle rosettes, and the kinetics of cellulose synthesis. Planta 160, 428–435.

Reisser, W., 1992. Green ciliates: principles of symbiosis formation between autotrophic and heterotrophic partners. In: Lewin, R.A. (Ed.), Origins of Plastids. Symbiogenesis, Prochlorophytes, and the Origins of Chloroplasts. Chapman & Hall, New York, pp. 27–43.

Remak, R., 1855. Untersuchungen ueber die Entwicklung der Wirbelthiere. G. Reimer, Berlin.

Ren, B., Robert, F., Wyrick, J.J., Aparicio, O., Jennings, E.G., Simon, I., Zeitlinger, J., Schreiber, J., Hannett, N., Kanin, E., Volkert, T.L., Wilson, C.J., Bell, S.P., Young, R.A., 2000. Genome-wide location and function of DNA binding proteins. Science 290, 2306–2309.

Renaud, F.L., Rowe, A.J., Gibbons, I.R., 1968. Some properties of the protein forming the outer fibers of cilia. J. Cell Biol. 36, 79–90.

Rensing, S.A., Lang, D., Zimmer, A., Terry, A., Salamov, A., Shapiro, H., Nishiyama, T., Perroud, P.-F., Lindquist, E., Kamisugi, Y., Tanahashi, T., Sakakibara, K., Fujita, T., Oishi, K., Shin-I, T., Kuroki, Y., Toyoda, A., Suzuki, Y., Hashimoto, S., Yamaguchi, K., Sugano, S., Kohara, Y., Fujiyama, A., Anterola, A., Aoki, S., Ashton, N., Barbazuk, W.B., Barker, E., Bennetzen, J., Blankenship, R., Cho, S.H., Dutcher, S.K., Estelle, M., Fawcett, J.A., Gundlach, H., Hanada, K., Hey, A., Hicks, K.A., Hughes, J., Lohr, M., Mayer, K., Melkozernov, A., Murata, T., Nelson, D., Pils, B., Prigge, M., Reiss, B., Renner, T., Rombauts, S., Rushton, P., Sanderfoot, A., Schween, G., Shiu, S.-H., Stueber, K., Theodoulou, F.L., Tu, H., de Peer, Y.V., Verrier, P.J., Waters, E., Wood, A., Yang, L., Cove, D., Cuming, A.C., Hasebe, M., Lucas, S., Mishler, B.D., Reski, R., Grigoriev, I.V., Quatrano, R.S., Boore, J.L., 2008. The genome of the moss *Physcomitrella patens* reveals evolutionary insights into the conquest of land by plants. Science 319, 64–69.

Rensink, W.A., Buell, C.R., 2005. Microarray expression profiling resources for plant genomics. Trends Plant Sci. 10, 603–609.

Renvoize, S., 1991. *Thamnocalamus spathaceus* and its hundred year flowering cycle. Curtiss Bot. Mag. 8, 185–194.

Reski, R., 1998a. Development, genetics and molecular biology of mosses. Bot. Acta 111, 1–15.

Reski, R., 1998b. *Physcomitrella* and *Arabidopsis*: the David and Goliath of reverse genetics. Trends Plant Sci. 3, 209–210.

Restrepo, M.A., Freed, D.D., Carrington, J.C., 1990. Nuclear transport of plant potyviral proteins. Plant Cell 2, 987–998.

Reumann, S., 2000. The structural properties of plant peroxisomes and their metabolic significance. Biol. Chem. 381, 639–648.

Reumann, S., 2004. Specification of the peroxisome targeting signals type 1 and type 2 of plant peroxisomes by bioinformatics analyses. Plant Physiol. 135, 783–800.

Reumann, S., 2011. Toward a definition of the complete proteome of plant peroxisomes: where experimental proteomics must be complemented by bioinformatics. Proteomics 11, 1764–1779.

Reumann, S., Babujee, L., Ma, C., Weinkoop, S., Siemsen, T., Antonicelli, G.E., Rasche, N., Lüder, F., Weckwerth, W., Jahn, O., 2007. Proteome analysis of *Arabidopsis* leaf peroxisomes reveals novel targeting peptides, metabolic and regulatory functions of peroxisomes. Plant Physiol. 150, 125–143.

Reumann, S., Bartel, B., 2016. Plant peroxisomes: recent discoveries in functional complexity, organelle homeostasis, and morphological dynamics. Curr. Opin. Plant Biol. 34, 17–26.

Reumann, S., Chowdhary, G., Lingner, T., 2016. Characterization, prediction and evolution of plant peroxisomal targeting signals type 1 (PTS1s). Biochim. Biophys. Acta 1863, 790–803.

Reumann, S., Heupel, R., Heldt, H.W., 1994. Compartmentation studies on spinach leaf peroxisomes. II. Evidence for the transfer of a reductant from the cytosol to the peroxisomal compartment via a malate shuttle. Planta 193, 167–173.

Reumann, S., Ma, C., Lemke, S., Babujee, L., August 27, 2004. AraPerox. A database of putative *Arabidopsis* proteins from plant peroxisomes. Plant Physiol. Preview.

Reumann, S., Maier, E., Benz, R., Heldt, H.W., 1995. The membrane of leaf peroxisomes contains a porin-like channel. J. Biol. Chem. 270, 17559–17565.

Reumann, S., Maier, E., Heldt, H.W., Benz, R., 1998. Permeability properties of the porin of spinach leaf peroxisomes. Eur. J. Biochem. 251, 359–366.

Reumann, S., Quan, S., Aung, K., Yang, P., Manandhar-Shrestha, K., Holbrook, D., Linka, N., Switzenberg, R., Wilkerson, C.G., Weber, A.P.M., Olsen, L.J., Hu, J., 2009. In-depth proteome analysis of Arabidopsis leaf peroxisomes combined with in vivo subcellular targeting verification indicates novel metabolic and regulatory functions of peroxisomes. Plant Physiol. 150, 125–114.

Reumann, S., Weber, A.P.M., 2006. Plant peroxisomes respire in the light: some gaps on the photorespiratory C2 cycle have become filled—others remain. Biochim. Biophys. Acta 1763, 1496–1510.

Revel, J.P., Karnovsky, M.J., 1967. Hexagonal arrays of subunits in intercellular junctions of the mouse heart and liver. J. Cell Biol. 33, C7–C12.

Reyes, J.C., Hennig, L., Gruissem, W., 2002. Chromatin- remodeling and memory factors. New regulators of plant development. Plant Physiol. 130, 1090–1101.

Reynolds, O., 1885. On the dilatancy of media composed of rigid particles in contact. With experimental illustrations. Philos. Mag. Ser. 5 (20), 469–481.

Reynolds, O., 1886. Dilatancy. Nature 33, 429–430.

Reynolds, O., 1892. Memoir of James Prescott Joule. Manchester Literary and Philosophical Society, Manchester.

Reynolds, O., 1901. An experimental investigation of the circumstances which determine whether the notion of water shall be direct or sinuous, and of the law of resistance in parallel channels (1881–1900). In: Reynolds, O. (Ed.), Papers on Mechanical and Physical Subjects, vol. II. Cambridge University Press, Cambridge, pp. 51–105.

Reynolds, A., 2007. The theory of the cell state and the question of cell autonomy in nineteenth and early twentieth-century biology. Sci. Context 20, 71–95.

Reynolds, A., 2018. In search if cell architecture. In: Matlin, K.S., Maienschein, J., Laubichler, M.D. (Eds.), Visions of Cell Biology: Reflections Inspired by Cowdry's General Cytology. University of Chicago Press, Chicago, pp. 46–72.

Rhea, R.P., 1966. Electron microscopic observations on the slime mold *Physarum polycephalum* with special reference to fibrillar structures. J. Ultrastruct. Res. 15, 349–379.

Rhoades, M.M., 1984. The early years of maize genetics. Ann. Rev. Genet. 18, 1–29.

Rhoades, D.M., McIntosh, L., 1993. The salicylic acid-inducible alternative oxidase gene aox1 and genes encoding pathogenesis-related proteins share regions of sequence similarity in their promoters. Plant Mol. Biol. 21, 615–624.

Rhoads, A., Au, K.F., 2015. PacBio sequencing and its applications. Genom. Proteom. Bioinform. 13, 278–289.

Rhodin, J., 1954. Correlation of Ultrastructural Organization and Function in Normal and Experimentally Changed Proximal Convoluted Tubule Cells of Mouse Kidney. Aktiebolaget Godvil, Stockholm.

Ricardi, M.M., González, R.M., Zhong, S., Domínguez, P.G., Duffy, T., Turjanski, P.G., Salgado Salter, J.D., Alleva, K., Carrari, F., Giovannoni, J.J., Estévez, J.M., Iusem, N.D., 2014. Genome-wide data (ChIP-seq) enabled identification of cell wall-related and aquaporin genes as targets of tomato ASR1, a drought stress-responsive transcription factor. BMC Plant Biol. 14, 29.

Rich, A.R., 1926. The place of R.-J.-H. Dutrochet in the development of the cell theory. Bull. Johns Hopkins Hosp. 39, 330–365.

Rich, A., 1997. Gamow and the genetic code. In: Harper, E., Parke, W.C., Anderson, D. (Eds.), George Gamow Symposium. ASP Conference Series, vol. 129, pp. 114–122.

Richards, E.J., 2006. Inherited epigenetic variation—revisiting soft inheritance. Nat. Rev. Genet. 7, 395–401.

Richards, T.W., December 6, 1919. Atomic Weights. Nobel Lecture.

Richardson, L.G.L., Paila, Y.D., Siman, S.R., Chen, Y., Smith, M.D., Schnell, D.J., 2014. Targeting and assembly of components of the TOC protein import complex at the chloroplast outer envelope membrane. Front. Plant Sci. 5, 269.

Richardson, A.O., Palmer, J.D., 2007. Horizontal gene transfer in plants. J. Exp. Bot. 58, 1–9.

Richardson, D.N., Simmons, M.P., Reddy, A.S., 2006. Comprehensive comparative analysis of kinesins in photosynthetic organisms. BMC Genom. 7, 18.

Richmond, P.A., Métraux, J.-P., Taiz, L., 1980. Cell expansion patterns and directionality of wall mechanical properties in Nitella. Plant Physiol. 65, 211–217.

Richmond, T.A., Somerville, C.R., 2000. The cellulose synthase superfamily. Plant Physiol. 124, 495–498.

Richter, J.B., 1792–1794. Anfangsgründe der Stöchiometrie; oder, Messkunst chemischer Elemente, 1968. G. Olms, Hildesheim. Reproduction of the edition by Ausg. Breslau und Hirschberg.

Richter, S., Anders, N., Wolters, H., Beckmann, H., Thomann, A., Heinrich, R., Schrader, J., Singh, M.K., Geldner, N., Mayer, U., Jürgens, G., 2010. Role of the GNOM gene in *Arabidopsis* apical-basal patterning—From mutant phenotype to cellular mechanism of action. Eur. J. Cell Biol. 89, 138–144.

Richter, L., Mason, H.S., Arntzen, C.J., 1996. Transgenic plants created for oral immunization against diarrheal diseases. J. Travel Med. 3, 52–56.

Ridge, R.W., Emons, A.M.C. (Eds.), 2000. Root Hairs. Cell and Molecular Biology. Springer, Tokyo.

Ridgway, E.B., Ashley, C.C., 1967. Calcium transients in single muscle fibers. Biochem. Biophys. Res. Commun. 29, 229–234.

Ridgway, E.B., Gilkey, J.C., Jaffe, L.F., 1977. Free calcium increases explosively in activating medaka eggs. Proc. Natl. Acad. Sci. U.S.A. 74, 623–627.

Rieder, C.L., 1991. Mitosis: towards a molecular understanding of chromosome behavior. Curr. Opin. Cell Biol. 3, 59–66.

Rieder, C.L., Alexander, S.P., 1990. Kinetochores are transported poleward along single astral microtubule during chromosome attachment to the spindle in newt lung cells. J. Cell Biol. 110, 81–95.

Rieder, C.L., Khodjakov, A., 2003. Mitosis through the microscope: advances in seeing inside live dividing cells. Science 300, 91–96.

Rieder, C.L., Khodjakov, A., Paliulis, L.V., Fortier, T.M., Cole, R.W., Sluder, G., 1997. Mitosis in vertebrate somatic cells with two spindles: implications for the metaphase/anaphase transition checkpoint and cleavage. Proc. Natl. Acad. Sci. U.S.A. 94, 5107–5112.

Riederer, M., Muller, C. (Eds.), 2006. Biology of the Plant Cuticle. Wiley-Blackwell, Oxford.

Riens, B., Lohaus, G., Heineke, D., Heldt, H.W., 1991. Amino acid and sucrose content determined in the cytosolic, chloroplastic, and vacuolar compartment and in the phloem sap of spinach leaves. Plant Physiol. 97, 227–233.

Riezman, H., Weir, E.M., Leaver, C.J., Titus, D.E., Becker, W.M., 1980. Regulation of glyoxysomal enzymes during germination of cucumber. 3. In vitro translation and characterization of four glyoxysomal enzymes. Plant Physiol. 65, 4046.

Rigal, A., Doyle, S.M., Robert, S., 2015. Live cell imaging of FM4-64, a tool for tracing the endocytotic pathways in *Arabidopsis* root cells. Methods Mol. Biol. 1242, 93–103.

Riley, M., Pardee, A.B., Jacob, F., Monod, J., 1960. On the expression of a structural gene. J. Mol. Biol. 2, 216–225.

Rim, Y., Huang, L., Chu, H., Han, X., Cho, W.K., Jeon, C.O., Kim, H.J., Hong, J.-C., Lucas, W.J., Kim, J.Y., 2011. Analysis of *Arabidopsis* transcription factor families revealed extensive capacity for cell-to-cell movement as well as discrete trafficking patterns. Mol. Cell 32, 519–526.

Rimington, C., 1936. The chemistry of the proteins and amino acids. Ann. Rev. Biochem. 5, 117–158.

Rincon, M., Boss, W.F., 1987. Myoinositol triphosphate mobilizes calcium from fusogenic carrot (*Daucus carota* L.) protoplasts. Plant Physiol. 83, 395–398.

Ringer, S., 1883. A further contribution regarding the influence of the different constituents of the blood on the contraction of the heart. J. Physiol. 4, 29–42.

Ris, H., 1943. A quantitative study of anaphase movement in the aphid *Tamalia*. Biol Bull. 85, 164–178.

Ris, H., 1949. The anaphase movement of chromosomes in the spermatocytes of the grasshopper. Biol. Bull. 96, 90–106.

Ris, H., 1961. Ultrastructure and molecular organization of genetic systems. Can. J. Genet. Cytol. 3, 95–120.

Ris, H., 1962. Interpretation of ultrastructure in the cell nucleus. In: Harris, R.J.C. (Ed.), The Interpretation of Ultrastructure, vol. 1. Academic Press, New York, pp. 69–88.

Ris, H., Chandler, B.L., 1963. The ultrastructure of genetic systems in prokaryotes and eukaryotes. Cold Spring Harbor Symp. Quant. Biol. 28, 1–8.

Ris, H., Mirsky, A.E., 1950. Quantitative cytochemical determination of desoxyribonucleic acid with the feulgen nucleal reaction. J. Gen. Physiol. 3, 125–146.

Ris, H., Plaut, W., 1962. Ultrastructure of DNA-containing areas in the chloroplast of Chlamydomonas. J. Cell Biol. 13, 383–391.

Ris, H., Witt, P., 1981. Structure of the mammalian kinetochore. Chromosoma 82, 153–170.

Risebrough, R.W., Tissières, A., Watson, J.D., 1962. Messenger-RNA attachment to active ribosomes. Proc. Natl. Acad. Sci. U.S.A. 48, 4300–4436.

Ritter, W.E., 1911. The controversy between materialism and vitalism: can it be ended? Science 33, 437–441.

Ritter, W.E., 1919. The Unity of the Organism, or the Organismal Conception of Life, vols. I and II. Gorham Press, Boston.

Ritter Jr., H., Inoué, S., Kubai, D., 1978. Mitosis in *Barbulanympha*. I. Spindle structure, formation, and kinetochore engagement. J. Cell Biol. 77, 638–654.

Ritzenthaler, C., Nebenführ, A., Movafeghi, A., Stussi-Garaud, C., Behnia, L., Pimpl, P., Staehelin, L.A., Robinson, D.G., 2002. Reevaluation of the effects of brefeldin A on plant cells using tobacco bright yellow 2 cells expressing Golgi-targeted green fluorescent protein and COPI antisera. Plant Cell 13, 237–261.

Roach, J.C., Boysen, C., Wang, K., Hood, L., 1995. Pairwise end sequencing: a unified approach to genomic mapping and sequencing. Genomics 26, 345–353.

Roach, M., Sauermann, H., 2017. The declining interest in an academic career. PLoS One 12, e0184130.

Robards, A.W., 1976. Plasmodesmata in higher plants. In: Gunning, B.E.S., Robards, A.W. (Eds.), Intercellular Communication in Plants: Studies on Plasmodesmata. Springer Verlag, Berlin, pp. 15–57.

Robards, A.W., Lucas, W.J., 1990. Plasmodesmata. Annu. Rev. Plant Physiol. Plant Mol. Biol. 41, 369–419.

Robatzek, S., 2007. Vesicle trafficking in plant immune responses. Cell Microbiol. 9, 1–8.

Robatzek, S., Chinchilla, D., Boller, T., 2006. Ligand-induced endocytosis of the pattern recognition receptor FLS2 in *Arabidopsis*. Genes Dev. 20, 537–542.

Robbins, J., Dilworth, S.N., Laskey, R.A., Dingwall, C., 1991. Two interdependent basic domains in nucleoplasmin nuclear targeting sequence: identification of a class of bipartite nuclear targeting sequence. Cell 64, 615–623.

Robbins, E., Gonatas, N.K., 1964. Histochemical and ultrastructural studies on HeLa cell cultures exposed to spindle inhibitors with special reference to the interphase cell. J. Histochem. Cytochem. 12, 704–711.

Robenek, H., Peveling, E., 1977. Ultrastructure of the cell wall regeneration of isolated protoplasts of *Skimmia japonica* Thumb. Planta 136, 135–145.

Roberson, R.W., Fuller, M.S., 1988. Ultrastructural aspects of the hyphal tip of *Sclerotium rolfsii* preserved by freeze substitution. Protoplasma 146, 143–149.

Roberts, H.F., 1929. Plant Hybridization before Mendel. Princeton University Press, Princeton, NJ.

Roberts, M., 1938. Bio-Politics. As Essay on the Physiology Pathology & Politics of the Social and Somatic Organism. J. Dent & Sons, London.

Roberts, R.B. (Ed.), 1958. Microsomal Particles and Protein Synthesis. Pergamon Press, New York.

Roberts, R.J., 1978. Restriction and modification enzymes and their recognition sequences. Gene 4, 183–193.

Roberts, D.M., 1989a. Detection of a calcium-activated protein kinase in *Mougeotia* by using synthetic peptide substrates. Plant Physiol. 91, 1613–1619.

Roberts, K., 1989b. The plant extracellular matrix. Curr. Opin. Cell Biol. 1, 1020–1027.

Roberts, R.M., 1989c. Serendipity. Accidental Discoveries in Science. John Wiley & Sons, New York.

Roberts, K., 1990. Structures at the plant cell surface. Curr. Opin. Cell Biol. 2, 920–928.

Roberts, K., 1994. The plant extracellular matrix: in a new expansive mood. Curr. Opin. Cell Biol. 6, 688–694.

Roberts, A.G., 2005. Plasmodesmal structure and development. In: Oparka, K.J. (Ed.), Plasmodesmata. Blackwell, Oxford, pp. 1–32.

Roberts, L., Baba, S., 1968. IAA induced xylem differentiation in the presence of colchicine. Plant Cell Physiol. 9, 315–321.

Roberts, I.M., Boevink, P., Roberts, A.G., Sauer, N., Reichel, C., Oparka, K.J., 2001. Dynamic changes in the frequency and architecture of plasmodesmata during sink-source transition in tobacco leaves. Protoplasma 218, 31–44.

Roberts, R.J., December 8, 1993. An Amazing Distortion in DNA Induced by a Methyltransferase. Nobel Lecture.

Roberts, S.K., Fischer, M., Dixon, G.K., Sanders, D., 1999. Divalent cation block of inward currents and low-affinity K^+ uptake in *Saccharomyces cervisiae*. J. Bacteriol. 181, 291–297.

Roberts, A.G., Oparka, K.J., 2003. Plasmodesmata and the control of symplastic transport. Plant Cell Environ. 26, 103–124.

Roberts, A.W., Roberts, E.M., 2004. Cellulose synthase (CesA) genes in algae and seedless plants. Cellulose 11, 419–435.

Roberts, A.W., Roberts, E.M., Delmer, D.P., 2002. Cellulose synthase (CesA) genes in the green alga *Mesotaenium caldariorum*. Eukaryot. Cell 1, 847–855.

Robertson, J.D., 1959. The ultrastructure of cell membranes and their derivatives. Biochem. Soc. Symp. 16, 3–43.

Robertson, J.D., 1964. Unit membranes: a review with recent new studies of experimental alterations and a new subunit structure in synaptic membranes. In: Locke, M. (Ed.), Cellular Membranes in Development. Academic Press, NewYork, pp. 1–81.

Robertson, J.D., 1987. The early days of electron microscopy of nerve tissues and membranes. Int. Rev. Cytol. 100, 129–201.

Robertson, M.P., Miller, S.L., 1995. An efficient prebiotic synthesis of cytosine and uracil. Nature 375, 772–774.

Robertson, D., Mitchell, G.P., Gilroy, J.S., Gerrish, C., Bolwell, G.P., Slabas, A.R., 1997. Differential extraction and protein sequencing reveals major differences in patterns of primary cell wall proteins from plants. J. Biol. Chem. 272, 15841–15848.

Robertson, E.J., Pyke, K.A., Leech, R.M., 1995. Arc6, an extreme chloroplast division mutant of *Arabidopsis* also alters proplastid proliferation and morphology in shoot and root apices. J. Cell Sci. 108, 2937–2944.

Robertson, E.J., Rutherford, S.M., Leech, R.M., 1996. Characterization of chloroplast division using the *Arabidopsis* mutant arc5. Plant Physiol. 112, 149–159.

Robinow, C., Angert, E.R., 1998. Nucleoids and coated vesicles of "*Epulopiscium*" spp. Arch. Microbiol. 170, 227–235.

Robins, R.D., Juniper, B.E., 1983. The secretory cycle of *Dionaea muscipula*. Ellis. I, II, III, IV, and V. New Phytol. 86, 279-296, 297–311, 313–327, 402–412, 413–422.

Robinson, R., 1955. The Structural Relations of Natural Products. Clarendon Press, Oxford.

Robinson, D.G., 1980. Dictyosome-endoplasmic reticulum associations in higher plant cells? A serial section analysis. Eur. J. Cell Biol. 23, 22–36.

Robinson, K.R., 1985. The responses of cells to electric fields: a review. J. Cell Biol. 101, 2023–2027.

Robinson, D.G., 1991a. What is a plant cell? The last word. Plant Cell 3, 1145–1146.

Robinson, T., 1991b. The Organic Constituents of Higher Plants. Their Chemistry and Interrelationships, sixth ed. Cordus Press, North Amherst, MA.

Robinson, D.G., 2003. Vesicle trafficking in plants. Zellbiologie aktuell 29 (2), 29–32.

Robinson, C.V., 2011. John Fenn (1917-2010). Nature 469, 300.

Robinson, D.G., Balusek, K., Depta, H., Hoh, B., Hostein, S.E.H., 1991. Isolation and characterization of plant coated vesicles. In: Hawes, C.R., Coleman, J.O.D., Evans, D.E. (Eds.), Endocytosis, Exocytosis and Vesicle Traffic in Plants. Cambridge University Press, Cambridge, pp. 65–79.

Robinson, D.G., Balusek, K., Freundt, H., 1989. Legumin antibodies recognize polypeptides in coated vesicles isolated from developing pea cotyledons. Protoplasma 150, 79–82.

Robinson, D.G., Brandizzi, F., Hawes, C., Nakano, A., 2015. Vesicles versus tubes: is endoplasmic reticulum-Golgi transport in plants fundamentally different from other eukaryotes. Plant Physiol. 168, 393–406.

Robinson, K.R., Cone, R., 1980. Polarization of fucoid eggs by a calcium ionophore gradient. Science 207, 77–78.

Robinson, R., December 12, 1947. Some Polycyclic Natural Products. Nobel Lecture.

Robinson, D.G., Depta, H., 1988. Coated vesicles. Ann. Rev. Plant Physiol. 39, 53–99.

Robinson, D.G., Ding, Y., Jiang, L., 2016. Unconventional protein secretion in plants: a critical assessment. Protoplasma 253, 31–43.

Robinson, D.G., Eisinger, W.R., Ray, P.M., 1976a. Dynamics of the Golgi system in wall matrix polysaccharide synthesis and secretion by pea cells. Ber. Dtsch. Bot. Ges. 89, 147–161.

Robinson, D.G., Glas, R., 1982. Secretion kinetics of hydroxyproline-containing macromolecules in carrot root disks. Plant Cell Rep. 1, 197–198.

Robinson, D.G., Grimm, I., Sachs, H., 1976b. Colchicine and microfibril orientation. Protoplasma 89, 375–380.

Robinson, D.G., Haschke, H.-P., Hinz, G., Hoh, B., Maeshima, M., Marty, F., 1996. Immunological detection of tonoplast polypeptides in the plasma membrane of pea cotyledons. Planta 198, 95–103.

Robinson, D.G., Herranz, A.-C., Bubeck, J., Pepperkok, R., Ritzenthaler, C., 2007. Membrane dynamics in the early secretory pathway. Crit. Rev. Plant Sci. 26, 199–225.

Robinson, D.G., Herzog, W., 1977. Structure, synthesis and orientation of microfibrils. III. A survey of the action of microtubule inhibitors on microtubules and microfibril orientation in *Oocystis solitaria*. Cytobiologie 15, 463–474.

Robinson, D.G., Hillmer, S., 1990. Endocytosis in plants. Physiol. Plant 79, 96–104.

Robinson, D.G., Hinz, G., 1996. Multiple mechanisms of protein body formation in pea cotyledons. Plant Physiol. Biochem. 34, 155–163.

Robinson, D.G., Hinz, G., 1997. Vacuole biogenesis and proton transport to the plant vacuole: a comparison with the yeast vacuole and the mammalian lysosome. Protoplasma 197, 1–25.

Robinson, K.R., Jaffe, L.F., 1973. Ion movements in a developing Fucoid egg. Dev. Biol. 35, 349–361.

Robinson, K.R., Jaffe, L.F., 1975. Polarizing fucoid eggs drive a calcium current through themselves. Science 187, 70–72.

Robinson, D.G., Jiang, L., Schumacher, K., 2008. The endosomal system in plants: charting new and familiar territories. Plant Physiol. 147, 1482–1492.

Robinson, K.R., Messerli, M.A., 2002. Left/Right, up/down: the role of endogenous electrical fields as directional signals in development, repair and invasion. BioEssays 25, 759–766.

Robinson, K.R., Muller, B.J., 1997. The coupling of cyclic GMP and photopolarization of *Pelvetia* zygotes. Dev. Biol. 187, 125–130.

Robinson, D.G., Preston, R.D., 1972. Plasmalemma structure in relation to microfibril biosynthesis in Oocystis. Planta 104, 234–246.

Robinson, D.G., Quader, H., 1981. Structure, synthesis, and orientation of microfibrils. IX. A freeze-fracture investigation of the *Oocystis* plasma membrane after inhibitor treatment. Eur. J. Cell Biol. 25, 278–288.

Robinson, D.G., Quader, H., 1982. The microtubule-microfibril syndrome. In: Lloyd, C.W. (Ed.), The Cytoskeleton in Plant Growth and Development. Academic Press, London, pp. 109–126.

Robinson, D.G., Rogers, J.C., 2000. Vacuolar Compartments. Sheffeld Academic Press and CRC Press, Boca Raton, FL.

Robinson, R.A., Stokes, R.H., 1959. Electrolyte Solutions. Butterworths Scientific Publications, London.

Rochaix, J.D., 2013. Redox regulation of thylakoid protein kinases and photosynthetic gene expression. Antioxid. Redox Signal 18, 2184–2201.

Rochaix, J.D., 2014. Regulation and dynamics of the light-harvesting system. Annu. Rev. Plant Biol. 65, 287–309.

Rockwell, N.C., Su, Y.-S., Lagarias, J.C., 2006. Phytochrome structure and signaling mechanisms. Annu. Rev. Plant Biol. 57, 837–858.

Rodrigo-Peiris, T., Xu, X.M., Zhao, Q., Wang, H.J., Meier, I., 2011. RanGAP is required for post-meiotic mitosis in female gametophyte development in *Arabidopsis thaliana*. J. Exp. Bot. 62, 2705–2714.

Rodriguez, E., Aregullin, M., Nishida, T., Uehara, S., Wrangham, R., Abramowski, Z., Finlayson, A., Towers, G.H.N., 1985. Thiarubrine A., a bioactive constituent of *Aspilia* (Asteraceae) comsumed by wild chimpanzees. Experientia 41, 419–420.

Rodriguez-Furlán, C., Miranda, G., Reggiardo, M., Hicks, G.R., Norambuena, L., 2016. High throughput selection of novel plant growth regulators: assessing the translatability of small bioactive molecules from *Arabidopsis* to crops. Plant Sci. 245, 50–60.

Roelfsema, M.R.G., Hedrich, R., 2002. Studying guard cells in the intact plant: modulation of stomatal movement by apoplastic factors. New Phytol. 153, 425–431.

Roelofsen, P.A., 1958. Cell wall structure as related to surface growth. Acta Bot. Neerl. 7, 77–89.

Roger, J., 1997. Buffon. A Life in Natural History. Cornell University Press, Ithaca, New York.

Rogers, H.J., 2005. Cytoskeletal regulation of the plane of cell division: an essential component of plant development and reproduction. Adv. Bot. 42, 69–111.

Rogers, S.L., Karcher, R.L., Roland, J.T., Minin, A.A., Steffen, W., Gelfand, V.I., 1999. Regulation of melanosome movement in the cell cycle by reversible association with myosin V. J. Cell Biol. 146, 1265–1276.

Rogers, M.M., McReynolds, A.W., Rogers, F.T., 1940. A determination of the masses and velocities of three radium B beta-particles: the relativistic mass of the electron. Phys. Rev. 57, 379–383.

Rogers, S.S., Waigh, T.A., Lu, J.R., 2008. Intracellular microrheology of motile *Amoeba proteus*. Biophys. J. 94, 3313–3322.

Roget, P.M., 1834. Animal and Vegetable Physiology, Considered with Reference to Natural Theology. W. Pickering, London.

Rojas, E.R., Hotton, S., Dumais, J., 2011. Chemically mediated mechanical expansion of the pollen tube cell wall. Biophys. J. 101, 1844–1853.

Rojas-Pierce, M., 2013. Targeting of tonoplast proteins to the vacuole. Plant Sci. 211, 132–136.

Rojo, E., Denecke, J., 2008. What is moving in the secretory pathway of plants? Plant Physiol. 147, 1493–1503.

Rolland, R., 1926. The Game of Love and Death. Translated by E. S. Brooks. Henry Holt and Co., New York.

Romagnoli, S., Cai, G., Cresti, M., 2003. In vitro assays demonstrate that pollen tube organelles use kinesin-related motor proteins to move along microtubules. Plant Cell 15, 251–269.

Römling, U., 2002. Molecular biology of cellulose production in bacteria. Res. Microbiol. 153, 205–212.

Ronaghi, M., Karamohamed, S., Pettersson, B., Uhlén, M., Nyrén, P., 1996. Real-time DNA sequencing using detection of pyrophosphate. Anal. Biochem. 242, 84–89.

Ronaghi, M., Uhlén, M., Nyrén, P., 1998. A sequencing method based on real-time pyrophosphate. Science 281, 363–365.

Ronceret, A., Sheehan, M.J., Pawlowski, W.P., 2007. Chromosome dynamics in meiosis. Plant Cell Monogr. 9, 103–124.

Roof, D.M., Meluh, P.B., Rose, M.D., 1992. Kinesin-related proteins required for assembly of the mitotic spindle. J. Cell Biol. 118, 95–108.

Roppolo, D., De Rybel, B., Tendon, V.D., Pfister, A., Alassimone, J., Vermeer, J.E.M., Yamazaki, M., Stierhof, Y.-D., Beeckman, T., Geldner, N., 2011. A novel protein family mediates Casparian strip formation in the endodermis. Nature 473, 380–383.

Roppolo, D., Geldner, N., 2012. Membrane and walls: who is master, who is servant? Curr. Opin. Plant Biol. 15, 608–617.

Rosamond, J., 1987. Structure and function of mitochondria. Int. Rev. Cytol Suppl. 17, 121–147.

Rose, A., 2007. Open mitosis: nuclear envelope dynamics. Plant Cell Monogr. 9, 207–230.

Rose, J.K., Bennett, A.B., 1999. Cooperative disassembly of the cellulose-xyloglucan network of plant cell walls: parallels between cell expansion and fruit ripening. Trends Plant Sci. 4, 176–183.

Rose, J.K.C., Cosgrove, D.J., Albersheim, P., Darvill, A.G., Bennett, A.B., 2000. Detection of expansin proteins and activity during tomato fruit ontogeny. Plant Physiol. 123, 1583–1592.

Rose, I., December 8, 2004. Ubiquitin at fox chase. In: Nobel Lecture.

Rose, J.K., Doms, R.W., 1988. Regulation of protein transport from the endoplasmic reticulum. Ann. Rev. Cell Biol. 4, 257–288.

Rose, J.K.C., Lee, S.J., 2010. Straying off the highway: trafficking of secreted plant proteins and complexity in the plant cell wall proteome. Plant Physiol. 153, 433–436.

Rosen, G., 1959. The conservation of energy and the study of metabolism. In: Brooks, C.McC., Cranefield, P.F. (Eds.), The Historical Development of Physiological Thought. Hafner Publishing Co, New York, pp. 245–263.

Rosen, W.G., 1968. Ultrastructure and physiology of pollen. Ann. Rev. Plant Physiol. 19, 435–462.

Rosen, W.G., Gawlik, S.R., Dashek, W.V., Siegesmund, K.A., 1964. Fine structure and cytochemistry of *Lilium* pollen tubes. Am. J. Bot. 51, 61–71.

Rosenbaum, J.L., Carlson, K., 1969. Cilia regeneration in *Tetrahymena* and its inhibition by colchicine. J. Cell Biol. 40, 415–425.

Rosenberg, J.L., 2004. The contributions of James Franck to photosynthesis research: a tribute. Photosynth. Res. 80, 71–76.

Rosenberger, R.F., Kessel, M., 1968. Nonrandom sister chromatid segregation and nuclear migration in hyphae of *Aspergillus nidulans*. J. Bacteriol. 96, 1208–1213.

Rosero, A., Žárský, V., Cvrčková, F., 2014. Visualizing and quantifying the in vivo structure and dynamics of the Arabidopsis cortical cytoskeleton using CLSM and VAEM. Methods Mol. Biol. 1080, 87–97.

Roshchina, V.V., 2001. Neurotransmitters in Plant Life. Science Publishers, Enfield, NH.

Roshchina, V.V., 2008. Fluorescing World of Plant Secreting Cells. Science Publishers, Enfield, NH.

Roshchina, V.V., Roshchina, V.D., 1993. The Excretory Function of Higher Plants. Springer-Verlag, Berlin.

Rosing, J., Slater, E.C., 1972. The value of $\Delta G°$ for the hydrolysis of ATP. Biochim. Biophys. Acta 267, 275–290.

Rossi, E., Azzone, G.F., 1970. The mechanism of ion translocation in mitochondria. 3. Coupling of K^+ efflux with ATP synthesis. Eur. J. Biochem. 12, 319–327.

Rőszer, T., 2012. The Biology of Subcellular Nitric Oxide. Springer, Dordrecht.

Roth, L.E., Daniels, E.W., 1962. Electron microscopic studies of mitosis in amebae. II. The giant ameba Pelomyxa carolinensis. J. Cell Biol. 12, 57–78.

Roth, T.F., Porter, K.R., 1964. Yolk protein uptake in the oocyte of the mosquito *Aedes Aegypti*. L. J. Cell Biol. 20, 313–332.

Rothberg, J.M., Leamon, J.H., 2008. The development and impact of 454 sequencing. Nat. Biotechnol. 26, 1117–1124.

Rothman, J.E., 1992. The reconstruction of intracellular protein transport in cell-free systems. Harvey Lect. 86, 65–85.

Rothman, T., 2003. Everything's Relative and other Fables from Science and Technology. John Wiley & Sons, Hoboken, NJ.

Rothman, J.E., 2010. The future of Golgi research. Mol. Biol. Cell 21, 3776–3780.

Rothman, J.E., December 7, 2013. The Principle of Membrane Fusion in the Cell. Nobel Lecture.

Rothman, J.E., Miller, R.L., Urbani, L.J., 1984a. Intercompartmental transport in the Golgi complex is a dissociative process: facile transfer of membrane proteins between two Golgi populations. J. Cell Biol. 99, 260–271.

Rothman, J.E., Urbani, L.J., Brands, R., 1984b. Transport of protein between cytoplasmic membranes of fused cells: correspondence to processes reconstituted in a cell-free system. J. Cell Biol. 99, 248–259.

Rouiller, C., Bernhard, W., 1956. "Microbodies" and the problem of mitochondrial regeneration in liver cells. J. Biophys. Biochem. Cytol. 2 (no. 4). Suppl. 355.

Rouiller, Bernhard, 1956. J. Biophys. Biochem. Cytol. 2 (4), 355.

Rounds, C.M., Hepler, P.K., Fuller, S.J., Winship, L.J., 2010. Oscillatory growth in lily pollen tubes does not require aerobic energy metabolism. Plant Physiol. 152, 736–746.

Rounds, C.M., Hepler, P.K., Winship, L.J., 2014. The apical actin fringe contributes to localized cell wall deposition and polarized growth in the lily pollen tube. Plant Physiol. 166, 139–151.

Rounds, C.M., Lubeck, E., Hepler, P.K., Winship, L.J., 2011a. Propidium iodide completes with Ca^{2+} to label pectin in pollen tubes and *Arabidopsis* root hairs. Plant Physiol. 157, 175–187.

Rounds, C.M., Winship, L.J., Hepler, P.K., 2011b. Pollen tube energetics: respiration, fermentation and the race to the ovule. AoB Plants plr019.

Rousseau, J.J., 1762. The Social Contract or Principles of Political Right. http://www.constitution.org/jjr/socon.htm.

Routier-Kierzkowska, A.-L., Smith, R.S., 2014. Mechanical measuremnts on living plant cells by micro-indentation with cellular force microscopy. Methods Mol. Biol. 1080, 135–146.

Roux, W., 1883. Ueber die Bedeutung der Kerntheilungsfiguren. Eine hypothetische Erörterung. Wilhem Engelmann, Leipzig.

Roux, W., 1888. Contributions to the developmental mechanics of the embryo. On the artificial production of half-embryos by destruction of one of the first two blastomeres, and the later development (postgeneration) of the missing half of the body, 1974. In: Willier, B.H., Oppenheimer, J.M. (Eds.), Foundations of Experimental Embryology, second ed. Hafner Press, New York, pp. 2–37.

Roy, S., Vian, B., 1991. Transmural exocytosis in maize root cap. Visualization by simultaneous use of a cellulose-probe and a fucose-probe. Protoplasma 161, 181–191.

Royo, J., Gómez, E., Hueros, G., 2007. Transfer cells. Plant Cell Monogr. 8, 73–89.

Ruan, Y.-L., Llewellyn, D.J., Furbank, R.T., 2001. The control of single-celled cotton fiber elongation by developmentally reversible gating of plasmodesmata and coordinated expression of sucrose and K^+ transporters and expansin. Plant Cell 13, 47–60.

Ruben, S., 1943. Photosynthesis and phosphorylation. J. Am. Chem. Soc. 65, 279–282.

Ruben, S., Kamen, M.D., 1940a. Radioactive carbon in the study of respiration in heterotrophic systems. Proc. Natl. Acad. Sci. U.S.A. 26, 418–422.

Ruben, S., Kamen, M.D., 1940b. Radioactive carbon of long half-life. Phys. Rev. 57, 549.

Ruben, S., Kamen, M.D., 1941. Long-lived radioactive carbon: C14. Phys. Rev. 59, 349–354.

Rubin, E.M., 2008. Genomics of cellulosic biofuels. Nature 454, 841–845.

Rubinstein, B., Luster, D.G., 1993. Plasma membrane redox activity: components and role in plant processes. Ann. Rev. Plant Physiol. Plant Mol. Biol. 44, 131–155.

Rubio, F., Gassmann, W., Schroeder, J.I., 1995. Sodium-driven potassium uptake by the plant potassium transporter HKT1 and mutations conferring salt tolerance. Science 270, 1660–1663.

Ruddle, F.H., 1984. The William Allan Award address: reverse genetics and beyond. Am. J. Hum. Genet. 36, 944–953.

Rüdiger, W., 2005. Prototropin phosphorylation. In: Wada, M., Shimazaki, K., Iino, M. (Eds.), Light Sensing in Plants. Springer, Tokyo, pp. 171–177.

Ruegger, M., Chapple, C., 2001. Mutations that reduce sinapoylmalate accumulation in *Arabidopsis thaliana* define loci with diverse roles in phenylpropanoid metabolism. Genetics 159, 1741–1749.

Ruegger, M., Meyer, K., Cusumano, J.C., Chapple, C., 1999. Regulation of ferulate-5-hydroxylase expression in *Arabidopsis* in the context of sinapate ester biosynthesis. Plant Physiol. 119, 101–110.

Ruestow, E.G., 1983. Images and ideas: Leeuwenhoek's perception of the spermatozoa. J. Hist. Biol. 16, 185–224.

Ruestow, E.G., 1996. The Microscope in the Dutch Republic. The Shaping of Discovery. Cambridge University Press, Cambridge.

Ruiz-May, E., Kim, S.-J., Brandizzi, F., Rose, J.K.C., 2012a. The secreted plant N-glycoproteome and associated secretory pathways. Front. Plant Sci. 3, 117.

Ruiz-May, E., Rose, J.K.C., 2013. Progress towards the tomato fruit cell wall proteome. Front. Plant Sci. 4, 159.

Ruiz-May, E., Thannhauser, T.W., Zhang, S., Rose, J.K.C., 2012b. Analytical technologies for identification and characterization of the plant *N*-glycoproteome. Front. Plant Sci. 3, 150.

Runnström, J., 1911. Untersuchungen über die Permeabilität des Seeigeleies für Farbstoffe. Arkiv. f. Zool. 7, 1–17.

Ruoslahti, E., 1988. Structures at the plant cell surface. Curr. Opin. Cell Biol. 4, 229–255.

Ruparel, H., Bi, L., Bai, X., Kim, D.H., Turro, N.J., Ju, J., 2005. Design and synthesis of a 3′-O-allyl photocleavable fluorescent nucleotide as a reversible terminator for DNA sequencing by synthesis. Proc. Natl. Acad. Sci. U.S.A. 102, 5932–5937.

Rupke, N., 1994. Richard Owen: Victorian Naturalist. Yale University Press, New Haven.

Rupke, N., 2009. Richard Owen: Biology without Darwin. Chicago University Press, Chicago.

Russ, V., Grolig, F., Wanger, G., 1991. Changes of cytoplasmic free Ca^{2+} in the green alga Mougeotia scalaris as monitored with indo-1, and their effect on the velocity of chloroplast movements. Planta 184, 105–112.

Russell, J.B., 1991. Inventing the Flat Earth. Columbus and Modern Historians. Praeger, New York.

Russinova, E., Borst, J.W., Kwaaitaal, M., Cano-Delgado, A., Yin, Y., Chory, J., de Vries, S.C., 2004. Heterodimerization and endocytosis of *Arabidopsis* brassinosteroid receptors BRI1 and AtSERK3. Plant Cell.

Rustom, A., Saffrich, R., Markovic, I., Walther, P., Gerdes, H.-H., 2004. Nanotubular highways for intercellular organellar transport. Science 303, 1007–1010.

Rutherford, E., 1909. Radioactive Transformations. Yale University Press, New Haven, CT.

Rutherford, 1936. Rutherford of Nelson, The Society for the protection of science and learning. Science 83, 372.

Rutherford, E., December 11, 1908. The Chemical Nature of the Alpha Particles from Radioactive Substances. Nobel Lecture.

Rutherford, S., Moore, I., 2002. The *Arabidopsis* Rab GTPase family: another enigma variation. Curr. Opin. Plant Biol. 5, 518–528.

Rutschow, H.L., Baskin, T.I., Kramer, E.M., 2014. The carrier AUXIN resistant (AUX1) dominates auxin flux into *Arabidopsis* protoplasts. New Phytol. 204, 536–544.

Rutter, B.D., Innes, R.W., 2017. Extracellular vesicles isolated from the leaf apoplast carry stress-response proteins. Plant Physiol. 173, 728–741.

Ruzin, S.E., 1999. Plant Microtechnique and Microscopy. Oxford University Press, Oxford.

Ryabov, E.V., van Wezel, R., Walsh, J., Hong, Y., 2004. Cell-to-Cell, but not long-distance, spread of RNA silencing that is induced in individual epidermal cells. J. Virol. 78, 3149–3154.

Ryan, C.A., 1990. Protease inhibitors in plants: genes for improving defenses against insects and pathogens. Annu. Rev. Phytopathol. 28, 425–449.

Ryan, C.A., Farmer, E.E., 1991. Oligosaccharide signals in plants: a current assessment. Annu. Rev. Plant Physiol. Plant Mol. Biol. 42, 651–674.

Rybicki, E.P., 2010. Plant-made vaccines for humans and animals. Plant Biotechnol. J. 8, 620–637.

Ryter, A., 1968. Association of the nucleus and the membrane of bacteria: a morphological study. Bacteriol. Rev. 32, 39–54.

Ryu, J.-H., Mizuno, K., Takagi, S., Nagai, R., 1997. Extracellular components implicated in the stationary organization of the actin cytoskeleton in mesophyll cells of *Vallisneria*. Plant Cell Physiol. 38, 420–432.

Ryu, J.-H., Takagi, S., Nagai, R., 1995. Stationary organization of the actin cytoskeleton in *Vallisneria*: the role of stable microfilaments at the end walls. J. Cell Sci. 108, 1531–1539.

Sabatini, D.D., Adesnik, M., 2013. Christian de Duve: explorer of the cell who discovered new organelles by using a centrifuge. Proc. Natl. Acad. Sci. U.S.A. 110, 13234.

Sabatini, D.D., Bensch, K., Barrnett, R.J., 1963. Cytochemistry and electron microscopy. The preservation of cellular ultrastructure and enzymatic activity by aldehyde fixation. J. Cell Biol. 17, 19–58.

Sabatini, D.D., Blobel, G., 1970. Controlled proteolysis of nascent polypeptides in rat liver cell fractions. II. Location of the polypeptides in rough microsomes. J. Cell Biol. 45, 146–157.

Sabatini, D.D., Louvard, D., Adesnik, M., 1991. Membranes. Editorial overview. Curr. Opin. Cell Biol. 3, 575–579.

Sabelli, P.A., Larkins, B.A., 2007. The endoreduplication cell cycle: regulation and function. Plant Cell Monogr. 9, 75–100.

Sabelli, P.A., Larkins, B.A., 2009. The development of endosperm in grasses. Plant Physiol. 149, 14–26.

Sabidó, E., Selevbek, N., Aebersold, R., 2012. Mass spectrometry-based proteomics for systems biology. Curr. Opin. Biotechnol. 23, 591–597.

Sablin, E.P., Case, R.B., Dai, S.C., Hart, C.L., Ruby, A., Vale, R.D., Fletterick, R.J., 1998. Direction determination in the minus- end-directed kinesin motor ncd. Nature 395, 813–816.

Sablin, E.P., Kull, F.J., Cooke, R., Vale, R.D., Fletterick, R.J., 1996. Crystal structure of the motor domain of the kinesin motor ncd. Nature 380, 555–559.

Sachs, J., 1882a. Text-book of Botany, second ed. Clarendon Press, Oxford.

Sachs, J., 1882b. Vorlesungen über Pflanzen-Physiologie. Verlag W. Engelmann, Leipzig.

Sachs, J. von, 1887. Lectures on the Physiology of Plants. Translated by H. Marshall Ward. Clarendon Press, Oxford.

Sachs, J. von, 1906. History of Botany. (1530–1860). Translated by H. E. F, Garnsey. Revised by I. B. Balfour. Clarendon Press, Oxford.

Sachs, T., 2006. How can plants choose the most promising organs? In: Baluska, F., Mancuso, S., Volkmann, D. (Eds.), Communication in Plants. Springer, Berlin, pp. 53–63.

Sachs, H., Grimm, I., Robinson, D.G., 1976. Structure, synthesis and orientation of microfibrils. I. Architecture and development of the wall of *Oocystis solitaria*. Cytobiologie 14, 49–60.

Sack, F.D., 1991. What is a plant cell? Continued Plant Cell 3, 844.

Sack, F.D., 1997. Plastids and gravitropic sensing. Planta 203, S63–S68.

Sack, F.D., 2004. Yoda would be proud: valves for land plants. Science 304, 1461–1462.

Sadava, D., Chrispeels, M.J., 1973. Hydroxyproline-rich cell wall protein (extension): role in the cessation of elongation in excised pea epicotyls. Dev. Biol. 30, 49–55.

Sadeghnezhad, E., Sharifi, M., Zare-Maivan, H., 2016. Profiling of acidic (amino and phenolic acids) and phenylpropanoids production in response to methyl jasmonate-induced oxidative stress in *Schrophularia striata* suspension cells. Planta 244, 75–85.

Saftner, R.A., Raschke, K., 1981. Electrical potentials in stomatal complexes. Plant Physiol. 67, 1124–1132.

Sagan, L., 1967. On the origin of mitosing cells. J. Theor. Biol. 14, 225–274.

Sager, R., Lee, J.-Y., 2014. Plasmodesmata in integrated cell signaling: insights from development and environmental signals and stresses. J. Exp. Bot. 65, 6337–6358.

Sager, R.E., Lee, J.-Y., 2018. Plasmodesmata at a glance. J. Cell Sci. 131, jcs209346.

Sager, R., Ryan, F.J., 1961. Cell Heredity. An Analysis of the Mechanisms of Heredity at the Cellular Level. John Wiley & Sons, Inc., New York.

Sagi, G., Katz, A., Guenoune-Gelbart, D., Epel, B.L., 2005. Class 1 reversibly glycosylated polypeptides are plasmodesmal-associated proteins delivered to plasmodesmata via the Golgi apparatus. Plant Cell 17, 1788–1800.

Saheki, Y., De Camillo, P., 2017. Endoplasmic reticulum-plasma membrane contact sites. Annu. Rev. Biochem. 86, 659–684.

Saiki, R.K., Gelfand, D.H., Stoffel, S., Scharf, S.J., Higuschi, R., Horn, G.T., Mullis, K.B., Erlich, H.A., 1988. Primer-directed enzymatic amplification of DNA with a thermostable DNA polymerase. Science 239, 487–491.

Saiki, R.K., Scharf, S., Faloona, F., Mullis, K.B., Horn, G.T., Erlich, H.A., Arnhein, N., 1985. Enzymatic amplification of β-globulin sequences and restriction site analysis for diagnosis of sickle cell anemia. Science 230, 1350–1354.

Saint-Jore, C.M., Evins, J., Batoko, H., Brandizzi, F., Moore, I., Hawes, C., 2002. Redistribution of membrane proteins between the Golgi apparatus and endoplasmic reticulum in plants is reversible and not dependent on cytoskeletal networks. Plant J. 29, 661–678.

Saito, S.Y., Karaki, H., 1996. A family of novel actin-inhibiting marine toxins. Clin. Exp. Pharma. Physiol. 23, 743–746.

Saito, C., Ueda, T., 2009. Functions of RAB and SNARE proteins in plant life. Int. Rev. Cell Mol. Biol. 274, 183–233.

Sakaguchi, S., Hogetsu, T., Hara, N., 1988a. Arrangement of cortical microtubules in the shoot apex of *Vinca major*. L. Planta 175, 403–411.

Sakaguchi, S., Hogetsu, T., Hara, N., 1988b. Arrangement of cortical microtubules at the surface of the shoot apex in Vinca major L: observations by immunofluorescence microscopy. Bot. Mag. (Tokyo) 101, 497–507.

Sakaguchi, S., Hogetsu, T., Hara, N., 1990. Specific arrangements of cortical microtubules are correlated with the architecture of meristems in shoot spices of angiosperms and gymnosperms. Bot. Mag. (Tokyo) 103, 143–163.

Sakai, A., 2014. Generative cells. In: Noguchi, T., Kawano, S., Tsukaya, H., Matsunaga, S., Sakai, A., Karahara, I., Hayashi, Y. (Eds.), Atlas of Plant Cell Structure. Springer, Japan, pp. 157–186.

Sakai, Y., Inoue, S., Harada, A., Shimazaki, K., Takagi, S., 2015b. Blue-light-induced rapid chloroplast de-anchoring in *Vallisneria* epidermal cells. J. Integr. Plant Biol. 57, 93–105.

Sakai, Y., Inoue, S., Harada, A., Shimazaki, K., Takagi, S., 2015b. Blue-light-induced rapid chloroplast de-anchoring in Vallisneria epidermal cells. J. Integr. Plant Biol. 57, 93–105.

Sakai, T., Kagawa, T., Kasahara, M., Swartz, T.E., Christie, J.M., Briggs, W.R., Wada, M., Okada, K., 2001. *Arabidopsis* nph1 and npl1: blue light receptors that mediate both phototropism and chloroplast relocation. Proc. Natl. Acad. Sci. U.S.A. 98, 6969–6974.

Sakai, H., Naito, K., Ogiso-Tanaka, E., Takahashi, Y., Iseki, K., Muto, C., Satou, K., Teruya, K., Shiroma, A., Shimoji, M., Hirano, T., Itoh, T., Kaga, A., Tomooka, N., 2015a. The power of single molecule real-time sequencing technology in the de novo assembly of a eukaryotic genome. Sci. Rep. 5, 16780.

Sakai, Y., Takagi, S., 2005. Reorganized actin filaments under high-intensity blue light anchor chloroplasts along the anticlinal walls of *Vallisneria* epidermal cells. Planta 221, 823–830.

Sakai, Y., Takagi, S., 2017. Roles of actin cytoskeleton for regulation of chloroplast anchoring. Plant Signal. Behav. https://doi.org/10.1080/15592324.2017.1370163.

Sakamoto, K., Briggs, W.R., 2002. Cellular and subcellular localization of phototropin I. Plant Cell 14, 1723–1735.

Sakamoto, T., Inui, Y.T., Yuraguchi, S., Yoshizumi, T., Matsunaga, S., Mastui, M., Umeda, M., Fukui, K., Fujiwara, T., 2011. Condensin II alleviates DNA damage and is essential for tolerance of boron overload stress in Arabidopsis. Plant Cell 23, 3533–3546.

Sakamoto, T., Inui, Y.T., Yuraguchi, S., Yoshizumi, T., Matsunaga, S., Mastui, M., Umeda, M., Fukui, K., Fujiwara, T., 2013. Condensin II alleviates DNA damage and is essential for tolerance of boron overload stress in *Arabidopsis*. Plant Cell 23, 3533–3546.

Sakamoto, Y., Takagi, S., 2013. Little nuclei 1 and 4 regulate nuclear morphology in *Arabidopsis thaliana*. Plant Cell Physiol. 54, 622–633.

Sakano, K., Tazawa, M., 1985. Metabolic conversion of amino acids loaded in the vacuole of *Chara australis* internodal cells. Plant Physiol. 78, 673–677.

Sakaue-Sawano, A., Kurokawa, H., Morimura, T., Hanyu, A., Hama, H., Osawa, H., Kashiwagi, S., Fukami, K., Miyata, T., Miyoshi, H., Imamura, T., Ogawa, M., Masai, H., Miyawaki, A., 2008. Visualizing spatiotemporal dynamics of multicellular cell-cycle progression. Cell 132, 487–498.

Sakiyama-Sogo, M., Shibaoka, H., 1993. Gibberellin A2 and abscisic and cause the reorientation of cortical microtubles in epicotyl cells of the decapitated dwarf pea. Plant Cell Physiol. 34, 431–437.

Sakmann, B., December 9, 1991. Elementary Steps in Synaptic Transmission Revealed by Currents Through Single ion Channels. Nobel Lecture.

Sakmann, B., Neher, E. (Eds.), 1983. Single-Channel Recording. Plenum Press, New York.

Sakurai, M., Pak, J.Y., Muramatsu, Y., Fukuhara, T., 2004. Integrin-like protein at the invaginated plasma membrane of epidermal cells in mature leaves of the marine angiosperm *Zostera marina* L. Planta 220, 271–277.

Saladino, R., Brucato, J.R., De Sio, A., Botta, G., Pace, E., Gambicorti, L., 2011. Photochemical synthesis of citric acid cycle intermediates based on titanium dioxide. Astrobiology 11, 815–824.

Salanoubat, M., Lemcke, K., Rieger, M., Ansorge, W., Unseld, M., et al., 2000. Sequence and analysis of chromosome 3 of the plant *Arabidopsis thaliana*. Nature 408, 820–822.

Salas Fernandez, M.G., Bao, Y., Tang, L., Schnable, P.S., 2017. A high-throughput, field-based phenotyping technology for tall biomass crops. Plant Physiol. 174, 2008–2022.

Sale, W.S., Satir, P., 1977. Direction of active sliding of microtubules in *Tetrahymena* cilia. Proc. Natl. Acad. Sci. U.S.A. 74, 2045–2049.

Saleh, A., Alvarez-Venegas, R., Avramova, Z., 2008. An efficient chromatin immunoprecipitation (ChiP) protocol for studying histone modifications in *Arabidopsis* plants. Nat. Protoc. 3, 1019–1025.

Salic, A., Waters, J.C., Mitchison, T.J., 2004. Vertebrate shugoshin links sister centromere cohesion and kinetochore microtubule stability in mitosis. Cell 118, 567–578.

Salisbury, F.B., 1993. Gravitropism: changing ideas. Hortic. Rev. 15, 233–278.

Salisbury, F.B., 1998. The discovery of biological clocks. Discov. Plant Biol. 1, 287–327.

Salisbury, J.L., Baron, A., Colling, D., Martindale, V., Sanders, M., 1986. Calcium-modulated contractile proteins associated with the eucaryotic centrosome. Cell Motil. Cytoskelet. 6, 193–197.

Salisbury, J.L., Baron, A.T., Sanders, M.A., 1988. The centrin-based cytoskeleton of *Chlamydomonas reinhartii*: distribution in interphase and mitotic cells. J. Cell Biol. 107, 635–641.

Salisbury, J.L., Baron, A., Surek, B., Melkonian, M., 1984. Striated flagellar roots: isolation and partial characterization of a calcium-modulated contractile organelle. J. Cell Biol. 99, 962–970.

Salisbury, J.L., Sanders, M.A., Harpst, L., 1987. Flagellar root contraction and nuclear movement during flagellar regeneration in Chlamydomonas reinhardtii. J. Cell Biol. 105, 1799–1805.

Salminen, A., Novick, P.J., 1987. A ras-like protein is required for a post-Golgi event in yeast secretion. Cell 49, 527–538.

Salmon, E.D., 1975a. Pressure-induced depolymerization of spindle microtubules. I. Changes in birefringence and spindle length. J. Cell Biol. 65, 603–614.

Salmon, E.D., 1975b. Pressure-induced depolymerization of spindle microtubules. II. Thermodynmics of in vivo spindle assembly. J. Cell Biol. 66, 114–127.

Salmon, E.D., 1989. Microtubule dynamics and chromosome movement. In: Hyams, J.S., Brinkley, B.R. (Eds.), Mitosis: Molecules and Mechanisms. Academic Press, Canada, pp. 119–181.

Salmon, M.S., Bayer, E.M.F., 2012. Dissecting plasmodesmata molecular composition by mass spectrometry-based proteomics. Front. Plant Sci. 3, 307.

Salmon, E.D., Ellis, G.W., 1975. A new miniature hydrostatic pressure chamber for microscopy. J. Cell Biol. 65, 587–602.

Salmon, E.D., Leslie, R.J., Saxton, W.M., Karow, M.L., Mclntosh, J.R., 1984c. Spindle microtubule dynamics in sea urchin embryos: analysis using a fluorescein-labeled tubulin and measurements of fluorescence redistri- bution after laser photobleaching. J. Cell Biol. 99, 2165–2174.

Salmon, E.D., McKeel, M., Hays, T., 1984a. Rapid rate of tubulin dissociation from microtubules in the mitotic spindle in vivo measured by blocking polymerization with colchicine. J. Cell Biol. 99, 1066–1075.

Salmon, E.D., Saxton, W.M., Leslie, R.J., Karow, M.L., Mclntosh, J.R., 1984b. Diffusion coefficient of fluorescein-labeled tubulin in the cytoplasm of embryonic cells of a sea urchin: video image analysis of fluores- cence redistribution after photobleaching. J. Cell Biol. 99, 2157–2164.

Salomon, M., Christie, J.M., Knieb, E., Lempert, U., Briggs, W.R., 2000. Photochemical and mutational analysis of the FMN- binding domains of the plant blue light receptor, phototropin. Biochemistry 39, 9401–9410.

Salomon, S., Grunewals, D., Stüber, K., Schaaf, S., MacLean, D., Schulze-Lefert, P., Robatzek, S., 2010. High-throughput confocal imaging of intact live tissue enables quantification of membrane trafficking in Arabidopsis. Plant Physiol. 1096–1104.

Salpeter, E.E., 1974. Dying stars and reborn dust. Rev. Mod. Phys. 46, 433–446.

Salpeter, E.E., 2002. A generalist looks back. Annu. Rev. Astron. Astrophys. 40, 1–25.

Saltman, P., Kunitake, G., Spolter, H., Stitt, C., 1956. The dark fixation of CO_2 by succulent leaves. The first products. Plant Physiol. 31, 464–468.

Salvato, F., Havelund, J.F., Chen, M., Rao, R.S.P., Rogowska-Wrzesinska, A., Jensen, O.N., Gang, D.R., Thelen, J.J., Møller, I.M., 2014. The potato tuber mitochondrial proteome. Plant Physiol. 164, 637–653.

Salzberg, S.L., Yorke, J.A., 2005. Beware of mis-assembled genomes. Bioinformatics 21, 4320–4321.

Sampathkumar, A., Lindeboom, J.J., Debolt, S., Gutrerrez, R., Ehrardt, D.W., Ketelaar, T., Persson, S., 2011. Live cell imaging reveals structural associations between actin and microtubule cytoskeleton in Arabidopsis. Plant Cell 23, 2302–2313.

Sampson, K., Pickett-Heaps, J.D., 2001. Phallacidin stains the kinetochore region in the mitotic spindle of the green alga *Oedogonium* spp. Protoplasma 217, 166–176.

Sampson, K., Pickett-Heaps, J.D., Forer, A., 1996. Cytochalasin D blocks chromosomal attachment to the spindle in the green alga *Oedogonium*. Protoplasma 192, 130–144.

Samuel, M., Bleackley, M., Anderson, M., Mathivanan, S., 2015. Extracellular vesicles including exosomes in cross kingdom regulation: a viewpoint from plant-fungal interactions. Front. Plant Sci.

Samuels, A.L., Bisalputra, T., 1990. Endocytosis in elongating root cells of Lobelia erinos. J. Cell Sci. 97, 157–165.

Samuels Jr., A.L., Giddings, T.H., Staehelin, L.A., 1995. Cytokinesis in tobacco BY-2 and root tip cells: a new model of cell plate formation in higher plants. J. Cell Biol. 130, 1345–1357.

Samuels, A.L., Staehelin, L.A., 1996. Caffeine inhibits cell plate formation by disrupting membrane reorganization just after the vesicle fusion step. Protoplasma 195, 144–155.

San Pietro, A., Lang, H.M., 1956. Accumulation of reduced pyridine nucleotides by illuminated grana. Science 124, 118–119.

Sanchez, A.M., Bosch, M., Bots, M., Nieuwland, J., Feron, R., Mariani, C., 2004. Pistil factors controlling pollination. Plant Cell 16, S98–S106.

Sanchez, R.A., Ferris, J.P., Orgel, L.E., 1966. Conditions for purine synthesis: did prebiotic synthesis occur at low temperatures? Science 153, 72–73.

Sanchez, R.A., Ferris, J.P., Orgel, L.E., 1968. Studies in prebiotic synthesis. IV. Conversion of 4-aminoimidazole-5-carbonitrile derivatives to purines. J. Mol. Biol. 38, 121–128.

Sanchez-Fernandez, T., Davies, G.E., Coleman, J.O.D., Rea, P.A., 2001. The *Arabidopsis* thaliana ABC protein superfamily, a complete inventory. J. Biol. Chem. 276, 30231–30244.

Sánchez-León, S., Gil-Humanes, J., Ozuna, C.V., Giménez, M.J., Sousa, C., Voytas, D.F., Barro, F., 2017. Low-gluten, non-transgenic wheat engineered with CRISPR/Cas9. Plant Biotechnol. J. 2017, 1–9.

Sandberg, C., 1994. The Red Dyes: Cochineal, Madder, and Murex Purple. Lark Books, New York.

Sandelius, A.S., Morré, D.J., 1987. Characteristics of a phosphatidylinositol exchange activity of soybean microsomes. Plant Physiol. 84, 1022–1027.

Sandelius, A.S., Penel, C., Auderset, G., Brightman, A., Millard, M., Morré, D.J., 1986. Isolation of highly purified fractions of plasma membrane and tonoplast from the same homogenate of soybean hypocotyls by free-flow electrophoresis. Plant Physiol. 81, 177–185.

Sanderfoot, A., 2007. Increases in the number of SNARE genes parallels the rise of multicellularity among green plants. Plant Physiol. 144, 6–17.

Sanderfoot, A.A., Ahmed, S.U., Marty-Mazars, D., Rapoport, I., Kischhausen, T., Marty, F., Raikhel, N.V., 1998. A putative vacuolar cargo receptor partially colocalizes with AtPEP12p on a prevacuolar compartment in *Arabidopsis* roots. Proc. Natl. Acad. Sci. U.S.A. 195, 9920–9925.

Sanderfoot, A.A., Assaad, F.F., Raikhel, N.V., 2000. The *Arabidopsis* genome: an abundance of soluble N-ethylmaleimide-sensitive factor adaptor proteins. Plant Physiol. 124, 1558–1569.

Sanders, D., 1990. Kinetic modeling of plant and fungal membrane transport systems. Annu. Rev. Plant Physiol. Plant Mol. Biol. 41, 77–107.

Sanders, B., 2016. In: Our Revolution: A Future to Believe. Thomas Dunne Books, New York.

Sanders, L.C., Wang, C.S., Walling, L.L., Lord, E.M., 1991. A homolog of the substrate adhesion molecule vitronectin occurs in four species of flowering plants. Plant Cell 3, 629–635.

Sandstrom, R.P., Cleland, R.E., 1989. Comparison of the lipid composition of oat root and coleoptile plasma membranes. Plant Physiol. 90, 1207–1213.

Sanford, J.C., 1990. Biolistic plant transformation. Physiol. Plant 79, 206–209.

Sanford, J.C., 2008. Genetic Entropy & The Mystery of the Genome, third ed. FMS Publications, Waterloo, New York.

Sanford, J.C., Johnston, S.A., 1985. The concept of parasite-derived resistance—Deriving resistance genes from the parasite's own genome. J. Theor. Biol. 113, 395–405.

Sanger, F., 1981. Determination of nucleotide sequences in DNA. Science 214, 1205–1210.

Sanger, F., 1988. Sequences, sequences, and sequences. Ann. Rev. Biochem. 57, 1–28.

Sanger, F., Air, G.M., Barrell, B.G., Brown, N.L., Coulson, A.R., Fiddes, J.C., Hutchison, C.A., Slocombe, P.M., Smith, M., 1977a. Nucleotide sequence of bacteriophage ΦX174 DNA. Nature 265, 687–695.

Sanger, F., Coulson, A.R., 1975. A rapid method for determining sequences in DNA by primed synthesis with DNA polymerase. J. Mol. Biol. 94, 441–448.

Sanger, F., Coulson, A.R., 1978. The use of thin acrylamide gels for DNA sequencing. FEBS Lett. 87, 107–110.

Sanger, F., Coulson, A.R., Barrell, B.G., Smith, A.J.H., Roe, B.A., 1980. Cloning in single-stranded bacteriophage as an aid to rapid DNA sequencing. J. Mol. Biol. 143, 161–178.

Sanger, F., Coulson, A.R., Hong, G.F., Hill, D.F., Peterson, G.B., 1982. Nucleotide sequence of bacteriophage λ DNA. J. Mol. Biol. 162, 729–773.

Sanger, F., December 8, 1980. Determination of Nucleotide Sequences in DNA. Nobel Lecture.

Sanger, F., December 11, 1958. The Chemistry of Insulin. Nobel Lecture.

Sanger, F., Donelson, J.E., Coulson, A.R., Kössel, H., Fischer, D., 1973. Use of DNA polymerase I primed by a synthetic oligonucleotide to a nucleotide sequence in Phage f1 DNA. Proc. Natl. Acad. Sci. U.S.A. 70, 1209–1213.

Sanger, F., Nicklen, S., Coulson, A.R., 1977b. DNA sequencing with chain-terminating inhibitors. Proc. Natl. Acad. Sci. U.S.A. 74, 5463–5467.

Sano, T., Katsuna, N., Higami, T., Oda, Y., Yoneda, A., Kumagai-Sano, F., Hasezawa, S., 2007. Cytoskeletal and vacuolar dynamics during plant cell division: approaches using structure- visualized cells. Plant Cell Monogr. 9, 125–140.

Sansom, C., 2006. The beginnings of bioinformatics. Biochemist 28 (6), 48–49.

Santarius, K.A., Heber, U., 1965. Changes in the intracellular levels of ATP, ADP, AMP and Pi and regulatory function of the adenylate system in leaf cells during photosynthesis. Biochim. Biophys. Acta 102, 39–54.

Santi, L., Batchelor, L., Huang, Z., Hjelm, B., Kilbourne, J., Arntzen, C.J., Chen, Q., Mason, H.S., 2008. An efficient plant viral expression system generating orally immunogenic Norwalk virus-like particles. Vaccine 26, 1846–1854.

Santuari, L., Hardtke, C.S., 2010. The case for resequencing studies of *Arabidopsis thaliana* accessions: mining the dark matter of natural genetic variation. F1000 Biol. Rep. 2, 85.

Sapala, A., Runions, A., Routier-Kierzkowska, A.-L., Das Gupta, M., Hong, L., Hofhuis, H., Verger, S., Mosca, G., Li, C.-B., Hay, A., Hamant, O., Roeder, A.H.K., Tsiantis, M., Prusinkiewicz, P., S Smith, R., 2018. Why plants make puzzle cells, and how their shape emerges. eLife 7, e32794.

Sapolsky, R.M., 1998. Why Zebras Don't Get Ulcers. Barnes & Noble Books, New York.

Sapp, J., 2007. Mitochondria and their host. In: Martin, W.F., Müller, M. (Eds.), Origin of Mitochondria and Hydrogenosomes. Springer, Berlin, pp. 57–83.

Saranak, J., Foster, K.W., 1994. The *in vivo* cleavage of carotenoids into retinoids in *Chlamydomonas reinhardtii*. J. Exp. Bot. 45, 505–511.

Saranak, J., Foster, K.W., 1997. Rhodopsin guides fungal phototaxis. Nature 387, 465–466.

Saranak, J., Foster, K.W., 2000. Reducing agents and light break an S-S bond activating rhodopsin *in vivo* in *Chlamydomonas*. Biochem. Biophys. Res. Commun. 275, 286–291.

Saraogi, I., Shan, S.-O., 2013. Co-translational protein targeting to the bacterial membrane. Biochim. Biophys. Acta 1843, 1433–1441.

Sardet, C., 2002. Special issue: microtrabecular concept of the cytoplasm revisited. Biol. Cell. 94, 553.

Sartres, J.-P., 1946. Existentialism is a Humanism. Translated by P. Mairet. https://www.marxists.org/reference/archive/sartre/works/exist/sartre.htm.

Sartres, J.-P., 2007. Existentialism is a Humanism. Translated by C. Macomber. Yale University Press, New Haven, CT.

Sarwar Khan, M., 2007. Engineering photorespiration in chloroplasts: a novel strategy for increasing biomass production. Trends Biotechnol. 25, 437–440.

Sastry, S.K., Horwitz, A.F., 1993. Integrin cytoplasmic domains: mediators of cytoskeletal linkages and extra-and intracellular initiated transmembrane signaling. Curr. Opin. Cell Biol. 5, 819–831.

Satell, G., April 15, 2016. The 3 Big Technologies to Watch Over the Next Decade—Genomics, Nanotechnology and Robotics. Forbes Magazine. https://www.forbes.com/sites/gregsatell/2016/04/15/the-3-big-technologies-to-watch-over-the-next-decade-genomics-nanotechnology-and-robotics/#17b05e734c0c.

Satiat-Jeunemaitre, B., Hawes, C., 1994. G.A.T.T. (A general agreement on traffic and transport) and brefeldin A in plant cells. Plant Cell 6, 463–467.

Satiat-Jeunemaitre, B., Martin, B., Hawes, C., 1992. Plant cell wall architecture is revealed by rapid-freezing and deep-etching. Protoplasma 167, 33–42.

Satiat-Jeunemaitre, B., Steele, C., Hawes, C., 1996. Golgi-membrane dynamics are cytoskeleton dependent: a study on Golgi stack movement induced by brefeldin A. Protoplasma 191, 21–33.

Satir, P., 1961. Cilia. Sci. Am. 204, 108–116.

Satir, P., 1965. Studies on cilia. II. Examination of the distal region of the ciliary shaft and the role of the filaments in motility. J. Cell Biol. 26, 805–834.

Satir, P., 1974. How cilia move. Sci. Am. 231, 45–50.

Satir, P., 1975. Ciliary and flagellar movement: an introduction. In: Inoué, S., Stephens, R.E. (Eds.), Molecules and Cell Movement. Raven Press, New York, pp. 143–149.

Satir, P., 1997a. Keith R. Porter and the first electron micrograph of a cell. Endeavour 21, 169–171.

Satir, P., 1997b. Keith Roberts Porter: 1912–1997. J. Cell Biol. 138, 223–224.

Sato, S., 1972. Mitochondria. Selected Papers in Biochemistry, vol. 10. University Park Press, Baltimore.

Sato, N., Albrieux, C., Joyard, J., Douce, R., Kuroiwa, T., 1993. Detection and characterization of a plastid envelope DNA- binding protein which may anchor plastid nucleoids. EMBO J. 12, 555–561.

Sato, H., Ellis, G.W., Inoué, S., 1975. Microtubular origin of mitotic spindle form birefringence. J. Cell Biol. 67, 501–517.

Sato, Y., Kadota, A., Wada, M., 1999. Mechanically induced avoidance response of chloroplasts in fern protonemal cells. Plant Physiol. 121, 37–44.

Sato, M., Schwarz, W.H., Selden, S.C., Pollard, T.D., 1988. Mechanical properties of brain tubulin and microtubules. J. Cell Biol. 106, 1205–1211.

Sato, Y., Wada, M., Kadota, A., 2001a. Choice of tracks, microtubules and/or actin filaments for chloroplast photo-movement is differentially controlled by phytochrome and a blue light receptor. J. Cell Sci. 114, 269–279.

Sato, Y., Wada, M., Kadota, A., 2001b. External Ca^{2+} is essential for chloroplast movement induced by mechanical stimulation but not by light stimulation. Plant Physiol. 127, 497–504.

Sato, M., Wong, T.Z., Allen, R.D., 1983. Rheological properties of living cytoplasm: endoplasm of *Physarum plasmodium*. J. Cell Biol. 97, 1089–1097.

Sattarzadeh, A., Franzen, R., Schmelzer, E., 2008. The Arabidopsis class VIII myosin ATM2 is involved in endocytosis. Cell Motil. Cytoskelet. 65, 457–468.

Sattarzadeh, A., Krahmer, J., Germain, A.D., Hanson, M.R., 2009. A myosin XI tail domain homologous to the yeast myosin vacuole-binding domain interacts with plastids and stromules in *Nicotiana benthamiana*. Mol. Plant 2, 1351–1358.

Sattarzadeh, A., Schmelzer, E., Hanson, M.R., 2011. Analysis of organelle targeting by DIL domains of the *Arabidopsis* myosin XI family. Front. Plant Sci. 2, 72.

Sattarzadeh, A., Schmelzer, E., Hanson, M.R., 2013. *Arabidopsis* myosin XI sub-domains homologous to the yeast myo2p organelle inheritance sub-domain target subcellular structures in plant cells. Front. Plant Sci. 4, 407.

Sattelle, D.B., Buchan, P.B., 1976. Cytoplasmic streaming in *Chara corallina* studied by laser light scattering. J. Cell Sci. 22, 633–643.

Satter, R.L., Morse, M.J., Lee, Y., Crain, R.C., Coté, G., Moran, N., 1988. Light and clock-controlled leaflet movements in *Samanea saman*: a physiological, biophysical and biochemical analysis. Bot. Acta 101, 205–213.

Sauer, H.W., 1982. Developmental Biology of Physarum. Cambridge University Press, Cambridge.

Sauer, G., Körner, R., Hanisch, A., Ries, A., Nigg, E.A., Silljé, H.W., 2005. Proteome analysis of the human mitotic spindle. Mol. Cell. Proteom. 4, 35–43.

Sauermann, H., Roach, M., 2016. Why pursue the postdoc path. Science 352, 663–664.

Saunders, J.A., 1979. Investigations of vacuoles isolated from tobacco—quantitation of nicotine. Plant Physiol. 64, 74–78.

Saunders, M.J., 1986a. Correlation of electrical current influx with nuclear position and division in *Funaria* caulonemal tip cells. Protoplasma 132, 32–37.

Saunders, M.J., 1986b. Cytokinin activation and redistribution of plasma membrane ion channels in *Funaria*. A vibrating-microelectrode and cytoskeleton-inhibitor study. Planta 167, 402–409.

Saunders, M.J., Hepler, P.K., 1981. Localization of membrane associated calcium following cytokinin treatment in *Funaria* using chlortetracycline. Planta 152, 272–281.

Saunders, M.J., Hepler, P.K., 1982. Calcium ionophore A23187 stimulates cytokinin-like mitosis in *Funaria*. Science 217, 943–945.

Sautter, C., 1986. Microbody transition in greening watermelon cotyledons. Double immunocytochemical labeling of isocitrate lyase and hydroxypyruvate reductase. Planta 167, 491–503.

Saveleva, N.V., Burlakovskiy, M.S., Yemelyanov, V.V., Lutova, L.A., 2016. Transgenic plants as bioreactors to produce substances for medical and veterinary uses. Russ. J. Genet. Appl. Res. 6, 712–724.

Sawin, K.E., LeGuellec, K., Philippe, M., Mitchison, T.J., 1992a. Mitotic spindle organization by a plus-end directed microtubule motor. Nature 359, 540–543.

Sawin, K.E., Mitchison, T.J., 1991a. Mitotic spindle assembly by two different pathways in vitro. J. Cell Biol. 112, 925–940.

Sawin, K.E., Mitchison, T.J., 1991b. Poleward microtubule flux in mitotic spindles. J. Cell Biol. 112, 941–954.

Sawin, K.E., Mitchison, T.J., Wordeman, L.G., 1992b. Evidence for kinesin-related proteins in the mitotic apparatus using peptide antibodies. J. Cell Sci. 101, 303–313.

Sawyer, R.M., Boulter, D., Gatehouse, J.A., 1987. Nuclease sensitivity of storage-protein genes in isolated nuclei of pea seeds. Planta 171, 254–258.

Sax, K., 1923. The association of size differences with seed-coat pattern and pigmentation in *Phaseolus vulgaris*. Genetics 8, 552–556.

Sax, K., 1944. Soviet biology. Science 99, 298–299.

Sax, K., O'Mara, J.G., 1941. Mechanism of mitosis in pollen tubes. Bot. Gaz. 102, 629–636.

Sax, H.J., Sax, K., 1935. Chromosome structure and behavior in mitosis and meiosis. J. Arnold Arbor. 16, 423–439.

Saxena, I., Brown Jr., R.M., 2005. Cellulose biosynthesis: current views and evolving concepts. Ann. Bot. 96, 9–21.

Saxton, M.J., Jacobson, K., 1997. Single-particle tracking: applications to membrane dynamics. Annu. Rev. Biophys. Biomol. Struct. 26, 373–399.

Sayre, A., 1975. Rosalind Franklin & DNA. W.W. Norton & Co., New York.

Sazuka, T., Keta, S., Shiratake, K., Yamaki, S., Shibata, D., 2004. A proteomic approach to identification of transmembrane proteins and membrane-anchored proteins of *Arabidopsis* thaliana by peptide sequencing. DNA Res. 11, 101–113.

Scala, J., Schwab, D., Simmons, E., 1968. The fine structure of the digestive gland of Venus's fly trap. Am. J. Bot. 55, 649–657.

Scanlon, M.J., Ohtsu, K., Timmermans, M., Schnable, P.S., 2009. Laser microdissection-mediated isolation and *in vitro* transcriptional amplification of plant RNA. Curr. Protoc. Mol. Biol. 87A, 25A.3.1-25A.3.15.

Schachtman, D.P., Schroeder, J.I., Lucas, W.J., Anderson, J.A., Gaber, R.F., 1992. Expression of an inward-rectifying potassuim channel by the *Arabidopsis* KAT1 cDNA. Science 258, 1654–1658.

Schaefer, D., Zrÿd, J.-P., 1997. Efficient gene targeting in the moss *Physcomitrella patens*. Plant J. 11, 1195–1206.

Schäfer, E.A., 1902. Essentials of Histology, sixth ed. Longmans, Green, and Co, New York.

Schaller, G.E., Sussman, M.R., 1988. Isolation and sequence of tryptic peptides from the proton-pumping ATPase of the oat plasma membrane. Plant Physiol. 86, 512–516.

Schattat, M., Barton, K., Mathur, J., 2011a. Correlated behavior implicates stromules in increasing the interactive surface between plastids and ER tubules. Plant Signal. Behav. 6, 715–718.

Schattat, M., Barton, K., Baudisch, B., Klosgen, R.B., Mathur, J., 2011b. Plastid stromule branching coincides with contiguous endoplasmic reticulum dynamics. Plant Physiol. 155, 1667–1677.

Schatten, G., Epel, D., 1996. In memorium. Daniel Mazia (1913–1996). Cell Motil. Cytoskelet. 34, 249–257.

Schatten, G., Epel, D., 1997. In memorium. Daniel Mazia (1913–1996). Exp. Cell Res. 231, 1–2.

Schatz, G., 1996. Biographical Memoir of Efraim Racker, vol. 70. National Academy of Sciences, Washington, DC, pp. 320–346.

Schatz, M.C., Delcher, A.L., Salzberg, S.L., 2010. Assembly of large genomes using second-generation sequencing. Genome Res. 20, 1165–1173.

Schatz, G., Dobberstein, B., 1996. Common principles of protein translocation across membranes. Science 271, 1519–1526.

Scheer, U., Benavente, R., 1990. Functional and dynamic aspects of the mammalian nucleolus. BioEssays 12, 343–346.

Scheer, U., Franke, W.W., 1972. Annulate lamellae in plant cells: formation during microsporogenesis and pollen development in *Canna generalis* Bailey. Planta 107, 145–159.

Schekman, R., 1985. Protein localization and membrane traffic. Ann. Rev. Cell Biol. 1, 115–143.

Schekman, R.W., 1996. Regulation of membrane traffic in the secretory pathway. Harvey Lect. 90, 41–57.

Schekman, R., 2005. Peroxisomes: another branch of the secretory pathway? Cell 122, 1–7.

Schekman, R., December 7, 2013. Genes and Proteins that Control the Secretory Pathway. Nobel Lecture.

Schekman, R., Orci, L., 1996. Coat proteins and vesicle budding. Science 271, 1526–1533.

Scheller, H.V., Ulvskov, P., 2010. Hemicelluloses. Annu. Rev. Plant Biol. 61, 263–289.

Schena, M., Shalon, D., Brown, P.O., Davis, R.W., 1995. Quantitative monitoring of gene-expression patterns with a complementary-DNA microarray. Science 270, 467–470.

Schenk, H.E.A., Bayer, M.G., Zook, D., 1987. Cyanelles. From symbiont to organelle. Ann. N.Y. Acad. Sci. 503, 151–167.

Scherer, G.F.E., von Drop, B., Schöllmann, C., Volkmann, D., 1992. Proton-transport activity, sidedness, and morphometry of tonoplast and plasma-membrane vesicles purified by free-flow electrophoresis from roots of *Lepidium sativum* L. and hypocotyls of *Cucurbita pepo*. L. Planta 186, 483–494.

Scheres, B., Benfey, P.B., 1999. Asymmetric cell division in plants. Annu. Rev. Plant Physiol. Plant Mol. Biol. 50, 505–537.

Scherp, P., Hasenstein, K.H., 2007. Anisotropic viscosity of the *Chara* (Characeae) rhizoid cytoplasm. Am. J. Bot. 94, 1930–1934.

Schibler, M.J., Pickett-Heaps, J.D., 1987. The kinetochore fibre structure in the acentric spindles of *Oedogonium*. Protoplasma 137, 29–44.

Schickore, J., 2018. Methodological reflections in general cytology in historical perspective. In: Matlin, K.S., Maienschein, J., Laubichler, M.D. (Eds.), Visions of Cell Biology: Reflections Inspired by Cowdry's General Cytology. University of Chicago Press, Chicago, pp. 73–99.

Schiebel, E., 2000. γ-tubulin complexes: binding to the centrosome, regulation and microtubule nucleation. Curr. Opin. Cell Biol. 12, 113–118.

Schiebel, W., Haas, B., Marinkovic, S., Klanner, A., Sänger, H.L., 1993a. RNA-directed RNA polymerase from tomato leaves. I. Purification and physical properties. J. Biol. Chem. 263, 11851–11857.

Schiebel, W., Haas, B., Marinkovic, S., Klanner, A., Sänger, H.L., 1993b. RNA-directed RNA polymerase from tomato leaves. II. Catalytic in vitro properties. J. Biol. Chem. 268, 11858–11867.

Schilperoort, R.A., 1970. Investigations on plant tumors, crown gall. On the biochemietry of tumor-induction by *Agrobacterium tumefaciens*. Demmenie N. V., Leiden.

Schimke, R.T., Doyle, D., 1970. Control of enzyme levels in animal tissues. Annu. Rev. Biochem. 39, 929–976.

Schiml, S., Puchta, H., 2016. Revolutionizing plant biology: multiple ways of genome engineering by CRIPR/Cas. Plant Methods 12, 8.

Schimper, A.F.W., 1883. Ueber die Entwickelung der Chlorophyllkörner und Farbkörper. Bot. Z. 41,105,121, 137,153.

Schimper, A.F.W., 1885. Untersuchungen über die Chlorophyllkörper und die ihnen homologen Gebilde. Jahr. Wiss. Bot. 16, 1–247.

Schindelman, G., Morikami, A., Jung, J., Baskin, T.I., Carpita, N.C., McCann, M.C., Benfey, P.N., 2001. The COBRA gene encodes a putative glycosylphosphatidyl-inositol anchored protein, which is polarly localized and necessary for oriented cell expansion in *Arabidopsis*. Genes Dev. 15, 1115–1127.

Schindler, M., 1995. Cell optical displacement assay (CODA)—measurements of cytoskeletal tension in living plant cells with a laser optical trap. Methods Cell Biol. 49, 71–84.

Schindler, M., Meiners, S., Cheresh, D.A., 1989. RGD-dependent linkage between plant cell wall and plasma membrane: consequences for growth. J. Cell Biol. 108, 1955–1965.

Schleiden, M.J., 1838. Contribution to Phytogenesis, 1867. Sydenhan Soc., London.

Schleiden, M.J., 1842. Grundzüge der wissenschaftlichen Botanik. W. Engelmann, Leipzig.

Schleiden, M.J., 1849. Principles of Scientific Botany or Botany as an Inductive Science. Translated by E. Lankester. Longman, Brown, Green, and Longmans, London.

Schleiden, M.J., 1853. Poetry of the Vegetable World. A Popular Exposition of the Science of Botany and its Relations to Man. Moore, Anderson, Wilstach & Keys, Cincinnati, OH.

Schleiden, M.J., 1969. Principles of Scientific Botany or Botany as an Inductive Science, 1849. Johnson Reprint Corp., New York. Translated by E. Lankester.

Schlesinger, G., Miller, S.L., 1983. Prebiotic synthesis in atmospheres containing CH_4, CO and CO_2. I. Amino Acids. J. Mol. Evol. 19, 376–382.

Schliephacke, M., Kremp, A., Schmid, H.P., Köhler, K., Kull, V., 1991. Prosomes (proteosomes) of higher plants. Eur. J. Cell Biol. 55, 114–121.

Schliwa, M., 1984. Mechanisms of intracellular organelle transport. In: Shay, J.W. (Ed.), Cell and Muscle Motility. Plenum, New York, pp. 1–82.

Schliwa, M., 1999. Myosin steps backwards. Nature 401, 431–432.

Schliwa, M., Enteneuer, U., Porter, K.R., 1987. Release of enzymes of intermediary metabolism from permeabilized cells: further evidence in siupport of a structural organization of the cytoplasmic matrix. Eur. J. Cell Biol. 44, 214–218.

Schliwa, M., van Blerkom, J., Porter, K.R., 1981. Stabilization and the cytoplasmic ground substance in detergent-opened cells and a structural and biochemical analysis of its composition. Proc. Natl. Acad. Sci. U.S.A. 78, 4329–4333.

Schmid, A.-M.M., 1996. Cytoskeleton-cell wall interactions in diatoms. In: Abstracts of the 1st European Phycological Congress. Aug. 11-18, 1996, Cologne, p. 7.

Schmid, V.H.R., Meindl, U., 1992. Microtubules do not control orientation of secondary cell wall microfibril deposition in *Micrasterias*. Protoplasma 169, 148–154.

Schmidt, A., 1924. Histologische Studien an phanerogamen Vegetationspunkten. Bot. Arch. 8, 345–404.

Schmidt, W.J., 1924. Die Bausteine des Tierköpers in Polarisiertem Lichte. Friedrich Cohen, Bonn.

Schmidt, W.J., 1928. Der submikroskopische Bau des Chromatins. Zool. Jbr. 45, 177–216.

Schmidt, C.L.A., 1932a. The chemistry of the amino acids and the proteins. Ann. Rev. Biochem. 1, 151–170.

Schmidt, W.J., 1932b. Die Doppelbrechung α-Thymonukleinsäure im Hinblick auf die Doppelbrechung des Chromosoms. Naturwissenschaften 20, 658.

Schmidt, C.L.A., 1933. The chemistry of the amino acids and the proteins. Ann. Rev. Biochem. 2, 71–94.

Schmidt, W.J., 1936. Doppelbrechung von Kernspindel und Chromosomen im lebenden, sich furchenden Ei von *Psammechinus miliaris* (Müll.). Ber. Oberhess. Ges. Natur- u. Heilk. 17, 140–144.

Schmidt, W.J., 1937. Die Doppelbrechung von Karyoplasma, Zytoplasma und Metaplasma. Gebrüder Borntraeger, Berlin.

Schmidt, W.J., 1939. Doppelbrechung der Kernspindel und Zugfasertheorie der Chromosomenbewegung. Chromosoma 1, 253–264.

Schmidt, A.L., Briskin, D.P., 1993. Energy transduction in tonoplast vesicles from red beet (*Beta vulgaris* L.) storage tissue: H^+/substrate stoichiometries for the H^+-ATPase and the H^+- PPase. Arch. Biochem. Biophys. 301, 165–173.

Schmidt, W.E., Ebel, J., 1987. Specific binding of a fungal glucan phytoalexin elicitor to membrane fraction from soybean *Glycine max*. Proc. Natl. Acad. Sci. U.S.A. 84, 4117–4121.

Schmidt, J.A., Eckert, R., 1976. Calcium couples flagellar reversal to photostimulation in *Chlamydomonas reinhardtii*. Nature 262, 713–715.

Schmidt, U.G., Endler, A., Schelbert, S., Brunner, A., Schnell, M., Neuhaus, H.E., Marty-Mazars, D., Marty, F., Baginsky, S., Martinoia, E., 2007. Novel tonoplast transporters identified using a proteomic approach with vacuoles isolated from cauliflower buds. Plant Physiol. 145, 216–229.

Schmidt, A., Jäger, K., 1992. Open questions about sulfur metabolism in plants. Ann. Rev. Plant Physiol. Plant Mol. Biol. 43, 325–349.

Schmidt, A.C., Lambert, A.M., 1990. Microinjected fluorescent phalloidin in vivo reveals the F-actin dynamics and assembly in higher plant mitotic cells. Plant Cell 2, 129–138.

Schmidt, M.H.W., Vogel, A., Denton, A.K., Istace, B., Wormit, A., van de Geest, H., Bolger, M.E., et al., 2017. De novo assembly of a new *Solanum pennellii* accession using nanopore sequencing. Plant Cell 29, 2336–2348.

Schmiedel, G., Schnepf, E., 1980. Polarity and growth of caulonema tip cells of the moss *Funaria hygrometrica*. Planta 147, 405–413.

Schmit, A.C., Lambert, A.M., 1987. Characterization and dynamics of cytoplasmic F-actin in higher plant endosperm cells during interphase, mitosis, and cytokinesis. J. Cell Biol. 105, 2157–2166.

Schmit, A.-C., Lambert, A.-M., 1988. Plant actin filament and microtubule interaction during anaphase-telophase transi- tion: effects of antagonist drugs. Biol. Cell. 64, 309–319.

Schmit, A.C., Lambert, A.M., 1990a. Microinjected fluorescent phalloidin in vivo reveals the F-actin dynamics and assembly in higher plant mitotic cells. Plant Cell 2, 129–138.

Schmit, A.C., Lambert, A.M., 1990b. Plant actin filament and microtubule interactions during anaphase-telophase transition. Effects of antagonist drugs. Biol. Cell 64, 309–320.

Schmit, A.-C., Nick, P., 2008. Microtubules and the evolution of mitosis. Plant Cell Monogr. 11, 233–266.

Schmitt, F.O., 1939. The ultrastructure of protoplasmic constituents. Physiol. Rev. 19, 270–302.

Schmitt, F.O., 1990. The Never-Ceasing Search. American Philosophical Society, Philadelphia, PA.

Schmitt, F.O., Bear, R.S., Ponder, E., 1936. Optical properties of the red cell membrane. J. Cell. Comp. Physiol. 9, 89–92.

Schmitt, F.O., Bear, R.S., Ponder, E., 1938. The red cell envelope considered as a Wiener mixed body. J. Cell. Comp. Physiol. 11, 309–313.

Schmitt, F.O., Livingston, R.B., 1972. Aharon Katzir-Katchalsky. Brain Res. 46, 427–434.

Schmitt, F.O., Livingston, R.B., 1973. Aharon (Katzir) Katchalsky. Annu. Rev. Biophys. Bioeng. 2, 1–7.

Schmitt, F.O., Palmer, K.J., 1940. X-ray diffraction studies of lipide and lipide-protein systems. Cold Spring Harb. Symp. Quant. Biol. 8, 94–101.

Schmitz, K., Kühn, R., 1982. Fine structure, distribution and frequency of plasmodesmata and pits in the cortex of *Laminaria hyperborea* and *L. saccharina*. Planta 154, 385–392.

Schmutzer, T., Bolger, M.E., Rudd, S., Chen, J., Gundlach, H., Arend, D., Oppermann, M., Weise, S., Lange, M., SPannagl, M., Usadel, B., Mayer, K.F.X., Scholz, U., 2017. Bioinformatics in the plant genomic and phenomic domain: the German contribution to resources, services and perspectives. J. Biotechnol. 261, 37–45.

Schnabl, H., Vienken, J., Zimmermann, U., 1980. Regular arrays of intramembranous particles in the plasmalemma of guard cells and mesophyll cell protoplasts of Vicia faba. Planta 148, 231–237.

Schnable, P.S., Ware, D., Fulton, R.S., Stein, J.C., Wei, F., Pasternak, S., Liang, C., Zhang, J., Fulton, L., Graves, T.A., Minx, P., Reily, A.D., Courtney, L., Kruchowski, S.S., Tomlinson, C., Strong, C., Delehaunty, K., Fronick, C., Courtney, B., Rock, S.M., Belter, E., Du, F., Kim, K., Abbott, R.M., Cotton, M., Levy, A., Marchetto, P., Ochoa, K., Jackson, S.M., Gillam, B., Chen, W., Yan, L., Higginbotham, J., Cardenas, M., Waligorski, J., Applebaum, E., Phelps, L., Falcone, J., Kanchi, K., Thane, T., Scimone, A., Thane, N., Henke, J., Wang, T., Ruppert, J., Shah, N., Rotter, K., Hodges, J., Ingenthron, E., Cordes, M., Kohlberg, S., Sgro, J., Delgado, B., Mead, K., Chinwalla, A., Leonard, S.,

Crouse, K., Collura, K., Kudrna, D., Currie, J., He, R., Angelova, A., Rajasekar, S., Mueller, T., Lomeli, R., Scara, G., Ko, A., Delaney, K., Wissotski, M., Lopez, G., Campos, D., Braidotti, M., Ashley, E., Golser, W., Kim, H., Lee, S., Lin, J., Dujmic, Z., Kim, W., Talag, J., Zuccolo, A., Fan, C., Sebastian, A., Kramer, M., Spiegel, L., Nascimento, L., Zutavern, T., Miller, B., Ambroise, C., Muller, S., Spooner, W., Narechania, A., Ren, L., Wei, S., Kumari, S., Faga, B., Levy, M.J., McMahan, L., Van Buren, P., Vaughn, M.W., Ying, K., Yeh, C.T., Emrich, S.J., Jia, Y., Kalyanaraman, A., Hsia, A.P., Barbazuk, W.B., Baucom, R.S., Brutnell, T.P., Carpita, N.C., Chaparro, C., Chia, J.M., Deragon, J.M., Estill, J.C., Fu, Y., Jeddeloh, J.A., Han, Y., Lee, H., Li, P., Lisch, D.R., Liu, S., Liu, Z., Nagel, D.H., McCann, M.C., San-Miguel, P., Myers, A.M., Nettleton, D., Nguyen, J., Penning, B.W., Ponnala, L., Schneider, K.L., Schwartz, D.C., Sharma, A., Soderlund, C., Springer, N.M., Sun, Q., Wang, H., Waterman, M., Westerman, R., Wolfgruber, T.K., Yang, L., Yu, Y., Zhang, L., Zhou, S., Zhu, Q., Bennetzen, J.L., Dawe, R.K., Jiang, J., Jiang, N., Presting, G.G., Wessler, S.R., Aluru, S., Martienssen, R.A., Clifton, S.W., McCombie, W.R., Wing, R.A., Wilson, R.K., 2009. The B73 maize genome: complexity, diversity, and dynamics. Science 326, 1112–1115.

Schnable, P.S., Wise, R.P., 1998. The molecular basis of cytoplasmic male sterility. Trends Plant Sci. 3, 175–180.

Schneider, S., Beyhl, D., Hedrich, R., Sauer, N., 2008. Functional and physiological characterization of *Arabidopsis* INOSITOL TRANSPORTER1, a novel tonoplast-localized transporter for myo-inositol. Plant Cell 20, 1073–1087.

Schneider, G.F., Kowalczyk, S.F., Calado, V.E., Pandraud, G., Zandbergen, H.W., Vandersypen, L.M.K., Dekker, C., 2010. DNA translocation through graphene nanopores. Nano Lett. 10, 3163–3167.

Schneider, M.V., Orchard, S., 2011. Omics technologies, data and bioinformatics principles. Bioinformatics for omics data: methods and protocols. Methods Mol. Biol. 719, 3–30.

Schneiter, K., Brügger, B., Sandhoff, R., Zellnig, G., Leber, A., Lampl, M., Athenstaedt, K., Hrastnik, C., Eder, S., Daum, G., Paltauf, F., Wieland, F.T., Kohlwein, S.D., 1999. Electrospray ionization tandem mass spectrometry (ESI-MS/MS) analysis of the lipid molecular species composition of yeast subcellular membrane reveals acyl chain-based sorting/remodelling of distinct molecular species en route to the plasma membrane. J. Cell Biol. 146, 741–754.

Schnell, D.J., Blobel, G., 1993. Identification of intermediates in the pathway of protein import into chloroplasts and their localization to envelope contact sites. J. Cell Biol. 120, 103–115.

Schnell, D.J., Kessler, F., Blobel, G., 1994. Isolation of components of the chloroplast protein import machinery. Science 266, 1007–1012.

Schnepf, E., 1969a. Sekretion und Exkretion bei Pflanzen. Protoplasmatologia VIII. Physiologie des Protoplasmas. 8. Sekretion und Exkretion bei Pflanzen.

Schnepf, E., 1969b. Über den Feinbau von Öldrüsen. I. Die Drüsenhaare von *Artium lappa*. Protoplasma 67, 185–194.

Schnepf, E., 1969c. Über den Feinbau von Öldrüsen. II. Die Ölgänge von *Solidago canadensis* und die Exkretschläuche von Arctium lappa. Protoplasma 67, 205–207.

Schnepf, E., 1986. Cellular polarity. Ann. Rev. Plant Physiol. 37, 23–47.

Schnepf, E., Hrdina, B., Lehne, A., 1982. Spore germination, development of the microtubule system and protonema cell morphogenesis in the moss, *Funaria hygrometrica*: effects of inhibitors and growth substances. Biochem. Physiol. Pflanzen 177, 461–482.

Schoberer, J., Runions, J., Steinkellner, H., STrasser, R., Hawes, C., Osterrieder, A., 2010. Sequential depletion and acquisition of proteins during Golgi stack disassembly and reformation. Traffic 11, 1429–1444.

Schoberer, J., Strasser, R., 2011. Sub-compartmental organization of Golgi-resident N-glycan processing enzymes in plants. Mol. Plant 4, 220–228.

Schoenheimer, R., 1942. The Dynamic State of Body Constituents. Harvard University Press, Cambridge, MA.

Schoenheimer, R., Ratner, S., Rittenberg, D., 1939a. Studies in protein metabolism. VII. Metabolism of tyrosine. J. Biol. Chem. 127, 333–344.

Schoenheimer, R., Ratner, S., Rittenberg, D., 1939b. Studies in protein metabolism. VX. The metabolic activity of body proteins investigated with *l*(-)-leucine containing two isotopes. J. Biol. Chem. 130, 703–732.

Schoenheimer, R., Rittenberg, D., 1938. The application of isotopes to the study of intermediary metabolism. Science 87, 221–226.

Scholes, T.A., Hinkle, P.C., 1984. Energetics of ATP-driven reverse electron transfer from cytochrome c to fumarate and from succinate to NAD in submitochrondria particles. Biochemistry 23, 3341–3345.

Scholes, J.D., Rolfe, S.A., 2009. Chlorophyll fluorescence imaging as tool for understanding the impact of fungal diseases on plant performance: a phenomics perspective. Funct. Plant Biol. 36, 880–892.

Schönbohm, E., 1972. Die Wirkung von SH-Blockern sowie von Licht und Dunkel auf die Verankerung der Mougeotia-Chloroplasten im cytoplasmatischen Wandbelag. Z. Pflanzenphysiol. 66, 113–132.

Schönbohm, E., 1975. Der Einblub von Colchicine sowie von Cytochalasin B auf fadige Plasmastrukturen, auf die Verankerung der Chloroplasten sowie aug die orientierte Chloroplastenbewegung. Ber. Dtsch. Bot. Ges. 88, 211–224.

Schönknecht, G., Brown, J.E., Verchot-Lubicz, J., 2008. Plasmodesmata transport GFP alone or fused to potato virus X TGBp1 is diffusion driven. Protoplasma 232, 143–152.

Schopf, J.W., 1992. The oldest fossils and what they mean. In: Schopf, J.W. (Ed.), Major Events in the History of Life. Jones and Bartlett Publishers, Sudbury, MA, pp. 29–63.

Schopfer, C.R., Hepler, P.K., 1991. Distribution of membranes and the cytoskeleton during cell plate formation in pollen mother cells of *Tradescantia*. J. Cell Sci. 100, 717–728.

Schott, D.H., Collins, R.N., Bretcher, A., 2002. Secretory vesicle transport velocity in living cells depends on the myosin-V lever arm length. J. Cell Biol. 156, 35–39.

Schrader, F., 1934. On the reality of spindle fibers. Biol. Bull. 67, 519–533.

Schrader, F., 1936. The kinetochore or spindle fiber locus in *Amphiuma tridactylus*. Biol. Bull. 70, 484–498.

Schrader, F., 1944. Mitosis. Columbia University Press, New York.

Schramm, D.N., Turner, M.S., 1998. Big-bang nucleosynthesis enters the precision era. Rev. Mod. Phys. 70, 303–318.

Schrecker, A.W., Kornberg, A., 1950. Reversible enzymatic synthesis of flavin-adenine dinucleotide. J. Biol. Chem. 182, 795–803.

Schreiber, L., 2010. Transport barriers made of cutin, suberin and associated waxes. Trends Plant Sci. 15, 546–553.

Schröder-Lang, S., Schwärzel, M., Seifert, R., Strünker, T., Kateriya, S., Looser, J., Watanabe, M., Kaupp, U.B., Hegemann, P., Nagel, G., 2007. Fast manipulation of cellular cAMP level by light *in vivo*. Nat. Methods 4, 39–42.

Schrödinger, E., 1946. What Is Life? Cambridge University Press, Cambridge.

Schrödinger, E., 1964. Statistical Thermodynamics. Cambridge University Press, Cambridge.

Schrödinger, E., 1996. Nature and the Greeks and Science and Humanism. Cambridge University Press, Cambridge.

Schroeder, J.I., 1988. K^+ transport properties of K^+ channels in the plasma membrane of *Vicia faba* guard cells. J. Gen. Physiol. 92, 667–683.

Schroeder, J.I., 1989. A quantitative analysis of outward rectifying K^+ channels in guard cell protoplasts from Vicia faba. J. Membr. Biol. 107, 229–235.

Schroeder, D.V., 2000. An Introduction to Thermal Physics. Addison-Wesley Longman, San Francisco.

Schroeder, J.I., Allen, G.J., Hugouvieux, V., Kwak, J.M., Waner, D., 2001. Guard cell signal transduction. Annu. Rev. Plant Physiol. Plant Mol. Biol. 52, 627–658.

Schroeder, J.I., Hagiwara, S., 1990. Repetitive increases in cytosolic Ca^{2+} of guard cells by abscisic acid activation of nonselective Ca^{2+} permeable channels. Proc. Natl. Acad. Sci. U.S.A. 87, 9305–9309.

Schroeder, J.I., Raschke, K., Neher, E., 1987. Voltage dependence of K^+ channels in guard-cell protoplasts. Proc. Natl. Acad. Sci. U.S.A. 84, 4108–4112.

Schroeder, J.I., Thuleau, P., 1991. Ca^{2+} channels in higher plant cells. Plant Cell 3, 555–559.

Schroeder, J.I., Ward, J.M., Grassman, W., 1994. Perspectives on the physiology and structure of inward-rectifying K^+ channels in higher plants: biophysical implications for K^+ uptake. Ann. Rev. Biophys. Biomol. Struct. 23, 441–471.

Schroer, T.A., 1991. Association of motor proteins with membranes. Curr. Opin. Cell Biol. 3, 133–137.

Schröter, K., Läuchli, A., Sievers, A., 1975. Mikroanalytische Identifikation von Bariumsulfat-Kristallen in den Statolithen der Rhizoide von *Chara fragilis* Desv. Planta 122, 213–225.

Schubert, M., Petersson, U.A., Haas, B.J., Funk, C., Schroder, W.P., Kieselbach, T., 2002. Proteome map of the chloroplast lumen of *Arabidopsis thaliana*. J. Biol. Chem. 277, 8354–8365.

Schultes, R.E., 1976. Hallucinogenic Plants. Golden Press, New York.

Schultes, R.E., 1988. Where the Gods Reign: Plants and Peoples of the Colombian Amazon. Synergetic Press, Oracle, AZ.

Schultes, R.E., Hofmann, A., 1979. Plants of the Gods: Origins of Hallucinogenic Use. McGraw-Hill, New York.

Schultes, R.E., Hofmann, A., 1980. The Botany and Chemistry of Hallucinogens, second ed. Thomas, Springfield, IL.

Schultes, R.E., Raffauf, R.F., 1990. The Healing Forest: Medicinal and Toxic Plants of the Northwest Amazonia. Dioscorides Press, Portland, OR.

Schultes, R.E., Raffauf, R.F., 1992. Vine of the Soul: Medicine Men, Their Plants and Rituals in the Colombian Amazonia. Synergetic Press, Oracle, AZ.

Schultz, A.M., Henderson, L.E., Oroszlan, S., 1988. Fatty acylation of proteins. Ann. Rev. Cell Biol. 4, 611–647.

Schultz, H.N., Jørgensen, B.B., 2001. Big bacteria. Annu. Rev. Microbiol. 55, 105–137.

Schultz, S.G., Solomon, A.K., 1961. Determination of the effective hydrodynamic radii of small molecules by viscometry. J. Gen. Physiol. 44, 1189–1199.

Schultze, M., 1863. Das Protoplasma der Rhizopoden und der Pfanzenzellen. W. Engelmann, Leipzig.

Schulz, A., 1992. Living sieve cells of conifers as visualized by confocal, laser-scanning fluorescence microscopy. Protoplasma 166, 153–164.

Schulz-Lessdorf, B., Hedrich, R., 1995. Protons and calcium modulate SV-type channels in the vacuolar-lysosomal compartment–channel interaction with calmodulin inhibitors. Planta 197, 655–671.

Schulze, W.X., 2010. Proteomics approaches to understand protein phosphorylation in pathway modulation. Curr. Opin. Plant Biol. 13, 280–287.

Schumaker, K.S., Sze, H., 1985. A Ca^{2+}/H^+ antiport system driven by the proton electrochemical gradient of a tonoplast H^+-ATPase from oat roots. Plant Physiol. 79, 111–1117.

Schumaker, K.S., Sze, H., 1986. Calcium transport into the vacuole of oat roots. Characterization of H^+/Ca^{2+} exchange activity. J. Biol. Chem. 261, 12171–12178.

Schumaker, K.S., Sze, H., 1987. Inositol 1,4,5-trisphosphate releases Ca^{2+} from vacuolar membrane vesicles of oat roots. J. Biol. Chem. 262, 3944–3946.

Schumaker, K.S., Sze, H., 1990. Solubilization and reconstitution of the oat root vacuolar H^+/Ca^{2+} exchanger. Plant Physiol. 92, 340–345.

Schuster, W., Brennicke, A., 1994. The plant mitochondrial genome: physical structure, information contents, RNA editing, and gene migration to the nucleus. Ann. Rev. Plant Physiol. Plant Mol. Biol. 45, 61–78.

Schuster, G., Dewit, M., Staehelin, L.A., Ohad, I., 1986. Transient inactivation of the thylakoid photosystem II light-harvesting protein kinase system and concomitant changes in intramembrane particle size during photoinhibition of Chlamydomonas reinhardtii. J. Cell Biol. 103, 71–80.

Schuster, W., Knoop, V., Hiesel, R., Grohmann, L., Brennicke, A., 1993. The mitochondrial genome on its way to the nucleus: different stages of gene transfer in plants. FEBS Lett. 325, 140–144.

Schwab, D.E., Simmons, E., Scala, J., 1969. Fine structure changes during function of the digestive gland of Venus's fly trap. Am. J. Bot. 56, 88–100.

Schwann, T., 1838. Ueber die Analogie in der Structur und dem Wachsthume der Thiere und Pflanzen. Neue. Not. Geb. Nat. Helk 5, 33–36.

Schwann, T., 1839. Microscopical Resarches into the Accordance in the Structure and Growth of Animals and Plants, 1867. Sydenhan Soc., London.

Schwann, T., 1847. Microscopical researches into the accordance in the structure and growth of animals and plants. In: Adventures in Molecular Biology. Harvey Lect., vol. 46. The Sydenham Society, London, pp. 3–44. Astbury, W.T., 1950/51.

Moments of discovery. In: Schwartz, G., Bishop, P.W. (Eds.), 1958. The Development of Modern Science, vol. II. Basic Books, New York.

Schwartz, R.B., Dayhoff, M.O., 1978. Origin of prokaryotes, eukaryotes, mitochondria and chloroplasts. Science 199, 395–403.

Schwartz, M.A., Ingber, D.E., Lawerence, M., Springer, T.A., Lechene, C., 1991. Multiple integrins share the ability to induce elevation of intracellular pH. Exp. Cell Res. 195, 533–535.

Schwartz, D.C., Li, X., Hernandez, L.I., Ramnarain, S.P., Huff, E.J., Wang, Y.-K., 1993. Ordered restriction maps of *Saccharomyces cerevisae* chromosomes by optical mapping. Science 262, 110–114.

Schwartz, A.W., Orgel, L.E., 1985. Template-directed synthesis of novel, nucleic acid-like structures. Science 228, 585–587.

Schwartz, M.A., Schaller, M.D., Ginsberg, M.H., 1995. Integrins: emerging paradigms of signal transduction. Ann. Rev. Cell Biol. 11, 549–599.

Schween, G., Egener, T., Fritzowsky, D., Granado, J., Guitton, M.-C., Hartmann, N., Hohe, A., Holtorf, H., Lang, D., Lucht, J.M., Reinhard, C., Rensing, S.A., Schlink, K., Schulte, J., Reski, R., 2005. Disruption libraries: production parameters and mutant phenotypes. Plant Biol. 7, 228–237.

Schweikert, M., Meyer, B., 2001. Characterization of intracellular bacteria in the freshwater dinoflagellate *Peridinium cinctum*. Protoplasma 217, 177–184.

Schwuchow, J.M., Kern, V.D., Sack, F.D., 2002. Tip-growing cells of the moss *Ceratodon purpureus* are gravitropic in high-density media. Plant Physiol. 130, 2095–2100.

Scofield, G.N., Beven, A.F., Shaw, P.J., Doonan, J.H., 1992. Identification and localization of a nucleoporin-like protein component of the plant nuclear matrix. Planta 187, 414–420.

Scott, P., Lyne, R.L., ap Rees, T., 1995. Metabolism of maltose and sucrose by microspores isolated from barley (*Hordeum vulgare* L.). Planta 197, 435–441.

Scudellari, M., 2011. Data deluge. Scientist 25 (10). http://www.the-scientist.com/?articles.view/articleNo/31212/title/Data-Deluge/.

Seagull, R.W., Falconer, M.M., Weedenburg, C.A., 1987. Microfilaments: dynamic arrays in higher plant cells. J. Cell Biol. 104, 995–1004.

Seaman, M.N.J., Burd, C.G., Emr, S.D., 1996. Receptor signalling and the regulation of endocytotic membrane transport. Curr. Opin. Cell Biol. 8, 549–556.

Searcy, D.G., 1975. Histone-like protein in the prokaryote *Thermoplasma acidophilum*. Biochim. Biophys. Acta 395, 535–547.

Searcy, D.G., 1987. Phylogenetic and phenotypic relationships between the eukaryotic nucleocytoplasm and thermophilic archebacteria. Ann. N.Y. Acad. Sci. 503, 168–179.

Searcy, D.G., Stein, D.B., 1980. Nucleoprotein subunit structure in an unusual prokaryotic organism: *Thermoplasma acidophilum*. Biochim. Biophys. Acta 609, 180–195.

Seaton, D.D., Graf, A., Baerenfaller, Stitt, M., Millar, A.J., Gruissem, W., 2018. Photoperiodic control of the Arabidopsis proteome reveals a translational coincidence mechanism. Mol. Syst. Biol. 14, e7962.

Seberg, O., Petersen, G., Davis, J.I., Pires, J.C., Stevenson, D., Chase, M.W., Fay, M.F., Devey, D.S., Sytsma, K.J., Yohan, P., 2012. Phylogeny of the Asparagales, based on three plastid and two mitochondrial genes. Am. J. Bot. 99, 875–889.

Seder, A.W., Wilson, D.F., 1951. Electron microscope studies on the normal and colchicinized mitotic figures of the onion root tip (*Allium cepa*). Biol. Bull. 100, 107–115.

Sedwick, C., 2014. Gohta Goshima: questing for answers on the mitotic spindle. J. Cell Biol. 206, 148–149.

Segal, L.A., 2001. Computing an organism. Proc. Natl. Acad. Sci. U.S.A. 98, 3639–3640.

Segel, I.H., 1968. Biochemical Calculations. John Wiley and Sons, New York.

Segev, N., 2001. Ypt an dRab GTPases: insight into functions through novel interactions. Curr. Opin. Cell Biol. 13, 500–511.

Seguí-Simarro, J.M., Austin II, R., White, E.A., Staehelin, L.A., 2004. Electron tomographic analysis of somatic cell plate formation in meristematic cells of *Arabidopsis* preserved by high-pressure freezing. Plant Cell 16, 836–856.

Seguí-Simarro, J.M., Otegui, M.S., Austin II, J.R., Staehelin, L.A., 2007. Plant cytokinesis—insights gained from electron tomography studies. Plant Cell Monogr. 9, 251–287.

Séguin, A., Lavoisier, A., 1789. Premier memoire sur la respiration des animaux. Mem de l'Acad. Sci. 566–584.

Seifert, G.J., Blaukopf, C., 2010. Irritable walls: the plant extracellular matrix and signaling. Plant Physiol. 153, 467–478.

Seifriz, W., 1920. Viscosity values of protoplasm as determined by microdissection. Bot. Gaz. 34, 307–324.

Seifriz, W., 1921. Observations of some physical properties of protoplasm by aid of microdissection. Ann. Bot. 35, 269–296.

Seifriz, W., 1924. An elastic value of protoplasm, with further observation on the viscosity of protoplasm. Brit. J. Exp. Biol. 2, 1–11.

Seifriz, W., 1928. How do the life processes work? Sci. Am. 138 (1), 18–21.

Seifriz, W., 1929. The structure of protoplasm. Biol. Rev. 4, 76–102.

Seifriz, W., 1930. The plasticity of protoplasm. J. Rheol. 1, 261–268.

Seifriz, W., 1931. The structure of protoplasm. Science 73, 648–649.

Seifriz, W., 1935. The structure of protoplasm. Bot. Rev. 1, 18–36.

Seifriz, W., 1936. Protoplasm. McGraw-Hill, New York.

Seifriz, W., 1938a. Recent contributions to the theory of protoplasmic structure. Science 88, 21–25.

Seifriz, W., 1938b. The Physiology of Plants. John Wiley and Sons, New York.

Seifriz, W., 1943a. Creative imagination and indeterminism. Philos. Sci. 10 (1), 25–33.

Seifriz, W., 1943b. Protoplasmic streaming. Bot. Rev. 9, 49–123.

Seifriz, W., 1945a. The physical properties of protoplasm. Ann. Rev. Physiol. 7, 35–60.

Seifriz, W., 1945b. The structure of protoplasm II. Bot. Rev. 11, 231–259.

Seifriz, W., 1953. Mechanism of protoplasmic movement. Nature 171, 1136–1138.

Seifriz, W., 1955a. The physical chemistry of cytoplasm. In: Ruhland, W. (Ed.), Handb. d. Pflanzenphysiol, 1, pp. 340–382.

Seifriz, W., 1955b. Poetic imagination necessary in rigid, scientific pursuit of truth. The Daily Pennsylvanian, p. 7 (Friday, January 14, 1955).

Seifriz, W., 2008. Knowledge and understanding. In: Congrès International de Philosophie – 24-30 Septembre 1944 DES PROBLÈMES DE LA CONNAISSANCE Travaux du Congrès international de philosophie consacré aux problèmes de la connaissance. "Moun" – Revue de philosophie Publiée deux fois l'an par les professeurs de l'Institut de Philosophie Saint François de Sales.

Seitz, K., 1964. Das Wirkungsspektrum der Photodinese bei *Elodea canadensis*. Protoplasma 58, 621–640.

Seitz, K., 1967. Kirkungsspektren für fie Starklichtbewegung der Chloroplasten, die Photodinese und die licht-abhängige Viskositätsänderrung bei Vallisneria spiralis ssp. torta. Z. Pflanzenphysiol. 56, 246–261.

Seitz, K., 1971. Die Ursache der Phototaxis der Chloroplasten: Ein ATP gradient? Z. Pfanzenphysiol. 64, 241–256.

Seitz, K., 1979. Light induced changes in the centrifugability of chloroplasts: different action spectra and different influence of inhibitors in the low and high intensity range. Z. Pflanzenphysiol. 95, 1–12.

Seitz, K., 1987. Light-dependent movement of chloroplasts in higher plant cells. Acta. Physiol. Plant. 9, 137–148.

Seki, M., Awata, J.-Y., Shimada, K., Kashiyama, T., Ito, K., Yamamoto, K., 2003. Susceptibility of *Chara* myosin to SH reagents. Plant Cell Physiol. 44, 201–205.

Seksek, O., Biwerski, J., Verkman, A.S., 1997. Translational diffusion of macromolecules-sized solutes in cytoplasm and nucleus. J. Cell Biol. 138, 131–142.

Senebier, J., 1788. Experiences sur l'action de la lumière solaire dans la végétation. Chez Briand, Genève.

Senn, G., 1908. Die Gestalts und Lageveraudderungen der Pflanzen-chromatophoren. Englemann, Leipzig.

Sentenac, H., Bonneaudm, N., Minet, M., Lacroute, F., Salmon, J.-M., Gaymard, F., Grignon, C., 1992. Cloning and expression in yeast of a plant potassium ion transport system. Science 256, 663–665.

Seo, T.S., Bai, X., Kim, D.H., Meng, Q., Shi, S., Ruparel, H., Li, Z., Turro, N.J., Ju, J., 2005. Four-color DNA sequencing by synthesis on a chip using photocleavable fluorescent nucleotides. Proc. Natl. Acad. Sci. U.S.A. 102, 5926–5931.

Sequeira, L., 1983. Mechanisms of induced resistance in plants. Annu. Rev. Microbiol. 37, 51–79.

Sequeira, L., 1984. Cross protection and induced resistance: their potential for plant disease control. Trends Biotechnol. 2, 25–29.

Sequeira, L., 1988. On becoming a plant pathologist: the changing scene. Annu. Rev. Phytopathol. 26, 1–13.

Sequeira, L., 2000. Legacy for the millennium: a century of progress in plant pathology. Annu. Rev. Phytopathol. 38, 1–17.

Serafini, T., Orci, L., Amherdt, M., Brunner, M., Kahn, R.A., Rothman, J.E., 1991. ADP-ribosylation factor is a subunit of the coat of Golgi-derived COP-coated vesicles: a novel rolle for GTP-binding protein. Cell 67, 239–253.

Seravin, L.N., 2001. The principle of counter-directional morphological evolution and its significance for construction the megasystem of protists and other eukaryotes. Protistology 2, 6–14.

Seren, Ü., Grimm, D., Fitz, J., Weigel, D., Nordborg, M., Borgwardt, K., Korte, A., 2017. AraPheno: a public database for *Arabidopsis thaliana* phenotypes. Nucleic Acids Res. 45, D1054–D1059.

Serio, T.R., Cashikar, A.G., Kowal, A.S., Sawicki, G.J., Moslehi, J.J., Serpell, L., Arnsdorf, M.F., Lindquist, S.L., 2000. Nucleated conformational conversion and the replication of conformational information by a prion determinant. Science 289, 1317–1321.

Serlin, B.S., Ferrell, S., 1989. The involvement of microtubules in chloroplast rotation in the alga, *Mougeotia*. Plant Sci. 60, 1–8.

Serlin, B.S., Roux, S.J., 1984. Modulation of chloroplast movement in the green alga *Mougeotia* by the Ca^{2+} ionophore A23187 and by calmodulin antagonists. Proc. Natl. Acad. Sci. U.S.A. 81, 6368–6372.

Serrano, R., 1988. Structure and function of proton translocating ATPase in plasma membranes of plants and fungi. Biochim. Biophys. Acta 947, 1–28.

Serrano, R., 1989. Structure and function of plasma membrane ATPase. Ann. Rev. Plant Physiol. 40, 61–94.

Serrano, R., Kielland-Brandt, M.C., Fink, G.R., 1986. Yeast plasma membrane ATPase is essential for growth and has homology with ($Na^+ + K^+$), K^+- and Ca^{2+}-ATPases. Nature 319, 689–693.

Servick, K., 2017. Circular RNAs hint at new realm of genetics. Science 355, 1363.

Sexton, R., Hall, J.L., 1974. Fine structure and cytochemistry of the abscission zone cells of Phaseolus leaves: I. Ultrastructural changes occurring during abscission. Ann. Bot. 38, 849–854.

Sexton, R., Jamieson, G.G.C., Allan, H.I.L., 1977. An ultrastructural study of abscission zone cells with special reference to the mechanism of enzyme secretion. Protoplasma 91, 369–387.

Seymour, R.S., 2004. Dynamics and precision of thermoregulatory responses of eastern skunk cabbage Symplocarpus foetidus. Plant Cell Environ. 27, 1014–1022.

Seymour, R.S., Bartholomew, G.A., Barhhardt, M.C., 1983. Respiration and heat production by the inflorescence of *Philodendron selloum*. Planta 157, 336–343.

Seymour, D.K., Becker, C., 2017. The causes and consequences of DNA methylome variation in plants. Curr. Opin. Plant Biol. 36, 56–63.

Shah, P., Powell, A.L.T., Orlando, R., Bergmann, C., Gutierrez-Sanchez, G., 2012. Proteomic analysis of ripening tomato fruit infected by Botrytis cinerea. J. Proteome Res. 11, 2178–2192.

Shakespeare, W., 1623. The Taming of the Shrew, 1992. Pocket Books, New York.

Shamu, C.E., Murray, A.W., 1992. Sister chromatid separation in frog egg extracts requires DNA topoisomerase II activity during anaphase. J. Cell Biol. 117, 921–934.

Shan, S.O., Walter, P., 2005. Co-translational targeting by the signal recognition particle. FEBS Lett. 579, 921–926.

Shanklin, S., Jabben, M., Vierstra, R.D., 1987. Red light-induced formation of ubiquitin-phytochrome conjugates: identification of possible intermediates of phytochrome degradation. Proc. Natl. Acad. Sci. U.S.A. 84, 359–363.

Shannon, T.M., Henry, Y., Picton, J.M., Steer, M.W., 1982. Polarity in higher plant dictyosomes. Protoplasma 112, 189–195.

Shapiro, R., 1999. Prebiotic cytosine synthesis: a critical analysis and implications for the origin of life. Proc. Natl. Acad. Sci. U.S.A. 96, 4396–4401.

Shapiro, A.L., Scharff, M.D., Maizel, J.V., Uhr, J.W., 1966. Polyribosomal synthesis and assembly of the H and L chains of gamma globulin. Proc. Natl. Acad. Sci. U.S.A. 56, 216–221.

Shapiro, A.L., Viñuela, E., Maizel Jr., J.V., 1967. Molecular weight estimation of polypeptide chains by electrophoresis in SDS-polyacrylamide gels. Biochem. Biophys. Res. Commun. 28, 815–820.

Shapley, H., 1926. Starlight. George H. Doran Co., New York.

Shapley, H., 1943. Galaxies. The Blakiston Co., Philadelphia, PA.

Shapley, H., 1958. Of Stars and Men. Human Response to an Expanding Universe. Beacon Press, Boston.

Shapley, H., 1967. Beyond the Observatory. Charles Scribner's Sons, New York.

Sharfman, M., Bar, M., Ehrlich, M., Schuster, S., Melech-Bonfil, S., Ezer, R., Sessa, G., Avni, A., 2011. Endosomal signaling of the tomato leucine-rich repeat receptor-like protein LeEix2. Plant J. 68, 413–423.

Sharma, A.K., Sharma, M.K., 2009. Plants as bioreactors: recent developments and emerging opportunities. Biotechnol. Adv. 27, 811–832.

Sharma, S., Shrivastava, N., 2016. Renaissance in pytomedicines: promising implications of NGS technologies. Planta 244, 19–38.

Sharp, L.W., 1914. Spermatogenesis in Marsilea. Bot. Gaz. 58, 419–431.

Sharp, L.W., 1921. An Introduction to Cytology. McGraw-Hill, New York.

Sharp, L.W., 1929. Structure of large somatic chromosomes. Bot. Gaz. 88, 349–382.

Sharp, L.W., 1934. Introduction to Cytology, third ed. McGraw-Hill, New York.

Sharp, P.A., December 8, 1993. Split genes and RNA splicing. Nobel Lecture.

Sharp, D.J., Kuriyama, R., Essner, R., Baas, P.W., 1997. Expression of a minus-end-directed motor protein induces Sf9 cells to form axon-like processes with uniform microtubule polarity orientation. J. Cell Sci. 110, 2373–2380.

Sharp, P.A., Sugden, B., Sambrook, J., 1973. Detection of two restriction endonuclease activities in *Haemophilis parainfluenzae* using analytical agarose-ethidium bromide electrophoresis. Biochemistry 12, 3055–3063.

Shaw, S., 2006. Imaging the live plant cell. Plant J. 45, 573–598.

Shaw, S.L., Kamyar, R., Ehrhardt, D.W., 2003. Sustained microtubule treadmilling in *Arabidopsis* cortical arrays. Science 300, 1715–1718.

Shaw, P., Rawlins, D., Higlett, M., 1993. Nuclear and nucleolar structure in plants. In: Heslop-Harrison, J.S., Flavell, R.B. (Eds.), The Chromosome. BIOS Scientific Publishers, Oxford, pp. 161–171.

Shea, E.M., Gilbeant, D.M., Carpita, N.C., 1989. Structural analyses of the cell walls regenerated by carrot protoplasts. Planta 179, 293–308.

Sheahan, M.B., McCurdy, D.W., Rose, R.J., 2005. Mitochondria as a connected population: ensuring continuity of the mitochondrial genome during plant cell dedifferentiation through massive mitochondrial fusion. Plant J. 44, 744–755.

Sheahan, M.B., Rose, R.J., McCurdy, D.W., 2004. Organelle inheritance in plant cell division: the actin cytoskeleton is required for unbiased inheritance of chloroplasts, mitochondria and endoplasmic reticulum in dividing protoplasts. Plant J. 37, 379–390.

Sheen, J., Hwang, S., Niwa, Y., Kobayashi, H., Galbraith, D.W., 1995. Green-fluorescent protein as a new vital marker in plant cells. Plant J. 8, 777–784.

Sheetz, M.P., 1989. Kinesin structure and function. In: Warner, F.D., McIntosh, J.R. (Eds.), Cell Movement, Kinesin, Dynein and Microtubule Dynamics, vol. 2. Alan R. Liss, Inc., New York, pp. 277–285.

Sheetz, M.P., Chasan, R., Spudich, J.A., 1984. ATP-dependent movement of myosin in vitro: characterization of a quantitative assay. J. Cell Biol. 99, 1867–1871.

Sheetz, M.P., Spudich, J.A., 1983a. Movement of myosin-coated fluorescent beads on actin cables in vitro. Nature 303, 31–35.

Sheetz, M.P., Spudich, J.A., 1983b. Movement of myosin-coated structures on actin cables. Cell Motil. 3, 485–489.

Sheffer, M., Fried, A., Gottlieb, H.E., Tietz, A., Avron, M., 1986. Lipid composition of the plasma-membrane of the halotolerant alga, Dunaliella salina. Biochim. Biophys. Acta 857, 165–172.

Shelanski, M.L., Taylor, E.W., 1967. Isolation of a protein subunit from microtubules. J. Cell Biol. 34, 549–554.

Shelden, E., Wadsworth, P., 1992a. Microinjection of biotin-tubulin into anaphase cells induces transient elongation of kinetochore microtubules and reversal of chromosome-to-pole motion. J. Cell Biol. 116, 1409–1420.

Shelden, E., Wadsworth, P., 1992b. Observation and quantification of individual microtubule behavior in vivo: microtubule dynamics are cell-type specific. J. Cell Biol. 120, 935–945.

Shelley, M., 1818, 1974. In: Rieger, James (Ed.), Frankenstein or, The Modern Prometheus. The 1818 Text. University of Chicago Press, Chicago.

Shemer, T.A., Harpaz-Saad, S., Belausov, E., Lovat, N., Krokhin, O., Spicer, V., Standing, K.G., Goldschmidt, E.E., Eyal, Y., 2008. Citrus chlorophyllase dynamics at ethylene-induced fruit color-break: a study of chlorophyllase expression, posttranslational processing kinetics, and in situ intracellular localization. Plant Physiol. 148, 108–118.

Shemi, A., Ben-Dor, S., Vardi, A., 2015. Elucidating the composition and conservation of the autophagy pathway in photosynthetic eukaryotes. Autophagy 11, 701–715.

Shen, W.-H., 2007. G1/S transition and the Rb-E2F pathway. Plant Cell Monogr. 9, 59–73.

Shen, Z., Collatos, A.R., Bibeau, J.P., Furt, F., Vidali, L., 2012. Phylogenetic analysis of the kinesin superfamily from *Physcomitrella*. Front. Plant Sci. 3, 230.

Shen, G., Kuppa, S., Venkataramani, S., Wang, J., Yan, J., Qiu, X., Zhang, H., 2010. Ankyrin repeat-containing protein 2A is an essential molecular chaperone for peroxisomal membrane-bound Ascorbate peroxidase 3 in *Arabidopsis*. Plant Cell 22, 811–831.

Shen, Z., Liu, Y.-C., Bibeau, J.P., Lemoi, K.P., Tüzel, E., Vidali, L., 2015. The kinesin-like protein, KAC1/2, regulate actin dynamics underlying chloroplast light-avoidance in *Physcomitrella patens*. J. Integr. Plant Biol. 57, 106–119.

Shen-Miller, J., Hinchman, R.R., 1974. Gravity sensing in plants: a critique of the statolith theory. BioScience 24, 643–651.

Shen-Miller, J., Miller, C., 1972. Distribution and activation of the Golgi apparatus in geotropism. Plant Physiol. 49, 634–639.

Shendure, J., Balasubramanian, S., Church, G.M., Gilbert, W., Rogers, J., Schloss, J.A., Waterston, R.H., 2017. DNA sequencing at 40: past, present and future. Nature 550, 345–353.

Shendure, J., Ji, H., 2008. Next-generation DNA sequencing. Nat. Biotechnol. 26, 1135–1145.

Shendure, J., Porreca, G.J., Reppas, N.B., Lin, X., McCutcheon, J.P., Rosenbaum, A.M., Wang, M.D., Zhang, K., Mitra, R.D., Church, G.M., 2005. Accurate multiplex polony sequencing of an evolved bacterial genome. Science 309, 1728–1732.

Shepherd, V.A., Beilby, M.J., 1999. The effect of an extracellular mucilage on the response to osmotic shock in the charophyte alga Lamprothamnium papulosum. J. Membr. Biol. 170,229–242.

Shepherd, V.A., Beilby, M.J., Heslop, D., 1999. Ecophysiology of the hypotonic response in the salt-tolerant charophyte alga *Lamprothamnium papulosum*. Plant Cell Environ. 22, 333–346.

Shepherd, V.A., Goodwin, P.B., 1989. The porosity of permeabilized *Chara* cells. Aust. J. Plant Physiol. 16, 231–239.

Sherman-Broyles, S., Bombarely, A., Doyle, J., 2017. Characterizing the allopolyploid species among the wild relatives of soybean: Utility of reduced representation genotyping methodologies. J. Systemat. Evol. 55, 365–376.

Sherrier, D.J., VandenBosch, K.A., 1994. Secretion of cell wall polysaccharides in *Vicia* root hairs. Plant J. 5, 185–195.

Sherrington, C.S., 1906. The Integrative Action of the Nervous System. Charles Scribner's Sons, New York.

Sheth, B.P., Thaker, V.S., 2014. Plant systems biology: insights, advances and challenges. Planta 240, 33–54.

Shi, X., Castandet, B., Germain, A., Hanson, M.R., Bentolila, S., 2017a. ORRM5, an RNA recognition motif-containing protein, has a unique effect on mitochondrial RNA editing. J. Exp. Bot. 68, 2833–2847.

Shi, X., Hanson, M.R., Bentolila, S., 2017b. Functional diversity of Arabidopsis organelle-localized RNA-recognition motif-containing proteins. Wiley Interdisp. Rev. RNA 8, e1420.

Shi, H., Ishitani, M., Kim, C., Zhu, J.-K., 2000. The *Arabidopsis thaliana* salt tolerance gene SOS1 encodes a putative Na^+/H^+ antiporter. Proc. Natl. Acad. Sci. U.S.A. 97, 6896–6901.

Shi, L.-X., Theg, S.M., 2010. A stromal heat shock protein 70 system functions in protein import into chloroplasts in the moss *Physcomitrella patens*. Plant Cell 22, 205–220.

Shi, L.-X., Theg, S.M., 2013. The chloroplast protein import system: from algae to trees. Biochim. Biophys. Acta 1833, 314–331.

Shibaoka, H., 1972. Gibberellin-colchicine interaction in elongation of azuki bean epicotyl sections. Plant Cell Physiol. 13, 461–469.

Shibaoka, H., 1974. Involvement of wall microtubules in gibberellin promotion and kinetin inhibition of stem elongation. Plant Cell Physiol. 15, 255–263.

Shibaoka, H., 1992. Cytokinesis in tobacco BY-2 cells. In: Shibaoka, H. (Ed.), Cellular Basis of Growth and Development. Proc. of the VII International Symposium in Conjunction with the Awarding of the International Prize of Biology. Toyonaka, Osaka, pp. 119–126.

Shibaoka, H., 1993. Regulation by gibberellins of the orientation of cortical microtubules in plant cells. Aust. J. Plant Physiol. 20, 461–470.

Shibaoka, H., 1994. Plant hormone-induced changes in the orientation of cortical microtubules: alterations in the cross-linking between microtubules and the plasma membrane. Annu. Rev. Plant Physiol. Plant Mol. Biol. 45, 527–544.

Shibaoka, H., Hogetsu, T., 1977. Effects of ethyl N-phenyl carbamate on wall microtubules and on gibberellin and kinetin- controlled cell expansion. Bot. Mag. (Tokyo) 90, 317–321.

Shibaoka, H., Nagai, R., 1994. The plant cytoskeleton. Curr. Biol. 6, 10–15.

Shibata, M., Oikawa, K., Yoshimoto, K., Kondo, M., Mano, S., Yamada, K., Hayashi, M., Sakamoto, W., Ohsumi, Y., Nishimura, M., 2013. Highly oxidized peroxisomes are selectively degraded via autophagy in *Arabidopsis*. Plant Cell 25, 4967–4983.

Shiina, T., Hayashi, K., Ishii, N., Morikawa, K., Toyoshima, Y., 2000. Chloroplast tubules visualized in transplastomic plants expressing green fluorescent protein. Plant Cell Physiol. 41, 367–371.

Shiina, T., Nishii, A., Toyoshima, Y., Bogorad, L., 1997. Identification of promoter elements involved in the cytosolic Ca^{2+}-mediated photoregulation of maize cab-m1 expression. Plant Physiol. 115, 477–483.

Shiina, T., Nishii, A., Toyoshima, Y., Bogorad, L., 1997. Identification of promoter elements involved in the cytosolic Ca^{2+}-mediated photoregulation of maize cab-m1 expression. Plant Physiol. 115, 477–483.

Shiina, T., Wayne, R., Tung, H.Y.L., Tazawa, M., 1988. Possible involvement of protein phosphorylation/dephosphorylation in the modulation of Ca^{2+} channel in tonoplast-free cells of Nitellopsis. J. Membr. Biol. 102, 255–264.

Shikama, K., Nakamura, K.-I., 1973. Standard free energy maps for the hydrolysis of ATP as a function of pH and metal ion concentration: comparison of metal ions. Arch. Biochem. Biophys. 157, 457–463.

Shimada, T.L., Ogawa, Y., Shimada, T., Hara-Nishimura, I., 2011. A non-destructive screenable marker, OsFAST, for identifying transgenic rice seeds. Plant Signal. Behav. 6, 1454–1456.

Shimamura, T., 1939. Cytological studies of polyploidy induced by colchicine. Cytologia 9, 486–494.

Shimaoka, T., Ohnishi, M., Sazuka, T., Mitsuhashi, N., Hara-Nishimura, I., Shimazaki, K., Maeshima, M., Yokota, A., Tomizawa, K., Mimura, T., 2004. Isolation of intact vacuoles and proteomic analysis of tonoplast from suspension-cultured cells of *Arabidopsis thaliana*. Plant Cell Physiol. 45, 672–683.

Shimazaki, K., Kinoshita, T., Nishimura, M., 1992. Involvement of calmodulin and calmodulin-dependent myosin light chain kinase in blue light-dependent H^+ pumping by guard cell protoplasts from *Vicia faba* L. Plant Physiol. 99, 1416–1421.

Shimmen, T., 1988a. Characean actin bundles as a tool for studying actomyosin-based motility. Bot. Mag. (Tokyo) 101, 533–544.

Shimmen, T., 1988b. Cytoplasmic streaming regulatedaby adenine nucleotides and inorganic phosphates in characeae. Protoplasma (Suppl. 1), 3–9.

Shimmen, T., 1996. Studies on mechano-perception in characean cells: development of a monitoring apparatus. Plant Cell Physiol. 37, 591–597.

Shimmen, T., 1997. Studies on mechano-perception in Characeae: effects of external Ca^{2+} and Cl^-. Plant Cell Physiol. 38, 691–697.

Shimmen, T., 2001a. Electrical perception of "death message" in *Chara*: involvement of turgor pressure. Plant Cell Physiol. 42, 366–373.

Shimmen, T., 2001b. Involvement of receptor potentials and action potentials in mechano-perception in plants. Aust. J. Plant Physiol. 28, 567–576.

Shimmen, T., 2002. Electrical perception of "death message" in *Chara*: analysis of rapid component and ionic process. Plant Cell Physiol. 43, 1575–1584.

Shimmen, T., 2003. Studies on mechano-perception in tha characeae: transduction of pressure signals into electrical signals. Plant Cell Physiol. 44, 1215–1224.

Shimmen, T., 2008. Electrophysiological characterization of the node in *Chara corallina*: functional differentiation for wounding response. Plant Cell Physiol. 49, 264–272.

Shimmen, T., Hamatani, M., Saito, S., Yokota, E., Mimura, T., Fusetani, N., Karaki, H., 1995. Roles of actin filaments in cytoplasmic streaming and organization of transvacuolar strands in root hair cells of *Hydrocharis*. Protoplasma 185, 188–193.

Shimmen, T., Ridge, R.W., Lambiris, I., Plazinski, J., Yokota, E., Williamson, R.E., 2000. Plant myosins. Protoplasma 214, 1–10.

Shimmen, T., Tazawa, M., 1977. Control of membrane potential and excitability of *Chara* cells with ATP and Mg^{2+}. J. Membr. Biol. 117, 201–221.

Shimmen, T., Tazawa, M., 1982a. Cytoplasmic streaming in the cell model of *Nitella*. Protoplasma 112, 101–106.

Shimmen, T., Tazawa, M., 1982b. Reconstitution of cytoplasmic streaming in characeae. Protoplasma 113, 127–131.

Shimmen, T., Tazawa, M., 1983. Control of cytoplasmic streaming by ATP, Mg^{2+} and cytochalasin B in permeabilized characeae cell. Protoplasma 115, 18–24.

Shimmen, T., Wakabayashi, A., 2008. Involvement of membrane potential in alkaline band formation by intermodal cells of *Chara corallina*. Plant Cell Physiol. 49, 1614–1620.

Shimmen, T., Yano, M., 1984. Active sliding movement of latex beads coated with skeletal muscle myosin on Chara actin bundles. Protoplasma 121, 132–137.

Shimmen, T., Yano, M., 1994. Active sliding movement of latex beads coated with skeletal muscle myosin on *Chara* actin bundles. Protoplasma 121, 132–137.

Shimmen, T., Yokota, E., 2004. Cytoplasmic streaming in plants. Curr. Opin. Cell Biol. 16, 68–72.

Shimomura, O., 1979. Structure of the chromophore of *Aequorea* green fluorescent protein. FEBS Lett. 104, 220–222.

Shimomura, O., December 8, 2008. Discovery of Green Fluorescent Protein, GFP. Nobel Lecture.

Shimomura, O., Goto, T., Hirata, Y., 1957. Crystalline *Cypridina* luciferin. Bull. Chem. Soc. Jpn. 30, 929–933.

Shimomura, O., Johnson, F.H., 1969. Properties of the bioluminescent protein aequorin. Biochemistry 8, 3991–3997.

Shimomura, O., Johnson, F.H., 1972. Structure of the light-emitting moiety of aequorin. Biochemistry 11, 1602–1608.

Shimomura, O., Johnson, F.H., Morise, H., 1974. Mechanism of the luminescent intramolecular reaction of aequorin. Biochemistry 13, 3278–3286.

Shimomura, O., Johnson, F.H., Saiga, Y., 1962. Extraction, purification and properties of aequorin, a bioluminescent protein from the luminous hydromedusan, *Aequorea*. J. Cell. Comp. Physiol. 59, 223–239.

Shimomura, O., Johnson, F.H., Saiga, Y., 1963. Further data on the bioluminescent protein, aequorin. J. Cell. Comp. Physiol. 62, 1–8.

Shimoni, E., Rav-Hon, O., Ohad, I., Brumfeld, V., Reich, Z., 2005. Three-dimensional organization of higher-plant chloroplast thylakoid membranes revealed by electron tomography. Plant Cell 17, 2580–2586.

Shin, S., Moore Jr., T.S., 1990. Phosphatidylethanolamine synthesis by caster bean endosperm. Plant Physiol. 93, 148–153, 154–159.

Shine, I., Wrobel, S., 1976. Thomas Hunt Morgan. Pioneer of Genetics. The University Press of Kentucky, Lexington, Kentucky.

Shinozaki, K., Ohne, M., Tanaka, M., Wakasugi, T., Chunwongse, N., Obokata, J., Yamaguchi-Shinozaki, K., Ohto, C., Torazawa, K., Meng, B.Y., Sugita, M., Deno, H., Kamogashira, T., Yamada, K., Kusuda, J., Takaiwa, F., Kato, A., Tohdoh, N., Shimada, H., Sugiura, M., 1986. The complete nucleotide sequence of the tobacco chloroplast genome: its gene organization and expression. EMBO J. 5, 2043–2049.

Shintani, T., Suzuki, K., Kamada, Y., Noda, T., Ohsumi, Y., 2001. Apg2p functions in autophagosome formation on the perivacuolar structure. J. Biol. Chem. 276, 30452–30460.

Shintomi, K., Hirano, T., 2011. The relative ratio of condensin I to II determines chromosome shapes. Gens Dev. 25, 1464–1469.

Shintomi, K., Inoue, F., Watanabe, H., Ohsumi, K., Ohsugi, M., Hirano, T., 2017. Mitotic chromosome assembly despite nucleosome depletion in *Xenopus* egg extracts. Science 356, 1284–1287.

Shintomi, K., Takahashi, T.S., Hirano, T., 2015. Reconstitution of mitotic chromatids with a minimum set of purified factors. Nat. Cell Biol. 17, 1014–1023.

Shiratake, K., Martinoia, E., 2007. Transporters in fruit vacuoles. Plant Biotechnol. 24, 127–133.

Shiva Kumar, N., Stevens, M.H.H., Kiss, J.Z., 2008. Plastid movement in the statocytes of the ARG1 (altered response to gravity) mutant. Am. J. Bot. 95, 177–184.

Shorter, J., Lindquist, S., 2004. Hsp104 catalyzes formation and elimination of self-replicating Sup35 prion conformers. Science 304, 1793–1797.

Showalter, A.M., 1993. Structure and function of plant cell wall proteins. Plant Cell 5, 9–23.

Shulaev, V., Chapman, K.D., 2017. Plant lipidomics at the crossroads: from technology to biology driven science. Biochim. Biophys. Acta 1862, 786–791.

Shuman, S., 2009. DNA ligases: progress and prospects. J. Biol. Chem. 284, 17365–17369.

Shy, G., Ehler, L., Herman, E., Galili, G., 2001. Expression patterns of genes encoding endomembrane proteins support a reduced function of the golgi in wheat endosperm during the onset of storage protein deposition. J. Exp. Bot. 52, 2387–2388.

Sibaoka, T., 1962. Excitable cells in Mimosa. Science 137, 226.

Sibaoka, T., 1966. Action potentials in plant cells. Symp. Soc. Exp. Biol. 20, 49–73.

Sibaoka, T., 1969. Physiology of rapid movements in plants. Annu. Rev. Plant Physiol. 20, 165–194.

Siegel, S.M., 1962. The Plant Cell Wall. Macmillin, New York.

Siekevitz, P., 1952. Uptake of radioactive alanine in vitro into the proteins of rat liver fractions. J. Biol. Chem. 195, 549–565.

Siekevitz, P., 1957. Powerhouse of the cell. Sci. Am. 197 (1), 131–140.

Siekevitz, P., Jamieson, J.D., 2012. Establishing the secretory pathway of the pancreatic exocrine cell. In: Moberg, C.L. (Ed.), Entering and Unseen World. Rockefeller University Press, New York, pp. 150–155.

Siekevitz, P., Palade, G.E., 1958a. A cytochemical study on the pancreas of the guinea pig. I. J. Biophys. Biochem. Cytol. 4, 203–218.

Siekevitz, P., Palade, G.E., 1958b. A cytochemical study on the pancreas of the guinea pig. II. J. Biophys. Biochem. Cytol. 4, 309–318.

Siekevitz, P., Palade, G.E., 1958c. A cytochemical study on the pancreas of the guinea pig. III. J. Biophys. Biochem. Cytol. 4, 557–566.

Siekevitz, P., Palade, G.E., 1959. A cytochemical study on the pancreas of the guinea pig. IV. J. Biophys. Biochem. Cytol 5, 1–10.

Siekevitz, P., Palade, G.E., 1960a. A cytochemical study on the pancreas of the guinea pig. V. J. Biophys. Biochem. Cytol. 7, 619–630.

Siekevitz, P., Palade, G.E., 1960b. A cytochemical study on the pancreas of the guinea pig. VI. J. Biophys. Biochem. Cytol. 7, 631–644.

Siever, R., 1979. The Earth. In: Folsome, C.E. (Ed.), Life. Origin and Evolution. Readings From Scientific American. W. H. Freeman and Co., San Francisco, pp. 25–31.

Sievers, A., 1963. Beteiligung des Golgi-Apparates bei der Bildung der Zellwand von Wurzelhaaren. Protoplasma 56, 188–192.

Sievers, A., 1966. Lysosomen–ähnliche Kompartment ein Pflanzenzelle. Naturwissenschaften 53, 334–335.

Signer, R., Caspersson, T., Hammarsten, E., 1938. Molecular shape and size of thymonucleic acid. Nature 141, 122.

Sikela, J.M., Auffray, C., 1993. Finding new genes faster than ever. Nat. Genet. 3, 189–191.

Sikora, J., Wasik, A., 1978. Cytoplasmic streaming with Ni^{2+} immobilized *Paramecium aurelia*. Acta Protozool. 17, 389–397.

Silk, W.K., 2000. A tribute to Paul Green. J. Plant Growth Regul. 19, 2–6.

Silva-Sanchez, C., Li, H., Chen, S., 2015. Recent advances and challenges in plant phosphoproteomics. Proteomics 15, 1127–1141.

Silver, R.B., 1986. Mitosis in sand dollar embryos is inhibited by antibodies directed against the calcium transport enzyme of muscle. Proc. Natl. Acad. Sci. U.S.A. 83, 4302–4306.

Silver, R.B., 1989. Nuclear envelope breakdown and mitosis in sand dollar embryos is inhibited by microinjection of calcium buffers in a calcium-reversible fashion, and by antagonists of intracellular Ca^{2+} channels. Dev. Biol. 131, 11–26.

Silver, P.A., 1991. How proteins enter the nucleus. Cell 64, 489–497.

Silver, R.B., Cole, R.D., Cande, W.Z., 1980. Isolation of mitotic apparatus containing vesicles with calcium sequestration activity. Cell 19, 505–516.

Silver, P., Goodson, H., 1989. Nuclear protein transport. Crit. Rev. Biochem. Mol. Biol. 24, 419–435.

Silverman-Gavrila, R., Forer, A., 2000. Evidence that actin and myosin are involved in the poleward flux of tubulin in metaphase kinetochore microtubules of crane-fly spermatocytes. J. Cell Sci. 113, 597–609.

Silverman-Gavrila, R., Forer, A., 2001. Effects of anti-myosin drugs on anaphase chromosome movement and cytokinesis in crane-fly primary spermatocytes. Cell Motil. Cytoskelet. 50, 180–197.

Simerly, C., Balczon, R., Brinkley, B.R., Schatten, G., 1990. Microinjected kinetochore antibodies interfere with chromosome movement in meiotic and mitotic mouse oocytes. J. Cell Biol. 111, 1491–1504.

Simkin, J.L., 1959. Protein biosynthesis. Ann. Rev. Biochem. 28, 145–170.

Simon, A.L., 1934. Champagne. Constable & Co., London.

Simon, S.M., 1993. Translocation of proteins across the endoplasmic reticulum. Curr. Opin. Cell Biol. 5, 581–588.

Simon, S., 2002. Translocation of macromolecules across membranes and through aqueous channels. Translocation across membranes. Struct. Dyn. Confined Polym. 2, 37–66.

Simon, R., Altschuler, Y., Ruben, R., Galili, G., 1990. Two closely related wheat storage proteins follow a markedly different subcellular route in Xenopus laevis oocytes. Plant Cell 2, 941–950.

Simon, S.M., Blobel, G., 1991. A protein-conducting channel in the endoplasmic reticulum. Cell 65, 371–380.

Simon, S.M., Blobel, G., 1992. Signal peptides open proteinconducting channels in *E. coli*. Cell 69, 677–684.

Simon, S.M., Blobel, G., Zimmerberg, J., 1989. Large aqueous channels in membrane vesicles derived from the rough endoplasmic reticulum of canine pancreas or the plasma membrane of *Escherichia coli*. Proc. Natl. Acad. Sci. U.S.A. 86, 6176–6180.

Simon, S.M., Peskin, S.C., Oster, G.F., 1992. What drives the translocation of proteins? Proc. Natl. Acad. Sci. U.S.A. 89, 3770–3774.

Simoni, R.D., Hill, R.L., Vaughan, M., 2002. Urease, the first crystalline enzyme and the proof that enzymes are proteins: the work of James B. Sumner. J. Biol. Chem. 277, e23.

Simons, P., 1992. The Action Plant. Blackwell, Oxford.

Simons, K., Ikonen, E., 1997. Functional rafts in cell membranes. Nature 387, 569–572.

Simons, K., Toomre, D., 2000. Lipid rafts and signal transduction. Nat. Rev. Mol. Cell Biol. 1, 31–39.

Simpson, D.J., 1978. Freeze-fracture studies on barley plastid membranes. I. Wildtype etioplast. Carlsberg Res. Commun. 43, 145–170.

Simpson, D.J., 1979. Freeze-fracture studies on barley plastid membranes. II. Location of the light-harvesting chlorophyll protein. Carlsberg Res. Commun. 44, 305–336.

Simpson, C., Thomas, C., Findlay, K., Bayer, E., Maule, A.J., 2009. An *Arabidopsis* GPI-anchor plasmodesmatal neck protein with callose binding activity and potential to regulate cell-to-cell trafficking. Plant Cell 21, 581–594.

Sineshchekov, O.A., Jung, K.H., Spudich, J.L., 2002. Two rhodopsins mediate phototaxis to low- and high-intensity light in *Chlamydomonas reinhardtii*. Proc. Natl. Acad. Sci. U.S.A. 99, 8689–8694.

Singer, S.J., 1975. Architecture and topology of biologic membranes. In: Weissmann, G., Claiborne, R. (Eds.), Cell Membranes. Biochemistry, Cell Biology & Pathology. HP Publishing Co., New York, pp. 35–44.

Singer, S.J., 1990. The structure and insertion of integral proteins in membranes. Ann. Rev. Cell Biol. 6, 242–296.

Singer, S.J., 1992. The structure and function of membranes. A personal memoir. J. Membr. Biol. 129, 3–12.

Singer, M.F., Heppel, L.A., Hilmoe, R.J., 1957. Oligonucleotides as primers for polynucleotide phosphorylase. Biochim. Biophys. Acta 26, 447–448.

Singer, M.F., Heppel, L.A., Hilmoe, R.J., 1960. Oligonucleotides as primers for polynucleotide phosphorylase. J. Biol. Chem. 235, 738–750.

Singer, S.J., Nicolson, G.L., 1972. The fluid mosaic model of the structure of cell membranes. Science 175, 720–731.

Singh, K.B., 1998. Transcriptional regulation in plants: the importance of combinatorial control. Plant Physiol. 118, 1111–1120.

Singh, M.K., Krüger, F., Beckmann, H., Brumm, S., Vermeer, J.E.M., Munnik, T., Mayer, U., Stierhof, Y.-D., Grefen, C., Schumacher, K., Jürgens, G., 2014. Protein delivery to vacuole requires SAND protein-dependent RabGTPase conversion. Curr. Biol. 24, 1383–1389.

Sinnott, E.W., 1960. Plant Morphogenesis. McGraw-Hill, New York.

Sinnott, E.W., 1961. Cell & Psyche. The Biology of Purpose. Harper & Row, New York.

Sinnott, E.W., Bloch, R., 1940. Cytoplasmic behavior during division of vacuolate plant cells. Proc. Natl. Acad. Sci. U.S.A. 26, 223–227.

Sinnott, E.W., Bloch, R., 1941. Division in vacuolate plant cells. Am. J. Bot. 28, 225–232.

Sinnott, E.W., Bloch, R., 1944. Visible expression of cytoplasmic patterns in the differentiation of xylem strands. Proc. Natl. Acad. Sci. U.S.A. 30, 388–392.

Sinnott, E.W., Bloch, R., 1945. The cytoplasmic basis of intercellular patterns in vascular differentiation. Am. J. Bot. 32, 151–156.

Sinnott, E.W., Dunn, L.C., Dobzhansky, Th, 1950. Principles of Genetics, fourth ed. McGraw-Hill, New York.

Sinsheimer, R.L., 1954. The action of pancreatic desoxyribonuclease. I. Isolation of mono- and dinucleotides. J. Biol. Chem. 208, 445–459.

Sinsheimer, R.L., 1955. The action of pancreatic desoxyribonuclease. II. Isomeric dinucleotides. J. Biol. Chem. 215, 579–583.

Sinsheimer, R.L., 1977. Recombinant DNA. Annu. Rev. Biochem. 46, 415–438.

Siqueira Jr., J.F., Fouad, A.F., Rôças, I.N., 2012. Pyrosequencing as a tool for better understanding of human microbiomes. J. Oral Microbiol. 4, 10743.

Sirault, X.R.R., James, R.A., Furbank, R.T., 2009. A new screening method for osmotic component of salinity tolerance in cereals using infrared thermography. Funct. Plant Biol. 36, 970–977.

Sirks, M.J., 1938. Plasmatic inheritance. Bot. Rev. 4, 113–131.

Sitte, P., 2005. Werner W. Franke: a retrospective to earlier years. Eur. J. Cell Biol. 84, 87–88.

Siwicki, K.K., Eastman, C., Petersen, G., Rosbash, M., Hall, J.C., 1988. Antibodies to the *period* product od *Drosophila* reveal diverse tissue distribution and rhythmic changes in the visual system. Neuron 1, 141–150.

Sjolund, R.D., Shih, C.Y., 1983. Freeze-fracture analysis of phloem structure in plant tissue cultures. I. The sieve element reticulum. J. Ultrastruc. Res. 82, 111–121.

Sjöstrand, F.S., Andersson-Cedergren, E., Dewey, M.M., 1958. The ultrastructure of the interculated discs of frog, mouse and guninea pig cardiac muscle. J. Ultrastruc. Res. 1, 271–287.

Sjöstrand, F.S., Hanzon, V., 1954. Ultrastructure of Golgi apparatus of exocrine cells of mouse pancreas. Exp. Cell Res. 7, 415–429.

Skibbins, R.V., Rieder, C.L., Salmon, E.D., 1995. Kinetochore mobility after severing between sister centromeres using laser microsurgery: evidence that kinetochore directional instability and position is regulated by tension. J. Cell Sci. 108, 2537–2548.

Skulachev, V.P., 1990. Power transmission along biological membranes. J. Membr. Biol. 114, 97–112.

Slade, K.M., Baker, R., Chia, M., Thompson, N.L., Pielak, G.J., 2009. Effects of recombinant protein expression on green fluorescent protein diffusion in *Escherichia coli*. Biochemistry 48, 083–5089.

Slater, E.C., 1953. Mechanism of phosphorylation in the respiratory chain. Nature 172, 975–978.

Slautterback, D.B., 1963. Cytoplasmic microtubules. I. Hydra. J. Cell Biol. 18, 367–388.

Slavkin, H.C. (Ed.), 1972. The Comparative Molecular Biology of Extracellular Matrices. Academic Press, New York.

Slavkin, H.C., Greulich, R.C. (Eds.), 1975. Extracellular Matrix Influences on Gene Expression. Academic Press, New York.

Sleep, N.H., Zahnle, K.J., Kasting, J.F., Morowitz, H.J., 1989. Annihilation of ecosystems by large asteroid impacts on the early Earth. Nature 342, 139–142.

Sleigh, M.A., 1962. The Biology of Cilia and Flagella. Macmillan, New York.

Sleigh, M.A. (Ed.), 1974. Cilia and Flagella. Academic Press, New York.

Slepchenko, B.M., Schaff, J.C., Carson, J.H., Loew, L.M., 2002. Computational cell biology: spatiotemporal simulation of cellular events. Annu. Rev. Biophys. Biomol. Struct. 31, 423–441.

Sluiman, H.J., Lokhorst, G.M., 1988. The ultrastructure of cellular division (autosporogenesis) in the coccoid green alga, *Trebouxia* aggregate, revealed by rapid freeze fixation and freeze substitution. Protoplasma 144, 149–159.

Smalley, R.E., December 7, 1996. Discovering the fullerenes. Nobel Lecture.

Smertenko, A., Assaad, F., Baluška, F., Bezanilla, M., Buschmann, H., Drakakaki, G., Hauser, M.-T., Janson, M., Mineyuki, Y., Moore, I., Müller, S., Murata, T., Otegui, M.S., Panteris, E., Rasmussen, C., Schmit, A.-C., Šamaj, J., Samuela, L., Staehelin, L.A., van Damme, D., Wastenys, G., Žárský, V., 2017. Plant cytokinesis: terminology for structures and processes. Trends Cell Biol. 27, 885–894.

Smertenko, A., Saleh, N., Igarashi, H., Mori, H., Hauser-Hahn, I., Jiang, C.-J., Sonobe, S., Lloyd, C.W., Hussey, P.J., 2000. A new class of microtubule-associated proteins in plants. Nat. Cell Biol. 2, 750–753.

Smirnova, E.A., Bajer, A.S., 1992. Spindle poles in higher plant mitosis. Cell. Motil. Cytoskelet. 23, 1–7.

Smirnova, E.A., Bajer, A.S., 1994. Microtubule converging centers and reorganization of the interphase cytoskeleton and the mitotic spindle in higher plant *Haemanthus*. Cell. Motil. Cytoskelet. 27, 219–233.

Smirnova, E.A., Bajer, A.S., 1998. Early stages of spindle formation and independence of chromosome and microtubule cycles in *Haemanthus* endosperm. Cell. Motil. Cytoskelet. 40, 22–37.

Smith, L.M., 1988. Fluorescence-based automated DNA sequence analysis. In: Setlow, J.K. (Ed.), Genetic Engineering. Genetic Engineering (Principles and Methods), vol. 10. Springer, Boston, MA, pp. 81–108.

Smith, P.R., et al., 1999. Anomalous diffusion of major histocompatibility complex class I molecules on HeLa cells determined by single particle tracking. Biophys. J. 76, 3331–3344.

Smith, L.G., 2001. Plant cell division: building walls in the right places. Nat. Rev. Mol. Cell Biol. 2, 33–39.

Smith, M., Brown, N.L., Air, G.M., Barrell, B.G., Coulson, A.R., Hutchison III, C.A., Sanger, F., 1977. DNA sequence at the C termini of the overlapping genes A and B in bacteriophage ΦX174. Nature 265, 702–705.

Smith, M., Butler, R.D., 1971. Ultrastructural aspects of petal development in Cucumis sativus with particular reference to the chromoplasts. Protoplasma 73, 1–13.

Smith, H.O., December 8, 1978. Nucleotide sequence specificity of restriction endonucleases. Nobel Lecture.

Smith, T.A., Kohorn, B.D., 1994. Mutations in a signal sequence for the thylakoid membrane identify multiple protein transport pathways and nuclear suppressors. J. Cell Biol. 126, 365–374.

Smith, J.A.C., Marigo, G., Lüttge, U., Ball, E., 1982. Adenine- nucleotide levels during crassulacean acid metabolism and the energetics of malate accumulation in *Kalanchoë tubiflora*. Plant Sci. Lett. 26, 13–21.

Smith, L.M., Sanders, J.Z., Kaiser, R.J., Hughes, P., Dodd, C., Connell, C.R., Heiner, C., Kent, S.B.H., Hood, L.E., 1986. Fluorescence detection in automated DNA sequence analysis. Nature 321, 674–679.

Smith, J.D., Todd, P., Staehelin, L.A., 1997. Modulation of statolith mass and grouping in white clover (*Trifolium repens*) grown in 1-g, microgravity and on a clinostat. Plant J. 12, 1361–1373.

Smith, E.F., Townsend, C.O., 1907. A plant-tumor of bacterial origin. Science 25, 671–673.

Smith, H.O., Wilcox, K.W., 1970. A restriction enzyme from *Hemophilus influenzae*. I. Purification and general properties. J. Mol. Biol. 51, 379–391.

Smithies, O., 2007a. Oliver Smithies—Biographical. https://www.nobelprize.org/nobel_prizes/medicine/laureates/2007/smithies-bio.html.

Smithies, O., December 7, 2007b. Turning Pages. Nobel Lecture.

Smithies, O., Gregg, R.G., Koralewski, M.A., Kurcherlapati, R.S., 1985. Insertion of DNA sequences into the human chromosomal β-globin locus by homologous recombination. Nature 317, 230–234.

Smoot, G., Davidson, K., 1993. Wrinkles in Time. HarperCollins, New York.

Smyth, R.D., Saranak, J., Foster, K.W., 1988. Algal visual systems and their photoreceptor pigments. Prog. Phycol. Res. 6, 255–286.

Sneader, W., 2000. The discovery of asprin. BJM 321, 1591–1594.

Snell, W.J., 1990. Adhesion and signalling during fertilization in multicellular and unicellular organisms. Curr. Opin. Cell Biol. 2, 821–832.

Snow, C.M., Senior, A., Gerace, L., 1987. Monoclonal antibodies identify a group of nuclear pore complex glycoproteins. J. Cell Biol. 104, 1143–1156.

Snyder, J.A., McIntosh, J.R., 1975. Initiation and growth of microtubules from mitotic centers in lysed mammalian cells. J. Cell Biol. 67, 744–760.

Søgaard, M., Tani, K., Ye, R.R., Geromans, S., Tempst, P., Kirchhausen, T., Rotheman, J.E., Söllner, T., 1994. A Rab protein is required for the assembly of SNARE complexes in the docking vesicles. Cell 78, 937–948.

Sohn, E.J., Kim, E.S., Zhao, M., Kim, S.J., Kim, H., Kim, Y.-W., Lee, Y.J., Hillmer, S., Sohn, U., Jiang, L., Hwang, I., 2003. Rha1, an *Arabidopsis* Rab5 homolog, plays a critical role in the vacuolar trafficking of soluble cargo proteins. Plant Cell 15, 1057–1070.

Soll, J., Schleiff, E., 2004. Protein import into chloroplasts. Nat. Rev. Mol. Cell Biol. 5, 198–208.

Söllner, T., Whiteheart, S.W., Brunner, M., Erdjument-Bromage, H., Geromans, S., Tempst, P., Rothman, J.E., 1993. SNAP receptors implicated in vesicle targeting and fusion. Nature 362, 318–324.

Solomon, A.K., 1992. Interactions between band 3 and other transport-related proteins. Prog. Cell Res. 2, 269–283.

Solomon, A.K., Vennesland, B., Klemperer, F.W., Buchanan, J.M., Hastings, A.B., 1941. The participation of carbon dioxide in the carbohydrate cycle. J. Biol. Chem. 140, 171–182.

Solomonson, L.P., Barber, M.J., 1990. Assimilatory nitrate reductase: functional properties and regulation. Ann. Rev. Plant Physiol. Plant Mol. Biol. 41, 225–253.

Somers, D.E., Devlin, P.F., Kay, S.A., 1998. Phytochromes and cryptochromes in the entrainment of the *Arabidopsis* circadian clock. Science 282, 1488–1490.

Somerville, C.R., 1986. Analysis of photosynthesis with mutants of higher plants and algae. Annu. Rev. Plant Physiol. 37, 467–507.

Somerville, C.R., 2001. An early *Arabidopsis* demonstration. Resolving a few issues concerning photorespiration. Plant Physiol. 125, 20–24.

Somerville, C., 2006. Cellulose synthesis in higher plants. Annu. Rev. Cell Dev. Biol. 22, 53–78.

Somerville, C.R., 2008. Developing Cellulosic Biofuels. http://www.ciw.edu/somerville_keynote.

Somerville, C., Bauer, S., Brininstool, G., Facette, M., Hamann, T., Milne, J., Osborne, E., Paredez, A., Persson, S., Rabb, T., Vorwerk, S., Youngs, H., 2004. Toward a systems approach to understanding plant cell walls. Science 306, 2206–2211.

Somerville, C.R., Ogren, W.L., 1979. A phosphoglycolate phosphatase-defiocient mutant of *Arabidopsis*. Nature 280, 833–836.

Somerville, C.R., Ogren, W.L., 1980. Photorespiration mutants of *Arabidopsis thaliana* deficient in serine-glyoxylate aminotransferase activity. Proc. Natl. Acad. Sci. U.S.A. 77, 2684–2687.

Somerville, C.R., Ogren, W.L., 1982. Mutants of the cruciferous plant *Arabidopsis thaliana* lacking glycine decarboxylase activity. Biochem. J. 202, 373–380.

Sommarin, M., Sandelius, A.S., 1988. Phosphatidylinositol and phosphatidylinositolphosphate kinases in plant plasma membranes. Biochim. Biophys. Acta 958, 268–278.

Sommer, H., Beltrán, J.P., Huijser, P., Pape, H., Lönnig, W.E., Saedler, H., Schwarz-Sommer, Z., 1990. Deficiens, a homeotic gene involved in the control of flower morphogenesis in *Antirrhinum majus*: the protein shows homology to transcription factors. EMBO J. 9, 605–613.

Sommer, A., Geist, B., Da Ines, O., Gehwolf, R., Schäffner, A.R., Obermeyer, G., 2008. Ectopic expression of *Arabidopsis thaliana* plasma membrane intrinsic protein 2 aquaprorins in lily pollen increases the plasma membrane water permeability of grain but not of tube protoplasts. New Phytol. 180, 787–797.

Son, O., Yang, H.S., Lee, H.J., Lee, M.Y., Shin, K.H., Jeon, S.L., Lee, M.S., Choi, S.Y., Chun, J.Y., Kim, H., et al., 2003. Expression of srab7 and SCaM genes required for endocytosis of rhizobium in root nodules. Plant. Sci. 165, 1239–1244.

Sonesson, A., Widell, S., 1993. Cytoskeleton components of inside-out and right-side-out plasma membrane vesicles from plants. Protoplasma 177, 45–52.

Song, G., Jia, M., Chen, K., Kong, X., Khattak, B., Xie, C., Li, A., Mao, L., 2016. CRISPR/Cas9: a powerful tool for crop genome editing. Crop J. 4, 75–82.

Songstad, D.D., Petolino, J.F., Voytas, D.F., Reichert, N.A., 2017. Genome editing of plants. Crit. Rev. Plant Sci. 36, 1–23.

Sonneborn, T.M., 1942. Sex hormones in unicellular organisms. Cold Spring Harb. Symp. Quant. Biol. 10, 111–125.

Sonnichsen, B., Fuellekrug, J., Van, P.N., Diekmann, W., Robinson, D.G., Mieskes, G., 1994. Retention and retrieval: both mechanisms cooperate to maintain calreticulin in the endoplasmic reticulum. J. Cell Sci. 107, 2705–2717.

Sonobe, S., Yamamoto, S., Motomura, M., Shimmen, T., 2001. Isolation of cortical MTs from tobacco BY-2 cells. Plant Cell Physiol. 42, 162–169.

Sopory, S.K., Oelmüller, R., Maheshwari, S.C. (Eds.), 2001. Signal Transduction in Plants. Kluwer Academic, New York.

Sørensen, S.P.L., 1909. Enzymstudien II. Biochem. Z. 21, 131–304.

Sørensen, S.P.L., 1925. Proteins. The Fleishmann Company, New York.

Sørensen, I., Fei, Z., Andreas, A., Willats, W.G.T., Domozych, D.S., Rose, J.K.C., 2014. Stable transformation and reverse genetic analysis of *Penium margaritaceum*: a platform for studies of charophyte green algae, the immediate ancestors of land plants. Plant J. 77, 339–351.

Sørensen, I., Willats, W.G., 2011. Screening and characterization of plant cell walls using carbohydrate microarrays. Methods Mol. Biol. 715, 115–121.

Sottomayor, M., Pinto, M.C., Salema, R., DiCosmo, F., Pedreoo, M.A., Ros Barcelo, A., 1996. The vacuolar localization of a basic peroxidase isoenzyme responsible for the synthesis of α-3′,4′-anhydrovinblastine in *Catharanthus roseus*. Plant Cell Environ. 19, 761–767.

Sottosanto, J.B., Gelli, A., Blumwald, E., 2004. DNA array analyses of *Arabidopsis thaliana* lacking a vacuolar Na^+/H^+ antiporter: impact of AtNHX1 on gene expression. Plant J. 40, 752–771.

Soukup, A., 2014. Selected simple methods of plant cell wall histochemistry and staining for light microscopy. Methods Mol. Biol. 1080, 25–40.

Southam, C.M., Erhlich, J., 1943. Effects of extracts of western red-cedar heartwood on certain wood-decaying fungi in culture. Phytopathology 33, 517–524.

Southern, E.M., 1975. Detection of specific sequences among DNA fragments separated by gel electrophoresis. J. Mol. Biol. 98, 503–517.

Southern, E.M., 1996. DNA chips: analysing sequence by hybridization to oligonucleotides on a large scale. Trends Genet. 12, 110–115.

Southern, E.M., Maskos, U., Elder, J.K., 1992. Analyzing and comparing nucleic acid sequences by hybridization to arrays of oligonucleotides: evaluation using experimental models. Genomics 13, 1008–1017.

Southwick, F.S., Purich, D.L., 1998. *Listeria* and *Shigella* actin-based motility in host cells. Trans. Am. Clin. Climatol. Assoc. 109, 160–173.

Sovonick, S.A., Geiger, D.R., Fellows, R.J., 1974. Evidence for active phloem loading in the minor veins of sugar beet. Plant Physiol. 54, 886–891.

Soyars, C.L., James, S.R., Nimchuk, Z.L., 2016. Ready, aim, shoot: stem cell regulation of the shoot apical meristem. Curr. Opin. Plant Biol. 29, 163–168.

Spalding, E.P., Harper, J.F., 2011. The ins and outs of cellular Ca(2+) transport. Curr. Opin. Plant Biol. 14, 715–720.

Spallanzani, L., 1769. Nouvelles Recherches sur les Découvertes Microscopiques et la Génération des Corps Organisés. Ches Lacombe, Londres & Paris.

Spallanzani, L., 1776. Observations et expériences faites sur les Animalicules des Infusions, 1920. L'École Polytechnique, Paris.

Spallanzani, L., 1784. Dissertations Relative to the Natural History of the Animals and Vegetables. J. Murray, London.

Spallanzani, L., 1803. Memorie su la Respirazione. Agnello Nobile, Milano.

Spanner, D.C., 1979. The electroosmotic theory of phloem transport: a final restatement. Plant Cell Environ. 2, 107–121.

Spanswick, R.M., 1972. Evidence for an electrogenic pump in *Nitella translucens*. I. The effects of pH, K^+, Na^+, lightand temperature on the membrane potential and resistance. Biochim. Biophys. Acta 288, 73–89.

Spanswick, R.M., 1974a. Evidence for an electrogenic pump in *Nitella translucens*. II. Control of the light-stimulated component of the membrane potential. Biochim. Biophys. Acta 332, 387–389.

Spanswick, R.M., 1974b. Symplastic transport in plants. Symp. Soc. Exp. Biol. 28, 127–137.

Spanswick, R.M., 2006. Electrogenic pumps. In: Volkov, A.G. (Ed.), Plant Electrophysiology—Theory & Methods. Springer-Verlag, Berlin, pp. 221–246.

Spanswick, R.M., Costerton, J.W.F., 1967. Plasmodesmata in *Nitella translucens*: structure and electrical resistance. J. Cell Sci. 2, 451–464.

Sparkes, I.A., 2010. Motoring around the plant cell: insights from plant myosins. Biochem. Soc. Trans. 38, 833–838.

Sparkes, I., Runions, J., Hawes, C., Griffing, L.R., 2010. Movement and remodeling of the endoplasmic reticulum in nondiving cells of tobacco leaves. Plant Cell 21, 3937–3949.

Sparkes, I.A., Teanby, N.A., Hawes, C., 2008. Truncated myosin XI tail fusions inhibit peroxisome, Golgi, and mitochondrial movement in tobacco leaf epidermal cells: a genetic tool for the next generation. J. Exp. Bot. 59, 2499–2512.

Spear, D.G., Barr, J.K., Barr, C.E., 1969. Localization of hydrogen ion and chloride ion fluxes in Nitella. J. Gen. Physiol. 54, 397–414.

Speksnijder, J.E., Miller, A.L., Weisenseel, M.H., Chen, T.-H., Jaffe, L.F., 1989. Calcium buffer injections block fucoid egg development by facilitating calcium diffusion. Proc. Natl. Acad. Sci. U.S.A. 86, 6607–6611.

Spemann, H., December 12, 1935. The organizer-effect in embryonic development. Nobel Lecture.

Spencer, H., 1860. The social organism. Westminst. Rev. lxxiii (90–121), 105–108.

Spencer, H., 1864. The Principles of Biology, vol. I. D. Appleton and Co., New York.

Spencer, H., 1880. First Principles. Reprinted from Fifth London Edition. A. L. Bur, New York.

Spencer, H., 1880. First Principles, fourth ed. H. M. Caldwell Co., New York.

Spencer, N., 2009. Darwin and God. Society for Promoting Christian Knowledge, London.

Spencer, M.W., Casson, S.A., Lindsey, K., 2007. Transcriptional profiling of the Arabidopsis embryo. Plant Physiol. 143, 924–940.

Spencer, D., Whitfield, P.R., 1969. The characteristics of spinach chloroplast DNA polymerase. Arch. Biochem. Biophys. 132, 477–488.

Spencer, D., Wildman, S.G., 1962. Observations on the grana- containing chloroplasts and a proposed model of chloroplast structure. Aust. J. Biol. Sci. 15, 599–615.

Spiegelman, S., 1956. On the nature of the enzyme-forming system. In: Gaebler, O.H. (Ed.), Enzymes: Units of Biological Structure and Function. Academic Press, New York, pp. 67–89.

Spiker, S., 1985. Plant chromatin structure. Ann. Rev. Plant Physiol. 36, 235–253.

Spiker, S., Murry, M., Thompson, W.F., 1983. DNase I sensitivity of transcriptionally active genes in intact nuclei and isolated chromatin. Proc. Natl. Acad. Sci. U.S.A. 80, 815–819.

Spiker, S., Thompson, W.F., 1996. Nuclear matrix attachment regions and transgene expression in plants. Plant Physiol. 110, 15–21.

Spiller, D.G., Wood, C.D., Rand, D.A., White, M.R., 2010. Measurement of single-cell dynamics. Nature 465, 736–745.

Spinoza, B., 1492. The ethics of Spinoza: the road to inner freedom, 1995. In: Runes. Carol Publishing Group, New York.

Spoehr, H.A., 1926. Photosynthesis. The Chemical Catalog Co., New York.

Sprague, S.G., Camm, E.L., Green, B.R., Staehelin, L.A., 1985. Reconstitution of light-harvesting complexes and photosystem II cores into galactolipid and phospholipid liposomes. J. Cell Biol. 100, 552–557.

Spreitzer, R.J., 1993. Genetic dissection of rubisco structure and function. Ann. Rev. Plant Physiol. Plant Mol. Biol. 44, 411–434.

Springer, M., 2006. Applied Biosystems: Celebrating 25 Years of Advancing Science. American Laboratory News. May.

Spudich, J., 2013. Memorie of Hugh E. Huxley (1924-2013). Mol. Biol. Cell 24, 2769–2771.

Spurck, T.P., Pickett-Heaps, J.D., 1987. On the mechanism of Anaphase A: evidence that ATP is needed for microtubule disassembly and not generation of polewards force. J. Cell Biol. 105, 1691–1705.

Spyrides, E.J., Lipmann, F., 1962. Polypeptide synthesis with sucrose gradient fractions of *E. coli* ribosomes. Proc. Natl. Acad. Sci. U.S.A. 48, 1977–1983.

Square, A., 1899. Flatland. A Romance of Many Dimensions. E.A. Abbott. Little, Brown, and Co., Boston.

Srb, A.M., Horowitz, N.H., 1944. The ornithine cycle in Neurospora and its genetic control. J. Biol. Chem. 154, 129–139.

Srb, A.M., Owens, R.D., 1952. General Genetics. W. H. Freeman and Co., San Franscisco.

Sreenivasulu, S., Usadel, B., Winter, A., Radchuk, V., Scholz, U., Stein, N., Weschke, W., Strickert, M., Close, T.J., Stitt, M., Graner, A., Wobus, U., 2008. Barley grain maturation and germination: metabolic pathway and regulatory network commonalities and differences highlighted by New MapMan/PageMan profiling tools. Plant Physiol. 146, 1438–1758.

Srere, P.A., 1985. The metabolon. Trends Biochem. Sci. 10, 109–110.

Srivastava, D.K., Bernhard, S.A., 1986. Metabolite transfer via enzyme-enzyme complexes. Science 234, 1081–1086.

Srivastava, A., Krishnamoorthy, G., 1997. Cell type and spatial location dependence of cytoplasmic viscosity measured by real time-resolved fluorescence microscopy. Arch. Biochem. Biophys. 340, 159–167.

Staal, M., Maathuis, F.J.N., Elzenga, T.M., Overbeek, J.H.M., Prins, H.B.A., 1991. Na^+/H^+ antiport activity in tonoplast vesicles from roots of the salt-tolerant Plantago maritima and the salt-sensitive Plantago media. Physiol. Plant 82, 179–184.

Stabenau, H., 1992. Peroxisomes in algae and evolutionary changes of enzymes in peroxisomes and mitochondria of green algae. In: Stabenau, H. (Ed.), Phylogenetic Changes in Peroxisomes of Algae. Phylogeny of Plant Peroxisomes. Proc. of an Int. Symp. on Phylogeny of Algal and Plant Peroxisomes. University of Oldenburg, Bad Zwischenahn, Germany, pp. 42–79. Sept. 18–20, 1991.

Stadelmann, E.J., Lee, O.Y., 1974. Inverse changes of water and non-electrolyte permeability. In: Bolis, L., Bloch, K., Luria, S.E., Lynen, F. (Eds.), Comparative Biochemistry and Physiology of Transport. North-Holland Publishing Co., Amsterdam, pp. 434–441.

Staden, R., 1977. Sequence data handling by computer. Nucleic Acids Res. 4, 4037–4051.

Staden, R., 1979. A strategy of DNA sequencing employing computer programs. Nucleic Acids Res. 6, 2601–2610.

Staden, R., 1980. A new computer method for the storage and manipulation of DNA gel reading data. Nucleic Acids Res. 8, 3673–3694.

Staden, R., 1983. Computer methods for DNA sequencers. Lab. Tech. Biochem. Mol. Biol. 10, 311–348.

Stadler, L.J., 1954. The gene. Science 120, 811–819.

Staehelin, L.A., 1983. Control and regulation of the spatial organization of membrane components by membrane-membrane interactions. Mod. Cell Biol. 2, 73–92.

Staehelin, L.A., 1986. Chloroplast structure and supramolecular organization of photosynthetic membranes. Encyl. Plant Physiol. Photosynth. III. 19, 1–84.

Staehelin, A., 1991. What is a plant cell? A response. Plant Cell 3, 553.

Staehelin, L.A., 2003. Chloroplast structure: from chlorophyll granules to supra-molecular architecture of thylakoid membranes. Photosynth. Res. 76, 185–196.

Staehelin, L.A., Arntzen, C.J., 1983. Regulation of chloroplast membrane function: protein phosphorylation changes the spatial organization of membrane components. J. Cell Biol. 97, 1327–1337.

Staehelin Jr., L.A., Giddings, T.H., Kiss, J.Z., Sack, F., 1990. Macromolecular differentiation of Golgi stacks in root tips of *Arabidopsis* and *Nicotiana* seedings as visualized in high pressure frozen and freeze substituted samples. Protoplasma 157, 75–91.

Staehelin, L.A., Giddings, T.H., Zhang, G.F., 1992. Immunocytochemical analysis of the functional compartmentalization of plant Golgi Stacks. In: Shibaoka, H. (Ed.), Cellular Basis of Growth and Development in plants. Proc. of the VII International Symposium in conjunction with the awarding of the international prize for Biology. November 26–28, 1991. Toyonaka, Osaka, pp. 45–53.

Staehelin, L.A., Hepler, P.K., 1996. Cytokinesis in higher plants. Cell 84, 821–824.

Staehelin, L.A., Kang, B.-H., 2008. Nanoscale architecture of endoplasmic reticulum export sites and of Golgi membranes as determined by electron tomography. Plant Physiol. 147, 1454–1468.

Staehelin, L.A., Kyle, D.J., Arntzen, C.J., 1982. Spillover is mediated by reversible migration of LHCP between grana and stroma thylakoids. Plant Physiol. 69, S69.

Staehelin, L.A., Moore, I., 1995. The plant Golgi apparatus: structure, functional organization and trafficking mechanisms. Annu. Rev. Plant Physiol. Plant Mol. Biol. 46, 261–288.

Staehelin., 1997. The plant ER: a dynamic organelle composed of a large number of discrete functional domains. Plant J. 11, 1151–1165.

Stafelt, M.G., 1955. The protoplasmic viscosity of terrestrial plants and its sensitivity to light. Protoplasma 55, 285–292.

Stafford, H.A., 1991. What is a plant cell? Plant Cell 3, 331.

Stafstrom, J.P., Staehelin, L.A., 1988. Antibody localization of extensin in cell walls of carrot storage roots. Planta 174, 321–332.

Staiger, C.J., Hussey, P.J., 2004. Actin and actin-modulating proteins. Ann. Plant Rev. 10, 21–80.

Staiger, C.J., Lloyd, C.W., 1991. The plant cytoskeleton. Curr. Opin. Cell Biol. 3, 33–42.

Staiger, C.J., Schliwa, M., 1987. Actin localization and function in higher plants. Protoplasma 141, 1–12.

Stairs, S., Nikmal, A., Bučar, D.-K., Zheng, S.-L., Szostak, J.W., Powner, M.W., 2017. Divergent prebiotic synthesis of pyrimidine and 8-oxo-purine ribonucleotides. Nat. Commun. 8, 15270.

Stanier, R.Y., Adelberg, E., Douderoff, M., 1963. The Microbial World, second ed. Prentiss-Hall, Engelwood Cliffs, NJ.

Stanley, W.M., 1935. Isolation of a crystalline protein possessing the properties of tobacco-mosaic virus. Science 81, 644–645.

Stanley, W.M., Valens, E.G., 1961. Viruses and the Nature of Life. E. P. Dutton & Co., New York.

Stanton, E.C., 1848. The senaca falls declaration. http://www.constitution.org/woll/seneca.htm.

Stark, J., 1938. The pragmatic and the dogmatic spirit in physics. Nature 141, 770–772.

Starr, D., 2016. When DNA is lying. Science 351, 1133–1136.

Starr, D.A., Fridolfsson, H.N., 2010. Interactions between nuclei and the cytoskeleton by SUN-KASH nuclear envelope bridges. Annu. Rev. Cell Dev. Biol. 26, 421–444.

Starr, D.A., Williams, B.C., Li, Z., Etemad-Moghadam, B., Dawe, R.K., Goldberg, M.L., 1997. Conservation of the centromere/kinetochore protein ZW10. J. Cell Biol. 138, 1289–1301.

Staudinger, H., 1961. From Organic Chemistry to Macromolecules A Scientific Autobiography Based on My Original Papers. Wiley-Interscience, New York.

Staverman, A.J., 1951. The theory of measurement of osmotic pressure. Recueil des Travaux Chimiques des Pays-Bas 70, 344–352.

Staves, M.P., 1997. Cytoplasmic streaming and gravity sensing in Chara internodal cells. Planta 203, S79–S84.

Staves, M.P., LaClair II, J.W., 1985. Nuclear synchrony in Valonia macrophysa (Chlorophyta): light microscopy and flow cytometry. J. Phycol. 21, 68–71.

Staves, M., Mescher, M., Shepherd, V., Brenner, E., van Volkenburgh, E., 2008. Symposia in plant neurobiology: a new venue for discussion of plant behavior and communication. Plant Sci. Bull. 54 (2), 56–57.

Staves, M.P., Wayne, R., 1993. The touch induced action potential in *Chara*: an inquiry into the ionic basis and the mechanoreceptor. Aust. J. Plant Physiol. 20, 471–488.

Staves, M.P., Wayne, R., Leopold, A.C., 1992. Hydrostatic pressure mimics gravitational pressure in characean cells. Protoplasma 168, 141–152.

Staves, M.P., Wayne, R., Leopold, A.C., 1995. Detection of gravity-induced polarity of cytoplasmic streaming in *Chara*. Protoplasma 188, 38–48.

Staves, M.P., Wayne, R., Leopold, A.C., 1997a. Cytochalasin D does not inhibit gravitropism in roots. Am. J. Bot. 84, 1530–1535.

Staves, M.P., Wayne, R., Leopold, A.C., 1997b. The effect of the external medium on the gravitropic curvature of rice (*Oryza sativa* Poaceae) roots. Am. J. Bot. 84, 1522–1529.

Staves, M.P., Wayne, R., Leopold, A.C., 1997b. The effect of the external medium on the gravitropic curvature of rice (Oryza sativa Poaceae) roots. Am. J. Bot. 84, 1522–1529.

Staves, M.P., Whitsit, K., Yeung, E., Wayne, R., 2005. The interaction between gravity and light in stimulating tropistic responses of *Chara* internodal cells. J. Gravit. Phys. 12, 158–160.

Stavolone, L., Lionetti, V., 2017. Extracellular matrix in plants and animals: hooks and locks for viruses. Front. Microbiol. 8, 1760.

Staxen, I., Pical, C., Montgomery, L., Gray, J.E., Hetherington, A.M., McAinsh, M.R., 1999. Abscisic acid induced oscillations in guard-cell cytosolic free calcium that involve phosphoinositide-specific phospholipase C. Proc. Natl. Acad. Sci. U.S.A. 96, 1779–1784.

Stearns, T., 1997. Motoring to the finish: kinesin and dynein work together to orient the yeast mitotic spindle. J. Cell Biol. 138, 957–960.

Stebbings, H., Hyams, J.S., 1979. Cell Motility. Longman, London.

Stebbins, G.L., 1982. Darwin to DNA, Molecules to Humanity. W. H. Freeman and Co., San Francisco.

Stebbins, G.L., 1985. Polyploidy, hybridization, and the invasion of new habitats. Ann. Missouri Bot. Gard. 72, 824–832.

Stebbins, G.L., Jain, S.K., 1960. Developmental studies of cell differentiation in the epidermis of monocotyledons. Dev. Biol. 2, 409–426 477–500.

Steck, T.L., 1974. The organization of proteins in the human red blood cell membrane. J. Cell Biol. 62, 1–19.

Stecker, E.K., Minkoff, B.B., Sussman, M.R., 2014. Phosphoproteomic analyses reveal early signaling events in the osmotic stress response. Plant Physiol. 165, 1171–1187.

Steckline, V.S., 1983. Zermelo, Boltzmann, and the recurrence paradox. Am. J. Phys. 51, 894–897.

Steele, E.J., Al-Mufti, S., Augustyn, K.A., Chandrajith, R., Coghlan, J.P., Coulson, S.G., Ghosh, S., Gillman, M., Gorczynski, R.M., Klycem, B., Louism, G., Mahanama, K., Oliver, K.R., Padron, J., Qu, J., Schuster, J.A., Smith, W.E., Snyder, D.P., Steele, J.A., Stewart, B.J., Temple, R., To,koro, G., Tout, C.A., Unzicker, A., Wainwright, M., Wallis, J., Wallis, D.H., Wallis, M.K., Wetherall, J., Wickramasinghe, D.T., Wickramasinghe, J.T., Wickramasinghe, N.C., Liu, Y., 2018. Cause of Cambrian explosion - terrestrial or cosmic? Prog. Biophys. Mol. Biol. 136, 3–23.

Steer, M.W., Steer, J.M., 1989. Pollen tube tip growth. New Phytol. 111, 323–358.

Stefano, G., Brandizzi, F., 2018. Advances in plant ER architecture and dynamics. Plant Physiol. 176, 178–186.

Stefano, G., Hawes, C., Brandizzi, F., 2014a. ER—the key to the highway. Curr. Opin. Plant Biol. 22, 30–38.

Stefano, G., Renna, L., Brandizzi, F., 2014b. The endoplasmic reticulum exerts control over organelle streaming during cell expansion. J. Cell Sci. 127, 947–953.

Steffens, J.C., 1990. The heavy metal—binding peptides of plants. Annu. Rev. Plant Physiol. Plant Mol. Biol. 41, 553–575.

Steger, D., Berry, D., Haider, S., Horn, M., Wagner, M., et al., 2011. Systematic spatial bias in DNA microarray hybridization is caused by probe spot position-dependent variability in lateral diffusion. PLoS One 6, e23727.

Stein, W.D., 1986. Transport and Diffusion Across Cell Membranes. Academic Press, San Diego.

Stein, J.C., Dixit, R., Nasrallah, M.E., Nasrallah, J.B., 1996. SRK, the stigma-specific S locus receptor kinase of *Brassica*, is targeted to the plasma membrane in transgenic tobacco. Plant Cell 8, 429–445.

Stein, J.C., Nasrallah, J.B., 1993. A plant receptor-like gene. The S-locus receptor kinase of *Brassica oleracea* L. encodes a functional serine-threonine kinase. Plant Physiol. 101, 1103–1106.

Stein, D.B., Searcy, D.G., 1978. Physiologically important stabilization of DNA by a prokaryotic histone-like protein. Science 202, 219–221.

Steinbach Ulbrich, M., 2016. The last one to go. Spiritsail 30 (2), 3–10.

Steinberg, D., Mihalyi, E., 1957. The chemistry of proteins. Ann. Rev. Biochem. 26, 373–4118.

Steiner, J., Pfanzagl, B., Ma, Y., Löffelhardt, W., 2002. Evolution and biology of cyanelles. Biol. Environ. Proc. Roy. Irish Acad. 102B, 7–9.

Steinhardt, J., 1945. The chemistry of the amino acids and proteins. Ann. Rev. Biochem. 14, 145–174.

Steinhardt, R.A., Epel, D., 1974. Activation of sea-urchin eggs by a calcium ionophore. Proc. Natl. Acad. Sci. U.S.A. 71, 1915–1919.

Steinmann, E., Sjöstrand, F.S., 1955. The ultrastructure of chloroplasts. Exp. Cell Res. 8, 15–23.

Stenbeck, G., 1998. Soluble NSF-attachment proteins. Int. J. Biochem. Cell Biol. 30, 573–577.

Stenger, V.J., 2007. God: The Failed Hypothesis. How Science Shows that God Does Not Exist. Prometheus Books, Amherst, NY.

Stent, G.S., 1968. That was the molecular biology that was. Science 160, 390–395.

Stent, G.S. (Ed.), 1980. The Double Helix. W. W. Norton & Co., New York.

Stephenne, N., Margoni, R., Laneve, G., 2009. From real time border monitoring to a permeability model. In: Jasani, B., Pesaresi, M., Schneiderbauer, S., Zeug, G. (Eds.), Remote Sensing from Space. Supporting International Peace and Security. Springer, pp. 239–259.

Stephenson, J.L.M., Hawes, C.R., 1986. Stereology and stereometry of endoplasmic reticulum during differentiation in the maize root cap. Protoplasma 131, 32–46.

Steppuhn, A., Gase, K., Krock, B., Halitschke, R., Baldwin, I.T., 2004. Nicotine's defensive function in nature. PLoS Biol. 2, 1074–1080.

Sterling, J.D., Quigley, H.F., Orellana, A., Mohnen, D., 2001. The catalytic site if the pectin biosynthetic enzyme α-1,4-galacturonosyltransferase is located in the lumen of the Golgi. Plant Physiol. 127, 360–371.

Stern, C., 1931. Zytologisch-genetische Untersuchungen als Beweise für die Morganische Theorie des Factorenaustauschs. Biol. Zbl. 51, 547–587.

Stern, D.B., Lonsdale, D.M., 1982. Mitochondrial and chloroplast genomes of maize have a 12-kilobase DNA sequence in common. Nature 299, 698–702.

Stern, D.B., Palmer, J.D., Thompson, W.F., Lonsdale, D.M., 1983. Mitochondrial DNA sequence evolution and homology to chloroplast DNA in angiosperms. Plant Mol. Biol. 1, 467–477.

Stetler, D.A., DeMaggio, A.E., 1972. An ultra-structural study of fern gametophytes during one to two-dimensional development. Am. J. Bot. 59, 1011–1017.

Steudle, E., Tyerman, S.D., 1983. Determination of permeability coefficients, and hydraulic conductivity of Chara corallina using pressure probe: effect of solute concentration. J. Membr. Biol. 75, 85–96.

Steuer, E.R., Wordeman, L., Schroer, T.A., Sheetz, M.P., 1990. Localization of cytoplasmic dynein to mitotic spindles and kinetochores. Nature 345, 266–268.

Stevens, N.E., 1932. The fad as a factor in botanical publication. Science 75, 499–504.

Steward, F.C., 1933. The absorption and accumulation of solutes by living plant cells. V. Observations upon the effects of time, oxygen and salt concentration upon absorption and respiration by storage tissue. Protoplasma 18, 208–242.

Steward, F.C., 1941. Salt accumulationtby plants—The role of growth and metabolism. Trans. Faraday Soc. 33, 1006–1016.

Steward, C.R., Boggess, S.F., Aspinall, D., Paleg, L.G., 1977. Inhibition of proline oxidation by water stress. Plant Physiol. 59, 930–932.

Steward, F.C., Mapes, M.O., Mears, K., 1958. Growth and organized development of cultured cells. Am. J. Bot 45, 705–713.

Steward, F.C., Mühlethaler, K., 1953. The structure and development of the cell-wall in the Valoniaceae as revealed by the electron microscope. Ann. Bot. N. S. 17, 295–316.

Stewart, K.D., Mattox, K.R., 1975. Comparative cytology, evolution and classification of the green algae with some considerations of the origin of other organisms with chlorophylls a and b. Bot. Rev. 41, 104–135.

Stewart, R.N., Norris, K.H., Asen, S., 1975. Microspectrophotometric measurement of pH and pH effect on color of petal epidermal cells. Phytochemistry 14, 937–942.

Stewart, B., Tait, P.G., 1890. The Unseen Universe. MacMillan and Co., London.

Sticher, L., Biswas, A.K., Bush, D.S., Jones, R.L., 1990. Heat shock inhibits α-amylase synthese in barley aleurone without inhibiting the activity of endoplasmic reticulum marker enzymes. Plant Physiol. 92, 506–513.

Stiles, W., 1925. Photosynthesis. The Assimilation of Carbon by Green Plants. Longmans, Green and Co., London.

Stillman, B., 1989. Initiation of eukaryotic DNA replication. Ann. Rev. Cell Biol. 5, 197–245.

Stocking, C.R., Gifford, E.M., 1959. Incorporation of thymidine into chloroplasts of Spirogyra. Biochem. Biophys. Res. Commun. 1, 159–164.

Stockwell, C.R., Miller, J.H., 1974. Regions of cell wall expansion in the protonema of a fern. Am. J. Bot. 61, 375–378.

Stoeckenius, W., Engelman, D.M., 1969. Current models for the structure of biological membranes. J. Cell Biol. 42, 613–646.

Stoger, E., Fischer, R., Molone, M., Ma, J.K.-C., 2014. Plant molecular pharming for the treatment of chronic and infectious diseases. Annu. Rev. Plant Biol. 65, 743–768.

Stokes, G.G., 1864. On the reduction and oxidation of the colouring matter of the blood. Proc. R. Soc. 13 (144), 355–364.

Stokes, G.G., 1891. Natural Theology. The Gifford Lectures of 1891. Adam and Charles Black, London.

Stokes, G.G., 1893. Natural Theology. The Gifford Lectures of 1893. Adam and Charles Black, London.

Stokes, G.G., 1922. On the effect of the internal friction of fluids on the motion of pendulums. In: Mathematical and Physical Papers, vol. III. Cambridge University Press, Cambridge.

Stokes, K.D., McAndrew, R.S., Figueroa, R., Vitha, S., Osteryoung, K.W., 2000. Chloroplast division and morphology are differentially affected by overexpression of FtsZ1 and FtsZ2 genes in *Arabidopsis*. Plant Physiol. 124, 1668–1677.

Stolc, V., Samanta, M.P., Tongprasit, W., Marshall, W.F., 2005. Genome-wide transcriptional analysis of flagellar regeneration in *Chlamydomonas reinhardtii* identifies orthologs of ciliary disease genes. Proc. Natl. Acad. Sci. U.S.A. 102, 3703–3707.

Stone, I., 1982. The Healing Factor. Vitamin C Against Disease Forwards By L. Pauling and A. Szent-Györgyi. Perigee Books, New York.

Storrie, B., Nilsson, T., 2002. The Golgi apparatus: balancing new with old. Traffic 3, 521–529.

Story, M., Congalton, R.G., 1986. Accuracy assessment: a user's perspective. Photogramm. Eng. Remote Sens. 52, 397–399.

Stout, S.C., Clark, G.B., Archer-Evans, S., Roux, S.J., 2003. Rapid and efficient gene expression in a single-cell model system, *Ceratopteris richardii*. Plant Physiol. 131, 1165–1168.

Stowe, B.B., Yamaki, T., 1959. Gibberellins: stimulants of plant growth. Science 129, 807–816.

Strader, L.C., Bartel, B., 2008. A new path to auxin. Nat. Chem. Biol. 4, 337–339.

Strain, H.H., 1958. Chloroplast Pigments and chromatographic analysis. In: Thirty-Second Annual Priestley Lecture. Pennsylvania State University, University Park, PA.

Strasburger, E., 1875. Ueber Zellbilbung und Zelltheilung. Hermann Dabis, Jena.

Strasburger, E., 1884. Neue Untersuchungen über den Befruchtungsvorgang bei den Phanerogamen als Grundlage für eine Theorie der Zeugung. Gustav Fischer, Jena.

Strasburger, E., 1897. Ueber Cytoplasmastrukturen, Kern- und Zelltheilung. Jahr. Wiss. Bot. 30, 375–405.

Strasburger, E., 1898. Die pflanzlichen Zellhäute. Jahr. Wiss. Bot. 31, 511–598.

Strasburger, E., Hillhouse, W., 1911. Handbook of Practical Botany, seventh ed. George Allen and Co., London.

Strasburger, E., Jost, L., Schenck, H., Karsten, G., 1912. A Text – Book of Botany. MacMillan, London.

Stratmann, J.W., Ryan, C.A., 1997. Myelin basic protein kinase activity in tomato leaves is induced systemically by wounding and increases in response to systemin and oligosaccharide elicitors. Proc. Natl. Acad. Sci. U.S.A. 94, 11085–11089.

Straub, F.B., 1981. From respiration of muscle to actin (1934 – 1943). In: Semenza, G. (Ed.), Of Oxygen, Fuels, and Living Matter. Part 1. John Wiley & Sons, Chichester, pp. 325–344.

Straus, W., 1953. Chromoplasts–development of crystalline forms, structure, state of the pigments. Bot. Rev. 19, 147–186.

Strauss, E.C., Kobori, J.A., Siu, G., Hood, L.E., 1986. Specific primer-directed DNA sequencing. Anal. Biochem. 154, 314–326.

Strepp, R., Scholz, S., Kruse, S., Speth, V., Reski, R., 1998. Plant nuclear gene knockout reveals a role in plastid division for the homolog of the bacterial cell division protein FtsZ, an ancestral tubulin. Proc. Natl. Acad. Sci. U.S.A. 95, 4368–4373.

Strnad, M., Caño-Delgado, A.I., Friml, J., Madder, A., Russinova, E., 2012. Fluorescent castasterone reveals BRI1 signaling from the plasma membrane. Nat. Chem. Biol. 8, 583–589.

Strother, P.K., Barghoorn, E.S., 1980. Microspheres from Swartkoppie formation: a review. In: Halvorson, H.O., van Holde, K.E. (Eds.), The Origins of Life and Evolution. Alan R. Liss, Inc., New York, pp. 1–18.

Strugger, S., 1935. Praktikum der Zell- und Gewebephysiologie der Pflanze. Gebrüder Borntraeger, Berlin.

Struhl, K., Stinchcomb, D.T., Scherer, S., Davis, R.W., 1979. High frequency transformation of yeast: autonomous replication of hybrid DNA molecules. Proc. Natl. Acad. Sci. U.S.A. 76, 1035–1039.

Struve, O., 1962. The Universe. MIT Press, Cambridge, MA.

Stubblefield, E., 1975. Analysis of the replication pattern of Chinese hamster chromosomes using 5-bromodeoxyuridine suppression of 33258 Hoechst fluorescence. Chromosoma 53, 209–221.

Sturbois-Balcerzak, B., Vincent, P., Maneta-Peyret, L., Duvert, M., Satiat-Jeunemaitre, B., Cassagne, C., Moreau, P., 1999. ATP- dependent formation of phosphatidylserine-rich vesicles from the endoplasmic reticulum of leek cells. Plant Physiol. 120, 245–256.

Sturtevant, A.H., 1913. The linear arrangement of sex-linked factors in Drosophila, as shown by their mode of association. J. Exp. Zool. 14, 43–59.

Sturtevant, A.H., 1925. The effects of unequal crossing over at the bar locus in Drosophilia. Genetics 10, 117–147.

Sturtevant, A.H., 1951. The relation of genes and chromosomes. In: Dunn, L.C. (Ed.), Genetics in the 20th Century. Macmillan, New York, pp. 101–110.

Styles, E.B., Founk, K.J., Zamparo, L.A., Sing, T.L., Altintas, D., Ribeyre, C., Ribaud, V., Rougemont, J., Mayhew, D., Costanzo, M., Usaj, M., Verster, A.J., Koch, E.N., Novarina, D., Graf, M., Luke, B., Muzi-Falconi, M., Myers, C.L., Andrews, B.J., 2016. Exploring quantitative yeast phenomics with single-cell analysis of DNA damage foci. Cell Syst. 3, 264–277.

Su, F., Gu, W., Zhai, Z.H., 1990a. The keratin intermediate filament-like system in maize protoplasts. Cell Res. 1, 11–16.

Su, F., Gu, W., Zhai, Z.H., 1990b. The keratin intermediate filament-like system in the plant mesophyll cells. Sci. Sinica. Ser. B 33, 1084–1091.

Su, P.-H., Li, H., 2010. Stromal Hsp70 is important for protein translocation into pea and *Arabidopsis* chloroplasts. Plant Cell 22, 1516–1531.

Su, S., Liu, Z., Chen, C., Zhang, Y., Wang, X., Zhu, L., Miao, L., Wang, X.C., Yuan, M., 2010. Cucumber mosaic virus movement protein severs actin filaments to increase the plasmodesmatal size exclusion limit in tobacco. Plant Cell 22, 1373–1387.

Suda, Y., Nakano, A., 2012. The yeast Golgi apparatus. Traffic 13, 505–510.

Sueoka, N., 1960. Mitotic replication of deoxyribonucleic acid in Chlamydomonas Reinhardti. Proc. Natl. Acad. Sci. U.S.A. 46, 83–91.

Suetsuga, N., Mittman, F., Wagner, G., Hughes, J., Wada, M., 2005. A chimeric photoreceptor gene, neochrome, has arisen twice during plant evolution. Proc. Natl. Acad. Sci. U.S.A. 102, 13705–13709.

Suetsuga, N., Wada, M., 2007. Phytochrome-dependent photomovement responses mediated by phototropin family proteins in cryptogam plants. Photochem. Photobiol. 83, 87–93.

Sugawara, O., Oshimura, M., Koi, M., Annab, L.A., Barrett, J.C., 1990. Induction of cellular senescence in immortalized cells by human chromosome 1. Science 247, 707–710.

Sugiura, M., 2003. History of chloroplast genomics. Photosynth. Res. 76, 371–377.

Suh, S., Moran, N., Lee, Y., 2000. Blue light activates depolarization-dependent K channels in flexor cells from Samanea saman motor organs via two mechanisms. Plant Physiol. 123, 833–843.

Sullivan, N., 1964. Pioneer Astronomers. E. M. Hale and CO., Eau Clare, WI.

Summers, K.E., Gibbons, I.R., 1971. Adenosine triphosphate- induced sliding of tubules in trypsin-treated flagella of sea urchin sperm. Proc. Natl. Acad. Sci. U.S.A. 68, 3092–3096.

Summers, K., Kirschner, M.W., 1979. Characteristics of the polar assembly and disassembly of microtubules observed in vitro by dark–field light microscopy. J. Cell Biol. 86, 402–416.

Sumner, J.B., 1926. The isolation and crystallization of the enzyme urease. J. Biol. Chem. 69, 435–441.

Sumner, J.B., 1927. Textbook of Biological Chemistry. MacMillan, New York.

Sumner, J.B., 1933. The chemical nature of enzymes. Science 78, 335.

Sumner, J.B., 1935. Enzymes. Annu. Rev. Biochem. 4, 37–58.

Sumner, J.B., 1937. The story of urease. J. Chem. Educ. 14, 255–259.

Sumner, J.B., December 12, 1946. The Chemical Nature of Enzymes. Nobel Lecture.

Sumner, J.B., Graham, V.A., 1925. The nature of insoluble urease. Proc. Soc. Exp. Biol. Med. 22, 504–506.

Sumner, J.B., Graham, V.A., Noback, C.V., 1924. The purification of jack bean urease. Proc. Soc. Exp. Biol. Med. 21, 551–552.

Sumner, L.W., Mendes, P., Dixon, R.A., 2003. Plant metabolomics: large-scale phytochemistry in th e functional genomics era. Phytochemistry 62, 817–836.

Sumner, J.B., Somers, G.F., 1943. Chemistry and Methods of Enzymes. Academic Press, New York.

Sun, G., Bailey, D., Jones, M.W., Markwell, J., 1989. Chloroplast thylakoid protein phosphatase is a membrane surface-associated activity. Plant Physiol. 89, 238–243.

Sun, T., Germain, A., Giloteaux, L., Hammani, K., Barkan, A., Hanson, M.R., Bentolila, S., 2013. An RNA recognition motif-containing protein is required for plastid RNA editing in *Arabidopsis* and maize. Proc. Natl. Acad. Sci. U.S.A. 110, E1169–E1178.

Sun, Z., Jin, X., Albert, R., Assmann, S.M., 2014. Multi-level modeling of light-induced stomatal opening offers new insights into its regulation by drought. PLoS Comp. Biol. 10, e1003930.

Sun, X., Jones, W.T., Rikkerink, E.H.A., 2012. GRAS proteins: the versatile roles of intrinsically disordered proteins in plant signalling. Biochem. J. 442, 1–12.

Sun, G., Ohya, Y., Anraku, Y., 1991. Half-calmodulin is sufficient for cell proliferation. J. Biol. Chem. 266, 7008–7015.

Sun, Y., Qian, H., Xu, X.-D., Han, Y., Yen, L.-F., Sun, D.-Y., 2000. Integrin-like proteins in the pollen tube: detection, localization and function. Plant Cell Physiol. 41, 1136–1142.

Sun, T., Shi, X., Friso, G., Van Wijk, K., Bentolila, S., Hanson, M.R., 2015a. A zinc finger motif-containing protein is essential for chloroplast RNA editing. PLoS Genet. 11, e1005028.

Sun, G.-H., Uyeda, T.Q.P., Kuroiwa, T., 1988. Destruction of organelle nuclei during spermatogenesis in Chara corallina examined by staining with DAPI and anti-DNA antibody. Protoplasma 144, 185–188.

Sun, T., Zhang, Y., Li, Y., Ding, Q., Zhang, Y., 2015b. ChIP-seq reveals broad roles of SARD1 and CBP60g in regulating plant immunity. Nat. Commun. 6, 10159.

Sun, Q., Zybailov, B., Majeran, W., Friso, G., Olinares, P.D.B., van Wijk, K.J., 2009. PPDB, the plant proteomics database at cornell. Nucleic Acids Res. 37, D969–D974.

Sundell, C.L., Singer, R.H., 1991. Requirement of microfilaments in sorting of actin messenger RNA. Science 253, 1275–1277.

Sunil Kumar, G.B., Ganapathi, T.R., Srinivas, L., Revathi, C.J., Bapat, V.A., 2005. Secretion of hepatitis B surface antigen in transformed tobacco cell suspension cultures. Biotech. Lett. 27, 927–932.

Sunkar, R., Zhu, J.-K., 2004. Novel and stress-regulated microRNAs and other small RNAs from *Arabidopsis*. Plant Cell 16, 2001–2019.

Surpin, M., Raikhel, N., 2004. Traffic jams affect plant development and signal transduction. Nat. Rev. Mol. Cell Biol. 5, 100–109.

Sussman, M.R., 1994. Molecular analysis of proteins in the plant plasma membrane. Annu. Rev. Plant Physiol. Plant Mol. Biol. 45, 211–234.

Sussman, M.R., Amasino, R.M., Young, J.C., Krysan, P., Austin-Phillips, S., 2000. The *Arabidopsis* knockout facility at the University of Wisconsin-Madison. Plant Physiol. 124, 1465–1467.

Sussman, M.R., Harper, J.F., 1989. Molecular biology of the plasma membrane of higher plants. Plant Cell 1, 953–960.

Sutera, S.P., Skalak, R., 1993. The history of Poiseuille's law. Annu. Rev. Fluid Mech. 25, 1–19.

Sutherland, E.W., December 11, 1971. Studies on the Mechanism of Hormone Action. Nobel Lecture.

Sutter, J.U., Campanoni, P., Blatt, M.R., Paneque, M., 2006. Setting SNAREs in a different wood. Traffic 7, 627–638.

Suwabe, K., Yano, K., 2008. Omics databases in plant science: key to systems biology. Plant Biotechnol. 25, 413–422.

Suzuki, L., Johnson, C.H., 2001. Algae know the time of day: circadian and photoperiodic programs. J. Phycol. 37, 933–942.

Suzuki, K., Kirisako, T., Kamada, Y., Mizushima, N., Noda, T., Ohsumi, Y., 2001. The pre-autophagosomal structure organized by concerted functions of APG genes is essential for autophagosome formation. EMBO J. 20, 5971–5981.

Suzuki, N.N., Yoshimoto, K., Fujioka, Y., Ohsumi, Y., Inagaki, F., 2005. The crystal structure of plant ATG12 and its biological implication inaautophagy. Autophagy 1, 119–126.

Svedberg, T., 1937. The ultra-centrifuge and the study of high- molecular compounds. Nature 139, 1051–1052.

Svedberg, T., May 19, 1927. The ultracentrifuge. Nobel Lecture.

Svedberg, T., Pedersen, K.O., 1940. The Ultracentrifuge. Clarendon Press, Oxford.

Svennelid, F., Olsson, A., Piotrowski, M., Rosenquist, M., Ottman, C., Larsson, C., Oecking, C., Sommarin, M., 1999. Phosphorylation of Thr-948 at the C terminus of the plasma membrane H^+-ATPase creates a binding site for the regulatory 14-3-3 protein. Plant Cell 11, 2379–2392.

Svitashev, S.J., Young, K., Schwartz, C., Gao, H., Falco, S.C., Cigan, A.M., 2015. Targeted mutagenesis, precise gene editing, and site-specific gene insertion in maize using Cas9 and guide RNA. Plant Physiol. 169, 931–945.

Svoboda, K., Block, S.M., 1994. Force and velocity measured for single kinesin molecules. Cell 77, 773–784.

Svoboda, K., Schmidt, C.F., Schnapp, B., Block, S.M., 1993. Direct observation of kinesin stepping by optical trapping interferometry. Nature 365, 721–727.

Swalla, B.J., Moon, R.T., Jeffrey, W.R., 1985. Developmental significance of a cortical cytoskeletal domain in *Chaetopterus* eggs. Dev. Biol. 111, 434–450.

Swaminathan, R., Hoang, C., Verkman, A., 1997. Photobleaching recovery and anisotropy decay of green fluorescent protein GFP-S65T in solution and cells: cytoplasmic viscosity probed by green fluorescent protein translational and rotational diffusion. Biophys. J. 72, 2015–2022.

Swann, M.M., 1951a. Protoplasmic structure and mitosis. I. J. Exp. Biol. 28, 417–433.

Swann, M.M., 1951b. Protoplasmic structure and mitosis. II. J. Exp. Biol. 28, 434–444.

Swann, M.M., 1952. The spindle. In: Hughes, A. (Ed.), The Mitotic Cycle. Academic Press, New York, pp. 119–133.

Swanson, C.P., 1951. Polyploidy in the pteridophytes. Q. Rev. Biol. 26, 281–282.

Swanson, C.P., 1957. Cytology and Cytogenetics. Prentice-Hall, Englewood Cliffs, NJ.

Swanson, S.J., Choi, W.-G., Chanoca, A., Gilroy, S., 2011. In vivo imaging of Ca^{2+}, pH, and reactive oxygen species using fluorescent probes in plants. Annu. Rev. Plant Biol. 62, 273–297.

Swanson, S.J., Jones, R.L., 1996. Gibberellic acid induces vacuolar acidification in barley aleurone. Plant Cell 8, 2211–2221.

Swanson, C.P., Merz, T., Young, W.J., 1967. Cytogenetics. Prentice-Hall, Englewood Cliffs, NJ.

Swanson, C.P., Webster, P.L., 1985. The Cell. Prentice-Hall, Englewood Cliffs, NJ.

Swatzell, L.J., Edelmann, R.E., Makaroff, C.A., Kiss, J.Z., 1999. Integrin-like proteins are localized to plasma membrane fractions, not plastids, in *Arabidopsis*. Plant Cell Physiol. 40, 173–183.

Swedlow, J.R., Agard, D.A., Dedat, J.W., 1993. Chromosome structure inside the nucleus. Curr. Opin. Cell Biol. 5, 412–416.

Sweeney, B.M., 1944. The effect of auxin on protoplasmic streaming in root hairs of *Avena*. Am. J. Bot. 31, 78–80.

Sweeney, B.M., 1987. Rhythmic Phenomena in Plants, second ed. Academic Press, San Diego.

Sweeney, B.M., Hastings, J.W., 1958. Rhythmic cell division in populations of Gonyaulax polyedra. J. Protozool. 5, 217–234.

Sweeney, B.M., Haxo, F.T., 1961. Persistence of a photosynthetic rhythm in enucleated Acetabularia. Science 134, 1361–1363.

Sweeney, B.M., Thimann, K.V., 1942. The effect of auxins on protoplasmic streaming III. J. Gen. Physiol. 25, 841–851.

Syn, N.L., Wang, L., Chow, E.K.-H., Lim, C.T., Goh, B.-C., 2017. Exosomes in cancer nanomedicine and immunotherapy: prospects and challenges. Trends Biotechnol. 35, 665–676.

Synge, J.L., 1951. Science: Sense and Nonsense. Cape, London.

Synge, J.L., 1970. Talking about Relativity. North-Holland Publishing Co., Amsterdam.

Synge, R.L.M., December 12, 1952. Applications of Partition Chromatography. Nobel Lecture.

Synge, J.L., Griffith, B.A., 1949. Principles of Mechanics, second ed. McGraw-Hill, New York.

Synge, R.L.M., William, E.F., 1990. Albert Charles Chibnall, 28 January 1894—10 January 1988. Biogr. Mems. Fell. R. Soc. 35, 55–96.

Sze, H., 1985. H^+ translocating ATPases: advances using membrane vesicles. Ann. Rev. Plant Physiol. 36, 175–208.

Sze, H., Geisler, M., Murphy, A.S., 2014. Linking the evolution of plant transporters to their functions. Front. Plant Sci. 4, 547.

Sze, H., Hodges, T.K., 1977. Selectivity of alkali cation influx across the plasma membrane of oat roots. Cation specificity of the plasma membrane ATPase. Plant Physiol. 59, 641–646.

Szent-Györgi, A., 1957. Bioenergetics. Academic Press, New York.

Szent-Györgi, A., 1966. In search of simplicity and generalizations (50 years Poaching in Science). In: Kaplan, N.O., Kennedy, E.P. (Eds.), Current Aspects of Biochemical Energetics. Fritz Lipmann Dedicatory Volume. Academic Press, New York, pp. 63–75.

Szent-Györgi, A.G., 1996. Regulation of contraction by calcium binding myosins. Biophys. Chem. 59, 357–363.

Szent-Györgi, A.G., Kalabokis, V.N., Perreault-Micale, C.L., 1999. Regulation by molluscan myosins. Mol. Cell. Biochem. 190, 55–62.

Szent-Györgyi, A., 1935. Über die Bedeutung der Fumarsäure für die tierische Gewebsatmung. Hoppe-Seyler's. Z. Physiol. Chem. 236, 1–20.

Szent-Györgyi, A., 1937. Studies on Biological Oxidation and Some of Its Catalysts. Eggenbergersche Buchhandlung Karl Rényi, Budapest.

Szent-Györgyi, A., 1939a. Biological oxidations and vitamins. Harvey Lect. 34, 265–279.

Szent-Györgyi, A., 1939b. On Oxidation, Fermentation, Vitamins, Health and Disease. Williams & Wilkins, Baltimore, MD.

Szent-Györgyi, A., 1941a. Towards a new biochemistry. Science 93, 609–611.

Szent-Györgyi, A., 1941b. Whatever a cell does. Chron. Bot. 6, 169–171.

Szent-Györgyi, A., 1943. Science needs freedom. World Digest Curr. Fact Comment 55 (50).

Szent-Györgyi, A., 1947. Chemistry of Muscle Contraction. Academic Press, New York.

Szent-Györgyi, A., 1948. Nature of Life: A Study of Muscle. Academic Press, New York.

Szent-Györgyi, A., 1949a. Free-energy relations and contraction of actomyosin. Biol. Bull. 96, 140–161.

Szent-Györgyi, A., 1949b. Muscle research. Sci. Am. 180, 22–25.

Szent-Györgyi, A., 1953a. Chemical Physiology of Contraction in Body and Heart Muscle. Academic Press, New York.

Szent-Györgyi, A.G., 1953b. Meromyosins, the subunits of myosin. Arch. Biochem. Biophys. 42, 305–320.

Szent-Györgyi, A., 1957. Science, ethics, and politics. Science 125, 225–226.

Szent-Györgyi, A., 1960. Introduction to a Submolecular Biology. Academic Press, New York.

Szent-Györgyi, A., 1963. Lost in the twentieth century. Ann. Rev. Biochem. 32, 1–14.

Szent-Györgyi, A., 1963a. Lost in the twentieth century. Ann. Rev. Biochem. 32, 1–14.

Szent-Györgyi, A., 1963b. Science, Ethics and Politics. Vantage Press, New York.

Szent-Györgyi, A., 1964. Teaching and the expanding knowledge. Science 146, 1278–1279.

Szent-Györgyi, A., 1970. The Crazy Ape. Written by a Biologist for the Young. Philosophical Library, New York.

Szent-Györgyi, A., 1971. Biology and pathology of water. Persepct. Biol. Med. Winter 239–249.

Szent-Györgyi, A., 1972. Dionysians and apollonians. Science 176, 966.

Szent-Györgyi, A.G., 2004. The early history of the biochemistry of muscle contraction. J. Gen. Physiol. 123, 631–641.

Szent-Györgyi, A.G., Bagshaw, C.R., 2012. A tribute to Annemarie Weber (1923-2012). J. Muscle Res. Cell Motil. 33, 301–303.

Szent-Györgyi, A., Banga, I., 1941. Adenosinetriphosphatase. Science 93, 158.

Szent-Györgyi, A., December 11, 1937. Oxidation, Energy Transfer, and Vitamins. Nobel Lecture.

Szilard, L., 1964. On the decrease of entropy in a thermodynamic system by the intervention of intelligent beings. Behav. Sci. 9, 301–310.

Szmidt, A.E., Aldén, T., Hällgren, J.-E., 1987. Paternal inheritance of chloroplast DNA in Carix. Plant Mol. Biol. 9, 59–64.

Szostak, J., 2009. Ray Wu, as remembered by a former student. Sci. China Series C Life Sci. 52, 108–110.

Szostak, J.W., Bartel, D.P., Luisi, P.L., 2001. Synthesizing life. Nature 409, 387–390.

Szostak, J.W., Orr-Weaver, T.L., Rothstein, R.J., Stahl, F.W., 1983. The double-strand-break repair model for recombination. Cell 33, 25–35.

Szostak, J., Wu, R., 1979. Insertion of a genetic marker into the ribosomal DNA of yeast. Plasmid 2, 536–554.

Szponarski, W., Sommerer, N., Boyer, J.C., Rossignol, M., Gibart, R., 2004. Large-scale characterization of integral proteins from *Arabidopsis* vacuolar membrane by two-dimensional liquid chromatography. Proteomics 4, 397–406.

Tabata, S., Kaneko, T., Nakamura, Y., Kotani, H., Kato, T., Asamizu, E., Miyajima, N., Sasamoto, S., Kimura, T., Hosouchi, T., Kawashima, K., Kohara, M., Matsumoto, M., Matsuno, A., Muraki, A., Nakayama, S., Nakazaki, N., Naruo, K., Okumura, S., Shinpo, S., Takeuchi, C., Wada, T., Watanabe, A., Yamada, M., Yasuda, M., Sato, S., de la Bastide, M., Huang, E., Spiegel, L., Gnoj, L., O'Shaughnessy, A., Preston, R., Habermann, K., Murray, J., Johnson, D., Rohlfing, T., Nelson, J., Stoneking, T., Pepin, K., Spieth, J., Sekhon, M., Armstrong, J., Becker, M., Belter, E., Cordum, H., Cordes, M., Courtney, L., Courtney, W., Dante, M., Du, H., Edwards, J., Fryman, J., Haakensen, B., Lamar, E., Latreille, P., Leonard, S., Meyer, R., Mulvaney, E., Ozersky, P., Riley, A., Strowmatt, C., Wagner-McPherson, C., Wollam, A., Yoakum, M., Bell, M., Dedhia, N., Parnell, L., Shah, R., Rodriguez, M., See, L.H., Vil, D., Baker, J., Kirchoff, K., Toth, K., King, L., Bahret, A., Miller, B., Marra, M., Martienssen, R., McCombie, W.R., Wilson, R.K., Murphy, G., Bancroft, I., Volckaert, G., Wambutt, R., Düsterhöft, A., Stiekema, W., Pohl, T., Entian, K.D., Terryn, N., Hartley, N., Bent, E., Johnson, S., Langham, S.A., McCullagh, B., Robben, J., Grymonprez, B., Zimmermann, W., Ramsperger, U., Wedler, H., Balke, K., Wedler, E., Peters, S., van Staveren, M., Dirkse, W., Mooijman, P., Lankhorst, R.K., Weitzenegger, T., Bothe, G., Rose, M., Hauf, J., Berneiser, S., Hempel, S., Feldpausch, M., Lamberth, S., Villarroel, R., Gielen, J., Ardiles, W., Bents, O., Lemcke, K., Kolesov, G., Mayer, K., Rudd, S., Schoof, H., Schueller, C., Zaccaria, P., Mewes, H.W., Bevan, M., Fransz, P., Kazusa DNA Research Institute, Cold Spring Harbor, Washington University in St Louis Sequencing Consortium, European Union Arabidopsis Genome Sequencing Consortium, 2000. Sequence and analysis of chromosome 5 of the plant *Arabidopsis thaliana*. Nature 408, 823–826.

Taber, G., 2007. To Cork or not to Cork: Tradition, Romance, Science, and the Battle for the Wine Bottle. Scribner, New York.

Taber, G., 2009. In: Search of Bacchus. Wanderings in the Wonderful World of Wine Tourism. Scribner, New York.

Tabor, S., Richardson, C.C., 1995. A single residue in DNA polymerases of the *Escherichia coli* DNA polymerase I family is critical for distinguishing between deoxy- and dideoxyribonucleotides. Proc. Natl. Acad. Sci. U.S.A. 92, 6339–6343.

Tachikawa, M., Mochizuki, A., 2015. Nonlinearity in cytoplasmic viscosity can generate an essential symmetry breaking in cellular behaviors. J. Theor. Biol. 364, 260–265.

Tadege, M., Wen, J., He, J., Tu, H., Kwak, Y., Eschstruth, A., Cayrel, A., Endre, G., Zhao, P.X., Chabaud, M., Ratet, P., Mysore, K., 2008. Large scale insertional mutagenesis using Tnt1 retrotransposon in the model legume *Medicago truncatula*. Plant J. 54, 335–347.

Tafolla-Arellano, J.C., Zheng, Y., Sun, H., Jiao, C., Ruiz-May, E., Hernández-Oñate, M.A., González-León, A., Báez-Sañudo, R., Fei, Z., Domozych, D., Rose, J.K.C., Tiznado-Hernández, M.E., 2017. Transcriptome analysis of mango (*Mangifera indica* L.) fruit epidermal peel to identify putative cuticle-associated genes. Sci. Rep. 7, 46163.

Taguchi, T., Uraguchi, A., Katsumi, M., 1999. Auxin- and acid-induced changes in the mechanical properties of the cell wall. Plant Cell Physiol. 40, 743–749.

Tague, B.W., Chrispeels, M.J., 1987. The plant vacuolar protein, phytohemagglutinin, is transported to the vacuole of transgenic yeast. J. Cell Biol. 105, 1971–1979.

Tait, P.G., Steele, W.J., 1878. A Treatise on Dynamics of a Particle, fourth ed. Macmillan, London.

Taiz, L., 1984. Plant cell expansion: regulation of cell wall mechanical properties. Ann. Rev. Plant Physiol. 35, 585–657.

Taiz, L., 1992. The plant vacuole. J. Exp. Biol. 172, 113–122.

Taiz, L., Jones, R.L., 1973. Plasmodesmata and an associated cell wall component in barley aleurone tissue. Am. J. Bot. 60, 67–75.

Taiz, L., Struve, I., Rausch, T., Bernasconi, P., Gogarten, J.P., Kibak, H., Taiz, S.L., 1990. The vacuolar ATPase: structure, evolution, and promoter analysis. In: Leonard, R.T., Hepler, P.K. (Eds.), Calcium in Plant Growth and Development. The American Society of Plant Physiologists Symposium Series, vol. 4. American Society of Plant Physiologists, Rockville, Maryland, pp. 55–59.

Taiz, S.L., Taiz, L., 1991. Ultrastructural comparison of the vacuolar and mitochondrial H^+-ATPases of Daucus carota. Bot. Acta 104, 117–121.

Taiz, L., Zeiger, E., 2006. Plant Physiology, fourth ed. Sinauer, Sunderland, MA.

Takacs, E.M., Li, J., Du, C., Ponnala, L., Janick-Buckner, D., Yu, J., Muehlbauer, G.J., Schnable, P.S., Timmermans, M.C.P., Sun, Q., Nettleton, D., Scanlon, M.J., 2012. Ontogeny of the maize shoot apical meristem. Plant Cell 24, 3219–3234.

Takagi, S., 1997. Photoregulation of cytoplasmic streaming: cell biological dissection of signal transduction pathway. J. Plant Res. 110, 299–303.

Takagi, S., 2003. Actin-based photo-orientation movement of chloroplasts in plant cells. J. Exp. Biol. 206, 1963–1969.

Takagi, S., Kamitsubo, E., Nagai, R., 1989. Light action on chloroplast behavior under centrifugal acceleration. In: Tazawa, M., Katsumi, M., Masuda, Y., Okamoto, H. (Eds.), Plant Water Relations and Growth under Stress. Yamada Science Foundation, Tokyo, Japan, pp. 455–457.

Takagi, S., Kamitsubo, E., Nagai, R., 1991. Light-induced changes in the behavior of chloroplasts under centrifugation in Vallisneria epidermal cells. J. Plant Physiol. 138, 257–262.

Takagi, S., Kamitsubo, E., Nagai, R., 1992. Visualization of a rapid, red/far-red light-dependent reaction by centrifuge microscopy. Protoplasma 168, 153–158.

Takagi, S., Kong, S.-G., Mineyuki, Y., Furuya, M., 2003. Regulation of actin-dependent cytoplasmic motility by type II phytochrome occurs within seconds in *Vallisneria* epidermal cells. Plant Cell 15, 331–345.

Takagi, S., Nagai, R., 1985. Light-controlled cytoplasmic streaming in Vallisneria mesophyll cells. Plant Cell Physiol. 26, 941–951.

Takagi, S., Nagai, R., 1988. Light-affected Ca^{2+} fluxes in protoplasts form Vallisneria mesophyll cells. Plant Physiol. 88, 228–232.

Takagi, M., Ogata, K., 1968. Direct evidence for albumin biosynthesis by membrane bound polysomes in rat liver. Biochem. Biophys. Res. Commun. 33, 55–60.

Takahashi, F., 2016. Blue-light-regulated transcription factor, Aureochrome, in photosynthetic stramenopiles. J. Plant Res. 129, 189–197.

Takahashi, S., Bauwe, H., Badger, M., 2007b. Impairment of the photorespiratory pathway accelerates photoinhibition of photosystem II by suppression of repair but not acceleration of damage processes in *Arabidopsis*. Plant Physiol. 144, 487–494.

Takahashi, F., Hishinuma, T., Kataoka, H., 2001. Blue light-induced branching in Vaucheria. Requirement of nuclear accumulation in the irradiated region. Plant Cell Physiol. 42, 274–285.

Takahashi, S., Takahashi, S., Bauwe, H., Badger, M., 2007. Impairment of the photorespiratory pathway accelerates photoinhibition of photosystem II by suppression of repair but not acceleration of damage processes in *Arabidopsis*. Plant Physiol. 144, 487–494.

Takahashi, F., Yamagata, D., Ishikawa, M., Fukamatsu, Y., Ogura, Y., Kasahara, M., Kiyosue, T., Kikuyama, M., Wada, M., Kataoka, H., 2007a. Aureochrome, a photoreceptor required for photomorphogenesis in stramenopiles. Proc. Natl. Acad. Sci. U.S.A. 104, 19625–19630.

Takano, J., Tanaka, M., Toyoda, A., Miwa, K., Kasai, K., Fuji, K., Onouchi, H., Naito, S., Fujiwara, T., 2010. Polar localization and degradation of *Arabidopsis* boron transporters through distinct trafficking pathways. Proc. Natl. Acad. Sci. U.S.A. 102, 12276–12281.

Takano, M., Wayne, R., 2000. ATP is synthesized by the alternative pathway in Chara corallina. Plant Physiol. Abstr. #751.

Takatsuka, C., Inoue, Y., Higuchi, T., Hillmer, S., Robinson, D.G., Moriyasu, Y., 2011. Autophagy in tobacco BY-2 cells cultured under sucrose starvation conditions. Isolation of the autolysosome and its characterization. Plant Cell Physiol. 52, 2074–2087.

Takeda, K., Shibaoka, H., 1981a. Changes in microfibril arrangement on the linear surface of the epidermal cell walls in the epicotyl of Vigna angularis Ohwi et Ohashi during cell growth. Planta 151, 385–392.

Takeda, K., Shibaoka, H., 1981b. Effects of gibberellin and colchicine on microfibril arrangement in epidermal cell walls of *Vigna angularis* Ohwi et Ohashi epicotyls. Planta 151, 393–398.

Takemiya, A., Kinoshita, T., Asanuma, M., Shimazaki, K., 2006. Protein phosphatase 1 positively regulates stomatal opening in response to blue light in *Vicia faba*. Proc. Natl. Acad. Sci. U.S.A. 103, 13549–13554.

Takemiya, A., Shimazaki, K., 2010. Phosphatidic acid inhibits blue lightinduced stomatal opening via inhibition of protein phosphatase 1 [corrected]. Plant Physiol. 153, 1555–1562.

Takemiya, A., Shimazaki, K., 2016. *Arabidopsis* phot1 and phot2 phosphorylate BLUS1 kinase with different efficiencies in stomatal opening. J. Plant Res. 129, 167–174.

Takemiya, A., Sugiyama, N., Fujimoto, H., Tsutsumi, T., Yamauchi, S., Hiyama, A., Tada, Y., Christie, J.M., Shimazaki, K., 2013a. Phosphorylation of BLUS1 kinase by phototropins is a primary step in stomatal opening. Nat. Commun. 4, 2094.

Takemiya, A., Yamauchi, S., Yano, T., Ariyoshi, C., Shimazaki, K., 2013b. Identification of a regulatory subunit of protein phosphatase 1 which mediates blue light signaling for stomatal opening. Plant Cell Physiol. 54, 24–35.

Takeshige, K., Baba, M., Tsuboi, S., Noda, T., Ohsumi, Y., 1992. Autophagy in yeast demonstrated with proteinase-deficient mutants and conditions for its induction. J. Cell Biol. 119, 301–311.

Takeshige, K., Shimmen, T., Tazawa, M., 1986. Quantitative analysis of the ATP-dependent H^+ efflux and pump current driven by an electrogenic pump in Nitellopsis obtusa. Plant Cell Physiol. 27, 337–348.

Takeshige, K., Tazawa, M., 1989a. Determination of the inorganic pyrophosphate level and its subcellular localization in *Chara corallina*. J. Biol. Chem. 264, 3262–3266.

Takeshige, K., Tazawa, M., 1989b. Measurement of the cytoplasmic and vacuolar buffer capacities in *Chara corallina*. Plant Physiol. 89, 1049–1052.

Takeshige, K., Tazawa, M., Hager, A., 1988. Characterization of the H^+ translocating adenosine triphosphatase and pyrophosphatase of vacuolar membranes isolated by means of a perfusion technique from *Chara corallina*. Plant Physiol. 86, 1168–1173.

Takeyama, N., Kiyono, H., Yuki, Y., 2015. Plant-based vaccines for animals and humans: recent advances in technology and clinical trials. Ther. Adv. Vaccines 3, 139–154.

Talbot, M.J., Wasteneys, G.O., Offler, C.E., McCurdy, D.W., 2007. Cellulose synthesis is required for deposition of reticulate wall ingrowths in transfer cells. Plant Cell Physiol. 48, 147–158.

Tallman, G., 1992. The chemiosmotic model of stomatal opening revisited. Crit. Rev. Plant Sci. 11, 35–57.

Talmadge, K.W., Keegstra, K., Bauer, W.D., Albersheim, P., 1973. The structure of plant cell walls. I. The macromolecular components of the walls of suspension cultured sycamore cells with a detailed analysis of the pectic polysaccharides. Plant Physiol. 51, 158–173.

Tamura, K., Fukao, Y., Iwamoto, M., Haraguchi, T., Hara-Nishimura, I., 2010. Identification and characterization of nuclear pore complex components in *Arabidopsis thaliana*. Plant Cell 22, 4084–4097.

Tamura, K., Goto, C., Hara-Nishimura, I., 2015. Recent advances in understanding plant nuclear envelope proteins involved in nuclear morphology. J. Exp. Bot. 66, 1641–1647.

Tamura, K., Hara-Nishimura, I., 2013. The molecular architecture of the plant nuclear pore complex. J. Exp. Biol. 64, 823–832.

Tamura, K., Hara-Nishimura, I., 2014. Functional insights of nucleocytoplasmic transport in plants. Front. Plant Sci. 5, 118.

Tan, B.C., Lim, Y.S., Lau, S.-E., 2017. Proteomics in commercial crops: an overview. J. Proteom. 169, 176–188.

Tanaka, K., 1987. Eukaryotes: scanning electron microscopy of intracellular structures. Int. Rev. Cytol. (Suppl. 17), 89–147.

Tanaka, K., December 8, 2002. The Origin of Macromolecule Ionization by Laser Irradiation. Nobel Lecture.

Tanaka, K., Murata, K., Yamazaki, M., Onosato, K., Miyao, A., Hirochika, H., 2003. Three distinct rice cellulose synthase catalytic subunit genes required for cellulose synthesis in the secondary wall. Plant Physiol. 133, 73–83.

Tanaka, M., Murata-Mori, M., Kadono, T., Yamada, T., Kawano, T., Kosaka, T., Hosoya, H., 2002. Complete elimination of endosymbiotic algae from Paramecium bursaria and its confirmation by diagnostic PCR. Acto Protozool. 41, 255–261.

Tanaka, K., Waki, H., Ido, Y., Akita, A., Yoshida, Y., Yoshida, T., 1988. Protein and polymer analysis up to m/z 100,000 by laser ionization time-of-flight mass spectrometry. Rapid Commun. Mass Spectrom. 2, 151–153.

Tanaka, T., Yamada, M., 1979. A phosphatidyl choline exchange protein isolated from germinated castor bean endosperms. Plant Cell Physiol. 20, 533–542.

Tanchak, M.A., Fowke, L.C., 1987. The morphology of multivesicular bodies in soybean protoplasts and their role in endocytosis. Protoplasma 138, 173–182.

Tanchak, M.A., Griffing, L.R., Mersey, B.G., Fowke, L.C., 1984. Endocytosis of cationized ferritin by coated vesicles of soybean protoplasts. Planta 162, 481–486.

Tanchak, M.A., Rennie, P.J., Fowke, L.C., 1988. Ultrastructure of the partially coated reticulum and dictyosomes during endocytosis by soybean protoplasts. Planta 175, 433–441.

Tanford, C., 1989. Ben Franklin Stilled the Waves. Duke University Press, Durham, NC.

Tang, W., Deng, Z., Oses-Prieto, J.A., Suzuki, N., Zhu, S., Zhang, X., Burlingame, A.L., Wang, Z.-Y., 2008. Proteomics studies of brassinosteroid signal transduction using prefractionation and two-dimensional DIGE. Mol. Cell. Proteom. 7, 728–738.

Tang, X.C., Han, Y.F., 1999. Pharmacological profile fo Huperzine A, a novel acetylcholinesterase inhibitor from Chinese herb. CNS Drug Rev. 5, 281–300.

Tang, S., Mikala, G., Bahinski, A., Yatani, A., Varadi, G., Schwartz, A., 1993. Molecular localization of ion selectivity sites within the pore of an human L-type cardiac Ca^{2+} channel. J. Biol. Chem. 268, 13026–13029.

Tang, S., Wong, H.-C., Wang, Z.-M., Huang, Y., Zou, J., Zhou, Y., Pennati, A., Gadda, G., Delbono, O., Yamg, J.J., 2011. Design and application of a class of sensors to monitor Ca^{2+} dynamics in high Ca^{2+} concentration cellular compartments. Proc. Natl. Acad. Sci. U.S.A. 108, 16265–16270.

Tang, H., Zhang, X., Miao, C., Zhang, J., Ming, R., Schnable, J.C., Schnable, P.S., Lyons, E., Lu, J., 2015. ALLMAPS: Robust scaffold ordering based on multiple maps. Genome Biol. 16, 3.

Tangl, E., 1879. Ueber oftene Communicationen zwichen den Zellen des Endosperms einiger Samen. Jahr. Wiss. Bot. 12, 170–190.

Tani, T., Shribak, M., Oldenbourg, R., 2016. Living cells and dynamic molecules observed with the polarized light microscope: the legacy of Shinya Inoué. Biol. Bull. 231, 85–95.

Tanigawa, G., Orci, L., Amherdt, M., Ravazzola, M., Helms, J.B., Rothman, J.E., 1993. Hydrolysis of bound GTP by ARF protein triggers uncoating of Golgi-derived COP-coated vesicles. J. Cell Biol. 123, 1365–1371.

Tanksley, S.D., 1993. Mapping polygenes. Annu. Rev. Genet. 27, 205–233.

Tanksley, S.D., Ganal, M.W., Prince, J.P., de Vicente, T.M., Giovannoni, J.J., Grandillo, S., Martin, G.B., Messeguer, R., Miller, J.C., Miller, L., Paterson, A.H., Pineda, O., Röder, M.S., Wing, R.A., Wu, W., Young, N.D., 1992. High density molecular linkage maps of the tomato and potato genomes. Genetics 132, 1141–1160.

Tanksley, S.D., Young, N.D., Paterson, A.H., Bonierbale, M.W., 1989. RFLP mapping in plant breeding: new tools for an old science. Nat. Biotechnol. 7, 257–264.

Tantama, M., Martínez-François, J.R., Mongeon, R., Yellen, G., 2013. Imaging energy status in live cells with a fluorescent biosensor of the intracellular ATP-to-ADP ratio. Nat. Commun. 4, 2550.

Tao, K.L.J., Jagendorf, A.T., 1973. The ratio of free to membrane- bound chloroplast ribosomes. Biochim. Biophys. Acta 324, 518–532.

Tapley, E.C., Starr, D.A., 2013. Connecting the nucleus to the cytoskeleton by SUN-KASH bridges across the nuclear envelope. Curr. Opin. Cell Biol. 25, 57–62.

Tariq, M., Habu, Y., Paszkowski, J., 2002. Depletion of MOM1 in non-dividing cells of *Arabidopsis* plants releases transcriptional gene silencing. EMBO Rep. 3, 951–955.

Tariq, M., Paszkowski, J., 2004. DNA and histone methylation in plants. Trends Genet. 20, 244–251.

Tasaka, M., Kato, T., Fukaki, H., 2001. Genetic regulation of gravitropism in higher plants. Int. Rev. Cytol. 206, 135–154.

Tashiro, S., 1917. A Chemical Sign of Life. Univeristy of Chicago Press, Chicago.

Tate, B.F., Schaller, G.E., Sussman, M.R., Crain, R.C., 1989. Characterization of polyphosphoinositide phospholipase C from the plasma membrane of *Avena sativa*. Plant Physiol. 91, 1275–1279.

Tate, J.A., Soltis, D.E., Soltis, P.S., 2005. Polyploidy in plants. In: Gregory, T.R. (Ed.), The Evolution of the Genome. Elsevier, Amsterdam, pp. 371–426.

Tateno, Y., Nei, M., Tajima, F., 1982. Accuracy of estimated phylogenetic trees from molecular data. J. Mol. Evol. 18, 387–404.

Tatout, C., Evans, D.E., Vanrobays, E., Probst, A.V., Graumann, K., 2014. The plant LINC complex at the nuclear envelope. Chromosome Res. 22, 241–252.

Tatsumi, K., Nishimura, O., Itomi, K., Tanegashima, C., Kuraku, S., 2015. Optimization and cost-saving in tagmentation-based mate-pair library preparation and sequencing. BioTechniques 58, 253–257.

Tatum, E.L., Beadle, G.W., 1942. Genetic control of biochemical reactions in Neurospora: an "Aminobenzoicless" mutant. Proc. Natl. Acad. Sci. U.S.A. 28, 234–243.

Tatum, E.L., December 11, 1958. A Case History in Biological Research. Nobel Lecture.

Taylor, F.S., 1953. The idea of the quintessence. In: Underwood, E.A. (Ed.), Science, Medicine and History, vol. 1. Oxford University Press, Oxford, pp. 247–265.

Taylor, F.J.R., 1987. An overview of the status of evolutionary cell symbiosis theories. "Max" Taylor Ann. N.Y. Acad. Sci. 503, 1–16.

Taylor, J.B., 2008a. My Stroke of Insight. http://www.ted.com/index.php/talks/jill_bolte_taylor_s_powerful_stroke_of_ insight.html.

Taylor, N.G., 2008b. Cellulose biosynthesis and deposition in higher plants. New Phytol. 178, 239–252.

Taylor, N.G., 2011. A role for *Arabidopsis* dynamin related proteins DRP2A/B in endocytosis; DRP2 function is essential for plant growth. Plant Mol. Biol. 76, 117–129.

Taylor Jr., J.H., December 8, 1993. Binary Pulsars and Relativistic Gravity. Nobel Lecture.

Taylor, H.S., Lawrence, E.O., Langmuir, I., 1942. Molecular Films The Cyclotron & The New Biology. Rutgers University Press, New Brunswick, NJ.

Taylor, C.E., Marshall, I.C.B., 1992. Calcium and inositol 1,4,5- triphosphate receptors: a complex relationship. TIBS 17, 403–407.

Taylor, N.L., Millar, A.H., 2015. Plant mitochondrial proteomics. Methods Mol. Biol. 1305, 83–106.

Taylor, N.L., Stroher, E., Millar, A.H., 2014. *Arabidopsis* organelle isolation and characterization. Methods Mol. Biol. 1062, 551–572.

Taylor, J.H., Woods, P.S., Hughes, W.L., 1957. The organization and duplication of chromosomes as revealed by autoradiographic studies using tritium-labeled thymidine. Proc. Natl. Acad. Sci. U.S.A. 43, 122–129.

Tazawa, M., 1957. Neue Methode zur Messung des osmotischen Wertes einer Zelle. Protoplasma 48, 342–359.

Tazawa, M., 1968. Motive force of the cytoplasmic streaming in *Nitella*. Protoplasma 65, 207–222.

Tazawa, M., 2003. Cell physiological aspects of the plasma membrane H^+ pump. J. Plant Res. 116, 419–442.

Tazawa, M., 2011. Sixty years research with characean cells: fascinating material for plant cell biology. Prog. Bot. 72, 5–36.

Tazawa, M., Kishimoto, U., 1968. Cessation of cytoplasmic streaming of *Chara* internodes during action potential. Plant Cell Physiol. 9, 361–368.

Tazawa, M., Okazaki, Y., 1997. Water channel does not limit evaporation of water from plant cells. J. Plant Res. 110, 317–320.

Tazawa, M., Sutou, E., Shibasaka, M., 2001. Onion root water transport sensitive to water channel and K^+ channel inhibitors. Plant Cell Physiol. 42, 28–36.

Tazawa, M., Wayne, R., 1989. Characteristics of hydraulic conductivity of the plasmalemma in characean cells. In: Plant Water Relations and Growth Under Stress. Proc. Yamada Conference XII. Osaka, Japan, pp. 237–244.

Teh, B.T., Lim, K., Yong, C.H., Ng, C.C.Y., Rao, S.R., Rajasegaran, V., Lim, W.K., Ong, C.K., Chan, K., Cheng, V.K.Y., Soh, P.S., Swarup, S., Rozen, S.G., Nagarajan, N., Tan, P., 2017. The draft genome of tropical fruit durian (*Durio zibethinus*). Nat. Genet. 49, 1633–1641.

Tehei, M., Franzetti, B., Wood, K., Gabel, F., Fabiani, E., Jasnin, M., Zamponi, M., Oesterhelt, D., Zaccai, G., Ginzburg, M., Ginzburg, B., 2007. Neutron scattering reveals extremely slow cell water in a Dead Sea organism. Proc. Natl. Acad. Sci. U.S.A. 104, 766–771.

Teichert, I., Wolff, G., Kück, U., Nowrousian, M., 2012. Combining laser microdissection and RNA-seq to chart the transcriptional landscape of fungal development. BMC Genom. 13, 511.

Teilhard de Chardin, P., 1966. Man's Place in Nature. The Human Zoological Group. Translated by R. Hague. Collins, London.

Tel-Zur, N.Y., Mizrahi, A., Cisneros, A., Mouyal, J., Schneider, B., Doyle, J.J., 2011. Phenotypic and genomic characterization of vine cacti collection (Cactaceae). Genet. Resour. Crop Evol. 7, 1075–1085.

Telling, G.C., Parchi, P., DeArmond, S.J., Cortelli, P., Montagna, P., Gabizon, R., Mastrianni, J., Lugaresi, E., Gambetti, P., Prusiner, S.B., 1996. Evidence for the conformation of the pathologic isoform of the prion protein enciphering and propagating prion diversity. Science 274, 2079–2082.

Telzer, B.R., Haimo, L.T., 1981. Decoration of spindle microtubules with dynein: evidence for uniform polarity. J. Cell Biol. 89, 373–378.

Telzer, B.R., Moses, M.J., Rosenbaum, J.L., 1975. Assembly of microtubles onto kinetochores of isolated mitotic chromosomes of HeLa cells. Proc. Natl. Acad. Sci. U.S.A. 72, 4023–4027.

Temin, H.M., December 12, 1975. The DNA Provirus Hypothesis. Nobel Lecture.

Temin, H.M., Mizutani, S., 1970. Viral RNA-dependent DNA polymerase: RNA-dependent DNA polymerase in virions of Rous sarcoma virus. Nature 226, 211–213.

Tepfer, M., Taylor, I.E.P., 1981. The permeability of plant cell walls as measured by gel filtration chromatography. Science 213, 761–763.

Terada, Y., Tatsuka, M., Suzuki, F., Yasuda, Y., Fujita, S., Otsu, M., 1998. AIM-1: a mammalian midbody-associated protein required for cytokinesis. EMBO J. 17, 667–676.

Terasaki, M., 1989. Fluorescent labeling of the endoplasmic reticulum. Methods Cell Biol. 29, 126–135.

Terasaki, M., Chen, L.B., Fujiwara, K., 1986. Microtubules and the endoplasmic reticulum are highly interdependent structures. J. Cell Biol. 103, 1557–1568.

Terasaki, M., Song, J., Wong, J.R., Weiss, M.J., Chen, L.B., 1984. Localization of endoplasmic reticulum in living and glutaraldehyde-fixed cells with fluorescent dyes. Cell 38, 101–108.

Terashima, I., Ono, K., 2002. Effects of $HgCl_2$ on CO_2 dependence of leaf photosynthesis: evidence indicating involvement of aquaporins in CO_2 diffusion across the plasma membrane. Plant Cell Physiol. 43, 70–78.

Terauchi, M., Nagasato, C., Motomura, T., 2015. Plasmodesmata of brown algae. J. Plant Res. 128, 7–15.

Terris, M. (Ed.), 1964. Joseph Goldberger on Pellagra. Louisiana State University Press, Baton Rouge, LA.

Terry, B.R., Robards, A.W., 1987. Hydrodynamic radius alone governs the mobility of molecules through plasmodesmata. Planta 171, 145–157.

Testori, A., Skinner, J.D., Murray, A.W., Burgoyne, L.A., 1991. Phosphorylation and chromatin mechanics. The central importance of substrate conformation in determining the patterns of HL-60 nuclear phosphorylation. Biochem. Biophys. Res. Commun. 180, 329–333.

Tewari, K.K., Wildman, S.G., 1967. DNA polymerase in isolated tobacco chloroplasts and nature of the polymerized product. Proc. Natl. Acad. Sci. U.S.A. 58, 689–696.

Tezuka, K., Hyayashi, M., Ishihara, H., Akazawa, T., Takahashi, N., 1992. Studies on synthetic pathway of xylose-containing N-linked oligosaccharides deduced from substrate specificities of the processing enzymes in sycamore cells (*Acer pseudoplantanos* L.). Eur. J. Biochem. 203, 401–413.

Thakare, D., Yang, R., Steffen, J.G., Zhan, J., Wang, D., Clark, R.M., Wang, X., Yadegari, R., 2014. RNA-Seq analysis of laser-capture microdissected cells of the developing central starchy endosperm of maize. Genom. Data 2, 242–245.

Thayer, W.S., Hinkle, P.C., 1975a. Kinetics of adenosine triphosphate synthesis in bovine heart submitochondrial particles. J. Biol. Chem. 250, 5336–5342.

Thayer, W.S., Hinkle, P.C., 1975b. Synthesis of adenosine triphosphate by an artificially imposed electrochemical proton gradient in bovine heart submitochondrial particles. J. Biol. Chem. 250, 5330–5335.

The 1001 Genomes Consortium, 2016. 1,135 genomes reveal the global pattern of polymorphism in *Arabidopsis thaliana*. Cell 166, 481–491.

The Arabidopsis Genome Initiative, 2000. Analysis of the genome sequence of the flowering plant *Arabidopsis thaliana*. Nature 408, 796–815.

The Gene Ontology Consortium, 2000. Gene ontology: tool for the unification of biology. Nat. Genet. 25, 25–29.

The Gene Ontology Consortium, 2008. The gene ontology project in 2008. Nucleic Acids Res. 36, D440–D444.

The Gene Ontology Consortium, 2013. Gene ontology annotations and resources. Nucleic Acids Res. 41, D530–D535.

The Gene Ontology Consortium, 2015. Gene ontology consortium: going forward. Nucleic Acids Res. 43, D1049–D1056.

The International Barley Genome Sequencing Consortium, 2012. A physical, genetic and functional sequence assembly of the barley genome. Nature 491, 711–716.

The International Brachypodium Initiative, 2010. Genome sequence analysis of the model grass *Brachypodium distachyon*: insights into grass genome evolution. Nature 463, 763–768.

The Tomato Genome Consortium, 2012. The tomato genome sequence provides insights into fleshy fruit evolution. Nature 485, 635–641.

Theodoulou, F.L., Job, K., Slocombe, S.P., Footitt, S., Holdsworth, M., Baker, A., Larson, T.R., Graham, I.A., 2005. Jasmonic acid levels are reduced in COMATOSE ATP-binding cassette transporter mutants. Implications for transport of jasmonate precursors into peroxisomes. Plant Physiol. 137, 835–840.

Theologis, A., Ecker, J.R., Palm, C.J., Federspiel, N.A., Kaul, S., et al., 2000. Sequence and analysis of chromosome 1 of the plant *Arabidopsis thaliana*. Nature 408, 816–820.

Theophrastus, 1916. Enquiry into Plants. Translated by Arthur Hort. William Heinemann, London.

Theorell, H., 1936. Keilen's cytochrome c and the respiratory mechanism of Warburg and Christian. Nature 138, 687.

Theorell, H., 1959. Award Presentation Speech. http://www.nobelprize.org/nobel_prizes/medicine/laureates/1959/press.html.

Theorell, H., December 12, 1955. The nature and mode of action of oxidation enzymes. Nobel Lecture.

Theriot, J.A., Mitchison, T.J., Tilney, L.G., Portnoy, D.A., 1992. The rate of actin-based motility of intracellular *Listeria monocytogenes* equals the rate of actin polymerization. Nature 357, 257–260.

Theurkauf, W.E., 1994. Premature microtubule-dependent cytoplasmic streaming in cappuccino and spire mutant oocytes. Science 265, 2093–2096.

Thieffry, D., Sarkar, S., 1998. Forty years under the central dogma. Trends Biochem. Sci. 23, 312–316.

Thiel, R., 1957. And There Was Light. The Discovery of the Universe. Alfred A. Knopf, New York.

Thiel, G., Battey, N., 1998. Exocytosis in plants. Plant Mol. Bio. 38, 111–125.

Thiel, G., Blatt, M.R., 1994. Phosphatase antagonist okadaic acid inhibits steady-state K^+ currents in guard cells of *Vicia faba*. Plant J. 5, 727–733.

Thimann, K.V., 1950. Autumn colors. Sci. Am. 83 (4), 40–43.

Thimann, K.V., 1977. Hormone Action in the Whole Life of Plants. University of Massachusetts Press, Amherst.

Thimann, K.V., Reese, K., Nachmias, V.T., 1992. Actin and the elongation of plant cells. Protoplasma 171, 153–166.

Thoen, M.P.M., Davila Olivas, N.H., Kloth, K.J., Coolen, S., Huang, P.-P., Aarts, M.G.M., Bac-Molenaar, J.A., Bakker, J., Bouwmeester, H.J., Broekgaarden, C., Bucher, J., Busscher-Lange, J., Cheng, X., Fradin, E.F., Jongsma, M.A., Julkowska, M.M., Keurentjes, J.J.B., Ligterink, W., Pieterse, C.M.J., Ruyter-Spira, C., Smant, G.,

Testerink, C., Usadel, B., van Loon, J.J.A., van Pelt, J.A., van Schaik, C.C., van Wees, S.C.M., Visser, R.G.F., Voorrips, R., Vosman, B., Vreugdenhil, D., Warmerdam, S., Wiegers, G.L., van Heerwaarden, J., Kruijer, W., van Eeuwijk, F.A., Dicke, M., 2017. Genetic architecture of plant stress resistance: multi-trait genome-wide association mapping. New Phytol. 213, 1346–1362.

Thomas, J., 1939. L'acide citrique dans la respiation musculaire. Enzymologia 7, 231–238.

Thomas, L., 1974. The Lives of a Cell. Notes of a Biology Watcher. Viking Press, New York.

Thomas, R., 1992. Molecular genetics under an embryologist's microscope: Jean Brachet, 1909-1988. Genetics 131, 515–518.

Thomas, T.D., 2010. GIGA: a simple, efficient algorithm for gene tree inference in the genomic age. BMC Bioinform. 11, 312.

Thomas, C.L., Bayer, E.M., Ritzanthaler, C., Fernandez-Calvino, L., Maule, A.J., 2008. Specific targeting of a plasmodesmal protein affecting cell-to-cell-communication. PLoS Biol. 6, 180–189.

Thomas, E., Davey, M.R., 1975. From Single Cells to Plants. Wykeham Publications, Ltd., London.

Thompson, D.W., 1959. On Growth and Form, second ed., vols. I and II. Cambridge University Press, Cambridge.

Thompson, W.F., Beven, A.F., Wells, B., Shaw, P.J., 1997. Sites of rRNA transcription are widely dispersed though the nucleolus in Pisum sativum and can comprise single genes. Plant J. 12, 571–582.

Thompson, A.E., Crants, J., Schnable, P.S., Yu, J., Timmermans, M.C.P., Springer, N.M., Scanlon, M.J., Muehlbauer, G.J., 2014. Genetic control of maize shoot apical meristem architecture. G3 4, 1327–1337.

Thompson, J.E., Froese, C.D., Madey, E., Smith, M.D., Yuwen, H., 1998. Lipid metabolism during plant senescence. Prog. Lipid Res. 37, 119–141.

Thompson, M.V., Wolniak, M., 2008. A plasma membrane- anchored fluorescent protein fusion illuminates sieve element plasma membranes in *Arabidopsis* and Tobacco. Plant Physiol. 146, 1599–1610.

Thomson, W., 1852. On a universal tendency in nature to the dissipation of mechanical energy, 1976. In: Kestin, J. (Ed.), The Second Law of Thermodynamics. Dowden, Hutchinson & Ross, Inc., Stroudsburg, PA, pp. 194–197.

Thomson, W., 1976. On a universal tendency in nature to the dissipation of mechanical energy, 1952. In: Kestin, J. (Ed.), The Second Law of Thermodynamics. Dowden, Hutchinson & Ross, Inc., Stroudsburg, PA, pp. 194–197.

Thomson, W.W., Lewis, L.N., Coggins, C.W., 1967. The reversion of chromoplasts to chloroplasts in Valencia oranges. Cytologia 32, 117–124.

Thomson, W., Tait, P.G., 1862. Energy. Good Words (Oct.) 601–607.

Thoreau, H.D., 1849. On the Duty of Civil Disobedience. www.constitution.org/civ/civildis.htm.

Thoreau, H.D., 1854. Walden; or, Life in the Woods. Tichnor and Fields, Boston.

Thoreau, H.D., 1947. Civil disobedience. In: Cummins, S., Linscott, R.N. (Eds.), The World's Greatest Thinkers. Man and the State. The Political Philosophers. Random House, New York, pp. 297–320.

Thorneycroft, A., Sherson, S.M., Smith, S.M., 2001. Using gene knockouts to investigate plant metabolism. J. Exp. Bot. 52, 1593–1601.

Thorsness, M.K., Kandasamy, M.K., Nasrallah, M.E., Nasrallah, J.B., 1991. A *Brassica* S-locus gene promoter targets toxic gene expression and cell death to the pistil and pollen of transgenic *Nicotiana*. Dev. Biol. 143, 173–184.

Thorsness, M.K., Kandasamy, M.K., Nasrallah, M.E., Nasrallah, J.B., 1993. Genetic ablation of floral cells in *Arabidopsis*. Plant Cell 5, 253–261.

Thoyts, P.J.E., Millichip, M.L., Stobart, A.K., Griffiths, W.T., Shewry, P.R., Napier, J.A., 1995. Expression and in vitro targeting of a sunflower oleosin. Plant Mol. Biol. 29, 403–410.

Thuleau, P., Graziana, A., Canut, H., Ranjeva, R., 1990. A 75- kDa polypeptide, located primarily at the plasma membrane of carrot cell-suspension cultures, is photoaffinity labeled by the calcium channel blocker LU 49888. Proc. Natl. Acad. Sci. U.S.A. 87, 10000–10004.

Tiainen, P., Myllyharju, J., Koivunen, P., 2005. Characterization of a second *Arabidopsis thaliana* proyl 4-hydroxylase with distinct substrate specificity. J. Biol. Chem. 280, 1142–1148.

Tian, G.-W., Chen, M.-H., Zaltsman, A., Citovsky, V., 2006. Pollen-specific pectin methylase involved in pollen tube growth. Dev. Biol. 294, 83–91.

Tian, G.-W., Mohanty, A., Chary, S.N., Li, S., Paap, B., Drakakaki, G., Kopec, C.D., Li, J., Ehrhardt, D., Jackson, D., Rhee, S.Y., Raikhel, N.V., Citovsky, V., 2004. High-throughput fluorescent tagging of full-length *Arabidopsis* gene products in planta. Plant Physiol. 135, 25–38.

Tibbs, J., 1957. The nature of algal and related flagella. Biochim. Biophys. Acta 23, 275–288.

Tice, M.M., Lowe, D.R., 2004. Photosynthetic microbial mats in the 3,416-Mya-old ocean. Nature 431, 549–552.

Tiezzi, A., Moscatelli, A., Cai, G., Bartalesi, A., Cresti, M., 1992. An immunoreactive homolog of mammalian kinesin in *Nicotiana tabacum* pollen tubes. Cell Motil. Cytoskelet. 21, 132–137.

Tikkanen, M., Aro, E.-M., 2011. Thylakoid protein phosphorylation in dynamic regulation of photosystem II in higher plants. Biochim. Biophys. Acta 1817, 232–238.

Till, A., Lakhani, R., Burnett, S.F., Subramani, S., 2012. Pexophagy: the selective degradation of peroxisomes. Int. J. Cell Biol. 2012, 512721.

Tilney, L.G., 1976. The polymerization of actin. J. Cell Biol. 69, 51–57, 73–89.

Tilney, L.G., 1983. Interactions between actin filaments and membranes give spatial organization to cell. Mod. Cell Biol. 2, 163–199.

Tilney, L.G., Conelly, P.S., Portnoy, D.C., 1990a. Actin filament nucleation by the bacterial pathogen Listeria monocytogenes. J. Cell Biol. 111, 2979–2988.

Tilney, L.G., Cooke, T.J., Connelly, P.S., Tilney, M.S., 1990b. The distribution of plasmodesmata and its relationship to morphogenesis on fern gametophytes. Development 110, 1207–1221.

Tilney, L.G., Cooke, T.J., Connelly, P.S., Tilney, M.S., 1991. The structure of plasmodesmata as revealed by plasmolysis, detergent extraction, and protease digestion. J. Cell Biol. 112, 739–747.

Tilney, L.G., DeRosier, D.J., Tilney, M.S., 1992a. How *Listeria* exploits host cell actin to form its own cytoskeleton. I. Formation of a tail and how that tail might be involved in movement. J. Cell Biol. 118, 71–81.

Tilney, L.G., DeRosier, D.J., Weber, A., Tilney, M.S., 1992b. How *Listeria* exploits host cell actin to form its own cytoskeleton. II. Nucleation, actin filament polarity, filament assembly, and evidence for a pointed end capper. J. Cell Biol. 118, 83–93.

Tilney, L.G., Harb, O.S., Connelly, P.S., Robinson, C.G., Roy, C.R., 2001. How the parasitic bacterium *Legionella pneumophila* modifies its phagosome and transforms it not rough ER: implications for conversion of plasma membrane to the ER membrane. J. Cell Sci. 114, 4637–4650.

Tilney, L.G., Portnoy, D.A., 1989. Actin filaments and the growth, movement, and spread of the intracellular bacterial parasite, *Listeria* monocytogenes. J. Cell Biol. 109, 1597–1608.

Time, 1937. Glorius Handful. Monday, December 6, 1937. www.time.com/time/magazine/article/0,9171, 758538,00.html.

Time, 1940. The Pulse of Protoplasm. Monday, November 25, 1940. http://www.time.com/time/magazine/article/0,9171, 884191,00.html.

Time, 1945. Appeal to the Goths. Monday December 10, 1945.

Timko, M.P., Kausch, A.P., Castresana, C., Fassler, J., Herrera-Estrella, L., van den Broeck, G., van Montagu, M., Schell, J., Cashmore, A.R., 1985. Light regulation of plant gene expression by an upstream enhancer-like element. Nature 318, 579–582.

Timmis, J.N., Ayliffe, M.A., Huany, C.Y., Martin, W., 2004. Endosymbiotic gene transfer: organelle genomes forge eukaryotic chromosomes. Nat. Rev. Genet. 5, 123–135.

Timmis, J.N., Ingle, J., 1973. Environmentally induced changes in rRNA gene redundancy. Nat. New Biol. 244, 235–236.

Tippit, D.H., Pickett-Heaps, J.D., Leslie, R., 1980. Cell division in two large pennate diatoms *Hantzschia* and *Nitzschia*. III. A new proposal for kinetochore function during prometaphase. J. Cell Biol. 86, 402–416.

Tirlapur, U.K., Dahse, I., Reiss, B., Meurer, J., Oelmüller, R., 1999. Characterization of the activity of a plastid-targeted green fluorescent protein in Arabidopsis. Eur. J. Cell Biol. 78, 233–240.

Tiselius, A., 1939. The chemistry of proteins and amino acids. Ann. Rev. Biochem. 8, 155–184.

Tiselius, A., 1946. The 1946 Nobel Prize in Chemistry Award Ceremony Speech.

Tiselius, A.W.K., December 13, 1948. Electrophoresis and Adsorption Analysis as Aids in Investigations of Large Molecular Weight Substances and their Breakdown Products. Nobel Lecture.

Titel, C., Woehlecke, H., Afifi, I., Ehwald, R., 1997. Dynamics of limiting cell wall porosity in plant suspension cultures. Planta 203, 320–326.

Titorenko, V.I., Mullen, R.T., 2006. Peroxisome biogenesis: the peroxisomal endomembrane system and the role of the ER. J. Cell Biol. 174, 11–17.

Titorenko, V.I., Rachubinski, R.A., 1998. Mutants of the yeast *Yarrowia lipolytica* defective in protein exit from the endoplasmic reticulum are also defective in peroxisome biogenesis. Mol. Cell Biol. 18, 2789–2803.

Titus, D.E., Becker, W.M., 1985. Investigation of the glyoxysome- peroxisome transition in germinating cucumber cotyledons using double-label immunoelectron microscopy. J. Cell Biol. 101, 1288–1299.

Titus, M.A., Warick, H.M., Spudich, J.A., 1989. Multiple actin-based motor genes in *Dictyostelium*. Cell Regul. 1, 55–63.

Tiwary, B.K., 2015. Next-generation sequencing and assembly of plant genomes. In: Barh, D., Sarwar Khan, M., Davies, E. (Eds.), PlantOmics: The Omics of Plant Science. Springer India, pp. 53–64.

To, L.P., 1987. Are centrioles semiautonomous? Ann. N.Y. Acad. Sci. 503, 83–91.

Toby, T.K., Fornelli, L., Kelleher, N.L., 2016. Progress in top-down proteomics and the analysis of proteoforms. Annu. Rev. Anal. Chem. 9, 499–519.

Todd, A.R., December 11, 1957. Synthesis in the Study of Nucleotides. Nobel Lecture.

Tolbert, N.E., 1971. Microbodies-peroxisomes and glyoxysomes. Annu. Rev. Plant Physiol. 22, 45–74.

Tolbert, N.E., 1981. Metabolic pathways in peroxisomes and glyoxysomes. Ann. Rev. Biochem. 50, 133–157.

Tolbert, N.E., Oeser, A., Kisaki, T., Hageman, R.H., Yamazaki, R.K., 1968. Peroxisomes from spinach leaves containing enzymes related to glycolate metabolism. J. Biol. Chem. 243, 5179–5184.

Tolley, N., Sparkes, I., Craddock, C., Eastmond, P., Runions, J., Hawes, C., Frigerio, L., 2010. Transmembrane domain length is responsible for the ability of a plant reticulon to shape endoplasmic reticulum tubules *in vivo*. Plant J. 64, 411–418.

Tolley, N., Sparkes, I., Hunter, P.R., Craddock, C.P., Nuttall, J., Roberts, L.M., Hawes, C., Pedrazzini, E., Frigerio, L., 2008. Overexpression of a plant reticulon remodels the lumen of the cortical endoplasmic reticulum but does not perturb protein transport. Traffick 9, 94–102.

Tolmach, L.J., 1951. Effects of triphosphopyridine nucleotide upon oxygen evolution and carbon dioxide fixation by illuminated chloroplasts. Nature 167, 946–948.

Tomé, F., Jansseune, K., Saey, B., Grundy, J., Vandenbroucke, K., Hannah, M.A., et al., 2017. rosettR: protocol and software for seedling area and growth analysis. Plant Methods 13, 13.

Tomenius, K., Clapham, D., Meshi, T., 1987. Localization by immunogold cytochemistry of the virus-coded 30 K protein in plasmodesmata of leaves infected with tobacco mosaic virus. Virology 160, 363–371.

Tominaga, M., Ito, K., 2015. The molecular mechanism and physiological role of cytoplasmic streaming. Curr. Opin. Plant Biol. 27, 104–110.

Tominaga, M., Kimura, A., Yokota, E., Haraguchi, T., Shimmen, T., Yamamoto, K., Nakano, A., Ito, K., 2013. Cytoplasmic streaming velocity as a plant size determinant. Dev. Cell 27, 345–352.

Tominaga, M., Kojima, H., Yokota, E., Nakamori, R., Anson, M., Shimmen, T., Oiwa, K., 2012. Calcium-induced mechanical change in the neck domain alters the activity of plant myosin XI. J. Biol. Chem. 287, 30711–30718.

Tominaga, M., Kojima, H., Yokota, E., Orii, H., Nakamori, R., Katayama, E., Anson, M., Shimmen, T., Oiwa, K., 2003. Higher plant myosin XI moves processively on actin with 35 nm steps at high velocity. EMBO J. 22, 1263–1272.

Tominaga, M., Morita, K., Sonobe, S., Yokota, E., Shimmen, T, 1997. Microtubules regulate the organization of actin filaments at the cortical region in root hair cells of *Hydrocharis*. Protoplasma 199, 83–92.

Tominaga, M., Nakano, A., 2012. Plant-specific myosin XI, a molecular perspective. Front. Plant Sci. 3, 211.

Tominaga, Y., Shimmen, T., Tazawa, M., 1983. Control of cytoplasmic streaming by extracellular Ca^{2+} in permeabilized Nitella cells. Protoplasma 116, 75–77.

Tominaga, Y., Tazawa, M., 1981. Reversible inhibition of cytoplasmic streaming by intracellular Ca^{2+} in tonoplast-free cells of *Chara australis*. Protoplasma 109, 103–111.

Tominaga, Y., Wayne, R., Tung, H.Y.L., Tazawa, M., 1987. Phosphorylation-dephosphorylation is involved in Ca^{2+}-controlled cytoplasmic streaming of characean cells. Protoplasma 136, 161–169.

Tominaga, M., Yokota, E., Sonobe, S., Shimmen, T., 2000a. Mechanism of inhibition of cytoplasmic streaming by a myosin inhibitor, 2,3-butanedione monoxime. Protoplasma 213, 46–54.

Tominaga, M., Yokota, E., Vidali, L., Sonobe, S., Hepler, P.K., Shimmen, T., 2000b. The role of plant villin in the organization of the actin cytoskeleton, cytoplasmic streaming and the architecture of the transvacuolar strand in root hair cells of *Hydrocharis*. Planta 210, 836–843.

Tomishige, M., Kusumi, A., 1999. Compartmentalization of the erythrocyte membrane by the membrane skeleton: intercompartmental hop diffusion of band 3. Mol. Biol. Cell 10, 2475–2479.

Tomishige, M., Sako, M., Kusumi, A., 1998. Regulation mechanism of the lateral diffusion of band 3 in erythrocyte membranes by the membrane skeleton. J. Cell Biol. 142, 989–1000.

Tomos, A.D., Leigh, R.A., 1999. The pressure probe: a versatile tool in plant cell physiology. Annu. Rev. Plant Physiol. 50, 447–472.

Tomos, A.D., Malone, M., Pritchard, J., 1989. The biophysics of differential growth. Environ. Exp. Bot. 29, 7–23.

Tompa, P., 2013. Hydrogel formation by multivalent IDPs. A reincarnation of the microtrabecular lattice? Intrinsically Disord. Proteins 1, e24068.

Tong, C.-G., Dauwlder, M., Clawson, G.A., Hatem, C.C., Roux, S.J., 1993. The major nucleoside triphosphatase in pea (*Pisum satiuum* L.) nuclei and in rat liver nuclei share common epitopes also present in nuclear lamins. Plant Physiol. 101, 1005–1011.

Tonkin, C.J., Foth, B.J., Ralph, S.A., Struck, N., Cowman, A.F., McFadden, G.I., 2008. Evolution of malaria parasite plastid targeting sequences. Proc. Natl. Acad. Sci. U.S.A. 105, 4781–4785.

Toriello, N.M., Douglas, E.S., Thaitrong, N., Hsiao, S.C., Francis, M.B., Bertozzi, C.R., Mathies, R.A., 2008. Integrated microfluidic bioprocessor for single-cell gene expression analysis. Proc. Natl. Acad. Sci. U.S.A. 105, 20173–20178.

Toriyama, H., 1955. The migration of potassium in the primary pulvinus. Cytologia 20, 367–377.

Toriyama, H., 1962. The migration of potassium in the petiole of *Mimosa pudica*. Cytologia 27, 431–441.

Toriyama, H., 1974. Tannins and the tannin vacuole in the motor organ of higher plants. Sci. Rep. Tokyo Womans Christian Coll. 31, 368–382.

Toriyama, H., Jaffe, M.J., 1972. The migration of calcium and its role in the regulation of seismonasty in the motor cell of *Mimosa pudica* L (in Japanese). Biol. Mem. Tokyo Woman's Christian Coll. 21, 215–235.

Toriyama, H., Satô, S., 1968a. Electron microscope observation of the motor cell of *Mimosa pudica* L. Proc. Jpn. Acad. 44, 702–706.

Toriyama, H., Satô, S., 1968b. Electron microscope observation of the motor cell of *Mimosa pudica* L. Proc. Jpn. Acad. 44, 949–953.

Toriyama, H., Satô, S., 1970. On the central vacuole in the *Mimosa* motor cell. Cytologia 36, 359–375.

Toro, E., Shapiro, L., 2010. Bacterial chromosome organization and segregation. Cold Spring Harb. Perspect. Biol. 2, a000349.

Torrens-Spence, M.P., Fallon, T.R., Weng, J.K., 2016. A workflow for studying specialized metabolism in nonmodel eukaryotic organisms. Methods Enzymol. 576, 69–97.

Torrens-Spence, M.P., Pluskal, T., Li, F.S., Carballo, V., Weng, J.K., 2017. Complete pathway elucidation and heterologous reconstitution of Rhodiola salidroside biosynthesis. Mol. Plant. https://doi.org/10.1016/j.molp.2017.12.007.

Towbin, H., 1979. Electrophoretic transfer of proteins from polyacrylamide gels to nitrocellulose sheets: procedure and some applications. Proc. Natl. Acad. Sci. U.S.A. 76, 4350–4354.

Towbin, H., 1988. Citation Classic: Electrophoretic Transfer of Proteins from Polyacrylamide Gels to Nitrocellulose Sheets: Procedure and Some Applications. http://bit.ly/NzqLpF.

Townsend, J.A., Wright, D.A., Winfrey, R.J., Fu, F., Maeder, M.L., Joung, J.K., Voytas, D.F., 2009. High-frequency modification of plant genes using engineered zinc-finger nucleases. Nature 459, 442–445.

Toyooka, K., Goto, Y., Asatsuma, S., Koizumi, M., Mitsui, T., Matsuoka, K., 2009. A mobile secretory vesicle cluster involved in mass transport from the Golgi to the plant cell exterior. Plant Cell 21, 1212–1229.

Toyoshima, C., Nomura, H., 2002. Structural changes in the calcium pump accompanying the dissociation of calcium. Nature 418, 605–611.

Traas, J.A., Braat, P., Emons, A.M.C., Meekes, H., Derksen, J., 1985. Microtubules in root hairs. J. Cell Sci. 76, 303–320.

Traas, J.A., Doonan, J.H., Rawlins, D.J., Shaw, P.J., Watts, J., Lloyd, C.W., 1987. An actin network is present in cytoplasm throughout the cell cycle of carrot cells and associates with the dividing nucleus. J. Cell Biol. 105, 387–395.

Tran, J.C., Zamdborg, L., Ahlf, D.R., Lee, J.E., Catherman, A.D., et al., 2011. Mapping intact protein isoforms in discovery mode using top-down proteomics. Nature 480, 254–258.

Traube, M., 1867. Experimente zur Theorie der Zellenbildung und Endosmose. In: Reichert, C.B., du Bois-Reymond, E. (Eds.), Archiv für Anatomie, Physiologie und wisswnachaftliche Medicin. Veit et Comp, Leipzig, 87–128, 129–165.

Travis, J., 2004. Plant biology. NO-making enzyme no more: cell, PNAS papers retracted. Science.

Trebst, A., 1974. Energy conservation in photosynthetic electron transport of chloroplasts. Ann. Rev. Plant Physiol. 25, 423–458.

Trefil, J.S., 1983. The Moment of Creation. Charles Scribner's Sons, New York.

Trefil, J., Morowitz, H.J., Smith, E., 2009. The origin of life. Am. Sci. 97, 206–213.

Tregunna, E.B., Krotkov, G., Nelson, C.D., 1966. Effect of oxygen on the rate of photorespiration in detached tobacco leaves. Physiol. Plant 19, 723–733.

Trelease, R.N., 1984. Biogenesis of glyoxysomes. Ann. Rev. Plant Physiol. 35, 321–347.

Trelease, R.N., Becker, W.M., Gruber, P.J., Newcomb, E.H., 1971. Microbodies (glyoxysomes and peroxisomes) in cucumber cotyledons. Correlative biochemical and ultrastructural study in light-and dark-grown seedlings. Plant Physiol. 48, 461–475.

Trelease, R.N., Lee, M.S., Banjoko, A., Bunkelmann, J., 1996. C-terminal polypeptides are necessary and sufficient for in vivo targeting of transiently-expressed proteins to peroxisomes in suspension-cultured plant cells. Protoplasma 195, 156–167.

Trench, R.K., 1991. Cyanophora paradoxa Korschikoff and the origin of chloroplasts. In: Margulis, L., Fester, R. (Eds.), Symbiosis as a Source of Evolutionary Innovation. Speciation and Morphogenesis. MIT Press, Cambridge, MA, pp. 143–150.

Trentmann, O., Haferkamp, I., 2013. Current progress in tonoplast proteomics reveals insights into the function of the large central vacuole. Front. Plant Sci. 4, 34.

Treub, M., 1878. Quelque researches sur le role du noyau dans la division des cellules vegetales. Natuurk. Verh. K. Akad. D. Wetensch. Amsterdam Deel 19, 1–33.

Treviranus, G.R., 1802. Biologi, oder Philosophie der lebenden Natur für Naturforscher und Aerzte. Röwer, Göttingen.

Trewavas, A., 2000. Signal perception and transduction. In: Buchanan, B., Gruissem, W., Jones, R. (Eds.), Biochemistry & Molecular Biology of Plants. American Society of Plant Physiologists, Rockville, MD, pp. 930–987.

Trewavas, A., 2002. Plant intelligence: mindless mastery. Nature 415, 841.

Trewavas, A., 2006. The green plant as an intelligent organism. In: Baluska, F., Mancuso, S., Volkmann, D. (Eds.), Communication in Plants. Springer, Berlin, pp. 1–18.

Trewavas, A., 2014. Plant Behavious & Intelligence. Oxford University Press, Oxford.

Trewavas, A., 2016. Plant intelligence: as overview. BioScience 66, 542–551.

Trewavas, A., Baluska, F., 2011. The ubiquity of consciousness. EMBO Rep. 12, 1221–1225.

Troglodyte, A., 1891. Riddles of the Sphinx. A Study in the Philosophy of Evolution. F. C. S. Schiller. Swan Sonnenschein & Co., London.

Troland, L.T., 1917. Biological enigmas and the theory of enzyme action. Am. Nat. 51, 321–350.

Trost, P., 2003. Plasma membrane redox systems. Protoplasma 221, 1.

Trowbridge, I.S., 1991. Endocytosis and signals for internalization. Curr. Opin. Cell Biol. 3, 634–641.

Trump, D.J., 2016. Great Again: How to Fix Our Crippled America. Threshold Editions, New York.

Trusty, J.L., Johnson, K.J., Graeme Lockaby, B., Goertzen, L.R., 2007. Bi-parental cytoplasmic DNA inheritance in Wisteria (Fabaceae): evidence from a natural experiment. Plant Cell Physiol. 48, 662–665.

Tsai, M.A., Frank, R.S., Waugh, R.E., 1993. Passive mechanical behavior of human neutrophils: power-law fluid. Biophys. J. 65, 2078–2088.

Tsai, M.A., Frank, R.S., Waugh, R.E., 1993a. Passive mechanical behavior of human neutrophils: effect of cytochalasin B. Biophys. J. 66, 2166–2172.

Tsai, M.A., Frank, R.S., Waugh, R.E., 1993b. Passive mechanical behavior of human neutrophils: power-law fluid. Biophys. J. 65, 2078–2088.

Tsai, M.A., Frank, R.S., Waugh, R.E., 1994. Passive mechanical behavior of human neutrophils: effect of cytochalasin B. Biophys. J. 66, 2166–2172.

Tsai, S.Q., Zheng, Z., Nguyen, N.T., Liebers, M., Topkar, V.V., Thapar, V., Wyvekens, N., Khayter, C., Iafrate, A.J., Le, L.P., Aryee, M.J., Joung, J.K., 2015. Guide-Seq enables genome-wide profiling of off-target cleavage by CRISPR-Cas nucleases. Nat. Biotechnol. 33, 187–197.

Ts'o, P.O.P., Bonner, J., Vinograd, J., 1956. Microsomal nucleoprotein particles from pea seedlings. J. Biophys. Biochem. Cytol. 2, 451–465.

Ts'o, P.O.P., Eggman, L., Vinograd, J., 1957. Physical and chemical studies of myxomyosin, an ATP-sensitive protein in cytoplasm. Biochim. Biophys. Acta 25, 532–542.

Ts'o, P.O.P., Sato, C.S., 1959. The incorporation of leucine-C14 into microsomal particles and other subcellular components of the peas epicotyl. J. Biophys. Biochem. Cytol. 5, 59–68.

Tschermak, E., 1900. Concerning artificial crossing in *Pisum sativum*, 1950 Genetics 35, 42–47.

Tse, Y.C., Mo, B., Hillmer, S., Zhao, M., Lo, S.W., Robinson, D.G., 2004. Identification of multivesicular bodies as prevacuolar compartments in *Nicotiana tabacum* BY-2 cells. Plant Cell 16, 672–693.

Tsien, R.Y., December 8, 2008. Constructing and Exploiting the Fluorescent Protein Paintbox. Nobel Lecture.

Tsubo, Y., Matsuda, Y., 1984. Transmission of chloroplast genes in crosses between *Chlamydomonas reinhardtii* diploids: correlation with chloroplast nucleoid behavior in young zygotes. Curr. Genet. 8, 223–229.

Tsugita, A., Fraenkel-Conrat, H., 1960. The amino acid composition and C-terminal sequence of a chemically evoked mutant of TMV. Proc. Natl. Acad. Sci. U.S.A. 46, 636–641.

Tsukada, M., Ohsumi, Y., 1993. Isolation and characterization of autophagy-defective mutants of *Saccharomyces cerevisiae*. FEBS Lett. 333, 169–174.

Tsukaya, H., 2014. Meristems. In: Noguchi, T., Kawano, S., Tsukaya, H., Matsunaga, S., Sakai, A., Karahara, I., Hayashi, Y. (Eds.), Atlas of Plant Cell Structure. Springer, Japan, pp. 187–202.

Tu, Y., 2011. The discovery of artemisinin (qinghaosu) and gifts from Chinese medicine. Nat. Medods 17, 1217–1220.

Tu, Y., December 7, 2015. Artemisinin—a Gift from Traditional Chinese Medicine to the World. Nobel Lecture.

Tucker, E.B., 1982. Translocation in the staminal hairs of *Secresea purpurea*. I. A study of cell ultrastructure and cell-to-cell passage of molecular probes. Protoplasma 113, 193–201.

Tucker, R.B., 1983. Ilya Prigogine on the Arrow of Time. Omni Magazine. http://www.omnimagazine.com/archives/interviews/prigogine/index.html.

Tucker, E.B., 1988. Inositol bisphosphate and inositol triphosphate inhibit cell-to-cell passage of carboxyfluorescein in staminal hairs of *Secresea purpurea*. Planta 174, 358–363.

Tucker, E.B., 1990. Calcium-loaded 1,2-bis(2-aminophenoxy) ethane-N, N,N',N'-tetraacetic acid blocks cell-to-cell diffusion of carboxyfluorescein in stammal hairs of Secresea purpurea. Planta 182, 34–38.

Tucker, E.B., Boss, W.F., 1996. Matoparan-induced intracellular Ca^{2+} fluxes may regulate cell-to-cell communication in plants. Plant Physiol. 111, 459–467.

Tucker, M.R., Laux, T., 2007. Connecting the paths in plant stem cell regulation. Trends Cell Biol. 17, 403–410.

Tucker, E.B., Lee, M., Alli, S., Sookhdeo, V., Wada, M., Imaizumi, T., Kasahara, M., Hepler, P.K., 2005. UV-A induces two calcium waves in *Physcomitrella patens*. Plant Cell Physiol. 46, 1226–1236.

Tucker, T., Marra, M., Friedman, J.M., 2009. Massively parallel sequencing: the next big thing in genetic medicine. Am. J. Hum. Genet. 85, 142–154.

Tucker, E.B., Tucker, J.E., 1993. Cell-to-cell diffusion selectivity in staminal hairs of *Setcreasea purpurea*. Protoplasma 174, 36–44.

Tugal, H.B., Pool, M., Baker, A., 1999. *Arabidopsis* 22-kilodalton peroxisomal membrane protein. Nucleotide sequence analysis and biochemical characterization. Plant Physiol. 120, 309–320.

Turcatti, G., Romieu, A., Fedurco, M., Tairi, A.P., 2008. A new class of cleavable fluorescent nucleotides: synthesis and optimization as reversible terminators for DNA sequencing by synthesis. Nucleic Acids Res. 36, e25.

Turgeon, R., 1991. Symplastic phloem loading and the sink-source transition in leaves: a model. In: Bonnemain, V.L., Delrot, S., Dainty, J., Lucas, W.J. (Eds.), Recent Advances in Phloem Transport and Assimilate Compartmentation. Quest Editions, Paris, pp. 18–22.

Turgeon, R., 2000. Plasmodesmata and solute exchange in the phloem. Aust. J. Plant Physiol. 27, 521–529.

Turgeon, R., 2010. The role of phloem loading reconsidered. Plant Physiol. 152, 1817–1823.

Turgeon, R., Beebe, D.U., 1991. The evidence for symplastic phloem loading. Plant Physiol. 96, 349–354.

Turgeon, R., Beebe, D.U., Gowan, E., 1993. The intermediary cell: minor-vein anatomy and raffinose oligosaccharide synthesis in the Scrophulariaceae. Planta 191, 446–456.

Turgeon, R., Gowan, E., 1990. Phloem loading in *Coleus blumei* in the absence of carrier-mediated uptake of export sugar from the apoplast. Plant Physiol. 94, 1244–1249.

Turgeon, R., Hepler, P.K., 1989. Symplastic continuity between mesophyll and companion cells in minor veins of mature *Cucurbita pepo* L. leaves. Planta 179, 24–31.

Turgeon, R., Medville, R., 2004. Phloem loading. A reevaluation of the relationship between plasmodesmatal frequencies and loading strategies. Plant Physiol. 136, 3795–3803.

Turing, A.M., 1992a. In: Ince, D.C. (Ed.), Mechanical Intelligence. North-Holland, New York.

Turing, A.M., 1992b. In: Saunders, P.T. (Ed.), Morphogenesis. North-Holland, New York.

Turing, A.M., 1992c. Collected Works of Turing, A. M. In: Saunders, P.T. (Ed.), Morphogenesis. North-Holland, Amsterdam.

Türktas, M., Yücebili Kurtoğlu, K., Dorado, G., Zhang, B., Hernandez, P., Ünver, T., 2015. Turk. J. Agric. For. 39, 361–376.

Turner, C.E., Burridge, K., 1991. Transmembrane molecular assemblies in cell-extra matrix interactions. Curr. Opin. Cell Biol. 3, 849–853.

Turner, F.R., Whaley, W.G., 1965. Intercisternal elements of the Golgi apparatus. Science 147, 1303–1304.

Tuskan, G.A., Difazio, S., Jansson, S., Bohlmann, J., Grigoriev, I., Hellsten, U., Putnam, N., Ralph, S., Rombauts, S., Salamov, A., Schein, J., Sterck, L., Aerts, A., Bhalerao, R.R., Bhalerao, R.P., Blaudez, D., Boerjan, W., Brun, A., Brunner, A., Busov, V., Campbell, M., Carlson, J., Chalot, M., Chapman, J., Chen, G.L., Cooper, D., Coutinho, P.M., Couturier, J., Covert, S., Cronk, Q., Cunningham, R., Davis, J., Degroeve, S., Dejardin, A., Depamphilis, C., Detter, J., Dirks, B., Dubchak, I., Duplessis, S., Ehlting, J., Ellis, B., Gendler, K., Goodstein, D., Gribskov, M., Grimwood, J., Groover, A., Gunter, L., Hamberger, B., Heinze, B., Helariutta, Y., Henrissat, B., Holligan, D., Holt, R., Huang, W., Islam-Faridi, N., Jones, S., Jones-Rhoades, M., Jorgensen, R., Joshi, C., Kangasjarvi, J., Karlsson, J., Kelleher, C., Kirkpatrick, R., Kirst, M., Kohler, A., Kalluri, U., Larimer, F., Leebens-Mack, J., Leple, J.C., Locascio, P., Lou, Y., Lucas, S., Martin, F., Montanini, B., Napoli, C., Nelson, D.R., Nelson, C., Nieminen, K., Nilsson, O., Pereda, V., Peter, G., Philippe, R., Pilate, G., Poliakov, A., Razumovskaya, J., Richardson, P., Rinaldi, C., Ritland, K., Rouze, P., Ryaboy, D., Schmutz, J., Schrader, J., Segerman, B., Shin, H., Siddiqui, A., Sterky, F., Terry, A., Tsai, C.J., Uberbacher, E., Unneberg, P., Vahala, J., Wall, K., Wessler, S., Yang, G., Yin, T., Douglas, C., Marra, M., Sandberg, G., Van de Peer, Y., Rokhsar, D., 2006. The genome of black cottonwood, *Populus trichocarpa* (Torr.&Gray). Science 313, 1596–1604.

Tuteja, N., 2003. Plant DNA helicases: the long unwinding road. J. Exp. Bot. 54, 2201–2214.

Twain, M., 1923. Dr. Loeb's incredible discovery. In: Europe and Elsewhere. Harper & Brothers, NY.

Twell, D., Park, S.K., Hawkins, T.J., Schubert, D., Schmidt, R., Smertenko, A., Hussey, P.J., 2002. MOR1/GEM1 has an essential role in the plant-specific cytokinetic phragmoplast. Nat. Cell Biol. 4, 711–714.

Twyman, R.M., Schillberg, S., Fischer, R., 2005. Transgenic plants in the biopharmaceutical market. Expert Opin. Emerg. Drugs 10, 185–218.

Tyerman, S.D., Bohnert, H.J., Maurel, C., Steudle, E., Smith, J.A.C., 1999. Plant aquaporins: their molecular biology, biophysics and significance for plant water relations. J. Exp. Bot. 50, 1055–1071.

Tyerman, S.D., Niemietz, C.M., Bramley, H., 2002. Plant aquaporins: multifunctional water and solute channels with expanding roles. Plant Cell Environ. 25, 173–194.

Tyers, M., Mann, M., 2003. From genomics to proteomics. Nature 422, 193–197.

Tyndall, J., 1870. The Scientific Use of the Imagination: A Discourse. Longmans, Green, and Co., London.

Tyndall, J., 1898. Fragments of Science. D. Appleton and Co., New York.

Tyson, J., 1878. The cell doctrine: its history and present state. For the Use of Students in Medicine and Dentistry. Also, a Copious Bibliography of the Subject. Lindsay & Blakiston, Philadelphia.

Tzfira, T., Citovsky, V., 2006. Agrobacterium-mediated genetic transformation of plants Biology and biotechnology. Curr. Opin. Biotechnol. 17, 147–154.

Uchida, K., Furuya, M., 1997. Control of the entry into S phase by phytochrome and blue light receptor in the first cell cycle of fern spores. Plant Cell Physiol. 38, 1075–1079.

Uddin, M.I., Qi, Y., Yamada, S., Shibuya, I., Deng, X.-P., Kwak, S.-S., Kaminaka, H., Tanaka, K., 2008. Overexpression of a new rice vacuolar antiporter regulating protein OsARP improves salt tolerance in tobacco. Plant Cell Physiol. 49, 880–890.

Ueda, K., 1966. Fine structure of *Chlorogonium elongatum* with special reference to vacuole development. Cytologia 31, 461–472.

Ueda, K., Matsuyama, T., Hashimoto, T., 1999. Visualization of microtubules in living cells of transgenic *Arabidopsis thaliana*. Protoplasma 206, 201–206.

Ueda, T., Mori, Y., Nakagaki, T., Kobatake, Y., 1988. Changes in cyclic AMP and cyclic GMP concentration. Birefringent fibrils and contractile activity accompanying UV and blue light photoavoidance in plasmodia of an albino strain of *Physarum polycephalum*. Photochem. Photobiol. 47, 271–276.

Ueda, M., Nakamura, Y., 2007. Chemical basis of plant leaf movement. Plant Cell Physiol. 48, 400–407.

Ueda, H., Tamura, K., Hara-Nishimura, I., 2015. Functions of plant specific myosin XI: from intracellular motility to plant postures. Curr. Opin. Plant Biol. 28, 30–38.

Ueda, T., Terayama, K., Kurihara, K., Kobatake, Y., 1975. Threshold phenomena in chemoreception and taxis in slime mold *Physarum polycephalum*. J. Gen. Physiol. 65, 223–234.

Ueda, T., Uemura, T., Sato, M.H., Nakano, A., 2004. Functional differentiation of endosomes in *Arabidopsis* cells. Plant J. 40, 783–789.

Ueda, H., Yokota, E., Kutsuna, N., Shimada, T., Tamura, K., Shimmen, T., Hasezawa, S., Dolja, V.V., Hara-Nishimura, I., 2010. Myosin-dependent endoplasmic reticulum motility and F-actin organization in plant cells. Proc. Natl. Acad. Sci. U.S.A. 107, 6894–6899.

Ueguchi-Tanaka, M., Fujisawa, Y., Kobayashi, M., Ashikari, M., Iwasaki, Y., Kitano, H., Matsuoka, M., 2000. Rice dwarf mutant d1, which is defective in the α subunit of the heterotrimeric G protein, affects gibberellin signal transduction. Proc. Natl. Acad. Sci. U.S.A. 97, 11638–11743.

Uehara, R., Goshima, G., 2010. Functional central spindle assembly requires de novo microtubule generation in the interchromosomal region during anaphase. J. Cell Biol. 191, 259–267.

Uehara, K., Hogetsu, T., 1993. Arrangement of cortical microtubules during formation of bordered pit in the trochoides of *Taxus*. Protoplasma 173, 8–12.

Uehlein, N., Kai, L., Kaldenhoff, R., 2017. Plant aquaporins and CO_2. In: Chaumont, F., Tyerman, S.D. (Eds.), Plant Aquaporins. From Transport to Signaling. Springer, Cham, Switzerland, pp. 255–265.

Uehlein, N., Lovisolo, C., Siefritz, F., Kaldenhoff, R., 2003. The tobacco aquaporin NtAQP1 is a membrane CO_2 pore with physiological functions. Nature 425, 734–737.

Uehlein, N.B., Otto, D.T., Hanson, M., Fischer, M., McDowell, N., Kaldenhoff, R., 2008. Function of *Nicotiana tabacum* aquaporins as chloroplast gas pores challenges the concept of membrane CO_2 permeability. Plant Cell 20, 648–657.

Uehlein, N., Sperling, H., Heckwolf, M., Kaldenhoff, R., 2012. The *Arabidopsis* aquaporin PIP1;2 rules cellular CO_2 uptake. Plant Cell Environ. 35, 1077–1083.

Uemura, M., 2001. Ten years with Peter L. Steponkus (our collaboration on plant cold hardiness and membrane cryostability). Cryo Lett. 22, 341–352.

Uemura, T., Ueda, T., 2014. Plant vacuolar trafficking driven by RAB and SNARE proteins. Curr. Opin. Plant Biol. 22, 116–121.

Uemura, T., Ueda, T., Ohniwa, R.L., Nakano, A., Takeyasu, K., Sato, M.H., 2004. Systemic analysis of SNARE molecules in *Arabidopsis*: dissection of the post-Golgi network in plant cells. Cell Struct. Funct. 29, 49–65.

Uetz, P., Giot, L., Cagney, G., Mansfield, T.A., Judson, R.S., Knight, J.R., Lockshon, D., Narayan, V., Srinivasan, M., Pochart, P., et al., 2000. A comprehensive analysis of protein-protein interactions in *Saccharomyces cerevisiae*. Nature 403, 623–662.

Uhl, C.H., 1996. Chromosomes and polyploidy in *Lenophyllum* (Crassulaceae). Am. J. Bot. 83, 216–220.

Uhlén, P., 2006. Visualization of Na,K-ATPase interacting proteins using FRET technique. Ann. N.Y. Acd. Sci. 986, 514–518.

Uhlenbeck, G.E., Goudsmit, S.A., 1926. Spinning electrons and the structure of spectra. Nature 117, 264–265.

Uhlmann, F., Lottspeich, F., Nasmyth, K., 1999. Sister-chromtid separation at anaphase onset is promoted by cleavage of the cohesin subunit Scc1. Nature 400, 37–42.

Uhlmann, F., Nasmyth, K., 1998. Cohesion between sister chromatids must be established during DNA replication. Curr. Biol. 8, 1095–1101.

Uhlmann, F., Wernic, D., Poupart, M.A., Koonin, E.V., Nasemyth, K., 2000. Cleavage of cohesin by the CD clan protease separin triggers anaphase in yeast. Cell 103, 375–386.

Ulam, S., 1972. Some ideas and prospects in biomathematics. Ann. Rev. Biophys. Bioeng. 1, 277–292.

Ullich, W.R., Rigano, C., Fuggi, A., Aparicio, P.J. (Eds.), 1990. Inorganic Nitrogen in Plants and Microorganisms. Uptake and Metabolism. Springer-Verlag, Berlin.

Ulmasov, T., Hagen, G., Guilfoyle, T.J., 1997. ARF1, a transcription factor that binds to auxin response elements. Science 276, 1865–1868.

Umekawa, M., Klionsky, D.J., 2012. The cytoplasm-to-vacuole targeting pathway: a historical perspective. Int. J. Cell Biol. 2012, 142634.

Underwood, M.R., Fried, H.M., 1990. Characterization of nuclear localizing sequences derived from yeast ribosomal protein L29. EMBO J. 9, 91–100.

Ungerleider, S., 2001. Faust's Gold. St. Martin's Press, New York.

Unlü, M., Morgan, M.E., Minden, J.S., 1997. Difference gel electrophoresis: a single gel method for detecting changes in protein extracts. Electrophoresis 18, 2071–2077.

Unwin, P.N.T., Milligan, R.A., 1982. A large particle associated with the perimeter of the nuclear pore complex. J. Cell Biol. 93, 63–75.

Uozumi, N., Gassmann, W., Cao, Y., Schroeder, J.I., 1995. Identification of strong modifications in cation selectivity in an *Arabidopsis* inward rectifying potassium channel by mutant selection in yeast. J. Biol. Chem. 270, 24276–24281.

Upadhyay, S.K., Kumar, J., Alok, A., Tuli, R., 2013. RNA-guided genome editing for target gene mutations in wheat. G3 3, 2233–2238.

Urban, S., 1829. Memoir of Sir Humphry Davy, Bart. Gentlemans Mag. 99, 9–16.

Urbanowicz, B.R., Rayon, C., Carpita, N.C., 2004. Topology of the maize mixed-linkage (1→3),(1→4)-ß-D-glucan synthase at the Golgi membrane. Plant Physiol. 134, 758–768.

Ure, A., 1819. An account of some experiments made on the body of a criminal immediately after execution, with physiological and practical observations. J. Sci. Arts 6, 283–294.

Urey, H.C., 1939. The separation of isotopes and their use in chemistry and biology. In: Baitsell, G.A. (Ed.), Science in Progress. Yale University Press, New Haven, CT, pp. 35–77.

Urey, H.C., 1952a. The Planets. Yale University Press, New Haven, CT.

Urey, H.C., 1952b. On the early chemical history of the Earth and the origin of life. Proc. Natl. Acad. Sci. U.S.A. 38, 351–363.

Urey, H., 1979. The origin of the Earth. In: Life. Origin and Evolution. Readings from Scientific American. W. H. Freeman and Co., San Francisco, pp. 18–23.

Urey, H.C., February 14, 1935. Some Thermodynamic Properties of Hydrogen and Deuterium. Nobel Lecture.

Url, W.G., 1964. Phasenoptische Untersuchungen and Inner-epidermen der Zweibelschuppe von *Allium* cepa L. Protoplasma 58, 294–311.

Ursache, R., Grube Andersen, T., Marhavy, P., Geldner, N., 2018. A protocol for combining fluorescent proteins with histological stains for diverse cell wall components. Plant J. 93, 399–412.

Usaj, M.M., Styles, E.B., Verster, A.J., Friesen, H., Boone an dB, C., Andrews, J., 2016. High-content screening for quantitative cell biology. Trends Cell Biol. 26, P598–P611.

Ussher, J., 1658. The Annal of the World. Printed by E. Tyler, London.

Ussher, J., 1864. The Whole Works of the Most Rev. James Ussher. Collected by C. R. Elrington. Hodges, Smith, and Co., Dublin.

Ussing, H.H., 1947. Interpretation of the exchange of radio-sodium in isolated muscle. Nature 160, 262–263.

Ussing, H.H., 1949. Transport of ions across cellular membranes. Physiol. Rev. 29, 127–155.

Ussing, H.H., Zerahn, K., 1951. Active transport of sodium as the source of electric current in the short-circuited isolated frog skin. Acta Physiol. Scand. 23, 110–127.

Uversky, V.N., 2016. Dancing protein clouds: the strange biology and chaotic physics of intrinsically disordered proteins. J. Biol. Chem. 291, 6681–6688.

Uyeda, T.Q.P., 1996. Ultra-fast *Chara* myosin: a test case for the swinging lever arm model for force production by myosin. J. Plant Res. 109, 231–239.

Vaeck, M., Reynaerts, A., Höfte, H., Jansens, S., De Beuckeleer, M., Dean, C., Zabeau, M., Van Montagu, M., Leemans, J., 1987. Transgenic plants protected from insect attack. Nature 328, 33–37.

Vahey, M., Scordilis, S.P., 1980. Contractile proteins from the tomato. Can. J. Bot. 58, 797–801.

Vahey, M., Titus, M., Trautwein, R., Scordilis, S., 1982. Tomato actin and myosin: contractile proteins from a higher land plant. Cell Motil. 2, 131–147.

Vaisberg, E.A., Koonce, M.P., McIntosh, J.R., 1993. Cytoplasmic dynein plays a role in mammalian mitotic spindle formation. J. Cell Biol. 123, 849–858.

Valaskovic, G.A., Kelleher, N.L., McLafferty, F.W., 1996. Attomole protein characterization by capillary electrophoresis-mass spectrometry. Science 273, 1199–1201.

Valdivia, E.R., Sampedro, J., Lamb, C., Chopra, S., Cosgrove, D.J., 2007a. Recent proliferation and translocation of pollen group 1 allergen genes in the maize genome. Plant Physiol. 143, 1269–1281.

Valdivia, E.R., Wu, Y., Li, L.-C., Cosgrove, D.J., Stephenson, A.G., 2007b. A group-1 grass pollen allergen influences the outcome of pollen competition in maize. PLoS One E154.

Vale, R.D., 1987. Intracellular transport using microtubule-based motors. Ann. Rev. Cell Biol. 3, 347–378.

Vale, R.D., Milligan, R.A., 2000. The way things move: looking under the hood of molecular motor proteins. Science 288, 88–95.

Vale, R.D., Reese, T.S., Sheetz, M.P., 1985a. Identification of a novel force-generating protein, kinesin, involved in microtubule-based motility. Cell 42, 39–50. https://www.ibiology.org/ibioeducation/making-discoveries/discovery-talk-looking-for-myosin-and-finding-kinesin.html. https://www.ibiology.org/ibiomagazine/discovering-kinesin.html.

Vale, R.D., Schnapp, B.J., Mitchison, T.J., Steuer, E., Reese, T.S., Sheetz, M.P., 1985b. Different axoplasmic proteins generate movement in opposite directions along microtubules in vitro. Cell 43, 623–632.

Vale, R.D., Schnapp, B.J., Reese, T.S., Sheetz, M.P., 1985c. Movement of organelles along filaments dissociated from the axoplasm of the giant squid axon. Cell 40, 449–454.

Vale, R.D., Toyoshima, Y.Y., 1988. Rotation and translocation of microtubules in vitro induced by dyneins from *Tetrahymena* cilia. Cell 52, 459–470.

Vale, R.D., Toyoshima, Y.Y., 1989. Microtubule translocation properties of intact and proteolytically digested dyneins from *Tetrahymena* cilia. J. Cell Biol. 108, 2327–2334.

Valenstein, E.S., 2005. The War of the Soups and Sparks: The Discovery of Neurotransmitters and the Dispute over How Nerves Communicate. Columbia University Press, New York.

Valerio, M., Haraux, F., Gardeström, P., Diolez, P., 1993. Tissue specificity of the regulation of ATP hydrolysis by isolated plant mitochondria. FEBS Lett. 318, 113–117.

Vallee, R.B., Wall, J.S., Paschal, B.M., Shpetner, H.S., 1988. Microtubule-associated protein 1C from brain is a two-headed cytosolic dynein. Nature 332, 561–563.

Vallon, O., Bulte, L., Dainese, P., Olive, J., Bassi, R., Wollman, F.A., 1991. Lateral redistribution of cytochrome b6\f complexes along thylakoid membranes upon state transitions. Proc. Natl. Acad. Sci. U.S.A. 88, 8262–8266.

Valster, A.H., Hepler, P.K., 1997. Caffeine inhibition of cytokinesis: effect on the phragmoplast cytoskeleton in living *Tradescantia* stamen hair cells. Protoplasma 196, 155–166.

Valster, A.H., Pierson, E.S., Valenta, R., Hepler, P.K., Emons, A.M.C., 1997. Probing the plant actin cytoskeleton during cytokinesis and interphase by profilin microinjection. Plant Cell 9, 1815–1824.

van Bel, A.J.E., 1989. The challenge of symplastic phloem loading. Bot. Acta 102, 183–185.

van Bel, A.J.E., Ehlers, K., 2005. Electrical signaling via plasmodesmata. In: Oparka, K.J. (Ed.), Plasmodesmata. Blackwell, Oxford, pp. 263–278.

van Bel, A.J.E., van Kesteren, W.J.P. (Eds.), 1999. Plasmodesmata. Structure, Function, Role in Cell Communication. Springer- Verlag, Berlin.

van Beneden, E., 1883. Recherches sur la maturation de l'oeuf, la fécondation et la division cellulaire. Arch. Biol. 4, 265–641.

Van Bortle, K., Corces, V.G., 2012. Nuclear organization and genome function. Annu. Rev. Cell Dev. Biol. 28, 163–187.

van Daele, Y., Gaill, F., Goffinet, G., 1991. Parabolic pattern of a peculiar striated body in the tunic of the ascidian, *Halocynthia papillosa* Tunicate Ascidiacea. J. Struct. Biol. 106, 115–124.

van de Meene, A.M.L., Doblin, M.S., Bacic, A., 2017. The plant secretory pathway seen through the lens of the cell wall. Protoplasma 254, 75–94.

Van den Berg, B., Clemons, W.M., Collinson, I., Modis, Y., Hartmann, E., Harrison Jr., S.C., Rappoport, T.A., 2004. X-ray structure of a protein-conducting channel. Nature 427, 24–26.

van den Bulcke, M., Baun, G., Castresana, C., Van Montagu, M., Vandekerckhove, J., 1989. Characterization of vacuolar and extra vacuolar ß(1,3)-glucanases of tobacco: evidence for a strictly compartmentalized plant defense system. Proc. Natl. Acad. Sci. U.S.A. 86, 2673–2677.

van der Honing, H.S., de Ruijter, N.C., Emons, A.M., Ketelaar, T., 2010. Actin and myosin regulate cytoplasm stiffness in plant cells: a study using optical tweezers. New Phytol. 185, 90–102.

van der Hoorn, R.A.L., 2008. Plant proteases from phenotypes to molecular mechanisms. Annu. Rev. Plant Biol. 59, 191–223.

van der Hoorst, G.T., Muijtjens, M., Kobayashi, K., Takano, R., Kanno, S., Takao, M., de Wit, J., Verkerk, A., Eker, A.P., van Leenen, D., Buijs, R., Bootsma, D., Hoeijmakers, J.H., Yasui, A., 1999. Mammalian Cry1 and Cry2 are essential for maintenance of circadian rhythms. Nature 398, 627–630.

van der Krol, A., Mur, R., Beld, M., Mol, J.N.M., Stuitje, A.R., 1990. Flavonoid genes in petunia: addition of a limited number of gene copies may lead to a suppression of gene expression. Plant Cell 2, 291–299.

van der Shoot, C., Dietrich, M.A., Storms, M., Verbecke, J.A., Lucas, W.J., 1995. Establishment of a cell-to-cell communication pathway between separate carpels during gynoecium development. Planta 195, 450–455.

Van der Wilden, W., Herman, E.M., Chrispeels, M.J., 1980. Protein bodies of mung bean cotyledons as autophagic organelles. Proc. Natl. Acad. Sci. U.S.A. 77, 428–432.

van der Woude, W.J., Morré, D.J., Bracker, C.E., 1971. Isolation and characterization of secretory vesicles in germinated pollen of *Lillium longiflorum*. J. Cell Sci. 8, 331–351.

Van der Zand, A., Braakman, I., Tabak, H.F., 2010. Peroxisomal membrane proteins insert into the endoplasmic reticulum. Mol. Biol. Cell 21, 2057–2065.

van der Zand, A., Gent, J., Braakman, I., Tabak, H.F., 2012. Biochemically distinct vesicles from the endoplasmic reticulum fuse to form peroxisomes. Cell 149, 397–409.

Van Emon, J.M., 2016. The omics revolution in agricultural research. J. Agric. Food. Chem. 64, 36–44.

Van Fleet, D.S., 1961. Histochemistry and function of the endodermis. Bot. Rev. 27, 165–220.

van Gelder, R.N., von Zastrow, M.E., Yool, A., Dement, W.C., Barchas, J.D., Eberwine, J.W., 1990. Amplified RNA synthesized from limited quantities of heterogeneous cDNA. Proc. Natl. Acad. Sci. U.S.A. 87, 1663–1667.

van Gisbergen, P.A.C., Bezanilla, M., 2013. Plant formins: membrane aanchors for actin polymerization. Trends Cell Biol. 23, 227–233.

van Helmont, J.B., 1683. Aufgang der Artzney-Kunst, 1971. Kösel Verlag, München.

van Hemelryck, M., Bernal, R., Rojas, E., Dumais, J., Kroeger, J.H., 2017. A fresh look at growth oscillations in pollen tubes: kinematic and mechanistic descriptions. In: Obermeyer, G., Feijó, J. (Eds.), Pollen Tip Growth. Springer International, pp. 369–389.

van Holde, K.E., 1989. Chromatin. Springer-Verlag, New York.

van Iterson, W., Hoeniger, J.F.M., van Zanten, E.N., 1967. A "microtubule" in a bacterium. J. Cell Biol. 32, 1–10.

van Lammeren, A.A.M., Bednara, J., Wallemse, M.T.M., 1989. Organization of the actin cytoskeleton during pollen development in *Gasteria verrucosa* (Mill.) H. Duval visualized with rhodamine-phalloidin. Planta 178, 531–539.

van Leeuwenhoek, A., 1677. Concerning little animals observed in rain-well-sea- and snow water; as also in water wherein pepper had lain infused. Philos. Trans. R. Soc. Lond. 12, 821–831.

van Leeuwenhoek, A., 1678. De natis E semine genitali animalculus. Philos. Trans. 12 (142), 1040–1046.

van Leeuwenhoek, A., 1699a. Concerning his answers to objections made to his opinions concerning the animalcula semine masculine. Philos. Trans. 21, 270–272.

van Leeuwenhoek, A., 1699b. Concerning the animalcula in semine humano, etc. Philos. Trans. 21, 301–308.

van Leeuwenhoek, A., 1700. Concerning his further observations on the animalcula in semine masculine. Philos. Trans. 22, 739–746.

van Leeuwenhoek, A., 1701. Concerning the spawn of codfish, etc. Philos. Trans. 22, 821–824.

van Meer, G., 1993. Transport and sorting of membrane lipids. Curr. Opin. Cell Biol. 5, 661–673.

van Niel, C.B., 1941. The bacterial photosyntheses and their importance for the general problem of photosynthesis. Adv. Enzymol. 1, 263–328.

Van Roermund, C.W.T., Schroers, M.G., Wiese, J., Facchinelli, F., Kurz, S., Wilkinson, S., Charton, L., Wanders, R.J.A., Waterham, H.R., Weber, A.P.M., Link, N., 2016. The peroxisomal NAD carrier from *Arabidopsis* imports NAD in exchange with AMP. Plant Physiol. 171, 2127–2139.

van Roessel, P., Brand, A.H., 2002. Imaging into the future: visualizing gene expression and protein interactions with fluorescent proteins. Nat. Cell Biol. 4, E15–E20.

van Spronsen, E.A., Sarafis, V., Brakenhoff, G.J., van der Voort, H.T.M., Nanninga, N., 1989. Three-dimensional structure of living chloroplasts as visualized by confocal scanning laser microscopy. Protoplasma 148, 8–14.

van West, P., Morris, B.M., Reid, B., Appiah, A.A., Osborne, M.C., Campbell, T.A., Shepherd, S.J., Gow, N.A.R., 2002. Oomycete plant pathogens use electric fields to target roots. Mol. Plant Microbe Interact. 15, 790–798.

van Wijk, K.J., 2000. Proteomics of the chloroplast: experimentation and prediction. Trends Plant Sci. 5, 420–425.

van Wijk, K.J., Baginsky, S., 2011. Plastid proteomics in higher plants: current state and future goals. Plant Physiol. 155, 1578–1588.

van Wijk, K.J., Friso, G., Walther, D., Schulze, W.X., 2014. Meta-analysis of *Arabidopsis thaliana* phospho-proteomics data reveals compartmentalization of phosphorylation motifs. Plant Cell 26, 2367–2389.

van Wijk, K.J., Kessler, F., 2017. Plastoglobuli: plastid microcompartments with integrated functions in metabolism, plastid developmental transitions, and environmental adaptations. Annu. Rev. Plant Biol. 68, 253–289.

Van't Hof, J., 1975. DNA fiber replication in chromosomes of a higher plant (Pisum sativum). Exp. Cell Res. 93, 95–104.

Van't Hof, J., 1985. Control points within the cell cycle. In: Bryant, J.A., Francis, D. (Eds.), The Cell Divison Cycle in Plants. Cambridge University Press, Cambridge, pp. 1–13.

Van't Hof, J., 1988. Functional chromosomal structure: the replicon. In: Bryant, J.A., Dunham, V.L. (Eds.), DNA Replication in Plants. CRC Press, Boca Raton, FL, pp. 1–15.

Van't Hof, J., Bjerknes, C.A., 1977. 18 μm replication units of chromosomal DNA fibers of differentiated cells of pea (Pisum sativum). Chromosoma 64, 287–294.

Van't Hof, J., Bjerknes, C.A., 1981. Similar replicon properties of higher plant cells with different S periods and genome sizes. Exp. Cell Res. 136, 461–465.

Van't Hof, J., Hernández, P., Bjerknes, C.A., Lamm, S.S., 1987a. Location of the replication origin in the 9 kb repeat size class of rDNA in pea (*Pisum sativum*). Plant Mol. Biol. 9, 87–96.

Van't Hof, J., Kuniyaki, A., Bjerknes, C.A., 1978. The size and number of replicon families of chromosomal DNA of *Arabidopsis thaliana*. Chromosoma 68, 269–285.

Van't Hof, J., Lamm, S.S., 1992. Site of initiation of replication of the ribosomal RNA genes of pea (Pisum sativum) detected by 2-dimensional gel electrophoresis. Plant Mol. Biol. 20, 377–382.

Van't Hof, J., Lamm, S.S., Bjerknes, C.A., 1987b. Detection of replication initiation by a replicon family in DNA of synchronized pea (*Pisum sativum*) root cells using benzoylated napthoylated DEAE-cellulose chromatography. Plant Mol. Biol. 9, 77–86.

Van't Hoff, J.H., 1874. A suggestion looking to the extension into space of the structural formulas at present used in chemistry and a note upon the relation between the optical activity and the chemical constitution of organic compounds. Archives neerlandaises des sciences exactes et naturelles 9, 445–454.

Van't Hoff, J., 1888. The function of osmotic pressure in the analogy between solutions and gases. Philos. Mag. Ser. 5 (26), 81–105.

Van't Hoff, J.H., 1898–1899. Lectures on Theoretical and Physical Chemistry. Translated by R. A. Lehfeldt. Edward Arnold, London.

Van't Hoff, J.H., 1903. Physical Chemistry in the Service of the Sciences. University of Chicago Press, Chicago.

Van't Hoff, J.H., 1967. Imagination in Science. Translated by G. F. Springer. Springer, Berlin.

Van't Hoff, J.H., December 13, 1901. Osmotic Pressure and Chemical Equilibrium. Nobel Lecture.

Vance, D.E., Vance, J.E. (Eds.), 2008. Biochemistry of Lipids, Lipoportins and Membranes, fifth ed. Elsevier Science, B. V.

VandenBosch, K.A., Becker, W.M., Palevitz, B.A., 1996. The natural history of a scholar and gentleman. A biography of Eldon H. Newcomb. Protoplasma 195, 4–11.

Vandenbrink, J.P., Kiss, J.Z., Herranz, R., Medina, F.J., 2014. Light and gravity signals synergize in modulating plant development. Front. Plant Sci. 5, 563.

Vanderhoef, L.N., 1985. Auxin-enhanced elongation. In: Hormonal Regulation of Plant Growth and Development, vol. II. Nijhoff/Junk Publishers, Dordrecht/Boston, pp. 37–44.

Vanderhoef, L.N., Lu, T.S., Williams, C.A., 1977. Comparison of auxin-induced elongation in soybean hypocotyl. Plant Physiol. 59, 1004–1007.

Vanderhoef, L.N., Stahl, C.A., 1975. Separation of two responses to auxin by means of cytokinin inhibition. Proc. Natl. Acad. Sci. U.S.A. 72, 1822–1825.

Vandré, D.D., Borisy, G.G., 1989. The centrosome cycle in animal cells. In: Hyams, J.S., Brinkley, B.R. (Eds.), Mitosis: Molecules and Mechanisms. Academic Press, London, pp. 39–75.

Vankudavath, R.N., Bodanapu, R., Sreelakshmi, Y., Sharma, R., 2012. High-throughput phenotyping of plant populations using a personal digital assistant. In: Normanly, J. (Ed.), High-Throughput Phenotyping in Plants. Methods and Protocols. Springer, New York, pp. 97–116.

Vanlerberghe, G.C., 2013. Alternative oxidase: a mitochondrial respiratory pathway to maintain metabolic and signaling homeostasis during abiotic and biotic stress in plants. Int. J. Mol. Sci. 14, 6805–6847.

Vannerum, K., Huysman, M.J., De, R.R., Vuylsteke, M., Leliaert, F., Pollier, J., Lutz-Meindel, U., Gillard, J., De, V.L., Goossens, A., Inzé, D., Vyverman, W., 2011. Transcriptional analysis of cell growth and morphogenesis in the unicellular green alga *Micrasterias* (Streptophyta), with emphasis on the role of expansin. BMC Plant Biol. 11, 128.

Vanzin, G.F., Madson, M., Carpita, N.C., Raikhel, N.V., Keegstra, K., Reiter, W.-D., 2002. The mur2 mutant of *Arabidopsis* thaliana lacks fucosylated xyloglucan because of a lesion in fucosyltransferase AtFUT1. Proc. Natl. Acad. Sci. U.S.A. 99, 3340–3345.

Varagona, M.J., Schmidt, R.J., Raikhel, N.V., 1991. Monocot regulatory protein opaque-2 is localized in the nucleus of maize endosperm and transformed tobacco plants. Plant Cell 3, 105–113.

Varagona, M.J., Schmidt, R.J., Raikhel, N.V., 1992. Nuclear localization signal(s) required for nuclear targeting of the maize regulatory protein opaque-2. Plant Cell 4, 1213–1227.

Varga, L., 1950. Observations on the glycerol-extracted musculus psoas of the rabbit. Enzymologia 4, 196–211.

Varki, A., Altheide, T.K., 2005. Comparing the human and chimpanzee genomes: searching for needles in a haystack. Genome Res. 15, 1746–1758. Errata: Genome Research 19, 2343.

Varner, J.E., Lin, L.-S., 1989. Plant cell wall architecture. Cell 56, 231–239.

Varner, J.E., Taylor, R., 1989. New ways to look at the architecture of plant cell walls–Localization of polygalacturonate blocks in plant tissues. Plant Physiol. 91, 31–33.

Vastag, B., July 7, 2012. U.S. pushes for more scientists, but the jobs aren't there. Wash. Post.

Vaughan, M.A., Braselton, J.P., 1985. Nucleolar persistence in meristems of mung bean. Caryologia 38, 357–362.

Vaughn, K.C., 1981. Organelle transmission in higher plants: organelle alteration vs. physical exclusion. J. Hered. 72, 335–337.

Vaughn, K.C., 1985. Physical basis for the maternal inheritance of triazine resistance in *Amaranthus* hybrids. Weed Res. 25, 15–19.

Vaughn, K.C., Bowling, A.J., 2008. Recovery of microtubules on the blepharoplast of *Ceratopteris* spermatogenous cells after oryzalin treatment. Protoplasma 233, 231–240.

Vaughn, K.C., DeBonte, L.R., Wilson, K.G., Schaeffer, G.W., 1980. Organelle alteration as a mechanism for maternal inheritance. Science 208, 196–198.

Vaughn, K.C., Sheiman, T.C., Renzaglia, K.S., 1993. A centrin homologue is a component of the multilayered structure in bryophytes and pteridophytes. Protoplasma 175, 58–60.

Vaughn, K.C., Talbot, M.J., Offler, C.E., McCurdy, D.W., 2007. Wall ingrowths in epidermal transfer cells of *Vicia faba* cotyledons are modified primary walls marked by localized accumulations of arabinogalactan proteins. Plant Cell Physiol. 48, 159–168.

Vaux, L.D., Tooze, J., Fuller, S.D., 1990. Identification by antiidiotypic antibodies of an intracellular membrane protein that recognizes a mammalian endoplasmic reticulum retention signal. Nature 345, 495–501.

Vega-Palas, M., Ferl, R.J., 1995. The *Arabidopsis* Adh gene exhibits diverse nucleosome arrangements within a small DNase I-sensitive domain. Plant Cell 7, 1923–1932.

Vejdovský, F., 1926–1927. Structure and Development of the "Living Matter". Royal Bohemian Society of Sciences, Prague.

Velasquez, S.M., Ricardi, M.M., Dorosz, J.G., Fernandez, P.V., Nadra, A.D., Pol-Fachin, L., Egelund, J., Gille, S., Harholt, J., Ciancia, M., Verli, H., Pauly, M., Bacic, A., Olsen, C.E., Ulvskov, P., Petersen, B.L., Somerville, C., Iusem, N.D., Estevez, J.M., 2011. O-glycosylated cell wall proteins are essential in root hair growth. Science 332, 1401–1403.

Velculescu, V.E., Zhang, L., Zhou, W., Vogelstein, J., Basrai, M.A., Bassett Jr., D.E., Hieter, P., Vogelstein, B., Kinzler, K.W., 1997. Characterization of the yeast transcriptome. Cell 88, 243–251.

Venning, F.D., 1949. Stimulation by wind motion of collenchyma formation in celery petioles. Bot. Gaz. 110, 511–514.

Venter, J.C., Remington, K., Heidelberg, J.F., Halpern, A.L., Rusch, D., Eisen, J.A., Wu, D., Paulsen, I., Nelson, K.E., et al., 2004. Environmental genome shotgun sequencing of the Sargasso Sea. Science 304, 66–74.

Venter, J.C., Smith, H.O., Hood, L., 1996. A new strategy for genome sequencing. Nature 381, 364–366.

Venverloo, C.J., Hovenkamp, P.H., Wedda, A.J., Libbenga, K.R., 1980. Cell division in *Nautilocalyx* explants. I. Phragmosome, preprophase band and plane of division. Z. Pflanzenphysiol. 100, 161–174.

Verbelen, J.-P., Vissenberg, K., 2006. Cell expansion: past, present and perspectives. Plant Cell Monogr. 5, 1–6.

Verchot-Lubicz, J., Goldstein, R.E., 2010. Cytoplasmic streaming enables the distribution of molecules and vesicles in large plant cells. Protoplasma 240, 99–107.

Vergara, C.E., Carpita, N.C., 2001. ß-D-Glycan synthases and the CesA gene family: lessons to be learned from the mixed- linkage $(1 \rightarrow 3),(1 \rightarrow 4)$ß-D-glucan synthase. Plant Mol. Biol. 47, 145–160.

Verger, S., Long, Y., Boundaoud, A., Hamant, O., 2018. A tension-adhesion feedback loop in plant epidermis. eLife 7, e34460.

Verlander, M.S., Lohrmann, R., Orgel, L.E., 1973. Catalysts for the self-polymerization of adenosine cyclic 2′,3′-phosphate. J. Mol. Evol. 2, 303–316.

Verma, D.P., 2001. Cytokinesis and building of the cell plate in plants. Annu. Rev. Plant Physiol. Plant Mol. Biol. 52, 751–764.

Vernadsky, V.I., 1944. Problems of geochemistry, the fundamental matter-energy difference between the living and the inert natural bodies of the biosphere. Trans. Conn. Acad. Arts Sci. 35, 487–517.

Vernon, J.A., 1960. Generalizations from sensory depravation to fallout shelters. Disaster Study 12. In: Symposium on Human Problems in the Utilization of Fallout Shelters. 11 and 12 February 1960. National Academy of Sciences. National Research Council, Washington, DC, pp. 59–66.

Vernoud, V., Horton, A.C., Yang, Z., Nielsen, E., 2003. Analysis of the small GTPase gene superfamily of *Arabidopsis*. Plant Physiol. 131, 1191–1208.

Véry, A.-A., Sentenac, H., 2003. Molecular mechanisms and regulation of K^+ transport in higher plants. Annu. Rev. Plant Biol. 54, 575–603.

Vesk, P.A., Vesk, M., Gunning, B.E.S., 1996. Field emission scanning electron microscopy of microtubule arrays in higher plant cells. Protoplasma 195, 168–182.

Vesper, M.J., Evans, M.L., 1979. Nonhormonal induction of H^+ efflux from plant tissues and its correlation with growth. Proc. Natl. Acad. Sci. U.S.A. 76, 6366–6370.

Vicedo, M., 1991. Realism and simplicity in the Castle-East debate on the stability of the hereditary units: rhetorical devices versus substantive methodology. Stud. Hist. Philos. Sci. 22, 201–221.

Vickery, H.B., 1950. The origin of the word protein. Yale J. Biol. Med. 22, 387–393.

Vickery, H.B., 1953. The behavior of the organic acids and starch of *Bryophyllum* leaves during culture in continuous light. J. Biol. Chem. 205, 369–381.

Vickery, H.B., 1972. The history of the discovery of amino acids. II. A review of amino acids described since 1931 as components of proteins. Adv. Prot. Chem. 26, 81–171.

Vida, T.A., Emr, S.D., 1995. A new vital stain for visualizing vacuolar membrane dynamics and endocytosis in yeast. J. Cell Biol. 128, 779–792.

Vidali, L., Augustine, R.C., Fay, S.N., Franco, P., Pattavina, K.A., Bezanilla, M., 2009. Rapid screening for temperature sensitive alleles in plants. Plant Physiol. 151, 506–514.

Vidali, L., Hepler, P.K., 1997. Characterization and localization of profilin in pollen grains and tubes of *Lilium longiflorum*. Cell Motil. Cytoskelet. 36, 323–338.

Vidali, L., McKenna, S.T., Hepler, P.K., 2001. Actin polymerization is essential for pollen germination and tube growth. Mol. Biol. Cell 12, 2534–2545.

Vidali, L., Rounds, C., Hepler, P.K., Bezanilla, M., 2009c. Lifeact-mEGFP reveals a dynamic apical F-actin network in tip growing plant cells. PLoS One 4, e5744.

Vidali, L., van Gisbergen, P.A.C., Guérin, C., Franco, P., Li, M., Burkart, G.M., Augustine, R.C., Blanchoin, L., Bezanilla, M., 2009b. Rapid formin-mediated actin-filament elongation is essential for polarized plant cell growth. Proc. Natl. Acad. Sci. U.S.A. 106, 13341–13346.

Vidali, L., Yokota, E., Cheung, A.Y., Shimmen, T., Hepler, P.K., 1999. The 135 kDa actin-bundling protein from Lilium longiflorum pollen is the plant homologue of villin. Protoplasma 209, 283–291.

Vierstra, R.D., 1993. Protein degradation in plants. Annu. Rev. Plant Physiol. Plant Mol. Biol. 44, 385–410.

Vierstra, R.D., 2003. The ubiquitin/26S proteasome pathway, the complex last chapter in the life of many plant proteins. Trends Plant Sci. 8, 135–141.

Vigil, E.L., 1970. Cytochemical and developmental changes in microbodies (glyoxysomes) and related organelles of castor been endosperm. J. Cell Biol. 46, 435–454.

Vigil, E.L., 1983. Microbodies. In: Hall, J.L., Moore, A.L. (Eds.), Isolation of Membranes and Organelles from Plant Cells. Academic Press, London, pp. 211–236.

Vigil, E.L., Ruddat, M., 1973. Effect of gibberellic acid and actinomycin D on the formation and distribution of rough endoplasmic reticulum in barley aleurone cells. Plant Physiol. 51, 549–558.

Villa-Komaroff, L., Efstratiadis, A., Broome, S., Lomedico, P., Tizard, R., Naber, S.P., Chick, W.L., Gilbert, W., 1978. A bacterial clone synthesizing proinsulin. Proc. Natl. Acad. Sci. U.S.A. 75, 3727–3731.

Villanueva, M.A., Schindler, M., Wang, J.L., 2005. The nucleocytoplasmic microfilament network in protoplasts from cultured soybean cells is a plastic entity that pervades the cytoplasm except the central vacuole. Cell Biol. Int. 29, 936–942.

Villanueva, M.A., Taylor, J., Sui, X., Griffing, L.R., 1993. Endocytosis is plant protoplasts; visualization and quantitation of fluid-phase endocytosis using silver-enhanced bovine serum albumin-gold. J. Exp. Bot. 44 (Suppl.), 275–281.

Villarejo, A., Burén, S., Larsson, S., Déjardin, A., Monné, M., Rudhe, C., Karlsson, J., Jansson, S., Lerouge, P., Rolland, N., von Heijne, G., Grebe, M., Bako, L., Samuelsson, G., 2005. Evidence for a protein transported through the secretory pathway en route to the higher plant chloroplast. Nat. Cell Biol. 7, 1224–1231.

Villiers, T.A., 1967. Cytolysomes in long-dormant plant embryo cells. Nature 214, 1356.

Vincentini, F., Matile, P., 1993. Gerontosomes, a multifunctional type of peroxisomes in scenescent leaves. J. Plant Physiol. 142, 50–56.

Viotti, C., 2014. ER and vacuoles: never been closer. Front. Plant Sci. 5, 20.

Viotti, C., Bubeck, J., Stierhof, Y.-D., Krebs, M., Langhans, M., van den Berg, W., van Dongen, W., Richter, S., Geldner, N., Takano, J., Jürgens, G., de Vries, S.C., Robinson, D.G., Schumacher, K., 2010. Endocytic and secretory traffic in *Arabidopsis* merge in the trans-Golgi network/early endosome, an independent and highly dynamic organelle. Plant Cell 22, 1344–1357.

Viotti, C., Krüger, F., Krebs, M., Neubert, C., Fink, F., Lupanga, U., Scheuring, D., Boutté, Y., Frescatada-Rosa, M., Wolfenstetter, S., Sauer, N., Hillmer, S., Grebe, M., Schumacher, K., 2013. The endoplasmic reticulum is the main membrane source for biogenesis of the lytic vacuole in *Arabidopsis*. Plant Cell 25, 3434–3449.

Virchow, R., 1859. Atoms and individuals. Translated by Rather, L.J. In: Disease, Life, and Man. Selected Essays by Rudolf Virchow. Stanford University Press, Stanford, CA, pp. 120–141.

Virchow, R., 1860. Cellular Pathology, seventh American ed. Robert M. de Witt, Publishers, New York.

Virgil, 1952. Georgic II, 30 bce. In: The Poems of Virgil Great Books of the Western World. Encyclopaedia Britannica, Chicago. Translated by J. Rhoades.

Virgin, H.I., 1954. Further studies of the action spectrum for light-induced changes in the protoplasmic viscosity of *Elodea densa*. Physiol. Plant 7, 343–353.

Virgin, H.I., 1981. The physical state of protochlorophyll(ide) in plants. Ann. Rev. Plant Physiol. 32, 451–463.

Virgin, H.I., 1987. Effects of red, far-red and blue light on the cytoplasm of wheat leaf cells. Physiol. Plant 70, 203–208.

Virgo, S.E., 1933. Loschmidt's number. Sci. Prog. 27, 634–649.

Virshup, D.M., 1990. DNA replication. Curr. Opin. Cell Biol. 2, 453–460.

Vischer, E., Chargaff, E., 1948. The separation and quantitative estimation of purines and pyrimidines in minute amounts. J. Biol. Chem. 176, 703–714.

Vishniac, W., Ochoa, S., 1951. Photochemical reduction of pyridine nucleotides by spinach grana and coupled carbon dioxide fixation. Nature 167, 768–769.

Visnovitz, T., Világi, I., Varró, P., Kristóf, Z., 2007. Mechanoreceptor cells on the tertiary pulvini of *Mimosa pudica* L. Plant Signal. Behav. 2, 462–466.

Visser, R.G.F., Somhorst, I., Kuipers, G.J., Ruys, N.J., Feenstra, W.J., Jacobsen, E., 1991. Inhibition of the expression of the gene for granule-bound starch synthase in potato by antisense constructs. Mol. Gen. Genet. 225, 289–296.

Vitale, A., Chrispeels, M.J., 1984. Transient N acetylglucosamine in the biosynthesis of phytohemagglutinin: attachment in the Golgi apparatus and removal in the protein bodies. J. Cell Biol. 99, 133–140.

Vitale, A., Pedrazzini, E., 2005. Recombinant pharmaceuticals from plants: the plant endomembrane system as a bioreactor. Mol. Interv. 5, 216–225.

Vitha, S., McAndrew, R.S., Osteryoung, K.W., 2001. FtsZ ring formation at the chloroplast division site in plants. J. Cell Biol. 153, 111–120.

Vitha, A., Yang, M., Sack, F.D., Kiss, J.Z., 2007. Gravitropism in the starch excess mutant of *Arabidopsis thaliana*. Am. J. Bot. 94, 590–598.

Voelckel, C., Gruenheit, N., Lockhart, P., 2017. Evolutionary transcriptomics and proteomics: insight into plant adaptation. Trends Plant Sci. 22, 462–471.

Voelker, T.A., Florkiewicz, R.Z., Chrispeels, M.J., 1986. Secretion of phytohemagglutinin by monkey COS cells. Eur. J. Cell Biol. 42, 218–223.

Voelkerding, K.V., Dames, S.A., Durtschi, J.D., 2009. Next-generation sequencing: from basic research to diagnostics. Clin. Chem. 55, 641–658.

Voeller, B.R., 1968. The Chromosome Theory of Inheritance. Appleton-Century-Crofts, New York.

Vogel, S., 1981. Life in Moving Fluids. Princeton University Press, Princeton, NJ.

Vogelmann, T.C., 1993. Plant tissue optics. Ann. Rev. Plant Physiol. Plant Mol. Biol. 44, 231–251.

Vogler, H., Felekis, D., Nelson, B.J., Grossniklaus, U., 2015. Measuring the mechanical properties of plant cell walls. Plants 4, 167–182.

Vogler, H., Shamsudhin, N., Nelson, B.J., Grossniklaus, U., 2017. Measuring cytomechanical forces on growing pollen tubes. In: Obermeyer, G., Feijó, J. (Eds.), Pollen Tip Growth. Springer International, pp. 65–85.

Volkin, E., Astrachan, L., 1956. Phosphorous incorporation in *Escherichia coli* ribonucleic acid after infection with bacteriophage T2. Virology 2, 149–161.

Volkmann, D., Baluška, F., 2000. Actin cytoskeleton related to gravisensing in higher plants. In: Staiger, C.J., Baluška, F., Volkmann, D., Barlow, P.W. (Eds.), Actin: A Dynamic Framework for Multiple Plant Cell Functions. Kluwer Academic Publishers, Dordrecht, The Netherlands, pp. 557–572.

Volkmann, D., Baluška, F., Menzel, D., 2012. Eduard Strasburger (1844-1912): founder of modern plant cell biology. Protoplasma 249, 1163–1172.

Volkov, A.G. (Ed.), 2006. Plant Electrophysiology—Theory & Methods. Springer-Verlag, Berlin.

Volkov, A.G., Adesina, T., Markin, V.S., Jovanov, E., 2008. Kinetics and mechanism of *Dionaea* muscipula trap closing. Plant Physiol. 146, 694–702.

Volkov, A.G., Foster, J.C., Baker, K.D., Markin, V.S., 2010a. Mechanical and electrical anisotropy in *Mimosa pudica* pulvini. Plant Signal. Behav. 5, 1211–1221.

Volkov, A.G., Foster, J.C., Markin, V.S., 2010b. Signal transduction in *Mimosa pudica*: biologically closed electrical circuits. Plant Cell Environ. 33, 816–827.

von Baeyer, A., 1870. Ueber die Wasserentziehung und ihre Bedeutung für das Pflanzenleben und die Gärung. Ber. d. Deutsch. Chem. Gesells 3, 63–73.

von Euler-Chelpin, H., May 23, 1930. Fermentation of Sugars and Fermentative Enzymes. Nobel Lecture.

Von Georgievics, G., Grandmougin, E., 1920. A Text-Book of Dye Chemistry. Scott, Greenwood & Son, London.

von Goebel, K., 1926. Wilhelm Hofmeister. The Work and Life of a Nineteenth Century Botanist. Ray Society, London.

von Goethe, J.W., 1952. Goethe's Botanical Writings. Translated by Bertha Mueller. University of Hawaii Press, Honolulu.

von Goethe, J.W., 1952. Faust, 1808, Translated by George Madison Priest. Great Books of the Western World. University of Chicago, Chicago.

von Hanstein, J., 1870. Die Entwickelung des Keimes der Monokotylen und der Dikotylen. Bot. Abhandl 1, 1–112.

von Hanstein, J., 1880. In: Frommel, W., Pfaff, F. (Eds.), Das Protoplasma als Träger der pflanzlichen und thierischen Lebensverrichtungen. Für Laien und Sachgenossen dargestellt. From Sammlung von Vorträgen für das deutsche Volk. Winter, Heidelberg.

von Hanstein, J., 1882. Einige Züge aus der Biologie des Protoplasmas. Botanische Abh. Herausgeg. v. Hanstein. Bd. 4./ Heft 2. Bonn.

von Heijne, G., 1990. The signal peptide. J. Membr. Biol. 115, 195–201.

von Helmholtz, H., 1881. Popular Lectures on Scientific Subjects. Longmans, Green and Co., London.

von Helmholtz, H., 1903. Hermann von Helmholtz. An autobiographical sketch. Second Series by H. von Helmholtz. In: Popular Lectures on Scientific Subjects. Longmans, Green, and Co., London, pp. 266–291.

von Hofmann, A.W., 1876. The Life Work of Liebig in Experimental and Philosophical Chemistry. MacMillan, London.

von Laue, M., December 12, 1915. Concerning the Detection of X-ray Interferences. Nobel Lecture.

von Lenard, P.E.A., May 28, 1906. On cathode rays. Nobel Lecture.

von Mohl, H., 1852. Principles of the Anatomy and Physiology of the Vegetable Cell. Translated by A. Henfrey. John van Voorst, London.

von Nägeli, C., 1844. Memoir on the nuclei, formation, and growth of vegetable cells, 1846. In: Reports and Papers on Botany, vol. 1. Ray Society, London, pp. 215–292.

von Nägeli, C., 1966. Bläschenförmige Gebilde im Inhalte der Pflanzenzelle [1844] Z. Wissenschaften Botan. 3, 94–128. Reprinted as Zeitschrift für wissenschaftliche Botanik von M. J. Schleiden und Carl Nägeli. Erster Band. Ersters Heft. (Tab. I-IV). The Sources of Science, No. 27. Johnson Reprint Corporation, New York.

von Nägeli, K., Cramer, C., 1855. Pflanzenphysiologische Untersuchungen. F. Schulthess, Zurich.

von Neumann, J., 1958. The Computer and the Brain. Yale University Press, New Haven, CT.

von Neumann, J., Burks, A.W., 1966. Theory of Self-Reproducing Automata. University of Illinois Press, Urbana.

von Sachs, J., 1887. Lectures on the Physiology of Plants. Translated by H. Marshall Ward. Clarendon Press, Oxford.

von Sachs, J., 1892. Physiologische Notizen. II. Beiträge zur Zellentheorie. Flora 75, 57–67.

von Sachs, J., 1906. History of Botany. (1530–1860). Translated by H. E. F, Garnsey. Revised by I. B. Balfour. Clarendon Press, Oxford.

von Schaewen, A., Storm, A., O'Neill, J., Chrispeels, M.J., 1993. Isolation of a mutant *Arabidopsis* plant that lacks N-acetyl glucosaminyl transferase I and is unable to synthesize Golgi-modified complex N-linked glycans. Plant Physiol. 102, 1109–1118.

von Smoluchchowski, M., 1917. Verrsuch einer mathematischen Theorie der Koagulationskinetic kolloider Lösungen. Z. Physikal. Chem. 92, 129–168.

von Weisäcker, C.F., 1949. The History of Nature. University of Chicago Press, Chicago.

Vonnegut, K., 1999. Unpaid consultant. In: Bamboo Snuff Box. Uncollected Short Fiction. G. P. Putman's Sons, New York, pp. 171–182.

Voragen, A.G.J., Coenen, G.-J., Verhoef, R.P., Schols, H.A., 2009. Pectin, a versatile polysaccharide present in plant cell walls. Struct. Chem. 20, 263–275.

Voros, S., 2014. The Promise of Panomics: Full-Spectrum Omics will Outshine the Clashing Hues of Complex Diseases. Genetic Engineering & Biotechnology News. http://www.genengnews.com/gen-exclusives/the-promise-of-panomics/77900162.

Vosshall, L.B., Price, J.L., Sehgal, A., Saez, L., Young, M.W., 1994. Block in nuclear localization of period protein by a second clock mutation, timeless. Science 263, 1606–1609.

Voznesenskaya, E.V., Edwards, G.E., Kiirats, O., Artyusheva, E.G., Franceschi, V.R., 2003. Development of biochemical specialization and organelle partitioning in the single celled C4 system in leaves of *Borszczowia aralocaspica* (Chenopodiaceae). Am. J. Bot. 90, 1669–1680.

Voznesenskaya, E.V., Franceschi, V.R., Edwards, G.E., 2004. Light-dependent development of a single cell C4 photosynthesis in cotyledons of *Borszczowia aralocaspica* (Chenopodiaceae) during transformation from a storage to a photosynthetic organ. Ann. Bot. 93, 177–187.

Voznesenskaya, E.V., Franceschi, V.R., Kiirats, O., Artusheva, E.G., Freitag, H., Edwards, G.E., 2002. Proof of C4 photosynthesis without Kranz anatomy in *Binertia cycloptera* (Chenopodiaceae). Plant J. 31, 649–662.

Voznesenskaya, E.V., Franceschi, V.R., Kiirats, O., Freitag, E.G., Edwards, G.E., 2001. Kranz anatomy is not essential for terrestrial C4 plant photosynthesis. Nature 414, 543–546.

Voznesenskaya, E.V., Koteyeva, N.K., Chuong, S.D.X., Edwards, G.E., Akhani, H., Franceschi, V.R., 2005. Differentiation of cellular and biochemical features of the single cell C4 syndrome during leaf development in *Bienertia cycloptera* (Chenopodiaceae). Am. J. Bot. 92, 1784–1795.

Vreeland, V., Morse, S.R., Robichaux, R.H., Miller, K.L., Hua, S.-S.T., Laetsch, W.M., 1989. Pectate distribution and esterification in *Dubautia* leaves and soybean nodules, studied with a fluorescent hybridization probe. Planta 177, 435–446.

Vreugdenhil, D., Spanswick, R.M., 1987. The effect of vanadate on proton-sucrose cotransport in Ricinus cotyledons. Plant Physiol. 84, 605–608.

Waadt, R., Hitomi, K., Nishimura, N., Hitomi, C., Adams, S.R., Getzoff, E.D., Schroeder, J.I., 2014. FRET-based reporters for the direct visualization of abscisic acid concentration changes and distribution in Arabidopsis. eLife 3, e01739.

Wacholtz, M.C., Patel, S.S., Lipsky, P.E., 1989. Leukocyte function-associated antigen I is an activation molecule for human T cells. J. Exp. Med. 170, 431–448.

Wacker, I., Quader, H., Schnepf, E., 1988. Influence of the herbicide oryzalin on cytoskeleton and growth of *Funaria hygrometrica* protonemata. Protoplasma 142, 55–67.

Wacker, I., Schnepf, E., 1990. Unequal distribution of nuclear pores in oryzalin-induced mini-nuclei in protonema cells of the moss *Funaria hygrometrica*. Influence of the nucleolus. Protoplasma 158, 195–197.

Wada, B., 1950. The mechanism of mitosis based on studies of the submicroscopic structure and of the living state of the *Tradescantia* cell. Cytologia 16, 1–26.

Wada, B., 1966. Analysis of Mitosis. Cytologia 30 (Suppl.), 1–158.

Wada, B., 1970. Mitotic cell studies based on in vivo observations IV. The raison d'etre for the spindle membrane and the atractoplasm being composed of fibrils. Cytologia 35, 483–499.

Wada, B., 1972. Mitotic cell studies based on in vivo observations. VI. Artifact problems of the mitotic figures under the electron microscope. Cytologia 37, 709–719.

Wada, M., 1992. Control of site and timing of pre-prophase band development in Adiantum protonemata. In: Shibaoka, H. (Ed.), Cellular Basis of Growth and Development in Plants. Proceedings of the VII International Symposium in Conjunction with the Awarding of the International Prize of Biology. Osaka University, Osaka, pp. 137–144.

Wada, M., 2005. Chloroplast photorelocation movement. In: Wada, M., Shimazaki, K., Iino, M. (Eds.), Light Sensing in Plants. Springer, Tokyo, pp. 193–199.

Wada, M., 2013. Chloroplast movement. Plant Sci. 210, 177–182.

Wada, M., 2016. Chloroplast and nuclear photorelocation movements. Proc. Jpn. Acad. Ser. B Phys. Biol. Sci. 92, 387–411.

Wada, Y., Anraku, Y., 1992. Genes for directing vacuolar morphogenesis in *Saccharomyces cerevisiae*. J. Biol. Chem. 267, 18671–18675.

Wada, Y., Anraku, Y., 1994. Chemiosmotic coupling of ion transport in the yeast vacuole: its role in acidification inside organelles. J. Bioenerg. Biomembr. 26, 631–637.

Wada, S., Ishida, H., Izumi, M., Yoshimoto, K., Ohsumi, Y., Mae, T., Makino, A., 2009. Autophagy plays a role in chloroplast degradation during senescence in individually darkened leaves. Plant Physiol. 149, 885–893.

Wada, B., Izutsu, K., 1961. Effects of ultraviolet microbeam irradiation on mitosis studied in *Tradescantia* cells *in vivo*. Cytologia 16, 1–26.

Wada, M., Kadota, A., 1989. Photomorphogenesis in lower green plants. Annu. Rev. Plant Physiol. Plant Mol. Biol. 40, 169–191.

Wada, M., Kadota, A., Furuya, M., 1981. Intracellular photoreceptive site for polarotropism in protonema of the fern *Adiantum capillus-venesis* L. Plant Cell Physiol. 22, 1481–1488.

Wada, M., Kadota, A., Furuya, M., 1983. Intracellular localization and dichroic orientation of phytochrome is plasma membrane and/or ectoplasm of a centrifuged protonema of fern *Adiantum capillus-venesis*. Plant Cell Physiol. 24, 1441–1447.

Wada, M., Kadota, A., Furuya, M., 1990a. Changes in microtubule and microfibril arrangement during polarotropism in Adiantum protonemata. Bot. Mag. (Tokyo) 103, 391–401.

Wada, Y., Kitamoto, K., Kanbe, T., Tanaka, K., Anraku, Y., 1990b. The SLP1 gene of *Saccharomyces cerevisiae* is essential for vacuolar morphogenesis and function. Mol. Cell Biol. 10, 2214–2223.

Wada, M., Kagawa, T., Sato, Y., 2003. Chloroplast movement. Annu. Rev. Plant Biol. 54, 455–468.

Wada, Y., Ohsumi, Y., Anraku, Y., 1992. Genes for directing vacuolar morphogenesis in *Saccharomyces* cerevisiae. J. Biol. Chem. 267, 18665–18670.

Wada, M., Staehelin, L.A., 1981. Freeze-fracture observations on the plasma membrane, the cell wall and the cuticle of growing protonemata of *Adiantum capillus-veneris* L. Planta 151, 462–468.

Wada, M., Tsuboi, H., 2015. Gene silencing by DNA interference in fern gametophytes. In: Mysore, K.S., Senthil-Kumar, M. (Eds.), Plant Gene Silencing: Methods and Protocols Methods in Molecular Biology, vol. 1287. Springer Science+Business Media, New York, pp. 119–127.

Waddington, C.H., 1942. The epigenotype. Endeavour 1, 18–20.

Waddington, C.H., 1956. Principles of Embryology. Allen & Unwin, London.

Waddington, C.H., 1961. Molecular biology or ultrastructural biology? Nature 190, 184.

Waddington, C.H., 1969. Some European contributions to the pre-history of molecular biology. Nature 221, 318–321.

Waddington, C.H., 1977. Tools for Thought. Basic Books, New York.

Wade, N., June 30, 2002. Erwin Chargaff, 96, Pioneer in DNA Chemical Research. New York Times. http://www.nytimes.com/2002/06/30/nyregion/erwin-chargaff-96-pioneer-in-dna-chemical-research.html.

Wadman, M., 2008. James Watson's genome sequenced at high speed. Nature 452, 788.

Wadsworth, P., Lee, W.-L., Murata, T., Baskin, T.I., 2011. Variations on theme: spindle assembly in diverse cells. Protoplasma 248, 439–446.

Wadsworth, P., Salmon, E.D., 1986. Analysis of the treadmilling model during metaphase of mitosis using fluorescence redistribution after photobleaching. J. Cell Biol. 102, 1032–1038.

Wadsworth, P., Shelden, E., Rupp, G., Rieder, C.L., 1989. Biotin-tubulin incorporates into kinetochore fiber microtubules during early but not late anaphase. J. Cell Biol. 109, 2257–2265.

Wagner, G., 1979a. Actomyosin as a basic mechanism of movement in animals and plants. In: Haupt, W., Feinleib, M.E. (Eds.), Encyclopedia of Plant Physiology. Physiology of Movements, vol. 7. Springer, Berlin, pp. 114–126.

Wagner, G.J., 1979b. Content and vacuole/extra vacuole distribution of neutral sugars, free amino acids and anthocyanins in protoplasts. Plant Physiol. 64, 88–93.

Wagner, G., 1983. Higher plant vacuoles and tonoplasts. In: Hall, J.L., Moore, A.L. (Eds.), Isolation of Membranes and Organelles from Plant Cells. Academic Press, London, pp. 83–118.

Wagner, G., Bellini, E., 1976. Light-dependent fluxes and compartmentation of calcium in the green alga *Mougeotia*. Z. Pflanzenphysiol. 79, 283–291.

Wagner, V.T., Brian, L., Quatrano, R.S., 1992. Role of a vitronectin like molecule in embryo adhesion of the known alga *Fucus*. Proc. Natl. Acad. Sci. U.S.A. 89, 3644–3648.

Wagner, G., Haupt, W., Laux, A., 1972. Reversible inhibition of chloroplast movement by cytochalasin B in the green alga *Mougeotia*. Science 176, 808–809.

Wagner, G.J., Hrazdina, G., 1984. Endoplasmic reticulum as a site of phenylpropanoid and flavonoid metabolism in Hippeastrum. Plant Physiol. 74, 901–906.

Wagner, G., Klein, K., 1981. Mechanism of chloroplast movement in *Mougeotia*. Protoplasma 109, 169–185.

Wagner, S., Nietzel, T., Aller, I., Costa, A., Fricker, M.D., Meyer, A.J., Schwarzländer, M., 2015. Analysis of plant mitochondrial function using fluorescent protein sensors. Methods Mol. Biol. 1305, 241–252.

Wagner, G.J., Siegelman, H.W., 1975. Large-scale isolation of intact vacuoles and isolation of chloroplasts from protoplasts of mature plant tissues. Science 190, 1298–1299.

Wagner, G., Valentin, P., Dieter, P., Marmé, D., 1984. Identification of calmodulin in the green alga Mougeotia and its possible function in chloroplast reorientational movement. Planta 162, 62–67.

Waigmann, E., Turner, A., Peart, J., Roberts, K., Zambryski, P., 1997. Ultrastructural analysis of leaf trichomes plasmodesmata reveals major differences from mesophyll plasmodesmata. Planta 203, 75–84.

Wait, L., Dundle, S., Wilkenson, F.E., Moreau, P., Safranski, K., Reust, T., Morré, D.J., 1990. Cell-free transfer of sterols from dictyosome-like structures to plasma membrane vesicles of guinea pig testes. Protoplasma 154, 8–15.

Wakefield, J., Rieder, C., Macgregor, H., 2011. Mitosis—The story. Conley Rieder of the Wadsworth Center, Albany, NY, interviewed at the University of Exeter, UK, by James Wakefield and Herbert Macgregor, October 2010. Chromosome Res. 19, 275–290.

Walczak, C.E., Heald, R., 2008. Mechanisms of mitotic spindle assembly and function. Int. Rev. Cytol. 265, 111–158.

Wald, G., 1955. The origin of life. In: The Physics and Chemistry of Life, Scientific American. Simon and Schuster, New York, pp. 3–26.

Wald, G., 1963. The origins of life. In: The Scientific Endeavor. Centennial Celebration of the National Academy of Sciences. The Rockefeller Institute Press, New York, pp. 113–134.

Wald, G., 1986. How the theory of solutions arose. J. Chem. Educ. 63, 658–659. Also see Science, September 17, 1982.

Walderhaug, M.O., Post, R.L., Saccomani, G., Leonard, R.T., Briskin, D.P., 1985. Structural relatedness of three ion-transport adenosine triphosphatases around their active sites of phosphorylation. J. Biol. Chem. 260, 3852–3859.

Waldeyer, W., 1888. Ueber Karyokinese und ihre Bezeihung zu den Befruchtungsvorgängen. Arch. Mikr. Anat. 32, 1–122.

Waldeyer, W., 1890. Karyokinesis and its relation to the process of fertilization. Q. J. Microsc. Sci. 30, 159–281.

Waldschmidt-Leitz, E., 1932. Enzymes Ann. Rev. Biochem. 1, 69–88.

Waldschmidt-Leitz, E., 1934. Enzymes Ann. Rev. Biochem. 3, 39–58.

Walk, R.-A., Hoch, B., 1978. Cell-free synthesis of glyoxysomal malate dehydrogenase. Biochem. Biophys. Res. Commun. 81, 636–643.

Walker, C.E., 1905. The Essentials of Cytology. Archibald Constable & Co, London.

Walker, C.E., 1907. The Essentials of Cytology: An Introduction to the Study of Living Matter. Archibald Constable & Co., London.

Walker, C.E., 1910. Hereditary Characters and their Modes of Transmission. Edward Arnold, London.

Walker, R.I., 1938. The effect of colchicine on somatic cells of *Tradescantia paludosa*. J. Arnold Arbor. 19, 158–162.

Walker, N.A., 1955. Microelectrode experiments on *Nitella*. Aust. J. Biol. Sci. 8, 476–489.

Walker, D.A., 1956. Malate synthesis in a cell free extract from a Crassulacean plant. Nature 178, 593–594.

Walker, N.A., 1960a. The electric resistance of the cell membranes in Chara and Nitella species. Aust. J. Biol. Sci. 13, 468–478.

Walker, W.S., 1960b. Th effects of mechanical stimulation and etiolation on the collenchyma of Datura stramonium. Am. J. Bot. 47, 717–724.

Walker, J.C.G., 1983. Possible limits on the composition of the Archaean ocean. Nature 302, 518–520.

Walker, D., 1992. Energy, Plants and Man, second ed. Oxygraphics, Brighton, England.

Walker, D.A., 1997a. Tell me where all the past years are. Photosynth. Res. 51, 2–26.

Walker, J.E., December 8, 1997b. ATP synthesis by rotary catalysis. Nobel Lecture.

Walker, R.P., Waterworth, W.M., Hooley, R., 1993. Preparation and polypeptide composition of plasma membrane and other subcellular fractions from wild oats (*Avena fatua*) aleurone. Physiol. Plant 89, 388–398.

Walko, R.M., Nothnagel, E.A., 1989. Lateral diffusion of proteins and lipids in the plasma membrane of rose protoplast. Protoplasma 152, 46–56.

Wall, M.K., Mitchenall, L.A., Maxwell, A., 2004. *Arabidopsis thaliana* DNA gyrase is targeted to chloroplasts and mitochondria. Proc. Natl. Acad. Sci. U.S.A. 101, 7821–7826.

Wallace, 1844. A Dissertation on the True Age of the World. Professor. Elder and Co., Smith.

Wallace, A.R., 1858. On the tendency of species to form varieties; and on the perpetuation of varieties and species by natural means of selection. III. On the tendency of varieties to depart indefinitely from the original type. J. Proc. Linn. Soc. Lond. 3, 53–62.

Wallace, A.R., 1869. Sir Charles Lyell on geological climates and the origin of species. running title for review of Principles of Geology (10th ed.), 1867-1868, and Elements of Geology (6th ed.), 1865, both by Sir Charles Lyell Q. Rev. 126, 359–394 (April 1869: no. 252). This article was written anonymously, but referred to in Wallace's My Life (1908) p. 211.

Wallace, A.R., 1870. Contributions to the Theory of Natural Selection. A Series of Essays. Macmillan and Co., London.

Wallace, A.R., 1889. Darwinism; an Exposition of the Theory of Natural Selection with Some of Its Applications. Macmillan and Co., London.

Wallace, A.R., 1905. Man's Place in the Universe. McClure, Phillips and Co., New York.

Wallace, A.R., 1908. My Life. A Record of Events and Opinions, New Edition. Chapman and Hall, London.

Wallace, A.R., 1911. The World of Life; A Manifestation of Creative Power, Directive Mind and Ultimate Purpose. Newer and cheaper edition. Chapman and Hall, London.

Wallach, O., 1901. Briefwechsel zwischen J. Bezelius und F. Wöhler. W. Engelmann, Leipzig.

Waller, F., Nick, P., 1997. Response of actin microfilaments during phytochrome-controlled growth of maize seedling. Protoplasma 200, 154–162.

Waller, F., Riemann, M., Nick, P., 2002. A role for actin-driven secretion in auxin-induced growth. Protoplasma 219, 72–78.

Wallin, I., 1922. On the nature of mitochondria. Am. J. Anat. 30, 203–339.

Wallin, I.E., 1927. Symbionticism and The Origin of Species. Williams & Wilkins Co., Baltimore.

Wallis, C., 2017. Is the U.S. Education System Producing a Society of "Smart Fools"? https://www.scientificamerican.com/article/is-the-u-s-education-system-producing-a-society-of-ldquo-smart-fools-rdquo/.

Walmsley, A.M., Arntzen, C.J., 2003. Plant cell factories and mucosal vaccines. Curr. Opin. Biotechnol. 14, 145–150.

Walsby, A.E., 1975. Gas vesicles. Annu. Rev. Plant Physiol. 26, 427–439.

Walsby, A.E., 1982. The elastic compressibility of gas vesicles. Proc. R. Soc. Lond. B. 216, 355–368.

Walsby, A.E., 1994. Gas vesicles. Microbiol. Rev. 58, 94–144.

Walsby, A.E., Hodge, M.J.S., 2017. Schrödinger's code-script: not a genetic cipher but a code of development. Stud. Hist. Philos. Biol. Biomed. Sci. 63, 45–54.

Walsh, J., Krull, H.R., 1987. Heteromerous interactions among glycolytic enzymes and of glycolytic enzymes with F-actin: effects of polyethylene glycol. Biochim. Biophys. Acta 952, 83–91.

Walter, P., 1997. Signal sequence recognition and protein targeting to the endoplasmic reticulum membrane. Harvey Lect. 91, 115–131.

Walter, P., Lingappa, V.R., 1986. Mechanism of protein translocation across the endoplasmic reticulum membrane. Ann. Rev. Cell. Biol. 2, 499–516.

Walter, W.J., Machens, I., Rafieian, F., Diez, S., 2015. The non-processive rice kinesin-14 OsKCH1 transports actin filaments along microtubules with two distinct velocities. Nat. Plants 1, 15111.

Walter, H., Stadelmann, E., 1968. The physiological prerequisites for the transition of autotrophic plants from water to terrestrial life. BioScience 18, 694–701.

Walters, M.S., 1958. Aberrant chromosome movement and spindle formation in meiosis of Bromus hybrids: an interpretation of spindle organization. Am. J. Bot. 45, 271–289.

Walton, P.A., Hill, P.E., Subramani, S., 1995. Import of stably folded proteins into peroxisomes. Mol. Biol. Cell 6, 675–683.

Waltz, E., 2016. Gene-edited CRISPR mushroom escapes US regulation. Nature 532, 293.

Wang, H., 1966. Further analysis of the reaction of *Paramecium* to cigarette paper ash solutions. Can. J. Microbiol. 12, 125–131.

Wang, Z., Benning, C., 2012. Chloroplast lipid synthesis and lipid trafficking through ER-plastid membrane contact sites. Biochem. Soc. Trans. 40, 457–463.

Wang, X., Bian, Y., Cheng, K., Gu, L.F., Ye, M., Zou, H., et al., 2013c. A large-scale protein phosphorylation analysis reveals novel phosphorylation motifs and phosphoregulatory networks in *Arabidopsis*. J. Proteom. 78, 486–498.

Wang, W., Cao, X.H., Miclăus, M., Xu, J., Xiong, W., 2017. The promise of agriculture genomics. Int. J. Genom. 9743749.

Wang, Y., Cheng, X., Shan, Q., Zhang, Y., Liu, J., Gao, C., Qiu, J.L., 2014b. Simultaneous editing of three homoeoalleles in hexaploid bread wheat confers heritable resistance to powdery mildew. Nat. Biotechnol. 32, 947–951.

Wang, X., Chung, K.P., Lin, W., Jiang, L., 2017d. Protein secretion in plants: conventional and unconventional pathways and new techniques. J. Exp. Bot. https://doi.org/10.1093/jxb/erx262.

Wang, H., Dittmer, T.A., Richards, E.J., 2013. Arabidopsis CROWDED NUCLEI (CRWN) proteins are required for nuclear size control and heterochromatin organization. BMC Plant Biol. 13, 200.

Wang, J.J., Elliot, E., Williamson, R.E., 2008. Features of the primary wall CESA complex in wild type and cellulose-deficient mutants of *Arabidopsis thaliana*. J. Exp. Bot. 59, 2627–2637.

Wang, Z., Gerstein, M., Snyder, M., 2009b. RNA-Seq: a revolutionary tool for transcriptomics. Nat. Rev. Genet. 10, 57–63.

Wang, P., Hawes, C., Hussey, P.J., 2017b. Plant endoplasmic reticulum-plasma membrane contact sites. Trends Plant Sci. 22, 289–297.

Wang, S., Hazelrigg, T., 1994. Implications for bcd mRNA localization from spatial distribution of exu protein in *Drosophila* oogenesis. Nature 369, 400–403.

Wang, Z., Hobson, N., Galindo, L., Zhu, S., Shi, D., McDill, J., Yang, L., Hawkins, S., Neutelings, G., Datla, R., et al., 2012c. The genome of flax (*Linum usitatissimum*) assembled de novo from short sequence reads. Plant J. 72, 461–473.

Wang, H., Jiang, L., 2017. Polar protein exocytosis: lessons from plant pollen tube. In: Obermeyer, G., Feijó, J. (Eds.), Pollen Tip Growth. Springer International, pp. 107–127.

Wang, X., Komatsu, S., 2016. Plant subcellular proteomics: application for exploring optimal cell function in soybean. J. Proteom. 143, 45–56.

Wang, C., Liu, C., Rooqueiro, D., Grimm, D., Schwab, R., Becker, C., Lanz, C., Weigel, D., 2015a. Genome-wide analysis of local chromatin packing in *Arabidopsis thaliana*. Genome Res. 25, 246–256.

Wang, M., Lu, Y., Botella, J.R., Mao, Y., Hua, K., Zhu, J.-K., 2017a. Gen targeting by homology-directed repair in rice using a Geminivirus-based CRISPR/Cas9 system. Mol. Plant 10, 1007–1010.

Wang, J., Morris, A.J., Tolan, D.R., Pagliaro, L., 1996. The molecular nature of the F-actin binding activity of aldolase revealed with site-directed mutants. J. Biol. Chem. 271, 6861–6865.

Wang, Q.Y., Nick, P., 1998. The auxin response of actin is altered in the rice mutant yin-yang. Protoplasma 204, 22–33.

Wang, N., Nobel, P.S., 1998. Phloem transport of fructans in the crassulacean acid metabolism species Agave deserti. Plant Physiol. 116, 709–714.

Wang, Y., Noguchi, K., Ono, K., Inoue, S., Terashima, I., Kinoshita, T., 2014c. Overexpression of plasma membrane H^+-ATPase in guard cells promotes light-induced stomatal opening and enhances plant growth. Proc. Natl. Acad. Sci. U.S.A. 111, 533–538.

Wang, T., Park, Y.B., Cosgrove, D.J., Hong, M., 2015b. Cellulose-pecin spatial contacts are inherent to never-dried *Arabidopsis* primary cell walls: evidence from solid-state nuclear magnetic resonance. Plant Physiol. 168, 871–884.

Wang, Z., Pesacreta, T.C., 2004. A subclass of myosin XI is associated with mitochondria, plastids, and the molecular chaperone subunit TCP-1 alpha in maize. Cell Motil. Cytoskelet. 57, 218–232.

Wang, F., Shang, Y., Fan, B., Yu, J.-Q., Chen, Z., 2014a. *Arabidopsis* LIP5, a positive regulator of multivesicular body biogenesis, is a critical target of pathogen-responsive MAPK cascade in plant basal defense. PLoS Pathog. 10, e1004243.

Wang, N., Tytell, J.D., Ingber, D.E., 2009. Mechanotransduction at a distance: mechanically coupling the extracellular matrix with the nucleus. Nat. Rev. Mol. Cell Biol. 10, 75–82.

Wang, C.X., Wang, L., Thomas, C.R., 2004. Modelling the mechanical properties of single suspension-cultured tomato cells. Ann. Bot. 93, 443–453.

Wang, Y., Wu, W.-H., 2013. Potassium transport and signaling in higher plants. Annu. Rev. Plant Biol. 64, 451–476.

Wang, J.P., Xu, Y.P., Munyampundu, J.P., Liu, T.Y., Cai, X.Z., 2016. Calcium-dependent protein kinase (CDPK) and CDPK-related kinase (CRK) gene families in tomato: genome-wide identification and functional analyses in disease resistance. Mol. Genet. Genom. 291, 661–676.

Wang, P., Xue, L., Batelli, G., Lee, S., Hou, Y.J., Van Oosten, M.J., et al., 2013b. Quantitative phosphoproteomics identifies SnRK2 protein kinase substrates and reveals the effectors of abscisic acid action. Proc. Natl. Acad. Sci. U.S.A. 110, 11205–11210.

Wang, J., Xue, X., Ren, H., 2012a. New insights into the role of plant formins: regulating the organization of the actin and microtubule cytoskeleton. Protoplasma 249, S101–S107.

Wang, T., Zabotina, O., Hong, M., 2012b. Pectin-cellulose interactions in the *Arabidopsis* primary cell wall from two-dimensional magic-angle-spinning solid-state nuclear magnetic resonance. Biochemistry 51, 9846–9856.

Wangensteen, O.W., 1947. Research and the graduate student. Am. Sci. 35, 107–113.

Warburg, O., 1920. Über die Geschwindigkeit der photochemischen Kohlensäurezersetzung in lebenden Zellen. Biochem. Z. 103, 188–217.

Warburg, O. (Ed.), 1931a. The Metabolism of Tumours. Richard R, Smith, New York.

Warburg, O., 1964. Prefatory chapter. Ann. Rev. Biochem. 35, 1–14.

Warburg, O., Christian, W., 1931. Über Aktivierung der Robison Hexose-mono-Phosphorsäure in roten Blutzellen und die Gewinnung aktivierender Fermentlösungen. Biochem. Z. 242, 206–227.

Warburg, O., Christian, W., 1936. Pyridin, der wasserstoffübertragende Bestandteil von Gärungsfermenten. (Pyridin- Nucleotide). Biochem. Z. 287, 291–328.

Warburg, O.H., December 10, 1931. The oxygen-transferring ferment of respiration. In: Nobel Lect.

Ward, W.W., Cody, C.W., Prasher, D.C., Predergast, F.G., 1989. Sequence of the chemical structure of the hexapeptide chromophore of *Aequoria* green-fluorescent protein. Photochem. Photobiol. 49, 62S.

Ward, J.L., Harris, C., Lewis, J., Beale, M.H., 2003. Assessment of ^1H NMR spectroscopy and multivariate analysis as a technique for metabolite fingerprinting of *Arabidopsis thaliana*. Phytochemistry 62, 949–957.

Ward, J.M., Mäser, P., Schroeder, J.I., 2009. Plant ion channels: gene families, physiology, and functional genomics analyses. Annu. Rev. Physiol. 71, 59–82.

Ward, J.M., Reinders, A., Hsu, H.-T., Sze, H., 1992. Dissociation and reassembly of the vacuolar H^+-ATPase complex form oat roots. Plant Physiol. 99, 161–169.

Ward, J.M., Sze, H., 1992a. Proton transport activity of the purified vacuolar H^+-ATPase from oats. Plant Physiol. 99, 925–931.

Ward, J.M., Sze, H., 1992b. Subunit composition and organization of the vacuolar proton ATPase from oat roots. Plant Physiol. 99, 170–179.

Ward, J.M., Sze, H., 1995. Isolation and functional reconstruction of the vacuolar H^+-ATPase. Methods Cell Biol. 50, 149–160.

Wardrop, A.B., 1962. Cell wall organization in higher plants. I. The primary cell wall. Bot. Rev. 28, 241–285.

Wardrop, A.B., 1983. Evidence for the possible presence of a microtrabecular lattice in plant cells. Protoplasma 115, 81–87.

Warner, J.R., 1990. The nucleolus and ribosome formation. Curr. Opin. Cell Biol. 2, 521–527.

Warner, F.D., Satir, P., Gibbons, I.R., 1989. Cell Movement. In: The Dynein ATPases, vol. 1. Alan R. Liss, Inc., New York.

Warren, G., 1993. Bridging the gap. Nature 362, 297–298.

Warren, C.R., 2008. Stand aside stomata, another actor deserves centre stage: the forgotten role of the internal conductance to CO_2 transfer. J. Exp. Bot. 59, 1475–1487.

Wasik, A., Sikora, J., 1980. Effects of cytochalasin B and colchicine on cytoplasmic streaming in *Paramecium bursaria*. Acta Protozool. 19, 103–110.

Wasteneys, G.O., Brandizzi, F., 2013. A glorious half-century of microtubules. Plant J. 75, 185–188.

Wasteneys, G.O., Collings, D.A., 2006. The cytoskeleton and coordination of directional expansion in a multicellular context. Plant Cell Monogr. 5, 217–248.

Wasteneys, G.O., Collings, D.A., 2007. The cytoskeleton and co-ordination of directional expansion in a multicellular context. In: Verbelen, J.-P., Vissenberg, K. (Eds.), The Expanding Cell. Plant Cell Monogr. (5). Springer-Verlag Berlin, pp. 217–248.

Wasteneys, G.O., Collings, D.A., Gunning, B.E.S., Hepler, P.K., Menzel, D., 1996. Actin in living and fixed characean internodal cells: identification of a cortical array of fine actin strands and chloroplast actin rings. Protoplasma 190, 25–38.

Wasteneys, G.O., Williamson, R.E., 1987. Microtubule orientation in developing internodal cells of Nitella: a qualitative analysis. Eur. J. Cell Biol. 43, 14–22.

Wasteneys, G.O., Williamson, R.E., 1989. Reassembly of microtubules in *Nitella tasmanica*: assembly of cortical microtubules in branching clusters and its relevance to steady-state microtubule assembly. J. Cell Sci. 93, 705–714.

Watanabe, K., Breier, U., Hensel, G., Kumlehn, J., Schubert, I., Reiss, B., 2016. Stable gene replacement in barley by targeted double-strand break induction. J. Exp. Bot. 67, 1433–1445.

Watanabe, K., Kobayashi, A., Endo, M., Sage-Ono, K., Toki, S., Ono, M., 2017. CRISPR/Cas9-mediated mutagenesis of the *dihydroflavonol*-4-*reductase*-B (DFR-B) locus in the Japanese morning glory *Ipomoea (Pharbitis) nil*. Sci. Rep. 7, 10028.

Waterhouse, P.M., Helliwell, C.A., 2002. Exploring plant genomes by RNA-induced gene silencing. Nat. Rev. Genet. 4, 29–38.

Waters, M.G., Blobel, G., 1986. Secretory protein translocation in a yeast cell-free system can occur post translationally and requires ATP hydrolysis. J. Cell Biol. 102, 1543–1550.

Waters, J.C., Cole, R.W., Rieder, C.L., 1993. The force-producing mechanism for centrosome separation during spindle formation in vertebrates is intrinsic to each aster. J. Cell Biol. 122, 361–372.

Waters, M., Serafini, T., Rothman, J.E., 1991. 'Coatomer': a cytosolic protein complex containing subunits of non-clathrin-coated Golgi transport vesicles. Nature 349, 248–251.

Waterworth, W.G., Bhardwaj, R.M., Jiang, Q., Bray, C.M., West, C.E., 2010. DNA ligase is an important determinant of seed longevity. Plant J. 63, 848–860.

Waterworth, W.M., Kozak, J., Provost, C.M., Bray, C.M., Angelis, K.J., West, C.E., 2009. DNA ligase I deficient plants display severe growth defects and delayed repair of both DNA single and double strand breaks. BMC Plant Biol. 9, 79.

Watson, J.D., 1963. Involvement of RNA in the synthesis of proteins. Science 140, 17–26.

Watson, J.D., 1968. The Double Helix. Atheneum, New York.

Watson, J.D., 1992. A personal view of the project. In: Kevles, D.J., Hood, L. (Eds.), The Code of Codes: Scientific and Social Issues in the Human Genome Project. Harvard University Press, Cambridge, MA, pp. 164–173.

Watson, J.D., 1995. Values from a Chicago upbringing. Ann. N.Y. Acad. Sci. 758, 194–197.

Watson, J.D., 2003. DNA: The Secret of Life. Alfred A, Knopf, New York.

Watson, J.D., Berry, A., 2003. DNA. The Secret Life. Knopf, New York.

Watson, J.D., Crick, F.H.C., 1953a. A structure for deoxyribose nucleic acid. Nature 171, 737–738.

Watson, J.D., Crick, F.H.C., 1953b. Genetical implications of the structure of deoxyribonucleic acid. Nature 171, 964–967.

Watson, J.D., Crick, F.H.C., 1953c. The structure of DNA. Cold Spring Harb. Symp. Quant. Biol. 18, 123–131.

Watson, J.D., December 11, 1962. The Involvement of RNA in the Synthesis of Proteins. Nobel Lecture.

Watson, J.D., Hopkins, N.H., Roberts, J.W., Steitz, J.A., Weiner, A.M., 1987. Molecular Biology of the Gene, fourth ed. Benjamin/Cummings, Menlo Park, CA.

Watts, I., 1801. The Improvement of the Mind: or a Supplement to the Art of Logic: Containing a Variety of Remarks and Rules for the Attainment and communication of Useful Knowledge in Religion, in the Sciences, and in Common Life. Printed for J. Abraham, London.

Watts, I., 1806. Logic, or the Right Case of Reason, for the Inquiry after Truth, a Variety of Rules to Guard against Error in the Affairs of Religion and Human Life, as Well as in the Sciences. Third American ed. Printed by Ranlet & Norris, Exeter.

Watts, I., 1811. Logic: The Right Use of Reason in the Inquiry after Truth: With a Variety of Rules to Guard against Error in the Affairs of Religion and Human Life, as Well as in the Sciences. Printed for Thomas Tegg, London.

Waugh, D.F., Schmitt, F.O., 1940. Investigations of the thickness and ultrastructure of cellular membranes by the analytical leptoscope. Cold Spring Harb. Symp. Quant. Biol. 8, 233–241.

Wayne, R., 1994. The excitability of plant cells: with a special emphasis on characean internodal cells. Bot. Rev. 60, 265–367.

Wayne, R., 2009. Light and Video Microscopy. Elsevier, Amsterdam.

Wayne, R., 2012a. A fundamental, relativistic and irreversible law of motion: a unification of Newton's second law of motion and the second law of thermodynamics. Afr. Rev. Phys. 7, 115–134.

Wayne, R., 2012b. Perspective: an attempt at identifying the position of genes on the chromosomes of maize using X-ray induced chromosomal deficiencies. In: Kass, L.R. (Ed.), Perspectives on Nobel Laureate Barbara McClintock's Publications (1926-1984): A Companion Volume. Internet-First University Press, Ithaca, NY, pp. I.83–I.87.

Wayne, R., 2012c. Perspective: identifying the individual chromosomes of maize. In: Kass, L.R. (Ed.), Perspectives on Nobel Laureate Barbara McClintock's Publications (1926-1984): A Companion Volume. Internet-First University Press, Ithaca, NY, pp. I.77–I.86.

Wayne, R., 2012d. Rethinking the concept of space-time in the general theory of relativity: the deflection of starlight and the gravitational red shift. Afr. Rev. Phys. 7, 183–201.

Wayne, R., 2012e. Taking the Mechanics Out of Space-Time and Putting it Back into Quantum Mechanics. http://fqxi.org/community/forum/topic/1402.

Wayne, R., 2015. Symmetry and the order of events in time: the thermodynamics of blackbodies composed of positive or negative mass. Turk. J. Phys. 39, 209–226.

Wayne, R., 2015b. Explanation of the perihelion motion of Mercury in terms of a velocity-dependent correction to Newton's Law of gravitation. Afr. Rev. Phys. 10, 185–194.

Wayne, R., 2015c. Radiation friction: shedding light on dark energy. Afr. Rev. Phys. 10, 361–364.

Wayne, R., 2015d. Symmetry and the order of events in time: the thermodynamics of blackbodies composed of positive or negative mass. Turk. J. Phys. 39, 209–226.

Wayne, R., 2016a. A reinterpretation of stimulated emission as spontaneous emission under non thermodynamic equilibrium conditions. Afr. Rev. Phys. 11, 17–22.

Wayne, R., 2016b. Evolution in real time. South Asian J. Multidiscipl. Stud. 2, 1–47.

Wayne, R., 2016c. The gravitational segregation of matter and antimatter. Afr. Rev. Phys. 11, 11–16.

Wayne, R., 2017. A push to understand gravity: a heuristic model. Afr. Rev. Phys. 12, 6–22.

Wayne, R., Hepler, P.K., 1984. The role of calcium ions in phytochrome mediated germination of spores of *Onoclea sensibilis* L. Planta 160, 2–20.

Wayne, R., Hepler, P.K., 1985a. Red light stimulates an increase in intracellular calcium in the spores of *Onoclea sensibilis*. Plant Physiol. 77, 8–11.

Wayne, R., Hepler, P.K., 1985b. The atomic composition of *Onoclea sensibilis* spores. Am. Fern J. 75, 12–18.

Wayne, R., Kadota, A., Watanabe, M., Furuya, M., 1991. Photomovement in *Dunaliella salina*: fluence rate-response curves and action spectra. Planta 184, 515–524.

Wayne, R., Mimura, T., Shimmen, T., 1994. The relationship between carbon and water transport in single cells of *Chara corallina*. Protoplasma 180, 118–135.

Wayne, R., November 20, 2015a. As Time Goes By and Albert Einstein. Do the Fundamental things still apply? The Lansing Star. http://www.lansingstar.com/around-town/12214-as-time-goes-by-and-albert-einstein.

Wayne, R., Staves, M., 1991. The density of the cell sap and endoplasm of *Nitellopsis* and *Chara*. Plant Cell Physiol. 32, 1137–1144.

Wayne, R., Staves, M.P., 1996a. Newton's law of gravitation from the apple's perspective; or a down to Earth model of gravisensing. Physiol. Plant 98, 917–921.

Wayne, R., Staves, M.P., 1996b. The August Krogh principle applies to plants. BioScience 46, 365–369.

Wayne, R., Staves, M.P., 1997. A down-to-Earth- model of gravisensing. Am. Soc. Grav. Space Biol. Bull. 10 (2), 57–64.

Wayne, R., Staves, M.P., 1998. Connecting undergraduate plant cell biology students with the scientists about whom they learn: a bibliography. Am. Biol. Teach. 60, 510–517.

Wayne, R., Staves, M.P., 2008. Model Scientists. Commun. Integr. Biol. 1, 1–7.

Wayne, R., Staves, M.P., Leopold, A.C., 1990. Gravity-dependent polarity of cytoplasmic streaming in Nitellopsis. Protoplasma 155, 43–57.

Wayne, R., Staves, M.P., Leopold, A.C., 1992. The contribution of the extracellular matrix to gravisensing in characean cells. J. Cell Sci. 101, 611–623.

Wayne, R., Tazawa, M., 1988. The actin cytoskeleton and polar water permeability in characean cells. Protoplasma (Suppl. 2), 116–130.

Wayne, R., Tazawa, M., 1990. Nature of the water channels in the internodal cells of Nitellopsis. J. Membr. Biol. 116, 31–39.

Weaver, W., 1970. Molecular biology: origin of the term. Science 170, 581–582.

Weber, H.H., 1955. Adenosine triphosphate and motility of living systems. Harvey Lect. 49, 37–56.

Weber, H.H., 1958. The Motility of Muscle and Cells. Harvard University Press, Cambridge, MA.

Weber, A., 1976. Synopsis of the presentations. Symp. Soc. Exp. Biol. 30, 445–456.

Weber, A., 1988. Reminiscences on Albert Szent-Györgyi. Biol. Bull. 174, 214–233.

Weber, A., Brybrook, S., Huflejt, M., Mosca, G., Routier-Kiierzkowska, A.L., Smith, R.S., 2015. Measuring the mechanical properties of plant cells by combining micro0indentation with osmotic treatments. J. Exp. Bot. 66, 3229–3241.

Weber, A., Franzini-Armstrong, C., 2002. Hugh E. Huxley: birth of the filament sliding model of muscle contraction. Trends Cell Biol. 12, 243–245.

Weber, B., Zicola, J., Oka, R., Stam, M., 2016. Plant enhancers: a call for discovery. Trends Plant Sci. 21, 974–987.

Webster, P.L., 1979. Variation in sister-cell cycle durations and loss of synchrony in cell lineages in root apical meristems. Plant Sci. Lett. 14, 13–22.

Weeden, N.F., 1981. Genetic and biochemical implications of the endosymbiotic origin of the chloroplast. J. Mol. Evol. 17, 133–139.

Weeds, A., 2013. Hugh Huxley (1924-2013). Nature 500, 530.

Weeds, A., Maciver, S., 1993. F-actin capping proteins. Curr. Opin. Cell Biol. 5, 63–69.

Wegener, A., 1966. The Origin of Continents and Oceans. Translated from the 4th revised German edition by B. Biram. Dover Publications, New York.

Wegst, U.G.K., 2006. Wood for sound. Am. J. Bot. 93, 1439–1448.

Wei, S., Bian, Y., Zhao, Q., Chen, S., Mao, J., Song, C., Cheng, K., Xiao, Z., Zhang, C., Ma, W., Zou, H., Ye, M., Dai, S., 2017. Salinity-induced palmella formation mechanism in halotolerant *Dunaliella salina* revealed by quantitative proteomics and phosphoproteomics. Front. Plant Sci. 8, 810.

Wei, T., Hibino, H., Omura, T., 2009. Release of *Rice dwarf virus* from insect vector involves secretory exosomes derived from multivesicular bodies. Commun. Integr. Biol. 2, 324–326.

Wei, C., Lintilhac, P.M., 2003. Loss of stability—a new model for stress relaxation in plant cell walls. J. Theor. Biol. 224, 305–312.

Wei, C., Lintilhac, P.M., 2007. Loss of stability: a new look at the physics of cell wall behavior during plant growth. Plant Physiol. 145, 763–772.

Wei, C., Lintilhac, L.S., Lintilhac, P.M., 2006. Loss of stability, pH, and the anisotropic extensibility of Chara cell walls. Planta 223, 1058–1067.

Wei, C., Lintilhac, P.M., Tanguay, J.J., 2001. An insight into cell elasticity and load-bearing ability. Measurement and theory. Plant Physiol. 126, 1129–1138.

Weier, T.E., 1931. A Study of the moss plastid after fixation by mitochondnial, osmium and silver techniques. I. The plastid during sporogenesis in *Polytrichum commune*. La Cellule 40, 261.

Weier, T.E., 1932a. A Study of the moss plastid after fixation by mitochondnial, osmium and silver techniques. II. The plastid during spermatogenesis in *Polytnichum commune* and *Catharinaea undulata*. La Cellule 41, 51.

Weier, T.E., 1932b. A comparison of the plastid with the Golgi zone. Biol. Bull. 62, 126–139.

Weier, T.E., 1938a. The structure of the chloroplast. Bot. Rev. 4, 497–530.

Weier, T.E., 1938b. The viability of cells containing granular and optically homogeneous chloroplasts. Protoplasma 31, 346–350.

Weier, T.E., 1961. The ultramicro structure of starch-free chloroplasts of fully expanded leaves of *Nicotiana rustica*. Am. J. Bot. 48, 615–630.

Weier, T.E., Thomson, W.W., 1962. Membranes of mesophyll cells of Nicotiana rustica and Phaseolus vulgaris with particular reference to the chloroplast. Am. J. Bot. 49, 807–820.

Weigel, D., Ahn, J.H., Blázquez, M.A., Borevitz, J.O., Christensen, S.K., Fankhauser, C., Ferrándiz, C., Kardailsky, I., Malancharuvil, E.J., Neff, M.M., Nguyen, J.T., Sato, S., Wang, Z.-Y., Xia, Y., Dixon, R.A., Harrison, M.J., Lamb, C.J., Yanofsky, M.F., Chory, J., 2000. Activation tagging in *Arabidopsis*. Plant Physiol. 122, 1003–1013.

Weihs, D., Mason, T.G., Teitell, M.A., 2006. Bio-microrheology: a frontier in microrheology. Biophys. J. 91, 4296–4305.

Weinberg, S., 1977. The First Three Minutes. Basic Books, New York.

Weiner, H., Burnell, J.N., Woodrow, I.E., Heldt, H.W., Hatch, M.D., 1988. Metabolite diffusion into bundle sheath cells from C4 plants. Relation to C4 photosynthesis and plasmodesmatal function. Plant Physiol. 88, 815–822.

Weinsberg, S., 1979. Is nature simple? In: Huff, D., Prewett, O. (Eds.), The Nature of the Physicsl Universe, 1976 Nobel Conference. John Wiley & Sons, New York, pp. 47–62.

Weise, S.E., Kiss, J.Z., 1999. Gravitropism of inflorescence stems in starch-deficient mutants of Arabidopsis. Int. J. Plant Sci. 160, 521–527.

Weisenberg, R.C., 1972. Microtubule formation *in vitro* in solutions containing low calcium concentrations. Science 177, 1104–1105.

Weisenberg, R.C., Borisy, G.G., Taylor, E.W., 1968. The colchicine-binding protein of mammalian brain and its relation to microtubules. Biochemistry 7, 4466–4478.

Weismann, A., 1891. Essays upon Heredity and Kindred Biological Problems, second ed., vol. 1. Clarendon Press, Oxford.

Weismann, A., 1892. Essays upon Heredity and Kindred Biological Problems. Clarendon Press, Oxford.

Weismann, A., 1893. The Germ-Plasm. A Theory of Heredity. Charles Scribner's Sons, New York.

Weiss, P., 1940. The problem of cell individuality in development. Am. Nat. 74, 34–46.

Weiss, W., 1967. The toxicity of tobacco smoke solutions for *Paramecium*: the influence of various forms of filtration. Arch. Environ. Health 10, 904–909.

Weiss, A., 1981. Replication and evolution of inorganic systems. Angew. Chem. Int. Ed. 20, 850–860.

Weiss, T.F., 1996. Cellular Biophysics. In: Transport, vol. 1. MIT Press, Cambridge, MA.

Weiss, T.F., 1997. Cellular Biophysics. In: Electrical Properties, vol. 2. MIT Press, Cambridge, MA.

Weiss, S.B., Gladstone, L., 1959. A mammalian system for the incorporation of cytidine triphosphate into ribonucleic acid. J. Am. Chem. Soc. 81, 4118–4119.

Weiss, B., Richardson, C.C., 1967. Enzymatic breakage and joining of deoxyribonucleic acid, I. Repair of single-stranded breaks in DNA by an enzyme system from *Escherichia coli* infected with T4 bacteriophage. Proc. Natl. Acad. Sci. U.S.A. 57, 1021–1028.

Weiss, W., Weiss, W.A., 1964. Effect of tobacco smoke solutions on *Paramecium*. Arch. Env. Health 9, 500–504.

Weiss, W., Weiss, W.A., 1966. Toxicity of cigarette solutions for *Paramecium*. Influence of variations in smoking technique. Arch. Env. Health 12, 227–230.

Weiss, W., Weiss, W.A., 1967. The gas phase of cigarette smoke for *Paramecium*. Arch. Env. Health 14, 682–686.

Weissbach, A., Horecker, B.L., Hurwitz, J., 1956. The enzymatic formation of phosphoglyceric acid from ribulose diphosphate and carbon dioxide. J. Biol. Chem. 218, 795–810.

Welburn, J.P., Cheeseman, I.M., 2008. Toward a molecular structure of the eukaryotic kinetochore. Dev. Cell 15, 645–655.

Welburn, J.P., Vleugel, M., Liu, D., Yates 3rd, J.R., Lampson, M.A., Fukagawa, T., Cheeseman, I.M., 2010. Aurora B phosphorylates spatially distinct targets to differentially regulate the kinetochore–microtubule interface. Mol. Cell 38, 383–392.

Welch, G.R., Easterby, J.S., 1994. Metabolic channeling verus free diffusion: transition-time analysis. TIBS 19, 193–197.

Welch, R.M., Epstein, E., 1968. The dual mechanism of alkali cation absorption by plant cells: their parallel operation across the plasmalemma. Proc. Natl. Acad. Sci. U.S.A. 61, 447–453.

Welch, M.D., Way, M., 2013. Arp2/3-mediated actin-based motility: a tail of pathogen abuse. Cell Host Microbe 11, 242–255.

Welcome Trust Case Control Consortium, 2007. Genome-wide association study of 14,000 cases of seven common diseases and 3,000 shared controls. Nature 447, 661–678.

Weldon, W.F.R., 1901. Mendel's Laws of alternative inheritance in peas. Biometrika 1, 228–254.

Weldon, W.F.R., 1902. On the ambiguity of Mendel's categories. Biometrika 2, 44–55.

Wellburn, A.R., 1982. Bioenergetic and ultrastructural changes associated with chloroplast development. Int. Rev. Cytol. 80, 133–191.

Wellburn, A.R., 1987. Plastids. Int. Rev. Cytol Suppl. 17, 149–210.

Wellensiek, S.J., 1939. The newest fad, colchicine, and its origin. Chron. Bot. 5, 15–17.

Weller, B., Zourelidou, M., Frank, L., Barbosa, I.C.R., Fastner, A., Richter, S., Jürgens, G., Hammes, U.Z., Schwechheimer, C., 2017. Dynamic PIN-FORMED auxin carrier phosphorylation at he plasma membrane controls auxin efflux-dependent growth. Proc. Natl. Acad. Sci. U.S.A. 114, E887–E896.

Wells, W.A., 2005. The discovery of tubulin. J. Cell Biol. 169, 552.

Wells, H.G., Huxley, J., Wells, G.P., 1934. The Science of Life. Cassell, London.

Wells, A.L., Lin, A.W., Chen, L.Q., Safer, D., Cain, S.M., Hasson, T., Carragher, B.O., Milligan, R.A., Sweeney, H.L., 1999. Myosin VI is an actin-based motor that moves backwards. Nature 401, 505–508.

Welti, R., Shah, J., Li, W., Li, M., Chen, J., Burke, J.J., Fauconnier, M.L., Chapman, K., Chye, M.L., Wang, X., 2007. Plant lipidomics: discerning biological function by profiling plant complex lipds using mass spectrometry. Front. Biosci. 12, 2494–2506.

Welti, R., Wang, X., 2004. Lipid species profiling: a high-throughout approach to identify lipid compositional changes and determine the function of genes involved in lipid metabolism and signaling. Curr. Opin. Plant Biol. 7, 337–344.

Weng, J.K., 2014. The evolutionary paths towards complexity: a metabolic perspective. New Phytol. 201, 1141–1149.

Weng, J.K., Akiyama, T., Bonawitz, N.D., Li, X., Ralph, J., Chapple, C., 2010. Convergent evolution of syringyl lignin via distinct biosynthetic pathways in the lycophyte *Selaginella* and flowering plants. Plant Cell 22, 1033–1045.

Weng, J.K., Banks, J.A., Chapple, C., 2008a. Parallels in lignin biosynthesis: a study in *Selaginella moellendorffii* reveals convergence across 400 million years of evolution. Commun. Integr. Biol. 1, 20–22.

Weng, J.K., Chapple, C., 2010. The origin and evolution of lignin biosynthesis. New Phytol. 187, 273–285.

Weng, J.K., Li, X., Stout, J., Chapple, C., 2008b. Independent origins of syringyl lignin in vascular plants. Proc. Natl. Acad. Sci. U.S.A. 105, 7887–7892.

Weng, J.K., Noel, J.P., 2012. The remarkable pliability and promiscuity of specialized metabolism. Cold Spring Harb. Symp. Quant. Biol. 77, 309–320.

Weng, J.K., Noel, J.P., 2013. Chemodiversity in *Selaginella*: a reference system for parallel and convergent metabolic evolution in terrestrial plants. Front. Plant Sci. 4, 119.

Weng, J.K., Philippe, R.N., Noel, J.P., 2012. The rise of chemodiversity in plants. Science 336, 1667–1670.

Werner, J.H., Cai, H., Jett, J.H., Reha-Krantz, L., Keller, R.A., Goodwin, P.M., 2003. Progress towards single-molecule DNA sequencing: a one color demonstration. J. Biotechnol. 102, 1–14.

Werner, P., Holmes, F.L., 2002. Justus Liebig and the plant physiologists. J. Hist. Biol. 35, 421–441.

Wernicke, W., Gunther, P., Jung, G., 1993. Microtubules and cell shaping in the mesophyll of *Nigella* damascena L. Protoplasma 173, 8–12.

Wesley, S.V., Helliwell, C.A., Smithy, N.A., Wang, M.B., Rouse, D.T., Liu, Q., Gooding, P.S., Singh, S.P., Abbott, D., Stoutjesdijk, P.A., Robinson, S.P., Gleave, A.P., Green, A.G., Waterhouse, P.M., 2001. Construct design foe efficient, effective and high-throughput gene silencing in plants. Plant J. 27, 581–590.

Whaley, W.G., Kephart, J.E., Mollenhauer, H.H., 1959. Developmental changes in the golgi-apparatus of maize root cells. Am. J. Bot. 46, 743–751.

Whaley, W.G., Mollenhauer, H.H., 1963. The Golgi apparatus and cell plate formation—a postulate. J. Cell Biol. 17, 216–221.

Whaley, W.G., Mollenhauer, H.H., Leech, J.H., 1960. The ultrastructure of the meristematic cell. Am. J. Bot. 47, 319–399.

Whatley, J.M., 1977a. The fine structure of Prochloron. New Phytol. 79, 309–313.

Whatley, J.M., 1977b. Variation in the basic pathway of chloroplast development. New Phytol. 78, 407–420.

Whatley, J.M., 1978. A suggested cycle of plastid developmental interrelationships. New Phytol. 80, 489–502.

Wheeler, J.J., Boss, W.F., 1987. Polyphosphoinositides are present in plasma membranes isolated from fusogenic carrot cells. Plant Physiol. 85, 389–392.

Wheeler, D.A., Srinivasan, M., Egholm, M., Shen, Y., Chen, L., McGuire, A., He, W., Chen, Y.-J., Makhijani, V., Roth, G.T., Gomes, X., Tartaro, K., Niazi, F., Turcotte, C.L., Irzyk, G.P., Lupski, J.R., Chinault, C., Song, X.-Z., Liu, Y., Yuan, Y., Nazareth, L., Qin, X., Muzny, D.M., Margulies, M., Weinstock, G.M., Gibbs, R.A., Rothberg, J.M., 2008. The complete genome of an individual by massively parallel DNA sequencing. Nature 452, 872–876.

Whelan, S., 2008. Inferring trees. Methods Mol. Biol. 452, 297–309.

Whelan, J., Murcha, M.W. (Eds.), 2015. Plant Mitochondria. Methods and Protocols. Humana Press, Springer, New York.

Whewell, W., 1858. Novum Organon Renovatum. John W. Parker and Son, London.

Whitaker, M., 1997. Calcium and mitosis. Prog. Cell Cycle Res. 3, 261–269.

White, A.D., 1877. The Warfare of Science. Henry S. King & Co., London.

White, A.D., 1913. A History of the Warfare of Science with Technology in Christendom. D. Appleton and Co., New York.

White, C., 2015. We, Robots: Staying Human in the Age of Big Data. Melville House, Brooklyn, NY.

White, J.W., Andrade-Sanchez, P., Gore, M.A., Bronson, K.F., Coffelt, T.A., Conley, M.M., Feldmann, K.A., French, A.N., Heun, J.T., Hunsake, D.J., Jenks, M.A., Kimball, B.A., Roth, R.L., Strand, R.J., Thorp, K.R., Wall, G.W., Wang, G., 2012. Field -based phenomics for plant genetics research. Field Crops Res. 133, 101–112.

White, R.G., Badelt, K., Overall, R.L., Vesk, M., 1994. Actin associated with plasmodesmata. Protoplasma 180, 169–184.

White, R.G., Barton, D.A., 2011. The cytoskeleton in plasmodesmata: a role in intercellular transport? J. Exp. Bot. 62, 5249–5266.

White, J.G., Borisy, G.G., 1983. On the mechanism of cytokinesis in animal cells. J. Theor. Biol. 101, 289–316.

White, R.G., Hyde, G., Overall, R.L., 1990. Microtubule arrays in regenerating Mougeotia protoplasts may be oriented by electric fields. Protoplasma 158, 73–85.

White, J.A., Scandalios, J.G., 1988. Molecular biology of intracellular protein trafficking. Physiol. Plant. 74, 397–408.

Whitehead, A.N., 1925. Science and the Modern World. MacMillan, New York.

Whiteheart, S.W., Griff, I.C., Brunner, M., Clary, D.O., Mayer, T., Buhrow, S.A., Rotheman, J.E., 1993. SNAP family of SNF attachment proteins includes a brain-sepecific isoform. Nature 362, 353–355.

Whiteman, S.A., Nühse, T.S., Ashford, D.A., Sanders, D., Maathuis, F.J., 2008a. A proteomic and phosphoproteomic analysis of *Oryza sativa* plasma membrane and vacuolar membrane. Plant J. 56, 146–156.

Whiteman, S.A., Serazetdinova, L., Jones, A.M., Sanders, D., Rathjen, J., Peck, S.C., Maathuis, F.J., 2008b. Identification of novel proteins and phosphorylation sites in a tonoplast enriched membrane fraction of *Arabidopsis thaliana*. Proteomics 8, 3535–3547.

Whitewoods, C.D., Cammarata, J., Nemec Venza, Z., Sang, S., Crook, A.D., Aoyama, T., Wang, X.Y., Waller, M., Kamisugi, Y., C.Cuming, A., Szövényi, P., Nimchuk, Z.L., K.Roeder, A.H., Scanlon, M.J., Harrison, J., 2018. *CLAVATA* was a genetic novelty for the morphological innovation of 3D growth in land plants. Curr. Biol. 28, 2365–2376.e5.

Whitfield, P.R., Bottomley, W., 1983. Organization and structure of chloroplast genes. Ann. Rev. Plant Physiol. 34, 279–310.

Whitlrow, G.J., 1972. The Nature of Time. Penguin, Harmondsworth, England.

Whitman, C.O., 1890. Specialization and organization, companion principles of all progress. The most important need in American Biology. In: Biological Lectures Delivered at the Marine Biological Laboratory at Woods Hole in the Summer Session of 1890. Ginn, Boston, pp. 1–26.

Whitman, C.O., 1894. The inadequacy of the cell-theory of development. In: Biological Lectures Delivered at The Marine Biological Laboratory of Wood's Hole in the Summer Session of 1893. Ginn & Co., Boston.

Whitney, S.E.C., Gidley, M.J., McQueen-Mason, S.J., 2000. Probing expansin action using cellose/hemicellulose composites. Plant J. 22, 327–334.

Whitney, H.M., Kolle, M., Andrew, P., Chittka, L., Steiner, U., Glover, B.J., 2009. Floral iridescence, produced by diffractive optics, acts as a cue for animal pollinators. Science 323, 130–133.

Whitrow, G.J., 1961. The Natural Philosophy of Time. Thomas Nelson and Sons, London.

Whittaker, R.H., 1969. New concepts of kingdoms of organisms. Science 163, 150–160.

Whittingham, C.P., 1974. The Mechanism of Photosynthesis. Elsevier, New York.

Wichterman, R., 1986. The Biology of *Paramecium*, second ed. Plenum Press, New York.

Wick, S.M., Cho, S.O., 1988. Higher plant spindle poles contain a protein that reacts with antibodies to the calcium-binding protein centrin. J. Cell Biol. 107, 455a (abstr).

Wick, S.M., Hepler, P.K., 1980. Localization of Ca^{++}-containing antimonide precipitates during mitosis. J. Cell Biol. 86, 500–513.

Wick, S.M., Hepler, P.K., 1982. Selective localization of intracellular Ca^{2+} with potassium antimonide. J. Histochem. Cytochem. 30, 1190–1204.

Wick, S.M., Seagull, R.W., Osborn, M., Weber, K., Gunning, B.E.S., 1981. Immunofluorescence microscopy of organized microtubule arrays in structurally stabilized meristematic plant cells. J. Cell Biol. 89, 685–690.

Wickner, W.T., 2011. Eugene Patrick Kennedy, 1919-2011. Proc. Natl. Acad. Sci. U.S.A. 108, 19122–19123.

Wickramasinghe, N.S., Lacey Jr., J.C., 1994. Catalytic roles of the AMP at the 3′ end of tRNAs. Mol. Cell. Biochem. 139, 117–121.

Wickramasinghe, N.S., Staves, M.P., Lacey Jr., J.C., 1991. Stereoselective, nonenzymatic, intramolecular transfer of amino acids. Biochemistry 30, 2768–2772.

Wickstead, B., Gull, K., 2007. Dyneins across eukaryotes: a comparative genomic analysis. Traffic 8, 1708–1721.

Wiebe, H.H., 1978. The significance of plant vacuoles. BioScience 28, 327–331.

Wiedenhoeft, B., Schmidt, G.W., Palevitz, B.A., 1988. Dissociation and reassembly of soybean clathrin. Plant Physiol. 86, 412–416.

Wieland, H., 1932. On the Mechanism of Oxidation. Yale University Press, New Haven, CT.

Wieland, T., Bodanszky, M., 1991. Structure elucidation. In: The World of Peptides. Springer-Verlag, Berlin, pp. 113–135.

Wienecke, K., Glas, R., Robinson, D.G., 1982. Organelles involved in the synthesis and transport of hydroxyproline-containing glycoproteins in carrot root disks. Planta 155, 58–63.

Wienkoop, S., Baginsky, S., Weckwerth, W., 2010. *Arabidopsis thaliana* as a model organism for plant proteome research. J. Proteom. 73, 2239–2248.

Wientjes, E., Drop, B., Kouřil, R., Boekema, E.J., Croce, R., 2013. During state 1 to state 2 transition in *Arabidopsis thaliana*, the photosystem II supercomplex gets phosphorylated but does not disassemble. J. Biol. Chem. 288, 32821–32826.

Wiercinski, F.J., 1989. Calcium, an overview—1989. Biol. Bull. 176, 195–217.

Wiese, C., Zheng, Y., 2006. Microtubule nucleation: gamma-tubulin and beyond. J. Cell Sci. 119, 4143–4153.

Wightman, R., Turner, S., 2010a. Trafficking of the cellulose synthase complex in developing xylem vessels. Biochem. Soc. Trans. 38, 755–760.

Wightman, R., Turner, S., 2010b. Trafficking of the plant cellulose synthase complex. Plant Physiol. 153, 427–432.

Wikstrom, M.K.F., 1977. Proton pump coupled to cytochrome *c* oxidase in mitochondria. Nature 266, 271–273.

Wildman, S.G., 1979. Aspects of fraction 1 protein evolution. Arch. Biochem. Biophys. 196, 598–610.

Wildman, S.G., 1998. Discovery of Rubisco. Discov. Plant Biol. 1, 163–173.

Wildman, S.G., 2002. Along the trail from fraction 1 protein to Rubisco (ribulose bisphosphate carboxylase-oxygenase. Photosynth. Res. 73, 243–250.

Wildman, S.G., Bonner, J., 1947. The proteins of green leaves. I. Isolation, enzymatic properties and auxin content of spinach cytoplasmic proteins. Arch. Biochem. Biophys. 14, 381–413.

Wildman, S.G., Hirsch, A.M., Kirchanski, S.J., Spencer, D., 2004. Chloroplasts in living cells and the string-of-grana-concept of chloroplast structure revisited. Photosynth. Res. 80, 345–352.

Wildman, S.G., Hongladarom, R., Honda, S.I., 1962. Chloroplasts and mitochondria in living plant cells: cine photomicrographic studies. Science 138, 434–436.

Wildman, S.G., Jope, C., Atchison, B.A., 1974. Role of mitochondria in the origin of chloroplast starch grains. Plant Physiol. 54, 231–237.

Wildman, S.G., Jope, C.A., Atchison, B.A., 1980. Light microscopic analysis of the three dimensional structure of higher plant chloroplasts. Position of starch grains and probably spiral arrangement of stroma lamellae and grana. Bot. Gaz. 141, 24–36.

Wileman, T., 2013. Autophagy as a defence against pathogens. Essays Biochem. 55, 153–163.

Wiley, S., 2008. Hypothesis-free? No such thing. Scientist 22 (5). http://www.the-scientist.com/?articles.view/articleNo/26330/title/Hypothesis-Free–No-Such-Thing/.

Wilhelm, B.T., Marguerat, S., Watt, S., Schubert, F., Wood, V., Goodhead, I., Penkett, C.J., Rogers, J., Bähler, J., 2008. Dynamic repertoire of a eukaryotic transcriptome surveyed at single-nucleotide resolution. Nature 453, 1239–1243.

Wilkins, M.H.F., December 11, 1962. The Molecular Configuration of Nucleic Acids. Nobel Lecture.

Wilkins, M.R., Sanchez, J.-C., Cooley, A.A., Appel, R.D., Humphrey-Smith, I., Hochstrasser, D.F., Williams, K.L., 1995. Progress with proteome projects: why all proteins expressed by a genome should be identified and how to do it. Biotech. Genet. Eng. Rev. 13, 19–50.

Wilkins, M.H.F., Stokes, A.R., Wilson, H.R., 1953. Molecular structure of deoxypentose nucleic acids. Nature 171, 738–740.

Wille, A.C., Lucas, W.J., 1984. Ultrastructural and histochemical studies on guard cells. Planta 160, 129–142.

Willey, D.L., Fischer, K., Wachter, E., Link, T.A., Flügge, U.I., 1991. Molecular cloning and structural analysis of the phosphate translocator from pea chloroplasts and its comparison to the spinach phosphate translocator. Planta 183, 451–461.

Williams, R.J.P., 1961. Possible functions of chains of catalysts. J. Theor. Biol. 1, 1–17.

Williams, R.J.P., 1974. Calcium ions: their ligands and their function. Biochem. Soc. Symp. 39, 133–138.

Williams, R.J.P., 1976. Calcium chemistry and its relation to biological function. Symp. Soc. Exp. Biol. 30, 1–17.

Williams, R.J.P., 1980. A general introduction to the special properties of the calcium ion and their deployment in biology. In: Carafoli, F.E, Kretsinger, R.H., MacLennan, D.H., Wasserman, R.H. (Eds.), Calcium-Binding Proteins. Structure and Function. Siegel Elsevier North Holland, Amsterdam, pp. 3–10.

Williams, R.J.P., 1993a. The history of proton-driven ATP formation. Biosci. Rep. 13, 191–212.

Williams, N., 2016. Irene Manton, Erwin Schrödinger and the puzzle of chromosome structure. J. Hist. Biol. 49, 425–459.

Williams, M., 2017a. Entire photodamaged chloroplasts are transported to the central vacuole by autophagy. Plantae. https://plantae.org/research/entire-photodamaged-chloroplasts-are-transported-to-the-central-vacuole-by-autophagy/?utm_source=TrendMD&utm_medium=cpc&utm_campaign=Plantae_TrendMD_0.

Williams, M., 2017b. Photodamaged chloroplasts are targets of cellular garbage disposal. Plantae. https://plantae.org/research/photodamaged-chloroplasts-are-targets-of-cellular-garbage-disposal/?utm_source=TrendMD&utm_medium=cpc&utm_campaign=Plantae_TrendMD_0.

Williams, R.J.P., Fraúto da Silva, J.J.R., 1996. The natural Selection of the Chemical Elements, The Environment and Life's Chemistry. Clarendon Press, Oxford.

Williams, B.A.P., Hirt, R.P., Lucocq, J.M., Embley, T.M., 2002. A mitochondrial remnant in the microsporidian *Trachipleistophora hominis*. Nature 418, 865–869.

Williams, R.J.P., May 26, 1993. Science and the Good Life. The J. D. Bernal Lecture 1993 delivered at Birkbeck College. London.

Williams, L.E., Schueler, S.B., Briskin, D.P., 1990. Further characterization of the red beet plasma membrane Ca^{2+}-ATPase using GTP as an alternative substrate. Plant Physiol. 92, 747–754.

Williams, T.C.R., Sweetlove, L.J., Ratcliffe, R.G., 2011. Capturing metabolite channeling in metabolic phenotypes. Plant Physiol. 157, 981–984.

Williamson, R.E., 1975. Cytoplasmic streaming in *Chara*: a cell model activated by ATP and inhibited by cytochalasin B. J. Cell Sci. 17, 655–668.

Williamson, R.E., 1990. Alignment of cortical microtubules by anisotropic wall stresses. Aust. J. Plant Physiol. 17, 601–613.

Williamson, R.E., 1991. Orientation of cortical microtubules in interphase plant cells. Int. Rev. Cytol. 129, 135–206.

Williamson, R.E., Ashley, C.C., 1982. Free Ca^{2+} and cytoplasmic streaming in alga *Chara*. Nature 296, 647–651.

Williamson, R.E., McCurdy, D.W., Hurley, U.A., Perkin, J.L., 1987. Actin of *Chara* giant internodal cells. Plant Physiol. 85, 268–272.

Williamson, R.E., Perkin, J.L., McCurdy, D.W., Craig, S., Hurley, U.A., 1986. Production and use of monoclonal antibodies to study the cytoskeleton and other components of the cortical cytoplasm of Chara. Eur. J. Cell Biol. 41, 1–8.

Willison, J.H.M., Brown Jr., R.M., 1978. Cell wall structure and deposition in *Glaucocystis nostochinearum*. J. Cell Biol. 77, 103–119.

Willstätter, R., 1927. Problems and Methods in Enzyme Research. Cornell University, Ithaca.

Willstätter, R., 1965. From My Life. The Memoirs of Richard Willstätter. Translated by Lilli S. Hornig. W. A. Benjamin, Inc., New York.

Willstätter, R., June 3, 1920. On plant pigments. Nobel Prize, 1915. Nobel Lecture.

Willstätter, R., Stoll, A., 1928. Investigations on Chlorophyll. Methods and Results. Translated by F. M. Schertz and A. R. Merz. The Science Press, Lancaster, PA.

Wilson, E.B., 1895. An Atlas of the Fertilization and Karyokinesis of the Ovum. MacMillan, New York.

Wilson, E.B., 1896. The Cell in Development and Inheritance, 1966. Johnson Reprint Corp, New York.

Wilson, E.B., 1899. The structure of protoplasm. Science 10, 33–45.

Wilson, E.B., 1923. The Physical Basis of Life. Yale University Press, New Haven, CT.

Wilson, E.B., 1925. The Cell in Development and Heredity, third ed. MacMillan, New York.

Wilson Jr., E.B., 1952. An Introduction of Scientific Research. McGraw-Hill, New York.

Wilson, S.B., 1980. Energy conservation by the plant mitochondrial cyanide-insensitive oxidase. Biochem. J. 190, 349–360.

Wilson, E.O., 1998. Consilience: The Unity of Knowledge. Knopf, New York.

Wilson, T.P., Canny, M.J., McCully, M.E., Lefkovitch, L.P., 1990. Breakdown of cytoplasmic vacuoles. A model of endoplasmic membrane rearrangement. Protoplasma 155, 144–152.

Wilson, R.K., Chen, C., Avdalovic, N., Burns, J., Hood, L., 1990a. Development of an automated procedure for fluorescent DNA sequencing. Genomics 6, 626–634.

Wilson, R.K., Chen, C., Hood, L., 1990b. Optimization of asymmetric polymerase chain reaction for fluorescent DNA sequencing. BioTechniques 8, 184–189.

Wilson, R.J., Denny, P.W., Preiser, P.R., Rangachari, K., Roberts, K., Roy, A., Whyte, A., Strath, M., Moore, D.J., Moore, P.W., Williamson, D.H., 1996. Complete gene map of the plastid-like DNA of the malaria parasite *Plasmodium falciparum*. J. Mol. Biol. 261, 155–172.

Wilson, K.L., Foisner, R., 2010. Lamin-binding proteins. Cold Spring Harb. Perspect. Biol. 2, a000554.

Wilson, D.E., Lewis, M.J., Pelham, H.R., 1993. pH-dependent binding of KDEL to its receptor in vitro. J. Biol. Chem. 268, 7465–7468.

Wilson, E.B., Pollister, A.W., 1937. Observations on sperm formation in the centrurid scorpions with especial reference to the Golgi material. J. Morphol. 60, 407–443.

Wilson, I.W., Schiff, C.L., Hughes, D.E., Somerville, S.C., 2001. Quantitative trait loci analysis of powdery mildew disease resistance in the *Arabidopsis thaliana* accession Kashmir-1. Genetics 158, 1301–1309.

Wilson, B.A., Schisler, J.C., Willis, M.S., 2010. Sir Hans Adolf Krebs: architect of metabolic cycles. Lab Med. 41 (6), 377–380.

Wilson, D.W., Wilcox, C.A., Flynn, G.C., Chen, E., Kuang, W.-J., Hensel, W.J., Block, M.R., Ullrich, A., Rothman, J.E., 1989. A fusion protein required for vesicle-mediated transport in both mammalian cells and yeast. Nature 339, 355–359.

Wilson, R.K., Yuen, A.S., Clark, S.M., Spence, C., Arakelian, P., Hood, L.E., 1988. Automation of dideoxynucleotide DNA sequencing reactions using a robotic workstation. BioTechniques 6, 776–787.

Wimmers, L.E., Ewing, N.N., Meger, D.J., Bennett, A.B., 1990. Molecular biology of plant P-type ion-translocating ATPases. In: Leonard, R.T., Hepler, P.K. (Eds.), Calcium in Plant Growth and Development. The American Society of Plant Physiologists Symposium Series, vol. 4, pp. 36–45.

Winderlich, R., 1950. Carl Friedrich Wenzel 1740-1793. J. Chem. Educ. 27, 56–59.

Winkel, B.S.J., 2004. Metabolic channeling in plants. Annu. Rev. Plant Biol. 55, 85–107.

Winkel-Shirley, B., 2002. Biosynthesis of flavonoids and effects of stress. Curr. Opin. Plant Biol. 5, 218–223.

Winship, L.J., Obermeyer, Geitmann, A., Hepler, P.K., 2010. Under pressure, cell walls set the pace. Trends Plant Sci. 15, 363–369.

Winship, L.J., Obermeyer, Geitmann, A., Hepler, P.K., 2011. Pollen tubes and the physical world. Trends Plant Sci. 16, 353–355.

Winter, H., Robinson, D.G., Heldt, H.W., 1994. Subcellular volumes and metabolite concentrations in spinach leaves. Planta 193, 530–535.

Winter, K., Smith, J.A.C. (Eds.), 1996. Crassulacean Acid Metabolism: Biochemistry, Ecophysiology and Evolution. Springer, Berlin.

Wirth, H.E., Rigg, G.B., 1937. The acidity of the juice of Desmarestia. Am. J. Bot. 24, 68–70.

Wise, D., 1988. The diversity of mitosis: the value of evolutionary experiments. Biochem. Cell. Biol. 66, 515–529.

Wise, D.A., Bhattacharjee, L., 1992. Anti kinetochore antibodies interfere with prometaphase but not anaphase chromosome movement in living PtK2 cells. Cell Motil. Cytoskelet. 23, 157–167.

Wise, D., Cassimeris, L., Rieder, C.L., Wadsworth, P., Salmon, E.D., 1986. Incorporation of tubulin into kinetochore microtubules—relation to chromosome congression. J. Cell Biol. 103, 412.

Wise, R.P., Moscou, M.J., Bogdanove, A.J., Whitham, S.A., 2007. Transcripts profiling in host-pathogen interactions. Annu. Rev. Phytopathol. 45, 329–369.

Wisecaver, J.H., Borowsky, A.T., Tzin, V., Jander, G., Kliebenstein, D.J., Rokas, A., 2017. A global coexpression network approach for connecting genes to specialized metabolic pathways. Plant Cell 29, 944–959.

Wiśiewska, J., Xu, J., Seifertova, D., Brewer, P.B., Růžička, K., Blilou, I., Rouquié, D., Benková, E., Scheres, B., Friml, J., 2006. Polar PIN localization directs auxin flow in plants. Science 312, 883.

Witztum, A., 1978. Transcellular chloroplast banding patterns in leaves of *Elodea densa* induced by light and DCMU. Ann. Bot. 42, 1459–1462.

Witztum, A., Wayne, R., 2012. Button botany: plasmodesmata in vegetable ivory. Protoplasma 249, 721–724.

Witztum, A., Wayne, W., 2014. Fibre cables in the lacunae of *Typha* leaves contribute to a tensegrity structure. Ann. Bot. 113, 789–797.

Witztum, A., Wayne, R., 2015. Variation in fiber cables in the lacunae of leaves in hybrid swarms of Typha ×glauca. Aquatic Bot. 124, 39–44.

Witztum, A., Wayne, R., 2016a. Fiber cables in leaf blades of *Typha domingensis* and their absence in *Typha elephantina*: a diagnostic character for phylogenetic affinity. Isreael J. Plant Sci. 63, 116–123.

Witztum, A., Wayne, R., 2016b. Lignified and nonlignified fiber cables in the lacunae of *Typha angustifolia*. Protoplasma 253, 1589–1592.

Wodzicki, T.J., Brown, C.L., 1973. Organization and breakdown of the protoplast during maturation of pine tracheids. Am. J. Bot. 60, 631–640.

Woese, C.R., 1967. The Genetic Code. The Molecular Basis for Genetic Expression. Harper & Row, New York.

Woese, C.R., 1977. Endosymbionts and mitochondrial origins. J. Mol. Evol. 10, 93–96.

Woese, C.R., 1998. The universal ancestor. Proc. Natl. Acad. Sci. U.S.A. 95, 6854–6859.

Woese, C.R., Fox, G.E., 1977. The concept of cellular evolution. J. Mol. Evol. 10, 1–6.

Wohlfarth-Bottermann, K.E., 1962. Weitreichende, fibrilläre Protoplasma differenzierungen und ihre Bedeutung für die Protoplasmaströmung. Protoplasma 54, 514–539.

Wolf, S., Deom, C.M., Beachy, R.N., Lucas, W.J., 1989. Movement protein of tobacco mosaic virus modifies plasmodesmatal size exclusion limit. Science 246, 377–379.

Wolf, S., Deom, C.M., Beachy, R., Lucas, W.J., 1991. Plasmodesmatal function is probed using transgenic tobacco plants that express a virus movement protein. Plant Cell 3, 593–604.

Wolf, S., Greiner, S., 2012. Growth control by cell wall pectins. Protoplasma 249 (Suppl. 2), S169–S175.

Wolf, S., Lucas, W.J., 1994. Virus movement proteins and other molecular probes of plasmodesmata function. Plant Cell Environ. 17, 573–585.

Wolf, L., Rizzini, L., Stracke, R., Ulm, R., Rensing, S.A., 2010. The molecular and physiological response of *Physcomitrella patens* to UV-B radiation. Plant Physiol. 153, 1123–1134.

Wolfe, J., Dowgert, M.F., Steponkus, P.L., 1986. Mechanical study of the deformation and rupture of the plasma membranes of protoplasts during osmotic expansions. J. Membr. Biol. 93, 63–74.

Wolfenstetter, S., Wirsching, P., Dotzauer, D., Schneider, S., Sauer, N., 2012. Routes to the tonoplast: the sorting of tonoplast transporters in *Arabidopsis* mesophyll protoplasts. Plant Cell 24, 215–232.

Wolins, N.E., Donaldson, R.P., 1994. Specific binding of the peroxisomal protein targeting sequence ot glyoxysomal membranes. J. Biol. Chem. 269, 1149–1153.

Wolken, J.J., Palade, G.E., 1952. Fine structure of chloroplasts in two flagellates. Nature 170, 114–115.

Wolken, J.J., Palade, G.E., 1953. An electron microscope study of two flagellates, chloroplast structure and variation. Ann. N.Y. Acad. Sci. 56, 873–889.

Woll, K., Borsuk, L.A., Stransky, H., Nettleton, D., Schnable, P.S., Hochholdinger, F., 2005. Isolation, characterization, and pericycle-specific transcriptome analyses of the novel maize lateral and seminal root initiation mutant *rum1*. Plant Physiol. 139, 1255–1267.

Wollaston, W.H., 1808. On Super-acid and Sub-acid Salts. Philos. Trans. R. Soc. 98, 96–102.

Wollman, E.L., Jacob, F., Hayes, W., 1956. Conjugation and genetic recombination in *Escherichia coli* K-12. Cold Spring Harb. Symp. Quant. Biol. 21, 141–162.

Wollman, R., Stuurman, N., 2007. High throughput microscopy: from Raw images to discoveries. J. Cell Sci. 120, 3715–3722.

Wolniak, S.M., 1987. Lithium alters mitotic progression in stamen hair cells of *Tradescantia* in a time-dependent and reversible fashion. Eur. J. Cell Biol. 44, 286–293.

Wolniak, S.M., Cande, W.Z., 1980. Physiological requirements for ciliary reactivation of bracken fern spermatozoids. J. Cell Sci. 43, 195–207.

Wolniak, S.M., Hepler, P.K., Jackson, W.T., 1980. Detection of the membrane-calcium distribution during mitosis in *Haemanthus* endosperm with chlortetracycline. J. Cell Biol. 87, 23–32.

Wolniak, S.M., Hepler, P.K., Jackson, W.T., 1981. The coincident distribution of calcium-rich membranes and kinetochore fibers at metaphase in living endosperm cells of *Haemanthus*. Eur. J. Cell Biol. 25, 171–174.

Wolosewick, J.J., 2002. Joining the trk with Keith up the Sepentine Road—the lattice from another perspective. Biol. Cell. 94, 557–559.

Wolosewick, J.J., Porter, K.R., 1979. Microtrabecular lattice of the cytoplasmic ground substance; artifact or reality. J. Cell Biol. 82, 114–139.

Wolpert, L., Richards, A., 1997. Passionate Minds: The Inner World of Scientists. Oxford University Press, Oxford.

Wolt, J.D., Wang, K., Yang, B., 2016. The regulatory status of genome-edited crops. Plant Biotechnol. J. 14, 510–518.

Wolters-Arts, A.M.C., van Amstel, T., Derksen, J., 1993. Tracing cellulose microfibril orientation in inner primary cell walls. Protoplasma 175, 102–111.

Wolyn, D.J., Borevitz, J.O., Loudet, O., Schwartz, C., Maloof, J., Ecker, J.R., Berry, C.C., Chory, J., 2004. Light response quantitative trait loci identified with composite interval and eXtreme array mapping in *Arabidopsis thaliana*. Genetics 167, 907–917.

Won, H., Renner, S.S., 2003. Horizontal gene transfer from flowering plants to *Gnetum*. Proc. Natl. Acad. Sci. U.S.A. 100, 10824–10829.

Woo, J.W., Kim, J., Kwon, S.I., Corvalán, C., Cho, S.W., Kim, H., Kim, S.-G., Kim, S.-T., Choe, S., Kim, J.S., 2015. DNA-free genome editing in plants with preassembled CRISPR-Cas9 ribonucleoproteins. Nat. Biotechnol. 33, 1162–1165.

Wood, H.G., 1982. The discovery of the fixation of CO_2 by heterotrophic organisms and metabolism of the propionic acid bacteria. In: Semenza, G. (Ed.), Of Oxygen, Fuels, and Living Matter, Part 2. John Wiley & Sons, Chichester, pp. 173–250.

Woodcock, C.L.F., 1973. Ultrastructure of inactive chromatin. J. Cell Biol. 59, 368a.

Woodcock, C.L.F., Frado, L.-L.Y., Rattner, J.B., 1984. The higher order structure of chromatin: evidence for a helical ribbon arrangement. J. Cell Biol. 99, 42–52.

Woodcock, C.L.F., McEweon, B.F., Frank, J., 1991a. Ultrastructure of Chromatin. II. Three-dimensional reconstruction of isolated fibers. J. Cell Sci. 99, 107–114.

Woodcock, C.L.F., Miller, G.J., 1973. Ultrastructural features of the life cycle of *Acetabularia mediterranea*. I. Gametogenesis. Protoplasma 77, 313–329.

Woodcock, C.L.F., Safer, J.P., Stanchfield, J.E., 1976. Structural repeating units in chromatin. I. Evidence for their general occurrence. Exp. Cell Res. 97, 101–110.

Woodcock, C.L.F., Woodcock, H., Horowitz, R.A., 1991b. Ultrastructure of chromatin. I. Negative staining of isolated fibers. J. Cell Sci. 99, 99–106.

Woodger, J.H., 1948. Observations on the present state of embryology. Symp. Soc. Exp. Biol. 2, 354.

Woodhouse, E.D., 1933. Sap hydraulics. Plant Physiol. 8, 177–202.

Woodrow, I.E., Berry, J.A., 1988. Enzymatic regulation of photosynthetic CO_2 fixation in C3 plants. Ann. Rev. Plant Physiol. Mol. Biol. 39, 533–594.

Woodruff, L.L., 1939. Microscopy before the nineteenth century. Am. Nat 73, 485–516.

Woodward, R.B., Ayer, W.A., Beaton, J.M., Bickelhaupt, F., Bonnett, R., Buchschacher, P., Closs, G.L., Dutler, H., Hanna, J., Hauck, F.P., Itó, S., Langemann, A., Le Goff, E., Leimgruber, W., Lwowski, W., Sauer, J., Valenta, Z., Volz, H., 1960. The total synthesis of chlorophyll. J. Am. Chem. Soc. 82, 3800–3802.

Woodward, A.W., Bartel, B., 2005. The *Arabidopsis* peroxisomal targeting signal type 2 receptor PEX7 is necessary for peroxisomal function and dependent on PEX5. Mol. Biol. Cell 16, 573–583.

Woodward, R.B., December 11, 1965. Recent Advances in the Chemistry of Natural Products. Nobel Lecture.

Woolfenden, H.C., Bourdais, G., Kopischke, M., Miedes, E., Molina, A., Robatzek, S., Morris, R.J., 2017. A computational approach for inferring the cell wall properties that govern gueard cell dynamics. Plant J. 92, 5–18.

Woollard, A.A.D., Moore, I., 2008. The functions of Rab GTPases inn plant membrane traffic. Curr. Opin. Plant Biol. 11, 610–619.

Wordeman, L., Cande, W.Z., 1987. Reactivation of spindle elongation *in vitro* is correlated with the phosphorylation of a 205 kD spindle-associated protein. Cell 50, 535–543.

Wordeman, L., Mitchison, T.J., 1995. Identification and partial characterization of mitotic centromere-associated kinesis, a kinesin-related protein that associates with centromeres during mitosis. J. Cell Biol. 128, 95–105.

Work, T.S., Burdon, R.K., 1983. Introduction. Lab. Tech. Biochem. Mol. Biol. 10, 1–29.

Worley, L.G., 1946. The Golgi apparatus—an interpretation of its structure and significance. Ann. N.Y. Acad. Sci. 47, 1–56.

Worman, H.J., Juan, J., Blobel, G., Georgatos, S.D., 1988. A lamin B receptor in the nuclear envelope. Proc. Natl. Acad. Sci. U.S.A. 85, 8531–8534.

Wragham, R., Rodriguez, E., 1989. Selection of plants with medicinal properties by wild chimpanzees. Fitoterapia 60, 378–380.

Wraight, C.A., 2014. Roderick K. Clayton: a life, and some personal recollections. Photosynth. Res. 120, 9–26.

Wright, S., 1931. Evolution in Mendelian populations. Genetics 16, 97–159.

Wright, S., 1941. The physiology of the gene. Physiol. Rev. 21, 487–527.

Wright, S., 1945. Physiological aspects of genetics. Ann. Rev. Physiol. 7, 75–106.

Wright, J.P., Fisher, D.B., 1981. Measurement of the sieve tube membrane potential. Plant Physiol. 67, 845–848.

Wright, P., July 2, 2002. Erwin Chargaff Obituary The Guardian. https://www.theguardian.com/news/2002/jul/02/guardianobituaries.obituaries.

Wright, A.V., Nuñez, J.K., Doudna, J.A., 2016. Biology and applications of CRISPR systems: harnessing nature's toolbox for genome editing. Cell 164, 29–44.

Wright, B.L., Salisbury, J., Jarvik, J.W., 1985. A nucleus-basal body connector in *Chlamydomonas reinhardtii* that may function in basal body localization or segregation. J. Cell Biol. 101, 1903–1912.

Wright, A.J., Smith, L.G., 2007. Division plane orientation in plant cells. Plant Cell Monogr. 9, 33–57.

Wright, D.A., Townsend, J.A., Winfrey Jr., R.J., Irwin, P.A., Rajagopal, J., Lonosky, P.M., Hall, B.D., Jondle, M.D., Voytas, D.F., 2005. High-frequency homologous recombination in plants mediated by zinc-finger nucleases. Plant J. 44, 693–705.

Wright, S.F., Upadhyaya, A., 1996. Extraction of an abundant and unusual protein from soil and comparison with hyphal protein of arbuscular mycorrhizal fungi. Soil Sci. 161, 575–586.

Wright, S.F., Upadhyaya, A., 1998. A survey of soils for aggregate stability and glomalin, a glycoprotein produced by hyphae of arbuscular mycorrhizal fungi. Plant Soil 198, 97–107.

Wright, S.F., Upadhyaya, A., 1999. Quantification of arbuscular mycorrhizal fungi activity by the glomalin concentration on hyphal traps. Mycorrhiza 8, 283–285.

Wrischer, M., 1989. Ultrastructural localization of photosynthetic activity in thylakoids during chloroplast development in maize. Planta 177, 18–23.

Wu, R., 1970. Nucleotide sequence analysis of DNA. I. Partial sequence of the cohesive ends of bacteriophage lambda and 186 DNA. J. Mol. Biol. 51, 501–521.

Wu, R., 1972. Nucleotide sequence analysis of DNA. Nat. New Biol. 236, 198–200.

Wu, R., 1993. Development of enzyme-based methods for DNA sequence analysis and their applications in the genome projects. Adv. Enzymol. Relat. Area Mol. Biol. 67, 431–468.

Wu, R., 1994. Development of the primer-extension approach: a key role in DNA sequencing. Trends Biochem. Sci. 19, 429–433.

Wu, M., 1998a. The discovery of chloroplast DNA replication origin in *Chlamydomonas reinhardtii*. Discov. Plant Biol. 2, 73–84.

Wu, R., 1998b. Discovery of transgenic plants. Discov. Plant Biol. 1, 115–129.

Wu, R., Donelson, J., Padmanabhan, R., Hamilton, R., 1972. Determination of primary nucleotide sequences in DNA molecules. Bulletin de L'Institut Pasteur 70, 203–233.

Wu, M., Eisen, J.A., 2008. A simple, fast, and accurate method of phylogenomic inference. Genome Biol. 9, R151.

Wu, S., Gallagher, K.L., 2011. Mobile protein signals in plant development. Curr. Opin. Plant Biol. 14, 563–570.

Wu, S., Gallagher, K.L., 2012. Transcription factors on the move. Curr. Opin. Plant Biol. 15, 645–651.

Wu, Y., Hiratsuka, K., Neuhaus, G., Chua, N.H., 1996. Calcium and cGMP target distinct phytochrome-responsive elements. Plant J. 10, 1149–1154.

Wu, L., Joshi, C.P., Chiang, V.L., 2001. A xylem-specific cellulose synthase gene from aspen (*Populus tremoides*) is responsive to mechanical stress. Plant J. 22, 495–502.

Wu, R., Kaiser, A.D., 1967. Mapping the 5′-terminal nucleotides of the DNA of bacteriophage λ and related phases. Proc. Natl. Acad. Sci. U.S.A. 57, 170–177.

Wu, R., Kaiser, A.D., 1968. Structure and base sequence in the cohesive ends of bacteriophage lambda DNA. J. Mol. Biol. 35, 523–537.

Wu, Y., Llewellyn, D.J., White, R., Ruggiero, K., Al-Ghazi, Y., Dennis, E.S., 2007. Laser capture microdissection and cDNA microarrays used to generate gene expression profiles of the rapidly expanding fibre initial cells on the surface of cotton ovules. Planta 22, 1475–1490.

Wu, S.S.H., Platt, K.A., Ratnayake, C., Wang, T.-W., Ting, J.T.L., Huang, A.H.C., 1997. Isolation and characterization of neutral-lipid-containing organelles and globuli-filled plastids from *Brassica napus* tapetum. Proc. Natl. Acad. Sci. U.S.A. 94, 12711–12716.

Wu, J., Prole, D.L., Shen, Y., Lin, Z., Gnanasekaran, A., Liu, Y., Chen, L., Zhou, H., Chen, S.R.W., Usachev, Y.M., Taylor, C.W., Campbell, R.E., 2014. Red fluorescent genetically encoded Ca^{2+} indicators for use in mitochondria and endoplasmic reticulum. Biochem. J. 464, 13–22.

Wu, S., Scheible, W.-R., Schindelasch, D., ven den Daele, H., de Veylder, L., Baskin, T.I., 2010. A conditional mutation in *Arabidopsis thaliana* separase induces chromosome non-disjunction, aberrant morphogenesis and cyclin B1;1 stability. Development 137, 953–961.

Wu, R., Taylor, E., 1971. Nucleotide sequence analysis of DNA. II. Complete nucleotide sequence of the cohesive ends of bacteriophage λ DNA. J. Mol. Biol. 57, 491–511.

Wu, R., Tu, C.D., Padmanabhan, R., 1973. Nucleotide sequence analysis of DNA. XII. The chemical synthesis and sequence analysis of a dodecadeoxynucleotide which binds to the endolysin gene of bacteriophage lambda. Biochem. Biophys. Res. Commun. 55, 1092–1098.

Wu, N., Veillette, A., 2011. Immunology: magnesium in a signalling role. Nature 475, 462–463.

Wu, X., Weigel, D., Wigge, P.A., 2002. Signaling in plants by intercellular RNA and protein movement. Genes Dev. 16, 151–158.

Wurtele, M., Jelich-Ottman, C., Wittinghofer, A., Oecking, C., 2003. Structural view of a fungal toxin acting on a 14-3-3 regulatory complex. EMBO J. 22, 987–994.

Wuttke, H.G., 1976. Chromoplasts in *Rosa rugosa*: development and chemical characterization of tubular elements. Z. Naturforsch. Teil C 31, 456–460.

Wyatt, H.V., 1972. When does information become knowledge? Nature 256, 86–89.

Wythe, J.H., 1880. The Science of Life; or, Animal and Vegetable Biology. Phillips & Hunt, New York.

Xi, X., Kim, S.H., Tittmann, B., 2015. Atomic force microscopy based nonoindentation study of onion abaxial epidermis walls in aqueous environment. J. Appl. Phys. 117, 024703.

Xicluna, J., Lacombe, B., Dreyer, I., Alcon, C., Jeanguenin, L., Sentenac, H., Thibaud, J.-B., Chérel, I., 2007. Increased functional diversity of plant K^+ channels by preferential heteromerization of the Shaker-like subunits AKT2 and KAT2. J. Biol. Chem. 282, 486–494.

Xie, K., Chen, J., Wang, Q., Yang, Y., 2014a. Direct phosphorylation and activation of a mitogen-activated protein kinase by a calcium-dependent protein kinase in rice. Plant Cell 26, 3077–3089.

Xie, Q., Michaeli, S., Peled-Zahavi, H., Galili, G., 2015. Chloroplast degradation: one organelle, multiple degradation pathways. Trends Plant Sci. 20, 264–265.

Xie, K., Zhang, J., Yang, Y., 2014. Genome-wide prediction of highly specific RNA spacers for CRISPR-Cas9-mediated genome editing in model plants and major crops. Mol. Plant 7, 923–926.

Xin, M., Wang, X., Peng, H., Yao, Y., Xie, C., Han, Y., Ni, Z., Sun, Q., 2012. Transcriptome comparison of susceptible and resistant wheat in response to powdery mildew infection. Genom. Proteom. Bioinform. 10, 94–106.

Xing, T., Higgins, V.J., Blumwald, E., 1997. Identification of G-proteins mediating fungal elicitor-induced dephosphorylation of host plasma membrane H^+-ATPase. J. Exp. Bot. 48, 229–237.

Xing, Y., Lawrence, J.B., 1991. Preservation of specific RNA distribution within the chromatin-depleted nuclear substructure demonstrated by *in situ* hybridization coupled with biochemical fractionation. J. Cell Biol. 112, 1055–1063.

Xiong, D., Huang, J., Peng, S., Li, Y., 2017. A few enlarged chloroplasts are less efficient in photosynthesis than a large population of small chloroplasts in *Arabidopsis thaliana*. Sci. Rep. 7, 5782.

Xoconostle-Cázares, B., Ruiz-Medrano, R., Lucas, W.J., 2000. Proteolytic processing of CmPP36, a protein from the cytochrome b(5)-reductase family, is required for entry into the phloem translocation pathway. Plant J. 24, 735–747.

Xoconostle-Cázares, B., Xiang, Y., Ruiz-Medrano, R., Wang, H.-L., Monzer, J., Yoo, B.C., McFarland, K.C., Franceschi, V.R., Lucas, W.J., 1999. Plant paralog to viral movement protein that potentiates transport of mRNA into the phloem. Science 283, 94–98.

Xu, M., Cho, E., Burch-Smith, T.M., Zambryski, P.C., 2012. Plasmodesmata formation and cell-to-cell transport are reduced in *decreased size exclusion limit* 1 during embryogenesis in *Arabidopsis*. Proc. Natl. Acad. Sci. U.S.A. 109, 5098–5103.

Xu, C., Fan, J., Riekhof, W., Froehlich, J.E., Benning, C., 2003. A permease-like protein involved in ER to thylakoid lipid transfer in *Arabidopsis*. EMBO J. 22, 2370–2379.

Xu, C., Jiao, C., Sun, H., Cai, X., Wang, X., Ge, C., Zheng, Y., Liu, W., Sun, X., Xu, Y., Deng, J., Zhang, Z., Huang, S., Dai, S., Mou, B., Wang, Q., Fei, Z., Wang, Q., 2017. Draft genome of spinach and transcriptome diversity of 120 Spinacia accessions. Nat. Commun. 8, 15275.

Xu, X., Liu, X., Ge, S., Jensen, J.D., Hu, F., Li, X., Dong, Y., Gutenkunst, R.N., Fang, L., Li, L.J., He, W., Zhang, G., Zheng, X., Zhang, F., Li, Y., Yu, C., Kristiansen, K., Zhang, X., Wang, J., Wright, M., McCouch, S., Nielsen, R., Wang, J., Wang, W., 2012. Resequencing 50 accessions of cultivated and wild rice yields markers for identifying agronomically important genes. Nat. Biotechnol. 30, 105–111.

Xu, X., Xie, Q., McClung, C.R., 2010. Robust circadian rhythms of gene expression in *Brassica rapa* tissue culture. Plant Physiol. 153, 841–850.

Xue, Y., Wang, Y., Shen, H., 2016. Ray Wu, fifth business or father of DNA sequencing. Protein Cell 7, 467–470.

Yadav, S.R., Yan, D., Sevilem, I., Helariutta, Y. Plasmodesmata-mediated intercellular signaling during plant growth and development. Front. Plant Sci. 5 (44).

Yadav, R.K., Girke, T., Pasala, S., Xie, M., Reddy, G.V., 2009. Gene expression map of the *Arabidopsis* shoot apical meristem stem cell niche. Proc. Natl. Acad. Sci. U.S.A. 106, 4941–4946.

Yadav, R.K., Tavakkoli, M., Xie, M., Girke, T., Reddy, G.V., 2014. A high-resolution gene expression map of the *Arabidopsis* shoot meristem stem cell niche. Development 141, 2735–2744.

Yaemsiri, S., Hou, N., Slining, M.M., He, K., 2010. Growth rate of human fingernails and toenails in healthy American young adults. J. Eur. Acad. Dermatol. Venereol. 24, 420–423.

Yaffe, M.P., Kennedy, E.P., 1983. Intracellular phospholipid movement and the role of intracellular phospholipid transfer proteins in animal cells. Biochemistry 22, 1497–1507.

Yahalom, A., Lando, R., Katz, A., Epel, B.L., 1998. A calcium-dependent protein kinase is associated with maize mesocotyl plasmodesmata. J. Plant Physiol. 153, 354–362.

Yahalom, A., Warmbrodt, R.D., Laird, D.W., Traub, O., Revel, J.-P., Willecke, K., Epel, B.L., 1991. Maize mesocotyl plasmodesmata proteins cross-react with connexin gap junction protein antibodies. Plant Cell 3, 407–417.

Yamada, M., Goshima, G., 2017. Mitotic spindle assembly in land plants: moelcules and mechanisms. Biology 6, 6.

Yamada, K., Lim, J., Dale, J.M., Chen, H.M., Shinn, P., Palm, C.J., Southwick, A.M., et al., 2003. Empirical analysis of transcriptional activity in the *Arabidopsis* genome. Science 302, 842–846.

Yamada, M., Tanaka-Takiguchi, Y., Hayashi, M., Nishina, M., Goshima, G., 2017. Multiple kinesin-14 family members drive microtubule minus end-directed transport in plant cells. J. Cell Biol. 216, 1705–1714.

Yamaguchi, T., Aharon, G.S., Sottosanto, J.B., Blumwald, E., 2005. Vacuolar Na^+/H^+ antiporter cation selectivity is regulated by calmodulin from within the vacuole in a Ca^{2+}- and pH-dependent manner. Science 102, 16107–16112.

Yamaguchi, T., Apse, M.P., Shi, H., Blumwald, E., 2003. Topological analysis of a plant vacuolar Na^+/H^+ antiporter reveals a luminal C terminus that regulates antiporter cation selectivity. Proc. Natl. Acad. Sci. U.S.A. 100, 12510–12515.

Yamaguchi, T., Fukada-Tanaka, S., Inagaki, Y., Saito, N., Yonekura-Sakakibara, K., Tanaka, Y., Kusumi, T., Iida, S., 2001. Genes encoding the vacuolar Na^+/H^+ exchanger and flower coloration. Plant Cell Physiol. 42, 451–461.

Yamaguchi, Y., Nagai, R., 1981. Motile apparatus in *Vallisneria* leaf cells. I. Organization of microfilaments. J. Cell Sci. 48, 193–205.

Yamaguchi, J., Nishimura, M., 1984. Purification of glyoxysomal catalase and immunochemical comparison of glyoxysomal and leaf peroxisomal catalase in germinating pumpkin cotyledons. Plant Physiol. 74, 261–267.

Yamaguchi, N., Winter, C.M., Wu, M.-F., Kwon, C.S., William, D.A., Wagner, D., 2014. PROTOCOL: chromatin immunoprecipitation from Arabidopsis tissues. Arabidopsis Book 12, e0170.

Yamamoto, K., 2007. Plant myosins VIII, XI, and XIII. In: Coluccio, L.M. (Ed.), Myosins. Springer, Berlin, pp. 375–390.

Yamamoto, K., 2008. In: Coluccio, L.M. (Ed.), Plant myosins, VIII, XI, and XIII. In: Myosins: A Superfamily of Molecular Motors. Springer, Dordrecht, The Netherlands, pp. 375–390.

Yamamoto, T., Burke, J., Autz, G., Jagendorf, A.T., 1981. Bound ribosomes of pea chloroplasts thylakoid membranes: location and release *in vitro* by high salt, puromycin, and RNAse. Plant Physiol. 67, 940–949.

Yamamoto, K., Kikuyama, M., Sutoh-Yamamoto, N., Kamitsubo, E., 1994. Purification of actin based motor protein from *Chara corallina*. Proc. Jpn. Acad. Sci. Ser. B. Phys. Biol. Sci. 70, 175–180.

Yamamoto, K., Kikuyama, M., Sutoh-Yamamoto, N., Kamitsubo, E., Katayama, E., 1995. Myosin from alga *Chara*: unique structure revealed by electron microscopy. J. Mol. Biol. 254, 109–112.

Yamamoto, R., Masuda, Y., 1971. Stress-relaxation properties of the *Avena* coleoptile cell wall. Physiol. Plant 25, 330–335.

Yamamoto, K., Shimada, K., Ito, K., Hamada, S., Ishijima, A., Tsuchiya, T., Tazawa, M., 2006. *Chara* myosin and the energy of cytoplasmic streaming. Plant Cell Physiol. 47, 1427–1431.

Yamamoto, R., Shinozaki, K., Masuda, Y., 1970. Stress-relaxation properties of plant cell walls with special reference to auxin action. Plant Cell Physiol. 11, 947–956.

Yamanishi, H., Kasamo, K., 1993. Modulation of the activity of purified tonoplast H^+-ATPase from mung bean (*Vigna radiata* L.) hypocotyls by various lipids. Plant Cell Physiol. 34, 411–419.

Yamanishi, H., Kasamo, K., 1994. Effects of cerebroside and cholesterol on the reconstitution of tonoplast H^+-ATPase purified from mung bean (*Vigna radiata* L.) hypocotyles in liposomes. Plant Cell Physiol. 35, 655−663.

Yamasaki, L., Kanda, P., Lanford, R.E., 1989. Identification of four nuclear transport signal-binding proteins that interact with diverse transport signals. Mol. Cell Biol. 9, 3028−3036.

Yamashita, M., Fenn, J.B., 1984. Electrospray ion source. Another variation on the free-jet theme. J. Phys. Chem. 88, 4451−4459.

Yamashita, H., Sato, Y., Kanegae, T., Kagawa, T., Wada, M., Kadota, A., 2011. Chloroplast actin filaments organize meshwork on the photorelocated chloroplasts in the moss *Physcomitrella patens*. Planta 233, 357−368.

Yamauchi, S., Takemiya, A., Sakamoto, T., Kurata, T., Tsutsumi, T., Kinoshita, T., Shimazaki, K., 2016. The plasma membrane H^+-ATPase AHA1 plays a major role in stomatal opening in response to blue light. Plant Physiol. 171, 2731−2743.

Yamazaki, D., Yoshida, S., Asami, T., Kuchitsu, K., 2003. Visualization of abscisic acid-perception sites on the plasma membrane of stomatal guard cells. Plant J. 35, 129−139.

Yan, W., Chen, D., Kaufmann, K., 2016. Efficient multiplex mutagenesis by RNA-guided Cas9 and its use in the characterization of regulatory elements in the AGAMOUS gene. Plant Methods 12, 23.

Yanagida, M., 1990. Higher-order chromosome structure in yeast. J. Cell Sci. 96, 1−2.

Yanagida, T., Kakase, M., Nishiyama, K., Oosawa, F., 1984. Direct observation of motion of single F-actin filaments in the presence of myosin. Nature 307, 57−60.

Yanagisawa, T., Hasegawa, S., Mohri, H., 1968. The bound nucleotides of the isolated microtubules of sea-urchin sperm flagella and their possible role in flagellar movement. Exp. Cell Res. 52, 86−100.

Yang, X., Boateng, K.A., Strittmatter, L., Burgess, R., Makaroff, C.A., 2009. *Arabidopsis* separase function beyond the removal of sister chromatid cohesion during meiosis. Plant Physiol. 151, 323−333.

Yang, W., Burkhart, W., Cavallius, J., Merrick, W.C., Boss, W.F., 1993b. Purification and characterization of phosphalidyl inositol 4-kinase activator in carrot cells. J. Biol. Chem. 268, 392−398.

Yang, T., Davies, P.J., Reid, J.B., 1996. Genetic dissection of the relative roles of auxin and gibberrellin in the regulation of stem elongation in intact light-grown peas. Plant Physiol. 110, 1029−1034.

Yang, F., Demma, M., Warren, V., Dharmawardhane, S., Condeelis, J., 1990. Identification of an actin-binding protein from *Dictyostelium* as elongation factor 1a. Nature 347, 494−496.

Yang, W., Duan, L., Chen, G., Xiong, L., Liu, Q., 2013. Plant phenomics and high-throughput phenotyping: accelerating rice functional genomics using multidisciplinary technologies. Curr. Opin. Plant Biol. 16, 1−8.

Yang, Y.D., Elamawi, R., Bubeck, J., Pepperkok, R., Ritzenthaler, C., Robinson, D.G., 2005. Dynamics of COPII vesicles and the Golgi apparatus in cultured *Nicotiana tabaccum* BY-2 cells provides evidence for transient association of Golgi stacks with endoplasmic reticulum exit sites. Plant Cell 17, 1513−1531.

Yang, J., Ellimore, P.T., Sather, W.A., Zhang, J.-F., Tsien, R.W., 1993a. Molecular determinants of Ca^{2+} selectivity and in permeation in L-type Ca^{2+} channels. Nature 366, 158−161.

Yang, Y., Franc, V., Heck, A.J.R., 2017. Glycoproteomics: a balance between high-throughput and in-depth analysis. Trends Biotechnol. 35, 598−609.

Yang, H., Murphy, A.S., 2009. Functional expression and characterization of *Arabidopsis* ABCB, AUX 1 and PIN auxin transporters in *Schizosaccharomyces pombe*. Plant J. 59, 179−191.

Yang, J., Osman, K., Iqbal, M., Stekel, D.J., Luo, Z., Armstrong, S.J., Franklin, F., Chris, H., 2012. Inferring the *Brassica rapa* interactome using protein-protein interaction data from *Arabidopsis thaliana*. Front. Plant Sci. 3, 297.

Yang, Y., Wang, Q., Chen, Q., Yin, X., Qian, M., Sun, X., Yang, Y., 2017b. Genome-wide survey indicates diverse physiological roles of the barley (*Hordeum vulgare* L.) calcium-dependent protein kinase genes. Sci. Rep. 7, 5306.

Yang, C., Xing, L., Zhai, Z., 1992. Intermediate filaments in higher plant cells and their assembly in a cell-free system. Protoplasma 171, 44−54.

Yang, C., Xing, L., Zhai, Z., 1992. Intermediate filaments in higher plant cells and their assembly in a cell-free system. Protoplasma 171, 44−54.

Yang, J.-P., Xu, Y.-P., Munyampundu, J.-P., Liu, T.-Y., Cai, X.-Z., 2016. Calcium-dependent protein kinase (CDPK) and CDPK-related kinase (CRK) gene families in tomato: genome-wide identification and functional analysis in disease resistance. Mol. Genet. Genom. 291, 661−676.

Yang, Y., Zhang, S., Howe, K., Wilson, D.B., Moser, F., Irwin, D., Thannhauser, T.W., 2007. A comparison of nLC-ESI-MS/MS and nLC-MALDI-MS/MS for GelC-based protein identification and iTRAQ-based shotgun quantitative proteomics. J. Biomol. Tech. 18, 226−237.

Yano, K., Matsui, S., Tsuchiya, T., Maeshima, M., Kutsuna, N., Hasezawa, S., Moriyasu, Y., 2004. Contribution of the plasma membrane and central vacuole in the formation of autolysosomes in cultured tobacco cells. Plant Cell Physiol. 45, 951−957.

Yano, H., Sato, S., 2002. Combination of electron microscopic *in situ* hybridization and anti-DNA antibody labelling reveals a peculiar arrangement of ribosomal DNA in the fibrillar centres of the plant cell nucleus. J. Electron Microsc. 51, 231−239.

Yano, K., Suzuki, T., Moriyasu, Y., 2007. Constitutive autophagy in plant root cells. Autophagy 3, 360−362.

Yano, K., Yanagisawa, T., Mukae, K., Niwa, Y., Inoue, Y., Moriyasu, Y., 2015. Dissection of autophagy in tobacco BY-2 cells under sucrose starvation conditions using vacuolar H^+-ATPase inhibitor concanamycin A and the autophagy −related protein Atg8. Plant Signal. Behav. 10, e1082699.

Yanofsky, M.F., Ma, H., Bowman, J.L., Drews, G.N., Feldmann, K.A., Meyerowitz, E.M., 1990. The protein encoded by the *Arabidopsis* homeotic gene *agamous* resembles transcription factors. Nature 346, 35−39.

Yassour, M., Kaplan, T., Fraser, H.B., Levin, J.Z., Pfiffner, J., Adiconis, X., Schroth, G., Luo, S., Khrebtukova, I., Gnirke, A., Nusbaum, C., Thompson, D.-A., Friedman, N., Regev, A., 2009. Ab initio construction of a eukaryotic transcriptome by massively parallel mRNA processing. Proc. Natl. Acad. Sci. U.S.A. 106, 3264−3269.

Yasuhara, H., Shibaoka, H., 2000. Inhibition of cell-plate formation by brefeldin A inhibited the depolymerization of microtubules in the central region of the phragmoplast. Plant Cell Physiol. 41, 300−310.

Yasuhara, H., Sonobe, S., Shibaoka, H., 1992. ATP-sensitive binding to microtubules of polypeptides extracted from isolated phragmoplasts of tobacco BY-2 cells. Plant Cell Physiol. 33, 601−608.

Yasui, K., 1939. On the cytokinesis in some angiosperms, with special reference to the middle lamella initial (MLI) formation and the phragmoplast. Cytologia 9, 557–574.

Yatsuhashi, H., Hashimoto, T., Wada, M., 1987a. Dichroic orientation of photoreceptors for chloroplast movement in *Adiantum* protonemata. Non-helical orientation. Plant Sci. 51, 165–170.

Yatsuhashi, H., Kadota, A., Wada, M., 1985. Blue- and red-light action in photoorientation of chloroplasts in *Adiantum* protonemata. Planta 165, 43–50.

Yatsuhashi, H., Wada, M., Hashimoto, T., 1987b. Dichroic orientation of phytochrome and blue-light photoreceptor in *Adiantum* as determined by chloroplast movement. Acta Physiol. Plant. 9, 163–173.

Yazaki, J., Galli, M., Kim, A.Y., Nito, K., Aleman, F., Chang, K.N., Carvunis, A.R., Quan, R., Nguyen, H., Song, L., Alvarez, J.M., Huang, S.S., Chen, H., Ramachandran, N., Altmann, S., Gutiérrez, R.A., Hill, D.E., Schroeder, J.I., Chory, J., LaBaer, J., Vidal, M., Braun, P., Ecker, J.R., 2016. Mapping transcription factor interactome networks using HaloTag protein arrays. Proc. Natl. Acad. Sci. U.S.A. 113, E4238–E4247.

Yazaki, J., Gregory, B.D., Ecker, J.R., 2007. Mapping the genome landscape using tiling array technology. Curr. Opin. Plant Biol. 10, 534–542.

Yazdanbakhsh, N., Fisahn, J., 2012. High-throughput phenotyping of plant shoots. In: Normanly, J. (Ed.), High-Throughput Phenotyping in Plants. Methods and Protocols. Springer, New York, pp. 21–40.

Ycas, M., 1974. On earlier states of the biochemical system. J. Theor. Biol. 44, 145–160.

Ycas, M., Vincent, W.S., 1960. A ribonucleic acid fraction from yeast related in composition to desoxyribonucleic acid. Proc. Natl. Acad. Sci. U.S.A. 46, 804–811.

Ye, J., Zheng, Y., Yan, A., Chen, N., Wang, Z., Huang, S., Yang, Z., 2009. *Arabidopsis* formin 3 directs the formation of actin cables and polarized growth in pollen tubes. Plant Cell 23, 3868–3884.

Yeats, T.H., Howe, K.J., Matas, A.J., Buda, G.J., Thannhauser, T.W., Rose, J.K.C., 2010. Mining the surface proteome of tomato (*Solanum lycopersicum*) fruit for proteins associated with cuticle biogenesis. J. Exp. Bot. 61, 3759–3771.

Yeh, K.C., Lagarias, J.C., 1998. Eukaryotic phytochromes: light- regulated serin/threonine protein kinases with histidine kinase ancestry. Proc. Natl. Acad. Sci. U.S.A. 95, 13976–13981.

Yennawar, N., Li, L.-C., Dudzinski, D.M., Tabuchi, A., Cosgrove, D.J., 2006. Crystal structure and activities of EXPB1 (*Zea* m 1), a beta-expansin and group-1 pollen allergen from maize. Proc. Natl. Acad. Sci. U.S.A. 103, 14664–14671.

Yildiz, A., Tomishige, M., Vale, R.D., Selvin, P.R., 2004. Kinesin walks hand-over-hand. Science 303, 676–678.

Yin, P., Fan, H., Yuan, X., Wu, D., Pang, Y., Yan, C., Li, W., Wang, J., Yan, N., 2009. Structural insights into the mechanism of abscisic acid signaling by PYL proteins. Nat. Struct. Mol. Biol. 16, 1230–1237.

Yin, X., Komatsu, S., 2016. Plant nuclear proteomics for unraveling physiological function. New Biotechnol. 33, 644–654.

Yin, H., Wang, M.D., Svoboda, K., Landick, R., Block, S.M., Gelles, J., 1995. Transcription against an applied force. Science 270, 1653–1657.

Yoder, T.L., Zheng, H.-Q., Todd, P., Staehelin, L.A., 2001. Amyloplast sedimentation dynamics in maize columella cells support a new model for the gravity-sensing apparatus of roots. Plant Physiol. 125, 1045–1060.

Yokota, E., McDonald, A.R., Liu, B., Shimmen, T., Palevitz, B.A., 1995a. Localization of a 170 kDa myosin heavy chain in plant cells. Protoplasma 185, 178–187.

Yokota, E., Mimura, T., Shimmen, T., 1995b. Biochemical, immunochemical and immunohistochemical identification of myosin heavy chains in cultured cells of *Catharanthus roseus*. Plant Cell Physiol. 36, 1541–1547.

Yokota, E., Muto, S., Shimmen, T., 1999a. Inhibitory regulation of higher-plant myosin by Ca^{2+} ions. Plant Physiol. 119, 231–240.

Yokota, E., Muto, S., Shimmen, T., 2000. Calcium-calmodulin suppresses the filamentous actin-binding activity of a 135-kilodalton actin-bundling protein isolated from lily pollen tubes. Plant Physiol. 123, 645–654.

Yokota, E., Shimmen, T., 1994. Isolation and characterization of plant myosin from pollen tubes of lily. Protoplasma 177, 153–162.

Yokota, E., Shimmen, T., 1999. The 135-kDa actin-bundling protein from lily pollen tubes arranges F-actin into bundles with uniform polarity. Planta 209, 264–266.

Yokota, E., Shimmen, T., 2011. Plant myosins. The plant cytoskelton. Adv. Plant Biol. 2, 33–56.

Yokota, E., Ueda, H., Hashimoto, K., Orii, H., Shimada, T., Hara-Nishimura, I., Shimmen, T., 2011. Myosin XI-dependent formation of tubular structures from endoplasmic reticulum isolated from tobacco cultures BY-2 cells. Plant Physiol. 156, 129–143.

Yokota, E., Ueda, S., Tamura, K., Orii, H., Uchi, S., Sonebe, S., Hara-Nishimura, I., Shimmen, T., 2009. AN isoform of myosin XI is responsible for the translocation of endoplasmic reticulum in tobacco cultured BY-2 cells. J. Exp. Bot. 60, 197–212.

Yokota, E., Vidali, L., Tominaga, M., Tahara, H., Orii, H., Morizane, Y., Hepler, P.K., Shimmen, T., 2003. Plant 115-kDa actin-filament bundling protein, P-115-ABP, is a homologue of plant villin and is widely distributed in plants. Plant Cell Physiol. 44, 1088–1099.

Yokota, E., Yukawa, C., Muto, S., Sonobe, S., Shimmen, T., 1999b. Biochemical and immunocytochemical characterization of two types of myosins in cultured tobacco Bright Yellow-2 cells. Plant Physiol. 121, 525–534.

Yokoyama, R., Nishitani, K., 2001. Endoxyloglucan transferase is localized in both the cell plate and in the secretory pathway destined for the apoplast in tobacco cells. Plant Cell Physiol. 42, 292–300.

Yokoyama, R., Nishitani, K., 2004. Genomic basis for cell-wall diversity in plants. A comparative approach to gene families in rice and *Arabidopsis*. Plant Cell Physiol. 45, 1111–1121.

Yokoyama, R., Rose, J.K.C., Nishitani, K., 2004. A surprizing diversity and abundance of xyloglucan endotransglucosylase/hydrolases in rice. Classification and expression analysis. Plant Physiol. 134, 1088–1099.

Yonath, A., Leonard, K.R., Wittiman, H.G., 1987. A tunnel in the large ribosomal subunit revealed by three dimensional image reconstruction. Science 256, 813–816.

Yoo, B.-C., Kragler, F., Varkonyi-Gasic, E., Haywood, V., Archer-Evans, S., Lee, Y.M., Lough, T.J., Lucas, W.J., 2004. A systemic small RNA signaling system in plants. Plant Cell 16, 1979–2000.

Yorimitsu, T., Sato, K., Takeuchi, M., 2014. Molecular mechanisms of Sar/Arf GTPases in vesicular trafficking in yeast and plants. Front. Plant Sci. 5, 411.

Yoshida, K., Kawachi, M., Mori, M., Maeshima, M., Kondo, M., Nishimura, M., Kondo, T., 2005. The involvement of tonoplast proton pumps and $Na^+(K^+)/H^+$ exchangers in the change of petal color during flower opening of morning glory, *Ipomoea tricolor* cv. Heaven. Blue Plant Cell Physiol. 46, 407–415.

Yoshida, K., Kondo, T., Okazaki, Y., Katou, K., 1995. Cause of blue petal colour. Nature 373, 291.

Yoshida, Y., Kuroiwa, H., Misumi, O., Nishida, K., Yagisawa, F., Fujiiwara, T., Nanamiya, F., Kuroiwa, T., 2006. Isolated chloroplast division machinery can actively constrict after stretching. Science 313, 1435–1438.

Yoshida, K., Toyama-Kato, Y., Kameda, K., Kondo, T., 2003. Sepal color variation *of Hydrangea macrophylla* and vacuolar pH measured with a proton-selective microelectrode. Plant Cell Physiol. 44, 262–268.

Yoshida, S., Uemura, M., 1986. Lipid composition of plasma membranes and tonoplasts isolated from etiolated seedlings of mung bean (*Vigna radiata*). Plant Physiol. 82, 807–812.

Yoshihisa, T., Anraku, Y., 1990. A novel pathway of import of α-mannosidase, a marker enzyme of vacuolar membrane, in *Saccharomyces cerevisiae*. J. Biol. Chem. 265, 22418–22425.

Yoshimoto, K., Hanaoka, H., Sato, S., Kato, T., Tabata, S., Noda, T., Ohsumi, Y., 2004. Processing of ATG8s, ubiquitin-like proteins, and their deconjugation by ATG4s are essential for plant autophagy. Plant Cell 16, 2967–2983.

Yoshimoto, K., Jikumaru, Y., Kamiya, Y., Kusano, M., Consonni, C., Panstruga, R., Ohsumi, Y., Shirasu, K., 2009. Autophagy negatively regulates cell death by controlling NPR1-dependent salicylic acid signaling during senescence and the innate immune response in *Arabidopsis*. Plant Cell 21, 2914–2927.

Yoshimoto, K., Shibata, M., Kondo, M., Oikawa, K., Sato, M., Toyooka, K., Shirasu, K., Nishimura, M., Ohsumi, Y., 2014. Organ-specific quality control of plant peroxisomes is mediated by autophagy. J. Cell Sci. 127, 1161–1168.

You, W., Abe, S., Davies, E., 1992. Cosedimentation of pea root polysomes with the cytoskeleton. Cell Biol. Int. Rep. 16, 663–673.

Young, T., 1807. A Course of Lectures on Natural Philosophy and the Mechanical Arts. Joseph Johnson, St. Paul's Church Yard.

Young, R.E., Bisgrove, S.R., 2011. Microtubule plus end-tracking proteins and their activities in plants. The Plant Cytoskeleton. Adv. Plant Biol. 2, 95–117.

Youngman, M.J., Aiken Hobbs, A.E., Burgess, S.M., Srinivasan, M., Jensen, R.E., 2004. Mmm2p, a mitochondrial outer membrane protein required for yeast mitochondrial shape and maintenance of mtDNA nucleoids. J. Cell Biol. 164, 677–688.

Yruela, I., Contreras-Moreira, B., Dunker, A.K., Niklas, K.J., 2018. Evolution of protein ductility in duplicated genes of plants. Front. Plant Sci. 9, 1216.

Yruela, I., Oldfield, C.J., Niklas, K.J., Dunker, A.K., 2017. Evidence for a strong correlation between transcription factor protein disorder and organismic complexity. Genome Biol. Evol. 9, 1248–1265.

Yu, H.-G., Hiatt, E.N., Dawe, R.K., 2000. The plant kinetochore. Trends Plant Sci. 5, 543–547.

Yu, J., Hu, S., Wang, J., Wong, G.K.-S., Li, S., Liu, B., Deng, Y., Dai, L., et al., 2002. A draft sequence of the rice genome (*Oryza sativa* L. ssp. *indica*). Science 296, 79–92.

Yu, Y., Lashbrook, C.C., Hannapel, D.J., 2007. Tissue integrity and RNA quality of laser microdissected phloem of potato. Planta 226, 797–803.

Yu, C.P., Lin, J.-J., Li, W.-H., 2016. Positional distribution of transcription factor binding sites in *Arabidopsis thaliana*. Sci. Rep. 6, 25164.

Yu, L., Moshelion, M., Moran, N., 2001. Extracellular protons inhibit the activity of inward-rectifying K channels in the motor cells of *Samanea saman* pulvini. Plant Physiol. 127, 1310–1322.

Yu, H.-G., Muszynski, M.G., Dawe, R.K., 1999. The maize homologue of the cell cycle checkpoint protein MAD2 reveals kinetochore substructure and contrasting mitotic and meiotic localization patterns. J. Cell Biol. 145, 425–435.

Yuan, Y., Bayer, P.E., Batley, J., Edwards, D., 2017. Improvements in genomic technologies: application to crop genomics. Trends Biotechnol. 35, 547–558.

Yuan, L., Yang, X., Makaroff, C.A., 2011. Plant cohesins, common themes and unique roles. Curr. Protein Pept. Sci. 12, 93–104.

Yuasa, T., Okazaki, Y., Iwasaki, N., Muto, S., 1997. Involvement of a calcium-dependent protein kinase in hypoosmotic tugor regulation in a brackish water Characeae *Lamprothamnium succintum*. Plant Cell Physiol. 38, 586–594.

Yubuki, N., Leander, B.S., 2013. Evolution of microtubule organizing centers across the tree of eukaryotes. Plant J. 75, 230–244.

Yue, J., Xu, W., Ban, R., Huang, S., Miao, M., Tang, X., Liu, G., Liu, Y., 2016. PTIR: predicted Tomato Interactome Resource. Sci. Rep. 6, 25047.

Yuen, Y.G., Crain, R.C., 1993. Deflagellation of *Chlamydomonas reinhardtii* follows a rapid transitory accumulation of inositol 1,4-5,-trisphosphate and requires Ca^{2+} entry. J. Cell Biol. 123, 869–875.

Yutin, N.M., Suzuki, T., Teeling, H., Weber, M., Venter, J.C., Rausch, D.B., Beja, O., 2007. Assessing diversity and biogeography of aerobic anoygenic phototropic bacteria in surface waters of the Atlantic and Pacific Oceans using the Global Ocean Sampling expedition metagenomes. Environ. Microbiol. 9, 1464–1475.

Zaar, K., 1977. The biogenesis of cellulose by Acetobacter xylinum. Cytobiologie 16, 1–15.

Zaar, K., 1979. Visualization of pores (export sites) correlated with cellulose production in the envelope of the gram-negative bacterium *Acetobacter xylinum*. J. Cell Biol. 80, 773–777.

Zabrouskov, V., Giacomelli, L., van Wijk, K.J., McLafferty, F.W., 2003. A new approach for plant proteomics: characterization of chloroplast proteins of *Arabidopsis thaliana* by top-down mass spectrometry. Mol. Cell. Proteom. 2, 1253–1260.

Zachariadis, M., Galatis, B., Apostolakos, P., 2000. Study of mitosis in root-tip cells of*Triticum turgidum* treated with the DNA-intercalating agent ethidium bromide. Protoplasma 211, 151–164.

Zacharias, E., 1881. Ueber die chemische Beschaffenheit des Zellkerns. Bot. Z. 39, 169–176.

Zacharias, H., 2001. Key word: chromosome. Chromosome Res. 9, 345–355.

Zacharias, R., 2008. The End of Reason. Zondervan, Grand Rapids, MI.

Zacherl, S., La Venuta, G., Müller, H.M., Wegehingel, S., Dimou, E., Sehr, P., Lewis, J.D., Erfle, H., Pepperkok, R., Nickel, W., 2015. A direct role for ATP1A1 in unconventional secretion of fibroblast growth factor 2. J. Biol. Chem. 290, 3654–3665.

Zachleder, V., van den Ende, H., 1992. Cell cycle events in the green alga *Chlamydomonas eugametos* and their control by environmental factors. J. Cell Sci. 102, 469–474.

Zaitlin, M., 1998. The discovery of the causal agent of the tobacco mosaic disease. In: Kung, S.D., Yang, S.F. (Eds.), Discoveries in Plant Biology. World Publishing Co., Hong Kong, pp. 105–110.

Zambryski, P., 2004. Cell-to-cell transport of proteins and fluorescent tracers via plasmodesmata during plant development. J. Cell Biol. 164, 165–168.

Zambryski, P., Crawford, K., 2000. Plasmodesmata: gatekeepers for cell-to-cell transport of developmental signals in plants. Annu. Rev. Cell Dev. Biol. 16, 393–421.

Zamecnik, P.C., 1960. Historical and current aspects of the problem of protein synthesis. Harvey Lect. 54, 256–281.

Zamecnik, P.C., 1969. An historical account of protein synthesis, with current overtones—a personalized view. Cold Spring Harb. Symp. Quant. Biol. 34, 1–16.

Zamecnik, P.C., Keller, E.B., 1954. Relation between phosphate energy donors and incorporation of labeled amino acids into proteins. J. Biol. Chem. 209, 337–354.

Zamenhof, S., Chargaff, E., 1949. Evidence of the existence of a core in desoxyribonucleic acids. J. Biol. Chem. 178, 531–532.

Zamenhof, S., Chargaff, E., 1950a. Dissymmetry in nucleotide sequence of desoxypentose nucleic acids. J. Biol. Chem. 187, 1–14.

Zamenhof, S., Chargaff, E., 1950b. Studies on the diversity and the native state of desoxypentose nucleic acids. J. Biol. Chem. 186, 207–214.

Zamenhof, S., Settles, L.B., Chargaff, E., 1950. Isolation of desoxypentose nucleic acid from human sperm. Nature 165, 756.

Zamir, A., Hollet, R.W., Marquisse, M., 1965. Evidence for the occurence of a common pentanucleotide sequence in the structures of transfer ribonucleic acids. J. Biol. Chem. 240, 1267–1273.

Zaug, A.J., Cech, T.R., 1986. The intervening sequence RNA of *Tetrahymena* is an enzyme. Science 231, 470–475.

Zavaliev, R., Ueki, S., Epel, B.L., Citovsky, V., 2011. Biology of callose (β-1,3-glucan) turnover at plasmodesmata. Protoplasma 248, 117–130.

Zehring, W.A., Wheeler, D.A., Reddy, P., Konopka, R.J., Kyriacou, C.P., Rosbash, M., Hall, J.C., 1984. P-element transformation with *period* locus DNA restores rhythmicity to mutant, arrhythmic *Drosophila melanogaster*. Cell 39, 369–376.

Zeiger, E., Hepler, P.K., 1976. Production of guard cell protoplasts from onion and tobacco. Plant Physiol. 58, 492–498.

Zeiger, E., Hepler, P.K., 1977. Light and stomatal function: blue light stimulates swelling of guard cell protoplasts. Science 196, 887–889.

Zeiger, E., Hepler, P.K., 1979. Blue light-induced, intrinsic vacuolar fluorescence in onion guard cells. J. Cell Sci. 37, 1–10.

Zeiger, E., Moody, W., Hepler, P., Varela, F., 1977. Light sensitive membrane potentials in onion guard cells. Nature 270, 270–271.

Zeigler, D.M., Linnane, A.W., Green, D.E., Dass, C.M.S., Ris, H., 1958. Studies on the electron transport system. Biochim. Biophys. Acta 28, 524–538.

Zeitlin, S.G., Shelby, R.D., Sullivan, K.F., 2001. CENP-A is phosphorylated by Aurora B kinase and plays an unexpected role in completion of cytokinesis. J. Cell Biol. 155, 1147–1157.

Zelitch, I., 1953. Oxidation and reduction of glycolic and glyoxylic acids in plants. II. Glyoxylic acid reductase. J. Biol. Chem. 201, 719–726.

Zelitch, I., 1955. The isolation of crystalline glyoxylic acid reductase from tobacco leaves. J. Biol. Chem. 216, 553–575.

Zelitch, I., 1957. α-Hydroxysulfonates as inhibitors of the enzymatic oxidation of glycolic and lactic acids. J. Biol. Chem. 224, 251–260.

Zelitch, I., 1959. The relationship of glycolic acid to respiration and photosynthesis in tobacco leaves. J. Biol. Chem. 234, 3077–3081.

Zelitch, I., 1964. Organic acids and respiration in photosynthetic tissues. Annu. Rev. Plant Physiol. 15, 121–142.

Zelitch, I., 1965. The relation of glycolic acid synthesis to the primary photosynthetic carboxylation reaction in leaves. J. Biol. Chem. 240, 1869–1876.

Zelitch, I., 1966. Increased rate of net photosynthetic carbon dioxide uptake caused by the inhibition of glycolate oxidase. Plant Physiol. 41, 1623–1631.

Zelitch, I., 1971. Photosynthesis. Photorespiration, and Plant Productivity. Academic Press, New York.

Zelitch, I., 1974. The effect of glycidate, an inhibitor of glycolate synthesis, on photorespiration and net photosynthesis. Arch. Biochem. Biophys. 163, 367–377.

Zelitch, I., 1975. Pathways of carbon fixation in green plants. Annu. Rev. Biochem. 44, 123–145.

Zelitch, I., 2001. Travels in a world of small science. Photosynth. Res. 67, 157–176.

Zelitch, I., Ochoa, S., 1953. Oxidation and reduction of glycolic and glyoxylic acids in plants. I. Glycolic acid oxidase. J. Biol. Chem. 201, 707–718.

Zelitch, I., Schultes, N.P., Peterson, R.B., Brown, P., Brutnell, T.P., 2009. High glycolate oxidase activity is required for survival of maize in normal air. Plant Physiol. 149, 195–204.

Zelman, A.K., Dawe, A., Gehring, C., Berkowitz, G.A., 2012. Evolutionary and structural perspectives of plant cyclic-nucleotide gated cation channels. Front. Plant Sci. 3, 95.

Zeng, Y., Li, Q., Wang, H., Zhang, J., Du, J., Feng, H., Blumwald, E., Yu, L., Xu, G., 2017. Two NHX-type transporters from Helianthus tuberosus improve the tolerance of rice to salinity and nutrient deficiency. stress. Plant Biotechnol. J. 16, 310–321.

Zeng, H., Xu, L., Singh, A., Wang, H., Du, L., Poovaiah, B.W., 2015. Involvement of calmodulin and calmodulin-like proteins in plant responses to abiotic stresses. Front. Plant Sci. 6, 600.

Zerzour, R., Kroeger, J., Geitmann, A., 2009. Polar growth in pollen tubes is associated with spatially confined dynamic changes in cell mechanical properties. Dev. Biol. 334, 437–446.

Zetter, B.R., Brightman, S.E., 1990. Cell motility and the extracellular matrix. Curr. Opin. Cell Biol. 2, 850–856.

Zewail, A., December 8, 1999. Femtochemistry. Atomic-scale dynamics of the chemical bond using ultrafast lasers. Nobel Lecture.

Zhai, Y., Kronebusch, P.J., Borisy, G.G., 1995. Kinetochore microtubule dynamics and the metaphase-anaphase transition. J. Cell Biol. 131, 721–734.

Zhang, Y., Akintola, O.S., Liu, K.J.A., Sun, B., 2016d. Detection bias in microarray and sequencing transcriptomic analysis identified by housekeeping genes. Data Brief 6, 121–123.

Zhang, Y., Akintola, O.S., Liu, K.J.A., Sun, B., 2016e. Membrane gene ontology bias in sequency and microarray obtained by housekeeping-gene analysis. Gene 575, 559–566.

Zhang, G., Annan, R.S., Carr, S.A., Neubert, T.A., 2010. Overview of peptide and protein analysis by mass spectrometry. Curr. Protoc. Protein Sci. 16.1.1–16.1.30.

Zhang, H.X., Blumwald, E., 2001. Transgenic salt tolerant tomato plants accumulate salt in the foliage but not in the fruits. Nat. Biotechnol. 19, 765–768.

Zhang, X., Buehner, N.A., Hutson, A.M., Estes, M.K., Mason, H.S., 2006. Tomato is a highly effective vehicle for expression and oral immunization with Norwalk virus capsid protein. Plant Biotechnol. 4, 419–432.

Zhang, D.H., Callaham, D.A., Hepler, P.K., 1990a. Regulation of anaphase chromosome motion in *Tradescantia* stamen hair cells by calcium and related signaling agents. J. Cell Biol. 111, 171–182.

Zhang, H., Dawe, R.K., 2011. Mechanisms of plant spindle formation. Chromosome Res. 19, 335–344.

Zhang, H., Dawe, R.K., 2012. Total centromere size and genome size are strongly correlated in ten grass species. Chromosome Res. 20, 403–412.

Zhang, G.F., Driouich, A., Staehelin, L.A., 1993. Effect of monensin on plant Golgi: re-examination of the monensin-induced changes in cisternal architecture and functional activities of the Golgi apparatus of sycamore suspension-cultured cells. J. Cell Sci. 104, 819–831.

Zhang, J., Gong, L., Liu, C., Huang, Y., Zhang, D., Yua, Z., 2016. Field phenotyping robot design and validation for the crop breeding. IFAC-PapersOnLine 49–16, 281–286.

Zhang, J., Hill, D.R., Sylvester, A.W., 2007. Diversification of the RAB guanosine triphosphatase family in dicots and monocots. J. Integr. Plant Biol. 49, 1129–1141.

Zhang, H.X., Hodson, J.N., Williams, J.P., et al., 2001. Engineering salt-tolerant Brassica plants: characterization of yield and seed oil quality in transgenic plants with increased vacuolar sodium accumulation. Proc. Natl. Acad. Sci. U.S.A. 98, 12832–12836.

Zhang, X.L., Jiang, L., Xin, Q., Liu, Y., Tan, J.X., Chen, Z.Z., 2015. Structural basis and functions of abscisic acid receptors PYLs. Front. Plant Sci. 6, 88.

Zhang, J., Ma, H., Feng, J., Zeng, L., Wang, Z., Chen, S., 2008. Grape berry plasma membrane proteome analysis and its differential expression during ripening. J. Exp. Bot. 59, 2979–2990.

Zhang, X., Madi, S., Borsuk, L., Nettleton, D., Elshire, R.J., Buckner, B., Janick-Buckner, D., Beck, J., Timmermans, M., Schnable, P.S., Scanlon, M.J., 2007b. Laser microdissection of narrow sheath mutant maize uncovers novel gene expression in the shoot apical meristem. PLoS Genet. 3 (6), e101.

Zhang, X.V., Martin, S.T., 2006. Driving parts of Krebs cycle in reverse through mineral photochemistry. J. Am. Chem. Soc. 128, 16032–16033.

Zhang, R., Miner, J.J., Gorman, M.J., Rausch, K., Ramage, H., White, H.P., Zuani, A., Zhang, P., Fernandez, E., Zhang, Q., Dowd, K.A., Pierson, T.C., Cherry, S., Diamond, M.S., 2016. A CRISPER screen defines a signal peptide proessing pathway required by flaviviruses. Nature 535, 164–168.

Zhang, D., Nicklas, R.B., 1995. Chromosomes initiate spindle assembly upon experimental dissolution of the nuclear envelope in grasshopper spermatocytes. J. Cell Biol. 131, 1125–1131.

Zhang, D., Nicklas, R.B., 1996. "Anaphase" and cytokinesis in the absence of chromosomes. Nature 382, 466–468.

Zhang, Y.-P., Oertner, T.G., 2007. Optical induction of synaptic plasticity using a light-sensitive channel. Nat. Methods 4, 139–141.

Zhang, G.F., Staehelin, L.A., 1992. Functional compartmentation of the Golgi Apparatus of plant cells. Immunocytochemical analysis of high-pressure frozen-and freeze-substituted sycamore maple suspension culture cells. Plant Physiol. 99, 1070–1083.

Zhang, D.H., Wadsworth, P., Hepler, P.K., 1990b. Microtubule dynamics in living plant cells: confocal imaging of microinjected fluorescent brain tubulin. Proc. Natl. Acad. Sci. U.S.A. 87, 8820–8824.

Zhang, D.H., Wadsworth, P., Hepler, P.K., 1992. Modulation of anaphase spindle microtubule structure in stamen hair cells of *Tradescantia* by calcium and related agents. J. Cell Sci. 102, 79–89.

Zhang, D., Wadsworth, P., Hepler, P.K., 1993a. Dynamics of microfilaments are similar but distinct from microtubules during cytokinesis in living dividing plant cells. Cell Motil. Cytoskelet. 24, 151–155.

Zhang, F., Wang, L.P., Boyden, E.S., Deisseroth, K., 2006a. Channel rhodopsin-2 and optical control of excitable cells. Nat. Methods 3, 785–792.

Zhang, H., Wei, C., Yang, X., Chen, H., Yang, Y., Mo, Y., Li, H., Zhang, Y., Ma, J., Yang, J., Zhang, X., 2017a. Genome-wide identification and expression analysis of calcium-dependent protein kinase and its related kinase gene families in melon (*Cucumis melo* L.). PLoS One 12, e0176352.

Zhang, S.-S., Yang, H., Ding, L., Song, Z.-T., Ma, H., Chang, F., Liu, J.-X., 2017b. Tissue-specific transcriptomics reveals an important role of the unfolded protein response in maintaining fertility upon heat stress in Arabidopsis. Plant Cell 29, 1007–1023.

Zhang, H., Zhang, J., Lang, Z., Botella, J.R., Zhu, J.-K., 2017b. Genome editing—principles and applications for functional genomics research and crop improvement. Crit. Rev. Plant Sci. 36, 291–309.

Zhang, Y., Zhang, F., Li, X., Baller, J.A., Qi, Y., Starker, C.G., Bogdanove, A.J., Voytas, D.Y., 2013. Transcription activator-like effector nucleases enable efficient plant genome editing. Plant Physiol. 161, 20–27.

Zhang, M., Zhang, R., Qu, X., Huang, S., 2016b. Arabidopsis FIM5 decorates apical actin filaments and regulates their organization in the pollen tube. J. Exp. Bot. 67, 3407–3417.

Zhang, H., Zhao, F.-G., Tang, R.-J., Yu, Y., Song, J., Wang, Y., Li, L., Luan, S., 2017a. Two tonoplast MATE proteins function as turgor-regulating chloride channels in. Arabidopsis 114, E2036–E2045.

Zhang, H., Zhou, H., Berke, L., Heck, A.J., Mohammed, S., Scheres, B., et al., 2013a. Quantitative phosphoproteomics after auxin-stimulated lateral root induction identifies an SNX1 protein phosphorylation site required for growth. Mol. Cell. Proteom. 12, 1158–1169.

Zhao, J., Last, R.L., 1996. Coordinate regulation of the tryptophan biosynthetic pathway and indolic phytoalexin accumulation in *Arabidopsis*. Plant Cell 8, 2235–2244.

Zhen, R.G., Baykov, A.A., Bakuleva, N.P., Rea, P.A., 1994. Amin omethylenediphosphonate: a potent type specific inhibitor of both plant and phototrophic bacterial H^+-pyrophosphatases. Plant Physiol. 104, 153–159.

Zheng, H., Fischer von Mollard, G., Kovaleva, V., Stevens, T.H., Raikhel, N.V., 1999. The plant vesicle-associated SNARE AtVTI1a likely mediates vesicle transport from the trans- Golgi network to the prevacuolar compartment. Mol. Biol. Cell 10, 2251–2264.

Zheng, H., Kunst, L., Hawes, C., Moore, I., 2004. A GFP-based assay reveals a role for RHD3 in transport between endoplasmic reticulum and Golgi apparatus. Plant J. 37, 398–414.

Zheng, M., Peng, C., Liu, H., Tang, M., Yang, H., Li, X., Liu, J., Sun, X., Wang, X., Xu, J., Hua, W., Wang, H., 2017. Genome-wide association study reveals candidate genes for control of plant height, branch initiation height and branch number in rapeseed (Brassica napus L.). Front. Plant Sci. 8, 1246.

Zheng, Y., Ren, N., Wang, H., Stromberg, A.J., Perry, S.E., 2009. Global identification of targets of the *Arabidopsis* MADS domain protein Agamous-Like15. Plant Cell 21, 2563–2577.

Zheng, H.Q., Staehelin, L.A., 2001. Nodal endoplasmic reticulum, a specialized form of endoplasmic reticulum found in gravity- sensing root tip columella cells. Plant Physiol. 125, 252–265.

Zheng, H., Staehelin, L.A., 2011. Protein storage vacuoles are transformed into lytic vacuoles in root meristematic cells of germinating seedlings by multiple, cell type-specific mechanisms. Plant Physiol. 155, 2023–2035.

Zheng, G.X.Y., Terry, J.M., Belgrader, P., Ryvkin, P., Bent, Z.W., Wilson, R., Ziraldo, S.B., Wheeler, T.D., McDermott, G.P., Zhu, J., Gregory, M.T., et al., 2017a. Massively parallel digital transcriptional profiling of single cells. Nat. Commun. 8, 14049.

Zhong, X., Ding, B., 2008. Distinct RNA motifs mediate systemic RNA trafficking. Plant Signal. Beyond Behav. 3, 1–2.

Zhong, X., Tao, X., Stombaugh, J., Leontis, N., Ding, B., 2007. Tertiary structure and function of an RNA motif required for plant vascular entry to initiate systemic trafficking. EMBO J. 26, 3836–3846.

Zhou, R., Basu, K., Hartman, H., Matocha, C.J., Sears, S.K., Vali, H., Guzman, M.I., 2017. Catalyzed synthesis of zinc clays by prebiotic central metabolites. Sci. Rep. 7, 533.

Zhou, A., Bu, Y., Takano, T., Zhang, X., Liu, S., 2016. Conserved V-ATPase c subunit plays a role in plant growth by influencing V-ATPase-dependent endosomal trafficking. Plant Biotechnol. J. 14, 271–283.

Zhou, R., Cheng, L.-L., Wayne, R., 2003. Purification and characterization of sorbitol-6-phosphate phosphatase from apple leaves. Plant Sci. 165, 227–232.

Zhou, X., Graumann, K., David, E.E., Meier, I., 2012. Novel plant SUN–KASH bridges are involved in RanGAP anchoring and nuclear shape determination. J. Cell Biol. 196, 203–211.

Zhou, X., Graumann, K., Meier, I., 2015a. The plant nuclear envelope as a multifunctional platform LINCed by SUN and KASH. J. Exp. Bot. 66, 1649–1659.

Zhou, X., Groves, N.R., Meier, I., 2015b. Plant nuclear shape is independently determined by the SUN-WIP-WIT2-myosin XI-I complex and CRWN1. Nucleus 6, 144–153.

Zhou, X., Groves, N.R., Meier, I., 2015c. SUN anchors pollen WIP-WIT complexes at he vegetative nuclear envelope and is necessary for pollen tube targeting and fertility. J. Exp. Bot. 66, 7299–7307.

Zhou, R., Guzman, M.I., 2014. CO_2 reduction under periodic illumination of ZnS. J. Phys. Chem. C 118, 11649–11656.

Zhou, R., Guzman, M.I., 2016. Photocatalytic reduction of fumarate to succinate on ZnS mineral surfaces. J. Phys. Chem. C 120, 7349–7357.

Zhou, Y., Liu, W., Xu, Y.-P., Cao, J.-Y., Braan, J., Cai, X.-Z., 2013. Genome-wide identification and functional analyses of calmodulin genes in Solanaceous species. BMC Plant Biol. 13, 70.

Zhou, X., Meier, I., 2013. How plants LINC the SUN to KASH. Nucleus 4, 206–215.

Zhou, R., Quebedeaux, B., 2003. Changes in photosynthesis and carbohydrate metabolism in mature apple leaves in response to whole plant source-sink manipulation. J. Am. Soc. Hort. Sci. 128, 113–119.

Zhou, R., Sicher, R.C., Quebedeaux, B., 2001. Diurnal changes in carbohydrate metabolism in mature apple leaves. Aust. J. Plant Physiol. 28, 1143–1150.

Zhou, R., Sicher, R.C., Quebedeaux, B., 2002. Apple leaf sucrose-phosphate synthase is inhibited by sorbitol-6-phosphate. Funct. Plant Biol. 29, 569–574.

Zhu, J., Adli, M., Zou, J.Y., Verstappen, G., Coyne, M., Zhang, X., Durham, T., Miri, M., Deshpande, V., De Jager, P.L., Bennett, D.A., Houmard, J.A., Muoio, D.M., Onder, T.T., Camahort, R., Cowan, C.A., Meissner, A., Epstein, C.B., Shoresh, N., Bernstein, B.E., 2013. Genome-wide chromatin state transitions associated with developmental and environmental cues. Cell 152, 642–654.

Zhu, C., Bao, G., Wang, N., 2000. Cell mechanics: mechanical response, cell adhesion, and molecular deformation. Annu. Rev. Biomed. Eng. 2, 189–226.

Zhu, Q., Clarke, M., 1992. Association of calmodulin and an unconventional myosin with the contractile vacuole complexes of *Dictyostelium discoideum*. J. Cell Biol. 118, 347–358.

Zhu, C., Dixit, R., 2011. Single molecule analysis of the Arabidopsis FRA1 kinesin shows that it is a functional motor protein with unusually high processivity. Mol. Plant 4, 879–885.

Zhu, C., Dixit, R., 2012. Functions of the *Arabidopsis* kinesin superfamily of microtubule-based motor proteins. Protoplasma 249, 887–899.

Zhu, C., Ganguly, A., Baskin, T.I., McClosky, D.D., Anderson, C.T., Foster, C., Meunier, K.A., Okamoto, R., Berg, H., Dixit, R., 2015. The FRA1 kinesin contributes to cortical microtubule-mediated trafficking of cell wall components. Plant Physiol. 167, 780–792.

Zhu, P., Gu, H., Jiao, Y., Huang, D., Chen, M., 2011. Computational identification of protein-protein interactions in rice based on the predicted rice interactome network. Dev. Reprod. Biol. 9, 128–137.

Zhu, Q., Liu, T., Clarke, M., 1993. Calmodulin and the contractile vacuole complex in mitotic cells *Dictyostelium discoideum*. J. Cell Sci. 104, 1119–1127.

Zhu, Y., Nam, J., Carpita, N.C., Matthysse, A.G., Gelvin, S.B., 2003. *Agrobacterium*-mediated root transformation is inhibited by mutation of an *Arabidopsis* cellulose synthase-like gene. Plant Physiol. 133, 1000–1010.

Zhu, T., Rost, T.L., 2000. Directional cell-to-cell communication in the *Arabidopsis* root apical meristem. III. Plasmodesmata turnover and apoptosis in meristem and root cap cells during four weeks after germination. Protoplasma 213, 99–107.

Zhu, T., Wang, X., 2000. Large-scale profiling of the *Arabidopsis* transcriptome. Plant Physiol. 124, 1472–1476.

Zhu-Salzman, K., Zeng, R., 2015. Insect response to plant defensive protease inhibitors. Annu. Rev. Entomol. 60, 233–252.

Ziegenhain, C., Vieth, B., Parekh, S., Reinius, B., Guillaumet-Adkins, A., Smets, M., Leohardt, H., Heyn, H., Hellmann, I., Enard, W., 2017. Comparative analysis of single-cell RNA sequencing methods. Mol. Cell 65, 631–643.

Ziegler, J., Facchini, P.J., 2008. Alkaloid biosynthesis: metabolism and trafficking. Annu. Rev. Plant Biol. 59, 735–769.

Zillig, W., 1987. Eukarytoic traits in archaebacteria. Could the eukaryotic cytoplasm have arisen from archaebacterial origin? Ann. N.Y. Acad. Sci. 503, 78–82.

Zimin, A.V., Puiu, D., Hall, R., Kingan, S., Clavijo, B.J., Salzberg, S.L., 2017. The first near-complete assembly of the hexaploid bread wheat genome, Triticum aestivum. $(GIGA)^n$. Science 6, 1–7.

Zimin, A., Stevens, K., Crepeau, M., Holtz-Morris, A., Korabline, M., Marçais, G., Puiu, D., Roberts, M., Wegrzyn, J., de Jong, P., Neale, D., Salzberg, S., Yorke, J., Langley, C., 2014. Sequencing and assembly of the 22-Gb loblolly pine genome. Genetics 196, 875–890.

Zimmer, H.-G., 2005. Sydney Ringer, serendipity, and hard work. Clin. Cardiol. 28, 55–56.

Zimmerman, S.B., Trach, S.O., 1991. Estimation of macromolecule concentrations and excluded volume effects for the cytoplasm of Escherichia coli. J. Mol. Biol. 222, 599–620.

Zimmermann, M.H., Brown, C.L., 1971. Trees—Structure and Function. Springer-Verlag, New York.

Zingen-Sell, I., Hillmer, S., Robinson, D.G., Jones, R.L., 1990. Localization of α-amylase isozymes within the endomembrane system of barley aleurone. Protoplasma 154, 16–24.

Zirkle, C., 1937. The plant vacuole. Bot. Rev. 3, 1–30.

Zirkle, C., 1949. Death of a Science in Russia. University of Pennsylvania Press, Philadelphia.

Zolman, B.K., Bartel, B., 2004. An *Arabidopsis* indole-3-butyric acid-response mutant defective in PEROXIN6, an apparent ATPase implicated in peroxisomal function. Proc. Natl. Acad. Sci. U.S.A. 101, 1786–1791.

Zolman, B.K., Nyberg, M., Bartel, B., 2007. IBR3, a novel peroxisomal acyl-CoA dehydrogenase-like protein required for indole-3-butyric acid response. Plant Mol. Biol. 64, 59–72.

Zolman, B.K., Silva, I.D., Bartel, B., 2001. The *Arabidopsis* pxa1 mutant is defective in an ATP-Binding cassette transporter-like protein required for peroxisomal fatty acid β-oxidation. Plant Physiol. 127, 1266–1278.

Zonia, L., 2010. Spatial and temporal integration of signaling networks regulating pollen tube growth. J. Exp. Bot. 61, 1939–1957.

Zonia, L., Cordeiro, S., Tupý, J., Feijó, J.A., 2002. Oscillatory chloride efflux at he pollen tube apex has a role in growth and cell volume regulation and is targeted by inositol 3,4,5,6-tetrakisphosphate. Plant Cell 14, 2233–2249.

Zonia, L., Müller, M., Munnik, T., 2006. Hydrodynamics and cell volume oscillations in the pollen tube apical region are integral components of the biomechanics of *Nicotiana tabacum* pollen tube growth. Cell Biochem. Biophys. 46, 209–232.

Zonia, L., Munnik, T., 2007. Life under pressure: hydrostatic pressure in cell growth and function. Trends Plant Sci. 12, 90–97.

Zonia, L., Munnik, T., 2008. Vesicle trafficking dynamics and visualization of zones of exocytosis and endocytosis in tobacco pollen tubes. J. Exp. Bot. 59, 861–873.

Zonia, L., Munnik, T., 2009. Uncovering hidden treasures in pollen tube growth mechanics. Trends Plant Sci. 14, 318–327.

Zonia, L., Munnik, T., 2011. Understanding pollen tube growth: the hydrodynamic model versus the cell wall model. Trends Plant Sci. 16, 347–352.

Zot, H.G., Doberstein, S.K., Pollard, T.D., 1992. Myosin I moves actin filaments on a phospholipid substrate-implications for membrane targeting. J. Cell Biol. 116, 597–602.

Zou, J., Song, L., Zhang, W., Wang, Y., Ruan, S., Wu, W.-H., 2009. Comparative proteomic analysis of *Arabidopsis* mature pollen and germinated pollen. J. Integr. Plant Biol. 51, 438–455.

Zouhar, J., Hicks, G.R., Raikhel, N.V., 2004. Sorting inhibitors (Sortins): chemical compounds to study vacuolar sorting in *Arabidopsis*. Proc. Natl. Adcad. Sci. U.S.A. 101, 9497–9501.

Zsigmondy, R., 1909. Colloids and the Ultramicroscope. Translated by J. John Wiley & Sons, Alexander.

Zsigmondy, R.A., December 11, 1926. Properties of colloids. Nobel Lecture.

Zsigmondy, R., Spear, E.B., 1917. The Chemistry of Colloids. John Wiley & Sons, New York.

Zufferey, M., Montandon, C., Douet, V., Demarsy, E., Agne, B., Baginsky, S., Kessler, F., 2017. The novel chloroplast outer membrane kinase KOC1 is a required component of the plastid protein import machinery. J. Biol. Chem. 292, 6952–6964.

Zuluaga, A.P., Vega-Arreguín, J.C., Fei, Z., Matas, A.J., Patev, S., Fry, W.E., Rose, J.K.C., 2016a. Analysis of the tomato leaf transcriptome during successive hemibiotrophic stages of a compatible interaction with the oomycete pathogen *Phytophthora infestans*. Mol. Plant Pathol. 17, 42–54.

Zuluaga, A.P., Vega-Arreguín, J.C., Fei, Z., Ponnala, L., Lee, S.J., Matas, A.J., Patev, S., Fry, W.E., Rose, J.K.C., 2016b. Transcriptional dynamics of *Phytophthora infestans* during sequential stages of hemibiotrophic infection of tomato. Mol. Plant Pathol. 29–41.

Zuo, R., Hu, R.B., Chai, G.H., Xu, M.L., Qi, G., Kong, Y.Z., et al., 2013. Genome-wide identification, classification, and expression analysis of CDPK and its closely related gene families in poplar (Populus trichocarpa). Mol. Biol. Rep. 40, 2645–2662.

Zurawski, G., Clegg, M.T., 1987. Evolution of higher-plant chloroplast DNA-encoded genes: implications for structure-function and phylogenetic studies. Ann. Rev. Plant Physiol. 38, 391–418.

Zurzycki, J., Lelatko, Z., 1969. Action dichroism in the chloroplast rearrangements in various plant species. Acta Soc. Bot. Pol. 38, 493–506.

Zwiewka, M., Friml, J., 2012. Fluorescence imaging-based forward genetic screens to identify trafficking regulators in plants. Front. Plant Sci. 3, 97.

Zwolak, M., Di Ventra, M., 2008. Colloquium: physical approaches to DNA sequencing and detection. Rev. Mod. Phys. 80, 141–165.

Index

Note: 'Page numbers followed by "f" indicate figures and "t" indicate tables.'

D

H

I

J

K

L

M

Printed in the United States
By Bookmasters